ACIDIC PRECIPITATION

Proceedings of the
International Symposium on Acidic Precipitation,
Muskoka, Ontario, September 15–20, 1985

Edited by

HANS C. MARTIN

Atmospheric Environment Service, Downsview, Ontario

PART 1

Reprinted from Water, Air, and Soil Pollution, Volume 30 (1986)

Springer-Science+Business Media, B.V.

Library of Congress Cataloging-in-Publication Data

International Symposium on Acidic Precipitation
(1985 : Muskoka, Ont.)
Acidic precipitation.

'Reprinted from Water, air, and soil pollution, volumes 30 and 31 (1986).'
Includes bibliographies and indexes.
1. Acidic precipitation (Meteorology)—Environmental aspects—Congresses. 2. Acid deposition—Environmental aspects—Congresses. 3. Acid pollution of river, lakes, etc.—Congresses. 4. Air—Pollution—Environmental aspects—Congresses. I. Martin, H. C. II. Title.
QC924.5.I57 1985 363.7'386 86-33883

ISBN 978-94-010-8019-4 ISBN 978-94-009-3385-9 (eBook)
DOI 10.1007/ 978-94-009-3385-9

Originally published by D. Reidel Publishing Company, Dordrecht, Holland in 1987
Softcover reprint of the hardcover 1st edition 1987

ACIDIC PRECIPITATION

Proceedings of the
International Symposium on Acidic Precipitation
Muskoka, Ontario, September 15–20, 1985

Guest Editor

H. C. MARTIN
Atmospheric Environment Service, Downsview, Ontario

PART 1

ACIDIC PRECIPITATION

Proceedings of the
International Symposium on Acidic Precipitation
Muskoka, Ontario, September 15–20, 1985

Guest Editor

H. C. MARTIN

Atmospheric Environment Service, Downsview, Ontario

PART 2

EFFECTS OF ACIDIC MELTWATER ON CHEMICAL CONDITIONS AT NEARSHORE SPAWNING SITES

J.M. Gunn[1] and W. Keller[2]

[1]Department of Zoology, University of Guelph, Guelph, Ontario N1G 2W1 and Ontario Ministry of Natural Resources, Sudbury, Ontario P3A 4S2, Canada
[2]Ontario Ministry of the Environment, Sudbury, Ontario, Canada P3E 5P9

ABSTRACT. Chemical conditions at lake charr (Salvelinus namaycush) spawning sites were monitored during snowmelt in low alkalinity lakes in Ontario, Canada. Embryos within the interstitial water of the spawning substrate were exposed to abrupt and potentially toxic levels of H^+ and inorganic Al as acidic runoff water inundated the shallow nearshore sites. Early runoff events, those that occurred while the lake ice was still snow-covered and under-ice water temperatures were <2°C, appeared to be most threatening, because of deep penetration of the runoff water. These highly site-specific events exhibit wide temporal and spatial variability.

1. INTRODUCTION

In north temperate areas with high deposition of acidic precipitation, snowmelt can cause large pH depressions in low alkalinity streams and lakes (Hagen and Langeland, 1973; Hultberg, 1977; Jeffries et al., 1979; Keller, 1983). However, despite widespread concern over the potential impact of acidic meltwater, there are few data on chemical changes at the specific sites inhabited by fish, or on the mechanisms which determine the resultant conditions in such habitats. This paper describes the effects of snowmelt on water chemistry at lake charr (Salvelinus namaycush) spawning sites. Lake charr are particularly vulnerable to damage from acidic runoff water because the sensitive embryonic stages are present in the shallow nearshore spawning sites during the melt period (Beggs et al., 1985).

Our findings are primarily from a 3-yr study of water quality changes at spawning sites and other nearshore areas on Whitepine Lake (47°17'N, 80°50'W), a lake with a history of lake charr recruitment failures (Gunn and Keller, 1984). Daily water samples were drawn from above and within the spawning site rubble throughout the melt period (30-40 d). Sampling methods are described in detail in Gunn and Keller (1984, 1985). The study was extended to include an extensive survey of nearshore chemistry during spring runoff on 20 other northeastern Ontario lake charr lakes. Changes in levels of H^+ and inorganic Al (determined by dialysis separation and graphite furnace atomic absorption spectrophytometry) are emphasized in this paper because of their importance as fish toxicants (Haines, 1981; Baker and Schofield, 1982).

Water, Air, and Soil Pollution 30 (1986) 545 552.

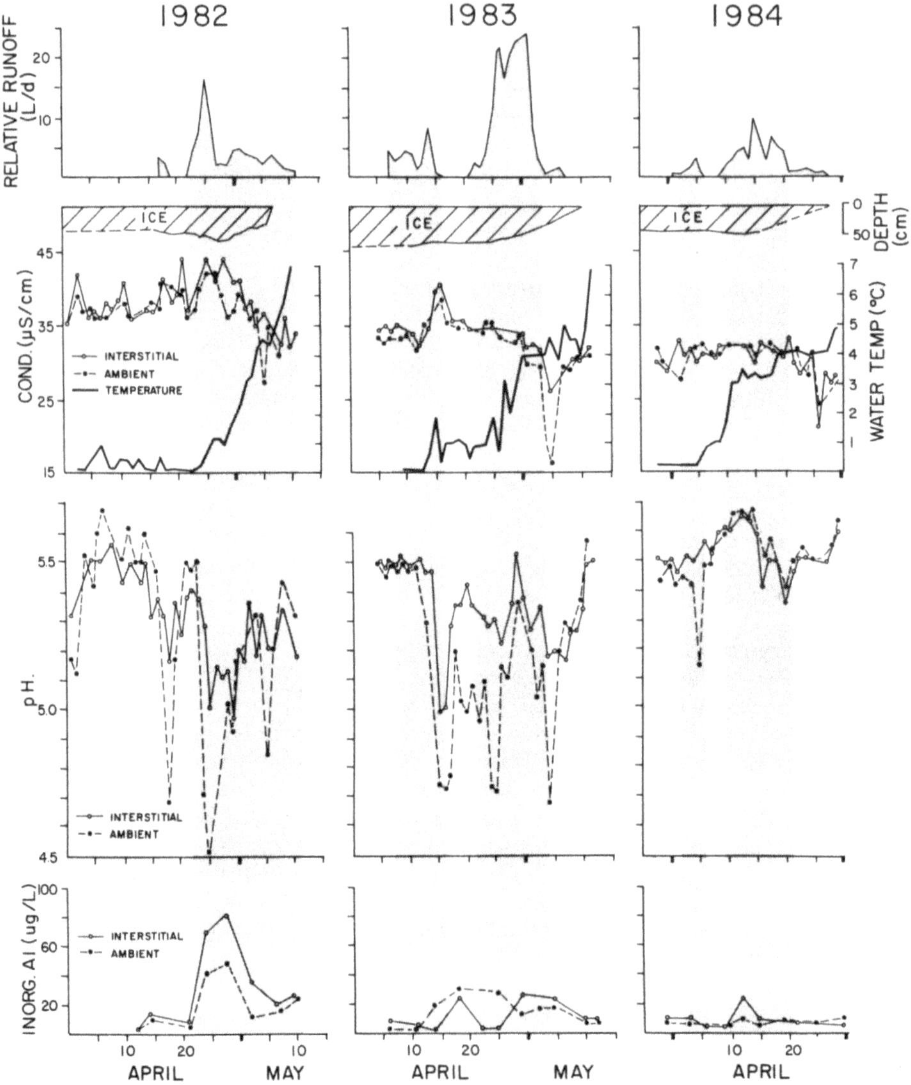

Fig. 1. Fluctuations in pH, specific conductance and inorganic Al concentrations within (interstitial) and above (ambient) the substrate of a lake charr spawning site (Whitepine Lake, site S), during the snowmelt period of three consecutive years. The shaded panels indicate the timing of the main runoff events, determined from the volume of water in a collector near the spawning site.

2. RESULTS

2.1 Intensive Monitoring

Snowmelt produced episodic pH depressions and elevations of inorganic Al within the substrate of nearshore spawning sites (Figure 1). At the main sampling site on Whitepine Lake, these episodic events were of 1 to 5 days duration (Figure 1). Considerable annual variability in pH and inorganic Al levels during snowmelt occurred at this site. Depressions in pH to 5.0 occurred during 1982 and 1983 but not in 1984. Maximum concentrations of potentially toxic inorganic Al varied annually from 24 to 83 $\mu g\ L^{-1}$.

Runoff water was enriched with several lithologically derived solutes before entering the lake. In the Whitepine Lake watershed, average concentrations of SiO3, Al and Fe were 115, 45, and 13 fold greater, respectively, in the runoff water, than in the winter snow pack (Gunn and Keller, 1985). Major pH depressions in nearshore waters were generally associated with increases in concentrations of these solutes. Short (1 to 2 day) pH depressions that occasionally occurred without elevations of Al, Fe and SiO3, appeared to result from direct inputs of rain and/or water from melting snow on the lake surface.

Major pH depressions did not always correspond with periods of maximum runoff from the terrestrial watershed. Penetration of acidic water to the substrate level, with resultant pH depressions, occurred only under cold water temperatures (<2 C). After disappearance of the surface snow cover and solar induced under-ice warming to >4 C, the less dense melt water formed thin layers (15 to 20 cm thick) that floated under the ice without affecting the interstitial conditions.

When snowmelt and cold water temperatures combined to allow penetration of runoff water to the substrate level, entrainment within the substrate inhibited mixing with epilimnetic water, and permitted maintenance of elevated concentrations of Al and other terrestrially derived solutes (Figure 1, 1982 data). A few sites exhibited depressed pH and elevated Al in the interstitial water without accompanying chemical changes in the water column. These sites appeared to be affected by subsurface seepage water moving from below or laterally through the substrate.

The conductivity of nearshore waters declined sharply during the melting of the ice pack (Figure 1). Dilution by melt water from the ice (specific conductance <10 μS/cm) caused a rapid drop in the concentration of most ions in the nearshore water, including H^+. After ice out, ambient chemical conditions returned quickly as nearshore and epilimnetic waters were mixed by the wind.

Individual sites along the shoreline of Whitepine Lake differed widely in the duration, mignitude and timing of pH depressions (Table I). Sites proximal to drainage channels usually showed more extended periods of pH depression and lower pH minima than other sites. Differences in slope, aspect and exposure of the drainage areas above these sites sppeared to be largely responsible for the asynchrony of runoff impacts.

Table I

Summary of pH conditions at 15 cm above the substrate at 10 shallow (< 1 m) nearshore sites on Whitepine Lake during the snow melt periods of 1982-1984. For each year, the reported data are for the samples taken daily during the 30 days prior to ice out.

	April 5-May 5, 1982				April 10-May 9, 1983				March 29-April 27, 1984			
Site[a]	x	No. days pH < 5.0	Minimum pH	Date of min.	x	No. days pH < 5.0	Minimum pH	Date of min.	x	No. days pH < 5.0	Minimum pH	Date of min.
1	5.1	6	4.6	17/4	5.3	0	5.0	15.4	5.3	4	4.9	7/4
2	5.2	4	4.4	17/4	5.2	3	4.4	15/4	5.3	3	4.3	1/4
3*	5.0	10	4.6	20/4	5.2	7	4.8	4/5	5.4	0	5.1	4/4
4*	4.8	10	4.3	19/4	4.6	19	4.3	14/4	5.2	3	4.7	29/3
5	5.1	5	4.2	29/4	5.1	5	4.7	14/4	5.2	3	4.7	30/3
6	5.0	9	4.3	17/4	5.0	9	4.5	26/4	5.2	4	4.5	31/3
7*	5.0	11	4.6	18/4	5.0	11	4.5	15/4	5.5	0	5.0	24/4
8	5.2	4	4.5	25/4	4.8	20	4.5	26/4	5.0	8	4.3	12/4
9*	4.6	20	4.1	20/4	4.4	26	4.1	23/4	4.7	20	4.2	7/4
10*	4.6	19	4.3	25/4	4.6	26	4.3	22/4	4.5	24	4.2	8/4

[a]The sites are located at 100 m intervals along the south and southeast shore of Whitepine Lake, a 67 ha lake charr lake with an annual average pH of 5.7. The sites are all located over granitic rubble and are representative of lake charr spawning sites in this lake.

*These sites are located near drainage channels and are usually influenced by greater amounts of terrestrial runoff water.

2.2 Extensive Surveys

Substantial pH depressions and elevations in Al concentrations were observed in the nearshore waters of most surveyed lakes during snowmelt (Table II). Wide site-specific differences were detected on individual lakes. For example, on Michaud Lake the sampled sites ranged in levels of total Al from 18 to 460 µg/L (inorganic Al 6 to 200 µg/L).

The degree of pH depression (ΔH^+ between late winter epilimnetic values and spring nearshore measurements) showed a significant negative correlation ($r = -0.304$, $P = 0.024$) with site depth, suggesting that shallow sites were more vulnerable to runoff impacts. Temperature was also negatively correlated ($r = -0.461$, $P < 0.001$) with degree of pH depression, indicating that at sites where under-ice warming had occurred, runoff water had a reduced impact at the substrate level.

3. DISCUSSION

Lake charr embryos within the spawning rubble can be exposed to severe pH depressions and elevated levels of inorganic Al during snowmelt. A conceptual model of some of the factors that influence spawning site chemistry is provided in Figure 2 and discussed below.

Table II

Comparison of chemical conditions of shallow (< 1 m) nearshore areas during spring runoff with ambient chemistry during late winter for 20 northeastern Ontario lake charr lakes. Nearshore chemistry is based on sampling 15 cm above the substrate at 2 to 4 locations per lake; ambient chemistry is based on 5m tube composite samples of offshore waters from 1 to 3 locations per lake.

Lake[a]	Latitude	Longitude	pH		Total Al (μg/L)		Conductivity (μS/cm)	
			Ambient (x)	Nearshore (Range)	Ambient (x)	Nearshore (Range)	Ambient (x)	Nearshore (Range)
Barbara	47°16'	84°20'	5.2	4.5-4.9	150	14-260	20	7-25
Bigwood	46°51'	81°05'	5.5	4.8-5.4	101	130-170	38	31-35
Black Beaver	47°24'	84°30'	5.4	4.7-5.1	80	33-180	19	10-21
Bluesucker	47°10'	80°37'	5.4	5.0-5.1	160	110-200	36	22-34
Burns	46°36'	83°07'	5.7	5.0-5.5	21	28-49	26	15-23
Florence	47°15'	80°33'	4.7	4.4-4.5	340	220-350	43	32-36
George	46°02'	81°24'	5.4	5.1-5.3	120	140-180	34	32-35
Grey Owl	47°16'	84°12'	5.2	4.6-4.9	153	30-200	21	14-22
Hubert	47°19'	84°27'	5.5	4.9	125	260-350	20	22-23
Iron	46°57'	80°21'	5.6	4.8-5.4	49	50-210	43	32-37
Jerry	47°22'	80°39'	5.0	4.3-4.8	220	160-880	37	17-41
Laundrie	47°07'	80°52'	5.2	5.3	106	83-92	36	27-30
Lulu	47°24'	80°45'	5.4	4.7-5.0	95	82-97	33	17-27
Marina	47°24'	80°40'	5.4	4.4-4.9	150	18-960	36	10-43
Michaud	46°49'	81°14'	5.3	4.3-4.5	137	18-460	35	22-44
North Hubert	47°20'	84°27'	5.5	4.9-5.1	100	160-210	19	22-23
Pilgrim	47°11'	80°41'	5.5	4.5-5.3	52	51-260	32	24-32
Regal	46°36'	83°05'	5.8	5.6-5.7	34	59-85	26	25
Seagram	47°06'	80°32'	5.2	5.1-5.2	145	170-290	40	35-36
Three Narrows	46°05'	81°27'	5.6	4.8	110	42-64	35	12

[a]All lakes were part of a test netting survey conducted to examine relationships between lake trout community status and degree of acidification (Beggs *et al.*, 1985).

The enrichment of dissolved Al and other metals that occurs in terrestrial watersheds increases the potential toxicity of runoff water to exposed biota. Pure meltwater may be toxic to fish (Grand et al., 1978), but field studies of fish mortalities during snowmelt have strongly implicated Al toxicity (Schofield and Trojnar, 1980; Gunn and Keller, 1984).

Terrestial runoff water is considerably diluted before it reaches the embryos in the substrate. In Whitepine Lake, the concentration of Al in the interstitial water rarely exceeded one tenth of that of the incoming runoff water (Gunn and Keller, 1985). Both ambient lake water and melt water from the lake surface ice and snow produce this dilution, but in our studies, the ambient lake water was more important because runoff usually preceded ice melt.

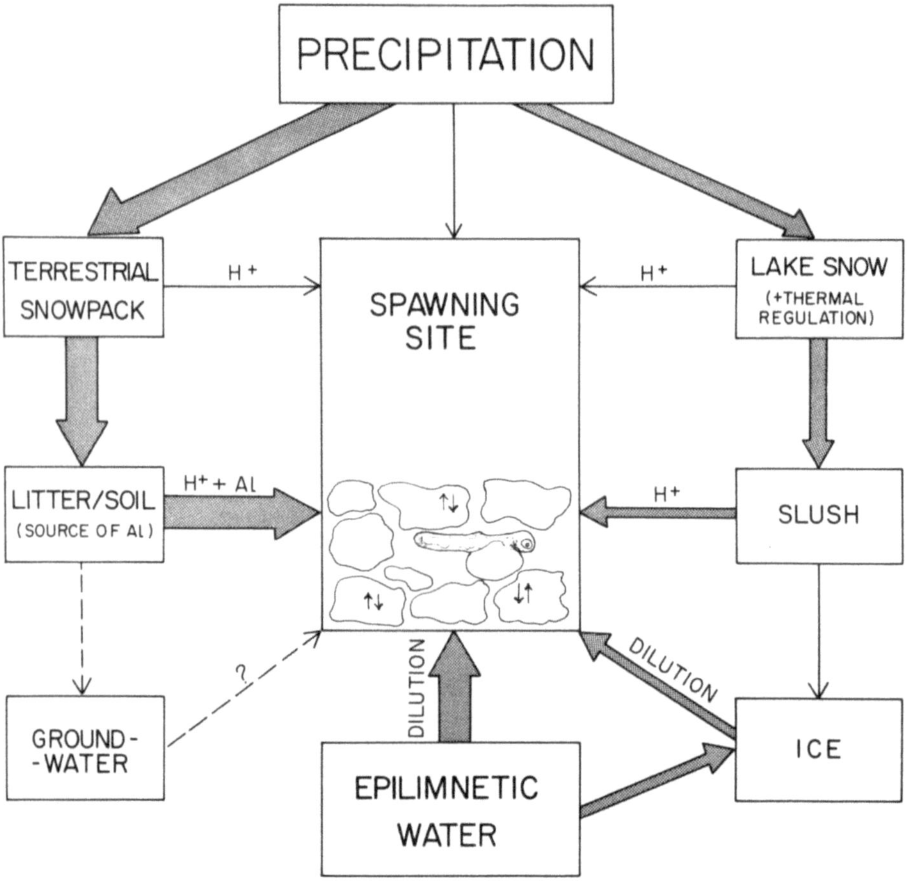

Fig. 2. Factors controlling spawning site water chemistry during the melt period. The width of the arrows indicates the relative contribution of H^+ or dilution water from the various sources.

Occasionally, the ice can produce, rather than dilute, acidic water. As H^+ and other ions freeze out of the upper portion of the ice, they collect in concentrated slush layers, usually at the interface of the granular white ice and the clear "black" ice (Hagen and Langeland, 1973; Jone and Ouellet, 1983). Some acidic layers can also be formed from runoff water that flows out on the ice and freezes before it can enter the water column (Gunn and Keller, 1985). These sources of acidic water may produce some of the pH depressions observed during periods of little or no terrestrial runoff (Figure 1), but they are not currently well documented.

The attenuating effect of surface snow on solar radiation (Maguire, 1975) controls under-ice thermal conditions, which in turn, through density differences, affect the mixing depth of incoming runoff water. We found that cold runoff water only penetrated to the substrate level when the lake was still snow covered and water temperatures were <2 C. Continuous temperature recordings at known spawning sites, or, on a more extensive basis, regional surveys of snow coverage on lake surfaces may therefore prove useful for predicting the timing of damaging pH depressions.

Weathering reactions between the acidic meltwater and the mineral substrate may increase the pH of the interstitial water (see Figure 1). In areas with calcareous substrates the weathering reactions would probably prevent toxic pH conditions from ever occurring. In the granitic substrate at our study sites, the slight neutralizing effect of the substrate may actually increase the toxicity of the interstitial water by affecting the form of Al present. Baker and Schofield (1982) found the inorganic Al was most toxic at pH 5.2 - 5.4.

Groundwater appears to exert little influence on chemical conditions at lake charr spawning sites, in contrast to its dominant role in the reproductive habitat of the closely related brook charr (*S. fontinalis*), a species that actively seeks groundwater upwelling sites for spawning (Webster and Eiriksdottir, 1975). Some of our monitored lake charr sites were periodically affected by acidic seepage water moving laterally through the substrate, but more typical groundwater, usually characterized by high conductivity and alkalinity because of its long contact time with surficial sediments and bedrock, was not detected.

SUMMARY

The chemistry of nearshore waters during snowmelt is controlled by many factors (Figure 2) and shows substantial spatial and temporal variability. Worst case conditions for lake charr embryos within the spawning substrate occur when substantial and sustained inputs of terrestrial runoff water reach shallow nearshore spawning areas before the disappearance of the snow cover from the ice. Least affected are offshore sites, in deeper water, in areas with little runoff and early disappearance of snow cover. Accurate predictions of runoff effects must take account of both the highly site-specific chemical trends and the patchy distribution of the reproductive habitat of the species.

ACKNOWLEDGEMENTS

P. Gale, V. Liimatainen and several student assistants helped with sample collection. Chemical analyses were carried out by Ministry of the Environment staff. G. Beggs, P. Dillon, D. Noakes and F. Tomassini provided helpful comments on an earlier version of the manuscript.

REFERENCES

Baker, J.P. and Schofield, C.L.: 1982, Water Air Soil Pollut. 18.289.

Beggs, G.L., Gunn, J.M. and Olver, C.H.: 1985, Ont. Fish. Tech. Rept. 17,1.

Grande, M., Muniz, I.P. and Andersen, S.: 1978, Int. Assoc. Theor. Appl. Limnol. Proc. 20,2076.

Gunn, J.M. and Keller, W.: 1984, Can. J. Fish. Aquat. Sci. 41,319.

Gunn, J.M. and Keller, W.: 1985, Annals. Glaciol. 7,208.

Hagen, A. and Langeland, A.: 1973, Environ. Pollut. 5,45.

Haines, T.A.: 1981, Trans. Am. Fish. Soc. 110,669.

Hultberg, H.: 1977, Water Air Soil Pollut. 7,279

Jeffries, D.S., Cox, C.M. and Dillon, P.J.: 1979, J. Fish. Res. Board Can. 36,640.

Jones, H.G. and Ouellet, M.: 1983, Eau du Quebec. 16,174.

Keller, W.: 1983, J. Great Lakes Res. 9,425

Maguire, R.I.: 1975, Environ. Canada. 91,1

Schofield, C.L. and Trojnar, J.R.: 1980, "Aluminum toxicity to brook trout (*Salvelinus fontinalis*) in acidified waters', In T.Y. Toribara, M.W. Miller, and P.E. Morrow (ed.) Polluted Rain. Plenum Press, New York, p.341.

Webster, D.A. and Eiriksdottir, G.: 1975, Trans. Amer. Fish. Soc. 91,168.

ASSOCIATION OF WETLAND ACIDITY WITH REPRODUCTIVE PARAMETERS AND INSECT PREY OF THE EASTERN KINGBIRD (*Tyrannus tyrannus*) NEAR SUDBURY, ONTARIO

V. Glooschenko, P. Blancher[1], J. Herskowitz[2], R. Fulthorpe[3], and S. Rang[4]
Wildlife Branch, Ontario Ministry of Natural Resources,
Toronto, Ontario, M7A 1W3
Canada.

ABSTRACT. A three year study (1981 to 1983) was conducted to determine possible influences of wetland acidification on the reproduction and growth of the Eastern Kingbird (*Tyrannus tyrannus*). This species uses emergent aquatic insect prey to feed its nestlings. Egg characteristics and nestling growth were monitored for birds nesting in 26 lakes in the Killarney region near Sudbury, Ontario (pH range 4.27 to 6.26). The major percentage of variation in kingbird reproductive factors including clutch initiation, egg weight loss and tarso-metatarsus bone growth was explained by differences between siblings and among nests on a single lake. A smaller percentage of the variance in reproductive parameters was explained by the variation in water chemistry among lakes. Canonical correlation showed that sulfate, nitrate/nitrite, alkalinity, pH, and several metals (Fe, Mn, Al) accounted for a minor but significant variance in growth characteristics. These water chemistry parameters are influenced by lake acidification. The relationship observed between lake acidity and kingbird reproduction is therefore measurable but minor compared to the influence of genetic and inter-nest variables. The study showed no food limitation for the kingbird attributed to lake acidity.

1. INTRODUCTION

The impact of acidic precipitation on aquatic organisms such as fish and benthos has been documented (Harvey, 1980); Wiederholm and Eriksson, 1977) and several studies have focused on the effect of

1. Canadian Wildlife Service, Ottawa, Ont.
2. Ministry of the Environment, Toronto, Ont.
3. Sierra Club of Ont., Toronto, Ont.
4. Inst. of Enviro. Studies, University of Toronto.

Water, Air, and Soil Pollution **30** (1986) 553 567.

acid precipitation on piscivorous waterfowl (Bendell et al.,1983; Nilsson and Nilsson, 1978). Whether the invertebrates available at a low pH (Confer et al., 1983) would provide adequate nutritional resources for successful waterfowl production has been under investigation (Stromberg and Longcore, 1983; Haseltine, 1985.; McNichol and Bendell, 1985.) Less information is available, however, on the effects of acidification on the growth and/or reproduction of other birds which use the aquatic environment as a food source.

Birds may accumulate metals and other contaminants in their diets (Custer and Mulhern, 1983; Haseltine et al., 1983) which can have effects on reproduction (Fimreite, 1971, 1974; Heinz, 1976, 1979). Nyholm and Myhrberg (1977) reported reproductive impairment in insectivorous birds breeding in Swedish Lapland. This was later attributed to possible Al interference with Ca and P metabolism in the female bird. A potential route of Al transport to the birds was presented and linked to acid precipitation by Nyholm (1981). Recent work has also shown that the abundance of breeding Dippers in Britain was correlated with pH (Ormerod, 1985; Ormerod et al., 1985). These authors suggested that the positive relationship between pH and the abundance of breeding Dippers was probably an indirect consequence of the abundance of their macroinvertebrate prey which tend to be scarce at low pH (Stoner et al., 1984).

The Eastern Kingbird often nests in or near wetlands or the littoral zones of lakes and feeds its nestlings on emerging aquatic insects. The purpose of this study was to determine: how much of the variance in egg characteristics and nestling growth of the Eastern Kingbird (*Tyrannus tyrannus*) was attributable to water chemistry differences between lakes; if the abundance of aquatic insect prey of the kingbird was correlated with lake acidity; and if metals in nestlings were associated with wetland acidity.

2. THE STUDY AREA

The Killarney region of Ontario (60 km southwest of Sudbury, Ontario; Figure 1) was selected for study because an impact on water chemistry is known to occur from influence of both long and short range atmospheric transport (Beamish and Harvey, 1972; Keller et al., 1980). Bedrock in the area is of low buffering capacity and includes the white quartzite mountains of the La Cloche range as well as granitic bedrock east of Killarney Provincial Park. There was a range in pH for the study sites from 4.27 to 6.26 although some wetlands were highly acidic (pH 4 to 5). Sulfates contribute far more than organic acids to the acidity of these waters (Glooschenko and Stevens, 1985).

Figure 1. Map of Study Area

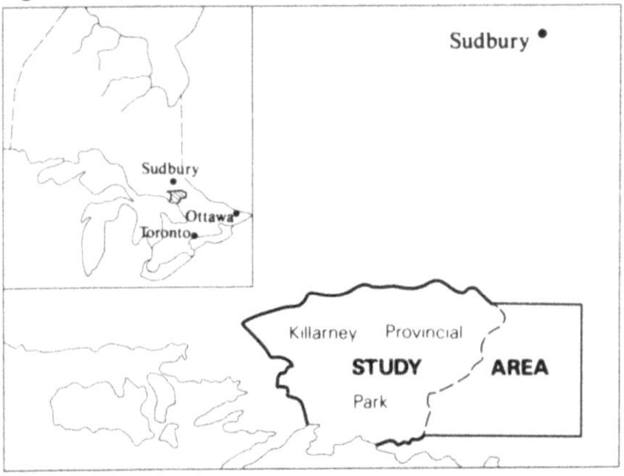

3. METHODS

Twenty-six wetlands were surveyed for kingbird nests in May and early June of 1981, 1982 and 1983. Nests were found at 56 locations. Over 3 yr, 145 nests with 404 eggs were examined. Sixty-three nests were successful, producing 215 young (Table I).

Table I. KINGBIRD BREEDING RECORDS IN THE KILLARNEY AREA

Year	Nests	Eggs	Total Young	Young used in Study *	Nests Producing Young
1983	40	110	78	69	22
1982	51	147	74	59	26
1981	54	147	63	51	15

* Includes only nestlings for which a complete set of measurements were available; these data used in statistical calculations.

Eggs were weighed and measured every four days to the nearest 0.01 g using a 5 g Pesola balance; length and width were measured to the nearest 0.05 mm using calipers. Nestlings were weighed and measured every four days until the birds fledged or disappeared and were presumed dead. Birds were weighed to the nearest 0.1 g using a 50 g Pesola balance. Wing chord length (distance between the upper wrist joint and first primary feather) of the right wing and tarso-metatarsus length (distance from the tarsal joint to the metatarsal joint) of the right leg were measured to the nearest 0.05 mm using calipers.

From these data egg volume, egg length-width ratio, egg water loss, growth rate of nestlings (i.e. weight increase), growth rate of tarso-metatarsus and growth rate of wings were calculated. Egg volume (V) was calculated from length (L) and width (W) measurements using the formula: $V = 0.507L \times W^2$ (Hoyt 1979). Surface area (A) was calculated as: $A = 4.951 \times V^{0.666}$ (Paganelli et al., 1974). Egg weight loss (E) was then calculated as:

$$E = \frac{\frac{\text{weight at time 1} - \text{weight at time 2}}{\text{time 2} - \text{time 1}}}{A}$$

Rate of weight gain, rate of bone growth and rate of wing growth were calculated using Ricklefs' (1967) graphical technique. Logistic curves were used for calculations of weight and tarsus growth rate and Gompertz curves for wing chord growth rate calculations.

A limited number (n=18) of eggs were collected and frozen, for determinations of eggshell thickness and composition. Nine eggs were collected from the Killarney area (mean alkalinity 2.9 mg l^{-1}) and, for comparative purposes, an additional nine were taken adjacent to wetlands remote from Sudbury on more buffered terrain. Seven of the latter were taken from the area of Lake Opinicon, 50 km north of Kingston, which is underlain by a mixture of granitic and marble bedrock (alkalinity 116 mg l^{-1}). The two remaining were collected from near Long Point on Lake Erie which is underlain by limestone bedrock (alkalinity 209 mg l^{-1}). In all cases eggs were taken from different nests. Eggshells were analyzed near the top (n=2), 0.75 cm from the top (n=4) and 2.0 cm from the top (n=4) for Na, Mg, Al, Si, P, S, K, Ca, Fe, Ba, Cr, Ni, Ag, Mn, Ti, Co, Hg, Zn, and Cs. Half-eggshells were coated with carbon and examined in a scanning electron microscope utilizing a static electron beam approximately 10 nm in diameter and an accelerating voltage of 15 KV for 60s under consistent lens conditions. The SEM was interfaced with a Traco Northern energy dispersive x-ray spectrometer and a TN2000 microanalyzer. Quantitative results were obtained with reference to polished plastic embedded mineral standards utilizing a least squares fitting routine with ZAF corrections. Eggshells were examined microscopically for differences in thickness.

Ten nestlings were collected in 1982 from ten different nests in the Killarney area. Nine nestlings were collected from the sites of higher alkalinity on granitic-marble (n=5) and on limestone bedrock (n=4) described earlier. Nestling metals analyses were conducted on freeze-dried liver tissue and air-dried (at room temperature) tarso-metatarsus bone and primary feathers. Neutron activation analysis was used to measure levels of Cl, Mn, As, Se, Ca, K, Mg, Al, and Na. Samples were dissolved overnight in HNO_3 acid plus perchloric acid in teflon bombs and diluted to 50 mL. Zinc, Fe, Ni, Cu and Pb were analysed by DCP atomic emission and the results checked by flame atomic absorption. Cold vapor atomic absorption was used to measure Cd. NBS bovine and oyster tissue, employed as external quality control standards, were measured with an accuracy of greater than 90% recovery for all certified elements except for cadmium in bovine (84%) and iron in oyster (79%) tissue. Values were near the borderline of the certified range for both elements and cadmium values were also near the limits of detection.

3.1 Aquatic Insects

Aquatic insects were collected in emergence traps in all 3 years (May to August). Triangular emergence traps, covered with plastic screening and covering a bottom surface area of 1 m^2, were placed in pairs along the littoral zones of the wetlands. Insects were collected every 4-8 days, frozen until they could be sorted, identified to family, enumerated, oven dried and total dry weight recorded.

3.2 Water Chemistry Analyses

Water samples were analyzed from 1981-1983 for the following: pH, T.I.P. alkalinity, conductivity, color (true), dissolved organic carbon (DOC), and the ions NO_3/NO_2, NH_4, SO_4, Cl, Ca, Mg, Na, K, Si, Fe, Mn, Cd, Pb, Zn and Al (total). Analyses for total PO_4 were carried out in 1981; in 1983 inorganic monomeric Al and Al^{+3} (in addition to total Al) were also measured. Major ions, aqueous metals and selected physico-chemical parameters (pH, conductivity, alkalinity, color) were analyzed by the Ontario Ministry of the Environment (MOE, 1975).

Kingbird egg characteristics and nestling growth were monitored from 1981 to 1983 at the study sites. If an egg did not hatch, or a nestling died before all growth characteristics could be measured, the data for that individual were not used in statistical calculations. A nested analysis of variance was performed to partition the variance in the population into three categories - within a single nest (among siblings), among nests within a single lake and among lakes (Snedecor and Cochran, 1967). The nested analysis of variance was done to determine if environmental differences among lakes exerted a significant influence on kingbird reproductive factors. Canonical correlation provides a way to measure the extent of shared variation between two sets of variables (Green, 1979). Here the technique was used to test if an overall relationship existed between reproductive and environmental variables.

Principal components analysis was used to reduce the volume of data handled and direct the study towards influential variables. PCA was also used here to determine which subset of water quality variables explained most of the water chemistry variation.

4. RESULTS

Results of water chemistry analyses from the study are summarized in Table II. Wetland names, pH, total number of successful nests/wetland for 1981-83 and egg growth parameters for kingbird nestlings raised on those lakes are given in Table III.

Simple correlations were performed to investigate relationships between egg characteristics and growth rate in the kingbird with some

Table II. WATER CHEMISTRY OF LAKES USED BY KINGBIRDS IN THE KILLARNEY AREA IN 1982 & 1983

Water Chemistry[1] Variable	Mean	Standard Deviation	Range
Alkalinity	2.9	2.5	0 - 11.1
pH (pH units)	5.13	0.55	4.27 - 6.26
DOC	3.4	2.6	<0.2 - 11.0
Sulphate	9.9	2.2	4.7 - 15.9
Nitrate/Nitrite	0.034	0.052	0.002 - 0.232
Ammonium	0.036	0.031	0.002 - 0.165
Iron	0.19	0.22	0.01 - 1.25
Manganese	0.10	0.07	0.01 - 0.27
Total Aluminum	0.218	0.205	0.038 - 1.100
Cadmium	0.000	0.0004	0.0002 - 0.0010
Lead	0.003	0.002	0.001 - 0.010
Zinc	0.011	0.009	0.001 - 0.039
Copper	0.006	0.006	0.001 - 0.036
Silicon	0.555	0.432	0.010 - 1.740
Chloride	1.17	2.30	0.05 - 9.90
Calcium	2.19	0.59	0.68 - 3.30
Magnesium	0.73	0.20	0.24 - 1.11
Sodium	1.13	1.35	0.20 - 5.82
Potasium	0.29	0.13	0.03 - 0.66
Kjeldahl Nitrogen	0.337	0.164	0.093 - 0.661
Conduct. (umhos/cm)	37.5	8.2	22.0 - 65.1
Colour (Hazen units)	14.0	10.9	5.0 - 40.0
Total phosphorus (1981 data)	0.002	0.002	0.001 - 0.006

[1] units are mg L^{-1} unless otherwise expressed

indices of water chemistry. Alkalinity, pH, DOC and Fe were positively correlated with nestling tarso-metatarsus growth rate ($p<0.05$); there was a negative relationship of this growth factor with SO_4, Mn, and Zn ($p<0.05$; Figure 2). Lower pH values were related to increases in egg water loss (Figure 3).

Figure 4 shows the results of the analyses of variance for two egg characteristics (egg volume and egg water loss) and one growth factor (tarso-metatarsus growth) for kingbirds (Note - for 1981 to 1983: for siblings in one nest, n = 24 lakes, 57 nests and 173 nestlings; for nests on a given lake, n = 12 lakes, 25 nests and 67 birds; for variation among lakes, n= 26 lakes, 63 nests and 179 nestlings). The major percentage of variation in kingbird egg and nestling growth characteristics was explained by differences among siblings and genetic plus environmental differences among nests on a single lake which are not discussed in this study. The variance among lakes was minor but significant for egg water loss ($p<0.05$) and marginally so for egg volume and tarso-metatarsus growth ($0.05<p<0.10$).

Table III. EGG AND GROWTH PARAMETERS OF KINGBIRDS NESTING ON 26 LAKES IN THE KILLARNEY AREA 1981-1983 (AVERAGE OF YEARLY MEANS)

Lake	pH	Successful Nests	Egg Volume (cm^3)	Length-Width Ratio	Egg Water Loss	Growth Rate (K) Weight Increase	Tarso-Metatarsus	Wings
					($mg\ day^{-1}\ cm^{-2}$)			
Johnnie Pond	4.33	1	3.88	1.34	2.82	0.40	0.28	0.26
Proulx	4.38	2	3.93	1.32	3.06	0.52	0.27	0.19
Johnnie Beaver	4.43	2	3.67	1.35	3.70	0.43	0.36	0.20
Ruth Roy	4.54	3	3.71	1.35	4.15	0.46	0.30	0.24
Hidden Bay	4.56	2	4.04	1.39	3.88	0.12	0.28	0.20
Chikanshing	4.75	2	4.08	1.31	3.57	0.50	0.27	0.23
Killarney 3	4.79	2	3.69	1.31	3.65	0.39	0.30	0.21
Killarney 2	4.79	2	4.01	1.42	3.42	0.44	0.32	0.21
Eucott	4.80	2	3.64	1.38	3.77	0.42	0.31	0.17
Muriel Pond	4.81	1	3.79	1.28	3.41	0.52	0.32	0.18
Killarney 1	4.91	1	4.12	1.27	3.38	0.39	0.23	0.15
Stumpy	4.99	2	4.12	1.36	3.23	0.44	0.31	0.21
Freeland	5.07	2	4.03	1.35	3.58	0.44	0.33	0.24
Double N.	5.10	4	4.33	1.28	4.53	0.47	0.33	0.21
Muriel Marsh	5.15	6	3.72	1.31	3.40	0.39	0.27	0.20
Double S.	5.16	2	4.09	1.36	4.63	0.48	0.32	0.21
Bell Swamp	5.30	1	3.85	1.25	4.05	0.46	0.29	0.20
Campsite	5.32	1	3.90	1.30	3.43	0.47	0.33	0.16
Tyson	5.48	5	3.88	1.34	3.53	0.43	0.32	0.20
Spoon	5.57	8	3.87	1.27	2.73	0.45	0.31	0.21
Bell Narrows	5.65	1	3.92	1.34	3.40	0.39	0.29	0.23
Bell Rd.	5.66	1	4.12	1.39	3.22	0.52	0.35	0.23
Heron Nest	5.75	3	3.78	1.36	3.62	0.47	0.34	0.21
Artist	5.90	3	4.00	1.37	3.79	0.44	0.37	0.25
Balsam	5.94	2	4.09	1.29	3.93	0.44	0.35	0.18
Broker	5.94	2	4.07	1.27	2.63	0.52	0.31	0.26

Figure 2. Correlation between kingbird nestlings tarso-metatarsus growth rate and water quality parameters.

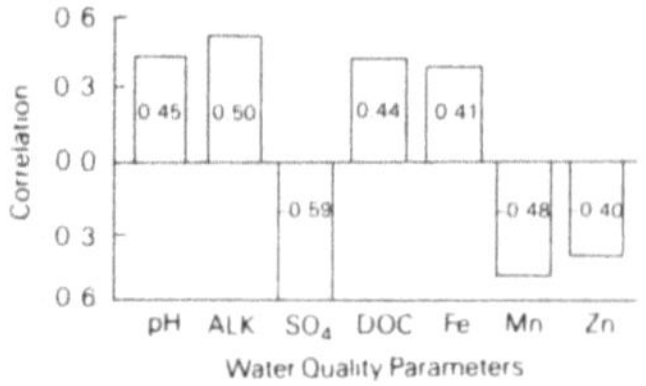

Figure 3. Correlations of kingbird reproductive parameters with water quality and insect metal concentrations.

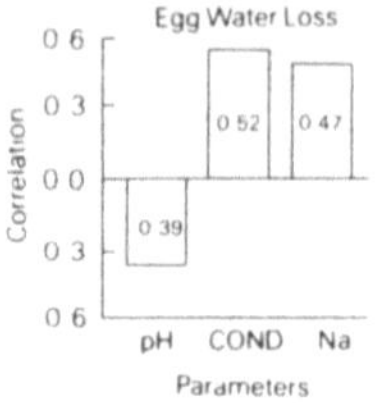

Figure 4. Analyses of variance components (%) for kingbird egg and nestling growth parameters.

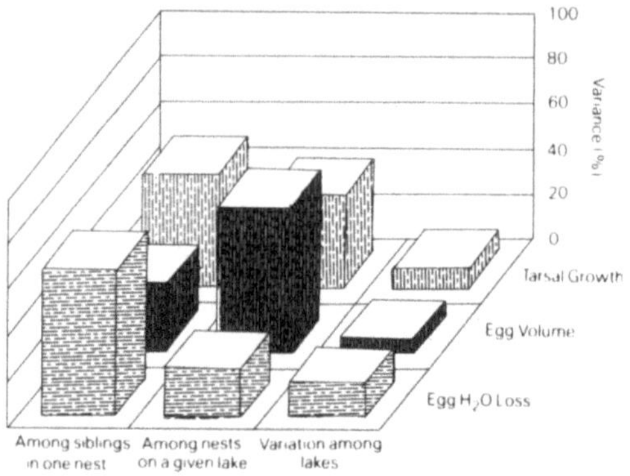

Environmental effects associated with the variation among lakes appeared therefore to exert a consistent but small influence on the reproduction and growth of the kingbird.

To investigate the effect of water chemistry differences on egg characteristics and nestling growth, canonical correlation was carried out on the average of yearly means for 1981, 1982 and 1983 (Table IV). The water chemistry factor explained 43% of the total nestling growth rate and the growth rate of wings ($p<0.004$). Variables related to acidity loaded onto the independent canonical variate. These were sulfate, nitrate/nitrite, pH, alkalinity and metals. Nestling growth rate (g day^{-1}) and growth rate of wings (mm day^{-1}) decreased as sulfates, nitrate/nitrite and some metals increased (Pb, Al, Zn, Mn) and as pH and alkalinity declined.

Three factors were identified by principal components analysis of water chemistry data from wetlands in the Killarney area. Factor 1 explained 35% of the variance (Table V). Strong positive factors include water chemistry variables related to lake acidity (pH, alkalinity) as well as those related to weathering (Mg, Na). Strong negative factors include SO_4, NO_3 and NO_2 which are related to lake acidity as well as the metals Al, Mn, Cd and Zn. All of the water chemistry variables which loaded strongly onto Factor 1 also showed variation with kingbird growth characteristics (Table IV). Factors 2 and 3 accounted for 16.4% and 12.1% of the variation observed and were associated with weathering and major ions.

Table IV. CANONICAL CORRELATION OF KINGBIRD GROWTH CHARACTERISTICS WITH WATER CHEMISTRY VARIABLES IN THE KILLARNEY AREA (Data used are the averages of yearly means - 1981,82,83)

Kingbird Growth Characteristic		Water Chemistry Variables	
Nestling Growth Rate ($g\ day^{-1}$)	-0.098	Nitrate/Nitrite	0.075
		SO_4	0.013
		Alkalinity	-0.496
Growth of Wings ($mm\ day^{-1}$)	-0.999	pH	-0.083
		DOC	0.205
		Fe	-0.121
		Mn	0.019
		Total Al	0.223
		True Color	0.310
		Cd	-0.894
		Pb	0.773
		Zn	0.074

Table V. FACTOR 1 - PRINCIPAL COMPONENTS ANALYSIS OF WATER CHEMISTRY OF KILLARNEY WETLANDS

Positive		Negative	
Alkalinity	* 0.875	SO4	* -0.800
DOC	* 0.779	Nitrite/ Nitrate	* -0.760
pH	* 0.762		
Colour	* 0.748	Al (total)	* -0.737
Fe	* 0.705	Mn	* -0.621
Na	0.574	Cd	* -0.615
Mg	0.553	Zn	-0.557

* $p<0.05$

Table VI. ELEMENTAL LEVELS OF EGGSHELLS FROM KILLARNEY AND LONG POINT/OPINICON AREAS ($mg\ L^{-1}$)

	No. of Eggshells	Na	Mg	P	S	K	Ca	Al
Killarney	9	0.735	0.762	2.657	2.867	1.001	37.407	0.42
Long Point/ Opinicon	9	0.038	0.234	0.998	1.360	0.248	46.734	0.33

Table VII. COMPARISON OF EGGSHELL THICKNESS DIFFERENCES AMONG PARTS OF EGGSHELL ($p<0.05$), BETWEEN GEOGRAPHIC AREAS ($p<0.001$)

Part of Eggshell	Killarney (n=9)	Buffered sites (n=9)	Thickness Ratio Killarney sites:Buffered sites
Tip	0.019	0.050	0.39
Mid	0.036	0.049	0.73
Bottom	0.021	0.050	0.42

Nestling crop analyses have shown that Odonates are the principal food of kingbird nestlings in the Killarney wetlands (Blancher and Glooschenko, 1986). Odonates comprised a major precentage of emergent insects captured in the emergence traps and Figure 5 shows that there was no prey limitation for the kingbird attributed to lake acidity.

Figure 5. Insect weight from emergence traps vs mean annual wetland pH (1981-3)

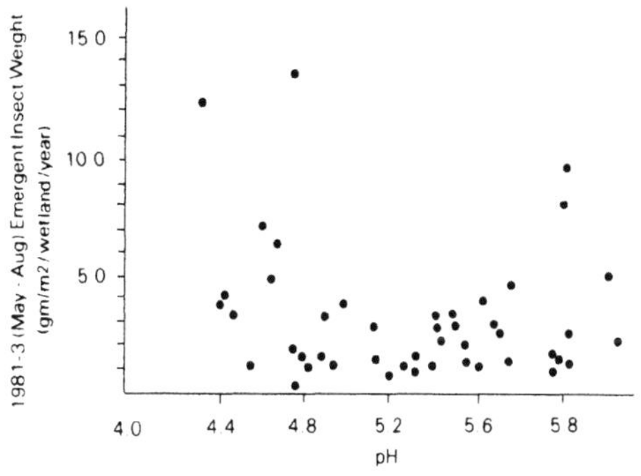

Differences in eggshell composition and thinning between the Killarney area and the two regions of higher alkalinity remote from Sudbury are shown in Tables VI and VII.

Levels of Hg found in Killarney nestlings were higher than those found in nestlings from buffered regions removed from the Sudbury influence (Table VIII).

5. DISCUSSION

This study shows a strong association of egg characteristics and growth rate of kingbirds with water chemistry. However, there was no proof of causality or evidence of serious reproductive impairment such as was reported for passerine birds in Sweden and linked to possible Al interference with Ca and P metabolism in the female bird (Nyholm, 1981). Despite the significance of the canonical correlation, water chemistry accounts for only a minor percentage of the total variance. More important sources of influence are related to natural variation (differences among siblings and among nests on a given lake).

Aquatic insects have been shown to be susceptible to the effects of low pH (Bell, 1971), and acidic watercourses are known to support diminished macroinvertebrate fauna in comparison with physiographically similar habitat which is less acidic (Sutcliffe and Carrick, 1973; Townsend et al., 1983). Several studies have shown relationships between food supply of avian species and their breeding densities (Enokson and Nilsson, 1983; Newton et al., 1977). Ormerod (1985) and Ormerod et al., (1985) have discussed the importance of ephemeropteran nymphs and trichopteran larvae to the diet of adult and nestling Dippers and the correlation of breeding Dippers with insect prey in higher pH streams. Schreiber and Vivellella (1985) have suggested that the food supply of other insectivorous birds, including kingbirds, may be limiting since the emergent insect prey of these birds is most sensitive to high acid levels at this life stage. However, the present study shows no limitation of insect prey of the kingbird attributable to wetland acidity.

Table VIII

ANALYSES OF VARIANCE FOR Hg ($\mu g\ g^{-1}$) IN NESTLINGS FROM THE KILLARNEY AND BUFFERED SITES

	Mean Killarney Level (n=10)	Standard Deviation	Mean Buffered Site Level (N=9)	Standard Deviation	Probability	Ratio of mean Killarney level/mean level at buffered sites
liver	0.516	0.474	0.174	0.044	0.05	3.0
feathers	0.950	0.267	0.490	0.198	0.001	1.9

Killarney eggshells were thinner than those from wetlands of higher alkalinity remote from Sudbury, but no effect on eggshell composition or characteristics was attributed to differences in acidity among lakes. Instead, Ca availability in the watershed may exert a major influence on eggshell composition. (Calcium concentration in the water from all Killarney wetlands was significantly lower ($p<.005$) than that from the Long Point and Lake Opinicon sites). However, as seen in Table IV, Ca was not strongly associated with kingbird growth characteristics. Total PO levels for the area wetlands noted from 1981 were low in comparison with nutrient input to other Sudbury basin lakes and wetlands (Glooschenko *et al.*, 1985). However the comparative data on eggshell composition indicates little or no P deficiency in eggshells from the Killarney area. Our study has shown that P content in the Killarney eggshells was twice as high as in eggshells from more buffered areas.

Effects of Al toxicity on avian growth have been seen in controlled studies when dietary P levels are lower than dietary Al levels of 2000 to 25000 ppm (Deobald and Elvehjem, 1935; Storer and Nelson, 1968; Steinborn *et al.*, 1957). Recent work on reproduction and growth of the ring dove has shown no toxic effects of total Al, at levels similar to those likely to occur in the environment (500 to 1500 ppm) when P intake was greater than Al (Carriere *et al.*, 1986). Levels of total Al in kingbird insect prey from the Killarney wetlands utilized in the present study were also in this range, varying from 17 to 1667 ppm (Glooschenko *et al.*, 1986). Although dietary P levels may have been lower than those of Al in insect prey from the Killarney area, P intake was not monitored. Consideration of the eggshell data suggests P would not appear to be limiting.

In comparison with other work, the present study shows Hg levels in the liver of kingbird nestlings to be similar to those found in pintail ducks (*Anas acuta*) from northern Ontario (Fimreite, 1971). Kingbird nestling Hg levels for liver and feathers were lower than those seen in mallard ducks (*Anas platyrhynchos*) fed a diet of 0.5 ppm methylmercury (1.62 ± 0.160, liver) which decreased reproductive success, but higher than those of control mallards (<0.1 ppm) (Heinz, 1979). However, Hg levels in kingbird nestlings from wetlands of higher alkalinity remote from Sudbury approached those of the control mallards. Future biomonitoring programs should include assessment of Hg body burdens in birds nesting near acidified areas.

ACKNOWLEDGEMENTS

The authors wish to extend thanks to those who have assisted with the completion of this study. Shelia Mackie, Barb Dowsley and Linda Graf gave excellent field assistance, and Mr. Pud Hunter of the Ministry of Natural Resources District Office in Aylmer, Ontario provided avian specimens. Mr. Cameron Ackerly, University of Guelph, analyzed eggshells with scanning electron microscopy. X-Ray Assay Laboratories, Toronto, performed tissue analysis of nestlings. Water quality analyses were performed by the Ontario Ministry of the Environment under Dr. Frank Tomassini. Mr. Clive Eastwood and Mr. Chris Wedeles performed statistical analyses and assisted in manuscript preparation. Dr. J. Rodrigues, University of Guelph, provided computer graphics and review. Useful suggestions were received from Dr. Thomas Nudds and Dr. Gerry Mackie, Zoology Department, University of Guelph, Dr. David Hussell and Dr. J. Maclean, Ontario Ministry of Natural Resources, and Dr. Bruce LaZerte and Dr. Peter Dillon, Ontario Ministry of the Environment. Typing assistance from Leslie Coo, Wanda Labanowicz and Sue Scott was much appreciated.

Jay Leather, Killarney Provincial Park, provided assistance and accomodation within the study area.

REFERENCES

Beamish, R.J. and Harvey, H.H.: 1972, J. Fish. Res. Bd. Can. **29**, 1131.

Bell, H. L.: 1971, Water Res. **5**, 313.

Bendell, B.E., McNicol, D.K. and Ross, R.K.: 1983, Effects of acidic precipitation on waterfowl populations in Northern Ontario Vol. II. LRTAP Program. Can. Wildl. Serv. Ont. Reg.

Blancher, P. and Glooschenko, V.: 1986, Association of pH with reproduction of insectivorous passerine birds. (in prep.).

Carriere, D, Peakall, D.B., Fischer, K.L. and Angehrn, P.: 1986, Can. J. of Zool. (in press).

Confer, J.L., Kaaret, J. and Likens, G.E.: 1983, Can. J. Fish. Aquat. Sci. **40**, 36.

Custer, T.W. and Mulhern, B.L.: 1983, Bull. Environm. Contam. Toxicol. **30**, 178.

Deobald, H.J. and Elvehjem, C.A.: 1935, Am. J. Physiol. **111**, 118.

Enokson, B. and Nilsson, S.G., 1983, J. Anim. Ecol. **52**, 927.

Fimreite, N.: 1971, Occasional Paper No. 9 Can. Wildl. Serv. 39 pp.

Fimreite, N.: 1974, J. Wildl. Manage. **38**(1), 120.

Glooschenko, V. and Stevens, W.: 1986, Sci. of the Tot. Environ. (in press).

Glooschenko, V., Blancher, P., Herskowirz, J., Fulthorpe, R. and Rang, S.: 1986, in prep.

Glooschenko, V., Weller, W., and Eastwood, C.: 1985, Technical Rep., Wildl. Br., Ont. Min. Nat. Res. Toronto, Ont. 30 pp.

Green, R.H.: 1979, John Wiley and Sons. 257 pp.

Harvey, H.H.: 1980, p. 83-98, In Drablos, D. and Tollan A. (eds), Proc. Int. Conf. Ecological Impact of Acid Precipitation. Sandefjord, Norway.

Haseltine, S.D., Fair, J.S., Sutcliffe, S.A. and Swineford, D.M.: 1983, p. 131-141, In Yahner, R. H. (ed.), Transactions of the Northeast Section, The Wildl. Soc., 40th Northeast Fish and Wildlife Conf., May 15-18, 1983.

Haseltine, S.D.: 1985, U.S.F.W.S., pers. comm.

Heinz, G.: 1976, J. Wildl. Manage. **40**(4), 82.

Heinz, G.: 1979, J. Wildl. Manage. **43**(2), 394.
Hoyt, D.F.: 1979, Auk **96**, 73.
Keller, W., Gunn, J.M. and Conroy, N.I.: 1980, In Drablos, D. and A. Tollan (eds) op cit.
McNicol, D.K. and Bendell, B.E.: 1985, Canadian Wildlife Service, pers. comm.
Newton, I., Marquiss, M., Wier, D.N. and Moss, D.: 1977, J. Anim. Ecol. **46**, 425.
Nilsson, S.C. and Nilsson, I.N.: 1978, Oikos **31**, 214.
Nyholm, N.E.I. and Myhrberg, H.E.: 1977, Oikos **29**, 336.
Nyholm, N.E.I.: 1981, Environ. Res. **26**, 363.
Ontario Ministry of the Environment: 1975, Handbook of Analytical Methods for Environmental Samples, 500 pp.
Ormerod, S.J.: 1985, Ibis **127**, 316.
Ormerod, S.J., Tyler, S.J. and Lewis, J.M.S.: 1985, Bird Study **32**, 33.
Paganelli, C.V., Olszowka, A. and Ar, A.: 1974, Condor **76**, 319.
Ricklefs, R.E.: 1967, Ecology **48**, 978.
Schreiber, R.K. and Vivillella, R.: 1985, p. 267-284. Proc. Air Pollut. Effects on Forest Ecosystems Sympos., St. Paul, MN. May 8-9, 1985.
Snedecor, G.W. and Cochran, W.G.: 1967, Statistical Methods, 6th Ed., Ames: Iowa State University Press. 593 pp.
Steinborn, K., Robbard, S. and Williams, C.: 1957, J. Appl. Physiol. **11**, 435.
Stoner, J.H., Gee, A.S. and Wade, K.R.: 1984, Environ. Pollut. **35**, 125.
Storer, N.L. and Nelson, T.S.: 1968, Poult. Sci. **47**, 244.
Stromberg, K.L. and Longcore, J.: 1983, Proc. Czechoslovak-American Sympos. on Toxic Effects of Chem. Environ. Contaminants upon Production and Reproduction Ability in Free-living Animals, Stroski Pleso. (in press).
Sutcliffe, D.W. and Carrick, T.R.: 1973, Freshwat. Biol. **3**, 437.
Townsend, C.R., Hildrew, A.G. and Francis, J.: 1983, Freshwater Biol. **13**, 521.
Wiederholm, T. and Eriksson.: 1977, Oikos **29**, 261.

THE EFFECTS OF LOW pH, LOW CALCIUM CONCENTRATIONS AND ELEVATED ALUMINIUM CONCENTRATIONS ON SODIUM FLUXES IN BROWN TROUT, *SALMO TRUTTA* L.

T. R. K. Dalziel, R. Morris, and D. J. A. Brown,
Department of Zoology,
University of Nottingham,
Nottingham NG7 2RD, England

Central Electricity Research Laboratories,
Ratcliffe-on-Soar,
Nottingham NG11 OEE, England

ABSTRACT. The effects of pH (c. 7.0, 5.4, 4.5 and 4.0), nominal Al levels (0 and 8 μmol L^{-1}) and Ca levels (10 and 50 μmol L^{-1}) on Na influx, efflux and netflux of brown trout have been investigated using artificial lake water of known composition. Low pH had little effect on influx, but tended to increase efflux, particularly in the low Ca treatments. A nominal addition of 8 μmol Al L^{-1} at pH 4.5 and 4.0 reduced influx significantly. Efflux was unaffected. Aluminium addition at pH c. 7.0 and 5.4 had no such effect. The measured Al concentrations at the end of the static 8 hr flux measuring experiments were markedly lower than the nominal amount of Al added to the start.

1. INTRODUCTION

Salmonid fishkills in Scandinavia (Leivestad and Muniz, 1976; Skogheim *et al*., 1983) have implicated the toxicity of Al in waters of low pH and low Ca concentration. Distressed and moribund fish found at fish-kill sites were found to have lower plasma Na and chloride levels than unaffected fish from the same geographical area. Muniz and Leivestad (1980), too, found significantly lower plasma chloride concentrations in fish exposed to elevated levels of Al at low pH. In addition they showed plasma salt depletion to be ameliorated when the external Ca concentration was increased.

It has been argued that the toxic mode of action of Al at low pH may be to impair ion exchange and so lead to lower blood salt levels (Muniz and Leivestad, 1980; Hesthagen and Skogheim, 1983). The purpose of this study was to measure the uptake (influx) and loss (efflux) of Na in brown trout exposed to acid media containing low concentrations of Ca with and without Al present.

2. MATERIALS AND METHODS

The experiments were conducted between October 1983 and March 1984 at the University of Nottingham. Brown trout (1 to 2 years old, c. 35 g)

Water, Air, and Soil Pollution **30** (1986) 569–577.

prior to experimentation were kept outside in a fibreglass bath (150 x 45 x 62 cm) with a constant supply of Nottingham tap water (pH 7.4 to 8.2; 0.78 mmol Ca L^{-1}; 0.67 mmol Na L^{-1}).

Two artificial freshwater media (A and B) simulating typical ionic concentrations where fishkills have occurred and differing only in Ca concentration were made up by adding the appropriate quantities of salts to deionized water (Table 1). Aluminium was added, where appropriate, as chloride. Five fish were tested at each of 16 combinations of pH (c. 7.0, 5.4, 4.5 and 4.0), Al concentration (0 and 8 μmol l^{-1}) and Ca concentration (10 and 50 μmol l^{-1}). Six fish were tested at any one time; the combinations used were selected at random and duplication avoided.

TABLE I. Concentration of ions in softwater media

ION	CONCENTRATION mmol L^{-1}		mg L^{-1}	
	MEDIUM A	MEDIUM B	MEDIUM A	MEDIUM B
Ca^{2+}	0.050	0.010	2.004	0.400
Na^{+}	0.050		1.150	
K^{+}	0.005		0.200	
Mg^{2+}	0.040		0.972	
Cl^{-}	0.100	0.060	3.545	2.127
SO_4^{2-}	0.065	0.045	6.244	4.323
NO_3^{-}	0.005		0.310	

Conductivity of Medium A = 26 μS cm^{-1}
Conductivity of Medium B = 17 μS cm^{-1}
(Ionic concentrations obtained by adding appropriate salts to deionized water).

Fish were not fed for at least 1 week prior to experimentation. At the start of an experiment the fish were individually placed in darkened plastic buckets fitted with lids and containing approximately 3 L of whichever Ca concentration medium they were to be tested in and aerated at 10°C ± 1°C. Twenty four hours later each fish was transferred to a darkened plastic flux chamber (8.0 x 8.5 x 25 cm) containing approximately 1 L of the correct medium and aerated at 10°C ± 1°C. Twenty four hours later the medium was siphoned from each chamber and replaced with 700 mL of fresh medium of appropriate pH and Al concentration. The volume was as low as possible to maximise the accuracy of both isotopic and ionic measurements. Two hours later 6 μCi of ^{22}Na was added to each chamber and after a further half hour (to allow distribution of isotope) the first samples were taken. Further sampling was carried out at two hourly intervals over the next 8 hr, giving a total of 5 sample times. At each sample time, 3 x 1 mL samples were taken by pipette from each chamber for eventual isotopic and ionic analysis.

Sulphuric acid (either 0.03 or 0.05 mol L^{-1} analytical grade) was added manually throughout the experiments to those chambers required to be maintained at low pH. pH was measured using a Radiometer PHM 82 meter with a Russell CWL-LCW pH electrode (specially for use in low conductivity waters) and was maintained to within 0.2 pH unit. pH control was not necessary in those chambers at circumneutral pH as the ammonia excreted by the fish in the relatively small volume of medium was sufficient to raise the pH to 6 or 7 over the experimental period.

At the conclusion of each experiment 49 mL of medium was removed from those chambers containing Al for total Al concentration analysis. The fish were individually removed, lightly dabbed dry with absorbent paper and weighed on a top-pan balance.

Samples were counted for gamma radiation using a Packard 500C gamma counter and the Na concentration measured by flame emission spectrophotometry (Pye Unicam SP 90). The drop in activity between successive samples was used to calculate influx of Na into the fish, and the change in Na concentration between successive samples the netflux of Na either into or out of the fish. Efflux values were calculated as the difference between influx and netflux.

Total Al concentrations were measured by the catechol-violet colorimetric method (Dougan and Wilson, 1974).

3. RESULTS

3.1. Aluminium Concentrations

Al concentration measurements made at the conclusion of each experiment are shown in Table II. In every case the original starting concentration was greatly reduced at the end of an experiment. The amounts of Al remaining in solution at pH c.7.0, 5.4 and 4.5 were similar but at pH 4.0 greater amounts of total dissolved Al were present at the end of the experiments. The standard errors were large due to considerable variation between results from different fish.

TABLE II. Total Al concentrations measured at the conclusion of experiments with a nominal starting concentration of 8 μmol L^{-1}.

pH	Ca Concentration/ μmol L^{-1}	Total Al concentration ($\pm$ S.E.) at end of experiment/μmol L^{-1}
c.7	50	1.63 $\pm$ 0.27
	10	2.84 $\pm$ 0.65
5.4	50	1.42 $\pm$ 0.11
	10	1.79 $\pm$ 0.58
4.5	50	1.31 $\pm$ 0.26
	10	1.19 $\pm$ 0.34
4.0	50	3.10 $\pm$ 0.39
	10	3.68 $\pm$ 0.74

3.2. Sodium Fluxes

Figures 1 and 2 show Na influx and efflux at each of the four pHs tested, with no Al present and with a nominal starting concentration of 8 μmol L^{-1}, respectively. (Note that within each time period (0 to 2 hr etc.) the flux results obtained in both Medium A and Medium B are shown.) In most cases a characteristic pattern of efflux rate becoming less with time is evident but influx rates appear to remain constant throughout the time periods. In the absence of Al (Figure 1) only at pH 4.0, especially at the lower Ca concentration, is there a greater efflux rate compared with the other pHs. At pH c.7.0, 5.4 and 4.5 influx rates appear to virtually mirror those of efflux such that a Na balance is maintained. At pH 4.0, however, this is not the case. Influx is insufficient to balance efflux and a net Na loss results. The higher Ca concentration tends to reduce Na efflux at pH 4.5 and 4.0. With a nominal concentration of 8 μmol Al L^{-1} present (Figure 2) efflux rates at each pH are of the same order as those which have no Al (Figure 1). At pH c.7.0 and 5.4 influx balances efflux but at pH 4.5 and 4.0 there is significant influx inhibition. Compared with the zero Al treatments the tendency for higher Ca concentration to reduce sodium efflux at low pH is less apparent.

4. DISCUSSION

4.1. Experimental

The fish were not cannulated in these experiments as this would have undoubtedly increased the risk of handling and experimental stress which can cause disturbances in ionic regulation primarily through increased urine output and consequently in ion loss through diuresis (Eddy, 1981). The most stressful time for the fish was likely to be the siphoning out of old medium before replacing with fresh and this is probably the reason for the higher efflux rates during the initial period.

4.2. Aluminium Loss from the Medium

The fact that so much Al was lost from solution (Table II) meant that it was impossible to state what concentration of Al inhibits Na influx at low pH. In these experiments only total Al was measured but classic Al solubility phase diagrams (cf. Burrows, 1977; Spry *et al*., 1981; Potts *et al*., 1985) show that at about pH 5.5 most Al is present in the form of insoluble polymeric $Al(OH)_3$ and that as the pH is reduced then the solubility increases, with the appearance of monomeric cationic forms, e.g. Al^{3+}, $Al(OH)^{2+}$, $Al(OH)_2{}^{+}$. Al has a strong tendency to bind to organic material (Burrows, 1977) and it is possible that some was removed from solution by becoming bound to the external surfaces of the fish and possibly also to the sides of the apparatus. Less Al was lost at pH 4.0 suggesting that the monomeric cationic forms adsorb less strongly than the more insoluble polymeric forms which predominate at higher pHs. Experiments to detect Al in the plasma have proved

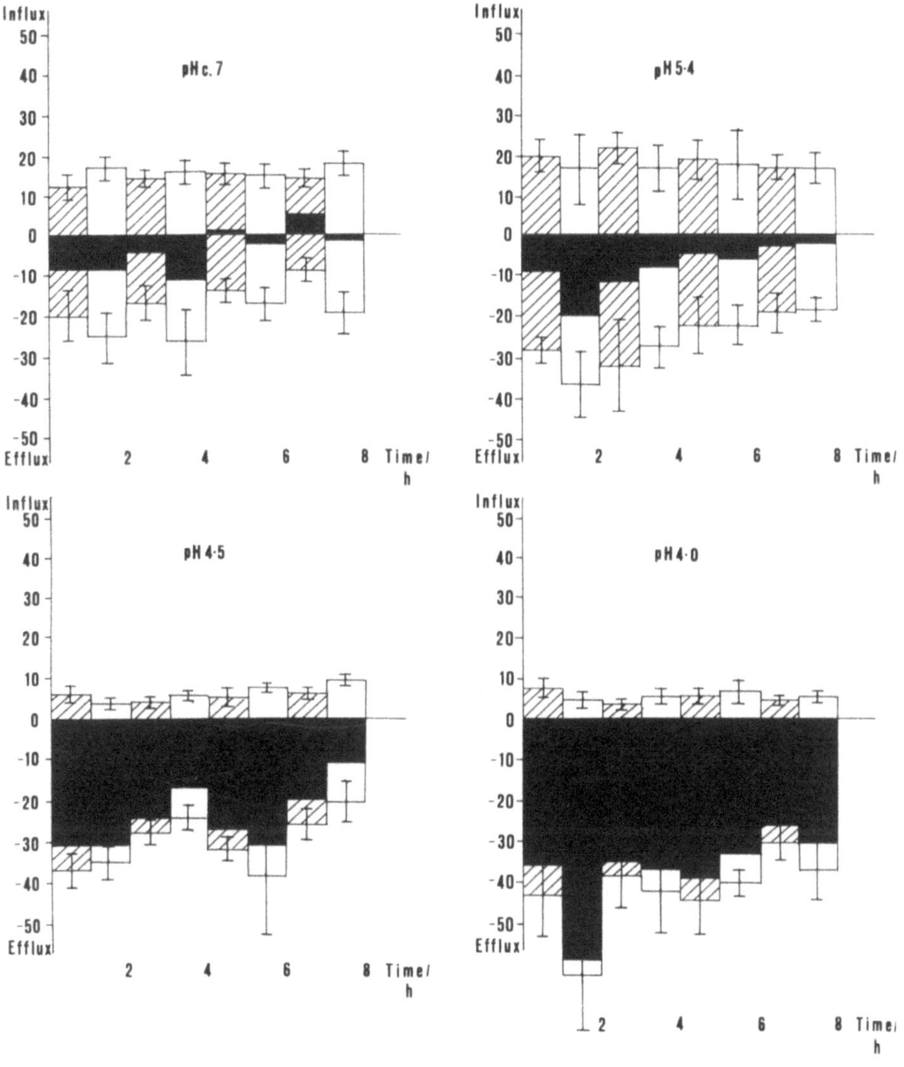

Fig.2 Sodium fluxes at different pHs with a nominal starting Al concentration of 8 µmol L^{-1} (S.E. shown)

(Ca) = 0·05 µmol L^{-1}

(Ca) = 0·01 µmol L^{-1}

shaded area indicates netflux
units: µmol Na^{+} $100g^{-1}$ hr^{-1}

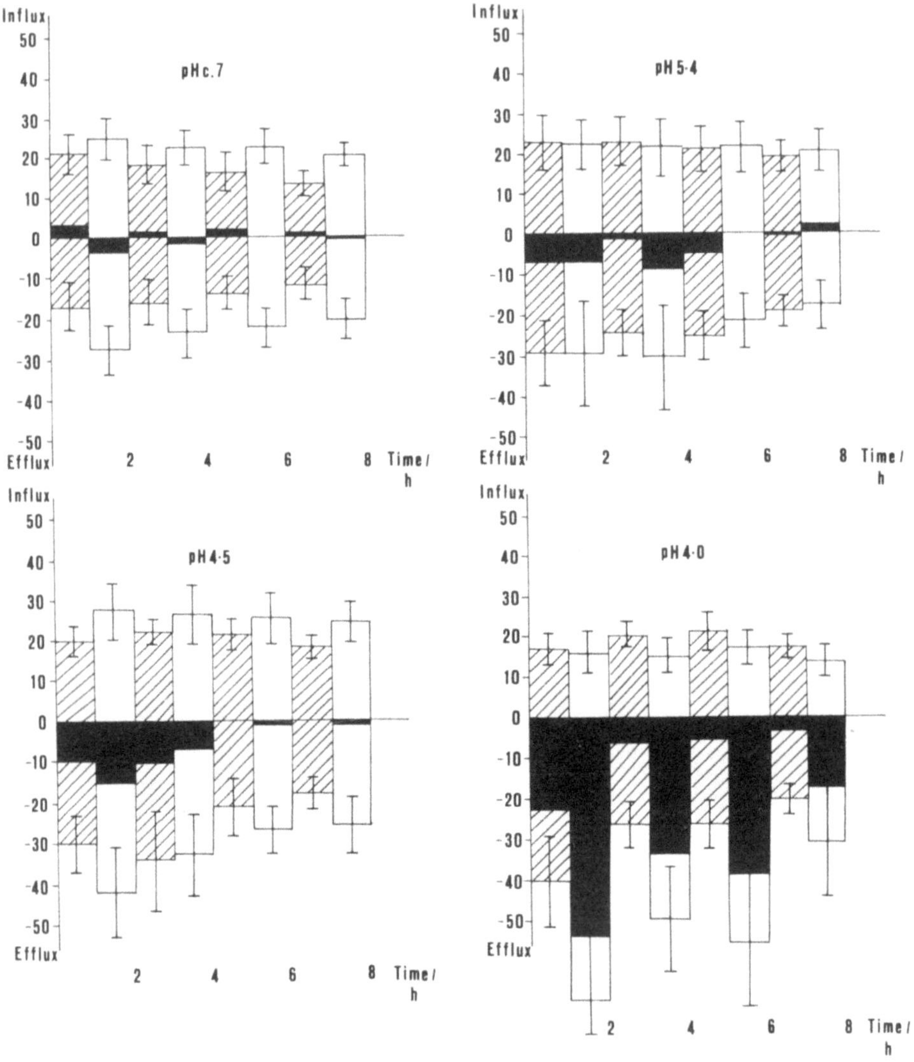

Fig.1 Sodium fluxes at different pHs with no aluminium added (S.E. shown)

(Ca) = 0·05 μmol L^{-1}

(Ca) = 0·01 μmol L^{-1}

shaded area indicates netflux
units: μmol Na^{+} 100g^{-1} hr^{-1}

negative, suggesting that Al does not actually enter the fish.

4.3. Sodium Fluxes

Acid exposure has three main effects at the gills according to Wood and McDonald (1981).

(i) passive H^+ entry along a concentration gradient;
(ii) inhibition of active Na^+/H^+ or NH_4^+ exchange;
(iii) increased passive permeability of branchial epithelium to Na^+, K^+ and Cl^- and perhaps other ions which diffuse out at increased rates along their concentration gradients.

An increase in external Ca concentration ameliorates the effect of acid by reducing the permeability of the branchial membrane, thus lessening the passive loss of ions down their respective electrochemical gradients (Cuthbert and Maetz, 1972; Eddy, 1975; McWilliams and Potts, 1978; McDonald et al., 1980). Ca also ameliorates Al toxicity (Muniz and Leivestad, 1980; Brown, 1983) and the inference has been made that the effect of Al on the gill is similar to that of H^+ stress (Muniz and Leivestad, 1980; Brown, 1983). In the present study the effect of pH per se on flux rates is small, with only the most extreme acid conditions (pH 4.0) producing an appreciable increase in Na and K loss. Even at this extreme pH, which is infrequent in natural conditions, there appears to be only slight non-significant ($p > 0.05$) evidence of hydrogen ions inhibiting Na uptake. Previous studies of the effect of low pH on Na fluxes (McWilliams and Potts, 1978; Goldthorp, 1982) have shown influx rates to be depressed and efflux rates increased as a result of exposure to acid conditions. The reasons why this was not found in this study are unclear but may be explained by the greater care taken over avoiding diuretic responses. There was a non-significant trend ($p > 0.05$) of Ca reducing the rate of passive loss of Na in most cases.

The presence of Al at low pH had no effect on Na efflux but significantly inhibited sodium influx when a nominal concentration of 8 µmol Al L^{-1} was presented at pH 4.5 and 4.0 although the actual concentration of total Al under these conditions was considerably less than the nominal (see Table II). It would seem, therefore, that the action of H^+ at the gill is different from that of Al and that because sodium influx inhibition was only apparent at pH 4.5 and 4.0, judging from the phase diagrams mentioned previously, it is among the monomeric cationic species that the inhibitory form of Al is to be found.

This study suggests that Ca does not ameliorate sodium influx inhibition by aluminium. As in previous studies there is evidence that Ca does reduce efflux at low pH and hence helps ion balance in this way. Further investigations using higher Ca concentrations may elucidate any Al/Ca interaction.

5. CONCLUSIONS

From the results of this study the need is identified to continue investigations into the effects of a range of concentrations of Al at low pH on Na balance in brown trout to establish more clearly (a) the actual concentration and species of Al which inhibits Na influx at low pH, (b) the type of mechanism involved in the toxic Al effect, (c) the kinetic relationship between the toxic Al concentration and the Na concentration of the medium, and (d) whether Ca levels can directly alter the inhibition of Na influx caused by Al at low pH.

REFERENCES

Brown, D.J.A.: 1983, Bull. Environ. Contam. Toxicol. 30, 582.

Burrows, W.D.: 1977, Critical Reviews in Environmental Control 7, 167.

Cuthbert, A.W. and Maetz, J.: 1972, J. Physiol. 221, 633.

Dougan, W.K. and Wilson, A.C.: 1974, Analyst 99, 413.

Eddy, F.B.: 1975, J. Comp. Physiol. 96, 131.

Eddy, F.B.: 1981, 'Effects of stress on osmotic and ionic regulation in fish'. In: Pickering, A.D. (Ed.), Stress and Fish. Academic Press, London and New York. pp 77-102.

Goldthorp, S.: 1982, 'The effect of acid water on osmoregulation in trout'. Ph.D. Thesis, University of Nottingham, England.

Hesthagen, T. and Skogheim, O.K.: 1983, 'High mortality of presmolt of Atlantic salmon, *Salmo salar* L. and sea trout, *Salmo trutta* L. during spring snowmelt 1982 in River Vikedalselva, Western Norway'. Manuscript.

Leivestad, H. and Muniz, I.P.: 1976, Nature 259, 391.

Muniz, I.P. and Leivestad, H.: 1980, 'Toxic effects of aluminium on the Brown trout (*Salmo trutta* L.).' In: Drablos, D. and Tollan, A. (Eds.), Proc. Int. Conf. Ecol. Impact Acid Precip., Norway. SNSF Project, pp 84-92.

McDonald, D.G., Hobe, H. and Wood, C.M.: 1980, J. Exp. Biol. 88, 109.

McWilliams, P.G. and Potts, W.T.W.: 1978, J. Comp. Physiol. 126, 277.

Potts, W.T.W., Hunt, D.T.E., Blake, S. and French, P.: 1985, 'Aluminium in acid waters - Chemistry and effects on fish.' Report prepared for Surface Water Acidification Programme.

Skogheim, O.K., Rosseland, B.O. and Sevaldrud, I.: 1983, Rep. Inst.

Freshw. Res., Drottningholm 61, 195-202.

Spry, D.J., Wood, C.M. and Hodson, P.V.: 1981, 'The effects of environmental acid on freshwater fish with particular reference to the soft-water lakes of Ontario and the modifying effects of heavy metals. A review.' Can. Tech. Rep. Fish. Aquat. Sci. 999, 145pp.

Wood, C.M. and McDonald, D.G.: 1981, 'Physiological mechanisms of acid toxicity to fish.' In: Acid Rain/Fisheries, Proceedings of an International Symposium on Acidic Precipitation and Fishery Impacts in North Eastern North America. Cornell University, Ithaca, New York.

On the prediction of acid precipitation events and their effects on fishes

Harold H. Harvey
Department of Zoology, University of Toronto
Toronto, Ontario M5S 1A1, Canada

Douglas M. Whelpdale
Atmospheric Environment Service
Environment Canada
Downsview, Ontario M3H 5T4, Canada

ABSTRACT. It is possible to predict acid rain events and melts of acid snow some 12 to 24 hr in advance, including estimation of the magnitude and duration of such events. This is sufficient notice to permit monitoring of stream chemistry and fish plasma and muscle ions before acid stress, and to continue this monitoring throughout and after specific events. Such a program has been in place for 2 yr in waters tributary to the Milford Bay Trout Hatchery, Ontario. During one snow melt in February 1984 surface waters showed a decline to pH 4 and associated negative ANC. Rainbow trout held in such water lost plasma Na and Cl rapidly and died within 28 hr. The hatchery water supply, consisting of a mixture of spring and surface water, showed a decline in alkalinity from 300 to 30 $\mu eq.L^{-1}$, and a pH change from 6.6 to 5.4, during snow melt. Total Al concentration increased from 42 to 222 $\mu g.L^{-1}$ during snow melt, with the "reactive" component increasing from 17 to 112 $\mu g.L^{-1}$. Rainbow trout held in this water did not show physiological stress. More rapid run-off of melt water could be expected to exhaust all of the alkalinity in the hatchery water supply permitting the pH to decline and Al concentration to rise to levels lethal to the hatchery stock of rainbow trout.

1. INTRODUCTION

The concern with the acidification of surface waters is based largely on the observed trends to lower alkalinity and/or pH, and with the loss of biota, especially fish populations, in regions of moderate to heavy acid loading. Such trends typically are based on annual weighted averages, for example, of pH of precipitation, lakes and streams. The biota, however, are more affected by episodic events, such as sudden pH depressions, than by long-term trends.

Water, Air, and Soil Pollution **30** (1986) 579–586.

In 1979-80 we purchased rainbow trout (Salmo gairdneri) from a privately owned hatchery at Milford Bay, Ontario, and in this context, analyzed the water supply of the hatchery. The alkalinity of the water, 300 $\mu eq.L^{-1}$, was high for a Precambrian Shield stream, the pH was 6.8, but total Al concentration was high. The possibility existed that this could represent a toxic concentration of Al at lower pH, or in combination with H^+, could be toxic at lower pH. Acid-related fish kills at hatcheries have been reported from Hallefors, Sweden (Dickson, 1978); Matre, Norway (Muniz, 1971) and Scotland (The Galloway News, October 28, 1982). Episodic fish kills in natural waters during pH depressions have been reported by Leivestad and Muniz (1976), Dickson (1978), and Harvey et al. (1982).

The stress of H^+ and/or Al toxicity on fishes results in the rapid loss of plasma Na^+ and Cl^-, and the loss of muscle K^+ and Cl^- (Packer and Dunson, 1970; Leivestad et al., 1980; McDonald and Wood, 1981; Fraser and Harvey, 1984). This ion loss is very large and can be measured readily in the field (Fraser and Harvey, 1984). Thus, stream water chemistry and the ionic composition of fishes held in such waters became the measures of the stresses of episodic acid loadings.

This assessment of episodic pH depressions required that both acid precipitation events and the snow-melt events be predictable in terms of time of commencement, magnitude, and duration of the event.

2. METHODS

The Milford Bay Trout Hatchery comprises a series of ponds fed by springs flowing continuously at 300 L min^{-1}, plus surface run-off varying from 0-1500 L min^{-1} (Figure 1). These sources differ in chemical composition and thus the combined flow entering the ponds varies in composition. Run-off from precipitation or snow melt both dilutes and reacts with the alkalinity of the ground-water component.

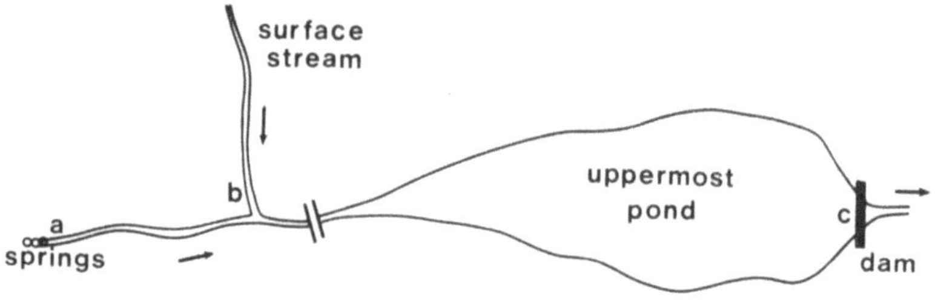

Figure 1. Map of the study site, Milford Bay, Muskoka, Ontario. Spring water, surface stream water, and the mixed flows were sampled at points a, b, and c, respectively.

Notice of a precipitation event was possible some 12 to 24 hr in advance. Based on the predicted movement of a storm system from, for example, the direction of the Ohio Valley, it was possible to estimate the likely acidity of a precipitation event plus the amount of precipitation. Similarly, warming trends, and especially rains during winter warm spells, could be predicted with sufficient accuracy to permit the monitoring of the magnitude and effects of such events. Precipitation was collected at the site during events, for calculation of deposition rate and acid loading.

Water samples were collected from the springs and surface stream, and at the outlet of the uppermost fish pond (Figure 1) and flow was measured at the three sites. (pH, alkalinity and dissolved oxygen were measured at 8-hr intervals. Alkalinity is reported in $\mu eq.L^{-1}$ as results are discussed in the context of solution electroneutrality; $\mu mol.L^{-1}$ = $\mu eq.L^{-1}$/charge). Detailed water chemistry was determined on samples taken at the beginning, mid-way and the end of events. Occasional water samples were analyzed for Al species, courtesy B. LaZerte, Ontario Ministry of Environment.

Rainbow trout, weighing from 96 to 179 g, were obtained from the Milford Bay Trout Farm. Twelve hours before commencing the experiment, five fish were sampled (plasma, muscle, gill) from the upper-most pond for evidence of stress. Three fish were sampled immediately following transportation from the pond cage to the holding tank (time zero) and served as a control on transport stress. Thereafter three fish were sampled every four hours. Blood and muscle samples were taken using the method described by Fraser and Harvey (1984). Gills were removed from the right side of each fish, weighed, dried to constant weight and analyzed for Al concentration, courtesy B. Loescher, Ontario Ministry of Environment.

3. RESULTS

Monitoring commenced with the first precipitation events of September, 1983. Initially, stream chemistry was monitored for every precipitation event, but as a knowledge of the stream chemistry response to atmospheric input was gained, monitoring was reduced to major events: heavy or very acid rains and rapid/prolonged snow melts. Six precipitation events monitored in the fall of 1983 (Figure 2) had pH of precipitation ranging from 3.8 to 4.3. The heavier rains showed the lower concentrations of $[H^+]$, and thus the acid loading was more similar from event to event than indicated by pH or mm of precipitation alone. No very heavy rains fell during the fall of 1983. Alkalinity of the hatchery ponds decreased from 290 to 360 to 110 to 150 $\mu eq.L^{-1}$ during three of the precipitation events, but it was not known if a precipitation event could reduce alkalinity to zero in the hatchery ponds.

Two snow melts were monitored during winter-spring 1984. On February 12, the day prior to commencement of the first melt, the hatchery ponds had a pH of 6.4 and an alkalinity of 292 $\mu eq.L^{-1}$; these declined to pH 5.3 and 50 $\mu eq.L^{-1}$ by February 15, a loss of 83% of initial alkalinity. In the second melt period, April 4-7, alkalinity in the ponds declined to 30 $\mu eq.L^{-1}$, or 10% of normal.

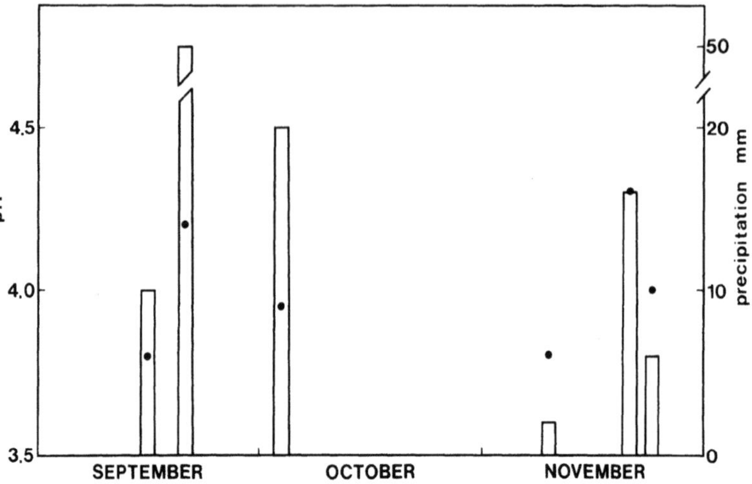

Figure 2. pH (solid circles) and amount of precipitation (bars) for six events monitored September-November, 1983.

The surface stream showed a pH of 6.0 and alkalinity of 78 ueq.L^{-1} on February 12, but by February 15, pH had declined to 4.8 and alkalinity was 0.5 (Figure 3). During the April 5 melt, the pH of the stream declined to 4.5 and alkalinity was -15.8 μeq.L^{-1}. During the February melt, the Al concentration of the surface stream doubled, and that of hatchery pond increased 5.5X (Figure 4).

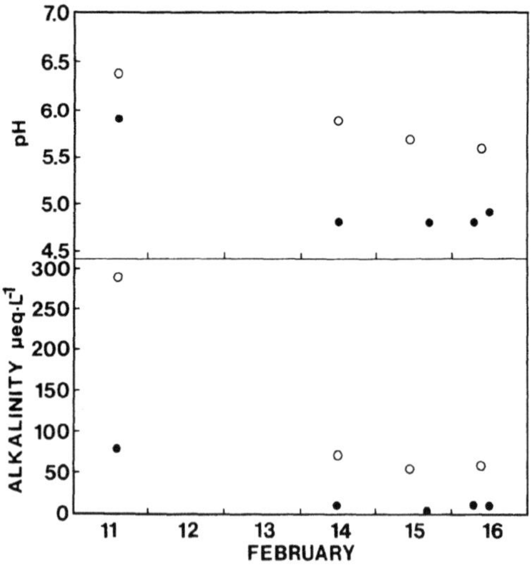

Figure 3. pH and alkalinity during a snow melt, February, 1983. Solid circles, surface stream (point b, Figure 1); open circles, mixed streams (point c).

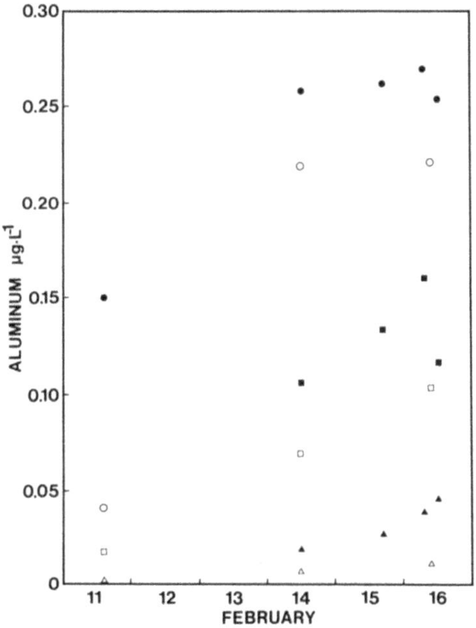

Figure 4. Aluminum concentration, including total (circles), organic (squares), and monomeric (triangles). Surface stream (point b, Figure 1) solid shapes; mixed streams (point c) open shapes.

The five rainbow trout sampled from the pond cage (-12 hr, Figures 5, 6) showed no evidence of acid stress. Fish transported to the stream site did not differ significantly in plasma or muscle ion concentrations. Trout exposed to the stream water, February 15-16, showed a rapid decline in plasma [Na^+] and [Cl^-] (Figure 5). Fish became moribund after 24-hour exposure and no fish survived beyond 28 hr. Muscle [Na^+] and [Cl^-] also declined (Figure 6) but [K^+] did not change significantly. An earlier study (Fraser and Harvey, 1984) established that such changes in concentration were the result of ion loss and not water uptake.

Coincident with this loss of Na^+ and Cl^- from plasma and muscle, there was a rapid accumulation of Al on the gills of the rainbow trout (Figure 7), from a concentration of 20 to 60 $\mu g.g^{-1}$ dry wt of tissue. This gill Al concentration was about 20 to 30% of that observed in rainbow trout held in Plastic Lake waters (Harvey and McArdle, 1986). In this experiment, the stress effects due to H^+ and Al could not be disaggregated.

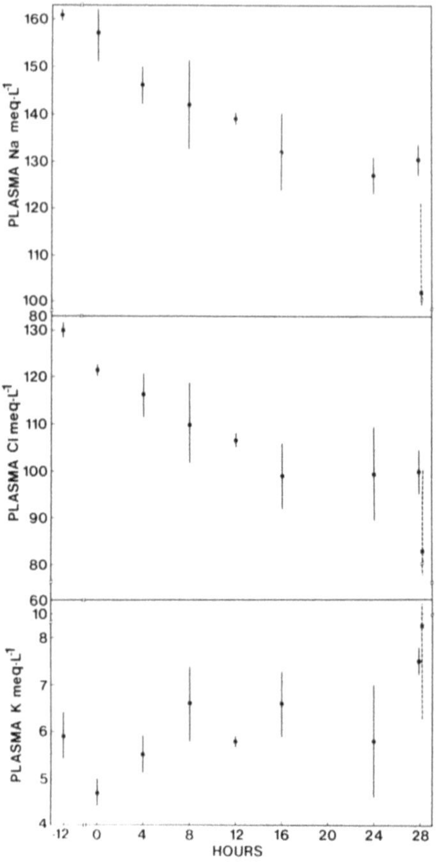

Figure 5. Concentrations of Na, Cl, K, in plasma of rainbow trout following exposure to surface stream water at point b, Figure 1. Mean (solid circle) and standard error (solid line) based on three fish; dashed line based on two moribund fish.

4. DISCUSSION AND CONCLUSIONS

Acid precipitation events and snow melts could be predicted sufficiently in advance to permit monitoring before and then during such events. During the period under study no severe episodes of acid deposition occurred. Moderately heavy rains resulted in the loss of one-half of the alkalinity of a stream that would have been considered only moderately sensitive (200 to 400 $\mu eq.L^{-1}$) to acid loading. Two snow melts resulted in the loss of 83 to 90% of alkalinity and pH depression from a range of 6.8 to 7.3, to a pH of 5.3. Under these conditions, rainbow trout did not show stress, as measured by ion loss from plasma and muscle.

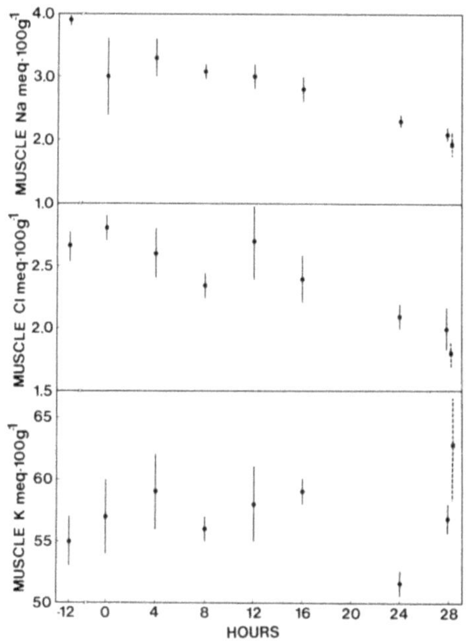

Figure 6. Concentrations of Na, Cl, K, in muscle (dry weight) of rainbow trout following exposure to stream water, point b, Figure 1. Mean (solid circle) and standard error (solid line) based on three fish; dashed line based on two moribund fish.

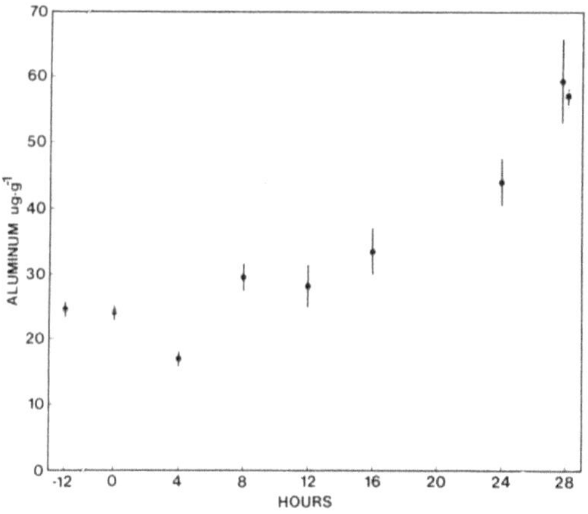

Figure 7. Concentrations of Al in gill tissue (dry weight) of rainbow trout following exposure to stream water, point b, Figure 1. Mean (solid circle) and standard error (solid line) based on three fish; dashed line based on two moribund fish.

In the surface-water stream, snow melts resulted in the rapid decline in alkalinity to zero and pH depression to 4.6. Under these conditions, loss of plasma ions was apparent within 4 hr and critical within 24 hr. These results raise the question of whether endemic fishes have/are/will suffer the same fate if surface waters elsewhere show comparable pH depression in response to episodic acid loading.

The early (February) and late (April) snow melts yielded similar results in terms of pH depression and exhaustion of alkalinity. There was, thus, no evidence that all the acid had been leached out of the snow pack in the first melt.

The inventory of rainbow trout at the Milford Bay Hatchery peaks at about 100,000 fish during hatching. That portion of the stock held in the series of ponds supplied with water as described here is at risk. It is considered likely that all of the alkalinity in the water supply could be exhaused via rapid snow melt, permitting a pH decline and Al toxicity increase to lethal levels. The Muskoka-Haliburton region of central Ontario experiences some of the most acidic deposition in eastern Canada. Combined with relatively frequent (~8 to 10 yr^{-1}) and often rapid melt periods, the potential for melts damaging to streams, fishes and other biota would appear to be quite high. It is considered possible that this could occur also during a major precipitation event.

The method for assessing risk, by holding "standard" rainbow trout in natural waters and measuring stress via ion loss, could be applied more widely in the evaluation of acid deposition damage and aquatic resources at risk.

5. REFERENCES

Dickson, W. 1978. Aluminum, pH and fish. The National Swedish Environment Protection Board Research Laboratory. 40 pp.

Fraser, G.A. and Harvey, H.H.: 1984, *Can. J. Zool.* **62**, 249.

Harvey, H.H., Dillon, P.J., Fraser, G.A., Somers, K.M., Fraser, P.E., and Lee, C.: 1982, *Amer. Chem. Soc., Div. Environ. Chem.* **22**, 438.

Harvey, H.H. and McArdle, J.M.: 1986, Physiological responses of rainbow trout (*Salmo gairdneri*) exposed to Plastic Lake inlet and outlet stream waters. Proceedings, International symposium on acidic precipitation, Muskoka '85. Submitted.

Leivestad, H. and Muniz, I.P.: 1976, *Nature (London)* **259**, 391.

Leivestad, H., Muniz, I.P., and Rosseland, B.O.: 1980, Acid stress in trout from a dilute mountain stream. In: Proceedings of the International Conference on the Ecological Impact of Acid Precipitation - Effects on forest and fish. D. Drabløs and A. Tollan, Ed. SNSF Project, 1432 Ås, Norway. pp. 318-319.

McDonald, D.G. and Wood, C.M.: 1981, *J. Exp. Biol.* **93**, 101.

Muniz, I.P.: 1971, personal communication.

Packer, R.K. and Dunson, W.A.: 1970, *J. Exp. Zool.* **174**, 65.

BASE ADDITIONS TO FLOWING ACIDIC WATER: EFFECTS ON SMOLTS OF ATLANTIC SALMON (*Salmo salar* L.)

O. K. Skogheim 1), B. O. Rosseland 1), E. Hoell 2) and F. Kroglund 1)
Directorate for Nature and Management,
Fish Research Division, Tungasletta 2, 7000 Trondheim, Norway
Present address: 1) Norsk Biotech A/S, P.O.Box 788 Krossen N-4301 Sandnes, Norway 2)NORCEM A/S, N-3950 Brevik, Norway

ABSTRACT. Acidic, Al-rich soft water (pH=5.1, Ca=1.0 mg L^{-1} and labile Al=160 µg Al L^{-1}) was treated by addition of soda (Na_2CO_3) and lime slurry ($CaCO_3$). Seven different water qualities of each type of treatment, covering the pH-range 5.1 to 8.2, were prepared in a flow-through system. Smolts of Atlantic salmon (*Salmo salar* L.) were used as test fish. In both types of treatment, mortality occurred at all pH-levels below 6.05. Above pH 6.05 no mortality occurred using lime slurry. Using soda, 10 % mortality occurred within 48 hr at pH above 7 due to the toxicity of aluminate at low levels of Ca. Plasma chloride levels indicated no physiological stress in the pH-range 6.45 to 7.0.

1. INTRODUCTION

Acidic soft water has been recognized as a serious problem in freshwater fish hatcheries in Norway since the 1920's (Dahl,1927). Rosseland and Skogheim (1984a) reviewed the various attempts to reduce effects of acidification on fisheries in Norway to date. In a previous short-term experiment with Atlantic salmon, NaOH, lime slurry and finegrained limestone were added to plastic enclosures placed in the epilimnion of an acidic lake (Skogheim *et al.*,1984). This experiment was of the batch-type and the water chemistry was unstable, simulating liming of lakes. In order to simulate effects of liming of running water, we conducted the present experiment in a flow-through system. Experimental objectives were to compare two neutralizing agents, both containing carbonate as the base but containing different cations; Ca (lime slurry) and Na (soda).

2. MATERIALS AND METHODS

2.1. Experimental design

Water from Lake Byglandsfjorden, southern Norway, was adjusted to pH=5.1 and labile Al=160 µg Al L^{-1} by pumping H_2SO_4 and $AlCl_3$ solutions into a continuous flow-system that supplied water to 15 fish tanks

Water, Air, and Soil Pollution **30** (1986) 587–592.

(30 L) with a flow rate of 1 L min^{-1} per tank. Neutralizing agents were pumped (Ismatec IP-12) into the water lines feeding each of the tanks. Neutralizing agents were: soda (Na_2CO_3, Merck p.a.) and lime slurry ($CaCO_3$, grainsize: 50 % < 2 µm).

2.2. Chemical analyses

In the acidified water (untreated) pH was monitored continuously with a Radiometer PHM 80 equipped with Radiometer electrode GK2401C. Temperature and pH were measured and water samples were taken 0.5, 13, 26, 34 and 47.5 hr after the start of the experiment. Aliquots were immediately prepared for Al analyses by the procedure described by Skogheim and Rosseland (1984). Acid reactive Al and non-labile Al were measured directly and labile Al was calculated as the difference. The analytical methods for determination of Ca and Na, alkalinity, sulphate and chloride, total fluoride and dissolved organic matter are given by Skogheim and Rosseland (1984).

Aluminium speciation was calculated using the computer program MINEQL (Westall et al.,1976). Inorganic monomeric species of aluminium were calculated from measured pH and concentrations of fluoride and sulphate. Ion strength was calculated from the measured ionic composition and varied between $4.1 \cdot 10^{-4}$ and $10.5 \cdot 10^{-4}$ M. For the equilibrium between aluminiumtrihydroxide and aquo Al :

$$Al(OH)_3 + 3\ H^+ = Al^{3+} + 3\ H_2O$$

a solubility constant corresponding to synthetic gibbsite (converted to the actual temperature) was used. The solubilty constant used here is based on earlier experience (Wright and Skogheim,1983; Rosseland and Skogheim,1984b). Equilibrium constants for the inorganic complexes used were identical to those noted by Driscoll (1980).

2.3. The test fish

The test fish were one year old smolts of Atlantic salmon (*Salmo salar* L.) of Imsa-strain. Mean ± standard deviation of length (cm) and weight (g) were 13.5 ± 2.5 and 21.5 ± 3.7, respectively, for 33 randomly sampled fish. Each tank contained 22 fish, which were not fed during the experiment. At intervals, blood samples were drawn by heart puncture. Samples were sentrifuged immediately (MSE Micro Hematocrit sentrifuge 346, 3 min) and analyzed for hematocrit and plasma chloride (Radiometer CMT 10 Chloride titrator). Cumulative mortality was calculated correcting for removals used in each blood sampling.

3. RESULTS

3.1. Water chemistry

Additions of neutralizing agents resulted in increased pH, alkalinity and concentrations of the respective base cation. The pH change induced changes in concentrations of non-labile Al and labile Al, whereas the level of acid reactive Al was constant (Table I).

TABLE I. Chemical components (mean ± SD, N = 5 unless otherwise indicated in parentheses) in untreated and treated waters sampled during the experiment.

TREATMENT	TANK NO	pH	Ca mg Ca L^{-1}	Na mg Na L^{-1}	Alkalinity meq L^{-1}	AL FRACTIONS Acid reactive (μg Al L^{-1})	Non-labile (μg Al L^{-1})	Labile (μg Al L^{-1})
	1	5.25±0.07	1.04±0.01	1.44(1)	0.000±0.000	256±4	110±21	146±26
	2	5.55±0.12	1.04±0.02	1.64(1)	0.003±0.005	226±4	116±12	110±13
SODA	3	5.65±0.06	1.05±0.01	1.84(1)	0.023±0.001	228±4	170± 9	58±12
	4	6.20±0.09	1.04±0.02	2.20(1)	0.036±0.002	231±3	191± 4	40± 4
(Na_2CO_3)	5	6.85±0.07	1.04±0.02	2.90(1)	0.063±0.003	232±4	202± 5	30± 4
	6	7.55±0.08	1.03±0.01	4.22(1)	0.125±0.005	228±6	178±12	50± 9
	7	8.20±0.04	1.03±0.04	4.92(1)	0.171±0.006	226±3	103±18	123±18
	8	5.50±0.07	1.27±0.04	1.41(1)	0.000±0.000	267±7	147± 8	120±16
LIME	9	5.65±0.05	1.40±0.10	1.29(1)	0.005±0.000	269±8	162± 2	107± 5
SLURRY	10	6.05±0.05	1.81±0.09	1.22(1)	0.025±0.002	267±7	197± 6	70±11
	11	6.45±0.20	2.17±0.10	1.32(1)	0.042±0.003	268±6	218± 5	50± 7
($CaCO_3$)	12	7.10±0.07	2.91±0.13	1.22(1)	0.082±0.005	264±11	238± 8	26± 9
	13	7.70±0.04	4.48±0.26	1.26(1)	0.159±0.002	269±8	207± 7	62±10
	14	8.05±0.05	5.86±0.25	1.24(1)	0.231±0.008	272±6	141±17	131±16
Untreated	15	5.10±0.01	1.00±0.05	1.25±0.02	0.000±0.000	245±3	89± 8	156± 8

The theoretical concentrations of inorganic monomeric Al-species, which can exist in equilibrium with a solid phase ($Al(OH)_3$) at each pH-level, were calculated (Figure 1).

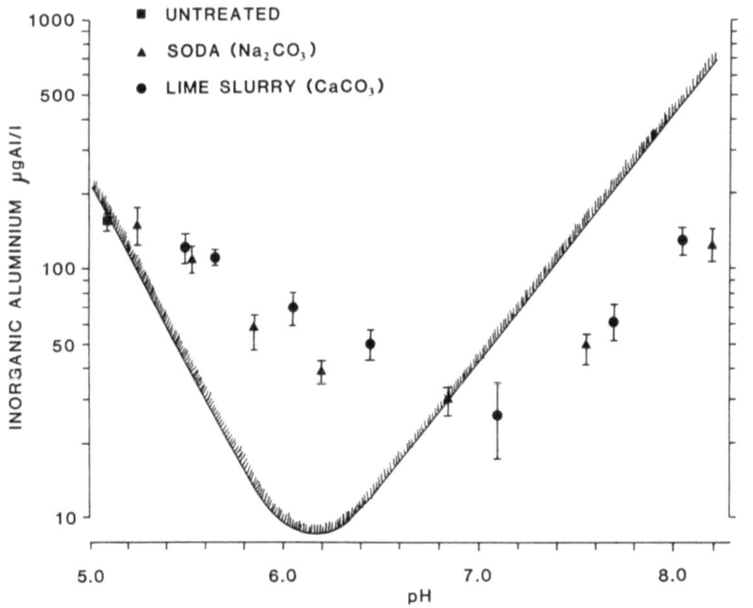

Fig. 1. Computed total inorganic monomeric Al in equilibrium with a solid phase of Al ($Al(OH)_3$, K_{so}=10 $^{9.35}$) in the pH-range 5.0 to 8.2. Mean ± SD (N=5) of the analyzed concentration of labile Al is shown.

The analyzed labile Al concentrations for the various treatments were plotted (as M±SD) on the solubility diagram (Figure 1). The fluoride concentration in the water used in this experiment was 46 µg L^{-1} (SD=2 µg L^{-1} for 10 random samples). In untreated water AlF^{2+} constituted 33% of the labile Al. Aquo Al, Al^{3+}, and the aluminiumhydroxides, $Al(OH)_2^+$ and $Al(OH)^{2+}$ constituted the remaining inorganic Al-species. At pH above 6.5, aluminate, $Al(OH)_4^-$, is the only significant inorganic Al species. The non-labile Al in newly neutralized water is thought to represent freshly precipitated $Al(OH)_3$ and polymeric forms of Al.

The concentration of dissolved organic matter was very low; UV-absorbance=0.040 cm^{-1} (SD=0.002, N=75), corresponding to 5 mg Pt L^{-1}. The concentrations (M±SD, N=10 random samples) of sulphate and chloride were 3.0 mg L^{-1} ± 0.4 and 2.3 mg L^{-1} ± 0.2 , respectively.

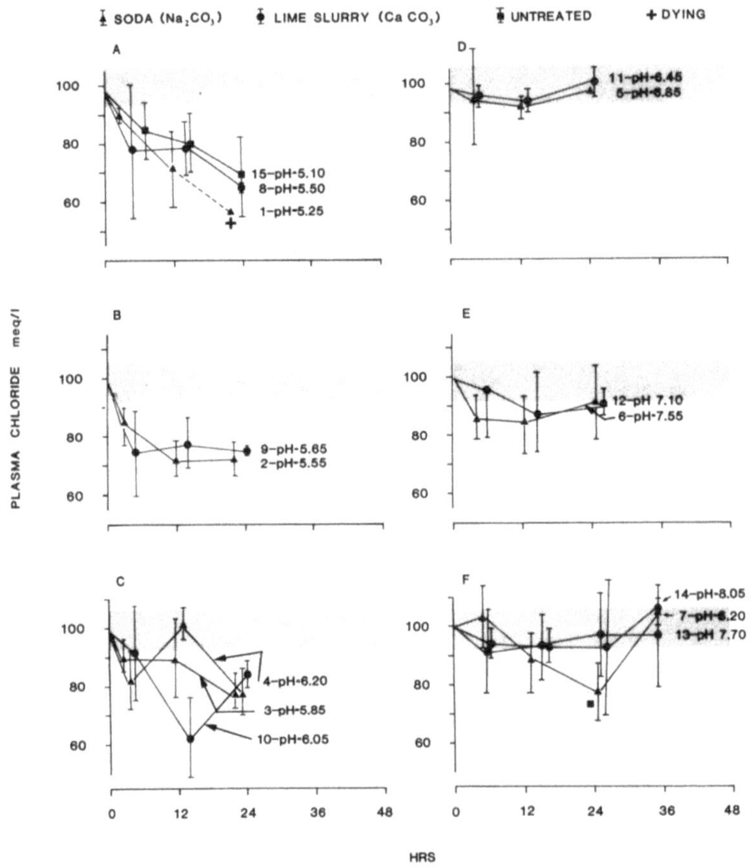

Fig. 2. Plasma chloride (mean ± SD, N=4) of smolts of Atlantic salmon exposed to treated and untreated water.

3.2. Fish responses

In untreated water at pH=5.10 the mortality at the end of the experiment (after 48 hr) was 100%. In waters treated with lime slurry the mortality was 100% at pH=5.50 and 5.60, and 70% at pH=6.05. At pH>6.45 there was no mortality. In waters treated with soda the mortality was 100% at pH=5.25, 90% at pH=5.55 and 25% at pH=5.85. At pH=6.20 and 6.85 there was no mortality, but at higher pH (pH=7.55 and 8.20) there was minor initial mortality (5 to 10%). The test fish were kept in limed water from Lake Byglandsfjord in large tanks ($4 m^2$) at the laboratory for 7 days prior to the experiments. Daily water samples revealed the following water quality (M$\pm$SD, N=7): pH=6.05$\pm$0.05, Ca=1.59$\pm$0.07 mg L^{-1} and labile Al=30 µg Al L^{-1}. The temperature varied in the range 4 to 5 °C. The smolts were sublethally stressed during this period. At the start of the experiment plasma chloride levels were 99$\pm$8 meq L^{-1} (N=17) and this level is taken as the reference level. Normal values for unstressed salmon of the same strain, size and stage are 120 to 130 meq L^{-1} (Rosseland and Skogheim, 1984b).

In untreated water, and in water with the lowest pH of both treatments, plasma chloride decreased rapidly and the fish died after ca. 24 hr. In water treated with soda at pH=6.85 and lime slurry at pH=6.45, plasma chloride levels were stable throughout the experiment and the mean values were within the variation observed among reference fish. In water treated with lime slurry at pH=7.70 and 8.05, plasma chloride levels were constant during the experiment. In water treated with soda at pH=8.20, plasma chloride decreased steadily during the first 24 hr, but after 36 hr plasma chloride had returned to the reference level (Figure 2).

4. DISCUSSION

The flow-through experimental system used in this study provided a constant water quality throughout the 48 hr exposure of fish in the various water treatments. When using soda and lime slurry as neutralizing agents, the rates of the chemical reactions are relatively rapid. These reactions are basically the acid neutralization and pH-change, and secondarily the changes in Al-speciation.

Comparing the theoretical equilibrium concentration with the actual analyzed concentration of labile Al, it appears that equilibrium exists in the untreated water at pH=5.1 and in treated water at pH=6.8 (Figure 1). For both neutralizing agents oversaturation exist in the pH-range 5.2 to 6.4 and undersaturation exists at pH>6.8. The disequilibrium between the dissolved and solid phase of Al-species was probably maintained throughout the experiment. In terms of the biological responses in the treatments, this chemical disequilibrium leading to precipitation of $Al(OH)_3$ should be of great concern as Dickson (1983) hypothesized the occurrence of a toxic zone caused by changes in Al species downstream the point of liming. Baker and Schofield (1982) found that Al was most toxic in oversaturated solutions at pH levels 5.2 to 5.4.

The rapid loss of plasma chloride and the 100% mortality in the untreated water is in accordance with results from several analogous experiments (Rosseland and Skogheim,1984a,b; Rosseland *et al.*,1986). Among the various treatments, only use of soda at pH=6.85 and lime slurry at pH=6.45 gave full survival and no loss of plasma chloride. In all other treatments (pH-range 5.25 to 6.25 and at pH>7.10) either mortality and/or loss of plasma chloride occurred. The complete survival and low stress at pH=6.45 and 6.85 is in accordance with the results from the full-scale liming of the acidic River Vaaraana by adding lime slurry to similar pH levels (Rosseland *et al.*,1986).

The fish response to oversaturation of inorganic Al is most dramatic at low pH and is moderate in the pH-range 5.50 to 6.20. Lower levels of labile Al and higher concentration of cations with increasing pH (Table I) within this range may explain this result. At pH>6.5 aluminate, $Al(OH)_4^-$, is in the predominant Al-species. Skogheim *et al.* (1984) found sublethal effects on smolt of Atlantic salmon at aluminate concentrations of 50 to 60 µg Al L^{-1} (pH=7.0, Ca=1.3 mg L^{-1}) using NaOH as neutralizing agent. When using limestone, aluminate concentrations as high as 90 µg Al L^{-1} at pH=8.15, Ca=5.25 mg L^{-1} did not induce effects on fish (Skogheim *et al.*,1984). These results are clearly verified in the current study, as no mortality or severe loss of plasma chloride occurred at pH≥6.45 using lime slurry. In contrast, the use of soda at pH≥7.10 lead to sublethal stress and a minor mortality.

The protective role of Ca at high levels of aluminate is clearly demonstrated. The results points to the importance of using Ca-containing bases for deacidification and of maintaining pH within the range 6.5 to 7.0 if non-Ca-containing based are used.

5. REFERENCES

- Baker, J.P. and Schofield, C.L.: 1982, Water,Air,Soil Pollut.**18**, 289
- Dahl, K.: 1927, Salmon and Trout Magazine **46**, 35
- Dickson, W.: 1983, Vatten **39**, 400
- Driscoll, C.T.: 1980, 'Chemical characterization of some dilute acidified lakes and streams in the Adirondack Region of New York State' Ph. D. Thesis, Cornell Univ., N.Y.
- Rosseland, B.O. and Skogheim, O.K.: 1984a, Fisheries **9**, 10
- Rosseland, B.O. and Skogheim, O.K.: 1984b, Rep. Inst. Freshw. Res. Drottningholm **61**, 186
- Rosseland, B.O., Skogheim, O.K., Abrahamsen, H. and Matzow, D.: 1986, Can. J. Fish. Aquat. Sci. **43**,
- Skogheim, O.K. and Rosseland, B.O.: 1984, Rep. Inst. Freshw. Res. Drottningholm **61**, 177
- Skogheim, O.K., Rosseland, B.O., Kroglund, F. and Hagenlund, G.: 1984, 'Addition of NaOH, limesluryy and finegrained limestone to acidified lake water and the effects on smolts of Atlantic salmon (*Salmo salar* L.)', Kalkingsprosjektet, rapport 19/84, 28 p.
- Westall, J.C., Zacjary, J.L. and Morel, F.N.M.: 1976,'MINEQL, a computer program for the calculation of chemical equilibrium composition of aqueous systems', Dep. of Civil Eng. MIT, Techn.note No. 18.
- Wright, R.F. and Skogheim, O.K.: 1983, Vatten **39**, 301

FIELD AND LABORATORY STUDIES OF EXPOSURES OF BROWN TROUT TO ACID WATERS

K.Sadler
Central Electricity Research Laboratories,
Freshwater Biology Unit,
Midlands Region Scientific Services Department,
Ratcliffe-on-Soar,
Nottingham, NG11 OEE,
England

A.W.H. Turnpenny
Central Electricity Research Laboratories, Marine Biology Unit,
Fawley,
Southampton, SO4 1TW,
England

ABSTRACT. Some recent work on the effects of acid waters on brown trout are presented. Laboratory bioassay experiments have demonstrated that yearling trout are relatively insensitive to pH >4.3. Aluminium is demonstrated to be extremely toxic with suppression of growth occurring at concentrations above 20 µg L^{-1} at pH 4.4 to 5.2. Aluminium toxicity is reduced at high pH (5.9 and 6.3). Field studies carried out on 61 acidic and circumneutral streams in upland areas of England and Wales showed a strong relationship between water quality and standing crop of 1+ brown trout. Measured pH levels per se were too high to be directly toxic. On the other hand, heavy metal and Al concentrations could account for low or zero brown trout biomass in the more acidic streams. A mobile bioassay laboratory has been developed to allow controlled bioassay experiments to be carried out in the field. Natural and synthesised waters can be tested concurrently in multi-factorial experiments with in situ determinations of pH, Ca, Al (total and monomeric) and other water quality characteristics.

1. INTRODUCTION

There is much evidence which indicates that the survival and growth of fish at low pH are dependent upon other water quality factors, principally Ca and Al (Muniz and Leivestad, 1980; Schofield, 1980; Schofield and Trojnar, 1980; Baker and Schofield, 1982; Driscoll et al., 1980; Brown, 1983; Fivelstad and Leivestad, 1984).
In order to understand the complex interactions between different aspects of water quality it is necessary to conduct carefully controlled bioassay experiments in which the exact composition of the water is known and it is possible to test independently variations in each constituent. There are, however, limitations

Water, Air, and Soil Pollution 30 (1986) 593–599.

to this bioassay approach when trying to predict responses in natural waters. It needs to be demonstrated that all of the toxic substances in the water or any detoxifying complexing agents have been considered. Also the effect of variable conditions in natural waters needs to be investigated. There is therefore a requirement for a number of approaches ranging from pure bioassay studies through experiments designed to simulate natural conditions, to field studies. This paper reports some of the findings from recent work carried out at the Central Electricity Research Laboratories which have adopted these different approaches.

2. MATERIALS AND METHODS

2.1. Field Studies

A survey of the water chemistry, and fish populations of 61 upland streams in Central and North Wales and the Peak District of England was conducted in 1982-3. These three areas are all characterized by hard, unreactive geologies (principally shales, mudstones or millstone grit) which give rise to oligotrophic waters, but differ in acid deposition rates (25 to 30 and >50 kg S ha^{-1} yr^{-1} for Welsh and Peak District sites, respectively; Martin, 1980).
Quantitative sampling was carried out by repeated electrofishing of 50 to 200 m stream reaches, and populations estimated by the 2-catch method of Seber and Le Cren (1967) or Zippin's (1958) multiple capture method. Water quality was sampled on five occasions over the 12-mo period preceding fish sampling. Variables measured were pH and conductivity (in the field), Ca, Mg, Al, Cu, Fe, Pb, Zn, S (by inductively coupled plasma emission spectrometry), NO_3 (UV spectrometry following removal of organic C by filtration through activated C), Cl (chloride ion selective electrode by titration with Ag NO_3) and K by flame emission spectrometry. Additional water sampling was carried out on two occasions in 1985 to measure the concentrations of labile monomeric Al, the fraction considered to be most toxic to fish (Driscoll et al., 1980). This was done using the catechol violet method and cation ion exchange resin by the method of Seip, et al. (1984).

2.2. Laboratory Bioassay

Yearling brown trout (Salmo trutta) from a hatchery in Derbyshire were exposed to various conditions in a flow-through system in which synthetic salt solutions can be maintained at constant pH levels. This experimental system consists of 12 tanks (capacity 100 L) supplied with water which has been recirculated through sand and activated C filters and ion exchange resins to which is added concentrated salt solutions of known composition. All experiments have been conducted at 12 ± 1.5 °K, with a nominal water composition of 1 mg L^{-1} Na, 1 mg L^{-1} Ca, 0.5 mg L^{-1} Mg, and 0.3 mg L^{-1}K, all salts added as chlorides. pH is controlled continuously by titration with H_2SO_4 or KOH as appropriate (pH of unregulated tank water is approximately 5.5). The fish were acclimatised to the system for 7 days and then marked individually by subcutaneous injection of latex dye. Growth rates were then measured over 6 weeks when fed on commercial trout pellets at a rate of 2% of biomass per day.

2.3. Field Experiments

To help to overcome the problem of distinguishing toxicological from ecological and other effects, a mobile bioassay laboratory was developed. This enables controlled laboratory toxicity studies to be carried out in the field, on the banks of rivers and lakes. This mobile laboratory is provided with a temperature-controlled section containing the experimental aquaria and dosing pumps and a section containing computerized control and analytical equipment including a continuous-flow autoanalyser for in situ determination of Al and Ca levels. The aquaria are arranged in four columns, each capable of holding a variety of tank combinations, from 8 x 10 L capacity to a maximum of 48 x 0.1 L tanks. The water supplied to each of the four columns can be treated in a variety of ways including carbon filtration, ion exchange and pH control. Computer-controlled dosing pumps supply individual tanks with any required ions. With this system, synthesized waters comparable to those used in conventional laboratory bioassays can be tested alongside natural waters.

This apparatus has been used in an experiment carried out at Loch Fleet, an acid lake in Galloway, SW Scotland. The lake has been devoid of brown trout for the past three decades and a study is now in progress to assess various treatment options that would restore trout fisheries to this and similarly affected waters. The main water quality problems preventing a trout fishery appeared to be low pH (4.0 to 4.5) and associated elevated Al levels (approximately 200 µg L^{-1} total) and low Ca levels (0.7 mg L^{-1}).
Brown trout swim-up fry of two different softwater strains were exposed for 9 days to various experimental treatments, including (a) raw lake water (pH 4.4, 210 µg total Al L^{-1}, 0.7 mg Ca L^{-1}), (b) the same lake water adjusted to the seasonal minimum pH of 4.0 by H_2SO_4 addition, (c) lake water adjusted to pH 5.4 by NaOH addition and (d) control: lake water, deionized and C filtered, Ca and Na added (pH 5.2, 0.7 mg Ca L^{-1}, 1 mg Na L^{-1}, 13 µg Al L^{-1}). The experiment was conducted at a temperature of 10 to 12°K.

3. RESULTS

3.1. Field Studies

A wide range of water chemistry conditions were found during the survey with pH 4.45 to 7.30, 1 to 27 mg L^{-1} Ca, 0.5 to 7 mg L^{-1} Mg, 3 to 12 mg L^{-1} Na, 0.3 to 5 mg L^{-1} K, 0 to 0.02 mg L^{-1} Cu, 0 to 1 mg L^{-1} Fe, 0 to 0.1 mg L^{-1}Zn, 2.5 to 44 mg L^{-1} SO_4, 6 to 22 mg L^{-1} Cl and 0.3 to 8 mg $L^{-1}NO_3$ (based on averages of 5 samples).
The fish populations were dominated by brown trout which was found at all of the fish containing sites. The catches tended to be mostly of the 1+ age group as smaller fish were not sampled very efficiently by the method used. The results are therefore presented for this age group to prevent any bias due to the occasional capture of large older fish on the biomass values.
pH, labile monomeric Al and heavy metal concentrations were the most highly correlated factors with the biomass of the 1+ age group of trout (Figure 1). The toxicity of heavy metals (Cu, Zn, Pb) was found to be well represented by the index MeTox calculated using the method of Brown (1968) based on the toxicity of these metals to rainbow trout (Salmo gairdneri Richardson).

3.2. Laboratory Bioassays

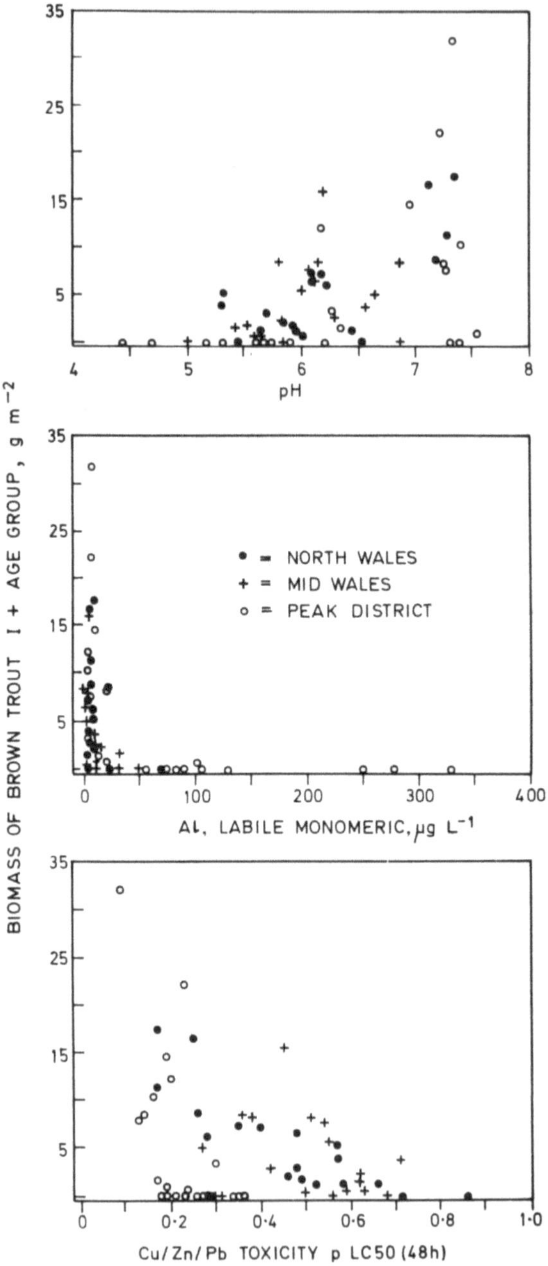

FIG. 1 BROWN TROUT BIOMASS IN RELATION TO WATER QUALITY

The results of an experiment in which brown trout were exposed to pH 4.4, 4.8 or 5.2, at varying additions of aqueous Al (added as Al chloride) are shown in Figure 2. In the absence of Al the fish all showed good growth and survival independent of pH but a threshold concentration of approximately 20 $\mu g\ L^{-1}$ Al was found above which growth rates are suppressed. The importance of pH to Al toxicity has been demonstrated by exposing fish to 50 $\mu g\ L^{-1}$ of Al at pH 4.3 to 6.2 (Figure 3). Growth rates of the fish at pH 4.3 in the absence of Al were clearly reduced and mortalities were high indicating that for the tested strain of brown trout there is a relatively sharp limit to their long term survival and growth at pH 4.3 to 4.4. The greatest toxicity attributable to Al per se occurred at pH 5.1 with relatively little suppression of growth at pH 5.9 and 6.3. This clearly indicates the differing toxicity of the various Al hydroxide complexes which predominate at different pH's. Preliminary analysis using thermodynamic equilibrium calculations to estimate the Al species present in the test conditions (Driscoll et al., 1984) would suggest that the $Al(OH)^{++}$ species is the most toxic form, as was also suggested by Fivelsted and Leivestad (1984), but this needs to be confirmed in further experiments.

3.3. Field Experiments

Figure 4 shows the results of the trout survival experiment conducted in Loch Fleet water (raw or adjusted to pH 4.0 or 5.4) and in optimal control conditions. The results for both stocks of fish indicate that the toxicity of the waters was decreased, by increasing pH from 4.4 to 5.4, to that found in control water lacking Al. Reducing the pH to 4.0 substantially increased toxicity, particularly to the Loch Dee stock.

4. DISCUSSION

The difficulties of interpreting field study results are illustrated by Figure 1 in which relationships with several environmental variables may be found, but there remains a large amount of variability which cannot be accounted for. Also, the environmental variables themselves may be intercorrelated.
The pH values exhibit an association with biomass, and 80% of streams (n = 10) with mean pH <5.5 were fishless compared with 19% of 51 streams with pH > 5.5. Population density and condition factor were also lower in the more acidic streams.
It is generally accepted that pH values above 5.0 per se are unlikely to be harmful to fish, and the bioassay studies indicate that good growth rates can be maintained down to pH 4.4 (Figure 2). Only two streams in the study had mean pH values below 5.0 but it is clear that streams of the type studied are susceptible to short-term (<24 hr) episodes of low pH which can only be adequately represen-ted by near-continuous chemical monitoring. Therefore, the relationship observed in Figure 1 may represent an underlying relationship with episodic pH minima.
However, Al and heavy metal concentrations associated with the more acid streams would account for much of the observed variation in biomass. Field studies of toxicity of heavy metals derived from natural outcrops, mine-tailings and also ammunition dumps on the River Mawddach system in West Wales, (Howells et al., 1983) showed that brown trout biomass was negatively correlated with

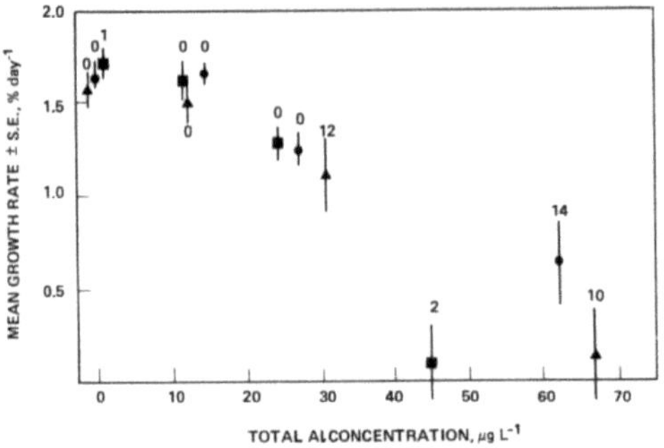

FIG. 2 MEAN GROWTH RATES OVER 6 WEEKS OF BROWN TROUT AT DIFFERENT AL CONCENTRATIONS. ▲ = pH 4.4 ● = pH 4.8 ■ = pH 5.2. NUMBERS REFER TO THE NUMBER OF MORTALITIES (ex 20) OBSERVED IN EACH TANK

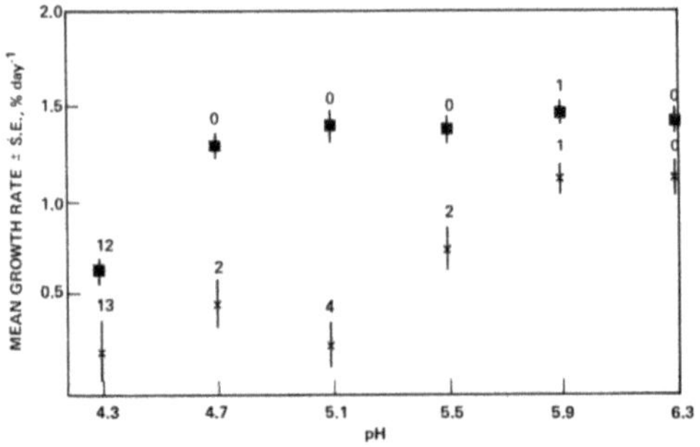

FIG. 3 MEAN GROWTH RATES OVER 6 WEEKS OF BROWN TROUT AT DIFFERENT pH LEVELS. ■ = 0 AL × = NOMINAL 50 $\mu g\ l^{-1}$ AL. NUMBERS REFER TO THE NUMBER OF MORTALITIES (ex 20) OBSERVED IN EACH TANK

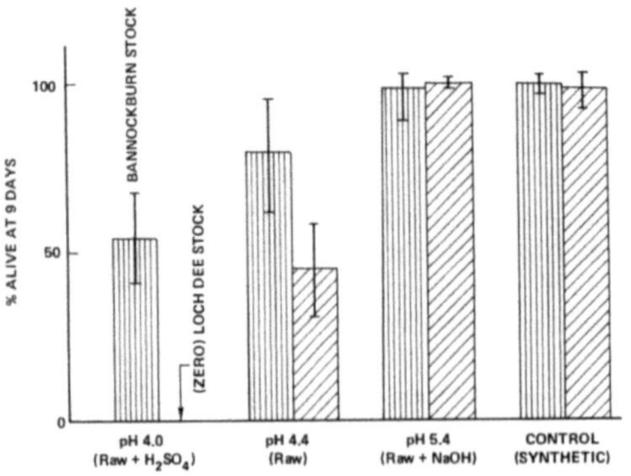

FIG. 4 LOCH FLEET BIOASSAY TEST: SURVIVAL OF TWO SOFT WATER STRAINS OF BROWN TROUT IN RAW AND pH–REGULATED LOCH WATER. CALCIUM LEVEL WAS 0.7 mg L^{-1}, TOTAL AL 210 $\mu g\ L^{-1}$. (BARS REPRESENT 95% CONFIDENCE INTERVALS)

MeTox, zero biomass occurring at around 0.56 of the 48h LC_{50} to rainbow trout. On inspection of Figure 1, it appears that trout biomass was probably controlled primarily by heavy metal toxicity in the Welsh streams, and by Al toxicity in the Peak District streams. Moreover, there is a threshold of approximately 40 µg L^{-1} labile monomeric Al above which streams were fishless which corresponds well with the threshold found in the bioassay experiments of reduced growth above 20 µg L^{-1} (Figure 2).

Experimental field studies in which natural conditions can be manipulated help to eliminate the ambiguities which arise in the interpretation of population surveys. Laboratory studies on the toxicity of Al to trout have shown that Al is most toxic at pH values in the range 5.0 to 5.5 (Figure 3; Schofield and Trojnar, 1980; Baker and Schofield,1982; Brown, 1983), hence suggesting that lime applications to Loch Fleet, by raising pH, might increase Al toxicity rather than having the desired ameliorative effect. In fact the toxicity was much reduced at pH 5.4 which suggests that most or all of the Al was not present in toxic form and that pH per se is the most important factor in this lake. In a single sample of the Loch Fleet water from the experiment, only about half of the Al (100 µg L^{-1}) at pH 4.5 was in labile monomeric form as estimated by fractionation through a cation exchange column (Driscoll et al., 1984). This alone would not account for the low toxicity attributable to Al and it could be that Al species of low toxicity were present within the labile monomeric fraction due, for example, to fluoride complexation (Driscoll et al.,1984; La Zerte, 1984). In future work the role of Al will be considered further by a combination of Al speciation studies and post mortem histochemical studies of exposed fish to establish whether or not the fish absorb Al and whether it is in a potentially toxic form.

5. REFERENCES

Baker, J.P. and Schofield, C.L.: 1982. Water Air and Soil Pollut., **18**, 289.
Brown, D.J.A.: 1983, Bull. Environm. Contam. Toxicol., **30**, 582.
Brown, V.M.: 1968, Water Res., **2**, 723.
Driscoll, C.D., Baker, J.P., Bisogni J.J. and Schofield, C.L.: 1980, Nature, **284**, 161.
Driscoll, L.D., Baker, J.P., Bisogni, J.J. and Schofield, C.L.: 1984, In Geological Aspects of Acid Deposition (Ed. Bricker, O.P.) Butterworth, **55**.
Fivelstad, S. and Leivestad, H.: 1984, Inst. Freshwater Res. Drottningholm, **61**, 69.
Howells, E.J., Howells, M.E. and Alabaster, J.S.: 1983, J.Fish Biol. **22**, 447.
La Zerte, B.D. : 1984, Can. J. Fish. Aquat.Sci., **41**, 766.
Martin, A.: 1980, Envir. Poll.(Series B) **1**, 177.
Muniz, I.P. and Leivestad, H.: 1980, Proc. Int.Conf.Ecol. Impact Acid Precip. (Eds. Drablos, D. and Tollan A.). Sandefjord, Norway SNSF Project, 320.
Schofield, C.L.: 1980, Atmospheric Sulphur Deposition Environmental Impact and Health Effects (Eds. Shriner, D.S., Richmond, C.R. and Lindberg, S.E.), Ann Arbor Science, 345.
Schofield, C.L. and Trojnar J.R.: 1980, In: Proc. Conf. Polluted Rain (Eds. Toribara, T.Y., Miller, M.W. and Morrow, P.F.,) Plenum, New York, 341.
Seber, G.A.F. and Le Cren, E.D.: 1967, J. Anim.Ecol., **36**, 631.
Seip, H.M., Muller, L. and Naas, A.: 1984, Water Air and Soil Pollut., **23**, 81.
Zippin, C.: 1958. J. Wildl. Mgmt., **22**, 82.

Fish Species Distribution in Relation to Lake Acidity in Ontario

D.L. Wales and G.L. Beggs
Ontario Ministry of Natural Resources
Queen's Park, Toronto
Ontario, Canada, M7A 1W3

ABSTRACT. Survey data from approximately 2,900 Ontario lakes were used to examine the distribution of 12 fish species with respect to lake pH. Yellow perch (Perca flavescens), white sucker (Catostomus commersoni), brook trout (Salvelinus fontinalis), pumpkinseed (Lepomis gibbosus), fathead minnow (Pimephales promelas) and redbelly dace (Phoxinus eos) proved to be tolerant of low pH (pH <6.4). Lake trout (Salvelinus namaycush), common shiner (Notropis cornutus), blacknose shiner (Notropis heterolepis), lake whitefish (Coregonus clupeaformis), walleye (Stizostedion vitreum vitreum) and northern pike (Esox lucius) showed limited distribution in low pH lakes (pH <6.4). The limited distribution of lake whitefish and northern pike likely relates to zoogeographic factors rather than their intolerance to low pH. Lake trout and common shiner occur in lakes susceptible to, and receiving high loadings of acidic deposition. However, without historical data, we cannot determine if their decreased distribution in low pH systems is due to intolerance to low pH.

1. INTRODUCTION

Acidification has been shown to be related to fisheries losses in the Sudbury area (Beamish and Harvey, 1972), with few cases documented elsewhere (Beggs et al., 1985a). Matuszek and Beggs (1986) using data from about 2,900 lakes in Ontario showed a decline in fish species number with increasing lake acidity. Multiple regression analysis of log (species number +1) on 19 physical and chemical variables showed that for all lakes with pH values below 6.4, pH was selected as the variable explaining the major portion of variation in species richness, followed by lake area, dissolved organic carbon, elevation, and total Al (total r^2 = 40, p = 0.0001). In this study, using the same data base as Matuszek and Beggs (1986), the distribution of selected fish species are examined in relation to lake pH, to determine which species are well distributed above and below pH 6.4. For those species showing limited distribution below 6.4, significant lake physical correlates of pH are examined to determine whether the observed distribution relates to zoogeographical factors rather than the species sensitivity to low pH.

Water, Air, and Soil Pollution **30** (1986) 601–609.

2. METHODS

Fish species presence, lake area and latitude data originated from the Aquatic Habitat Inventory Program, sampling methods are outlined in Dodge et al., (1985).

Water chemistry samples were collected primarily as tube composite samples (78%) and analyzed for pH, Ca and total Al according to methods specified by the Ontario Ministry of the Environment (1981). Only 12% of the samples were collected during the calendar months of March, April and May; therefore, very few samples would reflect worst case water chemical conditions usually present during spring melt (Jeffries et al., 1979).

To examine the distribution of each fish species with respect to lake pH, lakes in the data base were divided into pH intervals (0.1 units). The proportion of the total number of surveyed lakes containing each fish species was calculated for each pH interval (Figure 1 and 2). Lakes in pH intervals less than 5.1 or greater than 8.5 were eliminated from this analysis due to small sample size. The fish species data set is biased towards large lakes and lakes containing sport fish species (Minns, 1981).

Spearman correlation coefficients and univariate descriptive statistics (median, 5th and 95th percentiles) were calculated using SAS (SAS Institute Inc., 1982) (Table I, Table II).

3. RESULTS

The fraction of sampled lakes containing pumpkinseed and redbelly dace increases with decreasing pH over the pH interval 7.5 to 5.5 (Figure 1). The species fraction for brook trout increases with decreases in lake pH from pH 7.5 to 6.2 and remains constant from pH 6.2 to 5.1 (Figure 1). The species fraction for yellow perch, fathead minnow and common white sucker remain relatively constant from pH 7.5 to 5.6 (Figure 1). The above species were classified as being well distributed in low pH lakes (pH <6.4). Only at very low pH values are there signs of decreasing occurrence for these species.

For northern pike, walleye, lake whitefish and blacknose shiner, the species fraction decreased constantly with decreasing lake pH over the pH range 7.5 to 5.1 (Figure 2). Species fraction for common shiner and lake trout decrease with decreasing pH below 6.6 and 6.8 respectively (Figure 2). These species were classified as showing limited distribution in low pH lakes (pH <6.4).

Lake area, elevation, latitude, Ca and total Al are significantly ($p = 0.0001$) correlated with pH (Table I). In this paper, acidic deposition is represented by the variable latitude. The maximum deposition occurs in southern Ontario (mean annual weighted deposition of 45 kg wet sulphate ha^{-1} $year^{-1}$) and a southeast to northwest gradient of decreasing deposition is observed (Chan et al., 1983). Loadings of sulphate in northern Ontario are less than 10 kg ha^{-1} $year^{-1}$ (Chan et al., 1983).

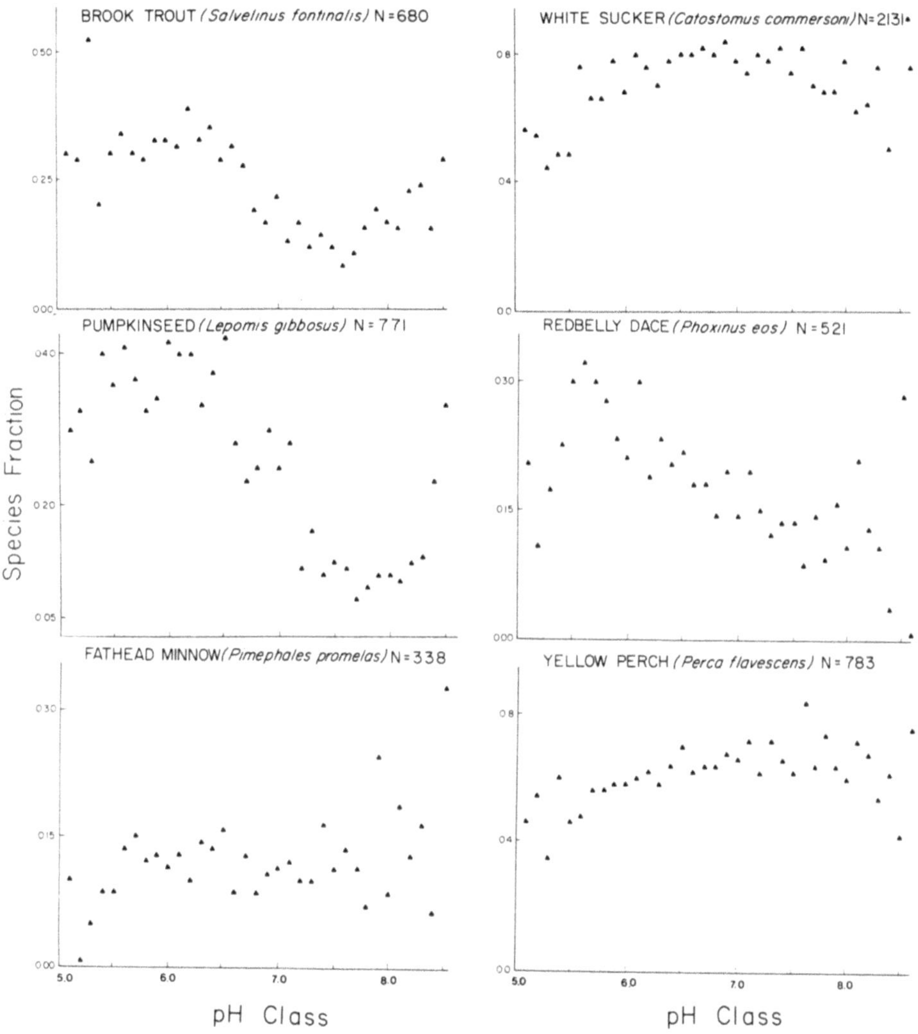

Figure 1. The fraction of sampled lakes containing selected fish species by pH interval. One pH interval spans 0.1 units (eg. interval 6.0 includes lakes with 6.0<=pH<6.1). Species subjectively classified as well distributed at low pH (pH<6.4).

Lake whitefish occur in the largest lakes with 50% of the sampled lakes having lake areas greater than or equal to 326 ha (Table II). Common shiner occur in the smallest lakes with 50% of the sampled lakes being less than 84 ha in area (Table II). Northern pike and walleye are distributed in northern Ontario, with 50% of their sampled lakes located north of latitude 48°36' with 48°29' respectively (Table II). Common shiner are largely distributed in southern Ontario with 50% of their sampled lakes located south of 46°02' (Table II).

Elevation was also significantly correlated with pH (Table I); however, it was not considered in this analysis due to the small variability in elevation observed between species.

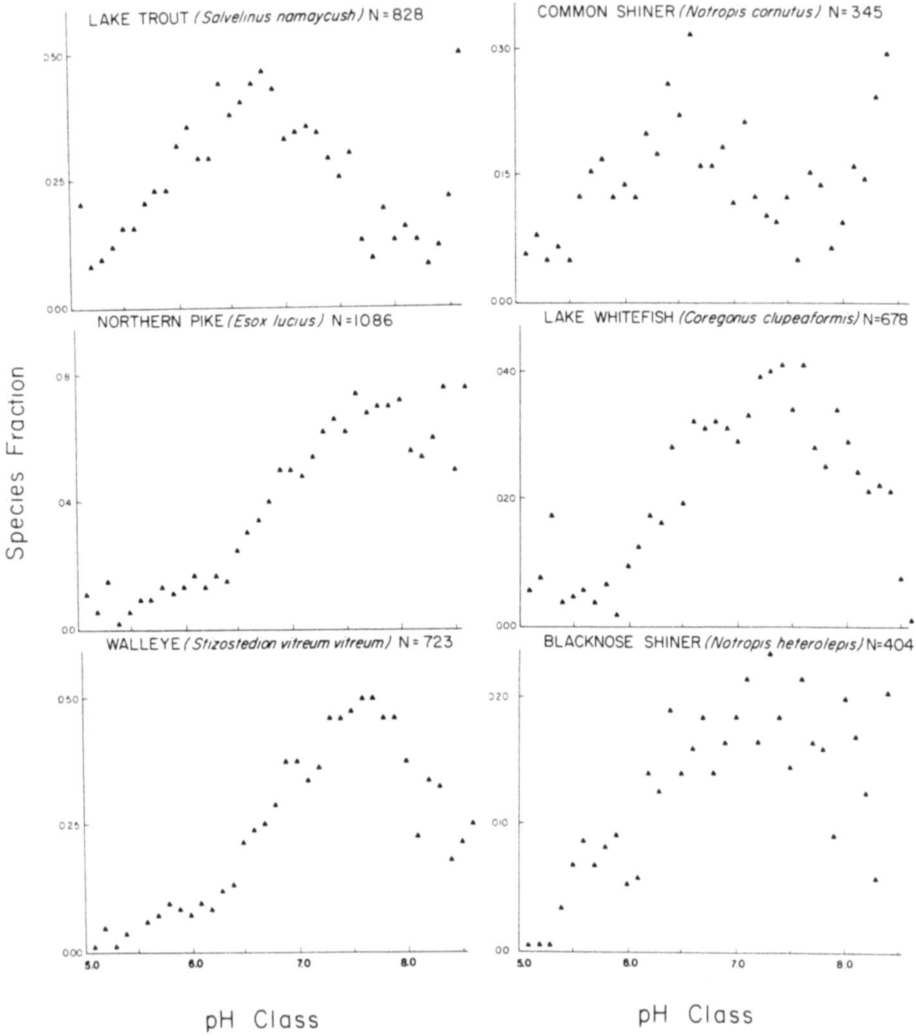

Figure 2. The fraction of sampled lakes containing selected fish species by pH interval. One pH interval spans 0.1 units (eg. interval 6.0 includes lakes with 6.0<=pH<6.1). Species subjectively classified as showing limited distribution at low pH (pH<6.4).

Table I. Spearman correlation coefficients for selected variables. Significance at the 0.0001 level is denoted by an asterisk.

	Elevation	Latitude	pH	Ca	Al
Lake area	-0.053	0.260*	0.226*	0.093*	-0.101*
Elevation		0.043	-0.221*	-0.347*	0.069
Latitude			0.482*	-0.484*	-0.354*
pH				0.843*	-0.630*
Ca					-0.441*

Table II. Descriptive statistics (median, 5th and 95th percentiles (P5 and P95) for latitude (degree, minutes) and lake area (ha) for all sampled lake trout, walleye, northern pike, lake whitefish, common shiner and blacknose shiner lakes, and all sampled lakes in Ontario.

	latitude			lake area (ha)		
	median	P5	P95	median	P5	P95
lake trout	46°33'	44°55'	50°02'	223	23	2401
walleye	48°29'	45°00'	50°20'	247	39	3700
northern pike	48°36'	45°09'	50°19'	212	19	2964
lake whitefish	48°05'	45°09'	50°21'	326	41	4267
common shiner	46°02'	45°03'	49°44'	84	7	2046
blacknose shiner	48°08'	44°56'	50°25'	166	12	2955
all lakes	46°48'	44°53'	49°57'	82	6	1399

4. DISCUSSION

Presence of any species is a function of many biotic and abiotic factors. If a species is found to occur at low pH, one can assume it is tolerant of low pH. However, if a species is not found at low pH or has limited distribution compared to higher pH, without historical data we cannot imply the species is sensitive to low pH or disappeared from these systems.

4.1 Species well distributed at low pH

Both laboratory and field studies have shown brook trout to be very tolerant of low pH with successful reproduction occurring in lakes with pH as low as 5.0 (Schofield, 1976; Beggs et al., 1986). Our study found brook trout to be well distributed at low pHs. Yellow perch, redbelly dace and pumpkinseeds were also well distributed across all pHs, coinciding with observations by other researchers (Beamish and Harvey, 1972; Rahel and Magnuson, 1983).

White suckers were common across the entire range of pH (Figure 1). Rahel (1986) found common white suckers to be a major species present in low alkalinity lakes in northern Wisconsin. Recruitment of white suckers ceased due to fry mortality only after lake 223 was experimentally acidified to pH 5.02 (Schindler et al., 1986), suggesting they are tolerant to low pH.

Fathead minnow showed fairly constant distribution over the pH range 5.6 to 8.0. This species failed to reproduce in lake 223 after the lake was experimentally acidified to pH 6.13 (Schindler et al., 1986). Rahel and Magnuson (1983) also reported this species to be very sensitive to low pH. Both these studies differ from the results of this study, in which we found fathead minnows in 47 lakes over the pH range 5.1 to 5.99. Given the short life span of the species, it is unlikely that fathead minnows are present in our lakes but not recruiting. Inlets and outlets may serve as a refuge from potentially toxic waters. Rahel and Magnuson (1983) reported that the acidic lakes in their study tended to be without inlets or outlets and, therefore, without a temporary refuge area for fish. Tolerance to acidic conditions is also thought to vary with acclimation and the strain of the fish (Flick et al., 1981). Since lake acidification due to deposition is likely to be a much slower process than experimental acidification, the differences in species tolerance between this study and fathead minnows in lake 223 may be related to acclimation. Alternatively, the difference may be due to genetically inherent strain characteristics.

4.2 Species with limited distribution in low pH waters

As pH is highly correlated with Ca and total Al (Table I), parameters known to affect survival in low pH waters (McDonald et al., 1980; Baker and Schofield, 1982), in the current study it is not possible to determine if pH is the only limiting factor. It may be acting in combination with correlates such as Ca and/or Al.

Lake trout show limited distribution in low pH lakes. Field studies have shown lake trout to be sensitive to low pH with recruitment failures occurring at pH <5.5 (Beggs et al., 1986; Schindler et al., 1986). They are also distributed in areas both sensitive to and receiving high acid loadings with 55% of the total Ontario lake trout resource located in the area of the province receiving the highest loadings of acidic deposition (Beggs et al., 1985b).

Common shiner are distributed in southern Ontario with 50% of the sampled lakes located south of latitude 46°02' (Table II), an area of the province receiving mean annual wet sulphate deposition of >20 kg ha^{-1} $year^{-1}$ (Chan et al., 1983). A sulphate deposition of below 20 kg ha^{-1} $year^{-1}$ has been proposed as a loading critical for protection of sensitive water bodies (MOI, 1983). Therefore, at least half of the sampled common shiner lakes are receiving deposition levels believed to damage fish populations. Common shiner are distributed in small lakes (Table II) which are more likely to be of low pH when compared to larger lakes (Table I). The species has been found to be sensitive to low pH, disappearing from lakes at pH levels ranging from 6.5 to 5.7 (Harvey, 1980; Rahel and Magnuson, 1983).

Walleye and northern pike were found to be poorly distributed at low pH. Field bioassay studies show high mortality of walleye eggs at pH 5.4 (Hulsman et al., 1983). In the laboratory, mortality of northern pike embryos occurred at pH 5.0 (Johanssan and Kihlstrom, 1975). Walleye and northern pike are generally distributed in northern Ontario with 50% of their sampled lakes occurring north of 48°29' and 48°36', respectively (Table II). These areas receive mean annual deposition of less than 12 kg sulphate ha^{-1} $year^{-1}$ (Chan et al., 1983). Acid lakes are generally found in areas receiving sulphate loadings in excess of 15 to 20 kg ha^{-1} $year^{-1}$ (MOI, 1983). Both species inhabit shallow productive waters of large size and high conductivity (Johnson et al., 1977); lakes generally not susceptible to acidification stress. Therefore, these species, by virtue of their geographic distribution and the physical characteristics of the lakes they inhabit, are distributed in lakes of higher pH.

Fifty percent of the sampled blacknose shiner and lake whitefish lakes are located north of 48°05' (Table II), an area of the province receiving sulphate deposition of <12 kg ha^{-1} (Chan et al., 1983). Of the species examined in this report, lake whitefish occur in lakes with the largest median area (325 ha) (Table II). Their limited distribution in low pH waters may, therefore, be a reflection of their distribution in northern lakes of large size rather than their sensitivity to low pH.

The median lake area for sampled blacknose shiner lakes (166 ha) show they inhabit lakes of intermediate size when compared to the total population of sampled lakes (Table II). Other researchers (Rahel and Magnuson, 1983) report an absense of blacknose shiner below pH 6.2. Their limited distribution in low pH lakes may, therefore, be a reflection of both their northern Ontario distribution and their chemical sensitivity to low pH.

Regional distribution of fish species data with respect to lake pH, used in combination with lake physical and geographic characteristics provide an indication of the species' tolerance to acidification stress. The absence of common shiner and lake trout at low pH could not be explained on a zoogeographic basis. The absense of walleye, northern pike and lake whitefish in low pH lakes is likely due to the distribution of the species in lakes and areas of the province not susceptible to acidification stress. The absense of blacknose shiner in low pH lakes may be due, at least in part to their distribution in northern lakes.

4. ACKNOWLEDGEMENT

Research funds were provided by the Ontario Ministry of Natural Resources. We thank L. Maki, B. Keller and P. Dillon of the Ontario Ministry of the Environment for providing some of the chemical data. J. Matuszek and D. Green conducted the data analyses. We thank M. McMurtry, J. Matuszek and J. Gunn for critically reviewing the manuscript.

5. REFERENCES

Baker, J.P. and Schofield, C.L.: 1982, Water, Air, Soil Pollut. 18,289.

Beamish, R.J. and Harvey, H.H.: 1972, J. Fish Res. Board Can. 40,3.

Beggs, G.L., MacLean, J.A., Gunn, J.M., Jones, M.L. and Minns, C.K.: 1985a, The case for effects of acid deposition on Ontario fisheries. U.S. Fish and Wildlife Tech. Report, in press

Beggs, G.L., Gunn, J.M., Olver, C.H.: 1985b, The sensitivity of Ontario lake trout (Salvelinus namaycush) and lake trout lakes to acidification. Ont. Fish. Tech. Rep. Ser. no. 17

Beggs, G.L., Gunn, J.M., Shuter, B.J. and Ihssen, P.E.: 1986, Water, Air, Soil Pollut., submitted

Chan, W.H., Tang, A.J. and Lusis, M.A.: 1983, Precipitation concentration and wet deposition fields of pollutants in Ontario, September 1980 to December 1981. Ontario Ministry of the Environment, Air Resources Branch, Toronto, Ontario. ARB-61-83 ARSP

Dodge, D.P., Goodchild, G.A., MacRitchie, I., Tilt, J.C. and Waldriff, D.G.: 1985, Manual of Instructions, Aquatic Habitat Inventory Surveys. Ontario Ministry of Natural Resources, Fisheries Branch, Toronto.

Flick, W.A., Schofield, C.L. and Webster, D.A.: 1981, Remedial Actions for Interim Maintenance of Fish Stocks in Acidified Waters. In: Acid Rain/Fisheries, R.E. Johnson (ed.), American Fisheries Society, Bethesda, MD.

Harvey, H.H.: 1980, Widespread and diverse changes in the biota of North American lakes and rivers coincident with acidification. In: Ecological Impact of Acid Precipitation, SNSF project, Drablos and Tollan (ed).

Hulsman, P.F., Powles, P.M. and Gunn, J.M.: 1983, Tran. Amer. Fish. Soc. 112,680.

Jeffries, D.S., Cox, C.M. and Dillon, P.J.: 1979, J. Fish. Res. Board Can. 36,640.

Johansson, N. and Kihlstrom, J.E.: 1975, Envir. Res. 9,12.

Johnson, M.G., Leach, J.H., Minns, C.K. and Olver, C.H.: 1977, Can. J. Fish. Aquat. Sci. 34,1592.

Matuszek, J.E. and Beggs, G.L.: 1986, Fish species richness in relation to pH in Ontario lakes. Ontario Ministry of Natural Resources, Fisheries Branch, Toronto, Ontario. Unpub. Report.

McDonald, D.G., Hobe, H. and Wood, C.M.: 1980, J. Exp. Bio. 88,109.

Minns, C.K.: 1981, Acid rain: a preliminary estimate of the risk to Ontario's inland fisheries. Can. MS Rep. Fish. Aquat. Sci. 1622

MOI: 1983, Canada-U.S.A. Memorandum of Intent Work Group 1 (Impact Assessment) Final Report

Ontario Ministry of the Environment: 1981, Outlines of Analytical Methods, A guide to the occurrence, significance, sampling and analysis of chemical and microbiological parameters in water, sediment, soil, vegetation and air. Toronto, Ontario. Unpublished report.

Rahel, F.J.: 1986, Can. J. Fish. Aquat. Sci. 43,124.

Rahel, F.J. and Magnuson, J.J.: 1983, Can. J. Fish. Aquat. Sci. 40,3.

SAS: 1982, User's Guide: Basics, SAS Institute Inc., Cary, North Carolina

Schindler, D.W., Mills, K.H., Malley, D.F., Findlay, D.I., Shearer, J.A., Davies, I.J., Turner, M.A., Linsey, G.A. and Cruikshank, D.R.: 1986, Science 228,1395.

Schofield, C: 1976, Ambio 5,228.

POPULATION DYNAMICS OF YELLOW PERCH (PERCA FLAVESCENS) AND PUMPKINSEED (LEPOMIS GIBBOSUS) IN TWO ACID-STRESSED LAKES

J. Sun
Institute of Hydrobiology
Academia Sinica
Wuhan, Hubei
People's Republic of China

H. H. Harvey*
Department of Zoology
University of Toronto
Toronto, Ontario M5S 1A1
Canada

* Author for all correspondence

ABSTRACT. Fish population parameters: abundance, age-class composition, and size-at-age, which respond to environmental stresses, were investigated in two lakes showing chemical and biological evidence of acidification. Plastic Lake, Muskoka District of Ontario, shows spring pH depression and Al toxicity in surface waters, with an associated annual fish kill. In Crosson Lake, pH 5.2 to 5.3, there has been a dramatic decline in abundance of the white sucker (Catostomus commersoni). In Plastic Lake the abundance of pumpkinseed (Lepomis gibbosus) has declined and that of rock bass (Ambloplites rupestris) has increased. In Crosson Lake size-at-age has increased in yellow perch (Perca flavescens), pumpkinseed abundance has increased but with no change in growth rate.

1. INTRODUCTION

Several attributes of fish populations have been shown to respond to the stresses of acidification: (1) fish abundance may be reduced; (2) age-class composition may shift to older fish with failed recruitment, or to younger fish with premature death; (3) size-at-age may be increased as a consequence of lessened competition for food, or reduced presumably in response to loss of acid-sensitive prey and/or physiological stress. The purpose of this study was to examine the major fish populations in an acidifying lake, Plastic (Dillon, 1985), and a lake of low pH, Crosson (Zimmerman and Harvey, 1979), for evidence of perturbation of these population attributes 1979 to 1984.

Water, Air, and Soil Pollution **30** (1986) 611–617.

2. METHODS

Fish were sampled by means of standardized gear, including survey gill nets, fine and coarse-mesh trap nets, baited and plastic traps, and seines. In 1984, large minnow traps were used also. Catch per unit effort was calculated separately for each gear; trap and gill net captures were used in comparisons of age-class distribution. The techniques of age determination were described in Beamish and Harvey (1969) and Ryan and Harvey (1977, 1980). Lake pH and alkalinity were determined at time of fish capture. Annual average pH data for the 2 lakes, 1979 to 1984 was obtained from Reid (1985). Water chemistry profiles were taken from Zimmerman and Harvey (1979) and Reid et al. (1983). Fish population estimates, based on mark-recapture were calculated separately for small fish by fin-clip: pumpkinseed (Lepomis gibbosus) and rock bass (Ambloplites rupestris), 6.5 to 10 cm; yellow perch (Perca flavescens) and white sucker (Catostomus commersoni), 7 to 15 cm; and by numbered tags for larger fish (Harvey and Lee, 1981). The advantages of comparing mean age at a specific size were discussed by Ryan and Harvey (1977, 1980).

3. RESULTS AND DISCUSSION

In Plastic Lake, the fish fauna has shown the following responses in recent decades: (1) Brook trout (Salvelinus fontinalis) were stocked in 11 yr, 1940 to 1959, with an average of 1,409 fingerlings per stocking, but only 2 brook trout, age ~5 yr, were captured in 1984. (2) Rainbow trout were stocked at an annual average of 1,218 fingerlings, with captures rare and the last capture in 1973. (3) The population of outlet-spawning white suckers has become extinct. (4) There are annual fish kills of endemic species, especially pumpkinseeds (Harvey et al., 1982). (5) Rainbow trout held in cages within Plastic Lake and tributary streams died with the symptoms of acid and Al stress (Harvey and McArdle, 1986).

Three endemic species showed some evidence of perturbation 1978 to 1984. Pumpkinseed abundance was reduced, based on a Petersen estimate of 8,896 fish in 1979 and comparison of CPUE for gill nets, the population was 27% smaller in 1984. The numbers of dead and moribund pumpkinseeds observed and recovered, during the annual counts at ice-out, were small relative to the estimated population. It is not known if the decline in pumpkinseed abundance was the direct result of the annual spring fish kills. Pumpkinseed of age 4 and 5 yr were significantly smaller in 1984, but age at 90 mm was not significantly different, 1979 to 1984 (Table I).

Yellow perch in Plastic Lake were sparse in 1979, at $6.2.ha^{-1}$, and based on comparison of CPUE for gill nets, the population increased 1.4X by 1984. Perch 4 and 5 yr old were significantly larger and mean age at 140 mm declined (Table II).

Rock bass in Plastic Lake were too rare to calculate a population estimated in 1979, but CPUE increased 13.5X by 1984, with recruitment of larger year classes into the population 1980 to 1982. The rock bass of Plastic Lake grew to 110 mm at an early age, only one of the 24 populations of Ryan and Harvey (1977) grew as quickly (Table III).

TABLE I

Lake		Mean back-calculated fork length (mm)* of pumpkinseed at the end of last full year of life of fish in each age group							Mean age at 90 mm
		1	2	3	4	5	6	7	(yr)
Plastic	1979	31.4 ±1.5 (20)	58.6 ±15 (60)	80.4 ±1.4 (89)	100.4 ± 1.6 (45)	126.6 ± 7.0 (3)	140.8 ±20.5 (3)		3.5
	1980	28.5 ±1.6 (15)	58.6 ±5.1 (7)	80.9 ±1.8 (49)	96.5 ±1.6 (55)	117.0 ± 2.5 (17)	146.3 ±18.4 (2)	166.3 (1)	3.6
	1984	30.1 ±1.4 (71)	56.1 ±2.2 (36)	78.3 ±2.3 (26)	97.7 ±1.7 (34)	112.4 ± 4.8 (7)			3.8
Crosson	1979	32.5 ±2.4 (18)	54.7 ±2.5 (14)	71.2 ±2.5 (23)	95.4 ±1.3 (124)	113.0 ± 2.1 (32)			3.8
	1984	27.2 ±1.5 (39)	55.7 ±1.6 (43)	74.2 ±1.7 (20)	89.6 ±2.7 (32)	114.8 ± 3.7 (16)	134.3 ± 6.6 (5)		3.9

* Shown below the mean are 95% confidence interval and sample size.

TABLE II

Lake		Mean back-calculated fork length (mm)* of yellow perch at the end of last full year of life of fish in each age group										Mean age at 140 mm
		1	2	3	4	5	6	7	8	9	10	(yr)
Plastic	1979	66.7	123.1	149.0	164.9	188.3		233.9				
		± 7.6	± 7.7	± 2.6	± 2.2							3.1
		(2)	(8)	(21)	(34)	(1)		(1)				
	1980			149.3	163.8	188.5	210.7	237.7		280.5	317.0	
					± 5.2	± 3.0	± 6.5	±66.1				2.9
				(1)	(9)	(22)	(3)	(2)		(1)	(1)	
	1984	69.8	130.5	166.3	196.9	208.8	222.8	240.5	259.8	295.1		
		± 3.1	± 4.9		± 1.7	± 5.6	±47.6	±31.8	±12.1	±129.6		2.6
		(13)	(15)	(1)	(3)	(4)	(2)	(2)	(2)	(2)		
Crosson	1979	66.1	86.9	110.7	130.8	152.0	171.6	201.9	217.9		260.5	
		±11.5	± 8.1	± 2.1	± 1.8	± 1.5	± 2.2	±10.4	± 7.9			4.4
		(3)	(4)	(24)	(48)	(41)	(26)	(3)	(4)		(1)	
	1984	70.7	93.9	112.7	137.3	169.5	190.7	216.6	240.9			
		± 1.8	± 1.3	± 1.7	± 3.0	±118.2	± 5.1	±62.3				3.9
		(39)	(37)	(42)	(22)	(2)	(4)	(2)	(1)			

* Shown below the mean are 95% confidence interval and sample size.

Crosson Lake has been stocked with rainbow trout as recently as 1978 with no record of capture, and with brook trout, of which three fish were caught in 1984. There has been a dramatic decline in the population of white suckers. The 1979 Petersen estimate yielded 764 mature fish, and catch per net set declined from 5.0 (1978) and 5.8 (1979) to 2.0 (1983) and 0.25 (1984). This decrease in abundance was not due to exploitation (sampling) (Trippel and Harvey, 1986), and the white sucker population of the control lake, Harp (pH 6.8), showed no change in abundance in response to a similar mark-recapture effort. There are no historic data on age-class distribution of the population, and young fish dominated the population 1979 to 1982 (Figure 1). White sucker reproduction could be expected to fail in the pH range 5.2 to 5.3, but the age-class composition did not indicate that this was the mechanism of action. The fish biomass estimates for Crosson Lake in 1979 were: white sucker 4.02 $kg.ha^{-1}$, pumpkinseed 30.7 $kg.ha^{-1}$, and yellow perch 78.7 $kg.ha^{-1}$. Thus the near extinction of white suckers could be expected to provide little additional resources, relative to the biomass of pumpkinseed and yellow perch, for the observed changes in these species, 1979 to 1984. The yellow perch population declined 40% in abundance, but showed an increase in size-at-age for all ages, and age at 140 mm declined significantly (Table II). The data for yellow perch, both lakes, were superimposed on the results of Ryan and Harvey (1980) for comparison (Figure 2). The two data sets show a trend to lower age at 140 mm for populations in acid lakes, but correlation coefficients were low: r=0.25, n=36 (Ryan and Harvey, 1980) and r=0.26, n=10, this study. Pumpkinseed trap net CPUE increased 6 fold, 1979 to 1984, but there were no significant changes in size at age or age at 90 mm (Table I).

TABLE III

Plastic Lake	Mean back-calculated fork length (mm)* of rock bass at the end of last full year of life of fish in each age group							Mean age at 110 mm
	1	2	3	4	5	6	7	(yr)
1979	36.2 ±5.4 (4)	90.1 ±5.1 (4)	132.2 ±12.4 (6)			215.1 (1)		2.8
1980	31.4 (1)	104.4 ±21.6 (4)		164.3 ±11.9 (5)				2.6
1984	33.8 ±1.3 (81)	80.0 ±4.6 (29)	135.8 ± 3.5 (39)	177.4 ± 7.0 (12)	202.3 ± 4.4 (8)	226.9 ± 1.7 (2)	245.5 ±33.1 (2)	2.7

* Shown below the mean are 95% confidence interval and sample size.

Lakes Plastic and Crosson are inhospitable to acid-sensitive fish species. The population dynamics of more acid tolerant species may be affected indirectly.

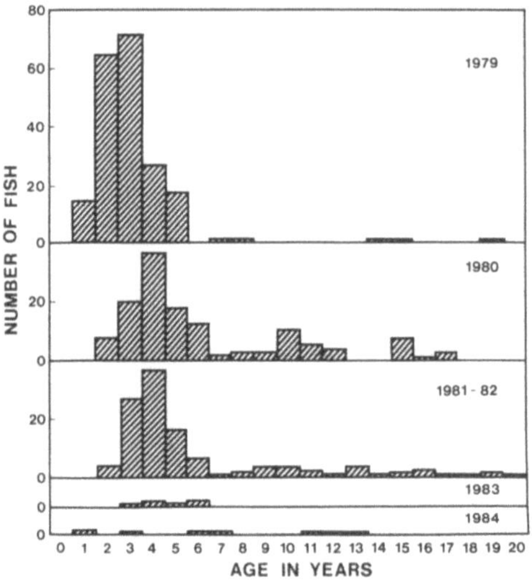

Figure 1. Age-class composition of white suckers sampled from Crosson Lake. Source: Harvey (1985).

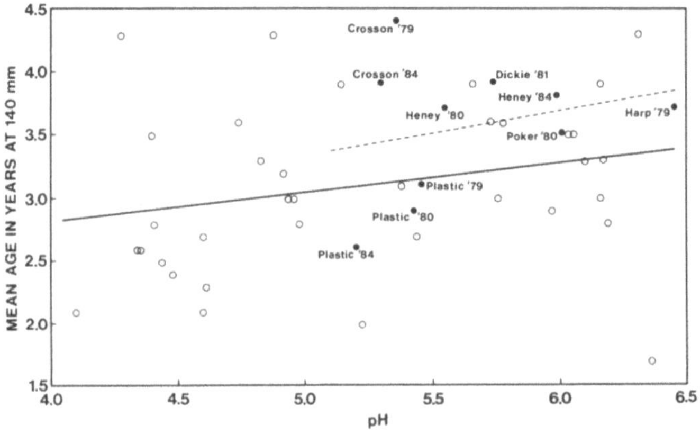

Figure 2. Mean age at 140 mm for yellow perch in relation to lake pH (at time of fish capture). Open circles and solid line (Y = 1.884 + 0.230X) from Ryan and Harvey (1980). Asterisks and dashed line (Y = 1.566 + 0.351X) based on the two study lakes plus unpublished data from lakes Dickie, Poker, Harp, and Heney.

4. REFERENCES

Beamish, R.J., and Harvey, H.H.: 1969, J. Fish. Res. Board Can. **29**, 1131.

Dillon, P.J.: 1985, Rates of alkalinity generation and loss in aquatic and terrestrial systems. Abstracts, Muskoka Conference '85, p. 162.

Harvey, H.H.: 1982, Population responses of fishes in acidified waters. In: Acid rain/fisheries, R.E. Johnson (ed.), American Fisheries Society, Bethesda, Maryland, pp. 227-242.

Harvey, H.H.: 1985, The biological evidence of acidification, mechanism of action and an attempt at predicting acidification effects. In: Symposium on the effects of air pollution on forest and water ecosystems, Helsinki, April 23-24, 1985, pp. 63-78.

Harvey, H.H., and Lee, C.: 1981, Fish populations in the calibrated watersheds of Muskoka-Haliburton. Conf. Proc., Technology Transfer Conf. No. 2, Ontario Ministry of Environment, pp. 316-370.

Harvey, H.H., Dillon, P.J., Fraser, G.A., Somers, K.M., Fraser, P.E., and Lee, C.: 1982. Elevated metals and enhanced metal uptake in fishes in acid-stressed waters. Extended Abstract, Am. Chem. Soc., Div. Environ. Chem., pp. 438-441.

Harvey, H.H., and McArdle, J.M.: 1986, Water Air Soil Pollut., in press.

Reid, R.A.: 1985, Personal communication.

Reid, R.A., Girard, R., and Locke, B. 1983, Oxygen profiles in the Muskoka-Haliburton study lakes (1976-1982), Ontario Ministry of Environment, Data Report DR 83/5. 62 pp.

Ryan, P.M., and Harvey, H.H.: 1977, J. Fish. Res. Board Can. 34, 2079.

Ryan, P.M., and Harvey, H.H.: 1980, Environ. Biol. Fish. **5**, 97.

Trippel, E.A., and Harvey, H.H.: 1986, Can. J. Fish. Aquat. Sci., submitted.

Zimmerman, A.P., and Harvey, H.H.: 1979, Sensitivity to acidification of waters of Ontario and neighboring states. Report to Ontario Hydro. 48 pp. + appendices.

FISH KILLS of ATLANTIC SALMON (*SALMO SALAR*) and BROWN TROUT (*SALMO TRUTTA*) in an ACIDIFIED RIVER of SW NORWAY

TRYGVE HESTHAGEN
Directorate for Nature Management
Fish Research Division
Tungasletta 2
N-7000 Trondheim, Norway

Abstract. This paper documents population responses of Atlantic salmon and brown trout to fish kills in River Vikedal, an acidified river in SW Norway. The angling statistics show that the adult population of Atlantic salmon in the river has decreased to low numbers during recent years, whereas catches of migrant brown trout have increased in the same period. A total of 659 dead salmon and trout were either collected or observed during spring snowmelt in the years 1982-1985. Parr suffered the highest mortality, and most large specimens. Only a few dead kelts were registered. In the springs of most unfavourable water quality and most severe episods of fish kills (1983 and 1984), salmon parr mortality were significantly higher than that of brown trout. Episodic death of parr in the spring is thought to be an important cause of the reduction in the adult stock of Atlantic salmon in the river during recent years.

1. Introduction

The acidification of freshwaters in Norway has eliminated populations of Atlantic salmon (*Salmo salar* L.) in all major rivers in the southernmost counties (Jensen and Snekvik, 1972; Leivestad *et al*., 1976; Muniz, 1981). The salmon populations in these rivers began to decline at the beginning of this century, and by the 1960's they became virtually extinct.

Eggs, alevins and fry are those life stages at which Atlantic salmon are considered most vulnerable to low pH (Jensen and Snekvik, 1972; Daye and Garside, 1977; Daye, 1980; Muniz and Leivestad, 1980; Haines, 1981; Lacroix, 1985; Lacroix *et al*., 1985). In some cases fish kills involving adult salmon have also been observed (Huitfeldt-Kaas, 1922; Rosseland, 1953; Snekvik, 1975; Skogheim *et al*., 1984).

A severe fish kill of young salmonids was observed in River Vikedal, an acidified river in SW Norway (Figure 1), during snowmelt in spring 1981 (Rosseland and Skogheim, 1982; State Pollution Control Authority, 1983). The angling statistics for

Water, Air, and Soil Pollution **30** (1986) 619–628

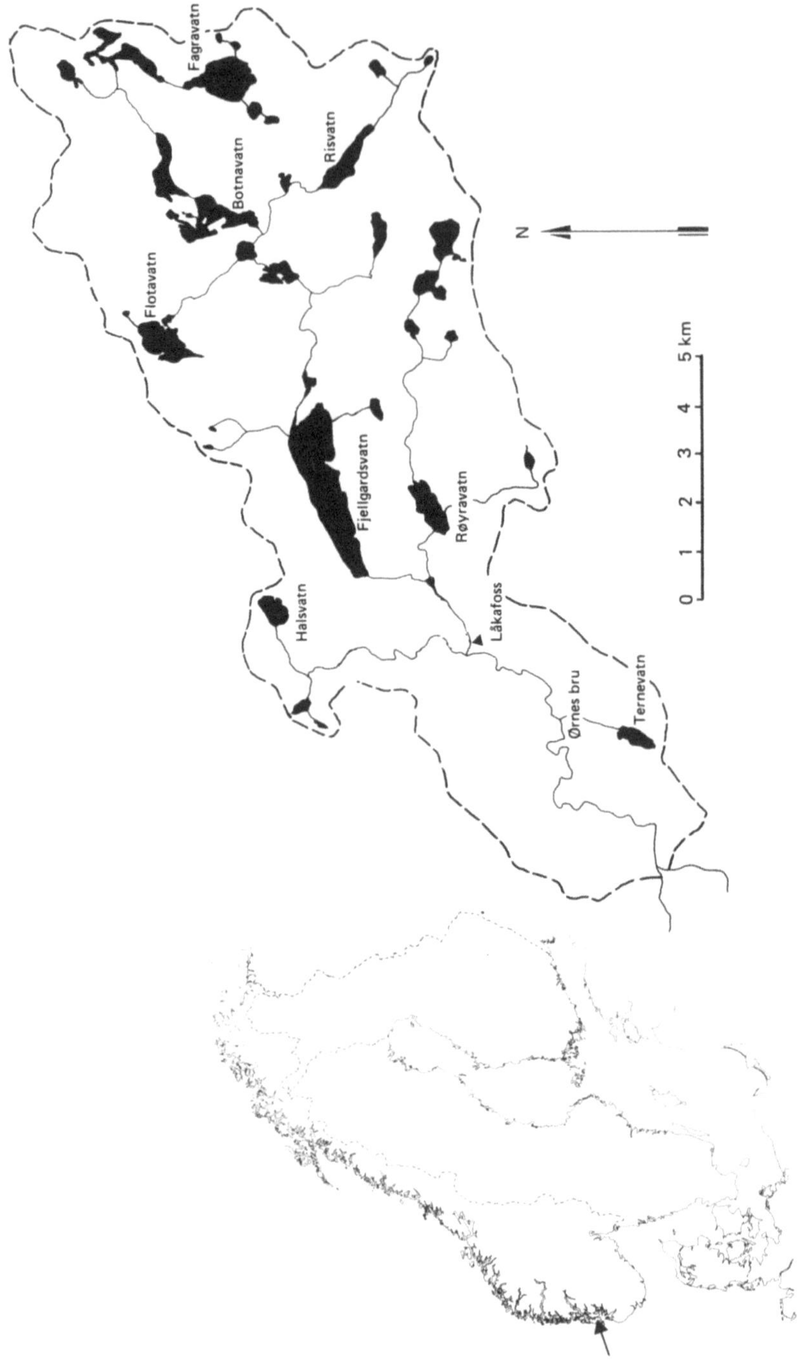

Fig. 1. Geographic location and catchment area of River Vikedal.

this river demonstrate a marked decline in catches of adult Atlantic salmon during recent years, whereas those of brown trout (Salmo trutta L.) have increased during the same period, Figure 2. Compared with the early 1979's, the pH level in the river Vikedal dropped markedly during the late part of this decade (State Pollution Control Authority).

A study which included the collection of dead fish and monitoring of water quality was initiated in spring 1982, in the event that new episodes of fish kill occurred. In addition, electrofishing has been carried out every fall since 1981 in order to assess any changes in the juvenile stocks of Atlantic salmon and brown trout.

Episodic fish kills have been registered every spring during 1982-1985, and based on continuous pH measurements there is evidence to link mortality to low pH effects (Hesthagen and Henriksen ms.). Experiments with captive Atlantic salmon parr in spring 1984, confirmed correlations between fish kill and water quality (Rosseland et al., 1986a).
The objectives of the present study were to determine the species and stages with the highest mortality during these episodes of fish kills, and to document related changes in species composition of juvenile Atlantic salmon and brown trout.

2. Study area

River Vikedal is located in SW Norway at 59°30'N, 5°53'E (Figure 1). The catchment area is 119 km^2, having an axial length of 22.6 km, and 29 lakes $\geq$ 2.4 ha with a total surface area of 812 ha (State Pollution Control Authority, 1984). The drainage area consists mainly of slowly weathering rocks (granite and gneisses), as well as some sparse glacilacustrine and glacifluvial deposits. In the lower part of the catchment, there are 87 farms with a total area of 4 km^2 cultivated land. Mean annual precipitation is 2650 mm. Weighted mean precipitation pH from April 1982 to November 1983 was 4.58, with the lowest value being 3.93. Based on monthly sampling during the last decade, the mean annual pH in the lower part of River Vikedal has ranged between 5.35 to 5.60 (State Pollution Control Authority, 1984). Average waterflow at the river mouth is 9 m^3 s^{-1}, and river width ranges between 20 to 30 m. In addition to Atlantic salmon and brown trout, River Vikedal supports stocks of European eel (Anguilla anguilla (L.)), lamprey (Lampetra planeri, (Bloch)), three-spined sticklebacks (Gasterosteus aculeatus L.) and sand flounder (Platichthys flesus (L)). Anadromous brown trout and Atlantic salmon ascend 10 km of the river system up to the waterfall Låkafossen (Figure 1). No tributary to the river is important with respect to fish production.

3. Methods

Surveys for collecting dead fish were generally carried out daily during spring snowmelt in 1982-1985. Three stream

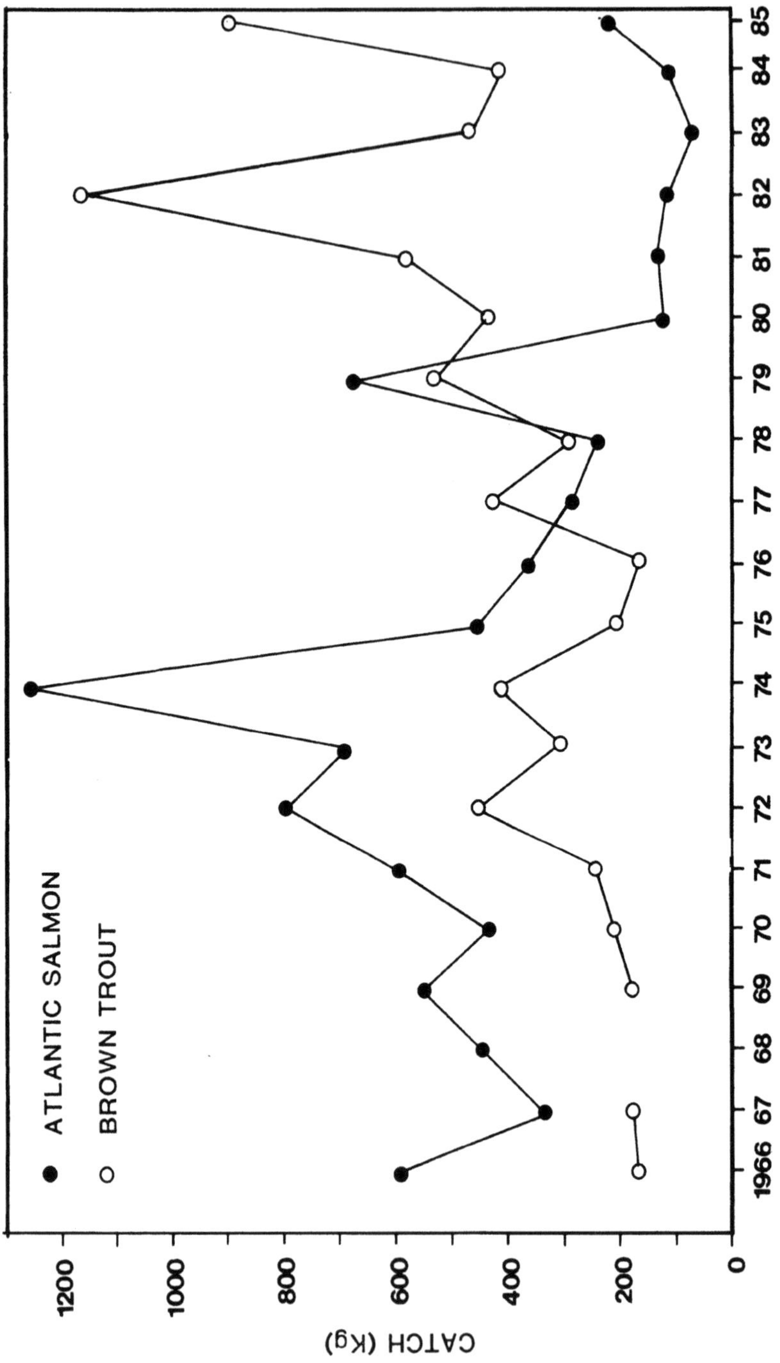

Fig. 2. Official angling statistics (kg) of Atlantic salmon and migrant brown trout caught in River Vikedal during 1966-1984.

stretches of the river and one stretch in the brachish area were included (Hesthagen and Henriksen ms.). Some of the dead fish (salmon or trout) observed could not be collected because they were located in deep areas of the river. Collected fish were frozen shortly after each trip, and the total length (0.1cm) was measured later. Dead salmon and trout with a total length < 20 cm cm were classified as parr, (Hesthagen and Garnås, 1984). Bigger specimens were classified as kelts if they showed sea-growth in their scales.

Juvenile fish density was calculated at three stations in fall 1981 (total area 585m^2), at five stations in 1982-1984 (642 m^2), and at 16 stations in fall 1985 (1712 m^2). Sampling was done with an electrofishing apparatus (1600 V, D.C. unloaded) at sites located in areas which included suitable rearing habitats for both fry and parr.

All fish caught during electrofishing were measured for total length (0.1cm).
Statistical analysis of difference in frequencies was carried out by testing equality of two percentages (Sokal and Rohlf, 1969).

4. Results

4.1. Size of fish found dead and caught by electrofishing

A total of 560 dead Atlantic salmon and brown trout were collected during the years 1982-1985. The mortality was highest in the spring of 1984 (172 salmon and 31 trout), and lowest in the spring of 1985 (1 salmon and 21 trout). In addition, 99 dead salmon and trout were observed during the same period, the majority being registered in 1984 (72 fish). The frequency of kelts among these specimens was high (42.5%).

Generally, juvenile fish between 4-19 cm in length were most abundant among dead fish for both salmon and trout (Figure 3). Only a few dead brown trout kelts were collected (8 specimens). The size of fish caught by electrofishing in the fall prior to springs fish kill showed that fry (age 0+) were dominant (length between 3 to 7 cm), indicating that large parr suffered a higher mortality than small parr. However, taking into account the growth between fall and spring, the size of dead brown trout in the springs of 1984 and 1985 seems merely to reflect the length frequency distribution in the population in the fall of 1983 and 1984, respectively.

4.2. Species composition for dead fish and for fish caught by electrofishing

Proportions of Atlantic salmon (68%) and brown trout (32%), found dead in the spring of 1982, did nor differ significantly from the salmon/trout ratio among live fish obtained by electrofishing in the previous fall (X^2=0.56, p>0.05). However, in the springs of both 1983 and 1984 salmon suffered the highest mortality of these species ($p < 0.05$, X^2=8.39 and

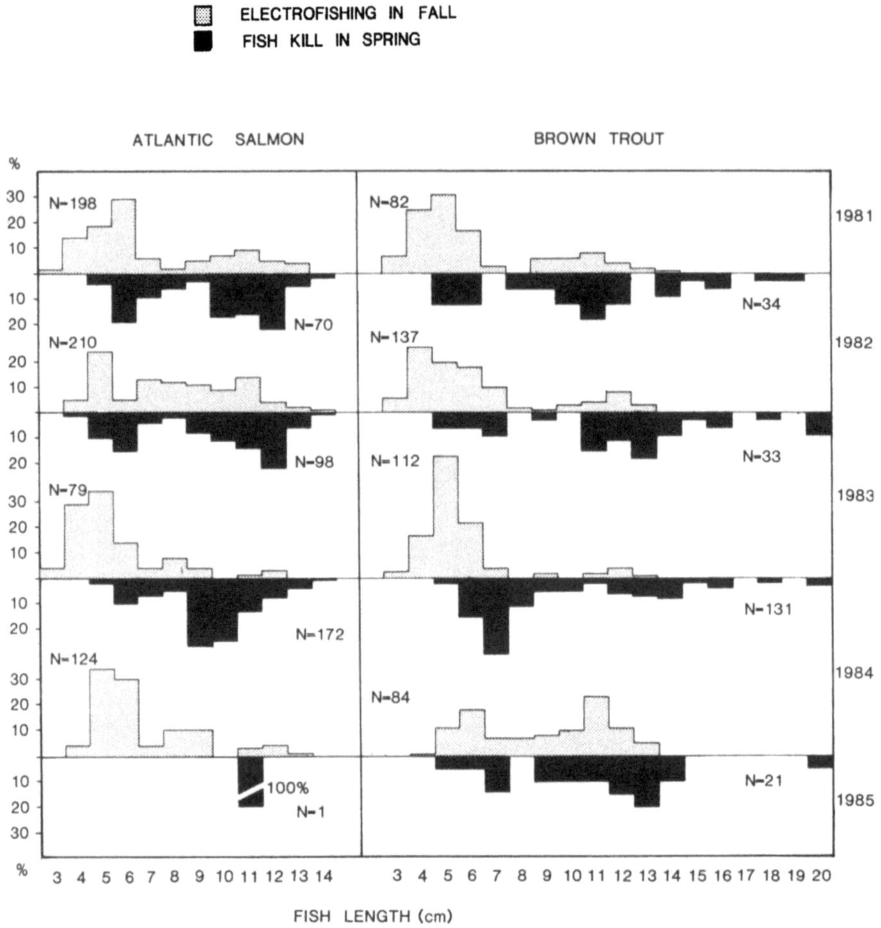

Fig. 3. Length-frequency distribution (%) of juvenile Atlantic salmon and brown trout caught by electrofishing in River Vikedal in fall and that of dead specimens collected next spring. Data from 1981-1985. N=Number of fish.

11.12, respectively). The proportion of juvenile salmon in the river decreased gradually from 1982 (75%) to 1984 (41%). However, in the fall of 1985 the number of young salmon was significantly higher (58%) than that of brown trout ($p < 0.05$).

5. Discussion

Fish kills involving mainly juvenile Atlantic salmon and brown trout were documented every spring from 1982-1985 in River Vikedal, SW Norway. The catches of adult Atlantic salmon have been reduced drastically during the 1980's, while those of migrant brown trout have increased during the same period.

Generally, large parr of both Atlantic salmon and brown trout suffered a significantly higher mortality than small parr. Some of the large parr are assumed to be potensial smolt that year (cf. Elson 1957). For Atlantic salmon, this difference in tolerance to acid water between fish in these two stages (parr and pre-smolt) has also been verified in laboratory experiments (Rosseland and Skogheim, 1982; 1984; Rosseland et al., 1986b. Furthermore, smoltification does not proceed normally in Atlantic salmon subjected to low pH (< 4.9) (Saunders et al., 1983). The lack of recaptures of released smolts of Atlantic salmon in acidic rivers in southernmost Norway also indicates that fish in this stage are sensitive to acidic water (Hansen, 1982).

For brown trout, the size dependent mortality was inconclusive. Laboratory experiments seems to confirm this result, showing that brown trout does not exhibit the same difference in sensitivity between parr and pre-smolts in acid Al-rich water (Rosseland and Skogheim, 1982). They found that two-year-old migrant brown trout (pre-smolts) were more sensitive than one-year-old fish in only one of four tank experiments.

Relatively few dead adults of Atlantic salmon/brown trout were observed in the river. These fishes are probably kelts because salmon and trout do not ascend River Vikedal before June.

Compared with other years the lowest pH levels were recorded in spring 1983 and 1984 (Hesthagen and Henriksen ms.), which coincide with the most severe episodes of fish kills. In these two years salmon parr mortality in spring was significantly higher than that of trout parr, Salmon parr densities in the following falls were low indicating that salmon is the most sensitive species to acid water (unpubl. data). Atlantic salmon was the first species to disappear from acidified Norwegian salmon rivers (Jensen and Snekvik, 1972; Leivestad et al., 1976). Data on physiological stress also confirm these results, showing that one- and two-year-old salmon were more sensitive to acidic, Al-rich water than both migrant- and resident brown trout of the same age (Fivelstad and Leivestad, 1984; Rosseland and Skogheim, 1984; Rosseland et al., 1986 b). In comparative studies on the egg and alevin stages, Atlantic salmon experienced also the highest mortality of these two species (Bua and Snekvik, 1972; Johansson et al., 1977; Grande et al., 1978; Skogheim and Rosseland, 1984).

Generally, the density of Atlantic salmon and brown trout fry in River Vikedal have been relatively constant during the study period (unpubl. data). However, for Atlantic salmon, the fry density was higher in 1985 compared with that in previous years. In the spring of 1984, River Vikedal was strongly acidic with pH values between 4.9 to 5.2 during most of the hatching and postemergent period compared with values between 5.3 to 5.6 in 1985 (Hesthagen and Henriksen ms.). Hatching experiments of Atlantic salmon conducted in the river in these two years showed higher survival in 1985 (unpublished data). Thus, some mortality on natural spawned eggs might occur during the alevin stage and hence reduced survival of Atlantic salmon fry (Bua and Snekvik 1972; Johansson et al., 1977; Grande et al.; 1978;

Skogheim *et al*., 1984). Hence, a partial failure of recruitment in addition to a high mortality of large parr might occur. However, the ultimate population effect (i.e. on smolt production) of such a mortality on young stages is difficult to judge because survival of juvenile Atlantic salmon is density dependent (Buck and Hay, 1984).

The lower return of adult Atlantic salmon compared with that of migrant brown trout may partly be due to difference in exploitation rates. Most Norwegian salmon stocks are heavily exploited in oceanic fisheries, while migrant brown trout are mainly restricted to the fjords, where they are less subjected to fishing. For the salmon in the neighbouring in River Imsa, the exploitation rate of both hatchery reared and wild salmon is estimated at 0.90-0.95 (Hansen, 1986). No such data are available on migrant brown trout, but compared with salmon it is considered to be low. There is also some evidence to suggest that an increase in exploitation for Norwegian salmon has occurred during recent years (Hansen, 1986).

Mortality of parr during spring might not be typical of acidified salmon rivers in other countries. The fish kills in River Vikedal were strongly episodic and occured after rapid changes in water quality due to increased water flow (heavy rainfall or snowmelting) (Hesthagen and Henriksen ms.). In some rivers of Nova Scotia in Canada, the salmon populations now have either disappeared or declined as a result of low pH (Farmer *et al*., 1980; Watt *et al*., 1983). These changes have been correlated to high mortality of eggs, alevins and postemergent stages (Lacroix, 1985; Lacroix *et al*., 1985). However, in these rivers, the minimum pH in spring occurred before the peak flow because snow was higher in pH than rain (Watt *et al*., 1983).

Acknowledgements

I am most grateful to Einar, Helge and Jonny Langhelle for their assistance during this study. Bror Jonsson gave valuable comments on a draft of this manuscript, and Mrs. Amy Lightfoot kindly corrected the English.

References

Bohlin, T.: 1981, *Rep.Inst.Freshw.Res.Drottningholm* 59, 5.
Bua, B. and Snekvik, E.: 1972, Klekkeforsøk med rogn av laksefisk 1966-1971. Virkning av Surhet og Saltinnhold i Klekkevannet. (Hatching Experiments with Roe from Salmonid Fishes 1966-1971. Effects of Acidity and Dissolved Salt in the Water) *Vann* 7, 86 (In Norwegian).
Buck, K.J. and Hay, D.: 1984, *J. Fish Biol*. 23, 1.
Daye, P.G. and Garside, E.T.: 1977, *Can. J. Zool*. 55, 1504.
Daye, P.G.: 1980, *Can. J. Fish. Aquat. Sci*. 37, 1035.
Elson, P.F.: 1957, *Can. Fish. Cult*. 21, 1.

Farmer, G.J., Goff, T.R., Ashfield, D., and Samant, H.S.: 1980, 'Some Effects of the Acidification of the Atlantic Salmon Rivers in Nova Scotia', Tech. Rep. Fish. Aquat. Sci., 972: viii, 13 p.

Fivelstad, S. and Leivestad, H.: 1984, Rep. Inst. Freshw. Res. Drottningholm 61, 69.

Grande, M., Muniz, I.P. and Andersen, S.: 1978, Verh. Int. Verein. Limnol. 20, 2076.

Haines, T.A.: 1981, 'Effects of Acid Rain on Atlantic salmon Rivers and Restoration Efforts in the United States', in Acid Rain and the Atlantic salmon, Proceedings of a conference held in November 22 - 23, 1980. Int. Atlantic Salmon Found. Spec. Publ. 10, 1981, pp. 57-63.

Hansen, L.P. 1982, 'Gjenfangster av Merket Laksesmolt Salmo salar L. Utsatt i to Sure Elver på Sørlandet. (Recaptures of Tagged Smolts of Atlantic Salmon Salmo salar L. Released on two Acid Rivers in southern Norway). Fauna 35, 145. (Summary in English).

Hansen, L.P.: 1986, 'The Salmon Catch Data available for Analysis in Norway', in D. Jenkins (ed.), The status of Atlantic salmon in Scotland, Symposium Int. Terr. Ecol., Aberdeen, Scotland (in press).

Henriksen, A., Skogheim, O.K., and Rosseland, B.O.: 1984, Vatten 40, 255.

Hesthagen, T. and Garnås, E.: 1984, Fauna norv. Ser. A, 5, 46.

Huitfeldt-Kaas, H.: 1922, 'Om Aarsaken til Massedød av Laks og Ørret i Frafjordelven, Helleelven og Dirdalselven i Ryfylke høsten 1920' (On the Cause of Mass Death of Salmon and Trout in Frafjordelven, Helleelven and Dirdalselven in Ryfylke, Autumn 1920). Norsk Jæger- og Fiskefor., tidsskrift, 1-2, 37. Oslo, Norway.

Jensen, K.W. and Snekvik, E.: 1972, Ambio 1, 223.

Johansson, N., Runn, P., and Milbrink, G.: 1977, Zoon 5, 127.

Lacroix, G.L.: 1985, Can. J. Fish. Aquat. Sci. 42, 292.

Lacroix, G.L., Gordon, D.J., and Johnston, D.J.: 1985, Can. J. Fish. Aquat. Sci. 42, 768.

Leivestad, H., Hendrey, G., Muniz, I.P., and Snekvik, E.: 1976, 'Effect of Acid Precipitation on Freshwater Organisms', in Brekke, F.H. (ed.), Impact of acid precipitation on forest and freshwater ecosystems in Norway, SNSF-project, FR 6/76, Aas, Norway, pp. 87-111.

Leivestad, H. and Muniz, I.P.: 1976, Nature 259, 391.

Muniz, I.P.: 1981, 'Acidification and the Norwegian Salmon', in Acid Rain and the Atlantic Salmon, Proceedings of a conference held in November 22-23, 1980. Int. Atlantic Salmon Found. Spec. Publ. 10, 1981, pp. 65-72.

Muniz, I.P. and Leivestad, H.: 1980, 'Acidification - Effects on Freshwater Fish', in D. Drabløs and A. Tollan (eds.), Ecological Impact of Acid Precipitation, SNSF-project, Aas, Norway, pp. 84-92.

Rosseland, L.: 1953, 'Om Virksomheten i 1948' (Annual report 1948) in Annual Report from the Fishery Inspector 1948-50, Department of Agriculture, Oslo, Norway.

Rosseland, B.O. and Skogheim, O.K.: 1982, 'Physiological Stress and Mortality of Atlantic Salmon, Salmo salar L., in

Acid Water with high Levels of Aluminium', Inter. Council for the Exploration of the Sea, C.M. 1982/M: 29, 16p.

Rosseland, B.O. and Skogheim, O.K.: 1984, Rep. Inst. Freshw. Res. Drottningholm 61, 185.

Rosseland, B.O., Skogheim, O.K. and Sevaldrud, I.H.: 1986b. Water, Air and Soil Pollut. in press.

Rosseland, B.O., Skogheim, O.K., Kroglund, F., and Hoell, E.: 1986b, Water, Air, and Soil Pollut. in press.

Saunders, R.L., Henderson, E.B., Harmon, P.R., Johnston, C.E., and Eales, J.G.: 1983, Can. J. Fish. Aquat. Sci. 40, 1203.

Skogheim, O.K., Rosseland, B.O., and Sevaldrud, I.: 1984, Rep. Inst. Freshw. Res. Drottningholm 61, 195.

Skogheim, O.K. and Rosseland, B.O.: 1984, Rep. Inst. Freshw. Res. Drottningholm 61, 177.

Snekvik, E.: 1975, 'Episoder med Fiskedød i forbindelse med Forsuring av Vassdrag i 1969' (Episodes of Fish Kills in relation to Acidification of Rivers in 1969). Unpubl. Report, Directorate for Wildlife and Freshwaterfish, Aas, Norway.

Sokal, R.R. and Rohlf, F.J.: 1969, Biometry. W.H. Freman & Company, San Francisco. 776 p.

State Pollution Control Authority.: 1983, 'National Program for Environmental Monitoring' Report 97/83. Oslo, Norway.

State Pollution Control Authority.: 1984, 'National Program for Environmental Monitoring' Report 123/84. Oslo, Norway.

State Pollution Control Authority.: 1985, 'National Program for Environmental Monitoring' Report 201/85. Oslo, Norway.

Watt, W.D., Scott, C.D., and White, W.J: 1983, Can. J. Fish. Aquat. Sci. 40, 462.

INCREASE IN THE ABUNDANCE OF CLADOCERA AT pH 5.1 IN EXPERIMENTALLY-ACIDIFIED LAKE 223, EXPERIMENTAL LAKES AREA, ONTARIO

D.F. Malley and P.S.S. Chang
Department of Fisheries and Oceans
Freshwater Institute
Winnipeg, MB R3T 2N6 Canada

ABSTRACT. Lake 223 in the Experimental Lakes Area, northwestern Ontario was experimentally-acidified with H_2SO_4 from 1976 to the present. Cladocerans increased in absolute and relative abundance and in total biomass when the pH of Lake 223 was lowered from 6.7 to 5.1. Comparing the cladoceran community at pH 6.7 (1974) with that at pH 5.1 (1981 to 1983), Daphnia galeata mendotae disappeared, Diaphanosoma birgei became rare, Holopedium gibberum increased in abundance and Daphnia catawba, recorded for the first time in 1980, became abundant. Bosmina longirostris became more abundant. Except for the decline in D. birgei, the changes in cladoceran species composition with acidification are consistent with known responses of these species to low pH. The increase in importance of cladocerans in 1981 to 1983 was at the expense of copepods. It was hypothesized to be caused by the decline in predation by small fish during 1981 to 1983 as acidification reduced minnow populations and inhibited reproduction in trout and sucker. The loss of the population of the invertebrate predator, the opposum shrimp, Mysis relicta, in 1979 had little effect on cladoceran biomass.

1. INTRODUCTION

Some Cladocera are among the most sensitive crustaceans to acidification. Species of Daphnia disappeared from Swedish lakes when the pH fell below 6.0 (Almer et al., 1974). In acidified lakes in south Norway, Raddum et al. (1980) noted that daphnids were intolerant to acidification, disappearing at around pH 5.0. In acidified lakes in North America, most species of Daphnia are rare or absent below pH 5.0 (Sprules, 1975; Roff and Kwiatkowski, 1977; Confer et al., 1983; Keller and Pitblado, 1984). Nevertheless, at least two species of Daphnia are tolerant to acid. Daphnia catawba and D. pulicaria (rare) occurred primarily in lakes in the La Cloche Mountains with pH below 6.0 (Sprules, 1975). Daphnia pulex was found to be abundant in some acidic Ontario lakes by Keller and Pitblado (1984).

Water, Air, and Soil Pollution **30** (1986) 629–638.

Although daphnids are generally sensitive to acid, other cladocerans are more tolerant. Some highly acidified lakes (pH 4.1 to 4.4) and some acidic metal-contaminated lakes have zooplankton communities dominated (up to 90% of ice-free season biomass) by the single cladoceran species, Bosmina longirostris (Yan and Strus, 1980), or by this species and the calanoid copepod Diaptomus minutus (NRCC, 1981). Numerous authors report that B. longirostris and/or Holopedium gibberum are tolerant to acid (Almer et al., 1974; Sprules, 1975; Roff and Kwiatkowski, 1977; Confer et al., 1983).

The experimental acidification of Lake 223, in the Experimental Lakes Area, northwestern Ontario, from 1976 to 1983 provided an opportunity to observe the responses of the zooplankton community to decreasing pH values from 6.7 to 5.1, including three annual cycles at pH 5.1. The most striking change in the zooplankton community over this pH range was the dominance of community biomass by cladocerans at pH 5.1 (Figure 1F, in Schindler et al., 1985). In this paper we report the seasonal changes in cladoceran biomass and in species composition and abundance with acidification of Lake 223 from pH 6.7 to 5.1 and examine possible causes of the increase in cladoceran biomass.

2. MATERIALS AND METHODS

Lake 223 is situated in the Experimental Lakes Area (ELA) at 49°42'N latitude and 93°42'N longitude. Surface area and maximum depth are 27.3 ha and 14.4 m.

Method of acidification with H_2SO_4 and results from the first several years of acidification are described by Schindler et al. (1980) and Schindler (1980). The first 8 yr of the experiment are summarized by Schindler et al. (1985). Total acid inputs are given by Cruikshank (1984).

Zooplankton were sampled at one station at the center of the lake. In 1974 a 28.7 L transparent trap (Schindler, 1969) was used to obtain a vertical series of samples. In 1977 a four-barrel non-closing sampler was hauled separately through the epilimnion, epilimnion and metalimnion and entire water column (Chang et al., 1980). Zooplankton counts in the metalimnion and hypolimnion were obtained by subtraction. From 1978 to 1983, zooplankton sampling was performed using a two-barrel closing sampler described by Chang et al. (1980). All gear were fitted with nets of 53 µm mesh. Results of test sampling to compare zooplankton abundances obtained by these three gear indicated there were no systematic differences in the gear for cladocerans and other planktonic groups (Chang and Malley, 1986). Sampling was biweekly or monthly. Number of sampling dates in each year was 1974, 7; 1977 to 1983, 10,8,12,16,16,17, and 15, respectively. Zooplankton sampling, identification and enumeration methods, numerical abundances of species and life stages and summaries of biomass on each sampling date for the 1974 to 1983 period are given by Chang et al. (1981) and Chang and Malley (1986). Data are not provided here for the years 1975 and 1976 because only the upper 2 m of the lake were sampled.

Conversion from abundance to biomass was made using dry weight estimates and length-dry weight relationships given by Lawrence *et al.* (1986).

3. RESULTS

Total zooplankton biomass in a volume-weighted column of water of 0 to 13 or 14 m was not reduced by acidification and may have increased during 1980 (Figure 1). Prior to acidification, mean ice-free season biomass was 269 mg m^{-2}, and during 1977 to 1980, it was 318, 218, 204 and 401 mg m^{-2}, respectively. During 1981 to 1983, mean ice-free season biomass was 279, 236 and 247 mg m^{-2}, respectively.

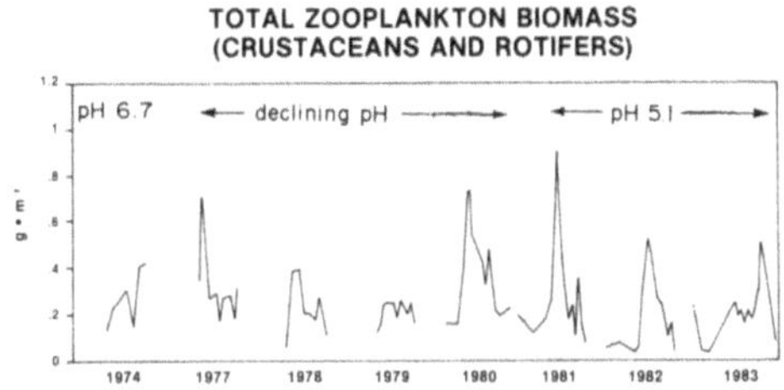

Figure 1. Total zooplankton biomass in Lake 223 during the pre-acidification year (1974), years of progressive acidification (1977 to 1980) and years of pH of approximately 5.1 (1981 to 1983).

Although total zooplankton biomass did not change greatly with acidification, Figure 2 shows that during 1981 to 1983 total biomass of cladocerans increased. This occurred at the expense of the

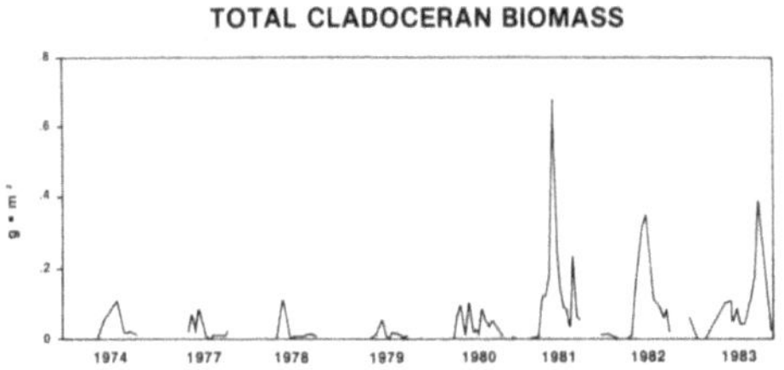

Figure 2. Total biomass of cladocerans in Lake 223 during the 1974 to 1983 period.

calanoid copepods and to a lesser extent the cyclopoid copepods (see Figure 1F, Schindler et al., 1985). In 1974, adults of D. minutus were present at mean ice-free season biomass of 33 mg m^{-2}. During 1981 to 1983 their biomass was 26, 17 and 24 mg m^{-2}, respectively.

Cladocerans also exhibited marked changes in species composition. The community in 1974 at pH 6.7 was dominated by Daphnia galeata mendotae, with Diaphanosoma birgei and Bosmina longirostris also present (Figure 3). Holopedium gibberum was rare. At pH 5.1, Daphnia catawba, H. gibberum and B. longirostris were all abundant. Diaphanosoma birgei and D. g. mendotae had disappeared or become rare (Figure 3).

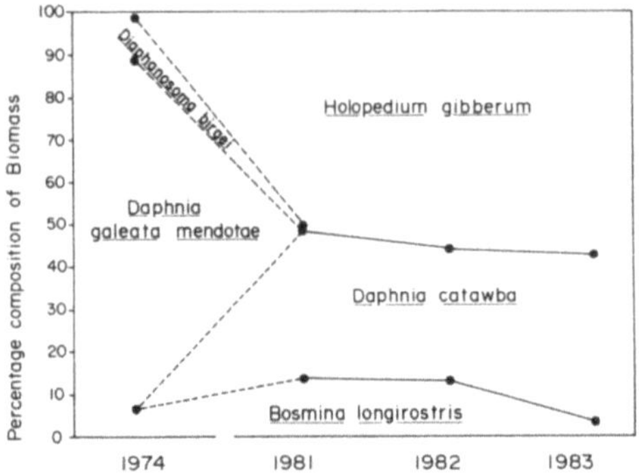

Figure 3. Contribution by various species to mean ice-free season biomass of cladocerans in Lake 223, prior to acidification and during the pH 5.1 period.

Figure 4A shows that D. g. mendotae, present but not abundant throughout the ice-free season in 1974 and 1977 to 1980, apparently disappeared by mid-summer 1981. Diaphanosoma birgei, changed little in mean ice-free season abundance from 1974 to 1980 but it became rare and was present only sporadically at pH 5.1 (Figure 4B). Although the less frequent sampling in 1974 and 1977 to 1979 may have missed population maxima, B. longirostris appeared to increase in number in 1980 to 1983. It was the most abundant cladoceran at pH 5.1 (Figure 4C). Holopedium gibberum, rare in 1974, apparently became more abundant between 1977 and 1980 and was abundant at pH 5.1(1981 to 1983) (Figure 4D). Daphnia catawba, recorded in the samples from Lake 223 for the first time in 1980, was the dominant and usually sole daphnid at pH 5.1 (Figure 4E). It replaced D. g. mendotae almost entirely during 1981 to 1983.

Chaoborus has been incidentally sampled along with zooplankton. There is no evidence to suggest that it changed markedly in abundance with acidification.

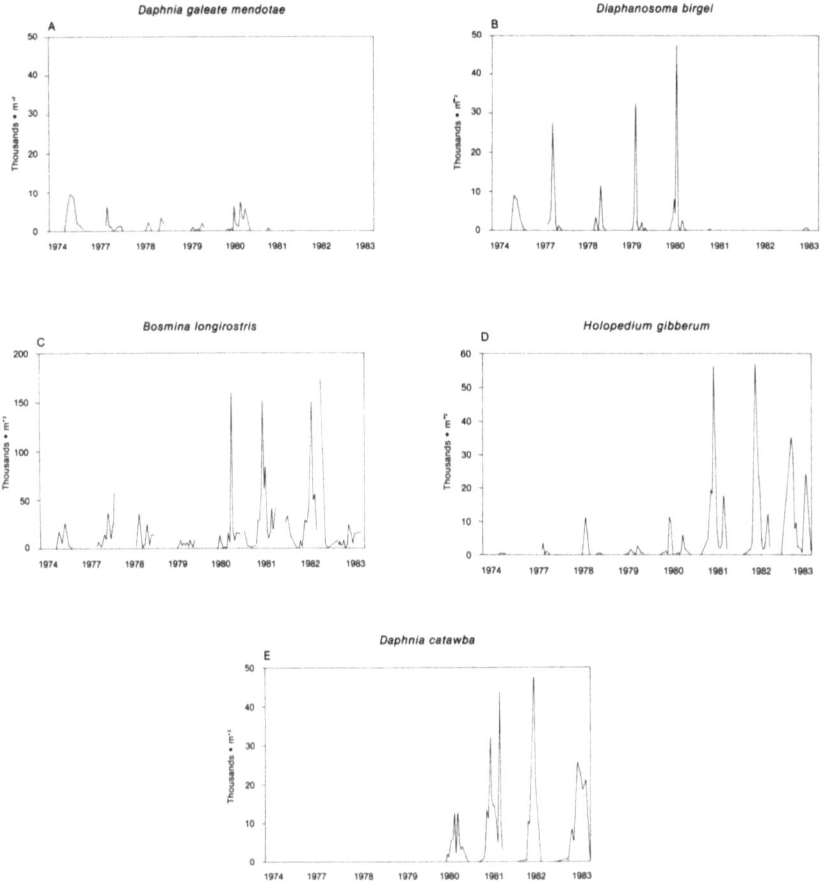

Figure 4. Seasonal abundances of A. Daphnia galeata mendotae, B. Diaphanosoma birgei, C. Bosmina longirostris, D. Holopedium gibberum, E. Daphnia catawba, prior to acidification (1974), during increasing acidification (1977 to 1980) and at pH 5.1 (1981 to 1983).

4. DISCUSSION

Some of the changes in species composition with acidification were expected as indicated by a review of the literature by Malley et al. (1982). Bosmina longirostris, H. gibberum and D. catawba are known to be acid tolerant (Sprules, 1975; Roff and Kwiatkowski, 1977; Dillon et al., 1984; Keller and Pitblado, 1984). Diaphanosoma sp. was found to be tolerant to about pH 4.1 (Sprules, 1975). D. g. mendotae was found in lakes in the Sudbury area which ranged in modal pH from 6.75 to 3.95 but it was rare in the most acidic ones (Roff and Kwiatkowski, 1977). The decline of D. birgei and D. g.

mendotae in Lake 223 at pH 5.1 may be explained by lower competitive abilities relative to the other species in pH 5.1, rather than their inability to survive at this pH level, but we have no direct evidence to support this.

Dominance of the zooplankton community biomass at pH 5.1 by cladocerans was not expected. Sprules (1975) reported that in lakes at pH 5.0 to 5.2 Diaptomus minutus contributed 40 to 99% to abundance. In only one of 9 cases were cladocerans more abundant than D. minutus. In lakes at pH 5.0 in the Sudbury, Ontario area, calanoids were more abundant than cladocerans even though cladocerans were more abundant at pH 5.0 than at pH 6.0 or 4.0 (Roff and Kwiatkowski, 1977). Diaptomus minutus comprised 74.3% of the crustacean abundance in the most acidic (mean pH 4.7), most oligotrophic lakes in a survey in northeastern Ontario (Keller and Pitblado, 1984).

Several hypotheses of cause and effect were examined to explain the dominance of cladocerans rather than D. minutus at pH 5.1 in Lake 223. Changes in relative proportions of zooplankton groups from copepod to cladoceran dominance is usually associated with eutrophication (Bradshaw, 1964; McNaught and Buzzard, 1973). McNaught (1975) hypothesizes that calanoids are more successful in oligotrophic waters because of superior filtering capacity and high ingestion rate and ingestion efficiencies at low cell densities. Cladocerans are successful under mesotrophic conditions when feeding efficiency is less important and when high intrinsic birth rate is an advantage. Nevertheless, it is probable that the increase in cladocerans in Lake 223 was not caused by increasing lake trophy and increased food availability for zooplankton. Schindler et al. (1985) show that the biomass of algae of edible size was a slightly lower proportion of total algal biomass in 1981 to 1983 than in 1974, even though phytoplankton productivity and standing algal biomass were higher than in 1974. Food availability for zooplankton therefore probably did not change greatly with acidification.

Acidification is normally associated with an increase in concentrations of metals. Diaptomus minutus appears to be intolerant of high Cu and Ni concentrations found in acidic Clearwater and Swan lakes near Sudbury (Yan and Strus, 1980). Because pure acid was added directly to Lake 223, the only metal which increased appreciably in Lake 223 with acidification was Mn, leached from the sediments. Aluminum increased only slightly (Malley et al., 1982). The slight increase in heavy metal concentrations in Lake 223 is discounted as a cause of the decline in D. minutus relative to cladocerans at pH 5.1.

A major change between 1974 and 1981 to 1983 was the loss of the invertebrate predator, the opposum shrimp, Mysis relicta, which disappeared in 1979 (Nero and Schindler, 1983). Although the food habits of M. relicta have not been studied in ELA lakes, an abundant literature indicates that it is an opportunistic feeder, ranging between herbivore, omnivore, detritivore, predator. It probably plays an important role in structuring zooplankton communities (Grossnickle, 1982). The decline of M. relicta in Lake 223 (1979) was poorly correlated in time with the increased importance of Cladocera (1981 to 1983) and would not appear to be its cause even though M. relicta undoubtedly fed at least in part on zooplankton. The invertebrate predator, Chaoborus, evidentally remained rare throughout the acidification process and is not implicated in the increase in cladocerans. The low abundance of Chaoborus before acidification is consistent with the presence of fish in Lake 223.

The biological events in Lake 223 which correlated best with the increase in cladocerans were the dramatic decline in abundance of small fish. Fry of lake trout were essentially not present in 1980 to 1983. Fry of the slightly more acid-tolerant white sucker were severely reduced in number in 1981 to 1983 (Mills et al., 1986). Fathead minnow was rare in 1979 and 1980 and absent in subsequent years. In 1982, its replacement, the pearl dace, had a very poor year class and the species was rare in 1983. In 1981 to 1983, therefore, few young-of-the-year of any fish species (except pearl dace in 1981) were observed (Schindler et al., 1985; Mills et al, 1986). We hypothesize that in Lake 223, enhancement of cladocerans, in general, and the large abundance of Daphnia catawba at pH 5.1, resulted from greatly reduced predation by small fish caused by acidification. Fish predation eliminates larger prey items (Brooks and Dodson, 1965; Northcote and Clarotto, 1975). It is not surprising, therefore, that release of vertebrate predation by acidification led to an increase in large planktonic crustaceans (D. catawba and H. gibberum).

It is not clear whether the release of fish predation alone was the reason for the dominance of cladocerans over D. minutus. It may be that D. minutus at ELA does not respond favourably to acid. At ELA D. minutus is on the edge of its geographical distribution and Patalas (1986) speculates that such organisms may be less able to withstand additional stress (e.g. acid) than those in the center of distribution.

5. TAXONOMIC NOTES

1. Diaphanosoma in Lake 223 are identified as D. birgei after Korinek (1981). It is probably the D. leuchtenbergianum of other authors. We referred to this population as D. brachyurum in Chang et al. (1981) and Malley et al. (1982). There is considerable confusion in the exact relationships between D. birgei, D. leuchtenbergianum and D. brachyurum as used by different authors.
2. Daphnia catawba was first recorded in Lake 223 in 1980. Malley et al. (1982) referred to it as D. catawba x schoedleri because of the interpretation that specimens tended to bear half of the taxonomic characters of D. catawba and half of those of D. schoedleri as described by Brooks (1957). From further work on this population by W. Geiling (Limnoservices Inc., Lansdowne, Ontario), it appears that these Daphnia are closer to D. catawba than to any other species description (Brooks, 1957).

3. Daphnia galeata mendotae is reported here to occur in Lake 223 from 1974 to 1980. Reexamination of some samples confirms this identification for Daphnia in 1974 but indicates that the Daphnia (other than D. catawba) in 1980 better fit the description for D. dubia. Until we complete the reexamination of 1974 to 1980 samples we continue to call these D. galeata mendotae.

6. ACKNOWLEDGMENTS

Zooplankton were sampled by field crews under the supervision of E. Matheson, J. Penny and D. Cruikshank. Identification and enumeration of zooplankton were performed by I. Delbaere, W. Findlay and W. Geiling (Limnoservices, Inc.). Data analysis was assisted by the computer programmers, G. Mueller, R. Scerbo and D. Schultz. Figures were prepared by the Freshwater Institute Graphics Services. L. Wilson and D. Laroque typed the manuscript. We thank P. Campbell, I. Davies, K. Patalas and D.W. Schindler for constructive reviews.

7. REFERENCES

Almer, B., Dickson, W., Ekström, C., Hörnström, E., and Miller, U.: 1974, Ambio **3**, 30.

Bradshaw, A.S.: 1964, Verh. Internat. Verein. Limnol., **15**, 700.

Brooks, J.L.: 1957, 'The Systematics of North American Daphnia'. Memoirs Conn. Acad. Arts Sci., **13**, 1-180.

Brooks, J.L., and Dodson, S.L.: 1965, Science, **150**, 28.

Chang, P.S.S., Malley, D.F., Findlay, W.J., Mueller, G., and Barnes, R.T.: 1980, 'Species Composition and Seasonal Abundance of Zooplankton in Lake 227, Experimental Lakes Area, Northwestern Ontario, 1969-1978', Can. Data Rep. Fish. Aquat. Sci., **182**, iv + 101 p.

Chang, P.S.S., Malley, D.F., Delbaere, I.L., and Mueller, G.: 1981, 'Species Composition and Seasonal Abundance of Zooplankton in Lake 223, Experimental Lakes Area, Northwestern Ontario: Before and During Acidification, 1974-1979', Can. Data Rep. Fish. Aquat. Sci., **290**, iv + 42 p.

Chang, P.S.S., and Malley, D.F.: 1986, 'Zooplankton in Lake 223, Experimental Lakes Area, Northwestern Ontario, 1974-1983', Can. Data Rep. Fish. Aquat. Sci., in press.

Confer, J.L., Kaaret, T., and Likens, G.E.: 1983, Can. J. Fish. Aquat Sci., **40**, 36.

Cruikshank, D.R.: 1984, 'Whole Lake Chemical Additions in the Experimental Lakes Area. 1969-1983', Can. Data Rep. Fish. Aquat. Sci., **449**, iv + 23 p.

Dillon, P.J., Yan, N.D., and Harvey, H.H.: 1984, CRC Critical Reviews in Environmental Control, **13**, 167.

Grossnickle, N.E.: 1982, Hydrobiologia, **93**, 101.

Keller, W., and Pitblado, J.R.: 1984, Wat. Air Soil Pollut., **23**, 271.
Korinek, V.: 1981, Can. J. Zool., **59**, 1115.
Lawrence, S.G., Malley, D.F., Findlay, W.J., MacIver, M.A., and Delbaere, I.L.: 1986, 'A Method for Estimating Dry Weight of Freshwater Planktonic Crustaceans from Measures of Length and Shape', Can. J. Fish. Aquat. Sci., in press.
Malley, D.F., Findlay, D.F., and Chang, P.S.S.: 1982, 'Ecological Effects of Acid Precipitation on Zooplankton', In D'Itri, F.M. (ed.), Acid Precipitation: Effects on Ecological Systems, Ann Arbor Science Publishers, Ann Arbor, MI, 297-327.
McNaught, D.C.: 1975, Verh. Internat. Verein. Limnol., **19**, 724.
McNaught, D.C., and Buzzard, M.: 1973, 'Changes in Zooplankton Populations in Lake Ontario (1939-1972)', Proc. 16th Conf. Great Lakes Res., 76-86.
Mills, K.H., Chalanchuk, S.M., Mohr, L.C., and Davies, I.J.: 1986, 'Responses of Fish Populations in Lake 223 to Eight Years of Experimental Acidification', Can. J. Fish. Aquat. Sci., in press.
National Research Council of Canada: 1981, 'Acidification in the Canadian Aquatic Environment: Scientific Criteria for Assessing the Effects of Acidic Deposition on Aquatic Ecosystems', NRCC, No. **18475**, 369 p.
Nero, R.W., and Schindler, D.W.: 1983, Can. J. Fish. Aquat. Sci., **40**, 1905.
Northcote, T.G., and Clarotto, R.: 1975, Verh. Intern. Verein. Limnol., **19**, 2378.
Patalas, K.: 1986, 'The Geographical Distribution of Mesocyclops edax (S.A. Forbes) in Lakes of Canada', Proc. Sec. Intern. Conf. on Copepoda, Ottawa, Canada, 13-17 August 1984, Nat. Mus. Can. Syllogeus, **58**, 400-408.
Raddum, G.G., Hobaek, A., Lomsland, E.R., and Johnsen, T.: 1980, 'Phytoplankton and Zooplankton in Acidified Lakes in South Norway', In Drablos, D., and Tollan, A. (eds), Ecological Impact of Acid Precipitaton, SNSF Project, Oslo, Norway, 332-333.
Roff, J.C., and Kwiatkowski, R.E.: 1977, Can. J. Zool., **55**, 899.
Schindler, D.W.: 1969, J. Fish. Res. Board Can., **26**, 1948.
Schindler, D.W.: 1980, 'Experimental Acidification of a Whole Lake: A Test of the Oligotrophication Hypothesis', In Drablos, D., and Tollan, A. (eds.), Ecological Impact of Acid Precipitation, SNSF Project, Oslo, Norway, 370-374.
Schindler, D.W., Wagemann, R., Cook, R.B., Ruszczynski, T., and Prokopowich, J.: 1980, Can. J. Fish. Aquat. Sci., **37**, 342.
Schindler, D.W., Mills, K.H., Malley, D.F., Findlay, D.L., Shearer, J.A., Davies, I.J., Turner, M.A., Linsey, G.A., and Cruikshank, D.R.: 1985, Science, **228**, 1395.
Sprules, W.G.: 1975, J. Fish. Res. Board Can., **32**, 389.
Yan, N.D., and Strus, R.: 1980, Can. J. Fish. Aquat. Sci., **37**, 2282.

LINKING LABORATORY AND FIELD RESPONSES OF FISH POPULATIONS TO ACIDIFICATION

Webster Van Winkle, Sigurd W. Christensen, and James E. Breck
Environmental Sciences Division
Oak Ridge National Laboratory
P.O. Box X
Oak Ridge, TN 37831 USA

ABSTRACT. To increase understanding of mechanisms by which fish populations respond to water chemistry changes resulting from acidification, it is necessary to link the results of laboratory studies of the effects of pH, Al, and Ca on the survival, growth, and reproduction of individual fish to responses of fish populations in the field. Our framework for achieving this goal is based on the types of data commonly available from both laboratory and field studies. One of the models (PHALCA) in this framework estimates the number of fish surviving as a function of time, given pH, Al, and Ca levels. A second model (FISHEGGS) evaluates the reduction in reproductive potential of a fish population attributable to the effects of pH, Al, and Ca on the survival of young-of-the-year and older fish and on fecundity. Preliminary results from these two models are presented, and the entire framework will be applied and tested over the next year using data for brook trout and lakes in the Adirondacks. The framework and models are a complementary alternative to the statistical analysis of survey data on water chemistry and on presence or absence of a fish species.

1. INTRODUCTION

To increase understanding of mechanisms by which fish populations respond to water chemistry changes resulting from acidification, it is necessary to link the results of laboratory studies of the effects of pH, Al, and Ca on the survival, growth, and reproduction of individual fish to responses of fish populations in the field (Baker 1984).

We have designed a framework for linking laboratory exposure studies to fish occurrence in the field, based on data commonly available from the laboratory and the field. One of the models (PHALCA) in this framework estimates the number of fish surviving as a function of time, given pH, Al, and Ca levels. A second model (FISHEGGS) evaluates the reduction in reproductive potential of a fish population attributable to the effects of pH, Al, and Ca on the survival of young-of-the-year fish, survival of older fish, and fecundity. Using data collected by the University of Wyoming as part of this project and

Water, Air, and Soil Pollution **30** (1986) 639–648.

those collected by other researchers on other projects, this framework will be applied and tested over the next year using data for brook trout and lakes in the Adirondacks.

The approach to date for estimating the probability of presence of a given fish species in surface waters sensitive to acidification has been to statistically analyze survey data on water chemistry and on presence/absence of the fish species of interest (Baker and Harvey, 1984; Reckhow et al., 1985). Consequently, this approach makes no use of results from laboratory or in situ bioassays, nor does it explicitly incorporate information on the life cycle of the fish species. The alternative framework we are proposing is new, as are the two models PHALCA and FISHEGGS. Our objective is to develop quantitative methods that can be applied using the types of data currently available and those becoming available from bioassays, basic life cycle biology, and field surveys of water chemistry and fish presence/absence. The framework and component models are designed to be flexible if new types of data (e.g., bioassay and field data on aluminum speciation and complexation) become available. Finally, the framework is designed to address questions on a spatial scale of a population of lakes in a region, rather than a single lake, and on a temporal scale of years to decades, rather than days to weeks as would be required to investigate the effects of episodic exposures during snowmelt.

The objective of this paper is to describe briefly the framework and our two models, PHALCA and FISHEGGS. Preliminary results from PHALCA and FISHEGGS are presented, but an application of the entire framework was not available at the time of the Muskoka Conference. Because the framework and models are unique, however, we want to make them more accessible to the general research community at this time, to stimulate discussion and to promote the synthesis of laboratory and field data useful in assessing the impact of acidification on fish populations.

2. DEVELOPING A FRAMEWORK FOR LINKING LABORATORY EXPOSURE STUDIES TO FISH OCCURRENCE IN THE FIELD

The objectives of our framework are to be able to (1) predict the probability of occurrence of a fish species in a lake, given data on pH, Al, and Ca levels in that lake; (2) predict the change in the probability of occurrence of a fish species in a lake, given a specified change in the pH, Al, and Ca levels in that lake; and (3) evaluate the effects of mitigation measures, such as liming, reduced fishing mortality, and stocking, on the probability of occurrence.

The steps involved in setting up the framework are indicated in Figure 1. Step 1 involves chronic laboratory bioassay experiments designed to delineate the consequences of different levels of acidification stress (pH, Al, and Ca) on mortality and fecundity. Extensive experiments of this type are being conducted on four fish species at the University of Wyoming as part of this LAF Project (Breck et al., in press; Ingersoll et al., in press). Step 2 involves

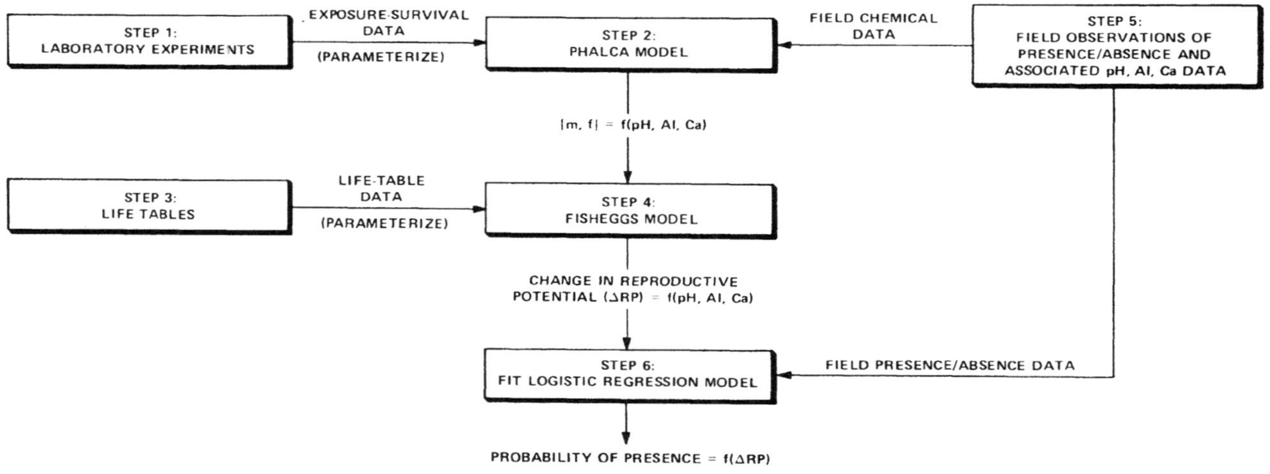

Figure 1. Steps involved in setting up the Lake Acidification and Fisheries framework for relating laboratory exposure studies to changes in the occurrence of a fish population in the field.

estimating parameters in the PHALCA model, using survival-time data from the laboratory experiments in Step 1. This model transforms survival-time data into estimates of conditional mortality rate (CMR) and will be expanded to translate fecundity effects into conditional reductions in fecundity. The PHALCA model is discussed further in Section 3.

Step 3 involves summarizing from the literature the life history characteristics of each fish species of interest, namely age-specific estimates of mortality, fraction of fish that are female, fraction of females that are sexually mature, and number of eggs produced per sexually mature female. Step 4 involves our FISHEGGS model, which integrates impacts on individual life stages over the entire life cycle. The input requirements for FISHEGGS are the basic life table data from Step 3 and estimates of the CMR and conditional fecundity effects from Step 2. The output is an estimate of change in reproductive potential (ΔRP) from an a priori control or baseline of pH, Al, and Ca to some specified combination of pH, Al, and Ca resulting in CMRs and fecundity effects greater than zero. The FISHEGGS model is discussed further in Section 4.

Step 5 requires results from a field survey of a set of lakes in a given region in which both water chemistry (pH, Al, and Ca) and presence/absence of the fish species of interest have been determined. Such data sets are available and others are becoming available, especially from New York State's Adirondack Lake Survey and EPA's National Surface Water Survey. Implicitly such presence/absence data incorporate community/ecosystem effects and compensatory phenomena. We have decided not to attempt to incorporate these effects explicitly into FISHEGGS since the data necessary to do this will never be available for large numbers of lakes. Step 6 involves statistically fitting a logistic regression model, using the presence/absence data from Step 5 as the dependent variable and the estimated change in reproductive potential (ΔRP) from Step 4 as the independent variable. Although the logistic regression model is not the only appropriate regression model to use when the dependent variable is binary, such as presence vs absence, there are theoretical reasons favoring this model (SAS, 1983) and there is precedent for its use in relating presence/absence of fish populations to water chemistry variables (Baker and Harvey, 1984; Reckhow et al., 1985). The estimated change in reproductive potential (ΔRP) is estimated for each lake in the field data set from Step 5 by transforming the pH, Al, and Ca for that lake through the PHALCA and FISHEGGS models (Steps 2 and 4). At this point, the framework is set up and ready to be applied or tested using a second set of field data.

The steps involved in applying or testing the framework are indicated in Figure 2. If the second set of field data includes data on pH, Al, and Ca only, then the framework can be applied to predict the probability of presence. Prior to this use of the framework, however, it is essential that the framework be tested on alternative data sets that include both water chemistry and presence/absence data so that the model predictions from the framework can be compared with the observed presence/absence data. We also envision other applications involving mitigation by liming and stocking.

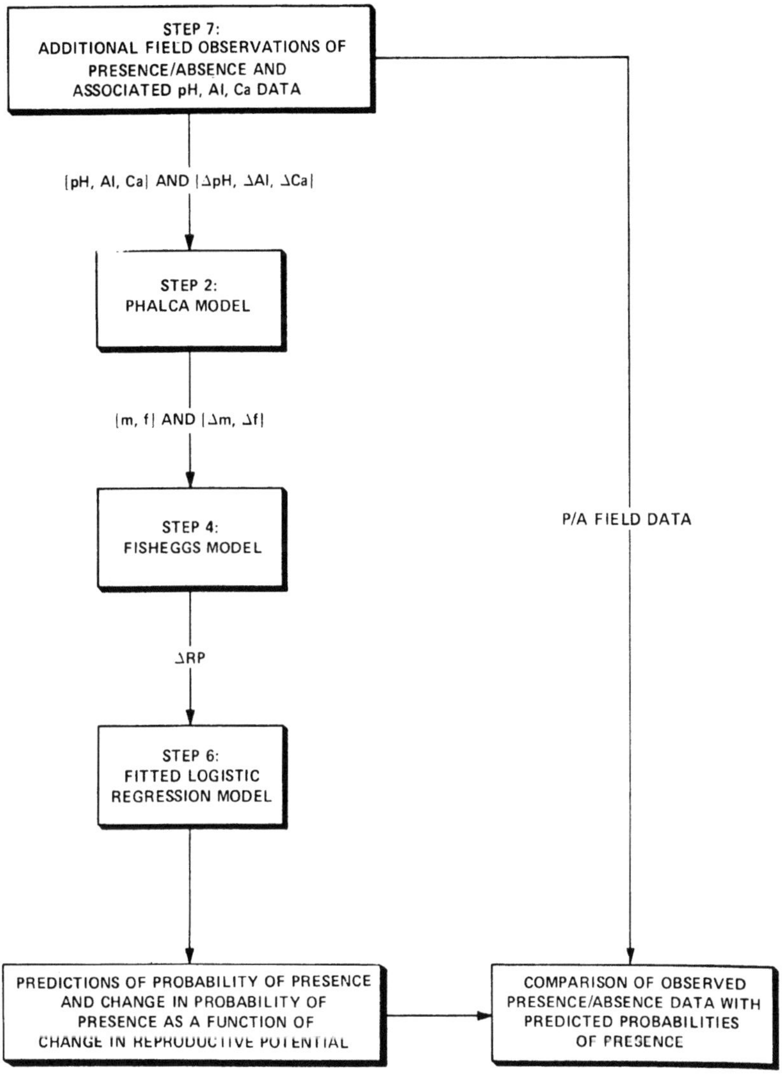

Figure 2. Steps involved in applying and testing the Lake Acidification and Fisheries framework.

3. APPLYING SURVIVAL-TIME ANALYSIS TO THE RESULTS OF LABORATORY EXPOSURE STUDIES TO ESTIMATE CONDITIONAL MORTALITY RATES (PHALCA)

Survival-time (or failure-time) analysis utilizes time-to-death, rather than the number of individuals alive after a fixed exposure time, as the measured response. We have developed the interactive computer model PHALCA to estimate the fraction surviving as a function of time, given the exposure history to pH, Al, and Ca. The CMRs (Ricker, 1975) are then calculated from values of fraction surviving. The advantages of survival-time analysis are that (1) it utilizes all of the daily survival data collected over the course of the experiment at all combinations of pH, Al, and Ca and not just survival on the final day, (2) it is possible to estimate survival at times earlier than the final observation time, and (3) it permits modeling survival as a function of fluctuating exposures. At present, the model uses the Weibull distribution (Weibull, 1951) to estimate the fraction surviving at time $\underline{t}$; this distribution is simple (two parameters) and flexible and has been applied to bioassay data (Christensen, 1984) and life table data (Pinder et al., 1978), as well as industrial failure-time data (Lee, 1980).

The Weibull function for the fraction surviving at time $\underline{t}$ is $S(\underline{t}) = \exp[-(\underline{a}\underline{t})^{\underline{b}}]$, where $\underline{a}$ and $\underline{b}$ are the two parameters to be estimated as a joint function of pH, Al, and Ca. In our fitting of the Weibull distribution to survival-time data we have obtained maximum likelihood estimators of these parameters, using stepwise multiple logistic regression, starting with a full quadratic model of the three independent variables. We are using this statistical approach to empirically approximate the response of the fish to these factors. An advantage of the multiple logistic regression approach is that the survival time can be estimated for combinations of pH, Al, and Ca not originally measured but within the range of the experimental conditions. Alternative equations can be used in cases where the approximation is judged to be inadequate. For example, the probit function can also be fit to cumulative frequency distributions having a sigmoid shape; the differences in fit are large only in the tails of the distributions.

Figure 3 illustrates sample output from PHALCA for early life stages of brook trout for fraction surviving (including 95% confidence intervals) and for CMR as a function of time for a single combination of pH, Al, and Ca. The CMR is calculated as $CMR = [S^*(\underline{t}) - S(\underline{t})]/S^*(\underline{t})$, where $S^*(\underline{t})$ is the estimated fraction survival on day $\underline{t}$ under the a priori control conditions of pH = 6.5, Al = 0.0 $\mu g\ L^{-1}$, and Ca = 8 mg L^{-1}, and $S(\underline{t})$ is the estimated fraction survival on day $\underline{t}$ under any other specified water chemistry condition. The CMR estimates are, in turn, used in the model FISHEGGS (see Section 4).

4. EVALUATING THE EFFECTS OF ACIDIFICATION STRESS ON THE REPRODUCTIVE POTENTIAL OF FISH POPULATIONS (FISHEGGS)

Scientists interested in performing laboratory or field bioassays to evaluate the effects of acidification or other stresses on fish populations usually want to focus their initial efforts on the most

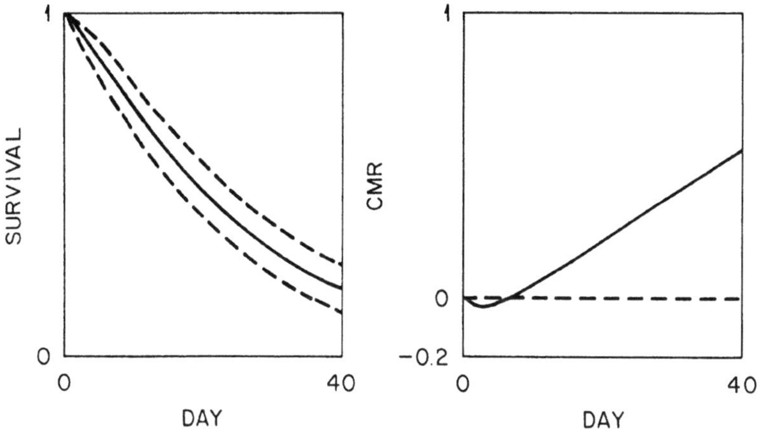

Figure 3. Sample output from the interactive computer model PHALCA of fraction surviving (and 95% confidence bounds as dashed lines) and conditional mortality rate as a function of exposure time. Results are based on data from experiments performed at the Red Buttes Fish Physiology and Toxicology Laboratory at the University of Wyoming involving continuous exposure of brook trout to a matrix of pH, Al, and Ca levels for 40 days, starting with eyed eggs. Results in this figure are for pH = 6.5, Al = 0.0 μg L^{-1}, and Ca = 0.5 mg L^{-1}.

critical life stage. Typical choices include studying (1) reduced survival and/or growth of young-of-the-year fish, (2) reduced survival and/or growth of yearling and older fish, and (3) impairment of the reproductive function of mature females. For fish populations that spawn repeatedly, age-specific effects of these three diverse types do not translate directly into consequences to the population. Rather, the population-level consequences of diverse acidification stresses depend on both the magnitudes of the stresses and on the pattern of survival and fecundity rates for the population.

The simplest mechanistic approach to translating these diverse effects on individual life stages into a "common denominator" that reflects population-level consequences is to evaluate the resulting change in reproductive potential. Example applications of the reproductive-potential approach are reported in Barnthouse (1984), Goodyear (1977), and Sadler (1983). The reproductive potential of a population is defined as the mean number of first-generation female eggs produced per present-generation female egg. In a population at equilibrium, the reproductive potential equals one, meaning that each female replaces herself. The reproductive potential will be less than one for a declining population and greater than one for an increasing population. Using this approach allows effects on different life stages and processes to be integrated into a single effect on a critical population parameter: the regenerative capacity, or reproductive potential, of the population.

FISHEGGS is a model designed for such an integration. Data needed to operate the model are age-specific estimates of survival, maturity, and fecundity, beginning with age 1 (yearlings). The user then specifies the relative magnitudes of effects (as percentage reductions) and the ranges of age over which the effects occur. The model evaluates the resulting percentage reduction in reproductive potential for the effects, singly as well as combined. Results are presented as tables and also as graphs (e.g., see Figure 4).

Results for brook trout show that relatively small but annually repeated water chemistry effects on survival of yearling and older fish can be more significant than larger effects on survival of young-of-the-year fish (Figure 4). In addition, repeated fixed reductions in survival of prereproductive and reproductive fish are more significant than equivalent annual reductions in fecundity (Figure 4). The ability to detect small effects on the survival of adults compared with young-of-the-year fish becomes increasingly important to identifying the critical stage(s) as reproduction is delayed and/or spread over a number of age groups and as exploitation of mature females is decreased.

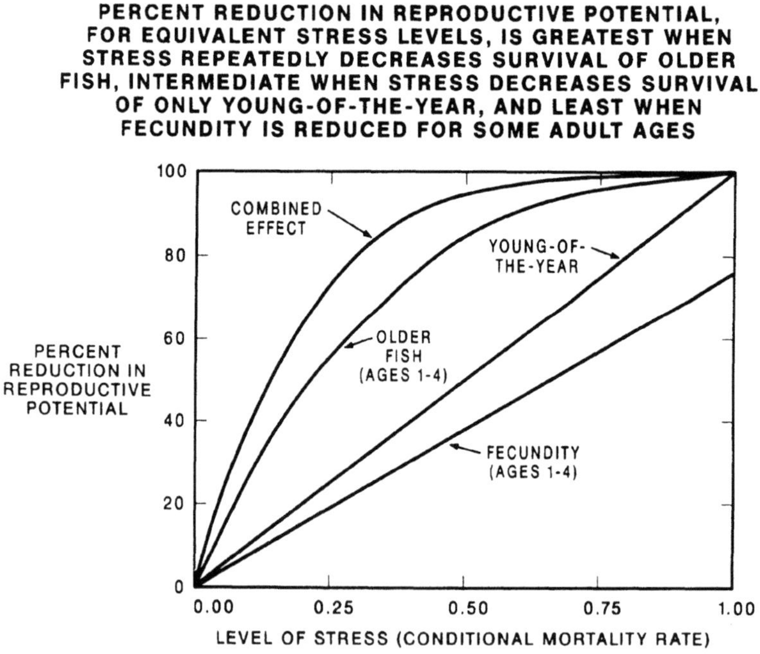

Figure 4. Sample output for brook trout from the interactive computer model FISHEGGS. These results show that percentage reduction in reproductive potential for equivalent stress levels is greatest when stress repeatedly decreases survival of older fish, intermediate when stress decreases survival of only young of the year, and least when fecundity is reduced for some adult ages.

5. DISCUSSION

The framework and the two models described above are designed to link the results from laboratory studies of the effects of acidification stress on physiological responses of individual fish to the responses of fish populations in the field. Thus, the framework and models can serve as assessment tools to help answer questions relating to current and future impacts of acidification on fishery resources in a given region (Figure 2). In addition, however, the framework and models can serve as research tools for setting priorities for alternative laboratory investigations. They can be used to evaluate threshold levels for pH, Al and Ca and to determine which life stages and life-cycle processes are most important in evaluations of population-level consequences of acidification.

We see this framework as complementing the approach of Baker and Harvey (1984) and Reckhow et al. (1985), which statistically relates presence/absence fisheries data directly to water chemistry data. The conceptual appeal of our approach is that it uses additional information on the life cycle of the fish species of interest; such information is recognized a priori as being relevant to any impact assessment. In our judgment, this increase in complexity is justified by the availability of the necessary life-cycle data and by the additional questions that can be addressed using our framework as opposed to the alternative approach. These additional questions relate to factors influencing the life cycle parameters (e.g., survival and fecundity) that are not directly a function of pH, Al, and Ca, such as changes in fishing effort and stocking.

6. ACKNOWLEDGMENTS

The authors acknowledge L. W. Barnthouse and D. L. DeAngelis for their helpful reviews of this manuscript. The Lake Acidification and Fisheries Project is sponsored by the Electric Power Research Institute, Palo Alto, California, under a contract with the University of Wyoming. The component of the project at Oak Ridge National Laboratory is sponsored by the University of Wyoming under Research Services Agreement ERD-83-317 with the U.S. Department of Energy. Oak Ridge National Laboratory is operated by Martin Marietta Energy Systems, Inc., for the U.S. Department of Energy under Contract No. DE-AC05-84OR21400. Publication No. 2725, Environmental Sciences Division, ORNL.

7. REFERENCES

Baker, J.P.: 1984, Fish. Section 5.6, pp. 5-74 to 5-132. IN R. A. Linthurst (ed.), The Acidic Deposition Phenomenon and Its Effects: Critical Assessment Review Papers, Volume II, Effects Sciences, EPA-600/8-83-016BF, U.S. Environmental Protection Agency, Washington, D.C.

Baker, J.P., and Harvey, T.B.: 1984, Critique of acid lakes and fish population status in the Adirondack Region of New York State. Draft final report to the U.S. Environmental Protection Agency. Acid Deposition Program, North Carolina State University, Raleigh, North Carolina.

Barnthouse, L.W.: 1984, Effects of contaminants on fish populations. Appendix B-4.1.4. IN C.S. Dudney and E.D. Copenhaver (eds.), Health and Environmental Effects Document for Direct Coal Liquefaction. Draft report prepared for the Office of Health and Environmental Research, U.S. Department of Energy, Washington, D.C.

Breck, J.E., Ingersoll, C.G., and Beauchamp, J.J. Modeling the mortality of early life stages of brook trout in response to fluctuating levels of pH, calcium, and aluminum. Environmental Toxicology and Chemistry (in press).

Christensen, E.R: 1984, Water Res. 18, 213.

Goodyear, C.P.: 1977, Assessing the impact of power plant mortality on the compensatory reserve of fish populations. pp. 186-195. IN W. Van Winkle (ed.), Assessing the Effects of Power-Plant-Induced Mortality on Fish Populations. Pergamon Press, New York.

Ingersoll, C.G., La Point, T.W., Breck, J.E., and Bergman, H.L. An early life stage brook trout (Salvelinus fontinalis) bioassay testing the independent and combined effects of pH, calcium, and aluminum in low conductivity water. Extended abstract. Fish and Wildlife Service special publication (in press).

Lee, E.T.: 1980, Statistical Methods for Survival Data Analysis. Wadsworth, Inc., Belmont, California. 557 pp.

Pinder, J.E., III, Wiener, J.G., and Smith, M.H.: 1978, Ecology 59, 175.

Reckhow, K.H., Black, R.W., Stockton, T.B., Jr., Vogt, J.D., and Wood, J.G.: 1985, Empirical models of fish response to lake acidification. Draft final report to the U.S. Environmental Protection Agency. Duke University, Durham, North Carolina.

Ricker, W.E.: 1975, Computation and Interpretation of Biological Statistics of Fish Populations. Fish. Board Can. Bull. 191. 382 pp.

Sadler, K.: 1983, Freshwater Biol. 13, 453.

SAS Institute Inc.: 1983, The LOGIST Procedure. pp. 181-202. IN SUGI Supplemental Library User's Guide, 1983 Edition. Cary, North Carolina.

Weibull, W.: 1951, J. App. Mechanics 18, 293.

LOCALIZATION OF ALUMINUM IN TISSUES OF FISH

C. Lee and H. H. Harvey
Department of Zoology
University of Toronto
Toronto, Ontario
Canada, M5S 1A1

ABSTRACT. The concentrations of Al in fish gills has been used as a measure of fish exposure to this metal in acidified waters. This experiment was designed to determine if other fish tissues also accumulate Al and thus possibly contribute to the cause of death. Rainbow trout (<u>Salmo gairdneri</u>) were exposed to the following four test conditions for 48 hr or until death: (1) pH 6.8, <0.001 $mg.L^{-1}$ Al (n=6); (2) pH 5.2, <0.001 $mg.L^{-1}$ Al (n=2); (3) pH 5.2, 1.0 $mg.L^{-1}$ Al (n=5); (4) pH 6.8, 1.0 $mg.L^{-1}$ Al (n=3). The trout were held in synthetic, low Ca water prior to, and during, experimentation. Esophagus-stomach, gonad, gall bladder, gill (left and right), heart, intestine, kidney, liver, muscle (epaxial), and spleen were digested in a 4:1 mixture of $HNO_3:HClO_4$ and analyzed by Inductively Coupled Plasma Emission Spectrophotometry. Elevated Al concentrations were found in gill and gastrointestinal tissues. Left and right gills of fish exposed to pH 5.2, 1.0 $mg.L^{-1}$ Al were the only tissues found to be significantly different ($p<0.01$) between the test conditions. The mean total Al concentrations of these test 3 fish, for the left and right gill were 3.61 and 4.33 $mg.g^{-1}$ Al dw. The Al concentration in the gastrointestinal tissues of the fish exposed to pH 6.8 at 1.0 $mg.L^{-1}$ Al was greater than that of the control fish, but not statistically significant. These results suggest that the analysis of whole gill remains an effective indicator of Al exposure in fishes at low pH.

1. INTRODUCTION

Although Al is the third most abundant element in the earth's crust, the concentration in fresh and salt waters usually is a fraction of a part per million. Aluminum hydrolyzes in water forming a series of compounds with hydroxide ion, which are amphoteric and thus yield soluble salts in acid solutions. Small changes in aqueous pH result in large differences in the concentration of free, monomeric Al in water. In the acidified lakes of Sweden, there has been a ten-fold increase in Al concentration (Dickson, 1975), and ground water acidified from pH 5.3 to 4.6 increased in Al concentration from 0.2 to 0.75 $mg.L^{-1}$ (Dickson, 1978). Aluminum

Water, Air, and Soil Pollution **30** (1986) 649–655.

concentration in the (stream) Austeråna increased from 0 to 58 μeq.L^{-1} during an acid precipitation event (Muniz and Leivestad, 1979). The toxicity of such water to fishes was first reported by Schofield (1977) when brook trout (Salvelinus fontinalis) held in lake water showed distress and mortality as pH declined from 7.2 to 5.9 and Al concentration rose from 20 to 320 μg.L^{-1}. Fish kills involving toxic concentrations of aqueous Al have now been reported from Norway, Sweden, United States, and Canada. Such fish show high concentrations of Al on the gills, as reported for example by (Grahn, 1980).

The physiological responses to Al exposure are similar to those induced by H^+ (Muniz and Leivestad, 1980), that is, the rapid loss of plasma Na and Cl. This may be due to the interference of Al with Na-K-ATPase in fish gills (Leivestad, 1985). It may also interfere with carbonic anhydrase in erythrocytes (Christensen and Tucker, 1976).

Aluminum is known to accumulate in a variety of mammalian tissues (Sorenson et al., 1974). High concentrations of Al have been identified in muscle of brown trout (Salmo trutta) but chronic uptake could have been from food or water (Hunter et al., 1980). For these reasons it was decided to determine whether Al would accumulate in other fish tissues during brief exposure to Al concentrations within the ranges reported for acidified waters (Wright et al., 1980).

2. METHODS

Rainbow trout of mean length 27.3 cm (SD ±1.2) and weight 212.9 g (SD ±6.6) were held for five days at 10°C on a 12 hr light-dark cycle, in low Ca water (Jansz, 1983) prior to experimentation. Prior to the introduction of trout, each tank was filled with 40 L of the synthetic, low Ca water, cooled to 10°C and Al was introduced into solution, where applicable, as $Al_2(SO_4)_3$. Due to the pH depression by the Al, pH was raised with (0.5M) KOH. Each tank was controlled to ±0.1 pH units using a Radiometer one-way pH stat system: pH meter (PHM 63), titrator (TTT 60), combination pH eletrodes (model GK2402C), a Bach-Simpson six channel automatic switch (model SAS2), and magnetic valves (MNV 2B) which controlled the flow of 0.1 N H_2SO_4. Mixing was provided by airstones and DO was measured every 12 hr by Winkler method (Anon., 1975).

On the sixth day, fish were introduced to each of the following test conditions: test 1 pH 6.8, <0.001 mg.L^{-1} Al (n=6); test 2 pH 5.2, <0.001 mg.L^{-1} Al (n=2); test 3 pH 5.2, 1.0 mg.L^{-1} Al (n=5); test 4 pH 6.8, 1.0 mg.L^{-1} Al (n=3). Test 5 represents a subsample of the fish held for 5 days in the low Ca water, pH 6.85, prior to experimentation (n=3). The experiment duration was 48 hr or until death. Death was determined as loss of equilibrium and lack of opercular movement. Replicate runs of tests 1, 3, and 5 were conducted to replace fish which escaped prior to the termination of the experiment.

Esophagus-stomach, cecae-intestine, heart, muscle, kidney, liver, gall bladder, spleen, left and right gill and gonad were removed, oven dried (60°C) to constant weight and digested in a 4:1 mixture of HNO_3:$HClO_4$. The tissues were brought to volume with 6% HNO_3 matrix compatible with the I.C.P.E.S. (Analytical Research Laboratories, model

34000). The samples were analyzed for total Al (Institute for Environmental Studies, University of Toronto). Consistent with analytical procedure, blanks and NBS Bovine liver samples were analyzed in conjuction with each series of fish tissues.

Water samples were taken for analysis at the beginning and end of each experiment. Procedures were as per the Ontario Ministry of the Environment (1980), and samples were stored at 4°C until analyzed. Pertinant chemistry is summarized in Table I.

TABLE I

Test	1*	2	3*	4	5*
pH	6.8	6.8	5.2	5.2	6.85
Nominal Al $\mu g.L^{-1}$	0.0	0.0	1,000	1,000	0.0
Al $\mu g.L^{-1}$ (total)	1.0	1.0	565	320	4.0
Cd	0.2	0.2	0.2	0.2	0.2
Cu	2.0	1.0	1.0	1.0	4.0
Mn	1.0	1.0	1.0	1.0	1.0
Ni	2.0	1.0	2.0	1.0	1.0
Zn	4.0	11.0	5.0	3.0	7.0
Ca $mg.L^{-1}$	2.2	2.6	2.4	2.0	2.6
Mg	4.80	4.82	4.80	4.52	4.32
Na	22.80	24.50	25.40	23.00	21.80
K	16.00	10.80	9.40	17.80	8.26
Cl	43.85	46.55	46.45	43.97	42.52
SO_4	34.28	41.38	33.71	34.77	25.51
DO	10.6	10.8	10.8	11.0	10.8
Temperature (°C)	10.2	10.2	10.2	10.2	10.4

* mean values of n=2.

Nominal Al concentration based on Al addition at commencement of experiment, all other metals and ligands are actual at termination of experiment. DO, temperature, pH, n=5; tests 1, 3, 5, n=2, and tests 2, 4, n=1 for all other parameters. Standard deviations: temperature ±0.0 except test 5, ±0.1; DO ±0.1, all tests.

Due to the small and unequal sample size, non parametric statistical analyses were conducted using the Statistical Analysis System (SAS) (Ray, 1982). For t-tests, the data were transformed using log(x+1). Means were ranked and analysed using the General Linear Models one-way ANOVA procedure. Sheffe's multiple range test was applied to determine which results differed significantly.

3. RESULTS

The synthetic water mixture, prepared from distilled-deionized water and analytical reagents, had an Al concentration of <0.001 mg.L^{-1} (Table I). In tests 1 and 2, Al concentration, did not reach limits of detection in initial or final water samples. In test 3, nominal Al 1.0 mg.L^{-1}, Al concentration declined to 0.320 mg.L^{-1} by the termination of the experiment. In test 4, pH 6.8, nominal Al 1.0 mg.L^{-1}, initial concentration of Al was 0.870 mg.L^{-1} and declined to 0.545 mg.L^{-1} by the end of the experiment. In the replication of tests, 1, 3 and 5, no differences were apparent among the water analyses.

All test 3 fish (pH 5.2, Al 1.0 mg.L^{-1}) died within 10 to 22 hr. The gills of these trout all showed an accumulation of white mucus between the filaments; this was not observed on the gills of the fish from the other test conditions. Trout of test 2 (pH 5.2, Al <0.001 mg.L^{-1}) appeared to be experiencing respiratory difficulties, as evidenced by gulping at the surface. The trout of tests 2 and 3 showed darkened pigmentation.

Tissue analyses yielded elevated concentrations of Al in/on the gills, stomach and intestine for tests 3 and 4, respectively (Table II). Test 3 trout had dry weight gill Al concentrations of 3.62 mg.g^{-1} (left) and 4.33 mg.g^{-1} (right) (Table II). These gills were the only tissues which showed a significant difference among the test fish ($p<0.01$). However, Sheffe's test failed to show a difference between tests, yielding a significant difference only between the test 3 and the holding tank (test 5) fish, for both right and left gill (Table III). There was no significant difference in Al concentration between the left and right gills (=0.05). Test 4 trout (pH 6.8, 1.0 mg.L^{-1} Al) showed significant ($p<0.001$) differences in the Al concentration between tissues. Sheffe's test differentiated only between the intestinal and gonadal tissues and again failed to discriminate between tissues for test 3, although significance was indicated ($p<0.001$). Test 5 (holding tank) and test 1 (control, pH 6.8, Al <0.001 mg.L^{-1}) fish showed only liver Al concentrations differed significantly ($p<0.01$).

4. DISCUSSION

At pH 5.2, rainbow trout rapidly accumulated Al on the gills (test 3). This rapid uptake and death in 10 to 22 hr is consistent with published reports, e.g., Harvey _et al_. (1982), Robinson and Deano (1985). No such accumulation of Al occurred in test 4, pH 6.8 and nominal Al of 1.0 mg.L^{-1}.

TABLE II

Test	1	2	3	4	5
pH	6.8	5.2	5.2	6.8	6.85
Nominal Al $\mu g.L^{-1}$	<1	<1	1,000	1,000	<1
Esophagus & Stomach	15.0± 27.0 (4)	15.1± 8.2 (2)	8.7± 3.6 (3)	1,339.0±2,073.0 (3)	20.7± 35.9 (3)
Cecae & Intestine	26.2± 44.9 (4)	21.7±23.5 (2)	12.7± 15.6 (2)	723.9± 978.1 (3)	3.3± 4.7 (2)
Heart	77.5±169.1 (5)	<1 (2)	11.7± 14.6 (3)	8.9± 15.5 (3)	260.0±450.3 (3)
Epaxial muscle	30.2± 48.5 (6)	10.0± 0.6 (2)	7.4± 7.7 (6)	11.8± 14.7 (4)	20.6± 26.1 (3)
Kidney	115.7±223.7 (5)	65.7±24.0 (2)	6.6± 4.9 (4)	7.2± 6.8 (3)	106.0±183.6 (3)
Liver	39.4± 28.9 (5)	39.3± 3.4 (2)	53.4± 55.4 (4)	30.5± 17.6 (3)	5.4± 9.4 (3)
Gall bladder	<1 (4)	3.9± 5.5 (2)	6.2± 9.5 (4)	<1 (3)	<1 (2)
Spleen	144.1±284.7 (5)	7.6 (1)	19.2± 22.7 (4)	3.7± 6.4 (3)	213.6±370.0 (3)
Left gill	37.3± 66.4 (4)	16.5± 1.8 (2)	3,615.3±1,586.7 (4)	27.2± 7.0 (3)	34.2± 59.2 (3)
Right gill	48.7± 90.9 (5)	19.9± 5.3 (2)	4,332.2±3,492.5 (4)	38.2± 15.6 (3)	34.7± 60.2 (3)
Gonad	185.6±311.2 (3)	25.0 (1)	45.1± 13.4 (2)	1.2 (1)	<1 (1)

Mean Al concentration $\mu g.g^{-1}$ dw ± SD, sample size in parentheses.

TABLE III

Test	1	2	3	4	5	Level of
Nominal Al $\mu g.L^{-1}$	0.0	0.0	1,000	1,000	0.0	Significance
pH	6.8	6.8	5.2	5.2	6.85	
Left gill	A	A	A	A		$p \leq 0.01$
	B	B		B	B	
Right gill	C	C	C	C		$p \leq 0.01$
	D	D		D	D	

Non-parametric analysis of variance of Al concentrations between test conditions; tissues with different letters are significantly different.

In test 4, all trout survived the 48 hr exposure to elevated Al. The pooled tissues, esophagus plus stomach, showed an Al concentration approximately 100X that of control groups 1 and 2; cecae and intestine had Al concentration 30X the control groups. A review of the literature failed to identify previous reports of this uptake of Al into the gastrointestinal tract by fishes at circum-neutral pH. At the higher pH of test 4, Al would be present in a polymeric form. It was not established that the Al measured was within the tissues, thus it was possible that colloidal Al was present merely within the gastrointestinal tract. No such gastrointestinal Al was evident in the test 3 fish, with similar total Al concentration in the water but at pH 5.2, where more of the Al would be present as monomeric Al.

In the remaining tissues, Al accumulation did not differ significantly between test 3 and 4 fish and controls. This is in contrast to reports of elevated Al in various tissues with prolonged exposure. Brown trout showed muscle Al concentrations of 475 $mg.kg^{-1}$ (Hunter _et al._, 1980), but the source could have been prey species (snails) and/or water. Kidney, liver and gills of rainbow trout showed elevated concentrations of Al (Buergel and Soltero, 1983) and these authors suggested such tissues may be the sites of Al storage or metabolism. In largemouth bass, gizzard shad, and channel catfish, Berg and Burn (1985) found significant differences in Al concentration between liver, kidney, brain, gill, and muscle, with brain concentration exceeding that of the gills.

Based on the 10 rainbow trout tissues examined, gill Al concentration remains the best indicator of Al exposure in acid waters. The cause of the high gastrointestinal Al concentration at circum-neutral pH warrants investigation.

5. REFERENCES

Anonymous: 1975, _Standard methods for the examination of water and wastewater_, 14th ed., American Public Health Association,

Washington, D.C. 1193 pp.
Berg, D.J. and Burns, T.A.: 1985, J. Freshw. Ecol. 3, 113.
Buergel, P.M. and Soltero, R.A.: 1983, J. Freshw. Ecol. 2, 37.
Christensen, G.M. and Tucker, J.H.: 1976, Chem.-Biol. Interactions **13**, 181.
Dickson, W.: 1975, Rep. Inst. Freshw. Res., Drottningholm **54**, 8.
Dickson, W.: 1978, Verh. Int. Verein. Limnol. **20**, 851.
Grahn, O.: 1980, Fishkills in two moderately acid lakes due to high Al concentrations. In: Conf. Ecol. Impact Acid Precip., Drabløs, D. and A. Tollan (eds.), Norway 1980, SNSF Proj., pp. 310-311.
Harvey, H.H., Dillon, P.J., Fraser, G.A., Somers, K.M., Fraser, P.E. and Lee, C.: 1982, Elevated metals and enhanced metal uptake in fishes in acid stressed waters, Extended Abstract, Am. Chem. Soc., Div. Environ. Chem., pp. 438-441.
Hunter, J.B., Ross, S.L., Tannahill, J.: 1980, Wat. Pollut. Contr. **79**, 413.
Jansz, G.F.: 1983, Some physiological consequences of exposure to and recovery from environmental acid stress with varied calcium concentrations in rainbow trout (Salmo gairdneri), M.Sc. thesis, University of Toronto.
Leivestad, H.: 1985, Aqueous speciation of Al and toxic effects on Atlantic salmon, Abstr. Int. Sym. Acidic Precip., Muskoka, September 1985, pp. 95-96.
Muniz, I.P. and Leivestad, H.: 1979, Langtidseksponering av fisk til surt vann, Forsøk med bekkerøye Salvelinus fontinalis Mitchill (Long term exposure of brook trout to acidic water), IR44/79, SNSF Proj., 1432 As, Norway. 32 pp.
Muniz, I.P. and Leivestad, H.: 1980, Toxic effects of Al on the brown trout, Salmo trutta L. In: Conf. Ecol. Impact Acid Precip., Drabløs, D. and A. Tollan (eds.), Norway 1980, SNSF Proj., pp. 320-321.
O.M.E.: 1980, Quality control report, Ontario Ministry of the Environment, Water Quality Section, Laboratory Services Branch, Rexdale, Ontario. 58 pp.
Ray, A.A.: 1982, SAS User's Guide: Statistics, 1982 ed., Cary, N.C., SAS Inst. Inc. 584 pp.
Robinson, J.W. and Deano, P.M.: 1985, J. Environ. Sci. Health **A20**, 193.
Schofield, C.L.: 1977, Acid snow-melt effects on water quality and fish survival in the Adirondack Mountains of New York State, Res. Tech. Completion Rept. Proj. No. A072-NY. 26 pp.
Sorenson, J.R.J., Campbell, I.R., Tepper, L.B. and Lingg, R.D.: 1974, Environ. Health Persp. **8**, 3.
Wright, R.F., Conroy, R., Dickson, W.T., Harriman, R., Henriksen, A. and Schofield, C.L.: 1980, Acidified lake districts of the world: a comparison of water chemistry of lakes in southern Norway, southern Sweden, souterhwestern Scotland, the Adirondack Mountains of New York and southeastern Ontario. In: Conf. Ecol. Impact Acid Precip., Drabløs, D. and A. Tollan (eds.), Norway 1980, SNSF Proj., pp. 377-389.

METAL CONCENTRATION AND CALCIFICATION OF BONE OF WHITE SUCKER (CATOSTOMUS COMMERSONI) IN RELATION TO LAKE pH

L. I. Bendell-Young and H. H. Harvey
Department of Zoology
University of Toronto
Toronto, Ontario M5S 1A1
Canada

ABSTRACT. The concentrations of 17 elements were determined in the bone of white suckers (Catostomus commersoni) netted from 5 acid (pH range 4.8 to 5.8) and 2 circumneutral (pH=6.2 and 6.3) lakes in south-central Ontario. The bone Ca:P dry weight ratios were similar (2.0:1) for all fish populations except those of George Lake (pH=4.8) which showed a significantly lower Ca:P ratio (1.9:1, $P < 0.05$). Magnesium was also lower in the bone of these fish and in fish from 2 other acid lakes. Only bone Ba and S concentrations in the 7 fish populations correlated significantly to lake pH (R=-0.9 and R=-0.8, respectively, $P < 0.05$). Bone Mn concentrations correlated to dissolved lake Mn concentrations (R=0.8, $P < 0.05$), and was 7 fold greater in the bone of fish from George Lake and 2 fold greater in King Lake (pH=5.0) fish, vs fish from the 2 circumneutral lakes. Bone Zn was significantly greater in white sucker from George Lake, and tended to be higher in this species from King Lake, compared to all other fish populations. Bone concentrations of Fe, Cu, Ni and Al showed no apparent trends among the 7 fish populations. Cd, Co, Cr, Mo, V and Be were not detected. The occurrence of a reduced Ca:P ratio coincident with the highest concentrations of Mn, Zn and Ba in the bone of fish from the most acidified environment suggests that increased metal concentrations which occur in surface waters coincident with lake acidification may affect bone calcification.

1. INTRODUCTION

The mechanisms by which acidification affect fishes in lakes of northeastern North America are complex. One possible mechanism is that acid stress interferes with the normal Ca dynamics, resulting in bone deformities (Beamish et al., 1975) and bone demineralization (Fraser and Harvey, 1982). However, recent field (Mills, 1984) and experimental (Rogers, 1984) evidence has indicated that conditions of chronic acid exposure alone are not sufficient to interfere with the normal Ca dynamics of fish.

Water, Air, and Soil Pollution 30 (1986) 657–664.

Coincident with lake acidification are increases in metal concentrations such as Cd and Zn (Dickson, 1975; Schindler et al., 1980). These metals can interfere with the Ca metabolism of bony tissues causing a reduction of bone Ca and P concentrations and a reduced Ca to P ratio coincident with increased bone metal concentrations (Muramoto, 1981; Sauer and Watabe, 1984), as well as inducing bone deformities (Bengtsson et al., 1985). It is possible therefore that increased metal concentrations in acid lakes may play a role in altering the Ca dynamics of fish. This may be apparent as a correlation between reduced bone Ca and P concentrations and a reduced Ca to P ratio and increased bone metal concentrations in the bone of white sucker from acid lakes. The objective of this study was to use a multielemental analysis of the bone of white sucker captured from acid and non-acid lakes to determine if such a relationship exists. A strong correlation may be indicative of a possible role of trace metals in acid waters interfering with the normal calcification of fish bone.

2. STUDY AREA AND METHODS

Seven lakes were chosen based on location, previous knowledge of the existing fish species and known water chemistry (Bendell-Young and Harvey, 1986). A detailed outline of the study area (location and lake morphology). Additional methods and water chemistry have been published (Bendell-Young and Harvey, 1986). Water and sediment chemistry for the 7 lakes are presented in Table I. The following elements were included in the analysis of the fish bone; Ca, P, Mg, S, Mn, Zn, Ba, Fe, Cu, Al, Ni, Cd, Co, Cr, Mo, V and Be. The 17 elements grouped as follows: macroelements; Ca, P, S and Mg; trace elements; Mn, Zn, Fe, Cu, Al, Ni and Ba and (post-analysis) non-detectable trace elements; Cd, Co, Cr, Mo, V, and Be. Accuracy of the bone analysis was measured by the inclusion of a National Bureau of Standards (NBS) bovine standard spiked with a 1,000 $\mu g.L^{-1}$ Ca standard to simulate the bone matrix. Experimentally determined values for Ca, P, S, Mn, Zn, Fe and Cu were within 10% of the NBS certified values. Determined values for Mg were within 15% of the certified values. Ba, Ni and Al were not detected in the NBS standard. Limits of detection for these 3 elements plus Cd, Co, Cr, V, Be and Mo were 0.005, 0.033, 0.027, 0.006, 0.007, 0.005, 0.003, 0.006 and 0.01 $\mu g.L^{-1}$, respectively.

Fish mean weight, fork length, bone elemental concentrations and the Ca to P dry weight ratios for the 7 white sucker populations are presented in Table II. Since only trace amounts of Ni were detected in the fish bone, ranges are presented. Elemental concentrations did not differ between sexes for the 7 fish populations (Students t-test; $P > 0.05$) and were not dependent on fish weight or fork length (Pearson's product correlation coefficient; $P > 0.05$). Since within-lake variability was not dependent on either fish sex or size, comparisons among the 7 fish populations were made. Homogeneity of group variances for the elemental concentrations in white sucker bone were verified through Levene's test prior to an one-way analysis of variance between elemental means. Where significant differences occurred,

Duncan's multiple comparisons were applied to indicate where they occurred. Pearson's product correlation coefficients were calculated to determine relationships between mean bone metal concentrations and 1) water chemistry parameters, 2) sediment chemistry parameters, and 3) mean bone elemental concentrations. Significant correlations are presented in Table III.

TABLE I

	WATER			SEDIMENT ± 1 SD						
Lake	Al	Zn ($\mu g.L^{-1}$)	Fe	Mn	Al ($mg.g^{-1}$)	Fe	Zn	Ni	Cu ($\mu g.g^{-1}$)	Ba[a]
George	180	30	50	3.3 ±0.7	27 ±4	48 ±22	256 ±102	113 ±16	106 ±24	830 ± 50
King	90	10	280	0.3 ±0.1	12 ±2	16 ± 2	123 ± 21	22 ± 7	23 ±11	634 ±180
McDonald	120	62	250	0.3 ±0.1	7 ±1	6 ± 2	97 ± 32	11 ± 3	15 ± 6	681 ±149
Crosson	50	10	430	0.3 ±0.1	14 ±2	26 ± 4	141 ± 63	16 ± 4	31 ± 8	319 ± 67
Chub	105	NA[b]	430	0.3 ±0.06	12 ±2	22 ± 8	142 ± 30	17 ± 4	28 ± 6	781 ±241
Red Chalk	15	10	180	2.3 ±1.5	15 ±5	45 ±26	210 ± 37	16 ± 1	64 ±40	393 ± 47
Harp	50	NA	130	3.2 ±1.0	15 ±5	40 ±19	165 ± 73	18 ± 6	33 ±20	1191 ±251

a: Ba sediment concentrations are "total" metal as compared to the "acid extractable" concentrations of the other metals.
b: Not available.

3. RESULTS AND DISCUSSION

3.1. Elemental concentrations in the bone of white suckers

Of the 11 elements that were detected in the fish vertebrae only Ba and S correlated to lake pH (R=-0.9 and R=-0.8, respectively; $P < 0.05$) (Table III). Barium is a large, highly electropositive cation and behaves differently than the transition metals such as Zn and Mn. For example, Schindler et al. (1980) reported uptake by seston and the rate of removal from the water column following acidification was the greatest for the transition metals and the least for the highly electropositive metals such as Ba. Similarly, Jackson et al. (1980) noted metal ions of large ionic radius, such as Ba, showed a greater affinity for the aqueous phase than did the smaller cations and attributed the observed effect to the low charge/radius ratio of Ba and the associated weak physical adsorption of this ion in an aqueous environment. These authors further suggested that under some circumstances Ba may be preferentially released from lake sediments to the interstitial water and overlying water column. This difference in the behavior of Ba as compared to the transition metals may explain in part the strong correlation that was found between lake pH and bone Ba concentrations in the 7 fish populations (Figure 1).

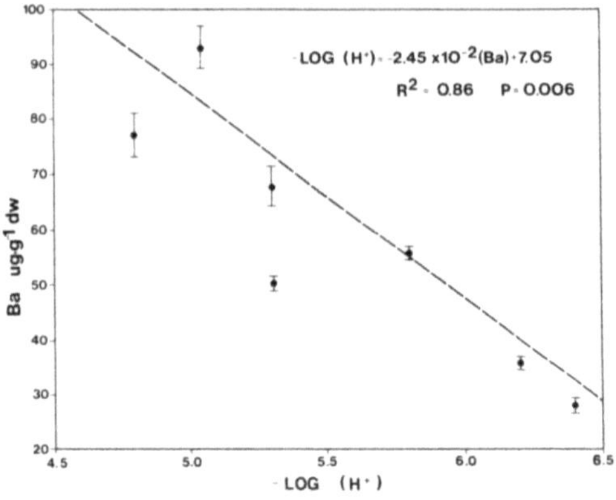

Figure 1. The regression of mean bone Ba concentration (± 1 SE) on lake pH for the 7 white sucker populations.

Bone S concentration was significantly greater in fish from 4 of 5 acid lakes as compared to fish from the 2 circumneutral lakes (Table II) and also correlated to lake pH (Table III). Neither the accumulation in nor toxicity of sulphate to fishes has received much study. Limited research has indicated that fish will accumulate S in skeletal tissue when exposed to increased water concentrations (Phillips et al., 1959). However, the sulphate anion passes through biological membranes with difficulty and must be taken up by a selected process with an expenditure of energy (Phillips et al., 1959).

Manganese was elevated 7 fold in the bone of white sucker from the most acid lake and 2 fold in the bone of fish from acid King Lake compared to white sucker from the remaining acid and circumneutral lakes (Table II). However, whereas white sucker bone Mn concentrations were strongly correlated to water Mn ($R=0.9$, $P < 0.05$), they were not related to lake pH (Table III). Similarly, Fraser and Harvey (1982) and Moreau et al. (1983) have noted that no simple relationship exists between lake pH and bone Mn concentrations. Recently, Bendell-Young and Harvey (1986) have shown that dissolved water and lake sediment concentrations of Mn are mediated strongly by the redox cycling of this metal as well as being influenced by changes in lake acidity. Bone concentrations of Mn in white sucker probably are correlated better with dissolved Mn

TABLE II

Lake	Wt g	FL cm	Ca	P mg.g^{-1}	Mg	S	Mn	Zn	Fe µg.g^{-1}	Al	Cu	Ni	Ca:P
George	555	33.5	208^{c}	109^{bc}	2.8^{d}	4.9^{b}	675^{a}	138^{a}	6^{ab}	15^{b}	2.3	(0.0-	1.9^{b}
(16)	±123	±3.2	±22	±13	±0.5	±0.3	±100	±25	±5	± 8	±1	6.6)	±0.08
King	651	34.4	242^{ab}	122^{a}	3.9^{a}	4.9^{b}	210^{b}	90^{bc}	8^{ab}	17^{ab}	8.9	(0.0-	1.98^{ab}
(19)	±125	±2.6	±29	±12	±0.4	±0.4	± 62	±11	±5	± 6	±1	4.2)	±0.07
McDonald	209	25.5	218^{c}	106^{c}	3.3^{cd}	4.6^{a}	62^{d}	80^{c}	8^{ab}	22^{a}	NA*	ND	2.05^{a}
(19)	± 80	±4.0	±26	±13	±0.7	±0.8	±15	±15	±5	±17			±0.07
Crosson	665	32.7	233^{bc}	117^{c}	3.2^{cd}	4.9^{b}	101^{c}	87^{c}	9^{a}	12^{ab}	6.4	ND	2.02^{ab}
(7)	±421	±6.3	±25	±13	±0.3	±0.4	± 58	±19	±7	± 7	±2		±0.1
Chub	669	35.1	243^{ab}	123^{a}	3.7^{a}	5.0^{b}	123^{c}	83^{c}	7^{ab}	22^{c}	9.4	(0.0-	1.96^{ab}
(19)	±115	±1.3	±33	±14	±0.3	±0.3	± 15	±13	±7	±11	±1	2.3)	±0.08
Red Chalk	1044	39.6	249^{a}	123^{a}	3.6^{ab}	4.6^{a}	147^{c}	83^{c}	11^{a}	15^{a}	9.6	(0.0-	2.02^{a}
(18)	±271	±4.1	±29	±14	±0.4	±0.2	± 65	±15	±7	± 9	±1	1.4)	±0.06
Harp	557	34.7	227^{b}	111^{b}	3.4^{bc}	4.3^{a}	118^{c}	64^{d}	3^{b}	8^{a}	NA	ND	2.05^{a}
(34)	±176	±4.0	±16	±10	±0.4	±0.3	± 33	±10	±3	± 4			±0.09

NA: Not available
ND: Not detected

TABLE III

II

	pH	ALK	COND	DOC	Mn	Al	Zn	Fe	Al	Fe	Ni	Cu	Ba	Zn	Mn	Ba	Mn	Zn	Fe	Al	Cu	Ni	Ca	P	Mg	S	Ca:P
	WATER								SEDIMENT							BONE											
pH	1																										
ALK	.8	1																									
COND	-	.76	1																								
DOC	-	-	-	1																							
Mn	-.73	-	-	-	1																						
Al	-.75	-	-	-	.8	1																					
Zn	-	-	-	-	-	-	1																				
Fe	-	-	-	-	-	-	-	1																			
Al	-	-	-	-.8	-	-	-	-	1																		
Fe	-	-	-	-.9	-	-	-	-	.8	1																	
Ni	-	-	-	-	.8	.73	-	-	.9	-	1																
Cu	-	-	-	-.9	-	-	-	-	.9	.8	.9	1															
Ba	-	-	.8	-	-	-	-	-	-	-	-	-	1														
Zn	-	-	-	-.9	-	-	-	-	.9	.9	-	.9	-	1													
Mn	.8	.9	-	-	-	-	-	-	-	-	-	-	-	-	1												
Ba	-.9	-.8	-	-	.75	-	-	-	-	-	-	-	-	-	-.8	1											
Mn	-	-	-	-	.9	-	-	-	.9	-	.9	.9	-	.8	-	-	1										
Zn	-	-	-	-	.9	-	-	-	.8	-	.9	.8	-	-	-	-	.9	1									
Fe	-	-	-.8	-	-	-	-	-	-	-	-	-	-.9	-	-	-	-	-	1								
Al	-	-	-	-	-	-	-	-	-	-	-	-	-	-	-	-	-	-	-	1							
Cu	-	-	-	-	-	-	-	-	-	-	-	-	-	-	-	-	-	-	-	-	1						
Ni	-	-	-	-	.9	-	-	-	-	-	.8	-	-	-	-	-	.9	.8	-	-	-	1					
Ca	-	-	-	-	-	-	-	-	-	-	-	-	-	-	-	-	-	-	-	-	.8	-	1				
P	-	-	-	-	-	-	-	-	-	-	-	-	-	-	-	-	-	-	-	-	.9	-	.9	1			
Mg	-	-	-	-	-	-	-	-	-	-	-	-	-	-	-	-	-	-	-	-	-	-	.9	.8	1		
S	-	-.8	-	-	-	-	-	-	-	-	-	-	-	-	-.8	-	-	-	-	-	-	-	-	-	-	1	
Ca:P	-	-	-	-	-.8	-	-	-	-	-	-.8	-	-	-	-	-	-.8	-.9	-	-	-	-.9	-	-	-	-	1

concentrations, where the latter are the result of both acid leaching and the geochemical cycling of this metal than with increased water concentrations resulting from pH depression alone.

Bone Zn concentration was significantly higher in white sucker from the most acid lake and tended to be elevated in this species from King Lake (Table II). Bone Zn concentrations were not correlated to either lake pH, or sediment concentrations (Table III). Since concentrations of dissolved Zn in 5 of the 7 study lakes were below detection, it was not possible to correlate lake Zn concentrations to white sucker bone Zn concentrations. Moreau et al. (1983) have reported the elevation of Zn in the opercula of brook trout (Salvelinus fontinalis) from 3 acidified as compared to 3 non-acidified lakes. Greater bone Zn concentrations in white sucker from lakes George and King are probably attributable to the increased mobilization and greater solubility of this metal in acid waters.

Although Fe, Al and Cu were detected in fish bone, there were no apparent trends in concentrations of these elements among the 7 fish populations studied (Table II). Bone Ni was only detected in trace amounts. Rather than the absence of significant differences being indicative of the availability of these 4 metals in acid and non-acid systems, it is more likely that the bone is not the best site for metal accumulation. For example, the liver has been shown to be the site of Cu storage and accumulation in the white sucker (Bendell-Young and Harvey, 1985) and gill accumulation of Al is indicative of its toxicity to fish (Schofield, 1976).

3.2. Elemental bone concentrations in relation to bone calcification

The mineralized portion of bone is composed of calcium triphosphate, which has a theoretical Ca:P weight ratio of 2.16:1 (Glimcher, 1959). Experimentally, this ratio has been shown to equal 2.11 in the scales of the mummichog (Fundulus heteroclitus) (Sauer and Watabe, 1984), and 2.05 in the vertebrae of carp (Cyprinus carpio) (Muramoto, 1981). Exposure of the mummichog to sublethal levels of Zn and the carp to sublethal levels of Cd resulted in bone demineralization and an overall reduction of the Ca:P ratio to 1.65 and 1.71 respectively, concurrent with an increase in bone metal concentrations.

In our study, fish from acid George Lake which contained the greatest bone metal concentrations of Mn, Zn and Ba also contained a significantly lower Ca:P ratio as compared to the 2 circumneutral fish populations (Table II). Bone Mg was also low in this population and tended to be low in the fish from 2 other acid lakes as compared to concentrations found in fish captured from 2 circumneutral lakes (Table III). It is possible that Zn, Mn and Ba are directly replacing Ca^{2+} in the hydroxapatite matrix, impairing bone calcification. However, direct ion replacement can only account for 7% of the noted reduction of bone Ca^{2+} in George Lake white suckers.

Rather than direct replacement of the metal ions in the bone matrix, one or a combination of these metals may be inhibiting the mineralization of the bone. For example, in addition to Cd and Zn inducing bone deformities in vivo, Mn and Zn have been shown to interfere with bone mineralization in vitro (Gluggenheim and Gaster, 1973). As well, even though it is a larger ion, Ba^{2+} can compete with Ca^{2+} for large anions (i.e. SO_4^{2-} and $ROSO_3^-$) because of favorable solvation factors. The complexes formed with such anions have low solubility, and in this way Ba may interefere with metabolic processes involving Ca (Williams, 1976).

4. REFERENCES

Beamish, R.J., Lockhart, W.J., VanLoon, J.C. and Harvey, H.H.: 1975, Ambio **4**, 98.

Bendell-Young, L.I. and Harvey, H.H.: 1985, unpublished data.

Bendell-Young, L.I. and Harvey, H.H.: 1986, Hydrobiologia 133, 117.

Bengtsson, B.E., Bengtsson, A. and Himberg, M.: 1985, Ambio **14**, 32.

Dickson, W.: 1975, Rep. Inst. Freshw. Res., Drottningholm **54**, 8.

Fraser, G.A. and Harvey, H.H.: 1982, Can. J. Fish. Aquat. Sci. **39**: 1289.

Glimcher, M.J.: 1959, Rev. Mod. Physics **31**, 359.

Gluggenheim, K. and Gaster, D.: 1973, The role of manganese, copper and zinc in the physiology of bones and teeth. In: Biological Mineralization, I. Zipkin (ed.), John Wiley and Sons, New York, pp. 444-458.

Jackson, T.A., Kipphut, G., Hesslein, R.H. and Schindler, D.W.: 1980, Can. J. Fish. Aquat. Sci. **37**, 387.

Mills, K.H.: 1984, Fish population response to experimental acidification of a small Ontario lake. In: Early Biotic Response to Advancing Lake Acidification, G.R. Hendrey (ed.), Acid Precipitation Series 6, pp. 117-131.

Moreau, G., Barbeau, C., Frenette, J.J., Saint-Onge, J. and Simoneau, M.: 1983, Can. J. Fish. Aquat. Sci. **40**: 1685.

Muramoto, S.: 1981, Environ. Pollut. **A24**, 125.

Phillips, A.A., Podiliak, H.A., Livingston, D.L., Duma, D.F. and Thresen, R.W.: 1959, Fish Res. Bull. **23**, 8.

Rogers, D.W.: 1984, Can. J. Fish. Aquat. Sci. **41**, 1774.

Schindler, D.W., Hesslein, R.H., Wagemann, R. and Broecker, W.S.: 1980, Can. J. Fish. Aquat. Sci. **37**: 373.

Schofield, C.L.: 1976, Ambio **5**, 228.

Sauer, G.R. and Watabe, N.: 1984, Aquat. Toxicol. **5**, 51.

Williams, R.J.P.: 1976, Calcium chemistry and its relation to biological function. In: Calcium in Biological Systems, Symp. Soc. Exp. Biol., 30th, London, Cambridge University Press, pp. 1-17.

OCCURRENCE OF SALMONID PARR AND EEL IN RELATION TO WATER QUALITY IN SMALL STREAMS ON THE WEST COAST OF SWEDEN

E. Degerman*, J-E Fogelgren*, B. Tengelin* and E. Thörnelöf**
* Inst. Freshwater Research ** Nat. Environmental Protec.Bd
Box 1302
S-170 11 Drottningholm S-171 25 Solna

ABSTRACT. In the summer of 1983 twelve streams on the west coast of Sweden were electrofished, and water samples were collected. In three of the streams water sampling was continued throughout the winter of 1983-84. Although summer pH showed little variation between stations there were large differences in fish abundance. The abundances of salmonid parr and eel were correlated with the alkalinity, and were significantly greater at stations with summer alkalinity above 0.25 $meq \cdot l^{-1}$. The winter water sampling programme showed drastic and rapid drops in the alkalinity during the snow-melt. Electrofishing data compiled from 7 streams during 1956-84 showed decreasing parr abundances in small streams (catchment area <1000 km^2) with low (<0.25 $meq \cdot l^{-1}$) alkalinity. It is suggested that the negative effect on fish is mainly due to toxic forms of Al and to a lesser extent on the direct toxicity of low pH levels.

1. INTRODUCTION

Since the beginning of the 1970s electrofishing surveys have revealed decreasing abundances of salmonid parr (brown trout - *Salmo trutta* - and Atlantic salmon - *S. salar*) in many of the smaller watercourses on the west coast of Sweden. Also declining catches of eel (*Anguilla anguilla*) were reported in streams on the west coast. It has been difficult to determine to what extent these changes depend on acidification. Electrofishing surveys are usually carried out in late summer and the water chemistry data collected at the same time usually show high pH values. However, it is well documented that the pH during snow-melts can decrease rapidly and remain low for periods of a few days to some weeks and during these episodes the concentrations of Al, Fe and Mn may increase (Bjärnborg, 1983; Dickson, 1983; Andersson and Nyberg, 1984).

The present study was undertaken in order to evaluate the extent of damage caused by acidification on the economically important salmonids and eels.

Water, Air, and Soil Pollution **30** (1986) 665–671.

2. METHODS

The project consisted of three different parts; electrofishing and a survey of the water chemistry in 12 streams (a total of 88 stations) in the summer of 1983, an intensive survey of the water chemistry in 2 streams (5 stations) in the winter of 1983/84 and finally, an evaluation of previously unpublished data from electrofishing investigations during the years 1956-84 in 7 streams in the region.

Electrofishing was conducted with standard gear (300 to 700 V D.C.) and performed two to four consecutive times to allow the total population of salmonids to be estimated with the removal method (Zippin, 1956). The eel population was not assessed by the removal method. Instead the abundances presented refer to all the eel caught at each station. The data for salmonids applies only to hard bottom biotopes, i.e. suitable salmonid reproduction areas. Water samples (one sample/station) were taken at 0.1 m depth and analysed for pH, alkalinity, color and conductivity. During the winter additional samples were collected and the water flow was estimated at one station in each of the two streams. Stations situated upstream from barriers to upstream-migrating salmonids are not included. No stockings have been performed at the investigated stations. Electrofishing data from 1956-84 was also compiled from 7 similar streams in the region. The statistical treatment was done using nonparametric tests; the Mann-Whitney U-test, Spearman Rank correlation and the Chi-Square test.

3. AREA INVESTIGATED

The streams are situated in the counties of Bohuslän and Halland on the Swedish west coast. The investigated stretches of the streams are situated below the former highest sea level, ca. 60 to 140 m above present sea level. The average annual precipitation is 700 to 1000 mm. The catchment areas are 100 to 1500 km^2 and the streams had an average width of 6.0 m and an average depth of 0.2 m. The water velocities were 0.05 to 1.5 m s^{-1}.

4. RESULTS

In the summer of 1983 the pH was above 6 at 89% of the stations and the alkalinity was greater than 0.1 meq$\cdot l^{-1}$ at 75% of the stations (Figure 1). The alkalinity showed a negative correlation with altitude ($p < 0.01$). The conductivity ranged from 5 to 33 mS m^{-1}. The water color was 10 to 300 mg Pt l^{-1}, with an average of 70 mg Pt l^{-1}.

During January 1984 a marked drop in alkalinity and pH occurred in connection with precipitation and temperatures above $0^{o}C$ (Figure 2). The alkalinity showed a significant negative correlation with water flow ($p < 0.001$). At some stations, especially at higher altitudes, the alkalinity reached low values and the buffering capacity disappeared at some stations where summer values had been 0.03 to 0.41 meq$\cdot l^{-1}$. At these stations the pH ranged from 4.5 to 5.5 during the snow-melt,

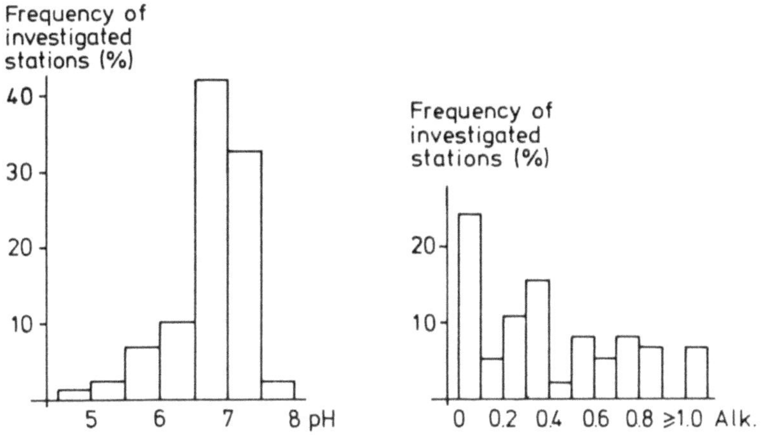

Figure 1. pH and alkalinity at the investigated 88 stations on the west coast of Sweden in the summer of Sweden in the summer of 1983.

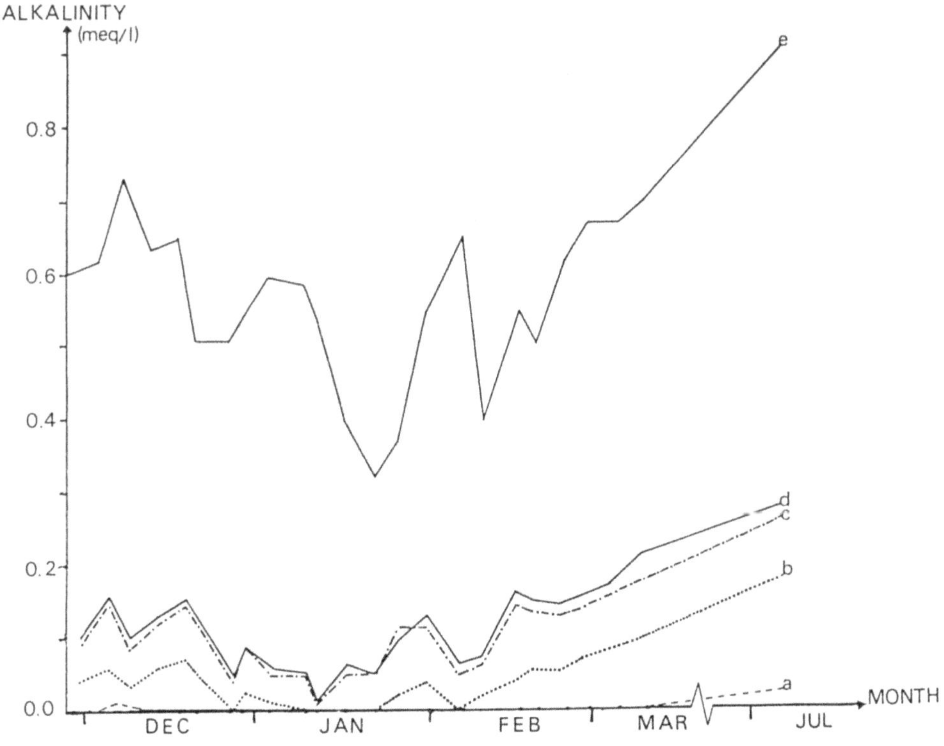

Figure 2. The alkalinity at five stations during December 1983 - July 1984. Stations a and e are from the upper and lower parts of stream Krogaån respectively. Stations b, c and d are from the upper and lower parts of stream Jörlandaån.

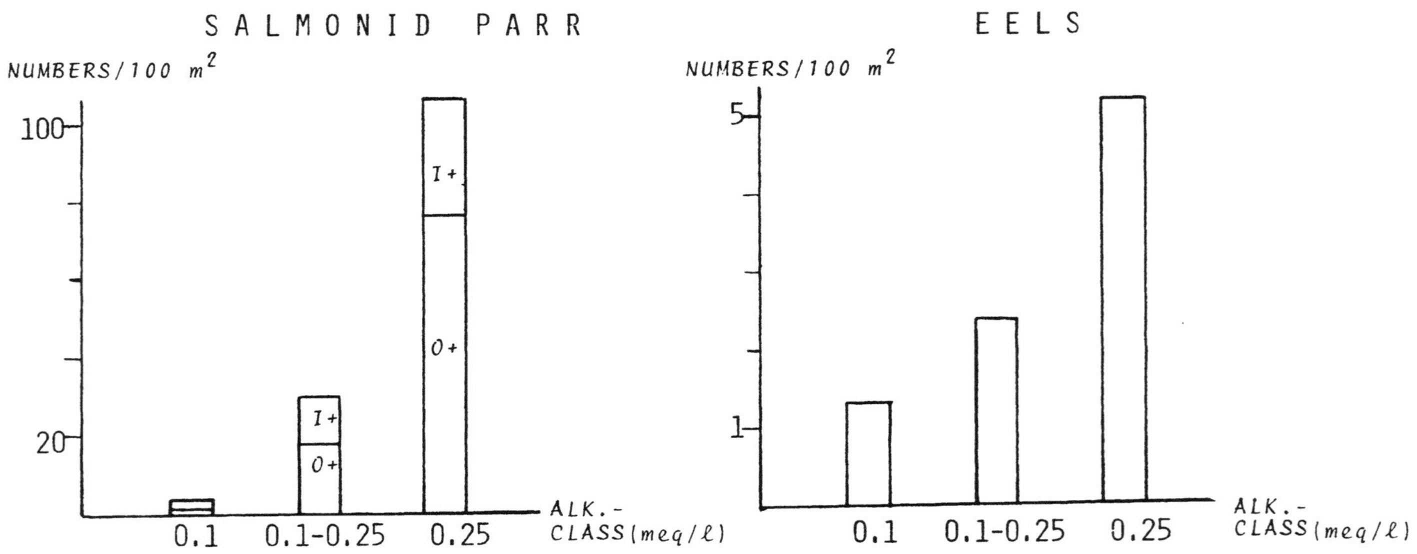

Figure 3. Abundance of salmonid parr (one- and two-summer old juveniles of brown trout and Atlantic salmon) and eel/100 m^2 at stations with different alkalinity (n=41 and n=45).

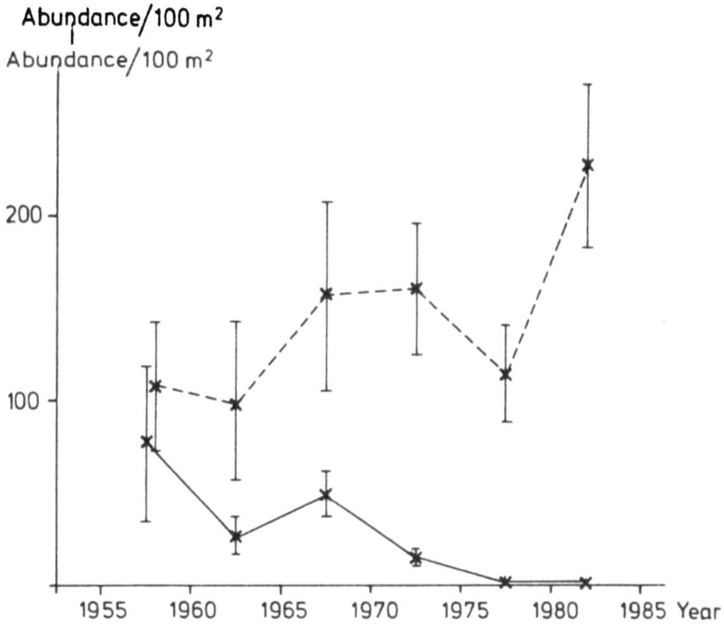

Figure 4. Mean abundance 100 m^2 of salmonid parr during five-year periods in seven streams on the west coast of Sweden during 1956-84. Solid line = streams with a summer alkalinity <0.25 meq·l^{-1} and a catchment area <1000 km^2 (n=2). Hatched line = streams with a summer alkalinity >0.25 meq.l^{-1} or an catchment area >1000 km^2 (n=5). The mean and standard errors are given.

whereas the values had been 5.7 to 6.7 in the summer water chemistry survey.

Brown trout occurred in all of the streams and Atlantic salmon was found in four streams. There was no significant correlation between the occurrence or abundance of salmonids and eel and the summer pH. The frequency of stations with salmonid parr was, however, significantly higher ($p < 0.05$) at alkalinities greater than 0.25 meq·l^{-1} as compared to the intervals 0 to 0.09 and 0.1 to 0.25 meq·l^{-1}. Eel were caught at 57 to 60% of the stations in the three alkalinity intervals mentioned above, i.e. there were no differences in occurrence correlated to alkalinity.

The abundance of salmonid parr (0+ and I+ of brown trout and Atlantic salmon combined) was significantly higher at stations with an alkalinity above 0.25 meq·l^{-1} as compared to stations with a lower alkalinity ($p < 0.05$) (Figure 3). There was a significant positive correlation between salmonid parr abundance and alkalinity ($p < 0.001$). No signifi-

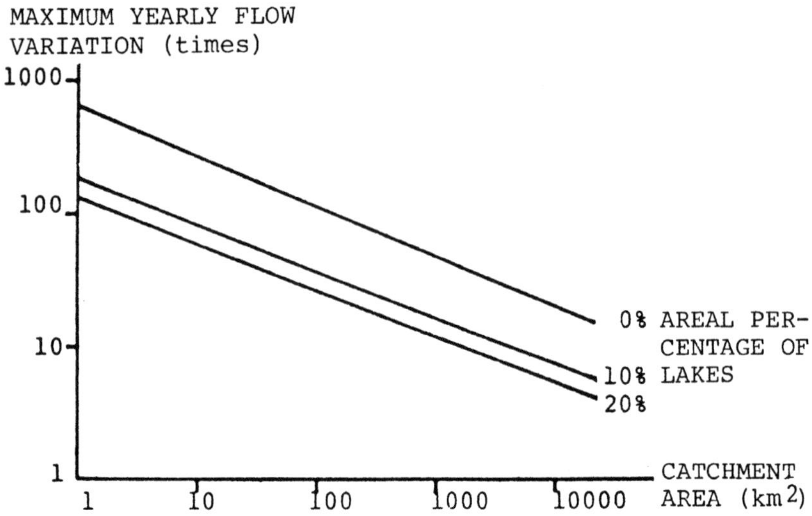

Figure 5. Maximum yearly flow variation (calculated as average yearly maximum flow divided by average minimum yearly flow) as a function of the catchment area and areal percentage of lakes. (Data from the Swedish Meteorological and Hydrological Institute.)

cant correlations were obtained between abundance and other parameters such as color, water velocity, stream width or conductivity. The latter parameter having been chosen as an index of the nutrient levels (P and N) in the streams. The abundance of eel showed the same pattern as for salmonid parr. It was significantly higher at alkalinities above 0.25 $meq \cdot l^{-1}$ ($p < 0.05$) and showed a significant positive correlation with the alkalinity ($p < 0.05$) (Figure 3).

Data compiled from previous electrofishing surveys in 7 similar streams in the region during 1956-84 showed yearly fluctuations in abundances of salmonid parr, but no significant trend over the period in the lower stretches of medium-sized streams (catchment area >1000 km^2) or in smaller streams with a summer alkalinity above 0.25 $meq \cdot l^{-1}$. In smaller rivers with low alkalinity, however, parr abundances at investigated stations decreased from the late 1950s onwards (Figure 4).

5. DISCUSSION

In spite of high pH values in summer, the abundances of salmonid parr and eel were low at some stations. During the spring thaw acid runoff may dilute the river water 100 to 600 times (Figure 5), giving low pH values. At stations with an alkalinity below 0.25 $meq \cdot l^{-1}$ such dilution results in the total loss of buffering capacity. The pH during winter at the investigated stations was 4.5-5.5 during a short period of only one week in January 1984. It is proposed that the formation of toxic Al complexes and to a lesser extent the short period of low pH caused salmonid mortality (e.g. Dickson, 1983; Rosseland and Skogheim, 1984). In the case of eel, which are known to be acid-tolerant (Almer et al., 1978), it has been shown that migrating juvenile eel prefer water of pH 7.5 as compared to 3.9 (Ask et al., 1971) and that high concentrations of Al cause mortality (Fjellheim et al., 1983). Miles (1968) has experimentally shown that eel prefer more productive water compared to less productive water. No correlation was found between the abundance of eel and the abundance of salmonid parr on all the investigated stations (hard and soft bottoms) and it is suggested that the distributions of the two species does not depend on any species interaction. The lower abundances of eel in waters with low alkalinity might thus rather be a combined effect of Al toxicity and avoidance reactions due to the low pH and productivity.

REFERENCES

Almer, B., Dickson, W., Ekström, C. and Hörnström, E.: 1978, 'Sulfur pollution in the aquatic ecosystem', p. 271-311. In Sulfur in the environment. Part II: Ecological impacts. Ed.: J.O. Nriagu. John Wiley & Sons, New York.

Andersson, P. and Nyberg, P.: 1984, 'Experiments with brown trout (Salmo trutta) during spring in mountain streams at low pH and elevated levels of iron, manganese and aluminium', Rep.Inst.Freshw. Res, Drottningholm 61:34-47

Ask, L., Berntsson, K.-E. and Holmberg, B.: 1971, 'Olika vattens attraktivitet på ålyngel', Inform.Inst.Freshw.Res., Drottningholm (14). 11 p. (In Swedish.)

Bjärnborg, B.: 1983, 'Dilution and acidification effects during the spring flood of four Swedish mountain brooks', Hydrobiologia 101: 19-26.

Dickson, W.: 1983, 'Liming toxicity of aluminium to fish', Vatten, Lund, Sweden 39:400-404.

Fjellheim, A., Raddum, G.G. and Sagen, T.: 1983, 'Effect of aluminium at low pH on the mortality of elvers (Anguilla anguilla L.)' PM Dept. Anim.Ecol., Zool.Mus., Univ.Bergen, Norway. 10 p.

Miles, S.G.: 1968, 'Rheotaxis of elvers of the American eel (Anguilla rostrata) in the laboratory from different streams in Nova Scotia', J.Fish.Res.Bd Canada 25:1591-1602.

Rosseland, B.O. and Skogheim, O.: 1984, 'A comparative study on salmonid fish species in acid aluminium-rich water. II. Physiological stress and mortality of one- and two-year-old fish', Rep.Inst.Freshw.Res., Drottningholm 61:163-169.

Zippin, C.: 1956, 'An evaluation of the removal method of estimating animal populations', Biometrics 12:163-189.

INFLUENCE OF ACIDIFICATION AND ALUMINUM ON THE DENSITY AND BIOMASS OF LOTIC BENTHIC INVERTEBRATES.

Martine Allard[1] and Guy Moreau
Dept. de biologie,
Universite Laval,
Quebec, Que. Canada G1K 7P4

Abstract. Experiments were carried out in plasticized wooden channels fed by a small creek in the Reserve des Laurentides, 80 km north of Quebec city. Channels were naturally colonized by invertebrates for 65 d before treatment. Treated channels were acidified in August with dilute H_2SO_4 only, or with acid plus a solution of Al sulfate (final concentration of 0.19 mg L^{-1}). The control channel received creek water only (pH 6.3 to 6.9). The addition of Al had no effect on invertebrate density and biomass. After 73 d of acidification, invertebrate densities were only one third the number found in the control channel. Invertebrate biomass was not different within channels, although biomass was generally higher in the two acidified channels. Difference in densities between acidified and non-acidified channels was attributed to lack of colonization and not to an increase in drift. Microtendipes, a large and resistant larva of Chironomidae constituted a large fraction of the biomass, largely outweighing numerous very small larvae. Effects on the density were attributed to the direct effect of low pH and not to indirect action through food limitations.

1. INTRODUCTION

Several studies have been done to assess the effects of acidification on freshwater ecosystems (see reviews by Haines, 1981; D'Itri, 1982; Haines and Johnson, 1982). At upper trophic levels, numerous detrimental effects have been observed on the physiology of fish and their population dynamics. The effects on benthic invertebrate communities also appear substantial. In several studies, the density, the biomass, the diversity, and the number of invertebrate species were reduced in acidified as compared to non-acidifed lakes or streams (Sutcliffe and Carrick, 1973; Wiederholm and Eriksson, 1977; Hall *et*

---

[1]present address: Institute for Environmental Studies, University of Toronto, Toronto, Ont. Canada M5S 1A4.

Water, Air, and Soil Pollution **30** (1986) 673–679.

al., 1980; Okland and Okland, 1980; Arnold et al., 1981; Zischke et al., 1983). However, Collins et al. (1981) did not find any differences in the densities, the biomass, or the the taxonomic composition between acidified and non-acidified lakes, while Crisman et al. (1980) did not observed differences in invertebrate densities and species numbers between acidified and non-acidified lakes in Florida. These contradictory results were tentatively attributed by Collins et al. (1981) to differences in Al concentrations. In Crisman et al. (1980) and Collins et al. (1981) studies, water pH were between 4.5 and 4.9, and Al concentration below 100 ug L^{-1}; while Al concentrations in Swedish and Norwegian lakes, where the most dramatic decreases in densities and number of species occurred, averaged between 200 and 250 ug Al L^{-1} in water of a mean pH of 4.7 (Wright et al., 1980).

Our objectives in this study was to test the influence of acidification and Al on the density and biomass of benthic invertebrates in streams.

2. MATERIALS AND METHODS

In nature, comparisons between acidified and non-acidified streams are difficult. The structure of benthic communities is often dependent on several physical and chemical characteristics of the sites. To distinguish between the influence of the sites features and the influence of the acidification or Al is rather difficult. For these reasons, we chose to work in experimental stream channels, in order to work with similar benthic communities before acidification. Any difference within the stream channels had to result from the acidification and/or Al, and not from differences in community structure between sites.

Experiments were carried out in the Foret Montmorency (Reserve des Laurentides, Quebec) on the Lake Laflamme outlet. The area is located on the Precambrian Shield at 47° 19' N and 71° 07' W at an altitude of 800 m. Creek water was diverted into plasticized wooden channels. Calibrated granitic pebbles contained in plastic baskets and unglazed ceramic tiles were utilized as artificial substrates for macroinvertebrates and algal colonization.

The experiment started in June 1982, at which time channels and substrates were cleaned. Organisms and sediment materials were introduced naturally by creek water into the channels. After 65 d of colonization, the acidification experiment was started (on August 8). Two channels were acidified with dilute H_2SO_4 to pH 4.0. One of these channels also received a solution of Al sulfate, such that the mean concentration of Al in the channel water was 0.19 mg L^{-1}. A third channel was fed with unaltered creek water (pH 6.3 to 6.9): this channel was the control. The experimental setup and the acidification procedure are described more precisely in Serodes et al. (1984) and Allard (1985).

Invertebrates were sampled one day before acidification (Aug. 8) and three times after acidification on August 20, September 14, and October 20, respectively. Sediments and periphytic algae were also sampled after acidification to monitor any change in the food resources for invertebrates. At each sampling time, in each channel, five pebble baskets were taken; invertebrates were completely sorted , counted and identified under a dissecting microscope at 12X. Invertebrates and sediments were dried separately at 50° C for 24 to 48 hr and weighed; sediments were ignited at 500° C for 2 hr and weighed to calculate their percentage of organic matter. Tiles were scraped to collect periphytic algae and chlorophyll *a* was extracted in 99% methanol (Marker, 1980). Water samples were also taken at each sampling time. Water analysis was done at the laboratories of the Quebec Ministry of the Environment within 24 hr. More details on water sampling and analysis are given in Allard (1985).

3. RESULTS AND DISCUSSION

Acidification resulted in only a few modifications of the water chemistry. Alkalinity was below detection limits, conductivity and sulfate were higher and total inorganic C was lower in acidified channels than in the control. Aluminum and metals (Cu, Zn, Cd, Cr, Ni, Pb) were not liberated from the sediments when the water was acidified (Table I). Therefore, any difference between the two acidified channels could be attributed to the effect of Al.

Table I. Chemical composition of the water[1]. Values are given as mean ± SD (N=3).

	Control	Acid	Acid + Al
pH	6.3 - 6.9	4.0	4.0
Alkalinity (mg $CaCO_3$ L^{-1})	5.3 ± 0.6	< 0.1	< 0.1
Conductivity (uS cm^{-1})	21.5 ± 1.3	42.5 ± 1.3	46.4 ± 1.2
Tot. Inorganic C (mg L^{-1})	1.1 ± 0.0	0.7 ± 0.3	0.7 ± 0.3
Tot. Organic C (mg L^{-1})	6.0 ± 1.0	5.7 ± 0.3	6.0 ± 0.5
SO_4 (mg L^{-1})	3.3 ± 0.3	10.5 ± 1.3	12.1 ± 0.8
Tot. dissolved Al (mg L^{-1})	0.07 ± 0.02	0.08 ± 0.03	0.19 ± 0.06

[1]Values for $N\text{-}NO_3$, $P\text{-}PO_4$, Ca, Mg, Na, K, Fe were similar for all channels, while Cu, Zn, Cd, Cr, Ni, Pb were always below detection limits, which were 5, 10, 1, 2, 20, 2 ug L^{-1} respectively.

The addition of Al to the water had no effect on the density and biomass of invertebrates (Table II). At a concentration of 0.2 mg L^{-1}, Al is toxic to fish (Baker, 1982). At similar and even higher concentrations (up to 30 mg L^{-1}), Al was not toxic to invertebrates (Allard and Moreau, 1985; Correa *et al.*, 1985; Havas and Likens, 1985).

However our experiment was conducted at pH 4.0 and it is possible that at higher pH, Al toxicity might be quite different.

Table II. Density and biomass of invertebrates, chlorophyll a in periphytic algae, dry weight and percentage organic matter of sediments before and after acidification. Channels are indicated as C for control, A for acid only, and AA for acid + Al. All values are given as mean ± SE, N=5 for sediment and invertebrates, N=3 for algal samples. Values followed by ** are significantly different ($p < 0.01$, Kruskal-Wallis).

		Before	After Acidification		
		Aug. 8 -1 d	Aug. 20 12 d	Sept. 14 37 d	Oct. 20 73 d
Density of invertebrates (no/0.01 m^2)	AA	535 ± 46	473 ± 50	429 ± 15	670 ± 58
	A	474 ± 25	566 ± 48	381 ± 47	635 ± 36
	C	450 ± 61	774 ± 102	904 ± 23 **	1701 ± 79 **
Biomass of invertebrates (mg d.w./ 0.01m^2)	AA	26.4 ± 6.0	25.2 ± 3.8	53.0 ± 4.8	70.0 ± 15.0
	A	22.2 ± 5.4	33.6 ± 6.6	42.8 ± 10.8	59.2 ± 16.6
	C	22.0 ± 4.4	23.0 ± 2.8	39.6 ± 7.8	39.6 ± 8.8
Algae (ug chloro a/cm^2)	AA	—	0.57 ± 0.11	2.18 ± 0.43	5.48 ± 2.8
	A	—	0.43 ± 0.06	1.81 ± 0.34	2.23 ± 3.9
	C	—	0.47 ± 0.18	1.61 ± 0.49	1.66 ± 0.6 *
Sediments (g d.w./ 0.01m^2)	AA	—	1.15 ± 0.22	2.14 ± 0.43	1.06 ± 0.14
	A	—	1.42 ± 0.25	1.62 ± 0.17	1.40 ± 0.31
	C	—	1.66 ± 0.43	2.63 ± 0.56	1.99 ± 0.21
Sediments (% organic matter)	AA	—	45.8 ± 1.4	45.9 ± 2.9	45.5 ± 1.1
	A	—	43.5 ± 2.5	43.5 ± 2.9	45.3 ± 2.0
	C	—	43.0 ± 1.8	42.3 ± 0.2	40.8 ± 0.4 **

The acidification itself reduced the density of invertebrates, no drift response was observed in the acidified channels after the starting of acidification (Allard and Moreau, 1984). After 37 d and 73 d of acidification, invertebrates were respectively two and three times less abundant in the acidified channels compared with the control channel. However, the invertebrate biomass did not show the same trend. Even though there was no significant difference in biomass among channels, the two acidified channels tended to have higher biomass than the control. The low density, high biomass in acidified channels and the high density, low biomass in the control could be easily explained based on observations of colonization. In September and October, colonization was intensive in the control channel; early instars of Chironomidae and small nymphs of Ephemeroptera were settling (Table III). Colonization in acidified channel, on the other hand, was restricted; densities at day 73 were only slightly higher than preacidification levels, and were due to the arrival of Chironomidae. On day 73, biomass in all channels was mainly due to larvae of Microtendipes (Chironomidae), which were big larvae which outweighed any other invertebrates. The early instars settling in the control, despite their high numbers, added relatively little to the biomass (Allard, 1985). This might explain why biomass was not significantly different between acidified and control channels.

Before acidification, densities and community structure were closely similar among channels (Table II and III). Chironomidae and Oligochaeta accounted for more than 75 % of the fauna. Soon after acidification (12 d), changes occurred in all channels; Chironomidae decreased while Nematoda increased (Table III). By the end of the experiment (73 d), the group Ephemeroptera was significantly lower in the acidified channels. The percentage of Chironomidae was similar among channels. However, acidified channels had less Chironomidae taxa (Allard, 1985). These changes in the community structure could have arisen from the acidification or from modifications in the food resources.

Table III. Percentage composition of the invertebrate community, before (day -1) and after acidification. Channels are indicated as C for control, A for acid only, and AA for acid + Al. Ch : Chironomidae, E : Ephemeroptera, Ol: Oligochaeta, N : Nematoda, C : Ceratopogonidae, O : others.

	CH	E	OL	N	C	O		CH	E	OL	N	C	O
day -1							day 12						
AA	41	10	33	10	4	2	AA	27	13	27	25	7	1
A	39	12	33	11	3	2	A	24	9	35	23	8	1
C	37	10	33	14	4	2	C	20	10	42	20	6	2
day 37							day 73						
AA	40	9	12	31	6	2	AA	49	7	9	27	5	3
A	43	9	12	27	7	2	A	46	6	14	26	7	1
C	25	14	29	25	7	0	C	44	12	13	22	8	1

It was concluded that acidification was acting directly on invertebrates and not indirectly through food limitation. The quantity of sediment/detritus available to collector-gatherers was not significantly different among channels (Table II). Chlorophyll _a_ was significantly higher in the two acidified channels after 73 d (Table II). Long filaments of algae were growing on the sides of the acidified channels, biomass averaging between 79 and 250 mg/0.01 m^2. Grazers and scrapers (which usually feed on periphytic algae) have been shown to be less abundant in acidified streams (Sutcliffe and Carrick, 1973; Hall _et al._, 1980; Otto and Svensson, 1983; Allard, 1985). In our experiments, there is no evidence that modifications of the algal communities could be responsible for their disappearence from the acidified channels. In fact, grazers, (mainly Orthocladiinae and Ephemeroptera) which were seen on top of the artificial substrates before acidification, disappeared as soon as the acidification started; grazers never returned foraging on the artificial substrates thereafter (Allard, 1985). Grazers left the acidified channels before any modifications in the algal community could have possibly taken place. Therefore, we conclude that they were responding to the increase in acidity and not to changes in the food supply.

On a long term basis, however, modification of the food supply could be more important than it was in our short-term experiment. Decomposition processes seemed to be lowered by acidification. This was confirmed by values of organic matter which were significantly higher in the two acidified channels than in the control. Lower microbial metabolism and leaf degradation were also observed in the acidified channels (Allard, 1985; Allard and Moreau, 1985). The consequence of a slower decomposition in acidified waters may be important, especially for shredders which rely upon allochthonous matter and its associated microbial population.

The consequences of the acidification along downstream reaches of streams could be numerous. Previous generalizations from the river continuum concept (Vannote _et al._, 1980) would be modified as a result of changes in the food web. Predictable modifications may include a shift from an allochthonous to an autochthonous energy source in headwater streams where production is not light limited. As a consequence, producers would be of greater importance higher in the stream gradient. Change in the relative importance of energy source would also modify macroinvertebrate distribution along the continuum. Shredders, which feed on coarse particulate organic matter (CPOM), would be abundant even in the lower part of the river as long as CPOM is not degraded. They could be replaced by filter feeders only when fine particulate organic matter (FPOM) is processed to ultrafine particulate organic matter (UFPOM). Detritivores such as Oligochaeta could also be favored throughout the system, especially if organic matter accumulates.

4. REFERENCES

Allard, M.: 1985, Etude experimentale des mecanismes d'action d'une acidification du milieu sur les communautes benthiques lotiques, Ph. D. thesis, Universite Laval, Quebec, pp 194.
Allard, M. and Moreau, G. : 1984, Verh. Int. Ver. Limnol. **22**; 1793.
Allard, M. and Moreau, G. : 1985, Can. J. Fish. Aquat. Sci. **42**, 1676.
Arnold, D.E., Bender, P.M., Hale, A.B., and Light, R.W.: 1981, 'Studies on infertile acidic Pennsylvania streams and their benthic communities', in Singer, R. (ed.) Effects of acidic precipitation on benthos, North American Benthological Society, Hamilton, NY, 15.
Baker, J.P.: 1982, 'Effects on fish of metals associated with acidification' in Haines, T.A. and Johnson, R.E. (eds) Acid rain/Fisheries, American Fishery Society, Bethseda, MA, 165.
Collins, N.C., Zimmerman, A.P. and Knoechel, R.: 1981, 'Comparisons of benthic infauna and epifauna biomasses in acidified and non-acidified Ontario Lakes', in Singer, R. (ed.) Effects of acidic precipitation on benthos, North American Benthological Society, Hamilton, NY, 35.
Correa, M., Coler, R.A. and Yin, C.M.: 1985, Hydrobiologia **121**, 151.
Crisman, T.L., Schulze, R.L., Brezonic, P.L. and Bloom, S.A.: 1980, 'Acid precipitation: the biotic response in Florida lakes', in Drablos, D. and Tollan, A. (eds) Ecological impact of acid precipitation, SNSF project, Oslo, Norway, 326.
D'Itri, F.M.: 1982, Acid precipitation, effects on ecological systems, Ann Arbor Science, Ann Arbor, MI, pp 506.
Haines, T.A.: 1981, Trans. Amer. Fish. Soc. **110**, 669.
Haines, T. A. and Johnson, R.E.: 1982, Acid rain/Fisheries, American Fisheries Society, Bethseda, MA, pp 357.
Hall, R.J., Likens, G.E., Fiance, S.B. and Hendrey, G.R.: 1980, Ecology **61**, 976.
Havas, M. and Likens, G.E.: 1985, Can. J. Zool. **63**, 1114.
Marker, A.F.H.: 1980, Arch. Hydrobiol. Beih. Ergebn. Limnol. **14**, 88.
Okland, J. and Okland, K.: 1980, 'pH level and food organisms for fish: studies of 1000 lakes in Norway', in Drablos, D. and Tollan, A. (eds) Ecological impact of acid precipitation, SNSF project, Oslo, Norway, 326.
Otto, C. and Svensson, B.S.: 1983, Archiv. Hydrobiol. **99**, 15.
Serodes, J.B., Moreau, G. and Allard, M.: 1984, Wat. Res. **18**, 95.
Sutcliffe, D.W. and Carrick, T.R.: 1973, Freshwat. Biol. **3**, 437.
Vannote, R.L., Minshall, G.W., Cummins, K.W., Sedell, J.R. and Cushing, C.E.: 1980, Can. J. Fish. Aquat. Sci. 37, 130.
Wiederholm, T. and Eriksson, L.: 1977, Oikos **29**, 261.
Wright, R.F., Conroy, N., Dickson, W.T., Harriman, R., Henriksen, A. and Schofield, C.L.: 1980, 'Acidified lake districts of the world: A comparison of water chemistry of lakes in southern Norway, southern Sweden, southwestern Scotland, the Adirondack Mountains of New-York and southeastern Ontario', in Drablos, D. and Tollan, A. (eds) Ecological impact of acid precipitation, SNSF project, Oslo, Norway, 337.
Zischke, J.A., Arthur, J.W., Nordlie, K.J., Hermanutz, R.O., Standen, D.A. and Henry, T.P.: 1983, Wat. Res. **17**, 47.

REACTION OF LOTIC PERIPHYTON TO EXPERIMENTAL ACIDIFICATION

D. Planas
Département de biologie
Université du Québec à Montréal
C.P. 8888, Succ. A
Montréal, Québec
H3C 3P8. Canada

G. Moreau
Département de biologie
Université Laval
Cité Universitaire
Québec, Québec
G1K 7P4. Canada

ABSTRACT. Two hypotheses that could explain increases in the biomass and production of lotic periphytic communities in acidified habitats were tested: 1) a greater bioavailability of P due to its greater release from sediments at low pH, and 2) utilization of S from the H_2SO_4 addition. The experiments were performed in semi-natural conditions by submitting the periphyton to either continuous or episodic acidification. The uptake of P and S by periphyton were determined at different times during the treatments. The increases in P and S uptake, in acidified habitats, are compatible with both hypotheses and could account for the increase in periphytic biomass and production.

1. INTRODUCTION

Grahn *et al.*, (1974) stated that the acidification of aquatic ecosystems could lead to the oligotrophication of lakes. This assertion for phytoplanktonic production has recently been questioned. Dillon *et al.* (1979) and Havens and DeCosta (1984) have experimentally demonstrated the control exerted by P on phytoplankton production and biomass at low pH as well as at neutrality. Other authors (Muller, 1980; Stockes, 1981; Schindler *et al.* 1985) have described prolific growths of periphyton in acidified lakes.

We have recently shown a strong increase in lotic periphytic production and biomass resulting from experimental acidification in channels (Parent *et al.*, 1986).

Three hypotheses can be put forth to explain this increase in biomass and production during acidification: 1) increase in P-availability by P-resolubilization from sediments at low pH, 2) S-SO_4 utilization by the algae, and 3) decrease in benthic grazing as a result of acidification.

The aim of this study is to test hypotheses one and two.

2. MATERIAL AND METHODS

Water, Air, and Soil Pollution **30** (1986) 681–686.

2.1. Experimental system

Experiments were performed in semi-natural conditions using two set-ups (Serodes et al., 1984) located on two similar first order streams of the Canadian Shield (Québec: 47° 19'N -71° 07'W).

2.2. Experimental procedure

After 3 to 4 weeks (from June to September) of algal colonization on unglazed ceremic tiles, the following experiments were performed in individual channels: 1) episodic acidification (E.A.): pH 4.0 to 4.5 every 3 days for 24 hr using 3.5 N H_2SO_4 (grade ACS), 2) continuous acidification (C.A.) during 3 mo at the same pH using the same acid, and 3) P enrichment (3 µM PO_4^{3-}) associated with E.A.

2.3. Chemical caracteristics

In each channel we measured: pH, conductivity, alkalinity, TIC, TOC, N (N-NH_4-NO_2-NO_3), P (MRP, PP, TP), SO_4, Cl, major cations (Ca, Mg, Na, K) and metals (Zn, Cd, CU, Fe, Mn, Hg, TAl and DAl). See Parent et al. (1986) for the methods used.

2.4. Radiotracer method

For P and S uptakes each colonized tile was put in a transparent Pyrex bottle, incubated in a shaker at constant temperature. For P uptake after the tracer was added subsampling, counting and rate constant calculations were done following the method of Levine (1975). This method was used since chemical analysis could not detect any variation in P concentrations after acidification. The S uptake samples were incubated during 4 to 5 hr. After incubation, the tiles were brushed out, filtered on 0.45 µm millipore membrane and counted on a scintillation counter as describe in Campbell and Baker (1978).

2.5. Periphyton biomass

Chlorophyll-a by methanol extraction (Marker, 1980) was used to monitor periphyton biomass.

3. RESULTS

In C.A. chlorophyll-a concentration in the acidified channel increased significantly ($p \leq 0.01$) and considerably (up to 4.5 fold) after 40 days of treatment with respect to the control (Figure 1a). This increase in biomass diminished with time and after approximately 3 mo the concentration of chlorophyll-a is not different in acidified and control channels. The uptake rate of P was significantly ($p \leq 0.01$) lower in the acidified channel than in the control from day 40 to the end of the experiment (Figure 1b), in spite of the biomass increase. The S uptake trend was inconsistent, it was significantly higher ($p \leq 0.05$) in the 1 mo of treatment and after 3 mo. Conversely, it was significantly ($p \leq 0.05$) lower than in the control after 40 days of treatment (Figure 1c).

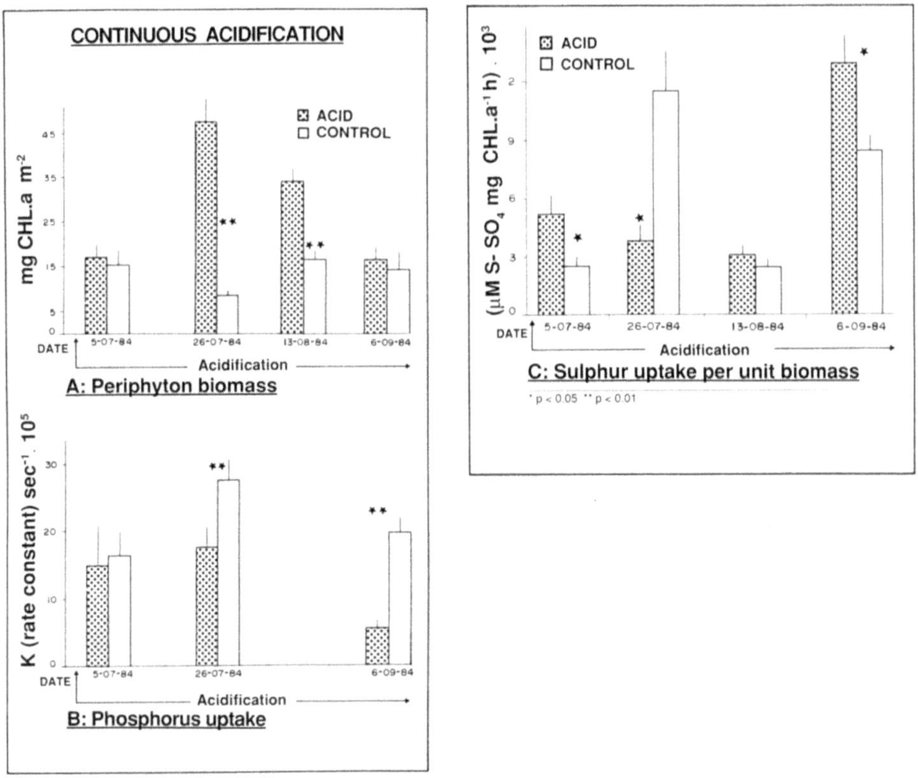

Figure 1. Evolution of: a) biomass, b) P uptake rate constant, c) S uptake per unit biomass in the continuously acidified channel in relation to the control.

In E.A. algal biomass was significantly ($p \leq 0.05$) lower in the acidified channel 24 hr after the start of acidification; identical in the acidified and control channels after 30 and 45 days of E.A. and significantly ($p \leq 0.01$) higher in the acidified channel after 64 days of treatment (Figure 2a). P enrichment produced a significant increase in biomass after about 1 mo. The rate of P uptake was never significantly different between the acidified and the control channels, although it was usually slightly lower in the former (Figure 2b). By contrast, P uptake rate was generally significantly higher in the acidified than in the P enriched channel where we measured the highest biomass of the three channels. During acidification, S uptake was always higher, usually significantly so, in the acidified channel; on the other hand immediately after acidification; S uptake was higher in the control (Figure 2c).

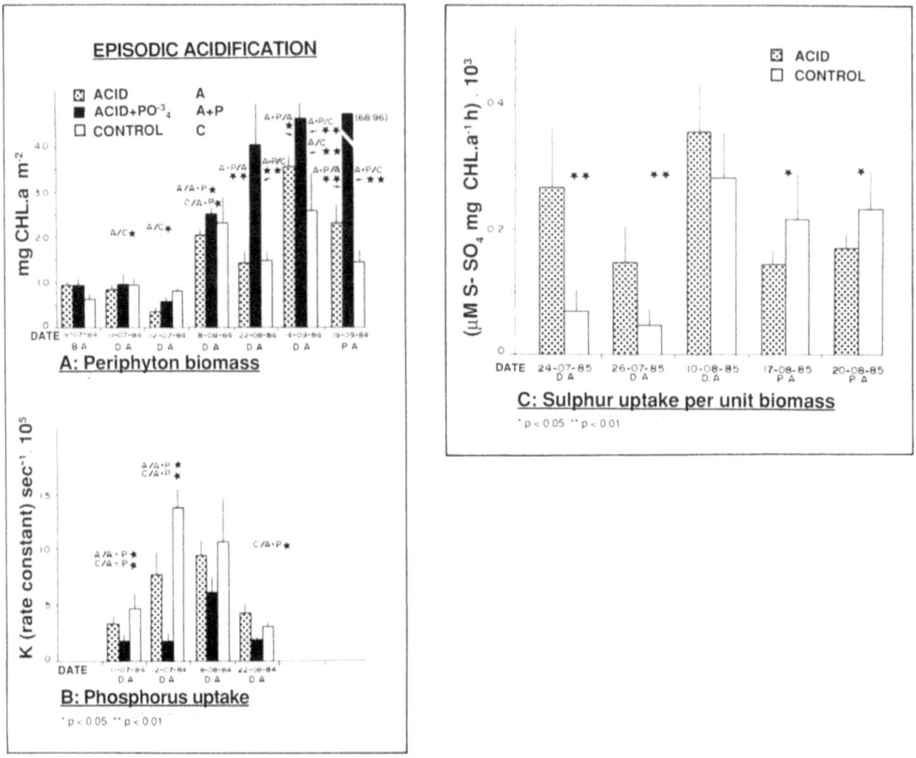

Figure 2. Evolution of: a) biomass, b) P uptake rate constant, c) S uptake per unit biomass in the episodic acidified channel in relation to the control. B.A. = before acidification; D.A. = during acidification; P.A. = post acidification; A/C = significant differences between acidified and control channels.

4. DISCUSSION

The lower P absorption constant in the P enriched habitat confirms, following Lean and Nalewajko (1979), that P uptake is an adequate measure of P availability. Based on these findings, the lower uptake constant in the acidified channel of the C.A. experiment suggests that more P was available in this channel than in the control.

The situation is not as clear in E.A. experiment. Although the uptake rate constant was always lower in the acidified channel, differences with the control were not statistically significant. Many things can explain this difference in response between the two experiments. We do not think that these differences could be related to the periphyton species composition, since control channels of both set-ups have similar dominant flora (mainly diatoms) and similar shifts occur in the acidified channels of both set-ups (changes in diatoms species and appearance of filamentous greens). The only difference was in the rate of change of species which occurs more quickly in the C.A. than in the E.A. A change in the physiological status of the algae in acidic condition is also possible but we do not have data to show this. It is more probable that this difference could be related to the rate of organic matter decomposition. Both experimental set-ups accumulated sediments very rich in organic matter (Allard and Moreau 1985a) and the degradation of these was slowed down in C.A. due to a decrease of microorganisms activity (Allard and Moreau 1985b). According to Mulholland et al. (1985) the P is normally fixed by microorganisms associated with CPOM (Coarse particulate organic matter). Consequently, in these lotic condition, it would result in a more permanent availability of P in C.A. than in E.A. where the input of P is only episodic.

That P is utilized by the algae in acidified milieu and increases its biomass seems evident from the P enriched channel. At the end of the experiment the biomass of this channel was 5 fold higher than in the control and nearly 3 times higher than the acidified.

With regards to S absorption, brought about by the addition of H2SO4 the response was much clearer in the E.A. experiment. Under these conditions S uptake is always significantly higher in the acidified channel. The fact that the rate of S absorption decreased significantly immediately after the end of experimental inputs seems to indicate, as has been shown in the laboratory by Nriagu and Wong (1976), that the absorption of S is of a "luxury" consumption type. This kind of absorption could explain why in the C.A. S uptake, initially higher under acid conditions, decreased once the algae had accumulated excess S. At the end of the experiment, the new period of elevated S absorption could possibly have coincided with a change in the specific composition of the algal community and the appearance of new species which had not yet accumulated S. On the other hand, one must note that the concentrations measured under natural conditions were very low (around 15 μeq^{-1}) and among the lowest of those observed for Canadian Shield waters (Lazerte and Dillon, 1984).

It is therefore possible to conclude that our results were consistent with formulated hypotheses. The increase in algal biomass in acidified conditions could be probably due to a greater availability of P, even though there is no detectable increase of this element in water, or to an increased utilization of S from the H2SO4 additions. The relative importance of these mechanisms certainly would depend on environmental conditions, such as the concentration of P and S in water and sediments.

These mechanisms however, do not exclude the third hypothesis, which was not tested in our experiments, a decrease in benthic grazing under acid conditions.

ACKNOWLEDGMENTS

The authors wish to thank FCAR, NSERC and the National Sportsman Fund for their financial support. Also, we are grateful to Environment Canada (Environment 2000) for providing funds for technical assistants. We also thank M. Allard, L. Parent and D. Houle for field assistance. C. Gascon helped in preparing the manuscript and R. Anderson provided helpful comments.

REFERENCES

Allard, M. and Moreau, G.: 1985a, Can. J. Fish Aquat. Sci. **42**, 1676.
Allard, M. and Moreau, G.: 1985b, Hydrobiologia in press.
Campbell, P.G.C. and Baker, J.H.: 1978, Can. J. Microbiol. **24**, 939.
Dillon, P.J., Yan, N.D., Scheider, W.A. and Conroy, N.: 1979: Arch. Hydrobiol. Beih. Ergebn. Limnol. **13**, 317.
Grahn, O., Hultberg, H. and Landner, L.: 1974, Ambio **3**, 93.
Havens, K.E. III and DeCosta, J.: 1984, Arch. Hydrobiol. **9**, 443.
Lazerte, B. and Dillon, P.J.: 1984, Can. J. Fish. Aquat. Sci. **41**, 1664.
Lean, D.R.S. and Nalewajko, C.: 1979, Arch. Hydrobiol. Beih. Ergebn. Limnol. **13**, 120.
Levine, S.: 1975, Verh. Internat. Verein. Limnol. **19**, 624.
Marker, A.F.H.: 1980, Arch. Hydrobiol Beih. Ergebn. Limnol. **16**, 88.
Mulholland, P.J., Newbold, J.D., Elwood, J.W., Ferren, L.A. and Webster, J.R.: 1985, Ecology **66**, 1012.
Muller, P.: 1980, Can. J. Fish. Aquat. Sci. **37**, 355.
Nriagu, J.O. and Wong, T.S.: 1976, Fractionation of sulphur isotopes by algae, pp. 734-738 In E.R. Klein and P.D. Klein (eds), U.S. Dept. Commerce, Technical Information Service, Springfield, Virginia.
Parent, L., Allard, M., Planas, D. and Moreau, G.: 1986, "The effects of short term and continuous experimental acidification on biomass and productivity of running water perithytic algae", pp. 28-41 In B.G. Isom, D. Denis and J.M. Bates (eds) Impact of Acid Rain and Deposition on Aquatic Systems. ASTM, STP 928, Philadelphia.
Schindler, D.W., Mills, K.H., Malley, D.F., Shearer, J.A., Davies, I.J., Turner, M.A., Linsey, G.A. and Cruikshank, D.R.: 1985, Science **228**, 4706.
Sérodes, J.B., Moreau, G. and Allard, M.: 1984, Water Res. **18**, 95
Stokes, P.M.: 1981, "Benthic Algal communities in acidic lakes", pp. 119-138 In R. Singer (ed), Effects of acidic Precipitation on Benthos. North American Benthological Society, Hamilton, N.Y.

PHYSIOLOGICAL RESPONSES OF RAINBOW TROUT <u>SALMO GAIRDNERI</u> EXPOSED TO PLASTIC LAKE INLET AND OUTLET STREAM WATERS

H. H. Harvey and J. M. McArdle
Department of Zoology
University of Toronto
Toronto, Ontario M5S 1A1, Canada

ABSTRACT. In Plastic Lake, Ontario, stocked rainbow trout (*Salmo gairdneri*) have failed to survive, one endemic fish species has become extinct and annual fish kills included up to five species, but especially pumpkinseeds (*Lepomis gibbosus*). The potential toxicity of Plastic Lake water was assessed by holding (hatchery) rainbow trout in the major inlet stream, the outlet, and in a portion of the outlet stream acidified to the pH of inlet No. 1. Stress on rainbow trout was assessed by measuring plasma and muscle concentrations of Na^+, Cl^-, and K^+, plus gill Al concentration. Trout held in Plastic Lake inlet No. 1 showed a rapid loss of plasma Na^+ from 138 to 85 $meq.L^{-1}$ and Cl^- from 120 to 75 $meq.L^{-1}$ within 24 hr. Gill Al concentration increased from 20 to 105 $\mu g.g^{-1}$ dry weight. Trout held in the outlet steam showed only slight loss of plasma Na^+ and Cl^- and no accumulation of Al on the gills. Trout held in the acidified outlet showed a significant loss of ions with plasma Na^+ depressed from 140 to 115 $meq.L^{-1}$ and plasma Cl^- from 125 to 95 $meq.L^{-1}$ over 24 hr. Gill Al concentration increased from 18 to 30 $\mu g.g^{-1}$ dry weight. The differences in stress response of rainbow trout held in the inlet and acidified outlet are likely due primarily to the difference in Al species concentrations in the two waters.

1. INTRODUCTION

The phenomenon of ionoregulation failure in fishes subject to acid stress is well documented under laboratory conditions (McDonald and Wood, 1981; Fraser and Harvey, 1984) and in brown trout *Salmo trutta*, from an acidified river (Leivestad and Muniz, 1976). To date, there has been no report of ionic imbalance in fishes from an acid lake. Acid stress appears to interfere with ion exchanges at the gill surface. Na^+ uptake in exchange for H^+, NH_4^+ loss; Cl^- uptake and HCO_3^- loss (Evans, 1975); the cause of death may be plasma hemoconcentration leading to circulatory failure (Wood and McDonald, 1982).

Aluminum in the free ionic form mimics the action of H^+, interfering with fish ionoregulation. One source of such Al is through the formation of weak acids Al^{+3}, $Al(OH)^{2+}$, and $Al(OH)_2^+$ during neutralization

Water, Air, and Soil Pollution **30** (1986) 687–694.

of excess strong acidity (Johnson, 1979). The lethality of Al in acid water has been documented (Schofield, 1977; Dickson, 1978; Muniz and Leivestad, 1980).

Following the first observed fish-kill in Plastic Lake (1978) experiments were conducted in which caged rainbow trout, Salmo gairdneri, were held within the inlet and outlet streams, and at sites within Plastic Lake (Harvey et al., 1982; Harvey, 1985). These experiments established the rate and extent of fish mortalities, and the rate of accumulation of Al on the gills.

The purposes of this experiment were to: define the stress of Plastic Lake waters on rainbow trout as measured by plasma and muscle concentrations of Na, Cl, K, and by gill Al; and to attempt to separate the effects of pH and Al by repeating these analyses on rainbow trout held in outlet stream water acidified to the same pH as Inlet No. 1, but at the lower Al concentration.

2. METHODS

Fish kills in Plastic Lake were investigated annually, 1978-1985, within a few days of ice cover leaving the lake, by collecting dead and dying fish around the perimeter of the lake in the depth zone 0 to 1 m. In three of the years, divers determined that dead and dying fish could be found to depths of several meters. Moribund fish were captured by dip-net, weighed, measured, a scale sample taken, bagged individually and frozen. Subsequently gills were removed, weighed, dried to constant weight and analyzed for Al.

The study was conducted on Plastic Lake inlet No. 1 (point a, Figure 1) June 19 to 20, 1982, on the lake outlet (point b, Figure 1) June 20 to 22, 1982; June 20 to 23, 1983, and on the acidified outlet (point b, Figure 1) June 20 to 23, 1983. For each experiment forty rainbow trout, 1^+ yr, 81 to 216 g, were purchased from the Milford Bay Trout Farm. The effect of transportation on the fish was assessed by conducting all blood, muscle and gill analyses on five fish at the hatchery (time = -3 hr) and on arrival at the field laboratory (time = 0 hr). No differences were found and the potential stress of transportation was discounted. The trout were held in a 600 L tank at a flow of 20 $L.min^{-1}$. At point b a 300 L trough was established for acidifying 20 $L.min^{-1}$ from pH 5.5 to nominal pH 4.1 (actual 4.2) and this water "acidified outlet" was fed to the 600 L tank.

Five trout from the holding tank were sampled every 4 hr. The methods of blood extraction and analysis, plus muscle removal and analysis were described previously (Fraser and Harvey, 1984). Gills were excised from the left side of each fish, weighed, dried to constant weight and analyzed for total Al (B. Loescher, Ontario Ministry of Environment, Rexdale, Ontario). Water pH and temperature were recorded hourly, D.O. every 4 hr, and alkalinity every 12 hr. Water samples were taken for Al speciation (B. Lazerte, Ontario Ministry of Environment, Dorset, Ontario).

Figure 1. Outline map of Plastic Lake with recovery sites: small circles, dead, and large circles, moribund pumpkinseeds; triangles, rainbow smelt; squares, frogs and tadpoles, based on collection from one day in April, 1981. Rainbow trout were held in inlet No. 1, point a, and lake outlet, point b.

3. RESULTS

The spring fish kill in Plastic Lake in 1981 was representative of the phenomenon in the years 1978 to 1985. The majority of the dead and dying fishes were pumpkinseeds, Lepomis gibbosus, but with occasional rainbow smelt, Osmerus mordax, brook stickleback, Culaea inconstans (not in 1981), yellow perch, Perca flavescens (not in 1981), bullfrog, Rana catesbiana, frog/tadpole sp., and crayfish sp. These organisms were observed most frequently near the inlet streams, and between Inlet No. 1 (south-west corner of the lake) and the outlet (Figure 1).

Moribund fish were analyzed, but chemical analyses were not performed on dead animals because of possible post-mortem changes in ionic concentration. In 1980, 11 pumpkinseeds had a mean concentration of 43.2±7.5 (S.E.) $\mu g\ Al.g^{-1}$ dry weight of gill tissue, and in 1981, nine pumpkinseeds yielded $\bar{x}$ = 37.0±16.5, and one rainbow smelt 350 $\mu g\ Al.g^{-1}$ dry weight.

The Al concentration of Inlet No. 1 was approximately 12X, and DOC 5X higher than the outlet (Table I). Aluminum speciation of the inlet water showed free Al 40% higher (5.9 vs 4.1 $\mu g.L^{-1}$) and fluoride-complexed Al 10X higher (72 vs 7.0 $\mu g.L^{-1}$) in inlet waters vs the outlet stream (Table I). Acidification of the outlet stream water altered substantially the proportions of the Al species present (Table I), reducing free Al and increasing the concentration of Al fluoride complexes.

Table I. Water chemistry at the three experimental sites, mean ± SE where n ≥ 4.

	Hatchery	Inlet No. 1	Outlet	Acidified Outlet
Temp °C	13	11.7 ± 0.3	16.2 ± 0.8	11.7 ± 1.4
pH	6.2	4.1 ± 0.1	5.4 ± 0.15	4.2 ± 0.14
		$\mu g.L^{-1}$		
Total Al	44	428	34.5 ± 2.3	37 ± 0.6
Polymeric Al	28	29	12.8 ± 2.4	7.6 ± 1.2
Organic Al	11	321	10.5 ± 2.6	15.4 ± 1.7
Inorganic Al	5	78	11.3 ± 2.5	14.0 ± 1.1
Al OH	4.99	5.9	4.1 ± 1.2	0.6 ± 0.1
Al F	0.01	72	7.0 ± 1.3	13.1 ± 1.2
Al SO_4		<0.001	0.03 ± 0.01	0.03 ± 0.01
		$mg.L^{-1}$		
DOC	11	11.0 ± 0.3	2.1 ± 0.05	2.7 ± 0.15
SO_4	8.7	5.6 ± 0.23	7.3 ± 0.15	9.02 ± 0.55
Ca	5.0	1.6 ± 0.00	2.25 ± 0.15	2.25 ± 0.15
Zn	0.008	0.016 ± 0.00	0.016 ± 0.005	0.010 ± 0.001
DO	8.3	9.5 ± 0.01	7.0 ± 0.7	8.4 ± 0.9

Rainbow trout held in Inlet No. 1 rapidly lost plasma ions (Figure 2), reaching a loss of 42% of Cl^- within 24 hr. Losses of this magnitude coincided with the fish becoming moribund, followed by death in a few hours. Trout in the outlet showed a temporary depression in $[Cl^-]$ from 125 to 100 $meq.L^{-1}$, followed by recovery. Trout held in the acidified outlet showed a 24% decline in plasma $[Cl^-]$, about half the loss of inlet fish, and without significant recovery. Plasma $[Na^+]$ was reduced 37% in inlet-held trout (Figure 2); outlet held fish showed a decline and recovery; acidified outlet fish showed a loss of 19% to a $[Na^+]$ of 117 $meq.L^{-1}$, without recovery.

Muscle $[Cl^-]$ and $[Na^+]$ (Figure 3) decreased rapidly in rainbow trout in the inlet water and showed no recovery over 25 hr. Muscle $[K^+]$ changed little over 20 hr but was lower (56 meq.100 g^{-1} dry weight) at 24 hr. Trout in the lake outlet showed a significant reduction in muscle $[Na^+]$, and no significant change in $[Cl^-]$ and $[K^+]$. Rainbow trout held in the acidified outlet showed temporary declines in muscle $[Na^+]$ and $[Cl^-]$, with minimum values at 8 hr and recovery by 24 hr.

Gill Al concentration increased rapidly from 20 to 105 $\mu g.g^{-1}$ dry weight for trout held in Inlet No. 1 (Figure 4). Rainbow trout in the outlet stream showed an initial increase in gill Al concentration, followed by a decline to less than control (hatchery) levels. Trout in the acidified outlet showed fluctuations in gill Al concentration, but no clear trend. The considerable variability in Al concentration for fish gills (Figure 4) may have been due to the sloughing of mucus containing Al. Mucus was observed being expelled from under the opercular flaps among fishes held in Inlet No. 1.

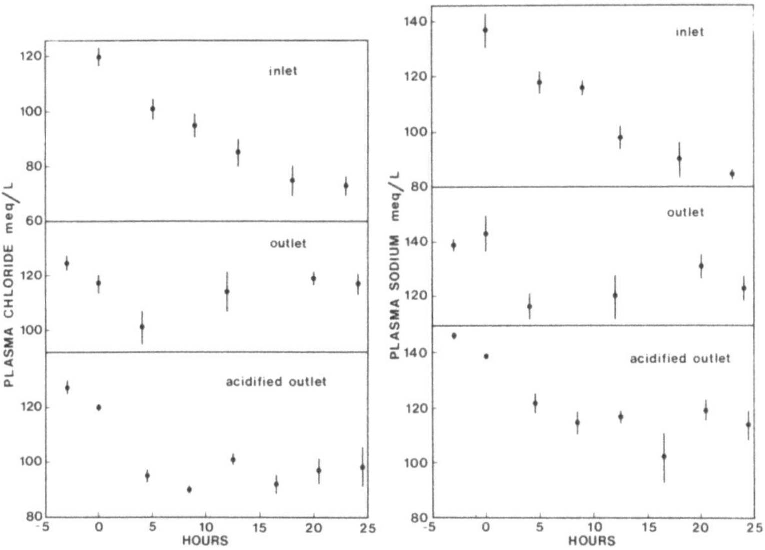

Figure 2. Plasma chloride and sodium concentrations of rainbow trout in the three test localities; solid circle, mean; vertical bar, standard error.

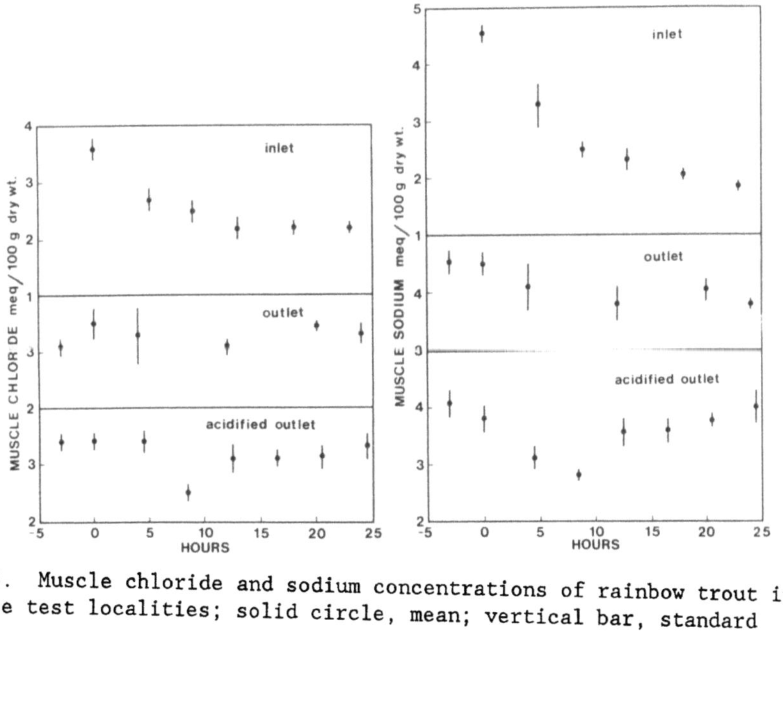

Figure 3. Muscle chloride and sodium concentrations of rainbow trout in the three test localities; solid circle, mean; vertical bar, standard error.

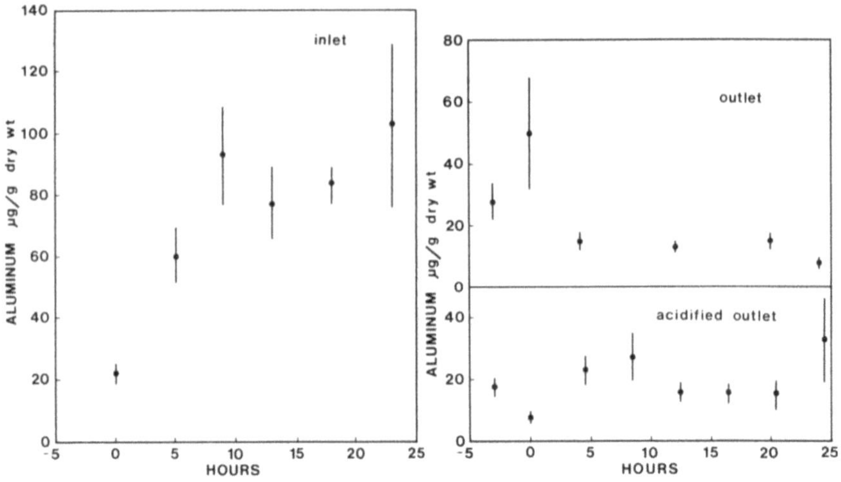

Figure 4. Gill concentrations of aluminum for rainbow trout held at the three test localities; solid circle, mean; vertical bar, standard error.

4. DISCUSSION

The first report of fish mortalities involving Al in the context of acid precipitation was by Schofield (1977), who related brook trout, *Salvelinus fontinalis* mortalities in Little Moose Lake to the increase in Al concentration from 20 to 320 $\mu g.L^{-1}$ with the first pulse of spring melt-water. The survival of stocked brook trout was related to Al concentrations in 53 Adirondack lakes (Schofield and Trojnar, 1980). Grahn (1980) reported high concentrations of Al on the gills of ciscoe, *Coregonus albula*, coincident with Al concentrations of 0.91 $mg.L^{-1}$ in lake Ransjon and 0.52 $mg.L^{-1}$ in lake Amten. In Plastic Lake, the 1981 fish-kill (Figure 1) was monitored in May at (total) Al concentrations of 210 $\mu g.L^{-1}$, declining over five days to 70 $\mu g.L^{-1}$, within the lake, and coincident with 280 $\mu g.L^{-1}$, declining to 200 $\mu g.L^{-1}$, in Inlet No. 1.

Rainbow trout held in inlet No. 1. showed a rapid increase in gill Al concentration from 20 to 90 $\mu g.L^{-1}$ within 8 hr (Figure 1) and no fish survived beyond 24 hr. These results were similar to those of 1981, when rainbow trout accummulated 65 - 80 $\mu g.L^{-1}$ gill Al before death (Harvey *et al*., 1982). In contrast, trout held within the lake at the mouth of the inlet, lived up to 5 days and accummulated 150 - 280 $\mu g.L^{-1}$ of Al on gill tissue, and trout held mid-lake experienced a 70% mortality at Al concentrations of 70 to 170 $\mu g.L^{-1}$ on gills (Harvey, 1985). From these results it may be inferred that Al accumulation on the gills of inlet-held fish is limited by the early death of the fish from the low pH and high Al concentration, whereas the longer survival of rainbow trout at pH 5.5 within the lake and the outlet stream makes possible the greater accummulation of Al on gill tissue. For such lake and outlet-held trout, Al toxicity may have been more important than H^+ toxicity. No other metal appearred to be present at toxic levels; Zn was highest in concentration at 16 $\mu g.L^{-1}$ in both inlet and outlet (Table I).

In this study rainbow trout were held in Inlet No. 1 at a total Al concentration of 428 $\mu g.L^{-1}$, of which 5.9 $\mu g.L^{-1}$ Al was in the reactive form. This was only slightly higher than the hatchery water (4.99 $\mu g.L^{-1}$) and Plastic Lake outlet (4.1 $\mu g.L^{-1}$). It is unlikely that the rapid stress response of inlet-held trout was due to this small difference. There remains the stress of $[H^+]$ and of Al fluorides. The acidified outlet waters were comparable in pH to inlet No. 1 (4.2 vs 4.1) but differed in concentrations of Al fluorides (13 vs 72 $\mu g.L^{-1}$) and free Al (0.6 vs 5.9 $\mu g.L^{-1}$). Thus the severe ionoreguation stress of inlet-held trout, was likely due to the combined effects of low pH plus the high concentrations of Al fluorides, given that the acidified-outlet trout showed a lesser stress to comparable $[H^+]$. These results are consistent with the observation of Baker and Schofield (1980) that the addition of fluoride reduced only partially the toxicity of free Al to brook trout fry.

It was shown previously that acid stress resulted in the loss of ions from both muscle and plasma, and that these changes in concentration could not have been due to water movement between tissues within the fish, as percent body water remained constant (Fraser and Harvey, 1984). Rainbow trout held in inlet No. 1 showed a severe loss of muscle Na and Cl concurrent with the loss of plasma ions, amounting to 38% loss of plasma Na and 37.5% loss of plasma Cl. These extreme losses are comparable to those of white suckers (Catostomus commersoni), 42% in 36 hr, and pumpkinseed, 38% in 19 days, at pH 4.0 (Fraser and Harvey, 1984). These intra- and extra-cellular compartments showed more independent responses in the trout held in the acidified outlet water. Plasma $[Na^+]$ and $[Cl^-]$ declined and remained low, vs muscle $[Na^+]$ and $[Cl^-]$ which declined and was followed by a partial recovery.

There remains the discrepancy between Al accumulation on the gills of endemic pumpkinseeds and the introduced rainbow trout. Al concentrations on the gills of the moribund pumpkinseeds were comparable to those of trout held in outlet and acidified-outlet waters, but such waters were not rapidly lethal to the rainbow trout. However, the duration of exposure of the pumpkinseeds to elevated Al concentrations in the lake was not known, thus the pumpkinseeds may have been responding to chronic Al exposure whereas the rainbow trout, 24-hr experiment, was based on acute exposure.

It is concluded that both low pH and high Al concentrations contributed to the ionoregulation stress of rainbow trout held in Plastic Lake inlet No. 1, and that the toxic Al was largely complexed to fluoride.

5. REFERENCES

Baker, J.P. and Schofield, C.L.: 1980, Aluminum toxicity to fish as related to acid precipitation and Adirondack surface water quality. pp. 292-293. In: Drabløs, D., and A. Tollan (Eds.) Ecological Impact of Acid Precipitation. Proc. Int. conf. ecol. impact acid precipitation, Norway. SNSF-project, Ås, Norway.

Dickson, W.: 1978, Aluminum, pH and fish. National Swedish Environment Protection Board Research Laboratory, 40 pp.

Evans, D.H.: 1975, Comp. Biochem. Physiol. **51**, 491.

Fraser, G.A. and Harvey, H.H.: 1984, Can. J. Zool. **62**, 249.

Grahn, O.: 1980, Fishkills in two moderately acid lakes due to high aluminum concentration. pp. 310-311. In: Drabløs, D., and A. Tollan (Eds.) Ecological Impact of Acid Precipitation. Proc. Int. conf. ecol. impact acid precipitation, Norway. SNSF-project, Ås, Norway.

Harvey, H.H., Dillon, P.J., Fraser, G.A., Somers, K.M., Fraser, P.E. and Lee, C.: 1982, Amer. Chem. Soc. Div. Environ. Chem., Extended Abstracts 22, 438.

Harvey, H.H.: 1985, The biological evidence of acidification, mechanisms of action, and an attempt at predicting acidification effects. pp. 63-78. In: Symposium on the effects of air pollution on forest and water ecosystems. Foundation for Research of Natural Resources in Finland, Helsinki, 143 pp.

Johnson, N.M.: 1979, Science **204**, 497.

Leivestad, H. and Muniz, I.P.: 1976, Nature **259**, 391.

McDonald, D.G. and Wood, C.M. 1981, J. Comp. Physiol. **126**, 277.

Muniz, I.P. and Leivestad, H. 1980, Acidification - effects on freshwater fish,. pp. 84-92. In: Drabløs, D., and A. Tollan (Eds.) Ecological Impact of Acid Precipitation. Proc. Int. conf. ecol. impact acid precipitation, Norway. SNSF-project, Ås, Norway.

Schofield, C.L.: 1977, Acid snow-melt effects on water quality and fish survival in the Adirondack Mountains of New York State. Res. Tech. Completion Report, Project No. A-072-NY, 26 pp.

Schofield, C.L. and Trojnar, J.R. 1980, Aluminum toxicity to brook trout (*Salvelinus fontinalis*) in acidified waters. pp. 341-362. In: Toribara, T.Y., M.W. Miller, and P.E. Morrow (Eds.) Polluted Rain. Plenum Publ. Corp.

Wood, C.M. and McDonald, D.G. 1982, Physiological mechanisms of acid toxicity to fish. pp. 197-225. In: Proc. Int. symposium on acidic precipitation and fishery impacts in North-eastern North America. Amer. Fish. Soc., Bethesda, MD.

PHYTOPLANKTON PRODUCTIVITY RESPONSES TO DIRECT ADDITION OF SULFURIC AND NITRIC ACIDS TO THE WATERS OF A DOUBLE-BASIN LAKE

J.A. Shearer and E.R. DeBruyn
Department of Fisheries and Oceans
Freshwater Institute
501 University Crescent
Winnipeg, MB R3T 2N6
Canada

ABSTRACT. The two basins of Lake 302, in the Experimental Lakes Area, northwestern Ontario, were separated by means of vinyl barriers. Beginning in 1982 and continuing through 1984, H_2SO_4 was added to the south basin during the ice-free months. Alkalinity was reduced rapidly and pH was lowered from near 6.6 to 5.6 by 1984. Transmission of light increased, resulting in a deeper euphotic zone. Over the same time period, HNO_3 was added to the north basin of Lake 302. Alkalinity was reduced much less in this basin and the pH dropped only to about 6.2 by 1984. Transmission of light did not increase. Mean epilimnetic phytoplankton production rates were greater in the HNO_3 basin than in the H_2SO_4 basin. However, in the basin receiving H_2SO_4, epilimnetic rates remained at or above those in reference lakes, and mean production rates below the mixed layer were greater than in the HNO_3 basin, probably because of increased light penetration. Mean euphotic zone production was similar in both basins and comparable to that measured in reference lakes. Phytoplankton community photosynthesis was not adversely affected by the additions of either HNO_3 or H_2SO_4.

1. INTRODUCTION

Direct acidification of small lakes can have major biological consequences, even during its early stages (Schindler et al., 1985). A whole lake acidification experiment, conducted on Lake 223 in the Experimental Lakes Area (ELA) in northwestern Ontario, demonstrated major changes in phytoplankton species composition during early stages of H_2SO_4 stress, but no major effects on phytoplankton community productivity (Schindler et al., 1985).

Another whole lake experiment was undertaken in the ELA to test whether the Lake 223 results were repeatable and, because the relative roles of H_2SO_4 and HNO_3 in lake acidification were not well understood, to determine whether HNO_3 would have similar effects to H_2SO_4. This paper discusses results relating to phytoplankton production during the early stages of this second acidification experiment.

Water, Air, and Soil Pollution **30** (1986) 695–702.

2. METHODS

Phytoplankton production was measured in both basins of L302, and in several non-manipulated, reference lakes, from 1980 through 1984. Whenever refinements of methodology were made, extensive intercalibration was conducted to ensure that the data from the differing techniques were comparable. The complete procedures for production measurement were described in detail by Shearer et al. (1985). They are summarized here.

Water samples, integrated over depth, were collected from each thermal stratum (epilimnion, metalimnion, hypolimnion), provided at least part of that stratum lay above the compensation depth (>0.5% of surface irradiance). The boundaries of these strata were determined from temperature profiles, and the compensation depth was determined from depth profiles of photosynthetically active radiation (PAR).

The water samples were returned to the laboratory and subsamples were incubated in an artificial light incubator (Fee, 1973a; Shearer et al., 1985) to determine C uptake rates over a range of irradiances. Incubations of 2 to 4 hr in duration, were carried out at in situ temperature and irradiance values. Following the incubations, C uptake was determined using acidification and bubbling (Schindler et al., 1972; Wessels and Birnbaum, 1979) and liquid scintillation spectrometry.

Whole lake production rates, integrated over time and depth (corrected for lake morphometry), were estimated using a numerical model developed by Fee (1973b, 1984). Simulated cloudless surface insolation was used when calculating whole lake production in order to standardize the solar input from year-to-year and eliminate a source of variation when comparing different years. Production rates in this paper are expressed as cloudless, mean daily values for the ice-free season (approximately May through early November). Because of low insolation and the shading effect of snow cover, winter production at the ELA is a negligible portion (<10%) of the annual total and was ignored in this study.

3. RESULTS

Mean, ice-free season values from both basins of L302 for selected physical, chemical and biological parameters for the years 1980 through 1984 are listed in Table I.

Before acidification (1980-1981), PAR at a depth of 4.0 m averaged 1.24 ± 0.06 times higher in L302S than in L302N (Table I; Figure 1). Coincident with the acid additions in 1982, transmission of PAR through the L302S (H_2SO_4) epilimnion increased relative to the L302N (HNO_3) epilimnion (Figure 1). The seasonal range of PAR values measured at 4.0 m for the H_2SO_4 basin was greater during the three years of acidification (1982-1984) because of the large increases in the maximum transmissions (Figure 1). The mean PAR at 4.0 m was 2.04 ± 0.27 times greater in the H_2SO_4 basin during these acidification years, and was negatively correlated with epilimnetic DOC

Lake 302 is a small, headwater, double-basin lake. It was previously the site of an eutrophication experiment (Schindler et al., 1980); however, monitoring during 1980 and 1981 revealed that no detectable effects of the nutrient additions remained. The two basins (north and south) are approximately equal in surface area (12.8 and 10.9 ha) and mean depth (5.7 and 5.1 m) (Brunskill and Schindler, 1971), and are connected by shallow channels on either side of a central island.

In May, 1981, reinforced vinyl barriers were installed between the L302 basins. Commencing in June, 1982, H_2SO_4 was added directly to the waters of the south basin (Cruikshank, 1984). Sufficient acid was added the first year to reduce the alkalinity of L302S but have minimal effect on pH. HNO_3 was added simultaneously to L302N, such that this basin received an approximately equivalent amount of H^+ (on an epilimnetic volume basis) as L302S. No attempt was made to predetermine the pH of the HNO_3 basin. Each subsequent year, following ice-out, the mean epilimnetic pH of the H_2SO_4 (south) basin was lowered to, and held near, a predetermined value (Table I). Equivalent HNO_3 was added to L302N. Epilimnetic pH values were determined several times per week during the ice-free period and acid was added as required to maintain the L302S target pH.

TABLE I
Summary of mean, ice-free season values for selected epilimnetic, physical, chemical and biological parameters, including estimates of epilimnetic and euphotic phytoplankton production (PP). All production rates are calculated for hypothetical "cloudless" weather conditions. Algal biomass (BMSS) is based on cell counts for the same water samples used for productivity measurement. Chlorophyll concentrations (CHL) were also determined from these water samples. The pH, H^+, alkalinity (ALKAL), dissolved inorganic C (DIC), dissolved organic C (DOC), total dissolved N (TDN) and total dissolved P (TDP) values were determined from water sampled independently. H^+ and ALKAL are in $\mu eq \cdot L^{-1}$; DIC and DOC in $mg \cdot L^{-1}$; TDN, TDL and CHL in $\mu g \cdot L^{-1}$; production rates in $mg \cdot C \cdot d^{-1}$; BMSS in $g \cdot m^{-3}$; PAR @ 4.0 m as % of surface irradiance.

YEAR	BASIN	TARGET pH	MEAN pH	H^+	ALKAL	DIC	DOC	TDN	TDP	CHL	EPILIMN PP m^{-3}	EPILIMN PP m^{-2}	EUPHOT PP m^{-3}	EUPHOT PP m^{-2}	BMSS	PAR
1980	North	Natural	6.51	0.31	N/A	0.81	6.29	289	4.0	4.05	64	245	60	319	1.18	8.5
	South	Natural	6.62	0.24	N/A	0.86	6.31	265	2.8	3.91	60	252	65	328	1.24	10.9
1981	North	Natural	6.64	0.23	56.0	0.92	6.84	253	4.0	3.38	81	312	78	431	2.24	10.6
	South	Natural	6.74	0.18	59.0	0.97	5.76	255	3.3	2.80	87	344	91	463	1.90	12.7
Pre-acid	North	Natural	6.58	0.27	56.0	0.87	6.57	271	4.0	3.72	72	279	69	375	1.71	9.6
	South	Natural	6.68	0.21	59.0	0.92	6.04	260	3.1	3.36	74	298	78	395	1.57	11.8
*1982	North	None	6.43	0.37	43.0	0.64	9.77	475	4.0	7.26	81	280	66	350	2.21	6.0
	South	**	6.21	0.62	35.0	0.65	7.88	276	3.1	4.09	49	189	52	263	1.52	11.4
*1983	North	None	6.35	0.45	36.0	0.60	6.08	610	5.0	3.40	60	218	45	259	2.09	7.0
	South	5.9	5.90	1.26	14.0	0.43	4.78	265	3.0	2.40	49	206	54	275	1.49	15.7
*1984	North	None	6.24	0.58	25.0	0.44	5.42	904	5.0	3.70	85	318	64	368	2.47	8.4
	South	5.6	5.60	2.51	4.5	0.29	4.21	270	4.4	4.40	78	348	79	402	2.58	14.5
Post-acid	North		6.34	0.47	34.7	0.56	7.09	663	4.7	4.79	75	272	58	325	2.26	7.1
	South		5.90	1.46	17.8	0.46	5.62	270	3.5	3.63	59	248	61	313	1.86	13.9

* Both basins acidified: North = HNO_3, South = H_2SO_4 ** Target was to reduce alkalinity

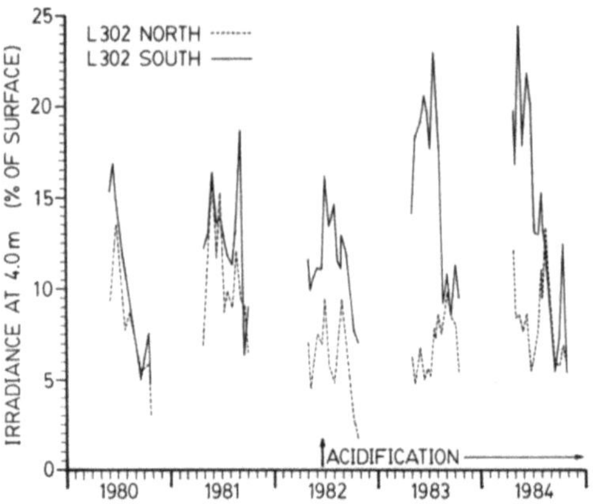

Fig. 1. Transmission of PAR (400-700 nm) vs time in the epilimnia of both basins of L302, expressed as the percentage of surface PAR reaching 4.0 m, the approximate depth of the mixed layer during August.

concentration ($r=-0.819$, $n=5$, $P<0.05$) in L302S. In L302N, where transmission of PAR did not increase, there was no correlation with DOC concentrations.

Production rates in both basins increased greatly between 1980 and 1981, although both years were pre-acidification. However, 1981 was also a high productivity year in reference basins (Figure 2).

During 1982, when L302S received the first additions of H_2SO_4, the phytoplankton production in this basin showed a sharp decline from pre-acidification values (Table I). In 1983, both epilimnetic and euphotic zone production rates remained almost the same as in 1982 (Table I). During the 1984 ice-free season, production in the H_2SO_4 basin increased sharply, both in the epilimnion and on a euphotic zone basis (Table I). The mixed layer deepened through 1983 and 1984, causing production per unit area to increase more rapidly.

In L302N, which received additions of HNO_3, the production per unit volume actually increased in the epilimnion during 1982. However, the production rates deeper in the water column were much lower and the euphotic zone rate decreased (Table I). Both epilimnetic and euphotic zone phytoplankton production in the HNO_3 basin declined noticeably in 1983 (Table I). Production in this basin then increased sharply in 1984 (Table I).

In summary, phytoplankton production in both basins varied from year-to-year but showed no clear trend during the period of study. The productivity fluctuations observed in the L302 basins were similar to those observed in nearby reference lakes (Figure 2). Mean daily phytoplankton production rates were not significantly correlated over time with mean epilimnetic pH in either basin, nor were they directly correlated with mean epilimnetic concentrations of total dissolved P, total dissolved N or chlorophyll a (Table I).

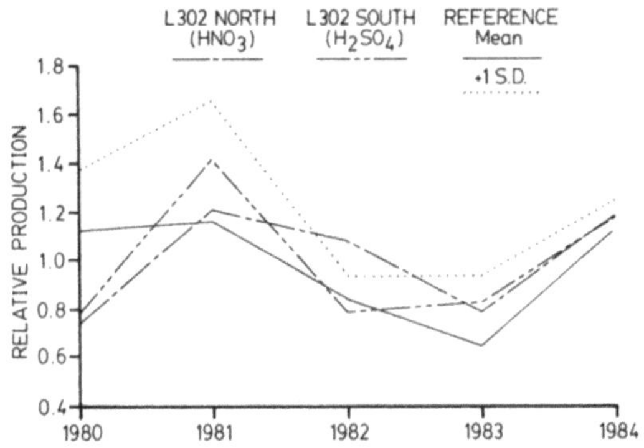

Fig. 2. Lake 302 production relative to the mean for pooled reference lakes, 1980-1984. A long-term mean value was calculated separately for each lake basin. The production for each year in each lake was then divided by the mean for that lake to obtain a relative value. These relative values were averaged for all reference lakes in a given year to produce the pooled reference values (solid line). The values for L302S and L302N are plotted separately. The dotted line represents the reference mean plus one standard deviaton. Using a Student t-test, only L302N, 1982, was found to be significantly different from the reference mean (n=7, P<0.05).

4. DISCUSSION

The Scandinavian literature (Grahn and Hultberg, 1974; Grahn et al., 1974; Hendrey et al., 1976; Leivestad et al., 1976; Andersson et al., 1980) suggests that lakes undergo "oligotrophication" in repsonse to acidification. However, few data exist from the early stages of acidification in Scandinavia. Algal productivity was often not measured and the hypothesis is based largely on the observations of low algal biomass and low species diversity. Also, many Scandinavian systems are severely impacted, and interpretations are complicated by high inputs of metals accompanying the acids. Many of these lakes are without fish. Increased zooplankton grazing pressures in such lakes may reduce algal biomass to give the impression of oligotrophication (Henrikson et al., 1980). Because the acids were added directly to the surface waters, the L302 study primarily examined the direct effects of acids. Acid interactions with the terrestrial watersheds were minimized. While Mn concentrations increased in both basins, and particularly in the south, after acidification began, other metals (Al, Cu, Zn) showed no clear trends in either basin and epilimnetic concentrations were generally less than 15 $\mu g \cdot L^{-1}$ (ELA staff, Freshwater Institute, Winnipeg, unpublished data).

Several other North American studies examined productivity of culturally acidified lakes. Kwiatkowski and Roff (1976), in lakes near Sudbury, Ontario, detected no reduced production at pH values above 5.5. At lower pH, there was evidence of reduced production per unit volume, although production per unit area declined only below pH 4.4. Dillon et al. (1979) reported no reduced productivity in Clearwater Lake, Ontario, at pH values of approximately 4.3. Havens and DeCosta (1984), in a series of enclosure experiments in Lake O' Woods, West Virginia, concluded that nutrient levels, not pH, were the prime controlling factor for phytoplankton production.

Transparency increases, in response to acidification, have been reported in various studies (Schofield, 1972; Almer et al., 1974; Yan, 1983; Schindler et al., 1985). This phenomenon has been attributed to the removal of dissolved organic C (DOC), in the form of yellow organic acids, possibly through precipitation with Al at low pH (Almer et al., 1978; Effler et al., 1985). The increased transmission of PAR in L302S seemed to be caused by a reduction in natural color, but mean Al concentrations were low (<15 $\mu g \cdot L^{-1}$) and there is no clear evidence of co-precipitation with DOC. The absence of a light transmission increase in the HNO_3 basin should not be interpreted immediately as a difference in response to HNO_3 vs H_2SO_4. In both L223 (Schindler et al., 1985) and L302S, the transparency increases were observed at pH values between 6.2 and 5.9, values lower than those reached in L302N.

Although this L302 experiment examined only the early stages of acidification, it confirmed the results from L223 (Schindler et al., 1985) and supported the phytoplankton productivity findings of other North American studies. At the pH values reached in L302S (5.6 and above), there were no apparent detrimental effects of H_2SO_4 acid on community phytoplankton productivity. Likewise, at pH values of 6.2 and above, there were no detrimental effects of HNO_3 on community phytoplankton productivity. While algal species composition and diversity may change rapidly in response to increased acidity (D.L. Findlay, Freshwater Institute, Winnipeg, unpublished data), phytoplankton productivity is not a sensitive, early indicator of acidification.

5. ACKNOWLEDGMENTS

Many ELA staff members contributed to this study. In particular, S. Meron, K. Tyson and K. Hallard measured C uptake. G. Linsey and his staff provided various chemical analyses. D. Findlay estimated phytoplankton biomass. D. Schindler coordinated the whole ecosystem study.

6. REFERENCES

Almer, B., Dickson, W., Ekstrom, C., Hornstrom, E. and Miller, U.:1974, Ambio 3, 30.

Almer, B., Dickson, W., Ekstrom, C. and Hornstrom, E.: 1978, 'Sulfur Pollution and the Aquatic Ecosystem", in Nriagu, J.O. (ed.),

Sulfur in the Environment: Part II, Ecological Impacts, John Wiley and Sons, New York, NY, 492 p.
Andersson, G., Gahnstrom, G. and Fleischer, S.: 1980, 'Rate of Oxygen Consumption and Decomposition in Sediment From Acid Lakes", in Drablos, D. and Tollan, A. (eds.), Proceedings of the International Conference on the Ecological Impact of Acid Precipitation, Sandefjord, Norway, March 11-14, 1980, SNSF Project, Oslo, 383 p.
Brunskill, G.J. and Schindler, D.W.: 1971, J. Fish. Res. Board Can. **28**, 277.
Cruikshank, D.R.: 1984, 'Whole Lake Chemical Additions in the Experimental Lakes Area, 1969-1983', Can. Data Rep. Fish. Aquat. Sci. **449**, iv + 23 p.
Dillon, P.J., Yan, N.D., Schneider, W.A. and Conroy, N.: 1979, Ergeb. Limnol. **13**, 317.
Effler, S.W., Schafran, G.C. and Driscoll, C.T.: 1985, Can. J. Fish. Aquat. Sci. **42**, 1707.
Fee, E.J.: 1973a, J. Fish. Res. Board Can. **30**, 1447.
Fee, E.J., 1973b, J. Fish. Res. Board Can. **30**, 1469.
Fee, E.J., 1984, 'Freshwater Institute Primary Production User's Guide', Can. Tech. Rep. Fish. Aquat. Sci. **1328**, v + 36 p.
Grahn, O.H. and Hultberg, H.: 1974, 'Forsurningseffekter pa Oligotrofa Sjoars Ekosystem - Integrerade Forandringar I Artssammansattning och Dynamik', Institutet for Vatten-och Luftvardsforskning, Gothenburg, Meddelande nr. **2**, 12 p.
Grahn, O., Hultberg, H. and Landner, L.: 1974, Ambio **3**, 93.
Havens, K.E., III and DeCosta, J.: 1984, Arch. Hydrobiol./Suppl. **69**, 443.
Hendrey, G.R., Baalsrud, K., Traaen, T.S., Laake, M. and Raddum, G.: 1976, Ambio **5**, 224.
Henrikson, L., Nyman, H.G., Oscarson, H.G. and Stenson, J.A.E.: 1980, Hydrobiologia **68**, 257.
Kwiatkowski, R.E. and Roff, J.C.: 1976, Can. J. Bot. **54**, 2546.
Leivestad, H., Hendrey, G., Muniz, I.P. and Snekvik, E.: 1976 'Effects of Acid Precipitation on Freshwater Organisms', in Braekke, F.H. (ed.), Impact of Acid Precipitation on Forest and Freshwater Ecosystems in Norway, Research Report 6/76, SNSF Project, NISK, **1432** As-NLH, p. 87-111.
Schindler, D.W., Schmidt, R.V. and Reid, R.A.: 1972, J. Fish. Res. Board Can. **29**, 1627.
Schindler, D.W., Ruszczynski, T. and Fee, E.J.: 1980, J. Fish. Res. Board Can. **37**, 320.
Schindler, D.W., Mills, K.H., Malley, D.F., Findlay, D.L., Shearer, J.A., Davies, I.J., Turner, M.A., Linsey, G.A. and Cruikshank, D.R.: 1985, Science **283**, 1395.
Schofield, C.O., Jr.: 1972, 'The Ecological Significance of Air-Pollution Induced Changes in Water Quality of Dilute Lake Districts in the Northeast', Trans. N.E. Fish Wildlife Conf., p. 98-112.
Shearer, J.A., DeBruyn, E.R., DeClercq, D.R., Schindler, D.W. and Fee, E.J.: 1985, 'Manual of Phytoplankton Pimary Production

Methodology', Can. Tech. Rep. Fish. Aquat. Sci. **1341**, iv + 58 p.
Wessels, C. and Birnbaum, E.: 1979, Limnol. Oceanogr. **25**, 360.
Yan, N.D.: 1983, Can. J. Fish. Aquat. Sci. **40**, 621.

ALUMINIUM IMPACT ON RESPIRATION OF LOTIC MAYFLIES AT LOW pH

Jan Herrmann and Kjell G. Andersson
Department of Animal Ecology
University of Lund
Ecology Building
S-223 62 Lund
Sweden

ABSTRACT. Three species of stream-living mayfly nymphs were exposed to total inorganic (labile) monomeric aluminium levels of 0, 0.5, and 2.0 mg L^{-1} at pH 4.0 and 4.8 for 10 days. Oxygen consumption rate of the nymphs was measured, using a closed-cylinder technique and micro-Winkler titration. The animals showed significant increases in respiration with raised Al concentrations. The mayfly species *Ephemera danica*, which is restricted to the less heavily acidified region of South Sweden, was most affected by Al. The response of the two *Heptagenia* species to elevated Al was less pronounced. The effect of pH, however, seemed less important. Two possible reasons for the respiration increase at high Al levels are: (1) A "chemical impact route", consisting of a compensatory mechanism due to decreased O_2 transport efficiency, because of impaired osmoregulation and ion transport; and/or (2) a "mechanical impact route" with Al hydroxide precipitation and mucus formation on the gills, causing lowered respiration efficiency. This stress of the increased respiration rate costs energy. Thus less will be available for growth and reproduction. A model of the proposed impact routes is presented.

1. INTRODUCTION

Acidifying precipitation has for many decades affected Sweden, and the chemical and biological effects has been especially severe in some southern and western areas with poorly buffered soils and surface waters (Johansson and Nyberg, 1981). Benthic stream invertebrate communities are known to become impoverished in acid conditions (Hall *et al.*, 1980; Singer, 1982; Sucliffe, 1983; Otto and Svensson, 1983), and especially nymphs of many mayfly species seem vulnerable (Engblom amd Lingdell, 1984). Detailed studies on the possible involved impact mechanisms for invertebrates are few, see however Sutcliffe (1983), Havas *et al.* (1984), Witters *et al.* and Appelberg (1985), and the review by Havas (1981). However, no studies of mayfly nymphs exist.

The impact of acid water on fish, a more prominent and commercially

Water, Air, and Soil Pollution **30** (1986) 703–709.

interesting component of the ecosystem, is much more well-known than invertebrates (e.g. Fromm, 1980; Muniz and Leivestad, 1980a, b; Baker and Schofield, 1982; Brown, 1983). Main effects of acid conditions on fish are reported to be impaired osmoregulation, leading to lowered content of Na^+, Cl^- and other ions; and precipitation of Al hydroxides and mucus on the gills, thus causing coughing and hyperventilation. To our knowledge, very little research has been conducted on the way oxygen consumption of invertebrates is affected with elevated acidity and Al impact.

This paper presents results on respiratory responses of a set of mayfly species exposed to different pH and Al conditions. Respiration reflects the extent of stress that the animal is experiencing, irrespective of its causation.

2. MATERIAL AND METHODS

Nymphs of three mayfly species were tested, all specimens being collected in fairly well buffered (pH 6 to 7) streams:

Heptagenia fuscogrisea Retz. This species is tolerant to low pH conditions and occurs almost everywhere in the acid parts of south Sweden. It can be found in streams as low as pH 4.2 (Engblom and Lingdell, 1984). The individual dry weight of the almost 1 yr old animals used was 1.5 to 4 mg.

H. sulphurea Müll. This species is more sensitive than its congener and is missing in streams with a pH less than 4.6 (Engblom and Lingdell, 1984). To judge from its total geographic range and wide tolerance limits with respect to other environmental factors, the species was probably more common some decades ago than nowadays (Svensson, B.S.: 1985, pers. comm.). Individual dry weight was 2 to 5 mg (almost 1 yr old).

Ephemera danica Müll. It occurs in less heavily acidified streams, and is fairly common at less acidic conditions (Engblom and Lingdell, 1984: lower limit at pH ca 5.5). Individual dry weight was 12 to 22 mg (almost 2 yr old).

A variety of regimes (see end of this chapter) were arranged in 2 L PVC jars. Filtrated stream water at pH 5.5 to 6.5, with very low total monomeric Al content ($\ll 0.07$ mg L^{-1}), and moderate contents of humic substance (20 to 40 mg Pt L^{-1}) and Ca (2.8 mg L^{-1}), was used for all experiments. The desired pH regimes were adjusted with H_2SO_4. Appropriate amounts of $Al_2(SO_4)_3$ were added and pH readjustments were made during two days before the experiments began.

Before the exposure began, the animals used were successively acclimatized towards their future pH conditions. The whole setup was kept at 8 to 10^oC, gentle air bubbling was added and pH adjusted daily. No food was added and no natural aubstrate was used, but a piece of nylon net was provided to which the animals could cling.

After an exposure time of 10 days, the respiration measurements were performed. First the O_2 bubbling was closed for 1 hr to let the water gas content stabilize. Then the O_2 content of the water was measured by means of a micro-Winkler method. Twenty mL of water was extracted with a syringe and the standard reagents were added to this syringe by means of 1 mL syringes. Titration with 0.01-M $Na_2S_2O_3$ was done from a microburette. Initial O_2 saturation was 75 to 85 %.

Subsequently, the experimental water and animals were gently transferred to small glass vials (ca 45 mL each). Change of O_2 saturation was considered to be very marginal. Lids with rubber gasket were mounted, so that no air could enter when submersed into a water bath at $10 \pm 1^{o}C$. To avoid O_2 stratification, each vial was equipped internally with a slow-speed propeller device.

After 4 to 6 hr, a micro-Winkler O_2 sample was taken directly from each vial. The animals were removed and dried at $60^{o}C$ overnight before being weighed. Content of monomeric Al of the experimental water was determined at the start and end of each 10 day period on a Varian AAS. Analyses included separation of Al into organic (non-labile) and inorganic (labile) fractions by means of a Dowex cation exchanger (for details see Nilsson and Bergkvist, 1983). In the following text, the denoted Al concentrations are total inorganic (labile) monomeric Al

The above procedure provided data for each vial that allowed calculation of the O_2 consumption per unit dry weight of animal and time. Each experimental regime comprised 5 to 10 vials (number of replicates) and each vial contained 1 to 4 animals, depending on animal size. Experiments were performed in May and June, and at six regimes, namely the Al contents of 0, 0.5 and 2.0 mg L^{-1} at both pH 4.0 and 4.8.

3. RESULTS AND DISCUSSION

In all series of experiments, comparisons between the respiration rate responses at Al concentrations 0 and 2.0 mg L^{-1} clearly showed that mayfly nymphs exhibit an increased respiration rate at a raised Al concentration (Figure 1). Also at Al = 0.5 mg L^{-1} the respiration rate levels of *H. sulphurea* and *E. danica* follows this raising pattern. For *H. fuscogrisea* we observed a "dip" at this concentration (Figure 1), which coincided with substantial problems in maintaining the correct Al levels.

Even if "the base level" (respiration rate in Al-free water) differed between the mayfly species used, all sets of data for the increase of respiration with raised Al levels showed a significant response (Table I). The effect of pH, however, seemed less important in these experiments (Figure 1).

The impact of Al on the respiration rate certainly puts a higher demand on the animal for aquiring energy, which can be at the expense of growth and reproduction. Havas (1980) suggested that the primary effect on aquatic invertebrates at acid conditions is that acid-base balance and osmoregulation are affected. She indicated that respiration effects would be less important, being less influenced at a moderately low pH levels, but also because it is a secondary effect, influenced by the earlier.

Two possible ways along which Al could physiologically affect mayfly nymphs are as follows (also Figure 2). (1). A layer of mucus is

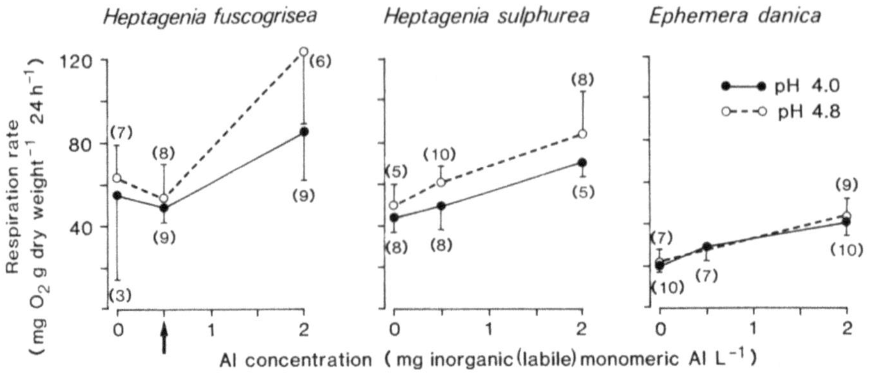

Fig. 1. Respiration rate (mg O_2 g dry weight^{-1} 24 hr^{-1}) of three species of mayfly nymphs, exposed to different Al concentrations (mg inorganic monomeric Al L^{-1}), at pH levels of 4.0 and 4.8. Number of experiments given within parentheses and 95 % confidence limits are indicated in one direction. Arrow indicates that Al levels decreased much below the denoted value.

formed, following impaired osmoregulation, and/or (2) hydroxide precipitates on the gills. Mucus or hydroxides prevents access to O_2 and

Table I. Significances (one-way ANOVA) for the data of increase in respiration rate, due to raised Al concentrations, presented in Figure 1. Also included, to the right, are the figures for the percentage increase of respiration rate, from Al = 0 mg L^{-1} to Al = 2.0 mg L^{-1}.

Mayfly species	pH level	F-value	df	significance	% increase
H. fuscogrisea	4.0	5.80	2/17	$p < 0.05$	54
	4.8	11.98	2/18	$p < 0.001$	95
H. sulphurea	4.0	9.24	2/18	$p < 0.01$	59
	4.8	7.21	2/20	$p < 0.01$	68
E. danica	4.0	16.27	2/24	$p < 0.001$	101
	4.8	19.98	1/13	$p < 0.001$	100

thus lowers the respiration efficiency. The animal would compensate this by increased respiration rate, thus reflecting a stressed situation. This response could be named the "mechanical impact route". However, a second type of response would include impaired osmoregulation and ion transport as well as other toxic effects, and could be called a "chemical impact route". This route would impede the O_2 transport and create a similarly increased compensatory respiration rate. In both cases, the effect would be an increased energy cost, resulting in less energy being available for growth and reproduction (cf. Sutcliffe, 1984). This model of proposed impact routes is inspired by and partly based on Muniz and Leivestad (1980a, b) and Havas (1981).

It should be added that a changed and disadvantageous behavior could be expected due to some of the impacts outlined above. Further, Al changes could also alter the bottom substrate composition in a stream, with respect to algae and humic substances (Dickson, 1978), as well as Al precipitates (McKnight and Feder, 1984), thereby indirectly affecting scraping invertebrates like mayfly nymphs.

The observed increase of O_2 consumption of the investigated mayfly nymphs at higher Al levels (Figure 1) may be due to mainly osmoregula-

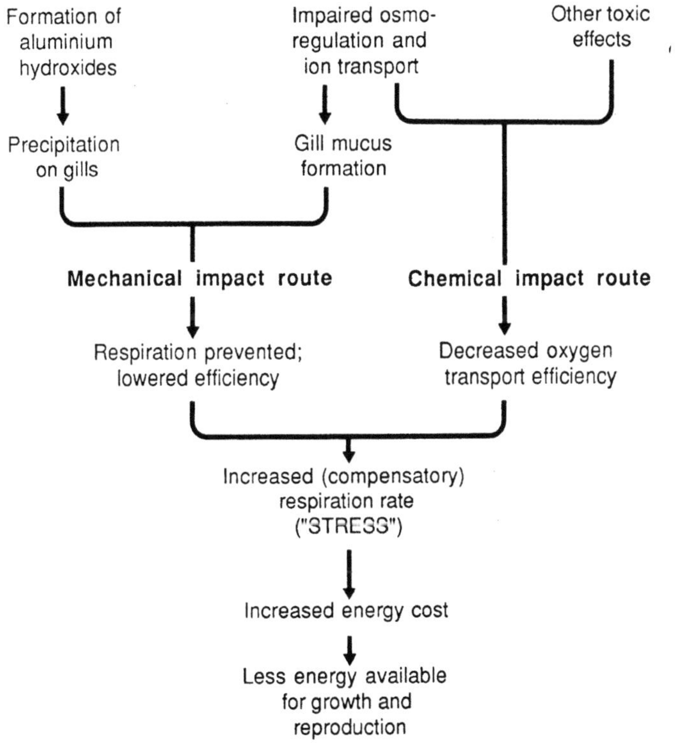

Fig. 2. A proposed model for the possible physiological impact routes for the effects of Al on respiration of mayfly nymphs at low pH conditions. See text for further explanation.

tory and ion transport problems (cf. Havas, 1981), thus lending support to the "chemical impact route" (Figure 2). Scattered observations of opaque material on gills and other parts of the mayfly nymphs during the experiments indicated that one or both of the "mechanical impact routes" also could be important. Hall *et al.* (1985) suggested that Al hydroxides could be toxic for invertebrates, but also that Al results in a disadvantageously decreased water surface tension.

Even if all series of experiments showed a significant response of increased respiration due to raised Al levels, there were differences in the degree of significance (Table I). Also, the percentage of increase was greater for *E. danica* (Table I). Thus, it is inferred that the respiration response was most pronounced and consistent for *E. danica*, but less so for *H. sulphurea* and *H. fuscogrisea*. This observation fits the fact that *E. danica* is mainly restricted to streams exhibiting relatively high pH and low Al levels. *H. fuscogrisea* is found in low pH/high Al streams, whereas *H. sulphurea* is intermediate in both these respects.

The exposure levels used, 0.5 and 2.0 mg Al L^{-1}, were not unrealistically high. In some streams in the investigated region of southern Sweden, stream water contents of 0.3 to 0.7 mg Al L^{-1} are not unusual at spring and autumn, but also levels > 1.0 mg Al L^{-1} has occasionally been measured. However, in the streams with Al levels of ca 1 mg L^{-1} (at pH around or below 4), no mayfly nymphs seem to exist, not even *Leptophlebia spp.* (Herrmann, unpubl.). Species of this genus are otherwise tolerant to low pH conditions (Otto and Svensson, 1983; Engblom and Lingdell, 1984; Herrmann, unpubl.).

Interstitial water often show a higher pH (and then probably lower Al levels) than that in the water column (Herrmann, unpubl.). Therefore, the conditions for the animals are usually "better" in or on the surface of the bottom sediment of an acid stream than above it, probably related to the buffering action of the organic and inorganic material in the sediment.

Havas and Hutchinson (1982) suggested that Al, besides the pH itself, is the key additional toxic factor for crustaceans in acid waters. They did not propose a particular mechanism, but Al-specific staining has indicated that this metal is related to the osmotic functions (Havas, 1986). Witters *et al.* (1984) have shown impaired Na fluxes at increased Al levels for corixids.

In the present study, the mayfly nymphs showed an evident respiration increase at raised Al levels. The main reason for this increase may be difficulty with ion regulation and transport. Further knowledge on the impact of Al and pH on mayfly osmoregulation is therefore needed.

ACKNOWLEDGMENTS

Our sincere gratitude to Mr F. Johansson for help with some of the respiration measurements and Mrs M. Gernersson for performing the Al analyses. Drs C. Enfield and B.S. Svensson, as well as two anonymous referees, kindly critisized an earlier draft of the manuscript. The study was financed by the Swedish National Environmental Protection Board.

REFERENCES

Appelberg, M.: 1985, *Hydrobiologia* **121**, 19.
Baker, J.P. and Schofield, C.L.: 1982, *Water, Air and Soil Pollut.* **18**, 289.
Brown, D.J.A.: 1983, *Bull. Environ. Contam. Toxicol.* **30**, 582.
Dickson, W.: 1978, *Verh. Internat. Verein. Limnol.* **20**, 851.
Engblom, E. and Lingdell, P.-E.: 1984, *Rep. Inst. Freshw. Res., Drottningholm* **61**, 60.
Fromm, P.O.: 1980, *Env. Biol. Fish.* **5**, 79.
Hall, R.J., Likens, G.E., Fiance, S.B. and Hendrey, G.R.: 1980, *Ecology* **61**, 976.
Hall, R.J., Driscoll, C.T., Likens, G.E. and Pratt, J.M.: 1985, *Limnol. Oceanogr.* **30**, 212.
Havas, M.: 1981, *Effects of Acidic Precipitation on Benthos*, R. Singer (ed.), North American Benthological Society, Springfield, p. 49.
Havas, M. and Hutchinson, T.C.: 1982, *Can. J. Fish. Aquat. Sci.* **39**, 890.
Havas, M., Hutchinson, T.C. and Likens, G.E.: 1984, *Can. J. Zool.* **62**, 1965.
Havas, M.: 1986, *Water, Air and Soil Pollut.*, this volume.
Johansson, K. and Nyberg, P.: 1980, *Inform. Inst. Freshw. Res., Drottningholm* **6**, 118 pp (In Swedish with English Summary)
McKnight, D.M. and Feder, G.L.: 1984, *Hydrobiologia* **119**, 129.
Muniz, I.P. and Leivestad, H.: 1980a, *Proc. Int. Conf. Ecol. Impact Acid Precip.*, D. Drabløs and A. Tollan (eds), Norway 1980, SNSF project, p. 84.
Muniz, I.P. and Leivestad, H.: 1980b, *Proc. Int. Conf. Ecol. Impact Acid Precip.*, D. Drabløs and A. Tollan (eds), Norway 1980, SNSF project, p. 320.
Nilsson, S.I. and Bergkvist, B.: 1983, *Water, Air and Soil Pollut.* **20**, 311.
Otto, C. and Svensson, B.S.: 1983, *Arch. Hydrobiol.* **99**, 15.
Singer, R.: 1982, *Acid Precipitation - Effects on Ecological Systems*, F.M. D'Itri (ed.), Ann Arbor Science, p. 329.
Sutcliffe, D.W.: 1983, *Rep. Freshw. Biol. Ass.* **51**, 30.
Sutcliffe, D.W.: 1984, *Freshw. Biol.* **14**, 443.
Witters, H., Vangenechten, J.H.D., van Puymbroeck, S. and Vanderborght, O.L.J.: 1984, *Bull. Environ. Contam. Toxicol.* **32**, 575.

Response of Lake Trout (*Salvelinus namaycush*) and Brook Trout (*S. fontinalis*) to Surface Water Acidification in Ontario

Gail L. Beggs and John M. Gunn

Ontario Ministry of Natural Resources
Toronto, Ontario, Canada M7A 1W3

ABSTRACT. Netting surveys of lakes varying in pH (4.4-7.1) showed that lake trout (Salvelinus namaycush) populations fail to recruit at pH <5.5 and are lost from lakes with pH<5.2. Brook trout (S. fontinalis) were extirpated in lakes with pH <5.0. In regional chemical surveys of Ontario lakes, it was found that 2% of sampled brook trout lakes and 2.5% of lake trout lakes were acidified (alkalinity <0 uEq L^{-1}). Threshold pH levels determined from fisheries assessments were used to estimate that 1% of lake trout and brook trout populations have been lost due to acidification.

1. INTRODUCTION

It has been well established that surface water acidification has resulted in the loss of fish populations in Scandinavia (Wright and Snekvik, 1978), the United States (Schofield, 1976) and Canada (Beamish and Harvey, 1972; Watt et al, 1983). Recently researchers have attempted to estimate the extent of such losses through the use of various mathematical models (Sadler, 1983; Baker and Harvey (cited in Haines and Baker, 1986); Somers and Harvey, 1984; Minns et al, 1986).

In this study we use empirical data on population status of lake trout and brook trout in lakes of differing pH (4.4 - 7.1) to determine a pH threshold for healthy populations. These thresholds are used in conjunction with regional chemical survey data to estimate the proportion of sampled Ontario lake trout and brook trout lakes lost and at risk from surface water acidification.

2. METHODS

2.1 Fish Population Assessment

Fish population and water chemistry data were collected on lakes known from reliable historic records to have supported reproducing populations of lake trout (n=36) and brook trout (n=23). Lakes were surveyed between May and September of 1982-1985 using 6 overnight sets of 76.2 m monofilament gillnets (15.2 m panels of 2.5, 3.8, 5.1, 6.3 and 7.6 cm

Water, Air, and Soil Pollution **30** (1986) 711–717.

stretched mesh). Nets were set to straddle the thermocline and well oxygenated depth strata. Catch-per-unit-effort (CUE), as number caught per gillnet set, was used as an index of relative abundance. Age composition, assessed by examining calcified structures (otoliths-lake trout, scales-brook trout), was used to determine whether recruitment of juveniles was occurring in the surveyed lakes. A 5 m tube composite sample of epilimnetic water was collected during the summer of the fisheries assessment and analyzed for pH, major ions and metals (MOE,1981).

2.2 Regional Surveys

Data on lake morphometry, water chemistry and fish species presence (determined by using gillnets, minnow traps, and seine nets according to the methods of Dodge et al (1985)) was obtained from an inventory survey of 2900 Ontario lakes (for further information see: Wales and Beggs,1986). Water samples for the extensive survey were primarily collected as 5 m tube composites during the fall or under ice, and were analyzed by methods described in MOE (1981). The extensive survey included water chemistry data on 38% (828) and 34% (680), respectively, of Ontario's lake trout and brook trout lakes. The data sets overlap for 199 lakes in which the species cohabit.

3. RESULTS AND DISCUSSION

3.1 Fish Population Assessment

The CUE for lake trout was reduced to 0 in all lakes of pH<5.2 (Figure 1). Some of the lakes with extinct lake trout populations contained large populations of acid tolerant species such as yellow perch (*Perca flavescens*), while lakes with pH below 4.7 were fishless. In lakes of pH 5.2 - 5.5, lake trout populations were almost entirely composed of

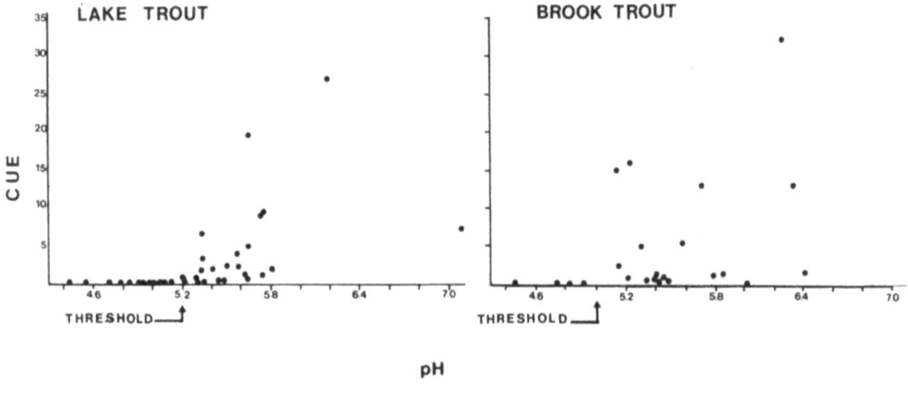

Fig. 1. Catch-per-unit effort (C.U.E.) (as numbers per overnight gillnet set) in relation to lake summer epilimnetic pH.

large old fish, indicating that successive recruitment failure was occurring. Populations existing within this transition stage (pH 5.2 - 5.5) appeared to be destined for extirpation unless pH increased.

Time series data from one of our assessment lakes, Whitepine Lake, supports the survey-derived threshold of approximately pH 5.5 for lake trout recruitment. Whitepine Lake is near Sudbury, Ontario and appears to be one of many lakes in that area with improved water quality as a result of recent reductions in smelter emissions (Keller and Pitblado,1986). In 1978 and 1980, when mean annual pH was 5.47 (range 5.26 to 5.85, mean inflection point alkalinity 1.2 uEq L^{-1}), the lake trout population of Whitepine Lake consisted of a small number of large, old (>15+) fish. Using the same assessment techniques, new recruits were first captured in 1982 when the mean pH was 5.63 (range 5.10 to 5.83, mean alkalinity 10.5 uEq L-1).

A similar threshold for lake trout recruitment, pH 5.6, was found by Schindler et al (1985) during the experimental acidification of Lake 223. Beamish (1976), using data from fisheries surveys of La Cloche Mountain lakes, also suggested lake trout populations are lost in lakes of pH less than 5.2 - 5.5.

Brook trout were more tolerant of low pH than lake trout. They were captured at a minimum pH of 5.1 (Figure 1). Above pH 5.0 there was no indication of reproductive failure. Below pH 5.0, populations were extinct. Populations consisting of older individuals, as was the case for lake trout, were not observed. The short life span of brook trout (5-8 years) may be responsible for such an apparently abrupt threshold.

In Joe Lake near Sudbury, Ontario, a reproducing brook trout population has developed coincident with improvements in water quality. Stocking was unsuccessful for several years when the lake pH was 4.8 - 5.0, but resulted in a self-sustaining population in 1983 when pH had risen to 6.2 (V. Liimatainen, Ontario Ministry of Natural Resources, Sudbury, pers. comm.). While our assessment of Joe Lake has not been intensive enough to isolate a pH threshold, we do know that reproduction occurred only after the lake pH was greater than 5.0.

The use of a pH threshold in this study does not imply that only H^+ is controlling the health of fish populations. The study lakes ranged in concentration of total aluminum, from 8 to 560 ug L^{-1}. Aluminum, particularly inorganic monomeric aluminum, has been shown to be toxic to fish at concentrations observed in acidic waters (Baker and Schofield,1982; Driscoll et al., 1980). As aluminum concentrations in natural waters are highly correlated with pH, our use of a pH threshold implies a corresponding range in total aluminum concentration (Figure 2). Similarily, other metals such as Cu (1-20 ug L^{-1}) and Zn (3-21 ug L^{-1}) are slightly elevated in some of these lakes, and have been found to increase the toxicity of low pH waters (Hutchinson and Sprague 1986). Many of the netting assessment lakes are $<$ 100 km from the metal smelters at Sudbury and continue to receive airborn inputs of Cu, Zn and other metals from this point source, even though current emissions of metals are far lower than in the past (Keller and Pitblado, 1986).

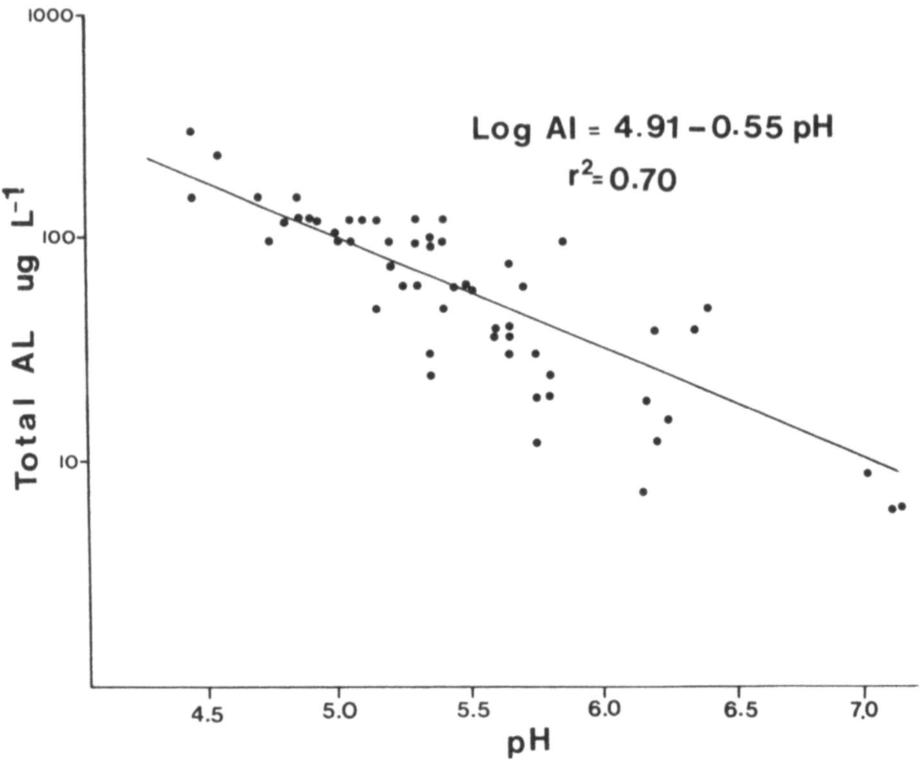

Fig. 2. The relationship between the concentration of total aluminum (ug L^{-1}) and pH in lake trout and brook trout lakes.

3.2 Regional Surveys

Brook trout lakes appear susceptible to acidification as they are predominantly clustered in the area of Ontario of high deposition (>20 kg SO_4 ha^{-1} yr^{-1}) (Figure 3), are relatively small in size (median surface area 30.2 ha), high in elevation and low in alkalinity (Table 1). Five (1%) of the surveyed lakes had a pH below the survival threshold of pH 5.0. Fifteen (2%) of the surveyed brook trout lakes had inflection point alkalinity of less than 0 uEq L^{-1} and are classified as acidified. As these systems have no capacity to buffer additional acidic input, brook trout populations existing in them can be considered at high risk from continued or increased acidic deposition. Further, 140 lakes or 29% of surveyed brook trout lakes in the area of high deposition have very low alkalinity ($\geq$0 and <40 uEq L^{-1}) and hence are classified as extremely sensitive to acidification.

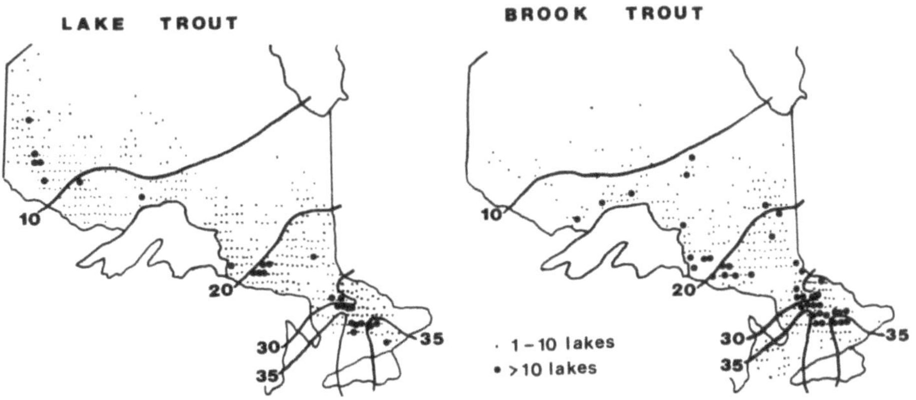

Fig. 3. The distribution of lake trout and brook trout lakes in Ontario and isopleths of wet sulphate deposits (1981). Isopleths from Chan et al. (1983).

Table 1
Selected physical and chemical characteristics of surveyed lake trout and brook trout lakes in Ontario

	Lake trout	Brook trout
Number surveyed	828	680
pH (min, med, max)	4.7, 6.8, 9.6	4.8, 6.5, 8.6
Alkalinity(uEq L^{-1}) (min, med, max)	-60, 115, 2834	-28, 77, 5850
Lake area (ha) (min, med, max)	1, 223, 27713	1, 30, 7235
Elevation (m) (min, med, max)	84, 371, 511	130, 396, 534
Number <pH threshold	8(<pH 5.2), 16(<pH 5.5)	5(<pH 5.0)
Number 0 <uEq L^{-1} alkalinity	20	15
Number 0 < alkalinity < 40 uEq L^{-1}	119	164
Number in area of highest deposition (>20 Kg ha^{-1} yr^{-1}) with 0 < alkalinity < 40 uEq L^{-1}	105	140

Although the physical and chemical characteristics of brook trout lakes render them susceptible to damage from acidic deposition, on the basis of our netting assessment results and toxicological data (Grande et al,1978), brook trout appear to be less sensitive to low pH than lake trout. Additionally, the ability of brook trout to detect and avoid acid water (Johnson and Webster, 1977; Gunn and Noakes, 1986), and the location of redds in the presence of groundwater upwellings

(Webster and Eiriksdottir, 1976), generally of higher alkalinity than surface water (Gunn, 1986), appears to protect early developmental stages (prior to emergence) of brook trout from acidic runoff and surface water.

Approximately 1% of sampled lake trout lakes have pH $<$5.2, the threshold for extirpation, 2% have pH $<$ 5.5, the threshold for successful recruitment, and 2.5% are acidified (alkalinity $<$ 0 uEq L^{-1}). In the area of Ontario receiving high deposition ($>$ 20 Kg wet sulphate ha^{-1} yr^{-1}) (Figure 3), 105 or 20% of sampled lakes have alkalinity levels between 0 and 40 uEq L^{-1} and are therefore considered extremely sensitive to continued high deposition. Although the majority (55%) of Ontario lake trout lakes are located in this region of high deposition (Beggs et al 1985), lake trout lakes are generally larger in area, somewhat lower in elevation and higher in alkalinity than brook trout lakes (Table 1), suggesting they are at less risk from acidic deposition. However, several behavioural traits may increase the susceptibility of lake trout populations to acidification damage. Lake trout spawn at nearshore sites, often in very shallow water (McMurtry unpublished, cited in Gunn, 1986). These locations are vulnerable to inundation by acidic potentially toxic meltwater (Gunn and Keller, 1984). Runoff coincides with the sensitive developmental period from hatching to emergence, increasing the probability of lake trout mortality. It is not known however, whether episodic mortality contributes to, or causes reductions or extirpation of populations.

ACKNOWLEDGEMENTS

Many colleagues assisted with the work presented here, including D. Green, V. Liimatainen, J. Matuszek, and D. Wales. W. Keller of Ontario Ministry of Environment provided the chemical data for the fisheries assessment lakes. This work was supported by the Ontario Ministry of Natural Resources Fisheries Acidification Program.

REFERENCES

Baker, J. P. and Schofield, C. L.: 1982, Water Air Soil Pollut. 18. 289.

Beamish, R. J.: 1976, Water Air Soil Pollut. 6. 501.

Beamish, R. J. and Harvey, H. H.: 1972, J. Fish. Res. Board Can. 40. 3.

Beggs, G. L., Gunn, J. M., and Olver, C. H.: 1985, Ont. Fish. Tech. Rept. 17. 1.

Chan, W. H., Tang, A. J. and Lusis, M. A.: 1983, Precipitation concentration and wet deposition fields of pollutants in Ontario, Sept. 1980 - Dec. 1981. Ontario Ministry of the Environment, Rept. ARB-61-83 ARSB, Toronto, 79 p.

Dodge, D. P., Goodchild, G. A., MacRichie, I., Tilt, J. A. and Waldriff, D. G.: 1985, Manual of instruction; aquatic habitat inventory surveys.

Ontario Ministry of Natural Resources, Toronto, 246 p.

Driscoll, C. T., Baker, J. P., Bisogni, J. J. Jr. and Schofield, C. L.: 1980, Nature 284. 161.

Grande, M., Muniz, I. P. and Andersen, S.: 1978, Verh. Internat. Verein. Limnol. 20. 2076.

Gunn, J. M. and Keller, W.: 1984, Can. J. Fish. Aquat. Sci. 41. 319.

Gunn, J. M.: 1986, Envir. Biol. Fish 15. (in press).

Gunn, J. M. and Noakes, D. L. G.: 1986, Water Air Soil Pollut. (this issue).

Haines, T. A. and Baker, J. P.: 1986, Water Air Soil Pollut. (this issue).

Hutchinson, N. J. and Sprague, J. B.: 1986, Can. J. Fish. Aquat. Sci. 43. 647.

Johnson, D. W. and Webster, D. A.: 1977, J. Fish. Res. Board Can. 34. 2215.

Keller, W. and Pitblado, J. R.: 1986, Water Air Soil Pollut. 29. 285.

Ministry of the Environment: 1981, Outline of Analytical Methods, unpublished report, Ontario Ministry of the Environment, Toronto, 246 p.

Minns, C. K., Kelso, J. R. M. and M. G. Johnson: 1986, Can. J. Fish. Aquat. Sci. 43. 900.

Sadler, K.: 1983, Freshwater Biol. 13. 453.

Schindler, D. W., Mills, K. H., Malley, D. F., Findlay, D. I., Shearer, J. A., Davies, I. J., Turner, M. A., Linsey, G. A., and Cruikshank, D. R.: 1985, Science 228. 1395.

Schofield, C.L.: 1976, Ambio 5. 228.

Somers, K. M. and Harvey, H. H.: 1984, Can. J. Fish. Aquat. Sci. 41. 20.

Wales, D. and Beggs, G. L.: 1986, Water Air Soil Pollut. (this issue).

Watt, W. D., Scott, C. D. and White, W. J.: 1983, Can. J. Fish. Aquat. Sci. 40. 462.

Webster, D. A. and Eiriksdottir, G.: 1976, Trans. Amer. Fish. Soc. 90. 168.

Wright, R. F. and Snekvik, E.: 1978, Verh. Internat. Verein. Limnol. 20. 763.

PHYTOPLANKTON COMMUNITY RESPONSES TO ACIDIFICATION OF LAKE 223, EXPERIMENTAL LAKES AREA, NORTHWESTERN ONTARIO

D.L. Findlay and S.E.M. Kasian
Department of Fisheries and Oceans
Freshwater Institute
501 University Crescent
Winnipeg, MB R3T 2N6
Canada

ABSTRACT. From 1976 to 1983 the pH of Lake 223 was artificially lowered by additions of H_2SO_4. From an initial level of 6.7, the pH was lowered at a rate of 0.5 pH units a year until it reached 5.0 and was held there for 3 yr. The decrease in pH caused major changes in the epilimnetic phytoplankton community in this lake. Biomass increased as pH decreased. Chlorophyte (Chlorella) abundance increased as pH decreased from 6.1 to 5.6 while Cyanophytes (Merismopedia and Chroococcus) and dinoflagellates (Gymnodinium and Peridinium) dominated once pH decreased below 5.6. Community diversities decreased because of these species shifts and a decrease in the number of species. The amount of edible biomass increased as the pH decreased from 6.7 to 5.6, then declined as pH decreased to 5.0.

1. INTRODUCTION

The response of phytoplankton communities to decreasing pH below 5.6 has been documented previously (Hörnström et al., 1973; Almer et al., 1974, 1978; Kwiatkowski and Roff, 1976; Yan and Stokes, 1978), but there are little data on the changes which occur above pH 5.6. In this paper we summarize the responses of the natural phytoplankton community in Lake 223, Experimental Lakes Area, northwestern Ontario, to decreasing pH from 6.7 to 5.0.

Based on the literature that does exist for the 5.0 to 6.0 pH range we were presented with the hypotheses that 1) biomass would not increase as pH levels decreased (Almer et al., 1978; Conroy et al., 1976), 2) species of chrysophytes, chlorophytes, and diatoms would be the first to disappear, being replaced by dinoflagellates (Yan and Stokes, 1978; Kwiatkowski and Roff, 1976; Almer et al., 1978), and 3) species diversity would decrease as pH decreased (Hornstrom et al., 1973; Yan and Stokes, 1978; Kwiatkowski and Roff, 1976). Data on Lake 223 for 1974 to 1983, 2 yr preacidification (1974-1975) and 8 yr acidification, are presented to test these hypotheses.

Water, Air, and Soil Pollution **30** (1986) 719–726.

2. METHODS

2.1. pH Manipulation

The pH was lowered to a targeted value by additions of H_2SO_4 immediately after ice-out, and maintained throughout the season with additions of acid as required (Table I). For details on amounts and methods of application of acid see Cruikshank, 1984.

TABLE I
Annual mean ice-free epilimnetic pH values for Lake 223

Year	1974	1975	1976	1977	1978	1979	1980	1981	1982	1983
pH	6.70	6.64	6.49	6.13	5.93	5.64	5.59	5.02	5.09	5.13

2.2. Sampling and Enumeration

Samples were collected during the ice-free season with an integrated sampler (Shearer, 1978). Subsamples of 125 mL were extracted and preserved in acid Lugol's solution for enumeration and identification. Samples were counted with a Zeiss inverted microscope using the Utermohl technique as modified by Nauwerck (1963) and identification was to the species level. For a more detailed description of counting procedures see Findlay and Saesura (1980). To compute wet weight biomass from cells L^{-1}, the standard method of using the geometric formula best fitted to the cell shape was used (Vollenweider, 1968), and a density displacement of 1 was assumed.

2.3. Diversity

Simpson's diversity index was chosen to summarize changes in the phytoplankton community structure. This index is biased towards the dominant species, and therefore changes in rare species may not be detected. In a review of biotic indices Washington (1984) demonstrated that Simpson's was the most suitable diversity index.

2.4. Edible Biomass

The effects of phytoplankton species changes were related to zooplankton by estimating the portion of phytoplankton biomass which was edible. Cells <20 µm in any dimension were considered to be edible. This is a conservative criteria compared to other literature values and may be an underestimate. Increasing the criterion to <35 µm increased the edible crop by a maximum of 12% in any year.

2.5. Statistical Comparison to Reference Lakes

To separate responses of acidification from temporal effects Lake 223 was compared to several reference lakes (222, 224, 239, 240, 302 south (1974-1981) and 382). A nested design analysis of variance (ANOVA) in the General Linear Models (GLM) procedure in the SAS (SAS, 1985) statistical package was performed on annual mean values. This procedure tests for a significant difference between the long-term trend of Lake 223 and the combined long-term trend of the reference lakes. For more detail on the analysis see Schindler et al. (1986).

3. RESULTS

3.1. Biomass and Species Composition

Significant changes occurred in the epilimnetic phytoplankton community of Lake 223 as pH decreased.

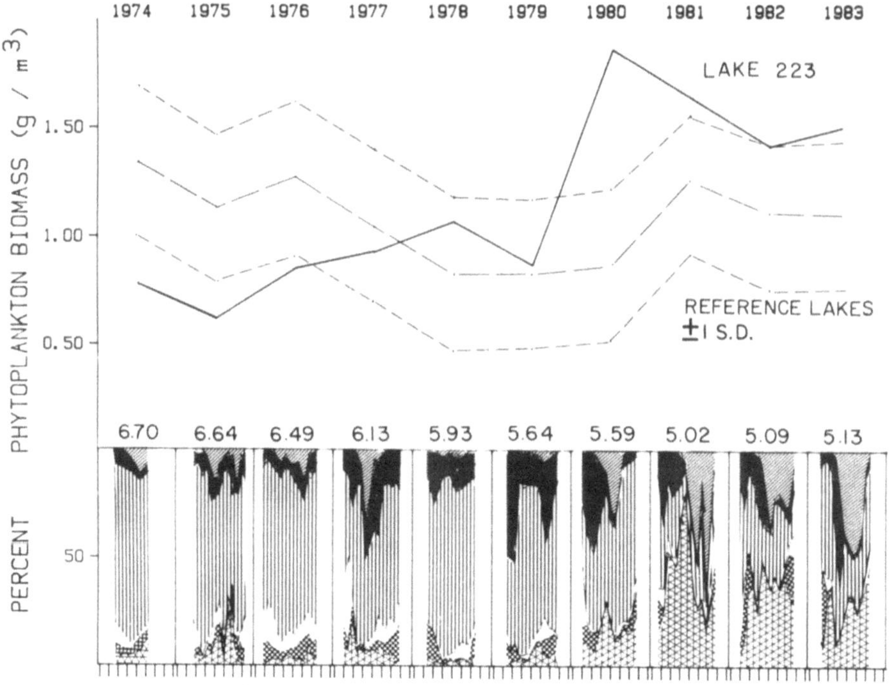

FIGURE 1. Mean annual epilimnetic biomass of Lake 223 vs reference lakes ±1 S.D. (upper panel) and seasonal species composition for ice-free epilimnion of Lake 223 (lower panel).
Cyanophytes, Chlorophytes, Chrysophytes, Diatoms, Cryptophytes, Dinoflagellates

Epilimnetic biomass increased slightly as pH declined from 6.7 to 5.6, then increased up to 40% when pH decreased below 5.6 (Figure 1). The nested ANOVA analysis (Table II) suggested a significant departure from the reference lake trend for 1976-1983 ($p = 0.039$). Also the increased biomass below pH 5.6 was higher than those recorded for any of the reference lakes for the same time period.

The most noticeable effects of acidification were the dramatic shifts in species composition (Figure 1, Table II). The first significant shift occurred in 1977 at pH 6.13. Chlorella mucosa (a small choccoidal chlorophyte) became abundant, eventually co-dominating with the chrysophytes (Findlay and Saesura, 1980). This species was previously not common to this area. The second significant shift occurred at pH 5.59 (1980). Dinobryon sertularia Ehrenberg, Mallomonas spp., Stichogloea spp., and Uroglena spp. increased in abundance and became the dominant chrysophycean species. As well, these species are large which caused biomass to increase to an annual maximum of 1875 mg m^{-3} (Figure 1). In addition to the increased chrysophyte abundance, there was an increase in the diatom Asterionella ralfsii W. Smith. This species was previously rare in the area but increased to 3.05×10^{6} cell L^{-1} (Findlay, 1984). Its rapid increase coincides with a decrease in silica (Schindler et al., 1985). The third significant shift in composition occurred when pH decreased below 5.6 (1981-1983) (Figure 1). The dinoflagellates, Gymnodinium sp. and Peridinium inconspicuum Lemmermann, and the cyanophyte, Chroococcus minutus (Keissler) Lemmermann, increased significantly and eventually became co-dominant. During the 3 yr when pH was held at 5.0, the phytoplankton assemblage was dominated by these species.

TABLE II
Probability that the trend in Lake 223 is equal to the trend in the control lakes (1976-1983).

Parameter mg m^{-3}*	Probability	Parameter mg m^{-3}	Probability
Biomass	0.0388	Chroococcus minutus	<0.0001
Diversity	<0.0001	Chlorella mucosa	<0.0001
Edible biomass	0.0083	Dinobryon sertularia	0.0985
Chrysophyceae	0.1291	Asterionella ralfsii	<0.0001
Cyanophyceae	<0.0001	Gymnodinium sp.	<0.0001
Chlorophyceae	0.0044	Peridinium inconspicuum	0.0030
Dinoflagellates	<0.0001		

*Except for diversity.

3.2. Diversity

The most significant change in the species diversity in Lake 223 was the increased range of values over a given season (Figure 2), especially below pH 5.6. These changes occurred during the summer when acid was being added and were due to a decrease in the number of species and frequent dominance of a few species. Using annual means there was a distinct decrease in diversity below pH 5.6. The nested ANOVA analysis (Table II) demonstrated a departure from the reference lake trend from 1976 - 1983 (P < 0.0001).

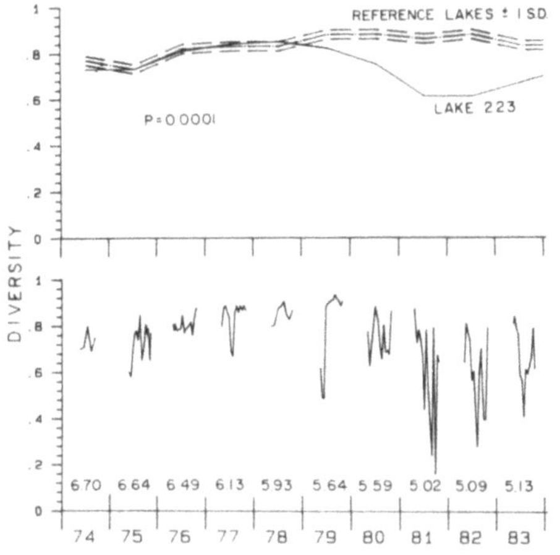

FIGURE 2. Mean annual epilimnetic species diversity of Lake 223 vs reference lakes (upper panel) and seasonal ice-free species diversity for the epilimnion of Lake 223 (lower panel).

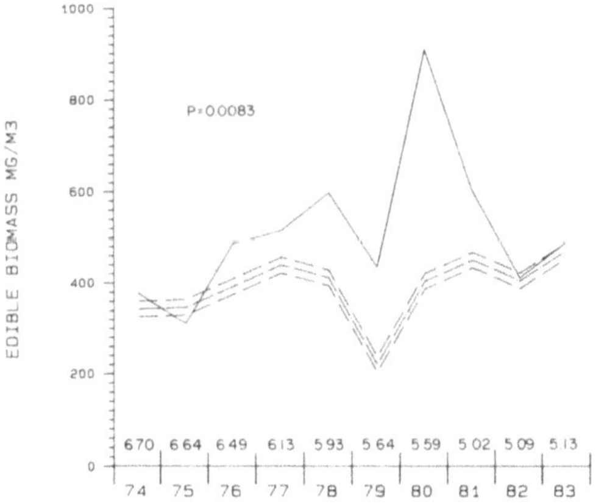

FIGURE 3. Mean annual epilimnetic edible biomass (cells <20 μ) of Lake 223 vs reference lakes.

3.3. Edible Biomass

The size structure of the phytoplankton community was altered by changes in species composition. The amount of biomass operationally considered edible (<20 μm in any dimension) increased as pH decreased from 6.7 to 5.6 (1974-1980) then decreased to estimates similar to preacidification when pH fell below 5.6 (1981-1983). This decrease was due to the abundance of large Gymnodinium sp. (35 to 40 μm diameter).

4. DISCUSSION

The epilimnetic phytoplankton community in Lake 223 experienced significant changes as pH decreased. Contrary to the hypothesis, biomass increased as pH decreased from 6.7 to 5.6 (Figure 1). There was no change in major nutrients (Schindler and Turner, 1982) and biomass was 2 × higher than that of the reference systems and there was a significant departure from the reference trend. This increase can be attributed to effects of acidification. Many of the species shifts which occurred were not referenced in the literature, although the final shift to dinoflagellate dominance was well documented in other acid affected areas. It appears that the majority of the chrysophytes, chlorophytes, cryptophytes and diatoms were sensitive to moderate to low pH levels. Those species which have thrived in Lake 223 below pH 6.0 may have several forms of protection against the acidic environment, which perhaps allow them to out compete other species. Species of Chlorella, Chroococcus and Gymnodinium are surrounded by a gelatinous sheath and Asterionella ralfsii has a thick siliceous casing. The decrease in the number of species and the increase in the abundance of a few species during summer has resulted in a decrease in diversity (Figure 2). As pH decreased, so did the number of species. Many of the smaller species were being replaced by larger species thus causing an increase in biomass. These changes in species composition have altered the food web. In the pH range 6.1 to 5.6, the assemblage was dominated by small chrysophytes, cyanophytes and chlorophytes. These species represented up to 60% of the total biomass and were a significant portion of the edible biomass as determined by size. Decreasing pH below 5.6 caused a species shift to large inedible Gymnodinium sp. causing a decrease in edible biomass (Figure 3). These changes may have had an effect on the zooplankton community in Lake 223. Malley and Chang (personal communication) hypothesized that the increase in edible biomass at pH levels of 6.0 to 5.6 was a factor in increased Cladoceran abundance.

5. CONCLUSION

The epilimnetic phytoplankton community in Lake 223 did respond to acidification. Most of the significant changes in community structure occurred below 5.6. In this experiment it appear that species composition and diversity are good indicators of early acidification whereas phytoplankton biomass appears to be less sensitive as an indicator. As well, Asterionella ralfsii, Chlorella mucosa, Chroococcus minutus and Peridinium inconspicuum also appear to be good indicators of acidification.

ACKNOWLEDGMENTS

We would like to thank E. R. DeBruyn and J.A. Shearer for sampling. Also, we would like to thank B. de March for her statistical advice and P. Campbell, D.W. Schindler, H.J. Kling, and K.H. Mills for their constructive criticisms of the manuscript.

REFERENCES

Almer, B., Dickson, W., Ekstrom, C. and Hörnström, E.: 1974, Ambio, **3**, 30.
Almer, B., Dickson, W., Ekstrom, C. and Hörnström, E.: 1978, 'Sulfur Pollution and the Aquatic Ecosystem', in J. Nriagu (ed.), Sulfur in the Environment, Part 2, John Wiley and Sons, New York, Chapter 7, 464 p.
Conroy, N., Hawley, K., Keller, W. and Lafrance, C.: 1976, Influences of the atmosphere on lakes in the Sudbury area. Atmospheric Contribution to the Chemistry of Lake Waters. J. Great Lakes Res. 2 (Suppl), p. 146-165.
Cruikshank, D.R.: 1984, 'Whole lake chemical additions in the Experimental Lakes Area. 1969-1983', Can. Data Rep. Fish. Aquat. Sci. **449**, iv + 23 p.
Findlay, D.L.: 1984, 'Effects on phytoplankton biomass, succession and composition in Lake 223 as a result of lowering pH levels from 5.6 to 5.2. Data from 1980 to 1982', Can. Manuscr. Rep. Fish. Aquat. Sci. **1761**, iv + 10 p.
Findlay, D.L. and Saesura, G.: 1980, 'Effects on Phytoplankton Biomass, Succession and Composition in Lake 223 as a Result of Lowering pH Levels from 7.0 to 5.6. Data from 1974 to 1979', Can. Manuscr. Rep. Fish. Aquat. Sci. **1584**, iv + 16 p.
Hörnström, E., Ekstrom, C., Miller, U., and Dickson, W.: 1973, Forsurningens Inverkan da Vastkustsjoar. Information fran Sotvattens-Laboratoriet, Drottningholm, No. **4**, 81 p.
Kwiatkowski, R.E. and Roff, J.C.: 1976, Can J. Bot. **54**, 2546.
Nauwerck, A.: 1963, Die Beziehungen zwischen Zooplantkon und Phytoplankton in See Erken. Symb. Bot. Upsal. **17(5)**, 163 p.
SAS Institute Inc.: 1985, SAS user's guide: statistics, Version 5 edition, SAS Institute Inc., Cary, NC. 956 p.
Schindler, D.W., DeMarch, B., Turner, M.A., and Playle, R.C.: 1986, 'Distinguishing Variation Due to Chemical Stressors from Natural Variation in Aquatic Ecosystems', in prep.

Schindler, D.W. and Turner, M.A.: 1982, Water, Air and Soil Pollution **18**, 259.
Schindler, D.W., Mills, K.H., Malley, D.F., Findlay, D.L., Shearer, J.A., Davies, I.J., Turner, M.A., Linsey, G.A., Cruikshank, D.R.: 1985, Science **228**, 1395.
Shearer, J.A.: 1978, 'Two devices for obtaining water samples integrated over depth', Can. Fish. Mar. Serv. Tech. Rep. **772**, iv + 9 p.
Vollenweider, R.A.: 1968, Scientific fundamentals of the eutrophication of lakes and flowing waters, with particular reference to nitrogen and phosphorus as factors in eutrophication, Tech. Rep. O.E.C.D. Paris Das/CSI/68.27, p. 1-182.
Washington, H.G.: 1984, Water Res. **18**, 653.
Yan, N.D. and Stokes, P.: 1978, Environ. Conserv. **5**, 653.

DISTRIBUTIONS OF ANURAN POPULATIONS IN CENTRAL ONTARIO RELATIVE TO HABITAT ACIDITY

Karen L. Clark
Biota Environmental Contractors
Box 21
Dorset, Ontario, Canada
POA 1EO

ABSTRACT. Adult anuran densities in 20 ponds in central Ontario were determined by using daytime visual censuses and by counting males calling during the breeding season. The pH of ponds varied from 4.55 to 6.37 and alkalinity from -1.9 mg L^{-1} to 23.2 $CaCO_3$. Bullfrog (Rana catesbeiana), small (2 to 3 cm snout-vent length) green frog (R. clamitans), and spring peeper (Hyla crucifer) densities were lower in more acidic ponds even though pH levels were not low enough to cause acute toxicity to eggs or larvae. The wood frog (R. sylvatica) was the only species whose density increased in more acidic ponds.

In a census of another 40 ponds for wood frog egg masses (an estimate of the number of females breeding in the pond), egg mass density was reduced as pond pH and alkalinity decreased even though the pH levels were not low enough to be acutely toxic to eggs or larvae (pH ranged from 4.44 to 6.63). These results suggest that long before the H^+ concentration in pond habitats increases to levels that cause acute toxicity, chronic sublethal effects in conjunction with indirect ecological effects of pond acidification (eg. changes in habitat structure, food resources, predator-prey relationships) may be affecting anuran populations.

1. INTRODUCTION

Many anuran species can tolerate high levels of acidity especially during the larval and adult lifestages (Gosner and Black, 1957; Saber and Dunson, 1978; Clark and Hall, 1985; Pierce, 1985). The embryonic stage is more sensitive to elevated acidity but some species eggs will hatch normally even at pH levels as low as 4.3 (Gosner and Black, 1957; Pierce et al., 1984). Most of the studies that have determined anuran sensitivity to increased acidity measured embryonic survival, or larval survival during short term exposures (Gosner and Black, 1957; Saber and Dunson, 1978; Clark and LaZerte 1985). Only one study examined chronic sublethal effects on larvae and they found that larvae had reduced growth rates at low pH (Freda and Dunson, 1985). Reduced larval growth rates can have ecological implications for anuran poulations as

Water, Air, and Soil Pollution **30** (1986) 727–734.

smaller larvae can have reduced ecological fitness (Wilbur, 1980; Crump, 1984).

Anuran populations are also subjected to indirect effects of pond acidification such as shifts in food resources (eg. changes in periphyton and zooplankton biomass and species composition) and alteration of habitat characteristics (eg. growth of Mougeotia). Chronic sub-lethal effects of acidification on anuran populations is difficult to measure. Reduced population sizes or shifts in species composition in ponds at sub-lethal pH levels compared to ponds with higher pH levels would suggest that there are indirect ecological effects of pond acidification on anuran communities. Studies have shown that some anuran species are absent in acidic ponds (Gosner and Black, 1957; Cooke and Frazer, 1976; Beebee and Griffin, 1977; Saber and Dunson, 1978; Hagstrom, 1980). Other studies have shown that habitat pH was not influencing the distribution of some species even in very acidic ponds (pH <5.0) (Pierce et al., 1984; Dale et al., 1985). However, these studies are based primarily on the presence or absence of a species. Long before a population is extirpated from an acidic pond, its population size should decline. Only one study, in Sweden, estimated anuran population sizes in ponds. It was found that the sizes of most anuran populations declined in more acidic habitats (Strijbosh, 1979).

Reductions in anuran populations could have far reaching implications for pond ecosystems since anurans can be major contributors to community biomass and are better sources of protein for tertiary consumers than either small mammals or birds (DeBenedictis, 1974; Burton and Likens, 1975; Cecil and Just, 1979).

The purpose of this study was to determine whether the population sizes of anurans in ponds in central Ontario are related to pond acidity.

2. METHODS

2.1 Adult Anuran Census

Twenty, ephemeral meltwater ponds in Muskoka and Haliburton counties in central Ontario were searched every 10 to 14 days from 15 May until 30 July 1980, for adult anurans. All ponds were temporary and dry by late summer. Ponds were small (surface area <0.075 ha) and the maximum water depth was between 0.25 and 1.5 m. Visual censuses for adults were conducted between 0800 and 1700 hr. When an individual was sighted the species and size class were recorded. Large differences in size that could be measured by eye occurred only for the bullfrog and green frog. The classes recorded were; small (2 to 3 cm snout-vent length), medium (3 to 5 cm) and large (>5 cm). Censuses of calling male anurans were conducted between 1800 and 0300 hr. pH, Ca concentration and alkalinity were determined weekly (Anonymous, 1975).

Since pond size varied greatly (3 to 753 m^2) and the number of individuals observed at each pond was related to pond surface area ($r = 0.47$, $p < 0.05$) any effects of pond size were reduced by using the total number of individuals of each species from all visits divided by the pond surface area. This value (the density of sightings of each species/m^2 of pond surface) was used to compare the population densities of adults of each species among the study ponds. Since pond chemistry was not correlated with pond size ($p > 0.05$) no potential effects of pond chemistry were obscured by using amphibian densities rather than total numbers.

2.2 Wood Frog Egg Mass Census

The same ponds were not censused in 1984 as those used in the 1980 census for adults. Similar criteria, however, were used to select the ponds.

Forty ponds were searched weekly between 15 April and 30 May 1984 for wood frog (*Rana sylvatica*) egg masses. Each female wood frog generally lays only one egg mass so the number of egg masses in a pond is an estimate of the size of the female breeding population. Pond size is positively correlated with the number of wood frog egg masses (Seale, 1982, this study $r = 0.65$, $p < 0.05$) but was not correlated with pond chemistry so the number of egg masses divided by pond surface area was used to compare population sizes among ponds. Hatching success was estimated by averaging the percentage of hatched eggs in two portions of each egg mass. pH, alkalinity, Ca, dissolved organic carbon (DOC), total Al and inorganic monomeric Al concentrations were measured weekly (Anonymous, 1975; LaZerte, 1984).

Correlation analysis was used to determine relationships between the number of sightings of anurans per unit area and wood frog egg mass density with pond water chemistry.

3. RESULTS AND DISCUSSION

3.1. Pond Water Chemistry

The water chemistry varied greatly between ponds. Mean pH varied from 4.44 to 6.63 (Table I). Total Al concentrations in the 1984 study ponds were high ($\bar{x} = 154$ µg L^{-1}) although only 8% of the ponds had inorganic monomeric Al levels that were within the range reported to be toxic to American toad (*Bufo americanus*) and wood frog embryos at pH <4.75 (Clark and LaZerte, 1985; Clark and Hall, 1985).

3.2. Adult Anuran Populations

Two species, the green frog and wood frog were very abundant and occurred in even the most acidic ponds (Tables I and II). The spring peeper (*Hyla crucifer*), bullfrog and American toad did not occur in ponds with pH <5.0 and the medium sized bullfrog was found only in ponds with pH >5.5. Other species that were observed much less frequently (<40% of the ponds) and are not included in any of the analyses are the chorus frog (*Pseudacris triseriata*), leopard frog (*Rana pipiens*), mink frog (*R. septentrionalis*), tree frog (*Hyla versicolor*), and the pickeral frog (*R. palustris*).

Table I. Mean and the range of anuran densities and water chemistry of the ponds censused in 1980 and 1984.

	Mean (standard error)		Range
1980 Pond Census (n = 20)			
Pond size (m^2)	1210	(430)	3-753
Alkalinity (mg L^{-1} $CaCO_3$)	6.69	(1.77)	0.4-37.7
pH	5.64		4.55-6.37
Ca (mg L^{-1})	3.3	(0.2)	0.4-37.7
Anuran density (no. ha^{-1})			
Green frog-total	5760	(1630)	140-24400
-small	1540	(520)	0-7700
-medium	4220	(1360)	30-23750
Bullfrog-total	2930	(1410)	0-21920
-small	1320	(760)	0-1300
-medium	1570	(780)	0-12700
-large	350	(180)	0-3300
Wood frog	940	(260)	0-3300
American toad	520	(330)	0-6670
Spring peeper	480	(150)	0-2090
1984 Pond Census (n = 40)			
Pond size (m^2)	55	(6)	1-185
Alkalinity (mg L^{-1} $CaCO_3$)	4.59	(1.00)	-1.90-23.18
pH	5.59		4.44-6.63
Ca (mg L^{-1})	3.2	(0.3)	0.9-10.3
Total Al (µg L^{-1})	154	(16)	15-428
Inorganic Al (µg L^{-1})	19	(3)	1-69
DOC (mg C L^{-1})	8	(1)	2-33
Wood frog egg mass density (no. masses m^2)	0.30	(0.07)	0-1.91

pH and alkalinity were correlated with the densities of individuals of several species (Table III). Alkalinity was more frequently correlated with species densities than pH probably because alkalinity is less temporally variable than pH and is an indicator of a pond's sensitivity to pH depressions.

Spring peeper, and bullfrog (especially the medium size bullfrog) densities declined in lower alkalinity ponds. Even though all size classes of green frogs were observed in all ponds, and the total number of green frogs was not related to pH or alkalinity, there were lower

Table II. The range of pH levels and the % of all ponds where each anuran species was observed.

Species	pH Min.	pH Max.	% of ponds (n = 20)
Green frogs-total	4.55	6.37	100
-small	4.55	6.37	85
-medium	4.55	6.37	100
Wood frogs	4.55	6.37	85
Bullfrogs -total	5.08	6.15	60
-small	5.08	6.15	55
-medium	5.50	6.15	50
-large	5.08	6.15	50
Spring peepers	5.08	6.37	50
American toad	5.08	6.03	45
Wood frog egg masses (1984 census, n = 40)	4.52	6.58	68

Table III. Spearman rank correlation coefficients between anuran densities and pond chemistry (n = 20).

Species	pH	Alkalinity	Ca
Green frogs-total	0.13	0.40	0.15
-small	0.31	0.62*	0.20
-medium	0.08	0.29	0.18
Bullfrogs -total	0.31	0.45*	0.30
-small	0.23	0.31	0.21
-medium	0.29	0.44*	0.20
-large	0.32	0.33	0.36
Wood frog	-0.48*	-0.55*	-0.01
Spring peeper	0.61*	0.69*	-0.17
American toad	-0.22	-0.13	0.01

*p <0.05

densities of small individuals in lower alkalinity ponds. In other parts of their range bullfrogs have been reported as being absent from acidic habitats where green frogs are common (Gosner and Black, 1957; Saber and Dunson, 1978).

The lower densities of spring peepers, bullfrogs or small green frogs in low alkalinity ponds is probably not due to direct toxicity of the increased H^+ concentrations. Adults are thought to be more tolerant to acidity than the egg or larval life stages and spring peeper, bullfrog and green frog embryos can tolerate pH levels as low as 4.3 (Gosner and Black, 1957; Pierce, 1985). It is highly likely that chronic sublethal effects of elevated H^+ levels (eg. reduced larval growth rates, changes in behaviour) as well as any ecological effects of low pH on other components of the pond community that have resulted in changes in food resources, habitat structure and predator-prey relations are influencing the distributions of these species.

Adult wood frogs occurred in all 20 ponds and this was the only species that was more common in the acidic, low alkalinity ponds. The wood frog in other parts of its range has been observed in acidic bogs with pH = 4.0 (Gosner and Black, 1957; Karns, 1983). Embryos can tolerate pH levels as low as 4.0 and tadpoles can tolerate pH 3.5 (Gosner and Black, 1957; Tome and Pough, 1982; Karns, 1983; Pierce et al., 1984; Clark and LaZerte, 1985). Adult wood frogs must benefit from any ecological changes due to pond acidification.

The American toad is terrestrial and very transient during the non-breeding season (Werner and McCune, 1979). It occurred almost exclusively along the shoreline of ponds and its distribution was not related to pond acidity.

3.3. Wood Frog Egg Mass Distributions

In the 1984 pond census, wood frog egg mass density was positively correlated wth pH ($r = 0.43$, $p < 0.05$) and alkalinity ($r = 0.38$, $p < 0.05$) and negatively correlated with inorganic monomeric Al (the toxic portion of the total Al concentration) ($r = -0.43$, $p < 0.05$). Acidic, low alkalinity ponds with high Al levels, therefore, supported smaller populations of breeding wood frogs (per unit area of pond) than less acidic ponds. The decline in egg mass densities at low pH was quite sharp and ponds with pH <5.5 had on average only 0.043 egg masses m^{-2} ($n = 20$) compared to 0.400 m^{-2} ($n = 20$) in ponds with higher pH levels pH levels (Mann Whitney μ test, $p < 0.05$).

None of the 40 ponds were acidic enough to cause direct embryonic or larval mortality and hatching success in most ponds was high (90 to 100%) and did not vary with pH ($p > 0.05$). However, 3 ponds with pH <4.75 did have inorganic monomeric Al concentrations that have been reported to be toxic to wood frog embryos (Clark and Hall, 1985; Clark and LaZerte, 1985). Two of the 3 ponds had no egg masses and the third had only 0.05 egg masses m^{-2}. Hatching success of these egg masses was high $\bar{x} = 98\%$) and so the pH could not have been low enough to cause Al toxicity (Clark and LaZerte 1985). Indirect effects of habitat acidity must have been influencing the distributions of the breeding populations.

4. CONCLUSIONS

In central Ontario the populations sizes of some anuran species are reduced in acidic temporary ponds even though the pH of the ponds is not low enough to be directly toxic to embryos and larvae. Field and laboratory bioassays to determine toxic levels of acidity are, therefore, not adequate to assess whether natural populations are being influenced by habitat acidification. Populations may decline long before local extinctions of anuran populations occur, or before larval or embryonic mortality can be observed in the field.

5. ACKNOWLEDGEMENTS

This study was funded as part of the Acid Precipitation in Ontario Study by the Ontario Ministry of the Environment and by the Ontario Ministry of Natural Resources. Many members of the staff at the Dorset Research Centre assisted with various aspects of the study. Thanks to Margot Shaw and Tim Raistrick for help with the field work. Constructive criticisms on the design and final manuscript were provided by Peter Dillon, Dave Euler, Bruce LaZerte and Ron Hall.

REFERENCES

Anonymous: 1975, Outline of analytical methods. Ont. Min. Environ. 130 pp.
Beebee, T.J.C. and Griffin, J.R.: 1977, J. Zool. Lond. **181**, 341.
Burton, T.M. and Likens, G.E.: 1975, Ecology **56**, 1068.
Cecil, S.G. and Just, J.J.: 1979, Copeia **1975**, 447.
Clark, K.L. and LaZerte, B.D.: 1985, Can. J. Fish. Aquat. Sci. **42**, 1544.
Clark, K.L. and Hall, R.J.: 1985, Can. J. Zool. **63**, 116.
Cooke, A.S. and Frazer, J.F.D.: 1976, J. Zool. Lond. **178**, 223.
Crump, M.L.: 1984, Herpetol. **40**, 265.
Dale, J.M., Freedman, B., and Kerekes, J.: 1985, Can. J. Zool. **63**, 97.
DeBenedictis, P.A.: 1974, Ecol. Monogr. **44**, 129.
Freda, J. and Dunson, W.A.: 1985, Copeia, **1985**, 415.
Gosner, K.K. and Black, I.H.: 1957, Ecology **38**, 256.
Hagstrom, T.: 1980, Reproductive strategy and success of amphibians in waters acidified by atmospheric pollution. European Herpetological Symposium Proceedings, Oxford.
Karns, D.R.: 1983, Toxic bog water in Northern Minnesota peatlands: ecological and evolutionary consequences for breeding amphibians. Univ. of Minnesota Ph.D. Thesis 300 pp.
LaZerte, B.D.:. 1984, Can. J. Fish. Aquat. Sci. **41**, 766.
MacDonald, D.G., Ozog, J.L., and Simons, B.P.: 1984, Can. J. Zool, **62**, 2171.
Pierce, B.A.: 1985, Bioscience **35**, 239.

Pierce, B.A., Hoskins, J.B., and Epstein, E.: 1984, J. of Herpet. **18**, 159.
Pough, F.H.: 1976, Science **192**, 68.
Saber, P.S., and Dunson, W.A.: 1978, J. Exp. Zoology **204**, 33.
Seale, D.B.: 1982, Copeia **1982**, 627.
Strijbosch, H.: 1979, Oikos **33**, 363.
Tome, M.E., and Pough, F.H.: 1982, Responses of amphibians to acid precipitation In T.A. Haines and R.E. Johnson (eds), Acid Rain/ Fisheries, American Fisheries Society, Bethesda, Maryland.
Werner, J.K. and McCune, M.B.: 1979, J. Herpetology **13**, 101.
Wilbur, H.M.: 1980, Ann. Rev. Ecol. Syst. **11**, 67.

A HEMATOXYLIN STAINING TECHNIQUE TO LOCATE SITES OF ALUMINUM BINDING IN AQUATIC PLANTS AND ANIMALS.

Magda Havas

Institute for Environmental Studies,
University of Toronto,
Toronto, Canada, M5S 1A4.

ABSTRACT

This paper describes a modified hematoxylin staining technique that can be used to locate sites of Al binding in freshwater plants and animals. This technique is fast, simple, and inexpensive to use. It is more reliable for organisms raised under controlled conditions, although it can be used on organisms isolated from the field. In the presence of Al, a purple stain appears which absorbs between 560 and 570 nm. This stain can be used as a live stain over the pH range of 3 to 9. The stain remains stable for at least 12 mo in alcohol-preserved specimens. Low concentrations of Fe (reddish-brown) and high concentrations of Pb (grey) and Cu (pink) can mask the reaction with Al (purple). Fluoride (10:1 molar ratio to Al), EDTA (1:1 ratio to Al), humic acid and PO_4 (1:10 ratio to Al) can prevent Al uptake or interfere with Al-hematein binding. Based on the technique described in this paper, sites of Al binding in aquatic plants and animals include: nucleus and cell wall of the green alga _Mougeotia_; cell wall of the aquatic moss _Leptodictyum riparium_; chloride cells, tip of the penis, and hind gut of the fairy shrimp, _Branchinecta paludosa_; and the anal papillae of the phantom midge, _Chaoborus_.

1. INTRODUCTION

For the past 70 yr, Al has been recognized as a major limiting factor on acidic soils due to its enhanced solubility at low pH (Hartwell and Pember, 1918). Within the past decade, elevated concentrations of Al have been found in acidifying freshwater ecosystems in both Scandinavia and North America (Dickson, 1980). Our understanding of the mechanism of Al toxicity can be substantially enhanced if sites of sorption and localization within the organism can be identified. The traditional methods of autoradiography and radioisotope identification are not possible since there are no suitable isotopes for Al. Highly sophisticated and sensitive methods of elemental microanalysis including Electron Energy Loss Analysis (EEL) and Energy Dispersive X-ray Microanalysis (EDX) can be used for Al, although sensitivity may not be adequate for some samples, and the effects of sample preparation (fixation, dehydration, embedding) on Al mobility need to be determined. In some cases, simple and considerably less expensive histochemical stains can be used as an alternative. One such stain for Al, a modified hematoxylin stain, is presented in this paper.

Water, Air, and Soil Pollution **30** (1986) 735–741.

Hematoxylin was first described as a useful biological stain in 1865 and rapidly became one of the most extensively used stains in the study of living organisms (Conn, 1929). Since its introduction, many versions of this stain have been described (see Cole, 1943; Baker 1962). In the traditional formula, high concentrations of Al (or Fe) are added to the stock solution to act as a mordant and bind hematein (oxidized hematoxylin) to plant or animal tissue. In the modified version, presented in this paper, the Al already on (or in) the organisms acts as the mordant. Both techniques, in effect, indicate sites of Al binding.

2. MATERIALS AND METHOD

2.1 Stock Solution

The stock solution consists of 2.0 g hematoxylin (anhydrous, certified) and 0.20 g sodium iodate dissolved in 1 L of distilled water. This gives half-oxidized hematoxylin and is a modification of the recipe presented by Gill et al. (1974). If anhydrous hematoxylin is unavailable, 2.36 g of crystalline hematoxylin ($C_{16}H_{14}O_6$ $3H_2O$) may be substituted. Smaller volumes (100 mL) of stock solution may be more economical, especially if only small volumes are required each week.

Before hematoxylin can act as a stain, it must be oxidized to hematein and mordanted. Sodium iodate is used as the oxidizing agent and aluminum on (or in) the organism acts as the mordant. The freshly made stock solution has a rich amber color, and continues to oxidize (ripen) and to change color with exposure to atmospheric O_2. Once the stock solution has a dull brown color it is no longer effective as a stain. A fresh solution should be prepared daily or weekly, as required. The rate of oxidation can be reduced by keeping the staining solution tighly capped and refrigerated. Stock solutions should be neutral or slightly acidic since alkalinity favors rapid oxidation and reduces its useful shelf-life. Alcohol or glycerin can also be added to increase the shelf-life of hematoxylin (Cole, 1943). Gill et al. (1974) added ethylene glycol (250 mL) and glacial acidic acid (20 mL) and reduced distilled water to 730 mL to impede oxidation.

2.2 Staining Procedure

The following method is ideal for organisms exposed to Al in laboratory bioassays (Fe should be omitted from the culture solution to prevent interference). It can also be used on organisms isolated from the field, although ligands and other metals may interfere with the stain if present at high concentrations (see next page).

Algae should be rinsed briefly with distilled water and then placed into the hematoxylin solution (full strength) for approximately 20 min. Larger organisms may need to be left longer to allow sufficient time for stain to penetrate. Some preliminary work is required for each type of organism. After staining, excess hematoxylin should be removed with distilled water. The specimen is now ready to be viewed under the microscope, or it can be preserved in alcohol (70% ethanol for zooplankton and macroinvertebrates, 95% for algae). Experiments with insect larvae and crustaceans showed that the Al-hematein stain had not faded after 12 mo of storage in alcohol.

Organisms were exposed for 24 hr to either 0.5 mg total Al L^{-1} (Mougeotia) or 1.0 mg total Al L^{-1} (remaining species). A synthetic soft water solution (pH 6.6, 2.5 mg Ca L^{-1}, 0.5 mg Mg L^{-1}, 1.8 mg Na L^{-1}, 0.5 mg K L^{-1}, 2.1 mg Cl L^{-1}, 5.3 mg SO_4 L^{-1}) adjusted to pH 5 was used for the bioassays (see Havas, 1985). The hematoxylin staining technique was tested successfully on freshwater filamentous algae, aquatic mosses, macrophytes, crustaceans, insect larvae, snail eggs, and fish gills. Results for just some of these are presented in this paper.

2.3 Spectral Absorbance and Color Development

100 uL of the hematoxylin stock solution were added to 20 mL of a solution containing 10 mg total Al L^{-1} adjusted to pH 5 with H_2SO_4. The spectral absorbance of this solution was recorded on a Beckman Model ACTA3 Scanning Spectrophotometer after 0, 10, 20, 30, 60 min and after 2 days to determine rate of color development (i.e. rate of Al-hematein binding). The spectral absorbance of hematoxylin over a range of Al concentrations (0, 0.5, 1.0, 2.5, 5.0 and 10.0 mg total Al L^{-1}) was also determined.

2.4 Interference

The interference of metals, anions, and organic compounds with the Al-hematein complex was tested since many potentially interferring ions may be present in culture solutions and in water samples collected from the field. Distilled water solutions adjusted from pH 5.0 to 5.5 with H_2SO_4 were prepared containing 10 mg L^{-1} of either the chloride or sulphate salts of one of each of the following metals: Al, Ba, Ca, Cd, Co, Cr, Cu, Fe, Hg, Mn, Pb, and Zn. 100 uL of the hematoxylin stock solution were added to 20 mL of each of the metal solutions. After 30 min spectral adsorption at the Al-hematein wavelength (560 to 570 nm) was recorded. Similarly, solutions containing molar ratios of 10:1, 1:1, and 1:10 of Al and one of EDTA, humic acid (weight basis), F, and PO_4 were tested for interference at 560 nm.

2.5 Energy Dispersive X-ray Microanalysis

I checked the hematoxylin staining technique using Scanning Electron Microscopy and X-ray Microanalysis at the Department of Zoology, University of Toronto. Selected specimens were air dried and mounted on Al-specimen stubs with double-sided tape and were coated with 20 to 30 nm of C (Edwards Vacuum Evaporator E12E). Specimens were examined in a Cambridge S180 Scanning Electron Microscope with an ORTEC 6230 Energy Dispersive X-ray Microanalysis System. X-rays were collected at a fixed take-off angle of 45°. X-ray spectra were recorded with a count time of 40 s at an accelerating voltage of 25 kV and a beam current of 6.5 x 10^{-9} A. The Al excitation window was set at 138 to 162 keV. Secondary electron images were recorded on Polaroid P/N 55 film.

3. RESULTS AND DISCUSSION

Unmordanted hematein in distilled water had a straw yellow color and absorbed between 380 and 390 nm. Aluminum-hematein had a purple color and a broad absorption band between 525 and 575 nm with peak absorption between 560 and 570 nm. Absorption was linear with increasing concentrations of Al up to (and possibly above) 10 mg L^{-1}. The color developed rapidly, and was fairly stable within 20 min at room temperature. Iron (less than 10 mg L^{-1}) interfered with the Al-hematein complex at 560 nm and gave a reddish-brown color. Lead (grey) and Cu (pink) also interfered visually when present at high concentrations (10 mg L^{-1}), although there was no spectral interference at the Al-hematein wavelength (560 nm). It was not possible to observe the Al-hematein complex in solutions containing humic acids due to the color of humic acids. This was true for all 3 ratios of humic acid to Al (10:1, 1:1, 1:10). EDTA interfered at a 1:1 molar ratio of EDTA:Al, but not at a 1:10 ratio. Phosphate interfered slightly (10%) at a ratio of 1:10 (PO_4:Al) and significantly (38%) at 1:1 (PO_4:Al), while F interfered at a ratio of 10:1 (F:Al). The exact hue of the Al-hematein stain was pH-dependent within the range of pH 3 to 9.

Energy Dispersive X-ray Microanalysis (EDX) and hematoxylin gave similar results for Al localization, although the EDX system was not as sensitive as hematoxylin for detecting low levels of Al and consequently did not provide good X-ray maps.

In the green alga, Mougeotia, (Plate 1) Al was detected as a deep purple stain (black in photograph) within the nucleus (N) and in the cell wall (CW). This strain of Mougeotia was extremely Al-sensitive (Stokes and Turner, 1985). The nucleus of an Al-tolerant strain of Mougeotia did not stain purple when exposed to the same and higher concentrations of Al, which suggests that exclusion may be involved in the Al tolerance of this alga.

In the submerged moss, Leptodictyum riparium, Al was observed as a purple stain primarily on the older leaves (not shown). Within the leaf, Al was found in the cell wall (CW) (Plate 2). Detoxification of Al by cell wall binding has been suggested by Clarkson (1969) for higher plants.

Fairy shrimp, Branchinecta paludosa, lack the protective carapace of other crustaceans and are relatively sensitive to Al (Havas, 1984). Aluminum was detected at the tip of the penis (P); in the hind gut near the anal opening (A), which is presumably due to anal drinking--a common feature among crustaceans; and on the leaf-like appendages on the thoracic legs which contain chloride cells (CC) and are responsible for the active uptake of salt from the environment (Plate 3). Experiments with Branchinecta showed that Al interfered with Na and Cl regulation (Havas, 1984), just as in Daphnia magna (Havas and Likens, 1985a), and that humic acids prevented Al bioaccumulation. Chloride cells in two other crustaceans, Daphnia and Holopedium were also stained purple following exposure to Al (not shown).

Plate 4 shows the posterior portion of the phantom midge, Chaoborus. The specimen on the left was exposed to Al while the one on the right was not. Both were placed into a hematoxylin solution.

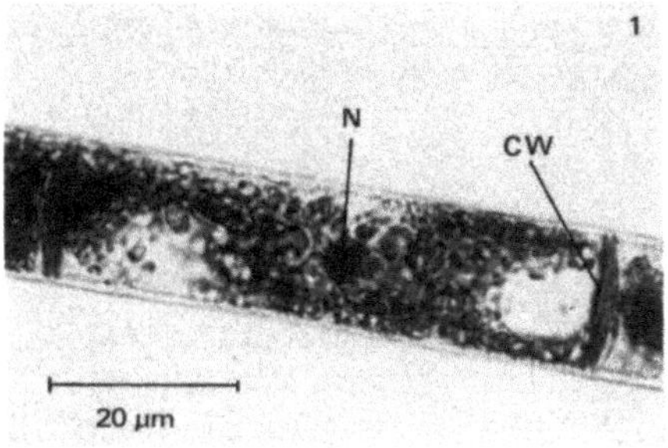

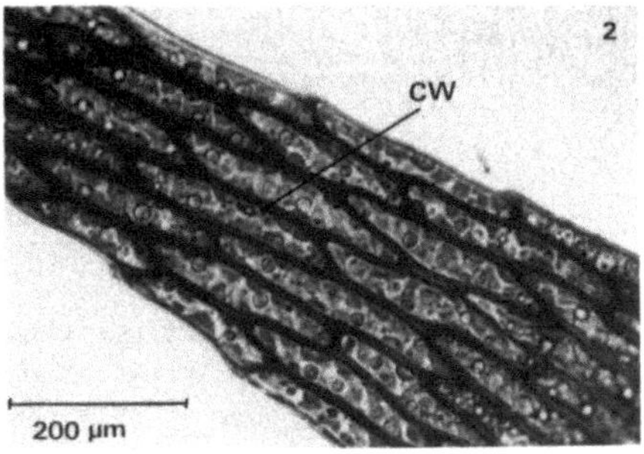

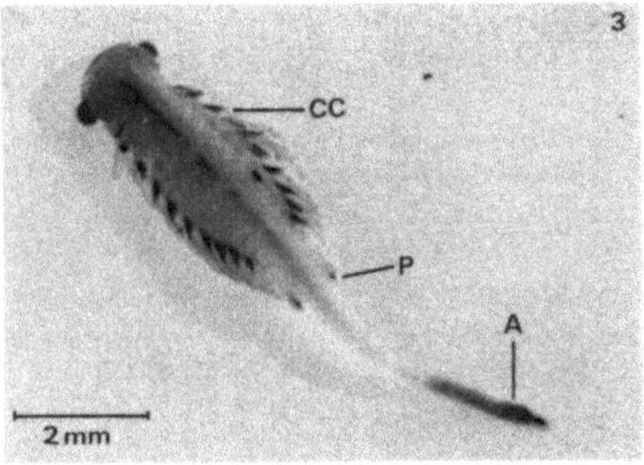

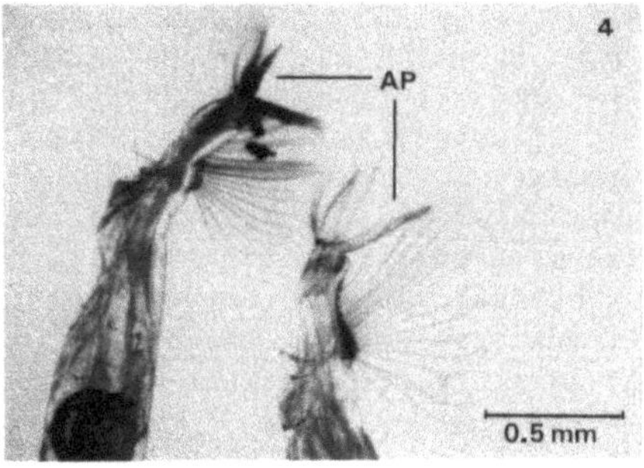

Plate 1. This strain of *Mougeotia* was extremely sensitive to Al and rapidly accumulated Al from the surrounding water. Aluminum was detected as a deep purple stain (black in photograph) in the nucleus (N) and in the cell wall (CW). Algae provided by P.M. Stokes and P.A.E. Turner.

Plate 2. Aluminum was found within the cell wall (CW) of the submerged moss, *Leptodictyum riparium*. Older fronds were stained more than younger fronds.

Plate 3. In the fairy shrimp, *Branchinecta paludosa*, Al was detected at the tip of the penis (P); in the hind gut near the anal opening (A), presumably due to anal drinking; and on leaf-like appendages on the thoracic legs which contain chloride cells (CC) and are involved with active salt uptake.

Plate 4. The posterior portion of two plantom midge larvae, *Chaoborus*, are shown. The specimen on the left was exposed to Al, the one on the right was not. Aluminum was found on the anal papillae (AP, left specimen), which are normally transparent (right specimen). The anal papillae are responsible for the active uptake of salt from the environment.

Aluminum can be seen as a purple stain (black in photograph) on the anal papillae (AP) of the specimen on the left. These papillae are normally transparent as can be seen on the specimen on the right. The anal papillae are the sites of salt uptake in diptera larvae just as are chloride cells in crustaceans (see Krogh 1965). The pupae showed no Al staining. Chaoborus is moderately Al-tolerant. Larvae can survive at (and probably above) 1 mg Al L^{-1} with no obvious ill effect (Havas and Likens, 1985b). Whether Al interferes with salt regulation in this species is not known.

The hematoxylin staining technique, presented in this paper, can be used to identify areas of Al binding in freshwater plants and animals. This same technique can also be used for terrestrial plants. Polle et al. (1978) and Wallace et al. (1982) used hematoxylin to quantify degree of Al-tolerance among different varieties of wheat seedlings by staining live roots. Aluminum-sensitive varieties showed more extensive staining than those which were Al-tolerant. Hutchinson et al. (1986) tested this stain using jack pine seedlings. Their results showed that Al was taken up by growing roots and was found in the outer cortex, especially in the cell wall, but not in vascular tissue.

Aluminum was detected in most of the aquatic species tested. Some species, however, excluded Al, which may be one mechanism (i.e. avoidance) of Al tolerance (Al-tolerant Mougeotia). Other species accumulated Al in the cell wall, presumably with less available for cytoplasmic uptake (Leptodictyum). Aluminum was found in the nucleus of Al-sensitive algae (Mougeotia), and on the chloride cells of Al-sensitive crustaceans (Branchinecta). The high concentrations of Al on the anal papillae of Chaoborus, which is a remarkably Al-tolerant species, has yet to be explained.

ACKNOWLEDGMENTS

I thank Tom Hutchinson who first suggested that this stain might be useful for Al localization in aquatic biota and who provided many useful references; Eric Lin for help with EDX; Verna Higgins for use of her scanning spectrophotometer, and Pamela Stokes and Peter Turner for supplying Mougeotia. Funding was provided by NSERC and a Connaught Fund from the University of Toronto.

4. REFERENCES

Baker, J.R.:1962, Quart. J. Micr. Sci. **103**; 493.

Clarkson, D.T.:1969, 'Metabolic aspects of aluminum toxicity and some possible mechanisms for resistance.' In: Rorison, I.H. (Ed.), Ecological Aspects of the Mineral Nutrition of Plants, Symp. Br. Ecol. Soc. 9, Sheffield, 1968, Blackwell Sci. Publ., Oxford, p. 138.

Cole, E.C.:1943, Stain Techn. **18**; 125.

Conn, H.J.:1929, Stain Techn. 4; 37.

Dickson, W.:1980, 'Properties of acidified waters.' In: Drablos, D. and Tollan, A. (Eds.), Proc. Int. Conf. Impact Acid Precipitation, Sandefjord, Norway, March 11-14, 1980, Norway, p. 75.

Gill, G.W., Frost, J.K, and Miller, K.A.:1974, Acta Cytologica **18**; 300.

Hartwell, B.L. and Pember, F.R.:1918, Soil Sci. **6**; 259.

Havas, M.:1984, unpublished data.

Havas, M.:1985, Can. J. Fish. Aquatic Sci. **42**; 1741.

Havas, M. and Likens, G.E.:1985a, Proc. Natl. Acad. Sci. **82**; 7345.

Havas, M. and Likens, G.E.:1985b, Can. J. Zool. **63**; 114.

Hutchinson, T.C., Munoz-Vega, G. and Bozic, L.E.:1986, Water, Air and Soil Pollut. (this volume).

Krogh, A.:1965, Osmotic Regulation in Aquatic Animals, Dover Publ. Inc., New York, 242 pp.

Stokes, P.M. and Turner, P.A.E.:1985, Institute for Environmetal Studies, University of Toronto, Toronto, Canada, M5S 1A4, personal communication.

Polle, E., Konzak, C.F., and Kittrick, J.A.:1978, Crop Sci. **18**; 823.

Wallace, S.U., Henning, S.J. and Anderson, I.C.:1982, Iowa State J. Research 57; 97.

INFLUENCE OF ACIDITY AND COMPETITION WITH FISH ON THE DEVELOPMENT OF DUCKLINGS IN QUEBEC

J.-L. DesGranges and J. Rodrigue
Canadian Wildlife Service
P.O. Box 10100
Ste-Foy, Quebec
CANADA G1V 4H5

ABSTRACT. In Quebec, in 1984 and 1985, a number of American Black Duck (Anas rubripes) and Common Goldeneye (Bucephala clangula) ducklings were reared on two acid lakes (pH 4.8 to 5.0) and one circumneutral lake (pH 6.0) in which the brook trout (Salvelinus fontinalis) populations had been manipulated. This allowed us to determine how acidity and competition with fish affected duckling feeding and growth. Strong acidity and the presence of fish were both required before the birds' development was restricted, the Black Duck being particularly affected. This result supports hypotheses on competition between fish and some waterfowl species.

1. INTRODUCTION

Acid rain can cause significant changes in lake chemistry, restricting fish reproduction and leading to profound changes in the aquatic community as a whole (Haines, 1981). Acidification also affects waterfowl by modifying the availability of their food supplies, and causing significant changes in the distribution of duck populations (Eriksson, 1984).

However, some duck species actually prefer to raise their young on acid lakes (DesGranges and Darveau, 1985). An experimental program was therefore devised in order to compare the growth and feeding activity of imprinted ducklings of two species on acid and non-acid lakes in which the fish populations were manipulated. The two species selected were the American Black Duck, a dabbler whose population is declining and which is known to avoid acid lakes, and the Common Goldeneye, a diving duck that seems to seek out acid lakes (DesGranges and Darveau, 1985).

2. STUDY AREAS AND METHODS

Studies were carried out on lakes situated on the Laurentian Shield approximately 75 km northwest of Quebec City (46°56'N, 71°41'W) were several lakes exhibit signs of advanced acidification (Dupont, 1984).

Water, Air, and Soil Pollution 30 (1986) 743–750.

The mean annual precipitation is about 100 cm, a third of which consists of acid snow (mean pH = 4.2) which cause significant acid episodes during springtime melt. Currently, about 25 kg ha^{-1} of wet sulphate is deposited annually.

Three similar lakes of 2 to 7 ha and with well developed riparian vegetation were selected in the Batiscan-Neilson controlled harvesting zone (ZEC) (DesGranges, 1985). While maximum depth varies considerably from lake to lake, each has fairly large sections of shallow water (< 1.5 m) and aquatic vegetation covers from 1.0 to 2.3 ha. Two of the lakes are acid: Lac au Cochon (pH = 4.8) was fishless during the first summer of field work (1984), while Lac du Rocher (pH = 5.0) supported a healthy brook trout population. The next spring, Lac au Cochon was stocked with 1500 brook trout, whilst nearly 750 trout were removed from Lac du Rocher using gill nets. The third lake, Lac Civens, was studied as a reference in one year. It contains brook trout and has a circumneutral pH (6.0).

To study duckling feeding ecology and weight growth, eggs were collected in the wild and incubated artificially. Ducklings were then imprinted to humans. Broods were first taken out on the lakes at the age of about one week and thereafter each day from approximatively 5:30 am to 8:00 pm. Time budgets were recorded throughout the day and were achieved by following the broods by canoe. Weight were recorded at the beginning and end of each day. The initial broods were made up of 6 to 8 ducklings from several different clutches, with a good mix of sexes. The mean weight of the ducklings when they first arrived at the lakes was not significantly different between broods. The disappearance or death of some individuals brought down the size of the broods mostly at the start of the study. The time involved in imprinting and rearing ducklings precluded the replacement of mortalities by bird of similar age to the experimental broods. Field observations were made from 13 June to 12 July in 1984 and from 16 June to 7 July in 1985. Separate broods of the same age were compared at the same time each year because of possible phenological or ontogenic effects.

The methods we used in this study were inspired by those employed by Hunter et al. (1985) in Maine. The morning weighings, measured while the ducklings' stomachs were empty (they had spent the night in heated shelters where only water was provided), yielded the measurements used in calculating growth curves. The ducklings' time budget was estimated as follows: at regular hourly interval over the course of the day, observers noted the number of ducklings in each brood (or the main group) that were either searching for food, moving from place to place, or resting. The distance travelled by the brood in 15 s was also recorded, although only movements associated with searching for food were noted. The ducklings' diet and the structure of the invertebrate and aquatic vegetation communities were also studied on a regular basis throughout the experiment (DesGranges, 1985, 1986).

the Goldeneye ducklings studied on Lac au Cochon in 1985 moved nearly twice as quickly as that studied the previous year when the lake contained no fish (5.0 VS 2.7 m min^{-1}; $t = 3.7$, d.f = 19 duckling-days, $p < 0.005$).

3.2 Resting time

In all cases, the Black Duck young spent less time resting than the Goldeneye ducklings (Figure 2). On the acid lakes, they spent 17 to 20% of the time resting when few fish were present which is a little more than on the circumneutral lake with fish (14%). When the acid lakes supported substantial fish populations, the Black Duck young tended to devote less time to resting (13% on acid lakes with fish VS 16% on "fishless" acid lakes) and more to searching for food, especially on Lac du Rocher ($t = 6.4$, d.f. = 4 ducklings, $p < 0.005$).

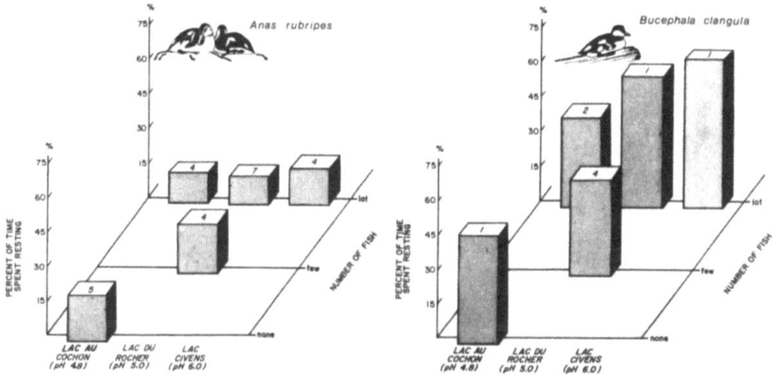

Figure 2. Influence of acidity and abundance of fish on resting time of Black Duck and Common Goldeneye ducklings.

The Goldeneye ducklings reared on the acid lakes spent more time searching for food (31.9 $\pm$ 3.6%, minimum = 14.5%, n = 7 ducklings) than did the single duckling released on the circumneutral lake (8.6%). On Lac au Cochon, the Goldeneye ducklings rested longer the summer the lake did not contain fish (45% VS 35%; $t = 1.8$, d.f. = 25 duckling-days, $p < 0.05$).

3.3 Growth rates

For each of the lakes, we drew weight growth curves using the mean

3. RESULTS

This study is based on 51 days of behavioral observations (total n = 19,342) and of growth measurements on 18 Black Duck ducklings and 7 Common Goldeneye ducklings. Mean and standard error for the various measurements are presented elsewhere (DesGranges, 1986). The Goldeneye results are handicapped by small sample size, especially the use of a single individual which survived during the study in 1984. This individual spent consecutive periods on each of the three lakes. Goldeneye results were nonetheless retained to show the basic similarity between the two species of ducklings. Inspection of the data for both species showed substantial ontogenic effects in the amount of time spent resting but not in the speed of movement (DesGranges, 1986).

3.1 Speed of movement

When they were feeding, the Black Duck young living on the two acid lakes covered approximately 9.3 m of shoreline a minute, while those on the circumneutral lake explored only about 6.1 m min^{-1} ($t = 6.0$, d.f. = 107 duckling-days, $p < 0.005$). Movements were 21% faster when the acid lakes were well stocked with fish ($t = 3.2$, d.f. = 78 duckling-days, $p < 0.005$).

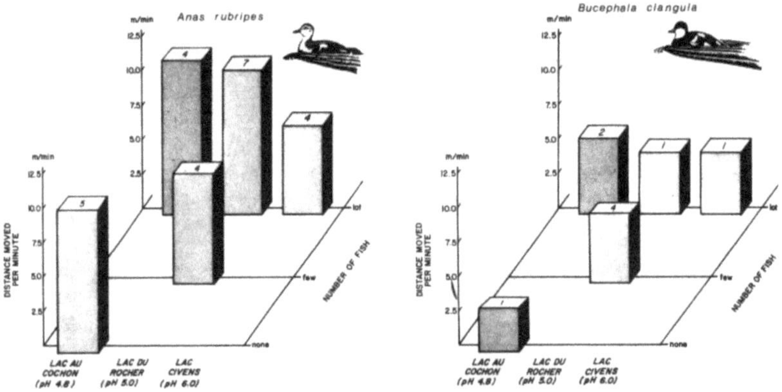

Figure 1. Influence of acidity and abundance of fish on movements of Black Duck and Common Goldeneye ducklings.

In general, average speed of surface movement by Common Goldeneye ducklings varies little as a function of lake acidity or abundance of fish (Figure 1). Nevertheless, it should be noted that weight of the ducklings in each brood for each day after their birth (DesGranges, 1985). Since examination of the individual weight growth curves showed that weight gain was nearly constant from the time the ducklings reached the lakes (i.e. 1 week), we applied the linear function (y = ax + b) to curve segments following this age to obtain slope measurements (i.e. mean growth rate in g d^{-1}) which were directly comparable from lake to lake and from species to species (Figure 3).

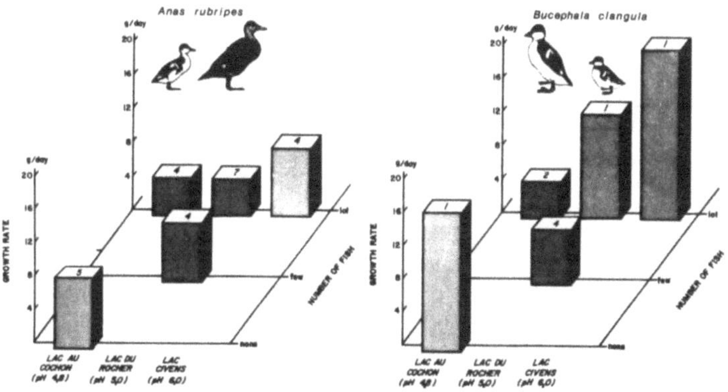

Figure 3. Influence of acidity and abundance of fish on growth rate of Black Duck and Common Goldeneye ducklings.

The Black Duck young reared on the well-stocked acid lakes grew 60% more slowly than when there were few or no fish (3.9 VS 6.4 g d^{-1}). The difference was particularly pronounced for Lac au Cochon (7.6 to 3.9 g d^{-1}; $t = 4.9$, d.f. = 5 ducklings, $p < 0.005$) (Table I). Meanwhile, on Lac du Rocher, the growth rate was higher (3.8 to 5.6 g d^{-1}; $t = 4.9$, d.f. = 4 ducklings, $p < 0.005$) when nearly 750 trout were removed from the lake. On the circumneutral lake, the growth rate was 7.6 g d^{-1} despite the presence of a healthy fish population, twice that exhibited by young Black Duck living on the well-stocked acid lakes (3.9 g d^{-1}; $t = 3.7$, d.f. = 8 ducklings, $p < 0.005$).

TABLE I

Influence of acidity and presence of fish on growth rates of Black Duck ducklings

	Lac au Cochon (pH: 4.8)	Lac du Rocher (pH: 5.0)	Lac Civens (pH: 6.0)
1984	7.6 ± 0.8 g d^{-1} *	3.8 ± 0.1 g d^{-1}	7.6 ± 1.2 g d^{-1}
	▽ Fish added ▽	▽ Fish removed ▽	
1985	3.9 ± 0.4 g d^{-1}	5.6 ± 0.2 g d^{-1}	

* Mean $\pm$ 1 standard error

The Common Goldeneye ducklings maintained a very high growth rate on both the circumneutral lake with fish (19.4 g d^{-1}) and the acid lake without fish (15.7 g d^{-1}). The single duckling reared on Lac du Rocher when it was well stocked with fish had been studied a week earlier on Lac du Cochon, which was then fishless. This duckling probably experienced difficulty feeding, since its growth rate dropped by one quarter (11.7 g d^{-1}) when it was transferred from Lac au Cochon to Lac du Rocher. The growth rate for the Goldeneye ducklings released on Lac du Rocher once many of the fish had been removed was more than one and a half that exhibited by the Goldeneye ducklings reared on Lac au Cochon after it had been stocked (5.5 VS 3.7 g d^{-1}; $t = 1.4$, d.f. = 4 ducklings, $p < 0.13$). The considerable drop (-500%) in the growth rate of the Goldeneye young on Lac au Cochon following its stocking with brook trout is particularly striking (3.7 $\pm$ 0.5 g d^{-1}, maximum = 4.2 g d^{-1}, n = 2 ducklings VS 15.7 g d^{-1}, n = 1 duckling).

4. DISCUSSION

Most of the results obtained to date suggest acid lakes with a good fish population (in the case of brook trout, at least) are detrimental to the development of Black Duck and Common Goldeneye ducklings. Young ducks living on such lakes usually move more quickly and devote more time to searching for food than those living on circumneutral lakes or fishless acid lakes. In general, their growth rate is also lowest when the presence of fish combines with strong acidity.

Epibenthic invertebrates are often less plentiful in acid lakes (Raddum, 1980; Collins _et al._, 1981) and their population levels may be too low to support both fish and ducks. In lakes having a high alkalinity (and therefore lower acidity), metamorphosis of some groups of insects is generally more successful (Bell, 1971), and the abundance of reproducing adults in the vicinity probably ensures the continued presence of large populations of aquatic insect larvae in these lakes despite predation by fish (Moyle, 1961). There should thus be enough food to allow ducklings to develop provided it is accessible (i.e. the lakes contain large sections of shallow water) and the other nesting requirements of the various duck species are met. If prey become scarce once the emergence period is over, the ducklings have the option of moving or feeding to a greater extent on plant matter if it is sufficiently abundant (Pehrsson, 1984).

Hunter et al. (1986) obtained results which seem partially incompatible with ours. The Black Duck ducklings they reared on two acidic ponds gained weight faster than ducklings on two circumneutral ponds. Differences in feeding behavior also implied that the acidic ponds had more duckling food and that ducklings on acidic ponds found food more readily. Hunter and we concluded that the Maine lakes were probably of generally lower quality as brood rearing habitat than the Quebec lakes. This would explain why the circumneutral, well-stocked lakes studied in Maine were not suited to the growth of Black Duck young (< 3.3 g d^{-1}) and why the Maine fishless acid lakes, for which the maximum growth rate was 5.3 g d^{-1}, were only slightly better duckling habitat. Therefore the threshold of invertebrate abundance at which fish competition becomes important was probably reached on the Maine circumneutral lakes but not on the Quebec circumneutral lake.

Pehrsson (1979, 1984) studied the feeding ecology of Mallard ducklings imprinted to domestic females. He noted that ducklings reared on fishless lakes found more food (more abundant and bigger preys) than ducklings reared on lakes containing fish. The study does not indicate whether the ducklings reared on lakes with fish were hungry, however, and growth rates were not measured. Nor does it provide information on the role of acidity; nearly all the lakes studied were acid.

A number of other studies have shown that some adult ducks prefer lakes with few or no fish. Examples include the Oldsquaw (Clangula hyemalis) (Pehrsson, 1974), the Common Goldeneye (Eriksson, 1979, 1983; Eadie and Keast, 1982; DesGranges 1985; DesGranges and Darveau, 1985), and sometimes the Mallard (yes, according to Pehrsson (1979, 1984); no, according to Eriksson (1983)). In most cases, lake acidity also appears to play a role, although no one has been able to demonstrate clearly its effect. Our study is a first step in this direction.

5. ACKNOWLEDGMENTS

By way of acknowledgment and thanks, we would like to stress the admirable work performed by all the "mother ducks" under the supervision of Alain Cossette, Serge Guimond, and Gaétan Morissette. We would also like to express our gratitude to Jacques Brodeur and Suzelle Giroux who processed some of the data, and Guy Michaud who prepared the figures. Finally, we thank M.O.G. Eriksson, M.L.

Hunter, Jr., and S.J. Ormerod for critical reviews of the manuscript. This work was partially funded by the federal Long-Range Transport of Airborne Pollutants (LRTAP) Program.

6. REFERENCES

Bell, H.L.: 1971, Water Research 5, 313.

Collins, N.C., Zimmerman, A.P., and Knoechel, R.: 1981, 'Comparison of benthic infauna and epifauna in acidified and nonacidified Ontario lakes'. In R. Singer (ed). Effects of acid precipitation on benthos. Proceedings of a regional symposium on benthic biology. North American Benthological Society. pp. 35-48.

DesGranges, J.-L. (ed): 1985, L'acidité des lacs et les canards: première étape. Technical report prepared under the "Environment 2000" program by Ducks Unlimited and the Canadian Wildlife Service. xii + 152 p.

DesGranges, J.-L. and Darveau, M.: 1985, Holarctic Ecology 8, 181.

DesGranges, J.-L.: 1986, in prep.

Dupont, J.: 1984, Nature et étendue de l'acidification de 97 lacs de la région de Portneuf. Service de la qualité des eaux, Ministère de l'Environnement du Québec. Publ. n° PA-10, 113 pp. + 6 annexes.

Eadie, J.M. and Keast, A.: 1982, Oecologia 55, 225.

Eriksson, M.O.G.: 1979, Oecologia 41, 99.

Eriksson, M.O.G.: 1983, Wildfowl 34, 27.

Eriksson, M.O.G.: 1984, Ambio 13, 260.

Haines, T.A.: 1981, Trans. Amer. Fish. Soc. 110, 669.

Hunter, M.L., Witham, J.W., and Jones, J.J.: 1985, Techniques for using the growth and behavior of imprinted ducklings to evaluate habitat quality. Maine Agricultural Station Technical Bull. N° 117. 18 pp.

Hunter, M.L., Jones, J.J., Gibbs, K.E., and Moring, J.R.: 1986, Oikos 47, in press.

Moyle, J.B.: 1961, Aquatic invertebrates as related to larger water plants and waterfowl. Minnesota Dept. Conserv. Investigational Report N° 233: 24 pp.

Pehrsson, O.: 1974, Int. Congr. Game Biol. 11, 259.

Pehrsson, O.: 1979, Viltrevy 10, 193.

Pehrsson, O.: 1984, J. Wildl. Manage. 48, 322.

Raddum, G.G.: 1980, 'Comparison of benthic invertebrates in lakes with different acidity'. In: D. Drablos and A. Tollan (eds). Ecological impact of acid precipitation. Proc. Int. Conf. Ecol. Impact Acid Precip. Norway, SNSF project. p. 330.

Mortality and physiological stress of year-classes of landlocked and migratory Atlantic salmon, brown trout and brook trout in acidic aluminium-rich soft water.

Rosseland, B.O.[1)], Skogheim, O.K.[1)], Kroglund, F.[1)] and E. Hoell[2)]
Directorate for Nature and Management, Fish Research Division, Tungasletta 2, N-7000 Trondheim, Norway.
1) Present address: Norsk Biotech A/S, P.O. Box 788, Krossen, N-4301 Sandnes, Norway. 2) Norcem A/S, N-3950 Brevik

ABSTRACT. Physiological stress, measured as changes in plasma chloride, and mortality were measured on different year-classes of landlocked and migratory Atlantic salmon, two strains of brown trout, and brook trout, in a flow-through system with acidic Al-rich soft water. The oldest year-classes of salmon were smolts. Water from Lake Byglandsfjord (pH = 5.9), was enriched with inorganic Al (as $AlCl_3$) and H_2SO_4 to pH = 5.1, total Al = 225 ug L^{-1}, and labile Al = 135 ug L^{-1}. As a reference, lake water was limed by means of a shellsand filter to pH = 6.2, increasing Ca-concentration from 1.0 to 1.5 mg L^{-1}. During the 83 hr experiment, neither mortality nor physiological stress occurred in any species or year-class in the limed water. In the acid water, no mortality occurred on any stage of brown trout or brook trout. Among the migratory and landlocked salmon, however, 5% of the alevins died after 49 and 70 hr, respectively. All smolts of both the landlocked and the migratory salmon died after 83 and 35 hr, respectively, the corresponding loss rate of plasma chloride was -0.76 and -1.26 meq Cl hr^{-1}. Brook trout, however, increased plasma ion concentration during the experimental period, and hence showed no stress.

1. INTRODUCTION

Between species and year-classes of salmonids, the smoltification period of migratory Atlantic salmon (*Salmo salar* L.) seems to be the most sensitive to acidic stress (Rosseland and Skogheim, 1984). The population of landlocked Atlantic salmon in the acidified Lake Byglandsfjorden, southern Norway, has declined during the last decades. In an experiment at Lake Byglandsfjorden, fingerlings of landlocked Atlantic salmon were found to be more sensitive than parr of brown trout (*Salmo trutta* L.) and brook trout (*Salvelinus fontinalis* Mitchill) (Skogheim *et al.*, 1984).

In this study, different year-classes of : 1) Atlantic salmon (landlocked and migratory), 2) two strains of brown trout (one strain being genetically selected towards a higher acid resistance ("Genetic", Gjedrem, 1980), and one strain from Lake Byglandsfjorden), and 3) brook

Water, Air, and Soil Pollution **30** (1986) 751–756.

trout, were exposed to acidic Al-rich soft water. Limed lake water served as reference. Comparison of smolts of the two salmon strains were given special attention. Mortality and changes in plasma chloride were recorded.

2. MATERIAL AND METHODS

The experiment was performed at the Bygland Fiskeanlegg, county of Aust-Agder, southern Norway, in the period May 8-11, 1984. Inorganic Al (as $AlCl_3$) and H_2SO_4 were added to water from Lake Byglandsfjorden (LW, pH = 5.9), by use of a peristaltic pump (Ismatech IP-12) in a continous flow-through system (acidified LW). The added chemicals were mixed by air flushing for 10 min prior to entering the fish tanks. Lake water limed by means of a shellsand filter served as reference water (limed LW).

Eyed eggs and 1-yr old smolts of migratory Atlantic salmon of River Imsa strain, were transported from the Research Station for Freshwater Fish, Ims, southwestern Norway, two weeks prior to the experiment. Eggs from the same strain had been fertilized and hatched in Lake Byglandsfjord waters, and were used as alevins in this experiment. The landlocked salmon, the two strains of brown trout, and the brook trout, had been raised from eggs in the Bygland Fiskeanlegg. The smolts of landlocked salmon were 2-yr old. Feeding was stopped two days before the experiment.

The fish were placed in a flow through system at 4°C, with black PVC-tanks with a volume of 30 L and retention time of 30 min. A semi-transparent black cover prevented visual disturbances. Separate tanks were used for mortality and blood sampling experiments. For mortality studies, 10 fish of each year-class of each species were placed in a single tank; for blood sampling, 20 fish per tank of each species were used. The 1-yr old brown trout was too small for blood sampling. The mean length (cm) ±SD, weight (g) ±SD and number of fish (n) in the different groups used for blood sampling were: migratory salmon 13.5 cm ± 2.5, 21.5 g ± 3.7 (33), landlocked salmon 14.6 cm ± 0.8, 25.5 g ± 5.4 (25), and brook trout 12.6 cm ± 6.0, 17.4 g ± 2.8 (31), respectively. Eggs and alevins were kept in floating net-cages made of plastic (mesh size 2 mm).

At preset intervals, tanks were checked for dead fish. Blood samples were drawn by heart puncture and analyzed as described by Rosseland and Skogheim (1984). Each fish gave rise to only one sample.

Water samples were taken on 16 occasions in the acidified water and on 10 occasions in the limed water. pH was measured immediately (Orion mod. 207), and aliquots prepared for Al analyses. All analytical methods are identical to those described by Skogheim and Rosseland (1984).

3. RESULTS

pH in the acidified lake water was fairly constant during the 83 hr experiment, with a mean value of 5.14 (Table I). Calcium concentration was low, 1.02 mg L^{-1}, with no bicarbonate alkalinity. The concentration

TABLE I. Mean value (M) and standard deviation (SD) for some chemical components in the acidified, Al-enriched lake water (LW), and in the limed lake water, N = number of observations. Dissolved organic matter (DOM) was estimated as absorbance of UV-light (at 254 nm).

		ACIDIFIED LW		LIMED LW	
COMPONENTS		M ± SD	N	M ± SD	N
pH		5.14 ± 0.03	16	6.18 ± 0.05	10
Ca	mg L^{-1}	1.02 ± 0.04	16	1.53 ± 0.05	10
Bicarbonate-alkalinity	ueq L^{-1}	0 ± 0	16	35	10
Acid-reactive Al	"	228 ± 5	16	111 ± 4	10
Non-labile Al	"	93 ± 4	16	92 ± 5	10
Labile Al	"	134 ± 8	16	20 ± 5	10
DOM	cm^{-1}	0.039 ± 0	16	0.039 ± 0	10

of acid-reactive Al was 228 ug L^{-1}, with 143 ug L^{-1} in the labile Al form. Fluoride was 46 ug L^{-1} for random samples.

In water which passed through the shellsand filter, pH increased to a mean of 6.18, Ca to 1.53 mg L^{-1}, and the bicarbonate alkalinity was 0.035 meq L^{-1} (Table I). The concentration of acid-reactive Al was 111 ug L^{-1} and 20 ug L^{-1} was in the labile Al form. The concentration of dissolved organic matter (DOM) was low and the same in the two waters. Water colour estimated against a Pt-standard was lower than 5 mg Pt L^{-1} in all samples.

The estimated equilibrium concentrations of labile Al with respect to a solid phase with similar solubility as synthetic gibbsite, indicate that in the acidified lake water there was a slight undersaturation (Table II). In the limed lake water, however, a slight oversaturation occurred. In the acidified water, the AlF^{2+} was the dominating Al-species, whereas aluminate ion ($Al(OH)_4^-$), Al hydroxide ($AL(OH)_2^+$) and Al trihydroxide ($Al(OH)_3^0$) were present in the limed water (Table II).

During the 83 hr experiment, no mortality occurred in any species or year-class in the limed water. In the acidified lake water, no mortality occurred on any stage or strain of brown trout or brook trout. However, both the migratory and landlocked salmon experienced a 5% mortality of alevins after 49 and 70 hr, respectively. One parr (10%) of the one-year old landlocked salmon died after 83 hr. Both the migratory and landlocked salmon smolts experienced 100% mortality after 35 hr (L_{T50} = 25 hr) and 83 hr ($L_{T\ 50}$ = 70 hr), respectively.

At the start of the experiment, plasma chloride concentration differed between groups; mean concentration ± SD in landlocked Atlantic salmon was 107 ± 1 meq L^{-1} (n = 3), versus 99 ± 8 meq L^{-1} (n = 8) for migratory salmon and 96 ± 9 meq L^{-1} (n = 4) for brook trout, respectively (Figure 1A).

Both the migratory and landlocked salmon lost plasma chloride at a rapid rate after exposure to the acidified lake water, Δ Cl = -1.22 and -0.76 meq hr^{-1}, respectively. Brook trout, however, increased plasma

TABLE II Analyzed labile Al and calculated Al-speciation in acidified lake water (LW) and limed LW. All concentrations in ug Al L^{-1}.

	ACIDIFIED LW	LIMED LW
Analyzed labile Al	134	20
Distribution of Al species		
AL^{3+}	22	0
$Al(OH)^{2+}$	28	0
$Al(OH)_2^{+}$	29	4
$Al(OH)_4^{-}$	0	7
AlF^{2+}	48	0
AlF_2^{+}	6	0
$AlSO_4^{+}$	1	0
Inorganic monomeric Al	134	11
$Al(OH)_3^{0}$	0	9
Inorganic Al	134	20
Equilibrium concentration of inorganic monomeric Al	161	11

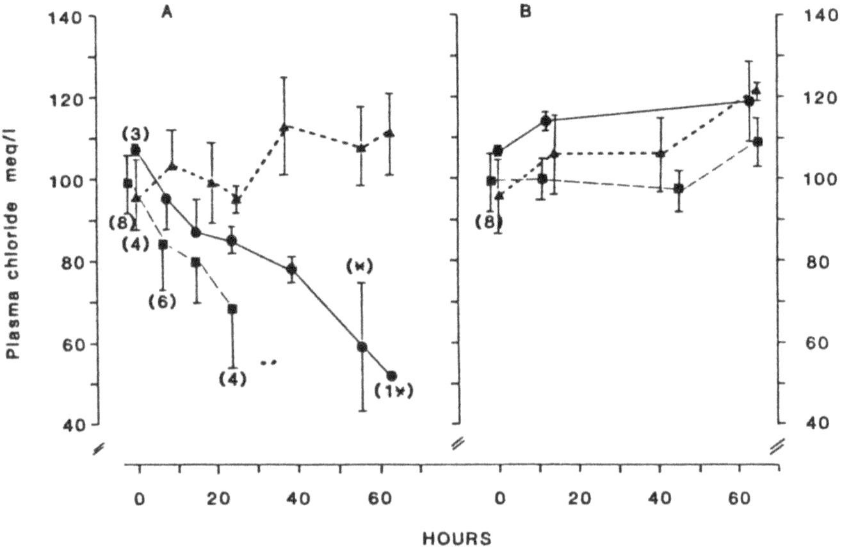

Fig. 1. Plasma chloride of migratory and landlocked Atlantic salmon and brook trout in A) acidified lake water and B) limed lake water. Mean (n=3 if not otherwise stated) and SD are given. Atlantic salmon: migratory (■---■), landlocked (●——●). Brook trout (▲---▲).

chloride concentration during the first 37 hr, $\Delta Cl = 0.32$ meq hr^{-1} and stabilized around a mean of 116 meq L^{-1} (Figure 1A). In the limed water, the levels of plasma chloride in all groups increased during the experiment (Figure 1B).

4. DISCUSSION

As to the equilibrium of labile Al, the acidified water was slightly undersaturated with respect to a solid phase with similar solubility as synthetic gibbsite, in contrast to the slight oversaturation in the limed lake water. In laboratory experiments, Baker and Schofield (1982) observed that oversaturated Al solutions were particularly toxic to salmonids. In connection with liming, however, such oversaturated solutions seems not to cause any stress (Rosseland et al., 1986).

No mortality occurred among the two strains of brown trout or the brook trout, whereas all stages (except eyed eggs) from alevins to parr and smolts of the two Atlantic salmon strains experienced different degrees of mortality. This species difference is in accordance with other studies (Grande et al., 1978; Rosseland and Skogheim, 1984; Skogheim et al., 1984; Fivelstad and Leivestad, 1984). The fact that 100% mortality only occurred in smolts supports the observation that salmon, either migratory or landlocked, are most sensitive during the smoltification period (Rosseland and Skogheim, 1982, 1984). The loss of plasma chloride of smolts in this experiment was more rapid than at comparable pH and labile Al-concentrations described by Rosseland and Skogheim (1984). The lower Ca- concentration in the experiment reported here (1.0 vs 2.0 mg Ca L^{-1}), is likely to be the explanation for this.

The species difference in sensitivity between brook trout and Atlantic salmon is clearly illustrated by the salmon loosing chloride rapidly whereas the brook trout gained chloride.

Handling and confinement have been shown to induce loss of plasma chloride in salmonids (Wedemeyer, 1972; Redding and Schreck, 1983). The fact that both salmon and brook trout plasma chloride concentration increased in the limed lake water, however, indicates that confinement in the experimental fish tanks are not likely to have caused any osmotic stress.

The migratory salmon had a subnormal level of plasma chloride at the experimental start, maybe due to stress caused by handling and transportation (Soivio and Virtanen, 1984). This might explain the more rapid ion loss and mortality compared to the landlocked salmon. Genetics, however, have been shown to play some role in the tolerance to acid waters in strains of brown trout (Gjedrem, 1980; McWilliams, 1982) and brook trout (Robinson et al., 1976). Acclimation, caused by raising of the landlocked salmon in the moderate acidified Lake Byglandsfjord, might also be of importance, as McWilliams (1980a,b), Schofield et al. (1981) and Fivelstad and Leivestad (1984) all found acclimation to be relevant within certain limits. Which of these factors, or combination were the cause of the observed difference in tolerance between landlocked and migratory salmon cannot be determined from this study.

REFERENCES

- Baker, J.P. and Schofield, C.L.,: 1982, Water, Air, Soil Pollut. 19, 289.
- Fivelstad, S. and Leivestad H,: 1984, Rep. Inst. Freshw. Res. Drottningholm 61, 69.
- Gjedrem, T.: 1980, 'Genetic variation in acid tolerance in brown trout', in: Drabløs, D. and Tollan A, (eds.), Ecological impact of acid precipitation, SNSF-project, pp. 308.
- Grande, M., Muniz, I.P., and Andersen S,: 1978, Verh.Int.Verein. Limnol. 20, 2076.
- McWilliams, P.G.: 1980a, J. Exp. Biol. 88, 259.
- McWilliams. P.G.: 1980b, J. Exp. Biol. 88, 267.
- McWilliams, P.G.: 1982, Comp. Biochem. Physiol. 72a, 515.
- Redding, J.M. and Schreck, C.B.,: 1983, Trans. Am. Fish. Soc. 112, 800.
- Robinson, G.D., Dunson, W.A., Wright, J.E., and Mamolito, G.E.,: 1976, J. Fish. Biol. 8, 5.
- Rosseland, B.O. and Skogheim, O.K.: 1982, 'Physiological stress and mortality of Atlantic salmon, Salmo salar L. in acid water with high levels of aluminium', Inter. Council for Exploration of the Sea C.M. 1982/M: 29, 15 p.
- Rosseland, B.O. and Skogheim, O.K.: 1984, Rep. Inst. Freshw. Res. Drottningholm 61, 186.
- Rosseland, B.O., Skogheim, O.K., Abrahamsen, H. and Matzow, D.: 1986, 'Limestone slurry reduces physiological stress and increases survival of Atlantic salmon (Salmo salar L.) in an acidic Norwegian river', Can. J. Fish Aquat. Sci. 43, in press.
- Schofield, C.L., Webster, D.A., Guthrie, G.A. and Flick, W.A.: 1981, 'Acclimation of brook trout to acidified waters', in: Management of acidified waters. Contr. C-164480, N.Y. State Dept. Environm. Cons. and Cornell Univ., N.Y.
- Skogheim, O.K. and Rosseland, B.O.: 1984, Rep. Inst. Freshw. Res. Drottningholm 61, 17.
- Skogheim, O.K., Rosseland, B.O., Hafsund, F., Kroglund, F. and Hagenlund, G.: 1984, 'Eksponering av bleke, aure og bekkerøye i Surt vann' (Exposure of landlocked salmon, brown trout and brook trout in acid water), Rapport fra Fiskeforskningen, 2/84, 14p. (In Norwegian with English summary).
- Soivio, A. and Virtanen, E.: 1984, 'Physiological effects of stocking stress on Baltic salmon (Salmo salar)', EIFAC Technical Paper No 42, Suppl. Vol. 1, 217.
- Wedemeyer, G.: 1972, J. Fish Res. Board Can. 29, 1780.

EFFECTS OF DIETARY ALUMINUM IN COMBINATION WITH REDUCED CALCIUM AND PHOSPHOROUS ON THE RING DOVE (*Streptopelia risoria*).

D. Carrière, K. Fischer, D. Peakall and P. Angehrn
Canadian Wildlife Service, National Wildlife Research Centre
Environment Canada, Ottawa, Canada, K1A 0E7.

ABSTRACT. Elevated levels of Al have been reported in surface waters and in certain aquatic organisms including aquatic insects in areas receiving acidic precipitation. It has been suggested that Al can affect avian reproduction. In this study, Ring doves were assigned to a control group fed a diet reduced in Ca and P and to a treated group fed the same diet supplemented with 0.1% Al. A four month dosing period did not result in any effect on egg production, fertility or hatchability. Eggshell permeability was decreased initially but subsequently recovered to a normal level. Dietary aluminum sulphate did not affect plasma Ca, P nor Mg levels of adults nor did it affect the pattern of growth nor the final weight of chicks fed on the same diet. There was a tendency for Al to accumulate in the sternum of growing doves fed levels up to 1500 ppm Al. Neither reproduction nor growth of this species were influenced by levels of Al equivalent to those occuring in areas receiving acid deposition. However, the results of a preliminary experiment suggest that decreasing the dietary levels of Ca and P below 0.9% and 0.5% respectively, in combination with excess Al (0.075%) caused significant impacts upon reproduction and growth. Similar experimental findings have been reported in the literature for domestic birds where levels of dietary Al exceeded those of Ca and P.

RESUME. Des niveaux élevé d'Al ont été détecté dans les eaux de surfaces et dans certains insectes aquatiques vivant dans des régions arrosée par les pluies acides. Certains travaux suggèrent que l'Al pourrait affecter la reproduction des oiseaux.

Nous examinons ici l'effet de l'Al (0.1 %) chez des tourterelles à collier placées sur un régime à teneure réduite en Ca et en P. Après 4 mois d'exposition à ce régime aucun effet sur la production d'oeuf, la fertilité ou le taux d'éclosion n'a été noté. La perméabilité de la coquille d'oeuf dans le groupe traité a d'abord diminué mais a ensuite rejoint le groupe contrôle.

Water, Air, and Soil Pollution **30** (1986) 757–764.

La présence d'Al dans le régime n'a pas affecte les niveaux de Ca, de P et de Mg dans le plasma des adultes et n'a pas affecté non plus la forme de la courbe de croissance et le poids final des poussins maintenus sur ce même régime. L'Al semble avoir été accumulé dans les os des jeunes tourterelles nourries à des niveaux allant jusqu'à 1500 ppm d'Al. Cependant, ni la reproduction ni la croissance n'a été affecté par l'Al dans la diète même à des niveaux similaires à ceux qu'entrainent les précipitations acides. Les résultats d'une étude préliminaire suggérent que l'Al (0.075%) administré avec une diête dont la teneur en Ca et en P est inférieure à 0.9% et 0.5% respectivement, pourrait affecter la reproduction et la croissance. Des résultats similaires ont été rapporté pour le poulet domestique nourrie d'une diête dont la teneure en Al dépasse la teneure en Ca et en P.

1. INTRODUCTION

Increased Al concentrations in surface waters due to acid precipitation have stimulated interest in the toxicity of this element to aquatic biota. Elevated levels of Al have been reported in emerging insects and in tissues of wild birds in Sweden (Nyholm and Myhrberg, 1977). Aluminum was found in the bone marrow of the pied flycatcher (Ficedula hypoleuca) nesting in areas adjacent to an acidified lake and was associated with impaired reproduction (Nyholm, 1981). The relative toxicity of dietary Al compared with other metals is low, the principal effect being the disturbance of P metabolism (Sorenson et al., 1974). Under normal conditions Al is little absorbed through the intestinal epithelium (Cam et al., 1976). In the alimentary canal, Al forms insoluble compounds with P resulting in an imbalance in the ratio of available Ca and P (Ondreicka et al., 1971). Absorption of Al along with P has been stimulated in chickens by injecting 1,25-$(OH)_2D_3$, a derivative of vitamin D (Long et al., 1980). In most vertebrates, the synthesis of this substance is mediated by the parathyroid hormone and normally produced by the kidneys when the need for Ca increases or when there is a shortage of dietary Ca. The production of 1,25-$(OH)_2D_3$ subsequently increases the synthesis of intestinal calcium-binding proteins which augment the absorption of Ca (Haussler and McCain, 1977; Norman, 1979). These proteins also increase the absorption of metals bound to the essential minerals including Pb (Shields and Mitchell, 1941; Six and Goyer, 1970) and Cd (Washko and Cousins, 1977). It has been previously noted that the absorption of many metals including Zn, Cd, Cs, and Co is enhanced when the level of vitamin D is high and when the parathyroid activity is increased (eg. during egg-laying) (Worker and Magicovsky, 1961a,b). Mayor et al., (1977) found a similar relationship between the level of parathyroid hormone and the gastrointestinal absorption of Al in the rat.

Birds stressed by a diet low in Ca and P may therefore be more sensitive to the presence of Al in the diet or may accumulate it to a greater extent. In this experiment we attempted to assess the effect of Al in concentrations similar to those likely to be encountered in prey of insectivorous birds from acid-stressed areas. The levels of dietary Ca and P were reduced from those available in standard commercial diets in an attempt to mimic the stress occuring in the environment. Sufficient Ca and P were provided to enable breeding and egg-laying.

2. MATERIALS AND METHODS

The measured parameters were chosen to identify the early signs of Al intoxication so that a means of monitoring the effects of Al stress in wild birds could be developed. Breeding pairs of Ring doves (*Streptopelia risoria*) were divided into two groups of twenty pairs each and both groups were fed a diet in which the Ca and P levels were set at 0.9 % and 0.5 % respectively. For the treated group, Al given as $Al_2(SO_4)_3.18H_2O$ was added to the diet in the concentration of 1000 ppm (0.1%). Aluminum sulphate was chosen because it is the most common form of Al in acidified waters (LaZerte, 1984) and soluble compounds of Al were found to be more toxic than insoluble forms to growing chicks (Storer and Nelson, 1968).

Both control and treated groups were divided into two equal subgroups of ten pairs each, in order to study in undisturbed conditions, the breeding productivity and growth in one subgroup and the egg permeability in the other subgroup.

In another study, also designed to examine the growth of juvenile ring doves exposed to dietary Al, a total of thirty 21 day-old chicks were divided into 4 groups. The control group was fed a purified diet modified to contain 0.6% Ca and 0.3% P. The other groups were given the same diet with respectively, 500 ppm, 1000 ppm and 1500 ppm Al mixed as $Al_2(SO_4)_3.18H_2O$.

3. RESULTS AND DISCUSSION

With the exception of a decline in moisture loss observed for the first eggs laid in the treated group (but followed by a recovery), none of the parameters of reproduction were significantly affected by the dosing levels of Al used here. No embryonic malformations were observed. The examination of two eggshells from each of the control and treated groups with a scanning electron microscope indicated that the number of pores and the pore area were similar between treatment groups, although one of the experimental shells had a low permeability index. Nevertheless, our findings do not confirm the field observations of Nyholm (*op. cit.*) of increased moisture loss and embryonic malformations.

Plasma Ca, P and Mg levels were not significantly affected by treatment. The higher levels of Ca and P in the plasma of females as compared with males were expected since they all had an egg in the oviduct when they were sacrificed. In our experiment, females from the

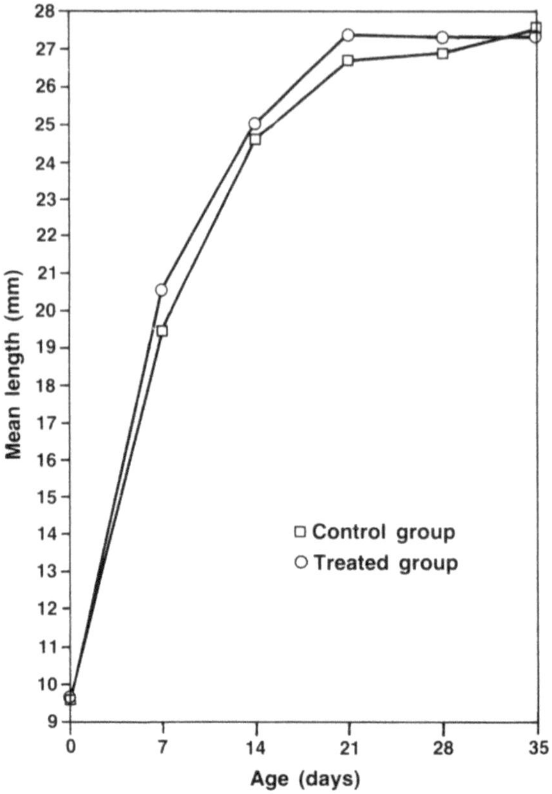

Fig.1
Growth curves of mean tarsus length of juvenile doves fed a control diet or 1000 ppm Al from 0 - 35 days.

treated group had plasma Ca levels slightly lower than control females.

Accumulation of Al occurred in at least the bones ($P<0.003$) and possibly in the kidneys of adult females (Table I). Aluminum concentrations in the pooled leg and wing bones of growing juvenile doves were not significantly affected by treatment ($P>0.05$) but a significant trend ($P<0.0136$) was found for the bones of the sternum to accumulate dietary Al (Table II). For the most part, Al was accumulated by bones of high P content. This was probably due to the tendency of Al to form stable aluminum phosphates (Nofre et al., 1963).

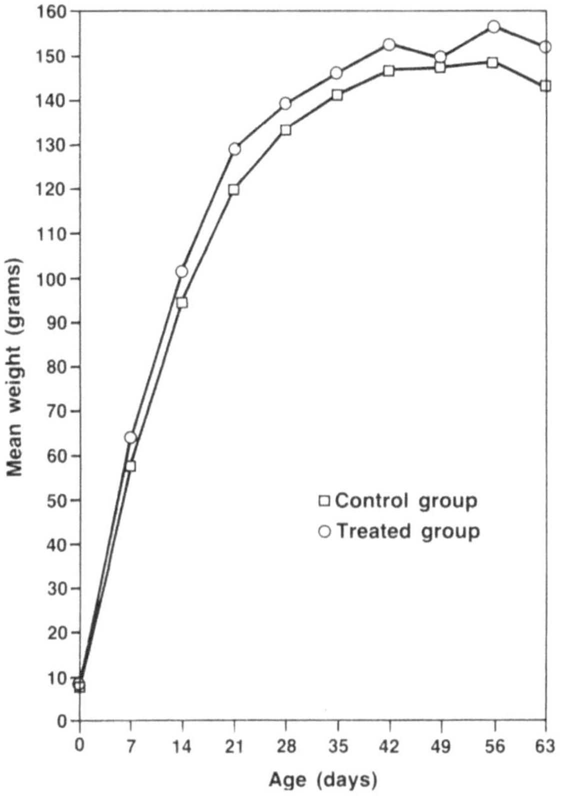

Fig.2
Mean weight increase in juvenile doves fed a control diet or 1000 ppm Al from 0 - 63 days.

In our experiment, Al treatment did not affect the growth of the doves (Figure 1). None of the dosing levels used affected the asymptotic weight or the rate of weight gain (Figure 2). For the juvenile doves fed various levels of Al, Ca and P, contents of bones were not significantly affected by treatment (Table II). The Ca:P ratio in bones was 2.1 for each of the 4 treatment groups in both the sternum and pooled leg and wing bones.

The levels of Ca and P administered in this experiment were higher than those of Al; it is considered that the effect of Al would have been greater if Ca and P conditions were deficient for growth and reproduction (Simkiss, 1961). In the course of a preliminary experiment with Ca, P and Al concentrations of 0.5%, 0.3% and 0.075%, respectively, eggshells were easily broken and crushed upon handling;

Table I. Aluminum content (µg g^{-1} on a dry weight basis) in femurae, kidneys and brain of adult ring doves.

Group	Tissue	Males	Females
Control	femurae	7.87 ± 2.51 (6/6)a	7.42 ± 2.14 (6/6)
	kidneys	0.75 (1/6)	0.97 (1/6)
	brain	n.d.	n.d.
Treated	femurae	6.75 ± 1.66 (6/6)	15.87 ± 10.30 (6/6)
	kidneys	0.92 ± 0.13 (5/6)	1.92 ± 0.47 (4/6)
	brain	n.d.	n.d.

a: mean ± s.d. (number above the detection limit out of 6 replicates)
n.d.: not detectable.

Table II. Calcium, P and Al levels (µg g^{-1} on a dry weight basis) in bones of growing ring doves administered four dietary levels of aluminum sulphate.

		Sternum	Pooled legs and wings bones
	Ca	169000 ± 17700 (12)a	137000 ± 16000 (12)a
Control	P	79800 ± 8700 (12)	64900 ± 7500 (12)
	Al	9.9 ± 3.3 (12)	6.6 ± 1.7 (12)
	Ca	158000 ± 21000 (6)	141000 ± 9300 (6)
500ppm	P	75700 ± 9300 (6)	67700 ± 7500 (6)
	Al	10.6 ± 4.2 (6)	7.0 ± 1.6 (6)
	Ca	160000 ± 29000 (7)	129000 ± 12500 (7)
1000ppm	P	76600 ± 13200 (7)	62000 ± 5300 (7)
	Al	11.3 ± 3.9 (7)	6.2 ± 1.4 (7)
	Ca	182000 ± 23000 (5)	143000 ± 17700 (5)
1500ppm	P	86400 ± 10600 (5)	69000 ± 8500 (5)
	Al	15.9 ± 6.7 (6)	6.7 ± 1.6 (5)

a: mean ± s.d. (n)

egg-laying was reduced and chick mortality was close to 100%. With the adoption of higher levels of Ca and P, no further effects were found, despite an increase in the Al added to the diet. The findings of this preliminary experiment are supported by other studies reported in the literature confirming the importance of Ca and P levels in determining Al absorption and toxicity to birds (Lipstein and Hurwitz, 1982).

4. CONCLUSIONS

No significant influence of dietary Al (0.1%) on reproduction or growth of Ring doves was observed in a laboratory experiment with adequate levels of Ca (0.9%) and P (0.5%) in the diet. The levels of Al in the diet were similar to those found in insect prey from areas impacted by acid precipitation (Hall and Likens, 1981). The difference between field observations and our experimental results may be due to species-to-species variation or the relative availability of Al and essential minerals in the diet.

Preliminary experiments with doves on diets with low levels of Ca (0.5%) and P (0.3%) in combination with Al (750 ppm) resulted in substantial effects on reproduction and growth. These findings are consistent with the literature which indicates that the ratio of Ca:P:Al is critical because of the importance of Ca and P in egg formation and chick development; the role of Al in binding P and displacing Ca in bone could have severe implications for reproduction and growth (Jones, 1938; Sorenson, op cit.).

Many areas subjected to acidification and mobilization of Al are also areas where the Ca and P levels in surface waters are low. Under these conditions, it is possible that the availability of Al could exceed those of the essential minerals in prey organisms and thereby cause reproductive impairment in avian predators.

ACKNOWLEDGEMENTS

We wish to thank G. Fox for his advice in setting up the experiment, B. Collins and B. Nagpal of the Canadian Wildlife Service for their assistance in the statistical analysis of the data; E. Hayakawa compiled and analyzed the data on reproductive performance; B. Loescher of the Ontario Ministry of Environment and F. Leung of the University Hospital of Western Ontario analyzed tissue samples; B. Petruk of the Department of Energy, Mines & Resources examined the eggshells using scanning electron microscopy; and finally we are grateful to D. Whitworth, A. Jacobsen, A. Harfenist and D. Jeffrey for their technical assisitance and M. Lis of the Canadian Wildlife Service for critically reviewing the manuscript.

REFERENCES

Cam, J.M., Luck, V.A., Eastwood, J.B. and H.E. de Wardener : 1976, Clin. Sci. Mol. Med. 51:407-414.

Hall, J.R. and G.E. Likens : 1981, Nature **292**:329-331.

Haussler, M.R. and McCain, T.A.: 1977, J. Engl. J. Med. **297**:974-983.

Jones, J.H. : 1938, Amer. J. Physiol. **124**:230-237.

LaZerte, B. D. : 1984, Can. J. Fish. Aquat. Sci. **41**:766-776.

Lipstein, B. and S. Hurwitz : 1982, Poult. Sci. **61**:951-954.

Long, J.F., Nagode, L.A., Kindig, O. and L. Liss : 1980, Neurotoxicology **1**:111-120.

Mayor, G.H., Keiser, J.A., Makdani, D. and P.K. Ku : 1977, Science **197**:1187-1189.

Nofre, C., Clement, J.M. and A. Cier : 1963, Pathol. Biol. **11**:853.

Norman, A.W. : 1979, Aust. N. Z. J. Med. **9**:9-16.

Nyholm, N.E. : 1981, Environ. Res. **26**:363-371.

Nyholm, N.E. and H.E. Myhrberg : 1977, Oikos **29**:336-341.

Ondreicka, R., J. Kortus and E. Ginter : 1971, p. 293-305 In: S.C. Skoryna and D. Waldron-Edward, eds. Intestinal Absorption of Metal Ions, Trace Elements and Radionuclides. Pergamon Press, New-York. 431 pp.

Shields, J.B. and H.H. Mitchell : 1941, J. Nutr. **21**:541-552.

Six, K.M. and R.A. Goyer: 1970, J. Lab. & Clin. Med. **76**:933-942.

Simkiss, K. : 1961, Biol. Rev. **36**:321-367.

Sorenson, J.R.J., I.R. Campbell, L.B. Tepper and R.D. Lingg : 1974, Environ. Health Perspect. **8**:3-95.

Storer, N.L. and T.S. Nelson : 1968, Poult. Sci. **47**:244-247.

Washko, P.W. and R.J. Cousins : 1977, J. Nutr. **107**:920-928.

Worker, N.A. and B.B. Migikovsky : 1961a, J. Nutr. **74**:490-494.

Worker, N.A. and B.B. Migikovsky : 1961b, J. Nutr. **75**:222-224.

BACKGROUND POLLUTION IN THE ARCTIC AIR MASS AND ITS RELEVANCE TO NORTH AMERICAN ACID RAIN STUDIES

L.A. Barrie
Atmospheric Environment Service, 4905 Dufferin Street, Downsview, Ontario, Canada, M3H 5T4

ABSTRACT. The Arctic air mass is a unique meteorological feature of the northern hemispheric atmosphere. Possessing well-defined meteorological characteristics, it occupies not only the polar region but also a large fraction of the Canadian and Eurasian land masses during the period November to April. Poor pollutant removal by precipitation and dry deposition within the air mass and a strong transport pathway between Eurasian mid-latitudinal sources and the north, result in elevated levels of acidic anthropogenic aerosols and gases in the air mass during winter. In summer, weak north/south transport and strong pollutant removal between the Arctic and mid-latitudes and within the Arctic, results in lower airborne concentrations of acidic pollutants. Due to the presence of the relatively polluted Arctic air mass, 'background' air concentrations of $SO_4^=$, SO_2 and total NO_3^- are elevated in western Canada during winter. Typical mean monthly concentrations from December to March are 0.8 to 2.1, 1.0 to 2.4 and 0.1 - 0.6 $\mu g\ m^{-3}$, respectively. In the absence of the neutralizing influence of alkaline soil dust, the acidity of snow forming in western Canada during winter is expected to range from 5 to 20 $\mu eq\ \ell^{-1}$.

1. INTRODUCTION

The North American airshed east of the Rockies between 30 and 60°N is a convergence zone for three air masses that originate in the Arctic; the Pacific and the Carribean (graphically illustrated in Figure 1). From the point of view of the acid deposition problem, these air masses provide the 'background' to man-made pollution on the continent. In meteorological terminology, an air mass is a large body of air which is approximately horizontally homogeneous in its moisture content and vertical distribution of temperature. The meteorological characteristics are acquired primarily from the source region (in this case the Arctic) and secondarily, from the modifying influence to which the air mass is subjected after leaving its source region. The concentration of trace constituents in this air is uniquely determined in the region of origin by a balance between natural and man-made

Water, Air, and Soil Pollution **30** (1986) 765–777.

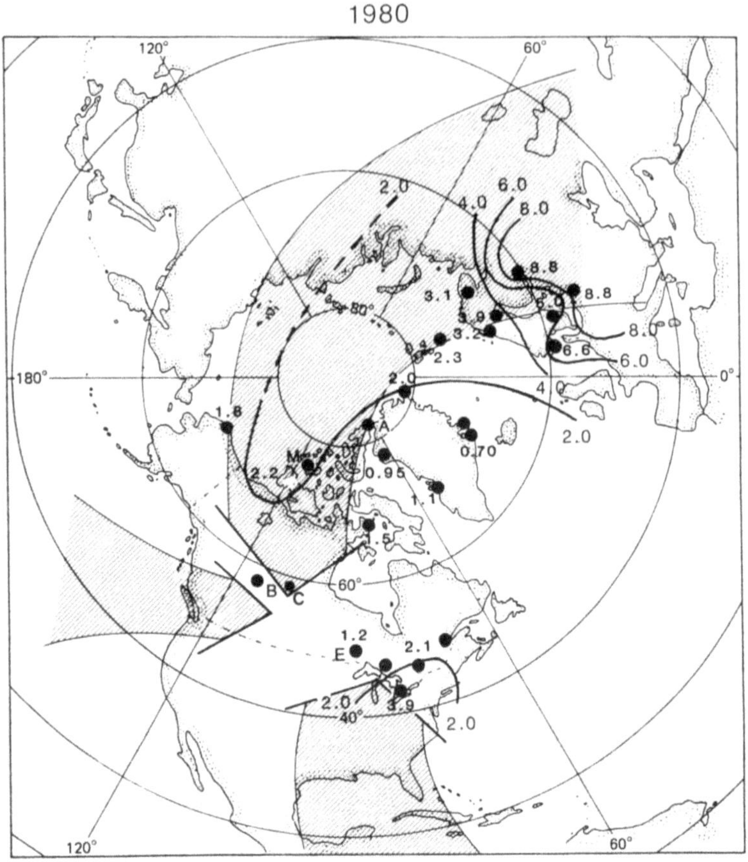

Figure 1. A schematic diagram illustrating the three main sources of air for the North American airshed (i.e. Arctic, Pacific, Carribean) superimposed on the spatial distribution of arithmetic mean $SO_4^{=}$ air concentrations ($\mu g\ m^{-3}$) for the period Jan.-April when the Arctic air mass is most polluted (from Barrie, 1986).

emissions on the one hand and removal processes and chemical reactions on the other hand. For a detailed review of our current knowledge of Arctic air pollution see Barrie (1986). This paper focuses on the Arctic air mass as a source of 'background air' into which North American pollution and in particular western Canadian pollution is released.

2. THE POLLUTION METEOROLOGY OF THE ARCTIC AIR MASS

The Arctic atmosphere is characterized by sub-zero temperatures much of the year, little precipitation and stable thermal stratification (especially in winter) that inhibits vertical mixing. During winter, it occupies, in addition to the polar region, a large fraction of the North American and Eurasian continents while in summer, it is restricted mainly to the polar region. Its mean position in January is shown in Figure 2 in relation to emissions of SO_2 in areas that are within the air mass at some time during the winter. Eurasian SO_2 sources that are potentially available to the Arctic in winter are more than double those in North America ($54x10^6$ vs $24x10^5$ tonne y^{-1}). When the predominant wind patterns are taken into account, the balance shifts even more towards Eurasia as the source of acidic air pollutants in the Arctic air mass. The Eurasia-Arctic-North America transport pathway is strong in winter and very weak in summer. Thus in summer, air from the North Pacific and the Carribean (see Figure 1) dominates most of North America below 60°N . Another winter/summer contrast of Arctic meteorology is the frequency of low

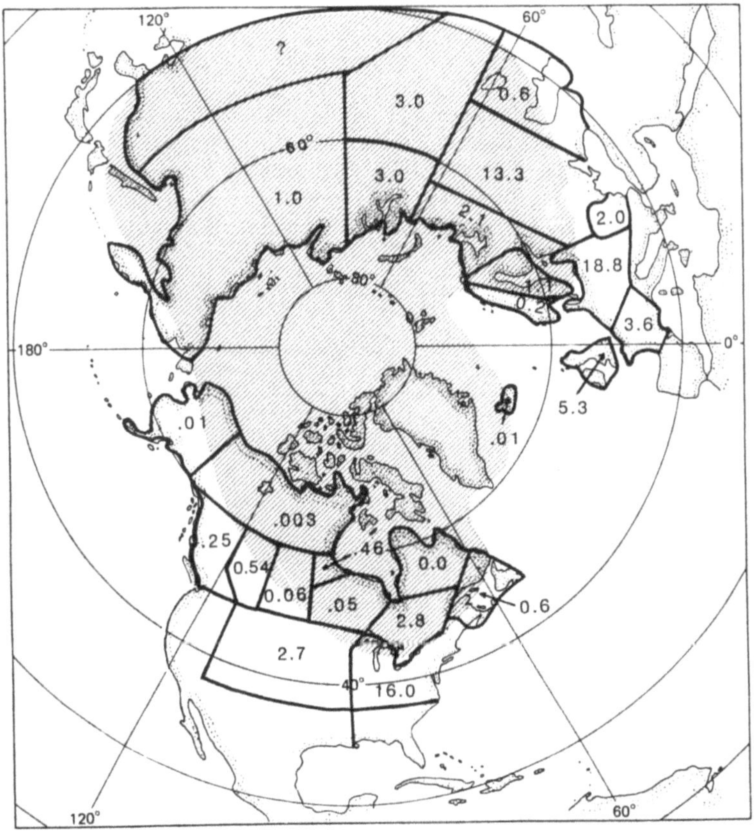

Figure 2. The mean position of the Arctic air mass in January (hatched area) from Barrie and Hoff (1984) and the annual SO_2 emissions (Tg) for areas in Arctic air at some time during the winter (from Barrie, 1986).

cloud cover and hence drizzle. In summer, low cloud occurs 50 to 70% of the time while in winter it appears less than 35% of the time. Thus, in winter not only is the transport of air pollution from the mid-latitudes to the Arctic strong but also the amount of cloud and precipitation that removes pollutants within the air mass is at a minimum. As will be seen in the next section, one manifestation of these pollution meteorological features is a strong seasonal variation in anthropogenic $SO_4^=$ and combustion-derived particulate elements such as V.

3. AIRBORNE ACIDIC POLLUTANTS IN THE ARCTIC AIR MASS

Throughout the Norwegian and North American Arctic during the winter half of the year, there are elevated levels of acidic particulate matter comprised of about 30% $SO_4^=$ (Barrie and Hoff, 1985). The seasonal variation in anthropogenic $SO_4^=$ and V (derived mainly from oil combustion) throughout the Arctic is typified by that at Mould Bay in the Canadian high Arctic between 1979 and 1983 (Figure 3). They are approximately 20 to 40 times higher in the winter than in the summer. In the high Arctic, airborne concentrations of total NO_3^- (particulate + HNO_3) are at least 7 times lower on a molar basis than those of $SO_4^=$ (Barrie et al., 1981; Barrie and Hoff, 1985). SO_2 concentrations do not peak for quite as long as $SO_4^=$ concentrations reaching a maximum in December to February. At its

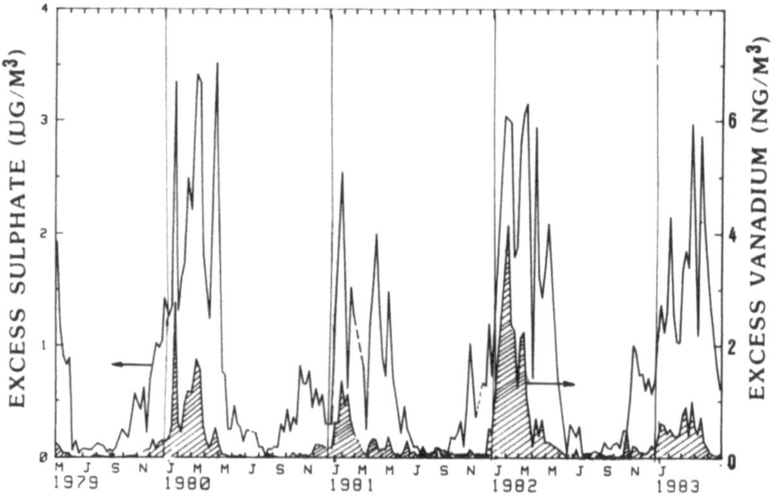

Figure 3. Seasonal variations of excess (non-sea salt) $SO_4^=$ and excess (non-soil) V in aerosols at Mould Bay, N.W.T. (M in Figure 1) that are typical of the high Arctic.

maximum, the abundance of SO_2 relative to $SO_4^=$ on a molar basis ranges from 1:1 to 3:1 (Barrie and Hoff, 1984, 1985). At other times of the year it is less abundant than $SO_4^=$.

The spatial distribution of the mean January to April ground level $SO_4^=$ concentration in 1980 is shown in Figure 1. It tends to decrease from the major source region in Eurasia into the Arctic where it is rather uniform throughout ranging from 1 to 2.3 $\mu g\ m^{-3}$. At present, there are few long term measurements of airborne $SO_4^=$ and other acidic substances available in the Canadian Arctic air mass south of the high Arctic. Those made at ELA-Kenora since 1979 and at Cree Lake, Saskatchewan since late 1982 (E and C in Figure 1, respectively) are the best available. In comparison to other Arctic sites, the mean wintertime $SO_4^=$ concentration at ELA was slightly lower than in the high Arctic in 1980 (Figure 1). In 1981, it was approximately the same as that in the high Arctic (Barrie, 1986). Since during much of winter, ELA and Cree Lake are in the Arctic air mass, it is of interest to: (i) find out if, at these sites a background $SO_4^=$ concentration can be detected above the interference of air pollution from North America and, if (i) is

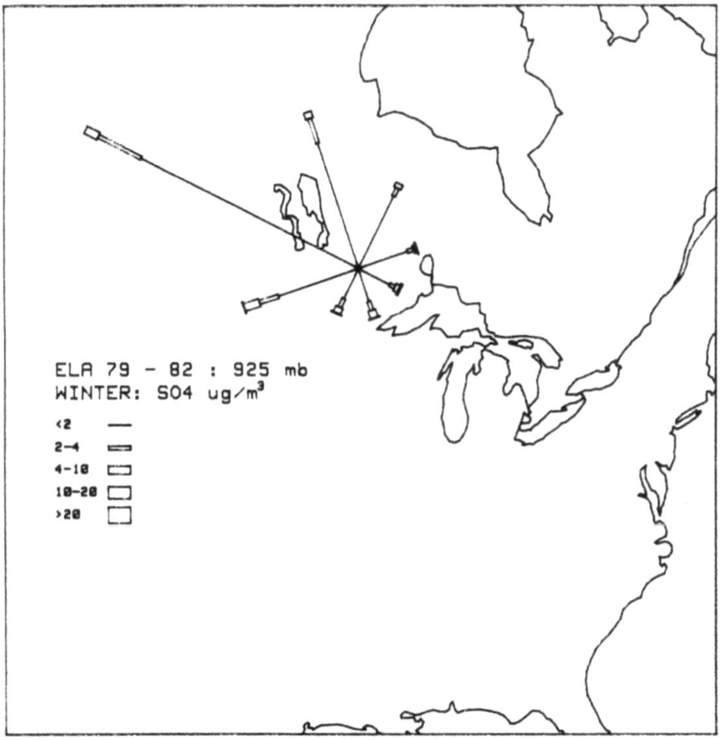

Figure 4. The mean-trajectory $SO_4^=$-air-concentration wind rose at ELA-Kenora (E in Figure 1) for the period Jan.-Mar. 1979-1982. 2-day, 925-mb back-trajectories were used.

affirmed, (ii) to determine what the relative magnitude and temporal variation of acidic substances are compared to the high Arctic.

In Figure 4, the frequency at which air comes to ELA from a particular direction is shown for the period Jan. to March 1979-1982. It is based on 2-day air parcel back-trajectories calculated at the 925 mb level. Also shown for each direction is the frequency of occurrence of a particular $SO_4^=$ concentration range. ELA receives air most frequently from the northern and westerly sectors. Furthermore, the majority of air arriving from these sectors has $SO_4^=$ concentrations less than 2 μg m^{-3}. Transport to ELA from the south and east that is known to bring higher concentrations from major sources in the lower Great Lakes region (Barrie et al., 1980) is infrequent. The highest concentration of $SO_4^=$ (i.e. > 4 μg m^{-3}) occurs solely in air arriving from the southwest to easterly sector. Thus, the influence of eastern sources at ELA is not frequent. It is even less prevlant at Cree Lake which is much further north and west. Cree Lake may occasionally be affected by emissions from the Oil Sands region to the west.

The seasonal variations of daily $SO_4^=$ concentrations at ELA and Cree Lake are shown in Figures 5 and 6. Several years of observations were used. The data set at ELA can be interpreted as having a large fraction of low concentrations associated mostly with background and a small fraction of higher concentrations associated

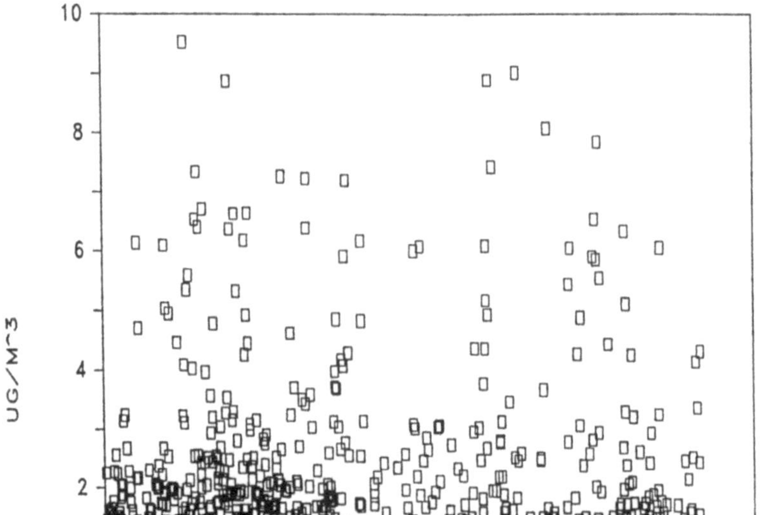

Figure 5. The seasonal variation of daily $SO_4^=$ air concentration at ELA Kenora (E in Figure 1) for 1979-1982.

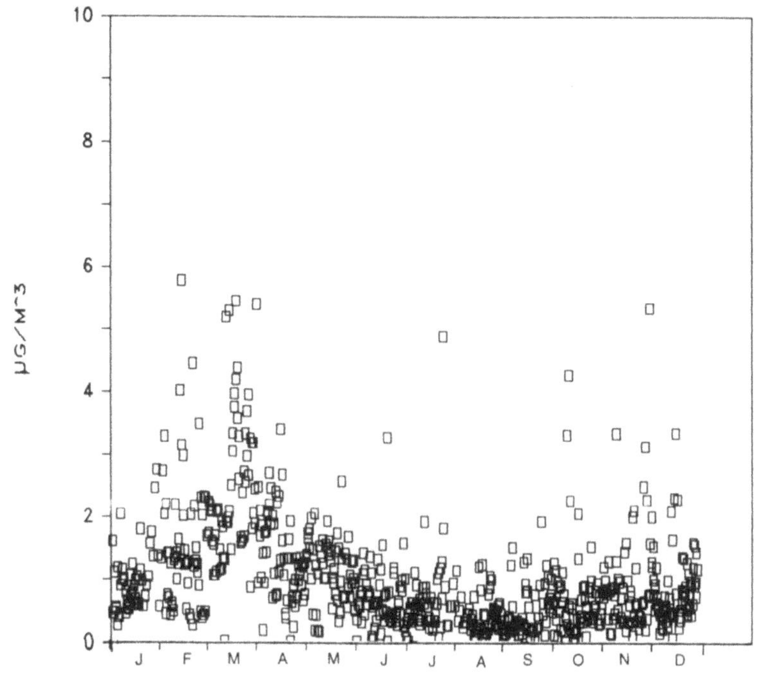

Figure 6. The seasonal variation of daily $SO_4^=$ air concentration at Cree Lake, Saskatchewan (C in Figure 1) for Nov. 1982 to Dec. 1984.

with the incursion of eastern North American pollution. This results in air concentrations being more log-normally distributed than normally distributed (Barrie et al. 1984). The seasonal variation and scatter of daily SO_2 and NO_3^- concentration was similar to those of $SO_4^=$ shown in Figure 5 and 6. Similarly, the observations at Cree Lake are a mixture of frequent 'background' concentrations and occasional higher concentrations from regional sources. Plots of SO_2 and NO_3^- data showed similar scatter indicating a more log-normal than normal distribution.

In choosing a quantitative measure of 'background' concentration, it is important that the infrequent high concentrations associated with regional sources are not given undue weight. Consequently, the monthly geometric mean was chosen. The seasonal variation of the means for $SO_4^=$, SO_2 and total NO_3^- is shown in Figure 7 for the two sites. A comparison of these results for $SO_4^=$ with the daily data plots (Figures 5 and 6) shows that the geometric mean fits well the dark region of high data density representing background.

The concentrations of all three acid-related substances peak in the winter months when the Arctic air mass is present at the sites. On a molar basis, $SO_4^=$ and SO_2 are much more abundant than NO_3^- just as they are in the high Arctic. Furthermore, the variations of $SO_4^=$ and SO_2 concentration is similar to that in

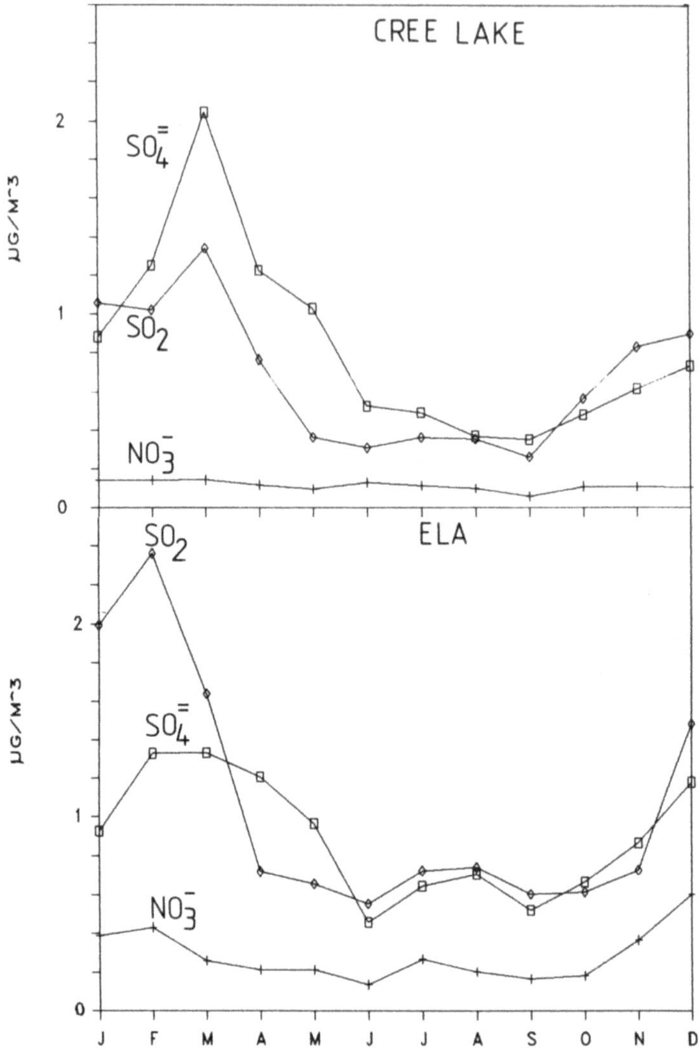

Figure 7. The seasonal variation of 'background' air concentrations at ELA-Kenora (1979-1982) and at Cree Lake (Nov. 1982 - Dec. 1984) represented by monthly geometric means.

the high Arctic. They begin to rise in November reaching a peak between January and April. Wintertime SO_2 and $NO_3^=$ were higher at ELA than at Cree Lake. This may be due to different sampling periods or may indicate that a fraction of the estimated 'background' at ELA for these substances is regional in origin. Mean $SO_4^=$ concentrations are similar at the two sites.

Thus, there is considerable evidence to suggest that the background concentration of acidic S compounds in western Canada undergoes a seasonal variation with a peak in the winter months when the Arctic air mass, in its most polluted state, covers much of the area. This in turn implies that Northern Pacific air contains lower concentrations of acid compounds than the winter Arctic air mass. There is additional support for this contention. Observations of the elemental content of aerosols were made by this author at Birch Mountain north of Fort McMurray, Alberta (B in Figure 1) under background northerly wind conditions in March 1976. The proximity of Birch Mountain to the periphery of the Arctic air mass means that both Pacific and Arctic air arrives there. The results are compared with similar measurements made by Rahn (1971) at Fort Smith in summer in Table I. Air-parcel back-trajectories were examined for the 3 sampling periods on Birch Mountain. It was found that the first period was distinct from the other two in that the air came predominantly from the North Pacific south of Alaska rather than from the Alaskan Arctic via the Mackenzie valley. Chemically, this manifests itself as lower aerosol V concentrations and higher Na and Cl concentrations indicative of sea salt. Higher V concentrations are expected in the polluted Arctic air than in the relatively unpolluted North Pacific air. In comparison to the summer values measured by Rahn (1971), V concentrations in the winter Arctic air were about 20 times higher. This is consistent with the seasonal variation in V observed in the Arctic air mass (see Figure 3). Another feature of the winter measurements at Birch Mountain is that, relative to Al (a crustal reference element), the following elements were enriched in the aerosol by more than 10-fold compared to soil: V, Sb, Zn, As, Cl, I, Br. This indicates that they are of non-crustal origin. the first four elements are likely of anthropogenic origin. Cl originates from the sea. Iodine and Br are either natural or anthropogenic in origin (Barrie and Hoff, 1985).

4. DEPOSITION OF ANTHROPOGENIC POLLUTANTS IN THE ARCTIC AIR MASS

In the high Arctic, there is strong evidence that in the absence of alkaline wind blown dust the acidity of the snow pack reflects the levels of anthropogenic pollution (Barrie et al., 1985). Thus, the acidity of snow deposited on the Agassiz ice cap on Northern Ellesmere Island (A in Figure 1) is a maximum in winter and a minimum in summer paralleling that of aerosol H^+ and $SO_4^=$ concentrations. Ice conductivity on Agassiz is strongly correlated with acidity. When a record of annual maximum conductivity of ice on Agassiz back to 1912 is compared to historical emissions of SO_2 and NO_x in Europe, one

Table I. A comparison of background aerosol composition (ng m^{-3}) measured at Birch Mountain 80 km northwest of Ft. Murray (B in Figure 1) under northerly flow conditions and at Fort Smith by Rahn (1971). Rahn's data represent the average of four two-week samples.

Element	Birch Mountain 1976 Sample 1 North Pacific	Sample 2 Alaskan Arctic	Sample 3 Alaskan Arctic	Fort Smith Summer 1970
Al	32.	43.	54.	54.
As	0.29	0.85	0.34	0.25
Br	3.6	1.6	1.4	0.44
Ca	15.	32.	31.	33.
Cl	142	42	47	7.
Cu	<.78	<1.0	<1.2	0.7
I	.40	.36	.42	0.16
K	25.	19.	27.	44.
Mg	24.	18.	20.	-
Mn	.51	.72	.78	1.2
Na	101.	59.	52.	15.
Sb	<.085	.13	<0.6	0.11
Sc	<.010	.0085 - .013	.0093 - .0160	.36
Ti	<3.8	5.2	6.1	4.1
V	.40	3.5	3.6	0.17
Zn	-	3.2 - 6.0	<4.4	3.1

finds the following (Barrie, et al., 1985). In the first half of this century, except for a short period after the 1912 eruption of Mt. Katmai in the Aleutians, the conductivity remained roughly constant paralleling that of SO_2 emissions in Europe. Thereafter, it increased by about 75% as SO_2 emissions doubled. The strong acidity of the glacial deposit at its winter-time maximum was approximately 8 $\mu eq\ \ell^{-1}$ in the first half of the century increasing thereafter to approximately 14 $\mu eq\ \ell^{-1}$. Thus, there is historical evidence that Arctic air pollution has been around since the turn of the century but has increased markedly since about 1956.

The presence of the Arctic air mass is evident in the air quality of western Canada and in glacial snow deposits in the high Arctic. But is its presence evident in precipitation and deposition in the area? At first glance, this may seem to be a trivial question with a straight-forward affirmative answer. However, because the days on which precipitation occurs in winter in western Canada are often associated with a perturbation of the Arctic front from its mean position and incursion of southern air (air from which precipitation draws its pollutants), the picture is complicated. This is especially true at locations such as ELA near the front (see Figure 1). At ELA, the fraction of days on which air arrives from the southwest to easterly sector is much greater for precipitation days than for all days (Barrie et al., 1984). Thus, unless one is far enough north from the Arctic front that intrusion of southern air does not influence precipitation, one cannot confidently identify Arctic air pollution as the source of precipitation pollutants. Thus, precipitation chemistry measurements in Manitoba and at ELA are difficult to interpret. However, there are some snowpack studies that have been made far enough north of possible frontal shifts that they do not suffer this drawback.

A snowpack chemistry survey in the Northwest Territories of Canada in March 1984 (Shewchuk, 1986) at sites on the pre-Cambrian Shield remote from wind blown dust, yielded H^+ concentrations in the range 2 to 10 $\mu eq\ \ell^{-1}$ (pH 5.0 to 5.7). $SO_4^=$ and NO_3^- concentrations were 7 to 18 $\mu eq\ \ell^{-1}$ and 5 to 7 $\mu eq\ \ell^{-1}$, respectively. Thus, the snowpack is slightly acidic and there is sufficient $SO_4^=$ and NO_3^- to explain much of the acidity. In another snowpack chemistry study in the Athabaska Oil Sands region in January 1978, Barrie (1980) found that at locations remote from the influence of the local emissions the snowpack had an H^+ concentration of 8 to 20 $\mu eq\ \ell^{-1}$ (pH 4.7 to 5.1) and the predominant anions were $SO_4^=$ and NO_3^-. This is considerably more acidic than precipitation in the region during summer. The levels of acidity found in both these snowpack studies are consistent with those found in the glacial snow of the Agassiz ice cap if one takes into account that levels of acidic $SO_4^=$ are somewhat lower in western Canada since it is more remote from Eurasian sources. Thus, there is some indication that during winter the presence of acidic air pollution in the Arctic air mass is seen in the acidity of deposition in western Canada.

5. CONCLUSION

The Arctic air mass is polluted in winter by substances released in Eurasia and, without much removal occurring, is transported into the polar region where it is dispersed and raises the 'background' concentrations of $SO_4^=$ and SO_2 during winter. When the air mass slips southward over North America, it carries this background pollution with it. It can be detected in short and long term air concentration observations made at ELA-Kenora and Cree Lake in western Canada. Typical background concentrations of $SO_4^=$, SO_2 and NO_3^- during December to March are 0.8 to 2.1, 1.0 to 2.4 and 0.1 to 0.6 $\mu g\ m^{-3}$, respectively. The influence of background Arctic air mass pollution on precipitation chemistry is restricted to those areas well north of the Arctic front in which precipitation obtains its pollutants from Arctic air. Further south closer to the front, precipitation often draws its air from intruding southern air. In the absence of the neutralizing influence of alkaline soil dust, the acidity of snow forming in Arctic air is expected to be in the range 5 to 20 $\mu eq\ \ell^{-1}$.

ACKNOWLEDGMENTS

Thanks are due to W. Sukloff, K. Anlauf and A. Smith for their assistance in preparing the material for this paper.

REFERENCES

Barrie, L.A.: 1986, Atmospheric Environment, 13, in press.

Barrie, L.A. and Hoff, R.M.: 1985, Atmospheric Environment, 12, 1995.

Barrie, L.A., Fishers, D. and Koerner, R.M.: 1985, Atmospheric Environment, 12, 205.

Barrie, L.A., Anlauf, K., Wiebe, H.A. and Fellin, P.: 1984, Acidic pollutants in air and precipitation at selected rural locations in Canada, in Deposition Both Wet and Dry, Ed. B.B. Hicks, Acid Precipitation Series, Vol. 4, Series Ed. J.I. Teasley, 15-35, Butterworths.

Barrie, L.A. and Hoff, R.M.: 1984, Atmospheric Environment, 18, 2711.

Barrie, L.A., Hoff, R.M. and Daggupaty, S.M.: 1981, Atmospheric Environment, 15, 1407.

Barrie, L.A., Wiebe, H.A., Anlauf, K. and Fellin, P.: 1980, Studies in Environmental Science, 8, 355.

Barrie, L.A.: 1980, The fate of particulate emissions from an isolated power plant in the Oil Sands area of western Canada, Ann. N.Y. Acad.

Sci., Aerosols: Anthropogenic and Natural Sources and Transport, 338, 434-452.

Rahn, K.A.: 1971, Sources of trace elements in aerosols - an approach to clean air, Ph.D. thesis, Univ. of michigan, Ann Arbor, Michigan, Xerox Univ. Microfilms (order No. 72-4956), 325 pp.

Shewchuk, S.R.: 1986, Acid deposition sensitivities within the Northwest Territories and current deposition to the snowpack and small lakes chemistry of a selected area of the Mackenzie District, Proc. Int. Conf. on Arctic Water Pollution Research: Applications of Science and Technology. Water, Science and Technology (in press).

BIOLOGICAL INDICATORS OF LAKE ACIDIFICATION

K.H. Mills and D.W. Schindler
Department of Fisheries and Oceans
Freshwater Institute
501 University Crescent
Winnipeg, MB R3T 2N6
Canada

ABSTRACT. Indicator taxa are identified, based on both synoptic surveys and whole lake acidification experiments, for lake acidification in the pH 6.0 to 5.0 range. Acidobiontic diatoms (e.g *Asterionella ralfsii*, *Fragillaria acidobiontica*, etc.), periphyton (*Mougeotia* and related species), macroinvertebrates (e.g. *Hyalella azteca*, *Orconectes* sp., etc.), leeches, and cyprinid fishes (e.g. *Pimephales promelas*, *Notropis cornutus*, etc.) are identified as target organisms during early phases of lake acidification.

1. INTRODUCTION

Our knowledge of fish population losses, and losses of other aquatic organisms, due to acidification has increased rapidly during the past 10 yrs. In many surveys the occurrences of taxa are correlated with lake pHs to infer pH tolerance of each species. These surveys generally show that many species are absent when lake pH is below 5.0, but changes in the pH 6.0 to 5.0 range are not as well understood (Hendrey, 1984). The purposes of this paper are to briefly summarize changes in selected species of aquatic groups (algae, zooplankton, benthos, and fishes) as lakes are acidified to pH 5.0, and to identify losses or appearances of indicator species which may be useful for future monitoring programs.

In this paper we will rely heavily on previous literature reviews of the susceptibility of aquatic organisms to acidification (Magnuson *et al.*, 1984; Haines, 1981), on surveys of presence or absence of species in lakes with different pH (Eilers *et al.*, 1984; Økland and Økland, 1980), and on results from the experimental acidification of a small lake trout lake in northwestern Ontario (Schindler *et al.*, 1985). An assumption in this approach is that species occurrences in lakes of different pH do reflect species tolerances. The potential problems of this assumption are listed in Eilers *et al.* (1984) and Magnuson *et al.* (1984), but results from both the experimental acidification of the small lake trout lake and laboratory studies support survey inferences.

Water, Air, and Soil Pollution **30** (1986) 779–789.

2. SURVEYS: PHYTOPLANKTON, PERIPHYTON, AND INVERTEBRATES

Altered species composition and reduced species number have been correlated with low pH levels in many lakes (Hendrey, 1982; Magnuson et al., 1984). Species losses occur in many taxonomic groups (Figure 1); diversity of most taxonomic groups decreases as lakes acidify. While changes occur in all taxonomic groups, specific changes of algae, crustacean zooplankton, benthic macroinvertebrates, and leeches are potential indicators of change in the early phases of lake acidification.

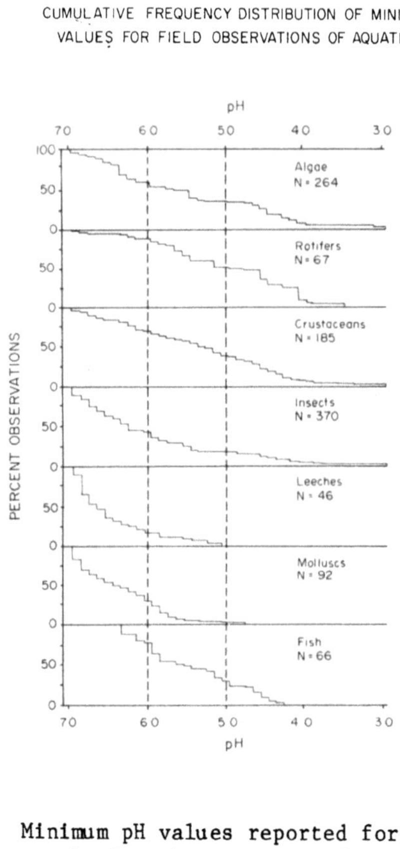

Figure 1. Minimum pH values reported for field observations of aquatic taxa, showing the percent reducton from pH 7.0 to pH 3.0. The range of pH from 6.0 to 5.0 is shown within the dotted lines. N is the number of species included in each panel. Redrawn from Eilers et al. (1984).

For algae, species losses occur in all taxonomic phyla, but in the pH 6.0 to 5.0 range the losses may be more dependent on nutrient status of a lake than pH (Almer et al., 1978; Dillon et al., 1979; NRCC, 1981). Acidobiontic diatoms, such as Asterionella ralfsii,

Fragillaria acidobiontica, Tabellaria quadriseptata, and Tabellaria binalis first occur when lake pH is lowered to approximately 5.5 (Davidson, 1984; Charles, 1985; Batterbee et al., 1985). The change from alkaliphilous to acidobiontic species is probably a good indicator of changing pH in a lake and is being used for the paleoreconstruction of pH status of some lakes (Renberg and Hellberg, 1982; Charles, 1984; Davidson, 1986).

Another characteristic change in the algal community of many acid lakes is the development of dense mats of periphyton (Almer et al., 1974; Stokes, 1981; Hendrey, 1982). Stokes (1981) characterized the types of mats which occur in acid lakes, but most of the observations and experimental research on periphytic algal growth have been conducted at very low pH (<5.0). While the type and extent of mats may vary among acid lakes, rapid development of Mougeotia-dominated Chlorophycean mats occurred in the experimentally acidified Lake 223 and the south basin of Lake 302 when pH was lowered below 6.0 (Schindler et al., 1985; Turner et al., 1986). These mats were very extensive and may be a widespread, obvious change that is characteristic of the early stages of lake acidification.

Acidic lakes have fewer crustacean zooplankton species than neutral lakes (Figure 1), but generally the species that are dominant in acidic lakes are important components of zooplankton communities in nearby non-acidified lakes (Sprules, 1975; Magnuson et al., 1984). Some Daphnia species are absent below pH 5.5 (Yan and Strus, 1980; Malley et al., 1982; Nilssen, 1984), but losses of species such as Mysis relicta (Nero and Schindler, 1983) may be more useful as indicators of lake acidification. A potentially confounding factor is that changes in upper trophic levels due to acidification could influence community composition of zooplankton (Erickson et al., 1980; Nilssen et al., 1984).

There is a general decline in species diversity of benthos as lakes are acidified (Magnuson et al., 1984). Potential indicators of lake acidification in the pH 6.0 to 5.0 range are losses of macroinvertebrates. Gammarus lacustris is widely distributed in North America (Holsinger, 1972) and is not found in Norwegian lakes with pH below 6.0 (Økland and Økland, 1985). Hyalella azteca is not present in Ontario lakes with pH less than 5.5 (Stephenson and Mackie, 1986). Gastropods and sphaeriids (e.g. Pisidium sp., Sphaerium sp., etc.) are usually not found in lakes with pH less than 6.0 (Økland and Økland, 1980). In addition, some species of mayflies (e.g. Baetos sp., Caenis sp., etc.) are very sensitive to pH change in the 6.0 to 5.0 range (Raddum and Fjellheim, 1984; Engblom and Lingdell, 1984). The ability of crayfish to tolerate acidic conditions varies between genera. Cambarus sp. are relatively acid tolerant (Warner, 1971), while Orconectes sp. apparently are not found in lakes that have pHs less than 5.5 (Berrill et al., 1985).

Based on a limited number of observations, leeches are a potentially valuable indicator group for acidification in the pH 6.0 to 5.0 range (Figure 1). Raddum (1980) and Raddum and Fjellheim

(1984) report that leeches are particularly sensitive in Norway, although there is little supporting data for North America (Magnuson et al., 1984). Leeches are not always common in oligotrophic lakes (Magnuson et al., 1984), but this is an obvious group for future research.

3. SURVEYS: FISHES

Because the losses of fish populations were often documented in initial acid rain research, there are many synoptic surveys of the presence or absence of fish species related to pH (Haines, 1981; Magnuson et al., 1984; Eilers et al., 1984). Many species are lost in the pH 6.0 to 5.0 range (Figure 1). In general, the occurrence of salmonid and centrarchid species does not change appreciably in the pH 6.0 to 5.0 range, but rapid changes occur for cyprinids and percids (Figure 2). The rapid decrease in presence of percids in the 6.0 to 5.0 range is due primarily to the loss of darters below pH 5.8 (Eilers et al., 1984). The greatest relative change in the 6.0 to 5.0 range occurs for cyprinids. Ironically, this family of important prey species is one of the least studied groups of fishes (Scott and Crossman, 1973). Although few centrarchid, salmonid, and percid species other than darters are lost in the 6.0 to 5.0 range, some important sport and commercial species such as lake trout (*Salvelinus namaycush*), walleye (*Stizostedion vitreum*), and smallmouth bass (*Micropterus dolomieui*) are lost (Magnuson et al., 1984).

As a taxonomic group, cyprinids are almost ideal indicators of the initial phases of lake acidification. Reproductive failures are the prime cause of fish population losses in acidic lakes (Haines, 1981; Magnuson et al., 1984). Minnow species have very short life spans (Tallman et al., 1984; Scott and Crossman, 1973); reproductive failures would be rapidly reflected in abundance declines. For longer-lived species, such as lake trout, walleye, and lake whitefish (*Coregonus clupeaformis*), the onset of reproductive failures might not be evident in catches for three or more years after the first occurrence. In addition, over-exploitation is usually not a confounding problem for cyprinids, and many species spawn in lake inflows during spring snow melt, a time of pH depression in many lakes. There are many small, shallow lakes which contain only cyprinids. These lakes often have extremely low alkalinities (<50 $\mu eq \cdot L^{-1}$) and rapid flushing rates (<1 yr). Such lakes probably would be affected by acidification prior to larger lakes, which often contain commerically important species.

The occurrence of some cyprinid species is restricted in the pH 6.0 to 5.0 range (Figure 3). Results of surveys by McNicol et al. (1986), Beamish et al. (1976), and our own survey data (Experimental Lakes Area staff, unpublished data) support this generalization. Common shiner (*Notropis cornutus*), blacknose dace (*Rhinichthys atratulus*), fathead minnow (*Pimephales promelas*), and creek chub (*Semotilus atromaculatus*) are usually not found in lakes with pH

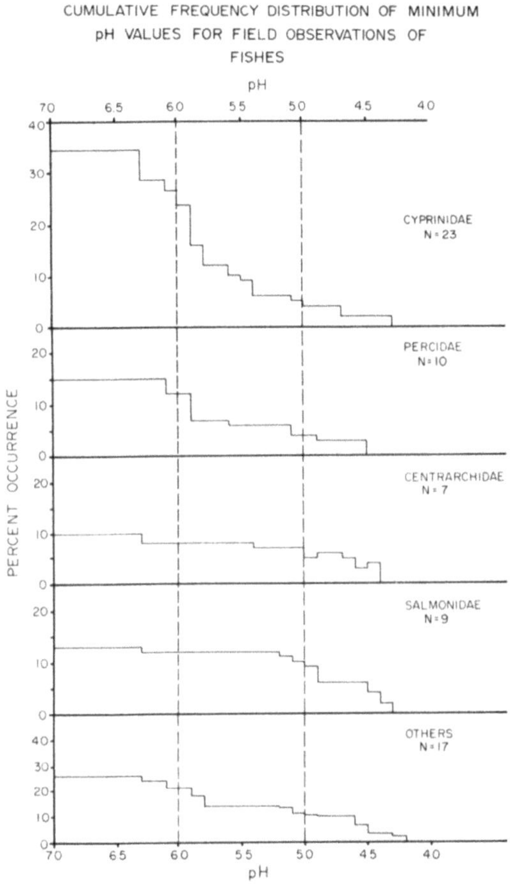

Figure 2. Minimum pH values reported for field observations of fishes, showing the percent reduction from pH 7.0 to pH 3.0. The range of pH from 6.0 to 5.0 is shown within the dotted lines. N is the number of species included in each panel. Redrawn from Eilers *et al*. (1984).

$\leq$5.8. Pearl dace (*Semotilus margarita*), finescale dace (*Phoxinus neogaeus*), and redbelly dace (*Phoxinus eos*) are found in lakes of lower pH, but usually are not present in lakes with pH $\leq$5.3 (Mills *et al*., 1986; Beamish *et al*. 1976; McNicol *et al*., 1986, Figure 3).

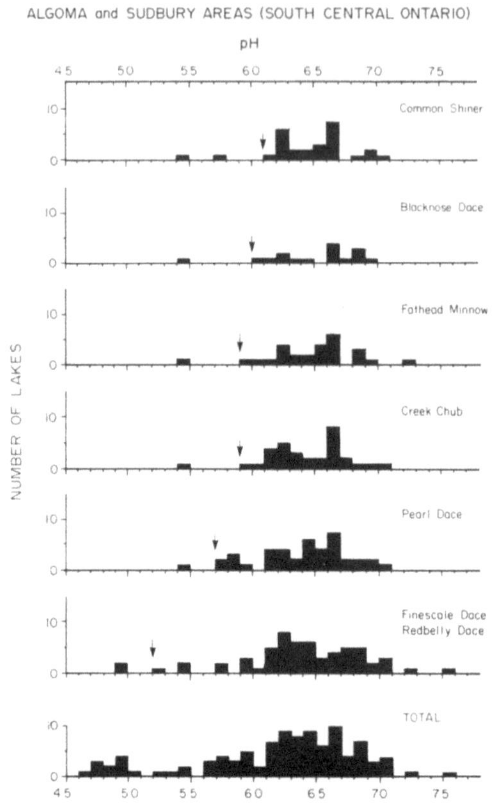

Figure 3. Presence of selected species of Cyprinidae in small lakes in the Sudbury and Algoma areas of Ontario. The bottom panel is the pH frequency distribution for all lakes that were sampled. Arrows represent approximate lower pH limit for each species. The single observations for each species at pH 5.4 are from one lake. Data were provided by D.K. McNicol, B.E. Bendell, and R.K. Ross, Canadian Wildlife Service, Hull.

4. THE LAKE 223 ACIDIFICATION

An important test of the relationship of field survey data to actual species pH tolerances is the detailed monitoring of those species during gradual lake acidification. With the exceptions of fish populations in a few lakes of the La Cloche Mountains, Ontario (Beamish and Harvey, 1972; Beamish 1974; Harvey and Lee, 1982), extensive temporal data for sequential losses of species from lakes undergoing gradual acidification are rare. One of the purposes of the experimental acidification of Lake 223, a small oligotrophic lake in northwestern Ontario, was to test predictions of species losses based on survey results. Changes in water chemistry and biota of Lake 223 are discussed extensively elsewhere (see Schindler et al., 1985 for a synopsis). As the lake epilimetic pH was lowered from 6.8 (1979) to 5.1 (1983), species losses and appearances occurred. In general, diversity of most algae, zooplankton, benthos, and fishes declined. Important lake trout prey (such as *Mysis relicta*, *Pimephales promelas*, etc.) were lost.

Table I
pH thresholds of biotic changes for selected Lake 223 organisms and thresholds for these changes inferred from synoptic surveys.

Biota	Lake 223 pH (year)	Survey pH
Mysis relicta abundance decline	5.93 (1978)	
Fathead minnow abundance decline	5.93 (1978)	5.8[a], 5.9[b], 6.7[c]
Mougeotia mats appear	5.64 (1979)	5.5[d], 5.8[e]
Asterionella ralfsii appears	5.64 (1979)	5.5[f]
Orconectes virilis abundance declines	5.59 (1980)	
Lake trout recruitment ceases	5.59 (1980)	5.5[g]
White sucker recruitment ceases	5.02 (1981)	4.9[c], 5.0[h]
Pearl dace abundance declines	5.09 (1982)	5.2[b], 5.4[a]
Leeches become rare	5.09 (1982)	
Mayfly *Hexagenia* disappears	5.13 (1983)	

[a] Beamish et al. 1976
[b] McNicol et al., 1986 (Fig. 3)
[c] Rahel and Magnuson, 1983
[d] Stokes, 1981
[e] Eloranta and Kumas, 1976
[f] Lowe, 1974
[g] Beggs et al., 1985
[h] Harvey, 1982
[i] Raddum and Fjellheim, 1984

Toxic thresholds for Lake 223 taxa confirmed many thresholds inferred from survey data (Table I). Survey data for other species which disappeared from Lake 223 are not available or are sparse. *Mysis relicta* is a widespread glacial relict in Canada, but no survey data are presently available to compare to the toxic threshold observed in Lake 223 by Nero and Schindler (1983). Despite frequent observations of *Mougeotia* mats in acid lakes, systematic surveys of its occurrence in lakes related to pH, similar to those for fishes, are not available. The data that are available support the threshold observed in Lake 223, although Mougeotia appeared during the acidification of Lake 302 at a somewhat higher pH (Turner et al., 1986). Survey data are not available for Orconectes virilis, but the toxic threshold observed for this species in Lake 223 is similar to pH thresholds inferred for other Orconectes species (Berrill et al., 1985). Comprehensive survey data are presently not available for leeches and Hexagenia mayflies. Toxic effects on lake trout and white sucker, two species that are often monitored in lake acidification surveys, occurred relatively late in the Lake 223 acidification. In addition, we could not confirm the recruitment failure for lake trout (which first occurred 1979) until 1981; age 0 and 1 lake trout are difficult to catch in lakes regardless of pH. Based on our Lake 223 experiment and survey data, a strong case can be made for use of (a) acidobiontic diatoms, (b) macroinvertebrates, and (c) non-sport or non-commercial fish species as indicators of the early phases of lake acidification.

REFERENCES

Almer, B., Dickson, W., Ekstrom, C., and Hornstrom, E.: 1974, Ambio, **3**, 30.

Almer, B., Dickson, W., Ekstrom, C., and Hornstrom, E.: 1978, 'Sulfur Pollution and Aquatic Ecosystems', In Nriagu (ed.), Sulfur in the Environment. Part II, John Wiley & Sons, New York, NY.

Batterbee, R.W., Flower, R.J., Stevenson, A.C., and Rippey, B.: 1985, Nature, **314**, 350.

Beamish, R.J.: 1974, Water Res., **8**, 85.

Beamish, R.J., Blouw, L.M., and McFarlane, G.A.: 1976, A Fish and Chemical Study of 109 Lakes in the Experimental Lakes Area (ELA), Northwestern Ontario, with Appended Reports on Lake Whitefish Ageing Errors and the Northwestern Ontario Baitfish Industry, Can. Fish. Mar. Ser. Tech. Rep., **607**, 116 p.

Beamish, R.J., and Harvey, H.H.: 1972, J. Fish. Res. Board Can., **29**, 1131.

Beggs, G.L., Gunn, J.M., and Olver, C.H.: 1985, The Sensitivity of Ontario Lake Trout (Salvelinus namaycush) and Lake Trout Lakes to Acidification, Ont. Fish. Tech. Rep. Ser., **17**, 24 p.

Berrill, M., Hollett, L., Margosian, A., and Hudson, J.: 1985, Can. J. Zool., **63**, 2586.

Charles, D.F.: 1984, Int. Ver. theor. angew. Limnol. Verh., **22**, 559.

Charles, D.F.: 1985, 'A New Diatom Species Fragilaria acidobiontica, From Acidic Lakes in Northeastern North America', In Smol, J., Batterbee, R., Davis, R., and Meriläinen, J. (eds.), Diatoms and Lake Acidity, W. Junk Publishers, The Hague, Netherlands.

Davidson, G.A.: 1984, Paleolimnological Reconstruction of the Acidification History of an Experimentally Acidified Lake, M.Sc. Thesis, University of Manitoba, Winnipeg, MB, 186 p.
Davidson, G.A.: 1986, Can. J. Fish. Aquat. Sci., **43**, in press.
Dillon, P., Yan, N., and Conroy, N.: 1979, Arch. Hydrobiol. Beih. Ergeb. Limnol., **13**, 317.
Eilers, J.M., Lien, G.L., and Berg, R.G.: 1984, Aquatic Organisms in Acidic Environments: A Literature Review, Wisc. Dep. Nat. Resour. Tech. Bull., **150**, 18 p.
Eloranta, P., and Kumas, S.: 1976, Biol. Res. Rep. Univ. Fjväskylä, **2**, 34.
Engblom, E., and Lingdell, P. E.: 1984, Rep. Inst. Freshwater Res. Drottningholm, **61**, 60.
Erickson, M.O.G., Henricksen, L., Nilson, B.-I., Nyman, G., Oscarson, H.G., and Stenson, A.E.: 1980, Ambio, **9**, 248.
Haines, T.A.: 1981, Trans. Am. Fish. Soc., **100**, 325.
Harvey, H.H.: 1982, 'Population Responses of Fish in Acidified Waters', in Johnson, R.E. (ed.), Acid Rain/Fisheries: Proceedings of an International Symposium on Acidic Precipitation and Fishery Impacts in Northeastern North America, American Fisheries Society, Bethesda, MD.
Harvey, H.H., and Lee, C.: 1982, 'Historical Changes Related to Surface Water pH Changes in Canada', In Johnson, R.E. (ed.), Acid Rain/Fisheries: Proceedings of an International Symposium on Acidic Precipitation and Fishery Impacts in Northeastern North America, American Fisheries Society, Bethesda, MD.
Hendrey, G.R.: 1982, 'Effects of Acidification on Aquatic Primary Producers and Decomposers', In Johnson, R.E. (ed.), Acid Rain/Fisheries: Proceedings of an International Symposium on Acidic Precipitation and Fishery Impacts in Northeastern North America, American Fisheries Society, Bethesda, MD.
Hendrey, G.R.: 1984, 'Preface to Volume 6', In Hendrey, G.R. (ed.), Early Biotic Response to Advancing Lake Acidification, Acid Precipitation Series, Vol. 6, Butterworth Publishers, Boston, MA.
Holsinger, J.R.: 1972, The Freshwater Amphipod Crustaceans (Gammaridae) of North America. Biota of Freshwater Ecosystems, Identification Manual No. 5, U.S. E.P.A. Water Pollution Control Research Series, **18050, ELO 04/72**, 89 p.
Lowe, R.L.: 1974, Environmental Requirements and Pollution Tolerance of Freshwater Diatoms, NTIS, U.S. Dept. Commerce, Springfield, PA.
Magnuson, J.J., Rahel, F.J., Singer, R., Peverly, J.H., Baker, J.P., Fischer, K., Driscol, C., and Schafarn, G.C.: 1984, 'Effects on Aquatic Biology', Chapter E-5 in Linthurst, R.A. (ed.), The Acidic Deposition Phenomenon and its Effects. Critical Assessment Review Papers. Vol. II, Effects Sciences, **EPA-600/8-83. 016 BF.**
Malley, D., Findlay, D.L., and Chang, P.: 1982, 'Ecological Effects of Acid Precipitation on Zooplankton', In D'Imtri, F. (ed.),

Acid Precipitation Effects on Ecological Systems, Ann Arbor Science, Ann Arbor, MI.
McNicol, D.K., Bendell, B.E., and Ross, R.K.: 1986, Studies of Acidification on Aquatic Wildlife in Canada: Vol. 3 - Waterfowl and Trophic Relationships in Headwater Lakes in Ontario. CWS Ocassional Paper, in prep.
Mills, K.H., Chalanchuk, S.M., Mohr, L.C., and Davies, I.J.: 1986, Can. J. Fish. Aquat. Sci., **43**, in press.
National Research Council of Canada (NRCC): 1981, Acidification in the Canadian Aquatic Environment: Scientific Criteria for Assessing the Effects of Acidic Deposition on Aquatic Resources, NRCC **18475**, Environmental Secretariat, Ottawa, ON.
Nero, R.W., and Schindler, D.W.: 1983, Can. J. Fish. Aquat. Sci., **40**, 1905.
Nilssen, J.P.: 1984, Rep. Inst. Freshwater Res. Drottningholm, **61**, 138.
Nilssen, J.P. Østdahl, T., and Potts, W.T.W.: 1984, Rep. Inst. Freshwater Res. Drottningholm, **61**, 148.
Økland, J., and Økland, K.A.: 1980, 'pH Level and Food Organisms for Fish: Studies of 1000 Lakes in Norway', In Drablos, D., and Tollan, A. (eds.), Ecological Impact of Acid Precipitation, Proceedings of an International Conference, Sandefjord, Norway, SNSF Project, Oslo.
Økland, K.A., and Økland, J.: 1985, Oecologia, **66**, 364.
Raddum, G.G.: 1980, 'Comparison of Benthic Invertebrates in Lakes with Different Acidity', In Drablos, D., and Tollan, A. (eds.) Ecological Impact of Acid Precipitation. Proceedings of an International Conference, Sandefjord, Norway, SNSF Project, Oslo.
Raddum, G.G., and Fjellheim, A.: 1984, Int. Verein. theor. angew. Limnol. Verh., **22**, 1973.
Rahel, F.J., and Magnuson, J.J.: 1983, Can. J. Fish. Aquat. Sci., **40**, 3.
Renberg, I., and Hellberg, T.: 1982, Ambio, **11**, 30.
Schindler, D.W., Mills, K.H., Malley, D.F., Findlay, D.L., Shearer, J.A., Davies, I.J., Turner, M.A., Linsey, G.A., and Cruikshank, D.R.: 1985, Science (Wash.), **228**, 1395.
Scott, W.B., and Crossman, E.J.: 1973, Freshwater Fishes of Canada, Bull. Fish. Res. Board Can., **184**, 966 p.
Sprules, W.: 1975, J. Fish. Res. Board Can., **32**, 389.
Stephenson, M., and Mackie, G.L.: 1986, Can. J. Fish. Aquat. Sci., **43**, 288.
Stokes, P.M.: 1981, 'Benthic Algal Communities in Acid Lakes', In Singer, R. (ed.), Effects of Acidic Precipitation on Benthos, North American Benthological Society, Hamilton, NY.
Tallman, R.F., Mills, K.H., and Rotter, R.G.: 1984, The Comparative Ecology of Pearl Dace (Semotilus marganta) and Fathead Minnow (Pimephales promelas) in Lake 114, The Experimental Lakes Area, Northwestern Ontario, with an Appended Key to the Cyprinids of the Experimental Lakes Area, Can. Manuscr. Rep. Fish. Aquat. Sci., **1756**, 27 p.

Turner, M.A., Jackson, M.B., Findlay, D.L., Graham, R.W., DeBruyn, and Vandermier, E.M.: 1986, Can. J. Fish. Aquat. Sci., in press.
Warner, R.W.: 1971, Ohio J. Sci., **71**, 202.
Yan, N.D., and Strus, R.: 1980, Can. J. Fish. Aquat. Sci., **37**, 2282.

The Acid-Base Balance of Peatlands: a short-term perspective

N. R. Urban
Environmental Engineering Program, University of Minnesota
122 CME Bldg, 500 Pillsbury Drive, Minneapolis, Mn 55455 USA

S. E. Bayley
Dept. of Botany, University of Manitoba, Winnipeg, Manitoba
R3T 2N2 Canada

ABSTRACT. Acidity budgets for an artificially acidified and an unacidified peatland in mid-continental North America are dominated by the production and export of organic acids. A fraction (about 50%) of these acids are neutralized by alkalinity from atmospheric deposition and upland run-off. The acid applied to the experimentally acidified site is neutralized by uptake of nitrate and sulfate. Short-term measurements show that nitrate is taken up rapidly in surface layers of moss while sulfate penetrates more deeply into the peat. Sulfate retention within the two sites ranges from 56 to 70% on an annual basis. Stripping of basic cations is shown to result from acid application to base-rich peat. These results suggest that short-term effects of acid deposition on peatlands will be slight, but long-term site acidification is possible.

I. Introduction

There are many important chemical links between peatlands and other systems. Globally peatlands are significant reservoirs of carbon (Clymo, 1986), and they may be important sources and sinks for N and S gases (Hemond, 1983; Urban, 1983). Wetlands greatly modify the composition of precipitation and upland or stream run-off flowing through them. They may consume alkalinity (Bayley and Schindler, 1986), remove trace metals (Eger et al., 1981), sequester nutrients (van der Valk et al., 1978), and neutralize acid mine drainage (Wieder and Lang, 1982). Perhaps most importantly, peatlands export large amounts of organic acids (Hemond, 1980; Urban et al., 1986) which have a profound impact on the acid-base balance of receiving waters (Gorham, 1957; Glover and Webb, 1979; Henriksen and Seip, 1980; Webb, 1982; Wright, 1983; Kerekes, 1984).

Hemond (1980) presented the first acidity budget for a bog and found that production of organic acids, primarily humic and fulvic acids arising from the decomposition of vegetation (McKnight et al., 1985),

Water, Air, and Soil Pollution **30** (1986) 791 800.

was the major source of acidity for Thoreau's Bog. Urban et al., (1986) in looking at internal processes generating and consuming acidity within a small bog in northern Minnesota also found that the acidity balance was governed largely by production of organic acids and inputs of alkalinity. A large survey of bogs in North America, Ireland, and England has shown that organic acid production is the dominant source of acidity in most bogs (Gorham et al., 1985; Urban et al., 1986). Only in severely polluted English bogs was mineral acidity (i.e., H_2SO_4) found. It is not known if organic acid production is affected by acid deposition (cf. Krug and Frink, 1983).

The relative importance of cation sequestration in peat as a source of acidity varies geographically, being greatest in maritime bogs and decreasing westward in North America (Urban et al., 1986). Acidity from uptake of base cations is low at Thoreau's Bog (Hemond, 1980) and fourfold higher in a Minnesota bog (Urban et al., 1986). Clymo (1986) suggests that acidity from ion exchange (cation sequestration) is likely to be greatest in sites receiving high annual precipitation and cation deposition. It is unknown if acid deposition will decrease cation retention in peatlands.

Except in severely polluted English sites (Lee et al., 1986) nitrate is efficiently retained in peatlands. Nitrate is generally undetectable in bog water (Gorham et al., 1985) and peat (Waughman, 1980). The loading at which nitrate changes from being stimulatory to detrimental to plant growth is not known.

Sulfate is more mobile than nitrate in peat. Typically only 60 to 90% of inputs are retained in bogs (Urban et al., 1986). Because of this mobility, deposition of H_2SO_4 may result in changes or damage in peatlands including toxicity to vegetation (Ferguson et al., 1978), cation leaching (Johnson and Cole, 1980), and site acidification. However, Hemond (1980) concluded that acid deposition at Thoreau's Bog was completely neutralized by nitrate and sulfate retention.

The objectives of this paper are two-fold: to compare the acidity budgets of the Marcell and ELA bogs and to examine short-term acidity fluxes associated with uptake and release of S and N. The sites are very similar in size and geographical location, but the ELA site is being acidified experimentally.

2. Experimental

2.1 Site Descriptions

Marcell bog watershed S-2 is located in northcentral Minnesota (47°32"N, 93°28"W). The peat deposit (3.24 ha) occupies a kettle-hole and is surrounded by 6.48 ha of mineral-soil upland. A 5 to 7 m-wide lagg surrounds the bog and funnels most upland run-off around it. The bog is forested with black spruce (*Picea mariana*) and an understory of ericaceous shrubs (*Ledum groenlandicum*, *Chamaedaphne calyculata*). Bryophytes (*Sphagnum magellanicum*, *Sphagnum angustifolium*) dominate the groundlayer vegetation. Important herbaceous species include *Carex trisperma*, *C. oligosperma*, and *Smilacina trifolia*. A hummock-hollow microtopography exists (Verry, 1984), but pools of standing water are absent. Limnic sediments seal the basin and have allowed the development of a perched bog water table (Verry and Timmons, 1982).

The ELA bog is located in the Experimental Lakes Area of northwestern Ontario (49°40"N, 93°43"W), approximately 300 km north of the Marcell site. Both the bog (3.67 ha) and the watershed (10.8 ha) are nearly identical in size to those at Marcell. The upland and a portion of the bog burned in 1974. Vegetation in the upland is a sparse regrowth of Jack pine (Pinus banksiana), aspen (Populus tremuloides), pin cherry (Prunus pennsylvanica), and alder (Alnus crispa) in thin soils (<10 cm) overlying granite bedrock. Vegetation in the bog (see Vitt and Bayley, 1984), consists of a tree layer of black spruce (Picea mariana), a shrub layer dominated by Labrador tea (Ledum groenlandicum) and leatherleaf (Chamaedaphne calyculata), a herb layer of Carex trisperma and Smilacina trifolia, and a ground layer of mosses (principally Sphagnum angustifolium and Sphagnum magellanicum).

A portion of the ELA site has been experimentally acidified for two and a half years. Once a month for 6 mo of the year, 2.7 ha of the bog are sprayed with a 1:1 (molar basis) mixture of HNO_3 and H_2SO_4 at a pH of 3. The acid application is designed to increase the acid-loading to the site 10-fold, which is equivalent to reducing the mean pH of precipitation from 4.9 to about 4 (Bayley et al., 1986). The unacidified portion of the site serves as a control and receives an equal volume of water without acid.

2.2 Methods

Details on sample collection and analysis for the construction of annual acidity budgets have been published elsewhere (Urban et al., 1986; Bayley and Schindler, 1986; Bayley et al., 1986). Only the 1984 acidity budget for the ELA site is discussed in this text. The budget presented for Marcell represents a long-term average and incorporates data from 1971 to 1984. Organic acids are measured by titration to pH 7. Mineral acidity is defined as negative alkalinity (measured by Gran titration for atmospheric deposition and upland runoff) or as the difference between cation and anion retention (ion uptake terms).

Throughfall was collected in the Marcell bog on an event basis from May through October in 1971-1973 and again in 1981-1983. Methods of collection and analysis appear in Verry and Timmons (1977) and Urban (1983). In order to determine the change in composition of rain as it passed through successive layers of moss and peat, plexiglass tubes (4 cm dia) were cut in half length-wise and inserted into the side of a hummock at various depths below the hummock top. Water samples obtained following a rain event were brought to the lab, filtered, and analyzed according to the same protocol as rain samples. Three profiles were obtained on each of three occassions. Anion analysis of all water samples was by ion chromatography (Dionex model 10).

3. Results and Discussion

3.1 Annual Acidity Budgets

Acidity budget components include acidity inputs (atmospheric deposition, run-off), outputs (streamflow), and any change in storage

Table I. Comparison of Annual Acidity Budgets for the ELA, Marcell, and Thoreau's Bogs.

		ELA	Marcell	Thoreau's Bog*
Mineral Acidity:				
Inputs and Sources				
Cation Uptake:	NH_4	16	20	5
	Ca,Mg,Na,K	70	55	11
Upland Run-off		-3	-44	0
Atmospheric Deposition		-7	-2	104
Acid Application		43	0	0
Total		**119**	**29**	**120**
Sinks				
Anion Uptake:	NO_3	46	16	24
	SO_4	64	24	95
Weathering		?	<76	0
Total		**110**	**116**	**119**
Mineral Acidity Export		**+9**	**-87**	**+1**
Organic Acidity:	Production	255	263	195
	Export	166	142	195
	Neutralized	**89**	**121**	**0**

All values as meq $m^{-2}yr^{-1}$.
*Hemond (1980). Organic acids were titrated to pH 8 (McKnight et al., 1985). Sulfate uptake has been adjusted from that originally reported to correct for organic S included in the measurement of sulfate.

(uptake or release) of anions and cations (Table I). Assimilation and ion exchange are not differentiated in uptake of cations. Anion uptake may be an assimilatory or dissimilatory process. Weathering, calculated by mass balances for Fe and Al species, refers to the dissolution and hydrolysis of Fe and Al inputs. Exported organic acidity is the titratable acidity in the outflow. Total production of organic acids equals exported acidity plus exported organic anions not balanced by H^+. All parameters listed were measured by element mass balances. The small discrepancy in the Marcell budget (i.e., the net mineral acidity is inadequate to account for all neutralized organic acids) may result from underestimating dry deposition or from loss of solutes through the bottom of the bog (deep seepage - Verry and Timmons, 1982). The discrepancy in the ELA budget suggests that weathering reactions may be important or that upland run-off is overestimated. No mineral acidity is exported from either the Marcell or ELA sites.

The three features which distinguish acidity budgets of bogs from those of other systems (viz., organic acid export, high net retention of cations and relatively high anion retention) are evident in each of the three budgets shown (Table I). The dominant feature is the large production and export of organic acids. In peatlands, unlike other terrestrial systems, there is no precipitation of humic and fulvic acids in mineral horizons so all acids produced are exported except those neutralized by alkalinity inputs or alkalinity generated in situ. Comparison of organic acid export from the 3 sites does not suggest that acid deposition has reduced the export from either the ELA site or Thoreau's Bog (cf. Krug and Frink, 1983).

Acidity from cation uptake is proportional to the magnitude of cation inputs. Cation inputs to Marcell are high due to its proximity to the prairie. Inputs to the ELA site are slightly higher due to the application of lake water as a medium for the acid application. Cation inputs and alkalinity from run-off are minimal at Thoreau's Bog because the system boundaries were established to exclude run-off from mineral soils (Hemond, 1980). Because approximately 90% of Ca, Mg, and Na in peat are exchangeable (Gore and Allen, 1956; Puustjarvi, 1957) it is base saturation not plant productivity which varies in proportion to cation inputs.

Alkalinity from anion uptake is proportional to anion inputs. All three sites are similar in exhibiting high retention efficiency of nitrate and sulfate. Over 98% of inputs of nitrate are retained at each site. Annual sulfate retention is 56% at Marcell and 74% at the ELA bog. At the Marcell and ELA bogs much of the SO_4 and NO_3 input is not acidic, so even incomplete retention offsets HNO_3 and H_2SO_4 inputs. At Thoreau's Bog anion uptake is adequate to offset acid deposition. However, because retention is not 100% the potential exists for acidification of peatlands by acid deposition. The upper limit for sulfate retention may be hydrologically controlled (Urban et al.,1986). Systems which are rapidly flushed are most prone to acidification. Preliminary evidence shows that water residence times in bogs decrease westward from maritime bogs (Urban et al., 1986). Thus bogs in the northeast currently receiving high loadings of acidic deposition may be most susceptible to acidification.

As long as anion uptake offsets acid deposition, the acidity balance of peatlands is determined by organic acid production and alkalinity inputs. The alkalinity input from upland run-off is high at Marcell because upland mineral soils are deep, slightly calcareous clay-loam tills (Verry and Timmons, 1982). Alkalinity inputs to Thoreau's bog are minimal because of the choice of system boundaries and minimal input of weatherable dust particles. Consequently, about half of the organic acids produced at Marcell and ELA are neutralized by alkalinity inputs while at Thoreau's Bog the entire organic acid production is exported.

Although on an annual basis acid deposition at rates extant in N. America may be neutralized by anion uptake in peatlands, it may have damaging effects on both shorter and longer time scales (Gorham et al., 1984). Short term effects might include temporary acidification if anion uptake is slow relative to acid deposition and surface acidification if anion uptake or reduction occurs deeper in the peat. Long-term damage may occur if basic cations are stripped by ion exchange from the surface moss and either flushed out of the site or re-absorbed by deeper layers of peat. It is also possible that increased rates of sulfate deposition may result in storage of labile forms of reduced S which may be oxidized later during dry periods when the water table is lowered. Short- and long-term damage may occur when, as in England, rates of sulfate and nitrate deposition are high enough to be toxic to vegetation (Tallis, 1964, Ferguson et al., 1978, Lee et al., 1986). The remainder of this paper will examine the potential for short-term effects by looking at the dynamics of alkalinity generation by sulfate and nitrate uptake.

3.2 Nitrate uptake

Vegetation of many bogs is nitrogen limited (Watt and Heinselmann, 1965; Moizuk and Livingston, 1966; Tilton, 1978) and rapid plant uptake of nitrate inputs is expected. This process begins in the tree canopy of forested bogs. Comparison of rainfall and throughfall composition at the Marcell bog consistently shows that throughfall is depleted in nitrate. Annually, approximately 30% of nitrate inputs are taken up within the canopy at this site (Verry and Timmons, 1977; Urban, 1983). Whether this nitrate uptake is by tree foliage or epiphytic organisms is unknown, but it acts to neutralize acidity present in rain.

Measurements at Marcell and ELA show that nitrate in rainfall and in the acid applications is quickly taken up in the surface layers of moss. Rainfall collected at various depths below the surface of

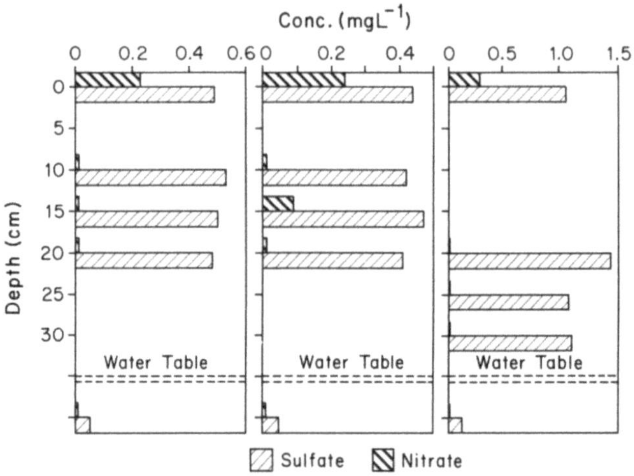

Figure 1. Removal of ions from rainwater percolating through a Sphagnum hummock. While nitrate is removed, sulfate is not taken up by the moss at any depth, but over time becomes depleted in anaerobic zones below the water table.

hummocks at Marcell (Figure 1) is completely depleted of nitrate by 10 cm depth. Porewater in the top 5 cm of peat showed no increase in nitrate concentration following four of six acid applications at the ELA site (Bayley et al., 1986). Nitrate in surface pools at the ELA bog is taken up rapidly following acid application. Within 24 hr following acidification concentrations return to low background levels (Figure 2). None of these measurements identifies the mechanism of nitrate uptake (assimilation, denitrification, or dissimilatory reduction to ammonium), but clearly nitrate uptake occurs rapidly at the bog surface. Consequently, current rates of atmospheric deposition of HNO_3 will not damage peatlands in North America by causing short-term or surface acidification, and may in fact act to fertilize these nutrient-poor ecosystems. This is in stark contrast to the situation in England where nitrate deposition is high enough to inhibit plant growth (Woodin et al., 1985).

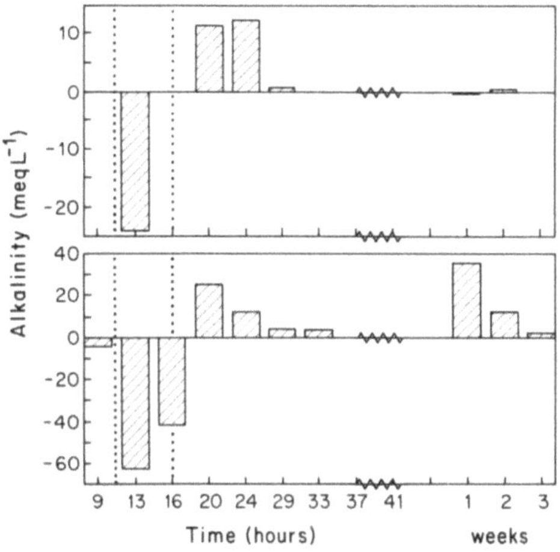

Figure 2. Alkalinity from anion uptake in surface pools of the ELA bog following acidification. Bars indicate the change in NO_3 (upper graph) and SO_4 (lower graph) concentration between two sampling times. Dotted lines denote the period of acid addition.

3.3 Sulfate Uptake.

There is conflicting evidence regarding the depth at which sulfate is removed in peat. At the Marcell bog, no depletion in sulfate concentration was noted in rain water percolating through hummocks (Figure 1). In contrast, Bayley et al. (1986) report that live moss removed most of the sulfate from the acid applications and pore water sulfate concentrations were seldom elevated following acidifications. Bayley et al. (1986) also report high rates of sulfate uptake by live and dead moss in laboratory experiments. Reasons for these differences between sites are not clear. It also is not clear if uptake by moss represents only temporary storage of sulfate which may be washed out later.

Alkalinity generation from sulfate uptake in surface waters at the ELA site occurs much more slowly than nitrate uptake (Figure 2). Sulfate concentrations remain elevated for over a week following acid application. This relatively slow rate of sulfate uptake allows sulfate loss from the bog to occur during periods of streamflow. Urban et al. (1986) noted a negative correlation between the fraction of sulfate inputs retained within bogs and the fraction of water inputs going to run-off. Thus temporary acidification from inputs of sulfuric acid is possible. There is also evidence suggesting that sulfate inputs are not permanently reduced within bogs, but may be reoxidized and released over longer time periods (Behr, 1985).

Mobility of sulfate may result in either acidification of bog waters or stripping of basic cations from moss and peat. In the ELA bog different processes occur in minerotrophic and oligotrophic areas (Figure 3). In oligotrophic, base-poor peat, addition of acid and lake water results in uptake of cations from the lake water and release of additional H^+ from the peat. Thus acid addition increases the acidity of the bog water. In contrast, in the minerotrophic area acid addition causes release of cations from the peat and a much smaller increase in acidity of the water. In both cases the pH change is buffered by the organic acids present. Similarly, oxidation and release of sulfate

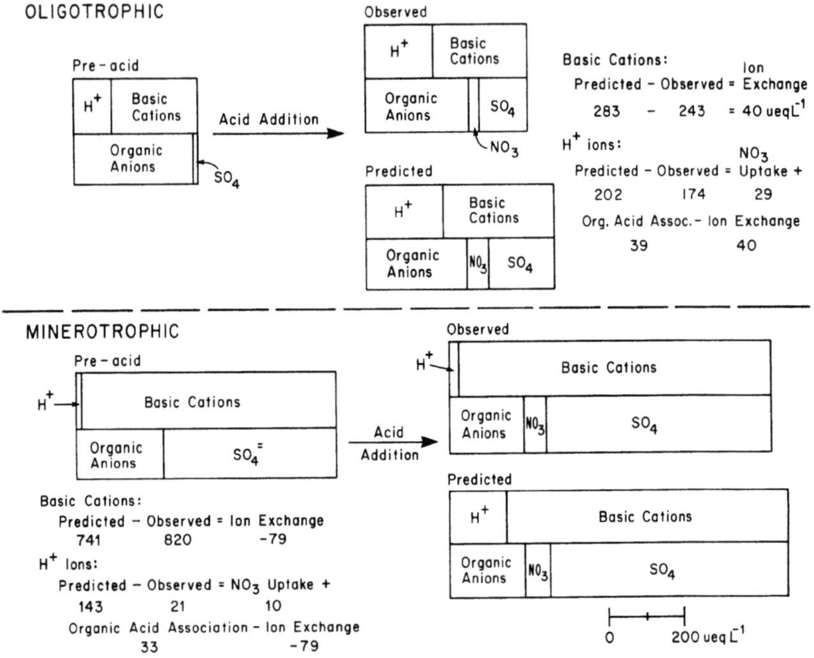

Figure 3. Changes in ionic composition of surface waters in the ELA bog following acid application are explained by processes of nitrate uptake, ion exchange, and protonation of organic anions. Predicted values would result if the applied acid were only diluted in the bog water and no chemical reactions occurred.

from minerotrophic peat following drought results in elevated Ca concentrations rather than reduced pH (Bayley, unpub.). Thus in the short-term, increased inputs of acid sulfate may both increase the acidity of bogwaters and strip basic cations from the moss and peat. When the base saturation of the peat is sufficiently lowered, export of mineral acids (i.e., H_2SO_4) can occur. This phenomenon already has been observed in highly polluted British bogs (Gorham et al., 1985; Urban et al., 1986).

4. Summary and Conclusions

The acidity balance of bogs is dominated by the production of organic acids. Production and export of organic acids from a small peatland in western Ontario is similar to that from a small bog in northern Minnesota and is unaffected by application of mineral acids. On an annual basis acidity applied to the Ontario bog is largely neutralized by uptake of the acid anions. Nitrate is taken up rapidly by surface moss so HNO_3 deposition is likely to have only a fertilizing effect on peatlands in North America. Sulfate is taken up more slowly and at greater depths in the peat. Reduced S may be reoxidized and released from peat following dry spells. From 26 to 44% of annual sulfate inputs are lost to run-off from the two sites examined. Hydrogen ions accompanying sulfate inputs are generally replaced by basic cations from minerotrophic peat. These observations suggest that short-term effects from acid deposition to peatlands will be small, but over longer time periods site acidification may occur due to sulfate mobility and base depletion by cation exchange. Clearly, peatlands have a large capacity to modify terrestrial drainage as well as precipitation via organic acid formation and alkalinity generation from sulfate and nitrate uptake.

Acknowledgments:
This work was funded by NSF (Grant No. DEB 7922142) and a Doctoral Dissertation Fellowship to N.U.

References:
Bayley, S.E. and Schindler D.S.: 1986, 'Sources of alkalinity in PreCambrian shield watersheds under natural conditions and after fire or acidification', in: Hutchinson T.C. and Meema, K.M. (eds), Effects of Air Pollutants on Forests, Agriculture and Wetlands, J.Wiley & Sons, New York

Bayley, S.E., Vitt, D.H., Newbury, R.W., Beaty, K.G., Behr, R. and Miller C, 1986, Can.J.Fish.Aquat.Sci, in press

Behr, R.S.: 1985, 'Sulfur dynamics in an experimentally acidified mire in northwestern Ontario', M.S.Thesis, University of Manitoba, 105 pp.

Clymo,R.S.: 1986, 'Interactions of Sphagnum with water and air' In: Hutchinson, T.C. and Meema, K.M. (eds), Effects of Air Pollutants on Forests, Agriculture and Wetlands, J.Wiley & Sons, N.Y.

Eger, P., Lapakko, K. and Otterson, P.: 1981,'Trace metal uptake by peat: Interaction of a white cedar bog and mining stockpile leachate, Proc. 6th Int. Peat Congr., Duluth, Mn, USA, pp 542-547

Ferguson, P., Lee, J.A. and Bell, J.: 1978, Envir. Poll., 16, 151,

Glover, G.M. and Webb, A.H.: 1979, Water Res., 13,781,

Gore, A.J. and Allen, S.E.: 1956, Oikos, 7,48,

Gorham, E.: 1957, Limnol. Oceanogr., 2, 12,

Gorham, E., Bayley. S,E. and Schindler, D.W.: 1984, Can.J. Fish. Aquat.Sci., **41**, 1256,
Gorham, E., Eisenreich, S.J., Ford, J., Santelmann, M.V.: 1985, 'The chemistry of bog waters', In: Stumm,W.(ed) Chemical Processes in Lakes, John Wiley and Sons, New York, p 339
Hemond, H.F.: 1980, Ecol.Monogr., **50**, 507,
Hemond, H.F.: 1983, Ecology, **64**, 99,
Henriksen, A. and Seip, H.M.: 1980, Water Res., **14**, 809,
Johnson, D.W. and Cole, D.W.: 1980, Envir. Internat., **3**, 79,
Kerekes, J.: 1984, 'Review of the regional aquatic program', Proc. of Atlantic Region LRTAP Monitoring and Effects Working Group, Bedford Inst. Oceanogr., Dartmouth, N.S.
Krug, E.C. and Frink, C.R.: 1983, Science, **221**, 520,
Lee, J.A., Press, M.C., Woodin, S. and Ferguson, P.: 1986, 'Responses to acidic deposition in ombrotrophic mires', in: Hutchinson, T.C. and Meema, K.M. (eds), Effects of Air Pollutants on Forests, Agriculture and Wetlands, J.Wiley & Sons, N.Y.
McKnight, D., Thurman, E., Wershaw, R., Hemond, H.: 1985, Ecology, **66**, 1339,
Moizuk, G.A. and Livingston, R.B.: 1966, Ecology, **47**, 942,
Puustjarvi, V.: 1957, Acta Agriculturae Scand., **7**, 190,
Tallis, J.H.: 1964, J. Ecol., **52**, 345,
Tilton, K.L.: 1978, J. Ecol., **66**, 499,
Urban, N.R.: 1983, 'The nitrogen cycle in a forested bog watershed in northern Minnesota', M.S. Thesis, Uni of Minnesota, 359 pp
Urban, N.R., Eisenreich, S.J. and Gorham, E.: 1986, 'Proton cycling in bogs: geographic variation in Northeastern North America', in: Hutchinson, T.C. and Meema, K.M. (eds), Effects of Air Pollutants on Forests, Agriculture and Wetlands, J.Wiley & Sons, N.Y.
van der Valk, A.G., Davis, C.B., Baker, J.L. and Beer, C.E.: 1978, 'Natural freshwater wetlands as nitrogen and phosphorus traps for land runoff', in: Greeson, P.E., Clark, J.R. and Clark, J.E. (eds.), Wetland Functions and Values: the State of our Understanding, pp.457-467
Verry, E.S.: 1984, 'Microtopography and water table fluctuation in a Sphagnum mire', In: Proc. 7th Internat. Peat Congress, Dublin
Verry, E.S. and Timmons, D.R.: 1977, Can. J. For.Res., **7**, 112,
Verry, E.S. and Timmons, D.R.: 1982, Ecology, **63**, 1456,
Vitt, D.H. and Bayley, S.E.: 1984, Can. J.Bot., **62**, 1485,
Watt, R.F. and Heinselman, M.L.: 1965, Ecology, **46**, 357,
Waughman, G.J.: 1980, J. Ecol., **68**, 1025,
Webb, A.H.: 1982, Water Res., **16**, 641,
Wieder, R.K. and Lang, G.E.: 1982, 'Modification of acid mine drainage in a freshwater wetland', in: Symposium on wetlands of the Unglaciated Appalachian Region, Morgantown, W.Virginia, pp. 43-53
Wright, R.: 1983, 'Predicting acidification of North American Lakes', Report 4, Acid Rain Res.Series, Norwegian Inst. for Water Res.,165 pp.
Woodin, S., Press, M.C. and Lee, J.A.: 1985, New Phytol., **99**, 381,

ON THE VARIABILITY OF SIMULATED SOURCE-RECEPTOR RELATIONSHIPS FOR SULFUR DEPOSITION

Perry J. Samson, Mark Fernau, and Patricia Allison
Department of Atmospheric and Oceanic Science
University of Michigan
Ann Arbor, Michigan 48109-2143

ABSTRACT. *The use of Lagrangian models to estimate source-receptor relationships for ambient $SO_4^=$ concentrations and S deposition has become fairly widespread over the past several years. This paper addresses the sensitivity of long-term simulations of a Lagrangian S transport and deposition model to actual variations in SO_2 emissions and meteorological conditions. The variations of predicted source-receptor relationships due to (1) the inclusion of day to day variations in emissions strength as opposed to the use of the annual average daily emission rate and (2) year-to-year variations in meteorological conditions were studied to identify causes of uncertainty in a Lagrangian model. The results suggested that adding information on day to day emission variations for a specific point source resulted in variations in estimated S wet deposition of the order of only 20% within 500 km of the source. Year-to-year variations in meteorological conditions, on the other hand, resulted in variations in predicted S wet deposition of the order of 50% for some receptors. The variation in estimated source-receptor relationships for a given source/receptor combination was found to range as high as 70% over a 5-yr modeling period.*

1. INTRODUCTION

The impact of sulfur dioxide and nitrogen oxide emissions on air and precipitation quality at receptors located hundreds of kilometers downwind of a source is still poorly understood. Several atmospheric transport and deposition models have been used to estimate these impacts but the modeling has been plagued with problems of uncertainty in both model formulation and in the necessary input data for emissions strengths and meteorology. Because the results of these models have been and will be used to aid in the design of costly emission control strategies it is important that we examine possible sources of uncertainty due both to input data quality and expected fluctuations of meteorological conditions. This paper examines the sensitivity of Lagrangian model simulations of airborne sulfate ($SO_4^=$) concentrations

Water, Air, and Soil Pollution **30** (1986) 801-813.

and wet sulfur (*S*) deposition to the use of seasonal, monthly and daily averaged emissions information instead of annually averaged emissions information. Additionally, the magnitude of expected variations in annual averaged ambient $SO_4^=$ concentrations and annual total *S* wet deposition due to year-to-year variations in meteorological conditions in eastern North America are estimated. Finally, the sensitivity of simulated source-receptor relationships to year-to-year variations in meteorological conditions are also explored.

The first part of this paper examines how much uncertainty in modeled estimates of source-receptor relationships results from ignoring temporal variations in sulfur dioxide (SO_2) emission rates. On a regional scale most modeling efforts have concentrated on representing meteorological variations accurately but have ignored temporal oscillations in SO_2 emission rates by using annually or, at best, seasonally averaged emissions to examine source-receptor relationships. Using daily emissions data from an electric generating facility located in southeastern Michigan, the relative effect of using emissions data representative of four different averaging periods was examined.

The second part of this paper illustrates the year-to-year changes expected in annual average $SO_4^=$ concentration and total annual *S* wet deposition at several locations in eastern North America due to changing meteorological conditions. This analysis was necessary to evaluate the natural "noise" upon which the signal of changing emissions impact will need to be discerned. In addition, variations in predicted source-receptor relationships were noted to assess the feasibility of being able to identify meaningful signals from targeted emission control scenarios.

The results of the model exercises are expressed as either the *natural potential* (also referred to as the "normalized source-receptor matrix elements") for contribution to a receptor or as an estimate of the *absolute contribution* to a receptor. The natural potential for contribution represents estimates of the probability of *S* emitted at a given location impacting a given receptor as SO_2, $SO_4^=$, *S* dry or *S* wet deposition due to the existing meteorological conditions and assumed chemical transformation and scavenging without regard to actual emission densities. The absolute contribution is the linear product of the natural potential field and the actual emissions density.

2. METHODS

The University of Michigan Atmospheric Contributions to Interregional Deposition (ACID) model calculates short-term (daily) SO_2 and $SO_4^=$ concentrations and *S* deposition due to regional-scale transport. Details of the model are presented elsewhere (Samson and Small, 1982). The ACID Model can be used in either of two modes. To investigate variations in downwind deposition patterns associated with varying emission averaging, the forward mode was used. To investigate changes in deposition at specific receptors due to the changes in meteorological conditions, the backward mode was used.

In order to estimate the transport path of the air leaving a source in the forward mode or arriving at a receptor in the backward mode trajectories have been calculated based on observed upper-air winds. A trajectory represents the center-of-mass of probability for air to be transported from a source to a receptor. The probability of a reactive tracer wet depositing at a point $\boldsymbol{x}$ at a time t, $W_s(\boldsymbol{x},t)$, can be expressed as

$$W_s(\boldsymbol{x},t) = \varepsilon_s(\boldsymbol{x},t) \int_{t-\tau}^{t} \int_{-\infty}^{\infty} \int_{-\infty}^{\infty} Q(\boldsymbol{x},t|\boldsymbol{x}',t')\, R(t|t') \times \times D_s(\boldsymbol{x}',t')\, \Lambda(\boldsymbol{x}',t')\, d\boldsymbol{x}'\, dt' \quad (1)$$

where $\varepsilon_s(\boldsymbol{x},t)$ is the fractional efficiency of wet deposition of the tracer at point $(\boldsymbol{x},t)$. $Q(\boldsymbol{x},t|\boldsymbol{x}',t')$ is the probability of an air parcel located at $\boldsymbol{x}'$ at time t' arriving at a receptor $\boldsymbol{x}$ at time t, $R(t|t')$ is the probability of the tracer not being chemically transformed to another species from time t' to time t, $D_s(\boldsymbol{x}',t')$ is the probability that the tracer will not be dry deposited at $(\boldsymbol{x}',t')$ and $\Lambda(\boldsymbol{x}',t')$ $(= 1-\varepsilon_s(\boldsymbol{x}',t'))$ is the probability that the tracer will not be wet deposited at $(\boldsymbol{x}',t')$. The integration is continued over a period τ which has been prescribed in this study as 72 hr (or less if the trajectory calculation terminated prematurely due to missing input data).

The product of the last three probability functions can be approximated for S by solving the coupled equations

$$ds/dt = -k_t s - k_d s - k_w s + E_s \quad (2)$$

$$dp/dt = +k_t s - \kappa_d p - \kappa_w p + E_p \quad (3)$$

where s is the probability of S occurring as SO_2, p is the probability of S occurring as $SO_4^=$, k_t is the linear rate of transformation of S from SO_2 to $SO_4^=$, k_d and k_w are the rates of dry and wet deposition, respectively, for SO_2, and κ_d and κ_w are the rates of dry and wet removal, respectively, for $SO_4^=$. Some fraction, E_p, (usually thought to be about 1% to 5%) of the initial probability of emitted S is assumed to be $SO_4^=$ while the majority of the initial probability is assumed to be SO_2. The chemical transformation rate, k_t and the SO_2 and $SO_4^=$ dry deposition rates are presumed to vary diurnally and seasonally. The wet deposition rates are assumed to be dependent upon precipitation rate following Scott (1982).

To test the sensitivity of simulated concentrations and deposition to emissions averaging the SO_2 emissions input data were varied to reflect daily, monthly, seasonally and annually averaged emissions for one source. All averaging periods longer than a day were derived from ensembles of daily emissions data. Seasons were defined as 3 month periods, beginning January, April, July and October. Model runs were made for all averaging times for each year from 1976 to 1981, except 1979, for which a complete year's worth of emissions data were not available. The model output included estimates of daily ambient SO_2 and $SO_4^=$ concentrations and daily total *S* dry and wet depositions in each cell of the 45 × 36 grid, ranging from 25°30'N to 51°21'N and from 60°88'W to 104°88'W. This grid dimensions were 1° longitude by 0.74° latitude (equal to the 1° longitude at the latitude of Ann Arbor, Michigan). Hence the area of the grid cells varied from the north (≈5700 km^2) to the south (≈8300 km^2). The model was run in the forward mode using four downwind (forward) trajectories per day.

To simulate the variations in concentrations, deposition and source-receptor relationships due to year-to-year variations in meteorological conditions the backward mode used four upwind (backward) trajectories per day to calculate the natural potential for contribution to $SO_4^=$ concentrations and *S* wet deposition at specific receptors. The potential was then multiplied by the emissions in each cell to estimate source contribution. In this study the potential was calculated for six years (1976 to 1981) and multiplied by an emissions inventory compiled for 1980. The model output includes estimates of contributions to daily ambient SO_2 and $SO_4^=$ concentrations and daily total *S* dry and wet depositions from each cell.

These estimates were integrated over the area of states and provinces to investigate modeled fluctuations in source-receptor relationships over the period.

3. INPUT DATA

3.1. Emissions Data

The emissions level for an electric generating facility varies from day to day in response to changing load demands, maintenance requirements and economic factors. Thus a wide distribution of daily emission strengths is possible. This variation in emission rates could, presumably, affect the estimation of annual impact on air and precipitation quality downwind of the facility in contrast to similar estimates made using only an annual averaged daily emission rate. In order to estimate the degree of variation in modeled deposition patterns incurred by ignoring day to day variations in emissions, the daily SO_2 emissions were obtained from one point source for five years: 1976, 1977, 1978, 1980, and 1981. Frequency distributions of daily emission rates normalized by the mean emission rate the whole period are shown in Figure 1. The figure shows that there was a wider distribution of emission rates in the earlier years studied than in later years. The greatest range in emissions occurred in 1977. The greatest standard deviation of daily emissions occurred in 1976 reflecting fluctuations in operating conditions during the source's first year of operation. In 1980 and 1981, there was a significant decrease in the peak emission and in the variation of the emissions from the year's mean.

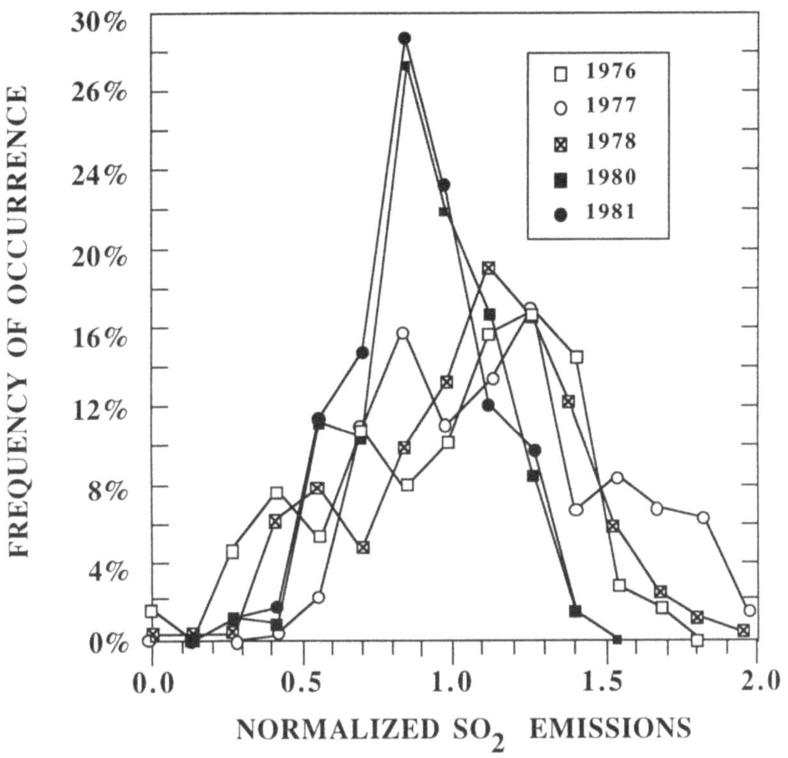

Figure 1. Frequency of occurrence of normalized SO_2 emissions from a major electric generating facility for five separate years. Emissions are normalized by the mean emission rate for the five-year period. The distribution for 1981 is shaded for reference.

The constancy of emissions from this source in later years is seen by plotting the coefficient of variation (the standard deviation divided by the mean annual emission rate) as a function of averaging time in Figure 2. The plot shows that in general the standard deviation of daily emissions was about 30% of the mean when sampled on a daily basis but decreases with longer averaging to zero when the data were averaged over the whole year.

There were year-to-year variations in the pattern, with the 3 early years showing the highest variation relative to the mean and the later 2 years the least. The importance of using day-by-day emissions information is presumably increased for those years with the highest coefficient of variation.

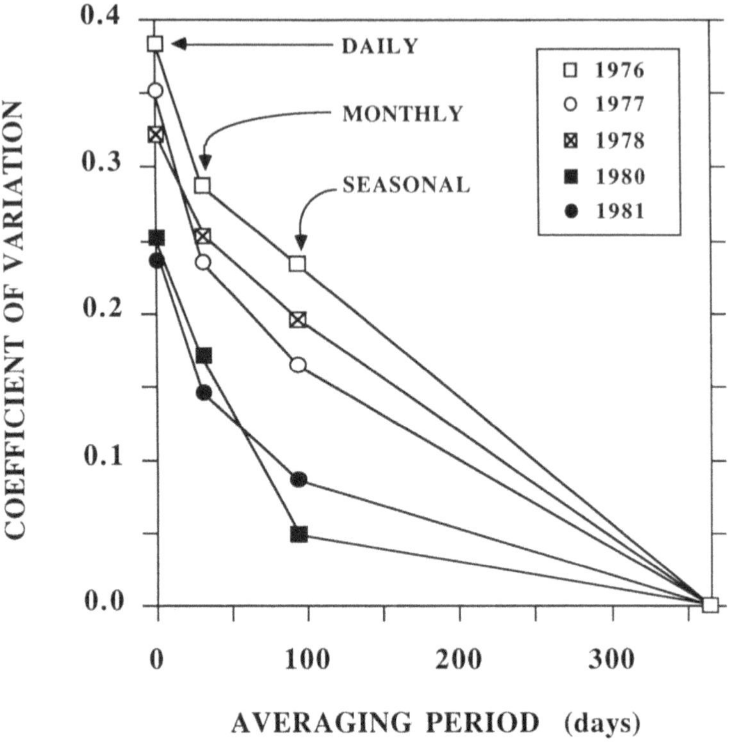

Figure 2. The coefficient of variation of SO_2 emissions shown in Fig. 1 for five separate years.

Emissions data for all of the United States and Canada were used to model the interannual variations in deposition resulting from changing meteorological conditions. The data used in this study were obtained as part of the International Sulfur Deposition Model Evaluation (ISDME) study (Clark, 1985). These data represent point and area source emissions for 1980.

3.2. Meteorological Data

The calculation of trajectories required the use of upper-air wind and temperature data. The data were obtained from the National Climatic Center in a format compatible with the NOAA-ATAD (Heffter, 1980) trajectory model. Precipitation information both at the receptors and upwind were derived from hourly precipitation data obtained from the National Climatic Center. These data were preprocessed to chronological files and then gridded into three-hour blocks as required by the ACID model. The gridding was based on ensemble averages of all reporting stations within each grid cell (see Keeler *et al.*, 1984). The precipitation amount from the grids was joined with the calculated trajectory location to provide an estimate of upwind pollutant removal.

4. RESULTS

4.1. The Effect of Temporal Averaging of SO_2 Emissions on Modeled Source-Receptor Relationships

The results from model exercises using the daily, monthly and seasonally averaged emissions data were compared to the results from model runs using the annually averaged data to identify the level of uncertainty introduced into modeling estimates when sub-annual fluctuations in deposition were ignored.

The estimated spatial distribution of *S* wet deposition potential resulting from a single point source presumed located in the Detroit Metropolitan area in 1976 is shown in Figure 3. These calculations suggest that the area of maximum exposure to wet *S* deposition extends to the east and north of this source region. The relatively long distances predicted between SO_2 emission and *S* wet deposition were presumably due to the slow conversion of SO_2 to $SO_4^{=}$, the latter being more efficiently scavenged by precipitation than the former. The preference for the *S* wet deposition potential to be highest to the northeast of the source was the result of the higher natural probability for southwest winds to be associated with precipitation.

The variability in estimated *S* wet deposition potential due to the choice of temporal averaging period of input emissions can be demonstrated by plotting the percent difference fields of daily, monthly and seasonally averaged emissions results versus those of the annually averaged emissions.

Since the difference was generally greatest between the annual and the daily averaged emissions output and since 1976 was the year with the greatest variability in the emissions, the percent difference for *S* wet deposition between the annually and daily averaged emissions results for 1976, shown in Figure 4, illustrates the expected worst-case variation.

The plot shows that *S* wet deposition estimates exhibited an area of difference of greater than 20% in northeastern Ohio and northern Pennsylvania. Another area of greater than 20% difference occurred in eastern Tennessee and Kentucky and western North Carolina, but the frequency of transport to these regions was estimated to be fairly low (Fig. 3).

Thus, lack of detailed emissions information lead to variations in the spatial distribution of potential annual deposition of the order of only 20%. However, the importance of using detailed emissions data will, presumably, increase with decreasing deposition averaging time. On the other hand, the variations in the emissions from an ensemble of sources within a grid cell (such as would be included in a regional-

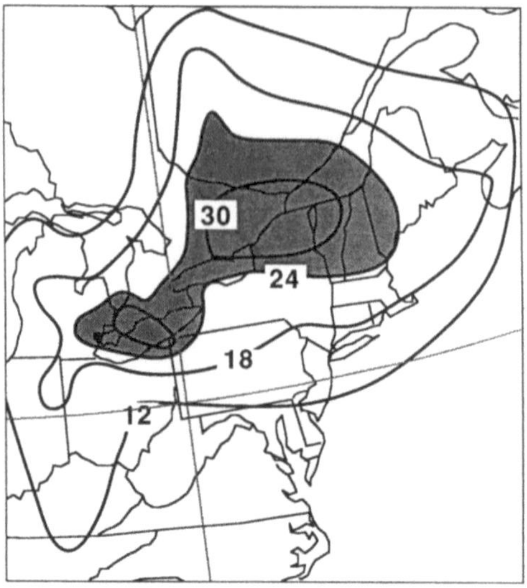

Figure 3. Probability of *S* wet deposition from a source located in Southeastern Michigan using 1976 annual average emissions and 1976 meteorology. Units are 10^{-6} km^{-2}.

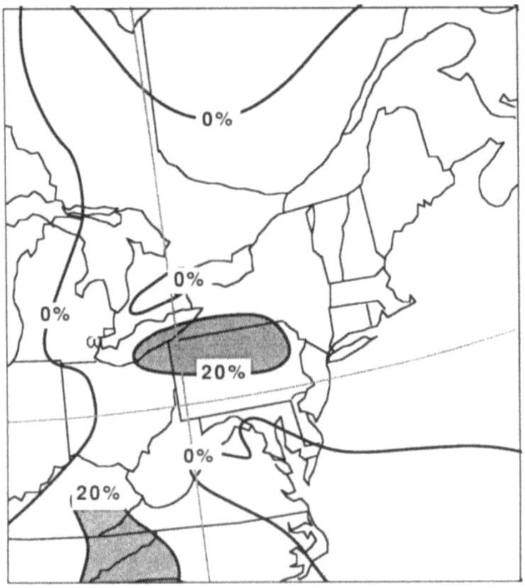

Figure 4. The percent difference between the probability of *S* wet deposition calculated for a source in southeast Michigan using 1976 meteorological conditions and daily emissions values and that obtained using the annual average emission rate.

scale transport and deposition model) would not be as pronounced as those for a single source.

4.2. Variations in Estimated *S* Wet Deposition in the Adirondacks due to Variations in Meteorological Conditions

The variation of estimated ambient $SO_4^=$ concentrations and *S* wet deposition due solely to variations in meteorological factors represents the "noise" upon which a signal of changing emissions is impressed. To evaluate the magnitude of this meteorological "noise", the ACID model was run in the backward mode for the ten receptors listed in Table I for each of six years using fixed 1980 annual SO_2 emissions. Figures 5 and 6 show the predicted annual average ambient $SO_4^=$ concentrations and S wet deposition, respectively, at each of the ten sites for each of the six years (1976-1981).

The range of predicted $SO_4^=$ concentrations were large at some sites and less at others. The variations in estimated annual average $SO_4^=$ concentrations had a range of variation of about -20% to +35% (ignoring Marcell, MN and Meade, NB where higher variations were predicted due to the low predicted values). The relative variation in ambient $SO_4^=$ was generally less than the relative range of the *S* wet deposition as would be expected because of the added variability in precipitation timing. The estimated annual *S* wet deposition varied about -45% to +45% from the deposition estimated for 1980. The range of predicted *S* wet deposition was relatively large for some of the receptors including Caldwell, Ohio; Kane Experimental Forest, Pennsylvania; Parsons, West Virginia; and Dorset, Ontario. Thus the model predicts that large variations in *S* wet deposition can be due to the variations in meteorology alone.

Table I. Receptor Site Locations for Multi-Year Modeling Analyses

SITE	LATITUDE	LONGITUDE
Whiteface Mountain, New York	44°23'	73°35
Kane Experimental Forest, Pennsylvania	41°36'	78°46'
Dorset, Ontario	45°13'	78°56'
Parsons, West Virginia	39°05'	79°40'
Hubbard Brook, New Hampshire	43°57'	71°42'
Georgia Station, Georgia	33°11'	84°24'
Marcell, Minnesota	47°32'	93°28'
Meade, New Brunswick	39°48'	81°32'
Caldwell, Ohio	39°48'	81°32'
Coweeta, North Carolina	35°04'	82°26'

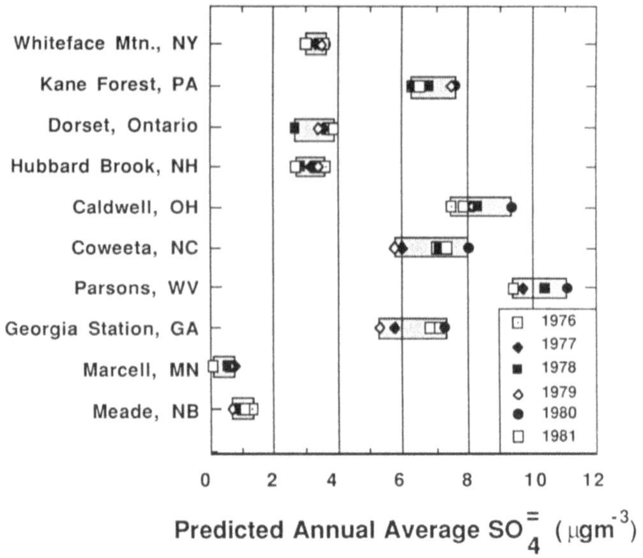

Figure 5. The predicted annual ambient $SO_4^=$ concentrations for each of the sites listed in Table I for each of six years.

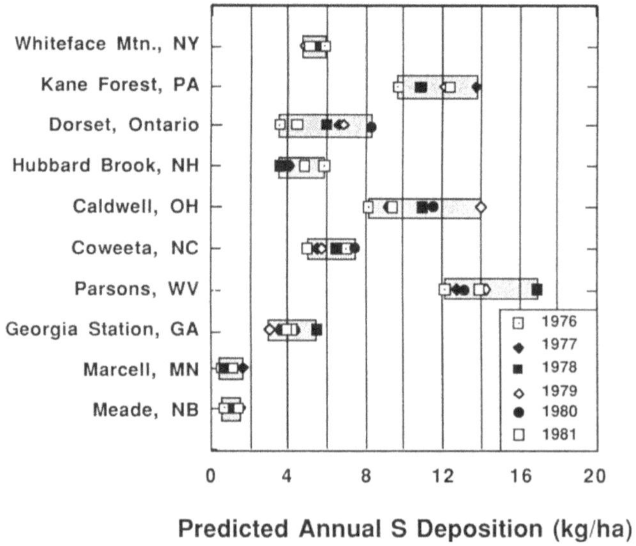

Figure 6. The predicted annual *S* wet depositions for each of the sites listed in Table I for each of six years.

4.3. Variations in Estimated Source-Receptor Relationships Due to Variations in Meteorological Conditions

The changing contributions of States and Provinces to annual average $SO_4^=$ concentrations and total annual *S* wet deposition at one of the receptors have been estimated using the ACID model in the backward mode. The changes in contribution reflect only the changing meteorological conditions from year-to-year.

Figure 7 shows the estimated absolute contribution of several States and/or Provinces to the annual average $SO_4^=$ concentrations at Whiteface Mountain, New York for each of six years modeled. Predicted absolute contributions changed considerably from year-to-year. The estimate of contribution from Ohio, for example, ranged from about 0.4 μgm^{-3} in 1981 to over 0.6 μgm^{-3} in 1979. Ontario, which was estimated to contribute the highest amount over the year was relatively stable in its contribution level. This variability could indicate the more episodic nature of transport from States in the Ohio River Basin.

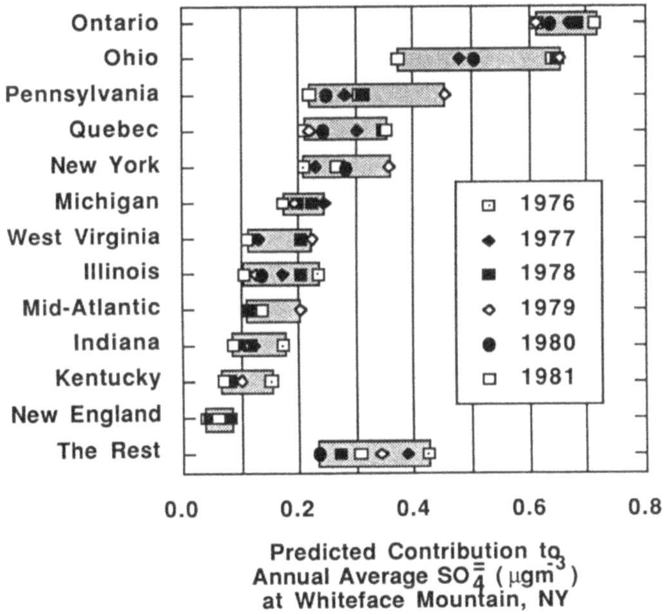

Figure 7. Variations in predicted contributions of various States and Provinces to annual ambient $SO_4^=$ concentrations at Whiteface Mountain, NY over a 6-yr period. It is assumed that SO_2 emissions for all years are equal to the 1980 emission rate, thus these numbers represent variations due to changing meteorological conditions only.

Figure 8 contains the estimated absolute contribution of several States and/or Provinces to the annual total S wet deposition at Whiteface Mountain, New York for each of the six years modeled. The figure shows that the predicted absolute contribution to S wet deposition can also change considerably from year-to-year. The estimate of contribution from Ohio ranged from about 0.9 kg ha^{-1} of S in 1979 to about 1.2 kg ha^{-1} in 1980. The increased importance of Ohio and Pennsylvania sources for S wet deposition reflect the bias in prevailing winds when precipitation occurred at the Whiteface Mountain site.

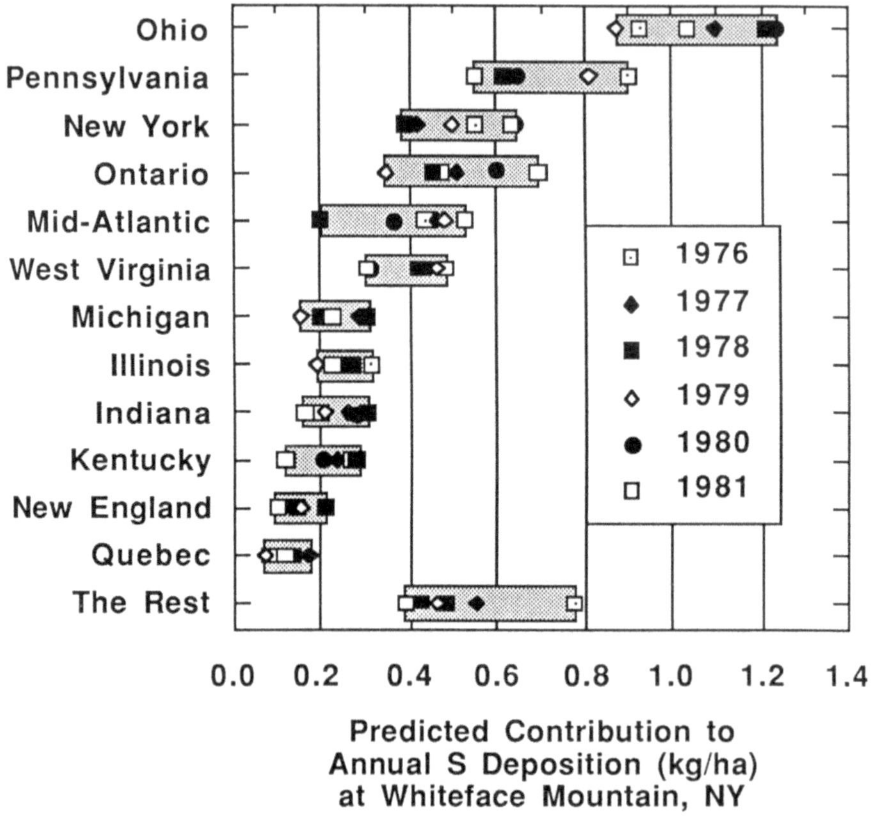

Figure 8. Variations in predicted contributions of various States and Provinces to S wet deposition at Whiteface Mountain, NY over a 6-yr period. It is assumed that SO_2 emissions for all years are equal to the 1980 emission rate, thus these numbers represent variations due to changing meteorological conditions only.

5. SUMMARY AND CONCLUSIONS

The failure to include sub-annual emissions variations in a Lagrangian regional-scale transport and deposition model with linear, first-order chemical conversion rates lead to variations in spatial patterns of source contributions of the order of 20% or less. While this variation is relatively small we have ignored possible non-linear interactions due to changing emissions of other pollutants such as nitrogen oxides and hydrocarbons.

Variations in meteorological conditions from year-to-year lead to variations in simulated $SO_4^=$ concentrations and *S* deposition independent of the influence of changing emissions. Estimates of deposition over six years varied from -20% to +35% of the $SO_4^=$ simulated for 1980. The variation in total annual *S* wet deposition was roughly ±45% of the amount simulated for 1980. The implication of these variations is that monitoring of trends in air concentrations and deposition due to changing emission values will need to be maintained for a sufficiently long time so that the trend could be discerned from the "noise" introduced by meteorology.

Similarly, the changes in simulated source-receptor relationships from year-to-year were substantial for certain State and Province source regions and the Whiteface Mountain location. The variations predicted suggest that the success of targeted emission control scenarios could be very difficult to evaluate.

REFERENCES

Clark, T.L.: 1985, private communication.

Heffter, J.L.: 1980, Air Resources Laboratories Atmospheric Transport and Dispersion Model (ARL-ATAD), NOAA Tech. Memo. ERL ARL-81, Air Resources Laboratories, Silver Springs, MD, 17 pp.

Keeler, G.J., Samson, P.J., and Small, M.J.: 1984, Representativeness of precipitation data in regional-scale acid deposition modeling, in *The Meteorology of Acid Deposition*, ed. P.J. Samson, Air Pollut. Cont. Assoc., Pittsburgh, PA, 225-240.

Samson, P.J. and Small, M.J.: 1984, Atmospheric trajectory models for diagnosing the sources of acid precipitation, in *Modeling of Total Acid Precipitation Impacts*, ed. J.L. Schnoor, Acid Precipitation Series, Vol. 9, Butterworth Publ., Boston, MA, 1-24.

Scott, B.C.: 1982, Theoretical estimates of the scavenging coefficient for soluable aerosol particles as a function of precipitation type, rate and altitude, *Atmos. Environ.*, **16**, 1753-1762.

ESTIMATION OF SOURCE-RECEPTOR MATRICES FOR DEPOSITION OF NO_x-N

Jack D. Shannon and Barry M. Lesht
Environmental Research Division
Argonne National Laboratory
Argonne, IL 60439 USA

ABSTRACT. The contributions of the anthropogenic sources of NO_x from various combinations of contiguous U.S. states or Canadian provinces to integrated deposition across selected states or provinces are estimated with the Advanced Statistical Trajectory Regional Air Pollution (ASTRAP) model. The model assumes linearity between emissions and deposition, and uses the same parameterization methods, although with different rates, as in simulations of transport and deposition of SO_x. Vertical distributions of emissions for the two classes of pollutants are substantially different in the gridded inventories used in simulations, with a weighted mean effective emission height of 160 m for NO_x and 310 m for SO_x. This might be expected to lead to an effective transport distance before deposition shorter for NO_x than for SO_x. However, the calculated fraction of NO_x emissions deposited within the contiguous United States and Canada south of 60 deg N (57%) is not greatly different from the fraction calculated for SO_x emissions (54%). This suggests that there may be compensating factors in the horizontal distribution of NO_x emissions, and in the lower dry deposition velocities for NO/NO_2 than for SO_2 in ASTRAP.

1. INTRODUCTION

Previous applications of linear models of long-range transport and deposition, particularly in regard to estimation of source-receptor transfer matrices, have primarily involved SO_x, as most proposed regulatory or policy actions have dealt with control of emissions of SO_2. In addition, the reactions of SO_2, particularly in clear air, have been found to be relatively slow and irreversible and thus amenable to the simplified parameterizations normally found in long-term regional models. Reactions of NO_x, on the other hand, have been studied primarily on the urban scale, where they are relatively rapid and reversible. However, it may be that for long-term, regional-scale modeling of transport and deposition, the competing reactions involving NO and NO_2 can be reasonably approximated with a linear, irreversible rate of transformation of combined NO/NO_2 to an imprecisely defined NO_3^-

Water, Air, and Soil Pollution **30** (1986) 815–824.

species, which would combine both HNO_3 and particulate nitrate. This is the approach we have taken here.

2. METHODS

The model that we have applied is the Advanced Statistical Trajectory Regional Air Pollution (ASTRAP) model (Shannon, 1981) A detailed discussion of the theoretical basis, assumptions, and limitations of the model can be found in Shannon (1985). Briefly, the ASTRAP model simulates monthly to seasonal average atmospheric concentrations and cumulative dry and wet deposition over scales from 100 km to continental distances. While the monthly or seasonal resolution is an inherent feature of the model, the minimum spatial resolution is determined by the resolution of the meteorological and emissions input. Although winds are resolved only to the approximate spacing of the upper-air network (300 to 400 km), while precipitation and emissions are resolved to 100 to 130 km, depending upon latitude, it is assumed that the influence of the winds on long-term deposition is dominated by synoptic-scale flow patterns and thus the lack of higher spatial resolution in the wind field is not critical to model resolution.

ASTRAP consists of three subprograms. One subprogram is a highly parameterized one-dimensional (vertical) numerical integration treating linear chemical transformation, dry deposition, diffusion, and slow leakage to the free troposphere, from each effective emission layer in turn. The parameterizations of gaseous and particulate dry deposition velocities, (Wesely and Hicks, 1977; Wesely and Shannon, 1984; Wesely et al., 1984) gas to particle transformation (Alkezweeny and Powell, 1977; Gillani et al., 1981; Meagher and Bailey, 1983), and profiles of vertical eddy diffusivity (Yamada, 1977; Hicks et al., 1981) vary diurnally and seasonally, but not daily nor horizontally. The deposition velocities and transformation rates peak near midday and have minimum values during the night, as they are all related, directly or indirectly, to the input of solar radiation. The seasonal variation of these parameters is approximated by scaling the basic diurnal curves, with some modification for the changing length of daylight hours. The amount of high quality field data on NO_x chemical transformations on scales appropriate to ASTRAP (seasonal and regional) is very limited. The rates used here have been selected primarily by analogy to S transformation parameterizations in ASTRAP, but are about 2.5 times larger. The diurnal pattern of vertical eddy diffusivity represents a typical cycle of nocturnal inversion formation, intensification, erosion from below and dissipation, with the maximum height of vigorous mixing and the timing of the diurnal pattern varying with season. Average, maximum and minimum values of parameterizations are given in Table I. Output of the subprogram includes statistics on the one-dimensional surface concentration, airborne burden, and dry deposition increment.

Horizontal dispersion and wet deposition statistics are produced in a second subprogram that calculates trajectories with wind and precipitation fields updated at 6-hour intervals. Mean trajectories, described by fitting bivariate normal distributions (puffs) to the time series of trajectory endpoint ensembles at 6-hr plume-age intervals, are calculated for a grid of virtual sources covering the U.S. and Canada. Wet removal of bulk pollutant (independent of speciation) is a function of the half-power of the precipitation per 6 hr. Amounts less than 1 mm are ignored, since the precipitation analysis procedure artificially increases precipitation frequency. A precipitation amount of 1 cm removes all pollutant from the mixed layer (2 cm is required for S removal in winter precipitation in the north); one half of the mass removed is deposited at the surface while the remainder of the calculated removal from the mixed layer is exported to the free troposphere where it is subject to later wet removal but is no longer subject to dry removal. This parameterization is an attempt to approximate the vertical redistribution of pollutant resulting from convective precipitation processes. Analysis of recent field data, particularly from the Processing of Emissions in Clouds and Precipitation (PRECP) program, may lead to better wet removal parameterizations; at this time parameterizations developed for bulk S removal (Hicks and Shannon, 1979) give reasonable results for NO_x-N.

TABLE I: MEAN VALUE AND DIURNAL RANGE OF ASTRAP PARAMETERIZATIONS

Parameter	Season	Average	High	Low
NO/NO_2 DRY DEPOSITION VELOCITY	Winter	0.20	0.45	0.08
	Spring	0.25	0.50	0.08
($cm\ s^{-1}$)	Summer	0.30	0.60	0.08
	Autumn	0.16	0.40	0.08
NO_3^- DRY DEPOSITION VELOCITY	Winter	0.40	0.80	0.15
	Spring	0.30	0.60	0.08
($cm\ s^{-1}$)	Summer	0.60	1.20	0.16
	Autumn	0.30	0.60	0.08
NO/NO_2 TO NO_3^- TRANSFORMATION RATE	Winter	1.0	2.8	0.4
	Spring	3.0	8.0	0.8
($\%\ hr^{-1}$)	Summer	4.0	10.0	1.0
	Autumn	1.8	4.8	0.6
MIXED LAYER DEPTH	Winter	400	1000	100
	Spring	800	1800	100
(m)	Summer	1000	1800	100
	Autumn	600	1400	100

The horizontal trajectory and vertical integration statistics are then used with an emission inventory in calculation of concentrations and deposition at specified receptor points. If a grid of receptor points representing cells of known area is used, the cells can be identified as to the state or province within which they lie, and average concentrations and total deposition across each state can then be generated. If the emission inventory is also identified as to the state or province of each source, source-receptor matrices can be produced.

In the results of ASTRAP simulations to follow, emission estimates for 1980 are used with each year of meteorology from 1976 through 1981 in turn to produce a 6-yr average. This should reduce the effects of climatological variability. The emission inventories of SO_x for both the U.S. and Canada and the U.S. inventory of emissions of NO_x are processed preliminary versions from the National Acid Precipitation Assessment Program, while the Canadian emission inventory for NO_x has been provided by the Atmospheric Environmental Service. The NO_x and SO_x emissions have been processed into a three-dimensional grid for each state or province, with the effective emission layers as defined in Table II and horizontal grid cells of 100 to 120 km on a side, increasing with latitude. The vertical distribution of the emissions has been estimated by applying the formula of Briggs (1971) to stack data, where ambient temperatures have been given a seasonal and latitudinal variation and gridded mean wind fields have been varied by a factor of two; most point source emissions are thus distributed among two or more layers. The horizontal location of the virtual source representing each grid cell has been computed from a weighted mean of the source locations within each cell, and initial plume spreads have been calculated from the standard deviations of sources about the

TABLE II: VERTICAL SUMMARY OF ASTRAP EMISSIONS INPUT

EMISSION LAYER (m)	SO_x (kT SO_2)	NO_x (kT NO_2)
0-100	4512	13284
100-200	7161	2816
200-300	4727	1845
300-400	4602	1530
400-600	5283	1625
600-800	3670	913

virtual source location. The entire NO_x emission field is represented by a total of 962 cells, although most cells do not contain emissions within all layers. As initially uniform pollutant distributions within each layer are assumed in ASTRAP, the vertical summation in Table II implies a mean effective emission height of 160 m for NO_x and 310 m for SO_x. The emissions of NO_x tend to be from much lower effective plume heights than is the case for SO_x because the transportation and industrial emission sectors are much more important for NO_x emissions. If the transformation and removal parameterizations for NO_x and SO_x were identical, this would lead to a shorter distance scale for NO_x dry and total deposition, since dry deposition is proportional to surface concentrations, which near the source would be higher for low-level emissions.

3. RESULTS AND DISCUSSION

Estimates of NO_x deposition are expressed in the tables to follow as equivalent nitrogen, in a manner analogous to deposition of SO_x in equivalent sulfur, because the wet removal parameterization is independent of speciation and because gas and particle are already lumped species. Nitrogen deposition calculations do not include any input from ammonia or ammonium.

The calculations used in producing source-receptor matrices are also represented in the NO_x dry and wet deposition fields shown in Figures 1 and 2. While no comparisons can be made with observations of dry deposition, the wet deposition field is generally consistent with observed values. Since monitoring networks are sited to minimize the effects of local sources, they should underestimate deposition peaks to some extent. On the other hand, part of the measured nitrate in precipitation may be produced by transformation of ammonium or by natural sources of N species. A comparison of ASTRAP simulations with observations during 1981 is shown in Figure 3. The coefficient of correlation is 0.86, explaining 74% of the observed variance.

A deposition budget for NO_x-N is given in Table III. Because of the large difference between the national emission inventories, emissions from the United States account for the larger share of wet and total deposition within Canada, as well as almost all deposition within the United States. For comparison, a deposition budget for SO_x is given in Table IV. Largely because the U.S./Canada SO_x emission ratio is only about 5:1 compared to 11:1 for NO_x, the estimated U.S. contribution to total S deposition in Canada is slightly less than one half, while the estimated contribution of U.S. sources to NO_x-N total deposition in Canada is almost two thirds.

Dry and wet NO_x-N deposition for aggregated sources and a few selected receptors are shown in Tables V and VI. It appears that an NO_x emission control strategy that applied only to the Northeast (Regions I-III) could have almost no effect upon deposition across Minnesota, North Carolina, or Ontario. The western states appear to be the dominant source of NO_x-N deposition in Minnesota, the southern states in North Carolina, and the central and western Great Lakes states dominate in Ontario. Emissions in the Northeast do appear to contribute one half of the NO_x-N deposition across New York and Maine.

From Tables III and IV, ASTRAP simulations calculate a total deposition of NO_x products over the U.S. and Canada that is only a slightly higher fraction of total NO_x emissions (57%) than the corresponding fraction for SO_x emissions (54%). This is a somewhat surprising result, since SO_x, on the average, is emitted from higher sources than is NO_x. For a specific pollutant, decreasing effective emission height increases local surface atmospheric concentrations and thus local dry deposition, and leaves less pollutant mass available for long-range transport. However, the effect in ASTRAP simulations for SO_x and NO_x-N deposition is confounded by differing parameterization rates for the two pollutant systems. In addition, different horizontal distributions of emissions of NO_x and SO_x across the United States and

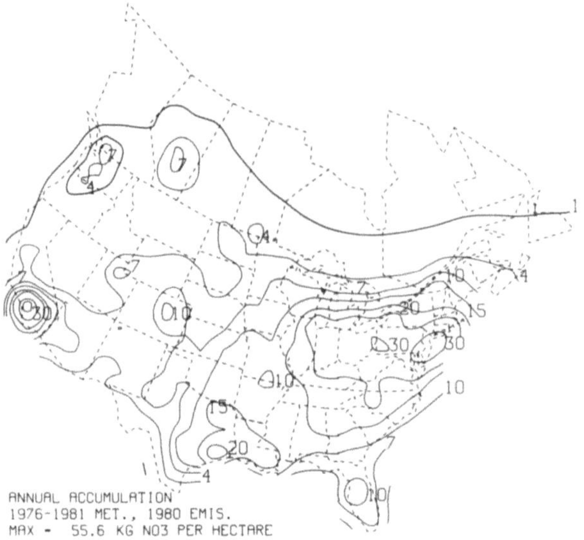

Fig. 1. ASTRAP simulation of the annual dry deposition of NO_x-N, in nitrate equivalent, for 1980 emissions and climatology for 1976 through 1981.

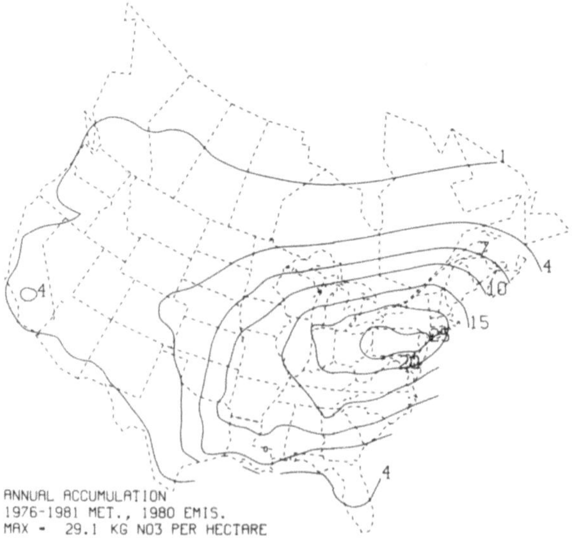

Fig. 2. ASTRAP simulation of the annual wet deposition of NO_x-N, in nitrate equivalent, for 1980 emissions and climatology for 1976 through 1981.

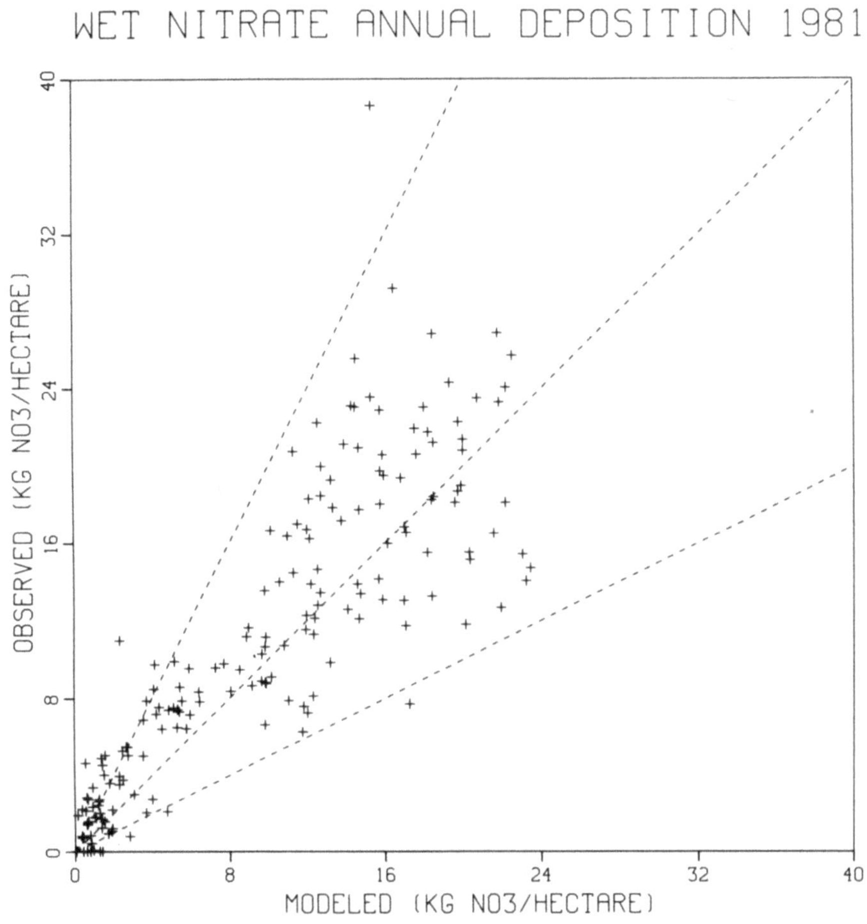

Fig. 3. Comparison of ASTRAP simulations with observed annual wet nitrate deposition during 1981. (Dashed lines are for a perfect fit and for factor of two range about a perfect fit.)

TABLE III: ANTHROPOGENIC NO_x-N DEPOSITION BUDGET (ktonne of equivalent N)

Source	Emissions	Dry		Wet		Total	
		U.S.	Canada	U.S.	Canada	U.S.	Canada
U.S.	6147	1706	108	1335	309	3041	417
Canada	555	65	146	25	85	90	231
Combined	6702	1771	254	1360	394	3131	648

TABLE IV: ANTHROPOGENIC S DEPOSITION BUDGET (ktonne of equivalent S)

Source	Emissions	Dry		Wet		Total	
		U.S.	Canada	U.S.	Canada	U.S.	Canada
U.S.	12580	3327	248	2625	685	5952	933
Canada	2400	163	593	77	394	240	987
Combined	14980	3490	841	2702	1079	6192	1920

TABLE V: SOURCE-RECEPTOR MATRIX FOR DRY DEPOSITION OF NO_x-N

Source	Emissions[1]	Receptor					
		MN	NC	NY	ME	ONT	NS
EPA Region I	162	*[2]	*	3.0[3]	3.9	0.2	0.8
" " II	309	*	0.3	20.6	2.3	1.9	0.6
" " III	676	*	8.7	10.1	1.7	2.5	0.5
" " IV	1111	0.1	37.8	1.4	0.2	1.2	0.1
" " V	1251	14.1	6.9	13.3	2.3	26.7	0.7
Western U.S.	2638	10.1	2.9	1.2	0.2	7.5	*
Atlantic Prov.	56	*	*	*	0.1	*	1.8
Quebec	101	*	*	0.7	3.7	0.9	0.6
Ontario	163	0.1	0.2	8.5	1.3	15.8	0.3
Western Canada	235	3.4	0.1	0.2	0.1	6.1	*
U.S. Total	6147	24.2	56.5	49.4	10.6	40.0	2.7
Canada Total	555	3.6	0.4	9.4	5.1	22.8	2.8
Total	6702	27.8	56.9	58.8	15.8	62.8	5.5

[1]Annual emission in equivalent N (ktonne)
[2]Asterisk indicates less than 0.05
[3]Annual deposition in equivalent N (ktonne)
Region I = New England; Region II = NY and NJ; Region III = PA, WV, DE, MD, VA, DC; Region IV = FL, MS, AL, GA, TN, SC, NC, KY; Region V = OH, IN, MI, IL, WI, MN

TABLE VI: SOURCE-RECEPTOR MATRIX FOR WET DEPOSITION OF NO_x-N

Source	Emissions[1]	Receptor					
		MN	NC	NY	ME	ONT	NS
EPA Region I	162	*[2]	0.1[3]	1.7	3.7	0.7	1.0
" " II	309	*	0.4	10.3	3.8	4.3	1.4
" " III	676	0.1	3.6	17.2	5.8	10.7	2.0
" " IV	1111	0.8	28.2	8.6	3.3	11.0	0.7
" " V	1251	8.4	4.1	19.5	6.1	53.6	2.0
Western U.S.	2638	18.8	6.2	5.3	0.5	29.7	0.4
Atlantic Prov.	56	*	*	*	0.2	*	0.8
Quebec	101	*	*	0.4	1.5	0.8	0.4
Ontario	163	0.1	0.2	3.7	1.5	10.7	0.5
Western Canada	235	1.5	0.1	0.2	0.1	5.6	*
U.S. Total	6147	28.2	42.6	62.6	23.2	110.2	7.5
Canada Total	555	1.6	0.3	4.3	3.2	17.0	1.8
Total	6702	29.8	42.9	66.9	26.4	127.2	9.3

[1]Annual emission in equivalent N (ktonne)
[2]Asterisk indicates less than 0.05
[3]Annual deposition in equivalent N (ktonne)
Region I = New England; Region II = NY and NJ; Region III = PA, WV, DE, MD, VA, DC; Region IV = FL, MS, AL, GA, TN, SC, NC, KY; Region V = OH, IN, MI, IL, WI, MN

Canada may result in shorter average travel times to the coasts for NO_x emissions as a whole.

It should be emphasized here that these estimates of source-receptor relationships are as yet unverified with field data, and that verification of ASTRAP NO_x simulations without regard to source attribution has been performed only for wet nitrate deposition. While model performance as depicted in Figure 3 is quite promising, a more rigorous comparison including air quality and dry deposition is necessary to quantify overall model uncertainties.

Acknowledgement: This work has been supported by the U.S. Environmental Protection Agency as part of the National Acid Precipitation Assessment Program. Because it has not been subject to Agency review, this paper does not necessarily reflect the views of the Agency. The preparation of meteorological data was conducted at the University of Michigan under the guidance of Prof. Perry Samson; the processing of emission inventories was done by Ms. Carmen Benkovitz at Brookhaven National Laboratory and Dr. Eva Voldner of the Atmospheric Environment Service of Canada. Dr. Brand Niemann of EPA provided a verification set for wet deposition. We gratefully acknowledge their contributions.

REFERENCES:

Alkezweeny A. S. and Powell, D. C.: 1977, Atmos. Environ. **11**, 179.

Briggs, G. A.: 1971, Some recent analysis of plume rise observations, Proc. 2nd International Clean Air Congress, England, H. M. and Baery, W. T. (ed.) Academic Press, New York.

Gillani, N. V., Kohli S., and Wilson, W. E.: 1981, Atmos. Environ. **15**, 2293.

Hicks, B. B. and Shannon J. D.: 1979, J. Appl. Meteor. **18**, 1415.

Hicks, B. B., Hess, G. D. Wesely, M. L., Yamada, T., Frenzen, P., Hart, R. L., Sisterson, D. L., Hess, P. E., Kulhanek, F. C., Lipschutz, R. C., and Zerbe, G. A.: 1981, The Sanagamon Field Experiments: Observations of the Diurnal Evolution of the Planetary Boundary Layer Over Land. Argonne National Laboratory Report ANL/RER/81-1.

Meagher, J. F. and Bailey, E. M.: 1983, J. Geophys. Res. **88**, 1525.

Shannon, J. D.: 1981, Atmos. Environ. **15**, 689-701.

Shannon, J.D.: 1985, User's Guide for the Advanced Statistical Trajectory Regional Air Pollution (ASTRAP) Model. U.S. Environmental Protection Agency Report EPA/600/8-85/016 (NTIS PB85-236784/XAB).

Wesely, M. L. and Hicks, B. B.: 1977, J. Air Poll. Control Assoc. **27**, 1110.

Wesely, M. L. and Shannon, J. D.: 1984: Environ. Progress **3**, 78.

Wesely, M. L., Cook, D. R., Hart, R. L., and Speer, R. E.: 1984, J. Geophys. Res. **90**, 2131.

Yamada, T.: 1977, Analyses of the Sangamon data. Argonne National Laboratory Report ANL/RER-77-65-IV, 49.

SOURCE-RECEPTOR RELATIONSHIPS FOR TRACE ELEMENTS IN NORTHERN EUROPE

Jozef M. Pacyna
Norwegian Institute for Air Research
P.O. Box 130, N-2001 Lillestrøm,
Norway

ABSTRACT. Measured air concentrations of several trace elements in southern Scandinavia have been related to long range transport and emissions of the same elements from sources in Europe using receptor modelling and statistical trajectory sector analysis. A good agreement, within a factor of 1.5 between estimates and measurements, has been obtained for Mn, Ni, Cu, and As. Significant differences were noted for V and Pb.

1. INTRODUCTION

An assessment of the contribution of source emissions to the ambient levels of air pollutants is a key requirement in air pollution studies. To obtain this information, data from emission surveys of air pollutants are related to the chemical composition of aerosols measured at remote areas, by applying various long-range transport models.

In the present paper, the emission data of several trace elements from various anthropogenic sources in Europe (Pacyna, 1984) are compared with measured air concentrations in Northern Europe (Pacyna et al., 1985), with the help of trajectory model calculations and trajectory sector statistics. Further, a group of trace elements has been selected to serve as indicators of long-range transport of air pollutants from Europe to Scandinavia.

2. ANTHROPOGENIC EMISSIONS OF TRACE ELEMENTS IN EUROPE

Emission estimates were based on: (1) trace element emission factors, estimated separately for all the sources considered, and (2) statistical information on the consumption of ores, rocks and fuel, and the production of various types of industrial goods. Fourteen trace elements: As, Be, Cd, Co, Cr, Cu, Mn, Mo, Ni, Pb, Sb, Se, V, and Zn were considered. The results of calculations are presented in Table I.

The amounts of the trace element discharges are affected by: (1) impurities in fuels, ores, and other raw materials, (2) physical-chemi-

Water, Air, and Soil Pollution **30** (1986) 825–835.

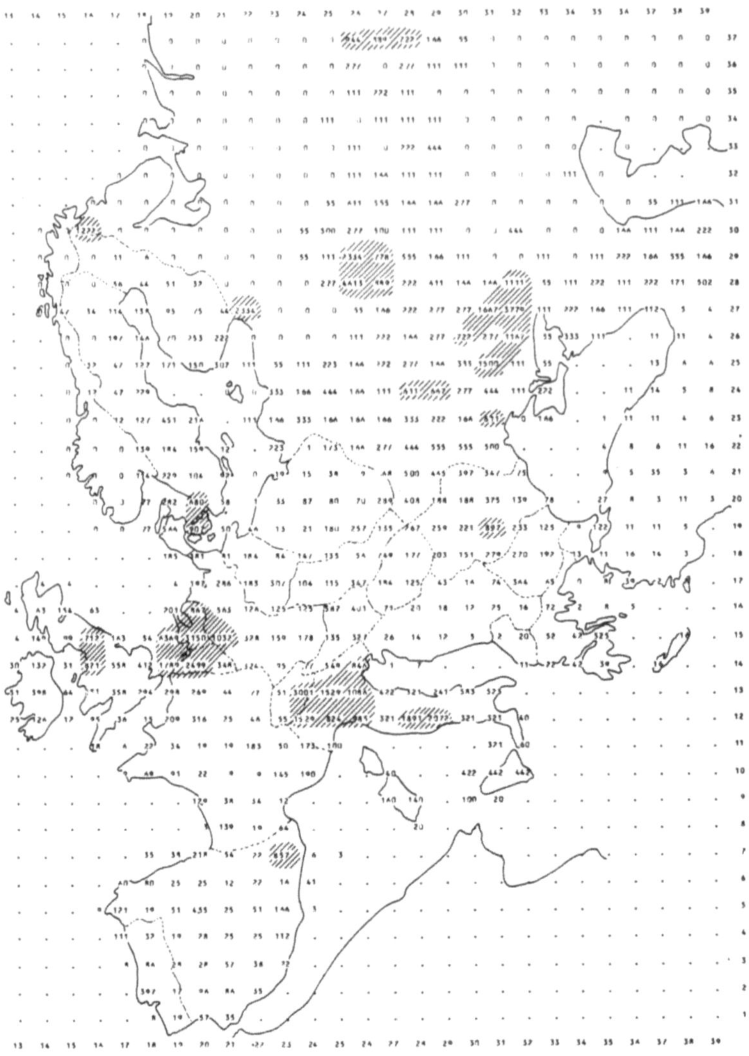

Fig. 1. Spatial distribution of vanadium emission in Europe in 1979 in 10^2 kg . yr^{-1}. The shaded areas represent locations with emissions higher than the average of the grid area.

cal properties of trace elements, (3) production technology employed, and (4) effectiveness of emission control devices (Pacyna, 1984).

All these factors have created a number of uncertainties in the emission estimates presented in Table I. Information on the content of trace elements in fuels and other raw materials, the efficiencies of control devices and the uses of different technologies to produce industrial goods is not always adequate. For some of the emission sources in Europe (e.g. waste incineration), data from the U.S. have been used to predict emission rates. Considering the substantial variations in the chemical composition of source materials and differences in the emission control technologies utilized, the use of U.S. data may not be always justified. However, relevant data from some European countries were difficult to obtain.

Source-receptor models require emission data to be spatially distributed. In this work, the spatial distributions of the trace elements were based on the location and capacities of power plants, smelters, mines, incinerators, steel and iron factories, and other industrial plants. In addition, any applicable European data from the literature were used, such as distributions of SO_2 emissions (Semb, 1978; Amble, 1981), distributions of NO_x emissions from fossil fuel combustion (Semb and Amble, 1981), and population density and energy consumption statistics. The EMEP grid system, with a resolution of 150 km, was used for the spatial distribution of the European emissions.

An example of distribution for V is given in Figure 1. Vanadium from oil combustion is emitted mainly in Western Europe and in the European part of the Soviet Union. The estimated spatial distributions of other elements are available in an other paper (Pacyna, 1984).

3. MEASUREMENTS OF ATMOSPHERIC TRACE ELEMENTS IN SCANDINAVIA

Time series of aerosol measurements at three EMEP (European Monitoring and Evaluation Programme) stations in Scandinavia, during February-June, 1980, were employed to relate estimates of the European emissions to measured ambient air concentrations.

The stations Birkenes in southern Norway, Rörvik near Gothenburg in Sweden, and Virolahti in southern Finland were selected for the study. With southerly winds, all these stations are exposed to long range transport of air pollutants from continental sources. Regular EMEP measurements of SO_2 and bulk aerosol, as well as aerosols in coarse and fine size fractions (< 2.5 μm EAD and 2.5 to 15.0 μm EAD) were used. The aerosol samples were analyzed by particle-induced X-ray emission analysis (PIXE).

Emission estimates can be related either to ambient concentrations measured during selected periods, (e.g. episodes of long-range transport of air pollutants, indicated by high concentrations of several trace elements), or to average concentrations over the entire period of sampling.

4. SOURCE-RECEPTOR RELATIONS FOR EPISODES OF LONG RANGE-TRANSPORT OF AIR POLLUTION

The origin of aerosols, measured during the episodes of long-range transport of air pollutants, can be analyzed with the help of a simple trajectory model. The model is based on the mass-balance equation:

$$\frac{dq}{dt} = (1 - \alpha) \frac{Q}{h} - k \cdot q \qquad (1)$$

where:

q = trace element concentration in air, ng m^{-3}.
t = time, s,
Q = trace element emission per unit area and time, ng $m^{-2} s^{-1}$, referred to the EMEP grid system,
h = height of mixing layer, m,
k = decay rate for the trace element considered (wet and dry deposition), s^{-1},
α = fraction of trace element emission deposited in the same grid element as it is emitted. (This local deposition supplements the deposition included in the decay rate, k.)

Dry deposition studies in the sorroundings of Cu and Pb smelters (e.g.,Glowiak et al., 1977) and coal-fired power plants (Pacyna, 1980) in Poland were used to assess the "local deposition" of the pollutants (coefficient "α" in Equation (1)). In these studies 11 to 15% of the trace element releases were found deposited in the emission area. These, however, may be overestimats for emissions from very tall stacks equipped with high efficiency control devices.

Model concentrations are inversely proportional to the mixing height "h". From radiosonde data for Europe (Eliassen and Saltbones, 1982), a constant mixing height of 1000 m can be assumed as a preliminary estimate.

Decay rates for trace elements were estimated from deposition velocities of the pollutants for constant mixing height.

Wet deposition of trace elements was not considered in calculating the decay rates. Thus, only episodes with no or very little precipitation between emission and study area were selected, possibly resulting in a slight overestimation of trace element concentrations.

The other simplification was the assumption of uniform distribution of concentrations within the mixing layer.

The trajectory model was used to calculate the trace element concentrations at Birkenes and Rörvik in southern Scandinavia.

4.1. Episodes in Scandinavia

During several periods, concentrations of many anthropogenic trace elements peaked at all three locations, particularly during 14-16 April, 1980, at Birkenes and Rörvik. The peaks appeared 2 days later at Virolahti. An examination of meteorological maps revealed transport of air masses from central Europe to Scandinavia during this period. The

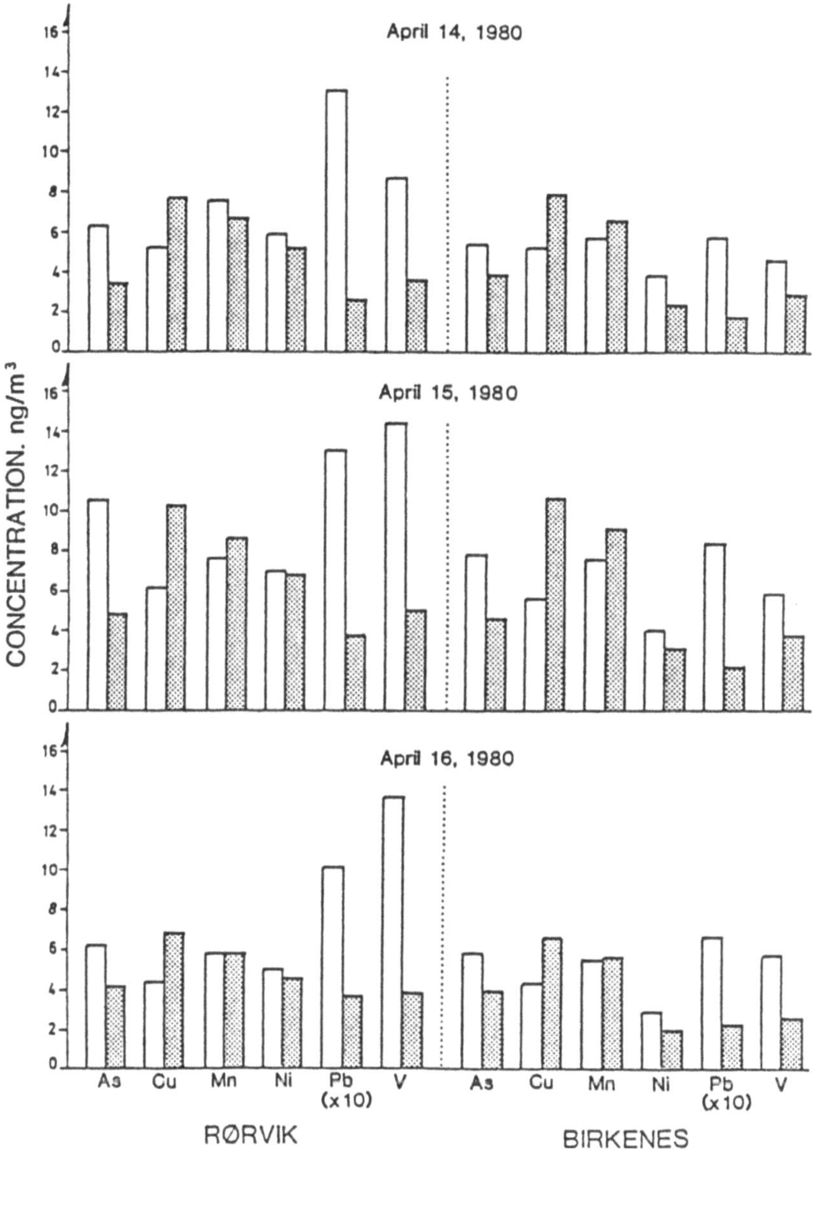

Fig. 2. Measured (open bars) and calculated concentrations (full bars) of trace elements at Rörvik and Birkenes in April 1980.

Table I: Annual emission of trace elements in Europe in 1979 (tonne y^{-1}) (Pacyna, 1984).

EUROPE	As	Be	Cd	Co	Cr	Cu	Mn	Mo	Ni	Pb	Sb	Se	V	Zn
1. Therm. power. pl.	284	21	101	787	1196	1377	1011	352	4580	1138	122	155	12600	1316
2. Indust.proc.comb.	394	29	155	1214	1580	2038	1378	493	7467	1652	154	218	21800	1824
3. Wood combustion	40		25			1500			375	562				4590
4. Gasoline comb.			31				92		1330	74300				
5. Mining			1			192	275		1640	1090		0.2		460
6. Primary n-f metal														
6.1 Cu-Ni	4490		595			7850				9250				2500
6.2 Zn-Cd	910		1550			440				7880		13		48800
6.3 Pb	300		8			120			140	10450				180
7. Second n-f metal														
7.1 Cu			2			61				55	1			660
7.2 Zn														2630
7.3 Pb			1							387				150
8. Iron, steel prod.			58		15400	1710	14770		340	14660				10250
9. Refuse incin.	11		84	4	53	260	114		10	804	100	32	19	5880
10. Phosphate fertil.			27			77			77	6				230
11. Cement prod.			15		663					746				
12. Industr. applic.	136													
13. TOTAL	6500	50	2700	2000	18900	15500	17600	850	16000	123000	380	420	34500	80000

850 mb air trajectories at Birkenes and Rörvik showed that the air masses had passed over the Federal Republic of Germany, the Netherlands, the German Democratic Republic, and Denmark.

The concentrations of the fine fraction of particles at Birkenes and Rörvik were compared with the model estimates for some selected trace elements, and the results are shown in Figure 2. The best agreement was obtained for Mn and Ni. Estimated concentrations were slightly higher for Cu, but lower for As and particularly for V and Pb. The elements Be, Cd, Co and Sb from Table I were not measured or difficult to analyse by PIXE. Thus, their emissions are not compared with measurements.

4.2. Possible reasons for differences

The major reasons for the differences between measurements and model predictions include: inadequate emission inventories, improper dry deposition velocities, disregard of wet deposition processes, wrong local deposition values, errors in the analytical procedures, and inaccuracies in trajectory calculations.

Uncertainties in anthropogenic emission inventories of trace elements have already been mentioned. The poor agreement between predictions and measurements for Pb and V at the Scandinavian sites is likely due to local combustion of residual oil and gasoline in Gothenburg (30 km from Rörvik) and Kristiansand located 25 km from Birkenes (e.g. Lannefors et al., 1983; SNV, 1982).

The most important natural source of trace elements is windblown dust. This, however, occurs mainly in desert areas, far from the present study region.

In the modelling the following dry deposition velocities for the aerosol-contained trace elements were assumed to be representative for the decay rate calculations: 0.1 cm s^{-1} for Cd and Se, 0.2 cm s^{-1} for As, Pb and Sb, 0.3 cm s^{-1} for V, 0.4 cm s^{-1} for Cr and Zn, and 0.5 cm s^{-1} for Mn and Ni (various papers by Davidson, e.g. Davidson, 1980).

It is extremely difficult to assess to what extend inaccuracies in trajectory calculations could have contributed to the total uncertainty of the model. Detailed discussions of all model uncertainties are available in other papers (Pacyna et al., 1984, 1985). The discrepancies between predicted and observed concentrations in our studies were typically within a factor of 2. Discrepancies of this magnitude are often encountered in even less complex situations, where the pollution sources have been better characterized, (e.g. Kowalczyk et al., 1982).

5. APPLICATION OF SECTOR STATISTICS

The impact of trace metal emissions in Europe on the ambient air concentrations measured in Scandinavia can be assessed not only from measurements during episodes of long range transport, but also from sector analysis of daily mean concentrations over a sampling period. Two approaches can be made. In the first, the average sectoral concentrations of trace elements in the fine fraction of particles are used to draw elemental diagrams for each sector at each station. In this study, the elemental concentrations, relative to V concentrations, were plotted for trace elements with more than 60% of mass in fine fractions. Then, the elements, that showed the largest differences between sectoral concentrations were selected to serve as elemental tracers for the European aerosol measured in Scandinavia. These elements included Cr, Cu, As, Se, Mo, and Sn, shown in Figure 3. To distinguish between the east and west European aerosols affecting receptors in Scandinavia, the concentrations of the above tracers were calculated separately for days with air pollution transport from western Europe, eastern Europe, and the Moscow and Urals areas to Scandinavia, deduced from 850 mb trajectories. The results are shown in Table II.

It should be noted that similar estimates of elemental tracers for total concentrations in this work have resulted in smaller differences between the east and west aerosols. Thus, total aerosols seem to be less suitable tracers than fine fraction aerosols.

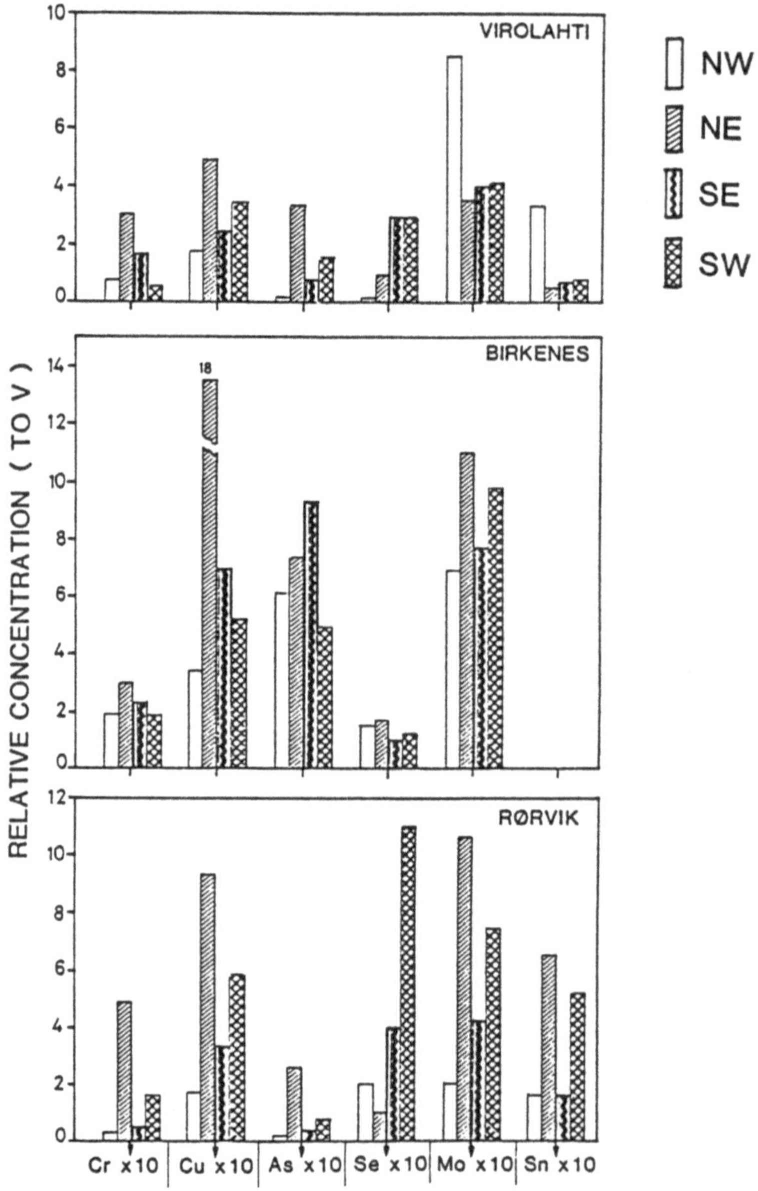

Fig. 3. The sectoral concentrations of trace elements relative to the V concentrations measured in fine fraction of particles at Virolahti, Rörvik, and Birkenes in April 1980.

Table II: Elemental ratios for the East and West European aerosol at at Birkenes and Rørvik

Ratio*[1]	West European*[2] 10 samples	East European*[3] 11 samples	Moscow + Urals 11 samples
Cr/V	0.23 - 0.37 0.30	0.12 - 0.17 0.14	0.5 - 0.8 0.70
Cu/V	0.70 - 1.10 0.92	0.07 - 0.20 0.13	1.2 - 1.4 1.1
As/V	0.80 - 1.10 0.91	0.15 - 0.37 0.28	3.4 - 4.1 3.8
Se/V	0.13 - 0.28 0.15	0.02 - 0.07 0.04	0.11 - 0.16 0.14
Mo/V	0.12 - 0.13 0.12	0.03 - 0.07 0.05	0.10 - 0.15 0.14
Sn/V	0.54 - 0.70 0.62	0.24 - 0.27 0.25	0.80 - 1.20 1.00

*[1] based on concentrations in particles < 2 µm EAD
*[2] including the United Kingdom, France, the Netherlands, Belgium, Luxemburg, and Federal Republic of Germany.
*[3] including German Democratic Republic, Poland, Czechoslovakia, Hungary and the following republics of the U.S.S.R: Lithuania, Latvia; Ukraine and White Russia.

In the second approach, the sectoral contributions of elemental concentrations at a given receptor point are related to the emissions within these sectors. The data in Table III show the ratio of element concentrations to element emission in a given sector, relative to the concentrations and emissions of V at the three stations in Scandinavia. Values less than one indicate that emissions have been overestimated relative to V, while values more than one that emissions were underestimated. Generally good agreement has been obtained for Mn, Ni, Cu, As, Se, and Pb, particularly for the southern European sectors.

6. CONCLUDING REMARKS

Emission estimates for trace elements from anthropogenic sources in Europe can be related to ambient air concentrations at remote locations in Scandinavia. To establish this source-receptor relationship it

Table III: Comparison of measured mean concentrations with modelled values, using V as the reference element*.

Element	S	Mn	Ni	Cu	Zn	As	Se	Pb	Cr
Birkenes									
NW	1.8	14.0	4.5	1.0	0.9	2.6	19	1.7	0.8
NE	1.1	23.0	4.1	4.1	15.0	3.7	3.3	6.4	0.8
SE	0.5	2.5	1.2	1.5	6.2	4.9	1.5	3.4	0.4
SW	1.0	2.8	1.7	1.3	3.9	2.6	3.4	2.0	0.3
Rørvik									
NW	0.9	1.8	1.3	0.7	3.5	1.0	3.5	1.9	0.3
NE	0.3	2.2	1.1	1.9	7.7	12.0	1.0	2.9	1.1
SE	0.2	1.1	0.9	0.6	3.0	1.7	0.6	1.6	0.1
SW	0.7	2.0	1.3	1.6	6.3	4.8	3.0	3.0	0.3
Virolahti									
NW	0.8	1.5	1.3	0.6	2.4	1.1	0.8	2.5	0.4
NE	0.2	4.2	1.0	1.0	7.0	1.6	0.2	2.2	0.7
SE	0.2	1.5	1.0	0.7	3.7	0.7	0.5	1.3	0.3
SW	0.4	1.4	1.1	0.7	1.8	0.7	0.7	1.0	0.1

*1 The data are in effect of

$$\frac{\dfrac{C_{i},\ \text{measured}}{C_{V},\ \text{measured}}}{\dfrac{E_{i,j},\ \text{calculated}}{E_{V}\ \text{calculated}}}$$

where C_i is a measured concentration of "i" and $E_{i,j}$, is a calculated emission of element "i" in sector "j",
V - vanadium

requires accurate emission surveys, air mass trajectories and information on the physical-chemical processes affecting the pollutants during their transport by air masses.

The assignment of the origins of remote aerosols, based upon the emission surveys, can be studied either for selected days with high concentrations and only slowly changing air trajectories, or for the entire sampling period. In the first case, a simple trajectory model can be used to assess the contribution of emissions from a certain source area to ambient concentrations of trace elements in the fine fraction of particles (i.e. <2 μm EAD), measured at remote receptors.

Sector statistics can also be used to assess the origin of aerosols at a given site from average sectoral concentrations for the whole period of measurements. In this method, a group of elements is selected as representative of certain emission area (s). These elements are often called tracers or indicators of the emission area (s). In this study for Scandinavian sites, the elements Cr, Cu, As, Se, Mo, and

Sn (relative to V) were found to be useful in differentiating the contributions of emissions from eastern and western Europe. It should be noted that the above indicators cannot be easily extended for source regions outside Europe due to possible differences in emission patterns.

REFERENCES

Amble, E.: 1981, Estimation of the spatial distribution of the population and the SO_2 emission in Europe and Turkey within the EMEP grid. Technical Report TR 8/81. The Norwegian Institute for Air Research, Lillestrøm, Norway.

Davidson, C.J.: 1980, Dry deposition of cadmium from the atmosphere. In: Cadmium in the Environment, J.O. Nriagu, ed., Wiley, New York, 115.

Eliassen, A., and Saltbones, J.: 1982, Modelling of Long Range Transport of Sulphur over Europe: A Two-Year Model Run and Some Model Experiments. EMEP/MSC-W Report 1/82. The Norwegian Meteorological Institute, Oslo, Norway.

Glowiak, B.J., Palczynski, R.J., and Pacyna, J.M.: 1977, Environmental lead and mercury in the area of a copper smelter. U.S. ERDA, Environmental Quality Report HASL-321, Health and Safety Laboratory, New York.

Kowalczyk, G.S., Gordon, G.E., and Rheingrover, S.W.: 1982, Envir. Sci. Technol., 16, 79.

Lannefors, H., Hanssen, H.-C., and Granat, L.: 1983, Atmos. Environ., 17, 87.

Pacyna, J.M.: 1980, Coal-fired power plants as a source of environmental contamination by trace metals and radionuclides. Habilitation thesis. Technical University of Wroclaw, Poland.

Pacyna, J.M.: 1984, Atmos. Environ., 18, 41.

Pacyna, J.M., Semb, A., and Hanssen, J.E.: 1984, Tellus, (Series B), 36B, 163.

Pacyna, J.M., Ottar, B., Hanssen, J.E. and Kemp, K.: 1985, The chemical composition of aerosols measured in Southern Scandinavia, Technical Report OR 66/84. The Norwegian Institute for Air Research, Lillestrøm, Norway.

Semb, A.: 1978, Atmos. Environ., 12, 455.

Semb, A., and Amble, E.: 1981, Emission of nitrogen oxides from fossil fuel combustion in Europe. Technical Report TR 13/81. The Norwegian Institute for Air Research, Lillestrøm, Norway.

SNV; 1982, Heavy metals and organic pollutants in the Swedish nature, Monitor 1982, Statens naturvårsverket, Solna, Sweden (in Swedish).

AMBIENT ACID AEROSOLS IN RURAL NOVA SCOTIA

Truis Smith-Palmer and Byron R. Wentzell
St. Francis Xavier University
Department of Chemistry
Antigonish, Nova Scotia
B2G 1C0

ABSTRACT. During the summer months, Nova Scotia is downwind of many of the large industrial areas of the eastern United States and central Canada. The studies described here were carried out in the summers of 1983 and 1984 in Antigonish, a small rural town in northeastern Nova Scotia with no local sources. Twenty-four hour samples were collected on Teflon filters using a dichotomous sampler, and were analyzed for sulphate, nitrate and chloride by ion chromatography and for strong acid using Gran titrations. Sulphate episodes (up to 27 $\mu g\ m^{-3}$) were frequent. They usually occurred in warm, hazy weather and were often accompanied by strong acid (up to 9 $\mu g\ m^{-3}$ as H_2SO_4).

1. INTRODUCTION

Nova Scotia (Figure 1) is in the vulnerable position of receptor site for pollutants from the southwest and west, i.e. it is downwind of most of the large industrial areas of North America (Shaw, 1979). Nova Scotian soils and water are particularly vulnerable because of their generally poor buffering capacity (Shaw, 1982). In addition, the majority of acid and sulphate particulate matter tends to be in the fraction of aerodynamic diameter less than 2 μm, which is also that fraction which can penetrate most deeply into the lungs. Recent studies suggest that relatively low concentrations of H_2SO_4 or even sulphate can aggravate already existing respiratory problems (Miller, 1974).

Sulphate episodes have previously been recorded in Nova Scotia (Shaw, 1979; Barrie et al., 1983), but the acidity of the ambient air has received little attention. This paper documents acid/sulphate episodes which occurred in a rural area of Nova Scotia during the summers of '83 and '84. The associated concentrations of various other species and certain metals were also investigated; further results will be reported in a subsequent paper.

2. SAMPLING STRATEGY

Water, Air, and Soil Pollution **30** (1986) 837–843.

Figure 1

Map of Nova Scotia showing the back surface trajectory from 1400Z, Julian Day 180, 1984 to 1200Z, Julian Day 181, 1984.

The sampling was conducted to measure the ambient concentration of inorganic pollutants to which the rural population may be exposed during summertime. Thus it was conducted in the backyard of a residence on one of the main streets of Antigonish, a town of approximately 5,000 in rural Nova Scotia (Figure 1), with no summertime local pollution sources. (In winter, university and domestic heating systems constitute a small but significant local pollution source.)

The sulphate, chloride and acid present in ambient aerosol were collected in a dichotomous sampler (cut-offs at 10 and 2.5 μm) on Teflon filters, which cause effectively no sulphate artifact and do not react with acid. Chloride is virtually all of marine origin, and was measured to allow us to estimate the amount of sulphate of marine origin.

3. EXPERIMENTAL PROCEDURE

Twenty-four hour samples (1200Z to 1200Z) were collected daily, Julian 230-253, (August 18 to September 11, 1983), by a Sierra-Anderson 241 dichotomous sampler with cut-offs at 10 and 2.5 μm, using 37 mm teflon filters with polypropylene rings and a 2 μm pore size (Ghia). Filters were sealed in petri dishes and stored at -14°C prior to analysis. Filters were cut exactly in half using a teflon coated razor on a teflon template. One half-filter was used to determine total acidity using a Gran Titration, the second half-filter was analyzed for SO_4^{2-}, NO_3^-, and Cl^- using Ion Chromatography (IC) (Dionex 12, fast anion column). Each half-filter was wetted with 0.1 mL ethanol to facilitate extraction by sonication in the appropriate solvent. The L.O.D. for SO_4^{2-}(3 x s. d. of blank) was 0.02 μg m^{-3} and for Cl^- was 0.04 μg m^{-3}.

The sulphate in each fraction was corrected for marine sulphate by assuming that all chloride was of marine origin, and that the marine SO_4^{2-}:Cl^- equivalent ratio 1:10.

The dichotomous sampler was also used as above from Julian 144-191, (May 23 to July 9, 1984), with the exception of the period 162-178 when samples were collected on alternate days.

Surface back trajectory analyses (24 hr), and surface weather analysis maps, were provided by the Atmospheric Environment Service (Bedford, NS).

4. RESULTS AND DISCUSSION

Sulphate and strong acid episodes were recorded in both sampling periods (Figures 2 and 3). These pollutants were concentrated in the fine fraction, with overall mean concentrations and standard deviations of 0.6±1.7 and 0.5±1.0 μg m^{-3} of acid (expressed as H_2SO_4) for the two studies, respectively, and 4.8±6.5 and 3.5±3.8 μg m^{-3} of sulphate, respectively. The coarse fraction was basic or neutral except during the largest acid episodes. On deposition rather than inhalation, some neutralization between fractions could occur and thus lessen the overall

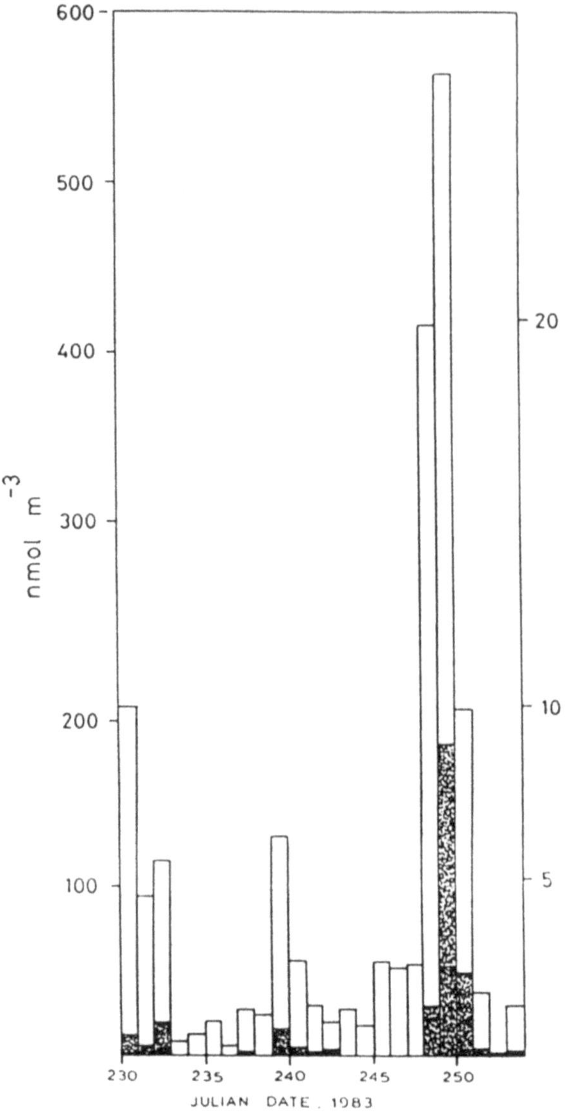

Figure 2

Sulphate and Strong Acid Concentrations (24h averages) for the first sampling period, Julian Days 230 - 253, 1983.

Sulphate □

Acid ▩

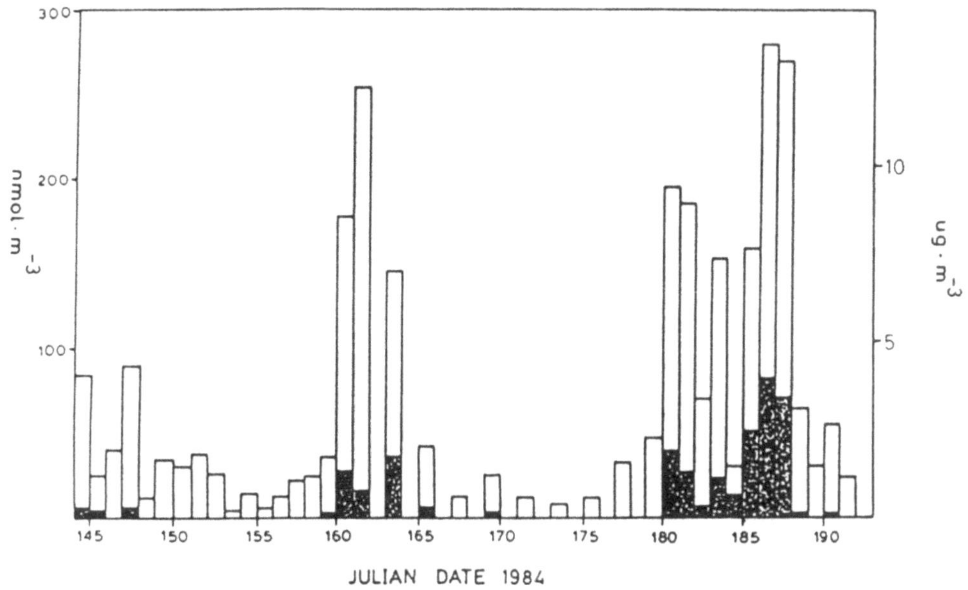

Figure 3

Sulphate and Strong Acid Concentrations (24h averages) for the second sampling period, Julian Days 144 - 191, 1984.

Sulphate □

Acid ▩

acidity. This is also the case if hi-Vol rather than dichotomous sampling is used. Sulphate pollution days (>10 $\mu g\ m^{-3}$ sulphate) occurred 10% of the sampling time, compared to a 5% frequency reported for Kejimkujik (southern Nova Scotia) over the 1977 sampling year (Barrie et al., 1983).

The sulphate values (Figures 2 and 3) were corrected for marine sulphate. The correction was significant for the fine fraction only on non-episodic days, but up to 75% of the coarse fractions were attributed to marine origin.

The ratio of acid/sulphate is an indicator of how much neutralization has occurred, and can be estimated by comparing the shaded to the unshaded areas of the histograms in Figures 2 and 3. The high acid episodes were associated with warm, hazy air and the air mass trajectories were usually over water for long distances (e.g. Figure 1), suggesting a lack of interaction with land-generated ammonia. There was a strong correlation ($r = 0.863$, $p < 0.001$) between acid and sulphate in both studies, supporting its identity as H_2SO_4 or NH_4HSO_4. Generally strong acids are present more frequently and in larger amounts in rural areas as opposed to urban areas, as neutralization can occur much faster over a city, although urban aerosols can also be acidic (Tanner et al., 1981).

Sampling on alternate days was initiated on day 162 of 1984 because a cold front from the northeast was seen to be moving in, suggesting that the sulphate episode was over. However, the front passed through within hours giving way to more hot hazy weather from the south and its associated sulphate. The last episode was the longest recorded in these studies, and was the result of several high pressure air masses moving in from the southwest. The low sulphate on day 182 was associated with general cleansing of the air mass by rain, while that on 184 was attributed to the passage of an intervening front with a westerly air flow.

The average sulphate concentrations are significantly higher than the average of 2.6±3.0 $\mu g\ m^{-3}$ found in an identically conducted 1983 winter study (Smith-Palmer et al., 1983), but the average winter acid is the same at 0.6±0.6 $\mu g\ m^{-3}$. The smaller standard deviation for the winter value reflects the fact that there were no very large episodes, but more frequent occurrences of small acid loadings. The mean acid concentrations are also comparable to that found for New Haven, Connecticut in a summer 1977 study (Tanner et al., 1981). Our maximum 24 hr acid value recorded was 9.1 $\mu g\ m^{-3}$ (as H_2SO_4), which is comparable to the 4 hr high of 11 $\mu g\ m^{-3}$ reported for a study in the Los Angeles Basin (Appel et al., 1982). The episodic and background sulphate values are also typical of those reported for studies elsewhere (Kelly et al., 1984; Kurtz et al., 1984).

5. SUMMARY AND CONCLUSIONS

Both sulphate and acid episodes occurred frequently in the summer months in Nova Scotia. Aerosol sulphate concentrations were typical of those reported previously in Nova Scotia, but in this study acid was also monitored and found to be present in episodic amounts during the major sulphate episodes. Changes in the acid and sulphate concentrations were related to weather patterns and episodes were associated with reduced visibility (haze) and warm air. Strong acid was highly correlated with sulphate and was attributed to H_2SO_4 and/or NH_4HSO_4; the path of the air masses over water from the southwest was thought to lessen the chance of neutralization.

REFERENCES

Appel, B. R., Hoffer, E. M., Tokiwa, Y. and Kothny, E. L.: 1982, Atmos. Environ. 16, 589.

Barrie, L. A., Anlauf, K. G., Wiebe, H. A. and Fellin, P.: 1983, Proc. Amer. Chem. Soc. Symp. on Acid Rain, March 1982. Ed. J. Teasley, Ann Arbor Science.

Kelly, T. J., Tanner, R. L., Newman, L., Galvin, P. J., and Kadlecek, J. A.: 1984, Atmos. Environ., 18, 2565.

Kurtz, J., Tang, A. J. S., Kirk, R. W., and Chan, W. H.: 1984, Atmos. Environ., 18, 387.

Miller, D. F.: 1974, Exploratory Study of FActors Affecting Aerosol Formation, PB-245-274.

Shaw, R. W.: 1979, Environ. Sci. Technol., 13, 406.

Shaw, R. W.: 1982, Atmos. Environ., 16, 337.

Smith-Palmer, T., Wentzell, B., Spengler, J. D., Turner, W.: 1983, Indoor-outdoor Levels of Nitrogen and Sulphur Species and their relation with Air Flow in Antigonish, Nova Scotia, unpublished report to Health and Welfare, Canada.

Tanner, R. L., Leaderer, B. P. and Spengler, J. D.: 1981, Environ. Sci. Technol., 15, 1150.

ANALYSIS AND ASSESSMENT OF PRECIPITATION CHEMISTRY AT CARIBOU, MAINE

R. S. Artz
Air Resources Laboratory/NOAA
Silver Spring, MD 20910 USA

Uri Dayan
Israel Atomic Energy Commission
Yavne 70600 Israel

ABSTRACT. Over the last 30 yr, Caribou, Maine has been the only U.S. collection site which has been a part of all four national precipitation chemistry networks. Due to its remote but strategically important location, the data from this site are very useful in evaluating transboundary transport of the major ions present in precipitation. This paper assesses the reliability of the Caribou data base and looks at the more recent data with the aid of the GAMBIT (Gridded Atmospheric Multilevel Backward Isobaric Trajectory) model. An examination of the historical data base indicates serious contamination problems in pre-1980 samples, particularly with $SO_4^=$ measurements. The trajectory climatology shows that the largest number of precipitation events track along the U.S. coast. Preliminary chemistry measurements indicate highest concentrations of H^+ and $SO_4^=$ occur from events associated with trajectories passing through southern Canada.

1. INTRODUCTION

During the past 30 yr, four major national precipitation networks have been established and operated over different periods of time up to the present. Due to its remote but important location, Caribou, Maine has been a precipitation chemistry collection site in all four networks. Although the record at Caribou has not been a continuous one, it does provide the longest intermittent series of measurements in one United States location.

The earliest network (Junge Network) that included Caribou was operated from June 1955 to June 1956 by the Air Force Cambridge Research Center. Samples were collected by event and processed by month. In this network, conductivity, SO_4^{-2}, NO_3^-, Cl^-, Ca^{+2}, Mg^{+2}, K^+, Na^+, and NH_4^+ were analyzed (Junge and Gustafson, 1956). The pH values were not measured. The second national network which included Caribou was operated from 1959 to 1966 by the Public Health Service

Water, Air, and Soil Pollution **30** (1986) 845–855.

and the National Center for Atmospheric Research (PHS/NCAR Network). Precipitation was collected after each event with an automated precipitation collector with a moisture-sensing grid. Toluene was added to minimize biological activity and pooled samples were analyzed on a monthly basis. The same species were analyzed as had been measured by the Junge network with the addition of pH (Lodge et al., 1968). The third network was operated by the National Oceanic and Atmospheric Administration (NOAA) and the Environmental Protection Agency (EPA) between 1972 and 1980, and was a part of the World Meteorological Organization's (WMO) regional and baseline monitoring program (NOAA/EPA/WMO network). The network used a Misco wet-only collector and analyzed for the same species as in the PHS/NCAR network. Several heavy metals were measured on an experimental basis (Miller and Highsmith, 1976). In 1980, the NOAA/EPA/WMO network joined the National Atmospheric Deposition Program (NADP) and switched to a weekly wet only sampling protocol using an Aerochem Metrics Model 301 Precipitation Collector. EPA ended its participation, but Caribou remained as a site operated by NOAA. Species analyzed are sample volume, pH, conductivity, Mg^{+2}, K^{+}, Na^{+}, NH_4^{+}, NO_3^{-}, Cl^{-}, SO_4^{-2}, and PO_4^{-3}. The network procedures are well documented (e.g. Bigelow, 1982).

2. RESULTS AND DISCUSSION

In Figure 1, annual precipitation weighted sulfate concentration means are plotted. Note that when network protocol changed in 1980, two values were plotted; the first value covers 6 mo of collection under NOAA/EPA/WMO protocol while the second value covers approximately 7 mo of collection under NADP protocol. Obvious problems were encountered in 1959 and 1972, the start-up years for the PHS/NCAR and NOAA/EPA/WMO networks. Causes for the very low PHS/NCAR concentrations and the very high NOAA/EPA/WMO values are not known as no quality assurance documentation for these periods is known to exist. NADP values for 1981 through 1983 were consistently lower than all previous networks, a fact that can probably be contributed to tighter lid seals, stricter protocols, and a generally reliable collector.

Annual deposition values (not shown) were calculated using NWS raingage measurements, for $SO_4^{=}$, NO_3^{-}, and H^{+} for the entire data set. NADP values tend to be lower than measurements from the other networks. The cause of this is not fully known but is probably related to evaporation of samples in the field prior to 1980. Other possible error sources include contamination by dust (which is highly probable for Ca^{+2} and Mg^{+2} but unlikely to be an important source of anthropogenic materials) and poor collector catch efficiency. Catch efficiency is a known problem of the Misco collector, used from 1972-1980, due to the tiny (144 cm^2) bucket orifice. No important siting changes are believed to have occurred since precipitation collection began in Caribou as the site has always been located near the National Weather Service station at the Caribou airport.

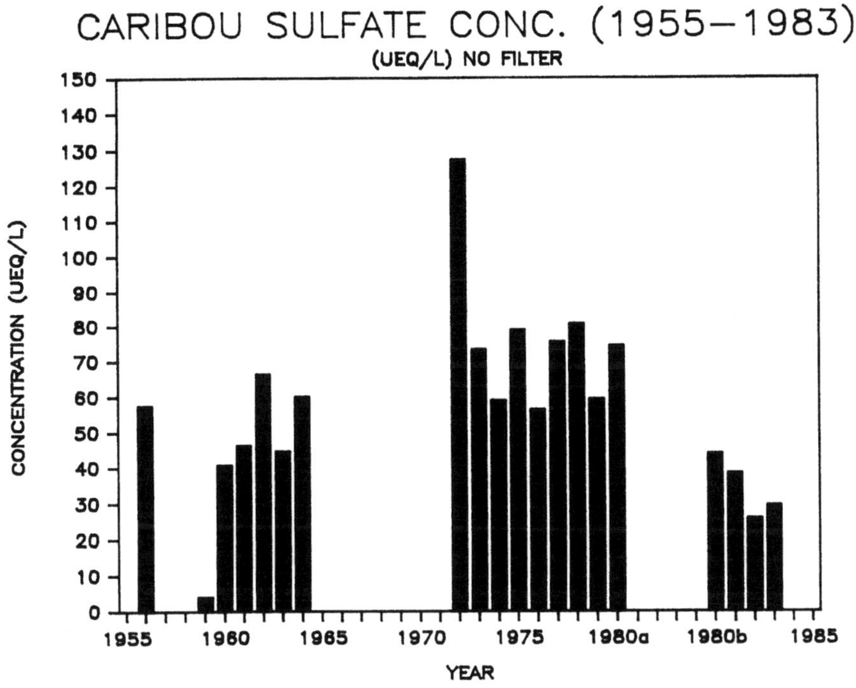

Fig. 1. Caribou, Maine annual weighted sulfate concentrations, 1955-1983.

Using data from the NOAA/EPA/WMO network, a gross estimation of concentration changes caused by evaporation of samples in the field can be made. In Figure 2, NOAA/EPA/WMO sulfate concentration data has been presented from 1972 to 1980 which shows apparent concentrations calculated without evaporation correction as well as annual values calculated using an evaporation correction ratio. Data are corrected by multiplying monthly samples by the ratio of the precipitation value from the precipitation chemistry samples to the value from the NWS raingage. It is seen from the figure that little change was observed from 1972 through 1974 as well as for 1980, a year of only partial collection. Large variations are observed from 1975 through 1979 indicating that evaporation was significant and caused up to a 100% increase in annual weighted concentrations during several years.

Figure 2 also provides estimates of the reasonableness of the evaporation correction. Data from 1972-1980 for the NOAA/NADP/WMO network are presented, as well as 1972-1980 Hubbard Brook, New Hampshire bulk sulfate data. (1980 represents an entire year of data for Hubbard Brook but only 6 mo for Caribou; due to the strong annual cycle in $SO_4^=$ concentration, 1980 Caribou data are probably too high due to the inclusion of all of the normally high events from the summer months and the loss of low winter values.) From wet-only isopleth maps (Barrie and Hales, 1984), it appears that Hubbard Brook sulfate values, especially for bulk samples, should always be higher in concentration than Caribou values. Figure 2 suggests, therefore, that sampling protocol probably changed after 1974 which resulted in increased evaporation. It should be stressed, however, that this is only speculation as no documentation for such a change is available.

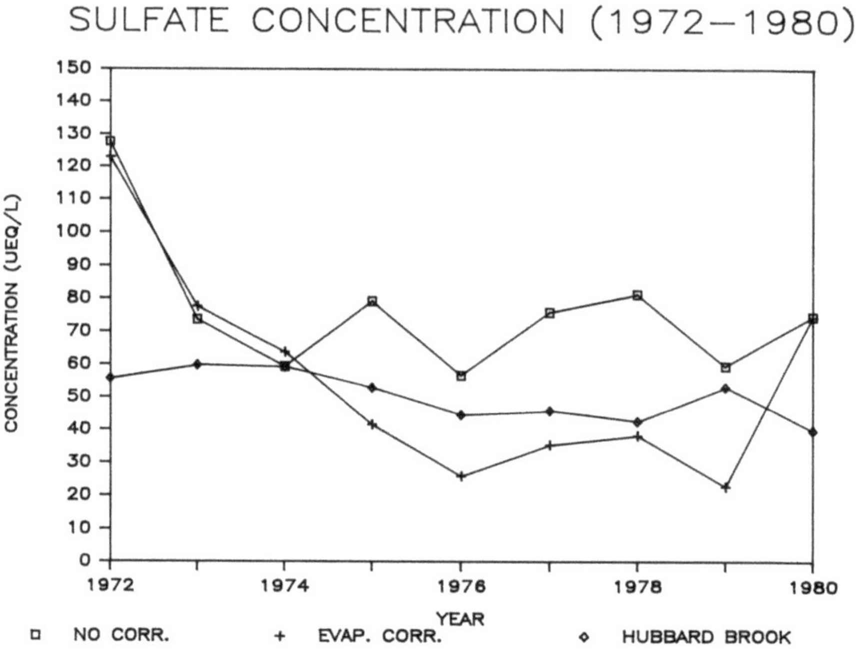

Fig. 2. Annual NOAA/EPA/WMO sulfate concentrations, with and without evaporation correction. Hubbard Brook, NH values are shown for comparison.

A 5 yr flow climatology of back trajectories was constructed in order to determine the influence of source regions on the precipitation amounts and precipitation chemistry at Caribou. The choice of trajectory method used to evaluate flow climatology is important, particularly for a highly baroclinic atmosphere, commonly observed in the winter and in northern latitudes. Unfortunately, however, highly specialized, physically accurate models are expensive to run and still are highly dependent upon available rawinsonde data. Of the available trajectory models (see, for example, Artz, 1982), the GMCC Gridded Atmospheric Multilevel Backward Isobaric Trajectory model (GAMBIT) (Harris, 1982) was chosen.

Details of the GAMBIT model have been described elsewhere (Harris, 1982) and its usefulness as a tool for constructing trajectory climatologies may be assessed from earlier studies such as

Miller and Harris (1985). Briefly, this model utilizes the National Meteorological Center gridded fields at 00Z and 12Z at standard pressure levels for input winds. Trajectories are calculated twice daily, are of 10 days duration, and were calculated at the 850 mb level for this study; trajectories were produced for all available 12-hr periods between 1979 and 1983.

Figure 3 shows the method of categorization used to construct the flow climatology. Three of the major sectors are Marine (MA) which includes the maritime Canadian provinces and the Atlantic Ocean, Eastern United States (E) which covers much of the heavily populated area of North America, and Pristine Canada (P) which includes only trajectories from Northern Canada. The Western United States (W) sector covers the remaining territory of the lower 48 states but requires trajectories to pass through the populated southern portion of Canada lying south of a line running from northern Lake Superior to the northernmost border shared by Maine and Canada. The Canada (C) sector covers all of Canada including the area encompassed by sector (P) in Figure 3, except for the maritime provinces. The (C) sector differs from the (P) sector in that trajectories are also required to pass through the populated southern portion of the country. The remaining three categories are missing (M), unable to classify (O) and stagnant (S) for trajetories moving very slowly in no particular sector for several days.

In the first evaluation of the climatology, a histogram of all 850 mb trajectories, both those associated and not associated with precipitation events at Caribou from 1979 through 1983 was constructed (Figure 4). Although the P category predominates, each of the first five categories is important. The two cleanest sectors, P and Ma, account for about 45% of all trajectories. Note that yearly flow appears quite reproducible with the possible exception of 1982 where relatively few Ma trajectories were observed and many trajectories were missing.

One of the objectives of the trajectory climatology was the determination of differences in precipitation chemistry at Caribou based upon geographical and political considerations. Each precipitation event greater than 0.1 cm as determined from official NMC reports was assigned to the appropriate back trajectory for a given 12-hr period. In Figure 5, the trajectory categories are plotted versus the number of wet-only trajectories associated with events exceeding 0.1 cm for the 5 yr period. The dominant categories, as would be expected, are Ma and E. Canadian trajectories are not an important source of precipitation. Note that while Ma and E trajectories remain the most important, annual variability, particularly for the Ma sector, may exceed 100%. In a separate exercise it was found that while Ma and E trajectories are recorded in the largest number of events, they also were associated with approximately 67% of the actual precipitation amounts.

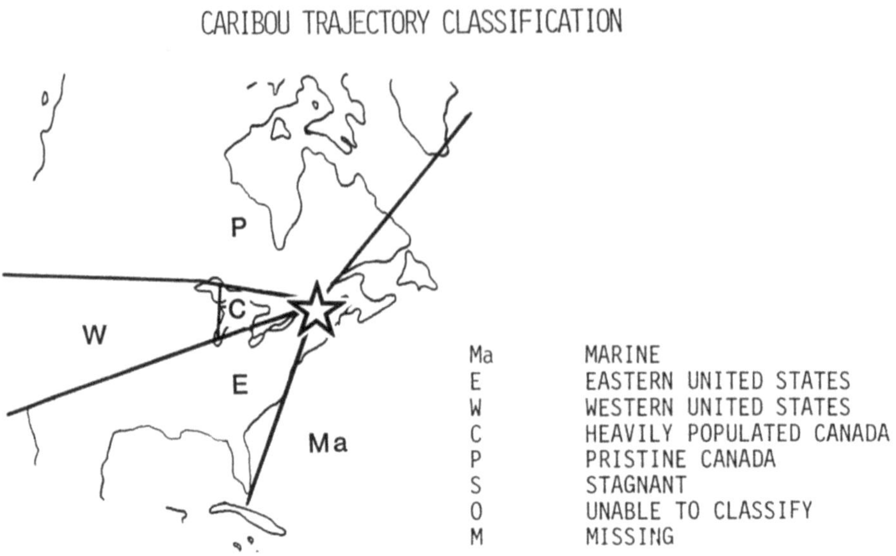

Fig. 3. Flow climatology classification diagram.

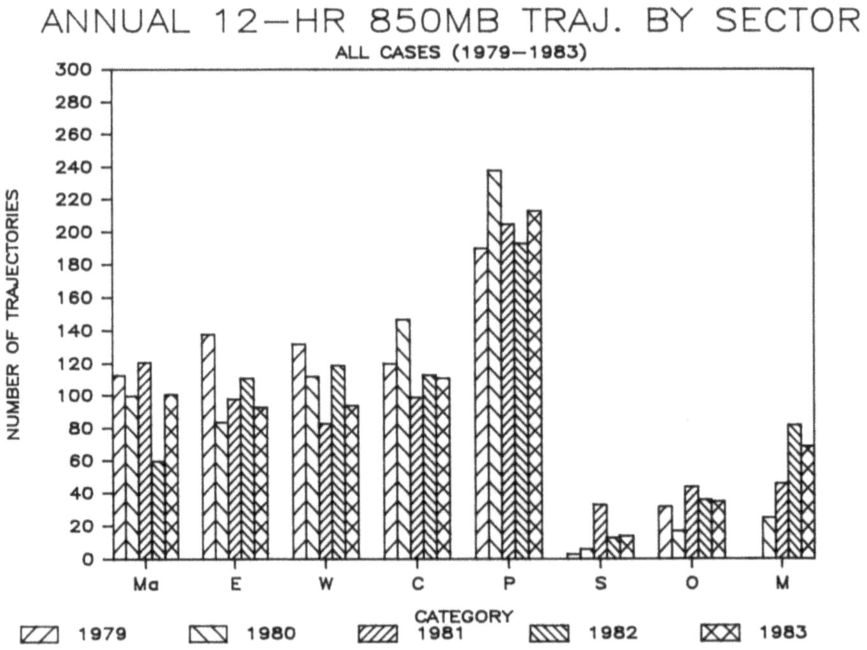

Fig. 4. Histogram of annual variations in 850 mb wet and dry flow patterns, 1979-1983.

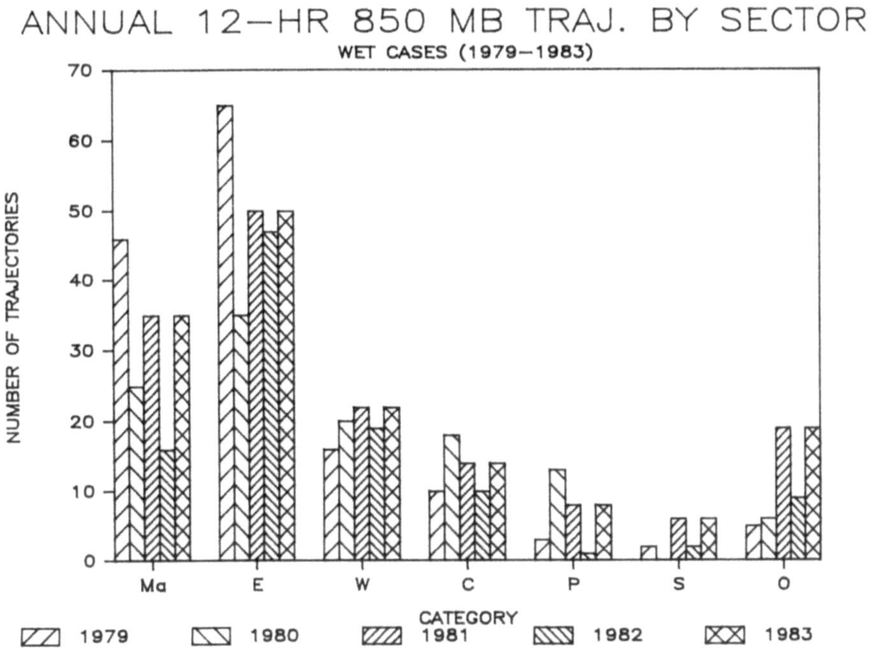

Fig. 5. 850 mb flow histogram for trajectories associated with Caribou precipitation events, 1979-1983.

In a final effort to determine the effect of wind flow on Caribou precipitation, the precipitation chemistry collected at Caribou through the National Atmospheric Deposition Program (NADP) was studied. The period studied covered May 1980 through December 1983. Unfortunately, NADP collection periods are weekly in duration making individual event chemistry and trajectory assessments very difficult. In order to assign appropriate 12 hr trajectories, only weekly measurements covering single events were considered, reducing the total number of individual events to 36 for the 3 yr period. All weeks with multiple events were discarded. Interestingly, while flow from categories Ma and E result in the largest quantity of precipitation (Figures 6a, b), flow from categories C and W results in precipitation events with highest concentrations of anthropogenic materials. Ca^{+2} and Na^{+} concentrations are reasonably stable across all sectors; Na^{+} is slightly higher from the marine sector. Owing to the very small data set, all values here must be viewed as preliminary; 36 total events do not result in a sufficient number of events in all categories to conduct thorough statistical tests.

Statistical category intercomparisons were computed using t-tests for H^{+}, NO_3^{-}, and $SO_4^{=}$. A special standard deviation equation was

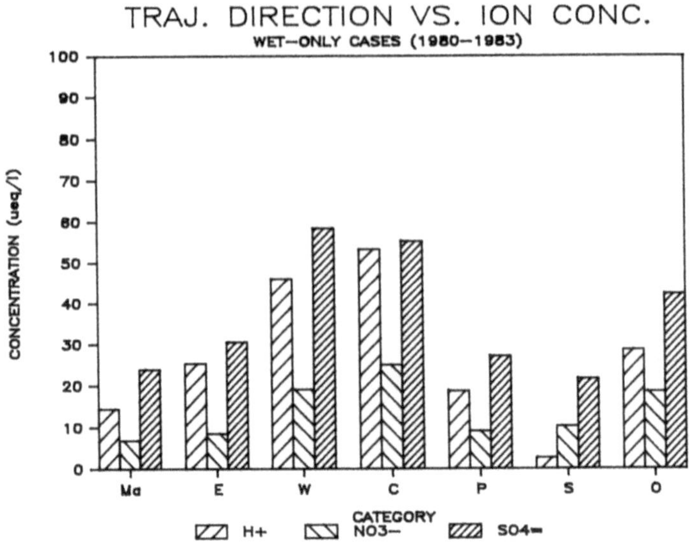

Fig. 6a. Histogram of H^+, NO_3^- and $SO_4^=$ concentration by trajectory sector for Caribou precipitation events, 1980-1983.

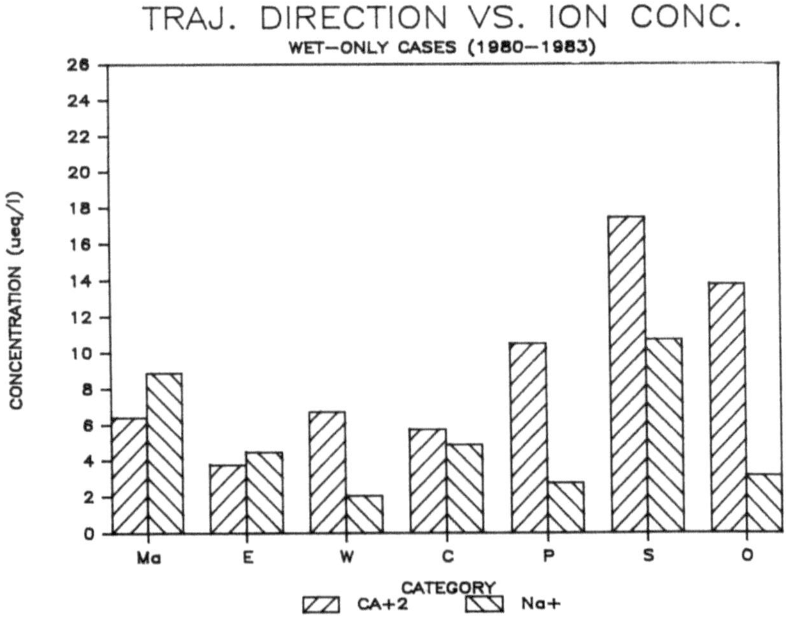

Fig. 6b. Same as 6a but for Ca^{++} and Na^+.

required for calculation of the t-tests in order to account for statistical arguments raised using weighted mean calculations. This formula has been thoroughly explained by Galloway et al. (1984). Results are shown in Figure 7. All tests are valid at the 95% level or greater. The purpose of these tests was to statistically determine if differences in chemical concentrations observed between sectors were significant or not. The large number of significant tests suggests that the geographical categorization is a valid method of describing groups of meteorological events for ions principally of anthropogenic origin. The three categories Ma, E and W show statistically significant differences from each other except for one test for NO_3^-. Statistical tests involving categories C and P are rarely significant although this may be due in part to the very small number of events available for the analysis. It should be pointed out, however, that no statistical differences were observed between categories W and C, indicating that they can probably be combined into one category.

H+	Ma	E	W	C	P
Ma	*	+	+	+	o
E		*	+	+	o
W			*	+	o
C				*	o
P					*

NO_3^-	Ma	E	W	C	P
Ma	*	o	+	+	o
E		*	+	+	o
W			*	o	+
C				*	o
P					*

$SO_4^=$	Ma	E	W	C	P
Ma	*	+	+	+	o
E		*	+	+	o
W			*	o	o
C				*	o
P					*

Category	# Samples
Ma	10
E	15
W	4
C	2
P	2
0	3

Fig. 7. T-test trajectory category intercomparisons for H^+, $SO_4^=$ and NO_3^-. A + indicates statistical significance at the 95% level while a o indicates that the test is not significant.

A word of caution should be raised concerning the validity of trajectories calculated during precipitation events. The model used in the preceding analysis is (as are all other operational trajectory models) subject to the least amount of error under dry, steady flow conditions, above the boundary layer, and not in the vicinity of frontal motion (Artz, 1982; Artz et al., 1985; Miller, 1986). Calculation of trajectories during precipitation events frequently involves strong vertical motion and wind shear associated with frontal movement which is inadequately resolved by the current North American rawinsonde network. For individual trajectories, the probability of large errors is very high. The strength of this analysis however, lies in the fact that a climatology was conducted, a procedure which should smooth out much of the error associated with individual trajectories; the more events represented in the climatology, the more faith that can be put in the results. In this particular analysis, enough data were available to ensure a reliable climatology of total precpitation amounts, although the chemical climatology associated with single event weeks of NADP data is less thorough than desired.

3. CONCLUSIONS

In summary, the following conclusions can be made:

- Four separate sampling protocols have been in effect in Caribou since 1956. Only data collected since 1980 are considered to be consistently reliable and of high quality. Pre-1980 data suffer from methodology and quality control problems and serious sample evaporation problems are suspected.

- The greatest amount of precipitation in Caribou is correlated with trajectories from over the ocean or across the eastern United States. Highest concentrations of $SO_4^=$ and NO_3^- are observed from trajectories out of Canada and the western United States; however, these areas are not an important source of precipitation amounts compared to the eastern U.S. and the Atlantic Ocean.

Acknowledgments - Special thanks are given to the staff at the Caribou, Maine National Weather Service Office for their diligence in collecting high quality precipitation samples over the past three decades. The authors would also like to thank Joyce M. Harris from the Geophysical Monitoring for Climatic Change Division, NOAA for help in providing Caribou trajectories. Funding for this project was through the National Acid Precipitation Assessment Program.

4. REFERENCES

Artz, R.S.: 1982, Comparison of methods of air parcel trajectory analysis for defining source-receptor relationships. M.S. Thesis, University of Virginia, Dept. of Environmental Sciences.

Artz, R.S., Pielke, R.A. and Galloway, J.: 1985, Atmos. Environ., 19, 47.

Barrie, L.A. and Hales, J.M.: 1984, Tellus, 36B, 333.

Bigelow, D.S.: 1982, NADP Instruction Manual, Site Operation, January 1982. Program Coordinator's Office, National Atmospheric Deposition Program, Natural Resource Ecology Laboratory, Colorado State University, Fort Collins, CO.

Galloway, J.N., Likens, G.E. and Hawley, M.E.: 1984, Science, 226, 829.

Harris, J.M.: 1982, The GMCC atmospheric trajectory program. NOAA Tech. Memo, ERL ARL-116, Air Resources Laboratory, Silver Spring, MD 30 pp.
Junge, C.E. and Gustafson, P.E.: 1956, Bul. Amer. Met. Soc., 37 244.
Lodge, J.P., Jr., Pate, J.B., Basbergill, W., Swanson, G.S., Hill, K.C., Lorange, E. and Lazrus, A.L.: 1968, Chemistry of United States precipitation. Final Report of the National Precipitation Sampling Network. Laboratory of Atmospheric Sciences, National Center for Atmospheric Research, Boulder, CO.
Miller, J.M.: 1985, The use of backward air trajectories in interpreting atmospheric chemistry data: A review. Proceedings, Muskoka Conference, Muskoka, Ontario, Canada.
Miller, J.M. and Harris, J.M.: 1985, Atmos. Environ., 19, 409.
Miller, J.M. and Highsmith, R.: 1976, The WMO Regional Precipitation Chemistry Network in the United States, Air Resources Laboratory, Silver Spring, MD.

A COMPARISON OF A MULTILAYER REGIONAL-SCALE ACID RAIN ASSESSMENT MODEL (MLRAPT) WITH A SINGLE-LAYER ASSESSMENT MODEL, RAPT

W. E. Davis and C. S. Glantz
Pacific Northwest Laboratory
P.O. Box 999
Richland, WA 99352 USA

ABSTRACT. This study, which is part of the 1985 Assessment on Current Damages of Acid Deposition, examines uncertainty due to treating transport processes using a single-layer model compared to a multilayer model. The Multilayer Regional Air Pollutant Transport Model, MLRAPT, was run in both modes, single layer and multilayer, using 224 sources during a 22-day period in July 1978 for the northeastern United States. Comparison between the two modes indicated that the average air concentrations were highly correlated (a correlation coefficient of 0.98). However, for wet deposition of $SO_4^=$, a correlation coefficient of 0.55 between the two predicted wet deposition fields was obtained. Results of the comparison of the two modes with the observed air concentration fields and with a limited comparison of observed wet deposition indicate that caution should be used when interpreting results of a single-layer model to develop source/receptor relationships.

1. INTRODUCTION

Long-term regional assessments are needed to evaluate the source/receptor effects from acid rain for the northeastern United States. These assessments will contribute to the development of emission control strategies, which policy makers will implement to minimize acid rain effects on certain receptors. A problem arises in that specific source/receptor relationships are limited by the uncertainties associated with their development. A failure to accurately estimate these relationships can result in the adoption of an erroneous control strategy. The problem then is to identify and quantify the uncertainties in source/receptor relationships so that policy makers can estimate the range of possible outcomes from the adoption of a selected strategy.

In particular, the neglect by long-term assessment models of shear and vertical motion increases uncertainty in source/receptor relationships. A previous short-term study over a 4-day period by Davis (1983b) demonstrated that significant air concentration differences can occur when results from a single-layer model are compared with results from a multilayer model that simulates vertical motion. This study addresses

Water, Air, and Soil Pollution **30** (1986) 857–866.

the question of whether there would still be significant differences between the models for longer time periods such as 20 days.

2. METHODS AND PROCEDURES

The multilayer and single-layer models were run for a 22-day period using 224 point and area pollution sources distributed across the eastern half of the United States and southeastern Canada. Model results are compared to evaluate the effects of vertical motion and wind shear on pollutant air concentrations and deposition patterns. Model predictions are also compared with field measurements. Data used by the models were collected during a SURE intensive period (Perhac, 1978) in July 1978. It was assumed, on the basis of the earlier trajectory end-point comparison (Davis, 1984) that the use of this summer month would produce the least amount of differences between the models.

The two models being compared are actually contained within a single Lagrangian regional-scale puff model, MLRAPT. The model has within its framework the option of being run in either a single-layer or multilayer mode. In the single-layer mode, the model reproduces RAPT (McNaughton et al., 1981), a regional-scale Lagrangian puff acid deposition assessment model. In the multilayer mode, the model is based both on RAPT and a previous multilayer model developed by Davis (1983a).

MLRAPT can be run in a number of different modes with a number of different options. In the multilayer mode, layer-averaged potential temperature is used to estimate vertical motion (Davis and Wendell, 1976; Davis, 1983a,b). The model assumes that puffs preserve their potential temperatures while residing in layers that undergo a diurnal temperature cycle. For this study (summer conditions), a temperature difference of more than 3°C between a puff and its layer's potential temperature will initiate a search for another layer with a potential temperature closer to that of the puff. If such a layer is found, the puff is moved to that layer. It is in this way that the multilayer model simulates vertical movement.

The models in this study assumed an hourly release of puffs, which are advected in hourly steps using gridded, layer-averaged u and v components. The layers used were 300 to 1500 m for the single-layer mode and 100 to 500, 501 to 1000, 1001 to 1500, 1501 to 2000, 2001 to 2500, and 2501 to 3000 m for the multilayer mode.

The following assumptions were made:

1. Horizontal diffusion was not used. The RAPT code cannot be run with horizontal diffusion. In order to eliminate differences due to horizontal diffusion, the MLRAPT was run with its option for horizontal diffusion turned off.
2. The depth of the daytime mixed layer is represented by a diurnal cycle. A daytime layer increases from a minimum depth at sunrise to a maximum depth in the afternoon. A nocturnal mixed layer builds after sunset.
3. Two vertical diffusion regimes are modeled: one above and one below the mixing height. Puffs within the daytime mixed layer and nocturnal mixed layer will expand in the layer according to the

current stability. An exception to this occurs when a puff has diffused to the maximum depth of the mixed layer. After this occurs, the puff diffuses as a fixed function of travel time. Puffs above the mixing height will diffuse as if the atmosphere is extremely stable.

4. Terrain effects are neglected.

For a more in-depth description of the RAPT mode, the reader is referred to the RAPT user's guide (McNaughton et al., 1981) and a report on results using the RAPT code (McNaughton, 1980).

Hourly gridded precipitation was used in both models. Since no precipitation for Canada was included, the results when discussed will not include wet deposition outside of the United States. The grid used for the hourly precipitation was a 1/10 NMC grid (38 km at 60°N) that encompasses the eastern half of the United States and southeastern Canada.

Twelve-hour rawinsonde and pibal data were used to compute fields of u and v layer-averaged components for both models on a 1/2 NMC grid (190 km at 60°N) (See Figure 1). Layer-averaged potential temperature and mixing ratios were also computed for the 1/2 NMC grid.

Table I lists the input data for the study period.

TABLE I. Model Input Data

Time Period of Study: July 1, 00Z - July 22, 12Z 1978
Maximum Mixing Height: 1500 m

Stability		Transformation Rate SO_2 to $SO_4^=$	
Hour (LST)	Stability Class	Hour (LST)	Rate (Fraction hr^{-1})
1-6	6	1-6	0.0025
7-11	4	7-18	0.02
12	3	19-24	0.0025
13	2		
14	3		
15-18	4		
19-24	5		

Dry Deposition Velocities		Wet Removal Coefficients	
SO_2	1.0 cm s^{-1}	SO_2	0.005P*
$SO_4^=$	0.1 cm s^{-1}	$SO_4^=$	$0.38P^{0.73}$*

* P ≡ precipitation measured in mm.

3. RESULTS

The single-layer and multilayer models were run using data for the first 22 days of July 1978. Figure 1 shows the 224 point and area sources used in the study. The size of the circle is proportional to source strength. Hourly puff releases were assumed at each source. The puffs from these sources were tracked until they left the advection grid.

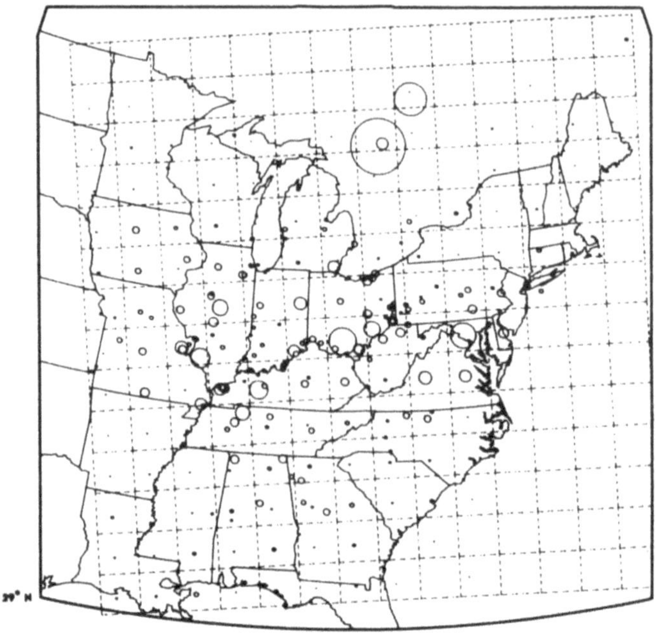

Figure 1. Sulfur emissions for point sources and area sources. Size of circle corresponds to source strength.

Table II shows the overall correlation coefficients based on a comparison between the values generated by both models at 53 receptor sites.

Model-predicted averages of SO_2 and $SO_4^{=}$ air concentrations at 53 locations were compared with measured values (Table II). In the case of SO_2, both models overestimate the average value of SO_2 in the Ohio River valley area while underestimating $SO_4^{=}$ air concentrations in the same area. This result indicates that a higher rate of transformation of SO_2 to $SO_4^{=}$ may be needed in the models.

TABLE II. Averages and Correlations*

	Correlations			Averages		
	Single & Multi	Multi & Obs	Single & Obs	Multi Layer	Single Layer	Observed
SO_2 Air Concentration	0.98	0.27	0.30	19.3	19.0	19.2
$SO_4^=$ Air Concentration	0.70	0.57	0.33	9.7	9.8	12.6
SO_2 Wet Deposition	0.94	-	-	4.7	4.5	-
$SO_4^=$ Wet Deposition	0.55	-0.08**	-0.06**	10.4	9.2	-

* Averages are the 22-day hourly average value at 53 receptors. Air concentrations are $\mu g\ m^{-3}$, deposition are $\mu g\ m^{-2}$.

** For only 5 sites (see Table III).

The correlations show that the two models are producing nearly the same SO_2 air concentrations and wet deposition patterns. The differences between the models begin to show in the $SO_4^=$ patterns. The $SO_4^=$ air concentration correlations show that the multilayer is significantly different than the single layer at the 95% level. The differences can be seen in the shift in the regions of maximum concentrations as shown in Figures 2 and 3. The multilayer was found to correlate better with the observed (Figure 4 and Table II).

Large differences occurred between the models' predicted $SO_4^=$ wet deposition fields (see Figures 5 and 6). The correlation coefficient between the models for $SO_4^=$ wet deposition for the 22-day period is 0.55. This value reflects the effect of wind shear and vertical motion on the deposition predictions by these model.

Model predictions of total $SO_4^=$ wet deposition for the 22-day period at five locations were compared with observed values. Both the single-layer and multilayer models of MLRAPT underestimate the $SO_4^=$ wet deposition by a factor of about 5 to 10 (see Table III). For this limited sample, the single-layer results appear to be slightly closer to observed values. However, an earlier study by Davis (1983b) indicated the multilayer results were closer to observed values.

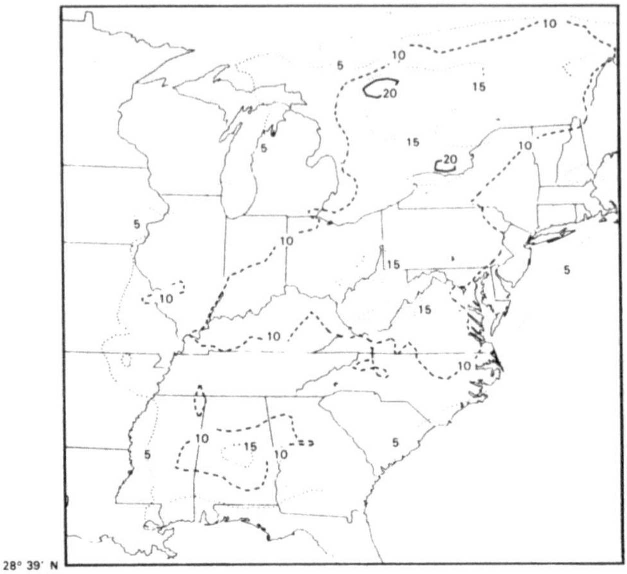

Figure 2. $SO_4^{=}$ air concentrations produced by the multilayer model ($\mu g\ m^{-3}$).

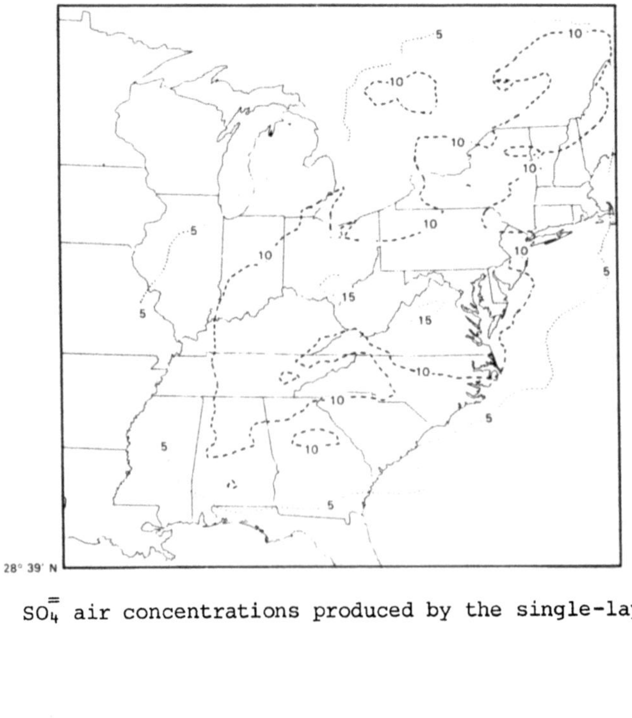

Figure 3. $SO_4^{=}$ air concentrations produced by the single-layer model ($\mu g\ m^{-3}$).

Figure 4. Observed $SO_4^{=}$ air concentrations ($\mu g\ m^{-3}$).

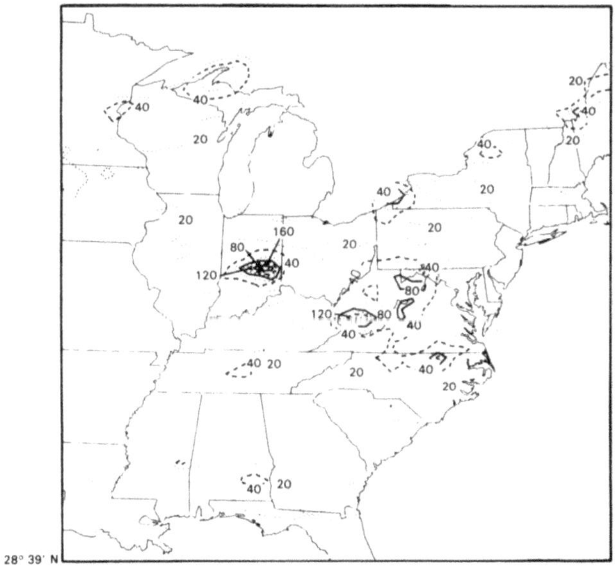

Figure 5. $SO_4^{=}$ wet deposition produced by the multilayer model ($10^3\ \mu g\ m^{-2}$).

Figure 6. $SO_4^{=}$ wet deposition produced by the single-layer model (10^3 µg m^{-2}).

TABLE III. Wet Deposition Comparisons (µg m^{-2})

Station	Observed	Single Layer	Multi-Layer
Whiteface Mountain, NY	$1.0*10^5$	$4.3*10^4$	$1.2*10^4$
Ithaca, NY	$3.1*10^5$	$1.8*10^4$	$9.8*10^3$
Champaign, IL	$4.5*10^5$	$3.4*10^4$	$8.2*10^3$
College Place, PA	$1.7*10^5$	$1.0*10^4$	$1.1*10^4$
Virginia	$3.1*10^5$	$2.0*10^4$	$1.2*10^4$

4. CONCLUSIONS

A comparison of predictions made by a single-layer and a multilayer model for a 22-day period in July 1978 revealed both striking similarities and significant differences. Minor differences occurred between single-layer and multilayer predictions of SO_2 air concentration and SO_2 wet deposition. Increased differences were obtained for $SO_4^{=}$ air concentrations. However, a slightly higher rate of transformation of SO_2 to $SO_4^{=}$ was indicated for the models.

The largest differences occurred between single-layer and multilayer predicted $SO_4^=$ wet deposition. The $SO_4^=$ wet deposition was underpredicted by both models when compared to observed values. This difference is thought to be in large part because of the lack of incloud conversion modeling within MLRAPT.

In summation, the two models predicted very similar SO_2 air concentrations during a study period characterized by limited frontal activity and infrequent precipitation. However, even with this limited activity, the simulation of vertical motion and wind shear by the multilayer model resulted in significant differences in the $SO_4^=$ air concentrations and wet deposition patterns from those computed by the single-layer model.

Additional studies during the fall, winter, and spring still need to be done to determine if the addition of a multilayer parameterization to RAPT will significantly improve model performance. The model needs to be run for periods with more frequent frontal activity and wet deposition measurements of $SO_4^=$. This will allow a more rigorous comparison between observed and modeled values during periods of increased wind shear effects.

Acknowledgments

The major source of funding for this research was provided by the U.S. Department of Energy. Additional support came as a part of the National Acid Precipitation Assessment Program by the U.S. Environmental Protection Agency Agreement EPA-DW930059, and the EPA Interagency Agreement DW89931201-01. The authors wish to thank Carl Berkowitz for his suggestions, A. R. Olsen for statistical work, Mike Kosorok for the use of his analysis program, and Elizabeth Owczarski for her help in preparing the manuscript. The Pacific Northwest Laboratory is operated for the U.S. Department of Energy under Contract DE-AC06-76RLO 1830 by Battelle Memorial Institute.

References

Berkowitz, C. M., Davis, W. E., and Knudson, D.: 1983, 'Potential Trajectory Errors in Single-Layer Wind Fields', paper appearing in The Meteorology of Acid Deposition, Perry Samson, Ed. APCA.

Davis, W. E.: 1983a, 'Preliminary Results of an Eight-Layer Regional Assessment Model Applied to the Problem of Acid Rain', In Energy and Environmental Chemistry: Acid Rain. Ed. L. H. Keith, Ann Arbor Science, Ann Arbor Michigan.

Davis, W. E.: 1983b, 'An Intercomparison of the Washout of $SO_4^=$ from a Single-Layer Regional Transport Model with a Multilayer Vertical-Motion Regional Transport Model', paper in Precipitation Scavenging, Dry Deposition, and Resuspension. Elsevier, New York.

Davis, W. E.: 1984, 'The Effect of Vertical Wind Shear on Regional-Scale Trajectories in Near-Frontal Conditions,' paper in the proceedings of Fourth Joint Conference on Applications of Air Pollution Meteorology with APCA. Portland, Oregon.

Davis, W. E., and Wendell, L. L.: 1976, 'Some Effects of Isentropic Vertical Motion Simulation in a Regional-Scale Quasi-Lagrangian Air Quality Model,' paper presented at the 3rd Symposium on Atmospheric Turbulence, Diffusion, and Air Quality, Raleigh, North Carolina.

McNaughton, D. J.: 1980, Atmos. Environ., **14**, 55.

McNaughton, D. J., Powell D. C., and Berkowitz C. M.: 1981, RAPT - The Pacific Northwest Laboratory Regional Air Pollutant Transport Model. PNL-3390, Pacific Northwest Laboratory, Richland, Washington.

Perhac, R. M.: 1978, Atmos. Environ. **12**, 641.

LOCAL PATTERNS OF SULPHUR DEPOSITION AROUND HALIFAX/DARTMOUTH, NOVA SCOTIA: 1979-83.

R.W. Shaw
Atmospheric Environment Service
Bedford Institute of Oceanography
P.O. Box 1006
Dartmouth, Nova Scotia, Canada
B2Y 4A2

ABSTRACT. Five years of sampling support earlier findings that SO_2 emissions from Halifax/Dartmouth, Nova Scotia (population 270,000) cause wet sulphate deposition within 40 or 50 km of the city to be twice regional values. The urban effect is greatest within the city itself. Mean monthly sulphate concentrations in precipitation are usually more highly correlated among stations than is monthly deposition; the highest correlations are among concentrations in and around the city. Month to month variations in sulphate concentration in the city and at Oyster Pond (50 km east of the city) appear to be affected mainly by urban influences; those at St. Margaret's Bay (25 km west of the city) appear to be affected by both urban and regional influences. The urban effect appears to have diminished somewhat during the 5 yr, possibly because the urban emissions have decreased by 35% since 1980.

1. INTRODUCTION

The review by Henderson and Shaw (1982) concluded that, in the long term, a few percent to about 20% of the S emitted from point and urban sources is deposited by precipitation within 50 or 100 km of the source. (Shaw, (1982) concluded that, in 1979, 50 to 60% of the wet sulphate deposition at St. Margaret's Bay, 25 km west of Halifax/Dartmouth, Nova Scotia (population 270,000) appeared to originate from emissions from the city. Following these results, sampling of individual precipitation events was continued at St. Margaret's Bay and started at Oyster Pond (50 km east of the city) and at Windsor Park (in the city proper) as shown in Figure 1. As in Shaw (1982), the observations have been compared with those at a regionally representative station at Kejimjujik, about 150 km west of Halifax/Dartmouth. This paper will present some preliminary results.

2. SAMPLING METHODOLOGY

Kejimkujik is relatively free of local influences; the other sites are, as will be seen, influenced to varying degrees by the urban emissions but (with the exception of Windsor Park which is 300 m from a heating plant) are free of deposition from nearby individual sources of SO_2.

As in Shaw (1982), Sangamo samplers expose a 20 cm diameter polyethylene bucket during precipitation only. At the end of the event the sample is stored in a refrigerator in a polyethylene bottle that has been rinsed with distilled, deionized water. Between events the sampler bucket is also rinsed with distilled, deionized water. Sulphate concentrations are

Water, Air, and Soil Pollution **30** (1986) 867–872.

measured using the method of ASTM (1970) with a detection limit of 0.5 mg L^{-1} and an accuracy of $\pm 10\%$. Chloride is used to make the sea salt correction.

3. MONTHLY CONCENTRATIONS AND DEPOSITIONS

In Table I, despite the different length of the sampling periods, the average monthly precipitation amount does not vary significantly among the stations. The precipitation-weighted mean sulphate concentrations, and mean monthly deposition of sulphate by precipitation are greatest at St. Margaret's Bay and Windsor Park and are least at Oyster Pond and Kejimkujik. (The significance of the differences will be discussed below.)

3.1. Mean Values and Standard Deviations

In Table II the 5 yr have been divided into three roughly equal periods corresponding to the times during which there was one, two or three stations operating around Halifax/Dartmouth. The standard deviation is from to 45 to 80% of the mean concentration, consistent with the factor 1.5 reported by Working Group 2 of the United States-Canada Memorandum of Intent on Transboundary Air Pollution (MOI, 1982).

During each period, the significance of the difference between the mean concentrations at the various stations is tested using the Wilcoxin Rank Sum non-parametric method (Malik and Malley, 1973). During January 1979 - August 1980, when only St. Margaret's Bay was operating, the mean monthly sulphate concentration there was significantly greater than that at Kejimkujik ($p < 0.005$). During the period September 1980 - August 1982 when St. Margaret's Bay and Oyster Pond were operating, the mean monthly concentrations at those stations were both greater than that at Kejimkujik, but the significance was less than that reported above. There was no significant difference between the mean concentrations at St. Margaret's Bay and Oyster Pond. During the most recent period, when the above stations were joined by Windsor Park in the city, only concentrations at Windsor Park were significantly greater than those at Kejimkujik. (They were also significantly greater than those at St. Margaret's Bay). There was no significant difference between mean monthly concentrations at St. Margaret's Bay and Oyster Pond.

In progressing from the earliest to most recent period, mean monthly concentrations decrease gradually, as well as the differences between concentrations at St. Margaret's Bay and Oyster Pond and those at Kejimkujik, suggesting a decrease in the urban effect. Emissions in Halifax/Dartmouth decreased by about 35% between 1980 and 1983, mainly due to a 55% decrease in electric utility emissions (J. Kozak, Environmental Protection Service, Dartmouth, Nova Scotia, private communication).

3.2. The Ratio of Monthly Mean Concentrations to those at Kejimkujik

In Table III, 69% of the monthly concentrations at St. Margaret's Bay are within a factor of 2 of those at Kejimkujik; 31% are greater than a factor of 2; and none is smaller than a factor of 2. Only 13% of monthly mean concentrations at Windsor Park, in the city proper, are within a factor of 2 of those at Kejimkujik; 81% are greater than a factor of 2 and 6% are smaller than a factor of 2.

In contrast, an overwhelming majority (81%) of the monthly concentrations at Oyster Pond are within a factor of 2 of those at Kejimkujik. More of them (16%) are greater than a factor of 2 than are less than a factor of 2 (3%), suggesting that, although there may be an influence of Halifax/Dartmouth on concentrations at Oyster Pond, it is much smaller than that at St. Margaret's Bay. This may also be deduced by comparing the mean concentrations at the two stations in Table II. Results similar to those in Table III were obtained from examining monthly depositions.

3.3. Linear Regressions among Monthly Concentrations, Rainfall Amounts and Deposition at the Stations

Figure 2 shows linear correlation coefficients (r^2) among the monthly sulphate concentration, rainfall amounts, and monthly mean deposition of sulphate by precipitation at the four stations during the three periods in Table II. (Values of 0.25, 0.20 and 0.16 are considered to be significant for 16, 20 and 24 pairs of values, respectively (Bailey, 1959)).

The highest correlations occur for monthly rainfall amounts, especially among the station in and around Halifax/Dartmouth. (The r^2 of 1.0 for HFX/OYS in Figure 2 results from using the same station, Shearwater Airport, for rainfall amounts.) Because deposition depends upon both rainfall amount and concentration, one might expect it to be less highly correlated among stations than the last two. However, in 4 of the 10 comparisons in Figure 2, this was not the case. This may be due to the episodicity of deposition; the above relationship may have held if correlations had been carried out for only those events contributing most of the deposition.

Mean monthly sulphate concentrations were most highly correlated among the stations in and around Halifax/Dartmouth, lending support to the likelihood of city emissions affecting these stations in a similar manner. Correlations with Kejimkujik were low for both Windsor Park and Oyster Pond, indicating that variations in sulphate concentrations at those two stations were affected more by local than by regional factors. Interestingly, monthly sulphate concentrations at St. Margaret's Bay were significantly correlated with Kejimkujik and the other two local stations; apparently variations at St. Margaret's Bay are affected by both local and regional influences.

Figure 3 shows the linear regression lines between the monthly concentrations at Windsor Park, St. Margaret's Bay, Oyster Pond and Kejimkujik for September 1982 - December 1983. Because concentrations at St. Margaret's Bay were the most highly correlated with those at the other stations, they were a logical candidate for the abscissa. For concentrations at St. Margaret's greater than 1.2 mg SO_4 L^{-1}, concentrations at Windsor Park are by far the greatest; those at Oyster Pond being less and those at Kejimkujik the least. For SO_4 concentrations at St. Margaret's Bay less than 1.0 mg L^{-1}, concentrations at Oyster Pond are slightly less than those at St. Margaret's Bay, possibly due to scatter in the data.

Similar relationships were found for the corresponding regression lines for monthly wet sulphate depositions, although the correlation coefficients were usually smaller.

5. CONCLUSIONS

(a) The 5 yr of sampling at St. Margaret's Bay and the addition of two sampling sites support the findings of Shaw (1982),i.e. that SO_2 emissions from Halifax/Dartmouth cause wet sulphate deposition within 40 or 50 km of the city to be as much as twice as great as those at a regionally representative station, Kejimkujik, 150 km away.

(b) The urban effect is greatest at Windsor Park within the city proper.

(c) Mean monthly sulphate concentrations in precipitation are usually more highly correlated among the stations than monthly wet sulphate deposition; the highest correlations are among the concentrations in and around the city.

(d) Month to month variations in sulphate concentrations at Windsor Park and Oyster Pond appear to be due more to urban than to regional effects; those at St. Margaret's Bay appear to be affected by both urban and regional influences.

(e) Lines of best linear fit for sulphate concentrations during the period September 1982 - December 1983 indicate that Windsor Park was most affected by urban emissions, followed by St. Margaret's Bay, with Oyster Pond a distant third.

(f) The urban effect appears to have diminished somewhat during the 5 yr, possibly because emissions within the city have decreased by 35% since 1980.

ACKNOWLEDGEMENTS

Sampling and analysis was carried out by Clark Wiseman, Don Walter, John Machell, Roger Percy and Bill Horne of the Environmental Protection Service (EPS), Environment Canada, Dartmouth, Nova Scotia.

REFERENCES

ASTM, 1970; Industrial Water Analysis Part 23. Am. Soc. Testing and Materials, Philadelphia, Pa., USA.

Bailey, M.T.J.; 1959, *Statistical Methods in Biology*, English Universities Press, London.

Henderson, D. and Shaw, R.W.; 1982, Local and Mesoscale Analysis Subgroup Report 2F-L, United States-Canada Memorandum of Intent on Transboundary Air Pollution. Atmospheric Environment Service, Downsview, Ontario M3H 5T4, 50 pp.

Malik, H.J. and Malley, K.; 1973, *A First Course in Probability and Statistics*. Addison-Wesley Publishing Co., Don Mills, Ontario.

MOI, 1982, Report of Working Group 2, United States-Canada Memorandum of Intent on Transboundary Air Pollution, Atmospheric Environment Service, Downsview, Ontario M3H 5T4, 185 pp.

Shaw, R.W.; 1982, *Atmospheric Environment* **16**, 337.

TABLE I: SUMMARY OF AVAILABLE EVENT DATA

Station	Sampling Period	No. of Samples	R(mm mo^{-1})	[SO4] (mg L^{-1})	SO4 DEP (g m^{-2} mo^{-1})
St. Margaret's Bay	Jan 79 - Dec 83	343 event	121	2.5	0.32
Oyster Pond	Sep 80 - Dec 83	136 event	123	1.2	0.17
Windsor Park (Halifax)	Sep 82 - Dec 83	71 event	102	2.6	0.39
Kejimkujik	Jan 79 - Dec 83	Monthly + event	119	1.2	0.15

TABLE II. Comparison of mean monthly SO4 concentrations in precipitation (mg L-1). Mean values designated as X_{av}; standard deviation as S, correlation coefficient as r^2.

Period	St. Margaret's		Oyster Pond		Windsor Park		Kejimkujik		r^2	Significant Difference?
	X_{av}	S	X_{av}	S	X_{av}	S	X_{av}	S		
Jan 79 - Aug 80	4.0	1.8	---	---	---	---	1.7	1.0	0.14	Yes $p < 0.005$
Sep 80 - Aug 82	2.0	1.6	---	---	---	---	1.3	0.7	0.37	Yes $p < 0.05$
	---	---	1.8	0.9	---	---	1.4	0.8	0.26	Yes $p < 0.025$
	2.2	1.7	1.7	0.9	---	---	---	---	0.46	No
Sep 82 - Dec 83	1.3	0.6	---	---	---	---	1.0	0.6	0.32	No
	---	---	1.1	0.4	---	---	0.9	0.5	0.03	No
	---	---	---	---	3.3	1.9	0.9	0.5	0.03	Yes $p < 0.025$
	1.2	0.5	1.1	0.4	---	---	---	---	0.28	No
	1.2	0.5	---	---	3.3	1.9	---	---	0.62	Yes $p < 0.025$
	---	---	1.1	0.4	3.3	1.9	---	---	0.41	Yes $p < 0.025$

TABLE III. Distribution of the ratio of monthly [S04] with that at Kejimkujik

Period	Station	%< 0.5	0.5<%<2.0	%> 2.0
Jan 79 - Dec 83	St. Margaret's Bay	0	69	31
Sep 80 - Dec 83	Oyster Pond	3	81	16
Sep 82 - Dec 83	Windsor Park	6	13	81

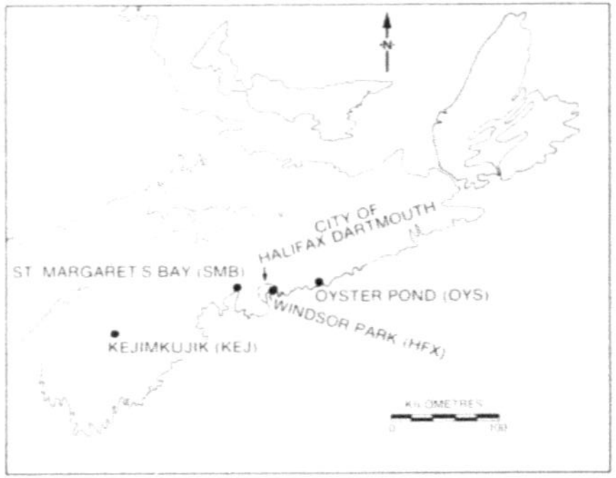

FIGURE 1: Location of sampling stations (heavy dots)

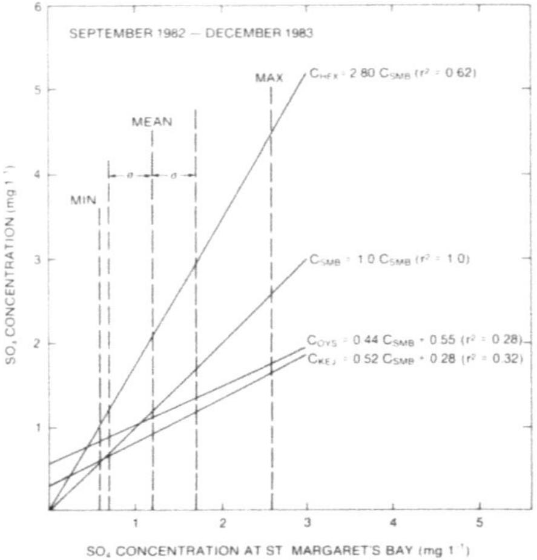

FIGURE 3: Lines of best fit for monthly sulphate concentration at Kejimkujik (KEJ), St. Margaret's Bay (SMB), Windsor Park (HFX) and Oyster Pond (OYS).

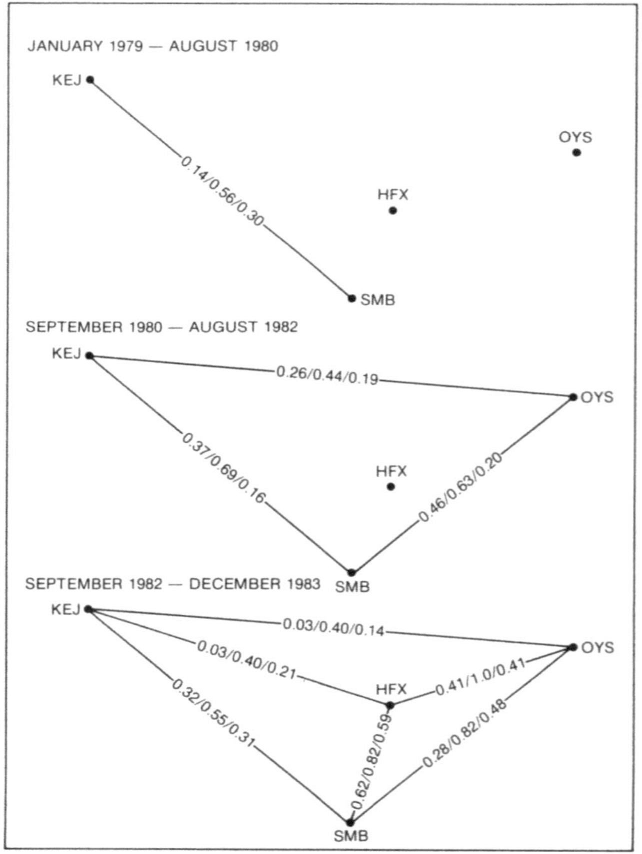

FIGURE 2: Correlations (r^2) of monthly sulphate concentrations, rainfall amounts and wet sulphate deposition among the sampling stations. Key is CONCENTRATION/RAINFALL/DEPOSITION.

METEOROLOGICAL ANALYSES OF ACIDIC PRECIPITATION IN ONTARIO

D. Yap and J. Kurtz
Air Resources Branch
Ontario Ministry of the Environment
125 Resources Road, Rexdale, Ontario, Canada
M9W 5L1

ABSTRACT. Meteorological analyses of precipitation and air quality data for the period 1976 to 1983 have been undertaken to infer potential source regions of acidity in Ontario. Techniques such as air parcel trajectory modelling and synoptic weather classification have been utilized. A meteorological data acquisition system has been implemented to provide historical and real-time trajectory statistics. This paper provides an overview of these studies. For areas of South and Central Ontario, precipitation events most commonly occur with pre-warm front and cyclonic situations, and trajectories from the south and southwest octants. Furthermore, most of the reported wet deposition of S and N is associated with trajectories from these octants, the direction of the Ohio Valley industrial source region and industrialized SW Ontario. High air concentrations of S and N are also associated with trajectories from these octants.

1. INTRODUCTION

As part of the Acidic Precipitation in Ontario Study (APIOS) program, a meteorological data acquisition system (MDAS), capable of providing support for special studies, as well as modelling activities, has been designed, developed and implemented. This computerized system is a package designed to receive weather teletype information from Canada's Atmospheric Environment Service (AES) and to store the incoming data in a form suitable for retrieval and further processing.

The area covered by MDAS extends approximately east of the Rockies and north of Florida. Air parcel trajectories, terminating or originating at a specific location, can be

Water, Air, and Soil Pollution **30** (1986) 873–878.

calculated by the system for interpreting precipitation events and air quality data. To date, trajectory analyses have been provided for events at various locations in Ontario and Eastern North America. MDAS is a multi-role system that assists the Air Resources Branch of the Ontario Ministry of the Environment in the performance of numerous functions, including long-term, episodic and real-time studies of acidic precipitation in Ontario. The package is currently implemented on a Data General Eclipse series mini computer. The major capabilities of the MDAS system consist of data acquisition, tabular output, graphics output, analysis programs, trajectories, archiving, macros and on-line documentation. A sample of the output produced by MDAS for acidic precipitation studies is shown in Figure 1.

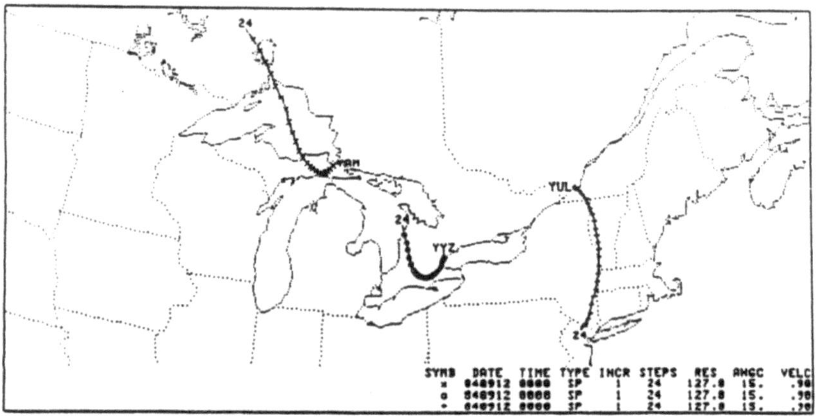

Figure 1: Three 24h forward surface trajectories calculated at one hour time steps.

In this paper, a summary of the meteorological studies for the period 1976 to 1983 is presented.

2. PRECIPITATION AND AIR QUALITY DATA

Air quality and precipitation chemistry concentration data for these studies were primarily obtained from the Acidic Precipitation in Ontario Study (APIOS) network for the period 1980 to 1983. Some earlier data (1976-1979) were collected for the Muskoka-Haliburton Region of Ontario as part of the Lakeshore Capacity Study. The province-wide APIOS networks (cumulative and event), operated by the Ontario Ministry of the Environment, were established in 1980 and 1981 to monitor both wet and dry deposition respectively (for details, see Chan et al., 1982). The event network stations, suitable for meteorological data analysis, are located in four clusters, but only one station for each cluster was used in this study, namely, Longwoods, Railton, Dorset and Fernberg (Figure 2).

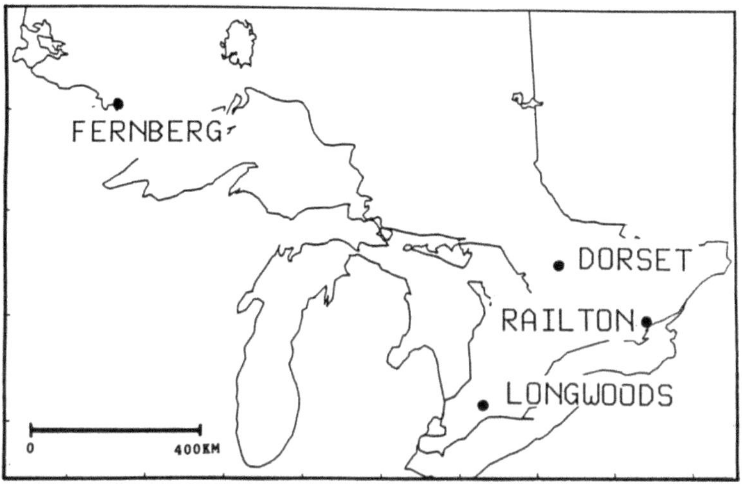

Figure 2: Location of the event network monitoring stations. Dorset is situated in the Muskoka-Haliburton Area of Ontario.

3. AIR PARCEL TRAJECTORY AND SYNOPTIC WEATHER SCHEMES

3.1. The Trajectory Model and Analysis Technique

Neglecting acceleration, representative horizontal winds (u and v components) in the atmospheric boundary layer are obtained from the surface and 850 mb geostrophic winds, with modifications for surface frictional effects. The task is accomplished with an objective analysis scheme using synoptic weather data from approximately 300 stations in a 30 x 30 grid system (grid length 127 km) for the surface level and using the height fields on a 11 x 11 grid system (grid length 381 km) for the 850 mb level. To obtain air parcel trajectories terminating or originating at a specific location, the three-hourly (typically) wind vector (surface) and the twelve-hourly wind vectors (850 mb) are integrated with backward or forward time-steps over the required periods, up to 2 to 3 days. Trajectories beyond this time period are not recommended since air flows then are likely to be subjected to significant curvature.

In these studies the possible source regions for wet deposition are inferred from the 850 mb (about 1500 m above the surface) and 'modified' surface geostrophic trajectories (see Smith and Jeffrey, 1975). The end-points of each of the back trajectories are classified according to "sector of origin" each sector being 45° in width and centered on one of the eight points of the compass. In addition, an unknown sector is included to handle cases where trajectories meander over two or more sectors or back trajectory end-points are very close to the receptor, etc.

3.2. Synoptic Weather Classification System

In order to study the relationship between synoptic weather patterns and measured pollutants, the eight classes defined by Heidorn (1979) to categorize the daily weather patterns were used. They are: Class 1: Post-cold front/pre-high pressure ridge. Class 2: High pressure ridge. Class 3: Post-ridge/pre-warm front. Class 4: Warm front. Class 5: Cyclonic. Class 6: Cold front. Class 7: Stationary front. Class 8: Weak pressure gradient.

4. SUMMARY OF FINDINGS

4.1. Long-term studies

Long-term studies of acidic precipitation in Ontario using meteorological analysis have only been conducted in recent years. One of the earliest works is that of Kurtz and Scheider(1981) which is based on measurements of pH, SO_4^{-2} and NO_3^- over a 33 mo period from August 1976 to April 1979 in the Muskoka-Haliburton region of Ontario. Sea-level geostrophic trajectories were used to determine the 'octant of origin' of the air mass associated with precipitation events. The study indicated that precipitation events most commonly occurred with trajectories from the south and southwest octants. In addition, the majority of H^+, SO_4^{-2} and NO_3^- deposited at Muskoka-Haliburton was associated with trajectories from the south and southwest. It was also noted that the average concentrations of H^+, SO_4^{-2} and NO_3^- in precipitation from the S, SW, W and NW octants are all substantial.

Since that initial study, air parcel trajectory analyses of precipitation data collected at locations in southwestern, central, eastern and northwestern Ontario have been conducted for the period July 1980-April 1983 in attempts to relate the wet deposition of various chemical species to octant of air mass origin and to study the effects of Sudbury emissions on precipitation quality (Kurtz, 1983; Yap and Kurtz, 1984). These studies suggest that southeast through southwest air flows are associated with about 60 to 80% of the wet deposition of sulphate and nitrate in Ontario. Loading contributions vary somewhat according to location and according to length of trajectory. Mean event loadings are also highest for airflows from the southerly sectors.

Application of this simple back-trajectory analysis to study the impact of the shutdown of Sudbury smelters on wet deposition in Ontario, however, gave, inconclusive results because of large meteorological induced variability in the precipitation chemistry and wet deposition data, and the fact that the smelter contribution was typically less than 15% at the receptors (Lusis, 1984).

These long-term trajectory studies suggest that southwesterly air reaching Southern and Central Ontario has previously passed over the industrialized Ohio Valley which is heavy in SO_x and NO_x emissions and has often traversed over industrialized regions of southwest Ontario as well. Such air is generally associated with more precipitation than air from other directions. This can be seen from synoptic meteorological consideration. For example, synoptic classes 3, 5 and 1 were most frequently associated with precipitation at the Dorset, Longwoods, Railton and Fernberg sites. These are, respectively: pre-warm front, cyclonic and post-cold front situations. The first two are associated with generally southerly winds and the last is associated with north or northwesterly winds behind the winter cold front. By far, most of the precipitation was associated with classes 3 and 5.

4.2. Episodic Study

During the period August 28-September 4, 1981, an acidic deposition episode occurred at Dorset in the ecologically sensitive Muskoka-Haliburton area of Ontario. The analysis of this event is reported by Kurtz et al., (1984). This eight day study period was characterized by 12% of the total 1981 precipitation, 28% of both the annual H^+ and SO_4^{-2} deposited at Dorset and 16% of the annual NO_3^- deposited at Dorset.

The analysis indicated that wet deposition of S and N was associated with air trajectories from the southwest, which is the direction of the Ohio Valley industrial source region, as well as industrialized southwestern Ontario. High air concentrations of S and N were also associated with trajectories from the southwest and appeared to be partly related to precipitation intensity. It was also suggested that episodic storms can have at least as great an impact on stream pH depression as snowmelt and spring runoff.

5. CONCLUSIONS

The potential source regions of acidity in Ontario have been investigated in recent years using a meteorological data acquisition system, air parcel trajectory techniques and a synoptic weather classification system. These meteorological studies indicate that most of reported wet deposition of S and N in South and Central Ontario is associated with trajectories from the south to southwest octants, the direction of the Ohio Valley and industrialized SW Ontario. In addition, high air concentrations of S and N are also associated with trajectories from these octants.

REFERENCES

Chan, W.H., Orr, D.B and Vet, R.J.: 1982a, "An Overview: The Event Wet/Dry Deposition Network", Ontario Ministry of the Environment Report ARB-11-82-ARSP.
Chan, W.H., Orr, D.B. and Vet, R.J.: 1982b, "An Overview: The Cumulative Wet/Dry Deposition Network", Ontario Ministry of the Environment Report ARB-15-72-ARSP.
Heidorn, K.: 1979, Water, Air and Soil Pollution, 11, 225.
Kurtz, J.: 1983, "Meteorological Analysis of Precipitation Event Sampling Data", Ontario Ministry of the Environment Report ARB-198-84-AQM.
Kurtz, J. and Scheider, W.A.: 1981, Atmos. Environ., 15, 1111.
Kurtz, J., Tang, A.J.S., Kirk, R.W. and Chan, W.H.: 1984, Atmos. Environ., 18, 387.
Lusis, M.: 1984, "Summary: Source Apportionment Analysis of Air and Precipitation Data to Determine the Contribution of the Sudbury Smelters to Atmospheric Deposition in Ontario", Ontario Ministry of the Environment Report ARB-194-84-ARSP.
Smith, F.B. and Jeffrey, G.H.: 1975, Atmos. Environ., 9, 643.
Yap, D. and Kurtz, J.: 1984, "Meteorological Studies to Quantify the Effects of Sudbury Emissions on Precipitation Quality and Air Quality During 1980-1983 with Emphasis on the Shut-Down Period", Ontario Ministry of the Environment Report ARB-192-84-AQM.

OPTIMIZATION OF EMISSION REDUCTIONS TO REDUCE ATMOSPHERIC SULPHUR DEPOSITION IN EUROPE: AN EXAMINATION OF A METHODOLOGY

R.W. Shaw
Atmospheric Environment Service
Bedford Institute of Oceanography
P.O. Box 1006
Dartmouth, Nova Scotia, Canada
B2Y 4A2

ABSTRACT. An examination has been made of the use of the LRT model of the Norwegian Meteorological Institute, which estimates the atmospheric exchanges of sulphur among 28 countries in Europe, to formulate optimized emission reduction scenarios. According to the methodology, emission reductions are needed in at least several countries to bring about a deposition reduction in a given country. However, it is most effective to reduce emissions in or near the country where deposition reductions are desired. To reduce deposition further, it is more effective to further reduce nearby emissions, rather than to seek additional emission reductions farther afield. Because of the complex distribution of strong source regions in Europe, an assessment of the advantages of an optimized emission reduction pattern over a uniform reduction scheme must await the better definition of deposition targets and the location where they are to be applied.

1. INTRODUCTION

The deposition of S in Europe to the surface from the atmosphere has been estimated to be as great as 5 to 10 g S m^{-2} a^{-1}, much greater than the proposed target of 0.3 to 0.5 g S m^{-2} a^{-1} that was proposed at the 1982 Stockholm Conference on Acidification of the Environment; the region of deposition greater than the above target extends from the Mediterranean to as far north as southern Scandinavia. Analysis of atmospheric trajectories, and modelling carried out during the OECD (1977) study, indicated that the S deposited in a given country is a result of emissions from both within the country itself and sources outside, some of which may be more than 1,000 km away.

Because of the complex atmospheric exchanges of S among the countries in Europe, and the high cost of removing S from stack gases, it is essential that any emission abatement program be optimized by concentrating efforts on those sources which cause the greatest deposition on sensitive receptors. Young and Shaw (1986) found that considerable economies could be realized in North America by using an optimized emission reduction scheme based upon an LRT model, rather than an across-the-board emission reduction. The LRT model developed at the Norwegian Meteorological Institute (OECD, 1977; Eliassen, 1978) has been used to estimate the deposition of S in each of 28 European countries resulting from emissions from a given country. This paper represents a preliminary attempt to assess the applicability of the methodology of Young and Shaw (1986) to the NMI/LRT model, and its potential for specifying the minimum emission reductions in Europe that would reduce deposition at sensitive receptor points to a given target value.

Water, Air, and Soil Pollution **30** (1986) 879–884

TABLE I: PARTIAL UNIT TRANSFER MATRIX FOR EUROPE SHOWING DEPOSITION (KG S HA^{-1} A^{-1}) AT RECEPTOR COUNTRY PER UNIT EMISSION (TG S A^{-1}) AT THE SOURCE COUNTRY.

RECEPTOR COUNTRY >>			9(DD)	10(D)	11(GR)
SOURCE COUNTRY					
1	Albania	ALB	0.0	0.0	5.5
2	Austria	A	0.0	1.8	0.0
3	Belgium	B	2.3	3.4	0.0
4	Bulgaria	BUL	0.0	0.0	6.2
5	Czechoslovakia	CS	4.1	1.5	0.2
6	Denmark	DK	1.5	0.8	0.0
7	Finland	SF	0.0	0.0	0.0
8	France	F	1.2	2.9	0.2
9	German Dem. Rep.	DD	27.6	2.8	0.1
10	Fed. Rep. Germany	D	5.2	14.9	0.2
11	Greece	GR	0.0	0.0	24.0
12	Hungary	HU	0.6	0.4	0.6
13	Iceland	IS	0.0	0.0	0.0
14	Ireland	IRL	0.0	0.0	0.0
15	Italy	I	0.2	0.5	0.6
16	Luxembourg	L	0.0	10.0	0.0
17	Netherlands	NL	2.8	4.6	0.0
18	Norway	N	0.0	0.0	0.0
19	Poland	PL	1.7	0.6	0.1
20	Portugal	PO	0.0	0.0	0.0
21	Romania	R	0.0	0.0	0.7
22	Spain	SP	0.0	0.3	0.2
23	Sweden	S	0.0	0.0	0.0
24	Switzerland	CH	0.0	5.0	0.0
25	Turkey	T	0.0	0.0	0.7
26	Soviet Union	SU	0.0	0.0	0.0
27	United Kingdom	UK	1.0	1.4	0.0
28	Yugoslavia	JU	0.5	0.5	1.7

TABLE II: ASSUMED PERCENTAGE SO_2 EMISSION REDUCTIONS IN EUROPE FROM 1980 TO 1994, FROM ECE (1985)

No.	Country	Identifier	% Emission Reduction
2	Austria	A	30
3	Belgium	B	30
5	Czeckoslavakia	CS	30
6	Denmark	DK	47
7	Finland	SF	30
8	France	F	50
10	Fed. Rep. Germany	D	50
15	Italy	I	30
17	Netherlands	NL	40
18	Norway	N	30
23	Sweden	S	60
24	Switzerland	CH	30
25	United Kingdom	UK	60

2. METHODOLOGY: DESCRIPTION AND EXAMPLE

The methodology has been described in Young and Shaw (1986), and makes use of the assumptions in linear LRTAP models that the deposition D_j at a given point may be described as:

$$D_j = \sum_{i=1} t_{ij} Q_i$$

where Q_i is the source strength of source i, N is the total number of sources and t_{ij} is the transfer coefficient quantifying the atmospheric linkage between source i and receptor j. It has the units of: (deposition per unit area per unit time)/(emission per unit time), and forms an ixj matrix called the "unit transfer matrix."

Using emission rates and the areas of each country, the S budget computed by the NMI model was converted to a unit transfer matrix for the period October 1978 - September 1980. Table I shows three columns of the unit matrix. Republic of Germany will be taken as an example in this paper for illustrating the methodology although it is recognized that the locations where deposition needs to be reduced in Europe are still undetermined.

In Table I, the country producing the largest deposition in the Federal Republic of Germany (D = receptor 10) per unit emission is the FRG itself (1.0 Tg S a^{-1} emission in the FRG results in 14.9 kg S ha^{-1} a^{-1} deposition). Therefore, if one were to reduce deposition in the FRG., it would be most effective to reduce emissions there, according to the model.

Once emissions in the F.R.G. have been reduced to a predetermined minimum (specified on a technological or other basis), emissions should be reduced in Luxemburg (L = source 16), the country with the next largest transfer element (1.0 Tg S a^{-1} emission in Luxemburg results in an estimated 10.0 kg S ha^{-1} a^{-1} deposition in the FRG). The next largest contributor is Switzerland (with a matrix element value of 5.0), followed closely by the Netherlands with a matrix element value of 4.6. It would not, according to the model, be effective to reduce emissions in Portugal or Romania (sources 20 and 21) because their source-receptor matrix elements are zero for the Federal Republic of Germany, indicating no atmospheric linkage.

If Greece (receptor 11) had been chosen as a target, Table I indicates that emission reductions would instead be required primarily in Greece, Bulgaria and Albania.

If it is desired to reduce deposition at more than one receptor point, two-dimensional ranking of source regions can be carried out as described in Young and Shaw (1986).

3. SAMPLE RESULTS

The optimized emission reductions will depend not only upon the source-receptor matrix but upon the constraints placed upon the emission reductions in each country, whether for technological or other reasons. Figure 1 shows the effect upon S deposition in the FRG of applying the above methodology for optimizing emission reductions, using constraints based upon the emission reductions in Table II that were committed or forecast by the Economic Commission for Europe (ECE, 1985). The predicted emission reductions, and the resulting reduction in deposition, are shown from left to right in Figure 1 with the most effective reductions on the left, although obviously the emission reductions could take place in any order.

The model predicts that the greatest deposition reduction in the F.R.G. (from 5.5 to 4.0 g S m^{-2} a^{-1}) will occur because of the 1.0 Tg S a^{-1} forecast to be removed from emissions in the FRG, Switzerland, the Netherlands and Belgium. The additional 2.3 Tg S a^{-1} removed from emissions in Austria, Czechoslovakia, Denmark, Italy and the United Kingdom will lead to only an additional 0.4 g S m^{-2} y^{-1} deposition reduction, to 3.6 g S m^{-2} a^{-1}. Emission reductions

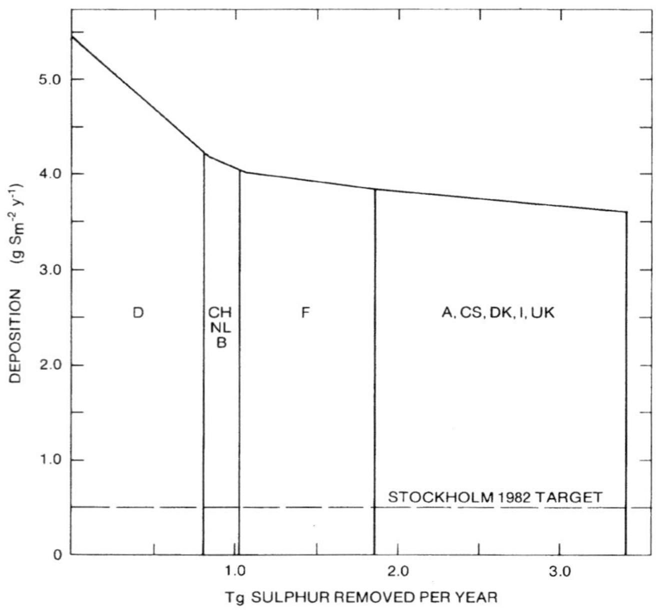

Figure 1. Reduction in deposition in the Federal Republic of Germany as a result of emission reductions in several countries as outlined in Table 2. The most effective emission reductions (i.e. 0.8 Tg S y^{-1} in the Federal Republic of Germany, "D") are shown on the left.

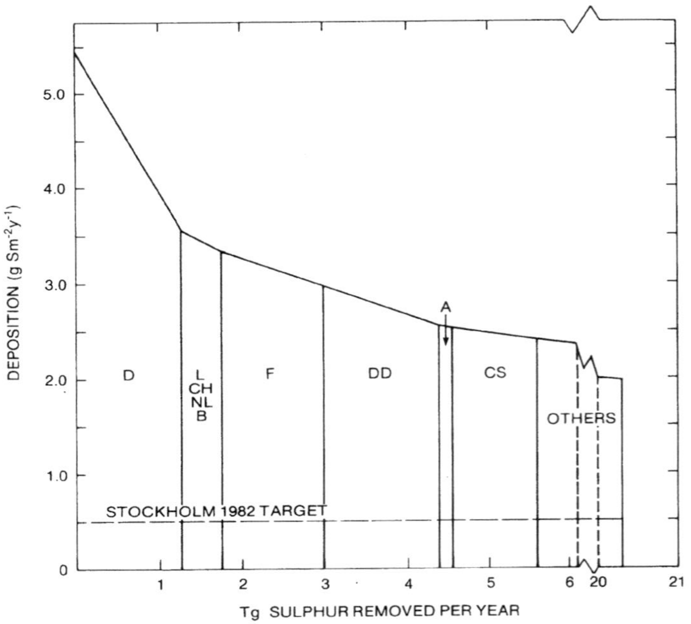

Figure 2. As in Figure 1, but assuming a greater emission reduction (70%) in each country.

in other countries will, according to the model, have little beneficial effect upon deposition in the FRG.

The results in Figure 1, therefore, indicate that it is most effective to reduce emissions in countries in and near the country where the deposition target is being applied, despite the transboundary atmospheric transport of S. Examinations of the atmospheric transfer matrix in Table I indicates that this result would be applicable to any target country.

To bring about further reductions in a given country it is more effective (according to the atmospheric model) to remove more S from emissions in countries in and near the location where the deposition target is being applied, than to reduce emissions further afield. Figure 2 shows the results of relaxing the constraints on emission reductions by increasing them to 70% in all countries covered by the model through applications of controls to the combustion of brown coal, anthracite, heavy fuel oil and diesel oil according to the analysis of Shaw (1983). In Figure 2, the removal of the first 2.0 Tg S from emissions will result in a deposition reduction of 2.1 g S m^{-2} a^{-1} compared to 1.4 g S m^{-2} a^{-1} in Figure 1. This is because more of the emission reductions in Figure 2 occur in countries that are closely linked via the atmosphere to the F.R.G.

4. UNCERTAINTIES AND LIMITATIONS OF THE METHODOLOGY

(a) Young and Shaw (1986) found that an estimated ±50% uncertainty in individual matrix elements had little effect upon the selection of the most important source regions for emission abatement. Because the highest ranking matrix elements for the European matrix have much greater values than the lower ranking elements, the above results are still expected to apply.

(b) Atmospheric transfer matrices are based upon meteorological data during a specified period. It is important that the period be representative of long term conditions.

(c) The methodology assumes that the chemistry and deposition processes affecting S are linear, i.e. that deposition changes are proportional to emission changes. This is believed to be not strictly true because of the interaction with the S chemistry of other pollutants such as oxides of N and oxidants. If a large enough geographical area is considered, the processes should be linear because what goes up must come down. Oppenheimer (1982) and Shaw and Young (1983) concluded that, while there are departures from linearity, they are not large. However, more work needs to be done to establish the bounds on linearity.

(d) It is recognized that other pollutants such as oxides of N and oxidants are very important in Europe, especially with respect to their potential for forest damage. Although the methodology described in this paper deals only with S deposition and may not be adaptable to control strategies for other pollutants because of their potentially greater non-linearity, S control strategies may still be important in their own right because, for example, the acidification of soil and the accompanying release of toxic metals may render forests more susceptible to stresses from other pollutants such as oxidants (Matzner and Murach, 1985; Zoettl and Huettl, 1985).

(e) In North America, the identification of strong source regions, sensitive receptors, and the geographical relationships among them are probably more clear-cut than they are in Europe. As a result, various workers have developed similar control strategies for North America, despite the use of different methodologies and LRT models (Streets *et al.*, 1983; Young and Shaw, 1986). In contrast, the situation in Europe appears to be more complex: strong source regions there are not "clustered" as much as they are in the industrialized Ohio Valley - Appalachia - Lower Great Lakes region in North America. Furthermore, sensitive target areas have been less completely defined to date than they have been in North America. The optimization methodology in this paper indicates that emissions should be reduced first in regions close to the target receptor(s); if it were found that deposition should be reduced at several separated points in Europe, the required emission reductions might be spread over a large area and, unlike North America, might not be too different from a uniform emission reduction. Therefore, the assessment of the applicability of the methodology in this paper to Europe must await the better definition of enviornmental targets.

5. CONCLUSIONS

(a) According to the atmospheric transfer matrix of the NMI/LRT model, which links S emissions in a given country in Europe with deposition in all 29 countries, emission reductions are needed in a large number of countries to bring about deposition reductions in a given country.

(b) Nevertheless, it is most effective to reduce emissions in and near the country where the deposition target is being applied.

(c) To bring about further deposition reductions, it is more effective to increase the amount of S being removed in countries in and near where the target is being applied, rather than to seek emission reductions farther afield.

(d) The applicability to Europe of the methodology in this paper must await the better definition of deposition targets, and the locations at which the targets will be applied.

REFERENCES

ECE, 1985: Strategies and Policies for the Abatement of Air Pollution. Report EB. AIR/R.1/Rev. 1. United Nations Economic Commission for Europe, Geneva.

Eliassen, A., 1978: *Atmospheric Environment* **12,** 479-487.

Matzner, E. and Murach, D., 1985: Soil acidification and its relationship to root growth in declining forest stands in Germany, Paper B-3(01), International Symposium on Acidic Precipitation, Muskoka, Ontario, Canada, September 1985.

OECD, 1977: The OECD Programme on Long Range Transport of Air Pollutants - Measurements and Findings. Organization for Economic Cooperation and Development. 1. Rue André Pascal, 757775, Paris CEDEX 16 France.

Oppenheimer, M., 1982: *Atmospheric Environment* **17,** 451-460.

Shaw, R.W., 1983: Optimization of Emission Reductions to Reduce Atmospheric Sulphur Deposition in Europe. Unpublished manuscript, 38 pp.

Shaw, R.W. and Young, J.W.S., 1983: *Atmospheric Environment* **17,** 2221-2229.

Streets, D.G., Knudson, D.A. and Shannon, J.D. 1983: *Environmental Science and Technology* **17,** 474A-485A.

Young, J.W.S. and Shaw, R.W., 1986: *Atmospheric Environment* **20,** 189-199.

Zoettl, H.W. and Huettl, R.F., 1985: Forest decline and status of nutrition, Paper B-3(05), International Symposium on Acidic Precipitation, Muskoka, Ontario, Canada, September 1985.

APPLICATION OF TRAJECTORY ANALYSIS TO THE ASSESSMENT OF LOCAL AND LONG-RANGE CONTRIBUTIONS TO ACIDIC DEPOSITION

R. G. de Pena*, G. D. Rolph, J. F. Takacs', J. O. Holian"
Department of Meteorology
503 Walker Building
The Pennsylvania State University
University Park, PA 16802

* Author for all correspondence.
' Present affiliation: Charles T. Main, Inc., Prudential Center, Boston, MA 02199 USA
" Present affiliation: NUS Corporation, 910 Clopper Road, Gaithersburg, MD 20878 USA

ABSTRACT. Two methods of back trajectories have been applied for 14 precipitation events. One method consisted of a combination of isentropic and constant level trajectories. The second method consisted only of the constant level trajectory. For each method a long range and a local contribution to the chemistry of the precipitation was calculated. The long range contribution was calculated by taking into account the SO_2 an NO_x emissions in the areas of the mixed layers traversed by the trajectories, the diffusion, and the dry and wet depositions along the trajectories path. For the local contribution, local emissions and the dry deposition were taken into account. From these contributions expected values of sulfate and nitrate in the precipitation were calculated and these values compared with the observed ones. Results indicate that: a) the isentropic-constant level model is much better in estimating the measured concentrations of sulfate and nitrate in precipitation than the constant level mode, and b) the long range transport of precursors contributed more to the total concentration than the local emissions.

1. INTRODUCTION

Increasing concern over acid rain has led many researchers to try to find methods of determining where the sources of precursor emissions are located. The question of the identity and location of particular sources is one of extreme sensitivity. If source-receptor relationships can be established there will be a firm basis for decisions such as the implementation of pollution control measures, changes in the fuel types, or perhaps even long-term geographic displacement of particular

Water, Air, and Soil Pollution 30 (1986) 885–896.

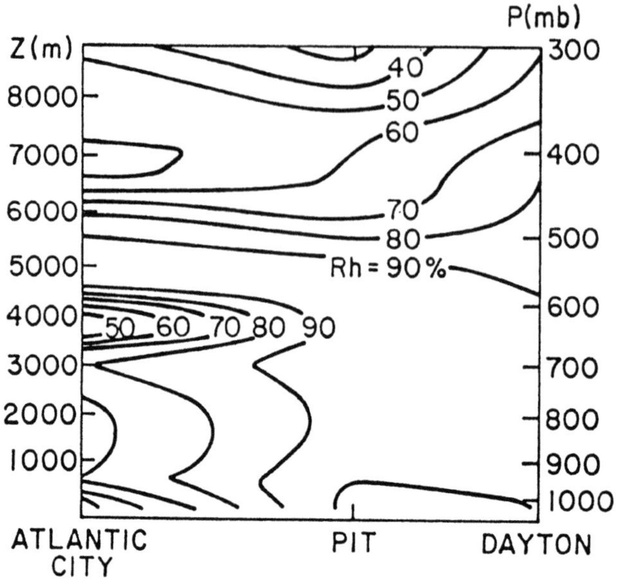

Fig. 1. Cross section of relative humidity along an East-West line from Atlantic City, New Jersey to Dayton, Ohio. Event May 2, 1985.

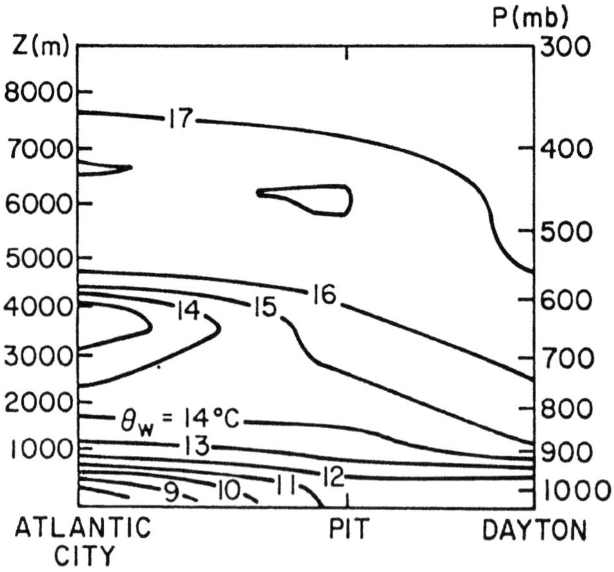

Fig. 2. Cross section of θ_w, moist potential temperature along an East-West line from Atlantic City, New Jersey to Dayton, Ohio. Event May 2, 1985.

industrial operations. Therefore, in an attempt to determine source-receptor relationships, two methods of back trajectories have been applied to precipitation samples collected on an event basis. The trajectories are used to trace the precursor emissions through synoptic systems that result in acidic deposition.

2. ISENTROPIC TRAJECTORY ANALYSIS

Insentropic trajectory analysis is based on the concept that the motion of air in the atmosphere proceeds in an adiabatic and reversible way and therefore the entropy or the potential temperature is conserved. Because we are dealing with precipitation cases and condensation occurs in the parcel of air the moist adiabatic potential temperature (θ_W), which is conserved in moist and dry processes, is the parameter chosen.

In the method applied, the isentropic trajectories are drawn on θ_W surfaces. To that end it is assumed that a particular weather system moves steadily without change in its velocity, shape or intensity (Carlson, 1980). This assumption is reasonable since sounding data are updated every 12 hr and most weather systems change little in such a small interval. Streamlines are determined on the θ_W surface and the locations of the parcel relative to the ground are obtained by subtracting the phase velocity of the system from the wind (de Pena et al., 1984).

The procedure has been automated, in the Penn State Department of Meteorology's microcomputer, to calculate trajectories back in time in 6-hr intervals.

The method requires the selection of the moist adiabtic potential temperature surface. To that end the precipitating cloud layer over the receptor site must be located. This is done by examining the upper-air soundings, vertical cross sections of relative humidity and moist potential temperature (Figures 1 and 2), and radar maps during the precipitation period. The sounding of the nearest upper-air station relative to the receptor region along with other nearby soundings are used to estimate the saturated layer. Since soundings are only taken every 12 hr, the sounding is chosen as close as possible to the time of precipitation.

Many times, however, the top and bottom of the precipitating cloud layer may not be evident from the sounding. In that case the cloud layer is determined through the use of vertical cross-sections of relative humidity. Precipitating cloud layers have been found to correspond roughly to a relative humidity of 90% or greater. Figure 1 is a vertical cross section of relative humidity along a line from Atlantic City, NJ to Dayton, OH which indicates a precipitating cloud layer centered over Pittsburgh, PA. In addition, radar maps can be used to find the height of the cloud top, and surface observations to find the height of the cloud base. Once the precipitating cloud layer has been found, the mean θ_w through this layer can be determined by overlaying a corresponding vertical cross section of relative humidity onto a corresponding vertical cross section of θ_w. In Figure 2 the mean θ_w is approximately 15°C (288°K).

3. ATL-ATAD METHOD

The ARL-ATAD model (Heffter, 1980) is based on the concept that a parcel of air in the mixed layer will be dispersed in the horizontal with the speed of the mean wind in that layer. The mean wind in a layer at a rawinsonde station is obtained by linearly weighting the winds with respect to the depth. In addition, the model calculates the height of the mixed layer according to the following criteria:

For the daylight hours (12Z to 18Z) the mixed layer is determined by the height of the lowest "critical inversion" defined by

$$\text{A)} \quad \Delta\theta/\Delta Z \geq 5 \text{ K/KM} \tag{1}$$

$$\text{B)} \quad \theta(\text{top}) - \theta(\text{base}) \geq 2 \text{ K} \tag{2}$$

where $\Delta\theta/\Delta Z$ is the potential temperature lapse rate and θ(top) and θ(base) refer to the potential temperatures of the top and base of a layer, respectively.

At night, when the mixed layer can become very shallow, the height of the mixed layer is assumed to be equal to twice the standard deviation of the vertical Gaussian pollution distribution and is calculated by

$$H = 2(2K_z t)^{1/2} \tag{3}$$

where K_z is the vertical coefficient of eddy diffusion taken as 1 m^2s^{-1} (Draxler, 1979) and t is the time of the trajectory segment (3 hours). The nighttime mixed layer is used for trajectories beginning at 00Z and 06Z, continuing until 12Z, thereafter, the daytime mixed layer is used.

The ARL-ATAD model calculates trajectories up to five days duration, starting every 6 hr moving either forward or backward in time. In our case, the backward mode was used exclusively.

4. SOURCE STRENGTH CALCULATIONS

For the source strength calculations (SS) two different sources are considered; one due to long-range transport and the other due to local sources.

Source strengths of sulfate and nitrate, respectively, are determined by first overlaying the complete trajectory onto a map of the United States and Southern Canada in which squares of 80 x 80 km have been marked off. Each square contains information on the number of tonne yr^{-1} of SO_2 and NO_x that was emitted in that square.* Typical emission values ranged from 0 to 562,700 tonne yr^{-1}. Only the squares traversed by the trajectory in the mixed layer are considered to contribute to the input of SO_2 and NO_x.

A box is assigned to each square with a volume equal to the surface of the square (80 x 80 km) multiplied by the height of the mixed layer at the time the trajectory passed through the square.

To calculate the contribution of each box to the pollutant transported by the trajectory, the emissions in the box, the dispersion, and the dry and wet deposition are taken into account. The contribution of pollutant from box n to box n + 1 (QP_n) is calculated according to the following equations:

$$Q_n = (QP_{n-1} + Q_{no})\left[1 - \frac{V_d t_n}{z_n}\right] e^{-\Omega t_{nr}} \tag{4}$$

$$QP_n = Q_n \frac{t_n}{t_{n+1}} \frac{V_{n+1}}{V'_{n+1}} \tag{5}$$

where Q_{no} is the emission in box n and QP_{n-1} the contribution from box n-1 to box n. The correction for dry deposition is given by the second term between brackets in Equation (4) where V_d is the dry deposition velocity taken as 0.5 cm s^{-1} for both SO_2 and NO_x, Z_n is the height of the mixed layer, t_n is the time the trajectory spends in the box, and t_{nr} is the rainfall time interval.

The correction for wet deposition is given by the last term in equation (4) with Ω, the washout ratio, equal to

$$\Omega_{SO_2} = 10 \times 10^{-5}\ J^{0.53} s^{-1}$$

$$\Omega_{NO_x} = 7.25 \times 10^{-5}\ J^{0.53}\ s^{-1}$$

where J is the precipitation rate (mm/hr) (McMahon and Dennison, 1979).

---

* This information was obtained from: Department of Natural Resources, State of Wisconsin, EPRI-SURE and Environment Canada.

The dispersion is taken into account in Equation (5) where V_{n+1} is the volume of box n+1 (V_{n+1} = 80 x 80 km x Z_{n+1}), V'_{n+1} is the expanded volume of box n + 1 (V'_{n+1} = 80 km x Z_{n+1} x (80 km + σ_y)), and t_{n+1} is the time the trajectory spends in box n+1.

If the last box traversed by the trajectory is m, then the concentration in it is:

$$X_m = \frac{Q_m t_m}{V_m} \tag{6}$$

and

$$Q_m = Q_{mo} + QP_{m-1}$$

X_m corrected for dry and wet deposition is the concentration at the site due to long range transport ($X_{\ell\ell}$).

The local contribution is given by

$$X_\ell = \frac{Q_{\ell 0} t_\ell}{V_\ell} \tag{7}$$

where $Q_{\ell 0}$ is the emission of the box corresponding to the site, t_ℓ is the time the trajectory spends in the local box and V_ℓ is the volume of the local box.

The concentration expected in the precipitation from the concentration in air is calculated by:

$$C = \varepsilon \frac{X}{L} \tag{8}$$

where C is the the concentration in precipitation, ε is the washout coefficient, L is the liquid water content, and X(total) = $X_\ell + X_{\ell\ell}$.

5. RESULTS

Two trajectories were calculated for each time period considered. One trajectory consisted of the isentropic method from the site backward in time until it intercepted the mixed layer. Once in the mixed layer, the ARL-ATAD model was used until a total trajectory duration of 36 hr had elapsed. The second trajectory consisted only of the ARL-ATAD method for a trajectory duration of 36 hr backward in time from the site. As an example, Figure 3 illustrates two trajectories obtained with the combined isentropic/ARL-ATAD model for the 2 May 1985 case. The solid line represents the ARL-ATAD trajectory segment and the dot-dashed line the isentropic trajectory segment. Figure 4 represents the same case with trajectories calculated by the ARL-ATAD method alone. Because vertical motions are present in mid-latitude weather systems, the isentropic/ARL-ATAD trajectories can, on occasion, differ substantially from those computed by the method of average-wind analysis (ARL-ATAD) alone. These differences were also found by Wisniewski and Sax (1979) when they compared an isentropic trajectory scheme developed by Danielsen (1961, 1967)[1] and the constant level model developed by Hefter and Taylor (1975).

Results of fourteen case studies are presented in Tables I and II, where actual SO_4 and NO_3 concentrations in precipitation are compared to the concentrations obtained by both types of trajectory analyses.

These results suggest that, in general, the values obtained by applying the isentropic/ARL-ATAD method are a better indicator of the actual concentration in the precipitation than those obtained by applying the ARL-ATAD method exclusively. This result is not surprising since the isentropic/ARL-ATAD method is able to represent the 3-dimensional motion of the air, particularly out of the mixed layer, whereas the ARL-ATAD method assumed the air remained in the mixed layer. Two obvious exceptions that can be noted in Tables I and II are Cases 1 and 8. These cases occurred when strong convection was present which was associated with thunderstorm activity. When convection is present, usually there are multiple surfaces with the same θ_w and following any particular θ_w surface is, at best, questionable. Therefore, in Cases 1 and 8 the ARL-ATAD model was considered to be the better of the two types of models. In addition, a breakdown of the local vs. long-range contributions are given in Tables I and II for the most representative trajectory analysis.

As Tables I and II show, the long-range contribution is much more important than the local contribution for both sulfate and nitrate in twelve out of fourteen cases. This contradicts an earlier study by Holian (1984), in which the nitrate concentration was determined to be mostly from local sources. However, this contradiction is not surprising because Holian considered a very high degree of dispersion which made the emissions from boxes located prior to the interception box (of the isentropic trajectory with the mixed layer) negligible.

---

[1] Danielsen (1961) used an objective method for computing isentropic trajectories on potential temperature surfaces which is based on the total-energy equation, whereas the authors of this paper used a graphical method for computing isentropic trajectories on moist potential temperature surfaces.

Table I. The case numbers, the concentrations measured in precipitation, the volume-weighted average concentrations expected in precipitation as calculated by the isentropic/ATAD and ATAD models, and the contributions to the expected concentrations in precipitation due to local and long-range sources. All the local and long-range source-strengths were computed by the isentropic/ATAD model except in cases 1 and 8 which were computed by the ATAD model (denoted by an asterisk).

		SO_4 Concentrations ($\mu mol \cdot L^{-1}$)			
Case	Precipitation	Isentropic & ARL-ATAD	ARL-ATAD	Local Contribution	Long-Range Contribution
1	39.0	0.3	39.2	0.3*	38.9*
2	2.7	5.9	997.6	1.4	4.5
3	7.4	16.2	87.4	1.1	15.1
4	30.7	41.6	81.3	1.1	40.5
5	19.0	8.7	42.8	1.7	7.0
6	42.0	15.6	111.3	1.0	14.6
7	10.7	19.1	313.8	2.8	16.3
8	25.1	1.3	111.0	0.9*	110.1*
9	10.3	2.19	132.3	1.0	20.9
10	14.1	13.9	246.0	1.6	12.3
11	9.1	14.5	66.4	0.8	13.7
12	7.5	54.1	94.1	1.3	52.8
13	14.5	3.2	219.6	3.08	0.15
14	4.9	2.3	112.0	2.3	0

Table II. Same as Table I but for nitrate.

NO_3 Concentrations ($\mu mol \cdot L^{-1}$)					
Case	Precipitation	Isentropic & ARL-ATAD	ARL-ATAD	Local Contribution	Long-Range Contribution
1	20.0	0.6	23.0	0.6*	22.4*
2	16.0	7.2	520.2	2.6	4.6
3	9.7	21.4	79.2	2.0	19.4
4	40.5	28.7	74.6	2.1	26.6
5	14.0	18.3	48.0	3.3	15.0
6	31.6	11.6	80.5	1.9	9.7
7	11.9	18.9	159.9	5.3	13.6
8	25.2	2.8	76.3	1.6*	74.6*
9	14.6	31.6	82.3	2.0	29.6
10	17.4	12.9	207.0	3.2	9.7
11	18.3	13.7	43.8	1.4	12.3
12	29.2	46.6	103.4	2.5	44.1
13	23.6	6.7	230.6	5.9	0.8
14	29.4	4.3	98.2	4.3	0

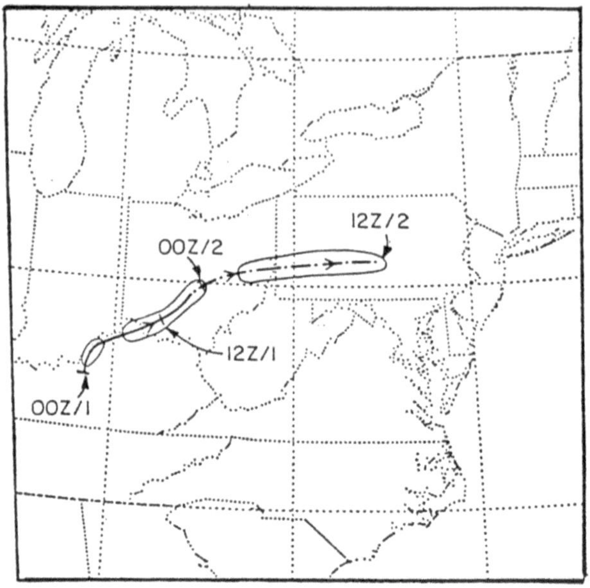

a)

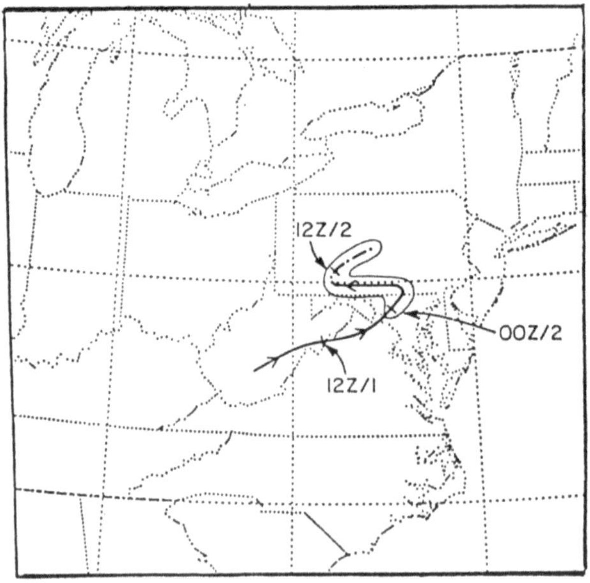

b)

Fig. 3. Combination of isentropic/ARL-ATAD trajectories for the event of May 2, 1985.

a) Trajectory starting at 1200 Z
b) Trajectory starting at 1800 Z

———————— ARL-ATAD segment
-·-·-·-·-·- Isentropic segment

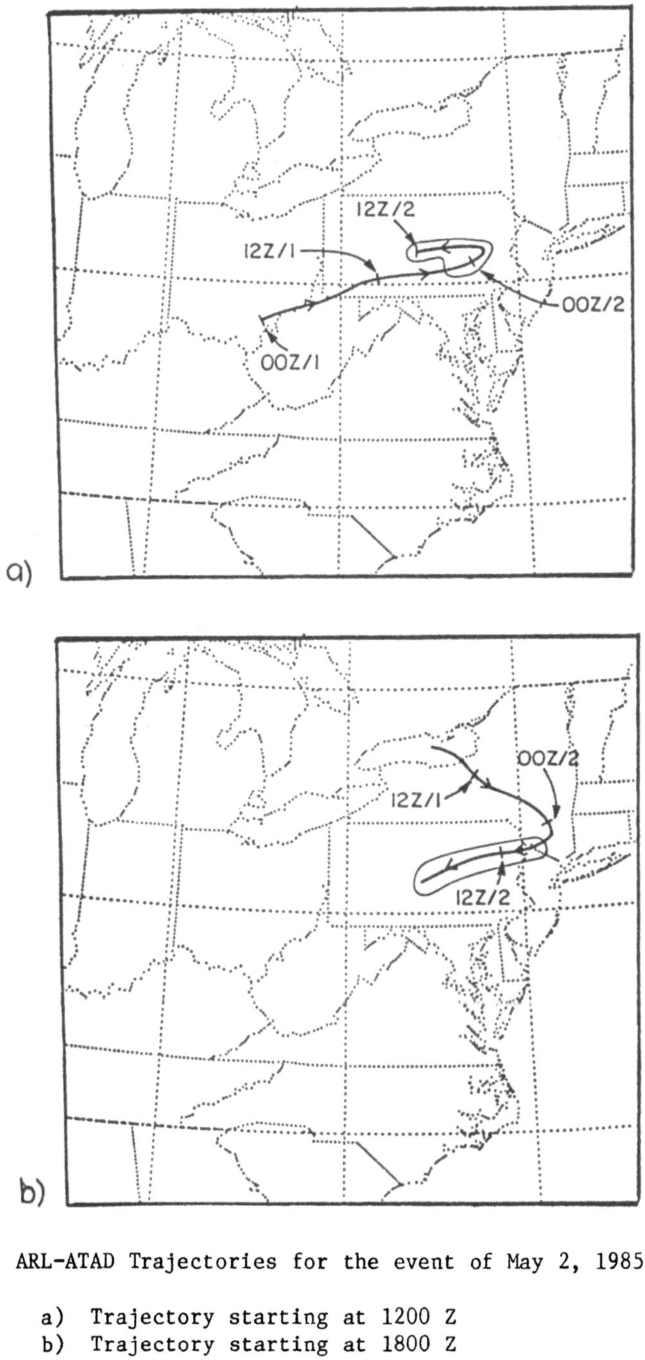

Fig. 4. ARL-ATAD Trajectories for the event of May 2, 1985.

a) Trajectory starting at 1200 Z
b) Trajectory starting at 1800 Z

ACKNOWLEDGEMENTS

This study was supported by NOAA Contract NA83RAC00087 and by DOE Grant DE-FG02-85ER60294.A000 for which we are grateful. The work reported here includes a portion of GDR's Master's thesis.

REFERENCES

Carlson, T.N.: 1980, Monthly Weather Review, 108, 1499.

Danielsen, E.F. and Bleck, R: 1967, Moist Isentropic Flow and Trajectories in a Developing Wave Cyclone, USAF Tech. Rept. No. 67-0617, Air Force Cambridge Res. Lab., 1-34 [NITS AD 670-847].

Danielsen, E.F.: 1961, Journal of Meteorology, 18, 479.

Draxler, R.R.: 1979, Atmospheric Environment, 13, 1559.

Hefter, J.L.: 1980, Air Resources Laboratories Atmospheric Transport and Dispersion Model, NOAA Technical Memorandum ERL ARL-81, U.S. Dept. of Commerce, 17 pp.

Hefter, J.L, Taylor, A.D.: 1975, A Regional-Continental Scale Transport, Diffusion, and Deposition Model, Part I: Trajectory Model, NOAA Technical Memorandum ERL-50, U.S. Dept. of Commerce, 16 pp.

Holian, J.O.: 1984, Acid Precipitation: Applying Isentropic Trajectories to Determine Source-Receptor Relationships. M.S. Thesis, Dept. of Meteorology, The Pennsylvania State University, University Park, PA 16802, 67 pp.

McMahon, T.A. and Dennison, P.J., 1979, Atmospheric Environment, 13, 571.

Pena, R.G. de, Carlson, T.M., Takacs, J.F. and Holian, J.O.: 1984, Atmospheric Environment, 18, 2665.

Wisniewski, J. and Sax, R.I.: 1979, Journal of Applied Meteor., 10, 1044.

SUDBURY SMELTER IMPACT ON ATMOSPHERIC DEPOSITION OF ACIDIC SUBSTANCES IN ONTARIO

M.A. Lusis, A.J.S. Tang, W.H. Chan, D. Yap, J. Kurtz, P.K. Misra
Ontario Ministry of the Environment
Air Resources Branch, 880 Bay Street, Toronto, Ontario,
Canada, M5S 1Z8
G. Ellenton
Eltek Environmental
218 Alexandra Avenue, Waterloo, Ontario, Canada N2L 1M7

ABSTRACT: During the period from June 1982 to March 1983, there was a prolonged shutdown of the INCO and Falconbridge smelters at Sudbury, Ontario. Taking advantage of this opportunity, the Air Resources Branch of the Ontario Ministry of the Environment undertook a detailed analysis of the impact that the resulting emission reductions had on the atmospheric deposition of acidity and SO_x in Ontario. This paper summarizes the various meteorological analyses and mathematical modeling estimates that were carried out in order to determine the contribution of the Sudbury smelters to atmospheric deposition in Ontario. For sulfates, the Sudbury contribution to the total wet deposition was typically less than 15% at the receptors studied. Furthermore, the smelters were found to contribute on the order of 10 to 20% of the total dry deposition of S compounds in central and northeastern Ontario, and less than 10% elsewhere.

1. INTRODUCTION

During the period from June 1982 to March 1983, there was a prolonged shutdown of the INCO and Falconbridge Smelters at Sudbury, Ontario. The former smelter is one of the largest point sources of SO_2 emissions in North America, with annual emissions falling in the range of 500 to 800 x 10^3 tonne in recent years, and issuing mostly from a 381-m "superstack". Falconbridge represents a considerably smaller source, with annual SO_2 emissions (mainly from a 93-m stack) since 1978 being in the vicinity of 100 x 10^3 tonne. In view of the significant contribution of the Sudbury smelters to the total emissions of SO_2 in eastern N. America, and especially in Ontario (in recent years these smelters have contributed about 50 to 60% of the provincial total), the Air Resources Branch of the Ontario Ministry of the Environment undertook a detailed analysis of the impact of these emission reductions on the atmospheric deposition of acidity and SO_x in Ontario. The analysis also included selected N compounds and trace metals.

Previous work, carried out in the Sudbury basin itself, had shown that it would be difficult to detect changes in atmospheric deposition due to changes in smelter operations (Ontario Ministry of the Environment, 1979; Scheider et al.,

Water, Air, and Soil Pollution **30** (1986) 897–908.

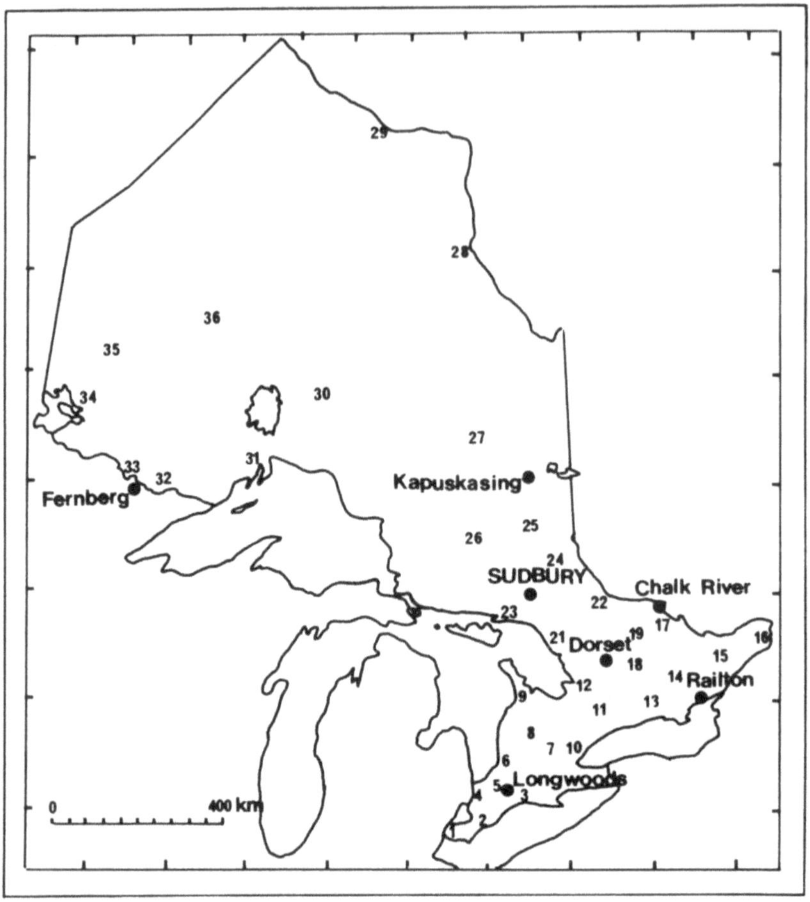

FIGURE 1: Location of Cumulative (numbered) and Daily (labeled) Monitoring Stations

1981). In fact, it has been estimated by aerometric analysis (Chan et al., 1984) that even within about 40 km of Sudbury, the smelters contribute on average only about 20% or less of the total wet deposition of most contaminants (including S) found in precipitation. The corresponding estimates for smelter contributions to the total dry deposition in the Sudbury basin were about 25% for S, and less than 20% for most trace metals other than Cu and Ni. Long-range transport from sources to the south was indentified as the major contributor to atmospheric deposition of SO_x and NO_x in the Sudbury basin.

The current analysis consisted largely of an examination of air and precipitation data collected during, and in the 2 yr prior to, the smelter shutdown in 1982/83, in an attempt to quantify the smelter contribution to wet and dry depositon at a number of receptors out to about 500 km from the source. Deposition maps and overall statistical comparisons were based on data from the Ministry's cumulative deposition network (Figure 1), which collects wet-only samples on a 28-day basis (these samples are analyzed for the major ions and a number of trace metals. - Chan et al., 1982). More detailed trajectory and meteorological analyses were based on data from daily air and precipitation samplers operated in Ontario by the Ministry (Chan et al., 1982), the Atmospheric Environment Service, and Ontario Hydro (Figure 1). The air samples were analyzed for sulfates, nitrates, ammonium and SO_2. The daily precipitation samples were analyzed for the major ions.

A Lagrangian long range transport model, which had been tested and found to successfully simulate monthly average wet deposition of sulfate at a number of receptors in eastern N. America over a 3 yr period, was also applied to determine the change in S deposition during the shutdown period of the Sudbury smelters.

It is not possible to give the full details of the above analyses here. Only the main analysis methods and results will be discussed. For details, the reader is referred to a series of Ministry of the Environment reports (Yap, 1984; Yap and Kurtz, 1984; Tang et al., 1984; Ellenton and Misra, 1984).

2. DISCUSSION OF ANALYSES AND RESULTS

2.1 Overall Monitoring Statistics (Tang et al., 1984)

Precipitation chemistry data obtained from stations within a radius of 250 km from Sudbury were grouped together on a seasonal basis. This allowed a comparison to be carried out, for the same season, of shutdown period average concentrations with the corresponding data collected during previous operating periods. This comparison was carried out for H^+, sulfates, nitrates and Cu in precipitation. The only consistent decrease during the smelter shutdown was found for Cu, which is a good tracer of smelter emissions. This is also the only one of the parameters examined for which a median 2-sample test of significance indicated a decrease (at the 95% confidence level) for most seasons during the shutdown period. Nitrates (for which smelter emissions are small), were not significantly different during the shut-down and operating periods. Sulfates and H^+ in precipitation showed no consistent, statistically significant decrease during the shut-down period.

2.2 Precipitation Concentration and Wet Deposition Contours (Tang et al., 1984)

Precipitation depth-weighted concentration and wet deposition contours of the above four parameters were plotted for both the shutdown period and the corresponding operating periods. It was found that the pattern in the decrease of SO_4 concentrations during the shutdown period resembled neither that for Cu, which is a good tracer for smelter emissions, nor that for NO_3 (NO_x are not emitted by the smelters in any appreciable amount). The observations seemed to be a resultant of both smelter emission changes, and a general decrease in SO_2 emissions from utilities in a number of states south of the Great Lakes during the study period, which between 1980 and 1982 amounted to more than 1×10^6 tonne on an annual basis (Yap, 1984). Concentration contours for H^+ did not display any readily recognizable patterns from which a smelter impact could be inferred. An examination of SO_4/NO_3 ratios, using NO_3 as a rough normalizing factor for meteorological variability (since nitrates are not emitted by the smelters), suggested an area of detectable smelter influence extending out to 200 to 400 km from Sudbury.

A comparison of wet deposition during the shutdown and operating periods proved even more difficult to interpret, since deposition is further complicated by year-to-year variations in the precipitation amount. Both SO_4 and NO_3 deposition were found to decrease during the shutdown periods at monitoring stations to the east of Sudbury suggesting that meteorological factors and/or changes in emissions in Southern Ontario and in states south of the Great Lakes may be largely responsible for these observations.

2.3 Meteorological Analysis of Daily Precipitation and Air Concentrations

A more detailed analysis was carried out, based on both back-trajectories and a consideration of the synoptic meteorology associated with each sample, to select sub-sets of samples for operating and shut-down periods where the air parcels had previously passed over Sudbury. A comparison of air and precipitation concentrations and deposition during the operating and shut-down periods then permitted an estimate of the Sudbury smelter contribution to atmospheric deposition at a number of receptor locations.

2.3.1 Precipitation Concentrations/Wet Deposition

2.3.1.1 Trajectory Analysis (Yap and Kurtz, 1984). For each of the six daily sampling sites shown in Figure 1, a "Sudbury sector" was defined as the octant of origin from the site which includes the Sudbury area. Surface geostrophic back-trajectories (with a frictional correction) were calculated from each site, the trajectory length being based on travel time of air parcels from the Sudbury area to the site. Thus, loading and concentration statistics for the Sudbury sector at Dorset were based on 6-hr northwesterly trajectories; for Chalk River, they were based on 6 hr west-northwesterly trajectories; for Longwoods, on 12-hr northerly trajectories; for Railton, on 12-hr northwesterly trajectories; for Fernberg, on 24-hr easterly trajectories; and for Kapuskasing, on 12-hr south-southeast trajectories. The details of the trajectory calculation method and sector assignment are given by Kurtz (1983).

The results of this analysis are summarized in Table I. As can be seen from this table, mean sulphate loadings were not depressed at all sites during the shutdown period but varied in general with precipitation amount. Mean sulphate concentrations, on the other hand, were typically lower at central, eastern and southeastern Ontario during the shutdown period. It should be added that mean nitrate concentrations in general were also low at central, eastern and southeastern Ontario during the shutdown period relative to values for the previous two years. The concentration reduction of nitrates cannot be attributed to the shutdown of the Sudbury sources, since the smelters are not significant emitters of NO_x.

Thus simple back-trajectory analysis gave inconclusive results regarding the impact of the smelters on wet deposition in Ontario. The following section

Table I
MEAN SULFATE LOADINGS, CONCENTRATIONS AND PRECIPITATION DEPTH DUE TO THE SUDBURY SECTOR+

Receptor	Period											
	Jul/80-Mar/81				Jul/81-Mar/82				Jul/82-Mar/83			
	n	P	L	C	n	P	L	C	n	P	L	C
Dorset	6	8	2.2	1.4	4	30	6.0	1.4	8	18	3.5	0.4
Longwoods	14	37	7.8	3.5	10	25	9.9	1.5	9	21	5.6	2.1
Railton	6	6	2.3	4.2	4	8	7.5	4.7	2	10	9.2	1.7
Fernberg	-	no data -			2	9	2.6	0.8	1	23	11.5	0.5
Chalk River	8	26	8.6	3.5	2	12	10.6	1.8	5	25	6.2	1.3
Kapuskasing	-	no data -			-	no data -			7	24	7.6	3.1

+ n = number of events; P = precipitation, mm; L = loading, mg m^{-2}; c = concentration, mg L^{-1}

describes another approach which was attempted, to assign an upper bound to the smelter contribution at a number of receptor sites.

2.3.1.2 Classification by Synoptic Meteorological and Back-Trajectory Patterns (Tang et al., 1984). Since the validity of simple surface geostrophic trajectories to accurately represent air parcel origins during precipitation can be criticized, the following approach was taken: a study was made of the synoptic pattern accompanying each precipitation event at a given monitoring site. Cases were identified for which source assignment by back-trajectory analysis could be made with some confidence. For Dorset, for example, the post-cold front, pre-high pressure ridge situation was selected for further study (the significant synoptic pattern for a Sudbury impact varies from site to site, and for some sites several patterns were studied). This sub-set of samples was subjected to further stratification, by requiring that the envelope of 48-hr geostrophic back-trajectories (which are calculated every 6 hr during the period of interest) should include Sudbury and should at all times be within a prescribed sector (for the Dorset receptor, for example, this was the 90° sector between west and north. In this way, we minimize the possibility of including cases where air parcels with recent history from the high emission density areas to south of Dorset have entered the "Sudbury sector").

The above stratification scheme resulted in a small set of precipitation samples, for which a potential Sudbury smelter impact could be expected, with a reasonable degree of confidence. For example, at Dorset, 9 such samples were identified during the July 80 - March 81 operating period; 3 during the July 81 - March 82 operating period; and 4 during the July 82-March 83 shutdown period. This sub-set was then subjected to further scrutiny, by applying the median 2-sample test of significance to the precipitation SO_4 and NO_3 concentrations. Only if the SO_4 concentrations during the shutdown period were significantly different from those of the operating period, while nitrate concentrations were not, did the analysis proceed further. Also, mean sulfate concentrations during the two operating periods had to be statistically the same.

While such a rigorous selection of events did result in a set of samples for which we could expect the smelter impact to be quantifiable by a comparison of operational and shutdown periods, serveral sites were eliminated due to insufficient data for a meaningful analysis, or because year-to-year changes in nitrate levels in the sample sub-set suggested the influence of source areas other than Sudbury. For the sites which were retained (Dorset , Railton and Chalk River), there was a further question to be addressed - viz. how to determine a representative Sudbury smelter contribution from such a small sub-set of data. The precentage of the total wet deposition associated with this sub-set was very small, and while we were reasonably confident of a potential Sudbury impact for this particular sub-set, there were clearly many other events and synoptic situations not included in the sub-set where the smelter could be contributing to the sulfate deposition.

The approach finally taken was to determine a precipitation concentration difference ("operating" minus "shutdown" period values of concentration) for this sub-set, which was attributed to Sudbury smelters. Then the Sudbury contribution to the total wet deposition at each site was determined by multiplying the above precipitation concentration difference by the total precipitation depth accumulated over all events where Sudbury may have been affecting the receptor area. For Dorset, for example, any event where at least one surface back-trajectory during the sampling period was in the quadrant between west and north was included in the determination of the above total depth. As a result, estimates of the smelter contribution are probably near the upper limits, since the criteria on precipitation depth are very loose (obviously not every event at Dorset, for which at least one of the four back-trajectories during the sampling period is within the 90° sector from west to north, will be influenced by Sudbury). However, since the sample size on which the source apportionment estimates are based is small, and since scavenging efficiency during the sub-set of precipitation events selected may not be representative of all events contributing to the wet deposition at the receptor of interest, caution is required with this interpretation.

The results of this calculation are shown in Column 2 of Table II. Out of interest, Column 3 also shows a rough upper limit estimate of the Subury contribution at Railton, Longwoods and Dorset based on simple back-trajectory analysis at these sites during the period from July 1980 to December 1981 (see Kurtz, 1983) when the smelters were in operation. The numbers shown in Column 3 represent the contribution to the total sulfate deposition, at each of the receptors, from three 45° sectors (the Sudbury sector and the two adjacent sectors) and can, therefore, also be considered as near-upper limit estimates, since the total sector covers 135° and sources within this sector other than Sudbury are also included.

The present analysis thus indicates that, for receptors out to about 400 km from the smelters, the smelter contribution to the wet sulfate deposition is expectd to be less than about 10 to 15% of the total. This is in agreement with previous work done within the Sudbury basin (Chan et al., 1984) which indicated a long-term smelter contribution to the wet sulfate deposition on the order of 10% of the total. Further out from these sources, the contribution is expected to be even less.

Table II
ESTIMATES OF SUDBURY SMELTER CONTRIBUTION TO THE SO_4 DEPOSITION AT VARIOUS RECEPTORS IN ONTARIO (% OF TOTAL)

Receptor	% Sudbury Smelter Contribution		
	By classification according to synoptic meteorology and back trajectories	From simple trajectory analysis (135° sector containing Sudbury)	Lagrangian model
Dorset	<12	< 8	15
Railton	< 2	<14	5
Longwoods	*	< 9	3
Chalk River	<12	-	10

* Insufficient data for analysis
\- Trajectory data not available for all sectors

2.3.2 Air Concentrations/Dry Deposition. To determine the impact of the Sudbury smelters on dry deposition in Ontario, air concentrations of sulfates, SO_2, nitrates and ammonium were compared during the shutdown and operating periods. We made the reasonable assumption that, if the same seasons for different years were compared, the change in the average dry deposition rate for a particular species would be proportional to the difference in its average air concentration during the two periods.

Daily air sample data were obtained from the daily monitoring stations shown in Figure 1. These included information on particulate sulfates, nitrates, and ammonium; HNO_3(sampled on nylon filter papers); and SO_2(sampled on filters impregnated with potassium carbonate). Two different approaches were used - one, a straightforward back-trajectory analysis (Yap and Kurtz, 1984); the other, a more detailed examination of trajectory and synoptic patterns, resulting in a sub-set of samples where the probability for a Sudbury impact was high (Tang et al., 1984). As in the case of the precipitation analysis, concentration differences attributable to Sudbury smelter operations were then used in the latter approach to estimate upper limits for the smelter impact on dry deposition at various receptor sites.

The method used for the back-trajectory analysis was the same as that described in Section 2.3.1.1. The results of this analysis are summarized in Table III. Note that at all monitoring stations, mean SO_2 concentrations within the Sudbury sector were depressed in the shut-down as compared to the operating period. Sulfate concentrations also decreased, except for the Longwoods and Fernberg sites. Although there was a decrease in SO_2 concentrations within the other sectors as well during the smelter shutdown period, the change was most pronounced within the Sudbury sector. The difference in mean SO_2 and SO_4 concentrations within the Sudbury sector during the smelter operating and shut-down periods was statistically significant at the 5% probability level, based on the median 2-sample test. On the other hand, for nitrate, nitric acid and particulate ammonium there was no signficiant difference.

Table III
MEAN SO_2 AND SULFATE AIR CONCENTRATIONS DUE TO THE SUDBURY SECTOR[++]

Receptor	Period							
	Jul/81-Mar/82				Jul/82-Mar/83			
	SO_2		SO_4		SO_2		SO_4	
	n	C	n	C	n	C	n	C
Dorset	23	8.7	22	2.3	17	1.3	18	0.7
Longwoods	38	5.0	39	2.6	31	3.0	33	3.0
Railton	39	4.7	40	1.5	19	2.4	19	1.3
Fernberg	10	2.2	10	0.7	12	1.2	13	1.1
Chalk River	23	8.2	23	2.0	13	1.9	18	0.8
Kapuskasing	11	7.9	11	4.4	9	1.8	9	1.9

[++] n = number of events; C = concentration (ug m^{-3})

Table IV
ESTIMATES OF SMELTER CONTRIBUTION (% OF THE TOTAL) TO SO_2 AND SO_4 DRY DEPOSITION IN ONTARIO

Location	Parameter	% Smelter Contribution		
		Trajectory	Trajectory/ Synoptic Meteorology	Lagrangian Model
Dorset	SO_2	11	< 37	20
	SO_4	4	< 18	9
Railton	SO_2	7	< 27	10
	SO_4	4	**	5
Longwoods	SO_2	3	< 5	5
	SO_4	*	**	4
Fernberg	SO_2	*	< 4	4
	SO_4	*	**	2
Kapuskasing	SO_2	18	< 47	20
	SO_4	7	< 46	5
Chalk River	SO_2	6	< 26	20
	SO_4	6	< 20	8

* Too small to quantify
** Insufficient data for analysis

The data in Table III were used (together with overall average air concentrations at each site and period of interest) to provide some preliminary estimates of the contribution of the Sudbury smelter to dry deposition of SO_2 and SO_4. These are shown in Table IV (Column 3).

The more detailed meteorological analysis proceeded along similar lines to what was done with the precipitation data (see Section 2.3.1.2), except that the smelter influence was generally more pronounced in the SO_2 and SO_4 air concentration data than in the precipitation data, and the number of samples available for assessing the smelter contribution was consequently larger. First, a sub-set of samples was selected based on a careful examination of the synoptic meteorology, 48-hr back-trajectories, and tests of statistical significance for the sulfur dioxide, sulfate and nitrate concentrations collected during the operating and shutdown periods. The contributions of the Sudbury smelters to the air concentrations was calculated for this sub-set, and was assumed to be representative of all air parcels coming to the receptor from a 90°-wide sector centered on Sudbury. The results, which are thought to be upper limits for the smelter contribution, are shown in Column 4 of Table IV. It should be noted that most of the smelter impact on sulfur dry deposition at the receptors studied here comes from SO_2, because the sulfur is emitted primarily as SO_2, and chemical conversion rates in the smelter plumes are small (Lusis and Wiebe, 1976; Chan et al., 1980).

2.4 Mathematical Modelling Estimates

The Lagrangian trajectory model used here has been described in detail by Ellenton et al. (1982, 1985). In the model, trajectories are initiated from each of 192 source centres in North America, and followed for 96 hr within the area from the Rocky Mountains in the west, to the east coast and from Florida in the south, to Hudson Bay. The transport of SO_2 emitted from each of the sources is computed along each trajectory, along with chemical transformations and removal to the earth's surface in rainy and dry conditions (assumed to be linear processes). This model was first used to successfully simulate the monthly wet deposition of sulfate at a number of receptors in Ontario and the northern United States during 1978-1980 (Ellenton and Misra, 1984), and to show that most of the month-to-month variability in wet deposition is explainable by meteorological factors, rather than emission fluctuations. It was then applied to estimate the impact of the Sudbury smelter on wet and dry sulfur deposition, by carrying out simulations, under 1980 meteorological conditions, with the smelters assumed to be operational; and non-operational. An example of the modelling results (for SO_4 wet deposition) is shown in Figure 2. There is reasonable agreement between the modelling and meteorological analysis source apportionment estimates, as can be seen from Tables II and IV.

3. SUMMARY

A source apportionment analysis was carried out to determine the contribution to dry and wet deposition at various receptors in Ontario, due to emissons from two large smelters at Sudbury. The analysis took advantage of a prolonged smelter shutdown during the latter half of 1982 and first quarter of 1983. Attention was focussed mainly on the SO_x, which are thought to be the most important contributors to acid deposition.

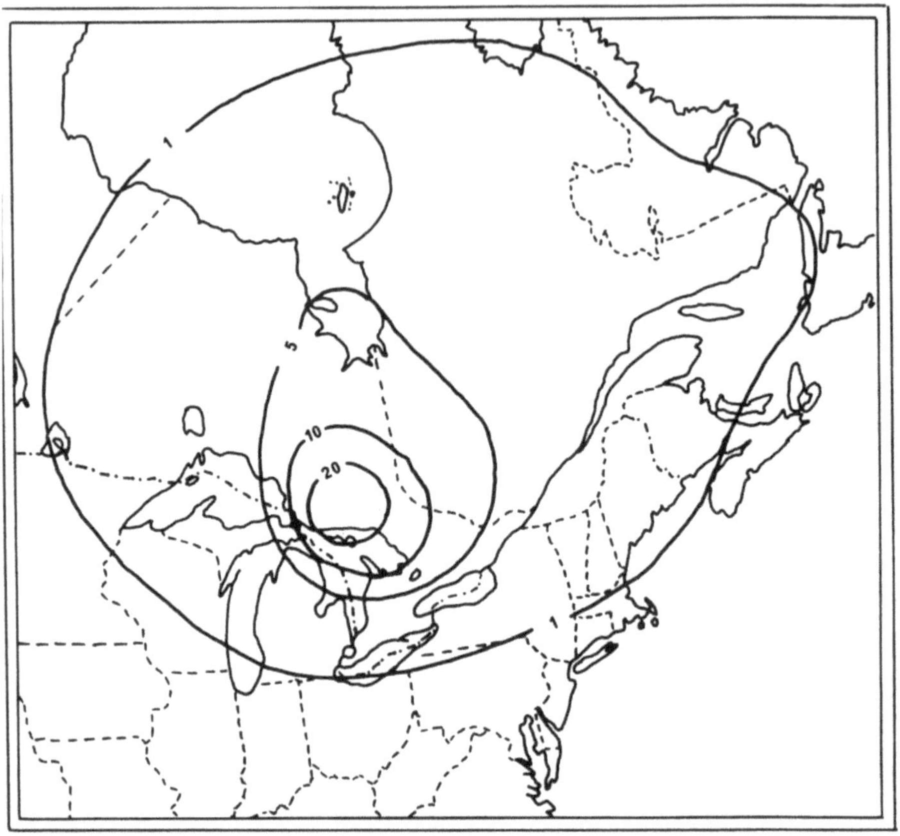

FIGURE 2: Model Estimate of Percentage Reduction in Wet Deposition of Sulfates Due to Shutdown of Sudbury Emissions (for 1980 Meteorological Conditions).

Despite the fact that the smelter shutdown resulted in a 60% decrease in the provincial SO_2 emissions, quantification of the smelter impact on S wet deposition by meteorological analysis of the data proved to be a difficult task, due to the very large meteorological variability in the precipitation chemistry and wet deposition data, and the fact that the smelter contribution constituted only 10 to 15% or less of the total at the receptors studied (these extended out to about 500 km from the source area). A statistical analysis of measurements within 250 km of the sources and an examination of the precipitation chemistry and wet deposition contours during the shutdown and operating periods failed to reveal any changes which could be clearly attributed to the smelters (except for Cu). Trajectory analysis also gave inconclusive results, with deposition from the Sudbury sector being dominated by variations in the precipitation amount, although at several receptor sites sulfate concentrations in the Sudbury sector did decrease during the smelter shutdown periods. A more detailed meteorological analysis (involving an examination of both back-trajectories and overall synoptic conditions accompanying each event) allowed a determination of the smelter contribution to wet deposition which is felt to approximate an upper limit and which was in rough agreement with similar upper estimates based on a simple trajectory analysis.

Changes in air concentrations and, therefore, dry deposition proved easier to detect. Generally, there was a readily detectable decrease in the average SO_2 levels within the Sudbury sector during the shut-down period. Changes in airborne sulfates were more difficult to quantify. Typical smelter contributions estimated for dry deposition of the airborne S at the receptors studied were about 20 to 30% of the total or less, with potentially larger contributions at receptors to the north of Sudbury (i.e., Kapuskasing).

The above results clearly have implications on the choice of parameters and the design of networks to monitor the impact of emission control strategies. On the basis of past experience with the Sudbury smelters, it would seem that, only if there are emission changes in a particular source area on the order of a million tonnes of SO_2 per year or more, would there be hope of quantifying downwind changes in the wet component of sulfur deposition with any degree of accuracy. At the same time, it must be pointed out that the air and precipitation monitoring network used in this study was not specifically designed to quantify the contribution of the Sudbury smelters. It is quite possible that a carefully-designed, source-oriented network would have a better chance of success. Furthermore, the use of tracers (in this case, As, Cu and In, for example) to assist in identifying the presence of air parcels which have passed over the smelters would be very useful in the sample stratification process, and the comparison of shut-down and operating period data.

Estimates of the Sudbury smelter contributions at the receptors of interest, made using the Lagrangian model, were in reasonable agreement with the results of the meteorological analyses, and reinforce the validity of the conclusions drawn in this study.

ACKNOWLEDGMENTS

Air and precipitation data for the Chalk River and Kapuskasing sites were kindly made available by the Atmospheric Environment Service and Ontario Hydro respectively.

REFERENCES

Chan, W.H., Vet, R.J., Lusis, M.A., Hunt, J.E. and Stevens, R.D.S.: 1980, Atmos. Environ., 14, 1159.

Chan, W.H., Orr, D.B. and Vet, R.J.: 1982a, "An Overview: the Event Wet/Dry Deposition Network" Ontario Ministry of the Environment Report ARB-11-82-ARSP.

Chan, W.H., Orr, D.B. and Vet, R.J.: 1982b, "An Overview: The Cumulative Wet/Dry Deposition Network", Ontario Ministry of the Environment Report ARB-15-82-ARSP.

Chan, W.H., Vet, R.J., Orr, D.B., Tang, A.J.S. and Lusis, M.A. (1984), Atmos. Environ., 18, 1001.

Ellenton, G., Ley, B. and Misra, P.K.: 1982, "Lagrangian Model of the Long Range Transport of Sulfur Oxides", Ontario Ministry of the Environment Report API 008/82.

Ellenton, G., and Misra, P.K.: 1984, "Examination of Monthly Wet Deposition by a Lagrangian Model and Its Application to Study the Effects of Source Control on Receptors", Ontario Ministry of the Environment Report ARB-198-84-AQM.

Kurtz, J.: 1983, "Meteorological Analysis of Precipitation Event Sampling Data", Ontario Ministry of the Environment Report AQM-83-5-1.

Lusis, M.A. and Wiebe, H.A.: 1976, Atmos. Environ. 10, 793.

Ontario Ministry of the Environment: 1979, "An Analysis of the Sudbury Environmental Study Network Precipitation Acidity Data, June-September, 1978" Report SES 011/82.

Scheider, W.A., Jefferies, D.S. and Dillon, P.J.: 1981, Atmos. Environ., 15, 945.

Tang, A.J.S., Chan, W.H. and Lusis, M.A.: 1984, "An Analysis of the Effects of the Sudbury Smelter Sources on Wet and Dry Deposition in Ontario", Ontario Ministry of the Environment Report ARB-124-84-ARSP.

Yap, D.: 1984, "Emission Inventory of Ontario and Eastern North America During 1980-1983 With Emphasis on the Sudbury Shut-down Period", Ontario Ministry of the Environment Report ARB-191-84-AQM.

Yap, D and Kurtz, J.: 1984, "Meteorological Studies to Qantify the Effects of Sudbury Emissions on Precipitation Quality and Air Quality During 1980-1983 With Emphasis on the Shut-down Period", Ontario Ministry of the Environment Report ARB-192-84-AQM.

A CRITIQUE OF MODELS FOR FRESHWATER AND SOIL ACIDIFICATION

John O. Reuss[1], Nils Christophersen[2] and Hans M. Seip[2]

[1] Colorado State University, Fort Collins, Colorado 80523, USA

[2] Center for Industrial Research P.O. Box 350 Blindern, 0314 Oslo 3, Norway

ABSTRACT. Four types of models quantifying effects of acid deposition on freshwaters are reviewed. These include Henriksen's empirical model, an adsorption isotherm model, soil-oriented charge balance models (Reuss-Johnson, Birkenes, MAGIC, ILWAS), and the Trickle Down model. Emphasis is on an assessment of critical assumptions; no attempt has been made to run the various models and compare results. The models range from simple to very complex and from empirical to highly process oriented. The various types have all proven useful and there has been a significant convergence concerning key processes. The importance of anion mobility, sulfate adsorption, ion exchange, dissolution of Al bearing minerals and weathering seems to be accepted by most workers. Future model improvement, however, relies to a large extent on further checking against observations.

1. INTRODUCTION

Models for acidification of soils and freshwaters are used to improve our understanding of important processes and mechanisms or to make predictions. The complexity of the models varies greatly. Simple models can concentrate on key processes and thus be very useful for improving our understanding. They may also be used for predictions, but at the risk of important processes being overlooked. Comparisons with observations are therefore essential. Instances where agreement is not obtained may yield useful information for improving the model. Complex models may be difficult to use, because they require too much data. Furthermore, if many parameters are calibrated it may be possible to reproduce the observations even if one or more basic assumptions are wrong. Calibration and verification should be carried out on separate data. Unfortunately, as data sets are often limited, in practice it may be difficult to strictly follow this rule.

In this paper we examine a number of models that have been proposed to describe or predict the effects of acid deposition on soils and waters. We believe that the usefulness and reliability of any such model

Water, Air, and Soil Pollution **30** (1986) 909–930.

ultimately depend on the validity of the assumptions on which it is based. We have, therefore, carefully examined the published accounts of these models and the equations that have been used to describe the system in an attempt to discern what are the critical assumptions determining its reponse. One of the problems in evaluating models is that those of most interest tend to be in a continuing state of evolution. Comments concerning published accounts may no longer be valid for current versions. For the most part we have tried to use versions that are available in the open literature; but where significant modifications have come to our attention through more recent manuscripts, we have included these in our discussion. The description of the models must necessarily be very sketchy in this paper. A more detailed description is also available (Reuss et al. 1986).

We have not attempted to run the various models and compare the outputs to each other or to some reference set of data. This might be a useful exercise, but we feel that a comparison of basic assumptions should come first. The number of models that can be dealt with in a short paper is limited, and the choice is necessarily somewhat arbitrary. We have selected models for which we have found fairly detailed descriptions in the open literature and that represent a number of different approaches.

Alkalinity or acid neutralizing capacity (ANC) is an important parameter in many models. For waters containing aluminum species and organic acids (RCOOH), at least two different definitions are used (see Reuss et al., 1986 for details). Using molar units the expressions are

$$\begin{aligned} ANC^1 = {} & [HCO_3^-] + 2[CO_3^{2-}] \\ & + [OH^-] + [RCOO^-] + [Al(OH)_4^-] \\ & - [H^+] - 3[Al^{3+}] - 2[Al(OH)^{2+}] - [Al(OH)_2^+] \end{aligned} \qquad (1.1)$$

$$\begin{aligned} ANC^2 = {} & [HCO_3^-] + 2[CO_3^{2-}] \\ & + [OH^-] + [RCOO^-] + [Al(OH)^{2+}] + 2[Al(OH)_2^+] \\ & + 3[Al(OH)_3^0] + 4[Al(OH)_4^-] \\ & - [H^+] \end{aligned} \qquad (1.2)$$

We have further

$$ANC^2 = ANC^1 + 3\ Al^t$$

where

$$\begin{aligned} Al^t = {} & [Al^{3+}] + [Al(OH)^{2+}] + [Al(OH)_2^+] + [Al(OH)_3^0] \\ & + [Al(OH)_4^-] \end{aligned}$$

For acid soil solutions or surface waters the concept that mineral acidity is equal to -ANC cannot be applied to ANC^2. ANC^1 is reasonably conservative with respect to the dissolution and precipitation of aluminum bearing minerals while ANC^2 is not. For this reason we suggest the use of ANC^1, particularly where ANC is to be tracked through the system.

2. EMPIRICAL ION BALANCE OR TITRATION MODELS

The most widely known model of this type is that proposed by Henriksen (1979, 1980) and further developed and applied by Henriksen (1982, 1984), Wright (1983) and Wright and Henriksen (1983). The model is used both to determine whether waters are "acidified" and to predict future changes in acidity. The approach is based on the ion balance principle and the concept of acid neutralizing capacity.

Henriksen (1980) defines acidification as:

$$\text{"acidification"} = ANC^0 - ANC \qquad (2.1)$$

where ANC^0 is the "preacidification ANC." (Henriksen actually uses the term Alk.) Simplifying the charge balance equation to include those components considered by Henriksen we obtain (concentrations are in eq. L^{-1} in this section)

$$[Ca^{2+} + Mg^{2+}]^* - [SO_4^{2-}]^* \approx [HCO_3^-] - [H^+] - \Sigma[Al^+] \qquad (2.2)$$

Where the asterisk is used for so-called non-marine concentrations and $\Sigma[Al^+]$ represents all positively charged Al-species. Either side of equation (2.2) may be used to define ANC (i.e. ANC^1, see Equation 1.1)

$$ANC \approx [Ca^{2+} + Mg^{2+}]^* - [SO_4^{2-}]^* \qquad (2.3a)$$

$$\text{or} \quad ANC \approx [HCO_3^-] - [H^+] - \Sigma[Al^+] \qquad (2.3b)$$

The expressions are, of course, simplified forms of ANC considering only a limited number of ions. What appear to be important relationships using one definition may prove to be trivial when the other is used.

In the original version of this approach, $[Ca^{2+} + Mg^{2+}]^*$ was assumed constant in time, and the present day value was used to estimate ANC^0. One relationship, obtained by regression using data from unimpacted areas, was:

$$ANC^0 = 0.93\ [Ca^{2+} + Mg^{2+}]^* - 14 \qquad (2.4)$$

The constants in this relationship apparently compensate for background non-marine sulfate and terms neglected in equation (2.2). In earlier work using this model, a catchment was considered to be similar to a solution in a beaker with respect to addition of sulfuric acid; ANC is reduced or titrated by the number of added sulfate equivalents. Henriksen (1980) seems to find support for this assumption using regressions between quantities derived from present day chemistry. Wright also uses relationships in which this assumption is implicit (Wright, 1983, chap. 3). These relationships support the model, however, only in the sense that they demonstrate the validity of the charge balance principle. They would occur regardless of how the lakes reached their present chemical status. Using this approach, an independent assessment of whether the increased sulfate has increased acidification (reduced ANC) or increased $[Ca^{2+} + Mg^{2+}]^*$, does not seem to be possible.

Changes in $[Ca^{2+} + Mg^{2+}]^*$ for increased $[SO_4^{2-}]^*$ are likely because the water comes into contact with vegetation, soils and sediments. To cope with this problem, Henriksen (1982) introduced the concept of a factor "F" defined as

$$F = \frac{\Delta[Ca^{2+} + Mg^{2+}]^*}{\Delta[SO_4^{2-}]^*} \tag{2.5}$$

F gives the fraction of the change in $[SO_4^{2-}]^*$ that is accompanied by a change in calcium and magnesium, leaving ANC unaffected (Equation 2.3a), the remainder going to change ANC. Critical assumptions in this modified approach seem to be:

a) Freshwater acidification is a result of acid deposition and is directly proportional to the increase in $[SO_4^{2-}]^*$. The proportionality factor (1-F) is assumed to be site-dependent, but in practice - if not in principle - constant over time and independent of the state of acidification.

b) Acidification is assumed to be reversible and the F factor for the acidification stage is assumed to apply during recovery as well.

c) The effect of acid deposition is assumed to be independent of or decoupled from the sea salt input. Only $[Ca^{2+} + Mg^{2+}]^*$ and $[SO_4^{2-}]^*$ are considered to contribute significantly to ANC (Equation 2.3a).

Acid precipitation is here assumed a priori to be the major cause of freshwater acidification. The most important question in using this model seems, however, to be the determination of the F factor. A satisfactory method for estimating F has proven difficult, particularly for lakes with alkalinity greater than zero. The calculation of a maximum value for F by assuming zero ANC^0 (Henriksen, 1982, 1984) seems to be reasonably valid for lakes that are presently acid. For lakes in southernmost Norway, Henriksen obtained F = 0.4-0.5. This estimate was corrected to F = 0.2 by a procedure with the implicit assumption that ANC^0 was equal to the background non-marine sulfate (= 20 μeqL^{-1}). We believe that F = 0.2 might be reasonable for susceptible areas of southern Norway as an average of preindustrial times to the present. The same value could possibly be used in other areas with similar geology and thin overburden. In less susceptible areas, larger values will be expected. Fortunately, values of F within a reasonable range - say, 0-0.4 - will in many cases not give too different estimates of freshwater pH for various depositions (Wright, 1983).

In spite of the inherent difficulties in estimating the F factor both the basic ion-balance principle involved and the F factor concept itself are very useful in helping us to understand the processes involved and the likely consequences of changing levels of acid deposition. One of the useful outputs of some of the more complex models to be discussed later are estimates of the F factor required by Henriksen's model.

3. ADSORPTION ISOTHERM MODELS

Arp (Arp 1983; Arp and Ramnarine, 1983) has developed a model which he describes as one "in which the total amount of acidity that can be retained by the soil is limited and qualitatively follows a Freundlich-type absorption (sic.) isotherm." According to Arp the objective has been to develop a simple yet quantitatively valid perspective for the extent of soil and lake acidification. Overall, the model first assumes that the base content of the soil will have reached some steady state, and that under acid deposition impact the system will approach some new steady state with lower base content.

For the natural or unaffected system Arp assumes that the variation in exchangeable bases, the base content B, with respect to time is given by

$$\frac{dB}{dt} = -k^0 B + W^0 \qquad (3.1)$$

where $k^0 B$ is the biological uptake, and W^0 is the net weathering input, prior to the onset of acid precipitation, defined as

$$W^0 = b^0_{atm} + b^0_w - b^0_{out} . \qquad (3.2)$$

b^0_{atm} is the atmospheric input of base cations, b^0_w the weathering input (including mineralization), and b^0_{out} the removal by leaching.

When acid deposition is introduced, it is assumed that weathering increases by an amount Δb^H:

$$W = W^0 + \Delta b^H \qquad (3.3)$$

The "base content" of the soil is defined by

$$B = B^{o} - \Delta HR \tag{3.4}$$

where the last term is the acid-cation retention.

The initial base content B^{0} may be taken directly as the content of exchangeable bases. However, following external H^{+} input, B no longer corresponds to the content of exchangeable bases, but represents a more abstract concept. The retention of acid cations ΔHR may, in fact, exceed the reduction in exchangeable bases due to factors such as SO_4^{2-} or NO_3^{-} adsorption or uptake by plants. The term "base content" is therefore unfortunate and we will refer to it as the B-variable.

The basic assumptions in this model seem to be:

a) The fraction by which the B-variable will be reduced before reaching a new steady state can be expressed as a Freundlich type adsorption isotherm, i.e.

$$\frac{B^{o} - B^{f}}{B^{o}} = (H/W)^{1/2} \tag{3.5}$$

where $B^{o}-B^{f}$ corresponds to the final amount of acid cation retention, ΔHR^{f}.

b) The rate of change in the B-variable may be expressed as:

$$\frac{dB}{dt} = -kB + W - H. \tag{3.6}$$

H is defined as the acid input, including wet and dry deposition and sometimes NH_4^{+}. H, W and k are independent of time. k is given by

$$k = \frac{W + (HW)^{1/2}}{B^{0}} \tag{3.7}$$

c) While net weathering (which includes mineralization and leaching) is assumed to increase as a result of acid deposition, it is assumed to be independent of B.

If B^{o}, W, and H are known, k and B^{f} can be calculated and the status at any time in the future can be predicted.

We are concerned with two aspects of the model. First, as recognized by Arp (1983), the model initially produces negative values for acid cation retention, *i.e.* base loss values in excess of acid inputs. Arp has chosen to regard the time at which the rate of acid cation loss becomes positive as the time of appearance of acid cations in the leachate, rather than the somewhat later time at which the accumulated acid cation loss becomes positive.

Secondly, we are concerned with assumption a) in that the acid cation retention is determined by the initial base content, the H^+ input and the weathering rate. It would seem to us that factors such as SO_4^{2-} adsorption and NO_3^- uptake would be major determinants of acid cation retention, and unlikely that these processes would be reflected in any meaningful way by $B^0(H/W)^{1/2}$.

The model is not sufficiently verified. The "observed" values used in the verification rely on additional assumptions and are often dependent on predicted values.

In spite of these limitations, we believe that the general approach may be useful and further work with this type of model may well be profitable. As demonstrated by Arp, many interesting and important parameters can be calculated from the derived equations.

4. SOIL-ORIENTED CHARGE BALANCE MODELS

In this section we will discuss some common features of a number of models currently in use, i.e. the models by Reuss and co-workers (Reuss, 1978, 1980, 1983; and Reuss and Johnson, 1985), "the Birkenes model" (Christophersen et al., 1982; Rustad et al., 1986), "ILWAS" (Chen et al, 1983; Goldstein et al., 1984; Gherini et al., 1985) and "MAGIC" (Cosby et al, 1985a,b). There are important differences among these models in scope, structure, and modelling philosophy. The models contain, however, very similar assumptions relative to a number of key soil chemical processes. They more or less explicitly incorporate the anion mobility concept (Seip, 1980) through the charge balance principle. These models require strong acid anion concentrations as input or calculated by a submodel. Basic common processes incorporated include: cation exchange processes, the carbonic acid - bicarbonate system, the dissolution of an aluminum mineral and/or a functional relationship between Al^{3+} and H^+ concentrations.

Chemical equilibria are generally assumed for these processes, but some models include kinetically controlled rate processes as well, and at least one (ILWAS) allows for equilibrium processes and rate limited processes within the same time step.

4.1 Anion submodels

The strong acid anion concentrations may be entered as inputs or calculated through the use of submodels. The SO_4^{2-} concentrations are particularly important, and a submodel incorporating a SO_4^{2-} isotherm is a common feature. Except in the ILWAS model, NO_3^- is generally ignored. Organic anions are considered in at least two of the models (ILWAS, Birkenes). Unless the model is applied only in situations where the soil solutions and leachates are consistently very acid, the HCO_3^- anion must be considered. It is commonly calculated using the equation:

$$(H^+)(HCO_3^-) = K_c (CO_2)_g \tag{4.1}$$

Here $(CO_2)_g$ is the partial pressure of CO_2 in the soil, and K_c is a temperature dependent constant. This, of course, requires that the CO_2 partial pressure be input or calculated through a submodel.

Sulfur in soils occurs both in organic and inorganic forms. In terms of total S present, the organic pool is generally dominant (Mitchell et al., 1983). Transformations between organic and inorganic sulfur species are important and the rates, which are not well known, will vary with season and meteorological conditions. The formation of sulfate through mineralization of organic S-compounds and desorption of inorganic sulfate may delay the effect of reduced deposition. None of the models treat mineralization in any detail, but it is taken into account to some extent in ILWAS and Birkenes.

The capacity of a soil to adsorb SO_4^{2-} is not a fixed quantity, but increases as the SO_4^{2-} concentration increases. The most common method of including this process is through use of an empirical relationship between the solution concentration and the adsorption capacity. Such "isotherms" must be experimentally determined and, unfortunately, may differ substantially among soils and among horizons within a given profile.

4.2 Cation exchange processes

These models utilize the fact that ion exchange is an important soil process, and that the reservoirs of cations such as Ca^{2+}, Mg^{2+}, Na^+, K^+, H^+, and Al^{3+} on exchange sites are much larger than the amounts in solution at any one time. The total charge associated with cations in solution will be determined by the anions in solution through the charge balance principle, while the relative amounts of each cation will be determined by ion exchange and/or solid solution equilibria. The ion exchange equilibria utilized are of the general form:

$$K_{a,b} = \frac{(A^{a+})^b}{(B^{b+})^a} \tag{4.2}$$

where $K_{a,b}$ is the activity ratio for two positively charged ions A and B of valence a^+ and b^+ respectively, and the brackets indicate ion activities in solution. This is a basic relationship in ion exchange theory (Bolt, 1967).

The value of $K_{a,b}$ is dependent on the fraction of each of the ions on the exchange complex, so that models using the activity ratio concept in the simple form of equation (4.2) are generally restricted to relatively short-term simulations where changes in the exchange fractions will not be significant (e.g. up to one or two decades). This limitation is removed by use of an exchange coefficient such as that of Gaines and Thomas (1953):

$$K_s = \frac{(A^{a+})^b\ (BX)^a}{(B^{b+})^a\ (AX)^b} \tag{4.3}$$

where K_s is a selection coefficient applicable to the two ions and the exchanger in question. AX and BX are the equivalent fractions of exchange sites occupied by the ions A and B respectively. Alternative formulations for exchange of ions of different charge such as the Gapon and Vanselow expressions also exist (Sposito, 1981).

This approach requires values for the selection coefficient but allows evaluation of the effect of changing amounts of an ion on the exchange complex. Accurate evaluation of these changes is often difficult in practice, due to uncertainties in the rate of cation replacement through weathering processes.

4.3 Aluminum buffering

One of the key processes in acid soils impacted by acid deposition is aluminum buffering. Increased acidity causes an increase in Al species in soils, particularly Al^{3+}. This process is normally included in the models by the assumption of an equilibrium reaction with a soil mineral, usually some form of $Al(OH)_3$:

$$Al(OH)_3 + 3H^+ = Al^{3+} + 3H_2O. \tag{4.4}$$

The solubility product expression for this reaction gives us

$$\frac{(Al^{3+})}{(H^+)^3} = K_{Al}. \qquad (4.5)$$

It is commonly assumed that the use of equation (4.5) limits the applicability of the models to those soils in which the solubility of aluminum is in the gibbsite range, i.e. log K_{Al} is in the range from about 8.0 to 10. This limitation has probably been overstated by both users and critics of the models. Equation (4.5) has the same form as equation (4.2), and can be used to describe Al^{3+} - H^+ exchange as well as the dissolution of $Al(OH)_3$ minerals. Thus, the use of equation (4.5) to describe the relationship between H^+ and Al^{3+} is not necessarily limited to soils where the solubility is controlled by $Al(OH)_3$ minerals.

4.4 Discussion of common assumptions

A set of simultaneous, non-linear equations usually including (4.1), (4.2) or (4.3), and (4.5) plus the charge balance constraint are solved to obtain the cation and bicarbonate concentrations. In addition to the various constants needed, $(CO_2)_g$ and the concentrations of the strong acids anions must be input or calculated by submodels. The critical assumptions generally inherent in these models are:

a) The ion-exchange processes are adequately described by the activity ratio equations (equation 4.2) or by equations of the type (4.3).

b) The aluminum concentrations are either determined by equilibrium with a soil mineral of known solubility, or the relationship between H^+ and Al^{3+} is adequately described by an ion exchange equation.

c) The concentration of strong acid anions, i.e. SO_4^{2-}, NO_3^-, Cl^-, and the partial pressure of CO_2 in the soil are known or can be calculated with acceptable accuracy.

Assumption a) implies that contact between the soil and the incoming water is sufficient so that the reactions may proceed to near equilibrium. Obviously, if a substantial portion of the acid deposition moves to the surface water with little or no contact with the soil, this assumption will be violated. Direct input to lakes, overland flow and so-called "macro-pore" flow represent such potential "by-pass" flows. The importance of these processes will be highly dependent on the characteristics of the catchment being simulated. In some studies, non-equilibrium situations are apparently observed (Driscoll et al., 1983; Seip et al., 1984). Inclusion of these processes generally requires a model including both hydrologic and soil-chemical processes and rather complex interactions between these modules.

The CO_2 partial pressure is an important input parameter for this modelling approach. So far, few measurements have been reported from forested catchments, and more data are clearly needed.

Our subjective judgement is that the approach represents a consistent framework for modelling of freshwater and soil acidification based on basic principles of soil chemistry. It should be emphasized, however, that the validity of the set of equations has not been sufficiently tested on observations from lysimeters or catchments. Data for more than 400 streamwater samples collected over several years in a small catchment in southernmost Norway gave some support to the model (Christophersen et al, 1984).

4.5 Basic equilibrium model

The model presented by Reuss (1978, 1980) represents the first attempt to use these principles in analyzing effects of acid deposition on soils and soil solutions. The basic relationships in this approach are essentially the mechanisms already described in Sections 4.2 and 4.3. The purpose of the modelling work has been to understand how these processes interact and not to simulate actual catchments. The ions included were SO_4^{2-}, HCO_3^-, Cl^-, Ca^{2+} (meant to include also Mg^{2+}), H^+ and Al^{3+}. Sulphate adsorption was described by a Langmuir isotherm, and ion-exchange was quantified utilizing the lime potential, LP (LP = pH + 1/2 log $[Ca^{2+}]$). When performing long-term simulations, changes in base saturation were related to changes in the lime potential through an empirical relationship.

Using the Gaines-Thomas selectivity coefficient for Al-Ca exchange and the gibbsite equilibrium assumption, Reuss and Johnson (1985) investigated the relationships between soil solution ANC and the soil CO_2 partial pressure, $(CO_2)_g$, at different strong acid anion levels. A major finding is that for a given $(CO_2)_g$, an increase in strong acid anions can result in a switch from positive to negative soil solution ANC. At the elevated $(CO_2)_g$ levels in the soil the pH of the soil solution will be only moderately affected, whereas the pH of the leachate (after CO_2 equilibration) will be strongly affected; i.e. a reduction of over one unit may occur. This mechanism could be of particular importance in explaining observed reductions in freshwater pH.

4.6 MAGIC (Model of Acidification of Groundwater In Catchments)

One of the main objectives seems to be to determine whether a lumped sum representation of the chemical reactions in soils gives a satisfactory description of gross chemical behaviour in catchments for use in long-term predictions. Cosby et al. (1985a) used an extended version of the Reuss-Johnson model (incorporating additional species such as Na^+, K^+, Mg^{2+}, Al-hydroxides and fluorides) on the White Oak Run catchment, Virginia. By including a sulphate submodel (Cosby et al. 1984) and estimating base cation weathering rates, a model suitable for long-term predictions (> 100 yrs.) of responses in water quality and soil properties in natural catchments was obtained (Cosby et al., 1985b).

Critical assumptions in present versions of MAGIC seem to be:

a) A single set of equilibrium equations of the Reuss-Johnson type adequately describes cation concentrations in soil solution and runoff both in individual samples and on an average (up to yearly) basis.

b) Long-term sulfate dynamics can be quantified using adsorption/-desorption isotherms of the Langmuir type. Nitrate is neglected.

Assumption a) is important in that the soils of the catchment are considered sufficiently homogenous so that a lumped description is valid. This implies that the routing of water within the catchment is of less importance, and a detailed hydrologic submodel is unnecessary; only the mass balance of water over the time step considered is needed. The single compartment model may be satisfactory for White Oak Run where most of the flow is always through one soil layer. In catchments where the flow changes between distinctly different soil layers depending on discharge rate the model is probably not adequate.

Weathering rates are input parameters that must be specified in MAGIC. The present streamwater chemistry of White Oak Run could only be reproduced by long-term simulations (140 yrs.) provided weathering rates were chosen within a fairly narrow range (Cosby et al., 1985b). This indicates the interesting possibility that weathering rates could be bracketed reasonably close by using current observations. Thus it may not be so critical to have independent estimates of these rates if basic assumptions in MAGIC are satisfied, and sources of H^+ in addition to acid precipitation are negligible or known.

The MAGIC model has a moderate data requirement and can fairly easily be applied to a new catchment. The model has been tested against observations to some extent (Cosby et al. 1985a; Wright et al., 1986) but the long-term predictions are, of course, hard to verify. There are many sources of errors, some of which have been investigated (Cosby et al., 1985d). In our opinion the sulfur dynamics in MAGIC may be too simple for reliable long-term predictions in many catchments . The neglect of mineralization (see Section 4.1) may result in serious errors in estimating the response to reduced deposition. A major use of MAGIC could be as an heuristic tool answering a series of "what if" questions based on a limited set of processes thought to be important (cf. Cosby et al., 1985c).

4.7 The Birkenes model

This model was originally developed to describe observed streamwater chemistry in the acidified Birkenes catchment in southernmost Norway (Christophersen and Wright, 1981; Christophersen et al., 1982), where streamwater chemistry is highly dependent upon flow. Highflow always gives low pH (4.2-4.5) and lowered Ca^{2+} levels, whereas lowflow results in pH ~ 5 and increased Ca^{2+} concentrations. A two-compartment hydrochemical model was developed to describe these typical variations. Recent versions also include a snow reservoir when appropriate. A later version of the model (Rustad et al., 1986) applied to inflow #4, Harp Lake, Ontario, includes organic anions in both reservoirs. The only aluminum species included is Al^{3+}.

Critical assumptions are:

a) Essentially the Reuss-Johnson equilibrium model is assumed to apply at highflow (for the upper reservoir), whereas lowflow cation chemistry is controlled by weathering.

b) Chloride, when significant, is assumed to be balanced by sodium, and both ions are neglected in this case. A linear sulfate adsorption term is assumed adequate, and nitrate is assumed negligible.

Other assumptions related to hydrology, sulfur dynamics, CO_2 pressure and treatment of organic compounds will not be discussed here.

Assumption a) implies that two soil reservoirs are sufficient to explain streamwater chemistry. The use of cation exchange in the upper reservoir and weathering in the lower one seems inconsistent. Though the importance of the processes may differ between soil layers, a better approach would be to include both cation exchange and weathering in both reservoirs.

The Birkenes model was originally aimed at explaining short-term (daily, seasonal) trends in streamwater chemistry. Present versions do not include a mass balance for exchangeable cations, i.e. the amounts of the exchangeable ions are assumed constant during a simulation run limiting direct use of the model to relatively short-term applications. The model is site-specific, and most parameters must be calibrated against observed streamwater chemistry for each new application. So far, the model has been applied to two catchments (Birkenes and Storgama) in Norway, to Harp Lake inflow #4, Canada, and to some catchments in Sweden (Grip, 1982; Grip et al., 1985). Though the success in predicting observed values varies, the results show that the processes incorporated are able to explain important trends in the observed streamwater chemistry at these rather different sites.

The model has tentatively been used for prediction (Christophersen et al., 1984; Rustad et al., 1986; Seip et al., 1986). To make it more useful in this respect, a mass balance must be kept for the exchangeable cations. As for MAGIC, however, predictions will be speculative and should be regarded as tentative.

4.8 The ILWAS model

The ILWAS model is the most comprehensive acid precipitation model developed to date (Chen et al., 1983 Goldstein et al., 1984; Gherini et al., 1985). It consists of a hydrological and a chemical part, both split into a number of submodels. The former simulates the hydrological processes of interception, throughfall, evapotranspiration, freezing and thawing of the soil, snow accumulation and melting, and soil and surface water hydraulics. The system may be divided into draining subcatchments. Each subcatchment contains compartments representing canopy, snowpack, and soil layers. The importance of hydrology (shallow versus deep flow) in determining the neutralization of incident precipitation, is emphasized by the authors (e.g. Gherini et al., 1985; Chen et al., 1984). The chemical models simulate concentrations of all major cations and anions, monomeric aluminum (including complexes with organic and inorganic ligands), organic acid analogues and dissolved inorganic carbon.

The processes considered comprise those in canopy, snowpack, soils and lakewater. Plant respiration, nutrient uptake, nitrification, and decomposition of litter may also be included. Nutrients are withdrawn from the soil solutions at rates which are specified by the user.

With respect to key soil processes, such as cation exchange and dissolution of aluminum, the ILWAS model is similar to the model described in Sections 4.1-4.4. Solution concentrations and exchangeable ions may be input and ion exchange coefficients calculated by a submodel. The ILWAS model has an option for non-equilibrium between aluminum in the aqueous phase and gibbsite. The carbonate system is also treated as described in section 4.1 with the CO_2 pressure being either input or simulated. The model simulates adsorption of sulfate by either a linear or a Langmuir type isotherm. A pH-dependent isotherm is used for organic acid adsorption. In water two types of organic compounds are considered, i.e. HR^1 and H_3R^2. The expression used for ANC corresponds to ANC^2 (Equation 1.2).

Assumptions in the ILWAS model pertinent to soil-water interactions are:

a) Relationships essentially as described in Section 4 determine the concentrations of cations. Weathering rates are assumed proportional to $[H^+]^a$, where $[H^+]$ is the acidity of the soil solution. Several soil reservoirs may be included.

b) A linear adsorption isotherm or a Langmuir type isotherm for sulfate is adequate.

A number of other assumptions related to the hydrological model and to the mathematical expressions used for various processes will not be discussed here.

The ILWAS model includes many more processes than any of the other models discussed. As the Birkenes model it is based on the philosophy that long-term calibration data are not available, so confidence in a model must be established by the ability to reproduce dynamics within season or a few seasons. The complexity of the ILWAS model may be an advantage, but may also divert attention from what could be the major processes over the long term. In principle, the model should be less site-specific than the Birkenes model. Whether this is true in practice remains to be seen. It seems unlikely that sufficient data will exist for utilizing the complete model in many cases without calibration. The material published on model verification is scarce.

5. ANC MASS BALANCE AND WEATHERING RATE MODELS

While several models incorporate weathering rates and acid neutralizing capacity (alkalinity) mass balance in one form or another, the models of Schnoor and co-workers rely heavily on these concepts. Here we will discuss the steady state "Trickle Down" model described by Schnoor and coworkers (Stumm et al., 1983; Schnoor and Stumm (1985) and Schnoor et al. 1986).

The model is described by Schnoor and Stumm (1985) as based on mass balance for ANC in the watershed and lake. The kinetics of the ANC supply is emphesized. ANC in the steady state is given by

$$ANC = \frac{I}{Q} L_{acy} (f-1) \qquad (5.1)$$

I = precipitation (inflow) volume per time ($m^3\ yr^{-1}$)
L_{acy} = total acidity of precipitation (dry + wet) ($\mu eq\ L^{-1}$)
Q = outflow or seepage rate ($m^3\ yr^{-1}$)
f = degree of acid neutralized in the watershed (-)

The f term is defined as

$$f = \frac{W}{R\,L_{acy}\,(0.1)} \qquad (5.2)$$

where W is weathering rate (eq ha^{-1} yr^{-1}) and R is precipitation input (cm yr^{-1}). The (0.1) arises from unit conversion.

L_{acy} includes the acidity input of both wet and dry deposition.

f is therefore the ratio of weathering (weathering here includes ion exchange) to acid loading (both expressed in eq.ha^{-1}). If f is < 1, the water will be acid, Equation (5.1). Note that this f is different from the F of Henriksen (Section 2), in that f is the total weathering or base supply from the catchment, expressed as a fraction of acid-loading, while Henriksen's F is the increase in base supply per unit of acid loading.

In the "Trickle Down" model the weathering or ANC supply rate, W, is assumed to be equal to the "rate in the absence of free acidity" (k_o) plus an acid input dependent term (Schnoor et al., 1986).

$$W = k_o + k_h \left(\frac{L_{acy}}{L_{ref}}\right)^m \qquad (5.3)$$

Here k_h is described as an "hydrolysis constant" and L_{ref} is a value of the input acidity, set by the modeler, such that when $L_{ref} = L_{acy}$, then $W = k_o + k_h$. The exponent m is between 0 and 1.0, but is generally taken as 0.5.

From equations (5.1), (5.2) and (5.3) an expression for ANC is obtained

$$ANC = \frac{I}{0.1RQ}\left(k_h\left(\frac{L_{acy}}{L_{ref}}\right)^m + k_o\right) - \frac{I}{Q}\cdot L_{acy} \qquad (5.4)$$

We consider the most critical assumptions of the steady state "Trickle Down" model to be:

a) The rate of supply of ANC to water within a catchment is considered to be kinetically controlled. The rate is the sum of a fixed or non-acid sensitive component, and a component which is fractional order with respect to the acidity of the deposition inputs.

b) The base export from exchange sites is assumed to be replenished so that we have a steady state with the net chemical weathering equal to the total base cation export.

c) Sulfate input-output balance is assumed.

We are concerned by the assumption that weathering kinetics are fractional order with respect to input acidity. While this might be the case if acid deposition were passing over a thin unweathered regolith, in most cases it will be entering soils which are buffered by several processes. The change in soil solution acidity will likely be much less than the change in the incoming precipitation. We would agree that weathering rates may well be fractional order with respect to soil solution acidity but fractional order with respect to incoming acidity does not seem appropriate.

Furthermore, as noted by Schnoor and Stumm (1985), the exchangeable base reservoir is commonly equivalent to perhaps 50-200 years of acid deposition inputs. This pool provides a substantial buffer of ANC supplying power, effectively delaying the attainment of the new steady state. The lag time associated with this buffer would seem to severely limit the usefulness of a model that relies strictly on the kinetics of weathering.

It is possible that our objections at least partly are due to semantics. Since increased acidity of the precipitation generally is correlated with increased concentrations of mobile anions, leaching of base cations is likely to increase with increasing L_{acy}. W in equation (5.3) should therefore perhaps be denoted "cation denudation rate".

Assumption c) is probably acceptable for many areas of the northern U.S., Canada, and Scandinavia, where sulfate adsorption capacities are relatively low. The papers we have seen do not discuss how nitrate inputs should be handled.

Potential model users should be aware that, due to differences in assumptions, in some cases this model may give quite different predictions than models based on exchange processes (or a combination of exchange and weathering). It seems that for sensitive systems (low base saturation or low value of k_h/L_{ref} in the Trickle Down model) all these models give a fairly linear response to input acidity. For less sensitive systems the predicted response over a range of input acidity will be quite different. More comparisons with observations are needed to find which models are most applicable.

A time-variable companion to the "Trickle Down" model that is specifically designed for seepage lakes, has been developed by Schnoor and coworkers (Schnoor et al., 1984, Lin and Schnoor, 1986). This model was briefly discussed by Reuss et al., 1986. Here we will just point out that weathering is treated quite differently in the two versions of the "Trickle Down" model.

6. CONCLUDING REMARKS

Modelling has proven to be a very useful tool in attempts to resolve the controversies concerning the effects of acid precipitation on freshwater chemistry and soil properties. The models that have been used to date represent a variety of approaches. They range from very simple to very complex, from dynamic to equilibrium, and from empirical to highly process oriented. There has been significant differences in the processes which the various modellers have accepted as critical in determining the system response. This is probably to be expected when modelling complex and poorly understood systems. Even so, substantial progress has been made. The empirical models have been useful in systemising large amounts of information and have led to a focus on important concepts. By using basic soil chemistry and quantifying key parameters essential insight as well as new ideas for experimental studies have been obtained. At their present stage, the models indicate that, under some conditions, acid deposition may have a strong effect on fresh water chemistry, particularly pH and aluminum concentration. The results have further clarified the role of other factors such as forest growth in contributing to water acidification.

We believe there has been a significant convergence of thought concerning the key processes. The importance of anion mobility, sulfate adsorption, ion exchange, and the dissolution of Al bearing minerals in determining the short-term solution chemistry response seems to be accepted by most workers, as does the role of mineral weathering for longer-term response.

There remains, however, a substantial divergence in modelling philosophy. The extremes of the approach are on the one hand a highly detailed process-oriented approach with its heavy demand for input data, computer time, and other resources. On the other hand is the "lumped parameter" approach which is much less detailed in reproducing short-term dynamics. There will probably never be a satisfactory resolution of this issue. Perhaps it is better so as we believe that both types have proven useful to date and will be more so in the future.

A catchment is an extremely complex system, and our understanding is certainly incomplete in important areas such as interplay between hydrology and chemistry, and aluminum chemistry in soils. Further development of nitrate models and better description of in-lake processes are also needed.

As stressed in earlier sections, the models must be checked against observations. There are at least three ways that observations should be made:

- Further studies in natural catchments, with more emphasis on details in flow pattern and soil chemistry. Long data series are of great importance for verification of models.

- Manipulation experiments with precipitation of varying chemistry (e.g. the RAIN project; Wright and Gjessing, 1986).

- Studies around point sources with changing emissions. The Sudbury area in Ontario provides good possibilities for such studies (LaZerte and Dillon, 1984).

One note of caution, however. While such studies are essential, a high degree of confidence would necessarily require long-term verification. It must be recognized that the ability to reproduce short-term dynamics does not guarantee accuracy of long-term prediction. Policy needs will not allow luxury of postponing predictive uses until long-term verification is accomplished. We must therefore accept some uncertainty in predictions made for this purpose.

ACKNOWLEDGEMENT

The work reported here was conducted while Professor Reuss visited Norway on a senior scientist fellowship from the Royal Norwegian Council for Scientific and Industrial Research. Support to the Center for Industrial Research to do this work was given by the above Council as well as SWAP (Surface Water Acidification Programme) administered by The Royal Society in London and the Academies in Norway and Sweden. Professor Reuss also received support from the Colorado Agricultural Experiment Station (Project 6232).

P.A. Arp, B.J. Cosby, A. Henriksen, S.A. Gherini, J.L. Schnoor and R.F. Wright have all received the parts of the manuscript pertinent to their work. Their comments are thankfully acknowledged. We have not found it possible to agree with all their suggestions, but hope our paper will lead to further fruitful discussions.

Arp, P.A.: 1983, Ecol. Model 19, 105.

Arp, P.A. and Ramnarine, S.: 1983, Ecol. Model. 19, 119.

Bolt, G.H.: 1967, Neth. J. Agric. Sci. 15, 81.

Chen, C.W., Gherini, S.A. Hudson, R.J.M., Dean, J.D.: 1983, The integrated lake-watershed acidification study. Vol. 1. Model principles and application procedures, EPRI EA-3221, Project 1109-5, Electric Power Research Inst., Palo Alto, Ca 94304.

Chen, C.W., Gherini, S.A., Peters, N.E., Murdoch, P.S., Newton, R.M., and Goldstein, R.A.: 1984, Water Resour. Res. 20, 1875.

Christophersen, N. and Wright, R.F.: 1981, Water Resour. Res. 17, 377.

Christophersen, N., Seip, H.M., and Wright, R.F.: 1982, Water Resour. Res. 18, 977.

Christophersen, N., Rustad, S., and Seip, H.M.: 1984, Phil. Trans. R. Soc. Lond. B305, 427.

Cosby, B.J., Hornberger, G.M., Wright, R.F. and Galloway, J.N.: 1984, A simple model of dissolved sulfate dynamics in soils. EOS Trans. AGU, 65 (16), 212.

Cosby, B.J., Wright, R.F., Hornberger, G.M. and Galloway, J.N.: 1985a, Water Resour. Res. 21, 51.

Cosby, B.J., Wright, R.F., Hornberger, G.M. and Galloway, J.N.: 1985b, Water Resour. Res. 21, 1591.

Cosby, B.J., Hornberger, G.M., Galloway, J.N., and Wright, R.F.: 1985c, Environ. Sci. Technol. 19, 1144.

Cosby, B.J., Hornberger, G.M., Wright, R.F., Rastetter, E.B., and Galloway, J.N.: 1985d, Estimating catchment water quality response to acid deposition using mathematical models of soil ion exchange processes. Proc. Mechanisms of Ion Transport in Soils. Swiss Federal Institute of Technology, 20-23 May, 1985.

Driscoll, C.T., Baker, J.P., Biscogni, J.J., and Schofield, C.L.: 1983, in Bricker, O.P. (ed.), Geological Aspects of Acid Deposition, Butterworth, pp. 55-75.

Gaines, G.L. and Thomas, H.C.: 1953, J.Chem.Phys. 21, 714.

Gherini, S.A., Mok, L., Hudson, R.J.M., Davis, G.F., Chen, C.W., and Goldstein, R.A.: 1985, Water Air Soil Pollut., 26, 425.

Goldstein, R.A., Gherini, S.A., Chen, S.W., Mok., L., and Hudson, R.J.M.: 1984, Phil. Trans. R. Soc. Lond. B305, 409.

Grip, H.: 1982, Water chemistry and runoff in forest streams at Kloten, UNGI Report No.58, Naturgeografiska Institutionen, Uppsala University, Sweden.

Grip, H., Jansson, P.E., Johnsson, H., and Nilsson, S.I.: 1985, in Andersson, F. and Olsson, B. (eds.), Lake Gårdsjön, Ecol. Bull. 37, pp. 176-192.

Henriksen, A.: 1979, Nature 278, 542.

Henriksen, A.: 1980, in Drabløs, D. and A. Tollan (eds.), Ecological Impact of Acid Precipitation, SNSF-prosject, Norwegian Inst. Water Research, Oslo, pp. 68-74.

Henriksen, A.: 1982, Preacidification pH-values in Norwegian rivers and lakes, Acid Rain Research Report 3/1982, Norwegian Inst. Water Research, Oslo.

Henriksen, A.: 1984, Verh. Internat. Verein. Limnol., 22, 692.

LaZerte, B.D. and Dillon, P.J.: 1984, Can. J. Fish. Aquat. Sci.: 41, 1664.

Lin, J.C. and Schnoor, J.L.: 1986, "An Acid Precipitation Model for Seepage Lakes," J.Environ.Eng. (in press).

Mitchell, M.J., David, H.B., and Morgan, C.R.: 1983, in Lebrun, P. et al. (eds.), New Trends in Soil Biology, Dieu-Brichart, pp. 75-85.

Reuss, J.O.: 1978, "Simulation of nutrient loss from soils due to rainfall acidity," Environ. Prot. Agency Rep. No. EPA-660/3-78-053, Corvallis, Oregon.

Reuss, J.O.: 1980, Ecol. Model: 11, 15.

Reuss, J.O.: 1983, J.Environ.Qual. 12, 591.

Reuss, J.O. and Johnson, D.W.: 1985, J. Environ. Qual. 14, 26.

Reuss, J.O., Christophersen, N. and Seip, H.M.: 1986, "A critique of models for freshwater and soil acidification", SI-Report 850137-1, Center for Industrial Research, Box 350, 0314 Oslo 3.

Rustad, S., Christophersen, N., Seip, H.M., and Dillon, P.J.: 1986, Can. J. Fish. Aquat. Sci. 43, 625.

Schnoor, J.L., Palmer, W.D., and Glass, G.E.: 1984, in Schnoor, J.L. (ed.), Modeling of Total Acid Precipitation Impacts, Vol. 9, Butterworth Publ. Inc., pp. 155-173.

Schnoor, J.L. and Stumm, W.: 1985, in Stumm, W. (ed.), Chemical Processes in Lakes, John Wiley & Sons, pp. 311-338.

Schnoor, J.L., Nikolaidis, N.P., and Glass, G.E.: 1986, J. Water Poll. Control Fed. 58, 139.

Seip, H.M.: 1980, in Drabløs, D. and Tollan, A. (eds.), Ecological Impact of Acid Precipitation, SNSF-project, Norwegian Inst. Water Research, pp. 358-366.

Seip, H.M., Müller, L., and Naas, A.: 1984, Water Air Soil Pollut. 23, 81.

Seip, H.M., Christophersen, N., and Rustad, S.: 1986, "Predicted changes in streamwater chemistry following reduced/increased sulphur deposition using the "Birkenes Model", Water Air Soil Pollut. (in press).

Sposito, G.: 1981, in Chemistry in the Soil Environment, ASA Spec. Publ. No. 40 (Amer. Soc. of Agronomy and Soil Science Soc. of America), pp.13-30.

Stumm, W., Morgan, J.J., and Schnoor, J.L.: 1983, Naturwissensch. 70, 216.

Wright, R.F.: 1983, Predicting Acidification of North American Lakes, Acid Rain Research Report 4/1983, Norwegian Inst. Water Research, Oslo.

Wright, R.F. and Henriksen, A.: 1983, Nature, 305, 422.

Wright, R.F. and Gjessing, E.: 1986, "Rain Project. Annual Report for 1985," Acid Rain Research Report 9/1986, Norwegian Inst. Water Research, Oslo.

Wright, R.F., Cosby, B.J., Hornberger, G.M. and Galloway, J.M.: 1986, Water Air Soil Pollut. (in press).

THE SIGNIFICANCE OF IN-LAKE PRODUCTION OF ALKALINITY

D.W. Schindler
Department of Fisheries and Oceans
Freshwater Institute
Winnipeg, MB R3T 2N6 Canada

ABSTRACT. Alkalinity production in terrestrial and aquatic ecosystems of Canada, the U.S.A., Norway and Sweden is calculated from either strong acid titrations or budgets for base cations and strong acid anions, using mass-balance budgets. Where alkalinity budgets for lakes and their catchments are calculated in acid-vulnerable geological settings, in-lake processes often contribute more to lake alkalinity than yield from terrestrial catchments. Nitrate and sulfate removal, and Ca exchange with sediments are the predominant alkalinity generating mechanisms in lakes.

Nitrate and sulfate removal rates increase as the concentrations of NO_3^- and SO_4^{2-} in lake water increase, so that in-lake acid neutralizing capacity increases as acid deposition increases. Both processes occur in sediments overlain by oxic waters, at rates which seem to be controlled primarily by diffusion.

1. INTRODUCTION

While the transfer of alkalinity from sediments to overlying waters has been known for decades (for example, Mortimer, 1941-42; Hutchinson, 1957), early authors did not quantitatively relate in-lake alkalinity production to other sources in a lake's watershed. Even so, it is curious that limnology texts (for example, Ruttner, 1959; Wetzel, 1975) and summaries of the effects of acid precipitation (for example, NRCC, 1981; Norton, 1982; Driscoll and Newton, 1985) have assumed that the alkalinity of lakes would be derived primarily from ion exchange and weathering of terrestrial soils and rocks.

In all of the above cases, standard weathering reactions were assumed to predominate in alkalinity production. The first illustrations of the importance of biological buffering in lakes by sulfate reduction in lakes were by Hongve (1978), Schindler (1978), and Schindler et al. (1980). Schindler and Turner (1982), Cook and Schindler (1983), and Cook et al. (1986) showed the quantitative

Water, Air, and Soil Pollution **30** (1986) 931–944.

importance of sulfate retention, and deduced that reduction and transfer of S to sediments was responsible. Schindler and Turner (1982), Kelly and Rudd (1984), and Cook et al. (1986) showed that sediments overlain by oxic water still reduced and retained sulfate. Kelly et al. (1982) showed that alkalinity generation was also related to biological productivity in lakes. Kilham (1982) found that sulfate removal and nitrate uptake completely neutralized H^+ input to Weber Lake, Michigan, and Hemond (1980) found similar results for Thoreau's Bog, Massachusetts. Baker et al. (1984), showed that alkalinity production was important in softwater lakes of Florida, Minnesota and Wisconsin. Carignan (1985) showed that fluxes of alkalinty from the sediments of Clearwater Lake in eastern Ontario could neutralize most of the acid inputs. Rudd et al. (1986a) showed that sulfate reduction and denitrification in epilimnetic sediments were important sources of alkalinity in low pH lakes from eastern Canada, the Adirondacks, and Norway. Schindler et al. (1985) and Schuurkes (1985) showed that the effect of N input on the alkalinity of freshwaters depended on the ionic form of N. van Breemen et al. (1982) show similar properties for terrestrial soils, and van Breemen et al. (1984) showed that acid-sensitive terrestrial watersheds yielded very little alkalinity. Several papers at the Muskoka '85 conference also have illustrated the importance of in situ alkalinity generation (Schiff and Anderson, 1985; Dillon and LaZerte, 1985; Jeffries et al., 1985). The topic was also discussed in detail at a pre-conference workshop sponsored by the Ontario Ministry of Environment.

This paper summarizes what is known about in-lake alkalinity generation from mass-balance studies, and compares the importance of in-lake processes with inputs to lakes for terrestrial catchments in determining the alkalinity balance of lakes.

2. METHODS

Studies of alkalinity generation in both lake and terrestrial ecosystems may be done by mass-balance, using any of three sets of measurements: 1) titratable alkalinity, 2) HCO_3^- plus H^+, or 3) budgets for individual base cations and strong acid anions. In method 3) it is assumed that the sum of all cations must equal the sum of all anions to maintain charge balance. In surface waters which typify acid sensitive areas:

$$H^+ + Ca^{2+} + Mg^{2+} + Na^+ + K^+ + NH_4^+ = HCO_3^- + SO_4^{2-} + Cl^- + NO_3^-. \quad (1)$$

Because alkalinity may be defined as:

$$Alk = HCO_3^- - H^+ \quad (2)$$

in typical acid sensitive waters alkalinity will equal the difference between basic cations and strong acid anions:

$$Alk = Ca^{2+} + Mg^{2+} + Na^{+} + K^{+} + NH_4^{+} - (SO_4^{2-} + Cl^{-} + NO_3^{-}). \qquad (3)$$

Aluminum (Al^{3+}) and organic anions (A^-) also contribute to alkalinity budgets in some lakes, but their role depends on the pH range of individual lakes and the pK_a values of different organic acids. Their entire role in the buffering process is not completely understood at present. Space does not allow their inclusion here, but calculations based on a few sites indicate that while Al^{3+} and A^- may be important in some lakes, their omission does not affect our general conclusions. ESSA (1986) review the problem.

If the inputs (I), outputs (O) and change in storage in the lake (ΔM) of the individual base cations and strong acid anions in Equation (3) are combined in mass balance calculations, the changes in a lake's alkalinity resulting from various ionic reactions can be calculated. Expressed as alkalinity production (P_a):
For base cations:

$$P_a = O + \Delta M - I \qquad (4)$$

and for strong acid anions:

$$P_a = I - (\Delta M + O). \qquad (5)$$

In general, release of base cations and removal of strong acid anions in an ecosystem are alkalinity producing reactions, exemplified by Ca^{2+} exchange, and removal of sulfate and nitrate. Conversely, removal of base cations such as ammonium, re-oxidation of reduced S and nitrification in ecosystems would consume alkalinity.

With the exception of uptake and release of nutrients by plankton, most *in situ* alkalinity production or consumption occurs in or near sediment surfaces (Carignan, 1985; Rudd *et al.*, 1986b; Schiff and Anderson, 1985, 1986). Both geochemical processes (precipitation, adsorption, dissolution, and desorption) and biological mechanisms (sulfate reduction, sulfate ester formation, denitrification, assimilation of nitrate and ammonium) can affect alkalinity. Both iron sulfides and organic sulfides are sinks for the reduced S, although the latter seem to be more important in the long-term (Landers *et al.*, 1983; Nriagu and Soon, 1985; Rudd *et al.*, 1986b). Reviews of pertinent aquatic reactions are given by Schnoor and Stumm (1985), Cook *et al.* (1986) and Rudd *et al.* (1986b). Analogous expressions for soils are given by van Breemen *et al.* (1982, 1984).

Terrestrial alkalinity production may be calculated in a matter analogous to that for freshwater lakes (Equation (4) or (5)). While ΔM is seldom known, it appears reasonable to disregard it for this comparison, because long-term fluxes of chemicals from catchments do not usually show uni-directional changes, unless they have been disturbed very recently (Likens *et al.*, 1977; Bayley and Schindler, 1986). Further discussion is given by van Breemen *et al.* (1984). The outflow of alkalinity from terrestrial catchments, called the yield in this paper, must be known in order to calculate the alkalinity budgets of lakes. I use the term production in the same way in both terrestrial and aquatic ecosystems, to designate the change in alkalinity occurring during passage of water through the system. In terrestrial ecosystems, this is yield-precipitation input. Similar studies were done by van Breemen et al. (1983, 1984) for terrestrial systems without reference to their contribution to lakes.

3. RESULTS

3.1. Terrestrial alkalinity production

Complete ion budgets for 33 terrestrial watersheds in acid-sensitive regions of Canada, Scandinavia and the U.S. were selected from the literature or contributed by scientists at the Muskoka '85 conference (see Table I for sources). North American sites produced from -20 to 193 meq m^{-2} yr^{-1} of alkalinity, with all but two sites producing less than 90 meq m^{-2} yr^{-1}. Scandinavian sites ranged from 9 to 112 meq m^{-2} yr^{-1}. There appeared to be no relationship between alkalinity production and deposition of H^+. This was not unexpected, because the reactivity of soils and bedrock in watersheds, the duration of interaction between water and soil, and the acidity of precipitation (i.e. its potential for weathering or ion-exchange reactions) should all affect the production of alkalinity. Furthermore, Stuanes (1980) has shown that increased yields of base cations from terrestrial plots did not occur until artificial acid precipitation had a pH less than 3. Fölster (1985) demonstrated that weathering rates in acid-sensitive areas receiving acid precipitation could not be distinguished from historical values. van Breemen et al. (1984) found a large variation in the ions affecting proton budgets of different watersheds.

Reactions involving release of Ca^{2+} or Mg^{2+}, or retention of NO_3^- were the most important mechanisms producing terrestrial alkalinity. Na^+ release was significant at a few sites. In addition, retention of SO_4^{2-} was significant in watersheds containing a high proportion of wetland, e.g., the ELA NE Subbasin, the Mersey River, Storgama STE03, and the Swedish watersheds. This retention is lower during periods of drought (Bayley et al., 1985).

Many of the watersheds appeared to export more sulfate than they received. The highest discrepancies were for watersheds near Sudbury, where recent reductions in sulfate emissions are likely to have caused non-steady state conditions. Negative values for other watersheds may indicate that input of S is underestimated, which appears to be a widespread problem (Johnson, 1985; Jeffries et al., 1984; Dillon and LaZerte, 1985). The high chloride losses from some sites (Sudbury and Scandinavia) are inexplicable, but deserve further research because they are too high and too numerous to be caused by errors in budget calculations. They represent important consumers of alkalinity, as shown in Table I.

TABLE I
Production of alkalinity by various ionic reactions in acid-sensitive terrestrial watersheds, as calculated from Equations (4) and (5). Values are in meq m^{-2} yr^{-1}. Aluminum and organic acids are ignored. Negative values indicate alkalinity removal.

	P_a due to: Release of					Removal of			ΣP_a by addition cations	ΣP_a by removal acid anions	ΣP_a
Description	Ca^{2+}	Mg^{2+}	Na^+	K^+	NH_4^+	SO_4^{2-}	NO_3^-	Cl^-			
CANADA											
ELA, Ontario[a]											
NW Subbasin	7	8	6	3	-15	-16	13	3	9	0	9
NE "	4	5	5	2	-15	12	15	5	1	32	33
E "	14	8	8	2	-15	- 1	14	4	17	17	34
Saqvaqjuac, N.W.T.[b]	58	23	34	5	not done	-37	not done	-38	120*	- 75*	45*
Muskoka, Ontario											
Harp Lake[c]											193***
Watershed 3[d]	54	34	11	2	-31	-28	29	-10	70	- 9	61
" 3A[d]	66	37	5	1	-31	-42	16	- 1	78	- 27	51
" 4[d]	44	28	7	2	-30	-27	34	- 2	51	5	56
" 5[d]	57	38	6	3	-30	-63	34	- 5	74	- 34	40
Plastic Lake[c]											-20***
Sudbury, Ontario[e]											
Clearwater 1	73	36	- 7	2	-25	-121	37	2	79	- 82	- 3
" 2	298	75	76	6	-25	-128	37	-341	430	-432	- 2
" 3	65	30	21	2	-26	-107	38	- 6	92	- 75	17
" 4	147	52	16	4	-22	-114	36	-124	197	-202	- 5
Middle	141	84	10	8	-28	-236	36	- 23	215	-223	- 8
Nelson 1	91	31	- 2	2	-22	-118	33	1	100	- 84	16
2	100	30	0	1	-22	-127	36	- 2	109	- 93	16
3	82	30	- 3	2	-22	- 88	34	- 1	89	- 55	34
4	129	54	5	5	-22	-171	35	0	171	-136	35
5	124	60	5	6	-18	-162	32	0	177	-130	47
6	110	88	4	5	-22	-123	34	6	185	- 83	102
Nova Scotia											
Mersey River[f]	12	12	21	1	-12	20	34	0	34	54	88
USA											
Hubbard Brook, NH[g]	58	21	24	2	-14	- 32	4	6	91	- 22	69
NORWAY											
Storgama STE01[h]	131	40	66	15	- 92	- 62	79	- 35	130	- 18	112
" STE03[h]	44	20	25	8	- 96	46	87	- 36	1	97	98
Birkenes[i]	59	25	54	2	- 48	- 60	45	- 52	92	- 67	25
Storgama[i]	32	11	19	5	- 25**	- 22	27	- 10	42	- 5	37
Gulen[i]	5	6	12	1	- 7	0	7	- 15	17	- 8	9
Todalen[i]	9	2	- 2	1	- 3	0	2	5	7	7	14
Längtjern 02[j]	31	7	5	- 1	no data	14	24	0	42*	38	80
" 03[j]	25	9	6	- 1	no data	13	24	0	39*	37	76*
SWEDEN											
Solmyren[k]	60	27	11	3	- 13	5	12	0	88	17	105
Vuoddasbäcken[k]	32	21	15	1	- 12	36	11	0	57	47	104
Lilla Tivsjön[k]	42	22	7	0	- 12	6	11	0	59	17	76

*excludes NO_3^- and/or NH_4^+
**estimated
***by Gran alkalinity mass-balance
[a]Schindler et al., 1986; [b]Welch, 1986; [c]Dillon and Lazerte, 1985; [d]NRCC, 1981; [e]Jeffries et al., 1984; [f]Kerekes, 1980; [g]Likens et al., 1977; [h]Braekke, 1981 (sub-basins of total Storgama watershed); [i]Overrein et al., 1980; [j]Wright, 1983; [k]Calles, 1983

To put the above acid-sensitive catchments in perspective, I calculated partial alkalinity budgets for calcareous catchments from data for Ca^{2+}, Mg^{2+}, Na^+, K^+ and SO_4^- which were summarized by Likens et al. (1977). As expected, alkalinity production in such watersheds was very high (up to 420 meq m^{-2} yr^{-1} at Carnation Creek, BC and 1313 meq m^{-2} yr^{-1} at Walker Branch, TN), and dominated by Ca^{2+} yield. Certain soils (i.e. Walker Branch) also retain sulfate, apparently by chemical adsorption rather than by sulfate reduction and precipitation as in lakes and bogs (Johnson and Todd, 1983). Either mechanism contributes to alkalinity production.

3.2. Terrestrial alkalinity yield

Because yield is the sum of terrestrial production and precipitation inputs, it will be lower than production in regions where precipitation has negative alkalinity, which is the case for all sites considered here. For the same acid-vulnerable watersheds summarized in Table I, the range in yield was -87 to 171 meq m^{-2} yr^{-1} (Table II). Watersheds in most high-deposition areas had alkalinity yields ranging from -50 to 50 meq m^{-2} yr^{-1}. ELA and Storgama also had low yields, despite low acid deposition. Values from Harp Lake, Ontario, watersheds in northern and central Sweden, and Saqvaqjuac, N.W.T. were over 100 meq m^{-2} yr^{-1}, perhaps indicating the presence of some calcareous material in the drainages.

3.3. Aquatic alkalinity production

Alkalinity production calculated from Gran alkalinity budgets for the lakes available varied by less than 3x with the exception of the Sault Ste. Marie sites, even though the regions encompassed receive deposition which varies by almost an order of magnitude in H^+, SO_4^{2-} and NO_3^-. Where both Gran alkalinity and ion-balance methods could be used, the agreement between methods was reasonably good, as one would expect if organic anions are relatively unimportant (Table III).

Ca^{2+} exchange, and removal of NO_3^- and SO_4^{2-} were the most important in situ processes producing alkalinity. Ca^{2+} exchange was much higher at the Sault Ste. Marie sites, which have small amounts of calcareous tills and lie in greenstone geology. Lake 302N, experimentally acidified with HNO_3, had a much higher rate of NO_3^- removal than the rest. Also, lakes 223 and 302S, to which H_2SO_4 was added, had the highest rates of SO_4^{2-} removal. SO_4^{2-} appeared to be exported, rather than retained in the Sault Ste. Marie lakes, although inputs may be underestimated due to high dry deposition in the area (Jeffries et al., 1984). Exchange for H^+ or biological uptake of other cations (Mg^{2+}, K^+, NH_4^+) may significantly affect alkalinity budgets, but are of lesser importance (Schindler et al., 1986; Cook et al., 1986; Jeffries et

TABLE II
The yield (Output) of various ions, from the terrestrial watersheds in Table 1. Σ alkalinity yield = total base cation yield − total strong acid anion yield. Data are in meq alkalinity m^{-2} yr^{-1}.

Description	Yield of Ca^{2+}	Mg^{2+}	Na^{+}	K^{+}	NH_4^{+}	SO_4^{2-}	NO_3^{-}	Cl^{-}	Σ base cation yield	Σ base acid anion yield	Σ alk. yield
CANADA											
ELA, Ontario											
NW Subbasin	22	15	11	5	0	39	2	4	53	45	8
NE "	20	12	10	3	0	11	0	2	45	13	32
E "	29	15	13	5	0	24	1	3	62	28	34
Saqvaqjuac, N.W.T.	150	24	41	5	not done	42	not done	45	220*	87*	133*
Muskoka, Ontario											
Harp Lake											125***
Watershed 3	84	41	26	6	2	99	8	2	159	109	50
" 3A	97	44	20	5	1	113	21	- 6	167	128	39
" 4	74	36	22	5	2	98	4	- 6	139	96	43
" 5	87	45	21	6	2	134	3	- 3	161	134	27
Plastic Lake											-87***
Sudbury, Ontario											
Clearwater 1	100	42	28	7	1	193	1	9	178	203	- 25
" 2	324	80	104	10	1	202	2	352	519	556	- 37
" 3	91	36	36	6	0	180	1	16	169	197	- 28
" 4	172	58	51	9	3	176	2	135	293	313	- 20
Middle	178	92	46	10	2	320	4	36	328	360	- 32
Nelson 1	114	38	25	6	2	187	2	9	185	198	- 13
2	123	36	28	4	1	197	0	6	192	203	- 11
3	106	37	24	6	2	156	2	6	175	164	11
4	152	60	32	8	1	239	1	8	253	248	5
5	163	66	32	10	4	230	4	8	275	242	33
6	132	94	32	8	1	192	2	14	267	208	59
Nova Scotia											
Mersey River	30	34	121	6	0	73	1	114	191	188	3
USA											
Hubbard Brook, NH	68	26	32	5	2	112	28	14	133	154	-21
NORWAY											
Storgama STE01	160	59	135	23	24	312	47	116	401	475	-74
" STE03	73	39	94	16	20	204	39	80	242	323	-81
Birkenes	72	43	133	8	5	164	8	133	261	305	-44
Storgama	39	15	33	7	not done	78	13	27	94	118	-24
Gjelen	10	22	78	3	1	29	2	91	114	122	- 8
Tödalen	13	0	23	2	1	9	1	25	47	35	12
Långtjern 02	38	11	15	3	15**	49	1	11	82	61	21
" 03	32	12	17	2	16**	50	1	11	79	62	17
SWEDEN											
Solmyren	82	34	25	6	1	14	1	8	148	23	125
Vuoddäsbacken	55	28	29	4	1	7	2	8	117	17	100
Lilla Tivsjön	59	27	15	2	1	8	1	6	104	15	89

*excludes NO_3^{-}, NH_4^{+}
**inputs are wet plus dry, rather than bulk ppt.
***by Gran alkalinity mass-balance

TABLE III
A summary of in-lake production of alkalinity and alkalinity generation by the three most commonly-important ionic reactions in acid-sensitive lakes. Data included are from mass-balance studies of either Gran alkalinity or ion balance, mesocosm mass balance studies, and pore-water studies. All data are in meq alkalinity m^{-2} yr^{-1}.

	Gran alkalinity	Ion balance alkalinity	Ca^{+2} exchange	NO_3^- removal	SO_4^{2-} removal
Whole-lake mass balance studies					
CANADA					
ELA - Lake 239[a]	118	92	29	19	39
Lake 223[b]	333	364	76	19	368
Lake 302S[c]	248	182	14	21	168
Lake 302N[c]	244	260	11	149	81
Muskoka					
Plastic Lake[d]	170	148	negligible	34	70
Harp Lake[d]	208	200	negligible	52	71
Sault Ste. Marie[e]	313	531	466	49	-110
So. Batchawana Lake	765	677	585	86	- 43
Turkey Lake					
USA					
Florida					
McCloud Lake[f]	-	-	-	17-20	26-34
NORWAY					
Langtjern[g]	-	111	43	17	57

[a] Schindler et al. (1986) - natural lake in low deposition area.
[b] Cook et al. (1986) - lake experimentally acidified to pH 5.1 with H_2SO_4. Data are 1981-83 average.
[c] Schindler et al. (1986 unpublished data) - Lake 302S was experimentally acidified to pH 5.6 with H_2SO_4, Lake 302N was experimentally acidified with the same amount per unit volume of HNO_3.
[d] P. Dillon and Lazerte (1985) - natural lakes in high deposition area.
[e] D. Jeffries et al. (1984) - natural lakes with some calcareous material, moderately acid deposition. Values may include input from groundwater seepage.
[f] Baker and Brezonik (1985a), Baker et al. (1984) - seepage lake, moderately acid deposition.
[g] Wright (1983) - natural lake in high deposition area.

al., 1985). Under anoxic conditions, diffusion of redox-sensitive species (Fe^{2+}, Mn^{2+}) from sediments may cause temporary increases in alkalinity, but this disappears at overturn, when the Fe^{2+} and Mn^{2+} are re-oxidized.

3.4. The role of terrestrial yield and *in situ* production in determining lake alkalinity

If *in situ* production were negligible, the alkalinity of a lake should be equal to the sum of average yield from the terrestrial basin and precipitation input to the lake surface, divided by the average outflow volume of the lake (Schindler et al., 1986; Dillon et al., 1985). Where the necessary measurements have been made, lakes always have much higher alkalinities than these theoretical calculations, illustrating the importance of *in situ* processes (Table IV). Similar conclusions could be reached for lakes in the Sudbury area, but because of recent reductions in regional sulfate

TABLE IV
Theoretical and observed alkalinity in acid-sensitive lakes, where theoretical alkalinity = (terrestrial yield + precipitation input) ÷ lake outflow volume. Alkalinity data in $\mu eq\ L^{-1}$.

Lake	τ, yr	Terrestrial drainage area / Lake area	Theoretical alkalinity	Observed alkalinity	Reference
Lake 239	6.2	6	73	150	Schindler *et al.*, 1986
Lake 223	8.7	6[a]	12	80	Schindler *et al.*, 1980; Cook *et al.*, 1986
Harp Lake	2.5	7	13	64	Dillon and Lazerte, 1985
Plastic Lake	3.0	3	- 75	8	Dillon and Lazerte, 1985
Langtjern	0.2	20	25	35	Wright, 1983
Lake McCloud	1.3	19	-133	0	Baker *et al.*, 1984

[a] Excluding drainage of upstream lakes.

emissions, these lakes are not at steady-state with respect to deposition (Dillon and LaZerte, 1985). Some have also been contaminated by road salt (Dillon *et al.*, 1986). We therefore have not included them in our comparison.

The role of *in situ* processes becomes still more important as lakes acidify. Addition of H_2SO_4 to ELA lakes 223 and 302S caused an almost linear increase in SO_4^{2-} reduction, with corresponding alkalinity production (Cook and Schindler, 1983; Schindler, 1985; Kelly and Rudd, 1984). In waters acidified with H_2SO_4, SO_4^{2-} removal became the dominant alkalinity-generating process, furnishing 92% of alkalinity production (Cook *et al.*, 1986; Schindler *et al.*, 1986). Likewise, addition of HNO_3 to Lake 302 North caused increased alkalinity production (Table III), largely due to increased denitrification (Rudd *et al.*, 1986a). While there are no data documenting long-term responses of terrestrial watersheds to increased acidification, it is unlikely that alkalinity export would increase, so that *in situ* processes will increase in relative importance in aquatic alkalinity budgets as lakes acidify.

4. DISCUSSION

In many acid-sensitive regions, terrestrial watersheds are unlikely to be the predominant alkalinity source for lakes (Table II). In some cases, they are actually consumers of alkalinity. For example, the average alkalinity yield from Hubbard Brook is 21 meq m^{-2} yr^{-1}, that from the Mersey River, NS is only +3, Norwegian watersheds range from -81 to +21, and 7 of 11 Sudbury watersheds have negative alkalinity yields. While terrestrial alkalinity yields from Sweden, Muskoka and ELA are slightly higher, in all cases aquatic alkalinity generation is too important to be ignored in budget calculations.

Sulfate retention appears to be a function of SO_4^{2-} concentration (Cook and Schindler, 1983; Schindler, 1985), which is in turn a function of sulfate input and water renewal (Baker and Brezonik, 1985b; Kelly et al., 1986). But the concentration of sulfate and nitrate alone does not appear to be sufficient to predict the rates of sulfate and nitrate reduction among lakes. The concentration of degradable organic substrate in near-surface sediments where sulfate reduction and denitrification occur is also important (Rudd et al., 1986a). Because of the dependence of a lake's acid neutralizing capacity on these biological reactions, it is presently impossible to predict the vulnerability of lakes to acidification from easily-measurable chemical, physical or geological characteristics alone. It follows that water column alkalinity is not necessarily a good indicator of the acid neutralizing capacity (ANC) of lakes, because alkalinity generated by sulfate and nitrate reduction increases when concentrations of these two ions are increased.

Clearly, alkalinity budgets in both terrestrial watersheds and lakes are partly a function of the composition of precipitation. For example, input and consumption of NH_4^+ can be expected to have the same effect on ecosystems as input of H^+, i.e. alkalinity consumption, (Vollenweider, 1963; Hunt and Boyd, 1981; Schindler et al., 1985; Schuurkes, 1985; van Breemen et al., 1982):

$$NH_4^+ \rightarrow \underset{\substack{\downarrow \\ \text{plants.}}}{NH_3} + H^+ \qquad (6)$$

A brief survey of the literature reveals that NH_4^+ concentrations in precipitation ranged from 17 to 173% of H^+ (see, for example, Likens et al., 1977). In most of the ecosystems studied, ammonium removal is nearly 100%. The sum of ammonium input plus H^+ input should be a better index of the acidification potential of precipitation on ecosystems than H^+ alone. Evaluating the effect of biologically-reactive anions in precipitation is more difficult, because of the wide variation in efficiencies of utilization by different ecosystems. Likens et al. (1977) present data for Hubbard Brook which suggest that the efficiency of terrestrial ecosystems at generating alkalinity may decrease as nitrate deposition increases, because nitrate retention by the terrestrial watershed decreased markedly in the 1960's and 1970's.

Lakes with larger terrestrial catchments will usually have higher rates of water renewal, which would be expected to reduce the retention of sulfate and nitrate as has been shown for nutrients in eutrophication (Dillon, 1975; Vollenweider, 1976; Schindler et al., 1978). This would result in a lower retention of alkalinity by the lake (Baker and Brezonik, 1985b; Kelly et al., 1986). Also, in systems where alkalinity yield from terrestrial basins is positive, lakes with larger terrestrial basins should have a higher proportion of their alkalinity originate outside the lake (Table V). However, even in Langtjern, with a terrestrial:lake surface ratio of 20 and a

TABLE V
A comparison of terrestrial and aquatic alkalinity production and terrestrial yield at several sites. All values are by Gran titration, in meq m^{-2} yr^{-1}.

	Terrestrial production	Terrestrial yield	Aquatic production
239	26	16	118
Harp	193	125	208
Plastic	-20	-87	170
Langtjern	93	19	57

water renewal time of 0.2 yr, 25% of the alkalinity in the system is generated within the lake. Clearly, even lakes with very short water renewal times can have a substantial part of their alkalinity generated in situ.

The number of data sets which are complete enough to allow calculation of separate alkalinity budgets for lakes and their watersheds is small, and the above conclusions must be regarded as tentative until checked more widely. However, studies of in situ alkalinity using other methods generally confirm my conclusions. Alkalinity exchange between sediments and overlying waters have been calculated by using flux-gradient calculations for ions in pore water (Baker et al., 1984; Carignan, 1985; Schiff and Anderson, 1985), and by changes in mass during periods when precipitation, inflow and outflow are negligible (Schindler and Turner, 1982). Alkalinity generation can also be measured in smaller scale experiments with isotopic tracers (Rudd et al., 1986a) and by mesocosm studies (Schiff and Anderson, 1986).

There were some exceptions worthy of discussion. Rudd et al. (1986a) found that reoxidation of S in sediments of Lake Hovvatn, Norway, destroyed much of the alkalinity which had been produced by sulfate reduction. However, high rates of nitrate reduction were important generators of alkalinity. Nitrate removal was very high in the Adirondacks, but negligible in Muskoka. Schiff and Anderson (1985) pointed out that while steep gradients of Fe^{2+}, Mn^{2+} and NH_4^+ usually occurred in lake sediments, they did not contribute greatly to the alkalinity of overlying water, because of reprecipitation of Fe and Mn at the sediment surface and biological uptake of NH_4^+. This point is not considered in earlier treatments.

Published watershed models to date have not included in-lake processes, although some modelling groups have indicated that recent revisions of models will include them. Such models clearly need to include in-lake processes in order to accurately predict changes in alkalinity for a wide variety of lake types.

ACKNOWLEDGMENTS

P.J. Dillon, L.A. Baker, D.S. Jeffries, S.E. Schiff and R.F. Anderson kindly supplied data for this review. Valuable comments on methods, the presentation at Muskoka, or the manuscript were provided by G.J. Brunskill, R.H. Hesslein, C.A. Kelly, J.W.M. Rudd, M.A. Turner and R.F. Wright, in addition to the above. This work was supported by the Canadian Department of Fisheries and Oceans.

REFERENCES

Baker, L.A., and Brezonik, P.L.: 1985a, 'Fate of Sulfate in a Soft-Water, Acidic Lake', in Caldwell, D.E., and Brierley, C.L. (eds.), Planetary Ecology, Van Nostrand Reinhold Co., New York, p. 297-306.

Baker, L.A., and Brezonik. P.L.: 1985b, Presentation at the Muskoka '85 Conference.

Baker, L.A., Peary, T.E., and Brezonik, P.L.: 1984, 'Neutralization of Acid Precipitation in Softwater Lakes', in Lake and Reservoir Management: Practical Applications. Proceedings 4th Meeting, North American Lake Management Society, Oct 14-16, 1984, McAfee, NJ, p. 356-360.

Bayley, S.E., Behr, R.S., and Kelly, C.A.: 1985, Presentation at the Muskoka '85 Conference.

Bayley, S.E., and Schindler, D.W.: 1986, 'Sources of Alkalinity in Precambrian Shield Watershed Streams Under Natural Conditions and After Stress (Fire or Acidification)', in Hutchinson, T.C., and Meema, K. (eds.), Effects of Acidic Deposition on Forests, Wetlands and Agricultural Ecosystems, Springer Verlag, NY, in press.

Braekke, F.H.: 1981,'Hydrochemistry of High Altitude Catchments in South Norway. 1. Effects of Summer Droughts and Soil-Vegetation Characteristics', Report Norwegian Forest Research Institute, **36.8**, 26 pp.

Calles, U.M.: 1983, Hydrobiologia **101**, 13.

Carignan, R.: 1985, Nature, **317**, 158.

Cook, R.B., and Schindler, D.W.: 1983, Ecol. Bull. (Stockholm), **35**, 115.

Cook, R.B., Kelly, C.A., Schindler, D.W., and Turner, M.A.: 1986, Limnol. Oceanogr., **31**, 134.

Dillon, P.J.: 1975, Limnol. Oceanogr., **20**, 28.

Dillon, P.J., and LaZerte, B.: 1985, Presentation at the Muskoka '85 Conference.

Dillon, P.J., Reid, R.A., and Girard, R.: 1986, Water, Air, Soil Pollut., **27**, in press.

Driscoll, C.T., and Newton, R.M.: 1985, Environ. Sci. Technol., **19**, 1018.

ESSA (Environmental and Social Systems Analysts, Ltd.): 1986, '"Brown Waters": Relative Importance of External and Internal Sources of Acidification on Catchment Biota', LRTAP Workshop No. 5, ESSA, Toronto, ON, 85.

Fölster, H.: 1985, 'Proton Consumption Rates in Holocene and Present-Day Weathering of Acid Forest Soils', in Driver, J.I. (ed.), The Chemistry of Weathering, D. Reidel Publ. Co., p. 197-209.
Hemond, H.F.: 1980, Ecol. Monogr., **50**, 507.
Henriksen, A.: 1982, 'Susceptibility of Surface Waters to Acidification', in Johnson, R.E. (ed.), Acid Rain/Fisheries: Proceedings of an International Symposium on Acidic Precipitation and Fishery Impacts in Northeastern North America, American Fisheries Society, Bethesda, MD, p. 103-121.
Hongve, D.: 1978, Verh. Internat. Verein. Limnol., **20**, 743.
Hunt, D.E., and Boyd, C.E.: 1981, Trans. Amer. Fish. Soc., **110**, 81.
Hutchinson, G.E.: 1957, A Treatise on Limnology, Vol. 1, John Wiley and Sons, 1015.
Jeffries, D.S., Semkin, R.G., Neureuther, P., and Seymour, M.: 1985, Presentation at the Muskoka '85 Conference.
Jeffries, D.S., Scheider, W.A., and Snyder, W.R.: 1984, 'Geochemical Interactions of Watersheds with Precipitation in Areas Affected by Smelter Emissions near Sudbury, Ontario', in Nriagu, J. (ed.), Environmental Impacts of Smelters, p. 195-241.
Johnson, D.W.: 1985, Presentation at the Muskoka '85 Conference.
Johnson, D.W., and Todd, D.E.: 1983, J. Soil Sci., **47**, 792.
Kelly, C.A., and Rudd, J.W.M.: 1984, Biogeochemistry, **1**, 63.
Kelly, C.A., Rudd, J.W.M., Cook, R.B., and Schindler, D.W.: 1982, Limmnol. Oceanogr., **27**, 868.
Kelly, C.A., Rudd, J.W.M., Hesslein, R.H., Schindler, D.W., Dillon, P.J., Driscoll, C.T., Gherini, S.A., and Hecky, R.E.: 1986, Biogeochemistry, in press.
Kerekes, J.J.: 1980, 'Preliminary Characterization of Three Lake Basins Sensitive to Acid Precipitation in Nova Scotia, Canada', in Drablos, D., and Tollan, A. (eds), Ecological Impact of Acid Precipitation, SNSF Project, Oslo, Norway, p. 68-74.
Kilham, P.: 1982, Limnol. Oceanogr., **27**, 856.
Landers, D.H., David, M.B., and Mitchell, M.J.: 1983, Anal. Chem., **14**, 245.
Likens, G.E., Bormann, F.H., Pierce, R.S., Eaton, J.S., and Johnson, N.M.: 1977, Biogeochemistry of a Forested Ecosystem, Springer-Verlag, New York, 146 pp.
Mortimer, C.H.: 1941-42, J. Ecol., **29**, 289 and **30**, 147.
Norton, S.A.: 1982, 'The Effects of Acidification on the Chemistry of Ground and Surface Waters', in Johnson, R.E. (ed), Acid Rain/Fisheries: Proceedings of an International Symposium on Acidic Precipitation and Fishery Impacts in Northeastern North America, American Fisheries Society, Bethesda, MD, p. 93-102.
NRCC (National Research Council Canada), 1981.: 'Acidification in the Canadian Aquatic Environment', NRCC No. **18475**, 369 pp.
Nriagu, J.O., and Soon, Y.K.: 1985, Nature, **303**, 692.
Overrein, L.N., Seip, H.M., and Tollan, A.: 1980, 'Acid Precipitation - Effects on Forest and Fish', SNSF Final Report, Oslo, 175 pp.
Rudd, J.W.M., Kelly, C.A., St. Louis, V., Hesslein, R.H., Furutani,

A., and Holoka, M.: 1986a, Limnol. Oceanogr., **31**, in press.
Rudd, J.W.M., Kelly, C.A., St. Louis, V., Hesslein, R.H., Furutani, A., and Holoka, M.H.: 1986b, Limnol. Oceanogr., **31**, in press.
Ruttner, F.: 1959, 'Fundamentals of Limnology', University of Toronto Press, Toronto, 295 pp.
Schiff, S., and Anderson, R.F.: 1985, Presentation at the Muskoka '85 Conference.
Schiff, S., and Anderson, R.F.: 1986, Can. J. Fish. Aquat. Sci., in press.
Schindler, D.W.: 1978, 'Chemical Budgets and Watershed Acidification', in Hendrey, G.R. (ed.), Limnological Aspects of Acid Precipitation, Brookhaven National Lab. Rep. **BML 51074 UC-11**, p. 13-14.
Schindler, D.W.: 1985, 'The Coupling of Elemental Cycles by Organisms: Evidence from Whole Lake Chemical Perturbations', in Stumm, W. (ed.), Chemical Processes in Lakes, John Wiley and Sons, NY, p. 225-250.
Schindler, D.W., Fee, E.J., and Ruszczynski, T.: 1978, J. Fish. Res. Board Can., **30**, 1511.
Schindler, D.W., and Turner, M.A.: 1982, Water Air Soil Pollut., **18**, 259.
Schindler, D.W., Turner, M.A., and Hesslein, R.H.: 1985, Biogeochemistry, **1**, 117.
Schindler, D.W., Turner, M.A., Stainton, M.P., and Linsey, G.A.: 1986, Science (Wash., DC), **232**, 844.
Schindler, D.W., Wagemann, R., Cook, R.B., Ruszczynski, T., and Prokopowich, J.: 1980, Can. J. Fish. Aquat. Sci., **37**, 342.
Schnoor, J.L., and Stumm, W.: 1985, 'Acidification of Aquatic and Terrestrial Systems', in Stumm, W. (ed.), Chemical Processes in Lakes, John Wiley and Sons, NY, p. 311-338.
Schuurkes, R.: 1985, Presentation at the Muskoka '85 Conference.
Stuanes, A.O.: 1980, 'Effects of Acid Precipitation on Soil and Forest', in Drablos, D., and Tollan, A. (eds), Ecological Impact of Acid Precipitation, SNSF Project, Oslo, Norway, p. 68-74.
van Breemen, N., Burrough, P.A., Velthorst, F.J., van Dobben, H.F., deWitt, T., Ridder, T.B., and Reijuders, H.F.: 1982, Nature, **299**, 548.
van Breemen, N., Mulder, J., and Driscoll, C.T.: 1983, Pl. Soil, **75**, 283.
van Breemen, N., Driscoll, C.T., and Mulder, J.: 1984, Nature, **307**, 599.
Vollenweider, R.A.: 1976, Mem. Ist. Ital. Idrobiol., **33**, 53.
Vollenweider, R.A.: 1963, Ist. Ital. Idrobiol. Dott. Marco De Marchi. Milan, **16**, 21 (in Italian).
Welch, H.E.: 1986, personal communication.
Wetzel, R.G., 1975, Limnology, W.B. Saunders Co., Philadelphia, 743 pp.
Wright, R.F.: 1983, Hydrobiologia, **101**, 1.

Effects of nitrogen deposition on the acidification of terrestrial and aquatic ecosystems

Peringe Grennfelt and Hans Hultberg
Swedish Environmental Research Institute
P.O. Box 5207, S-402 24 Göteborg, Sweden

Abstract

The concentration of ammonium and nitrate in precipitation has increased during this century. The deposition of N compounds (wet + dry) is reaching 30 to 40 kg $ha^{-1}yr^{-1}$ in many areas in Central Europe and above 20 kg in the southern parts of Scandinavia. In extreme situations throughfall data indicate depositions above 60 kg $ha^{-1}yr^{-1}$ in Central Europe and above 40 kg $ha^{-1}yr^{-1}$ in south Sweden. Very high depositions are observed on slopes at forest edges and adjacent to areas with animal farms and manure spreading.
In areas with low N deposition almost all deposited N (>95%) will be absorbed in the tree canopies or in the soil. In areas with high deposition an increased outflow is observed which in some cases reach 10 to 15 kg $ha^{-1}yr^{-1}$. The increased output is an indication of N saturation of the ecosystem and it leads to acidification effects in soils, soilwater, groundwater and surface waters.

1. Introduction

The concentration of N compounds in the precipitation over North America and Europe has successively increased (N.A.S., 1983; Galloway and Dillon, 1983; Galloway and Likens, 1981; Söderlund and Granat, 1982; Rodhe, 1982). Data from the European Air Chemistry Network (EACN) show a significant upward trend in ammonia and nitrate deposition during the last decades (Söderlund and Granat, 1982; Rodhe, 1982; Kallend et al., 1983). A doubling in nitrate concentrations from the end of the 50's to the beginning of the 70's has been shown (Rodhe, 1982). The upward trend for nitrate, however, is not observed after 1975 (Rodhe, 1985). The ratio of nitrate to sulphate in precipitation has also changed during the last decades. In south Scandinavia NO_3^-/SO_4^{2-} was 0.27 (on an equivalent basis) for the period 1955-59 and 0.47 for the period 1975-79 (Persson, 1982).

The upward trend in nitrate concentrations in precipitation corresponds well with the trend in NO emissions (W.S.L., 1983; Jernelöv and Lövblad, 1985). No similar study of the changes in ammonia emissions during this century has been made.
Since the wet deposition of nitrate and ammonium has increased both in absolute terms as well as in relation to sulphate it is important to evaluate the magnitude of this deposition, its interactions with the ecosystem and its possible effects. In this paper we will discuss these terms with respect to a coniferous forest ecosystem.

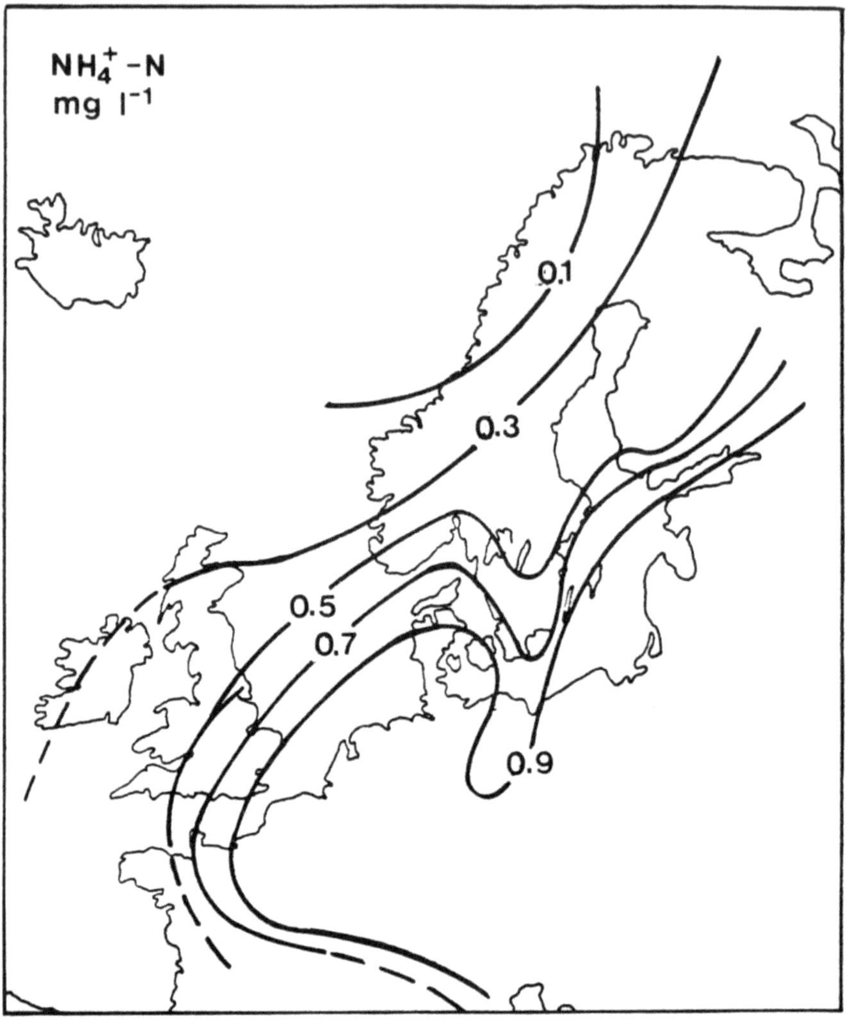

Fig. 1 Mean concentration of NH_4^+ in precipitation in Europe. Data from CCC/EMEP (SNV, 1984).

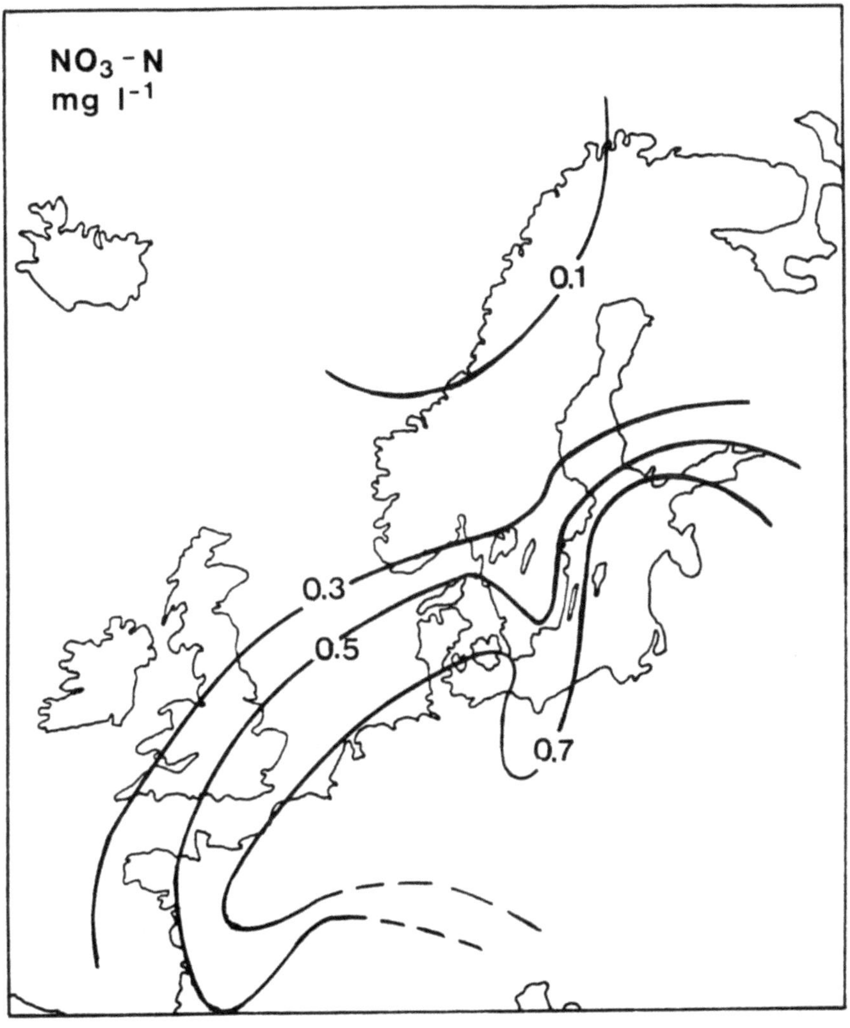

Fig. 2 Mean concentration of NO_3^- in precipitation.
Data from CCC/EMEP (SNV, 1984).

2. Deposition of nitrogen species

The flow of N compounds from the atmosphere to the ground and further into the different compartments of a forest ecosystem is a complicated process and it is only known partially. The main reasons why it is difficult to describe qualitatively as well as quantitatively are:

i) Anthropogenic N compounds occur in many different chemical forms.

For long-range transport and deposition the most important compounds are NO_2, gaseous HNO_3, particulate nitrate, and particulate ammonium.

ii) Nitrogen is a nutrient and deposited N compounds will, to a large extent, be metabolised and retained in terrestrial and aquatic ecosystems.

iii) Natural exchange processes between atmosphere and soil/vegetation, for example fixation of N from air and denitrification in the soil, may influence the net flow of N as well as be influenced by the anthropogenic deposition.

The distribution of wet deposition of ammonia and nitrate in Europe is investigated within the EACN and the European Monitoring and Evaluation Programme (EMEP) networks. The precipitation in the most polluted areas in Europe has today a mean concentration of 0.9 mg L^{-1} or more of NH_4^+-N and 0.7 mg L^{-1} or more of NO_3^--N. The concentration decreases successively when approaching the less polluted areas and it is near 0.30 mg L^{-1} of NH_4^+-N and 0.15 to 0.30 mg L^{-1} of NO_3^--N in the north of Great Britain and in the middle of Scandinavia (SNV, 1984) (Figures 1 and 2).

Besides the wet deposition of ammonium and nitrate, a deposition of several kilograms ha^{-1} of soluble organic N has been found (Rosén and Lundmark, 1985; Matzner et al., 1982). The origin of this N input is so far unknown; it might be of anthropogenic origin or it might be natural. For the N budget it might in many cases be necessary to include this input.

Dry deposition of oxidized N compounds will essentially occur in the form of NO_2, gaseous HNO_3 and particulate nitrates. The deposition of these species goes along different pathways and different mechanisms regulate the deposition. For NH_3 and NO and sometimes also for NO_2 a "negative" deposition is observed indicating that the vegetation and the soil might be a source of N compounds.

In the Lake Gårdsjön project we have estimated the dry deposition essentially based on atmospheric concentration monitoring at Gårdsjön or adjacent sites on the Swedish west coast and of the best available estimates on dry deposition velocities (Grennfelt et al., 1985; Hultberg and Grennfelt, 1986). The estimates indicate that the dry deposition input is of the same order of magnitude as the wet deposition both for oxidized and for reduced N compounds at this site (Table I).

Table I.

Dry and wet deposition of N compounds to a forest area within the Lake Gårdsjön watershed. The dry deposition estimates are based on monitored atmospheric concentrations at the site or elsewhere on the Swedish west coast and literature data on deposition velocities.

		Deposition kg N ha^{-1} yr^{-1}
Dry		
	NO_2 + HNO_3	3.6 - 5.1
	particulate NO_3^-	0.3 - 1.1
	NH_3	?
Wet	particulate NH_4^+	1.4 - 5.5
	NO_3^-	4.6
	NH_4^+	6.3
		16.2 - 22.6

3. Extremes in impact of nitrogen compounds

Given that the effect of atmospheric pollution on an individual tree is likely to depend mainly on the deposition rate in the immediate vicinity, it is reasonable to ask whether there are locations which receive particularly high deposition and whether this effect is of practical importance.

Slopes and tops of hills at high elevations which are exposed to polluted air masses are shown to receive much larger quantities of pollutants than flat land due to interception of cloud droplets with high concentrations of pollutants (Lovett et al., 1982; Scherbatskoy and Bliss, 1983).

Exposed forest edges adjacent to open fields and clearcuts is another deposition situation where one might expect increased deposition. In south Sweden this deposition has been studied by means of throughfall along parallel lines from the forest edge into the closed forest (Hasselrot and Grennfelt, 1985). For many ions the throughfall at the forest edge will be 2 to 4 times higher compared to the closed forest (Figure 3.). The results often show much more pronounced edge effects for nitrate and ammonium than for sulphate, indicating that specific deposition mechanisms sometimes will dominate. Besides the cloud droplet incerception, the gaseous deposition of HNO_3 and ammonia might be very large at the edges.

Another situation where very large N deposition might be expected is close to ammonia emission sources from the ground or low

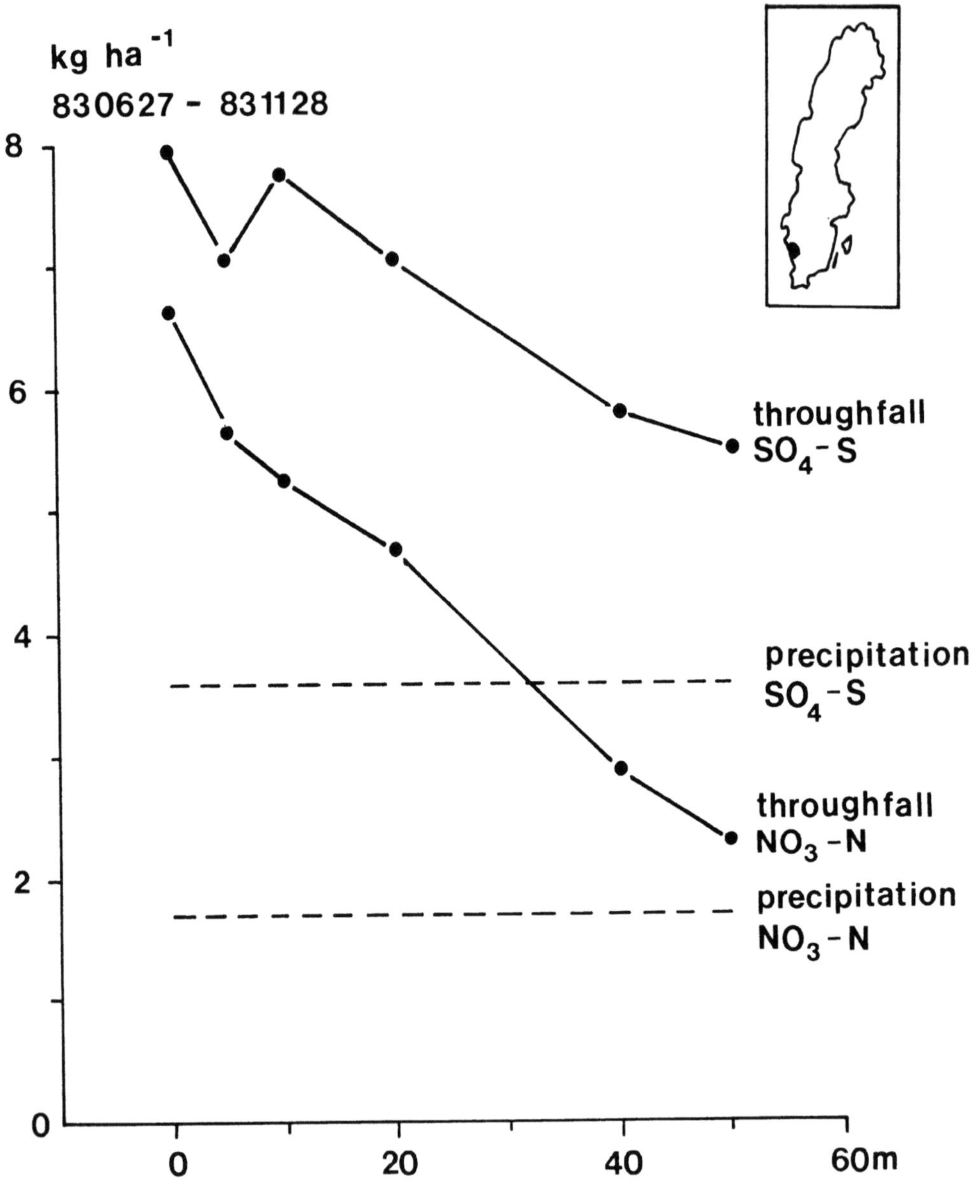

Fig. 3 Throughfall of nitrate and sulphate at a forest edge of a pine forest in south west Sweden (Hasselrot and Grennfelt, 1985).

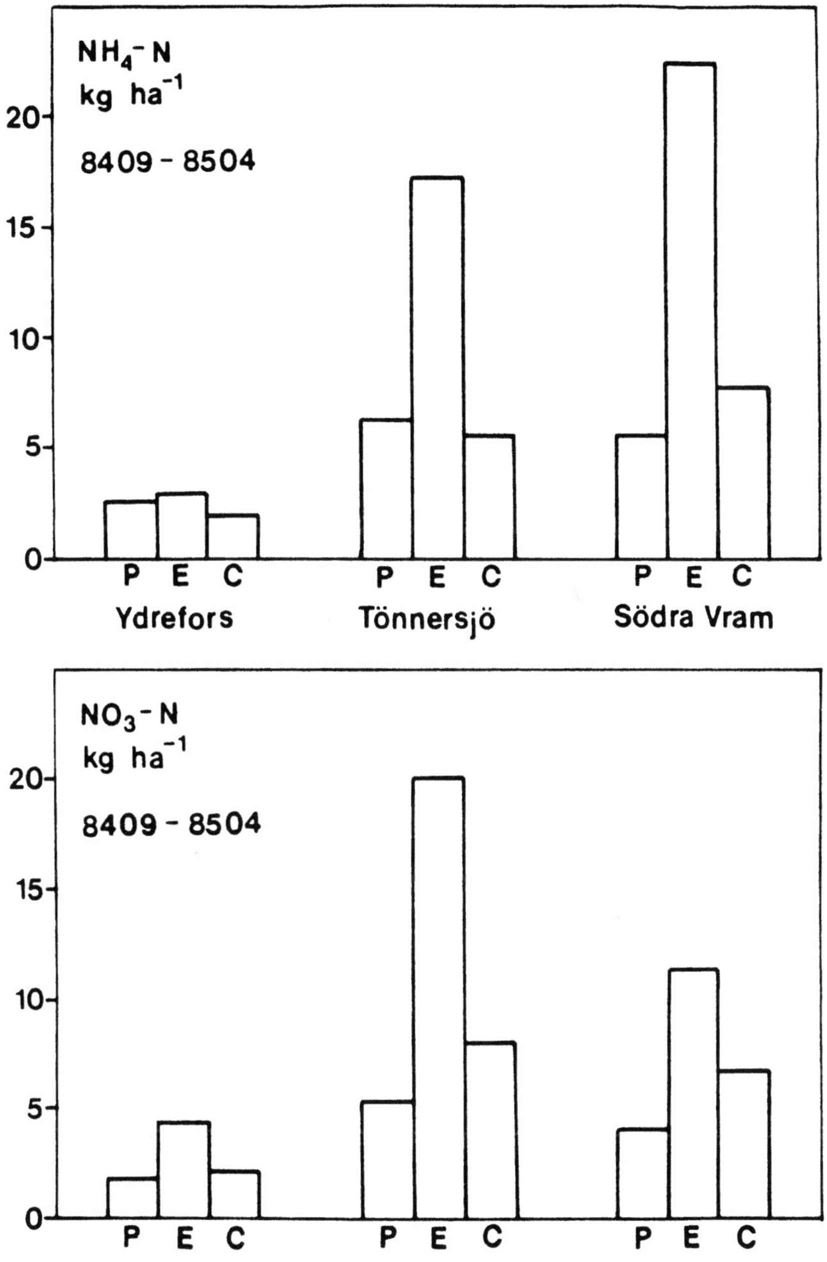

Fig. 4 Ammonium and nitrate flow in precipitation (P) and throughfall at the forest edge (E) and within the forest (C) of mature spruce forests at three sites in south Sweden. Data from 8 mo (Sept. 84 - April 85). Ydrefors is situated in the central part of south Sweden, Tönnersjö at the SW border of the south Swedish highlands, and Södra Vram in the agricultural lowlands in the very south of Sweden.

heights. Chicken farms and manure spreading are such sources. In the Netherlands very high ammonium depositions as well as ammonium concentrations in the soil have been shown (Roelofs et al., 1985; van Breemen and Jordens, 1983; van Breemen et al., 1982). A total input of inorganic N of 64 kg ha^{-1} yr^{-1} has been monitored of which 75% was in the form of NH_4^+ and 25% in the form of NO_3^- (van Breemen et al., 1982).

Forest edge studies in areas with a high density of cattle in south Sweden also indicate a very large influence by ammonia (Figure 4). From an ongoing study data from 11 mo showed a throughfall flow of 26.5 kg NH_4^+-N ha^{-1} and 14.9 kg NO_3^--N ha^{-1} at a forest edge, while the corresponding figures for the wet deposition (open gauges) were 6.9 and 4.9 kg ha^{-1}, respectively (Hasselrot and Grennfelt, 1985).

4. The interaction of nitrogen with tree canopies

Studies in areas with fairly low deposition of N show that the nitrate and ammonium flow beneath the canopy is smaller than the precipitation (Rosén and Lundmark, 1985; Horntvedt et al., 1980; Richter and Granat, 1978). If the dry deposition of N is then added one might assume a substantial absorption of the deposited N in the canopy. Fertilizing experiments show that N (ammonium and nitrate) applied to vegetation surfaces will easily be taken up and metabolized (Verry and Timmons, 1977). The same will happen when ammonium and nitrate are deposited to vegetation surfaces with dry processes as well as when the precipitation passes the canopy. From throughfall studies in Sweden and Norway it is shown that in the summertime more than 70% of the ammonium flow and more than 50% of the nitrate flow in precipitation may be absorbed in the canopy (Horntvedt et al., 1980; Rosén and Lundmark, 1985).

In more polluted areas in south Sweden and on the European continent these relations are changed and higher nitrate flow and sometimes also higher ammonium flow is measured in the throughfall than by precipitation. In these areas the deposition to the canopies is larger than the trees are able to absorb. The soil may also supply the trees with N in such quantities that it will affect the uptake. This effect was shown in an experiment where throughfall was studied in a stand of Corsican pine before and after fertilization of the soil (Miller, 1984).

The interaction of the canopy on the N flow is also dependent of the time of the year. In winters, when the biological activity is small, there seems to be no uptake in the canopy. This effect is clearly seen in Figure 5, where monthly throughfall data of ammonium at a forest edge is presented. Observe that in the summer (May-July) ammonium in throughfall is less than in the open precipitation while it in March is five times that in precipitation.

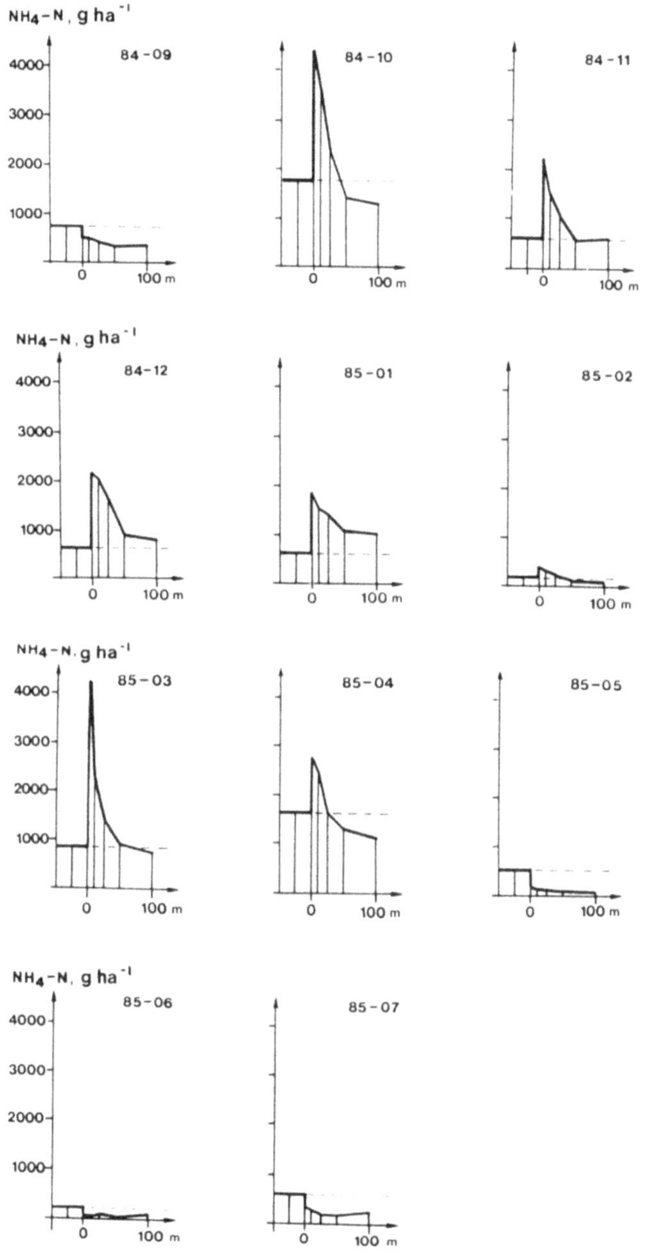

Fig. 5 The monthly variation in throughfall of ammonium at a forest edge of a spruce forest (Tönnersjöheden) in south west Sweden (Hasselrot and Grennfelt, 1985).

5. Nitrogen and ammonium interactions with the soil

Forest growth is generally limited by N deficiency and N deposition will have a fertilizing effect. The retention of N is therefore high in forest ecosystems. High N deposition may, however, cause a N saturation which in turn may cause a considerable output of nitrate (Matzner et al., 1982; Hauhs, 1984) along with leaching of basic cations and/or Al and H^+ from the soil. This will in turn cause a decrease in base saturation of the soil (soil acidification) and/or increased acidity in runoff water.

The acid-base relationships of the N cycle in forest systems are quite well known. Oxidation of ammonium to nitrate produce acid and H^+ is released by root uptake of ammonium. Ammonium formation from amino-groups during mineralisation of organic nitrogen and uptake of nitrate either consume H^+ or release OH^- ions.

The chemical and biochemical reactions of N, from the acidification point of view, are normally simplified to the H^+-producing and H^+-consuming reactions (Reuss, 1976; van Breemen, 1985):

Consuming H^+	$NO_3^- + H^+ \rightarrow$	bio-organic N
"	$NO_3^- + H^+ \rightarrow$	N_2, N_2O (denitrification)
Producing H^+	$NH_4^+ \rightarrow$	$NO_3^- + 2H^+$ (nitrification)
"	$NH_4^+ \rightarrow$	org.N + H^+.

If we assume that these processes are specific for the N-deposition effect on the H^+ budget of a soil the process is simplified to the following equation representing the net flow:

$$\Delta H^+ = (NH_4^+{}_{(in)} - NH_4^+{}_{(out)}) + (NO_3^-{}_{(out)} - NO_3^-{}_{(in)}).$$

Small errors in the estimate of reactions involving H^+ in the N cycle may cause serious errors in H^+ budget. Reuss (1982) therefore suggests that the acidification of soils due to acid deposition is quantified through reduction in soil base saturation. Nitrogen uptake increases biomass production which increases uptake of base cations and reduces base saturation (see also Hultberg, 1985). Nitrate leaching (primarily NO_3^-) occur along with an equivalent amount of cations, which at high base saturation are Ca, Mg, Na and K, but with decreasing base saturation Al-species and H^+ increase.

A consequence of this approach is that the acidifying effect is independent of the ionic form of the N deposition. One mole of ammonium nitrate have twice the potential to acidify the soil as one mole of HNO_3. One mole ammonium sulphate will have twice the acidifying effect as one mole of H_2SO_4. The potential to reduce base saturation by N and S inputs is therefore independent of form.

With this approach the potential acidity can be much larger than that obtained from the H^+ activity. The potential acidity from ammonium sulphate deposition in a woodland in the Netherlands was estimated to 7.14 kmol ha^{-1} while actual acidity in throughfall and stemflow was about 3% (pH = 4.6) of the potential (van Breemen and Jordens, 1983). The N-deposition reached at this site 64 ± 15 kg ha^{-1} yr^{-1} and both calcareous and acid soils leached NO_3 ions along with primarily Ca (calcareous soils) and Al (acid soils) as dominating cations. In the acid soils 65% of the nitrate in soil solution came from ammonia in throughfall (30% in the calcareous soil).

Wiklander (1983) stresses that N from deposition which is retained in the soils may increase nitrification after clear-cutting and induce high nitrate leaching along with decrease in soil base saturation and leaching of Al.
The increased Al-levels in soil solutions may result in precipitation of Al-P compounds in the soil (Broberg and Persson, 1985) which may cause P deficiency to trees and decreased catchment transport by runoff to lakes and streams. Decreased output of P and increased output of Al from the terrestrial system due to increased N inputs are two effects which have the most severe consequenses, oligotrophication and toxicity to fish and other organisms, in the aquatic ecosystem (Hultberg, 1985).

6. Mass flow of nitrogen compounds

Mass flow of N including precipitation and catchment runoff are studied at many forest sites in Europe. In some cases throughfall is also monitored but only few studies exist where the dry deposition is estimated (Table II). These input/output studies may be separated into three different categories

1. studies in low-polluted areas, mostly in central Sweden and Norway,

2. studies in moderately polluted areas in south Sweden and at some areas on the European Continent,

3. studies in highly polluted areas in central Europe.

The first category is characterized by a very high absorption of N in the canopy and a low runoff of N (<0.3 kg ha^{-1} yr^{-1}).

The second category is an intermediate phase where the N deposition affects the N circulation in the ecosystem. The load of N to the ground by throughfall reaches 5 to 10 kg N ha^{-1} but the runoff is still very low.

The third category is when nitrate leaching from the soil is elevated. The input at these places is larger than the ecosystem will be able to absorb.

Abrahamsen (1980) presented a regession between N-input by precipitation and catchment runoff of N, showing that 30% of the wet deposition was leached from the terrestrial system.
Runoff data from Central Europe (Table II), however, show in five cases that the output of nitrate is 53 to 92% of the input by precipitation when this input exceeds 13 kg inorganic N ha^{-1}. Nitrogen saturation may already have occurred in these forest ecosystems (Figure 6).

Table II.

Mass flows of NO_3-N and NH_4-N in some forested catchments kg $ha^{-1}yr^{-1}$

Place	Country	Wet depositon		Throughfall		Outflow		Reference and comments
		NH_4-N	NO_3-N	NH_4-N	NO_3-N	NH_4-N	NO_3-N	
Jergul	N	0.35	0.42			-	0.04	Wright, 1982
Birkenes	N	7.4	7.8			<1.0	<1.0	-"-
Hälsingland	S	1.04	1.05			0.05	0.08	Rosén and Lundmark, 1985
Jädraås	S	1.3	1.3	0.8	1.1	0.05	0.03	Bringmark, 1980 (Data from May-Nov)
Kloten	S	2.7	3.2			0.03	0.1	Grip, 1983
Gårdsjön	S	6.3	4.6	4.7	5.7	0.023	0.24	Grennfelt et al., 1985, Hultberg, 1985
Söderåsen	S	~7.0	~5.0	~10	~9	-	5.0	Wiklander, 1983, beech and 1/2 year budget of spruce stand
Langen Bramke	D	14.8	8.9	10.0	9.7	0.2	1.9	Hauhs, 1984
Dicke Bramke	D	(14.8)	(8.9)			0.3	6.4	-"-
Wintertal	D	(14.8)	(8.9)			0.3	16.0	-"-
Solling	D	12.6	8.1	16.3	16.0	0.18	11.9	Matzner et al., 1982
Waroneu	B	12.4	11.0	17.9	16.5	0.8	12.0	Buldgen and Remack, 1984
Robinette	B	12.4	11.0	19.5	21.5	0.8	16.5	- " - - " -
Vysoca Pec	CZ	7.5	5.5			0.0	12.0	Paces, 1985

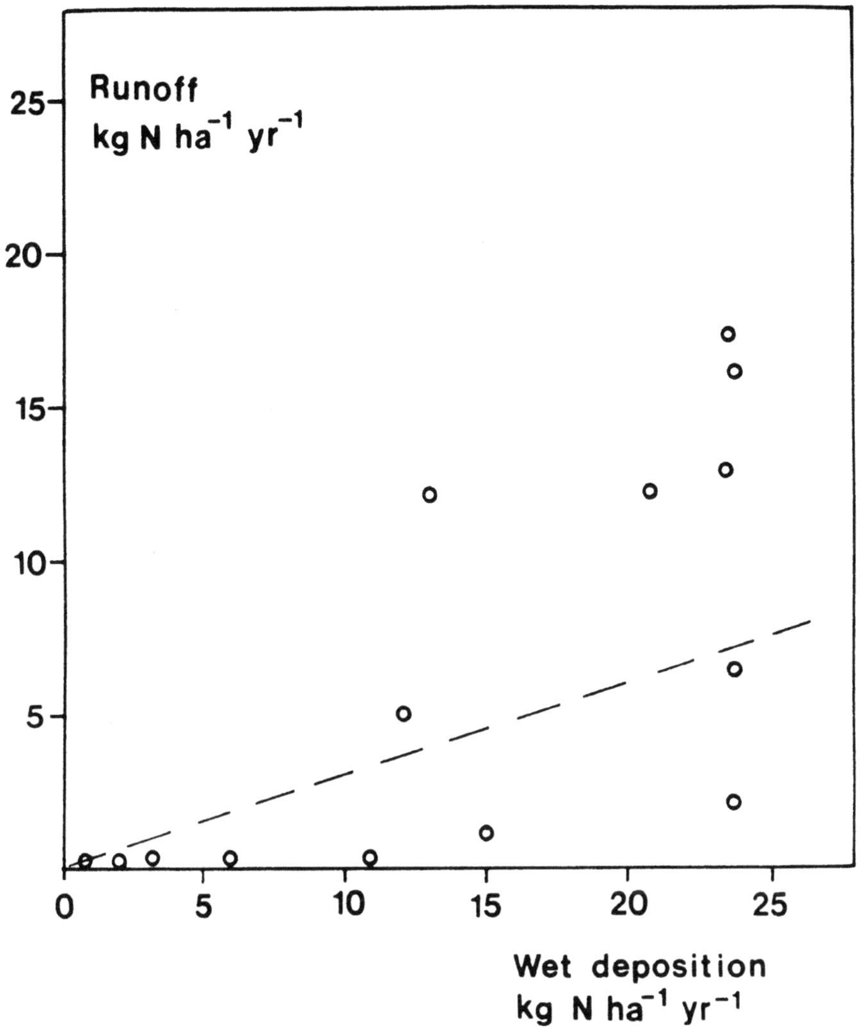

Fig. 6 Annual impact of $NH_4^+ + NO_3^-$ (kg ha^{-1}) by precipitation versus the leaching losses from watersheds and lysimeters in coniferous forests. Data from north and Central Europe. The line indicates the regression line obtained by Abrahamsen (1980).

7. Effects on forests of nitrogen deposition

Investigations of pine forests in the Netherlands show that needles of pine take up ammonia and excrete K, Mg and Ca which often results in K and/or Mg deficiencies. The high levels of N in needles are also correlated to fungal diseases. Field studies of forest damage to pine stands were correlated to increased ratios of NH_4 to K, Mg, and Ca in the soil solution (Roelofs *et al.*, 1985).

Nihlgård (1985) also stresses that the ammonium deposition is an additional explanation to forest decline. In his hypothesis, he assumes that high uptake of ammonia stimulates biomass production which may cause relative deficiencies in Mg, K, P, Mo, B and water, and accumulation of non-protein N may increase. Increased non-protein N and waste products from assimilation may cause the tree to release the leaf or the needle. High levels of N in leaves and needles cause a decrease in frost hardiness and an increased susceptibility to insect attacks, fungi, bacteria and viruses.

8. Effects of nitrogen deposition in aquatic ecosystems

Generally, sulphate is the most important strong acid anion in acidified lakes and streams and on a yearly basis only <5 to 15% of the acidity in nutrient-poor lakes is explained by nitrate (Tirén, 1980).

The nitrate concentration has, however, increased during the last decades in both lakes and streams (Dickson, 1980) and in deep nutrient-poor lakes this progressive increase has been shown to cause a decline in pH and alkalinity. Figure 7 show that the acid Lake Gårdsjön have had an increase in both the lake inlet and lake outlet during the last years.

Acid surges in spring co-incides with elevated concentrations of nitrate in runoff from forested catchments (Likens *et al.*, 1977) and in lakes (Galloway *et al.*, 1980). During late autumn and winter when no nitrate uptake by vegetation occur, nitrate in autumn rains and/or accumulation in snowpack may be transported from the catchments. High concentrations may also occur in streams draining catchments with forest dieback (Paces, 1983) along with high concentrations of Al. A positive correlation between Al and NO_3^- have been demonstrated by Schofield (1984) in Woods Lake outlet and inlet during snowmelt.

9. Conclusions

The deposition of N compounds to forest ecosystems is in most cases higher than previously considered. Deposition estimates and throughfall data indicate that the atmospheric input in many areas in Central Europe reaches 30 to 40 kg N $ha^{-1}yr^{-1}$ and in extremes above 60 kg N $ha^{-1}yr^{-1}$. In south Sweden the deposition exceeds 20 kg N $ha^{-1}yr^{-1}$ and in extremes above 40 kg $ha^{-1}yr^{-1}$.

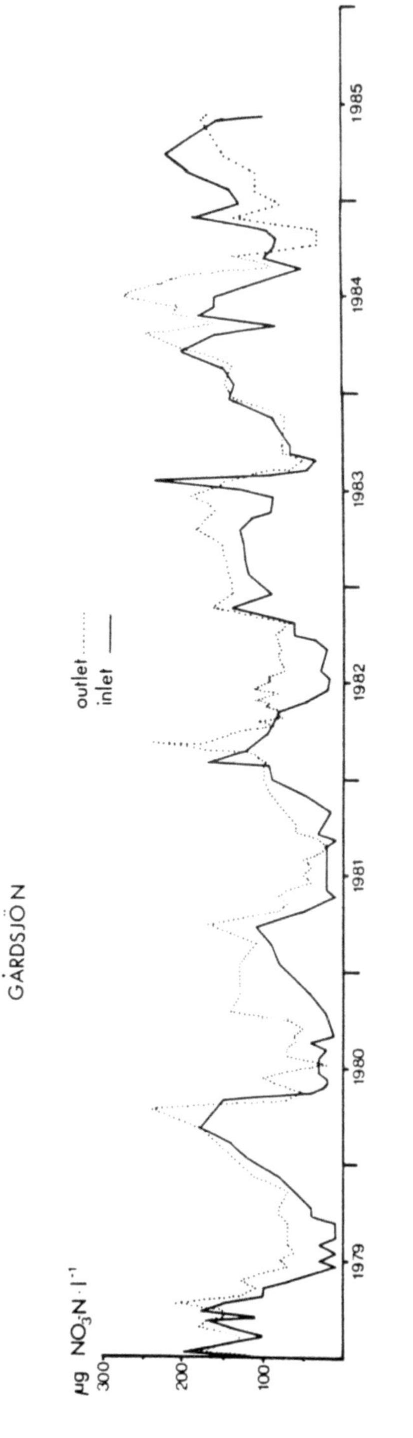

Fig. 7 Changes in the concentration of nitrate in Lake Gårdsjön in south west Sweden 1979 - mid 1985.

The consequence of an increased N input with respect to N utilization and forest growth was analyzed by Ågren (1983). With a theoretical C/N model, he found that with a deposition of 30 kg N ha^{-1} N saturation will be reached in fertile soils within 20 to 25 yr. For poor soils this saturation level will be reached within approx. 50 yr.

Our opinion is that many sites in Central Europe have had a yearly deposition of 30 kg N ha^{-1} for at last 25 yr and according to the model, these sites should be nitrogen saturated. The high leaching of nitrate (5 to 16 kg N $ha^{-1}yr^{-1}$) from many soils with a high N loading is probably the best evidence of this N saturation.

Moreover, there are indications of ecosystem effects related to high loadings of N, especially such loadings that exceed the absorption capacity of the ecosystem. Together with S deposition the N deposition may be an essential factor for soil acidification and forest damages as well as for acidification of lakes and streamwaters during snowmelt episodes.

References

Abrahamsen, G.: 1980, Acid Precipitation, Plant Nutrients and Forest Growth. In: Drabløs, D. and Tollan, A. (Eds.) Ecological Impacts of Acid Precipitation, 58-63. SNSF-project Oslo-Ås, Norway.

Bringmark, L.: 1980, Ecol. Bull. (Stockholm) 32, 341.

Buldgen, P. and Remack, J.: 1984, Sulphur budget in forested watersheds in Haute Ardenne region (Belgium). In: Belgian Research on Acid Deposition and the Sulphur Cycle (Ed. O.L.J. Vanderborght), p. 163-172. SCK/CEN Mol Belgum.

Dickson, W.: 1980, Properties of Acidified Lakes. In: Drabløs, D. and Tollan, A. (Eds.) Ecological Impact of Acid Precipitation pp. 75-83. SNSF-project, Oslo-Ås, Norway.

Galloway, J.N. and Dillon, P.J.: 1983, Effects of acid deposition the importance of nitrogen. In: Ecological Effects of Acid Deposition. National Swedish Environmental Protection Board Report PM 1636, p. 145-160.

Galloway, J.N. and Likens, G.E.: 1981, Atmos. Environm. 15, 1081.

Galloway, J.N., Schofield, C.L., Hendrey, G.R., Peters, N.E. and Johannes, A.J.: 1980, Sources of acidity in three lakes acidified during snowmelt. In: Drabløs, D. and Tollan, A. (Eds.) Ecological Impact of Acid Precipitation. Proc. Int. Conf. on Acid Precipitation - Effects on Forests and Fish, 264-265. Oslo-Ås, Norway.

Grennfelt, P., Larsson, S., Leyton, P. and Olsson, B.: 1985, Atmospheric deposition in the Lake Gårdsjön area, SW Sweden. Ecol. Bull. (Stockholm) 37, 101.

Grip, H.: 1983, Water chemistry and runoff in forest streams at Kloten. Department of Physical Geography Report No 58, Uppsala, Sweden.

Hauhs, M.: 1984, Wasser- und Stoffhaushalt im Einzugsgebiet der Langen Bramke (Harz). Department of Forestry, University of Göttingen.

Hasselrot, B. and Grennfelt, P.: 1985, Skogsfakta 8, 36.

Horntvedt, R., Dollard, G.J. and Joranger, E.: 1980, Effects of acid precipitation on soil and forest. 2. Atmosphere-vegetation interactions. In: Drabløs, D. and Tollan, A. (Eds.) Proc. Int. Conf. Ecological Impact of Acid Precipitation, 192-193, Oslo-Ås, Norway.

Hultberg, H.: 1985, Budgets of base cations chloride, nitrogen and sulphur in the acid Lake Gårdsjön, SW Sweden. Ecol. Bull. (Stockholm) 37, 133.

Hultberg, H. and Grennfelt, P.: 1986, Gårdsjön Project: lake acidification, chemistry in catchment runoff, lake liming and microcatchment manipulations. Water, Air, and Soil Poll. ...., ... (this conf.)

Jernelöv, M. and Lövblad, G.: 1985, Utsläpp till luft i Sverige 1982. National Swedish Environmental Protection Board, Solna, Rep. PM 1973 (In Swedish).

Kallend, A.S., Marsh, A.R.W., Pickles, J.H. and Proctor, M.W.: 1983, Atm. Environ., 17, 127.

Likens, G.E., Bormann, F.H., Pierce, R.S., Eaton, J.S. and Johnson, N.M.: 1977, Bio-geo-chemistry of a forested ecosystem, Springer-Verlag, New York.

Lovett, G.M., Reiners, W.A. and Olson, R.K.: 1982, Science 218, 1303.

Matzner, E., Kharna, P.K., Meiwes, K.J., Lindheim, M., Prenzel, J. and Ulrich, B.: 1982, Elementflüsse in Waldökosystemen im Solling. Datendokumentation. Göttinger Bodenkundliche Berichte 71. Institut für Bodenkunde und Waldernährung, Göttingen.

Miller, H.G.: 1984, Phil. Trans. R. Soc. Lond. B 305, 339.

N.A.S.: 1983, Acid deposition: Atmospheric Processes in Eastern North America. National Academy of Sciences. National Academy Press, Washington D.C., 375 pp.

Nihlgård, B.: 1985, The ammonium hypothesis - an additional explanation to the forest dieback in Europe. Ambio 14, 2.

Paces, T.: 1983, The pH and concentrations of sulphate, nitrate and aluminum in the runoff from representative catchments in the Elbe River basin. In: Ecol. Effects of Acid Deposition. National Swedish Environment Protection Board, Solna Sweden. Report PM 1636, 161-173.

Paces, T.: 1985, Nature 315, 31.

Persson, G (Ed.): 1982, Acidification Today and Tomorrow. Department of Agriculture, Sweden, 1-232.

Persson, G. and Broberg, O.:, 1985, Nutrient concentration in the acidified Lake Gårdsjön: The role of transport and retention of phosphorus, nitrogen and DOC in watershed and lake. Ecol. Bull. (Stockholm) 37, 158.

Reuss, J.O.: 1976, Chemical and biological relationships relevant to the effect of acid rainfall on the soil-plant system. In: Dochinger, L.S. and Seliga, T.A. (Eds.) Proc. of the First Int. Symposium on Acid Precipitation and the Forest Ecosystem. USDA Forest Service General Technical Rept. NE-23, p. 791.

Reuss, J.O.: 1982, Effect of nitrogen in acid precipitation on soil: An integrated approach. In: Simulation Models of Water Acidification. Documentation of a workshop in Uppsala, Sweden 16-18 June 1982, pp 130-137. Dep. of Ecology and Environmental Research. Swedish University of Agricultural Sciences, Uppsala.

Richter, A. and Granat, L.: 1978, Pine forest canopy throughfall measurements. Report AC 43. Dep. of Meteorology, University of Stockholm, Stockholm.

Rodhe, H.: 1982, Deposition of plant nutrients from the air. Skogs- o Lantbr.-akad. Tidskr. Suppl. 14, 32-36 (In Swedish)

Rodhe, H.: 1985, personal communication.

Roelofs, J.G.M., Kempers, A.J., Houdijk, A.L.F.M. and Janssen, J.: 1985, Plant and Soil 84, 45.

Rosén, K. and Lundmark, J.-E.: 1985, Kemiska förändringar i nederbörd vid passage av kronskiktet i en mellansvensk barrblandskog. Rapport i Skogsekologi och skoglig marklära 55 (In press).

Scherbatskoy, T. and Bliss, M.: 1983, Occurrence of acidic rain and cloud water in high elevation ecosystems in the green mountains of Vermont. Proc. of the APCA speciality meeting, "The Meteorology of Acidic Deposition", Hartford CT, Oct. 16-19, 1983.

Schofield, C.L.: 1984, Surface water chemistry in the ILWAS basins in Goldstein and Gherini (Eds.) The Integrated Lake Watershed Acidification Study, Vol. 4: Summary of Major Results 6.1-6.32 EPRI Report, Palo Alto CA.

SNV: 1984, Long Range Transport of Air Pollutants. Monitor 1984. National Swedish Environmental Protection Board, Solna, 1-134 (In Swedish).

Söderlund, R. and Granat, L.: 1982, Ammonium (NH_4^+) in precipitation - A presentation of data from the European Air Chemistry Network. Department of Meteorology, University of Stockholm and the International Meteorological Institute in Stockholm. Report CM 59.

Tirén, T.: 1980, Den interna svavel- och kväveomsättningens betydelse för försurning av limniska ekosystem. National Swedish Environmental Protection Board, Solna, Report PM 1376 (In Swedish).

van Breemen, N. and Jordens, E.R.: 1983, Effects of atmospheric ammonium sulphate on calcareous and non-calcareous soils of woodlands in the Netherlands. In: Ulrich and Pankrath. Effects of Accumulation of Air Pollutants in Forest Ecosystems, 171-182. D. Reidel Publ. Company.

van Breemen, N., Borrough, P.A., Velthorst, E.J., van Dobben, H.F., de Wit, T., Ridder, T.B. and Reijnders, H.F.R.: 1982, Nature 299, 548.

van Breemen, N.: 1985, Transformation of nitrogen and sulphur,and acidification of soils and waters. In: Johansson, I. (Ed) Hydrological and hydrogeochemical mechanisms and model approaches to the acidification of ecological systems. IHP Workshop, Uppsala, 15-16 September 1984. Nordic Hydrological Programme Report No 10, p. 131-136, Uppsala, Sweden.

Verry, E. and Timmons, D.R.: 1977, Can. J. For. Res. 7, 112.

Wiklander, G.: 1983, Skogs- o Lantbr.-Akad. Tidskrift, 122, 311.

Wright, R.F.: 1982, Gauged Catchments in Norway. Norwegian Institute for Water Research. Report.

W S L: 1983, Acid deposition in the United Kingdom. Warren Spring Laboratory, Stevenage, Herts., UK.

Ågren, G.: 1983, Model analyses of some consequences of acid precipitation on forest growth. In: Ecological Effects of Acid Deposition. National Swedish Environmental Protection Board Report PM 1636, 233-244.

SULFUR CYCLING IN FIVE FOREST ECOSYSTEMS

Dale W. Johnson
Daniel D. Richter
Environmental Sciences Division
Oak Ridge National Laboratory
P. O. Box X
Oak Ridge, TN 37831 USA

Helga Van Miegroet
Dale W. Cole
College of Forest Resources AR-10
University of Washington
Seattle, WA 98195 USA

J. M. Kelly
Tennessee Valley Authority
Division of Air and Water Resources
Muscle Shoals, AL 35660 USA

ABSTRACT. The cycling and retention of sulfur were studied in five forest ecosystems: a chestnut oak and yellow poplar stand on Walker Branch Watershed, Tennessee; a mixed oak stand on Camp Branch Watershed, Tennessee; and a red alder and Douglas-fir stand at the Thompson site, Washington. Calculations from foliage sulfur turnover indicate that about one-half of total sulfur input was dry in the Tennessee sites, whereas only one-tenth was dry in the Washington sites. Atmospheric sulfur inputs exceeded forest sulfur requirements in all cases, but three sites (chestnut oak, mixed oak, and red alder) showed a net ecosystem retention of atmospherically deposited sulfur. Net ecosystem sulfur retention was consistent with laboratory-determined sulfate adsorption isotherms within a given location (Walker Branch, Thompson site) but not between locations because of differing deposition histories and consequent differing degrees of soil sulfate saturation. No consistent relationships between soil sulfate adsorption capacity and other soil properties (pH, base saturation, iron, and aluminum oxides) were found.

1. INTRODUCTION

The relatively recent interest in the effects of atmospheric sulfur deposition has greatly increased the interest in sulfur cycling in forest ecosystems. Most nutrient-cycling research in forests has focused on the cycles of the other major nutrients--nitrogen

Water, Air, and Soil Pollution **30** (1986) 965–979.

phosphorus, calcium, potassium, and magnesium -- because sulfur is seldom growth-limiting and analytical techniques for sulfur are more difficult. There has been considerable research on tree sulfur nutrition in sulfur-deficient areas of Australia (Kelly and Lambert, 1972) and the northwestern United States (Youngberg and Dryness, 1966), however, and several researchers have reported a consistent molar ratio of sulfur to nitrogen of 0.03 in the foliage of both coniferous and deciduous species (Kelly and Lambert, 1972; Turner et al., 1980; Johnson et al., 1982). This is consistent with the ratio of sulfur to nitrogen in the three sulfur-bearing amino acids, cystine, cysteine, and methionine (Kelly and Lambert, 1972). Sulfur present in excess of that prescribed by this ratio accumulates as sulfate, and foliar sulfate is used as an index of tree sulfur status (Kelly and Lambert, 1972; Turner et al., 1977).

The relationships between organic nitrogen and organic sulfur in soils is less consistent because of the presence of nonprotein organic sulfur. Soil sulfur is commonly fractionated into the major categories of carbon-bonded sulfur (presumably including protein sulfur), ester sulfate, and inorganic soluble and adsorbed sulfate (Bettany et al., 1980; David et al., 1982). In most cases, carbon-bonded sulfur and ester sulfate are the predominant forms of soil sulfur, but adsorbed sulfate can account for > 50% of total soil sulfur in some instances (Meiwes and Khanna, 1981; Johnson et al., 1982).

Recent concerns over acid deposition effects have stimulated considerable research into mechanisms controlling the retention or release of atmospherically deposited sulfur in forest ecosystems. Specifically, mechanisms controlling the retention and release of sulfate have been intensely studied because of the role of this anion in leaching base cations and aluminum from terrestrial to aquatic ecosystems (Johnson and Cole, 1977; Johnson et al., 1979, 1982; Likens et al., 1977; Richter et al., 1983; Swank et al., 1984; Seip, 1980; Turner et al., 1980; Galloway et al., 1983; Meiwes and Khanna, 1981). Thus, the focus of sulfur-cycling research has shifted away from tree sulfur nutrition, which is most certainly adequate if not excessive in polluted regions (Johnson, 1985), to the identification of the key mechanisms by which atmospherically deposited sulfur is retained by forest ecosystems.

Sulfur uptake by forest vegetation can be a significant proportion of atmospheric sulfur input and can cause seasonal effects on ecosystem sulfur balances by converting SO_4^{2-} into organic forms. However, net annual sulfur accumulation in biomass (which is usually only 5 to 10% of uptake, the remainder being recycled) can account for only a fraction of the net annual accumulation of sulfur in those forest ecosystems that show a net ecosystem sulfur retention (i.e., in the southeastern United States and central Europe) (Shriner and Henderson, 1978; Johnson et al., 1982; Johnson, 1985; Kelly, 1984a; Meiwes and Khanna, 1981).

The soil is the major repository of sulfur in most forest ecosystems, and soil organic matter typically (but not always) contains most of the soil sulfur. Soil adsorbed sulfate is of varying importance, ranging from 6 to 66% of total soil sulfur in a variety of forest ecosystems (Johnson, 1985). Many researchers find that soil sulfate adsorption and desorption are the major controlling mechanisms for sulfate leaching in soils (Johnson and Cole, 1977; Johnson et al., 1979, Richter et al., 1983; Galloway et al., 1983). Others emphasize the potentially great importance of organic sulfur reactions in soils (David et al., 1982; Swank et al., 1984).

In this paper, we report the results of a comparative sulfur-cycling study in three deciduous forests in eastern Tennessee subject to relatively high atmospheric sulfur input and one coniferous (Douglas fir) and one deciduous (red alder) forest in western Washington subject to lower sulfur inputs. It was hypothesized that the older, more highly weathered Tennessee soils would have higher iron and aluminum oxide contents and, consequently, a greater sulfate adsorption capacity than the younger Washington soils. It was also hypothesized that the enhanced nitrogen status (caused by nitrogen fixation) in the red alder stand would cause greater retention of sulfur in organic forms than in the nitrogen-deficient Douglas fir or mixed deciduous stands.

2. SITES

2.1. Walker Branch Watershed, Tennessee

Walker Branch Watershed is situated in the ridge-and-valley physiographic province near Oak Ridge, Tennessee (Grigal and Goldstein, 1971). The climate type is humid mesothermal, with moderate summer and winter temperatures. Mean annual temperature is 14.3°C, and precipitation averages 130 cm/year, mostly as rain (Henderson et al., 1977). Underlying the watershed is Knox dolomite, a cherty, dense to coarsely crystalline rock of late Cambrian to early Ordovician age.

The chestnut oak stand on Walker Branch is located on a ridgetop on the Fullerton series soil, a Typic Paleudult. Vegetation type is chestnut oak, described by Grigal and Goldstein (1971). Species consist primarily of chestnut oak (Quercus prinus L.), hickory (Carya spp.), red maple (Acer rubrum L.), white oak (Quercus alba L.), black oak (Q. velutina Lam.), blackgum (Nyssa sylvatica Marsh.), sourwood (Oxydendrum arboreum L.), and yellow poplar (Liriodendron tulipifera L.), with occasional dogwood (Cornus florida L.) and northern red oak (Q. rubra L.). Understory is absent.

The yellow poplar stand is located in a sinkhole depression on the Tarklin series soil, a Typic Fragiudult. Vegetation type is yellow poplar, described by Grigal and Goldstein (1971). Species consist primarily of yellow poplar with occasional red maple, white oak, black locust (Robinia pseudoacacia L.), persimmon (Diospyros virginiana L.), black cherry (Prunus serotina Ehrh.), and Carolina buckthorn (Rhamnus caroliniana D.C.). Understory consists of dogwood, occasional eastern red cedar (Juniperus virginiana L.), and Japanese honeysuckle (Lonicera japonica).

2.2. Thompson Site, Washington

The Thompson site is located in the western foothills of the Cascade Mountains near Landsburg, Washington. The climate is maritime, with cool, dry summers and wet, moderate winters. Mean annual temperature is 9.8°C, and mean annual precipitation is 136 cm, most of which falls as rain between October and March. The soil at the study site belongs to the Alderwood series, a Dystric Entic Durochrept, and is composed of loosely consolidated ablation till of varying depth over a compacted basal till.

The research plots are located in a naturally regenerated 48-year-old stand of red alder (Alnus rubra Bong.) and, immediately adjacent to it, in a 48-year-old plantation of Douglas-fir (Pseudotsuga menziesii Bong. Carr.). The understory in the red alder stand is dominated by sword fern (Polystichum munitum Kaulf. Presl.) intermixed with occasional western hemlock (Tsuga heterophylla Raf. Sarg.), huckleberry (Vaccininum parviflorum Smith), vine maple (Acer circinatum Pursh.), Oregon grape (Berberis nervosa Pursh. Nutt.), and elderberry (Sambucus callicarpa Pursh.). Understory in the Douglas-fir stand is much less predominant and consists of salal (Gualtheria shallon Pursh.), Oregon grape (Berberis nervosa Pursh. Nutt.), huckleberry (Vaccininum spp.), and mosses.

2.3. Camp Branch Watershed, Tennessee

The Camp Branch Experimental Watershed is located on the Cumberland Plateau within the boundaries of Fall Creek Falls State Park in Bledsoe County, Tennessee. The climate of the Cumberland Plateau is temperate and continental. Winters are moderate with short cold periods, and the summers are mild to hot. Precipitation is well distributed over the year. Average annual precipitation is 144 cm, mostly in the form of rain. Mean annual temperature is 12°C.

The study site was located on an upper slope position on the Lily series soil, a Typic Hapludult derived from acid sandstone. Vegetation consisted of second growth mixed oak forest in which oak species constituted ~ 70% of the basal area. Quercus, coccinea, Q. stellata, Q. alba, and Q. velutina are dominant (Ramseur and Kelly 1981).

3. METHODS

At each plot, bulk precipitation and throughfall were collected by means of plastic funnels fitted to polyethylene bottles; litterfall was collected in screened litter traps of known area; and soil solutions were collected beneath litter, A horizons, and B horizons (at rooting depth) by means of tension lysimeters (Cole 1968). Collections were made on approximately a monthly basis. Aboveground biomass (foliage, branch, and bole) was determined from regression equations based on diameter (Harris et al., 1973; Gholz et al., 1979). Litter was sampled with randomly spaced 0.25-m^2 rings, and soils were sampled from lysimeter sites. Foliage, branch, bole, roots, litterfall, and forest floor samples were analyzed for total nitrogen by Kjeldahl digestion, total sulfur by a nitric-perchloric acid digestion procedure, and sulfate-sulfur by the method of Richter and Johnson (1983).

Soil samples were analyzed for percent gravel, bulk density (by determining pit volume with backfilling and weighing), percent nitrogen (Kjeldahl digestion, followed by Autoanalyzer analysis for NH_4^+), soluble SO_4^{2-} (by shaking with distilled water, followed by Autoanalyzer analysis), adsorbed SO_4^{2-} [by shaking with 0.032 M NaH_2PO_4 in a 1:5 soil:solution ratio for 1 h, followed by barium chloranilate analysis (Bertolacini and Barney, 1957)] and total sulfur by a nitric-perchloric acid digestion, followed by turbidometric analysis of SO_4^{2-}.

Sulfate adsorption isotherms were determined by treating 5 g of air-dried soil with successive 50-mL aliquots of 0.5, 1.0, 2.5, and 5.0 meq • L^{-1} of $CaSO_4$ in a mechanical vacuum extractor (Concept Engineering, Inc.) at a 1-h extraction time setting. Each column (2.5-cm diam by 13.2-cm length) was prepacked with 1 g of ashless filter pulp, and two columns were left without soils (as blanks). Adsorption or desorption of anions was calculated by the disappearance or appearance of anions in the eluent solution (after accounting for interstitial water and blank adsorption-desorption).

All waters were analyzed by pH; conductivity; Ca^{2+}, Mg^{2+}, K^+, and Na^+ (the latter four by atomic absorption), NH_4^+, NO_3^-, PO_4^{3-}, Cl^-, and SO_4^{2-} (the latter five by Technicon Autoanalyzer); and alkalinity (by titration). Checks of analyses were made by charge balance between cations and anions (except where organic anions were suspected) and by comparing calculated with measured conductivity (Golterman and Clymo 1969).

Nutrient requirement R is normally defined as that amount of nutrient used to produce new plant tissue as foliage, bole and branch wood, and roots (Cole and Rapp, 1981). Part of this requirement may be met by translocation T of nutrients stored in older tissue to new tissue during growth, the rest being satisfied by uptake U. Thus,

$$R = T + U \quad , \text{and}$$

$$U = R - T \quad .$$

Translocation is estimated as the difference among current year live foliage content FC and foliage leaching FL and leaf litterfall LL (in deciduous forests), or older foliage FO (in coniferous forests):

$$T = FC - FL - LL \text{ deciduous}$$

$$T = FC - FL - FO \text{ coniferous}$$

Due to uncertainties in the estimation of FL (see Sect. 4), we have assumed in these cases that U = R and, thus, T = 0. Unfortunately, root data were unavailable for the Washington sites, so only aboveground uptake and requirement are presented here.

4. RESULTS AND DISCUSSION

4.1. Sulfur Distribution

The total sulfur distributions at these five sites exhibited some basic similarities, as well as some specific differences. Although wood concentrations were much lower than foliage concentrations, the preponderance of wood biomass resulted in wood, not foliage, being the major aboveground sulfur pool in trees at all sites. The understory constituted a large sulfur pool in the red alder stand, perhaps due in part to its advanced stage of stand development (i.e., the alder overstory is beginning to break up). The understory is less important at the Douglas-fir site and virtually absent at the Tennessee sites.

Forest floor sulfur content (in midsummer) was roughly equal to aboveground tree sulfur content in the chestnut oak and yellow-poplar sites but more than tree content at the mixed oak and red alder sites. Forest floor sulfur content was very large at the red alder site, primarily as a result of sword fern litter caused in part by its advanced stage of stand development.

The soil was the major sulfur pool at all sites, constituting 88 to 97% of total ecosystem sulfur content (Figures 1-3). The red alder

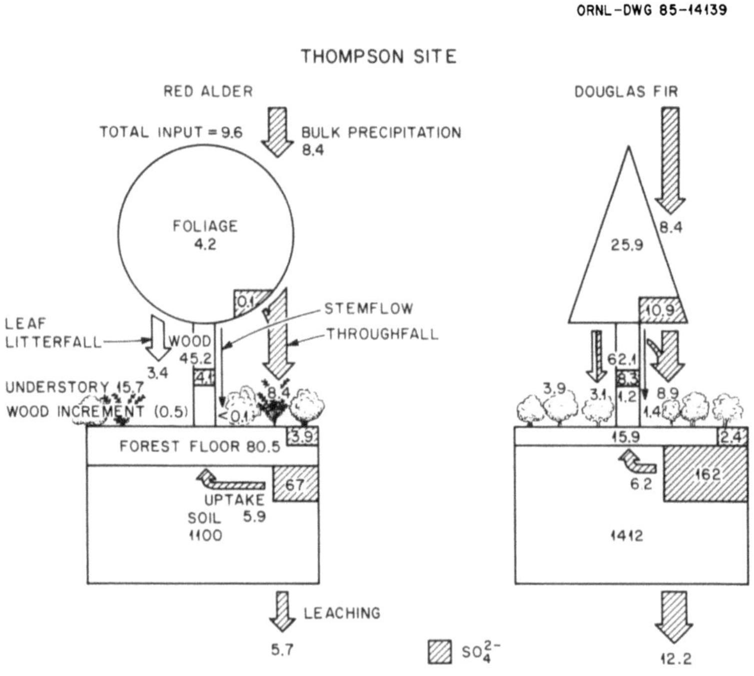

Figure 1. Total sulfur and sulfate-sulfur cycling in a chestnut oak and yellow poplar stand on Walker Branch Watershed, Tennessee (shaded areas represent SO_4^{2-}-S).

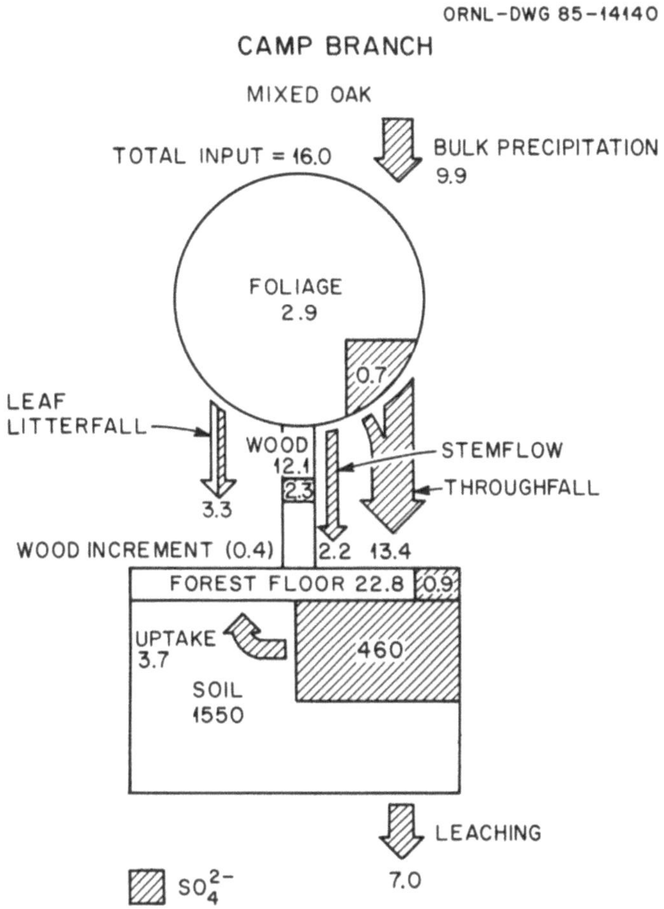

Figure 2. Total sulfur and sulfate-sulfur cycling in a mixed oak stand on Camp Branch Watershed, Tennessee (shaded areas represent SO_4^{2-}-S) (kg•ha^{-1} and kg•ha^{-1}•year^{-1}).

soil had the lowest total sulfur and sulfate-sulfur pool of the five soils studied here. To a large extent, this is due to the high gravel content of the red alder soil (60-78% vs 46-50% in the Douglas-fir soil and 8-30% in the 3 Tennessee soils), since total S concentration (<2 mm fraction) was actually highest in the red alder soil (Table I).

The SO_4^{2-} distributions in these five sites paralleled the total sulfur distributions in many ways but differed in specific ways related to soil SO_4^{2-} adsorption properties and nitrogen-sulfur

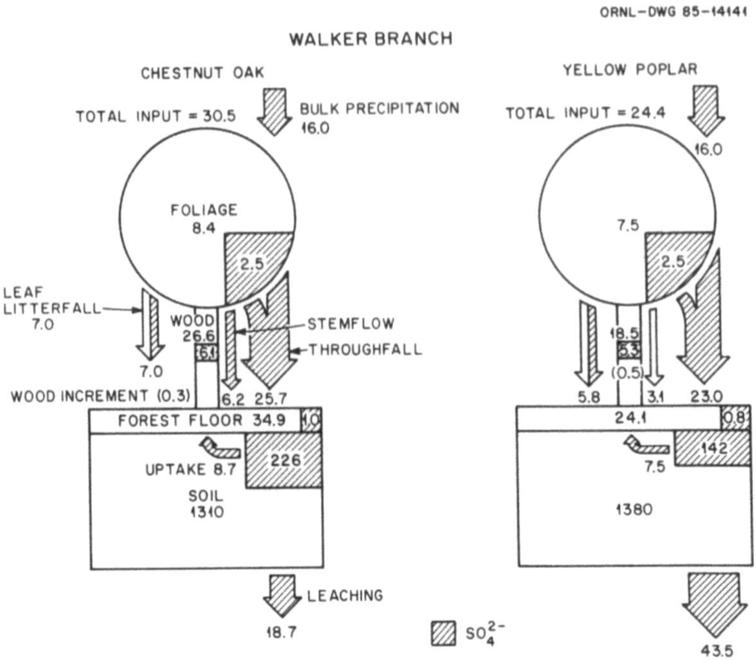

Figure 3. Total sulfur and sulfate-sulfur cycling in a red alder and Douglas fir stand at the Thompson site, Washington (shaded areas represent SO_4^{2-}-S) ($kg \bullet ha^{-1}$ and $kg \bullet ha^{-1} \bullet year^{-1}$).

interactions. The Lily (Camp Branch) soil had the highest soil SO_4^{2-} content followed by the Fullerton (chestnut oak), Alderwood (Douglas-fir), Tarklin (yellow-poplar) and Alderwood (red alder), respectively. (Again, the low SO_4^{2-} contents in the Alderwood soils are in part due to high gravel content, but in part they are also due to relatively low soil SO_4^{2-} concentration; Table I). In contrast to this pattern, the soil sulfate adsorption isotherms for selected horizons from the five sites indicated that the Alderwood (red alder) and Lily (mixed oak) soils had the greatest sulfate adsorption capacities, followed by the Alderwood (Douglas-fir), Fullerton (chestnut oak), and Tarklin (yellow poplar) soils, respectively. These adsorption capacities bear little relationship to native soil sulfate content, citrate-dithionite-extractable iron or aluminum, pH, cation exchange capacity (CEC), or base saturation (BS) when all sites are compared (Table I). However, the differences in sulfate adsorption capacity between the Fullerton and Tarklin soils are consistent

TABLE I. Some chemical properties of selected soil horizons from sulfur-cycling studies (after Johnson et al., 1981, Van Miegroet; 1986 and Richter, unpubl. data).

Site and soil	Horizon	pH[a]	CEC[b]	%BS[c]	Fe	Al[d]	Total S	SO_4^{2-}-S[e]
			(meq/100g)		----- %	-----	----- mg/kg	-----
Walker Branch								
Chestnut oak (Fullerton)	A1	4.6	6.2	61	0.94	0.39	243	7
	B2	4.8	13.6	9	2.49	0.33	129	57
Yellow poplar (Tarklin)	A1	4.6	7.6	74	0.51	0.15	204	10
	B2	4.7	3.9	34	0.75	0.17	144	14
Camp Branch								
Mixed oak (Lily)	A1	4.2	3.5	12	0.72	0.24	204	43
	B2	4.6	3.9	16	1.60	0.30	179	64
Thompson								
Red alder (Alderwood)	A1	4.6	10.0	15	1.31	0.84	977	10
	B2	5.2	5.5	19	1.00	0.42	295	30
Douglas fir (Alderwood)	A1	5.0	12.3	27	1.30	0.80	304	13
	B2	5.2	5.2	10	1.06	0.37	271	40

[a] 1:1 soil:water ratio.
[b] 1m-NH_4Cl extraction.
[c] Percent base saturation.
[d] Citrate-dithionite extraction.
[e] NaH_2PO_4 extraction.

with differences in citrate-dithionite iron and base saturation on Walker Branch. Similarly, the differences in sulfate adsorption capacity between the Alderwood A horizons are consistent with differences in pH. In a related study, Johnson, et al. (1986) noted a relationship between oxalate-extractable iron and aluminum and both sulfate and phosphate adsorption capacity in the Fullerton, Tarklin, and both Alderwood soils. Further research is needed to determine the soil properties lending Lily soils their high sulfate adsorption capacities.

In any event, the differences in SO_4^{2-} content between the Fullerton (chestnut oak) and Tarklin (yellow poplar) soils are apparently a result of differences in SO_4^{2-} adsorption capacity (Figures 1 and 4), which in turn are due to differences in iron + aluminum oxide content (Johnson et al., 1981). The differences in SO_4^{2-} content between the Alderwood (Douglas-fir) and Alderwood (red alder) soils are more likely due to differences in nitrogen status because the Alderwood (red alder) soil had a greater SO_4^{2-} adsorption capacity yet a lower proportion of total S as SO_4^{2-}-S (6%) than the adjacent Douglas-fir soil (11%) (Figure 4).

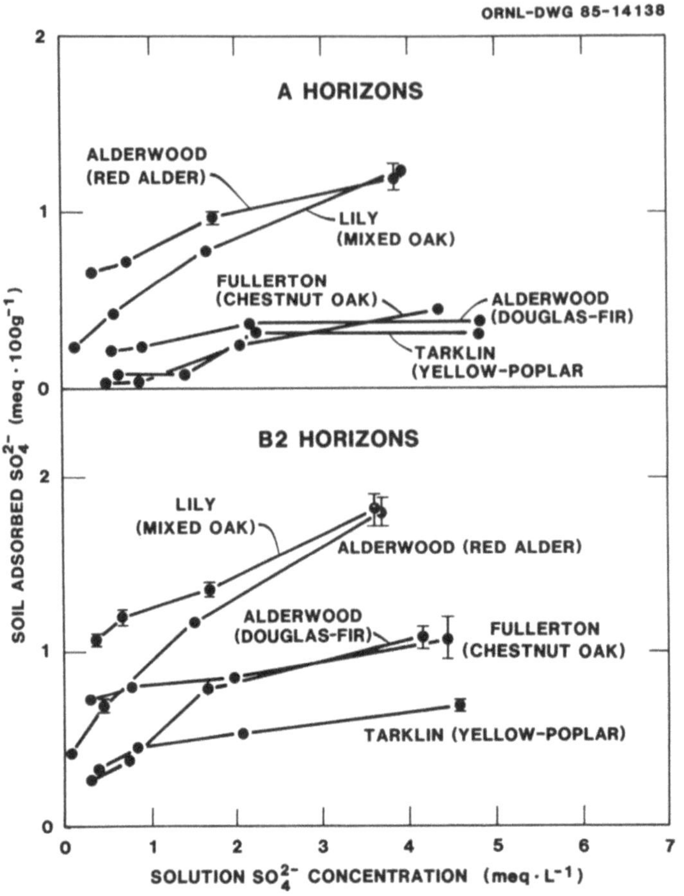

Figure 4. Soil sulfate adsorption isotherms for the five study sites.

4.2. Sulfur Fluxes

At first glance (Figures 1-3), it would appear that only the mixed oak site at Camp Branch, Tennessee, and the red alder site at Thompson, Washington, show a (slight) net ecosystem sulfur retention, whereas the other sites show net ecosystem losses of sulfur (chestnut oak and yellow poplar at Walker Branch, Tennessee, and Douglas fir at Thompson, Washington). However, independent estimates of atmospheric input to the chestnut oak and yellow poplar sites at Walker Branch (Lovett et al., 1985) and the mixed oak site at Camp Branch (Kelly 1984b) show that bulk precipitation underestimates total (wet plus dry) sulfur input by about 50%. The fact that bulk precipitation gives a low estimate of total sulfur input is evident when comparing foliage sulfur content with net sulfur fluxes from the crown (i.e., leaf litterfall plus crownwash + stemflow). Dividing net sulfur flux content by foliage sulfur content from the crown gives the number of times the foliar sulfur pool would have to turn over each year to account for net sulfur fluxes from the crown if net sulfur flux from the crown was assumed to be part of the internal sulfur cycle (i.e., as foliar leaching). For the chestnut oak, yellow poplar, and mixed oak stands, foliage sulfur contents equal 8.4, 7.5, and 2.9 $kg \bullet ha^{-1}$; whereas net sulfur fluxes are 22.9, 15.9, and 9.0 $kg \bullet ha \bullet yr^{-1}$ giving foliage sulfur turnover rates of 2.7, 2.1, and 3.1 per year (Figures 1 and 2). No other nutrient studied shows such an unrealistically high cycling rate, and thus, most net SO_4^{2-} flux from the crown can be assumed to be due to dry deposition at these sites. At the Washington sites the situation is far less clear because the Douglas fir stand retains 3 to 4 years of needles, confounding somewhat the foliar turnover calculation. It is noteworthy that in the red alder stand, the foliar sulfur turns over about once a year, while no net foliar leaching by crownwash can be observed (Figure 3). This implies either a lack of dry deposition or rapid uptake of dry-deposited sulfur and little internal SO_4^{2-} leaching. The latter is consistent with the observation that the red alder stand has virtually no foliar SO_4^{2-} -S pool (Figure 3). This is no doubt due to the enriched nitrogen status at the red alder site because increased nitrogen status would lead to synthesis of S-bearing proteins. By contrast, the adjacent Douglas fir stand has fully 42% of its foliar sulfur in SO_4^{2-} form (Figure 3). The Tennessee sites have only 24 to 33% of their foliar sulfur as SO_4^{2-}, a somewhat surprising result since sulfur inputs are higher than those at the Washington sites (Figures 1-2).

Assuming only one turnover of foliage sulfur per year, one can make an indirect estimate of total ecosystem sulfur input (I) from throughfall TF, stemflow SF, foliage content FC, and leaf litterfall LL at the deciduous forest sites (Meiwes et al., 1981). Dry-deposited S that is incorporated into foliage will be accounted for either in throughfall or litterfall in this budget unless it is translocated back into woody tissues prior to leaf senescence. We assume translocation is negligible, and also that the source of foliage leaching is the leaf itself rather than the xylem stream. Given these assumptions, foliar leaching FL must equal FC minus LL (FL = FC LL).

Total input I can then be assumed to be TF plus SF minus FL; that is,

$$I = (TF + SF) - FL$$

$$= (TF + SF) - (FC - LL).$$

Using these assumptions, total sulfur inputs at the chestnut oak, yellow poplar, mixed oak, and red alder sites are 30.5, 24.4, 16.0, and 9.6 $kg \bullet ha^{-1} \bullet yr^{-1}$ (Figures 1-3). Given these larger, more realistic estimates of sulfur input, it appears that all deciduous sites except the yellow poplar site retain a substantial amount (40 to 55%) of incoming sulfur. We have not estimated the foliage leaching from the Douglas fir site, and, thus, total input is not known; however, it would be < 10.3 $kg \cdot ha^{-1} \cdot yr^{-1}$ (i.e., the sum of SF plus TF).

A major assumption in this calculation is that foliage leaching occurs from the leaf itself rather than the xylem stream. If the xylem stream were a major source, foliage content could indeed turn over more than once per year, and, therefore, our input estimates would be overinflated.

Atmospheric sulfur inputs greatly exceed net annual sulfur increment in wood at all five sites (Figures 1-3). Aboveground sulfur uptake accounts for a significant proportion of atmospheric sulfur input, but most of this uptake is recycled in litterfall. Thus, the retention of sulfur in the chestnut oak, scarlet oak, and red alder sites appears to be occurring in the soil rather than by vegetation uptake and accumulation. Soil sulfate adsorption is one likely mechanism for sulfur accumulation in these ecosystems.

The extent to which a site accumulates SO_4^{2-} (by soil adsorption) from atmospheric deposition depends on both the SO_4^{2-} adsorption capacity of the soil (or alternatively, the slope of its adsorption isotherm) and the input SO_4^{2-} concentration. Thus, the differences in SO_4^{2-} accumulation between the red alder (accumulating SO_4^{2-}) and the Douglas-fir (not accumulating) sites are consistent with differences in soil SO_4^{2-} adsorption properties (Figures 3 and 4). Similarly, the differences between the chestnut oak (accumulating SO_4^{2-}) and the yellow poplar (not accumulating) sites are consistent with differences in soil SO_4^{2-} adsorption properties in the B horizon, at least at the low concentration ranges (Figures 1 and 4). Laboratory SO_4^{2-} adsorption isotherms cannot be directly compared with ecosystem sulfur retention in the chestnut oak and Douglas-fir sites, for example, because they have very different sulfur deposition histories.

5. SUMMARY AND CONCLUSIONS

The first of the two hypotheses tested in this study--that the older, more highly weathered Tennessee soils would adsorb more sulfate than the younger, less-weathered Washington soils--was not supported by the results of laboratory soil sulfate adsorption studies. The sulfate adsorption capacities were greatest in the Lily (mixed oak, Camp Branch, Tennessee) and Alderwood (red alder, Thompson, Washington) which are greater than Alderwood (Douglas-fir, Thompson, Washington) and Fullerton (chestnut oak, Walker Branch, Tennessee) which are greater than Tarklin (yellow poplar, Walker Branch, Tennessee). There were no consistent relationships between soil sulfate adsorption capacity and net ecosystem retention of atmospherically deposited sulfur when all sites were compared, but within a given location the soil sulfate adsorption capacities were consistent with ecosystem sulfur retention (i.e., the chestnut oak site had greater sulfate adsorption and greater ecosystem sulfur retention than the yellow poplar site at Walker Branch, and the same was true of the red alder vs Douglas-fir sites at

Thompson). Similarly, there were some relationships among certain soil properties (pH, base saturation, iron and aluminum hydrous oxides) and sulfate adsorption capacity within a location, but these relationships were not consistent when all soils were considered together.

The inconsistencies between soil sulfate adsorption capacity and ecosystem sulfur retention among all sites are thought to be due to differences in sulfur deposition histories (which do not exist within a given location such as Walker Branch or Thompson), which in turn offset the degree to which these soils have come into equilibrium with current atmospheric sulfur inputs.

The second hypothesis tested in this study - that the enhanced nitrogen status of the red alder stand would cause greater retention of sulfur in organic forms than the other nitrogen-deficient stands - was supported by these results. The red alder stand had lower vegetation and soil sulfate content than any of the other stands. Foliar sulfate levels were trace (indicating potential sulfur deficiency) in the red alder stand, but soil sulfate pools appeared quite adequate to supply tree sulfur requirements.

A nutrient-cycling approach (comparing foliage sulfur pools with sulfur return via litterfall and throughfall) allowed rough estimates of dry deposition of sulfur in the deciduous forest sites. These estimates indicate that about one-half of total sulfur input (24.4 to 30.5 $kg \cdot ha^{-1} \cdot yr^{-1}$) was due to dry deposition in the two Walker Branch, Tennessee, sites (a figure close to independent estimates by Lovett et al., 1985, at that site); about 40% of total deposition (16 $kg \cdot ha^{-1} \cdot yr^{-1}$) was dry at the Camp Branch, Tennessee, site; and about 13% of total deposition (9.6 $kg \cdot ha^{-1} \cdot yr^{-1}$) was dry at the Thompson, Washington, site.

These studies demonstrate the insights that can be gained into the modes of sulfur input (wet vs dry) and mechanisms of ecosystem sulfur retention by taking a nutrient-cycling approach. While it is convenient to use a particular ecosystem property (such as soil sulfate adsorption) for extrapolating results to a regional scale, a basic underpinning of site-intensive nutrient-cycling studies is necessary to examine and either validate or refute assumptions made in more simplistic regional approaches.

ACKNOWLEDGMENTS

Technical assistance by D. E. Todd (ORNL), R. Gonyea (University of Wyoming) and Alan Mays (TVA) is greatly appreciated. Research sponsored by the Electric Power Research Institute (RP-1813-1) and the Office of Health and Environmental Research, U.S. Department of Energy, under Contract No. DE-AC05-84OR21400 with Martin Marietta Energy Systems, Inc. Publication No. 2705, Environmental Sciences Division, ORNL.

REFERENCES

Bettany, J.R., Saggar, S., and Stewart, J.W.B.: 1980, Soil Soc. Am. J. 44, 70.

Bertolacini, R.J., and Barney, J.E.: 1957, Anal. Chem. 29, 281.

Cole, D.W.: 1968, Water Resour. Res. 4, 1127.

Cole, D.W., and Rapp, M.: 1981, 'Elemental Cycling in Forest Ecosystems,' pp. 341-409 in Dynamic Properties of Forest Ecosystems, ed. D.E. Reichle Cambridge University Press, London.

David, M.B., Mitchell, M.J., and Nakas, J.P.: 1982, Soil Sci. Soc. Am. J. 46, 847.

Galloway, J.N., Norton, S.A., and Church, M.R.: 1983, Environ. Sci. Technol. 17, 541.

Gholz, H.L., Grier, C.C., Campbell, A.G., and Brown, A.P.: 1979, Equations for Estimating Biomass and Leaf area of Plants in the Pacific Northwest, Forest Research Laboratory Paper No. 41, Oregon State University, Corvallis, Oregon.

Golterman, H.L., and Clymo, R.S., eds.: 1969, Methods for Chemical Analyses of Fresh Waters, IPB Handbook No. 8, International Biological Programme, London.

Grigal, D.F., and Goldstein, R.A.: 1971, J. Ecol. 59, 481.

Harris, W.F., Goldstein, R.A., and Henderson, G.S.: 1973, Analysis of Forest Biomass Pools, Annual Primary Production, and Turnover of Biomass for a Mixed Deciduous Forest Watershed,' pp. 41-64 IUFRO Biomass Studies, Mensuration, Growth and Yield, in ed. Harold Young, University of Maine Press, Orono.

Henderson, G.S., A.E. Hunley, and Selvidge, W.J.: 1977, Nutrient Discharge from Walker Branch Watershed. pp. 307-320. In Watershed Research in Eastern North America. A Environmental Studies, ed. D. L. Correll, Smithsonian Institution, Edgewater, Maryland.

Johnson, D. W.: 1985, Biogeochemistry 1,29.

Johnson, D.W., and Cole, D.W.: 1977, Water, Air, Soil Pollut. 7, 489.

Johnson, D.W., Cole, D.W., and Gessel, S.P.: 1979, Biotropica 11, 38.

Johnson, D.W., Cole, D.W., Horng, F.W., Van Miegroet, H., and Todd, D.E.: 1981, Chemical Characteristics of Two Forested Ultisols and Two Forested Inceptisols Relevant to Anion Production and Mobility. ORNL/TM-7646, Union Carbide Corp. Nuclear Div., Oak Ridge Natl. Lab.

Johnson, D.W., Cole, D.W., Van Miegroet, H., and Horng, F.W.: 1986, Factors affecting anion movement and retention in four forest soils. Soil Sci. Soc. Amer. J. (in press).

Johnson, D.W., Henderson, G.S., Huff, D.D., Lindberg, S.E., Richter, D.D., Shriner, D.S., Todd, D.E., and Turner, J: 1982, Oecologia 54, 141.

Kelly, J.M.: 1984a, J. Environ. Qual. 13, 405.

Kelly, J.M.: 1984b, 'Sulfur Input, Output and Distribution in Two Oak Forests,' pp. 265-89 in Forest Soils and Treatment Impacts, ed. E.L. Stone University of Tennessee, Knoxville.

Kelly, J., and Lambert, M.J.: 1972, Plant Soil 37, 395.

Likens, G.E., Bormann, F.H., Pierce, R.S., Eaton, J.S., and Johnson, N.M.: 1977, Biogeochemistry of a Forested Ecosystem. Springer-Verlag, New York.

Lovett, G.M., et al.: 1985, 'The Effects of Acid Deposition on Cation Leaching from a Deciduous Forest Canopy,' Can. J. For. Res., in press.
Meiwes, K.J., and Khanna, P.K.: 1981, Plant Soil 60, 369.
Ramseur, G.S., and Kelly, J.M.: 1979, 'J. Tenn. Acad. Sci. 56(3), 99.
Richter, D.D., and Johnson, D.W.: 1983, Soil Sci. Soc. Amer. J. 42, 522.
Richter, D.D., Johnson, D.W., and Todd, D.E.: 1983, J. Environ. Qual. 12, 263.
Seip, H.M.: 1980, 'Acidification of Freshwaters: Sources and Mechanisms,' pp. 358-66 in Ecological Impacts of Acid Precipitation, eds. D. Drablos and A. Tollan, Johs Grefslie Trykkeri, Mysen, Norway.
Shriner, D.S., and Henderson, G.S.: 1978, J. Environ. Qual. 7, 392.
Swank, W.T., Fitzgerald, J.W., and Ash, J.T.: 1984, Science 223, 182.
Turner, J., Johnson, D.W., and Lambert, M.J.: 1980, Oecol. Plant. 15, 27.
Turner, J., Lambert, M.J., and Gessel, S.P.: 1977, Can. J. For. Res. 7, 476.
Van Miegroet, H: 1986, 'Role of N status and N transformations in H+ budget, cation loss, and S retention mechanisms in adjacent Douglas-fir and red alder forests.' Ph.D. Thesis, University of Washington, Seattle, WA.
Youngberg, D.T. and Dryness, C.T.: 1966, Soil Sci. Soc. Amer. Proc. 29, 182.

A QUALITY ASSURANCE PROGRAM AND QUALITY ASSESSMENT OF THE ACIDIC PRECIPITATION IN ONTARIO STUDY (APIOS) DEPOSITION MONITORING NETWORKS

W.S. Bardswick, W.H. Chan and D.B. Orr
Ministry of the Environment, Air Resources Branch,
880 Bay St., Toronto, Ontario M5S 1Z8, Canada

ABSTRACT. The Air Resources Branch of the Ontario Ministry of the Environment currently operates two deposition monitoring networks designed with the general objective of understanding wet and dry deposition across the Province of Ontario. In both the daily (used interchangeably with the term event) and cumulative (28-day) networks, both wet-only precipitation and air filter samples are collected. Some of the major components of the quality assurance (QA) program of these networks are discussed. Preliminary quality control (QC) results obtained during 1983-84 are presented.

1. INTRODUCTION

The Air Resources Branch of the Ontario Ministry of the Environment, under the auspices of the Acidic Precipitation in Ontario Study (APIOS), currently operates two deposition monitoring networks (daily and cumulative) designed with the general objective of understanding wet and dry deposition across the Province of Ontario.

The daily precipitation network is comprised of 15 wet-only monitoring sites. Precipitation is collected in disposable polyethylene/nylon sampling bags inserted into Aerochem Metrics samplers. The daily air sampling network consists of 4 monitoring sites which collect air filter samples on a daily basis using Metrex SAS8-25 sequential air samplers and modified Millipore Swinnex 2-stage filter packs employing teflon, nylon and impregnated (K_2CO_3/glycerol) Whatman-41 filters. Dry deposition rates are estimated by using the airborne concentration and appropriate deposition velocity values.

The cumulative precipitation network is comprised of 36 monitoring sites which employ M.I.C. Type "A" (formerly called Sangamo) wet-only precipitation collectors (utilizing polyethylene/nylon sampling bag inserts) and precipitation gauges. Located at 24 of these cumulative precipitation sites are Metrex AS-2 low-volume air samplers using modified Millipore Swinnex 2-stage filter packs employing Whatman 40, nylon, and impregnated (K_2CO_3/glycerol) Whatman 41 filters to collect air filter samples for determining estimates of dry deposition.

Water, Air, and Soil Pollution **30** (1986) 981–990.

The collecting and reporting of data of a known quality is of utmost importance if historical trends and proper abatement strategies are to be made based on precipitation and air sampling network data. Quality assurance (QA) therefore, has been designated as an integral part of the APIOS monitoring program. A formal QA program has been implemented to ensure that the collection and reporting of data are complete, accurate, precise and representative. A QA Plan (Bardswick, 1984) has been prepared to implement the activities within this program. Quality assurance responsibilities and quality control (QC) sampling procedures associated with each QA activity are described in detail in a QA Manual (Ontario Ministry of the Environment, 1985a).

The focus of this paper is on field QA and QC procedures. Discussion of laboratory and data management QA/QC can be found in other published reports (Ontario Ministry of the Environment, 1983, 1985b).

2. QUALITY ASSURANCE PROGRAM AND RESULTS

2.1. Network Representativeness

Monitoring sites are documented in detail in terms of location (coordinates, topographic map), instrumentation, site logistics (road access, power), local pollutant sources, topography, land use, human and vehicular activity; each site is photographed from four compass points. Well-established siting criteria and site documentation were used to evaluate all sites to determine their representativeness for monitoring (Ontario Ministry of the Environment, 1985a). This evaluation resulted in sites qualitatively ranked in one of the following categories: Excellent, Good, Marginal or Poor. Sites ranked Marginal or Poor were either improved, or removed from the network. Data collected from sites that are considered marginal are closely scrutinized by network and database personnel.

Some monitoring sites have been relocated since network inception. To ensure that existing sites continue to meet major siting criteria, a re-evaluation of every site is undertaken every two years. Any site change occurring between these evaluations is documented using a notification of site change form. An evaluation of this change on data quality is carried out and documented.

2.2. Sample Collection and Instrumentation

Instrumentation is one of the most important factors in meeting the completeness objective in the monitoring program. All precipitation collectors, air samplers, and other monitoring equipment are tested for acceptance prior to field use. Equipment is calibrated, installed and maintained by following standard guidelines. Preventive maintenance and routine calibration procedures are documented and scheduled for all equipment. Instrumentation log books are maintained to provide records of all calibrations, preventive maintenance or required repairs. Supplies, parts and spare equipment are maintained both at the head office and field offices. Inventories and supply requests are carried out every six months to ensure that sufficient supplies and spare equipment are available for both networks.

The monitoring program relies heavily upon on-site operators to collect samples. Technical staff working out of regional offices are responsible for maintaining the instrumentation at each site and ensuring that sample collections are carried out properly by site operators. Training of site operators and technicians is therefore of utmost importance to ensure that samples are collected and processed for analyses following standard operating procedures. A formalized training program for operators and technicians has been implemented. A Technical and Operating Manual (Bardswick, 1983) documents all network procedures in detail from initial sample collection to final data validation and reporting.

Various QC activities have been implemented into the networks to estimate the precision and accuracy of the network's sample collection techniques. Table I summarizes some of the routine QC samples and procedures presently implemented in both networks.

2.2.1. Co-located Sampling

Overall sampling precision is determined by co-located instrumentation at selected monitoring sites. Discussions of co-location precipitation chemistry results over 28 day and daily sampling intervals are presented in detail in papers by Chan et al., (1985a,b). Reproducibility (100 - median absolute % difference) was found to be better than 88% for the acid-base parameters (total H^+, free H^+, $SO_4^=$, NO_3^-) for both the daily and cumulative precipitation networks.

In the cumulative and daily air sampling networks, $SO_4^=$, NO_3^- and HNO_3 reproducibility is better than 90% (based on 27 cumulative and 115 daily co-located samples) while SO_2 reproducibility is 75% or better. In the daily air sampling network, NH_4^+ reproducibility is better than 94%.

2.2.2. Triplicate Split Samples - Daily Precipitation Network

In the daily precipitation network, precipitation collected in bags is transferred to a polystyrene bottle (weighed by a digital balance to determine volume) with any excess volume decanted into a graduated volumetric cylinder and measured. Once a month, on average, a large precipitation sample is poured into three polystyrene bottles and submitted for triplicate measurements.

Maximum concentration percentage differences were calculated by dividing the maximum difference between the three triplicate results by the average of the three triplicate samples. The medians of these maximum percentage differences are presented in Table II.

Reproducibility (100 - median maximum % difference) calculated from triplicate sampling is better than 95% for field and laboratory pH, total H^+, $SO_4^=$ and NO_3^-. The ions, NH_4^+, K^+ and Mg^{++} are reproducible to 93% or better. The least reproducible ions are Cl^-, Na^+ and Ca^{++} with reproducibilities of 89%, 88% and 81% respectively.

Table II also shows the differences of reproducibility between co-

TABLE I. ROUTINE QC SAMPLES AND PROCEDURES

QC Sample/Procedure	QC Parameter Measured	Network Type/ Frequency
Co-located Sampling	Overall Precision	Daily Precip.-4 sites Cum. Precip.- 5 sites Daily Air - 1 site Cum. Air - 5 sites
Triplicate Split Sampling	Analytical Precision & Sample Handling	Daily Precip.- 1 event/site/month
Field Blank	Accuracy (field component)	Daily Precip.- 1 bag/site/month
Composite Precip. Handling	Accuracy (field component)	Daily Precip.- quarterly/site Cum. Precip. - semi-annual/site
Bag Blanks (from field offices)	Accuracy (field component)	Daily Precip.- quarterly/region Cum. Precip.- quarterly/region
Bag Procurement	Accuracy (container contamination)	9 bags per sample bag type (3 DDW, 3 DDW + HNO_3, composite)
Filter Pack Handling Blanks	Accuracy (field component)	Daily Air- quarterly/region Cum. Air- quarterly/region
Filter Blanks	Accuracy (filter contamination)	5 filters per filter type for every batch

Cum. - Cumulative
DDW - Deionized distilled water

TABLE II. RESULTS OF TRIPLICATE SAMPLING IN THE DAILY PRECIPITATION NETWORK

Parameter	n*	Median Maximum % Difference	Precision Comparison**(%)
Field pH	84	.51	-
Lab pH	121	.25	-
H_t^+	119	4.46	.05
$SO_4^=$	119	1.65	1.74
NO_3^-	121	2.59	.40
Ca^{++}	113	18.75	-5.36
Cl^-	108	10.71	1.10
Mg^{++}	115	6.52	3.53
K^+	105	6.12	16.38
Na^+	102	12.00	3.48
NH_4^+	120	5.36	.86

* n is the number of samples
** Median Absolute % Difference (Co-located) - Median Maximum % Difference (Triplicates)

located and triplicate sampling. Since co-located sampling is an overall estimate of the reproducibility of the total measurement system, triplicate reproducibilities should be better than co-located reproducibilities. These differences between the two types of samples (co-located and triplicate) can be attributed primarily to the precipitation collector and sampling bag variabilities and possibly a bias due to precipitation event volumes (since triplicates are only submitted if sufficient sample volume is available while co-located samples are always submitted, regardless of volume).

The precision differences (reproducibilites) between co-located and triplicate sampling ranged from less than ½ % (total H^+) to 16 % (K^+), while Ca^{++} showed a 5 % loss of precision. The loss of Ca^{++} reproducibility is speculated to have originated from particles in the samples. The non-homogeneity of large volume samples split into sub-samples may result in some of the split samples having more soluble Ca^{++}

TABLE III. DAILY PRECIPITATION NETWORK FIELD BLANK RESULTS, 1983-84

	Parameter							
	$SO_4^{=}$	NO_3^-	Ca^{++}	Cl^-	Mg^{++}	K^+	Na^+	NH_4^+
n	61	61	61	61	60	61	61	61
Results < w	53	56	38	45	57	47	43	49
Results < T	0	2	14	5	1	6	9	0
Measurable Results*	8	3	9	11	2	8	9	12
Mean (mg L^{-1})	0.15	0.04	0.02	0.06	0.015	0.019	0.029	0.016
Stand. Dev.	0.07	0.01	0.03	0.04	0	0.009	0.023	0.010
Mean Field Concentration Southeastern Ontario Sites (1982) mg L^{-1}	3.29	0.68	0.25	0.23	0.040	0.055	0.105	0.396

* - includes results with < T codings (for calculation of the mean, a value equal to ½ of the detection limit was used)
<W - "zero", value reported is minimum measurable amount
<T - less than detection limit

particulate matter than others. There was no similar loss of reproducibilty for Mg^{++} however. The increase (16%) in reproducibility of K^+ may be a result of sampling bag desorption (Chan et al., 1985a, b).

Total H^+, $SO_4^{=}$, and NO_3^- show small reproducibility changes (less than 2%) between co-located and triplicate results indicating that the precipitation collector and sampling bag variabilities are small for these parameters. This conclusion, however, is based on large volume samples only. Smaller volume event sample reproducibilities are probably influenced much more by the precipitation collector.

2.2.3. Field Blanks - Daily Precipitation Network

Since sample bags in the daily precipitation network are only removed after a precipitation sample has been collected, dry bags (bags that have been inserted into the collector but not exposed to precipitation) are randomly submitted to the laboratory as field blanks. The length of time of the field blanks in the precipitation collector ranged from 1 to 7 days with an average time of 3 days. The analyses of these field blanks are used to quantify dry deposition that is collected in daily wet deposition samples during days of no precipitation. Initially, 1000 mL of distilled deionized water was added to the dry bag, however, 250 mL was implemented as a more representative dilution for daily samples.

Table III shows the results of 61 field blank analyses from 14 monitoring sites during 1983-84. Approximately one half of these dry bag results used 250 mL as the dilution volume. Dry deposition of $SO_4^{=}$, NO_3^-, Mg^{++}, and NH_4^+ is negligible. Approximately 15% of all field blanks do show minor contributions of Cl^-, K^+, Ca^{++}, and Na^+. These contributions are usually small relative to field sample concentrations.

2.2.4 Composite Precipitation Handling - Daily and Cumulative Precipitation Networks

Composite precipitation samples of a known chemical concentration are handled by site operators in both precipitation networks (quarterly in the daily network and semi-annually in the cumulative network). Aliquots (250 mL) of a composite solution were prepared and shipped out to monitoring sites. Site operators poured an aliquot into a new sample bag in the collector and immediately removed the sample following standard operating procedures. Three aliquots were initially analyzed prior to shipment to the monitoring sites and three were stored and later analysed with the "handled" aliquots from the field sites. Standard network operating procedures were followed at all times in terms of sample handling, storage and shipment to the laboratory.

The results of three of these composite handling submissions are shown in Table IV, for both the cumulative and daily networks. For each submission at least 10 monitoring sites participated.

There are noticeable concentration changes for some parameters in these submissions (particularly Mg^{++}, Na^+ and K^+); however, the changes occur randomly for both "handled" and final laboratory samples. Because of some of the laboratory initial and final concentration inconsistencies, the use of composites as precipitation standards needs to be investigated further (especially long term storage and stability).

2.2.5. Bag Blanks - Daily and Cumulative Precipitation Networks

Polyethylene/nylon laminated bags are used for the collection of precipitation in both networks. The integrity of newly purchased sample bags is always checked by submitting random bags for laboratory analyses. However, to ensure that sampling bags maintain their integrity once shipped and stored in field offices, blank bags are chosen randomly from every regional office and analysed after a dilution with 250 mL of deionized distilled water.

During 1984, 29 sample bags were submitted for blank analysis. For $SO_4^{=}$, NO_3^-, Ca^{++}, Mg^{++}, K^+ and NH_4^+, all blank bag results were at or

TABLE IV. COMPOSITE PRECIPITATION HANDLING, 1983-84

$mg\ L^{-1}$	Sub. #	Lab Initial	Daily Network Mean (S.D.) Handling	Lab Final	Lab Initial	Cumulative Network Mean (S.D.) Handling	Lab Final
Lab pH	1	4.14	4.10 (.03)	4.06	4.41	4.42 (.06)	4.43
	2	4.32	4.41 (.05)	4.18	3.27	3.29 (.01)	3.27
	3	4.37	4.39 (.08)	4.36	4.37	4.35 (.03)	4.34
H_t^+	1	1.00	.930 (.025)	.979	.065	.065 (.010)	.061
	2	.067	.057 (.005)	.078	.572	.575 (.008)	.595
	3	.085	.060 (.006)	.067	.065	.065 (.003)	.064
$SO_4^=$	1	2.80	2.79 (.05)	2.80	2.45	2.55 (.03)	2.43
	2	3.30	3.36 (.23)	3.30	3.40	3.51 (.01)	3.48
	3	2.65	2.79 (.29)	2.58	2.42	2.44 (.02)	2.43
NO_3^-	1	.56	.57 (.01)	.48	.35	.37 (.01)	.33
	2	.35	.41 (.02)	.33	.51	.49 (.003)	.41
	3	.34	.34 (.02)	.25	.06	.09 (.03)	.01
Cl^-	1	.21	.21 (.02)	.19	.24	.22 (.01)	.18
	2	.16	.17 (.05)	.15	.19	.17 (.01)	.17
	3	.23	.24 (.06)	.22	.13	.14 (.02)	.14
Ca^{++}	1	.15	.18 (.05)	.16	.29	.25 (.02)	.25
	2	.48	.46 (.04)	.47	.32	.34 (.05)	.34
	3	.33	.32 (.06)	.29	.17	.25 (.07)	.20
Mg^{++}	1	.022	.015 (.006)	.013	.040	.032 (.004)	.038
	2	.075	.090 (.008)	.078	.045	.040 (.015)	.028
	3	.045	.049 (.012)	.035	.005	.024 (.013)	.005
Na^+	1	.112	.128 (.007)	.125	.248	.257 (.013)	.243
	2	.023	.038 (.010)	.030	.080	.076 (.011)	.072
	3	.185	.195 (.024)	.170	.010	.028 (.009)	.010
K^+	1	.015	.012 (.007)	.010	.045	.043 (.009)	.028
	2	.035	.033 (.009)	.008	.090	.089 (.024)	.001
	3	.040	.047 (.017)	.015	.005	.005 (0)	.005

Note: All initial and final laboratory sample results are means based on 3 aliquots; except for the 1st cumulative and 3rd daily submission where the mean is only based on 2 aliquots. NH_4^+ results are omitted because of chemical stability problems.

S.D.- Standard Deviation

less than the detection limits for all parameters. Blank bag results for Na^+ and Cl^- were at or less than the detection limit for 22 out of 29 results, with the remaining 7 result concentrations slightly above the detection limits.

2.2.6. Filter Pack Handling Blanks - Daily and Cumulative Air Networks

Blank filters from every batch purchased (teflon, nylon, Whatman-40) or prepared (impregnated Whatman-41) are submitted for blank analyses prior to deployment in the air sampling field programs. Three filters of each type from every lot number are analysed. Handling blanks are also submitted on a quarterly basis from each region. One loaded filter pack of each network type is stored in a Whirlpak bag at a monitoring site for a representative time period (cumulative -28 days, daily - 7 days). The pack is unloaded and shipped for analyses following standard operating procedures. Results of these handling blanks (15 blanks during 1984) have all been at or below laboratory detection limits.

2.3. External and Internal Audits

To independently assess the quality of data collected in the monitoring networks, external audits have been carred out in the monitoring program. The overall objective of these audits has been to verify that the program is effectively producing data that are complete, accurate, precise and representative. To meet this objective, auditors have evaluated the following components: 1) representativeness of sites 2) instrumentation and operator sample collection procedures at sites 3) data management system, and 4) QA/QC program to ensure it meets stated objectives.

Thus far, three external audits have been carried out with reports describing the results of site and operator assessments, laboratory assessments, and data management reviews. Weaknesses, strengths, and recommendations are also reported (Concord Scientific, 1983, 1984 a,b).

Internal audits have been introduced in 1985 to verify that the networks continue to operate according to network protocols. Each field office is audited annually to ensure that sites remain representative, and that operators and technicians continue to follow standard operating procedures.

3. SUMMARY

The QA program of the APIOS monitoring networks was initiated in 1982 and became fully operational by the end of 1983. Preliminary QC results from this program have indicated the following:

(1) Overall reproducibility in both the daily and 28 day precipitation networks determined by co-located sampling is better than 88% for total H^+, free H^+, $SO_4^=$ and NO_3^-. Reproducibility in both the daily and 28 day air sampling networks is better than 90% for $SO_4^=$, NO_3^- and HNO_3, and 75% for SO_2. In the daily air sampling network, NH_4^{3+} reproducibility is better than 94%.

(2) Reproducibility from split samples in the daily precipitation network is better than 95% for pH, total H^+, $SO_4^=$, and NO_3^-.
(3) Field blank analyses in the daily precipitation network indicated no contribution from dry deposition during days with no precipitation for $SO_4^=$, NO_3^-, Mg^{++} and NH_4^+; while Cl^-, K^+, Ca^{++} and Na^+ showed some minor contribution for approximately 15% of all field blank samples.
(4) Composite precipitation handling results indicated inconsistent agreement between the initial and final lab values for some parameters suggesting sample degradation. The use of composites as precipitation standards needs further investigation.
(5) Bag blank results in both precipitation networks were usually at or less than detection limits for all parameters. Cl^- and Na^+ results indicated some minor contributions on an irregular basis.
(6) Filter pack handling results were at or below detection limits for all samples.

REFERENCES

Bardswick, W.S.: 1983 "APIOS Technical and Operating Manual", Ontario Ministry of the Environment, Air Resources Branch, April, 1983.

Bardswick, W.S.: 1984, "Quality Assurance Plan - APIOS Deposition Monitoring Program", Ontario Ministry of the Environment Report #ARB-051-85-AQM.

Chan, W.H., Tang, A.J., Orr, D.B., Bardswick, W.S., and Lusis, M.A.: 1985a, "An Evaluation of the Precision and Accuracy of Daily Sampling for Wet Deposition of Trace Contaminants", a paper presented at the International Symposium on Acidic Precipitation, Muskoka, Ontario, September 15-20, 1985.

Chan, W.H., Tang, A.J., Orr, D.B., Bardswick, W.S., and Lusis, M.A.: 1985b, "An Evaluation of the Precision and Accuracy of Cumulative Sampling for Wet Deposition of Trace Contaminants", a paper presented at the International Symposium on Acidic Precipitation, Muskoka, Ontario, September 15-20, 1985.

Concord Scientific Corporation: 1983, "A Performance and Systems Audit of the APIOS Monitoring Networks". Ontario Ministry of the Environment Report #ARB-69-83-ARSP.

Concord Scientific Corporation: 1984a, "The Phase I Report of the Audit of the APIOS Deposition Monitoring Networks". Ontario Ministry of the Environment Report # ARB-228-84-ARSP.

Concord Scientific Corporation: 1984b, " The Phase II Report of the Audit of the APIOS Deposition Monitoring Networks". Ontario Ministry of the Environment Report # ARB-084-85-AQM.

Ontario Ministry of Environment: 1983, "Handbook of Analytical Methods for Environmental Samples - Volumes 1 and 2", Laboratory Services Branch.

Ontario Ministry of the Environment: 1985a, "Quality Assurance Manual - APIOS Deposition Monitoring Networks". Ontario Ministry of the Environment Report #ARB-051-85-AQM.

Ontario Ministry of the Environment: 1985b, "1982-84 Performance Report", Water Quality Section - Volumes 1 and 2, Laboratory Services Branch.

CONTEMPORARY CHEMISTRY OF NOVA SCOTIAN LAKES

J. K. Underwood*
Nova Scotia Department of the Environment
P.O. Box 2107
Halifax, Nova Scotia, Canada B3J 3B7

J. G. Ogden, III
Department of Biology
Dalhousie University
Halifax, Nova Scotia, Canada B3H 4J1

D. L. Smith
Nova Scotia Department of the Environment
P.O. Box 2107
Halifax, Nova Scotia, Canada B3J 3B7

ABSTRACT. Water samples were collected from 234 Nova Scotian lakes during the summers of 1981-1984. Samples from near-surface and mid-depth were composited to yield a single descriptive analysis for each lake. Median pH for the lakes was 5.37 (arithmetic mean 5.7), and ranged from 4.4 to 7.7. Conductivities <133 μS cm^{-1} were found for 96% of the lakes with higher values for lakes influenced by road salt, proximity to the sea, or geology. The most dilute lake (19.8 μS cm^{-1}) was approximately the salinity of rain water. Sodium, the most abundant cation (216 μeq L^{-1}) was 2.5x more concentrated than Ca. Among anions, Cl (228 μeq L^{-1}) exceeded SO_4 by 2.7 times. Alkalinity averaged 48 μeq L^{-1}, and 16% of the lakes (n = 37) had no measurable alkalinity at a detection limit of 2 μeq L^{-1}. Non-marine fractions of ions were calculated from the ratio of Cl to SO_4 in standard seawater, and ranged from 9.8% for Na to 100% for eight other ions. As with Na, Mg is strongly influenced by sea spray, while K is apparently influenced almost equally by marine and terrestrial sources. Sulfate and Ca are both strongly influenced by terrestrial sources. Color in the study lakes was as high as 300 T.C.U., and the implicit role of organic anion (ca. 50 μeq L^{-1}) in the ion sum is reviewed. Interactions between selected ion pairs are reviewed to examine the role of non-marine ions (e.g. SO_4^*, Ca^*, K^*) in lake chemistry. Results for Al, Fe, Mn, and alkalinity are also considered.

* Author for all correspondence.

Water, Air, and Soil Pollution **30** (1986) 991–999.

1. INTRODUCTION

Published reports of acid stress on riparian systems in Nova Scotia (Watt, 1981; NRCC, 1981), confirm the loss of salmonids, especially in the southwestern part of the province. There are no previous synoptic studies of the conditions in lakes throughout the province. There is chemical evidence of acidification of a number of lakes around Halifax (Watt et al., 1979), and Vaughan et al. (1982, 1983) describe diatom evidence of acidification in Halifax lakes and in a few colorless rural lakes.

As part of a continuing Lake Inventory program being conducted by the Nova Scotia Department of the Environment and the Department of Lands and Forests, 234 lakes were spot-sampled for water chemistry during the summer months of 1981 to 1984. The lakes were also surveyed for fish species using 23-hr net sets (Smith et al., 1986). Location of the lakes sampled in this study are shown in Figure 1.

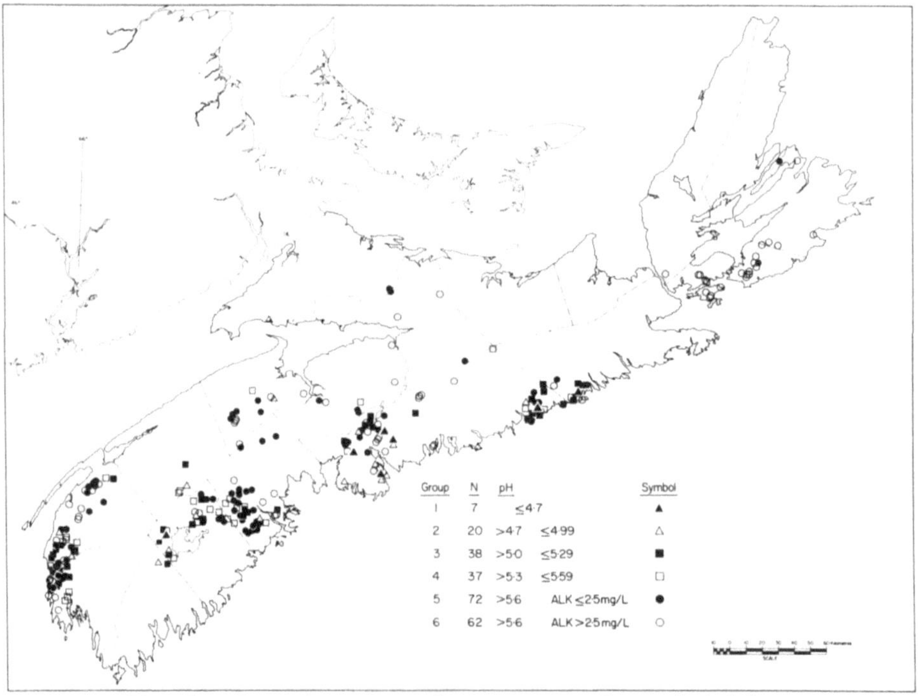

Fig. 1. Location map and pH groups of sampled lakes.

2. METHODS

One-litre lake water chemistry samples were obtained by Kemmerer sampler from a boat at 0.5m and mid-depth at each lake. Samples were analyzed separately and averaged to provide a mean estimate for each lake. Analyses were conducted at the Environmental Chemistry Laboratory of the Nova Scotia Department of Health. Atomic absorption was used for Na, K, Ca, and Mg (the latter two following Lanthanum correction). Colorimetric AutoAnalyzer techniques were used for most other analyses. Nitrate Nitrogen was reduced in Cd columns, and converted to azo dye; NH_4 was treated with alkaline phenol and sodium hypochlorite; alkalinity was determined by titration of a known volume of potassium acid phthalate buffer in a solution containing bromophenol blue; mercuric thiocyanate indicator was used in Cl measurements. DOC was by UV digestion and phenolphthalein indicator; PO_4 determination followed reaction with ammonium molybdate/ascorbic acid; and SO_4 was performed by methylthymol blue indicator on aliquots subjected to UV and H_2O_2 oxidation. Hellige color comparator discs were used for color measurements, and pH was determined with a Radiometer 026 using a combination electrode. Organic anion [A^-] was calculated from DOC and pH according to Oliver _et al._ (1983). A graphite furnace was used for Fe, Al, and Mn. Speciation was not attempted for these three metals, and ion budgets were calculated by assuming complete ionization. We are aware of the literature on metal speciation, but we decided that speculative assignment of some fractional values for ion species was as likely to introduce unknown errors as assuming complete ionization. Furthermore, despite including total Fe, Al, and Mn, ion balances reveal a small cation deficiency (C:A = 0.980). Detailed methods and validation procedures are discussed in Underwood (1984).

The lakes reflect influences of a variety of geologic and edaphic conditions, so the data set was subdivided into six categories, based primarily on the pH tolerance of Atlantic salmon (cf. Watt, 1981). Other classification schemes are possible, but we selected a technique based on known biological responses in Nova Scotia surface waters. The lake groupings and pH ranges are shown in Figure 1. Water chemistry data were converted to microequivalents and analyzed by SPSS (Nie _et al._, 1975).

3. RESULTS AND DISCUSSION

Mean values for cations and anions for all lakes are summarized in Table I, which includes determination of non-marine contribution calculated from ion ratios relative to Cl in seawater. Other chemical and physical data are also reviewed. The common occurrence (in Nova Scotia) of colored waters is of note (mean value 44.5 T.C.U.). Regressions of Color and DOC (mg L^{-1}) yielded a strong relationship as shown below:

$$DOC = 0.005\ Color + 3.26\ (r^2 = 0.72,\ n=234).$$

Sea salt influence in both precipitation and lakes in Nova Scotia is generally the major source for total ions present (Underwood, 1984). For lakes described in Table I, marine contributions account for 63% of the ions. Marine influence on individual ions varies considerably, from negligible for H^+, NH_4, Fe, Al, Mn, A^-, Alk, and NO_3, to over 90% for Na and Cl. Sulfate is mostly non-marine (28%

Table I. Summary water chemistry of 234 Nova Scotian lakes. Ionic units are in μeq L^{-1}.

Valence	Ion	Mean	S.D.	Median	Min.	Max.	Non-marine conc.	%
1+	Na	216.	128.	178.	69.6	740.	21.2	9.8
1+	K	9.09	5.28	7.9	0	33.5	4.8	52.8
2+	Ca	85.5	83.1	62.1	7.48	703.	76.8	89.8
2+	Mg	60.4	30.9	52.2	13.2	197.	16.1	26.7
1+	H	4.27*	5.96	1.74**	0.02	37.2	4.27	100.
1+	NH_4	0.66	2.12	0	0	21.4	0.66	100.
2.5+	Fe	11.2	15.8	8.28	0	90.3	11.2	100.
3+	Al	11.5	10.5	4.54	0	58.9	11.5	100.
2+	Mn	1.78	2.81	0.88	0	30.4	1.78	100.
	Total Cations	400.		316.				
1?-	A-	48.5	27.6	41.4	5.8	169.	48.5	100.
2-	Alkalinity	47.0	72.0	22.0	0	519.	45.7	97.2
2-	SO_4	83.7	44.9	72.9	35.4	448.	60.3	72.
1-	NO_3	0.52	3.0	0	0	35.7	0.52	100.
1-	Cl	228.	151.	182.	56.4	790.	--	--
	Total Anions	408.		318.				
	Cation:Anion	0.980		0.994				
	pH (arithmetic)	5.7	0.57	5.76	4.4	7.7		
	Conductivity (μS cm^{-1})	46.9	23.8	41.3	19.6	131.		
	Color (T.C.U.)	44.5	54.6	25.0	0	300.		
	DOC (mg L^{-1})	5.73	3.57	4.65	0.60	22.0		
	PO_4-P (mg L^{-1})	0.001	0.001	0	0	0.013		
	Turbidity (N.T.U.)	1.05	0.60	0.93	0.14	4.70		

(Ionic units in μeq L^{-1})
* = pH 5.37
** = pH 5.76

in lakes attributed to sea spray).

Individual mean values for the 6 pH groups are shown as histograms in Figure 2. Direct comparisons between and among groups 1 to 5 are readily made as their ion sums are quite similar. Na and Cl are present in similar proportions throughout the five groups. The other monovalent cation, K, is also present in similar proportions from group to group. As acidity increases towards Group 1, however, di- and tri-valent ions show either decreases (for Ca and Mg), or increases (Al, Fe). Sulfate is almost constant in Groups 2 to 5, but shows a slight increase in Group 1. Group 1 is a small data base (n = 7), so it is possible that values for SO_4 are not representative of this lake class in the province. Alternatively, it is possible that the increase in acidity in Group 1 is a result of H_2SO_4, and not a function of A^- which shows only a very minor increase with increasing H^+. It should be noted that A^- is present in all lake classes, and contributes 10 to 15% of the anion sums. Nitrogen forms a quantitatively negligible fraction of the ion sums, and was omitted from the histograms.

In Group 6, total ions are about 2/3 more concentrated than in the other groups. It is readily apparent that Ca and alkalinity have been enriched in this group, and if concentrations are determined from the ion sums and percentage scales, SO_4 shows enrichment as well. Bedrock and soils in several areas of the province are strongly

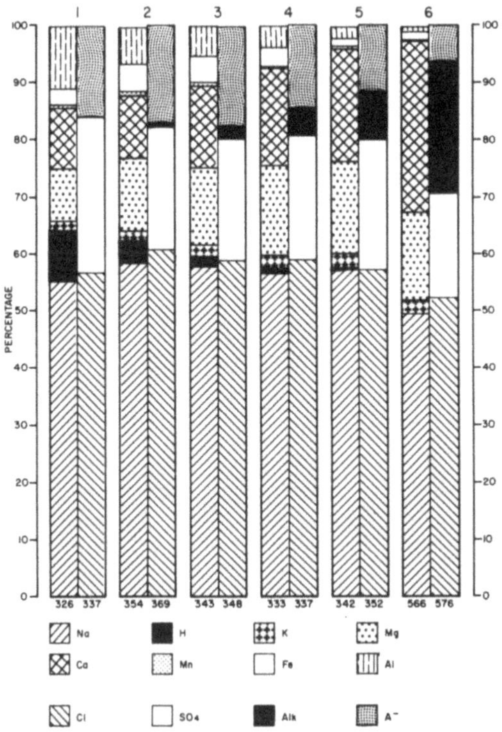

Fig. 2. Histograms of ion species in study lake groups. Cation and anion sums (μeq L^{-1}) are shown at the bottom of each histogram.

influenced by limestone and gypsum, which exist in sufficient quantities to make mineral exploitation economically viable.

Frequency distributions for selected ions are plotted in Figure 3. H^+ is biased by the pH selection criteria used to establish the groupings, and this should be noted in interpreting the results. Calcium shows an obvious shift towards lower concentrations as pH decreases. Alternatively, Al increases in concentration at the lower pH values. Mn shows a slight increase in concentration, starting in Group 3 (pH <5.3). The trend does not appear to be very strong. DOC appears to be more pronounced in Group 1 than Group 6, although A^- does not behave as strongly. Finally SO_4 shows a suggestive bimodality, with gypsum influence at high pH, and an alternative source of SO_4 at low pH. The lakes in this study were not selected according to bedrock types, and there are pyritic outcrops in a number of locations along the eastern shore of the province. The possibility of pyrite influence on some of the lakes should be borne in mind, although correlations between Fe and SO_4^* were negative in all six classes. Kerekes (1986), and Ogden, (1982), have found seasonality in SO_4

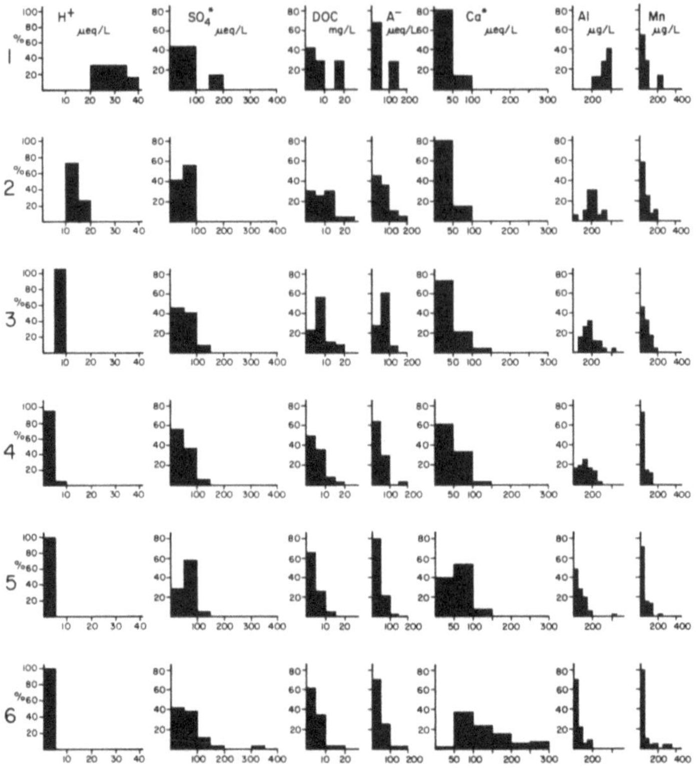

Fig. 3. Frequency distributions of selected ions in Nova Scotia lake waters.

concentrations in colored lakes, with lower than average concentrations during summer months. It should therefore be recognized that summer SO_4 concentrations of some lakes may not be representative of annual conditions.

Stepwise regression analyses were performed for several combinations of linear variables in each group. Logarithmic transformations were reviewed, but r^2 values were less consistent than among linear relationships. Table II shows multiple regression results for the lake groups, using both H^+ and SO_4^* as the dependent variables. The independent variables were selected from sets which included SO_4^*, H^+, A^-, Ca^*, Fe, Mn, and Al.

When H^+ was used as a dependent variable, regressions generally yielded low r^2 values. In the group of All lakes, H^+ was significantly influenced by Ca^*, Al, Fe, and A^-. Calcium and Al were also important in Group 6, while in Group 5, Fe was most strongly related to H^+. Manganese most strongly influenced H^+ in Group 4. No significant regressions were found for Groups 1, 2, and 3. Furthermore, no group showed a relationship between H^+ and SO_4^*.

Table II. Multiple regressions for H^+ and SO_4^* for all lakes (N = 234) and by Groups (1 to 6).

Group	n.	H^+ (all values μeq L^{-1})
All	234	$H^+ = 0.43 Al - 0.011 Ca^* - 0.06 Fe - 0.26A^- + 1.93$; $r^2 = 0.51$
1	7	Non significant
2	20	Non significant
3	38	Non significant
4	35	$H^+ = 0.29 Mn + 3.44$; $r^2 = 0.25$
5	72	$H^+ = 0.64 Fe + 1.08$; $r^2 = 0.18$
6	62	$H^+ = -0.0014 Ca^* + 0.0022 Al + 0.70$; $r^2 = 0.25$

Group	n.	SO_4^* (all values μeq L^{-1})
All	234	$SO_4^* = 0.36 Ca^* - 0.68 Fe + 1.28 Al - 0.22 A^- + 33.46$; $r^2 = 0.57$
1	7	$SO_4^* = 43.3 Mn - 3.27 Ca^* + 61.4$; $r^2 = 0.89$
2	20	$SO_4^* = 0.90 Ca^* - 0.62 Fe + 36.3$; $r^2 = 0.51$
3	38	$SO_4^* = 0.70 Ca^* - 0.35 A^- + 44.4$; $r^2 = 0.41$
4	35	$SO_4^* = 0.53 Ca^* + 26.2$; $r^2 = 0.19$
5	72	$SO_4^* = 0.60 Ca^* + 21.9$; $r^2 = 0.63$
6	62	$SO_4^* = 0.45 Ca^* + 4.93$; $r^2 = 0.61$

First through fifth order partial correlation analysis did produce a few significant relationships between H^+ and SO_4^{*+} for the 234 lakes. Significant relationships ($p < .001$) resulted from controlling Ca^* ($r = 0.25$), Ca^* and Fe ($r = 0.29$) and Ca^*, Fe, and A^- ($r = 0.30$).

In Group 1 some very strong partial correlations were produced, but caution is necessary because of the small number of lakes in this class. The following shows the results obtained:

H^+ vs SO_4^*	r
Zero order	0.07
First order (by Mn)	0.66
Second order (by Mn, Al)	0.85
Third order (by Mn, Al, Ca)	0.87
Third order (by Mn, Al, A^-)	0.86

An intriguing feature of this result is the implied importance of Mn. Since Mn is reputedly more toxic to fish than Al (McNeely et al., 1979), it is not unreasonable to suggest that present or future biological studies of metal toxicity should include Mn. No significant relationships were found by partial correlations for Groups 2 to 6.

Regressions with SO_4^* as the dependent variable yielded slightly more significant results (Table II). The relationship between SO_4^* and Ca^* was generally positive, presumably a result of gypsum influence. The sign did shift in Group 1, however, and may be related to leaching properties in the most acidic watersheds. Sulphate did show a small negative relationship with A^- in groups All and 3. This may be due in part to seasonality, as noted previously.

Significant correlations were found between H^+ and A^- in Groups 5 (r = 0.37), 6 (r = 0.31), and All (r = 0.23) although not in the more acidic groups (1 to 4). Partial correlations were not performed to further examine relationships between H^+ and A^-.

4. SUMMARY AND CONCLUSIONS

Summer (1981-1984) water chemistry results from 234 lakes in Nova Scotia show a mean pH of 5.7 with a range from 4.4 to 7.7. Approximately 10% of the lakes have pH <5.0 (n = 27). 134 of the sampled lakes (57%) had summer pH values >5.6.

All of the lakes reflect a strong marine influence (no lake in Nova Scotia is more than 50 km from the sea), yet all lakes demonstrate non-marine and non-geological sources of SO_4. Concentrations of NO_3 were negligible in all samples. Organic anions [A^-] contribute 10 to 15% of the ion sums.

Although water chemistry results showed the anticipated marked increase in Al at the lower pH levels, the observation that the first order partial correlation of H^+ vs SO_4^* is strongly affected by Mn (accounting for more than 80% of the variance) may imply a more important role for Mn in the biological response of lakes to acidification than previously considered.

ACKNOWLEDGEMENTS

Gratitude is extended to Mr. B. Sabean and staff, Nova Scotia Department of Lands and Forests, for water samples, and to Dr. R. McCurdy and staff for analytical services. Mr. D. Hirtle and Mrs. D. McConnell assisted with the preparation of the manuscript. Funding was provided by the Nova Scotia Department of the Environment.

REFERENCES

Kerekes, J.J. : 1986, Water Air Soil Pollut. this volume.

McNeely, R.N., Neimanis, V.P. and Dwyer, L. : 1979, Environment Canada, Inland Waters Directorate Report EN 37-54/1979, ix + 88p.

Nie, N.N., Hull, C.H., Jenkins, J.G., Steinbrenner, K. and Bent, D.H. : 1975, Statistical Package for the Social Sciences, McGraw-Hill, New York, xiv + 675p.

NRCC, National Research Council of Canada : 1981, Publ. NRCC No. 18475 of the Environmental Secretariat, Ottawa, K1A 0R6.

Ogden, III, J.G. : 1982, Water Air Soil Pollut. **17**, 119.

Oliver, B.E., Thurman, E.M. and Malcolm, R.L. : 1983, Geochim. Cosmochim. Acta **47**, 2031.

Smith, D.L., Underwood, J.K., Ogden, III, J.G. and Sabean, B. : 1986, Water Air Soil Pollut. this volume.

Underwood, J.K. : 1984, An analysis of the chemistry of precipitation in Nova Scotia, 1977-1980, Unpubl. Ph.D. thesis, Technical University of Nova Scotia, Halifax, xvi + 264p.

Vaughan, H.H., Underwood, J.K. and Ogden, III, J.G. : 1982, Water Air Soil Pollut. **18**, 353.

Vaughan, H.H., Underwood, J.K., Ogden, III, J.G. and Kerekes, J.J. : 1983, Nova Scotia Dept. Environ. Tech. Report, 18p.

Watt, W.W. : 1981, in Sochasky, L. ed. Acid Rain and the Atlantic Salmon Int. Atlantic Salmon Foundation Spec. Publ. Ser. no 10 : 39.

Watt, W.W., Scott, C.D. and Ray, S. : 1979, Limnol. Oceanogr. **24**, 1154.

EXAMINATION OF THE OLIGOTROPHICATION HYPOTHESIS: PHOSPORUS CYCLING IN AN ACIDIC FLORIDA LAKE

R.W. Ogburn III, Breedlove, Dennis & Associates, 2412 Forsyth Rd, Orlando, FL 32807 USA, and P.L. Brezonik, Dept. of Civil & Mineral Engrg., Univ. of Minnesota, Minneapolis, MN 55455 USA

ABSTRACT. Phosphorus cycling processes were studied in an acidic lake to test the oligotrophication hypothesis. P storages and inputs were measured; *in situ* and laboratory experiments measured pH effects on planktonic uptake and release, macrophyte release, and sediment-water exchange of SRP. Planktonic rates showed little response to pH; release from macrophytes initially was slowest at highest pH (5.5), but after 225 days was not affected over the pH range 3.7 to 5.5. Maximum SRP sorption by sediments occurred at pH 4.0 to 5.5; decreasing sorption above pH 5.5 provides some support for the oligtrophication hypothesis.

1. INTRODUCTION

Based on observations that acidic Scandinavian lakes had greater transparency and lower chlorophyll *a* (chl *a*) than non-acidic lakes, Grahn *et al*. (1974) hypothesized that acidification induces an "oligotrophication" process in lakes. Reduced rates of organic matter decomposition and coverage of sediments by algal mats and *Sphagnum* were thought to reduce nutrient recycling and primary production. Some surveys in North America have shown similar pH-related trends in transparency, nutrients, and chl *a* (e.g. Hendrey *et al*., 1976; Hendry and Brezonik, 1984), but others have not (Dillon *et al*., 1979). Experimental acidification of ELA Lake 223 pH to 5.0 did not reduce decomposition or nutrient concentrations (Kelly *et al*., 1984; Schindler et al., 1985). Much other conflicting evidence has been reported regarding the hypothesis, and its validity is questionable.

We studied P cycling in McCloud Lake to test effects of pH on (i) planktonic uptake and release of soluble reactive P (SRP); (ii) SRP release from decomposing macrophytes; and (iii) sediment-water exchange of SRP. The lake's P inputs and storages also were measured. Experimental details were given by Ogburn (1984). McCloud Lake is acidic, soft, uncolored, and oligo- to mesotrophic (Table I). Its small, uninhabited watershed in the sand-hills of north Florida is sparsely forested (longleaf pine/turkey oak). The lake has no surface drainage, and > 90% of its water is directly from rainfall. Because the soils are very permeable, it receives no overland runoff. Water residence time is ~10 yr; surface area is ~5 ha; z_{max} = 5 to 6 m, and z = 2.7 m.

Water, Air, and Soil Pollution **30** (1986) 1001–1006.

2. TP INPUTS AND STORAGES

Budget. Monthly wet-only precipitation and P concentrations were measured from August 1981 to July 1982; dryfall of TP was estimated from data of Brezonik et al. (1983). TP inputs from groundwater, which constitutes < 10% of the water budget, are negligible. The estimated TP budget for August 1981-July 1982 was 2.2 to 2.9 kg. The areal TP loading (44 to 58 mg m^{-2} yr^{-1}) is above criteria for mesotrophic lakes but below criteria for eutrophy. The mean TP storage in the lakewater was 1.54 kg for the period, and the TP residence time was ~0.6 yr. Maximum TP occurred in spring and summer, when chl a and zooplankton abundance were highest, but there were no major seasonal trends in TP or SRP over the period October 1980-September 1982. TP was higher in 1982 than 1980-81, which was a drought period with falling water levels. Higher TP in 1982 probably reflects increased inputs from rainfall and rising lake levels.

Sedimentation. Mid-lake sedimentation was measured during two winter and three summer months. Sedimentation TP fluxes ranged from 10 to 48 mg m^{-2} mo^{-1}. Estimated annual mid-lake sedimentation was 430 g m^{-2} yr^{-1} (seston dry wt) and 370 mg m^{-2} yr^{-1} (TP). The former value agrees fairly well with results of ^{210}Pb dating, which indicate net sediment accumulation of ~300 g m^{-2} yr^{-1} (dry wt) (Binford, 1984). Gross mid-lake sedimentation of TP is 6 to 8X the lake's annual areal TP loading, but sediment accumulation is limited to ~50% of the lake bottom. Based on the TP content of mid-lake surface sediment (0.034%) and sedimenting seston (0.085%), about 60% of the sedimenting TP is recycled to the water column.

Macrophytes. The dominant submergent plants in the lake (Eleocharis sp. and Websteria sp.) covered 26% of the bottom in September 1982. Coverage by Websteria was small (2% of the bottom), and the plants were small and patchy. The biomass of Eleocharis contained ~3.8 kg TP, (2.5X the lake's mean storage of TP and 1.3 to 1.7X the annual loading). The macrophytes probably obtain their P from sediments and do not compete with algae for water column P. Indeed, macrophytes may be an important mechanism for returning sediment P to the water column when they decompose.

Effect of pH on P release from decomposing Eleocharis was studied in the laboratory. Duplicate polyethylene bottles containing 0.50 ± 0.01 g intact plants and 4 L filtered lake water were incubated in the dark for 7 mo at pH 3.7, 4.1, 4.6, or 5.6. Bottles were aerated and pH adjusted weekly. Samples were collected 18 times over 225 days to determine release of SRP, soluble organic P (SOP), and particulate organic P (POP). Initial levels were < 5 ug L^{-1}, and SOP and POP remained < 10 ug L^{-1} during the study. About 40-58% of the plant TP (~800 ug TP) was released to the water mostly as SRP in 225 days; ~80% of the release occurred in the first 140 days. Final SRP levels were 80 to 115 ug L^{-1}, and no differences were observed among pH treatments on the last two sampling dates (ANOVA, $p > 0.05$). However, initial SRP release was fastest at pH 3.7, and a definite lag occurred at pH 5.5. Average release rates over days 0 to 51 (except pH 5.5: days 10 to 51) decreased with increasing pH, but differences were not statistically significant (ANOVA, $p > 0.05$):

TABLE I. SUMMARY OF LIMNOLOGICAL CONDITIONS IN McCLOUD L., 1980-1982.

Parameter	Value	Nutrients	X ± SD (n=24) ($\mu g\ L^{-1}$)
pH	4.5-4.6	TP	10 ± 6
Conductance ($\mu S\ cm^{-1}$)	42	SRP	4 ± 2
Color (PCU)	< 5	NH_4^+-N	59 ± 45
Transparency (m)		NO_3^--N	59 ± 30
Summer	1.7-2.5	Chl *a*	5.2 ± 3.4
Winter	to bottom	TN:TP	45 (wt:wt)
$Ca^{2+} + Mg^{2+}$ ($\mu eq\ L^{-1}$)	150		
SO_4^{2-} "	140		

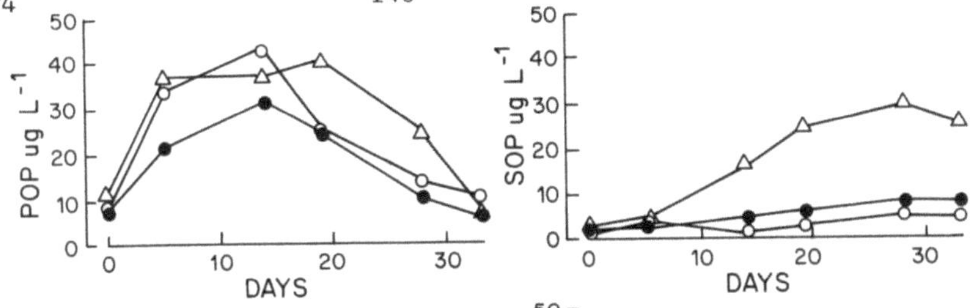

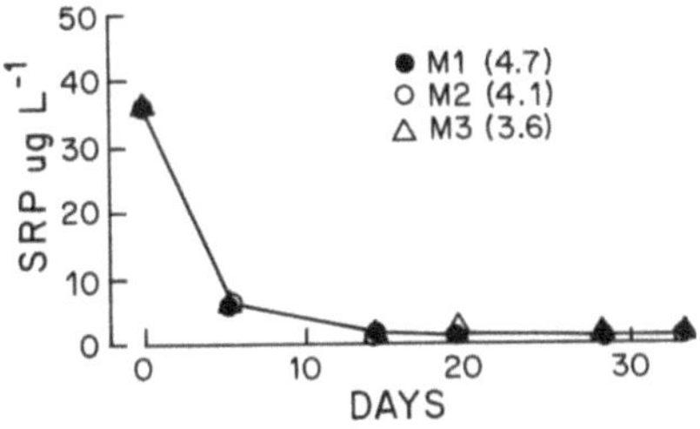

Figure 1. Temporal variations in particulate organic P (POP), soluble organic P (SOP), and soluble reactive P after SRP additions to mid-lake enclosures.

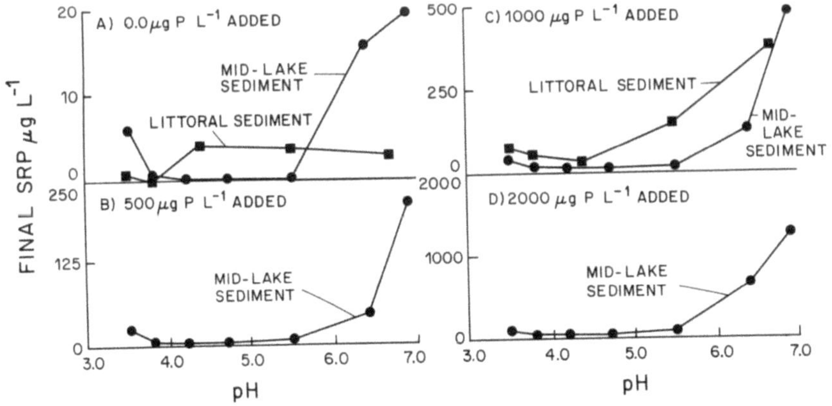

Figure 2. Effect of pH on SRP release (A) or sorption (B-D) by littoral and mid-lake sediments from McCloud Lake in well-mixed flasks.

pH	3.7	4.1	4.4	5.5
Rate (ug L^{-1} day^{-1})	1.08	1.03	0.63	0.38

Studies of perennial macrophytes in softwater lakes of northern Wisconsin indicate that biomass turnover time is ~1 to 2 yr (Boston, 1985). If steady-state conditions exist (production = loss), macrophyte TP production in McCloud L. would be ~1.9 to 3.8 kg yr^{-1}. Based on the decomposition results, most of the "labile" P (~50% of total plant content) is released in less than a year. Decomposition of submerged macrophytes thus may contribute ~0.9 to 1.9 kg yr^{-1} of SRP to the water column (which is 33-86% of the atmospheric TP loading). We conclude that macrophyte decomposition is an important mechanism for recycling P from sediments to the lakewater and that SRP release from macrophytes is insensitive to pH.

3. COMMUNITY PHOSPHORUS CYCLING

Phosphorus cycling in littoral and mid-lake communities was studied in laboratory and _in situ_ enclosures adjusted to several pH values. Rates of SRP uptake and release were measured with ^{32}P. We installed three littoral enclosures (4 m dia, open to the bottom) at 1 m depth in March 1981. pH was adjusted over 4 weeks to 3.6 (H_2SO_4), 5.1 (NaOH), or left ambient (4.6). A large SRP peak (max = 22 ug L^{-1}) occurred at pH 3.7 during pH adjustment, and a smaller peak (max = 9 ug L^{-1}) occurred at pH 5.1. These peaks may reflect transient responses of biota to pH change. No pH-related trends were noted in SRP or TP after initial adjustment.

Six 0.9 m dia, 2.2 m long polyethylene tubes closed at the bottom were floated at mid-lake in July 1982. Duplicate tubes were left at ambient pH (4.7) or adjusted to pH 4.1 or 3.6 (H_2SO_4) over 4 weeks. Ten weeks later, one tube at each pH was enriched ~10X with SRP and NH_4NO_3 (TIN) (1:10 P:N ratio). SRP was low in all tubes (< 2 ug L^{-1}) before nutrient addition, but TP was significantly lower at pH 3.7 than at higher pH (ANOVA, $p < 0.05$). SRP loss rates were the same at all 3 pH's after nutrient enrichment (Figure 1). POP responded similarly at all 3 pH's but remained elevated longer at pH 3.6. SOP increased gradually to ~30 ug L^{-1} at pH 3.6 but remained < 10 ug L^{-1} at pH > 4.0, suggesting reduced ability of biota to decompose SOP at pH 3.6. SRP uptake rates in the 6 tubes were measured by ^{32}P on 6 dates from August to November. Uptake rate coefficients (h^{-1}) were divided by algal biomass (mg L^{-1}) to normalize values. These coefficients tended to increase with decreasing pH, but the trend was not statistically significant and within-pH variance was large. Values probably were more affected by P availability than pH. Seston from mid-lake tubes was labeled with ^{32}P in lab incubations ; SRP release rates were comparable to uptake rates, but no pH trend was found.

Three 12-L carboys filled with depth-composited lake water (May 1983) were aerated in the lab on a 14/10 h L/D cycle. Two weeks after pH was adjusted to 3.6, 4.0, and 4.6 (ambient), the microcosms were enriched with SRP and TIN (1:10 ratio). The same pattern of SRP uptake occurred at all 3 pH's. Uptake was linear from day 5 to day 15, and 80 to 90% of the SRP was assimilated by day 15. No pH effect was observed. In summary, laboratory and in situ studies show that SRP uptake by McCloud L. phytoplankton is not inhibited at pH 3.6 compared to ambient pH. The only pH trend we observed was slower conversion of SOP to SRP at pH 3.6.

4. PHOSPHORUS SORPTION BY SEDIMENTS

We studied pH effects on SRP sorption and release in well-mixed sediment-water slurries and undisturbed sediment-water cores from littoral and mid-lake areas. Littoral sediment had higher sand (18.3%) and lower organic content (11.6%) than flocculent mid-lake sediments (0.0% sand, 77.7% organic [by volatile solids]). Mid-lake sediment had pH_{ZPC} = 3.6 (based on potentiometric titrations); ambient pH of the sediment was 5.5.

Littoral sediments (25 g wet wt, ~8 g dry wt) mixed with 100 mL filtered lake water for 5 days released very little SRP (Figure 2). Maximum release occurred at pH 4.4; release declined slowly above and rapidly below that value. Maximum sorption of 1 mg L^{-1} SRP added to the lake water occurred at pH 4.4 (96% sorbed) and decreased slightly at lower pH. Much less SRP was adsorbed at higher pH (only 55% at pH 6.6). Similar studies with mid-lake sediments (25 g wet wt, 1.7 g dry wt) showed no release in 100 mL filtered lake water equilibrated 48 h at pH 3.8 to 5.5 (Figure 2), but increasing amounts were released at higher pH (20 ug L^{-1} at pH 6.9). SRP sorbed from lake water with 0.5 to 2.0 mg L^{-1} SRP added showed trends similar to the littoral sediments, and percent sorbed closely followed percent SRP present as $H_2PO_4^-$. Sorption by mid-lake sediments fit the Langmuir equation, and fitted parameters varied with pH:

pH	3.5	3.8	4.2	5.5	6.9
Q_o (mg g^{-1})	0.63	0.68	0.65	0.12	0.07
b (L mg^{-1})	2.26	5.88	6.95	54.8	1.50
r^2	0.943	0.997	0.950	0.980	0.988

Binding capacity (Q_o) was nearly the same for pH 3.5 to 4.2 and decreased markedly at pH > 5. Binding energy (related to b) increased up to pH 5.5 but was lowest at pH 6.9. Our results generally agree with the adsorption envelope theory of Hingston et al. (1967). An SRP adsorption maximum also occurs between pH 4 and 5 with kaolinite (Edzwald, 1977), a common clay mineral in the area's soils.

Four 4-cm dia cores with 10 cm sediment and 30 cm water were incubated at each of 4 pH values (3.7 to 6.0) in the dark with overlying water aerated slowly. After 30 days at the pH-adjusted values, 1 mg L^{-1} SRP was added to duplicate cores. Duplicate cores with no added SRP and tubes with SRP added to lake water (sediment-free) served as controls. Little uptake or release occurred in all controls during 30 day incubations. SRP uptake by undisturbed sediment in amended cores followed first order kinetics. Uptake was fastest at pH 4.1 and slowest at 6.0; 66 to 92% of the added SRP was lost in 30 days. Uptake rate constants had the following pH pattern:

pH	4.1	>	3.7	=	4.7	>	6.0*	*by ANCOVA
k (day^{-1})	0.090		0.061		0.063		0.036	**regression fit to
r^2	0.86		0.93		0.97		0.99**	first order model

Thus kinetic trends (undisturbed cores) and equilibrium results (well-mixed systems) showed the least uptake of SRP at highest pH values.

Sediment sorption/release of SRP was the only P cycling process showing a significant pH effect. SRP sorption was maximum at pH 4.0 to 5.5 and decreased greatly at higher pH and slightly at lower pH. The trend agrees with previous studies on SRP sorption by soils and clay minerals and follows the percent SRP present as $H_2PO_4^-$. Changes in sediment surface charge also may affect the trend. These results suggest that acidification of circumneutral lakes to pH 5.0 to 5.5 could lead to decreased cycling of P because of enhanced binding of SRP by sediments. This offers some support for the oligotrophication hypothesis. Further acidification of McCloud L. would have little effect on P binding by sediment, however.

Acknowledgments. This study was performed at the University of Florida and was supported by a grant from the U.S. EPA-North Carolina State University cooperative research program on acid precipitation.

REFERENCES

Binford, M.: 1984, personal communication, Univ. of Florida, Gainesville.

Boston, H.: 1985, personal communication, Univ. of Wisconsin, Madison.

Brezonik, P.L., Hendry, C.D., et al.: 1983, Rep. EPA 600/3-83-004, U.S.EPA, Corvallis, Oregon.

Dillon, P.J., Yan, N.D., et al.: 1979, Arch. Hydrobiol. Beih., **13**, 317.

Edzwald, J.: 1977, in Suffet, I.H. (ed.), Fate of Pollutants in the Air and Water Environments, Pt. 2, Wiley-Interscience, New York.

Grahn, O., Hultberg, H., and Landner, L.: 1974, Ambio, **3**, 93.

Hendrey, G.R., Baalsrud, K., et al.: 1976, Ambio, **5**, 224.

Hendry, C.D., and Brezonik, P.L.: 1984, Water Resources Bull., **20**, 75.

Hingston, F.J., et al.: 1967, Nature, **215**, 1459.

Kelly, C.A., Rudd, J.W.M., et al.:1984, Limnol. Oceanogr., **29**, 687.

Ogburn, R.W., III: 1984, Ph.D. thesis, Univ. of Florida, Gainesville.

Schindler, D.W., Mills, K.H., et al.: 1985, Science, **228**, 1395.

METAL SPECIATION IN ACIDIFIED MOUNTAIN STREAMS IN CENTRAL SWEDEN

Hans Borg
The National Environmental Protection Board
Chemical Laboratory
Box 1302
S-171 25 SOLNA
Sweden

ABSTRACT. The study was performed in mountain streams in Central Sweden where a dramatic decrease in fish populations has been observed during the latest years. The spring runoff caused acid surges of pH 4.4 to 5.5 and high concentrations of Fe, Mn and Al occured in the stream waters. Water samples were collected during fish cage experiments performed during spring flood of 1985. The metal fractions were separated with filtration, in-situ dialysis and ion-exchange technique and determined using atomic absorption spectroscopy. The elements analyzed included Fe, Mn, Al, Zn, Cu, Pb and Cd. During spring flood, when a 100 % fish mortality was observed, particulate forms of Fe, Al and Mn dominated and were probably responsible for the toxic action. The high leaching of Fe, Mn and Al at spring flood during the recent years is a new phenomenon and seems to be caused by the increased acid deposition to the area.

INTRODUCTION

The acidic precipitation on Scandinavia and North America has resulted in increased concentrations of many metals in fresh waters (Dickson, 1975, 1980; Borg, 1983), including some of the more abundant elements in the earths crust, such as Al, Mn and occasionally also Fe. High peak values of these metals have been recorded in mountain streams during spring runoff (Bjärnborg, 1983). These elements, considered to be relatively harmless to aquatic life, compared to trace metals like Hg and Cd, now often occur in acidified waters in concentrations which can be toxic to fish. So far, Al has received most of the attention concerning effects on fish in acid waters (see for example Erichsen Jones, 1939; Baker and Schofield, 1980; Driscoll et al., 1980; Muniz and Leivestad, 1980; Skogheim et al., 1984.) The decreased pH in the waters generally promotes a larger fraction of most metals to occur in dissolved forms (Driscoll et al., 1980; Borg and Andersson, 1984).

This investigation continues the fish cage experiments performed during spring flood in 1982 and 1983 in a remote area in central

Water, Air, and Soil Pollution **30** (1986) 1007–1014.

Sweden, when acute toxic effects on fish were recorded (Andersson and Nyberg, 1984). The aim was to try to explain the observed effects on fish by separating the metal fractions by filtration, in situ dialysis and ion exchange technique. This paper will focus on the results of the chemical measurements, while the effects on fish will be reported in detail elsewhere.

STUDY AREA

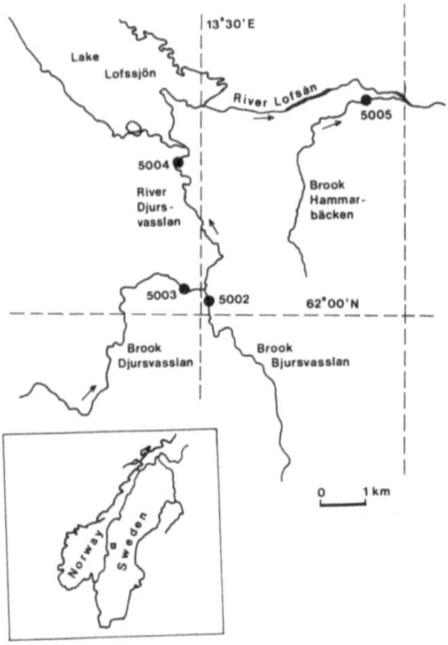

Figure 1. Map of the investigation area.

The investigation was performed in four mountain streams in Lofsdalen, central Sweden, at about 500 to 800 m a s l (Figure 1).The drainage areas are 8 to 35 km^2. Bogs and swamps are frequent in the upper parts, as a cause of a former lake basin from the glacial period, with large deposits of Fe, Mn and Al. Before 1978, the pH of the snow in the area was around 5.0 and trout and other fish populations inhabited the streams. Since then, the snow pH has been 4.2 to 4.7 and the spring thaws have caused acid episodes with a dramatic decrease in the stream invertebrate fauna and fish populations as a consequence (Andersson and Nyberg, 1984). Some parts of the drainage areas of Brook Hammarbäcken (5005) and Brook Bjursvasslan (5002) were limed in the autumn of 1983 and 1984, respectively. A low dose ($CaCO_3$, 200 kg ha^{-1} drainage area) was applied with helicopter, primarily on the swamps.

MATERIALS AND METHODS

Three fish cage experiments were performed with brown trout (Salmo trutta L.) before, during and after the spring flood in 1985 at four stations (Figure 1) as described in Andersson and Nyberg (1984).

The discharge was calculated using data from a water level gauge in river Djursvasslan. Water samples have been taken at least monthly since 1981 and during some periods even daily. During the fish cage experiments in 1985, water samples were taken at eleven occasions. Polyethylene bottles were used for general water chemical parameters and acid washed polypropylene bottles were used for metal analyses.

pH, color and alkalinity were measured both in the field and after arrival at the laboratory which always was within 24 hr after the sampling. Immediately after arrival, the samples for metal speciation were filtered through polycarbonate membranes (Nuclepore, 0.4 um) by pressure filtration with N_2 in a polycarbonate apparatus. Inorganic Al and Fe were separated from the filtered samples by cation exchange (Amberlite IR 120 resin, Na-form) according to Driscoll (1980). The non-exchangeable forms, which should approximate the levels of metal-organic complexes (Campbell et al., 1983 a,b; Driscoll, 1984), were collected and analysed for Al and Fe content. The dissolved fractions of metals were also collected by in situ dialysis (Benes and Steinnes, 1974; Borg and Andersson, 1984). The dialysis membranes (pore-size 2.4 nm), were allowed to equilibrate in the streams for 6 to 7 days. All the collected metal fractions were preserved with ultrapure HNO_3 (2 mL L^{-1}).

The analyzed metal fractions included total, dialysable (all elements), filterable (Fe, Mn, Al) and filterable-organic (Fe, Al). The labile inorganic fraction was calculated as the difference between filterable and filterable-organic fractions. All metals were determined using AAS with flame (Fe, Mn) or Zeeman graphite furnace (Al, Zn, Cu, Pb, Cd). pH, color, conductivity, major ions, alkalinity and total organic C (TOC) were determined according to Swedish Standard Methods (1985) whenever possible.

RESULTS

Large fluctuations in pH, alkalinity, water color, Fe, Mn and Al were observed during the recent years as examplified in Figures 2 to 3 (Brook Bjursvasslan).During low flow periods, alkalinity sometimes reached above 1 meq L^{-1}, whereas during spates, zero alkalinity and low levels of major ions generally was obtained (Na, K, Mg, Cl, 0.01 to 0.02, Ca 0.02 to 0.11 meq L^{-1}). The area is not influenced by specific S-rich deposits. The sulphate level in the streams was only about 0.02

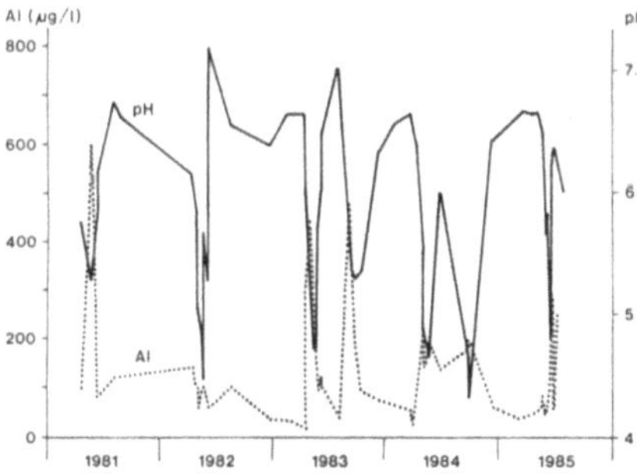

Figure 2. Temporal variations in pH and Al at station 5002, Brook Bjursvasslan.

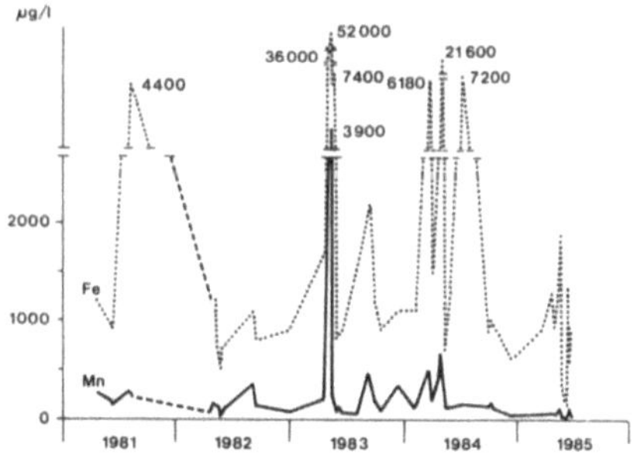

Figure 3. Temporal variations in Fe and Mn at station 5002, Brook Bjursvasslan.

to 0.05 meq L^{-1}, which is less than or equal to what is derived from atmospheric deposition. The sum of inorganic anions was generally 0.04 to 0.08 meq less than the sum of cations, indicating the presence of organic anions. pH drops and peak values of metals were typically associated with periods of high discharge, primarily spring flood.

The results from the spring 1985 are exemplified with station 5003 (Figures 4a, b, c). The discharge was low until May 7, when it started to increase and reached the top in May 21. The discharge then decreased again until May 25 when the snowmelt in the higher areas led to an even more intensive spring flood, but of shorter duration.

The pH-values decreased during the spring flood, from around 7.3 to 4.4. The Ca concentration decreased from 1.0 to 0.02 meq L^{-1} and the alkalinity from 1.0 to 0.01 meq L^{-1} but had increased again to 0.39 meq L^{-1} at the beginning of June. The TOC content increased with the discharge, from around 2.5 to 12 mg L^{-1} and was 8 to 10 mg L^{-1} after the spring flood.

The spring flood caused a dramatic increase of all three metals at both peak discharge occasions. Iron (Figure 4a), which reached 7000 µg L^{-1}, occurred almost entirely in the particulate form. After the spring flood Fe decreased to about the same level as before the thaw (around 300 µg L^{-1}) but with an even larger portion as particulate and organic Fe.

Before snow melt, total Al was around 10 µg L^{-1} (Figure 4b). During the spring flood it reached 175 to 230 µg L^{-1} of which 100 to 175 was bound to particles. When the discharge declined, total Al decreased to 40 to 50 µg L^{-1}. The labile inorganic Al concentration was in the range 0 to 10 µg L^{-1} through the whole sampling period.

Before the thaw Mn (Figure 4c) was in the range 15 to 40 µg L^{-1}, of which almost everything was dialysable. During spring flood it reached 300 and 420 µg L^{-1}, respectively, of which the most was particulate. The dialysable fraction then was only about 60 µg L^{-1}. After the spring flood Mn decreased to 20 to 60 µg L^{-1}, of which the most was filterable.

In spite of the large variations in pH, Fe, Al and Mn, the concentrations of the trace metals Zn, Pb, Cu and Cd did not fluctuate that much. However, the temporal variation looked much the same, with maximum levels at periods of high discharge. The only element which increased significantly over what could be considered as background level (Borg, 1983) was Pb, which reached 9.0 µg L^{-1} at station 5003 in May 1985. Pb was however primarily bound to particles and only 1 to 23% was dialysable, compared to 20 to 100% for Zn, 23 to 75% for Cd and 32 to 100% for Cu (Table I).

At station 5003, 100% trout mortality was recorded within 14 to 25 hr at two different tests during the first discharge peak. Equally short survival times were observed at station 5002 at pH 5.0 to 5.5 and at station 5004 at pH 5.3 to 5.5 However, the liming of the watershed gave decreased concentrations of metals in those streams, compared to previous years. Fe reached only 150 to 350, Al 40 to 65 and Mn 10 to 110 µg L^{-1}. In the other limed watershed, (stn 5005), where the pH was above 6.0 during the spring flood, no fish mortality was observed

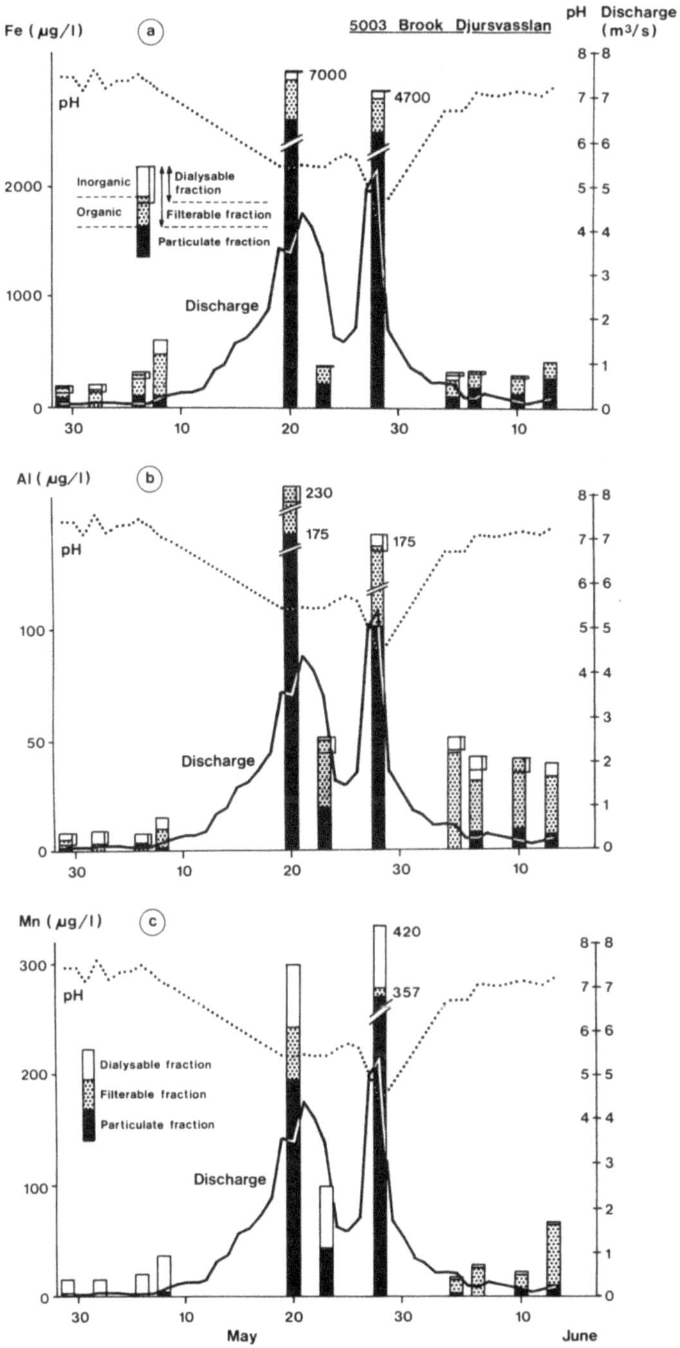

Figure 4a,b,c. Discharge, pH and different forms of Fe (a), Al (b) and Mn (c) during spring snowmelt, 1985.

(Olofsson, 1985). The metal levels on that site during high flow were lower than at any other site; 57 to 210 (Fe), 6 to 37 (Mn) and 29 to 34 $\mu g\ L^{-1}$ (Al).

Table I. The range of total metal concentrations and percentage dialysable fractions in the brooks Bjursvasslan (Stn 5002) and Djursvasslan (Stn 5003) during spring flood.

Stn		Zn	Cu	Pb	Cd
			Total concentration $\mu g\ L^{-1}$		
5002	May 1984	5.1-5.9	0.18-0.28	0.29-0.57	0.027-0.066
	May 1985	1.1-2.3	0.06-0.11	0.13-0.24	0.004-0.009
5003	May 1984	1.1-4.9	0.19-0.40	0.26-1.2	0.011-0.040
	May 1985	3.0-10	0.20-0.52	0.32-9.0	0.024-0.058
			Dialysable fraction (%)		
5002	May 1984	100	72	23	75
	May 1985	20	100	6	29
5003	May 1984	100	41	6	36
	May 1985	20	32	1	23

DISCUSSION

For practical reasons, the ion-exchange separations were performed in the laboratory after a transportation time of the samples of 24 hr as a maximum. The pH was however determined both in the field and in the laboratory. The laboratory-pH was about 0.2 pH units higher than the field values at pH 4.5 to 5.0 and around 0.1 units higher at 6.5 to 7.0. As these changes were relatively small, the transportation time of the samples has probably not influenced the metal speciation in any substantial way.

The dialysis membranes used had a pore-size of 2.4 nm, which would exclude all particulate and colloidal fractions. However, the low molecular weight fractions of fulvic acids might pass through this membranes. When checking the organic C content inside the dialysis tubing I generally found up to 20 % of the levels outside the tubing in these streamwaters. LaZerte (1984), using a 1000-mwco membrane, reported an organic penetration of 6 to 12 %, based on absorbance at 330 nm. Consequently, the dialysable fraction consists not only of inorganic forms, but includes probably some low molecular weight organic substances. The _in situ_ dialysis separation used here , gives an integrated picture of the dialysis period. In these streams however, very fast changes of pH and metal flows often occur, and as the diffusion of ions through the membrane might take a longer time than the changes in the stream, sometimes even less metal was found as dialysable than as inorganic labile metal (Figures 4a,b). Another source of error when dialysis methods are used in very humic waters is the Donnan disequilibrium which can occur if the organic anions make up a substantial fraction of the total amount of anions (LaZerte, 1984). However, as some organically complexed metal probably passed the membrane and the content of organic C in these waters was in the range 10 to 12 mg L^{-1} at the most, the Donnan disequilibrium was probably of minor importance. In spite of the potential errors discussed above, the general agreement between the dialysable and the labile inorganic metal fraction was relatively good. As expected, the dialysable fraction was mostly a bit larger or equal to the labile inorganic fraction. Consequently, it did not seem to be any problem with the filtration used prior to the ion exchange - such as clogging of the filter pores - especially as the filtered volume was only 100 mL.

The fish mortality recorded in 1982, was probably caused by Fe and possibly Mn while in 1983, also Al might have contributed (Andersson and Nyberg, 1984). These results further indicated that the metals were not only toxic when present in free ionic form but also in particulate form, which is strongly confirmed by the results reported here. The drastic fish mortality occurred when Fe, Al and Mn were present almost entirely as particulate bound forms. The metal species involved might be precipitates of Fe-hydroxides possibly associated with colloidal organic matter. Fe and Mn probably enter the streams as reduced forms, at least during periods of low discharge, when a large portion of the input comes via groundwater.Consequently, reduced metal forms might contribute to fish toxicity at some periods.

After entering the stream water, Fe and Mn are oxidized and precipitated further downstream. Thick layers of metal precipitates have also been observed on the streambeds during periods of low discharge (Olofsson, 1985). Resuspension of such precipitates by flushing contributes to the high concentrations of particulate bound metals, found at spring flood (Figure 4).

It is not likely that the drastic effect on fish were caused by the H^+ ions alone, as brown trout generally can tolerate the pH-levels which were recorded during the cage experiments (Grande et al., 1978; Leivestad et al., 1980).

The transport of large amounts of metals from these drainage areas during the recent years, is not a natural phenomenon but is strongly related to the observed increase in acid loading on the area, causing an increased mobility of the metals from the soil. The causes of the observed toxic effects on fish apparently are complex, but particulate metal forms seem to play an important role. In this case the toxic action was not due to labile inorganic Al, which otherwise often has been reported to be of major importance for fish mortality in acidified waters. The concentrations of the trace metals Zn, Cu, Pb and Cd were probably far too low to be of importance for the acute toxic effects on fish.

ACKNOWLEDGMENTS

I am most indepted to Miss Elisabeth Melin and Mr Erik Olofsson who performed the field experiments, water sampling and water chemical measurements in the field.

REFERENCES

Andersson, P. and Nyberg, P.: 1984, Rep. Inst. Freshw.Res., Drottningholm **61**, 34.

Baker, J.P. and Schofield, C.L.: 1980, Aluminium toxicity to fish as related to acid precipitation and Adirondack surface water quality. In: Drabløs, D. and Tollan, A. (eds), Ecological impact of acid precipitation. SNSF-project, Oslo-Ås, Norway, 292.

Beneš, P. and Steinnes, E.: 1974, Water Res. **8**, 947.

Bjärnborg, B.: 1983, Hydrobiologia **101**, 19.

Borg, H.: 1983, Hydrobiologia **101**, 27.

Borg, H. and Andersson, P.: 1984, Verh. Int. Verein.Limnol. **22**, 725.

Campbell, P.G.C., Bougie, R., Tessier, A. and Villeneuve, J-P.: 1983a, Verh. Int. Verein. Limnol. **22**, 371.

Campbell, P.G.C., Bisson, M., Bougie, R., Tessier, A. and Villeneuve, J-P.: 1983b, Anal. Chem. **55**, 2246.

Dickson, W.: 1975, Rep.Inst. Freshw. Res., Drottningholm **54**, 8

Dickson, W.: 1980, Properties of acidified waters. In: Drabløs, D. and Tollan, A. (eds), Ecological impact of acid precipitation. SNSF-project Oslo-Ås, Norway.

Driscoll, C.T.: 1980, Chemical characterization of some dilute acidified lakes and streams in the Adirondack region of New York State, Ph.D.Thesis, Cornell University, 1980.

Driscoll, C.T., Baker, J.P., Bisogni, J.J. and Schofield, C.L.: 1980, Nature, **284**, 161.

Driscoll, C.T.: 1984, Intern. J.Environ. Anal. Chem, **16**, 267.

Erichsen Jones, J.R.: 1939, J. Exp. Biol. **16**, 425.

Grande, M., Muniz, I.P. and Andersen, S.: 1978, Verh.Internat.Verein. Limnol. **20**, 2076.

La Zerte, B.D.: 1984, Can. J. Fish. Aquat.Sci. **41**, 766.

Leivestad, H., Muniz, I.P. and Rosseland, B.O.: 1980, Acid stress in trout from a dilute mountain stream. In: Drabløs, D. and Tollan, A. Ecological impact of acid precipitation. SNSF-project, Oslo-Ås, Norway, 318.

Muniz, I.P. and Leivestad, H,: 1980, Toxic effects of aluminium on the brown trout, Salmo trutta L. In: Drabløs, D. and Tollan, A. Ecological impact of acid precipitation. SNSF-project, Oslo-Ås, Norway, 320.

Olofsson, E.: 1985, personal communications.

Skogheim, O., Rosseland, B.O. and Sevaldrud, I.: 1984, Rep.Inst.Freshw. Res., Drottningholm **61**, 195.

Swedish Standard Methods (Svensk standard) 1985, SS 0281 series.

SNOW CHEMISTRY IN THE CENTRAL SIERRA NEVADA, CALIFORNIA

Neil Berg
Pacific Southwest Forest and Range Experiment Station
P.O. Box 245
Berkeley, CA 94701

ABSTRACT. Precipitation and snowpack basal outflow event samples were analyzed for pH and NO_3 and SO_4 concentrations at a site near Lake Tahoe, California. The seasonal minimum pH from basal outflow was synchronous with a rain-on-snow event in 1984. Outflow pH reflected the low ionic concentrations of precipitation at the site. There was no evidence for a pulse of acidified meltwater at this site either during a mid-February melt induced by high air temperatures in 1985 or during 1984 or 1985 main spring melt periods. During the mid-February 1985 melt, however, an alkalinity pulse occurred in basal outflow.

1. INTRODUCTION

In areas of the world receiving acidic deposition, the leaching of solutes stored in snow has been demonstrated by sudden increases of acidity in surface waters during spring snowmelt periods (Galloway *et al.*, 1980). These have been accompanied by elevated concentrations from outflow at the base of the snowpack (Colbeck, 1981). Most of the chemical load is released in the first third of the melt volume, but in some situations as much as 50% of the total H^+ and SO_4 present can leach from the snowpack in the first 10% of water content loss (Johannessen *et al.*, 1980).

Findings in Scandinavia, Japan, and the northern United States may not be applicable to the Sierra Nevada in California. Strong maritime influences on Pacific Coast meteorologic processes can produce winter rain storms as high as the crest of the central portion of the range and cause significant periods of near-freezing snow temperatures, and snowpack accumulations are greater than those described elsewhere. Typically, the snowpack forms a sequence of variable density layers, some of which may impede the vertical movement of rain or melt water (Berg, 1982). These characteristics of Sierra Nevada snowpacks affect the disposition of chemical constituents resident in the snowpack and also influence the timing and frequency of outflows to the streams.

Trade names or products are mentioned solely for information. No endorsement by the U.S. Department of Agriculture is implied.

Water, Air, and Soil Pollution **30** (1986) 1015–1021.

This paper analyzes precipitation and snowpack basal outflow chemistry at one site in the central Sierra Nevada of California for two winters. It assesses evidence for spring surges in snow meltwater acidity and describes chemical changes associated with mid-winter rain-on-snow and warm weather snowmelt events.

2. EXPERIMENTAL

Precipitation and snowpack basal outflow samples were collected from a 0.5-ha forest clearing at the Central Sierra Snow Laboratory (CSSL), a U.S. Forest Service facility located slightly below and west of the crest of the Sierra Nevada 25 km northwest of Lake Tahoe, California (2100 m, 39° 19' 30" N. Lat., 120° 22' W. Long.). High winter precipitation (1000 to 2500 mm) causes snowpack accumulations up to 5 m depth by early April. While snowfall predominates, rain and mixed snow and rain also occur at CSSL. Storms move toward CSSL from the west or southwest and potentially entrain pollutants from northern California's San Francisco Bay Area and Sacramento metropolitan areas as well as agricultural operations in the Central Valley (Roth et al., 1985).

During winter 1983-1984, information on the pH, NO_3 and SO_4 solute concentrations of precipitation and snowpack basal outflow was gathered. Stream discharge, precipitation amount and timing, and the rate of water outflow from the base of the snowpack were also monitored (Figure 1). Melt, rainfall, and mixed rain and snow precipitation events were identified and related to periods of elevated basal outflow and stream discharge. Particular emphasis was placed on identification of fractionation events. Similar analyses were made in water year 1985, but NO_3 and SO_4 concentrations were not determined.

3. RESULTS AND DISCUSSION

Changes in pH, NO_3, and SO_4 concentrations during the first 10 to 25% of the outflow period imply fractionation. Although acidification of surface waters can have appreciable ecological impacts, alkaline impurities in snow may also be released during melt periods (Colbeck, 1981; R.A. Sommerfeld, per. comm., 1986). Potential changes in meltwater chemistry depend upon the type and magnitude of the impurities. "Clean" precipitation ($5.0 < pH < 6.0$; Roth et al., 1985) has relatively few pollutants available for fractionation. Under this condition, if the impurities are acidic pH depression during melt is likely to be 0.2 to 0.3 units below the mean. More acidic precipitation can be expected to show pH minima greater than one pH unit below the mean during the first 10% of the outflow. At the least, relative pH minima should be associated with outflow events. Alkaline impurities may cause elevated pH in outflow waters.

Mean snowfall pH values, weighted by precipitation amount, were 5.3 in 1984 and 5.2 in 1985, and were within the range of natural occurrence (Table I). Other precipitation types were generally more acidic than snow or mixed rain and snowfall in 1984. The 1984 and 1985 weighted mean H^+ concentrations, calculated from pH, were 5.4 and 6.4 μeq L^{-1}, respectively, with standard deviations equaling 0.9 and 3.3 μeq L^{-1}. The 1984 and 1985 annual precipitation totals of 1880 and 1118 mm, respectively, correspond to H^+ loadings at CSSL of approximately 0.10 and 0.08 kg ha^{-1}.

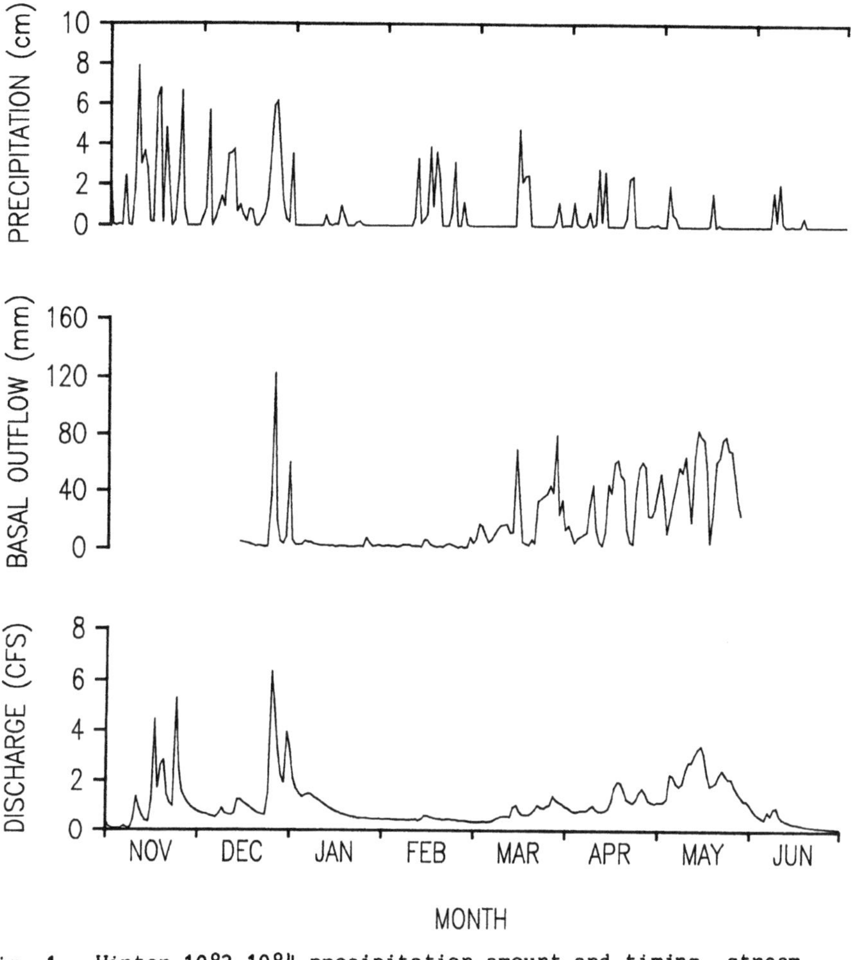

Fig. 1. Winter 1983-1984 precipitation amount and timing, stream discharge, and outflow from the base of the snowpack at the Central Sierra Snow Laboratory, north-central California.

In 1984, soluble NO_3 and SO_4 annual loadings approximated 1.2 and 4.5 kg ha^{-1}, respectively, at CSSL. Variability in the CSSL concentrations for SO_4, in particular, was high. The Pearson product moment correlation between pH and NO_3 was moderate and equaled -0.43 (n=80) and -0.79 (n=7) for snowfall and rainfall, respectively. SO_4:pH correlations were lower and equaled -0.10 (n=81) and -0.43 (n=7) for snowfall and rainfall, respectively.

TABLE I

Annual summary statistics for nitrate and sulfate solute concentrations ($\mu eq\ L^{-1}$) and pH of precipitation and basal outflow, Central Sierra Snow Laboratory, California.

	Precipitation 1984												1985	Outflow 1984						1985
	Rain			Mixed			Snow			Hail			Snow	Pan 1			Pan 2			Pan 1
	NO_3	SO_4	pH	NO_3	SO_4	pH	NO_3	SO_4	pH	NO_3	SO_4	pH	pH	NO_3	SO_4	pH	NO_3	SO_4	pH	pH
n[1]	7	7	7	13	13	13	80	80	89	3	3	5	32	21	21	31	23	23	22	75
M[2]	2.4	27	4.9	0.6	4.0	5.4	0.9	4.7	5.3	7.4	55	4.7	5.2	1.6	4.4	5.3	1.2	5.3	5.6	5.7
SD[3]	1.5	52	0.3	0.4	6.8	0.6	1.0	6.2	0.1	4.6	60	0.6	0.2	1.1	2.6	0.4	0.5	7.1	0.3	5.7
Max	5.0	143	5.2	1.6	22	6.4	6.4	28	5.7	12	124	5.7	5.9	3.3	11	6.4	2.7	25	6.5	6.8
Min	0.7	0	4.5	0.2	0	4.6	0	0	4.5	2.6	19	4.2	4.7	0.1	0	4.9	0.3	0	5.1	5.0

[1] Sample size. [2] Mean precipitation values weighted by precipitation amount.
[3] Standard Deviation.

Evidence of fractionation should be most obvious in changes in the basal outflow chemistry when snow metamorphic processes alone affect the water chemistry. Nevertheless, several factors confound a simple association between outflow chemistry and fractionation. Variations in the chemistry of atmospheric impurities may inhibit a straightforward identification of fractionation. If the pH of the impurity is similar to that of typical snowfall, fractionation may be masked. Stratigraphic features in the snowpack, such as high density horizontal layers, may direct liquid laterally, potentially away from a ground-sited melt pan (Berg, 1982). Vertical conduits channel flow so that some sections of the snowpack are saturated while adjacent sections are dry (Kattelmann, 1985). Such conduits were observed during both spring melt seasons. Depending upon the existing flow network and changes in its structure over time, water release from the base of the snowpack can vary widely in space and time. The optimum time to identify fractionation in the field under maritime conditions on the West Coast may well be during the initial outflow events, before the confounding effects accumulate. The chemistry of early and mid-winter outflows therefore take on added significance.

Basal outflow sampling began on December 30, 1983. Two mid-December outflow events were not monitored. Mean pH values from the pans equalled 5.3 and 5.6, and were within the general range of snowfall pH. Both pans showed elevated pH values during March 1984 and absolute or near-absolute minima during February. Mid-February pH values from one pan were 0.2 units lower than any other period, with February 12 and 13 values equalling 4.8 (Figure 2). The mid-February outflow pH's, associated with a rain-on-snow event, were the lowest of the year, and were consistent with the fractionation hypothesis. However, outflow pH's during the second week in February, a period of base level outflow and stream discharge, were only marginally higher than the mid-February recordings. Moreover, during a mixed rain and snowfall event in mid-March, outflow pH was appreciably above the mean and reached the seasonal high (6.49) for pan 2. Likewise, the March 20-22 and 26-27 outflow events were associated with pH values greater than the mean.

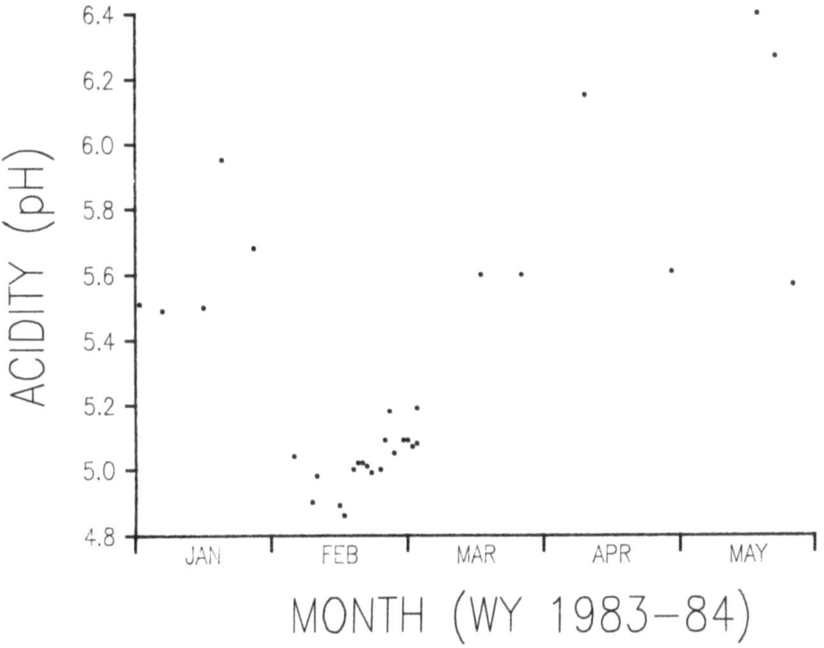

Fig. 2. Values for pH of basal outflow, Central Sierra Snow Laboratory 1983-1984.

These findings suggest that fractionation may have occurred in early winter; by the time the snow was inundated with new percolating water in March, it had already fractionated. This interpretation is supported by stream water pH data. The seasonal pH stream minimum occurred in mid-December, synchronous with a rain-on-snow event that caused elevated stream discharges.

The 1984 NO_3 and SO_4 solute concentrations from the basal outflow were slightly greater than the corresponding snowfall values (Table I). The third highest NO_3 concentration was recorded on February 12 in pan 1 and the highest on March 26 in pan 2 during an outflow event. But during other major outflow events, both NO_3 and SO_4 concentrations approximated or were lower than the mean value. In both pans, maximum SO_4 concentrations were recorded in May. Correlation between pH and NO_3 was relatively high and reached -0.63 (n=25) for one melt pan.

No outflow events were associated with rain on snow during 1985. Basal outflow increased dramatically, however, above ground melt rates of 0.5 mm day^{-1} during warm periods in mid-February and mid-March. Basal outflow pH showed no evidence of acidification for these periods or the main spring melt. An abrupt pH decrease early in the event was not evident and in fact the opposite occurred. The mid-February 1985 pH trend implies an alkalinity pulse; pH increased with basal outflow but diminished much faster than the basal outflow discharge (Figure 3). For the winter as a whole, outflow pH was generally greater than snow precipitation pH. This may have been due to an influx of alkaline dry deposition but information is not readily available to test this hypothesis.

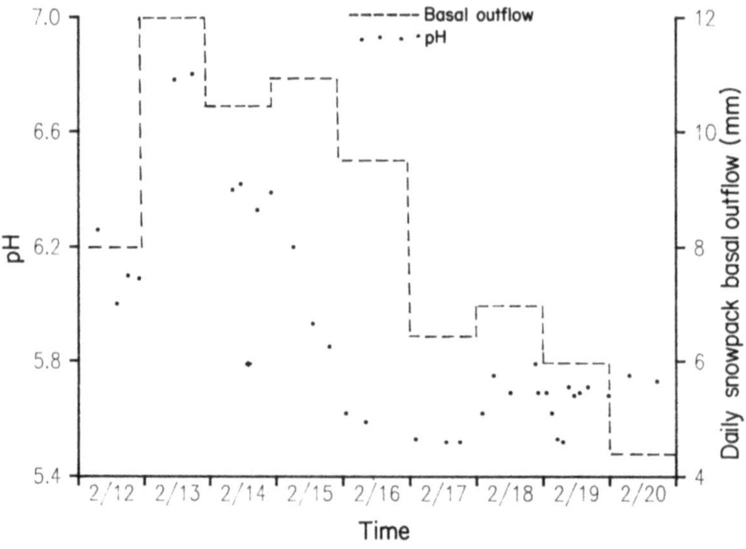

Fig. 3. Snowpack basal outflow pH and rate at the Central Sierra Snow Laboratory, mid-February 1985.

4. SUMMARY AND CONCLUSIONS

- During 2 spring melt periods, no evidence was found for acidification of melt water leaving the base of the snowpack at the Central Sierra Snow Laboratory.
- Fractionation may be occurring prior to spring melt. Lowered pH values for stream and basal outflow water were synchronous with rain or mixed rain and snowfall events during early and mid-winter when streamflow and basal outflow volumes increased.
- Evidence for an alkalinity pulse was observed. During a mid-February melt event, basal outflow increased by 0.6 pH units initially and thereafter decreased much faster than outflow discharge.
- Weighted mean precipitation pH varied with precipitation type and was lowest at the sampled site for hail and graupel (4.7), followed by rain (4.9), snow (5.3), and mixed rain and snow (5.4). Values of pH for snowfall did not vary appreciably between 1984 and 1985.

- Weighted mean NO_3 and SO_4 concentrations were in the range of 0.5 to 1.0 and 4 to 5 μeq L^{-1} for mixed rain and snow and snowfall, respectively. Rainfall and hail and graupel had concentrations elevated appreciably above these levels. SO_4 and NO_3 in the sampled basal snowpack outflow approximated the mean concentrations of snowfall.
- Annual H^+ loadings ranged from 0.07 to 0.1 kg ha^{-1}. NO_3 and SO_4 loadings approximated 1.2 and 4.5 kg ha^{-1} during 1984.
- Correlations between pH and NO_3 and SO_4 were poor. This may have been due partially to the existence of a small number of outliers, often up to one order of magnitude greater than the median NO_3 or SO_4 concentrations. NO_3 and SO_4 data were not as useful in this study as pH data because of their high variability.
- Ratios of median SO_4 to NO_3 concentrations varied from 2.6 to 3.8 for the 4 precipitation types, and from 2.3 to 2.8 for basal outflow.

Future studies on the disposition of impurities in the snowpack should emphasize monitoring of precipitation chemistry by type. Flow-proportional sampling of basal outflow and stream water is recommended to identify chemical changes associated with precipitation events, like rainfall onto the snow surface, which may trigger basal outflow. Layer or snowpack core sampling during rain or wet snowfall is needed to provide the optimum means of identifying chemical fractionation signatures. Information on the variability of precipitation and stream and lake water chemistries on a basin scale is also needed.

ACKNOWLEDGMENTS

This research was partially supported by the Division of Atmospheric Resources Research, Bureau of Reclamation, U.S.D.I. Sheri Woo deserves particular thanks for her assistance in both the lab and field.

REFERENCES

Berg, N.H.: 1982, 'Layer and Crust Development in a Central Sierra Nevada Snowpack: Some Preliminary Observations', Proc. West. Snow Conf., 180.

Colbeck, S.C.: 1981, Water Resources Res. **17**, 1383.

Galloway, J.N., C.L. Schofield, G.R. Hendrey, N.E. Peters and Johannes, A.H.: 1980, 'Sources of Acidity in Three Lakes Acidified During Snowmelt', Proc. Internat. Conf., Ecological Impact of Acid Precipitation, SNSF Proj. Rept. FA 65/80, 264.

Johannessen, M., A. Skartveit and Wright, R.F.: 1980, 'Streamwater Chemistry Before, During and After Snowmelt', Proc. Internat. Conf., Ecological Impact of Acid Precipitation, SNSF Proj. Rept. FA 65/80, 224.

Kattelmann, R.C.: 1985, Ann. Glaciol. **6**, 272.

Roth, P.H., C. Blanchard, J. Harte, H. Michaels, and El-Ashry, M.: 1985, 'The American West's Acid Rain Test', Research Report 1, World Resources Institute, Washington, D.C., 50 pp.

ALUMINUM SPECIATION IN RUNNING WATERS ON THE CANADIAN PRE-CAMBRIAN SHIELD: KINETIC ASPECTS.

P.G.C. Campbell, D. Thomassin and A. Tessier
Université du Québec, INRS-Eau
C.P. 7500, Sainte-Foy
Québec, Canada G1V 4C7

ABSTRACT. To evaluate the extent to which chemical disequilibrium may exist in stream waters following naturally occurring short-term pH changes, surface water samples were collected from two streams in the Lake Laflamme watershed north of Quebec City, placed in large polyethylene containers and stirred continuously at the *in situ* temperature; changes in Al speciation were followed with time for up to 24 hr. The original samples were collected during the spring snow-melt period and during summer low flows. In this latter case, a pH depression was induced by artificial acidification (H_2SO_4; pH 4) and changes in Al speciation with time were followed in both the acidified and the unacidified samples. Aluminum speciation as operationally defined did respond to artificial acidification (pH lowered; non-exchangeable and non-extractable Al decreased), and these pH-induced changes occurred rapidly (< 4 hr). During storage of natural water samples at *in situ* T for up to 24 hr, little change in Al speciation was noted, even in the case of samples collected during spring snow-melt; no clear cases of chemical disequilibrium existed among the studied samples. The distribution of Al among its various physico-chemical forms thus appears to be established within the stream itself; at a given stream sampling point the speciation of Al will not necessarily reflect its geochemical origins, but rather the prevailing *in situ* conditions (e.g. pH, ligand concentrations).

1. INTRODUCTION

Aluminum introduced into surface waters in response to environmental acidification will be partitioned among various possible physico-chemical forms. As these forms may well exhibit different chemical reactivities, it follows that measurement of the total concentration of Al will provide little indication of the metal's potential interactions with the abiotic or biotic components present in the environment; the corollary is that knowledge of the speciation of Al will be necessary for predicting its subsequent geochemical behavior and its biological availability.

Water, Air, and Soil Pollution **30** (1986) 1023–1032.

Running waters on the Canadian Pre-Cambrian Shield, and indeed in other temperate regions of similar hydrogeology, may exhibit short-term (hour to day) pH fluctuations in the pH range 7 to 4, e.g. during spring snow-melt (Jeffries et al., 1979; Johannessen et al., 1980) or autumn spates (Hendershot et al., 1984). Thermodynamic equilibrium calculations indicate that Al speciation should be particularly sensitive to changes in $[H^+]$ in this pH range (pH ↓, degree of complexation ↓). On the other hand, laboratory kinetic studies on synthetic aqueous solutions of Al(III) suggest that some Al hydrolysis reactions may be slow, notably in the presence of polynuclear hydroxy-Al species (Baes and Mesmer, 1976), with the result that equilibrium conditions are approached only very slowly.

In the present study we have considered the sensitivity of Al speciation in natural waters to pH changes and the rate at which any shift in speciation occurs after a change in pH. The study objective was to evaluate the extent to which chemical disequilibrium may exist in natural waters following naturally occurring short-term pH changes. All but one of the samples studied were collected during the spring snow-melt period, i.e. under conditions most likely to favor a state of chemical disequilibrium (changing physico-chemical conditions; T = 0°C, reaction rates slowed; high flow rates, short transit times).

2. MATERIALS AND METHODS

2.1 Sampling

Surface water samples were collected from two streams in the Lake Laflamme watershed, located on the Canadian Shield north of Quebec City (47° 19' N, 71° 07' W; see Jones et al. (1986) for a description of the Lake Laflamme area).

One or more samples were collected each week from the Lake Laflamme inlet (ET-9) during the 1985 snow-melt period (March - May) and changes in the filterable Al concentration and in Al speciation were followed as a function of stream flow and water quality. On three occasions (3 April; 22 April; 14 May) an additional volume of stream water was collected (ca. 20 L), placed in a large polyethylene container and stirred continuously in the field at the in situ temperature. Subsamples were removed over the first 24 hr of storage and subjected to a number of rapid analytical determinations (pH; filterable Al, $[Al]_F$; filterable extractable Al, $[Al]_{FE}$); precautions were taken to minimize sample warming during these experimental manipulations.

In addition, in an experiment designed to evaluate the sensitivity of Al speciation to pH changes, samples from the Lake Laflamme outlet were collected during the summer low flow period and a pH depression was simulated by artificially acidifying one sample with H_2SO_4 (initial pH = 6.7, final pH = 4.0); changes in Al speciation with time were followed in both the acidified and unacidified samples.

2.2 Analysis

Methods used to determine Al speciation usually involved a preliminary size separation step (filtration or equilibrium dialysis) followed by measurements of the chemical reactivity of the Al species thus separated (e.g., chelation-extraction or ion-exchange procedures).

Filtration was carried out with an acid-washed polycarbonate filtration apparatus fitted with a polycarbonate membrane (Nuclepore, 0.4 µm porosity; see Campbell et al. (1983) for details), whereas samples destined for equilibrium dialysis were placed in polyethylene graduated cylinders (sample volume 1 L) and dialysis tubing was suspended in the sample for 24 hr (dialysate volume ca. 40 mL; Spectrapor 6 dialysis tubing, nominal 1000 molecular weight cut-off; 20°C; see LaZerte (1984) for details).

The chelation-extraction procedure involved the following steps (Barnes, 1975; LaZerte, 1984) : (1) addition of ammonium hydroxide and 8-hydroxyquinoline (oxine) to the sample (final pH 8.3); (2) immediate addition of methyl isobutyl ketone (MIBK), followed by vigorous shaking of the sample (total elapsed time 15 s); (3) separation of the two immiscible phases and determination of the concentration of extractable Al, $[Al]_E$, in the organic solvent or, in some cases, the concentration of non-extractable Al remaining in the aqueous phase, $[Al]_{NE}$.

The ion-exchange procedure (Campbell et al., 1983) involved treatment of the sample with Chelex-100 ion-exchange resin (H^+-form, fractionally loaded with Ca and Mg; stirred batch process, 30 min contact time) to remove exchangeable Al, followed by determination of the residual non-exchangeable Al, $[Al]_{NEX}$.

All Al determinations were performed by atomic absorption spectrophotometry with electrothermal atomization (graphite tube). Certain additional water quality parameters were measured: pH, F (ion-selective electrode); SO_4 (ion chromatography); dissolved organic carbon (DOC : combustion + infrared detection CO_2). Standard procedures were used for all these complementary analyses (U.S. EPA, 1979; APHA, 1980).

2.3 Data treatment

The equilibrium speciation of inorganic aluminum was calculated with the MINEQL equilibrium model (Westall et al., 1976), after revision of certain of the original thermodynamic data (see Campbell et al. (1983) for the stability constants that were employed and the original references). The MINEQL program was run in either of two modes: allowing precipitation of $Al(OH)_3(s)$ (as microcrystalline gibbsite, $\log K_{so} = -32.65$) or suppressing this reaction.

3. RESULTS

3.1 Changes in Al speciation and stream water quality during spring snow-melt

The inlet stream ET-9 to Lake Laflamme experienced a notable pH depression (ca. 1.5 pH units) during the snow-melt period (Figure 1).

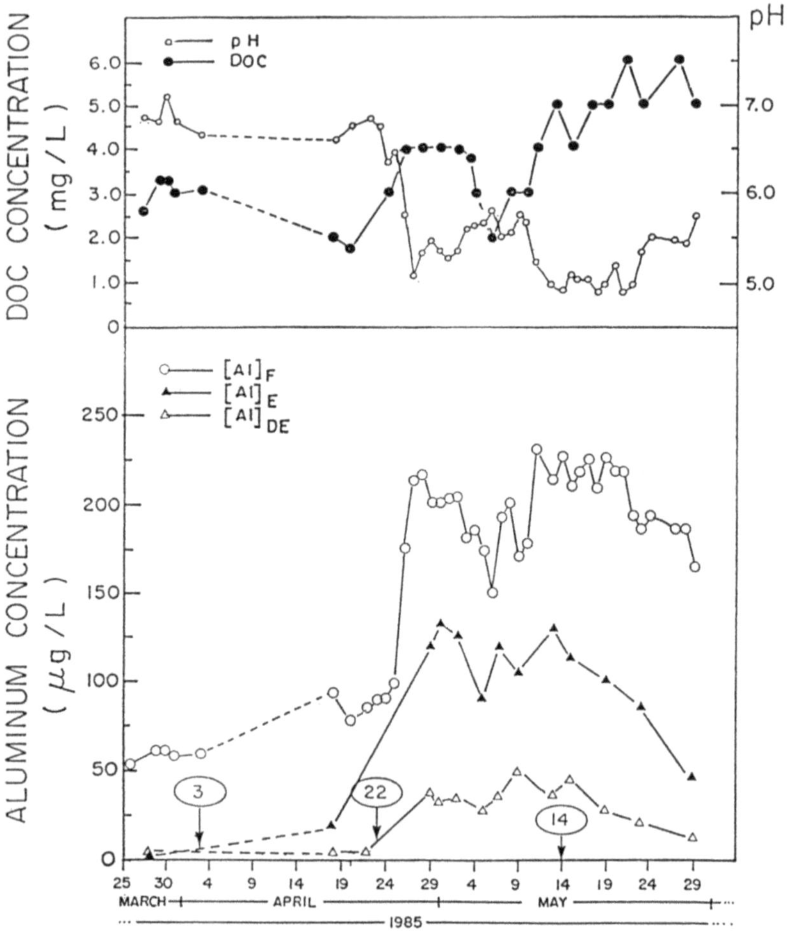

Figure 1: Changes in pH, DOC, filterable Al ($[Al]_F$), extractable Al ($[Al]_E$) and dialysable extractable Al ($[Al]_{DE}$) in the stream ET-9 during spring runoff. Arrows indicate collection dates for the samples used in the 24 hr time-course experiments.

From initial values in the range 7.0 to 6.5, the pH dropped to ca. 5.1 between April 24 and 26, the period that corresponded to the onset of massive snow-melt and spring rains. After a period of 2 weeks during which the stream water pH slowly rose to ca. 5.7, a second drop in pH occurred around May 11 to values < 5.0. Organic C levels also showed considerable variation during spring runoff; from initial values of 2 to 3 mg C L^{-1}, concentrations increased with the onset of snow-melt, decreased again as the stream slowly recovered from the initial pH depression, and then increased steadily over the last 3 weeks to the end of May, attaining values in the range 5 to 6 mg C L^{-1}.

During this same period of pH and DOC changes, the filterable Al concentrations, $[Al]_F$, increased markedly (ca. 2x) despite the increase in stream flow[1]. Such increases in Al concentration with increased flow, though not invariably observed, have been reported for streams on the Canadian pre-Cambrian Shield, in Ontario (LaZerte, 1984) and Quebec (Hendershot et al., 1984), and in the Adirondacks region of New York State (Driscoll et al., 1980). The relative proportions of the different Al forms also changed quite markedly over the study period. Before extensive snow-melt (i.e. before 24 April), concentrations of oxine-extractable Al, $[Al]_E$, and dialysable extractable Al, $[Al]_{DE}$, were very low (5 to 20 µg L^{-1} and ca. 5 µg L^{-1} respectively); these two forms represented < 20% of $[Al]_F$. With the initial pH depression, however, concentrations of both these forms increased abruptly. Levels remained elevated and relatively constant for about 3 weeks ($[Al]_E$ ca. 110 µg L^{-1}, $[Al]_{DE}$ ca. 35 µg L^{-1}), and then declined towards the end of the sampling period.

3.2 Stability of Al speciation after sampling

Large volumes of stream water were collected before and during the snow-melt period (dates indicated by arrows on Figure 1) and stored at the in situ water temperature (0°C); changes in Al speciation were followed with time (Table I).

For all three samples, pH values remained approximately constant over the 24 hr storage period (variability < ± 0.1 pH unit) and no discernable time trend was noted. Similarly, concentrations of filterable Al remained constant after sampling, with no evidence of progressive losses by precipitation or adsorption during storage: 3 April - $[Al]_F$ = 75 ± 1 µg L^{-1}; 22 April - $[Al]_F$ = 105 ± 4 µg L^{-1}; 14 May - $[Al]_F$ = 255 ± 4 µg L^{-1}. In all cases, the coefficient of variation among the 7 successive measurements was comparable to the analytical precision (< 5%). Thermodynamic equilibrium calculations (with $[Al]_{DE}$ as input data) indicated that samples collected during the snow-melt period were undersaturated with respect to micro-crystalline gibbsite (cf. LaZerte, 1984), and thus precipitation

[1] The stream ET-9 was unguaged, but runoff as calculated with a hydrological model (Jones et al., 1986) varied by more than two orders of magnitude during the study period.

Table I: Temporal evolution of pH and Al concentrations during sample storage at *in situ* temperature (Lake Laflamme inlet ET-9).

Date	Parameter	Sub-sampling interval 1	2	3	4	5	6	7
3 April	t(hr)	0.1	1.3	1.8	2.3	3.3	7.8	22.5
	pH	6.60	6.62	6.47	6.45	6.45	6.68	6.67
	$[Al]_F$ (μg L^{-1})	76 ±1	73 ±2	75 ±2	74 ±2	77 ±1	74 ±1	74 ±2
	$[Al]_{FE}$ (μg L^{-1})	32 ±2	26 ±2	26 ±2	28 ±2	26 ±3	27 ±4	24 ±5
22 April	t(hr)	0.2	0.8	1.7	2.3	3.5	8.0	26
	pH	6.36	6.45	6.45	6.49	6.49	6.48	6.48
	$[Al]_F$ (μg L^{-1})	109 ±9	105 ±4	107 ±3	105 ±5	99 ±5	110 ±10	102 ±5
	$[Al]_{FE}$ (μg L^{-1})	48 ±3	40 ±7	39 ±3	43 ±11	42 ±8	34 ±1	30 ±2
14 May	t(hr)	0.1	1.0	1.5	2.0	3.3	7.0	23
	pH	4.88	4.87	4.98	4.95	4.93	4.94	4.78
	$[Al]_F$ (μg/L^{-1})	253 ±7	254 ±2	250 ±10	260 ±3	258 ±6	260 ±15	252 ±7
	$[Al]_{FE}$ (μg L^{-1})	244 ±2	231 ±8	240 ±17	228 ±4	238 ±2	275 ±5	263 ±7

reactions involving $Al(OH)_3(s)$ would not have been anticipated.

The proportion of filterable extractable aluminum differed markedly from one sampling date to another ($[Al]_{FE}/[Al]_F$: 36% before snow-melt, 95% during snow-melt), but its concentration remained approximately constant after sampling : 3 April - $[Al]_{FE}$ = 27 ± 3 μg L^{-1}; 22 April - $[Al]_{FE}$ = 39 ± 6 μg L^{-1}; 14 May - $[Al]_{FE}$ = 246 ± 17 μg L^{-1}. The coefficients of variation for the 7 successive measurements were < 10%, somewhat higher than for $[Al]_F$, but nevertheless comparable to the analytical reproducbility of the chelation-extraction procedure. Only for the sample collected on 22 April did successive determinations show an apparent trend, a slight decrease in $[Al]_{FE}$ with time, but the differences were not statistically significant.

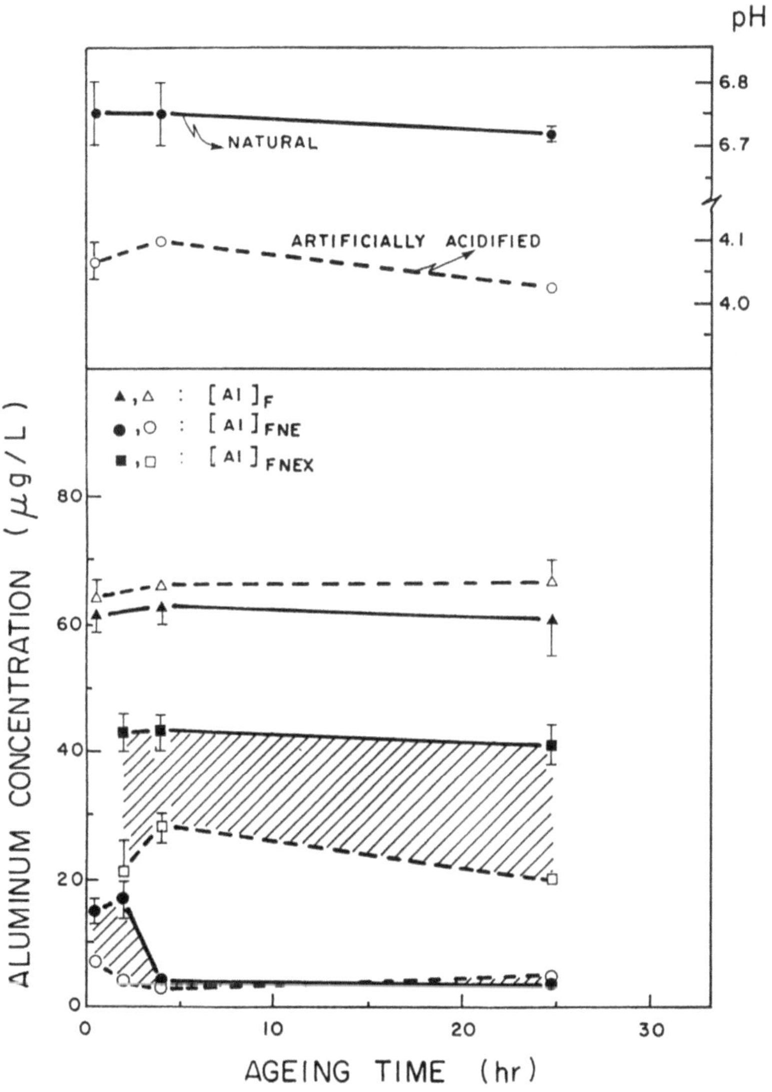

Figure 2: Evolution over 24 hr of a water sample collected during summer low flows (August 28) from the Lake Laflamme outlet: pH, filterable Al ($[Al]_F$), filterable non-extractable Al ($[Al]_{FNE}$), and filterable non-exchangeable Al ($[Al]_{FNEX}$). Solid symbols (Δ ● ■) correspond to the natural water; open symbols (Δ ○ □) refer to an artificially acidified sample.

As for the single sample collected from the Lake Laflamme outlet during stable summer low flows, it too showed no significant changes in pH, filterable Al or filterable non-exchangeable Al on storage (Figure 2): pH 6.74 ± 0.02 (n=3); $[Al]_F$ 62 ± 1 μg L^{-1} (n=3); $[Al]_{FNEX}$ 42 ± 1 μg L^{-1} (n=3). Only for filterable non-extractable Al, $[Al]_{FNE}$, was an apparent time-dependent change noted (see below).

When this sample was acidified to pH 4.0, the concentration of filterable non-exchangeable Al decreased significantly (ca. 50% ; 43 → 21 μg L^{-1}). This shift in speciation was noted as early as the first ion-exchange measurement (< 2 hr), and the difference between the acidified and the natural samples then remained constant over the next 24 hr. The artificial acidification also had a slight effect on the concentrations of filterable non-extractable Al; levels of $[Al]_{FNE}$ were initially slightly lower in the acidified sample, but this difference between acidified and natural samples effectively disappeared after 4 hr, due not to changes in the acidified sample but rather to a decrease in $[Al]_{FNE}$ in the natural samples, as noted above. It seems likely that the original sample contained some unstable polynuclear hydroxy-Al species that converted to monomeric (extractable) form on standing (< 4 hr); the acidification to pH 4.0 accelerated this process, such that the Al speciation in the two samples differred over the first several hours of their parallel storage.

4. DISCUSSION

The results from the induced pH depression demonstrated that Al speciation in a given water sample, as operationally defined, does respond to acidification (pH decreases; $[Al]_{FNEX}$ and $[Al]_{FNE}$ decrease), and that these pH-induced changes occur rapidly. The only comparable results of which we are aware are those recently reported by Driscoll (1984) who bufferred a stream water sample from the Adirondacks at various pH values and noted that the concentration of non-exchangeable Al decreased progressively below pH 5.0; the kinetics of this apparent decomplexation were not reported.

During storage of natural water samples at in situ temperature for up to 24 h, concentrations of both total filterable Al and extractable Al (i.e., $[Al]_F$ and $[Al]_{FE}$) almost invariably proved stable with time, even in the case of samples collected during spring snow-melt. No clear cases of chemical disequilibrium existed among the studied samples.

The distribution of Al among its various physico-chemical forms thus appears to be established within the stream itself. In other words, at a given stream sampling point the speciation of Al will not necessarily reflect its geochemical origins (cf. Hendershot et al., 1984), but rather the prevailing in situ conditions (pH, ligand concentrations, etc.). Consider, for example, the apparent increase in extractable organic Al during the snow-melt period (see Figure 1: $[Al]_E$ - $[Al]_{DE}$ = extractable organic Al; LaZerte, 1984); it would be tempting to relate this increase to the hydrological behavior of the catchment during snow-melt and the flowpath of the melt-water from the snow-pack into the stream (e.g., by involving lateral flow through the upper organic soil horizons). Note, however, that even if the aluminum were derived from the stream bottom (Hall et al., 1980) or from a mineral soil horizon, an increase in organic Al would be anticipated in any case, in response to the increase in DOC concentration (cf. Campbell et al., 1983: effect of addition of fulvic acid to a natural water sample). In other words, the original source of the spring Al pulse cannot be inferred unequivocally from the speciation results. The corollary of these results, provided they prove to be of general applicability, is that the reactions governing Al speciation in surface waters are reasonably rapid and that Al speciation in such systems should be amenable to thermodynamic modelling.

5. ACKNOWLEDGMENTS

The authors thank M. Allard, H.G. Jones and G. Moreau for help in sampling. Financial support for this research was provided by Fisheries and Oceans Canada, and by the Natural Sciences and Engineering Research Council of Canada.

6. REFERENCES

APHA: 1980, Standard methods for the examination of water and wastewater, 15th ed., American Public Health Assoc., Washington, D.C.

Baes, C.F., and Mesmer, R.E.: 1976, The hydrolysis of cations, John Wiley and Sons, New York, N.Y.

Barnes, R.A.: 1975, Chem. Geol., **15**, 177.

Campbell, P.G.C., Bisson, M., Bougie, R., Tessier, A., and Villeneuve, J.P.: 1983, Anal. Chem., **55**, 2246.

Driscoll, C.T.: 1984, Internat. J. Environ. Anal. Chem., **16**, 267.

Driscoll, C.T., Baker, J.P., Bisogni, J.J. and Schofield, C.L.: 1980, Nature, **284**, 161.

Hall, R.J., Likens, G.E., Fiance, S.B., and Hendry, G.R.: 1980, Ecology, **61**, 976.

Hendershot, W.H., Lalande, H., and Dufresne, A.: 1984, Water Pollut. Res. J. Canada, **19**, 11.

Jeffries, D.S., Cox, C.M., and Dillon, P.J.: 1979, J. Fish. Res. Board Canada, **36**, 640.

Johannessen, M., Skartveit, A., and Wright, R.F.: 1980, in Drablos, D. and Tollan, A. [Eds.], Ecological impact of acidic precipitation, Proc. Internat. Conference, Sandefjord, Norway: 224.

Jones, H.G., Sochanska, W., Stein, J., Roberge, J., and Charette, J.Y.: 1986, Water Air Soil Pollut. (this volume).

LaZerte, B.D.: 1984, Can. J. Fish. Aquat. Sci., **41**, 766.

U.S. EPA: 1979, Methods for chemical analysis of water and wastes, U.S. Environmental Protection Agency, Report No. EPA-600 14-79-020, Environ. Monitoring and Support Lab., Office of Research and Development, Cincinnati, Ohio.

Westall, J.C., Zachary, J.L., and Morel, F.M.M.: 1976, MINEQL, a computer program for the calculation of the chemical equilibrium composition of aqueous systems, Massachusetts Institute of Technology, Dept. Civil Eng., Tech. Report No. 18, Cambridge, Massachusetts.

INFLUENCE OF ATMOSPHERIC DEPOSITION ON LAKE MASS BALANCES IN THE TURKEY LAKES WATERSHED, CENTRAL ONTARIO

D.S. Jeffries, R.G. Semkin, R. Neureuther, M. Seymour
National Water Research Institute
Aquatic Ecology Division
Burlington, Ontario, Canada L7R 4A6
J.A. Nicolson
Great Lakes Forest Centre
Sault Ste Marie, Ontario, Canada P6A 5M7

ABSTRACT. Ion mass budgets were measured for 2 water yr (June-May, 1981-83) for a high and a low elevation lake and their associated catchments. The lakes are located in the Turkey Lakes Watershed (TLW) in central Ontario, Canada, which is an undeveloped basin located on the Canadian Shield, 50 km north of Sault Ste. Marie. The ionic budgets of the lakes show that atmospheric deposition directly to the lakes' surfaces is the principal input pathway for H^+ and NH_4, whereas basic cations, SO_4, NO_3, and probably alkalinity are supplied primarily by inflow from the surrounding terrestrial basin and/or upstream lake. The lakes strongly retain H^+ (i.e. output $\ll$ input), weakly retain the N species, and are in balance (i.e. output = input) for other ions except Ca and alkalinity which show an excess output compared to measured + estimated inputs. We hypothesize that an input of groundwater and/or seepage accounts for most of the Ca and alkalinity imbalance although the existence of within-lake alkalinity generation is probable also.

1. INTRODUCTION

A "calibrated basin" study in the Turkey Lakes Watershed (TLW) employs whole ecoystem mass balances to determine the most important biogeochemical factors that regulate the degree and rate of surface water acidification. Dillon et al. (1982) demonstrated the utility of this approach whereas Schindler et al. (1976), Likens et al. (1977), and Galloway et al. (1984) used similar studies to infer the course of specific biogeochemical processes occurring within aquatic or terrestrial ecosystems. In particular, Harvey et al. (1981) have reviewed the results of mass budget studies conducted in terrain sensitive to acidic deposition.

This paper will present major ion mass budgets for a high and low elevation lake and their associated terrestrial catchments located within the TLW. The study period spans 2 yr. The relative importance of differing input sources, acidifying species, and basin supply of alkalinity will be discussed.

Water, Air, and Soil Pollution **30** (1986) 1033–1044.

2. STUDY SITE

The TLW is an undeveloped basin located on the Canadian Shield, 50 km north of Sault Ste Marie, Ontario (Figure 1). It is completely forested (mixed hardwoods), is 10.5 km^2 in area, has over 300 m of relief, and contains a chain of 5 lakes that exhibit a gradient in both water quality (Table I) and response to acidic inputs. The basin is predominantly underlain by metamorphosed basalt (greenstone) and overlain by glacial till derived from the granites that dominate to the north (Semkin and Jeffries, 1983; Kusmirski and Cowell, 1983). The till is of variable thickness and contains a small amount (1 to 2%) of $CaCO_3$, particularly in the thicker deposits located in the lower portion of the watershed. Batchawana Lake is the headwater within the watershed and is separated into 2 distinct sub-basins; mass budgets for the southern sub-basin will be presented here. Batchawana L. South has a mean depth of 3.3 m, maximum depth of 10.9 m, lake area of 5.8 ha, and a total drainage basin area of 85.6 ha. Turkey L. is the last and largest lake in the TLW (mean depth = 12.2 m, max depth = 37.0 m, lake area = 52 ha, total drainage basin area = 803 ha). Water renewal times are 0.30 and 0.94 yr for Batchawana L. South and Turkey L., respectively (Jeffries et al., 1984b). Median major ion concentrations for the lake outflows given in Table I show that there is a 2-fold increase in Ca and >3-fold increase in alkalinity in the downstream direction; SO_4 concentrations are uniform. Further detail concerning watershed and lake characteristics is given by Jeffries and Semkin (1982).

TABLE I Median ion concentrations (ueq L^{-1}) for all samples (n=200) collected during 1980 to 1983 at the outflows of Batchawana Lake South and Turkey Lake.

	Batchawana Lake South	Turkey Lake
H	1	0
Ca	138	271
Mg	40	44
Na	23	27
K	7	7
NH_4	4	3
Alk	47	176
SO_4	123	134
Cl	10	10
NO_3	14	16

3. METHODS

Atmospheric deposition was measured as bulk deposition using a 0.25 m^2 teflon coated, stainless steel collector with a 1 week sampling time; daily measurements of precipitation quantity were obtained using standard rain gauges in summer and Nipher shielded snow gauges in winter (Semkin and Jeffries, 1985).

There are 26 stream stations in the TLW outfitted for continuous monitoring of stream flow. For this study, data from stations S0, S1, S3, S4, F34, F37, and F47 to F49 (Figure 1) were used

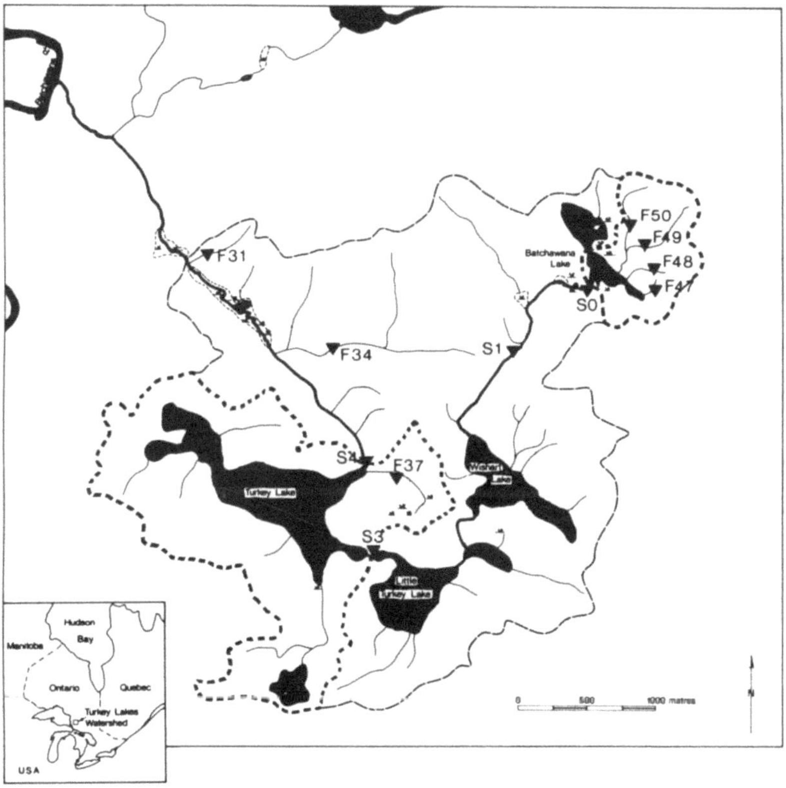

Figure 1. Location map of the Turkey Lakes Watershed showing basin boundary (fine dashed line) and drainage system. Triangular symbols give the location of stream sampling sites mentioned in the text. Boundary of the terrestrial basin immediately surrounding the 2 study lakes is indicated by the bold dashed line.

to determine the sub-basin runoff, inputs to, and outputs from the lakes. In the event that flow data from a station were missing, estimated values were generated from the records of complete data sets using the best available linear regression equation. No flow measurements from Batchawana L. North to Batchawana L. South were available; these were estimated on a monthly basis from the S1 record using the appropriate sub-basin area ratio. Lake Evaporation was calculated from the energy budget (Jeffries, 1983) using daily meteorological and solar radiation data and weekly measurements of lake temperature. Small changes in lake storage were determined from the difference in lake level on May 31 of each year.

Lake and stream samples for chemical analysis were collected on a variable schedule dependent on the hydrological flux within the system; hence, sampling frequency varied from once a month during mid-winter low flows to daily during spring melt peak flows. All chemical variables (H^+, Ca, Mg, Na, K, NH_4, SO_4, Cl, NO_3) were determined using standard techniques (Department of the Environment, 1979) except alkalinity which was analyzed using a Metrohm Model 636 Titroprocessor which depends on second derivative detection of the titration inflection point. Rigorous checking of this procedure has shown that the results obtained are equivalent to those determined using the F_1 Gran function (Stumm and Morgan, 1981). Stream loadings were calculated as the sum of the products of daily stream flow with the less frequent chemical data using the optimal methodology of Scheider et al. (1979). Input loading to Batchawana L. South from the outflow of Batchawana L. North was calculated as the sum of products of monthly estimated outflow (see above) with monthly average surface water and/or epilimnion composition.

Finally, the loading to the lakes from the remaining ungauged portion of the basins was estimated by areal scaling of the mean annual unit export from the gauged sub-basins associated with each of the lakes. While F31 and F34 do not drain into Turkey Lake, water yield and ion export information from these 2 low elevation streams was averaged with that from F37 when estimating the ungauged inputs to Turkey Lake. The sub-basins used to calculate mean annual unit export were selected so as to cover the range of soil and bedrock conditions present in each lake basin and, therefore, provide as reliable a loading estimate as possible. Nevertheless, there is substantial uncertainty in the input loadings derived from these values (see later discussion).

4. RESULTS AND DISCUSSION

4.1 Water Budgets

The water budget for each water year for both lakes is given in Table II. The gauged inflow component of the water budgets include all that associated with monitored inflowing headwater streams plus the outflow from the next higher lake. The ungauged inflow component was estimated by first determining the mean water yield for the

TABLE II Water budgets ($m^3 \times 10^5$) for Batchawana Lake South and Turkey Lake for 06/1981-05/1982 and 06/1982-05/1983.

Input Component	Batchawana Lake South		Turkey Lake	
	1981/82	1982/83	1981/82	1982/83
Precipitation	0.76	0.80	6.8	7.1
Gauged Inflow	2.51	2.44	42.9	47.2
Ungauged Inflow	3.05	3.21	18.5	18.4
Groundwater/Seepage	?	?	?	?
Total	6.32	6.45	68.2	72.8
Output Component				
Outflow	6.13	6.92	66.6	73.7
Evaporation	0.27	0.24	2.8	1.8
Storage	0.00	0.08	-0.9	1.0
Total	6.40	7.24	68.5	76.5
Out/In	1.01	1.12	1.00	1.05

monitored inflowing headwater streams associated with each lake and then extrapolating this value to cover the ungauged area.

Excluding any estimate for groundwater input or an elevation related increase in precipitation quantity (applicable to Batchawana L. only), the water budgets presented in Table II balance within 1% for both lakes in the first water year and to within 12% (Batchawana L. south) and 5% (Turkey L.) in the second water year. In all cases, the water budgets showed excess ouput over input.

Total precipitation for the 2 water yr (06/81-05/82 and 06/82-05/83) was 1304 and 1369 mm, respectively, which makes this a relatively wet environment compared to most of central Ontario (nearby Sault Ste Marie has a mean annual precipitation of 940 mm). This situation arises from the TLW's highland position in the lee of Lake Superior. Semkin and Jeffries (1985) have shown that precipitation quantity even varies up to 15% with elevation within the watershed; however, this factor has not been considered in the following analysis.

The pattern of precipitation was different between the 2 water yr. The first year experienced several heavy rainfall events in June 1981 followed by a relatively dry summer and autumn, a normal winter, and finished with a late spring melt. In contrast, the second water year experienced a very wet autumn, and a warm winter with multiple spring melt episodes. The effect of these differences is clearly evident in the stream hydrographs (e.g. Figure 2). If the "base flow"

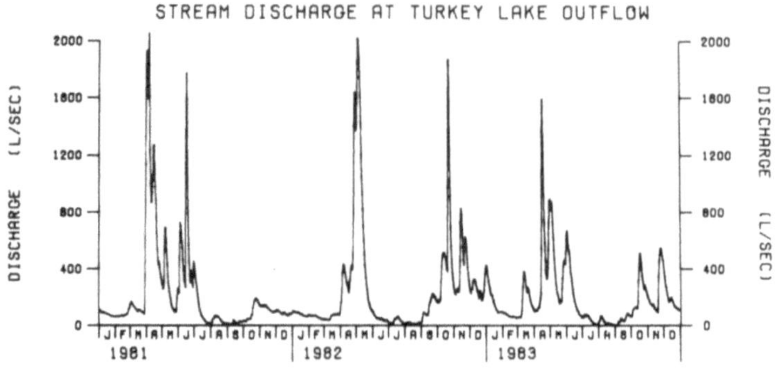

Figure 2. Hydrograph for the Turkey Lake outflow stream spanning the study period.

component of the hydrograph is simplistically defined as that associated with the sustained flow (i.e. that below the short-term peaks associated with specific precipitation or snowmelt events), then it is clear from Figure 2 that the second water year experienced a greater base flow water yield than the first year. It follows from this that the potential for the direct input of groundwater and/or seepage to the lakes was also greater during the second year, a factor important for interpretation of the chemical budgets that follow.

4.2 Chemical Budgets

The ionic budgets (in keq) for both water years are presented in Table III (Batchawana L. South) and Table IV (Turkey L.). In these tables the inputs have been separated into three components; namely, 1) bulk deposition to the lakes' surfaces, 2) terrestrial basin export, and 3) input from the higher lake in the chain (the major inflow for both lakes). The lakes' surfaces constitute 6.8 and 6.5% of the total drainage area for Batchawana L. South and Turkey L., respectively. Likewise, the terrestrial basin draining directly into the lakes is 65.3 and 32.4% of the total area, while the area represented by the inflow from the next higher lake corresponds to 28.0 and 61.2% of the total, respectively. The tables also include the ratio of total output to total input as an indicator of the degree of budgetary balance for each chemical parameter. We recognize that estimation of atmospheric deposition by bulk deposition may be inadequate for certain parameters, particularly for S and N species, due to the large uncertainties associated with collection of the dry component; nevertheless, we have proceeded using this estimate since its percentage of the total input to the lakes was generally small for all parameters except H^+ and NH_4. While a rigorous error analysis is incomplete at this time, we expect that the error in inflow and outflow components of the budets is on the order of ±10%. The

TABLE III Mass budgets (keq) for Batchawana Lake South for 06/1981-05/1982 and 06/1982-05/1983.

Ion	Period	Inputs			Total Input	Total Output	Out/In
		Bulk Deposition	Terrestrial Basin	Major Inflow			
H	81/82	3.1	1.6	0.3	5.1	0.7	0.14
	82/83	4.3	1.4	0.4	6.1	1.2	0.19
Ca	81/82	1.2	48.9	16.0	66.1	81.9	1.24
	82/83	0.5	42.2	20.8	63.4	91.9	1.45
Mg	81/82	0.5	13.6	5.2	19.3	23.3	1.21
	82/83	0.3	12.8	7.4	20.6	27.4	1.33
Na	81/82	0.3	8.3	3.3	11.9	13.5	1.13
	82/83	0.2	8.3	4.3	12.8	15.3	1.08
K	81/82	0.2	2.4	1.3	2.4	4.1	1.20
	82/83	0.1	2.5	1.9	2.5	5.1	1.11
NH_4	81/82	1.6	0.5	1.3	3.4	1.9	0.57
	82/83	1.5	0.4	0.8	2.7	2.3	0.85
Alk	81/82	-3.1*	8.4	5.5	10.7	25.0	2.34
	82/83	-4.3*	4.4	6.0	6.1	28.3	4.68
SO_4	81/82	4.1	53.3	17.5	74.8	77.6	1.04
	82/83	4.5	46.5	20.3	71.3	81.3	1.14
Cl	81/82	0.4	4.1	1.6	6.0	6.5	1.09
	82/83	0.2	3.4	1.9	5.5	7.7	1.39
NO_3	81/82	2.3	9.7	2.0	14.0	9.8	0.70
	82/83	2.3	6.3	2.3	10.8	9.3	0.86

*Alk demand on lake due to bulk deposition taken as negative H.

potential error in the terrestrial basin input (as a whole) is larger since it is composed of the sum of both measured export from gauged headwater sub-basins and the estimated export from the remaining ungauged area (see method description above). For Batchawana L. South, the ungauged component of the input from the terrestrial basin is 36% (or 23% of the total drainage area), while for Turkey L. it is 95% (or 31% of the total drainage area).

The mean annual gross export (meq m^{-2}) and standard deviation for Ca and alkalinity for terrestrial sub-basins associated with Batchawana L. South (i.e. F47 to F50) in the 81/82 and 82/83 water years were as follows: Ca = 85±13 and 74±11; alkalinity = 14±7 and 8±9. Similarly, for sub-basins associated with Turkey L. (F31, F34, and F37), the export values were: Ca = 193±72 and 191±62, alkalinity = 90±17 and 93±9. Jeffries et al. (1984a) have summarized the export of ions measured at several Shield locations in eastern North America. The export of Ca from the terrestrial sub-basins surrounding Batchawana L. South is comparable to that measured at Harp L. (south-central Ontario), generally, less than that measured at Sudbury (where total deposition is higher), and much greater than that found at Rawson L. (northwestern Ontario) or Hubbard Brook (New Hampshire).

TABLE IV Mass budgets (keq) for Turkey Lake South for 06/1981-85/1982 and 05/1982-06/1983.

Ion	Period	Inputs			Total Input	Total Output	Out/In
		Bulk Deposition	Terrestrial Basin	Major Inflow			
H	81/82	28	0	1	29	1	0.04
	82/83	38	0	1	40	0	0.00
Ca	81/82	11	511	929	1451	1638	1.13
	82/83	4	504	1043	1551	1972	1.27
Mg	81/82	4	96	177	277	282	1.02
	82/83	3	99	205	307	342	1.11
Na	81/82	3	60	100	163	167	1.02
	82/83	2	65	119	185	204	0.10
K	81/82	1	20	29	50	49	0.97
	82/83	1	22	36	59	57	0.96
NH_4	81/82	14	2	3	19	15	0.77
	82/83	14	2	3	18	10	1.57
Alk	81/82	-28*	242	504	718	1043	1.45
	82/83	-38*	249	580	788	1260	1.60
SO_4	81/82	37	302	563	902	894	0.99
	82/83	40	262	582	883	936	1.06
Cl	81/82	3	22	45	70	64	0.92
	82/83	2	22	47	71	72	1.02
NO_3	81/82	20	67	79	167	119	0.72
	82/83	20	53	78	152	111	0.73

*Alk demand on lake due to bulk deposition taken as negative H.

Export of Ca from sub-basins at the bottom of the TLW is higher than all of these locations (1 exception at Sudbury). Similarly, gross export of alkalinity at the top of the TLW compares favorably to that observed for Harp L., is greater than that at Sudbury, and less than that at Hubbard Brook; while at the bottom of the watershed, alkalinity export exceeds the levels measured at all the other locations. The high Ca and alkalinity export found in the TLW arises from the $CaCO_3$ present in the tills as noted above. Hence, the lakes in the TLW, particularly those at lower elevations, are only moderately susceptible to the impact of acidic deposition (at least in the short run).

Of the 10 chemical parameters in Tables III and IV, only H^+, NH_4, and NO_3 exhibited net retention by the lakes (i.e. Out/In < 1). Direct atmospheric deposition to the lakes' surfaces consitutes the major input pathway of H^+ and NH_4 due to strong retention of these ions within the terrestrial watershed. That is, annual unit output flux of these 2 parameters was generally <10% of the bulk deposition input flux for all gauged terrestrial sub-basins for both water years. Nitrate is not as strongly retained in the terrestrial basin of the TLW (unit output flux was 24 to 82% of bulk deposition input flux depending on sub-basin and year) as it is in other locations in south-central Ontario (Jeffries et al., 1984a) although the situation is similar to that found at Hubbard Brook, New Hampshire (Likens et al., 1977). Therefore, drainage from the surrounding basin is an important input source of NO_3 for the lakes. Of the total input to the lakes, 81 to 100% of the H^+, and only 14 to 43% of the N species are retained. Both N species exhibit a seasonal pattern in stream concentration (i.e. higher in winter and spring and lower in summer and autumn, presumably due to biological utilization). The large flux associated with spring runoff assures the relatively low N retention demonstrated by these lakes. Retention of NH_4 by the terrestrial basin through oxidation and incorporation in the biomass (or any other process involving oxidation) is an acidifying process while retention of NO_3 (which involves reduction) is an acid neutralizing process (Reuss, 1975). Hence, the strong retention of NH_4 and the smaller retention of NO_3 show that both N species as well as H^+ are playing a significant role in acidifying the terrestrial catchment and lakes of the TLW.

All remaining ions exhibited approximately equal or greater output than input for both lakes, and the dominating source was the terrestrial basin for Batchawana L. South and the major inflow for Turkey Lake. Dominance by differing input components for the 2 lakes simply reflects the drainage area associated with each component. The imbalance of output to input is sufficiently small for Mg (Turkey L. only), Na, K, SO_4, and Cl (1 exception) to assume that the lakes are in a steady state for these parameters.

Since the output from the lakes is reasonably precise (i.e. direct measurement involved; no estimation or assumptions required), the apparent excess output over input of Ca and alkalinity (both lakes) and Mg (Batchawana only) must be due to errors in the estimation of input components in Tables III and IV and/or the existence of an unaccounted input source. Estimation of the input from ungauged portions of the terrestrial basin contains substantial uncertainty since both the annual unit discharge and unit export of Ca and alkalinity (means and standard deviations given above) varied greatly among the gauged sub-basins. For Batchawana L. South, this potential error applies to 36% of the terrestrial basin; even if the resultant error was 100%, the "extra" Ca and alkalinity input would still be insufficient to balance the mass budgets for this lake except for Ca in 81/82. Since the proportion of ungauged area within the terrestrial basin is much greater for Turkey L. than it is for Batchawana, the potential error in terrestrial input may account for the Ca imbalance at Turkey L. However, this is probably not the case for alkalinity since variability in unit export among the 3 low elevation sub-basins was relatively smaller. It should be noted that input errors of this type would be expected to influence both ions in the same manner, i.e. either increasing or decreasing the supply of both at the same time, not increasing one while decreasing the other.

Another possible (and unmeasured) input source is groundwater entering the lake either by emergence at nearshore (and therefore ungauged) terrestrial locations, or as seepage upwards through bottom sediments. While the water flux associated with such an input source could be small, the enhanced chance for groundwater interaction with the $CaCO_3$ in the tills may result in a significant chemical flux of Ca and HCO_3. Support for this hypothesis is found in the fact that the excess output was greater in the second year compared to the first when a greater groundwater influence is apparent in the hydrographs (see discussion above); furthermore, the excess ouput of Ca approximately matches the excess alkalinity output for Batchawana L. South for both water years. Also, the difference in the Ca and alkalinity imbalances between years is greater than the difference in Mg imbalances. Note that the potential influence of groundwater input and errors made in estimating input from ungauged terrestrial sources (above) are not necessarily mutually exclusive.

Finally, it is possible that the excess Ca and alkalinity output was derived in part from a within-lake alkalinity generating mechanism. Schindler (1986) has reviewed the processes which contribute to within-lake alkalinity generation. Nitrate and SO_4 removal at the sediment interface, and Ca exchange with the sediments are the predominating mechanisms. Calcium exchange, like unmeasured groundwater seepage or erroneous terrestrial output discussed above, would increase the apparent lake output of Ca and alkalinity by approximately an equivalent amount; independent evidence exists that this mechanism is operating in the TLW (Schiff and Anderson, 1986). With the data presently available, it is impossible to distinguish the relative importance of the sedimentary source of Ca and alkalinity compared to groundwater seepage.

The excess alkalinity output for Turkey L. is greater than the excess Ca ouput, particularly for 81/82. This observation suggests the existence of still another source of alkalinity that is independent of Ca input, most probably related to SO_4 and/or NO_3 reduction. If SO_4 reduction does contribute to the excess alkalinity output in these lake systems, then it must do so within the confines of the error of the apparently balanced SO_4 budget, or else there is some other additional source of SO_4 such as groundwater seepage. As discussed by Schindler (1986), the relative importance of NO_3 and SO_4 reduction mechanisms for internal alkalinity generation is less important for lakes with short residence times such as these in the TLW.

Taken altogether, i.e. budgetary uncertainties and additional external and internal sources, these factors may explain the excess Ca and alkalinity outputs for the lakes in the TLW. Calculations of budgets for more years and other lakes in the chain, and consideration of other related information may help to verify which mechanisms are most important.

5. CONCLUSIONS

Mass budget studies in the TLW have yielded the following conclusions:

(1) Lake outputs approximately equal measured plus estimated plus inputs for Na, K, SO_4, Cl, and Mg (sometimes); lake concentrations for these ions are in a steady state;

(2) Almost all H^+ but only a part of the NH_4 and NO_3 entering the lakes are retained. The dominant input pathway for the former 2 species is via atmospheric deposition directly to the lakes' surfaces, while the dominant pathway for NO_3 is via surface runoff. All 3 species play an acidifying role in the TLW.

(3) Excess ouput of Ca and alkalinity over input suggests the presence of unaccounted sources for these ions. Potential sources include groundwater seepage (probable), Ca exchange with the sediments, and perhaps SO_4 reduction.

6. ACKNOWLEDGMENTS

Personnel from the Water Survey of Canada were responsible for converting continuous stream water level measurements into useable daily flow data. Chemical analyses were performed by D. Kurylo and co-workers in the GLFRC LRTAP laboratory, Sault Ste Marie. Personnel from the Data Management Section, NWRI supplied the information and some of the programs necessary for efficient storage and subsequent processing of the vast amount of data used in this paper. Finally, 2 anonymous referees took the time to provide excellent reviews of this paper; their comments have resulted in much improved product. We appreciate their efforts.

7. REFERENCES

Department of the Environment: 1979, Analytical Methods Manual. Inland Waters Directorage, Water Quality Branch, Ottawa, Canada.

Dillon, P.J., Jeffries, D.S., and Scheider, W.A.: 1982, Wat. Air Soil Pollut., **18**, 241.

Galloway, J.N., Altwicker, E.R., Church, M.R., Cosby, B.J., Davis, A.O., Hendry, G., Johannes, A.H., Nordstrom, K.D., Peters, N.E., Schofield, C.L., and Tokos, J.: 1984, 'The Integrated Lake-Watershed Acidification Study Volume 3: Lake Chemistry Program'. Report No. EA-3221, **3**, Electric Power Research Institute, Palo Alto, California.

Harvey, H.H., Pierce, R.C., Dillon, P.J., Kramer, J.R., and Whelpdale, D.M.: 1981, 'Acidification in the Canadian Aquatic Environment'. National Research Council Canada, NRCC Rep. No. 18475, Ottawa, Canada, 369 pp.

Jeffries, D.S.: 1983, 'Lake evaporation and energy budgets in the Turkey Lakes Watershed (1980-98)'. Turkey Lakes Watershed Unpublished Report No. 83-04, 57 pp.

Jeffries, D.S., Scheider, W.A. and Snyder, W.R.: 1984a, 'Geochemical interactions of watersheds with precipitation in areas affected by smelter emissions near Sudbury, Ontario'. In Nriagu, J.O. (ed.), Advances in environmental science and technology, Vol. 15, Environmental impact of smelters. John Wiley and Sons, New York, 195-241.

Jeffries, D.S., Semkin, R.G., and Neureuther, R.: 1984b, 'Data Report: Lake outflow hydrology (1980-1983)'. Turkey Lakes Watershed Unpublished Rep. No. 84-01, 30 pp.

Likens, G.E., Bormann, F.H., Pierce, R.S., Eaton, J.S., and Johnson, N.M.: 1977, Biogeochemistry of a forested ecosystem. Springer-Verlag, New York, 164 pp.

Reuss, J.S.: 1975, 'Chemical and biological relationships relevant to ecological effects of acid rain'. US EPA 660/3-75-032, Corvallis, Oregon.

Scheider, W.A., Moss, J.J., and Dillon, P.J.: 1979, 'Measurement and uses of hydraulic and nutrient budgets'. Proc. Conf. on Lake Restoration, August 22-24, 1978, US EPA 440/5-79-001, Washington, D.C.

Schiff, S.L. and Anderson, R.F.: 1986, Wat. Air Soil Pollut., in press.

Schindler, D.W.: 1986, Wat. Air Soil Pollut., in press.

Schindler, D.W., Newbury, R.W., Beaty, K.G., and Campbell, P.: 1976, J. Fish Res. Board Can., **33**, 2526.

Semkin, R.G. and Jeffries, D.S.: 1983, 'Rock chemistry in the Turkey Lakes Watershed'. Turkey Lakes Watershed Unpublished Rep. No. 83-03, 9 pp.

Semkin, R.G. and Jeffries, D.S.: 1985, 'Bulk deposition of ions in the Turkey Lakes Watershed'. Turkey Lakes Watershed Unpublished Rep. No. 85-01, 24 pp.

Stumm, W. and Morgan, J.J.: 1981, Aquatic Chemistry 2nd Edition, John Wiley and Sons, New York, 780 pp.

ACID NEUTRALIZING CAPACITY OF LAKES IN THE NORTH CASCADES AREA OF WASHINGTON STATE

D.F. Brakke and T.J. Loranger
Institute for Watershed Studies
Western Washington University
Bellingham, WA 98225

ABSTRACT: Thirty three lakes were surveyed in 1983 in the North Cascades area of Washington State and 27 additional lakes were surveyed in 1984 to characterize lake chemistry in the area. Lakewater ANC was less than 100 ueq l^{-1} for 37% of the lakes and less than 200 ueq l^{-1} for 68% of the lakes. The North Cascades are very similar to unimpacted sensitive areas of North America and Northern Europe in mean alkalinity and (Ca + Mg). Bedrock geology appears to have a significant influence on surface water ANC with lakes on granitic and metamorphic bedrocks having the lowest ANC.

1. INTRODUCTION

Acidification of low alkalinity surface waters has been documented for several areas in North America (U. S. - Canada MOI). Until recently, research on surface waters sensitive to acidification has been restricted mainly to Eastern North America. This has also been the case for other aspects of atmospheric deposition research.

Research on acidic deposition in the Pacific Northwest was reviewed by Vong and Larson (1983). They reported scattered chemical information from Cascade Lakes from a number of sources. They pointed out difficulties at present in evaluating some of the information because of different and changing methodologies. Subsequently, two reviews (Tetra Tech, 1985; Roth et al., 1985) have drawn attention to the very low ionic strengths of Cascade lakes and their proximity to emissions sources.

We conducted a survey of 33 Cascade Lakes in Northwestern Washington, in late summer-early fall of 1983 to characterize the chemistry of the lakes. The lakes ranged in elevation from 853 to 1676 m. The lakes selected were low color, oligotrophic, and predominantly located on granitic or gneissic bedrock. These results have been reported by Brakke (1984) and Brakke and Waddell (1985). In 1984, our sampling included 27 additional lakes and re-surveyed others to give broader geographic coverage and estimates of year-to-year and seasonal variability.

Water, Air, and Soil Pollution **30** (1986) 1045–1053.

2. METHODS

The chemical methods and parameters selected for analysis were derived from the draft "Sampling and Analysis Protocol for Long-Term Monitoring Sites" developed by the National Acid Precipitation Assessment Program (1983) and the Methods Manual for the National Surface Water Survey (EPA, 1985). The parameters and methods of analysis are given in Table I, along with the EPA method number where applicable. For some of the methods no EPA accepted method was available: acid neutralizing capacity (ANC), ion chromatography (IC) analyses of anions, and Al. Analyses of ANC followed Stumm and Morgan (1982). Manufacturer's instrument instructions and the National Surface Water Survey Methods Manual were followed for IC (Dionex).

Ten percent of the analyses were replicated for all parameters, except P, which was run in triplicate for each sample. The laboratory also ran routinely external standards from U.S. EPA for nutrients and minerals.

Table I. Methods of Analysis

Parameter	Instrumentation	EPA Method No.[1]
pH	Orion pH meter	150.1
Acid Neutralizing Capacity	Gran plot	---
Specific Conductance	Wheatstone bridge	120.1
Temperature	Thermistor	170.1
Color	Pt-Co comparator	110.2
Ca	AA flame	215.1
Mg	AA flame	242.1
Na	AA flame	273.1
K	AA flame	258.1
SO_4	IC, autoanalyzer	375.2
NO_3-N	Autoanalyzer	352.1
Cl	Ion-sel. electrode, IC	---
SiO_2	Autoanalyzer	370.1
TP	Autoanalyzer	365.4
SRP	Autoanalyzer	365.3
NH_3-N	Autoanalyzer	350.1
Fe	AA flame	236.1
Mn	AA flame	243.1

[1]Methods for Chemical analyses of Water and Wastes (USEPA 1983)

3. DESCRIPTION OF STUDY LAKES

The location of the 33 lakes sampled in 1983 and the 28 lakes sampled in 1984 is shown in Figure 1. With the exception of two lakes, the extensive survey in 1984 covered an area to the east of the 1983 survey.

The lakes were small (2 to 66 ha; 40 were < 10 ha) and ranged in elevation from 1130 to 2300 m. Seventeen lakes sampled in 1984 were > 1500 m, whereas most of the 1983 lakes were between 1200 to 1500 m.

Tertiary granites and gneisses were the most common bedrock types. Three lakes were on Quarternary volcanic rocks and two of the lakes were on dunite (olivine).

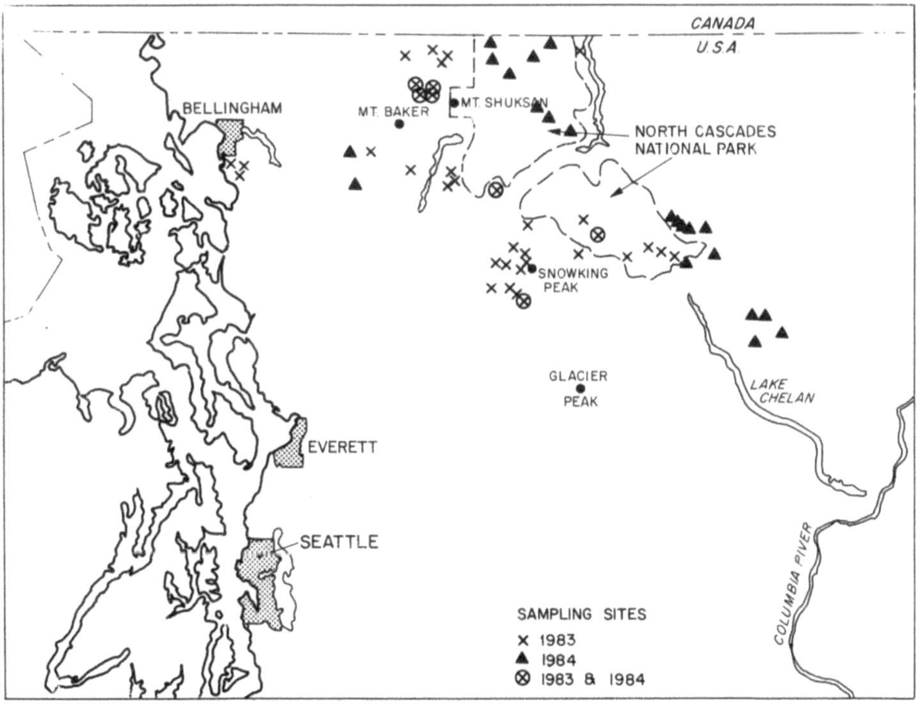

Figure 1. Location of lakes sampled in 1983 and 1984.

Table II. Geographic and physical characteristics of study lakes

Lake	Latitude	Longitude	Altitude (m)	Lake Area (ha)	Geology[1]
Blue	48°38"	121°47"	1220	5.3	CP
Bearpaw Mt.	48°56"	121°50"	1356	2.6	Jks
Big Watson	48°39"	121°35"	1330	18.5	Pjgs
Little Watson	48°39"	121°35"	1360	7.3	
Iceberg	48°50"	121°44"	1460	14.8	QV
Hayes	48°50"	121°44"	1460	5.3	QV
Upper Bagley	48°50"	121°41"	1292	3.6	QV
Lower Bagley	48°50"	121°42"	1280	4.5	QV
Ann	48°48"	121°39"	1520	2.4	Pjgs
Tomyhoi	48°58"	121°41"	1128	33.2	Cp
Upper Twin	48°56"	121°39"	1585	6.9	Cp
Lower Twin	48°56"	121°39"	1580	8.1	Cp
Anderson	48°39"	121°35"	1364	1.6	Pjgs
Doubtful	48°28"	121°03"	1641	12.1	Pjgn
Trapper	48°26"	121°00"	1269	59.4	Pjgn
Moraine	48°31"	121°05"	1372	21.0	Pjgn
Greenview	48°27"	120°54"	1663	16.3	Pjgn
Monogram	48°33"	121°17"	1478	11.3	Pjgn
Hozomeen	48°57"	121°03"	853	44.9	Cp
Thornton	48°40"	121°20"	1360	22.7	Pjgn
Pine	48°39"	122°28"	479	2.7	Tkc
Cedar	48°39"	122°28"	466	1.7	Tkc
Fragrance	48°38"	122°28"	312	2.6	Tkc
Boulder	48°19"	121°19"	1430	22.3	Pjgn
Slide	48°25"	121°20"	1006	15.0	Tkg
Louise	48°23"	121°27"	1524	2.4	Pjs
Cliff	48°22"	121°20"	1463	2.0	Mzg
Jug	48°23"	121°21"	1158	8.1	Mzg
Granite 3	48°27"	121°19"	1372	15.5	Pjs
Whale	48°27"	121°17"	1402	20.3	Tkg
Martin	48°27"	121°25"	1417	5.0	Tkg
Upper Falls	48°27"	121°24"	1372	9.1	Tkg
Arrowhead	48°25"	121°17"	1372	6.0	Tkg
Rainbow	48°24"	120°45"	1676	5.0	Tk
Waddell	48°26"	120°49"	1503	4.2	Tk/Pjgn
Elbow	48°41"	121°56"	1036	2.0	Cp
Heart	48°37"	121°57"	1250	6.1	TD
Wiseman	48°41"	121°57"	1130	7.5	TD
Copper	48°54"	121°27"	1700	3.2	Tg
Kwahnesom	48°60"	121°27"	1550	6.5	Tg
Silver	48°59"	121°14"	2050	66.4	Tg
Bear	48°56"	121°18"	1750	9.3	Pjgn
Tapto	48°52"	121°22"	1750	4.1	Tg
Luna	48°48"	121°17"	1500	6.9	Pjgn

Table II (cont.)

Lake	Latitude	Longitude	Altitude (m)	Lake Area (ha)	Geology[1]
Azure	48°45"	121°15"	1240	36.0	Pjgn
Sourdough	48°45"	121°08"	1400	13.5	Pjgn
Ann	48°30"	120°46"	1700	7.8	Tkg
Rainy	48°30"	120°45"	1460	21.9	Tkg
Blue	48°30"	120°40"	1920	6.2	Tkg
Wing	48°30"	120°49"	2100	6.1	Tkg
Lewis	48°30"	120°47"	1750	3.6	Tkg
Mcalister	48°25"	120°41"	1700	6.2	Tkg
Louis	48°24"	120°35"	1650	11.1	Tkg
Middle Oval	48°16"	120°27"	2040	8.6	Mzg
West Oval	48°16"	120°27"	2070	7.3	Mzg
Surprise	48°14"	120°27"	2000	19.3	Mzg
Libby	48°15"	120°19"	2300	3.9	Mzg

[1]Cp = Carboniferous-Permian sedimentary and volcanic rock
Pjs = metamorphic (high to medium grade)
Pjgn = gneiss
Pjgs = low to medium grade metamorphic
Qv = Quaternary volcanic
Tk, Tkg, Tg = Tertiary granitic
Tkc = Chuckanut sandstone
TD = dunite
Mzg = Mesozoic granitic

4. RESULTS AND DISCUSSION

ANC for the lakes sampled in 1984 ranged from 12 to 281 ueq l^{-1}, of which six lakes were < 40 ueq l^{-1} and 15 were < 200 ueq l^{-1}. The pH values ranged from 6.19 to 7.53, with eight lakes having pH < 7.0. Of the 57 lakes sampled in 1983 and 1984, nine lakes had ANC < 50 ueq l^{-1} and 38 of 57 were < 200 ueq l^{-1}.

Specific conductance was low (4 to 34 uS) as were nutrient concentrations. Nitrate values were not detectable using ion chromatography, while total phosphorus (TP) and soluble reactive phosphorus (SRP) concentrations were < 2 ug l^{-1}.

Calcium was normally the dominant cation, with the exception of two lakes (Weisman and Hart), which are located on dunite, a nonfelspathic plutonic rock that is rich in Mg and contains little if any Ca. If Ca and Mg are summed and plotted against ANC, they yield a highly significant regression equation with a slope slightly less than 1.0.

One primary area of low ANC in the northern portion of the North Cascades National Park Complex. In this area, six of eight lakes had ANC < 40 ueq l^{-1}, with the two other lakes at 110 and 130 ueq l^{-1}. In other areas, the variation in ANC between sites was considerable. In the Mt. Baker - Mt. Shuksan area, ANC varied from 5 to 550 ueq l^{-1}. For the Snowking Peak area, ANC concentrations were 50 to 370 ueq l^{-1}.

Within the southern portion of the North Cascades Park complex ANC was 30 to 250 ueq l^{-1}, whereas it was 120 to 280 ueq l^{-1} in the Washington and Rainy Pass area just to the east. The four lakes sampled in the Lake Chelan area ranged from 70 to 200 ueq l^{-1}.

The distributions of ANC were related to bedrock geology. Lakes on Carboniferous-Permian sedimentary and volcanic rocks had the highest ANC values. The lowest values were on granites, granitic gneisses and metamorphic equivalents (Figure 2). The lake survey of 1984 included lakes in the northern section of North Cascades National Park and had lower ANC values than the lakes samples in 1983. Elevation and lake size did not correlate with ANC. Those lakes with the lowest ANC values ranged greatly in area, whereas the smaller lakes (< 15 ha) had a wide range of ANC, indicating again the pronounced influence of bedrock geology.

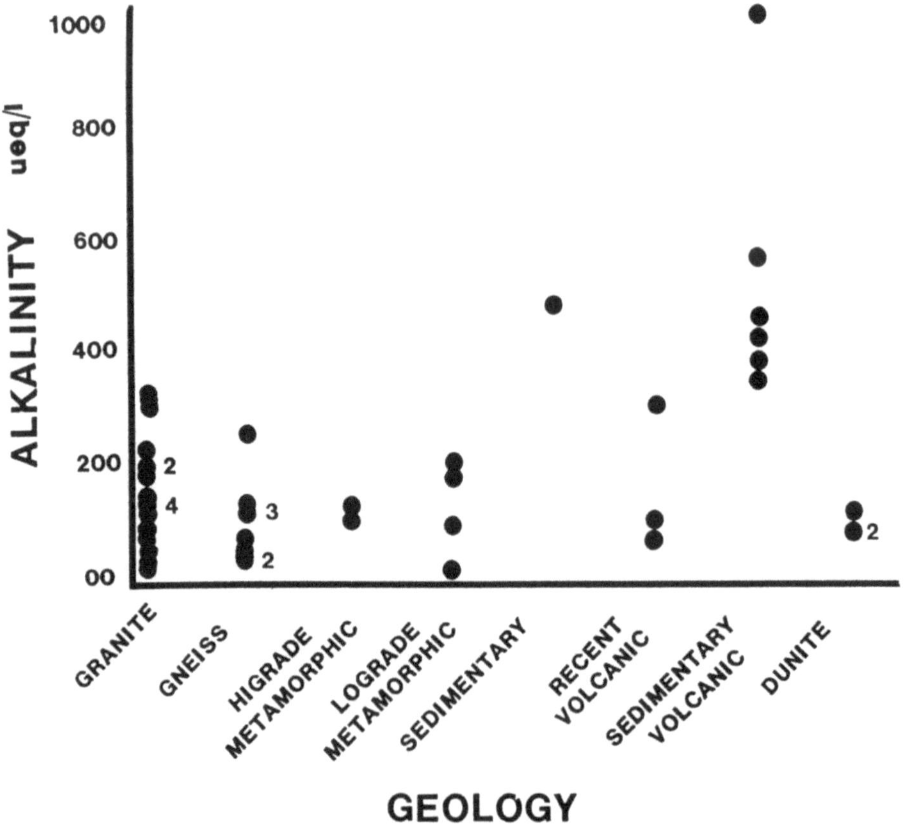

Figure 2. Alkalinity versus geology for North Cascades Lakes.

Regional water chemistry surveys have been conducted in many areas of North America and Scandinavia. These have been useful in characterizing natural background concentrations of alkalinity and other constituents in sensitive regions as well as for developing empirical relationships of loss of alkalinity due to acidic inputs.

Table III summarizes alkalinity data for lakes in sensitive regions of the Western U.S. The North Cascades lakes survey did not show any acidic lakes (pH 5.0), as occurs in regions affected by acidification (Table 5). The pH distribution for the Cascade lakes was similar to background areas (e.g. Northern Norway) where no effects due to acidification have been observed.

Table III. Summary of ANC distributions by area in the Western United States

Percent of lakes in ANC range (ueq/1)

Area	< 50	50-100	101-200	> 200	n	Source
North Cascades, WA	8	28	28	36	36	Brakke, 1984
North Cascades, WA	29	10	33	28	21	Brakke, 1985
Central Cascades, WA	20	50	17	13	30	Duncan, 1984
Alpine Lakes Wilderness, WA	86	14	0	0	7	Welch et al, 1984
Central, Southern Cascades, WA	35	19	12	35	26	Duncan, 1985
Central Cascades, WA	42	45	6	7	31	Logan et al, 1982
Olympics WA	0	0	29	71	7	Welch et al, 1985
Cascades OR	40	11	24	26	63	Nelson and Deluriche, 1983
Yellowstone WY	4	6	15	75	109	
Siskiyou CA	7	21	7	65	14	Wilson & Wood, 1984
Marble Mtns CA	0	0	66	33	3	Wilson & Wood, 1984
Scott-Salmon CA	11	26	37	26	19	Wilson & Wood, 1984
Trinity Divide, CA	14	29	14	43	14	Wilson & Wood, 1984
Caribow Wilderness, CA	0	33	33	33	6	Wilson & Wood, 1984
Trinity Alps Wilderness, CA	44	26	15	15	27	Wilson & Wood, 1984
Flat Tops Wilderness, CO	0	8	18	74	27	Turk & Adams, 1983

Table III (cont.)

Area	50	50-100	101-200	200	n	Source
Rocky Mt. NP CO	50	25	25	0	4	Baron, 1983
Big Horn Wilderness, WY	33	33	33	0	6	EPA Reg VIII
Uintas Wilderness, UT	20	70	10	0	20	EPA Reg VIII
San Juan-Weminuche Wilderness	33	0	8	58	12	EPA Reg VIII

Within the Western U.S., the distribution of ANC varies considerably between areas (Table III). Several portions of the Cascades, particularly the Alpine Lakes Wilderness, have a high proportion of lakes with ANC < 50 ueq l^{-1}. The variability between and within the areas is probably related to bedrock and soils. The North Cascades contains areas of low alkalinity, but also has considerable variability due to bedrock type and exposure.

The sum of non-marine Ca + Mg vs alkalinity for the North Cascades area also indicates that normal carbonation reactions are controlling the production of alkalinity from the watersheds. This regression is the basis of the empirical modelling approach of Henriksen (1980) and can be used to infer acidification or predict response to acidic loadings (Henriksen 1982). The North Cascades are similar to other unimpacted sensitive areas in terms of mean alkalinity and (Ca + Mg). The other areas compiled by Henriksen (1980b) are from various places, but have in common granitic or other metamorphic bedrock that weathers slowly, generating little alkalinity from the watersheds.

5. CONCLUSION

1. The results from a survey of 57 lakes in the North Cascades during 1983 and 1984 indicate that lakewater ANC was less than 100 ueq l^{-1} for 36 to 38% of the lakes and less than 200 ueq l^{-1} for 64 to 72% of the lakes sampled.

2. The North Cascades are similar to unimpacted sensitive areas of North America and Northern Europe in mean alkalinity and (Ca + Mg).

3. Bedrock geology had a significant affect on surface water ANC.

6. ACKNOWLEDGEMENTS

Funding for this project was provided by the U.S. EPA Region 10, Seattle, Washington and the Washington State Department of Ecology.

7. REFERENCES

Baron, J.:1983, Wat. Res. Bull. **19**, 897.

Brakke, D.F.:1983, Environmental Conservation **10**, 75.

Brakke, D.F.:1984, Final Report to Washington State Dept. of Ecology.

Brakke, D.F. and Waddell, D.F.:1985, Env. Canada Sp. Symp. (in press).

Duncan, L.D.:1985, Report to Wash. Dept. Ecology. 19p.

Henriksen, A.:1980, SNSF project, Oslo.

Henriksen, A.:1982, NIVA, Oslo.

Logan, R.M., Derby, J.C. and Duncan, L.C.:1982, Env. Sci. Technol. **16**, 771.

Nelson, P.O. and Delwiche, G.K.:1983, WWRI, OSU, Corvallis. 56p.

Roth, P., Blanchard, C., Harte, J., Michaels, H. and El-Ashry, M.T.:1985, World Resources Institute Res. Report **1**. 50p.

Tetra Tech.:1985, Report for Wash. State Dept. of Ecology.

Turk, J.T. and Adams, D.B.:1983, Wat. Resources **19:34**, 6-350.

U.S. EPA:1983, Methods for chemical analysis of water and wastes. EPA-600/4-79-020.

U.S. EPA:1985, Methods manual for National Surface Water Survey. Phase IA1. (in press).

Vong, R.J. and Larson, R.V.:1983, Puget Sound Power & Light, Report 1983-1.

Welch, E.B. and Spyridakis, D.E.:1984, Final report to Wash. Dept. of Ecology. 16p.

Welch, E.B. Spyridakis, D.E. and Smayda, T.:1985, Final report to Wash. Dept. of Ecology. 22p.

Wilson, D. and Wood, R.:1984, Dept. of Fish and Game.

MIX
Papier aus verantwortungsvollen Quellen
Paper from responsible sources
FSC® C105338

If you have any concerns about our products,
you can contact us on
ProductSafety@springernature.com

In case Publisher is established outside the EU,
the EU authorized representative is:
Springer Nature Customer Service Center GmbH
Europaplatz 3, 69115 Heidelberg, Germany

Printed by Libri Plureos GmbH
in Hamburg, Germany

ACIDIC PRECIPITATION

ACIDIC PRECIPITATION

Proceedings of the
International Symposium on Acidic Precipitation,
Muskoka, Ontario, September 15–20, 1985

Edited by

HANS C. MARTIN

Atmospheric Environment Service, Downsview, Ontario

PART 1

Reprinted from Water, Air, and Soil Pollution, Volume 30 (1986)

Springer-Science+Business Media, B.V.

Library of Congress Cataloging-in-Publication Data

CIP

International Symposium on Acidic Precipitation
(1985 : Muskoka, Ont.)
Acidic precipitation.

'Reprinted from Water, air, and soil pollution, volumes 30 and 31 (1986).'
Includes bibliographies and indexes.
1. Acidic precipitation (Meteorology)—Environmental aspects—Congresses. 2. Acid deposition—Environmental aspects—Congresses. 3. Acid pollution of river, lakes, etc.—Congresses. 4. Air—Pollution—Environmental aspects—Congresses. I. Martin, H. C. II. Title.
QC924.5.I57 1985 363.7'386 86-33883

ISBN 978-94-010-8019-4 ISBN 978-94-009-3385-9 (eBook)
DOI 10.1007/ 978-94-009-3385-9

Originally published by D. Reidel Publishing Company, Dordrecht, Holland in 1987
Softcover reprint of the hardcover 1st edition 1987

ACIDIC PRECIPITATION

Proceedings of the
International Symposium on Acidic Precipitation
Muskoka, Ontario, September 15–20, 1985

Guest Editor

H. C. MARTIN
Atmospheric Environment Service, Downsview, Ontario

PART 1

ACIDIC PRECIPITATION

Proceedings of the
International Symposium on Acidic Precipitation
Muskoka, Ontario, September 15–20, 1985

Guest Editor

H. C. MARTIN

Atmospheric Environment Service, Downsview, Ontario

PART 2

INTRODUCTION

Since the beginning of the 1980's, research has continued to clarify and improve our understanding of the problem of the ecological impact of acidic precipitation. From September 15-20, 1985, Canada convened the the Muskoka Symposium on Acidic Precipitation to review progress and help chart the direction of future studies. The Conference was held in central Ontario, Canada, a region of heavy sulphate deposition where aquatic effects are evident. Approximately 700 people attended, over 500 of whom were research scientists. Over 400 papers were presented. Nearly 200 of the papers are included in these proceedings, which together give the present state of the art of acid rain research.

The Conference focused on atmospheric science and aquatic and terrestrial effects. I have made some general observations on progress in clarifying issues and linkages between these areas of research. Outstanding issues which require a great deal more work are identified. These are my views, influenced by personal bias and limited by my expertise.

Atmospheric Sciences

Three aspects were covered: measurement techniques, the actual measurements, and source-receptor relationships.

The collection and analysis of precipitation are routine and generally of good quality although some problems remain with snow fall measurements. On the other hand, the direct measurement of dry deposition continues to pose serious technical problems. Fortunately, there has been considerable progress made in the indirect evaluation of dry deposition, and hence, towards the determination of total deposition. Recent progress has been made in techniques for measuring cloud water, H_2O_2, and atmospheric tracers. It is expected that these methods will gain wide usuage and significantly further our understanding of the atmosphere.

On the monitoring side, the spatial distribution of wet deposited material, and in some cases, dry, is available and has been linked clearly to emission regions. Background pollution levels at remote locations have been measured and we have a good sense of the relative importance of local and distant sources. The change in the sulphate and nitrate ratio with season is now well defined. Some progress has been made in the determination of O_3 concentrations on the regional scale. This atmospheric constituent is of great interest to foresters concerned with forest decline. In some areas, particularly where agriculture is dominant, base cations, in the form of atmospheric dust, can control air and precipitation chemistry. Some studies are beginning to clarify the importance of this atmospheric constituent. Over the past years highly polluted precipitation episodes have been measured on many

Water, Air and Soil Pollution **30** (1986) 1–3.

occasions. Further characterization of episodes will provide great assistance to researchers of aquatic effects where episodes may be the primary cause for degredation in some instances.

The modeling of the chemistry and physics of the atmopshere using the Lagrangian approach has progressed well, coincident with better emission and deposition inventories. Good qualitative estimates of deposition fields can be produced. Eulerian models, incorporating the subtleties of chemical and physical processes, have also evolved rapidly and will offer a greatly enhanced degree of modeling sophistication in the future.

Atmospheric researchers have yet to complete several important tasks:

(a) the reliable measurement of dry deposition;
(b) the chemical characterization of fog and low-level clouds in the context of forest research;
(c) the determination of the role of ice and super-cooled water in cloud processes;
(d) the clarification of the role of H_2O_2 in the oxidation processes (a matter of great importance in the development of control programs);
(e) the evaluating of models, especially the new generation of complex Eulerian models under development.

Aquatic Effects

The status of biological communities in aquatic systems has been under study longer than most aspects of the acidification issue. This is not surprising since irregularities in fish populations were among the first signals that something was seriously wrong with the natural environment. The effects were eventually linked to acidification. Despite declines in SO_2 emissions for the past 10 yr and some signs of lake recovery in moderately impacted areas, lakes in areas of high deposition continue to decline in biological quality. It is clear that the main acidifying agent is sulphate and that humic acids play a minor role, even during episodes. Land use practices, notably aforestation, do not account for acidification. Because of the increasing load of N on aquatic systems, there is a potential for elevated Al concentrations, marine eutrophication, as well as acidification.

There remain many unanswered questions. Primary among these are:

(a) What is the extent of the in-lake alkalinity production?
(b) What is the role of ammonia/ammonium in acidification?
(c) What is the consequence of ever-increasing emissions of NO_x, particularly in altering the chemistry of episodes?
(d) What are the best biological indicators of early-stage acidification?

Forest Effects

In the past 5 yr the perception of a link between air pollution and forest health has changed from passive interest to grave concern. Forest decline, a symptom of general ecosystem decline, has caught the scientific community by surprise. If raising an issue to high visibility fraught with controversy is progress, then we have made great progress over the past 5 yr. We note prophetic statements occurred at Sandefjord, although these were definitely in the minorities. A forest-effects problem was not generally acknowledged. However, Dr. Bernard Ulrich had a different view: 'The following ecological risks are noted: Enrichment of the mor layer with Cu and Pb close to toxic levels for microbial processes; and increasing levels of Al-toxicity in the mineral soil root zone. Severe consequences for the growth and the re-establishment of the forests are expected' (Sandefjord Abstracts, p. 50). Our understanding of the consequences of air pollutants on terrestrial ecosystems is rapidly improving. Current emphasis is on forests. It is clear that S and N compounds disrupt the integrity of protective foliar waxes and leach foliar nutrients. Ambient pH levels in precipitation negatively affect pollen germination, placing at risk the natural regeneration of forests. Enhanced H^+ concentrations in soils lead to base cation leaching. This loss of base cations can place stress on forest stands on nutrient-poor soils and lead to enhanced soil acidification. The decrease in precipitation pH can lead to the mobilization in soils of toxic metals which are available to roots or can move into aquatic systems.

The primary need in forest decline research is to clarify the relative importance of stress due to air pollution (including acidic precipitation) and other stresses, including disease, insects, weather extremes, and climate. The task involves differentiating between acidic precipitation effects on soils and normal soil acidification processes, establishing the rate of change of soil characteristics, and determining soil-root and -mycorrhizae interactions. The effect of H^+ loading on weathering rates is not well established, yet these rates are fundamental if we are to determine acceptable critical loadings for ecosystems (both terrestrial and aquatic) and if we expect to model these systems. Above ground, many of the direct impacts are understood but their importance has not been established adequately for different forest species and regions. The effects of O_3, which is N derived, compared with those of H^+ and air pollutants are not clear. This matter has great importance in the consideration of pollution control programs.

In many ways, the acidification process is a new kind of environmental problem. Cause-effect relationships cross traditional scientific discipline barriers. I refer to barriers because there is a resistance to working across disciplines. We are learning to overcome old traditions although we must move quickly to strengthen not only remote links (forest biologists and cloud chemists), but also closer links (aquatic biologists and aquatic chemists). Acidification may well be a training ground for the scientific community in preparation for more complex interdisciplinary environmental problems. I suggest that over the coming decades, our success in dealing with all toxic chemicals in the environment will be a good indication of what we have learned.

H. C. MARTIN
Guest Editor

LRTAP Liaison Officer
Atmospheric Environment Service
4905 Duffering Street
Downsview, Ontario
M3H 5T4

ATMOSPHERIC DEPOSITION IN FENNO-SCANDIA:
CHARACTERISTICS AND TRENDS

Arne Semb and Harald Dovland
Norwegian Institute for Air Research
P.O. Box 130, N-2001 Lillestrøm,
Norway

ABSTRACT. Chemical analyses of daily precipitation samples from "background" stations in Europe are discussed together with measurements of airborne SO_2 and sulphate aerosol, and trends in energy usage and SO_2 emissions. Emission sources contributing to the major part of the concentrations of sulphate and nitrate in precipitation are mostly 500 to 1000 km from the receptor area. Although there are no general statistically significant trends in the precipitation chemistry data, minor changes point to an effect of reduced SO_2 emissions in some areas. The daily data can be used to infer general conclusions with respect to precipitation scavenging efficiency.

1. INTRODUCTION

The purpose of this presentation is to give an overview of the present situation with respect to long-range transport and deposition of S and N compounds in Northern Europe. The discussion will be based on known emission trends and changes in fossil fuel consumption patterns, and a record of precipitation chemistry and air quality data for background sampling stations since the early 1970s.

Most of the data have been obtained through the OECD Programme on Long Range Transport of Air Pollutants (OECD, 1977), the Norwegian SNSF-project (Overrein *et al.*,1980) and the present Co-operative programme on monitoring and evaluation of long-range transmission of air pollutants in Europe (EMEP) organized under the United Economic Commission for Europe.

2. EMISSIONS

The SO_2 emissions inventory developed for the OECD Programme was updated in 1978 and is currently being revised. No major change in energy consumptions has occurred in this period. The energy crisis in 1973 initially seems to have stabilized the energy consumption. Since about 1978, however, use of liquid fuel has declined, particularly the use of resi-

Water, Air and Soil Pollution **30** (1986) 5 16.

dual fuel oil for electricity generation and industrial heat production. Use of fuel has declined generally in the industrial sector, while consumption of fuel for motor vehicles has increased slightly. It is seen from Figure 1 that only minor changes occurred between 1970 and 1979, after which the consumption of liquid fuel, particularly residual fuel oil, has decreased markedly.

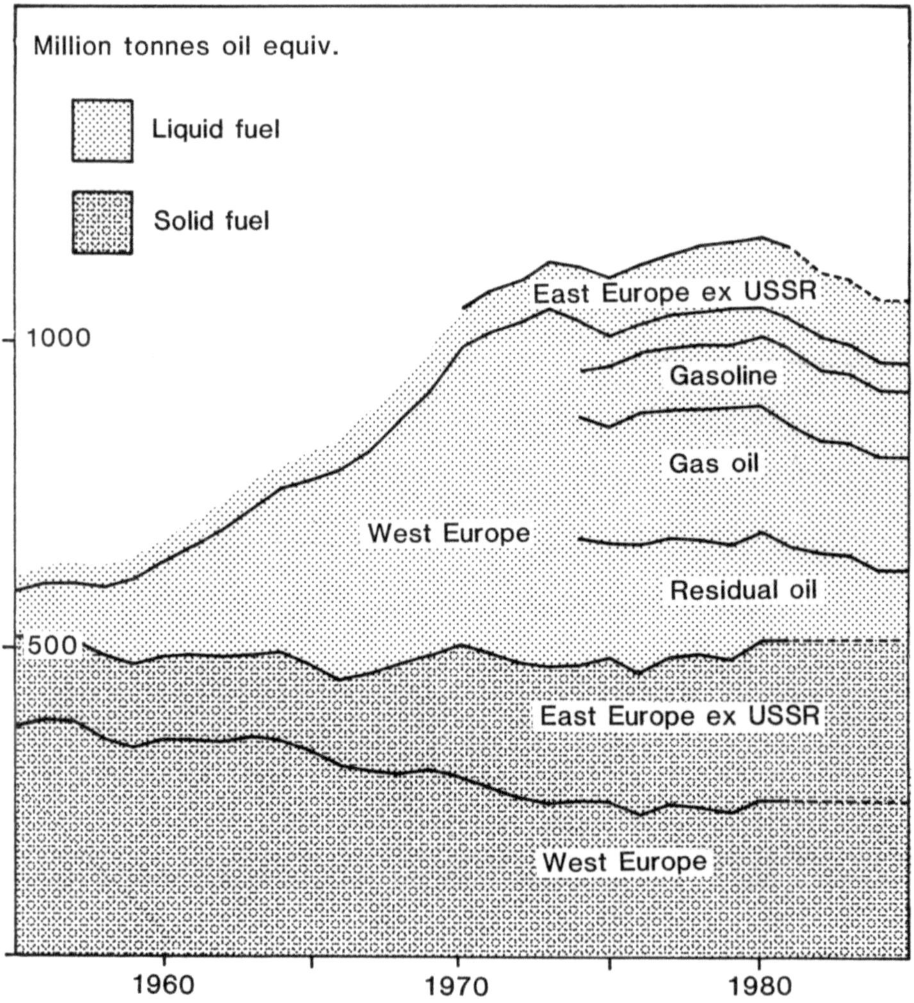

Figure 1: European consumption of solid and liquid fossil fuel, in energy equivalents, constructed from available international statistics.

It has been estimated that this changing pattern of oil consumption resulted in a reduction of SO_2 emissions from oil combustion of 38%, or 3.8 Mt as S, between 1979 and 1982 (CONCAWE, 1984).

Table I gives figures for total SO_2 emissions from some European countries, received from authorities in the different countries.

TABLE I. Sulphur dioxide emissions in 12 countries in north-western Europe (kg/yr as S).

Country	OECD/ LRTAP (~ 1973)	EMEP (~ 1978)	EMEP (~ 1983)	Δ 1973 - 83 (%)
Austria	221	215	147	- 33
Belgium	449	380	305	- 39
Denmark	250	228	152	- 39
Finland	274	270	179	- 35
France	1616	1800	1945	+ 20
Federal Rep. of Germany	1964	1815	1500	- 24
Ireland	80	87	70	- 13
Netherlands	290	240	170	- 41
Norway	91	75	51	- 44
Sweden	415	275	156	- 62
Switzerland	76	58	45	- 41
United Kingdom	2803	2490	1845	- 34
Σ 12 countries	8579	7933	6565	- 23

Few countries have systematic inventories of NO_x emissions, and reliable estimates are difficult to make in the absence of adequate emission factors. This applies both for fossil fueled thermoelectric power plants and for motor vehicles. A preliminary survey by Semb and Amble (1981) indicated that the total emission of NO_x Mtonne as NO_2 as compared with 20 Mtonne of SO_2-S. This was based on 1975 fuel consumption figures, and did not include emissions of NO from agricultural areas. Extrapolating from field measurements (Slemr and Seiler, 1984; Johansson and Granat, 1984), indicates that this emission source may be approach 1 Mtonne/yr for Europe as a whole. An important point in this respect is the consumption of N fertilizer in Europe, which was doubled between 1970 and 1980 (United Nations, 1983). This increase in N fertilizer consumption, and the resulting agricultural surplus, has probably had an even larger influence on the emissions of ammonia.

3. OBSERVED CONCENTRATIONS AND DEPOSITIONS

3.1 Deposition of acidic substances

The EMEP provides data on precipitation chemistry and/or ambient concentrations of SO_2 and sulphate aerosol from more than 80 "background stations" in 24 European countries. The objective of EMEP is to provide governments with information on the deposition and concentration of air pollutants and on the fluxes across boundaries. Thus, the measurements are also important in order to validate the model calculations which provide the integrated information that is required for policy decisions.

The results from the first 5 yr of the EMEP (1979.1982) have been used to give isolines for mean concentrations of sulphate, nitrate, and ammonium ions in precipitation in Europe. The main features of these maps (Figure 2 are essentially the same, with the highest concentrations near the largest emissions.

The pH in precipitation is largely determined by the balance between sulphate and nitrate on one side, and ammonium and other soil-derived components on the other. In general, however, the areas with the highest emissions of SO_2 and NO_x also have intensive agricultural production, so that there are only small anomalies in the pH-pattern as compared with the concentration of sulphate or nitrate in precipitation.

The deposition of S and N occur as dry deposition or with snow and rain. The wet deposition processes are the most important in Scandinavia, particularly in areas with heavy precipitation, where the amounts brought down approach 15 ha^{-1} of SO_4-S and 7 kg ha^{-1} of both NO_3-N and NH_4-N. This can be rather variable, because of the variations in precipitation amounts with topographic elements. It is our experience, however, that the chemical composition of precipitation is less variable than the amount of precipitation. Knowledge of the precipitation pattern can therefore be used to extrapolate from a limited number of precipitation sampling stations.

The wet deposition pattern for sulphate in Scandinavia has been constructed using data for a number of stations operated outside the EMEP system to obtain adequate deposition data (Figure 3). Daily, weekly, and even monthly sample collection periods are used. It has been found that by selecting proper sampling sites, the chemical composition is not critical dependent on sampling period. However, evaporation from the gauges is a problem in the summer months.

There is still not a satisfactory method available for determining the dry deposition to an actual, complex area. For a discussion of the present state of the art, reference is made to Pruppacher *et al.*, (1983) and a workshop proceeding edited by Grennfelt (1983). One result of these reconsiderations appears that the annual average dry deposition velocity for SO_2 in Scandinavia may be considerably less than 0.8 cm s^{-1}, chiefly because the highest concentrations occur in the winter period, with little vegetation activity and snow covered ground.

Dry deposition of NO_2 is also small, partly for the same reason, but may be more significant for HNO_3, judging from the very limited ambient

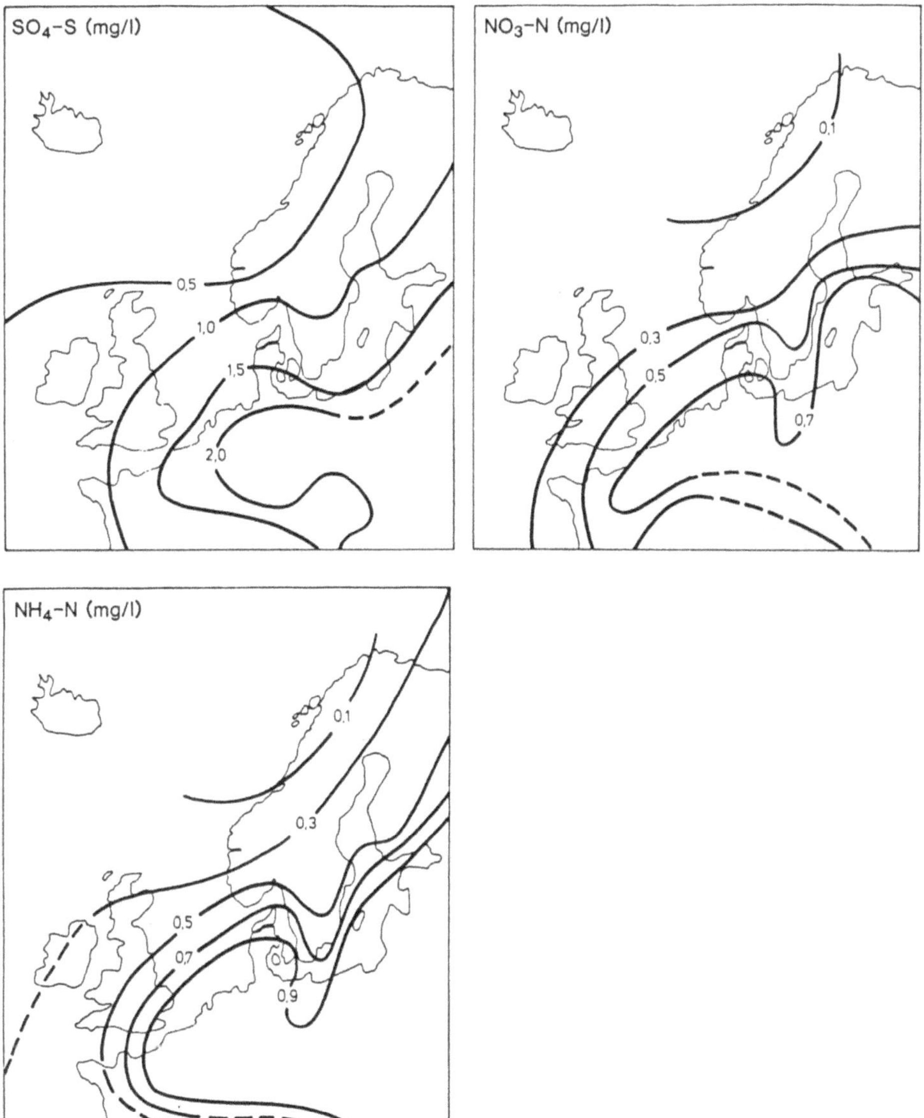

Figure 2. Mean weighted concentrations of non-marine sulphate, nitrate, and ammonium in precipitation at background stations in Europe 1978-1982.

Figure 3. Deposition of sulphate in Scandinavia by precipitation. Based on EMEP data and other observations for 1983 (Joranger, Granat and Järvinen, priv. commun.).

air concentration data.

The deposition of aerosols has aroused much controversy because of the difficulty in determining important parameters such as size distribution, surface area, and the complicated topography of coniferous forest canopies.

Recent studies (Wiman and Ågren, 1985; Wiman, 1985), emphasize the inhomogenity in aerosol deposition. Over-all deposition rates of sulphate aerosols are still difficult to quantify, but the deposition velocity for a forested area may be in the range 0.1 to 0.5 cm s^{-1}.

Adding together the dry depositions result in a contribution of S and N up to 30% of the deposition by precipitation in Scandinavia. Closer to the source areas, dry deposition is much more important, and will be the major deposition process over most of Europe.

3.2 Trends

The record of measurements at EMEP sites are from 1978 to 1984. In time, these are expected to give information of changing ambient concentrations and precipitation chemistry, which may be compared with emission data. For sites in Scandinavia, where weather and atmospheric transport conditions vary strongly from one year to the next, clear trends cannot be expected from short data series. For most of the stations there is a clear trend for lower concentrations of SO_2 in air (Figure 4). Sulphur dioxide concentration is, of course, more influenced by "near-by" indigenous sources than the more long-lived transformation product.

The precipitation chemistry data presented in Figure 5 are precipitation-weighted half-yearly mean concentrations. In general, summer half-years have higher concentrations of both sulphate and ammonium ions, while nitrate is not so well correlated. Declining trends in the concentration of sulphate in precipitation is most pronounced at Eskedalemuir in Scotland, which fit well with the reduction of SO_2 in the United Kingdom. A similar trend was observed for precipitation chemistry data from Pitlochry by Harriman and Wells (1984).

Apparent reductions of the concentration of sulphate in precipitation seem also to have occurred at Tange in Denmark and Ähtäri in Finland. There is no clear trend at Birkenes, for the 10-yr periode 1974-1984. It should be mentioned that Rodhe and Granat (1984), in examining data from the European Air Chemistry Network, found increasing concentrations of sulphate in precipitation in "Central Scandinavia" for successive 5-yr periods up to 1970-1974, then decreasing.

3.3 A precipitation event

Episodes with large deposition of sulphate, nitrate and ammonium ions are generally associated with heavy rainfall. It is not unusual to find that 10 to 20% of the annual deposition occurred on a single day. Figure 6 shows the deposition in Southern Norway during a 5-day period in September 1981. The very large amounts of precipitation and the precipitation pattern is typical of orographic precipitation, which occurred in conjunction with a front moving slowly from the West.

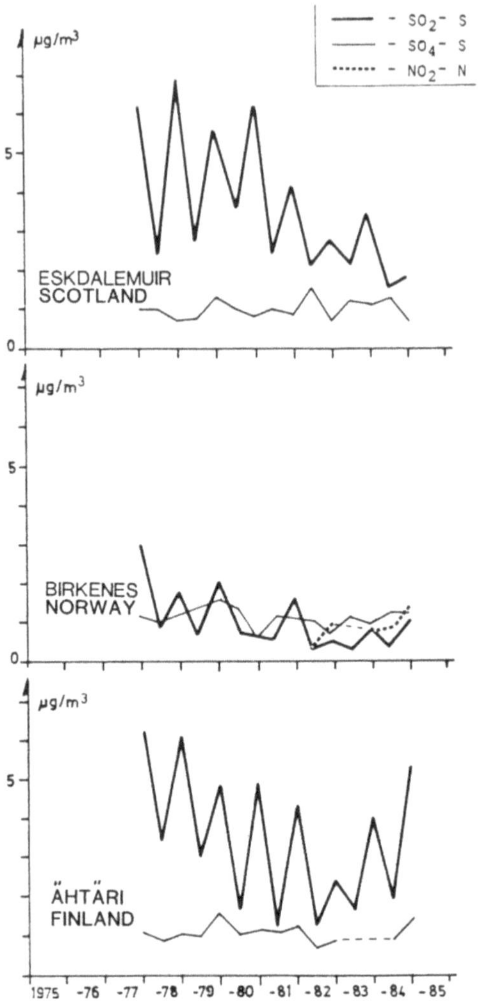

Figure 4. Mean half-yearly concentrations of sulphur dioxide and sulphate in air at 3 EMEP stations.

It is apparent from the precipitation chemistry data that concentrations of water-soluble ions are considerably reduced as the air passes through the precipitation field. This is more pronounced for nitrate than for the other components, possibly indicating that below-cloud scavenging of gaseous HNO_3 may be of some importance. However, sulphate and ammonium ion concentrations are also considerably reduced, which is not surprising in view of the precipitation intensity and persistence.

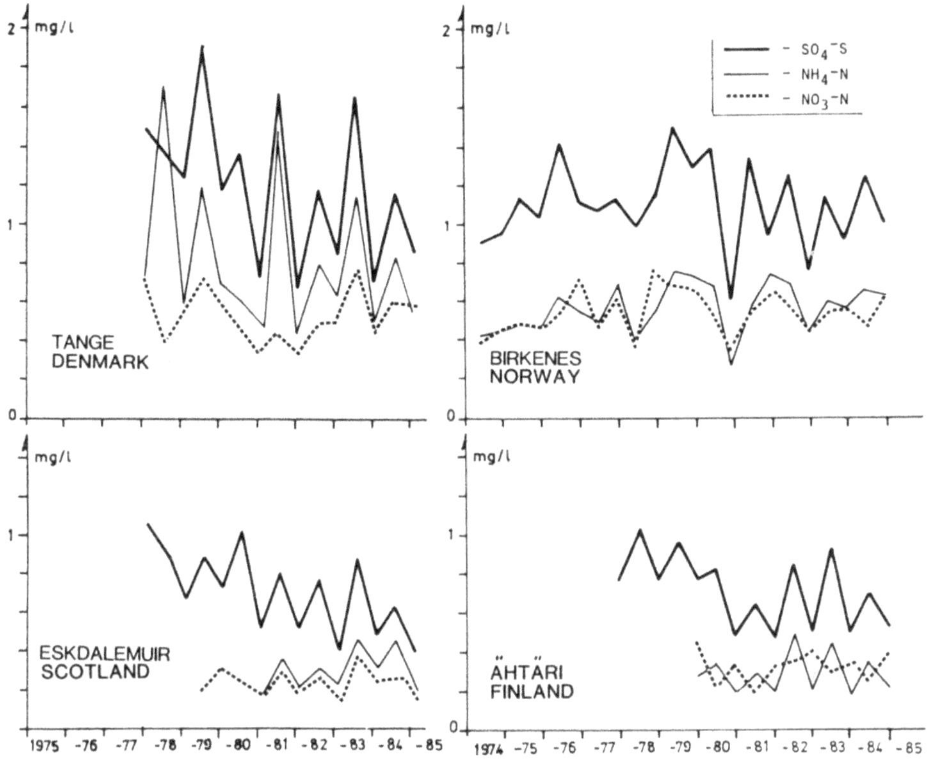

Figure 5. Mean precipitation-weighted half-yearly concentrations of sulphate, nitrate, and ammonium ions in precipitation at 4 EMEP stations.

The rate of deposition on 18 September, in mg m^{-3} d^{-1}, was about one order of magnitutde higher than the average emission rate per unit area in Central Europe. Air parcels passed over areas with high rates of emission for 1000 km transport distance, then for 500 km over areas with small or negligible emission rates. Typically, over half of the emitted SO_2 is deposited *en route*, and most of the remaining converted to sulphates before entering the orographically enhanced precipitation area, where virtually all of the pollutants become deposited within 50 to 100 km. Convergence undoubtedly increase the amounts of water and pollutants processed by the precipitation system. Intensive field studies are needed to resolve all the complex meteorological and chemical processes taking place during such events.

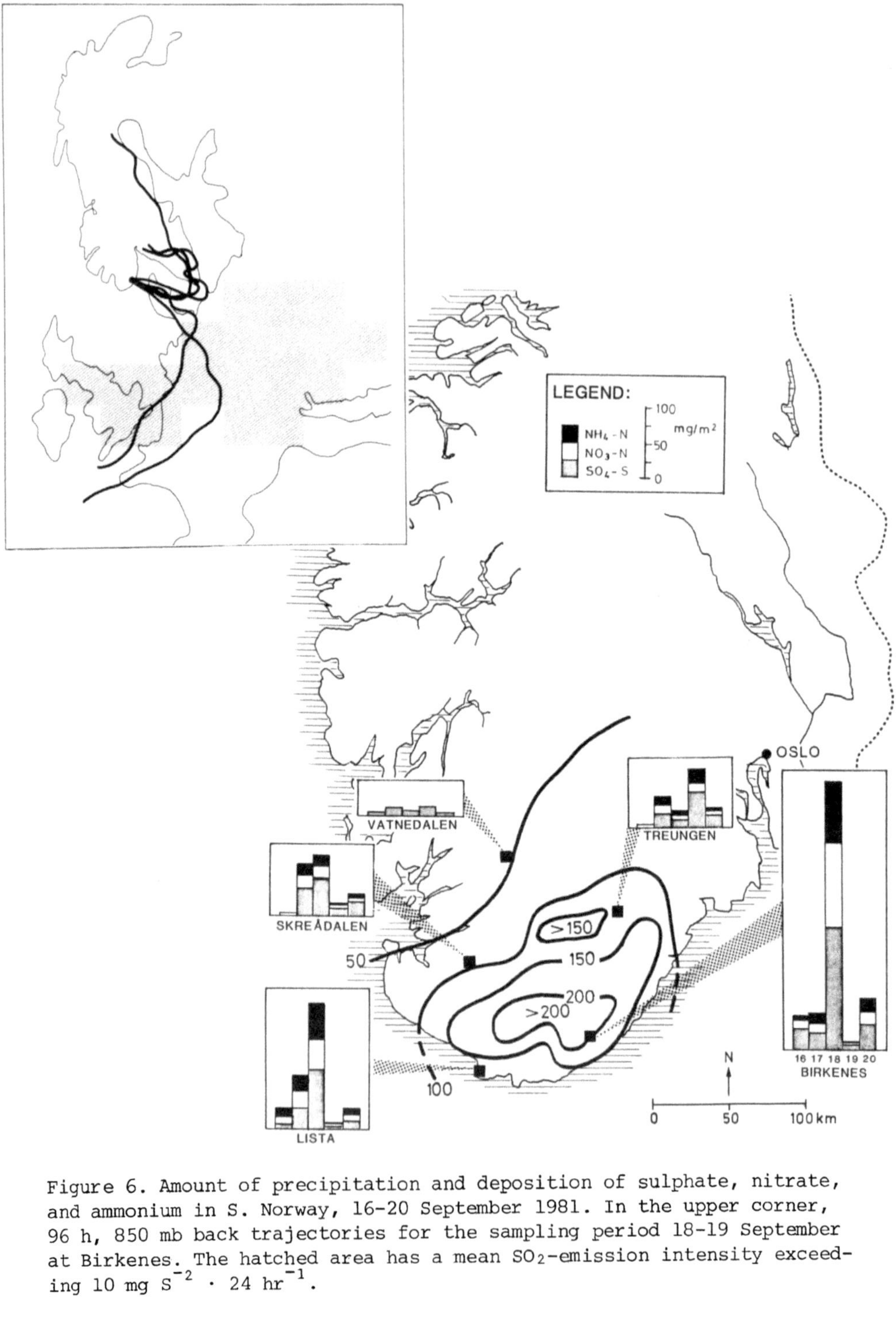

Figure 6. Amount of precipitation and deposition of sulphate, nitrate, and ammonium in S. Norway, 16-20 September 1981. In the upper corner, 96 h, 850 mb back trajectories for the sampling period 18-19 September at Birkenes. The hatched area has a mean SO_2-emission intensity exceeding 10 mg $S^{-2} \cdot 24\ hr^{-1}$.

3.4 Photochemical oxidants

During the last decade, photochemical oxidants and their precursors have received increasing attention, mainly because of their potential effects on crops and forest (e.g., Grennfelt and Schjoldager, 1984; Eliassen *et al.*, 1982). The formation of O_3 and other photochemical oxidants in the troposphere, involves emissions of NO_x and volatile organic compounds. Although research on photochemical oxidants seems to be at an earlier stage of development than research on acid deposition, there is evidence that large-scale formation and transport of O_3 is important. Thus, photochemical oxidants is another environmental problem of international dimensions.

Available O_3 measurements show that higher concentrations occur episodically in southern Scandinavia. Due to the variable summer climate in northern Europe, the frequency of episodes vary substantially. This may be illustrated by the occurrence of O_3 episodes in southern Norway during the last 8 yr (Table II). An O_3 episode has been tentatively defined as a day with O_3 concentrations above 120 $\mu g\ m^{-3}$ over a large area (i.e. measured at several stations simultaneously) or concentrations above 200 $\mu g\ m^{-3}$ at one station.

TABLE II. Occurrence of O_3 episodes in southern Norway (see text for further explanation).

Year	1977	1978	1979	1980	1981	1982	1983	1984
Number of episode days	2	3	5	75	17	27	9	5

The most pronounced oxidant episodes in southern Scandinavia are assosicated with sunny weather with light winds over northern Europe, and transport of the polluted air masses towards Scandinavia from the major source areas in northern Europe.

4. CONCLUDING REMARKS

Measurements of air and precipitation quality at rural sites in northern Europe may vary significantly from one year to another due to meteorological factors. Trend evaluations based on relatively short data series are therefore uncertain unless meteorological data are taken into account. A simple evaluation of available measurements indicate, however, decreasing concentrations of S in air and precipitation at some stations. Less data are available on N compounds, and no conclusion can be drawn. Continued measurements are important in order to facilitate future trend analysis. A better understanding an dquantitative knowledge of deposition processes is an important task for future research. Such studies will most likely require extensive field studies.

At present, photochemical oxidants seem to play an increasing role in relation to large-scale effects on the environment. This should be reflected in our national and international monitoring programmes. As a first approximation, acidic deposition and photochemical oxidants may be considered independently. However, in future scientific studies, it may be of importance to consider both these envionmental problems as an entity.

5. REFERENCES

CONCAWE: 1984, Sulphur dioxide emissions from oil refineries and combustion of oil products in western Europe in 1979 and 1982. Report No. 10/84, CONCAWE, den Haag, Netherlands.

Eliassen, A., Hov, Ø., Isaksen, I.S.A., Saltbones, J. and Stordal, F.: 1982, Journal of Applied Meteorology, 21, 1645.

Grennfelt, P. (ed.): 1983, Torrdeposition av svavel- och kväveföroreningar. SNV PM 1973 Swedish Environment Protection Board, Solna, Sweden.

Grennfelt, P. and Schjoldager, J.: 1984, Ambio, 13, 61.

Harriman, B. and Wells, D.E.: 1984, Causes and effects of surface water acidification in Scotland. The Institute of Water Pollution Control, Maidstone, Kent, England.

Johansson, C. and Granat, L.: 1984, Tellus, 36B, 26.

OECD: 1977, The OECD Programme on long range transport of air pollutants. Measurements and Findings. OECD, Paris, France.

Overrein, L.N., Seip, H.M. and Tollan, A.: 1980, Acid Precipitation - effects on forest and fish. Final report on the SNSF-project. SNSF-project FR 19/80. Norwegian Institute for Water Research, Oslo.

Pruppacher, H.R., Semonin, R.G. and Slinn, W.G.N.: 1983, Precipitation Scavenging, Dry Deposition and Resuspension. Proceedings of the Fourth International Conference, Santa Monica, California. Elsevier, New York.

Rohde, H. and Granat, L.: 1984, Atmospheric Environment, 18, 2627.

Semb, A. and Amble, E.: 1981, Emission of nitrogen oxides from fossil combustion in Europe. NILU TR 13/81. Norwegian Institute for Air Research, Lillestøm.

Slemr, F. and Seiler, W.: 1984, J. Atmospheric Chemistry, 2, 1.

United Nations: 1983, 1981 Statistical Yearbook. New York.

Wiman, B.L.B.: 1985, Aerosol dynamics in coniferous forests. Department of Plant Ecology, University of Lund, Sweden.

Wiman, B.L.B. and Ågren, G.I.: 1985, Atmospheric Environment, 19, 335.

ACIDIC DEPOSITION AND ITS EFFECTS ON THE FORESTS OF NORDIC EUROPE

Prof Folke Andersson
Section of Systems Ecology
Dept of Ecology and Environmental Research
P.O. Box 7072
Swedish University of Agricultural Sciences
S-750 07 UPPSALA, Sweden

ABSTRACT. Air pollution effects on the forest of Nordic Europe are analyzed with respect to the situation before and after 1980. No adverse effects were reported up to 1980. Inventories of forest vitality in terms of needle loss or crown density as well as tree growth have been made in parts of Norway and the whole of Sweden. For Norway spruce, needle loss was found to increase with tree age and climatic stress with increasing latitude and altitude. However, decreased crown density was also reported for SW Sweden. Needle loss is a non-specific signal not reacting to air pollution alone. Direct effects of gases such as SO_2 are not likely to occur, but high episodic levels of O_3 have been measured at a level possibly leading to chronic injuries. Increasing soil acidity is reported from forests in SW Sweden and also from an alpine locality in southern Norway. A differentiation of biological and deposition-dependent acidification has been attempted. A soil acidification of up to 1 pH unit has occurred in both humus and mineral soil layers. An increased release of Al and heavy metals combined with an increased loss of basic cations is now occurring. Soil acidification probably also results in less available amounts of phosphorus and is suggested to interact with N as limiting forest growth.

1. INTRODUCTION

In Nordic Europe the recent history of acidic deposition and its effects also has a focus on the forest environment. It began with the hypotheses that forests of southern Sweden should have a decreased forest growth and that changes of forest soils are a consequence of an increased load of acidity by the increased airborne pollution. These hypotheses were presented in the Swedish contributions to the UN Conference in Stockholm, 1972 (Anonymous, 1972). This was an impulse

Water, Air and Soil Pollution **30** (1986) 17–29.

for the active Norwegian research on 'Acid precipitation - Effects on forest and fish' (Drabløs and Tollan, 1980). In that project, as well as in Sweden, analyses of forest growth development were made. In addition, long-term experiments on artificial soil acidification and tree growth were established.

Finnish research activities into the effects of air pollution on the forest environment started around 1970 (Havas, 1971; Huttunen, 1985). From the beginning the problem focussed around the effects of different types of industrial pollutants. In these ecophysiologically dominated works the interpretation of the nature of damage required a consideration of the interaction between natural and air pollution stress factors. In particular the stress phenomenon of the harsh northern climate was dealt with. In the late 1970s a more direct interest was paid to regional problems connected to the acidic deposition on forest growth and soils.

Danish activities started with an interest on lichens as indicators to S pollution. Recently a wider interest has been demonstrated dealing with acidification problems of soils and vitality of forests.

This new interest in Denmark is shared with all the Nordic countries and is easily explained by the circumstances found in central Europe - the forest decline. An increased inventory and research activity started during 1983, when a presumed forest decline similar to the one in central Europe occurred in the southern part of the Nordic countries.

The aim of this plenary paper is:

- to summarize the understanding and status of the Nordic forests as a consequence of acidic deposition up to 1980-82;and
- to discuss present new evidences or indications of forest decline, soil changes and changes of forest growth.

This paper will only deal with conditions in remote areas far from emission sources and with descriptive research rather than evaluating ongoing experimental efforts. The air pollution situation in the Nordic countries is described by Semb and Dovland (1986).

2. PRE-1980 SITUATION

Very few or no field evidences demonstrating adverse effects of the acidic deposition on the forest of Nordic Europe were available at the Sandefjord meeting in 1980 (Drabløs and Tollan, 1980) or at the Stockholm Conference in 1982 (Anonymous, 1982). The Swedish case study from 1972 (Anonymous, 1972) generated some major hypotheses including decreased forest growth with a decreased base saturation of soils as a result of the increased input of acidic compounds. The assumed effects, if any, should mainly be indirect ones through the soil. In particular, nutritional questions were discussed (Abrahamsen, 1980). Besides analysis of forest growth based on year-ring analysis (Jonsson and Sundberg, 1972, Jonsson and Svensson, 1982; Strand, 1980) experimental efforts in the field and the laboratory were started with accelerated acidification (SNSF-project, 1980; Farrell <u>et al</u>., 1980; Popovic, 1980; Tamm, 1984; Tamm and Wiklander, 1980). The results from these experiments have formed a solid base for our understanding and analysis of the long-term effects. Nonetheless, our understanding of the long-term effects of acidic deposition on soils and also the forest ecosystem as a whole remains unsatisfactory.

On the basis that visible injuries were not reported and that the SO_2-levels generally were lower than those producing injuries, direct effects were not considered as a major concern (Abrahamsen, 1980).

In conclusion, the Sandefjord meeting in 1980 and the Stockholm Conference in 1982 did not provide any specific information on evidence of observed effects of air pollutants in the Nordic forests as growth decline or soil changes. However, Troedsson (1980) presented a study from central Sweden on changes over a 10-yr period of the humus layer of forests. A decrease of basic cations was demonstrated, correlated with an increased age of the forests. A distinction between biological and atmospheric dependent acidification was not possible.

3. POST-1980 SITUATION

The great concern of the acidification situation of the terrestrial environment has been an awareness of Nordic scientists ever since the end of the 1960's. However, as mentioned, evidence of changes has not been found. Experimental work with accelerated acidification has shown that a decrease of base saturation of the soil occurs (Farrell et al., 1980). However the soil acidification and changes of forest growth in these experiments have not shown a consistent pattern compared to the natural situation (Tamm and Wiklander, 1980). Calculation of surface water acidification indicates and also requires an assumption where a change of the soil conditions occurs (Christophersen, 1982).

The development during the 1980s of the acid deposition and air pollution situation in central Europe has given a new focus on forests and soils. Since 1983 several activities have started in the Nordic countries dealing with inventories of forest vitality. In the following we shall discuss signs of forest decline, soil changes and finally changes in forest growth and associated nutritional and metal solubility phenomena.

3.1 Forest vitality, crown thinning, needle loss

With symptoms of forest decline occurring in 1983 in Southern Sweden, similar to those found in Central Europe the need for regional inventories was obvious.

The decline was registered as needle loss and is understood as an indicator of vitality of trees as is tree growth.

Discussions between the Nordic countries took place early in 1984. In Norway and Sweden inventories were performed during the 1984 season. During the same period Denmark has made limited observations of stand vitality of Norway spruce. Finland has done introductory work on a regular inventory system of forest vitality and soil changes starting from 1985.

3.1.1 Norwegian inventory 1984. In connection with the Norwegian National Forest Survey an inventory covering a south-north transect in six counties from 58° to 65°N was made during 1984. A systematic clustered sample scheme was applied. Approximately 10000 sample trees of Norway spruce (6500 trees) and Scots pine (3500 trees) were classified. The crown density was judged or the needle loss in relation to a fully dense crown was estimated. Only the upper half of the tree was evaluated (Tveite, 1985).

A decreased tree vitality or an increased needle loss from south to north was demonstrated. There was also a general decline of vitality with increasing age and a decreased vitality or crown density was also observed with increasing altitude.

3.1.2 Swedish inventory 1984. In 1984 the Swedish National Forest Survey 1984 also made additional observations of tree vitality interpreted as percentage of needle loss (Bengtsson, 1985a and b). The inventory covers the whole of Sweden. The observations include - as also in Norway - both Norway spruce (6000 trees) and Scots pine (4000 trees). The design of the inventory is a systematic survey with a combination of observations on permanent and temporal tracts with clustered sample plots.

The results of the inventory describe the general situation, not details in different counties. In the following only results of Norway spruce are considered.

The Swedish results confirm those presented from Norway. Needle loss or crown thinning increases with age (Figures 1 and 2). However, a needle loss >20 % is common not only in parts of Northern Sweden but also in Southwestern Sweden (Figure 3). As in Norway the needle loss increases with altitude (Figure 4).

3.1.3 Conclusions. From these Norwegian and Swedish inventories it can be concluded that needle loss is a non-specific variable. Natural factors like climatic stress and aging over-shadow possible effects of air pollutants on tree vitality. Although not well documented earlier periods with drought have been found to result in an increased needle loss of conifers. Such an example was the episode of sub-top dying in 1975 in southernmost Sweden. There seems to be a variation over years in crown density or needle loss. Therefore observations over longer periods are required for a better interpretation of the present observations.

With the assumption generally stated in Central Europe it is contradictory to find that an increasing crown thinning or needle loss occurs towards north where decreasing levels of dry and wet deposition prevail. However the occurrence of needle loss in old forests of Norway spruce is frequent also in southwestern Sweden, an area with a higher load of air pollution than in northern parts. The forests here are also younger than in north Sweden. This observation supports the idea that an impact of air pollutants on trees is more probable in southernmost Nordic Europe.

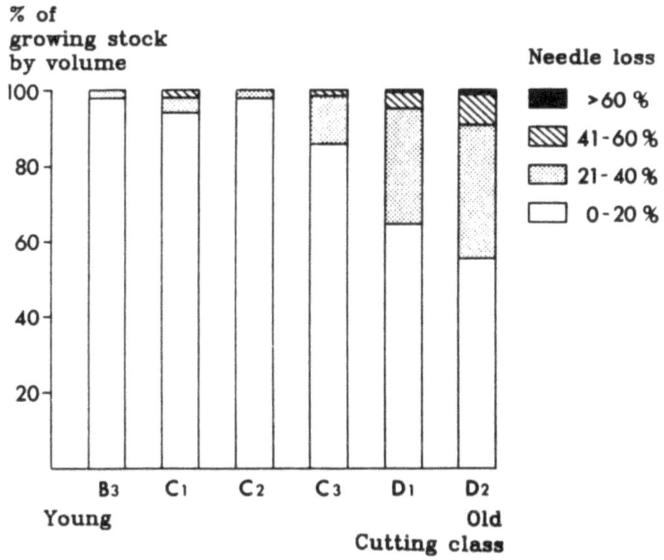

Figure 1. The age dependence of needle loss of Norway spruce in Sweden 1984 (Bengtsson, 1985a). From region 5, southern Sweden.

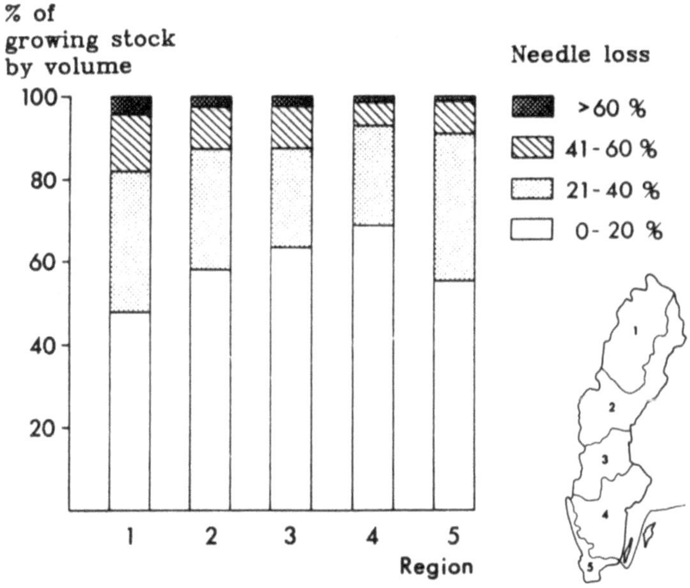

Figure 2. Needle loss of Norway spruce in different Swedish inventory regions 1984. Cutting class D2 = old trees (Bengtsson, 1985a).

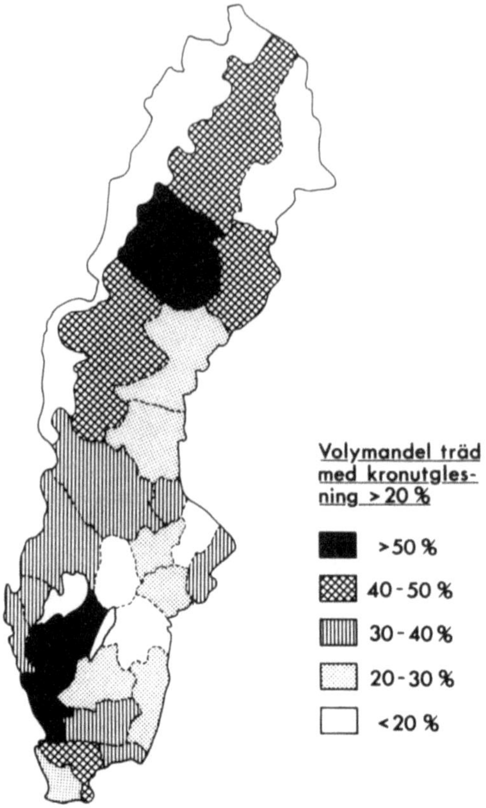

Figure 3. Percentage by volume of standing stock of old Norway spruce forest with a needle loss of more than 20% in different regions in Sweden, 1984 (Bengtsson, 1985a).

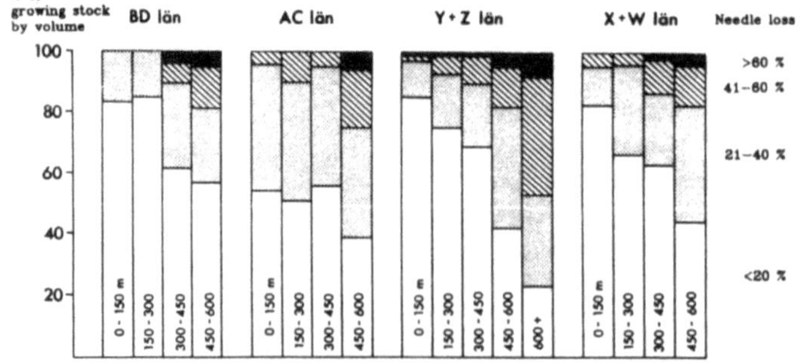

Figure 4. Altitude and needle loss of Norway spruce for four northern regions in Sweden 1984 (Bengtsson, 1985a).

3.2 Soil changes

Chemical changes of soil acidity have been documented in the literature. An increasing soil acidity has often been associated with an ageing forest, differences between tree species or as a consequence of local emissions. Recent investigations in Central Europe indicate, however, both changes in humus and mineral soil layers after a period varying from 12 to 30 yr (cf. literature cited by Hallbäcken and Tamm, 1986).

In Nordic Europe, recent information on soil changes has become available. It has been derived from areas which previously have been investigated and well documented. Now they have been subjected to a re-examination. We have now not only access to material covering a 30-yr-period from southernmost Sweden but also to material from the Swedish west coast covering a period of appr. 60 years. The latter material is unique as the recent use of the forests is well documented.

3.2.1 South Swedish information. During the 1940's, Linnermark (1960) investigated podsols and grey-brown forest soils in the southernmost part of Sweden, Skåne. These soils as well as the humus layer of the deciduous forests have now been re-examined (Tyler et al., 1985) using identical methods.

The humus layer of 230 deciduous forests in the southernmost parts of Sweden, Skåne, has been re-examined after a first inventory in 1953. Decreases in the pH of the humus layer of appr. 0.7 units are reported. The change in mull soils was larger than that in mor soils. The pH-decrease of the humus layer of 0.7 pH-unit was found to compare a change of base satuation of appr. 50 %. This means that half of the sum of the basic cations has been replaced by H^+ and Al during the period 1949/53 to 1984.

A change is also found in the B- and C-horizons (Figure 5). Considering that small changes have occurred in vegetation it seems difficult to explain the changes only by successional processes. Supported by other investigations, Tyler et al. (1985) concludes that "acid deposition is to a large extent accountable for the observed pH changes in forest soils of Skåne, south Sweden".

3.2.2 Southwest Swedish information. A unique material has become available, making it possible to follow soil changes under various canopies under a period from 1927 to 1984 (Hallbäcken and Tamm, 1986).

The re-examined area is situated on the west coast of southern Sweden, in the province of Halland. The acidity changes occurring in stands of beech and Norway spruce demonstrate comparable changes to those shown from the province of Skåne. The beech forest represents a continuous beech forest canopy for 300 yr while the Norway spruce stands represent mainly first and second generations on previous heathland.

By examining the pH of different soil layers of stands of different age of Norway spruce in 1927 and 1982-83, respectively, it has been possible to demonstrate the natural biological acidification;

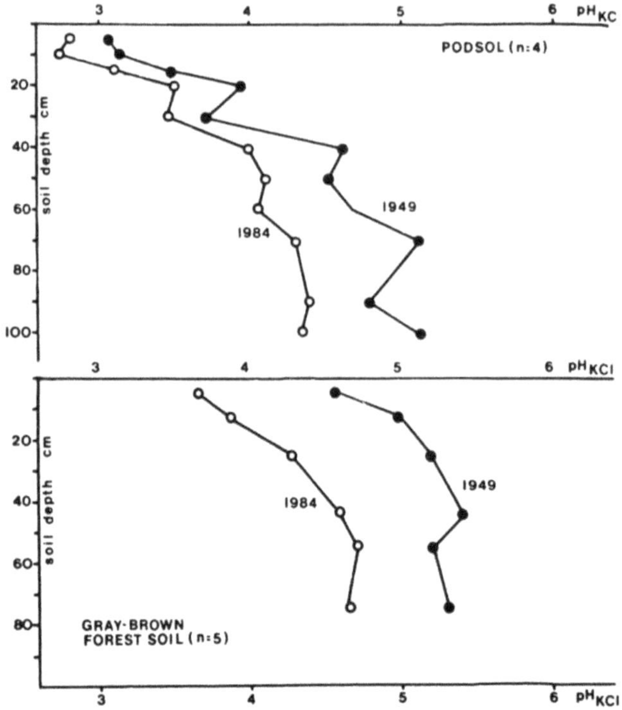

Figure 5. The pH changes at different depths between 1949 and 1984 in forest soils of southernmost Sweden. Means of four podsols (above) and five grey-brown forest soils (below). (Tyler et al., 1985).

an increasing acidity with stand age. In the C horizon the influence of the biological acidification is hardly discernable (Figure 6). Still, a general decrease of pH is detected after 60 yrs. This is interpreted as a most likely effect of the input of strong acids by the anthropogenic deposition to the forest stands.

3.2.3 Conclusions. These findings and indications of an acidification of forest soils in south Sweden are also supported by Dahl, (pers. communication) with results of a re-examination of soil studies above the timber line in Rondane, southern Norway. Here the changes in vegetation are insignificant or small. This supports the idea that the change might be attributed to the atmospheric input of H^+, S and N compounds. On the basis of the reported findings it can be concluded that the deposition of acidic compounds over the last 50 yr have contributed to an accelerated change of natural soils in the southern part of the Nordic region.

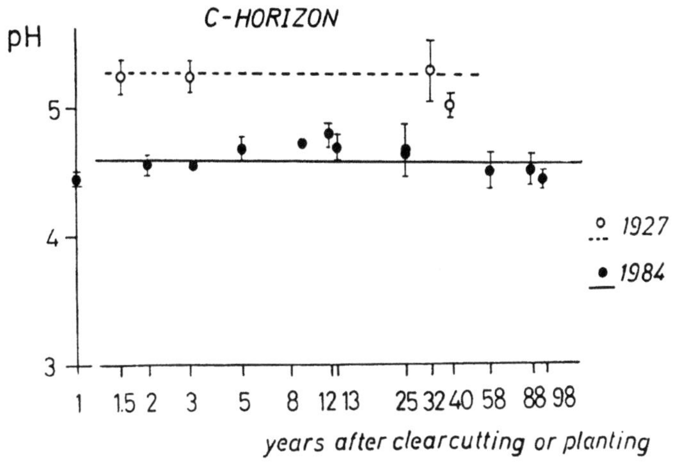

Figure 6. pH of the C-horizon plotted against arithmetic age for different Norway spruce stands at Tönnersjöheden's experimental forest, Halland, Sweden (Hallbäcken and Tamm, 1986).

3.3 Nutritional and metal solubility aspects of increased deposition and soil acidification

3.3.1 Availability of mineral nutrients. The growth of Nordic forests on mineral soils is N-limited (cf. Abrahamsen, 1980). With an increasing input of N an increased growth rate is assumed to be additional to the one gained by, e.g. improved forest management. However, it has not been possible to isolate the N component of this growth increase. Others consider that the increased content of the atmosphere of CO_2 also plays a role for increased growth of forests (Arovaara et al., 1984). An increased growth rate in stands of conservation forests has been attributed to the N deposition (Hari and Raunemaa, 1985).

With the N input and increased tree growth the demand for other elements increase. Leaching losses following the input of strong acids will deplete the soil of available elements. Magnesium has been considered as such a critical growth element. So far Mg has not been shown as a minimum element for southern Sweden. Some recent needle analyses of Norway spruce in damaged stands from southern Sweden indicate low levels of P (Aronsson, 1985). An increased soil acidification, as previously discussed, may cause deficiency situations as P is less available to plants in acid soils. With time a change between the different elements and their availability to tree growth can be expected.

By comparing the elemental composition of needle litter of Scots pine sampled during the period 1959-1979 from south and north Finland, different patterns have been obtained (Raunemaa et al., 1982), which might reflect a changing situation with increasing deposition of air pollutants.

3.3.2 Soil acidification and metal mobility. As demonstrated by Tyler et al. (1985) Al and some heavy metals will occur in higher concentrations in the soil solution after acidification. The pH of the mineral soil layers has decreased drastically during the last 30 to 50 yr of the soil profiles studied, which implies a pH decrease of the humus and mineral soil layers down to a present-date pH of 4 or less. This means that the soil is now within the Al buffer range.

Soil and root conditions have been studied in stands of Norway spruce in southernmost Sweden with signs of needle loss or decreased crown density. A low soil pH results in low amounts of available Ca and Mg and high amounts of Al and H^+. The total Al fraction in the soil solution varies between 1 and 15 mg Al L^{-1}. The root studies indicate an increasing portion of dead coarse roots with depth. This result is not meant to provide any evidence of toxicity. The information is given to indicate some present conditions in an area where soil acidification has occurred (Puhe et al., 1986).

3.4 Tree growth

In previous studies based on year-ring analysis no increase or decrease of forest growth can so far be associated with the pattern of air pollutants in Nordic Europe (Tveite, 1985 and literature cited there). The preliminary Swedish studies of stands of Norway spruce, as presented earlier, has not shown any correlation between amount of needle loss and tree growth (Eriksson, 1985). These studies are still limited and more concentrated efforts will be made in areas exposed to pollution along the coastal regions of Sweden. It should be expected that trees with a high degree of needle loss, will show decreased growth.

4. CONCLUSIONS

At present, it is the opinion of Danish, Finnish, Norwegian and Swedish inventory authorities that these natural types of damage dominate and that climatic factors are a most probable cause of many of the observed phenomena, like needle loss.

From southernmost Sweden, Skåne, an area with high pollution exposure, reports on damage to beech (Fagus sylvatica) have been received indicating stunting shoots, early yellowing of leaves, early leaf fall and bark damage. Young stands of planted Norway spruce less than 40 yr old often show unexplained injuries to needles as well as luxurious growth of green algae. At present it must be concluded that insufficient work is done on forests in areas exposed to air pollution. More work is needed in order to examine these unexplained findings, including possible direct damage by gases and acid deposition on plant surfaces.

Nordic Europe represents an area from high to low exposure of air pollutants. Areas of local influence occur. The vitality of the forest in terms of crown density (needle loss) of conifers over larger regions may vary mainly due to natural conditions such as drought and low temperature.

An interaction between air pollutants and plants in the southernmost part occurs today. The direct effects of gases such as SO_2, seem insignificant although effects of O_3 cannot be excluded.

The demonstrated soil acidification, in particular in southernmost Sweden, means that the forest ecosystems here are exposed to an air pollution stress which affects the forests. The recently observed crown thinning is most likely triggered by a drought condition, but it cannot be excluded that the changed conditions of the soil result in an additional stress. Although increased growth is registered, the forests in certain regions can be regarded as more sensitive to critical situations such as drought and frost due to a decreased vitality. The air pollution stress is here superimposed on the natural stress factors.

REFERENCES

Arovaara, H., Hari, P. and Kuusela, K.: 1984, Commun. Inst. For. Fenn. **122**, 1.

Abrahamsen, G.: 1980, 'Acid precipitation, plant nutrients and forest growth'. In: Drabløs, D, & Tollan, A. (eds) Ecological impact of acid precipitation. Proc. of an Int. Conf. Sandefjord, Norway, March 11-14, 1980, 58-63. SNSF-project, Ås, Norway.

Aronsson, A.: 1985, 'Trädens växtnäringstillstånd i områden med skogsskador'. ('Nutritional status of trees in areas with forest decline'). In: Skogsfakta Konferens 8. Uppsala. Sweden

Anonymous: 1972, Air pollution across national boundaries. The impact on the environment of sulfur in air and precipitation. Sweden's case study for the United Nations conference on the human environment. Stockholm. Royal Ministry for Foreign Affairs.

Anonymous: 1982, Acidification - today and tomorrow. A Swedish study prepared for the 1982 Stockholm Conference on the Acidification of the Environment. Ministry of Agriculture. Stockholm.

Bengtsson, G.: 1985a, 'Skogsskador i Sverige 1984 - en riksinventering'. (Forest damage in Sweden 1984 - a national inventory). Skogsfakta Konferens 8. Uppsala, Sweden.

Bengtsson, G.: 1985b, 'Approach and methodology for monitoring of the health of forests in Sweden'. Paper presented at IUFRO-Conf. Inventorying and Monitoring Endangered Forests, 19-24 August 1985, Zürich, Switzerland.

Christophersen, N.: 1982, Freshwater acidification - a modelling approach. Doctoral thesis. Inst. of Informatics, Univ. of Oslo, Norway.

Drabløs, D and Tollan A.: 1980, (Eds.) Ecological impact of acid precipitation. Proc. of an International Conference, Sandefjord, Norway, March 11-14, 1980. Oslo, Norway.

Eriksson, H.: 1985, 'Skogstillväxt i områden med skogsskador'. (Forest growth in areas with forest decline). Skogsfakta Konferens 8. Uppsala, Sweden.

Farrell, E.P., Nilsson, I., Tamm, C.O. and Wiklander, G.: 1980, 'Effects of artificial acidification with sulphuric acid on soil chemistry in a Scots pine forest'. In: Drabløs, D, & Tollan, A. (eds) Ecological impact of acid precipitation. Proc. of an Int. Conf. Sandefjord, Norway, March 11-14, 1980, 186-187. SNSF-project, Ås, Norway.

Hallbäcken, L. and Tamm, C.O.: 1986, Scand. J. For. Res. 1:219-232.

Hari, P. and Raunemaa, T.: 1985, 'The effect of changes in atmospheric composition on forest growth in Finland'. In: Symp. on the effects of air pollution on forest and water ecosystems, Helsinki April 23-24, 1985, 17-23. Helsinki, Finland.

Havas, P.: 1971, Acta For. Fennica 121:1.

Huttunen, S.: 1985, 'Ecophysiological effects of air pollution on conifers'. In: Symp. on the effects of air pollution on forest and water ecosystems, Helsinki April 23-24, 1985, 17-23. Helsinki, Finland.

Jonsson, B. and Sundberg, R.: 1972, Has the acidification by atmospheric pollution caused a growth reduction in Swedish forests? Dept of Forest Production. Reports and essays, 20. Stockholm, Sweden.

Jonsson, B. and Svensson, L-G.: 1982, A study of the effects of air pollution and forest yield. A follow-up of the report of Johnson and Sundberg 1972 and a new study based on forest types. Section of forest mensuration and managment. Swedish University of Agricultural Sciences, Umeå, Sweden.

Linnermark, N.: 1960, 'Podsol och brunjord I-II'. (Podsol and brown soil I-II). Publ. Inst. of Mineralogy, Paleontology and Quaternary Geology, Univ, of Lund, Sweden 75:1-233.

Popovic, B.: 1980, Acidification effect of urea and sulphur application on the forest soil in a percolation experiment. In: (Drabløs, D. & Tollan, A. (eds) Ecological impact of acid precipitation. Proc. of an Int. Conf. Sandefjord, Norway, March 11-14, 1980, 182-3. SNSF--project, Ås, Norway.

Puhe, J., Persson, H. and Börjesson, I.: 1986, 'Wurzelwachstum und Wurzelschäden in skandinavischen Nadelwäldern'. Allgemeine Forst Zeitschrift, 20.

Raunemaa, A., Hautojärvi, A., Kaisla, K., Gerlander, M., Erkinjuntti, R., Tuomi, T., Hari, P., Kellomäki, S. and Katainen, H-S.: 1982, Can. Journal of Forest Research. Vol. 12:384.

Semb, A. and Dovland, A.: 1985, 'Atmospheric deposition in Fenno--Scandia: characteristics, trends and transport modelling'. Proceedings of the Muskoka conference September 1985. WASP.

SNSF project: 1980, 'Effects of acid precipitation on soil and forest, SNSF-project, Norway'. In: (Drabløs, D. & Tollan, A. (eds) Ecological impact of acid precipitation. Proc. of an Int. Conf. Sandefjord, Norway, March 11-14, 1980, 190-211. SNSF-project, Ås, Norway.

Strand, S.: 1980, 'The effect of acid precipitation on tree growth'. In: Drabløs, D. and Tollan, A. (eds) Ecological impact of acid precipitation. Proc. of an Int. Conf. Sandefjord, Norway, March 11-14, 1980, 64-67. SNSF-project, Ås, Norway.

Tamm, C.O.: 1984, 'Is field experimentation a relevant method to study the effects of change in chemical climate on forest ecosystems'. In: Ågren, G. (ed.) State and change of forest ecosystems - indicators in current research. Swed. Univ. Agric. Sci. Dept Ecology & Environmental Research, Report nr 13, 251-259.

Tamm, C.O. and Wiklander, G.: 1980, 'Effects of artificial acidification with sulphuric acid on the growth in Scots pine forest'. In: Drabløs, D. & Tollan, A. (eds) Ecological impact of acid precipitation. Proc. of an Int. Conf. Sandefjord, Norway, March 11-14, 1980, 188-89. SNSF-project, Ås, Norway.

Troedsson, T.: 1980, 'Ten years acidification of Swedish forest soils'. In: Drabløs, D. and Tollan, A. Ecological impact of acid precipitation. Proc. of an Int. Conf. Sandefjord, Norway, March 11-14, 1980, 188-89. SNSF-project, Ås, Norway.

Tveite, B.: 1985, Air pollution and forest damage in Norway. Paper presented at the NATO-workshop in Toronto, Ontario, Canada. May 1985.

Tyler, G., Berggren, D., Bergkvist, B., Falkengren-Grerup, U., Folkeson, L. and Rühling, Å.: 1985, Soil acidification and metal solubility in forests of south Sweden. The NATO advanced research institute workshop: Effects of air pollutants, especially acidic deposition on forests, agriculture and wetlands; Toronto, May 12-17, 1985.

Gårdsjön Project: lake acidification, chemistry in catchment runoff, lake liming and microcatchment manipulations

HANS HULTBERG and PERINGE GRENNFELT
Swedish Environmental Research Institute
P.O. Box 5207
S-402 24 Göteborg
Sweden

ABSTRACT. The Gårdsjön Project is an ecosystem approach in studying acid deposition and its effects on terrestrial parts and lakes within a catchment. The study is an attempt to quantify the processes causing the acidification as well as an analyses of the chemical and biological conditions before liming of the lake and manipulations in microcatchments. This paper gives a brief overview of some studies already performed and recently started experiments within the project, such as lake liming, lime application on land, forest fertilization, clearcutting and applications of acid and neutral sulphate on land.

1. Introduction

A geological map of Northern Europe shows that areas underlain by lime-poor bedrock predominate in the Scandinavian countries: while most parts of Central Europe have few areas with low weathering bedrock. The Scandinavian countries have large numbers of lakes predominantly in areas with crystalline bedrock while only few lakes are found in Central Europe. The limited neutralizing capacity of soils and bedrock explains why reports on fish decline and dramatic changes in ecosystem structure due to increased acidity in streams and lakes first appeared in Sweden and Norway (Hultberg and Stenson, 1970; Jensen and Snekvik, 1972; Grahn et al., 1974; Almer et al., 1974; Muniz and Leivestad, 1980).

Water, Air and Soil Pollution **30** (1986) 31–46.

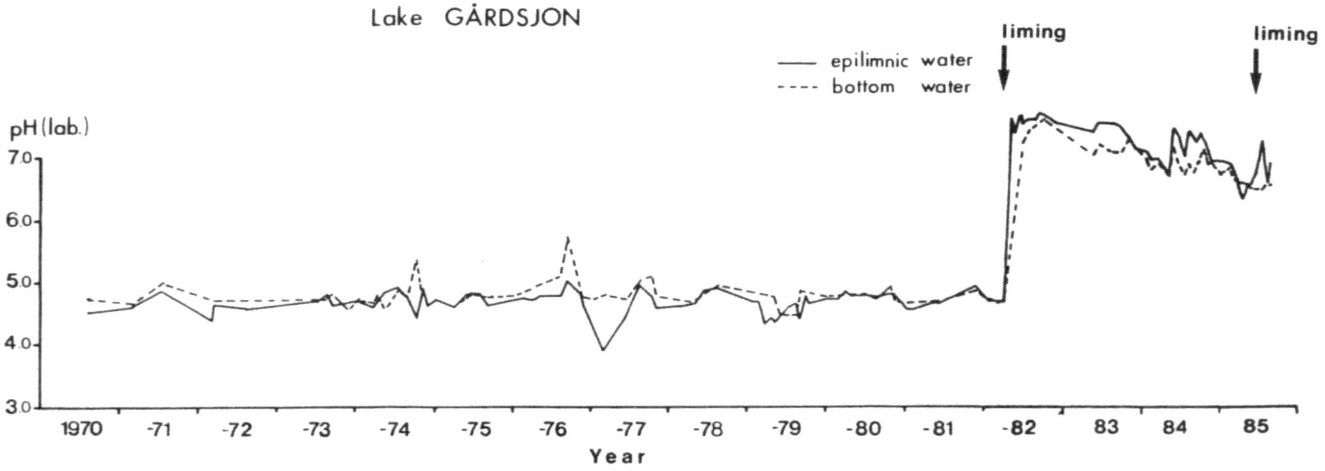

Fig. 1 pH (laboratory measurements) in epilimnic and bottom water (1 m above bottom at 18.5 m depth) in Lake Gårdsjön during the years 1970 to 1985.

2. Changes in Lake Gårdsjön acidity and fish populations

In Lake Gårdsjön and other neighboring lakes fish decline was observed since late 1940's and became a major problem during the 1960's. Selfsustaining populations of perch (Perca fluviatilis), roach (Leuciscus rutilus), Northern pike (Esox lucius), and European eel (Anguilla anguilla) were documented in 1900. By 1973 only 47 perch were left in this 31 ha large lake (Hultberg, 1985a).

pH-measurements are available from late 1940's when summer pH was measured to 6.25. During the years 1961 and 1962 monthly pH-measurements at six lakes and streams near to Lake Gårdsjön showed a pH >6.0 during summertime but sharp pH-drops to 4.3 to 5.5 during snowmelt and autumn rains. By 1970 pH at these sites had dropped to 4.5 to 5.2 during the whole year (Hultberg, 1985a). Epilimnic water in Lake Gårdsjön never exceeded pH 4.9 over the period 1970 to 1982 when the lake was limed (Figure 1).

The development of acidification was reconstructed by the diatom stratification of the lake sediment (Renberg and Wallin, 1985). A slow natural decline in pH (from ≈7 to ≈6) had occurred since the ice sheet finally retreated in 12500 B.D. After 1950 a rapid acidification from pH 6.0 to 4.7 has occurred. Soot particles from coal and oil burning appeared in the lake sediments from mid 1800 but increased rapidly after the 1940's (Renberg and Wik, 1985). The increased burning of oil and coal in the north European countries after the second world war caused severe air pollution of S compounds and soot particles. This coincides with the rapid acidification of Lake Gårdsjön and other waters in Sweden, which was demonstrated by the sediment analyses, actual pH-measurements and fish decline.

3. Chemistry of the Lake Gårdsjön catchment

3.1 Long-term chemical monitoring

Characteristic of the catchment is the thin Fe podsol soils (average soil depth ≈50 cm). The vegetation of the Lake Gårdsjön catchment consists of mature mixed coniferous forests and 10 to 20-yr old clear fellings. The total catchment is divided into five lake subcatchments and one stream subcatchment. The lakes are oligotrophic with pH 4.5 to 5.0 and clear water with Secchi depths >6 m.

Monitoring of the water chemistry in Lake Gårdsjön started in early 1970's. During the period 1971 to 1977 the sulphate concentration ranged 12 to 15 mg$\cdot$L^{-1} but dropped to 9 to 11 mg$\cdot$L^{-1} during 1980 to 1984 (Figure 2). This ≈20% drop may coincide with reductions in S emissions in Sweden and other European countries during the last 5 to 8 years. Other factors such as clear fellings and climate may have influenced this trend.

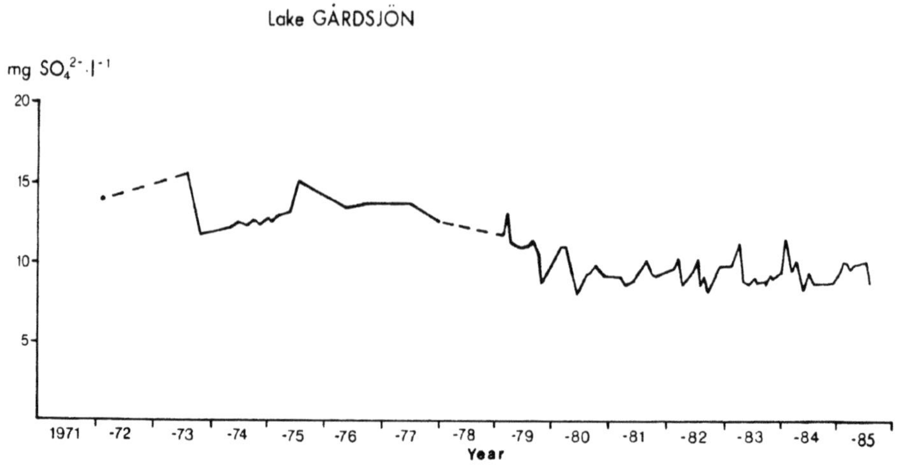

Fig. 2 Sulphate concentration in epilimnic water in Lake Gårdsjön during the years 1971 to 1985.

Nitrate increased, however, both in the lake inlet and outlet and in runoff from land during the years 1981 to 1985 (Grennfelt and Hultberg, 1986). Phosphorus decreased from 9 to 3 $\mu g \cdot L^{-1}$ in the lake during 1972 to 1978 (Hultberg 1985a). This trend continued during the period 1979 to 1982 (Persson and Broberg, 1985).

3.2 Quantitative studies of acidification processes

Two of the scientific tools used to quantify the acidification of Lake Gårdsjön are: mass balance studies in the 5 lake subcatchments and mass balance studies in three microcatchments (F1, F2, F3, Figure 3). This joint effort into fluxes of cations and anions through the catchment is also the basis to evaluate the importance of soil acidification processes for the lake acidification (Nilsson, 1985a).

The atmospheric input by wet- and dry deposition have been monitored by deposition collectors, SO_2, NO_2 and $SO_4{}^{2-}$-particulate monitoring and throughfall measurements (Grennfelt et al., 1985). Hydrological monitoring was performed at the lake outlet and at each of the four microcatchments (Johansson and Nilsson, 1985). A detailed soil and forest chemistry inventory was performed in one of the microcatchments. Lysimeters including the A-, B- and C-horizons, respectively, were installed in the same catchment (Nilsson and Bergqvist, 1983). Runoff from lake subcatchments and microcatchments was monitored on a weekly or biweekly basis (Hultberg, 1985b).

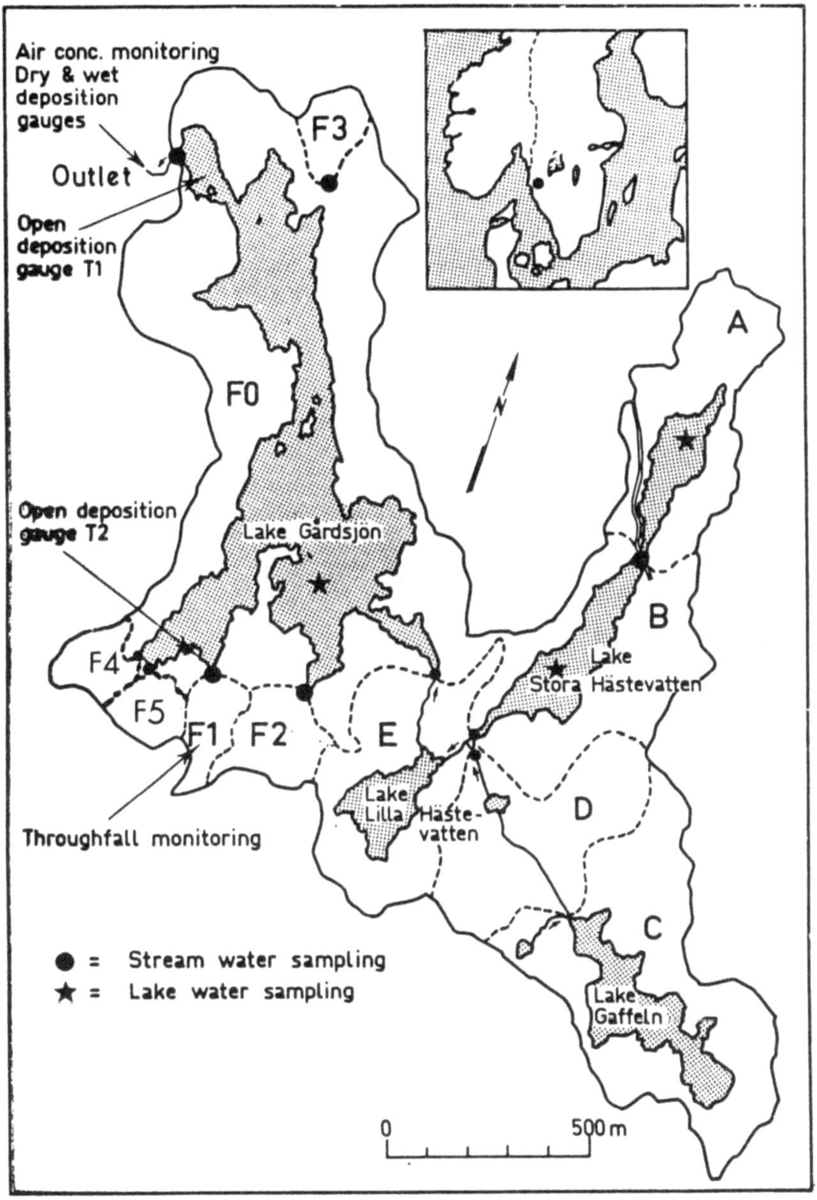

Fig. 3 Map showing location of the Lake Gårdsjön catchment site at the Swedish west coast and sampling sites for water chemical and deposition monitoring. Microcatchments are labelled F1 - F5 and terrestrial parts of lake subcatchment A - F0.

The yearly wet deposition of S was 11.4 kg•ha^{-1} during the investigation period, Oct. 1979 to Sept. 1981, and the dry S-deposition was estimated to 13.7 to 19.6 kg•ha^{-1} to a mature Norway spruce stand. Throughfall S deposition was measured to 25 kg•ha^{-1} which is comparable to the estimated atmospheric input of 25 to 31 kg•ha^{-1}. The flux of S by run-off from the forested microcatchment was 28.8 kg•ha^{-1}•yr^{-1} indicating that the S output from the catchment was similar to the atmospheric input. The long-term high S-deposition over the last 30 to 40 yr may therefore have resulted in S saturation of the thin soil cover in the catchment. Similar estimates of other major elements show that also Na and Cl deposited as seasalt in the catchment are transported through the system, while ammonium and nitrate were retained in the forest by canopy and root uptake and soil retention by >90% (Hultberg, 1985b). Protons were retained by 24% in the canopy (throughfall pH ≈3.7) and another 32% was retained in the soil.

The H^+ retention in the soil resulted in a small excess leaching of Ca and Mg (Hultberg 1985b) but soil- and groundwater was acid, with pH 4.5 to 5.0 in soil water in the B- and C-horizons. A major portion of the H^+ retention in soils resulted in leaching of Al (Nilsson, 1985b). The Al-concentration was lower in runoff compared to that in soil- and groundwaters and up to 55% was in the form of organic Al complexes (Lee, 1985). Aluminum uptake by humus in outflow parts along the stream and release of an equivalent amount of H^+, may explain the increased acidity in runoff (pH: 4.1 to 4.3) compared to soil- and groundwater.

The net annual tree uptake of base cations and sulphate were obtained from tissue concentrations and estimates of net primary production. The base cation depletion rate weathering + mineralization) was estimated as base cation output by runoff plus base cation accumulation in tree biomass minus the atmospheric deposition (wet + dry). The uncertainty in dry non-marine deposition of Ca and K is reflected by a range of the estimates (Hultberg, 1985b). The derived sequence was Ca>Mg>Na>K.

The annual sum of atmospheric input and average base cation depletion is compared to the sum of tree biomass accumulation and output by runoff from the terrestrial catchment in Figure 4. With the above mentioned assumption, atmospheric deposition is equally important as the soil and bedrock for the yearly fluxes of Ca, Mg and K while the source of the S flux is atmospheric deposition. On a yearly basis accumulation of Ca and K in tree biomass is as high as the total input by atmospheric deposition and output by runoff, while Mg is retained to a lesser extent. Compared to the deposition, almost no Na or S is retained in the trees.

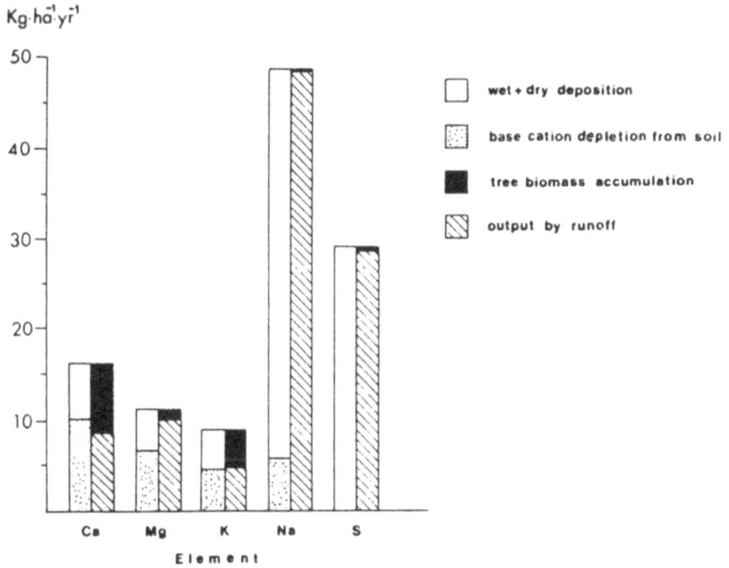

Fig. 4 Element flux of Ca, Mg, K, Na and S in kg•ha^{-1}yr^{-1} in a Norway spruce forested microcatchment (F1) in the Lake Gårdsjön catchment.

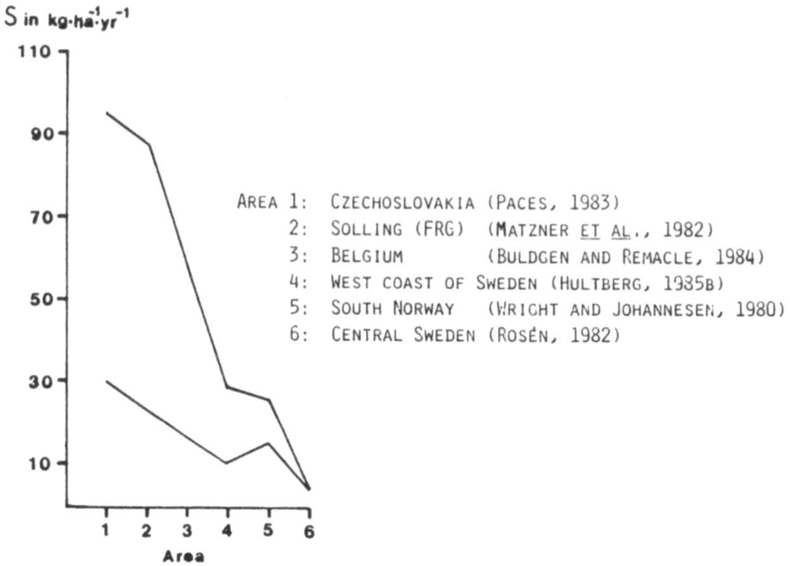

Fig. 5 Sulphur transport by runoff (upper curve) and wet deposition (lower curve) from six coniferous forested catchments in Central Europe, Sweden and Norway. Data originate from Paces (1983), Matzner et al. (1982), Buldgen and Remacle (1984), Hultberg (1985b), Wright and Johannesen (1980), and Rosén (1982).

Figure 5 shows the output of S compared to wet S-deposition from coniferous forested catchments in three countries in Central Europe and three catchments in Sweden and Norway (Hultberg 1983). Assuming the input/output of S are at steady-state in these catchments the dry S-deposition may be calculated as output by runoff minus wet input. Such estimates show that the dry S-deposition is 63 to 73% of the total in 4 sites including the Lake Gårdsjön catchment. The dry S-deposition is underestimated if S retention occur in the soils. Higher rainfall due to oreographic precipitation in the south Norwegian site gives a proportionally lower dry deposition of S (38%). The site in central Sweden is situated far from the major source areas of S emissions which results in a dry S-input of ≈0%. The gradient in both wet and total S deposition from high S emission areas in Central Europe to low S emission areas in central Sweden is sharp and linked to the SO_2 and ammonium sulphate particulate concentrations in the air mass.

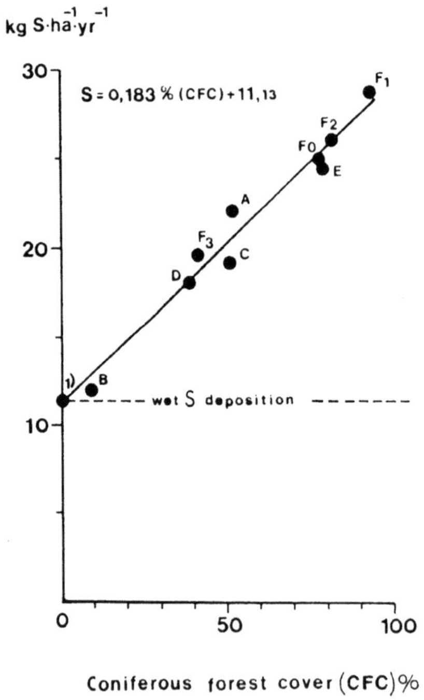

Fig. 6 Sulphur transport ($kg{\cdot}ha^{-1}yr^{-1}$) from terrestrial parts of lake subcatchments, microcatchments (and minicatchments 1)) in relation to coniferous forest cover (%). Location within the Lake Gårdsjön catchment labelled F1 - F3 and A - F0, are found in Figure 3. Minicatchment runoff from rock outcrops are labelled 1 (see Hultberg, 1985b).

The importance of the tree canopies for the total deposition of S is further illustrated by the measured and estimated S output from microcatchments and terrestrial parts of each lake subcatchment within the Lake Gårdsjön watershed (Hultberg, 1985b). The regression in Figure 6 (see also Figure 3) shows that the yearly S-transport from a catchment without a forest cover is 11.1 $kg \cdot ha^{-1} \cdot yr^{-1}$ which is close to the measured average wet deposition of S (11.4 $kg \cdot ha^{-1} \cdot yr^{-1}$) and indicates that dry S-deposition, SO_2 and $(NH_4)_2SO_4$ particulates, are directly related to the exposed needle surface of the coniferous trees.

The S-transport from a fully forested catchment with mature stands of Norway spruce and Scots pine was estimated to 29.4 $kg \cdot ha^{-1} \cdot yr^{-1}$.

3.3 Acidity in runoff from land

The acidity in runoff from calibrated catchments with long-term monitoring of chemistry in Europe and North America, was compared by use of a charge balance model (Hultberg, 1985b). The model is a simple linear regression based on non-marine (and non-roadsalt) concentrations ($\mu eq \cdot L^{-1}$) and the pH-range covered was 4.15 to 5.5 (Figure 7)

$$H^+ = 34.6 \cdot \frac{SO_4{}^{2-}(+NO_3{}^-)}{Ca^{2+} + Mg^{2+}} - 10.71,\ n = 13,\ r^2 = 0.93.$$

As shown in the regression sulphate is the most important anion to explain the acidity in runoff from terrestrial parts of catchments underlain by low weathering bedrock and lime-poor soils. The low nitrate concentrations in runoff have no significant effect on the regression. Recent observations of high NO_3-transport have, however, been shown from forested catchments in Central Europe (Grennfelt and Hultberg, 1986). When data from these sites are used nitrate must be included in the regression. Other cations or anions including organic anions seem to be less important to explain the acidity in runoff from coniferous forested catchments in the studied pH-range: 4.15 to 5.5. The weighted annual average pH (4.13 to 4.16) measured in two of the microcatchments at Lake Gårdsjön are among the most acid reported in the literture. Occasionally pH is as low as 3.8 to 3.9 for shorter periods (days).

Low leaching of base cations has resulted in high export of Al (4.8 to 6.0 $kg \cdot ha^{-1} \cdot yr^{-1}$) from all terrestrial parts in the Lake Gårdsjön catchment (Nilsson, 1985b). High concentrations of inorganic Al in the soil- and groundwater may precipitate P, which cause low transport of P from land to the Lake Gårdsjön watermass (Persson and Broberg, 1985).

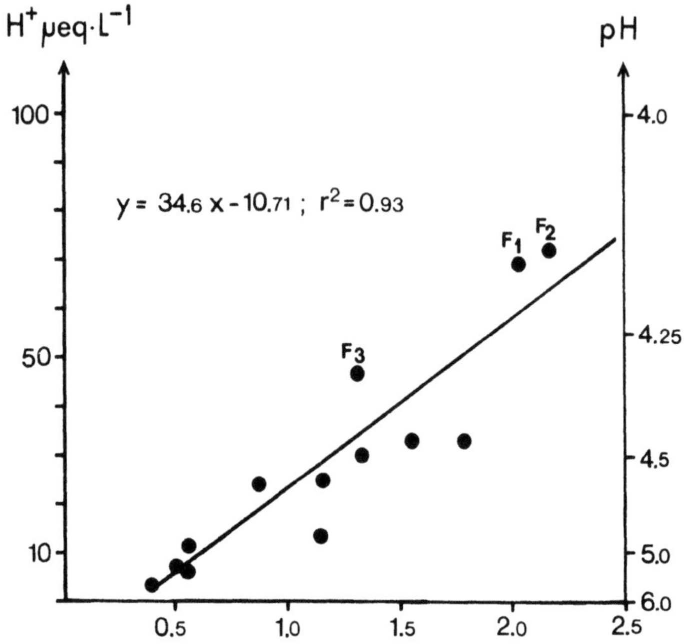

Fig. 7 Charge balance model of the ratio $\frac{SO_4^{2-}(+NO_3^-)}{Ca^{2+} + Mg^{2+}}$ in $\mu eq \cdot L^{-1}$ vs H^+ -concentration in runoff from long-term monitored catchments in Europe and North America. Data originate from Wright and Johannesen (1980, 1982), Henriksen and Wright (1977), Rosén (1982), Jefferies et al. (1984), Schindler et al. (1976), Likens et al. (1977), and Hultberg (1985b).

The increased concentrations of inorganic Al in the lake water may also reduce the P turnover rate in the pelagic system. Both these effects result from the soil interaction with acid deposition and may be the cause of oligotrophication of acid lakes (Grahn et al., 1974).

4. Liming of Lake Gårdsjön

110 tonne of finely ground limestone (0 to 0.2 mm) was spread in Lake Gårdsjön in April 1982. The total amount of Ca spread in the lake was 34.9 tonne. During a two and a half year period after liming a total of 29.9 tonne (84%) of the spread lime was dissolved.

A sharp increase in pH from 4.7 to 7 occurred in the whole water column and was only slightly depressed during

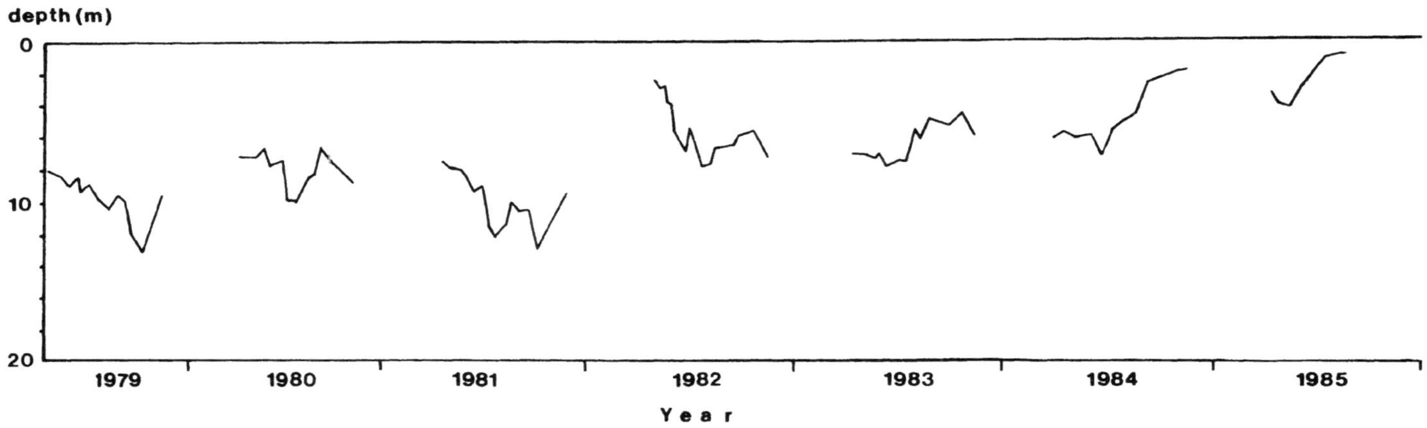

Fig. 8 Secchi-disc transparency (m) in Lake Gårdsjön from 1979 to 1985. Data originate from Larsson and Leyton (1985).

the snowmelt 1984 (Figure 1). The short retention time of water (1.1 to 1.3 yr) has reduced the initial alkalinity of 0.5 meq$\cdot L^{-1}$ to ≈0.3 in 1983, ≈0.13 in 1984 and <0.1 in 1985. This fast loss in alkalinity was reversed by a new liming of 30 tonne (9 tonne Ca) in October 1985. The lake had high concentrations of Ca during the years 1982, 1983 and 1984.

The inorganic Al concentration dropped from an average of 0.3 to 0.4 mg$\cdot L^{-1}$ to almost 0 during summer but input from surrounding land and upstream parts of the catchment increased the inorganic Al during autumn rain periods and at spring snowmelt. Beside these effects on the lake chemistry, no significant changes occurred in other major cations and anions.

Phytoplankton species composition and biomass changed, however, already during 1982 and from summer 1984 a dramatic change has occurred with prolific growth and extreme biomass of primarily Cosmocladium perissum, a small green algae. Consequently the Secchi-disc transparency was reduced from 9 to 12 m during acid conditions to late summer transparencies of <2 m during 1984 and 1985 (Figure 8). A fish population (Salmo trutta) was successfully introduced during autumn 1985.

5. Experimental treatments in microcatchments

5.1 Effects on acidity, Al and S in runoff by clearcutting

50% of the Norway spruce stand was cut in microcatchment F3 in 1976, and the runoff from this catchment had higher pH, lower Al-concentration and lower S-concentration compared to the fully forested F1 and F2 (see Figure 6). As the coniferous trees are extremely good collectors of dry S-deposition an experimental total clear felling of microcatchment F3 was decided to study if a reduction in acidity, Al-leaching and S-flux occur by the reduced dry S-deposition.

The Norway spruce trees were felled in April 1984 and each tree was individually examined of signs of forest decline symptoms. Figure 9 shows that a reduction of >20% of the needle biomass of the upper $^1/_3$ of the trees were observed on 46% of the stand (Westman and Bergholm, 1986). The dead fine root biomass increased and the living fine root biomass decreased with increased loss of needle biomass of the trees (Persson et al., 1985). The results from this detailed inventory of forest decline in southwest Sweden may be representative of a large area as a later inventory showed similar results in south Sweden.

Changes in chemistry of runoff water from microcatchment F3 are not evaluated in detail, but pH in runoff water show a steady increase to pH 4.8 in summer 1985 since the

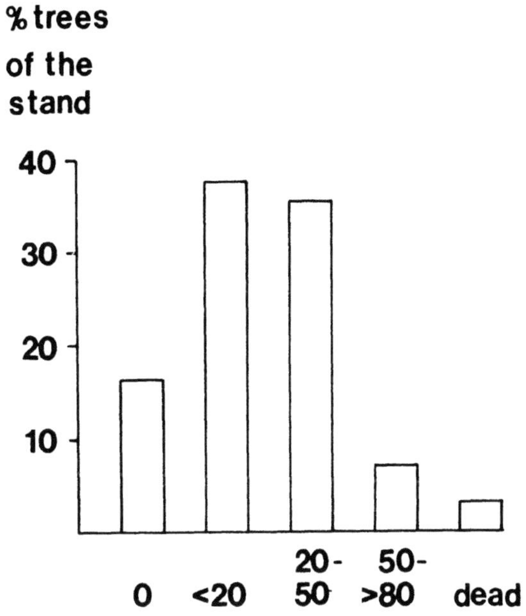

Fig. 9 Needle biomass loss inventory on a Norway spruce stand in microcatchment F3 in 1984.

clear felling. This is the highest pH recorded since 1979 when the water chemistry monitoring started. The reference catchment F1 does not show this trend. No trends are so far observed in Ca, Mg, Al or SO_4^{2-}. Increase of K and NH_4^+ may explain the increase in pH in runoff from F3. NH_4^+ may have formed in the soils of the catchment as a result of reducing conditions (depletion of O_2) due to increased groundwater table. Further monitoring during the following years are necessary to evaluate future trends in pH, NH_4^+, NO_3^-, SO_4^{2-}, Ca, Mg, and Al which all may be important for the evaluation of the effects of the clear felling.

5.2 Effects on acidity, Ca, Mg, and Al after liming

Microcatchment F2 was limed by 1500 kg finely ground (0 to 0.2 mm) dolomite in June 1984. Monitoring of pH, Ca, Al, and SO_4^{2-} have so far not shown any significant changes in runoff. Increased transport of Mg, NO_3^-, and humic acids (DOC) have, however, occurred. These changes are small and a repeated liming may be decided for spring 1986.

5.3 Effects on runoff chemistry of forest fertilization

Forest fertilization by ammonium-nitrate (AN) is used to increase forest growth on nutrient-poor soils since many years in Sweden. During the last few years no or only small growth increases have been observed after fertilization of Norway spruce stands in south Sweden. This may be a result of increased wet and dry deposition of NH_4^+ and NO_3^- which fertilizes the forests.

The AN application to microcatchment F4 (Figure 3) resulted in a lower pH in runoff after the treatment in August 1984. NH_4^+ and NO_3^- increased to extreme concentrations during the whole winter 1984/85 but dropped to lower concentrations from spring 1985.

Parallel treatments with limestone plus fertilizer and similar transports of NH_4^+ and NO_3^- were observed by this treatment. These experimental treatments are part of a long-term study, where new fertilizing techniques will be tested and the effects on tree growth and water chemistry will be studied.

5.4 Effects on runoff chemistry of neutral and acid sulphur

In order to further understand the influence of a S deposition on the soil and runoff chemistry two microcatchments were treated with neutral and acid sulphate (ca 100 kg S $ha^{-1} \cdot yr^{-1}$), respectively. As neutral sulphate Na_2SO_4 was used, and as acid sulphate elemental S, which slowly oxidize and produces H_2SO_4. These studies have started during 1984 and applications of Na_2SO_4 and elemental S were performed in October 1985. Rains after the treatments induced high concentrations of Na and SO_4^{2-} from the Na_2SO_4 treatment and a sharp drop from pH 4.2 to 3.7. The elemental S application has not resulted in any change in pH during the first weeks after treatment. Inorganic Al, however, increased in both treated catchments.

6. Conclusions

The studies described above are at several stages of development where some are finished and reported in Ecological Bulletin 37 and some are in an early stage with few results so far. We believe that these studies both on biological and chemical effects in the limed Lake Gårdsjön and the experimental treatments in forested microcatchments will be of general value for the understanding of the acidification processes as well as on abatement strategies.

Acknowledgment

We wish to thank all scientists involved in The Gårdsjön Project for valuable information on results not yet published. The studies on chemistry within The Gårdsjön Project was funded by the National Swedish Environment Protection Board.

References

Almer, B., Dickson, W., Ekström, E., Hörnström, E. & Miller, U.: 1974, Ambio, 3, 30.
Buldgen, P. and Remacle, J.: 1984, Acid deposition and the sulphur cycle. Proc. SCOPE Belgium, Vanderborght, O.L.J. (ed.), p. 163-172.
Grahn, O., Hultberg, H. and Landner, L.: 1974, Ambio, 3, 93.
Grennfelt, P., Larsson, S., Leyton, P. and Olsson, B.: 1985, Ecol. Bull. (Stockholm), 37, pp. 101.
Grennfelt, P. and Hultberg, H.: 1986, Water, Air, and Soil Pollut. (this conference), __, ...
Henriksen, A. and Wright, R.F.: 1977, Nordic Hydrol., 8, 1.
Hultberg, H. and Stenson, J.: 1970, Fauna och Flora, 1, 11.
Hultberg, H.: 1983, 'Effects of acid deposition on aquatic ecosystems', in Ott, H. and Stangel, H. (eds.), Acid Deposition - A Challenge for Europe, Commission of the European Communities, Brussels, p. 167-185.
Hultberg, H.: 1985a, Ecol. Bull. (Stockholm), 37, pp. 64.
Hultberg, H.: 1985b, Ecol. Bull. (Stockholm), 37, pp. 133.
Jefferies, D.S., Schneider, W.A. and Snyder, W.R.: 1984, 'Geochemical Interactions of Watersheds with Precipitation near Sudbury, Ontario', in Nriagu, J.O. (ed.), Environmental Impact of Smelters, Wiley & Sons.
Jensen, K.W. and Snekvik, E.: 1972, Ambio, 1, 223.
Johansson, S. and Nilsson, T.: 1985, Ecol. Bull., 37, pp. 86
Larsson, and Leyton,: 1985, personal communication.
Lee, Y.H.: 1985, Ecol. Bull. (Stockholm), 37, pp. 109.
Likens, G.E., Bormann, F.H., Pierce, R.C., Eaton, J.S. and Johnson, W.M.: 1977, 'Biochemistry of a Forested Ecosystem', Springer, N.Y., 146 pp.
Matzner, E., Khanna, P.K., Meiwes, K.J., Lindheim, M., Prenzel, J. and Ulrich, B.: 1982, 'Elementflüsse in Waldökosystemen im Solling - Datendokumentation', Göttinger Bodenkundliche Berichte, 71, pp. 267.

Muniz, I.P. and Leivestad, H.: 1980, Acidification-effects on freshwater fish, in Drabløs, D. and Tollan, A.(eds.), Ecological Impact of Acid Precipitation, Proc. Intern. Conf. Sandefjord, Norway, SNSF-project, Oslo-Ås, 84-92.

Nilsson, I.S. and Bergqvist, B.: 1983, Water, Air, and Soil Pollut., 20, pp. 311.

Nilsson, I.S.: 1985a, Ecol. Bull. (Stockholm), 37, pp. 311.

Nilsson, I.S.: 1985b, Ecol. Bull. (Stockholm), 37, pp. 120.

Paces, T.: 1983, 'The pH and concentration of sulphate, nitrate and aluminum in the runoff from representative catchments in the Elbe River basin', in Ecological Effects of Acid Precipitation, SNV Report (Stockholm), PM 1636.

Persson, G. and Broberg, O.: 1985, Ecol. Bull. (Stockholm), 37, pp. 158.

Persson, H., Börjesson, I. and Puhe, J.: 1985. Roots and mychorrhiza in areas with forest decline. Air Pollution and Forest Decline, Swedish University of Agricultural Sciences, Uppsala, Skogsfakta, 8, pp. 55.

Renberg, I. and Wallin, J.E.: 1985, Ecol. Bull., 37, pp. 47.

Renberg, I. and Wik, M.: 1985, Ecol. Bull., 37, pp. 53.

Rosén, K.: 1982, 'Supply, loss and distribution of nutrients in three coniferous forest-watersheds in Central Sweden, Swed. Univ. of Agricultural Sciences, Uppsala, Rep. 41.

Schindler, D.W., Newburg, R.W., Beaty, K.G. and Campbell, P.: 1976, J. Fish. Res. Board Can., 33, 2526.

Westman, and Bergholm,: 1986, to be published.

Wright, R.F. and Johannesen, M.: 1980, 'Input and output budgets of major ions at gauged catchments in Norway', in Drabløs, D. and Tollan, A. (eds.), Ecological Impact of Acid Precipitation, SNSF-proj., Oslo-As, pp 250-251.

Wright, R.F. and Johannesen, M.: 1982, personal communication.

PROJECT RAIN: CHANGING ACID DEPOSITION TO WHOLE CATCHMENTS.

THE FIRST YEAR OF TREATMENT

R.F. Wright[1], E. Gjessing[1], N. Christophersen[2], E. Lotse[3], H.M. Seip[2], A. Semb[4], B. Sletaune[5], R. Storhaug[6], and K. Wedum[6]

[1]Norwegian Institute for Water Research, Box 333, 0314 Oslo, Norway
[2]Center for Industrial Research, Box 350, 0314 Oslo, Norway
[3]Department of Soil Science, Swedish University for Agricultural Sciences, 75007 Uppsala, Sweden
[4]Norwegian Institute for Air Research, Box 130, 2001 Lillestrøm, Norway
[5]Norwegian Water Resources and Electricity Board, Box 5091, 0301 Oslo, Norway
[6]Aquateam A/S, Box 6593, 0501 Oslo 5, Norway

ABSTRACT. Project Rain (Reversing Acidification In Norway) is a 5-yr international research project aimed at investigating the effect on water and soil chemistry of changing acid deposition to whole catchments. The project comprises 2 parallel large-scale experimental manipulations -- artificial acidification at Sogndal and exclusion of acid rain at Risdalsheia. Treatment at Sogndal commenced April 1984 with the acidification of the snowpack by addition of H_2SO_4 (SOG2) and a 1:1 mixture of H_2SO_4 and HNO_3 (SOG4). Preliminary results indicate rapid and significant response in runoff chemistry to the acid treatment; pH decreased (to as low as 4.1 during snowmelt in 1984); SO_4, NO_3, and labile Al increased. Response during snowmelt 1985 was modest relative to 1984. At Risdalsheia treatment began in June 1984 with the mounting of the transparent panels on the roofs at KIM catchment (treatment by deacidified rain) and EGIL catchment (control with ambient acid rain). Preliminary data for the first year indicate that most runoff samples from KIM contain much lower NO_3 concentrations, about 20 to 30% lower SO_4 levels and pH 0.1 to 0.3 units higher than runoff from EGIL catchment. The treatments continue in 1985-87. Project RAIN provides experimental evidence bearing on target loading, reversibility of acidification, and the processes linking acid deposition, soil acidification and freshwater acidification.

Water, Air and Soil Pollution **30** (1986) 47–63.

1. INTRODUCTION

Vigorous efforts to obtain reductions in the emissions of acidifying compounds SO_2 and NOx to the atmosphere are in part based on the premise that such reductions will restore acidified waters. The magnitude and rate of response of natural ecosystems to changes in acid loading is, however, not well known, largely because such effects have been difficult to document in the absence of large-scale reductions. We have now begun widespread manipulations of natural headwater catchments in Norway. Project RAIN (Reversing Acidification In Norway), a 5-yr international research project, comprises two parallel experiments in which the response of soil and runoff chemistry to changes in loading of strong acids from the atmosphere are studied (Wright 1985). The RAIN project builds on short-term pilot-scale experiments conducted in Norway by Seip _et al._(1979) and Christophersen _et al._, (1982). The project provides information on reversibility of acidification, rate of response and target loadings.

At Sogndal, a "clean" area in western Norway, we are acidifying two pristine catchments by addition of H_2SO_4 (SOG2) and a 1:1 mixture of H_2SO_4 and HNO_3 (SOG4), respectively (Figure 1, Table I). At Risdalheia, an acidified area in southernmost Norway, we have excluded acid precipitation from a small catchment (KIM) by means of a roof and are watering with clean precipitation beneath the roof.

Project RAIN is a 5-yr study and began in June 1983. The project plan calls for approximately 1 yr of pre-treatment data, 3 yr of treatment and 1 yr of post-treatment recovery data. The first year was devoted to selection of sites, collection of pre-treatment data, and design, construction and installation of roofs, watering systems, weirs and sampling devices (Wright, 1985). The scientific program currently includes measurements of precipitation volume and chemical composition, microclimate beneath the roofs, soil chemistry, and runoff volume and chemical composition.

2. MATERIALS AND METHODS

2.1. Site descriptions

Sogndal. For the acidification experiments we selected 4 pristine, headwater catchments at Sogndal. The site is located above treeline at about 900 m above sealevel. Vegetation is alpine Soils are thin (average depth about 30 cm) and poorly-developed (US classification:

Table I. RAIN project. Overview of the experimental catchments and treatments.

Sogndal. Acid addition experiments.

Catchment	Treatment	Area
SOG1	control	96300 m^2
SOG2	H_2SO_4	7220 m^2
SOG3	control	43200 m^2
SOG4	$H_2SO_4+HNO_3$	1940 m^2

Risdalsheia. Acid exclusion experiments.

Catchment	Treatment	Area
KIM	roof, clean rain	860 m^2
EGIL	roof, acid rain	400 m^2
ROLF	no roof, acid rain	220 m^2

Lithic Haplumbrept, sandy, siliceous, frigid). Soil pH (H_2O) is 4.5 to 5.5. Bedrock is siliceous gneiss (Lotse and Otabbong, 1985). Precipitation at Sogndal is relatively clean with a volume-weighted average pH of 4.9 and excess sulfate concentration of 14 µeq L^{-1} (Table II). Ambient loading of excess sulfate is about 9 kg^{-1} ha^{-1} yr. Runoff chemistry from the 4 catchments prior to treatment indicates extremely sensitive systems with pH 5.5 to 6.0, Ca concentrations 10 to 30 µeq L^{-1}, SO_4 15 to 25 µeq L^{-1}, and alkalinity 5 to 15 µeq L^{-1} (Table II).

Risdalsheia. For the de-acidification experiments we selected 3 natural headwater catchments at Risdalsheia, located about 5 km from the long-term calibrated catchment at Birkenes. The site is 300 m above sealevel. Vegetation at Risdalsheia is similar to that at Sogndal except that here pine (*Pinus sylvestris L.*) and birch (*Betula pubescens L.*) are also present. The soils are similar -- thin (average depth 15 cm), patchy and poorly-developed. A major difference is that here soil

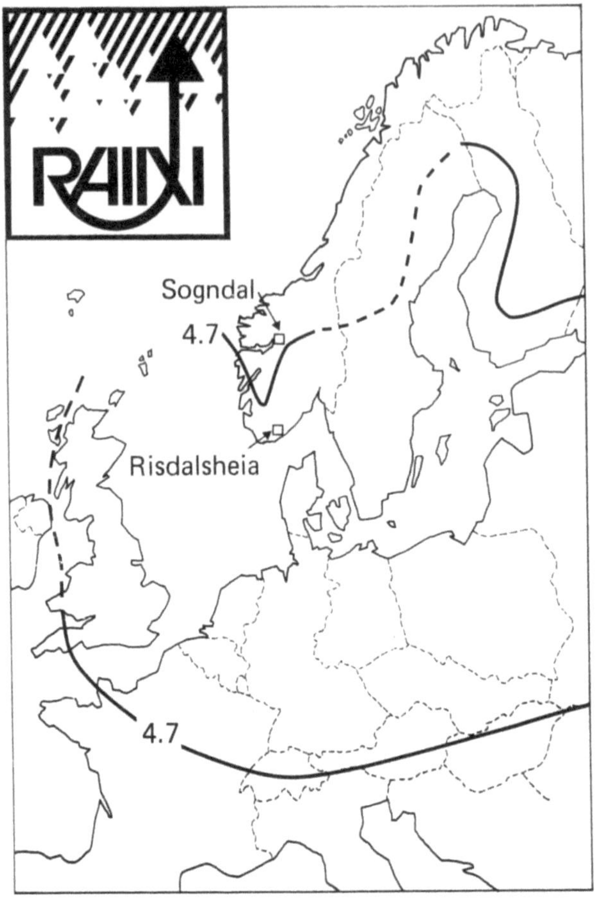

Figure 1. Location of the experimental catchments in Project RAIN. Areas within the pH 4.7 isoline receive precipitation with a yearly-weighted average pH below 4.7.

pH (H_2O) is about one unit lower (pH 3.9 to 4.5). Bedrock is biotite granite. Risdalsheia lies in the zone of maximum deposition of acid components in Norway. Data from the nearby Birkenes station show a long-term volume-weighted average pH of 4.2 in wet precipitation (Table III) (Christophersen and Wright, 1981). Deposition of sulfate is 70 kg ha^{-1} yr^{-1} (50 wet plus 20 dry). Runoff from the 3 catchments prior to treatment was highly acidic (pH 3.9 to 4.2) with high levels of sulfate and inorganic labile Al (Table III).

Table II. Volume-weighted mean concentrations of major ions in precipitation (in) and runoff (out) for the first year of treatment (15 November 1983 to 13 November 1984) at Sogndal. SOG2 receives H_2SO_4, SOG4 a 1:1 mixture of H_2SO_4 and HNO_3, and SOG1 and SOG3 are untreated controls. Units: water in mm, concentrations in μeq L^{-1}.

	SOG1 control		SOG2 H_2SO_4		SOG3 Control		SOG4 HNO_3+H_2SO_4	
	In	Out	In	Out	In	Out	In	Out
H_2O	1126	1026	1186	1086	1126	1026	1186	1086
H^+	17	2	75	8	17	3	75	4
(pH)	(4.8)	(5.7)	(4.1)	(5.1)	(4.8)	(5.5)	(4.1)	(5.6)
Na	62	88	62	59	62	87	62	51
K	5	11	5	4	5	2	5	2
Ca	8	20	8	22	8	21	8	21
Mg	14	15	13	13	14	18	13	11
Al^A	0	1	0	3	0	1	0	2
NH_4	8	6	8	3	8	1	8	1
NO_3	8	2	8	4	8	2	37	4
SO_4	27	28	84	40	27	24	55	27
Cl	70	89	69	60	70	83	69	57
HCO_3	0	15	0	5	0	2	0	6
Σ^+	114	145	171	112	114	133	171	97
Σ^-	97	134	161	109	97	111	161	94

[A] includes Al^{+3}, and inorganic complexes with OH^-, SO_4^{-2}, and F^-.

2.2. Measurements and treatments

Sogndal. At Sogndal 3 of the 4 catchments are equipped with weir and level recorder for continuous gauging of discharge. Samples of bulk precipitation for chemical analysis are collected weekly at Haukås farm located about 3 km east of the site at 500 m above sealevel. Parameters measured include volume, pH, conductivity, Na, K, Ca, Mg, NH_4, NO_3, SO_4, and Cl. Runoff samples for chemical analysis are collected weekly at the control catchments and at the treated catchments hourly during watering episodes, daily following episodes, weekly during the summer and fortnightly during the winter. Parameters measured include pH, conductivity, Na, K, Ca, Mg, NH_4, NO_3, SO_4, Cl, F, alkalinity, total-Al, acid-reactive Al, non-labile Al, total organic C (TOC), and SiO_2. Details are given by Wright (1985).

Soils at Sogndal were sampled in August 1983 and again in August 1984 (Lotse and Otabbong, 1985). Vegetation samples were collected in August 1984. In addition to description of soil profiles the analyical program includes pH, loss on ignition, C, N, S, exchangeable acidity, exchangeable base cations (Na, K, Ca, Mg, NH_4, and Al), base saturation, cation exchange capacity, and sulfate adsorption characteristics.

Treatment at Sogndal commenced in April 1984 with the addition of acid to the snowpack. About 0.02 mm of pH 1.9 acid was spread evenly over catchments SOG2 and SOG4. Prior to treatment bulk snow had a pH of about 4.9; after acid addition pH was 4.2. A commercial sprinkling system was used to apply acid in four separate episodes during August-October 1984. The acid was added to lakewater from SOG1 (Table II); 11 mm of pH 3.2 water was added. Before and after each treatment 2 mm of pH 5.6 lakewater was added to wet up and wash the vegetation, respectively. Watering intensity is 2 mm hr^{-1}. No direct damage to the vegetation was observed. Details are given by Wright (1985). Treatment continued with acid addition to the snowpack in April 1985 and watering in June 1985. A total of 106 keq H^+ km^{-2} has been added during the period April 1984-June 1985.

Risdalsheia. At Risdalsheia the experiment includes 3 headwater catchments (Table I). Two are covered by roofs (KIM treatment; EGIL control) and one (ROLF) serves as untreated reference. The roofs extend 2 to 3 m beyond the catchment boundaries.

Bulk precipitation is collected weekly at Risdalheia to supplement data from the nearby station at Birkenes, at which daily samples for air- and precipitation chemistry are available. Runoff volume is gauged continuously at the 3 catchments (ROLF catchment began October 1984).

Soils at KIM and EGIL catchments were first sampled in October 1983. Soils and vegetation at all three catchments were sampled in August 1984. Sampling is conducted such that integrated values for each catchment are obtained. The analytical program is similar to that at Sogndal (Lotse and Otabbong, 1985).

Table III. Volume-weighted average precipitation volume and chemistry at Birkenes (1972-78) (Christophersen and Wright 1981), under the roofs at EGIL and KIM catchments (29 October 1984 - 12 June 1985), and runoff volume and chemistry at EGIL and KIM catchments during the first year of treatment (13 June 1984 - 12 June 1985). Data for reference catchment ROLF are not shown, because runoff data prior to October 1984 are not available. Units: water in mm, concentrations in µeq L^{-1}.

	Birkenes	KIM acid exclusion		EGIL control	
	In	In	Out	In	Out
H_2O	1310	714[B]	620	900[B]	807
H^+	47	11	82	59	90
(pH)	(4.3)	(5.0)	(4.1)	(4.2)	(4.0)
Na	51	66	80	60	86
K	-	3	7	6	8
Ca	12	5	12	12	14
Mg	14	15	16	15	23
Al[A]	0	0	11	0	14
NH_4	36	0	8	58	16
NO_3	35	1	8	43	34
SO_4	65	13	88	87	111
Cl	59	86	111	72	109
Σ^+	160	101	216	210	251
Σ^-	159	100	207	202	254

[A] includes Al^{+3}, and inorganic complexes with OH^-, SO_4^{-2}, and F^-.

[B] volume for period 13 June 1984 - 12 June 1985.

Treatment at KIM and EGIL catchments entails collection of incoming precipitation by means of gutter and cistern systems. At KIM the water is sent through ion-exchange columns. Seawater is added to give the natural levels of seawater salts. The resulting precipitation has a chemical composition equivalent to that at Birkenes except that H^+, excess SO_4, NO_3, and NH_4 are removed. The water is pumped back out to a sprinkler system mounted beneath the roof. Watering proceeds at 2 mm hr^{-1}. The system at EGIL catchment is similar except that here the water is not treated but merely recycled back beneath the roof. The entire system is automatic and regulated by the water level in the cisterns. The system capacity is about 50 mm day^{-1}. During the winter we shut down the watering systems, lower side panels to prevent snow from blowing into the catchments, and make artificial snow using commercial snow-making equipment. Details are given by Wright (1985).

3. RESULTS

3.1. Sogndal

At Sogndal both catchments responded immediately and dramatically to the acid treatments (Figure 2). During snowmelt in 1984 runoff at SOG2 was acidified to pH 4.1 and contained high levels of sulfate and labile Al. Due to technical difficulties no samples from SOG4 were collected during snowmelt in 1984. During the 4 episodes of acid addition in August-October 1984 pH levels fell to 4.7 at SOG2 and 4.3 at SOG4. The dominant anion was sulfate at SOG2 and approximately equivalent concentrations of sulfate and nitrate at SOG4 (Figure 2). At both catchments acidic runoff contained high concentrations of labile Al.

Whereas SOG4 responded and returned to background levels within hours, the response at SOG2 is considerably damped. Runoff chemistry exhibits a step-like pattern with increasingly higher acidity, sulfate and labile Al concentrations following each of the four treatments (Figure 2). The small pond at the bottom of SOG2 provides a hydrologic buffer that apparently acts to smooth out sudden changes in runoff chemistry such as observed at SOG4.

Of the 40 keq SO_4 km^{-2} in the 1984 snowpack of SOG2 approximately 45 % left in runoff. Only 20 % (7 keq km^{-2}) of the H^+ left in runoff, however, with the difference accounted for by increased labile Al and base cations and decreased bicarbonate alkalinity. Mass transport budgets for the 6-week period in 1984 during which the 4 batches of 11 mm acidified rain were applied indicate that at SOG2 only about 20 % of the 48 keq km^{-2} SO_4 applied left in runoff. At SOG4 about 50 % of the 26 keq km^{-2} SO_4 and 30 % of the 23 keq km^{-2} NO_3 left the catchment in runoff. At both catchments about 3 % of the applied H+ (45 keq km^{-2}) left in runoff.

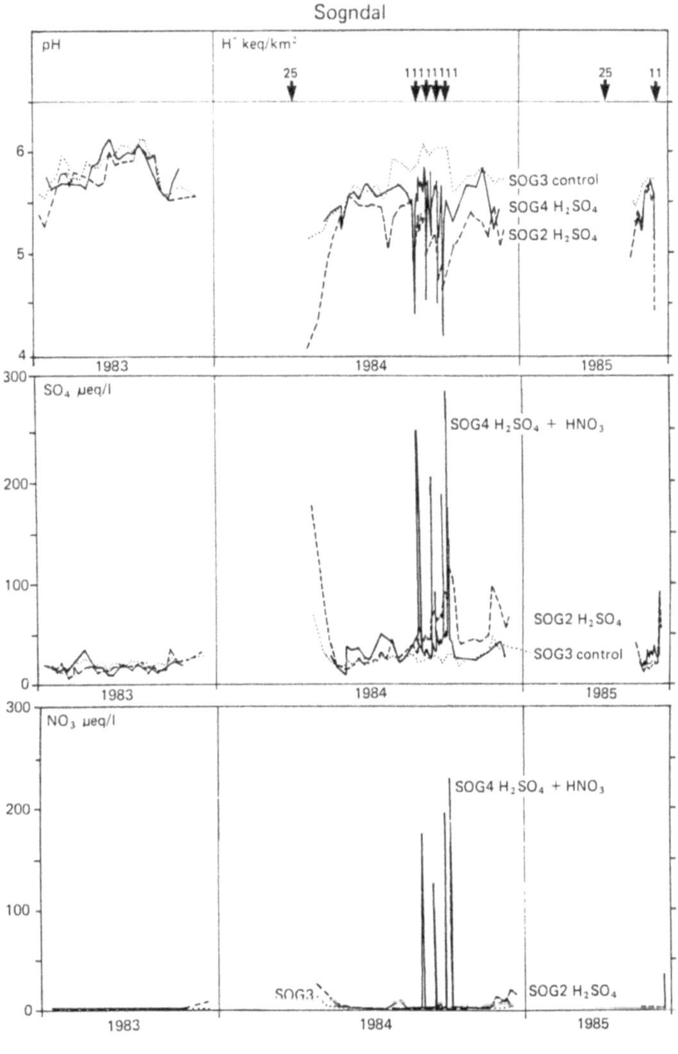

Figure 2. pH and concentrations of sulfate and nitrate in runoff from catchments SOG2, SOG3 and SOG4 at Sogndal over the period June 1983 - June 1985. Treatments began in April 1984 with the acid addition to the snowpack. Acid additions (either 25 or 11 keq km^{-2}) are indicated by arrows. Control catchment SOG1 (not shown) exhibits chemistry similar to SOG3.

Snowmelt in 1985 again resulted in acidification of runoff at SOG2 and SOG4 relative to the control catchments (Figure 2). SOG2 apparently did not fully recover from the acid treatments the previous autumn and exhibited slightly lower pH levels and higher sulfate levels in baseflow throughout the winter. Minimum measured pH in runoff from SOG2 during snowmelt in 1985 was about 5.0, substantialy higher than the pH 4.1 measured in the first snowmelt of 1984.

The first spraying of acid onto the soil in 1985 was carried out in June, only a few weeks following the snowmelt. The response at both catchments SOG2 and SOG4 was similar to that observed following application of acid the previous summer and autumn, except that while the pH levels again dropped to about pH 4.4, the concentrations of the acid anions sulfate and nitrate were considerably lower than at similar pH levels the previous autumn. The runoff in June 1985 was considerably more dilute and contained lower concentrations of the base cations, probably due to the flushing of the soil by large volumes of dilute snowmelt water only a few weeks previously.

Risdalsheia

Nitrate concentrations in runoff at KIM catchment declined promptly in response to treatment (Figure 3). During the first 10 weeks following beginning of treatment in mid-June 1984 there was very little rain or runoff. With the onset of the autumn rains in September nitrate concentrations in runoff decreased promptly from about 10 to 20 $\mu eq\ L^{-1}$ to below detection limit and remained low until late autumn. Sulfate concentrations in runoff began to deviate relative to levels at EGIL and ROLF catchments in mid-October, at which time about 350 mm of precipitation had been ion-exchanged. At this time pH levels began to go up and levels of labile Al decreased (Figure 3). By mid-December 1984 at the onset of winter sulfate concentrations at KIM were about 80 $\mu eq\ L^{-1}$ as compared to 100 to 120 $\mu eq\ L^{-1}$ at EGIL and ROLF, and pH levels were about 0.2 units higher.

This trend was resumed in the snowmelt. Whereas runoff at KIM exhibited sulfate levels about 80 to 100 $\mu eq\ L^{-1}$ during the snowmelt period, runoff at EGIL and ROLF repeated the pattern observed during the pre-treatment year 1983 with first high levels (160 $\mu eq\ L^{-1}$) followed by lower levels (100 to 120 $\mu eq\ L^{-1}$). Nitrate was again much lower at KIM (0 to 10 $\mu eq\ L^{-1}$) than at EGIL and ROLF (20 to 40 $\mu eq\ L^{-1}$), and pH levels were also slightly higher (Figure 3).

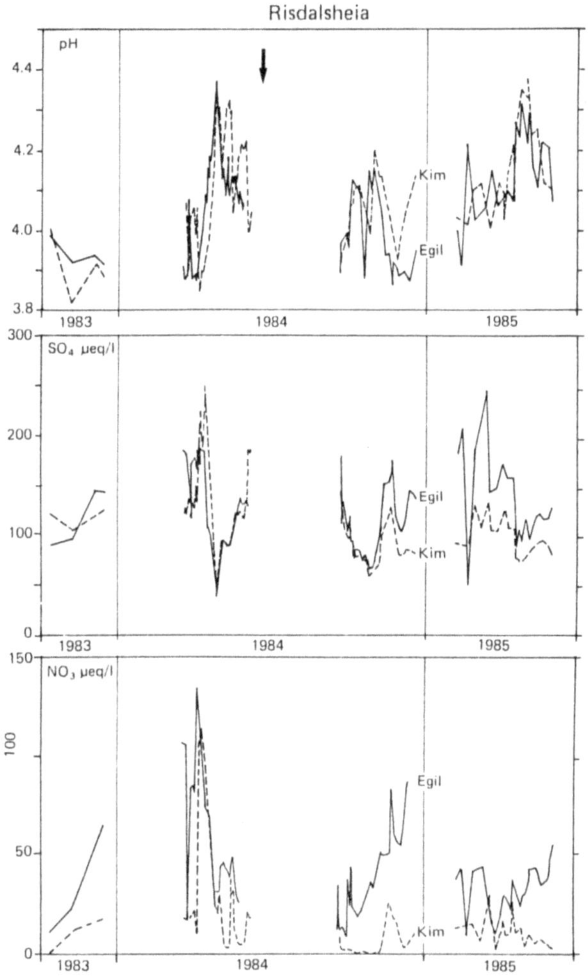

Figure 3. pH and concentrations of sulfate and nitrate in runoff from KIM and EGIL catchments at Risdalsheia over the period October 1983 - June 1985. ROLF catchment (not shown) exhibits chemistry similar to EGIL. Treatment began in mid-June 1984 (arrow).

During the first year of treatment (15 June 1984 - 14 June 1985) 715 mm of clean precipitation (600 mm rain and 115 mm snow) had been added to KIM catchment, and 900 mm of acid precipitation (785 mm as rain and 115 mm as snow) added to EGIL catchment. The difference in applied volume is due largely to technical problems encountered in autumn 1984 at KIM catchment. Precipitation volume at Birkenes over this period was about 1500 mm of which 350 mm came as snow.

3.2. Aluminum fractions

Aluminum concentrations in runoff at Sogndal and Risdalsheia are a function largely of pH and concentration of TOC. Inorganic Al (LAL) levels increase with decreasing pH (Figure 4). The samples from Risdalsheia and those from Sogndal cluster into two distinct groups indicating that LAL concentrations at these sites are controlled by different solid phases. At Risdalheia LAL levels give a pK (expressed as 3 pH-pAl^{+3}) for $Al(OH)_3$ of about 6.8, whereas at Sogndal a value of about 9.1 is indicated (Figure 4). Organic Al levels (NLAL) increase with increasing concentrations of TOC. Linear regression of NLAL on TOC for 193 samples from Risdalsheia gives NLAL (µg L^{-1}) = 43.5 + 12.2 TOC (mg C L^{-1}) with r = 0.75.

4. DISCUSSION

We explain the difference in export of applied acid between snowmelt and autumn 1984 at SOG2 by the probably shorter and less complete contact of meltwater with soil during snowmelt; a larger fraction of the applied acid reaches the weir without undergoing neutralization in the soil.

The difference between SOG2 and SOG4 during the 4 autumn waterings in 1984 is more difficult to explain, but is probably related to hydrologic differences between the catchments rather than differences in soil characteristics. SOG2 contains a small pond that provides a hydrologic buffer. Response at SOG2 is thus considerably damped relative to the spiky response at SOG4 (Figure 2). Response at SOG2 requires chemical changes in the pond water by flushing with runoff from the surrounding area.

Response during snowmelt in 1985 was different than during snowmelt in 1984. pH levels in runoff from SOG2 declined to 4.1 in 1984 but only 5.0 in 1985 despite similar loading of acid to the snowpack. The difference may be related to snowpack volume or nature of the snowmelting.

At Risdalsheia exclusion of wet deposition of acid resulted in immediate decline in nitrate concentrations in runoff at KIM catchment (Figure 3). These terrestrial catchments are unable to retain all the incoming nitrate in precipitation, and thus nitrate mobility contributes significantly to runoff acidity. These data suggest that at Risdalsheia the catchments are saturated with respect to N.

The first year of treatment at KIM catchment caused significant reduction in sulfate concentrations in runoff (Figure 3). The normal seasonal pattern of sulfate concentration in runoff is one of maximum levels during the first phases of snowmelt and in the first runoff following the onset of autumn rains and minimum levels at the end of snowmelt in May and at the end of autumn in November (Figure 3). At the nearby gauged catchment at Birkenes and at the Storgama catchment about 50 km further inland this pattern is well-explained by a simple model for hydrology and sulfate in which water-soluble sulfate accumulates in the snowpack and during the relatively dry summers and then is washed out during periods of high flow (Christophersen and Wright, 1981; Christophersen et al. 1983). The peak concentrations are governed largely by the amount of accumulated and mineralized sulfate while the minimum concentrations and regulated by a process analogous to sulfate adsorption. Treatment at Risdasheia has now largely eliminated the peaks while the minimum level of 80 to 100 µeq L^{-1} is retained.

KIM catchment is bleeding sulfate; concentrations in rain and snow are now about 13 µeq L^{-1} while runnoff still contains 80 to 100 µeq L^{-1} (Table III). Some of this difference may be accounted for by an unknown amount of dry deposit that is not excluded by the roof. At Birkenes dry deposition of sulfate as SO_2 gas and SO_4 particles is estimated at 20 kg ha^{-1} yr^{-1}. Distributed over the 700 mm runoff at KIM catchment 1984-85 this amounts to about 50 µeq L^{-1}. Prior to treatment dry deposition at Risdalsheia was certainly less than at Birkenes. Birkenes is heavily forested and interception will be large whereas at Risdalsheia the sparse forest offers much lower surface area for interception of gases and particles. Furthermore the roof will act to exclude a fraction of the dry deposit. Although dry deposit under the roofs can theoretically account for up to 50 µeq L^{-1} SO_4 in runoff, we estimate this contribution to be much lower.

From measurements of the sulfate adsorption characteristics of the soils at Risdalsheia (Lotse and Otabbong, 1985) we estimate that sulfate levels in runoff should begin to decline from the present minimum plateau of 80 to 100 μeq L^{-1} during the second year of treatment. If, however, a significant net conversion of other forms of S in the catchment such as organic S to water-soluble sulfate occurs, then further decline in sulfate concentrations can be substantially delayed. At present we have no measure of the rate of conversion to water-soluble sulfate in these catchments.

At Risdalheia pH levels in runoff during late autumn 1984 and spring 1985 from KIM catchment have been 0.1 to 0.3 units higher than in runoff from EGIL catchment since mid-October 1984. pH levels at all 3 catchments reach maximum levels during the last phase of snowmelt. This can be explained simply by dilution of the strong acid solution by meltwater of relatively low ionic strength. At KIM catchment after one year of treatment sulfate is still the dominant anion in runoff. pH levels remain below pH 4.4, the range in which aluminum buffering plays an important role. As the treatment continues and if sulfate levels decline further, inorganic Al concentrations will decline and pH will increase with the result that pH levels may become governed by weak organic acids. With the high levels of TOC (10 to 20 mg C L^{-1}) in runoff from the Risdalsheia catchments pH levels may ultimately not go above 4.6 to 5.0.

Acid additions at Sogndal caused increases in labile Al from natural levels of <50 μg Al L^{-1} to as much as 1000 μg Al L^{-1} (Figure 4). Our data indicate that an $Al(OH)_3$ solid phase with pK of about 9.1 can account for labile Al concentrations at Sogndal (Figure 4). This solubility corresponds to that of synthetic gibbsite at 10^0C or natural gibbsite at 20^0C (Driscoll et al., 1984, Seip et al., 1984). Similar LAL-pH relationships are found elsewhere in Norway (Wright and Skogheim, 1983; Seip et al., 1984) and in North America (Driscoll, 1980; Driscoll et al., 1984; Johnson et al., 1981).

At Risdalsheia labile Al levels can be explained by an Al hydroxide phase with pK of 6.8, much lower than at Sogndal (Figure 4). This site difference may be related to differences in soil chemistry, in particular soil pH (Lotse and Otabbong, 1985). A possible explanation is that at Sogndal with soil pH in the 4.6 to 5.4 range, Al mobilized in the surface soil layers at low pH will be precipitated out again in the lower soil layers at higher pH or in the stream channel itself as the soil solution degasses and looses CO_2 to the atmosphere. This process is described quantitatively by Reuss and Johnson (1985). Addition of strong acid to these soils would then result in re-solution of this previously precipitated Al hydroxide phase, presumably an easily-soluble phase. At Risdalsheia, on the other hand, both soil pH

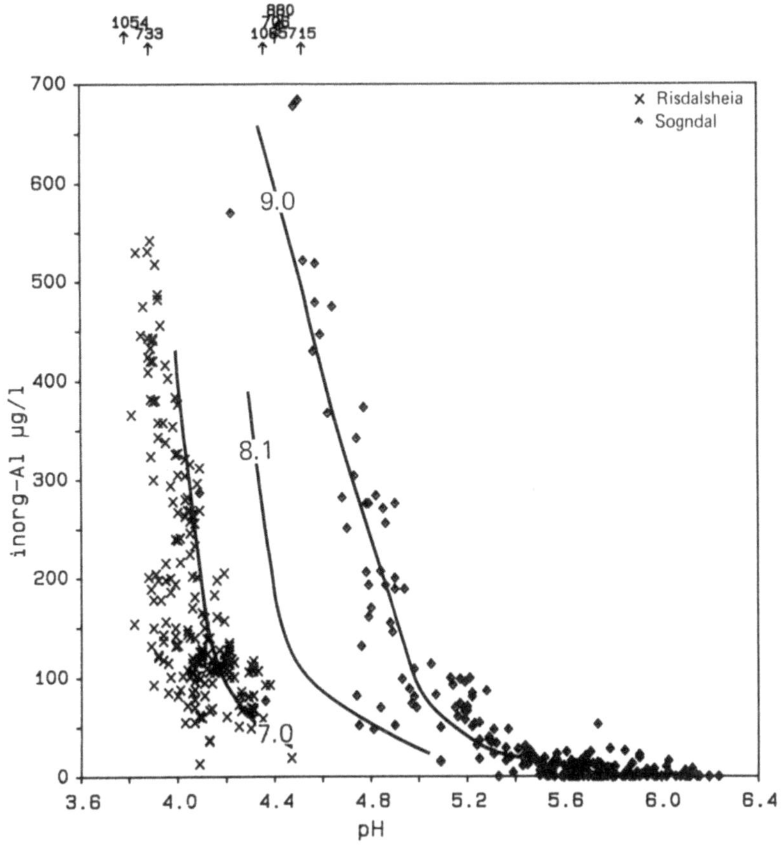

Figure 4. Scatter diagram of pH and concentration of inorganic-Al (labile Al) in runoff samples collected October 1983 - October 1984 at Sogndal (n=187) and Risdalsheia (n=193). Also shown are the lines of equilibrium with $Al(OH)_3$ with pK (expressed as $3pH-pAl^{+3}$) of 7.0, 8.1, and 9.0.

and runoff are in the range 3.8 to 4.2, and Al dissolved in soil solution is simply not precipitated before leaving in runoff. The solubility product of 6.8 may reflect dissolution of Al from a primary mineral such as feldspar or from secondary clay minerals, rather than from an easily-soluble Al hydroxide phase. On the other hand the observed LAL-pH relationships could be simply due to differences in ion-exchange characteristics of these two soils.

Already at the one-year mark the RAIN project provides information about the response of whole catchments to changes in loading of acids from the atmosphere. Although many of the results presented here are preliminary, we are hopeful that the RAIN project will help us understand the complex relationship between soil and water acidification and provide information useful for the establishment of target loadings and goals for future levels of emissions of acidifying compounds.

ACKNOWLEDGMENTS

We thank S. Andersen and R. Høgberget for technical assistance, and N. Knagenhjelm, Sogn Televerk and Arendal Televerk for permission to use private roads. N. Dalaker, H. Haukås, A. Meltveit and A. Risdal provided local assistance.

Financial support comes from the Norwegian Ministry of Environment, The Royal Norwegian Council for Scientific and Industrial Research, The Ontario Ministry of the Environment, Environment Canada, The Swedish National Environmental Protection Board, the Surface Water Acidification Programme (SWAP) (The Royal Society, the Norwegain Academy of Science and Letters, and the Royal Swedish Academy of Sciences), and in-house funds from the participating institutes.

REFERENCES

Christophersen, N., Dymbe, L.H., Johannessen, M., and Seip, H.M.:1983, Ecol. Modelling 21, 35.

Christophersen, N., Stuanes, A.O., and Wright, R.F.: 1982, Nordic Hydrol. 13, 115.

Christophersen, N., and Wright, R.F.: 1981, Water Resour. Res. 17, 377.

Driscoll, C.T.: 1980, 'Chemical characterization of some dilute acidified lakes and streams in the Adirondack Region of New York State', Thesis, Cornell Univ., Ithaca, NY.

Driscoll, C.T., Baker, J.P., Bisgoni, J.J., and Schofield, C.L.: 1984, 'Aluminum speciation and equilibria in dilute acidic surface waters of the Adirondack Region of New York State', p.55-76, In Bricker, O.P. (ed.), Geological Aspects of Acid Deposition, Butterworth, Boston.

Johnson, N.M., Driscoll, C.T., Eaton, J.S., Likens, G.E., and McDowell, W.H.: 1981' , Geochim. Cosmochim. Acta 45,1421.

Lotse, E., and Otabbong, E.: 1985, 'Physiochemical properties of soils at Risdalsheia and Sogndal: RAIN project', Acid Rain Res. Rept. 8/85 (Norwegian Inst. Water Research, Oslo, Norway), 48pp.

Ruess, J.O., and Johnson, D.W.: 1985, J.Environ.Qual. 14, 26.

Seip, H.M., Müller, L., and Naas,A.: 1984, Water, Air, Soil Pollut. 23, 81.

Seip, H.M., Gjessing, E.T., and Kamben, H.: 1979,' Importance of the composition of the precipitation for the pH in runoff-experiments with artificial precipitation on partly soil-covered "mini-catchmnts"'. Internal Report IR 47/79, (SNSF-project, 1432 Ås, Norway), 34 pp.

Wright, R.F.: 1985, 'RAIN project. Annual report for 1984', Acid Rain Res. Rept. 7/85 (Norwegian Inst. Water Research, Oslo, Norway), 39pp.

Wright, R.F., and Skogheim, O.K.: 1983, Vatten 39, 301.

ACID DEPOSITION AND EFFECTS IN NORDIC EUROPE.
DAMAGE TO FISH POPULATIONS IN SCANDINAVIA CONTINUE TO APACE.

Rosseland, B.O.[1], Skogheim, O.K.[1] and Sevaldrud, I.H.[2]
Directorate for Nature and Management, Fish Research Division, Tungasletta 2, N-7000 Trondheim , Norway.

ABSTRACT. Although the decline in fish populations due to acidic water in Norway started as early as in the 1920's the most rapid losses appeared during the 1960-70's. Until 1978, the populations of Atlantic salmon had disappeared from the southernmost part of Norway, and in these areas, more than half of the brown trout populations had been lost. Today, in spite of no increase in acid depositions, the fishery problems seems to continue at the same speed. Data based on interviews of the local fish authority shows that lakes still holding a fish population in the late 70's, have experienced a 30% loss of brown trout populations and a 12% loss of perch in the period 1978-1983. This trend have been confirmed by testfishing in lake systems having long data series. Salmon rivers on the western coast of Norway have experienced several episodes of fish kills due to rapid changes in water quality. These fish kills have mainly affected smolts of Atlantic salmon. Spawning migrating salmon on entering their acidified home river have also been affected. In Sweden, several salmon populations along the western coast have been lost due to acidification with no positive trends so far in the 1980's. Areas in central Sweden and in some high mountain areas are still experiencing a continuous and increasing acidification with detrimental effects on invertebrates and fish.

In Finland, an increase in acidic deposition during the last decades have occurred, leading to acidification in the most sensitive freshwater systems. Although some acidified freshwater lakes are reported to have lost their fish stocks, few data on fish population effects are available.

INTRODUCTION

The losses of freshwater fish populations in Scandinavia due to acidic precipitation have been well documented (Jensen and Snekvik, 1972; Leivestad et al., 1976; Overrein et al., 1980; Haines, 1981; Sevaldrud et al., 1980). Atlantic salmon (Salmo salar L.) was the first species to

Present address: 1) Norsk Biotech A/S, P.O. Box 788, Krossen, N-4301 Sandnes, Norway. 2) N-2710 Bjoneroa, Norway.

Water, Air and Soil Pollution **30** (1986) 65–74.

be affected (Huitfeldt-Kaas, 1922; Dahl, 1927), but an increasing acidification have further affected, populations of brown trout (Salmo trutta L.), roach (Rutilus rutilus L.) and perch (Perca fluviatilis L.) (Jensen and Snekvik, 1972; Almer et al., 1974; Leivestad et al., 1976). In Norway, the losses of brown trout populations were especially rapid throughout the 1960's, and by the mid 1970's an area of 33 000 km^2 were affected (Sevaldrud et al., 1980; Sevaldrud and Muniz, 1980; Muniz et al., 1984). At that time, no tendency for the trend to level off was observed. Based on this trend, Sevaldrud et al. (1980) predicted a 80% loss of brown trout populations in the most affected counties by the end of the 1980's.

During the last decade, the amounts of acidic precipitation transported to Norway and Sweden seems to have leveled off (SFT 1984a). This paper describes the effects on fish populations caused by acidification in Scandinavia sofar in the 1980's.

1. NORWAY

When using the interview-method (Sevaldrud et al. 1980; Sevaldrud and Muniz, 1980) in the county of Buskerud, an increase in affected areas by 20% from the 1970's compared to 1982 was found (SFT, 1983a).

The catchment area of River Gjerstad, county of Aust Agder, was updated in 1982. Timetrends for population changes of brown trout are available for 41 lakes. By 1977, 46% of the populations were reported lost, 27% had a declining population, whereas 27% of the lakes were reported to have a good stock of trout. In the 5- years period 1977 - 1982, however, 75% of the former good populations were reported declining. Of the 11 sparsely populated lakes 18% had become barren, 36% were still declining whereas 45% had a sparse but unchanged population (SFT, 1983a). In contrast to most barren lakes in southern Norway who lost their fish in the 1960's (Sevaldrud et al., 1980), the main losses of fish populations in Gjerstad occurred in the late 1970's and in the 1980's. Affected lakes are closely related to the geology in the area. Main losses have occurred in the area with granitic and poorly buffered bedrock (Telemark-formation), and only in recent years in areas with more easily weathered bedrock (Bamle-formation) (SFT, 1983a).

Data from Gjerstad initiated a larger updating of lakes in the two counties Aust- and Vest Agder. Lakes reported to have a fish population by the end of the 1970's, was checked for fishery status, and water samples were taken. In the period 1978 - 1983, 30% of the brown trout populations and 12% of the perch populations had disappeared (Sevaldrud and Skogheim, 1986). If this trend is representative for the relative few populations of trout and perch that were not updated, 71% of the brown trout populations and 43% of the perch populations have been lost in these two counties by 1983. Changes in fishery status reflects the general water chemistry; barren lakes having the lowest pH and calcium, and highest concentration of labile aluminium (Sevaldrud and Skogheim, 1986).

Two areas within these counties, River Tovdal and Farsund, have been given a special concern since the mid 1970's. Lake Tveitvatn, a part of River Tovdal, once had a stunted brown trout population (Haabesland, 1972). After a massiv fish kill in 1975 (Leivestad and Muniz, 1976), the population rapidly declined by both recruitment failure and postspawning mortalities (Rosseland et al., 1980). After 1980, the population continuously declined (SFT, 1981, 1982) and by 1982 the brown trout in lake Tveitvatn had disappeared (SFT, 1983a). Weekly water samples from the outlet of lake Tveitvatn showed periodes in autumn and spring 1977 - 1978, with pH down to 4.5 and a total Al-concentration up to 590 ugAl L^{-1} (SFT, 1981). Episodic changes in water chemistry seems thus to be the obvious explanation to the losses of fisheries in this river.

In the area of Birkeland, lower part of the River Tovdal catchment area, 6 lakes formerly having mixed populations of perch and brown trout, were updated by testfishing in 1984. In 1976, two of the lakes had mixed and stunted populations of perch and trout, whereas the other four had either both populations sparsely represented, or one of them being lost (Rosseland et al., 1980, 1981). In 1984 no changes had occurred in the two lakes with stunted populations. In the other four, three of them had become barren whereas the last had a sparse population of perch, a population being good in 1976 (Lake Djupedalsvatn) (SFT,1985). These results confirms the trend in the updating performed by Sevaldrud and Skogheim (1986).

Fish populations in two lakes close to Farsund, county of Vest Agder, lake Gjervollstadvatn and lake Saudlandsvatn, have been monitored since 1977. Both lakes once had a stunted brown trout population, but since 1977 high mortality of eggs (90 - 100%) have occurred in the spawning brooks (Muniz and Leivestad, 1980; SFT, 1982, 1983a, 1984a). Water chemistry have been monitored on a weekly basis. During the winter 1982/83, pH in the two brooks were in periodes 4.72 - 4.98 and 4.89 - 5.14 in lake Gjervollstadvatn and lake Saudlandsvatn respectively (SFT, 1984a). Testfishing every second year have demonstrated in both lakes a decrease in catch per effort and an increased condition factor. In lake Gjervollstadvatn, a slight increase in growth have been observed. The changes occurred earlier in lake Gjervollstadvatn, reflecting the smaller catchment area, shorter retention time and more fluctuating water chemistry in this lake compared to lake Saudlandsvatn (Rosseland and Sevaldrud, unpublished; SFT, 1984a).

Areas on the western coast of Norway seems today to have the largest potential for increasing acidification problems (SFT, 1984a). Lakes in the catchment area of River Ogna, county of Rogaland, south western Norway, were investigated during 1982 - 1983 (Hesthagen et al., 1982; SFT, 1983a, 1984a). Of 44 lakes with brown trout, 20 (46%) had become barren whereas 10 (22%), were reported declining. The losses appeared to have been mainly in the 1950 - 1960's. The salmon population in River Ogna have experienced several episodic fish kills in connection to acid waters both on younger year-classes and on spawning migrating salmon (Snekvik, 1975; Hesthagen et al., 1982; Nordland, 1983; Skogheim et al., 1984).

Water quality in River Ogna have been monitored since 1970. During the winter 1981/82 an extensive fish kill on the younger yearclasses of salmon occurred (Hesthagen et al., 1982). This fish kill coincided with periodes of very low pH (4.6) in the main river. Skogheim et al. (1984) describes an episodic fish kill in August 1982 downstream of an acid tributary. At least 50 spawning migrating salmon from 3 - 10 kg, still infected with sea lice (Lepeophteirus spp.), were observed dead or dying in the acidic river water (pH = 4.8, labile Al = 317 ug L^{-1}). Bloodsamples showed osmoregulatory disturbances, and gills of dead and dying salmon were enriched with aluminium in the order of 10 compared to reference fish (Skogheim et al., 1984).

In October 1983, after a long period of heavy rain with low pH, another fish kill in the upper part of River Ogna occurred. In the period 15-25 October, weighted mean pH of precipitation at Lista was 4.51 (amount of precipitation 94.7 mm) with episodic rainfall of pH 4.15 coming from Western Europe and U.K. (October 15, amount of precipitation 22 mm) (SFT, 1984a; E. Joranger, NILU pers. com.). On October 21, after the worst flood had ceased, a farmer observed the first dying salmon. During October 23 - 24 another 17 salmon and 5 brown trout were found dead. Except for one parr, the salmon were spawners from 3 - 5 kg (Nordland, 1983). A number of spawners had escaped from the toxic water by migrating into a small brook, too small to provide spawning facilities. Watersamples taken on October 24 showed pH 5.65 and 84 ug labile Al L^{-1} in the main river, whereas the pH was 6.09 in the brook with 15 ug labile Al L^{-1}. The salmon had obviously tried to escape by actively seeking better water quality.

No fish kills have been reported since, but in 1984, the abundance of salmon parr in the lower part of the river (downstream the acid tributary) was still low compared to the area upstream. Hatching experiments showed 80 - 100% mortality downstream. Close to the estuary, the mortality was 10% (SFT, 1985). This indicates yearly variations in water chemistry and thus population effects, illustrating the long term aspect when fish populations decline.

River Vikedal, situated 100 km north of River Ogna, county of Rogaland, was once a typical salmon river and one of the best in the district. The 22 lakes in the catchment area had populations of brown trout, two of them with mixed populations of trout and Arctic char (Salvelinus alpinus L.). Population changes occurred mainly during the 1960's, and today 7 lakes (32%) are barren, 5 lakes (23%) are reported to have a declining population whereas 10 lakes (46%) have a good or unchanged brown trout population. One of the char populations has vanished as well. In autumn 1982, the barren lakes had a pH between 4.8 - 5.1 (SFT, 1984b).

The Atlantic salmon is restricted to the lower 10 km of the river. The first fish kill, was observed in mid April 1981, after a flood caused by spring snowmelt and rain. Large amounts of smolts were observed floating dead in the estuary (J. Langhelle, Vikedal, pers. com.). Taxation two weeks later when the mass mortality had ceased, resulted in 29 dead salmon and 25 trout, 28% of the salmon being smolts (Rosseland and Sevaldrud, unpublished,; SFT, 1983b). Electrofishing the same autumn, however, showed normal abundance of parr. These mortalities in connection to spring snow melt, occurred yearly between 1981 - 1984. Although most yearclasses of these two species were represented, two year old smolts of salmon dominated (SFT, 1984b). Although daily watersamples were taken during spring 1982 - 83, the fish kill could not be related to any given episode because of the time between the kill and the observations. A monitor recording pH, conductivity and temperature demonstrated in 1983 large pH fluctuations related to changes in waterflow. During one episode in August, pH dropped from 5.55 to 4.90 within 3 hours, but no recordings of any fish mortality were performed (SFT, 1984b; Henriksen et al., 1984).

To relate water chemistry to fish kill directly, the monitor was installed in connection to fish tanks in November 1983, and daily water samples for analyzes of aluminium speciation were taken. During the next 3 weeks, 3 episodes occurred with a rapid decrease in pH and calcium , and increase in labile Al in connection to increases in waterflow due to rain and snow melt. Of salmon eggs, parr and presmolt, only the presmolt died (Henriksen et al., 1984). In spring 1984, a small episode occurred in March, whereas the major snowmelting period came in mid April to mid May. The high mortality of smolts (100%) in the fish tanks (Skogheim, Rosseland and Henriksen, MS) coincided with the mortality in the main river (SFT, 1985; Hesthagen, 1986). Thus, the field laboratory studies demonstrated both the higher sensitivity of salmon during smoltification, as described by Rosseland and Skogheim (1982, 1984), and that the fish kill in the main river were attributed to episodic changes in water quality often to rapid to be picked up by daily watersampling.

River Vikedal is concidered to be the next river to loose its natural stock of Atlantic salmon. The rod catch statistics, and the fact that the rod catch by mid July 1985, a normally very good fishing season, were only 2 salmon, points to the fact that this prediction might in the near future come through.

2. SWEDEN

By 1980, water chemistry data from about 8000 lakes showed that most acid lakes were situated in the southwestern part of the country. In these lakes, about 75% had a pH-value lower than 5.5 during at least some part of the year with 33% of the lakes having a pH in summer below 5.0 (Johansson and Nyberg, 1981). In the southwestern part of Sweden, roach populations were concidered to have declined in about 2000 lakes, representing aproximately 50% of the lakes in the area. In the south and middle part of Sweden, another 2500 lakes had been damaged with further 6500 lakes assumed to have declining fish populations (Johansson and Nyberg, 1981).

Degerman et al. (1985) report results from electrofishing in 12 different watercourses on the Swedish west coast. The abundance of eel and parr of brown trout and Atlantic salmon were significantly greater at stations having an alkalinity above 0.25 mekv L^{-1}. At lower alkalinity, water chemistry was the single most important density regulating factor. Comparing water chemistry data and electrofishing results from 7 rivers for the years 1955 and 1984, showed that in the lower part of the small rivers (catchment areas less than 1000 km^2), as in the larger, water chemistry and salmonid abundance seemed unchanged. In the middle and upper parts of the small rivers, however, the alkalinity and fish abundance had significantly decreased, and in some cases fish had disappeared (Degerman et al., 1985).

These results are in accordance with the losses of brown trout from small streams in Delsbo, central Sweden reported by Andersson and Andersson (1984). A few decades ago, all the 21 streams had trout whereas electrofishing in 1978 revealed no trout in streams having summer pH below 6.0. Watersamples collected monthly since 1980, have demonstrated a spring pH less than 4.8 in the barren streams, whereas streams with summer and spring pH of $\leq$6.5 and $\leq$5.1, respectively, had a reduced abundance from one to four compared to high pH streams (Andersson and Andersson, 1984).

Johansson and Nyberg (1981) pointed to the lack of information from fish populations in mountain areas of central and northern Sweden. Andersson and Nyberg (1984) describes the decline of populations and mortality of brown trout and roach caused by severe drops in pH during the spring thaw in area in central Sweden. Experiments with caged brown trout in these streams demonstrated high mortality as pH dropped from 6.1 - 7.2 to 4.5. High concentrations of iron, manganese and aluminium in connection to the low pH seemed to explain the mortality, iron concidered to be the most important (Andersson and Nyberg, 1984). This area experience today a very rapid and continuing acidification and metal-contamination, where streams with water alkalinity less than 1.2 mekv L^{-1} now are barren (P. Nyberg, pers. com.)

3. FINLAND

During the last few decades, the atmospheric supply of sulphur and the acidity of precipitation have conciderably increased in Finland, and diatome-analyzes indicate acidification in many lakes during the last three decades (Kämäre, 1984). Although many lakes are acid due to humic acids, many clearwater lakes have reached a pH below 5.5, and concentrations of zink, titanium, cadmium, copper and mercury increases in the sediments.

Large coastal areas of Finland have problems with clay and soils enriched with sulphur, leading to large acidification problems and fish population declines (Sevola, 1980; Sevola et al., 1982; Håstbacka, 1984). However, several studies have concluded that the water chemistry outside such areas also have a decreasing pH and alkalinity, both in rivers (Alasaareta and Heinonen, 1984) and in lakes (Pätilä, 1982, 1984; Helminen, 1984; Laaksonen and Malin, 1984). Few studies, however, gives documentation of declining fish populations or barren lakes caused by acid rain. Of 52 lakes in northern Espoo, 20% had an alkalinity from zero to less than 0.05 mmol L^{-1} in summer 1984. Although no large damage to fish populations were found, a few small lakes have thus experienced fish mortality (Pätila, 1984).

The coastal area of Nyland are the area of Finland receiving the highest inputs of acidic precipitation. Of 107 lakes, nearly 50% had an alkalinity less than 0.05 mmol L^{-1} many with a pH below 5.0. However only one lake was extreme acid (pH 4.4), and only a few were concidered by the locals to have lost their fish populations due to acid rain

(Helminen, 1984).

With few exceptions, data on fish populations from acidfication areas are lacking. An exception is a study reported by Rask (1984) on a perch population in a small extremely humic forest lake in Evo, Lammi. The population had a reproduction failure (egg mortality) in the springs caused by low pH. In the spring 1981, many adult perch died probably due to the low pH and high aluminium concentration in the water (Rask, 1984).

4. CONCLUSION

In Norway and Sweden, no positive trend can be seen in the fishery status of lakes and rivers in the most affected areas. On the contrary, new areas especially in the high mountains in Sweden and on the Norwegian west coast are affected by acidification. Large amounts of acidic precipitation, short retention time in the lake systems and a bedrock not easily weathered often with thin soils in the catchment area, are common characteristics for these recent acidified areas.

Fishery status based on interviews or testfishing, is a result of a long term influence. Therefore, the fish status gives the result of past and recent perturbation on the environment. The lag-phase between the cause and its effects should be cept in mind. Especially in these acidification areas, year to year variations in precipitation and lake water chemistry are of importance, as are possible refuge areas for fish survival in the catchment area.

Several authors have tried to calculate the reestablishment of fish populations after an increase in lake pH caused by reductions in sulphur emissions (Brown and Sadler, 1981; Muniz and Seip, 1982; Wright and Henriksen, 1983; Muniz et al., 1984). However, water chemistry data used in all these papers are based on single samples in most cases taken at the lake outlet. The temporal and spatial variations in lake water chemistry or the even more important condition in brooks or tributaries used as spawning areas are thus not taken into account. The practical value of these "theoretical" predictions are thus questioned.

In spite of the difficulties, in predicting trends, rate of reestablishment and quantification of fish population effects, reductions in sulphur emissions are the only solution to the still increasing acidification problem in Scandinavia. When such reductions appears, chemical and biological management in the environment will significantly shorten the recovery period.

REFERENCES

-Alasaarela, E. and Heinonen, P.: 1984, 'Alkalinity and chemical oxygen demand in some Finnish rivers during the periods 1911 - 1931 and 1962 - 1979', Publication of the Water Research Institute, National Board of Waters, Finland, No 57, 3 - 13.

-Almer, B., Dicksen, W., Ekstrøm, C., Hörnström, E. and Miller, U.: 1974, Ambio 3, 30.

-Andersson, B. and Andersson, P.: 1984, Rep. Inst. Fresh. Res. Drottningholm 61, 28.
-Andersson, P. and Nyberg, P.: 1984, Rep. Inst. Freshw. Res. Drottningholm 61, 34.
-Brown, D.J.A. and Sadler, K.: 1981, J. Appl. Ecol. 18, 433.
-Dahl, K.: 1927, Salmon and Trout Magazine 46, 35.
-Degerman, E., Fogelgren, J.E. Tengelin, B. and Thörnelöf, E.: 1985, 'Förekomst och täthet av havsörig, lax och ål i försurade mindre vattendrag på svenska Västkusten' ("Occurrence of brown trout, Atlantic salmon and eel in small acidified water courses on the west coast of Sweden, Inf. Inst. Freshw. Res. Drottningholm 1/1985, 84 pp.
-Haines, T.: 1981, Trans. Am. Fish. Soc. 110, 669.
-Helminen, O.: 1984, 'Försuringssituasjonen i den västernyländska Kusttraktens sjöar', ("The acidification of lakes in the Vastnylandska coastal area), Västra Nylands Vattensskyddsförening r.f. Småskrift 5, 14 pp.
-Henriksen, A., Skogheim, O.K. and Rosseland, B.O.: 1984, Vatten 40, 255.
-Hesthagen, T.: 1985, ' Fish kill of Atlantic salmon (Salmo salar) and brown trout (Salmo trutta) in a acidified river in south western Norway', MS Muskoka Conference '85.
-Hesthagen, T., Sevaldrud, I.H. and Skogheim, O.K.: 1982, ' Ogna og Helgåvassdraget i Rogaland. Vannkvalitet og fiskestatus' (River Ogna and River Helgå in Rogaland. Water quality and fish status), Rapp. fra Fiskeforskningen 4/82, 34 p. (In Norwegian).
-Huitfeldt-Kaas, H.: 1922, 'Om aarsaken til massedød av laks og ørret i Frafjordelven, Helleelven og Dirdalselven i Ryfylke høsten 1920', (On the cause of mass death of salmon and trout in River Frafjord, River Helle and River Dirdal in Ryfylke, autumn 1920), Norsk Jæger- og Fiskefor. tidsskrift, 37-44. (In Norwegian).
-Hästbacka, H. 1984. En katastrof som vi redan drabbats av: Försuringen som lurar i marken "A catastrophy already hit us: The acidification that hides in the soil", Finsk Natur 6, 25 - 27.
-Haabesland, K.K.: 1972, 'Alders- og tilvekstforhold hos ørret (Salmo trutta L.) i et overbefolket sørlandsvassdrag' (Age and growth in brown trout (Salmo trutta L.) in a stunted river in Sørlandet), Thesis, University of Trondheim, 87 p. + tables. (In Norwegian)
-Jensen, K.W. and Snekvik, E.: 1972, Ambio 1, 223.
-Johansson, K. and Nyberg, P.: 1981, 'Försuringen av svenska ytvatten-effekter och omfattning 1980', (The acidification of Swedish lakes, effects and extension 1980), Inf. Inst. Freshw. Res. Drottningholm. 6, 118 pp.
-Kämäri, J.: 1984, 'National review of water quality aspects: acid precipitation in Finnish waters', Vannet i Norden 3, 46 -64.
-Laaksonen, R. and Malin, V.: 1984, 'Changes in ion distribution in Finnish lake water during the period 1968 - 1983', Publication of the Water Reseach Institute, National Board of Waters, Finland No. 57, 59 - 76.
-Leivestad, H., Hendrey, G., Muniz, I.P. and Snekvik, E.: 1976, 'Effect of acid precipitation on freshwater organisms', in Braekke, F.H.(ed.), Impact of acid precipitation on forest and fresh-

water ecosystems in Norway, 87-111, SNSF-project, FR 6/76.
-Leivestad, H. and Muniz, I.P.: 1976, Nature, 259, 391.
-Muniz, I.P. and Leivestad. H.: 1980, 'Acidification-effects on freshwater fish', in, Drabløs, D. and Tollan, A. (eds.), Ecological impact of acid preciptation, p. 84-92, SNSF-project.
-Muniz, I.P. and Seip, H.: 1982, 'Possible effects of reduced Norwegian sulphur emissions on the fish populations in lakes in southernmost Norway', Report No. 810313-2, Central institute for Industrial Research, Oslo.
-Muniz, I.P., Seip, H.M. and Sevaldrud, I.H.: 1984, Water, Air and Soil Pollut. 23, 97.
-Nordland, J.: 1983, 'Fiskedød i Ognaelva oktober 1983' (Fish kill in River Ogna October 1983), Note from Rogaland Environmental Protection Authority, (In Norwegian).
-Overrein, L., Seip, H.M. and Tollan, A.: 1980, 'Acid precipitation effects on forest and fish', SNSF-project, FR 19/80, 175 p.
-Pätilä, A.: 1982 'Changes in water chemistry due to acidification - study of 23 lakes in southern Finland', Luonnon Tutkija 86, 19 - 22. (In Finnish with English summary).
-Pätilä, A. 1984,: 'Sensitivity to acidificatoin in lakes in northern Espoo in 1984. Espoon ympäristönsurjelulautakunnan julkaisu 8/84, 31 pp. + tables (In Finnish with English summary)
-Rask, M.: 1984, Ann. Zool. Fennici 21, 15.
-Rosseland, B.O. Sevaldrud, I.H. Svalastog D. and Muniz, I.P.: 1980, 'Studies of freshwater fish populations - effects of acidification on reproduction, population structure, growth and food selection', in, Drabløs, D. and Tollan, A. (eds.) Ecological impact of acid precipitation, p. 336-337, SNSF-project.
-Rosseland, B.O. Sevaldrud, I.H. Svalastog, D. and Muniz, I.P.: 1981, 'Bestandsundersøkelser på fiskebestander fra forsuringsområdene i Aust- Agder fylke 1976' (Studies on fish populations in the acidification area of Aust- Agder county), Rapport fra Fiskeforskningen 1981/4, 78 p. (In Norwegian).
-Rosseland, B.O. and Skogheim, O.K.: 1982, 'Physiological stress and mortality of Atlantic salmon, Salmo salar L. in acid water with high levels of aluminium', Inter. Council for Exploration of the Sea C.M. 1982/M:29,15 p.
-Rosseland, B.O. and Skogheim, O.K.: 1984, Rep. Inst. Freshw. Res. Drottningholm 61, 186.
-Sevaldrud, I.H. Muniz, I.P. and Kalvenes, S.: 1980, 'Loss of fish populations in Southern Norway. Dynamics and magnitude of the problem', in, Drabløs, D. and Tollan, A. (eds.) Ecological impact of acid precipitation, pp. 350-351, SNSF-prosject.
-Sevaldrud, I.H. and Muniz, I.P.: 1980, 'Sure vatn og innlands fisket i Norge. Resultater fra intervjuundersøkelsene 1974-1979' (Acid waters and inland fisheries in Norway. Results from interviews 1974 - 1979), SNSF-prosjektet IR 77/80, 92 pp + tab.
-Sevaldrud, I.H. and Skogheim, O.K.: 1986, 'Changes in fish populations in southernmost Norway during the last decade', Water, Air and Soil Pollut. (In press)
-Sevola, P.: 1980, 'Forntidshavets bottenävjor - Østerbottens

problemleror', (Ancient seabottoms, - The clay problems of Østerbottens), Finsk Natur 2, 42 - 44. (In Finnish).
-Sevola, P., Hudd, R. and Hildén, M.: 1982, 'Acidification of watercourses in sulphid clay areas', Luonnon Tutkija 86, 62 - 64. (In Finnish with English summary).
-SFT: 1981, 'Overvåking av langtransportert forurenset luft og nedbør. Årsrapport 1980' (Monitoring of long range transported pollutants in air and precipitation, Annual report 1980), Statlig program for forurensningsovervåking. Rapp. 26/81, 143p., Statens forurensningstilsyn, Oslo. (In Norwegian).
-SFT: 1982,'Overvåking av langstransportert forurenset luft og nedbør. Årsrapport 1981' (Monitoring of long range transported pollutants in air and precipitation. Annual report 1981), Statlig program for forurensningsovervåking. Rapp 64/82, 176 p., Statens forurensningstilsyn, Oslo. (In Norwegian).
-SFT: 1983a, 'Overvåking av langtransportert forurenset luft og nedbør. Årsrapport 1982' (Monitoring of long range transported pollutants in air and precipitation. Annual report 1982), Statlig program for forurensningsovervåking. Rapp. 108/83,228 p, Statens forurensningstilsyn, Oslo. (In Norwegian).
-SFT: 1983b, 'Vannkjemisk og fiskeribiologiske undersøkelser i Vikedalsvassdraget 1981 - 1982. Vannkvalitet og fiskedød våren 1982' (Water chemistry and fish-biological investigations in River Vikedal 1981 - 1982. Waterquality and fish kill spring 1982), Statlig program for forurensningsovervåking, Rapp. 97/83, 64 p., Statens forurensningstilsyn, Oslo. (In Norwegian).
-SFT: 1984a, 'Overvåking av langtransportert forurenset luft og nedbør. Årsrapport 1983', (Monitoring of long range transported pollutants in air and precipitation. Annual report 1983), Statlig program for forurensningsovervåking. Rapp. 162/84, 198 p., Statens forurensningstilsyn, Oslo. (In Norwegian).
-SFT: 1984b, 'Vikedalsvassdraget. Nedbør-, vannkjemiske og biolog iske undersøkelser i 1981-1983', (River Vikedal. Precipitation, water chemistry and biological investigations 1981 - 1983), Statlig program for forurensningsovervåking. Rapp. 123/84, 160 p. Statens Forurensningstilsyn, Oslo. (In Norwegian).
-SFT: 1985, 'Overvåkning av langtransportert forurenset luft og nedbør. Årsrapport 1984' (Monitoring of long range transported pollutants in air and precipitation. Annual report 1984), Statlig program for forurensnings overvåking. Rapp. 85, Statens forurensningstilsyn, Oslo. (In Norwegian).
-Skogheim, O.K. Rosseland, B.O. and Sevaldrud, I.: 1984, Rep. Inst. Freshw. Res. Drottningholm 61, 195.
-Snekvik, E.: 1975, 'Episoder med fiskedød i forbindelse med forsuring av vassdrag i 1969', (Episodic fish kills in relation to acidification of rivers in 1969). Notat fra Fiskeforskningen, 9 p. (In Norwegian).
-Wright, R.F. and Henriksen, A.: 1983, Nature 305, 422.

MEASURING DRY DEPOSITION: A RE-ASSESSMENT OF THE STATE OF THE ART

B. B. Hicks
NOAA/ERL Air Resources Laboratory
Atmospheric Turbulence and Diffusion Division
Post Office Box E, Oak Ridge, TN 37831, U.S.A.

ABSTRACT. The exchange process known as dry deposition encompasses the dynamic exchange of trace gases and aerosol particles, and the gravitational settling of large particles. Except for particles large enough that their sedimentation velocity exceeds turbulent velocities, the rate of deposition is mostly determined by surface properties, such as roughness, stickiness, and wetness, and by atmospheric stability. Thus, it is difficult to interpret results obtained using collection devices having surfaces different from those of nature. Other methods for measuring dry deposition exist, mostly micrometeorological, but these are sufficiently complicated that routine application as in a monitoring network is not yet feasible. For some chemical species and in some locations, inferential methods offer considerable promise. These methods measure atmospheric concentrations (C) of the relevant chemical species, and derive relevant deposition velocities (v_d) on a site-specific, species-dependent, and time-evolving basis. The dry deposition rates of interest are then evaluated as the product $v_d.C$. A major goal of current research programs is to provide the knowledge necessary to evaluate v_d. Experimental methods are reviewed, and potential sources of error are examined, for both the research methods and the "concentration monitoring" methodologies presently being advocated for use in numerical models as well as for routine monitoring.

1. INTRODUCTION

The aura of uncertainty associated with the wide range of atmosphere/surface exchange processes collectively known as "dry deposition" continues to be aggravated by an inability to make direct measurements except in very limited circumstances. The general conclusions of earlier discussions of different experimental techniques (e.g. Hicks et al., 1980) remain largely unaffected by more recent developments. However, there has been a continuing gradual improvement in measurement capabilities, both in the context of routine monitoring and intensive measurement.

Water, Air and Soil Pollution **30** (1986) 75–90.

There is little argument concerning the importance of dry deposition as a mechanism for transferring atmospheric particles and trace gases to underlying surfaces. Atmospheric budget calculations (e.g. Galloway et al., 1984) indicate that the ratio of dry to wet deposition averages about 0.4 across the eastern portion of North America. Model assessments (e.g. Shannon, 1981) suggest that the ratio is about 1.8. Such differences are often found in the dry deposition literature. They will not be discussed further here. Instead, the present discussion will focus on the causes of the differences that exist among the wide variety of measurement methods that are currently being employed. The relevance is more on the development of monitoring methods and model parameterizations than on the assessments resulting from application of these tools once they are developed.

2. ACCUMULATION ON SURFACES AND IN VESSELS

2.1 Surrogate surfaces and vessels

For many years, wet/dry bucket samplers have been used to investigate deposition of atmospheric particulate pollutants. Experience obtained in studies of radioactive fallout suggested that such wet/dry samplers yielded results which agreed with known inputs and measured atmospheric inventories (see Volchok et al., 1970; Volchock and Graveson, 1976). However, exploratory application of these techniques to non-radioactive particulate pollutants has underlined severe limitations due to contamination by birds, wind-blown dust, etc. (see Feely et al., 1985).

Such wet/dry collectors are but one of many forms of artificial collection devices used to measure the deposition of atmospheric particles. Dustfall collectors are sufficiently well accepted that exploratory measures to standardize collection methods have been made. Flat plates and shallow pans of various configurations, constructed of a wide range of materials have also been employed.

Some of the conceptual problems associated with the use of any of these artificial devices were reviewed by Hicks et al. (1980). In general, the devices gain credibility as particle sizes increase. For substances like sulfate, for which Altshuller (1983) reports relatively little material in the super-μm size range at urban and rural sites, gravitational settling may be relatively unimportant. It seems likely that the proportion of atmospheric sulfate in the super-μm size fraction will depend on the distance from sources and meteorological conditions, as well as on emission characteristics. Thus, the correlation between measurements of sulfate deposition by any artificial collection device and deposition to the natural surface over which the device is exposed is likely to vary with location and time.

In the case of rapidly falling particles, it is clear that collection devices will provide accurate measures of the surface deposition; the devices merely intercept falling particles before they arrive at the ground. However, questions arise regarding the performance of surrogate-surface devices used to measure deposition of particles small enough to be influenced by turbulence. An upwards-facing collection surface mounted above the natural surface might well collect particles being carried by downdrafts, but not those being carried by updrafts. Dasch (1983) presents comparisons between deposition data obtained using several different surrogate surfaces, mounted with both upward- and downward-facing exposure. The upward-facing surfaces always collected more material, but the ratio of the upward to downward-facing results varied from about 2 to 5. It is not obvious how such data should be combined in order to estimate the net flux to the underlying surface. Workers familiar with turbulent exchange might rely on the difference between upward-facing and downward-facing surfaces, whereas those accustomed to working with two-sided foliar surfaces might take the sum of the two results.

The above discussion lends some guidance concerning the performance of surrogate-surface devices. The major conceptual problems should be minimized for particles large enough that the gravitational sedimentation speed exceeds turbulent vertical velocities. In daytime, vertical velocities associated with the turbulent exchange process are typically of the order of 1 m s^{-1}; hence a surrogate surface exposed in daytime might be expected to yield unequivocal results for particles larger than about 100 μm diameter. At night, vertical velocities are normally much less, and thus good results should be obtained for particles possibly as small as 30 μm diameter. However, most current interest focuses on S- and N-bearing particles, mostly formed by gas-to-particle conversion and hence typically $<$ 1 μm diameter (see Altshuller, 1983). A further difficulty concerns contamination by wind-blown dust and local biological activity (birds, insects, foliage, etc.; see Hicks, 1982).

Surrogate surface devices for measuring sub-μm particulate sulfate deposition continue to be investigated, mainly because they offer the only method by which a chemical sample can be obtained with convenience approaching that of wet deposition collection. Data obtained in the 1981/2 comparison studies conducted at Champaign/ Urbana, Illinois (Dolske and Gatz, 1984, 1985), are summarized in Figure 1. Data are presented as a plot of effective deposition velocity (the ratio of flux density per unit horizontal area to the atmospheric concentration) against the collection area index (the total amount of inside surface area per unit horizontal area, somewhat like the leaf area index of agricultural and forest meteorology). The 1981 data provide intriguing evidence of some order among the collection devices. The 1982 data show similar behavior for most of the devices, but appear to differ substantially for the dry bucket collections. The data obtained using micrometeorology (Wesely _et al._, 1985) are in general agreement with the results obtained using surfaces and vessels, provided the appropriate surface area index of the natural vegetation is assumed to be low. However, the agreement may be fortuitous since the micrometeorological methods favor measurement of sub-μm particles whereas the collection devices work best for super-μm. Moreover, Dasch (1983) reports a different ordering among deposition surfaces, and finds evidence of a dependence on the chemical composition of the surface rather than on the area index.

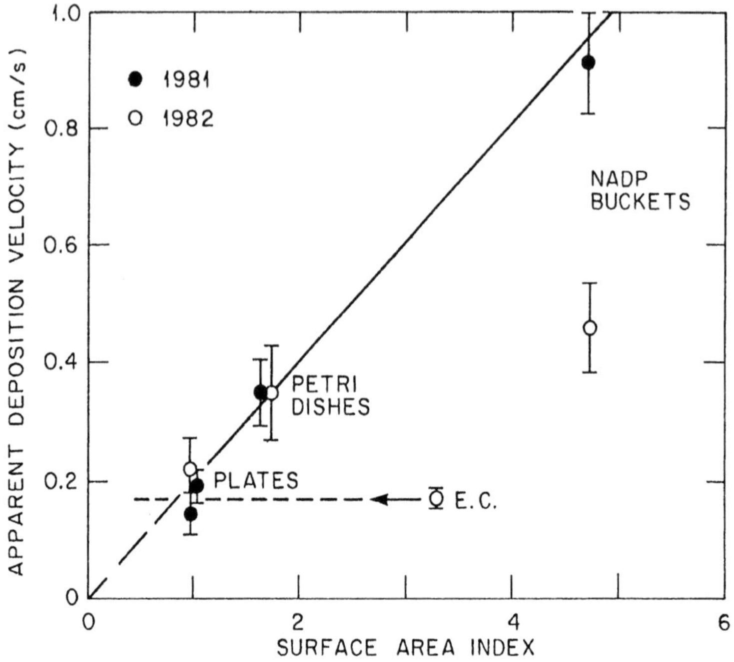

Figure 1. Apparent deposition velocities (calculated on the basis of horizontal area) derived using a variety of surfaces and vessels as described by Dolske and Gatz (1985), plotted against the internal area of the collection devices per unit horizontal area. Data are derived from Dolske and Gatz (1984), Davidson et al. (1985), and Feely et al. (1985). The average of micrometeorological measurements by Wesely et al. (1985) is also shown, (identified as E.C.).

The inability of surrogate surface devices or collection vessels to simulate the intricacies of gas exchange between the atmosphere and biological systems has tended to focus attention on alternative deposition measurement methods in recent years.

If deposited material can be removed from the natural surface on which it deposits, or if the surface can be analyzed sequentially so as to determine the rate of change of accumulation with time, then clearly the conceptual problems of surrogate-surface sampling can be circumvented. The measurement of the rate of accumulation of material on an exposed natural surface, although practical only in a few circumstances, constitutes the most direct method for determining dry deposition.

Snow surfaces have been most frequently studied. Lead and S deposition to snow was reported by Dovland and Eliassen (1976), Ibrahim et al. (1983), and Cadle et al. (1985). The last of these papers also addresses gaseous exchange, for SO_2 and HNO_3. Ibrahim et al. compare snow data with results obtained using artificial collectors. In general, the conclusions of these studies fall in line with the results of other techniques. For HNO_3, surface uptake appears to be exceedingly rapid, such that the surface resistance can be neglected. For SO_2, the surface resistance depends on surface temperature and pH. For submicron particles, the low deposition velocities typical of such studies appear similar to the predictions of theory.

Many surfaces lend themselves to periodic washing in order to remove and analyze dry-deposited particles, thus constituting a direct evaluation of dry deposition rates. If interest is in deposition to surfaces of buildings (masonry, stone, metal, glass, etc.), then clearly the most representative measurements will be made by studying either these surfaces themselves or materials with identical physical, chemical, and biological characteristics.

The washing of foliage is of special interest. Studies conducted at an oak-hickory forest near Oak Ridge, Tennessee, have combined leaf washing with studies of throughfall and of deposition to surrogate surfaces (see Lovett and Lindberg, 1984; Lindberg and Lovett, 1985), in the overall context of pollutant cycling (Johnson et al., 1982). Lindberg and Harriss (1981) report a deposition velocity of 0.13 cm s^{-1} (verified by Lindberg and Lovett, 1985) for the deposition of sulfate particles to inert surfaces, from which they estimated the canopy deposition velocity (as viewed from aloft) to be about 0.5 cm s^{-1}*. Lovett and Lindberg (1984) derived an independent estimate of the overall canopy v_d for sulfate, from throughfall studies: 0.6 cm s^{-1} for the growing season, and 0.4 cm s^{-1} for the dormant season. Substantially different results were obtained for other particulate species, possibly because of the different size range involved.

It is tempting to compare the conclusions of Figure 1 with the results derived here. The leaf area index of the Oak Ridge forest canopy was about 11 (double-sided). Extrapolation of the line drawn in Figure 1 would suggest a far greater deposition velocity than has been demonstrated by any of the field studies considered here.

Dasch (1985) found that deposition velocities for sulfate particles to leaves of trees varied with the distance from the edge of an oak/elm stand. Values were substantially different than for the Oak Ridge study, perhaps because of differences in the particle size spectrum at the two locations but also possibly because of biological differences of the trees involved (the elms appeared better sulfate collectors than the oaks, for example).

The results of such recent studies have emphasized the need for great care when interpreting data from such surrogate surface and individual-leaf studies. Most workers express the data obtained in terms of the area of the collecting surface, in which case the leaf area index (either single- or double-sided, depending on the circumstance) might be used to derive a first-order estimate of whole-canopy

* But Lindberg and Lovett (1985) point to a difference in results between rimmed and rimless artificial collecting surfaces when mounted over grass, that could possibly influence these conclusions.

behavior. Further work is necessary to answer questions concerning the roles of (1) the shape of surrogate-surface collectors, (2) the chemical composition of the surface, (3) differences in surface structure, and (4) differences in the particle size distribution.

3. MICROMETEOROLOGICAL METHODS

Measurement of the atmosphere/surface turbulent exchange of sensible heat, water vapor, and momentum is a central problem of experimental meteorology. Dry deposition is a related turbulent exchange process. Many of the methods initially developed to measure the more standard meteorological fluxes are now being used to measure atmosphere/surface exchange of trace gases and small particles.

Eddy correlation methods provide direct measurements of the turbulent transport of material across a plane defined by local streamline flow. Gradient methods provide indirect measurements of the same quantity, estimated from knowledge of the appropriate eddy diffusivity. In practice, neither approach is either simple in comparison to the other or suitable for use with more than a very limited range of chemical species.

Most often, eddy correlation methods use tower-mounted instruments with time response of the order of one second. Recent technological advances now also permit measurement using more rapidly-responding sensors aboard aircraft (e.g. see Lenschow et al., 1980). Tower data provide a detailed time history of atmosphere-surface exchange above a specific area, whereas aircraft tend to provide spatial (or line) averages at specific times. Depending on the goal of a particular study, either aircraft or tower-based systems may be preferred. However, there is often benefit in coupling aircraft and tower-based studies.

Turbulent exchange is statistical in nature, being the consequence of poorly-correlated atmospheric motions that must be measured with accuracy in order to derive the net exchange. It is relatively simple to compute the statistical uncertainty associated with the average eddy flux computed from (say) second-by-second experimental evaluations of the instantaneous vertical exchange rate. Figure 2 shows the time trend of SO_2 fluxes computed using a flame photometric S sensor, fitted with a filter to remove particles from the sampled air stream, and operated in conjunction with a three-dimensional sonic anemometer system. The sensors were mounted on a scaffolding tower at about 6 m height, at an airfield near Linkenheim, FRG during March, 1985. The surface was dormant grass. Standard error bands associated with the computed eddy fluxes are shown. (Note that negative eddy fluxes indicate transport towards the surface, i.e. deposition, and that no significant occasions of gaseous S emission are apparent).

Figure 3 illustrates results obtained by two independent measurement systems (one as already illustrated in Figure 2). The differences between the two measurement systems are about the same as the errors that are expected based on the statistics of the measurement.

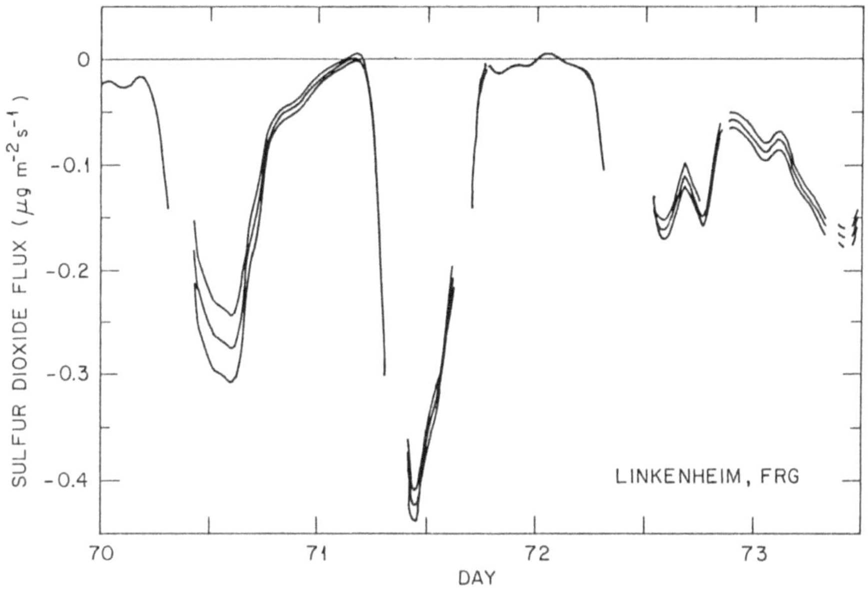

Figure 2. A time sequence of SO_2 eddy flux measurements, obtained during 1985 over a dormant grass surface near Linkenheim, FRG. The upper and lower curves show the statistical bounds associated with the average fluxes. Data have been smoothed using a running mean (1-2-4-2-1 weighting). Figure provided by R. T. McMillen.

Questions also arise about the validity of "S" measurements as reported by flame photometers such as were used in the Linkenheim study. A series of tests was conducted to quantify the magnitude of the artifact term, resulting from the effect of contamination by H_2O, CO_2, etc. First, a SO_2 denuder was inserted into the air stream of one S sensor, without imposing any other change on the experimental set-up. The eddy flux then deduced was 0.003 μg $m^{-2}s^{-1}$, about 1% of the daytime maxima evident in Figures 2 and 3. The experiment was repeated, using the second S sensor, with similar results.

Fairall (1984), Wesely and Hart (1984), and Lenschow and Kristensen (1985), have addressed related questions from a more theoretical direction. Webb et al. (1980) point to the need to apply corrections to raw micrometeorological flux data in order to account for effects mainly of Stefan flow and density fluctuations. The former correction arises as a consequence of displacement of air at the surface if evaporation is occurring (q.v. Hicks, 1983). The latter correction is necessary if a sensor responds to density fluctuations of the species of interest instead of the mixing ratio. Stefan flow usually amounts to less than 0.05 cm s^{-1}, and must be considered whenever fluxes such as

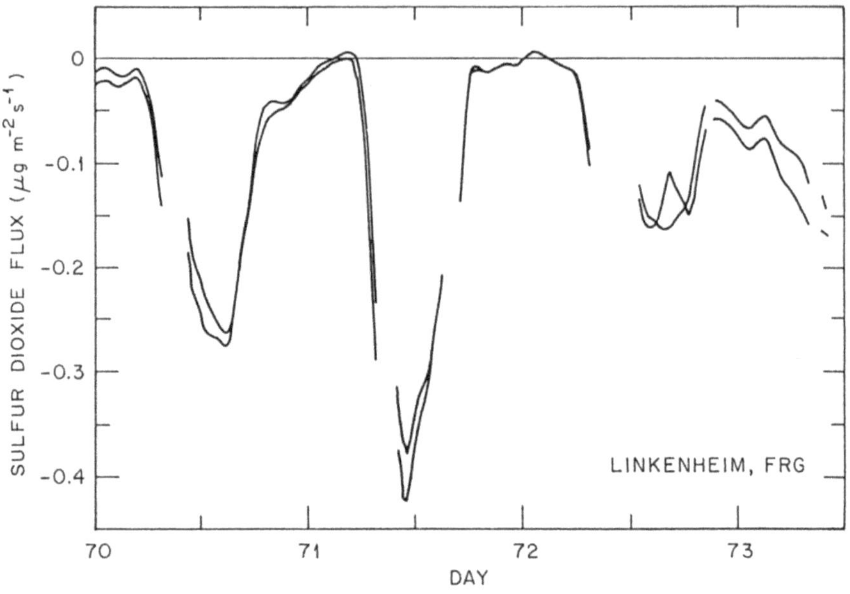

Figure 3. A comparison between eddy fluxes of SO_2 as derived from two independent, adjacent, flame photometer systems: Figure provided by R. T. McMillen.

are illustrated in Figures 2 and 3 are used to compute deposition velocities. The flame photometers used to produce the records illustrated here measure S in a constant-temperature environment; hence, they are not greatly susceptible to errors arising from the density effects that are the main source of concern in the context of the Webb et al., analysis. In general, however, it appears that eddy correlation methods should not be expected to resolve deposition velocities with a single-determination precision less than about 0.05 cm s^{-1}. This does not mean that the method is inappropriate in the case of slowly-depositing pollutants, but that ensemble averages must then be used to reduce the effects of the various errors that arise.

A variation on conventional eddy correlation methods is the so-called eddy accumulation technique. This replaces the usual requirement for a fast-response chemical sampling system by a fast-response sampling system which accumulates an air sample at a rate proportional to the measured vertical velocity. Two sample accumulators are used, one each for updrafts and downdrafts. In theory, the true eddy flux can be deduced from measurements of concentrations in the two accumulated samples. In practice, the technique has two major limitations --- (i) the sampling system must operate with rapid response over a wide dynamic range, and (ii) very small differences in concentration must be measured with accuracy, (see Hicks and McMillen,

1984). In practice, the technique appears most suitable for studies of rapidly-depositing pollutants. Work continues on its development, even though little success has yet been reported and even though the application of the method will be limited to micrometeorologically acceptable sites whenever such development is successful.

The discussion above focuses on the statistical nature of turbulent exchange and its consequences on measurement of eddy fluxes. Similar arguments arise in the interpretation of gradient data. First, it is evident that vertical concentration differences are exceedingly difficult to measure with the accuracy necessary to determine fluxes. The statistical variability imposed by turbulence means that it is inappropriate to simply take periodic measurements of concentration at various heights and then take the difference of averages; scatter would then be too great. Moreover, the differences being measured are so small that the use of independent sensors at different levels cannot generally be recommended. An optimal configuration might be to employ identical sampling lines from the heights of interest, with high flow rate and with stilling chambers to impose the desired time-averaging, and then to sample these different air streams using a single instrument. Even when this procedure is adopted, it is sometimes not possible to derive a consistent measurement of the vertical gradient.

Interpretation of concentration gradient measurements requires knowledge of eddy diffusivities. Over acceptably uniform pasture, or other surfaces with no significant depth of vegetation relative to the height of gradient measurement, standard micrometeorological relations can be applied to estimate eddy diffusivity. However, over tall crops or forests, questions arise about the co-location of relevant sources and sinks within the canopy, and hence there is sometimes doubt as to the applicability of diffusivities derived from consideration of heat exchange, or moisture. In practice, gradient methods should be used only when alternatives are not practical.

Micrometeorological flux-measurement methods yield data on the rate of turbulent exchange across a quasi-horizontal plane at some experimentally convenient level above the surface. To interpret such measurements, it is necessary that the fluxes be nearly constant with height, or that corrections be made if flux divergences are significant. Flux divergences arise whenever conditions are changing with time, over non-uniform surfaces, or in situations in which individual plumes from identifiable point sources dominate the transfer process. A recent experimental determination of the importance of local concentration gradients and time trends on the measurement of eddy fluxes at about 5 m above a forest indicated that the resulting errors would rarely exceed 0.1 cm s^{-1} in the deduced deposition velocity (see Hicks et al., 1986). Clearly, such matters as flux divergences are of greatest importance in studies of slowy-depositing substances. For studies of particle exchange, for example, it may be necessary to construct ensemble averages from a large number of individual observations in order to clarify the overall situation.

The matter of extrapolation of fluxes measured above the surface to the surface itself is obviously of greatest importance in the context of aircraft operations. In practice, studies of eddy fluxes using tower-mounted equipment are most common. Although the results are derived from point measurements, they are indicative of an upwind area of considerable extent. If this upwind area is uniform, then the point-measurement data are indeed representative of it provided the standard site requirements are also satisfied.

The case of particle fluxes to a natural surface continues to be of special interest, because of the frequent experimental finding that fluxes are not always directed towards the surface (Wesely et al., 1982). Recent studies at the Oak Ridge forest research site (Hicks et al., 1986) suggest that the surface can be a sink for particles of atmospheric origin (i.e. sulfates, etc.) while reactions occurring near the surface cause it to also be a source of other particles (perhaps hydrocarbons). At this time, there is an inadequate experimental basis for extended discussion. However, it is clear that the frequent detection of particle fluxes out of canopies should not be viewed as an indication of problems of the measurement systems but more as a feature of the exchange process that remains to be explored.

4. INFERENCE OF DEPOSITION USING CONCENTRATION DATA

The aim of conventional studies of micrometeorological turbulent fluxes is usually to investigate the nature of turbulent exchange. In the present context, the goal is usually to quantify and formulate the resistances that are of importance in an overall transfer model, as discussed by O'Dell et al. (1977), for example. The micrometeorological methods are analogous to laboratory studies, whose benefit is that they can be conducted in more controlled conditions and using more sensitive equipment than is appropriate in other situations. The results obtained from such studies are not greatly limited by the constraints imposed by the operation of the micrometeorological equipment required to make the measurements.

The micrometeorological methods used in research programs, as discussed above, are sufficiently complicated that they are rarely appropriate for routine monitoring. Instead, most monitoring programs make use of inferential methods, by which dry deposition rates are calculated from measurements of the air concentrations of the gaseous and particulate chemical species of interest, using one of many different levels of sophistication of the deposition velocity calculation.

In much early work, estimates of the deposition flux were derived by applying an assumed-constant deposition velocity (typically 1 cm s^{-1} for SO_2 and 0.1 cm s^{-1} for sulfate particles) to concentration data. Methods based on characterization of land use were then introduced to account for the average diurnal and seasonal variability, as a function of surface vegetation and geographical location (see Sheih et al., 1979). At present, methods for deriving species-dependent, site-specific, and time-evolving deposition velocities from on-site measurements are being tested in trial networks (Hosker and Womack, 1985; Hicks et al., 1985).

Figure 4 illustrates the average diurnal cycle of deposition velocities computed by application of a deposition velocity model to data measured at the same forest site as was discussed earlier, at Oak Ridge (Tennessee), for one week during January 1985. Initial tests of this system have emphasized the importance of two major considerations. First, the deposition velocity of many trace species appears to be controlled by surface moisture. Surface wetness cannot yet be measured well; the estimates of its effects as seen in Figure 4 are based on measurements using a wetness sensor that is still under development, and assume that the entire foliar surface is wet when the sensor is wetted. Clearly, a better method for accounting for the effects of surface wetness is required. Second, the assumption that the flux is always directed towards the surface appears adequate for O_3, HNO_3 and SO_2, but is clearly not always appropriate for trace gases that can have surface sources, such as H_2S, NH_3, and NO_x. Thus, even if accurate measurements of NO_x were available, and even if an appropriate resistance model were at hand, then surface deposition fluxes could still not be computed from atmospheric concentration data with confidence. In brief, the atmospheric concentrations that are measured may be the consequence of emissions from the surface instead of being the cause of deposition to it.

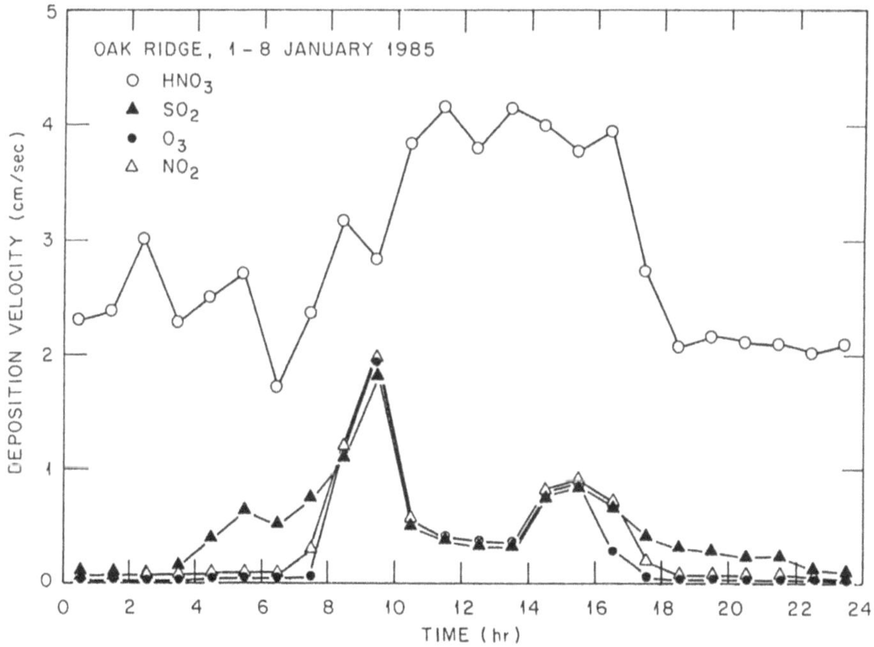

Figure 4. Average diurnal cycles of deposition velocities for HNO_3, SO_2, NO_2, and O_3, deduced from seven days of data obtained for a deciduous forest at Oak Ridge, Tennessee. Note the early morning peak in SO_2 deposition velocity, associated with surface wetness.

5. THEORY VERSUS MEASUREMENTS

Attention is frequently directed to differences perceived between the results of theories and experiments, especially for particle deposition. Much of the experimental basis for this debate has been reviewed above, and a number of difficulties and possible misconceptions have been identified.

There is considerable insecurity associated with many field investigations of particle deposition. This lack of certainty is due to the low deposition velocities which it is required to measure, and to the need to account for fluxes from the surface as well as deposition to it. For example, consider the apparent disagreement between micrometeorological results and the conclusions of studies of the deposition of natural radioactive particles. Micrometeorological measurements made over a conifer canopy have yielded about 1.5 cm s^{-1} for the daytime surface resistance to both sub-μm sulfate particle transfer and small particles in the 0.05 to 0.1 μm size range (Wesely and Hicks, 1979); hence, daytime deposition velocities of about 0.5 cm s^{-1} should be expected. On the other hand, studies of atmospheric radioactivity deposition to a similar canopy have indicated deposition velocities of about 0.02 cm s^{-1} (Russell et al., 1981). The results do not necessarily disagree. The eddy-flux results refer to the canopy as a whole during daytime, whereas the radioactivity results refer to a unit area of leaf within the canopy, averaged over time. If the leaf area index were five, and if deposition at night were near to zero (both of which are reasonable assumptions), then the apparent differences are resolved. There remain, however, significant differences between the predictions of deposition models of forest canopies, built up from understanding of deposition to individual surfaces (e.g. Slinn, 1982) and the results of field studies. These differences appear to diminish as the canopy becomes smoother. Recent field experiments conducted over grass, for example, tend to verify the relatively low deposition velocities predicted by theory (Wesely et al., 1985).

Discussion of dry deposition of particles is also sometimes confused by different physical pictures of the important features of the phenomenon. There is considerable interest in questions regarding local sources, resuspension, and saltation. In this instance, the characteristic feature of the relevant equilibrium boundary layer near the surface is that concentrations are maintained at any given height above the surface, as a balance between fluxes from the surface below and deposition from layers above. However, most modern concerns are related to questions of atmospheric emission, transport, and subsequent deposition. In this case, the surface at some location distant from relevant sources is better viewed as a net sink for the material in question, as air parcels move across the terrain of interest. Atmospheric turbulence is usually taken to be in equilibrium with the roughness and thermal characteristics of the underlying surface. In the absence of local sources of the particles under consideration, there is then near-constancy of vertical fluxes with height. The discussion of field data presented above suggests that a new approach combining features of both resuspension-layer and constant-flux layer derivations may be desirable.

6. CONCLUSIONS

If all other factors are equal, then dry deposition will be greatest in areas with highest concentrations in air. It is usual to evaluate dry deposition rates from air concentration data by applying a pollutant (or particle-size) -specific deposition velocity to concentration data referring to air within the lowest few metres of the surface. These deposition velocities are known to vary with the texture of the surface, its biological properties, and the prevailing meteorological conditions. However, several important conclusions are now evident concerning the general applicability of the method.

The inferential, "concentration-monitoring" approach is fundamentally incapable of producing information on the direction of the transfer. Thus, it is inappropriate for general studies of the exchange of trace gases and particles that may have a surface source.

Knowledge of appropriate deposition velocities is essential. The necessary information is most conveniently summarized in terms of a resistance model of the overall exchange process. The formulations included in such a model are the products of intensive field and laboratory studies. At the present stage of development, multiple-resistance models are only applicable when aerodynamic transfer is by turbulent exchange and gravity. Complex terrain presents difficulties that are presently unresolved.

Most intensive studies of deposition rely on surface-accumulation methods and micrometeorology. Within each category, both direct and indirect methods can be identified. Foliar extraction, snowpack accumulation, etc., constitute direct evaluations of dry deposition to natural surfaces. Measurements of deposition to surrogate surfaces can be considered indirect; extension to the case of a natural surface is not trivial. Eddy correlation provides a direct measurement of the turbulent transfer through the air, whereas gradient data provide an opportunity to infer fluxes only when appropriate eddy diffusivities are known.

In practice, all micrometeorological methods are subject to scatter caused by the effects of atmospheric turbulence as well as instrumental and analytical errors. Errors associated with the measurement of deposition velocities using micrometeorological techniques appear to be of the order of 0.05 cm s^{-1} for any individual 30 to 60 min average.

Differences between experimental results and theoretical predictions appear to be diminishing with time. Remaining inconsistencies are greatest for the case of particle exchange, for which it now appears necessary to consider the surface as an active participant in the exchange process.

Acknowledgments.

This work was conducted with support of the National Oceanic and Atmospheric Administration and the U. S. Department of Energy, as a contribution to the National Acid Precipitation Assessment Program. The experimental results used here were generated as part of the NAPAP "CORE" dry deposition research program, also involving Argonne National

Laboratory, Pennsylvania State University, and the EPA Atmospheric Sciences Research Laboratory. The micrometeorological data shown here were obtained by D. R. Matt, R. T. McMillen, and J. D. Womack.

References

Altshuller, A. P.: 1983, The Acidic Deposition Phenomenon and its Effects. Critical Assessment Review Papers. Chapter A-5, EPA Report Number EPA-600/8-83-016A.

Cadle, S. H., Dasch, J. M., and Mulawa, P. A.: 1985, General Motors Research Laboratories Publication GMR-4918, ENV-196, 30 pp.

Dasch, J. M.: 1983, Precipitation Scavenging, Dry Deposition, and Resuspension, Elsevier Press, New York, p 883.

Dasch, J. M.: 1985, General Motors Research Laboratories Publication GMR-5019, ENV-203, 28 pp.

Davidson, C. I., Lindberg, S. E., Schmidt, J. A., Cartwright, L. G., and Landis, L. R.: 1985, J. Geophys. Res., **90**, 2123.

Dolske, D. A., and Gatz, D. F.: 1984, Deposition Both Wet and Dry, Butterworth Publishers, Stoneham, MA, p 121.

Dolske, D. A., and Gatz, D. F.: 1985, J. Geophys. Res., **90**, 2076.

Dovland, H., and Eliassen, A.: 1976, Atmos. Environ., **10**, 783.

Fairall, C. W.: 1984, Atmos. Environ., **18**, 1329.

Feely, H. W., Bowen, D. C., Nagourney, S. J., and Torquato, C. C.: 1985, J. Geophys. Res., **90**, 2161.

Galloway, J. N., Whelpdale, D. M., and Wolff, G. T.: 1984, Atmos. Environ., **18**, 2595.

Hicks, B. B., Wesely, M. L., and Durham, J. L.: 1980, Critique of Methods to Measure Dry Deposition, Workshop Summary, EPA Report Number EPA-600/9-80-050, NTIS PB81-126443.

Hicks, B. B.: 1982, Proceedings, Symposium on Monitoring and Assessment of Airborne Pollutants with Special Emphasis on Long-Range Transport and Deposition of Acidic Materials, National Research Council of Canada Report NRCC/CNRC No. 20642, p 245.

Hicks, B. B.: 1983, The Acid Deposition Phenomenon and its Effects, Critical Assessment Review Papers. Chapter A-7, EPA Report Number EPA-a600/8-83-016A.

Hicks, B. B., and McMillen, R. T.: 1984, J. Climate and Appl. Meteorol., 23, 637.

Hicks, B. B., Hutchison, B. A., Matt, D. R., Fairall, C. W., de Pena, R. G., Thomson, D. R., Wesely, M. L., Sisterson, D. L., Durham, J. L., Knapp, K. T., and Ellestad, T. G.: 1986, Preliminary Operations and Plans of "CORE" Dry Deposition Monitoring Stations, ATDL Contribution Number 84/25.

Hicks, B. B., Baldocchi, D. D., R. P. Hosker, Jr., B. A. Hutchison, D. R. Matt, R. T. McMillen, and L. C. Satterfield, 1985, On the Use of Monitored Air Concentration to Infer Dry Deposition (1985), NOAA Tech Memo., ERL/ARL-141.

Hosker, R. P. H., Jr., and Womack, J. D.: 1985, Proceedings, 5th. Annual Symposium on Recent Advances in the Measurement of Air Pollutants, U.S. EPA, in press.

Ibrahim, M., Barrie, L. A., and Fanaki, F.: 1983, Atmos. Environ., 17, 781.

Johnson, D. W., Henderson, G. S., Huff, D. D., Lindberg, S. E., Richter, D. D., Shriner, D. S., Todd, D. E., and Turner, J.: 1982, Oecologia (Berl), 54, 141.

Lenschow, D. H., Delany, A. C., Stankov, B. B., and Stedman, D. H.: 1980, Boundary-Layer Meteorol., 19, 249.

Lenschow, D. H., and Kristensen, L.: 1985, J. Atmos. and Ocean. Tech., 2, 68.

Lindberg, S. E., and Harriss, R. C.: 1981, Water, Air and Soil Pollut., 16, 13.

Lindberg, S. E., and Lovett, G. M.: 1985, Environ. Sci. and Technol., 19, 238.

Lovett, G. M., and Lindberg, S. E.: 1984, J. Appl. Ecology, 21, 1013.

O'Dell, R. A., Toeheri, M., and Kabel, R. L.:1977, J. Air Pollut. Contr. Assoc., 27, 1104.

Russell, I. J., Choquette, C. E., Fang, S-L., Dundulis, W. P., Pao, A. A., and Pszenny, A. A. P.: 1981, J. Geophys. Res., 86, 5247.

Shannon, J. D.: 1981, Atmos. Environ., 15, 689.

Sheih, C-M., Wesely, M. L., and Hicks, B. B.: 1979, Atmos. Environ., 13, 1361.

Slinn, W. G. N.: 1982, Atmos. Environ., 16, 1785.

Volchok, H. L., Feiner, M., Simpson, H. J., Broecker, W. S., Noshkin, V. E., Bowen, V. T., and Willis, E.: 1970, J. Geophys. Res., 75, 1084.

Volchok, H. L., and Graveson, R. T.: 1976, Proceedings, Second Federal Conference on the Great Lakes, Great Lakes Commission, p 259.

Webb, E. K., Pearman, G. I., and Leuning, R.: 1980; Quart. J. Roy. Meteorol. Soc., **106**, 85.

Wesely, M. L., and Hicks, B. B.: 1979, Proceedings, Fourth Symposium on Turbulence, Diffusion, and Air Quality, (Am. Meteorol. Soc., Boston), p 510.

Wesely, M. L., Cook, D. R., Hart, R. L., Hicks, B. B., Durham, J. L., Speer, R. E., Stedman, D. H., and Tropp, R. J.,: 1982, Precipitation Scavenging, Dry Deposition, and Resuspension, Elsevier Press, New York, p 943.

Wesely, M. L., and Hart, R. L.: 1984, The Forest Atmosphere Interaction, D. Reidel, p 591.

Wesely, M. L., Cook, D. R., Hart, R. L., and Speer, R. E.: 1985, J. Geophys. Res., **90**, 2131.

POLLUTANT WET DEPOSITION MECHANISMS IN PRECIPITATION AND FOG WATER

L.A. Barrie and R.S. Schemenauer
Atmospheric Environment Service
4905 Dufferin Street
Downsview, Ontario, Canada
M3H 5T4

ABSTRACT. Wet deposition of acid-related substances takes place by two processes: precipitation scavenging and fog water impaction/sedimentation on natural surfaces. The relative importance of each deposition pathway depends on the frequency of occurrence of precipitation or fog, the magnitude of the event and the efficiency of pollutant removal by each mechanism. The latter, in turn, is governed by the type of cloud or fog, complex precipitation formation mechanisms and cloud-surface interactions. These factors are examined in the light of our current knowledge. Particular emphasis is placed on how cloud micro-physical as well as air and precipitation measurements, made aloft by aircraft and at the ground, have been used to further our knowledge of wet deposition mechanisms. Future research is needed to quantify the importance of the fog-water deposition pathway in eastern North America to better understand the interaction of gaseous pollutants with cloud and fog-water and to improve our knowledge of pollutant scavenging processes in mesoscale and synoptic weather systems.

1. INTRODUCTION

Particles and gases are removed from the atmosphere by precipitation and fog-water deposition (wet deposition) as well as by direct uptake at the earth's surface (dry deposition). The deposition of relatively clean dew or frost (Wisniewski, 1982) is expected to lead to little direct pollutant deposition. Indirectly, however, by altering surface roughness and wetness it may enhance dry deposition of particles and gases. The wet deposition process is a major pathway of acidic pollutants from the the atmosphere to the biosphere. In eastern North America, approximately 71% of the total S reaching the earth annually is wet deposited by precipitation (Galloway and Whelpdale, 1980). Four years of air and precipitation chemistry observations in the southern part of eastern Canada show that wet deposition by precipitation accounts for 70 to 80% of the total annual atmospheric deposition of S and N (Barrie and Sirios, 1986). The wet

Water, Air and Soil Pollution **30** (1986) 91–104.

deposition process involves complex chemical transformation of SO_2 and NO_2 to H_2SO_4 and HNO_3 in a cloud, as well as the processes of cloud formation, precipitation formation and, in the case of fog water deposition, the interaction of fog droplets with the Earth's surface. In this paper, a brief review is given of the mechanisms of wet deposition and the methods that are used to gain a better understanding of them.

Precipitation at mid-latitudes is usually associated with extra-tropical cyclones (low pressure zones) which have cloud patterns similar to that shown by the satellite photo in Figure 1. These cyclones typically involve the intrusion of a tongue of warm southern air northward from the Gulf of Mexico or Caribbean that overrides colder northern air. This in turn is undercut from the west-northwest by colder northern air. A warm front marks the boundary between advancing warm air and the colder air over which it is rising. A cold front lies between the advancing cold air in the northwest and the

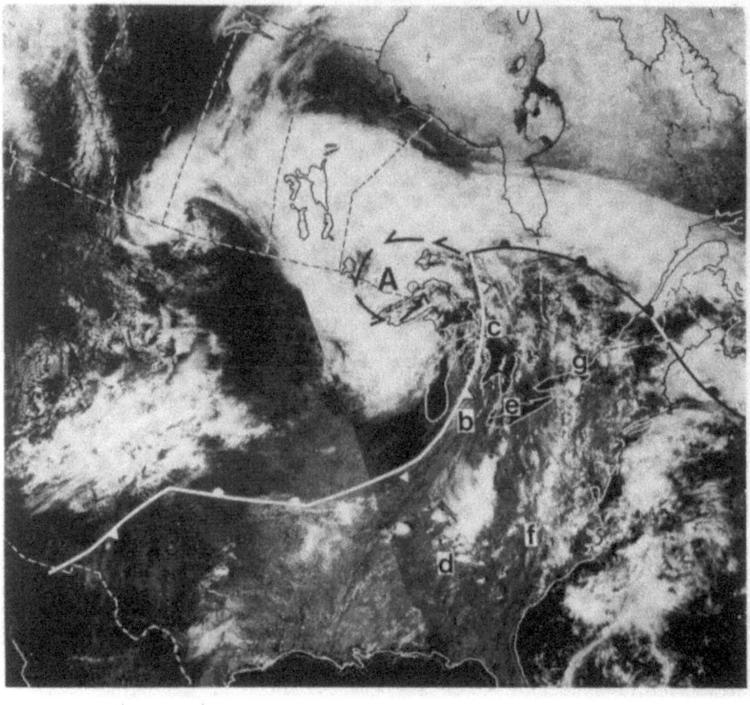

Figure 1: A satellite photo in the visible, illustrating the typical cloud and frontal structure associated with an extra-tropical cyclone occurring May 12, 1985. Δ marks a cold front, a warm front. Convective cloud bands are visible in the warm sector between b and c, e and d as well as g and f. Warm frontal cloud is present in central Quebec and northwestern Ontario.

warm air. Different pollutant removal situations arise depending on location relative to these fronts in the cyclone. In the vicinity of the cold front and in the warm sector ahead of the cold front, convective clouds of 2 to 20 km diameter form in clusters. Each cloud draws air from the near-surface boundary layer below. Ahead of the surface warm front the situation is quite different. Clouds form over large areas as a result of widespread uplift. The air in which they form originates from the near-surface layer several hundreds of kilometres to the south. Precipitation falls from the overriding warm air into cooler air below.

The nature of precipitating clouds in extra-tropical cyclones has been extensively studied using satellite photography, radar, and aircraft (Parsons and Hobbs, 1983). Precipitation tends to occur from organized bands of convective cells (see examples in the warm sector of Figure 1). This is true even in the warm frontal sector. The nature of a precipitation event associated with the passage of an extra-tropical cyclone depends on the intensity of the cyclone, its speed and the location of the sampling site relative to it. A climatology of storm features in the northeastern United States (Thorp and Scott, 1982) shows that, in general, there is a marked difference in storm types between winter and summer. In winter, storms move twice as fast across the region (15 m s^{-1} versus 7 m s^{-1}) but their extent is such that the average duration is much longer (26 hr) than in summer (2.5 hr). Precipitation intensity is lower in winter than in summer (0.9 versus 2.5 mm hr^{-1}) . The fraction of storms that are isolated convective bands is higher in summer than in winter.

Fogs can be simply regarded as clouds in contact with the Earth's surface. In some cases, this is literally true, for example when a layer of cloud passes over a mountain. In other cases, the fog is formed by processes (different from those of cloud formation) that take place near the Earth's surface. If air is cooled sufficiently, the water vapour in it will condense into liquid drops. The cooling can be produced by several different meteorological situations leading to different types of fogs. At night, the earth loses heat through longwave radiation. Sufficient cooling leads to 'radiation fog'. Light winds ($<$ 10 km hr^{-1}) promote the development by mixing cool, moist, near-surface air to moderate heights. At higher wind speeds, mixing with the drier air aloft tends to inhibit fog formation. When warm moist air moves over a cold surface, the cooling that results can produce an 'advection fog'. Most coastal fogs are of this type. Air that is forced to rise over a topographical barrier experiences decreasing atmospheric pressure that produces expansion and cooling of the air. These conditions can generate 'upslope fog'. Other types of fog are less important vehicles of wet deposition. 'Steam fog' is produced by the condensation of water vapour from a relatively warm water surface in the cooler air above. 'Ice fog' is produced at temperatures below -30°C when water vapor sublimes directly onto ice nucleii (particles) thus forming tiny ice crystals. This illustrates a characteristic of fogs (and clouds) that is important in the deposition of pollutants, namely, they remain liquid (containing supercooled droplets) at temperatures as low as ≈ -20°C and

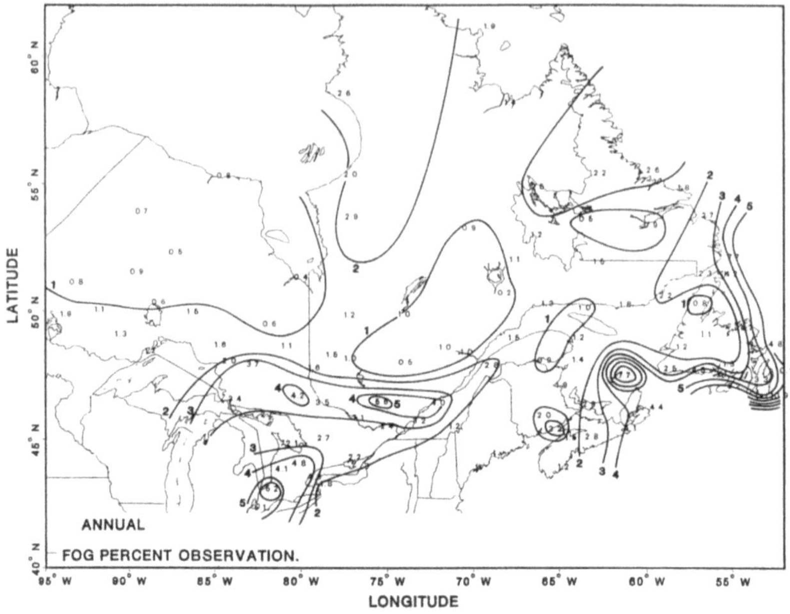

Figure 2a: The percentage of hourly observations reporting fog in eastern Canada. The data represent a 30 yr annual average (1953-1982) (Joe, 1985).

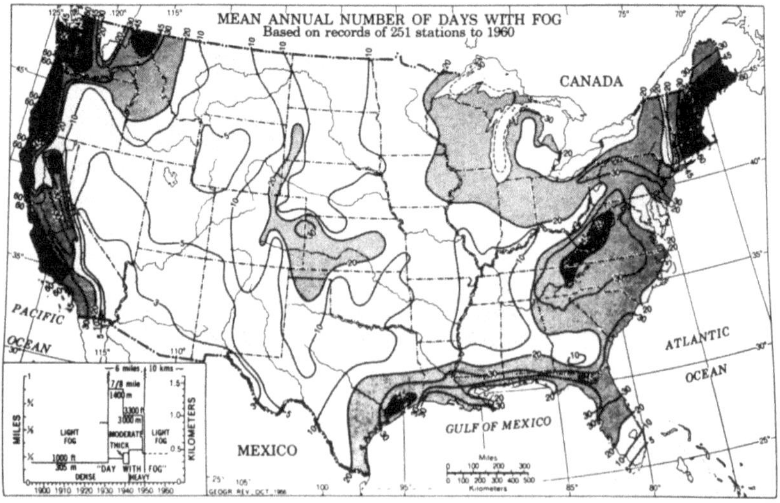

Figure 2b: The mean annual number of days with fog in the United States (Court and Gerston, 1966).

occasionally to much colder temperatures.

At most North American locations fog occurs on 0.5 to 5 day mo^{-1}. Figure 2 shows the spatial distribution of fog frequency in the United States and eastern Canada, respectively.

2. MECHANISMS OF POLLUTANT WET DEPOSITION

2.1 Precipitation Scavenging

At mid-latitudes, pollutants are scavenged from the atmosphere by precipitation via the sequence of processes depicted in Figure 3.

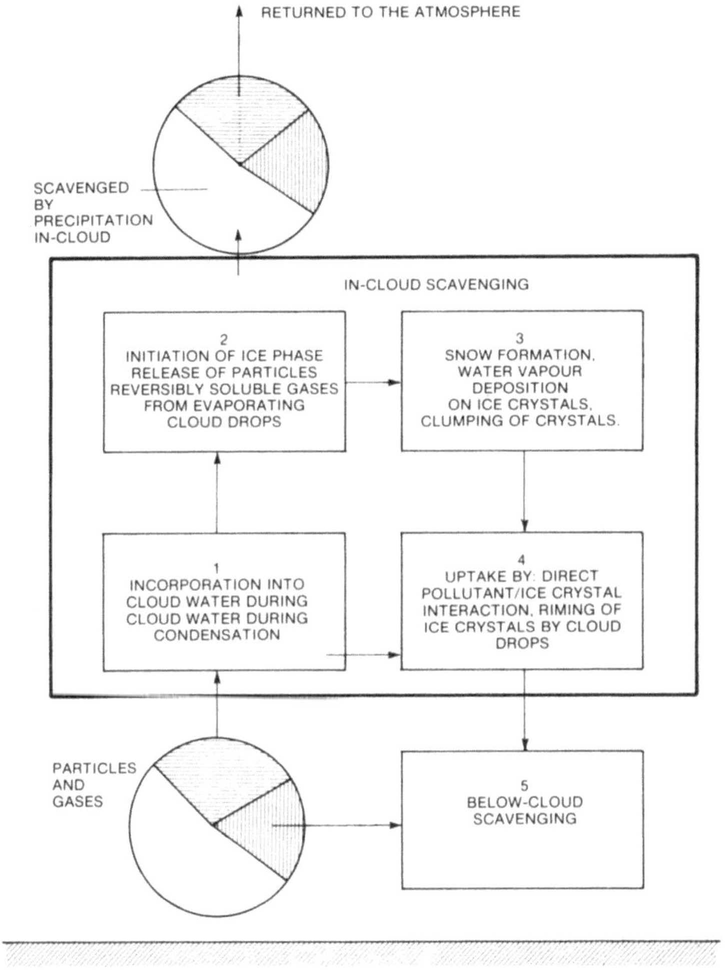

Figure 3: The precipitation scavenging process.

Condensation of water on hygroscopic aerosols entering a cloud as well as dissolution of soluble gases such as HNO_3 and SO_2 (step 1) leads to supercooled cloud drops containing inorganic ions, insoluble particulate matter and organic constituents. Typically in precipitating clouds, condensation takes place on particles of diameter greater than 0.2 μm (Pruppacher and Klett, 1978). The greater the supersaturation in a cloud the lower the minimum particle size. In precipitating clouds 50 to 100% of the particle mass is usually incorporated into cloud water at this stage. A fraction of SO_2 and NO_2 may also be converted to $SO_4^=$ and NO_3^- at this point or shortly thereafter. Indirect evidence based on an analysis of ground level observations (Barrie, 1985) indicates that 20 to 100% of SO_2 entering rain storms in eastern Canada is oxidized to $SO_4^=$ and enters cloud water while a much smaller fraction is converted in snow storms. There is also evidence that considerable amounts of cloud water NO_3^- are formed by NO_2 oxidation in both winter and summer. This is confirmed by aircraft observations (Leaitch et al., 1985).

Two things can happen to the cloud water: (i) if ice is present in-cloud, cloud water can evaporate. Water vapor, driven by the difference in saturation vapor pressure of water and ice, diffuses from the drops to ice crystals (steps 2 and 3). The gaseous and particulate matter in cloud water is then returned to the atmosphere in a form that is not necessarily the same as when it entered the water (ii) cloud water can be captured by large snow crystals (diameter > 200 μm). This process is called riming. It is the collection by snow of supercooled cloud drops 2 to 50 μm diameter by inertial impaction and interception as the heavier snow crystals fall past the small drops. The cloud drops freeze upon contact. Pollutants trapped in cloud water thereby enter precipitation and fall from the cloud.

Particles and gases that do not enter cloud water during condensation (step 1) or that are released back into the air during vapor growth of the ice phase (step 2) can be incorporated into ice crystals (steps 3 and 4). This occurs by Brownian diffusion, electrical attraction and inertial impaction. Gases may react with ice surfaces. HNO_3 is more reactive with ice than SO_2. Thus precipitation leaving cloud base can consist of rimed ice, unrimed ice or if the temperature at cloud base is above 0°C, water drops. Scott (1981) has found that snow storms with mostly unrimed snow crystals (i.e. scavenging steps 1-2-3-4) remove SO_x from the atmosphere much less efficiently than storms with rimed snow crystals (i.e. scavenging steps 1-2-3-4 and 1-4). Particles and gases that are not scavenged from the cloud by the precipitation processes discussed above leave the cloud at the evaporating edges. As a result of chemical reactions in cloud water, the chemical composition of air at the outflowing edges can be different than that at the inflow. Furthermore, the size distribution of particulate matter can be changed as a result of in-cloud chemical and physical transformations.

Rain and snow falling below the cloud base scavenge particles and reactive gases from the atmosphere by the same mechanisms as they do within cloud with the exception of the riming process and with the addition of phoretic forces caused by evaporation in the sub-saturated air beneath the cloud. A quantitative description of the interaction between drops or ice crystals and pollutant gases or particles is a complex one. Even in the relatively simple case of a plate-like ice crystal scavenging particles, the collision efficiency depends on crystal and aerosol size distributions, on the degree of charging and on evaporation (Martin et al., 1980). Add to this an extremely variable ice crystal shape and one is left with a complicated situation to handle theoretically, experimentally or observationally.

2.2 Fog Deposition

The means by which gases and particles are incorporated into fog droplets are the same as those discussed above for cloud droplets. The ice phase is usually less important in fogs than in precipitating clouds and the riming process is essentially non-existent due to the small sizes of any ice crystals. Therefore, a simplified version of Figure 3 applies to pollutant deposition mechanisms in fogs. Pollutant-bearing fog droplets can be: deposited on the Earth's surface by sedimentation, impacted on surface roughness elements such as tree needles or leaves, and evaporated either at the edges of the cloud or internally. The first two processes constitute pollutant deposition pathways. They are dependent on the physical/chemical properties of a fog, namely: the size distribution of fog drops; the chemical composition of fog drops as a function of drop size; as well as on characteristics of the earth's surface and meteorological factors. Juisto (1981) conducted an extensive review of the current state of knowledge of drop size distributions and of the visual range in fogs. All these properties are extremely variable in time and in space as is seen below. Mean drop diameters of surface generated fogs are typically 10 (inland) to 20 μm (coastal). Mean number concentrations range from 10 to 200 drops cm^{-3} and the maximum liquid water content is approximately 0.3 g m^{-3}. With the exception of mountain top fogs, the number concentrations and liquid water contents of most fogs are lower than one typically finds in clouds.

The microstructure of fogs is strongly influenced by the characteristics of the condensation nucleii (Hudson, 1980). Thus, polluted air masses with high concentrations of condensation nucleii generally produce fogs with high cloud drop number concentrations and smaller drop sizes. Sea fogs generally have lower total droplet concentrations with larger particles. Knowledge of the chemical composition of fog water as a function of size is important because the collection efficiency of foliage and other objects is dependent on droplet size. Thus, wet deposition to an object may include only droplets in a specific size range and not the entire droplet spectrum. Unfortunately, reliable size dependent chemical composition data is not available for fogs (nor for clouds).

Aspects of the fogwater deposition process other than net pollutant deposition are also important in determining the impact on vegetation. Figure 4 is a schematic showing the deposition processes on a conifer needle. The fog droplets contain particulates such as soil derived dust, pollen, fly ash, $SO_4^=$ and NO_3^- as well as H^+, SO_2, oxidants (O_3, H_2O_2, etc.) and other dissolved gases. When a droplet contacts the needle, damage can result directly from its acidity. But even if only the waxy surface of the needle is destroyed, this will leave it prone to subsequent damage. Evaporation of the deposited droplet can lead to even higher droplet acidity, and hence, even greater damage. Conversely, the small supersaturations that exist in fogs can cause condensation and hence dilution of the drop. The deposited droplet can also migrate over the needle surface encountering material that has been dry deposited. Like evaporation, this increases the chemical concentration of trace substances in the drop.

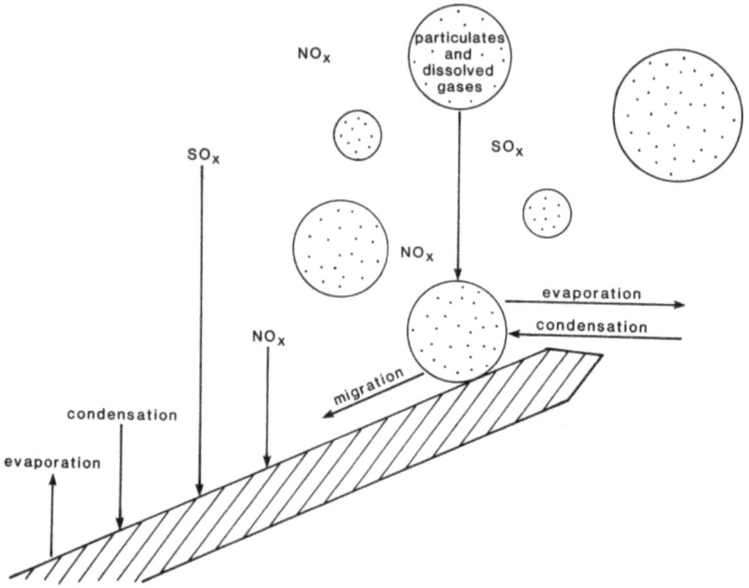

Figure 4: A schematic of the deposition of fog water drops on a conifer needle and factors that influence their composition after deposition.

3. THEORETICAL AND OBSERVATIONAL APPROACHES TO UNDERSTANDING WET DEPOSITION MECHANISMS

The ultimate goal of wet deposition research is to develop the capability of understanding the influence of a precipitation or fog system on atmospheric trace constituents in its vicinity. Thus, from meteorological observations of storm or fog characteristics and models, one would like to know what fraction of a pollutant entering a system is removed, the redistribution of remaining pollutant in the atmosphere and the physical/chemical transformation it has undergone.

An important step in attaining this goal is to understand wet deposition mechanisms. This involves three simultaneous activities: making in situ atmospheric observations of processes, developing theoretical descriptions of them and whenever possible testing them in the laboratory. The mechanisms of interest include the following: (i) nucleation scavenging of particles (ii) the oxidation steps associated with $SO_4^=$ and NO_3^- formation in cloud or fog water from SO_2 and NO_2 (iii) the in-cloud transfer of cloud water pollutants to precipitation (iv) below-cloud scavenging of pollutants by precipitation and (v) the interaction of fogs with the Earth's surface.

The nucleation scavenging efficiency (F) is defined as the fraction of total particulate mass incorporated into cloud (or fog) water during the condensation growth stage. F can be measured from aircraft by measuring the concentration of particulate matter in air entering a cloud and the air equivalent concentration in cloud water just inside the cloud (Leaitch et al., 1985; Hegg et al., 1984a). The latter measurement involves the cloud liquid water content (L in kg m^{-3} of air). An alternate approach is to measure the partitioning of mass between the gas and liquid phase just after condensation (Daum et al., 1984). Hegg et al. (1984b) report an F-value of 0.8 ± 0.2 for $SO_4^=$ bearing particles in non-precipitating cumulus, stratocumulus and stratus clouds. On the basis of measurements in stratus and stratocumulus clouds, Daum et al. (1984) conclude that the mass of aerosol in the air between the droplets is generally low but not negligible compared to the mass in cloud water. Theoretical calculations of F can be made if the composition and mass size distribution of the condensation nuclei as well as cloud supersaturation are known (e.g. Jensen and Charlson, 1984). These tend to agree with observations.

Measurements of SO_2 and NO_2 oxidation in-cloud are done from aircraft by one of two methods: (i) a mass balancing method (Hegg et al., 1984a; Lazrus et al., 1983; Leaitch et al., 1985) (ii) detection of changes in the relative composition of trace constituents in air just outside the cloud and in cloud water just inside the cloud (Daum et al., 1984). In the mass balancing method, measurement of an increase in the total amount of $SO_4^=$ or NO_3^- in air entering a cloud is attempted. It has proven to be an effective method in detecting large amounts of conversion (60% of existing $SO_4^=$ or more). Its limitation is the large uncertainty involved in calculating small differences in large numbers. The other method involves measuring the ratio of $SO_4^=$ or NO_3^- to another particulate trace constituent that does not have a gaseous precursor inside and outside the cloud. To date this method has not been widely used. Theoretical models of in-cloud oxidation of SO_x and NO_x have proven useful in guiding and explaining observations (e.g. Chameides, 1984).

3.1 Precipitation Scavenging

Aircraft studies of the transfer of cloud water pollutants to precipitation and of below-cloud scavenging have been relatively scarce. It is particularly hard to conduct observations within precipitating convective bands because of the difficult flying conditions. Considering the complexity of the microphysical processes involved in precipitation scavenging, the variability in macroscopic features of storms and practical difficulties in flying in precipitating regions of extra-tropical cyclones, it is not surprising that other ways of using ground level observations to gain insight into scavenging processes have been sought (Slinn, 1984; Hidy and Countess, 1984; Barrie, 1985). For instance, for modelling purposes or in cases where air concentrations but no precipitation measurements are available, one often needs to estimate a substance's concentration in precipitation knowing its concentration in the atmosphere, the precipitation amount and microphysical features such as whether the riming process was active in the storm. The scavenging ratio (W) defined as the time-average concentration of substances in rain (C_R) divided by the time-average concentration in air entering a cloud (C_A) is one parameter which has proven useful in this regard. Its advantages and disadvantages are discussed by Barrie (1985). Ground level daily W values measured in Canada were used to gain insight into SO_2 oxidation in precipitating systems. Results indicate that W which is calculated from airborne particulate $SO_4^=$ concentrations is, on average, lower on days with relatively low SO_2 (SO_2 < 40% total S) than on days with relatively high SO_2 (SO_2 > 70% total S). This suggests that SO_2 oxidation is taking place. From a fit of a physical model to data, it was concluded that, on average, in eastern Canada, in-cloud oxidation accounts for approximately 60% of the $SO_4^=$ observed in rain and for almost none of the $SO_4^=$ observed in snow. A similar analysis for NO_3^- predicts that much of the NO_3^- observed in precipitation originates from in-cloud NO_2 oxidation regardless of precipitation type. This is consistent with recent aircraft observations (Leaitch et al., 1985) and with the observation that the molar ratio of $SO_4^=$ to NO_3^- is much higher in precipitation than in air (Summers and Barrie, 1986).

Theoretical simulations of pollutant removal by storm systems associated with extra-tropical cyclones can be used to understand scavenging processes. There are two types of models: diagnostic ones and simulative ones. The first type relies heavily upon observations in addition to a theoretical formulation of cloud processes to yield insight into the net scavenging of pollutants by a 'real' storm (e.g. Easter and Hales, 1983; Hegg et al., 1984b). The second type, is a dynamic storm model with chemistry incorporated (e.g. Tremblay and Leighton, 1986). Provided storm dynamics and microphysics resemble those of actual storms insight can be gained into scavenging processes.

3.2 Fog Deposition

In some areas, the amount of water deposited in fogs can be a substantial fraction of the total liquid water deposited (Nagel, 1956; Lovett, 1984; Lovett and Reiners, 1986). Values of fog-water deposition rates ranged from ≈ 0.1 to 4 mm hr^{-1} and are supported by other observations (e.g. Merriam, 1973). These are comparable to precipitation rates. Falconer and Falconer (1980) report cloud water pH measured on Whiteface Mountain, N.Y. and review the existing

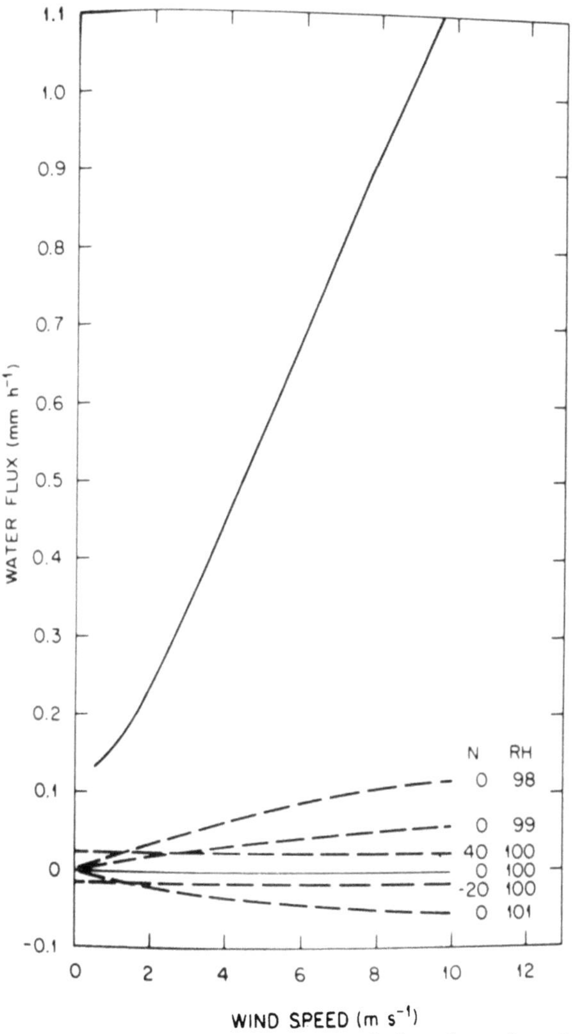

Figure 5: Simulated cloud water deposition (——) and evaporation (- - -) rates plotted against canopy-top wind speed. Evaporation rates are given for several combinations of net radiation (N) in W m^{-2} and relative humidity (R.H.) in per cent. Cloud liquid water content = 0.4 g m^{-3} modal droplet diameter = 10 μm and temperature = 10°C. Net cloud deposition is the difference between gross cloud deposition and evaporation (from Lovett, 1984).

literature. Nearly 90% of all Whiteface fog water (cloud moving over the mountain) pH observations lie in the range 2.7 to 4.7. In contrast, the most acidic rain on the mountain seldom has a pH < 4. Low fog water pH values have also been documented recently by Wisniewski (1982), Fuzzi et al. (1983) and Hoffmann (1985). In areas where fog frequency is high (coastlines or mountainous areas), the H^+ deposition from fog is comparable to that from precipitation. The same is likely true for $SO_4^=$ and NO_3^-.

Several models of vegetation-fog interactions have been developed. Shuttleworth (1977) looked at fogwater capture and the evaporation-condensation process. His theoretical description was quite restricted applying only to wind driven fog in areas of extensive uniform vegetation (i.e. mountain forests rather than lowlands with radiation fogs). Lovett (1984) used a more complex model to study a similar situation. The deposition of cloud droplets to a balsam fir forest whose canopy was divided into 1 m strata each containing seven canopy component types was modelled. Three drop-size classes were used. Theory and observations at a sub-alpine site in New England agreed reasonably well with fogwater deposition of the order of a few tenth's of millimetres per hour. The deposition velocity of fog water ranged from 1 to 80 cm s^{-1}. Figure 5 shows the marked effect on the water flux resulting from an increase in canopy top wind speed.

The mass flux of fog water to a forest canopy can be calculated with some confidence provided fog microphysical and surface boundary layer meteorological characteristics are known. In principle, if the distribution of pollutant mass as a function of drop size is known, the net deposition of pollutant by fog can be calculated. In practice, however, the measurement of composition as a function of size is extremely difficult. Field programs have concentrated on developing passive and active (fan driven) fog water samplers designed to obtain samples over as large a droplet size range as possible. This in itself is a major challenge given the wide range of wind, temperature and fog conditions encountered.

4. CONCLUSION

Wet deposition of pollutants takes place in complex precipitating cloud and fog systems. In some mountainous and coastal areas, fog water deposition is at least as important as deposition in precipitation. Thus, previous estimates of the impact of acidic deposition which did not take into account fog water (and to a lesser extent dew) are actually underestimates. Insight into precipitation scavenging processes is gained by making measurements quite different from those used to understand fog water deposition. Precipitation scavenging is studied using ground level air and precipitation chemistry measurements and cloud microphysical and chemical observations from aircraft. Deposition in fogs requires a knowledge of the fog microphysical and chemical properties, details of the structure of the Earth's surface and the micrometeorology of the boundary layer. From available evidence, one can conclude that the acidity of cloud or fog water originates mainly from particulate sulphate and nitrate as well as gaseous SO_2, HNO_3 and NO_2. Furthermore, the acidity of deposited fog water is generally higher than that of precipitation, whereas the amount of fog water deposited is generally less than that of precipitation. The relative contribution of precipitation and fog water to wet deposition depends on the nature and frequency of fog, total precipitation amount and the nature of the Earth's surface. These factors are so variable spatially that an accurate assessment of the role of fog water deposition has not been made on a regional basis.

Models of wet deposition due to fog water, although relatively primitive, show basic agreement with observations. Improvements are needed in: the characterization of fog droplet chemistry as a function of size; the description of wind flow around mountains and in the parameterization of deposition to a variety of vegetative surfaces. Instrumentation and techniques to measure fog water deposition at the Earth's surface are still in a developmental stage. In contrast, the methodology of making a precipitation deposition measurement is well developed and relatively easily used. Precipitation scavenging has been modelled for a much longer time than fog water deposition. Despite this, a capability to understand the pathways that pollutants travel through storms has not been sufficiently developed. Consequently, much more work is needed before one has a non-empirical formulation of precipitation scavenging to use in chemical transport models.

REFERENCES

Barrie, L.A.: 1985, J. Geophys. Res., **90** , 5789.

Barrie, L.A. and Sirois, A.: 1986, J. Water Air Soil Poll., submitted.

Chameides, W.L.: 1984, J. Geophys. Res., **89** , 4739.

Court, A. and Gerston, R.D.: 1966, Geograph. Rev., **56** , 1966.

Daum, P.H., Schwartz, S.E., and Newman, L.: 1984, J. Geophys Res., **89** , 1147.

Easter, R.C. and Hales, J.M.: 1983, Interpretations of the OSCAR data for reactive gas scavenging, Precipitation Scavenging, Dry Deposition and Resuspension, eds. Pruppacher, Semonin, Slinn, Elsevier, New York, pp 649-662.

Falconer, R.E. and Falconer, P.D.: 1980, J. Geophys. Res., **85** , 7465.

Fuzzi, S., Orsi, G., and Mariotti, M.: 1983, J. Aerosol. Sci., **14** , 135.

Galloway, J.N. and Whelpdale, D.M.: 1980, Atmospheric Environment, **14** , 409.

Hegg, D.A., Hobbs, P.V., and Radke, L.F.: 1984a, Atmospheric Environment, **18** , 1939.

Hegg, D.A., Rutledge, S.A., and Hobbs, P.V.: 1984b, J. Geophys. Res., **89** , 7133.

Hidy, G.M. and Countess, R.: 1984, Combined analysis of air quality and precipitation chemistry data, Chapt. 3 in Deposition: Both Wet and Dry, Acid Precipitation Ser., Vol 4, ed. by H.H. Hicks, 37-62, Butterworth, Boston.
Hoffman, M.R.: 1985, American Chemical Society abstracts, 190th National Meeting, AMS Division of Environmental Chemistry, Chicago, Ill., **25** , 10-11.
Hudson, J.G.: 1980, J. Atmos. Sci., **37** , 1854.
Jensen, J.B. and Charlson, R.J.: 1984, Tellus, **36B** , 367.
Jiusto, J.E.: 1981, Fog structure. From, Clouds: Their Formation, Optical Properties, and Effects, Academic Press, edited by P.V. Hobbs and A. Deepak, 187-235.
Joe, P.I.: 1985, private communication.
Lazrus, A.L., Haagensen, P.L., Kok, G.L., Huebert, B.J., Kreitzberg, C.W., Likens, G.E., Mohnen, V.A., Wilson, W.E., and Winchester, J.W.: 1983, Atmospheric Environment, **17** , 581.
Leaitch, W.R., Strapp, J.W., Wiebe, H.A., Anlauf, K.G., and Isaac, G.A.: 1985, J. Geophys. Res., submitted.
Lovett, G.M.: 1984, Atmospheric Environment, **18** , 361.
Lovett, G.M. and Reiners, W.A.: 1986, J. Water Air Soil Poll., 1986.
Martin, J.J., Wang, P.K., and Pruppacher, H.R.: 1980, J. Colloid and Interf. Sci., **78** , 44.
Merriam, R.A.: 1973, Water Resour. Res., **9** , 1591.
Nagel, J.F.: 1956, Quart. J. Roy. Meteor. Soc., **82** , 452.
Parsons, D.B. and Hobbs, P.V.: 1983, J. Atmos. Sci., **40** , 2377.
Pruppacher, H.R. and Klett, J.D.: 1978, Microphysics of clouds and precipitation, Reidel, Dordrecht, Boston, London, 714 pp.
Scott, B.C.: 1981, J. Appl. Met., **20** , 619.
Shuttleworth, W.J.: 1977, Bound.-Layer Meteor., **12** , 463.
Slinn, W.G.N.: 1984, Precipitation scavenging in Atmospheric Sciences and Power Production, edited by D. Randerson, Chapt. 12, US DOG Tech. Inf. Centre, Oak Ridge, Tennessee, 533-572.
Summers, P.W. and Barrie, L.A.: 1986, J. Water Air Soil Poll., 1986.
Tremblay, A. and Leighton, H.: 1986, J. Climate Appl. Meteor., in press.
Thorp, J.M. and Scott, B.C.: 1982, Atmospheric Environment, **16** , 1763.
Wisniewski, J.: 1982, Water, Air, Soil Pollut., **17** , 361.

Atmospheric Measurements of Nitrogen Dioxide with a Sensitive Luminol Instrument

H.I. Schiff, G.I. Mackay, C. Castledine, G.W. Harris and Q. Tran
Unisearch Associates Inc., 222 Snidercroft Rd., Concord, Ontario, Canada, L4K 1B5

ABSTRACT A highly sensitive, lightweight, portable instrument has been developed for continuous monitoring of NO_2 in air. It operates by detecting the chemiluminescence produced when NO_2 encounters a surface wetted with a specially formulated solution containing luminol. Unlike other chemiluminescent instruments it measures NO_2 directly and does not require prior conversion of NO_2 to NO. It does not respond to H_2O_2, NO, HNO_3, NH_3, CO, CO_2, SO_2 or organic nitrates. The only interferences encountered to date are from O_3 and PAN with the response to O_3 being less than 0.2% of its response to NO_2. The response time of the instrument is less than 1 s and its sensitivity is better than 5 pptv. The instrument shows a negative temperature dependence of about 2% $°C^{-1}$ which is compensated electronically. Measurements have been made with the instrument during a number of field missions in polluted urban and relatively clean rural air. The relative influences of photochemistry and transport from local sources has been observed. Simultaneous measurements made with this instrument and the unequivocal tunable diode laser absorption spectroscopy method have shown excellent agreement.

1. INTRODUCTION

The measurement of NO_2 in air is of paramount importance for a number of reasons. Its presence in the atmosphere, whether from natural or human sources, initiates atmospheric chemistry. NO_2 is the only gas in the troposphere which can be photodissociated by sunlight to produce O_3. The atmospheric chemistry resulting from its dissociation leads to the oxidation of the reduced gases in the atmosphere including H_2S, SO_2, CO and the hydrocarbons. For example, model calculations by Fishman et al. (1985) have shown that NO_2 has, by far, the most pronounced effect on the ozone budget and its height profile in the atmosphere.

Thus NO_2 contributes indirectly to acid deposition by affecting the oxidant level in the atmosphere which converts S(IV) to S(VI). It

Water, Air and Soil Pollution **30** (1986) 105–114.

contributes directly by being itself converted to HNO_3, mainly by reaction with OH radicals and by dry deposition.

Evaluation of models developed to attack the problems of photochemical oxidation and acid deposition require accurate measurements of this important species. Monitoring NO_2 at the large number of sites needed for evaluation of these models require instrumentation which is inexpensive, sensitive and easy to operate. It is also important that they can provide real time measurements with good time resolution. Filter pack methods do not meet these requirements. Instruments based on the chemiluminescent NO/O_3 reaction can provide the desired time response but depend upon prior conversion of NO_2 to NO. Most of the conversion methods also reduce other nitrogen species to NO which can lead to serious interference problems. In addition, the weight and power requirements of the instrument make it difficult to use in field monitoring stations.

This paper describes an instrument which meets the requirements for field monitoring and has sufficiently fast time response to be of interest for dry deposition studies. It responds to NO_2 directly and, unlike the conventional chemiluminescent instrument, does not require prior conversion to NO. In addition, its small size and ability to operate with rechargeable batteries make it highly suitable for indoor and outdoor air quality measurements. The paper also describes a number of field measurements made with this instrument.

2. PRINCIPLE OF THE METHOD

The instrument operates by detecting the chemiluminescence produced when NO_2 encounters a surface wetted with a specially formulated, alkaline, luminol solution. Unlike other chemiluminescent instruments, it measures NO_2 directly and does not require prior conversion to NO. Although many oxidizers can produce chemiluminescence with luminol most do so in the liquid phase and require the presence of catalysts such as metal ions. In contrast, chemiluminescence with NO_2 appears to be a surface reaction not requiring the presence of metal ions. Confining the reaction to a surface and using a solution made with deionized water avoids interferences with most other gases.

The method was first described by Maeda et al. (1980). The sampled air was passed over a pool of the luminol solution which was viewed with a photomultiplier. The surface area of this system was not, however, well defined and splashing of the solution up the sides of the reservoir changed the sensitivity.

Wendall et al. (1983) overcame this difficulty by flowing the solution down a filter paper, thus providing a defined surface to be viewed by the photomultiplier.

We have made further modifications and have investigated the parameters of the instrument performance with the aim of optimizing its sensitivity and freedom from interferences.

3. DESCRIPTION OF THE INSTRUMENT

A block diagram of the instrument, trademarked LUMINOX is shown in Figure 1. The sampled air is drawn through the instrument by a micro pump at a constant flow rate of about 1.5 standard liters per minute (SLM). The response of the instrument increases only slowly with flow rate above about 1 SLM. The air passes along the 2 cm length of a fabric wick wetted with the luminol solution. The solution is continuously replenished at the top of the wick and removed at the bottom by a small, two-stage, peristaltic pump.

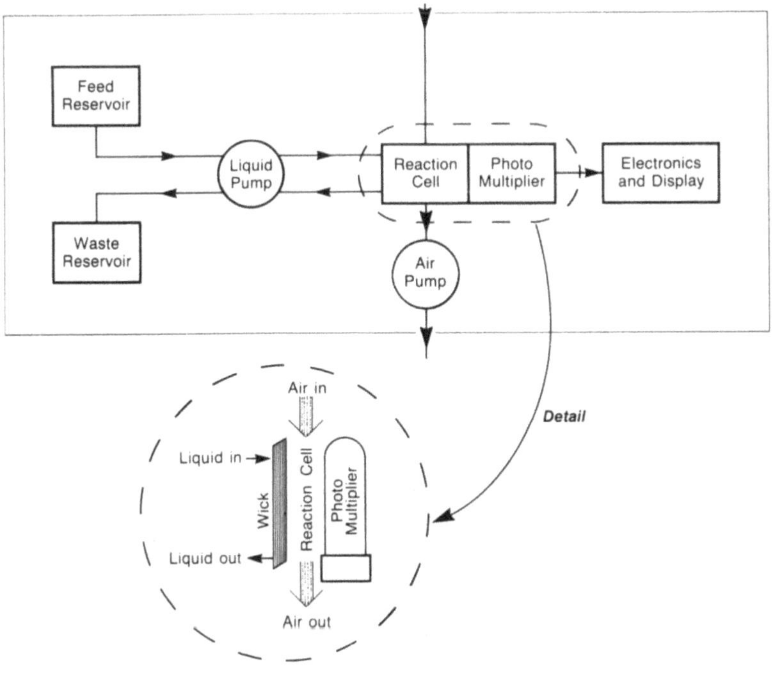

Fig. 1 Block diagram of the instrument.

The flow rate of the solution is about 0.05 mL sec^{-1}. A bottle, containing 250 mL of the solution located in the instrument provides for 80 hr of measurement. The used liquid is pumped into another 250 mL bottle, also located in the instrument. If longer operation is desired without replenishing the solution, provision is made for attaching larger reservoirs external to the instrument.

A photomultiplier views the wick and its signal provides a measure of the NO_2 mixing ratio. The signal is displayed on a 4.5 digit LCD display. The analogue signal is also presented to BNC connectors at the front and back of the instrument for connection to a chart recorder or a data acquisition device. Switches permit the LCD display to read the stand-by battery voltage and the photomultiplier voltage. One potentiometer permits zeroing the instrument and another permits adjustment to absolute NO_2 readings when the instrument is attached to a calibration sources. The instrument is calibrated by adding a known concentration of NO_2 to the air stream.

The instrument is compact and lightweight, having dimensions of 38 x 20 x 22 cm and a weight of 7 kgm. It has been designed to be very rugged and to prevent any leakage of the solution. The instrument can also operate for at least 3 hr from its rechargeable battery. When the line cord is plugged in the battery is charged automatically.

The solution contains luminol (5-amino-2, 3-dihydro-1, 4 phthalazine dione), NaOH, Na_2SO_3 and alcohol, dissolved in deionized water. The presence of Na_2SO_3 increases its response to NO_2 while decreasing its response to O_3. The addition of one of a number of alcohols was found to enhance its response to NO_2.

Studies were made on the effect of the concentration of each of these components on the response of the instrument to NO_2 and to potential interfering gases. In general, the response towards NO_2 was found to increase with concentration of each component, reach a broad maximum and then decrease. A selection was made for a formulation which appears to optimize the sensitivity towards NO_2 while minimizing its response to other gases, particularly O_3.

Possible interferences were investigated when a number of gases was added to laboratory air containing a few ppbv of NO_2. No signal changes above noise level were observed for additions of CO and CO_2 at concentrations up to 1000 ppmv and for SO_2, NH_3, H_2O_2, HNO_3, and CH_3NO_3 at concentrations below 100 ppbv. Addition of 80 ppbv of NO reduced the signal from NO_2 by about 5%. The only positive interferences encountered were from O_3 and PAN. For the luminol solution selected the response of the instrument to O_3 was 750 times less than to the same concentration of NO_2. For measurements of low levels of NO_2 in ambient air where the concentration O_3 is greater than 750 larger it may be necessary to preferentially scrub the O_3 or to make the appropriate correction.

The response of the instrument to PAN is still somewhat uncertain and appears to depend upon the nature and the concentration of the alcohol used in the solution. However, the agreement of NO_2 measurements made with this instrument and the unequivocal tunable diode laser absorption spectroscopic (TDLAS) method (Hastie et al. 1983) under polluted conditions, where the PAN concentration were expected to be high, suggest a relatively low response to PAN. Further work is being conducted to characterize this interference.

The noise level of the signal, largely due to the photomultiplier, corresponds to an NO_2 concentration of less than 5 pptv. This does not mean, however, that the detection limit is this low since we have not demonstrated that there are not interferences from other gases which can contribute to the signal at this level. The zero signal was found to remain within this degree of variance for periods of several hours.

The response time of the instrument is demonstrated in Figure 2 when 22 ppbv of NO_2 is suddenly added to or removed from the sampled air. The response time is less than 1 s for a 20% change and less than 10 s for a 99% change in NO_2 mixing ratio.

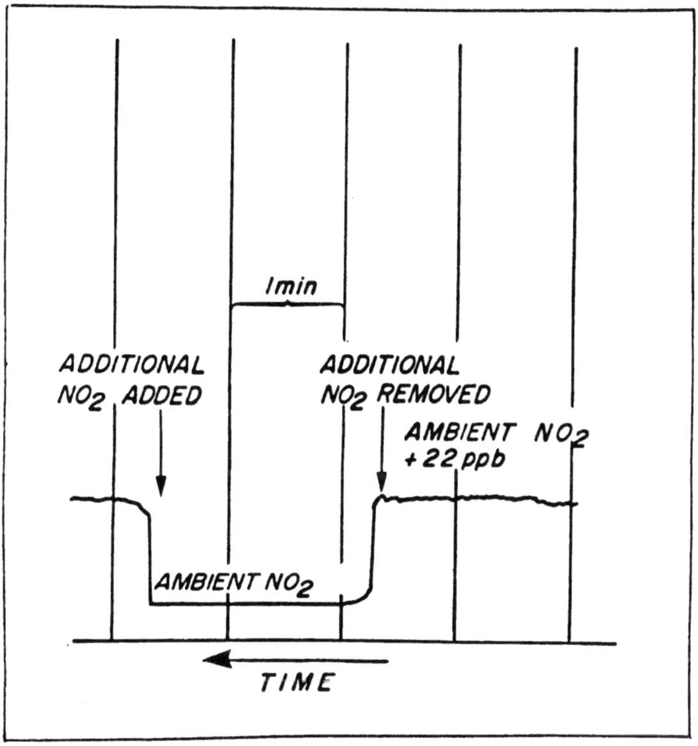

Fig. 2 Response of the instrument as a function of time.

Figure 3 shows that the response of the instrument to NO_2 mixing ratio is linear to within ±5% up to 40 ppbv. Figure 4 shows the dependence of the instrument response as a function of temperature over the range of 5 to 45°C. The response decreases linearly at a rate of about 2% per deg which is independent of the NO_2 mixing ratio. An electronic circuit is being designed to compensate for this temperature dependence.

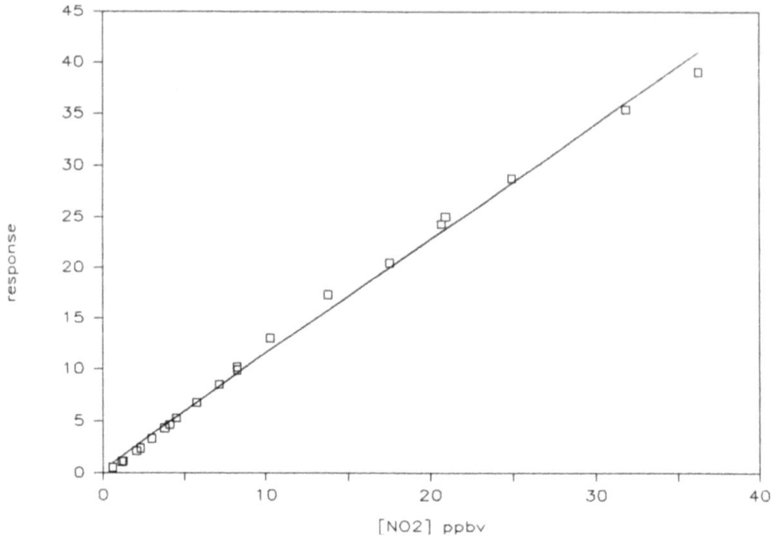

Fig. 3 Laboratory test of the response of the instrument as a function of NO_2 mixing ratio based on a permeation tube source and dynamic double dilution.

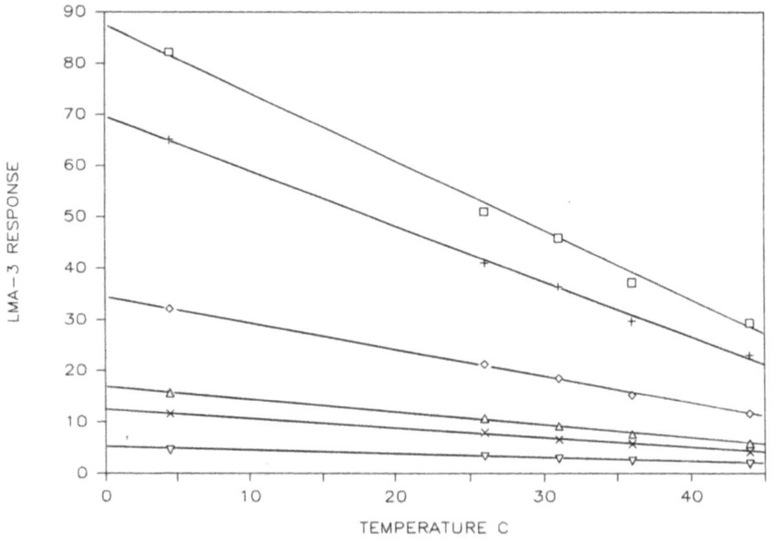

Fig. 4 Response of the instrument as a function of temperature for a number of different NO_2 mixing ratios:
(□) 52.8 ppbv (+) 41.0 ppbv (◇) 20.1 ppbv
(△) 10.1 ppbv (x) 8.0 ppbv (▽) 3.9 ppbv

4. MEASUREMENT RESULT

The instrument has been used to measure NO_2 in urban and rural air and in smog chamber experiments. On a number of occasions intercomparisons were made of simultaneous measurements made with the instrument and the tunable diode laser absorption spectroscopic (TDLAS) method (Hastie et al. 1983). Figure 5 shows an example of such a comparison made in Metropolitan Toronto which shows typical agreement within ±10% between the two methods.

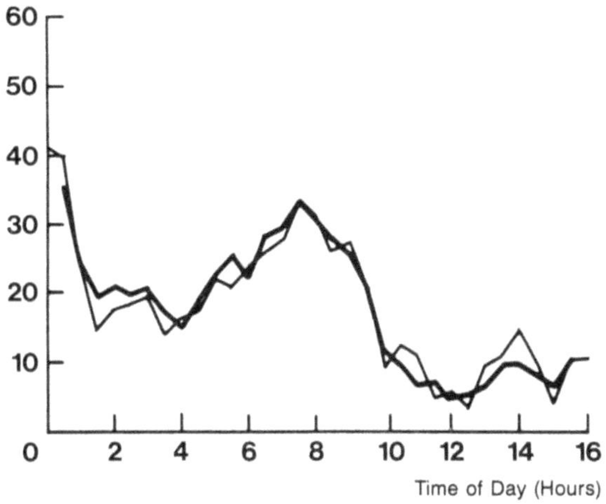

Fig. 5 Comparison between 30 min average measurements made by the luminol instrument (light line) and the TDLAS system (dark line) made in Metropolitan Toronto, June 1984.

A series of measurements were made during the summer of 1985 at Cold Creek, Ontario, about 30 km northwest of Toronto. This is a relatively clean site with prevailing winds from the S.W. and N.W. quadrants, which do not contain nearby major sources. Under prevailing wind conditions the NO_2 mixing ratios were usually less than 3 ppbv and showed characteristic diurnal behavior as exemplified in Figure 6. On the basis of supporting measurements, the NO_2 behavior at the site seems to reflect the joint influence of transport and photochemistry. After sunset the NO_2 is observed to increase, in part due to oxidation of NO by O_3. The NO_2 subsequently decreases and this decrease may be associated with the further oxidation by O_3 to yield NO_3 and N_2O_5. After sunrise the NO_2 mixing ratios increase. NO_3 and N_2O_5 are rapidly converted to NO_2 and photochemistry results in the conversion of any fresh NO arriving at the site to NO_2.

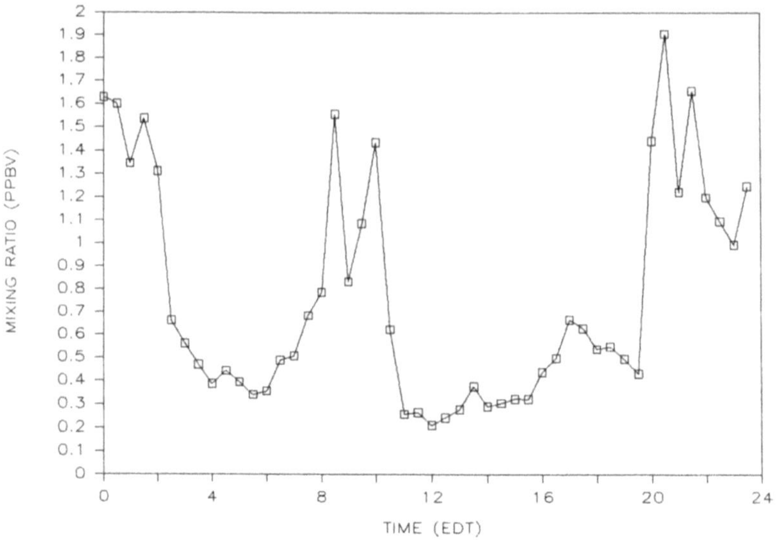

Fig. 6 30 mn average NO_2 measurements made at Cold Creek, Ontario on June 24, 1985.

Much higher concentrations were observed when the wind direction was from Metropolitan Toronto. Figure 7 shows an example of this condition where both the luminol instrument and the TDLAS system were used to make NO_2 measurements. Again the agreement between the two methods is excellent. Values as high as 80 ppbv are observed with somewhat different diurnal behavior reflecting the influence of air transport from the polluted source region superimposed upon the photochemistry.

During September, 1985 a measurement program was undertaken at Claremont, CA in relatively polluted air. Figure 8 shows hourly averaged values obtained by the two methods. Mixing ratios higher than 100 ppbv were encountered while the diurnal behaviour shows the mixed influences of photochemistry and local sources. Morning and evening maxima were observed but the evening maximum is delayed somewhat relative to the observations at Cold Creek. The agreement between the two methods is generally good except for the evening maxima on the night of September 13 and 14 when the luminol instrument gave values some 35% higher than the TDLAS measurements. On the other hand good agreement was found for the morning maxima on those and other days. This suggests that luminol instrument is also responding to some other species which may have been present in abnormally high concentrations during those polluted evening events.

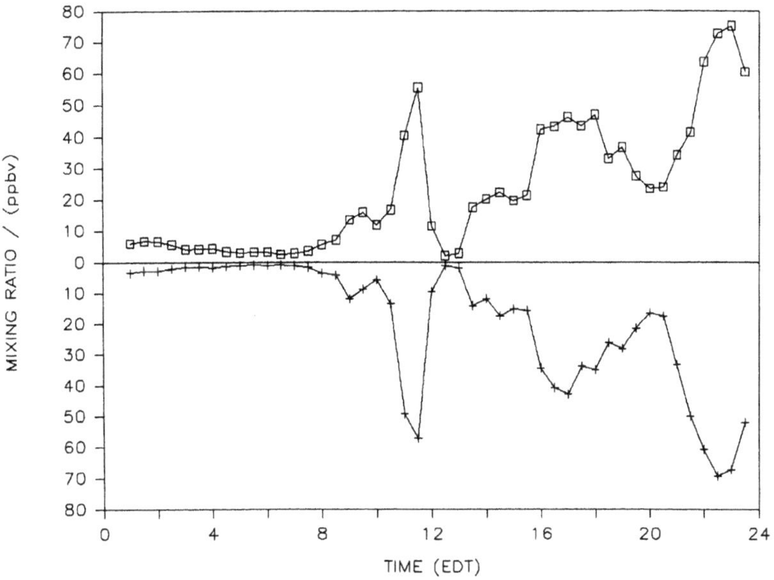

Fig. 7 Comparisons between 30 min average measurements made by the luminol instrument (+) and the TDLAS system (□) at Cold Creek, Ontario June 21, 1985.

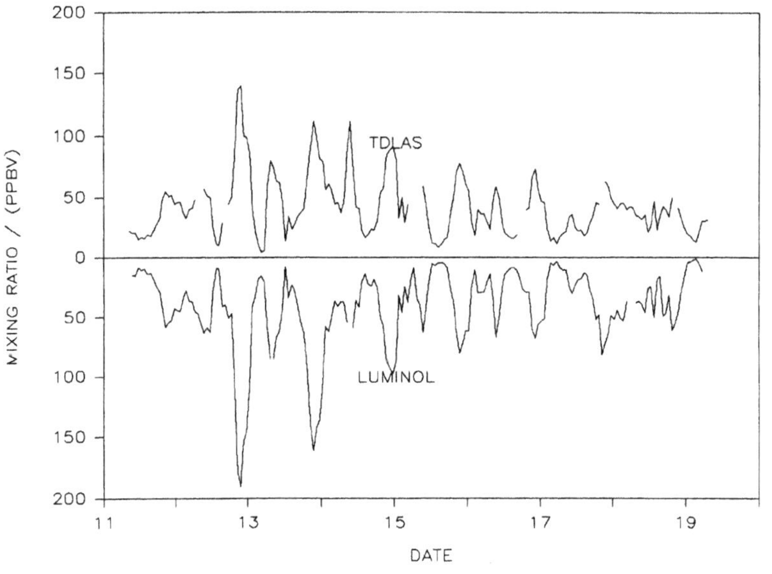

Fig. 8 Comparison between one hr average measurements made simultaneously by the luminol instrument and the TDLAS system at Claremont, CA, during September, 1985.

5. CONCLUSIONS

Chemiluminescence with a luminol solution provides a direct method for measuring NO_2 in ambient air with no necessity to convert the gas to NO as in other chemiluminescent methods. An instrument based on this method has been constructed which is lightweight, rugged, portable and is very simple to operate.

It has a sensitivity better than 5 ppbv and a response time less than 1 sec. A formulation has been found for the solution which minimizes interferences from other gases, optimizes the sensitivity and provides near-linear response to NO_2 over a considerable range.

No interferences were found for H_2O_2, NO, HNO_3, NH_3, CO, CO_2, SO_2 or organic nitro compounds at concentrations normally found in air. Interferences with O_3 have been reduced to less than 0.2% of the equivalent concentration of NO_2 by suitable adjustments in the composition of the solution. The interference from PAN is still under investigation.

The instrument response was found to have a negative temperature dependence of about 2% $°C^{-1}$ and a circuit has been installed to compensate for this effect electronically.

The instrument appears to be well suited for indoor and outdoor monitoring of NO_2 in ambient air. Its size, weight and portability makes it attractive for providing the network measurements required to evaluate models directed to the problems of acid deposition and photochemical oxidation.

Acknowledgments

The authors are grateful to the Atmospheric Environment Service of Canada and the Environment Protection Agency of the U.S.A. for financial assistance in developing the instrumental method.

References

Fishman, J., Vukovich, F.M., and Browell, E.V., 1985, J. Atmos. Chem. 3, 299.

Hastie, D.R., Mackay, G.I., Iguchi, T., Ridley, B.A., and Schiff, H.I., 1983, Environ. Sci. & Technol., 17, 352A.

Maeda, Y., Aoki, and Munemori, M., 1980, Anal. Chem., 52, 307.

Wendell, G.J., Stedman, D.H., and Cantrell, C.A., 1983, Anal. Chem. 55., 937.

RAIN, SNOW AND LAKE WATER CHEMISTRY ON AND NEAR THE PRECAMBRIAN SHIELD OF WESTERN CANADA

S.R. Shewchuk
Saskatchewan Research Council
15 Innovation Blvd.
Saskatoon, Saskatchewan, Canada
S7N 2X8

ABSTRACT. Regional surveys of rain, snow and lake water chemistry have been conducted in several locations of western Canada which are considered sensitive to acidic deposition and are within several hundred kilometres of major Canadian point source emitters (664,000 tonne SO_2 in 1981). Snow and rain in the western Precambrian Shield is acidic with pH values generally ranging between 5.0 and 6.4. However, there is little evidence to suggest that the area is receiving significant amounts of acidic deposition. Total deposition of sulphate is approximately 2 kg ha^{-1} between the months of April to October in 1983. Environmental impacts at present are minimal with most small lakes in the region having pH values near nuetrality and low concentrations of metals such as Al.

1. INTRODUCTION

At several locations in western Canada the Precambrian Shield is sensitive to acidic deposition (Shewchuk _et al._, 1981). There are significant point source emitters of SO_x and NO_x within and near this area. An emissions inventory for these sources is given in Still Waters (1981). Total emissions for 1981 were 664,000 tonne SO_2.

Preliminary studies of S levels within the atmosphere near the Precambrian Shield in Saskatchewan were initiated at the Saskatchewan Research Council (SRC) in 1979 (Shewchuk, 1979). The study site selected was La Loche, (Figure 1). Aerosol and gaseous S were studied with dual filter pack procedures outlined by Lusis _et al._ (1977). Daily average gaseous S concentrations were in the range of 0.11 to 1.00 $\mu g(S)\ m^{-3}$. After this study a major baseline review of sensitivities to acidic deposition aquatic, soil and surficial geological systems was conducted for Alberta and Saskatchewan (Shewchuk _et al._, 1981 and Shewchuk, 1982a).

A series of projects were subsequently undertaken to measure sulphate, nitrate and H^+ concentrations in the snow, rain and lake waters of selected areas of western Canada. Baseline results for snow, rain and lake water chemistry are available in several SRC reports (Shewchuk,

Water, Air and Soil Pollution **30** (1986) 115–123.

1982b, 1983a, 1984, 1985). Since 1981, event precipitation collection has been conducted within the Saskatchewan Precambrian Shield area at Southend, Saskatchewan (Shewchuk, 1982c). A regional network consisting of six stations was set up by Saskatchewan Environment in 1982. The 1982 and 1983 results from this network are reported in Shewchuk (1985).

A large area of the bedrock of the Northwest Territories (NWT) also consists of Precambrian Shield, so a review of acidic deposition sensitivities of lakewater and terrestrial systems was conducted in 1983 (Shewchuk, 1983b). This was supplemented by a field project on the monitoring of snow and lake water quality within the District of MacKenzie (Shewchuk, 1984). The purpose of this report is to identify current levels of acid deposition to the snowpack. The chemistry of the lake water was also monitored at this time.

2. PROJECTS SUMMARY

Figure 1 is a regional map of the area of western Canada where SRC has conducted environment studies. Only the sites that are named are considered for this presentation. Data are available for all sites shown in this Figure, but the original reports as outlined by Shewchuk in this paper must be consulted for further information. In all of

Table I. **Chemistry of snow cores from lakes from western Canada. Saskatchewan, Alberta and Manitoba data were collected in 1983. Northwest Territories data were collected in 1984. Cores represent approximately 100 days of accumulation.**

Lake	pH	Sulphate SO_4 (mg L^{-1})	Nitrate NO_3 (mg L^{-1})	Conductivity (μmho cm^{-1})	Ca (mg L^{-1})
a) Precambrian Shield Lakes					
Wollaston, SK	5.0	0.51	0.34	5.3	0.06
Cairns, SK	5.1	0.73	0.65	7.7	0.22
Haultain, SK	5.1	0.47	0.61	6.4	0.20
Reeve, SK	5.0	0.51	0.43	6.0	0.28
Forrest, SK	5.3	0.89	0.65	6.5	0.24
Sid, NWT	5.0	0.72	0.43	4.0	0.11
Snowbird, NWT	5.2	0.34	0.34	6.0	0.06
Lady Grey, NWT	6.4	0.21	0.30	4.0	0.09
b) Fringe or off Shield Lakes					
Anglin, SK	6.0	1.67	1.04	9.7	0.29
Athapapuskow, MN	5.6	0.80	0.87	5.5	0.26
Gordon, AB	5.5	0.78	0.87	41.3	2.10

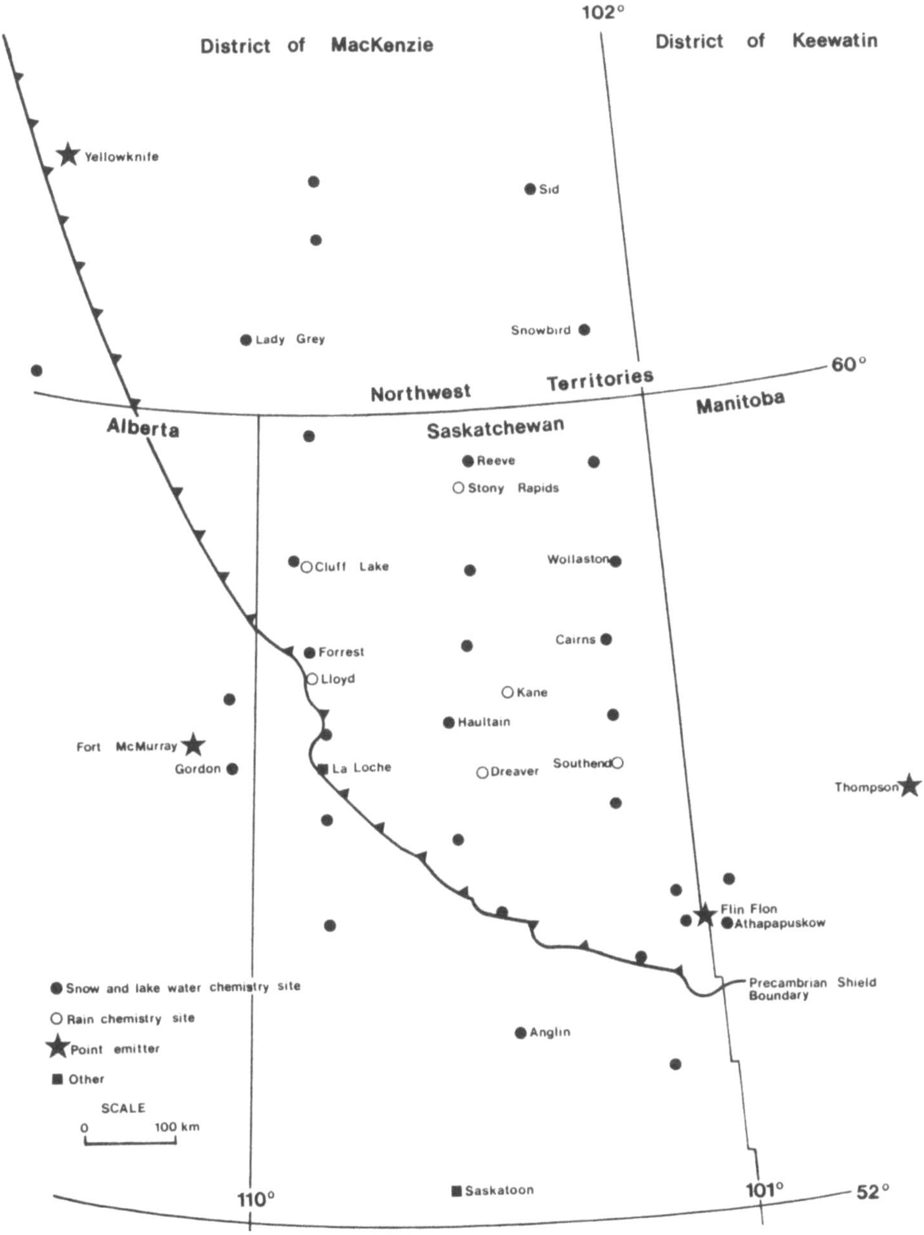

Figure 1. Locations within Western Canada where major baseline studies on acid deposition impacts have been conducted by the Saskatchewan Research Council.

Table II. Deposition of sulphate (SO_4) and nitrate (NO_3) to the snowpack at several locations during the winter season (ca. 100 days). All samples collected in late March 1983.

Location	Equivalent Water Depth (cm)	Sulphate deposition (kg ha^{-1})	Nitrate deposition (kg ha^{-1})
a) on Shield			
Wollaston, SK	6.9	0.35	0.26
Cairns, SK	4.9	0.36	0.30
Haultain, SK	10.4	0.49	0.09
Reeve, SK	5.1	0.26	0.22
Forrest, SK	5.2	0.46	0.35
Sid, NWT	8.4	0.61	0.35
Snowbird, NWT	5.3	0.18	0.17
Lady Grey, NWT	6.6	0.14	0.57
b) off or fringe Shield			
Anglin, SK	6.0	1.00	0.61
Athapapuskow, MB	4.5	0.36	0.39
Gordon, AB	5.1	0.40	0.43

this work the concentrations of sulphate, nitrate, H^+ and some metals in lakewaters and precipitation were the primary focus of attention.

This area of Canada is ideal for conducting snow chemistry studies. The mean daily temperature drops below freezing in mid to late November, and does not thaw until mid to late March. Thus there is a period of approximately one hundred days during which little melting of a snowpack can occur so that pollutants deposited on the snowpack remain trapped until the spring melt (Colbeck, 1981). Table I is a summary of the quality of the snow collected from several of the lake surface sites. All snow cores were collected with an acrylic tube and kept frozen until analysis. Snow cores were analyzed at SRC laboratories with appropriate quality control procedures for dilute solutions. Snow samples and field bottles were precleaned before each field trip. Sulphates were determined by ion chromatography, nitrates were determined by Cd reduction and pH values were determined by sensitive electrodes.

The deposition summary of sulphate and nitrate to the snowpack at several sites is shown in Table II. Deposition values are quite similar both for off-Shield and on-Shield locations. For the approximate 100 day period, sulphate deposition totaled 1.00 to 0.14 kg ha^{-1}.

Event rain only chemistry has been studied in northern Saskatchewan at six sites (see Figure 1). Results for pH and sulphate are seen

Table III. Summary of event rain chemistry data from the Saskatchewan Environment network of 1983, for sulphate. The precipitation volume weighted pH and concentration is also given. In collection period was from April 1 to October 31 (214) days for liquid precipitation only.

Site	Precipitation volume weighted pH	Precipitation volume weighted means for SO_4 (mg L^{-1})	Sulphate Deposition (kg ha^{-1})
Kane Lake	5.2	0.49	1.5
Dreaver Lake	5.3	0.63	1.9
Cluff Lake	5.2	0.47	1.2
Lloyd Lake	5.7	0.88	2.4

Table IV. Summary of event rain chemistry data from the Saskatchewan Environment network of 1983, for nitrate. The precipitation volume weighted mean concentrations and wet April 1 to October 31 (214 days) for liquid precipitation only.

Site	Precipitation weighted means for nitrate NO^3 (mg L^{-1})	Nitrate Deposition (kg ha^{-1})
Kane Lake	0.31	0.89
Dreaver Lake	0.49	1.51
Cluff Lake	0.18	0.44
Lloyd Lake	0.53	1.37

in Table III. Concentrations reported have been volume weighted from all rain chemistry determined from event samples. The pH values were determined by back calculations from H^+. Samples were frozen in the field and melted prior to analysis at the laboratory. Volume weighted concentrations and depositions of NO_3 are shown in Table IV. The wet deposition of nitrate for the Precambrian Shield area ranged from 0.44 to 1.51 kg NO_3 ha^{-1} for rain for the period April to October 1983.

Table V. Summary of water chemistry for small lakes in western Canada.

Nearest large lake	pH	Sulphate (mg L^{-1})	Alkalinity as $CaCO_3$ (mg L^{-1})	Conductivity (µmho cm^{-1})	Ca (mg L^{-1})
a) Precambrian Shield Lakes					
Wollaston, SK	7.3	2.5	10	28	2.5
Cairns, SK	6.9	2.9	12	39	3.6
Haultain, SK	7.6	1.1	27	56	5.7
Reeve, SK	7.4	2.6	14	37	4.9
Forrest, SK	6.6	1.3	7	19	2.1
Sid, NWT	7.7	1.4	7	19	2.5
Snowbird, NWT	7.5	7.5	18	49	8.8
b) Fringe or off Shield Lakes					
Anglin, SK	8.0	46.0	309	610	68
Athapapuskow, MN	8.2	10.0	262	482	58
Gordon, AB	8.0	0.8	115	213	34
Lady Grey, NWT	7.6	1.0	124	320	59

Water chemistry of small lakes in several of the sites within western Canada is shown in Table V. Names of the nearest major lake systems are given in this table. Lakewater was collected from one metre level below the ice surface with a standard Kemmerer bottle (Boyd, 1979). Most of the lakes showed Al concentrations consistently below 0.05 mg L^{-1}. Additional results are in Shewchuk (1983a, 1984). The pH, conductivity and alkalinity of the Precambrian Shield lakes indicated that they were significantly more dilute than off-Shield lakes. According to criteria for pH, conductivity and alkalinity given by Altshuller and McBean (1979), the Precambrian Shield lakes in this area of Canada are highly sensitive to acid deposition. Since pH values were in the range 6.6 to 7.6 few acidification impacts are evident (or suspected).

3. DISCUSSION

The Precambrian Shield in western Canada has been identified as being sensitive to acid deposition. Since in this area of Saskatchewan and the Northwest Territories there is a low and reasonably uniform deposition of sulphate and nitrate, this suggests that natural deposition processes are dominating the precipitation chemistry. The deposition of sulphate onto the snowpack of Norway has been studied by

Johannessen *et al.*, (1977). The reported sulphate deposition to the snowpack ranged from 1 to 6 kg ha^{-1}. Remote areas inland had deposition values less than 1 kg ha^{-1}. This latter value was considered background. Consequently, I contend that sulphate deposition in this area of Canada may be considered as representative of background values.

Wet deposition of sulphate from rain for this area was in the order of 2 kg ha^{-1}. The rain was quite dilute and acidic. It is thought that the pH values of natural rainwater is controlled by the equilibrium with atmospheric CO_2 at a pH value of 5.6 (McQuaker *et al.*, 1983). There are many reasons for the decrease of pH below this value due to natural sources such as, natural organic acids (Galloway *et al.*, 1982). A preliminary study by Iribarne and Shewchuk, (1985) has shown that organic acids are, in part, a cause for the decrease in pH for snow from this area of Canada.

The Precambrian Shield has major freshwater lake systems that are highly sensitive to acid deposition based on lack of buffering capacity. There is little evidence at present to show that significant acidification has occurred. This is clearly seen in the alkalinity and pH data. The pH and metals observed in most Precambrian Shield lakes we studied also support this observation. The area of western Canada as outlined in this presentation has point source emitters of SO_x and NO_x. These sources likely do contribute to the acid deposition of this area. However, precipitation chemistry is so dilute that deposition to the environment of acid anions is significantly less that in areas reporting impacts (Bangay and Riordan, 1983). In addition, the ion chemistry of the dilute Precambrian Shield lakes in this region indicates little acidification has taken place.

ACKNOWLEDGMENTS

Much of this research work was undertaken through contracts received form Alberta Environment, Environment Canada, Government of the Northwest Territories and Saskatchewan Environment. I gratefully acknowledge the support of all of these agencies, particularly Saskatchewan Environment, in the programs of the Environment Division of the Saskatchewan Research Council. All of the chemistry analysis reported in this paper were conducted by Dr. W. Yuen of the Saskatchewan Research Council's Analytical Chemistry Laboratories.

REFERENCES

Altshuller, A.P. and McBean, G.A.: 1979, The LRTAP problem in North America, a preliminary overview. United States-Canada Research Consultation Group on the Long-Reange Transport of Air Pollutants. Atmospheric Environment Service, Downsview, Ontario.

Bangay, G.E. and Riordan, C.: 1983, Memorandum of Intent on Transboundary Air Pollution, Impact Assessment Work Group I, Final Report.

Boyd, C.E.: 1979, Water quality in warm water fish ponds. Auburn University. Craft Master Printers, Opelika, Alabama.

Collbeck, S.C.: 1981, Water Resources Research 17, 1383.

Galloway, J.N., Likens, G.E. and Miller, J.M.: 1982, J. Geoph. Res. 87, 8771.

Iribarne, J.V. and Shewchuk, S.R.: 1985, On the origin of strong acidity in the snowpack of northern Saskatchewan, SRC Technical Report No. E-906-39-B-85.

Johannessen, M., Dale, T., Gjessing, E.T., Henriksen, A. And Wright, R.F.: 1977, Acid Precipitation in Norway: The Regional Distribution of Contaminants in Snow and the Chemical Concentration Processes During Snowmelt. Proc. Int. Symp. Isotopes and Impurities in Snow and Ice, Int. Assoc. Hydrol. Sci. Grenoble, France, 28-30 August, 1975. IAHS-AISH Publication No. 118.

Lusis, M.A., Barrie, L.A., Wiebe, H.A. and Anlauf, K.G.: 1977, Development and testing of a filter pack for simultaneous sampling of gaseous and particulate sulphur compounds in the Atmosphere. Report ARQA, Atmos. Chemistry, Criteria and Standards Div., Atmos. Environ. Service, Downsview, Ontario.

McQuaker, N.R., Kluckner, P.D. and Sandberg, D.K.: 1983, Environ. Sci. Technol., 17, 431.

Shewchuk, S.R.: 1979, A Background study of Potential Sulphur Stress in West Central Saskatchewan. SRC Technical Report P79-6.

Shewchuk, S.R., Abouguendia, Z.M., Atton, F.M., Dublin, J., Godwin, R.C., Holowaychuk, N., Hopkinson, R., Liaw, W.K., Maybank, J., Padbury, G.A. and Schreiner, B.T.: 1981, Transport of Acid Forming Emissions and Potential Effects of Deposition in Northeastern Alberta and Northern Saskatchewan: A problem Analysis. SRC Technical Report No. 122.

Shewchuk, S.R.: 1982a, Water, Air and Soil Pollut 18, 413.

Shewchuk, S.R.: 1982b, A snowpack chemistry study of small lakes in Northern Saskatchewan. SRC Technical Report No. 137.

Shewchuk, S.R.: 1982c, A Precipitation Study by Event Collection from Reindeer Lake, Saskatchewan. SRC Technical Report No. 137.

Shewchuk, S.R.: 1983a, A snowpack and surface water chemistry survey of small lakes in Northern Saskatchewan. SRC Technical Report No. 155.

Shewchuk, S.R.: 1983b, An Acid Deposition Perspective for the Northwest Territories. Dept. Of Information, Government of the Northwest Territories. ISBN 0-7708-8116.

Shewchuk, S.R.: 1984, A Snowpack and Water Quality Survey of Several Small Lakes in the Southwest Precambrian Shield area of the Northwest Territories: An Acid Deposition Approach. SRC Technical Report No. 171.

Shewchuk, S.R.: 1985, An Air Mass Back Trajectory and Precipitation Chemistry Analysis with Northern Saskatchewan, SRC Technical Report No. 173.

Still Waters: 1981, the chilling reality of acid rain. Parliamentary Committee, Minister of Supply and Services Canada ISBN 0-662-11723-9 pp 150.

ACIDIC PRECIPITATION IN WESTERN NORTH AMERICA: TRENDS, SOURCES, AND ALTITUDE EFFECTS IN NEW MEXICO 1979-1985

Carl J. Popp
Donald K. Brandvold
Department of Chemistry
New Mexico Institute of Mining & Technology
Socorro, New Mexico 87801 USA

Austin Long
Lisa Warneke
Laboratory of Isotope Geochronology
Department of Geoscience
University of Arizona
Tucson, Arizona 85721 USA

ABSTRACT. Volume-weighted pH values in central New Mexico have averaged 3.8 to 5.1 during the period 1979-1985. Samples collected at a high altitude site (3200m) have lower pH values than found for low altitude samples (1400m). Both dry deposition and event-averaged pH values have been higher than the volume-weighted averages due to neutralization by terrestrial material. During the period 1980-1984, changes in pH values and wet sulfate loading have correlated to SO_2 emissions from regional non-ferrous smelters. Sulfur isotope analyses of sulfate extracted from regional rain samples yielded a $\delta^{34}S_{CD}(^{o}/oo)$ of $+3.91 \pm 1.1$,

indicating very little regional differentiation which, in turn, suggests that the regional atmospheric sulfate scavenged by precipitation is well-mixed and relatively homogeneous.

INTRODUCTION.

Determination of the presence of acidic precipitation in the Rocky Mountain (Lewis and Grant 1981, Grant and Lewis 1982; Huebert et al. 1982) and southwestern desert regions (Popp et al. 1982, 1984) of the western United States has only recently received attention in the acid rain literature. Because of the low population density, lack of industrialization and urbanization, and alkaline nature of much of the soil, it has generally been assumed that acidic precipitation, even if present, would probably not have deleterious effects. However, the volume-weighted pH values often average less than 5 and, therefore, may be cause of concern, specifically at high altitudes.

Studies of the sources of regional haze using the Fine Particle Collection Network have suggested that Cu smelters in Arizona, New

Water, Air and Soil Pollution **30** (1986) 125–133.

Mexico, and Mexico may be major contributors to visibility problems in the Rocky Mountain West and may also, therefore, contribute acid precursors in large quantities (Ashbaugh et al. 1984). Evidence supporting a linear response of acidic deposition to smelter emissions in the western United States has recently been presented by Oppenheimer et al. (1985), who linked wet sulfate loading at National Acid Deposition Program stations in Utah, Colorado, and Arizona to non-ferrous smelter SO_2 emissions. Lewis et al. 1984 using a bulk atmospheric deposition network in Colorado found precipitation acidity gradients related to, 1) power plant and urban area locations and prevailing winds, 2) differential transport of acids and alkaline, and 3) amount of wet precipitation.

In this study, we summarize pH, sulfate, and nitrate loading using precipitation data from two sites in central New Mexico from 1979 to 1985. The proximity of the sites coupled with the large altitude difference provides an opportunity to determine the relative differences between acidic deposition at high mountain and desert sites. In addition, we wish to report stable isotope S analysis results obtained from sulfate extracted from regional precipitation samples and from coal samples from the Four Corners region of New Mexico.

2. METHODS AND MATERIALS

2.1 Study Areas

The Langmuir Laboratory for Atmospheric Research at 3200 m altitude is in a spruce-fir climate while Socorro, NM at 1400 m in altitude is in an Upper Sonoron life zone and provides an opportunity to determine the effects of altitude (high desert vs. mountain) at sites which are geographically close (27 air km apart) (Popp et al. 1982, 1984).

2.2 Sample Collection and Handling

The large, 1 m diameter funnels used to allow collection of large volumes of samples from the relatively infrequent and small precipitation events have been previously described (Popp et al. 1982). Precipitation samples were collected on an event basis and the funnels were either covered between events to exclude dry deposition or were rinsed with distilled water immediately prior to an event. The samples were stabilized shortly after collection by filtration through 0.2 μm membrane filters and stored in HNO_3/water rinsed polyethylene or polypropylene bottles.

Dry deposition was collected by rinsing the funnel with 100 to 500 mL of distilled water after 7 to 10 days with no precipitation. The samples were allowed to stand for 15 min and then filtered to stabilize, as above.

2.3 Analysis Procedures

Analytical procedures were verified with U.S. Environmental Protection Agency (USEPA) standards as controls. Approximately 10% of the samples were run as duplicates and spikes and blanks were routinely analyzed. The pH values were determined using a glass electrode and pH meter calibrated with two buffers (pH 4 and 7) and two "artificial rain" solutions which were made with dilute nitric acid and titrated accurately at pH values approximately 3 to 4.

Nitrate and sulfate were analyzed by ion chromatography using a Wescan 2 mL sample loop, anion column (Wescan 269-101), and conductivity detector (Wescan 213A) coupled with a Beckman Model 110A high performance liquid chromatograph (Buchholz et al., 1982). Early in the study, nitrate and sulfate were determined using USEPA methods 353.3 (NO_3^-) and 375.4 (SO_4^{2-}) respectively (USEPA, 1979).

2.4 Sulfur Isotope Analysis

Sulfate was extracted from precipitation by ion exchange using Amberlite CG-400 anion exchange resin. The ion exchange columns were backflushed with 0.3 molar sodium chloride and sulfate was precipitated as barium sulfate using 10% barium chloride. Coal samples were digested in HNO_3 and H_2O_2 and the sulfate was precipitated with 10% barium chloride. The barium sulfate samples were treated with dilute HCl to remove traces of carbonate before conversion to SO_2 used in the mass spectrometric determination.

Sulfur dioxide for mass spectrometric analysis was obtained by heating a mixture of 25 mg $BaSO_4$, 200 mg SiO_2, and 200 mg Cu_2O in a 6 mm O.D. Vycor boat at 1100°C for 15 min under vacuum. The SO_2 was separated from H_2O and CO_2 cryogenically and analyzed in a V.G. Micromass 602-D mass-spectrometer. Laboratory accuracy and precision is +/- 0.25 per mil and the values are referenced to Canon Diablo troilite.

3. RESULTS AND DISCUSSION

3.1 Hydrogen Ion, Sulfate, and Nitrate Trends and Loading

The volume-weighted (VW) and event-averaged (EA) pH values are summarized in Table I. The VW pH values were always lower than the EAs, except for the 1981 samples from Langmuir Laboratory. The VW

Table I. Volume and event averaged pH values at the Socorro and Langmuir sites and volume-weighted sulfate and nitrate concentrations (mgL^{-1}) at the Socorro site. More than 90% of the events were sampled.

	Socorro						Langmuir[a]			
		Vol.Wt.	Vol.Wt.	Vol.Wt.	Event			Vol.Wt.	Event	
Year	Vol (mm)	pH	SO_4^{2-}	NO_3^-	pH	Events	Vol (mm)	pH	pH	Events
1979[c]	76	4.7	2.0	---	5.0	8	71	5.1	5.4	7
1980	236	4.8	1.0	0.2	5.2	21	196	4.6	5.0	10
1981	113	4.1	4.9	1.7	5.0	30	72	4.2	4.1	16
1982	153	4.7	3.1	2.6	5.3 (6.2)[b]	32	74	4.5	4.7	7
1983	161	4.1	3.9	3.0	4.9	41	---	---	---	---
1984	201	4.4	1.7	1.6	5.3	42	214	3.8	4.1	13
1985[d]	65	4.7	1.8	---	5.5 (6.6)[b]	21	112	4.4	4.9 (6.0)[b]	24

[a]June - Sept. samples ~60% of yearly precipitation

[b]Values in parentheses are for dry deposition

[c]Data July through December, 1979

[d]Data through July, 1985

pH values averaged 0.7 pH units lower than the EA pH values at the Socorro (lower altitude) site and 0.2 pH units lower at the Langmuir (mountain) site. These data clearly show the neutralizing effects of the presence of alkaline terrestrial material having a greater influence at the desert site (Popp et al. 1984). The general pH trends at Socorro and Langmuir mimic each other from one year to the next and the values are generally higher at the valley site.

The pH values for dry deposition in contact with distilled water (Table I) are considerably higher than either the volume-weighted or event-averaged pH's, again indicating the neutralizing affect of local soil material (Popp et al. 1982, 1984, Lewis et al. 1984). Also, the pH of dissolved dry deposition for 1985 at the valley site (6.6) is greater than that determined at the mountain site (6.0) for the same period.

The volume weighted mean (VWM) sulfate and nitrate concentrations in precipitation from the Socorro sample site from 1979-1985 are shown in Table I. Sulfur dioxide emissions from non-ferrous smelters in Arizona and New Mexico are shown in Table II. These data may be more applicable to our sampling site than using the four state regional data compiled by Oppenheimer et al. (1985) because of the proximity of the site location in Central New Mexico to the smelters and also because the Arizona and New Mexico non-ferrous smelter sulfur dioxide emissions constitute a majority of such total regional

Table II. SO_2 Emissions (metric tonne yr^{-1} x 10^{-3}) from Non-ferrous Smelters in Arizona and New Mexico (Oppenheimer et al., 1985). Data for 1984 was obtained from Niemann (1985) and the New Mexico Air Quality Bureau (1985).

State	1980	1981	1982	1983	1984
Arizona	579	794	374	446	560
New Mexico	122	145	129	181	100
Total	701	939	503	627	660

emissions. The pH data for 1979 are for the last six months of 1979 and for 1985 are for the first six months and do not represent data from an entire year, so trends are examined only from 1980-84. As shown in Figure 1, the pH values are inversely proportional to both SO_2 emissions and wet sulfate loading for the 1980-84 period and support the correspondence proposed by Oppenheimer, at least with respect to SO_2 emissions and wet sulfate deposition. However, nitrate loading varied considerably during this period (although not in proportion to sulfate loading) and must certainly be a factor in the variation of pH. The sample site at Socorro is the only collection site during this study period which was located in New Mexico and in proximity to the path of Arizona and New Mexico SO_2 emissions proposed by Ashbaugh et al. (1984) and should be the most likely to show relationships with SO_2 emissions.

In order to account for variations in yearly precipitation amount, wet sulfate loading at Socorro is also shown in Figure 1. These data generally agree with sulfate loading reported in Colorado by Lewis and Grant (1982). During this period (1980-84) the Socorro site received its greatest rainfall (236 mm) in 1980 and its least (113 mm) in 1981, yet much more wet sulfate loading occurred in 1981. The data from 1980-1983 rise and fall with non-ferrous SO_2 emissions shown in Table II. In 1984, the Arizona SO_2 smelter emissions went up but the New Mexico emissions fell (Table II) for a slightly greater 1984 total than occurred in 1983. Yet, the volume weighted pH was lower and sulfate and nitrate values were higher in 1983 than in 1984, so it is clear that the SO_2 emissions alone do not tell the entire story.

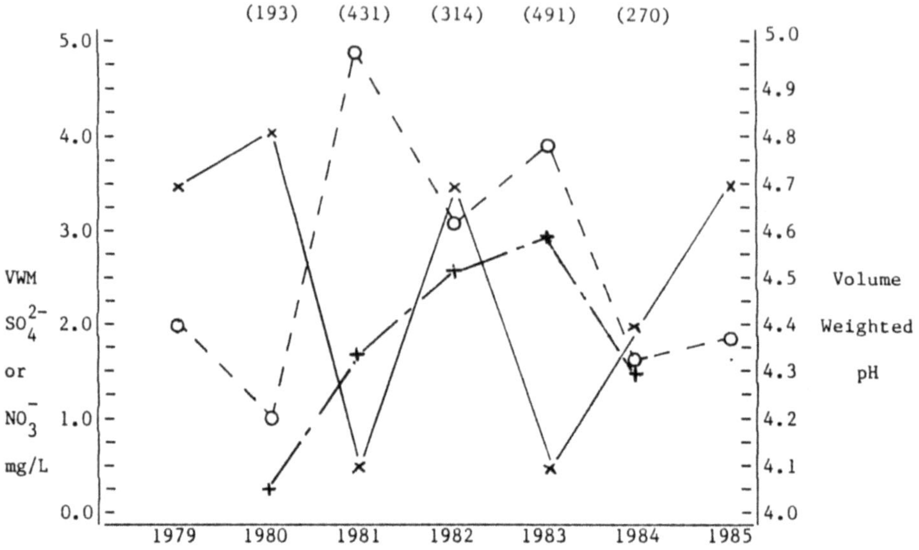

Figure 1. VWM sulfate and nitrate concentrations and pH values at Socorro: 1979-1985. Wet sulfate deposition in mg m^{-2} is shown in parentheses. O = sulfate ✕ = pH + = nitrate

3.2 Sulfur Isotopes

Sulfur isotope values ($\delta^{34}S$ $^o/oo$) for sulfate extracted from precipitation in 1984 and 1985 and for selected Four Corners coal samples are summarized in Table III. The values for precipitation sulfate show little variation while

Table III. $\delta^{34}S_{CD}$ ($^o/oo$) for Precipitation and Coal-Derived Sulfate in New Mexico

Site	$\delta^{34}S(^o/oo)$
Socorro	+2.9 ± 1.0 (11 values)
Langmuir Lab	+3.9 ± 1.2 (6 values)
Chama	+4.8 (1 value)
Coal (Four Corners)	-16.7,+2.7,+8.3,+1.8,-0.4,-12.7
Sulfide Ores	~-5 to +5
Evaporites (Local)	~+20

the coal samples vary widely. In addition, porphyry Cu deposits in Utah (Nielson, 1974) and Arizona (Esstoc, 1986) exhibit ranges in $\delta^{34}S$ ($^{o}/oo$) from approximately +5 $^{o}/oo$ to -5 $^{o}/oo$. Nielson (1974) reported ^{34}S values of -2 to +5 $^{o}/oo$ in rain in the Salt Lake City area which he attributed to Cu smelter emissions. Because of the greater distance of the New Mexico sampling sites from possible point sources (as compared to Salt Lake City and Bingham, Utah) a greater amount of atmospheric mixing is probably occurring leading to a more homogeneous distribution of $\delta^{34}S$ values in the precipitation.

The contribution of terrestrial sulfate to the total wet sulfate loading is difficult to determine experimentally. Much of the surface sulfate in New Mexico occurs in evaporite deposits in which $\delta^{34}S$ values have been found to range from +9 to +27 $^{o}/oo$ (Hoefs, 1980); so if a large amount of sulfate in precipitation were derived from entrained terrestrial material, the $\delta^{34}S$ should be much higher than actually found (Table III). These studies of regional sulfur isotope ratios in precipitation, dry deposition and SO_2 gas are continuing.

4. CONCLUSIONS

Changes in pH and wet sulfate loading at the Socorro, New Mexico sample site have varied inversely from 1980-1984 suggesting that H_2SO_4 is a primary acid responsible for acidic precipitation in the region. Wet nitrate loading also varied during the study period, but did not give as good a correspondence with pH as sulfate did. The pH values are generally lower at high altitude sites than at low altitude sites and dry deposition is considerably more basic than wet, mitigating the acidity in wet deposition.

Preliminary S isotope data suggest a reasonably homogeneous sulfate component in precipitation which could be derived from regional, but distant point sources such as Cu smelters and coal burning power plants. The relatively low $\delta^{34}S$ for precipitation-derived sulfate also suggests that entrainment of evaporite sulfates does not contribute large quantities to the total wet sulfate deposition.

ACKNOWLEDGMENTS

The authors wish to thank Mr. Charles Moore, Director of the Langmuir Laboratory for allowing us access to the mountain laboratory facilities. In addition, Frank Campbell of NMBMMR provided us with coal samples from the Four Corners region for analysis. The S isotope studies were partially supported by the New Mexico State Mining and Mineral Resources Research Institute under Project #9C-EF-10-9.

REFERENCES

Ashbaugh, L.L., Myrup, L.O., and Flocchini, R.G.: 1984, 'A Principal Component Analysis of Sulfur Concentrations in the Western United States', Atmos. Environ., **18**, 783-791.

Buchholz, A.E., Verplough, C.I., and Smith J.L.: 1982, 'A Method for the Simultaneous Measurement of Less Than a Part-Per-Million of Chloride, Nitrate, and Sulfate in Aqueous Samples by Nonsuppressed Ion Chromatography', J. Chromatogr. Sci., **20**, 449-501.

Esstoc, C.J.: personal communication.

Grant, M.C. and Lewis, Jr., W.M.: 1982, 'Chemical Loading Rates from Precipitation in the Colorado Rockies', Tellus, **34**, 74.

Hoefs, J.: 1980, Stable Isotope Geochemistry, 2nd Ed., Springer-Verlag, NY, p. 44.

Huebert, B.J., Norton, R.B., Bollinger, M.J., Parrish, D.D., Hahn C., Bush, Y.A., Murphy, P.C., Fehsenfeld, F.C., and Albritton, D.L.: 1982, 'Gas Phase and Precipitation Acidities in the Colorado Mountains', In Proc. Int. Sympo.on Hydrometeorology, American Water Resource Association, 17.

Lewis, Jr., W.M. Grant, M.C., and Saunders, III, J.F.: 1984, 'Chemical Patterns of Bulk Atmospheric Deposition in the State of Colorado', Water Resource Res., **20**, 1691.

Lewis, Jr., W. M. and Grant M.C.:1981, 'Acid Precipitation in the Western United States', Science, **207**, 176.

New Mexico Air Quality Bureau: 1985, Department of Health and Environment, Santa Fe, New Mexico.

Nielson, H.: 1974, 'Isotope Composition of the Major Contributors to Atmospheric Sulfur', Tellus, **26**, 213.

Niemann, B.L.: 1985, 'Critical Analysis of Regional Acid Deposition Source-Receptor Relationships for Smelters in the Intermountain West'. Presented at the International Conference on Acidic Precipitation, Muskoka, Ontario.

Oppenheimer, M., Epstein, C.B., and Yuhnke, R.E.: 1985, 'Acid Deposition, Smelter Emissions, and the Linearity Issue in the Western United States', Science, **229**, 859-862.

Popp, C.J., Jensen, C.M., Brandvold, D.K. and Brandvold, L.A.: 1982, 'Precipitation Analysis in Central New Mexico', In Energy and the Environment, II, Ann Arbor Science Press, Ann Arbor, Michigan.

Popp, C.J., Brandvold, D.C., Ohline, R.W., and Brandvold, L.A.: 1984, 'Nature of Precipitation and Atmospheric Particulates in Central and Northern New Mexico', In Deposition, Both Wet and Dry, Ann Arbor Science Press, Ann Arbor, Michigan.

USEPA: 1979, 'Methods for the Analyses of Water and Wastes', EPA-600014-79-020, U.S. EPA, Cincinnati, OH.

PATTERNS OF ACID DEPOSITION TO A DANISH SPRUCE FOREST

N. E. v. Freiesleben, C. Ridder and L. Rasmussen
Laboratory of Environmental Sciences and Ecology
Technical University of Denmark
Building 224, DK 2800 Lyngby
Denmark

ABSTRACT. Bulk precipitation and bulk throughfall was collected during the period September to November 1984 in a Danish spruce forest. Samples were analyzed for all major anions and cations as well as strong and total acidity. The acid load to the forest ecosystem was estimated adding the throughfall fluxes of protons (79 eq ha^{-1} mo^{-1}), ammonium (99 eq ha^{-1} mo^{-1}) and a calculated estimate of the protons buffered by exchange processes in the canopy (75 eq ha^{-1} mo^{-1}). This is still a minimum estimate but it exceeds the proton load determined by pH measurement in bulk throughfall and bulk precipitation by factors 3 and 6, respectively. Throughfall fluxes of all major cations and anions except ammonium decreased with distance from the trunk.

1. INTRODUCTION

The chemical composition of precipitation is substantially altered on passage through a tree canopy due to downwash of particles and gases deposited on the plant surface as well as uptake and release of substances by the tree (Mayer and Ulrich, 1978; Miller and Miller, 1980; Sehmel, 1980; Grennfelt *et al.*, 1980; van Breemen *et al.*, 1982; Lovett and Lindberg, 1984). Significant spatial variation in throughfall chemistry must be expected.

Deposition of particles and gases substantially increases the acid input to the forest ecosystem, but part of the acid can be buffered by exchange with Ca, Mg or K in the leaf tissue (Ulrich, 1983; Adams and Hutchinson, 1984; Evans *et al.*, 1980). Buffering reactions of this kind can even raise pH in throughfall compared to precipitation (Cronan *et al.*, 1980). However, the protons removed from the throughfall flux by this means are transferred to an internal tree flux and ultimately reaches the soil in exchange with nutrient ions taken up by the roots in order to compensate the canopy loss. Another part of the protons deposited in the canopy is neutralized by reaction with ammonia on particles or aerosols in the atmosphere or on the plant surface itself. Ammonium entering the forest ecosystem is either taken up by the plant in exchange for a proton or oxidized to nitrate in the soil, rendering two pro-

Water, Air and Soil Pollution **30** (1986) 135–141.

tons for each ammonium ion, one of which is consumed if nitrate is taken up by the tree. In order to calculate the acid load on the forest ecosystem it is necessary to sum the throughfall fluxes of H^+ and ammonium ions and add a contribution from nutrient ions exchanged with protons in the canopy.

The present study was undertaken to estimate the acid load on a Danish spruce forest as well as to study the dependence of throughfall chemistry on distance from the nearest tree trunk.

2. MATERIALS AND METHODS

This investigation was conducted in Gribskov Forest (The Strødam Reserve) in North Zealand, Denmark. The forest is situated in a farming area 5 km NW of the town of Hillerød with 24,000 inhabitants, and 30 km NW of the Copenhagen area. The distance from The Sound is 15 km. The vegetation at the study area consists of even-aged forty years old Norway Spruce (*Picea abies*), canopy cover 63% of ground surface area. Yearly precipitation was 551 mm of which 350 mm reached the forest floor as throughfall.

Bulk precipitation and bulk canopy throughfall was collected in the period September to November 1984, one collection per month. Standard polyethylene funnel collectors (diameter 200 mm) were used in all sampling. Bulk precipitation was collected in an open pasture with 50 m to the nearest edge of the forest. The study area was divided into 16 squares of 25x25 m, in each of which a throughfall collector was placed in varying distance from the nearest trunk (range 23 to 55 cm: 8 collectors, range 120 to 135 cm: 5 collectors, range 200 to 250 cm: 3 collectors). The precipitation deposition obtained by bulk sampling includes wet deposition plus precipitation of particles according to gravity (Ulrich, 1983). Samples being visually contaminated was discarded.

Determination of strong and total acidity was accomplished by titration with NaOH using Grans' procedure (Gran, 1952). All pH measurements were made in quiescent solution and special care was taken to avoid error potentials (McQuaker *et al.*, 1983). Weak acidity was obtained as the difference between total and strong acidity. Water samples were analyzed for pH and acidity on the day of collection, stored at 4°C for a few days and subsequently analyzed for organic and inorganic constituents. Organic content was determined by TOC-analyzer. Analysis of Ca, Mg, K and Na was performed by AAS in an acetylene-air flame and ammonium was analyzed using an ammonia electrode, all on unfiltered samples. Chloride, nitrate and sulphate was determined by ion chromatography (Dogan and Hærdi, 1981) after filtering (0.5 μm).

3. RESULTS AND DISCUSSION

Precipitation chemistry in Gribskov Forest, Denmark, is dominated by H^+, ammonium and Na as the major cationic components, and sulphate, chloride and nitrate as the major anions (Table I). The relative abundance of Na, chloride and ammonium compaired to similar studies (Likens *et al.*, 1976; Cronan, 1980; Gorham *et al.*, 1984) reflects the influence of sea spray

TABLE I

Ion fluxes in bulk precipitation and bulk canopy throughfall. Monthly means September to November 1984 at Gribskov Forest, Denmark.

Ionic species	bulk precipitation	bulk[1] precipitation	bulk[2] throughfall	interception[3] deposition	canopy[3] leaching
pH	4.0	4.1	3.6		
Strong acidity			79 ± 95%		
Weak acidity			109 ± 42%		
H^+	40	34	79 ± 89%		
NH_4^+	41	32	99 ± 74%	> 58	< 0
Ca^{2+}	11	12	86 ± 46%	23	52
Mg^{2+}	8	9	50 ± 43%	19	23
K^+	2	3	65 ± 30%	4	59
Na^+	43	38	133 ± 30%	90	0
Total cations	145	128	512		
SO_4^{2-}	48	49	256 ± 50%		
NO_3^-	32	34	57 ± 49%	> 25	< 0
Cl^-	41	36	149 ± 34%	108	0
Total anions	121	119	462		
Anion deficit	24	9	50		

Fluxes: Eq ha^{-1} mo^{-1} (1 mo = 30 day).
1. Monthly mean during May 1984 to May 1985.
2. Mean ± s.d. for 16 collectors. Standard deviation within distance groups was 10-20%.
3. Calculated, see text.

(NaCl) and agricultural activities (ammonium from fertilizers and animal wastes). The primary source of H^+, sulfate and nitrate are anthropogenic combustion processes (Galloway *et al.*, 1984). Ca, Mg and K from particles of different origin (soil, sea spray or anthropogenic emissions) contribute to a lesser extent to the sum of cation equivalents. The precipitation chemistry in the autumn 1984 (September to November) does not differ substantially from the yearly mean (May 1984 to May 1985), but minor increases are observed in the deposition of Na and chloride as a result of increased storm frequency during autumn as well as increased ammonium deposition caused by agricultural activities.

As precipitation passes through the canopy, the flux of all ions increases by the following factors: K (34.3), Ca (8.0), Mg (6.1), sulphate (5.3), chloride (3.6), Na (3.1), ammonium (2.4), H^+ (2.0), nitrate (1.8), and anion deficit (2.1). The observed anion deficit is presumed to represent organic anions (Galloway *et al.*, 1984; Cronan, 1980), which was confirmed by TOC-measurement showing a linear correlation with the anion deficit (r = 0.82).

The enrichment in bulk throughfall compaired to precipitation is caused by interception deposition (impaction of aerosols plus deposition of gases) as well as release and uptake of substances by the canopy. Neither interception deposition nor canopy leaching is easily quantified, but for elements without gaseous components (e.g. Na, Ca, Mg and K) the respective contributions can be separated assuming that these elements show the same ratio between precipitation deposition and interception deposition for each individual sample (Ulrich, 1983). This ratio can be estimated using Na,which is generally believed not to be noticeably affected by canopy exchange (Cronan, 1980), except for some minor leaching from scenescent needles (Ulrich, 1983). The Na enrichment in throughfall is hence interpreted solely as downwash of particles, aerosols or droplets deposited in the canopy. Especially K being the most biologically active cation is subject to canopy exchange (Table I).

The increased sulphate flux in throughfall is mainly caused by interception and gaseous deposition (Eaton et al., 1978; Grennfelt, 1983), though some minor sulphate leaching is to be expected. In spruce forest in Solling (FRG) canopy leaching of sulphate has been quantified to approximately 10 eq ha^{-1} mo^{-1} (Meiwes and Khanna, 1981). The dry deposition of sulphate by sea spray is 11 eq ha^{-1} mo^{-1} assuming that all dry deposited Na originates from sea spray.

Nitrogen compounds being important nutrients can be assimilated directly by the foliage (Nihlgård, 1985). This explains why ammonium and nitrate fluxes in throughfall are less elevated than the Na flux, even though gaseous deposition is supposed to occur.

The throughfall enrichment by protons is relatively small, and still it is a wellknown fact that, apart from particle deposition, a substantial amount of protons is produced on the plant surface and in the plant tissue as a result of gas (SO_2 and NO_x) deposition and subsequent formation of acids (Grennfelt et al., 1980). The acid on the plant surface can be neutralized by cations (Ca, Mg or K) leached from exchangeable pools in the leaf or from salts deposited on the leaf surface (Adams and Hutchinson, 1984). Leaching of cations involves exchange reactions on the leaf surface, in which cations on exchange sites of the cuticle and cell walls are exchanged for H^+. Ca, Mg and K might as well be leached from scenescent needles or throughout the growing season due to metabolic processes (K). In an acid rain environment leaching of Ca and Mg are dominated by exchange processes with H^+ (Ulrich, 1983). Canopy buffering has the effect of removing protons from the throughfall flux, but these protons are still posing a stress on the ecosystem. In order to compensate the canopy loss, Ca and Mg ions are taken up by the roots. This process release a proton load to the soil, especially concentrated in the vicinity of the fine roots.

Ammonium is an acidifying agent because it is either taken up by the roots in exchange with protons or nitrified, a process producing two protons, only one of which is taken up together with the nitrate anion.

As a consequence the acid load on the forest soil can be expressed as the summed equivalents of H, ammonium and canopy exchanged Ca and Mg. The acid load in the present study is 253 eq ha^{-1} mo^{-1}, which exceeds the proton load determined by pH measurement in bulk throughfall and bulk precipitation by factors 3 and 6, respectively.

The calculated acid load is still a minimum in which the following processes have not been taken into account: 1) Gas absorption into the plant cells give rise to acid production. This acidification is buffered by increased root uptake of Ca and Mg extruding protons to the soil. 2) Part of the canopy leaching of K is probably the result of exchange processes with protons. Ca, Mg and K ions are either exchanged with H^+ or leached together with anions. These anions must be sulphate (11 eq ha^{-1} mo^{-1} estimated according to Meiwes and Khanna, 1981) or organic anions (50-24=26 eq ha^{-1} mo^{-1}). The total exchange of Ca, Mg and K with H^+ then amounts to 52+23+59-11-26=97 eq ha^{-1} mo^{-1}. By this calculation the acid load increase to 275 eq ha^{-1} mo^{-1} indicating that also K is involved in cation exchange with H^+. 3) If all nitrate is leached following nitrification the acid contribution might amount to 99 eq ha^{-1} mo^{-1}. 4) The contribution of weak organic acids extruded from the living plant canopy, which can be estimated by subtracting the ammonia throughfall flux from the measured weak acidity. In this study, organic acids contributes 10 eq ha^{-1} mo^{-1} or only 9% of the weak acidity, ammonium making up the rest. Furthermore, the major part of the organic acids are non-protolyzed in acid forest soil, and ultimately they are decomposed to CO_2 and water.

The calculated acid load can be accounted for by summing wet acid deposition with an estimate of gas deposition of SO_2 and NO_x. The proton input with wet deposition equals H^+ (40 eq ha^{-1} mo^{-1}) plus NH_4^+ (41 eq ha^{-1} mo^{-1}). Dry deposition of SO_2 can be estimated as the total sulphate flux in throughfall minus wet deposition, dry salt deposition and sulphate leaching (256-48-11-10=187 eq ha^{-1} mo^{-1}). The minimum NO_x deposition is the nitrate flux minus wet deposition (57-32=25 eq ha^{-1} mo^{-1}). In this calculation nitrate assimilation by the canopy has not been taken into account. Wet acid deposition plus dry deposition of SO_2 and NO_x then equals 293 eq ha^{-1} mo^{-1}.

The large standard deviations in throughfall chemistry (Table I) reflects a great spatial variability which was found to be correlated to distance from the nearest tree trunk (Figure 1). The proton flux shows a dramatic decrease with distance from the trunk, which can not be explained by differences in particle deposition, since elements not being subject to canopy exchange are deposited in amounts either slightly decreasing with distance (chloride) or showing no decrease on the first 135 cm (Na). The observed variation in the proton flux is probably caused by a combination of several effects: 1) A large SO_2 deposition in the tree top, i.e. close to the trunk, produces H_2SO_4. This is in agreement with the radial decrease in the sulphate flux. 2) Buffering in the canopy. The buffering agent is primarily ammonia, but cation exchange processes contribute to buffering as well. Increased buffering with distance can be caused by increased residence time of precipitation in the canopy due to passage of larger amounts of needle mass than close to the trunk where only the top branches carry needles. Comparing throughfall fluxes with bulk precipitation shows that throughfall at all distances is enriched with respect to all ions except H^+, reflecting the buffering capacity of the canopy.

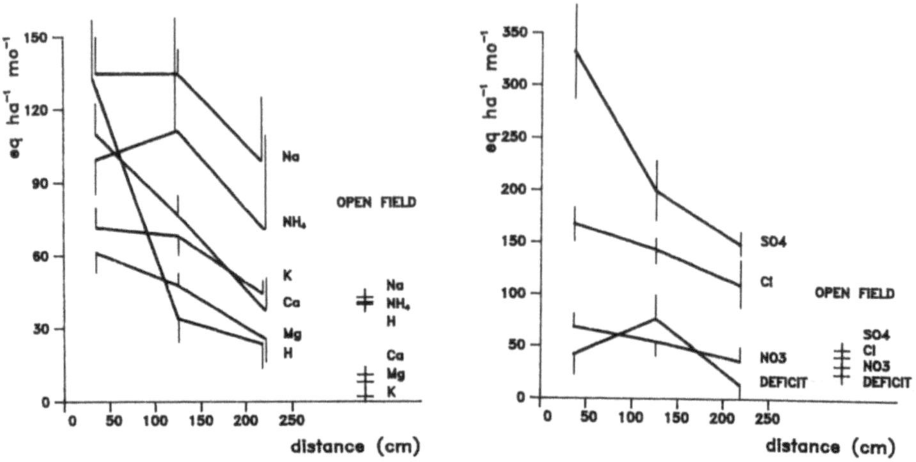

Figure 1. Ion fluxes in throughfall as a function of distance from the tree trunk. Means of distance groups are: 36 cm (range 23-55 cm), 125 cm (range 120-135 cm) and 217 cm (range 200-250 cm). Vertical bars indicate ± standard deviation.

REFERENCES

Adams, C. M. and Hutchinson, T. C.: 1984, New Phytologist 97, 463.
Cronan, C. S.: 1980, Oikos 34, 272.
Cronan, C. S., Reiners, W. A. and Reynolds, Jr., R. C.: 1980, in Drabløs, D. and Tollan, A. (eds.), Proc. Int. Conf. Ecol. Impact Acid Precip., Norway 1980, SNSF Project, p 158.
Dogan, S. and Hærdi, W.: 1981, Chimia 35, 339.
Eaton, J. S., Likens, G. E. and Bormann, F. H.: 1978, Tellus 30, 546.
Evans, L. S., Curry, T. M. and Lewin, K. F.: 1981, New Phytologist 88, 403.
Galloway, J. N., Likens, G. E. and Hawley, M. E.: 1984, Science 226, 829.
Gorham, E., Martin, F. B. and Litzau, J. T.: 1984, Science 225, 407.
Gran, G.: 1952, Analyst 77, 661.
Grennfelt, P., Bengtson, C. and Skärby, L.: 1980, in Hutchinson, T. C. and Havas, M. (eds.),'Effects of acid precipitation on terrestrial ecosystems', Plenum Press, New York, p 29.
Grennfelt, P. (ed.): 1983, snv pm 1763, Naturvårdsverket, Solna, Sweden.
Likens, G. E., Johnson, N. M., Galloway, J. N. and Bormann, F. H.: 1976, Science 194, 643.
Lovett, G. M. and Lindberg, S. E.: 1984, J. of Applied Ecology 21, 1013.
Mayer, R. and Ulrich, B.: 1978, Atmos. Environ. 12, 375.
McQuaker, N. R., Kluckner, P. D. and Sandberg, D. K.: 1983, Environ. Sci. Technol. 17, 431.
Meiwes, K. J. and Khanna, P. K.: 1981, Plant and Soil 60, 369.
Miller, H. G. and Miller, J. D.: 1980, in Drabløs, D. and Tollan A. (eds.)

Proc. Int. Conf. Ecol. Impact Acid Precip., Norway 1980, SNSF project, p 33.
Nihlgård, B.: 1985, Ambio 14, 2.
Sehmel, G. A.: 1980, Atmos. Environ. 14, 983.
Ulrich, B.: 1983, in Ulrich, B. and Pankrath, J. (eds.), 'Effects of Accumulation of Air Pollutants in Forest Ecosystems', D. Reidel Publishing Company, Dordrecht, p 33.
Van Breemen, N., Burrough, P. A., Velthorst, E. J., Van Dobben, H. B., de Wit, T., Ridder, T. B. and Reijnders, H. F. R.: 1982, Nature 299, 548.

CHEMICAL COMPOSITION OF PRECIPITATION AT LONG ISLAND, NY

Yin-Nan Lee, Ji Shen[1], and Paul J. Klotz
Environmental Chemistry Division
Department of Applied Science
Brookhaven National Laboratory
Upton, NY 11973
U.S.A.

ABSTRACT. Chemical analyses were performed on 387 sequential precipitation samples, both rain and snow, collected at Brookhaven National Laboratory, Long Island, New York, on 96 different days during the period from October 1983 to June 1985. The collection time for each individual sample was typically 30 min. The chemical species determined and the volume-weighted average and maximum concentrations (in parentheses) in units of μM were: H_2O_2:7(120), H_2CO:6(60), H^+:51(500), NO_3^-:30(350), SO_4^{2-}:25(270), NH_4^+:18(250), Ca^{2+}:3(48), Na^+:32(400), and Cl^-:37(380). A strong seasonal dependence was observed for H_2O_2 concentration, the maximum concentration in summer (120 μM) being 6 times greater than in winter. In addition, H_2O_2 concentration also exhibited a strong diurnal variation, with a maximum peaking in the afternoon and a minimum after midnight. These observations suggest that the production of H_2O_2 may be governed by photochemical activity. Nitrate concentration also displayed a strong diurnal variation, having a maximum around noon, but did not show a seasonal dependence. Relationships between various species and their possible sources are briefly discussed.

1. INTRODUCTION

In recent years much attention has been focused on the "acid rain" phenomenon because the acidic precipitation is believed to be responsible for adverse environmental effects such as modification of lake ecology and damage to forests (Glass et al., 1982; NRC, 1983). In order to gain a better understanding of the processes that play major roles in acid production as well as of wet deposition with respect to species and concentrations, chemical composition of precipitation has to be characterized.

[1] Visiting Scientist from the Institute of Environmental Chemistry, Academia Sinica, Beijing, China

Water, Air and Soil Pollution **30** (1986) 143–152.

One of the major issues in acid precipitation research concerns the identification of aqueous-phase reaction pathways for the oxidation of SO_2 and NO_x (Calvert, 1984). Among the various aqueous-phase oxidants of S(IV), H_2O_2 is considered to be by far the most important one, mainly because of its high solubility (Martin and Damschen, 1981) and fast reaction kinetics in the pH range characteristic of rain and cloudwater (Hoffmann and Edwards, 1975; Penkett et al., 1979). Despite the facts that atmospheric H_2O_2 is produced by the recombination reaction of HO_2 radicals, and that its concentrations in the ppb range have been predicted by model calculations (Kleinman, 1984; Calvert and Stockwell 1983) there have been only a few measurements of this species (e.g., Kelly et al., 1985; Heikes et al., 1985). Conseqeuntly, the contribution of the aqueous H_2O_2-S(IV) reaction to oxidation of SO_2 in the atmosphere remains highly uncertain.

In this work we report the determination of H_2O_2 concentration in collected rainwater samples. Although such an approach may be indirect as a means of determining atmospheric H_2O_2 concentrations, nonetheless, these measurements should provide information for H_2O_2 regarding its variability, its relative abundance with respect to SO_2, and its possible sources. Formaldehyde (HCHO) was also measured in an effort to relate H_2O_2 to other major photochemical species coexisting in the precipitation samples. In addition, the concentrations of H^+, NO_3^-, SO_4^{2-}, NH_4^+, Na^+, Cl^-, and Ca^{2+} were measured. Mn^{2+} was measured for a small fraction of the samples ($\leq$ 10%). This work was carried out at Brookhaven National Laboratory, Long Island, New York. We make no distinction between the terms precipitation and rain, and that snow samples are included in these data. When discussing seasonal dependence, October-March is referred to as winter and April-September as summer. Average concentration means volume-weighted.

2. EXPERIMENTAL SECTION

Rainwater collection was made in an open area with a 44-cm diameter funnel and 1-L bottles, both of high density polyethylene. The collector was placed in the field only after precipitation had actually begun in order to minimize possible contamination from dry deposition. The duration of collection for individual samples varied form 10 to 60 min depending on rainfall rate, typical collection time being less than 30 min. After measurement of volume, pH, $[H_2O_2]$, and conductivity, samples were transferred into polyethylene bottles and stored in a refrigerator for further analysis.

The acidity of the solutions was measured at room temperature by a pH meter (Beckman, Model 4500) equipped with a combination electrode (Orion, Ross electrode). The solution was continuously stirred during the measurement using a magnetic stirrer.

Hydrogen peroxide concentration was determined using an enzyme-fluorescence technique (Guilbault et al., 1968) as modified by Lazrus and coworkers (1985). The detection limit (LOD) of this method is ~0.02 μM. HCHO was determined by a colorimetric method using p-dinitrophenyl-hydrazine as a coupling agent (Kuntz et al., 1980), in conjunction with an uv-vis spectrophotometer (Beckman, DU-7). The uncertainty associated with this technique was ±3 μM.

The concentrations of other ions were measured by the standard methods: SO_4^{2-} and NO_3^- by ion chromatography (Dionex), NH_4^+ and Cl^- by autoanalyzer technique (Technicon), and Na^+, Ca^{2+}, and Mn^{2+} by atomic absorption (Instrument Laboratories).

All of the chemical analyses were carried out within the shortest time possible after collection. The maximum elapsed time were: 30 min for initial pH measurement (pH_1, see below) and H_2O_2; 1 day for HCHO; 3 days for NH_4^+, NO_3^-, and SO_4^{2-}; and 2 weeks for the rest of the ions.

3. RESULTS AND DISCUSSION

For the period from October 1983 to June 1985 a total of 387 samples were collected on 96 different days. Measurements of pH and H_2O_2 were performed on all of the samples; ionic species were carried out for nearly 90% of the samples (~10% of the sample had insufficient volume). Measurement of HCHO commenced in June 1984.

3.1 pH Measurement

In order to examine the uncertainty in pH measurement, the pH of precipitation samples was measured twice, once immediately after collection (pH_1) and the second time at ~3 days later (pH_2), with the use of identical equipment and the same calibration procedure. Comparison of these two values should provide information on the precision of the measurement as well as the possibility of systematic drift of the acidity of the solutions arising form either chemical reactions of rainwater constituents or artifacts due to the method of storage.

Plotting pH_2 against pH_1 shows that the data points spread evenly about the ideal 1:1 line. A least-squares fit to the data having a slope of unity is displaced slightly above the ideal line by ~0.04 pH unit; the difference is insignificant in view of the rather large measurement uncertainty as indicated by the scatter of the data i.e., ~0.05 + 0.08(pH-3.4) pH unit (1 sigma). This observation therefore shows the absence of any appreciable systematic change of sample pH upon storage. Furthermore, it may be noted that such an uncertainty in pH may be representative of field experiments of similar nature.

3.2 Hydrogen Peroxide

The concentration of H_2O_2 was measured within 30 min of the sample collection. For some of the samples containing higher levels of H_2O_2 (> 10 μM) analysis was also made to determine the concentration of organic peroxides. Using the catalase addition method (Lazrus et al., 1985), we found no evidence of the existence of organic peroxides within the uncertainty of the analysis (± 1%). This observation is consistent with current knowledge of the atmospheric concentrations of organic peroxides and their aqueous solubilities (Gaffney and Senum, 1984) and it may be concluded that organic peroxides, despite their potential importance in photooxidant formation (Hanst and Gay, 1983), may not contribute significantly to the aqueous-phase oxidation of dissolved SO_2.

The $[H_2O_2]$ determined is presented in Figure 1 as a function of the time of the year. A seasonal dependence showing a summer maximum and a winter minimum is indicated. The maximum observed for the summer (120

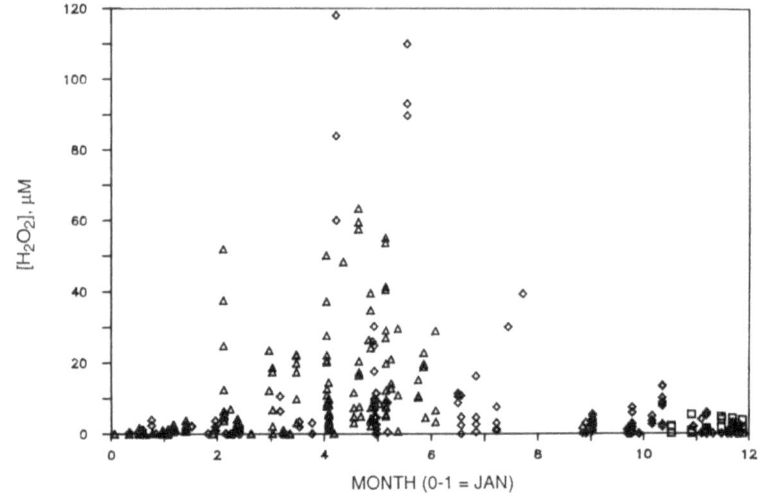

Figure 1. Plot of $[H_2O_2]$ as a function of time of the year of the collection. Legends: □,1983(n = 27); ◇, 1984 (n = 166); Δ, 1985 (n = 183).

µM), is six times greater than that observed for the winter. In Figure 2 $[H_2O_2]$ is plotted as a function of the time of day of sample collection. A diurnal variation in which the H_2O_2 concentration peaks in the afternoon and reaches a minimum during the night is indicated and suggests, in conjunction with the seasonal dependence, that the production of atmospheric H_2O_2 may be primarily governed by photochemical activity. The fact that samples having high concentration of H_2O_2 do not always contain high levels of SO_4^{2-} and NO_3^- indicates that sources of H_2O_2 may not be necessarily coupled to polluted air masses. A recent study has suggested that clouds (thereby precipitation) can acquire H_2O_2 from the free troposphere (Heikes et al., 1985), and therefore points out the importance of atmospheric mixing in the overall process of SO_2 oxidation.

3.3 Formaldehyde

Formaldehyde is present in the atmosphere as a reaction product of photochemical oxidation of hydrocarbons (NAS, 1976). It is also believed to be a precursor of H_2O_2 because it photolyzes to form HO_2 and reacts with various free radical species to produce H_2O_2 (Bufalini et al., 1979). Although the analytical method used in this work is not

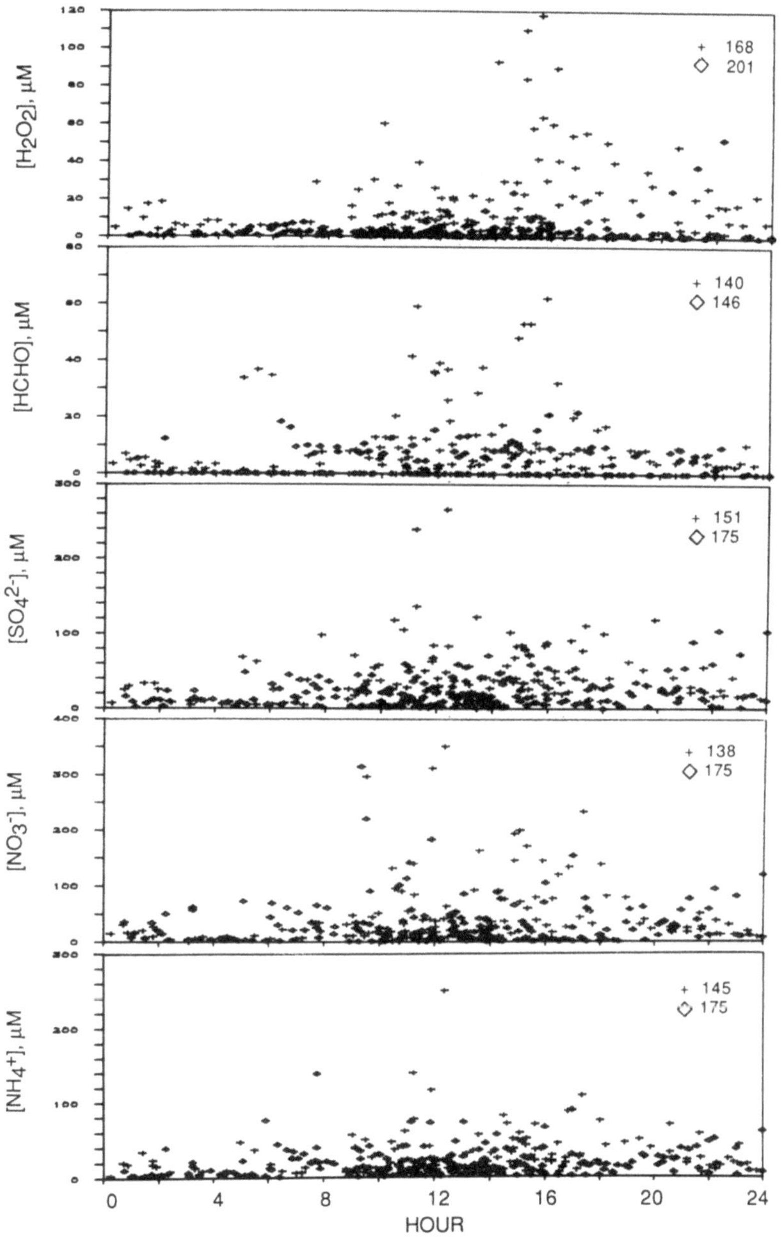

Figure 2. Diurnal variation of the concentrations of H_2O_2, HCHO, SO_4^{2-}, NO_3^-, and NH_4^+ of the precipitation samples. Legends: +, summer; ◇, winter.

specific to HCHO but also responds to other aldehydes, the fact that the higher aldehyde homologues, such as acetaldehyde, are much less soluble in water (by a factor of > 100) than HCHO makes it possible to attribute the signal almost entirely to HCHO.

The [HCHO] is plotted as a function of the time of day in Figure 2. The maximum concentration observed was 60 μM. The corresponding gas-phase concentration of 10 ppb, calculated using a Henry's law solubility of 6×10^3 M atm^{-1} (Blair and Ledbury, 1925) assuming gas-liquid equilibrium, is in good agreement with gas-phase HCHO concentrations reported for the same location (Tanner and Meng, 1984). Although HCHO exhibits a relatively clear diurnal variation having a maximum at around noon for the summer data, similar to H_2O_2, there exists no significant correlation between H_2O_2 and HCHO (correlation coefficient = 0.18). It does not seem possible to substantiate any precursor-product relationship between HCHO and H_2O_2 from this work. However, HCHO is found to be significantly correlated with sulfate ($r = 0.63$), which may be interpreted to mean that HCHO is frequently associated with polluted air masses.

3.4 Nitrate and Sulfate

Concentrations of nitrate and sulfate plotted as a function of the hour of day are also shown in Figure 2. A strong diurnal variation is displayed in the nitrate data showing a maximum centered around noon. The same set of data, on the other hand, does not seem to show a noticeable seasonal dependence. The latter behavior seems to suggest that either there exist possible additional pathways for nitrate generation (Richards, 1983) or $[HNO_3]$ may be higher in the winter season (arising from, e.g., less efficient vertical mixing). Relative to SO_4^{2-}, NO_3^- also shows a higher winter/summer ratio of average concentration (0.67 vs. 0.51).

The sulfate data displayed a weaker seasonal and diurnal variation than that of nitrate (when corrected for sea salt sulfate by Na^+, the seasonal dependence became more discernable). However, it is interesting to note that the average sulfate concentration is nearly a factor of 2 higher in summer (32 μM) than in winter (17 μM). The molar concentrations of NO_3^- are higher than that of SO_4^{2-} in samples of high concentrations. These two species, however, have nearly identical average concentrations for the summer ($[NO_3^-]$ = 33 μM); for the winter season, nitrate (22 μM) is higher than sulfate. This difference may be attributed entirely to the high nitrate/sulfate ratio of the snow (5.4) which accounts for ~6% of the precipitation. It should be noted that the winter rain, however, showed no difference in nitrate/sulfate ratio from that of summer rain (0.95). Similar findings have been reported by Raynor and Hayes (1983).

3.5 Ammonium and Other Ions

The concentration of NH_4^+ is plotted as a function of the time of day of collection (Figure 2). The maximum value is ~250 μM, which coincides with that of sulfate. Despite an identifiable diurnal variation, with a maximum at around noon, seasonal dependence appears to be absent. This observation is in qualitative agreement with the fact that the difference between the average concentration for summer (21 μM) and for winter (15 μM) is rather small comparing to other species.

The maximum and volume-weighted average concentration of Ca^{2+} was 48 and 3.2 μM, respectively. Mn^{2+} was never found to be greater than our limit of detection, i.e., 2 μM.

Sodium and chloride were found to be present in the precipitation samples in rather high concentrations. The maximum and average concentrations were 400 and 32 μM for Na^+ and 380 and 37 μM for Cl^-. It is interesting to note that these ions exhibit a seasonal dependence totally opposite to the rest of the chemical species studied in this work. Instead of a low-in-winter and high-in-summer pattern, Na^+ and Cl^- were found to have winter/summer average concentrations of 54/15 and 56/20 (in μM). The fact that summer data show an excess of Cl^- and winter data show an excess of Na^+ (over seawater) and the concentrations of these ions are much higher in winter than in summer indicate that winter precipitation chemistry may be more significantly modified by air masses of marine origin than summer (Lebowitz and de Pena, 1985). As such, the lower concentration of SO_4^{2-} and NO_3^- observed in winter precipitation may be attributed, at least partially, to the dilution by cleaner marine air, and cannot be ascribed entirely to diminished chemical activities.

3.6 Ion Concentrations and Their Ratios

Table I lists the average and maximum concentrations of the species determined in this work along with some of those observed by Raynor and Hayes (1984) for the same location. It is noted that although the maximum concentrations of SO_4^{2-} and NO_3^-, and even H^+, are fairly variable, the average concentrations of these species do not differ significantly; the most noticeable case is pH: the three average pH values of 4.27, 4.30, and 4.29 for 1977, 1981, and 1983-1985, respectively, are essentially identical. It seems that either there has been no

TABLE I.
Volume-Weighted Average (and Maximum) Concentration Found in Precipitation, in Units of μM

Species	This Work	1977*	1981*
H^+	51.2(500)	53.5(562)	50.2(631)
NH_4^+	17.9(250)	16.9(260)	19.6(286)
$SO_4^=$	24.7(270)	62.1(338)	50.4(358)
NO_3^-	27.9(320)	24.7(431)	28.0(575)
Ca^{++}	3.2(48)		
Na^+	32.4(400)	44.3(2544)	22.9(461)
Cl^-	37.2(380)	51.5(1459)	33.3(564)
H_2O_2	7.0(120)		
H_2CO	5.9(60)		

*Raynor and Hayes, 1984

significant change in emission and deposition pattern over that 9-yr period, or the average concentrations do not provide enough detail to reflect possible long-term trend in precipitation chemistry.

In order to address the extent to which H_2SO_4 and HNO_3 are responsible for the total strong acids found in precipitation, we examined the ratio of the concentration of total free acid, i.e., H^+ ion, to the sum of that of SO_4^{2-} and NO_3^-. Since a substantial fraction of the free acid is normally neutralized by ammonia, the total free acid should include NH_4^+. Figure 3 presents histograms for this ratio, with separate summer and winter entries. In addition to the fact that summer data show a distribution much narrower than that of the winter data, the two distributions are also noted to center around slightly different values, summer being at unity and winter at 0.9. These observations suggest that, relative to summer samples, winter samples may contain more metal cations associated with either nitrate or sulfate. If strong acids other than H_2SO_4 and HNO_3 are also present in the precipitation samples the ratio should become greater than unity. The fraction having a ratio of 1.2 observed for the summer data is ~20%. This may be interpreted as that 20% of the summer precipitation samples contain 20% more acid than can be accounted for by H_2SO_4 and HNO_3 alone. Even so, the amount of unidentified strong acids constitutes only a minor fraction of the total acidity, viz., in the order of ~5%. This argument is consistent with the fact that the sum of the average concentrations of H^+ (62 μM) and NH_4^+ (20 μM) is similar to that of SO_4^{2-} (2x32 μM) and NO_3^- (33 μM), within the uncertainty of the measurements (e.g., that of H^+).

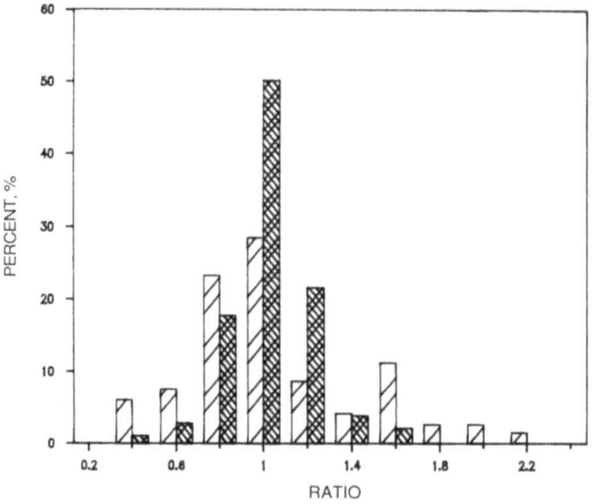

Figure 3. Histograms of volume-weighted ratio of $[NH_4^+] + [H^+]$ to $2x[SO_4^=] + [NO_3^-]$. Legends: darkhatched, summer; light-hatched, winter.

ACKNOWLEDGMENTS

This work was supported by the National Acid Deposition Assessment Program through the PRECP program funded by the U.S. Department of Energy under contract No. DE-AC02-76CH00016.

REFERENCES

Bufalini, J.J., Lancaster, H.T., Namie, G.R., and Gay, B.W., Jr., J. Environ. Sci. Health, **A14**, 135. Calvert, J.G., ed., 1984, SO_2 and NO_2 oxidation mechanisms: Atmospheric considerations, Butterworth, Boston.

Calvert, J.G. and Stockwell, W.R., 1983, Environ. Sci.Technol. **17**, 428A.

Gaffney, J.S. and Senum, G.I., 1984, Peroxides, peracids, aldehydes and PANs and their links to natural and anthropogenic organic sources, in Proceedings of the Conference on Gas-Liquid Chemistry of Natural Waters, Brookhaven National Laboratory, Upton, New York, April, Paper No. 5.

Glass, N.R., Arnold, D.E., Galloway, J.N., Hendrey, G.R., Lee, J.J., McFee, W.W., Norton, S.A., Powers, C.F., Rambo, D.L., and Schofield, C.L., 1982, Environ. Sci. Technol., **16**, 162A.

Guilbault, G.G., Brignac, P., and Juneau, M., 1968, Anal. Chem., **40**, 1256.

Hanst, P.L. and Gay, B. Jr., 1983, Atmos. Environ., **17**, 2259.

Hoffmann, M.R. and Edwards, J.O., 1975, J. Phys. Chem., **79**, 2096.

Heikes, B.G., Lazrus, A.L., and Kok, G.L., 1985, Measurements of H_2O_2 in the lower troposphere, presented at the 17th International Symposium on Free Radicals, Granby, Colorado, August 18-23, 1985.

Kelly, T.J., Daum, P.H., and Schwartz, S.E., 1985, J. Geophys. Res. **90**, 7861.

Kleinman, L.I., 1984, Atmospheric formation of gas phase H_2O_2 and other oxidants in the eastern U.S., Report BNL-33367, Brookhaven National Laboratory, Upton, NY.

Kuntz, R. Lonneman, W., Namie, G., and Hull, L.A., 1980, Anal. Letts. **13**, 1409.

Lazrus, L.L., Kok, G.L., Gitlin, S.N., and Lind, J.A., 1985, Anal. Chem., **57**, 917.

Lebowitz, L.G., and de Pena, R.G., 1985;, J. Geophys. Res., **90**, 8149.

Martin, L.R. and Damschen, D.E., 1981, Atmos. Environ., **15**, 1615.

National Research Council, 1983, Acid deposition: Atmospheric processes in Eastern Northern America, Environmental Studies Board, National Academy Press, Washington, DC.

National Academy of Sciences, 1976, Vapor-Phase Organic Pollutants, Committee on Medical and Biological Effects of Environmental Pollutants, Washington, DC.

Penkett, S.A., Jones, B.M.R., Brice, K.A., and Eggleton, A.E.J., 1979, Atmos. Environ., **13**, 123.

Raynor, G.S. and Hayes, J.V., 1983, Differential rain and snow scavenging efficiency implied by ionic concentration differences in winter precipitation. In Precipitation, Scavenging, Dry Deposition, and Resuspension, Pruppacher, H.R., Semonin, R.G., and Slinn, W.G.N., eds. Elsevier, New York, 1983.
Raynor, G.S. and Hayes, J.V., 1984, The sequential precipitation chemistry sampling program at Brookhaven National Laboratory with an analytical summary of 1976-1982 data. Brookhaven National Laboratory report, No. 51819.
Richards, L.W., 1983, Atmos. Environ., **17**, 397.
Tanner, R.I., and Meng, Z., 1984, Environ. Sci. Technol., **18**, 723.

A COMPARISON OF SUMMER AND WINTER MEASUREMENTS OF ATMOSPHERIC NITROGEN AND SULPHUR COMPOUNDS

K.G. ANLAUF, J.W. BOTTENHEIM, K.A. BRICE AND H.A. WIEBE
Atmospheric Environment Service
4905 Dufferin St.,
Downsview, Ont., Canada M3H 5T4

ABSTRACT. Summer and winter concentrations of acidic atmospheric species and their precursors were measured in central Ontario. Large seasonal differences in concentrations were observed for some species. The concentrations of primary emission species, SO_2 and NO_x, were much larger in winter, whereas O_3 and aerosol $SO_4^=$ were much reduced. HNO_3, PAN and aerosol NO_3^- showed little seasonal change; PAN represented about 30% of the total oxidized nitrate. During pollution episodes and in summer, the molar ratio of nitrate (aerosol NO_3^- plus HNO_3) to aerosol $SO_4^=$ was about 0.1 to 0.2; however, during winter, this molar ratio was about 1 to 2.

1. INTRODUCTION

Past studies (e.g. see Barrie and Hales, 1984) have shown that annual mean sulphate molar concentrations were about the same as nitrate in precipitation in eastern North America. However, on a seasonal basis, more recent studies have shown large variations in their relative contributions to precipitation acidity (e.g. Summers and Barrie, 1986). The relative contributions of these species to the acidity in precipitation are controlled by the ambient gas-phase and aerosol chemical make-up. Even though photochemical reactions are thought to be the driving force behind the formation of these atmospheric acidic compounds, it is not expected that atmospheric chemical reactivity is negligible in winter (Calvert et al., 1985). Rather, there may be a shift in the relative contribution of N and S species to atmospheric acidity.

An analysis of the 1980 emission inventory by Summers and Barrie (1986) has shown that the molar ratio of SO_x to NO_x emissions was about 1.1 for the northeastern U.S. (This ratio may have decreased slightly since 1980). Furthermore, there was less than 5% seasonal variability in this ratio. For central Ontario, a similar ratio is expected (see also Barrie and Hales, 1984). Hence, any seasonal differences in atmospheric concentrations in these geographical regions cannot be attributed to changes in emissions but to changes in atmospheric processes.

This paper presents some of the results of two field studies

Water, Air and Soil Pollution **30** (1986) 153–160.

conducted in central Ontario, Canada. The measurements were made near Powassan, a village about 30 km south of North Bay, Ontario for the summer period June 29 to July 29, 1982 and the winter period January 15 to February 24, 1984. Some of the summer-winter differences will be examined.

2. EXPERIMENTAL PROCEDURE

A triple filter pack was used to collect 6-hr samples (except for occasional shorter sampling periods during day-time) of atmospheric aerosol (analyzed for the major inorganic ions), gaseous HNO_3 and SO_2. O_3 and NO_x were measured using chemiluminescence analyzers and peroxyacetylnitrate (PAN) was measured using conventional gas chromatography employing electron capture detection (details of these methods are given by Anlauf et al.,1985a,b).

3. RESULTS AND DISCUSSION

The measurement results are shown in Figures 1 to 3. For both summer and winter periods, the frequent occurrences of high pollutant concentrations were invariably associated with air masses of southern origin. Calculated 4-day back-trajectories (1000, 925 and 850 mb) showed that southerly air masses were being sampled during the following periods: (1) in 1982, July 5-7, 11-12, 15-18 and 25; (2) in 1984, Jan. 22-25, 27, 29, Feb. 1-5, 9-15, 18-19 and 22-24 (the latter period was unseasonably warm). These air masses had passed over source regions rich in SO_2 and NO_x emissions (see Barrie and Hales, 1984, for maps of spatial distributions of the emission source strengths).

3.1. Ozone

The summer O_3 concentrations were much higher than those of winter and they exhibited the typical large diurnal variation (reflecting the importance of solar intensity) and peaked in the range 100 to 150 ppb during southerly pollutant episodes. During winter, O_3 was relatively constant at about 25 to 35 ppb with little diurnal variation and no clear correlation with air mass origin was evident (except for the unique Feb. 22-24 period). The measured summer-winter difference in ambient O_3 concentrations is not unexpected. Ambient O_3 is produced by the reaction of O (from solar UV photolysis of NO_2) with O_2. Hence, in winter, the lower solar UV intensity would result in less O_3 formation.

3.2. NO_x, HNO_3 and aerosol nitrate (p-NO_3)

For the summer period, ambient NO_x (mostly NO_2 with some NO) was less than 5 ppb (detection limit of the commercial analyzer); whereas in winter, NO_x concentrations ranged from 10 to 30 ppb during pollutant episodes. This winter increase in NO_2 concentrations may be due to two factors: (1) the winter decrease in ambient OH radicals (Calvert et al., 1985) markedly reduces the gas-phase oxidation rate of NO_2, and (2) the NO_2 deposition velocity is expected to be lower in winter because of the greater atmospheric stability (this is the case for other pollutants; Voldner and Sirois, 1986). Both these factors contribute to the transport of NO_2 over greater distances, resulting in higher concentrations at Powassan which is located several hundred kilometres downwind of major sources.

Figure 1 shows the summer and winter concentrations of p-NO_3 and HNO_3. The episodes of high nitrate concentrations occurred when southerly polluted air masses were being sampled at the measurement site. It is obvious that the production of atmospheric nitrates has not been diminished in winter. During the summer episodes, p-NO_3 peaked at about 0.3 to 0.7 ppb and HNO_3 at about 1 to 3 ppb; the corresponding ratio of p-NO_3 to (HNO_3 + p-NO_3) was, on average, about 0.15. In winter, p-NO_3 peaked at about 0.5 to 2 ppb (excepting the Feb. 22-24 episode) and HNO_3 at about 1 to 5 ppb; the corresponding winter ratio

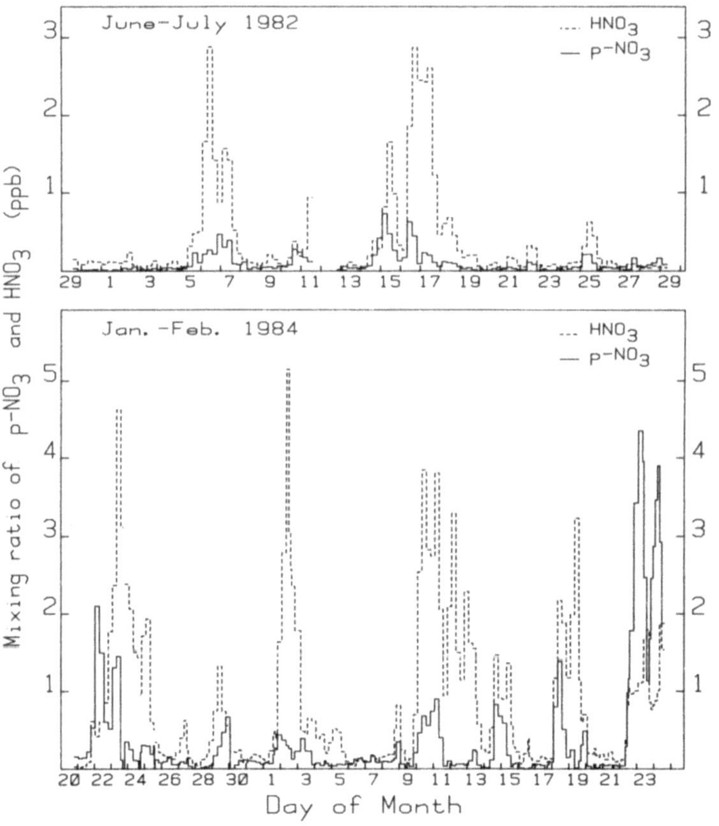

Figure 1. Concentration of particulate nitrate (p-NO_3) and HNO_3 at Powassan, Ontario for the periods June-July, 1982 and January-February, 1984.

of p-NO_3 to (HNO_3 + p-NO_3) was, on average, 0.2 to 0.3 but was highly variable.

Presently, it is thought that there are two major mechanisms for formation of atmospheric nitrate. Firstly, the reaction of NO_2 with the OH radical produces gaseous HNO_3. Secondly, NO_2 reacts with O_3 to form the NO_3 radical; the latter reacts with more NO_2 to form N_2O_5 which, in turn, hydrolyzes to HNO_3 in the liquid phase (wet aerosol or cloud droplets).

The similarity in summer and winter concentrations of gaseous HNO_3 may be due to several factors: (1) the winter decrease in OH concentration (Bottenheim et al.,1984; Calvert et al.,1985) is compensated to some extent by the increase in ambient winter NO_2 concentration; (2) HNO_3 formed in the ambient acidic aerosol may be volatilized; (3) similar to other pollutants, it is expected that the dry deposition loss would be smaller in winter (Voldner and Sirois, 1986).

For p-NO_3, the factors that could result in a winter increase are: (1) higher local NO_2 concentration, (2) lower deposition velocity, (3) more night-time hours available for the formation of the unstable intermediate NO_3 radical and (4) greater thermal stability of the intermediate N_2O_5. However, these factors would be offset to some extent by the decrease in ambient O_3 concentration. The measurements show that, if anything, there was a slight increase in winter p-NO_3 concentrations.

3.3. Peroxyacetylnitrate

PAN is another product indicative of atmospheric photochemical activity. It is formed by the reaction of OH radicals with hydrocarbons and NO_2; however, PAN is subject to thermal dissociation and hence may form a source of NO_2 far downstream of NO_x sources.

During the summer period, PAN was measured only from July 23-27. This included a brief period of southerly synoptic flow at our measurement site, July 25. The peak concentration on July 25 was about 0.7 ppb; for the remainder of the measurement period PAN was less than 0.2 ppb. For the full 5-day period, the average ratio of PAN to total nitrate (t-NO_3 = PAN + HNO_3 + p-NO_3) was 0.34 (for more details on this period, see Bottenheim et al., 1984).

During the winter study period, similarly to the HNO_3 and p-NO_3 discussed earlier, there was a significant production of PAN and again the episodes of high PAN concentrations coincided with the periods of southerly air flow. The PAN concentrations peaked at about 1 to 2.5 ppb during these periods and the ratio of PAN / t-NO_3 was about 0.3 ±0.1. During northerly air flows, this ratio was 0.4 to 0.7.

3.4. SO_2 and aerosol sulphate (p-SO_4)

Figure 2 shows the concentration of SO_2 along with the ratio of p-SO_4 / (p-SO_4 + SO_2) for the two study periods. Figure 3 shows the concentration of p-SO_4. It is seen that the winter SO_2 concentrations were much larger than those of summer and conversely for p-SO_4. The episodes of high concentrations in winter were associated with southerly air masses as discussed earlier. Also, Figure 2 shows that the molar ratio of p-SO_4

to total sulphur ($p\text{-}SO_4$ + SO_2) was about 0.6 to 0.8 during summer pollutant episodes, whereas in winter, this ratio was generally less than 0.2 (in fact, frequently about 0.05).

There are two important mechanisms for oxidation of SO_2 to $p\text{-}SO_4$. These are the gas-phase reaction of SO_2 with OH radicals and the liquid-phase reactions of SO_2 with H_2O_2 and O_3 in cloud water. In winter, as discussed earlier, the ambient OH is expected to be at least 4 times lower than in summer and the ambient O_3 is also very much reduced. Furthermore, measurements of H_2O_2 in winter-time cloud samples showed trace or no measureable amounts (results will be reported at a later

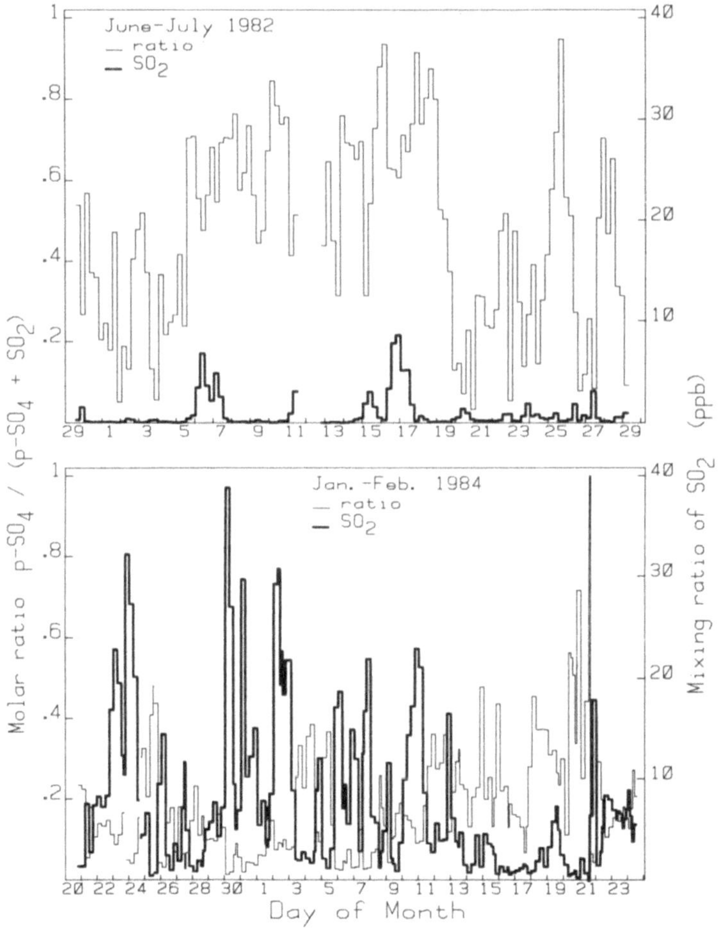

Figure 2. Concentration of SO_2 and the ratio $p\text{-}SO_4/(p\text{-}SO_4 + SO_2)$ at Powassan, Ontario for the periods June-July, 1982 and January-February, 1984.

date). This observation was expected because H_2O_2 is formed from the recombination of two HO_2 radicals and, in turn, their concentration is dependent on the ambient photochemical reactivity. The sharp decrease in these oxidizing species would markedly reduce the winter-time production of p-SO_4. For both SO_2 and p-SO_4, the winter deposition velocities are about a factor of 3 to 4 lower than those of summer (Voldner and Sirois, 1986). For SO_2, this would add to the higher winter concentrations measured at Powassan. For p-SO_4, however, the decreased winter deposition velocity would enhance the long-range transport of p-SO_4 that is formed en route from sources. Our measurements show that winter p-SO_4 concentrations were much lower than those of summer. This would

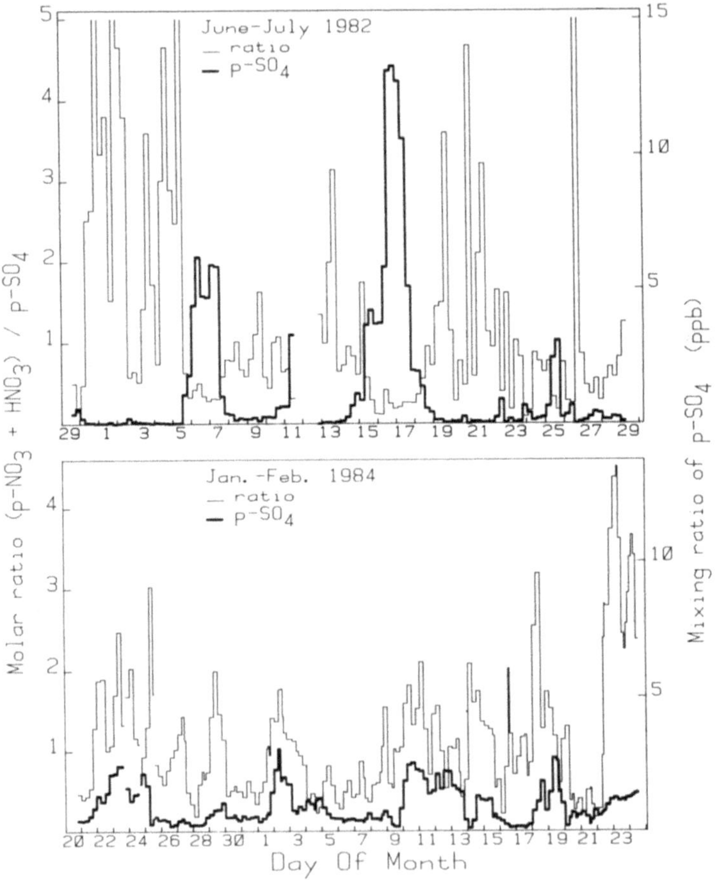

Figure 3. Concentration of particulate sulphate (p-SO_4) and the ratio of (p-NO_3 + HNO_3) / p-SO_4 at Powassan, Ontario for the periods June-July, 1982 and January-February, 1984.

imply that the decrease in the winter oxidation rate is the dominant factor influencing ambient p-SO_4 concentrations at our measurement site. This is in agreement with studies on power plant plumes; these have shown that the net oxidation rate of SO_2 is very much reduced in winter (as low as 0.2% hr^{-1} in winter and as high as 1-4% hr^{-1} in summer; e.g., see Meagher et al., 1981; Anlauf et al., 1982).

3.5. The ratio (p-NO_3 + HNO_3) / p-SO_4

A major goal of the field study was to measure the relative contribution of atmospheric nitrate and sulphate to potential atmospheric acidity. In Figure 3 are plotted the molar ratios (p-NO_3 + HNO_3) / p-SO_4 for summer and winter. During summer pollutant episodes, this ratio was 0.2 to 0.4. However, during winter, it was generally in the range 1 to 2. Hence, as a result of the marked decrease in winter oxidation of SO_2 to p-SO_4 and yet little change in the net oxidation of NO_2 to atmospheric nitrates, the contribution of nitrates to potential winter atmospheric acidity has increased 5-fold (on a molar basis) relative to that of sulphate.

4. SUMMARY AND CONCLUSIONS

Summer and winter measurements were made of primary emission pollutants (NO, NO_2 and SO_2), their oxidation products (aerosol NO_3 and SO_4, HNO_3, PAN) and O_3 in central Ontario, at a site several hundred kilometres downwind of major pollutant emission regions. The major findings were: (1) Episodes of large pollutant concentrations were associated with southerly air masses that had passed over the SO_2 and NO_x source regions; (2) Large seasonal differences were found between winter and summer concentrations of SO_2, NO_2, O_3 and p-SO_4. SO_2 and NO_2 concentrations were much larger in winter, and conversely for O_3 and p-SO_4; (3) The oxidized N species, HNO_3, PAN and p-NO_3 showed little seasonal difference with possibly slightly larger concentrations during the winter study period. For both seasons, p-NO_3 was a smaller fraction of the total acidic nitrate (p-NO_3 + HNO_3); in summer, p-NO_3 constituted about 15% and, in winter, it was about 20 to 30%. PAN comprised about 30% of the total measured nitrate (PAN + HNO_3 + p-NO_3); (4) In summer, the molar ratio of nitrate (p-NO_3 + HNO_3) to p-SO_4 was about 0.2 to 0.4 during the periods of southerly pollutant episodes; however, in winter, this ratio was about 1 to 2. This has important consequences with respect to the relative importance of the contribution of atmospheric nitrate to sulphate to winter dry and wet deposition. Elsewhere in this volume, Summers and Barrie (1986) showed that the winter molar ratio of nitrate to sulphate in wet deposition in NE America was about 2 and, in summer, it was about 0.7. This is complementary to the present measurements which show the importance of atmospheric N-chemistry during winter.

5. ACKNOWLEDGEMENTS

The authors wish to express their gratitude to Drs. P. Daum and T. Kelly of Brookhaven National Laboratories, USA, for making available the Jan.-Feb. 1984 NO_x data, and to P. Fellin, O. Hunt and D. MacTavish for their expert assistance and advice.

6. REFERENCES

Anlauf, K.G., Fellin, P., Wiebe, H.A., and Melo, O.T.: 1982, Atmospheric Environment 16, 455.

Anlauf, K.G., Fellin, P., Wiebe, H.A., Schiff, H.I., Mackay, G.I., Braman, R.S., and Gilbert, R.: 1985a, Atmospheric Environment 19, 325.

Anlauf, K.G., Bottenheim, J.W., Brice, K.A., Fellin, P., Wiebe, H.A., Schiff, H.I., Mackay, G.I., Braman, R.S., and Gilbert, R.: 1985b, Atmospheric Environment 19, 1859.

Barrie, L.A., and Hales, J.M.: 1984, Tellus 36B, 333.

Bottenheim, J.W., Brice, K.A., and Anlauf, K.G.: 1984, Atmospheric Environment 18, 2609.

Calvert, J.G., Lazrus, A., Kok, G.L., Heikes,B.G., Walega, J.G., Lind, J., and Cantrell, C.A.: 1985, Nature 317, 27.

Meagher, J.F., Stockburger, L., Bonanno, R.J., Bailey, E.M., and Luria, M.: 1981, Atmospheric Environment 15, 749.

Summers, P.W., and Barrie,L.A.: 1986, J. Water Air Soil Poll. (submitted).

Voldner,E.C., and Sirois,A.: 1986, J. Water Air Soil Poll. (submitted).

LICHEN SULPHUR AND LEAD LEVELS IN RELATION TO DEPOSITION PATTERNS IN EASTERN CANADA

E.M. Zakshek, K.J. Puckett
Atmospheric Environment Service
4905 Dufferin Street, Downsview, Ontario, M3H 5T4, Canada and
K.E. Percy,
Maritime Forest Research Centre, Canadian Forestry Service,
Fredericton, New Brunswick, E3B 5P7, Canada

ABSTRACT. Marked differences exist between the S and Pb concentration in Cladina rangiferina collected from eastern Canada and the Northwest Territories, Canada. These differences reflected the differing emission/deposition rates in the two regions. The spatial distribution of S and Pb in the lichen from eastern Canada is described in detail. Mean S concentrations ranged from 329 to 959 $\mu g\ g^{-1}$; Pb concentrations varied in the range 3 to 21 $\mu g\ g^{-1}$. A regional gradient in lichen S concentrations was evident with the highest concentrations being found in central and north-central Ontario and the lowest in Newfoundland. A regional gradient for Pb was apparent but was not as well defined as that for S. The regional distribution of S illustrated by the lichen agreed with another indirect measurement of S deposition. Lichen S concentrations correlated with measured wet sulphate deposition.

1. INTRODUCTION

The atmosphere is an important pathway for the dispersion of potentially toxic elements. Large spatial and temporal variations in element deposition occur making it difficult to define trends in deposition on a regional scale. Consequently, various plant types which accumulate these elements have been used to monitor these trends. Both lichens and mosses have been used extensively to monitor deposition because they are dependent to a large extent on wet and dry deposition to the plant surface for their mineral nutrients. The high retention capacity of these plants results in plant element concentrations in excess of their expected physiological needs. This high retention capacity together with their perennial nature has led to the extensive use of these plants as long term integrators of atmospheric elemental deposition.

Ideally, the only element source should be deposition from the atmosphere and hence, the use of plants in contact with mineral substrates complicates the interpretation of measured lichen element concentrations. Ombrotrophic peat bogs receive their nutrients

Water, Air and Soil Pollution **30** (1986) 161–169.

primarily from the atmosphere and so the Sphagnum species which constitute such bogs or lichens colonizing the bog surfaces are good receptors for the study of the deposition of pollutants to terrestrial surfaces.

Lichens and mosses have been used to monitor local deposition around point and area emission sources, but their use has now been extended to describe regional deposition in Scandinavia (Gydesen et al., 1983) and North America (Groet, 1976; Percy, 1982). Comparing different areas of Canada, Zakshek et al. (1985) found that S and Pb were the only elements which showed consistently higher values in lichens from eastern Canada when compared to the Northwest Territories (NWT) and suggested that these differences were due to differing deposition loadings. In this paper the spatial distribution of S and Pb in lichens from eastern Canada is described in more detail and lichen S concentrations are compared to direct and indirect measurements of S deposition.

2. METHODS AND MATERIALS

In 1980, the lichen Cladina rangiferina (L.) Harm. was collected from the 14 sites in eastern Canada shown in Figure 1. Lichen samples were collected from ombrotrophic peat bogs. The lichen was cleaned of debris

Fig. 1. Map of eastern Canada showing lichen sampling sites.

and the dead and decaying portions of the podetia removed. The podetia were ground in liquid nitrogen, sieved and the fraction which passed through a mesh 70 sieve retained. After oven-drying (75°C for 24hr) the powder was made into pellets for S and Pb analyses by X-ray fluorescence analysis. Details of pellet construction and the X-ray fluorescence technique are given in Tomassini (1976) and Tomassini et al. (1976). The data are expressed as $\mu g\ g^{-1}$ and are the mean of 5-10 replicates ± 1SD.

In 1981, the lichen was collected extensively in eastern Canada for additional Pb and S analyses. The collection and processing procedures were the same as those described above except that only 5 replicates/site were analyzed. A sampling grid (127x127 km) was superimposed on eastern Canada The grid is the same used for computing the long range transport and deposition of S and its use facilitates the comparison of predicted S deposition, measured deposition and lichen S concentrations. Suitable ombrotrophic peat bogs were located in 85 grid squares.

3. RESULTS AND DISCUSSION

The Pb and S concentrations in C. rangiferina from eastern Canada were compared to concentrations in the lichen collected from the NWT (Puckett, 1978). Lead and S concentrations were consistently higher in eastern Canada, with up to 8 and 4 fold differences respectively. Sulphur concentrations were relatively low in Northwestern Ontario (ELA, ER, RR, and WR), rose to peak values (Figure 2) in north-central and central Ontario (EL, FR, D, O) and dropped back to low values in Québec, Labrador and Nova Scotia (FC, W, SI, PM). A similar pattern was observed for Pb; low values in the west and north of the sampling area and higher values in central regions (Figure 3).

Higher S and Pb concentrations in C. rangiferina from eastern Canada are probably the result of qualitative and quantitative differences in the sources of these elements between the two regions. Natural sources differ between the regions, both in their nature and in strength. Natural emissions of S are three times greater in eastern Canada than in the NWT (Environmental Protection Service, 1980). Anthropogenic emissions differ greatly between the areas; man-made emissions in eastern Canada in 1976 were 3.8 x 10^6 tonne compared to 3000 tonne in the NWT (Environmental Protection Service, 1981). These large difference in emissions are reflected in both air concentration and deposition values for S. Daily air concentration data for SO_2 in the NWT are limited but Barrie and Hoff (1984) report peak values of 4.3 $\mu g\ m^{-3}$ while peak values for southern Ontario frequently exceed 15 $\mu g\ m^{-3}$ (Barrie et al., 1984). Similarly, sulphate levels in air and precipitation are usually higher in eastern Canada than in the NWT (Barrie et al., 1981; Barrie and Hales, 1984). Lead emissions in the NWT were 478 tonne in 1978 compared to 8119 tonne in eastern Canada (Environmental Protection Service, 1983). Thus, the lichen appears to be qualitatively reflecting the differences in S and Pb availability in the two regions.

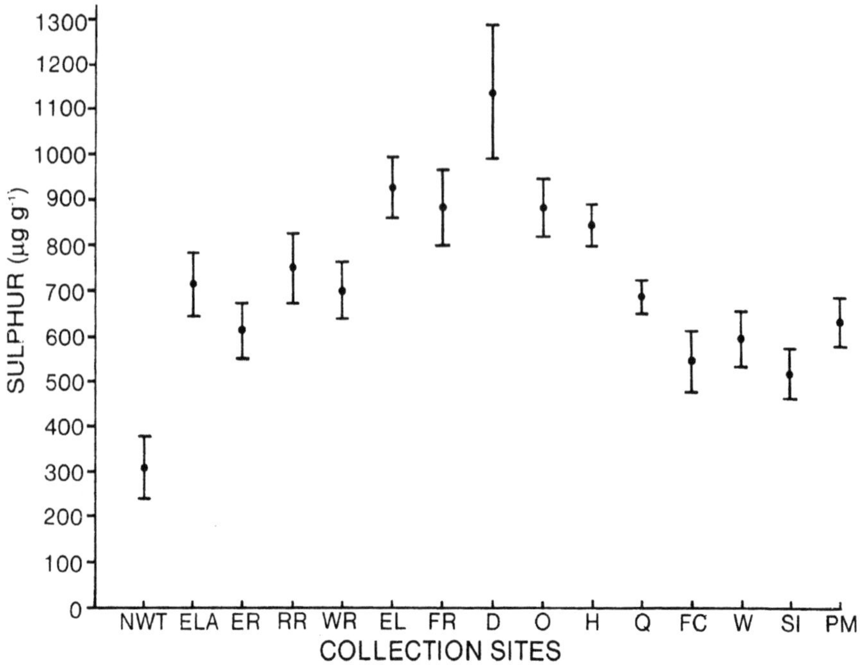

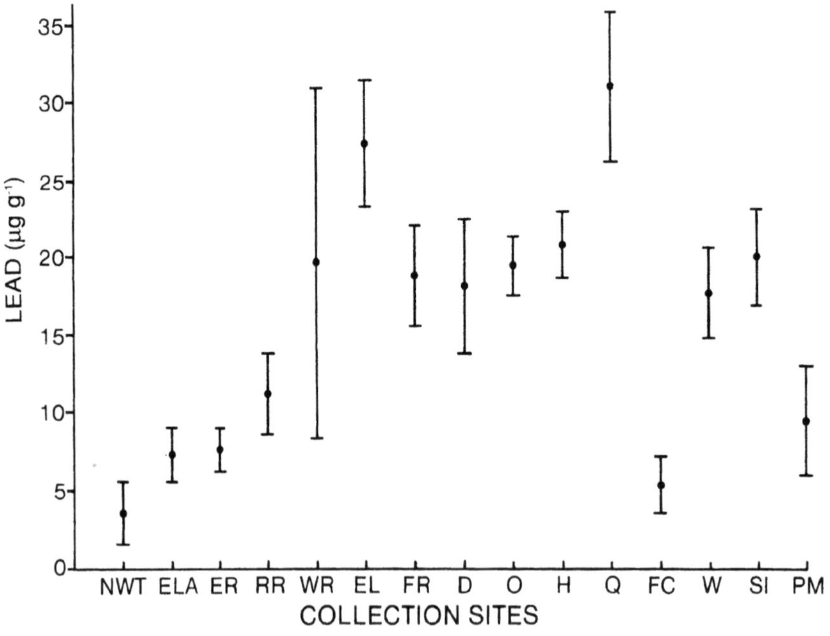

Figs.2. & 3. Sulphur and lead concentrations in C. rangiferina from eastern Canada and the Northwest Territories. Northwest Territories data from Puckett (1978).

As a result of the more intensive sampling in eastern Canada, well defined regional gradients for S could be described (Figure 4). The highest concentrations were found again in north central and central Ontario with the highest concentrations (~1000 µg g^{-1}) being found in the area immediately adjacent to the smelting complex in Sudbury, Ontario. With increasing distance to the east, the S values became progressively less with the lowest concentrations (<350 µg g^{-1}) being being found in Newfoundland. This regional gradient is illustrated further in Figure 5. Sulphur concentrations along a south-west/north-east transect (seen in Figure 4) are plotted as a function of distance along the transect. Initially, the S concentrations increase, peak and then decrease gradually to values much lower than the initial values and approach the concentrations seen in the Northwest Territories.

Regional gradients for the Pb concentrations (Figure 6) were not as well defined as those for S. In general, uniform and high concentrations (15-20 µg m^{-1}) were found throughout southern Ontario and southern Québec with the lower concentrations (< 10 µg g^{-1}) in the more remote areas (Newfoundland).

Lichen S levels were compared with another indirect measure of S deposition. Thompson and Hutton (1985) have examined the sulphate levels in lakes in eastern Canada and have calculated sulphate yields (concentrations x runoff) which are shown in Figure 7 together with the lichen S concentration from the most appropriate grid square. Both

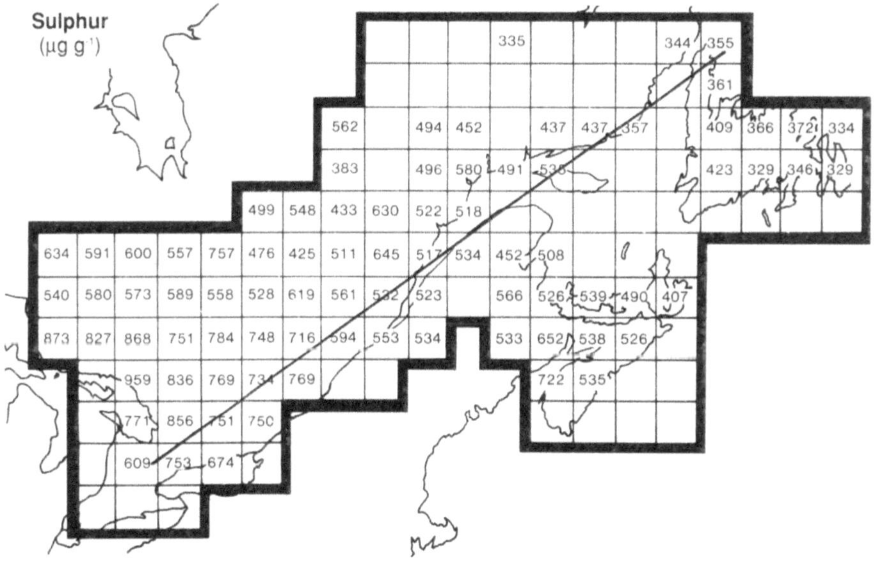

Fig. 4. Sulphur concentrations in C. rangiferina in eastern Canada. Data are the mean value of 5 replicates.

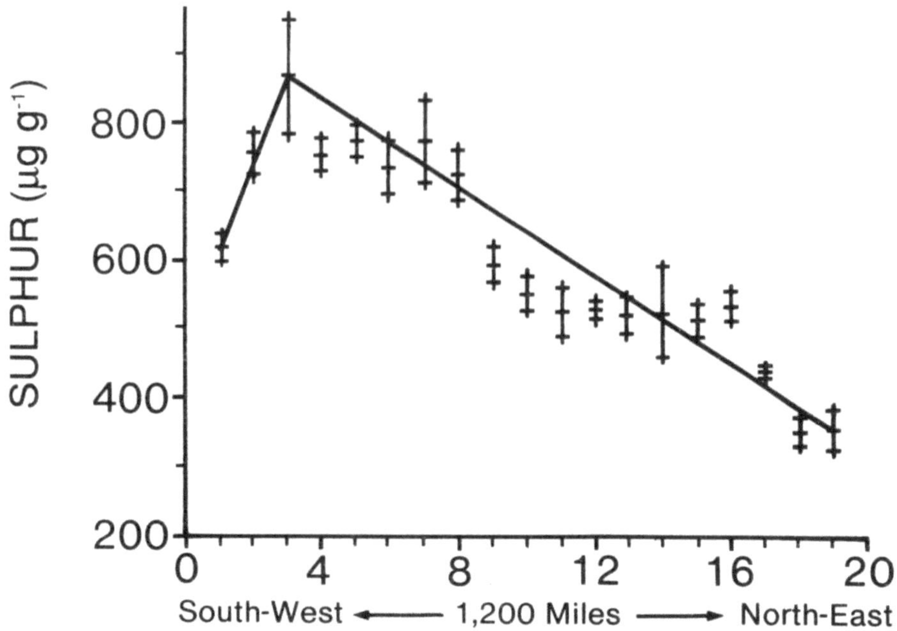

Fig. 5. Sulphur concentrations in C. rangiferina along a SW-NE transect through eastern Canada. The gradations on the X-axis represent individual grid squares along the transect.

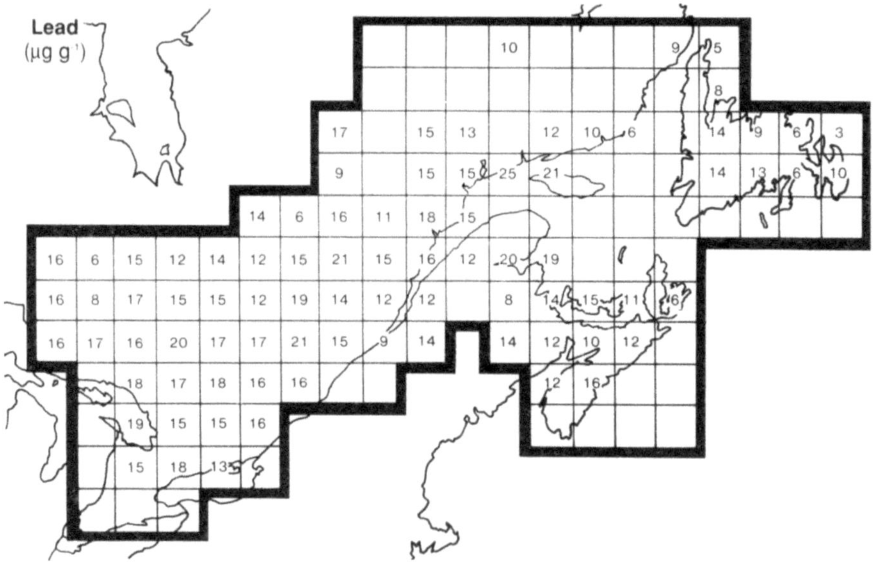

Fig. 6. Lead concentrations in C. rangiferina in eastern Canada. Data are the mean value of 5 replicates.

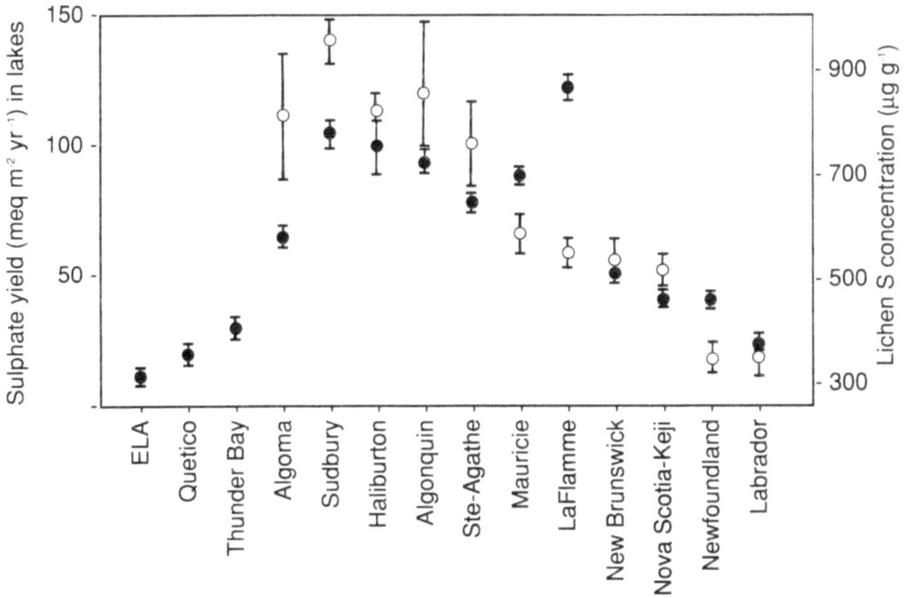

Fig. 7. A comparison of S concentrations in C. rangiferina and sulphate yields in lakes. ●, SO_4 in lakes; ○, S in lichens.

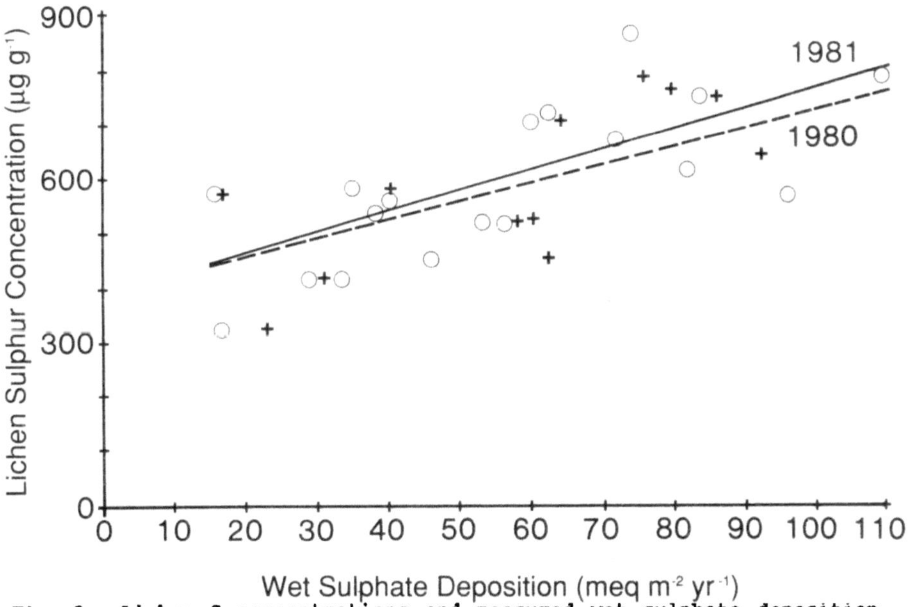

Fig. 8. Lichen S concentrations and measured wet sulphate deposition (1980, 1981) in eastern Canada. +, 1980; ○, 1981.

indirect measurements of S deposition show the same pattern; the lowest lake S yields/lichen concentrations are found in the more remote sites in the east and both receptors show higher values in central Ontario and Québec.

The acceptability of using lichens to describe regional gradients in Pb and S deposition rests, in part, on the agreement between the biomonitoring data and more direct measurements of element deposition. A more quantitative relationship between lichen S levels and S deposition can be described. Annual estimates of wet sulphate deposition for 1980 and 1981 have been reported for 12 and 18 sites respectively (Barrie and Sirois, 1982; Sirois, 1983). A comparison of these measured deposition values with lichen S concentrations from the most appropriate grid square is shown in Figure 8. In general, increasing sulphate deposition is mirrored by increasing lichen S concentrations. The correlation between the two variables although significant ($r=0.6$, $p < 0.01$, $n=18$) does not account for a large component of the variation. There are several reasons for the lack of a more complete agreement. Firstly, the lichen and precipitation collection sites are not co-located; some precipitation sites were close to urban areas whereas the lichen sites were not. Secondly, sulphate deposition is corrected for sea salt contribution whereas the lichen values are not. More important perhaps in explaining the incomplete agreement is that the lichen will reflect both wet and dry deposition to the thallus whereas only wet deposition was compared with the lichen S concentration.

4. REFERENCES

Barrie, L.A. and Sirois, A.: 1982, 'An analysis and assessment of precipitation measurements made by CANSAP (The Canadian Network for Sampling Precipitation): 1977-1980'. Report AQRB-82-003-T. Atmospheric Environment Service, Environment Canada, Downsview, Ontario, M3H 5T4.

Barrie, L.A., Hoff, R.M. and Daggupaty, S.M.: 1981, Atmos. Environ. 15, 1407.

Barrie, L.A. and Hoff, R.M.: 1984, Atmos. Environ. 18, 2711.

Barrie, L.A. and Hales, J.M.: 1984, Tellus, 36, 333.

Barrie, L.A., Anlauf, K.G., Wiebe, H.A., and Fellin, P.: 1984, 'Acidic pollutants in air and precipitation at selected rural locations in Canada.' In: Deposition both wet and dry (B.B. Hicks, ed.). Acid precipitation series - Vol. 4, Butterworth Publishers, London.

Environmental Protection Service: 1980, 'National inventory of natural sources and emissions of sulphur compounds'. Economic and Technical Review Report EPS 3-AP-79-2, Environment Canada, Ottawa.

Environmental Protection Service: 1981, 'A nationwide inventory of emissions of air contaminants (1976). Economic and Technical Review Report EPS 3-AP-80-1, Environment Canada, Ottawa.

Environmental Protection Service: 1983, 'National inventory of sources and emissions for lead (1978). Economic and Technical Review Report EPS 3-EP-83-6, Environment Canada, Ottawa.

Groet, S.S: 1976, Oikos, 27, 445.

Gydesen, H., Pilegaard, K., Rasmussen, L., and Ruhling, A.: 1983, 'Moss analysis used as a means of surveying the atmospheric heavy metal deposition in Sweden, Denmark, and Greenland'. Statens naturvaradsverk 1670, 44pp.

Percy, K.E.: 1982, Water, Soil and Air Pollution, 19, 341.

Puckett, K.J.: 1978, 'Element levels in lichens from the Northwest Territories'. Report ARQA-56-78. Atmospheric Environment Service, Environment Canada, Downsview, M3H 5T4.

Sirois, A.: 1983, 'The Canadian Network for Sampling Precipitation (CANSAP) annual report'. Atmospheric Environment Service, Environment Canada, Downsview, Ontario, M3H 5T4.

Thompson, M.E. and Hutton, M.B.: 1985, Water, Air and Soil Pollution, 24, 77.

Tomassini, F.D.: 1976, 'The measurement of photosynthetic ^{14}C fixation rates and potassium efflux to assess the sensitivity of lichens to sulphur dioxide, and the adaptation of x-ray fluorescence to determine the elemental content in lichens'. M.Sc. Thesis, Laurentian University, Sudbury, Ontario, Canada.

Tomassini, F.D., Puckett, K.J., Nieboer, E., Richardson, D.H.S., and Grace, B.: 1976, Can. J. Bot., 54, 1591.

Zakshek, E.M., Percy, K.E., and Puckett, K.J.: 1985, 'Deposition patterns of lead, sulphur and other elements in eastern Canada as monitored by the lichen Cladina rangiferina'. In: Proceedings of the International Conference on Heavy Metals in the Environment. Athens, Vol. 2, 578-580.

THE DISTRIBUTION OF POLLUTANTS NEAR A FRONTAL SURFACE: A COMPARISON BETWEEN FIELD EXPERIMENT AND MODELING

C.M. Banic and G.A. Isaac
Atmospheric Environment Service
Downsview, Ontario, Canada. M3H 5T4

H.R. Cho and J.V. Iribarne
University of Toronto
Toronto, Ontario, Canada. M5S 1A7

ABSTRACT. A number of frontal systems passed through the experimental area during the 1984 North Bay Acid Snow Study. Two different cold frontal systems were examined using aircraft instrumented for cloud and air chemistry, and cloud microphysics measurements. A distinct maximum in aerosol particle number concentration was detected near both frontal surfaces. In at least one case, a peak in NO_x concentrations was observed simultaneously with the aerosol particle maximum. Using a model of frontogenesis, the aerosol particle layer can be explained in terms of air circulations associated with a front. The model also demonstrates how better flight plans might be designed for air chemistry studies using aircraft.

1. INTRODUCTION

Frontal systems are responsible for much of the precipitation, often quite acidic, that falls on the acid sensitive areas of North America. The distribution of pollutants near the frontal surface is of interest since fronts form the boundary between different air masses, which often have very different pollutant concentrations. This study uses a numerical frontal model to explain observations made during a field project conducted near North Bay, Ontario, from 16 January to 24 February 1984. There is clear meteorological evidence that the project aircraft penetrated the frontal zone of two cold front systems. These cases, both weak cold fronts in relatively unpolluted air masses, were chosen because there was an indication of interesting results in terms of the aerosol particle distribution near the frontal surface.

2. METHOD

A National Aeronautical Establishment Twin Otter and a Canada Centre for Remote Sensing DC-3, were used to obtain profiles of temperature,

Water, Air and Soil Pollution **30** (1986) 171–177.

dewpoint, pressure, air motion, gaseous N species, and cloud droplet and aerosol particle number densities. In addition, samples of cloud water, snow and aerosol particles were collected. The aerosol particle concentrations used in this paper were measured with a Particle Measuring Systems Active Scattering Aerosol Spectrometer Probe which detects particles in the 0.2 to 2 µm size range (see Isaac **et al.**, 1986). The aircraft NO_x detector used is a specially designed high sensitivity instrument (Tanner **et al.**, 1983).

An overview paper on the North Bay field project describing the instrumentation in more detail has been prepared by Isaac and Daum (1986).

3. RESULTS AND DISCUSSION

3.1 Meteorology and Operations:

The first system of interest passed through the study area on 4 February. It was associated with a weak low pressure centre situated to the southwest of North Bay. The frontal system was quasi-stationary, but locally had many of the dynamic features of a cold front. Its passage at the surface stations in the study area was characterized by a small perturbation in temperature and by a shift in surface wind direction from south westerly to north easterly.

The results presented in this paper for 4 February were taken on a flight to the northwest of North Bay. Four frontal crossings were made. Aerosol filter samples were taken in the warm sector both at 0.9 km and at 3.1 km. Snow samples were collected along the flight path as well as cloud water samples when in cloud.

On 9 February the second system, a cold front, moved through the project area from the northeast. Its passage at the surface stations was characterized by a drop in temperature, a rise in pressure, and a veering of the surface wind from south westerly to easterly or south easterly.

The flight on 9 February was to the south of North Bay, intersecting the cold front both on ascent and descent. One aerosol filter and two cloud water samples were collected in the warm sector.

For both studies the flight tracks were almost perpendicular to the front.

3.2 Field Observations:

Vertical profiles measured by the aircraft are shown in Figure 1. Just as at the surface, the frontal zone was characterized by a shift in temperature and by a veering of the wind. Associated with the frontal zone was a definite peak in aerosol particle number concentration. On 4 February the aerosol layer was closely associated with the region where the winds were changing direction, and the maximum vertical gradient in temperature was at a slightly lower altitude. The vertical aerosol particle number concentration profile for 9 February is very complex but the aerosol maximum again occurs near the same altitude as the wind shift.

The NO_x levels were low throughout both flights. However there is some indication of a peak in NO_x associated with the aerosol particle maximum of 4 February.

3.3 Modeling of the Aerosol Particle Layer:

To simulate the formation of the aerosol layer, a model of frontogenesis was developed, based on the uniform potential vorticity model described in Hoskins and Bretherton (1972). The model simulates

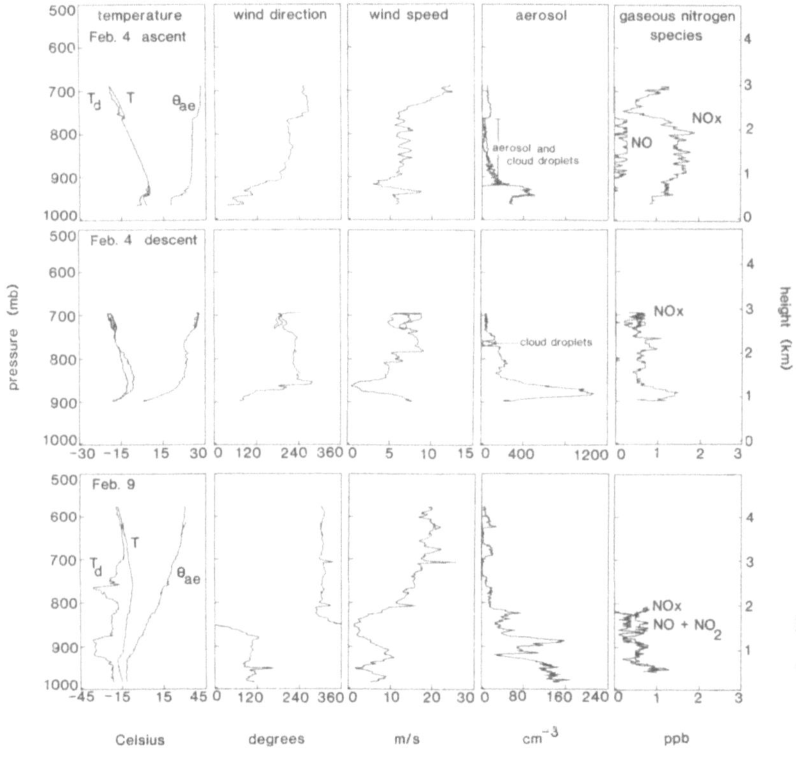

Fig. 1: Profile summary for the flights of 4 and 9 February 1984. The curves shown for temperature are dew point (left), dry bulb (centre), and adiabatic equivalent potential (right) temperatures. For wind direction 0° is north. In the fourth frame the curve is for aerosol particles unless otherwise indicated. [NO] was measured for the descent of February 4, and was found to be ≈ 0. The ascent and descent profiles for February 4 were taken ≈ 200 km apart.

the formation of a front in two spatial dimensions. The orientation of the front is perpendicular to the x-axis of the model. The z-axis is in the vertical direction. The assumption is made that the frontogenesis is caused by a deformation field at the synoptic scale with a rate of deformation $\alpha = 10^{-5}\ s^{-1}$. When such a flow field is forced upon a region of the atmosphere with large horizontal temperature contrast a self-accelerating contraction is induced which leads to the rapid formation of a front.

The initial potential temperature field of the participating air masses is assumed to be

$$\theta = \theta_o + \frac{N^2\ \theta_o}{g}\ z + \frac{\Delta\theta}{\pi} \tan^{-1} (x/L) \qquad (1)$$

where N is the Brünt-Väisälä frequency, g the gravitational acceleration, x the horizontal distance and L the horizontal scale. The values of the parameters used are

$$\theta_o = 300\ K,\ N = 10^{-2}\ s^{-1},\ \Delta\theta = 24K,\ L = 1000\ km. \qquad (2)$$

The initial distribution of aerosol particle number concentration is assumed to be

$$A(x,z) = a_o(x)\ \exp(-z/D) \qquad (3)$$

with D = 1 km. The aerosol concentration decreases exponentially with height. Two horizontal profiles are assumed for $a_o(x)$. In the interior of the warm and cold air masses the aerosol concentration at the 975 mb level is assumed in both cases to be 1200 and 200 cm^{-3}, respectively. These were based on the observed concentrations reported by Isaac and Daum (1986). In the region between the two air masses, a linear transition is assumed. In the first case (distribution I) the transition takes place over a distance of 1000 km in a region where most of the variation in temperature takes place. In the second case (distribution II) the transition takes place over a distance of 500 km, so that the slope of the fractional variation is comparable to that of the temperature field in the main region of the transition zone.

The simulated distributions of aerosol concentrations when the front is fully formed (t = 49 h) is shown in Figure 2 for the second aerosol distribution used. The solid curves are contours of constant aerosol concentration. The dashed curves are constant potential temperature contours. For both initial distributions used an aerosol layer formed along the frontal surface. At a fixed position along the x-axis, aerosol concentration first decreased with height, reaching a minimum near the level where the vertical gradient of potential temperature appeared to be the steepest. Above this level a second maximum in aerosol concentration was located.

Comparing the two cases, one finds that the aerosol layer has a more distinct structure when the initial horizontal gradient of aerosol concentration is steeper (distribution II). In the first simulated case, although an elevated aerosol layer may be defined, the maximum concentration in the layer is much lower than that near the ground at the same location. Furthermore, no second maximum of aerosol concentration at an elevated level can be found within about 100 km of the position of the surface front. In the second case, a well-defined aerosol layer can be found along the entire frontal surface below the 2.5 km level. The maximum concentration in the layer is comparable to that near the ground at most places.

3.4 Comparison of Modeling Results and Field Observation:

The results of the model agree well with the observations of the aerosol layer. The minimum in aerosol concentration is located at the region of maximum vertical potential temperature gradient as predicted by the model. The aerosol concentration minima predicted by the model are not as low as those observed, and the maxima are broader. However the model does predict sharper maxima for the more abrupt variation in aerosol concentration between the two air masses (distribution II). As well, the model is stopped upon formation of the front. Additional dynamic effects as the frontal system moves have not been considered.

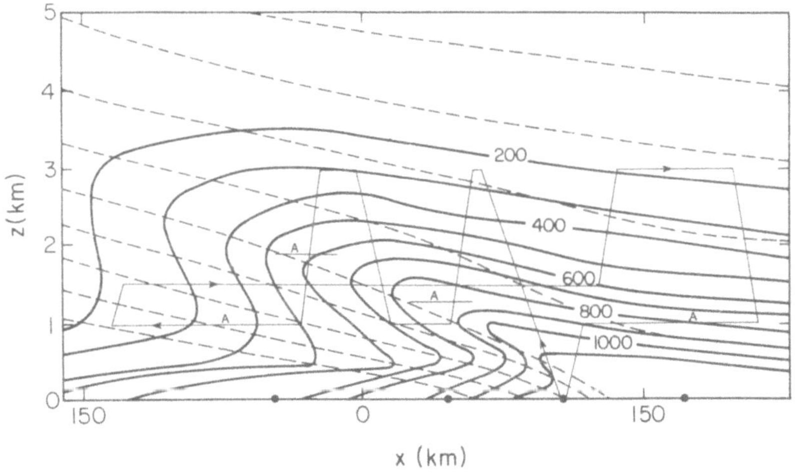

Fig. 2: Model results for initial aerosol distribution II. The dashed lines are intervals of θ where $\Delta\theta = 2.6$ K and the heavy solid lines are the aerosol concentrations. The light solid line shows an idealized flight plan for sampling the chemical constituents of the air masses and frontal zone. Appropriate ground sampling sites are indicated by ● .

3.5 Relationship Between Aerosol Composition and Cloud Chemistry:

If it can be established that the aerosol collected on a filter is representative of the aerosol in the air feeding the cloud, then the composition of constituents of the aerosol can be compared with the composition of the cloud water. In studies of frontal situations, it is difficult to determine where to sample the aerosol. The conditions at the ground may not be representative of those aloft where the clouds are forming because the air above and below the front has different origin. In addition, the air forming the cloud may have travelled long distances after leaving the ground source of pollutants, and its composition in terms of primary and secondary pollutants may have changed.

In consideration of the location of the aerosol layer relative to the frontal surface, and the conservation of particle number density (aerosol + cloud droplets) observed in these field studies, the indication is that the clouds observed on 4 and 9 February may be formed by condensation about aerosol particles such as in the aerosol layer. A substantial fraction (60 to 80%) of the aerosol particles are incorporated into the cloud water as cloud condensation nuclei since the cloud droplet concentrations were equal to or greater than the aerosol concentration in cloud.

A preliminary analysis of the cloud chemistry data shows the pH of the cloud water collected on 4 February to be 3.6. The aerosol composition on the ground before and after the frontal passage, and as measured in the aerosol layer (at 0.9 km) was very similar. (There need not be a difference in aerosol particle concentration across a front. This would depend on the origin of the air masses.) If one half of the aerosol particles, either ground based or in the layer, were to be dissolved in the cloud water a pH of 3.7 would be realized. The cloud water acidity is well correlated with the aerosol acidity. The HNO_3 concentration measured aloft is low and has not been included in the acidity. If the HNO_3 measured at the ground was to be included in the cloud water acidity the pH would be 3.4.

For 9 February the cloud water pH is 4.8. If one-half the ground based aerosol was to be dissolved the cloud water pH would be 3.9. The cloud water acidity cannot be explained by the ground-based aerosol. Any dissolved HNO_3 would increase this discrepancy.

Aerosol filter samples should be taken in the front associated pollutant layer aloft during field experiments. It is operationally sensible (relatively high concentration) and it would be useful to compare the composition of this aerosol with that at the ground, and with the composition of cloud water.

The model results suggest the flight plan shown in Figure 2. The aircraft would take-off shortly after passage of the surface front and make up to six frontal crossings while sampling air both in the warm and cold air masses. The aircraft would fly in a holding pattern at the levels indicated by A to collect aerosol samples. The low altitude sections should be done as low as operationally possible. It may be necessary to use more than one aircraft to make these measurements, both for operational reasons, and because the front is moving. Measurements of aerosol concentration and filter samples should be made at selected sites along the surface.

4. SUMMARY

There is a layer of relatively high aerosol particle number concentration associated with cold front surfaces. A model of frontogenesis can be used to predict the formation of this layer. Development of this layer depends on the contrast of pollutant concentrations between the two air masses, the vertical variation of the pollutant concentration in each air mass, and the developmental stage of the front.

Since the model can follow the aerosol, it can also be used to follow pollutant gases. If chemistry and moisture are included in the model, predictions of the interaction between pollutants and cloud droplets may be made. Field experiments could then be designed to sample along profiles suggested by the model.

5. ACKNOWLEDGMENTS

The North Bay 1984 field program was a cooperative effort between the Atmospheric Environment Service (AES), the National Aeronautical Establishment, the Canada Centre for Remote Sensing, the Ontario Ministry of Environment, Ontario Hydro, and the Brookhaven National Laboratory (BNL) of the United States. The authors thank the many scientists and technicians from AES and BNL who collected the data and provided advice for this paper, with special mention given to J.W. Strapp and W.R. Leaitch of AES.

6. REFERENCES

Hoskins, B.J. and Bretherton, F.P.: 1972, J. Atm. Sci. **29**, 11.

Isaac, G.A. and Daum, P.H.: 1986, 'Winter Cloud Chemistry Studies Near North Bay, Canada'. Paper to be presented at Conference on Cloud Physics, Sept. 22-26, 1986, Colorado.

Isaac, G.A., Leaitch, W.R., Strapp, J.W. and Analuf, K.G.: 1986, Atmos. Environ. **20**, 157.

Tanner. R.L., Daum, P.H., and Kelly, T.J.: 1983, Int. J. Envir. Analyt. Chem. **13**, 323.

MONTHLY MEAN SPATIAL VARIATIONS OF DRY DEPOSITION VELOCITIES OF OXIDES OF SULPHUR AND NITROGEN

E.C. Voldner and A. Sirois
Environment Canada
4905 Dufferin Street,
Ontario, M3H 5T4, Canada

ABSTRACT. Dry deposition processes play an important role in delivering acidic SO_x and NO_x to the surface. Modeled regional S and N budgets in North America indicate that about half of the total deposition is dry deposited. Thus, there is advantage to be gained in more accurately representing dry deposition in 'acid deposition' model calculations. A scheme, based on a resistance analogy, for estimating areal-average deposition velocities of gaseous and particulate oxides of S and N is used. From a literature review of observations, estimates of surface resistances for gases and near surface resistances for particles are obtained. Boundary layer theory and meteorological data are used to estimate aerodynamic resistances for gases and particulates as well as near surface resistance for gases. Coupled with information on land-use, monthly mean dry deposition velocity fields are obtained. This paper briefly reviews the parameterization schemes and the required data bases. It discusses spatial and temporal variation in dry deposition velocities of the oxides.

1. INTRODUCTION

Dry deposition processes play an important role in removing acidic oxides of S and N from the atmosphere. Modeled S and N budgets in North America indicate that about half of the total deposition is dry deposited (Galloway and Whelpdale, 1980; Voldner and Shannon, 1983; Logan, 1983). Thus, there is advantages to be gained in more accurately representing dry deposition in 'acid deposition' model calculations. In this study the effort is focussed on North America.

Dry deposition is most commonly estimated as the product of the measured or model predicted air concentration of the pollutant near the ground and a dry deposition velocity. The deposition velocity is the inverse of the total resistance to pollutant uptake.

$$V_z^{-1} = r_t = r_a + r_b + r_s$$

Water, Air and Soil Pollution **30** (1986) 179–186.

where r_a is the aerodynamic resistance, r_b the near-surface resistance and r_s the surface resistance. V_z depends on surface characteristics and meteorological conditions (see e.g. Chamberlain, 1980; Hicks, 1984), which vary both spatially and temporally.

Temporal variation in dry deposition velocity is generally considered in operational long-range transport (LRT) models, but owing to lack of information, the spatial variation is usually ignored. This was recognized as a shortcoming in current modeling efforts at a resent workshop on 'Sources and Evaluation of Uncertainty in Long Range Transport Models' (A.M.S., 1985).

We have estimated the monthly mean dry deposition velocity fields of SO_x and NO_x with a spatial resolution of about 100 km, compatible with most LRT models. This is an extension to and improvement of earlier work on SO_x by Sheih et al. (1979) and Masse and Voldner (1983). In this paper we summarize the methodology and give examples of temporal and spatial variations in dry deposition velocities of the acidic oxides.

2. APPROACH

Dry deposition velocities of SO_x and NO_x are functions of surface characteristics and meteorological conditions. The surface conditions depend on the physical and physiological state of the ground cover, which are influenced by the season. In this study the progression of seasons are approximated from information on climatological snow boundary, ice-cover and near-surface temperature. Thus winter prevails north of the snow boundary, spring/fall prevails in the region between the snow boundary and the 10°C surface isotherm and summer prevails in areas south of the 10°C isotherm. Climatological information on formation and break-up of ice cover determines if a water body is frozen or open.

The aerodynamic resistance (r_a) associated with turbulent eddy transport in the surface boundary layer is calculated from the micro-meteorological relationships of Wesely and Hicks (1977) and Sheih et al. (1979). The calculations require information on atmospheric stability and associated wind speeds, which were derived from hourly weather observations in North America (Sirois and Voldner, 1985). Information on surface roughness is also required. Values of roughness over land were obtained from a literature review (Voldner et al., 1985), while over open water bodies the relationship by Hicks and Liss was used (1976).

The near-surface layer resistance (r_b) is more difficult to parameterize. For gaseous deposition to all surfaces except snow the expressions, based on molecular diffusivity, by Wesely and Hicks (1977), Sheih et al. (1979) and Hicks (1984) were used, while for snow a relationship derived from experiments by Barry (1965) was deemed more appropriate. The choice of parameterization is discussed by Voldner et al. (1985). The expressions are less suitable for transfer of particles across the near-surface layer, since the transfer is particle-size dependent (Slinn and Slinn, 1980; Slinn,

1982; Ibrahim et al., 1983). Except for over water, where we follow suggestions by Slinn and Slinn (1980), we rely on field observations.

The surface resistance for particles less than 10 μm in diameter (Hicks and Garland, 1983) is negligibly small to all surfaces (i.e. bounce-off does not occur), while for gases, it depends on the chemical and biological reactivity of the gases with the surface. The resistance is difficult to parameterize and hence we have relied on field experiments. The choice of r_b for particles and of r_s for gases is well documented in a literature review by Voldner et al. (1985). In essence, 'average' resistance values are found for each season and land-use type. Thus the choice of values reflects, for example, a typical change in an agricultural region, where the surface is snow covered in the winter, is a mixture of bare soil and active crop in the spring, mature crop in summer and a mixture of bare soil and senescent or stubbed vegetation in fall.

3. SYNTHESIS OF RESISTANCES INTO MONTHLY MEAN DRY DEPOSITION VELOCITIES

The following procedure was used in estimating a monthly grid-square average V_z:

i) the frequency distribution of occurrence of seven stability classes spanning the range of atmospheric stability and the associated mean horizontal wind speed were derived from hourly weather observations (Sirois and Voldner, 1985).

ii) a best estimate of surface roughness for each of the seven surface types in the land-use data bank (Sirois and Voldner, 1984) was made for each season (for water the expression by Hicks and Liss (1976) was used).

iii) for each surface type and for each stability class, for which a mean wind speed was known, V_z was calculated from r_a, r_b and r_s

iv) the known frequency distribution of stability classes for the month of interest was used to calculate the mean monthly deposition velocity for a particular surface

v) to obtain the grid-square mean deposition velocity, the surface area mean V_z was calculated from the fractional surface cover and the mean monthly V_z of each surface type obtained in (iv).

4. DISCUSSION

Monthly mean dry deposition velocity fields were computed for SO_2, NO_x, sulphate, nitrate and HNO_3. These will be discussed in detail elsewhere. Here we merely wish to illustrate the results with a few selected examples.

The monthly mean dry deposition velocities of SO_2 for January and July are shown in Figure 1a and b with a spatial resolution of 127 x 127 km. The solid line shows the snow boundary, north of which winter is assumed. The ice boundary, illustrated by the dash-dot

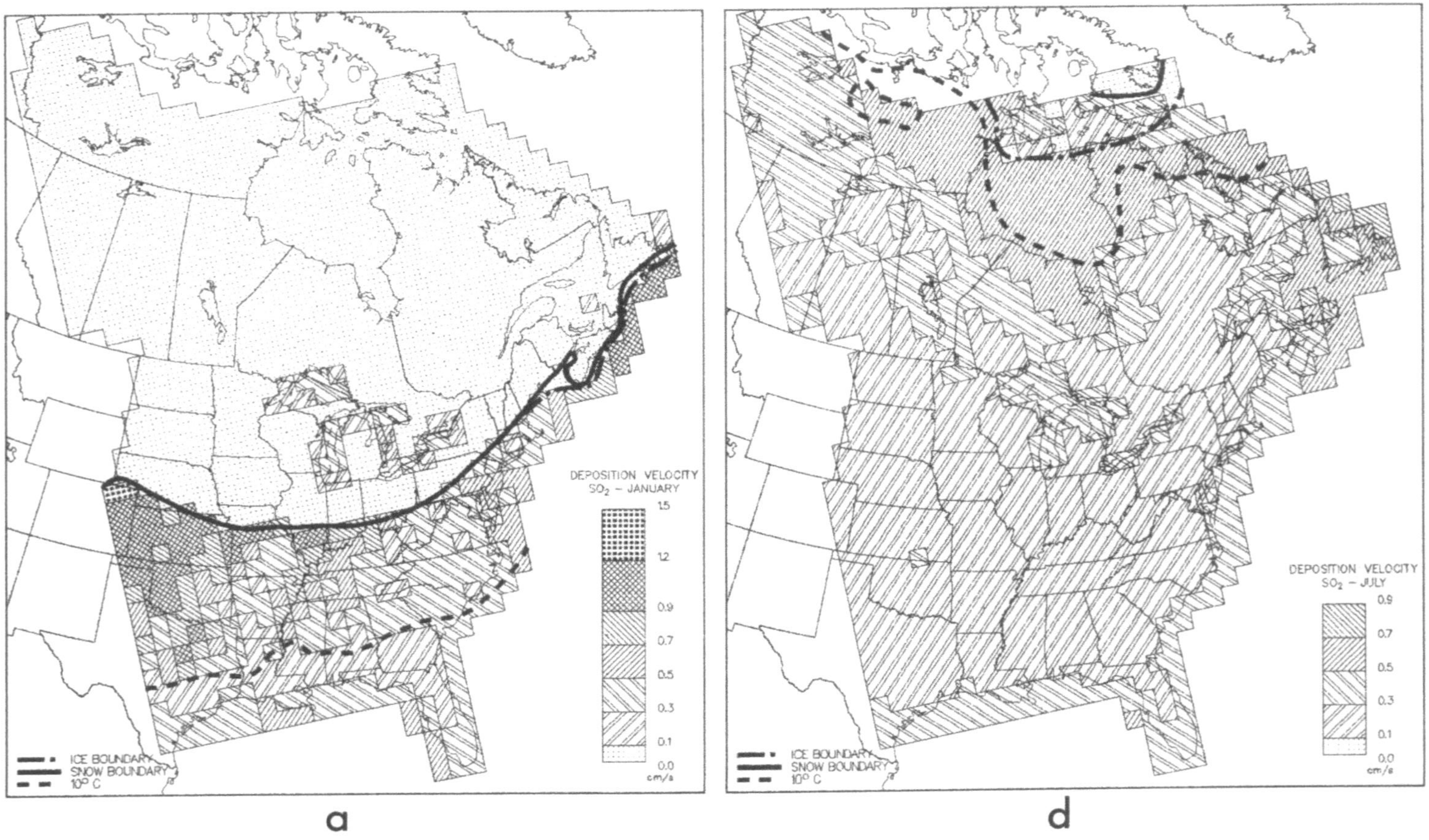

Figure 1: Monthly mean dry deposition velocity fields of SO_2 for a) January and b) July.

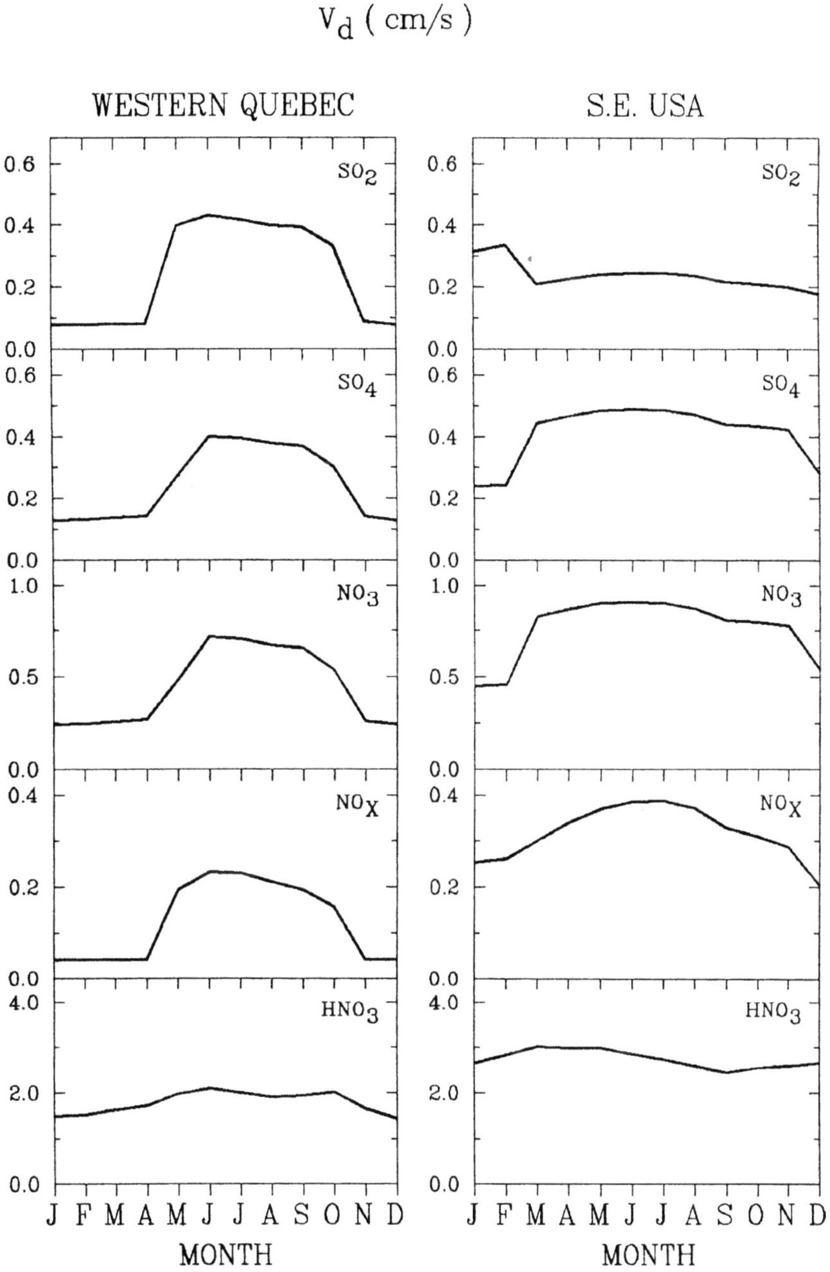

Figure 2: Temporal variation in the dry deposition velocity of SO_x and NO_x at two locations in North America.

line, runs south of the snow boundary in Figure 1b, indicating that the water bodies have a significant ice cover even in regions classified as spring. In Figure 1a, snow and ice boundaries essentially coincide. The Great Lakes, however, remain ice free. The 10°C isotherm, drawn as a dashed line, approximates the boundary between spring and summer. Thus, in July most of the continent experiences summer-like conditions.

The importance of accounting for the spatial and temporal variation in the dry deposition velocity of SO_2 is clearly illustrated in Figure 1a. On the average, the dry deposition velocity in the northern snow-covered regions is less than 0.1 cm s^{-1}. The highest dry deposition velocities are found in the springtime over the mainly agricultural region in the southwest, where the average values reaches 1 cm s^{-1}. During summer, the dry deposition velocities are fairly homogeneous over the region with monthly mean values in the range 0.2 to 0.5 cm s^{-1}.

Next we would like to illustrate the temporal variation in dry deposition velocities of SO_2, NO_x, sulphate, nitrate and HNO_3 at two grid locations in eastern North America; viz at the Ontario/Quebec border and in North Carolina (S.E. U.S.A.). At the western Quebec site the dominant land-use categories are decidous forest ($\sim$20%), coniferous forest ($\sim$20%) and swamp ($\sim$50%), while at the U.S. site the decidous forest contributes more than 45%, coniferous about 30% and cultivation 20%. The deposition velocities for all species, except HNO_3 are low at the Quebec site during winter, due to the dormancy of the forest and low uptake over frozen swamp, increase during spring, peak in the summer and decrease in the fall. This is caused primarily by the change in metabolic activity and in leaf area index of the foilage as well as thawing of the swamp land. The dry deposition velocity of the nitrate particles is roughly twice that of sulphate particles, since the nitrate particles generally are larger than the sulphate particles. The dry deposition velocity of HNO_3 is fairly flat, reaching values of about 2 cm s^{-1} in the summer. The HNO_3 profile essentially reflects the aerodynamic resistance at the sites, since it is assumed that the surface resistance for HNO_3 is low (0.1 cm s^{-1}).

The variation in dry deposition velocities at the southern site is less pronounced, mainly due to the long summer-like conditions. The dry deposition velocity of SO_2 dioxide is out of phase with the other species. Thus the deposition velocity is higher during January - February, i.e. during "fall and spring". The high values are attributed to the low resistance to uptake of SO_2 to bare soil. On the average, the aerodynamic resistance at the southern site is lower than at the northern. This is mainly due to the higher solar intensity and therefore higher frequency of unstable and neutral conditions in the south.

5. SUMMARY

The paper gives a brief review of an effort to estimate dry

deposition velocities of acidic oxides in North America in a format suitable for input to LRT models and for interpretation of air quality measurements. The necessity of this kind of work was recognized at the Woods Hole workshop.

It should be emphasized that the focus has been on estimating representative monthly means rather than following short term fluctuations in dry deposition velocity over a specific surface. Thus, the values reported here may deviate from instantaneous values measured in field experiments.

REFERENCES

American Meteorological Society: 1985, Workshop on Uncertainty in Long-Range Transport Modelling, Woods Hole, Mass. Sept. 18-21, 1984. Proceedings to be published in AMS Bulletin.

Barry P.J.: 1965, The use of radioactive tracer gases to study the rate of exchange of water vapour between air and natural surfaces, Symp. on Isotope Techniques in the Hydrologic Cycle, University of Illinois, Champaign-Urbana, Illinois, Atomic Energy of Canada, Report 2576.

Chamberlain A.C.: 1980, Dry deposition of sulfur dioxide, Chap. 22, Atmospheric Sulfur Deposition, eds. Shriner, Richmond, Lindberg, Ann Arbor Science, 187-197.

Galloway J.N. and Whelpdale D.M.: 1980, Atmospheric Environment, 14, 409.

Hicks B.B.: 1984, The acidic deposition phenomenon and its effect A-7 Dry deposition processes, Critical Assessment Review-Papers, Vol. 1, Atmospheric Sciences, EPA-600/8-83-016AF, 7-1 - 7-70.

Hicks, B.B. and Garland J.A.: 1983, Overview and suggestions for future research on dry deposition, Precipitation Scavenging, Dry Deposition and Resuspension, eds. Pruppacher, Semonin and Slinn, 1429-1432, Elsevier.

Hicks B.B. and Liss P.S.: 1976, Tellus, 28, 348.

Ibrahim M., Barrie L.A. and Fanaki F.: 1983, Atmospheric Environment, 17, 781.

Logan J.: 1983, Nitrogen oxides in the tropospheric: global and regional budgets, J. Geophys. Res., 88, 785 and 807.

Masse C. and Voldner E.C.: 1983, Estimation of dry deposition velocities of sulfur over Canada and United States east of the Rocky Mountains, Precipitation Scavenging, Dry Deposition, and Resuspension, eds. Pruppacher, Semonin and Slinn, 991-1002, Elsevier.

Sheih C.M., Weseley M.L. and Hicks B.B.: 1979, A guide for estimating dry deposition velocities of sulfur over the eastern United States and surrounding regions, Argonne National Laboratory, Rep. ANL/RER-79-2, 1-56.

Sirois A. and Voldner E.C.: 1985, Estimation of dry deposition velocities in North America part II: The meteorological data banks, Canadian Meteorological Research Report - accepted for publication.

Sirois A. and Voldner E.C.: 1984, Estimation of dry deposition velocities in North America part I: The land-use data bank, Canadian Meteorological Research Report CMRR 1/84, pp. 28.
Slinn, W.G.N.: 1982, Atmospheric Environment, 16, 1785.
Slinn S.A. and Slinn W.G.N.: 1980, Predictions for particle deposition on natural waters, Atmospheric Environment, 14, 1013-1016.
Voldner E.C. and Shannon J.D.: 1983, Evaluation of predicted wet deposition fields of sulfur in eastern Canada, Transactions: the Meteorology of Acid Deposition, an APCA Specialty Conference, 387-399.
Voldner E.C., Barrie L.A. and Sirois A.: 1985, A review of dry deposition of oxides of sulphur and nitrogen with emphasis on long range transport modelling in North America, Submitted to Atmospheric Environment.
Wesely M.L. and Hicks B.B.: 1977, J. Air Poll. Contr. Assoc., 27, 1110.

EDDY CORRELATION MEASUREMENTS OF DRY DEPOSITION FLUXES USING A TUNABLE DIODE LASER ABSORPTION SPECTROMETER GAS MONITOR

G.C. Edwards and G.L. Ogram
Ontario Hydro Research Division
800 Kipling Avenue
Toronto, Ontario
Canada, M8Z 5S4

ABSTRACT. An instrument package, based on a tunable diode laser absorption spectrometer and comprehensive micrometeorological instrumentation, has been successfully developed for making dry deposition flux measurements, using eddy correlation techniques. Preliminary results from the field testing and evaluation of the instrument package at a rural location are presented. Low deposition velocities were measured for NO_2 and SO_2 to snow.

1. INTRODUCTION

Over the past 2 yr an instrument package based on a tunable diode laser absorption spectrometer (TDLAS) and a comprehensive micrometeorological instrument package has been developed for making dry deposition flux measurements using eddy correlation techniques, Edwards and Ogram (1983). The overall objective of the research program is to investigate the physical processes that control dry deposition and to quantify dry deposition for various chemical species, surface types and meteorological conditions. An important objective of the first phase of the research was to develop the potential of the TDLAS to make fast response, high sensitivity in-field measurements of trace gas concentrations.

Diode laser absorption spectroscopy has, in the past, been successfully used to monitor atmospheric trace gases at ppbv levels, (Reid et al., 1978a; Schiff et al., 1983). The technique has the advantage of high sensitivity along with high selectivity. Because of the narrow linewidth of the laser it is possible to unambiguously measure a particular gas species without the problem of interference from other background gases. The technique is also versatile in that it can be applied to many trace gases of interest. In these experiments the time response of the TDLAS has been improved to allow the fast time response necessary for eddy correlation.

Water, Air and Soil Pollution **30** (1986) 187–194.

The following sections discuss the instrumentation, experimentation and results of the preliminary field tests, along with information on present activities.

2. INSTRUMENT PACKAGE

The reliable measurement of pollutant fluxes by the eddy correlation technique requires the measurement of key surface layer quantities u,v,w,T,q and C, sensitively and with fast response. The vertical wind speed and the pollutant fluctuating components w' and c', respectively, determines the pollutant flux F by,

$$F_c = \overline{w'c'} + \overline{w}\,\overline{c}$$

where the overbar indicates averaging time. The measurement of the horizontal wind speed u facilitates the calculation of the momentum flux and the friction velocity. Temperature and humidity measurements allow determination of sensible and latent heat fluxes.

The principal instrumentation used to measure the variable for calculating the fluxes were two sonic anemometers (Kaijo Denki Dat 310, 61B omnidirectional), two Lyman-alpha hygrometers (Atmospheric Instrumentation Research Inc), and the TDLAS. (Mention of a commercial product does not constitute approval by the research organizations involved, to the exclusion of other commercial products suitable for the same purposes).

Instruments were tower-mounted at the 4- and 8-m levels providing measurements of u', v', w', T' at these levels, and pollutant concentration of C' at one level. This configuration was used to test the instrument package.

To be useful in eddy correlation measurements the TDLAS must operate with a response time and sensitivity of <1 s and ~1 ppbv, respectively. The demonstrated sensitivity of the instrument is already sufficient for eddy correlation measurements of many species, see Table I. However, the time response to fluctuations in ambient concentration has been limited by the exchange time of the sampling cell. Fast exchange times were achieved here by rapidly pumping the sampling cell.

The principle of the TDLAS technique is the measurement of the extent of laser absorption over a long path giving a measure of the number of pollutant molecules in this path. The system is built around a Laser Analytics diode laser system and multipass cell. The diode laser is mounted in a cold head cooled by a cryogenic He refrigerator. The center frequency of the laser emission is controlled by the cold head temperature. Each diode is temperature-tunable over typically 100 cm^{-1}.

Diodes may be obtained to cover the region of the infrared spectrum from about 2 to 30 cm^{-1}, a region in which many trace gases of interest have strong absorption lines. The laser beam is passed through the multipass cell which enables the beam path length through the gas sample to be set at values ranging from 4 to 100 m. The air sample is flowed rapidly through the cell at a pressure of 10 Torr or less. The narrow linewidth of the diode laser (~10^{-4} cm^{-1}) ensures that the individual absorption lines are easily resolved without interference from neighbouring lines due to other absorbers and thereby enabling the detection without ambiguity of a particular species. The sample exchange time in the cell provides the upper limit on detection response time. The relatively large volume of the cell (~10 L) requires a fast pumping speed to achieve exchange times of less than one second. A Bush R5 super series rotary pump with a pumping speed of ~80 L s^{-1} at 10 Torr was used. For the flows under consideration, sample introduction time delays can be maintained near 0.1 s while maintaining sufficient gas conductance with an inlet tube of a few meters in length. This enables the multipass cell to be placed a reasonable distance from the sampling point. After passage through the multipass cell the beam is detected with a HgCdTe IR detector.

Second harmonic detection is used to enhance signal to noise and has the benefit of making the instrument relatively insensitive to vibration. The components of the system are reasonably compact and portable.

TABLE I ATMOSPHERIC TRACE GASES DETECTED USING DIODE LASERS

Species	Wavelength Region (cm^{-1})	Detection Limit (ppbv)	Reference
SO_2	1140	3	
	1370	0.2-1	Reid (1979)
NO_2	1600	0.1	
	1600	0.5	Reid (1980)
NO	1850	0.5	Shiff (1983)
HNO_3	1690-1720	0.35	
O_3	1050	0.5	El-Sherbiny (1979)
NH_3	1140	0.1	Reid (1978a)
N_2O	1140	1	Reid (1978b)

3. FIELD EXPERIMENTATION AND RESULTS

The instrument package was field-tested and evaluated at the University of Guelph Elora Agricultural Research Station located 5 km south of Elora, Ontario and 16 km NW of Guelph, Ontario, Canada. The site was selected because the topography and ground cover provided acceptable horizontal homogeneity and fetch conditions.

The eddy correlation experiments were located at the northern midpoint of a relatively flat region of agricultural land. The fall experiments were carried out over a field of alfalfa stubble. The stubble was from 6 to 10 cm high with some leaf attached. The surface roughness length was estimated to be from 5 to 10 cm. The uniform upwind fetch for these conditions was approximately 600 m for wind directions of ESE clockwise through SW, before the surface cover changed. Fields of maize surrounded the alfalfa; however, the maize was cut before the field experiments were carried out.

During winter, the entire region was snow covered (estimated roughness length of 0.01 to 0.1 cm) providing a uniform upwind fetch of approximately 1500 m or more for wind directions from ESE clockwise through NW. Although ground cover was uniform during winter, the topography beyond 1500 m was slightly rolling.

No major point sources of pollutants were identified in the immediate area nor were there any major highways nearby. The closest industrial area was Guelph.

The Elora field tests were carried out from mid-September, 1983, to mid-February, 1984. As the objective was to test and demonstrate the field package assembled, the primary goal of the field experiments was to gain experience with and evaluate the performance of the systems. The collection of dry deposition data was secondary. For this reason field studies were carried out during daylight hours and suitable meteorological conditions (ie, southerly flows, clear skies) only. This constraint limited the amount of data obtained. Another limitation was that, on many occasions, pollutant concentrations were too low for measurement. However, three days of pollutant concentration data were reduced and analyzed using standard signal processing, statistical and micrometeorological techniques (Bendat and Piersol, 1971; Kaimal et al., 1972). Where needed corrections were applied to the data; to account for time lags due to the length of the laser sampling tube (ie ~0.2 s), and for the effects of interferences by H_2O vapour and CO_2, Wesley, (1983) and Webb et al. (1980). The laser sampling line was colocated with the sonic anemometer. Care was taken to position the sampling tube so that the air volume within the anemometer array was not disturbed (ie, intake rate ~ 2.5 L s^{-1}). Tests confirmed the sonic measurements were not influenced by the presence of the intake tube. The sonic sampling volume exchange rate is typically many times faster than the laser intake sampling rate.

On the three days, a total of approximately 9 hr of deposition data were obtained. However, some data were lost due to data logger malfunction. Data analysis to date has been limited to the hours when dry deposition data were obtained and to the 4 m tower level, where pollutant concentrations were measured.

Insufficient deposition data were obtained to observe trends (eg, diurnal variations, etc) or to quantify deposition rates, statistically, for a particular gaseous species to various surfaces under a range of meteorological conditions.

Table II presents a summary of the preliminary dry deposition data obtained during the Elora field tests. Table III presents the supporting measured and derived parameters. Standard micrometeorological data handling procedures were applied to the data. Coordinate rotation, test for stationarity, trend removal where approprite etc.

Whenever a full data set were available, standard tests of the micrometeorological measurements were carried out. That is, do drag coefficients, roughness lengths and energy balances conform to expectations? Not all data presented in the tables represented a full data set; however, for those cases when all data were available results were within expectation. Roughness parameter values and friction coefficients were within predicted ranges. Unfortunately, only rough estimates of net radiation were available for energy budget closure. However, without estimates of storage terms, reasonable balance is seen in Table III between the estimated net radiation and sensible and latent heat fluxes.

The half-hourly deposition velocities measured for NO_2 ranged from near zero to 1.9 cm s^{-1} under various meteorological conditions and over alfalfa and snow surfaces. These values are within the range reported in the sparse literature of NO_2 measurements, Sehmel (1980); MCMahon and Denison (1979), and Wesley et al. (1982) over alfalfa and snow. It is important, however, to point out that most of the previously reported measurements were for NO_x (ie, NO + NO_2) and, where NO_2 was measured, it is not clear that NO_2 was measured unambiguously (Huebert et al., 1983).

For the NO_2 measurement over snow, an upward flux of 0.015 cm s^{-1} was measured. For the experiment on February 2, 1984, no large fluctuations in NO and NO_2 concentrations were apparent from the period 10:00 to 15:00 EST with approximate NO and total NO_x levels of 13 and 16 ppbv, respectively, measured using a chemiluminescence nitrogen oxides monitor, implying NO_2 levels of ~3 ppbv. NO_2 has low solubility in water at low concentrations, therefore, given a ~3 ppbv ambient concentration of NO_2 over a frozen snow surface as was the case here, no deposition would be expected. The small upward flux measured is considered to be anomalous, possibly indicating the noise of the technique for this day, given the low ambient concentation.

Unfortunately, the ambient SO_2 concentrations were sufficiently high to allow measurement of the deposition velocity on only one occasion during the field studies. The half-hour of good quality data obtained shows a deposition velocity to snow under stable conditions of ~0.03 cm s^{-1}. As was the case with NO_2, there is a paucity of data for SO_2 deposition to a snow surface (Sehmel, 1980; Whelpdale and Shaw, 1974).

To the authors' knowledge, no study has measured fluxes to snow directly by eddy correlation. Whelpdale and Shaw, (1974) using a gradient technique, inferred mean deposition velocities to snow, under stable conditions, on the order of 0.05 cm s^{-1}. As well, chamber studies by Granat and Johansson, (1983) indicate that uptake of SO_2 by snow at temperatures below -3^oC is slow, inferring deposition velocities of less than 0.1 cm s^{-1}.

4. CONCLUSIONS

The development of a tunable diode laser absorption spectrometer as a high sensitivity, fast response gas monitor suitable for eddy correlation measurements of dry deposition fluxes was successfully undertaken. The technique has the additional advantages of high selectivity and versatility in trace gas detection. As well, suitable micrometeorological instrumentation was assembled and reliable field measurements of dry deposition rates were carried out.

As very few unambiguous, direct measurements of NO_2 and SO_2 deposition to snow exist, the few results obtained have significance to the deposition literature, however preliminary these may be. The low deposition velocities of NO_2 and SO_2 to snow reinforce the concept of very long-range transport (VLRT) under winter conditions (Melo, 1984).

The NO_2 measurements to alfalfa show large variability. To properly interpret these fluxes under all conditions, future experiments should consider the effect of chemical reactions involving the species NO, NO_2 and O_3 (Fitzjarrald and Lenschow, 1983; Duyzer et al., 1983).

ACKNOWLEDGMENTS

This work is being conducted under contract to the Canadian Electrical Association, reference 187 G 294. Co-funding is provided by Ontario Hydro.

REFERENCES

Bendat, J.S., and Piersol, A.G.: 1971, 'Random Data: Analysis and Measurement Procedures,' Wiley-Interscience, Toronto.

Duyzer, J.H., Meyer, G.M., and Van Aalst, R.M.: 1983, Atm. Env. 17, 2117.

Edwards, G.C., and Ogram, G.L.: 1983, 'Development of an Instrument Package for the Measurement of Dry Deposition Fluxes by Eddy Correlation,' Proceedings of the Fourth International Conference on Preciitation Scavenging, Dry Deposition and Resuspension, Vol. 2, Elsevier, Amsterdam.

El-Sherbiny, M., Ballik, E.A., Shewchun, J., Garside, B.K., and Reid, J.: 1979, Appl. Opt. 18, 1198.

Fitzjarrald, D.R., and Lenschow, D.H.: 1983, Atm Env 17, 2505.

Granat, L., and Johansson, C.: 1983, Atm. Env. 17, 600.

Huebert, B.J., Wesely, M.L., and Stedman, D.H.: 1983. Atm. Env. 17, 600.

Kaimal, J.C., Wyngaard, J.C., Izumi, X., Cote, O.R.: 1972, Quart J.R. Met-Soc 98, 564.

MCMahon, J.A., Denison, P.J.: 1979, Atm. Env. 13, 571.

Melo, O.T., '1984 Meeting of the CEGB/OH/TVA Working Group on Long Range Transport and Atmospheric Chemistry,' Ontario Hydro Research Division Report C84-94-K.

Reid, J., Shewchun, J., Garside, B.K., and Ballik, E.A.: 1978a, Opt. Eng. 17, 56.

Reid, J., Shewchun, J., Garside, B.K. and Ballik, E.A.: 1978b, Appl. Opt. 300.

Reid, J., Garside, B.K., Shewchun, J., El-Sherbiny, M., and Ballic, E.A.: 1978c, Appl. Opt. 17, 806.

Reid, J., Garside, B.K., Shewchun, J.: 1979, Opt. Quant. Electronics 11, 385.

Reid, J., El-Sherbiny, M., Garside, B.K., and Ballik, E.A.: 1980, Appl. Opt. 19, 3349.

Schiff, H.I., Hastie, D.R., Mackay, G.I., Iguchi, T., and Ridley, B.A.: 1983, Env. Sci. and Technol. 17, 352A.

Sehmel, G.A.: 1980, Atm. Env. 14, 983.

Webb, E.K., Pearman, G.I., and Leunig, R.: 1980, J.R. Meteorol. Soc., 106, 85.

Wesley, M.L., Eastman, J.A., Stedman, D.H., Yalvac, E.D.: 1982, Atm. Env. 16, 815.

Wesley, M.L.: 1983, Turbulent Transport of Ozone to Surfaces Common in the Eastern Half of the United States, in Trace Atmospheric Constituents: Properties, Transformations, and Fates, edited by S.E. Schwartz, Chap. 8, John Wiley, New York.

Whelpdale, D.M., and Shaw, R.W.: 1974, Tellux XXVI, 196.

EFFECT OF A STRATUS CLOUD ON THE REDISTRIBUTION AND TRANSFORMATION OF POLLUTANTS

H.-R. Cho, J.V. Iribarne
Department of Physics
University of Toronto
Toronto, Ontario, Canada M5S 1A7

T.A. Kavassalis,* O.T. Melo, Y.T. Tam
Ontario Hydro Research Division
800 Kipling Avenue
Toronto, Ontario, Canada M8Z 5S4

W.J. Moroz
Ontario Hydro Design and Development Division
700 University Avenue
Toronto, Ontario, Canada M5G 1X6

ABSTRACT. A schematic bidimensional model has been developed to describe an extended stratus formed by warm, moist air in a frontal system. Parcel trajectories are followed through the cloud, and the formation of precipitation is considered on the basis of simple assumptions. To the dynamic description, a chemical module has been coupled, which includes the oxidation of SO_2 in aqueous-phase (by H_2O_2, O_3 and radicals scavenged from interstitial air) and that of NO and NO_2 in gaseous phase with subsequent dissolution of HNO_2 and HNO_3. The pH of precipitation is calculated. Sensitivity tests have been run to study the influence of various dynamic, microphysical and chemical parameters, such as cloud base temperature, cloud thickness, vertical velocities, initiation temperature and distribution of precipitation in the cloud, and the initial concentrations of different chemical species.

1. INTRODUCTION

The role of clouds in the transport and transformation of air pollutants is one of the main problems in acid rain research. A great deal of research, both experimental and theoretical, is being done to improve our understanding of the role of clouds. This paper describes the results of a model study designed specifically to examine the role of a stratus cloud system in redistributing and transforming pollutants. The model includes a set of chemical reactions appropriate for an extended stratus cloud system. In order

*Author for all correspondence

Water, Air and Soil Pollution **30** (1986) 195–203.

to simplify the problem, a schematic model of stratiform cloud systems is adopted in which the air flow field is specified according to the main observed features of such a cloud system (Palmen and Newton, 1969). The model is used to examine the relative importance of environmental conditions and cloud processes to the transformation of pollutants.

2. THE SCHEMATIC MODEL

The model (see Figure 1) is a time-independent, bidirectional representation of the cloud system formed by rising air in the warm sector of an extratropical cyclone. Curved air parcel trajectories are described in the model by their straight line projections on the x-p plane. It is assumed that at x=0 a layer of air between p_b and p_t becomes saturated and, as the air proceeds to increasing x at a constant horizontal velocity v, it rises with a specified vertical velocity w, forming a stratus cloud system. At a distance X, given by the intersection of the warm front with the 500 mb isobar, the air parcels start to subside and the cloud disappears. The cloud region is thus represented in two-dimensions by the area between x=0, x=X, the cloud base and the top trajectory. It is further assumed that the cloud base remains at a constant level p_b up to the point where this isobar meets the frontal surface. Beyond this point the base of the cloud coincides with the frontal surface. The frontal surface (frontal line in this bidirectional model) is selected to give a linear variation between lnp and x.

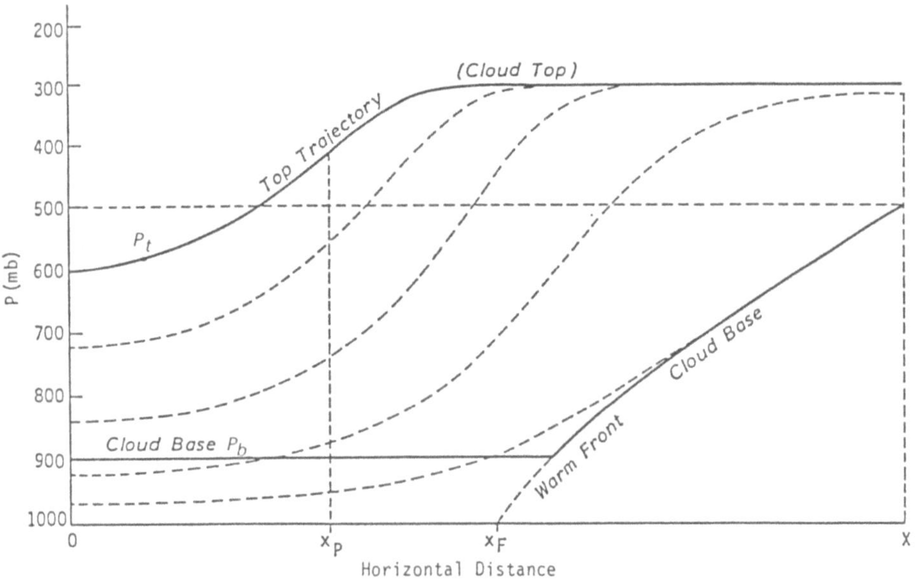

Figure 1. Schematic stratiform cloud model. The parameters used are described in the text.

The vertical velocity is assumed to have a symmetric parabolic profile with zero values at the ground (1000 mb) and the 300 mb levels. The local maximum velocity, w_m, is further assumed to be zero at x=0 and x=X and to have an overall maximum, w_F, at $x=x_F$.

At x=0, the vertical temperature profile is assumed to be that of the standard atmosphere. Beyond x=0, air parcels below the base of the cloud follow dry adiabatic expansion. Within the cloud, air parcels follow a saturated adiabatic process. Expressions for these temperature profiles may be found in Iribarne and Godson (1981).

Two precipitation schemes were considered (Mason, 1971). In the first, or Bergeron-Findeisen process, precipitation occurs when the top of the cloud reaches temperatures of $\sim$-20°C. Ice particles formed at these temperatures fall through the cloud, picking up water by sublimation and accretion of supercooled droplets. The second process sets in as the cloud ages and the ice phase gradually fills the entire cloud region above the 0°C isotherm. In this glaciated region, the water vapor condensed by the adiabatic ascent becomes deposited on the ice crystals. A region of liquid cloud water can exist below this glaciated zone, where rain droplets grow by accretion of liquid cloud water.

A very simple approach was adopted for calculating precipitation rate. This was to assume a "local precipitation efficiency" α, defined as the fraction of the vapour condensed that is converted into precipitation. Thus, in a steady-state situation, the precipitation rate through any horizontal plane at a height z, P(z), is given by the integration of this conversion rate from the cloud top to the height z. The value of   was assumed constant throughout the cloud.

3. THE CHEMISTRY MODEL

A large number of chemical species and reactions affect the chemistry of cloud water and precipitation. After carefully assessing the relative importance of these, it was decided that, for the purposes of this schematic model, it would be adequate to consider only the production of HNO_2 and HNO_3 in the gas phase and the oxidation of SO_2 by O_3, H_2O_2 and free radicals in the liquid phase. The influence of other species, such as NH_3, HCl, $CaCO_3$ and $MgCO_3$, is described through a parameter, N_A, termed initial acidity.

The gas phase production of HNO_2 and HNO_3 is described in the model by the two reactions:

$$NO + OH \xrightarrow[k_{NO}]{M} HNO_2 \tag{1}$$

$$NO_2 + OH \xrightarrow[k_{NO_2}]{M} HNO_3 \quad . \tag{2}$$

These acids, along with SO_2, O_3, H_2O_2 and free radicals dissolve in cloud water. The time constant for this dissolution when cloud droplets of $\sim 10\ \mu m$ diameter are involved, is at most 100 s, considerably shorter than droplet or cloud lifetimes. Equilibrium between the liquid and gas phase is thus assumed.

$$[x] = H_x\ P_x. \tag{3}$$

If the chemical species ionize in solution, then a modified Henry's law coefficient, H_x^*, is used. Of course, if the condensed phase is ice, no dissolution is allowed. When freezing of droplets occurs, volatile species, such as dissolved SO_2, are released to the gas-phase. This is described in the model by retention coefficients, Γ_x.

Precipitation drops, because of their larger size, take longer to equilibrate. This effect is handled in the model by assuming that the concentration of highly soluble species (e.g., H_2O_2, SO_2 at pH > 5.5) in precipitation is unaltered as the drops fall through the cloud. The concentration of less soluble species (e.g., SO_2 at pH < 5.5) is assumed to be in solubility equilibirum at all times.

In the pH range of interest (3-6), HSO_3^- is the main S(IV) species in solution. The oxidation of S(IV) in solution is thus described by

$$HSO_3^- + H_2O_2(aq) \xrightarrow[k_{H_2O_2}]{} H^+ + SO_4^{2-} + H_2O(aq) \tag{4}$$

$$HSO_3^- + O_3(aq) \xrightarrow[k_{O_3}]{} H^+ + SO_4^{2-} + O_2(aq) \tag{5}$$

$$HSO_3^- + OH(aq) \xrightarrow[k_{OH}]{} H^+ + SO_4^{2-} + \ldots \quad . \tag{6}$$

The H_2O_2 in solution is scavenged from the gas phase and produced in solution from HO_2 radicals scavenged from the gas phase. The strength of this second source of H_2O_2 (aq) is called S_{HO_2}. The pH of precipitation at ground level is calculated in the model by using a precipitation weighted average of the acidity parameter and the contribution due to the local SO_2 concentration.

Model calculations involved a range of temperatures. The temperature dependence of the various equilibria and rate constants was estimated using current literature values.

4. RESULTS

The model equations were solved using a Lagrangian parcel method with ten vertical levels of resolution. Typical values of the parameters which enter into the model are listed in Table I. Model predictions for this typical or "base case" parameter set were made and the sensitivity of the model to the various parameters was studied. Some of these results are presented here.

The cloud microphysics and precipitation mechanism can have a striking effect on the efficiency of conversion of SO_2 and NO_x to acidic species in precipitation. Figure 2 shows isopleths of SO_2 for the two precipitation schemes. From Table II we can see that, for precipitation scheme 1, the fraction of SO_2 that is deposited as acidic H_2SO_4 in precipitation is relatively unaffected by the temperature at ground level. In this model the oxidation of SO_2 occurs in the liquid droplets even after they have supercooled. The oxidation/deposition efficiency can therefore be related to the precipitation efficiency of the cloud.

In Table III we can see that, for precipitation scheme 2, the oxidation/deposition efficiency of SO_2 and NO_2 varies significantly with temperature. At a ground temperature of 5^oC we have the situation where the entire cloud is glaciated and consequently no aqueous phase oxidation of SO_2 occurs. The gas phase oxidation of NO and NO_2 is not affected by

TABLE I

Parameter Values for the Base Case

T_o	15^oC	C_{NO}	0.1 ppb
P_b	900 mb	C_{O_3}	50 ppb
w_F	-0.005 mb s^{-1}	N_A	0.3×10^{-7} moles/kg^{-1} air
v	10 m s^{1}	Γ_{SO_2}	0.25
α	0.5	$\Gamma_{H_2O_2}$	0.25
C_{SO_2}	10 ppb	$\Gamma_{SO_4^=}$	1.0
$C_{H_2O_2}$	2 ppb	$\Gamma_{NO_3^-}$	1.0
C_{NO_2}	0.1 ppb	Γ_{acid}	1.0

TABLE II

TEMPERATURE SENSITIVITY FOR PRECIPITATION SCHEME 1

Temperature at Ground Level °C	Average $[SO_4^{=}]$ M	Average $[NO_3^-]$ M	Fraction of SO_2 oxidized and deposited in precipitation %	Fraction of NO_2 oxidized and deposited in precipitation %
5	0.34×10^{-3}	0.19×10^{-5}	64	35
15	0.12×10^{-3}	0.80×10^{-6}	62	41
25	0.71×10^{-4}	0.49×10^{-6}	73	50

TABLE III

TEMPERATURE SENSITIVITY FOR PRECIPITATION SCHEME 2

Temperature at Ground Level °C	Average $[SO_4^{=}]$ M	Average $[NO_3^-]$ M	Fraction of SO_2 oxidized and deposited in precipitation %	Fraction of NO_2 oxidized and deposited in precipitation %
5	0.86×10^{-6}	0.44×10^{-7}	0	2
15	0.64×10^{-4}	0.35×10^{-6}	57	31
25	0.63×10^{-4}	0.36×10^{-6}	92	53

the cloud becoming glaciated but the acids produced are not taken up by snow crystals. Note that in this model, the average sulfate and nitrate concentrations do not vary significantly with temperature in going from $T = 15^\circ C$ to $T = 25^\circ C$.

The sensitivity of the two models to the initial SO_2 concentration is indicated in Figure 3. The liquid phase oxidation of SO_2 proceeds to completion for all initial values of the SO_2 concentration with the first microphysics and precipitation scheme and hence the absence of any non-linearity. In this case, an oxidation/deposition efficiency of roughly

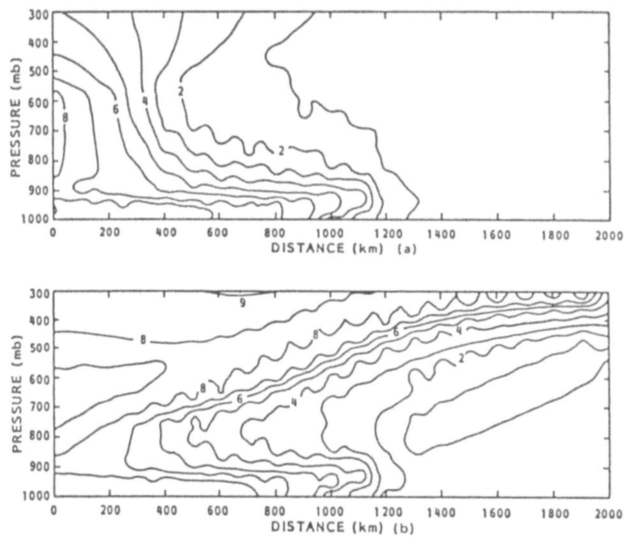

Figure 2. Isopleths of SO_2 concentration in ppb for (a) Precipitation Scheme 1 and (b) Precipitation Scheme 2.

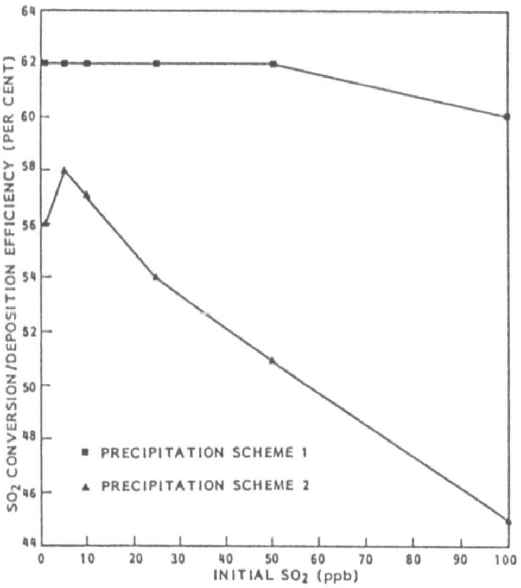

Figure 3. Fraction of initial SO_2 oxidized and deposited in precipitation as a function of the initial SO_2 concentration.

60% indicates that on the order of 60% of the water condensed has fallen as precipitation. With the second precipitation scheme the SO_2 oxidation is often incomplete before a given air parcel becomes glaciated. This leads to a slight nonlinear relationship between the initial SO_2 concentration and the concentration of H_2SO_4 in the precipitation. The gas phase oxidation of NO_x proceeds via a pseudo-first order mechanism in our model and therefore nonlinearities are not expected between the initial NO_x concentration and the corresponding acid concentration in precipitation.

A number of other sensitivity studies were performed on both models. Space limitations prevent us from presenting those results here.

5. SUMMARY AND CONCLUSIONS

The purpose of this study was to examine the transformation and redistribution of the gas phase pollutants SO_2, and NO_2 by stratiform cloud systems. The stratiform cloud model we used was schematic but contained all of the salient features of a stratus formed by warm moist air in midlatitude cyclones. By implementing two different precipitation and microphysics schemes we have demonstrated the importance of cloud microphysics in cloud/pollutant interactions. We have shown that cloud microphysical processes can determine the spatial distribution of acidic precipitation as well as the efficiency with which SO_2 is converted to H_2SO_4.

It appears therefore that proper source/receptor relationships, for a region where in-cloud oxidation of SO_2 and NO_x are important, can only be established when detailed information is known about the dominant microphysical and precipitation processes.

In future work the present schematic cloud models will be replaced with dynamical models which represent self-consistent solutions of the dynamical equations. An improved description of cloud microphysics will be incorporated in these dynamical models.

ACKNOWLEDGMENTS

This work is being conducted under contract to the Canadian Electrical Association, reference 403 G 348. Co-funding is provided by Ontario Hydro. One of us, T.A. Kavassalis, would like to thank the Natural Sciences and Engineering Research Council of Canada for financial support in the form of an Industrial Research Fellowship.

REFERENCES

Iribarne, J.V. and Godson, W.L.: 1981, Atmospheric Thermodynamics, 2nd ed. D. Reidel Publ. Co.

Mason, B.J.: 1971, The Physics of Clouds, 2nd ed. Clarendon Press, Oxford.

Palmen, E. and Newton, C.W.: 1969, Atmospheric Circulation Systems. Academic Press, New York.

MEASUREMENT OF DRY DEPOSITION TO VEGETATIVE SURFACES

Jean Muhlbaier Dasch
Environmental Science Department
General Motors Research Laboratory
Warren, MI 48090 USA

ABSTRACT. The importance of dry deposition was assessed to oak and pine trees, in a hardwood and coniferous canopy, respectively. In both canopies, ambient particle and gas concentrations on the interior were decreased relative to perimeter concentrations due to dry deposition scavenging by the canopy. Deposition of all species was higher to perimeter leaves than to interior leaves, due both to the lower concentrations and the lower wind speeds in the sheltered interior. The deposition velocities measured to the pine needles were similar to the values measured to the oak leaves.

1. INTRODUCTION

During two successive summers, dry deposition was measured to forests: a hardwood canopy in 1983 and a pine canopy in 1984 (Dasch, 1985a, b). In both studies air concentrations and dry deposition were measured at both the perimeter and interior locations of the canopy. Dry deposition was measured by washing the deposited material from the leaves as has been done in other studies (Lindberg and Lovett, 1985; Sickles et al., 1983). The results of the two studies will be compared in this paper.

2. EXPERIMENTAL

Sampling was conducted at a stand of deciduous trees (50 ha) in Warren, MI from June through October, 1983. Warren is a suburban area located seven km north of the Detroit city line. Dry deposition was sampled from two oak trees, one on the southern perimeter of the wooded area and the other on the interior, 27 m north of the perimeter tree. From July through November, 1984, sampling was conducted in a 9-ha stand of Austrian pine trees. The stand is located in rural Lapeer County, MI which is 65 km north of Detroit. Dry deposition was sampled from two pine trees, one on the western perimeter of the stand and the other on the interior, 80 m east of the perimeter tree. The prevailing wind direction from June to November is from the southwest so the perimeter trees at each site were exposed to maximum air flow.

Water, Air and Soil Pollution **30** (1986) 205–210.

During both studies, ambient particles and gases were collected at the perimeter and interior sites. Dry deposition was also collected at the perimeter and interior sites by washing the deposits from the leaves. The main differences between the two studies were the tree type, the suburban vs. the rural location, and the distance from the perimeter to the interior tree (27 m for the hardwood canopy and 80 m for the pine canopy). The same deposited species were measured on the leaves (Pb^{++}, Ca^{++}, NO_3^-, and SO_4^{--}) with the addition of Mg^{++} and NH_4^+ during the pine study.

Particles and gases were collected under each of the four trees. Air was sampled at 10 L min^{-1} through a triple-stack filter: a 1-μm pore-size Teflon filter collected particles, a 1-μm pore-size nylon filter collected HNO_3 or volatilized NO_3^-, and double cellulose nitrate filters impregnated with a 25% K_2CO_3, 10% glycerol solution collected SO_2. The Teflon filter was extracted in 50 mL of a leachate solution consisting of deionized water acidified to pH 4 with hydrofluoric acid and containing 100 μL of toluene as a biological inhibitor. After extraction, the liquid was divided in half. One half was further acidified to pH 1 with Ultrex HNO_3 for Pb^{++} analysis and the other half was saved for NO_3^-, SO_4^{--}, NH_4^+, Ca^{++} and Mg^{++} analyses. The nylon filter was extracted in 50 mL of the same bicarbonate-carbonate eluant used in the ion chromatograph and analyzed for NO_3^-. The SO_2 filters were extracted in 100 mL of a 0.2% H_2O_2 solution and analyzed for SO_4^{--}.

At the start of a dry period, one section of a branch on each tree and a control branch were cleaned to remove deposits. The control branch was then sheltered from dry deposition by enclosing it in a transparent polyethylene bag while hydrocarbon-free air flowed through the bag to exclude ambient air. The other cleaned branches on each tree were marked as the test branches. At the end of each dry period (varying from 1 to 9 days), the control and test branches were extracted with leachate solution to remove dry deposited material. Complete sampling details can be found in Dasch (1985a,b).

Wet deposition was collected on an event basis in a polyethylene bucket in an Aerochem-Metric collector which protected the precipitation from dry deposition. The water was filtered through 0.4 μm pore-size filters and refrigerated until analysis.

All samples were divided in two aliquots; one was acidified to pH 1 with Ultrex HNO_3 for Pb^{++} analysis by graphite furnace atomic absorption spectroscopy. The other aliquot was analyzed for NO_3^- and SO_4^{--} by ion chromatography, Ca^{++} and Mg^{++} by flame atomic absorption spectroscopy and NH_4^+ by the indophenol colorimetric method.

3. RESULTS

3.1. Ambient Concentrations

Particle and gas concentrations in the ambient air were measured near the perimeter and interior trees. The average concentrations are shown in Table I. The sampling period varied from 24 to 199 hr. For

Table I
Ambient Concentrations ($\mu g\ m^{-3}$)

	1983 Oak Study			1984 Pine Study		
	Perimeter	Interior	Ratio	Perimeter	Interior	Ratio
Pb^{++}	0.114	0.108	0.92	0.024	0.025	1.03
Ca^{++}	1.25	1.14	0.89	0.61	0.53	0.86
NO_3^-	1.29	1.12	0.94	1.86	1.50	0.72
SO_4^{--}	8.81	8.15	0.98	5.79	5.26	0.97
Mg^{++}	ND*	ND	ND	0.11	0.097	0.88
NH_4^+	ND	ND	ND	2.97	2.59	0.94
HNO_3	3.16	2.36	0.77	1.65	1.00	0.60
SO_2	11.2	10.7	0.98	7.37	6.28	0.89

ND: Not determined

all species but Pb^{++} to the pine canopy, interior concentrations were lower than perimeter concentrations indicating scavenging by the canopy. The median ratio of interior/perimeter concentrations for each time period is also shown in Table I. The ratio is high for SO_4^{--} and SO_2 indicating little scavenging by the canopy. HNO_3 has a low ratio indicating that it is efficiently removed by passage through the canopy. Removal of HNO_3 is consistently higher than for SO_2, indicating a higher deposition velocity for HNO_3 than SO_2. There was an excellent correlation between perimeter concentrations and interior concentrations with correlation coefficients varying between 0.72 and 0.96 (all significant at <0.01% level). Other studies have also noted lower particle concentrations within a canopy than outside the canopy (Wiman and Lannefors, 1985; Hofken et al., 1983).

The ambient concentrations were higher in Warren during the oak study than in Lapeer during the pine study. Due to the proximity of the urban area to Warren, ambient concentrations are, in general, higher in Warren than in Lapeer (Dasch and Cadle, 1984). The ratios of interior/perimeter concentrations are generally lower in the pine study, particularly for HNO_3 and SO_2. This is presumably due to the greater distance between perimeter and interior trees in the pine study than in the oak study with a greater opportunity for dry removal.

3.2. Dry Deposition

Direct measurements of dry deposition to leaves were made by washing the deposited particles from the trees. The foliar component was determined from the control branches which were protected from deposition. This foliar component was then subtracted from the extracted material on the test branches to arrive at a dry deposition component. The average dry deposition fluxes to the leaves are shown in Table II based on 22 tests to the oak leaves and 15 tests to the pine needles.

The tests were made throughout the sampling period with each test representing a new set of leaves. The flux is based on the upper and lower surface area of the oak leaves and the total surface area of the pine needles.

The effect of the location within the canopy on dry deposition can

Table II
Mean Dry Deposition Fluxes to Leaf Surfaces (neq cm^{-2} day^{-1})

	Perimeter Oak	Interior Oak	Perimeter Pine	Interior Pine
Pb^{++}*	1.7±2.5	1.3±2.8	0.39±0.58	0.15±0.71
Ca^{++}	2.3±2.3	0.4±1.5	1.7±1.7	0.31±0.80
NO_3^-	0.91±0.76	0.56±0.50	1.7±1.1	0.51±0.51
SO_4^{--}	0.43±0.53	0.28±0.67	0.34±0.45	0.15±0.50
Mg^{++}	ND	ND	0.50±0.67	0.09±0.48
NH_4^+	ND	ND	0.72±0.92	0.31±0.74

*peq cm^{-2} day^{-1}

be seen by comparing the perimeter and interior location. The perimeter dry deposition flux is greater than the interior flux for all species and for both types of forest. The higher dry deposition flux to the perimeter is consistent with the higher ambient concentrations at the perimeter location. However, the differences in dry deposition flux at the two locations are greater than the differences between ambient concentrations. Presumably, the additional factor that decreases the flux to the interior leaves is the lower wind speeds encountered in the canopy.

The only significant difference in deposition to the oak and pine trees is for Pb^{++} which is much higher to the oak leaves. This is probably due to the higher traffic density in suburban Warren compared to rural Lapeer and is in agreement with the five-fold higher ambient Pb^{++} levels in Warren.

3.3. Deposition Velocity

Deposition velocities (Vd) were determined for each time period; mean values and standard deviations are shown in Table III. The deposition velocity was determined from the ratio of the dry deposition flux to the ambient particle concentration. The effect of gaseous species is not included. The values in Table III represent Vd's to individual surfaces. A canopy Vd would be this value multiplied by the leaf area index.

Pb^{++} has the lowest Vd of the measured species as expected since it has a small mass median diameter (MMD). SO_4^{--} and NH_4^+ are next with Vd's of about 0.04 cm s^{-1} on the perimeter. Ca^{++} and Mg^{++}, the large particle species, are next with Vd's of about 0.6 cm s^{-1}. The anomalous specie is NO_3^- which has a very high Vd but has a smaller MMD than Ca^{++} or Mg^{++}. The most likely explanation is that gaseous HNO_3 is contributing to the measured deposition.

Table III
Mean Deposition Velocities to Leaf Surfaces (cm s^{-1})

	Perimeter Oak	Interior Oak	Perimeter Pine	Interior Pine
Pb^{++}	0.02±0.03	0.01±0.02	0.03±0.05	0.01±0.06
Ca^{++}	0.41±0.23	0.07±0.31	0.62±0.45	0.12±0.36
NO_3^-*	1.1±1.7	0.73±0.93	1.3±1.0	0.47±0.51
SO_4^{--}*	0.047±0.071	0.026±0.058	0.045±0.037	0.015±0.040
Mg^{++}	ND	ND	0.57±0.53	0.37±0.82
NH_4^+*	ND	ND	0.056±0.068	0.010±0.042

*Based on ambient particle concentrations, not gases.

3.4. Wet vs. Dry Deposition

The total input to the trees from wet and dry deposition are compared in Table IV. The values are in terms of deposited material per cm^2 of ground area. To obtain the dry values, the deposition flux from Table II was multiplied by the leaf area index (12 for the oak tree to include the two sides of the leaf and 9 for the pine tree). The dry

Table IV
Wet vs. Dry Deposition (neq cm^{-2} day^{-1})

	Wet	Dry to Oak Perimeter	Dry to Pine Perimeter
Pb^{++}	0.020	0.020	0.0035
Ca^{++}	5.2	28	15
NO_3^-	11	11	15
SO_4^{--}	21	5.2	3.1
Mg^{++}	2.1	ND	4.5
NH_4^+	11	ND	6.5

deposition flux in each case is for the perimeter tree. Dry deposition is more important than wet deposition for the large-particle species, Ca^{++} and Mg^{++}. It is less important for the small-particle species, NH_4^+ and SO_4^{--}. Dry deposition of Pb^{++} was important near the urban area but not at the rural location. These values can only represent crude estimates. The dry deposition values are biased by the sampling position; sampling was conducted at one height on trees exposed to the air flow, thus receiving maximum dry deposition impacts. No additional influence from the top of the canopy was considered.

4. SUMMARY

The importance of dry deposition was assessed to trees on the perimeter and the interior of forest canopies. Ambient concentrations and dry deposition were measured to pine and oak leaves. Ambient particle and gas concentrations on the interior were decreased relative to perimeter concentrations due to dry deposition scavenging by the canopy. Deposition of all species was higher to perimeter leaves than to interior leaves for both the oak leaves and the pine leaves, due both to the lower concentrations and the lower wind speeds in the sheltered interior. The deposition velocities measured to the pine needles were similar to the values measured to the oak leaves. Dry deposition was most important compared to wet deposition for the large particle species and least important for the small particles species.

ACKNOWLEDGMENTS

I would like to thank Ken Kennedy for his help with sample collection and analysis and Pat Mulawa and Rene Vandervennet for their analytical assistance.

REFERENCES

Dasch, J.M.: 1985a, 'Measurement of dry deposition to a deciduous canopy', GMR Research Publication GMR-5019.

Dasch, J.M.: 1985b, 'Measurement of dry deposition to a pine canopy', GMR Research Publication GMR-5153.

Dasch, J.M. and Cadle, S.H.: 1984, Atmos. Environ., 18, 1009.

Hofken, K.D., Meixner, F.X., Ehhalt, D.H.: 1983, 'Deposition of atmospheric trace constituents onto different natural surfaces', Precipitation Scavenging, Dry Deposition, and Resuspension, Pruppacher, H.R., Semonin, R.G., Slinn, W.G.N. (eds), Elsevier Press, 8325-836.

Lindberg, S.E. and Lovett, G.M.: 1985, Environ. Sci. Tech, 19, 238.

Sickles, J.E., Bach, W.D., Spiller, L.L.: 1983, 'Comparison of several methods for estimating dry deposition flux' Precipitation Scavenging, Dry Deposition, and Resuspension, Pruppacher, H.R., Semonin, R.G., Slinn, W.G.N. (eds), Elsevier Press, 979-990.

Wiman, B.L.B. and Lannefors, H.O.: 1985, Atmos. Environ., 19, 349.

MODELING OF THROUGHFALL CHEMISTRY AND INDIRECT MEASUREMENT OF DRY DEPOSITION

A. H. Johannes, Y. L. Chen, K. Dackson, and T. Suleski
School of Chemical Engineering
Oklahoma State University
Stillwater, OK 74078-0537 USA

ABSTRACT. Precipitation over a forested watershed is altered by interaction with plant surfaces which act as a filter for airborne gases and particles. This results in a major transfer to the forest floor of materials captured, washed, and leached from the forest canopy. Sequential samples of wetfall and sequential samples of throughfall under deciduous and coniferous trees were collected and chemically analyzed for major anions and cations. A simple washoff, mixing model based on leaf area index was used to simulate throughfall chemistry and to decouple foliar exudation from dry deposition. Model results gave excellent predictions of the measured sequential throughfall using estimated values of dry deposition. The model can also be used to calculate dry deposition, if the sequential throughfall data and wetfall data are used as input variables.

1. Introduction

Dry deposition of airborne gases and particles is a major atmospheric removal process estimated to be of similar magnitude with wet deposition. Trees and other vegetation act as natural filters in the environment by providing high surface area for dry deposition. Precipitation over a forested watershed is chemically altered as it cascades to the forest floor by a brief but significant interaction with plant surfaces. The result is a major transfer to the forest floor of chemicals captured, washed and leached from the forest canopy. The chemically altered precipitation is termed throughfall if it falls from the leaves and branches of the tree or stem flow if it runs down the tree trunk.

Throughfall and stemflow studies have shown major enrichment in acid anions and base cations. (Eaton et al., 1973; Likens et al., 1977; Lindberg and Harriss, 1981). Hydrogen ion concentration was found to be dependent on tree species with coniferous canopies decreasing pH and deciduous increasing pH (Johannes et al., 1985).

Chemical changes in throughfall can result from at least four major processes: washoff and dissolution of dry deposition, chemical reaction in the liquid phase, leaching from the vegetation, and

Water, Air and Soil Pollution **30** (1986) 211–216.

chemical uptake by the vegetation. Controversy exists as to which mechanism is dominant and whether competing mechanisms exist for different chemical species.

Event throughfall data while providing information about net changes, masks details on the initial portion of an event where washoff of dry deposition is most likely to have the greatest effect. To study the dynamics during an event, sequential samples of throughfall and wetfall were collected and chemically analyzed. The resulting data were studied using a simple mixing model.

2. Methods

Samples were collected in the Adirondack Mountains of New York State near Big Moose, NY (43°49'N, 74°55'W) in September and early October 1981. A total of 9 events were sampled but events 6-9 were of limited value since the trees began to defoliate in early October. Sequential samples were collected approximately every 60 mL. During the later part of low intensity, protracted events, the collection frequency dropped to hourly intervals.

Throughfall collectors consisted of two 19 cm polyethylene funnels connected with tygon tubing to a 500 mL polyethylene collection bottle. A Whatman 500 filter was placed inside each funnel to remove any canopy debris prior to collection. Under each tree sampled and out in the open, a total event collector, sequential collector, and rain gauge were located. Dryfall was collected between rain events using a white polyethylene bucket.

Samples were transported in the sealed collection bottles to the field laboratory where sample volume, field pH, and conductivity were measured. The remainder of the sample was then filtered through a 0.4µm Nucleopore filter and stored in a clear polyethylene bottle. Samples were refrigerated at 3°C without preservatives until chemically analyzed. Dryfall samples were extracted with 250 mL of distilled-deionized water and processed in the same manner.

Samples were analyzed for pH, SO_4^{-2}, NO_3^-, Cl^-, NH_4^+, Ca^{+2}, Mg^{+2}, Na^+ and K^+. The cations (except NH_4^+) were measured using a Perkin Elmer Model 403 Atomic Absorption Spectrophotometer. Anions and NH_4^+ were analyzed using Technicon Auto Analyzers. Specific details of the chemical analyses are reported elsewhere (Johannes et al., 1981).

To determine differences between deciduous and coniferous throughfall chemistry, collectors were located under an American Beech tree (deciduous) and a Red Spruce tree (coniferous) of similar diameter. Leaf surface area and leaf area index were calculated for each tree using regression equations and the diameter at breast height (Whittaker et al., 1974). Calculated leaf area index for the American Beech and Red Spruce was 2.7 and 5.9, respectively.

3. Results and Models

Typical sequential data showed highest ion concentrations occurred during the initial portion of an event and decreased during the event. For most ions (except NH_4+) coniferous throughfall contained much higher concentrations than the deciduous throughfall. Deciduous throughfall in the latter portion of an event had anion concentrations which were similar to wetfall. Cation concentrations in throughfall were one or two orders of magnitude higher than wetfall. Although most cation concentrations decreased during an event, the later concentrations for both deciduous and coniferous throughfall remained much higher than wetfall. Organic acids were not measured during the study, but since most of the samples were slightly colored (yellow or brown) and total anion to cation ratios were less than unity, organic acids were assumed to be present.

The cascade of water through a tree follows many different flow paths depending on leaf and branch structure. Contact time on each leaf surface is rather short and is dependent on many variables. The tree was therefore conceptualized in this preliminary model as a series of stages or levels where dissolution, leaching and mixing occurred. The number of stages for each tree was based on total leaf surface area divided by the projected area of the tree. The result is leaf area index and this value rounded to the next higher integer was used to determine the number of stages per tree. Dry deposition was assumed to be present on each stage at the start of an event.

The relationship between wetfall and throughfall volume under the American Beech was obtained by linear regression. Throughfall volume was found to be a linear function of wetfall volume (Figure 1). A least squares fit of the sequential data yielded the following regression equation for the American Beech (volume in L):

$$[\text{Throughfall Volume}] = 0.711\ [\text{Wetfall Volume}] - 0.028.$$

An interpretation of this equation is that a tree has a unique storage capacity (generally termed interception or holdup) which must be exceeded before any throughfall is observed. A holdup parameter may be derived from the preceding equation by setting the throughfall volume equal to zero and solving for the wetfall volume required (or 0.0392 L in this case). Since stemflow was not measured during this study but was required for the mass balance of water, existing data from a previous study was used as an estimate.

The model calculates throughfall volume and ion concentrations from one stage to the next with respect to time. Initially, each compartment is assumed to contain only dry deposition. The incoming throughfall (or wet deposition in the first stage) enters a stage and completely mixes with any water already present. Stages must be completely filled before any stemflow or overflow to the next compartment occurs.

Leaching parameters were determined by linear regression between ion concentrations in throughfall and wetfall. Ions were grouped using Duncan's test. Two primary groupings were indicated with several overlap ions. Group one was comprised of Na^+, Ca^{2+}, Mg^{2+} and K^+. Group two was comprised of NO_3^-, Cl^-, and SO_4^{2-}. Overlap ions included Cl^-, Mg^{2+} and Na^+. Maximum estimates based on wetfall concentrations and on estimates of dryfall, factored by the leaf area index, showed that group one ion concentrations had to result from foliar exudation rather than dry deposition. No evidence was found for the group two ions to support enrichment by exudation.

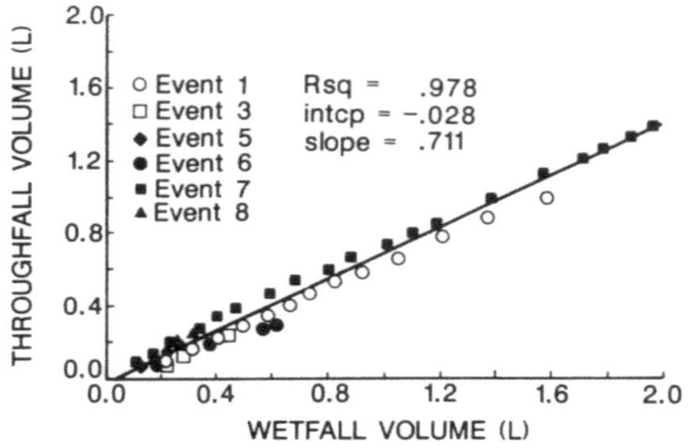

Figure 1. Relationship between event throughfall and wetfall volumes.

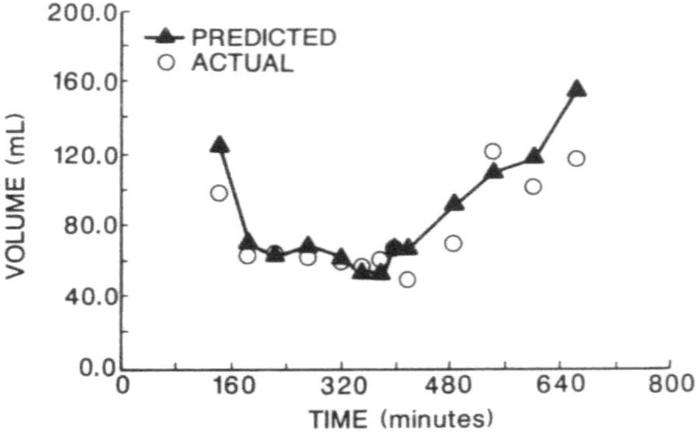

Figure 2. Predicted and measured volumes during model calibration phase.

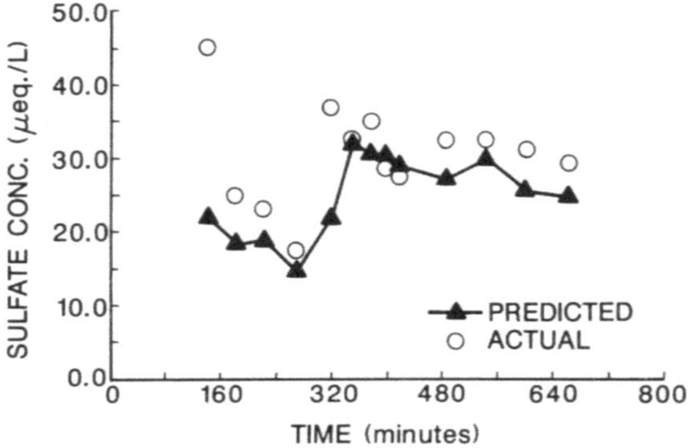

Figure 3. Predicted and measured sulfate concentrations during model calibration (no leaching or dry deposition).

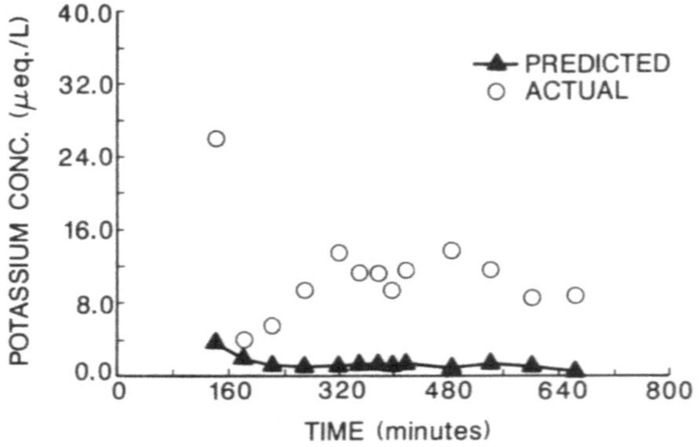

Figure 4. Predicted and measured potassium concentrations during model calibration (no leaching or dry deposition).

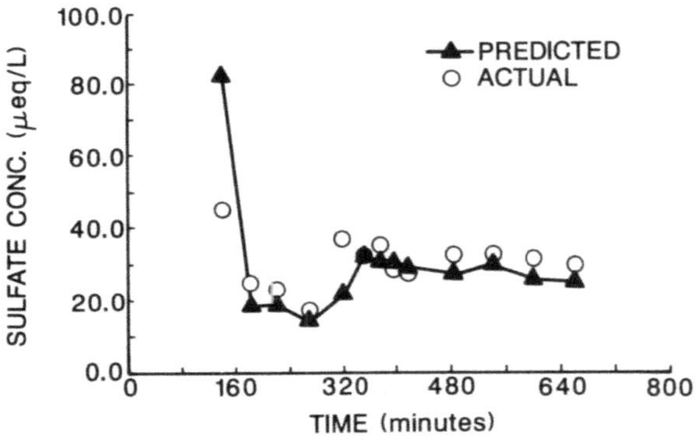

Figure 5. Predicted and measured sulfate concentrations during model calibration (dry deposition included).

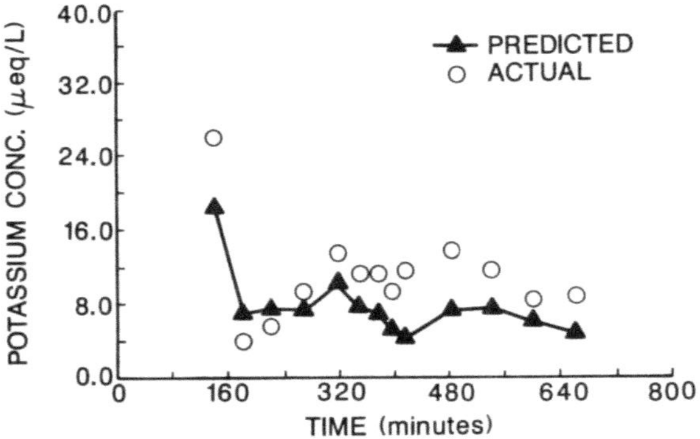

Figure 6. Predicted and measured potassium concentrations during model calibration (leaching and dry deposition included).

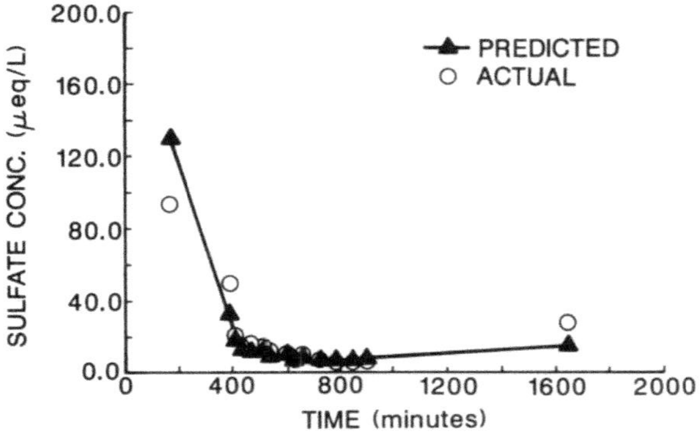

Figure 7. Predicted and measured sulfate concentrations during model verification phase.

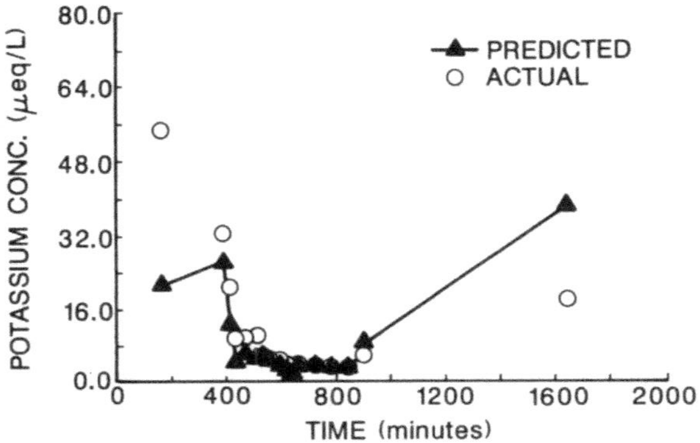

Figure 8. Predicted and measured potassium concentrations during model verification phase.

The model was first run in a calibration mode with no leaching or dry deposition. The calculated water balance (Figure 2) showed excellent agreement with the measured values. Predicted SO_4^{2-} and K^+ concentrations (Figures 3 and 4) gave low values, as expected, with little correlation for K^+ with measured values. Inclusion of dry deposition in the model did not significantly change the K^+ concentrations but gave significantly better sulfate concentrations (Figure 5). Leaching coefficients based on mean concentrations were than introduced into the model. Potassium ion concentrations (Figure 6) increased significantly and gave reasonable estimates of measured values.

The model was then rerun on a new and untested event using the same leaching factors derived during calibration. Water balance, sulfate concentrations (Figure 7) and K^+ concentrations (Figure 8), showed excellent agreement with the measured results. Similar agreement was found for the other anions and cations with the exception of NH_4^+.

4. Conclusions

Ionic species were found to be grouped into 3 sets: Washoff controlled (SO_4^{2-}, NO_3^- and Cl^-,); leaching controlled (K^+, Ca^{2+}, Na^+ and Mg^{2+}); and washoff-leaching competitive (NH_4^+). For the leaching controlled ions, dry deposition was of only minor importance, and then only in the very early stages of an event. Leaf area index was found to be the most important modeling parameter. Leaching coefficients, although determined empirically during the calibration phase, gave excellent agreement during the comparison phase. In this preliminary study, the mass residue left on the leaves at the end of an event was not taken into account and resulted in somewhat lower concentrations during the initial portion of the event. In spite of these deficiencies, the model yields reasonable predictions of throughfall concentrations.

5. References

Eaton, J. S., Likens, G. E., and Bormann, F. H.: 1973, J. Ecology 61, 495.

Johannes, A. H., Altwicker, E. R., and Clesceri, N. L.: 1981, Characterization of Acidic Precipitation in the Adirondack Region, Electric Power Research Institute, EA-1826.

Johannes, A. H., Altwicker, E. R., and Clesceri, N. L.: 1985, Water, Air, and Soil Pollut. 26, 339.

Likens, G. E., Bormann, F. H., Pierce, R. S., Eaton, J. S., and Johnson, N. M.: 1977, Biogeochemistry of a Forested Ecosystem, Springer-Verlog, NY.

Lindberg, S. E., and Harriss, R. C.: 1981, Water, Air and Soil Pollut. 16, 13.

Whittaker, R. H., Bormann, F. H., Likens, G. E., and Siccama, T. G.: 1974, Ecological Monographs 44, 233.

THE COMPOSITION OF PRECIPITATION AT SNOQUALMIE PASS AND STEVENS PASS IN THE CENTRAL CASCADES OF WASHINGTON STATE

L. Clint Duncan*

E. B. Welch#

Wally Ausserer*

*Chemistry Department
Central Washington University
Ellensburg, WA 98926 USA

#Department of Civil Engineering
University of Washington
Seattle, WA 98195 USA

ABSTRACT. Bulk precipitation samples were collected weekly during the winters of 1984 and 1985 at two high elevation sites in the Washington Cascades. The Puget basin, west and upwind of the Cascade sites, includes an urban industrialized region with a variety of area and point sources of acidic oxides. The precipitation was found to be low in contaminant content with $SO_4^=$ and NO_3^- concentrations as low or lower than those reported for remote pristine stations in the US or elsewhere. It is suggested that the reduced contaminant levels result from an interplay of the meteorological and washout factors unique to the region.

1. INTRODUCTION

The Cascade range in Washington state is east and downwind of populated urban areas that serve as non-point NO_x and SO_2 sources (Everett-Seattle-Tacoma) and is also downwind of three relatively large acid oxide point sources, the ASARCO smelter (SO_2), the Centralia Power Plant (SO_2 and NO_x), and Mount St. Helens (SO_2), (Logan et al., 1982; Vong, 1982; PSACPA, 1985; SWAPCA, 1985; USGS 1985).

Water, Air and Soil Pollution **30** (1986) 217–229.

The results of studies accomplished to date indicate that the precipitation falling on the western portion of Washington State including the Cascades is contaminated with strong acids. Moist marine air sweeps westerly over the coast, then over the urban and point sources depositing 'acid rain' in the urban Puget sound region and then in the Cascades. (Larson et al., 1975; Deither, 1979; Logan et al., 1982; Ausserer and Duncan, 1984; Welch et al., 1984; Vong et al., 1985).

The average precipitation composition of samples collected at the Hoh National Atmospheric Deposition Program collection site 35 km east of the Pacific coast in Washington (NADP, 1985) indicates little contaminate introduction occurs before the air masses reach the urban areas. The annual volume-weighted average pH was 5.5 at Hoh with volume weighted average $SO_4^=$ and NO_3^- concentrations of 8.2 and 1.3 µeq L^{-1} respectively. Half the $SO_4^=$ at that site is believed to be "sea salt" $SO_4^=$ (Vong et al., 1985). The pH value is near that expected for a solution saturated with atmospheric CO_2 at room temperature.

Precipitation in the Puget Sound urban areas contains excess strong acid. For example, an urban Seattle collection study (Vong et al., 1985) of 1983 indicated a volume weighted average pH of 4.40 (39.8 µeq L^{-1} H^+) with average $SO_4^=$ and NO_3^- concentrations of 29.6 and 7.2 µeq L^{-1}, respectively. The same investigators carried out a collection study at Bellingham Washington (242 km north of Seattle yet west of the Cascades) in 1983 and reported an average pH of 4.57 (26.9 µeq L^{-1} H^+) with $SO_4^=$ and NO_3^- concentrations of 18.5 and 5.3 µeq L^{-1}, respectively.

The foothills east of the Puget Sound basin receive a greater volume of precipitation than the urban Puget Sound region (Logan et al., 1982). At the same time the precipitation contaminant concentrations are somewhat diminished compared to values observed in the urban Puget Sound basin. For example, the volume weighed average pH at the Tolt reservoir (48 km N.W. of Seattle) was 4.59 (25.9 µeq L^{-1} H^+) and $SO_4^=$ and NO_3^- concentrations were 17.1 and 3.7 µeq L^{-1} respectively (Vong et al., 1985).

Precipitation falling on the Cascades contains lesser contaminant concentrations yet. Also the composition remains relatively uniform over the mountains from west to east. Logan et al.,(1982) sampled bulk precipitation at five sites traversing the Cascades winter-spring 1981 and reported volume weighted concentration average of pH = 4.67 to 4.72 (19.0 to 21.3 µeq L^{-1} H^+), $SO_4^=$ = 15 to 18 µeq L^{-1}, and NO_3^- = 9 to 11 µeq L^{-1} over the Cascade ridge.

Generally WNW winds sweep over western Washington urban and point source locations before the air masses interact with the Cascades (Pennell et al., 1985). The average direction of air flow over southwest Washington is such that the projected shadow of the urban plume would fall on the Cascades somewhat north of Seattle (Vong, 1982) if it was not diverted. The Snoqualmie Pass (transect) previously studied, lies south of Seattle (Logan et al., 1982). This report describes the composition of precipitation collected (winter 1983-84 and winter 1984-85) at three sites along a Cascade mountain transect

(Stevens Pass) located approximately 33 km north of Snoqualmie Pass. Precipitation samplers were also co-collected at one location on the Snoqualmie Pass for purposes of spatial comparisons.

2. EXPERIMENTAL

Deposition Sampling: Precipitation samples of this study were collected at sites as shown on Figure 1 commencing fall 1983. Two bulk 640 cm^2 collector buckets and one Aerochem Metric "wet/only collector" were placed approximately 3m apart in a triangular configuration at the Skykomish site. Bulk collectors were used at the other sites. The bulk collectors were supported at least 2m above the ground in all cases. Before use, all buckets were thoroughly scrubbed with a brush using sodium bicarbonate as an abrasive agent, soaked with 2N HCl, then with deionized water. The final rinse was allowed to set in the bucket for several hours after which the conductivity of that rinse water was determined. If the conductivity exceeded 1.0 $\mu S\ cm^{-1}$, the process was repeated until a desired rinse conductivity of less than 1.0 $\mu S\ cm^{-1}$ was achieved. The samples were collected weekly, promptly transported to the laboratory, thawed if necessary, then transferred to new rinsed polyethylene bottles and refrigerated near 4 °C until analysis.

Analysis: A tabulation showing quality control data and the methods used is shown in Table I. The pH values were determined in the laboratory on 100 mL samples that had come to room temperature and constant pH. The pH of those samples was observed to increase with time With stabilization occurring after 15 to 30 min. Acidities (AC) were determined after pH measurement by potentiometric titration with 0.01000 N NaOH. The titrant was added using precalibrated 25.0 and 100.0 µL eppendorf pipets. End points were detected using the Gran method (McQuaker et al., 1983). A Corning 125 pH meter equipped with a Corning combination electrode was used for the titration and pH determination after a two point calibration with pH 7.00 and 4.01 buffers. The pH of a 100.0 $\mu eq\ L^{-1}$ HCl solution was determined after each calibration to insure the electrode was functioning properly. The sample conductivities were determined using a YSI Model 31 conductivity bridge and a detection cell with K = 0.100. Cell constants were determined using 1.000 x 10^{-4} M KCl (APHA, 1976). Cl^-, NO_3^- and $SO_4^=$ concentrations were determined using a Dionex 2000i Ion Chromatograph equipped with a HPIC-As4 anion column and an AFS-1 anion suppression system (EPA, 1984). Samples were prefiltered through a .4 um Nucleopore filter before analyses. Ca^{++}, Mg^{++}, Na^+, and K^+ concentrations were determined using flame atomic absorption techniques (EPA, 1979). An Instrumentation Laboratory model 251 spectrophotometer with a single slot burner was used. The phenate method was used to determine ammonium ion concentrations (APHA, 1976). Absorbances were measured using a Bausch and Lomb Spectronic 20 spectrophotometer.

The average ion balance defined (Sisterson et al., 1985) as: B = [(anions - cations) / (anions + cations)] x 100% was -6.1% with a std. dev. of 14% for 210 precipitation samples. The measured sample conductivities were compared with those calculated using analytical ion

Table I

ANALYTICAL METHODS AND QUALITY CONTROL

Parameter	Method	Average (a) Concentration	Std. Dev. (b)
Ammonia (Std. Meth., 1976)	Phenate	17.4	0.88
Acidity (McQuaker *et al.*, 1983)	Titr.	10.4	0.62
Ca (EPA, 1979)	A.A.	5.4	0.22
Chloride (EPA, 1984)	I.C.	10.69	1.04
Conductivity (EPA, 1979)	Probe	21.7 μS cm^{-1}	0.6 μS cm^{-1}
Mg (EPA, 1979)	A.A.	5.1	0.39
Nitrate (EPA, 1984)	I.C.	10.27	0.45
pH (McQuaker *et al.*, 1983)	Probe	4.06	0.05
K (EPA, 1979)	A.A.	5.0	0.33
Na (EPA, 1979)	A.A.	5.1	0.33
Sulfate (EPA, 1984)	I.C.	9.88	0.46

(a) Trial sample concentration μeq L^{-1} (except pH)
(b) Standard deviation of replicates of trial sample

concentrations and limiting equivalent ionic conductances. The result expressed in terms of the ratio,

$$C = \frac{\text{(calculated conductivity - measured conductivity)}}{\text{(calculated conductivity + measured conductivity)}} \times 100\%$$

was calculated for each sample. The average for 195 samples was C = 6% with a std. dev. of 9%.

3. RESULTS AND DISCUSSION

Collector characteristics: bulk precipitation collectors are used at the mountain pass sites because of the difficulties associated with operation of a wet/only collector in the extreme weather conditions at those sites. Most of the precipitation in the Cascades occurs in the

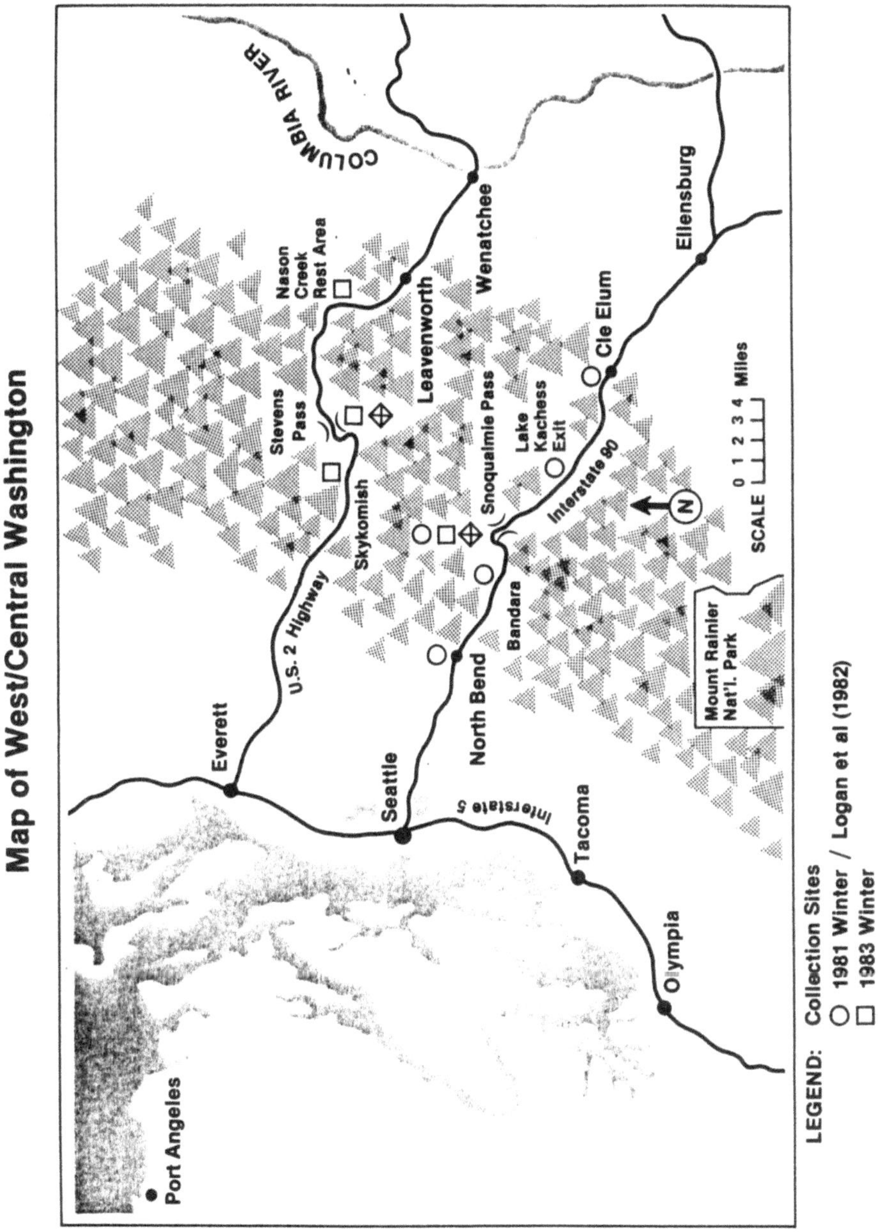
Map of West/Central Washington
Port Angeles
Everett
Seattle
Tacoma
Olympia
Interstate 5
U.S. 2 Highway
Skykomish
Stevens Pass
Nason Creek Rest Area
Leavenworth
Wenatchee
COLUMBIA RIVER
North Bend
Bandara
Snoqualmie Pass
Interstate 90
Lake Kachess Exit
Cle Elum
Ellensburg
Mount Rainier Nat'l. Park
N
SCALE 0 1 2 3 4 Miles
LEGEND: Collection Sites
1981 Winter / Logan et al (1982)
1983 Winter
1984

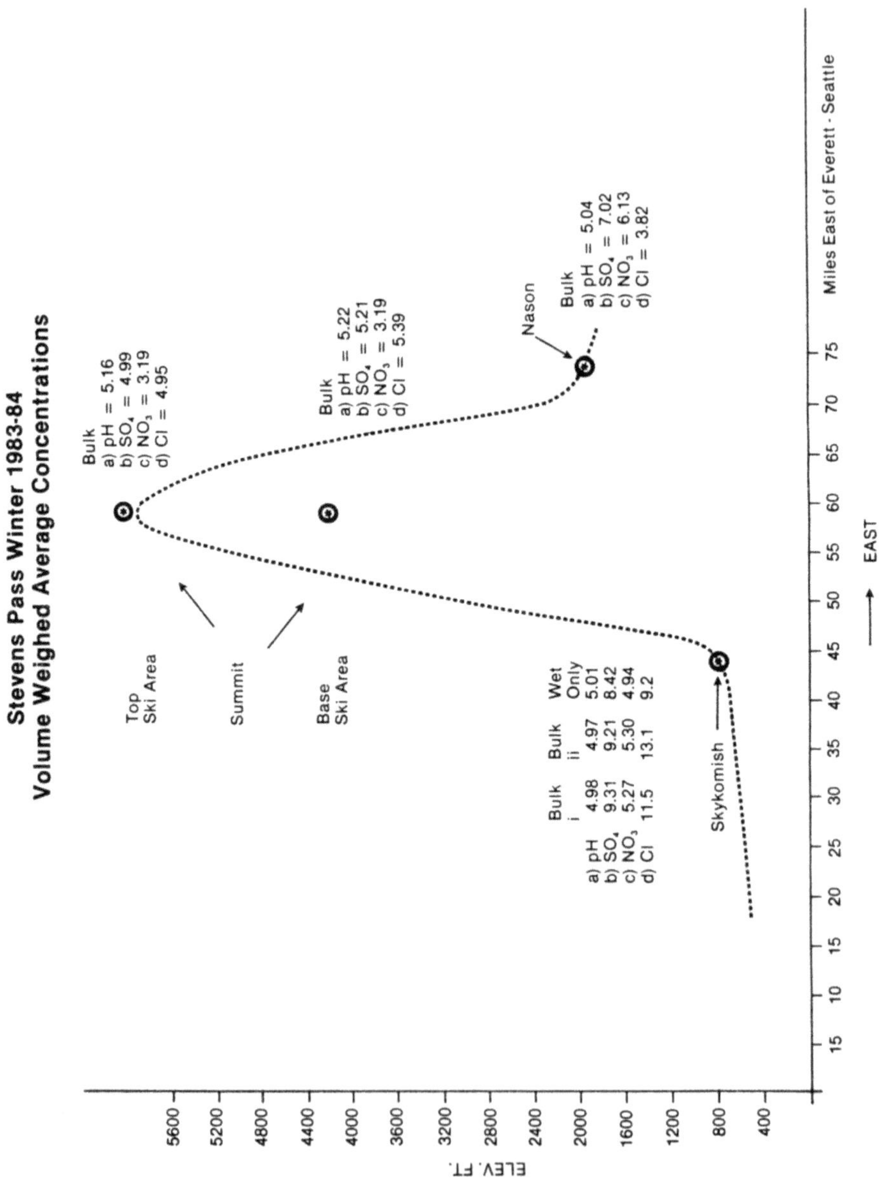
Stevens Pass Winter 1983-84
Volume Weighed Average Concentrations
Top Ski Area
Summit
Base Ski Area
Bulk
a) pH = 5.16
b) SO4 = 4.99
c) NO3 = 3.19
d) Cl = 4.95
Bulk
a) pH = 5.22
b) SO4 = 5.21
c) NO3 = 3.19
d) Cl = 5.39
Nason
Bulk
a) pH = 5.04
b) SO4 = 7.02
c) NO3 = 6.13
d) Cl = 3.82
Bulk i Bulk ii Wet Only
a) pH 4.98 4.97 5.01
b) SO4 9.31 9.21 8.42
c) NO3 5.27 5.30 4.94
d) Cl 11.5 13.1 9.2
Skykomish
ELEV. FT.
5600 5200 4800 4400 4000 3600 3200 2800 2400 2000 1600 1200 800 400
15 10 15 20 25 30 35 40 45 50 55 60 65 70 75
Miles East of Everett - Seattle
EAST

fall-winter months and a majority of that above 610 m occurs as snow (Logan et al., 1982). Samples collected in wet/only collectors differ from those obtained in bulk collectors (Chan et al., 1984). The differences in wet/only and bulk collection data must arise in part because of local dry deposition characteristics. The magnitude of those influences was determined by comparing collector sample characteristics at one site. The most easterly collection site at Skykomish (Figure 1) receives the least snow of any of the Stevens Pass transect sites so was used to observe the bulk collector, wet/only collector relationships over a period of 25 weeks winter 1983-84. The results are shown on Tables II and III.

The bulk collection data differ from wet/only collector data for several parameters. The wet/only collector misses at least 8% of the precipitation volume, presumably at the start of each event (Chan et al., 1984). Bulk and wet/only data were considered different if the mean % difference (Table II) of Bulk 1 - wet/only exceeded the mean % difference of Bulk 1 - Bulk 2 by a magnitude greater than 1.96 x [std. dev. (Bulk 1 - Bulk 2)] x $28^{-1/2}$. The pH of precipitation collected in the wet/only container was the same as that of deposition collected in the bulk container while $SO_4^=$ and all cation concentrations were higher in the bulk samples. Nitrate concentrations in wet/only and bulk were not significantly different. Our observation of elevated $SO_4^=$ yet unchanged pH content in the bulk collector is consistent with that of Dasch (1985) who suggested that the bulk samples absorb more acidifying SO_2 than protected rain samples, but that the resulting excess acidity is more than made up for by the neutralizing effect of dry deposition. In summary the volume weighted pH values of the samples collected were essentially independent of mode of collection. Overall the bulk collector data overestimates wet $SO_4^=$ deposition. An added uncertainty is introduced at the higher elevation sites in that the bulk collectors do not collect all of the snow that falls in some events. If there is a consistent

Table II

Compositional data bulk vs wet/only collectors winter 1983-84 weekly collection, Skykomish, WA

	V(l)	pH	$SO_4^=$	NO_3^-	Cl^-	Ca^{++}	Mg^{++}	Na^+	K^+
Bulk 1 vs Bulk 2									
D%	0.57	0.08	0.74	1.1	3.9	2.1	33.	10.	9.
std. dev.	0.33	1.03	2.3	1.2	27.	67.	109.	43.4	90.3
N=25									
Bulk 1 vs Wet/Only									
D%	7.8	0.43	15.	0.1	21.	46.	53.	45.	27.
std. dev.	22.	1.6	20.	35.	32.	101.	132.	69.	89.
N=24									

$$D\% = \text{mean of } 2 \times \frac{\{\text{value 1st collector} - \text{value 2nd collector}\}}{\{\text{value 1st collector} + \text{value 2nd collector}\}} \times 100$$

std. dev. = std. dev. of D%

Table III

Precipitation weighed mean concentrations of major species Stevens Pass transect winter 1983-84, Stevens Pass winter 1984-85, Snoqualmie Pass winter 1984-85, Global Precipitation Network Stations 1980-81, Selected NADP stations 1982.

	Volume Weighted Average Concentrations µeq L^{-1}									
	pH	NH_4^+	Ca^{++}	Mg^{++}	Na^+	K^+	$SO_4^=$	NO_3^-	Cl^-	Sa
Winter 1983-84										
Skykomish										
Bulk 1	4.98	*	0.4	0.2	10.5	0.2	9.3	5.3	11.5	14.6
Bulk 2	4.97	*	0.5	0.3	9.9	0.1	9.2	5.3	13.1	14.5
Wet/Only	5.01	*	0.3	0.3	7.5	0.1	8.4	4.9	9.2	14.5
Stevens Lo	5.22	*	0.4	0.2	4.0	0.2	5.2	3.2	5.0	11.1
Stevens Hi	5.16	*	0.3	0.3	4.2	0.1	5.0	3.2	5.4	11.0
Stevens Nason	5.04	*	0.7	0.3	2.2	0.1	7.0	6.2	3.8	13.4
Snoqualmie Pass	5.14	1.6	0.5	0.6	7.1	0.7	8.3	5.1	8.7	12.6
Winter 84-85										
Stevens Hi	5.18	1.3	1.3	0.7	3.2	0.4	7.4	4.7	5.6	9.0
Snoqualmie Pass	5.04	3.4	1.5	1.3	5.8	0.5	13.	7.4	8.1	17.8
Amsterdam Is#	4.92	2.1	7.4	39.	177.	4.	(8.8)	1.7	208.	
Poker Flat AK#	4.96	1.1	0.1	0.2	1.	0.6	(7.2)	1.9	2.6	
Katherine Aust.#	4.78	2.0	2.5	2.	7.	0.9	(5.5)	4.3	11.8	
San Carlos Venz.#	4.81	2.3	0.3	0.5	1.8	0.8	(2.7)	2.6	2.5	
St. Georges Ber.#	4.79	3.8	9.7	34.	147.	4.3	(18.)	5.5	175.	
Hopeland CA$	5.37	3.3	2.	4.3	18.	0.5	6.9	3.9	21.	
Sequia CA$	5.3	3.3	1.5	1.	2.7	0.4	6.0	3.4	7.3	
Mana Loa HI$	5.34	1.7	1.	0.6	1.2	0.1	5.6	1.1	1.7	
Hoh River WA$	5.53	1.1	2.0	5.3	21.4	1.9	7.5	1.1	25.	

* = below analytical scatter
Sa = titrable strong acid
\# = Global Precipitation Network Station (Galloway et al.,1982)
$ = NADP, (1985)
( ) = Calculated excess sulfate

Collection periods [Julian dates]: 1983-84, Skykomish 338/83-127/84; Stevens Lo 303/83-127/84; Stevens Hi 303/83-127/84; Stevens Nason 288/83-127/84; Snoqualmie Pass 22/84-127/84. Collection periods 1984-85, Stevens Hi 319/84-156/85; Snoqualmie Pass 23/85-135/85.

trend in compositional change during a snow event the bulk snow concentrations will show a bias. The volume weighted averages reported here are based on the volume of precipitation collected.

The precipitation composition across the Stevens pass transect: The volume weighted average concentrations obtained over a 25-week period commencing November 1983 are shown in Figure 2. The average pH varied from 5.0 to 5.2 (6.3 to 10 µeq L^{-1}) across the transect while the bulk $SO_4^=$ concentrations ranged from a low of 5.0 µeq L^{-1} at the highest elevation to 9.2 µeq L^{-1} at the most westerly collection site. Nitrate concentrations varied from 3.2 µeq L^{-1} at the highest elevation to 6.3 µeq L^{-1} at the most easterly collection site.

The $SO_4^=$ concentrations at the Stevens Pass site winter 1983-84 and winter 1984-85 were low (Table III). In fact they were as low or lower than the $SO_4^=$ concentrations observed (1980-81) at four of the five remote collection sites of the Global Precipitation Chemistry Project (Galloway et al., 1982). The data from those sites has been used in an attempt to define a working 'background' precipitation $SO_4^=$ level (Galloway et al., 1984).

The Stevens Pass NO_3^- concentrations exceeded the concentrations reported (Galloway et al., 1982) for Amsterdam Island and Poker Flat, Alaska but were less than the average concentrations reported for the St. Georges Bermuda, Katherine Australia, and San Carlos Venezuela sites. Only four of the 65 NADP sites discussed in the 1982 annual data summary (NADP 1985) showed $SO_4^=$ and or NO_3^- concentrations lower than those reported here. Those stations were located at Hopeland California, Mauna Loa Hawaii, Olympic National Park Washington and Sequoia National Park California. The first three of those sites are Coastal while the third is inland at higher elevation.

The observation of low $SO_4^=$ and low NO_3^- concentrations in the mountain precipitation is significant in view of the sensitivity of the watershed source lakes which lie along the Cascade crest (Logan et al., 1982; Welch and Spyridakis, 1985).

The observation of a dramatic dropoff in contaminant concentrations in the mountain precipitation as compared to that of precipitation in the adjacent Puget Sound basin is consistent with a description of the meteorological factors at play that was proposed by Pennell and co-workers (1985). They considered the marine origin of the air masses sweeping over the Puget Sound basin, the contaminant sources (See Table IV for Southwest Washington Acid Oxide Inventory), and the usual storm trajectories and suggested that precipitation events could be characterized by an initial transient 'purge' phase in which oxidized SO_x and NO_x were removed from the air mass (most often over the Puget Sound basin). They suggested that precipitation contaminant concentrations would depend on the history of the air mass and could be relatively high during the 'purge' phase. Following that 'purge' phase precipitation composition is governed by the rate at which primary unreacted emissions are oxidized and scavenged while the supporting air masses move on to and interact with the mountains. Precipitation contaminant concentra-

Table IV

S.W. Washington Acid Oxide
Emission Estimates Tonne

	I SWAPCA		II PSAPCA		III St. Helens	Sum I,II,III kilo tonne		Equivalent Ratio
	SO_2	NO_x	SO_2	NO_x	SO_2 (tonne)	SO_2	NO_x	$SO_2/NO_{1.5}$
1979	P59,950	P37,600	P99,637 *36,558 136,195	P12,532 *78,780 91,312	?	196	129	
1981	P55,900	P34,950	P101,532 *36,558 138,080	P11,635 *78,780 90,415	105,130	309	125	2.84
1982	P37,600	P35,000	P88,982 *36,558 125,540	P11,809 *78,780 90,589	45,130	210	126	1.96
1983	P37,500	P26,900	P86,499 *36,558 123,057	P10,952 *78,780 89,732	38,720	192	117	2.02
1984	P51,900	P31,300	P70,326 *36,558 106,884	?11,000 *78,780 89,780	26,713	185	121	1.81

P = Point sources
* = Area source estimate (1980)
? = Estimate
I = Southwest Air Pollution Control agency data (includes coal powered generating plant)
II = Puget Sound Air Pollution Control agency data (includes Seattle–Tacoma urban area and the ASARCO copper smelter)
III = USGS data

tions were predicted to be lower during this 'quasi-steady-state' regime. Most often the cascades would receive precipitation of this type.

There may be a washout effect coupled with the meteorological factors cited above which serves to decrease contaminant concentrations in the Cascades even further. There is a lesser washout efficiency of snow for $SO_4^=$ as compared to rain (Scott, 1981). Most of the high Cascade precipitation occurs as winter snow so this factor would serve to further attenuate the effect of Puget basin SO_x injections. Snow seems to be an efficient HNO_3 absorber (Chang, 1984) so seasonal changes in NO_3^- content related to precipitation type would not be expected. Logan et al.(1982) did report that $SO_4^=$ concentrations increased winter-spring while NO_3^- concentrations did not vary with time. Late spring 1985 the ASARCO smelter ceased operation. The collection periods described in this report extend beyond the closure period for approximately one and one-half months. The data did not indicate dramatic changes in $SO_4^=$ concentration in that time period.

4. CONCLUSION

The chemical composition of the precipitation collected at Stevens Pass and Snoqualmie Pass, Washington, winters 1983-84 and 1984-85, differs little in NO_3^-, $SO_4^=$ and H^+ concentration from the composition of precipitation collected at remote stations in the world that were sited because of their pristine locations. The Cascade sites are down wind and are within 60 km of significant acidic oxide point and area sources. The relative insensitivity of precipitation composition to near-by source inventory may, in part, arise because of the storm meterology in the Puget basin. The inefficiency of snow as a $SO_4^=$ collector must also have an effect. Whatever the cause, the sensitive Cascade crest receptor waters are receiving input that differs little from 'background levels'. Future studies will define the effect, if any, of the ASARCO closure on precipitation composition.

ACKNOWLEDGMENT. The research on which this paper is based was financed, in part by the United States Department of the Interior as authorized by the Water Research and Development Act of 1975 (P.L. 95-467) and made available through the State of Washington Water Research Center. Partial support was also supplied by the Washington State Department of Ecology. We are thankful for the help and advice of many people including the management at the Snoqualmie Pass and Stevens Pass ski areas, many people of the Wenatchee National Forest USFS, and Craig Wilbur, Washington State Department of Transportation (avalanche control).

References

APHA:1976, American Public Health Association, Washington, D.C.

Ausserer, W. and Duncan, L.C.:1984, 'Precipitation Chemistry at Skykomish Washington', report to Washington State DOE, Olympia, Washington.

Chan, W.H., Lusis, M.A., Stevens R.D.S., and Vet R.J.:1984, Water Air and Soil Pollut. **23**, 1.

Chang, T.Y.:1984, Atmos. Environ., **18**, 191.

Dasch, J.M.:1985, Environ. Sci. Technol. **19**, 721.

Dethier, D.P.:1979, Water Resour. Resh. **15**, 757.

EPA:1984, 'Determination of Inorganic Ions by Ion Chromatography', Method 300.0 EPA-600/4-84-017: Office of Research and Development, U.S. Environ. Prot. Agency, Cincinnati, Ohio.

EPA:1979, 'Methods for Chemical Analysis of Water and Wastes, EPA-600/4-79-020; Office of Research and Development, U.S. Environ. Prot. Agency, Cincinnati, Ohio.

Galloway, J.N., Likens, G.E. and Hawley, M.E.:1984, Science **226**, 829.

Galloway, J.N., Likens, G. E., Keene, W.C., and Miller, J.M.:1982, J. Geophys. Res. **87**, 8771.

Larson, T.V., Charlson, R.J., Knudson, E.J., Christian G.D., and Harrison, H.:1975, Water Air and Soil Pollut. **4**, 319.

Logan, R.M., Derby, J.C., and Duncan L.C.:1982, Environ. Sci. Technol. **16**, 771.

McQuaker, N.R., Kluckner, P.D., and Sandberg, D.K.:1983, Environ. Sci. Technol. **17**, 431.

NADP:1985, 'NADP Annual Data Summary, Precipitation Chemistry in The United States 1982', J.H. Gibson program coordinator, Colorado State Univ. Fort Collins, Colorado.

Pennell, W.T., Hales, J., and Scott, B.C.:1985, 'Wet Scavenging Processes in Puget Sound Storms', Proceedings of the Acid Rain Symposium Annual Conference of the Northwest Association for Environmental Studies - University of Victoria, Victoria, B.C. Canada.

PSAPCA:1985, Written communications from Puget Sound Air Pollution Control Agency, Seattle, Washington.

Scott, B.C.:1981, J. Appl. Meteor. 20, 619.
Sisterson D.L., Wurfel, B.E., and Lesht, B.M.:1985, Atmos. Environ. 19, 1453.

SWAPCA:1985, Written communications from Southwest Air Pollution Control Agency, Vancouver, Washington.

USGS:1985, Communications from United States Geological Survey, Vancouver, Washington.

Vong, R. J.:1982 'Puget Sound Transport Winds with Application to Sulfate Wet Deposition', MSE Thesis, Univ. of Washington, Seattle, Washington.

Vong, R.J., Larson, T.V., Covert, A.P. and Waggoner, A.P.:1985, Water Air and Soil Pollut. 26, 71.

Welch, E.B., Chamberlain, W.H., and Spyridakis, D.E.:1984, Northwest Science 58, 85.

Welch, E.B. and Spyridakis, D.E.:1985, 'Chemical Content of Washington Cascade Mountain Lakes and Indications of Acidification.' Abstracts, Int. Symp. Acid Precip. Muskoka, Canada.

PRECIPITATION CHEMISTRY MEASUREMENT IN ALBERTA

Henry L. Bertram, Nirmal C. Das
Alberta Environmental Centre, P.O. Bag 4000,
Vegreville, Alberta, Canada T0B 4L0
and
Yan K. Lau
Alberta Environment, Air Quality Control Branch,
9820 - 106 St. Edmonton, Alberta, Canada T5K 2J6

ABSTRACT. Precipitation samples in Alberta were collected and analyzed monthly from six Alberta Environment stations. Samples were collected with Sangamo samplers and analyzed for the major ions, pH and acidity. The data were tabulated and analyzed for spatial distribution, seasonal variation, temporal trends, ionic character and wet sulphate deposition. The major ionic species in Alberta precipitation are Ca^{2+}, SO_4^{2-}, NH_4^+ and NO_3^-. The spatial distribution shows a slight decrease in pH from southern Alberta (pH 6.0) to northern Alberta (pH 5.4). The seasonal variation shows higher hydrogen ion content in the summer months (pH 5.4 in summer and pH 5.8 in winter). Temporal trends are not apparent over the five year period investigated. The five year average wet sulphate deposition rate in Alberta is 9.1 kg ha^{-1} yr^{-1}.

1. INTRODUCTION

The Alberta and Federal Governments undertook the monitoring of precipitation in Alberta in the year 1977-78. From 1978 to 1984, the CANSAP (Canadian Network for Sampling Precipitation) network consisted of five monitoring sites and the Alberta Environment network consisted of six monitoring sites. This paper examines the 1980 to 1984 data from the Alberta Environment network. For the spatial distribution of pH CANSAP data were included.

The purpose of the Alberta Environment network is to determine the quality of precipitation in the province and the temporal trends caused by population growth and industrialization of the Province.

2. SAMPLING AND ANALYSIS

The Sangamo Type A wet-only precipitation sampler, which only opens during precipitation events, was used. The accumulated sample was collected on a monthly basis.

Analyses consisted of the major cations: H^+, Ca^{2+}, Mg^{2+}, NH_4^+, Na^+, K^+, and the major anions: SO_4^{2-}, NO_3^-, Cl^-, and PO_4^{3-}. Due to the nature of precipitation, which is a nonequilibrium system at low concentration, pH and acidity measurements are difficult to determine accurately and require special procedures. pH was measured by

Water, Air and Soil Pollution **30** (1986) 231–237.

calibrating the pH meter with pH 4.0 and pH 7.0 buffers and then checking the calibration with a strong acid (quality control standard of low ionic strength and pH 4.5). The sample was then stirred for 1 minute, allowed to become quiescent and then measured. Atomic absorption (AA) was used for Ca^{2+} and Mg^{2+}, and ion chromatography (IC) for the other ions. These techniques are well suited for the low concentrations encountered in precipitation. Bicarbonate was determined by calculation.

The quality control procedure consisted of spiking every 5th sample and running every 10th sample as a duplicate, as well as participating in interlaboratory quality control studies. Samples were reanalyzed for those parameters which exceeded their predetermined upper range. Specific conductance ratio, measured versus calculated, and ion balances were determined for each sample to check on the internal consistency of the analytical results. As part of the quality assurance program, blind samples were analyzed occasionally.

3. RESULTS AND DISCUSSION

3.1 Spatial Distribution

The spatial distribution of hydrogen ion concentration, as indicated by pH, and the locations of the Alberta Environment and CANSAP stations, are illustrated in Figure 1. The average pH decreases from pH 6.1 in the south to pH 5.2 in the north. Stations in higher rainfall forested areas such as Beaverlodge (485 mm yr^{-1} of precipitation) have the lowest pH while those in cultivated lower rainfall areas such as Suffield (335 mm yr^{-1} of precipitation) have the highest pH.

The five year average SO_4^{2-} concentration, precipitations volume, and wet sulphate deposition (Alberta Environment Stations) are tabulated in Table I. In general southern Alberta has a higher sulphate concentration and wet sulphate deposition rate. The wet sulphate deposition and acid precipitation are not closely related in Alberta.

The ions Ca^{2+}, SO_4^{2-}, NO_3^-, and NH_4^+ all show decreases in their average concentration from south to north (Lau, 1985). The northern forest and forested areas of the eastern slopes of the Rockies would contribute less soil originated dust components when compared to the drier southern prairies. The northeast corner of Alberta, which is located on the Precambrian Shield, is the area of the province most susceptible to the effects of acid rain. In this area shallow top soil and insufficient lake sediments lack the ability to buffer acid precipitation for very long. This is also an area in which more acidic precipitation is occurring.

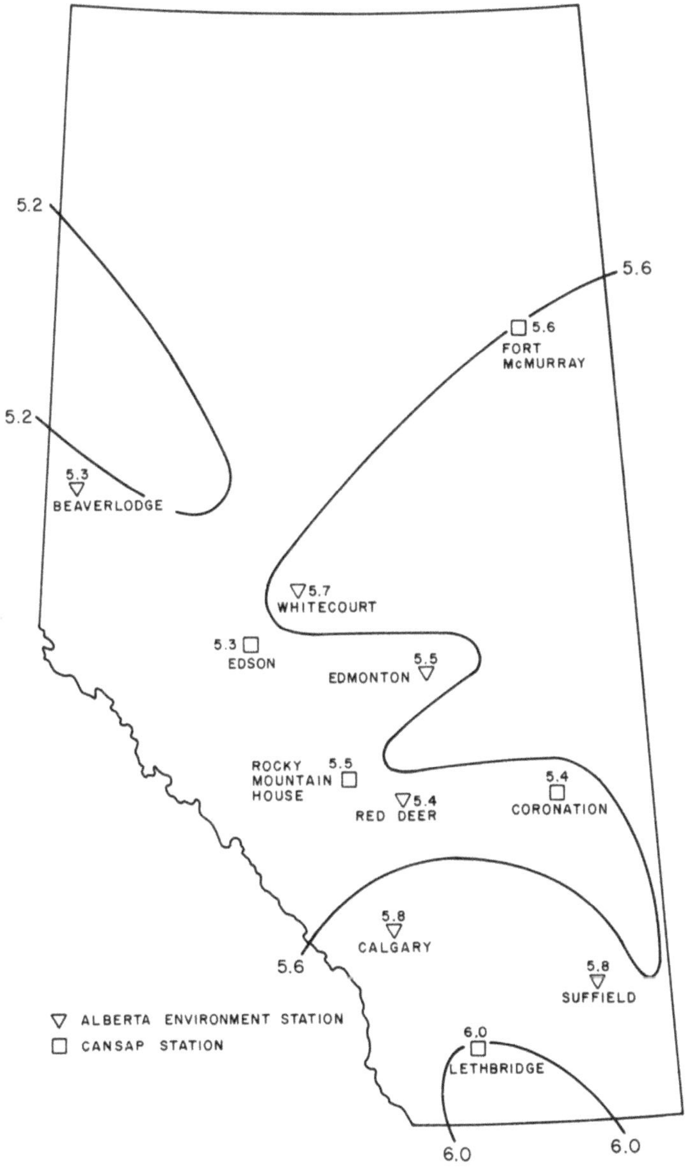

Fig. 1. Average pH of precipitation in Alberta 1980-1984

TABLE I

Wet Sulphate Deposition in Alberta (1980-1984)

Station	Volume-Weighted Average Sulphate $mg\ L^{-1}$	Average Annual Precipitation mm	Average Wet Sulphate Deposition $kg\ ha^{-1}\ yr^{-1}$
Beaverlodge	1.19	485	5.8
Whitecourt	2.06	571	11.8
Edmonton	1.96	472	9.3
Red Deer	1.88	488	9.2
Calgary	2.90	409	11.9
Suffield	2.04	335	6.8
Alberta Average			9.1

3.2 Seasonal Variation

The seasonal variation of major ions is illustrated in Figure 2. H^+ ion concentration is highest in the summer months, June to September. This is the opposite to what one would expect considering local anthropogenic sources of acid precursors, because the gas processing industry emissions of SO_2 (major source) are approximately 1.5 times higher in winter than in summer (Alberta Environment 1982; 1984). The concentrations of the other major ions Ca^{2+}, SO_4^{2-}, and NO^{3-} are lower in the summer months, June to September, and higher the rest of the year. Although the anthropogenic sources of alkaline earth components such as road dust and farming activity would be expected to contribute more in summer months (Environment Canada, 1981), the higher amount of precipitation in summer dominates the precipitation character by diluting the ions present. The seasonal differences in ion concentrations (H^+, Ca^{2+}, SO_4^{2-}, NO_3^-) were determined to be significant by means of the Smith-Satterthwaite t-test on the lognormal distributions (Lau, 1985; Chan et al, 1984). Ammonium ion does not have any pronounced seasonal variation. Anthropogenic sources of ammonia would be highest in the summer. The fact that the high Ca^{2+} ion concentrations are associated with the low H^+ ion concentrations (Figure 2) suggests that neutralization of the acidity is taking place in the precipitation. The seasonal variations are best explained by the quantity of precipitation in each season, which is much higher in the summer than in the winter.

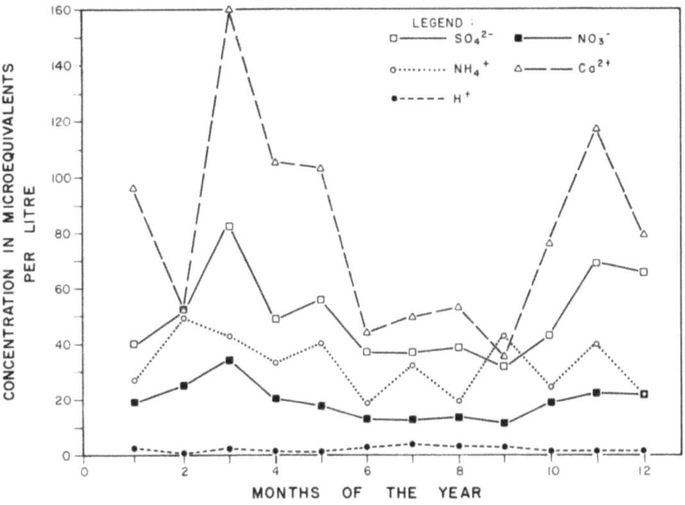

Fig. 2. Seasonal variation of precipitation quality in Alberta for the period 1980-84, Alberta Environment network

3.3 Temporal Trends

The annual volume-weighted average H^+ ion concentration has shown little variation (3.24 to 2.76 μeq L^{-1}) over the five-year period and no pronounced trend is apparent. This is the same whether one uses the overall provincial average (Table II) or individual station averages (Lau, 1985). Over the five-year period Ca^{2+} exhibits the widest variation of any ion in its yearly average concentration (33.8 to 105.4 μeq L^{-1}). As calcium originates largely from windborne dusts, such factors as the occurrence of a dry year and the extent of forest fires, etc. would be expected to contribute to this variation. Sulphate, nitrate and ammonium do not show as pronounced a variation in their yearly averages. The values reflect the wide variability in the nature of precipitation in Alberta. There does not appear to be any trend towards increasing or decreasing values over the five year period. Over the five year period, the relative abundance of the major ions has remained the same, that is $Ca^{2+} > SO_4^{2-} > NH_4^+ > NO_3^- > H^+$.

3.4 Ionic Character

Precipitation in Alberta is less acidic (average pH 5.5) than in eastern North America (average pH 4.1 to 4.4) (National Research Council, 1983) or northern Europe (average pH 4 to 4.5) (Swedish Ministry of Agriculture, 1982) where forest and lake damage has been observed. In Alberta only 5% of the samples have a pH of less than

5.0, with the lowest recorded pH (in the CANSAP and Alberta Environment network) being 4.3. This is illustrated in the pH frequency distribution (Figure 3). The concentration of various ions are in the following order: $Ca^{2+} > SO_4^{2-} > NH_4^+ > NO_3^- > Na^+ > Mg^{2+} > HCO_3^- > K^+ > Cl^- > H^+$. The average wet sulphate deposition rate in Alberta is 9.1 kg ha^{-1} yr^{-1}.

The dominant ionic species in Alberta precipitation are Ca^{2+}, SO_4^{2-}, NH_4^+ and NO_3^-, (see Table II) whereas in eastern North America and northern Europe the dominant ions are SO_4^{2-}, H^+ and NO_3^-. There is no significant correlation between hydrogen ion and sulphate ion in Alberta precipitation, as occurs in eastern North America.

TABLE II
Data Summary For Alberta Environment Network

Volume weighted average concentration in μeq L^{-1}										
ION YEAR	H^+	NH_4^+	Na^+	K^+	Ca^{2+}	Mg^{2+}	SO_4^{2-}	NO_3^-	Cl^-	HCO_3^-
1980	3.24	44.40	48.34	14.23	105.36	22.71	53.72	17.65	4.86	17.65
1981	3.11	33.19	6.68	5.63	67.03	12.40	52.34	19.12	4.26	9.06
1982	2.82	22.26	8.82	4.96	35.83	12.33	31.49	11.12	5.01	6.40
1983	3.00	24.14	6.34	3.75	36.57	9.98	37.36	15.62	4.17	9.22
1984	2.76	29.95	5.62	7.62	70.09	17.91	41.58	15.22	4.81	21.01
80-84 Ave.	2.98	30.79	15.16	7.24	62.98	15.06	43.30	15.75	6.63	12.67
80-84 Ave. pH 5.53										

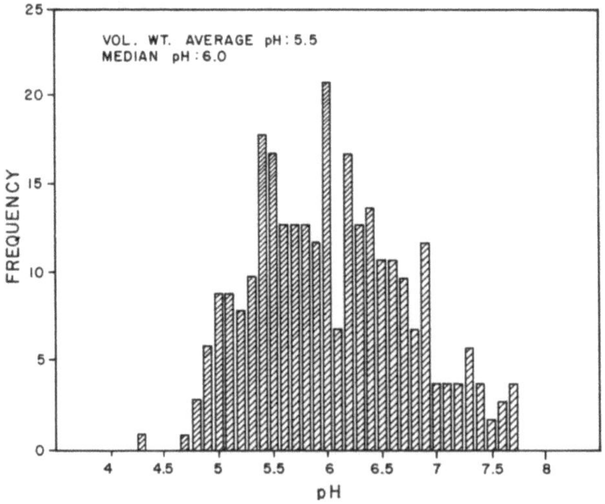

Fig. 3. Alberta Precipitation Quality - pH frequency distribution, Alberta Environment network - 1980 to 1984

4. CONCLUSIONS

We conclude from the investigation of Alberta precipitation that:

1) The ionic character is: $Ca^{2+} > SO_4^{2-} > NH_4^+ > NO_3^- > Na^+ > Mg^{2+} > HCO_3^- > K^+ > Cl^- > H^+$.
2) The spatial distribution shows a decrease in pH (from 6.0 to 5.2) from south to north in the province.
3) Seasonal variations show higher H^+ concentrations in the summer months (pH 5.4 in summer, pH 5.8 in winter).
4) Temporal trends are not apparent over the five year period investigated.
5) The five year average wet sulphate deposition in Alberta is 9.1 kg ha^{-1} yr^{-1}.

ACKNOWLEDGMENTS

The authors wish to acknowledge Dr. H. S. Sandhu and Bonnie Magill of Alberta Environment, Research Management Division, and John Torneby of Alberta Environment, Air Quality Control Branch for their helpful comments on the drafts. The five years data used in this paper have been produced by a number of people. The authors would like to thank collectively, those involved in sampling at Air Quality Control Branch, and those involved in laboratory analyses at the Alberta Environmental Centre, for their diligent efforts.

REFERENCES

Alberta Environment: 1982, Industrial Sulphur Emissions For Alberta 1977-81, Pollution Control Division, Edmonton, Alberta.

Alberta Environment: 1984, Nitrogen Oxides Emissions For Alberta 1978-1982, Pollution Control Division, Edmonton, Alberta.

Chan, W. H., Lusis, M. A., Stevens, R. D. and Vet, R. J.: 1984, 'A Precipitation Sampler Intercomparison', Water, Air and Soil Pollution 23, 1.

Environment Canada: 1981, National Inventory of Natural Sources and Emissions of Primary Particulates, Air Pollution Control Directorate, Ottawa, Canada.

Lau, Y. K.: 1985, A Five Year (1978-1982) Summary of Precipitation Chemistry Measurements in Alberta, Alberta Environment, Air Quality Control Branch, Edmonton, Alberta.

National Research Council: 1983, Acid Deposition Atmospheric Processes in Eastern North America, National Academy Press. Washington, D.C., p. 14.

Swedish Ministry of Agriculture Environment' 82 Committee: 1982, Acidification Today and Tomorrow, 1982 Stockholm Conference, Risbergs Tryckeri AB, Uddevalla, Sweden.

RAINFALL ACIDITY IN NORTHERN BRITAIN - EXPLORING THE DATA

J.N. Cape and D. Fowler
Institute of Terrestrial Ecology
Bush Estate
Penicuik
Midlothian
EH26 OQB
UK

ABSTRACT. Daily rainfall pH measurements from a site in south Scotland were found to have a bimodal distribution. Similarly, monthly data from a network of bulk rain collectors in northern Britain gave bimodal distributions. Sites with an annual average pH of 4.7 had occasional monthly pH values of pH 4 or less. The annual average pH was a poor indicator of the frequency of such acidic periods, and may not be a good indicator of potential effects. Data from a daily and monthly collector at the same site are compared and the discrepancies discussed. Daily pH measurements gave significantly greater acidities than monthly pH measurements, but these differences were small for months with rain-weighted pH $<$ 5.

1. Introduction

There are many different ways of collecting precipitation to measure its acidity; collectors may be continuously-open (bulk) or wet-only, and may be sampled by event, daily, weekly or monthly. Even with identical collectors, the sampling period necessarily affects the distribution of measured pH and may also affect the precipitation-weighted average. This study investigates the geographical variation in monthly pH over 4 years, and the comparison of identical daily and monthly collectors at the same site.

2. Methods

Full details of site locations, methods and chemical analyses for the monthly network may be found in Cape et al (1984). Daily samples were made at Bush (10 km south of Edinburgh) from a bulk collector identical to that used for the monthly network.

3. Results

The data from the daily collector have been used to describe the

Water, Air and Soil Pollution **30** (1986) 239–244.

statistical nature of acid rain 'events' and associated meteorology of episodes (Fowler and Cape, 1984a). They showed a bimodal frequency distribution of daily rain pH, which allowed classification into two populations, although rainfall amount was log-normally distributed. The bimodal behavior of daily data is shown by approximation to two intersecting straight lines (Figure 1). Results from the daily and adjacent monthly collector were also used to estimate the contribution of dry deposition to sulphate collected by the bulk gauges (Fowler and Cape, 1984b).

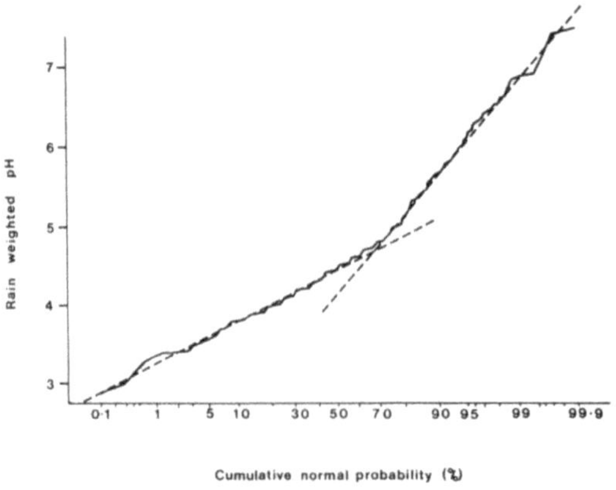

Figure 1. Bimodal normal distribution of daily rain pH at Bush, 1978-1981.

2.1 Monthly data

Locations of sites in the monthly network are shown in Figure 2. The frequency distribution of monthly rainfall-weighted pH, if plotted on a normal probability graph, should show a straight line for a single normally distributed population. When plotted in this way, data over 4 yr (1978-1981) for all sites from the monthly network showed a bimodal type of frequency distribution for monthly average pH. Two examples are shown in Figure 3; the first relates to Bush, where the daily measurements are made, and the second (Inverpolly) to a site on the north-west coast of Scotland. When straight-line segments are fitted to the distributions, each site may be assigned two classes of pH; a low pH class representing 'polluted' rain, and a high pH class representing 'unpolluted' rain. The means of the low pH class for each site are shown in Figure 2, with the % of time that each site gives samples in the low-pH class. Annual average pH changes by 0.5 unit (a factor of 3 in acidity) from west to east over northern Britain (Fowler et al., 1982), with highest pH in the west. The separation of monthly data into the two pH classes shows the inadequacies of representing pH data by annual means.

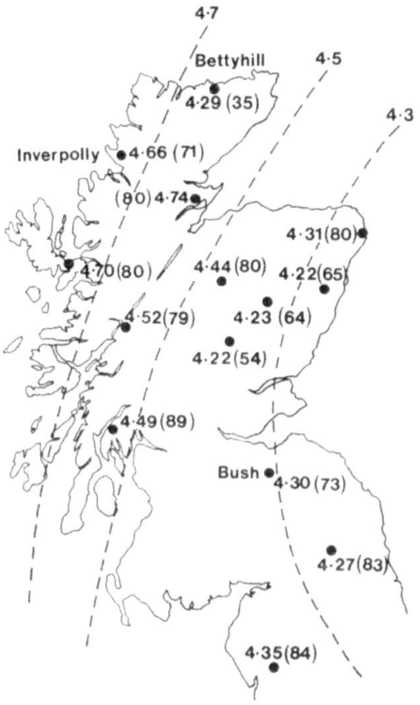

Figure 2. Map of northern Britain showing locations of the monthly network sites. The figures at each are the mean pH of the 'polluted' class of rain, and the % frequency of occurrence in this class. The lines show the distribution of the precipitation-weighted annual mean pH from 1978-1981.

(i) At all sites, even those in the north-west with annual mean pH of 4.7, there are months with average pH less than 4. The difference between west and east is generated by the proportion of rain in each pH class.

(ii) Sites which have very similar annual average pH may have very different distributions of monthly pH; for example, the site at Inverpolly and the site at Bettyhill on the north coast of Scotland both have an annual average pH around 4.7, yet the site at Bettyhill experiences a much lower pH in its 'polluted' rain (mean 4.29) than Inverpolly (mean 4.66).

For all sites in the network, the 'unpolluted' rain pH class has a mean value between pH 4.7 and pH 6, and a much larger standard deviation (slope in Figure 3) than for the 'polluted' rain pH class.

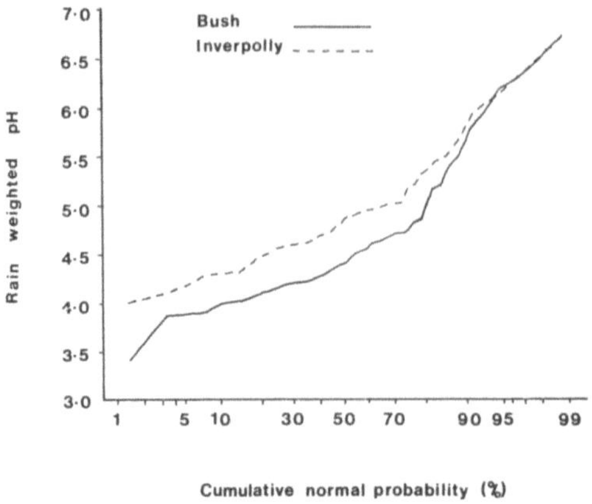

Figure 3. Distribution of monthly precipitation-weighted pH for two of the sites in the monthly network. Bush is in SE Scotland (mean pH of 'polluted' class 4.30) and Inverpolly on the NW coast (mean pH of 'polluted' class 4.66).

3.2 Comparison of monthly and daily data.

The range of variation of daily pH is much greater than that for monthly samples, as there is no 'dilution' of extreme events by rain during the rest of the month.

The comparison of daily and monthly pH data may be confounded by contamination of samples. The rôle of dry deposition of S-containing material has been discussed elsewhere (Fowler and Cape, 1984b). In theory, the relationship between rainfall amount collected daily or monthly over the same period would be exact. In practice, however, there may be differences arising, for example, from evaporation from a monthly gauge. Figure 4 shows the frequency distributions of monthly rainfall totals at Bush as measured in the monthly gauge or calculated by summing collections from the daily gauge. As expected, the differences are small, and comparison of the paired data points over 4 yr gave a correlation coefficient of 0.958 and a constant factor between amounts from the daily gauge relative to the monthly gauge of 0.99 ± 0.02.

In contrast, Figure 5 shows the frequency distributions of the monthly rainfall-weighted pH as measured from the monthly sample or as calculated from daily samples. In this case the possible sources of error are legion: daily samples may be too small to measure an accurate pH, the amount of dry deposition of both acidifying and neutralizing materials is likely to be different, a contamination of one daily sample out of a month would contaminate a whole monthly sample, etc. Comparison of the paired data points (Figure 6) shows that the major differences arise for those samples which are only weakly acidic. It is in these conditions that biological activity would be expected to be greatest (Galloway and Likens, 1976) and where small additional amounts of contamination have the greatest potential for changing the pH by a larger value. If data from the monthly gauge with pH > 5 are excluded, the relationship between monthly measured pH and pH calculated from daily data is surprisingly good (Figure 6) ($r = 0.79$; $n = 37$; factor 'daily' : 'monthly' = 0.98 ± 0.01). On average, the monthly pH from daily samples was less than that from the monthly samples (ie. more acidic). This difference is in the same direction as found by de Pena et al., (1985) and, in view of the different techniques and locations used, suggests that common processes are operating.

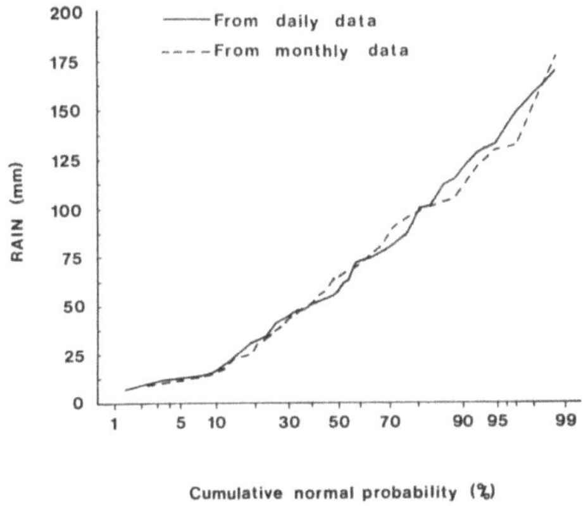

Figure 4 Distribution of monthly precipitation at Bush (1978-1981) - comparison of daily and monthly collectors.

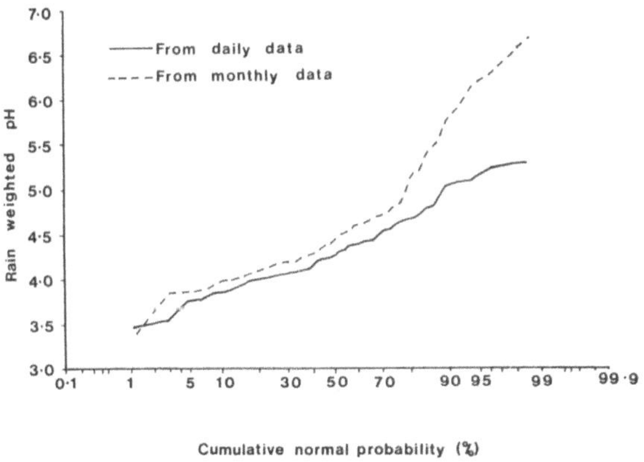

Figure 5 Distribution of monthly precipitation-weighted pH at Bush (1978-1981) - comparison of daily and monthly collectors.

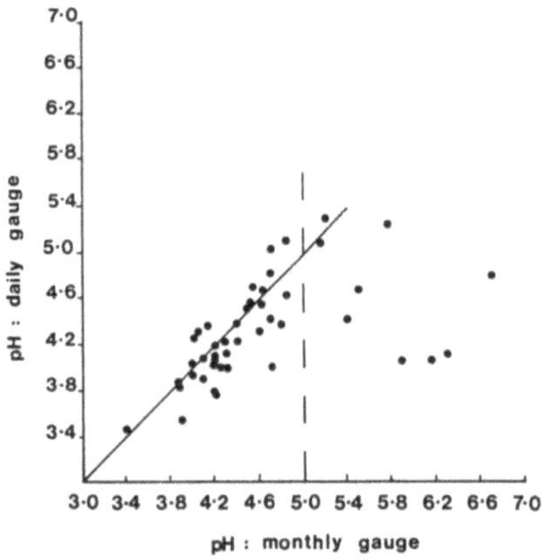

Figure 6 Relationship between precipitation-weighted pH from monthly and daily collectors at Bush for each month.

4. Conclusions

Even monthly bulk rainfall measurements give an indication of the range of pH which may be experienced at a given site. The extent to which data may be split into two pH classes (or the skewness of the overall distribution) gives valuable information on the frequency of low pH rainfall, which may be different for sites with a similar average precipitation-weighted pH.

Comparison with daily data suggests that monthly collections underestimate the acidity (overestimate pH), particularly for samples with large pH values (pH > 5).

References

Cape, J.N., Fowler, D., Kinnaird, J.W., Paterson, I.S., Leith, I.D. and Nicholson, I.A.: 1984, Atmospheric Environment, **18**, 1921.

Fowler, D., Cape, J.N., Leith, I.D., Paterson, I.S., Kinnaird, J.W. and Nicholson, I.A.: 1982, Nature, **297**, 383.

Fowler, D. and Cape, J.N.: 1984a, Atmospheric Environment, **18**, 1859.

Fowler, D. and Cape, J.N.: 1984b, Atmospheric Environment, **18**, 183.

Galloway, J.N., Likens, G.E., Keene, W.C. and Miller, J.M.: 1982, J. Geophys. Res., **87**, 8871.

Galloway, J.N. and Likens, G.E.: 1976, Calibration of collection procedures for the determination of precipitation chemistry. In: Proc. 1st Int Symp. Acid Precipitation and Forest Ecosystems, Dochinger, L.S. and Seliga, T.A. (eds), pp 137-155, USDA Forest Service Report NE-23.

Larsen, R.I.: 1974, J. Air Pollut. Control Assocn., **24**, 551.

de Pena, R.G., Walker, K.C., Lebowitz, L. and Micka, J.G.: 1985, Atmospheric Environment, **19**, 151.

THE ROLE OF ALKALINE MATERIALS IN PRECIPITATION CHEMISTRY: A BRIEF REVIEW OF THE ISSUES

Donald F. Gatz, William R. Barnard, and Gary J. Stensland
Illinois State Water Survey
2204 Griffith Drive
Champaign, Illinois 61820 USA

ABSTRACT. The purpose of this paper is to provide a brief overview of the relatively unexplored role of alkaline materials in precipitation chemistry. Clearly, they can play fully as important a role as acidic materials in determining pH. The first major issue concerns sources. Comparison of Ca/K and Ca/Mg ratios in precipitation and dry deposition with those in likely sources indicates that both unpaved roads and soils make important contributions. Elemental emissions fluxes have been derived from literature estimates of mass emissions fluxes and element abundances in the important sources, but are subject to large uncertainties owing to a lack of adequate data. It is quite clear, however, that conventional (smokestack) sources are minor compared to open sources such as those already identified. Another major issue concerns interactions between alkaline aerosols and water in the atmosphere. Reactions involving suspended solids that lead to removal of H-ions from solution include ion exchange and mineral weathering. A simplified model of acid buffering indicates that NH4, Ca, Mg, K, and Na buffer between 25 and 50% of the potential acids in U.S. precipitation everywhere east of the Mississippi River. The third major issue concerns wet and dry deposition fluxes of alkaline materials. Wet deposition fluxes are currently being measured adequately by a nationwide network of weekly samplers. There is no agreed-upon method for monitoring dry deposition, but available information suggests that dry deposition accounts for somewhat more than half of the Ca deposition. A list of research and data needs is also provided.

1. INTRODUCTION

The purpose of this paper is to discuss the major issues connected with the role of alkaline materials in precipitation chemistry, and thus to focus attention on the alkaline constituents of precipitation, which are commonly overlooked. Nonetheless, the pH of precipitation is determined by its content of both acids and bases.

Dust from the earth's surface is commonly assumed to be alkaline; it is commonly reported in the literature as a constituent of the atmosphere and of precipitation. There is an extensive literature on

Water, Air and Soil Pollution **30** (1986) 245–251.

long-distance transport of desert dust from various locations. Alkaline and alkaline earth cations in precipitation are typically attributed to such sources as "soil dust", "terrestrial dust", or "crustal dust." Specific sources include soil, from which dust may be raised by wind erosion and tilling operations, and also unpaved roads, from which dust is raised by vehicle traffic. Several authors have used chemical element balance methods to estimate the contributions of various sources to the constituents of precipitation. Most results indicate that terrestrial sources account for up to about 30% of the mass or equivalents of precipitation impurities. Neutralization of acids by alkaline materials in precipitation has been estimated (Munger and Eisenreich, 1983) to range between about 50 and 90% in various regions of the U.S. and Canada.

We discuss a variety of issues related to alkaline materials in precipitation under three main headings: source issues, aerosol-water interactions issues, and deposition issues. We end with some suggestions of research and data needed to advance knowledge in this subject area.

2. SOURCE ISSUES

2.1. Source Identification

The primary issue related to sources of alkaline materials in precipitation is their identification and apportionment. Factor analysis of precipitation chemistry data usually identifies a crustal or terrestrial source having high loadings for such elements as Ca, Mg, and K, but multiple sources have not been distinguishable (e.g., Gatz, 1984). Another approach is to compare element ratios in likely source materials with those observed in precipitation, and also in dry deposition. An example appears in Table I, where Ca/Mg and Ca/K ratios in soils and unpaved road dust are compared with ratios observed in wet and dry deposition (bucket collections, which some would call dryfall) in the eastern U.S. High ratios in both forms of atmospheric deposition indicate that unpaved roads must contribute a substantial portion of these cations to both wet and dry deposition. However, if there are unknown, but important, sources of Ca, or if Ca from soils is enriched to a great degree by fractionation during mobilization, then unpaved road sources may be less important than indicated in Table I. Comparison of the Ca/K ratios in both sources (Table I) with the spatial distributions of the Ca/K ratio in both wet and dry deposition in the U.S. (not shown) confirms the overall conclusion above, and suggests that unpaved roads are major sources of crustal cations in large sections of the Midwest and West. It is thus of interest to examine further the source strengths of the major crustal source types, as well as those of the individual alkaline elements.

Table I. Comparison of element ratios in crustal sources and wet and dry deposition in the 31 eastern states, from National Atmospheric Deposition Program/National Trends Network data.

	Ca/Mg	Ca/K
Crustal sources* (soluble, ≤10 um)		
Wind erosion	2.30	0.89
Tilling	2.25	0.78
Unpaved roads	7.74	17.8
Precipitation (Ratios of site medians, 75 NADP/NTN sites)		
Minimum	1.41	2.01
20th percentile	2.75	3.93
50th percentile	4.34	6.00
80th percentile	5.45	8.40
Maximum	8.48	16.31
Dry deposition (Ratios of site medians, 65 NADP/NTN sites)		
Minimum	1.3	0.9
20th percentile	2.9	2.4
50th percentile	4.2	3.9
80th percentile	5.6	8.5
Maximum	14.0	13.9

* Ratios of element emissions from Gatz *et al.* (1985).

2.2. Mass and Element Emissions

Since element emissions are computed as the product of the total mass emissions of the major sources of alkaline materials and their respective element abundances, it is clear that accurate estimates of both source strengths and source compositions are required. Emissions from open sources, such as soil wind erosion, agricultural tilling, and unpaved roads, far exceed conventional (smokestack) emissions of alkaline materials (Gatz *et al.*, 1985). Estimates of mass emissions from these sources are available for the U.S. (Evans and Cooper, 1980; Heisler *et al.*, 1985). Unfortunately these estimates vary considerably for the major sources. We believe that the smaller estimates of Evans and Cooper are closer to being correct. Even these may be too high in some cases, however, judging from mass budget comparisons, by element, of total U.S. emissions with total estimated wet and dry deposition.

The major source compositions required are those for soils and unpaved road materials. For soils, a global mean value (e.g., Bowen, 1966) is sometimes used, but more localized compositions, such as are available from the data of Boerngen and Shacklette (1981), are preferable. From this latter reference, the mean abundances of the four major alkaline elements in the 31 eastern U.S. states are: Na, 0.70%; Mg, 0.30%; K, 1.30%; and Ca, 0.36%.

The situation with respect to the elemental composition of dust from unpaved roads is not as good. Few or no data are available in the literature. Mean percent abundances (± standard errors) from our own recent measurements (Barnard, et al., 1985) of 12 samples from crushed limestone roads in 6 states are: Na, 0.22 ± 0.03; Mg, 2.46 ± 0.46; K, 0.66 ± 0.09, and Ca, 24.38 ± 1.07. Similar measurements on 10 washed gravel samples from 7 states gave: Na, 0.71 ± 0.14; Mg, 2.56 ± 0.53, K, 1.38 ± 0.14, and Ca, 7.83 ± 1.72. The third major category of unpaved road surfaces, soil surfaces, appears from very limited data to have approximately the same composition as local soils. Of particular note in the data given above are the high values of Ca abundance in unpaved road surface materials, especially limestone. Also note that these measurements were made on sieved (< 53 um) materials from the road surface, rather than on road-derived aerosols, which may have somewhat different abundances if fractionation takes place.

Combining mass emissions and element abundances for the major sources of alkaline aerosols yields the element emissions fluxes shown in Table II. Combined element fluxes for five categories of conventional source emissions are shown for comparison. (Fluxes from the individual categories, and the assumed source compositions, are given in the paper referenced in the table.)

Table II. Total mass and estimated cation emissions from major open and conventional sources in the 31 "eastern" states. Emission estimates for open sources are limited to soluble materials ≤ 10 um; these limitations do not apply to the estimates of emissions from conventional sources.

	Thousands of tonnes/yr (1980)*				
Source category	Total mass	Na	Mg	K	Ca
OPEN**					
Wind erosion	37,500	107	47.4	123	109
Tilling	7,420	18.9	8.28	23.7	18.7
Unpaved roads	117,000	172	531	231	4,110
Sum	162,000	297	587	378	4,240
CONVENTIONAL***	8,410	109	119	71.6	274

* Rounded to three significant figures.

** Total mass emissions are from Evans and Cooper (1980); element emissions are based on the work of Stensland et al. (1985).

*** Sum of separate emissions estimated for the following categories: Fuel Combustion, Industrial Processes, Solid Waste Disposal, Transportation, and Miscellaneous (less unpaved roads.) Total mass emissions are from U.S. EPA (1982); element abundances used to calculate emissions are from Gatz et al. (1985).

3. AEROSOL-WATER INTERACTIONS ISSUES

The nomenclature of these interactions deserves some initial comment. "Buffering" is a broad term that can include all forms of H-ion removal from solution, but is often used specifically to refer to removal in a potentially reversible reaction, such as ion exchange. "Neutralization" is a narrower term, referring specifically to removal of H-ion in an irreversible reaction such as formation of water in a reaction of a strong base with a strong acid. For convenience, we shall use the popular term "acid neutralizing capacity" (ANC) as the general term for removal of H-ion from solution.

The major H-ion removal reactions in cloud or precipitation water include: 1) ion exchange, in which cations adsorbed by clay minerals or other substrates are exchanged for H-ions in solution, and 2) mineral weathering (dissolution). Cation exchange reactions are essentially instantaneous, reaction rates being usually diffusion limited. Mineral weathering, on the other hand, may be relatively rapid, as in the case of finely divided $CaCO_3$ particles, or quite slow (days to weeks), as in the case of silicate minerals.

The general situation is that the ANC attributed to a "base cation" in precipitation is less than or equal to the soluble portion of the element (since the element can also be introduced as a neutral salt), This value in turn is less than or equal to the total concentration of the element in precipitation, since elements can also occur in essentially insoluble phases. The insoluble fraction is generally not analyzed in precipitation, but a few measurements in the literature (e.g., Cawse, 1974) and some of our own unpublished measurements suggest that Na and Ca are about 95 to 100% soluble in precipitation (although in some cases Na can be is much less), Mg about 80%, and K near 50%. It is usually not possible to estimate the ANC of these ions from their concentrations in precipitation, so we must then consider appropriate measurements of source materials.

Source measurements should ideally be made on aerosols derived from bulk source materials, but often the practical first step is to analyze the bulk material. We need measurements of both solubility and ANC of the source materials, but such measurements must be done with proper attention to insuring that the samples analyzed properly simulate the particles likely to encounter cloud droplets or precipitation. Specifically, experiments must match ambient conditions with respect to particle size distribution, solvent pH, contact time, and solid/liquid ratio.

It is possible to estimate the approximate ANC due to ammonium and the crustal cations in precipitation by applying a set of assumptions from the literature (Asman et al., 1981; Munger and Eisenreich, 1983) to data from the NADP/NTN network. Spatial distributions of percent neutralization by the sum of NH_4, K, Mg, and Ca indicate that these cations could neutralize 25 to 50% of the potential acids in precipitation everywhere east of the Mississippi River, on an annual basis. It appears that NH_4 and the sum of Mg, K, and Ca each account for about half of the ANC, with Ca having the largest effect among the three crustal elements.

4. DEPOSITION ISSUES

The issues here concern the relative magnitudes and spatial patterns of the wet and dry deposition fluxes to the surface. Wet fluxes are readily measured at the present time by the NADP/NTN national network of precipitation samplers. Dry deposition measurements are far more controversial, since there is no agreed-upon method for routine monitoring. However, for crustal aerosols, which occur principally as relatively large particles, bucket measurements may contain useful information. Plots of wet and dry deposition of Ca (based on dry bucket data) indicate that dry deposition accounts for somewhat more than half of the total Ca deposition. Maximum wet deposition of Ca occurs in a broad swath across the Great Plains, while the maximum dry deposition flux occurs in the Midwest, from Illinois to West Virginia.

5. RESEARCH AND DATA NEEDS

We see three key areas where research is needed to advance our knowledge in the area of alkaline substances in precipitation chemistry. The first of these is research to improve methods of estimating aerosol emissions, especially for the major sources of unpaved road emissions, soil wind erosion, and agricultural tilling. Crustal element emissions appear to be generally overestimated, by comparison with deposition, and a key need is for improvement in methods of estimating total mass emitted. A related data need is for additional measurements of the element abundances in source materials, particularly in aerosols derived from these same major open sources. Implied in this statement is the strong possibility that element fractionation occurs during the mobilization of particles from surface sources. The second perceived research need is for more investigation of mobilization mechanisms, particularly as they relate to element fractionation. The third key research need is the development of methods for measuring the solubility and neutralizing capacity of crustal source materials in precipitation. The problem involves careful matching of source sample specifications to typical ambient conditions, especially such parameters as particle size distributions and aerosol/solvent mass ratios. A related data need is for additional measurements of the solid/liquid partitioning of alkaline elements in precipitation.

ACKNOWLEDGMENTS

This work has been funded as part of the National Acid Precipitation Assessment Program by the National Oceanic and Atmospheric Administration, under Contract NA82RAC00153. Additional support for this work came from the National Science Foundation Atmospheric Chemistry Program, under Grant ATM 83-16643, and from the State of Illinois.

REFERENCES

Asman, W. A. H., Slanina, J., and Baard, J. H.: 1981, Water, Air, Soil Pollut., 16, 159.

Barnard, W. R., Stensland, G. J., and Gatz, D. F.: 1985, 'Alkaline Materials Flux from Unpaved Roads: Source Strength, Chemistry, and Potential for Acid Neutralization. Paper prepared for presentation at the International Symposium on Acidic Precipitation, Muskoka, Ontario, Canada, September 15-20.

Boerngen, J. G., and Shacklette, H. T.: 1981, 'Chemical Analyses of Soil and Other Surficial Materials of the Conterminous United States', Open File Report 81-197, Geological Survey, U.S. Dept. of Interior, Denver, CO.

Bowen, H. J. M.: 1966, Trace Elements in Biochemistry, Academic Press, New York, 241 pp.

Cawse, P. A.: 1974, 'A Survey of Atmospheric Trace Elements in the U.K., 1972-1973', AERE-R7669, HSMO, London.

Evans, J.S., and Cooper, D. W.: 1980, J. Air Pollut. Control Assoc., 30, 1298.

Gatz, D. F.: 1984, Atmos. Environ. 18, 1895.

Gatz D. F., Barnard, W. R., and Stensland, G. J.: 1985, 'Dust from Unpaved Roads as a Source of Cations in Precipitation', Paper 85-6B.6, Proceedings of the 78th Annual Meeting, Air Pollution Control Association, Pittsburgh, PA.

Heisler, S. L., Young, J., Lurmann, F., Hayden, P., Collins, H., Collins, J., Yeransian, T., Kringel, D., and Hay, G.: 1985, 'The EPRI 1982 National Air Pollutant Emission Inventory', Paper 85-4.3, Proceedings of the 78th Annual Meeting, Air Pollution Control Association, Pittsburgh, PA.

Munger, J. W., and Eisenreich, S. J.: 1983, Environ. Sci. Technol., 17, 32A.

Stensland, G. J., Barnard, W. R., and Gatz, D. F.: 1985, 'The Flux of Ca, Mg, K, and Na into the Atmosphere for Acid Rain Neutralization', Paper presented at the Air Pollution Control Association Acid Rain Symposium, Boulder, CO, January 30 - February 1, 1985.

U.S. EPA: 1982, '1980 National Emissions Report', National Air Data Analysis Division, Office of Air Quality Planning and Standards, Research Triangle Park, NC 27711, EPA-450/4-82-010.

SNOW CHEMISTRY IN THE FLIN FLON AREA OF MANITOBA, 1981-1984

S. Floyd Phillips, David L. Wotton, and Dale B. McEachern
Environmental Management, Manitoba Department of Environment and Workplace Safety and Health
Bldg. 2, 139 Tuxedo Avenue,
Winnipeg, Manitoba, Canada, R3N 0H6

ABSTRACT. The March snowpack was sampled at seven sites along two 40 km transects southeast and south southeast of the Hudson Bay Mining and Smelting Co. smelter in Flin Flon, Manitoba, Canada, from 1981 to 1984. The purpose was to determine deposition patterns of smelter emissions to use in environmental impact assessment. Zinc deposition decreased with distance from the smelter according to an inverse curvilinear relationship, thus implicating the smelter as the source. Prevailing northwest winds resulted in higher deposition in the southeast than in the south southeast direction. Snowpack monitoring responded effectively to the reduction in emissions which occurred when a new electrostatic precipitator was installed. There is no evidence that sulphate-S in the snow originated from smelter emissions even though there were up to 800 tonne day^{-1} of SO_2 emitted. The pH of the snow decreased with distance from the smelter and was positively correlated with heavy metals. pH was not correlated with sulphate deposition. Snow monitoring was an effective and sensitive tool for determining snowpack chemistry and air pollutant dispersion around the smelter at Flin Flon.

1. INTRODUCTION

Precipitation monitoring has been used extensively in both North America and Europe to monitor the "acid rain" phenomenon (Shewchuk, 1982; Barrie and Walmsley, 1978; Thornton and Eisenreich, 1982; Seip, 1978). Precipitation monitoring was previously conducted in the area of Flin Flon, Manitoba, Canada in 1976 and 1977 to determine air pollutant fallout from the Zn/Cu smelting complex (Franzin et al., 1979). They found that analyses of the naturally deposited snow pack was an effective method of monitoring the dispersion of air pollutants. Air pollutants can be deposited to the snow profile via three processes, (i) gravitational fallout of particulates (ii) scavenging of pollutants from the air during precipitation events and (iii) adsorption/absorption of gaseous pollutants to the surface of fallen snow (Barrie and Whelpdale, 1977).

Since 1976, the Environmental Management Division of Manitoba has

Water, Air and Soil Pollution **30** (1986) 253–261.

been undertaking studies in the vicinity of primary smelters in Northern Manitoba to assess the environmental impacts of heavy metals and S contained in the stack emissions. Monitoring programs had two basic objectives; to determine the deposition pattern of emissions, and to determine the effects of emissions on environmental components, including soils, vegetation and wildlife. Annual snow sampling surveys have been carried out in the Flin Flon area since 1979 to monitor pH and the deposition patterns of heavy metal and S stack emissions from the Zn/Cu smelter operated by Hudson Bay Mining and Smelting Company. Results of the 1979 and 1980 snow sampling surveys have been reported previously (Hogan and Wotton, 1980). The present report deals with the results of the 1981 to 1984 annual snow surveys.

2. METHODS

2.1. Sampling Method

Snow samples were collected in the Flin Flon area at seven sites located in the southeast quadrant and within a 40 km radius of the Hudson Bay Mining and Smelting stack. The sample stations were located on ice covered ponds or sheltered bays of lakes to minimize drifting and also to provide a substrate which was free of soil organic matter contamination.

All samples were collected before 10 March to avoid the spring melt which can begin as early as March 15. March sampling was also prudent because it provides a measure of total winter-time accumulation. Six replicate snow samples were collected at each sampling site per survey, three for heavy metal analyses (Zn, Cu, Fe, Pb and Cd) and three for pH and major ion analyses (pH, acidity, conductivity, Ca, Mg, K, Na, SO_4, NO_3, NH_3 and Cl). The samples for heavy metal analysis were collected in 1 L linear polyethylene bottles which were pre-washed with HCl and HNO_3. The pH/major ion samples were collected in new heavy-weight plastic bags. A subsample of the plastic bags was tested previously to insure that they were free of chemical contaminants.

Glass distilled water (200 mL) was placed in 20 bags and the bag was sealed. Aliquots were removed and analyzed for all major ions as well as pH, conductivity. In addition, empty bags were taken to the field with the sample bags on each survey that was conducted and after the addition of distilled water one was analyzed with each batch of samples.

Core samples of snow were collected with an acrylic plastic tube (i.d. 3.7 cm). The tube was pressed through the snow pack to the ice, which forced the snow up into the tube. A plastic plunger was used to compact the snow to keep it from falling out as the sampler was pulled out of the snow pack. The compacted snow cylinder was ejected into the sample bottle or plastic bag by applying pressure with the plastic plunger. Cores were sampled until each bottle was filled or alternately until a sufficient sample was accumulated in each bag. The number of cores taken for each replicate varied from 6 to 15 depending on the snow depth. Screw tops were used to close the bottles while the plastic bags were folded over and stapled. The samples were kept frozen by placing them in insulated coolers with freeze packs; they were transferred to a freezer until analyzed. All samples were weighed, and replicates for heavy metal analyses were acidified with 5 mL of concentrated HNO_3. The samples were analyzed by the W. M. Ward Technical Services Laboratory, Manitoba Department of Environment and Workplace Safety and Health.

The detection limits for Zn, Cu, Fe, Pb and Cd were 0.01, 0.005, 0.02, 0.005 and .001 μg L^{-1}, respectively. The pH readings were accurate to 0.1 pH units, the lower detection limit for acidity was 1 mg L^{-1} and conductivity readings had a coefficient of variation of 2.6%. Detection limits for major ions were 0.2 mg L^{-1} for SO_4, 0.1 mg L^{-1} for Cl, 0.02 mg L^{-1} for Ca and Na, 0.01 mg L^{-1} for K and NO_3, 0.005 mg L^{-1} for Mg and 0.002 mg L^{-1} for NH_3.

2.2. Data Analysis

Results of the metal and sulphate analyses were reported from the laboratory as mg L^{-1}. Using the area sampled and the volume of sample, these data were standardized to mg m^{-2}.

The mean deposition by site and survey was calculated for each metal and major ion. Hydrogen ion concentrations were used in the calculation of mean pH.

The data were generally consistent between the three replicate samples but a few analyses were found to be excessively high or excessively low relative to the other two replicates. These outliers were excluded from the calculations.

3. RESULTS AND DISCUSSION

3.1. Heavy Metals

The metals which are emitted as particulates from the Hudson Bay Mining and Smelting stack include but are not restricted to Zn, Cu, Fe, Pb and Cd (Hudson Bay Mining and Smelting, 1982). Of these, Zn comprises the largest fraction in the emissions (25%) and was also the dominant metal in the snow samples (Table I). Since the distribution pattern was similar for all metals, this discussion deals only with the Zn data.

Deposition of Zn to the snow was inversely related to distance, that is, highest close to the smelter, and decreasing as distance increased. The correlation between Zn deposition and distance is best described by the power curve, $Zn = aD^b$; where Zn is the Zn deposition in mg m^{-2} and D is the distance from the smelter in kilometres. All correlation coefficients were highly significant (Table II). This is a curvilinear relationship wherein the greatest decline in Zn occurred over the first 10 to 15 km from the smelter. Beyond 15 km, amounts were considerably lower but continued to decline gradually toward background. Zinc deposition 38 km southeast of the smelter was slightly elevated compared with background for northern Manitoba (0.8 μg g^{-1}). The correlation between metal deposition

TABLE I

Mean heavy metal deposition (mg m^{-2}) to snow, Flin Flon, Manitoba, 1981 to 1984

Year	Site	Dist.(km)	Zn ± s.d.	Cu	Fe	Pb	Cd
1981	1	5.1	2548.9 ± 337.1	98.9	63.1	45.1	0.78
	5	5.8	5386.5 ± 76.6	162.5	80.7	82.1	0.34
	2	8.9	2094.1 ± 109.3	67.9	40.7	36.2	1.25
	6	14.6	558.9 ± 94.6	19.6	30.0	18.0	1.34
	3	15.5	685.6 ± 17.2	18.2	15.1	15.5	1.37
	7	35.7	96.9 ± 6.8	5.5	12.6	5.3	0.48
	4	38.1	84.0 ± 2.9	4.6	9.2	5.4	0.42
1982	1	5.1	2813.9 ± 185.6	94.4	69.8	47.7	8.23
	5	5.8	2249.4 ± 56.6	84.9	61.7	32.9	4.50
	2	8.9	1860.4 ± 517.7	65.7	54.8	36.4	4.77
	6	14.6	209.5 ± 57.9	31.0	34.8	11.3	1.91
	3	15.5	217.9 ± 32.1	13.3	15.3	7.4	0.97
	7	35.7	71.6 ± 4.8	17.3	12.0	5.4	0.83
	4	38.1	38.0 ± 4.2	3.4	9.7	2.5	0.33
1983	1	5.1	55.6 ± 2.2	19.0	12.1	3.6	0.26
	5	5.8	83.5 ± 4.9	11.9	7.3	4.9	0.40
	2	8.9	54.7 ± 5.6	7.1	7.6	2.7	0.28
	6	14.6	11.5 ± 1.4	2.5	20.8	1.1	0.22
	3	15.5	13.2 ± 0.7	2.4	2.2	1.1	0.22
	7	35.7	8.8 ± 0.7	1.5	2.3	0.9	0.44
	4	38.1	4.2 ± 0.2	0.7	4.2	0.6	0.35
1984	1	5.1	103.6 ± 36.0	13.5	9.5	2.8	0.29
	5	5.8	343.1 ± 159.5	42.0	23.4	11.9	0.94
	2	8.9	70.0 ± 16.2	12.3	9.0	4.1	0.52
	6	14.6	21.9 ± 9.5	4.0	45.2	1.1	0.22
	3	15.5	36.8 ± 15.3	4.6	6.2	1.2	0.22
	7	35.7	4.9 ± 1.5	2.3	3.7	0.6	0.10
	4	38.1	3.0 ± 0.2	1.1	1.2	0.5	0.10

and distance is consistent with that found by Franzin _et al_. (1979). This provides evidence that the smelter was the source of the Zn in the snow.

Furthermore, these data show that heavy metal deposition was generally highest at site 5, 5.8 km southeast of the smelter. That was most clearly demonstrated in the 1981 data but also occurred in the 1983 and 1984 results (Figure 1). Wind direction at Flin Flon during the winter months of November to March is predominantly from the northwest (18 to 23%) with relatively strong west (19 to 14%) and north (7 to 13%) components. The plume was therefore carried in the

TABLE II

Power curve relationships between Zn deposition and distance from the smelter at Flin Flon, Manitoba.

Year	a	b	r^2	Prob.
1981	6596.2	-0.82	0.41	<0.005
1982	99,599.1	-2.12	0.94	<0.0005
1983	632.6	-1.32	0.88	<0.0005
1984	5142.1	-1.98	0.88	<0.0005

southeast direction a greater percentage of the time. This could explain why more metals (except Cd) were deposited at site 5, 5.8 km southeast than at site 1, 5.1 km east southeast.

The data in Figure 1 show that Zn emissions decreased over the 1981 to 1983 period, with the largest reduction (20 fold) occurring in 1983. Even though no monitoring was done in the SW, NW or NE quadrants, we do not believe that the reduction in deposition was the result of a change in the distribution pattern of the emissions. Wind direction was similar in all years; the combined frequency of wind from the W, NW and N directions was 41.8% in 1981, 45.5 % in 1982 and 41.1% in 1983. It is possible that the installation of a new electrostatic precipitator on the Zn process could have caused a significant reduction in particulate emissions(Hudson Bay Mining and Smelting, 1982). It is anticipated that the reduction of heavy metal deposition will benefit the environment in the Flin Flon area by slowing the accumulation rate of heavy metals.

Zinc deposition increased marginally in 1984; a three fold increase occurred at site 5, 5.8 km southeast of the smelter while other sites were slightly higher (Table I). The combined frequency of wind from the W, NW and N directions was similar to that in 1983 (44%) and it is unlikely that the difference could be attributed to changes in the dispersion pattern. Hudson Bay Mining and Smelting Co. report that the efficiency of the old baghouse has been declining. Because the two precipitators operate in series, this could have been the reason for higher particulate emissions in that year.

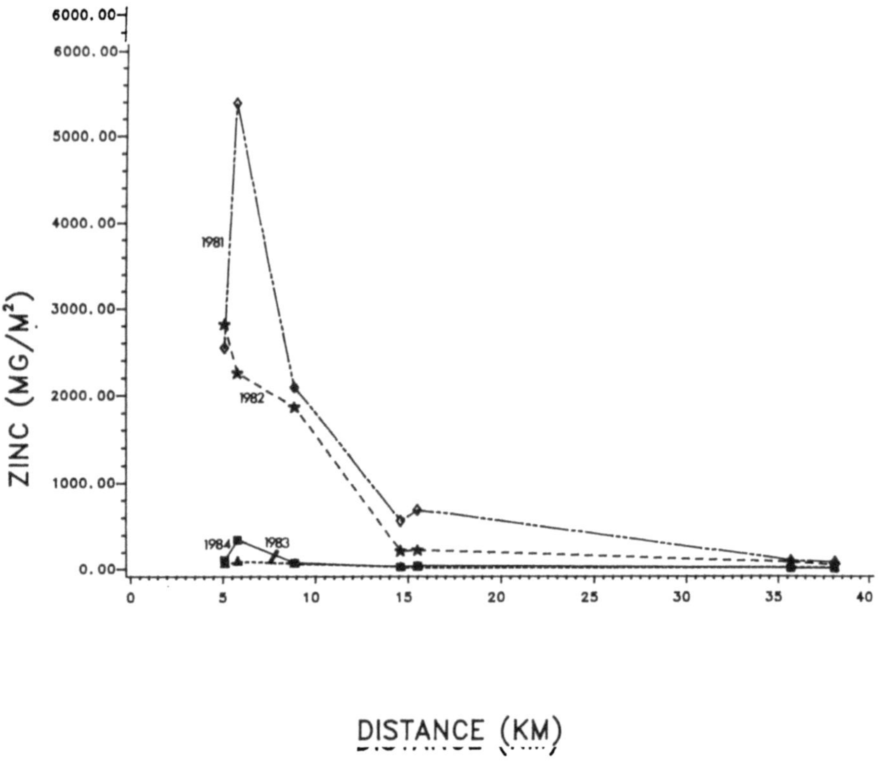

Fig. 1. Mean Zn deposition to snow at Flin Flon, Manitoba, 1981 to 1984. Decrease from 1981/82 to 1983/84 resulted from installation of a new electrostatic precipitator.

3.2. Sulphate-Sulphur

In 1982, sulphate deposition was higher at close-in sites suggesting that smelter emissions may have been a contributing factor (Figure 2). However, except for site 5 in 1984, the 1983 and 1984 sulphate data do not support this trend. Although it would be reasonable to assume that this smelter which emits between 450 and 800 tonne of SO_2 day^{-1}, should increase the local deposition of sulphur, there is insufficient evidence to prove that it was a factor. Local deposition could be lower than expected because long range transport of SO_2 emissions from this tall stack is probably occurring.

Franzin et al. (1979) speculated that the sulphate ion may be deposited in the form of zinc sulphate, copper sulphate, etc. Our

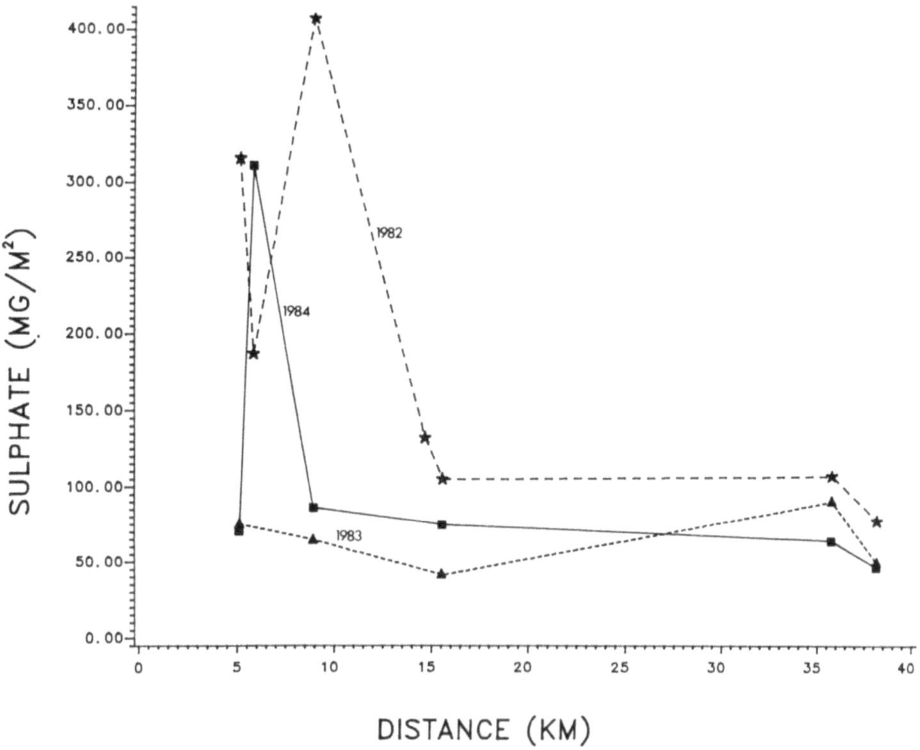

Fig. 2. Mean sulphate deposition to snow at Flin Flon, Manitoba, 1982 to 1984.

data seem to lend some support to that theory in that there was a coincident reduction in sulphate deposition in 1983 along with the reduction in heavy metals. However, the obvious lack of correlation between heavy metal deposition and sulphate deposition to snow in the Flin Flon area, suggests that metallic sulphates may not be major contributors to the total sulphate deposition.

Sulphur deposition to snow can be detrimental to the environment because it can result in acidic meltwater. This does not seem to be the case in the Flin Flon area because the pH of the snow has not decreased at close-in sites where sulphate deposition was highest.

3.3. Snow pH

Except for 1983, the pH of the snow was generally above the accepted figure of pH 5.6 which has been used as the upper limit in the definition of acid precipitation (Figure 3). However, there was no evidence of an inverse relationship between sulphate deposition and pH, even though SO_x are the most likely acid precursors in the smelter emissions. Instead, the pH of the snow tended to decrease with distance from the smelter which supports the findings of Franzin et al. (1979), who observed a positive correlation with metal levels. One theory which might be used to explain decreasing pH as a function of distance, relates to medium-range transport plus time dependent chemical reactions on acid precursors.

On the other hand, this could occur if the particulate emissions were basic. At close-in sites, they would be dominant, thus masking the acidic components in the emissions. Since the particulate load in the smelter plume decreases with distance, there would be progressively fewer bases to buffer the acidic components. Reduced particulates (lower heavy metal deposition) could explain the reason for the lower pH's observed in 1983. However, in 1984, the pH was much higher than in 1983 even though metal levels had increased only slightly. Detailed characterization of the chemistry of the smelter

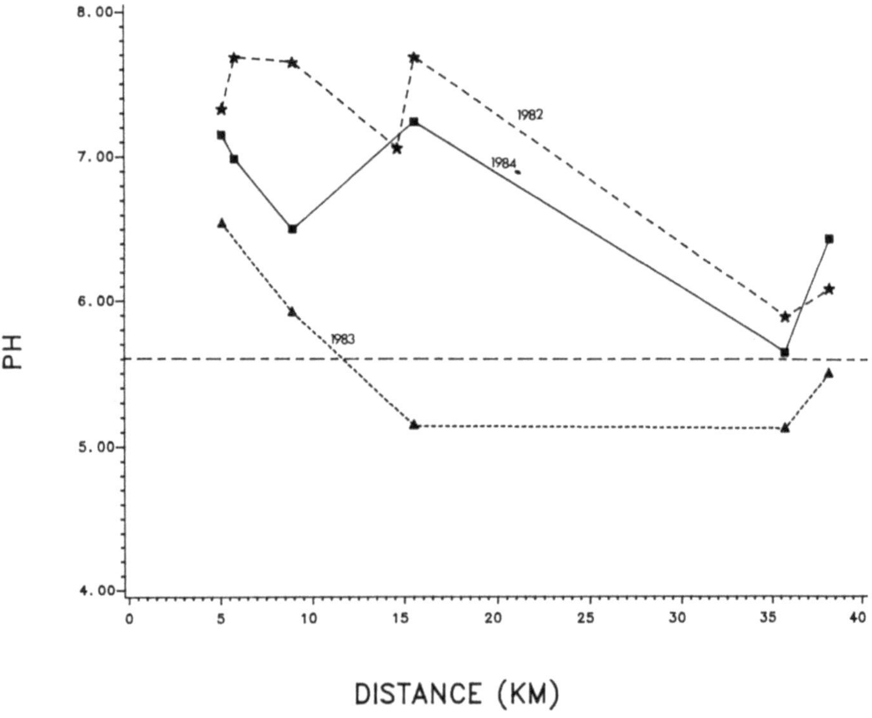

Fig. 3. Mean pH of snow at Flin Flon, Manitoba, 1982 to 1984.

emissions as well as that of the air masses impinging on the area would be required to try to explain this anomaly.

4. CONCLUSIONS

Zinc deposition to the snowpack was inversely correlated with distance from the Hudson Bay Mining and Smelting Co. stack, thus implicating smelter emissions as the source of the metal bearing particulates. Prevailing northwest winds resulted in higher deposition in the southeast direction than in the east southeast. Snowpack monitoring responded effectively to fluctuations in smelter emissions as demonstrated by the 20 fold decrease in Zn deposition after a new electrostatic precipitator was installed.

In spite of high SO_2 emissions from the smelter, there is no conclusive evidence that the accumulated sulphate in the snowpack derived from those emissions.

The pH of the snow was usually higher than 5.6 the currently accepted pH of acid precipitation. There was a trend of decreasing pH with distance from the smelter. Higher pH readings seemed to be related to higher heavy metal deposition, but there was no relationship between sulphates and pH.

Overall, snow monitoring proved to be an effective and sensitive tool with which to determine snowpack chemistry and emission deposition patterns around a point source such as the Hudson Bay Mining and Smelting Co. smelter at Flin Flon.

REFERENCES

Barrie, L. A. and Walmsley, J. L.: 1978, Atmos. Environ., 12, 2331.

Barrie, L. A. and Whelpdale, D. M.: 1977, 'Wintertime Sulphur Deposition Measurements in the Athabasca Oil Sands Area", Proc. Alberta Sulphur Gas Workshop, University of Alberta, Edmonton, Nov. 17-18, 306.

Franzin, W. G., McFarlane, G. A., and Lutz, A.: 1979, Environ. Sci. & Technol., 13(12), 1513.

Hogan G. and Wotton, D.: 1980, 'A Preliminary Report on The Effects of Atmospheric Effluents from Mining and Smelting Industries on Forest Vegetation and Soils', Submission to the Clean Environment Commission, Province of Manitoba.

Hudson Bay Mining and Smelting Company Limited: 1982, 'Manitoba Government Stack Emission Compliance Test, Letter to Environmental Management dated November 18, 1982.

Seip, H. M.: 1978, 'Acid Snow - Snowpack Chemistry and Snowmelt', In: Effects of Acid Precipitation on Terrestrial Ecosystems, Edited by T. C. Hutchinson and M. Haras. Plenum Press New York, 654 pp.

Shewchuk, S. R.: 1982, 'A Precipitation Study by Event Collection from Reindeer Lake, Saskatchewan', Sask. Res. Council Tech. Report No. 135, 23 pp.

Thornton, J. D. and Eisenreich, S. J.: 1982, Atm. Environment 16(8), 1945.

SPATIAL AND TEMPORAL PATTERN OF SULFATE AND NITRATE WET DEPOSITION IN ONTARIO

A.J.S. Tang, W.H. Chan, D.H.S. Chung and M.A. Lusis Ontario Ministry of the Environment
Air Resources Branch, 880 Bay Street, Toronto, Ontario, Canada M5S 1Z8

ABSTRACT. The spatial patterns of the precipitation-weighted mean concentration and total deposition of sulfate and nitrate in Ontario from 1981 to 1983 were determined using the Modified Simple Kriging method. The 3-yr averaged spatial patterns of sulfate and nitrate concentrations were similar to the sulfate and nitrate 3-yr averaged deposition patterns in Ontario. The overall spatial variability of nitrate was slightly higher than sulfate. The seasonal variation of nitrate concentrations and deposition were less regular than sulfate.

1. INTRODUCTION

In response to the need to estimate long term wet deposition amount, as well as the pattern of air pollutants in Ontario, the Ontario Ministry of the Environment has set up a cumulative network in the fall of 1980 (Chan et al., 1984). Before the end of 1981, samples were collected over a monthly period. Starting from January 1982, a 28-day sampling period was adopted. The 36 sampling sites have been distributed more densely in southern Ontario than in northern Ontario. In this study the basic instrument used was the Sangamo (M.I.C.) wet-only collector. Precipitation samples were collected in polyethylene bags inserted into a sampling container. At each site the precipitation collector was colocated with a storage precipitation gauge which serves as the primary standard of precipitation depth during the sampling period. Aside from the acid and base-related contaminants, other selected major ions, nutrients and trace metals were also determined.

Our intention is to present the spatial and temporal variations of precipitation concentration and wet deposition of sulfate and nitrate in Ontario from 1981 to 1983. During the period of June 1982 to March 1983, there were extensive shutdowns of the Sudbury INCO and Falconbridge smelters, which had contributed approximately one-half of the total Ontario SO_2 emissions. The impact of the Sudbury smelters on wet deposition of sulfate at a number of receptors in central and southern Ontario was found to be small - typically less than 15% (Tang et al., 1984; Yap et al., 1984). Hence, the spatial patterns of precipitation concentration and wet deposition given here should still be representative of the situation with more typical provincial SO_2 emissions.

Water, Air and Soil Pollution **30** (1986) 263–273.

2. METHODS

The spatial patterns of the 3-yr. averaged precipitation concentration and wet deposition for sulfate and nitrate (1981 to 1983) were determined using the Modified Simple Kriging Method (Tang et al., 1985a), while the averaged spatial variabilities were estimated by the Standard Deviation function S.

$$S=\left\{\frac{1}{N^h}\sum_{i=1}^{N^h}\left[Z\left(X_i+h\right)-Z\left(X_i\right)\right]\right\}^{\frac{1}{2}}$$

$Z(X_i)$ is the 3-yr. averaged precipitation concentration or wet deposition; h is the distance between a pair of stations, and N^h is the number of pairs of stations that are seperated by h. The standard deviation function represents the averaged deviation of measurement between two stations with a separation h.

For studying the temporal variation of precipitation concentration and wet deposition, Ontario was divided into four regions (see Figure 1) according to the spatial correlation of precipitation sulfate concentration (Tang et al., 1985b). The median values of the precipitation-weighted seasonal mean concentration and wet deposition of the stations in each region were calculated and plotted in Figures 5 and 6.

3. RESULTS

3.1. Spatial Variability

In general, the 3-yr. averaged spatial patterns of sulfate concentration and deposition (Figures 2a and 2b), as well as, nitrate concentration and deposition (Figures 3a and 3b) in Ontario were similar and show a north to south gradient. The averaged sulfate and nitrate concentrations and total depositions in southern Ontario were higher then those in northern Ontario by a factor of three and four times, respectively.

The standard deviation functions of the 3-yr. averaged sulfate concentration and total deposition were given in Figure 4. In general, the overall spatial variabilities of nitrate were slightly higher than sulfate for a separation less than 1150 km (see slope of the fitted lines), beyond which distance the variabilities of sulfate and nitrate were the same. The averaged deviations of the 3-yr. averaged concentration and the total wet deposition increased linearly as the separation inceased up to 1150 km. When the separations were greater than 1150 km, the averaged deviation of concentration decreased linearly, whereas the averaged deviation of deposition remained at a constant level.

3.2. Temporal Variability

In general, the temporal patterns of the median of seasonal sulfate and nitrate concentration and deposition were similar in the southeastern and southwestern regions. Sulfate concentrations were higher in the spring than the other seasons in all regions, however, sulfate concentrations in southern Ontario were high in both spring and summer (see Figures 5 and 6). The seasonal variation of nitrate concentrations was less regular than sulfate concentrations. In southern Ontario, nitrate concentrations were higher in the spring than in the other seasons in 1981 and 1982, whereas in 1983, nitrate concentrations were relatively high in winter and summer. In the northern and central regions, nitrate concentrations were higher in winter.

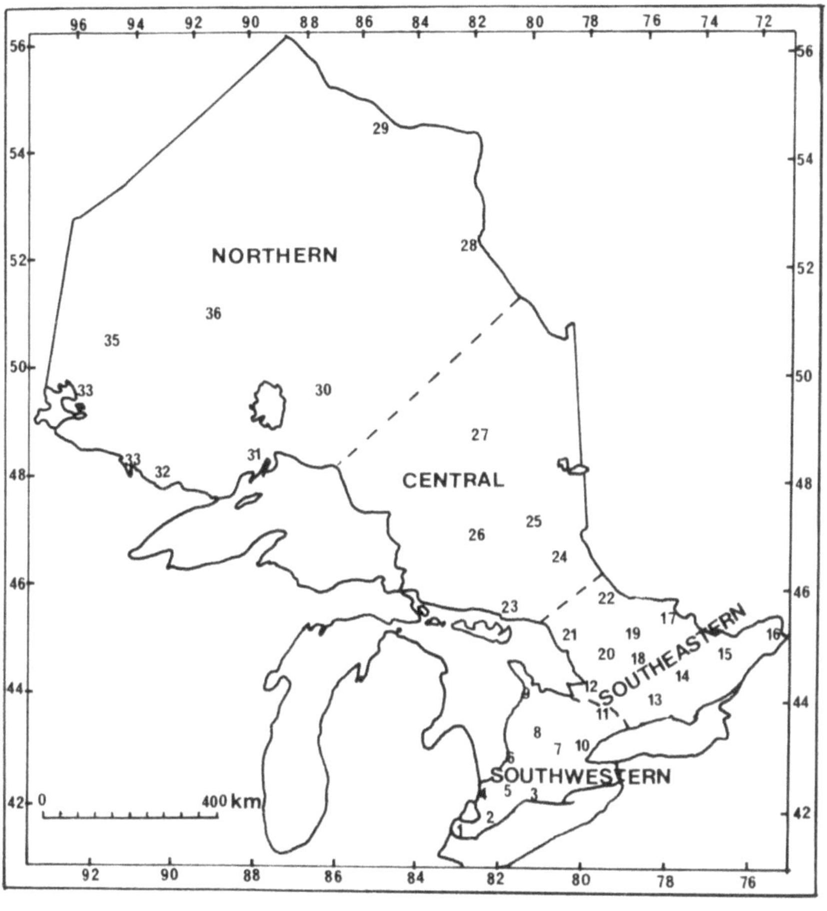

Figure 1. Four defined regions in Ontario in temporal variability study.

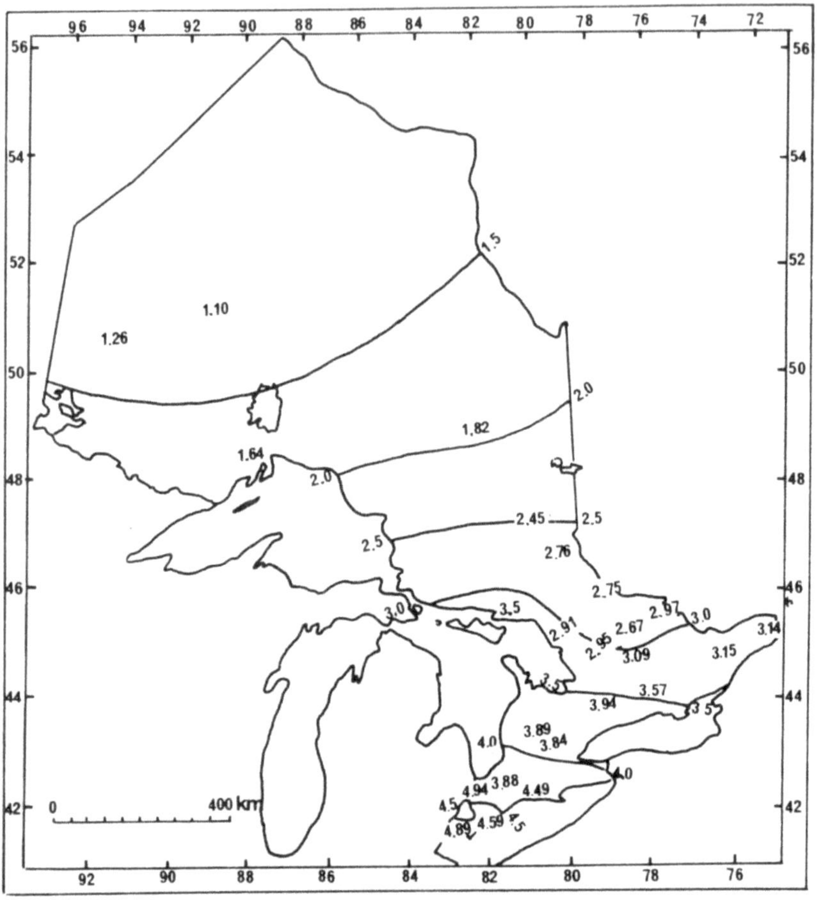
96
94
92
90
88
86
84
82
80
78
76
74
72
56
54
52
50
48
46
44
42
1.5
1.10
1.26
2.0
1.82
1.64
2.0
2.45
2.5
2.5
2.76
2.75
3.0
3.5
2.91
2.95
2.67
2.97
3.0
3.14
3.09
3.15
3.57
3.94
3.5
3.89
3.84
4.0
4.0
4.94
3.88
4.49
4.5
4.89
4.59
4.5
0
400 km
92
90
88
86
84
82
80
78
76

Figure (2a)

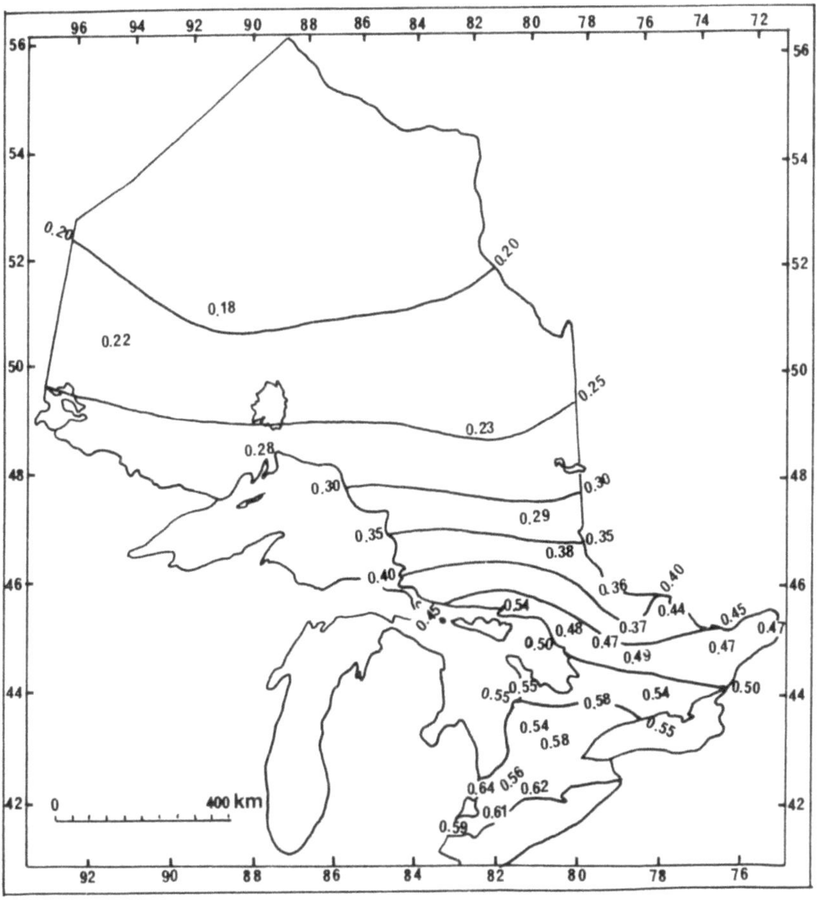

Figure (2b)

Figure 2. Precipitation-weighted mean concentrations (mg L^{-1}) in Ontario from 1981 to 1983.

a. SO_4

b. $N\text{-}NO_3$

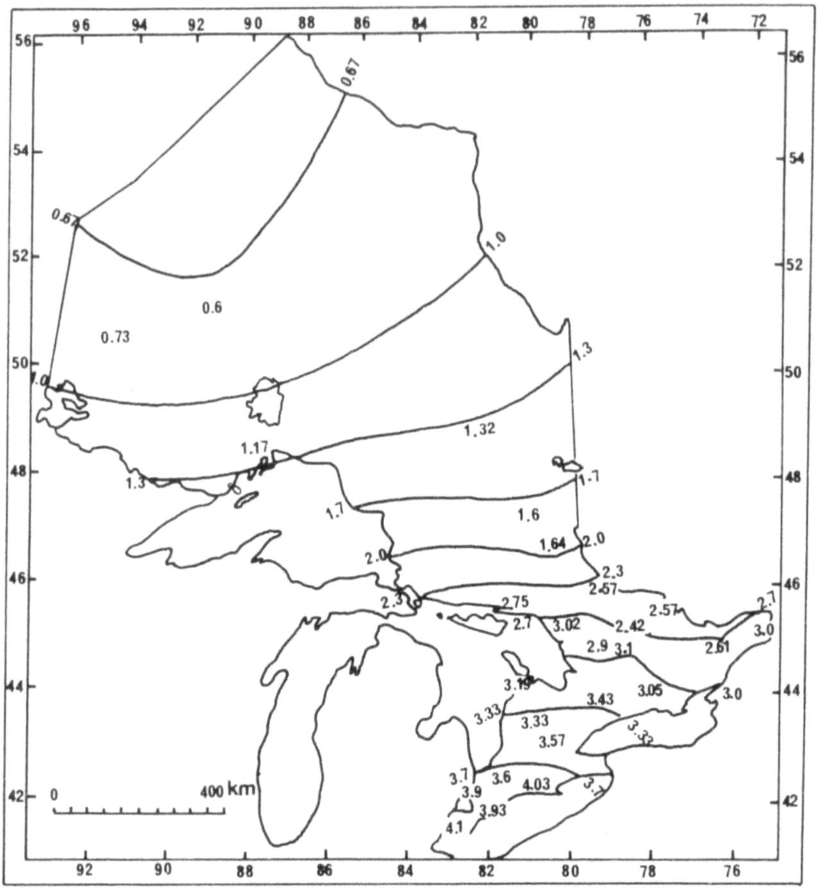
0.67
0.67
1.0
0.6
0.73
1.3
1.32
1.17
1.3
1.7
1.6
1.7
2.0
1.64
2.3
2.57
2.75
2.3
2.7
3.02
2.42
2.57
2.7
3.0
2.9
2.61
3.43
3.05
3.0
3.33
3.33
3.57
3.7
3.6
4.03
3.9
3.93
4.1
0
400 km

Figure (3a)

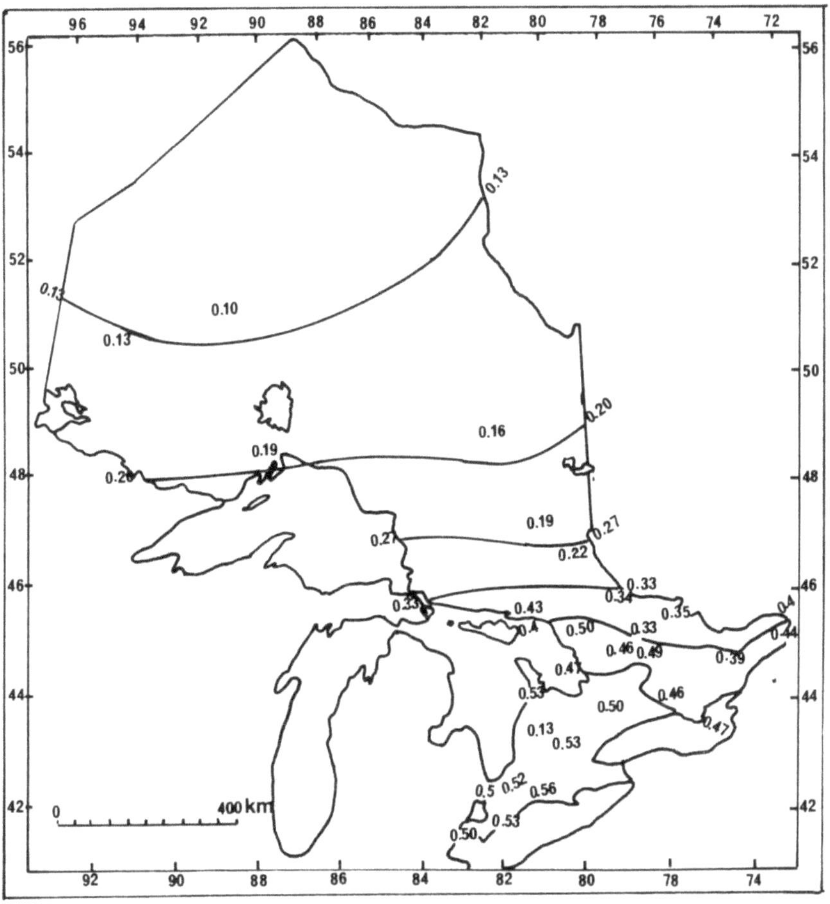

Figure (3b)

Figure 3. Annual averaged wet deposition (g m^{-2}) in Ontario from 1981 to 1983.

a. SO_4

b. $N\text{-}NO_3$

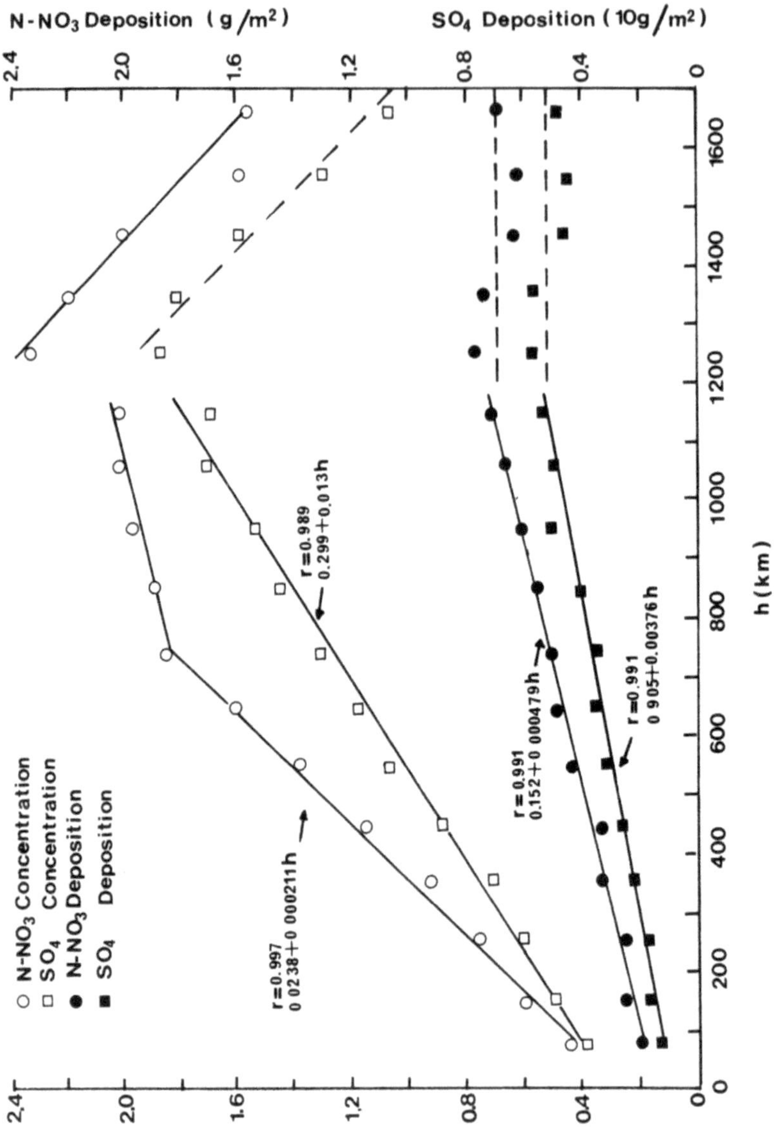

Figure 4. The standard deviation functions.

Figure (5a)

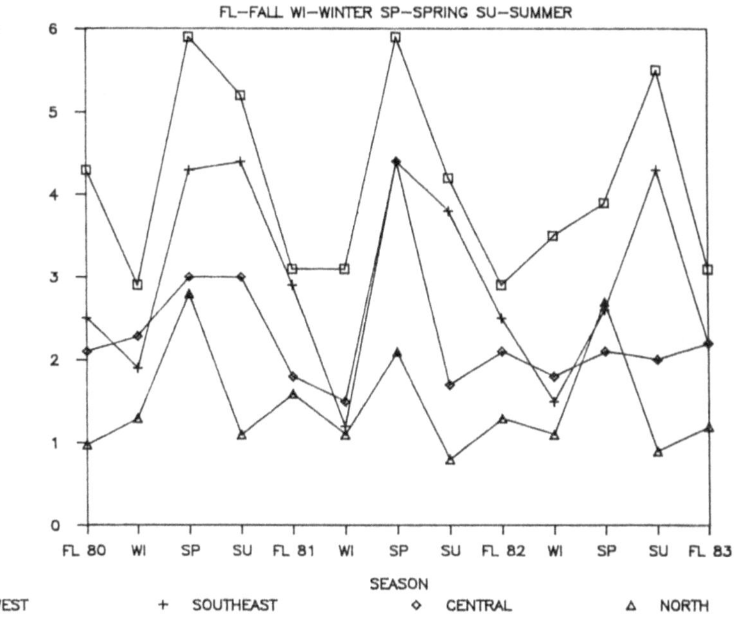

Figure (5b)

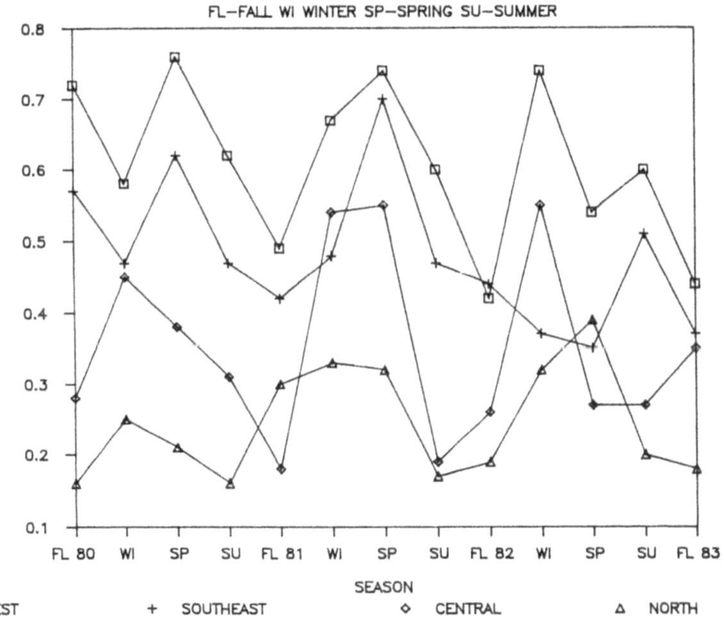

Figure 5. Median of the precipitation-weighted mean seasonal concentrations in Ontario. a. SO_4 b. N-NO_3

Figure (6a)

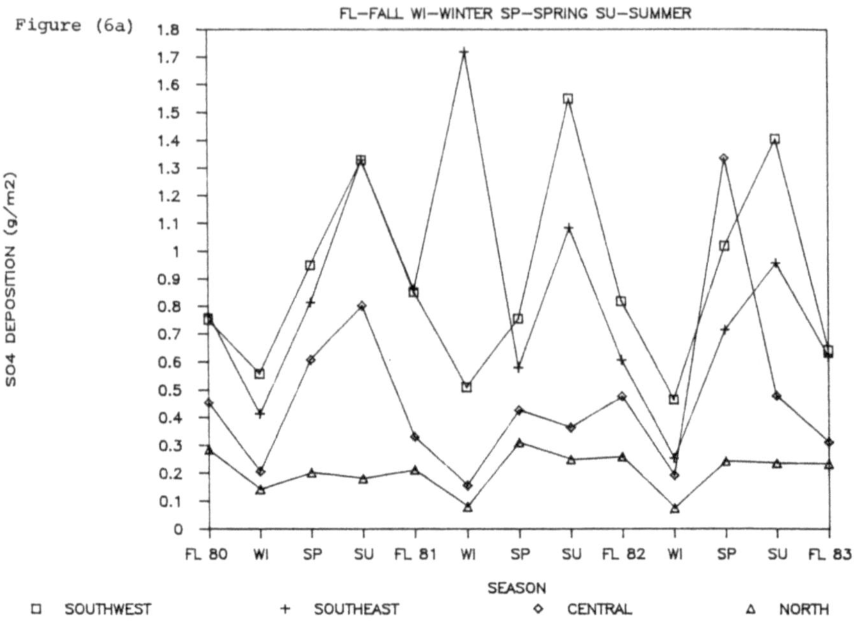

Figure (6b)

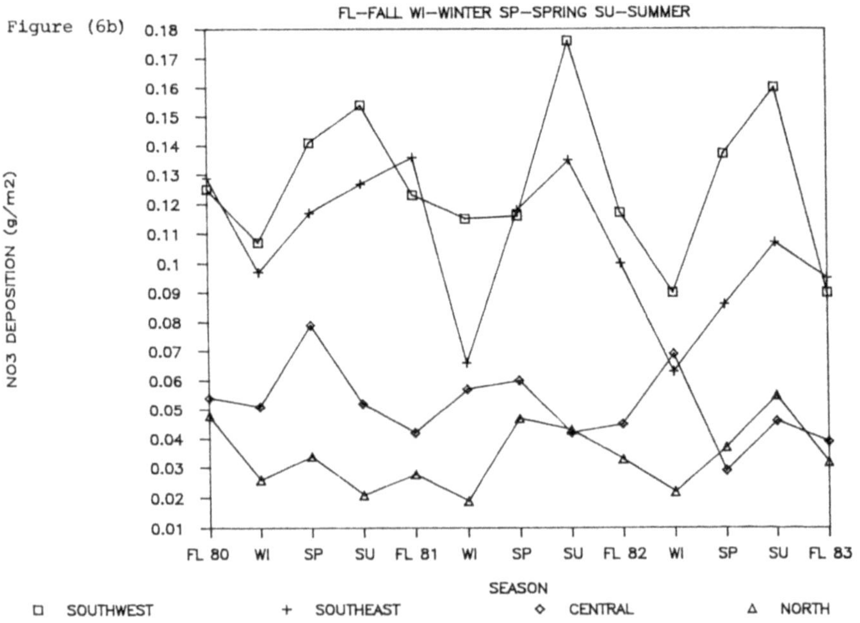

Figure 6. Median of the seasonal wet depositions in Ontario.

a. SO_4 b. $N\text{-}NO_3$

Sulfate deposition was higher in the summer in the southern and central regions but higher in the spring in the northern region. Nitrate deposition pattern was less regular than sulfate deposition pattern. In the southern region, nitrate deposition was higher in summer, whereas in the northern and central regions, nitrate deposition was higher in spring of 1981 and 1982. In 1983, nitrate deposition was higher in the winter and summer in the central and northern regions, respectively.

4. CONCLUSIONS

The spatial and temporal variabilities of precipitation concentration and wet deposition of acid pollutants in Ontario were examined using the cumulative network data. The major conclusions were:

(1) the 3-yr. averaged spatial patterns of sulfate and nitrate concentrations were similar to the sulfate and nitrate 3-yr. averaged deposition patterns in Ontario;
(2) the overall spatial variability of nitrate was slightly higher than sulfate;
(3) the median seasonal sulfate concentration was higher in spring, in all regions of Ontario, however, in the southern region, it was high in both spring and summer;
(4) the median seasonal nitrate concentration was higher in spring, in the southern region (except in 1983) but in the central and northern regions is was higher in winter;
(5) the seasonal sulfate deposition was higher in summer, in the southern and central regions, and in spring, in the northern region;
(6) the seasonal nitrate deposition was higher in summer, in the southern region, and in spring, in the central and northern regions (except in 1983).

REFERENCES

Chan, W.H. Orr, D.B., and Vet, R.J.: 1984, 'Acidic Precipitation in Ontario Study. An overview: The cumulative wet/dry deposition network' (1st revised edition), Ontario Ministry of the Environment Report ARB-164-83-ARSP.

Tang, A.J.S., Chan, W.H., and Lusis, M.A.: 1984, 'An Analysis of the Effects of the Sudbury Smelter Sources on Wet and Dry Deposition in Ontario'. Ontario Ministry of the Environment Report ARB-14-84-ARSP.

Tang, A.J.S. and Chan, W.H.: 1985a, 'Spatial Trend Analysis and Uncertainty Estimates of Acid Precipitation Data in Ontario', the 78th APCA Annual Meeting and Exhibition paper no. 85-6A.6.

Tang, A.J.S., Chan, W.H., and Chung, D.H.S.: 1985b, published results.

Yap, D. and Kurtz, J.: 1984, "Meteorological Studies to Quantify the Effects of Sudbury Emissions on Precipitation Quality and Air Quality During 1980-1983 With Emphasis on the Shut-down Period", Ontario Ministry of the Environment Report ARB-192-84-AQM.

THE SPATIAL AND TEMPORAL VARIATION OF THE SULPHATE TO NITRATE RATIO IN PRECIPITATION IN EASTERN NORTH AMERICA

Peter W. Summers and Leonard A. Barrie
Atmospheric Environment Service
Environment Canada
4905 Dufferin Street
Downsview, Ontario M3H 5T4, Canada

ABSTRACT. Four years of precipitation chemistry data for eastern North America were used to investigate seasonal and geographical variations in $SO_4^=/NO_3^-$ ratio. Several distinct regimes occur. One, in the region of heaviest acidic deposition extending from the states south of the Great Lakes across New England and southeastern Canada, has a very strong seasonal variation in the $SO_4^=/NO_3^-$ molar ratio in deposition. The ratio ranges from about 1.5 in summer to about 0.5 in winter. Another, in the smaller area of Texas and surrounding states, shows the reverse seasonal pattern. Yet another, in the high plains states, has a double maximum in the ratio in Spring and Fall. The remainder of the region has an irregular seasonal pattern. Insight into the cause of $SO_4^=/NO_3^-$ variations was obtained using a simple chemical transport box model. It showed that the chemical transformation of SO_2 and NO_x in the atmosphere is a major factor. A comparison of model predictions and observations indicate that in the vicinity of mid-western American sources the molar ratio of amount of SO_2 oxidized in-cloud to that of NO_2 is 0.5 in winter and 1.5 in summer.

1. INTRODUCTION

Precipitation chemistry observations in eastern North America in the late 1970's indicated that, when averaged over a year, the molar concentrations of $SO_4^=$ and NO_3^- were approximately equal (i.e. the potential contribution to the acidity from $SO_4^=$ was roughly twice that from NO_3^-). Not much attention was given to intra-annual variations.

Indications that there was a seasonal variation in $SO_4^=/NO_3^-$, first emerged from snowpack chemistry studies. Samples from the Adirondack Park Region of New York State collected in 1977 (Galvin and Cline, 1978) had an equivalent ratio of $SO_4^=$ to NO_3^- of only 0.52. Similar measurements in January 1981 in the Canadian Shield of eastern Canada (Barrie and Vet, 1984) yielded a ratio of 0.63. The implications of a high proportion of NO_3^- relative to $SO_4^=$ in the snowpack was discussed by McLean (1981). Precipitation chemistry data

Water, Air and Soil Pollution **30** (1986) 275–283.

confirmed the seasonality in $SO_4^=/NO_3^-$. At Pennsylvania State University over a 2 yr period the molar ratio of $SO_4^=/NO_3^-$ ranged from 1.5 in July to less than 0.5 in the winter (Bowersox and de Pena, 1980).

To date, the most extensive analysis of the relative behavior of S and N in the atmosphere was made by the National Academy of Sciences (1983). It showed that for the year 1980, in a large area of eastern North America (especially where the deposition rates are high), the annual average molar ratio $SO_4^=/NO_3^-$ is close to unity. In this paper, the above analysis will be extended using 4 yr of data (1980 to 1983) from the U.S. NADP and Canadian CANSAP networks to look at the temporal and spatial variations in the ratio at a much larger number of sites in the whole of eastern North America. Barrie and Hales (1984) briefly discuss the operation of these networks and indicate that $SO_4^=$ and NO_3^- concentration behave consistently across the border. Only those stations with 3 or 4 yrs of record and at least 75% of the monthly values were used. The average ratio of $SO_4^=/NO_3^-$ can be calculated several ways. Because of the importance of the total loadings over a period of a season or longer, the averaging method used in this paper was to divide the precipitation-amount weighted mean concentrations of $SO_4^=$ and NO_3^- calculated separately for the time period in question.

The ratio of sulphur to nitrogen in emissions can only be expressed in molar units and so, to facilitate comparisons in the following text, S/N will denote the molar ratio of $SO_4^=$ to NO_3^- in precipitation. However, when considering the potential effects on the ecosystem, the ion equivalent ratio is more relevant. Therefore, both will be used as appropriate, noting the simple conversion:

$$(S/N)_{equivalents} = 2\ (S/N)_{molar}.$$

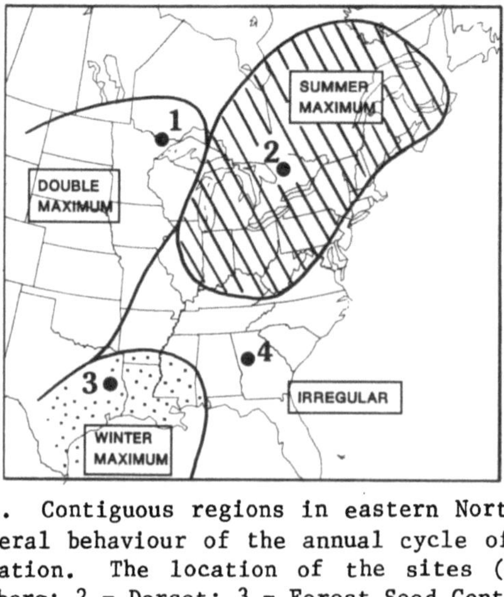

Figure 1. Contiguous regions in eastern North America which show the same general behaviour of the annual cycle of the S/N molar ratio in precipitation. The location of the sites (shown in Figure 2) are: 1 - Fernberg; 2 - Dorset; 3 - Forest Seed Centre; 4 - Georgia Station

2. RESULTS

The geographical distribution of the average values of S/N in precipitation for the 4 yr period since 1980 is almost identical to that produced for the year 1980 by the National Academy of Sciences (1983) and modified by Barrie and Hales (1984). Within the region of high acidity (i.e. pH less than 4.5) in the northeast U.S.A. and southeast Canada (Barrie and Hales, 1984), the annual mean S/N varies from 1.25 to 0.85. That is the ionic equivalent ratio varies from 2.5 to 1.7.

The annual cycle of S/N at each site showed one of four distinctive patterns: (i) a strong summer maximum and winter minimum in S/N; (ii) the opposite to (i) i.e., a winter maximum and summer minimum; (iii) a

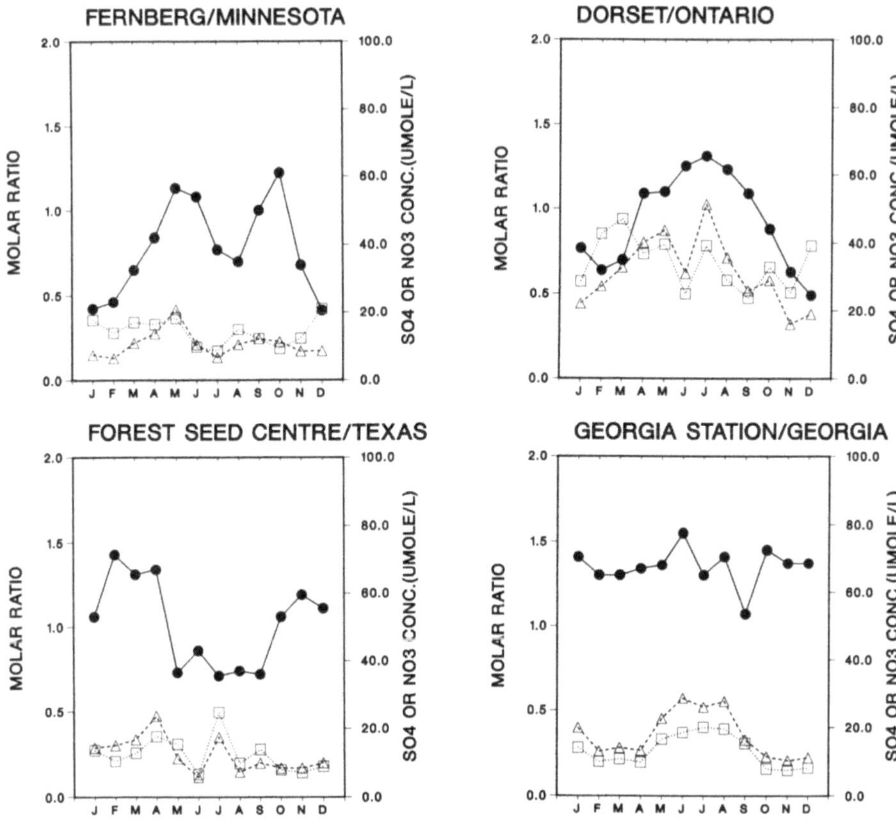

Figure 2. The monthly variation of the S/N molar ratio in precipitation (●——●) and the concentrations of $SO_4^{=}$ (Δ----Δ) and NO_3^- (□........□) at the locations indicated in Figure 1, showing the typical pattern for each region.

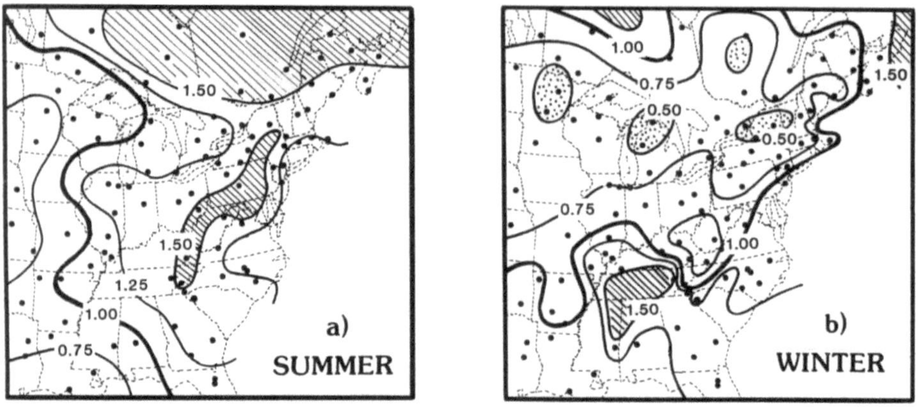

Figure 3. The average seasonal value of the S/N molar ratio in precipitation a) summer (June to August); b) winter (December to February).

double (Spring and Fall) maximum and double (Summer and Winter) minimum in the ratio; (iv) an indeterminate pattern with no systematic maxima or minima. Contiguous regions in eastern North America associated with one of these regimes are shown in Figure 1. Typical examples of the annual cycle of S/N for each region are shown in Figure 2. Also shown is the annual cycle of the $SO_4^{=}$ and NO_3^- concentrations separately. Although not shown in Figure 2, the range of values of the three or four months making up each data point show (in all but the irregular region), little or no overlap between seasons, thus indicating significant seasonal variations.

In the summer season, S/N is greater than 1 over the whole of the eastern United States and Canada (Figure 3a). In a smaller region extending along the Appalachian Mountains, the ratio reaches a maximum of about 1.5. Thus on an ionic equivalents basis, the potential contribution of $SO_4^{=}$ to acidity is between two and three times as much as that of NO_3^-. This occurs in a season when the ecosystem is most active biologically and when the maximum amount of total acid is deposited.

From summer to winter (Figure 3b), there is a dramatic reversal of the S/N ratio in the northeast whilst in the southeast it stays essentially the same. In the region of maximum deposition and lowest winter pH, the S/N is less than 1.0 with a large area of less than 0.75 in the central Appalachians, the Great Lakes, New England, southern Ontario and southwestern Quebec. Within this area, there are regional pockets where S/N drops below 0.5 (i.e. where on an ion equivalent basis nitrate contributes more to the acidity than sulphate). The regions where the nitrate contributes between 40 to 55% of the acidity corresponds to the region with accumulating snowpack during the winter season. Thus, nitrate plays a major role in the acidity of the spring runoff with this melting snow.

An interesting feature of the patterns in Figure 3 is illustrated in the annual variation shown for the Forest Seed Centre in Texas (Figure 2). The cycle of the ratio in the Texas, Louisiana area has a pronounced Summer minimum, which is completely out of phase with that in the northeast. In the Spring and Fall seasons, the geographical patterns are intermediate between those of Summer and Winter.

4. DISCUSSION OF RESULTS

There are many factors that control the ratio of sulphate to nitrate concentrations in wet deposition. These include (i) the ratio of SO_x to NO_x in the upwind emissions; (ii) the relative rates of oxidation of the gaseous oxides to secondary products; (iii) the relative efficiencies of the dry and wet removal of the gaseous and particulate S and N compounds. Both the overall geographical variations and the observed seasonal changes in the ratio must reflect the geographical and seasonal changes in the above three factors. The latter, in turn, are dependent on several meteorological parameters.

4.1 The Impact of Emissions

An emissions inventory for the 1980 base year, prepared for the U.S. National Acid Precipitation Assessment Program (Toothman and Yates, 1984), contains State by State emissions on a seasonal basis for both SO_x and NO_x. The states corresponding approximately to the boundaries shown in Figure 1 can, therefore, be grouped together and the annual averages and the inter-seasonal variability of S/N in emissions calculated, as shown in Table I. Since 1980, the SO_x emissions have generally decreased slightly and the NO_x emissions have increased slightly, therefore, the ratios in Table I are a slight overestimate of the average values for the whole period 1980-1983 for which the deposition data are valid. No comparable data are available for eastern Canadian emissions but the seasonal variability should not be too different from that in the NE region. Within a given region there is

Table I: The annual average SO_x/NO_x molar ratio in emissions and variability, expressed as % difference between highest and lowest seasons, for the regions in the U.S.A. corresponding approximately to the zones shown in Figure 1.

	NE	N. Central	SE	S. Central
Annual Mean	1.16	0.71	1.06	0.29
Seasonal Difference (%)	5	11	8	15

very little change in the molar ratio in emissions between seasons (less than 5% in the northeast). Thus we conclude that the large seasonal change in the wet deposition ratio is not driven by emissions. There is a marked difference in the molar ratio of emissions between regions. Values are very low in the southwest and just above 1.0 in the region east of the Mississippi Valley, and again there does not appear to be any direct correspondence with the annual value of the ratio in wet deposition.

4.2 Impact of Transport, Chemical Transformation and Selective Removal

In order to investigate the dependence of S/N in precipitation on factors other than source strengths, a simple line-source chemical-transport model was developed. Consider the case of air flowing at a uniform speed (U) into the climatologically favored wind sector to the north and northeast (see trajectory wind roses of Barrie et al., 1984) from the Ohio source region into southern Ontario and central Québec. Pollutant sources between the Ohio River Valley and southern Canada are relatively small and therefore are neglected (see emissions distribution, Barrie et al., 1984). The fraction of total S and N emissions comprised of sulphates and nitrates (P_s and P_n, respectively) was also taken into account. Pollutants are assumed to be well mixed to a height H. Before encountering precipitation, NO_2 and SO_2 are transformed to secondary products (NO_3^-, HNO_3, $SO_4^=$) at a linear rate (K_n and K_s, respectively) and removed by dry deposition at a rate proportional to their airborne concentrations (dry deposition velocities for gaseous (g) and particulate (p) components: V_s^p, V_s^g, V_n^p, V_n^g). When pollutants encounter precipitation, a fraction of airborne SO_2 and NO_2 (F_s and F_n, respectively) is assumed to be oxidized in-cloud and to contribute to the S and N in precipitation. The concentration of S and N in rain (C_r) was calculated from concentration in air (K_a) using the empirical scavenging ratio relationships of Barrie (1985) as follows:

$$C_r = WK_a$$

where

for $SO_4^=$
$$W = W_p^{SO_4^=}\left[1 + F_s K_a^{SO_2} \Big/ K_a^{SO_4^=}\right]$$

for NO_3^-
$$W = W_p^{T\text{-}NO_3^-}\left[1 + F_n K_a^{NO_2} \Big/ K_a^{T\text{-}NO_3^-}\right]$$

It was assumed that $W_p^{SO_4^=}$ and $W_p^{T\text{-}NO_3^-}$ were equal. The first term in brackets represents nucleation scavenging while the second term represents in-cloud production by oxidation of SO_2 and NO_2.

Table II: Values of model parameters (defined in the text) used in the estimation of the ratio of S to N in precipitation to that in emission (RF) as a function of downwind distance.

	P_s	P_n	U	H	K_s	K_n	V_s^p	V_n^p	V_s^g	V_n^g
	%	%	m s^{-1}	km	%hr^{-1}	%hr^{-1}	cm s^{-1}			
Summer	3	0.02	5	1.3	1	4	0.3	1.0	0.3	0.1
Winter	3	0.02	10	0.9	0.1	1	0.3	0.6	0.2	0.05

Thus using a simple line-source box model with uniform advection and the removal parameters discussed above, the ratio (R) of the molar ratio S/N in precipitation to that in emissions was calculated as a function of downwind distance (D) from Ohio. Values of all input variables except F_s and F_n are shown in Table II for summer and for winter. Seasonal values of H consistent with those for eastern Canada (Portelli, 1977) were used. Deposition velocities were taken from Voldner and Sirois (1986). V_n was chosen to represent the dry deposition of a 50 to 50% mix of NO_3^- and HNO_3. The model showed that close to the source region R is mainly a function of the relative oxidation of SO_2 and NO_2 in-cloud; that is of the ratio F_s/F_n. This strong dependence was prevalent within 100 km of the source in summer and within 500 km in the winter. Such a result is not unexpected since P_s and P_n are small and there is little time for clear-air oxidation to occur. In other words, the results are insensitive to P_s and P_n. A comparison of observed R values in the Ohio region (calculated from Figures 3 and 4, and Table I) with model predictions leads to the conclusion that the ratio of in-cloud oxidation of SO_2 to that of NO_2 is 1.5 and 0.5 in summer and winter, respectively.

In winter, the dependence of R on D was weak for: (i) D up to 2000 km, (ii) F_s/F_n of 0.5 and (iii) F_s ranging from 5 to 20%. This agrees with observed ratios of S/N in precipitation during winter along a band running northeastward out of Ohio (see S/N in Figure 3). In contrast, during summer the dependence of R on D was stronger. Sensitivity tests showed that while close to the source, R was most dependent on F_s/F_n, remote from the source it was a strong function of the transformation rates K_s and K_n. This is illustrated in Figure 4. Three model-predicted R versus D relationships are shown for K_s and K_n values one half of -, equal to -, and double the values in Table II. Also shown are observed R values calculated from observed S/N values (Figure 3b) along a band running northeastward out of the Ohio River Valley and from the S/N ratio in emissions. Model predictions showed a better fit to observations for K_s and K_n of 0.5 and 2% hr^{-1} (curve 2) rather than for the other oxidation rates (curves 1

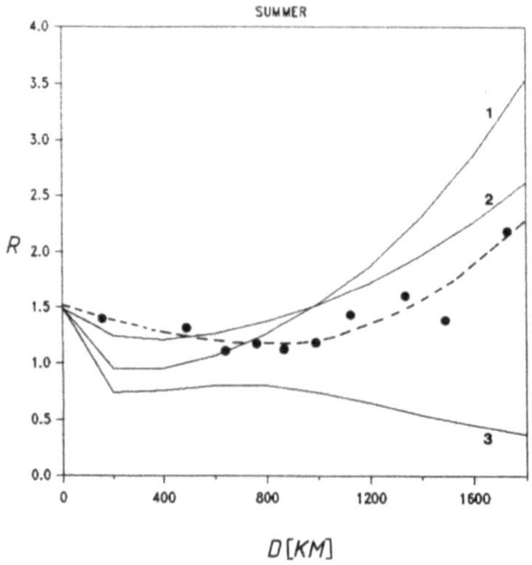

Figure 4. A comparison of the observed (dashed line) and model (solid line) predicted dependence of the ratio (R), of S/N in precipitation to that in emissions, on downwind distance (D) from Ohio in summer. The model parameters are given in Table II with F_s=40%, F_n=28%. For curve 1, K_s=1%hr^{-1}, K_n=4%hr^{-1}; for curve 2, K_s=0.5%hr^{-1}, K_n=2%hr^{-1}; for curve 3, K_s=2%hr^{-1}, K_n=8%hr^{-1}.

and 3). The comparison indicates clearly that mean oxidation rates K_s and K_n as great as 2% and 8% hr^{-1} (curve 3), respectively are unrealistic. Thus, the general features of the observed variations of R with distance downwind of Ohio can be explained by a relatively simple model that embodies the main processes affecting SO_x and NO_x in the atmosphere.

5. CONCLUSIONS

Four years of observations have been used to show that the relative abundance of $SO_4^=$ and NO_3^- in precipitation in eastern North America varies seasonally and spatially. The ratio of S/N peaks in summer in the U.S. northeast and eastern Canada, has a double maximum centred on May and October in north central U.S./central Canada and peaks in winter within and downwind of Texas. No significant seasonal variations are observed in the U.S. southeast. Over most of the region of highest deposition, nitrates contribute almost as much to the acidity of snow (on an ion equivalent basis) as sulphates, and at a few locations nitrates contribute even more than sulphates.

The variation of S/N in precipitation within down-wind distances up to 2000 km from major mid-western U.S. sources is small in winter. However, in summer S/N increases beyond 1200 km. In the northern half of eastern North America the summer maximum in S/N is mainly due to a summer maximum in $SO_4^=$ concentration. That is, NO_3^- concentrations do not vary as much as $SO_4^=$. The strong seasonality of S/N in precipitation is not driven by variations in emissions. A simple chemical transport model indicates that chemical transformation of SO_2 and NO_2 is a major controlling factor of S/N in precipitation. In particular, it is estimated that the ratio of the amount of SO_2 oxidized in-cloud to that of NO_2 is 1.5 in summer and 0.5 in winter. In summer, the mean clear-air oxidation rate of SO_2 and NO_2 is definitely less than 2 and 8% hr^{-1}, respectively. Model predictions fit observations best for respective rates of 0.5 to 1 and 2 to 4% hr^{-1}.

6. REFERENCES

Barrie, L.A., Anlauf, K.G., Wiebe, H.A., and Fellin, P., 1984: Acidic pollutants in air and precipitation at selected rural locations in Canada, Chapter 2 in Deposition Both Wet and Dry, ed. B.B. Hicks, Acid. Precip. Series, Vol. 4, Series Ed. J.I. Teasley, Butterworth Publishers, 15-35.

Barrie, L.A. and Hales, J.M., 1984: Tellus, 36B, 333.

Barrie, L.A. and Vet, R.J., 1984: Atmos. Environ., 18, 1459.

Barrie, L.A., 1985: J.G.R., 90, D3, 5789.

Bowersox, V.C. and de Pena, R.G., 1980: J.G.R., 85, 5614.

Galvin, P.J. and Cline, J.A., 1978: Atmos. Environ., 12, 1163.

McLean, R.A.N., 1981: J. Air Poll. Control Assoc., 31, 1184.

National Academy of Sciences: 1983, Acid Deposition - Atmospheric Processes in Eastern North America, Nat. Academy Press, Washington, D.C., 375 pp.

Portelli, R.V., 1977: Mixing heights, wind speeds and ventilation coefficients for Canada. Climatological Studies No. 31, Atmospheric Environment Service, Downsview, Ont. Canada, 87 pp.

Toothman, D.A. and Yates, J.C.: 1984:, Status report on the development of the NAPAP emission inventory for the 1980 base year and summary of preliminary data, EPA Report EPA-600/7-84-091.

Voldner, E.C. and Sirois, A., 1986: J. Water Air Soil Poll. submitted.

ALKALINE MATERIALS FLUX FROM UNPAVED ROADS: SOURCE STRENGTH, CHEMISTRY AND POTENTIAL FOR ACID RAIN NEUTRALIZATION

W. R. Barnard
Gary J. Stensland
Donald F. Gatz
Illinois State Water Survey
2204 Griffith Dr.
Champaign, IL 61820 U.S.A.

ABSTRACT. This paper presents data on the alkaline element chemistry of unpaved road surface materials and the potential flux of these elements into the atmosphere. Additionally, we discuss the potential acid neutralizing capacity of unpaved road surface materials in precipitation. The relative contribution of unpaved roads to the total open source flux of alkaline materials is also examined. Preliminary calculations indicate that unpaved roads contribute more than 90% of all open source Ca and Mg in 37 and 33 of the 48 contiguous states, respectively.

1. INTRODUCTION

Most research efforts dealing with precipitation acidity have been concerned with attempts to understand the influence of acid materials or chemical reactions leading to the formation of acids in precipitation. As a result, the role of alkaline substances in determining precipitation acidity has been largely overlooked. Precipitation acidity is a function of its contents of both acid and bases and any attempt to understand the processes causing acid precipitation must deal with the potential acid neutralizing capacity of alkaline materials. The major cations in precipitation associated with alkaline compounds (other than ammonia) are Ca, Mg, K and Na. Ammonia and ammonium compounds are not considered in this paper because of their different sources and behavior.

Alkaline materials in precipitation have usually been attributed to crustal sources but can be derived from either conventional (industrial) or open sources. Unpaved roads are one of the main open source contributors to total suspended particulate levels (Carpenter and Weant, 1978; Evans and Cooper, 1980) but their role in precipitation acidity has not been fully explored. This paper presents evidence that unpaved roads can be a significant source of alkaline materials to precipitation.

Water, Air and Soil Pollution **30** (1986) 285–293.

2. MATERIALS AND METHODS

2.1 Sampling and Analysis

Unpaved road surface material was collected by sweeping the material from several areas of the road (within 10 to 20 m of each other) into a dustpan and then emptying the dustpan into a sample bag. Usually 4 to 5 areas were sampled on each road with 2 to 4 of the sample areas taken from the tracked portion of the road and the remainder from the edge of the road. Sample size was typically in the range of 0.2 to 4 kg depending upon the availablity of loose material on the road. Once the samples were collected, they were transported back to the lab for cataloging and physical and chemical analyses.

Each sample's grain size distribution was determined by standard dry sieving methods. The <53 μm fraction was utilized for chemical analysis because it represents the smallest upper size limit for particles classified as silt. The silt content of unpaved roads has been identified as an important parameter in the estimation of particulate mass emissions from travel on unpaved roads (Cowherd et al., 1974b; Dyck and Stukel, 1976; McCaldin, 1979). The silt fraction of the samples were analyzed using the $LiBO_2$ fusion/atomic absorption spectrometry method described by Gatz et al. (1984).

2.2 Estimating the Elemental Natural Emission Fluxes by State for Ca, Mg, Na, and K.

All elemental emission flux equations are of the following general form:

$$F = \frac{(E.F.) \times (S.E.) \times (E.A.) \times (S.C.F.) \times (S.F.) \times (N.F.) \times (F.F)}{Area} \quad (1)$$

where

F = Emission Flux (tonne km^{-2} yr^{-1} for a particular state)
E.F. = Emission Factor (tonne yr^{-1}
S.E. = Source Extent - for a state)
E.A. = Elemental Abundance (fraction by weight)
S.C.F. = Size Cut Factor (fraction of mass)
S.F. = Solubility Factor (fraction)
N.F. = Neutralization Factor (fraction)
Area = Area of State (km^2)
F.F. = Fractionation Factor (enrichment or impoverishment).

The values used for the product of the first two terms, (E.F. x S.E.) in equation (1) were those reported by Evans and Cooper (1980) for agricultural wind erosion, agricultural tilling and unpaved roads (derived primarily from the research of Woodruff and Siddoway, 1965; Cowherd et al., 1974a,b).

Soil elemental abundance values for agricultural tilling and wind erosion emissions in each state were taken as the median value for each element in each state from USGS data (Boerngen and Shacklette, 1981). F.F. values were set to 1 because no data were available.

Elemental abundance values (E.A. in Equation (1)) for unpaved roads emissions are not readily available in the literature. The values used in this paper for a given element in a given state were calculated according to the following formula:

E.A. = [(fraction of vehicle miles traveled on roads of Type D and Type E) x (elemental abundance for Type D and E roads)] + [(fraction of vehicle miles traveled on roads of Type A, B and C) x (USGS soil elemental abundance value)] (2)

where

Type A roads = primitive nonsurfaced
Type B roads = unimproved nonsurfaced
Type C roads = graded and drained nonsurfaced
Type D roads = soil surfaced
Type E roads = slag, gravel and stone surfaced.

Cowherd et al. (1974a) estimated that the average U.S. ratio of vehicle miles traveled on Type D and E roads to vehicle miles traveled on Type A, B and C roads was 3. This means that 75% of vehicle miles traveled are on Type D and E roads. From personal experience (based on gathering unpaved road surface materials in twenty states) we feel that at most, 20% of the Type D and E roads are Type D and thus at least 80% are Type E. Type D roads are assumed to have an elemental abundance similar to soils. This seems to be a valid assumption based on the data presented in Table I. Table I compares actual chemical data on four soil surfaced road materials compared to the USGS mean, median, 10th and 90th percentile values for surficial soils in that state. Except for Ca and Na in Vermont the road material values are similar to the soils. The high Ca and Na values found in Vermont are probably due to the addition of $CaCl_2$ to the roads as a binder and to the addition of NaCl for road deicing (since the sample was obtained in the winter).

The elemental abundance values for Type E roads were somewhat more difficult to derive. Gatz et al. (1984) have presented elemental concentration values for unpaved road materials near Champaign, Illinois. These road materials were either crushed limestone or washed gravel. The data in Table II shows the chemical composition of limestone, gravel and soil surfaced roads for 20 samples from 15 states determined in this study. We are aware of no other data on the elemental abundance of unpaved road surface material. We are assuming that 50% of Type E roads are slag or gravel surfaced and 50% are crushed stone surfaced and that these surface types have elemental abundance values equal to the mean values determined from the data in Table II and that given by Gatz et al. (1984) for washed gravel and crushed limestone, respectively. Therefore, our overall elemental abundance values for all unpaved roads represents a mixture based on usage (vehicle-miles) and the assumptions listed above. Equation (2) can be rewritten as follows:

E.A. = (0.40 x USGS state soil elemental abundance) + (0.30 x washed gravel elemental abundance) + (0.30 x crushed limestone elemental abundance). (3)

Table I. Comparison of alkaline elemental chemistry of surficial soils (Boerngen and Shacklette, 1981) and soil surfaced unpaved road materials (all values in wt %).

State	Soil (S) or Road (R)	Ca	Mg	K	Na
CO	S-10th Percentile	.52	.23	1.29	.52
	S-90th Percentile	8.10	1.50	2.90	2.00
	S-Median	1.40	.70	2.12	1.00
	S-Mean (N=72)	2.94	.80	2.14	1.22
	R-Surface Material (N=1)	1.02	.49	2.33	1.17
NE	10th Percentile	.60	.07	1.50	.70
	90th Percentile	4.43	1.00	2.50	1.00
	Median	.75	.50	1.99	.70
	Mean (N=19)	1.35	.46	1.95	.79
	Road Surface Material (N=1)	1.43	.40	2.48	1.32
NM	10th Percentile	.70	.20	.90	.50
	90th Percentile	9.50	1.50	2.81	2.00
	Median	2.20	.70	1.80	1.00
	Mean (N=59)	3.58	.82	1.81	1.22
	Road Surface Material (N=1)	1.76	.63	2.35	1.45
VT	10th Percentile	.50	.70	.75	.70
	90th Percentile	.80	1.50	2.80	1.00
	Median	.60	1.00	1.50	.70
	Mean (N=3)	.63	1.07	1.68	.80
	Road Surface Material (N=1)	2.01	1.33	1.73	1.43

Table II. Unpaved road surface chemistry (all values in wt %).

Sample # Site	Ca	Mg	K	Na	Mn	Al	Fe	Si	Surface Material Type
IA-06	24.62	3.10	.53	.24	.06	1.61	1.32	9.24	CL
IN-02	27.57	1.33	.42	.09	.11	.94	1.08	6.63	CL
KY-02	26.20	4.06	1.19	.08	.04	1.69	.75	6.84	CL
IN-03	19.77	5.41	.56	.30	.06	1.32	.99	11.83	CL
TN-03	26.25	4.38	.46	.08	.02	1.35	1.27	7.01	CL
PA-01	29.21	1.40	.66	.23	.03	1.62	1.03	8.53	CL
PA-04	18.12	1.25	1.35	.20	.03	3.60	2.11	17.55	CL
PA-07	1.52	.48	1.31	.27	.06	4.56	3.34	22.60	WG
MN-01	11.65	4.10	1.19	1.14	.08	3.69	2.06	20.36	WG
SD-02A	12.03	5.36	.71	.39	.61	2.41	11.15	11.69	WG
VA-03	13.55	3.58	1.39	.22	.07	3.39	2.02	17.45	WG
NC-02	1.47	.86	2.30	1.53	.07	6.64	4.05	30.93	WG
IL-10	14.60	1.75	.85	.40	.07	2.00	1.38	15.64	WG
SD-06	9.08	3.29	1.35	.86	.58	3.92	2.80	22.96	WG
VA-01	2.98	1.72	1.80	1.20	.09	6.27	4.75	28.35	WG
OH-01	1.16	.68	1.55	.52	.07	4.78	3.96	24.26	WG
NE-01	1.43	.40	2.48	1.32	.06	5.08	1.50	32.22	S
VT-04	2.01	1.33	1.73	1.43	.11	6.65	4.80	27.87	S
CO-04	1.02	.49	2.33	1.17	.04	5.59	2.80	29.41	S
NM-01	1.76	.63	2.35	1.45	.05	5.84	2.87	27.20	S

WG = Washed Gravel; CL = Crushed Limestone; S = Soil Surfaced

The first term in this equation represents road types A, B, C and D.

Size cut factors (S.C.F. in Equation (1)) were used to determine the fraction of particles ≤10 μm in diameter of the ≤30 μm diameter fraction on a mass basis. Evans and Cooper's (1980) total particulate emissions from open sources represent the ≤30 μm diameter fraction since their values are derived from Cowherd et al. (1974a), whose emissions factors are based on hi-vol data assumed to have an effective sampling cutoff diameter of 30 m. The 10 μm diameter cutoff was chosen because 1) it represents a size that can easily be transported long distances before settling out (thereby increasing its potential for neutralizing acid precipitation on a regional rather than only a local scale), and 2) it represents the smaller size mode of the typical bimodal soil aerosol distribution.

The S.C.F. chosen for wind erosion is 0.40. This was derived by plotting cumulative % mass vs diameter for soil aerosol data presented by Environmental Applications Group Ltd. (1982) and determining the mass fraction ≤10 μm. The size cut factor for agricultural tilling is assumed to be the same as that for wind erosion.

Cowherd et al. (1977) measured the particle size distribution of emissions from two unpaved roads in Kansas. One road was a gravel surfaced road and the other was a dirt road. The average ≤10 μm diameter cumulative mass fraction in this study was 0.35 and this was adopted as the S.C.F. for unpaved road emissions. This value may range from 0.24 to 0.45. The ability of the air to support particles of a given size depends on the vertical turbulent fluctuations at the time and therefore will be a function of time. For practical purposes in determining flux values however, a constant factor was assumed. The U.S. EPA (1983) recommends 0.45 as the reduction factor for ≤10 μm diameter emissions from unpaved roads. However, unpublished road emissions data (Thompson Pace, personal communication) from Lincoln, Nebraska using dichotomous samplers with a 15 m diameter cutoff showed a dichotomous to high-vol mass ratio of 0.24. The fraction ≤10 μm would be even less.

The solubility factor (S.F. in Eq. 1) is derived from two soluble/insoluble data sets; daily bulk data from St. Louis and 1 year of weekly sampling data for wet-side precipitation samples from Glen Ellyn, Illinois (Gatz et al., 1984). The solubility factor for Ca is 0.94, Mg is 0.74, K is 0.55 and Na is 0.90.

The neutralization factor for each element is assumed to be 1, i.e., all of the soluble element participates in the neutralization of precipitation acidity. Although no data is available yet on the actual neutralizing capacity of unpaved road surface materials our flux estimates thus represent the maximum potential neutralizing capacity of alkaline materials from unpaved roads.

3. RESULTS

Table III presents data on a state-by-state basis for alkaline emissions from unpaved roads, agricultural tilling and agricultural wind erosion. These values represent the fluxes of the four crustal elements

Table III. State-by-state summary of elemental fluxes of alkaline materials from various open sources (all values in tonnes/km^2-yr x 100) and percent of total elemental flux due to unpaved road emissions.

State	Wind Erosion				Agricultural Tilling				Unpaved Roads				Total				Unpaved Roads/Total (%)			
	Ca	Mg	K	Na	Ca	Mg	K	Na	Ca	Mg	K	Na	Ca	Mg	K	Na	Ca	Mg	K	Na
AL	.03	.01	.03	.02	.01	.01	.01	.01	115.60	14.21	4.84	3.68	115.64	14.23	4.88	3.71	99.97	99.85	99.23	99.10
AR	.06	.05	.14	.06	.13	.10	.31	.31	111.40	13.87	5.65	3.65	111.59	14.02	6.10	3.83	99.83	98.96	92.57	95.39
AZ	2.11	1.08	1.77	1.32	.95	.49	.80	.80	120.09	17.23	9.99	7.46	123.14	18.80	12.56	9.37	97.52	91.66	79.57	79.64
CA	10.12	4.36	5.70	7.95	3.67	1.58	2.07	2.07	310.31	44.01	22.61	24.66	324.10	49.95	30.38	35.50	95.74	88.11	74.43	69.48
CO	8.42	3.25	7.24	5.64	3.80	1.47	3.27	3.27	87.79	11.85	7.14	5.48	100.01	16.56	17.64	13.66	87.78	71.56	40.46	40.09
CT	.20	.09	.21	.33	.05	.02	.05	.05	86.21	11.36	6.08	7.11	86.46	11.48	6.33	7.52	99.72	99.03	95.93	94.58
DE	.26	.10	1.37	.54	.04	.02	.23	.23	52.88	6.45	3.74	2.14	53.19	6.57	5.33	2.77	99.43	98.21	70.06	77.36
FL	.12	.01	.03	.02	.00	0.00	.00	.00	103.39	12.37	3.90	2.89	103.52	12.38	3.93	2.91	99.88	99.90	99.20	99.24
GA	.10	.06	.11	.08	.01	.01	.01	.01	150.13	18.65	6.62	4.91	150.24	18.72	6.74	5.00	99.92	99.62	98.24	98.28
IA	16.06	7.84	17.47	13.34	3.36	1.64	3.65	2.79	105.13	13.89	7.30	5.53	124.55	23.36	28.43	21.66	84.41	59.44	25.69	25.53
ID	9.34	4.03	4.53	3.93	3.86	1.67	1.88	1.62	92.46	13.57	6.87	5.58	105.65	19.27	13.28	11.13	87.51	70.41	51.75	50.15
IL	4.02	1.52	6.27	4.32	1.67	.63	2.61	1.80	173.91	22.05	12.87	9.21	179.60	24.21	21.74	15.34	96.83	91.09	59.19	60.07
IN	2.01	1.25	3.15	2.12	.74	.46	1.16	.78	151.83	20.19	11.32	8.04	154.57	21.90	15.63	10.95	98.23	92.21	72.43	73.48
KS	24.91	13.52	41.41	32.90	4.68	2.54	7.78	6.18	84.26	11.16	6.99	5.40	113.85	27.23	56.18	44.48	74.01	41.00	12.44	12.14
KY	.18	.18	.59	.30	.06	.07	.21	.11	80.29	10.34	5.33	3.39	80.54	10.58	6.12	3.80	99.69	97.66	87.05	89.41
LA	.09	.05	.17	.09	.13	.06	.23	.12	116.38	14.71	7.47	4.89	116.60	14.82	7.86	5.10	99.81	99.28	95.05	95.90
MA	.27	.37	.32	.35	.02	.03	.03	.03	33.05	5.01	2.14	1.94	33.35	5.42	2.49	2.32	99.11	92.51	85.90	83.55
MD	.09	.05	.34	.14	.03	.02	.11	.05	143.65	17.64	8.63	5.26	143.77	17.70	9.08	5.44	99.91	99.66	95.01	96.66
ME	.04	.06	.08	.08	.01	.01	.02	.02	25.47	3.80	1.91	1.65	25.51	3.88	2.00	1.74	99.81	98.05	95.24	94.72
MI	.29	.09	.45	.39	.15	.05	.23	.20	139.11	17.29	9.13	7.41	139.55	17.43	9.81	8.00	99.68	99.20	93.14	92.59
MN	26.93	9.44	19.78	22.95	3.03	1.06	2.22	2.58	98.56	12.86	6.60	6.22	128.52	23.36	28.60	31.75	76.69	55.07	23.07	19.59
MO	1.43	.81	3.30	2.76	.65	.37	1.51	1.26	112.12	14.20	7.60	6.01	114.20	15.38	12.41	10.03	98.18	92.33	61.27	59.91
MS	.05	.04	.11	.05	.08	.06	.15	.07	81.89	10.51	5.03	3.14	82.02	10.60	5.29	3.25	99.84	99.11	95.16	96.49
MT	22.79	10.58	14.32	12.87	3.19	1.48	2.00	1.80	21.84	3.11	1.62	1.35	47.82	15.18	17.94	16.02	45.67	20.51	9.05	8.41
NB	9.51	4.89	14.45	8.32	5.10	2.62	7.76	4.47	58.35	7.72	4.73	3.08	72.96	15.23	26.94	15.86	79.98	50.70	17.57	19.39
NC	.22	.11	.21	.34	.03	.02	.03	.04	179.15	22.45	8.50	8.24	179.40	22.58	8.74	8.63	99.86	99.44	97.30	95.53
ND	207.25	100.59	152.16	143.92	18.64	9.05	13.69	12.95	64.70	9.05	4.72	4.04	290.59	118.68	170.57	160.91	22.26	7.62	2.77	2.51
NH	.05	.03	.08	.09	.23	.11	.33	.41	221.01	29.19	17.89	18.27	221.30	29.33	18.30	18.77	99.87	99.54	97.78	97.35
NJ	.12	.07	.24	.28	.04	.02	.07	.08	297.55	37.29	17.43	15.97	297.70	37.37	17.74	16.34	99.95	99.77	98.25	97.77
NM	13.16	3.23	6.17	5.61	.13	.03	.06	.05	98.05	12.84	7.09	5.93	111.33	16.09	13.32	11.59	88.07	79.75	53.23	51.16
NV	.77	.36	.44	.43	1.22	.56	.69	.68	29.92	4.57	2.59	2.32	31.91	5.48	3.72	3.43	93.78	83.25	69.56	67.46
NY	.21	.22	.48	.38	.09	.10	.21	.16	49.39	6.64	3.46	2.64	49.68	6.95	4.15	3.18	99.40	95.44	83.33	83.05
OH	10.45	7.49	23.81	15.95	.45	.32	1.03	.69	187.83	24.57	14.22	10.02	198.73	32.29	39.05	26.66	94.51	75.87	36.40	37.59
OK	2.29	1.01	5.24	4.29	.58	.25	1.32	1.08	105.00	13.13	7.18	5.63	107.86	14.39	13.74	10.99	97.35	91.24	52.26	51.17
OR	1.08	.35	.36	.64	.40	.13	.13	.24	81.58	11.32	5.11	6.34	83.06	11.80	5.60	7.23	98.22	95.91	91.22	87.80
PA	.14	.24	.62	.36	.06	.10	.27	.15	360.37	47.63	25.01	16.65	360.57	47.97	25.90	17.17	99.95	99.29	96.57	97.00
RI	.04	.02	.03	.05	.04	.02	.03	.05	70.49	9.63	4.84	5.76	70.56	9.67	4.90	5.85	99.90	99.63	98.83	98.40
SC	.09	.02	.02	.08	.01	.00	.00	.01	64.31	7.80	2.39	2.11	64.40	7.83	2.41	2.20	99.85	99.68	99.23	95.88
SD	59.38	29.13	52.57	50.61	7.62	3.74	6.75	6.49	39.75	5.43	2.91	2.51	106.75	38.29	62.23	59.61	37.24	14.18	4.67	4.21
TN	.04	.03	.08	.04	.15	.10	.27	.14	120.58	15.30	6.92	4.62	120.77	15.43	7.27	4.81	99.84	99.18	95.24	96.16
TX	3.25	1.04	1.55	1.27	.71	.23	.34	.28	84.77	11.02	4.84	3.76	88.73	12.29	6.73	5.31	95.54	89.66	71.91	70.88
UT	3.43	.52	.65	.63	2.38	.36	.45	.44	83.55	10.51	5.24	4.55	89.35	11.39	6.33	5.62	93.50	92.29	82.69	81.00
VA	.23	.06	.19	.10	.10	.02	.08	.04	73.24	9.05	4.15	2.77	73.57	9.13	4.42	2.91	99.55	99.13	93.87	95.04
VT	.23	.29	.33	.25	.08	.10	.11	.09	94.08	13.99	6.59	4.99	94.38	14.38	7.03	5.32	99.68	97.27	93.77	93.72
WA	5.03	1.86	1.73	4.53	.21	.08	.07	.19	134.53	18.86	8.15	12.97	139.77	20.81	9.95	17.70	96.25	90.67	81.88	73.31
WI	2.07	.58	4.32	2.47	.67	.19	1.40	.80	81.84	10.15	6.69	4.35	84.59	10.92	12.41	7.63	96.75	92.95	53.91	57.03
WV	.03	.04	.12	.05	.02	.03	.09	.04	188.36	24.32	12.64	7.62	188.40	24.39	12.86	7.72	99.98	99.73	98.34	98.79
WY	3.19	1.05	1.74	1.05	1.24	.41	.68	.41	20.33	2.77	1.53	1.02	24.76	4.22	3.94	2.47	82.10	65.63	38.74	41.24

into the atmosphere capable of neutralizing acids in precipitation. The last four columns in Table III show the percentage of the total fluxes of Ca, Mg, K and Na that are due to unpaved roads emissions. For Ca, 37 of the 48 contiguous states have 90% or more of their total alkaline emissions from unpaved roads. Only the central and high plains states (ID, MT, WY, CO, NM, ND, SD, NB, KS, MN and IA) have less than 90% of their Ca emissions from unpaved roads. These states have significant mass emissions due to wind erosion and therefore the wind erosion contribution is significant despite the 5 to 6 fold differences in concentration between the Ca content of soils and road surface materials. This same effect is seen when examining the data for Mg. In addition to the eleven central and high Plains states mentioned above, NV, CA, TX and OH have less than 90% of their total Mg flux due to unpaved roads emissions. This is also the result of significant wind erosion mass emissions overcoming the concentration difference between soil and road surface material. In fact, OH, CA, and TX rank 7th, 9th and 11th in wind erosion mass emissions, respectively (ND, SD, KS, MN, MT, IA, NB and NM rank 1-6, 8th and 10th) (Evans and Cooper, 1980). The unpaved roads contributions to K and Na fluxes are less than 90% in 27 and 28 states, respectively, because soils and road surface materials are much more similar in their K and Na concentrations than their Ca and Mg concentrations.

4. CONCLUSIONS

The data presented here show that unpaved roads can be a significant source of alkaline materials to the atmosphere and have the potential to neutralize a significant fraction of the acids in precipitation. Initial estimates of fluxes of Ca and Mg show that more than 90% of the total open source fluxes are due to unpaved roads emissions in 37 and 33 states, respectively. The dominance of roads over soils is mainly caused by the high concentrations of these elements in road surface materials relative to soil concentrations.

The fluxes presented in this paper are based on equations and data available in the literature that are very imprecise. Our initial estimates showed that the emissions flux of alkaline materials from natural sources exceeded wet and dry deposition of these elements by a factor of 2 to 3. Therefore, the estimates given in this paper (for all sources) are probably biased high (Stensland et al., 1985).

Estimates of open source alkaline emissions fluxes could be improved in several ways. The estimates presented here are based on the analyses of 20 unpaved road surface material samples. We currently have an inventory of approximately 100 samples from 20 states which are awaiting analysis. Additional research will be conducted.

5. ACKNOWLEDGMENTS

This work was funded as part of the National Acid Precipitation Assessment Program under NOAA Contract NA82RAC 00153. Special thanks also go to Jean Dennison for manuscript preparation.

6. REFERENCES

Boerngen, J. C. and Shacklette, H. T.: 1981, 'Chemical analyses of soils and other surficial materials in the coterminous United States', U.S. Geological Survey Report 81-197.

Carpenter, B. H. and Weant, G. E., III: 1978, 'Particulate control for fugitive dust,' U.S.E.P.A., Research Triangle Park, NC, EPA 600/7-78-071, April 1978.

Cowherd, C. C., Guenther, C. M., and Wallace, D. D.: 1974a, 'Emissions inventory of agricultural sites. U.S. Environmental Protection Agency, Research Triangle Park, NC, EPA 450/3-74-085.

Cowherd, C. C., Axetell, C. K., Guenther, C. M., and Jutze, G. A.: 1974b, 'Development of emission factors for fugitive dust sources.' U.S. Environmental Protection Agency, Research Triangle Park, NC., EPA 450/3-74-037.

Cowherd, C. C., Maxwell, C. M., and Nelson, D. W.: 1977, 'Quantification of dust entrainment from paved roadways.' U.S. Environmental Protection Agency, Research Triangle Park, NC. EPA 450/3-77/027.

Dyck, R. I. J. and Stukel, J. J.: 1976, Environ. Sci. Technol., 10, 1046.

Environmental Applications Group, Limited: 1982, 'National inventory of natural sources and emissions of alkaline particulates,' Prepared for Atmospheric Environment Service, Environment Canada. Contract No. OSE81-00614.

Evans, J. S. and Cooper, D. W.: 1980, J. Air Pollut. Control Assoc., 30, 1298.

Gatz D. F., Stensland, G. J., Miller, M. V., and Chu, L.-C.: 1984, 'Alkaline Aerosols: An initial investigation of their role in determining precipitation acidity,' Final Rept. NSF Grant ATM 77-24294. Illinois State Water Survey Contract Report No. 343.

McCaldin, R. O.: 1977, 'Fugitive Dust Study for Pima County Air Quality Control District, Tucson, Arizona,' Pima County Health Department, Air Quality Control District Pub., AQ-91A.

Pace, Thompson: 1985, personal communication.

Stensland, J. G., Barnard, W. R., and Gatz, D. F.: 1985, 'The Flux of Ca, Mg, K, and Na into the Atmosphere for Acid Rain Neutralization.' Paper presented at the Acid Deposition Symposium, Boulder, CO.

U.S. Environmental Protection Agency: 1983, 'Section 11.2, Fugitive Dust Sources. An AP-42 Update of Open Source Fugitive Dust Emissions,' EPA Report No. EPA-450/4-83-010, prepared by Midwest Research Institute.

Woodruff, N. P. and Siddoway, F. H.: 1965, 'A wind erosion equation,' Soil Sci. Soc. Am. Proc., 29:5, 602-608.

UNCERTAINTIES IN ESTIMATING AREAL MEANS: WITH APPLICATIONS TO NADP/NTN DATA

Chin-Fei Hsu[1]
Illinois State Water Survey
2204 Griffith Drive
Champaign, IL 61820 USA

ABSTRACT. The National Atmospheric Deposition Program (NADP) and National Trends Network (NTN) have been gathering data to estimate mean deposition of various constituents of precipitation over areas of varying size and time periods of varying length. The mean values are estimated mainly by sampling at various locations and at discrete times. Such estimations introduce uncertainty about the means since the samplers represent only a minute fraction of the total area to be estimated. When two or more such means are compared, some estimation of uncertainties is required to derive the statistical significance. This paper examines two methods capable of estimating such uncertainty -- areal-temporal correlation and kriging approaches. These methods are applied to the NADP/NTN data, and the estimated uncertainties are compared for concentrations of nine constituents, plus conductivity and pH of the rainfall. The estimated root mean square errors (RMSE) of both methods are smaller than the non-weighted standard deviation. The RMSE of the correlation approach is smaller than that of kriging for K, NH_4, NO_3, pH, and precipitation volume for the 3 U.S. sub-regions. Mixed results occur for other ions.

1. INTRODUCTION

Areal means of depositions are usually estimated by sampling a finite number of sites at discrete times. Such estimation introduces uncertainty about the mean because the samplers cover only a minute fraction of the total area. In order to properly interpret the meaning of the areal mean, it is necessary to know the associated uncertainty. Furthermore, if two or more such means are to be compared, the uncertainty of each of the means is needed to obtain meaningful statistical significance.

---

[1]The work was done while the author was at the Illinois State Water Survey. Now the author is at AT & T Bell Laboratories, Middletown, NJ 07748.

Water, Air and Soil Pollution **30** (1986) 295–302.

The National Atmospheric Deposition Program (NADP) and National Trends Network (NTN) networks have been gathering data appropriate for estimating mean deposition of various constituents of precipitation over areas of varying size and time periods of varying length since 1978. However, the uncertainty associated with mean deposition has not received enough attention. This paper compares two methods of estimating the uncertainty associated with a spatial-temporal mean deposition.

2. METHODS FOR ESTIMATING AREAL MEAN

Stochastic methods have been applied to the problem of modeling spatial-temporal rainfall processes (see Hall and Barclay, 1975, for a review.) Some of these methods have an appealing feature in that the error associated with the estimator can be derived. Therefore by comparing the magnitudes of error, it is possible to select an optimum estimation method with the smallest error. In principle, estimation of the uncertainty of a spatial-temporal mean for rainfall is analogous to that of estimating rainwater impurities. Deposition of rainwater impurity is, however, different from that of rainfall alone, in that impurity deposition includes rainfall as a factor when expressed as a product of rainfall and impurity concentration. Thus, the statistical methods used for rainfall must be examined to determine their appropriateness for application to deposition.

Some of the statistical methods for estimating the uncertainties of spatial-temporal means have been applied to rainwater impurity deposition, for example, Karol and Myatch (1972), Gatz and Naiman (1980), and Finkelstein (1984). However, it appears that there has been no comprehensive comparison of the methods on a common data set. This paper presents the results of such a comparison using the NADP/NTN data. The methods selected for comparison include kriging and the spatial-temporal correlation approach. These methods have been shown to be superior to the simple arithmetic average.

2.1. Kriging

The method of kriging was first used by D. G. Krige to estimate ore reserves, and was subsequently developed by Matheron (e.g., 1971) and others at the Ecole des Mines de Paris. In general, kriging has been used to find the relationship between spatial correlation and precision of interpolation, and/or to estimate means of regionalized variables. Usually, only a 'single realization' of the underlying process (Z) is available. When the estimation is over an area, the method is called block kriging. In addition, Z can be assumed to satisfy, in order of complexity of the underlying process, spatial (weak) stationarity, intrinsic hypothesis, or a drift (Delhomme, 1978).

The mean of Z over an area A is usually estimated by a weighted average of N points. The estimator is derived so that it is unbiased, and minimizes the mean square error (MSE) of the mean estimation by solving a 'kriging system' of linear equations. If the estimator satisfies the condition of 'generalized increment,' its variance can be shown to depend only on the spatial distance between the points (Delhomme, 1978). A jackknife procedure as proposed by Delhomme (1978) is used in this paper to find the order of generalized covariance function.

2.2. The correlation approach

A model for representing the temporal-spatial covariance function of a surface rainfall process was proposed by Rodriguez-Iturbe and Mejia (1974). Gatz and Naiman (1980) used this approach to investigate the spatial variability of a number of ions in rainwater. The model assumed that the stochastic process is weakly stationary in time and space, and that the covariance structure of the process is separable into a temporal component and a spatial component. The temporal covariance function has been approximated by an exponential form. In addition, by assuming the process to be isotropic, the spatial covariance function can be taken to be either a negative exponential function or a first-order modified Bessel function. The exponential function is used in this paper.

Rodriguez-Iturbe and Mejia (1974) showed that the MSE of a weighted areal mean can be expressed as a product of 3 terms: a point variance, a variance reduction factor due to temporal sampling, and another due to spatial sampling. Bras and Colon (1978) extended the estimation procedure to allow for the covariance function to be non-separable by adopting a stationary (multivariate) autoregressive model for the underlying process Z. In this paper, the point variance is estimated by pooling values from all the sites at all time intervals. The overall auto-correlation is taken to be the average of auto-correlations calculated for each site. An estimate of the spatial-temporal mean is computed to be a double-weighted mean, namely, a weighted average of site means. (The site mean is computed using only non-missing values at each site.)

3. DATA

The NADP/NTN network began collecting data in the summer of 1978, and stations are still being added. This paper used data through the end of 1983. The collecting sites used in these analyses are shown in Figure 1a. The data consist of weekly (Tuesday to Tuesday) values of ion concentrations, conductivity, pH, and rain volume. In the analyses, 33 sites with total samples for less than 52 weeks are not included. Most of these sites began operation in 1983. Questionable samples, as recorded in field and laboratory notes, were not used, except for values from irregularly sampled periods, which are converted to 7-day values whenever possible. No attempt was made to estimate the missing values.

4. RESULTS OF THE PRELIMINARY ANALYSES

For each site included, the arithmetic mean and standard deviation of pH, calculated over all non-missing samples, are shown in Figure 1. It can be seen that the pH values in the eastern half of the U.S. are less than 5.0, except those of the Gulf coast and Florida. An obvious NE-SW gradient of pH values exists in this half. A minimum of pH values less than 4.3 is found over the upper Ohio River Valley. The standard

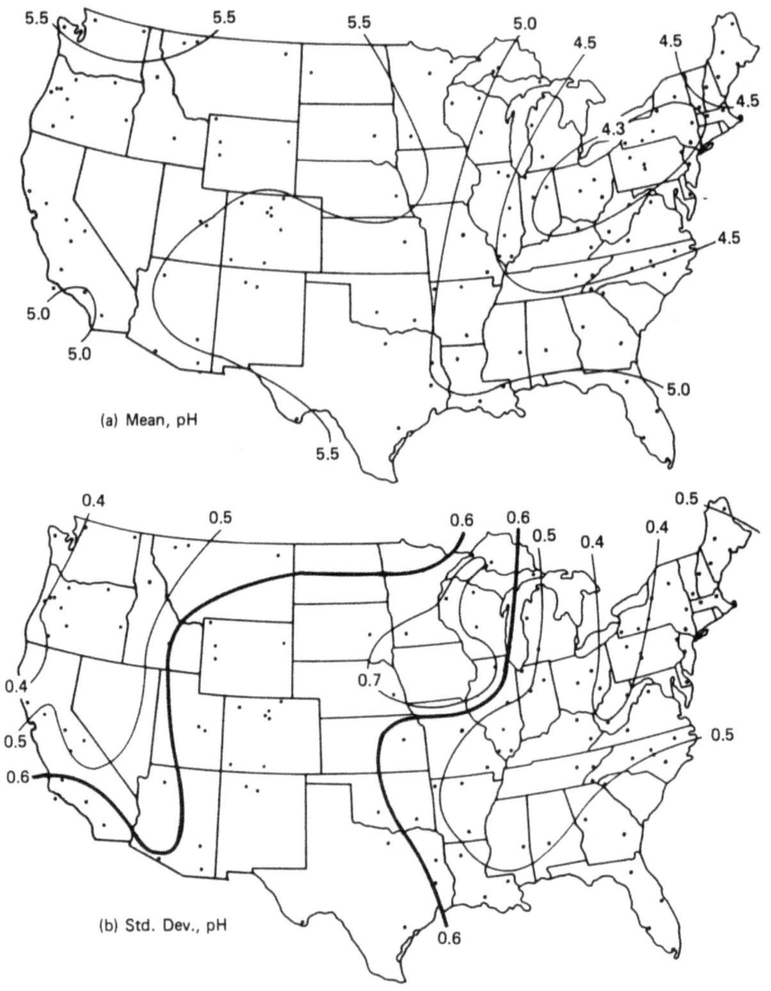

Fig. 1. Time-average pH values, (a) arithmetic mean, (b) standard deviation.

deviation of pH at all sites has a range of 0.4 to 0.7. A noticeable area of high standard deviation (greater than 0.7) occurs over Iowa and parts of Minnesota and Wisconsin. The coefficient of variation (defined as standard deviation divided by mean) is very close to 0.1 for most sites.

Since the assumption of homogeneous variance over the area of study is needed in the methods of estimating areal mean, the entire U.S. is sub-divided into 3 regions bounded by the 0.6 standard deviation line (as shown in Figure 1b) for the analyses. Essentially, the 3 regions thus divided represented the eastern one-third, the central, and the western one-fourth of the U.S., respectively.

The arithmetic (areal) mean and standard deviation for nine ions, conductivity, and pH over the 3 regions and the entire U.S. are shown in Table I. The mean pH is 20% smaller in the eastern region than in the western region. The overall mean pH for the continental U.S. is 4.95. The arithmetic means of conductivity, NO_3, and SO_4 have a noticeable east-west gradient with the largest values in the east region. The standard deviations, however, do not show such a gradient. A reverse east-west gradient with the largest values in the west region is observed for Cl, pH, and PO_4. Values that are largest in the central region include those for Ca, K, Mg, Na, and NH_4.

Table I. Arithmetic Mean and Standard Deviation.

	Region*			
	East	Central	West	All
No. of Sites	43	44	10	97
Ca (mgL^{-1})	0.365/0.09	0.670/0.13	0.225/0.16	0.489/0.07
Cl (mgL^{-1})	0.471/0.11	0.493/0.11	0.596/0.26	0.494/0.07
Conductivity	33.0/2.8	20.3/2.2	9.3/3.2	24.8/1.6
K (mgL^{-1})	0.057/0.04	0.091/0.05	0.072/0.09	0.073/0.03
Mg (mgL^{-1})	0.088/0.05	0.131/0.06	0.088/0.10	0.108/0.03
Na (mgL^{-1})	0.321/0.09	0.437/0.10	0.415/0.21	0.383/0.06
NH_4 (mgL^{-1})	0.362/0.09	0.462/0.10	0.137/0.12	0.385/0.06
NO_3 (mgL^{-1})	2.12/0.22	1.80/0.20	0.58/0.25	1.82/0.14
pH	4.48/0.33	5.28/0.35	5.53/0.78	4.95/0.23
PO_4 (mgL^{-1})	0.010/0.02	0.012/0.02	0.017/0.04	0.012/0.01
SO_4 (mgL^{-1})	3.41/0.28	2.39/0.24	0.82/0.30	2.68/0.17
Volume (L)	1.47/0.19	1.02/0.15	1.84/0.45	1.30/0.12

* Refer to Fig. 1b for regions defined. The 0.6 s.d. value is used to sub-divide the U.S. into three regions.

Results of the block kriging are shown in Table II. A drift is allowed in the model-building process. In general, the kriging weighted means are slightly larger than the arithmetic means except for NH_4, NO_3, and PO_4. The kriging (square root) MSE is smaller than the corresponding standard deviations of K, Mg, and pH in all regions, and is larger than that of SO_4.

The MSE of an estimator for areal mean depends not only on the temporal structure of the process, but also on the density of the collector network and the shape of the region. A square is used in this paper to enclose each of the three regions and the entire U.S. in order to derive a characteristic correlation distance (Rodriguez-Iturbe and Mejia, 1974) and a characteristic correlation. The estimated characteristic distance (shown in Table III) is a measure of the average

Table II. Results of Block Kriging.

Order of Gen. Cov. Fn. Areal Mean/Sq rt MSE	Region: East	Central	West	All
Ca (mgL^{-1})	1 0.352/0.10	1 0.695/0.15	1 0.257/0.12	1 0.585/0.15
Cl (mgL^{-1})	1 0.579/0.39	0 0.493/0.05	1 0.605/0.12	1 0.717/0.37
Conductivity	0 32.9/0.95	1 21.7/6.8	0 9.3/0.40	1 22.4/4.8
K (mgL^{-1})	0 0.057/0.02	0 0.093/0.03	1 0.084/0.04	1 0.084/0.03
Mg (mgL^{-1})	1 0.088/0.00	1 0.131/0.00	1 0.092/0.02	1 0.133/0.04
Na (mgL^{-1})	0 0.321/0.02	1 0.543/0.20	3 1.135/0.00	1 0.517/0.19
NH_4 (mgL^{-1})	1 0.357/0.08	1 0.449/0.15	0 0.137/0.04	1 0.352/0.14
NO_3 (mgL^{-1})	1 1.989/0.24	1 1.889/1.1	0 0.582/0.09	1 1.631/0.51
pH	1 4.52/0.09	1 5.28/0.28	1 5.55/0.14	1 5.12/0.19
PO_4 (mgL^{-1})	1 0.007/0.01	1 0.010/0.02	1 0.025/0.02	1 0.009/0.02
SO_4 (mgL^{-1})	1 3.17/0.42	1 2.51/0.60	0 0.82/0.09	1 2.43/0.55
Volume (L)	1 1.46/0.23	1 1.09/0.14	0 1.84/0.18	1 1.23/0.22

distance between two randomly selected points in the region. The results of the correlation approach are shown in Table III. (Data which can not be analyzed because they do not satisfy underlying assumptions are left as blanks in the table.) It can be seen that the variance

Table III. Results of Correlation Approach, Using Exponential Functions in Spatial and Temporal Components.

	Region			
	East	Central	West	All
Char. Corr. Dist. (km)	570	923	316	1012
Char. corr.	0.20	0.16	0.29	0.16
Scaling factor (h*)	0.003	0.002	0.004	0.002
Auto-corr. (pooled) Var. Red.-temp./spat. Areal Mean/Sq rt MSE				
Ca (mgL^{-1})	0.17	0.21	0.22	0.27
	0.005/0.153	0.006/	0.007/0.174	0.006/
	0.264/0.35	/	0.184/0.05	/
Cl (mgL^{-1})	0.40	0.14	0.09	0.33
	0.008/0.154	0.005/0.163	0.005/0.175	0.007/0.158
	0.304/0.06	0.837/0.08	0.530/0.10	0.679/0.08
Conductivity	0.34	0.19	0.14	0.35
	0.007/0.172	0.005/0.168	0.006/0.171	0.007/0.165
	21.3/1.5	21.2/1.3	8.6/0.72	18.8/1.5
K (mgL^{-1})	0.12	0.12	0.09	0.16
	0.004/0.133	0.005/0.171	0.005/0.172	0.005/0.156
	0.040/0.01	0.087/0.02	0.056/0.02	0.078/0.01
Mg (mgL^{-1})	0.29	0.47	0.23	0.48
	0.006/0.153	0.010/0.168	0.007/0.116	0.010/160
	0.062/0.01	0.150/0.02	0.076/0.01	0.118/0.01
Na (mgL^{-1})	0.30	0.25	0.09	0.33
	0.007/0.161	0.006/0.159	0.006/0.040	0.007/0.159
	0.195/0.05	0.634/0.08	0.372/0.05	0.505/0.07
NH_4 (mgL^{-1})	0.22	0.17	0.13	0.30
	0.005/0.166	0.005/0.165	0.006/0.161	0.006/0.164
	0.265/0.03	0.420/0.08	0.150/0.02	0.346/0.06
NO_3 (mgL^{-1})	0.19	0.11	0.18	0.25
	0.005/0.173	0.004/0.169	0.006/	0.006/0.164
	1.42/0.12	1.66/0.13		1.40/0.14
pH	0.43	0.48	0.35	0.53
	0.009/0.173	0.010/0.169	0.010/0.165	0.011/0.165
	3.29/0.02	5.70/0.05	5.25/0.04	5.03/0.04
SO_4 (mgL^{-1})	0.35	0.28	0.13	0.44
	0.007/0.174	0.006/0.164	0.006/0.157	0.009/0.165
	2.29/0.20	2.32/0.20	0.80/0.09	2.12/0.23
Volume (L)	0.14	0.28	0.49	0.17
	0.005/0.173	0.006/0.157	0.013/0.165	0.005/0.153
	1.01/0.06	0.92/0.06	1.51/0.17	0.97/0.01

reduction factor due to time is smaller than that due to space. The weighted mean calculated in the correlation approach is generally smaller than the arithmetic average except for a few values in the central region and the entire U.S. Almost all of the (square root) MSE's are smaller than the corresponding standard deviations.

Comparing Tables II and III, it can be seen that the MSE of the correlation approach is smaller than that of kriging for K, NH_4, NO_3, pH, and volume for all regions. Mixed results occur for other ions. The weighted mean calculated by the correlation approach is smaller than that by kriging.

5. DISCUSSION

The kriging technique requires less stringent assumptions than the correlation approach. The technique allows for spatial non-stationarity and anisotropic covariance structure and is appropriate in estimating long-term areal mean deposition. Finkelstein (1984) preferred kriging over the correlation approach or the time series approach. But kriging assumes that the observed total deposition at each site is a reasonably 'good' representation of the true (long-term) total deposition, an assumption which might require many years of deposition data at many points to satisfy. Recent kriging theory allows for processes with a time component, and thus provides a more realistic fitting to data.

Acknowledgments. Part of the research in this paper was supported by a U.S. Geological Survey Grant INT 14-08-0001-G-964. The opinions expressed here are those of the author, and do not necessarily reflect USGS's views.

The programming help of Jim Angel and the assistance of Jose Guerrero are appreciated.

6. REFERENCES

Bras, R. L and Colon R.: 1978, Water Res. Research, 14, 878.

Delhomme, J. P.: 1978, Advances in Water Resources, 1, 251.

Finkelstein, P. L.: 1984, J. Clim. and Appl. Meteor. 23, 52.

Gatz, D. F Naiman D. Q.: 1980, Spatial variability of rain water impurities in mesoscale events. Symp., Interm. Range Atmos. Transp. Proc. and Tech. Assess., 1-3 Oct., Gatlinburg, TN, 6 pp.

Hall, A. J. and Barclay P. A.: 1975, Methods of determining areal rainfall from observed data. Predic. in Catchment Hydro., T. G. Chapman and F. X. Dunin (eds.), Australia, 47-57.

Karol, I. L. and Myatch L. T.: 1972, Tellus, 26, 421.

Matheron, G.: 1971, The theory of regionalized variables and its applications. Cahiers du Centre de Morphologie Math., 5, Ecole des Mines, Fantainebleau, France, 211 pp.

Rodriguez-Iturbe, I. and Mejia J. M.: 1974, Water Res. Research, 10 713.

WET AND DRY DEPOSITION OF SULPHATES AND NITRATES IN EASTERN CANADA: 1979-1982

L.A. Barrie and A. Sirois
Atmospheric Environment Service
4905 Dufferin Street, Downsview, Ontario
Canada M3H 5T4

ABSTRACT. Daily air and precipitation chemistry observations at six rural locations in eastern Canada were analyzed to obtain wet and dry deposition. Dry deposition was calculated from air concentrations using deposition velocities originating from a recent literature review and synthesis exercise involving land use types. Total annual deposition ranges for $SO_4^=$ from 10 to 86 mmol m^{-2} and for NO_3^- excluding NO_2 contributions to dry deposition from 13 to 62 mmol m^{-2}. Dry deposition accounts for an estimated 22 and 21% of the total $SO_4^=$ and NO_3^- deposition, respectively. For NO_3^-, this fraction increases to 30% if NO_2 concentration to dry deposition is included. There is a marked seasonal variation in total $SO_4^=$ deposition but not in that of NO_3^-. Both wet and dry deposition are episodic. 20% of daily events deliver between 47 and 70% of the deposition.

1. INTRODUCTION

Deposition of acidic sulphates and nitrates from the atmosphere to aquatic and terrestrial ecosystems in Canada takes place via three processes: direct uptake of particles and gases at the earth's surface (dry deposition), removal by rain and snow and fogwater deposition. The latter two processes constitute wet deposition. In late 1978, the Air and Precipitation Monitoring Network (APN) was established in Canada with the aim of measuring atmospheric pollutant inputs to sensitive ecosystems. Fogwater deposition observations were not made. Henceforth, wet deposition will mean deposition in rain and snow. For a discussion of the importance of fog water deposition see Barrie and Schemenauer (1986). Between 1979 and 1982, measurements were made at 6 rural sites (Figure 1). Various aspects of the data have been analyzed and reported elsewhere (Barrie et al., 1980, 1984; Anlauf et al., 1980; Barrie, 1982; Whelpdale and Barrie, 1982) but not until recently has it been possible to address quantitatively the absolute value and relative magnitude of wet and dry deposition for an extended period of time. In this paper the results of 4 yr. of measurements from 1979-1982 are summarized in terms of wet and dry deposition of $SO_4^=$ and NO_3^- to an area of 100 km radius around each

Water, Air and Soil Pollution **30** (1986) 303–310.

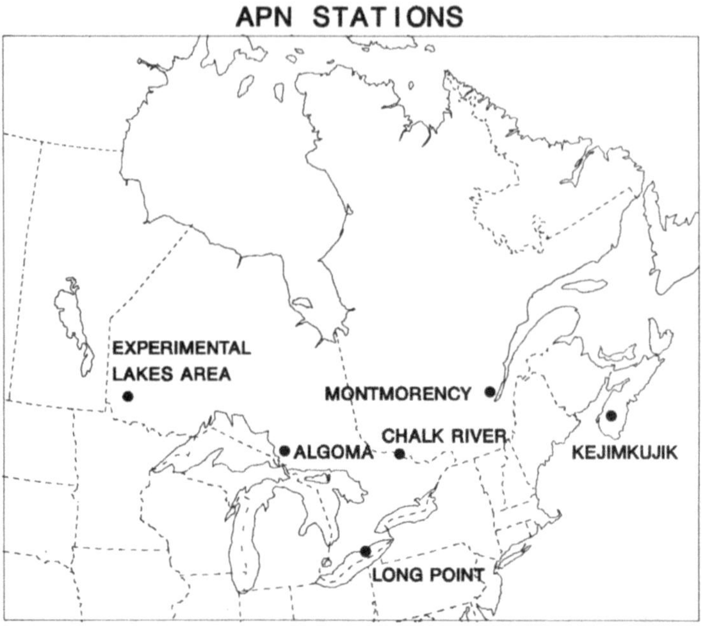

Figure 1. The location of APN sites at which air and precipitation chemistry measurements were made.

site. A detailed presentation and discussion of results can be found elsewhere (Sirois and Barrie, 1986).

2. OBSERVATIONS

A detailed description of the sampling equipment, procedures and analytical techniques is published elsewhere in a series of data reports (Data). Both air and precipitation chemistry measurements were made on a daily basis.

In estimating dry deposition of nitrate, the assumption was made that on a monthly mean basis the total nitrate we observe is partitioned between particulate NO_3^- and HNO_3 in the same proportion as they were in 1983 when a three filter teflon-nylon-Whatman 41 system was used allowing us to separate gaseous and particulate NO_3^-. Using this system, there is a tendency to underestimate the fraction of NO_3^- that is HNO_3 because of the volatilization of particulate NO_3^- from the teflon filter. However, recent observations have led us to believe that the errors thus incurred are less than 20 to 35% on a monthly basis (Anlauf et al., 1985; Cadle, 1985). In the following estimates of NO_3^- dry deposition, we do not include the contribution due to NO_2 and PAN because air concentration measurements are not available. However, based on available NO_2 and PAN observations, Sirois and Barrie (1986) estimated that the contribution of NO_2 to the total nitrate dry deposition is about 50% and that of PAN is negligible. Unless otherwise mentioned, reference to NO_3^- deposition below does not include NO_2 dry deposition.

3. THE CALCULATION OF WET AND DRY DEPOSITION FROM OBSERVATIONS

Daily wet deposition was calculated as the product of the $SO_4^=$ or NO_3^- concentration from the wet-only precipitation collector and the standard precipitation amount. Monthly wet deposition was calculated as the product of the standard-precipitation-amount-weighted concentration and the standard precipitation amount. If not all precipitation events were analyzed during a month, we applied the criterion that those available must represent at least 50% of the total precipitation. Annual deposition was calculated provided 8 or more monthly concentration values and 12 mo. of precipitation amount were available. For a given surface and chemical compound, daily dry deposition (D) is calculated as the product of the daily mean air concentration (C), the dry deposition velocity (V) and the averaging time (t). The monthly-mean areal- average D to an area 100 km in radius about a station is the product of the monthly arithmetic mean concentration (C), the areal mean monthly dry deposition velocity ($\bar{V}$) and T. If daily air concentration measurements are missing in a month, the criterion was applied that at least 50% must be available to estimate C, otherwise the deposition is reported as missing. $\bar{V}$ is calculated from the monthly mean V values for each surface type in the area about a station by weighting the values according to fractional surface coverage. Surface type and site specific deposition velocities estimated by Voldner et al., (1986) were used. Annual means of the monthly-mean areal-average deposition velocities of SO_2, $SO_4^=$, HNO_3 and NO_3^- were in the range 0.19 to 0.30, 0.25 to 0.37, 1.09 to 2.93 and 0.35 to 0.65, respectively. Annual mean dry deposition was calculated if 9 or more monthly values were available. Missing data were obtained by interpolation.

4. RESULTS AND DISCUSSION

The annual deposition of $SO_4^=$ and NO_3^- and the annual mean fraction of total deposition that was dry are shown in Tables I and II, respectively. Observations in Ontario by Chan et al., (1986) are also included. Due to sampling problems, the uncertainty in air concentrations and hence in the dry deposition estimates is higher at Chalk River than at the other locations. Ranked in order of decreasing S deposition the stations are Long Point, Algoma, Chalk River, Montmorency, Kejimkujik and ELA. Sulphur deposition is not markedly different at Montmorency and Kejimkujik. When ranked according to NO_3^- deposition, the stations appear in the same order as above with the notable difference that deposition at Montmorency is markedly higher by 40% than that at Kejimkujik. The 4-yr. average

TABLE I: The annual total deposition (T) of $SO_4^=$ (mmol m^{-2}) and the fraction that is comprised of dry deposition (D) as measured at APN sites for 4-yr. in comparison to results measured by Chan et al. (1985) in Ontario. Note that T does not include fog-water deposition.

	APN										ONTARIO	
	1979		1980		1981		1982		4-YR MEAN		1982	
	T	D(%)	T	D(%)	T	(D%)	T	D(%)	T	D(%)	T	(D%)
LPT	71*	43	62	41	65	37	86*	26	71	37	58	22
ALG	NM		ID		40*	16	38	17	39	16	27#	20#
CH. R.	43	27	36*	23	38	20	34	27	38	24	24	22
MONT	NM		ID		25	16	27	12	26	14		
KEJI	ID		28	22	28	16	24	26	27	21		
ELA	13	22	8	31	10	27	14	22	11	26	12	13

TABLE II: The annual total deposition (T) of NO_3^- (mmol m^{-2} and the fraction that is comprised of dry deposition (D) as measured at APN sites for 4-yr in comparison to results measured by Chan et al. (1985) in Ontario in 1982. Note that T does not include fog water deposition or NO_2 dry deposition.

	APN										ONTARIO	
	1979		1980		1981		1982		4-YR MEAN		1982	
	T	D(%)	T	D(%)	T	(D%)	T	D(%)	T	D(%)	T	(D%)
LPT	55*	24	53	22	53	20	62	9	56	19	67	35
ALG	NM		ID		44*	22	41	22	42	22	39#	45#
CH. R.	39	28	37*	20	36	15	34	24	37	22	29	25
MONT	NM		ID		26	20	28	15	27	17	-	-
KEJI	ID		20	18	19	14	16	20	19	18	-	-
ELA	15	25	ID		13	34	16	25	15	28	19	37

ID - insufficient data
NM - no measurement
* - precipitation amount estimated
# - Ontario sampling sites more than 120 km from Algoma

fraction of total S deposition comprised of dry deposition ranges from 14 to 37%. For nitrate, the highest dry deposition fraction is at ELA (28%) while it is 17 to 22% at the other stations. Based on the observed deposition at APN sites, we estimate that in the southern part of Canada east of Manitoba 22% and 21% of the total $SO_4^=$ and NO_3^- deposited comes down as dry deposition, respectively. However, the fraction of dry deposition for nitrate increases to 30% if NO_2 contribution is added. Using earlier measurements in the literature, Galloway and Whelpdale (1980) estimated that in all of eastern Canada 29% of $SO_4^=$ deposition was dry. Similarily, N budget estimates by Logan (1983) indicate that in all of eastern Canada dry deposition of NO_3^- is 17 to 18% of the total.

A comparison of the seasonal variation of total monthly deposition and its wet and dry components at the sites is made in Figure 2 for $SO_4^=$ and NO_3^-, respectively. At all locations, there is a marked seasonal variation in wet deposition. It peaks in the summer and is a maximum in winter. In contrast, the seasonal variation in dry deposition is not the same at all sites. Dry deposition varies seasonally at all sites but ELA where no marked variation is apparent. This has to do with differences in the seasonal cycles of SO_2 and $SO_4^=$ at ELA and the other other sites (Barrie et al., 1984). There is also a distinct seasonal variation in the fraction of total dry deposition falling as SO_2. It peaks in the winter and is a minimum in summer. The variation is much less pronounced at ELA than at the other sites. Seasonal variations in total NO_3^- deposition are much less pronounced than those of $SO_4^=$ (Figure 2). This is due to the off-setting influence of wet and dry deposition cycles that are in anti-phase. Dry deposition peaks in the winter and wet deposition peaks, if at all, in the spring, summer and fall. This is in general true for all stations. The fraction of total $SO_4^=$ and NO_3^- that is dry deposited undergoes a marked seasonal variation at ELA, Long Point and at Chalk River ($SO_4^=$ only) but not at the others. At ELA for both constituents, it is a minimum in November to February of about 20 to 25% and 55 to 70% for $SO_4^=$ and NO_3^-, respectively. At Long Point it has two relative maxima (in December-January and in April- May) and a minimum in February-March. At Chalk River, the dry deposition fraction of $SO_4^=$ is higher in January and February (40 to 50%) than in the rest of the year (20 to 30%).

From the point of view of effects of acidic deposition that is episodic in nature, it is useful to know what fraction of total deposition is delivered by what fraction of events or, in other words, the episodicity. The fraction of total wet or dry deposition delivered by 20% of the daily events (E) ranges from 47 to 70%. For $SO_4^=$, E of wet deposition is higher than for dry, especially at Kejimkujik (higher by 6% absolute E=units), Algoma (10%) and ELA (14%). The reverse is true for NO_3^-, especially so at Algoma (lower by 13%) and Chalk River (16%). Intersite differences were apparent. For both $SO_4^=$ and NO_3^-, E of wet and of dry deposition was $10 \pm 5\%$ higher at the more remote sites than at Chalk River and Long Point closer to sources (the exception to this was E of NO_3^- for dry deposition at Chalk River which was similar to that at other sites).

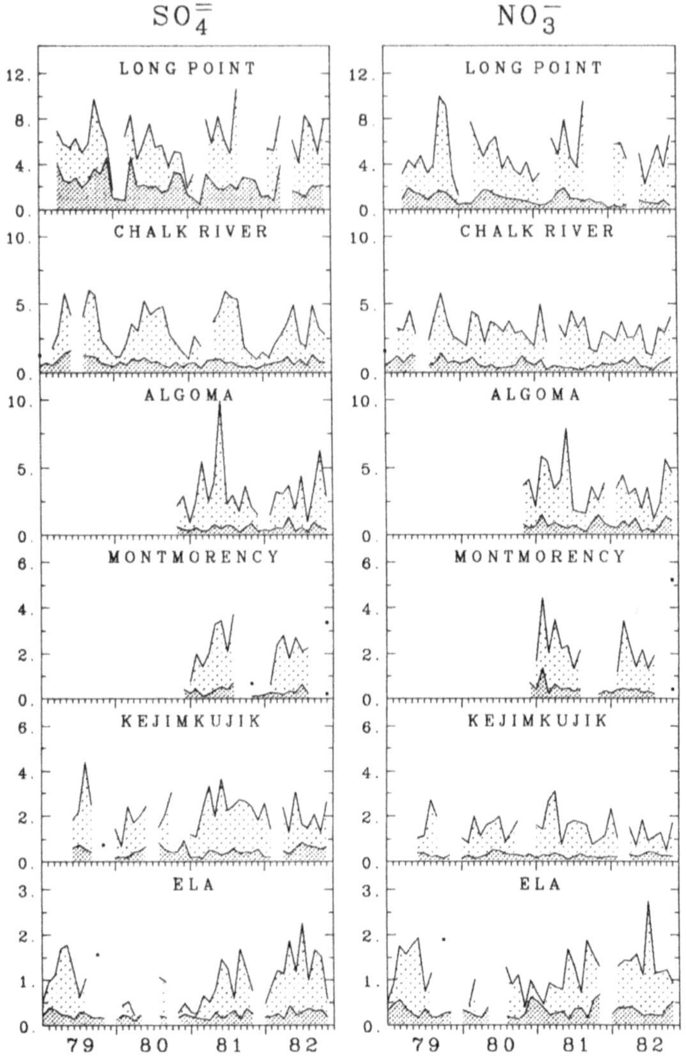

Figure 2. The temporal variation of total monthly $SO_4^=$ and NO_3^- deposition and its dry component.

This is expected since the closer a site is to major sources the more uniform the air pollutant concentrations are likely to be.

5. CONCLUSION

Four years of air and precipitation chemistry measurements at six locations in eastern Canada have been used to calculate the wet and dry deposition of $SO_4^=$ and NO_3^- to an area 100 km in radius about each site. The mean annual total deposition of $SO_4^=$ and NO_3^- in eastern Canada ranged from 8 to 86 and 13 to 62 mmol m^{-2}, respectively. The highest deposition occurred at Long Point on the north shore of Lake Erie and the lowest at ELA in northwestern Ontario. At all sites but ELA, annual $SO_4^=$ wet deposition was greater than the target loadings of 21 mmol m^{-2} (20 kg-$SO_4^=$ yr^{-1}) considered to be dangerous to aquatic ecosystems (MOI, 1984). In the southern half of eastern Canada dry deposition accounted for 22 and 21% of the total annual deposition of $SO_4^=$ and NO_3^- (excluding NO_2 dry deposition) respectively. For NO_3^-, the inclusion of NO_2 dry deposition raises the fraction dry from 21 to 30%. These total deposition estimates do not include the contribution of fogwater which in some areas can be substantial (Barrie and Schemenauer, 1986). There was a marked seasonal variation in the total deposition of $SO_4^=$ but not in that of NO_3^-. $SO_4^=$ deposition peaked in the summer. This means that there is a seasonal variation in the ratio of $SO_4^=$ and NO_3^- deposition; a fact that has been noted elsewhere (Summers and Barrie, 1986). The episodicity of wet and dry deposition of $SO_4^=$ and NO_3^- in eastern Canada is such that 20% of the daily events deliver 47 to 70% of the deposition. For $SO_4^=$, the episodicity of wet deposition is higher than for dry, particularly at remote sites, while the reverse is true for NO_3^-.

ACKNOWLEDGMENTS

The authors thank J. Kovalick for his invaluable assistance in data preparation and analysis.

REFERENCES

Anlauf, K., Barrie, L.A., Wiebe, H.A. and Fellin, P.: 1980, Major ion concentrations in atmospheric particulates sampled in eastern Canada, Canadian Research, Feb., 49-53.

Anlauf, K.G., Fellin, P., Wiebe, H.A., Schiff, H.I., Mackay, G.I., Braman, R.S. and Gilbert, R.: 1985, Atmospheric Environment, 19, 325.

Barrie, L.A., Wiebe, H.A., Anlauf, K. and Fellin, P.: 1980, The Canadian Air and Precipitation Monitoring Network (APN) in Studies in Environmental Science, Vol .8, 355-360, Elsevier Press, Amsterdam.

Barrie, L.A.: 1982, Environment Canada's long range transport of atmospheric pollutants program: atmospheric studies, Acid Precipitation Effects on Ecological Systems, Ed. F.M. D'Itri, Ann Arbor Science, Michigan, 141-161.

Barrie, L.A., Anlauf, K., Wiebe, H.A. and Fellin, P.: 1984, Acidic pollutants in air and precipitation at selected rural locations in Canada, in Deposition Both Wet and Dry, Ed. B.B. Hicks, Acid Precipitation Series, Vol. 4, Series, Ed. J.I. Teasley, Butterworth, London, Toronto.

Barrie, L.A. and Schemenauer, R.S.: 1986, Water, Air and Soil Pollution, this volume.

Cadle, S.H.: 1985, Atmospheric Environment, 19, 181.

Chan, W.H., Tang, A.J.S., Chung, D.H.S., Kirk, R.W. and Lusis, M.A.: 1986, Atmospheric Environment, submitted.

Data, APN internal data reports Barrie et al, Rep. #'s AQRB-80-002-T, AQRB-81-026-, AQRB-82-009-T, AQRB-84-005-T, Atmospheric Environment Service, 4905 Dufferin Street, Downsview, Ontario M3H 5T4.

Galloway, J.N. and Whelpdale, D.M.: 1980, Atmospheric Environment, 14, 409.

Logan, J.A.: 1983, J. of Geophys. Res., 88, 10785.

MOI: 1983, Final Report of the United States Canada Memorandum of Intent on Transboundary Air Pollution Impact Assessment Work Group I. Chairmen Bangay Riordam.

Sirois, A. and Barrie, L.A.: 1986, Tellus. submitted.

Summers, P.W. and Barrie, L.A.: 1986, Water, Air and Soil Pollution, this volume.

Voldner, E.C., Barrie, L.A. and Sirois, A.: 1986, Atmospheric Environment, in press.

Whelpdale, D.M. and Barrie, L.A.: 1982, Water, Air and Soil Pollution, 18, 7.

OVERVIEW OF HISTORICAL AND PALEOECOLOGICAL STUDIES OF ACIDIC AIR POLLUTION AND ITS EFFECTS

Ronald B. Davis
Department of Botany & Plant Pathology
and Institute for Quaternary Studies
University of Maine
Orono, ME 04469
U.S.A.

Pamela M. Stokes
Department of Botany and Institute for Environmental Studies
University of Toronto
Toronto, Ontario
M5S 1A4
Canada

ABSTRACT. Historical evidence of acid deposition and its effects which was presented at Muskoka, Ontario in September, 1985 is summarized. This evidence consists of written records of the past chemistry and biology of atmospheric, terrestrial, and aquatic systems; it also includes evidence from archived collections which were "revisited", from tree rings, and from the chemical and biological "records" in lake sediment and peat from which histories of airborne contaminants and ecosystem responses to these contaminants were inferred.

1. INTRODUCTION

Historical and paleoecological studies have contributed significantly to our understanding of acidic precipitation and its effects. In this overview, we give selected highlights of the papers that were included in the section on "Historical Perspectives of Acidification" at the International Symposium on Acidic Precipitation in Muskoka, Ontario in 1985.

2. FOREST DAMAGE

Johnson (1986) describes the widespread red spruce decline in northeastern United States since ca. 1960 and analyzes its possible causes. He also presents historical information on red spruce declines in the 1870's and 1880's. Air pollutant concentrations were certainly

Water, Air and Soil Pollution **30** (1986) 311–318.

very much lower in 1870-1890 than since 1960, and he argues that pollutants were not the cause of the early declines but that natural causes were. As the causes of the post-1960 decline have not definitely been shown to be anthropogenic, Johnson's finding of earlier, natural declines strengthens the argument that natural factors, if not solely responsible, could at least be having an amplifying effect. By study of affected populations, climatic records, and tree rings, he shows that the recent decline could be a function of anomalously cold winter and spring weather. He argues that while there is a correlation between the recent decline and high concentrations of air pollutants, particularly at high elevation, it is unlikely that pollutants alone are responsible for so widespread a phenomenon.

3. FISHES, OTHER AQUATIC BIOTA, AND WATER CHEMISTRY

Haines and Baker (1986) review historical fisheries and pH data for lakes in certain areas of the United States east of the Great Lakes, and find that the Adirondack Mts., New York, data set remains the strongest for evaluating the role of acidification in decline of fisheries. These authors systematically evaluate alternative explanations for the decline and conclude that acidification is the most likely cause. Fisheries declines associated with acidity have also been found in Pennsylvania and Massachusetts, but the quality and/or quantity of the data are lower than in the Adirondacks. Data from New Hampshire and Vermont suggest limited effects on fisheries, and for Maine no effects have yet been demonstrated.

Losses of fish populations in southern Norway are documented by Sevaldrud and Skogheim (1986), based on interviews. Interview results held up well for reliability when checked by test fishing at a subsample of lakes. These authors report that pre-1978 trends of declining brown trout and perch populations have continued in 1983. Lakes with decreased or exterminated populations of either species had lower Ca concentrations, lower pH, and higher labile Al concentrations than lakes whose populations remained unchanged between 1978 and 1983.

Leuven et al. (1986) report that at least 67% of the Netherlands' lentic soft waters (<1 meq L^{-1}) have been acidified. This conclusion is based on historical chemical and biological data, inference from water chemistry models of acidification, and paleoecological studies. These authors report losses of fish stocks in the soft waters, based on historical records and on inference from the present distributions of fishes and water chemistry.

Stewart et al. (1986) review the evidence for loss of fisheries in Ontario lakes. The authors demonstrate loss of fisheries that are apparently related to large point sources of acidifying emissions such as Sudbury. Stewart et al. (1986) find only one Ontario lake where acidification is not related to such a point source. This lake, located in the south-central part of the province, shows evidence of fishery loss apparently resulting from acidification by regional acid deposition. The authors stress that the finding of only one such site may be due to inadequate data. Evidence for decreased quality of spawning habitat for lake trout, in terms of increased episodic pH depression and increased Al is presented for sites both near to and distant from large point sources.

By a comparison of invertebrates found in 1937-1942 with those found in 1984 in the same Ontario streams, Hall and Findeis (1985) inferred that water acidification had taken place there. Acid sensitive species such as certain mayflies, stoneflies and caddisflies had "disappeared" from streams that in 1984 experienced strong pH depressions (to pH 4.9) but the species were still present in streams whose pH remained $\geqslant$6.0.

4. CHEMISTRY OF LAKE SEDIMENT AND PEAT

Norton (1986) discusses the chemical record of air contaminants in lake sediments and attempts to link this record to air pollution history and the acidification of lake water. He demonstrates the difficulties in making these linkages. The stratigraphy of airborne contaminants in lake sediments cannot be interpreted simply as a direct quantitative chronologic record of atmospheric deposition on the lake. The sedimentary record reflects, in addition to direct atmospheric loading, effects of the terrestrial watershed in "processing" deposits from the atmosphere (affecting fluxes to the lake), variable distributions of sediment constituents within the lake, and post-depositional modifications of the sediments. Norton shows that it is possible to strengthen the interpretation of the chemical sedimentary record in terms of airborne contaminants and their effect by the analysis of multiple sedimentary components in the same core, and by sorting through multiple working hypotheses to explain correlated changes.

Stratigraphic records of S in lake sediments in the Adirondacks, northern New England, northern Great Lake States, and northern Florida were described by Mitchell et al. (1985). The data were collected as part of the Paleoecological Investigation of Recent Lake Acidification (PIRLA) Project (Charles et al., 1986). Sulfur concentrations and accumulation rates in pre-1900 sediment are similar in all regions. Sulfur enrichment of post-1900 compared to pre-1900 sediment was observed in the four regions. Enrichment factors greater than 2 were most consistently found in the Adirondacks, as would be predicted from patterns of atmospheric deposition (Altwicker and Johannes, 1985). Stratigraphic patterns were most difficult to interpret in the Florida lakes. There and elsewhere, the use of sedimentary stratigraphy of S as a chronologic record of deposition of atmospheric S must be approached with great caution because of confounding factors including diagenesis and mobility of S in sediments. This problem is also discussed by Norton (1986).

The widespread increase of Hg starting in sediment dated at mid- to late- 1800's, such as that reported by Megar (1986) for northern Minnesota, supports the belief that long range atmospheric transport of

Hg has resulted in contamination of sediments and watersheds remote from point sources. Megar also proposes that acidic precipitation could enhance the transport of Hg from lake watersheds. Increases since 1800 of several other elements associated with acidic precipitation and other air pollution complexes are estimated by peat stratigraphic study in the Appalachians by Schell *et al.* (1986). In Pennsylvania, enrichment factors greater than 100 are reported for Cl, N and S, and 10 to 40 for Pb, Br, Ca and Sb.

Increases of certain trace metals including Pb and Zn start in lake sediment dating to the mid- to late-1800's in northern New England and the Adirondacks (Norton, 1986). In the upper part of sediment profiles in acidic (pH $\leq$ 5.5) lakes, Zn accumulation rates decrease as a result of increased leaching due to acidification and/or sediment diagenesis. Decreases in sedimentary accumulations of Ca and Mn relative to Ti probably result from increased leaching of Ca and Mn by acidified waters. Accumulation rates of V start increasing in sediment dating 1920-1940, reflecting the upswing in fuel oil consumption.

5. RECONSTRUCTION OF LAKE pH BY ANALYSIS OF SEDIMENTARY REMAINS OF DIATOMS

Battarbee and Charles (1986) describe pH reconstructions based on diatoms from Scotland, Scandinavia, continental Europe, Canada and U.S.A., and demonstrate that acidification, probably due to acidified precipitation, has occurred in certain lakes in all these regions. Geographic distributions of the lakes shown to have been acidified and the chronology of acidification in these lakes are consistent with geographic patterns of precipitation chemistry and with what is known of the history of acidic emissions. The authors point out that the diatom technique is still the only reliable one for accomplishing detailed reconstruction of lake pH. Charles *et al.* (1986) imply that sedimentary remains of silica-scaled chrysophytes may soon be used for this purpose, and Norton (1986) indicates that sediment chemistry can be used for inferring the crossing of certain pH thresholds.

Taylor and Duthie (1986) indicate a relationship between (1) abundance of sedimentary remains of the diatoms *Cyclotella stelligera* and *C. kutzingiana* and (2) depth of shield lakes in Ontario, but no relationship with pH. In other geographic regions, investigators have found relationships between the distributions of these species and pH and have used the species for paleolimnologic reconstruction of pH. Taylor and Duthie's (1986) results demonstrate that the extrapolation of diatom pH calibrations from one region to another for use in reconstruction of lake pH is a risky procedure.

6. PRECIPITATION CHEMISTRY

Altwicker and Johannes (1985) characterized the differences in precipitation chemistry between the "North East Region" (including

Adirondacks and northern new England areas) and the "North Central Region" (including the northern Great Lakes region) in terms of the ratio of equivalents of (1) "Sum(+)" i.e. $Ca^{++} + Mg^{++} + NH_4^{+} + K^{+} + Na^{+}$ to (2) "Sum(-)", i.e. $SO_4^{=} + NO_3^{-} + Cl^{-}$. The authors pointed out a clear difference between the "non-acidic (pH > 4.5)" "North Central Region" with a ratio of ca. 1:1 and the "acidic (pH < 4.5)" "North East Region" with 1:>2. A comparison was made of ratios from the 1950's (Junge sites) with ratios from the late 1970's and early 1980's for these two regions and also for the "Midwest" and "East Central" regions. Decreased pH was inferred from increases of "Sum(-)" in all but the "North Central Region." This is consistent with results of the PIRLA Project (Charles et al., 1986) and of fishery/water chemistry studies (Haines and Baker, 1986; Stewart et al., 1986). The methodological approach of Altwicker and Johannes (1985) is useful for discerning spatial and temporal trends of acidity of precipitation when reliable measurements of pH are not available, but because of possible problems with the quality of older data on cations and anions, the approach still needs to be used with caution.

7. INTEGRATIVE PALEOECOLOGICAL APPROACHES, AND MODELLING

Charles et al. (1986) describe the PIRLA Project and give some of the early results of the project. They demonstrate the effectiveness of multiparameter and multidisciplinary approaches in paleoecological studies. The sedimentary contents may be used in two ways: first, as a record of changes outside the lake, both in the atmosphere and in the terrestrial watershed, and second, as a record of responses of the lake to atmospheric and terrestrial changes. Charles et al. (1986) show that the evaluation of both aspects of the sedimentary record is complementary for the understanding of the history and effects of acidic precipitation. The PIRLA Project covers four acid sensitive regions of North America where lake sediment cores have been analyzed for biological indicators (including diatom and chrysophyte remains) of past water chemistry. Inferred chemical trends are interpreted by comparison with accumulations in the same cores of airborne contaminants associated with acid deposition such as polycyclic aromatic hydrocarbons, soot particles, and trace metals. An important goal of the project is the estimation of errors associated with the paleoecological reconstructions. This is being accomplished by replicate determinations for several of the parameters at each major step in the procedure, including analyses of multiple cores from at least one lake in each region. The PIRLA investigators have demonstrated that certain lakes now with pH 4.4 to 5.0 in the Adirondack Mts., N.Y., and in northern New England have been acidified (by acidic precipitation) by $\leq$ 1 pH unit starting in the 1920's to 1950's depending on the particular lake. These lakes were all naturally less than pH 6 prior to the recent acidification. By historical study of changes in the watersheds (e.g., logging, forest fire, forest succession) it has been possible for several lakes to exclude watershed changes as causes of the acidification. Study lakes now with pH greater than 6.0 in these areas show no evidence of recent acidification. Lakes studied in the northern Great Lakes region (one of which is pH 4.4) show no evidence of recent acidification. This probably results from the different precipitation chemistry in this region compared with the Adirondacks and New England.

Wright et al. (1986) reconstruct water pH back to pre-pollution times in four lakes using a model (MAGIC) that accepts inputs for historical trends in acidic air pollutants as well as data on the capacity of watershed soil to buffer and otherwise alter such pollutants. These authors demonstrate that terrestrial processes "damp, delay, and moderate" the responses of lakes to acid deposition. Major responses of lakes have occured over only 1 to 3 decades compared to the longer term, slower increase of acid deposition that started earlier. The time-series lake pH responses generated by the model are in reasonable agreement with reconstructions of pH based on sedimentary diatom remains.

8. CONCLUDING COMMENTS

Any retrospective study is likely to give an incomplete and only roughly quantitative picture of chemical and biological changes relating to acidification. Nevertheless, we believe that historical and paleoecological studies such as those presented at Muskoka and highlighted here can contribute, if used with appropriate care, to our information base on trends of acidification. Ingenuity and persistence in uncovering old records, and the development of increasingly sophisticated techniques for determining trends from these records and from lake sediment and peat stratigraphies are contributing to our understanding of acidic precipitation and its effects. Particularly encouraging is the agreement now emerging among the results of widely different approaches such as the modelling of trends in acidification, study of spatial and temporal patterns of precipitation chemistry, comparison of new and old biological and chemical data for terrestrial and aquatic systems, analyses of sediment/stratigraphic "records" of airborne contaminants, and environmental inference based on sedimentary diatom remains and other sedimentary indicators of past conditions. For example, the contrasting acidification history of the heavily impacted Adirondacks and the lightly impacted northern Great Lakes region (except for point source areas like Sudbury) has been demonstrated by historic studies of precipitation chemistry and fisheries as well as sediment studies. Because of the heavy impacts in the Adirondacks, and therefore the relative ease in demonstrating these impacts, the area is the best historically documented one in North America for demonstrating the effects of regional acid deposition including lake acidification and fishery losses. For other regions, less comprehensive but similar historic data demonstrate lake acidification, loss of fisheries and other effects of acidic deposition.

As a final comment, we point to what we perceive as a major difficulty in interpreting historical and paleoecological data. This difficulty derives from the co-occurrence of natural and anthropogenic processes and the diverse ways that airborne contaminants may affect terrestrial and aquatic systems. The difficulty may be partly overcome by increasing our understanding in general of ecosystem function and watershed biogeochemistry. Notably in short supply is good comparative information on the variation in response of ecological systems to natural perturbations. Such information can be obtained from long-term ecological study and monitoring in relatively unpolluted areas, and by sophisticated paleoecological study. Support for such fundamental and long-term studies is at least as important as for studies of pollution effects themselves, because without knowledge of non-anthropogenic variation it is impossible to place anthropogenic variation in a meaningful context.

ACKNOWLEDGMENTS

We thank S.A. Norton for his review and helpful comments on drafts of this paper.

REFERENCES

Altwicker, E.R. and Johannes, A.H.: 1985, 'Acidic Precipitation: A Graphical Presentation of Historical and Spatial Perspectives in Eastern North America', Abstracts, Intern. Acidic Prec., Muskoka, Ont., 84.

Battarbee, R.W. and Charles, D.F.: 1986, *Water, Air and Soil Pollut.*, this volume.

Charles, D.F., Whitehead, D.R., Anderson D.S., *et al.*: 1986, *Water, Air and Soil Pollut.*, this volume.

Haines, T.A., and Baker, J.P.: 1986, *Water, Air and Soil Pollut.*, this volume.

Hall, T.A., and Findeis, J.; 1985, 'Biological Effects of Acidification. V. Recent Changes in Stream Invertebrate Communities: Is Acidification the Cause?' Abstracts, Intern. Symp. Acidic Prec., Muskoka, Ont., 86.

Johnson, A.H.: 1986, *Water, Air and Soil Pollut.*, this volume.

Leuven, R.S.E.W., Kersten, H.L.H., Schuurkes, J.A.A.R., Roelofs, J.G.M., and Arts, G.H.P.: 1986, *Water, Air and Soil Pollut.*, this volume.

Meger, S.A.: 1986, Water, Air and Soil Pollut., this volume.

Mitchell, M.J., Owen, J.S., Schindler, S.C., and David, M.B.: 1985, Comparison of Sulfur Sediment Profiles in Selected Lakes of the United States and Their Relationship to Acidic Deposition', Abstracts, Intern. Symp. Acidic Prec., Muskoka, Ont., 87.

Norton, S.A.: 1986, Water, Air and Soil Pollut., this volume.

Schell, W.R., Sanchez, A.L., Granlund, C., and Underhill, D.W.: 1986, Water, Air and Soil Pollut., this volume.

Sevaldrud, I.H., and Skogheim, O.K.: 1986, Water, Air and Soil Pollut., this volume.

Stewart, T.J., MacLean, J.A., Hicks, F.J., Liimatainen, V.A., and Beggs, G.L.: 1986, Water, Air and Soil Pollut., this volume.

Taylor, M.C., and Duthie, H.C.: 1986, Water, Air and Soil Pollut., this volume.

Wright, R.F., Cosby, B.J., Hornberger, G.M., and Galloway, J.N.: 1986, Air and Soil Pollut., this volume.

Recent and Historic Red Spruce Mortality: Evidence of Climatic Influence[1]

Arthur H. Johnson
Department of Landscape Architecture and Regional Planning
and Department of Geology/D4
University of Pennsylvania
Philadelphia, PA 19104 USA

Andrew J. Friedland, Jonathan G. Dushoff
Department of Geology
University of Pennsylvania
Philadelphia, PA 19104 USA

Abstract

A period of declining growth and extensive mortality of red spruce (*Picea rubens* Sarg.) has been in progress for the past two decades in the montane forests of New York, Vermont and New Hampshire. Those areas are subject to particularly high levels of pollutant input and there is concern that airborne chemicals have destabilized the forest ecosystems. Field reports indicate that major episodes of mortality also occurred prior to the era of widespread industrial pollution, particularly in the 1870's and 1880's. The present and past periods of mortality have similarities as well as differences, and whether the primary causes are the same cannot be confidently determined with the information currently available. One similarity is that recent and past episodes of red spruce mortality tended to occur during or immediately after periods which are climatically unfavorable for this species.

1. INTRODUCTION

Several recent studies have reported extensive mortality of red spruce (*Picea rubens* Sarg.) in the montane forests of the northern Appalachians (Siccama *et al.*, 1982; Johnson and Siccama, 1983; Foster and Reiners, 1983; Carey *et al.*, 1984; Scott *et al* 1984), without unusual

---

[1]This work was supported by a grant from the A.W. Mellon Foundation and by the U.S. EPA through Interagency Agreement no. DW 89931334-01-0 to U.S. DOE and Oak Ridge National Laboratory. This manuscript has not been subject to EPA review and does not reflect the views of EPA. No official endorsement should be inferred.

Water, Air and Soil Pollution **30** (1986) 319–330.

mortality of balsam fir (Abies balsamea L. Mill.), or paper birch (Betula payrifera var. cordifolia Marsh [Regel]) the other major species. Two decades of sharply declining ring-widths are associated with the spruce mortality (Johnson and McLaughlin, 1986; Cook, 1986; Hornbeck and Smith, 1985).

Recent studies of the deposition of airborne chemicals in high elevation New England forests suggest that the input of acidic compounds and heavy metals is approximately 3 to 8 times higher in the coniferous forests (> 1000 m elevation) than in the adjacent, lower-elevation hardwood forests (<800m) (Lovett et al., 1982; Scherbatskoy and Bliss, 1984, Friedland et al.,1984a). The higher deposition rates result from orographic effects and in particular from the interception of cloud water, which has comparatively high concentrations of airborne chemicals. The estimates for deposition of H^+ (3.9 Kg ha^{-1} yr^{-1}), NH_4-N (16 Kg ha^{-1} yr^{-1}), NO_3-N (28 to 40 Kg ha^{-1} yr^{-1}), SO_4-S (68 Kg ha^{-1}, yr^{-1}) and Pb 500 to 700 g ha^{-1} yr^{-1}) in bulk precipitation and in cloudwater (Lovett et al., 1982, Scherbatskoy and Bliss, 1984, Friedland et al., 1984a) suggest that the high elevation forests of New England are subject to higher rates of acid and metal deposition than any other major forest ecosystem in eastern North America except for areas near smelters.

Ozone monitoring at Whiteface Mountain, NY indicates that average monthly values during the growing season fluctuate between 40 and 60 ppb with short-term excursions to about 100 ppb (Burgess et al., 1984; J. Kadlecek, unpublished data, State University of New York, Albany). Recent concern that O_3 is at least in part responsible for the decline of forests in the Federal Republic of Germany and the documented detrimental effects of ambient O_3 levels in the field (i.e. Miller et al.,1982) and in the laboratory (Reich and Amundson, 1985) has furthered the idea that chemicals of anthropogenic origin may have promoted the observed deterioration of spruce populations.

These findings, in combination with the fact that no obvious biotic causes for red spruce decline have been discovered as yet, have focused attention on the possiblity that the current episode of spruce deterioration is caused by long-distance transport of acidic compounds and other pollutants.

On the other hand, there were episodes of red spruce mortality in the 19th century, and based on temporal trends in S emissions in eastern North America (Husar, 1986) industrial pollution can be ruled out as a contributor to those episodes of mortality. It is generally accepted that climate change or severe climatic events can play important roles in the occurrence of tree disease (Manion, 1981). Accordingly, the response of red spruce to climate as determined from dendroclimatic studies is reviewed here, and the nature and timing of well-documented periods of spruce mortality are summarized. Finally, the association between periods of spruce decline and unfavorable climatic conditions is discussed.

2. CHARACTERISTICS OF RED SPRUCE AND THE RECENT DECLINE

Red spruce is a long-lived, shade-tolerant species which, in montane environments of the Northeast, reaches it maximum importance on the mid to upper slopes. Longevity and shade tolerance allow red spruce to compete successfully with its more vigorous but shorter lived competitors, balsam fir and white birch. The characteristics of the high elevation coniferous forests has been reviewed elsewhere (Siccama, 1974; McIntosh and Hurley, 1964; Whittaker, 1956). Red spruce in all mature size classes may show symptoms of disease or injury in the montane forests of New York, Vermont and New Hampshire (e.g. Johnson and Siccama, 1983; Friedland et al., 1984b). Red spruce show needle discoloration and death, crown thinning, an altered branching pattern and the eventual death of branches, individuals and large patches of mature trees. Typically (but not uniformly) the crowns of mature, declining red spruce appear to deteriorate from the top down and from the outside inward (Johnson and Siccama, 1983). Root studies in the northern Appalachians indicate that Armillaria mellea, a common root pathogen, plays a secondary role in the observed spruce deterioration (Carey et al., 1984). No primary biotic agents have been discovered to date, but thorough, systematic studies are still much needed in this regard. Primary pathogens, should they be present, escaped detection in at least one thorough study (Hadfield, 1968).

The deterioration of red spruce populations in the northern Appalachians is occurring to the greatest degree near its upper elevational limit (about 1200 m). Johnson and McLaughlin (1986) have reviewed data from the northern Appalachians on mortality, canopy deterioration, stand ages, basal area values, quantitative comparisons over the past 20 yr. and tree ring chronologies. They reported that (1) large reductions (40 to 70%) in live spruce basal area have occurred at some sites over the past two decades; (2) abrupt substantial and sustained reductions in breast-height tree ring widths began about 20 to 25 yr. ago, and; (3) that the degree of spruce deterioration is independent of present stand basal area, and disturbance history, with some increase in mortality in the larger size classes (>30 cm dbh).

3. HISTORICAL EVIDENCE OF SPRUCE ABUNDANCE AND SPRUCE MORTALITY

It is reasonably clear that red spruce were more abundant in the pre-settlement forests of Vermont and New Hampshire than they are at present. Siccama (1971) reviewed information on the pre-1800 forests of Chittenden County, VT using data from town line surveys. From 1750 to 1960, at high elevation (> 700 m), spruce decreased from 52% to 12.8% of the total number of stems, and at low elevation (< 700 M), the change was from 18.5% to 4.5% in non-agricultural "stony land".

In a review of red spruce population changes in Virginia and other parts of the southern Appalachians, Pielke (1979) concluded that there was a substantial loss of red spruce over the last century from elevations above 900 m. The cause for the loss of spruce was thought to be logging, which gave rise to unfavorable microclimatic and site conditions for seedling establishment. Fire, which became more frequent due to large amounts of slash that remained after logging, also contributed to the loss of spruce. There are suggestions that climatic warming might explain the decrease of red spruce in undisturbed areas (Hamburg and Cogbill, 1986).

Since red spruce was an important timber species in northern New England during the last century, it attracted attention whenever it died in uncharacteristic numbers. More than 40 reports of sizeable areas of dead and dying red spruce are traceable, dating back to the early 1800's (Figure 1). While some reports from the last century are not well documented, there is sufficient evidence to conclude that there have been at least two periods of extensive mortality in New England and the Adirondacks: a recent period from at least 1962 to the present, and one which occurred during the period 1871 to 1885 (Hopkins, 1901). A period of mortality was recorded in the Adirondacks during the late 1930's which extended into the 1940's (Howard, 1938; McIntyre, 1938; Smith, 1948)[1].

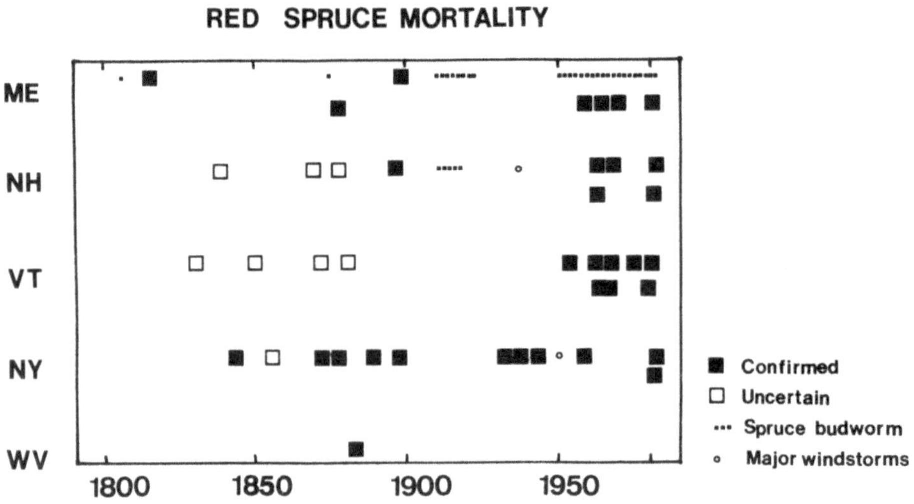

Fig. 1. Reports of dead and dying red spruce in the eastern U.S. (Hopkins et al., 1891, 1901; Weiss et al.,1985; unpublished reports from W.G. Howard, 1938; H.L. McIntyre,1938; W. Smith, 1948; Mook and Eno, 1956; Hadfield, 1968; Wheeler, 1965; Kelso, 1965; Tegethoff, 1964; Siccama, 1974; Siccama et al.,1982; Johnson and Siccama,1983; Scott et al.,1985; unpublished notes on red spruce mortality and winter injury, D. Stark, State of Maine Dept. of Entomology Laboratory, Augusta).

---

[1] W.G. Howard (1938), H.L. McIntyre (1938), and W. Smith (1948). Unpublished letters collected by M. Bermingham, New York State Department of Environmental Protection, Albany, NY.

While there are many reports of dead and dying red spruce, most are qualitative. In assessing the presence or absence of mortality through time, systematic records of forest conditions at fixed locations would comprise the most rigorous and reliable data set but we are unable to locate any such records. The information from New York (figure 1) is probably the most reliable since there are more or less continous records of diseases or other problems with timber back to about 1911 with considerable formal reporting dating from 1885[1].

By the late 1880's it was estimated that "one-third to one-half of the fully grown spruce timber" was killed in the Adirondacks (Seventh Report of the Forest Commission, Albany, NY, 1891) and other accounts from Vermont, New Hampshire and Maine (Hopkins, 1901) also suggest that a substantial percentage of the red spruce were killed during this period. Accounts of the blight of the late 19th century offered several possible causes. Mortality was attributed to the spruce beetle (Dendroctonus rufipennis Kirby, called D. piceaperda by Hopkins, 1901) (Weiss et al.,1985) with many suggestions that there were other factors which predisposed the trees to beetle attack (Hopkins, 1891,1901; 2nd Ann. Report of the New York State Entomologist, 1885). Drought, slash from heavy logging promoting a build-up of the spruce beetle, extreme changes in winter temperature and many other factors were suggested as contributing to or causing spruce mortality (Hopkins, 1891; First Report of the Forest Commission [NY], 1885).

Descriptions of the symptoms indicate greater mortality among older trees, mortality at high and low elevation and on wet and dry sites, dieback beginning with the smaller branches, red foliage in late summer, and "wormholes through the sapwood".

Throughout the period 1915-1980, there are reports of other types of insect problems, including spruce budworm (Choristoneura fumiferana Clem.) outbreaks and damage from the European spruce sawfly (Diprion hercyniae Hartig.) but as far as is known, those insects were not responsible for the major episodes of mortality delineated on figure 1 except as noted (Annual Reports of the New York State Conservation Commission, 1923-1963). In the period of red spruce mortality in the Adirondacks described by Howard, 1938, McIntyre, 1938 and Smith, 1948 (op.cit.), light spruce beetle infestations were suspected along with the mortality.

Except in the budworm-affected areas, the current episode of red spruce mortality appears to have been in progress at least since 1962 when reports of dead and dying red spruce on the mountains of Vermont, New Hampshire and western Maine were frequent (e.g. Tegethoff, 1964; Wheeler, 1965; Stark 1986; Kelso 1965; Hadfield, 1968; Siccama, 1974; Weiss et al., 1985). In addition to the sighting of dead and dying spruce, there was widespread and severe damage from winter dessication or freezing which resulted in red needles and bud damage in early spring (Figure 2). Most of the mortality was attributed to severe winter conditions and summer drought by the reporting scientists when recognizable symptoms of biotic disease were not observed. Secondary pathogens, but no primary pathogens or beetles were reported in association with the winter damage. By 1982 it was recognized that mortality of high-elevation red spruce was a regional phenomenon which appeared to be

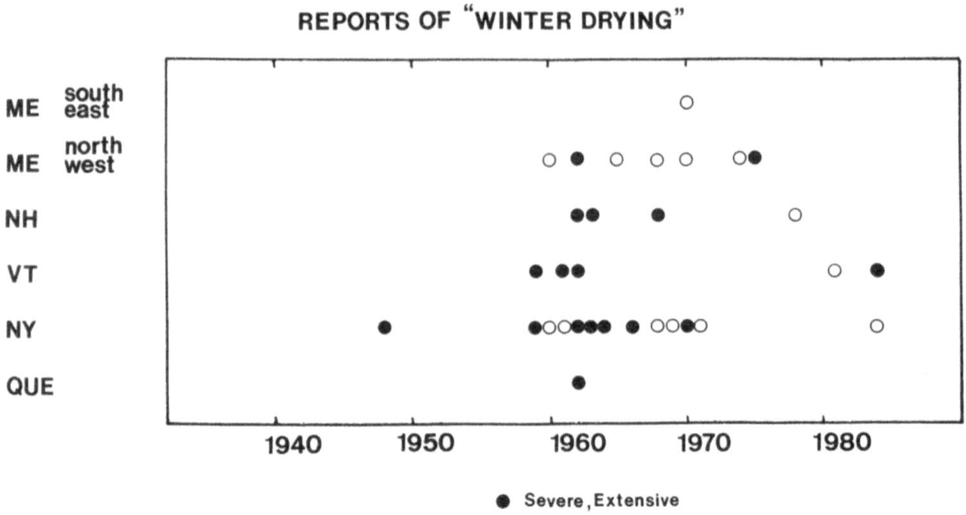

Fig. 2. Reports of "winter drying" or winter injury (dessication) to red spruce in New England. Year designated represents the spring of the year in which red needles were observed. The best coverage is from Maine from 1955-1976, and in New York (New York State Forest Pest Reports, 1971-1977; New York State Forest Insect and Disease Control Ann. Repts.,1968-1976; Annual Reports of the New York Conservation Commission, 1950-1963; Biennial Report of the Department of Forests and Parks (VT) 1959-60, 1961-62; Friedland et al., 1984b; Weiss et al.,1985; Kelso, 1965; Wheeler, 1965; Tegethoff, 1964; D. Stark, op.cit.

reasonably synchronous (Johnson and Siccama, 1983; Foster and Reiners, 1983) and accompanied by a sharp decrease in radial growth (Johnson and McLaughlin, 1986).

4. TREE-RING-CLIMATE-RELATIONSHIPS

Since red spruce is a long-lived species and shows sensitivity to temperature variations, it has been examined in tree ring research for use in reconstructing past climate. A few long-term chronologies (i.e. 250 to 300 yr) and some shorter-term chronologies (130 yr) from the northeastern U.S. have been analyzed (Roman and Raynal 1980; Cook, 1986; Conkey, 1979, 1984 a,b). Those findings were summarized by Johnson and McLaughlin (1986) and tend to show clear anomalies in ring width after about 1960 which coincides with the recent period of mortality. The current, anomalous change in ring width is reasonably synchronous across New York and New England suggesting some large-scale controlling mechanism.

By modeling ring-width response using average monthly precipitation and temperature values, the tree-ring studies have indicated that high-elevation red spruce ring widths are negatively correlated with 1) warm temperatures in late summer prior to the year of ring formation and; 2) cold temperatures during the December or January prior to the growing season. For high elevation old growth red spruce at three sites in Maine, Conkey (1979) found that the relationship between ring width and previous summer (July, August) temperatures was negative and that December and January temperature were positively correlated with ring width. Cook (1986) found that for high elevation red spruce at Lake Arnold (Adirondack Mountains, NY), prior July (-), August (-), and December (+) temperatures were variables most strongly correlated with ring width in the following growing season. While the physiological bases for these relationships are not established, warm late summers and cold early winters might be related to unfavorable effects on C budgeting

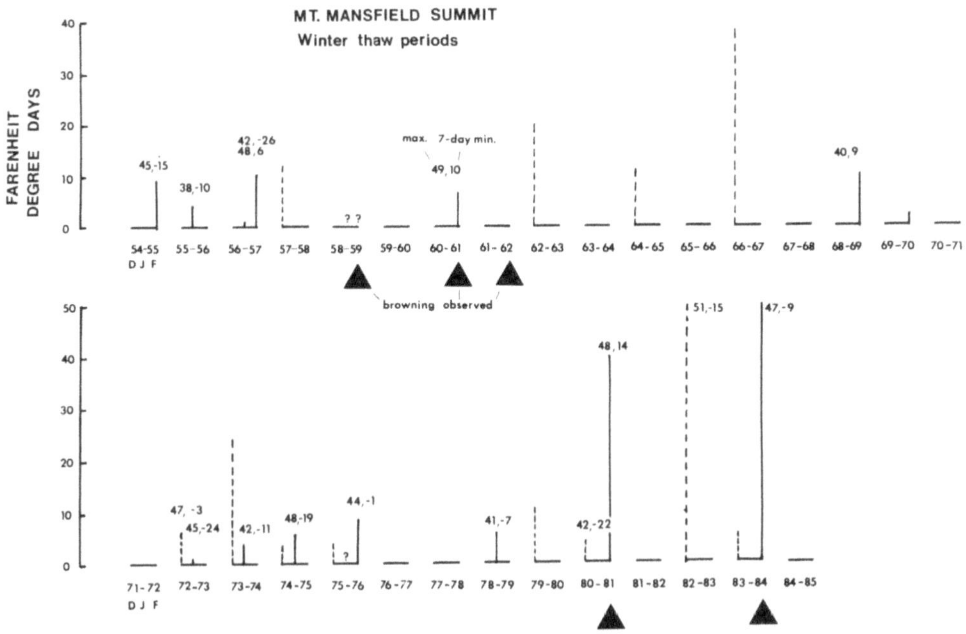

Fig. 3. Degree days for winter thaw periods at the summit of Mt. Mansfield, VT. Bar heights show the degree days associated with thaw events which occurred from 1955-1984 with the maximum temperature recorded during the thaw, and the minimum temperature recorded within seven days of the end of above-freezing temperatures. Based on tree ring - climate relationships, warm Decembers are probably not adverse, and December thaw periods are shown as dashed bars. Darts show years in which brown needles signifying dessication or freezing injury were observed.

and on winter hardiness. Extended mid-winter thaws have been associated with bud and needle death observed in spring (Curry and Church, 1952; Friedland et al., 1984b). However, based on temperature records from the weather station on Mt. Mansfield, VT (elevation 1300 m), it does not appear that mid-winter thaw periods were sufficient in the late 1950's and early 1960's to produce the observed damage (Figure 3).
From long-term temperature records at Burlington, VT, Lake Placid, NY, Albany, NY, Watertown, NY, and Cooperstown, NY, monthly temperatures were reconstructed for the period 1820 to 1984 (E.R. Cook, Lamont-Doherty Geological laboratory and A.H. Johnson, unpublished data). Figure 4 shows the co-occurrence of above-average August temperatures and below-average December temperatures. It is interesting to note that several documented periods of red spruce mortality, 1871-1885, 1938-1945, and 1962- present followed periods which were characterized by unfavorable temperatures.

Another climatic anomaly, the drought of the mid-1960's may have been unfavorable for red spruce, but owing to the fact that the peak drought occurred well after mortality was first observed and because the high elevation forest may have received adequate moisture even during the driest years (Siccama, 1974), drought is judged to be less likely than unfavorable temperatures as the critical, initiating factor in the current spruce decline. However, since the drought of the mid-1960's was particularly severe (Cook and Jacoby, 1977), and since observers in Vermont (Biennial Report of the Department of Forests and Parks (VT), 1965-66) noted what they believed to be drought effects in several tree species, we do not reject drought as a factor which may have contributed to the spruce decline.

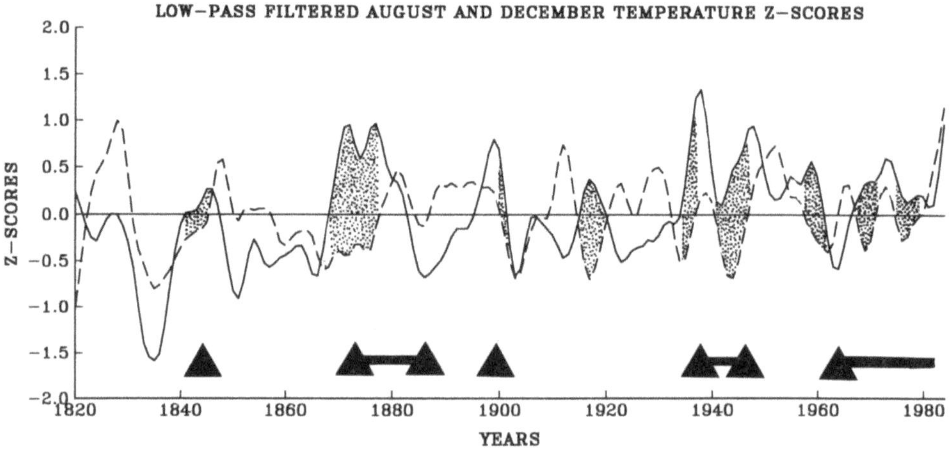

Fig. 4. Trends in August (solid) and December (dashed) temperatures in the Adirondack Mountains, New York (Northern Plateau Climatic Division). Periods when colder then average December coincide with warmer than average Augusts are shaded. The solid triangles show periods of region-wide red spruce mortality.

5. SUMMARY

Tree ring analyses suggest that red spruce are adversely affected by warm late-summer and cold early-winter temperatures. The timing of recorded episodes of red spruce mortality suggests that spruce decline is promoted by unfavorable climatic periods. This possibility warrants further investigation through analyses of weather records, tree rings and the physiological basis for the response to warm late summers and cold early winters. It seems unlikely that the influence of airborne chemicals alone could cause a synchronized reduction in ring-width over such a wide area as described by Johnson and McLaughlin (1986), Cook (1986) and Hornbeck and Smith (1985). The occurrence of widespread and severe winter damage to buds and foliage and winter and summer temperature anomalies at the onset of recent ring width decreases suggest that climatic fluctuations have at least a synchronizing effect in the current episode of spruce decline and possibly in prior episodes of red spruce decline. The mechanisms by which climatic events could cause spruce decline are speculative at present but might involve adverse effects of temperature on carbon budgets, and/or on winter hardiness. Other stress factors such as pathogens, drought, spring frosts, and airborne chemicals may be involved. We suggest that research designed to evaluate those as causal factors should take into account the effects of temperature on basic physiological processes.

REFERENCES CITED

Burgess, R.L., David, M.B., Manion, P.D., Mitchell, M.J., Mohen, V.A., Raynal,D.J., Schaedle, M., White, E.H. :1984, Effects of Acidic Deposition on Forest Ecosystems in the Northeastern United States: An Evaluation of Current Evidence. New York State College of Environmental Science and Forestry, Syracuse, NY.

Carey, A.C., Miller, E.A., Geballe, G.T., Wargo, P.M., Smith, W.H. and Siccama, T.G.: 1984, Plant Disease **68**: 794.

Conkey, L.E.: 1984a, Dendrochronology and forest productivity: red spruce wood density and ring width in Maine. In M.M. Harris and A.M. Spearing, (eds.). Research in Foresst Productivity, Use and Pest Control. USDA Forest Service, Gen. Tech. Report. NE-90 p. 69-75.

Conkey, L.E.: 1984b, X-ray denstometry: Wood density as a measure of forest productivity and disturbance. In D.D. Davis ed. Air Pollution and Productivity of the forest. Izack Walton League of America, Washington, D.C. p. 287-296.

Conkey, L.E.: 1979, Tree Ring Bull. **19**, 29.

Cook, E.R.: 1986, The use and limitations of dendrochronology in studying effects of air pollution on forests. In T. Hutchinson (ed.) Proceedings of a NATO Advanced Research Workshop, Toronto, Ontario, May 8-15, 1895. (in press).

Cook, E.R.and Jacoby, G.C.:1977, Science **198**,399.

Curry, J.R.and Church, T.W.: 1952, J. For. **50**, 114.

Foster, J.R. and Reiners, W.A.: 1983, Vegetation patterns in a virgin subalpine forest, Crawford Notch, White Mountains, New Hampshire. Bull. Torrey Bot. Club.

Friedland, A.J., Johnson, A.H. and Siccama, T.G.: 1984a, Water, Air, Soil, Pollut. **21**, 161.

Friedland, A.J., Gregory, R.A., Karenlampi, L. and Johnson, A.H.: 1984b, Can. J. For. Res. **14**, 963.

Hadfield, J.S.: 1968, Evaluation of diseases of red spruce on the Chamberlin Hill Sale, Rochester Ranger District, Green Mountain National Forest. File report A-68-8 5230. USDA- Forest Service Northeastern Area, State and Private Forestry Amherst FPC Field Office, Amherst, MA. 7p.

Hamburg, S., and Cogbill, C.V.: 1986, unpublished.

Hopkins, A.D.: 1901, Insect enemies of the spruce in the Northeast. U.S. Dept Agriculture Div. of Entomol. Bull. no. 28 new series p. 15-29.

Hopkins, A.D. : 1891, Forest and shade tree insects. II. W. Virgina Experiment Station Third Ann. Report. p. 171-180.

Hornbeck, J. and Smith, R.B.: 1985, Can. J. For. Res. **15**:199.

Husar, R.B.: 1986, Emissions of sulfur dioxide and nitrogen oxides and trends for eastern North America. Report of the Committee on Monitoring and Trends in Acidic Deposition. National Research Council. National Academy Press, Washington, D.C. p. 48-92.

Johnson, A.H. and Siccama, T.G.: 1983, Environ. Sci. Technology **17**, 294a.

Johnson, A.H. and McLaughlin, S.B.: 1986, The nature and timing of the deterioration of red spruce in the northern Appalachians. Report of the Committee on Monitoring and Trends in Acidic Deposition. National Research Council. National Academy Press, Washington, D.C. p. 200-230.

Kadlecek, J. : 1985, personal communication.

Kelso, E.G.: 1965, Memorandum 5220,2480 of July 23, 1965. U.S. Forest Service Northern FPC Zone 6 Main Street Amerst, MA.

Lovett, G.M., Reiners, W.R. and Olson, R.K.: 1982, Science **218**: 1303-1304.

Manion, P.D.: 1981, Tree Disease Concepts. Prentice Hall Inc., Inglewood

McIntosh, R.P. and Hurley, R.T.: 1964, Ecology **45**:314.

Miller, P.R., Taylor, O.C. and Wilhour, R.G.: 1982, Oxidant air pollution effects on a western coniferous forest ecosystem. EPA 660/D- 82 - 276.

Mook, P. V. and Eno, H.G.: 1956, Relation of heart rots to mortality of red spruce in the Green Mountain National Forest. USDA Forest Service. Northeastern Forest Experiment Station. Forest Res. Note 59. Upper Darby, PA.

New York State. 1891, Seventh Report of the Forest Commission. Albany, New York.

New York State. Forest Pest Reports, 1971-1977. Forest Insect and Disease Control Ann. Reports 1968-1976. Ann. Report of the State New York Conservation Commission 1911-1963. Albany, NY.

Pielke, R.A.: 1979, The Distributuion of Spruce in West-Central Virginia Before Lumbering. Report no. UVA-ENVSCI-MESO-1979-1. University of Virginia.

Reich, P.B., and Amundson, R.G.: 1985, Science **230**:566.

Roman, J.R. and Raynal, D.J.: 1980, Effects of acid precipitation on vegetation. In Raynal, D.J., A.L. Leaf, P.D. Manion and C.J.K. Wang (eds.). Actual and potential effects of acid precipitation on a forest ecosystem in the Adirondack Mountains. New York State ERDA Report 80-28, Albany, NY p. 4-1-4-63.

Scherbatskoy, T. and Bliss, M. : 1984, Occurrence of acidic rain and cloud water in high elevation ecosystems in the Green Mts. of Vermont. In Transactions of the APCA Specialty Conference: The Meteorology of Acidic Deposition. P. J. Samson (ed.) Hartford, CT, Oct. 16-19, 1983, p 449-463.

Scott, J.T., Siccama, T.G., Johnson, A.H. and Breisch., A.R. :1984, Bull. Torrey. Bot. Club **111**: 438.

Siccama, T.G.: 1971, American Midland Naturalist **85**: 153.

Siccama, T.G.: 1974, Vegetation, soil and climate on the Green Mountains of Vermont. Ecol. Monogr. **44**:325.

Siccama, T.G., Bliss, M. and Vogelmann, H.W.: 1982, Bull. Torrey Bot. Club **109**: 163.

Tegethoff, A.C.: 1964, Memorandum 5220, Sept. 25, 1964. White Mt. Nat. Forest. High Elevation Spruce Mortality. U.S. Forest Service Northern FPC Zone, 6 Main St. Amherst, MA. 2 p.

Weiss, M.J., McCreery, L.R., Millers, I., O'Brien, J.T. and Miller-Weeks, M.: 1985, Cooperative Survey of Red Spruce and Balsm Fir Decline and Mortality in New York, New Hampshire and Vermont - 1984. Interim Report. USDA Forest Service, Forest Pest Management, P.O. Box 640, Durham, NH. 130p.

Wheeler, G.S.: 1965, Memorandum 2400, 5100 July 1, 1965. White Mountain National Forest, Laconia, NH.

Whittaker, R.H.: 1956, Ecol. Mongr. **26**: 1.

A REVIEW OF THE CHEMICAL RECORD IN LAKE SEDIMENT OF ENERGY RELATED AIR POLLUTION AND ITS EFFECTS ON LAKES

Stephen A. Norton
Department of Geological Sciences
110 Boardman Hall
University of Maine
Orono, ME 04469 USA

ABSTRACT. Inter-lake variation in accumulation rates of energy-related elements is a function of gross sedimentation rate (sediment focusing), position on pollution gradients, and water chemistry. Accumulation rates of Pb in lake sediments from profundal area cores in the Adirondack Mountains of New York and northern New England range from 0.1 to 0.2 $ug/cm^2/y$ (pre-1800 A.D.) to as much as 2 $ug/cm^2/y$ (recent sediment). Rates increase from the late 1800's to nearly the present, in parallel. Accumulation rates for V remain at background values which range up to 0.5 $ug/cm^2/y$ and increase 5 to 10 x background in Adirondack Mountain lakes. Chronically acidic lakes have a subsurface maximum. Of these three metals, only Pb has elevated deposition rates in high elevation lakes in the Rocky Mountains of Colorado. Mining activity is believed responsible for the implied air pollution there.

I. INTRODUCTION

Air pollution of urban environments has been recognized for at least 130 years (e.g., Smith, 1852). In rural settings, industrial activities such as smelting of sulfide ores has been recognized for over 50 yrs as a factor in biological effects (e.g., Tyler, 1984). As early as the 1920's, effects on the fisheries of southern Norway, remote from point sources of pollution, were related to acidification of surface waters (Dahl, 1927), almost certainly due to regional air pollution. Since that time awareness of regional air pollution has grown slowly. Seminal works include those of Gorham (1958) in England in the 1950's, Oden (1968) and coworkers in Sweden in 1960's, and the rest of the world in the 1970's and more recently.

Husar (1986) has suggested that, for the northeastern United States, SO_x emissions started increasing about 100 yrs ago and may have remained fairly constant within 50% for the last 55 to 65 years. NO_x emissions have increased until the early 1970's as a result of a variety of industrial and cultural activities. Concurrent with these emissions is the injection into the atmosphere of many elements such as

Water, Air and Soil Pollution **30** (1986) 331–345.

Pb, Zn, Cd, Hg, and V from solid (coal) or liquid (oil) fuels; chemical compounds formed by combustion, e.g., polycyclic aromatic hydrocarbons (PAHs); and solid products such as soot, and glassy or magnetic spheres. Smelting of sulfide ores and automotive consumption of tetraethyl lead contribute significantly to the atmospheric burden of metals.

Qualitative and quantitative characterization of the emissions are poor; characterization and quantification of subsequent deposition are infrequent, not spatially well distributed, and for many substances not analyzable because of the lack of technology.

Many of these emitted substances are deposited in lake sediment but generally not at a rate equivalent to the rate of atmospheric deposition on the lake surface. This lack of correspondence is caused by a variety of processes including chemical mobilization, terrestrial processing of pollutants, sediment focusing, bypassing of sediment due to flushing, and post-depositional alteration of sediment. The variety of energy-related materials deposited in sediment vary with time and space. This is caused by a variable mix of emissions with time, the changing distribution of sources, the technology of emission controls, and the airshed for lake receptors. Some pollutants have multiple sources, e.g., Pb comes from the smelting of sulfide ores, consumption of coal and oil, and gasoline additives. Some materials come primarily from a particular source (e.g., PAHs from coal burning; V from oil).

This paper presents a review of the record of changing atmospheric deposition as preserved in lake sediment and focuses on trace metals. Many of the data are from the northeastern U.S. but similar types of data have been gathered in Canada, Great Britain, and Fenno-Scandinavia.

2. MATERIALS AND METHODS

Reconstruction of atmospheric deposition from the lake sediment record requires a lake with a drainage basin with minimal and known disturbance, an absence of *in situ* disturbance of sediment, cores with undisturbed stratigraphy (see e.g., Renberg, 1979; Davis and Doyle, 1969), and an excellent absolute chronology. A variety of chronological evidence and techniques has been and is used in this regard. Researchers attempt to use as many dating techniques as possible. Pollutants themselves may be used as chronological tools, after they have been properly calibrated. [Spherules-magnetic or glassy (Oldfield *et al.*, 1984); Soot (Goldberg *et al.*, 1981); PAHs (Gschwend and Hites, 1981); Trace metals (Norton, 1986); Isotopic ratios - Pb, S (Shirahata *et al.*, 1980); DDT, PCBs (Rapaport *et al.*, 1985); and Magnetism (Bloemendal *et al.*, 1979).] Varves provide the best and continuous dating (see, *e.g.*, Renberg, 1979). However, varved sediments are relatively rare in the lakes of North America. There, ^{210}Pb dating is more commonly used although it is not without problems (see, e.g., Imboden and Stiller, 1982; Evans and Rigler, 1983). In most cases where ancillary dating is available, ^{210}Pb dates appear to give satisfactory (internally consistent) chronology

(Appleby et al., 1979). Under optimum condition, ^{210}Pb dating can yield reliable dates for sediment up to about 150 yrs old. Continuous dating control down core is necessary for the determination of accumulation rates. If the latter are based solely on chronostratigraphic markers, assumptions of constancy of sedimentation rates must be made using interpolation. Absolute chronostratigraphic markers include ^{137}Cs (Pennington et al., 1973; Davis et al., 1984); $^{239/240}Pu$ (Jaakkola et al., 1983); Fire-related charcoal (e.g., Charles, 1984); DDT and other organics (Rapaport et al., 1985); Sedimentation events (Battarbee and Flower, 1984); and Pollen (Davis et al., 1983). Techniques of chemical analysis of sediment range from the determination of bulk chemistry (Buckley and Cranston, 1971) to metal speciation studies (Rapin et al., 1986). Many of these techniques and analysis of particulate materials are reviewed in Norton (1984).

3. RESULTS

Recent paleolimnological interpretations have recognized the importance of the multiplicity of evidence for the reconstruction of the history of atmospheric pollution. Many elements have been reported as having had an increase in atmospheric deposition rate as evidenced from cores of lake sediment, soil studies, and cores of ombrogenic peat. These include most commonly Pb, Zn, Cu, Cd, As, Sb, Hg, and V and less commonly Fe, Mn, Ni and Co. Products of combustion may be incorporated directly into sediment (e.g., PAHs) or indirectly into sediment (e.g., SO_x). Solid particles derived from combustion, including soot, charcoal, and magnetic or glassy spheres have also been reported as having stratigraphic and temporal significance. The most comprehensive study to date is the PIRLA (Paleolimnological Investigation of Recent Lake Acidification) study. In that study numerous biological, chemical, and physical properties of well-dated sediment have been analyzed from the same core (Charles et al., 1985). A wide variety of pollutants appear to be related to the period of industrialization in eastern North America. However different constituents increase at different times (e.g., Pb and V), presumably because of different source terms with different chronologies. Similarly, the deposition of some pollutants (e.g., PAHs) peaks and then decreases as a result of changes in the technology of burning and emission controls (see e.g. Charles and Norton, 1986). The concentrations of several constituents of sediment (e.g., Zn and S) do not have a straightforward relationship to atmospheric emission and deposition. For these two elements, there may be a direct linkage between atmospheric emissions and sediment accumulation. However, additional processes related to pre- or post-depositional mobilization or diagenetic deposition of Zn and S and certain other elements make it impossible to unambiguously infer atmospheric deposition rates from sediment accumulation rates (Carignan and Tessier, 1985).

The faithfulness with which a sedimentary record reflects atmospheric deposition is a complex function of gross sedimentation rates, focusing of sediment containing the constituent of interest,

TABLE I
Concentration and accumulation rate for Pb derived from sediment cores from 10 lakes in the Adirondack Mountains, NY

	Pb Concentration (ppm)				Pb Accumulation Rate ($\mu g cm^{-2} yr^{-1}$)			
	Background	Maximum	Year of Max.	Maximum / Background	Background	Maximum	Year of Max.	Maximum / Background
Brooktrout	27	558	1982	21	0.14	2.06	1968	15
Deep	25	640	1973	14	0.06	2.01	1973	34
Fourth	15	270	1976	18	0.10	2.55	1976	26
Jerseyfield	26	675	1979	26	0.12	3.42	1979	29
Merriam	20	459	1980	23	0.07	3.70	1980	53
Panther	40	498	1967	12	0.25	5.5	1981	22
Sagamore	65	335	1979	5	0.8	4.82	1983	6
(BS) Silver	26	307	1978	12	0.12	2.50	1980	21
T	37	299	1982	8	0.16	2.54	1974	16
Upper Wallface	24	759	1979	32	0.09	3.36	1979	37

pre- and post-depositional chemical and mechanical mobility of sediment (resuspension, bioturbation, etc.), and diagenetic alteration of sediment.

Certain elements are chemically relatively immobile after deposition, including Pb (Kahl and Norton, 1983; Davis et al., 1982) and V. Typically, accumulation rates for Pb in lakes start to increase above pre-industrial values in sediment dated between 1850-1900. The depositional history of Pb integrates emissions due to coal burning, sulfide ore smelting, and tetraethyl lead additives to gasoline and their recent reductions. Pre-industrial accumulation rates range widely depending on the chemistry of the watershed soils and the gross sedimentation rate at the coring site. Thus lakes of a region have variable accumulation rates for airborne pollutants (Table 1). The variability in modern (post-1850) accumulation rates for a single lake may be due to a number of causes including errors in sediment chronology, local pollution sources, and signal to noise ratio. Considerable deep water sediment is derived from resuspension of older shallower water sediment, diluting (and delaying the recognition of) any atmospheric signal such as increased atmospheric deposition of Pb. Similarly, the recent decrease in atmospheric deposition of Pb will not be directly reflected in a quantitative or proportional decrease in net sediment accumulation for Pb due to dilution by recently deposited Pb-rich sediment. Bioturbation and resuspension complicate further the stratigraphic record. Thus, a lag in the response of the sediment chemistry to changes in atmospheric deposition of pollutants may exist. Peak accumulation rates occur in near-recent sediments but differ in magnitude (Table 1); consequently, total anthropogenic Pb cm^{-2} for the last 100 yrs varies by over 200%. This conclusion is similar to that reached by Dillon and Evans (1982) who examined whole lake burdens, using multiple cores from single lakes. Lakes with short water residence times typically have lower accumulation rates for all constituents including pollutants. For example we have observed a strong correlation between low integrated ^{210}Pb and low total anthropogenic Pb, both of which are largely atmospherically derived. Similar depositional history trends are found over wide areas (Figure 1) although accumulation rates vary significantly.

The accumulation histories for V for two lakes in the Adirondack Mountains are shown in Figure 2. The onset of increased accumulation occurs in the 1920-1940 period, corresponding to increased use of oil. Peak values are reached in near-recent sediment.

The total flux of Pb, V, and other elements is, in part, controlled by the flux of these elements which is indigenous to the drainage basin. Normalization of the accumulation rate of a pollutant (e.g., Pb) to that of a chemically, immobile element derived principally from the drainage basin (e.g., TiO_2) enables one to evaluate or isolate the influence of increased atmospheric deposition (compare Figures 1a and 3a). The V/TiO_2 ratio versus time for Silver Lake shows only a slight change in slope since 1900 due to a large increase in the TiO_2 accumulation rate which accompanies the increase in the V accumulation rate. The latter is comprised of two components, V from the atmosphere and the drainage basin particulates.

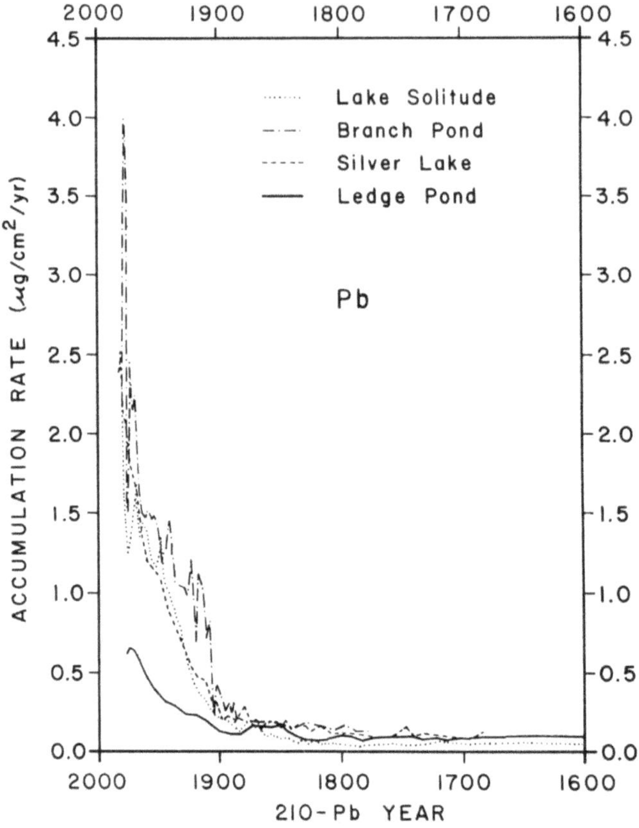

Figure 1: Accumulation rates of Pb in sediment cores from (a) Silver Lake, NY, (b) Branch Lake, VT, (c) Solitude Pond, NH, and (d) Ledge Pond, ME. Dates prior to ca. 1800 are extrapolated.

The atmospheric influx of Pb clearly increases by the late 1800's in all northern New England and New York lakes (ca. 40 lakes) and in the 1920-1940 period for V (Figures 3a and b). The V/TiO_2 ratio versus time for Silver Lake shows only a slight change in slope since 1900 due to a large increase in the TiO_2 accumulation rate, which accompanies the increase in the V accumulation rate. The latter is comprised of two components, V from the atmosphere and the drainage basin particulates.

An element with greater chemical mobility and subject to diagenetic processing such as Zn (Figure 4) shows less correspondence

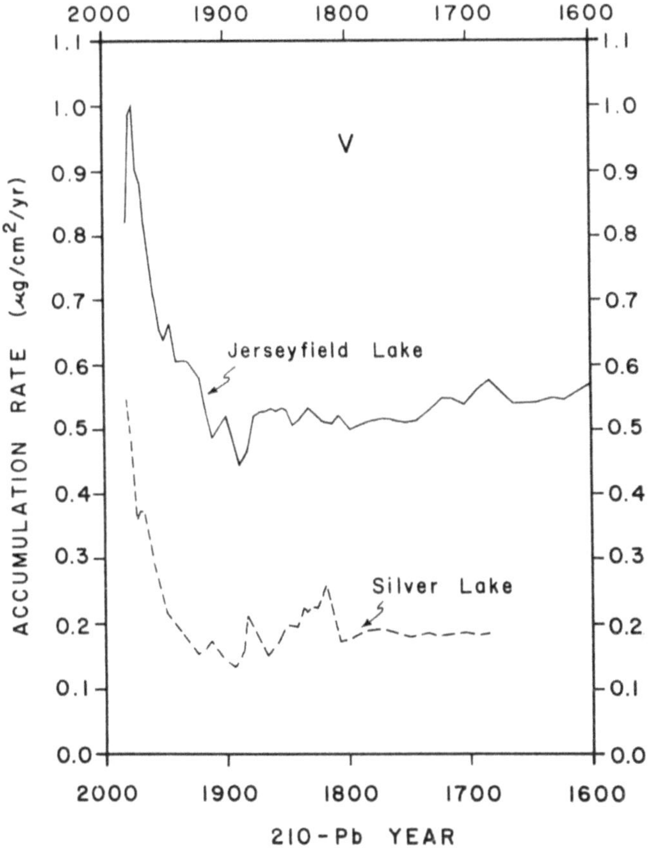

Figure 2: Accumulation rates of V in sediment from (a) Silver and (b) Jerseyfield Lakes, NY. Dates prior to ca. 1800 are extrapolated.

from lake to lake. Many lakes have an upcore decreasing accumulation rate at, or near, the top of the sediment column. Lower lake water pH is related to the decrease in the recent net accumulation rate of Zn. This is consistent with the observations that Zn is leached from sediment as overlying water becomes acidified (whole lake experiments - Schindler et al., 1980; microcosm experiments - Kahl and Norton, 1983). It is also consistent with a model of increased $SO_4^{=}$ reduction beneath the sediment-water interface, formation of interstitial H_2S and precipitation of Zn from interstitial water (Carignan and Tessier, 1985). The result is migration of Zn from the water column into

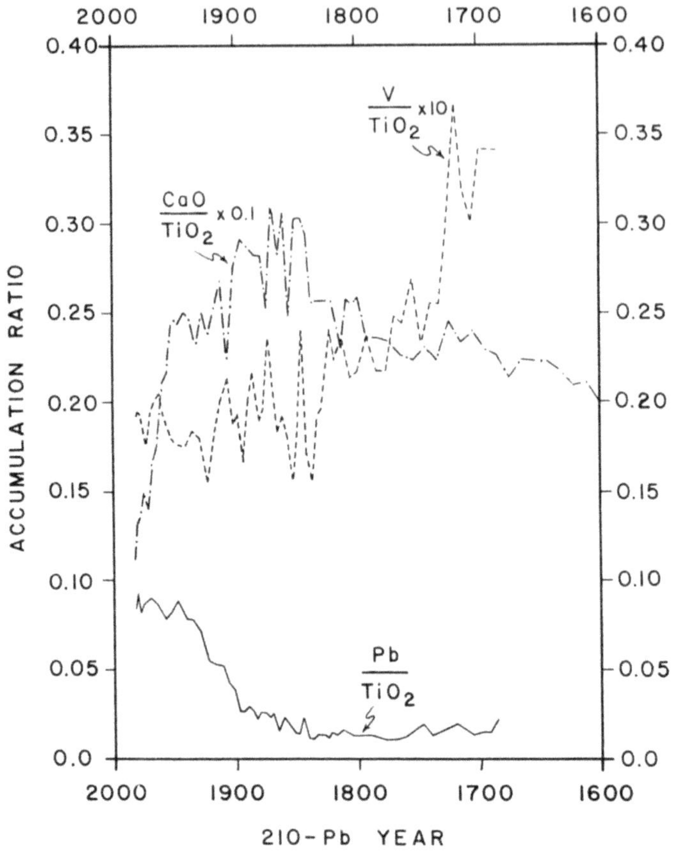

Figure 3: Accumulation rate ratios (thus, concentration ratios) of (a) Pb/TiO_2 and (b) V/TiO_2 for Silver Lake, NY, and CaO/TiO_2 for Jerseyfield Lake, NY.

already deposited sediment. Evans et al. (1983) have found that retention of Zn is positive in acidified lakes but less than for circumneutral lakes. It is likely that all these processes are important in determining concentrations and accumulation rates in lake sediment of certain chalcophile metals such as Zn, Cd, Ni, Co, and Cu.

Although the accumulation rate of S in sediments in eastern North America has been elevated over the last 100 yrs (by inference from Husar, 1986), temporal trends vary considerably from lake to lake (see, e.g., Mitchell et al., 1985). Sulfur may be delivered to the sediment as particulate organic S, adsorbed $^=$on particulate material, or be

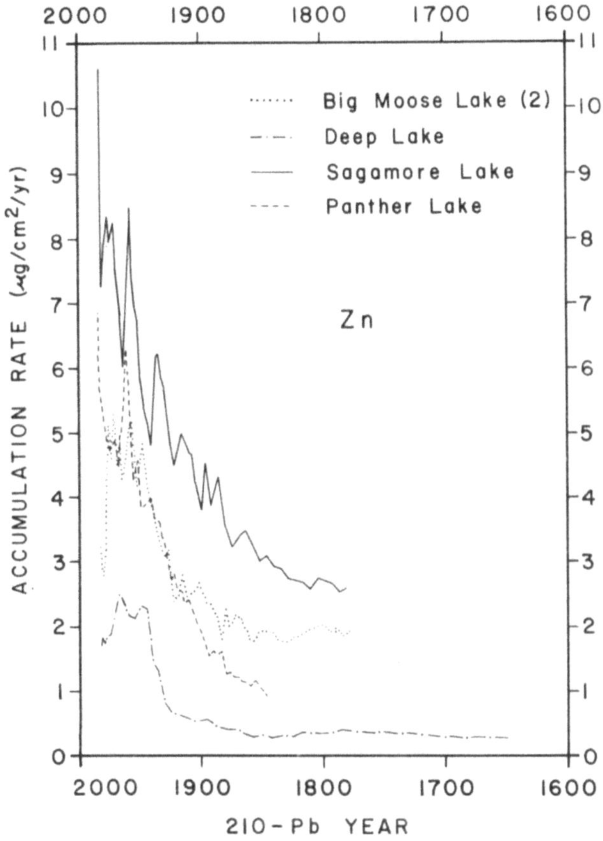

Figure 4: Accumulation rates of Zn in sediment from (a) Sagamore, (b) Big Moose, (c) and Deep Lakes, and (d) Panther Pond, NY. Dates prior to ca. 1800 are extrapolated.

incorporated into sediment via $SO_4^=$ reduction and precipitation of native S or metal sulfides. Transport of atmospheric S from the terrestrial system to the lake may be decreased by bioaccumulation in aggrading terrestrial systems, or because of $SO_4^=$ adsorption by soils (Johnson and Cole, 1980). Consequently, S accumulation rates will not, in most cases, parallel changes in atmospheric deposition.

Lake acidification is most precisely documented in paleolimnological analysis by changes in diatom (Charles and Norton, 1986) and chrysophyte (Smol et al., 1984) assemblages. The evidence from diatoms is slightly confounded by imperfectly understood

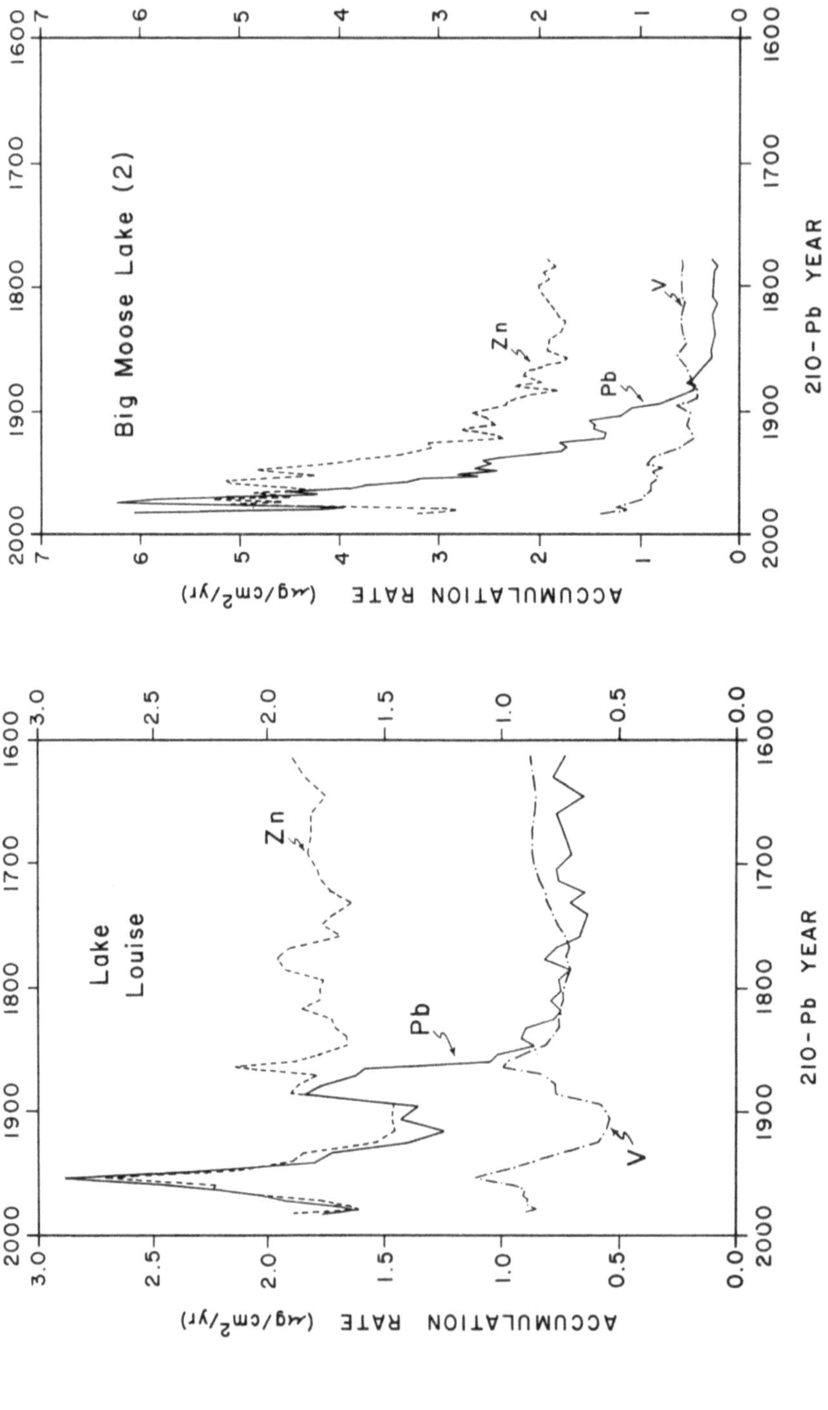

Figure 5: Accumulation rates of Pb, Zn, and V in sediment from (a) Lake Louise in Rocky Mountain National Park, CO and from (b) Big Moose Lake, NY. Dates prior to ca. 1800 are extrapolated.

relationships between water chemistry parameters, other than pH, and diatom taxa (Davis and Anderson, 1985; Davis et al., 1985). Resuspension of older sediment followed by deposition with modern sediment confuses the calibration of diatoms and Chrysophyceae, as well as the chemical record. The autecology of Chrysophyceae is poorly understood; consequently pH reconstructions are not yet possible, only pH trends. This biological evidence for acidification commonly occurs in younger sediment than does chemical evidence of water acidification, possibly because benthic diatoms which dominate in acidic waters live in a slightly better buffered and more nutrient rich environment than planktonic diatoms. Declines in concentrations and accumulation rates of Mn, Ca, and Zn in near surface sediments in presently acidic lakes have been interpreted as due, at least in part, to water column acidification (see, e.g., Kahl and Norton, 1983). The latter would cause leaching of these elements from already-deposited sediment or reduced adsorption on sedimenting particles. Thus chemical effects of acidification could affect sediments which pre-date the acidification of the water column. This selective depletion can best be seen in the CaO/TiO_2 accumulation rate (or concentration ratio) for sediment (Figure 3c). However, the accumulation of CaO in the sediment is also linked to changes in the accumulation rate of organic matter (Engstrom and Wright, 1984); as much as 25% of the ash content of organic matter is CaO. Davis et al. (1983) have noted that chemical effects of acidification on diatoms appear to lag behind chemical effects of metal pollution and acidification in time equivalent sediment for some lakes. This may be due in part to the in situ leaching of older sediment or diagenesis of S-bearing phases. The MnO/TiO_2 accumulation rate ratio or the Zn/TiO_2 ratio commonly reinforces the interpretation of acidification based on CaO/TiO_2 ratios. However, diagenetic changes unrelated to acidification can affect both MnO and Zn concentrations and accumulation rates in the sediment.

Interpretation of the chemical record preserved in sediment can be best accomplished by analysis of many components; each single component, interpreted by itself, typically yields multiple working hypotheses. For example, Figure 5 displays accumulation rates for Pb, Zn, and V for Lake Louise in Colorado and Big Moose Lake in New York. Comparisons of the Pb accumulation rates for Colorado and New York suggest a comparable history of atmospheric pollution. However the lack of accelerated deposition of Zn or V for the Rocky Mountain area, elements associated with the burning of coal and oil, respectively, indicates a different history of pollution, notably different sources of pollution for the two areas. The erratic behavior of V over the last 200 yrs is matched by that of TiO_2. It appears that Pb increases relative to TiO_2 but Zn and V do not.

4. SUMMARY

Lake sediments in the northeastern United States record an increase in the accumulation rate of Pb which starts in the period 1850-1900, with most lakes falling in the period 1860-1880. Peak values are typically reached in the 1970's. The values are more a function of local lake morphometry and gross sedimentation rate than of position along a pollution gradient.

A similar early initiation of increased atmospheric deposition of Zn is indicated but the accumulation rate is only on the order of 20 to 30% of the atmospheric deposition rate. In acidic lakes (pH $\leq$ 5.5) accumulation rates for Zn decline relative to Pb in recent sediment. This may be related to acidification and sediment diagenesis. V accumulation rates increase from 1920-1940 to nearly the present, generally paralleling oil consumption.

Acidification of lake waters is indicated by recent (last 100 yrs) loss of Ca, Mn, and Zn from sediment relative to TiO_2. These chemical trends pre-date biological fossil evidence for water column acidification but post depositional leaching of sediments may be a cause of this. High elevation lakes in Rocky Mountain National Park have received increased loading of Pb from the atmosphere from the mining of sulfide ore in the region, not from fossil fuel emissions. Multiple lines of evidence need to be assessed to properly interpret the history of atmospheric pollution and its effects on lake water and sediment chemistry.

5. ACKNOWLEDGMENTS

This research has been supported by grants from the U.S. National Science Foundation, Department of the Interior, Environmental Protection Agency, and Department of Energy; The National Research Council; The Electric Power Research Institute; and the Norwegian Institute for Water Research. Data collection for Branch Lake (VT), Solitude Pond (NH), and Ledge Pond (ME) was funded by joint research grants to R.B. Davis and myself. Ronald B. Davis, Terry A. Haines, Arne Henriksen, Richard F. Wright, David F. Brakke, and Jeffrey S. Kahl have provided continuous counsel during my studies of lake sediment. Nearly all of the data appearing under my name has been generated and quality assurred under the close scrutiny of Marilyn Morrison and Geneva Blake.

References

Appleby, P.G. and Oldfield, F., 1978, Catena, 5, 1.

Appleby, P.G., Oldfield, F., Thompson, R., Huttunen, P., and Tolonen, K., 1979, Nature, 280, 53.

Battarbee, R.W. and Flower, R.J., 1984, Limnol. Oceanogr., 29, 1325.

Bloemendal, J., Oldfield, F., and Thompson, R., 1979, Nature, 280, 50.

Buckley, D.E. and Cranston, R.E., 1971, Chem. Geol., 7, 273.

Carignan, R. and Tessier, A., 1985, Science, 228, 1524.

Charles, D.F., 1984, Verein. Limnol., 22, 559.

Charles, D.F. and Norton, S.A., 1986, Direct comparison of diatom and chemical data: Paleolimnological evidence for trends in atmospheric deposition of acids and metals in Monitoring and Assessment of Trends in Acid Deposition: U.S. National Academy of Sciences. 402.

Charles, D.F., Whitehead, D.R., Engstrom, D.R., Fry, B.D., Hites, R.A., Norton, S.A., Owen, J., Roll, L.A., Schindler, S., Smol, J.P.,Uutala, A.J., White, J.R., and Wise, R.J., 1986 (in press), Paleolimnological evidence for recent acidification of Big Moose Lake, Adirondack Mountains, N.Y. (U.S.A.): Biogeochemistry.

Dahl, K., 1927, Salmon and Trout Magazine, 46, 35.

Davis, A., Galloway, J.N., and Nordstrom, D.K., 1982, Limnol. Oceanogr., 27, 163.

Davis, R.B. and Doyle, R.W., 1969, Limnol. Oceanogr., 14, 643.

Davis, R.B., Norton, S.A., Hess, C.T., and Brakke, D.F., 1983, Hydrobiologia, 103, 113.

Davis, R.B., Hess, C.T., Norton, S.A. Hanson, D.W, Hoagland, K.D., and Anderson, D.S., 1984, Chemical Geology, 44, 151.

Davis, R.B. and Anderson, D., 1985, Hydrobiologia, 120, 69.

Davis, R.B., Anderson, D., and Berge, F., 1985, Nature, 316, 436.

Dillon, P.J. and Evans, R.D., 1982, Hydrobiologia, 91, 121.

Engstrom, D.R. and Wright, H.E., Jr., 1984, Chemical stratigraphy of lake sediments as a record of environmental change in Haworth, E.Y. and Lund, J.W.G., eds., Lake Sediments and Environmental History. Studies in Paleolimnology and Paleoecology. Leicester University Press, 11.

Evans, H.E., Smith, P.J., and Dillon, P.J., 1983, Canad. Jour. Fish. Aquat. Sci., 40, 570.

Evans, R.D. and Rigler, F.H., 1983, Canad. Jour. Fish. Aquat. Sci., 40, 506.

Goldberg, E.D., Hodge, V.F., Griffin, J.J., Koide, M., and Edginton, D.N., 1981, Envir. Sc. Tech., 15, 466.

Gorham, E., 1958, Phil. Trans. R. Soc. Lon., 47, 147.

Gschwend, P.M. and Hites, R.A., 1981, Geoch. Cosmoch. Acta. 45, 2359.

Husar, R.B., 1986, Manmade SO_x and NO_x emissions and trends for eastern North America in Monitoring and Assessment of Trends in Acid Deposition: U.S. National Academy of Sciences, 48.

Imboden, D.M. and Stiller, M., 1982, Jour. Geophys. Res., 87, 557.

Jaakkola, T., Tolonen, K., Huttunen, P., and Leskinen, S., 1983, Hydrobiologia, 103, 15.

Johnson, D. and Cole, D.W., 1980, Envir. Intern., 3, 79.

Kahl, J.S. and Norton, S.A., 1983, Metal input and mobilization in two acid-stressed lake watersheds in Maine: Land and Water Resources Center, Univ. ME, U.S.O.W.R.T. Compl. Rept. A-053, 70 p.

Mitchell, M.J., David, M.B., and Uutala, A.J., 1985, Hydrobiologia, 121, 121.

Norton, S.A., 1986, The chemical stratigraphy of lake sediments in Monitoring and Assessment of Trends in Acid Deposition: U.S. National Academy of Sciences, 369.

Norton, S.A. (ed), 1984, Proceedings of a Workshop on Paleolimnological Studies of the History and Effects of Acidic Precipitation: U.S.E.P.A. Project Rept. CR-811631-01-0, 452 p.

Oden, S., 1968, The acidification of air and precipitation and its consequences in the natural environment: Ecology Comm. Bull. 1, National Science Res. Council, Stockholm. Arlington, VA. Translation Consultants Ltd. Tr-1172.

Oldfield, F., Bloemendal, J., Barker, L., Hunt, A., Jones, J.M., Jones, M.D.H., Maxted, R., Richardson, N., Sahota, J., and Tolonen, K., 1984, Magnetic measurements of atmospheric particulates and ombrotrophic peat: A review in Norton, S.A. (ed.), Proceedings of a Workshop on Paleolimnological Studies of the History and Effects of Acidic Precipitation, U.S.E.P.A. Project Rept. CR-811631-01-0, 452 p.

Pennington, W., Cambray, R.S., and Fisher, E.M., 1973, Nature, 242, 324.

Rapaport, R.A., Capel, P.D., Eisenreich, S.J., and Robbins, 1985, Current trends and biogeochemistry of DDT in the Great Lakes region (ext. abs.): Am. Chem. Soc., Envir. Div., Chicago, 25, #2, 134.

Rapin, F., Tessier, A., Campbell, P.G.C. and Carignan, R., 1986, Potential artifacts in the determination of metal partitioning in sediments by a sequential extraction procedure: Env. Sci. and Tech., in review.

Renberg, I., 1979, Environmental monitoring by chemical, physical and biological analyses of annually-laminated lake sediments - the possibilities of the method in the use of ecological variables in environmental monitoring: The National Swedish Environment Protection Board, Report PM 1151, 318.

Schindler, D.W., Hesslein, R.H., and Wagermann, R., 1980, Canad. Jour. Fish. Aquat. Sci., 37, 373.

Shirahata, H., Elias, R.W., Patterson, C.C., and Koide, M., 1980, Geochim. Cosmoch. Acta, 44, 149.

Smith, R.A., 1852, Air and Rain: The beginnings of a chemical climatology: Longmans, Green, London.

Smol, J.P., Charles, D.F., and Whitehead, D.R., 1984, Canad. Jour. Bot., 62, 911.

Tyler, G., 1984, Ambio, 13, 18.

DIATOM-BASED pH RECONSTRUCTION STUDIES OF ACID LAKES IN EUROPE AND NORTH AMERICA: A SYNTHESIS

R. W. BATTARBEE
Palaeoecology Research Unit
University College London
26 Bedford Way
London WC1H OAP, U.K.

and

D. F. CHARLES
Department of Biology
Indiana University
Bloomington
In.47405, U.S.A.

ABSTRACT. The analysis of diatom assemblages in lake sediments is being increasingly used to reconstruct past trends in the pH of acid lakes. This paper briefly reviews the data that are now available for Europe and North America and concludes that the evidence is consistent with the acid deposition hypothesis for recent lake acidification.

1. INTRODUCTION

There are few acid lakes for which reliable long term pH records are available. For this reason attention has focussed on the use of fossil diatom assemblages in lake sediments as indicators of past pH trends and as a means of reconstructing past pH values. The essential features of the approach were developed prior to the recent interest in "acid rain" (cf. Hustedt, 1937-39; Nygaard, 1956; Meriläinen, 1967) but in the last few years further advances have been made (Renberg and Hellberg, 1982; Davis and Anderson, 1985; Charles, 1985, Smol et al., 1986). In most cases an attempt is made to establish a quantitative relationship between contemporary diatom assemblages and pH for a series of lakes along a pH gradient and to use this relationship to predict the pH represented by a fossil assemblage. There are many problems and assumptions in using this technique that have been pointed out by Battarbee (1984a, b; Charles and Norton (1986) and Davis and Smol (1986) but in most cases, where sufficient taxonomic care has been taken, good results can be achieved. Many pH reconstruction studies have been carried out and in this paper we present a brief synthesis of the results now available, either published or in press, from Europe and North America.

Water, Air and Soil Pollution **30** (1986) 347–354.

2. EUROPE: NORWAY, SWEDEN AND FINLAND

The first attempts to trace the history of recent lake acidification using diatom analysis were carried out in Sweden by Miller (Miller, 1973; Almer *et al.*, 1974), and in Norway by Berge (1975). More recent work in Northern Europe includes Berge (1979, 1985), Davis and Berge (1980), Renberg and Hellberg (1982), Davis *et al.* (1983), Tolonen and Jaakkola (1983), Renberg and Wahlin (1985), Tolonen *et al.* (1985), and Simola *et al.* (1986). In most cases where the diatom data indicate that recent acidification has taken place the diatom changes involve a decline in plantonic taxa, especially *Cyclotella kützingiana*, a decline in circumneutral non-planktonic taxa such as *Anomoeoneis vitrea* and *Achnanthes minutissima* and an increase in the acidobiontic *Tabellaria* species, *T. binalis* and *T. quadriseptata* (cf. Flower and Battarbee, 1985).

Although sediment cores from some of the study lakes have not been reliably dated it is apparent that acidification in Norway began for different lakes at different times between 1890 and 1930, whereas in Sweden and Finland the acidification appears to be confined to the post-1945 period. For example in Norway pH decreases of between 0.6 and 0.8 units are inferred for Nedre Målmesvatn, Hovvatn, and Holmvatn beginning in about 1890, 1918, and 1927, respectively (Davis *et al.*, 1983). For Gårdsjön in Sweden Renberg and Wahlin (1985) argue for a pH decline of approximately 1.3 to 1.5 units since 1950 whilst for Hauklampi in southern Finland Tolonen and Jaakkola (1983) describe a decrease of about 1 unit since 1960.

The full significance of these differences is not yet clear, but the dates are consistent with those expected on the basis of S emission history within Europe and support the view that the eastern parts of Scandinavia and Finland receive most of their acid deposition from Central and Eastern Europe, whereas Norway is more affected by emissions from Western Europe and the United Kingdom.

In all three countries a continuing program of diatom-based pH reconstruction work is being carried out. In Norway and Sweden much of it is organized within the framework of the Royal Society Surface Waters Acidification Project (SWAP) (cf. Battarbee and Renberg, 1985), and is designed to assess the causes of acidification rather than to look for evidence of acidification *per se*.

3. EUROPE: U.K., THE NETHERLANDS, WEST GERMANY AND AUSTRIA

In Western and Central Europe diatom analysis of sediment cores from acid lakes has been carried out in the U.K. (Flower and Battarbee, 1983; Battarbee, 1984a; Battarbee and Flower, 1985; Battarbee *et al.*, 1985; Haworth, 1985a, b); The Netherlands (Dickman *et al.*, 1986), West Germany and Austria (Arzet *et al.*, 1985, 1986).

Published work from the U.K. stems from sites in the Galloway region of S.W. Scotland. Although the species changes are broadly similar to those described above for Northern Europe the timing is different, with the first changes taking place in the middle of the nineteenth century, somewhat earlier than Norway and considerably earlier than Sweden and Finland. This is a difference that might be expected in light of the longer history of fossil fuel combustion of the U.K. compared with Northern Europe.

Work in progress is extending this approach to sites in acid sensitive regions in the rest of Scotland, Wales, and England in order to test alternative hypotheses to the acid deposition hypothesis, especially to evaluate the effects of recent afforestation, and to assess the regional extent of the acidification problem. Preliminary data show that acidified lakes occur in West Wales (Fritz, 1985), Cumbria (Haworth, 1985b), and the Grampian region of Scotland (Flower, 1986), as well as in Galloway.

In West Germany Arzet has analyzed sediment cores from a number of lakes including Grosser and Kleiner Arbersee in the Bavarian Forest, Herrenwiesersee, a lake in the Black Forest, and from Pinnsee, in the Lauenburg Lake Area (Steinberg et al., 1984; Arzet et al., 1986). All sites have modern values <pH 5.0 and the diatom data indicate that all have been acidified, although Herrenwiesersee appears to have had a pH below 5.0 for a very long time. The precise timing of acidification is not clear since the chronology is based on ^{137}Cs dating, a method not always reliable in acidic waters (cf. Davis et al., 1984a, b). However, from ^{137}Cs dates and from other circumstantial evidence Arzet et al. suggest that this acidification has taken place within the last 30 to 40 yr and that the prime cuase has been an increase in acid deposition.

Arzet et al. (1986) have also studied four small lakes in the Austrian Alps that are located on acid bedrock at more than 2000 m above sea level and where the pH of precipitation is approximately 4.9 to 5.0. Here the diatom data show no trends to lower water pH although there are increases in heavy metal concentrations towards the sediment surface (Steinberg, 1985).

In other parts of Europe without deep water lakes, van Dam has been using early twentieth century diatom samples and diatoms associated with old macrophyte samples to reconstruct water pH. Although this technique does not provide a continuous sequence of data through time and is restricted by the availability of samples it can provide excellent information on water quality in situations where sedimentary records are absent as in the case of streams, rivers and small pools (van Dam et al., 1981). In this way van Dam has established that many surface waters have become more acid over the last 50 yr or so in The Netherlands, Belgium, Denmark, and parts of West Germany. Most recently Dickman et al. (1986) have shown that shallow moorland pools in The Netherlands do have accumulating sediment and that the sediment record substantiates the acidification inferences previously based on diatoms in herbarium collections.

4. NORTH AMERICA

Several palaeoecological studies of lake acidification are now completed or in progress in the United States and Canada. The PIRLA (Palaeoecological Investigation of Recent Lake Acidification) project (Charles and Whitehead, 1986; Charles et al., 1986) is the largest.

Results of studies performed by several investigators on 31 North American lakes in the Adirondacks (NY), New England, Eastern Canada and the Rocky Mountains have been recently summarized (Charles and Norton, 1986). The data indicate that weakly buffered lakes in some regions have become more acidic in recent decades (Del Prete and Galloway, 1983; Charles, 1984; Davis et al., 1984a). Of 11 Adirondack lakes studied six of the seven lakes with a current pH 5.0 or less show clear evidence of recent pH decreases (Big Moose L., Deep L., Honnedaga L., L. Arnold, Upper Wallface P., Woods L. and Little Echo Pond). No substantial pH decrease is indicated for any of the four lakes with a current pH >5.2. On the basis of ^{210}Pb dating, most rapid pH decreases (0.5 to 1.0 pH units) occurred between 1930 and 1970. In all cases the primary cause of recent acidification appears to be atmospheric deposition of strong acids, although watershed disturbance may play a role at some sites.

Diatom stratigraphies and reconstructions have been completed for 13 New England lakes (Davis et al., 1983; Norton, et al.; Whiting et al., 1985; Ford, 1986; Davis, 1985). At four sites small decreases in pH were observed but these were only about the magnitude of the standard error of the mean for the regression (c. 0.2 to 0.3 pH units). Two of these lakes are presently pH 4.5 to 4.8 and two 5.5 to 6.0. Three other lakes with current pH 4.5 to 4.8 have decreased in pH by 0.5 to 0.7 units. For all seven lakes the decreases started between 1920 and 1950. The lakes with the earliest and largest pH decreases are situated in areas of greatest acid loading, suggesting deposition of strong acids as the major cause of acidification. In two additional lakes with current pH values of 5.0 to 5.5 and with the same inferred pH in the 1700's as during recent decades pH was 0.3 to 0.5 units higher in the 1800's when heavy lumbering of the catchments was taking place. The four remaining lakes with present pH values of 4.5 to 5.0 show no change in diatom inferred pH since the 1770's.

Recent studies of three low pH (4.4 to 5.6), low alkalinity (-38 to 9 $\mu eq.^{-1}$) lakes in Michigan (L. Andrus and L. McNearney) and Wisconsin (L. Otto Mielke) indicate no recent pH decrease (Kingston et al., 1985; Charles et al., 1986) and similar conclusions have been reached for Rocky Mountain lakes that are dilute (conductivities 10 to 25 $\mu S\ cm^{-1}$ and poorly buffered (alk. 28 to 129 $\mu eq.^{-1}$) (Baron, 1983; Beeson, 1984).

In Canada data are available for only a few lakes and mostly for sites with current pH values above pH 5.5 to 6.0. However, recent diatom inferred pH decreases have been determined for lakes B and CS in the Wawa region (Dickman et al., 1983) and are perhaps related to the influence of local emission sources.

A number of other studies are in progress in several regions. In Eastern Canada surface sediment samples and sediment cores from a range of low alkalinity lakes are under investigation (Duthie et al., 1985; Nicholls, 1985; Dixit and Evans, 1985; Smol, 1985; Davidson, 1985). In the United States similar work is being carried out in the Pocono Mts. PA (Sherman, 1985), and in the Sierra Nevada, CA (Holmes et al., 1985).

5. CONCLUSIONS

The use of diatoms in reconstructing pH has evolved rapidly during the last 10 yr and has reached a point where the technique is well tested. It offers the only means whereby the pH history of a lake can be reliably reconstructed, both in terms of trends and in terms of past pH values. As such it is the best way of assessing the regional distribution of acidified lakes within countries and within continents. In the next 5 yr results from studies already in progress should reveal a definitive distribution pattern of acidified lakes both in space and in time. Already expected patterns are emerging. Lakes in sensitive bedrock areas within the regions of high acid deposition have been acidified whereas similar sites outside these regions have not, and sensitive sites located close to regions of early industrial development began to be acidified before sites in countries where industrialisation has occurred more recently. All the data so far available show a time and space pattern that is consistent with the acid deposition hypothesis.

ACKNOWLEDGMENTS

This review has been put together with the generous help of many colleagues who have allowed us to use their unpublished data. In particular we should like to thank Frode Berge, Ingemar Renberg, Heikki Simola, Pertti Huttunen, Jouko Meriläinen, Kimmo Tolonen, Herman van Dam, Klaus Arzet, Elizabeth Haworth, Hamish Duthie, Mark Taylor, Sushil Dixit, Ken Nicholls, John Sherman, John Kingston, Keith Camburn, Russell Kreis, Roger Sweets, Ronald Davis, Dennis Anderson, Mark Whiting, Robert Holmes, Geoffery Davidson, David Beeson, and Michael Dickman. Ron Davis and Roger Flower made valuable comments on the manuscript. RWB would also like to thank Pam Stokes and the organizing committee of the Muskoka Conference for financial assistance.

REFERENCES

Almer, B., Dickson, W., Ekstrom, C. Hornstrom, E., and Miller U.: 1974, Ambio 3, 30.

Arzet, K., Krause-Dellin, D., and Steinberg, C.: 1986, 'Acidification of four lakes in the Federal Republic of Germany as reflected by diatom assemblages, cladoceran remains and sediment chemistry', in Smol, J.P., Battarbee, R.W., Davis, R.B., and Meriläinen, J. (Eds.) Diatoms and Lake Acidity, W. Junk, The Netherlands, p. 227.

Arzet, K., Steinberg, C., Psenner, R., and Schulz, N.: 1985, 'Diatom distribution and diatom inferred pH in the sediment of four high altitude, alpine lakes', Abstract for the IVth International Symposium on Paleolimnology, Ossiach, Austria.

Baron, J.: 1983, Water Resources Bulletin, 19, p. 897.

Battarbee, R.W.: 1984a, Phil. Trans. R. Soc. 305, p. 451.

Battarbee, R.W.: 1984b, 'Diatom analysis', in Norton, S.A. (Ed.) Proceedings of a Workshop in Palaeolimnological Studies of the History and Effects of Acidic Precipitation, Rockland, Maine, p. 275.
Battarbee, R.W., and Flower, R.J.: 1985, 'Palaeoecological evidence for the timing and causes of lake acidification in Galloway, Southwest Scotland', Working Paper No. 8, Palaeoecological Research Unit, University College London, 79 pp.
Battarbee, R.W., and Renberg, I.: 1985, 'Royal Society Surface Water Acidification Project (SWAP) Palaeolimnology Programme', Working Paper No. 12, Palaeoecology Research Unit, University College London, 18 pp.
Battarbee, R.W., Flower, R.J., Stevenson, A.C., and Rippey, B.: 1985, Nature, 314, p. 350.
Beeson, D.R.: 1984, Bull. Ecol. Soc. Am., 65, p. 135.
Berge, F.: 1975, 'pH-forandringer og sedimentasjon av diatomeer i Langtjern', Norges Teknisk-Naturvitenskapelige Forskningsråd, Intern. rapport 11, p. 1.
Berge, F.: 1979, 'Kiselalger og pH i noen insjoer i Agder og Hordaland', SNSF Prosjektet IR 42/79, 64 pp.
Berge, F.: 1985, 'Relationships of diatom taxa to pH and other environmental factors in Norwegian soft-water lakes', Ph.D. Dissertation, University of Maine, Orone, Maine, USA.
Charles, D.F., 1984, Verh. Int. Verein. Limnol. 22, p. 559.
Charles, D.F.: 1986, 'The PIRLA Project (Palaeoecological Investigation of Recent Lake Acidification): Preliminary results for the Adirondacks, New England, N. Great Lakes States, and N. Florida', Water, Air and Soil Pollution (this volume).
Charles, D.F., and Norton, S.A.: 1986, 'Paleolimnological evidence for trends in atmospheric deposition of acids and metals', in Acid Deposition: Long Term Trends, National Academy Press, Washington, D.C., USA, p. 335.
Charles, D.F. and Whitehead, D.R.: 1986: 'The PIRLA Project: Palaeoecological Investigation of Recent Lake Acidification', Abstract for the IVth International Symposium on Palaeolimnology, Ossiach, Austria.
Davidson, G.: 1985, personal communication.
Davis, R.B.: 1985, personal communication.
Davis, R.B. and Anderson, D.S.: 1985, Hydrobiologia 120, p. 69.
Davis, R.B., Anderson, D.S., Charles, D.F., and Galloway, J.N.: 1984a, 'Two hundred year pH history of Woods, Sagamore, and Panther lakes in the Adirondack Mountains, New York State, USA', Completion Report, Electrical Power Research Institute, Palo Alto. California, USA.
Davis, R.B., Hess, C.T., Norton, S.A., Hanson, D.W., Hoagland, K.D., and Anderson, D.S.: 1984b, Chem. Geol., 44, p. 151.
Davis, R.B., and Berge, F.: 1980, 'Atmospheric deposition in Norway during the last 300 years as recorded in SNSF lake sediments II. Diatom stratigraphy and inferred pH', in Drablos, D. and Tollan, A. (Eds.) Ecological impact of acid precipitation, Proceedings of an International Conference, Sandefjord, Norway. SNSF Project, Oslo, Norway, p. 270.
Davis, R.B., Norton, S.A., Hess, C.T., and Brakke, D.F.: 1983, Hydrobiologia, 103, p. 113.
Davis, R.B., and Smol, J.P.: 1986, 'The use of sedimentary remains of siliceous algae for inferring past chemistry of lake water - problems,

potential and research needs', in Smol, J.P., Battarbee, R.W., Davis, R.B., and Meriläinen, J. (Eds.) Diatoms and Lake Acidity, W. Junk and Sons, The Hague, The Netherlands, p. 291.

Del Prete, A., and Galloway, J.N.: 1983, 'Temporal trends in the pH of Woods, Sagamore, and Panther lakes as determined by an analysis of diatom populations', in The integrated Lake-Watershed Acidification Study: Proceedings of the ILWAS Annual Review Conference, Electrical Power Research Institute, Palo Alto, California, USA.

Dickman, M., Dixit, S., Fortescue, J., Barlow, R., Dixit, S., and Evans, R.: 1985, personal communication.

Terasmae, J.: 1983, Water Air and Soil Pollution, 21, p. 227.

Dickman, M.D., van Dam, H., van Geel, B., Klink, A.G., and van der Wijk, A.: 1986, Arch. Hydrobiol. (in press).

Duthie, H.: 1985, personal communication.

Flower, R.J., and Battarbee, R.W.: 1983, Nature 305, p. 130.

Flower, R.J., and Battarbee, R.W.: 1985, Br. Phycol. J. 20, p. 69.

Flower, R.J.: 1986, unpublished data.

Ford, M.S.: 1986, 'The recent history of a naturally acidic lake (Cone Pond, N.H.)', in Smol, J.P., Battarbee, R.W., Davis, R.B., and Meriläinen, J. (Eds.): Diatoms and Lake Acidity, W. Junk and Sons, The Hague, The Netherlands, p. 131.

Fritz, S.: 1985, unpublished data.

Haworth, E.Y.: 1985a, 'The highly nervous system of the English lakes: aquatic ecosystem sensitivity to external changes, as demonstrated by diatoms', FBA Annual Report, 53, p. 60.

Haworth, E.Y.: 1985b, personal communication.

Holmes, R.: 1985, personal communication.

Hustedt, F.: 1937-39, Arch Hydrobiol. (suppl.) 15 & 16.

Kingston, J.C., Kreis, R., and Camburn, K.: 1985, personal communication.

Meriläinen, J.: 1967, Ann. Bot. Fenn. 4, p. 51.

Miller, U.: 1973, Statens Naturvardsverk Publikationer, 7, p. 43.

Nichols, K.: 1985, personal communication.

Norton, S.A., Davis, R.B., and Anderson, D.S.: 1985, 'The distribution and extent of acid and metal precipitation in northern New England', Completion Report, U.S.F. and W.S., Department of Interior, Project 14-16-0009-79-040, Washington D.C., USA, 136 pp.

Nygaard, G.: 1956, Fol. Limnol. Scand. 8, p. 32.

Renberg, I., and Hellberg, I.: 1982, Ambio 11, p. 30.

Renberg, I., and Wahlin, J-E.: 1985, Ecol. Bull. 37, p. 219.

Sherman, J.: 1985, personal communication.

Simola, H., Kenttamies, K., and Sandman, O.: 1986, Aqua Fennica, in press.

Smol, J.P., Battarbee, R.W., Davis, R.B., and Meriläinen, J. (Eds.): 1986, 'Diatoms and Lake Acidity: reconstructing pH from siliceous algal remains in lake sediments', W. Junk, The Hague, The Netherlands, 307 p.

Smol, J.P.: 1985, personal communication.

Steinberg, C.: 1985, personal communication.

Steinberg, C., Arzet, K., and Krause-Dellin, D.: 1984, Naturwissenshaften 71, p. 631.

Tolonen, K., and Jaakkola, T.: 1983, Ann. Bot. Fenn., 20, p. 57.
Tolonen, K., Liukkonen, M., Harjula, R., and Patila, A.: 1985, 'Acidification of small lakes in Finland documented by sedimentary diatom and chrysophycean remains' in Smol, J.P., Battarbee, R.W., Davis, R.B., and Merilainen, J. (Eds.), Diatoms and lake acidity, W. Junk and Sons, The Hague, The Netherlands, p. 169.
Van Dam, H., Survmond, G., and ter Braak, C.J.F.: 1981, Hydrobiologia 83, p. 425.
Whiting, M.C., Anderson, D.S., and Davis, R.B.: 1985, 'Further paleolimnological evidence for possible lake acidification in northern New England', Abstract for 48th Annual Meeting of the American Society of Limnology and Oceanography, Minneapolis, MN, USA, p. 24.

THE PIRLA PROJECT (PALEOECOLOGICAL INVESTIGATION OF RECENT LAKE ACIDIFICATION): PRELIMINARY RESULTS FOR THE ADIRONDACKS, NEW ENGLAND, N. GREAT LAKES STATES, AND N. FLORIDA

Donald F. Charles,[1] Donald R. Whitehead,[1] Dennis S. Anderson,[2] Raymond Bienert,[3] Keith E. Camburn,[4] Robert B. Cook,[5] Thomas L. Crisman,[3] Ronald B. Davis,[2] Jesse Ford[6], Brian D. Fry,[1] Ronald A. Hites,[7] Jeffrey S. Kahl,[8] John C. Kingston,[4] Russell G. Kreis, Jr.,[4] Myron J. Mitchell,[9] Stephen A. Norton,[8] Linda A. Roll,[7] John P. Smol,[10] P. Roger Sweets,[4] Allen J. Uutala,[9] Jeffrey R. White,[7] Mark C. Whiting,[11] Robert J. Wise.[1]

1. Department of Biology, Indiana University, Bloomington, IN 47405 USA.
2. Department of Botany and Plant Pathology, University of Maine at Orono, Orono, ME 04469 USA.
3. Department of Environmental Engineering, University of Florida, Gainesville, FL 32611 USA.
4. Department of Geology, University of Minnesota, Duluth, MN 55812 USA.
5. Department of Geoscience, Indiana University of Pennsylvania, Indiana, PA 15705 USA.
6. Ecosystems Research Center, Cornell University, Ithaca, NY 14853 USA.
7. School of Public and Environmental Affairs and Department of Chemistry, Indiana University, Bloomington, IN 47405 USA.
8. Department of Geological Sciences, University of Maine at Orono, Orono, ME 04469 USA.
9. Department of Environmental and Forest Biology, SUNY College of Environmental Science and Forestry, Syracuse, NY 13210 USA.
10. Department of Biology, Queens University, Kingston, Ontario K7L 3N6 CANADA.
11. Department of Botany and Plant Pathology, University of Maine at Orono, Orono, ME 04469 USA. Present address: Department of Biology, Indiana University, Bloomington, IN 47405, USA.

Water, Air and Soil Pollution **30** (1986) 355–365.

ABSTRACT. The PIRLA project is an interdisciplinary paleoecological study designed to provide reconstructions of the recent acidification histories of a representative set of lakes in four acid-sensitive regions in North America. We are trying to determine if lakes in the study regions have acidified, and if so, to what extent, over what time period and why. Sediment cores from 5 to 15 lakes in each region are being analyzed for several characteristics. Diatoms and chrysophytes are being used to reconstruct lakewater pH. Results for three Adirondack lakes with current pH of 4.8 to 5.0 indicate a decrease in pH beginning in the 1930's-1950's. Increased atmospheric deposition of strong acids appears to be the primary factor responsible for the pH decline. Two lakes (pH 4.4 and 4.7) in New England show clear evidence of acidification probably due to acidic deposition. Preliminary reconstructions for two lakes in Michigan (pH 4.4 and 5.6), one in Wisconsin (pH 5.3), and one in Minnesota (pH 6.8) suggest no recent pH decrease. For the one Florida lake (pH 4.4) analyzed, inferred pH decreases by about 0.5 unit, beginning in the 1950s; the cause has not been determined.

1. INTRODUCTION

The PIRLA project is designed to reconstruct acidification histories of lakes in four acid sensitive regions in North America (Figure 1).

Major questions addressed by the study are: Have lakes in acid sensitive regions acidified? If so, what are the relative roles of (1) natural, long-term acidification, (2) watershed disturbance followed by vegetation regrowth, and (3) atmospheric deposition of strong acids? The questions are evaluated by examining data for individual lakes and individual regions, and by making comparisons among regions. Lake-watershed interactions are similar within regions but each region has its own characteristic hydrology, land use patterns, and strong acid loadings (Table I). The hypothesis being evaluated is that the lakes in the regions receiving greatest atmospheric loading should be most strongly and/or most rapidly acidified in the recent past. Also, if changes in historical pH are consistently observed within more than one region, the relative amounts of acidification should be proportional to regional acid loading.

The purpose of this paper is to review briefly the initial results of the PIRLA project for each study region. More detailed description of the project, results, and methods are given elsewhere (Charles and Whitehead, 1986a,b,c).

The PIRLA project is funded by the Electric Power Research Institute. It involves more than 30 investigators at nine academic institutions. Scheduled duration of the project is from June 1983 through December 1986.

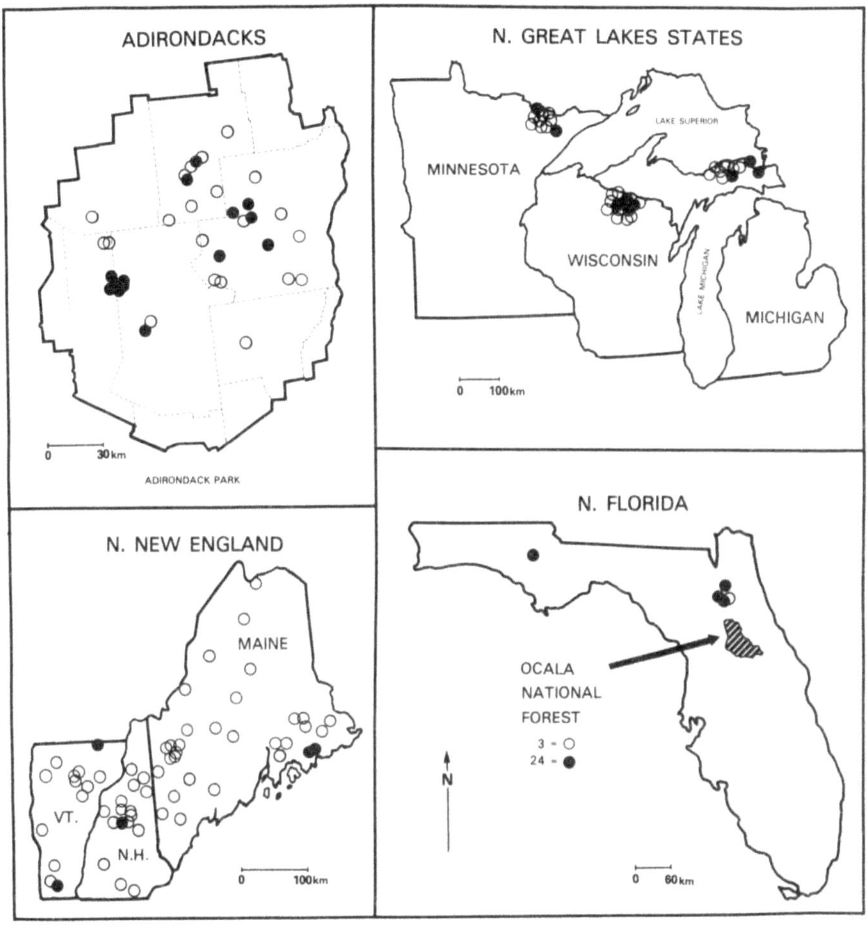

Fig. 1. Location of PIRLA study lakes in the Adirondack Mountains, N.Y., northern New England (Maine, New Hampshire and Vermont), northern Great Lakes States (Michigan, Wisconsin, and Minnesota), and northern Florida. The number of lakes in each category in the Ocala National Forest is indicated on the figure. All circles indicate lakes which comprise the diatom surface sediment--water chemistry calibration data set. Solid dots represent lakes cored for stratigraphic analysis. The latter lakes are listed in Battarbee and Charles 1986, along with current measured lakewater pH.

TABLE I

Wet precipitation pH and ion deposition for PIRLA study regions. Numbers are rough estimates for the regions, based on maps in Barrie and Hales (1984) and Glass and Loucks (1985). NGLS = Northern Great Lakes States.

Region	Deposition pH	H^+	meq·m^{-2}·y^{-1} $SO_4^{=}$	NO_3^-
Adirondack Park, N.Y.	4.2	60	60	30
N. New England	4.3-4.4	45	40	20
N. Michigan (NGLS)	4.4	26	37	19
N. Florida	4.6	30	40	20
N. Wisconsin (NGLS)	4.7	17	31	17
N. Minnesota (NGLS)	4.9	9	22	12

2. METHODS

Sediment cores from 5 to 15 small (typically 5 to 50 ha, Z_m = 10 to 20 m) lakes are being studied in each region. Stratigraphic analyses of several sediment characteristics (diatoms, chrysophytes, chironomids, pollen, charcoal, total metals, total S, S-isotopes, polycyclic aromatic hydrocarbons (PAH), and soot particles) provide historical information on trends in lake pH, watershed disturbances, and atmospheric deposition from many sources, including the combustion of fossil fuels. Sediments are dated using ^{210}Pb and chronostratigraphic markers such as pollen and charcoal.

Diatom assemblages are being used to reconstruct lakewater pH (e.g., Battarbee and Charles, 1986). Predictive equations (e.g., derived from indices α and B, and multiple regression) have been calculated for each region based on surface sediment (0 to 1 cm) diatom assemblage and water chemistry data for at least 35 lakes in each region (Charles, 1985; Davis and Anderson, 1985; Kreis _et al._, 1986; Sweets _et al._, 1986). Measurements of air-equilibrated pH for the calibrations were generally made six to ten times for each lake over a period of 2 yr although the sampling frequency for Florida lakes was less. Standard errors for the best prediction of pH from diatom assemblages are between ± 0.20 to 0.40 pH units for all regions.

3. RESULTS

3.1 _Adirondacks_

Reconstructions of lakewater pH have been completed for five Adirondack PIRLA lakes (Big Moose, pH 4.9; Upper Wallface, pH 4.9; Deep, pH 4.7; Arnold, pH 4.8; Little Echo, pH 4.3). All of these lakes have negative alkalinity, and sediment data for all indicate recent acidification.

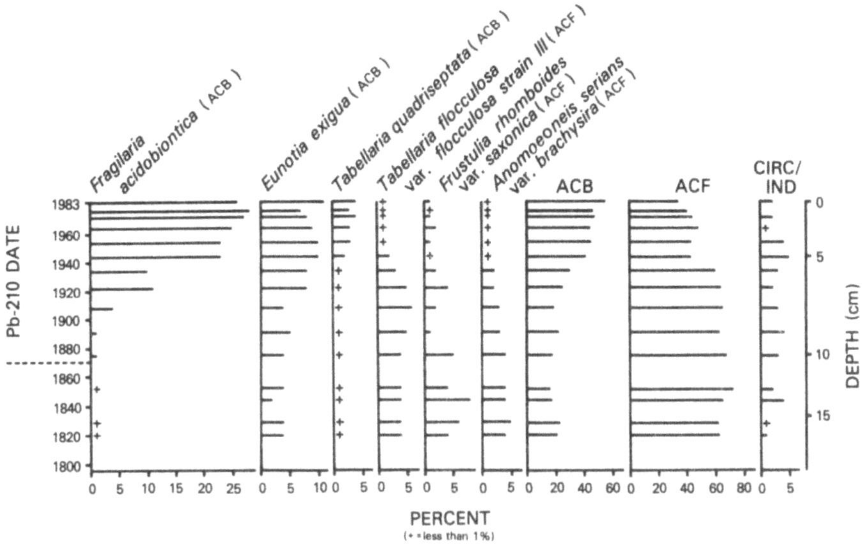

Fig. 2. Dominant diatom taxa in U. Wallface P., Adirondack Park, N.Y., core 1. The pH category for each taxon is indicated: ACB = acidobiontic; ACF = acidophilic; IND = indifferent/circumneutral. Alkaliphilic taxa were rare and occurred in only three intervals within the core. Accumulation rate of valves was nearly constant throughout the core.

Diatom inferred (DI) pH values of Adirondack lakes (studied by other investigators) that have a current pH above 5.5 do not decrease toward the surface (Charles and Norton, 1986), most probably because these lakes have higher acid neutralizing capacity.

Results for all lakes (Charles, 1984; Charles et al., 1986; Charles and Norton, 1986) are similar except for Little Echo P., a bog lake. Complete data sets are available for the first three lakes and only these lakes are discussed here. In all lakes, (e.g., Upper Wallface, Figure 2) Fragilaria acidobiontica Charles increases toward the surface. This taxon and Navicula tenuicephala Hustedt are generally found only in lakes with pH less than 5.0 to 5.5 and with Al concentrations greater than 200 $\mu g \cdot L^{-1}$ (Charles, 1986). Thus, these trends may indicate increasing Al concentrations, which are now about 200 to 650 $\mu g \cdot L^{-1}$. Percentages of indifferent/circumneutral taxa decrease toward the surface of the cores. The abundances of Anomoeoneis serians var. brachysira (Breb. ex Kutz.) Hust. and Frustulia rhomboides var. saxonica (Rabh.) DeT. also generally decrease. This trend may be indicative of decreased concentrations of total organic C (Davis et al., 1985) and probably of organic acids as well. Diatom inferred (DI) pH decreases by about 1.0 pH unit for Big Moose Lake, and about 0.3 to 0.7 pH units for the other four lakes. The onset of pH decreases ranges from about 1930 for U. Wallface Pond (Figure 3) to the 1950's.

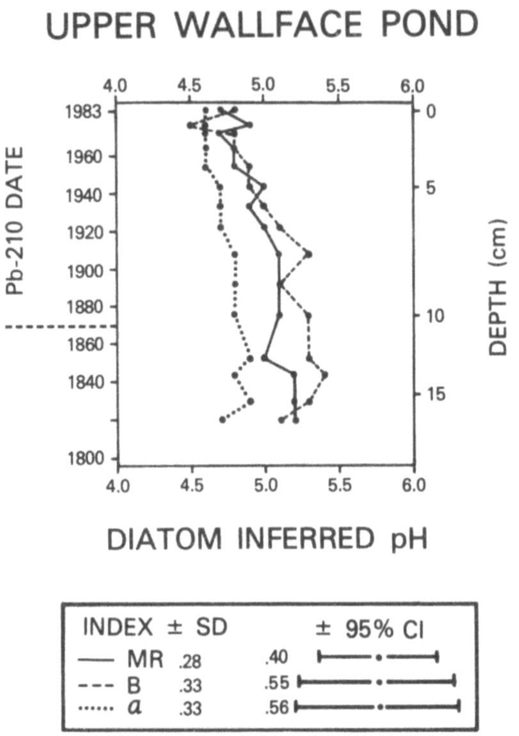

Fig. 3. Diatom inferred pH for Upper Wallface P., Adirondack Park, N.Y., core 1, based on equations in Charles, 1985. Standard deviation (SD) and 95% confidence intervals for single DI pH values are based on predictive relationships derived from the Adirondack lake calibration data set (Charles, 1985).

Chrysophyte data also indicate declining pH in all lakes but Little Echo (U. Wallface, Christie and Smol, 1986; Deep, Smol et al. 1984; Big Moose, Charles et al., 1986). Both diatom and chrysophyte data have been used to reconstruct the long-term history of U. Wallface Pond (Whitehead et al., 1986; Christie and Smol, 1986). The DI pH of the lake declined from above 7 to ≈ 5 from about 12,000 yr to 8,000 yr ago. The DI pH remained about 5 throughout the Holocene until the recent acidification phase. Thus, the change since about the 1930's represents a recent acidification phase of a magnitude which has not occurred previously.

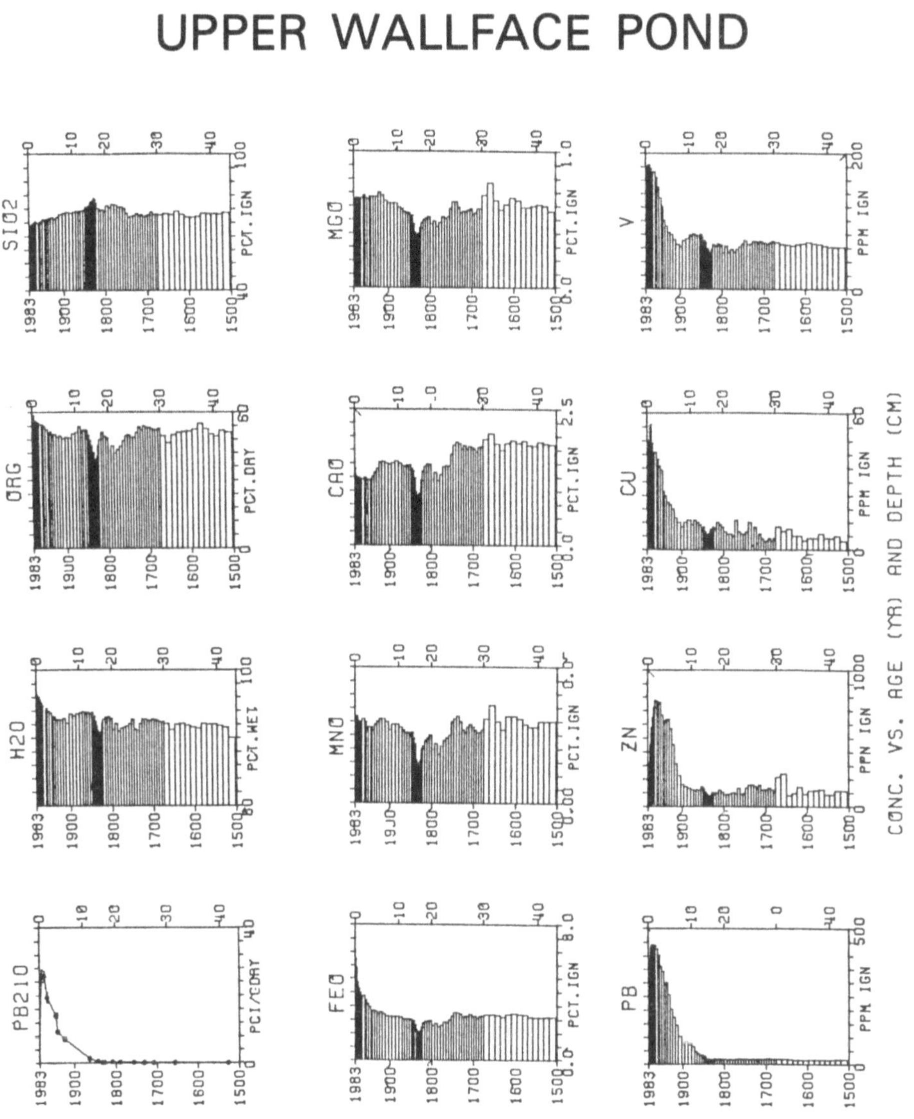

Fig. 4. Concentrations of total metals and ^{210}Pb in U. Wallface P., core 1. Dates are based on the ^{210}Pb CRS model. Each bar represents analysis of one sediment interval.

Stratigraphic analysis of chironomid remains in the Deep Lake (Uutala, 1985) and U. Wallface Pond and Big Moose Lake cores and of Cladocera in the Deep Lake and Wallface Pond cores (Paterson, 1985) is consistent with other data indicating lake acidification. Although sequential extractions of metals such as Al from Big Moose Lake sediment cores show a complex pattern, increases in labile Al fractions correspond with the decrease in DI pH observed in that lake.

The most reasonable explanation for the recent decrease in lakewater pH in the lakes is increased atmospheric deposition of strong acids, though watershed disturbance and natural acidification may also play minor roles. Lakewater concentrations of SO_4 are high (110 to 145 $\mu eq \cdot L^{-1}$) due to atmospheric inputs of S, and are primarily responsible for the current low lakewater pH. Increased inputs of total S to the sediment (Mitchell et al., 1986) and shifts in S isotope ratios (Fry, 1986) both suggest the most reasonable explanation for the decline in pH is acid deposition. The DI pH decreases follow changes in these sediment characteristics, and in others that also indicate input of materials associated with fossil fuel combustion, such as the metals Pb, Cu, V, and Zn (Figure 4), PAH, and soot particles (e.g., Charles et al., 1986). Watershed disturbance has probably not been an important cause of recent acidification in these lakes (Charles and Norton, 1986). There is no record of fire or logging in the U. Wallface Pond watershed. These disturbances have occurred in the Big Moose Lake and Deep Lake watersheds, but acidification trends are not logically related to the disturbances. For example, the DI pH decreases begin occurring before the disturbances, or they occur after recent (1950's) logging, but not after earlier logging (1890's-1910's). Natural, long-term acidification due to leaching of base cations from the soil was probably the dominant acidification process until the mid-1900's when the rate of acidification increases, as indicated by the sudden change in diatom and chrysophyte flora.

3.2 Northern New England

Preliminary DI pH reconstructions for cores of three New England PIRLA study lakes show that all three lakes had a pH between 5.0 and 5.5 prior to European settlement (ca. late 1700's), and that Mud Pond (current pH 4.7), Haystack Pond, and perhaps Little Long Pond, have decreased in pH since 1920 (Whiting et al., 1985). The most likely cause for acidification of Mud Pond and Haystack Pond appears to be acidic deposition. Sediment accumulation rates of metals released by combustion of fossil fuels and industrial activity begin increasing in the late 1800's, and continued to rise until the last two decades. Declining concentrations of Ca and Mn at the surface of the sediment may indicate increasing acidification of the lake/watershed system. Concentrations of total S and PAH also increase toward the core surface, beginning prior to the time DI pH begins to decline.

3.3 Northern Great Lakes States (NGLS)

PIRLA study lakes in the NGLS are different as a group from those in other regions in that a large percentage are seepage lakes and are very dilute (conductivities as low as 9 $\mu S \cdot cm^{-1}$ at 25°C), with low alkalinity (-38 to 80 $\mu eq \cdot L^{-1}$). Often 40 to 75% of the basic cations (Ca, Mg, Na, K) in the water column originate from precipitation (Cook et al., 1986). Concentrations of Al are generally low (less than 50 $\mu g \cdot L^{-1}$).

Preliminary pH reconstructions are complete for four lakes: McNearney, MI; Andrus, MI; Otto Mielke, WI; and Hustler, MN. No recent pH changes are indicated in these lakes. There are often significant shifts in diatom species composition in the cores, but these usually occur within single diatom-pH categories; therefore no DI pH shift is observed. Explanations for the recent changing diatom floristics may include responses to recent land-use events that changed nutrient inputs to the lakes, and effects of fire.

Concentrations of PAH, S, Pb, Cu, V, and Zn increase toward the top of the McNearney Lake core, and indicate increased atmospheric loadings of these materials. However, accumulation rates for PAH are typically an order of magnitude lower than those seen in the northeastern USA (e.g., Gschwend and Hites, 1981).

3.4 Florida

More than 60% of the Florida PIRLA study lakes are characterized by a pH < 5.0 (Garren et al., 1986). They are situated on thin, sandy soils in areas of low relief. Color varies, even in the most acidic lakes, from clear to highly colored (1 to 265 Pt-Co units true color). Most of the lakes contain viable fish populations (e.g., largemouth bass). Lakewater concentrations of SO_4 are sometimes higher than in the Adirondacks (35 to 256 $\mu eq.L^{-1}$; ave. = 92 $\mu eq.L^{-1}$), but Cl has consistently higher values (64 to 624 $\mu eq.L^{-1}$; ave. = 218 $\mu eq.L^{-1}$). Further analysis is necessary to confidently assess the sources of acidity in N. Florida lakes.

Most of the initial work has concentrated on Lake Barco, which is very clear (5 Pt-Co units true color) and the most acidic of the cored lakes (pH = 4.4). Diatom inferred pH indicates a decrease of 0.4-0.5 pH units. The corresponding floristic changes occur abruptly in the late 1950's. Total S and PAH increase in concentration toward the surface of the sediment, suggesting increased input of materials related to fossil fuel combustion since the 1930's.

Despite this evidence, and until the other analyses (particularly geochemistry) are complete, our preliminary working hypothesis for the more acidic lakes assumes that pHs have been < 5.5 since at least pre-industrial times. Long-term acidification by several possible processes (e.g., oxidation of marine deposits, evapoconcentration) has yet to be ruled out. In particular, responses following controlled burning in the area of Lake Barco may be a factor in any pH decrease in the 1900's.

4. SUMMARY

Analyses of 13 cores from the four PIRLA study regions indicate that at least some low alkalinity lakes in the Adirondacks and New England have acidified recently, primarily due to acid deposition. Diatom inferred pH for the one Florida core analyzed indicates a recent pH decrease, but the possible causes have not yet been fully examined. There was no evidence to indicate acidification of low pH (pH < 5.5) lakes in the Northern Great Lakes states.

ACKNOWLEDGMENTS

This is contribution number 10 of the PIRLA project, which is funded by the Electric Power Research Institute (RP-2174-10).

For all regions except Florida, ^{210}Pb was analyzed by G. Blake, and total metals by M. Morrison, both at the University of Maine at Orono. M. Binford did the ^{210}Pb dating for the Florida lakes.

REFERENCES

Barrie, L.A. and Hales, J.M: 1984, Tellus 36B, 333.

Battarbee, R.W. and Charles, D.F.: 1986, Water, Air, and Soil Poll. (this volume).

Charles, D.F.: 1984, Verh. Internatl. Verein. Limnol. 22, 559.

Charles, D.F.: 1985, Ecology 66, 994.

Charles, D.F.: 1986, 'A New Diatom Species, Fragilaria acidobiontica, from Acidic Lakes in Northeastern North America', in Smol, J.P., Battarbee, R., Davis, R.B., and Merilainen,J. (eds.), Diatoms and Lake Acidity, W. Junk, Dordrecht, The Netherlands, 35.

Charles, D.F. and Norton, S.A.: 1986, 'Paleolimnological Evidence for Trends in Atmospheric Deposition of Acids and Metals', in Acid Deposition: Long-term Trends, National Academy Press, Washington, D.C., 335.

Charles, D.F. and Whitehead, D.R., eds.: 1986a, Paleoecological Investigation of Recent Lake Acidification (PIRLA): Interim Report, Electric Power Research Institute, Palo Alto, CA (in press).

Charles, D.F. and Whitehead, D.R.: 1986b, Paleoecological Investigation of Recent Lake Acidification (PIRLA): Methods and Project Description, Electric Power Research Institute, Palo Alto, CA (in press).

Charles, D.F. and Whitehead, D.R.: 1986c, 'The PIRLA Project: Paleoecological Investigation of Recent Lake Acidification,' in Loffler, H. (ed.), Proceedings of the IVth International Symposium on Paleolimnology, Developments in Hydrobiology Series, W. Junk, Dordrecht, The Netherlands, (submitted).

Charles, D.F., Whitehead, D.R., Engstrom, D.R., Fry, B.D., Hites, R.A., Norton, S.A., Owen, J.S., Roll, L.A., Schindler, S.C., Smol, J.P., Uutala, A.J., White, J.R., and Wise, R.J.: 1986, Biogeochem., (in press).

Christie, C.E. and Smol, J.P.: 1986, 'Recent and Long-term Acidification of Upper Wallface Pond (N.Y.) as Indicated by Mallomonadacean Microfossils', in Loffler, H., (ed.), Proceedings of the IVth International Symposium on Paleolimnology, Developments in Hydrobiology Series, W. Junk, Dordrecht, The Netherlands (submitted).
Cook, R.B., Kelley, C.A., Kreis Jr., R.G., and Kingston, J.C.: 1986, 'Chemical Limnology of Lakes in the Northern Great Lakes States', in Charles, D.F. and Whitehead, D.R. (eds.) Paleoecological Investigation of Recent Lake Acidification: Interim Report, Electric Power Research Institute, Palo Alto, CA, (in press).
Davis, R.B. and Anderson, D.S.: 1985, Hydrobiologia 12, 69.
Davis, R.B., Anderson, D.S., and Berge, F.: 1985, Nature 316, 436.
Fry, B.: 1986, Biogeochem., (in press).
Garren, R.A., Clarkson, C.L., and Crisman, T.L.: 1986, 'Chemistry of Lakes in the Ocala National Forest, Florida', in Charles, D.F. and Whitehead, D.R. (eds.), Paleoecological Investigation of Recent Lake Acidification: Interim Report, Electric Power Research Institute, Palo Alto, CA, (in press).
Glass, G.E. and Loucks, O.L.: 1986, Environ. Sci. Technol. 20, 35.
Gschwend and Hites, R.A.: 1981, Geochim. Cosmochim. Acta 45, 2359.
Kreis, Jr., R.G., Kingston, J.C., Camburn, K.E., and Cook, R.B.: 1986, 'Diatom--pH Relationships in the Northern Great Lakes Region for Predicting Past Lake Acidity' in Charles, D.F. and Whitehead, D.R. (eds.). Paleoecological Investigation of Recent Lake Acidification: Interim Report, Electric Power Research Institute, Palo Alto, CA, (in press).
Mitchell, M.J., Schindler, S.C., Owen, J.S., and Norton, S.A.: 1986, J. Environ. Qual., (submitted).
Paterson, M.J.: 1985, Paleolimnological Reconstruction of Cladoceran Response to Presumed Acidifcation of Four Lakes in the Adirondack Mountains (New York). Masters thesis. Department of Biology, Indiana University, Bloomington, IN.
Smol, J.P., Charles, D.F., and Whitehead, D.R.: 1984, Nature 307, 628.
Sweets, P.R.: 1986, 'The Relationship Between Surface Sediment Diatoms and pH in Northern Florida' in Charles, D.F. and Whitehead, D.R. (eds.), Paleoecological Investigation of Recent Lake Acidification: Interim Report, Electric Power Research Institute, Palo Alto, CA, (in press).
Uutala, A.: 1986, Entomol. Scand., (submitted).
Whitehead, D.R., Charles, D.F., Reed, S.E., Jackson, S.T., and Sheehan, M.C.: 1986, 'Late-glacial and Holocene Acidification History of Adirondack (N.Y.) Lakes', in Smol, J.P., Battarbee, R., Davis, R.B., and Merilainen, J. (eds.). Diatoms and Lake Acidity, W. Junk, Dordrecht, Netherlands, 251.
Whiting, M.C., Anderson, D.S., and Davis, R.B.: 1985, 'Further Paleolimnological Evidence for Possible Lake Acidification in Northern New England', Amer. Soc. Limnol. Oceanogr. Meeting, Minneapolis, MN, June, 1985 (abstract).

COMPARISON OF PALEOLIMNOLOGICAL WITH MAGIC MODEL RECONSTRUCTIONS OF WATER ACIDIFICATION

R.F. Wright[1], B.J. Cosby[2], G.M. Hornberger[2], and J.N. Galloway[2]

[1]Norwegian Institute for Water Research, Box 333, 0314 Oslo, Norway
[2]Department of Environmental Sciences, University of Virginia, Charlottesville, VA 22903, USA

ABSTRACT. In sensitive areas receiving acidic deposition, paleolimnolgical data indicate changes in lake pH over 1 to 3 decades during the past century. Estimates of deposition of SOx and NOx over this same period suggest that deposition rates changed (1) earlier and (2) more slowly than did changes in lake chemistry. Clearly chemical and biological processes in the terrestrial catchment damp, delay, and moderate the response of surface water pH to deposition of acidifying compounds. This response is controlled by key terrestrial processes that include chemical weathering, sulfate adsorption, cation exchange, dissolution and precipitation of Al compounds, and dissolution and dissociation of inorganic C. MAGIC (Model of Acidifcation of Groundwater In Catchments) provides a tool by which these processes can be simultaneously and quantitatively linked to examine the impact of acid deposition on surface water chemistry. We have applied MAGIC to 4 lakes from which paleolimnological reconstructions are available - Big Moose Lake in the Adirondacks, Loch Grannoch in Scotland, Lake Gårdsjøn in Sweden, and Lake Hovvatn in Norway. The results indicate that the processes linked in MAGIC can account for temporal trends in pH and alkalinity such as those obtained from paleolimnological data.

1. INTRODUCTION

Lakes and streams in large areas of eastern North America and Europe have been acidified during the past several decades (Likens et al., 1979; Overrein et al., 1980). Historical records of water chemistry and fish populations as well as paleolimnological data such as diatom remains preserved in lake sediments allow reconstruction of the history of acidification of individual lakes. Such reconstructions in

Water, Air and Soil Pollution **30** (1986) 367–380.

turn provide independent evidence by which the relationships between acid deposition, soil acidification and surface water acidification can be tested.

We have recently developed an intermediate-complexity process-oriented model for soil and water acidification (MAGIC - Model of Acidification of Groundwater In Catchments) (Cosby et al., 1985a,b). We report here application of MAGIC to 4 lakes from which historical or paleolimnological data are available - Big Moose Lake in the Adirondack Mountains of New York, USA (Charles, 1984), Loch Grannoch in the Galloway Hills of southwestern Scotland (Flower and Battarbee, 1983), Lake Gårdsjøn on the west coast of Sweden (Renberg and Hellberg, 1982), and Lake Hovvatn in southernmost Norway (Davis et al.,1985).

2. DATA REQUIREMENTS FOR MAGIC

MAGIC consists of (1) a soil-soil solution chemical equilibria section in which the concentrations of major ions are assumed governed by simultaneous reactions involving sulfate adsorption, cation exchange, dissolution and precipitation of Al, and dissolution of inorganic C and (2) a mass-balance section in which the flux of major ions to and from the soil is assumed governed by atmospheric inputs, chemical weathering inputs, net uptake in biomass and loss to runoff (Figure 1). At the heart of MAGIC is the size of the pool of exchangeable base cations in the soil; as the fluxes to and from this pool change over time due to changes in acid deposition, the chemical equilibria between soil and soil solution shift to give changes in surface water chemistry. The degree and rate of change in surface water acidity thus depends both on flux factors and on the inherent characteristics of the impacted soils.

Application of MAGIC requires data for (1) soil physical and chemical characteristics, (2) hydrologic fluxes, (3) precipitation chemistry covering the period from the onset of acid deposition to the present, and (4) estimate of weathering rate. Unfortunately for no catchment do there exist sufficient data such that a mathematically unique solution is obtained. We are thus forced to fit MAGIC to present-day soil and water chemistry data for each catchment. We use a trial and adjustment procedure in which we chose a set of weathering rates for each of the 4 base cations and an initial base saturation such that when run for the 140 yr from 1844 to 1984 the model arrives at values for streamwater chemistry and soil chemistry that agree with the measured present-day values. Our version of the model takes the

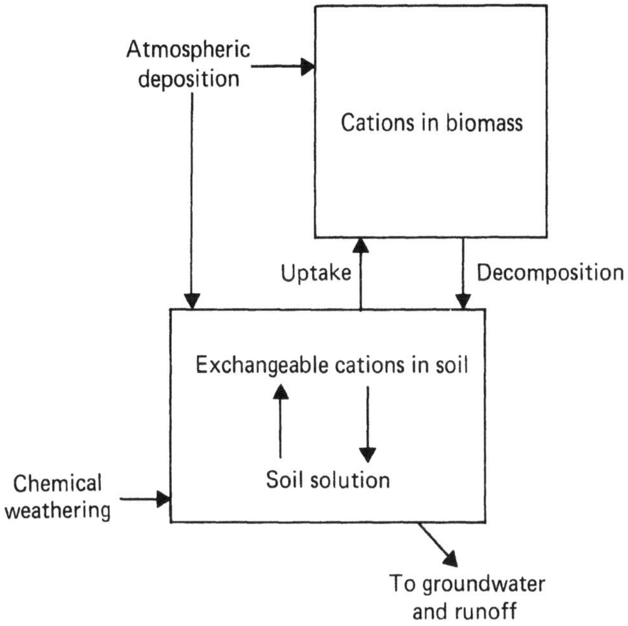

Figure 1. Schematic view of MAGIC (Model for acidification of A Groundwater In Catchments) (Cosby et al., 1985a,b). MAGIC is comprised of 2 parts - fluxes to and from the soil and a series of chemical equilibria between soil and soil solution.

year 1844 as a "pre-acidification" starting point and assumes that at that time the catchment was in steady state with respect to fluxes of major ions into and out of the soil. We perform the MAGIC recontructions "blind", that is without reference to the paleolimnological data. MAGIC is designed to simulate streamwater chemistry; in applying MAGIC to these lakes we assume that no chemical changes occur in the lake ecosystem.

For Big Moose Lake we use water chemistry data from Driscoll (1983), soils data from the B horizon at nearby Woods Lake (Cronan, 1985), and precipitation volume and chemistry data from Johannes and Altwicker (1980). For Loch Grannoch we use water chemistry data from Harriman et al. (1986), soils data from the B horizon at Birkenes in Norway (Frank, 1980), and precipitation volume and chemistry data from the station at Cairnsmore of Fleet (Burns et al., 1984). For Lake Gårdsjøn we use water chemistry data from Hultberg (1985b), soils data from the B horizon (Melkerud, 1983), and precipitation volume and chemistry data and hydrologic fluxes from Hultberg (1985b). For Lake Hovvatn we use water chemistry data from Wright (1985), soils data from the Storgama area (Seip et al., 1979), precipitation chemistry data from the Treungen station (Christophersen et al., 1984), and hydrologic fluxes from Wright (1985). Precipitation chemistry was modified to account for the higher seasalt deposition at Hovvatn. The remainder of the necessary data we estimated from ranges reported from other catchments or simply "best guesses" (sulfate adsorption characteristics, mean soil temperature, mean soil CO_2 partial pressure, mean soil depth, and the solubility of Al hydroxide) (Table I).

From the diatom-reconstructed pH values we estimate alkalinity. We define alkalinity as $HCO_3 - H^+ - \Sigma Al^+$ where HCO_3 ($\mu eq\ L^{-1}$) is obtained from the relationship $\log HCO_3 = -5.93 + 1.22$ pH, and sum of positively-charged Al species ($\mu eq\ L^{-1}$) is obtained from the relationship $\log \Sigma Al = 12.7 - 2.56$ pH (Wright and Henriksen, 1983)

Historical sequences of precipitation chemistry for the 140 yr period 1844-1984 are also required to run MAGIC. We assume deposition has paralleled emissions of SO_2 and NOx. For Big Moose Lake we use the emission sequence for the Adirondack region given by OTA (1984), for Loch Grannoch the sequence for the UK given by Warren Spring Laboratory (1983), and for Lakes Gårdsjøn and Hovvatn the sequence for SO_2 in Europe estimated by OECD (1976). The deposition is scaled such that the present-day deposition agrees with the measured precipitation chemistry and volume at each site.

3. RESULTS

3.1. Big Moose Lake

The pH reconstruction obtained from the diatom stratigraphy at Big Moose Lake suggests that from a "pre-acidification" pH of about 5.9 the lake declined slowly to pH 5.6 in 1960 and then broke sharply to pH 4.6 by 1980 (Charles, 1984) (Figure 2). Measured pH in the 1980's is 5.1 (Driscoll, 1983). The pH 5.1 value is consistent with the remaining major-ion chemistry. The pH curves give the appearance of rather large differences between the two methods. Much of the discrepancy occurs in the pH range 5.0 to 5.5 at which small changes in alkalinity give large

Table I. Parameter values used for MAGIC reconstructions. Data are either measured or estimated (see text for data sources).

	Big Moose Lake	Loch Grannoch	Lake Gårdsjön	Lake Hovvatn
Catchment characteristics				
precipitation (m yr^{-1})	1.0	2.0	1.1	1.5
runoff (m yr^{-1})	0.7	1.5	.07	1.2
dry deposition factor	1.2	1.2	1.6	1.2
Soil characteristics				
mean depth (m)	2.0	1.5	0.8	0.3
mean T (^{0}C)	8	8	8	5
pCO_2 (atm)	0.02	0.02	0.02	0.02
soil mass (kg m^{-2})	2650	1990	1400	400
sulfate absorption				
half-saturation (μeq L^{-1})	100	100	100	100
maximum capacity (meq kg^{-1})	0.3	0.3	1	1
cation exch. capacity (meq $100g^{-1}$)	8	13	10	31
selectivity coefficients				
log S (AlCa)	2.3	0.5	1.9	1.0
log S (AlMg)	3.7	2.9	4.1	2.2
log S (AlNa)	-0.4	0.0	0.7	-0.5
log S (AlK)	-3.0	-2.8	-1.0	-4.0
$Al(OH)_3$ solubility				
log K (soil)	9.1	8.6	8.6	8.0
log K (stream)	9.1	8.6	8.6	8.6
weathering rates				
Ca (meq m^{-2} yr^{-1})	13	10	10	3
Mg (meq m^{-2} yr^{-1})	10	6	12	1
Na (meq m^{-2} yr^{-1})	14	0.5	3	0.5
K (meq m^{-2} yr^{-1})	2	2	2	0.5

changes in pH. Comparison of reconstructed alkalinity reveals rather small differences.

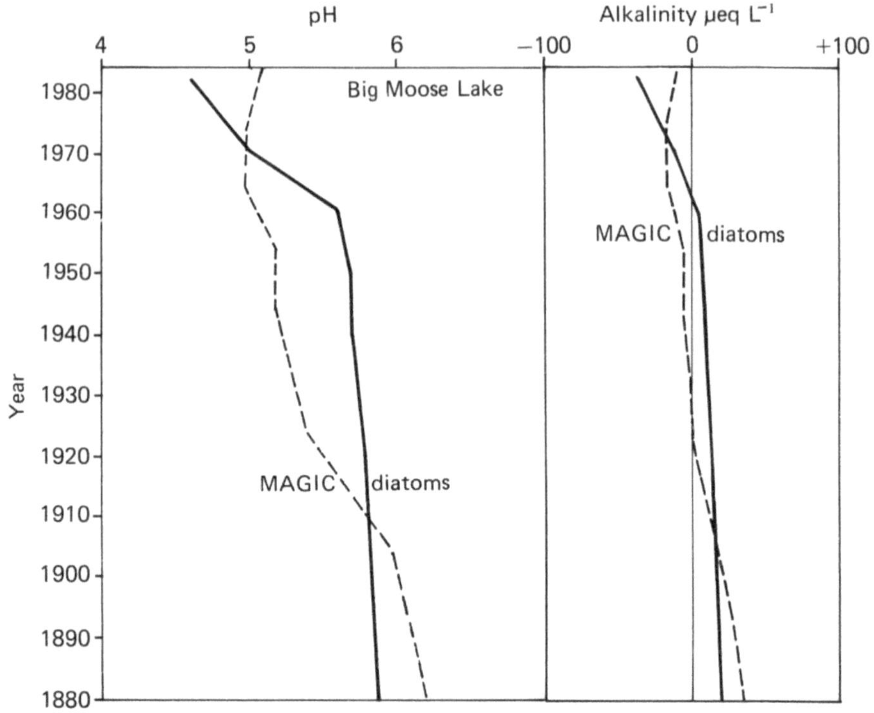

Figure 2. Reconstructed pH and alkalinity for Big Moose Lake, Adirondack Mountains, New York, USA, from the diatom stratigraphy (Charles, 1984) and from MAGIC.

3.2. Loch Grannoch

The diatom reconstruction indicates a slow decline from pH levels of about 5.6 in the late 1800's to about 4.8 in 1970 and then to present-day pH 4.5 (Figure 3). The diatom stratigraphy at Loch Grannoch is complicated during the period 1950-70 by the influx of terrestrial diatoms following disturbance in the catchment (Flower and Battarbee, 1983). MAGIC indicates that the pH at Loch Grannoch declined sharply around 1880-1900 and then more slowly to present-day levels of pH 4.5. The reconstructed alkalinity curves also differ in timing of maximum rate of decrease.

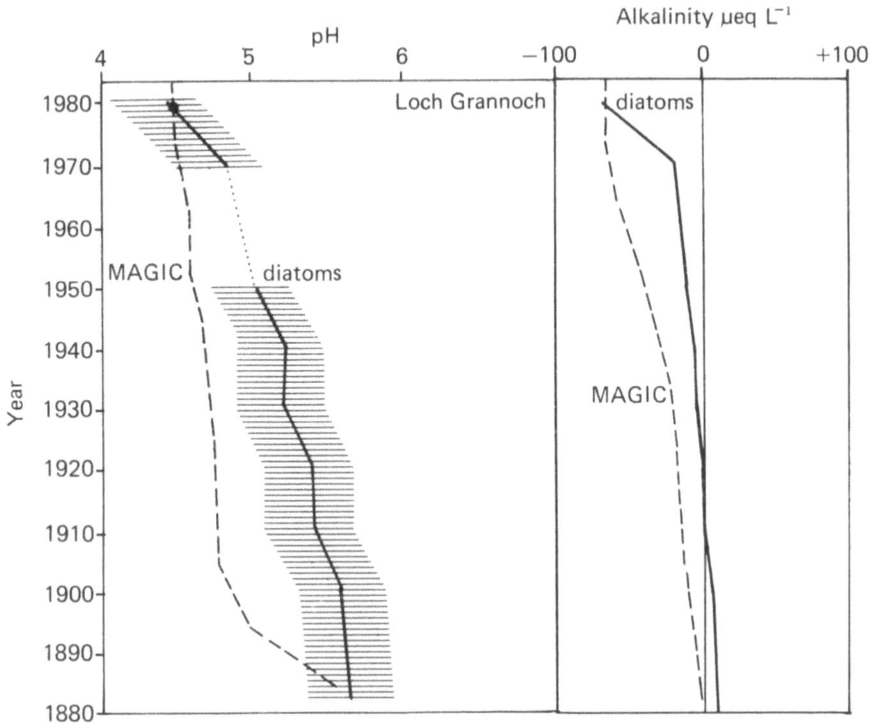

Figure 3. Reconstructed pH and alkalinity for Loch Grannoch, Galloway Hills, southwestern Scotland, UK, from the diatom stratigraphy (Flower and Battarbee, 1983) and from MAGIC. The shaded region about the diatom curve is the estimate of uncertainty indicated by Flower and Battarbee (1983).

3.3. Lake Gårdsjøn

At Lake Gårdsjøn the diatom stratigraphy again points to a slow decline from pH about 6.0 in 1960 followed by a sharp break to pH 4.5 in 1980 (Renberg and Hellberg, 1982) (Figure 4). MAGIC indicates a more gradual pH decline with the final drop to levels below pH 5 occurring around 1960, the period in which fish populations were affected (Hultberg, 1985a). The diatom stratigraphy indicates a somewhat lower pH today than measured, but this difference is negligible. Reconstructed alkalinity curves indicate a parallel development over time.

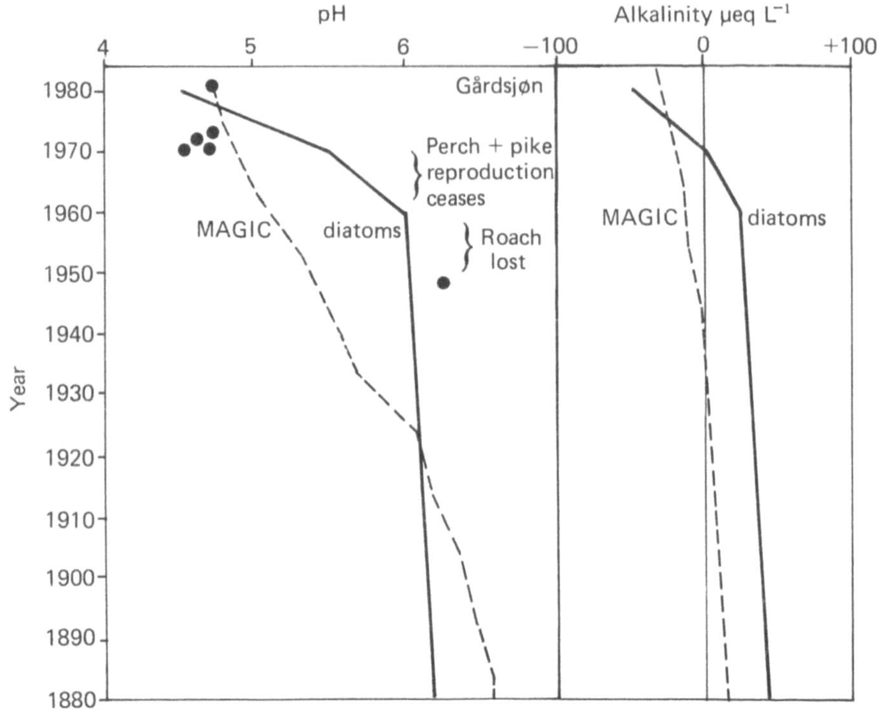

Figure 4. Reconstructed pH and alkalinity for Lake Gårdsjøn, west coast of Sweden, from the diatom stratigraphy (Renberg and Hellberg, 1982) and from MAGIC. Also shown are historical pH measurements and fisheries status (Hultberg, 1985a).

3.4. Lake Hovvatn

The diatom curve for Hovvatn (Davis et al., 1985) suggests that the lake originally had pH of only 5.0 and that the acidification to present-day levels of 4.3 to 4.5 occurred as early as 1910 (Figure 5). MAGIC indicates a similar pH history except that the original pH level was somewhat higher. Historical fisheries data from Hovvatn indicate that the native brown trout population may have begun to experience reproduction failure as early as the 1920's and the last fish was caught in the lake in 1947 (Wright, 1985). This information is consistent with the historical pH reconstructions. Reconstructed alkalinity curves are quite similar.

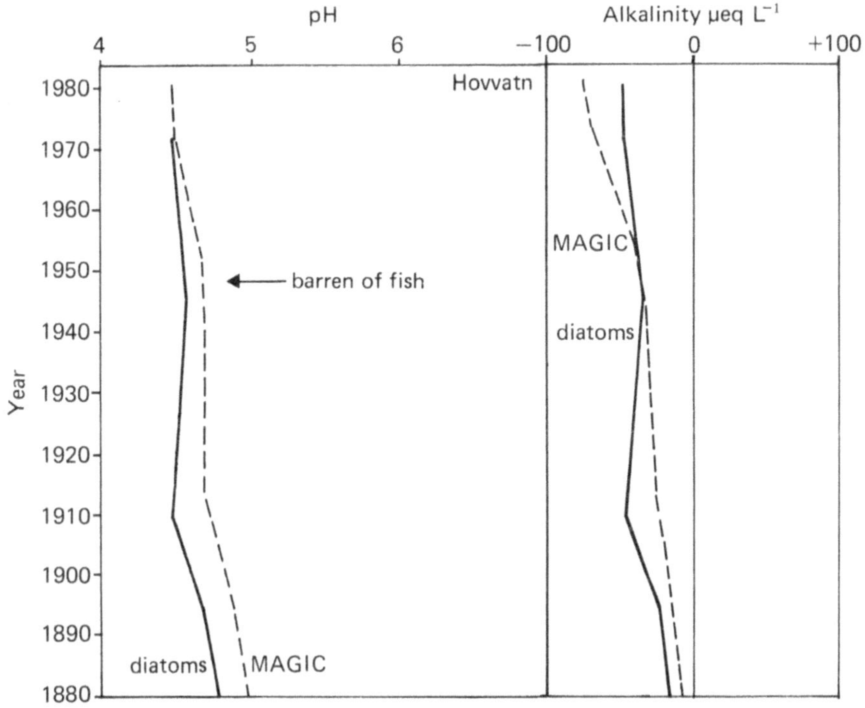

Figure 5. Reconstructed pH and alkalinity for Lake Hovvatn, southernmost Norway, from the diatom stratigraphy (Davis et al., 1985) and from MAGIC.

4. DISCUSSION

In all four cases the MAGIC model applied "blind" to independent data yields pH and alkalinity reconstructions that are most probably within the range of uncertainties inherent in both MAGIC and the diatom methods. It is difficult to assess the uncertainities in the MAGIC reconstructions despite the fact that MAGIC is a relatively simple model. Several of the input parameters are interrelated such that an acceptable range for one parameter acts to restrict the possible ranges for other parameters (Cosby et al., 1985a). For example, if the weathering rate is set too low, then the base saturation chosen for pre-acidification conditions must be high such that after 140 yr of acid deposition the system passes through the present-day soil chemistry and water chemistry "windows". Furthermore the uncertainty in the deposition sequence used to drive the system is difficult to estimate. These problems are not solved by turning to more complicated models such as the ILWAS model (Gherini et al., 1985), because independent data on a catchment scale for many of the key parameters such as weathering rates simply do not exist. Complex models tend to obscure the uncertainties and interrelationships between processes and parameters. We are continuing our work with MAGIC in an effort to assess the model's uncertainties.

The difference between pH history at Loch Grannoch (Figure 3) and Lake Gårdsjøn (Figure 4) can be largely explained by differences in deposition amounts and histories at these 2 sites. Whereas MAGIC indicates that both had initially about the same pool of available base cations to neutralize acid inputs, already in 1910 acid deposition levels at Loch Grannoch had reached levels similar to present-day levels at Lake Gårdsjøn. Thus Loch Grannoch "broke through" early whereas at Gårdsjøn pH levels remained above pH 5 until 1960.

Delay in catchment response due to sulfate adsorption in the soils is minimal for all of these 4 lakes. The soils are all young, podsolic soils developed in post-glacial times on granitic bedrock. Such soils have typically little ability to adsorb sulfate (Johnson and Todd 1983). The choice of parameter values for the sulfate adsorption characteristics for these 4 sites is not of critical importance. In all cases the sulfate flux out of the catchments is approximately equal to the sulfur flux into the catchments from the atmosphere.

These applications of MAGIC all assume that the weathering rate of base cations is constant and is not a function of soil pH during the 140-yr period. We have tested the importance of this assumption for several sites including Lake Gårdsjøn and Lake Hovvatn. Although changes in the pH of surface waters at these sites is large, changes in the pH of soil solution are small. Thus introducing a pH-dependent weathering rate has very little influence on the reconstructed lake pH and alkalinity.

MAGIC provides reconstruction of not only pH but of all the major ions. The results for the 4 lakes can thus be placed in Henriksen's (1980) empirical nomograph and compared with reconstructions using the Wright and Henriksen (1983) approach. The key uncertainty in these reconstructions is the value set for the "F-factor", which Henriksen (1980) defines as the change in concentrations of Ca^* and Mg^* for a given change in the concentration of SO_4^*. The asterisks denote the non-marine fraction. The MAGIC reconstructions indicate that the F-factors for these 4 lakes calculated over the period from 1844

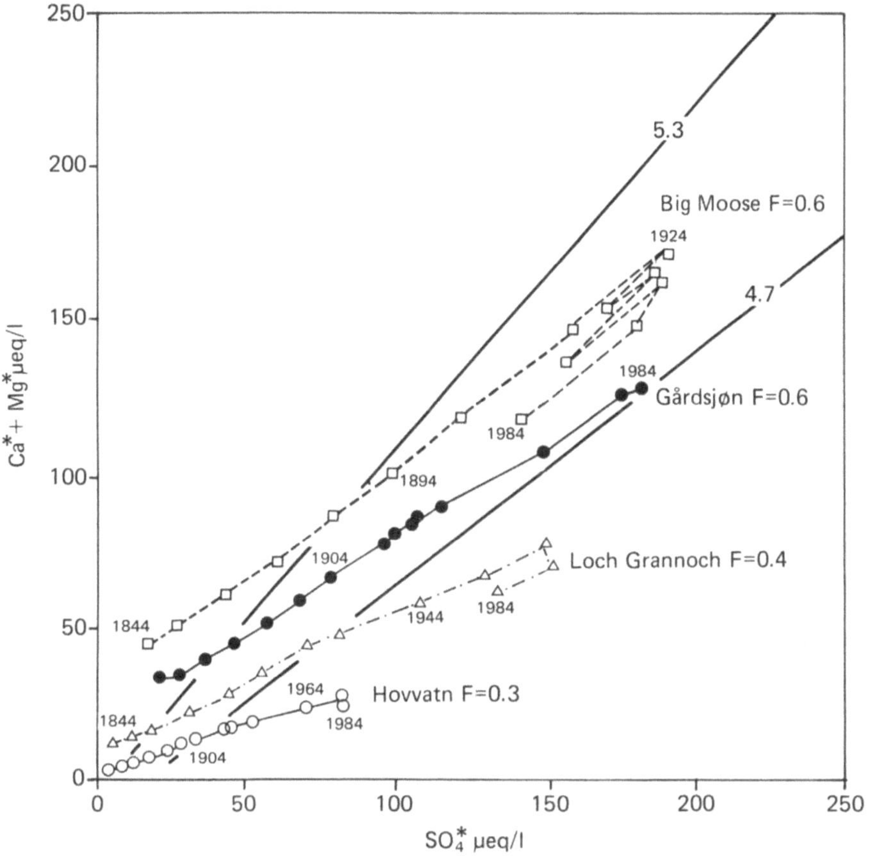

Figure 6. Reconstructed chemistry of 4 lakes using MAGIC and plotted in Henriksen's (1980) empirical nomograph for lake acidification. This nomograph relates the sum of concentrations of calcium and magnesium on the y-axis to the concentration of sulfate on the x-axis. The asterisks denote the non-marine fraction. The pH 5.3 and 4.7 lines separate lakes into 3 pH classes - bicarbonate lakes lie above the 5.3 line, transitional lakes lie between the 5.3 and 4.7 lines, and acid lakes lie below the 4.7 line. The trajectories run from 1844 to 1984.Each point represents a 10-yr interval. Also indicated are the F-factors for the entire period 1844-1984.

(pre-acidification) to 1984 vary between 0.3 for Hovvatn to 0.6 for Big Moose Lake (Figure 6). This range of F-factor is somewhat higher than that found by Henriksen (1984) and Wright (1983) for lakes in large regions of Europe and North America. The trajectories of each lake over time in the Henriksen nomograph indicates that the F-factor varies over time and that the greatest change occurs in lakes with the highest F-factor.

The general agreement between the pH reconstructions from historical and paleolimnological data and those from MAGIC indicates that the processes and assumptions contained in our model can account for historical trends in surface water chemistry. The next step is now to use MAGIC to predict soil and water acidification in response to future changes in acid deposition and to set target loadings for individual sites and regions.

ACKNOWLEDGMENTS

We thank C. T. Driscoll and H. Hultberg for unpublished data. This work was supported in part by the Surface Water Acidification Programme (SWAP) (The Royal Society, the Norwegian Academy of Science and Letters, and the Royal Swedish Academy of Sciences), and the Norwegian Institute for Water Research.

REFERENCES

Burns, J.C., Coy, J.S., Tervet, D.J., Harriman, R., Morrison, B.R.S. and Quine, C.P.: 1984, Fish. Mgmt. 15, 145.

Charles, D.F.: 1984, Verh. Internat. Verein. Limnolog. 22, 559.

Christophersen, N., Dymbe, L.H., Johannessen, M., and Seip, H.M.: 1984, Ecol. Modelling 21, 35.

Cosby, B.J., Hornberger, G.M., Galloway, J.N., and Wright, R.F.: 1985a, Water Resour. Res. 21, 51.

Cosby, B.J., Wright, R.F., Hornberger, G.M., and Galloway, J.N.: 1985b, Water Resour. Res. 21, 1591.

Cronan, C.S.: 1985, Water Air soil Pollut. 26, 355.

Davis, R.B., Andersen, D.S., and Berge, F.: 1985, Nature 316, 436.

Driscoll, C.T.: 1983, personal communication.

Flower, R.J., and Battarbee, R.W.: 1983, Nature 305, 130.

Frank, J.: 1980, 'Soil survey at Birkenes, a small catchment in Aust-Agder county, Southern Norway' Tech. Rept. TN 60/80 (SNSF-project, NISK, 1432 Ås, Norway), 41pp.(in Norwegian).

Gherini, S.A., Mok, L., Hudson, R.J.M., Davis, G.F., Chen, C.W., and Goldstein, R.A.: 1985, Water Air Soil Pollut. 26, 425.

Harriman, R., Morrison, B.R.S., Caines, L.A., Collen, P., and Watt, A.W.S.: in press,' The acidification of surface waters in Galloway, South-west Scotland, in relation to changes in their fishery status', Water Air Soil Pollut.

Henriksen, A.: 1980, 'Acidification of freshwaters -- a large scale titration' p.68-74, In Drabløs, D., and Tollan, A. (eds.) Ecol. Impact of Acid Precip. (SNSF-project, NISK, 1432 Ås,Norway) 383pp.

Henriksen, A.: 1984, Verh. Internat. Verein. Limnol. 22, 692.

Hultberg, H.: 1985a, Ecol.Bull.(Stockholm) 37, 64.

Hultberg, H.: 1985b, Ecol.Bull.(Stockholm) 37, 133.

Johannes, A.H., and Altwicker,E.R.: 1980, 'Atmospheric inputs to three Adirondack lake watersheds', p.256-257 In Drabløs, D., and Tollan, A. (eds.) Ecol. Impact Acid Precip. (SNSF-project, NISK, 1432 Ås, Norway) 383 pp.

Johnson, D. W.,and Todd, D.E.: 1983, Soil Sci. Soc. Am. J. 47, 792.

Likens, G.E., Wright, R.F., Galloway, J.N.,and Butler, T.J.: 1979, Sci. Am. 241(10), 43.

Melkerud, P-A.: 1983, 'Quaternary deposits and bedrock outcrops in an area around Lake Gårdsjøn, south-western Sweden, with physical, mineralogical and geochemical investigations', Rept. Forest Ecol. Forest Soils 44, (Swedish University - of Agricultural Sciences, Uppsala) 87 pp.

OECD: 1976, 'The OECD programme on long range transport of air pollutants', (Organisation of Economic Co-operation and Development, Paris).

OTA: 1984, 'Acid rain and transported air pollutants: implications for public policy', OTA-O-204 (U.S. Congress Office of Technology Assessment, Washington, D.C.)

Overrein, L.N., Seip, H.M., and Tollan, A.: 1980, 'Acid precipitation - effects on forest and fish. Final report of the SNSF-project 1972-1980', Research Rept. FR 19/80 (SNSF-project, NISK, 1432 Ås, Norway) 175 pp.

Renberg, I., and Hellberg, T.: 1982, Ambio 11,30.

Seip, H.M., Abrahamsen, G., Gjessing, E.T., and Stuanes, A.: 1979, 'Studies of soil-, precipitation- and runoff-chemistry in six small natural plots ("mini-catchments")', Internal Rept. 46/79 (SNSF-project, NISK, 1432 Ås, Norway), 62pp.

Warren Spring Laboratory: 1983, 'Acid deposition in the United Kingdom', U.K. Review Group on Acid Rain (Warren Spring Laboratory, Stevenage, UK, SG1 2BX).

Wright, R.F.: 1983, 'Predicting acidification of North American lakes', Acid Rain Res. Rept. 4/83 (Norw. Inst. Water Research, Oslo, Norway),165pp.

Wright, R.F.: 1985, Can. J. Fish. Aquat. Sci. 42, 1103.

Wright, R.F., and Henriksen, A.: 1983, Nature 305, 422.

CHANGES IN FISH POPULATIONS IN SOUTHERNMOST NORWAY DURING THE LAST DECADE

I. H. Sevaldrud and O. K. Skogheim 1)
Directorate for Nature and Management,
Fish Research Division, Tungasletta 2, 7000 Trondheim, Norway
Present adress: 1) Norsk Biotech A/S, P. O. Box 788 Krossen,
N-4301 Sandnes, Norway

ABSTRACT. Up to 1978 51% of brown trout populations and 27% of perch populations in the four southernmost counties of Norway were lost. During 1983 the fish status of lakes in the two southernmost counties which in the period 1974 to 1978 still had fish were updated.During the period 1978 to 1983, 30% of the remaining brown trout populations and 12% of the perch populations were lost. By 1983 71% of brown trout populations and 43% of perch populations in this area have been lost. In October 1983, 623 (77%) of the lakes were sampled for analyses of water quality. The status and change in fish populations during the period from 1974-78 to 1983 were highly related to water quality status.

1. INTRODUCTION

Brown trout (Salmo trutta L.) and perch (Perca fluviatilis L.) are the most widespread species in the acidified areas in the four southernmost counties in Norway. Up to 1978 51% of the brown trout populations and 27% of the perch populations in this area were lost. Several investigations in the 1980's indicate that the deterioration of fish populations in acidified areas in southern Norway continues despite of the constant level of acidic precipitation during the last decade (SFT, 1983,1984; Andersen et al.,1984; Rosseland et al.,1985).The aim of this investigation was to update the fish status in lakes in the counties Aust-Agder and Vest-Agder which in 1978 were reported to have fish. The lakes were also surveyed in order to establish basic relationship between water quality and fish population status and changes.

2. MATERIALS AND METHODS

Sevaldrud and Muniz (1980) reported fish status for about 3400 lakes (surface area >5 ha) in southern Norway. This investigation was carried out during the years 1974 to 1978. In order to derive the trend in fish status changes, lakes in the two most severely damaged counties (Figure 1) were selected for reexamination of fish status. During 1983 fish

Water, Air and Soil Pollution 30 (1986) 381–386.

status for 591 (83%) of the trout populations and 195 (78%) of perch populations were updated by the same method as used by Sevaldrud and Muniz (1980). The time span between these two investigations is for short refered to as "the last decade".

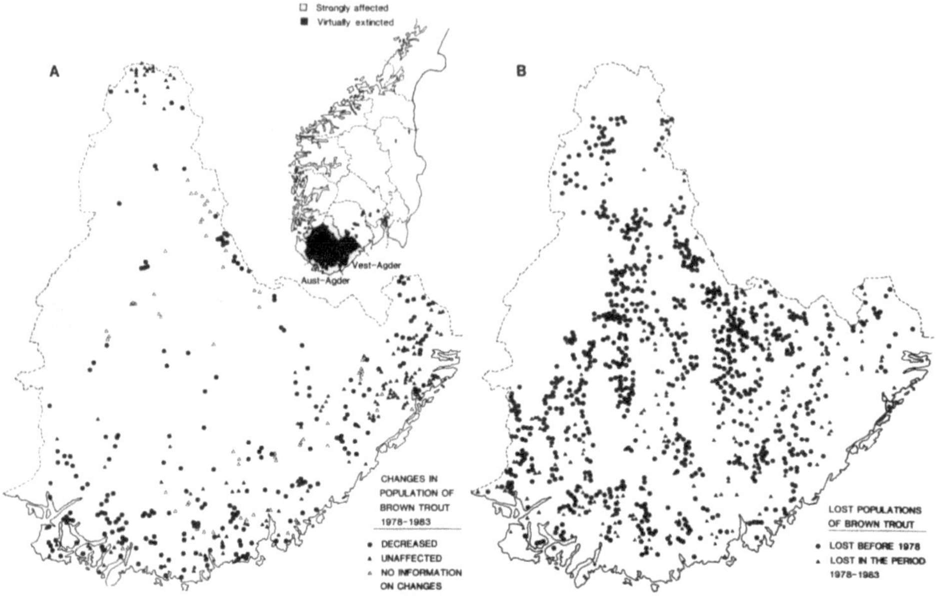

Fig. 1. A: Location of brown trout populations lost up to 1974-78 and populations lost in the period 1974-78 to 1983. Inserted: Regional distribution of areas with affected fish populations in southern Norway (Muniz and Sevaldrud,1980). B: Location of brown trout populations lost and decreased in the period 1974-78 to 1983. Unaffected and not updated populations are marked. For shaded areas no information were available.

During one week in October 1983 water samples were collected from 623 (77%) of the lakes. The samples were taken at or near the outlet of the lakes. In the field laboratory pH was measured by using Radiometer PHM29 or PHM64 equipped with a Radiometer GK2401C combination electrode and water for analyses of non-labile Al was ion-exchanged (Dowex-50,Na^+ form) and stored in polyethylene tubes whereas samples for acid-reactive Al were stored in tubes without any pretreatment. In the laboratory Al was analyzed by the method of Dougan and Wilson (1974) adapted to flow-injection-analyses technique(FIA). To samples for analyses of Ca $LaNO_3$ /HCl were added and stored in polyethylene tubes and analyzed by AAS(Perkin-Elmer mod.603). Dissolved organic matter was estimated by UV-absorbance (UVICORD LKB 8300) at 254 nm.

3. RESULTS

3.1. Status and changes in fish populations

Based on data from 591(83%) lakes inhabited by brown trout according to the former investigation, 30% of the populations were lost, 39% had decreased, while 31% were still unaffected(Figure 1A,B and 2). By 1983 a total of 71% of brown trout populations had been lost. Updating of 195 perch populations in 1983(78% of the remaining populations) showed that 12% were lost, 20% were decreased while 62% were unaffected (Figure 2). By 1983 a total of 43% of the perch populations had been lost.

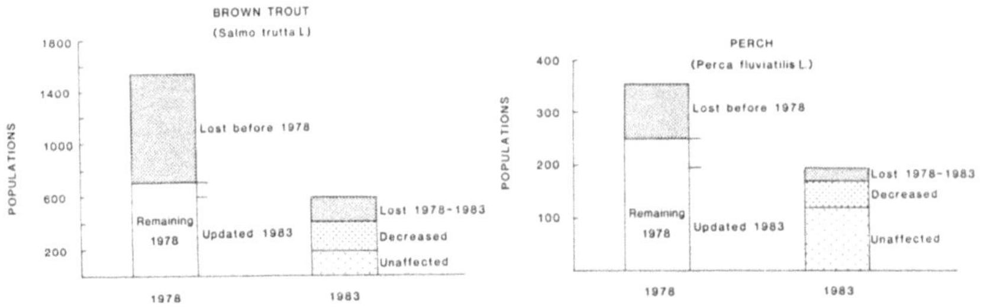

Fig. 2. Fish population status in Aust-Agder and Vest-Agder up to 1974-78 and change in the period 1974-78 to 1983.

3.2. Relationship between water quality and fish population status and changes.

For each fish species both status (good, sparse, lost) of population in 1983 and change (unaffected, decreased, lost) during the last decade were used as data sorting criteria. For each category of status/change, descriptive statistics of water quality components were calculated. Three components were selected because of their experimental documented detrimental effects on fish in acidic water: pH (Grande et al.,1978), Ca (Brown,1982), and Al (Schofield,1977; Driscoll et al.,1980). Median value, 25% and 75% quantiles of pH and mean value ± standard deviation for Ca and labile Al were selected as descriptive statistics of each fish population category (Figure 3). Total Al is separated into two fractions; labile Al and non-labile Al, generally found to be correlated to pH and dissolved organic matter, respectively (Driscoll,1980). These correlations were confirmed by the following linear regressions:

Labile Al = 577 - 93.0 pH $r = 0.710$, $n = 539$ $p<0.001$

Non-labile Al = 33.4 - 516 ORG $r = 0.772$, $n = 557$ $p<0.001$

Aluminium is given in μg Al l^{-1} and dissolved organic matter (ORG) is estimated as absorbance of UV-light (cm^{-1}). Multiple regression analyses including pH in the non-labile Al-ORG regression did not improve this correlation significantly.

4. DISCUSSION

The reported changes in fish population status during the last decade

(Figure 1 and 2) are regarded as significant. The applied interview method has been confirmed by test-fishing (Rosseland et al.,1980) and the accuracy is estimated to exceed 90% (Rosseland, 1985). This continued loss of fish populations may be due to both abiotic and biotic factors. The delay between the increased acidic deposition in the 1960's and an eventually change in water quality may be highly variable from lake to lake due to different catchment reactivity. Abiotic factors like heterogenity of hydrological, hydrochemical and physical characters of a lake-ecosystem may have vital importance for survival of fish populations in acidified areas. Further, factors like gradually desimation of the population, reduction of reproduction potential, adaption and genetic selection should be taken into account.

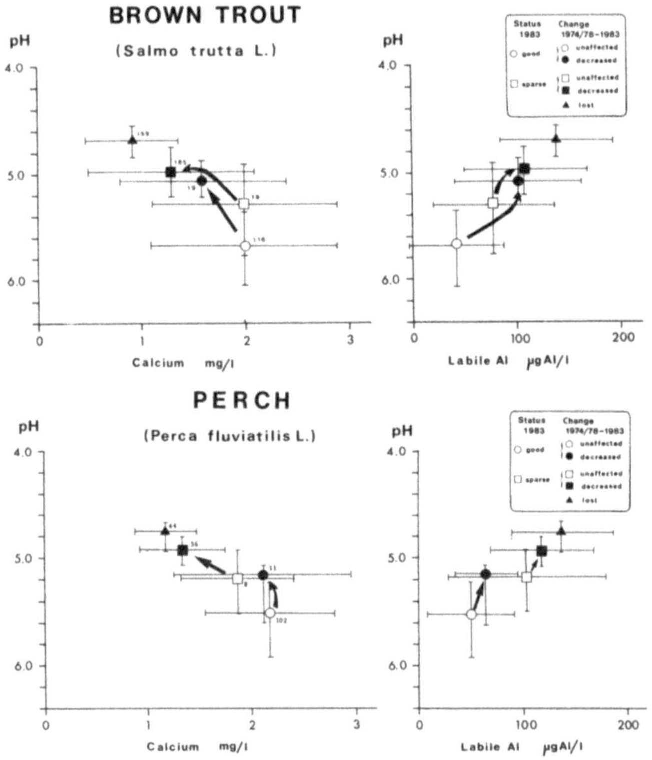

Fig. 3. Descriptive statistics for chemical components for five categories of fish populations status and change. Mean value and standard deviation for Ca and labile Al are shown. For pH median value and 25% and 75% quantiles are marked. Numbers of populations in each category are marked.

The brown trout populations affected during the last decade are located to areas at lower altitudes and near the coast(Figure 1). Brown trout populations in areas at higher altitude were recorded as virtually extinct during the former investigation. The unaffected populations, however, are restricted to an area (Bamble formation) near the eastern coast and to a small northern area with bedrock deviating from the characteristic granites and felsic bedrock elsewere in this region.

Several authors have discussed relationships between water quality and fish status (Wright and Snekvik,1978; Sevaldrud and Muniz,1980; Brown and Sadler,1981; Sadler,1983; Muniz et al.,1984). The water quality data have with few exceptions been restricted to one sample per lake, thus no spatial nor temporal variations have been taken into consideration. Muniz et al.(1984) gave several relevant objections against this use of single water samples from regional surveys and claims that such data sets have limited value to predict biological events. Despite of these obvious objections, quantitative analyses of relationships between water quality and fish status have nevertheless been undertaken. These empirical relationships are then used to predict changes in fish population status for given changes in water quality (Brown and Sadler, 1981; Sadler,1983; Wright and Henriksen,1983; Muniz et al.,1984). The value of these quantitative predictions is thus strongly limited.

The water quality data analyzed in this paper were sampled during a synoptic survey in a period of stable meteorological and hydrological conditions. Being aware of the limitations of these data, only some basic empirical relationships between water quality and fish status will be established. Despite of the large variations of the chemical components within each category, very clear patterns emerge when plotting descriptive statistics of the various categories of the pH-Ca and pH-labile Al diagrams (Figure 3). For both fish species considered population status in the order good, sparse and lost moves towards lower pH, lower concentration of Ca and higher concentration of labile Al. Likewise, if the population status have changed (e.g. decreased or lost) it moves in the same direction in the diagrammes. The pattern of fish population status in the pH-Ca diagram confirms the basic relationship between these factors and status and change in fish populations (Wright and Snekvik,1978; Sevaldrud and Muniz,1980). Wright and Snekvik (1978) did not find a general correlation between fish status and total Al concentration. Using labile Al, however, a clear pattern appears (Figure 3).

The basic relationship between water quality and fish population status and change illustrated in Figure 3 should not be expressed in quantitative terms. On the contrary, the large variation within each category should be focused on. Crucial water quality limits can be established for fish species, whereas establishment of limits for populations must include a number of abiotic and biotic factors reflecting local conditions. Quantitative predictions on fish population status and change based on simple water quality relationships are thus of limited value.

ACKNOWLEDGMENTS. We gratelully acknowledge Hans Kalleberg, Dag Matzow and Ørnulf Haraldstad for cooperation during this study. Frode Kroglund and Gøsta Hagenlund are acknowledged for water sample preparation and analysis at the field laboratory. We are also indebted to Sigmund Kalvenes and Jan Ruud Hansen for data prosessing, and Svein Fjellanger for drawing maps. The staff at Fish Research Division at Aas, especially Marit Dahl is also gratefully acknowledged for all kind of assistance. Bjørn Olav Rosseland and Dr. Hans Martin Seip are gratefully acknowledged for constructive criticism of the manuscript.

5. REFERENCES

-Andersen, R., Muniz, I.P. and Skurdal, J.: 1984, Rep. Inst. Freshw. Res. Drottningholm **61**, 5.

-Brown, D.J.A. and Sadler, K.: 1981, J. Appl. Ecol. **18**, 433.

-Brown, D.J.A.: 1982, Water, Air, Soil Pollut. **18**, 343.

-Dougan, W.K. and Wilson, A.L.: 1974, Analyst **99**,413.

-Driscoll, C.T.: 1980, 'Chemical characterization of some dilute acidified lakes and streams in the Adirondack Region of New York State', Ph. D. Thesis, Cornell Univ., N.Y.

-Driscoll, C.T., Baker, J.P., Bisogni, J.J.Jr. and Schofield, C.L.: 1980, Nature **284**, 161.

-Grande, M., Muniz, I.P. and Andersen, S.: 1978, Verh. Int. Verein. Limnol. **20**, 2076.

-Muniz, I.P., Seip, H.M. and Sevaldrud, I.H.: 1984, Water, Air, Soil Pollut. **23**, 97.

-Rosseland, B.O., Sevaldrud, I.H., Svalastog D. and Muniz, I.P.: 1980, 'Studies of freshwater fish populations - effects of acidification on reproduction, population structure, growth and food selection', in, Drabløs, D. and Tollan, A. (eds.) Ecological impact of acid precipitation, p. 336-337, SNSF-project.

-Rosseland, B.O., Skogheim, O.K. and Sevaldrud, I.H.: 1985, 'Acid deposition and effects in nordic Europe. Damage to fish populations in Scandinavia continue to apace', MS Muskoka Conference' 85.

-Rosseland, B.O.:1985, personal communication.

-Sadler, K.: 1983, Freshw. Biol. **13**, 453.

-Schofield, C.L.: 1977, 'Acid snow-smelt effects on water quality and fish survival in the Adirondack Mountains of New York State',Res.Proj. Tech.Comp.Rep. Project A-072-NY. Cornell University,Ithaca, N.Y. 27 p.

-Sevaldrud, I.H. and Muniz, I.P.: 1980, 'Sure vatn og innlandsfisket i Norge. Resultater fra intervjuundersøkelsene 1974-1979', (Acid waters and inland fisheries in Norway. Results from interviews 1974 - 1979), SNSF-prosjektet IR 77/80, 92 pp + tab.

-SFT: 1983, 'Overvåking av langtransportert forurenset luft og nedbør. Årsrapport 1982'(Monitoring of long range transported pollutants in air and precipitation. Annual report 1982), Statlig program for forurensningsovervåking.Rapp.108/83, Statens forurensningstilsyn,Oslo.

-SFT: 1984, 'Overvåking av langtransportert forurenset luft og nedbør. Årsrapport 1983',(Monitoring of long range transported pollutants in air and precipitation. Annual report 1983), Statlig program for forurensningsovervåking.Rapp.162/84, Statens forurensningstilsyn,Oslo.

-Wright, R.F. and Henriksen, A.: 1983, Nature **305**, 422.

-Wright, R.T. and Snekvik, A.: 1978, Verh. Int. Verein. Limnol. 20,765.

EVIDENCE FOR RECENT ACIDIFICATION OF LENTIC SOFT WATERS IN THE NETHERLANDS

R.S.E.W. Leuven, H.L.M. Kersten, J.A.A.R. Schuurkes, J.G.M. Roelofs and G.H.P. Arts.
Laboratory of Aquatic Ecology, Catholic University, Toernooiveld, 6525 ED Nijmegen, The Netherlands.

ABSTRACT During 1983-1984 about 90% of the Dutch soft waters appeared to be acid, and 35% of them showed pH-values below 4.0. The alkalinity of most waters (70%) was less than 0.1 $meq.L^{-1}$. Evidence for recent acidification is derived from temporal trends in water quality, fish population status, shifts in aquatic biota and application of current empirical models for lake acidification. It has been concluded that at least 59% of the Dutch soft waters have been recently acidified. This is a conservative estimate. Depending on the criterion used for assessment of acidification this percentage might be as high as 96%.

1. INTRODUCTION

Although The Netherlands are situated in the center of an acid rain area the impact of acidifying deposition on aquatic ecosystems has not been recognized at an early stage. Undoubtedly an important reason is that most surface waters are well buffered to acid additions. However, the pleistocene sandy soils in the southern and eastern parts of the country and the coastal dunes north of Bergen are highly weathered and exhibit a low acid-neutralizing capacity. Particularly in these areas many soft waters are found which seem to be highly susceptible to acidifying precipitation (Leuven and Schuurkes, 1985). Qualitative information about water acidification in The Netherlands was first obtained by comparisons of old data on diatom assemblages and macrophyte communities with recent ones (van Dam *et al.*, 1981; Roelofs, 1983).

In the present paper spatial and temporal comparisons of water quality and fish population status are described. Furthermore, several types of evidence available for assessment of recent acidification of Dutch soft waters are discussed in detail.

2. MATERIALS AND METHODS

During 1978-1985, water samples have occasionally been taken from a thousand surface waters. Since 1983 a group of 187 poorly buffered waters (alkalinity < 2 $meq.L^{-1}$) has been more regularly sampled, with a frequency of about two times per year. The sampling procedures and analytical methods, and the water quality, geographical position and some geomorphological data of the sampling sites are given by Kersten (1985). In 54 permanent, isolated soft waters the fish assemblages have been

Water, Air and Soil Pollution **30** (1986) 387–392.

investigated. The fishing methods, collected fish species and water quality data have been described by Leuven and Schuurkes (1985). Old data concerning fishery status and historical physico-chemical data have been obtained from several publications and archives. The results of archive activities, including references and analytical methods, have been summarized by Kersten (1985) and Leuven and Schuurkes (1985).

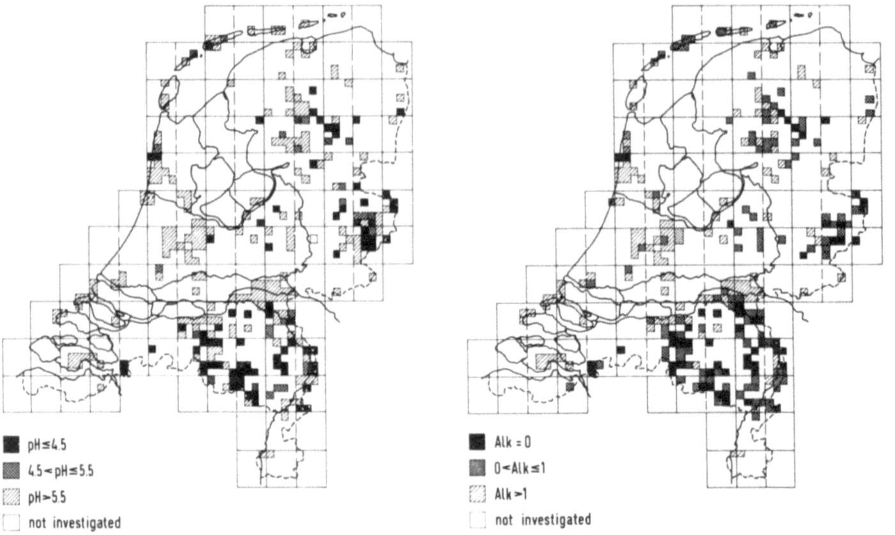

Fig. 1. The lowest pH and alkalinity values of surface waters in The Netherlands per grit of 25 km^2, measured during the period 1978-1985.

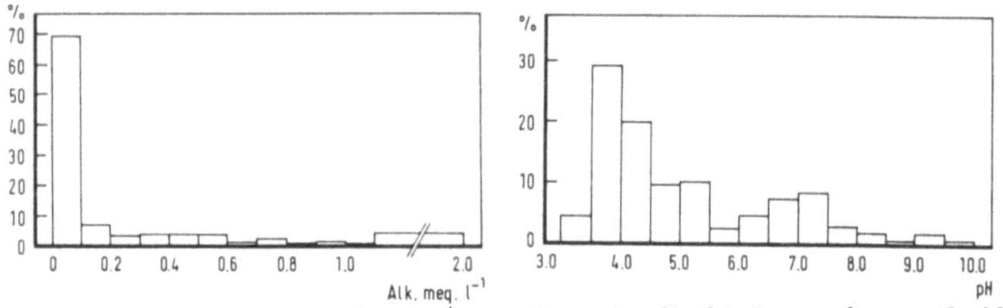

Fig. 2. The relative distribution of pH and alkalinity values of 187 lentic soft waters in The Netherlands (average values 1983-1984).

3. RESULTS

Areas sensitive to acidification can be largely identified on the basis of geological information and water quality data. The existing pH and alkalinity data of a thousand waters are summarized in Figure 1. The locations of surface waters with low pH and alkalinity correspond rather well with the areas which are practically devoid of lime.

Figure 2 shows the relative frequency of the pH and alkalinity values of a randomly selected group of soft waters, mainly moorland pools dune pools and some small lakes. These waters are generally small (< 10 ha), shallow (mean depth 0.5 m), fully mixed, and with periodically fluctuating water levels. The sampled waters reflect the variation in pH and alkalinity due to local differences in hydrology, buffering capacity of soil and water, acidifying deposition and eutrophication. About 90% of the sampled waters appeared to be acid, and 35% was extremely acid (pH < 4.0). The highest frequency of waters was recorded as being within the pH-class 3.5-4.0. The alkalinity of most waters (70%) was less than 0.1 meq.L^{-1}. The extremely acid waters were hydrologically isolated and mainly dependent on rain water. Those waters having a moderately low or high pH often received buffered ground- and/or inlet water. Some of these waters were surrounded by fertilized agricultural land and had been eutrophicated to some extent.

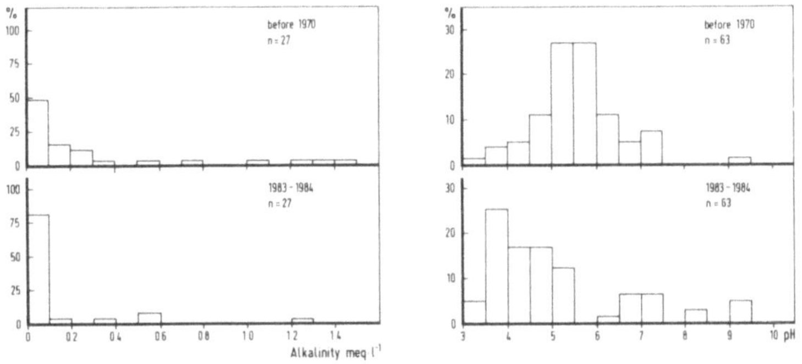

Fig. 3. Frequency distribution of pH and alkalinity levels measured in Dutch soft waters during two time spans (n: number of waters).

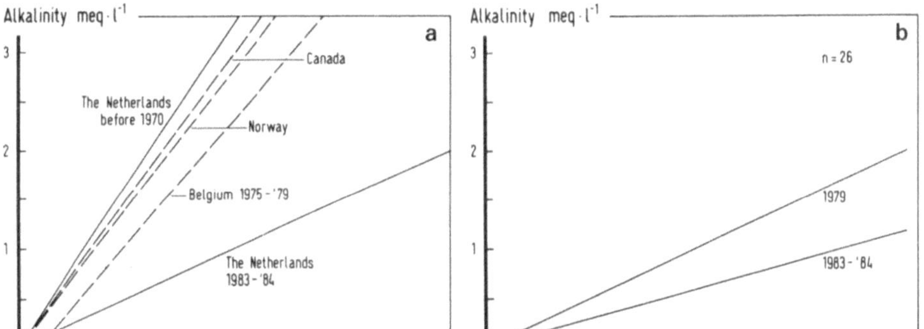

Fig. 4a. Relations between the alkalinity and Ca concentrations for 187 poorly and moderately buffered waters in The Netherlands during 1983-1984 ([Alk.] = 0.44[Ca^{2+}], r=0.80, p<0.001) for 9 soft waters in The Netherlands measured before 1970 ([Alk.] = 1.65[Ca^{2+}] - 0.14, r=0.98, p<0.005) for 59 not acidified oligotrophic lakes in Norway ([Alk.] = 1.32[Ca^{2+}] - 0.029) for 98 lakes in Canada ([Alk.] = 1.42[Ca^{2+}] - 0.032) and for the moderately acid pools and alkaline lakes in the northern part of Belgium ([Alk.] = 1.13[Ca^{2+}] - 0.23). Equations for lakes in Belgium, Canada and Norway are listed by Vangenechten (1980).

Fig. 4b. Relations between the alkalinity and Ca concentrations for 26 soft waters sampled in 1979 ([Alk.] = 0.49 [Ca^{2+}] + 0.005, r=0.5, p<0.005) and again during 1983-1984 ([Alk.] = 0.29[Ca^{2+}] + 0.05, r=0.77, p<0.005).

The results of our archive studies and literature reviews are presented in Figure 3. Reliable data on pH and alkalinity of Dutch soft waters before 1950 are scarce. In order to minimize errors associated with natural fluctuations, different sampling methods and pH measurement procedures,"average" pH and alkalinity data over a relatively large time span (1900-1970) were used as reference values. In that way a relatively large number of observations could be used and the effects of extreme values and unreliable measurements are partly eliminated. In spite of the methodical problems, comparisons of old and recent data clearly indicate acidification. Before 1970 most soft waters showed slightly acid or circumneutral pH values. A decrease in pH and alkalinity has apparently occurred in 75% and 93% of the waters, respectively. In most waters pH decreased 0 to 2 units.

The relations between alkalinity and Ca content for several groups of surface waters are illustrated in Figure 4a. In all cases the alkalinity shows a strong positive correlation with the Ca concentration. During 1983-1984 the Dutch soft waters exhibited a strong deficiency of alkalinity in comparison with unaffected lakes in Canada and Norway or older data of similar systems in The Netherlands or Belgium. Figure 4b shows alkalinity-Ca relationships for the same group of surface waters during two periods. The data were obtained by using identical sampling- and analytical methods and also indicated progressive acidification of the pools and lakes. During the last 5 yr the overall decrease in alkalinity was up to 60%.

The fishery status of 54 soft waters is summarized in Figure 5. Well-developed fish populations occur frequently in slightly acid and alkaline systems. The percentage of waters with large and species-rich fish populations decreases with increasing acidity. Most waters with a pH below 5.0 are without fish. However, historical data show that at least 67% of the nowadays acid waters formerly harbored fish populations. So, comparisons of historical data concerning fish occurrence in susceptible pools and lakes also give strong evidence for recent acidification of these waters.

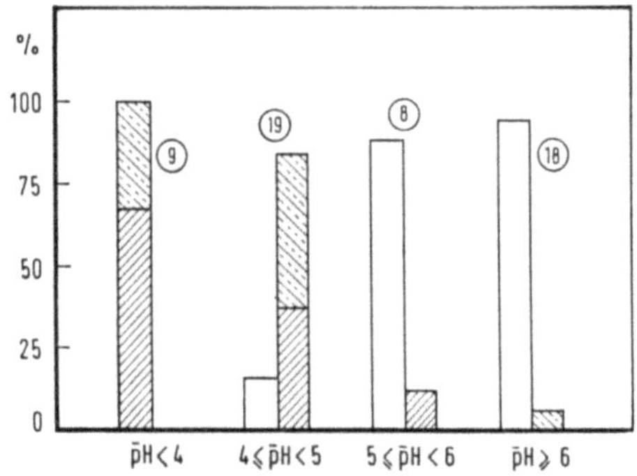

Fig 5. Fishery status of 54 poorly buffered waters in The Netherlands during 1983-1984. [open box] good fish populations, [hatched box] lost fish populations, [dotted hatched box] fishless and no historical data available.

4. DISCUSSION

In The Netherlands, several thousand pools and small lakes occur on soils susceptible to acidifying substances. These waters show remarkable geomorphological differences with sensitive waters in North America and Scandinavia. The acid-loading of Dutch surface waters is very high in comparison with other European countries (Leuven and Schuurkes, 1985; Schuurkes, 1986).

During 1983-1984 most lentic soft waters appeared to be acid and exhibited extremely low alkalinities. It is likely that these ecosystems have been altered as a result of continuous high loading with acidifying substances. This study provides evidence for their recent acidification. The results are briefly summarized in Table I. Temporal trends in water quality data clearly indicate recent acidification. However, Kramer and Tessier (1982) described that the accuracy of old data is often doubtful. Indeed many factors (e.g. differences in sampling methods, instrumentation and techniques) may be involved in biases and must be considered carefully. Nevertheless, temporal trends in recent data sets (Figure 4b) strongly support the historical reconstructions and indicate, beyond doubt, progressive acidification. Detailed time series of water quality data have been drawn up for several systems (Kersten, 1985; Leuven and Schuurkes, 1985).

Several empirical relationships have been suggested to estimate the degree of acidification or the susceptibility of surface waters to acidifying substances (Henriksen, 1982; Kramer and Tessier, 1982). The Calcite Saturation Index (CSI) of a lake indicates how well it assimilates H^+ ions without suffering a decrease in pH (Conroy *et al.*, 1974). CSI values greater than 3 represent acid waters and systems that may be affected by an input of acidifying substances. The CSI's for Dutch soft waters (average values 1983-1984) show that 77% of these waters may be affected by acid loading. Henriksen (1982) pointed out that plots of pH against Ca concentrations of lakes should distinguish between acidified and unacidified waters. Leuven and Schuurkes (1985) presented a pH-Ca scatter diagram of poorly and moderately buffered waters in The Netherlands. Most of the plotted values (96%) are found above the empirical "acidification indicator". The degree of acidification of lakes (AC-index) can be calculated according to Henriksen (1982). During 1983-1984, most Dutch soft waters (96%) reflected a positive AC-index (Leuven and Schuurkes, 1985). Although the applied models are derived from simple chemical considerations, several assumptions must be made (Kramer and Tessier, 1982). The empirical relations have been obtained from waters in a limited geographical area and generalizations of observations can be accompanied by errors.

Acidification of aquatic ecosystems may also be established by studying changes in plant and animal communities. Van Dam *et al.* (1981) compared old and recent diatom assemblages from pristine moorland pools in The Netherlands. Shifts in pH spectra indicated a considerable increase in the acidity of 69% of the waters. Roelofs (1983) investigated sixty-eight waters which were known to be dominated by *Littorella uniflora* (L.) Aschers. In 59% of the systems *L. uniflora* had been replaced by submerged *Juncus bulbosus* L. and/or *Sphagnum* spp. These changes have been caused by acidification. Comparisons of historical and recent data concerning fishery status clearly indicate recent acidification too. Arts (Unpublished data) made historical reconstructions of macrophyte communities in pristine soft waters. Preliminary results show that acidification of many systems has started already at the beginning of the 20th century, whereas most deleterious changes in aquatic biota have occurred after 1950. These results are in agreement with temporal trends in water quality data. The presented data show that at least 59% of the Dutch soft waters have recently become acidified. Depending on the method used for assessment of recent acidification, this percentage might be as high as 96% (Table I).

A major problem with respect to acidification of surface waters is to establish in what way and to what extent precipitation contributes to this process. For Dutch soft waters, the AC-index was strongly correlated with the excess of sulfate, i.e. the sulfate concentration corrected for sea spray (Leuven and Schuurkes, 1985). The excess of sulphate in Norwegian lakes is strongly correlated with the excess of sulphate or H^+ concentration in precipitation (Henriksen, 1982). In this way it has been established that H_2SO_4 in precipitation may be a major contributor to the decrease of alkalinity in surface waters. The latter is queried by Kramer and Tessier (1982). Where geological conditions allows it sulphide or sulphate weathering may also contribute to excess sulphate in surface waters. Furthermore, rain water composition in The Netherlands differs markedly from that in Norway and apart from H_2SO_4 also airborne NH_4^+ deposits play an important role in acidification processes (Leuven and Schuurkes, 1985; Schuurkes, 1986). To put this matter beyond all doubt, simulated ecosystems have been exposed to synthetic rain waters of different quality. The results of these experiments unambiguously show that in regions susceptible to acid loading the present rain water quality causes strong acidification of soft water ecosystems (Schuurkes, 1986).

Table I. Estimations of the percentage of soft waters in The Netherlands which has recently become acidified.

Method	n	%
Temporal trends in water quality		
- pH data (before 1970 - 1983/84)	63	74
- Alkalinity data (before 1970 - 1983/84)	27	93
Current empirical models		
- Calcite saturation index (CSI > 3)[1]	187	77
- Calcium - pH plots[1]	187	96
- Acidification index (AC > 0)[1]	187	96
Comparison of historical data on aquatic biota		
- Changes in diatom spectra (van Dam *et al.*,1981)	16	69
- decline of macrophyte communities (Roelofs, 1983)	69	59
- lost fish populations	54	67

n: number of waters. [1]Calculated according to Henriksen (1982).

5. REFERENCES

Conroy, N., Jeffries, D.S., and Kramer, J.R.: 1974, Proc. 9 th Canadian Symp. Wat. poll. Res. Can., 45.
Dam, H. van, Suurmond, G., and Braak, C.J.F. ter: 1981, *Hydrobiologia* **83**, 425.
Henriksen, A.: 1982, 'Susceptibility of Surface Waters to Acidification', In R.E. Johnson (Ed.), Proc. Int. Symp. Acid. Precip. Fish. Imp., American Fisheries Society, Bethesda, Maryland, p. 103.
Kersten, H.L.M.:1985,'*Fysisch-chemische Gegevens van zwak gebufferde Wateren*', Rep.158 Lab. Aquat. Ecol., Nijmegen, The Netherlands, 278 pp.
Kramer, J. and Tessier, A.: 1982, *Environ. Sci. Technol.* **16**, 606.
Leuven, R.S.E.W., and Schuurkes, J.A.A.R.: 1985. Ministerie V.R.O.M., Publikatiereeks Lucht 47, 131 pp.
Roelofs, J.G.M.: 1984, *Aquat. Bot.* **17**, 139.
Schuurkes, J.A.A.R.: 1986, *Experientia* **42**, 351.
Vangenechten, J.H.D.: 1980, *Arch. Hydrobiol.* **90**, 265.

NEW DATA FROM PEAT BOGS MAY GIVE A HISTORICAL PERSPECTIVE ON ACID DEPOSITION

W.R. Schell, A.L. Sanchez, C. Granlund
University of Pittsburgh, Graduate School of Public Health
Pittsburgh, PA 15261 U.S.A.

ABSTRACT. Dynamic changes in certain bog ecosystems together with environmental trace elements may be utilized to reconstruct the history of atmospheric chemicals deposited on the terrestrial biosphere. Cores collected from ombrotrophic bogs in eastern New York, western Pennsylvania and western Virginia were dated using the Pb-210 method and analyzed for 34 elements by neutron activation analysis. The fallout Cs-137 profiles were utilized to cross check the Pb-210 dates assigned for each layer and to estimate the transport of mobile elements similar to Cs. A comparison has been made between the relative enrichment factors of the elements accumulated at different times since 1800 on the dated peat bog layers and the crustal abundance of elements compared to Al. At the PA site, relative enrichments of greater than 100 times are found for Cl, N, S and Br; relative enrichments of 10 to 40 times are found for Pb, Ca and Sb. At the NY site, generally lower relative enrichments are found with values of 10 to 40 times for Cl, Cr, and Mn and less than 10 times for N and S.

1. INTRODUCTION

The global problem of man's input of toxic chemicals on ecosystems has shown up in recent years in fish kills, reduction and loss of forests, and destruction of buildings, roads and monuments. The question we should ask is not only what the levels of toxic chemicals are today, but how have we arrived at these levels at different regions of the world? We need historical evidence of the amounts of toxic chemicals deposited. Recent studies indicate that ombrotrophic peat bogs may have maintained such a detailed history of fallout chemicals from the atmosphere to the Earth's surface at many latitudes. Such material can complement data from polar ice sheets which have been shown to be excellent collectors of radionuclides from nuclear weapon tests starting in 1952.[1,2,3,4] The long-lived radionuclides produced naturally and from nuclear weapons tests can be used as particle tracers at the atmosphere - lithosphere boundary for latitudes other than polar using bogs and swamps. This study uses the radioactive tracers to evaluate chemicals deposited on bog ecosystems of the Appalachian Mountains of Pennsylvania (40° 7' N, 79° 11' W) and Virginia (37° 19' N, 80° 30' W), and a plateau area of Western New York (42° 19' N, 75° W).

Water, Air and Soil Pollution **30** (1986) 393–409.

2. MATERIALS AND METHODS

The samples were collected by driving a 1.5 m long, 10 cm diameter, polyvinyl chloride pipe into a low lying bog hollow away from tree roots and hummocks which could disturb the profile. The cores were stored vertically at $4^{o}C$ in a refrigerator or frozen until measured. Treatment of the core included sectioning into 1 to 2 cm layers, weighing and drying at $70^{o}C$, homogenizing by grinding in a mortar and pestle, and wet ashing with H_2O_2 - HNO_3. Three to five gram aliquots of the dry material were usually sufficient for Cs-137, Pb-210 and Ra-226 gamma radioactivity measurements using a calibrated Ge (Int) detector and pulse height analysis system. For measurement of Po-210 (used for Pb-210 dating by alpha particle detection of the daughter radionuclide),(5,6) a chemical yield tracer, Po-208, was added. The average radionuclide counting errors and their ranges were: Po-210 - 3% (2-6), Cs-137 - 14% (6-25).

For stable elements, the methods included: neutron activation analysis, plasma emission spectrometry (Spectroscan Model 3A), atomic absorption spectrometry (Perkin Elmer Models 4000 and 603), sulfur analysis with Hewlett Packard Model 78668, and carbon, hydrogen and nitrogen analyses - using Hewlett Packard Model 9909. The quality of the measurements was checked by NBS Standard Reference material and through the participation in the EPA quality assurance program for radionuclides in water.

3. RESULTS

Measurements of Pb-210 (half-life of 22.35 years) are used to determine the age for each layer of the core. Lead-210 is produced from Rn-222 (half life of 3.8 days) which decays in the atmosphere after emanating from the Ra-226 present in soil and rocks. Because of the short half life of Rn-222, a continuous source of Pb-210 is available for deposition on the earth. Since all soil contains some of the parent Ra-226, the amount of Pb-210 in excess of the amount in equilibrium, i.e. unsupported, is used to date the layers.

Figure 1 gives the unsupported Pb-210 deposition (pCi cm^{-2} yr^{-1} 0.037 Bq cm^{-2} yr^{-1}) vs depth and year for cores collected at ombrotrophic bogs in: a) New York, (NY) b) Pennsylvania, (PA) and c) Virginia, (VA). The exponential decay of the measured values vs depth shows that the 22.35 year decay half-life of Pb-210 can be used to determine the date of each layer by the constant rate of supply or CRS Model.(7,8) The CRS Model assumes that the Pb-210 flux from the atmosphere is constant and that the Pb-210 does not migrate. Any change in the matrix material, i.e., loss of mass (g) due to microbial decomposition or increase of mass due to erosion containing no Pb-210, would not affect the atmospheric Pb-210 flux whereas it would affect the dating based on the constant initial concentration or the CIC Model.

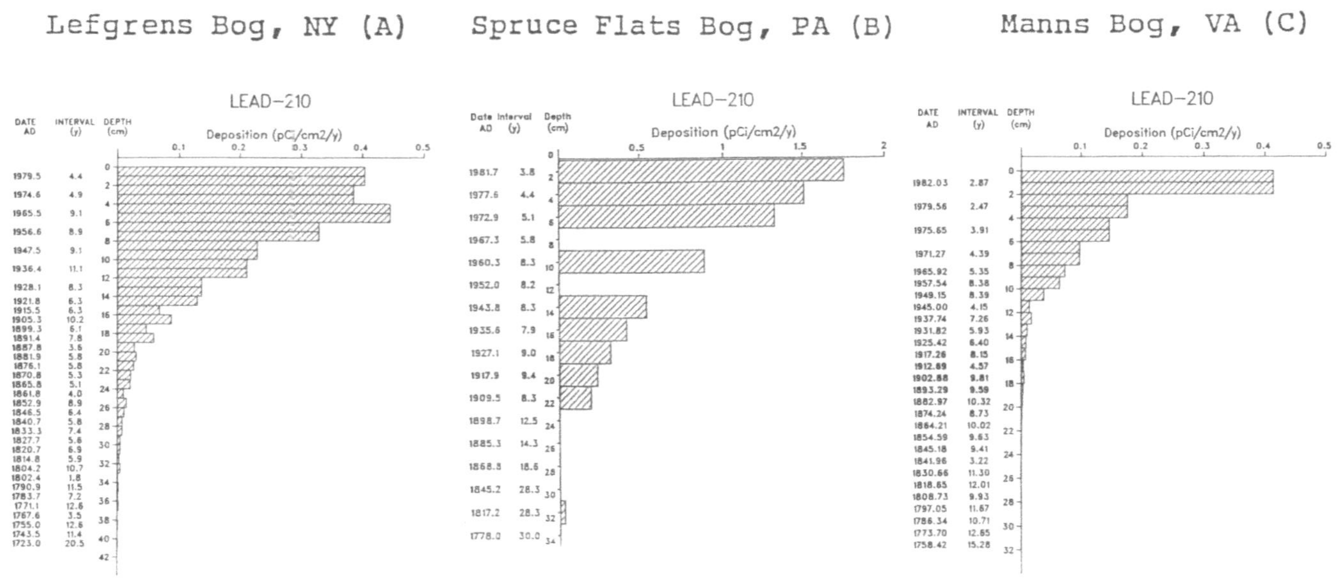

Figure 1. Deposition of unsupported Pb-210 (pCi cm^{-2} yr^{-1}) in peat bog cores collected in: A) New York (EFC-1); B) Pennsylvania (LRC-1), and C) Virginia (MAC-5).

Measurements of Cs-137 deposition on the three bogs with depth and date are shown in Figures 2A, B, and C, compared with what was

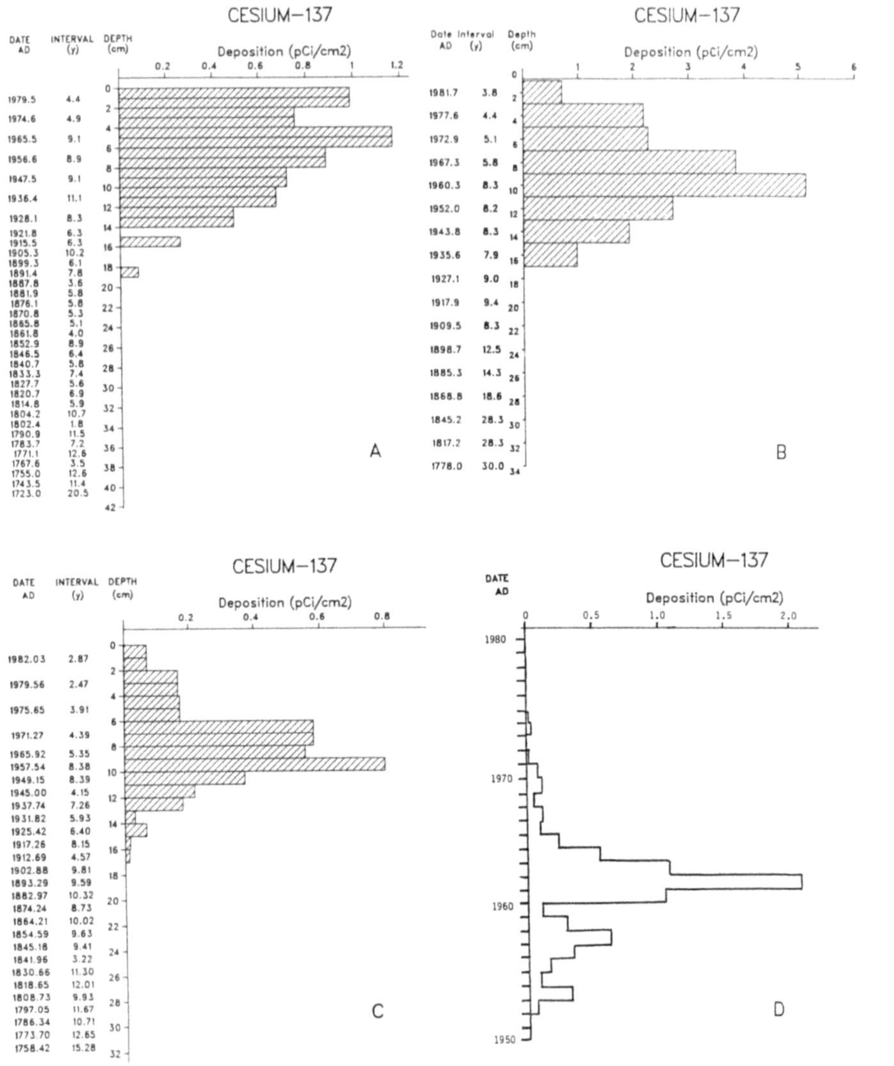

Figure 2. Deposition of Cs-137 (pCi cm^{-2}) with depth in peat bog cores collected in A) New York (EFC-1); B) Pennsylvania (LRC-1); C) Virginia (MAC-5); and D) Global Fallout.

actually deposited during the years when nuclear testing produced the fallout radionuclides (Figure 2D). The maximum 1963 fallout peak is evident in each core but the smaller 1959, 1956 and 1954 peaks are missing. This would be expected if Cs-137 has diffused over the past 30 yr since deposition. At the NY bog, Cs-137 is transported (upward and downward) to the 1890 layer, whereas, for the PA and VA bogs downward diffusion occurs to the 1920 layer.

The major constituent of a bog is the carbon content of the plant material which is usually 40% to 60% of the total dry mass. It is with this matrix that other elements must interact. Figure 3 shows the C net accumulation rate* of the three bogs studied; the H values mirror the C values. The C accumulation rate fluctuates in each section by a factor of two at the NY and VA bogs indicating the inhomogeneity of the growing peat. The PA bog appears to be more homogeneous with depth than the other two bogs, having a surface C accumulation rate of almost three times higher (30 mg cm^{-2} yr^{-1} vs ~ 10 mg cm^{-2} yr^{-1}). However, unlike the NY bog, the C in the PA bog decreases with depth reaching a value of 1% at 30 cm. The loss of C has resulted from microbial decay processes.

The net accumulation rate of S in the three bogs is shown in Figure 3. The S accumulation rate appears to have reached a maximum at the NY bog of 0.1 to 0.13 mg cm^{-2} yr^{-1} at about 1890, a peak at 1966 of 0.08 mg cm^{-2} yr^{-1} and then has decreased to the present rate of 0.02 mg cm^{-2} yr^{-1}. At the PA bog, the S reached a maximum in 1972 of 0.18 mg cm^{-2} yr^{-1} and has remained constant to the present. At the VA bog, an early peak in 1842 of 0.12 mg cm^{-2} yr^{-1} is found with peaks of 0.17 mg cm^{-2} yr^{-1} in 1949, 0.16 mg cm^{-2} yr^{-1} in 1971; the present value has decreased to 0.03 mg cm^{-2} yr^{-1}.

The net accumulation rate of N in the three bogs is shown in Figure 3. The N accumulation appears to have reached an all time maximum of 0.43 mg cm^{-2} yr^{-1} in 1804 at the NY bog with several peaks observed over time including the present maximum value of 0.31 mg cm^{-2} yr^{-1}; the minimum values have been 0.1 mg cm^{-2} yr^{-1}. At the PA bog, the N accumulation rate apparently has reached a peak value of 1 mg cm^{-2} yr^{-1} in 1972 and has remained constant to the present; the minimum values have been 0.02 mg cm^{-2} yr^{-1} before 1845. At the VA bog, peaks of 0.44 mg cm^{-2} yr^{-1} have been observed in 1949 and 1971; the present value and the minimum values observed over time are about 0.06 mg cm^{-2} yr^{-1}.

The deposition of Br is shown in Figure 3 for the PA and NY bogs; the VA bog has not yet been measured. An exponential increase in deposition at both bogs is observed after 1860 from about 0.1 μg cm^{-2}

*The net accumulation rate is the difference between the total amount of the element deposited initially minus the amount lost by aerobic or anaerobic decay processes between the time of active growth and the measurement.

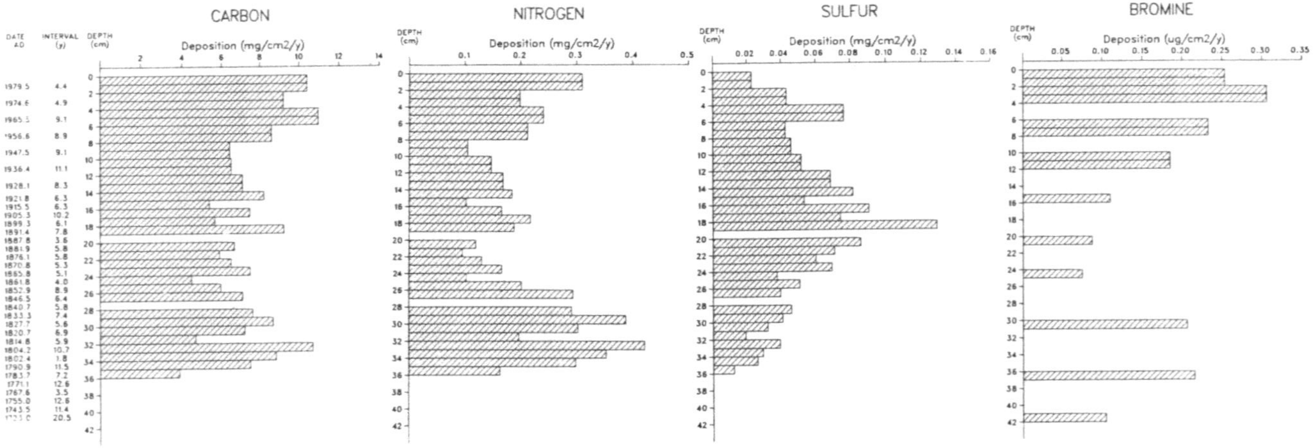

Figure 3 A. Net deposition of C, S, N, and Br in a peat bog core collected in New York, (EFC-1).

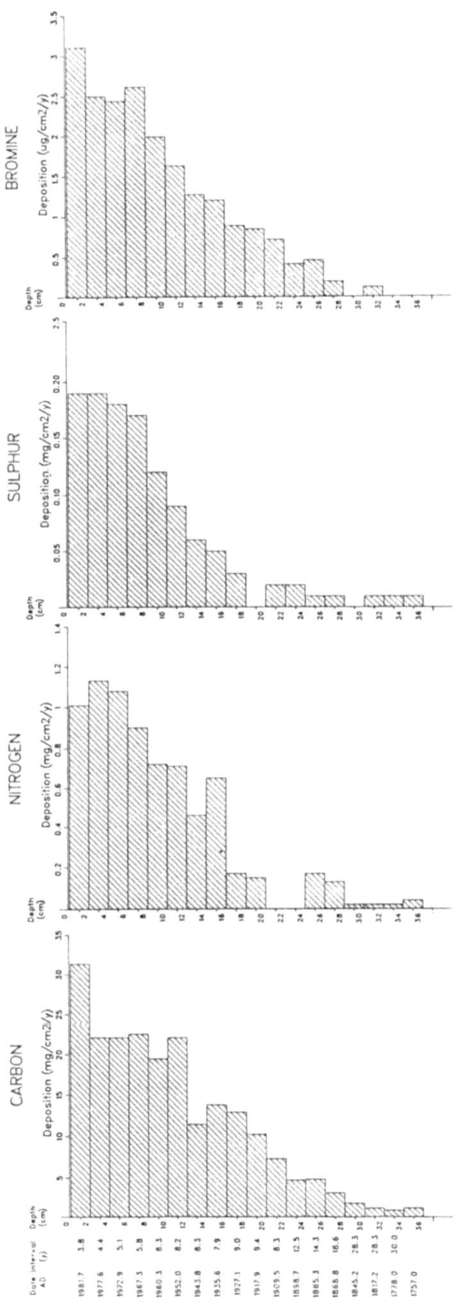

Figure 3 B. Net deposition of C, S, N, and Br in a peat bog core collected in Pennsylvania, (LRC-1).

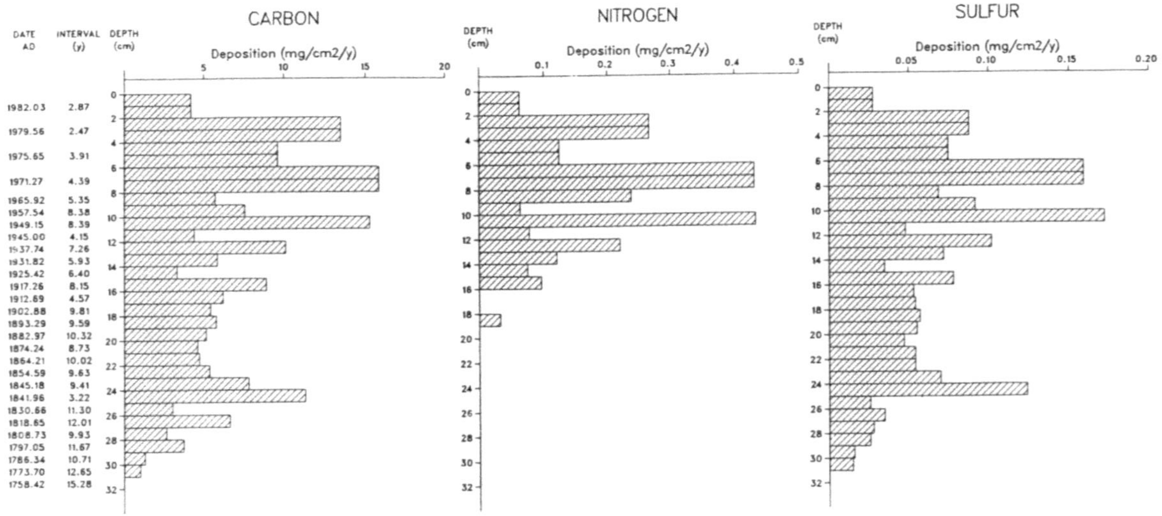

Figure 3 C. Net deposition of C, S, N, and Br in a peat bog core collected in Virginia, (MAC-5).

yr^{-1} to the present values of 3.0 $\mu g\ cm^{-2}\ yr^{-1}$ for PA and 0.3 $\mu g\ cm^{-2}\ yr^{-1}$ for NY.

Other elements measured in the PA bog are shown in Table I which gives the enrichment factor based on terrestrial abundances using Al or Sc as the normalizing factor:[9,10]

$$\text{Enrichment Factor} = Ef_x = \frac{\dfrac{\text{Element X}}{\text{Al (Sc)}}\ (\text{time t})}{\dfrac{\text{Element X}}{\text{Al (Sc)}}\ (\text{Terrestrial abundance}).}$$

At the PA bog, no relative enrichment over time, $[Ef_x(\text{present})/Ef_x(1817)]$, is observed for the rare earth metals, alkali metals, and for Mg, Ta, U, V. The elements Co, Cr, Se, W and Ba show less than a factor of two relative enrichment. The group As, Cd, Zn, P, Fe and Mn have relative enrichment factors of 2 to 10 times. The relative enrichment of 10 to 40 times is found for Pb, Ca, and Sb. The greatest relative enrichment of more than 100 times is found for Cl, N, S and Br.

For the NY bog (Table II), relative enrichment of 0 to 2 times from 1800 to the present is shown by rare earth elements, Mg and N; relative enrichments of 2 to 10 times are shown for As, Br, Co, Cs, Fe, K, Na, S, Sb, Se, Ta, U and Zn; relative enrichments of 10 to 40 times are shown for Cl, Cr and Mn.

4. DISCUSSION

Ombrotrophic peat bogs receive their nutrients and moisture from the atmosphere. Atmospheric particles are intercepted by vegetation and subsequently deposited and retained by the bog. If appropriate site locations for sample collection are selected, many chemicals present in the atmosphere and which have deposited on the bogs over the past one to two centuries can be determined. To understand the extent to which the airshed has influenced the deposition of chemicals on the bogs, a natural radioactive tracer of atmospheric particles, Pb-210, has been utilized.

The PA bog shows an example of the deposition of large amounts of particles carried by prevailing winds originating from across the great plains of North America. Here, at the first mountain range east of the Great Plains, comparatively large deposition rates of 0.89 $pCi\ cm^{-2}yr^{-1}$ for Pb-210 have been measured. At regions where the airshed is predominately from over the ocean, much lower Pb-210 deposition rates are found (e.g. Seattle, WA, 0.19 $pCi\ cm^{-2}yr^{-1}$[11]). The airshed of

Table I. Enrichment factors for elements deposited compared to Al in Spruce Flats Bog (PA), (LRC-1).

Date (y)	Interval (years)	Depth (cm)	AlConc (mg/g)	AlDep (mg/cm^2y)	As	Ba	Br	C	Ca	Cd
1981.7	3.8	2	14.2	1.05	179.97	0.78	95.86	12069	0.39	57.33
1977.6	4.4	4	31.6	2.15	23.44	0.68	37.77	4168	0.12	15.44
1972.9	5.1	6	31.9	2.22	23.22	0.69	35.88	4052	0.13	16.57
1967.3	5.8	8	51.2	5.39	12.26		15.82	1699		
1960.3	8.3	10	49.8	6.55	9.34	0.67	9.86	1208	0.02	5.71
1952.0	8.2	12								
1943.8	8.3	14	51.2	6.59	6.53	0.61	6.28	702	0.02	7.15
1935.6	7.9	16								
1927.1	9.0	18								
1917.9	9.4	20	47.8	8.74	5.39		3.16	475		
1909.5	8.3	22								
1898.7	12.5	24	47.8	8.29	3.35		1.60	230		
1885.3	14.3	26								
1868.8	18.6	28								
1845.5	28.3	30								
1817.2	28.3	32	41.2	4.54	2.85	0.42	0.83	102		
1787.0	30.0	34								
1757.0	30.0	36								

Date (y)	Ce	Cl	Co	Cr	Cs	Eu	Fe	H	Hf	K
1981.7	2.18	1239.30		19.16	2.12		2.47	147.4	1.99	0.55
1977.6	2.12	17.93	0.21	1.43	2.74	1.65	1.02	49.1	3.26	0.48
1972.9	1.87	5.51	0.18	1.28	2.81	1.28	1.05	47.7	3.08	0.60
1967.3	2.10		0.15	0.96	2.99	1.68	0.47	22.9	3.27	0.56
1960.3	2.14		0.11	1.01	3.35	1.65	0.35	16.8	4.04	0.57
1952.0										
1943.8	2.51		0.14	1.15	4.13	1.93	0.42	10.7	5.02	0.63
1935.6										
1927.1										
1917.9	2.36		0.10	1.02	3.01	1.74	0.24	7.8	5.80	0.65
1909.5										
1898.7	2.50		0.10	1.02	3.35	2.20	0.22	4.5	6.56	0.59
1885.3										
1868.8										
1845.5										
1817.2	2.60		0.12	0.96	2.39	1.76	0.21	3.7	7.70	0.67
1787.0										

the NY bog may include the Great Lakes to the west; and the smaller amount of forest vegetation present compared to the PA bog may have resulted in a lower deposition rate of 0.68 pCi cm^{-2}yr^{-1} for Pb-210. Similarly, the VA bog, which may be influenced partly by marine air, has an even lower deposition rate for Pb-210 of 0.43 pCi cm^{-2}yr^{-1}. The forest soils of New Hampshire, for comparison, have Pb-210 deposition rates of (0.35-1.07) pCi cm^{-2}yr^{-1}, and the Eastern Continental coastal deposition at New Haven, CT, is 0.45 pCi cm^{-2}yr^{-1}.[12] The actual deposition rate of Pb-210 at a collection site and of the tropospheric particles it traces depends upon the airshed as well as on the

Table I. (cont'd).

Date (y)	La	Mg	Mn	N	Na	P	Pb	Rb	S	Sb
1981.7	1.91	2.94	0.48	3898.7	0.07	7.10	71.01	1.38	57.33	3058.78
1977.6	1.81	2.37	0.14	2135.4	0.06	4.41	39.38	1.76	27.71	66.64
1972.9	1.87	2.99	0.14	1987.9	0.07	4.13	35.09	1.64	25.49	43.45
1967.3	1.82	2.96	0.10	674.9	0.06			1.80	9.77	21.44
1960.3	1.92	2.37	0.08	448.9	0.06	2.33	12.18	2.10	5.65	15.35
1952.0										
1943.8	2.02	2.70	0.08	285.8	0.06	1.51	7.57	2.39	3.05	19.05
1935.6										
1927.1										
1917.9	2.04	2.49	0.07	68.0	0.06			2.23		7.31
1909.5										
1898.7	1.95	2.73	0.07		0.06			2.65	0.65	4.59
1885.3										
1868.8										
1845.5										
1817.2	2.16	2.50	0.06	19.7	0.06	0.56	3.04	1.88	0.46	5.62
1787.0										
1757.0										

Date (y)	Sc	Se	Sm	Ta	U	V	W	Zn
1981.7	0.65	97.47	1.34			0.82		10.07
1977.6	0.63	100.85	1.34	0.83	1.37	0.81	4.96	4.12
1972.9	0.63	115.20	1.35	0.78	2.11	0.92	6.20	3.42
1967.3	0.63	85.11	1.32	0.93	1.65	0.76	4.62	2.50
1960.3	0.65	35.59	1.40	1.09	1.66	0.79	5.29	2.50
1952.0								
1943.8	0.80	58.43	1.49	1.23	1.90	0.79	4.16	2.40
1935.6								
1927.1								
1917.9	0.66	51.71	1.53	1.25	1.66	0.76	3.78	1.26
1909.5								
1898.7	0.66	68.03	1.71	1.39	1.60	0.77	2.52	
1885.3								
1868.8								
1845.5								
1817.2		59.99	1.68	1.27	1.74	0.71	1.82	
1787.0								
1757.0								

elevation, rainfall, types and amounts of vegetation and the impaction of cloud and fog droplets on foliage.[13]

In a similar way, the deposition of global fallout from nuclear weapons tests varies with latitude and with the same parameters mentioned above for Pb-210. The PA bog has the largest deposition of Cs-137, 18.8 pCi cm^{-2}, followed by NY, 6.02 pCi cm^{-2} and VA at 3.25 pCi cm^{-2} when measured in 1983-1984. The decay corrected values for Cs-137 are: NY-128 mCi km^{-2}, PA-340 mCi km^{-2} and VA-58 mCi km^{-2} which can be compared with the value estimated at Stockholm, Sweden, 66 mCi

Table II. Enrichment factors for elements deposited compared to Al in Lefgrens Bog (NY), (EFC-1).

Date (y)	Interval (years)	Depth (cm)	AlConc (mg/g)	AlDep (mg/cm^2y)	As	Br	Cr	Cl	Co	Cr
1979.5	4.4	2	1.35	0.044	50	294	3.27	12554	1.37	4.03
1974.6	4.9	4	1.15	0.041	104	459	2.46	5116	1.98	9.03
1965.5	9.1	6								
1956.6	8.9	8	1.12	0.020	84	396	1.73	2100	1.81	2.29
1947.5	9.1	10								
1936.4	11.1	12	1.86	0.022	85	250	3.29		1.00	3.28
1928.1	8.3	14								
1921.8	6.3	15								
1915.5	6.3	16	2.38	0.034	103	146	3.86	753	0.68	1.77
1905.3	10.2	17								
1899.3	6.1	18								
1891.4	7.8	19								
1887.8	3.6	20								
1881.9	5.9	21	1.16	0.016	112	193	1.96	619	1.39	1.78
1876.1	5.8	22								
1870.8	5.3	23								
1865.8	5.1	24								
1861.8	4.0	25	1.35	0.021	60	212	2.85	352	0.83	1.00
1852.9	8.9	26								
1846.5	6.4	27								
1840.7	5.7	28								
1833.3	7.4	29								
1827.7	5.7	30								
1820.7	6.9	31	0.94	0.017	12	531	1.38	396	0.72	0.83
1814.8	5.9	32								
1804.2	10.7	33								
1802.4	1.8	34								
1790.9	11.5	35								
1783.7	7.1	36								
1771.1	12.6	37	1.10	0.016	12	457	3.34	656	0.58	0.69
1767.6	3.5	38								
1755.0	12.7	39								
1743.5	11.4	40								
1723.0	20.5	41								
		42	1.03	0.008	13	446	1.68	594	0.72	0.42

km^{-2}(14). Thus Cs-137 measured at the PA bog is approximately 5 times greater than the deposition value for Sweden. As stated before, the deposition of Pb-210 and other elements are also greater than at the NY and VA bogs, indicating that atmospheric fallout is more concentrated at the PA site. No systematic study has yet been conducted to identify local differences in radioactive fallout deposition based on the aforementioned parameters which have been identified for Pb-210.

Carbon compounds present in the peat include organically bound N, S, and P, and are part of the mass which is lost. Aerobic and anaerobic microbes must be responsible for decomposition of the humus producing the volatile compounds CO_2, CH_4, $(CH_3)_2S$, H_2S, etc. The PA bog site must be particularly well suited for micro-organisms to decompose the large amount of C at a rapid rate; the NY and VA bogs have less decomposition of organic matter. Because of this loss of

Table II. (cont'd).

Date (y)	Cs	Fe	H	Hf	K	La	Mg	Mn	N	Na
1979.5		1.07	1906	4.40	3.56	2.33		4.62	44866	0.33
1974.6	7.41	1.82	2492	7.41	3.05	2.29		3.46	37148	0.31
1965.5										
1956.6		1.67	2324	2.40	1.99	2.50		3.62	45126	0.30
1947.5										
1936.4	2.90	0.80	1515	3.37	1.06	2.42		1.43	24709	0.26
1928.1										
1921.8										
1915.5	2.57	0.59	1244	4.65	1.09	2.34	2.43	0.37	16937	0.34
1905.3										
1899.3										
1891.4										
1887.8										
1881.9		1.14	2736	4.01		4.25	2.95	0.31	32912	0.33
1876.1										
1870.8										
1865.8										
1861.8		0.69	2241	5.44	1.23	2.37	2.43	0.12	36133	0.41
1852.9										
1846.5										
1840.7										
1833.3										
1827.7										
1820.7	3.47	0.66	3291	2.38	0.77	4.13		0.25	97319	0.20
1814.8										
1804.2										
1802.4										
1790.9										
1783.7										
1771.1	1.36	0.48		2.08	0.44	2.37		0.15		0.17
1767.6										
1755.0										
1743.5										
1723.0										
		0.49	2980	1.71		4.26	2.83	0.20	72153	0.11

mass, elements would exhibit a concentration which would be greater than that which was originally deposited. This increase in concentration would be proportional to the decomposition of the C and would not reflect the actual rate of element deposition from the atmosphere.

The accumulation rates of S and N appear to have varied extensively over the past 200 yr in the three bogs. Recent experiments by S. Bailey (1985) indicate that sulfate and nitrate artificially added to a peatland are fixed rapidly. Of the total S found in hardwood and conifer soils in the Adirondak Mountains (NY), 93% was present as organic bound S; 74% as C bound sulfur and 18% as ester sulfate.[15,16] Thus, organic bound S would not be mobilized but would accumulate in organic rich soils from atmospheric input. In evaluating the atmospheric input hypothesis, additional historical evidence is needed which can show that the regional air shed influencing the NY bog

Table II. (cont'd).

Date (y)	S	Sb	Sc	Se	Sm	Ta	V	Zn
1979.5	255	1518	1.09		1.99		3.52	56.69
1974.6	626	1327	1.09	603	1.99		3.93	74.70
1965.5								
1956.6	700	371	1.04	1360	1.87		3.99	88.90
1947.5								
1936.4	673	304	1.18	607	2.06	2.34	2.53	47.61
1928.1								
1921.8								
1915.5	684	299	1.29	765	2.14	1.59	1.72	24.49
1905.3								
1899.3								
1891.4								
1887.8								
1881.9	1831	378	0.55	954	1.86		0.57	63.02
1876.1								
1870.8								
1865.8								
1861.8	1019	169	1.13	444	1.91	1.48	1.07	21.77
1852.9								
1846.5								
1840.7								
1833.3								
1827.7								
1820.7	795		3.76	456	1.38		0.77	30.82
1814.8								
1804.2								
1802.4								
1790.9								
1783.7								
1771.1			1.09		1.72	0.67	0.71	13.34
1767.6								
1755.0								
1743.5								
1723.0								
	485		1.90	353	1.59		0.58	14.57

carried extensive S in 1891, and excess N in 1800-1827 for deposition. For the VA bog, S peaks in 1842, 1949, 1971 and N peaks in 1949 and 1971 must be explained by sources from its regional air shed. For the PA bog, the air shed has been influenced by regional mineral extraction and by clear cut logging and steel production industries. The net accumulation of S and N increased from about 1880 and reached a maximum at about 1972. However, these values may underestimate the actual deposition since microbial decomposition reduced the C and, in all probability, the net accumulation of N and S.

A better acidic deposition indicator, and perhaps an analogue element for S and N may be Br. In contrast to S and N, Br does not participate in the same anaerobic decay processes which would be responsible for losses of S and N in the bog ecosystem. The Br deposition has increased exponentially in both the PA and NY bogs after 1860 with a doubling rate of 35 yr. The source of Br can be coal (bituminous coal 43 ppm, NBS, SRM 1632a), oil and leaded gasoline.

Other trace elements which have been transported in the atmosphere and deposited on the bog over the amount present in the earth crust is shown by the enrichment factor EF_x.[9] Both the bog and the crustal amounts are normalized to a single abundant element such as Al(Sc). The history of the enrichment factors illustrates the change in element deposition due to man's activities. At a given site the enrichment factor before man's activities can be used as a baseline for the enrichment of elements at the present. This method probably is the most accurate means of determining relative enrichment factors of elements.

The differences in the relative enrichments of the two bogs may be partly due to the lower atmospheric input at NY and to the greater mobility of elements at the NY bog compared to the PA bog.

5. CONCLUSIONS

This study illustrates that certain mountain top bog and marsh areas can be used to determine the history of chemicals deposited on the earth. The natural and nuclear weapons produced radionuclides can be used to measure the ages of the layers deposited during the growing season of each year. However, possible migration of mobile elements such as Cs-137 must be considered in interpreting the data. Lead-210 and Pb do not appear to be mobilized during the time period studied. The organic matter is decomposed at different rates at the three sites. Thus, to evaluate the change in element deposition over time, the decomposition and loss of volatile forms of C, H, N, and S must be considered. Deposition or net accumulation rates and not concentration values should be used to identify the change in deposition over time. Carbon decomposes about 50% each 40 yr for the PA bog and about 50% for the first 40 yr for the NY and VA bogs, but little decomposition after this time. The rate of decomposition depends on the microorganisms and nutrients present.

The PA bog has the largest deposition of Pb-210 and fallout radionuclides followed by NY and VA. The PA bog also has the greatest deposition rate of trace metals and nutrients reflecting the airshed which includes the heavy industrial and mining sources upwind of the bog. Of particular interest at the NY bog is the apparent large increase in net accumulation of S in 1890, N in 1840, and Br in 1820 which was larger than the present values. It is possible that agricultural and industrial activities at these early dates upwind of the bog sites may have produced more S and N for deposition locally than the present anthropogenic inputs. Further time-series measurements are needed to determine if regional rather than long-range sources predominate in the amounts of chemicals deposited from the atmosphere.

References

1. Herron, M.M.J.: 1982, J.Geophys. Res. 87, 3052.

2. Weiss, H., Bertine, K. Koide, M.E., Goldberg, E.D.: 1975, Geochim. et Cosmochim. Acta, 39, 1.

3. Oeschger, H.: 1982, 'The Contribution of Radioactive and Chemical Dating to The Understanding of The Environmental System.,' Nuclear and Chemical Dating Techniques, L.A. Currie, Ed., 5-42, American Chemical Society, ACS Symposium Series 176, Washington, DC.

4. Koide, M., Bertine, K.K., Chow, T.J., Goldberg, E.D.: 1985, Earth and Planet Sci Lett 72, 1-8. 2.

5. Schell, W.R.: 1982, 'Dating Recent (200 Years) Events in Sediments from Lakes, Estuaries and Deep Ocean Environments Using Lead-210,' Nuclear and Chemical Dating Techniques, L.A. Currie, Ed., 331-361, American Chemical Society, ACS Symposium Series 176, Washington, DC.

6. Schell, W.R. and Barnes, R.S.: 1986, 'Environmental Isotope and Anthropogenic Tracers of Recent Lake Sedimentation,' Handbook of Environmental Isotope Geochemistry, Vol. B, The Terrestrial Environment, J.C. Fontes, P. Fritz, Eds., Elsevier, Amsterdam, in press.

7. Appleby, P.G., Oldfield, F.: 1978, Catena 5, 1.

8. Oldfield, F., Appleby, P.G., Battarbee, R.W.: 1978, Nature 271, 339.

9. Rahn, K.A.: 1976, "The Chemical Composition of the Atmospheric Aerosol," Technical Report, Graduate School of Oceanography, University of Rhode Island.

10. Rahn, K.A.: 1984, Science, 223, 132.

11. Nevissi, A.E.: 1985, Health Phys. 48, 169.

12. Graustein, W.C., Turekian, K.K.: 1986, in "Proceedings of the 4th International Symposium on Precipitation Scavenging, Dry Deposition and Resuspension," Elsevier, Amsterdam, in press.

13. Lovett, G.M., Reiners, U.A. and Olsen, R.K.: 1982, Science, 218, 1303.

14. DeGeer, L.E., Arntsing, R., Vintersued, I., Sisefsky, J., Jakobsson, S., Engstrom, J.A.: 1978, in: "Forsvarets for Skingsanstalt," FOA Report C40089-T2 (Al) 1978, and in "Environmental Quarterly," Environmental Measurements Laboratory, EML-349, 149-161, U.S. Department of Energy, New York, NY.

15. Bailey, S.: 1985, Personnel Communication.

16. David, M.B., Mitchell, M.S., and Nakas, J.P.: 1982, Soil. Sci. Soc. Am. J. **46**, 847.

POLLUTED PRECIPITATION AND THE GEOCHRONOLOGY OF MERCURY DEPOSITION IN LAKE SEDIMENT OF NORTHERN MINNESOTA

Steven A. Meger
Group for the South Fork, Inc.
P.O. Box 569
Bridgehampton, NY 11932
USA

ABSTRACT. Elevated Hg levels in game fish from wilderness lakes in northern Minnesota led to the present study of sediment cores from two lakes to ascertain the source and history of Hg deposition. Natural background levels of Hg were found to range from 0.03 to 0.06 $\mu g\ g^{-1}$, with cultural levels as high as 0.16 $\mu g\ g^{-1}$. Reconstructed geochronologies reveal a dramatic two-fold increase in Hg flux, from 0.008 to 0.017 $\mu g\ cm^{-2}\ yr^{-1}$, occuring after the year 1880, suggesting an anthropogenic influence. No industrial or geologic source of Hg is found in the study watersheds. The entire historical increase in Hg flux can be accounted for by atmospheric loading provided that 1/5 of all the Hg presently supplied to the watershed via precipitation is ultimately deposited in lake sediment. Hg levels in fish are not correlated with Hg levels in lake sediment, although there is a link to acid-sensitivity of lake water, amount of acid-neutralizing geologic material exposed in the watershed, and watershed area/lake volume ratio. Thermal stratification of lake water and a complexation-adsorption mechanism are proposed to account for variations in Hg levels observed in sediment collected from different sites.

1. INTRODUCTION

Widespread Hg contamination of game fish from northern Minnesota lakes led state health officials to issue fish consumption advisories, although the source of Hg remained unknown (Glazer and Bohlander, 1978). The present study was undertaken to investigate the history of Hg deposition as recorded by lake sediment in order to evaluate the relative influence of natural and cultural inputs. Two wilderness lakes located within the Canadian shield along the Minnesota-Ontario border, Crane Lake and Kabetogama Lake were selected for study on the basis of Hg levels detected in game fish (Figure 1). Crane Lake is recognized as containing Hg contaminated gamefish with mean Hg levels of 1.18 $\mu g\ g^{-1}$ in Northern pike (Esox lucius), and 0.79 $\mu g\ g^{-1}$ in Walleye (Stizostedion vitreum) the highest levels reported to date. Kabetogama Lake was selected for purposes of comparison, having yielded

Water, Air and Soil Pollution **30** (1986) 411–419.

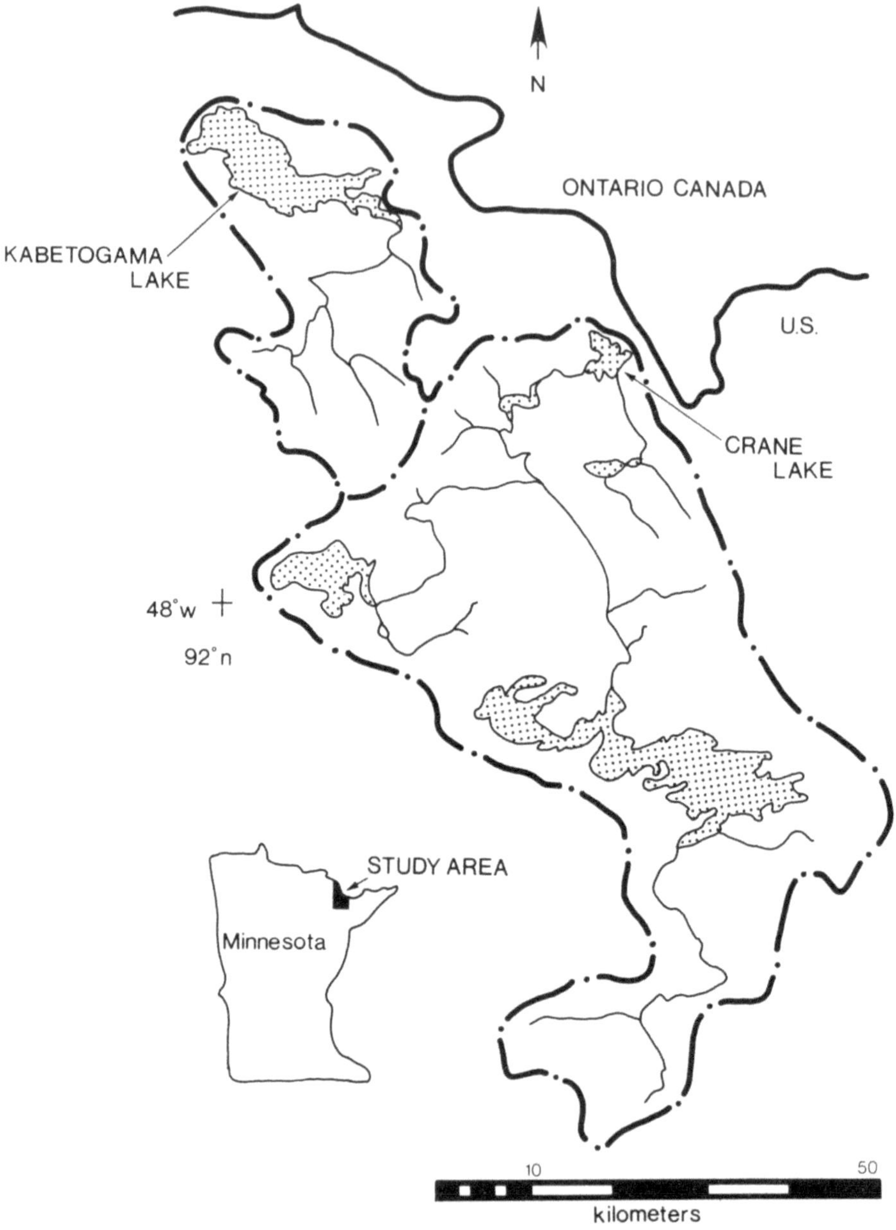

Fig. 1. Location of study lakes.

the lowest mean Hg levels detected, 0.16 µg g^{-1}, for both Northern pike and Walleye.

The study lakes show marked physical differences in lake basin and watershed morphology. Table I indicates Crane Lake is a relatively small, deep lake with a large catchment area, in contrast to Kabetogama Lake which is a larger, shallower lake with a smaller watershed.

Table I. Morphometric Parameters of Study Lakes

Lake	Surface Area (10^3 ha)	Volume (10^8 m^3)	Mean Depth (m)	Watershed Area (10^4 ha)
Crane	1.22	1.9	8.95	27.20
Kabetogama	9.38	5.18	5.53	7.25

The study watersheds lie adjacent to each other in an area characterized by a Precambrian bedrock - glacial drift complex overlain by thin acid soils derived from grantic rock. The study lakes are located in a wilderness setting which has remained essentially undisturbed by man with the exception of the lumbering of pine trees which commenced in 1893 and concluded in the mid-1930's. For purposes of this study, the year 1880 is assumed to represent the beginning of man's influence on the environment.

2. MATERIALS AND METHODS

Three lake sediment cores, each approximately 80 cm in length, were retrieved using a piston corer from a small boat. Two cores, C-1 and C-2, were collected from Crane Lake at water depths of 7 m and 21 m, respectively. Core K-1 was taken from Kabetogama Lake at a water depth of 10 m. Temperature profiles of the water column were recorded at each core site. Sediment cores were partitioned at 1-cm intervals and analyzed for total Hg by flameless atomic absorption spectrophotometry (Hatch and Ott, 1968) yielding a detection limit of 0.01 µg g^{-1}. Sediment samples were also analyzed for clay content (<2 µm) and organic C content (weight loss on ignition). Sedimentation rates for each core were ascertained by Cs-137 radionuclide dating.

3. RESULTS

All sediment cores consisted of fine-grained, olive-brown copropel. Sedimentation rates were found to be 0.28 cm yr^{-1}, 0.19 cm yr^{-1}, and 0.23 cm yr^{-1} for cores C-1, C-2, and K-1, respectively. Hg levels displayed no significant correlation with respect to either clay or organic C content.

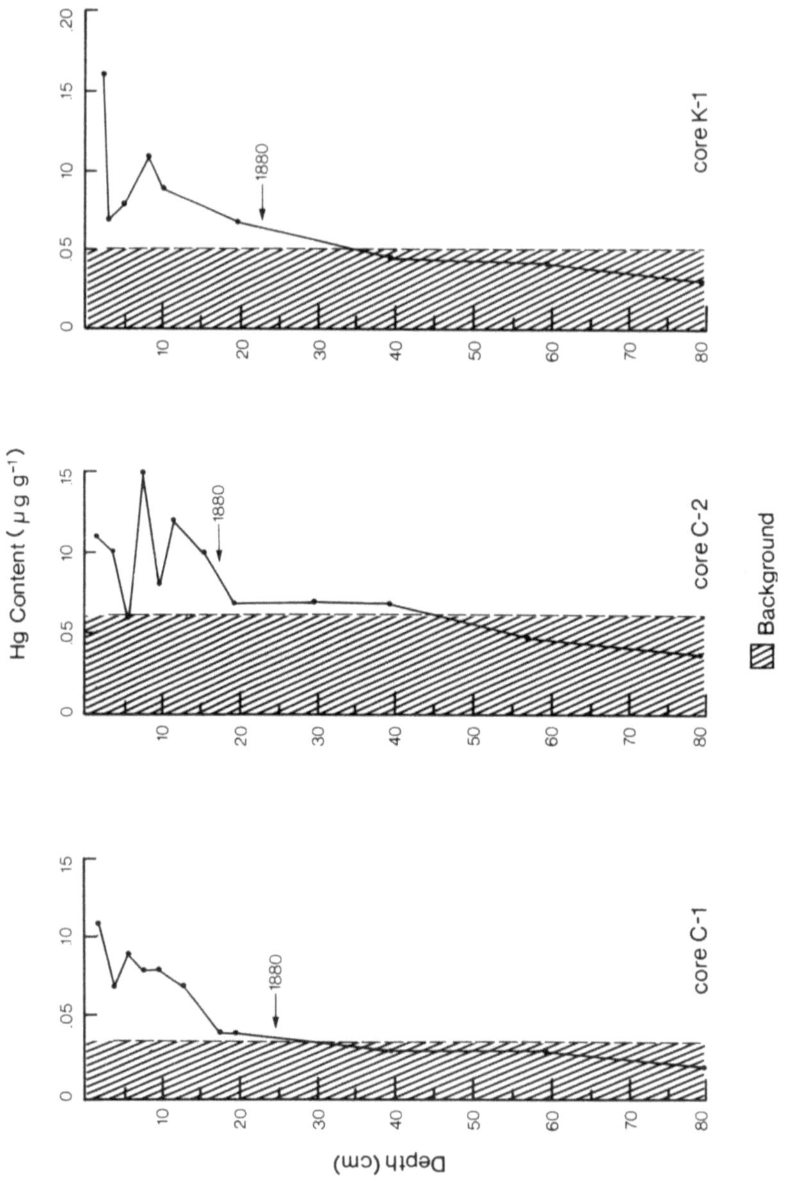

Fig. 2. Hg profiles of lake sediment cores.
Note the historical increase since the year 1880.

Figure 2 illustrates a discernible increase in Hg observed in all cores within the top ±25 cm of sediment. Natural background levels of Hg in pre-1880 sediment were found to be 0.03, 0.06, and 0.05 µg g^{-1} dw, with a two-fold enrichment in post-1880 sediment of 0.06, 0.10, and 0.09 µg g^{-1} for cores C-1, C-2, and K-1, respectively. Table II displays the increase in Hg flux observed in the post-1880 sediment of all cores.

Table II. Historical Comparison of Hg Flux in Lake Cores

(expressed in µg cm^{-2} yr^{-1})

Core	Mean Hg Flux pre-1880 (natural)	Mean Hg Flux post-1880 (cultural)	Anthropogenic Loading	Enrichment Factor
C-1	0.008	0.020	0.012	2.5X
C-2	0.007	0.013	0.006	1.9X
Crane Lake (overall)	0.008	0.017	0.009	2.1X
K-1	0.006	0.007	0.001	1.2X

Comparison of pre-1880 and post-1880 Hg flux values provides an estimate of antropogenic loading and the degree of enrichment. These Hg enrichment factors are comparable to a factor of 1.4X reported by Johnson (1985) for the Turkey Lakes system in Ontario, Canada.

4. DISCUSSION

No geological source of Hg is known to exist in the study area (Glazer and Bohlander, 1978; Siegel, 1979; Southwick, 1982). Granite and diorite rock samples collected near Crane Lake failed to yield Hg levels above the detection limit of 0.01 µg g^{-1}. Industrial sources are not found the study watersheds which lie between protected wilderness areas of Voyageurs National Park and the Boundary Waters Canoe Area. Hg from gasoline leakage of recreational boats was found to be insignificant (<0.01% of post-1880 Hg flux). A dramatic enrichment in Hg flux observed in sediment deposited since the turn of the century suggest an anthropogenic influence.

Glass and Loucks (1980) estimate the atmospheric Hg flux in northern Minnesota to be at most 0.013 g ha^{-1} per snow season. However, this estimate accounts for only one-third of the annual precipitation. Thus, the total annual atmospheric Hg flux is more on the order of 0.020 g ha^{-1} yr^{-1}. The entire increase in Hg flux observed in post-1880 sediment of Crane Lake can be accounted for by an atmospheric source, provided that 1/5 of all the Hg supplied to the watershed via precipitation is ultimately deposited in the lake bottom. This fraction of 1/5 compares favorably with the fraction of 1/7 inferred from the Hg budget presented by Bloomfield et al. (1980) for a similar wilderness lake located in the Adirondacks, New York. Atmospheric loading of Hg has been reported for remote lakes in Michigan, Switzerland, Sweden, the Lake Superior region, and Canada (D'Itri et al., 1971; Vernet and Thomas, 1972; Bjorkland et al., 1984; Glass et al., 1984; Johnson, 1985).

The inability of the Crane Lake watershed to retain Hg in the terrestrial environment may be a direct result of acidic precipitation. Hydrogen ions supplied by acid rain and snow may enhance the mobility of Hg and its transformation into biologically available forms (Suns et al., 1985). Reports of precipitation acidity in the study region range from mean pH 4.3 (Verry, 1983) to mean pH 4.82 in snow (Glass and Loucks, 1980). The absence of a large soil volume to buffer lakes coupled with a general historical decrease of lake water alkalinity in northeastern Minnesota lakes (Heiskary et al., 1981) suggests the region has been receiving acidic precipitation which may serve as a mechanism to mobilize Hg in the study watersheds.

Given that the source of Hg in the study area is atmospheric loading, it is curious that fish in Crane Lake display Hg levels several times larger than nearby Kabetogama Lake. This variation, however, can be accounted for by differences in lake water pH, the buffering capability of watersheds, and the watershed area/lake volume ratio.

A link between acidic lake water and elevated Hg levels in fish is recognized in the literature (Brouzes et al., 1977; Suns et al., 1980; Tomlinson et al., 1980; Wren and MacCrimmon, 1983). Accordingly, Crane Lake, yielding elevated Hg levels in fish, is more acidic and more suspectible to acidification (mean pH 7.2, mean alkalinity 380 $\mu eq\ L^{-1}$), than is Kabetogama Lake (mean pH 7.6, mean alkalinity 816 $\mu eq\ L^{-1}$). This variation in lake water chemistry appears to be related to a difference in the amount of acid-neutralizing material exposed in the two watersheds. Both watersheds contain Lake Agassiz Ca bicarbonate clays capable of buffering lake water acidity. However, the Kabetogama Lake watershed contains significantly more Lake Agassiz clay per unit surface area than does the Crane Lake watershed, 34% and 4% of the total watershed area, respectively.

Crane Lake also displays a larger watershed area/lake volume ratio than does Kabetogama Lake, 25.0 and 1.6 $m^2\ m^{-3}$, respectively. This trend has also been reported by Suns et al. (1980) for several lakes in Ontario where elevated Hg levels in fish are explained by the influence of watershed area and lake basin morphology acting upon Hg derived from atmospheric origins.

Although Hg levels in fish are higher in Crane Lake, this trend is not reflected by the Hg levels found in sediment. The lack of a direct relationship between Hg levels in sediment and fish is discussed by Brouzes et al. (1977).

A complexation-adsorption mechanism may account for the

variation of Hg in sediment observed between cores C-1 and C-2 in Crane Lake. Comparison of these cores reveal that both pre-1880 and post-1880 Hg levels are about twice as high in C-2. Core C-2 was retrived at a water depth of 21 m, one of deepest parts of Crane Lake. This site displayed a well-defined thermal stratification at the time of coring in mid-August. Because anaerobic conditions are likely to occur in hypolimnion waters, Hg compounds which settle into the profundal reaches of the C-2 site are likely converted to insoluble HgS. Soluble Hg complexes would tend to be adsorbed to suspended organic particulate settling downward through the water column. In contrast, core C-1 was collected at a water depth of 7 m, a site which displayed no thermal stratification at the time of coring. Aerobic conditions, which are likely to be more prevalent at this site, would tend to oxidize S to sulfate and thus limit the precipitation of HgS. Organic matter would also tend to be oxidized resulting in less particulate matter available for the adsorption and subsequent deposition of soluble Hg complexes. D'Itri et al. (1971) observe a similar trend in two lakes of northern Michigan where bottom sediment from a thermally stratified lake showed higher Hg levels than sediment collected from a lake possessing year-around aerobic conditions.

The degree to which increasing Hg flux to lake sediment is due to atmospheric loading, or to the enhanced mobility of Hg in the watershed by acidification, remains uncertain. This question could be resolved by collecting sediment cores from both poorly and well-buffered lakes and comparing trends in Hg flux. Cores could also be profiled for metals which can be attributed to natural weathering (Al, Rb, Sr, Ti, Zr) and compared to pH dependent elements (Ca, Mg), as well as to metals (Cd, Cr, Cu, Ni, Pb, Zn) whose environmental levels may be chiefly enriched by antropogenic influences.

ACKNOWLEDGMENTS

This research was made possible through the cooperation of the University of Delaware, Lamont-Doherty Geological Observatory, University of Minnesota, Minnesota Pollution Control Agency, U.S. Geological Survey - St. Paul, U.S. Environmental Protection Agency - Duluth, National Park Service, and Group for the South Fork. Special thanks is extended to my father, Martin Meger, for his technical ingenuity and overall support.

REFERENCES

Bloomfield, J.A., Quinn, S.O., Scrudato, R.J., Long, D., Richards, A., and Ryan, F.: 1980, 'Atmospheric and watershed inputs of mercury to Cranberry Lake, St. Lawrence County, New York', in Toribara, T.Y., Morton, M.W., and Morrow, P.E. IV, (eds.), Polluted Rain, Plenum Press, p. 175-210.

Bjorkland, I., Borg, H., and Johansson, K.: 1984, Ambio, 13, 118.

Brouzes, R.J. P., McLean R.A.N., and Tomlinson, G.H.: 1977, 'The link between pH of natural water and the mercury content of fish', Domtar Research Centre, Senneville, Quebec.

D'Itri, F.M., Annett, C.S., and Fast, A.W.: 1971, Marine Technology Society Journal, 5-6, 10.

Glass, G.E. and Loucks, O.L.: 1980, 'Impacts of airborne pollutants on wilderness areas along the Minnesota-Ontario border', U.S. Environmental Protection Agency, EPA-600/3-80-44.

Glass, G.E., Leonard, E.N., Chan, W.H., and Orr, D.B.: 1984, 'Airborne mercury in precipitation in the Lake Superior region', paper presented at the 27th Conference of the Lake Superior Research Program, Brock University, St. Catherines, Ontario, Canada.

Glazer, R. and Bohlander, D.: 1978, 'Mercury levels in fish from eleven northeastern Minnesota Lakes, 1977', Minnesota Department of Natural Resources, Division of Fish and Wildlife, Report 355.

Hatch, W.R., and Ott, W.L.: 1968, Analytical Chemistry, **40**, 2085.

Heiskary, S.A., Hora, M.E. and Thornton, J.D.: 1981, 'Acid precipitation impact assessment in Minnesota derived from current and historical data', Minnesota Pollution Control Agency, Division of Water Quality.

Johnson, M.G.: 1985, 'Temporal and spatial trends in metal loadings to sediment of the Turkey Lakes', in Abstracts of the International Symposium on Acidic Precipitation, Muskoka '85, Ontario, Canada.

Siegel, D.I.: 1979, 'Mercury in lakes and streams in and adjacent to Voyageurs National Park - a project proposal', U.S. Geological Survey, St. Paul, Minnesota, unpublished.

Southwick, D.L.: 1982, personal communication.

Suns, K., Curry, C., and Russell, D.: 1980, 'The effects of water quality and morphometric parameters on mercury uptake by yearling yellow perch, Ontario Ministry of the Environment, Technical Report LTS 80-1.

Suns, K., Hitchin, G., Russell, D., and Pastorek, E.: 1985, 'Metal residue accumulations in fishes from Muskoka-Haliburton lakes', in Abstracts of the International Symposium on Acidic Precipitation, Muskoka '85, Ontario, Canada.

Tomlinson, G.H., Brouzes, R.J.P., McLean, R.A.N., and Kadlecek, J.: 1980, 'The link between acid precipitation, poorly buffered waters, mercury and fish', in Drables, D. and Tollan, A., (eds.), Proceedings of an International Conference on the Ecological Impact of Acid

Precipitation, SNSF Project, Sandefjord, Norway.

Vernet, J.P. and Thomas, R.L.: 1972, Eclogae Geologicae Helvetiae, 65, 293.

Verry, E.S., 1983, Water Resources Research, 19, 454.

Wren, C.D., and MacCrimmon, H.R.: 1983, Canadian Journal of Fisheries and Aquatic Sciences, 40, 1737.

ECOLOGICAL EFFECTS OF ACIDIFICATION ON PRIMARY PRODUCERS IN AQUATIC SYSTEMS

Pamela M. Stokes
Department of Botany and
Institute for Environmental Studies
University of Toronto
Toronto, Ontario M5S 1A1 Canada

ABSTRACT. For each of the aquatic communities of the phytoplankton, the periphyton and the macrophytes, recent studies of community structure related to acidification are reviewed. While biomass and productivity do not generally decrease, there are rather consistent decreases in species richness and profound changes in species composition as pH decreases. It is unclear whether these changes result from the effect of H^+, lowered concentrations of nutrients, change in chemical form of nutrients, toxic metals, or alterations in grazing pressure, but evidence for each of these is considered. We have most information on the plankton, less on periphyton and little on macrophytes. The massive invasion of *Sphagnum* and the resulting oligotrophication hypothesized to have occurred in Swedish lakes does not seem to be occurring in other regions under acid stress. Implications for the aquatic ecosystem as a result of these structural changes include loss of resilience in the plant communities, habitat effects related to excessive growth of periphyton and macrophytes, modification of nutrient cycling by the benthic mats, and alteration of the food base for grazers. More studies are required on functional changes in primary producer communities related to acidification; these will lead to more testable hypotheses concerning mechanisms.

1. INTRODUCTION

While a great deal of attention has been paid to the relationship between decreasing pH of surface waters and loss or decline of fish populations, more recently some emphasis has also been given to the effects of acidification on communities and populations of lower trophic levels. Several recent reviews on these subjects have been published (NRCC, 1981; Haines, 1981; Dillon *et al.*, 1984; Linthurst and Altshuller, 1984) and for the most part, a similar pattern emerges for all communities. Whether based upon spatial surveys, experimental acidification or the (rare) temporal sequences, species richness decreases as pH decreases, and eventually community biomass decreases.

Water, Air and Soil Pollution **30** (1986) 421–438.

TABLE I. Studies which show increased biomass or cell numbers of dinoflagellates with decreasing pH.

Lake/Region	Type of Study	Observations/Results	References
Swedish Lakes	Survey of lakes, pH 4.6-6.8.	Biomass and cell numbers of Pyrrophyta increased at pH below 6.0; Diatomae and Cyanophytes decreased.	Dickson *et al.*, 1975
Carlyle Lake, La Cloche Mountains Ontario, Canada	Experimental acidification in enclosures.	In the pH 4.5 cylinder, increased biomass of Pyrrophyta, mainly due to *Peridinium limbatum*, and decrease of Chrysophyceae and Diatomae.	Yan and Stokes, 1978
La Cloche Mountain Lakes, Ontario Canada	Survey of nine lakes pH 4.2-7.0.	General increase of Pyrrophyta biomass with decreased pH, accompanied by decrease in Crysophyceae and Diatomae biomass.	Stokes and Yung, 1985
Experimental Lakes area, N.W. Ontario, Canada	Experimental acidification from pH 6.8 to pH 5.0 over 8 years.	Gradual decrease in Chrysophyceae, and Diatomae, increases in Peridineae and Cyanophyceae. The latter two more evident below pH 5.6.	Findlay and Saesura, 1980; Findlay, 1984
N. Florida	Thirteen lakes, pH 4.5-7.0, monitored, 1978-1979.	Loss of Cyanophyceae and increase of Chlorophyceae between pH 6 and 7; highly acidic lakes (4.5-5.0) dominated by greens and *Peridinium*.	Crisman *et al.*, 1980
Sudbury, Ontario, Canada	1973-77 study of Clearwater L, pH 4.1-4.3 and metal contaminated, compared with twelve other softwater lakes, pH 4.3-6.7.	Percentage of phytoplankton biomass of Dinophyceae negatively correlated with pH reverse true for combined Chrysophyceae and Bacillariophyceae biomass. In Clearwater L, 30-55% of phytoplankton biomass was *Peridinium inconspicuum*.	Yan, 1979
S. Central Ontario, Canada	Plastic L., pH 5.6 compared with St. Norah L., pH 6.5 over 1982-83 including ice cover.	Increase (up to 90%) in Pyrrophyta biovolume in Plastic L., compared with St. Nora with 25%. Great decrease in Bacillariophyceae in Plastic L.	O'Mahoney, 1985

The latter response is by no means as uniform or consistent as the former, and the critical pH at which it occurs is also quite variable.

In this study, the recent literature (mainly post-1980) on the response(s) of primary producers to acidification is reviewed and considered in terms of (1) community response (taxonomic composition, biomass, and functional aspects) to acidification, (2) individual species response to lowered pH, and (3) proposed or demonstrated mechanisms to explain pH related changes in (1) and (2). Three community types are recognized: the phytoplankton, the attached algae, and the macrophytes (including bryophytes). Implications of these responses and changes in primary producer communities for aquatic ecosystem are considered.

2. PHYTOPLANKTON - STRUCTURAL STUDIES

Species richness declines with declining pH. Community structure tends to show a rather consistent pattern with decreasing pH, in that the relative biomass of Chrysophyceae and Bacillariophyceae decreases, and Pyrrophyta (Peridineae) biomass increases at low pH. While this is by no means universal, the examples listed in Table I exemplify this trend for spatial surveys and experimental studies in Scandinavia, eastern Canada and Florida. When acidified lakes are limed, the trend is reversed (Yan and La France, 1984; Hultberg and Andersson, 1982).

Total biomass of the phytoplankton shows much less consistent trends with declining pH. Early studies involving a small number of lakes suggested a decline in biomass with acidification. For a range of lakes of pH 4.05 to 7.15 in the La Cloche Mountains, Kwiatkowski and Roff (1976) showed a significant non-linear correlation between chlorophyll *a* and Secchi disc transparency; the latter also showed a non linear correlation with pH. Bleiwas (1983) found a significant correlation between pH and phytoplankton biomass for nine lakes of pH 4.3 to 7.2 in the same geographical area, but when the data for the lake of lowest pH were removed from the calculation, there was no longer a significant correlation. In a survey of nine Adirondack lakes of pH 4.4 to 6.9 (Roberts *et al.* 1985) there was no relationship between chlorophyll *a* and pH, and for 20 New Hampshire lakes of pH 4.5 to 7.3 (Boylen *et al.*,1984), total phytoplankton biomass was not correlated with pH or acid neutralizing capacity. Similarly, Raddum *et al.* (1980) showed no correlation for Southern Norway lakes between pH and biomass. During the experimental acidification of Lake 223, as the pH decreased, there were increases in phytoplankton biomass, and even though concurrent increases were observed in nearby reference lakes, by the 4th year of acidification (1979) whole lake phytoplankton biomass in Lake 223 was much greater; the percentage increase since pre-acidification was also much greater than for reference lakes (Findlay and Saesura, 1980; Findlay, 1984).

Phytoplankton biomass in softwater lakes is independent of pH (Table II). Almer et al. (1978) and Yan (1979) showed for lakes of a range of pH from acidic to neutral, a correlation between total P and phytoplankton biomass, but no correlation with pH, and it is now generally agreed that "phytoplankton biomass is independent of lake pH among oligotrophic lakes" (Dillon et al., 1984). Nevertheless, (See Section 3) at extremely low pH there may be a loss of P from the water column which could result in decreased biomass.

Occurrences of individual algal species in the plankton have been related to acidification. The blue green non nitrogen-fixing *Merismopedia* is frequently dominant in terms of cell numbers in acidic lakes (Kwaitkowski and Roff, 1976; Yan and Stokes, 1978), but species of the dinoflagellate *Peridinium* - *P. inconspicuum* and *P. limbatum* are most

TABLE II. Range of phytoplankton biomass in selected studies of softwater Swedish and North American lakes.

Lake	Biomass $mg.m^{-3}$	Reference
Swedish lakes	< 1.0	Almer et al., 1974
Labrador softwater lakes	35-533	Duthie and Ostrofsky, 1974
Ontario softwater lakes		
pH 5.5-7.8	100-830	Dillon et al., 1979
pH 4.2-5.1	260-730	Dillon et al., 1979
ELA Lake 223		
at pH 5.6 Epilimnion	650-825	Schindler, 1980
Hypolimnion	950-2100	
Carlyle Lake, Killarney	50-60	Yan and Stokes, 1978
Killarney, Ontario softwater lakes, 1976	31-259	Stokes and Yung, 1985
Plastic Lake, Ontario	40-390	O'Mahoney, 1985
St. Nora Lake, Ontario pH 6.5	40-700	O'Mahoney, 1985

frequently cited as dominating the biomass at low pH (Yan and Stokes, 1978; Yan and La France, 1984; Dickson et al., 1975; Crisman et al., 1980; O'Mahoney, 1985; Havens, 1984). The dinoflagellate, Gymnodinium became important as Lake 223 was acidified (Schindler et al., 1985). In contrast, the acidic Ruth Roy lake (pH 4.3 to 4.5) in the La Cloche Mountains of Ontario did not show a dominance of dinoflagellates (Bleiwas, 1983). The diatom Asterionella ralfsii, known as an acidobiontic diatom and one of many diatom species used as indicators by palaeolimnologists (e.g., Charles, 1984), became important for the first time in Lake 223 when the pH dropped to 5.2 to 5.6 (Findlay, 1984)

Another planktonic species of interest in acidifying lakes is the prymnesiophyte Chrysochromulina breviturrita. This organism has been associated with odour-causing blooms and is typically found in poorly buffered moderately acidic (pH 5.6 to 6.2) lakes (Nicholls et al., 1982). It has been collected from over 80 such lakes and has recently been studied in pure culture by Wehr et al. (1985). Its pH tolerance in the laboratory is rather broad, with abrupt declines in growth rate and yield at pH 3.5 and 6.9, respectively.

3. PHYTOPLANKTON: FUNCTIONAL AND MECHANISTIC STUDIES

Few attempts have been made to measure primary productivity in a comparative way for acidic lakes, but studies indicate that productivity, like biomass, is unrelated to pH. In a comparison between two lakes in Ontario, over the period September 1982 to June 1983, O'Mahoney (1985) obtained values for ^{14}C fixation of 1 to 12.5 ug C $L^{-1}h^{-1}$ in Plastic lake (pH 5.6) and 0.3 to 4.5 in St. Nora Lake (pH 6.5). Shearer et al. (1986) showed that ^{14}C fixation rates in Lake 223 did not decrease with acidification.

Phosphorus is the most common limiting nutrient for planktonic algae and it has been suggested that as lakes acidify P chemistry is altered (eg. Dickson, 1978). Because Al concentrations increase at low pH, P may be removed from water through adsorption or precipitation reactions with Al. Actual demonstration that P concentrations or P availability decreased at low pH are not available; where surveys have been made there is no consistent pattern of P with pH. As pointed out by Schindler et al. (1985), Lake 223 differs from lakes acidified by atmospheric deposition, in its much lower levels of Al, resulting from lack of terrestrial acidification. The effect of acidification on production in Lake 223 may therefore be atypical, because of the low Al in the experiment, and caution is required in extrapolating the results for primary production from this situation to lakes acidified by atmospheric deposition. Even so, the species composition of Lake 223 resembles other acidic lakes in quite a striking manner. Most studies agree that there is no consistent decrease in productivity as pH declines (Linthurst and Altshuller, 1984). This is not really surprising; if organisms that occur at the lower pHs are acid tolerant, they should be able to fix C efficiently at low pH.

Although studies are scarce, there are some suggestions that other nutrients may partly explain the pH preference of certain acidobiontic or acidophilic species. For example, the appearance of *Asterionella ralfsii* in Lake 223 at pH 5.2 to 5.6 coincided with a sharp decrease in Si (Findlay, 1984). It has been shown that variation in Si uptake kinetics may confer competitive advantage on certain diatom species (Tilman and Kilham, 1976), and it is possible that *A. ralfsii* may be more efficient at transporting or storing Si than other diatoms. In Lake 223, the diatoms that existed at higher pH and higher Si were lost as *S. ralfsii* increased. Other studies have shown a decrease in Si as pH decreases (e.g., Davis, 1985) but have not yet shown correlated changes in diatom species. Culture studies are required to elucidate this and other mechanisms.

The laboratory studies of Wehr *et al.* (1985) on *Chrysochromulina breviturrita* indicate that the upper pH limit is probably determined by bicarbonate, since alkalinities above 84 µeq L^{-1} were inhibitory, and since additions of bicarbonate definitely decreased growth and yield. If *C. breviturrita* cannot use HCO_3, but only CO_2, as a C source then, above pH 6.0, bicarbonate alkalinity would rapidly affect growth. The authors could not, however, explain the lower pH limit in terms of C source, since they demonstrated that there was no C limitation at pH 3.5. They also suggested that ammonium was the preferred N source, since yields were lower on nitrate than on ammonium salts, but N preference did not appear to be the major factor controlling growth at different pHs.

Metal toxicity is a potential problem for algae in acidic lakes. Even in lakes remote from smelter sources, Al, Mn and Zn concentrations increase as pH decreases (e.g., Bailey and Stokes, 1984), and Al is present at concentrations toxic to fish (Baker and Schofield, 1982). In acidic lakes close to smelters, concentrations of additional metals may reach toxic concentations. Phytoplankton in several acidic lakes polluted by smelters with elevated Cu and Ni (as well as high Al, Mn and Zn) do not differ very much from lakes of comparable pH without the smelter metals (Yan and Dillon, 1984). Possibly the algae have become metal tolerant (Yan, 1979) or the metals at low pH are not taken up by the algae (Campbell and Stokes, 1985). For Al, Mn and Zn, which covary with H+, it is not possible to separate the H+ effect from the metal effect in field studies. To date, no culture studies on planktonic algae have clearly resolved the respective roles of H+ and Al in limiting the number of species that can occur in low pH lakes.

Other possible mechanisms that could control the phytoplankton composition and biomass in acidic lakes relate to second order effects of acidification. Of these, grazing is the most frequently cited. A number of authors have demonstrated the complex effects that herbivore grazing can have on phytoplankton, including depletion of biomass by cropping, changing composition by selective grazing or differential assimilation of species, alteration of efficiency of production through nutrient regeneration, reduction of productivity by intensive grazing or even enhancement of certain species by removal of competitors (e.g., Redfield, 1980; Porter, 1976; Lynch and Shapiro, 1981). Few studies have attempted to evaluate the effects of lowering pH on the effects of grazers on phytoplankton. However, pH affects the size distribution and species composition of zooplankton (Sprules, 1975; Bleiwas, 1983) and therefore could affect grazing pressure by any or all of the mechanisms listed above. The small crustacean herbivores, e.g., Bosmina sp. and Diaptomus sp., which come to dominate the herbivorous zooplankton in acidic lakes, are probably not capable of grazing larger algae, and the success of the dinoflagellates in acidic lakes may be related to this size change in the herbivores. In a study of an acidic lake near Sudbury, Ontario, gut content analysis revealed no Peridinium in the herbivores, although the alga was abundant in the lake (Yan and Strus, 1980). However, Peridinium is probably not grazed in neutral lakes either, so this is unlikely to be the complete explanation. Filtering rates in acidic and circumneutral lakes were measured by Bleiwas (1983) but she found no significant relationship between community grazing rates and pH. Yan and Strus (1980) discussed the possible impact of phytoplankton on zooplankton in the acidic Clearwater Lake near Sudbury. They concluded that the volume of algae filtered by the zooplankton at the measured density of the latter in the lake would be relatively small. This would also have meant that there was only very limited grazing pressure on the phytoplankton as a whole.

In summary, no clear support for any of the hypotheses relating to phytoplankton response to acidification has yet emerged.

4. PERIPHYTON: STRUCTURAL STUDIES

One of the striking, visible plant responses to acidification is the development of benthic or epiphytic mats or clouds of filamentous green algae or (mainly on sediments) mats of blue-green algae (Singer, 1983; Boylen et al., 1984; Turner et al., 1986; Jackson, 1985). The dominant algae in the green mats are almost invariably species of Mougeotia and Zygnema (Jackson, 1985), but Spirogyra can also occur at moderately low pH (Turner et al., 1986). Several workers have also documented blue-green dominated mats (Lazarek, 1983; Turner et al., 1986; Singer, 1983). Species richness, especially of diatoms in the mats, tends to decrease with decreasing pH (Stokes, 1981).

Müller (1980) described the species composition and productivity, for periphytic algae in artificially acidified cylinders in a lake, and found the now familiar phenomenon of Mougeotia dominance described earlier by Hendrey (1976). In a recent study (Turner et al., 1986), biomass was estimated by mapping the percentage of shoreline affected for 3 acidified lakes: Lake 223, Lake 203N and 203S, as well as a reference neutral lake, Lake 239, in the Experimental Lakes area of northwestern Ontario. Mougeotia sp. dominated the shoreline of L223 with 99-100% coverage of rocks, while Mougeotia with Zygnema dominated the S and acid basin of Lake 302 (pH 5.6) with 100% rock cover in 1984.

The HNO_3 acid basin of Lake 302 (pH 6.2) was dominated by Spyrogyra and Zygnema but with only 7% cover on rocks. There was no shoreline growth of filamentous green algae for Lake 239 (pH 6.8). For the HNO_3 acid site, it may be that the pH is not yet sufficiently low to trigger the growth of the filamentous greens, or it may be that HNO_3 acid per se does not elicit the same response as H_2SO_4.

In several other studies, standing stock of attached algal biomass has been estimated variously by chl a, cover, or dry weight, and seems to be consistently increased at low pH (Boylen et al., 1984; Muller, 1980; Parent et al., 1985; Stokes, 1984). In addition, there is descriptive or anecdotal information from a number of areas in eastern North America, Scandinavia and southwestern U.S.A. which confirms the presence of a conspicuous green algal mat in many acidic systems.

The basis for comparison among lakes in the context of standing stock or biomass of the attached algal communities is quite unsatisfactory because of the lack of standardized methods for measurement. All that can really be said is that the phenomenon is widespread in lakes and streams of pH 6.0 or less, with more dramatic visible effects below pH 5.5. Although widespread, the phenomenon does not occur universally in acidic systems nor is it confined to acidic systems; occasionally large Zygnematacean mats are found in neutral waters. A recent observation by Jackson (1985) on Bowland Lake, Ontario, described how the existing algal mat was almost completely eliminated when the pH was raised to > 6.0 by liming. This occurred over a single year.

5. PERIPHYTON: FUNCTIONAL AND MECHANISTIC STUDIES

The increased standing stock of periphytic algae at low pH has led us to question whether these algae are in fact more productive at low pH, or whether they grow at essentially the same rate across their pH range, while potential removal processes are altered at low pH. Relatively few studies have been made to date on production rates of periphyton. Hendrey (1976) considered that ^{14}C fixation, normalized to chl a, was greater at the low pH than at neutral pH in his acidified stream channels. Müller's (1980) study of periphyton in acidified and control enclosures showed considerable variation in production rates and also seasonal differences, but in summary, production based on unit biomass was 6>4.5=6.5 in June, while in August the ranking was pH 5>6.5>4.5>6. These results may have to re-interpreted in light of problems with methodology (Turner et al., 1986).

Production in the epilithic periphyton of the middle littoral zone in the acidified Lake 223, Lake 203S and 203N was studied by Turner et al. (1985). Below the wave zone, net photosynthesis in the algal mats (composed mainly of blue-greens with some greens and diatoms), decreased with decreasing pH although the community structure and biomass was in general unchanged from pH 6.9 to 5.6. It should be noted that Turner et al. did not include communities within the wave zone, i.e., those dominated by filamentous greens. As well as showing pH dependence, net photosynthetic rates of the haptobenthos, at ambient dissolved inorganic carbon (DIC) compared with rates at elevated DIC, showed a significant increase at the higher DIC. For all the Lakes, 223, 302S, and 302N, the authors suggest that the community is C limited at low pH.

Growth rates of Mougeotia spp. isolated from acidic lakes and grown in the laboratory under controlled conditions revealed a distinct low pH preference for two isolates of Mougeotia from Plastic Lake, Ontario (pH 5.6) (Jackson, 1985). Turner and Stokes (1985) compared growth rates of two different Mougeotia isolates from Ruth Roy Lake (pH 4.5) and Chub Lake (pH 5.8) and found that the isolate from the less acidic site grew more poorly at pH 4.5 than did the isolate from the acidic site.

Turner et al. (1986) have proposed adaptation to C limitation as a mechanism to explain the success of Mougeotia and other filamentous green algae in acidic lakes. For the experimentally acidified lakes to pHs $\geq$ 5.6 , the filamentous algae are predominantly restricted to the wave zone. These authors suggest that under conditions of limiting DIC, the growth habit of these algae (i.e., filamentous, flexible and loosely attached) enables them to maximize uptake of dissolved inorganic carbon. Their surface area is increased and limitation of gas transport across the boundary layer is decreased by the constant movement that they experience in the wave zone. The compacted blue-green mat, on the other hand, is less well-adapted to DIC limitation. The authors also suggest that below the wave zone, the blue-greens may prevent attachment of the filamentous greens. Neither of these hypothesis has yet been tested.

Although Turner et al. have not yet seen the filamentous green community below the wave zone, this is not true for other studies. Several authors describe green algal mats in fairly deep quiet waters. These are always in very clear water (Stokes, 1981; Hendrey and Vertucci, 1980; Singer, 1983). The role of light as a factor for the expansion of the green algal community has not been elucidated. Mats only occur deeper than the littoral zone in very clear lakes, and increased clarity is related to acidification (Almer et al., 1978; Schindler, 1980; Singer, 1983); however there has been no systematic survey to correlate the two and no evidence exists for a causal relationship.

Proposed mechanisms to explain the periphyton acidification phenomenon include:

(i) The algae have a low pH preference, i.e., are acidobiontic

(ii) The herbivores that normally graze and limit the biomass are absent below pH 5.6 and certainly begin to decline as pH decreases below 6.0.

(iii) Microbial decomposition is decreased at low pH, permitting more biomass to accumulate.

(iv) Competition from other algae and other members of the aufwuchs (periphyton) is decreased at low pH.

The studies described in this section relate only to hypothesis (i). As for the other proposals, there is only circumstantial evidence for (ii) and no work to date has been published for the periphytic community relating to (iii) or (iv).

6. MACROPHYTES: STRUCTURAL ASPECTS

In an early study of lake acidification, Grahn and Hultberg (1974) described a process termed "oligotrophication", in which the community of the bryophyte *Sphagnum* played a key role in the biological and chemical changes that occurred as a lake acidified. For a number of softwater lakes in Sweden, these workers noted invasions of *Sphagnum* moss which spread over the lake bottom, apparently eliminating the normal isoetid macrophyte communities. They proposed that once established, *Sphaghnum* actually contributed to acidification by exchanging nutrient cations for H^+. Since that time, Grahn (1985) has described a time series of *Sphagnum* invasion in the acidic Lake Gärdsjön in Sweden. In 1980, Hendrey and Vertucci reported on extensive *Sphagnum* mats in Lake Colden in New York; the moss appeared to have increased its cover since the earlier part of the century. However, the *Sphagnum* invasion has not yet been described elsewhere in eastern North America or in other parts of Europe. For lakes ranging from neutral to 3.9 in the Netherlands, the studies of Roelofs (1983) and Roelofs *et al.* (1984) identify losses of isoetid species, notably *Littorella*, and replacement by *Juncus bulbosus*, with some *Sphagnum*, but the latter was not described as forming extensive mats. The surveys of Roberts *et al.* (1985) for nine Adirondack lakes, of Boylen *et al.* (1984) for 20 New Hampshire lakes, and of Yan *et al.* (1985) for 30 Ontario lakes, each identify certain macrophyte species as acid tolerant or acidobiontic; however, none of these studies reported extensive Sphagnum mats.

Species richness was determined by Yan *et al.* (1985) for 30 lakes of pH 7.0 to 4.0. They found no significant correlation with pH for tracheophyte richness, or for any macrophytes, but they did find a highly significant negative correlation for pH and bryophyte richness, and a positive correlation for charophytes. Charophyte and tracheophyte richness was significantly correlated with calcium concentration over the range 2.0 to 5.0 mg L^{-1} Ca. The Roberts *et al.* study found a decrease in species richness of macrophytes for nine lakes in the Adirondacks, over the range 6.9 to 4.4. But Boylen *et al.* (1984) found no correlation between species richness and pH for 20 New Hampshire lakes, pH 7.3 to 4.5, using the same methods as the previous study.

Macrophyte biomass was measured in several of the studies noted above but only for Ontario are there data on a series of lakes across a gradient of acidity, even for softwater lakes, ranging in g m^{-2} from 0.085 to 0.55 (Wilson, 1935, cited in Hutchinson, 1975) through 8.1 (Moeller, 1975) to 20.0 (Grahn, 1985) to 40.0 (Eriksson, 1973, cited in Wile et al., 1985) to 59-326W (Wile et al., 1985). Macrophyte biomass is extremely variable. This variation may partly be a reflection of the methods used. However, the survey of Wile et al. (1985) used the same method for all the lakes and showed that the lake of lowest pH, which was Clearwater Lake in Sudbury, had the highest macrophyte biomass of their set; in fact, Clearwater Lake had the highest macrophyte biomass for any softwater lake on record: 3260 g m^{-2}!

Functional and mechanistic studies of macrophytes in acid-stressed lakes are lacking except for two examples. Production estimates of *Sphagnum* growth rates described by Grahn (1985) for a single lake indicated high rates for *Sphagnum* compared with other macrophytes in the acidic Lake Gardsjon. Grahn suggested that this was related to the fact that *Sphagnum* can assimilate free CO_2 exclusively and grow in shade, and to the fact that mosses are not limited by hydrostatic pressure and can grow at great depths. In this context, Singer (1983) noted that *Utricularia geminiscapa*, which grows in 18 m of water in the acidic Silver Lake, was clearly not affected or limited by hydrostatic pressure.

Roelofs et al. (1984) suggested that the success of *Juncus bulbosus* in acidic Lakes and its competitive advantage over *Littorella* were related to its ability to tolerate and use CO_2 from the water column, while most isoetids can only obtain their inorganic C from sediment. Roelofs supported this experimentally and also suggested that ammonium-N was preferred over nitrate-N by the acidobiontic *Juncus bulbosus*.

7. SUMMARY, RESEARCH NEED AND IMPLICATIONS FOR AQUATIC ECOSYSTEMS

Information on the three major primary producer communities with respect to acidification decreases from phytoplankton, to periphyton to macrophytes. For all studies, with the possible exception of recent ones on macrophytes, it has been shown that species composition changes and species richness declines with pH but that biomass does not follow this pattern. The effects of these changes on higher trophic levels are probably minimal. For the producer communities it is likely that the simplification resulting from decrease in richness may make the community less resilient to additional stresses of a natural or anthropogenic nature. This was suggested by Marmorek (1984), based on Yan's data from Clearwater Lake and on Marmorek's data from artificially acidified enclosures. The temporal variability in the acidified (simplified) communities was much greater than in the neutral richer communities. It is expected that the amplitude of the movement from the average or midpoint when an additional perturbation occurs will be larger when fewer species are present because there is less redundancy in the acidified systems.

The increase in benthic vegetation, whether in filamentous algae or in macrophytes, which can occur with acidification, is accompanied by an increased clarity of water. However, no causal link has been made between the former and the latter phenomena. Because planktonic primary production does not decrease as water clarity increases, it is unlikely that nutrient demand by phytoplankton declines as pH decreases. The primary mechanism for increased benthic biomass remains unexplained at present, and the Sphagnum invasion described for some Swedish lakes has not been seen elsewhere to date.

The implications of changes in benthic plant communities in terms of habitat modification are not trivial. Suggestions have been made that fish spawning might be affected (either adversely or positively) by the benthic plant mats, and Schindler et al. (1985) have recently observed that the expansion of the algal mats in the littoral zone of Lake 223 coincided with changes in fish spawning. Habitats for benthic epi- and in-fauna are also subject to change from the same processes; to date no direct effects have been demonstrated, but benthic mats may in fact ameliorate low pH and high Al at the sediment water interface.

The loosely attached algal mats are unlikely to affect nutrient cycling except as they themselves absorb nutrients, but the more compact blue-green mats and certain moss mats may influence the normal microbial and physical processes related to nutrient release. Again, this has not yet been directly demonstrated. If such nutrient cycling effects were operating, one would expect to find decreased nutrients in acid lakes, but such a decrease has rarely been observed. Various mechanisms related to nutrient availability or nutrient uptake as affected by acidification have been proposed or demonstrated and these are summarized in Table III. To date, most evidence relates to C. The potential for P to influence plant communities as pH is decreased requires much more investigation, as does the recent observation relating decline in Si with acidification to shifts in diatom species composition.

The actual species composition of the producers may be important for herbivores, but to date there is no evidence that herbivorous communities in acidifying lakes are changing because of alterations in the food base. The opposite may be true, and experimental manipulation of grazers appears to be a promising area of research in this context.

In general, the descriptive phase (except for macrophyte communities) has been adequately covered. The main need is for functional studies on primary producers and for tests of hypotheses concerning mechanisms, including direct and higher order interactions.

TABLE III. Possible mechanisms related to algal nutrition for changes in primary producers with acidification.

Factor	Change/Response	Examples	References
Phosphate	Decrease in concentration of P as a result of increased Al concentrations.	None available.	Stumm and Morgan, 1981
Silica	Si concentration decreases as decreases, limiting for diatoms. High affinity for Si would be advantageous.	L. 223. *Asterionella ralfusu* increased significantly and other diatoms decreased at pH 5.2-5.6, coincident with a sharp drop in Si concentration.	Findlay, 1984
Carbon	CO_2/bicarbonate ratio changes with pH; acid tolerant algae use CO_2.	Above pH 6.9, *Chrysochromulina breviturrita* fails to grow; alkalinity greater than 84 ueq L^{-1} is inhibitory. The alga appears incapable of using DIC as bicarbonate.	Wehr *et al.*, 1985
Carbon	Net photosynthesis significantly correlated with DIC; DIC limiting at low pH.	Haptobenthos in acidified systems has pH dependent production rates; comparing ambient with elevated DIC, significant increase in net photosynthesis.	Turner *et al.*, 1985
Carbon	Plants which dominate at low pH can only use CO_2, not bicarbonate.	Suggestion that *Sphagnum* uses only CO_2; other aquatic macrophytes used CO_2 or bicarbonate.	Grahn, 1985

8. ACKNOWLEDGEMENTS

I wish to thank Norman Yan and Michael Jackson of the Ontario Ministry of the Environment, Michael Turner, David Findlay and John Shearer of the Freshwater Institute, Winnipeg, John Wehr and Lewis Brown from the University of Western Ontario, Olle Grahn from Sweden, Debbie Roberts, Charles Boylen and Bob Singer, from Rensselaer and many others, who were most generous in providing data, slides, reports and manuscripts, sometimes in advance of publication, which enabled me to keep the information as up-to-date as possible. Judy Eichmanis assisted in literature searches and Jennifer Stoikos, Penny Ashcroft Moore and Brenda Missen typed the manuscripts, tables and references. I appreciate their help.

References

Almer, B., Dickson, W., Eckstrom, C., and Hornstrom, E.: 1978, 'Sulfur pollution and the aquatic ecosystem', Sulfur in the Environment Part II. Ecological Impacts, John Wiley & Sons, New York. p. 271-311.

Baker, J.P. and Schofield, C.L.: 1982, Water, Air, Soil Pollut. **18**, 289.

Bailey, R.C., and Stokes, P.M.: 1984, 'Evaluation of filamentous algae as biomonitors of metal accumulation in softwater lakes: A multivariate approach', Aquatic Toxicology and Hazard Assessment: Seventh Symposium, ASTM STP 854, American Society for Testing and Materials, Philadelphia, 1985, pp. 5-26.

Bleiwas, A.S.H.: 1983, 'Zooplankton grazing rates in some Ontario lakes with a range of pH's, 1979-1982', M.Sc. Thesis, Department of Botany, University of Toronto, 211p.

Boylen, C.W., Singer, R. and Roberts, D.A.: 1984, 'Biological and chemical field survey of small New Hampshire Lakes sensitive to acid deposition', EPA/NCSU Acid Deposition Program Contract Final Report No. APP-0313-1983. Freshwater Institute Tech. Report No. 85-1, Troy, New York.

Campbell, P.G.C., and Stokes, P.M.: 1985, Can. J. Fish. Aquat. Sci. **42**, in press.

Charles, D.F.: 1984, Verh. Internat. Verein. Limnol. **22**, 559.

Crisman, T.L., Schulze, R.L., Brezonik, P.L., and Bloom, S.A.: 1980, 'Acid precipitation: the biotic response in Florida lakes', Proc. Int. Conf. Ecol. Impact. Acid Precip., SNSF Project.

Davis, R.B.: 1985, Department of Botany, University of Maine at Orono, personal communication.

Dickson, W.: 1978, Verh. Internat. Verein. Limnol. **20**, 851.

Dickson, W., Hornstrom, E., Eckstrom, C., Almer, D.: 1975, 'Rodinsjoar Soder Om Dalaben', Information fran Drottingholm, #7, 140p.

Dillon, P.J., Yan, N.D., and Harvey, H.H.: 1984, CRC Crit. Rev. in Env. Cont. **13**, 167.

Dillon, P.J., Yan, N.D., Scheider, W.A. and Conroy, N.: 1979, Arch. Hydrobiol. Beih., Ergebr. Limnol. **13**, 317.

Duthie, H.C., and Ostrofsky, M.L.: 1974, J. Fish. Res. Bd. Can. **31**, 1105.

Findlay, D.L., and Saesura, G.: 1980, 'Effects of phytoplankton biomass succession and composition in Lake 223 as a result of lowering pH levels from 7.0 to 5.6. Data from 1974 to 1979', Can. M/S Report of Fish. Aquat. Sci. No. 1585: iv + 16p.

Findlay, D.L.: 1984, 'Effects on phytoplankton biomass, succession and composition in Lake 223 as a result of lowering pH levels from 5.6 to 5.2. Data from 1980-1982', Can. M/S Report of Fish. Aquat. Sci. No. 1761: 10p.

Grahn, O.: 1985, 'Macrophyte biomass and production in Lake Gardsjon - an acidified lake in southwestern Sweden', Swedish Environmental Research Group, unpublished manuscript.

Grahn, O., and Hultberg, H.: 1974, Ambio **3**, 93.

Haines, T.A.: 1981, Trans. Amer. Fish. Soc. **10**, 669.

Havens, K.E.: 1984, 'Freshwater plankton response to acidification, Ph.D. Diss. W. Virginia University.

Hendrey, G.R.: 1976, 'Effects of pH on the growth of periphytic algae in artificial stream channels', Sur. Nedbors Virkning pa Skog og Fisk Project IR 25/76.

Hendrey, G.R., and Vertucci, F.: 1980, 'Benthic plant communities in acidic Lake Colden, New York: Sphagnum and the algal mat, Ecological Impacts Acid Precipitation, SNSF Project, 314-315pp.

Hultberg, H. and Andersson, I.B.: 1982, Water, Air, Soil Pollut. **18**, 311.

Hutchinson, E.: 1975, A Treatise on Limnology, John Wiley and Sons Inc., **2**, 1115 pp.

Jackson, M.B.: 1985, 'Filamentous algae in Ontario softwater lakes', Int. Symp. on Acidic Precipitation, Muskoka Canada, Sept. 15-20, 1985, p. 117-118 (Abstract).

Kwiatkowski, R.E. and Roff, J.C.: 1976, Can. J. Bot. **54**, 2546.

Lazarek, S.: 1983, 'Structural and functional aspects of epiphytic and benthic algae in the acidified Lake Gardsjon, Sweden', Ph.D. Dissertation, University of Lund, 112p.

Linthurst, R.A. and Altshuller, A.P. (Ed.): 1984. 'The Acidic Deposition Phenomenon and Its Effects: Critical Assessment Review Papers', **2** EPA/600/19. 748pp.

Lynch, M., and Shapiro, J.: 1981, Limnol. Oceanogr. **26**, 86.

Marmorek, D.: 1984, 'The effects of lake acidifiction on zooplankton community structure and phytoplankton-zooplankton interactions: an experimental approach', M.Sc. Thesis, University of British Columbia.

Moeller, R.E.: 1975, Verh. Internat. Verein. Limnol. **19**, 1004.

Muller, P.: 1980, Can. J. Fish. Aquat. Sci. **37**, 355.

NRCC: 1981, 'Effects of acidification on the Canadian environment', Associate Committee on Scientific Criteria for Environmental Quality, NRCC 18475.

Nicholls, K.H., Beaver, J.L. and Estabrook, R.H.: 1982, Hydrobiologia **96**, 91.

O'Mahoney, M.A.: 1985, 'A seasonal study of an acidifying and circumneutral shield lakes: effects of pH shock on phosphorus concentration and kinetics and photosynthesis', M.Sc. Thesis, Botany Department, University of Toronto, Toronto, Ontario, Canada, 188p.

Parent, L., Allard, M., Planas, D., and Moreau, G.: 1985, 'The effects of short term and continuous experimental acidification on biomass and productivity of running water periphytic algae', Proc. ASTM Symposium, 1985.

Porter, K.G.: 1976, Science **192**, 1332.

Raddum, G.G., Hobaek, A., Lomsland, E.R., and Johnsen, T.: 1980, 'Phytoplankton and zooplankton in acidified lakes in South Norway', Proc. Int. Conf. Ecol. Impact Precip, SNSF Project.

Redfield, G.W.: 1980, Hydrobiologia **70**, 217.

Roelofs, J.G.M, Shuurkes, J.A.A.R., and Smits, A.J.M.: 1984, Aquatic Botany **18**, 389.

Roelofs, J.G.M.: 1983, Aquat. Bot. **17**, 139.

Roberts, D.A., Singer, R., and Boylen, C.W.: 1985, Aquat. Bot. **20**, in press.

Schindler, D.W., Mills, K.H., Malley, D.F., Findlay, D.L., Shearer, J.A., Davies, I.J., Turner, M.A., Linsey, G.A., and D.R. Cruikshank: 1985, Science **228**, 1395.

Schindler, D.W.: 1980, 'Experimental acidification of a whole lake: a test of the oligotrophication hypothesis', Ecological Impact of Acid Precipitation, SNSF Project, 370-374pp.

Shearer, J.A., Fee, E.J., DeBruyn, E.R., and DeClercq, D.R.: 1986, Can. J. Fish. Aquat. Sci. **43**, 000-000.

Singer, R.: 1983, Aquatic Bot. **16**, 49.

Sprules, G.: 1975, J. Fish. Res. Bd. Can. **32**, 382.

Stokes, P.M.: 1981, 'Benthic algal communities in acidic lakes', Effects of Acidic Precipitation on Benthos, Proc. Symp. Acidic Precipitation on Benthos, 1980, North American Benthological Society, Hamilton, N.Y., pp. 119-138.

Stokes, P.M.: 1984, 'pH-related changes in attached algal communities of softwater lakes', Early Biotic Responses to Advancing Lake Acidification, Acid Precipitation Series 6, Butterworth, 43p.

Stokes, P.M., and Yung, Y.K.: 1985, 'Phytoplankton in selected LaCloche Lakes, Ontario pH 6.2-7.0', Silicous Algae as Indicators of Lake Acidification, Developments in Hydrobiology, (Hydrobiologia) 57.

Stumm, W., and Morgan, J.J.: 1981, Aquatic chemistry. An introduction emphasizing chemical equilibria in natural waters, John Wiley and Sons Inc., New York, N.Y., 780p.

Tilman, D. and Kilham, S.S.: 1976, J. Phycol. **12**, 375.

Turner, M., Jackson, M., Findlay, D., Schindler, D., Graham, R., DeBruyn, E., and Vandermeen, L.: 1986, 'Responses of haptobenthos to experimental whole lake additions of sulphuric and nitric acids', Can. J. Fish Aquat. Sci. in press.

Turner, P.A.E. and Stokes, P.M.: 1985, 'The growth of filamentous algae (Mougeotia spp.) isolated from acidic lakes grown over pH gradient 4.5 to 6.9', Int. Symp. on Acidic Precipitation, Sept. 15-20, 1985, Muskoka, Canada, p. 120, (Abstract).

Wehr, J.D., Brown, L.M., and O'Grady, K.: 1985, 'Physiological ecology of the bloom-forming algae Chrysochromulina breviturrita Nich. (Prymnesiophyceae) from lakes influenced by acid precipitation', unpublished manuscript.

Wile, I., Miller, G.E., Hitchin, G.G. and Yan, N.D.: 1985, Can. Field-Natural in press.

Yan, N.D.: 1979, Water, Air, Soil Pollution **11**, 43.

Yan, N.D. and Stokes, P.M.: 1978, Environ. Conserv. **5**, 93.

Yan, N.D. and Strus, R.: 1980, Can. J. Fish. Aquat. Sci. **37**, 2282.

Yan, N.D. and Dillon, P.J. 1984: 'Experimental neutralization of lakes near Sudbury, Ontario', Environmental Impacts of Smelters, John Wiley and Sons Inc., New York, N.Y.. 417-456.

Yan, N.D. and Lafrance, C.: 1984, 'Responses of acidic and neutralized lakes near Sudbury, Ontario to nutrient enrichment', Environmental Impacts of Smelters, John Wiley and Sons, New York, N.Y., 458-521pp.

Yan, N.D., Miller, G.E., Wile, I., and Hitchin, G.G.: 1985, Aquat. Bot., in press.

THE INFLUENCE OF ACIDIC POND WATER ON AMPHIBIANS: A REVIEW

Joseph Freda

Program in Ecology, The Pennsylvania State University, University Park, PA 16802 USA

Abstract: Acidic pond water may influence the reproduction of amphibians by causing direct mortality of embryos and larvae, and/or by disrupting trophic relationships between amphibians and other aquatic organisms. The embryo is the most sensitive stage of development and may abort soon after exposure to very low pH or eventually fail to hatch at a more moderate pH. Larvae are killed by disruption of Na and Cl balance. The toxicity of pond water is governed by complex interactions of pH, temperature, and the concentrations of Al, Ca, and organic acids. The reduction in recruitment associated with acidification of ponds affects both the local distribution and abundance of amphibians. The relative importance of different mechanisms of acidification of amphibian breeding sites is unknown. Amphibian breeding ponds are generally small, low in buffering capacity, and darkly stained by humic substances, making it difficult to discriminate between the contribution of acid from atmospheric input and natural sources (such as sphagnum and organic acids). It is however quite clear that temporary ponds are extremely vulnerable to acidification by rain and that they require continued monitoring in order to detect future changes in biological and/or chemical characteristics.

1. INTRODUCTION

The toxic nature of surface waters low in pH and the resulting influence on the distribution and abundance of aquatic organisms have been recognized for several decades (Jewell and Brown, 1924; Gosner and Black, 1957; Smith, 1960). These early studies were conducted within the naturally acidic sphagnum and cedar bogs of Michigan and the New Jersey pine barrens. Pough (1976) was the first to report a possible effect of acidic precipitation on amphibians. He observed high mortality of embryos of the spotted salamander *(Ambystoma maculatum)* in acidic ponds near Ithaca, New York. These observations paralleled results of laboratory bioassays at similar pH's, implicating low pH as the source of toxicity. In the years following these published observations, an explosion of research investigating possible effects of low pH water on amphibians has occurred. The main objective of this review is to consolidate and evaluate all current information describing the effects of low pH water on amphibians. Special emphasis will be placed on the possible effects of acidic precipitation on amphibian populations.

2. THE TOXICITY OF ACIDIC WATER TO AMPHIBIANS

Inter-specific Differences in Embryonic Tolerance - Amphibian development passes through a series of discrete life stages: embryo, larva, and adult. Most species (in temperate North America) require water for oviposition and larval development, but adults can be aquatic, semi-aquatic, or terrestrial. As with other aquatic organisms, reproductive failure is the major effect of acidic water on amphibians (Haines, 1981). Schlichter (1981) reported that water with a pH below 6.3 decreased sperm motility and

Corresponding Address: Dept. Biology, McMaster University, Hamilton, Ontario, L8S 4K1 Canada

Water, Air and Soil Pollution **30** (1986) 439–450.

Table I. Summary of lethal and critical pH's of embryos of 26 species of amphibians determined under laboratory conditions. Lethal pH causes 100% mortality of embryos; the point where hatching success declines below levels in neutral water is termed the critical pH.

Species	Lethal pH	Critical pH	Investigator(s)
FROGS *(Anura)*			
Hylidae			
Acris gryllus	4.1	4.6	1
Hyla andersoni	3.4-3.7	3.8	1, 15
Hyla crucifer	3.8-4.2	4.2	1, 8, 11
Hyla versicolor	3.8	4.3	1
Pseudacris nigrita	3.8	4.1	1
Pseudacris triseriata	4.2	-	8
Bufonidae			
Bufo americanus	3.8-4.2	4.0-4.2	8, 11, 12, 13
Bufo boreas	-	4.0	3
Bufo bufo	4.25	4.5	2, 16
Bufo calamita	4.5	4.75	2
Bufo woodhousei	4.0	4.2	13, 15
Pipidae			
Xenopus laevis	3.0-4.1	3.5-4.5	5, 6, 7, 11
Ranidae			
Rana arvalis	3.5	4.0	16
Rana catesbeiana	3.9	4.3	1, 5
Rana clamitans	3.7-3.8	4.1	1, 15
Rana esculenta	4.0	4.5	16
Rana palustris	4.0-4.5	4.3-4.5	1, 11
Rana pipiens	4.5	-	8, 13
Rana sylvatica	3.5-4.0	3.9-4.25	1,7,8,10,11,12,13,14
Rana temporaria	4.25	4.5	2, 16
Rana utricularia	3.7	4.1	1
Rana virgatipes	3.4	3.8	1
SALAMANDERS *(Urodela)*			
Ambystomatidae			
Ambystoma jeffersonianum	4.0-4.5	4.5-5.0	4, 13
Ambystoma laterale	4.2	-	8
Ambystoma maculatum	4.0-4.5	4.5-5.0	4, 11, 13
Ambystoma texanum	4.0	5.0	9

Investigators:(1) Gosner and Black, 1957, (2) Beebee and Griffin, 1977, (3) Porter and Hakanson, 1976, (4) Pough and Wilson, 1977, (5) Saber and Dunson, 1978, (6) Dunson and Connell, 1982, (7) Tome and Pough, 1982, (8) Karns, 1983, (9) Punzo, 1983, (10) Pierce *et al.*, 1984, (11) Dale *et al.*, 1986, (12) Clark and Lazerte, 1985, (13) Freda and Dunson, 1985b, (14) Pierce and Sikland, 1985, (15) Freda and Dunson, 1986, (16) Leuven *et al.*, 1986.

fertilization success in the leopard frog *(Rana pipiens)*, but these results are misleading because toxic sodium acetate/acetic acid buffers were used. Dale *et al.* (1986) observed complete mortality of embryos of the African clawed frog *(Xenopus laevis)* in solutions buffered by sodium acetate/acetic acid. Karns (1983) reported 100% fertilization of eggs of the wood frog *(Rana sylvatica)*, the american toad *(Bufo americanus)*, and the leopard frog *(R. pipiens)* in pH 4.2 bog water. These embryos subsequently died from developmental abnormalities induced by the continued exposure to bog water. Similarly, most embryonic mortality in acidic ponds occurs during the latter stages of development, indicating little effect of acidic pond water on fertilization.

The majority of studies investigating toxic properties of acid water have focused on embryonic development because this life stage is the most sensitive to low pH (Gosner and Black, 1957; Pierce *et al.*, 1984; Dale *et al.*, 1986). The results of laboratory bioassays conducted at low pH on 26 species of North American and European amphibians are presented in Table I. I have assigned two values to each species: 1) the pH at which hatching success decreased below levels normally occurring in neutral water (critical pH), and 2) the pH at which complete mortality occurred (lethal pH). Large interspecific differences (> 1.0 pH unit) in tolerance to low pH exist and are generally correlated with habitat acidity. For example, the pine barrens treefrog *(Hyla andersoni)* and the carpenter frog *(Rana virgatipes)* are the most tolerant species tested thus far and are endemic to the naturally acidic coastal plain of the eastern United States. Documentation of inter-specific differences is also essential in predicting which species of amphibians are most vulnerable to acidic precipitation.

Differences in breeding behavior among species may also contribute to inter-specific variability in hatching success of embryos. Intra- and inter-pond variability in pH can be exploited by species able to discriminate between toxic and non-toxic oviposition sites. The pH within a pond can exhibit considerable spatial variation (> 1.0 unit; Beebee and Griffin, 1977; Pierce *et al.*, 1984), and can fluctuate weekly as well as increase gradually during spring and summer (Freda and Dunson, 1985c). The span between the critical and lethal pH of embryos of many species is only 0.2 to 0.5 units (Table I). Therefore microhabitat placement of egg masses and timing of oviposition are critical in ponds near the edge of a species' tolerance range. Salamanders can differentiate between acidic and neutral substrates, but whether they can resolve relatively small differences (< 0.5 pH units) is not known (Mushinsky and Brodie, 1975). Beebee and Griffin (1977) observed that populations of natterjack toads *(Bufo calamita)*, native to naturally acidic heathland regions, refused to breed in low pH ponds lethal to embryos. Individuals transplanted to heathland areas from neutral coastal regions failed to discriminate and oviposited in lethally acidic ponds. *Rana sylvatica* and *B. americanus* were also able to discriminate between odors of high pH fens and acidic bogs in olfactometer experiments (Karns, 1983). A behavioral trait of *R. sylvatica* which may alter survival of embryos in acidic ponds is the construction of communal egg rafts containing several dozen egg masses. Hatching success in acidic bog water was higher in large egg rafts (> 65 egg masses) in comparison to small rafts (< 10 egg masses; Karns, 1983).

Intra-specific Variation in Embryonic Tolerance - A high degree of intra-specific variation in tolerance to low pH has also been documented. At the most basic level, considerable variability in tolerance can exist within a single egg mass. Pough (1976) noted that within a single pond, some egg masses were viable while 100% mortality occurred in others. Pierce and Sikand (1985) were the first to examine whether these inter-clutch differences had a genetic basis. They crossed 4 *R. sylvatica* females to 5 different males, resulting in 20 different male/female combinations. At pH 3.75 and 4.0, no significant differences in hatching of embryos derived from crosses of different males to a single female were found. However highly significant differences occurred between embryos taken from different females. This maternal effect may have a genetic basis or reflect environmental influences such as diet, physiological condition of female, and/or maturity of eggs.

Inter-populational variability in embryonic tolerance to low pH was first hypothesized by Cook (1983). He observed little mortality of *Ambystoma* salamanders in ponds with a pH of 4.2 in Massachusetts, whereas Pough (1976) observed high mortality

Table II. Percent mortality of wood frog embryos *Rana sylvatica* measured by different investigators under laboratory conditions.

Investigator(s)	Mortality (%) at pH 3.5	Mortality (%) at pH 4.0
Gosner and Black, 1957	100%	0%
Tome and Pough, 1982	45%	10%
Karns, 1983	-	89%[c]
Pierce et al., 1984	100%	0%
Clark and Lazerte, 1985	-	19%[b]
Freda and Dunson, 1985b	-	100%
Dale et al., 1986	-	88%
Pierce and Harvey, 1986	58%[a]	13%

a at pH 3.75
b at pH 4.14
c at pH 4.20

in ponds with pH's below 6.0. Unfortunately, toxic elements such as Al in pond water were not measured in these studies. Table II summarizes the results of controlled laboratory studies which have tested the low pH tolerance of *R. sylvatica* with the intent of making comparisons between geographic regions. Large intra-specific differences in lethal pH (> 0.5 pH units) are immediately apparent. Drawing conclusions from these comparisons is also risky because experimental protocols differed considerably. Under carefully controlled conditions, Pierce and Harvey (1986) tested whether this inter-populational variation had a genetic basis by making half-sibling crosses of *R. sylvatica* as previously described, but using individuals from different populations. They found that different populations were different in tolerance to low pH, but as before, the degree of tolerance was not genetically determined and could be attributted to maternal factors. Tolerance of embryos from different ponds was significantly different but was not correlated with pond pH (Pierce and Harvey, 1986). The pH tolerance of embryos of *A. maculatum* in New York also was not correlated with pond pH (Tome and Pough, 1982).

The investigations of Tome and Pough (1982), and Pierce and Harvey (1986) were all conducted in the laboratory under controlled but somewhat different conditions. Differences in mortality of embryos under field conditions at a specific pH are difficult to interpret since the toxicity of pond water is governed by more than just the concentration of H^+. Under field conditions, the complex interactions of pH, temperature, and the concentrations of Al, Ca and organic compounds may ultimately determine hatching success. Comparisons among laboratory bioassays which have employed vastly different experimental protocols should also be made cautiously. The concept that several independent variables interact with the physiological and biochemical adaptations of each species to produce a unique toxicological response is the major theme of this review and the subject of the next section.

Physiological and toxicological responses of embryos - Amphibian embryos killed by low pH display characteristic morphological traits depending on pH, species, and developmental stage. Embryos stop developing and die within 12 hr of exposure to water less than 0.25 pH units below the pH of 100% lethality (Pough and Wilson, 1977; Freda and Dunson, 1985b). A mottled appearance of the surface layer of cells and a shrunken, rubbery texture of egg jelly and membranes are characteristic of mortality of young embryos at these extremely low pH's. The most commonly observed abnormality at slightly higher but still lethal pH (within 0.25 units of 100% lethal level) is a distinct curling of embryos within the vitelline membrane; termed the "curling defect" (Tome and Pough, 1982; Freda and Dunson, 1985b). The vitelline membrane normally lifts off the embryo at neurulation, creating the perivitelline space (Salthe, 1965; Carroll and Hedrick, 1974). At low pH, the perivitelline space never enlarges, forcing the elongating embryo to fold over itself. The degree of curling can be moderate, only causing a slight curvature, or severe enough to compress the embryo into a solid ball (Freda and Dunson, 1985b). Embryos do not curl or die if the vitelline membrane is removed (Dunson an Connell, 1982; Karns, 1983). This result suggests that mortality associated with curling results from vitelline membrane dysfunction rather than a general toxic effect on the embryo itself. It has been hypothesized that H^+ invade the perivitelline space and deactivate the hatching enzyme (Gosner and Black, 1957; Dunson and Connell, 1982). The hatching enzyme of *X. laevis in vitro* is totally inhibited at pH 4.0; the pH of perivitelline fluid in fish embryos parallels environmental pH (Urch and Hedrick, 1981; Peterson *et al.*, 1982). The hatching enzyme is responsible for the changes in the vitelline membrane which allow it to expand in response to the osmotically driven influx of water (Salthe, 1965; Carroll and Hedrick, 1974). The digestive action of the hatching enzyme also facilitates the rupture of the vitelline membrane by muscular contractions of the embryo at hatching (Carroll and Hedrick, 1974; Yoshizaki, 1978). Curled embryos that do manage to hatch often have curved spines. These abnormal larvae are only capable of swimming in tight circles with a doubtful prospect for survival. Staining techniques have revealed that vertebrae of larvae with bent spines are actually fused (Dale *et al.*, 1986). Other less commonly observed abnormalities induced by acidic water include swelling of the thoracic region, stunted gills, failure to retract the yolk plug, and deformation of the posterior trunk (Pough and Wilson, 1977; Karns, 1983; Freda and Dunson, 1986a).

In acidic ponds (< pH 5.0), the solubility of Al increases to toxic levels (Clark and Hall, 1985; Freda and Dunson; 1986a). Dissolved Al can lower hatching success of fish and amphibians at very low concentrations (0.05 to 0.2 mg L^{-1}) and its toxicity varies with pH and the inorganic and organic complexes that it forms (Baker and Schofield, 1980; Clark and Lazerte, 1985). The toxicity of Al to fish is attributable to Al^{+3} and $Al(OH)_3$ while AlF is intermediate and the organically bound form is harmless (Driscoll *et al.*, 1980). Clark and Lazerte (1985) have discovered a completely different relationship between pH and Al toxicity in amphibians. *Rana sylvatica* and *B. americanus* embryos did not respond differently to Al^{+3}, $Al(OH)_3$, or AlF. At pH 4.75, concentrations of inorganic monomeric Al up to 100 µg L^{-1} had no effect. However, when pH was lowered to 4.32 or 4.14, a dual threshold phenomenon was observed. Small but statistically significant increases in mortality (19 to 20%) occurred when Al was increased from 0 to 10 or 20 µg L^{-1}, whereas no further significant increases resulted unless the concentration was elevated to 200 µg L^{-1} (Clark and Lazerte, 1985). It is unfortunate that the toxicity of organically bound Al has not been determined because a large percentage of amphibian breeding ponds contain high concentrations of organic compounds (Freda and Dunson, 1986a). Attributing the mortality of amphibian embryos in field situations to a specific toxic element is difficult because pH and the concentrations of Al and organic compounds (DOC) are correlated. To circumvent these inter-relations, Clark and Hall (1985) artifically acidified a stream and were able to manipulate pH and Al while DOC remained constant. Hatching of *B. americanus* at pH 4.3 was reduced from 72 to 11% when inorganic monomeric Al increased from 35 to 46 µg L^{-1}, whereas only small non-significant declines in hatching of the more low pH tolerant *R. sylvatica* occurred (91 to 83%). The physiological effects of Al have not been studied in amphibians, but this element does increase the severity of the curling defect (Clark and Lazerte, 1985). Acidified lakes can also have elevated concentrations of other heavy metals (Cu, Fe, Mn, Zn, etc.). These heavy metals have been shown to be a major source of toxicity to amphibians in acidic mine drainage and leachates from pyritic formations (Porter and Hakanson, 1976; Mathews and Morgan, 1982). Glooschenko *et al.*, (1985) have measured toxic levels of Fe, Mn, and Al in acidic amphibian breeding ponds near Sudbury, Ontario.

Low pH and high concentrations of Al are the major causes of mortality in acidic water, but several other physical and chemical variables can modify their action. Calcium, Mg and Na have a species-specific influence on hatching success at low pH (Dale *et al.*, 1986; Freda and Dunson, 1985b). In all species tested, these cations (especially Ca) at concentrations above 10 mg L^{-1} block the inhibition of development that occurs at very low pH's. Cations alone, however, can inhibit the hatching enzyme and intensify curling in acidic water (Salthe, 1985; Urch and Hedrick, 1981; Katagiri, 1975; Freda and Dunson, 1985b). A species-specific response to the combined effects of low pH and high Ca (> 10 mg L^{-1}) results from a differential ability of embryos to hatch when curled. For example, Ca reduces mortality during early development at low pH in both *R. sylvatica* and *A. maculatum*, but *R. sylvatica* never hatches when severely curled whereas *A. maculatum* is able to do so (Freda and Dunson, 1985b). The concentrations of single cations needed to influence hatching success (> 10 mg L^{-1}) rarely occur in acidic ponds, but preliminary evidence suggests that the interaction of several cations at realistic natural concentrations (1 mg L^{-1}) can influence mortality at low pH (Freda and Dunson, 1985b, 1986a). Calcium also has an important role in reducing the toxicity of Al and other heavy metals in fish (Sprague and Ramsey, 1965; Brown and Lynam, 1982).

Organic compounds which commonly occur at very high concentrations in temporary forest ponds may also influence embryonic mortality at low pH. Dunson and Connell (1982) reported that the critical and lethal pH for embryos of *X. laevis* were 0.4 pH units higher in darkly stained bog water as compared to clear creek water, but these media were different in ionic composition. *Rana clamitans*, *B. woodhousei*, and *H. andersoni* also experienced high mortality in acidic bog water as compared to laboratory solutions with a similar pH and chemical composition (Freda and Dunson, 1986a).

Water temperature can also influence hatching of embryos in acidic ponds. Species-specific temperature optima exist, and a departure from optimal temperatures increases mortality at all pH's but to a higher degree at low pH (Pough and Wilson, 1977; Dale *et al.*, 1986). Mortality of early spring breeding species increases sharply below 5 C and above 15 C. Hatching success of embryos of *R. sylvatica* in a Minnesota bog was several fold higher during a warm spring (5 to 27 C) than in a cold spring (3 to 14 C), whereas pH was similar (Karns, 1983).

Physiological effects of low pH on larvae - Inter-specific differences in tolerance of embryos to low pH carry over into the larval stage, accompanied by a general increase in tolerance. For example, embryos of *R. pipiens* can not hatch below pH 4.5, but the lethal pH for larvae is pH 4.0 (Freda and Dunson, 1984, 1985b). In comparison, the lethal pH for embryos and larvae of the more low pH tolerant *R. sylvatica* are 4.0 and 3.5, respectively (Pierce *et al.*, 1984). Low pH tolerance of larvae also increases with age, but it is not known whether this is due to an increase in size or physiological changes during development (Pierce *et al.*, 1984, Freda and Dunson, 1985a). In contrast with studies of embryos, inter-populational differences in low pH tolerance of *R. sylvatica* larvae are genetically based and inversely correlated with pond pH (Pierce and Harvey, 1986).

Because larvae can withstand lower pH's than embryos, direct mortality of larvae in acidic ponds may not be a major limiting factor. For early spring breeding species, the pH experienced by embryos is also usually lower than for larvae, because pond pH increases during spring and summer (Freda and Dunson, 1985c). Direct mortality of larvae would therefore only occur after episodic declines in pH resulting from rainfall or snowmelt (Gascon and Bider, 1985). However, both Karns (1983) and Freda and Dunson (1986a) observed successful hatching of embryos in acidic ponds but failed to find any larvae even though pond pH remained the same or increased.

The primary physiological effect of low pH on larvae is the disruption of Na and Cl balance (Freda and Dunson, 1984, 1986b; McDonald *et al.* 1985). Acute exposure to lethal pH's inhibits active uptake of Na and Cl at the gills and promotes massive losses of these ions. The stimulation of Na efflux in fish at low pH is thought to result from displacement of bound Ca from inter-cellular diffusion channels (in gills) by H^+ (McDonald, 1983). The depression of sodium influx results from a disruption of Na^+/H^+ exchange mechanisms. The inhibition of influx is quantitatively insignificant and death ensues when 50% of the body Na content is lost (Freda and Dunson, 1984). Intra- and inter-specific differences in sensitivity to low pH are correlated with the magnitude o Na efflux. Increasing the concentration of Ca in acid water increases the survival time of larvae or prevents mortality because this ion slows the lethal loss of body Na (Freda and Dunson, 1984).

Chronic sublethal exposure to low pH also reduces body sodium concentration by 10 to 20% and compensatory mechanisms which maintain ion balance are stimulated (Freda and Dunson, 1984). Sodium influx (in pH 4.5 test water) in *R. pipiens* tadpoles pre-exposed (7 d) to pH 4.5 was higher than in pH 5.8 maintained animals in pH 5.8 test water. Sodium efflux was also reduced by low pH pre-exposure (7 d; Freda and Dunson, 1984). In field studies, *R. sylvatica* tadpoles from a low pH pond (4.1 to 4.9) were 10 to 20% lower in body Na and Cl concentration than tadpoles collected from a high pH pond (5.7 to 6.4). Tadpoles from both ponds also lost Na at a faster rate when placed in the acidic pond than in the neutral pond (Freda and Dunson, 1985a)

3. THE INFLUENCE OF POND PH ON AMPHIBIAN POPULATIONS

Gosner and Black (1957) were the first to report an association between a species' tolerance to low pH and its abundance in acidic ponds. *Hyla andersoni* and *R. virgatipies* have a lethal pH of 3.4 and commonly breed in the lowest pH ponds (< 4.0) in the naturally acidic New Jersey Pine Barrens. *Rana catesbeiana* (lethal pH of 3.9) was restricted to isolated areas containing high pH water. Saber and Dunson (1978) reported a lowered species diversity within a sphagnum bog in central Pennsylvania (pH 4.2). *Rana clamitans* was the most abundant species whereas *R. catesbeiana* was noticeably absent. Laboratory and field exposures of embryos and larvae of both species to bog water (pH 4.2) resulted in complete survival for *R. clamitans*, and 100% percent mortality for *R. catesbeiana* (Saber and Dunson, 1978). In Great Britain, the natterjack toad *(B. calamita)* did not breed in ponds with a pH below 5.0, supporting laboratory tests which revealed high embryonic mortality below this pH (Beebee and Griffin, 1977). Cooke and Frazier (1976) reported that smooth and palmate newts *(Triturus vulgaris* and *T. helvetica*, respectively) were never found in ponds with pH's below 3.9. Smooth newts were also rarely encountered in ponds below pH 6.0. Numbers of breeding adults, egg masses, larvae, and metamorphosing frogs, and embryonic hatching success were significantly lower in acidic bogs in comparison to neutral fens (Karns, 1983). These previously mentioned investigations all dealt with naturally acidic ecosystems. In 1976, Pough reported that declines in the pH of amphibian breeding ponds were having a detrimental effect on *Ambystoma* salamanders in New York. Clark and Euler (1982) surveyed 20 streams and 21 ephemeral meltwater ponds in central Ontario for pH and species of amphibians present. Principal component analysis revealed negative associations between densities of *R. catesbeiana, H. crucifer*, and *R. sylvatica* and habitat acidity. Gascon and Planas (1986) conducted similar surveys of *R. sylvatica* in Quebec, but also measured hatching success of embryos. Acidity and DOC were inversely correlated with density of egg masses and percent hatching was directly correlated with pond pH. In the Netherlands, Leuven *et al.* (1986) found pond pH to have a large influence on the distribution and abundance of amphibians. In contrast, Cook (1983) found no correlation between hatching success of *Ambystoma* salamanders and pond pH, since little mortality occurred even in the most acidic ponds. Dale *et al.* (1985) also found no correlation between presence of an amphibian species and several chemical variables in 169 locations in Nova Scotia. However, in this study, calling adults were used as an indicator of presence, which does not necessarily indicate oviposition or recruitment (Freda and Dunson, 1986a).

Another powerful approach to studying the exclusion of amphibians from acidic ponds is to transplant embryos of missing species into uninhabited ponds low in pH. Low pH or Al toxicity can be confirmed by examining dead embryos for the characteristic curling defect. In central Pennsylvania, *A. jeffersonianum* did not breed in ponds with a pH below 4.6, whereas the more low pH tolerant *R. sylvatica* was found in most pond regardless of pH (Freda and Dunson, 1986a). Embryos of both species were transplanted into the 12 ponds surveyed. Hatching of *R. sylvatica* was not correlated with pond pH. However mortality of *A. jeffersonianum* did increase with declining pond pH and high mortality (62 to 100%) occurred in 5 of 6 ponds uninhabited by this species.

Direct mortality of embryos and larvae of amphibians in acidic ponds has an obvious effect on reproductive success, but other more subtle mechanisms may be operating. Changes in temporary pond food chains may also influence recruitment. Community structure and populations of benthic invertebrates and zooplankton on which larval salamanders feed and the phytoplankton on which tadpoles are dependent can be dramatically affected by acidification (see review by Haines, 1981). Water low in pH can also directly reduce the growth rate of tadpoles and salamander larvae (Freda and Dunson, 1985a, 1986a; Ling *et al.*, 1986). Slower growing individuals in acidic temporary ponds may not metamorphose before the time of pond drying. Smaller larvae are also more susceptible to gape-limited predators such as newts and salamander larvae (Wilbur, 1984). The outcome of intra- and inter-specific competition among amphibian larvae is heavily dependent on body size (Wilbur, 1984).

In summary, acidic water influences the distribution and abundance of amphibians through complex interactions of several chemical and physical variables (pH, temperature, Al, Ca, organic acids) and the physiological, biochemical, behavioral, and ecological attributes of each species. Predictions of embryonic mortality in acidic ponds are further complicated by temporal variation in both the chemical environment and sensitivity of embryos.

4. SOURCES OF ACIDITY IN AMPHIBIAN BREEDING PONDS

I have summarized chemical surveys of amphibian breeding sites in North America and Europe (Table III). Ponds were categorized into 3 groups based on their lowest observed pH: below pH 4.5, between pH 4.5 and 5.0, and above pH 5.0. The relative distribution of pond pH's in each geographic region varies considerably. For example, 81% of ponds in the naturally acidic New Jersey Pine Barrens are below 4.5 in pH, whereas other regions range from 5 to 29%. These inter-regional differences in general levels of acidity have not been addressed but probably involve differences in geology, soils, vegetation type, and H^+ loading from acidic precipitation and/or natural chemical and biological processes. In most non-agricultural regions, a large percentage of ponds have a pH below 5.0 (15 to 87%) indicating that toxicity of pond water could be a major force structuring amphibian communities.

Multivariate toxicity studies and detailed biological and chemical surveys of amphibian habitat have been conducted but no study has rigorously ascertained the source of acidity in ponds of low pH. Acidic precipitation has been identified as the major source of hydrogen ions in many lakes and streams and may also be responsible for the acidification of temporary ponds. However, the same characteristics that make temporary ponds vulnerable to acidic precipitation (small volume and low alkalinity) are also associated with vulnerability to acidification by natural processes. Sphagnum moss can acidify water by exchange of endogenous H^+ for environmental Ca (Clymo, 1967; Kilham, 1982). In several studies, the ponds lowest in pH contained or were surrounded by abundant sphagnum (Freda and Dunson, 1985c; Ling *et al.*, 1986; Gascon and Planas, 1986). The sphagnum, however, may have simply invaded ponds acidified by rain or somehow increased the vulnerability of ponds to acidic precipitation (ie. lower Ca conc.). The high concentrations of organic acids in temporary ponds also can not be ignored (Dale *et al.*, 1985; Freda and Dunson, 1986a). These compounds are capable of influencing the pH of surface water and are the primary source of H^+ in many naturally acidic aquatic habitats (Patrick *et al.*, 1979; Hemond, 1980). Many acidic temporary ponds also have a pH well below that of rain or snowmelt, thus lending support to the idea that natural sources of acidity are a contributing factor (Dale *et al.*, 1985; Freda and Dunson, 1985c; Gascon and Planas, 1986). Because no historical data for temporary ponds are available, one approach to assessing the influence of acid rain on pond chemistry is to measure pond pH before and after rainstorms. Pough (1976) and Cook

Table III. Summary of the distribution of pond pH in 13 regions of North America and Europe.

Location	Total no. of Ponds Surveyed	The % of Ponds with a pH < 4.5	4.5-5.0	> 5.0	Source
NORTH AMERICA					
Massachusetts	14	28%	43%	28%	Cook, 1983
Nova Scotia	153	5%	10%	85%	Dale and Freedman, 1983
C. Pennsylvania	39	18%	18%	64%	Freda and Dunson, 1985c
N.E. Pennsylvania	36	11%	6%	83%	Freda and Dunson, 1985c
S. New Jersey	16	81%	6%	13%	Freda and Dunson, 1986
E. Ontario	49	10%	14%	76%	Glooschenko et al., 1985
C. Ontario	20	5%	15%	80%	Clark, 1984
Quebec	15	7%	20%	73%	Gascon and Planas, 1986
EUROPE					
England					
Hampshire	52	37%	6%	57%	Cooke and Frazer, 1976
South, Agricultural	81		2%[a]	98%	Beebee and Griffin, 1977
N.W., Coastal Dune	31		10%[a]	90%	
South, Heathland	93		48%[a]	52%	
The Netherlands	96	29%[b]	32%[a]	39%	Leuven et al., 1986

a pH 4.0-5.0
b pH < 4.0

(1983) observed large declines in pH (0.25 to 0.75) after rainstorms, but Freda and Dunson (1985c) and Pierce (1985) found no correlation between weekly pond pH and atmospheric H^+ loading and did not observe significant declines in pH after rainstorms. This type of assessment is, however, too simplistic because percolation of rain water through peat deposits or organically enriched, water saturated soils can acidify ponds (Hemond, 1980). The fact that ponds are lowest in pH early in spring is not, however, diagnostic of acidification by snowmelt because the pH of naturally acidic ponds also increases during spring and summer due to higher temperatures and biological activity. The major point of this discussion is to illuminate possible difficulties in rigorously determining the relative contribution of acidic precipitation to the acidity of temporary ponds. Detailed bio-geochemical studies measuring ion and water budgets of a large series of ponds are needed before accurate conclusions can be drawn. It is however quite clear that temporary ponds are extremely vulnerable to acidification because of their very low alkalinity (0 to 2 mg L^{-1} $CaCO_3$; Clark and Euler, 1982; Dale *et al.*, 1985; Freda and Dunson, 1985c; Gascon and Planas, 1986). Due to the pervasive character of acidic precipitation and the finite buffering capacity of soils and waters, future declines in pH should be anticipated. Even further modest declines in pH (0.2 units) may have substantial effects on amphibian populations because large percentages of ponds are already below pH 5.0, close to the lethal pH of many species.

Much more information is needed if we are to detect and deal with current and future acidification of amphibian breeding ponds. For example, many extensive areas of North America which experience acidic precipitation have not been surveyed. In addition, virtually all studies have narrowly focused on amphibians which breed in temporary ponds. Plethodontid salamander communities are equally at risk because many species within this taxonomic group live and breed in mountain headwater springs and seeps, which experience drastic episodic declines in pH after snowmelt and rainstorms (Corbett and Lynch, 1982; DeWalle *et al.*, 1984).

In conclusion, pond acidity is currently having a major effect on the local distribution and abundance of temporary pond amphibians. Continued monitoring is necessary to determine the precise role of natural and man-made processes in pond acidification.

LITERATURE CITED

Baker, J.P. and Schofield, C.L.: 1980, *Water Air Soil Pollut.* **18** , 289.

Beebee, T.J.C. and Griffin, J.R.: 1977, *J. Zool., Lond.* **181** , 341.

Brown, D.J.A. and Lynam, S.: 1982, *Bull. Envir. Contam. Toxicol.* **30** , 582.

Carroll, E.J. and Hedrick, J.L.: 1974, *Devel. Biol.* **38** , 1.

Clark, K.L.: 1984, unpublished observations.

Clark, K.L. and Euler, D.: 1982, *The Importance of pH and Habitat Disturbance in Amphibian Distributions in Central Ontario.* Technical Report of the Wildlife Habitat Component, Lakeshore Capacity Study. Ontario Ministry of Natural Resources, Queen's Park, Toronto.

Clark, K.L. and Hall, R.J.: 1985, *Can. J. Zool.* **63** , 116.

Clark, K.L. and Lazerte, B.D.: 1985, *Can. J. Fish. Aquat. Sci.* **42** , 1544.

Clymo, R.S.: 1967, 'Control of cation concentrations, and in particular of pH, in *Sphagnum* dominated communities'. p. 273-284. In: *Chemical Environment in the Aquatic Environment.* H.L. Golterman and R.S. Clymo eds. N. V. Noord-Hollandsche Uitgevers Maatschappij - Amsterdam 1967.

Cook, R.P.: 1983, *Biol. Cons.* **27** , 77.

Cooke, A.S. and Frazer, J.F.D.: 1976, *J. Zool., Lond.* **178** , 223.

Corbett, E.S. and Lynch, J.A.: 1982, 'Rapid fluctuations in streamflow, pH, and associated water quality parameters during a stormflow event'. p. 461-464. *Proc. Inter. Sympo. Hydrometerol.* Am. Water Resour. Assoc. June 13-17, Denver, CO.

Dale, J. and Freedman, B.: 1983, *Effects of Acidity on Nova Scotia Amphibians.* First Annual Report for the Canadian Widlife Service and Institute for Resource and Environment Studies. Dalhousie University, Halifax, Nova Scotia.

Dale, J.M., Freedman, B. and Kerekes, J.: 1985, *Can. J. Zool.* **63** , 97.

Dale, J.M., Freedman, B. and Kerekes, J.: 1986, *Proc. N.S. Inst. Sci., in press.*

DeWalle, D.R., Sharpe, W.E., Leibfried, R.T., Dinicola, R.S., Kimmel, W.G. and Sherwin, L.S.: 1984, *J. Environ. Qual.* **13** , 619.

Driscoll, C.T., Baker, J.P., Bisogni, J.J. and Schofield, C.L.: 1980, *Nature* **284** , 161.

Dunson, W.A. and Connell, J.: 1982, *J. Herp.* **16** , 314.

Freda, J. and Dunson, W.A.: 1984, *Physiol. Zool.* **57** , 435.

Freda, J. and Dunson, W.A.: 1985a, *Copeia* **1985(2)** , 414.

Freda, J. and Dunson, W.A.: 1985b, *Can. J. Zool.* **63** , 2649.

Freda, J. and Dunson, W.A.: 1985c, *The Effect of Acidic Precipitation on Amphibians breeding in Temporary Ponds in Pennsylvania.* U.S. Fish and Wildlife Service, Eastern Energy and Land Use Team. Biol. Report 80(40.21), Box 705, Kearneyville, WV, 25430.

Freda, J. and Dunson, W.A.: 1986a, *Copeia* **1986(2)** , 454.

Freda, J. and Dunson, W.A.: 1986b, *J. Comp. Physiol. B, in press.*

Gascon, C. and Bider, J.R.: 1985, *Can. Field-Nat.* **99** , 259.

Gascon, C. and Planas, D.: 1986 *Can. J. Zool.* **64** , 543.

Glooschenko, V., Weller, W. and Eastwood, C.: 1985, *Amphibian Abundance and Distribution with Respect to Environmental Contaminants and pH Stress Adjacent to Sudbury, Ontario.* Technical report of Wildlife branch, Ontario Ministry of Natural Resources, Queen's Park, Toronto.

Gosner, K.L. and Black, I.H.: 1957, *Ecology* **38** , 256.

Haines, T.A.: 1981, *Trans. Am. Fish. Soc.* **110** , 669.

Hemond, H.F.: 1980, *Ecol. Monogr.* **50** , 507.

Jewell, M.E. and Brown, H.W.: 1929, *Ecology* **10** , 427.

Karns, D.R.: 1983, *Toxic Bog Water in Northern Minnesota Peatlands: Ecological and Evolutionary Consequences for Breeding Amphibians.* Ph.D. Thesis, University of Minnesota, Minneapolis, MN.

Katagiri, C.: 1975, *J. Exp. Zool.* **193** , 109.

Kilham, P.: 1982, *Michigan Botanist* **21** , 159.

Leuven, R.S.E.W., den Hartog, C., Christiaans, M.M.C and Heijligers, W.H.C.: 1986, *Experentia, in press.*

Ling, R.W., VanAmberg, J.P. and Werner, J.K.: 1986, *Ambystoma maculatum* and *Rana sylvatica* in Upper Michigan'. *J. Herpetol., in press.*

Mathews, R.C. and Morgan, E.L.: 1982, *J. Environ. Qual.* **11** , 102.

McDonald, D.G.: 1983, *Can. J. Zool.* **61** , 691.

McDonald, D.G., Ozog, J.L. and Simons, B.P.: 1985, *Can. J. Zool.* **62** , 2171.

Mushinsky, H.R. and Brodie, E.D.: 1975, *Amer. Midl. Nat.* **93** , 440.

Patrick, R., Matson, B. and Anderson, L.: 1979, 'Streams and lakes in the Pine Barrens'. p. 169-193. In: *Pine Barrens: Ecosystems and Landscape.* R.T.T. Forman, ed. Academic Press, N.Y.

Peterson, R.H., Daye, P.G., Lacroix, G.L. and Garside, E.T.: 1982, 'Reproduction in fish experiencing acid and metal stresses'. p. 177-196. In: *Acid Rain / Fisheries.* R.E. Johnson, ed. Am. Fish. Soc., Bethesda, MD.

Pierce, B.A.:1985, Unpublished observations.

Pierce, B.A. and Harvey, J.M.: 1986, *Copeia, in press.*

Pierce, B.A., Hoskins, J.B. and Epstein, E.: 1984, *J. Herp.* **18** , 159.

Pierce, B.A. and Sikand, N.: 1985, *Can. J. Zool.* **63** , 1647.

Porter, K.R. and Hakanson, D.E.: 1976, *Copeia* **1976(2)** , 327.

Pough, F.H.: 1976, *Science* **192** , 68.

Pough, F.H. and Wilson, R.E.: 1977, *Water Air Soil Pollut.* **7** , 307.

Punzo, F.: 1983, **31** , 467.

Saber, P.A. and Dunson, W.A.: 1978, *J. Exp. Zool.* **204** , 133.

Salthe, S.N.: 1965, *Physiol. Zool.* **38** , 80.

Schlichter, L.C.: 1981, *Can J. Zool.* **59** , 1693.

Smith, R.F.: 1960, *An Ecological Study of an Acid Pond in the New Jersey Coastal Plain.* Ph.D. Thesis, Rutgers University, New Brunswick, N.J.

Sprague, J.B. and Ramsey, B.A.: 1965, *J. Fish. Res. Bd. Canad.* **22** , 425.

Tome, M.A. and Pough, F.H.: 1982, 'Responses of amphibians to acid precipitation'. p. 245-254. In: *Acid Rain / Fisheries.* R.E. Johnson, ed. Amer. Fish Soc. Bethesda, MD.

Urch, U.A. and Hedrick, J.L.: 1981, *Arch. Biochem. Biophys.* **206** , 424.

Wilbur, H.M.: 1984, 'Complex life cycles and community organization in amphibians'. p. 195-224. In: *A New Ecology - Novel Approaches to Interactive Systems.* P.W. Price, C.N. Slobodchoff, and W.S. Gaud, eds. John Wiley and Son, N.Y.

Yoshizaki, N.: 1978, *J. Exp. Zool.* **203** , 127.

ECOLOGICAL EFFECTS OF ACIDIFICATION ON TERTIARY CONSUMERS. FISH POPULATION RESPONSES.

Rosseland, B.O. *
Directorate for Nature and Management,
Fish Research Division,
Tungasletta 2, N-7000 Trondheim, Norway.

ABSTRACT. Acidification effects fish populations in various ways, dependant on several factors of biotic and abiotic nature. The most important abiotic factor is water quality and its variability through season and during episodes. Among important biotic factors are fish species, development stages and spawning strategy. Also the specimens residence at time of toxic water and their ability to seek areas of better quality (refuges), are of main importance. For inland populations, reproduction failure seems to be the most important factor for population losses. Which lifestage being most affected might, however, differ from one population to another, even among the same species. Eggs and alevins have been considered to be the most sensitive stages, although high mortality of postspawners have been reported from several lakes. Anadromous species, like Atlantic salmon, have stages of development that are particularly sensitive to changes in water quality. These stages are connected to physiological changes when preparing for change in salinity (smoltification and return from sea). Unfortunately these physiological changes coincides with the most critical environmental periodes; spring melt and autumn rain.

1 INTRODUCTION

The historical and present data on disappearance of fish populations in acidified areas, leaves no doubt to the reason; low pH, elevated levels of Al and low concentrations of ions as Ca and other factors which can ameliorate the toxic effects. Although all throphic levels in aquatic ecosystems are affected during acdification, the disappearence of fish populations have been given greatest concern (Beamish and Harvey, 1972; Almer et al., 1974, 1978; Leivestad et al., 1976; Schofield, 1976; Overrein et al.,1980; Haines,1981).

In Scandinavia, fish kills in connection with acidic waters have been observed since the turn of the century (Huitfeldt - Kaas, 1922). At

* Present address: Norsk Biotech A/S, P.O. Box 788, Krossen, N-4301 Sandnes, Norway

Water, Air and Soil Pollution **30** (1986) 451–460.

first, populations of Atlantic salmon (Salmo salar), later on brown trout (Salmo trutta) and other freshwater species were affected. By the end of the 1970's the fish populations in an area 33,000 km^2 had been affected in Norway (Sevaldrud et al., 1980). Today the loss of fish populations still continuous in Scandinavia (Sevaldrud and Skogheim, 1986; Rosseland et al., 1986a), with an increasing concern also in the North Eastern U.S.A. and Canada (Kelso and Minns, 1982; Johnson, 1982).

To understand the various modes which acidification can act on fish populations, the mechanisms involved must be clarified: a) species and developmental stage sensitivity, including differences between stationary and anadromous populations, b) spawning periodes, sites and strategy, and c) differences in water quality in respect to a chronic versus episodic change, and thus the susceptibility of a lake versus river/brook population.

2. REPRODUCTION FAILURE, AGEING.

The data on reproduction failure as the main cause for disappearance of fish populations are comprehensive (Jensen and Snekvik, 1972; Rosseland et al., 1980; Harvey, 1982; Beamish et al., 1975; Haines, 1981; Leivestad et al., 1976; Schofield, 1976; Muniz and Leivestad,1980; Peterson et al., 1982). The early life history stages at which the acidification acts, however, can differ. At least three different developmental stages are important for reproduction : 1) the oogenesis 2) the development from fertilization to hatching, through alevin stage, and during and shortly after swim up, and 3) the survival of spawners.

Craig and Baksi (1977) and Peterson et al. (1982) pointed to the oogenesis and fertilization periode as sensitive to low pH. Failure in producing viable eggs owing to a reduced serum Ca level have been reported from Canada (Beamish, 1976). Wiener (1983) found a depleted plasma Ca concentration in maturing white sucker (Catostomus commersoni) compared to fish from neutral lakes. Failure in normal Ca metabolism in acid waters is also supported by Fraser and Harvey (1982) showing a depletion of Ca from bone in white sucker in relation to lake acidification. An increased number of females with unshed eggs have also been reported from acid lakes (Beamish et al., 1975). On the other hand several studies have concluded that the females seems to develop eggs and spawn normally (Rosseland et al., 1980; Kelso and Gunn, 1984; Rask, 1984b). This is in accordance with the results of Rodgers (1984), who did not find any significant reduction in axial skeleton and visceral tissue Ca in Brook trout (Salvelinus fontinalis) in low pH and low Ca waters, thus indicating no severe disturbances in the Ca metabolism.

After fertilization, the embryo are susceptible through the whole developmental periode although the time shortly after fertilization and prior to and during hatching seems to be of major concern (Peterson et al., 1980; Runn et al., 1977; Skogheim and Rosseland, 1984; Rask, 1984a) Three factors are of main importance determine the sensitivity and thus toxicity at hatching: the eggshell hardening, reduced fry mobility and reduced hatching enzyme activity at low pH (Peterson et al., 1980; Runn et al., 1977; Haya and Waiwood, 1981). In general, the pH seems to be of major importance for the toxicity at the egg stage, whereas the role of Al as a toxic factor increases after hatching (Baker and Schofield, 1980; Baker, 1982).

Eggs and alevins lying in the gravel are to some extent protected from rapid changes in pH. This is due to the longer retention time of the interstitial water and thus a better buffered microenvironment (Gunn and Keller, 1984b; Lacroix et al., 1985). Nevertheless they can experience high mortality during acid run off, as the concentration of Al in the interstitial water in some cases can exceed that of the ambient water in spite of a higher pH (Gunn and Keller, 1984a). When the interstitial water is neutralized by mixture of limestone in the gravel, however, the mortality decreases (Gunn and Keller, 1980, 1984b; Rosseland and Skogheim, 1984a).

Rask (1983) found increasing mortality and development time for eggs of perch (Perca fluviatilis) in waters of pH 5.0 to 4.0 compared to eggs developed in pH 6.4. Skogheim and Rosseland (1984) demonstrated that acidic conditions at the alevin stage affected the resorption of the yolk sac. At pH 4.9 to 5.5 the number of degree-days from hatching to initiation of feeding increased, by 80, 45, 45 and 18% for Atlantic salmon, sea trout, brown trout and brook trout respectively, when compared to pH 6.7. In other words, the retardation being largest for the most sensitive species. This phenomenon might be an important respons under natural conditions, as a prolongation of the resorption periode decreases the probability that post swim-up fry are exposed to acid episodes during snow melt (Skogheim and Rosseland, 1984).

Lacroix et al. (1985) demonstrated a high mortality of post swim up fry of Atlantic salmon in river water of pH 5.0. As the fry body ion reserve was not significantly different from fry in neutralized river water, the observed weight reduction, inactivity and high mortality was suggested to be caused by starvation, leading to declining and lost salmon populations in Novia Scotia rivers.

The role of aluminum as a toxic factor after hatch (Baker and Schofield, 1980), have been clearly demonstrated by Hulsman et al. (1983). One hundred % alevins of rainbow trout (Salmo gairdneri) died in a clearwater creek at pH 5.4, whereas only 1% died in a humic creek at pH 5.5. Both creeks had a rather similar concentration of total Al, 137 vs 167 ug Al L^{-1}, but since the color (in Hazen units) were 2 vs 13, more Al was assumed to be chelated and thus detoxified in the more humic creek.

Mortality in the early life stage history will lead to a reduction in or even total lack of younger year classes (ageing), depending on year to year variation in water chemistry in the spawning areas (Harvey, 1982; Rosseland et al., 1980; Andersen et al., 1984; Rask, 1984b; Beggs et al., 1985).

2. GROWTH AND CONDITION FACTOR

In the early phases of ageing, the remaining fish often respond by increased condition factor and growth, probably caused by a reduced competition for food (Rosseland et al., 1980,; Andersen et al. ,1984;

SFT, 1984,; Andersson and Andersson ,1984; Jensen and Snekvik, 1972; Almer et al., 1978,; Mossberg and Nyberg ,1976; Lessmark, 1976). On the other hand, an improved condition factor does not necessarily have to be linked to an increased growth (Beamish et al., 1975; Rosseland et al., 1980; Harvey, 1982). Tank experiments with brook trout exposed to acid but sublethal waters demonstrated a slower growth in spite of feeding ad lib compared to fish of same origin kept in neutralized waters (Muniz and Leivestad, 1979). The same result was obtained using rainbow trout and Arctic char (Salvelinus alpinus) (Edwards and Hjeldnes, 1977). Brown trout exposed to sublethal and lethal acidic brook waters have been shown to increase their standard metabolism, probably in connection to increased cellular activity trying to restore the ionic balance (Rosseland, 1980).

Fish populations in acidified areas will through most of the year live under unfavorable conditions. Habitats like brooks, rivers and lakes with a short retention time, will have large seasonal fluctuations in water quality. These periods will therefore require much energy (increased metabolism) for surviving only. The period from late spring to early autumn is favorable both because of a normally surplus food supply and good water quality. In this period, a weight increment leading to a high condition factor is reasonable. On the yearly energy budget, however, the unfavorable period might outclass the favorable one, thus the surplus energy necessery for growth might be lacking. This might explain the lack of growth respons often observed in declining fish populations in spite of the reduced competion for food (Rosseland et al., 1980; Ryan and Harvey, 1980; Beamish, 1974; Kelso and Gunn, 1984). This might also be the main explanation for the observed two fold growth in brook trout compared to brown trout in a mixed population in an acidified river system (SFT, 1982), as brook trout have shown to be more physiological tolerant than brown trout to acid waters (Grande et al., 1978; Rosseland, 1980; Rosseland and Skogheim, 1984b). No evidence are given for a food shortage being responsible for fish population losses during acidification (Beamish, 1974; Rosseland et al., 1980).

3. MORTALITY OF ADULT FISH, POSTSPAWNING MORTALITY, JUVENILIZATION

Although an increasing mortality on the young life history stages are the most common cause for fish population extinction, mortality of adult fish do occur. In the cases where these mortalities have been observed, it has been in connection with episodic changes in water quality, during spring snowmelt or heavy autumn rain (Huitfeldt - Kaas, 1922; Dahl, 1927; Rosseland, 1953; Leivestad and Muniz, 1976; Schofield, 1977; Skogheim et al., 1984; Henriksen et al., 1984; Rask, 1984b; Andersson and Nyberg, 1984; Harvey and Lee, 1982; Snekvik, 1975; Rosseland et al., 1986a). As this period is characterized by high water flow, however, the chance of observing dead fish is limited and often based on coincidence (Leivestad and Muniz, 1976). A sudden lack of older fish (juvenilization) in a population might therfore be an indirect observation of such a phenomenon (Rosseland et al., 1980; Andersen et al., 1984; Frenette and Dodson, 1984; Kelso and Gunn, 1984; Rask, 1984b).

Tveitvatn, southern Norway, will be used as an example. In the 1960's the stunted brown trout population was a part of a feeding experiment, trying to increase growth (Dannevig, 1966). After a massive fish kill during snow melt in spring 1975 (Leivestad and Muniz, 1976), the brown trout population rapidly decreased (Rosseland et al., 1980), and by 1982 the population had disappeared (SFT, 1984). Typically for this population was the high postspawning mortality, especially on females, a phenomenon also found in other brown trout populations, and to a less extent in populations of Arctic char in Norway (Andersen et al., 1984), perch in Finland (Rask, 1984b) and in lake trout (Salvelinus namaycusk Walbaum) and brook trout populations in Canada (Kelso and Gunn, 1984; Frenette and Dodson, 1984). Brook and river spawners normally enters the spawning areas on high waterflow, in connection with heavy autumn rain. As this normally is a period of low pH (Henriksen and Wright, 1977), the chemical conditions can be sublethal or lethal. Although spawners of brown trout seems to be relative physiological tolerant to acid waters (Muniz et al., 1986), postspawners have shown to be more sensitive than juveniles in early spring (Rosseland and Skogheim unpublished).

4. REFUGES AS SELECTION PARAMETRES

In some cases, observations of fish selecting or avoiding waters of a certain quality have been conducted. Muniz et al. (1978) describes how physiologicaly stressed adult brown trout gathered outside the outlet of a small brook, too small to house the fishes itself. pH in the brook was 6.0 compared to pH 4.7 in the main river. By using the method of Webster and Eiriksdottir (1976), Johnson and Webster (1977) described how brook trout avoided spawning sites with pH below 5. During episodes of extreme toxic concentrations of pH and Al, spawning migrating Atlantic salmon have either been observed drifting or swimming downstream towards better water quality (Skogheim et al., 1984) or been found hiding in a small brook with better water quality while other salmon died in the more acid main river (Rosseland et al., 1986a). A similar phenomenon is described in Andersson and Nyberg (1984) where brown trout, and roach (Rutilus rutilus L.) tried to escape toxic water during spring thaw by downward migration.

Although the genetic and physiological constituents of the individual fish sets the major limitations to acid tolerance, the ability of fishes to activly seek and find refuges in periods of marginal water quality in their main habitat, might nevertheless be an important selection parameter in nature. Thus differences in this respect might as well be an important factor in determin which species to disappear first within the same lake system. In addition, differences in spawning sites (lake or brook spawners) and spawning periods (spring vs autumn) might as well add to the complexity. This is why some lakes, having mixed populations of brown trout and perch, have lost its trout populations but still holds its perch, and vice versa (Rosseland et al., 1980). Having this in mind, one should be very carefull to use field observations of e.g. water chemistry of the lake outlet and fish status, and draw conclusion as to the tolerance of the different species. This is emphasized by Haines (1981).

5. SUPERSENSITIVE STAGES.

In recent years, the smoltification period have been focused as a extremely sensitive stage in the life cycle of the anadromous Atlantic salmon (Rosseland and Skogheim, 1982, 1984b; Rosseland et al., 1986b; Saunders et al., 1983; Johnston et al., 1984). The loss of Atlantic salmon populations in acidified areas in Norway (Jensen and Snekvik, 1972; Leivestad et al., 1976) may be understood by this "supersensitive" smolt stage.

River Vikedal, southwestern Norway, was once a typical salmon river. Since 1981, however, yearly mortalities of smolts of Atlantic salmon and sea trout in connection to spring snowmelt have been observed (SFT, 1983b, 1984; Hesthagen, 1986; Rosseland et al., 1986a). These mortalities seems to have been selective towards the 2-yr old smolts, as the abundance of both summerold and one year old salmon measured by electrofishing in autumn, have been normal in these years. Field laboratory experiments in River Vikedal have verified these observations (Henriksen et al.,1984).

Variations in life history may be a safeguard against loss of small salmon stocks through several successive years of reproductive failure, since nonspawning inividuals in the river or sea could spawn in subsequent years (Saunders and Schom, 1985). Nevertheless, when the number of descending smolts decreases due to high spring mortality, and the exploitation in the sea is ca. 90% (Hansen ,1985), the chance for a successful salmon run is rapidly decreasing. On return from sea, however, another critical period appears. Spawning migrating salmon caught in sea outside the river estuary, have shown to be extremely sensitive to low pH when immediately exposed to fresh water (Potts et al., 1985). This high sensitivity can lead to severe mortality if the upward migration coincides with an acid episode in their native river (Skogheim et al., 1984).

The time-lag between a complete stop in smoltproduction to a stop in the run of spawners depends on the typical sea age of the salmon stock (1 to 3 yr). As the variations in water chemistry between years can be great, the decrease in the salmon population will be gradual. In such a situation, a more tolerant species like sea trout (Rosseland and Skogheim, 1984b) will normally expand. This has been the case in the now barren salmon rivers at Sørlandet (Jensen and Snekvik, 1972), and is now the case in River Vikedal, which today can be characterized as a typical sea trout river (Hesthagen, 1986).

ACKNOWLEDGMENT. I greatly acknowledge the comments given to the manucript by my colleagues Odd K. Skogheim and Iver Sevaldrud.

REFERENCES

-Almer, B., Dickson, W., Ekstrøm, C., Hörnström, E. and Miller, U.: 1974, Ambio 3, 30.

-Almer, B., Dickson, W., Ekstrøm, C. and Hörnström, E.: 1978, 'Sulfur pollution and the aquatic ecosystem,' in, Nriagu, J.O. (ed), Sulfur in the environment: Part II, Ecological impacts. pp. 271 - 311, John Wiley & Sons. 1978.

-Andersen, R., Muniz, I.P. and Skurdal, J.: 1984, Rep. Inst. Freshw. Res. Drottningholm 61, 5.

-Andersson, B. and Andersson, P.: 1984, Rep. Inst. Fresh. Res. Drottningholm 61, 28.

-Andersson, P. and Nyberg, P.: 1984, Rep. Inst. Freshw. Res. Drottningholm 61, 34.

-Baker, J.P. and Schofield, C.L.: 1980, 'Aluminium toxicity to fish as related to acid precipiation and Adirondack surface water quality', in, Drabløs, D. and Tollan, A. (eds)., Ecological impact of acid precipitation, pp 292-293, SNSF-project.

-Baker, J.P.: 1982, 'Effects on fish of metals associated with acidification', in, Johnson, R.E. (ed), Acid rain/fisheries (Bethesda, MD, Am. Fish. Soc. 1982), pp. 45 - 55.

-Beamish, R.J. and Harvey, H.H.: 1972, , J. Fish Res. Board Can. 29, 1131.

-Beamish, R.J.: 1974, J. Fish Res. Board Can. 31, 49.

-Beamish, R.J., Lockhart, W.L., Van Loon, J.C. and Harvey, H.H.: 1975, Ambio 4, 98.

-Beamish, R.J.: 1976', Water Air Soil Pollut. 6, 501.

-Beggs, G.L., Gunn, J.M. and Olver, C.H.: 1985, 'The sensitivity of Ontario lake trout (Salvelinus namaycush) and lake trout lakes to acidification', Ont. Fish Tech. Rep Ser. No 17, iii & 24 p.

-Craig, G.R. and Baksi, W.F.; 1977, Water Res. 11, 621.

-Dahl, K.: 1927, Salmon and Trout Magazine 46, 35.

-Dannevig, G.: 1966, 'Kulturarbeidet i Tovdalselva. Noen retningslinjer og resultater', (Fish culturing in River Tovdal. Some procedures and results), Statens Biologiske Stasjon, Arendal. Stens. Rapport, 6 p. (In Norwegian).

-Edwards, D.J. and Hjeldnes, S.: 1977, 'Growth and survival of salmonids in water of different pH', SNSF-project, FR 10/12, 12 p.

-Fraser, G.A. and Harvey, H.H.: 1982, Can. J. Fish. Aquat. Sci. 39, 1289.

-Frenette, J.J. and Dodson, J.J.: 1984, Can. J. Fish. Aquat. Sci, 41, 865.

-Grande, M., Muniz, I.P., and Andersen, S.: 1978, Verh. Int. Verein. Limnol, 20, 2076.

-Gunn, J.M. and Keller, W.: 1980, Can J. Fish Aquat. Sci. 37, 1522.

-Gunn, J.M. and Keller, W.: 1984, Can. J. Fish. Aquat. Sci. 41, 319.

-Gunn, J.M. and Keller, W.: 1984b, Fisheries 9, 19.

-Haines, T.: 1981, 'Effect of acid rain on Atlantic salmon rivers and restoration effects in the United States', Proc. Comf. Acid Rein and the Atlantic salmon. Nov. 22 - 23, 1980. International Atlantic salmon foundation special publication, 10, 57 - 63.

-Hansen, L.P.: 1985, 'Salmon catch data available for analyzes in Norway', in, Jenkins, D. (ed.), The status of Atlantic salmon in Scotland, J.T.E. symposium (In press.)
-Harvey, H.H. and Lee, C.: 1982, 'Historical fisheries changes related to surface water pH changes in Canada', in, Johnson, R.E. (ed.), Acid rain/Fisheries (Bethesda, MD, Am. Fish. Soc. 1982) pp. 45 - 55.
-Harvey, H.H.: 1982, 'Population responses of fishes in acidified waters', in Johnson, R.E. (ed.), Acid rain/fisheries (Bethesda, MD, Am. Fish. Soc. 1982), pp. 227 - 242.
-Haya, K.and Waiwood, B.A.: 1981, Bull Environm. Contam. Toxicol. 27, 7.
-Henriksen, A. and Wright, R.F.: 1977, Nordic Hydrol. 8, 1.
-Henriksen, A., Skogheim, O.K. and Rosseland, B.O.: 1984, Vatten 40, 255.
-Hesthagen, T.: 1986, 'Fish kill of Atlantic salmon (Salmo salar) and brown trout (Salmo trutta) in an acidified river in south western Norway', MS Muskoka Conference '85.
-Huitfeldt-Kaas, H.: 1922, 'Om aarsaken til massedød av laks og ørret i Frafjordelven, Helleelven og Dirdalselven i Ryfylke høsten 1920', (On the cause of mass death of salmon and trout in River Frafjord, River Helle and River Dirdals in Ryfylke, autumn 1920), Norsk Jæger- og Fiskefor. tidsskrift, 37-44. (In Norwegian).
-Hulsman, P.F., Powles, P.M. and Gunn, J.M.: 1983, Trans. Am. Fish. Soc. 112, 680.
-Jensen, K.W. and Snekvik, E.: 1972, Ambio 1, 223.
-Johnson, D.W. and Webster, D.A.: 1977, J. Fish. Res. Board Can. 34, 2215.
-Johnson, M.G.: 1982, 'Acid rain and fisheries - an overview', in: Johnson, R.E. (ed), Acid Rain/Fisheries (Bethesda, M.D, Am.Fish.Soc. 1982) pp. 329 - 339.
-Johnston, C.E., Saunders, R.L., Henderson, E.B., Harmon, P.R. and Davidson, K.: 1984, 'Chronic effects of low pH on some physiological aspects of smoltification on Atlantic salmon (Salmo salar)', Can.Tech. Rep. Fish. Aquat. Sci. 1294,iii+7p.
-Kelso, J.R.M. and Minns, C.K.: 1982, 'Current status of lake acidification and its effect on the fishery resources of Canada', in, Johnson, R.E. (ed.) Acid Rain/Fisheries. (Bethesda, M.D, Am. Fish. Soc. 1982) pp. 69 - 82.
-Kelso, J.R.M. and Gunn, J.M.: 1984,' Responses of fish communities to acidic water in Ontario', in, Hendrey, G.R. (ed.) Early biotic responses to advancing lake acidification. Butterworth Publishers. Stoneham, Maryland, p. 105 - 115.
-Lacroix, G.L., Gordon, D.J. and Johnston, D.J.: 1985, Can. J. Fish. Aquat. Sci 42, 768.
-Leivestad, H., Hendrey, G., Muniz, I.P. and Snekvik, E.: 1976, 'Effect of acid precipitation on freshwater organisms', in Braekke, F.H.(ed.), Impact of acid precipitation on forest and freshwater ecosystems in Norway, 87-111, SNSF-project, FR 6/76.
-Leivestad, H. and Muniz, I.P.: 1976, Nature, 259, 391.
-Lessmark, O.: 1976, 'Försurningens inverkan på fiskefaunan i några småländska sjöar', (Effects of acidification on the fish fauna of some

lakes in Småland), Inf. Inst. Freshw. Res. Drottningholm 7/1976, 25 p. (In Swedish).
-Mossberg, P. and Nyberg, P.: 1976, 'Försurningseffekter på bottenfauna och fisk i Västra Skjälsjön', (Effects of acidification on bottom fauna and fish in Lake Västra Skjälsjön), Inf. Inst. Freshw. Res. Drottningholm 9/1976, 23 p. (In Swedish).
-Muniz, I.P. Leivestad, H. and Rosseland, B.O.: 1978, 'Stressmålinger på fisk i sure vassdrag'. (Stress measurements on fish in acidic rivers), Nordforsk,publ. 1978/2, 233-247.
-Muniz, I.P. and Leivestad, H.: 1979, 'Langtidseksponering av fisk til surt vann. Forsøk med bekkerøye Salvelinus fontinalis Mitchill' (Longterm exposure of fish to acid water. Experiment with brook trout Salvelinus foutinalis Mitchill), SNSF-project IR 44/79, 32p.
-Muniz, I.P. and Leivestad. H.: 1980, 'Acidification-effects on freshwater fish', in, Drabløs, D. and Tollan, A. (eds.), Ecological impact of acid preciptation, p. 84-92, SNSF-project.
-Muniz, I.P., Andersen, R. and Sullivan, T.J.: 1986, 'Physiological respons of brown trout (Salmo trutta L.) spawners and postspawners exposed to acidic aluminium-rich stream water', MS, Muskoka Conference'85.
-Overrein, L., Seip, H.M. and Tollan, A.: 1980, 'Acid precipitation effects on forest and fish', SNSF-project, FR 19/80, 175 p.
-Peterson, R.H. Daye, P.G. and Metcalfe, J.L.: 1980, Can. J. Fish. Aquat. Sci. 37, 770.
-Peterson, R.H., Daye, R.G., Lacroix, G.L., and Garside, E.T.: 1982, 'Reproduction in fish experiencing acid and metal stress', in, Johnson, R. (ed.), Acid rain/fisheries. (Bethesda, M.D. Am. Fish. Soc. 1982) pp.177-195. Am. Fish. Soc.
-Potts, W.T.W., Talbot, C. and Eddy, F.B.: 1985, J. Exp. Biol., 118, 455.
-Rask, M.: 1983, Ann. Zool. Fennici 20, 73.
-Rask, M.: 1984a, Ann. Zool. Fennici 21, 9.
-Rask, M.: 1984b, Ann. Zool. Fennici 21, 15.
-Rodgers, D.W.: 1984, Can. J. Fish. Aquat. Sci. 41, 1774.
-Rosseland, B.O. Sevaldrud, I.H. Svalastog D. and Muniz, I.P.: 1980, 'Studies of freshwater fish populations - effects of acidification on reproduction, population structure, growth and food selection', in, Drabløs, D. and Tollan, A. (eds.) Ecological impact of acid precipitation, p. 336-337, SNSF-project.
-Rosseland, B.O.: 1980, 'Effects of acid water on metabolism and gill ventilation in brown trout, Salmo trutta L., and brook trout, Salvelinus fontinalis Mitchill', in, Drabløs, D. and Tollan, A. (eds.) Ecological impact of acid precipitation, pp. 348-349, SNSF-project.
-Rosseland, B.O. and Skogheim, O.K.: 1982, 'Physiological stress and mortality of Atlantic salmon, Salmo salar L. in acid water with high levels of aluminium', Inter. Council for Exploration of the Sea C.M. 1982/M:29,15 p.
-Rosseland, B.O. and Skogheim, O.K.: 1984, Fisheries 9, 10.
-Rosseland, B.O. and Skogheim, O.K.: 1984, Rep. Inst. Freshw. Res. Drottningholm 61, 186.
-Rosseland, B.O., Skogheim, O.K. and Sevaldrud, I.H.: 1986a, 'Acid deposition and effects in nordic Europe. Damage to fish populations in

Scandinavia continue to apace', Water, Soil and Air Pollut..
-Rosseland, B.O., Skogheim, O.K., Kroglund, F. and Hoell, E.: 1986b, 'Mortality and physiological stress on different year-classes of landlocked and migratory Atlantic salmon,brown trout and brook trout in acidic aluminium-rich soft water', Water, Air and Soil Pollut.
-Rosseland, L.: 1953, 'Om virksomheten i 1948. I: Fiskeriinspektørens årsmelding for årene 1948, 1949, 1950' (Annual Report 1948. in, Annual Report on Freshwaterfisheries 1948, 1949, 1950), Directorate for Wildlife and Freshwater Fish, Ås. (In Norwegian).
-Runn, P. Johansson, N. and Milbrink, G.: 1977, Zoon 5, 115.
-Ryan, D.M. and Harvey, H.: 1980, Environ. Biol. Fish.5, 97.
-Saunders, R.L. Henderson, E.B. Harmon, P.R. Johnston, C.E. and Eales, J.G.: 1983, Can. J. Fish. Aquat. Sci. 40, 1203.
-Saunders, R.L. and Schom, C.B.: 1985, Can. J. Fish. Aquat. Sci. 42, 615.
-Schofield, C.L.: 1976, Ambio 5, 228.
-Schofield, C.L.: 1977, 'Acid snow-smelt effects on water quality and fish survival in the Adirondack Mountains of New York State', Res. Proj. Tech. Comp.Rep. Project A-072-NY. Cornell University, Ithaca, N.Y. 27 p.
-Sevaldrud, I.H. Muniz, I.P. and Kalvenes, S.: 1980, 'Loss of fish populations in Southern Norway. Dynamics and magnitude of the problem', in, Drabløs, D. and Tollan, A. (eds.) Ecological impact of acid precipitation, pp. 350-351, SNSF-project.
-Sevaldrud, I.H. and Skogheim O.K.: 1986, 'Changes in fish populations in southernmost Norway during the last decade', Water, Soil and Air Pollut.
-SFT: 1982,'Overvåking av langstransportert forurenset luft og nedbør. Årsrapport 1981' (Monitoring of long range transported pollutants in air and precipitation. Annual report 1981), Rapport 64/82, 176 p., Statens forurensningstilsyn, Oslo. (In Norwegian).
-SFT: 1983, 'Vannkjemisk og fiskeribiologiske undersøkelser i Vikedalsvassdraget 1981 - 1982. Vannkvalitet og fiskedød våren 1982' (Water chemistry and fish biological investigations in River Vikedal 1981-1982. Waterquality and fish kill spring 1982), Rapp. 97/83, 63 p. Statens forurensningstilsyn, Oslo (In Norwegian).
-SFT: 1984, 'Monitoring of long range transported pollutants in air and precipitation. Annual report 1983', Rapp. 162/84, 198 p., Statens forurensningstilsyn, Oslo (In Norwegian).
-Skogheim, O.K., Rosseland, B.O. and Sevaldrud, I.: 1984, Rep. Inst. Freshw. Res. Drottningholm 61, 195.
-Skogheim, O.K. and Rosseland, B.O.: 1984, Rep. Inst. Freshw. Res. Drottningholm 61, 177.
-Snekvik, E.: 1975, 'Episodic fish kills in relation to acidification of rivers in 1969'. Notat fra Fiskeforskningen, 9 p.(In Norwegian).
-Webster, D.A. and Eriksdotir, G.: 1976, Trans. Am. Fish. Soc. 105, 416.
-Wiener, J.G.: 1983, 'Comparative analyses of fish populations in naturally acidic and circumneutral lakes in northern Wisconsin', Fish and Wildlife Service, Eastern Energy and Land Use Team, FWS/OBS - 80/40.16, 107 p.

FISH RESPONSES TO ACIDITY IN QUÉBEC LAKES: A REVIEW

J.-J. Frenette, Y. Richard
Ministère de l'Environnement du Québec, Direction des relevés aquatiques, 3900 rue Marly, 5ième étage, Sainte-Foy, Québec. G1X 4E4. Canada.
and G. Moreau
Département de biologie, Université Laval, Cité Universitaire, Sainte-Foy, Québec. G1K 7P4. Canada.

ABSTRACT. To establish the impact of acidity on fish populations, studies were conducted in 37 Québec lakes located in four regions; the réserve des Laurentides and Portneuf and the Charlevoix and Témiscamingue regions. Density (catch per unit effort) of brook charr (Salvelinus fontinalis) decreases with increasing acidity. Moreover, in the Charlevoix region, this species has disappeared from three acid lakes (4.6 $\leqslant$pH$\leqslant$5.1) with low Ca levels. Unlike growth, condition demonstrates a close relationship to acidity in brook charr populations. The total Al concentration in gills decreases with increasing size and pH. Lake acidity and sensitivity to acidification introduces problems in gamefish management. A survey of 17 lakes of the Témiscamingue region reveals that species diversity and total fish biomass are much lower in acid lakes than non acid lakes. In addition, two acid lakes are devoid of fish. Cyprinidae and Johnny darters (Etheostoma nigrum) are abundant in lakes with a pH level of 5.9 to 7.0 but are absent in lakes with a pH lower than 5.2. The yellow perch (Perca flavescens) is the only fish that appears to be tolerant to a wide pH range. This species, however, is in poor condition in acid lakes as compared with non acid lakes.

1. INTRODUCTION

Since 1981, several studies have been conducted by the Direction des relevés aquatiques du ministère de l'Environnement du Québec, to evaluate the impact of acidity on brook charr populations (Salvelinus fontinalis) and fish communities from 37 lakes in Québec. The purpose of this paper is to identify through a synthesis of these studies the major modifications induced by acidity on diversity, density, biomass, growth, condition and survival of fish populations. The extent of these modifications may have an impact on fish exploitation.

Water, Air and Soil Pollution **30** (1986) 461–475.

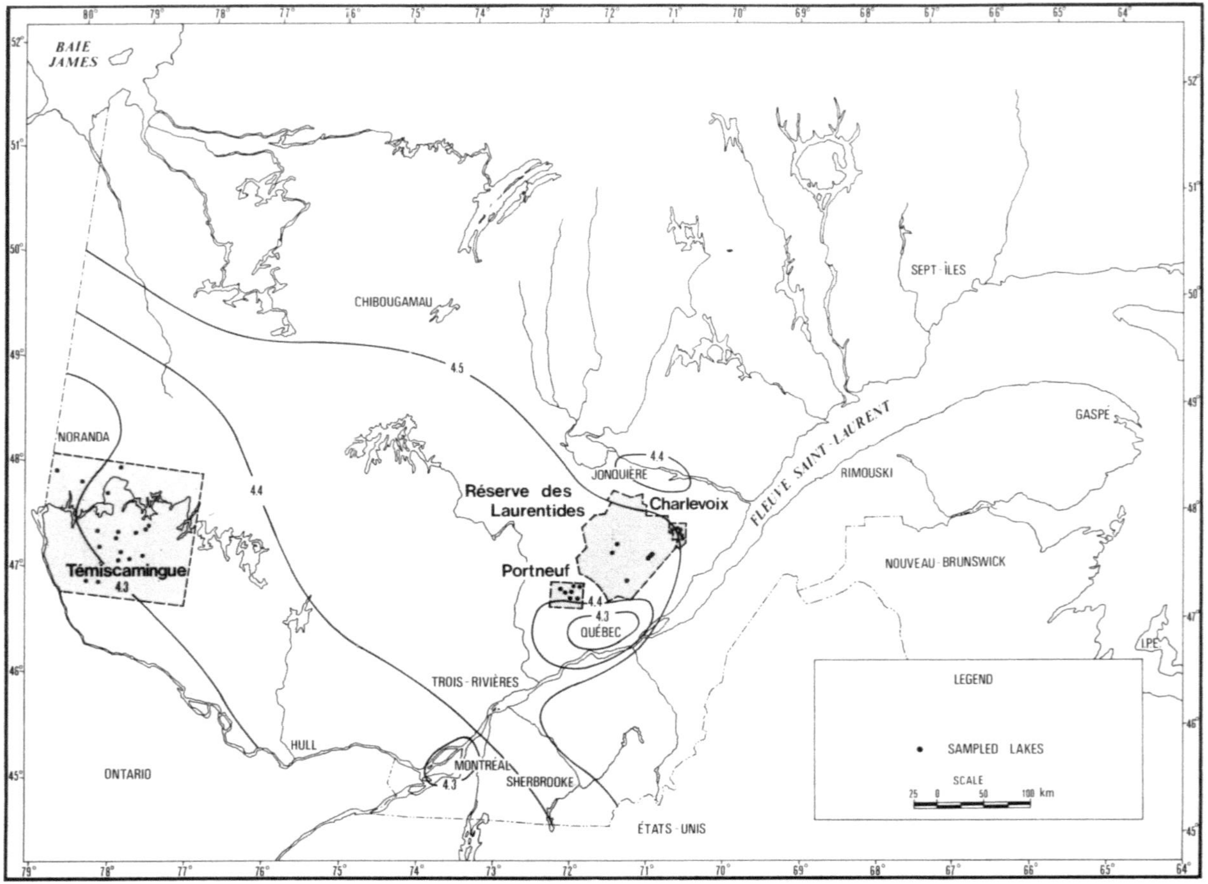

FIGURE 1. Study areas and precipitation weighted mean pH value for southern Québec, 1982-1983. (after Grimard, 1984).

2. MATERIALS AND METHODS

Study area - The weighted mean pH values varied in Québec precipitation between 4.3 to 4.9. The lowest pH are found in southern and southwestern Québec and more specifically near Noranda, Québec and Montréal (Figure 1). This gradient was used to select target regions for the study of fish populations. These are the réserve des Laurentides and the regions of Portneuf (réserve de Portneuf and ZEC Batiscan-Neilson) and Charlevoix (ZEC des Martres) for allopatric brook charr populations, and the region of Témiscamingue for multispecies populations (Figure 1).

Lake selection - Physico-chemical sampling was conducted in April 1981, -82, -83 and -84 in the réserve des Laurentides and the Portneuf, Charlevoix and Témiscamingue regions respectively. A total of 286 lakes were visited. Based on the results of these samplings, 37 lakes were retained for the comparative study of their fish populations. These lakes were representative of the acidity gradient specific to each region (Table 1).

Fish sampling - Fish were sampled in June 1981, -82, -83 and -84 in order to minimize variations due to season. Sampling was realized with experimental gill nets composed of six panels (25, 38, 51, 63, 76 and 89 mm) 7.6 m wide and 1.8 m high. All nets were deployed during the night for a minimum of 12 hr. Minnow traps and trap nets were also used for sampling and analysis are described in Richard (1982, 1985a,b, 1986), Moreau et al. (1984) and Frenette (1986a,b).

3. RESULTS AND DISCUSSION

3.1. Allopatric Brook Charr Populations

Density - Catch per net per 12 hr (c.p.u.) of brook charr decreases with increasing acidity (Figure 2). In addition no fish were captured from three Charlevoix lakes where the pH varied from 4.6 to 5.1 (Table 1). These results are in agreement with the findings of several studies which generally concede that for pH $\leqslant$ 5.0, brook charr reproductive success is greatly diminished (Daye and Garside, 1975; Menendez, 1976; Johansson et al., 1977; Grande et al., 1978; Harvey, 1980; and Schofield and Trojnar, 1980), and population density is reduced following a decrease in recruitment of young age-classes (Jensen, 1971). At these pH values, the species has completely disappeared from many Adirondacks lakes (Schofield, 1982) and the Sudbury region (Conroy et al., 1978).

Calcium levels appear to be a better predictive factor of population density (Figure 2) since c.p.u. was more highly correlated to Ca^{++} concentration that to pH. Field studies have in fact demonstrated that for a given pH, the percentage of lakes without fish is always higher for lakes with low Ca levels than for those with high Ca levels (Brown, 1982). The disappearance of brook charr from lakes Robbé, Freddo and Premier Lac Paul may be due to this factor since the Ca levels present

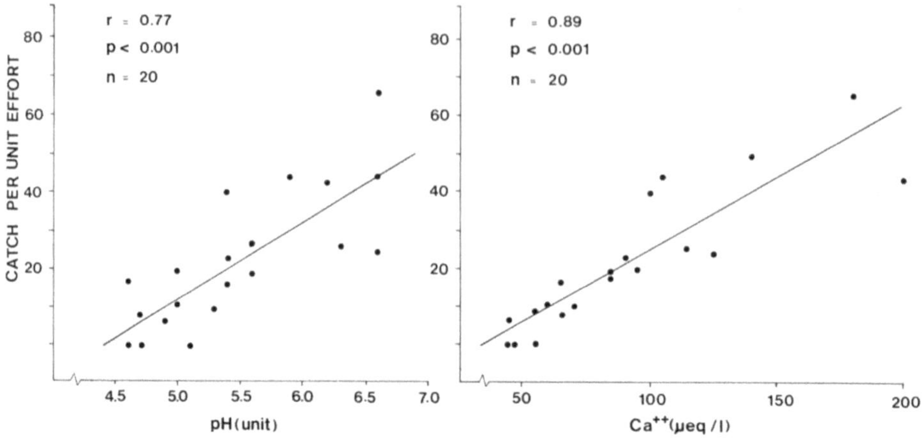

FIGURE 2. Density (catch per unit effort) of brook charr in relation to pH and Ca^{++} concentrations in lakes.

in these lakes are lower than those of lakes with similar pHs in the réserve des Laurentides and the Portneuf region (Table 1). Even a small increase in Ca levels reduces acid stress. Calcium facilitates osmotic regulation by allowing better Na^+ and Cl^- ion retention, thus enhancing survival of fish (Leivestad, 1982; Wood and McDonald, 1982).

Growth - The slopes of growth curves for the brook charr (Figure 3) demonstrate that fish in acid lakes from the Portneuf (Poliquin, Morasse) and Charlevoix (Côte-à-Côte, Deuxième Lac Paul) regions grow faster than those from other lakes (covariance analysis, $p < 0.001$). However, the situation is reversed in the réserve des Laurentides where brook charr grow more slowly.

The better growth of some fish populations in acid lakes may be attributed to an increase in food availability following a decrease in fish density, as a result of diminished intra and/or interspecific competition (Harvey, 1982). This hypothesis has been discussed by Frenette and Dodson (1984) who argue that the relationship between greater food supply and increased growth rate must take into account the maintenance requirement of fish and the potential growth capacity (influenced by growth hormone) which set the limit for voluntary food intake and hence growth rate (Brett, 1979). Therefore, the relation between increased food supply and increased growth rate cannot be readily assumed. In addition to this study, many authors have demonstrated different inter and intraspecific responses in growth to acidity as reviewed by Harvey (1982). Growth may be affected by many other factors such as climate, morphometry, nutrient status (Ryder, 1982), exploitation (Johnson, 1976; Power, 1978), selective mortality (Frenette and Dodson, 1984) and genetic strain (Webster and Flick, 1975).

TABLE I. CHEMICAL AND MORPHOMETRIC CHARACTERISTICS OF THE QUÉBEC STUDY LAKES (1981-1984). DASH INDICATES NO DATA.

LAKE	pH (unit)	ALCALINITY (μeqL^{-1})	CONDUCTIVITY (μScm^{-1})	Ca (μeqL^{-1})	SULPHATE (μeqL^{-1})	COLOR (U. HAZEN)	FILTRATED Al (μgL^{-1})	MEAN DEPTH (m)	AREA (ha)
RÉSERVE DES LAURENTIDES (APRIL 1981)									
OVIDE	4.7	22.0	19.0	64.9	95.8	17	220	4.7	12.9
HAMON	5.0	6.0	16.0	59.9	87.4	24	240	—	20.7
MONETTE	5.3	8.0	16.0	69.9	87.4	28	190	3.1	14.4
MIGNON	6.6	154.0	25.0	124.8	41.6	7	50	—	23.3
OPALIN	6.6	280.0	37.0	199.6	62.5	18	70	1.5	15.5
NOIR	6.6	218.0	33.0	179.6	54.1	14	50	4.5	23.3
PORTNEUF (APRIL 1982)									
POLIQUIN	4.6	-35.0	26.5	84.8	124.9	42	390	11.9	26.0
MORASSE	5.0	-12.0	23.0	94.8	104.1	21	310	7.0	13.0
WILLIAM	5.4	11.9	21.5	89.8	72.8	11	180	4.9	34.0
MOULINEAU	5.4	9.9	22.0	99.8	93.7	4	140	1.7	14.0
ÉDITHE	5.6	31.0	17.0	84.8	83.3	11	140	5.6	21.0
DES ÎLES	5.9	37.2	20.0	104.8	124.9	10	100	6.6	28.0
MAIN DE FER	6.2	107.0	26.0	139.7	41.6	19	50	4.0	16.0
CHARLEVOIX (APRIL 1983)									
ROBBÉ	4.6	-26.0	21.5	44.9	62.5	35	370	7.4	8.3
FREDDO	4.7	-21.0	17.2	44.9	52.1	22	300	12.8	14.1
CÔTE-À-CÔTE	4.9	-10.0	13.0	44.9	52.1	5	20	1.5	15.0
PREMIER LAC PAUL	5.1	4.0	15.5	54.9	62.5	18	300	6.9	3.7
DEUXIÈME LAC PAUL	5.4	26.0	14.5	64.9	62.5	19	200	1.3	8.8
DU BOUCHON	5.6	26.0	13.0	54.9	31.2	20	190	5.4	16.0
GUGY	6.3	102.0	21.5	114.8	72.9	19	40	1.2	14.0
TÉMISCAMINGUE (JUNE 1984)									
14	4.8	-16.0	18.0	44.9	93.7	19	80	1.9	29.4
3	4.9	-13.0	25.0	69.9	156.1	5	120	6.2	28.9
4	4.9	-11.0	23.0	69.9	124.9	20	130	3.9	34.8
16	5.0	- 9.0	26.8	84.8	177.0	10	120	4.5	43.6
7	5.1	- 7,0	19.7	69.9	114.5	15	60	5.0	32.4
1	5.2	- 6.0	25.5	99.8	166.6	22	140	6.1	48.3
2	5.2	- 5.0	21.3	74.8	135.3	5	70	4.1	21.1
17	5.2	2.0	27.0	79.8	177.0	5	120	4.3	29.9
5	5.9	10.0	20.5	94.8	135.3	3	10	7.1	33.7
6	6.1	30.0	24.7	99.8	156.1	15	50	5.7	16.9
9	6.1	32.0	30.3	159.7	177.0	23	140	2.3	19.6
11	6.1	24.0	22.7	99.8	145.7	8	50	4.0	37.5
15	6.1	28.0	34.0	144.7	218.6	10	80	2.7	46.1
10	6.2	26.0	26.8	119.8	177.0	4	10	6.4	34.9
8	6.4	38.0	24.9	119.8	145.7	7	20	7.9	51.3
12	6.9	38.0	33.8	179.6	177.0	5	10	8.6	34.0
13	7.0	30.0	33.9	164.7	177.0	4	10	7.0	42.2

Condition - In the three regions (Figure 4), brook charr populations inhabiting acid lakes were in better condition that those in less acid lakes (covariance analysis, $p < 0.001$). These differences were more pronounced in cases where pH was lower than 5.0. The closer relationship between fish condition and acidity appears to be more representative of the effects of acid stress than that of growth and acidity (Figure 3). These results confirm those obtained by Rosseland et al., (1980), Harvey (1982) and Light (1983), who showed that a change in condition due to

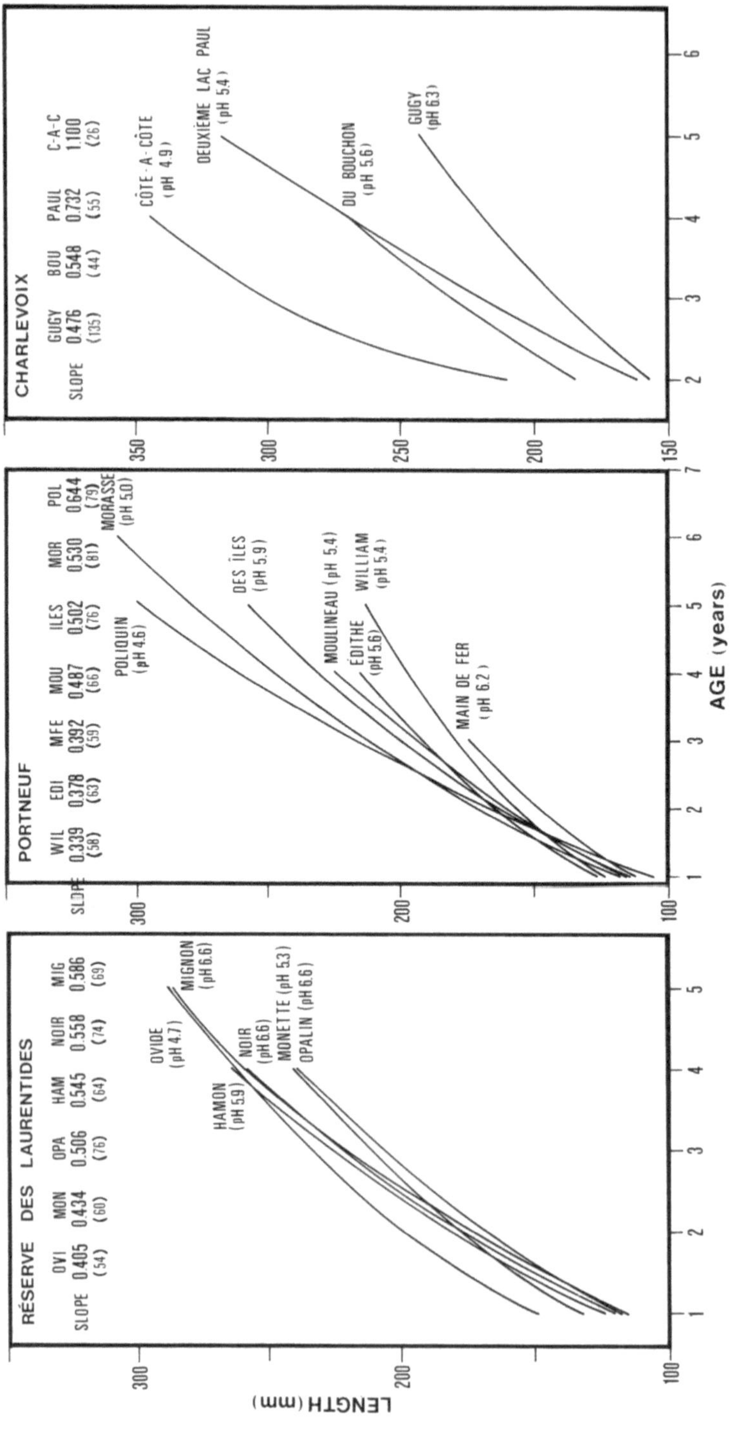

FIGURE 3. Growth of brook charr captured in the réserve des Laurentides and the Portneuf and Charlevoix regions. Curves were modeled according to log (length)= a + b log (age) equation. ( ) indicates the sample size.

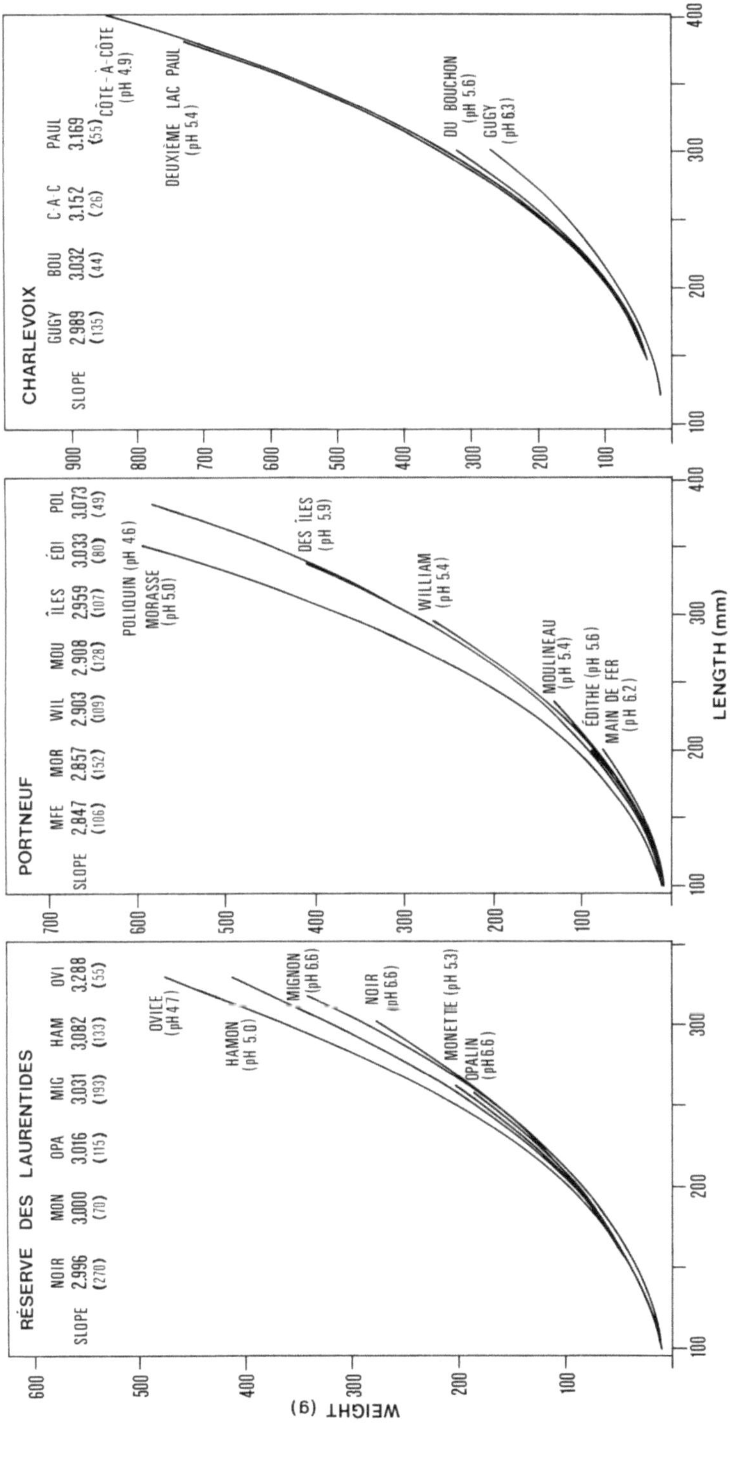

FIGURE 4. Weight - length relationship among brook charr captured in the réserve des Laurentides and the Portneuf and Charlevoix regions. Curves were modeled according to log (weight)= a + b log (length). ( ) indicates sample size.

TABLE II. Mean Al concentrations ($\mu g g^{-1}$) in brook charr gills for each length-class. ( ) sample size; [ ] standard deviation; common letter indicates non-significant differences (p>0.05) between lakes (Duncan test).

LAKE	pH (unit)	LENGTH - CLASS (mm) 90-140	141-190	191-240	>240
POLIQUIN	4.6	151^{a} (7) [166.7]	74^{c} (7) [21.6]	78^{e} (9) [77.1]	67 (5) [52.9]
ÉDITHE	5.6	86^{a} (7) [57.1]	48^{c} (10) [50.5]	16^{e} (13) [7.9]	
MAIN DE FER	6.2	12^{b} (11) [16.2]	9^{d} (17) [10.7]	5^{f} (2) [0.9]	

acid stress may occur independent of changes in growth. Light (1983) attributes the changes in fish condition to an increase in food supply following a decrease in density. The hypothesis that increased food supply, is identical to the hypothesis presented to explain increased growth. If both hypotheses are correct, then growth and condition should be positively correlated. This however is not the case. The difference observed between growth and condition emphasizes the ambiguity of the phenomenon and one of these hypotheses must be incorrect. The increased availability of food to surviving fish has never been demonstrated thus all hypotheses relating growth and condition to food availability are suspect.

Aluminum in the gills - Brook charr from lakes Poliquin, Édithe and Main de Fer (Portneuf), exhibit a decrease in total Al concentration in the cens) and the northern pike (*Esox lucius*) are found in four other lakes, and only one lake supports four species in association. The situation was different for non acid lakes where eight lakes out of a total of nine exhibited community structures including three to nine species in association. The remaining lake supports only two species. The low number of species in acid lakes was significantly different from non acid lakes (Mann Whitney U-test, $p < 0.001$). In addition, the number of species is positively correlated to pH ($r = 0.76$, $p < 0.05$; Frenette, 1986b). By taking into account morphoedaphic factors such as mean depth and area, explanatory variables of the number of species (Barbour and Brown, 1974; Harvey, 1978, 1981; Magnuson *et al.*, 1984), partial regression analyses demonstrate that acidity was the determining factor in explaining the difference (Frenette, 1986b).

Cyprinidae, with the exception of fallfish (*Semotilus corporalis*), and Johnny darters (*Etheostoma nigrum*) were very abundant in lakes with a pH of 5.9 to 7.0 but were absent from lakes with a pH $\leqslant$ 5.2. These fishes are generally known to be very sensitive to acidity (Harvey, 1980; Rahel and Magnuson, 1983; Magnuson *et al.*, 1984). Predation by

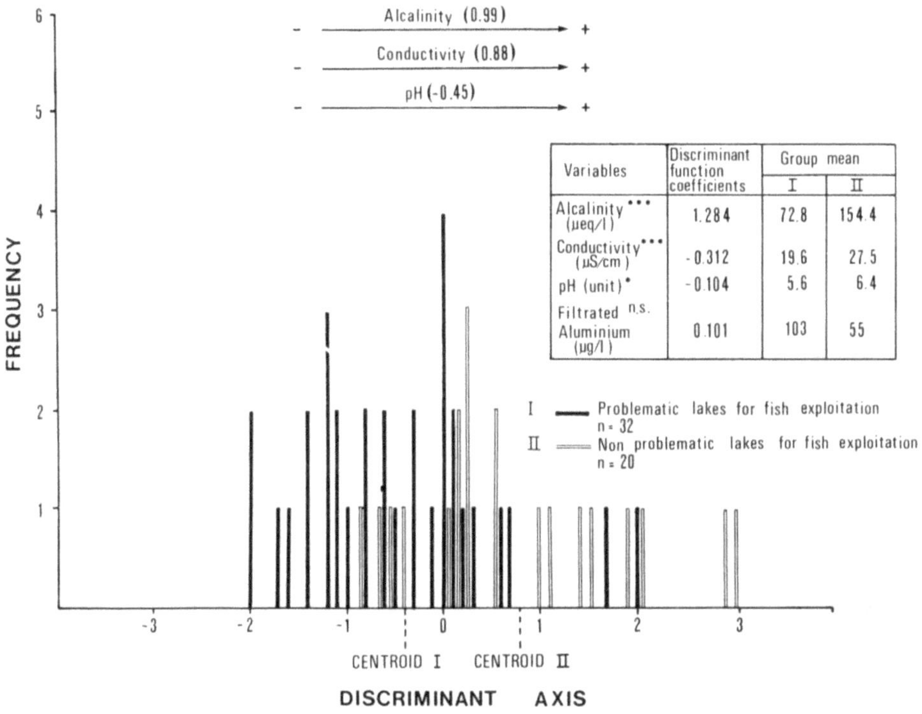

FIGURE 5. Frequency distribution of problematic and non problematic lakes for gamefish management after discriminant analysis. Values between brackets indicate the correlation between the discriminant variable and the discriminant axis.***: p< 0.001,**:p< 0.05, N.S.: p>0.05, Student's t test.

the yellow perch and northern pike such as mentioned by Somers and Harvey (1984) cannot explain their disappearance since these species were also present in non acid lakes.

Other species such as the white sucker (Catostomus commersoni) and the northern pike are poorly represented in acid lakes. The white sucker is known to be sensitive to acidity. At pH lower than 5.0, this species often disappears (Magnuson et al., 1984). However, northern pike have been found in lakes with a pH as low as 4.0, although their reproduction appears to be adversely affected starting at pH 4.5 (Magnuson et al., 1984). In fact, only the yellow perch can tolerate a low pH. This species is highly resistant to acidity and is found in waters with a pH as low as 4.2 (Harvey, 1982).

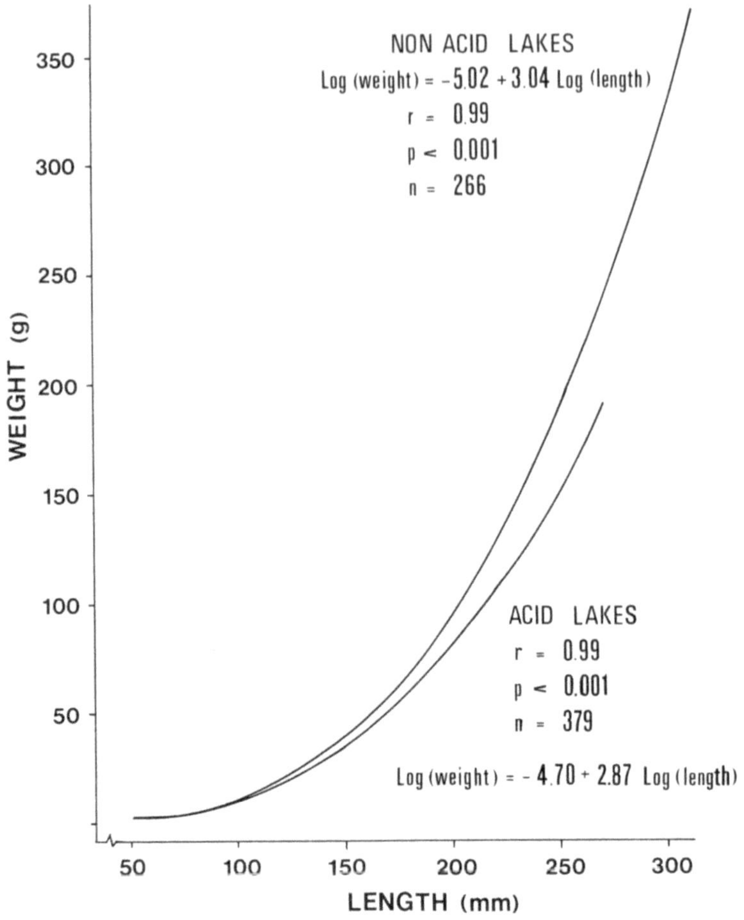

FIGURE 6. Weight - length relationship among yellow perch from acidic and non acidic lakes in the Témiscamingue region.

The presence of a species in an acid environment does not mean that it experiences optimum conditions for survival. In fact, the weight-length relationship (Figure 6) demonstrates a poor condition among yellow perch living in acid lakes as compared with those in non acid lakes (covariance analysis, $p < 0.001$).

Two to three-year old yellow perch are considered as piscivorous (Scott and Crossman, 1973). The absence of prey in the majority of acid

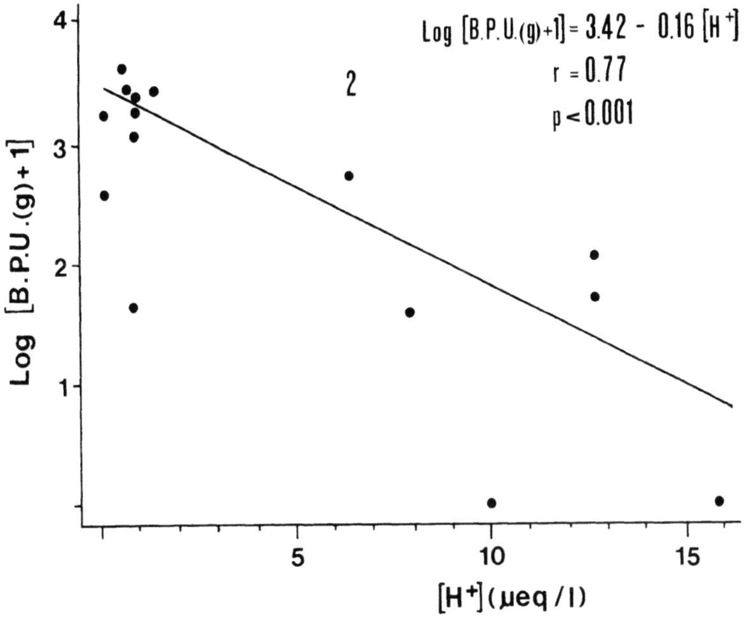

FIGURE 7. Relation between the total fish biomass per unit effort and the H^+ concentration in Témiscamingue lakes.

lakes can therefore act as a limiting factor and, in combination with an increase in metabolic needs, decrease the species condition.

Biomass - The relationship between total biomass per unit effort (biomass sampled by three types of gear during 12 hr) and the concentration in H^+ ions demonstrates that the biomass decreases with an increase in acidity (Figure 7). However, the productive capacity of an oligotrophic lake is dependent on its morphometric characteristics and its nutrient status (Ryder, 1982). In this study, the mean depth, area, conductivity (equivalent to the total amount of dissolved substances, Ryder et al., (1974)) and the mean depth-conductivity ratio are not correlated to the biomass (Frenette, 1986b). Hence, acidity appears to be responsible for the decrease in biomass observed in lakes of the Témiscamingue region.

The catches per unit effort for a given species were not significantly correlated to acidity (Frenette, 1986b). The absence of relationship may be attributed to biotic interaction between several species such as competition and predation which probably vary from lake to lake. These results confirm those by Richard (1985a) who observed a relationship between density and acidity levels in allopatric populations of the Portneuf region which was absent in sympatric populations of brook charr.

Table III. Numbers of fish and catch per unit effort ( ) of different species collected with experimental gill nets, minnow traps and trap nets in 17 Témiscamingue lakes during 1984.

Species	ACID LAKES								NON ACID LAKES								
Lake number	14	3	4	16	7	2	17	1	5	9	11	6	15	10	8	12	13
pH	4.8	4.9	4.9	5.0	5.1	5.2	5.2	5.2	5.9	6.1	6.1	6.1	6.1	6.2	6.4	6.9	7.0
Yellow perch (*Perca flavescens*)	0	37 (5.0)	38 (4.4)	0	237 (28.8)	0	208 (24.3)	2 (0.6)	45 (5.7)	16 (2.0)	15 (1.8)	10 (1.1)	219 (25.6)	10 (1.2)	1 (0.1)	4 (0.6)	0
Northern pike (*Esox lucius*)	0	0	0	0	0	17 (2.3)	0	0	8 (1.0)	17 (12.2)	9 (1.1)	0	0	3 (0.4)	28 (3.6)	0	0
White sucker (*Catostomus commersoni*)	0	0	0	0	0	0	0	71 (7.7)	0	0	0	12 (1.3)	36 (4.3)	27 (3.1)	0	18 (2.4)	28 (3.4)
Brook stickleback (*Culaea inconstans*)	0	0	0	0	0	0	8 (0.9)	0	0	0	0	0	11 (1.2)	0	0	0	0
Lake whitefish (*Coregonus clupeaformis*)	0	0	0	0	0	0	0	4 (0.4)	0	0	0	0	0	0	0	0	0
Fallfish (*Semotilus corporalis*)	0	0	0	0	0	0	0	13 (1.4)	0	0	0	0	0	0	0	0	0
Blacknose shiner (*Notropis heterolepis*)	0	0	0	0	0	0	0	0	0	25 (3.3)	5 (0.6)	46 (5.1)	1 (0.1)	1 (0.1)	0	17 (2.2)	0
Northern redbelly dace (*Chrosomus eos*)	0	0	0	0	0	0	0	0	0	0	0	114 (13.1)	72 (9.7)	0	0	1005 (130.9)	788 (95.3)
Golden shiner (*Notemigonus crysoleucas*)	0	0	0	0	0	0	0	0	2 (0.3)	0	0	25 (2.8)	120 (14.3)	0	0	1 (0.1)	0
Pearl dace (*Semotilus margarita*)	0	0	0	0	0	0	0	0	0	0	0	202 (23.0)	31 (3.7)	0	0	31 (3.9)	29 (3.5)
Common shiner (*Notropis cornutus*)	0	0	0	0	0	0	0	0	0	0	0	0	0	0	0	5 (0.7)	86 (10.4)
Johny darter (*Etheostoma nigrum*)	0	0	0	0	0	0	0	0	7 (0.9)	0	0	7 (0.8)	0	0	0	11 (1.4)	19 (2.3)
Pumpkinseed (*Lepomis gibbosus*)	0	0	0	0	0	0	0	0	0	0	0	0	0	0	0	1 (0.1)	0
Burbot (*Lota lota*)	0	0	0	0	0	0	0	0	0	0	0	0	0	1 (0.1)	0	0	0
Mottled sculpin (*Cottus bairdi*)	0	0	0	0	0	0	0	0	1 (0.1)	0	0	0	0	0	0	0	0
Slimy sculpin (*Cottus cognatus*)	0	0	0	0	0	0	0	0	2 (0.2)	0	0	0	0	0	0	0	0
Total number of fish	0	37	38	0	237	17	216	90	65	58	29	416	490	42	29	1093	950
Total number of species	0	1	1	0	1	1	2	4	6	3	3	7	7	5	2	9	5

ACKNOWLEDGMENTS

We thank Dr. Julian J. Dodson who critically reviewed this manuscript.

REFERENCES

BAKER, J., and SCHOFIELD, C.: 1980, Alumininium toxicity to fish as related to acid precipitation and Adirondack surface water quality, p. 292-293. *In* Drablos and Tollan, 1980.
BAKER, J.P.: 1982, Effects on fish of metals associated with acidification, p. 167-176. *In* R. Johnson (ed.). Acid rain/fisheries. Northeastern Division, American Fisheries Society, Bethesda, MD. 357 p.
BARBOUR, C.D., and BROWN, H.H.: 1974, *Am. Nat.* **108**, 473.
BRETT, J.R.: 1979, Environmental factors and growth, p. 599-677. *In* W.S. Hoar, D.J. Randall, and J.A. Brett (ed.) Fish physiology. Vol. 8. Bioenergetics and growth. Academic Press, New York, NY. 786 p.
BROWN, D.J.A.: 1982, *Water, Air, and Soil Pollution* **18**, 343.
CHEVALIER, G., GAUTHIER, L., et MOREAU, G.: 1985, *Can. J. Zool.* **63**, 2062.
CONROY, M.I., HAWLEY, R., and KELLER, N.: 1978, Extensive monitoring of lakes in the Greater Sudbury area 1974-1976, Sudbury Experimental Study Report, Ontario Ministry of the Environment, 187 p.
CRONAN, C.S., and SCHOFIELD, C.L.: 1979, *Science* (Wash, DC) **204**, 304.
DAYE, P., and GARSIDE, E.: 1975, *Can. J. Zool.* **53**, 639.
DRABLOS, D., and TOLLAN, A., editors, 1980, Proceedings of the international conference on the ecological impact of acid precipitation, *In* Acid precipitation - Effects on forest and fish project, Aas, Norway.
FRENETTE, J.-J., and DODSON, J.J.: 1984, *Can. J. Fish. Aquat. Sci.* **41**, 865.
FRENETTE, J.-J.: 1986a, Étude du niveau d'acidité de 39 lacs du Témiscamingue, Direction des relevés aquatiques, ministère de l'Environnement du Québec, Publication no. PA-24, Envirodoq no. 860151, 70 p.
FRENETTE, J.-J.: 1986b. Effet de l'acidité sur les communautés piscicoles de 17 lacs du Témiscamingue, Direction des relevés aquatiques, ministère de l'Environnement du Québec, Publication no. PA-23, Envirodoq No. 860150, 102 p.
GRANDE, M., MUNIZ, I., and ANDERSEN, S.: 1978, *Verh. Int. Ver. Limnol.* **20**, 2076.
GRIMARD, Y.: 1984, Réseau d'échantillonnage des précipitations du Québec: Sommaire des données de la qualité des eaux de précipitations 1981-1982-1983, Service de la Qualité des eaux, ministère de l'Environnement du Québec, rapport PA-11, Envirodoq 840595, 163 p.
HARVEY, H.H.: 1978, *Verh. Int. Ver. Limnol.* **20**, 2031.
HARVEY, H.H.: 1980, Wildespread and diverse changes in the biota of North American lakes and rivers coincident with acidification, p. 93-98. *In* Drablos and Tollan, 1980.
HARVEY, H.H.: 1981, *Verh. Int. Ver. Limnol.* **21**, 1222.
HARVEY, H.H.: 1982, Population responses of fishes in acidified waters, p. 227-242. *In* R. Johnson (ed.) Acid rain/fisheries. Northeastern Division, American Fisheries Society, Bethesda, MD. 357 p.

JENSEN, A.L.: 1971, *Trans. Amer. Fish. Soc.* 100, 456.
JOHANSSON, N., RUNN, D., and MILBRINK, G.: 1977, *Zoom* 5, 127.
JOHNSON, L.: 1976, *J. Fish. Res. Board Can.* 33, 2459.
LEIVESTAD, H.: 1982, Physiological effects of acid on fish, p. 157-164. *In* R. Johnson (ed.) Acid rain/fisheries. Northeastern Division, American Fisheries Society, Bethesda, MD. 357 p.
LIGHT, R.W.: 1983, Growth and reproductive ecology of the brook trout, *Salvelinus fontinalis*, in streams of differing vulnerability to acid atmospheric deposition. Ph.D. Thesis, the Pennsylvania State University, Ann Arbor, Ml 48106, 198 p.
MAGNUSON, J.J., BAKER, J.P., and RAHEL, F.J.: 1984, *Phil. Trans. R. Soc. Lond.* B305, 501.
MENENDEZ, R.: 1976, *J. Fish. Res. Board Can.* 33, 118.
MOREAU, G., BARBEAU, C., FRENETTE, J.-J., LÉVESQUE, F., SAINT-ONGE, J., et SIMONEAU, M.: 1984, Influence des précipitations acides sur les populations d'omble de fontaine (*Salvelinus fontinalis*) de la réserve des Laurentides, rapport publié pour la Direction générale des inventaires et de la recherche, ministère de l'Environnement du Québec, Publication no. PA-7, Envirodoq No. 8404671, 124 p.
MUIR, B.S.: 1969, *J. Fish. Res. Board Can.* 26, 165.
POWER, G.: 1978, *J. Fish. Res. Board Can.* 35, 53.
RAHEL, F.J., and MAGNUSON, J.J.: 1983, *Can. J. Fish. Aquat. Sci.* 40, 3.
RICHARD, Y.: 1982, Relations entre le niveau d'acidité de 158 lacs du parc des Laurentides et l'évolution de leurs statistiques de pêche au cours des neuf dernières années, Service de la qualité des eaux, ministère de l'Environnement du Québec, publication no. PA-5, Envirodoq No. 03921, 115 p.
RICHARD, Y.: 1985a, Influence du niveau d'acidité des milieux lacustres sur la faune ichtyologique de la région de Portneuf, Québec, Direction des relevés aquatiques, ministère de l'Environnement du Québec, Publication no. PA-15, Envirodoq No. 850237, 197 p., 10 annexes.
RICHARD, Y.: 1985b, Niveau d'acidité de 54 lacs de la région de Charlevoix, ZEC des Martres, Québec, Direction des relevés aquatiques, ministère de l'Environnement du Québec, Publication no. PA-22, Envirodoq no. 850741 p., 1 annexe.
RICHARD, Y.: 1986, Effets de l'acidité sur les populations ichtyologiques de la ZEC des Martres, Charlevoix, Québec, Direction des relevés aquatiques, ministère de l'Environnement du Québec, Publication no. PA-22, Envirodoq no. 850741, 47 p., 1 annexe.
ROBINSON, G.D., DUNSON, W.A., WRIGHT, J.W., and MAMOLITO, G.E.: 1976, *J. Fish Biol.* 8, 5.
ROSSELAND, B.O., SEVALDRUD, I., SVALASTOG, D., and MUNIZ, I.P.: 1980, Studies on freshwater fish populations - effects of acidification on reproduction, population structure, growth and food selection, p. 336-337. *In* Drablos and Tollan, 1980.
RYDER, R.A., KERR, S.R., LOFTUS, K.H., and REGIER, H.A.: 1974, *J. Fish. Res. Board Can.* 31, 663.
RYDER, R.A.: 1982, *Trans. Amer. Fish. Soc.* 111, 154.
SCHOFIELD, C., and TROJNAR, J.R.: 1980, Aluminium toxicity to brook trout (*Salvelinus fontinalis)* in acidified waters, p. 341-365. *In* Polluted rain. T.Y. Toribara, M.W. Miller, and P.E. Morrow, Plenum Press, New York, N.Y.

SCHOFIELD, C.L.: 1982, Historical fisheries changes in the United States related to decrease in surface water pH, p. 57-67. *In* R. Johnson (ed.) Acid rain/fisheries, Northeastern Division, American Fisheries Society, Bethesda, MD. 357 p.
SOMERS, K.M., and HARVEY, H.H.: 1984, *Can. J. Fish. Aquat. Sci.* **41**, 20.
SCOTT, W.B., and CROSSMAN, E.J.: 1973, Freshwater fishes of Canada. *Fish. Res. Board Can. Bull.* **184**, 966 p.
SWARTS, F.A., DUNSON, W.A., and WRIGHT, J.E.: 1978, *Trans. Am. Fish. Soc.* **107**, 651.
WEBSTER, D.A., and FLICK, W.A.: 1975, Species management. *In* Wild Trout Management. Proc. Yellowstone Nat. Park Symposium, 1974. Trout unlimited: 40-47.
WOOD, C.M., and McDONALD, D.G.: 1982, Physiological mechanisms of acid toxicity to fish, p. 197-226. *In* R. Johnson (ed.) Acid rain/fisheries. Northeastern Division, American Fisheries Society, Bethesda, MD. 357 p.

FISH SPECIES DISTRIBUTION IN RELATION TO WATER CHEMISTRY IN SELECTED MAINE LAKES

Stanislas J. Pauwels
Zoology Department
University of Maine
Orono, ME 04469 USA

Terry A. Haines
U.S. Fish and Wildlife Service
Columbia National Fisheries Research Laboratory
Field Research Station
Zoology Department, University of Maine
Orono, ME 04469 USA

ABSTRACT. We examined the possible effects of acidity on the number of fish species in 22 lakes in Maine, ranging in pH from 4.4 to 7.0 (mean values). We caught no fish in three lakes with pH <5.0, but collected 1 to 9 species in the remaining 19 lakes (pH 5.4 to 7.0). Brook trout (*Salvelinus fontinalis*), golden shiners (*Notemigonus crysoleucas*), and white suckers (*Catostomus commersoni*) were ubiquitous, but no common shiners (*Notropis cornutus*) or creek chubs (*Semotilus atromaculatus*) were collected from lakes with pH below 6.0 and 5.9, respectively. The fishless lakes differed from the others primarily in water chemistry variables related to acidity, i.e., pH, Al, and concentration of divalent cations. Among lakes that contained fish, the factors related to the number of species collected were lake surface area and maximum depth, which may be related to habitat quantity and diversity. Cluster analysis identified two distinct fish species groups - depauperate and cyprinid-sucker - but multiple comparison analysis failed to relate any measured chemical or physical variable to these two groups, probably because water chemistry was suitable for reproduction by these species.

1. INTRODUCTION

The mean annual volume-weighted pH of the precipitation currently falling on the northeastern United States is between 4.2 and 4.4 (Interagency Task Force on Acid Precipitation Program, 1984). Large areas of New England, including Maine, are underlain by bedrock that is low in acid neutralizing capacity (Hendrey *et al.*, 1980). A

Water, Air and Soil Pollution **30** (1986) 477–488.

previous study (Haines and Akielaszek, 1983) showed that in about 8% of the lakes surveyed in northern New England pH was <5.0, and in about 40% the acid neutralizing capacity (ANC) was <100 μeqL^{-1}. However, no data were collected during that study to ascertain the relation between fish populations and lake acidity. In the present study, we attempted to measure the possible effects of acidity on fish species assemblages in Maine lakes that were not subject to local perturbation.

2. MATERIALS AND METHODS

We surveyed and ranked 25 lakes in Maine according to the following six criteria: (1) no history of fish stocking or chemical reclamation, (2) presence of a permanent outlet but no permanent inlet, (3) lack of significant human disturbance in the watershed, (4) lack of bog characteristics (low color), (5) surface area less than 50 ha, and (6) reasonable accessibility. The lakes were sufficiently small and remote that angling was judged to be minimal. Three lakes were

Table I. Morphological characteristics of the 22 study lakes.

Name	Latitude	Longitude	Surface Area (ha)	Drainage Area (ha)	Elev. (m)	Maximum Depth (m)
Mud	44°37'50"	68°05'38"	2	68	110	14
Cloud	45°24'59"	69°21'26"	8	34	734	11
Crater	45°31'13"	69°07'01"	6	18	771	8
Lost	45°25'41"	69°16'58"	6	60	325	3
Garcock	44°16'10"	69°21'03"	1	26	120	6
Horseshoe	45°17'30"	69°22'46"	26	194	162	4
Browns	43°54'27"	70°43'21"	39	311	159	2
Unnamed	45°41'00"	69°10'00"	7	27	382	2
Eighteen	45°19'00"	69°27'42"	6	34	229	9
Little Hurd	45°49'22"	69°01'30"	24	1984	222	27
Davis	44°33'33"	69°59'58"	7	70	253	11
Fogg	45°35'27"	70°04'01"	9	197	532	6
Johnston	45°35'50"	69°03'42"	24	264	416	18
Knox	45°01'19"	67°53'41"	21	337	104	6
Little Fisher	45°46'04"	69°16'40"	14	179	363	2
Twin	45°47'47"	68°49'33"	24	158	241	26
Bear	45°38'51"	69°11'40"	5	163	418	4
Porter	45°09'32"	68°13'02"	23	161	140	4
Big Fisher	45°46'41"	69°17'18"	23	153	351	3
Upper Pug	45°18'59"	68°05'34"	20	236	153	5
Trout	45°22'32"	67°53'12"	2	41	241	3
Green	45°00'51"	68°03'20"	26	114	139	4

Table II. Limnological characteristics of the 22 study lakes. Values shown are the arithmetic means of the water chemistry variables of all the surface and bottom samples combined.

Name	pH (units)	Cond ($\mu S cm^{-1}$)	Al ($\mu g L^{-1}$)	Alk	Ca	Na	Mg	K ($\mu eq L^{-1}$)	NO_3	SO_4	Cl	Sum	Color (units)
Mud	4.4	26	393	-37	33	96	29	8	8	86	109	211	20
Cloud	4.8	21	271	-14	34	26	25	9	9	97	27	112	10
Crater	4.8	16	296	-10	13	21	17	6	4	76	18	81	0
Lost	5.4	18	163	6	67	34	24	10	6	78	22	135	30
Garcock	5.5	20	63	18	94	36	30	7	36	76	26	156	40
Horseshoe	5.6	17	51	7	70	40	28	6	23	83	26	140	30
Browns	5.8	21	123	22	73	77	29	5	2	63	49	171	40
Unnamed	5.9	17	66	23	73	34	28	11	5	45	26	135	60
Eighteen	6.0	14	42	23	47	39	27	7	12	60	24	118	10
Little Hurd	6.0	20	83	31	94	43	23	12	12	97	22	165	10
Davis	6.1	26	129	64	100	61	35	10	7	80	40	199	60
Fogg	6.1	13	33	19	46	28	44	9	4	64	18	120	10
Johnston	6.1	15	39	28	52	38	39	8	2	71	22	134	10
Knox	6.2	22	83	43	77	84	29	13	1	72	47	194	40
Little Fisher	6.3	25	128	46	123	41	60	8	3	117	14	231	40
Twin	6.3	22	24	57	102	44	28	11	3	93	23	185	10
Bear	6.4	21	60	82	102	65	26	9	3	28	25	183	60
Porter	6.4	20	40	46	75	76	26	13	1	69	32	175	20
Big Fisher	6.5	22	42	81	119	34	46	8	3	66	24	196	20
Upper Pug	6.7	25	42	94	126	70	41	13	1	73	28	233	20
Trout	6.8	28	45	90	107	74	40	8	5	83	21	222	10
Green	7.0	25	28	112	88	92	31	21	7	54	40	226	10

[a]Sum=Cations+Anions

eliminated from further consideration and 22 were selected for intensive study of water chemistry and fish populations (Tables I and II). Each lake was sampled three times for water chemistry. Sampling occured at least once before and at least once after fall overturn. The water sample collection and analysis followed protocols and quality assurance procedures specified by the National Acid Precipitation Assessment Program (1985). In calculating organic anions from water color, we followed the procedure described by Haines and Akielaszek (1983).

Fish were sampled once in each lake during summer months of 1982 or 1983. In addition, six lakes were resurveyed in 1984. For the 1982 and 1983 survey seasons, the fish sampling protocol consisted of an overnight set (average fishing time, 19 hr) of one nylon experimental gill net, 60 or 75 m long, set perpendicular to the shoreline, and four metal minnow traps baited with commercial dry dog food. The gill nets were 2 m deep, and consisted of four panels of different mesh sizes (25, 35, 50, and 63 mm stretched measure). The 75 m net contained an additional panel of 75 mm mesh.

All fish captured were identified to species and enumerated. Inasmuch as pumpkinseeds (*Lepomis gibbosus*) and redbreast sunfish (*L. auritus*) hybridize extensively (Scott and Crossman 1973) and the specimen of both species that we caught were invariably small (5 to 10cm long) and difficult to identify to species, we combined the two species into one category, namely sunfish, in the statistical analyses.

Six lakes were resurveyed in the summer of 1984 to assess whether or not the sampling effort of the previous 2 yr had been adequate to accurately describe the fish populations. Three acidic lakes (Garcock, Unnamed and Eighteen Pond) and three circumneutral lakes (Knox Pond, Trout Pond, and Green Lake) were selected based on the survey data from 1982 and 1983. These lakes were judged to be lower in number of species than other lakes in the data set with similar pH. The fishing protocol for 1984 increased the sampling effort by a factor of more than two over the initial surveys.

3. RESULTS

The repeat sampling of six lakes with increased effort produced no additional species in the three acidic lakes, and a total of nine additional species populations in the three circumneutral lakes. Moreover, three species populations sampled during the 1982-1983 survey in the three circumneutral lakes were not recaptured during the 1984 survey.

No fish were collected from Crater Pond (mean pH 4.8), Cloud Pond (mean pH 4.8), or Mud Pond (mean pH 4.4). Golden shiners (*Notemigonus crysolucas*) were caught in 11 lakes, and brook trout (*Salvelinus fontinalis*) and white sucker (*Catostomi commersoni*) were caught in 10 (Figure 1). No single species was taken in all 19 lakes, but all lakes of pH 5.4 and above contained one or more fish species. White suckers and sunfish were the most widely distributed species with respect to pH, occurring over the entire range of pH in lakes containing fish; they were closely followed by brook trout (Figure 1). Of the fish species caught in five or more lakes, creek chubs (Semotilus atromaculatus) occurred only in waters of pH 5.9 and above, and common shiners (Notropis cornutus) only in those of pH 6.0 and above (Figure 1).

A stepwise forward multiple regression analysis of the nine morphological and chemical variables on number of fish species in the 22 study lakes resulted in an r^2 of 0.86 for a six factor model (Table III). However, three variables (divalent cation concentration, surface area, and maximum depth) accounted for 96% of this total. A stepwise forward multiple regression of the same variables on the number of fish species in the 19 lakes that contained fish resulted in a r^2 of 0.68 for a four factor model (Table III). The same three variables accounted for most of the regression.

A clustering analysis in which we used Fager's similarity coefficient and the single linkage clustering strategy (Boesch, 1977) produced two distinct lake clusters (Table IV). Lakes in the first cluster yielded few species (mean 2.6; range = 1 to 4). The most common species in this group were the golden shiner and brook trout. With the exception of the ubiquitous golden shiner, and three lakes each containing one species of minnow or dace, no cyprinids, sunfish, or white suckers were present in these lakes. The one exception was Bear Pond, in which we collected both common shiners and creek chubs and which therefore had a different biological community. As a result, we arbitrarily assigned Bear Pond to the next cluster. On the basis of the scarcity of fish species in the first cluster, we termed this group of lakes depauperate. Most lakes in this group were relatively deep, lacked extensive inshore areas, and had little or no aquatic vegetation.

All lakes in the second cluster contained sunfish, white suckers, or cyprinids, or a combination of these species. The number of species caught in the lakes of this group, termed cyprinid-sucker lakes, ranged from 3 to 9 (mean 5.5). The cyprinid-sucker lakes tended to be shallow or have extensive shallow areas with emergent and submergent vegetation. The third group of lakes, not shown in Table IV, consisted the three fishless lakes. All three were relatively deep and had pH <5.0.

Nonparametric Kruskall-Wallis analysis of variance tests (Zar, 1984) indicated that pH, divalent cation concentration, Al concentration, and number of fish species per lake differed significantly among the three groups ($p < 0.05$). Nonparametric multiple comparison tests (Zar, 1984) showed that the fishless lakes differed from the other lakes with respect to water chemistry. The depauperate and cyprinid-sucker lake groups differed between themselves in the number of fish species but not in any of the measured lake morphological or water chemistry variables (Table V).

```
                                         pH
 5.4  5.5  5.6  5.7  5.8  5.9  6.0  6.1  6.2  6.3  6.4  6.5  6.6  6.7  6.8  6.9  7.0
                                                                                     a
 <---------------------- White sucker ------------------------------------------>(10)

 <------------------------ Sunfish ---------------------------------------------> (7)

 <---------------------- Finescale dace ---------------->                          (3)

      <----------------- Golden shiners ---------------->                         (11)

           <----------------- Brook trout ---------------------------->           (10)

           <----------------- Banded killifish^b ------------------------>         (2)

           <------------------ Brown bullhead ---------->                          (2)

           <-------------- Fathead minnow^b --->                                   (2)

           <-------------- Blacknose dace^b --->                                   (2)

          . Rosyface shiner^b                                                      (1)

                     <----------- Yellow perch^b ------------------->               (2)

                     <----------- Chain pickerel ------------------>               (5)

                          <-------- N. redbelly dace --->                          (3)

                          <-------- Creek chub --------->                          (6)

                          <-------- Pearl dace --------->                          (3)

                               <--------------- Common shiner -------------------> (5)

                                               <-------- Fallfish^b ------>         (2)

                                              . Arctic char^b                      (1)

                                                              . White perch^b      (1)

                                                       Rainbow smelt^b.            (1)
```

[a] Frequency of occurrence
[b] Banded killifish *Fundulus diaphanus*, fathead minnow *Pimephales promelas*, blacknose dace *Rhinichthys atratulus*, rosyface shiner *Notropis rubellus*, yellow perch *Perca flavescens*, fallfish *Semotilus corporalis*, Arctic char (landlocked) *Salvelinus alpinus*, white perch *Morone americana*, rainbow smelt *Osmerus mordax*.

Figure 1. Distribution of fish species by lake pH.

Table III. Stepwise forward multiple regression analyses with number of fish species as the dependent variable and lake physical and chemical factors as the independent variables.

All lakes			
Step	Variables	r^2	p
1	Divalent cation concentration	0.58	0.0001
2	Surface area	0.74	0.0001
3	Maximum depth	0.80	0.0001
4	Aluminum concentration	0.83	0.0001
5	Drainage area	0.84	0.0001
6	Total ionic strength	0.86	0.0001

Lakes that yielded fish			
Step	Variables	r^2	p
1	Surface area	0.27	0.021
2	Divalent cation concentration	0.47	0.006
3	Maximum depth	0.60	0.003
4	Total ionic strength	0.68	0.002

4. DISCUSSION

The results of the 1984 resurvey do not suggest that certain species were systematically missed during the previous survey, although some unknown factors seem to influence fish capture from one year to another. Despite a doubling in sampling effort, three species populations caught in the three circumneutral lakes during the 1982-1983 survey seasons were not recaptured during the 1984 sampling season. Of the nine additional species populations collected during the second survey, only one occurred in more than one lake: golden shiner in Green Lake and Trout Pond. The eight remaining populations were only caught once. Moreover, only two of the nine species (golden

Table IV. Lake group and fish species occurrence based on the cluster analysis. Only fish species caught in three or more lakes were included in the analysis and are shown in this figure.

Lake	Fish Species[a]										
	CS	CC	SF	WS	GS	BT	BB	CP	PD	RD	FD
Depauperate Lakes											
Garcock Pond					X						
Browns Pond					X		X	X			
Eighteen Pond					X			X			
Johnston Pond					X	X					
Fogg Pond						X			X		
Davis Pond						X		X			
Twin Lake					X	X					
Trout Pond						X					
Cyprinid-Sucker Lakes											
Little Fisher Pond		X		X		X	X			X	X
Big Fisher Pond	X	X		X	X	X	X		X	X	X
Unnamed Pond		X		X	X				X	X	
Green Lake	X		X	X							
Knox Pond			X	X				X			
Lost Pond			X	X							X
Little Hurd Pond	X	X	X	X	X	X					
Porter Lake	X	X	X	X	X						
Upper Pug Lake			X	X				X			
Horseshoe Pond			X	X	X	X	X				
Bear Pond	X	X			X	X					

[a] Fish Species codes: BT=brook trout BB=brown bullhead Ictalurus nebulosus CC=creek chub CP=chain pickerel CS=common shiner FD=finescale dace Phoxinus neogaeus GS=golden shiner PD=pearl dace Semotilus margarita RD=northern redbelly dace SF=sunfish WS=white sucker

Table V. Results of Tukey's nonparametric multiple comparison analysis for fish species groups with respect to the significant variables identified by Kruskall-Wallis analysis. Groups connected by a solid line are not significantly different ($p<0.05$). Groups connected by a dotted line are marginally different ($0.05<p<0.10$).

Variable	Group		
Number of fish species	Fishless	Depauperate	Cyprinid-Sucker
Aluminum	Depauperate	Cyprinid-Sucker	Fishless
pH	Fishless	Cyprinid-Sucker	Depauperate
Divalent cations	Fishless	Cyprinid-Sucker	Depauperate

shiner and northern redbelly dace Phoxinus eos) occurred in any appreciable numbers (97 and 14, respectively). The remaining seven species populations were represented by one or two fish each and were apparently quite rare in those lakes.

In our study, white suckers and sunfish were the most widely distributed species with respect to acidity, being collected at pH 5.4, closely followed by golden shiners (pH 5.5), and brook trout (pH 5.6). Creek chubs (pH 5.9) and common shiners (pH 6.0) were distributed over the narrowest pH range. These findings are similar to those from surveys conducted in other areas. Brook trout have been widely reported to occur in acidic lakes (Altshuller and Linthurst, 1984). In a survey of Vermont lakes similar to the ones surveyed by us, Langdon (1983) collected brook trout from a lake of pH 4.67. Harvey (1980) found white suckers reproducing in lakes of pH 5.0, and Langdon (1983, 1984) collected them at pH 5.48. Schindler et al. (1985) found that reproduction in white suckers ceased at about pH 5.1 in a lake acidification experiment. Golden shiners were collected in lakes of pH 5.1 by Harvey (1980) and Langdon (1983, 1984). Of the two less acid tolerant species, creek chubs were not collected below pH 6.1 (Rahel and Magnuson, 1983) and pH 5.8 (Langdon, 1983, 1984). Harvey (1980) and Rahel and Magnuson (1983) reported that common shiners were lacking in lakes of pH below 5.7 and 6.2, respectively. The observed distribution of these two species may partly be a reflection of their greater sensitivity to low pH.

Multiple regression analysis indicated that lake divalent cation concentration accounted for most of the variation in number of fish species when all 22 lakes were included in the data set. The importance of Ca in determining the status of brown trout (*Salmo trutta*) in 700 Norwegian lakes was shown by Wright and Snekvik (1978). Moreover, Wright and Snekvik (1978) reported that fishless lakes predominated at low pH (pH <4.5) and low Ca concentrations (Ca <20 μeqL^{-1}). The average Ca concentration in the three acidic, fishless lakes we surveyed was 27 vs 78 μeqL^{-1} in the depauperate lakes and 92 μeqL^{-1} in the cyprinid-sucker lakes.

Water chemistry variables related to acidity (i.e. pH, Al) appear to be important in determining the presence or absence of fish (Schofield, 1977; Baker and Schofield, 1980; Driscoll *et al*., 1980). Both these variables differed significantly between the fishless lakes and the lakes that yielded fish (Table V). Low pH and Al concentrations as low as 100 μgL^{-1} have been shown to be toxic to fish (Muniz and Leivestad, 1980; Rosseland and Skogheim, 1984; Schofield and Trojnar, 1980). The average Al concentration for the three fishless lakes was 320 vs 68 μgL^{-1} for the remaining lakes. We measured only total Al, which includes the inorganic monomeric form that is considered to be the most toxic to fish (Driscoll *et al*., 1980). Color (dissolved organic C) levels were very low in the fishless lakes, and recent measurements indicated that over half of the total Al concentration in these lakes was in the labile, monomeric form (Haines, 1985).

For lakes supporting fish, the multiple regression analysis indicated that surface area, divalent cation concentration, and maximum depth accounted for most of the variability in the number of fish species among lakes (Table III). The influence of lake surface area on number of fish species has previously been noted (Harvey, 1975, 1978; Rahel and Magnuson, 1983; Tonn and Magnuson, 1982; Eadie and Keast, 1984). This is a problem as it may be difficult to distinguish the influence that surface area has on the number of species from that of other variables. Maximum depth and surface area may reflect habitat quantity and diversity.

The processes that are responsible for producing the species assemblages described in the lakes supporting fish are not known. The hypothesis of unequal accessibility, which could result in different species assemblages (Rubec, 1975), seems unlikely. All our lakes were drained lakes, with permanent outlets, and lake elevation was not a significant variable in determining species richness. The lack of minnows and sunfish in the depauperate lakes could be due to the fact that most of these lakes tended to be deep, unproductive, and devoid of inshore areas. Extensive shallow and weedy areas, habitats preferred by both cyprinids and centrarchids (Moyle, 1973; Werner *et al*., 1978), were lacking. The paucity of species in the depaurerate lakes may also be a reflection of the lack of structural heterogeneity and adequate food supplies (Menge and Sutherland, 1976; Werner *et al*., 1978; Eadie and Keast, 1984; Rahel, 1984). Tonn and Magnuson (1982) noted the absence of cyprinids from small Wisconsin lakes with suitable habitats and attributed this to the presence of large piscivores such as largemouth bass (Micropterus salmoides) and northern pike (Esox lucius). In the depauperate lakes, only chain pickerel (Esox niger) and brook trout were piscivorous. But chain pickerel were only caught in three of those lakes and only in small numbers, and also occurred in two of the cyprinid-sucker lakes. Brook trout were ubiquitous in the cyprinid-sucker lakes, all of which contained several species of minnows and sunfish.

Finally, although no strong relationship could be established between number of fish species and water chemistry in the lakes that supported fish populations, we do not imply that such a relationship does not exist. Certain minnow species were absent from lakes with pH below 6. The reason a more pronounced influence was not detected in this study was probably because only three of the 19 lakes that yielded fish had a pH below 5.8 and thus had marginal water quality for fish (Table II).

ACKNOWLEDGMENTS

This research was funded as part of the National Acid Precipitation Assessment Program by the U.S. Environmental Protection Agency. Additional funding was supplied by the U.S. Fish and Wildlife Service and the University of Maine. We thank Charles Jagoe for his help in collecting the field samples, and Stephen Norton and Jeffrey Kahl for performing laboratory analyses.

REFERENCES

Altshuller, A., and Linthurst, R. (ed): 1984, EPA-600/8-83-016BF, Washington, D.C.

Baker, J.P., and Schofield, C.L.: 1980, 'Aluminum toxicity to fish as related to acid precipitation and Adirondack surface water quality' in Proceedings of an International Conference on the Ecological Impact of Acid Precipitation, D. Drablos and A. Tollan (ed.), Aas, Norway, p. 292-293.

Boesch, D.: 1977, EPA-600/3-77-033.

Driscoll, C.T., Jr., Baker, J.P., Bisogni, J.J., Jr., and Schofield, C.L.: 1980, Nature (Lond.) 284, 161.

Eadie, J. McA., and Keast, A.: 1984, Can. J. Zool. 62, 1689.

Haines, T.A.: 1985, unpublished data.

Haines, T.A., and Akielaszek, J.J.: 1983, FWS/OBS-80/40.15. Kearneysville, W.V.

Harvey, H.H.: 1975, Verh. Int. Verein. Limnol. 19, 2405.

Harvey, H.H.: 1978, Verh. Int. Verein. Limnol. 20, 2031.

Harvey, H.H.: 1980, 'Widespread and diverse changes in the biota of North American lakes and rivers coincident with acidification' in Proceedings of an International Conference Ecological Impact of Acid Precipitation, D. Drablos and A. Tollan (ed.), Aas, Norway, p. 93-98.

Hendrey, G., Galloway, J., Norton, S., Schofield, C., Shaffer, P., and Burns, D.: 1980, EPA-600/3-80-024. Washington, D.C.

Interagency Task Force on Acid Precipitation: 1984, National Acid Precip. Assessment Prog., Washington, DC.

Langdon, R.: 1983, Vermont Agency of Environmental Conservation, Department of Water Resources and Environmental Engineering, 69 p.

Langdon, R.: 1984, Vermont Agency of Environmental Conservation, Department of Water Resources and Environmental Engineering, 45 p.

Menge, B.A., and Sutherland, J.P.: 1976, Am. Nat. 110, 351.

Moyle, P.B.: 1973, Trans. Am. Fish. Soc. 102, 794.

Muniz, I.P., and Leivestad, H.: 1980, 'Toxic effects of aluminum on the brown trout, *Salmo trutta* L.' in Proceedings of an International Conference on the Ecological Impact of Acid Precipitation, D. Drablos and A. Tollan (ed.), Aas, Norway, 1980, p. 320-321.

National Acid Precipitation Assessment Program: 1985, Working protocol for sampling, sample analysis and quality assurance and quality control for the U.S. E.P.A. Long Term Surface Water Monitoring Program. Draft Report. Interagency Task Force on Acid Precipitation, Env. Res. Lab., Corvallis, Oregon 20 p.

Rahel, F.J.: 1984, Ecology 65, 1276.

Rahel, F.J., and Magnuson, J.J.: 1983, Can. J. Fish. Aquat. Sci. 40, 3.

Rosseland, B., and Skogheim, O.: 1984, Rep. Inst. Freshw. Res. Drottningholm 61, 186.

Rubec, P.J.: 1975, Can. Field Nat. 89, 389.

Schindler, D., Mills, K., Malley, D., Findlay, D., Shearer, J., Davies, I., Turner, M., Linsey, G., and Cruikshank, D.: 1985, Science 228, 1395.

Schofield, C.L. 1977. Research Project Completion Report. Project No. A-072-NY.

Schofield, C.L., Trojnar, J.R.: 1980, T. Toribara, M. Miller, and P. Morrow (ed): Polluted Rain, Plenum Press, New York, p. 341-366.

Scott, W.B., and E.J. Crossman: 1973, Freshwater Fishes of Canada, Bull. 184, Fisheries Research Board of Canada, Ottawa.

Tonn, W.M., and Magnuson, J.J.: 1982, Ecology 63, 1149.

Werner, E.E., Hall, D.J., and Werner, M.D.: 1978, Environ. Biol. Fishes 3, 163.

Wright, R.F., and Snekvik, E.: 1978, Verh. Int. Verein. Limnol. 20, 765.

Zar, J.H. 1984. Biostatistical analysis. 2nd. ed. Prentice-Hall, NJ.

FISH SPECIES DISTRIBUTION AND WATER CHEMISTRY IN NOVA SCOTIA LAKES

D. L. Smith and J. K. Underwood*
Nova Scotia Department of the Environment
P.O. Box 2107
Halifax, N.S., Canada B3J 3B7

J. G. Ogden, III
Department of Biology
Dalhousie University
Halifax, N.S., Canada B3H 4J1

B. C. Sabean
Nova Scotia Department of Lands and Forests
120 Exhibition Street
Kentville, N.S., Canada B4N 4E5

ABSTRACT. During the summers of 1981-1984, 19,714 fish (23 species) were netted in 234 Nova Scotian lakes. Surface and mid-depth water samples were also analyzed for major ions, metals, and DOC. Lakewater pH varied from 4.4 to 7.7, including eight lakes which produced no fish in standard 23-hr net sets. Fish data were partitioned into 6 pH groupings for analysis. Stepwise multiple regressions of fish species vs. H+, SO_4, Al, Fe, and Mn showed little predictive power. Productive lakes ranged up to 530 μg L^{-1} Al, 1680 μg L^{-1} Fe, and 836 μg L^{-1} Mn. Apart from pH, fish distribution and abundance showed no significant relationships with water chemistry data. We note, however, that the more acidic lakes had fewer species of fish.

1. INTRODUCTION

The sensitivity and loss of fish species from acidic deposition in lakes has been documented by Schofield (1976), Overrein et al. (1980), Haines (1981), Harvey (1982) and for southwestern Nova Scotian rivers and streams by Watt (1981). Increasing acid levels affect fish directly through H+ toxicity (Daye and Garside, 1977), and indirectly through reproductive and recruitment failure, histopathologic changes in tissues (Daye and Garside, 1980), degradation of spawning sites, disruption of the food chain (Haines, 1981), and Al toxicity (Baker,1982). This paper focusses on relationships between various species of fresh water (and anadromous) fish and various water chemistry parameters in

*Author for all correspondence

Water, Air and Soil Pollution **30** (1986) 489–496.

Nova Scotian lakes. Livingstone (1951) documented the fresh water fish of Nova Scotia, but we are unaware of any previous survey of fish distribution in relation to water chemistry in Nova Scotia. Although our data set is limited to 234 lakes from a resource that includes approximately 10,000 lakes (Smith, 1963), this report provides an initial overview of fish relative to water chemistry from one of the most heavily impacted acidic deposition areas in North America.

2. METHODS

During the summers of 1981-1984, 234 provincial lakes were sampled by the Nova Scotia Department of Lands and Forests (Figure 1). Fish surveys were conducted with 2m x 30m gill nets, each of which had four 2m x 7.4m panels of different mesh sizes (1.3, 1.9, 2.5, and 3.8 cm/side). The nets were set for approximately 23 hr (cf. Haines et al., 1986). Water samples from near-surface and mid-depth were also collected and analyzed for major ions, organic C, and metals (Underwood et al., 1986).

The lakes were divided into 6 groups (Figure 1) based on pH, alkalinity, and tolerance of the Atlantic salmon (Watt, 1981). Data were analyzed using SPSS (Nie et al., 1975).

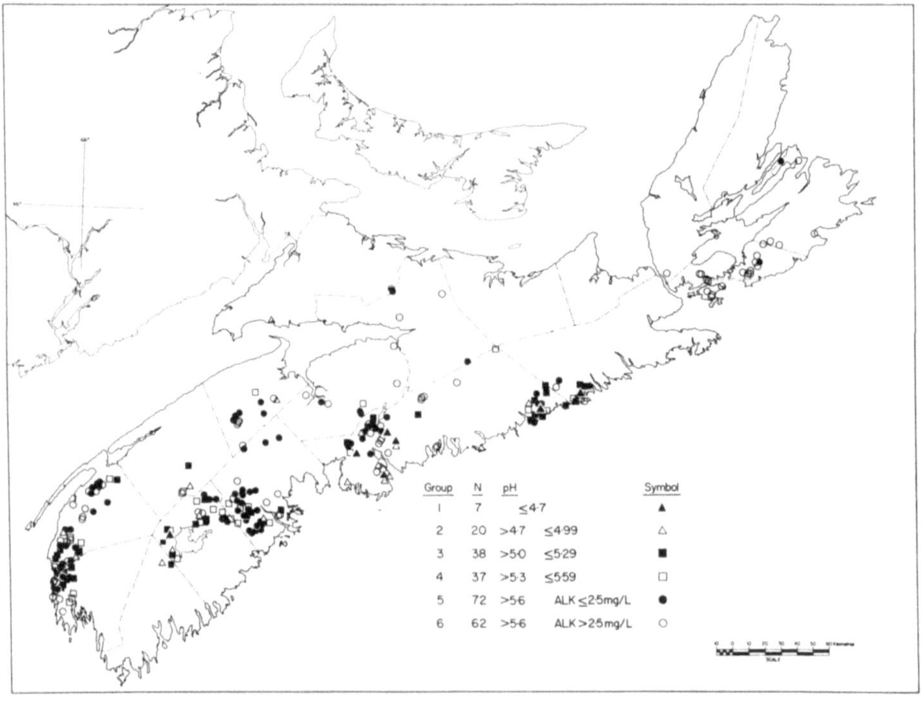

Fig. 1. Location map of sampled lakes and pH class groupings.

Table I. Species list of fish caught. Mean and (maximum) values (µg L^{-1}) for Al, Fe, and Mn shown at right of table. Minimum values were < 10 µg L^{-1}, and usually 0.

Species No.	Common Name	Scientific Name	No. Caught	No. Lakes	Mean/(Maximum) µg L^{-1} Al	Fe	Mn
1.*	Atlantic Salmon	Salmo salar	10	7	88 (175)	120 (260)	44 (138)
2.*	Brook Trout	Salvelinus fontinalis	483	91	95 (335)	176 (1380)	51 (836)
4.*	Rainbow Trout	Salmo gairdneri	6	2	15 (30)	54 (74)	19 (25)
5.*	Brown Trout	Salmo trutta	2	2	78 (120)	124 (165)	37 (51)
6.*	Smallmouth Bass	Micropterus dolmieui	14	5	55 (132)	43 (91)	31 (70)
7.*	White Perch	Morone americana	3599	99	98 (530)	217 (1380)	51 (836)
8.*	Striped Bass	Morone saxatilis	2	1	15	10	8
9.*	Yellow Perch	Perca flavescens	3792	114	105 (530)	248 (1390)	48 (499)
10.*	Chain Pickerel	Esox niger	4	3	147 (190)	245 (410)	16 (22)
11.*	Brown Bullhead	Ictalurus nebulosus	2301	119	104 (530)	254 (1680)	41 (296)
15.*	American Eel	Anguilla rostrata	58	32	104 (270)	194 (1030)	74 (836)
16.*	White Sucker	Catostomus commersoni	1739	134	100 (530)	215 (1326)	54 (836)
18.*	Lake Chub	Couesius plumbeus	230	24	85 (270)	251 (1100)	53 (209)
20.*	Creek Chub	Semotilus atromaculatus	124	15	55 (143)	113 (581)	35 (125)
22.*	Redbelly Dace	Chrosomus eos	389	9	46 (130)	75 (175)	18 (51)
23.*	Golden Shiner	Notemigonus crysoleucas	829	85	105 (440)	258 (1680)	59 (836)
24.*	Common Shiner	Notropis cornutus	30	7	36 (145)	150 (484)	54 (138)
26.*	Threespine Stickleback	Gasterosteus aculeatus	12	4	36 (85)	50 (103)	18 (51)
27.*	Fourspine Stickleback	Apeltes quadracus	26	4	110 (320)	161 (307)	27 (67)
28.*	Ninespine Stickleback	Pungitius pungitius	195	29	111 (390)	182 (698)	56 (296)
29.*	Banded Killifish	Fundulus diaphanus	5582	122	98 (530)	218 (1390)	55 (836)
30.*	Mummichog	Fundulus heteroclitus	1	1	14	40	12
32.*	Gaspereau	Alosa psuedoharengus	286	25	113 (360)	199 (420)	47 (160)
3.**	Lake Trout	Salvelinus namaycush	--	--	--	--	--
12.**	Smelts	Osmeridae	--	--	--	--	--
13.**	Lake Whitefish	Coregonus clupeaformis	--	--	--	--	--
14.**	Atlantic Whitefish	Coregonus canadensis	--	--	--	--	--
17.**	Blacknose Dace	Rhinicthys astralatus	--	--	--	--	--
19.**	Fallfish	Semotilus corporalis	--	--	--	--	--
21.**	Pearl Dace	Semotilus margarita	--	--	--	--	--
25.**	Blacknose Shiner	Notropis hererolepis	--	--	--	--	--
31.**	Sea Lamprey	Petromyzon marinus	--	--	--	--	--
33.**	American Shad	Alosa sapidissima	--	--	--	--	--
		TOTAL	19,714				

* - Species caught in survey

** - Species not netted but known from other surveys.

3. RESULTS AND DISCUSSION

Species caught in the survey (23) or known for the lakes but not recovered in this survey (10), are listed in Table I together with the number of fish caught, lakes of occurrence, and the mean and maximum values of Al, Fe, and Mn (μg L^{-1}) for each species recovered in the survey. Figure 2 shows the number of lakes (abscissa) in which each species was caught by pH group. For example, the ninespine stickleback (Fish sp. no. 28) was found in two lakes in Group 1 (pH < 4.7), 1 lake in Group 2, seven lakes in Group 3, three lakes in Group 4, five lakes in Group 5, and three lakes in Group 6 (pH>5.6, alk. >2.5 mg L^{-1}).

Figure 3 presents the percentage of lakes within which each fish species was caught, e.g., fish species no. 28 was found in 35% of the Group 1 lakes, 4% of Group 2 lakes, 12% of Group 3 lakes, 5% of Group 4 lakes, 15% of Group 5 lakes, and 6% of Group 6 lakes. Percentages were calculated to permit comparison since each Group includes differing numbers of lakes.

Six species (ninespine stickleback, fourspine stickleback, yellow perch, brook trout, white sucker, and banded killifish) occur in Group 1 (pH <4.7) lakes. American eel, golden shiner, white perch, and brown bullhead first occur in Group 2 lakes. Five more species (gaspereau, common shiner, Atlantic salmon, creek chub, and lake chub) first appeared in Group 3 lakes. Three additional species were found in groups

Fig. 2. Fish species distribution.

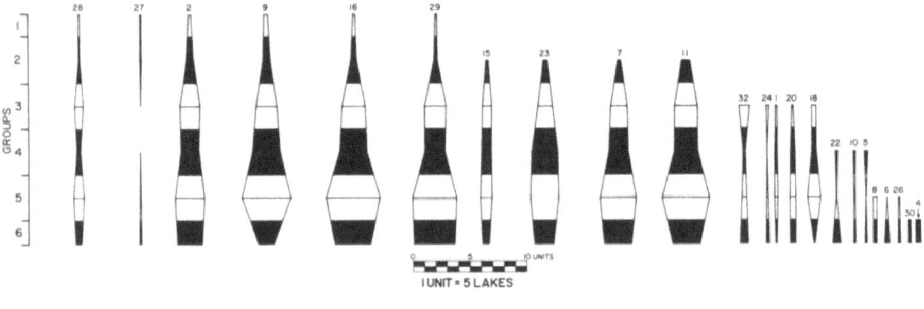

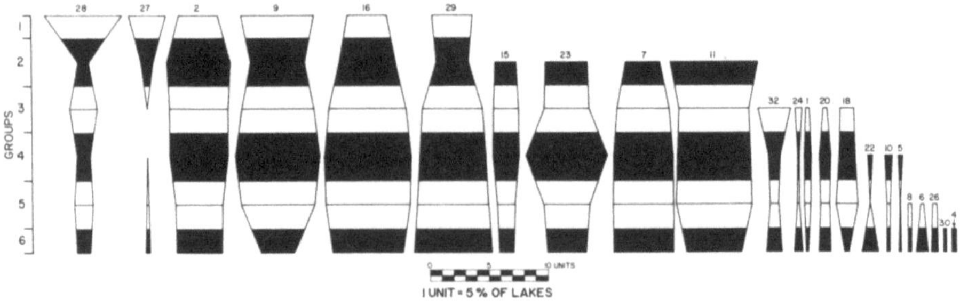

Fig. 3. Relative distribution of species in study lake groups.

4 (redbelly dace, chain pickerel, and brown trout) and 5 (striped bass, smallmouth bass, and threespine stickleback). Two species (mummichog and rainbow trout) were found only in Group 6 lakes.

Figures 2 and 3 do not show any trends in the abundance of fish species. With the exception of No. 27 (Fourspine stickleback), once a species appears in the figure, however, they are found in all consecutively higher pH groups. The lowest pH of occurrence for 11 species is shown in Table II. Most of the species listed are within the ranges shown in Table II for extinction or reproductive failure. The minimum pH for rainbow trout, brown trout, and lake chub are higher for Nova Scotia than shown by the references noted in Table II. Rainbow and

Table II. Minimum pH values for 10 fish species in Nova Scotia.

Species	minimum pH Nova Scotia	minimum pH other surveys	(References)
Yellow perch	4.43	4.2 to 4.8	(1,2)
White sucker	4.65	4.5 to 5.2	(1,2,3)
Brook trout	4.65	4.5 to 5.0	(4,5,6)
Brown bullhead	4.84	4.5 to 5.2	(1,2,3,7)
Golden shiner	4.84	4.8 to 5.2	(2)
Lake chub	5.09	4.5 to 4.7	(1)
Common shiner	5.27	<5.7	(2)
Atlantic salmon	5.27	5.0 to 5.5	(5,6)
Brown trout	6.27	4.5 to 5.5	(5,6,8,9)
Rainbow trout	6.56	5.5 to 6.0	(5,6)

(1) Beamish (1976); (2) Harvey (1980); (3) Beamish, et al. (1975); (4) Schofield (1976); Grande, et al. (1978); Peterson, et al. (1982); (7) Wood and McDonald (1982); (8) Jensen and Snekvik (1972); (9) Wright and Snekvik (1978).

brown trout are introduced in Nova Scotia and are restricted to unspecified lakes in various portions of the province. Our survey apparently included only a few of these lakes (e.g., Table I, spp. 4 and 5 found in only 2 lakes) and we therefore attach no particular significance to their occurrence because of restricted sample size. Lake chub, (N=230 in 24 lakes) however, are widely distributed in Nova Scotian lakes and streams (Scott and Crossman, 1973) and our minimum pH value (5.09) may therefore imply a decreased tolerance to acidity in Nova Scotian lakes.

We note that the minimum pH values at which common shiner is found in this study is lower than that cited by Mills and Schindler (1986). Similarly, White suckers (N=1739 in 134 lakes) were found at pH levels below those found to inhibit reproduction (Schindler et al., 1985). Additional age/pH class structure information will be required to determine if the existing populations are viable at these lower pH values. Rahel and Magnuson (1980) hypothesize that some species are genetically pre-adapted to naturally occurring acidity. The total number of fish species in each group ranged from 6 for Group 1 through

10, 15, 17, 21, and 25 for Groups 2 through 6, respectively. The more acidic lakes therefore, are inhabited by fewer species. Eight lakes were found in which no fish were captured. Three of these lakes were from Group I and 5 from Group II.

Pearson Product Moment and Spearman Rank correlations were determined between all fish species and H+, conductivity, colour, turbidity, Na, Ca, Mg, NH_4, alkalinity, Cl, SO_4, NO_3, PO_4, DOC, Al, Fe, and Mn. Of the few significant ($p < .05$) relationships found, none had an r >0.4. Stepwise multiple regressions of fish species vs. H+, SO_4, Al, Fe, and Mn revealed little predictive power from these variables.

Schindler, et al. (1985) have noted that acidic deposition may have deleterious effects on the condition of animals, impairing growth and other physical or physiological properties short of mortality. In the present survey, data on physical condition was not generally noted, but length and weight were recorded for brook trout (Figure 4). There does not appear to be any trend in length and weight related to pH group. The distribution of length is skewed to the left within the size range for the species as given by Scott and Crossman (1973).

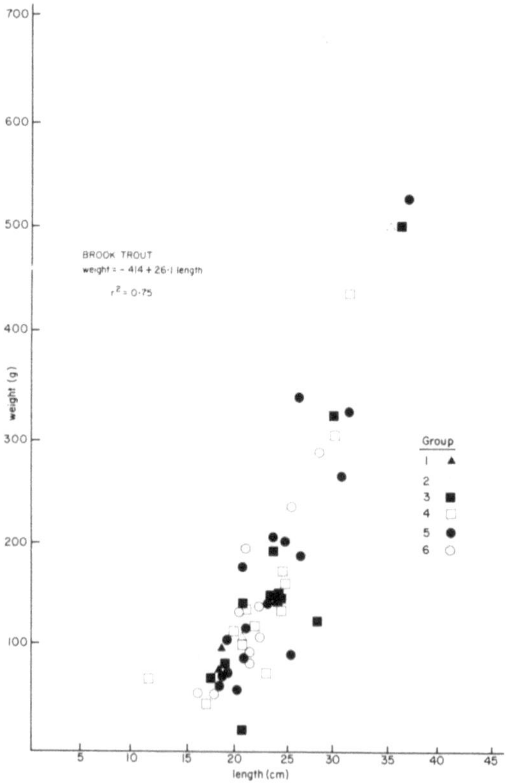

Fig. 4. Brook trout length vs. weight.

4. SUMMARY AND CONCLUSIONS

A total of 23 species (19,714 individuals) of fish were found in 234 Nova Scotian Lakes sampled by 23-hr net sets (60 m^2), and their distribution and abundance is compared with near-surface and mid-depth water chemistry samples collected during the summer months, 1981-1984. Data from an additional 8 species recovered in other surveys are included. Apart from pH, fish distribution and abundance showed no significant relationships with water chemistry data. pH tolerance levels are inferred from the lowest pH of occurrence. The more acidic lakes had fewer species of fish. Future studies should include age/pH class structure and a detailed analysis of habitat conditions within the lakes.

ACKNOWLEDGMENTS

Fish samples were collected and analyzed by staff of the Nova Scotia Department of Lands and Forests. Funding for water chemistry analyses was provided by the Nova Scotia Department of the Environment.

The authors are indebted to D. Hirtle for drafting the illustrations.

REFERENCES

Baker, J. P.: 1985, p. 165 in R. E. Johnson (ed.), Acid Rain/Fisheries, American Fisheries Society, Bethesda, Maryland, 357 pp.
Beamish, R.: 1976, Water, Air, and Soil Pollut. **6**, 501.
Beamish, R., Lockhart, W. VanLoon, J., and Harvey, H.: 1975, Ambio 4, 98.
Daye, P.G. and Garside, E.T.: 1977, Can. J. Zool. **55**,1504.
Daye, P. G. and Garside, E.T.: 1980, Can. J. Zool, **58**, 27.
Grande, M., Muniz, I., and Andersen, S.: 1978, Verh. Internat. Verein. Limnol., **20**,2076.
Haines, T.A.: 1981, Trans. Amer. Fish. Soc. **110**, 669.
Haines, T. A., Pauwels, S. J., and Jagoe, C. H.: 1986, Biol. Report 80 (40.23), U.S. Fish and Wildlife Service, Eastern Energy and Land Use Team, 139pp.
Harvey, H. H.: 1980, p. 93 in D. Drabløs, and A. Tollan (eds.), Ecological Impact of Acid Precipitation, Proc. Int. Conf. Ecol. Impact Acid Precip., Sandefjord, Norway, 383 pp.
Harvey, H. H.: 1982, p. 227 in R. E. Johnson (ed.), Acid Rain/Fisheries, American Fisheries Society, Bethesda, Maryland, 357pp.
Jensen, K., and Snekvik,E.: 1978, Ambio, **1**, 223.
Livingstone, D. A.: 1951, Proc. N.S. Inst. Sci. **23**, 90pp.
Mills, K. H., and Schindler, D.W.: 1986, Water, Air, and Soil Pollut. (this volume).
Nie, N.H., Hul, C.H., Jenkins, J. G., Steinbrenner, K., and Bent, 2ed, McGraw-Hill, New York, 675.
Overrein, L.N., Seip, H.M., and Tollan, A.: 1980, Acid precipitation-effects on forest and fish. Final report, SNSF project, 1971-1980, Reclamo, Oslo, 175pp.

Peterson, R. H., Daye, P.G., Lacroix, G. L., and Garside, E. T.: 1982, p. 177 in R.E. Johnson (ed.), Acid Rain/Fisheries, American Fisheries Society, Bethesda, Maryland, 357 pp.

Rahel, F. J., and Magnuson, J.J.: 1980, 'Fish in naturally acidic lakes of northern Wisconsin, U.S.A.'; in D. Drabløs, and A. Tollan (eds.) Ecological Impact of Acid Precipitation, Proc. Int. Conf. Ecol. Impact Acid Precip., Sandefjord, Norway, 334-335.

Schindler, D.W., Mills, K.H., Malley, D.F., Findlay, D.L., Shearer, J.A., Davies, I.J., Turner, M.A., Linsey, G.A., and Cruikshank, D.R.: 1985, Science **228**, 1395.

Schofield, C. L.: 1976, Ambio 5,228.

Scott, W. B., and Crossman, E.J.: 1973, Freshwater fishes of Canada. Fisheries Research Board of Canada, Bull. 184, 996.

Smith, M.W.: 1963, 'The Atlantic Provinces of Canada', in D. G. Frey (ed.), Limnology in North America, Univ. Wisconsin Press, Milwaukee.

Underwood, J.K., Ogden, III, J. G., and Smith, D.L.: 1986, Water, Air, and Soil Pollut., (this volume).

Watt, W. D.: 1981, pp. 39-45 in L. Sochasky (ed.), Acid Rain and the Atlantic Salmon, International Atlantic Salmon Special Publication Series no. 10, St. Andrews, New Brunswick.

Wood, C. M., and McDonald, D. G.: 1982. p. 197 in R. E. Johnson (ed.), Acid Rain/Fisheries, American Fisheries Society, Bethesda, Maryland, 357 pp.

Wright, R., and Snekvik, E.: 1978, Limnol. Oceanog. **23**, 487.

Avoidance of Low pH and Elevated Al Concentrations by Brook Charr (*Salvelinus fontinalis*) Alevins in Laboratory Tests

John M. Gunn and David L.G. Noakes

Department of Zoology, University of Guelph,
Guelph, Ontario, Canada N1G 2W1

ABSTRACT. Laboratory studies were conducted to test the ability of brook charr (Salvelinus fontinalis) alevins, the earliest free-swimming life stage of the species, to detect and avoid toxic levels of H^+ and inorganic Al. Alevins were tested in steep gradient choice tanks using a range of H^+ (pH 4.0 to 5.5) and Al (0 to 500 μg L^{-1}) concentrations in low Ca (2.0 mg L^{-1}) water. The young brook charr actively avoided acidic water with a pH < 5.0. Aluminum additions of 500 μg L^{-1} increased the avoidance response. The observed behavioral response of alevins to low pH and elevated levels of Al, may be of significant adaptive advantage in systems undergoing acidification.

1. INTRODUCTION

Near the end of the yolk absorption period, brook charr (Salvelinus fontinalis) emerge from the gravel of the redd, fill their gas bladder and commence exogenous feeding. This shift from embryonic to juvenile life style not only involves morphological and physiological changes in the fish, but potentially exposes alevins to abrupt physiochemical changes in the water as they move from the interstitial to the surface environment. In areas receiving acidic precipitation, the surface waters are often far more acidic than the upwelling groundwater that percolates through the redd (Trojnar, 1977; Gunn, 1985). These chemical differences may be further exaggerated if emergence coincides with the pH depressions in surface waters that often accompany snowmelt.

While the sensitivity of the early life stages of fish to acidic water is well known (Haines, 1981) and many recognize the potential damage caused by episodic pH depressions (Jeffries et al., 1979), there is little information on the role that avoidance reactions may play in protecting fish from the deleterious chemical conditions typical of acid stressed systems. It has been shown that fish can sense minute concentrations of many other toxic chemicals and will often avoid contaminated waters at levels well below lethal conditions (Giattina and Garton, 1983).

In this study we tested the ability of brook charr alevins to

Water, Air and Soil Pollution **30** (1986) 497–503.

avoid toxic concentrations of H+ and inorganic Al, the factors generally considered responsible for fish mortality in acidified lakes and streams (Haines, 1981; Baker, 1982). Tests were conducted using prepared solutions as well as water collected from an acidic lake.

2. MATERIAL AND METHODS

Brook charr eggs were obtained from 1) brook stock at Ontario Ministry of Natural Resources Fisheries Laboratory at Maple, Ontario, and 2) from wild fish in Dickson Lake (45°47'N, 78°12'W), Ontario. Shortly after hatching, embryos were acclimated to the control conditions that were used throughout this experiment: temperature 4.8 (± 0.2)°C ($\bar{X}$(± 95%CI)), pH 6.9(± 0.1), specific conductance @ 25 C 23.0(± 1.2) uS cm^{-1}, hardness 7.2(± 0.5) mg L^{-1} as $CaCO_3$, Ca^{++} 2.0(± 0.1) mg L^{-1}, Na^{++} 1.4(± .03) mg L^{-1}, and inorganic Al <0.001 mg L^{-1}. The avoidance experiments were run using fish from the start of swim-up to the end of yolk absorption, a period of approximately 3 weeks. The mean total length and dry weight of alevins used in the avoidance tests were 23.6 ± 0.3 mm and 13.4 ± 0.2 mg respectively.

Avoidance reactions were tested in steep gradient choice tanks (Figure 1) with a flow rate of 400 ml min^{-1} per side. Dye tests confirmed that sharp separation of the test waters occurred at the center drain in these tanks.

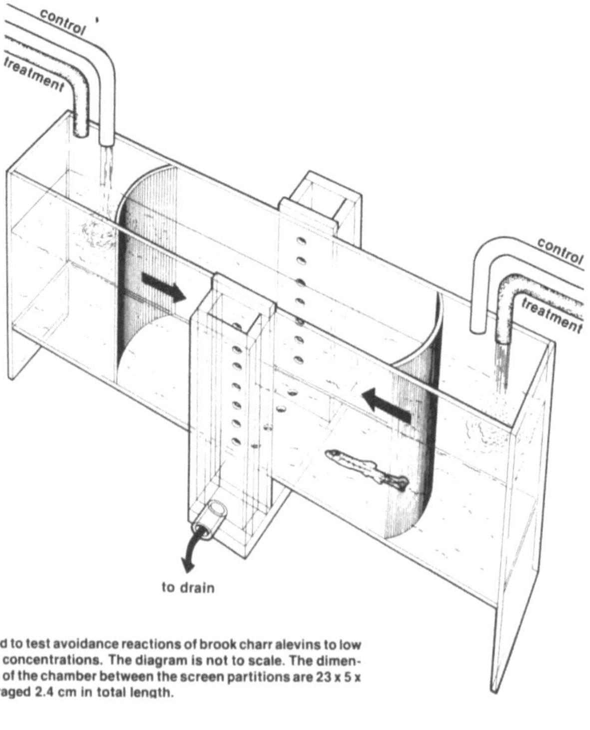

Fig. 1. Chamber used to test avoidance reactions of brook charr alevins to low pH and elevated Al concentrations. The diagram is not to scale. The dimensions of the portion of the chamber between the screen partitions are 23 x 5 x 10 cm. Alevins averaged 2.4 cm in total length.

A single alevin was placed in the tank while control water (pH 6.9, 0 Al) flowed from both sides. After an acclimation period of 5 min, the control water on one side was replaced by treatment water and the time spent on the left and right side of the center drain was recorded as the fish swam back and forth for 15 min. The trip durations were recorded manually by an observer using a Esterline Angus event recorder. Three chambers, each separated by opaque barriers were run simultaneously with the fish being observed using an overhead mirror. The start of a test run was defined as the point of first entry of the alevin into the treatment side water. There were approximately 12 fish used per treatment, with each fish used only once. Only the data from actively swimming fish (min of 2 return trips) were used in later analysis. The side used for the treatment water was reversed between each run to eliminate chamber effects. Control runs were conducted during which control water was maintained on both sides throughout the test period. All runs were conducted in dim light (1 mE m^{-2} s^{-1} at the water surface).

The avoidance reactions of alevins were tested under the following conditions:

1) a range of pH levels (4.0 to 5.5)
2) low pH (4.5, 5.0) + Al (0 to 500 µg L^{-1})
3) in "natural" water collected from an acidic lake

Synthetic test solutions were prepared in 200 L polyethylene barrels by adding $Al_2(SO_4)_3$ and H_2SO_4 to the control water mixture of deionized and dechlorinated tap water (98/2 mixture). Vigorous aeration (to drive off CO_2) and repeated titrations with 0.02N H_2SO_4 were used to produce stable pH conditions. Solutions were cooled to 5 C (using plastic emersion coils attached to a heat exchanger), aerated, pumped to insulated headtanks, and passed through calibrated flow valves to the avoidance chambers. Solutions were in contact with only the plastic components of the apparatus.

Water was collected from Ruth-Roy Lake (46°06' N, 81°14' W), an acid lake near Sudbury, Ontario to test the response of alevins to chemical conditions in "natural" water. In preliminary tests, using static bioassay procedures, the Ruth-Roy Lake water (pH 4.6, cond. 33 uS cm^{-1}, Ca 1.3 mg L^{-1}, Na 0.5 mg L^{-1}, inorg. Al 496 µg L^{-1}, Zn 26 µg L^{-1}, Cu 2 µg L^{-1}, Ni 17 µg L^{-1}) proved to be lethal to brook charr alevins in < 48 hr. For the avoidance tests, the lake water was handled in the laboratory with the same procedures outlined above.

Duplicate water samples were collected from the barrels for analysis of pH (Radiometer PM82), specific conductance (Radiometer CDM3), and hardness (EDTA titration). Samples were submitted to the Ontario Ministry of Environment (MOE) laboratory for major ion and metal analysis by procedures outlined in MOE (1981). Inorganic monomeric aluminum was analyzed using dialysis separation followed by graphite furnace atomic absorption spectroscopy. (On average the measured total and inorganic Al for each of the 4 nominal additions were: 50 - 38, 33 µg L^{-1}; 100 - 105, 72 µg L^{-1}; 200 - 205, 149 µg L^{-1}; 500 - 508, 389 µg L^{-1}). Samples were collected directly from the water flowing into the avoidance chambers for analysis of DO and free CO_2 (calculated from acid titration curve). In the first set of runs (using the Maple stock of fish) temperature was measured in the barrels at the start of a run. In the second set (Dickson L. fish) temperatures of the control and treatment solutions were measured continuously within the avoidance chambers using a YSI thermister calibrated against a NBS standard thermometer (accurate to 0.1°C).

The duration of time spent in the treatment and control ends of the tank were compared using nonparametric tests (Kruskal-Wallis, Wilcoxon, $p<.05$) after logit (deMarch and Scherer, 1980) and arcsine transformations of the data.

3. RESULTS AND DISCUSSION

Brook charr alevins avoided acidic water in the pH range of 4.0-5.0 (Figure 2), indicating that this early life stage can detect and will attempt to avoid the pH levels that other investigators (Baker and Schofield, 1982; Ingersoll _et al._, 1985) have found to be lethal to brook charr alevins in similar low conductivity, low Ca^{++} waters. The upper pH limit for the avoidance reaction was not well defined in this experiment, but most alevins appeared unaffected by pH > 5.0.

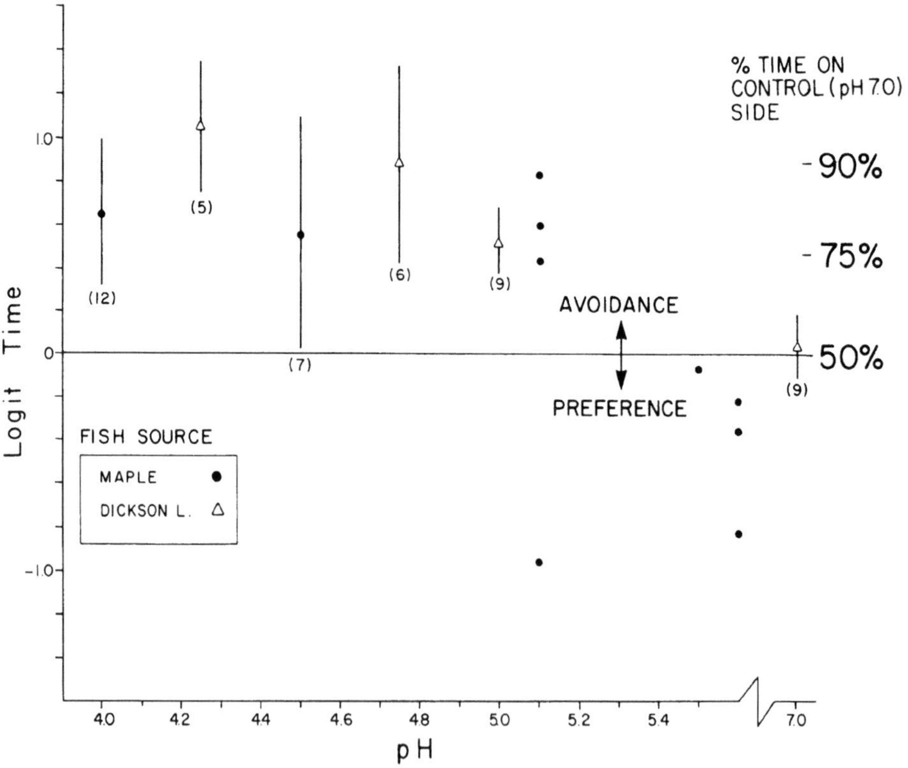

Fig. 2. Results from avoidance tests using brook charr alevins exposed to various low pH levels. Mean, 95% conf. interval, and number of runs are indicated. Data for individual fish are shown for pH levels for which there were too few values to construct confidence intervals. The location of some equivalent values for % time on the control side are indicated.

The procedures were designed to test the avoidance of H^+ ions alone. The use of soft water and excessive aeration kept free CO_2 to less than 5.0 mg L^{-1}. Hoglund and Hardig (1969) have shown that free CO_2, evolved through the acidification of test solutions, is more strongly avoided than the H^+, a fact that limits the usefulness of data on avoidance reactions to low pH in water with a high carbonate concentrations (Wells, 1915; Bishai, 1962; Ishio, 1965). The reagent additions increased the concentration of sulfate ions in the test solutions, but these were not expected to produce significant behavioral reactions (Maciorowski et al., 1977).

There were no observed differences between the response of alevins to acid lake water, and the prepared solutions with similar pH and Al concentrations (Figure 3). Both were strongly avoided.

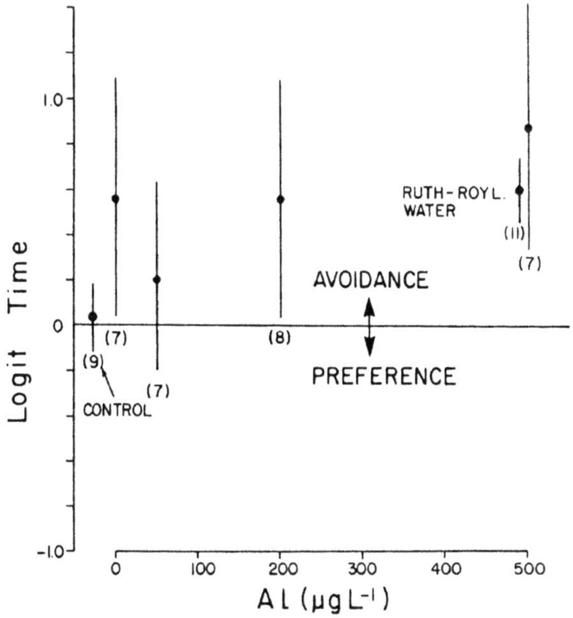

Fig. 3. Avoidance reaction to low pH (in all cases pH is 4.5) plus increasing concentrations of Al. Mean, 95% C.I., and number of runs are indicated. Aluminum levels are nominal values (see text for measured values).

There was an increase in the avoidance reaction (i.e. greater time spent on the control side) with the addition of Al to low pH (4.5) water (Figure 3), however, only with the highest Al concentration (500 μg L^{-1}) was the difference statistically significant ($P < .01$). Similarly, when alevins were confronted with low pH on both sides of the chamber, they avoided the side with the Al (Table I), but, again the enhanced avoidance reaction occurred only with the largest Al additions (500 μg L^{-1}). Van Coillie et al. (1983) conducted similar studies with adult brook charr (age 2+) and reported a threshold of approximately 100 μg L^{-1} of total AL to produce avoidance at pH 5.6.

Table I

Combined effects of low pH and aluminium on the avoidance reaction of brook charr alevins

	Control Side		Treatment Side		% Time on Control Side	Logit[a] Time
	pH	Al (μL^{-1})	pH	Al (μL^{-1})	x (95% C.I.)n	x (95% C.I.)n
Control						
	7.0	0	7.0	0	47.0(15.7) 9	.036 (.146) 9
Treatment						
	4.5	0	4.5	200	50.8(27.3) 6	.013 (.550) 6
	4.5	200	4.5	500	55.2(24.2) 7	.188 (.618) 7
	4.5	0	4.5	500	72.9(12.8) 7	.473 (.285) 7
	5.0	0	5.0	100	62.8(15.5) 9	.318 (.396) 9
	5.0	0	5.0	500	80.7(13.1) 8*	.720 (.356) 8*

[a] log $\left(\frac{\text{time spent on control side}}{\text{time spent on treatment side}}\right)$

*significantly different ($p \leq 0.05$) than control

Avoidance responses have been inferred from the distribution of fish relative to areas of low pH (Muniz and Leivestad, 1980), but to date, there are no detailed in situ studies of avoidance reactions to acidic water for any fish species. Fish respond to a variety of abiotic and biotic factors in their natural environment. It is probably difficult to predict the response to a single factor, such as pH, in isolation, but these laboratory findings suggest that avoidance reactions may be of significant adaptive advantage to even the earliest free-swimming life stage of brook charr.

4. ACKNOWLEDGEMENT

Laboratory space and equipment was kindly provided by J. Bishop, J. Lee, J. Raltson and G. Westlake of Ontario Ministry of the Environment. Funds were provided by Ontario Ministry of Natural Resources and the National Research Council of Canada (contract DSS 20ST 31048--4-200, Dept. of Supplies and Services). W. Kerr and R. Angelow provided technical assistance. G. Beggs, C. Neville, J. Sprague, G. Westlake and N. Yan offered constructive criticisms of an earlier draft of the manuscript.

5. REFERENCES

Baker, J.P. 1982, 'Effects on fish of metals associated with acidification', In: Johnson, R.E. (ed.) Acid Rain/Fisheries, Am. Fish. Soc. Publ., Bethesda, Maryland, p. 165.

Baker, J.P. and Schofield, C.L. 1982, Water Air Soil Pollut. 18, 289.

Bishai, H.M. 1962, J. Cons., Con. Int. Explor. Mer. 27, 18.

De March, B.G.E. and Scherer, E. 1980, Can. Tech. Rep. Fish. Aquat. Sci. 975, 171.

Giattina, J.D. and Garton, R.R. 1983, Res. Rev. 87, 43.

Gunn, J.M. 1986, Envir. Biol. Fish. (in press).

Haines, T.A. 1981, Trans. Am. Fish. Soc. 110, 669.

Hoglund, L.B. and Hardig, J. 1969, Rep. Inst. Freshwater Res. Drottingholm. 49, 76.

Ingersoll, C.G., La Point, T.W., Breck, J. and Bergman, H.L. 1985, U.S. Fish. Wildl. Rept. 80 (40.21) No. 21, 42.

Ishio, S. 1965, Adv. Water Pollut. Res. Proc. Int. Conf. 2nd. 1, 19.

Jeffries, D.S., Cox, C.M. and Dillon, P.J. 1979, J. Fish. Res. Board Can. 36, 640.

Maciorowski, H.D., McV. Clarke, R. and Scherer, E. 1977, 'The use of avoidance-preference bioassays with aquatic invertebrates', In: Parker, W.R., et al. (ed.) Proc. 3rd Annual Aquat. Tox. Workshop, Halifax, N.S., Can. Environ. Prot. Serv. Tech. Rept. EPS-5AR-77-1, p. 49.

Ministry of the Environment (MOE). 1981, 'Outline of analytical procedures', MOE Tech. Rept., Toronto, Ontario, Canada. 246 p.

Muniz, I.P. and Leivestad, H. 1980, 'Acidification - effects on freshwater fish', In: Drablos, D. and Tollan, A. (ed.). Proc. Inter. Conf. Ecol. Impact Acid Precip., Sandefjord, Norway. p. 84.

Trojnar, J.R. 1977, J. Fish. Res. Board Can. 34, 574.

Van Coillie, R., Thellen, C., Campbell, P.G.C. and Vigneauit, Y. 1983, Can. Tech. Rept. Fish. Aquat. Sci. 1237, 1.

Wells, M.M. 1915, Biol. Bull. (Woods Hole, Mass.) 29, 221.

BIOLOGY AND CHEMISTRY OF THREE PENNSYLVANIA LAKES: RESPONSES TO ACID PRECIPITATION

P. T. Bradt, J. L. Dudley (Ctr. Marine & Environ. Studies, Chandler #17, Lehigh Univ., Bethlehem, PA 18015 USA), M. B. Berg (Dept. Biol. Sci., Univ. Notre Dame, Notre Dame, IN 46556 USA), and D. S. Barrasso (Waksman Inst. Microbiology, Piscataway, NJ 08854 USA)

ABSTRACT. The biology and chemistry of three northeastern Pennsylvania lakes was studied from summer 1981 through summer 1983 to evaluate lakes with different sensitivities to acidification. At the acidified lake (total alkalinity $\leq$ 0.0 µeq L^{-1}) there were fewer phytoplankton and zooplankton species than at the moderately sensitive lakes. The most numerous plankton species in all three lakes are reportedly acid tolerant. Among the benthic macro- invertebrates (BMI) there were more acid tolerant Chironomidae at the acidified lake, but more acid intolerant Ephemeroptera and Mollusca and a higher wet weight at the least sensitive lake. There were no differences among the lakes' BMI mean total numbers or mean number of taxa. The fish community at the acidified lake was dominated by stunted *Lepomis gibbosus*, but *L. machrochirous* were most abundant in the other lakes. Principal component analysis suggested a shift in all three lakes over the sampling period toward combined lower pH, alkalinity, specific conductance, Ca and Mg and higher Al and Mn. Such chemical changes have been associated with acidification. The rate and extent of acidification appeared to be controlled by geological and hydrological characteristics of the drainage basins.

1. INTRODUCTION

Northeastern Pennsylvania receives some of the most acidic precipitation in North America (1982 mean weighted pH = 4.17; Lynch and Corbett, 1983) and contains surface waters sensitive to acidification (OTA, 1984). Soils are generally acidic and many surface waters are low in dissolved minerals (Arnold *et al.*, 1980; Bradt and Berg, 1983; Arnold *et al.*, 1985). To our knowledge previous studies in the area (Barker, 1978; Bradt and Berg, 1983) have recorded only one time chemical analyses with no time-trend studies of either biology or chemistry.

Water, Air and Soil Pollution **30** (1986) 505–513.

Given the acid precipitation in the area and the sensitive conditions, some waters could be demonstrating biological and/or chemical changes associated with acidification. The purposes of this study were to determine 1) if there were any biological or chemical differences among certain sensitive lakes and 2) if acid precipitation was affecting lake biota or chemistry.

We studied three lakes with different sensitivities to acidification. Long Pond (1981 total alkalinity = 187 μeq L^{-1}) and Lake Lacawac (1981 total alkalinity = 56 μeq L^{-1}) were moderately sensitive and Deep Lake was acidified (1981 total alkalinity $\leq$ 0.0 μeq L^{-1}) according to U.S./Canada (1981) criteria.

2. METHODS

Phytoplankton, zooplankton and benthic macroinvertebrates (BMI) were sampled once each season (excluding winter) from summer 1981 through summer 1983. Phytoplankton were sampled with a Kemmerer sampler (three equal increments between Secchi depth and surface), samples pooled and biovolumes calculated by the methods of APHA (1976). Macro and microzooplankton were sampled (at least 3 vertical tows from bottom) with a Wisconsin net (0.06 mm mesh) and the samples pooled. BMI were at sampled at 3 to 7 sites (3 replicates per site) with an Ekman grab sampler (0.23 m^2) in the boulder-free littoral zone. Fish were sampled annually by electrofishing and trap, hoop, and seine nets. Water chemistry was sampled at the same time as the biology (spring, summer, fall) and through the ice in winter. The number of biological and chemical samples (Tables I and II) taken was determined by the lake's size. Total alkalinity was determined by double end-point titration (APHA 1976). Methods of other chemical analyses are outlined in APHA (1976) and Bradt and Berg (1983). Al was extracted by the method of Barnes (1975).

Surficial geology and woody vegetation were reported by Dudley (1985). The following hydrological calculations were made: overland travel time, average detention time, and ground water flow-time (Lindsley et al., 1982). Precipitation chemistry was obtained for the area (EPRI, 1983; Lynch and Corbett, 1983; EPRI, 1984). Statistical analyses were performed using the SPSS Program (Nie et al., 1975; Hull and Nie, 1981). See Bradt et al. (1984) for all detailed methodologies.

3. RESULTS AND DISCUSSION

Phytoplankton species richness was lower at Deep Lake than at the other lakes (Table I). Chrysophyceae dominated algal numbers and biovolume at Deep Lake and biovolume at Long Pond (Table I). Cyanobacteria were most abundant at the other two lakes and dominated the biovolume at Lake Lacawac. Mean cell numbers appeared highest at Long Pond (Table I), while biovolume appeared highest at Deep Lake,

TABLE I - BIOLOGICAL CHARACTERISTICS OF STUDY LAKES

	DEEP LAKE	LAKE LACAWAC	LONG POND
PHYTOPLANKTON			
No. Midlake Sample	15	15	16
Numerical Dominant	Dinobryon sertularia	Merismopedia Microcystis	Anabaena circinalis
Biovolume Dominant	D. sertularia	Merismopedia glauca	D.sertularia
Mean No. Taxa mL^{-1}	8.4*	12.5	12.5
Mean No. Cells mL^{-1}	675	1060	2776
Mean Biovolume u3 mL^{-1}	$2.5x10^7$	$2.9x10^6$	$8.3x10^6$
ZOOPLANKTON			
No.Midlake Samples	14	14	21
Numerical Dominant			
Crustacea	Diaptomus minutus	Mesocyclops edax	M. edax
Rotifera	Keratella taurocephala	Keratella crassa	Keratella cochlearis
Total No. Species Crustacea+Rotifera	4+1	7+7	6+9
BENTHIC MACROINVERTEBRATES**			
No. of Samples	66	71	75
Numerical Dominant	Chironomidae	Chironomidae	Amphipoda
Mean per sample	74	40	34
Wet Weight Dominant	Oligochaeta	Odonata	Mollusca
Mean No. Taxa	12.8	13.6	14.0
Mean No. Organisms	104	93	107
Mean Wet Weight(mg)	205	232	629*
FISH			
Numerical Dominant	Lepomis gibbosus	L. macrochirous	L. macrochirous
No. Dominants Collected	989	143***	260
Total Number of Taxa	3	7	14

* Significantly different from other means, Oneway ANOVA ($p<0.01$) followed by Student-Neuman Keuls (SNK) *a posteriori* test

** Per Ekman grab sample (0.23 m^2 approx.) *** 1981 and 1982 only probably due to numerous large Mougeotia cells in 1981. Because of large variations in the samples no statistical differences were detected among the lakes in either algal cell numbers or biovolume.

Dinobryon was an important phytoplankton species in all three lakes, it was found in acid lakes by Kelso et al. (1982). The most numerous phytoplankton group, the Cyanobacteria, at Lake Lacawac and Long Pond has been reported as acid sensitive by some re seachers (Almer et al., 1978; Yan and Stokes, 1978) and acid tolerant by others (Kwiatowski and Roff, 1976). In this division sensitivity to acidification appears to be species dependent.

Fewer total number of zooplankton species were collected at Deep Lake than at the other lakes (Table I). The most numerous species in acidified Deep Lake have been reported from acid waters (Ahlstrom, 1943; Sprules, 1975; Confer et al., 1983). Mesocyclops edax was the only species found in all three lakes.

A total of 21,541 BMI were collected. There were no differences among the lakes in mean total BMI numbers or mean number of taxa, but mean BMI wet weight was highest at Long Pond (Table I). Numerically there were more Ephemeroptera, Gastropoda and Pelecypoda at Long Pond, but more Chironomidae at Deep Lake. Amphipoda dominated at Long Pond, were numerous at Lake Lacawac (30.2% of total numbers) and are considered acid sensitive (Okland and Okland, 1980). Only one amphipod was collected at Deep Lake. The occurrence of more acid sensitive Ephemeroptera, Gastropoda and Pelecypoda and a higher wet weight at Long Pond suggests a productive fauna of a circumneutral lake. The dominant BMI family, Chironomidae (Diptera) at Deep Lake and Lake Lacawac often dominates in acid lakes (Wiederholm and Eriksson, 1977; Mossberg and Nyberg, 1979; Raddum and Saether, 1981; Walker et al., 1985).

The fish community at Deep Lake was dominated by stunted Lepomis gibbosus (Table I) in poor condition. Harvey (1982) reports these fish as acid tolerant. Growth rate of L. gibbosus at Deep Lake was lower (mean 5 yr length = 90mm) than at the other lakes (5 yr = 190mm). Lepomis macrochirus at Lake Lacawac and Long Pond were in good condition.

Table II shows the significant water chemistry differences among the lakes over the sampling period. Principal component (PC) analyses of chemical data indicated that 43% of the variance in the data set was correlated with pH, soluble Al, Ca, total alkalinity, Mg, specific conductance and Mn (Figure 1). This component represented a chemical gradient from sites with low pH, alkalinity, specific conductance, Ca and Mg concentrations and high Al and Mn to sites with high pH, alkalinity, specific conductance, Ca and Mg and low Al and Mn. All lakes differed significantly (ANOVA $p<0.01$) with respect to this gradient described by the first PC. Chronological positions of the individual samples along the first PC suggested a shift in all three lakes over the sampling period toward a combination of lower pH, alkalinity, specific conductance, Ca and Mg and higher soluble Al and Mn (Figure 1). These chemical changes have been associated with decreasing pH and increasing acidification (Henrickson, 1980; 1982).

TABLE II - SURFACE WATER CHEMISTRY - 1981 TO 1983
DIFFERENCES AMONG STUDY LAKES - MEANS (STANDARD DEVIATION)

	TOTAL ALKALINITY $\mu eq\ L^{-1}$	CONDUCTIVITY $\mu S\ cm^{-1}$	SOLUBLE ALUMINUM $\mu eq\ L^{-1}$	MAGNESIUM $\mu eq\ L^{-1}$	CALCIUM $\mu eq\ L^{-1}$	SULFATE $\mu eq\ L^{-1}$
			DEEP LAKE			
	(n = 19, 2 surface sites each sampling date)					
	0.0*	24	7.6*	29*	74*	242
SD	(0.0)	(3.6)	(2.6)	(11.5)	(28.8)	(58)
			LAKE LACAWAC			
	(n = 19, 2 surface sites each sampling date)					
	47.4*	29	1.3*	45*	180*	276
SD	(58.6)	(6.6)	(1.3)	(12.4)	(98)	(281)
			LONG POND			
	(n = 21, 2 or 3 surface sites each sampling date)					
	190.0*	43*	0.1*	62*	296*	263
SD	(76.0)	(18.2)	(0.4)	(12.8)	(40.7)	(56)

* Means significantly different, Oneway ANOVA & Student-Neuman-Keuls *a posteriori* Test

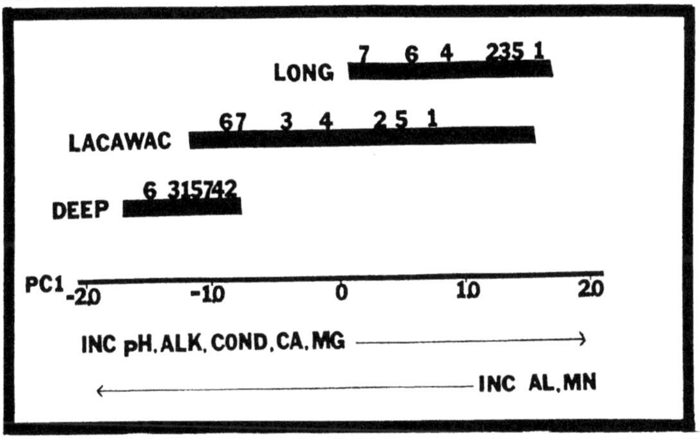

Figure 1. Position of Mean Principal Component Scores for Lake Water Chemistry (by season) along the First Principal Component (PC 1). 1 = Summer 1981, 2 = Autumn 1981, 3 = Spring 1982, 4 = Summer 1982, 5 = Autumn 1982, 6 = Spring 1983, 7 = Summer 1983. Solid bar = range of principal component scores. ALK = total alkalinity, COND = specific conductance.

Table III lists the lakes and drainage basin characteristics. Dominant woody vegetation differed in the three drainage basins. Acidified Deep Lake is the smallest lake (Table III) and is the most severely impacted of the three lakes by acidification. Historical records indicate an alkalinity of 0 to 100 $\mu eq\ L^{-1}$ in 1979 (Powlus,

1980) so the exhaustion of alkalinity is apparently recent. The pH decreased from a high of 7.2 (summer 1981) to a low of 4.0 (spring 1982) and remained below 4.4 for the remainder of the study. Historical records indicate a pH near 6.0. The unusually high pH in summer 1981 may have been caused by high primary productivity (Morgan, 1985) and/or sparse rainfall (Hultberg, 1985). Precipitation in 1981 was approximately 18% below normal. Precipitation obtains little neutralization from the drainage basin, entering the lake relatively unchanged chemically - a "precipitation dominated lake" (Eilers *et al.*, 1983).

TABLE III - DRAINAGE BASIN AND LAKE CHARACTERISTICS

	DEEP	LACAWAC	LONG
Drainage Basin (ha)	16	69	372
Surface Area (ha)	3.0	21	33
Mean Depth (m)	4.1	8.1	4.0
Maximum Depth (m)	6.8	13.5	7.0
Volume (m^3)	122×10^5	168×10^6	126×10^6
Retention Time (Days)	325	1220	200
Altitude (m)	613	480	453
Source	Springs	Springs	2 Streams, Springs
Dominant Woody Vegetation (% importance value*)	*Sassafras albidum*(29.4%) *Acer rubrum* (25.4%) *Quercus prinus*(15.7%)	*Tsuga canadensis*(26.7) *Rhododendron maxima*(14.3) *Fagus grandifolia*(10.2)	*Acer saccharum*(33.6) *Fagus grandifolia*(16.9) *Acer rubrum*(11.8)

* See Dudley (1985) for methodology

Deep Lake, however, is not devoid of plants, invertebrates or fish. The large phytoplankton biomass at Deep Lake suggests that biomass (here expressed as biovolume) may be independent of lake pH (Dillon *et al.*, 1984). Numbers of Chironomidae increased (ANOVA $p<0.01$) over the study period and the most numerous chironomid ,*Procladius*, (Tanypodinae) is evidently tolerant of acid conditions (Singer, 1982; Walker *et al.*, 1985).

Lake Lacawac contains the largest water volume and the longest retention time (Table III). Acid precipitation does not appear to be affecting Lake Lacawac as severely as Deep Lake. An adjacent palustrine wetland and underlying carbonate-cemented shale provide some neutralizing capacity within the drainage basin. However, decreases in both specific conductance (ANOVA $p<0.03$) and alkalinity (ANOVA $p<0.04$) over the sampling period suggest acidification.

Long Pond has a larger drainage basin , thicker soils and a longer ground water flow-time (an order of magnitude longer than the other lakes). Precipitation stays longer in the basin and may be partially neutralized by soil and ground water interactions before entering the lake (Newton and April, 1982). A decrease in specific conductance (ANOVA $p<0.02$) suggests that acidification may be occurring.

4. CONCLUSIONS

Biological differences among the study lakes were 1) absence of acid sensitive organisms and presence of acid tolerant groups in acidified Deep Lake, 2) fewer phytoplankton, zooplankton and fish species in the acidified lake, 3) increase in acid tolerant Chironomidae in the acidified lake, 4) presence of some acid sensitive and some acid tolerant organisms in moderately sensitive Lake Lacawac, and 5) presence of many acid sensitive organisms and a higher BMI wet weight in circumneutral Long Pond.

All three study lakes showed some chemical evidences of acidification though they differed in sensitivities. The extent of acidification was greatest in Deep Lake where there was minimal neutralizing capacity within the drainage basin. The small amount of buffering capacity has evidently been exhausted by the continued influx of acid precipitation. The other lakes had capabilities within their drainage basins to neutralize some acid precipitation so acidification has been slower. The extent and rate of acidification appears to be partially controlled by geological and hydrological characteristics within the drainage basins. If acid precipitation continues more biological and chemical changes associated with acidification may be expected not only in the acidified lake, but also in the moderately sensitive lakes.

ACKNOWLEDGMENTS

Pennsylvania Power and Light Co. provided the financial resources for this research. H. G. Pfeiffer and R. Domermuth gave counsel and guidance. The following people cooperated in this research: D. E. Arnold (fish), P. B. Myers and J. Cotter (geology), R. N. Weisman and P. Sitkowski (hydrology), C. Williamson and S. Daubney (zooplankton) and H. Reinert (statistical analyses). The Lehigh University Center for Social Research and the Computing Center provided computer expertise.

REFERENCES

Ahlstrom E. H.: 1943, *Bull. Amer. Mus. Nat. Hist.* **80**, 411.

Almer, B., Dickson, W., Ekstrom, C., and Hornstrom, E.: 1978, Sulfur pollution and the Aquatic Ecosystem. John Wiley, New York. In J. O. Nriagu (ed.). *Sulfur in the Environment, Part II:* Ecological Impacts. pp 271-311.

American Public Health Association (APHA): 1976, Standard Methods for the Examination of Water and Waste Water. American Public Health Assoc., Washington, D.C. 14th ed., 1149 pp.
Arnold D. E., Light R. W., and Dymond, V. J.: 1980, Probable Effects of Acid Precipitation on Pennsylvania Waters. Technical Report. 600/3-80-012. U.S. Environmental Protection Agency, Corvallis, Oregon. 24 pp.
Arnold, D. E., Light, R. W., and Paul E. A.: 1985, Vulnerability of Selected Lakes and Streams in the Middle Atlantic Region to Acidification: A regional survey. Technical Report. FWS/OBS-80/40.19. U.S. Fish and Wildlife Service, Eastern Energy and Land Use Team. Kearneysville, W.V. 133 pp.
Barker, J. L.: 1978, Characteristics of Pennsylvania Recreational Lakes. Pennsylvania Department of Environmental Resources in Cooperation with U.S. Geological Survey, Harrisburg, PA.226 pp.
Barnes, R. A.: 1975, Chem. Geol. **15**, 177.
Bradt, P. T. and Berg, M. B.: 1983. Proc. PA Acad. Sci. **57**, 190.
Bradt, P. T., Berg, M. B., Barrasso, D. S. and Dudley, J. L.: 1984, The Biological and Chemical Impact of Acid Precipitation on Pocono Mountain Lakes. Technical Report. Pennsylvania Power and Light, Allentown, PA 215 pp.
Confer, J. L., Kaaret, T. and Likens, G. E.: 1983, Can. J. Fish. Aquat. Sci. **40**, 36.
Dillon, P. J., Yan, N. D. and Harvey, H. H.: 1984, CRC Critical Reviews in Environmental Control. **13**, 167.
Dudley, J. L.: 1985, Proc. Penna. Acad. Sci. **59**, 33.
Eilers, J. M., Glass, G. E., Webster, K. E. and Rogalla, J.: 1983, Can. J. Fish. Aquat. Sci. **40**, 1896.
Electric Power Research Institute (EPRI): 1983. Precipitation Data Displays for January 1, 1979-June 30, 1982. Vols. 1 & 2. Technical Report. 103. UAPSP. Vol. 1: 204 pp. Vol. 2: 184 pp.
Electric Power Research Institute (EPRI): 1984. Precipitation Data for January 1, 1983 to June 30, 1983. Palo Alto, CA.
Harvey, H. H.: 1982, Population Responses of Fish in Acidified Waters. R. E. Johnson (Ed.) Acid Rain/Fisheries. Am. Fish. Soc. American Fisheries Society, Bethesda, MD pp. 227-242.
Henriksen, A.: 1982, Susceptibility of Surface Waters to Acidification. In R. E. Johnson (Ed.) Acid Rain/Fisheries. Amer. Fish Soc. Bethesda, MD pp. 103-121.
Henriksen, A.: 1980, Acidification of Freshwaters- a large scale Titration. In D. Drablos & A. Tollan (eds.) Ecological Impact of Acid Precipitation. SNSF Proj. Oslo, Norway pp. 68-74.
Hull, C. H. and N. H. Nie.: 1981, SPSS-UPDATE 7-9. McGraw-Hill Book C., New York. 402 pp.
Hultberg, H.: 1985, Ecol. Bull. **37**, 133.
Kelso, J. R. M., Love, R. J., Lipsit, J. H. and Dermott, R.: 1982, Chemical and Biological Status of Headwater Lakes in the Sault Ste. Marie District, Ontario. In F. D'Itri (Ed.) Acid Precipitation. Ann Arbor Sci. Ann Arbor, MI pp. 165-208.
Kwiatkowski, R. E. and Roff, J. C.: 1976, Can. J. Bot. **54**, 2546.

Lindsley, R. K. Jr., Kohler, M. A. and Paulhus, J. L. H.: 1982, Hydrology for Engineers. Third Ed. McGraw-Hill Book Co., New York. 508 pp.

Lynch, J. A. and Corbett, E. S.: 1983, Atmospheric Deposition: Spatial and Temporal Variation in Pennsylvania 1982. Instit. Land and Water Resources, Penna. State University, University Park, PA 204 pp.

Morgan, M. D.: 1985, Hydrobiologia. **128**, 239.

Mossberg, P. and Nyberg, P.: 1979, Inst. Freshwater Res. Drottningholm. **58**, 77.

Newton, R. and April, R.: 1982, Northeast. Env. Sci. 1, 143.

Nie, N. H., Hull, C. H., Jenkins, J. G., Steinbrenner, K. and Bent, D. H.: 1975, Statistical Package for the Social Sciences. McGraw-Hill Book Co., New York, NY 675 pp.

Office of Technology Assessment (OTA): 1984. Acid Rain and Transported Air Pollutants: Implications for Public Policy. U. S. Congress, OTA-0-204. Washington, D.C. 323 pp.

Okland, J. and Okland, K.'A.: 1980, Acidification threatens Trout Diet. Research in Norway 1980. Harald Lyche and Co., A. S., Norway, pp 21-26.

Powlus, R. D.: 1980, The Morphology, Seasonal Abundance, and Spatial Distribution of Diaptomus Minutus Nauplii in Deep Lake. Master's thesis. East Stroudsburg State University, East Stroudsburg, PA 53 pp.

Raddum, G. G. and Saether, O. E.: 1981, Verh. Int. Verein. Limnol. **21**, 367.

Singer, R.: 1982, Effects of Acidic Precipitation on Benthos. In F. D'Itri (Ed.) Acid Precipitation. Ann Arbor Science. Ann Arbor, MI pp 329-363.

Sprules W.G.: 1975, J. Fish. Res. Bd. Can. **32**, 389.

U.S.-Canada: 1981, U.S.-Canada Memorandum of Intent on Transboundary Air Pollution. Impact Assessment Working Group I, phase II interim working paper. Technical Report. Impact Assessment Working Group I. Washington, DC and Ottawa, Ontario, Canada. 345 pp.

Walker, I. R., Fernando, C. H., and Paterson, C. G.: 1985, Hydrobiologia. **120**, 11.

Wiederholm, T. and Eriksson, L.: 1977, Oikos. **29**, 261.

Yan, N. D. and Stokes, P.: 1978, Environ. Conserv. **5**, 93.

BONE CONCENTRATION OF MANGANESE IN WHITE SUCKER (*CATOSTOMUS COMMERSONI*) FROM ACID, CIRCUMNEUTRAL AND METAL-STRESSED LAKES

H. H. Harvey, G. A. Fraser and J. M. McArdle
Department of Zoology
University of Toronto
Toronto, Ontario M5S 1A1
Canada

ABSTRACT. Four hundred and thirty-five white suckers were collected from 24 lakes in 7 regions (Wawa, La Cloche, Parry Sound, Algonquin, Muskoka, Haliburton, North Bay) of Ontario, concurrent with measurement of pH, alkalinity and 18 metals and ligands. Lakes Summit and Moran have been contaminated by the dumping of mine tailings. One hundred and seventy-eight fish of six additional species were collected from two of the lakes. Vertebral centra were analyzed by neutron activation analysis.

Manganese concentration in lake water was correlated with $[H^+]$, $Y = 0.024 + 5914X$, $r^2 = 0.79$, $p < 0.0001$, $n = 22$. In vertebral centra, [Mn] was similar among white suckers from the same lake, but with large differences between lakes (range of population means, 0.223 to 2.42 $meq.100g^{-1}$ dw). Bone [Mn] was correlated positively with lake [Mn] ($r^2 = 0.55$, $p < 0.0001$) and negatively with DOC ($r^2 = 0.36$, $p < 0.01$); lake [Mn] and DOC were not correlated. In stepwise multiple regression, lake [Mn] + DOC + alkalinity accounted for 87% of the variance in bone [Mn]. Concentration factors, calculated as bone [Mn] ÷ lake [Mn] were 1,000 to 4,000X for many fish populations, with the lowest values from the two tailings-contaminated lakes, and the highest C.F.'s from lakes of lowest [Mn] in water, implying some degree of regulation of Mn uptake by the fish. Bone [Mn] of pelagic and littoral fishes were significantly lower than those found in the benthic-feeding white suckers.

1. INTRODUCTION

Manganese concentration has been reported to be higher in lakes of low pH: west-coast Sweden (Dickson, 1975, 1980); the La Cloche lakes (Harvey *et al.*, 1982); the Wawa lakes (Somers and Harvey, 1984); Sudbury and Muskoka lakes (Harvey *et al.*, 1981), in comparison with the ELA lakes (Beamish, 1976). Most of these acid lakes are known to have acidified within recent decades. This increased lake concentration of Mn in part is due to the increased export of cations from watersheds in response to acid loading, and to the reduction in the rate of transformation of Mn^{2+}

Water, Air and Soil Pollution **30** (1986) 515–521.

to insoluble MnO_x (Harvey et al., 1981). Manganese accumulation in sediments is governed by redox processes, and thus sediment concentration is a function of organic carbon availability and depth in the lake (Davison et al., 1985). In acidified Lake 223, Mn loss to the sediment was reduced (Schindler et al., 1980), and in Gårdsjon, Swedish west coast, Mn concentration in surface sediments is lower than from deeper in the core, implying solubilization of sediment manganese during the recent acidification of the lake (Reuther et al., 1983).

The purpose of this study was to relate [Mn] in fishes to [Mn] in lakes. [Mn] in water is highly pH dependent and the bioavailability of Mn may be affected by dissolved organics. The approach was thus to attempt to correlate [Mn] in vertebral centra of white suckers with lake chemical variables, especially Mn, pH, alkalinity and DOC (dissolved organic carbon).

2. METHODS

Lake water samples were collected at 0.5 m and analyzed for pH, alkalinity (Gran titration) and O_2 in the field, plus Na, K, Cl, SO_4, Si, P, Tot. N, NH_4, NO_3, DOC, DIC, Fe, Mn, Al, As, Be, Ca, Cd, Co, Cr, Cu, Mg, Mo, Ni, Pb, V, Zn, in the laboratory (OME, 1980). An attempt was made to capture 20 white suckers from each lake (Table I). Vertebrae #9-13 were removed, cleaned and dried to constant weight, and Mn determined by neutron activation analysis, SLOWPOKE-1 reactor, University of Toronto (Fraser and Harvey, 1982). The two tailings contaminated lakes were omitted from calculations where n = 22.

3. RESULTS

Manganese concentration in lake water varied directly with $[H^+]$ (Figure 1); the line of best fit, $Y = 0.024 + 5{,}914X$ showed an r^2 of 0.79, $p < 0.0001$, $n = 22$.

[Mn] in the vertebral centra of white suckers was very similar among suckers from the same lake, as shown by the standard error on the mean (Table I). The means of the fish samples differed greatly between lakes, ranging from 0.223 to 2.42 meq.100 g^{-1} dw. In a between-lake analysis, centrum [Mn] was correlated strongly with lake [Mn] (Figure 2), $r^2 = 0.55$, $p < 0.0001$.

Bone [Mn] was correlated negatively with DOC, $r^2 = 0.36$, $p < 0.01$ (Figure 3). Internal correlations (r) were calculated among the principal chemical variables (Table II). Lake [Mn] and [DOC] were not correlated and thus could be treated as independent variables. Centrum [Mn] was correlated negatively with lake pH, $r = -0.53$, $p < 0.01$, but lake pH and [Mn] were correlated strongly (-0.78) and so could not be considered as independent variables. No other chemical variables were correlated significantly with bone [Mn].

TABLE I

Lake		Bone Mn meq.$100g^{-1}$ dw Mean	SE	n	Water pH	Alkalinity $\mu eq.L^{-1}$	Mn $mg.L^{-1}$	Ca $mg.L^{-1}$	DOC $mg.L^{-1}$
George	*	2.42	0.15	20	4.70	-3.20	0.148	2.7	1.4
Round		2.157	0.114	8	4.90	0.0	0.100	1.7	0.9
Summit		1.631	0.084	15	7.30	397	0.400	21.0	5.6
Wallace		1.388	0.070	20	7.63	735	0.034	18.0	4.9
Little Coon		0.997	0.065	10	5.30	9.5	0.075	2.0	3.0
East		0.757	0.038	32	6.49	60.9	0.010	1.74	3.8
Black Trout		0.752	0.047	20	5.99	144	0.018	5.6	4.0
King	*	0.70	0.05	16	5.08	-3.5	0.093	2.8	5.6
Blue Chalk	*	0.59	0.05	27	6.59	64.1	0.024	2.9	1.8
Moran		0.570	0.036	11	7.30	639	0.330	43.0	1.9
Bigwind		0.498	0.045	24	6.25	5.62	0.024	3.0	3.7
Fogal		0.449	0.036	10	5.06	-4.4	0.040	1.7	6.9
Harp	*	0.39	0.02	19	6.45	116.0	0.028	3.4	4.5
Oldfield		0.382	0.018	16	5.45	2.6	0.058	2.2	6.2
Red Chalk		0.364	0.047	14	6.32	6.60	0.029	2.8	3.0
Dickie		0.360	0.022	18	5.70	4.64	0.048	2.6	5.4
Brush		0.359	0.049	20	6.55	31.2	0.020	2.6	2.5
Chub		0.358	0.016	20	5.76	4.63	0.045	2.9	4.8
Opeongo	*	0.35	0.02	20	6.37	69.0	0.010	3.5	4.7
Gullfeather		0.345	0.016	20	5.88	5.45	0.048	2.8	6.6
Crosson	*	0.32	0.020	17	5.36	0.30	0.028	2.5	4.0
Lovell		0.275	0.036	20	7.49	427	0.035	11.0	7.9
Bentshoe		0.244	0.015	20	5.64	14	0.029	2.1	4.3
McDonald		0.223	0.012	18	5.2	10.9	0.040	2.2	3.8

* Data from Fraser and Harvey (1982).

TABLE II

pH	1.0				
alkalinity	0.74*	1.0			
Mn	-0.64*	-0.21*	1.0		
Ca	0.70	0.99*	-0.15	1.0	
DOC	0.16	0.24	-0.27	0.25	1.0
	pH	alkalinity	Mn	Ca	DOC

* $p < 0.01$

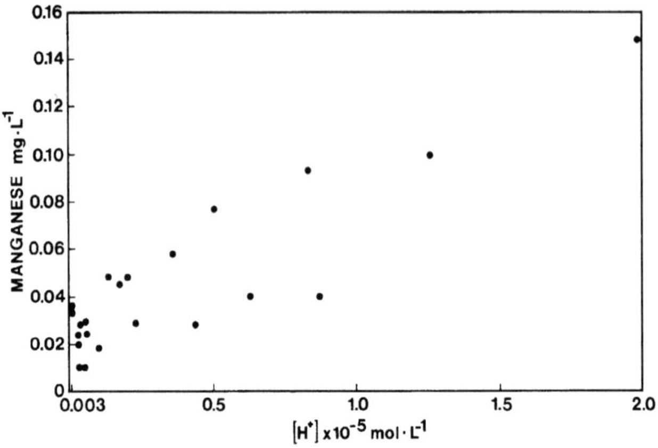

Figure 1. [Mn] in lake water in relation to $[H^+]$. n = 22.

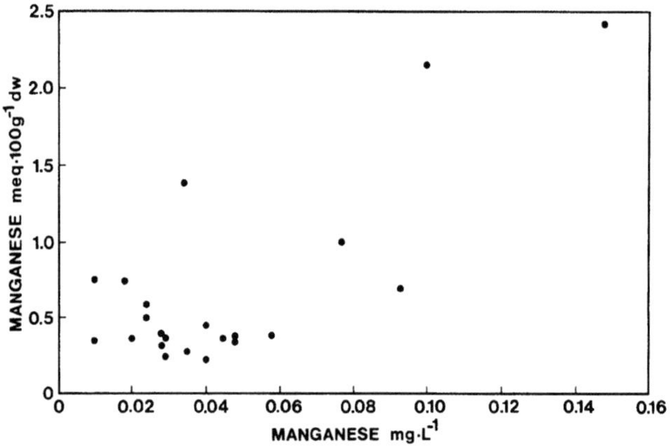

Figure 2. [Mn] in white sucker vertebral centra vs [Mn] in lake water. n = 22.

In stepwise multiple regression (SMR), 66% of the variance in bone [Mn] between lakes was explained by lake [Mn], and the inclusion of DOC raised this to 84%. Both variables were significant ($p < 0.001$). The inclusion of lake alkalinity in SMR accounted for an additional 3% of the variance to 87%, but at a slightly reduced significance ($p < 0.005$). Subsequent additions of lake variables did not change the significance of the regression.

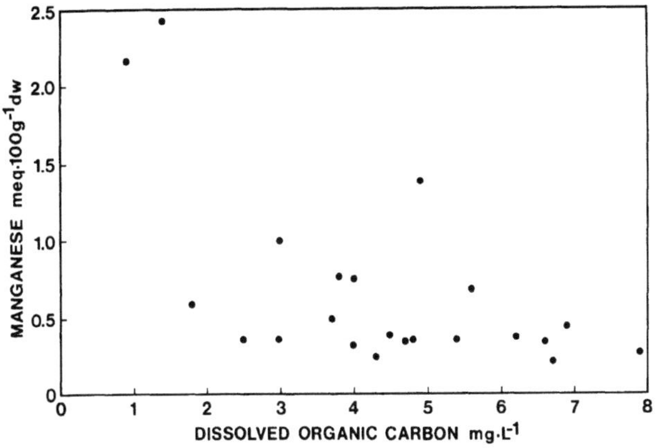

Figure 3. [Mn] in white sucker vertebral centra vs [DOC] in lake water. n = 22.

The comparison of bone [Mn] between species for lakes Round and George showed some important differences. White sucker [Mn] were similar, but yellow perch (Perca flavescens) and pumpkinseed (Lepomis gibbosus) from George Lake had twice the bone [Mn] of Round Lake fish, and brown bullhead (Ictalurus nebulosus) from Round Lake showed the higher [Mn] (Table III).

TABLE III

Species	George Lake Mean	SE	n	Round Lake Mean	SE	n
White sucker	2.42	0.15	20	2.157	0.114	8
Pumpkinseed	1.445	0.115	23	0.653	0.040	22
Yellow perch	1.416	0.169	29	0.637	0.053	19
Brown bullhead	1.392	0.134	12	1.850	0.062	20
Rock bass	1.292	0.205	22	-	-	-
Lake trout	0.411	0.092	7	-	-	-
Lake herring	0.269	0.021	23	0.436	-	1

4. DISCUSSION

Direct uptake of manganese from water has been determined using ^{54}Mn and ^{55}Mn isotopes, with yellow perch reaching equilibrium state in 8.3 days (Kearns and Vetter, 1982) and the minnow (Phoxinus laevis) 7 days (Nicholas and Kirchman, 1974). Based on such a short biological half-life, several authors have postulated a regulatory system for control of

[Mn] in fishes (Kneip and Laver, 1973; Kwasnik et al., 1978). Fish show limited ability to bioconcentrate Mn. Concentration factors (ratio of ^{54}Mn in the animal to the concentration in medium water) varied from 11 for yellow perch (Kearns and Vetter, 1982), 22.6 for fathead minnow (*Pimephales promelas*) (Kwasnik et al., 1978) to 50 for *P. laevis* (Nicholas and Kirchman, 1974).

In this study, concentration factors (CF) were calculated from Mn meq.kg^{-1} in bone/Mn meq.L^{-1} in water. The mean for 22 white sucker populations was 5,039 ± 899 SE. The lowest CF's were in fishes from the tailings-contaminated lakes Moran (475X) and Summit (1,117X), where [Mn] in the water was 10X the mean (0.045 mg.L^{-1}) of the other 22 lakes. Three of the four white sucker populations showing the highest CF's, East Lake 18,925X, Wallace Lake 11,567X, Black Trout 10,743X, Opeongo 8,750X had the lowest [Mn] in lake water (Table I). These extreme responses would support the hypothesis that [Mn] in fishes is regulated to some extent. Thus for the 24 lakes, water [Mn] ranged over 40X, lowest to highest, but bone [Mn] was 10.9X. This lesser range of bone [Mn] does not obviate its use in comparisons of fish populations either within lakes over time, or between lakes.

In both simple correlation and in stepwise multiple regression, DOC was correlated negatively with vertebral centrum [Mn], and was second in importance only to [Mn] in lake water. Stumm and Morgan (1970), however, cautioned against such environmental correlations being interpreted as direct evidence of Mn complexation to organics. Manganese is known to complex weakly with dissolved organics, especially at low pH. In the hypothetical metal-ligand model of Stumm and Bilinski (1973) only 7% of Mn was complexed to organics, and the model of Wilson (1978) predicted humic-Mn complexes of 3% at pH 7.5. There was little association of Mn with organic matter (Beck et al., 1974), 5.1% of Mn was complexed to humics (Mantoura et al., 1978) and for high-organic Okefenokee water (pH 3.5, DOC 35 ppm) only 30% of Mn was likely complexed to dissolved organics (Giesy and Briese, 1977). Thus there is no basis for concluding that organic complexation of Mn was reducing the availability of the metal to the white suckers despite the strong negative correlation between DOC in lake water and [Mn] in bone.

The comparison of [Mn] in vertebral centra of all large fish species in lakes George and Round (Table III) revealed significant differences between species in the same lake and within species between lakes. From this it is concluded that Mn CF's will need to be determined from a larger number of fish populations before comparisons can be made of CF's in pelagic and littoral species with those of the benthic-feeding white sucker.

Investigation of [Mn] in hard tissues of fishes is continuing: (1) with respect to Mn input, cycling and biological uptake in lakes, and (2) by use of archival material to measure changes in [Mn] in the same populations over time. Because [Mn] in water is highly pH dependent, acidification of surface waters can be expected to increase [Mn] in hard tissues of fishes.

5. REFERENCES

Beamish, R.J.: 1976, Water Air Soil Pollut. **6**, 501.
Beck, K.C., Reuther, J.H., and Purdue, E.M.: 1974, Geochim. Cosmochim. Acta **38**, 341.
Bendell-Young, L.I., and Harvey, H.H.: 1985, Hydrobiologia **133**, 117.
Davison, W., Hilton, J., Lishman, J.P., and Pennington, W.: 1985, Environ. Sci. Technol. **19**, 356.
Dickson, W.: 1975, Rep. Inst. Freshw. Res., Drottningholm **54**, 8.
Dickson, W.: 1980, Properties of acidified waters. In: Proc. Int. Conf. Ecological Impact Acid Precipitation, Norway, 1980, D. Drabløs and A. Tollan (eds.), pp. 75-83.
Fraser, G.A., and Harvey, H.H.: 1982, Can. J. Fish. Aquat. Sci. **39**, 1289.
Giesy, J.P., Jr., and Briese, L.A.: 1977, Chem. Geol. **20**, 109.
Harvey, H.H., Pierce, R.C., Dillon, P.J., Kramer, J.R., and Whelpdale, D.M. 1981, Acidification in the Canadian aquatic environment, NRCC No 18745, National Research Council of Canada, Ottawa. 369 pp.
Harvey, H.H., Dillon, P.J., Fraser, G.A., Somers, K.M., Fraser, P.E., and Lee, C.: 1982, Elevated metals and enhanced metal uptake in fishes in acid-stressed waters, Extended Abstract, Am. Chem. Soc., Div. Environ. Chem., pp. 438-441.
Kearns, P.K., and Vetter, R.J.: 1982, Hydrobiologia **88**, 277.
Kneip, T.J., and Laver, G.J.: 1973, Trace metal concentration factors in aquatic ecosystems. In: Progress in Analytical Chemistry, Vol. 5, Plenum Press, New York.
Kwasnik, G.M., Vetter, R.J., and Atchison, G.J.: 1978, Hydrobiologia **59**, 181.
Mantoura, R.F.C., Dickson, A., and Riley, J.P.: 1978, Estuarine Coastal Mar. Sci. **6**, 387.
Nicolas, C., and Kirchman, R.: 1974, Contamination of freshwater fish by [Mn]-54 and Co-60, EUR-5176E: 36-52, Centre D'Etude De L'Energie Nuclearie, Mol, Belgium.
O.M.E.: 1980, Quality control report, Ontario Ministry of the Environment, Water Quality Section, Laboratory Services Branch, Rexdale, Ontario.
Reuther, R., Andersson, J.B., and Forstner, U.: 1983, Effect of pH changes on the distribution of chemical forms of heavy metals in sediment cores from Swedish lakes, Int. Conf., Heavy Metals in the Environment, Heidelberg, Vol. 2, pp. 868-871.
Schindler, D.W., Hesslein, R.H., and Wagemann, R. 1980, Can. J. Fish. Aquat. Sci. **37**, 373.
Somers, K.M., and Harvey, H.H.: 1984, Can. J. Fish. Aquat. Sci. **41**, 20.
Stumm, W., and Bilinski, H.: 1973, Trace metals in natural waters; difficulties of interpretation arising from our ignorance on their speciation. In: Advances in Water Pollution Research, Jenkins, S.H. (ed.), Pergamon Press, Oxford, pp. 39-49.
Stumm, W.E., and Morgan, J.J.: 1970, Aquatic Chemistry, Wiley-Interscience, New York. 583 pp.
Wilson, D.E.: 1978, Limnol. Oceanogr. **23**, 499.

ANNOUNCEMENTS

Indoor Air '87

The 4th International Conference on Indoor Air Quality and Climate (INDOOR AIR '87), which will take place in Berlin (West), 17–21 August 1987, is intended to serve this purpose. INDOOR AIR '87, organized by the Institute for Water, Soil and Air Hygiene, will create a forum for a fruitful exchange of scientific information concerning the measurement and characterisation of factors affecting indoor air quality, health effects associated with the indoor environment, the reduction and prevention of indoor air pollution and other problems related to the well-being of humans in the indoor environment. For additional information contact:

INDOOR AIR '87
c/o CPO Hanser Service GmbH
Schaumburgallee 12
D – 1000 Berlin 19 / Germany

Second International Conference on New Frontiers for Hazardous Waste Management

Gaithersburg, Maryland – NUS Corporation announced today that papers and technical posters are being accepted for the Second International Conference on New Frontiers for Hazardous Waste Management. The conference will be held September 27–30, 1987 in Pittsburgh, Pennsylvania, U.S.A. The purpose of the conference is to examine the current state of technology for the disposal of hazardous waste. Emphasis will be on original, unpublished papers that summarize important new technological developments or concepts with broad international application.

Abstracts should be submitted to Ms. Lynne Caster, Conference Co-Director, NUS Corporation, Park West Two, Cliff Mine Road, Pittsburg, PA 15275, U.S.A.

Papers Needed for 11th Symposium on Aquatic Toxicology & Hazard Assessment

Philadelphia, PA – Papers are needed for the 11th Symposium on Aquatic Toxicology and Hazard Assessment, to be held 10–12 May, 1987 in Cincinnati, Ohio. The symposium is sponsored by ASTM Committee E-47 on Biological Effects and Environmental Fate and its Subcommittee E47.01 on Aquatic Toxicology.

The science of aquatic toxicology and hazard assessment deals with transport, transformation, and effects of chemicals in aquatic ecosystems. Relevant tools of the science include laboratory and field experiments, field monitoring, and statistical and mathematical models. The eleventh symposium will address all of these activities as they contribute to decision making for environmental releases of chemicals.

For more information contact Glenn Suter, Symposium Chairman, Environmental Sciences Division, Oak Ridge National Laboratory, P.O. Box X, Oak Ridge, TN 37831, U.S.A., 615/574-7306 (FTS 624-7306); or Michael Lewis, Co-Chairman, Environmental Safety Department, The Procter and Gamble Company, Ivorydale Technical Center, Cincinnati, OH 45217, U.S.A., 513/627-6010.

Vulnerability of Soil and Groundwater to Pollutants

The National Institute of Public Health and Environmental Hygiene of The Netherlands is organising an international conference on the vulnerability of soil and groundwater to pollutants (VSGP) March 30–April 3, 1987. Contact:

VSGP '87 c/o KIvI
P.O. Box 30424
2500 GK The Hague
The Netherlands
Telephone: (0)70-919900

Technical Seminar on Chemical Spills

The Fourth "Technical Seminar on Chemical Spills", sponsored by Environment Canada, will be held at the Sheraton Centre, Toronto, Ontario, February 10th, 11th, and 12th, 1987.

This seminar will provide a forum for the exchange of technical information, regarding chemical spill control, among scientists, engineers, technically oriented operatives and technological innovations in the field of chemical spill control.

For further information contact:

Michel Cloutier
Program Administrator
Technology Development and Technical Services Branch
Environment Canada, Conservation and Protection
Ottawa, Ontario
K1A 1C8
Telephone: (819)953-1189

Radioactivity and Oceanography

This symposium will be held at Cherbourg (France), June 2–5, 1987, on the use of natural and artificial radionuclides for oceanographic studies. For addition information write to:

SFEN – 'Radioactivity and Oceanography'
48 rue de la Procession, F-75724, Paris, France

International Conference: Acid Rain

Athens, Greece; 30th August – 4th September, 1987

The conference will address physico-chemical, biological and technological aspects of acid rain in its widest sense.

Major themes are as follows:

1. Sources of Pollutants (including Natural Sources) and Atmospheric Chemistry
2. Atmospheric Transport Processes and Modelling
3. Deposition Processes
4. Catchment and Soil Processes
5. Effects – Aquatic Ecosystems
6. Effects – Gaseous Pollutants and Terrestrial Plants
7. Effects – Acid Precipitation, Terrestrial Plants and Soils
8. Effects – Buildings and other Materials
9. Aerosols and Visibility
10. The Role of Acid Rain in Forest Decline
11. Direct and Indirect Effects on Human Health
12. Control Strategies – Technological and Economic Implications

Abstracts (200–300 words; 6 copies) of submitted papers should be received by 1st March 1987.

Further details available from:

Dr. Roy M. Harrison
Institute of Aerosol Science
Department of Chemistry,
University of Essex,
Colchester CO4 3SQ, U.K.

42nd Annual Purdue Industrial Waste Conference West Lafayette, Indiana, May 12–14, 1987

Address:
Purdue Industrial Waste Conference
School of Civil Engineering
Purdue University
West Lafayette, IN 47907

For information contact:
Cynthia S. Dalton
Conference Secretary
(317)-494-2194

Heavy Metals in the Environment

The 6th International Heavy Metals in the Environment Conference will be held in New Orleans on 15–18 September, 1987.

This year papers have been invited for submission on the main themes of environmental pathways, fate and effects, exposure and health effects, disposal and control strategies and legislation.

Contact:
David Cowan
Technical Media Services
62 Kelvingrove Street
Glasgow
G3 7SA

Telephone: 041 332 6636
Telefax: 041 221 5318

Chemical Accidents

This event will be held in Rome on 7–10 July, 1987.

For the first time, on a world scale, industrial and academic concerns will gather to examine the subject of Chemical Accidents as outlined by the themes of the conference. Particularly highlighting accidents occurring at stationary facilities (either confined within the area or spreading out to the general environment), during chemical transportation, misuse, waste disposal, or involving food chain contamination.

In addition to these themes, attention will also focus on the contingency planning for radiological accidents, such as Chernobyl, as they require the same responsiveness and manoeuvring of the emergency services.

Contact:
David Cowan
Technical Media Services
62 Kelvingrove Street
Glasgow
G3 7SA

Telephone: 041 332 6636
Telefax: 041 221 5318

CALL FOR PAPERS

WATER RESOURCES MANAGEMENT

A new journal: first issue Spring 1987

D. Reidel Publishing Company are pleased to announce that in the Spring of 1987 they will commence publication of *Water Resources Management*, for the European Committee of Water Resources Management.

Water Resources Management aims to become an international multidisciplinary forum for the presentation of original contributions and the exchange of knowledge and experiences on water resources management. Contributions on the following areas are welcome:

(i) Water resources assessment, development, conservation, and control with emphasis on the policies and strategies);

(ii) Planning and design of water resource systems, and

(iii) Operation, maintenance, and administration of water resource systems.

As closely related to the above, the following areas are also of interest to the Journal: hydrological and technological aspects; water demand and consumption; applied surface and groundwater hydrology; water management techniques; simulation and modelling of water resource systems; forecasting and control of quantity and quality of water; economic and social aspects of water use; legislation and water resources protection.

The Journal is supported scientifically by the European Committee for Water Resources Management, a scientific and technical committee working within the framework of the International Association of Hydrological Sciences (IAHS). Therefore, the Journal is expected to play a major role in achieving the purposes of the committee which are to promote the application of scientific knowledge to practical engineering activities, and enhance European and international cooperation of water resources management.

Research advances in all aspects of Water Resources Management may be presented in the form of *research papers* which should be as concise as possible with a maximum length of 8000 words or equivalent.

Technical communications (up to 5000 words) are intended for describing applications of new models and technologies as well as case studies of particular scientific and engineering interest.

Research notes should be very brief letters (up to 2000 words) which report advance information on current research in progress prior to a more complete contribution subsequently, or short papers on limited facets of an investigation. Occasionally, the Journal will publish *extended reviews* by well known scientists. Extended reviews are invited papers which should not exceed 16000 words or equivalent.

Apart from the above, the Journal welcomes *discussions* on the published articles, *book reviews, conference reports*, and *announcements* and *new book announcements*. Discussion on papers will be accepted for eight months following publication and should be as brief as possible.

Manuscripts intended for publication should be written in English and should be submitted in triplicate to the Editor at the following address:

Water Resources Management
P.O. Box 60082
15310 Agia Paraskevi
Athens
Greece.

Detailed Information for Authors is available from The Journals Department, D. Reidel Publishing Company, P.O. Box 17, 3300 AA Dordrecht, Holland, (Telephone: (31)78-334208. Telex: 29245 reid nl.)

Subscription information is available from the same address, also sample copies.

COMPOSITION OF THE BENTHOS IN RELATION TO pH IN THE LaCLOCHE LAKES

H. H. Harvey and J. M. McArdle
Department of Zoology
University of Toronto
Toronto, Ontario M5S 1A1
Canada

ABSTRACT. In the LaCloche Mountains of Ontario, 11 lakes spanning the pH range 4.8 to 6.7 were sampled for benthic organisms. Sixty sediment cores were collected from each lake, organisms >0.5 mm were separated and 25 taxa were identified. The number of taxa present was correlated negatively with $[H^+]$: $Y = 23.2 - 4.27 \times 10^5 X$, $r^2 = 0.61$, $p < 0.005$. The density of all organisms was not correlated with lake pH. Six taxa: gastropods, pelecypods, daphnids, ephemeropterans, amphipods and ceratopogonids, either were reduced in abundance or absent from the more acid lakes. Bosminids and acari were more abundant in lakes of pH 5.0 to 5.5. The densities of the remaining taxa, notably chironomids and oligochaetes, appeared to be independent of pH over the range 4.4 to 6.7. These results are discussed in relation to the hypotheses which have been advanced to explain the differences observed in benthos distribution between lakes in Europe and in North America.

1. INTRODUCTION

The purpose of this study was to determine the presence-absence and density of 25 taxa of benthic animals in order to test current hypotheses on the distribution of lake benthos in relation to pH. Results are presented for presence-absence and density of benthic organisms; standing crop estimates will be reported in a subsequent larger publication.

There are several reports in the literature of organism presence in or absence from the sediments of acid lakes. Typically such reports are based on an examination of 1 to 5 sediment samples per lake. The adequacy of such sampling is unclear. Allison and Harvey (1980) quantified the benthos of a lake by means of stratified sampling totalling 147 cores, and based on information gain with each additional core, calculated the optimum number of cores to be 60. For this reason, 60 sediment cores were examined from each of the 11 lakes.

The literature review in this paper was confined to comparison of field survey results and the testing of such hypotheses as have been advanced to explain the observed differences in lake benthos with respect to pH. Thorough reviews have been published recently of

Water, Air, and Soil Pollution **30** (1986) 529–536.

laboratory pH tolerance (Nilsson et al., 1984), field experiments (Ravera, 1986) and interactions (Økland and Økland, 1986).

2. METHODS

Benthos was sampled by divers collecting 10 sediment cores and by counting the larger macrobenthos in a 1 m wide band along each of 6 transect lines laid on the lake bottom (Allison and Harvey, 1981). Organisms were separated both by flotation and by passing the sediment through a 0.5 mm sieve. Organisms were identified into 25 taxa and counted. The whole-lake mean density of each taxon was calculated from: total taxon count ÷ total area of cores m^{-2}.

3. RESULTS AND DISCUSSION

The number of taxa in each of the 11 lakes was regressed on $[H^+]$ (Figure 1) as $Y=23.2-4.27 \times 10^5 X$, $r^2=0.61$, $p<0.005$. The regression was not significant if the most acid lake was deleted from the calculation. The 6 lakes of pH 5.9 to 6.7 had a mean of 22.8±1.5 taxa/lake, the 4 lakes of pH 5.0 to 5.4 had a mean of 20.5±1.3 taxa, whereas Lumsden Lake at pH 4.4 yielded only 14 taxa. Thus, Lumsden Lake lacked 10 taxa and the lakes of pH 5.0 to 5.4 lacked 3 to 5 taxa (Table I). These results were comparable to those of Mossberg and Nyberg (1979) for 7 acid Swedish lakes, where taxa declined from 31 at pH 5 to 19 at pH 4.2. In contrast, Crisman et al. (1980) found the number of taxa and mean abundance were not significantly different between 20 lakes classified into 5 pH intervals. For all taxa, the density of organisms varied little between lakes, 1 to 4 x $10^4.m^{-2}$ (Table I), and was not related to lake pH.

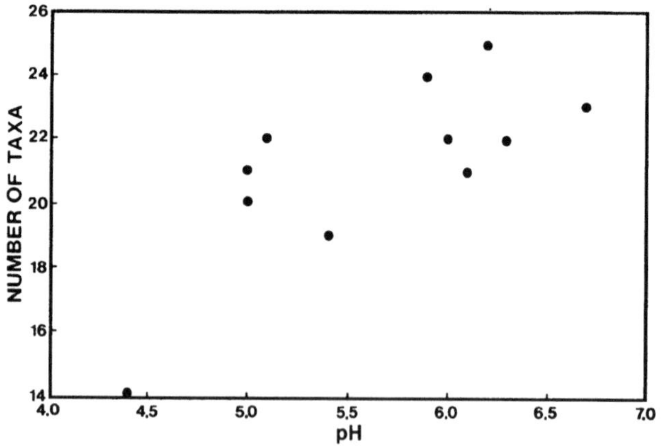

Figure 1. Number of taxa vs lake pH for 11 LaCloche lakes.

TABLE I

Lake	Lumsden	A.Y. Jackson	Johnnie	Carlyle	George	Kakakise	Ishmael	Florence	Evangeline	Hele	ALexander
pH	4.4	5.0	5.0	5.1	5.4	5.9	6.0	6.1	6.2	6.3	6.7
Oligochaeta	538	1,061	290	442	285	738	340	156	484	364	393
Nemata	57	403	26	297	1,105	202	99	0	52	349	15
Bosminidae	0	276	79	187	155	105	57	0	17	0	15
Chydoridae	722	1,550	290	1,163	1,606	720	2,604	835	2,366	1,674	990
Daphniidae	0	21	9	17	0	615	226	42	207	102	58
Holopedidae	0	0	53	76	0	9	14	14	17	0	0
Macrothricidae	1,019	21	580	526	933	518	892	495	889	73	2,096
Sididae	42	42	114	297	86	272	665	127	354	422	233
Ostracoda	127	2,463	79	289	52	1,414	708	439	2,168	320	422
Calanoida	0	211	228	2,395	630	1,230	42	85	173	87	15
Cyclopoida	920	7,473	1,142	3,380	4,706	11,279	3,085	2,109	5,725	1,572	3,101
Harpacticoida	42	85	53	127	639	9	297	28	345	87	44
Amphipoda	0	105	0	0	35	343	1,345	410	138	277	553
Elmidae	0	0	0	8	0	35	0	0	0	0	87
Ceratopogonidae	57	275	360	603	501	650	948	693	354	1,572	189
Chaoborinae	85	594	35	238	26	88	0	0	458	0	15
Chironomidae	5,944	15,900	10,954	18,231	6,606	12,078	17,945	7,260	14,378	14,790	12,300
Ephemeroptera	0	0	0	0	60	9	57	42	147	189	29
Neuroptera	57	403	26	297	1,105	202	99	0	52	349	15
Odonata	0	0	0	25	0	18	0	57	17	29	15
Trichoptera	28	658	18	17	9	53	71	57	52	73	15
Acari	99	8,322	791	5,562	2,323	193	212	42	311	146	116
Pelecypoda	0	42	18	93	95	255	2,321	382	475	2,722	73
Gastropoda	0	0	0	0	0	44	127	113	52	218	0
Miscellaneous	269	403	176	17	328	290	0	28	0	44	15

$Numbers.m^{-2}$ whole lake area.

No gastropods were found in lakes below pH 5.8; and whereas pelecypods were sparse at low pH, they were absent from only the two most-acid lakes (Figure 2). These results are consistent with reports from Europe where gastropods and pelecypods have been reported as sparse (J. Økland, 1969, pH 4.4-4.6; Grahn et al., 1974, pH 4.4-5.4; Raddum, 1980), or absent (Hagen and Langeland, 1973, pH 5.7-6.2; Wiederholm and Eriksson, 1977, pH 3.7-4.6). Amphipods were absent from 3 acid lakes (Figure 2), and daphnids were sparse in or absent from the 5 most acid lakes. In a Swedish lake, Lepidurus arcticus was not found in lakes of pH <6.1 (Borgstrom et al., 1976), Gammarus lacustris was never found in (Norwegian) lakes of pH below 6.0 (Økland, 1969), and amphipods were absent from (lake) Flovatn, pH 5.7-6.2 (Hagen and Langeland, 1973). North American malacostracan, Mysis, was reported as acid sensitive (Nero and Schindler, 1983). Bosminids were more abundant in most lakes of low pH (Figure 2), but none were taken from the cores of Lumsden Lake, the most acid, and the only fish-free lake. The densities of the remaining crustaceans: chydorids, Holopedium, macrothricids, sidids, calanoids, cyclopoids, harpacticoids, were unrelated to pH over the pH range of the study lakes. The 10-fold greater abundance in the cladoceran Latona, reported by Collins et al. (1981) in three acidified lakes, was not observed. The group "miscellaneous" which was predominantly isopods, had its greatest abundance in 5 of the lakes of pH <6.0 (Table I).

The lake benthos was chironomid dominated but density was not pH related (Figure 2). Similarly Chaoborous density was not correlated with lake pH. The third dipteran group, the Ceratopogonidae, showed lower densities in acid lake (Figure 2). Mossberg and Nyberg (1979) quantified the benthos in seven acid lakes and reported the increasing abundance of chironomids (86% of fauna in the most acid lake), at the expense of Phaenopsectra with decreasing pH. Ephemeropterans were absent from the 3 most-acid lakes (Figure 2), consistent with the literature: absent at pH 5.7-6.2 (Hagen and Langeland, 1973), less species at pH 4-4.5 vs >6.5 (Hendrey et al., 1976), and reduced abundance (Grahn et al., 1974). Trichopterans, neuropterans, plecopterans, and odonates showed no obvious relationship between density and lake pH (Table I), such as has been reported (Hagen and Langeland, 1973; Grahn et al., 1974; Hendrey et al., 1976).

Oligochaete densities were not pH dependent (Figure 2) which agrees with the report of Raddum (1980) that oligochaete abundances were similar between acid and non-acid lakes within acidified regions of Norway. Wiederholm and Eriksson (1977) in contrast, found oligochaetes sparse in lake Trestickeln, pH 3.9-4.6, a lake more acid than any in this study. In the 7 acid Florida lakes, oligochaetes were replaced by chironomids as the dominant benthic organisms (Crisman et al., 1980).

Nematode densities were independent of pH (Table I) in contrast to Leivestad et al. (1976) who noted high numbers of nematodes in the littoral of acid lakes.

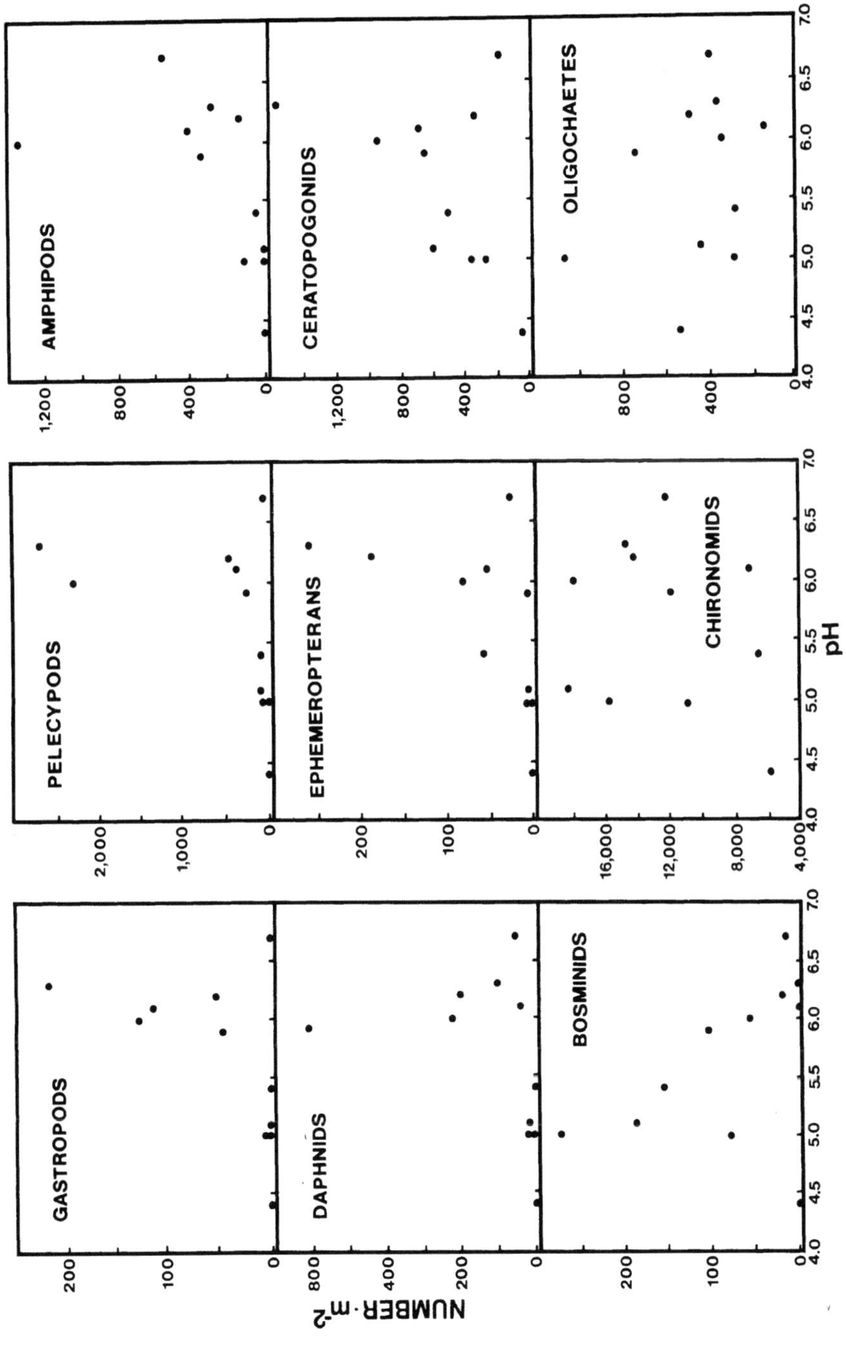

Figure 2. Nine taxa illustrating reduced, increased and independent densities in relation to lake pH.

Water mites were much more abundant in 4 of the 5 most-acid lakes (Figure 2), possibly in response to reduced fish predation. This is consistent with the Norwegian experience (Leivestad et al., 1976) of high densities of acari inshore, but in contrast to the report of Collins et al. (1981) that water mite densities were not elevated in three acidified lakes, and the report of Eriksson et al. (1980) that abundance was decreased.

The LaCloche lakes are ultra oligotrophic: lakes George and Carlyle both had total P values of $\bar{x}$=3 and range 1 to 5 $\mu g.L^{-1}$ with chlorophyll-a values of 0.3 and 0.6 $\mu g.L^{-1}$, respectively (Zimmerman et al., 1983). These very low concentrations are in contrast to the 27 $\mu g.L^{-1}$ and 9 $\mu g.L^{-1}$ total P plus 18 $\mu g.L^{-1}$ and 1.3 $\mu g.L^{-1}$ chlorophyll-a in lakes Drain (pH 4.0) and Springfield (pH 3.6) reported by Kerekes et al. (1984). Thus nutrient status in the LaCloche lakes can be expected to be less important in benthos diversity than it was in acid-eutrophic Springfield Lake.

Several North American reports show that the benthic fauna was not diminished in lakes of low pH or that the reduction was not significant (Roff and Kwiatkowski, 1977; Hendry et al., 1980; Crisman et al., 1980; Collins et al., 1981; Kelso et al., 1982; Dermott, 1985). Four hypotheses have been advanced to explain the discrepancy in biological response, European vs North American lakes (Økland and Økland, 1986). (1) The decrease in low-pH tolerant species is compensated for by an increase in acid-tolerant insects which are normally susceptible to fish predation. This might explain constancy of biomass, but not major taxa. Nor is it clear how this biological response would generate the apparent differences between European and North American lakes. (2) Acid lakes have a higher proportion of organisms living in the sediments (vs pelagic). Again there is no evidence that this is more true of North American lakes. (3) The concentration of Al is lower in North American lakes. The data presented by Wright et al. (1980) do not support this: Adirondack lakes showed much higher Al concentrations than lakes in Norway, Sweden and Scotland. Regional comparisons of Al concentrations may be possible, but international ones are unrealistic. (4) North American benthos has a longer history of adaptation to extreme environments. This requires the demonstration that the high $[H^+]$ was one of those extremes.

In summary, the analysis of the benthos from the 11 LaCloche lakes revealed absence or sparseness of taxa at low pH largely consistent with European reports. It may be premature to speculate as to the cause of differences in benthos response between North American and European lakes.

4. REFERENCES

Allison, W.R., and Harvey, H.H.: 1980, unpublished data.

Allison, W.R., and Harvey, H.H.: 1981, Methods for assessing the benthos of acidifying lakes. In: Effects of Acidic Precipitation on Benthos, R. Singer (ed.), N. Am. Benthol. Soc. 154 pp.

Beamish, R.J., and Harvey, H.H.: 1972, J. Fish. Res. Board Can. **29**, 1131.

Borgström, R., Brittain, J., and Lillehammer, A.: 1976, Evertebrater og surt vann. Oversikt over innsamligings-lokaliteter, Sur Nedbors Virkning Pa Skog og Fisk SNSF IR 21/76. 33 pp.

Collins, N.C., Zimmerman, A.P., and Knoechel, R.: 1981, Comparisons of benthic infauna and epifauna biomass in acidified and non-acidified Ontario lakes. In: Effects of Acidic Precipitation on Benthos, R. Singer (ed.), N. Am. Benthol. Soc. 154 pp.

Crisman, T.L., Schulze, R.L., Brezonik, P.L., and Bloom, S.A.: 1980, Acid precipitation: the biotic response in Florida lakes. In: Ecological Impact of Acid Precipitation, D. Drabløs and A. Tollan (eds.), SNSF Project, Ås, Norway. 383 pp.

Dermott, R.: 1985, Hydrobiologia **128**, 31.

Eriksson, M.O.G., Henrikson, L., Nilsson, B.-I., Nyman, G., Oscarson, H.G., and Stenson, A.E.: 1980, Ambio **9**, 248.

Fryer, G.: 1980, Freshw. Biol. 10, 41.

Grahn, O.H., Hultberg, H., and Landner, L.: 1974, Ambio 3, 93.

Hagen, A., and Langeland, A.: 1973, Environ. Pollut. **5**, 45.

Harvey, H.H.: 1980, Widespread and diverse changes in the biota of North American lakes and rivers coincident with acidification. In: Ecological Impact Acid Precipitation, D. Drabløs and A. Tollan (eds.), SNSF Project, Ås, Norway, pp. 93-98.

Harvey, H.H.: 1982, Population responses of fishes in acidified waters, Proc. Int. Conf. Acid Rain/Fisheries, Am. Fish. Soc., Cornell, pp. 45-55.

Harvey, H.H., Pierce, R.C., Dillon, P.J., Kramer, J.R., and Whelpdale, D.W.: 1981, Acidification in the Canadian aquatic environment, NRCC No. 18745, National Research Council of Canada, Ottawa. 369 pp.

Harvey, H.H., Dillon, P.J., Fraser, G.A., Somers, K.M., Fraser, P.E., and Lee, C.: 1982, Elevated metals and enhanced metal uptake in fishes in acid-stressed waters, Extended Abstract, Am. Chem. Soc., Conf. Proc., pp. 438-441.

Hendrey, G.R., Baalsrud, K., Traacn, T., Laake, M., and Raddum, G.: 1976, Ambio 5, 224.

Hendrey, G.R., Yan, N.D., and Baumgartner, K.J.: 1980, Responses of freshwater plants and animals to acidification, Proc. Int. Symp. Inland Water and Lake Restoration, pp. 457-466.

Kelso, J.R.M., Love, R.J., Lipsit, J.H., and Dermott, R.: 1982, Chemical and biological status of headwater lakes in the Sault Ste. Marie district, Ontario. In: Acid Precipitation. Effects on Ecological Systems, F.M. D'Itri (ed.), Ann Arbor Science.

Kerekes, J., Freedman, B., Howell, G., and Clifford, P.: 1984, Water Pollut. Res. J. Can. **19**, 1.

Klie, W.: 1937, Arch. Hydrobiol. **31**, 85.

Leivestad, H., Hendrey, G., Muniz, I.P., and Snekvik, E.: 1976, Effects of acid precipitation on freshwater organisms. In: Impacts of Acid Precipitation on Forest and Freshwater Ecosystems in Norway, F.H. Bracekke (ed.), SNSF FR 6/76, SNSF Project, Ås, Norway. 111 pp.

Mossberg, P., and Nyberg, P.: 1979, Rep. Inst. Freshw. Res., Drottningholm **58**, 77.
Nero, R.W., and Schindler, D.W.: 1983, Can. J. Fish. Aquat. Sci. 40, 1905.
Nilssen, J.P., Østdahl, T., and Potts, W.T.W.: 1984, Rep. Inst. Freshw. Res., Drottningholm **59**, 148.
Økland, J.: 1979, Malacologia **9**, 143.
Økland, J., and Økland, K.A.: 1986, Experientia 42, 471.
Økland, K.A.: 1969, Norw. J. Zool. **17**, 111.
Raddum, G.G.: 1980, Comparisons of benthic invertebrates in lakes with different acidity. In: Ecological Impact of Acid Precipitation, D. Drabløs and A. Tollan (eds.), SNSF Project, Ås, Norway. 383 pp.
Ravera, O.: 1986, Experientia 42, 507.
Roff, J.C., and Kwiatkowski, R.E.: 1977, Can. J. Zool. **55**, 899.
Stenson, J.A.E.: 1979, Rep. Inst. Freshw. Res., Drottningholm **58**, 166.
Wiederholm, T., and Eriksson, L.: 1977, Oikos **29**, 261.
Zimmerman, A.P., Noble, K.M., Gates, M.A., and Paloheimo, J.E.: 1983, Can. J. Fish. Aquat. Sci. **40**, 1788.

RESPONSES OF PERCH, PERCA FLUVIATILIS L., FROM AN ACIDIC AND A NEUTRAL LAKE TO ACIDIC WATER

M. Rask
Lammi Biological Station
University of Helsinki
SF - 16900 Lammi
Finland

E. Virtanen
Division of Physiology
Department of Zoology
University of Helsinki
Arkadiankatu 7, Helsinki 10
Finland

ABSTRACT. Responses to low pH of perch, *Perca fluviatilis*, from a naturally acid and a neutral lake were compared by 24 hr exposures to pH 4.6, 4.1 and 3.8 and by 72 hr exposures to pH 4.5. Plasma osmolality and plasma concentrations of Na and chloride decreased in fish from both lakes during acid exposures. Significant differences between the populations were observed at pH 4.1 and 4.5. Hematocrits of the fish from the acid lake increased rapidly and at higher pH compared with those of fish from the neutral lake. This was interpreted as an adaptation to their normal acidic environment, connected with the maintenance of red cell oxygen affinity. The perch from the acid lake maintained their muscle water balance at lower pH better than did the fish from the neutral lake.

1. INTRODUCTION

The decrease of plasma osmolality, Na and chloride are well known responses of salmonids to environmental acidity (McWilliams, 1980; Lee *et al.*, 1983; Giles *et al.*, 1984; Barton *et al.*, 1985). Less is known about the adaptation to acid conditions of fish populations from lakes with initially different acidity. However, differences in the pH tolerance between strains of a fish species have been reported (Robinson *et al.*, 1976; Swarts *et al.*, 1978; Gjedrem, 1980). With a few exceptions (Rahel, 1983; Lyons, 1982), these studies have concentrated on salmonid fish species.

In this study we compared the responses to low pH of perch from a naturally acidic lake (pH 4.9 to 5.2) and from a neutral lake (pH 6.3 to 7.2) to see whether there are differences between the two popula-

Water, Air, and Soil Pollution **30** (1986) 537–543.

tions in ability to adapt to acidic conditions. Perch is tolerant to low pH (Runn et al., 1977; Magnuson et al., 1984) and it is common in both Europe and North America in regions which are vulnerable to acid precipitation (According to Thorpe (1977), *Perca fluviatilis* L. and *Perca flavescens* Mitchill can be considered the same species). In southern Finland, where acidification of small oligotrophic lakes also is taking place (Tolonen and Jaakkola, 1983), perch is the most common fish species.

2. MATERIALS AND METHODS

Perch were captured with traps (1 cm square mesh) from naturally acid L. Opiston Valkjärvi (K. Tolonen, personal communication) and from neutral L. Valkea Mustajärvi. Both lakes are groundwater seepage lakes and their physicochemical properties are quite similar except for pH and alkalinity (Table I). For a more detailed description of water

TABLE I. The properties of water in the study lakes and in the experimental water of Evo Station

	Opiston Valkjärvi	Valkea Mustajärvi	Evo Station
Surface area (ha)	4.2	13.9	
Maximum depth (m)	8	11	
pH	5.1	6.5	6.0
Alkalinity (meq/l)	0.01	0.07	0.08
Conductivity (μS/cm 20 °C)	16	23	40
Colour (mg Pt/l)	20	30	150
Total P (μg/l)	19	13	50
Total N (μg/l)	390	410	890
Na (mg/l)	0.4	1.0	1.6
K (mg/l)	0.6	0.8	0.8
Ca (mg/l)	0.9	2.2	4.5
Mg (mg/l)	0.2	0.5	1.1
Fe (mg/l)	0.07	0.07	0.5
Al (mg/l)	0.09	0.09	0.16
SO_4 (mg/l)	3.5	4.9	9.9

analyses, see Rask et al. (1986). The experimental fishes from both lakes were transferred to Evo Inland Fisheries and Aquaculture Research Station, where they were acclimated for three weeks in 700 L tanks (Table I) with water pH 5.8 to 6.2 and fed with crustacean zooplankton and fish larvae.

Two series of exposures were performed. In the first, 5 to 10 perch from both lakes were exposed to pH 3.8, 4.1, 4.6 and 6.0 (control) for 24 hr. In the second series perch from both lakes (n = 50 + 50) were transferred to 200 L basins with mean water pH 4.5 (range

4.1 to 4.8) and sampled before the transfer (control) and after 2, 24 and 72 hr (5 to 10 fish/sample). The pH in both series was adjusted with analytical grade H_2SO_4 before starting the experiments. The fish were stunned with a blow on the head and the blood samples were taken immediately from the caudal vein with 1 ml heparinized syringes. Hematocrit values were determined immediately and plasma samples were deepfrozen after centrifugation (3 min, 10000 rpm). For determination of muscle water content samples from lateral white muscle were taken and frozen in tightly closed polypropylen tubes. The analyses of plasma osmolality, plasma Na^+ and Cl^- concentrations and muscle water content were performed according to Soivio and Virtanen (1980).

The statistical significance of the differences between the two perch populations was determined with Student´s t-test. The symbols (NS=not significant, NS°=P<0.1, * =P<0.05, ** =P<0.01, *** =P<0.001) are included in the figures.

3. RESULTS

Both perch populations exhibited a decrease in plasma osmolality and plasma concentrations of Na^+ and Cl^- when exposed to low pH (Figure 1). The values for perch from the acidic lake were initially somewhat higher than those for perch from the neutral lake. Differences could also be seen after the 24 hr exposures and these were significant at pH 4.1 in plasma osmolality and plasma chloride. The 72 hr exposure to pH 4.5 showed that the osmolality, Na^+ and Cl^- decreased during the first hours of the experiment. A partial recovery followed, except that in fish from the neutral lake Cl^- still remained relatively low after 72 hr.

The hematocrit values of perch from both populations increased after exposure to low pH, but the course of increase was strikingly different between the populations (Figure 2). In the perch from the acidic lake the hematocrits were initially higher and increased evenly with decreasing pH, whereas the perch from the neutral lake maintained their hematocrits constant at pH 4.6 and 4.1. Only at pH 3.8 was a clear increase observed. A corresponding difference between the populations was also observed in the 72 hr experiment at pH 4.5: the fish from the acidic lake showed a sharp increase in Hct during the first 2 hr, a response which did not occur so strikingly in the fish from the neutral lake.

The water content of the muscle tissue was initially slightly higher in the perch from the neutral lake (Figure 3) and it increased evenly with decreasing pH. The fish from the acidic lake maintained their muscle water content constant at pH 4.6 and 4.1, but at 3.8 a clear increase was observed. The 72 hr exposure showed the same trend: a sharp but transient increase in muscle water content of perch from the neutral lake but no remarkable alterations in the fish from the acid lake.

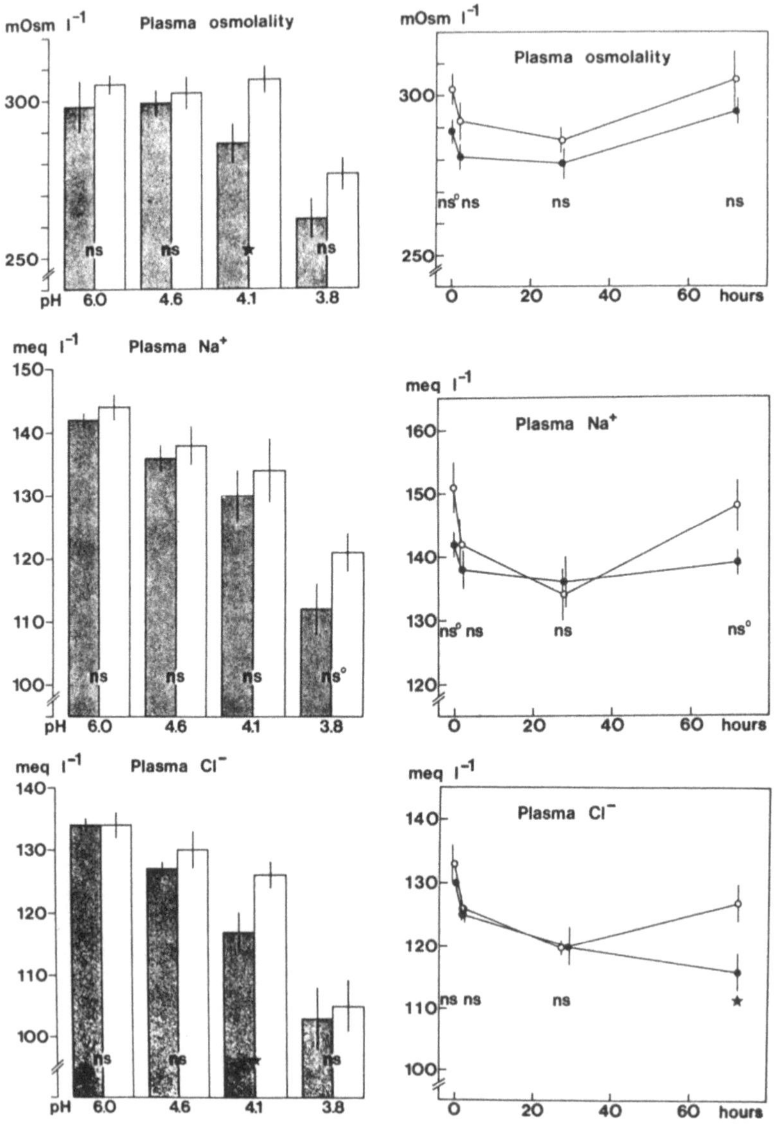

Figure 1. Plasma osmolality (top) and plasma concentrations of Na^+ (middle) and Cl^- (bottom) of perch after 24 hr exposures at pH 6.0, 4.6, 4.1 and 3.8 (left) and 72 hr exposures at pH 4.5 (right). White columns and open circles refer to perch from the acid lake and shaded columns and closed circles to those from the neutral lake.

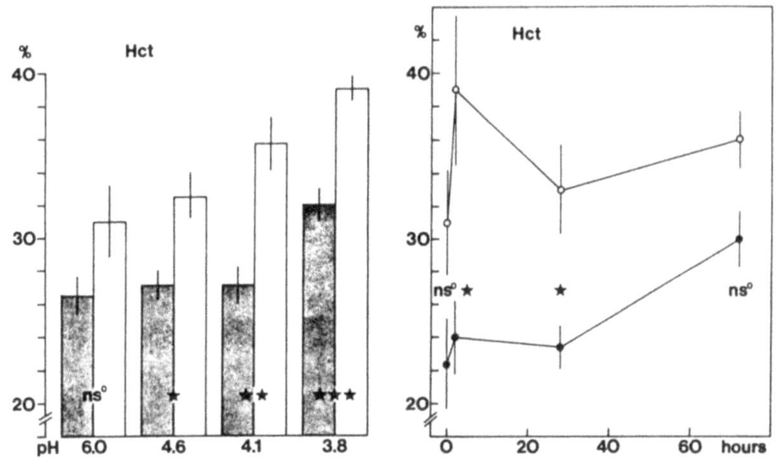

Figure 2. The hematocrit values of perch from the two populations after 24 hr exposures (left) and 72 hr exposures (right). White columns and open circles refer to perch from the acid lake, shaded columns and closed circles to those from the neutral lake.

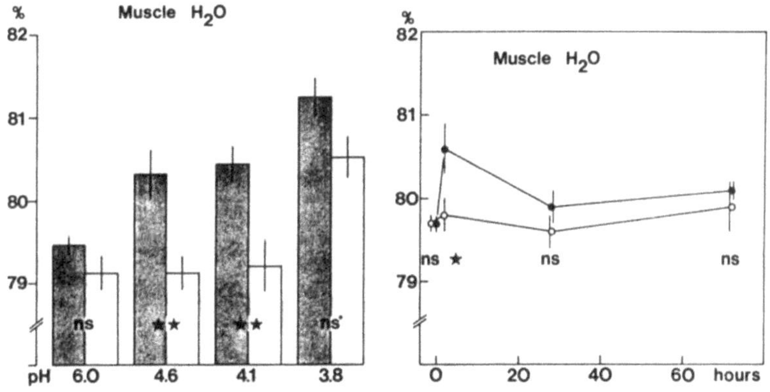

Figure 3. The muscle water content (%) of perch from the two populations after 24 hr (left) and 72 hr (right) exposures. White columns and open circles for the perch from the acid lake, shaded columns and closed circles for those from the neutral lake.

4. DISCUSSION

According to our results pH levels of 4.0 to 4.5 seem to be critical for the maintenance of plasma ion balance in perch. Reproductive failures of perch at the same pH range (Rask et al., 1986) support this conclusion. Therefore we believe that significant differences between the two perch populations in plasma chloride and osmolality at pH 4.1 indicate a better adaptation of ionic regulation to low pH in the perch from the acid lake. Maybe the choice of a lethal pH level of 3.2 was the reason why Lyons (1982) did not find any differences in the rates of decrease of plasma Na between two populations (one from an acidic and the other from an alkaline lake) of yellow perch, *Perca flavescens*. However, perch from the same brown water lakes survived differently in experimental acid exposures (Rahel, 1983).

The increase in hematocrits that was observed here as in other studies (Lee et al., 1983; Barton et al., 1985) is attributed to swelling of red blood cells (Milligan and Wood, 1982). At constant extracellular pH swelling improves the oxygen affinity of red blood cells by diluting hemoglobin and organic phosphates in the cells and by increasing intracellular pH (Soivio and Nikinmaa, 1981), and may also help maintaining oxygen loading in the face of extracellular acidification. Thus, the higher hematocrits in the perch from the acidic lake and the increase at higher pH than in fishes from the neutral lake, can be seen as an advantageous adaptation to low pH. Chloride accumulates in red cells during swelling (Baroin et al., 1984). Therefore the decrease in plasma chloride can be due to ion loss from the fish (McDonald and Wood, 1981; Lee et al., 1983; Giles et al., 1984) or due to accumulation of chloride into red cells, or both.

Disturbances in the ionic balance of fish as observed during acid exposures, are often followed by increased muscle water content (Lee et al., 1983). The ability of perch from the acid lake to maintain their muscle water content at pH levels of 4.6 and 4.1 is another indication of their better adaptation to acidic conditions.

ACKNOWLEDGMENTS

We thank Leena Rask for help during the exposures, Hannele Virtanen for performing plasma analyses, Mr. M. Nikinmaa for valuable comments on the manuscript and Mr. R.I. Jones for correcting our language. The work was financially supported by the Maj and Tor Nessling Foundation.

REFERENCES

Baroin, A., Garcia-Romeu, F., Lamarre, T. and Motais, R.: 1984, J. Physiol. 350, 137.
Barton, B.A., Weiner, G.S. and Schreck, C.B.: 1985, Can. J. Fish. Aquat. Sci. 42, 710.
Giles, M.A., Majewski, H.S. and Hobden, B.: 1984, Can. J. Fish. Aquat. Sci. 41, 1686.
Gjedrem, T.: 1980, 'Genetic variation in tolerance of brown trout to acid water', SNSF-project, FR 5/76, 11 pp.
Lee, R.M., Gerking, S.D. and Jezierska, B.: 1983, Env. Biol. Fish. 8, 115.
Lyons, J.: 1982, Comp. Biochem. Physiol. 73A, 437.
Magnuson, J.J., Baker, J.B. and Rahel, F.J.: 1984, Phil. Trans. R. Soc. Lond. B305, 501.
McDonald, D.G. and Wood, C.M.: 1981, J. exp. Biol. 93, 101.
McWilliams, P.G.: 1980, J. exp. Biol. 88, 269.
Milligan, C.L. and Wood, C.M.: 1982, J. exp. Biol. 99, 397.
Rahel, F.J.: 1983, Can. J. Zool. 61, 147.
Rask, M., Heinänen, A., Salonen, K., Arvola, L., Bergström, I., Liukkonen, M. and Ojala, A.: 1986, Arch. Hydrobiol.
Robinson, G.D., Dunson, W.A., Wright, J.E. and Mamolito, G.E.: 1976, J. Fish Biol. 8, 5.
Runn, P., Johansson, N. and Milbrink, G.: 1977, Zoon, 5, 115.
Soivio, A. and Virtanen, E.: 1980, 'Methods for physiological experiments on fish', Ekotoxicologiska Metoder för Akvatisk Miljö, Nordforsk, Helsinki, Report no. 16, 15 pp.
Soivio, A. and Nikinmaa, M.: 1981, 'The swelling of erythrocytes in relation to the oxygen affinity of the blood of the rainbow trout, Salmo gairdneri Richardson', In: Pickering, A.D. (ed.), Stress and Fish, 103-119, Academic Press, London.
Swarts, F.A., Dunson, W.A. and Wright, J.E.: 1978, Trans. Amer. Fish. Soc. 107, 651.
Thorpe, J.E.: 1977, J. Fish. Res. Board Can. 34, 1504.
Tolonen, K. and Jaakkola, T.: 1983, Ann. Bot. Fennici 20, 57.

MIX
Papier aus verantwortungsvollen Quellen
Paper from responsible sources
FSC® C105338

If you have any concerns about our products,
you can contact us on
ProductSafety@springernature.com

In case Publisher is established outside the EU,
the EU authorized representative is:
Springer Nature Customer Service Center GmbH
Europaplatz 3, 69115 Heidelberg, Germany

Printed by Libri Plureos GmbH
in Hamburg, Germany

ACIDIC PRECIPITATION

ACIDIC PRECIPITATION

Proceedings of the
International Symposium on Acidic Precipitation,
Muskoka, Ontario, September 15–20, 1985

Edited by

HANS C. MARTIN
Atmospheric Environment Service, Downsview, Ontario

PART 2

Reprinted from Water, Air, and Soil Pollution, Volume 31 (1986)

Springer-Science+Business Media, B.V.

Library of Congress Cataloging-in-Publication Data

CIP

International Symposium on Acidic Precipitation
(1985 : Muskoka, Ont.)
Acidic precipitation.

'Reprinted from Water, air, and soil pollution, volumes 30 and 31 (1986).'
Includes bibliographies and indexes.
1. Acidic precipitation (Meteorology)—Environmental aspects—Congresses. 2. Acid deposition—Environmental aspects—Congresses. 3. Acid pollution of river, lakes, etc.—Congresses. 4. Air—Pollution—Environmental aspects—Congresses. I. Martin, H. C. II. Title.
QC924.5.I57 1985 363.7'386 86-33883
ISBN 978-94-010-8019-4 ISBN 978-94-009-3385-9 (eBook)
DOI 10.1007/ 978-94-009-3385-9

Originally published by D. Reidel Publishing Company, Dordrecht, Holland in 1987
Softcover reprint of the hardcover 1st edition 1987

ACIDIC PRECIPITATION

Proceedings of the
International Symposium on Acidic Precipitation
Muskoka, Ontario, September 15–20, 1985

Guest Editor

H. C. MARTIN

Atmospheric Environment Service, Downsview, Ontario

PART 2

ACIDIC PRECIPITATION

Proceedings of the
International Symposium on Acidic Precipitation
Muskoka, Ontario, September 15–20, 1985

Guest Editor

H. C. MARTIN

Atmospheric Environment Service, Downsview, Ontario

PART 1

IONIC COMPOSITION OF ACID LAKES IN RELATION TO AIRBORNE INPUTS AND WATERSHED CHARACTERISTICS

Gary E. Glass*
Environmental Research Laboratory-Duluth
U.S. Environmental Protection Agency
6201 Congdon Boulevard
Duluth, MN 55804 USA

John A. Sorensen, Barbara W. Liukkonen and George R. Rapp, Jr.
College of Science and Engineering
University of Minnesota-Duluth
Duluth, MN 55812 USA

Orie L. Loucks
Holcomb Research Institute
Butler University
Indianapolis, IN 46208 USA

ABSTRACT. Present acid forming emissions to the atmosphere have the potential to alter significantly the chemistry of rain, snow, and surface water of weakly buffered lakes in the Upper Midwest. Average precipitation pH from field measurements during 1979-1983 declined from west to east from 4.8, 4.6, and 4.3 along a cross-section of sites in Minnesota, Wisconsin, and Michigan respectively where 990 lake and stream sampling sites were studied. Measurements of weakly buffered lakes show a parallel decline in lake water pH with the lowest values measured, 5.1, 4.6 and 4.4, respectively in the same regions. Correspondingly, the percentage of lakes sampled with little or no acid neutralizing capacity (ANC) was found to increase from 0 to 4 and 13%, respectively. The geographic patterns in ionic composition of airborne acids and bases, and the resultant surface water concentrations are compared. The acid forming capacity (AFC) from airborne inputs is calculated using mass balance and in-lake processes. Stoichiometric acid-base reactions are used to balance the observed chemical differences between airborne inputs and surface water composition considering nitrification, denitrification, other oxidation-reduction reactions, and the evaporation concentration process. Microbial activity in surface water can result in a net decrease in ionic strength from the conversion of most of the ammonium and nitrate to neutral compounds and biomass, but only a partial reduction of about 20% of the sulfate inputs to weakly buffered lakes. The resulting AFC of airborne inputs are calculated to range from 30 to 60, 50 to 90, and

Water, Air, and Soil Pollution **31** (1986) 1–15.

80 to 130 μeq H^+L^{-1}, respectively, in northeastern Minnesota-Ontario, northcentral Wisconsin and northern Michigan-Ontario. The differences in AFC of airborne inputs from west to east, and differences in in-lake processes explain the observed acidity of weakly buffered lakes across the region.

1. INTRODUCTION

One of the major unresolved questions about the acidification of lakes continues to be whether observed lake chemistry is or is not primarily the result of chemicals from atmospheric inputs after watershed processing. Simple stoichiometric and mass balance modeling on a population of lakes could be used to evaluate the extent of correspondence between input chemistry and in-lake concentrations. This paper applies established principles of mass balance, evaporative concentration and biological processing of chemicals to determine whether precipitation inputs influence lake responses.

The compounds contributing most to the differences in precipitation chemistry across three Great Lakes States (GLS) of Minnesota, Wisconsin, and Michigan, offer the potential for understanding regional and local variation in pH, ionic compositions, acid-forming precursors, their reactive products (SO_4^{2-}, NO_3^-, H^+, and NH_4^+), and subsequent impacts on surface water quality (Glass and Loucks, 1986). Soil- and agriculturally-derived ions (Thornton and Eisenreich, 1982; Munger, 1982; Guiang et al., 1984) also need to be considered as inputs to surface water relating to geographic variations in rainfall chemistry (Glass and Loucks, 1980; Stensland and Semonin, 1982; Cogbill and Likens, 1982; Likens and Butler, 1981). The objectives of this paper are, therefore: to evaluate the significance of geographic patterns in the ion composition of acidic and basic reactants in precipitation and surface water across the GLS region and to compare the concentrations of acid and base forming compounds as atmospheric inputs with the resulting surface water chemistry.

2. METHODS*

The precipitation data reported in this analysis are summarized, in part, from a complete reporting and analyses by Glass and Loucks (1986). Sampling and analyses of 990 surface water sites in the study region began in the fall of 1978 in Minnesota, 1979 in Wisconsin, 1981 in Michigan and Ontario and continued through the summer of 1983 (see Table I). The water bodies were chosen using the sensitivity criteria developed by Glass and Loucks (1980) with an approximately equal number of sensitive and non-sensitive lakes across each region with a 5 to 10% sampling frequency of the total lake populations. In-situ measurements of water quality and water samples were collected at mid-lake at a

*Manufacturers and trade names are given for identification and no endorsement by the USEPA is intended or should be implied.

Table I. General distribution of the 990 lake and stream sampling sites for assessment of surface water sensitivity in the Upper Midwest States and a summary of the number of lakes and streams, and the frequency distribution of these sites in five ANC (sensitivity) classes (Glass, 1984).

Region		Acid Neutralizaton Capacity Interval (μeqL^{-1})					Total
		<0	0-39	40-199	200-499	500+	Sites
Minnesota-	%	-	0.4	41	37	22	46
Ontario	Count	0	2	186	166	100	454
Wisconsin	%	4	23	23	24	28	45
	Count	20	100	99	105	121	445
Wisconsin[1]							
Total	%	1	6.8	30.6	61.6		100
Total	Count	142	967	4,369	8,783		14,261
Michigan-	%	13	10	19	17	42	9
Ontario	Count	12	9	17	15	38	91
Totals	%	3.2	11.2	30.5	28.9	26.2	100
	Count	32	111	302	286	259	990

[1]Total inventory of lakes was available for all of Wisconsin from the Wisconsin DNR (1984).

depth of 1 m during spring and fall turnovers when the lakes were well mixed and during summer and winter stratification as well. Detailed sampling procedures and analytical methods are given in Rapp et al. (1985).

The lake chemistry data have been subjected to quality assurance procedures for single parameter measurements and inter-parameter relationships to identify outliers and evaluate possible systematic errors in measurement or manipulation using charge balance (APHA, 1980), calculated pH-alkalinity (Stumm and Morgan, 1981), and comparison of measured vs calculated conductivity (Quint and Viallard, 1978). The precision and accuracy (based on standard additions and USEPA and NBS standard samples) of the measurements are, on average, within 5% of expected values. Calculated conductivity for 11 ions was usually within 5% of measured conductivity and variations outside of 15% were considered as outliers. The water chemical data are given as mean values averaged from all samples taken over the 5 yr period of study. The mean for pH values was determined by averaging H^+ concentrations.

Watershed and lake morphometric characterization, water renewal times, and hydrologic evaporative concentration factors were determined using pre-existing surveys, USGS topographic maps and long-term data from water atlases (Geraghty et al., 1973). A detailed description of these methods and the statistical methods of analysis are found in Eilers et al. (1983) and Rapp et al. (1985, 1986). Precipitation chemistry and atmospheric inputs were determined for individual watersheds using linear interpolation between the precipitation monitoring sites. Evaporative concentration by watersheds was also calculated using the ratio of chloride concentrations in precipitation to those of lake water. Sulfate concentration ratios were computed similarly using the wet plus dry fall values. The lakes with the lowest acid neutralizing capacities (ANC ≤ 0) (lowest terrestrial inputs to lake water) were selected for detailed study in this analysis to identify and examine characteristics controlling lake water quality. They are identified in Table III with relevant descriptive data.

3. RESULTS AND DISCUSSION

Geographic Patterns in the Ionic Composition of Airborne Acid and Base Forming Substances: Since the potential for adverse effects to aquatic resources is controlled by the total dose of polluting substances (concentration(s) x rate of uptake x time; U.S. EPA, 1976), the average concentrations as well as volume weighted averages of atmospheric inputs (loadings) are of interest (Table II). The proton concentration results from numerous reactions between acid-forming precursors, ion exchange, and neutralization reactions that occur with the specific mix of gaseous and particulate materials washed out of the atmosphere during a given precipitation event. Average precipitation pH from field measurements made during 1979-1983 declines from 4.8 in northeastern Minnesota, to 4.6 in northcentral Wisconsin and to 4.3 in the Upper Peninsula of Michigan. The average concentrations are important in evaluating the maximum chemical potential for chemical and biological reactions while the volume-weighted values are important in determining loadings and mass balance calculations. The major anion concentrations are indicative of strong acid precursors in the atmosphere across the region and the site averages are given in Table II and plotted in Figure 1A. The major cations, in addition to H^+ in descending order of average concentration, are NH_4^+, Ca^{2+}, Mg^{2+}, K^+ and Na^+. The gradient in acidic reactants and products is consistent with the regional pattern of sulfur and nitrogen oxide emissions from sources in the region (Altshuller, 1980). The sum of $SO_4^{2-} + NO_3^-$ closely parallels the sum of $H^+ + NH_4^+$, also indicating the anthropogenic sources of strong atmospheric acidity (Brosset, 1980; Gorham et al., 1984; Calvert, 1984; NRC/NAS, 1983).

The ammonium (NH_4^+) ion is an important reaction product of acid aerosols with gaseous ammonia, $H_2SO_{4(s)} + 2NH_{3(g)} \longrightarrow (NH_4)_2SO_{4(s)}$ (Munger, 1982; Brosset, 1980). Ammonia neutralizes H^+ in precipitation, but the ammonium ion serves as a major contributor to the acidic inputs to surface water due to microbial activity (nitrification) that occurs in surface water: $NH_4^+ + 2O_2 \longrightarrow 2H^+ + NO_3^- + H_2O$ (Schnoor and

Table II. Average and volume-weighted ion concentrations in precipitation for the 3-4 year period from July 1979 through November 1983 at three NADP sites.

Location	No. Samples	Precip-itation cm yr^{-1}	Average and/Volume Weighted Average Ion Concentration, μeqL^{-1}										
			H^+	NH_4^+	SO_4^{2-}	NO_3^-	Cl^-	Ca^{2+}	Mg^{2+}	K^+	Na^+	(+)	(-)
Fernberg Road,	119	75	12	23	39	25	6.0	17.8	9.9	1.2	7.2		
Minnesota			11	15	25	14	2.9	8.9	3.6	0.9	2.8	42.4	41.9
Trout Lake,	161	83	21	25	49	30	5.5	21.1	7.0	1.2	5.9		
Wisconsin			22	20	37	21	2.8	10.6	3.2	0.7	2.5	59.0	60.8
Douglas Lake,	180	71	42	30	67	40	5.7	22.4	8.1	1.4	6.7		
Michigan			39	19	51	28	3.4	11.8	4.1	0.8	4.0	78.7	82.4

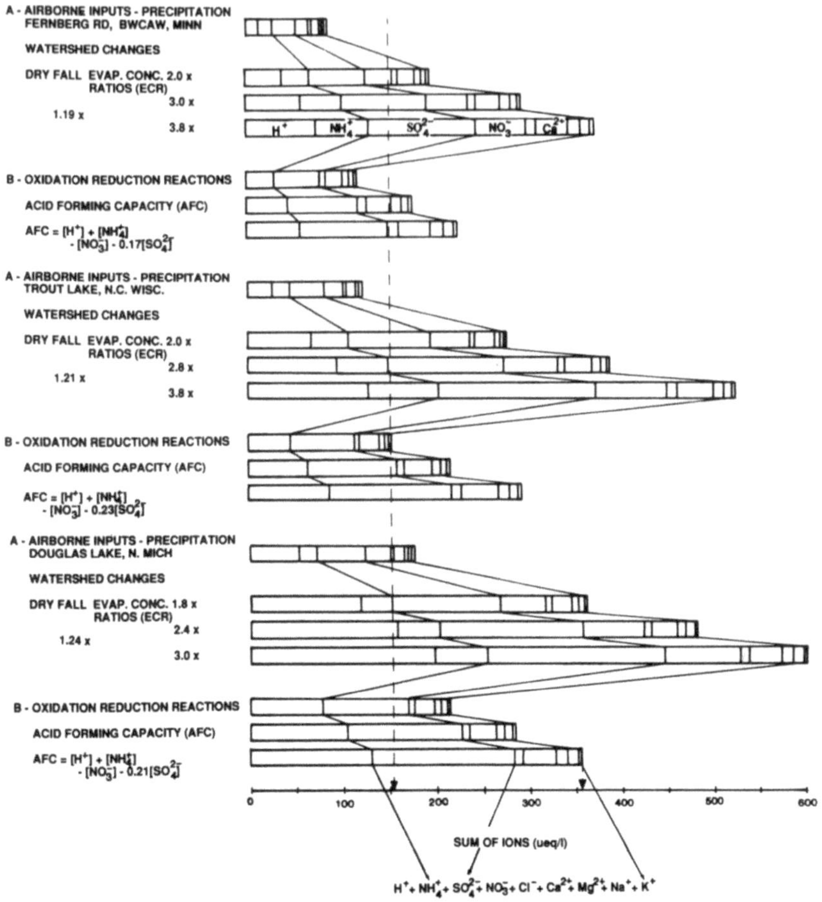

Fig. 1. Annual average ion concentrations in precipitation at monitoring sites showing the expected concentrations in unbuffered surface water following dry deposition and concentration by evaporation and redox reactions. Upper: the Boundary Waters Canoe Area Wilderness - Quetico Provincial Park regions of Minnesota and Ontario; Middle: the Nicolet and Ottawa National Forest regions of northcentral Wisconsin; and Lower: the Eastern Hiawatha National Forest - Pictured Rocks National Lakeshore region in Upper Michigan and near the Batchawana Mountain Lake Superior Provincial Park region of Ontario.

Stumm, 1984). In Table II, the ammonium and H^+ ions are the dominant cations, 67%. These, with the base cations (Ca^{2+}, Mg^{2+}, Na^+, and K^+) are in ionic balance with sulfate, nitrate and chloride anions. The volume-weighted average concentration of the sum of base cations is 19 ± 3 μeqL^{-1} for all sites (Table II) and results from ion exchange and acid neutralization reactions with H^+ and from the wash-out and weathering of soil particles (Verry, 1983).

Dry deposition of gases and particulates can increase the sulfate and nitrate deposition onto lakes and their watersheds by varying amounts depending upon many factors (Glass and Brydges, 1982). For the years 1981-1982, Chan et al. (1985) measured low level gaseous and particulate deposition and concluded that the dry components add an additional 19 to 24% SO_4^{2-} and 60 to 90% NO_3^- to the wet deposition values in northeastern Minnesota and northern Michigan, respectively. These values were used to estimate total acidic deposition to the study watersheds assuming that equivalent sulfate, nitrate, and H^+ are added.

Ion Composition and Potential Acid Forming Capacity (AFC) from Airborne Inputs to Surface Water: The changes from the chemistry of water as precipitation to that of surface water following input processes are complex and involve many physical, chemical and biological processes (Stumm et al., 1983; Glass and Brydges, 1982). For this study (focused on hydrologically isolated seepage lakes) we considered three major processes causing the changes in concentrations using a mass balance: evaporative concentration, nitrification-denitrification, and sulfate reduction.

Concentration of all ionic components in precipitation occurs by evaporative processes from water and from various terrestrial components within the watershed. The calculated results of predicted concentrations considering watershed evaporative concentration on atmospheric inputs (low, average and high values) are shown in Figure 1A together with the annual average (volume weighted) composition of precipitation in the three study regions.

Microbial and mediation reactions are known to occur for the more reactive components in precipitation that have nutrient properties (Schnoor and Stumm, 1984). These compounds can be modeled by stoichiometry of two oxidation-reduction reactions involving NH_4^+ and NO_3^-. The first is the nitrification process adding acidity:

$NH_4^+ + 2O_2 = NO_3^- + H_2O + 2H^+$ and the second is the denitrification process reducing acidity:

$$4NO_3^- + 5CH_2O + 4H^+ = 5CO_2 + 2N_2 + 7H_2O.$$

The complete reaction of one equivalent of each ion is assumed to result in no overall change in acidity or alkalinity content over long reaction times of several years, considering the water renewal times of the lakes (see Table III) (Schnoor and Stumm, 1984; Stumm et al., 1983; Schnoor et al., 1983). We modeled the reaction stoichiometry in precipitation in the short term and for correlation analyses to be $[H^+] + [NH_4^+] - [NO_3^-]$ since both NH_4^+ and NO_3^- reactions are essentially complete over the water renewal time of these seepage lakes, as indi-

Table III. Site, morphometric and watershed data for acidic lakes in the Upper Midwest study region.

Lake Name	County, State-Prov.	Lake Type (1)	pH	ANC μeqL^{-1}	Color Pt-Co	Evap. Conc. Ratio	Lake Area, ha	Depth Max. m	Water Renew. yr(2)
McNearney	Chippewa, MI	1	4.4	-38	3	2.2	50	7.5	41
Airhole	Langlade, WI	1	4.6	-26	115	2.7	34	3.0	25
Avery	Chippewa, MI	1	4.6	-27	45	1.9	6	3.0	11
Bear	Chippewa, MI	1	4.6	-24	54	2.2	36	3.6	7
Garlinghouse	Chippewa, MI	1	4.6	-27	7	3.4	17	3.7	20
Morgan	Florence, WI	1	4.6	-26	43	1.9	10	2.4	33
Denton	Vilas, WI	1	4.7	-12	110	2.6	10	10.1	28
Johnson	Chippewa, MI	1	4.8	-14	6	2.0	16	4.5	19
Stuart	Luce, MI	1	4.8	-17	6	2.2	30	6.9	25
Summit	Langlade, WI	4	4.8	-9	133	2.7	114	7.9	39
West Branch	Alger, MI	4	4.9	-14	70	2.1	15	2.7	12
Denton	Oneida, WI	1	4.9	-6	55	2.6	19	6.7	27
Malby	Oneida, WI	1	5.0	-7	6	2.7	18	4.0	16
Nineweb	Vilas, WI	1	5.0	-11	7	2.7	15	7.9	21
Bastille	Forest, WI	1	5.0	-12	10	2.5	18	7.0	40
Sand	Oneida, WI	1	5.0	-10	4	2.8	16	5.8	22
Silverbass	Oneida, WI	1	5.1	-8	5	2.7	11	3.4	20
Shallow	Oneida, WI	1	5.1	-4	9	2.7	11	2.1	16
Gordon	Vilas, WI	1	5.1	-7	10	3.1	21	5.5	23
McGrath	Oneida, WI	1	5.2	-5	7	3.3	20	7.3	21
Kelly	Luce, MI	1	5.2	-5	8	2.0	12	9.0	27
Clear	Langlade, WI	1	5.2	-1	9	3.1	35	7.0	42
Mill	Oneida, WI	1	5.2	-2	17	2.9	21	1.8	12
Long	Oneida, WI	5	5.2	1	10	1.8	8	2.0	8
JR08	Algoma, Ont.	1	5.2	-5	7	2.9	43	9.4	34
Harmony	Vilas, WI	2	5.2	3	11	2.6	31	4.0	19
Camp 12	Vilas, WI	1	5.3	-0	6	2.7	34	8.2	26
JR05	Algoma, Ont.	5	5.3	-2	30	1.8	8	3.0	10
Cusino	Schoolcft, MI	1	5.3	-1	22	2.1	53	10.8	19
Torch	Vilas, WI	1	5.3	0	7	2.7	16	4.3	17
Burrows	Oneida, WI	1	5.3	4	24	2.7	60	7.9	26
Deadmans	Luce, MI	1	5.4	-5	10	2.0	11	16.5	38
Sugar Camp	Oneida, WI	1	5.4	-0	4	3.8	210	11.6	32
Lumen	Oneida, WI	1	5.4	1	7	3.1	19	5.8	24
Great Bass	Oneida, WI	1	5.4	-1	15	3.0	45	4.6	17
Hildebrand	Lincoln, WI	1	5.4	1	7	2.8	23	7.3	34
Joyce	Vilas, WI	1	5.5	4	7	2.8	11	10.1	26
Scotchman	Oneida, WI	1	5.5	3	7	2.7	13	6.4	21
Luna	Forest, WI	1	5.5	5	5	3.0	26	4.8	25
Buteau	Lincoln, WI	1	5.6	2	6	3.2	21	15.2	46
Ike Walton	Vilas, WI	3	5.6	5	11	3.2	574	18.6	15

(1) Hydrologic type 1, seepage; 2, inflow; 3, drained; 4, drainage; 5, headwater; (Eilers, et al., 1983).
(2) Calculated from lake surface precipitation inputs.

cated by low lake water concentrations of these ions.

Sulfate reduction was measured using a simple mass balance comparison of present airborne inputs to lake water concentrations. Considering wet inputs alone, sulfate retention averages 77, 79, and 83% for the 20 most weakly buffered lakes in Wisconsin, Michigan, and Minnesota, respectively (see Figure 1). This difference was used to approximate the amount of sulfate reduction and associated H^+ consumption, and assumes lake sulfate to be in steady state with present inputs. The reduction reaction considered for sulfate is:

$$SO_4^{2-} + 2CH_2O + 2H^+ = 2CO_2 + H_2S + H_2O.$$

This reaction leads to an assumed upper limit on H^+ consumption since the lake data are really a time series and are dependent upon the watershed history preceding sample collection (Rapp et al., 1985).

Thus, the net result of oxidation-reduction reaction and mass balance stoichiometry gives the potential acid forming capacity for volume-weighted precipitation inputs:

$$[H^+], AFC = [H^+] + [NH_4^+] - [NO_3^-] - 0.2\ [SO_4^{2-}].$$

These results are shown graphically in Figure 1B where the expected maximum yields of H^+ are calculated considering the four processes described above. When all variations in evaporative concentration, wet and dry inputs, and oxidation-reduction reactions are calculated, total potential H^+ concentrations are calculated to be: 30 to 60, 50 to 90, and 80 to 130 μeqL^{-1}, respectively, for the resultant surface water across each state's (MN, WI, MI) study region on an annual basis. These results are shown in Figure 1B with the other ions resulting from airborne inputs as the resultant predicted concentrations for low, average and high evaporative concentration ratios (ECR). Lines connecting NH_4^+, SO_4^{2-} and K^+ are for comparative purposes for the range of watershed characteristics studied.

Additional reactions can occur, such as weathering, ion exchange, and synthesis/decomposition of biomass. These reactions can "consume" H^+ and produce increased concentrations of Ca^{2+}, Mg^{2+}, Na^+, K^+, Al(III), and SiO_2 (Schnoor and Stumm, 1984). The differences between predicted and measured lake composition can be seen by comparing Figure 1 with Figure 2 for 32 acid lakes in the study region.

Expected vs Observed Ionic Concentrations in Lake Water: The measured ionic concentrations of the most acid lakes listed in Table III are presented in bar-graph form in descending order of $[H^+]$ (Figure 2). The ionic concentrations in lake water are ordered along the bars to facilitate comparison with the concentrations for atmospheric inputs shown in Figure 1. Increased cations, Al, and organic anions generally account for the decreased H^+ that is expected considering atmospheric inputs. The broad pattern of surface water chemistry that corresponds with that of the precipitation chemistry can be seen in pH and ANC in surface water from 990 lake (861) and stream (129) sampling sites summarized in Table I. The lowest pH value from a lake in Minnesota was 5.1, in Wisconsin it was 4.6 and in Michigan it was 4.4. All of the

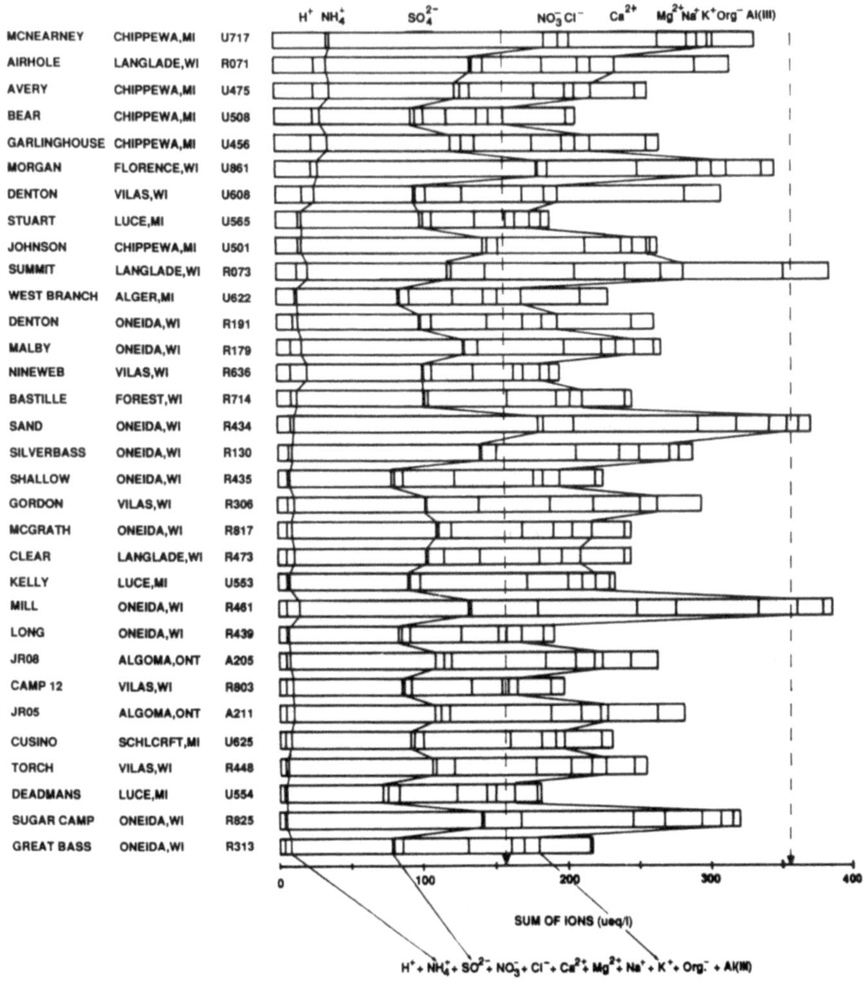

Fig. 2. Average ion concentrations for 32 acid lakes in the Upper Midwest study region. Dotted lines indicate the expected minimum and maximum resultant ion sums from atmospheric inputs and watershed factors given in Figure 1.

lakes in Minnesota/western Ontario showed positive measurable ANC, with all values greater than 40 μeqL^{-1}, except two. The 32 lakes showing strongest acidity (ANC <0) are found in Wisconsin and northern Michigan/central Ontario. This corresponds to 4% of the lakes in northcentral Wisconsin and 13% of the lakes in the eastern U.P. of Michigan.

Differences in lake water concentrations of ions from those predicted by atmospheric input can occur as a result of individual characteristics of the lakes, and processes induced in the lake basin by different reaction kinetics, ion exchange, dissolution of soil and sediment components, and cultural inputs (Glass and Brydges, 1982). The potential for these processes to yield varying concentrations of base cations, Al and corresponding bicarbonate, chloride and organic anions has been evaluated in a series of related studies (e.g., Rapp et al., 1985). Correlations obtained between watershed properties and strong acidity in these lakes show $[H^+]$ to be most strongly related to atmospheric acid forming inputs (indicating sources) and morphometric factors where weathering and ion exchange reactions are consuming H^+, Table IV. Conservative ion concentrations show positive correlations with evaporative concentration ratio and terrestrial anthropogenic influences, while the more reactive components, NH_4^+ and NO_3^- show negative correlations.

These results indicate that the most sensitive lake basins in the study region must annually supply a quantifiable amount of acid-neutralizing substances and/or ionic exchange substrate. This is to neutralize acid forming substances from atmospheric inputs due to differences in inputs and watershed characteristics in amounts of excess acid forming inputs ranging from 30 to 140 μeL^{-1} H^+, increasing from west to east across the region. Acidic waters and reduced resource value will result in basins that are unable to neutralize excess acidic inputs and maintain a lake pH above 6.5, the lowest pH value considered safe for an abundance of freshwater aquatic life (USEPA, 1976).

4. CONCLUSIONS

In a regional assessment of relationships between precipitation and surface water chemistry, care must be taken to consider a wide variety of potentially significant variables (Glass and Brydges, 1982; Rapp et al., 1985, 1986; Glass and Loucks, 1986). Physical and biological factors play key roles in the continuing dynamic changes in water chemistry as rock, soil, and sediment weathering contribute solutes; micro and macro organisms grow and cycle nutrients; and insoluble and particulate matter precipitate and accumulate as sediments in the basins. In this paper, local characteristics of precipitation and surface water composition have been considered in detail in relation to watershed processes transforming precipitation into surface water.

Two conclusions can be drawn that are relevant to the aquatic resources at risk from acidic deposition in northern Michigan, Wisconsin, and Minnesota. First, ions (H^+ + NH_4^+) from acid-forming gases (SO_x and NO_x) are observed in deposition with the resultant oxi-

Table IV. Selected correlation coefficients for lake water chemistry and selected atmospheric inputs and watershed variables for 41 acidic lakes.

Variable	Lake pH	Lake ANC	Lake Color	Lake Dep. Max.	L.Wat. Ren. Time	Evap. Conc. Ratio	Coni. For. %	N. Lat.	W. Long.	Prec. Depth	SO_4^{-2} Conc. Prec.	AFC
Lake ANC	+0.94 **											
Lake Color	-0.50 **	-0.34 *										
Lake Depth, Max.	+0.44 **	+0.33 *	-									
Lake Water Ren. Time	-	-	-	+0.51 **								
Evap. Conc. Ratio	+0.40 **	+0.36 *	-	+0.27	-							
Coniferous Forest, %	-0.31 *	-0.28	+0.44 **	-	-	-						
N. Latitude	-	-	-	-	-0.46 **	-0.75 **	-					
W. Longitude	+0.43 **	+0.48 **	-	-	-	+0.83 **	-0.26	-0.86 **				
Precip. Depth	+0.45 **	+0.47 **	-	+0.29	-	-	-	-	+0.44 **			
SO_4^{-2} Conc. in Precip.	-0.49 **	-0.54 **	-	-	-	-0.79 **	-	+0.72 **	-0.98 **	-0.57 **		
AFC; wet $[H^+] + [NH_4^+]$	-0.51 **	-0.56 **	-	-	-	-0.76 **	-	+0.65 **	-0.95 **	-0.62 **	+1.00 **	
AFC - total, wet & dry $[H^+] + [NH_4^+] - [NO_3^-] - 0.2 [SO_4^{-2}]$	-0.51 **	-0.57 **	-	-	-	-0.75 **	-	+0.84 **	-1.00 **	-0.46 **	+0.98 **	+0.96 **

**significance <0.01; *significance <0.05 and $\geq$0.1; no designation <0.10 and $\geq$0.05

dation products (SO_4^{2-} and NO_3^-) ranging in potential concentrations (volume weighted) in surface water from 110 to 190 μeqL^{-1} and 50 to 80 μeqL^{-1}, respectively, from west to east across the region. After considering stoichiometric oxidation reduction reactions of NH_4^+, NO_3^-, and SO_4^{2-}, and evaporative concentration, the calculated AFC of acidifying substances in $\mu eq\ H^+\ L^{-1}$ of lake inflow water can range from 30 to 60, 50 to 90, and 80 to 130, respectively, in the study watersheds of Minnesota, Wisconsin and Michigan. The pattern in AFC must be considered with $[NH_4^+]$, $[NO_3^-]$, and $[SO_4^{2-}]$ in assessing anthropogenic acid-precursor sources and implications for acidification of hydrologically isolated surface waters such as seepage lakes.

Second, present-day atmospheric inputs of acid forming compounds show strong relationships with acidity of lakes (after considering watershed factors) in northern Michigan/Ontario (12 acid lakes, 13% of the sampling sites) and northcentral Wisconsin (20 acid lakes, 4% of the sampling sites). Inputs of AFC to these lakes appear to have exceeded the ability of the watersheds to neutralize precipitation inputs and the resulting lake water remains acidic. These mass balance calculations show that acidic atmospheric inputs can dominate lake chemistry, forcing close chemical correspondence between precipitaton inputs and acidic surface water composition. For the acid lakes studied in this analysis, the variations in the resultant lake water H^+ concentration were most strongly related to acid forming inputs and in-lake processes. Additional studies of similarities and differences in the receiving lake basins are required to determine whether the present atmopheric inputs are in equilibrium with the observed surface water chemistry or whether these inputs have induced changes in the aquatic and terrestrial resources of these regions.

ACKNOWLEDGEMENTS

The data used in this report represents the combined efforts of many individuals from the University of Minnesota-Duluth, the USEPA Environmental Research Laboratory-Duluth and Corvallis, Wisconsin Department of Natural Resources, Minnesota Pollution Control Agency, Michigan Department of Natural Resources, U.S. Forest Service, U.S. Geological Survey, U.S. Fish and Wildlife Service, Ontario Ministry of the Environment, Environment Canada, NADP-Natural Resources Ecology Laboratory, Ft. Collins, Colorado and NADP site investigators. We acknowledge their hard work and thank them for their initiative and endurance. This work was supported by the USEPA National Acid Precipitation Assessment Program through Coöperative Agreements CR809412 with the University of Minnesota, Duluth and CR809484 with the Wisconsin Department of Resources, and Interagency Agreement AD12F3581 with the USDA Forest Service, Duluth, MN.

REFERENCES

Altshuller, A.P.: 1980, Environ. Sci. Technol. 14, 1337.

APHA: 1980, Standard Methods for the Examination of Wastewater. 15th ed. Am. Public Health Assoc., Washington, D.C., pp. 30-31.

Brosset, C.: 1980, 'Atmospheric Sulfur Deposition', Shriner, D.S., Richmond, C.R., Lindberg, S.E., Eds., Ann Arbor Science Publishers, Inc.: Ann Arbor, Michigan, pp. 145-152.

Calvert, J.G., Ed.: 1984, SO_2, NO and NO_2 Oxidation Mechanisms: Atmospheric Considerations, Vol. 3, Acid Precipitation Series, Butterworth Publishers: Boston, Massachusetts, 254 pp.

Chan, W.H., Tang, A.J.S., Chung, D.H.S., and Lusis, M.A.: 1985, Air Concentration and Dry Deposition Fields in Ontario - 1982, Ontario MOE, Toronto, Canada, Report No. ARB-001-85-AQM.

Cogbill, C.V., and Likens, G.E.: 1982, Water Resources Res. 10, 1133.

Eilers, J.M., Glass, G.E., Webster, K.E., and Rogalla, J.A.: 1983, Can. J. Fish. Aquat. Sci. 40, 1896.

Geraghty, J.J., Miller, D.W., Van der Leeden, F., and Troise, F.L.: 1973, Water Atlas of the U.S., Water Information Center Pub.: New York, 128 pp.

Glass, G.E.: 1984, Acid Rain Progress Report: Technical Findings and Research Issues. Testimony to the Northeast-Midwest Congressional Coalition, U.S. House of Representatives, James L. Oberstar and Frank Horton, Co-Chairmen. Hearing held March 30, 1984, Minneapolis, MN. 25 pp.

Glass, G.E., and Brydges, T.G.: 1982, 'Acid Rain/Fisheries', Proc. Symp. on Acid Precipitation and Fishery Impacts in Northeastern North America, Johnson, R.E., and Haines, T.A., Eds., Cornell University: Ithaca, New York, pp. 265-286.

Glass, G.E., and Loucks, O.L.: 1986, Environ. Sci. Technol., 20, 35.

Glass, G.E., and Loucks, O.L.: 1980, U.S. Environ. Prot. Agency, EPA-600/3-80-044: Duluth, Minnesota, 186 pp.

Gorham, E., Martin, F.B., and Litzau, J.T.: 1984, Science 225, 407.

Guiang, S.F., Krupa, S.V., and Pratt, G.C.: 1984, Atmos. Environ. 18, 1677.

Likens, G.E., and Butler, T.J.: 1981, Atmos. Environ. 15, 1103.

Munger, J.W.: 1982, Atmos. Environ. 16, 1633.

National Research Council/National Academy of Science: 1983, Acid Deposition: Atmospheric Processes in Eastern North America, A Review of Current Scientific Understanding, National Academy Press: Washington, D.C., 375 pp.

Quint, J., and Viallard, A.: 1978, J. Solution Chem. 7, 533.

Rapp, G., Jr., Allert, J.D., Liukkonen, B.W., Ilse, J.A., Loucks, O.L., and Glass, G.E.: 1985, Environ. Internatl. 11, 425.

Rapp, G., Jr., Liukkonen, B.W., Allert, J.D., Sorensen, J.A., Glass, G.E., and Loucks, O.L.: 1986, Environ. Geol., in press.

Schnoor, J.L., Palmer, W.D., Jr., and Glass, G.E.: 1983, 'Modeling of Total Acid Precipitation Impacts, Schnoor, J.L., Ed., Ann Arbor Science Publishers, Inc.: Ann Arbor, Michigan, pp. 155-173.

Schnoor, J.L., and Stumm, W.: 1984, Chemical Processes in Lakes, Stumm, W., Ed., Wiley Interscience: New York.

Stensland, G.J., and Semonin, R.G.: 1982, Bull. Am. Meteorol. Soc. 63, 1277.

Stumm, W., and Morgan, J.J.: 1981, Aquatic Chemistry, Wiley Interscience, New York, NY.

Stumm, W., Morgan, J.J., and Schnoor, J.L.: 1983, Naturwissenschaften 70, 216.

Thornton, J.D., and Eisenreich, S.J.: 1982, Atmos. Environ. 8, 1945.

U.S. Environmental Protection Agency: 1976, Quality Criteria for Water, Office of Water and Hazardous Materials, USEPA: Washington, D.C., 256 pp.

Verry, E.S.: 1983, Water Resources Res. 19, 454.

Wisconsin Department of Natural Resources: 1984, Distribution and Inventory of Softwater Lakes in Wisconsin. Progress Report CR809484, Rhinelander, Wisconsin, March, 20 pp.

THE CATION DENUDATION RATE MODEL - ITS CONTINUED VALIDITY

Mary E. Thompson
Aquatic Physics and Systems Division
National Water Research Institute
Canada Centre for Inland Waters
Burlington, Ontario, Canada
L7R 4A6

ABSTRACT. The rationale of the cation denudation rate (CDR) model is reviewed, and the history of the application of the model is described. The CDR model was first applied to rivers in Nova Scotia, some of which are strongly colored and carry significant amounts of organic matter. Analytical problems encountered in strongly colored coastal rivers in the determinations of sulfate ions, and their effects upon the application of the model are discussed. The methyl thymol blue (MTB) method for sulfate incorporates part of the organic anion, here called soluble reactive organic anion, or SRA^-. Charge balance is obtained between MTB 'sulfate' and major ionic species. The CDR model requires charge balance; therefore MTB 'sulfate' yields are used for the model. This is equivalent to using sulfate plus SRA^- yields for the model. Comparing the status of these rivers in the early 1970s and in the early 1980s, we find that the excess sulfate yields are lower in the latter period, and the mean pHs are generally higher. The model is used to predict the mean pH of these rivers to be expected in the absence of SRA^-, or of excess sulfate, or of both, that is, for simple cation-bicarbonate waters.

1. INTRODUCTION

The CDR model was developed during 1979 and 1980, being first applied to data for rivers in Nova Scotia and Newfoundland, for which relatively long chemical records of data exist. The CDR model compares the rate of atmospheric deposition of acid-forming species, especially sulfate, with the rate of chemical weathering of base cations, or, more specifically, the rate at which rivers carry off the four major cations, Ca^{++}, Mg^{++}, Na^+ and K^+. In the absence of information on rates of atmospheric deposition, the river yields of excess sulfate are used as a proxy. Some of these rivers, especially those in southern Nova Scotia, are strongly

Water, Air, and Soil Pollution **31** (1986) 17–26.

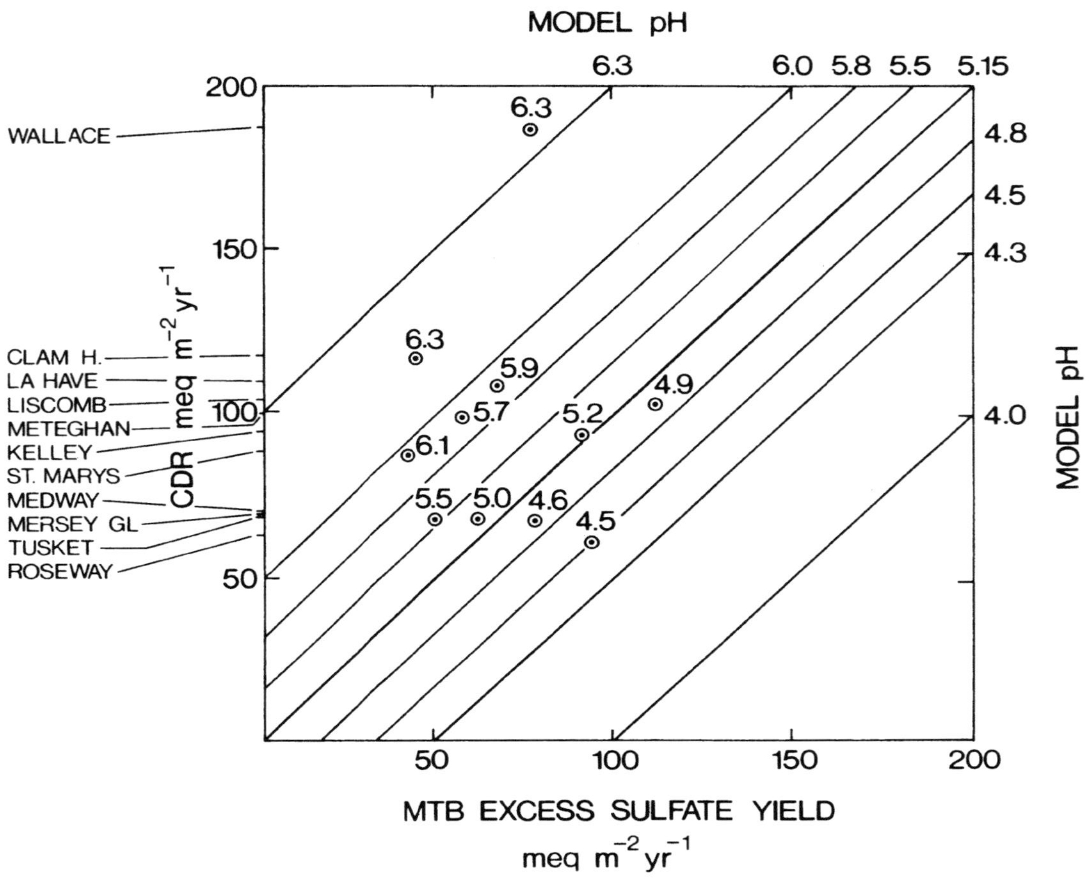

Figure 1. The CDR model plot for eleven rivers in Nova Scotia, 1981 - 1983. Mean observed pHs are shown on the plot, model predicted pHs on the outside. (Mersey GL indicates the Mersey River sampled at George Lake.)

tea-colored, and carry significant concentrations of organic anions. The exact nature, and seasonal variations in the properties of these organic compounds have yet to be sorted out, but the influence of the organic anion on the colorimetric (MTB) method for sulfate analysis was not yet appreciated in 1980, although the model worked well, even on the more strongly colored rivers.

The rationale of the CDR model is reviewed, and the total organic anion, A^-, calculated by the method of Oliver et al. (1983) is subdivided into reactive (SRA^-) and unreactive (SUA^-) fractions, which explain why the model works well. Some data on pH and the relative concentrations of DOC, A^-, SRA^-, SUA^- and [Al +Fe] are presented. Sulfate yields, by methods that are not affected by organics, are compared for the two periods 1971-1973 and 1981-1983. Finally, the model is used to predict the mean pH to be expected in the absence of one or both of the acidifying anions, sulfate and SRA^-.

The analytical data are supplied by the Water Quality Branch Laboratory, Atlantic Region, Environment Canada, at Moncton, New Brunswick. The data are stored in NAQUADAT, Canada's national water quality data storage system (Demayo, 1970). The eleven rivers in Nova Scotia to be discussed here are gauged by Water Survey of Canada.

2. DESCRIPTION OF THE MODEL

The CDR model plot has been used to predict the change in mean annual pH to be expected if the excess sulfate yields (ESYs) are reduced and if CDRs are constant (United States - Canada Memorandum of Intent, 1983). CDRs, calculated as mean discharge-weighted, seasalt-corrected sum of cations (DWSU+) times runoff, actually vary from year to year because runoff varies. There is little evidence that mean DWSU+ varies significantly in most of these rivers, if a sufficiently large number of samples are used to calculate the mean. This sufficiently large number is commonly about 35; for which the standard error of the mean is commonly 1 to 2 $\mu eq\ L^{-1}$, therefore, the data are averaged over three years or more, because the rivers are sampled only about 12 times a year.

The CDR model plot is shown (Figure 1) for data averaged over the period November 1981 to December 1983, and the data used for the model plot are listed in Table I. The cation denudation rate (CDR) model is based upon two simple principles:

At pHs between 7.0 and 5.2, a charge balance should exist among these major ion species

$$Ca^{++} + Mg^{++} + Na^{+} + K^{+} = Cl^{-} + SO_4^{--} + HCO_3^{-} \qquad (1)$$

where all species are in μeq L^{-1}, and

$$pH = pK + pP_{CO_2} - pHCO_3^{-} \qquad (2)$$

where pP_{CO_2} is the negative logarithm of the CO_2 pressure, $pHCO_3^-$ is the negative logarithm of the bicarbonate ion concentration, and pK is the negative logarithm of the equilibrium constant for the reaction:

$$H_2O_{aq} + CO_{2g} = H^{+}_{aq} + HCO_3^{-}{}_{aq} \qquad (3)$$

where aq stands for aqueous, i.e., species dissolved in water, and g stands for gaseous.

This reaction is used because it is written in terms of all but one of the variables measured or to be calculated; the equilibrium constant for this reaction has the advantage of being relatively insensitive to temperature over the temperature range of interest, pK being 7.8 at 25° and 7.7 at 0° (Garrels and Christ, 1965). As temperature data are not generally available for application of the model, a temperature-insensitive constant is desirable.

A special case of Equation (1) above occurs in acid waters, where the charge balance includes H^+ but not bicarbonate ion

$$Ca^{++} + Mg^{++} + Na^{+} + K^{+} + H^{+} = Cl^{-} + SO_4^{--} \qquad (4)$$

and $pH = -\log [H^+]$.

3. SULFATE AND ORGANIC ANIONS

During 1981, it was found that sulfate measured in strongly colored waters by the colorimetric methyl thymol blue (MTB) method was reported as falsely high (Kerekes et al., 1984; Cheam et al., 1985). Thus, for the strongly colored rivers of southern Nova Scotia, the excess sulfate yields were overestimated. It is now known that a large part of the acid state of those rivers is due to the presence of acid organic anions (Oliver et al., 1983).

One reason that the data fit the model well is that charge balance between $Ca^{++} + Mg^{++} + Na^{+} + K^{+}$ and $SO_4^{--} + Cl^{-} + HCO_3^{-}$ was, and still is, more closely obtained using MTB sulfate than using ion chromatographic (IC) sulfate. That is because the MTB sulfate method incorporates some of the organic anions,

TABLE I. **Runoff, discharge-weighted sum of cations, CDR, MTB SO_4^{--}, yield, mean measured pH and CDR model-predicted pH are shown for the period November 1981 to December 1983**

River	Latitude Degrees N	Longitude Degrees W	Runoff $m\ yr^{-1}$	DWSU+ $\mu eq\ L^{-1}$	CDR $meq\ m^{-2}\ yr^{-1}$	MTB SO_4^{--} Yield $meq\ m^{-2}\ yr^{-1}$	Mean pH	Model pH
Meteghan	44-12-58	66-07-02	0.724	137.0	99.0	58.2	5.7	5.9
Kelley	45-35-10	64-27-05	1.17	79.4	93.0	91.4	5.2	5.2
Wallace	45-40-42	63-33-35	1.23	150.0	186.0	78.0	6.3	6.3
Tusket	43-55-24	65-52-12	0.783	87.2	68.3	78.4	4.6	5.0
Roseway	42-50-18	65-22-12	1.00	61.2	61.5	92.6	4.5	4.5
Mersey GL	44-20-00	65-12-16	1.00	69.2	69.3	60.9	5.0	5.2
Medway	44-10-24	64-39-36	0.860	80.7	69.4	49.8	5.5	5.6
La Have	44-26-48	64-35-30	0.874	125.0	109.0	68.4	5.9	5.9
Liscomb	45-00-54	62-05-45	1.32	77.5	103.0	112.2	4.9	5.0
St. Marys	45-10-24	61-58-54	0.824	105.0	86.6	41.6	6.1	6.0
Clam Harbour	45-28-06	61-27-36	0.877	133.0	116.0	44.9	6.3	6.2

TABLE II For 1983, the number of samples, n, the mean pH, the mean dissolved organic carbon (DOC) in mg L^{-1}, the mean organic anion, A^-, calcuated from pH and DOC by the method of Oliver et al. (1983), the mean SRA^- calculated as the difference between MTB and IC sulfates, the mean SUA^- calculated as the difference between A^- and SRA^-, the mean Al^{+3} + Fe^{+3}, and the mean IC sulfate. The last five species are in μeq L^{-1}.

	n	pH	DOC	A^-	SRA^-	SUA^-	[Al + Fe]	IC SO_4^{--}
			mg L^{-1}	μeq L^{-1}	μeq L^{-1}	μeq L^{-1}	μeq L^{-1}	μeq L^{-1}
Roseway	10	4.4	13.4	86	53	33	39	39.5
Tusket	10	4.6	11.0	75	39	36	44	56.7
Liscomb	11	4.8	9.2	69	39	30	40	34.5
Mersey	27	5.0	8.4	63	24	39	31	38.8
Medway	12	5.4	6.7	57	21	36	27	36.4
Kelley	11	5.6	6.4	57	19	38	27	54.2
Meteghan	17	5.7	9.1	81	26	55	39	55.7
La Have	11	5.9	6.4	58	16	42	27	58.8
St. Marys	10	6.0	4.6	42	14	28	20	36.4
Clam H.	9	6.1	6.0	55	11	44	15	45.5
Wallace	9	6.5	3.8	37	6	31	15	57.3

perhaps the more reactive anions, referred to here as soluble reactive organic anion, or SRA^-.

The concentration of organic anion, A^-, may be calculated from pH and dissolved organic carbon (DOC), using the equations of Oliver et al. (1983). A^- calculated in this way is always higher in concentration than A^- calculated as the difference between MTB and IC sulfates. It is now generally agreed that IC sulfate is sulfate ion and that MTB sulfate is sulfate plus some organic anion (Cheam et al., 1985). The MTB method apparently incorporates SRA^- because during the procedure, in the presence of excess Ba^{++}, the pH is adjusted to 11 or 12, at which pH part of the organic matter reacts with Ba^{++}. Schnitzer and Khan (1972) describe the use of $Ba(OH)_2$ to measure the acid content of humic material. The MTB method for sulfate, therefore, probably incorporates the acid part of the soluble organic anion; the remainder (the difference between A^- calculated by the method of Oliver et al. (1983) and SRA^-, or soluble unreactive A^- (SUA^-), is very likely complexed by Fe^{+++} and Al^{+++} and possibly other metals. Thus, when MTB sulfate yields are plotted on the CDR Model plot, actually sulfate ion plus SRA^- yields are plotted, and the data, as shown in Table I and on Figure 1, fit the model well.

Table II shows data for 1983, the number of samples, the mean pH, the mean DOC in mg L^{-1}, the mean A^- calculated by the method of Oliver et al. (1983) from pH and DOC, mean SRA^- calculated as the difference between MTB and IC sulfates, mean SUA^- calculated as the difference between A^- and SRA^-, the mean [Al + Fe] concentrations, and the mean IC sulfate concentrations for the eleven rivers in Nova Scotia. The last five species are all in μeq L^{-1}. The data show that DOC, SRA^- and [Al + Fe] are consistently lower in concentration in rivers with higher mean pH values. SUA^-, on the other hand, is more or less constant in concentration in all of the rivers. In the lower pH rivers, there is more than enough [Al + Fe] to account for all the SUA^- (assuming complexation), but, in the higher pH rivers, there is insufficient [Al + Fe] to complex with SUA^- on a one to one basis. Further explanation of the data shown in this table awaits a better evaluation of the accuracy in the calculation of A^- and in the measurement of the association constants between Al and Fe with A^-.

4. DECREASES IN SULFATE YIELDS

Over the period 1971 to mid 1973, sulfate analyses were done by the Thorin method, a titration carried out at low pH, which, if carefully done, yields results similar to those of the ion chromatograph (Cheam et al., 1985). Figure 2 shows excess sulfate yields (ESYs) (Thorin method) 1971 - 1973 compared with

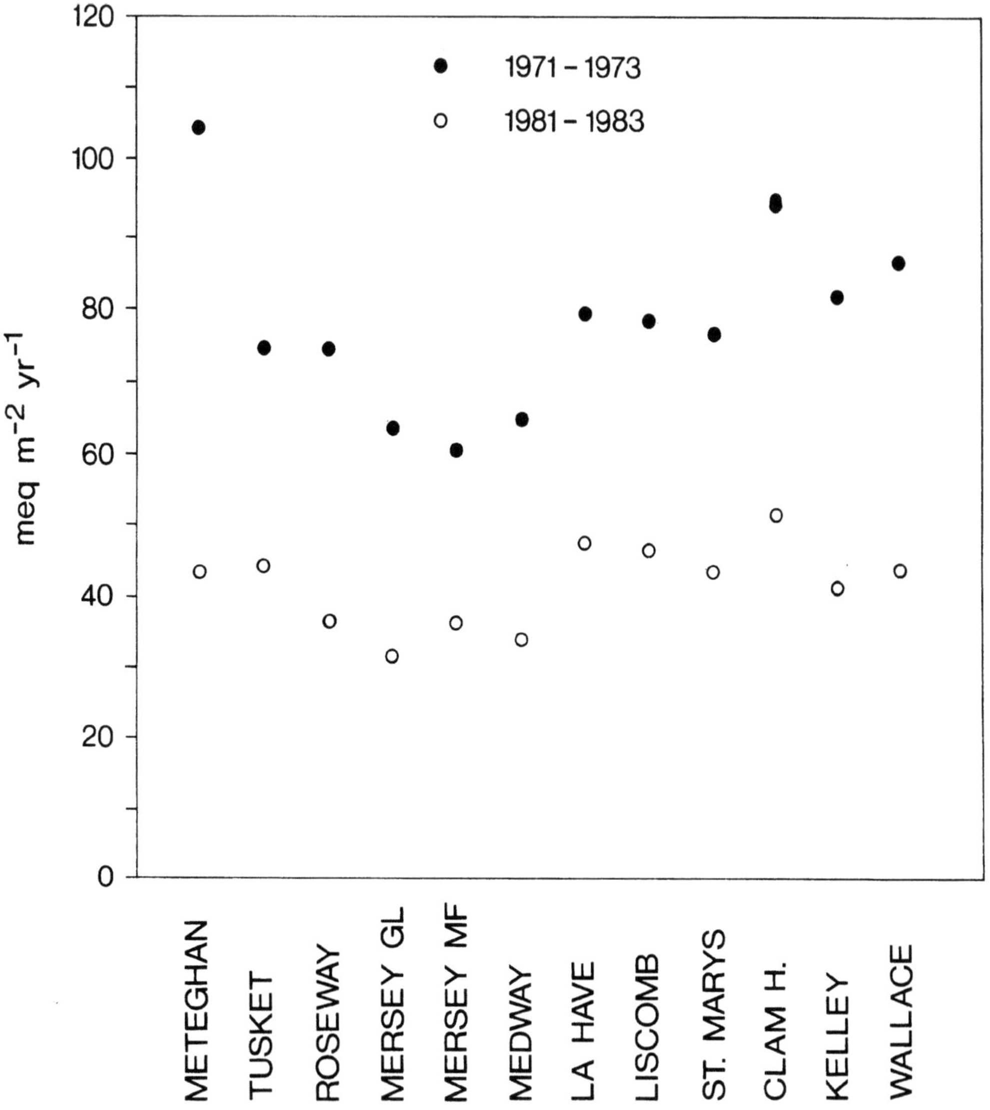

Figure 2. A comparison of excess sulfate yields of eleven rivers in Nova Scotia, 1971 - 1973 (solid circles), and 1981 - 1983 (open circles). (Mersey MF indicates the Mersey River samples at Mill Falls.)

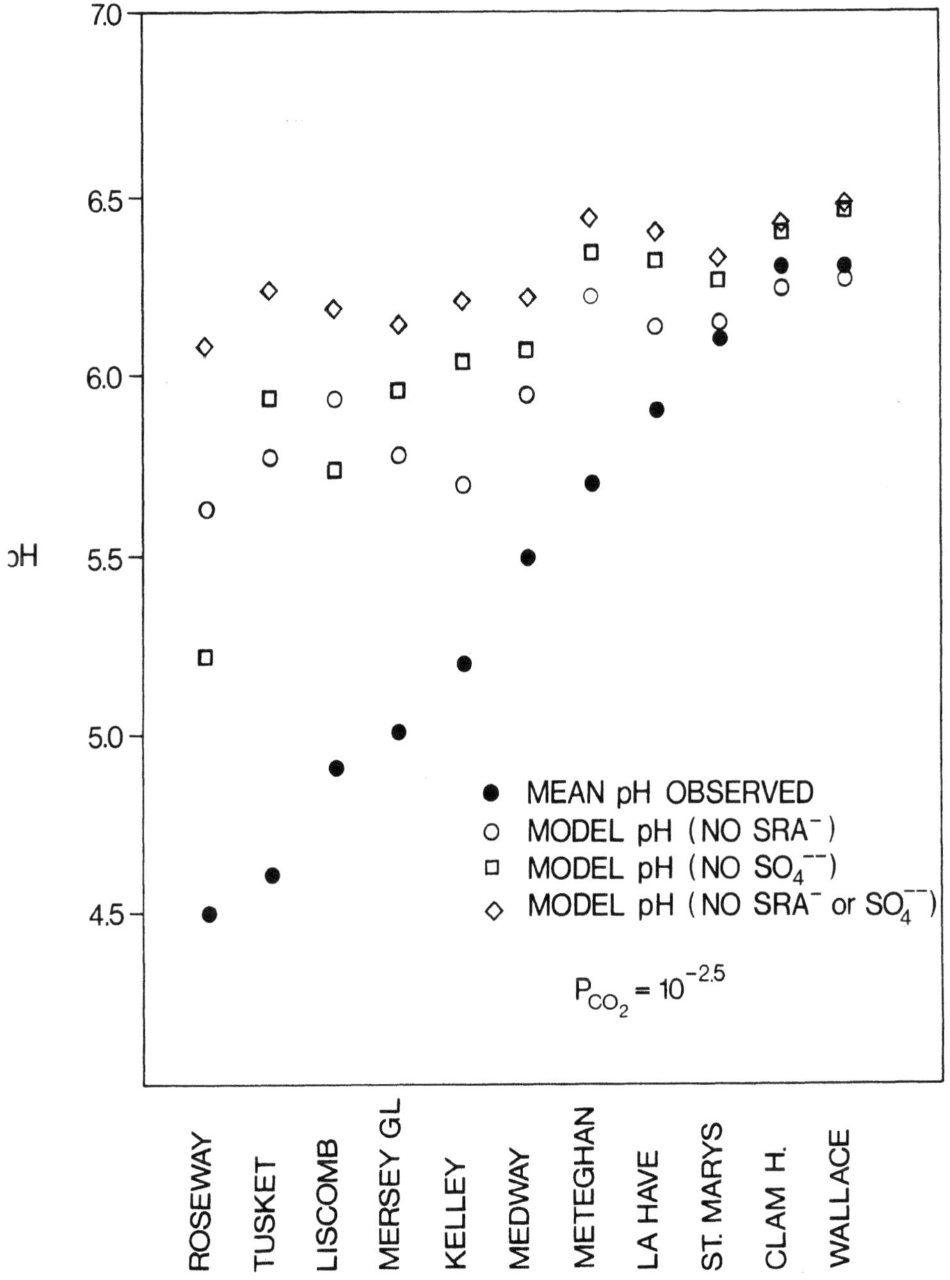

Figure 3. Mean pH, 1983, for eleven rivers in Nova Scotia (solid circles), model predicted mean pH in the absence of SRA^- (open circles), in the absence of sulfate (open squares), and in the absence of both, that is, for simple cation - bicarbonate waters (open diamonds).

ESYs 1981 - 1983 (IC method). The yields are consistently lower in the latter period, and the mean pHs, as would be expected, are somewhat higher for most of the rivers. The pH rise is not as pronounced as the decrease in ESYs, because of the influence of organic anions, especially in the more strongly colored rivers.

5. MODEL pH PREDICTIONS

As has been mentioned, the CDR model can be used to predict changes in pH if there is a change in ESYs or of yields of SRA^-. The relative impact of SRA^- and of IC sulfate can be seen on Figure 3. Figure 3 shows the 1983 mean pH, the rivers being arranged in order of increasing mean pH, the model-predicted pH in the absence of SRA^-, the model-predicted pH in the absence of IC SO_4^{--}, and the model-predicted pH in the absence of both SRA^- and IC SO_4^{--}, that is, for simple cation bicarbonate waters. For all but two of the rivers, the removal of SO_4^{--} has a more beneficial effect on pH than the removal of SRA^-. For these soft water rivers, mean pH cannot be expected ever to rise above 6.5; for the three hardest water rivers, the St. Marys, the Clam Harbour, and the Wallace, the removal of the acid anions would not significantly affect the pH.

REFERENCES

Cheam, V., Chau, A.S.Y. and Todd, S.: 1985, Sulfate in colored waters: Investigation on methodologies, data reliability and approaches for salvaging historical colorimetrical data. National Water Research Institute Contribution No. 85-95.

Demayo, A.: 1970, A storage and retrieval system for water quality data. Department of Energy, Mines and Resources, Inland Waters Branch, Report Series No. 9, Ottawa.

Garrels, R.M. and Christ, C.L.: 1965, Solutions, Minerals and Eqilibria. Harper & Row, New York, p. 89.

Kerekes, J., Howell, G. and Pollock, T.: 1984, Verh. Internat. Verein, Limnol., **22**, 1811.

Oliver, B.G., Thurman, E.M. and Malcolm, R.L.: 1983, Geoch. Cosmoch. Acta., **47**, 2031-2035.

Schnitzer, M. and Khan, S.U.: 1972, Humic substances in the environment. Marcel Dekker, Inc., New York, pp. 37-39.

Thompson, Mary E.: 1982, Water, Air and Soil Poll. **18**, 215.

United States-Canada Memorandum of Intent of Transboundary Air Pollution: 1983, Impact Assessment, Work Group 1, Final Report, January 1983, pp 3-206, 3-211.

A PREMILINARY ASSESSMENT OF THE CHEMICAL AND HYDROLOGICAL INTERACTION OF ACIDIC SNOWMELT WATER WITH THE TERRESTRIAL PORTION OF A CANADIAN SHIELD CATCHMENT

M.C.English[1], D.S. Jeffries[1], N.W.Foster[2], R.G.Semkin[1], and P.W.Hazlett[2]

1. National Water Research Institute, Canada Centre for Inland Waters, Environment Canada, 867 Lakeshore Rd. Burlington, Ontario, L7R 4A6, Canada
2. Great Lakes Forest Centre, Canadian Forestry Service, Agriculture Canada, 1219 Queen St E., Sault Ste Marie, Ontario, P6A 5M7, Canada

ABSTRACT. This study examines certain aspects of the hydrological and chemical interaction of springmelt runoff with the terrestrial portion of a Canadian Shield catchment. Though NO_3^- is the dominant anion in the snowpack for most of the winter, SO_4^{-2} deposition increases at the end of the winter season (March) such that it is the dominant anion in snowpack meltwater runoff. Pronounced formation of ice lenses in the snowpack are responsible for notable movement of H^+, NO_3^- and SO_4^{-2} ions downslope prior to contact with the ground. Subsurface meltwater flow through the lower portion of the slope is much more acidic (pH 5.18) than that flowing within the subsurface in the upper slope (pH 6.44).

1.Introduction

Reports summarizing research conducted on acidic snowmelt runoff conclude that in order to understand the significance of the event to terrestrial ecosystems it is imperative to understand the details of meltwater runoff both within the snowpack and through the ground (Seip,1980;Overrein et al.,1981).

Two factors combine to increase the significance of within-snowpack downslope transfer of water and ions. The first is that with prolonged runoff the effectiveness of the ice lenses to divert flow downslope prior to contact with the ground will decrease (Colbeck,1977). Secondly, during the initial melt period when downslope diversion of water is most efficient there is a disproportionate exsolving of ions from the snowpack (Seip,1980).

Till structure is such that the lower hillslope will receive a greater proportion of water than the upper portion. This hydrological pattern implies that weathering rates in the lower slope may become progressively greater from the top to the bottom of a hillslope.

The two purposes of this research are: i) to quantify the downslope transfer of water and ions along ice lenses in the snowpack prior to ground contact and ii) to examine whether elevational differences in soil percolate chemistry occur along a hillslope.

2.Methods

The study is ongoing at the Turkey Lakes Watershed (TLW) located 50 km north of Sault Ste Marie, Ontario, Canada (47°2' 14" N, 84° 25' 30" W).The study site is a

Water, Air, and Soil Pollution **31** (1986) 27–34.

hillslope (E.S.E. aspect) draining directly into Little Turkey Lake by subsurface seepage. The 20°slope is forested with maple (Acer saccharum Marsh.) and yellow birch (Betula lutea Michx. f.). The soil is an orthic ferro-humic podzol. The till structure consists of an upper ablation till (silty-loam) overlying a compact basal till (sand-loam),(Cowell and Wickware,1983). At the study site the ablation till ranges in thickness from < 0.50 m at the top of the slope to 1.2 m at the base of the slope.

2.1 Snowpack

Estimations of water equivalence and snowpack chemistry were made by conducting snow surveys adjacent to each subplot using an NRC snow kit (Adams and Barr,1974) and composite snow samples extracted with a clean 0.15m diameter plexiglass tube. Surveys were completed just prior to springmelt. Rain and snowfall after this date (27 March 1985) were sampled at the TLW meteorological station and at the study site, respectively.

2.2 Snowmelt runoff

At the study site a large scale snowmelt runoff plot was constructed. A fence, 1.0m in height delineates the perimeter of the 100.0m (upslope) by 10.0m structure (Figure 1). Within the plot, small trees and shrubs were cut to ground level and removed from the site. Larger trees and ground vegetation (< 0.10 m)) were left intact.

The runoff plot was divided into 5 subplots, each approximately 20.0m (upslope) by 10.0m. At the downslope end of each subplot a diversion wall (0.15m in height) was constructed. The entire plot including the fence and diversion walls were covered in clear 6 mil plastic just after leaf fall and prior to formation of the snowpack. The diversion walls channel meltwater reaching the plastic (ground) surface within each subplot to a 0.10m plastic tile at the side of the subplot. The meltwater from each subplot drains into a 10° V-notch weir box equipped with a water level recorder and a gas heater. Water samples for chemical analyses were taken frequently from the discharging tile at each subplot. The water level recorders were calibrated each time a sample was taken.

2.3 Mineral soil percolate

To examine the chemistry of percolate at the ablation-basal till boundary 6 lysimeters were placed at each of the three sampling locations along the hillslope (Figure 1).

Each lysimeter drains into a 15.0 L carbouy equipped with a syphon which automatically drains the carbouy once it has filled. A counter is connected to each carbouy and is activated each time the container fills up. In this manner it was possible to register the flow per unit of time. Samples for chemical analyses were taken daily.

2.4 Laboratory

The pH of all samples were read at the field laboratory within 12 hr of sampling. Analyses of SO_4^{-2} and NO_3^- were done using standard methods. Detection limits for SO_4^{-2} and NO_3^- are 2.08 x 10^{-3} meq·L^{-1} and 3.23 x 10^{-4} meq·L^{-1} respectively.

3. Results and Discussion

3.1 Snowpack

Just prior to the initiation of thaw events (mid March) standing water in the

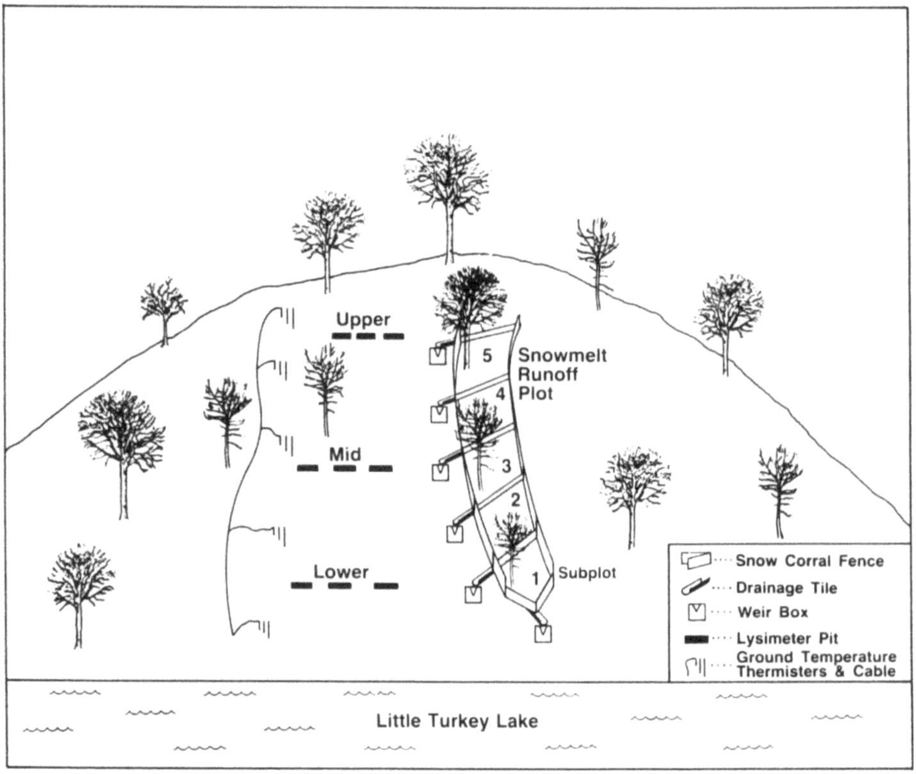

Figure 1: Schematic illustration of instrumentation on hillslope study site.

snowpack ranged between 0.15 m (upper hillslope, depth: 0.62 m, n= 15) and 0.27 m (lower hillslope, depth: 1.17 m, n = 15).

The SO_4^{-2} and NO_3^- content of the snowpack changed in mid March when warm southerly air invaded the region. Prior to mid March NO_3^- was the dominant anion in winter precipitation. With this change SO_4^{-2} became the dominant anion in the snowpack and later in snowpack runoff.

Although the dominant anion in the snowpack prior to mid March was NO_3^-, the difference between $[NO_3^-]$ and $[SO_4^{-2}]$ in ion equivalents at this time is not as significant as that reported by Barrie and Vet (1984) for snowpacks in southern Ontario and Quebec. Barrie and Vet report a $[NO_3^-]$ to $[SO_4^{-2}]$ ratio of approximately 3:1; the cold snowpack (prior to mid March) ratio determined in this study is approximately 1.2:1. During the same period Semkin and Jeffries (1985) report a ratio for the TLW snowpack of 1.7:1. The sampling sites reported for Barrie and Vet (1984) and Semkin and Jeffries (1985) are unforested open sites . The sampling sites for this study are all in a mature forest. Fowler (1980) attests to the effectiveness of vegetation as a trap for aerosols. Horntvedt et al. (1980) report enrichment of SO_4^{-2} under trees at Birkenes, Norway; they attribute this to dry deposition.

The dissimilar ratios of $[NO_3^-]$ to $[SO_4^{-2}]$ both recorded at TLW may be attributable to the presence or absence of above snowpack vegetation. The large difference noted between the results of Barrie and Vet(1984) and the TLW data are probably due to geographical differences ; the sites of the former study located closer to larger population centres.

3.2 Snowmelt runoff

Figure 2 illustrates SO_4^{-2}, NO_3^- and H^+ ion mass and snow water equivalence within the entire snowmelt runoff plot (included are snow and rainfall through the melt period). For comparison, total water and individual ion masses discharging from the separate subplots are shown.

Figure 2 indicates that downslope movement of meltwater within the snowpack occurs. The estimated water content of subplots 2 to 5 (inclusive) prior to melt is, in each case, greater than the corresponding measured output. At the lower end of the hillslope subplot 1 shows the opposite pattern; a significantly greater volume of water discharges from the plot than was estimated in the plot.

The water balance pattern recorded in the subplots is attributed to the effectiveness of ice lenses in the snowpack to divert melt-water downslope. Flow of meltwater into the subplots from outside the perimeter fence is thought minimal as pathways around the perimeter of the runoff plot were made throughout the winter and spring

In Figure 2 the mean water volume estimated within the entire snowmelt runoff structure exceeds the total amount of runoff measured from the 5 subplots. This error may in part be attributable to rainfall quantity measurement differences between the meteorological sampling site and the study site (2 km apart). It is probable that the difference between total water content and total runoff is not as pronounced as illustrated.

The mass of ions received in subplot 1 at the base of the hillslope is disproportionally greater than the volume of additional meltwater it receives from upslope sources (Figure 2). This is attributed to the pronounced exsolving of ions out of the snowpack during the early part of the melt when downslope diversion of meltwater by ice lenses would be most efficient.

The mean concentrations of SO_4^{-2}, NO_3^- and H^+ within subplot 1 were estimated to be .047, .035, and .054 (pH 4.27) meq·L$^-$, respectively. During the runoff period $[SO_4^{-2}]$, $[NO_3^-]$ and $[H^+]$ in the discharge from subplot 1 were 0.105, 0.065, and

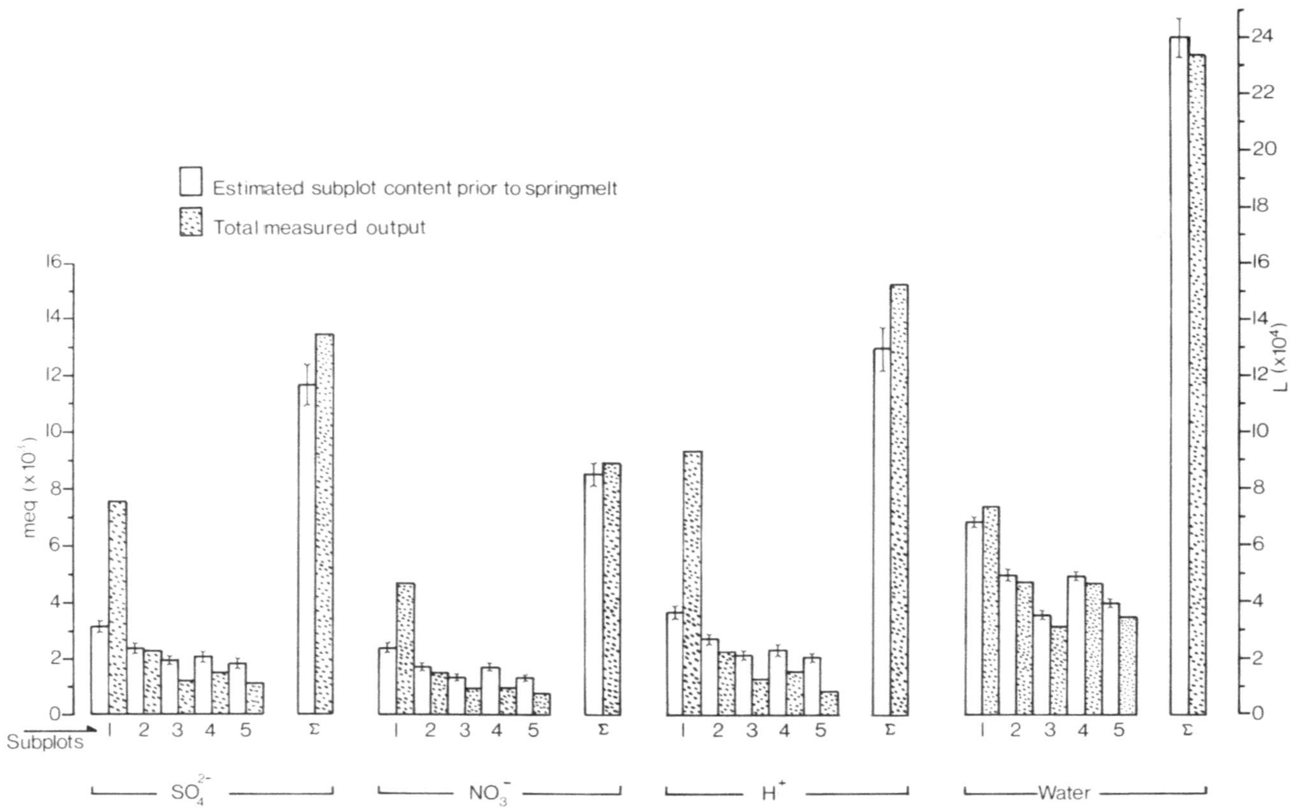

Figure 2: Estimated water, SO_4^{-2},NO_3^-, and H^+content of snowmelt runoff plots prior to runoff and that measured during the springmelt period. Range in the estimated plot values indicates expected maximum and minimum variation attributed to σ_x of the water and chemistry estimates.

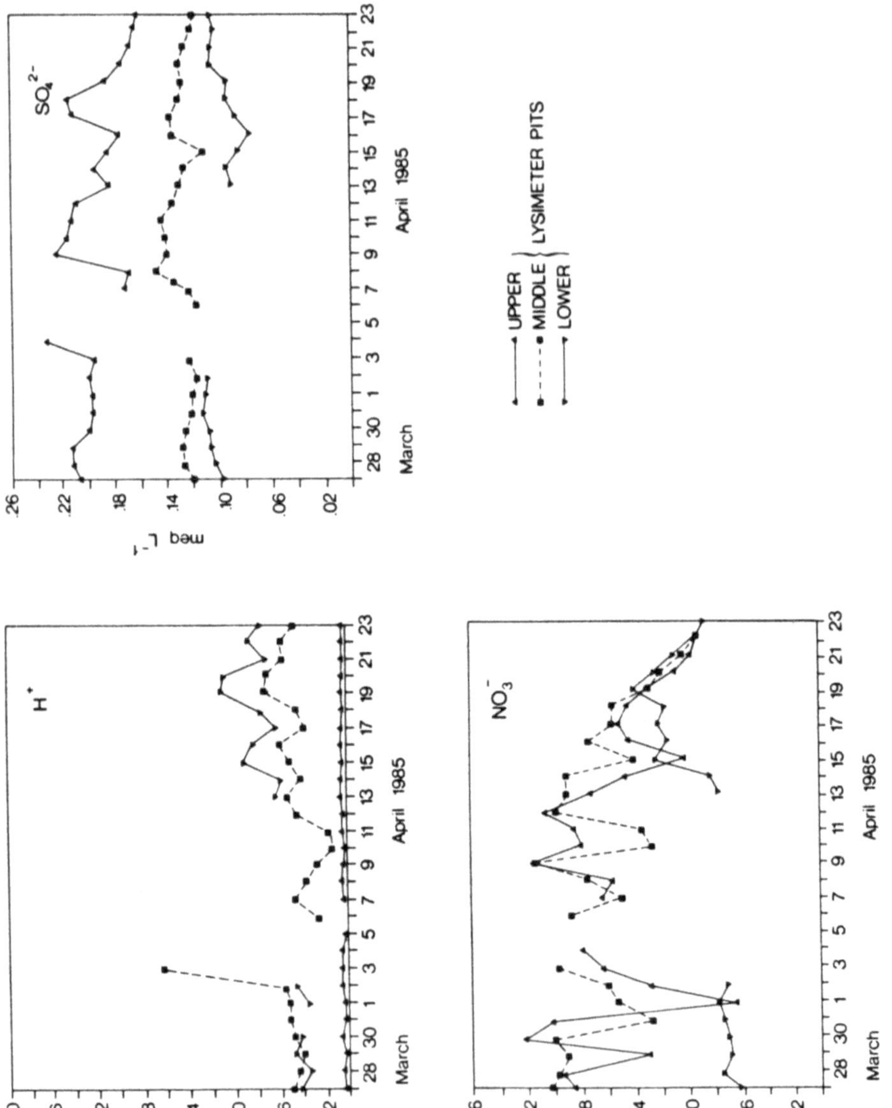

Figure 3: Mean SO_4^{-2}, NO_3^-, and H^+ concentrations in upper, middle and lower slope mineral soil lysimeters, spring 1985.

0.120 (pH 3.92) meq·L^{-1} respectively. For all ions the discharge in subplot 1 exceeds the estimated content of the plot. With the exception of SO_4^{-2} in subplot 2 the upslope subplots all show a reverse pattern; a mass reduction in discharge over the estimated content.

3.3 Subsurface runoff chemistry

The differences in concentration at the lower, middle and upper lysimeter sites for: i) H^+ and ii) SO_4^{-2} are statistically significant (t-test, 99% Confidence Interval).

At the upper lysimeter site meltwater infiltrating the ground surface had a mean $[H^+]$ of 2.34 x 10^{-2} meq·L^{-1}(pH 4.63), the mean $[H^+]$ of the sample discharging from the lysimeters was 3.63 x 10^{-4} meq·L^{-1}(pH 6.44). The subsurface water at the mid and lower hillslope sites had mean $[H^+]$ of 4.89 x 10^{-3} meq·L^{-1} (pH 5.31) and 6.59 x 10^{-3} meq·L^{-1}(pH 5.18). The corresponding mean $[H^+]$ of meltwater runoff from the snowpack at the respective slope locations was 3.16 x 10^{-2} meq·L^{-1}(pH 4.50) and 0.13 meq·L^{-1} (pH 3.89).

The pattern of $[H^+]$ in the subsurface waters (Figure 3) along the hillslope is thought to be, in part, a function of hydrology. Runoff is largely diverted downslope along the contact zone between the ablation and basal tills. Consequently the lower slope receives significantly more subsurface flow than the upper slope. In addition, snow accumulation (greater at the base of the slope) and the reported downslope diversion of meltwater within the snowpack all contribute to an increased loading of H^+ into the lower slope.

The $[SO_4^{-2}]$ is highest (mean: 0.19 meq·L^{-1}) in the upper hillslope and progressively lower downslope; midslope site,mean: 0.12 meq·L^{-1} and the lower hillslope values, mean: 0.09 meq·L^{-1}. The differences are attributed to two factors; both relate to the disproportionate volume of water the lower portions of the hillslope receive during periods of runoff. The initial factor is dilution. The second factor is scouring efficiency. It is assumed that the scouring of the soil-ablation till matrix is most efficient at the base of the slope. In the lower slope the $[SO_4^{-2}]$ of snowmelt water entering the soil, (mean: 0.11 meq· L^{-1}), is essentially equal to that determined for the mineral soil percolate, (mean: 0.12 meq·L^{-1}). The differences are greater the further upslope the samples are extracted. For example, the mean $[SO_4^{-2}]$ in the mineral soil percolate is 0.19 meq· L^{-1} while the mean concentration in the infiltrating snowmelt water (subplot 5) is 0.03 meq· L^{-1}.

Of note in Figure 3 is the increasing $[NO_3^-]$ in the lower slope at a time when discharge is increasing (14-20 April 1985) and the concentrations in the middle and upper hillslope site lysimeter percolate are decreasing. The subsurface and within-snowpack transfer of NO_3^- downslope accounts for the increased $[NO_3^-]$ in percolate at the base of the hillslope.

4. Conclusions

1. Though NO_3^- is the dominant anion in the forest snowpack prior to the infusion of warm southerly air masses in March, contribution of SO_4^{-2} from the vegetation (presumably as dry deposition) and the dominance of SO_4^{-2} over NO_3^- in the precipitation beginning in March results in SO_4^{-2} as the dominant anion in snowpack runoff.
2. The diversion of ions downslope within the snowpack results in a disproportionate loading of ions to different portions of the hillslope; reduced ion contribution to the upper portion of the hillslope and enriched loading in the lower section. This process reduces the interaction of snowpack source ions with the soil-till matrix, thus the potential buffering of the terrestrial sustem to atmospheric

source acids during a significant portion of the annual precipitation input to the terrestrial system is reduced.

3. It is thought the disproportionate quantity of water flowing through the upper ablation tills in the lower portions of the hillslope has resulted in increased weathering and therefore a decreased potential to buffer incoming mineral acids.

Acknowledgments

The authors acknowledge the assistance of M.Griffin, S. Schiff, R. Neureuther, M. Seymour and E. Walker, the Technical Operations Division at the Canada Centre for Inland Waters, Burlington, Ontario and the personal at the LRTAP laboratory at Great Lakes Forest Centre, Sault Ste Marie, Ontario. Jim English critically edited the manuscript.

Bibliography

Adams,W.P. and Barr,D.R.:1974, Occ. Paper No. 3 Dept. Geography, Trent University. 11-26.

Barrie,L.A. and Vet,R.J.: 1984, Atmospheric Environment.**18**(7):1459-1469.

Colbeck,S.C.:1977, J. Glaciology. **19**: 571-588.

Cowell,D.W. and Wickware,G.M.:1983, Turkey Lakes Watershed Study Unpublished Report Series. TLW-83-08. 25p.

Fowler,D.:1980, Ecological impact of acid precipitation,Proceedings of an international conference, Sandefjord,Norway. 22-32.

Horntvedt,R., Dollard,G.J., and Joranger,E.:1980, Ecological impact of acidic precipitation. D.Drablos and A.Tollan (eds.) Sandefjord, Norway. 192-193.

Overrein,L.N., Seip,H.M. and Tollan,A.:1981, Acidic precipitation. Effects on forest and fish.Final report of the SNSF project. 1972-1980. 175 p.

Seip, H.M.:1980, In: Effects of acidic precipitation on terrestrial ecosystems. T.C. Hutchinson and M. Havas (eds.) NATO-API series, Plenum Press, NYC.

Semkin,R.G. and Jeffries,D.S.:1985, Int. Symposium on acidic precipitation. Minette, Ontario.

TEMPORAL CHEMICAL VARIABILITY IN ACID SENSITIVE HIGH ELEVATION LAKES

E. B. Welch, D. E. Spyridakis, and T. Smayda
Department of Civil Engineering
University of Washington
Seattle, WA 98195 U.S.A.

ABSTRACT. Chemical indicators suggest that slight, but discernable acidification occurs during smowmelt in some highly sensitive Cascade Mountain Lakes (mean alkalinity 20 µeq L^{-1}). Although some SO_4 in the lakes (mean 13 µeq L^{-1}) comes from local geologic sources, several considerations suggest that some also comes from atmospheric deposition and anthropogenic sources. If sampling is stratified, the relatively low lake-to-lake and year-to-year variability in chemical constituents demonstrates that these highly sensitive lakes represent excellent indicators of acidification. The contention that precipitation pH >4.6 is needed for protection of sensitive lakes is supported. However, the slight but temporary acidification effect currently detectable during snowmelt, suggests that for adequate protection of these highly sensitive lakes, precipitation pH should be >4.7 to 4.8.

1. INTRODUCTION

Washington's Cascade Mountain lakes are some of the most sensitive waters in the United States (Omernik and Powers, 1982). Their sensitivity is due to the resistent bedrock, thin soils and relatively small steep-sided catchments. Thus, their response to increased acid deposition would be rather quick and extensive.

The Cascades receive precipitation that is acidified by anthropogenic sources of strong acid. An annual mean pH of precipitation as low as 4.7 was determined in 1981 (Logan et al., 1982). Subsequent monitoring in 1984 showed a mean pH of 5.1 (Assere and Duncan, 1984). Although a precipitation pH of 4.7 could be within a range conceivably caused by natural sources of S (Charlson and Rodhe, 1982), precipitation and cored snow from the Cascades and the Olympic Peninsula, where anthropogenic impact on precipitation is slight, indicate that some of the Cascade precipitation is man-caused (Welch et al., 1984; Vong and Waggoner, 1983). While that level is low compared to areas where damage from acidification has been reported, it is nevertheless near that considered critical (4.6) for sensitive receptors (Evans et al., 1981).

Water, Air, and Soil Pollution **31** (1986) 35–44.

Important questions regarding potential acidification in Cascade lakes are: (1) What is the distribution of the most sensitive lakes; 2) Is man-caused acidification occurring and where; and 3) Are there seasonal trends or episodes in lake chemical constituents that would indicate acidification?

The first question can be most accurately answered from a large, statistically designed survey such as the one conducted by EPA for the fall of 1985. In an attempt to answer questions 2 and 3, seven lakes in the Alpine Lakes Wilderness Area were sampled six times during 1984 and twice during 1983 to study seasonal changes in relatively fast- and slow-flush lakes. The three slow-flush lakes were also sampled twice in 1982. The general location of these lakes is shown in Figure 1. For morphometric and location descriptions of these lakes, see Welch and Spyridakis (1984).

The fast-flush lakes, Rock, Lunker, Hi Lo and Quartz, have watershed:lake volume ratios ranging from 6 to 11. Angeline, Big Heart and Copper are slow-flush lakes with watershed:lake volume ratios less than 0.1. Although water inputs to determine actual flushing rates are unknown, an assumption of constant precipitation and evaporation among the watersheds would yield flushing rates proportional to the watershed:lake volume ratios.

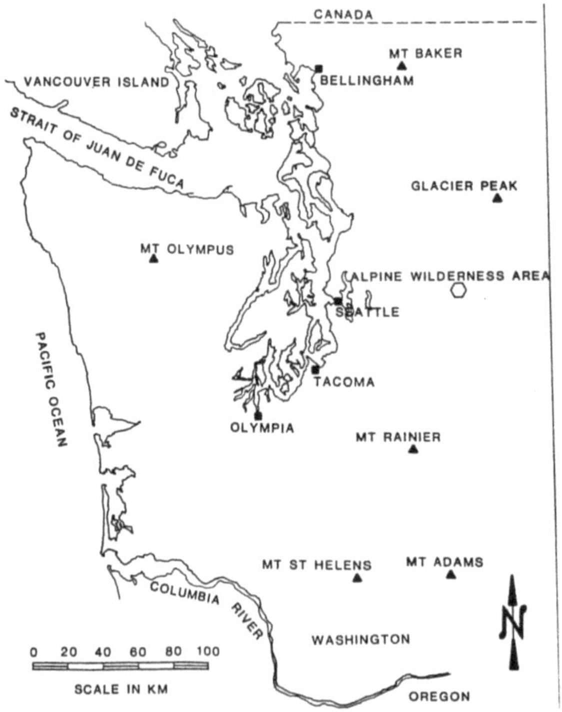

Figure 1. Location of Alpine Wilderness Area lakes in the Washington Cascade mountains.

2. METHODS AND MATERIALS

2.1 Sampling

Water samples were collected six times at about 0.5 m below the surface at a distance of about 20 m from shore. A hand line was used with a float secured 0.5 m from the weighted bottle. On one occasion samples were collected at 20 m intervals in the deep lakes. Bottles were acid washed and multiply, distilled-water rinsed in the laboratory and subsequently rinsed before filling. Upon return to the laboratory, one fraction was filtered through a rinsed 0.45 µm Millipore filter and the filtrate frozen for determination of cations and anions while the unfiltered fraction was refrigerated.

2.2 Analysis

The anions chloride, sulfate and nitrate were determined with a Dionex QIC ion chromatograph fitted with an AS-4 column. Analytical precision for Cl, SO_4 and NO_3 was ± 2.9%, ± 3.3%, and ± 50% of the mean, respectively. These analyses were performed by C. Duncan, Central Washington University.

Major cations (Na, Ca, Mg and K) were determined by atomic absorption spectrometry with an IL model 551 using the methodology of EPA (1979). Precision for the metals ranged from ± 1.1% to ± 6.5% of the mean, and the accuracy error ranged from 0.4% to 3.8% of the true value.

Alkalinity was determined using the conductometric method (Golterman, 1969, pp. 32 and 141). Analytical precision was ± 10% of the mean, and accuracy error was within 1% of the true value. Comparison with Gran titration showed nearly identical results.

The pH of samples was determined in the laboratory with a Corning Digital 112 meter equipped with a high quality glass electrode after a 48-hour equilibration to room temperature (20° C). Determinations were made under quiescent conditions after instrument drifting had ceased (5 to 30 min). Following calibration against pH 4 and 7 buffers, a dilute HCl buffer of pH 5.0 ± 0.01 (calibrated against a Na_2CO_3 primary standard solution) was used as a validity check. Determinations were precise to 0.5 to 1.0% of the mean.

Equilibrium pH was calculated by assuming that HCO_3 is equal to alkalinity and $P_{CO2} = 10^{-3.40}$ ATM. Agreement between measured and calculated pH were generally good (mean difference 0.27 ± 4.5%); however, for a few samples from lakes with alkalinity <20 µeq L^{-1}, calculations exceeded measured pH by about 0.5 units. This may be due in part to oxidation of reduced sulfur following sample collection or inaccurate pH measurements in low-alkalinity samples.

Specific conductance was determined at 25 ± 0.5^{o}C with a Sybron-Barnstead model PM 70 CB conductivity bridge. Precision was ± 1.1% of the mean and the accuracy error was within 1.1% of the true value. Color was determined with an Hellige Aqua tester.

Ammonia nitrogen was determined with the phenate method (APHA, 1980 and EPA, 1979). Lakes with low alkalinity contained ammonia near the detection limit of the method (1.0 μg L^{-1}).

Except for about 10 samples from 4 lakes in 1984, the charge balance showed an anion-cation difference of <15%; for all samples the mean difference in % was 15 ± 14.7. The mean difference in % between measured and calculated specific conductance was 6.2 ± 12. The mean difference in % between cations and anions in 1983 was 12 ± 8.

2.3 Acidification Indices

Two indices were used to determine if acidification is presently occurring in any Cascade lakes. The first index originates from Henriksen's (1980) model:

$$Alk = 0.91\ (Ca^{*} + Mg^{*}) - Alk$$

where 0.91 $(Ca^{*} + Mg^{*})$ represents preacidification alkalinity based on a relationship of alkalinity to $Ca^{*} + Mg^{*}$ (* indicates sea salt correction) in unacidified lakes of the world. Alk is current alkalinity of the lake in question. Ca and Mg are assumed not to change appreciably with acidification. Evidence for that has been presented elsewhere (Henriksen, 1980; Wright 1983; and Wright and Henriksen, 1983). As SO_4 replaces alkalinity in the acidification process, Alk increases. Example areas showing positive acidification are as follows (Wright, 1983): 30 lakes in the Rocky Mountains, 9 μeq L^{-1}; 206 in the Adirondacks, 119 μeq l^{-1}; 208 at Sudbury, Ontario, 192 μeq L^{-1}.

The ratio of Alk/(Ca + Mg + Na + K) is the second index that has been used to detect acidification (Burns et al., 1981). Essentially the same principle holds with this index as with Henriksen's model. This ratio decreases as SO_4 replaces Alk and the cations remain relatively stable, or if cations increase,
the increase is less than SO_4. Burns et al. found this ratio to range from 0.78 to 0.94 in areas receiving no or little acid deposition, while average values from areas receiving acid deposition were as follows: 44 lakes in Scandinavia, 0.064; 35 in North Carolina, 0.56; 38 in New Hampshire, 0.31.

3. RESULTS

3.1 Lake Chemical Characteristics

These Alpine Lakes Wilderness Area (ALWA) lakes are some of the most sensitive in the Cascades. They have clear water with low color content (5 Pt units). Alkalinity in the seven studied lakes averaged 23.3 ± 12 μeq L-1 for two sampling dates in 1983 and 20.3 ± 14 μeq L-1 for five dates in 1984. Quartz Lake had the lowest alkalinity (7.3 ± 1.0 μeq L-1) and Copper Lake had the highest (55 ± 15 μeq L-1). The pH was infrequently less than 6.0 and was never below 5.5. SO4 was consistent among the lakes with lake means ranging from 9.5 to 17.2 μeq L-1 with a six-lake mean of 13 ± 3.1 μeq L-1, except for a rather high and constant level in Quartz Lake (average 33.4 ± 5.9 μeq L^{-1}).

3.2 Snowmelt Effects on Lake Chemical Content

Distinct seasonal trends were observed in three of the four fast-flush lakes in the ALWA. These lakes (Rock, Hi Lo, Lunker and Quartz) are considered to flush relatively fast because of their high watershed area:lake volume ratios. Figure 2 shows that alkalinity increased by about four-fold in Rock Lake following the one to two month snowmelt period (May through June), while that variable about doubled in Hi Lo and Lunker. Alkalinity in Quartz was unaffected by runoff.

The recovery of alkalinity in the fast-flush lakes during summer may represent a gradual replacement of low alkalinity snowmelt water with groundwater. The greater response in Rock Lake was partly due to entrainment of deeper, higher alkalinity water. Replacement of lake water with snowmelt water can produce episodes of low alkalinity, as well as low pH, if SO_4, NO_3 and H^+ in the snowpack are elevated (Henriksen et al.. 1984).

The question is, does some of the lowered lake alkalinity and pH (6.1 to 6.2) during snowmelt result from anthropogenic acid deposition? Snowcores from 100 km north of the (ALWA) contained 11.7 ± 4.9 μeq L^{-1} of SO_4 and 7.3 ± 3.4 μeq L-1 of NO_3, which were, respectively, 2- and 7-fold greater than the levels determined in Olympic Mountain snow and had a pH of 5.3 (Welch et al., 1984). Vong and Waggoner (1983) also showed higher SO_4 downwind from sources in the Puget Sound basin than in the Olympics.

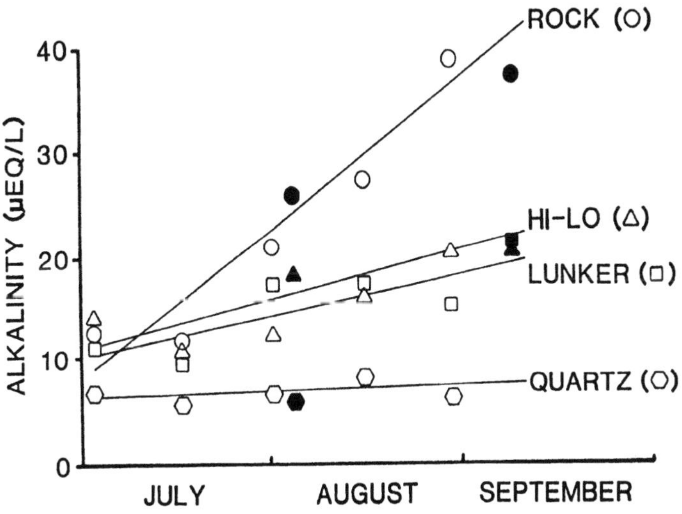

Figure 2. Seasonal change in alkalinity in the surface water in fast-flush lakes in the Alpine Wilderness Area. Solid symbols are values from 1983 and open symbols from 1984. Lines are calculated.

The described indices were used to detect acidification. The index Alk was higher in July than in August and the difference was greatest for the fast-flush lakes (Table I). The Alk/cations ratio was lower in July than in August and the difference was also greatest for the fast-flush lakes.

The ratio of Alk/specific conductance was used to determine if the low alkalinities following snowmelt were simply a straight dilution of lake water with low-alkalinity snowmelt water, with no loss of alkalinity by replacement with SO_4 and NO_3. This ratio was proportionately lower during July than in August (Table I). If the alkalinity reduction were a simple dilution, the ratios should have remained similar throughout the summer. This suggests that snowmelt water not only diluted the lake water but also preferentially depleted lake alkalinity as a result of its elevated SO_4 and NO_3 content

The lower alkalinities during snowmelt resulted in lower Alk/(SO_4 + NO_3) ratios (Figure 3). SO_4 showed no significant change between July and August and NO_3 decreased in Rock and Lunker (3.7 to 0.2 μeq L^{-1} and 2.9 to 0.3 μeq L^{-1}, respectively), probably due to phytoplankton uptake. The minimum ratio occurred later in Hi Lo than in Rock and Lunker and was associated with a slight increase in SO_4. The latter two are in the same watershed. Quartz Lake had a constant ratio due to its very high SO_4 (up to 38.6 μeq L^{-1}).

The slow-flush lakes did not show a pronounced seasonal trend in alkalinity. Lack of such a trend in those lakes has been observed previously (Welch et al., 1984). A low alkalinity value was, however, observed in Big Heart (Angeline was not sampled) on the first sampling date. The indices listed in Table I also suggest lesser effects of snowmelt in these lakes.

Table I. Indices of acidification (mean ± SE) in three fast-flush (Hi Lo, Lunker, and Rock) and two slow-flush lakes (Angeline and Big Heart) during July and August, 1984

		Alk	Alk/cations	Alk/sp cond
fast-flush,	July	9.6 + 2.2	0.27 + 0.03	2.2 + 0.06
	August	3.5 + 2.7	0.38 + 0.06	3.1 + 0.06
slow-flush,	July	3.5 + 3.3	0.57 + 0.15	3.7 + 0.00
	August	2.6 + 1.1	0.62 + 0.17	3.2 + 0.14

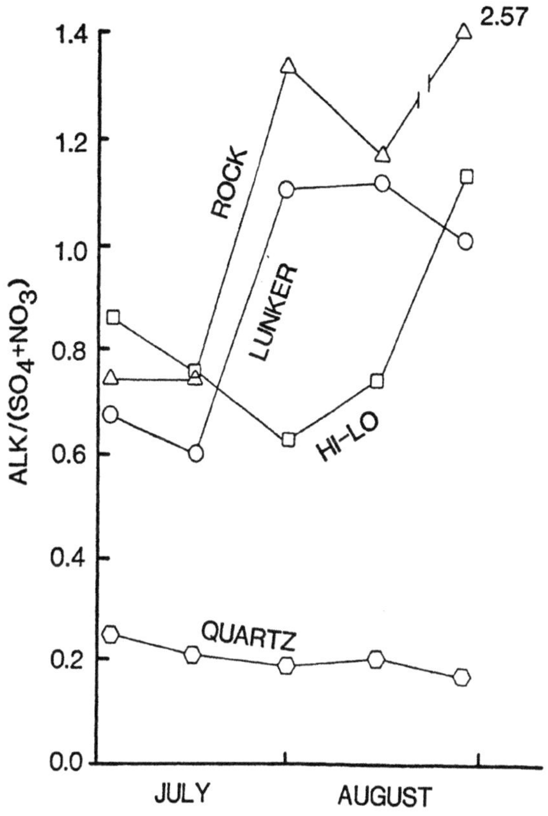

Figure 3. Seasonal change in the ratio of Alk/(SO4 + NO3) in four fast-flush lakes in the Alpine Wilderness Area during 1984.

The slow-flush lakes are deep (two >100 m), and the effects of snowmelt water may be dampened by entrainment of higher alkalinity water from depth. The profiles of chemical content (Table II) show that alkalinity increases by about 50% at 100 m, compared to the surface concentrations, in Angeline and Big Heart and the increase is even more marked in Copper. Ca and Mg increase proportionately, although Na and K are nearly constant with depth in Big Heart and Angeline. However, anions SO_4 and Cl are constant with depth in Angeline and Big Heart, although they roughly double in Copper.

The profiles suggest that these lakes may not mix completely during the year and that higher alkalinity water, possibly similar to groundwater without snowmelt dilution, accumulates in the deep waters. Also, the near constant SO_4 and Cl with depth may suggest that the main sources for these anions are not groundwater, but atmospheric deposition.

Table II. Chemical and physical characteristics in surface samples from vertical profiles of Alpine Lakes Wilderness Area Lakes, during 1984.

	(m) Depth	(°C) Temp	Ca^{2+} (µeq L^{-1})	Mg^{2+}	Na^+	K^+	SO_4^{2-}	Cl^-	NO_3^-	Alk	(lab) pH	(µmoh/cm) Spec. Cond.	(Pt-units) Color
Angeline (8-16-84)	1/2	16.2	13.2	2.9	10.7	1.4	9.9	12.2	2.0	15.2	6.79	5.3	5
	10	6.2	20.0	3.8	12.9	1.5	11.1	9.0	2.0	12.4	6.19	5.1	5
	20	4.2	20.6	3.3	53.6	11.1	8.7	8.6	2.3	16.1	6.21	6.7	5
	30	4.0	21.4	5.5	44.4	10.2	8.7	8.3	2.6	11.9	6.18	5.2	5
	40	4.0	19.2	3.3	15.9	2.4	11.3	36.9	5.9	20.2	6.39	6.1	5
	50	4.0	26.6	7.0	11.0	1.5	8.4	9.2	2.6	20.9	6.43	5.8	5
	60	4.0	22.8	3.9	16.6	2.0	8.6	8.5	2.9	19.1	6.26	5.1	5
	70	4.0	22.0	5.1	10.6	1.0	8.3	11.9	3.4	20.7	6.39	5.8	5
	100	4.0	23.8	5.4	11.7	1.7	8.7	8.2	3.6	22.5	6.51	6.1	5
Big Heart (8-29-84)	1/2	12.0	21.8	5.7	14.9	0.6	9.0	10.1	1.9	21.1	6.71	6.2	5
	10	8.0	24.0	5.9	12.4	1.5	8.6	11.3	3.0	22.8	6.42	5.8	5
	20	5.1	26.4	6.4	14.8	2.1	10.7	12.5	0	24.9	6.53	6.6	5
	30	4.2	31.4	7.4	19.7	4.1	8.6	6.9	3.0	26.5	6.61	6.9	5
	40	4.3	28.6	6.8	18.2	3.0	8.1	8.0	2.4	26.9	6.62	6.9	5
	60	4.0	29.6	6.9	17.3	3.0	8.5	8.6	3.1	21.9	6.46	7.0	5
	80	4.0	31.0	7.1	13.3	1.5	9.6	9.4	3.7	27.7	6.60	6.0	5
	100	4.0	35.6	7.9	14.2	2.1	8.8	10.8	4.4	31.7	6.49	7.5	5
Copper (8-30-84)	1/2	13.9	55.7	13.6	14.7	2.2	10.1	6.3	1.0	56.6	6.97	10.4	5
	10	12.6	56.2	13.4	31.7	8.8	10.3	6.6	1.1	58.6	6.95	10.2	5
	20	5.0	110.0	27.6	21.8	3.6	15.7	8.4	1.0	149.1	7.28	17.8	5
	30	4.0	125.5	32.4	24.4	4.6	12.7	7.3	1.0	165.0	7.31	20.6	5
	40	4.4	243.0	56.5	36.1	8.2	16.5	9.7	2.3	323.0	7.54	38.3	5
	50	4.0	275.0	69.5	43.5	10.8	19.4	11.1	2.5	435.7	7.67	49.7	5
	60	4.0	318.0	79.8	46.6	11.4	20.3	10.3	2.5	531.1	7.89	59.7	5

3.3 Constituent Variability

If data during snowmelt (through mid July) are omitted, the variability in alkalinity and SO_4 from year-to-year and lake-to-lake is rather small. If all samples are pooled, standard errors are only 3 and 4% of the means for SO_4 and about 9% for alkalinity (Table III). The stability of constituents in the ALWA lakes illustrates their value as indicator lakes.

Table III. Mean alkalinity and sulfate (+SE) in surface samples (n) collected at the surface of six lakes in the Alpine Lakes Wilderness Area from mid July through August during 1983 to 1984. Data from 1982 included for the slow-flush lakes Welch et al., 1984).

	Slow Flush	Fast Flush
Alkalinity*	19.7 + 1.7 (16)	20.3 + 1.9 (17)
SO_4	10.3 + 0.3 (25)	17.1 + 0.7 (17)

*Copper Lake omitted due to high values.

4. DISCUSSION

The two indices of acidification, plus the ratios of Alk/sp. cond. and Alk/(SO_4 + NO_3), suggest that some of the lowered alkalinity following snowmelt may be due to acidification rather than totally a result of dilution. The magnitude is so low, however, that it is difficult to demonstrate convincingly. Nevertheless, the rather constant SO_4 content with depth in Angeline and Big Heart Lakes, and its similarity with precipitation SO_4, indicate strongly that much of the SO_4 in the lakes comes from atmospheric deposition. SO_4 concentrations in precipitation from Snoqualmie Pass, which is immediately south of these lakes, have ranged from 8.3 to 14.8 μeq L^{-1} (Logan et al., 1982; Asserer and Duncan, 1984). Also, SO_4 and NO_3 are higher in Cascade precipitation, downwind from anthropogenic sources in the Puget Sound Basin. Therefore, some of the lake SO_4 is no doubt anthropogenic in origin. The steep, granite basins of these lakes, especially Angeline and Big Heart, probably result in minimal modification of the acid inputs during snowmelt. Dividing SO_4 deposition by expected water runoff indicates that from 40 to 85% of slow-flush lake SO_4 may come from precipitation.

The high levels of SO_4 in Quartz Lake, relative to concentrations in precipitation, and evidence of sulfide bearing rocks in the watershed, indicate local geologic sources. Thus, most of the low alkalinity in Quartz Lake is probably due to natural acidification. The higher SO_4 content of the fast-flush lakes (exclusive of Quartz) may also reflect local sources of sulfur.

Calculation of net SO^*_4 (Wright, 1983) results in a mean of 1.1 ± 2.0 μeq L^{-1} for the five Alpine Wilderness Area lakes. By this approach, although possibly not appropriate for this area, nearly all, or 11.9 μeq L^{-1} (13-1.1), of the SO^*_4 in these lakes may come from local sources.

The relatively low level of acid deposition, even though the lakes are highly sensitive, suggests that detection of a doubling or halving in acid deposition would be difficult. This is so because seasonal and annual variability among lakes is usually considered rather great. However, if data from the these lakes are stratified for flushing rate and sensitivity, the variability becomes relatively small. The pooled, 3-yr data set indicates for the slow-flush lakes, that an alkalinity change of about 4 μeq L^{-1} and a SO_4 change of about 2 μeq L^{-1} would be detectable at the 95% level. Changes would need to be slightly higher in the fast-flush lakes to be detectable; about 10 μeq L^{-1} for alkalinity and 3.6 μeq L^{-1} for SO_4. Therefore, these lakes clearly represent sensitive indicators of change, not only because of their very low alkalinity, but also because variability in constituents among lakes with similar flushing rate and among years is relatively low.

Although slight acidification is indicated during snowmelt, these highly sensitive Cascade Mountain Lakes are not receiving enough acid deposition for the slight effect from anthropogenic sources to be easily separated from local effects of watershed weathering. However, the general consensus that a mean annual pH >4.6 will protect sensitive areas (Evans et al., 1981) may not be sufficient for these lakes.

Because of their very low alkalinity and the possible existence of slight effects with precipitation pH of 4.7 to 5.1, precipitation pH should probably be > 4.7 to 4.8 (20 to 16 μeq L^{-1} H^+) for adequate protection.

REFERENCES

ALPHA: 1980, Standard methods for the examination of water and wastewater. Amer. Public Health Assoc., 14th edition.
Asserer, W. and Duncan, L.C.: 1984, Precipitation chemistry at Skykomish, Washington, Report for WA State Dept. of Ecology.
Burns, D.A., Galloway, J.N., and Hendrey, G.R.: 1981, Water, Air, and Soil Pollut. **16**, 275.
Charlson, R.J. and Rodhe, H: 1982, Nature **295**, 683.
EPA: 1979, Methods for chemical analysis of water and wastes.
Evans, L.S., Hendrey, G.R., Stensland, G.J., Johnson, D.W., and Frances, A.J.: 1981, Water, Air and Soil Pollut. **16**, 469.
Golterman, H.L.: 1969, Methods for chemical analysis of fresh waters. IBP Handbook No. 8, Blackwell Scientific Publ., Oxford and Edinburgh, 166 pp.
Henriksen, A.: 1980, Acidification of freshwaters--a large-scale titration, in D. Drablos and A. Tollan, eds., Int. Conf. Ecol. Impac Acid Precip., SNSF Proj. Rep., Norway, pp.68-74.
Henriksen, A., Skogheim, O.K., and Rooseland, B.O.: 1984, Vatten **40**, 255.
Logan, R.M., Derby, J.C., and Duncan, L.C.: 1982, Environ. Sci. and Tech. **16**, 771.
Omernik, J.M., and Powers, C.F.: 1982, Total alkalinity of surface waters--a national map, EPA-600/D-82-333.
PSPCA: 1980, Puget Sound Pollution Control Agency, a report.
Vong, R. and Waggoner, A.P.: 1983, The chemical composition of Western Washington rainwater. Report to USEPA, Reg. X, Dept. Civil Engr., University of Washington, 54 pp.
Welch, E.B. and Spyridakis, D.E.: 1984, Acidification potential in mountain lakes in Washington, Final report to State of WA, Dept. of Ecology, 16 pp.
Welch, E.B., Chamberlain, W.H., and Spyridakis, D.E.: 1984, Northwest Science **58**, **85**.
Wright, R.F.: 1983, Predicting acidification of North American lakes, Report 4/83, Norwegian Inst. Water Res. (NIVA), Oslo, 165 pp.
Wright, R.F. and Henriksen, A.: 1983, Nature **303**, 422.

MODELLING HYDROLOGY AND SULPHATE CONCENTRATION IN SMALL CATCHMENTS IN CENTRAL ONTARIO

E. deGrosbois, P.J. Dillon,
Dorset Research Centre
Ontario Ministry of the Environment
Box 39
Dorset, Ontario, Canada POA 1E0

H.M. Seip, R. Seip
Department of Chemistry
University of Oslo
Box 1033, Blindern
0315 Oslo 3, Norway

ABSTRACT. The Birkenes-Storgama model of hydrology and streamwater chemistry which was modified by Seip et al. (1985) to simulate data for the Harp Lake #4 sub-catchment was applied to the Harp Lake #3, #4, and #6 sub-catchments for the 1980 to 1984 period. The model was roughly calibrated using a readily available optimization procedure on a relatively limited suite of model parameters. The calibration period was much shorter (8 mo for the hydrology sub-model and 12 mo for the sulphate sub-model) and the confirmation period longer (about 3 yr) than used previously. A statistical assessment indicates that the hydrology sub-model performs with good reliability over the confirmation period, and the sulphate sub-model performs fairly well. The need for modification to the sulphate sub-model may be indicated. An important result is that the Birkenes-Storgama model can be applied at new temporal and spatial locations without extensive calibration, and produce useful simulations.

1. INTRODUCTION

An important concern in the development of a mathematical model of a complex system is its performance when applied to other systems with similar characteristics. A model may predict very well for the subject system on which it was developed, but an assessment of its general usefulness requires an evaluation of its performance on other systems with similar features.

Recently, the Birkenes-Storgama model of stream hydrology and stream chemistry (Christophersen et al. 1982) was transported from its subject system in Norway to operate on a sub-catchment (#4) at Harp Lake, Ontario, Canada. Modifications to the model included the introduction of "piston flow" and a term for sulphate reduction in deeper soil horizons. The model worked with a good degree of success for the period May 1977 to April 1982 (Seip et al. 1985, Rustad et al., 1986).

This paper examines the performance of the same model, without major modification, on three streams in the Harp Lake basin. For each catchment, the hydrologic sub-model is calibrated over an 8-mo period

Water, Air, and Soil Pollution **31** (1986) 45–57.

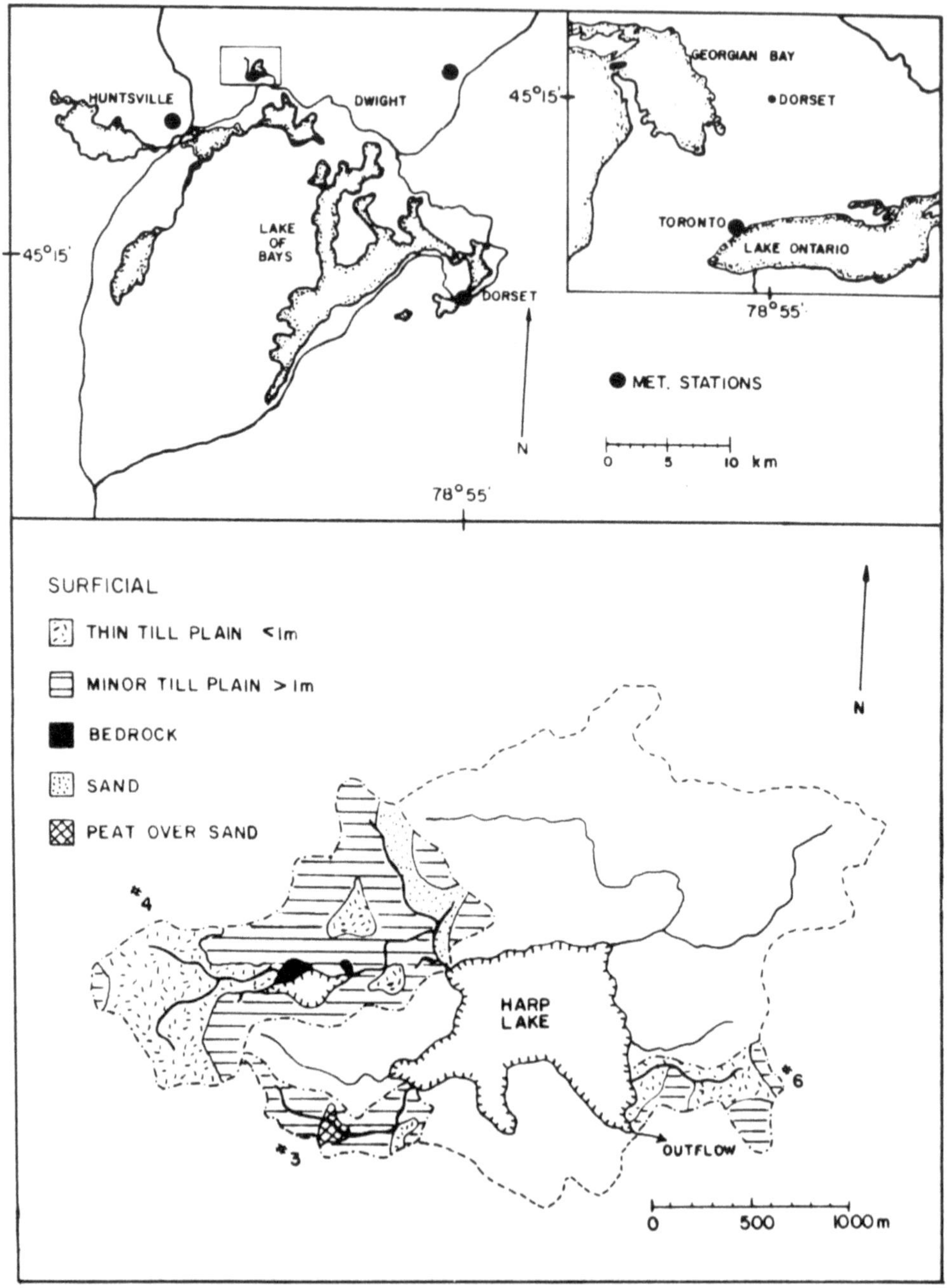

Figure 1 Location of Harp Lake and surficial geology of the #3, #4, and #6 sub-catchments (from Jeffries and Snyder, 1983).

on a limited suite of parameters. Remaining data are used as confirmation of the model. Since dry deposition of S-compounds can be assumed identical for all catchments, the sulphate sub-model was used with minor modifications only and was calibrated only on Harp Lake sub-catchment #4. Lack of detailed knowledge of the soil properties makes it difficult to compare the optimized parameters with catchment characteristics. In this paper we focus on statistical aspects of model performance.

2. SITE DESCRIPTION

Inflows #3, #4 and #6 are tributaries of Harp Lake, 8 km northeast of Huntsville, Ontario (Figure 1).

Inflow #3, with a catchment area of 30.0 ha, is underlain by granitized biotite and hornblende gneiss which is covered by a minor till plain with some peat over sand deposits. The catchment is principally covered with deciduous forest.

The Harp #4 catchment has an area of 119.5 ha and is almost completely covered with deciduous forest. There are several very small beaver ponds (total area ~3.8 ha) within the watershed. The bedrock underlying most of the subcatchment is amphibolite and schist, with a smaller area of biotite and hornblende gneiss. The overburden consists of a minor till plain (sandy loam to sand deposits >1m thick), with thin deposits (<1m) and rock ridges covering most of the remainder of the subcatchment.

The Harp #6 catchment has an area of 9.8 ha and is covered with mixed deciduous and conifer canopy. The bedrock is mainly diorite with granitized biotite and hornblende gneiss underlying the northern boundary. Most of the overburden consists of thin till and rock ridges with small deposits of peat over sand and minor till plain.

3. DATA BASE

Meteorological data required for the model include mean daily air temperature and daily precipitation. These parameters and sulphate concentration in precipitation are assumed to be identical for all three Harp sub-catchments. No corrections are made for gauge efficiency. Mean daily air temperatures were estimated as a distance-weighted average of the values at the nearby Dwight and Huntsville stations. Daily precipitation was obtained from the nearest neighbour to Harp Lake (Huntsville), or, if missing, from the next nearest neighbour (Dwight). This is a realistic situation insofar as it may be desirable to apply the model at locations for which little data are specifically available, yet it must also be viewed as a source of error when there are notable episodic differences between Dwight and Huntsville. Thunderstorms in the area can produce highly sporadic precipitation in summer months, the snow cover is quite variable, and temperature gradients can vary by about two Celsius degrees between the two stations.

Sulphate in precipitation was measured using precipitation collectors located in the Harp Lake catchment. Concentration of sulphate in the collector was assigned to all precipitation previous to that collection and since the last collection. Dry deposition of S was was estimated to be 0.4 g S m^{-2} yr^{-1} (Chan et al. 1985). All other non-calibrated parameters were used at the same levels as reported by Seip et al. (1985).

Sulphate in streamwater was measured bi-weekly using ion chromatography.

4. MODEL DESCRIPTION

Basically, the model consists of three hydrologic reservoirs: S, a snow reservoir, and A and B, a shallow soil and deep soil reservoir, respectively. Precipitation is allocated as rain or snow between S and A, and routed between S, A, B, and the stream on the basis of dynamic flow rates mainly governed by water depths in each reservoir. Discharge from the A and B reservoirs is given by

$$Q_a = A_k\ (A-A_{min})$$

$$Q_b = B_k\ (B-B_{min})$$

In these equations A and B represent the amount of water in each reservoir. The fraction of Q_a routed to the B-reservoir is

$$A_{sig} = K_{as}\left(1 - \frac{B-B_{min}}{B_{max}-B_{min}}\right)$$

which is slightly different from the expression used by Seip et al. (1985)

The sulphate concentration in Q_a is

$$C_a = F_a K_a = (F_a{}'-AC_a)K_a$$

where F_a is the amount of water-soluble sulphate in the solid phase in reservoir A and $F_a{}'$ is the amount of water-soluble sulphate both in solution and in the solid phase. (Note that the value given for K_a in the Appendix of Seip et al. 1985, is a misprint. The correct value is 0.00167 m^2 L^{-1}). In the upper reservoir sulphur compounds are oxidized to sulphate during dry periods. The parameter determining this mineralization rate is D_{miner}. The concentration of sulphate in discharge from reservoir B depends on a parameter D_{red} governing sulphate reduction (or adsorption) during summer and fall, and K_b which is rate constant for a first order reaction towards an equilibrium value (see Seip et al. 1985 for details).

The hydrologic sub-model was calibrated for each sub-catchment over an 8-mo period. For sub-catchment #3, the calibration period was October 1, 1980 to May 31, 1981 and confirmed over the three subsequent years. Because of a large number of missing values in the first year, sub-catchments #4 and #6 were calibrated for the period October 1, 1981 to May 31, 1982. Confirmation for these catchments was done on the preceding year and on the two subsequent years so that the hydrology of all three catchments were simulated over the period July 1980 to May 1984. Calibration was done by fitting model predictions to observations using non-linear least squares optimization (Marquardt, 1963). The parameters calibrated were the starting depths for A and B

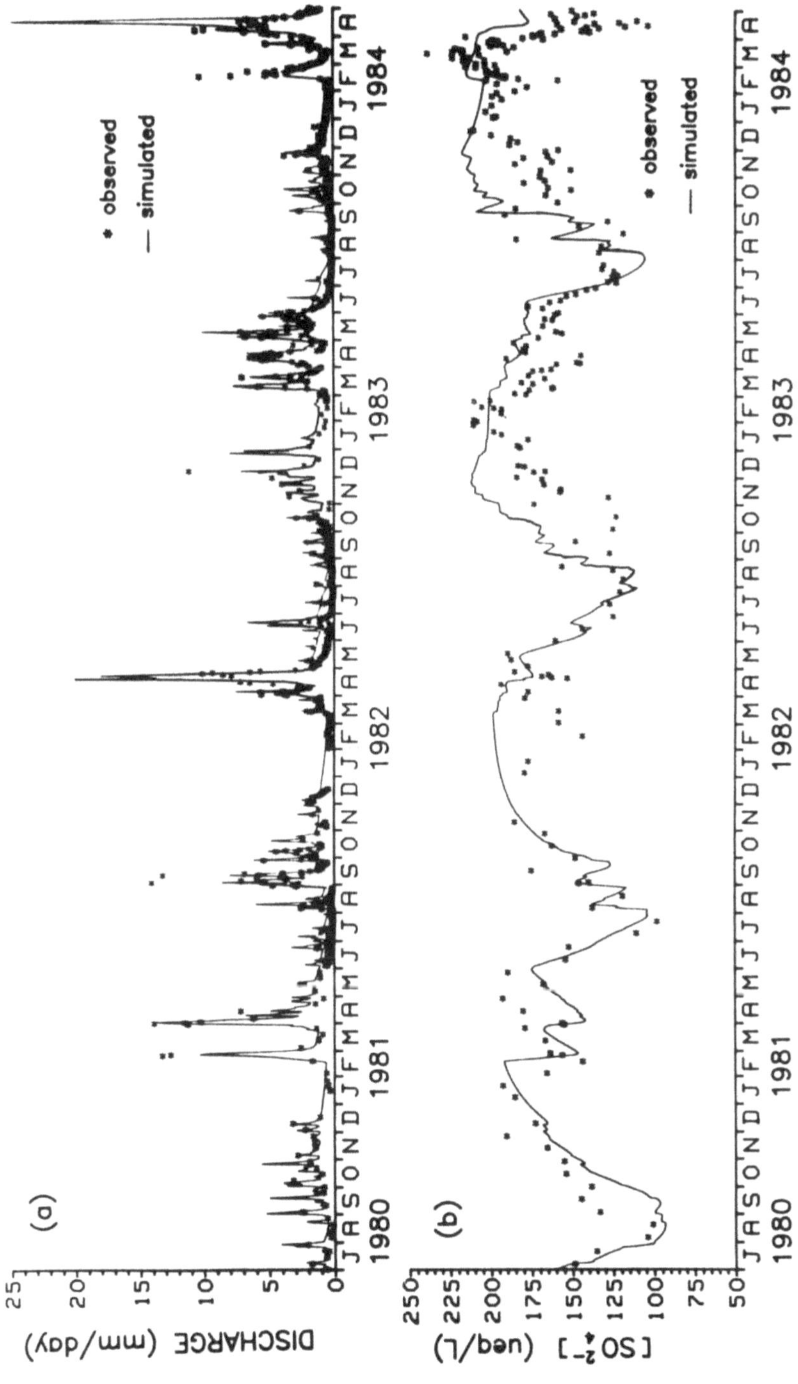

Figure 2 Harp sub-catchment #4: July 1980 to April 1984.
(a) Observed and simulated discharge. Calibration period was October 1981 to May 1982.
(b) Observed and simulated sulphate concentrations. Calibration period was July 1980 to June 1981.

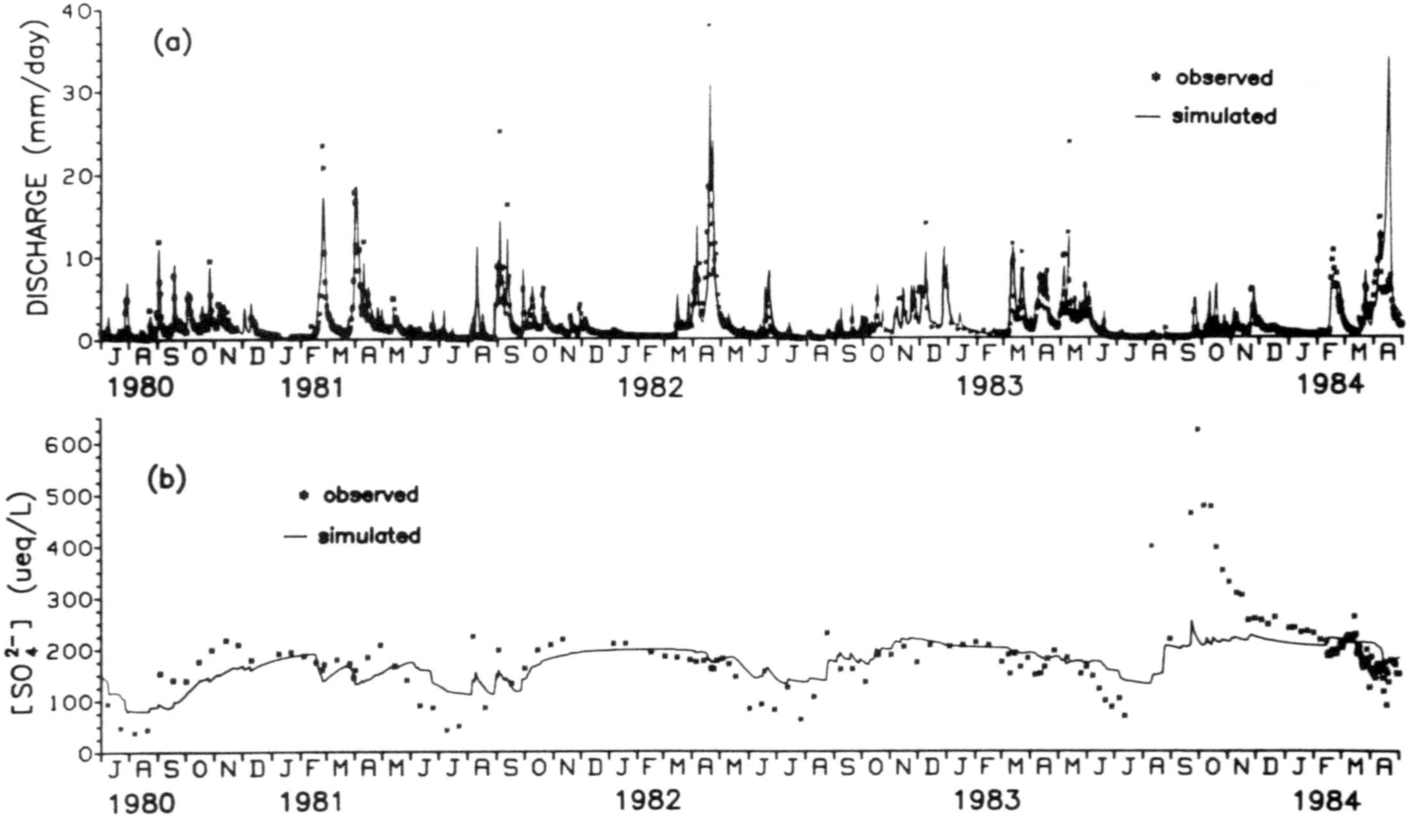

Figure 3 Harp sub-catchment #3: July 1980 to April 1984.
(a) Observed and simulated discharge. Calibration period was October 1980 to May 1981.
(b) Observed and simulated sulphate concentrations.

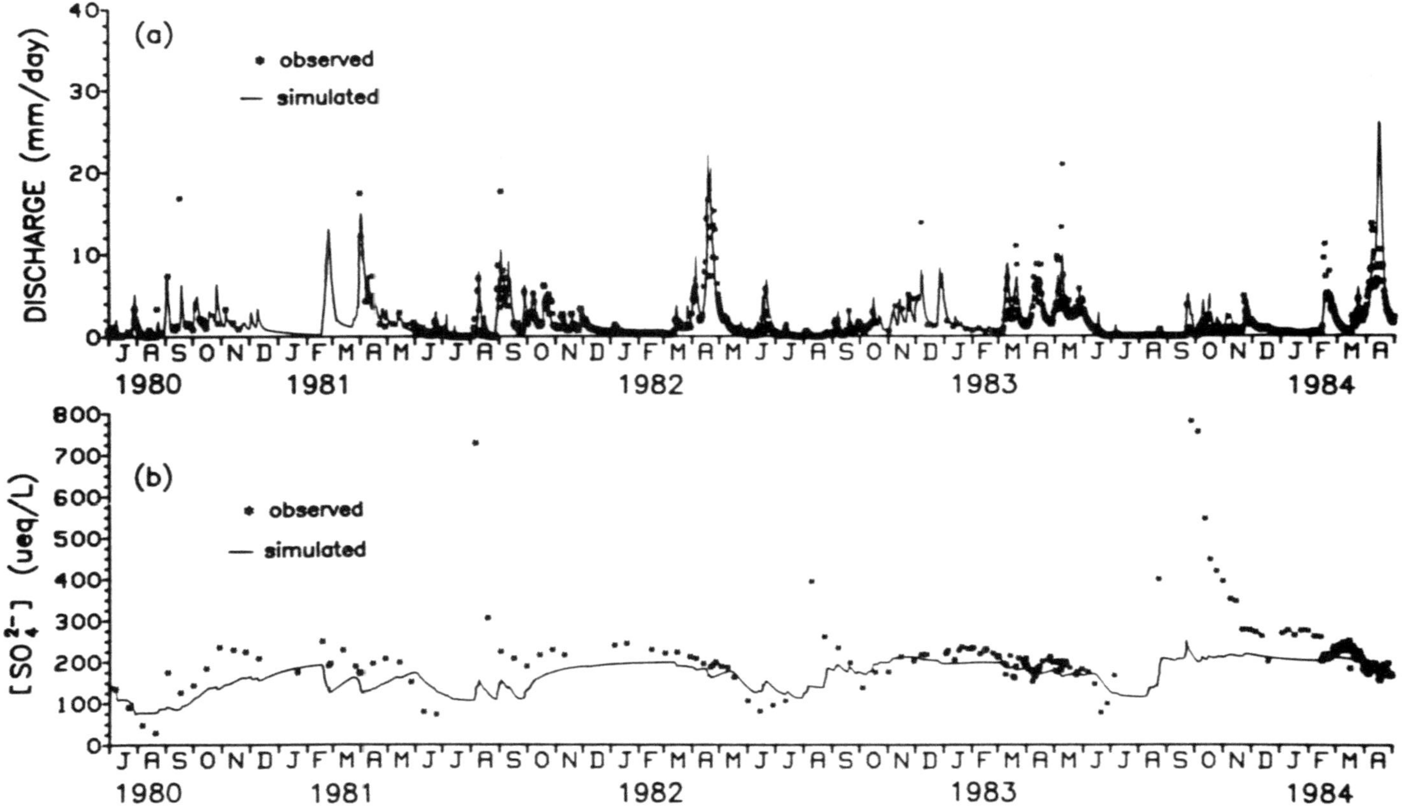

Figure 4 Harp sub-catchment #6: July 1980 to April 1984.
(a) Observed and simulated discharge. Calibration period was October 1981 to May 1982.
(b) Observed and simulated sulphate concentrations.

reservoirs, the flow rates A_k and B_k, K_{as}, and reservoir characteristics A_{min}, A_{max}, B_{min}, and B_{max}.
All other parameters were set at values described by Seip et al. (1985). The sulphate sub-model was calibrated on sub-catchment #4 only for the 1-yr period July 1, 1980 to June 30, 1981. The parameters calibrated were: D_{miner}, D_{red}, the starting value $F_a'(0)$, and the constant determining the concentration in the A reservoir K_a. For sub-catchments #3 and #6, some adjustments were made to sulphate parameters on a trial and error basis.

Model performance was assessed over the calibration period and the confirmation period for the hydrological sub-model on each catchment, and for the sulphate sub-model on sub-catchment #4; and, assessed over the full 4-yr period for the sulphate sub-model on #3 and #6. Three different statistics are used to assess the performance of the model. The correlation (Pearson's r) between observed and predicted values is an indicator of episodic coincidence, and the slope ($\beta 1$) and intercept ($\beta 0$) of the regression line are indications of over-prediction or under-prediction by the model. A P statistic (used by Seip et al. 1985) indicates the relative improvement over some benchmark model. Nash and Sutcliffe (1970) have used a mathematically similar statistic, E_f, to indicate forecasting efficiency. In our case, the benchmark model used was the mean observed value.

$$P = \frac{\sum_{i=1}^{n} (x_b - x_i)^2 - \sum_{i=1}^{n} (\hat{x}_i - x_i)^2}{\sum_{i=1}^{n} (x_b - x_i)^2}$$

where x_i and $\hat{x}_i$ are observed and simulated values respectively, and $i=1$ to n,
and x_b is a benchmark: a scalar (ie. the mean) or a vector (ie. values from another model)

As P approaches 1.0, it indicates a 1:1 correspondence between simulations and observations, whereas an r^2 of 1.0 does not.

5. RESULTS AND DISCUSSION

From a visual inspection of Figures 2 to 4, we may conclude that the model is reasonably successful in predicting daily hydrology in all three streams over the 3-yr period. The simulation shows a large runoff event during April 1984 which is not corroborated by observation. It seems likely this may be the result of a poor estimation of the meteorologic data for Harp Lake at this point in time. It should be noted that the optimization procedure used represents only a preliminary attempt at calibrating the Birkenes model for a new system. Only a limited suite of parameters are included, and the criteria for optimum are change of 1% or less in the parameter values or the total sum of squares, or 5 iterations, whichever comes first. Clearly more rigorous and comprehensive calibrations are possible.

The performance statistics for the hydrologic sub-model are presented in Table I. The largest sub-catchment, Harp #4, also has the largest B reservoir. All three streams indicate a very good episodic coincidence in both calibration and confirmation periods (Pearson's r greater than 0.75) and in all cases the simulated hydrology represents a good improvement over the benchmark model (the mean observed value in this case).

The performance statistics for the sulphate sub-model are presented in Table II. It is interesting to note the changes in the optimized parameter values for Harp #4. The mineralization rate in the A reservoir, D_{miner}, increased to 2500 μeq $m^{-2} \cdot d^{-1}$ and the fraction of sulphate in solution that is reduced in the B reservoir was decreased to 0.0185. $F_a'(0)$, the amount of water soluble sulphate both in solution and in the solid phase in A, reduced to 16,150 μeq m^{-2}. This value is almost certainly too low and indicates the necessity for caution in using uninformed optimization procedures. It probably resulted from optimizing the initial conditions for the calibration period as well as being positively correlated with sulphate reduction. The constant determining the concentration in the upper reservoir, K_a increased to 0.00167 m^2 L^{-1}. The rate constant K_b increased to 0.022 (day^{-1}).

There is a tendency for larger variations in the observed concentrations than in the simulated ones during snowmelt. Due to fractionation of the ions in snow during melting the sulphate concentration in meltwater must decrease considerably as the melting proceeds. The observed streamwater concentrations do not show a consistent pattern of this kind. It is likely that the observed variation in stream sulphate concentration reflects changes in flow paths more than decrease in meltwater concentration.

For Harp #3 and #6 sub-catchments, we set K_b to 0.028 and 0.030 (day^{-1}), respectively. We also used a higher reduction fraction for D_{red} (0.045) and a slightly lower rate of mineralization (D_{miner} = 2400 μeq m^{-2} d^{-1}) and $F_a'(0)$ was set to yield satisfactory starting values.

The hydrologic sub-model for sub-catchment #4 calibrated at P = 0.77 (77% better than the benchmark model). The 95% confidence interval on the regression line (r^2 = 0.77) includes a slope of 1.0 ($0.99 \leq \beta 1 \leq 1.18$) and an intercept of 0 ($-0.50 \leq \beta 0 \leq 0.04$). For the confirmation period, P = 0.41, or a 41% improvement over the benchmark. The 95% confidence intervals on the regression line indicates an intercept that includes 0 ($-0.02 \leq \beta 0 \leq 0.20$) but a slope less than 1.0. This indicates that the model is slightly over-predicting during the confirmation period.

The calibration of the sulphate sub-model for Harp #4 was fairly successful with P = 0.25 or a 25% improvement over the mean observed value. The regression equation indicates that high values are over-predicted and low values are under-predicted by the model. This indicates that a decrease in the mineralization rate and an increase in the reduction fraction might be required. The sulphate sub-model performs only slightly poorer during the confirmation period for Harp #4.

TABLE I: Hydrologic sub-model: parameter values and performance

Parameter Values:	Sub-catchment #4	Sub-catchment #3	Sub-catchment #6	Seip et al. (1985)
A_k (d^{-1})	0.54	0.75	0.41	0.7
B_k (d^{-1})	0.0095	0.042	0.025	0.04
A_{min} (mm)	6.5	8.19	6.80	15.0
A_{max} (mm)	30.6	64.0	122.	80.0
B_{min} (mm)	49.4	80.1	48.0	100.0
B_{max} (mm)	417.	268.	286.	160.0
K_{as}	0.946	0.60	0.60	N/A
Calibration Period Performance:				
Pearson's r	0.88	0.89	0.78	
P statistic	0.77	0.78	0.60	
Regression Slope	1.09	0.91	0.91	
95% C.I. upper limit	1.18	0.97	1.01	
95% C.I. lower limit	0.99	0.84	0.82	
Regression Intercept	-0.23	0.18	0.43	
95% C.I. upper limit	0.04	0.45	0.80	
95% C.I. lower limit	-0.50	-0.09	0.06	
Confirmation Period Performance:				
Pearson's r	0.80	0.76	0.79	
P. Statistic	0.41	0.47	0.56	
Regression slope	0.64	0.65	0.70	
95% C.I. upper limit	0.67	0.69	0.73	
95% C.I. lower limit	0.60	0.62	0.66	
Regression intercept	0.09	0.47	0.23	
95% C.I. upper limit	0.20	0.59	0.35	
95% C.I. lower limit	-0.02	0.34	0.11	

TABLE II: Sulphate sub-model: parameter values and performance

Parameter Values:	Sub-catchment #4	Sub-catchment #3	Sub-catchment #6	Seip *et al.* (1985)
D_{miner} (μeq m^{-2} d^{-1})	2500.	2400.	2400.	1050.
D_{red}	0.0185	0.045	0.045	0.03
K_a (m^2 L^{-1})	1.67 10^{-3}	1.67 10^{-3}	1.67 10^{-3}	8.33 10^{-4}
$F_a'(0)$ (μeq m^{-2})	16,150.	8000.	10,000.	235,000.
K_b (d^{-1})	0.022	0.028	0.030	235,000.
Calibration Period Performance:				
Pearson's r	0.76	N/A	N/A	
P statistic	0.25	N/A	N/A	
Regression Slope	0.65	N/A	N/A	
95% C.I. upper limit	0.87	-	-	
95% C.I. lower limit	0.43	-	-	
Regression intercept	62.5	N/A	N/A	
95% C.I. upper limit	96.0	-	-	
95% C.I. lower limit	29.1	-	-	
Confirmation Period Performance:				
Pearson's r	0.69	0.50	0.36	
P. Statistic	0.14	0.25	-0.01	
Regression slope	0.70	1.16	1.08	
95% C.I. upper limit	0.79	1.46	1.47	
95% C.I. lower limit	0.60	0.87	0.69	
Regression intercept	39.5	-31.3	17.6	
95% C.I. upper limit	58.0	23.9	89.0	
95% C.I. lower limit	20.9	-86.5	-53.9	

For sub-catchment #3, the hydrologic sub-model calibrated at P = 0.78 and during the confirmation period shows a 47% improvement over the benchmark (P = 0.47). The sulphate sub-model performs fairly well over the 4-yr period (Figure 3). Although the parameter determining the sulphate reduction D_{red}, was increased compared to the value used for Harp #4 (Table II), the observed sulphate concentrations tend to be lower than the simulated ones during summer indicating that an even higher value should have been chosen. The simulations indicate a 25% improvement over the benchmark (P = 0.25). The 95% confidence interval on the regression line includes a slope of 1.0 and an intercept of 0. These results are good considering that the sulphate sub-model for both Harp #3 and #6 was not mathematically calibrated.

The hydrology sub-model for sub-catchment #6 calibrated at P = 0.60. For the confirmation period on sub-catchment #6, the simulation represents a 56% improvement over the benchmark (P = 0.56).

The sulphate sub-model for Harp #6 was less successful. The simulation did not represent an improvement over the benchmark mean-as-model. The occurrence of a very large sulphate peak (over 750 $\mu eq/L^{-1}$) that occurred in September 1983 after a particularly dry period during the late summer is not reproduced. This phenomenon was also observed in Harp #3. The performance of both Harp #3 and Harp #6 sulphate simulations is drastically improved if we consider only the first 3 yr of the confirmation (i.e. excluding the sulphate spike). This argues strongly for additional modifications to the model; in particular, a mineralization term that would come into effect whenever the previously saturated anoxic soil reservoirs are dry and strongly oxidizing conditions prevail.

Clearly, the Birkenes-Storgama model can produce good simulations of hydrology and sulphate chemistry when applied at different temporal and spatial locations. Although this study undertakes only a preliminary calibration of model parameters over a relatively short period (in this case, 1 yr), the efficacy of the model is readily apparent. Without extensive prior knowledge of the new systems, the model can be calibrated to yield useful results by using a readily available optimization algorithm. It is important to note that whereas by definition mathematical optimization will produce better model performance and artificially better estimates of parameter values, these values are, in the present case, reasonable and fall well within the expected range. The performance of the model relative to changes in parameter values will indicate areas both where enhancements are necessary and where additional empirical data are required.

ACKNOWLEDGEMENTS

This work was supported in part by a NATO grant to H.M.S.

REFERENCES

Chan, W., and Lusis, M.: 1985, "Deposition fields in Ontario, 1981-82". (unpublished manuscript)

Christophersen, N., Seip, H.M., and Wright, R.F.: 1982, Water Resour. Res. 18, 977.

Jeffries, D.S., and Snyder, W.R.: 1983, "Geology and geochemistry of the Muskoka-Haliburton study area". Data Report DR 83/2, Ontario Ministry of the Environment.

Marquardt, D.W.: 1963, J. Soc. Ind. Appl. Math. 2, 431.

Nash, J.E., and Sutcliffe, J.V.: 1970, J. Hydrol. 10, 282.

Rustad, S., Christophersen, N., Seip, H.M., and Dillon, P.J.: 1986, Can. J. Fish. Aquat. Sci. 43, 625-633.

Seip, H.M., Seip, R., Dillon, P.J., and deGrosbois, E.: 1985, Can. J. Fish. Aquat. Sci. 42, 927.

CHANGES IN THE CHEMISTRY OF LAKES NEAR SUBURY, ONTARIO FOLLOWING REDUCTIONS OF SO_2 EMISSIONS

P.J. Dillon, R.A. Reid and R. Girard
Dorset Research Centre
Ontario Ministry of the Environment
P.O. Box 39, Dorset, Ontario, Canada
POA 1EO

ABSTRACT. Emissions of SO_2 in the Sudbury area declined from an estimated average of 1.41 X 10^6 tonne yr^{-1} in 1973-78 to 0.68 X 10^6 tonne yr^{-1} in 1979-85. As a result, SO_4 concentrations of lakes in the area have decreased, and the pH of each of the acidic lakes that was studied has increased. Aluminum, Cu, Ni and Zn concentrations have also decreased; however, the latter three metals have probably declined because of reductions in emissions of metals from the smelters rather than because of the pH changes in the lakes.

1. INTRODUCTION

Nickel and copper-containing sulphide ores have been mined and smelted in the Sudbury area for almost 100 yr. As a result of these activities, acidification and trace metal contamination of lakes and flowing waters, contamination of soils with trace metals, and devastation of forests in and around Sudbury have occurred.

In the early 1900s open roasting methods were used to smelt the ores; this had pronounced effects on the local terrain (Linzon, 1978; Gorham and Gordon, 1960). In the 1930s, the open roasting methods were replaced by smelters which reduced the localized damage; however, fumigations still prevented recovery in the local environment, and extended the zone of environmental effects (Dillon and Smith, 1984).

The construction of a single, 381 m stack at Copper Cliff in 1972 in place of the shorter stacks resulted in greater dispersion of the SO_2 emissions although a decrease in total SO_2 emissions in the Sudbury basin occurred concurrently. Emissions of SO_2 from all of the smelters in the Sudbury area averaged 2.20 X 10^6 tonne yr^{-1} between 1950 and 1972, but declined to an average of 1.41 X 10^6 tonne yr^{-1} in 1973-77 following the introduction of the tall stack. Since 1978, average annual SO_2 emissions at Sudbury have decreased an additional 54% to 0.68 tonne yr^{-1} (average of 1979-85: Figure 1).

In this paper we assess the effect of these reductions on the chemistry of 5 acidified lakes in the area.

Water, Air, and Soil Pollution **31** (1986) 59–65.

2. STUDY AREA

We have monitored the chemistry of four lakes in the Sudbury area continuously since 1973. These have been experimentally manipulated, two with base additions ($CaCO_3$ and $Ca(OH)_2$) and P (Middle, Hannah), and one with base alone (Lohi). The latter was re-acidified to pH ~4.8 by 1978 while the former two lakes have maintained pH's greater than 6.5.

The fourth lake, Clearwater Lake, has served as a control for the entire study period. The lakes and their catchments are described in detail by Jeffries et al. (1984). The sampling programme and the data analysis procedures are outlined in Yan and Miller (1984) and Yan and Dillon (1984). The fifth lake, Swan Lake, was sampled in only 1977 and 1982 (LaZerte and Dillon, 1984).

3. RESULTS AND DISCUSSION

3.1 Clearwater Lake

The sulphate concentration in Clearwater Lake averaged 545 $\mu eq\ L^{-1}$ in the period 1973-77. Beginning in 1978, the SO_2 emission levels of the Sudbury smelters declined (Figure 1), resulting in a decrease in the local S deposition rate (Scheider et al., 1981). This decreased deposition resulted in changes in the chemistry of Clearwater Lake. The sulphate concentration in the lake began to decline (Figure 2), in 1978, and by 1984 had reached 370 $\mu eq\ L^{-1}$, a drop of 32%. Since the water replenishment time of the lake averages 3 to 4 yr (Scheider 1984), further changes in concentration reflecting the decrease in emissions may continue for several additional years.

Between 1973 and 1977, the average pH (calculated as the arithmetic mean H^+ concentration) of Clearwater Lake was 4.23. As the SO_4 concentration has decreased, the pH has increased, reaching a mean value of 4.61 in 1984. This relative change in H^+ activity (59%) has been even greater than the change in sulphate concentration. The dissolved organic carbon (DOC) content of the lake has not changed over the course of the study, but has remained very low, averaging 0.4 mg C L^{-1} in 1981-84 compared to 0.6 mg C L^{-1} in 1973 (LaZerte and Dillon, 1984). The estimated contribution of the dissociated organic acid functional groups is too small (2 to 3 $\mu eq\ L^{-1}$; calculated using the method of Oliver et al., 1983) to appreciably affect lake pH.

It is well-known that acidification of lakes and streams results in elevation of Al concentrations. The total Al level of Clearwater Lake, which averaged 430 $\mu g\ L^{-1}$ in 1973-77, dropped to 190 $\mu g\ L^{-1}$ by 1982-84 (Figure 3). The concentration appears to have stabilized at this level.

There is a substantial atmospheric deposition of other metals, notably Cu, Ni and Zn, in the Sudbury area (Scheider et al., 1981; Jeffries, 1984). The levels of these metals in Clearwater Lake have also decreased following reduction in SO_2 emissions; Ni by 24%, Cu by 48% and Zn by 52% (Figure 3). These reductions almost certainly

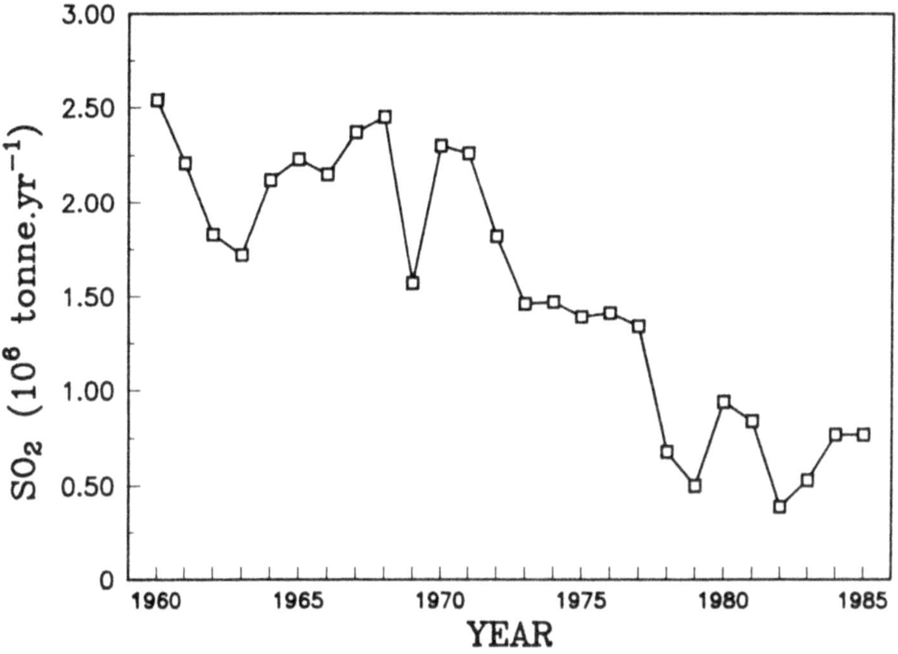

Figure 1 SO_2 emissions (10^6 tonne yr^{-1}) in the Sudbury basin, 1960-1985.

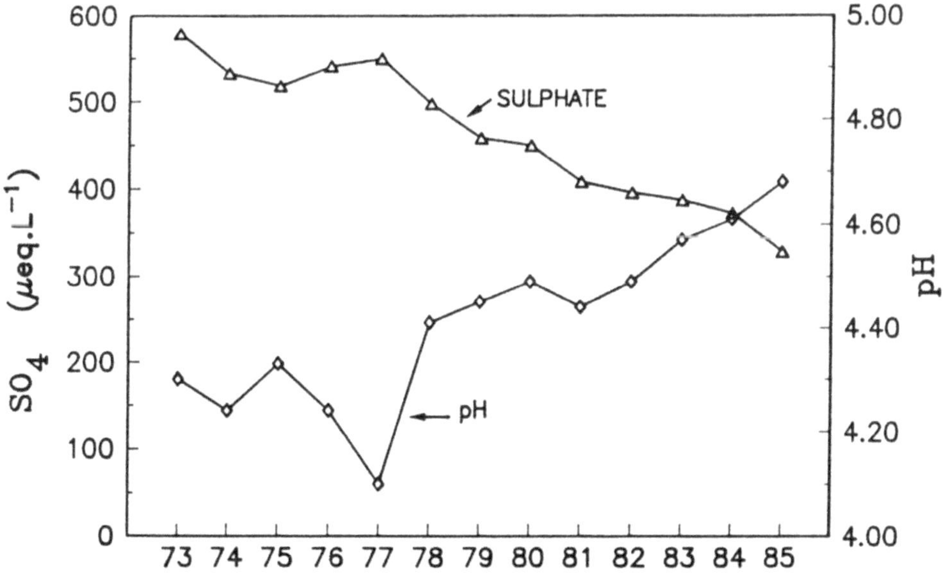

Figure 2 SO_4 concentration (μeq L^{-1}) and pH in Clearwater Lake, 1973-85. Results are whole-lake annual averages.

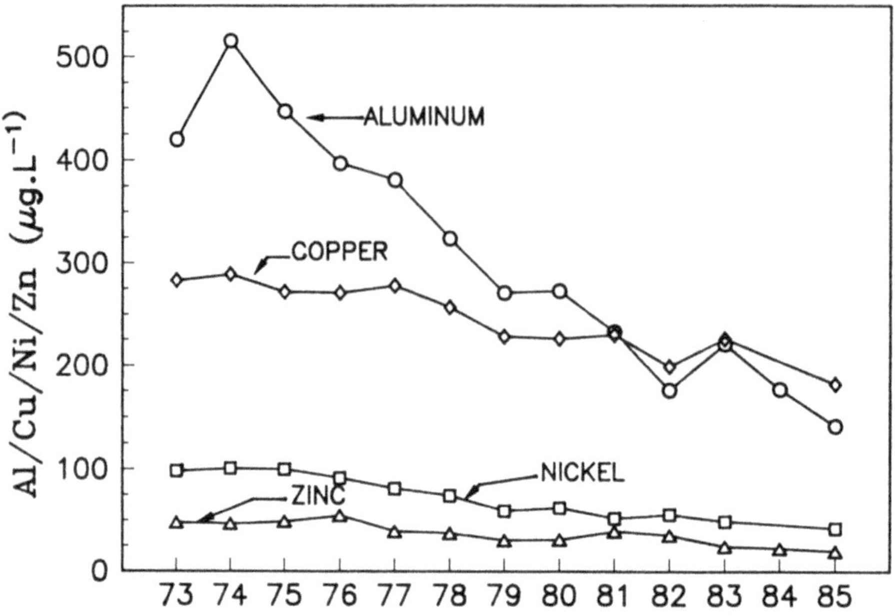

Figure 3 Aluminum, copper, nickel and zinc concentrations (µg L^{-1}) in Clearwater Lake, 1973-85. Results are whole-lake annual averages.

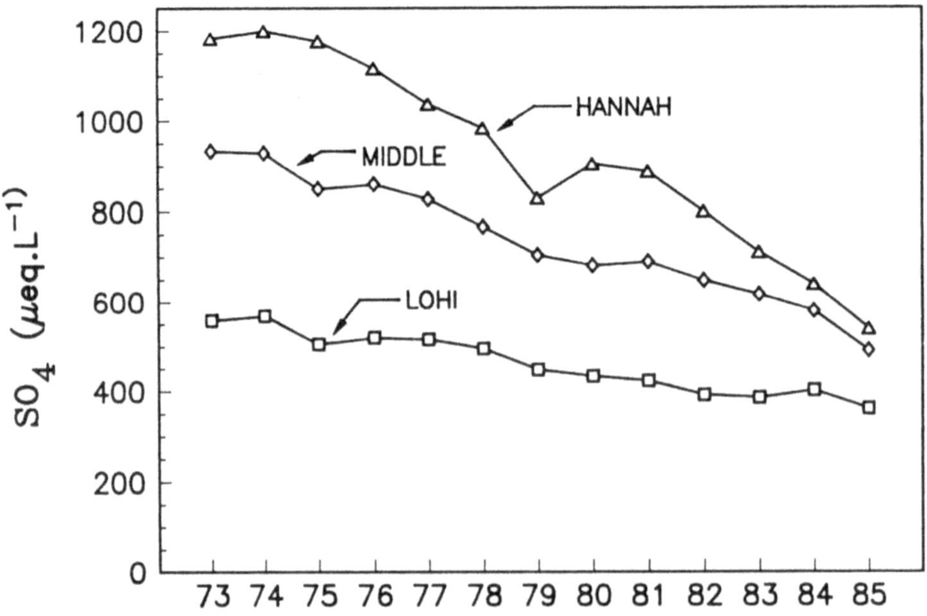

Figure 4 SO_4 concentration (µeq L^{-1}) in Lohi, Middle and Hannah Lakes, 1973-85.

reflect reduced metal emissions rather than a secondary effect of pH changes.

3.2 Lohi, Hannah and Middle Lakes

Lohi, Hannah and Middle Lakes were experimentally manipulated during the course of this study by the addition of bases ($CaCO_3$ and $Ca(OH)_2$) and, in the case of the latter two, the algal nutrient P (Yan and Lafrance, 1984). As a result, changes in pH and alkalinity in these lakes reflect these manipulations in addition to the local decrease in SO_2 emissions. However, like Clearwater Lake, the SO_4 concentrations have dropped in all 3 lakes (Figure 4), the measured decreases (1973-77 vs. 1984) being 25% in Lohi Lake, 34% in Middle Lake and 44% in Hannah Lake.

There are at least two factors resulting in different relative decreases in SO_4 concentration in the 4 lakes. Clearwater and Hannah Lakes are headwater lakes draining into Lohi and Middle Lakes, respectively. Since much of the SO_4 supply to Lohi and Middle Lakes consists of input from the upstream lake (Dillon, 1984), chemical response in the 'downstream' lakes must necessarily be slower than in the headwater lakes. Once a new steady state is reached (approx. 3 to 4 times the water replenishment time of the upstream lakes), the relative decrease in SO_4 concentration in Lohi Lake should be the same as in Clearwater Lake, and that in Middle, the same as in Hannah Lake.

On the other hand, the relative decrease in SO_4 concentration in Hannah and Middle Lakes is greater than that in Clearwater and Lohi Lakes. This result is attributed to the location of the former 2 lakes; Hannah and Middle Lakes, are closer to the major source of SO_2, the Copper Cliff smelter. Scheider et al. (1981) and Dillon et al. (1982) demonstrated that there was an appreciable S deposition gradient within 12 km of the Copper Cliff smelter prior to the decrease in SO_2 emissions. This has almost certainly resulted in a greater decrease in SO_2 deposition at Middle and Hannah Lakes than at Clearwater and Lohi Lakes because the component of deposition attributable to long-range transport has not changed appreciably.

3.3 Swan Lake

Swan Lake was studied much less intensively than the other 4 lakes. Its pH increased from 3.96 in 1977 to 4.80 in 1982, an 85% decrease in H^+ concentration. In 1977, its SO_4 concentration was 580 $\mu eq\ L^{-1}$; by 1982 it had dropped to 220 $\mu eq\ L^{-1}$, a decline of 62% (Table I). This decline was greater than that observed in any of the other study lakes, probably because a change in the O_2 regime in Swan Lake occurred between 1977 and 1982. In the latter year, the bottom waters were anoxic; this was not the case in 1977 (W. Keller, unpub. data). The anoxia probably resulted in increased rate of SO_4 reduction.

The 4 major study lakes received significant quantities of Na and/or Ca as a result of the applications of salts (NaCl, $CaCl_2$) to roads in their catchments during the course of these studies (Dillon,

TABLE I: Chemistry of Swan Lake in 1977 and 1982 (in μeq L^{-1} except pH).

Parameter	1977	1982	Δ
pH	3.96	4.80	
H	110	16	94
SO_4	580	220	360
Cl	73	130	-57
Al	32	8	24
Ca	319	210	109
Mg	115	69	46
Na	65	52	13
K	15	11	4
Base cations	514	342	172

1984). Swan Lake, however, received relatively little Na or Ca from salt treatments, as evidenced by the change in Cl concentration (Table I). It is therefore possible to compare the change in SO_4 with the change in both base cations and alkalinity. The decrease in base cation concentration, when corrected for the change in Cl, accounted for 64% of the drop in SO_4 concentration, while only 36% of the SO_4 decrease resulted in a change in alkalinity. This indicates that the so-called 'F-factor', the proportion of acid supplied to a system that meant to increase base cation content as opposed to decrease alkalinity, (Henriksen, 1982) for Swan Lake is 0.64, a value substantially greater than Henriksen reported for other locations.

4. SUMMARY

Reduced SO_4 emissions at Sudbury resulted in a rapid decrease in SO_4 concentration in all lakes studied. The pH of lakes that had not been experimentally manipulated increased as SO_4 decreased. Total Al concentration decreased. Base cation content of the one lake that received little Na or Ca from road treatments decreased as SO_4 concentration decreased. These results are strong evidence for a direct link between S deposition and lake acidification. They also demonstrate that a decrease in S deposition resulted in a rapid improvement in the chemistry of acidified lakes.

ACKNOWLEDGEMENTS

We would like to thank W. Keller for the use of unpublished Swan Lake data and N. Yan, B. LaZerte and D. Jeffries for review of the manuscript.

REFERENCES

Dillon, P.J.: 1984, The use of mass balances and mass balance models for quantification of the effects of anthropogenic activities on lakes near Sudbury, Ontario. p. 283-347, in "Environmental Impact of Smelters", J. Nriagu (ed.), Advances in Environmental Science Series, John Wiley and Sons, Inc.

Dillon, P.J., Jeffries D.S., and Scheider W.A.: 1982, Wat. Air Soil Pollut. 18: 241.

Dillon, P.J., and Smith, P.J.: 1984, Trace metal and nutrient accumulation in the sediments of lakes near Sudbury, Ontario. p. 375-426, in "Environmental Impact of Smelters", J. Nriagu (ed.), Advances in Environmental Science Series, John Wiley and Sons, Inc.

Gorham, E. and Gordon, A.G.: 1960, Can. J. Bot. 38, 447.

Henriksen, A.: 1982, Changes in base cation concentrations due to freshwater acidification, Norwegian Institute for Water Research, Report 1/1982.

Jeffries, D.S.: 1984, Atmospheric deposition of pollutants in the Sudbury Area, p. 117-154, in "Environmental Impact of Smelters", J. Nriagu (ed.), Advances in Environmental Science Series, John Wiley and Sons, Inc.

Jeffries, D.S., Scheider, W.A., and Snyder, W.R.: 1984, Geochemical interactions of watersheds with precipitation in areas affected by smelter emissions near Sudbury, Ontario, pp. 195-241, in "Environmental Impacts of Smelters". J. Nriagu (ed.), Advances in Environmental Science Series, John Wiley and Sons, Inc.

LaZerte, B.D. and Dillon, P.J.: 1984, Can. J. Fish. Aquat. Sci. 41, 1664.

Linzon, S.N.: 1978, Effects of airborne sulphur pollutants in plants, p. 109-162, in "Sulphur in the Environment, Part 2, Ecological Impacts", J. Nriagu (ed.), John Wiley and Sons, Inc.

Oliver, B.G., Thurman, E.M., and Malcolm, R.L.: 1983, Geochim. Cosmochim. Acta 47, 2031.

Scheider, W.A.; 1984, Lake water budgets in areas affected by smelting practices near Sudbury, Ontario. p. 155-194, in "Environmental Impact of Smelters", J. Nriagu (ed.), Advances in Environmnetal Science Series, John Wiley and Sons, Inc.

Scheider, W.A., Jeffries, D.S., and Dillon, P.J.: 1981, Atmos. Environ. 15, 945.

Yan, N.D., and Dillon, P.J.: 1984, Experimental neutralization of lakes near Sudbury, Ontario. p. 417-456, in "Environmental Impacts of Smelters", J. Nriagu (ed.), Advances in Environmental Science Series, John Wiley and Son, Inc.

Yan, N.D. and Lafrance, C. 1984. Responses of acidic neutralized lakes near Sudbury, Ontario to nutrient enrichment. p. 457-521, in "Environmental Impacts of Smelters", J. Nriagu (ed.), Advances in Environmnetal Science Series, John Wiley and Son, Inc.

Yan, N.D. and Miller, G.E. 1984. Effects of deposition of acids and metals on chemistry and biology of lakes near Sudbury, Ontario. p. 243-282, in "Environmental Impacts of Smelters", J. Nriagu (ed.), Advances in Environmental Science Series, John Wiley and Son, Inc.

ROLE OF PRECIPITATION CHEMISTRY VERSUS OTHER WATERSHED PROPERTIES IN WISCONSIN LAKE ACIDIFICATION

Orie L. Loucks
Butler University, Holcomb Research Institute
4600 Sunset Ave., Indianapolis, IN., USA

Gary E. Glass
U.S. Environmental Protection Agency
Duluth, MN., USA

John A. Sorensen, Barbara W. Liukkonen,
James Allert and George Rapp, Jr.
College of Science and Engineering
University of Minnesota
Duluth, MN., USA

ABSTRACT. Data for over 100 watershed properties, including aspects of topography, hydrology, geology, soils, vegetation, lake morphometry and input precipitation chemistry, have been developed since 1980 for 316 watersheds in northern Wisconsin. The hypothesis being evaluated for this lake population is that the observed water chemistry, can be accounted for as a function of antecedent water and chemical inputs, after considering exchange processes in the lake and watershed and the lake/groundwater interactions. The variables found by regression analysis to explain observed variability in color, sulfate, and acid neutralizing capacity (ANC) levels in Wisconsin lakes are: for color, vegetative characteristics, mean depth, and water renewal times; for sulfate, precipitation concentration of sulfur, evaporative concentration, and lake water renewal time; ANC appears to be controlled by the size of the watershed, lake depth or water renewal time, and the intensity of anthropogenic inputs and cultural developments in the watershed. These results differ from previous studies in Wisconsin and nearby areas of Michigan and Minnesota by indicating that in some lakes acidity may not be in equilibrium with current precipitation chemistry.

1. INTRODUCTION

The northeastern portion of Minnesota, north central Wisconsin, and the Upper Peninsula of Michigan are areas with several characteristics of acid sensitivity in common: all three regions support a large number of lakes with recreational importance to their respective states, they are all located on bedrock or surficial deposits of low buffering

Water, Air, and Soil Pollution **31** (1986) 67–77.

capacity, and they are all characterized by a mixed conifer-hardwood vegetation with a similar history of logging and forest recovery during the twentieth century. There are differences from one state to the other, however, in total rainfall and evaporation, in the total inputs of acidic substances (Glass and Loucks, 1986), in cultural development, and in present-day lake chemistry, a response pattern that could be attributable either to the differences in precipitation chemistry or the differences in land use. This paper utilizes data on the chemical deposition patterns and on watershed and land use variables to evaluate the role of watershed characteristics, particularly land use, hydrology and groundwater, in controlling the concentration of chemicals in a population of Wisconsin lakes.

The study utilizes the watershed approach found effective in previous research (Likens et al. 1977; Chen et al. 1982) and extends its application to entire populations of watersheds and subtending lakes in the upper midwest (see Rapp et al. 1986 for details). The study considers the possibility of defining a dose/response relationship utilizing both known causal relationships (linkages with-in the framework of a single lake/watershed system), and population-based relationships recognizable through correlations describing the association between lake water chemistry and watershed characteristics.

The research framework has been described in previous papers (Glass and Loucks, 1986; Rapp et al., 1986). Briefly stated, it utilizes the conceptual abstraction of a watershed as a receptor system with the following major processes influencing the inputs and outputs of water and other substances: runoff; recharge; evaporation; ionic neutralization and mobilization; biological uptake; and anthropogenic interactions. Specific properties of the lake and terrestrial watershed can be quantified to represent the potential for each of the above processes to have a significant influence on chemical exchange within the watershed and, therefore, on the concentrations observed in the subtending lake. The hypotheses to be tested are essentially that watershed factors can each have a significant influence on reactions following the input of chemicals, producing a multiple factor explanation of the observed lake chemistry. Accordingly, the principal methodology is a combination of single-factor relationship and matrix correlation studies.

2. THE STUDY AREA AND DATABASE

The principal geomorphological characteristics of the study area have been determined by glacial processes. The resulting topography consists of areas of pitted outwash with lakes in deep deposits of outwash sands and sandy morainal till. The coarse texture of the sandy outwash and moraines allows much of the drainage from one basin to another to take place as seepage below the surface of the soil. The result is that a relatively large proportion (locally, as much as 40%) of the lakes are "seepage lakes" with neither inlet nor outlet streams. Most of the larger watersheds, however, have sufficient flow to maintain a surface water stream which controls the hydrology and chemistry of subtending lakes.

Overlying the physical configuration of the lakes in the upper midwest is a significant gradient in the concentration of acidifying substances in rain and snow (Glass and Loucks, 1986). The principal features of this deposition gradient, as characterized by representative sites in northeastern Minnesota, northcentral Wisconsin, and the eastern portion of the Upper Peninsula, can be summarized by noting that wet sulfate deposition ranges from a low of 9 kg ha^{-1} yr^{-1} in western Minnesota, to an average of 15 kg ha^{-1} yr^{-1} in central Wisconsin, and a maximum of 17 kg ha^{-1} yr^{-1} in the eastern part of the Upper Peninsula of Michigan.

The data for the study have been developed for a representative sub-sample of lakes identified for study between 1979 and 1983. The goal was to sample up to 10% of the total lake population in the designated study areas of each state (Glass et al. 1986). Subsequently, complete characterization of over 100 watershed properties was carried out for each of the watersheds supporting a study lake (Rapp et al., 1985; 1986). Many of the topographic, morphometric, geological, vegetative, and anthropogenic characteristics were measurable directly from maps and aerial photos and were viewed as primary cartographic data. However, there are other variables that must be derived from regional data such as the precipitation and evaporation at each of the watershed/lake sites.

One of the important variables that must be interpolated regionally is the sulfur concentration in precipitation at each watershed sites. The computation utilized the results of precipitation monitoring at the three primary NADP sites in the region (Glass and Loucks, 1985), and at adjacent sites to the north and south, sufficient to allow calculation of precipitation chemistry between the various monitoring stations using a linear interpolation technique. In addition, the monitored evaporation in the region (pan evaporation corrected for the differences between the pan and natural open-water surfaces), as well as total precipitation, have been obtained for each site by linear interpolation techniques. Total wet deposition of substances was calculated for each site as a product of the interpolated values of total preciptation and annual average ionic concentrations.

A key process in the hydrology of this sub-humid region, however, is evaporation and the resulting increase in concentration of chemicals in surface water. A term describing the result of this process is evaporative concentration ratio (ECR), the multiplier by which the ion components in the precipitation are increased in concentration in the remaining water after evaporation.

A second important aspect of lake/watershed system hydrology is the flushing or renewal of the lake water through annual inflow. Water renewal time is defined as the length of time required for a volume of inflow water to have entered (or left) the lake equal to the volume of water in the lake. For many small lakes, renewal times are about one year, but can range widely. However, in the upper midwest, a relatively large proportion of the lakes are "seepage" lakes, having no aboveground inlet or outlet, and this "hydrologic type" was found by Eilers et al., (1983) to be a good predictor of apparent sensitivity to acidic deposition. Seepage lakes, however, range from perched "recharge" lakes which drain outward to the surrounding soil (and have no groundwater inflow), to groundwater dominated seepage lakes where a below-ground drainage passes through the seepage lake. The former, with inputs of only precipitation, less evaporation, have very long water renewal times, while the latter have short renewal times.

Physical determination of the water budgets for these two classes of seepage lake, and the many intermediates, is nearly impossible. However, the recent work by Stauffer (1985) shows that the relative similarity of Mg concentrations, on the one hand to precipitation inputs (evaporation concentrated), or to concentrations in surrounding groundwater on the other hand, provides a chemical indicator of precipitation dominated vs groundwater controlled seepage lakes. Numerous examples of isolated, perched lakes show that both Cl^- and Mg^+ are present in the lakes at the concentration expected from precipitation X ECR. Accordingly, these lakes are assumed to have only precipitation directly on the lake, which after evaporation, is the only inflow (generating a corresponding groundwater outflow) that determines water renewal. As Mg concentrations in the lake approach that of Mg in the surrounding groundwater, most of the surface watershed around and above the lake is assumed to feed groundwater to the lake, in addition to the precipitation input (Stauffer, 1985). Thus, total lake water renewal time, based on Mg as an indicator, is calculated using only lake surface precipitation input for the low Mg lakes (ca. 0.3 $mg\ell^{-1}$), scaling up to a larger proportion of the surface watershed, until a concentration of 3.0 $mg\ell^{-1}$ is encountered. At Mg concentrations above this level, as is observed in larger watersheds, one can assume the equivalent of a "drainage" lake system with ground water making a major contribution.

In addition, there are important in-lake processes which act on nitrate and sulfate to breakdown the anion form, through reduction processes, yielding ANC (Schnoor and Stumm, 1984). The possibility exists that a local source of sulfate may be present in the watershed, either from sulfide minerals (Rapp et al., 1985), or as a waste product of local anthropogenic origin. Accordingly, both the conservative ion, chloride, and evaporation concentration ratio, were used to predict sulfate concentration in lake water solely as a result of precipitation input (excluding all watersheds having roads that could be salted during the winter). From this, the ratio of observed lake sulfate concentration over the expected lake sulfate concentration gives a S retention index (SRI) suggesting local sources when the ratio exceeds 1.0, and indicating sulfate reduction in the lake or a non-equilibrium with precipitation inputs when the ratio is appreciably below one.

Also important is the potential alteration of lake water quality through development of summer residences and feeder road systems around the lake shores. Counts of these summer or year-round residences are readily available, but a few on a small lake can have relatively more impact than the same number on a large lake. Few lakes in Wisconsin have no residences. Chloride, S and Na are all relatively conservative indicators of wastes from septic systems, and show generally small influences at a density of less than one dwelling unit per three hectares of lake surface area. Accordingly, the principal measures of local anthropogenic influence used here is the dwelling count/lake area ratio (DC/LA) along with total road length. The DC/LA ratio ranges from 0 to as much as 10; a cutoff of 0.35 units per ha has been used in the analysis to select for a pristine or lightly altered lake population.

3. RESULTS

The characteristics of lake populations are best represented initially by considering the number and percentage of lakes in each of the principal alkalinity classes used in lake acidification surveys (see Table II in Glass et al., 1986). The Wisconsin study population has 4% of the total set in a negative alkalinity status (ANC <0 $\mu eq \ell^{-1}$), while 12% of the Michigan/Ontario population is in this class and none of the Minnesota/Ontario lakes. The question being addressed in this paper, then, is whether, when all watershed factors are considered, the total acid and base forming inputs from precipitation explain the modest level of acid lakes in Wisconsin, or alternatively, whether natural processes independent of human activity provide a sufficient explanation.

First, it is useful to consider the range in the watershed factor data for the Wisconsin sample in terms of simple means, medians and standard deviations observed (Table I). The population is presented in three successively more restrictive subsets (A, B, and C): first, all lakes with watershed areas less than 1000 ha.; secondly, a subset of the first where, in addition, the dwelling count/lake area ratio is <0.35; and thirdly, a subset of the first in which only lakes with ANC <100 $\mu eq \ell^{-1}$ are considered. A key watershed input variable, wet SO_4^{2-} deposition, averages 15.0 kg Ha^{-1} yr^{-1} with only a few sites below 14 or above 16 in all three populations.

The great differences due to density of summer residences are evident in the lower average Cl^-, SO_4^{2-} and Na^+ for the set with DC/LA ratio <0.35 (B), compared with all watersheds of <1000 ha (A). Also of potential significance are the differences in indicated lake water renewal time, shifting from an average of 18 yr for both (A) and (B), to 26 yr for the narrowly defined "sensitive" set (ANC <100). The lower average color value in the sensitive set (C), corresponds with lower nutrient status, but has higher SO_4 concentrations than the A or B populations. The amount of chlorophyl A is not greatly influenced by removing high residence-density lakes (B), but it is lower in the low ANC lakes where NO_3^- Ca^+ and Mg^{2+} are significantly lower. The ion ratio results, and the good correspondence of calculated versus measured conductivity, indicate the very high level of closure for all chemical analyses, thus reducing measurement error as a source of variability in these data. The slight departure of the ion ratio on the + side is due to organic anions not included in the anion sum at the present time.

One step in the statistical analysis of over 100 variables at up to 233 sites was to prepare a correlation matrix in which all of the

TABLE I. MEANS, MEDIANS, AND STANDARD DIVATIONS OF WATERSHED AND LAKE RESPONSE VARIABLES FOR THREE SETS OF WISCONSIN LAKES

Measured Variable	(A) All lakes with watersheds <1,000 ha			(B) Watersheds <1,000 ha & DC/LA Ratio <0.35			(C) Watersheds <1,000 ha, ANC <100 $\mu eq\ell^{-1}$		
	Mean	Med.	Stand. Dev.	Mean	Med.	Stand. Dev.	Mean	Med.	Stand. Dev.
		(N = 233)			(N = 111)			(N = 78)	
Wet SO_4^{2-} Dep.	15.0	15.0	0.52	15.0	15.0	0.50	15.0	15.0	0.53
T. Renewal-Mg (yrs)	17.5	15.6	16.0	17.8	16.6	16.4	25.9	23.8	15.4
Decid. For. (%)	59.8	60.8	19.0	60.4	63.3	20.8	61.5	63.1	19.0
Dwelling/Lk. Area R.	0.55	0.36	0.86	0.13	0.13	0.11	0.49	0.32	0.55
Tot. Rd. Length (km)	2.17	1.50	2.12	1.37	0.70	1.79	1.67	1.30	1.66
Color (Pt-Co Units)	33.3	13.5	52.9	37.5	15.9	56.2	18.5	9.0	27.3
Acid. Neutr. Cap. ($\mu eq\ell^{-1}$)	224	91	260	209	65	270	32.5	27.7	28.1
pH (pH units)	6.75	6.80	0.87	6.67	6.62	0.89	6.07	6.10	0.54
SO_4 ($mg\ell^{-1}$)	3.50	3.38	1.14	3.27	3.28	0.97	3.89	3.77	1.15
Cl ($mg\ell^{-1}$)	1.09	0.46	1.66	0.79	0.30	1.14	0.70	0.39	0.91
Na ($mg\ell^{-1}$)	1.06	0.60	1.11	0.88	0.45	0.92	0.58	0.37	0.68
Chlor-A ($\mu gm\ell^{-1}$)	6.56	4.20	6.84	6.78	4.20	7.70	4.04	3.20	3.17
NO_3 ($mg\ell^{-1}$)	0.17	0.08	0.25	0.16	0.09	0.20	0.11	0.09	0.09
Mg ($mg\ell^{-1}$)	1.34	0.75	1.09	1.26	0.65	1.12	0.56	0.58	0.17
Ca ($mg\ell^{-1}$)	3.91	1.99	3.60	3.69	1.73	3.78	1.33	1.28	0.41
+/- (ion ratio)	1.09	1.07	0.13	1.12	1.09	0.13	1.07	1.06	0.12
C/M (Calc./Meas. Cond.)	1.00	0.99	0.06	0.99	0.98	0.06	0.97	0.97	0.05

watershed variables and all of the lake response measures can be evaluated in a cross-correlation with each other. While the primary concern of the study was to evaluate correlations between watershed factors and the resultant lake chemistry, the correlation matrix also allows study of intercorrelation between inputs and watershed factors themselves, and among lake chemistry variables. The results in Table II show the significant correlations for each of the major categories of input and watershed influence. Results for color are shown because color can be an important interactive property of lakes, and because previous studies (Rapp et al., 1985, 1986) have shown our approach provides a high level of explanation for color.

A wide range of factors contributes to both the in-lake sulfate concentrations and acid neutralizing capacity (ANC). The significant correlation between wet sulfate deposition (which has only a small gradient within Wisconsin) and observed lake SO_4^{2-} for the 233 watersheds of <1000 ha must be judged also in relation to the large correlation of SO_4^{2-} with water renewal time (estimated by Mg), and apparently, to S retention (SRI). The lakes with long renewal times (25 yr and up) may not be in equilibrium with modern precipitation inputs of S, a phenomenon that is indistinguishable at present from possible sulfur reduction in those lakes. The apparent importance of SO_4^{2-} deficiencies (as indicated by SRI<0) therefore must be viewed as indicative of two processes, reduction and hydrologic lag time. The fact that measured in-lake SO_4^{2-} concentration is one of the terms in the S retention ratio means that the high correlation of SO_4^{2-} with SRI should be considered only parenthetically. Relatively large correlations also occur between observed ANC and total lake water renewal time, sulfate retention index, watershed area, and anthropogenic influences such as total road length.

Part B of Table II shows results for the smaller population obtained when lakes with relatively high anthropogenic influences are removed. The factors influencing color now have higher correlation values, as these lakes average slightly higher values of color. Correlations for SO_4^{2-} are not greatly different, but the influence of mean depth, watershed area and water renewal time on ANC have all increased. Residence density is now a negligable influence on SO_4^{2-}, and road length is a negligible source of explanation for variation in ANC.

The results in Part C show relationships for a population of low alkalinity lakes (ANC <100 $\mu eq \ell^{-1}$). Water renewal time is not a major factor controlling SO_4^{2-} in this set because the population already is selected for long renewal times (see Table I) and has higher than average SO_4^{2-}. Summer residence density emerges as an important factor in SO_4^{2-} concentrations, as does the local gradient in precipitation concentration of SO_4^{2-}. The latter also appears to influence observed ANC.

Overall, atmospheric deposition of SO_4^{2-} can be seen as an important factor, along with surficial, morphometric and hydrologic factors such as renewal time. As already noted, the correlation of in-lake SO_4^{2-} concentration with water renewal time may also reflect 20-98 changes in precipitation chemistry due to the long averaging time represented by the long-renewal times of some lakes. The gradient in

TABLE II. SELECTED CORRELATION COEFFICIENTS FOR THREE SUCCESSIVE POPULATIONS OF WISCONSIN WATERSHEDS

Independent Variables	Dependent Variables: Color	Sulfate	Acid Neut.Cap.
A. 233 Watersheds of <1,000 ha.			
Precip. Conc. of SO_4^{2-}	+.24**	+.22**	
Total watershed area	+.23**		+.44**
Evap. conc. ratio	-.36**	+.20**	
Total renewal time	-.43**	+.26**	-.62**
Mean depth	-.44**	+.20**	-.20**
S. retention index	.14*	+.95**	-.45**
% Marsh	+.22**	-.22**	
Upland deciduous	-.35**		
Dwellings/lake surf. ratio		+.20**	+.16*
Total road length	-.20**		+.20**
B. 111 Watersheds of <1,000 ha. and DC/LA Ratio <.35.			
Precip. Conc. of SO_4^{2-}	+.41**	+.23**	
Total watershed area			+.51**
Evaporative concentration ratio	-.41**	+.22**	
Total renewal time	-.41**	+.25**	-.71**
Mean depth	-.50**	+.28**	-.34**
S. retention index		+.94**	-.41**
% Marsh	+.24*		
Upland deciduous	-.40**		
Dwellings/lake surface ratio			+.17*
Total road length	-.28**		
C. 120 Watersheds of <1,000 ha. and ANC <100 $\mu eq \ell^{-1}$.			
Precip. Conc. of SO_4^{2-}		+0.30**	-.22*
Total watershed area			
Evaporative concentration ratio	-.28**	+0.27**	
Total renewal time			
Mean depth	-.33**		-.24**
S. retention index		+.94**	-.44**
% Marsh	+.19*		
Upland deciduous	-.32**		
Dwellings/lake surface ratio	-.22*	+.36**	
Total road length	-.24**		

Statistical Significance: ** $p<0.01$; *p0.01 to <.050.

wet deposition of SO_4^{2-} correlates with observed variation in ANC in only one of the populations, but the very large correlation of water renewal time with ANC (particularly in the low residence density set) could be the result of changes in precipitation SO_4^{2-} concentration over time. This question is still unanswered at this time and investigation is continuing.

4. DISCUSSION AND CONCLUSIONS

A comparison can now be made of these Wisconsin results with the analyses presented by Eilers _et al._ in 1983, as well as results from Minnesota (Rapp _et al._, 1985) and from the U.P. of Michigan (Rapp _et al._, 1986). In both Minnesota and Michigan, as in these Wisconsin results, annual wet deposition of SO_4^{2-} has had an important correlation with the observed [SO_4^{2-}] in lake water. On the other hand, the correlations of ANC with acidic inputs and various watershed variables are only modest in all three states. Thus, one reaches the conclusion that watershed variables, including precipitation inputs, contributing to high explanation are:

- For color - Water renewal time and local organic soils;
- For SO_4^{2-} - Precipitation SO_4^{2-} concentration consistently; other variables such as water renewal time, evaporative concentration and bedrock chemistry vary from region to region;
- For ANC - Bedrock or surficial materials, local land use, and wet sulfate deposition.

These results provide a stong contrast with the findings of Eilers _et al._ (1983) in their study of Wisconsin lakes. Their findings emphasized "lake hydrology" as the factor determining sensitivity to acidic deposition, as this paper does, but the variable used was "lake type", rather than a measure of hydrology. The seepage type was shown to be strongly associated with a measure representing the greatest sensitivity of the lakes to acid inputs. Here we show that within the seepage lake type, differences in water renewal time are important in differentiating sensitive from less sensitive lakes. High precipitation inputs of SO_4^{2-} correlate with elevated concentrations in the lakes, but the high correlation of water renewal time with SO_4^{2-} concentration (lakes with renewal times over 25 yr having the lowest SO_4^{2-} concentrations) suggests that atmospheric inputs and watershed properties may be inter-connected in previously unrecognized ways.

Two other conclusions can be drawn from these results. It is clear that anthropogenic development, including roads and summer residences, can exert a profound influence on the chemistry of the Wisconsin lake population as a whole. This influence extends across a number of response variables, including in-lake chloride, Ca and Na concentrations, and both SO_4^{2-} and ANC. When a lake population is chosen so as to remove the anthropogenic influence, the importance of lake morphometry, hydrology and acidic inputs is more evident.

The results also show the importance of lake water renewal time in determining the extent to which present precipitation chemistry appears to influence present lake chemistry. The very slow rates of addition of water volume to the precipitation dominated lakes lead to the observation that the present lake chemistry may be more characteristic of precipitation inputs averaged over the past 50 yr than of input chemistry during the past two or three years. These water renewal time results place in question the widespread assumption of many acid deposition assessment programs that the presently observed response in the lake can be thought of as representing an equilibrium with present precipitation chemistry.

ACKNOWLEDGMENTS

The data used in this report represent the combined effort of many individuals from the University of Minnesota-Duluth, Environmental Research Laboratory-Duluth and Environmental Research Laboratory-Corvallis, Wisconsin, Department of Natural Resources, Minnesota Pollution Control Agency, Minnesota Department of Natural Resources, USDA Forest Service-Duluth, U.S. Geological Survey, U.S. Fish & Wildlife Service, Ontario Ministry of the Environment, Environment Canada, NADP Natural Resources Ecology Laboratory, Ft. Collins, Colorado and NADP site investigators. This work was supported by the U.S. EPA National Acid Precipitation Assessment Program through Cooperative Agreements with the University of Minnesota-Duluth (CR809412), Wisconsin Department of Natural Resources (CR-809484), with the USDA Forest Service-Duluth, Minnesota (AD12F3581), and a subvention by the Holcomb Research Institute, Butler University, Indianapolis, Indiana.

REFERENCES

Chen, C.W., Gherini, S.A., Hudson, R.J.M., and Dean, J.D.: 1983, *The integrated lake-watershed acidification study: Model principles and application procedures.* Electric Power Research Institute, EPRI EA-3221, Palo Alto, CA.

Eilers, J.M., Glass, G.E., Webster, K.E., and Rogalla, J.A.: 1983, *Can. J. Fish. Aquat. Sci.* **11** 1896.

Glass, G.E., and Loucks, O.L.: 1986, *Environ. Sci. Technol.*, **20**, No.1, pp. 35–43.

Glass, G.E., Sorensen, J.A., Liukkonen, B.W., Rapp, G. Jr., Loucks, O.L.: 1986, This volume.

Likens, G.E., et al.: 1977, *Bio-Geo-Chemistry of a Forested Ecosystem.* Springer-Verlag, N.Y., Berlin, Heidelberg, pp. 148.

Rapp, G., Jr., Allert, J.D., Liukkonen, B.W., Ilse, J.A., Loucks, O.L., and Glass, C.E.: 1985, *Environ. International* **11** 425.

Rapp, George, Jr., Liukkonen, Barbara W., Allert, James D., Glass, Gary E., and Loucks, O.L.: 1986, *Geologic and Atmospheric-Input Factors Affecting Watershed Chemistry in Upper Michigan*. In Press, Environmental Geology.
Schnoor, J.L., and Stumm, W.: 1984, "Acidification of Aquatic and Terrestrial Systems." *In Chemical Processes in Lakes*. W. Stumm (Ed.), Wiley Interscience, New York, N.Y.
Stauffer, R.E.: 1985, *Envionr. Sci. Technol.*, **19** 405.

DISCLAIMER

Although the research described in this article has been funded wholly or in part by the U.S. EPA under Cooperative Ageement #CR809412 to the University of Minnesota–Duluth/Butler University, it has not been subjected to the agency's peer and administrative review and therefore may not necessarily reflect the views of the agency and no official endorsement should be inferred.

ADIRONDACK HEADWATER LAKE CHEMISTRY RELATIONSHIPS WITH WATERSHED CHARACTERISTICS

Carolyn T. Hunsaker,* John L. Malanchuk,** Richard J. Olson,*
Sigurd W. Christensen,* Robert S. Turner*

*Environmental Sciences Division, Oak Ridge National Laboratory, P.O. Box X, Oak Ridge, TN 37831

**Acid Deposition Assessment Staff, U.S. Environmental Protection Agency, 401 M Street, S.W., Washington, DC 20460

ABSTRACT. The Adirondack Region of New York State has been identified as having surface waters sensitive to acidic deposition and as receiving large annual inputs of acidic deposition. The large amount of data available for this region makes a quantitative study of the region possible. Compiled from a variety of sources, the Adirondack Watershed Data Base (AWDB) contains information on lake chemistry; lake elevation, area, and volume; and associated watershed data, such as size, slope, aspect, elevation, vegetation and wetland types, beaver activity, fire and logging history, and soils data. Bivariate and multivariate procedures were used to examine relationships between watershed attributes and lake chemistry. Because the variables in the data base are being refined and modified, the current relationships should be considered preliminary. Preliminary results indicate that wet deposition, lake elevation, and forest cover are the principal variables that are associated with variance in the data for lake pH and acid neutralizing capacity (ANC) in the Adirondacks. For headwater lakes in the Adirondacks, we estimate approximately 50% have a total ANC $\leq$ 40 μeq L^{-1} and 40% have a pH $\leq$ 5.5.

1. INTRODUCTION

The Adirondack Region of New York State has been identified as having surface waters sensitive to acidic deposition and as receiving large annual inputs of acidic deposition. Quantitative analysis of data available for this region should increase our knowledge of why some lakes acidify in a region and others do not. The primary objective of this Adirondack study is to examine, on a regional scale, watershed attributes that account for variability and change in lake pH and acid neutralizing capacity (ANC). Secondarily, the relationships developed by statistical analyses are used to predict the acidity of additional Adirondack lakes for which chemistry data do not exist. A model using

Water, Air, and Soil Pollution **31** (1986) 79–88.

readily available data that can reasonably predict the acidity of a lake will help assess the status of the nation's lakes without performing costly and lengthy data collection programs. Initial analyses involve only headwater lakes (the upper most lake in a watershed), but work with complex lake systems is ongoing.

2. DATA AND METHODS

2.1. Data

The Adirondack Watershed Data Base (AWDB) includes general data (e.g., lake name, location, size, elevation) for all 2759 lakes in the Adirondack Ecological Zone (AEZ, approximately a 300-m elevation contour) (Baker et al., 1984) and most watershed attribute data for 463 headwater lakes. AWDB watershed variables were selected based on their presumed relative importance to lake chemistry, including the acidic deposition issue, and on their regional availability. The AWDB links data on a watershed's morphology, physiography, bedrock, soils, land cover, land use, wetlands, disturbances (presence of cabins, fires, and logging), beaver activity, and atmospheric deposition with data on lake water chemistry and fishery status. Watershed data were compiled by several organizations from maps, aerial photographs, and regional monitoring networks (Table I). The Fish Information Network (FIN) Data Base (Baker et al., 1984) provided water chemistry and fishery status data compiled from many surveys. The AWDB will be documented and available from the Acid Deposition Data Network (ADDNET) at Oak Ridge National Laboratory (ORNL) (Rosen et al., 1986).

Although the AWDB contains some data for thousands of lakes in the Adirondacks, extensive watershed attribute data were available for only the 453 headwater lakes used in this study. In developing models for pH and for ANC, all lakes were used which had measurements of both the candidate predictor variables (Table II) and the response variable from FIN: 237 lakes for pH and 210 lakes for ANC. Predictions of pH or of ANC could then be made for lakes having no chemistry data but with the necessary predictor variables selected for the final models: 216 lakes for pH and 242 lakes for ANC. These 453 lakes constitute the majority of Adirondack headwater lakes; a relatively small number of headwater lakes in the southwestern part of the AEZ were not included because wetland data were unavailable. In the National Surface Water Survey (NSWS), 48 of our headwater lakes have both pH and ANC data. Initial analyses were limited to headwater lakes because we believed relationships would be most apparent for these simple watershed systems.

2.2. Methods

Techniques for identifying relationships among environmental variables included both bivariate and multivariate analyses using SAS (1982) (Figure 1). For each lake, the water chemistry was represented by averaging all surface observations for each individual water chemistry variable for the summer months (May-September) from 1974 through 1983,

TABLE I. Types of data in the Adirondack Watershed Data Base and their sources.

Data type	Source	Compiled by[†]
Morphology	U.S. Geological Survey's topographic maps; New York Department of Environmental Conservation's records; aerial photos	SUNY/P
Physiography	Defense Mapping Agency's TOPOCOM digital topography data	ORNL
Bedrock	1982 National Atmospheric Deposition Programs's State-Level Atlas	ORNL
Soils	1974 Soil Conservation Service's mesoscale maps, Soils-5, and Pedon Data Base; U.S. Environmental Protection Agency's Pilot Soil Survey	APA/ORNL
Land cover	1978 LANDSAT imagery	APA/ORNL
Wetlands	1982 Wetland map	SUNY/P
Cabins	1978-1983 Aerial photos	SUNY/P
Fire, logging (historic)	1916 New York State map	SUNY/P
Beavers	1978-1981 Aerial photos	SUNY/P
Deposition	1951-1981 Precipitation norms 1980-1982 Deposition monitoring	ORNL
Land use	APA Park Plan	APA
Water chemistry	Fish Information Network-FIN (assembled from several sources)	NCSU
Fish status	FIN	NCSU

[†]SUNY/P - State University of New York at Plattsburgh.
ORNL - Oak Ridge National Laboratory.
APA - Adirondack Park Agency.
NCSU - North Carolina State University.

TABLE II. Candidate variables selected for possible inclusion in multivariate analyses, significant Spearman rank order correlations for pH and alkalinity from the FIN subset of lakes, and the a priori expectation of the direction of relationship between a predictor variable and a response variable. The variables were selected based on knowledge of watershed processes and simple correlations with lake chemistry.

Predictor Value	Spearman Correlations for pH (-)	Spearman Correlations for pH (+)	Spearman Correlations for Alkalinity (-)	Spearman Correlations for Alkalinity (+)	Expected Direction of Relationship with pH or Alkalinity
<u>Morphology</u>					
Lake area		***[a]		**	+
Watershed area		***		**	+
Watershed to lake area ratio				*	+
Lake development ratio		*			+
<u>Physiology</u>					
Lake elevation	*****		****		-
Slope >15% (% watershed)		**		**	-
Slope >25% (% watershed)		**		**	-
<u>Geology</u>					
Rock buffering, none-low + low-medium (% watershed)	**		**		-
<u>Soils</u> (% watershed)					
Rock outcrops		*		*	-
Stoney soils		*		*	either
Moderately steep soils	**		**		-
Very steep soils	**		**		-
Moderate runoff potential		**		**	+
High runoff potential	**		**		-
Base saturation, <20%					-
Soil pH, <5.0	*		*		-
Cation exchange capacity, <10%		**		*	-
Exchangeable bases, <6		*			-
Organic Matter, >10%				*	-
<u>Land Cover</u> (% watershed)					
Conifer and wet conifer forest cover	****		****		-
Hardwood forest cover		**		**	+
Nonforest cover		***		***	+
<u>Wetlands</u>					
Total wetland (% perimeter)	**		**		either
Very acid wetland (% perimeter)	*		**		-
Nonacid wetland (% perimeter)					either
Total wetland (% watershed)	*		*		either
Very acid wetland (% watershed)	*		*		-
Nonacid wetland (% watershed)					either
<u>Cabins</u>					
1978 Cabins to lake ratio		***		**	+
<u>Disturbances</u> (% watershed)					
Burned		**		*	+
<u>Deposition</u>					
Average annual H^+ deposition	*****		*****		-
Sulfate, wet deposition	*****		*****		-
Nitrate, wet deposition	*****		*****		-
Sulfate-nitrate, wet deposition	*****		*****		-

[a]Each asterisk (*) represents a magnitude of approximately 0.1; therefore, ** indicates r=0.2. These results are presented in this manner because of their preliminary nature.

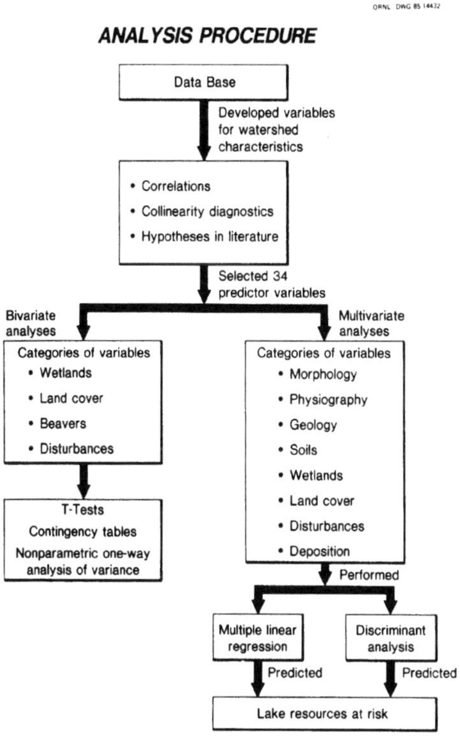

Figure 1. Procedure used for variable selection and statistical analyses to determine significant relationships between variables and to predict lake resources at risk.

excluding questionable data from 1979. Correlations between watershed variables and lake pH and ANC, as well as hypotheses and data in the literature on watershed influences on surface water chemistry, were instrumental in the selection of 34 predictor variables (Table II). Collinearity diagnostics reduced this number to 23 and 28 variables which were analyzed for the NSWS and FIN subsets of lakes, respectively (Hunsaker et al., 1986).

Bivariate statistical techniques were used to determine relationships between individual watershed characteristics and lake chemistry. Spearman rank correlations, t-tests, contingency tables, and nonparametric one-way analyses of variance were the statistical techniques used to examine current hypotheses about the influence of wetlands, forest cover, beaver activity, and land cover disturbances on lake water chemistry (Hunsaker et al., 1986).

Stepwise multiple linear regression (MLR) and stepwise discriminant analysis techniques were used to determine multivariate relationships between lake pH or ANC and watershed attributes, using 23 to 28 selected variables depending on the subset of lakes. The MLR analyses were performed on two subsets of lakes: the NSWS subset of 48 lakes and the FIN subset of 237 (pH) or 210 (ANC) lakes. For the discriminant analysis, two classes were used for both pH (acid, pH $\leq$ 5.5; basic, pH > 5.5) and ANC (acid, ANC $\leq$ 40 μeq L^{-1}; basic, ANC > 40 μeq L^{-1}). For the purpose of illustration, we present only the results of the Spearman rank correlations, the multiple linear regression, and the discriminant analysis; the reader is referred to Hunsaker et al. (1986) for more details on these and other analyses.

3. RESULTS

3.1. Bivariate Analysis

The magnitude and direction (positive or negative) of the Spearman rank correlations between a watershed variable and lake chemistry were similar for lake pH and ANC (Table II). Among the variables selected for input to multivariate analyses, the strongest correlations were between lake chemistry and average annual wet deposition (H^+, sulfate, and nitrate ions), percentage of watershed in coniferous forest cover, and lake elevation (r = -0.4 to -0.5). As expected, these were all negative correlations. Weaker positive correlations exist between lake chemistry and lake surface area, watershed area, percentage of watershed having a slope > 15%, percentage of watershed with soils having a moderate runoff potential or a cation exchange capacity of < 10%, percentage of watershed with hardwood forest cover or unforested, and percentage of watershed in which there were fires in the early 1900s (r = 0.1 to 0.3). Moderate negative correlations (r = -0.1 to -0.3) with lake pH and ANC were found for watershed variables representing the percentage of the watershed with no or low-to-medium bedrock buffering capacity; with soils that are moderately steep, very steep, or that have a high runoff potential; or with wetlands. Of these variables, the percentage of watershed having a slope > 15% and the percentage of watershed with soils having a cation exchange capacity of < 10% have correlations that are opposite to what one would expect based on theory.

3.2 Multivariate Analyses

Three watershed variables were selected consistently by both the MLR and discriminant analyses as being the strongest predictors of both lake pH and ANC: wet H^+ deposition, lake elevation, and percentage of watershed in conifers. Wet H^+ deposition was a surrogate for acidic deposition because all atmospheric deposition variables were collinear. Lake area, percentage of watershed having a slope > 25%, and percentage of watershed in rock outcrops were frequently significant variables in the multivariate analyses. Multiple linear regression predictions were similar to observed pH and ANC values (Figure 2). Discriminant analysis predictions were very similar to the MLR predictions. Existing MLR models can account for 38 to 74% of the variation in lake pH and alkalinity, depending on the subset of

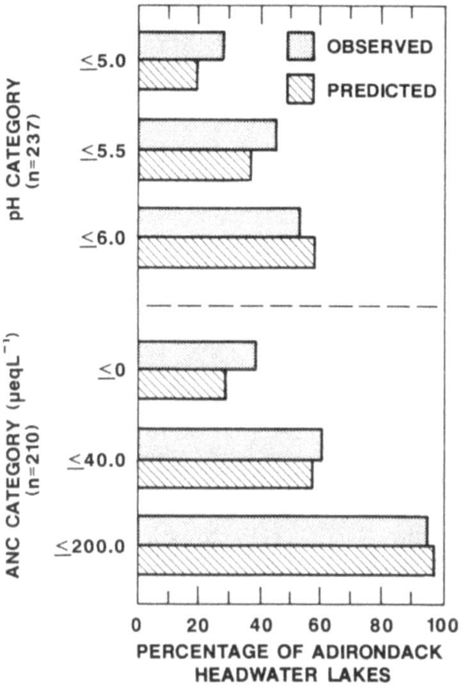

Figure 2. Comparison of the actual percentage of Adirondack headwater lakes from the FIN subset with model predictions for pH and ANC cumulative categories. Observed percentage in each category is based on actual measurements. Predicted percentage in each category is based on the multiple linear regression models developed from the FIN subset.

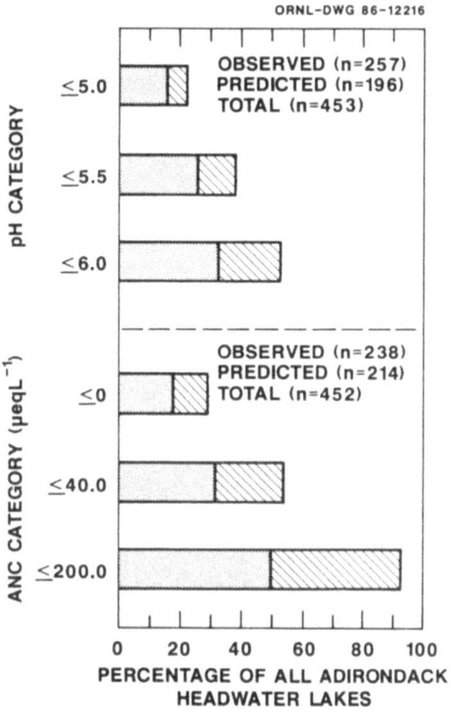

Figure 3. Estimates of the total percentage of all Adirondack headwater lakes in pH and ANC cumulative categories. These estimates are based on a combination of actual observed data (solid bar), when available from NSWS or FIN, and predictions (lined bar) based on the multiple linear regression models developed from FIN data.

watersheds used to fit the model. The four-variable models derived from the NSWS subset always accounted for more variability but are based on a very small sample size. Our predictions of the number of lakes having an average ANC or pH in a specified category were within 10% of the observed values using the six-variable MLR model developed from FIN data; however, the MLR tended to underpredict the number of lakes in lower pH and ANC categories (Figure 2). Discriminant functions using eight variables and two classes correctly classified 81 to 84% of the lakes with observed pH and ANC measurements. As the data base is expanded and modified these results may change slightly, but we expect the basic conclusions concerning the importance of various predictors and the ability of the models to classify correctly to remain essentially unchanged.

The empirical relationships (based on the FIN subset) developed for watershed characteristics and lake chemistry were used to predict individual lake pH and ANC values for the 216 and 242 headwater lakes for which no chemical data exist. Using measured values for lakes having data and MLR predictions for lakes without data, our analysis indicated that approximately 20% of headwater lakes have a pH of $\leq$ 5.0 and 30% have an ANC of $\leq$ 0 µeq L^{-1} (Figure 3).

4. CONCLUSIONS

The AWDB contains data on watershed attributes for 463 headwater lakes and water chemistry for more than 200 lakes, depending on the variables of interest. Results presented in this paper are preliminary because we are currently correcting data, such as latitude and longitude coordinates; revising a number of variables; adding new variables; and using a different method to remove collinear variables. However, we do not expect our final conclusions to be very different from those presented here. Results from the bivariate analyses were sometimes consistent with current hypotheses. For example, the watersheds with a high percentage of conifers are more likely to have lakes with low pH and ANC. Also, wetlands do appear to be associated with differences in lake chemistry. Further analyses on complex lake systems will help substantiate these results. The MLR and discriminant analysis techniques select similar predictor variables - wet deposition; lake elevation; lake area; and percentage of watershed in conifers, having soils with a slope > 25%, and in rock outcrop. Wet deposition (H^+) is the strongest predictor variable in all multivariate models. The degree to which relationships determined by this preliminary research may hold for other regions of the country is not known but is the subject of future work. However, these relationships suggest some factors that should be considered important in assessing the acidification of surface waters.

For headwater lakes in the Adirondacks (combining observed, NSWS data was preferred over FIN data, with predicted data), approximately 50% have a total ANC of $\leq$ 40 µeq L^{-1} and 40% have a pH of $\leq$ 5.5 (levels believed to be detrimental to some fish species). Headwater lakes have a much smaller average size and a slightly higher average elevation than the average of all lakes in the Adirondacks (Hunsaker _et al._, 1986); therefore, one expects these lakes to tend to have lower pH and ANC than the entire population of Adirondack lakes (Baker and Harvey, 1984). With such information as currently exists for headwater lakes in the Adirondacks, we estimate that approximately 30% of these lakes have an ANC of $\leq$ 0 µeq L^{-1} and 20% have a pH of $\leq$ 5.0. Our estimates appear to be low for these categories (Figure 2). The FIN subset of lakes may not be completely representative of Adirondack headwater lakes, and this may account for our apparent underprediction of the number of lakes in the $\leq$ 5.0 and $\leq$ 0 µeq L^{-1} categories.

Our estimates are similar to those developed by Baker and Harvey (1984) and Schnoor et al. (1985) for all Adirondack lakes, using different techniques and/or data. The appropriateness of the MLR models, developed for headwater lakes in the Adirondacks, to nonheadwater lakes is currently being investigated.

ACKNOWLEDGEMENTS

Many people have made this study possible. The authors thank J. J. Beauchamp of the Engineering Physics and Mathematics Division at ORNL for his statistical advice. We also acknowledge the data compilation and mapping efforts of R. C. Durfee, P. R. Coleman, D. L. Wilson, and F. E. Latham of the Computing and Telecommunications Division at ORNL, and G. K. Gruendling, D. J. Bogucki, K. B. Adams, and E. Allen of the State University of New York at Plattsburgh. We thank as well J. P. Baker of Kilkelly Environmental Associates and R. P. Curran of the Adirondack Park Agency for supplying us with data.

This research was sponsored as part of the National Acid Precipitation Assessment Program (NAPAP) by the U.S. Environmental Protection Agency (EPA), and by the Office of Environmental Analysis, U.S. Department of Energy, under EPA Interagency Agreement No. 40-1488-84 and under Contract No. DE-AC05-84OR21400 with Martin Marietta Energy Systems, Inc. The research described in this report has not been subjected to EPA's or NAPAP's required peer and policy review and does not necessarily reflect the views of these organizations, therefore no official endorsement should be inferred. Publication No. 2632, Environmental Sciences Division, ORNL.

REFERENCES

Baker, J. P., and Harvey, T. B.: 1984, Critique of acid lakes and fish population status in the Adirondack Region of New York state. Draft final report to the U.S. Environmental Protection Agency.

Baker, J. P., Harvey, T. B., and Nicolette, J. P.: 1984, Compilation of available data on the status of fish populations in regions of the northeastern U.S. susceptible to acidic deposition. Draft final report to the U.S. Environmental Protection Agency.

Hunsaker, C. T., Olson, R. J., Christensen, S. W., Turner, R. S., and Beauchamp, J. J.: 1986, Empirical relationships between watershed attributes and lake chemistry in the Adirondack Region, ORNL/TM-9838, Oak Ridge National Laboratory, Oak Ridge, Tennessee, in press.

Rosen, A. E., Olson, R. J., Durfee, R. C., Wilson, D. L., Coleman, P. R., Brandt, C. C., Turner, R. S., and Hunsaker, C. T.: 1986, An Adirondacks Watershed Data Base: Attribute and mapping information for regional acidic deposition studies, ORNL/TM-10144, Oak Ridge National Laboratory, Oak Ridge, Tennessee, in press.

SAS Institute, Inc.: 1982, SAS User's Guide: Statistics, Cary, North Carolina.

Schnoor, J. L., Nair, D. R., Kuhns, M. A., Lee, S. J., and Nikolaidis, N. P.: 1985, Lake resources at risk due to acid deposition in the northeastern United States. Draft final report to the Environmental Protection Agency. Cooperative agreement CR-811257-01.

MODEL OF INTERNAL ALKALINITY GENERATION: SULFATE RETENTION COMPONENT

L. A. Baker, P. L. Brezonik
Dept. Civil and Mineral Engineering
500 Pillsbury Dr.
University of Minnesota
Minneapolis, MN 55455 USA

C. D. Pollman
KBN Engineering and
Applied Sciences
P. O. Box 14288
Gainesville, FLA 32604

ABSTRACT. Internal alkalinity generation is modeled by an input-output approach in which equations to describe budgets for sulfate, nitrate, ammonium, and base cations are linked to an alkalinity budget equation. Calibration of the sulfate model using ion budgets for 14 softwater lakes shows that the sulfate sink coefficient is reasonably uniform (mean = 0.46 m yr^{-1}) and can be used to predict sulfate retention. Model predictions show that internal sulfate sinks are needed to correctly predict lakewater $[SO_4^{2-}]$ and that in-lake sulfate sinks can account for over 50% of input. For experimentally acidified Little Rock Lake, Wisconsin, the sulfate model predicts 90% recovery of $[SO_4^{2-}]$ 13 years after acid additions stop.

1. INTRODUCTION

There is a growing awareness that in-lake (internal) alkalinity generation (IAG) is important in regulating alkalinity in softwater lakes. Schindler et al. (1986a) concluded that IAG exceeds terrestial alkalinity production when the watershed: lake ratio is < 8:1, and IAG is important in regulating lakewater alkalinity in seepage lakes common in Wisconsin, Minnesota, Ontario, and central Florida (See Baker et al., 1986). Despite the significance of IAG, regional lake acidification models do not explicitly consider internal processes, although the Trickle Down Model (Schnoor, 1985) includes a 0th order term for IAG. This paper presents the framework for a simple model of IAG and develops the sulfate submodel.

2. THEORETICAL DEVELOPMENT

A model of IAG can be developed using input/output modeling concepts analogous to those used in nutrient loading models (Vollenweider, 1975) and alkalinity/ electroneutrality concepts (Cook et al., 1986). It consists of five linked differential equations that describe mass balances for alkalinity and for ions whose production or consumption affects alkalinity:

Water, Air, and Soil Pollution **31** (1986) 89–94.

$$d[alk]/dt = (1/V)\{J_{alk} - S_o[alk] + k_{SO4}A[SO_4^{2-}] + + k_{NO3}[NO_3^-]V + WA - k_{NH4}[NH_4^+]V\} \quad (1)$$

$$d[SO_4^{2-}]/dt = (1/V)\{J_{SO4} - [SO_4^{2-}](S_o + k_{SO4}A)\} \quad (2)$$

$$d[NO_3^-]/dt = (1/V)\{J_{NO3} - [NO_3^-](S_o + k_{NO3}V\} \quad (3)$$

$$d[NH_4^+]/dt = (1/V)\{J_{NH4} - [NH_4^+](S_o + k_{NH4}V)\} \quad (4)$$

$$d[\sum M^+]/dt = (1/V)\{J_m - [\sum M^+]S_o + WA\} \quad (5)$$

where [i] = concentration of i, meq m^{-3} ($[\sum M^+]$ = sum of cations produced by in-lake weathering),
V = lake volume, m^3,
A = lake area, m^2,
S_o = outflow, m^3 yr-1,
J_i = loading of constituent i, meq yr^{-1},
k_{SO4} = first order loss term for SO_4^{2-} sink, m yr^{-1},
k_{NO3}, k_{NH4} = first order loss constants for NO_3^- and NH_4^+, yr^{-1},
W = in-lake weathering rate, meq m^{-2}-yr.

Change in alkalinity is the sum of alkalinity inputs plus IAG minus outflows and internal alkalinity consumption (Equation (1)). IAG occurs by sulfate retention (assimilatory and dissimilatory reduction), nitrate retention (assimilatory reduction or denitrification), and cation production; internal alkalinity consumption results from ammonium retention. Equations describing budgets for these ions include inputs, outputs, sink/source terms (Equations (2) to (5)). Justification for the form of these equations is given elsewhere (Baker and Brezonik, 1986). This approach could readily be expanded to include other ions, but current evidence (See Schindler, et al., 1986a) suggests the ions included are most important in alkalinity production and consumption.

Loading terms in these equations represent total inputs from all sources (wet + dry atmospheric inputs plus stream and groundwater inputs). Inputs must be obtained from field measurements or a separate watershed model. To predict IAG, four model coefficients (k_{SO4}, k_{NH4}, k_{NO3} and W) are needed. The remainder of this paper shows calibration and application of the sulfate submodel (Equation (2)); calibration of other coefficients will be presented later (Baker and Brezonik, 1986).

3. CALIBRATION OF SULFATE SUBMODEL.

In order to use the sulfate submodel to make predictions, one needs to obtain a value for k_{SO4}. A first-order expression was chosen because areal sulfate reduction rates appear to be directly proportional to $[SO_4^{2-}]$ in whole-lake studies (Cook et al., 1986) and in mesocosm and microcosm experiments (Perry et al., 1986). Calibration was based on ion budgets for 14 softwater lakes. At steady-state ($d[SO_4^{2-}]/dt = 0$), Equation (2) can be rearranged to calculate k_{SO4}:

$$k_{SO4} = R*z/t_w/(100 - R) \qquad (6)$$

where R = retention of SO_4^{2-}, as % input,
z = mean depth, m, and
t_w = water residence time (V/S_o), yr.

The mean value of k_{SO4} was 0.46 m yr^{-1}, with a standard deviation of 0.30 m/yr. Values ranged from 0.11 m yr^{-1} (L. Gardsjon, Sweden) to 1.28 m yr^{-1} (Magnolia L., Florida) but for 12 of the 14 k_{SO4} was between 0.3 and 0.8 m yr^{-1}). This uniformity among most lakes suggests a common physical limitation for sulfate reduction, namely, diffusion into the sediments (for most of these lakes, anoxic hypolimnia are small or absent; sulfate reduction occurs in sediments where the overlying water is oxic).

The mean k_{SO4} was used to predict sulfate retention (R_p) for lakes in the calibration data set by rearranging Equation (6) to calculate predicted retention ($R_p = k_{SO4}*100/(k_{SO4} + z/t_w)$. As seen in Figure 1, predicted sulfate retention agrees well with measured retention (R_m); a regression of R_p vs. R_m produced the equation:

$$R_p = 7.0 + 0.89*R_m. \qquad (7)$$

The slope of this equation is close to 1.0 and the r^2 value is high (0.86); thus the model is useful to predict sulfate retention.

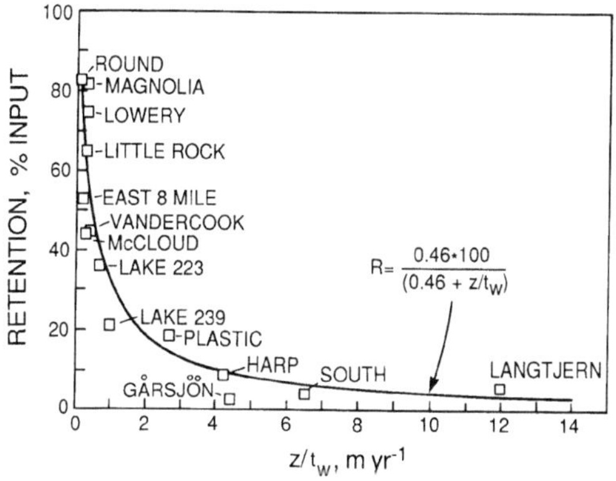

Figure 1. Sulfate retention as a function of z/t_w. Points represent measured retention. Data sources are: McCloud (Baker et al., 1986; Lowery and Magnolia (Baker et al., 1985); Lake 239 (Schindler et al., 1986b); Lake 223 (Cook et al., 1986); Vandercook (Schnoor, 1986), Round and East Eight Mile (Wentz et al., 1986); Langtjern (Wright, 1983); South (Mitchell, et al., 1985); Plastic and Harp (Dillon, 1986); Gardsjon (Hultberg, 1985). Line is calculated using k_{SO4} = 0.46 m yr^{-1}.

The model shows that hydrologic differences can account for much of the variability in observed sulfate retention. Thus, in several lakes with high values of z/t_w (Langtjern, South, and Gardsjon), predicted and observed sulfate retention is low. On the other hand, in lakes with z/t_w < 0.5 m yr^{-1} (all seepage lakes), predicted and observed sulfate retention is over 40%. These lakes are characterized by a lack of stream inflows or outflows; water inputs are derived primarily from direct precipitation onto the lake surface.

4. APPLICATION

Equations to predict steady-state $[SO_4^{2-}]$ and areal sulfate flux can be derived from Equation (2):

$$[SO_4^{2-}] = L_{SO4}/(z/t_w + k_{SO4}) \quad (8)$$

$$F_{SO4} = k_{SO4}*[SO_4^{2-}] \quad (9)$$

where L_{SO4} = areal sulfate loading, meq m^{-2} yr^{-1},
F_{SO4} = areal flux (internal consumption rate), meq m^{-2} yr^{-1}.

Predicted $[SO_4^{2-}]$ was calculated from Equation (8) for L_{SO4} = 30 meq m^{-2} yr^{-1} and z = 5 m using the calibrated k_{SO4} (0.46 m yr^{-1}) and a k_{SO4} of 0 m yr^{-1} (i.e., no sulfate retention). As seen in Fig. 2A, internal sinks are important in controlling lakewater $[SO_4^{2-}]$ in lakes with long residence times. For example, with t_w = 10 yr, predicted $[SO_4^{2-}]$ is 60 meq m^{-3} with k_{SO4} = 0.0 m yr^{-1}, but only 31 meq m^{-3} with k_{SO4} = 0.46 m yr^{-1}. For lakes with T_w < 1 yr, internal sulfate sinks have little effect on $[SO_4^{2-}]$.

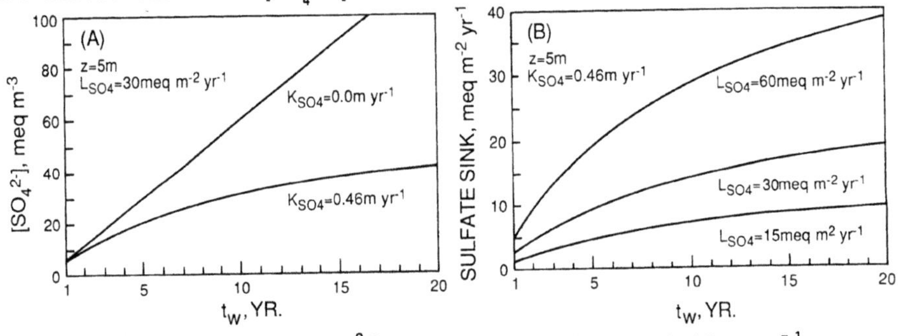

Figure 2. (A) Predicted $[SO_4^{2-}]$ vs. t_w using k_{SO4} = 0.46 m yr^{-1} (calibrated value) or 0.0 m yr^{-1} (no sulfate retention). (B) Predicted sulfate flux vs. t_w for various values of L_{SO4}.

Internal sulfate fluxes are calculated as a function of t_w and L_{SO4} with lakes of z = 5 m in Figure 2B. For lakes with t_w > 10 yr, internal sulfate consumption accounts for > 50% of input. Because F_{SO4} varies in direct proportion to $[SO_4^{2-}]$, internal consumption increases as L_{SO4} increases. Thus, for a lake with t_w = 10 yr and z = 5 m, a doubling of L_{SO4} from 30 meq m^{-2} yr^{-1} to 60 meq m^{-2} yr^{-1} increases F_{SO4} from 15 meq m^{-2} yr^{-1} to 31 meq m^{-2} yr^{-1}. Conversely, a decrease in L_{SO4} results in a decrease in F_{SO4}. Since internal sulfate sinks generate alkalinity, it follows that IAG changes direct response to changes in L_{SO4}. The model also shows that IAG supplied by sulfate consumption processes is a signficant fraction of typical areal H^+ deposition rates (10-80 meq/m^2-yr^{-1} in most of North America), so for lakes that receive most of their water and ions from direct precipitation, sulfate retention is an important H^+-neutralization mechanism.

The sulfate model has also been used to predict the rate of recovery for Little Rock Lake, a softwater lake in northern Wisconsin that is currently the subject of an EPA-funded whole-lake acidification experiment (Brezonik et al., 1986). Current plans are to reach pH 4.5 (pre-acidification pH = 6.1) by 1990. Sulfate concentrations during recovery were predicted using a non-steady state solution of Equation (2) and a k_{SO4} of 0.46 m yr^{-1} (determined before acidification). Predicted $[SO_4^{2-}]$ decreases to within 50% of and 10% of pre-experimental levels within four years and 13 years, respectively, after experimental acidification ceases (compared with t_w = 16 years) (Figure 3).

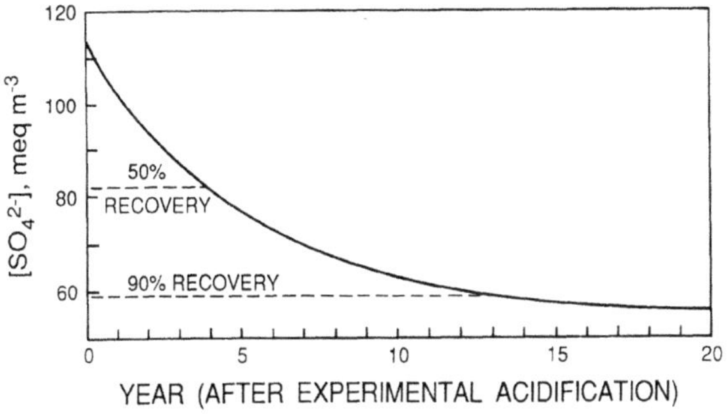

Figure 3. Predicted $[SO_4^{2-}]$ in Little Rock Lake following experimental acidification. Initial post-acidification $[SO_4^{2-}]$ was estimated from titration of lakewater to pH 4.5 from an pre-acidification alkalinity = 25 ueq L^{-1} and $[SO_4^{2-}]$ = 56 ueq L^{-1}. For acidified basin, z = 4.15 m, t_w = 16 yr, L_{SO4} = 40.4 meq m^{-2} yr^{-1} (non-experimental input), k_{SO4} = 0.46 m yr^{-1} (determined before experimental acidification).

4. CONCLUSIONS

The proposed model provides the basis for predicting IAG by linking simple models for each alkalinity-generating or consuming process to an alkalinity budget equation. Because of its simplicity and modest data requirements, the model is suitable for application on a regional basis. We have shown that k_{SO4} is reasonably uniform in a variety of lakes and that the model can be used to predict sulfate retention. Calibration of other components of the model (Equations (3)-(5)) is underway in order to complete model development (Baker and Brezonik, 1986).

ACKNOWLEDGMENTS

Concepts in this paper were presented at the Workshop on Modeling Lake Acidification, held in Dorset, Ontario, Sept. 13-14, 1985. The authors thank D. Schindler, M. Turner, J. Schnoor, P. Dillon, C. Kelly, and P. Garrison for their constructive suggestions and for sharing pre-publication data on several lake budgets. Data for several lakes became available after initial review and were added in the final draft.

REFERENCES

Baker, L. A., Pollman, C. D., and Brezonik, P. L. : 1985, 'Development of Interpretive Models for Lake Acidification in Florida', in Florida Acid Deposition Study, Phase IV, Environmental Sciences and Engineering, Inc., ESE 83-152-0602-0120.

Baker, L. A., and Brezonik, P. L.: 1986, 'Model of Internal Alkalinity Generation for Softwater Lakes', in preparation.

Baker, L. A., and Brezonik, P. L., and Edgerton, E. S.: 1986, Water Resources Research 22,175.

Brezonik, P.L., Baker, L., Eaton, J., Frost, T., Garrison, P., Kratz, T. Magnuson, J., Perry, J., Perry, T., Rose, W., Swenson, W., Watras, C., and Webster, K.: 1986, Water, Air, Soil Pollut., in review.

Cook, R. B., Kelly, C. A., Schindler, D. W., and Turner, M. A.: 1986, Limnol. Oceanogr. 31, 134.

Dillon, P. J: 1986, personal communication.

Hultberg, H.: 1985, Ecological Bulletin (Stockholm) 37, 133.

Mitchell, M.J., David M.B., and Uutala, A.J.: 1985, Hydrobiologica 121, 121.

Perry, T. E., Baker, L. A., and Brezonik, P. L.: 1986, 'Comparison of Sulfate Reduction in Microcosms, Mesocosms, and in situ at Little Rock Lake', Proc. Seventh N. Am. Lake Mgmt. Soc. Meeting, in press.

Schindler, D. W., Rudd, J. W. M, Kelly, C. A., and Turner, M. A.: 1986a, Water, Air, Soil Pollut., in review.

Schindler, D. W., Turner, M. A., Stainton, M. P., and Linsey, G. A.: 1986b, Science 232, 844.

Schnoor, J. L.: 1985, 'Acidification of Aquatic and Terrestial Systems' in Stumm, W. and Schnoor, J. L. (eds.), Chemical Processes in Lakes, John Wiley and Sons, 311.

Schnoor, J. L.: 1986, personal communication.

Vollenweider, R. A.: 1975, Scheiz. Z. Hydrol. 37, 53.

Wentz, D. A., Garrison, P. J., and Bockheim, J. G.: 1986. 'Chemical input-output budgets for Round and East Eight Mile lakes', report to Electric Power Research Institute.

Wright, R.F.: 1983, Hydrobiologica 101, 1.

EMPIRICAL MODELS FOR LAKE ACIDIFICATION IN THE UPPER GREAT LAKES REGION

J.A. Rogalla, P.L. Brezonik, Dept. of Civil & Mineral Engrg., Univ. of Minnesota, Minneapolis, MN 55455, USA, and G.E. Glass, U.S. EPA, 6201 Congdon Blvd., Duluth, MN 55108, USA

ABSTRACT. A large data base on inland lakes in the Upper Great Lakes Region (UGLR) was used to evaluate assumptions and relationships of empirical acidification models. Improved methods to calculate background alkalinity and background SO_4^{2-} are reported; SO_4^{2-} enrichment factors indicate that terrestrial SO_4^{2-} sources and watershed or lake sinks must be considered for site-specific backgound SO_4^{2-} estimates. Significant relationships were found between lake acidification estimated as change in SO_4^{2-} and precipitation acidity but not between changes in lake alkalinity and precipitation acidity in this lightly impacted region.

1. INTRODUCTION

Several empirical models have been developed to predict lake acidification based on titration and ion balance (electroneutrality) concepts. The simplest description of long-term lake acidification is the titration of alkalinity by atmospheric inputs of H_2SO_4. Thus acidification can be measured in two ways: loss of alkalinity (ΔAlk) and increase in sulfate(ΔSO_4). Quantification of these terms requires knowledge of the original lake conditions, i.e. alkalinity and sulfate concentrations prior to enhanced deposition of sulfate from anthropogenic sources.

Criticisms of existing models stem from assumptions used to estimate these parameters. Original alkalinity (Alk°) was estimated by Henriksen (1980) as a function of Ca or (Ca+Mg). This implies that weathering of other base cations contributes negligibly to Alk production or is constant among all watersheds. Original (background) sulfate ($SO_4{}^{o}{}_1$) has been ignored or assumed constant in a region (e.g. Henriksen, 1980). This assumes terrestrial S sources and site-specific SO_4^{2-} sinks (e.g. biological reduction and soil adsorption) are unimportant.

We used a large data base of chemistry, hydrologic, and land-use data for lakes across the UGLR in conjunction with a precipitation chemistry data base to evaluate these assumptions. This paper describes improved methods to calculate lake-specific estimates of Alk° and $SO_4{}^{o}{}_1$ by use of a complete electroneutrality statement. The importance of local sources and sinks for sulfate in lakes of this region is demonstrated; preliminary atmospheric loading-lake response relationships are summarized.

Water, Air, and Soil Pollution **31** (1986) 95–100.

2. THE UPPER GREAT LAKES DATA BASE

The lake chemistry data base includes 1547 lakes in Minnesota, northern Wisconsin, and Upper Michigan. Data were obtained from the U.S. EPA, Minnesota Pollution Control Association (MPCA), and U.S. Forest Service (USFS). EPA data represent averages for 837 lakes sampled across the region from 1978-1983. Parameters include hydrologic type (Eilers et al.,1983), pH, alkalinity, color, major ions and nutrients. The MPCA and USFS data are averages for 710 Minnesota lakes sampled during the same time period. Parameters include pH, alkalinity, color, and major ions. Annual average wet deposition data for 21 sites across the region were obtained from NADP, MPCA, and Ontario monitoring networks. Variables include pH, SO_4^{2-}, and Cl^-. Annual deposition and annual average concentration of precipitation components were interpolated for each lake (Pratt, 1985). Acidic components increase from west to east across the region; H^+ ranged from 6 to 49 μeq L^{-1} and SO_4^{2-} from 20 to 68 μeq L^{-1}.

Several criteria were used to select data for model evaluation and development (Table I). Data were inspected for consistency; lakes with reasonable ion balances ($0.80 < \sum+/\sum- < 1.50$) were included, but lakes with $Cl^- > 5$ mg L^{-1} (an indicator of cultural influence) were deleted. Frequency distributions by alkalinity and hydrologic class (Figure 1) illustrate regional distributions of acid-sensitive lakes and the relative importance of various water sources. According to Eilers et al. (1983), the most acid-sensitive lakes in the region are precipitation-dominated seepage lakes. Nearly 40% of the lakes in the data base are seepage lakes, and most of these are acid-sensitive (Alk < 200 μeq L^{-1}). Overall, one-third of the lakes are acid-sensitive by this criterion.

To determine whether unmeasured organic anions explain the anion deficits found in most ion balances, we computed anion contributions from organic color for 511 lakes by the method of Oliver et al. (1983). Comparison between initial and corrected ion balances showed that organic anions were insignificant at color < 50 PCU (n = 414), but significant contributions were found at color > 50 PCU (n = 97). To maintain consistency with data used for previous models and minimize uncertainty in calculations, we selected lakes with Alk < 500 μeqL^{-1} and color < 50 PCU for model evaluation and development of loading-response relationships. This included 505 lakes with complete major ion data (Table II).

3. BACKGROUND SULFATE AND ORIGINAL ALKALINITY

Background sulfate in lakewater ($SO_4{}^{\circ}{}_l$) is naturally occurring sulfate from atmospheric or terrestrial sources. This parameter is needed to compute both Alk° and ΔSO_4. Our estimates were calculated from background precipitation estimates ($SO_4{}^{\circ}{}_p$) and evaporative concentration factors. A range for $SO_4{}^{\circ}{}_p$ was estimated as follows. $SO_4{}^{\circ}{}_p$ in remote areas was estimated to be 5 μeq L^{-1} (Galloway, 1985). This probably is a global minimum value, and $SO_4{}^{\circ}{}_p$ in many terrestrial areas is likely influenced by natural S sources: volcanoes, S-bearing minerals, H_2S

Table I. Lake Selection Criteria ($N_{initial}$ = 1547)

Criterion	$\sum+/\sum-$	$Cl^- < 5.0$ mg L^{-1}	Color < 50 PCU	Alk < 500 μeq L^{-1}	Incomplete ion analysis
N lakes remaining	1509	1472	1155	705	505

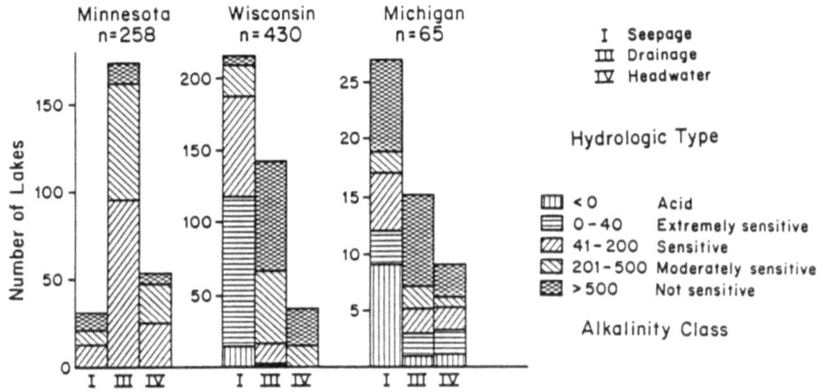

Figure 1. Alkalinity ranges for hydrologic classes. A majority of UGLR lakes with Alk < 200 μeq L^{-1} are precipitation-dominated seepage lakes.

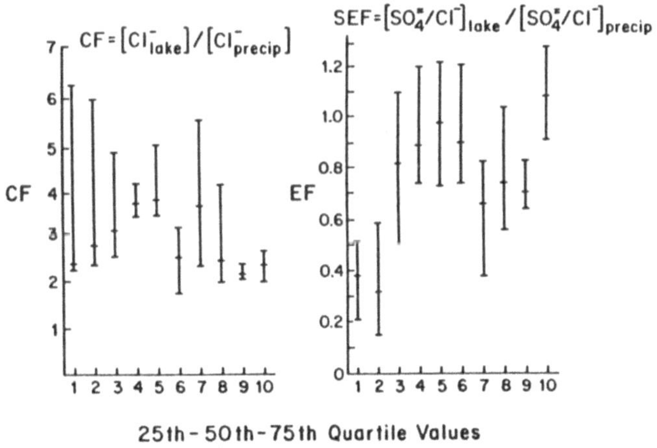

Figure 2. 25th, 50th and 75th quartile values of CF and SEF for counties across the UGLR (west to east). 1-5, Minnesota counties; 6-8, Wisconsin counties; 9-10, Michigan counties. CF range indicates that precipitation ions should be concentrated ca. 2-4X in lakes; EF range illustrates importance of watershed processes in determining lake $[SO_4^{2-}]$.

TABLE II. DESCRIPTIVE STATISTICS FOR UGLR LAKES USED IN ANALYSIS (N = 505)

Parameter		Mean	Std. Dev.	Max	Min
Alkalinity	μeq L^{-1}	163	134	500	-38
pH		6.67	0.64	9.00	4.44
Conductance	μS cm^{-1}	30	14	71	9
Color	PCU	18	13	50	0
Cl^-	μeq L^{-1}	13	14	117	1
SO_4^{2-}	"	81	30	189	1
Ca^{2+}	"	142	90	485	25
Mg^{2+}	"	74	48	297	21
Na^+	"	32	19	121	1
K^+	"	15	10	72	4
$\sum^+/\sum^-$	"	263/261	143/139	704/702	64/60

emissions from wetlands, and sulfate-rich soils. Precipitation in the UGLR is influenced by all but the first source: prairie soils to the west are high in $CaSO_4$; wetlands are abundant in the region; localized sulfide mineral deposits occur, especially in NE Minnesota. Although we cannot quantify these sources from present data, it is reasonable to assume that $SO_4{}^\circ{}_p$ in the UGLR is above the global minimum. An upper bound for $SO_4{}^\circ{}_p$ can be obtained from an empirical model (Pratt, 1985) that relates pH and ion concentrations in precipitation. Background estimates calculated for the UGLR by this model are 23 to 29 μeq L^{-1}. Because recent data were used for these estimates, they likely are higher than historical levels. A reasonable range of $SO_4{}^\circ{}_p$ for the UGLR is 10 to 20 μeq L^{-1} (between the global minimum and estimates from Pratt's model).

If we assume there are no local sources and that SO_4^{2-} behaves conservatively, we can estimate $SO_4{}^\circ{}_1$ as the product of $SO_4{}^\circ{}_p$ and evaporative concentration. Assuming Cl^- is conservative, we can estimate the latter from the ratio $CF = [Cl^-]_1/[Cl^-]_p$, which acts as a surrogate for hydrologic data. Atmospherically-derived $SO_4{}^\circ{}_1$ thus is: $SO_4{}^\circ{}_1 = SO_4{}^\circ{}_p xCF$. Q_{25}-Q_{75}, the first to third quartile range for CF (Figure 2a), is 2.3 to 4.4, and the median is 3.3; this implies that $SO_4{}^\circ{}_1$ should be ca. 2-4X higher than $SO_4{}^\circ{}_p$, or 20-80 μeq L^{-1}.

Regional variations in $SO_4{}^\circ{}_1$ result from variations in CF and from SO_4^{2-} sources or losses in watersheds, the importance of which can be evaluated from SO_4^{2-} enrichment factors (SEF): $\{[SO_4^=]/[Cl^-]\}_1/\{[SO_4^=]/[Cl^-]\}_p$. SEF > 1 indicates watershed sources; SEF = 1 indicates conservative behavior; and SEF < 1 indicates lake or watershed sinks. A bare majority of lakes in the UGLR exhibit nearly conservative behavior (0.75 < EF < 1.25) (Figure 2b). Estimates of $SO_4{}^\circ{}_1$ based on $SO_4{}^\circ{}_p$ and CF should be valid for these lakes. Some lakes, especially in NE Minnesota, have significant terrestrial sources; SO_4^{2-} reduction or other sinks are important in nearly 40% of the lakes. Watershed sources of S can be estimated from a lake's SEF and present $[SO_4^{2-}]$, but because the influence of watershed processes on Alk and base cation levels in such lakes is uncertain, we did not use them to develop acidification relationships.

Existing models estimate Alk° from Ca or (Ca + Mg). This assumes contributions from Na and K are negigible or regionally constant (Kramer and Tessier, 1982). For UGLR lakes with Alk < 200 μeq L^{-1}, the average (Na + K) contribution to total cations ($\sum BC$) is 21%, Q_{25}-Q_{75} is 16 to 23%, and the total range is 8 to 62%. Weathering of Na and K thus contributes varying but significant amounts of Alk to acid-sensitive lakes. Errors can be reduced by using all base cations to estimate Alk°, which is best defined by the electroneutrality condition:

$$Alk^{\circ} = \sum BC^{\circ} - Cl^{-} - SO_4{}^{\circ}{}_1 = \sum BC - \Delta\sum BC - Cl^{-} - SO_4{}^{\circ}{}_1 \quad (1A)$$
$$= \sum BC - F\Delta\, SO_4 - Cl^{-} - SO_4{}^{\circ}{}_1 \quad (1B)$$

Superscript ° denotes background values; Cl^{-} is the present value, (assumed constant); $\Delta\sum BC = F\Delta SO_4$ is the change in $\sum BC$ resulting from increased $SO_4{}^{2-}$ loading; F is the cation enhancement factor (e.g. Henriksen, 1980). We found no statistical evidence for a non-zero F in the UGLR. For the present, we assume F = 0. This may be reasonable for seepage lakes, which have little interaction with watersheds. Thus the equation, $Alk^{\circ} = \sum BC - Cl^{-} - SO_4{}^{\circ}{}_1$, provides lake-specific estimates of Alk°, uses all available data and maintains the ion balance.

4. LAKE ACIDIFICATION AND LOADING-RESPONSE ESTIMATES

Lake acidification (estimated as change in alkalinity or sulfate) can be calculated from measured lake chemistry and estimated $SO_4{}^{\circ}{}_1$ and Alk°: $\Delta Alk = Alk^{\circ} - Alk^{m}$, and $\Delta SO_4 = SO_4{}^{m} - SO_4{}^{\circ}{}_1$. Superscript m denotes present values. A 1:1 relationship is expected between ΔAlk and ΔSO_4. The two terms are not independent, however, because $SO_4{}^{\circ}{}_1$ is common to both. Consequently, the relationship is a restatement of the electroneutrality condition. Regression analysis of the 505 lakes yielded the expected trend, with slope near 1 and intercept near 0 for the estimated range of $SO_4{}^{\circ}{}_p$: $\Delta Alk = 11.7 + 0.92\Delta SO_4$, $p < .0001$, (for $SO_4{}^{\circ}{}_p = 10$ μeq L^{-1}); and $\Delta Alk = 8.4 + 1.01\Delta SO_4$, $p < 0.001$ (for $SO_4{}^{\circ}{}_p = 20$ μeq L^{-1}).

Negative or very high ΔAlk and ΔSO_4 were found for some lakes; as indicated by SEF values, these lakes have local sources or sinks of $SO_4{}^{2-}$ (Rogalla, 1986). In the latter case, estimates of $SO_4{}^{\circ}{}_1$ are too high, causing low or negative ΔAlk or ΔSO_4. Conversely, $SO_4{}^{\circ}{}_1$ is underestimated for lakes with terrestrial sources of S, and ΔAlk and ΔSO_4 are too high. For lakes with roughly conservative $SO_4{}^{2-}$ behavior (0.75 < EF < 1.25) and $SO_4{}^{\circ}{}_p = 10$ or 20 μeq L^{-1}, respective Q_{25}-Q_{75} ranges for ΔAlk are 46 to 83 μeq L^{-1} and 22 to 55 μeq L^{-1}; coresponding values for ΔSO_4 are 42 to 61 μeq L^{-1} and 16 to 36 μeq L^{-1}. ΔALk and ΔSO_4 estimates for these lakes should represent change by atmospheric inputs, and these were used develop loading-response relationships.

Relationships between precipitation acidity and acidification estimates were developed by regression analysis. Table III summarizes results for seepage and drainage lake subsets. The ΔSO_4-$H^{+}{}_{prec}$ relationships are statistically significant and not different from each other. Intercepts

TABLE III. LOADING-RESPONSE RELATIONSHIPS FOR LAKES WITH EF ≈ 1.

	F	p	R^2 (%)
Seepage Lakes (N=83)			
$\Delta SO_4(10)^* = 22.3 + 1.61\ H^+_p$	13.7	< 0.01	14.5
$\Delta SO_4(20) = -15.8 + 2.32\ H^+_p$	43.8	< 0.01	35.1
$\Delta Alk(10) = 56.8 + 0.10\ H^+_p$	< 1	>> 0.10	0
$\Delta Alk(20) = 18.7 + 0.80\ H^+_p$	1.4	> 0.10	1.7
Drainage Lakes (N=60)			
$\Delta SO_4(10) = 24.2 + 1.78\ H^+_p$	11.4	< 0.01	16.4
$\Delta SO_4(20) = -17.7 + 2.32\ H^+_p$	22.5	< 0.01	28.0
$\Delta Alk(10) = 34.6 + 1.16\ H^+_p$	< 1	>> 0.10	0.6
$\Delta Alk(20) = 76.5 + 0.62\ H^+_p$	1.2	>> 0.10	2.0

* Values in parentheses represent assumed $SO4^{o}_{p}$ in μeq L^{-1}.

for $SO_4{}^{o}{}_p$ = 10 μeq L^{-1} are positive, while those for 20 μeq L^{-1} are negative, which suggests that the true value of $SO_4{}^{o}{}_p$ is in this range. (An intercept of 0 would be expected for lakes at steady state with respect to atmospheric inputs.) The ΔAlk relationships are not statistically significant, probably because local (lake and watershed) processes that produce or consume Alk are sufficiently variable to mask regional relationships in this area of relatively low loading of atmospheric acidity.

ACKNOWLEDGMENT. This work was supported by the Minnesota Pollution Control Agency, University of Minnesota Water Resources Research Center, and the U.S. EPA. This paper was not reviewed by the latter agency and may not reflect its views. No official endorsement should be inferred.

5. REFERENCES

Eilers, J.M., Glass, G.E., Webster, K.E., and Rogalla, J.A.: 1983, Can. J. Fish. Aquatic Sci., 40, 1896.

Galloway, J.N.: 1985, 'Precipitation Composition in Remote Areas', Presented Int. Conf. on Acidic Precipitaion, Muskoka, ON, Sept., 1985.

Henriksen, A.: 1980, 'Acidification of Freshwaters - A Large-scale Titraation', In Proc. Int. Conf. Ecol. Impact Acid Precip., SNSF Proj., Oslo.

Kramer, J. and Tessier, A.: 1982, Environ. Sci. Technol., **16**, 606a.

Oliver, B.G., Thurman, E.M., and Malcolm, R.L.: 1983, Geochim. Cosmochim. Acta., **47**, 2031.

Pratt, G.C.: 1985, unpub. ms., Minn. Poll. Contr. Ag., Roseville, MN

Rogalla, J.A.: 1986, M.S. Thesis, Univ. of Minnesota, Minneapolis.

RETENTION AND RELEASE OF S FROM A FRESHWATER WETLAND

S. E. Bayley, R. S. Behr and C.A. Kelly
Department of Botany
University of Manitoba
Winnipeg, MB R3T 2N2
Canada

Department of Microbiology
University of Manitoba
Winnipeg, MB R3T 2N2
Canada

ABSTRACT. A freshwater wetland at the Experimental Lakes Area in northwestern Ontario stored most of the SO_4^{2-} received annually from precipitation, runoff and experimental additions. The S budget was determined for a small fen spray irrigated with H_2SO_4 and HNO_3. Annual S retention was greatest during the first year of experimental addition of H_2SO_4 (73% of input in 1983). Retention was lowest (22%) in 1984, a year of lower than average precipitation, with a long hot summer. During years with hot, dry summers, SO_4^{2-} was produced from the reoxidation of reduced S compounds in the peat and released to surface waters. The autumn SO_4^{2-} pulse was accompanied by the release of Ca and Mg but was not accompanied by a H^+ release as has been detected in eastern Ontario and southern Norway, areas which receive more acidic precipitation.

1. INTRODUCTION

Many freshwater wetlands appear to retain S on an annual basis. Braekke (1981) found a 35% retention of SO_4^{2-} in the peatland-dominated Storgama watershed in S. Norway during dry years and only a 4% retention during wet years. Urban et al. (1986) found retention of SO_4^{2-} averaged 56% in Marcell Bog in northern Minnesota, an area relatively unaffected by anthropogénic atmospheric deposition. Calles (1983) found a 25 to 64% retention of SO_4^{2-} in Swedish watersheds which contained from 8 to 33% bogs. Hemond (1980) calculated that approximately 77% of the input SO_4^{2-} was retained in Thoreau's Bog in the northeastern US, in spite of high atmospheric S deposition.

Other studies have documented the decrease in SO_4^{2-} concentration as water passes through a wetland (Vitt and Bayley, 1984; Kerekes et al., 1985; Wieder and Lang, 1984). Wieder and Lang observed that acidic, sulfate-rich mine drainage was neutralized as it passed through a *Sphagnum*-dominated bog in West Virginia.

Retention of SO_4^{2} in wetlands would be expected to be lower in areas with a large snowpack and subsequent rapid springmelt or in

Water, Air, and Soil Pollution **31** (1986) 101–114.

wetlands with high rates of water flow, because there would be little time for the water and SO_4^{2-} to remain in contact with the reduced substrate. Lazerte and Dillon (1984) found that a fen in Dorset, Ontario released high concentrations of SO_4^{2-} during the springmelt but retained SO_4^{2-} during the summer. It appears that SO_4^{2-} is retained more effectively when the hydrologic conditions maximize the contact time between the water, vegetation and anaerobic peat.

In contrast to most wetlands studied, terrestrial watersheds in acid sensitive areas release most of the SO_4^{2-} received in precipitation (Braekke, 1981; Likens et al., 1977; Bayley and Schindler, 1986; Schindler et al., 1976; Wright, 1983). Wetlands located at the outflow of terrestrial watersheds can modify the acid pulse and total SO_4^{2-} received by streams and lakes even in acid impacted areas (Braekke, 1981). However, during hot dry seasons, the water level in many wetlands drops, allowing O_2 to penetrate further into the peat and reduced sulfides are oxidized to sulfate. With subsequent rainstorms, SO_4^{2-} is flushed from the wetland to downstream waters. This autumn SO_4^{2-} pulse has been observed at ELA (this study), Norway (Braekke, 1981), Dorset, Ontario (Lazerte and Dillon, 1985), England (Rippon et al., 1980; Brown, 1980), and West Virginia (Wieder, 1985).

This study presents SO_4^{2-} budgets under natural conditions as well as under a regime of increasing SO_4^{2-} additions. We present evidence that a poor fen in northwestern Ontario retains SO_4^{2-} on an annual basis, even with a 50% increase in SO_4^{2-} loading from experimental acidification. Secondly, we present evidence that wetlands can lose a large portion of the stored SO_4^{2-} if the hydrologic conditions favor oxidation of reduced S compounds. The data show that SO_4^{2-} release is accompanied by release of Ca and Mg while the H^+ concentration does not increase. We hypothesize that the autumn SO_4^{2-} pulses in wetlands are not accompanied by a decrease in pH in the early stages of acid deposition (prior to Ca and Mg exhaustion), or in wetlands with high cation loading.

While the previously mentioned SO_4^{2-} mass balance studies of wetlands have documented SO_4^{2-} retention, some of these studies have methodological problems which this study does not have. This study area has very low dry deposition of SO_4^{2-} (Barrie, 1982) and good control of the hydrology. The basin is bound by granitic bedrock and there is little ground water seepage in or out of the basin. The low SO_4^{2-} deposition coupled with known experimental addition of H_2SO_4 permitted greater accuracy in measurement of SO_4^{2-} retention than is possible under circumstances where the major input is atmospheric only.

2. MATERIALS AND SITE DESCRIPTION

The experimental wetland (3.67 ha) (Figure 1) is part of the Northeast watershed (10.58 ha) of L.239, located at the Experimental Lakes Area (ELA) in northwestern Ontario, Canada (49°40'N and

93°44'W). The upland portion of the watershed (6.91 ha) is covered with a thin layer of mineral soil derived from the underlying granite bedrock. The terrestrial portion of the watershed burned in 1980 and has been quickly re-vegetated with Pinus banksiana, Populus tremuloides and other boreal species. The peatland is dominated by black spruce (Picea mariana) and three Sphagnum species, S. magellanicum, S. angustifolium and S. fuscum. See Vitt and Bayley (1984) for a detailed vegetation description. The peatland is composed of an oligotrophic center area and a more minerotrophic edge area (Figure 1). Runoff from the watershed flows directly into the peatland. The single outflow stream from the peatland is gauged with a weir.

Water samples were collected weekly during the ice free season at various stations within the fen during 1984. Outflow samples were collected weekly since 1971 but the outflow data used in the mass balance calculations is from 1981 through 1984. Precipitation was collected after each event. SO_4^{2-} concentration was measured with an ion chromatograph (Dionex) using filtered samples. Ca and Mg concentration were measured by atomic absorption (Stainton et al., 1977).

Sulfate budgets for the fen were calculated using the approach outlined in Schindler et al. (1976). Annual runoff volume into the fen was calculated from the water balance of the fen (i.e Runoff = Outflow + Evapotranspiration - Precipitation on fen ± Δ Storage). Evaporation was derived from the traditional mass transfer relationship as described by Newbury and Beaty (1980) for small ELA lakes. Storage in the fen was assumed to be zero, based on water level measurements. Concentrations of SO_4^{2-} from a pool in the upland watershed were multiplied by the runoff volume to calculate SO_4^{2-} loading to the fen due to runoff.

Total S and total inorganic S in peat cores were measured as described in Behr (1985). Total S was measured on a LECO sulfur analyser, model SC-132. Total inorganic S was measured as the sum of acid volatile sulfides (AVS), elemental S (S°) and pyrite (FeS_2). AVS released from sediment by HCL was trapped with a zinc acetate trap and the ZnS precipitate measured by iodine titration. S° and FeS_2 were measured by the chromium (II) reduction technique (Howarth and Merkel, 1984).

In 1983, experimental acidification of the fen began. The purpose of the experiment was to simulate lowering the average pH of precipitation falling on the fen from pH 5 to pH 4. To achieve this, H_2SO_4 and HNO_3 (on a equinormal basis) were added to lake water (pH 6.56; Ca 127 μEq L^{-1}) and sprayed once each month during the ice-free season. An experimental part of the fen received the acid plus lake water (2.66 ha) while an upslope part of the fen (0.85 ha) was used as a control and received only lake water (Figure 1).

The fen can be divided into an edge or lagg zone, which is a minerotrophic area receiving runoff from the surrounding uplands and an oligotrophic center area (Figure 1) which does not receive upland runoff. In this paper data from the minerotrophic pools and outflow

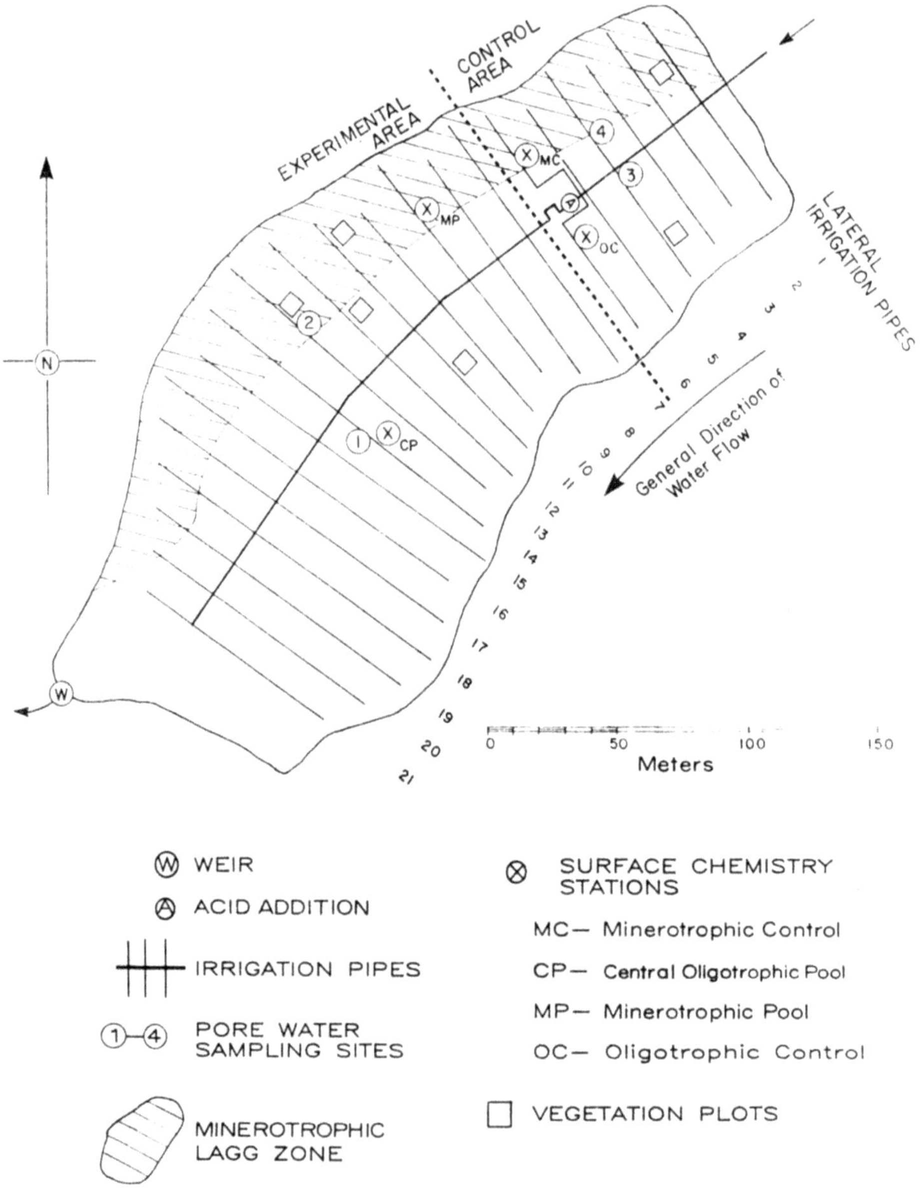

Figure 1. Map of NE Fen showing acidified area and control area. Hatched area is the minerotrophic edge.

as well as the whole fen mass balance are presented. The minerotrophic experimental site received acid and lake water applications and runoff from the surrounding watershed (which was not irrigated). The minerotrophic control pool received lake water applications plus runoff from the surrounding watersheds. For further details of the acidification procedure see Bayley et al. (1986).

3. RESULTS AND DISCUSSION

3.1. Wetlands as a Sink for S

On an annual basis, the fen at ELA was a sink for S. From 1981-1984, SO_4^{2-} retention ranged from 22 to 73% of inputs (Table I). The annual storage ranged from 21 to 51 meq SO_4^{2-} m^{-2} yr^{-1} (0.33 to 0.82 g S m^{-2} yr^{-1}). Higher absolute and percent S storage was detected after experimental acidification in 1983 but not in 1984. Both were years of lower than average precipitation. Pakarinen (1981) and Pakarinen and Tolonen (1980) measured approximately 0.87 to 1.08 g S m^{-2} yr^{-1} retained in South Finnish bogs. Brown (1985) calculated a rate of S accumulation in peat of 4.76 to 6.06 g m^{-2} yr^{-1} in an area with high S deposition. Gaseous losses of S from the fen were not measured and would reduce calculated retention if these processes were important. However, Brown and Macqueen (1985) concluded that volatilization of H_2S from peat cores was negligible.

TABLE I

Annual SO_4^{2-} budgets in the experimental wetland (1981-1984) in meq SO_4 m^{-2} yr^{-1}.

	1981	1982	1983	1984
Precipitation	23	44	25	25
Direct runoff into wetland	40	53	25	31
Experimental input	0	0	20	37
Total inputs	63	97	70	93
Outflow	23	53	19	72
Retention	40	44	51	21
% retention	64%	46%	73%	22%
Annual precipitation (mm)	643	708	565	613

3.2. Storage of S in Peat

Atmospheric bulk deposition of SO_4^{2-} at ELA has averaged 22.3 mEq m^{-2} SO_4^{2-} over the last 11 years (Linsey et al., 1986). While the deposition at ELA can be considered low, there has been sufficient input from the atmosphere and runoff for the peat to store up to 155 μmol S g^{-1} dry matter (1.4%) in the 10 to 15 cm depths (Figure 2), and over 90% of the S was organic S (Behr, 1985). Pakarinen and Tolonen (1980) found peak concentrations of 109 μmol S g^{-1} dry matter in a South Finnish bog.

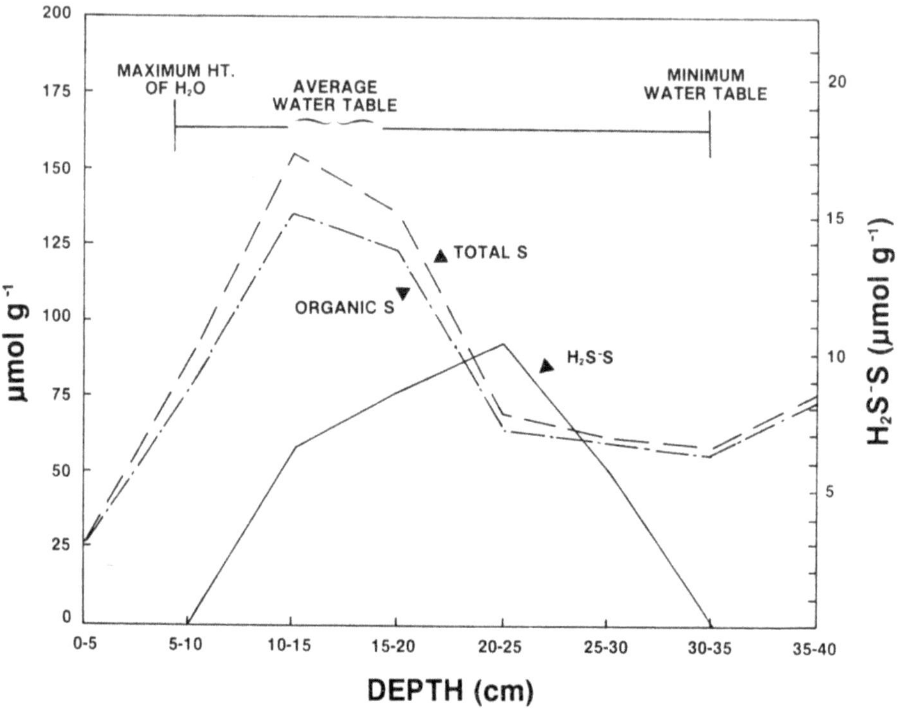

Figure 2. Concentration of total and organic S in a peat core. Concentration of H_2S-S in porewater on 29 June 1984. Average water table fluctuation is from 10-20 cm below moss surface. The minimum water level was at 32 cm during August 1985.

Because the fen retained a large percentage of SO_4^{2-} inputs, and because SO_4^{2-} reduction has been shown to occur there. (Behr, 1985), it seems reasonable to believe that the peak in S concentration at 10 to 15 cm was due to downward diffusion of SO_4^{2-} and its reduction there. Also, SO_4^{2-} reduction can lead to formation of organic S (Behr, 1985; Rudd et al., 1986), probably as a result of H_2S reactions with organic molecules (Casagrande et al., 1979; Brown, 1986). The organic S formed in this way in lake sediments has been shown to be partly lost during winter when O_2 penetrates further into the sediments (Rudd et al., 1986). Loss of metal sulfides due to oxidation is well known. Thus, any change in the O_2 penetration depth in the peat (controlled in large part by changes in water level) could result in remobilization of reduced S from both inorganic and organic forms.

The water level in the fen generally fluctuated between 10 and 20 cm below the surface of the moss which is coincident with the zone containing the highest S concentration (Figure 2). Hydrogen sulfide concentration was also highest in that zone (Figure 2) indicating that O_2 was depleted to within a few cm of the H_2O surface, resulting in conditions suitable for SO_4^{2-} reduction. During the 1984 summer drought the water level dropped 30 cm below the moss surface and O_2 probably penetrated to that depth, oxidizing the reduced S. Then, when autumn rain came, this oxidized S was flushed out as described below.

3.3. Autumn SO_4^{2-} pulse

With the first storms after the summer drought, there was an increase in the concentration of SO_4^{2-} in the minerotrophic control pool starting 12 September 1984 (Figure 3a). The concentration increased from a summer low of 17 to 731 μEq L^{-1} in mid September. The abrupt increase was detected after a 13 mm rainfall, followed by a 15 mm application of lake water. The concentration remained higher than spring and summer values for the remainder of the ice-free season (Figure 3). The SO_4^{2-} concentration at the control station (Figure 3a) was higher than the SO_4^{2-} concentration in the minerotrophic experimental station (Figure 3b) which had received 758 μEq L^{-1} of SO_4^{2-} as H_2SO_4 during the acidification experiment. This may be because the control receives more runoff than does the experimental station. The concentration of SO_4^{2-} in the outflow of the fen also showed an increase after the summer drought (data not shown).

3.4. Oxidation of reduced S

We believe the sharp increase in SO_4^{2-} concentration in September in the minerotrophic control pool (Figure 3a) was caused primarily by the oxidation of reduced S during the preceding dry period, not by runoff of SO_4^{2-} from the upland watershed. This is suggested because the SO_4^{2-} increase also occurred in August in the same pool with just the addition of lake water (Figure 3a). August 1984 was the driest August on record (Beaty, 1986) and there was no upland runoff.

Another line of evidence suggesting that oxidation of reduced S compounds caused the increase in surface water concentration was the decrease in total inorganic sulfides in peat cores from spring 1984 to autumn 1984 (Figure 4). There was an average 36% reduction in

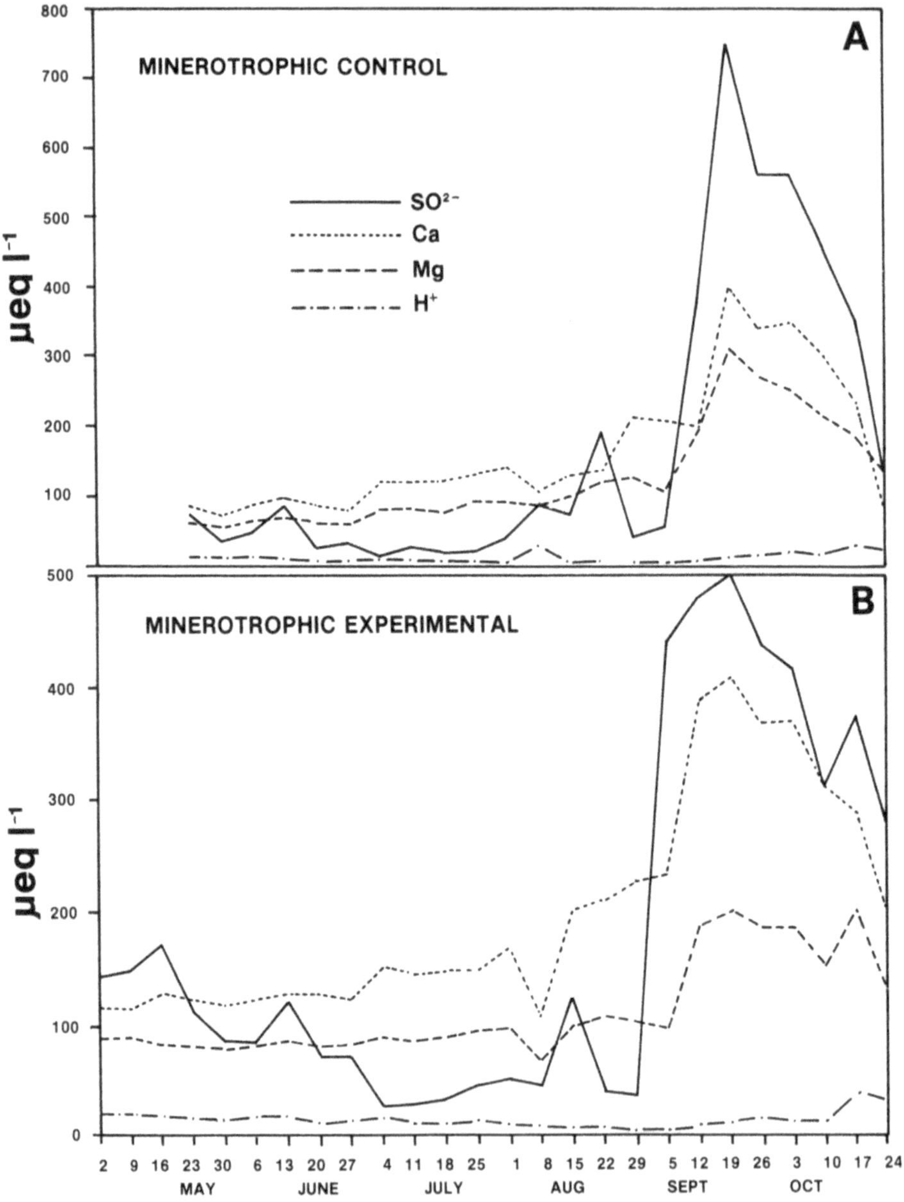

Figure 3. a. Weekly concentrations of SO_4^{2-}, Ca, Mg and H^+ during the 1984 ice-free season in the minerotrophic control site. b. Weekly concentrations of ions in the minerotrophic experimental site.

concentration per gram of peat over the dry summer with most of that loss in the 10 to 20 cm depths (Figure 4).

Another contributor to the observed autumn SO_4^{2-} pulse might be enhanced mineralization of S from organic peat by heterotrophic decomposer activity in the dry aerobic peat.

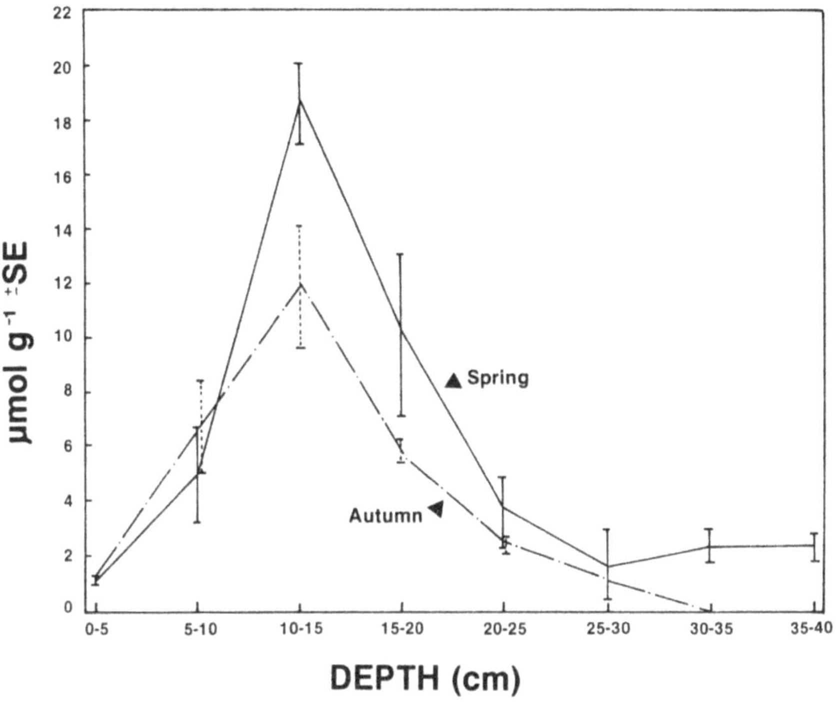

Figure 4. Inorganic S concentration in peat cores collected 31 May 84, 2 June 84 (spring), 1 Oct 84 and 2 Oct 84. Data in µEq g^{-1} dry wt ± SE, n = 4.

3.5. Release of Ca and Mg

The release of SO_4^{2-} after summer drought was accompanied by the release of Ca and Mg in both control and experimental stations (Figures 3a and b). Ca and Mg (data not shown) also increased in the outflow of the fen. Sphagnum moss releases cations with the addition of anions (Bayley, 1986).

The higher autumn concentration of Ca, Mg and SO_4^{2-} could be due to evaporation, drying of the moss and concentrating of the water during the dry summer. With autumn rains, the Ca and Mg salts of sulfate would be dissolved and released in the water. Assuming that sodium sulfate would dry and redissolve in a similar fashion as $CaSO_4$ and $MgSO_4$, then Na can be used as a predictor to account for the evaporative drying of Ca, Mg and SO_4^{2-}. The ratio of spring Na concentration to autumn Na concentration is assumed to represent evaporative concentration of all the cation salts of sulfate. Based on this ratio, the autumn concentrations of Ca, Mg and SO_4^{2-} were much higher than that predicted based on evaporative drying alone. In fact, Ca and Mg concentrations were 2 times and SO_4^{2-} 5 times higher than what would be expected from evaporative drying.

It is unlikely that the Ca and Mg release could be from the lake water or acid application since lake water concentrations of Ca and Mg averaged 127 μEq L^{-1} and 58 μEq L^{-1}, respectively. In September after the first rains and fresh water application, the Ca concentration reached 476 μEq L^{-1} and Mg 375 μEq L^{-1} in the minerotrophic control pool. While storm runoff into the fen could increase the Ca and Mg concentrations to some extent, it is unlikely that runoff was the immediate source of the increased Ca and Mg. Comparison of minerotrophic experimental surface pool Ca and Mg concentrations before acidification in August 1984 with concentrations after acidification show a similar release of Ca and Mg as the SO_4^{2-} concentration increased (Figure 5a), and since there was virtually no rainfall in August, the Ca and Mg must have come from the Sphagnum mosses and peat.

In September 1984 natural reoxidation of S in the control site produced higher concentrations of S than did acidification of the experimental site (Figure 5b and c). In these minerotrophic sites, increases in SO_4^{2-} were not accompanied by increases in H^+ concentration (Figure 3a and b, 5a, b and c).

In a fen in Dorset, Ontario, during a dry autumn, the SO_4^{2-} rose to 1000 μEq L^{-1}, Ca increased to 675 μEq L^{-1}, while the H^+ rose to 140 μeq L^{-1} (Lazerte and Dillon, 1985). Similarly, in the Storgama peatlands of Norway, Braekke (1981) detected a September increase in SO_4^{2-}, Ca and H^+ during dry years.

The most apparent difference in the ELA site and the two sites mentioned above is that the fen at ELA has received acid for only 2 yrs while the Dorset, Ontario and Norwegian sites have had high atmospheric loadings for decades. Another explanation for the lower H^+ concentration at ELA could be that the cation (primarily Ca and Mg) supply rates to the fen are higher than the cation supply at the Dorset, Ontario or Storgama Norwegian sites. However, Table II indicates that the ELA fen had lower cation supply in runoff than the Norwegian site (Braekke, 1981) and similar cation supply to the Dorset, Ontario site (NRC, 1981). Therefore, differences in cation supply are not likely to be causes of the lower H^+ concentrations at the ELA site. The release of H^+ at the Dorset, Ontario and Norwegian sites during the autumn SO_4^{2-} pulse could be due to reduction of Ca and Mg neutralization capacity under the longer and higher SO_4^{2-} deposition conditions, however, both of those sites still release Ca and Mg during the SO_4^{2-} pulse.

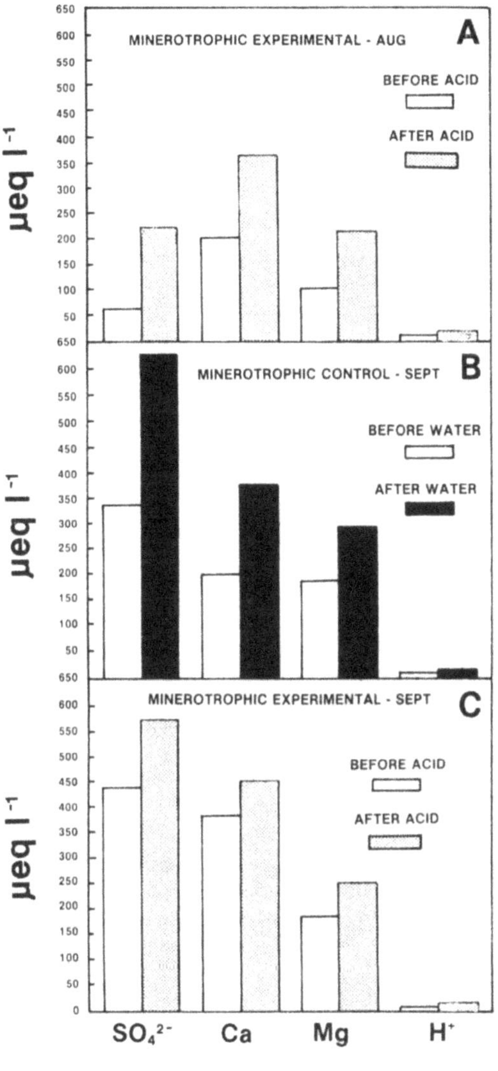

Figure 5. Effect of autumn rains on ion concentrations. a. Concentrations of ions in the experimental site before and after acidification in August 1984 during the summer drought.
b. Concentrations of ions in the control site before and after addition of lakewater in September 1984, after the summer drought.
c. Concentrations of ions in the experimental site before and after acidification in September 1984, after the summer drought.

TABLE II

Runoff loadings into wetlands. Data in meq m^{-2} yr^{-1}.

		NW Ontario	Dorset, Ontario				Storgama
		ELA NE fen	Harp Lake watersheds				Norway
			3	3A	4	5	STE 01
SO_4	Total	108* (51)	99	113	98	134	312
Ca	Total	90 (20)	84	97	74	87	160
Mg	Total	42 (9)	41	44	36	45	59

*Total input at ELA is the sum of natural runoff plus the experimental input. Experimental input alone is in brackets. At Dorset and Storgama, the total input is runoff from the watersheds.

4. CONCLUSION

Annual SO_4^{2-} retention by the fen is high (mean 51%) in the 4 yr of study, including 2 yr of elevated SO_4^{2-} inputs from experimental acid additions. Release of SO_4 was greatest during the first complete year of experimental acidification, however, the release of SO_4 also occurred in the control areas after the long dry summer, indicating that the acidification was not the dominant factor in the release of SO_4.

When the water table declined during dry summers, reduced S compounds were oxidized to SO_4^{2-} and flushed from the fen causing an autumn SO_4^{2-} pulse. This pulse was locally important to downstream waters even though the fen still retained some of its SO_4^{2-} inputs. The hydrology of the wetland (depth and frequency of flooding) and the total amount of precipitation may be the most important factors determining annual S retention.

5. ACKNOWLEDGMENTS

This work was funded by the Canadian Department of Fisheries and Oceans and NSERC grant A2671. We thank M. Stainton, G. Linsey, M. Turner, K. Beaty, and R. Newbury for chemical and hydrologic data. We appreciate the critical reviews of R. Hesslein, D.W. Schindler, G. Brunskill and P. Campbell.

6. REFERENCES

Bayley, S.E.: 1986, unpublished data.

Bayley, S.E., and Schindler, D.W.: 1986, 'Sources of Alkalinity in Precambrian Shield Watersheds under Natural Conditions and after Fire or Acidification' in Hutchinson T. and Meema K. (eds.), Effects of Acid Deposition on Forests, Wetlands and Agricultural Ecosystems, Proceedings of the Workshop. Springer-Verlag, New York, in press.

Bayley, S.E., Vitt, D.H., Newbury, R.W., Beaty, K.G., Behr, R., and Miller, C.: 1986, Can. J. Fish. Aquat. Sci., in press.

Beaty, K.G.: 1986, unpublished data.

Behr, R.S.: 1985, 'Sulfur Dynamics in an Experimentally Acidified Mire in Northwestern Ontario', M.Sc. Thesis, University of Manitoba, MB, various pagination.

Braekke, F.H.: 1981, 'Hydrochemistry of High Altitude Catchments in South Norway, 1. Effects of Summer Droughts and Soil-Vegetation Characteristics', Rep. Norw. For. Res. Inst., **36.8**, 26 pp.

Brown, K.A.: 1980, 'The Distribution of Sulfur Compounds in a Peat Bog in Relation to Stream Water Chemistry'. British Report of Central Electricity Research Laboratories, RD/L/N 150/80, Central Electricity Generating Board. Surrey, England, 18 pp.

Brown, K.A.: 1985, Soil Biol. and Biochem., **17**, 39.

Brown, K.A.: 1986, Soil Biol. and Biochem., **18**, 131.

Brown, K.A., and MacQueen, J.F.: 1985, Soil Biol. and Biochem., **17**, 411.

Calles, U.M.: 1983, Hydrobiologia, **101**, 13.

Casagrande, D.J., Idowu, G., Friedman, A., Rickert, P., Siefert, K., and Schlenz, D.: 1979, Nature, **282**, 599.

Hemond, H.F.: 1980, Ecol. Monogr., **50**, 507.

Howarth, R.W., and Merkel, S: 1984, Limnol. Oceanogr., **29**, 598.

Kerekes, J., Beauchamp, S., Jordon, W., and Vaughan, H.: 1985, 'Sources of Sulfate Ions and Acidity in Wetlands and Lakes in Nova Scotia'. Abstract in International Symposium on Acidic Precipitation, Muskoka, ON, p. 237.

Lazerte, B.D., and Dillon, P.J.: 1984, Can. J. Fish. Aquat. Sci., **41**, 1664.

Lazerte, B.D., and Dillon, P.J.: 1985, 'Chemical Transformations. V. Chemical Transformations of Stream and Soilwater Flowing Through Sphagnum Conifer Swamps of Central Ontario', Abstract in International Symposium on Acidic Precipitation, Muskoka, ON, p. 207.

Likens, G.E., Bormann, F.H., Pierce, R.S., Eaton, J.S., and Johnson, N.M.: 1977, 'Biogeochemistry of a Forested Ecosystem', Springer-Verlag, New York, 146 pp.

Linsey, G.A., Schindler, D.W., and Stainton, M.P.: 1986, Can. J. Fish. Aquat. Sci., in press.

National Research Council Canada.: 1981, 'Acidification in the Canadian Aquatic Environment', NRCC No. **18475**, 369 pp.

Newbury, R.W., and Beaty, K.G.: 1980, Can. J. Fish, Aquat. Sci., **37**, 335.

Pakarinen, P.: 1981, Ann. Bot. Fenn., **18**, 275.
Pakarinen, P., and Tolonen, K.: 1980, Suo, **31**, 1.
Rippon, J.E., Skeffington, R.A., Wood, M.J., Brown, K.A., and Brown, D.J.A.: 1980, 'Hydrogen, Sulphur and Nitrogen Budgets in Soils and Catchments', p. 276-277, in Drablos, D., and Tollan, A. (eds.) Ecological Impact of Acid Precipitation, SNSF Project, Oslo, Norway.
Rudd, J.W.M., Kelly, C.A., Furutani, A.: 1986, Limnol. Oceanogr., in press.
Schindler, D.W., Newbury, R.W., Beaty, K.G., and P. Campbell.: 1976, J. Fish. Res. Board Can., **33**, 2526.
Stainton, M.P., Capel, M.J., and Armstrong, F.A.J.: 1977, 'The Chemical Analysis of Fresh Water', 2nd edition, Can. Fish. Mar. Serv. Misc. Spec. Publ., **25**, 166 pp.
Urban, N.R., Eisenreich, S.J., and Gorham, E.: 1986, 'Proton Cycling in Bogs: Geographic Variations in Northeastern North America, in Hutchinson, T., and Meema, K. (eds.), Effects of Acid Deposition on Forests, Wetlands and Agricultural Ecosystems, Proceedings of the Workshop, Springer-Verlag, New York.
Vitt, D.H., and Bayley, S.E.: 1984, Can. J. Bot., **62**, 1485.
Wieder, R.K.: 1985, Biogeochemistry, **1**, 277.
Wieder, R.K., and Lang, G.E.: 1984, Water Air Soil Pollut., **23**, 381.
Wright, R.F.: 1983, Hydrobiologia, **101**, 1.

EXPERIMENTAL ACIDIFICATION OF LITTLE ROCK LAKE, WISCONSIN

P.L. Brezonik[1], L.A. Baker[1], J.R. Eaton[2], T.M. Frost[3], P. Garrison[4], T.K. Kratz[3], J.J. Magnuson[3], W.J. Rose[6], B.K. Shephard[7], W.A. Swenson[8], C.J. Watras[3,4], and K.E. Webster[4]

[1]Dept. Civil & Mineral Engrg., Univ. of Minnesota, Minneapolis, MN 55455, U.S.A.; [2]U.S. EPA, Duluth, MN; [3]Univ. of Wisconsin, Madison; [4]Wisconsin Dept. of Nat. Resour.; [5]Univ. of Minnesota, St. Paul; [6]U.S. Geol. Surv., Madison, WI; [7]Iowa State Univ., Ames; Univ. of Wisconsin, Superior

ABSTRACT. The controlled acidification of a two-basin lake is described. The lake was divided by a vinyl curtain in 1984; acidification of one basin began in 1985. Target pH values of 5.5, 5.0 and 4.5 are planned for 2-yr increments. Biotic and chemical responses and internal alkalinity generation are being studied. Baseline studies, initial results at pH 5.5, and predictions of lake responses to acidification are described.

1. INTRODUCTION

The effects of acid deposition on lake ecosystems are still poorly understood despite more than a decade of research and monitoring studies. This situation is mainly a consequence of the methods commonly used for such studies: statistical analysis of lake survey data, comparison of historical and recent lake data, and laboratory experiments. Surveys are useful to generate hypotheses, but their deficiencies in proving cause-effect relationships are well-known. Historical data rarely are adequate to quantify pH and alkalinity changes, let alone biotic changes in lakes. Laboratory experiments are a valuable source of information on the direct sensitivity of specific organisms and processes to pH-related parameters, but indirect effects resulting from changes in community linkages and trophic level interactions cannot be established in the laboratory.

Few acidification experiments have been performed at the whole ecosystem level. To our knowledge, such experiments on whole lakes have been conducted in North America only at the Experimental Lakes Area of northwestern Ontario (Schindler et al., 1985). Studies on Lake 223, a cold water system acidified with H_2SO_4, documented dramatic effects on lake trout and their food sources. Major biogeochemical processes were relatively insensitive to pH over the range 6.5 to 5.0, and primary production did not decrease, as hypothesized by Grahn et al., 1974). Although much information was gained from this study, many questions remain unanswered.

Water, Air, and Soil Pollution **31** (1986) 115–121.

Similar studies should be conducted on acid-sensitive lakes with cool and warm water fish communities in other geologic and climatic regions. As part of the U.S. National Acid Precipitation Assessment Program (NAPAP), we have begun an acidification experiment on a lake in northern Wisconsin.

2. SITE DESCRIPTION AND EXPERIMENTAL DESIGN

Planning of the study began in fall 1982. Site selection was based on six criteria: (i) undisturbed watershed with limited access; (ii) suitable water chemistry (low alkalinity and color); (iii) representative biota for acid-sensitive lakes in the region; (iv) multiple basin morphometry; (v) proximity to a field laboratory; and (vi) location in a state where permission for the manipulation could be obtained. Little Rock Lake (Vilas County, Wis.) meets all these criteria. Located on state-owned land, its uninhabited watershed is wholly forested. Lake access (by unpaved road) is restricted to project staff. The Wisconsin Legislature passed a law in May 1984 that permits acidification of the lake.

Little Rock Lake has two main basins connected by a narrow constriction. The south basin (8.3 ha, z_{max} = 6.5 m) does not stratify thermally; the north basin (9.9 ha, z_{max} = 10.5 m) does, and its small hypolimnion (~8% of the basin volume) becomes anoxic. Like many acid-sensitive lakes in this region, the lake has no surface inlet or outlet and receives nearly all its water as precipitation. As a result, it has very low ionic content, organic color, and nutrient levels (Table I). Specific conductance is about the same as that of local precipitation. With an alkalinity of only 25 ueq L^{-1}, the lake is highly sensitive to acidification. Chlorophyll _a_ typically is 1 to 3 ug L^{-1}, and inorganic P is undetectable (< 1 ug L^{-1}) in the epilimnion during summer. Nonetheless, the lake has an abundant fish assemblage dominated by yellow perch; small populations of largemouth bass, rock bass and black crappie also are present.

Sampling to characterize the water, sediments, and biota of both basins began in August, 1983. After one year of monitoring, a curtain of dacronfiber-reinforced polyvinyl sheeting was installed across the 70 m constriction connecting the two basins. Acidification of the north basin with electroytic grade H_2SO_4 began in May 1985, shortly after ice-out. Present plans are to acidify the north basin over a 6 yr period beginning in spring 1985 (2 yr each at pH 5.5, 5.0, and 4.5). According to this plan, the south basin will remain as a control. This schedule was selected to allow time for response at each pH within the constraints of acheiving results at pH < 5.0 in time for the NAPAP 1989 assessment. The schedule and pH increments will be determined, however, by the biotic responses observed during the study. Except for fish and macrophytes, all species have longevities of less than two years.

An alternative experimental design still being considered would acidify both basins on a staggered schedule. Acidification of the south basin would start 2 yr after that of the north basin; the basins thus could always be compared at 0.5 pH differences. This design has the advantage

TABLE I. CHEMICAL CHARACTERISTICS OF LITTLE ROCK LAKE*

Variable	North Basin X	North Basin Range	South Basin X	South Basin Range
Sp. cond. uS/cm	11.3	10.6-17.4	12.6	10.3-16.8
Color Pt-Co units	9	2-20	12	3-27
pH	6.04	5.76-6.33	6.04	5.93-6.27
Alkalinity ueq/L	25.2	13.9-31.1	25.8	21.3-28.7
Ca^{2+} mg L^{-1}	0.89	0.76-1.04	0.86	0.71-1.02
Mg^{2+} "	0.30	0.28-0.41	0.29	0.27-0.41
K^{+} "	0.57	0.49-0.65	0.55	0.48-0.64
Al ug L^{-1}	5.2	2.6-7.2	7.1	3.4-9.5
Cl^{-} mg L^{-1}	0.27	0.20-0.30	0.28	0.20-0.33
F^{-} "	0.02	0.01-0.03	0.01	<0.01-0.02
SO_4^{2-} "	2.56	2.30-3.14	2.61	2.30-3.44
$NH_4^+ + NO_3^-$ ug N L^{-1}	15.6	<1-109	11.8	<1-63
TN ug/L	190	90-390	210	60-300
SRP "	1.2	0-2	1.1	0-2
TP "	8.5	5-20	9.0	5-24
DOC mg L^{-1}	2.9	2.3-3.7	2.8	2.2-3.6
Chl _a_ ug L^{-1}	2.5	0.4-5.4	2.2	1.1-6.6

*Values are for surface water samples from mid lake station in each basin for period Sept. 1983 to Dec.1984.

of replicating the experiment at little added cost and increases the number of treatment-control comparisons that could be made. It does eliminate contemporaneous comparisons between pH 4.5 and 6.0, however. We are aware of the potential problem of "pseudoreplication" (Hurlbert, 1984) in both designs and in nearly all large ecosystem manipulations, where replication of both controls and treatments is seldom possible. We have the advantage of access to data on other lakes to help interpret results from pH manipulations on Little Rock Lake. A long-term ecological research program is monitoring 7 nearby lakes by the same procedures used at Little Rock Lake (Magnuson _et al_., 1984a). These data will be useful to evaluate interlake and interbasin variability on a multi-year basis.

3. BACKGROUND STUDIES AND INITIAL FINDINGS

Hydrologic studies indicate that groundwater near the lake flows northwesterly and that the lake surface is above the water table except at the east end of the south basin, where some inseepage occurs. Recharge occurs elsewhere in the lake. The north basin was selected for acidification because of its down-gradient position in the flow system. Inseepage is a minor component of the water budget, but it may be more important in ion budgets because groundwater has a higher ionic content than rain.

Extensive pre-acidification monitoring verified initial impressions that the two basins are very similar in water chemistry and biota. Statistical analyses (paired difference t-test, F-test) showed no significant differences between the basins in means and variances of all measured variables (Table I). The basins have similar temporal patterns of pH, alkalinity, major ions, chlorophyll a, and nutrients. A pH depression during late winter apparently is caused by a buildup of dissolved CO_2 under the ice rather than from mineral acids in snowpack or subsurface flow through acidic litter (Kratz et al., 1986). This conclusion is supported by several findings: alkalinity was constant under the ice; pH depression was greater near the bottom, where respiratory production of CO_2 is greater; and the depression disappeared soon after ice-out. Concentrations of major ions show only small temporal variations except during snow melt and ice-out; the latter variations probably reflect initial inputs from melting snowpack and later dilution from melting ice. Inorganic N and P forms and dissolved SiO_2 accumulate under the ice but are rapidly removed from the water column by plankton after ice-out.

Species distribution and abundance in the major trophic levels are similar in the two basins. Dominant phytoplankton taxa include: Asterionella, Fragilaria, Dinobryon, Synura, Euglena, and Merismopedia. The zooplankton community is diverse (> 40 recorded species). It is dominated numerically by rotifers, including Conchilus unicornis, Keratella cochlearis, K. taurocephala, and Polyarthra remata. Bosmina longirostris, Holopedium gibberum, Cyclops bicuspidatus, Epischura lacustris, and Diaptomus minutus are the most abundant crustaceans. The benthic macroinvertebrate community is dominated by chironomid larvae (esp. Procladius and Tanytarsus), but it contains representatives of most other insect orders, as well as amphipods and sphaeriid clams. Three species of sponges are common on the littoral bottom. Small macrophytes (primarily Myriophyllum tenellum, Isoetes, Gratiola, and Eliocharis) cover much of the littoral zone of both basins. Patches of filamentous algae (primarily Oedogonium spp. and Mougeotia spp.) were observed in both basins in 1983 and 1984; a much more extensive mat (primarily Mougeotia) covered up to 25% of the littoral zone (z < 3 m) in the north basin during summer 1985, after the basin pH was lowered to 5.5. Extensive measurements of fish populations indicate that yellow perch populations are similar in the two basins. Surveys of largemouth bass populations in summer 1985 showed significant interbasin differences, however, and some adults and fry were transferred to distribute populations more evenly.

Experiments were conducted in large in situ enclosures during 1984 and 1985 to obtain preliminary information on responses of biota and biogeochemical processes to acidification. Results were used to design the whole lake treatment and to make predictions of lake responses described below. Ten littoral enclosures (4 m dia; 1.2-1.5 m deep; open to the bottom) were placed in the north basin in May 1984. Pairs were acidified to pH 5.5, 5.0, and 4.5 with H_2SO_4 and 5.0 with HCl; one pair was left at pH 6.0. Responses were studied extensively during the summer. Results from 1985 experiments at a pelagic site are still being evaluated.

Al in littoral enclosures increased with decreasing pH but only from 6 to ~25 ug L^{-1} at pH 4.5. Zn increased from 1-2 ug L^{-1} at pH 5.5 and 6.0 to

6-9 ug L^{-1} at pH 5.0 and 4.5, and Cd increased from ~0.1 ug L^{-1} at pH 6.0 to ~0.45 ug L^{-1} at pH 4.5. Total N and P levels, chlorophyll a, and primary production were highest at pH 5.0, but values at other pH's were not different from the controls. Consistent with published reports (e.g. Muller, 1980), Mougeotia replaced Oedogonium as the dominant periphyton at pH < 5.0. Some changes in zooplankton species were noted; Diaptomus minutus decreased and Keratella taurocephala increased with pH. In situ bioassays suggested that most of the lake's 14 major species of zooplankton tolerate pH 5.0, but half the species have LC_{50} values in the range pH 4.5 to 5.0. Decomposition of oak and birch leaves decreased significantly over the pH range, but benthic invertebrates were unaffected.

Because of internal (sediment) alkalinity generation, weekly to biweekly additions of acid were needed to maintain target pH values. The pH 5.0 enclosures treated with H_2SO_4 required greater additions than those receiving HCl; SO_4^{2-} reduction occurred at a rate of 0.59 meq m^{-2} day^{-1} in the pH 5.0 (H_2SO_4) enclosures and accounted for 59% of the generated alkalinity (Perry, 1986). Exchange of sediment Ca^{2+} and Mg^{2+} accounted for the remaining alkalinity generation. Sulfate reduction increased with decreasing pH in the enclosures because $[SO_4^{2-}]$ was higher at lower pH; reduction rates were first order in $[SO_4^{2-}]$.

4. PREDICTION OF WHOLE-LAKE RESPONSES TO ACIDIFICATION

Predictions of ecosystem responses to stress are based on four types of information: (i) baseline data on the ecosystem of concern; (ii) mesoscale experiments (in limno-enclosures); (iii) published information, which for pH effects on lakes is based primarily on laboratory studies and lake surveys; and (iv) laboratory and in situ tests with indigenous species. Evaluation of present abilities to predict ecosystem-level effects of acidification from these four types of information is important because ecosystem manipulations are costly, and time-consuming.

Before acidification began (spring 1985), we developed an extensive list of predictions concerning the lake's responses to acidification. Predictions were based on all four types of information described above. The list includes over 100 predictions of chemical changes, biotic responses (qualitative and quantitative), and changes in biogeochemical process rates. The list was sent to other experts for review and is available from the authors (Brezonik et al., 1986). Table II lists predictions we consider most important. While it was easy to make broad, qualitative predictions, it was more difficult to make quantitative predictions for specific processes or taxa. For example, it is likely that Al will increase with decreasing pH, but predicting the actual increase at a given pH is much more difficult. Based on our enclosure studies and review of the literature (e.g. Haines, 1981; Hendrey, 1984; Magnuson et al., 1984b) we predicted only small changes in ecosystem structure and process rates at pH 5.5, but the subtle structural changes expected to occur above pH 5.0 are more difficult to predict than the dramatic responses expected at lower pH. Predicted changes in species composition are thought to be

TABLE II. PREDICTIONS OF LAKE RESPONSES TO ACIDIFICATION

pH Interval	Predictions
6.0 | 5.5	Epischura lacustris wil not be present (2)*; increased abundance of filamentous green algae, esp. Mougeotia, Oedogonium, Zygnema (2) Phytoplankton biomass and production constant or slight increase (1) No difference in mean length of age 0 fish (all 4 major species) between basins, but greater variance in treatment side because of reduced food conversion efficiency (2); no changes in abundance, population structure, survival, and recruitment (1) Ca increases 25-50%, steady-state sulfate, $[SO_4^{2-}]_{ss}$, of 4 mg L^{-1} reached in 1 yr, no increase in Al(1); no change in TP and TN (2)
5.5 | 5.0	Few zooplankton species changes at pH > 5.0; some changes in population dynamics (2); leeches (3) and amphipod C.gracilis (2)disappear No change in largemouth and rock bass abundance; slight decline in condition of 4 major species; black crappie recruitment failure (1) Ca increases 75-100%, $[SO_4^{2-}]_{ss}$ of 6 mg L^{-1} reached in 1 yr (1); Al increases to 9-20 ug L^{-1} (2); aqueous Cd, Zn, Mn increase (2)
5.0 | 4.5	C. unicornis, P. remata, P. vulgaris disappear (3); K. taurocephala, B. longirostris, D. minutus dominate zooplankton (2); periphyton almost exclusively Mougeotia (1); phytoplankton biomass and production decline (2); flagellated and large algal species dominate (3); loss of Ephemeroptera, Sphaeriidae, and caddisfly Limnephilus (2) Largemouth and rock bass abundance and production decline because of changes in diet, lower survival, lower food conversion efficiency; yellow perch condition and growth decline but survival does not (1) Ca increases 100-125%; $[SO_4^{2-}]_{ss}$ of 9 mg L^{-1} reached in 1 yr; SiO_2 increases because of decline in diatoms (2); Al increases to 20-45 ug L^{-1}; small increases in aqueous Cd, Mn, Zn; increased bioaccumulation of Cd, Cu, Pb, and Zn (2)
6.0 | 4.5	Small decline in water column TP and TN at pH 4.5 (2); dominant filamentous algae shift from Oedogonium (pH 5.5-6.0) to Mougeotia (pH 4.5) (2); litter decomposition significantly lower at pH < 5.0 (1); no decrease in P release from leaf litter at pH > 4.5 (2) Alkalinity generation (primarily SO_4^{2-} reduction) increases with decreasing pH (1); lake requires 2X stoichiometric amount to maintain target pH in first yr of pH decrease, and less in second yr (2) SRP sorption by sediment stronger at pH 4.5-5.0 than 5.5-6.0, but this is not correlated with SRP changes in lakewater (2) Sensitivity of reproductive processes and/or pre-juvenile stages exceed later stages and cause recruitment failure by direct (water chemistry) effects for some species at pH > 4.5 (2); loss of prey for fish larvae change some fish populations at pH > that directly affecting fish (3); increased aqueous levels induce elevated Cd, Zn, and Hg in fish organs and muscle (1); changes in blood osmolality and gill morphology useful to indicate pH stress (2)

*Confidence level: 1 = highest; 2 = medium; 3 = lowest confidence more reliable than predictions about population density.

The abundant information published on aquatic effects of acidity is based almost entirely on synoptic surveys or experiments on isolated components of lake ecosystems (e.g. single species). Such information may not allow accurate (quantitative) predictions of the responses of intact ecosystems (e.g. changes in process rates and population densities) widely reported to occur below pH 5 or the more subtle effects expected in the pH range 5.0 to 6.0. Whole ecosystem experiments like we are conducting at Little Rock Lake will greatly improve our understanding of the causes and mechanisms of acid-related changes in lakes and greatly increase our confidence in predicting the consequences of anthropogenic changes in lake pH.

Acknowledgment. Work described in this paper was supported by cooperative agreements between the U.S. EPA (Environmental Research Laboratory - Duluth) and the institutions associated with the authors.

REFERENCES

Brezonik, P.L. (ed.): 1986, 'Background Studies on Little Rock Lake, Wisconsin, and Predictions of its Responses to Artificial Acidification'. Publ. 114, Wat. Resources Res. Cent., Univ. Minnesota, St. Paul, MN.

Grahn, O., Hultberg, H., and Landner, L.: 1974, Ambio, 3, 93.

Haines, T.: 1981, Trans. Am. Fish. Soc., **110**, 669.

Hendrey, G.R. (ed.): 1984, 'Early Biotic Responses to Advancing Lake Acidification'. Vol. 6, Acid Precip. Series, Butterworth Publ., Boston, MA.

Hurlbert, S.H.: 1984, Ecol. Monogr., **54**, 187.

Kratz, T.K., Brezonik, P.L., Bowser, C.J., and Cook, R.B.: 1986, in prep.

Magnuson, J.J., Bowser, C.J., and Kratz, T.K.: 1984a, Verh. Internat. Verein. Limnol., 22, 533.

Magnuson, J.J., Baker, J.P., and Rahel, E.J.: 1984b, Phil. Trans. R. Soc. Lond., **B 305**, 501.

Muller, P.: 1980, Can. J. Fish. Aquat. Sci., **37**, 355.

Perry, T.E.: 1986, M.S. thesis, Univ. Minnesota, Minneapolis.

Schindler, D.W., Mills, K., Malley, D., Findlay, D., Shearer, J., Davies, I., Turner, M., Linsey, G., and Cruikshank, D.: 1985, Science, **228**, 1395.

TEMPORAL VARIABILITY OF LAKE WATERS IN THE NORTH CASCADES MOUNTAINS (WASHINGTON, U.S.A.)

T.J. Loranger, D.F. Brakke, M.B. Bonoff, and B.F. Gall, Institute for Watershed Studies, Western Washington University, Bellingham, WA 98225 USA

ABSTRACT. Five lakes in the North Cascades were sampled at regular time and depth intervals during the open water period in 1984. Surface water ANC was depressed during snowmelt and then increased as summer and fall progressed. Shallow lakes circulated completely in the fall, whereas deep lakes ($>$ 75 m) did not. Sulfate and NO_3 concentrations in the lakes increased during the snowmelt period and then decreased until ice-over. No significant differences were found between surface ANC measurements during fall circulation in successive years.

1. INTRODUCTION

The North Cascades of Washington State are characterized by granitic, volcanic and metamorphic bedrocks that weather slowly generating little buffering capacity in surface waters. The watersheds often have steep slopes with thin soil profiles, thus limiting contact time of incoming precipitation with the soil and bedrock components of the watershed.

North Cascade lakes were surveyed for chemical characteristics in 1983 and 1984. In both years single surface water samples were collected during fall circulation in the lakes. Out of a total of 54 lakes surveyed only 14 lakes had acid neutralizing capacity (ANC) values $>$ 200 ueq l^{-1}, whereas 21 lakes had ANC values $<$ 100 ueq l^{-1}(Brakke and Loranger, 1986). Distribution of bedrock types appears to be the controlling factor on surface water ANC. Lakes on gneissic and granitic bedrock typically have lower ANC whereas lakes on sedimentary bedrock have greater ANC (Brakke and Waddell, 1985). None of the lakes showed apparent acidification.

This study was designed to determine the extent of variability of water chemistry in North Cascades lakes and evaluate major factors contributing to this variability. Five lakes were sampled at regular time and depth intervals during ice free conditions. Major processes examined were, period and extent of ANC depletion in lakes of varying depth and watershed area to lake area ratio (wa/la), year-to-year variability of ANC in lakes and snowmelt influence on lake anion levels.

Water, Air, and Soil Pollution **31** (1986) 123–129.

2. METHODS

2.1. Lake selection and sampling schedule

Lakes were selected based on the survey results from 1983 (Brakke and Waddell, 1985). All study lakes had ANC< 150 ueq l^{-1}. The lakes selected represent a wide range of basin size and shape, watershed area to lake area ratio, and watershed geology (Table I).

Table I. Lake Physical Data

Lake	Area (ha)	Elev. (m)	Geol.[1]	Depth (m)	Watershed Area: Lake Area
Doubtful	12.1	1641	PJgn	19	20
Thornton	22.0	1351	PJgn	92	10
Iceberg	14.8	1463	QV	30	4
Bagley	3.6	1292	QV	6	40
Boulder	22.0	1524	K^s	80 (est.)	7

[1]QV = volcanic (Mt. Baker formation)
PJgn = gneiss
K^s = Skagit Metamorphic Suite

All lakes were sampled three times during summer and fall, 1984, with the first sampling date at ice-out or shortly thereafter. Ice free conditions varied with the aspect and elevation of each lake. The second sampling took place in September at the time of thermal stratification and the final date was in October during fall circulation. Sampling stations were established at a central location on all the lakes and on each sampling date two to four samples were taken depending upon lake depth. Doubtful Lake was sampled more intensively because of its low ANC and accessibility. The lake was sampled at biweekly intervals until the lake was no longer accessible due to snow conditions. Replicate samples were collected at the lake outlet beginning in June before ice free conditions and continuing through the fall. The lake water column was sampled in replicate at depths of 0, 1.5, 5, 10, and 17 m.

2.2 Chemical analyses

All water and snow samples were collected and stored in Nalgene bottles that had been acid washed with 2N HCl, rinsed six times with distilled water (specific conductance < 1 uS) and then rinsed with sample water. Aliquots prepared for the analysis of SiO_2, NO_3-N, Cl, SO_4, NH_3-N, total phosphate (TP) and soluble reactive phosphate (SRP) were filtered through 0.45 um Nucleopore filters that were pre-rinsed with 100 mL distilled water.

The chemical methods and parameters selected for analysis were derived from the draft "Sampling and Analysis Protocol for Long-term Monitoring Sites" developed by the National Acid Precipitation Assessment Program (1983) and the methods manual for the National Surface Water Survey (EPA, 1985). Total P, SRP, NH_3-N and SiO_2 were analyzed colorimetrically with a Technicon II Autoanalyzer. Calcium, Mg, Na, and K were analyzed by atomic absorption spectroscopy. Sulfate, Cl, and NO_3 were measured on a Dionex ion chromatograph.

Turbidity, specific conductance, ANC and pH were analyzed within 24 hr of collection. pH was measured under quiescsent conditions with samples warmed to buffer temperature. The pH was recorded at the first stabilization period, usually about 20 min after electrode emersion. ANC was determined by a Gran Titration (Stumm and Morgan, 1982). A minimum of 10% of all analyses were replicated for all parameters except for TP, SRP and NH_3-N which were run in duplicate because of the sensitivity of the analyses.

Relative mean deviation of analyses (precision) was generally less than 5%. For the entire parameter set only NH_3 (14.1%) and turbidity (7.5%) had relative mean deviation values over 5%. Mean cation:anion ratios for all study lakes were near 1.0. Coefficients of variation (cv) of cations/anions were near 10% in Bagley, Hayes, and Iceberg Lakes and near 30% in Doubtful and Boulder Lake.

3. RESULTS AND DISCUSSION

3.1. Mean ion concentrations and variability of ANC

Mean total cationic strength of the six study lakes ranged from 65 ueq l^{-1} cations in Thornton Lake to 200 ueq l^{-1} of cations in Boulder Lake. Calcium was the dominant cation in all of the lakes. In all lakes except Doubtful Lake, which had instrusive pyritic deposits (FeS_2), HCO_3 was the dominant anion. For Bagley Lake, Boulder Lake, Iceberg Lake and Thornton Lake, $HCO_3 > SO_4 > Cl > NO_3$. In Doubtful Lake, $SO_4 > HCO_3 > Cl > NO_3$.

Because of a stoichiometric increase in ANC with base cations (C_B) when carbonation reactions predominate within a watershed, ANC is a good indicator of total weathering products in solutions. Differences in ANC with the depth are also used here to represent chemical stratification of ions .

For surface waters during the open water period (August - October), ANC generally increased over time, with the exception of Thornton Lake, where it remained constant (Figure 1). Dilution from snowmelt runoff in August apparently caused the lowest ANC values. As greater contact time with soils and bedrock occurred, ANC increased. Over the same period, NO_3^- concentrations were depleted, indicating that phytoplankton uptake may have also generated alkalinity.

Bagley Lake was sampled at 1.5 and 5 m. In early August (mo 1 in Figure 1) mean ANC values were at a maximum at both depths increasing through September (mo. 2) and October (mo. 3). Chemical stratification did not occur in this lake during the 1984 sampling season. In Iceberg

Lake, the 1.5 m samples had an ANC only half that of the 10 and 20 m samples during August. The surface samples showed the greatest dilution from snowmelt. Surface water ANC increased over time and concentrations at depth decreased. By October, the water column was in circulation (Figure 2).

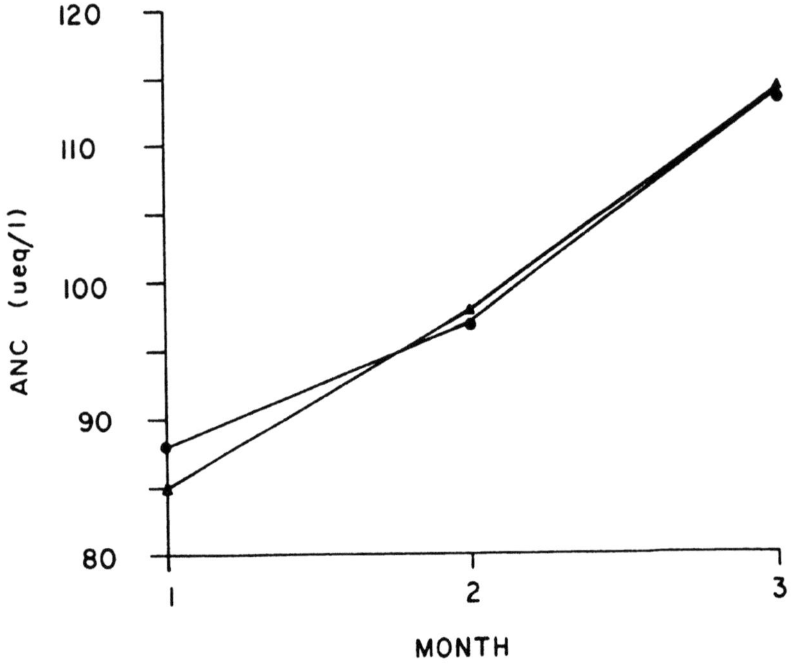

Figure 1. ANC values at 1.5 m (o) and 5 m (Δ) in Bagley Lake (August, September, October, 1984).

Thornton and Boulder Lake were sampled at 1.5, 10, 20, and 45 m (Figures 3 and 4). Both lakes stratified during the sampling season. Thornton Lake had similar ANC values at 1.5m, 10m, and 20m on all 3 sampling dates. The 45 m depth was sampled on the second and third sampling dates, and ANC values were considerably higher than at the shallower depths. Boulder Lake became ice free later and froze earlier than the other lakes, consequently the lake was sampled on only 2 dates. The lake showed chemical differences at each depth on both sampling dates. It appears that neither lake circulates completely, maintaining higher ANC values at depth throughout the open water period, and presumably under ice cover as well. Boulder lake had chemical differences at each depth in September, whereas Thornton Lake was circulating to 20 m. Undoubtably the greater exposure of Thornton Lake promotes surface water mixing.

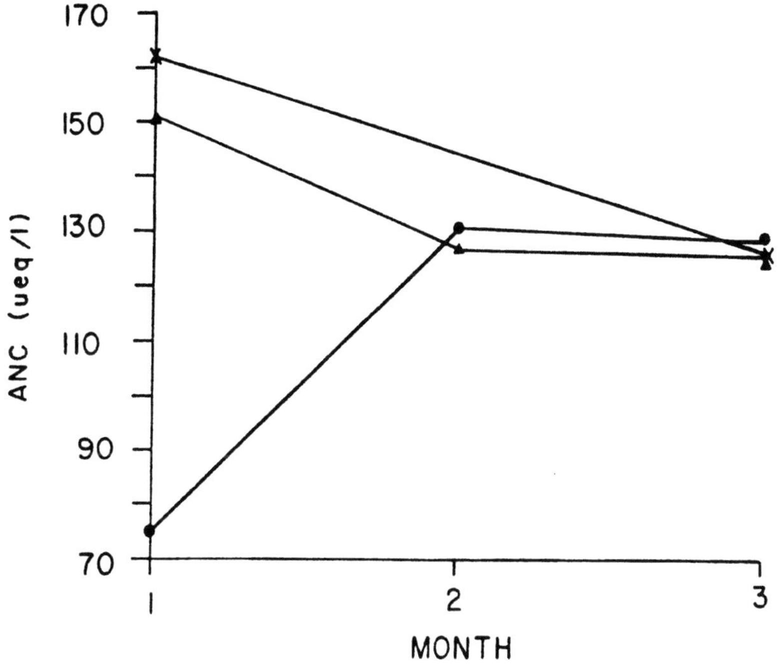

Figure 2. ANC values at 1.5 m (o), 10 m (△), and 20 m (*) in Iceberg Lake (August, September, October, 1984).

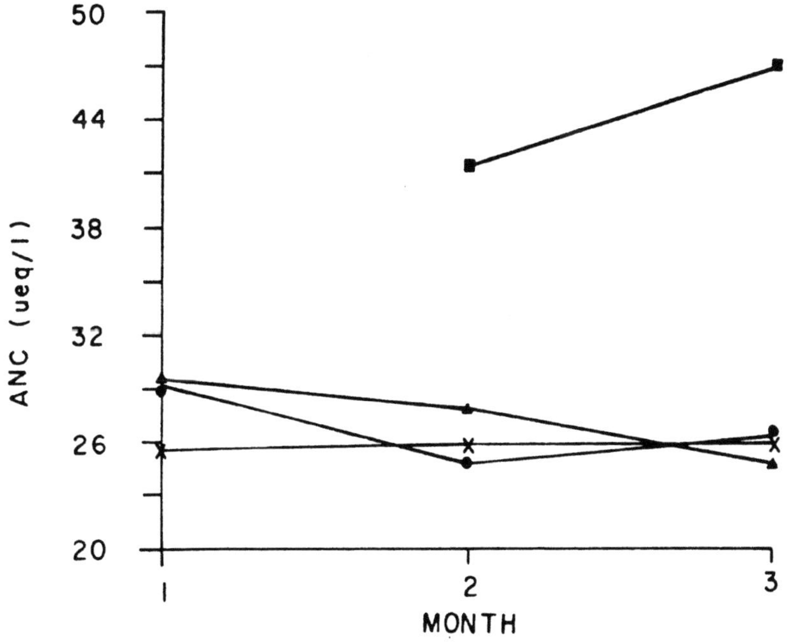

Figure 3. ANC values at 1.5 m (o), 10 m (△), 20 m (*), and 45 m (□) in Thornton Lake (August, September, October, 1984).

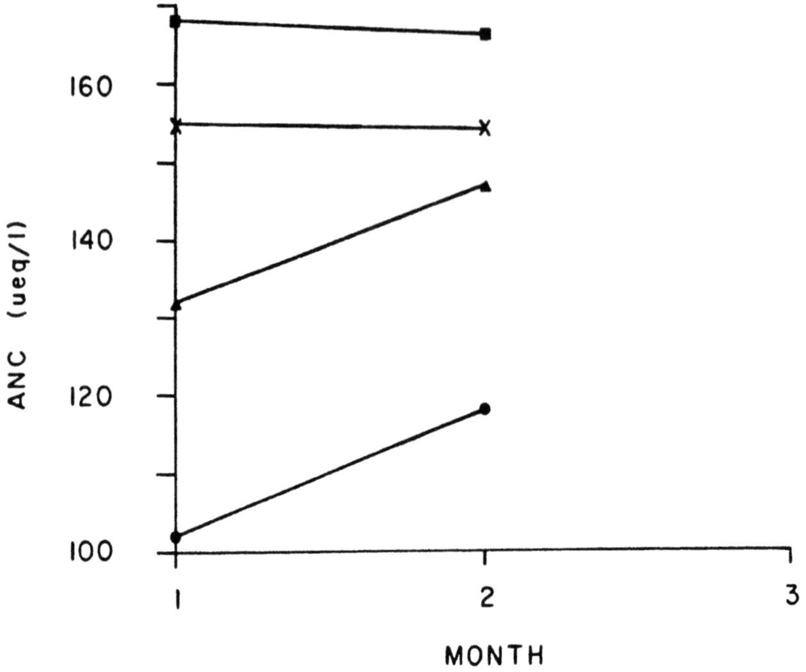

Figure 4. ANC values at 1.5 m (o), 10 m (△), 20 m (*), and 45 m (□) in Boulder Lake (August, September, 1984).

Replicate outlet samples collected in Doubtful Lake in July before ice free conditions in the lake had the lowest mean ANC value of all samples taken from the lake. Immediately after ice out, mean ANC (n=11) within the lake was 29 ueq l^{-1} dropping to 26 ueq l^{-1} on the third sampling date and rising steadily with decreasing lake discharge to 30 ueq l^{-1} on the final sampling date.

Year-to-year variability of ANC was generally not significant in the study lakes. For example, Doubtful Lake had an ANC of 29 ueq l^{-1} on 20 September 1983 and 18 September 1984. Bagley Lake was 86 ueq l^{-1} 17 August 1983 and 88 ueq l^{-1} on 6 August 1984, and Boulder Lake was 117 ueq l^{-1} during equivalent periods in both years. ANC in Iceberg Lake was 100 ueq l^{-1} in August 1983, but only 75 ueq l^{-1} in 1984. This may have been due to the later snowmelt period in 1984, because the September value was about 130 ueq l^{-1}. The ANC measured in 1983 at Thornton Lake was 17 ueq l^{-1}, which was less than any period in 1984.

3.2. Anion concentrations

Sulfate levels in Doubtful Lake were significantly higher ($p < 0.05$) the three weeks following ice out. Concentrations stayed near 85 ueq l^{-1}. On the final two sampling dates, SO_4 concentrations were near 40 ueq l^{-1}, very similar to concentrations before ice out. The source of the SO_4 appears to be a pyrite seam running through part of the basin.

Maximum SO_4 levels at 1.5 m occurred at the first sampling date in all lakes except Doubtful. Bagley and Boulder had SO_4 concentrations near 55 ueq l^{-1} following ice out, dropping steadily to 35 ueq l^{-1} and 40 ueq l^{-1}, respectively, at the third and final sampling in October. Iceberg Lake, Hayes Lake and Thornton Lake had SO_4 concentrations of 25, 18, and 18 ueq l^{-1}, respectively, once ice free, with SO_4 also decreasing steadily through October.

Nitrate levels paralleled the trends in SO_4. Concentrations of 4.5 to 7.5 ueq l^{-1} were recorded in all lakes at the first ice free sampling. All study lakes had nondetectable levels of NO_3 by the second and third sampling. The snowpack, then, seemed to be a source of SO_4 and NO_3 to the lakes and concentrations decreased over time.

4. SUMMARY

Snowmelt runoff resulted in ANC depletion immediately following ice-out in the study lakes. Lake depth, wa/la and rate of snowpack melt interacted to mediate period and extent of ANC depletion.

The two shallowest lakes (<20 m) with large wa/la, had ANC depletion throughout the water column. No chemical stratification occurred during the open water period.

A lake of moderate depth (30 m) had only surface water ANC depletion. The lake was chemically stratified until fall mixing as a result of the dilute surface water.

Two deep lakes (> 75 m) had ANC depletion in surface water, with depth of depletion governed by lake exposure. Chemical stratification within these deep lakes persisted through fall and apparently under ice cover as well.

Year-to-year variability of ANC was not significant in the study lakes. Any differences can be associated with persistence of snowpack in the summer.

Sulfate and NO_3 concentrations were highest immediately following ice-out. Snowpack was an apparent source of SO_4 and NO_3 to the lakes.

5. ACKNOWLEDGEMENTS

Funding for this project was provided by the U.S. EPA Region 10, Seattle, Washington, the Washington State Department of Ecology, the Northwest Pulp and Paper Association, and the Weyerhauser Company.

6. REFERENCES

Brakke, D.F. and Loranger, T.J.:1986, Water Air and Soil Pollution.

Brakke, D.F. and Waddell, D.A.:1985, Env. Canada Special Symp. (in press).

Duncan, L.C.:1985, Report to Washington Dept. of Ecology. 19 p.

Omernick, J.M. and Griffith, G.E.:1985, U.S. EPA (in press).

Welch, E.B., Spyridakis, D.E., and Smayda T.:1985, Final report to Wash. Dept. of Ecology. 22 pp.

SNOWMELT ACIDIC SHOCK STUDY IN SOUTH CENTRAL ONTARIO

B.E. Goodison, P.Y.T. Louie and J.R. Metcalfe
Atmospheric Environment Service
4905 Dufferin St., Downsview, Ontario
Canada M3H 5T4

ABSTRACT. A two part project to study snowmelt acidic shock potential in south central Ontario has been undertaken by the Atmospheric Environment Service. One part involved development of a snowmelt acidic shock model suitable for application to specific basins to provide time series of snowmelt rate, snowpack and melt water chemistry. The U.S. National Weather Service River Forecast System's conceptual snowmelt algorithm (HYDRO-17) was chosen as the basis for the snowmelt acidic shock model. Routines to handle the acid concentration in the snowpack and melt water have been incorporated into the model. The second part of this project involved a field study to collect data on melt rate, snowpack and melt water chemistry for model development and verification. Two 3 x 3 m snowmelt lysimeters were constructed for this purpose at Dorset, Ontario. An overview of the project is presented along with descriptions of the acidic snowmelt model, the field experimental design and discussion of the results from two field seasons.

1. INTRODUCTION

In regions of Canada and other northern countries characterized by seasonal snow cover and subject to acid precipitation, sudden and intense doses of acidic melt waters can be discharged into lakes and streams during late winter and/or early spring. This so called snowmelt acidic shock phenomenon results from the accumulation of air pollutants in the winter snowpack which are then released in high concentration during the first melt events. The impact to lake ecosystems is severe, since these melt events can occur at the worst possible time for fish and amphibians--spawning time (Leivestad *et al.*, 1976; Jeffries *et al.*, 1979).

The pathway for air pollutants from local and distant sources through the atmosphere and snowpack to the lake and streams is very complex. Air pollutants accumulate in the snowpack directly from precipitation containing pollutants incorporated during its formation and scavenged from the atmosphere. There is also a contribution from dry atmospheric fallout onto the snow surface. The processes which affect the ultimate concentration of the pollutants both in the snowpack and the initial melt water include snowpack metamorphism, snow sublimation, freeze-thaw cycles prior to initial runoff, non-uniform distribution of pollutants in the snowpack, channelling of melt water flow down distinct pathways through the snowpack and the possibility of pollutant migration into the underlying soil during the accumulation period. Once the melt water leaves the snowpack, it is subject to more interactions with the soil surface and leaf litter and this can have a significant effect on the final pollutant concentration reaching the streams and lakes.

A recent review of acidic snowmelt processes and current related research activities was conducted by Marmorek *et al.* (1984). It was apparent, however, that the interactions of many of the above mentioned factors are still not fully understood and that spatial and temporal variations could

Water, Air, and Soil Pollution **31** (1986) 131–138.

be significant. This study combines model development with field investigation to assess the accumulation of pollutants in the snowpack and their release in the melt water at the base of the pack, exclusive of the effects of the soil-vegetation complex. It is intended to complement earlier studies in the region which focussed on snowpack and stream chemistry (e.g. Jeffries and Snyder, 1981), and to provide measurements of snowpack and melt water concentrations which can be used to validate climatological models of snowmelt and ultimately be used to help understand the relative role of precipitation, snowpack and soil processes on the runoff chemistry. This paper provides an overview of the study and discusses initial field and model results.

2. MODEL DESCRIPTION

The empirical acidic snowmelt shock potential model developed in this study provides a simple simulation of the snowpack and pollutant accumulation and depletion. The snowmelt algorithm is based largely on the U. S. National Weather Service River Forecast System (NWSRFS) snow accumulation and ablation model (Anderson, 1973). This algorithm may be described as a psuedo-energy budget of the snowpack as it considers most of the relevant physical processes involved in snowmelt. The algorithm's applications, limitations and calibration requirements are well documented.

A simple chemistry budget, similar to the one described by Wilson and Barrie (1981), is used to accumulate the pollutants in the snowpack and to apportion the concentration of pollutants to be leached from the pack by each effective melt event. The melt algorithm only requires readily available meteorological observations while the basic inputs for the chemical budget are dry and wet deposition rates of H^+ derived from deposition measurements at stations in the Canadian Air and Precipitation Monitoring Network. Wet deposition H^+ rates are based on direct pH measurements of daily precipitation samples taken with a Sangamo sampler. The dry deposition H^+ rates are based on daily concentrations of $[SO_2]$, $[SO_4^{2-}]$, $[HNO_3]$ and $[NO_3^-]$ determined from measurements using a three filter sequential air sampling system and using appropriate deposition velocities. Barrie and Sirois (1985) provide a more detailed description of the deposition measurement, the analytical procedures and their relative accuracies.

The model maintains an account of the pollutant concentration once a snowpack is established. For each time step, pollutants from dry deposition and wet deposition (if precipitation has occurred) are accumulated and added to the snowpack pollutant concentration. Initially in this study, it has been assumed that pollutants accumulate without neutralization over the cold season. The melt model then determines if there is an effective melt event, i.e. when the amount of snowmelt and rain is greater than the liquid water holding capacity of the snowpack. The model proceeds to the next time step if no effective melt has occurred. For each melt event, the model computes the amount of pollutant to be removed by the melt water. This amount is proportional to the melt water removed, thus:

$$dAp/Ap = n(dAs/As) \tag{1}$$

where:

dAp = amount of pollutants removed (ueq l^{-1})

Ap = total pollutant concentration in snowpack before melt (ueq l^{-1})

dAs = amount of snowpack removed as melt water (mm)

As = total snowpack water equivalent before melt (mm)

n = constant of proportionality.

Integration of Equation (1) provides the following relationship:

$$Ap_f/Ap_i = (As_f/As_i)^n \tag{2}$$

where:

Ap_i = initial snowpack pollutant concentration before melt (ueq l^{-1})

Ap_f = final snowpack pollutant concentration after melt (ueq l^{-1})

As_i = initial snowpack water equivalent before melt (mm)

As_f = final snowpack water equivalent after melt (mm).

Since,

$$As_f = As_i - ME$$

where:

ME = effective melt water (mm)

the final snowpack pollutant concentration after an effective melt can be computed by:

$$Ap_f = Ap_i[(As_i - ME)/As_f]^n \tag{3}$$

and the concentration of the melt water, Cm, is given by:

$$Cm = (Ap_iAs_i - Ap_fAs_f)/ME. \tag{4}$$

The constant, n, determines the magnitude of the shock event and it is likely to depend on the site and snowpack conditions. Previous field and laboratory observations of the snowmelt enrichment process (Johannessen and Henriksen, 1978; Colbeck, 1981) have suggested that 50 to 80% of the total snowpack pollutant is removed by the first 30% of the melt water which gives a range for the value of n from 1.9 to 4.5. These values provide some guidance for choosing a value for n for model testing, but ultimately, model calibration will require field observations of melt rate and melt water concentration to determine n.

3. FIELD INVESTIGATION

The field investigations were conducted near the Ontario Ministry of Environment's (OME) Dorset Research Station to complement other studies in the area investigating the problem of long range transport of air pollutants and to take advantage of available climatological and precipitation chemistry data. This region of the Canadian Shield is characterized by a multitude of lakes surrounded by mixed deciduous and coniferous forest. Jefferies and Snyder (1983) provide further details of the basic geology of the study area. Continuous winter snow cover normally begins in early December and ends in late March or early April. Snow depth may reach 100 cm and the water equivalent can exceed 200 mm. Snowmelt can occur anytime during the winter and melt events during the January to March period are commonly accompanied by rain.

Snowmelt lysimeters, one in the open and one in a nearby maple-birch forest, were constructed to measure snowmelt. Their design combined the most practical features used by Goodison (1977) and Price (1984) and basically consisted of a collector, a flow-measuring device and a conduit linking the two. The collector was a 3 x 3 m plot, enclosed by 50 cm high styrofoam lined wooden walls. The entire collector was lined with 6 mil plastic sheeting in order to contain all melt and to isolate the snowpack from the underlying soil. Both plots were situated on gentle slopes allowing natural drainage of the melt water through the outlet into a dual tipping bucket recording system. The flow rate through each tipping bucket was controlled to minimize measurement error which is characteristic of this type of measuring system. A correction curve, dependent on the rate of melt, was developed and used in the calculation of hourly melt. Meteorological observations at the open site, including air temperature, humidity, wind speed, radiation, precipitation and snowpack temperature were logged on-site and can be remotely accessed via a satellite link. Temperature, humidity and wind speed measured at the forest site were logged on-site only. A more complete description of the snowmelt plot system and meteorological observing program is provided in Goodison and Metcalfe (1984).

Snowcore and melt water samples were taken at selected intervals during the winter season. Ontario Ministry of Environment laboratories at Dorset and Toronto analyzed these samples for pH, alkalinity, conductivity, NO_3^-, NH_4^+, Al^{3+}, Ca^{2+}, Na^+, K^+, Fl^-, Cl^-, SO_4^{2-}, Si^{2+} and DOC. Melt water samples were collected manually in 500 mL plastic sample jars as the water flowed out of the PVC conduit from the melt plot before entering the tipping buckets. Snow core samples for chemical analysis were taken using a 10 cm diameter acrylic plastic tube. Several vertical samples were combined into a bulk sample to obtain enough water volume for analysis and to minimize potential errors related to point variability in chemical constituents. Prior to the initial melt, samples of each stratigraphic layer were taken to determine the vertical profile of pollutants in the snowpack. Basic snowpack information recorded throughout the winter included depth, water equivalent and profiles of temperature and stratigraphy which would support interpretation of snowpack and melt water chemistry.

Procedures followed in obtaining samples and the laboratory methods used for chemical analyses are described in Ontario Ministry of Environment (1983). Chemical analyses of all water samples were completed as soon as possible after collection and usually within 24 hr for non-conservative parameters. The electrometric method (use of a H^+ sensitive glass electrode in conjunction with a suitable reference electrode) was used to determine pH.

4. RESULTS AND DISCUSSION

Maximum snow depths for the 1983/84 and 1984/85 seasons were 74 cm and 75 cm, respectively. Figure 1 shows snowpack depth and the measured snowpack H^+ concentration for the open and forest sites during the 1984/85 season. The H^+ concentration of the forest site snowpack was consistently lower than the open site snowpack. The mean snowpack H^+ concentration was observed to be lower than the accumulated H^+ concentration measured in snowfall precipitation samples.

Over the two seasons melt water pH values ranged from 3.53 to 5.35. The lowest pH was measured during the initial melt in February of each season rising slowly with each successive melt event. All significant melts were associated with rain-on-snow events thus the chemistry of the rain water will have an important impact on the melt water chemistry--the worst scenario being a highly polluted rainfall with an early season melt event. Such a condition was not observed during the two seasons of observations. The pH of rainfall associated with melt events ranged from 4.02 to 4.54. The pH of the initial melt water for each event was lower than the rainfall pH.

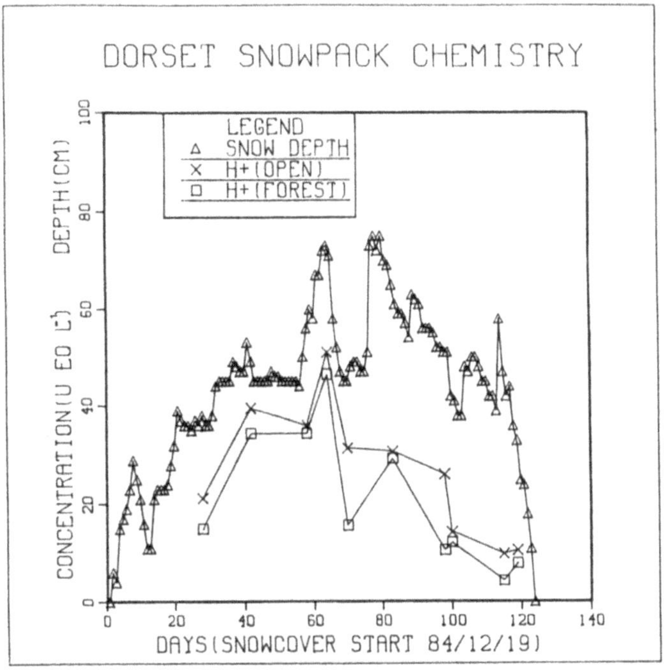
DORSET SNOWPACK CHEMISTRY
LEGEND
△ SNOW DEPTH
× H+(OPEN)
□ H+(FOREST)
CONCENTRATION(U EQ L⁻¹) DEPTH(CM)
0
20
40
60
80
100
120
140
DAYS(SNOWCOVER START 84/12/19)

Fig. 1.

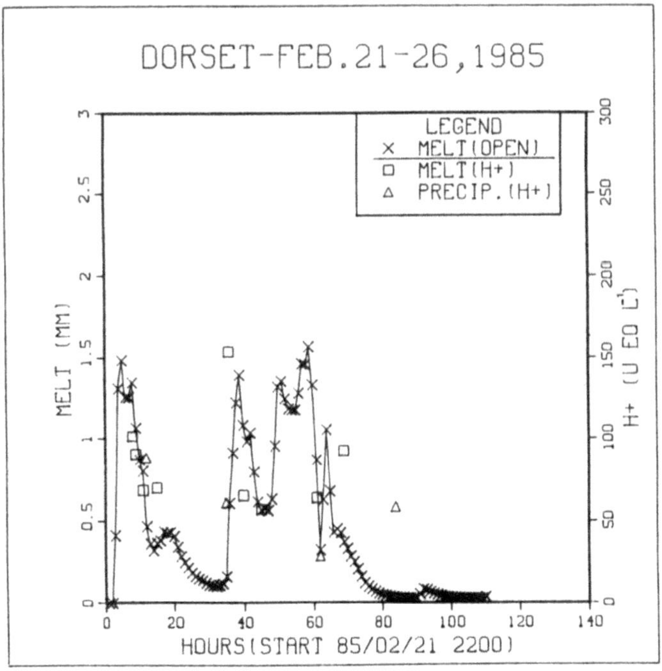
DORSET-FEB.21-26,1985
LEGEND
× MELT(OPEN)
□ MELT(H+)
△ PRECIP.(H+)
MELT (MM)
0
0.5
1
1.5
2
2.5
3
H+ (U EQ L⁻¹)
50
100
150
200
250
300
HOURS(START 85/02/21 2200)

Fig. 2.

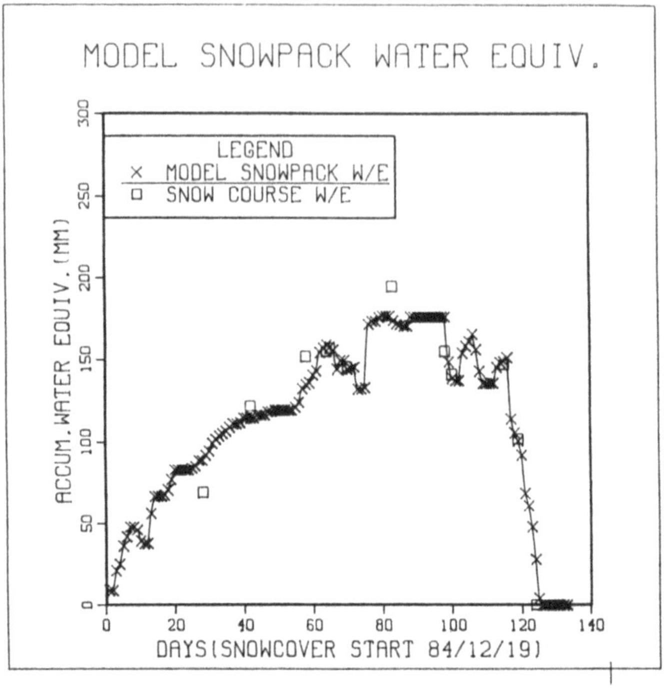
MODEL SNOWPACK WATER EQUIV.
LEGEND
× MODEL SNOWPACK W/E
□ SNOW COURSE W/E
ACCUM.WATER EQUIV.(MM)
0
50
100
150
200
250
300
0
20
40
60
80
100
120
140
DAYS(SNOWCOVER START 84/12/19)

Fig. 3.

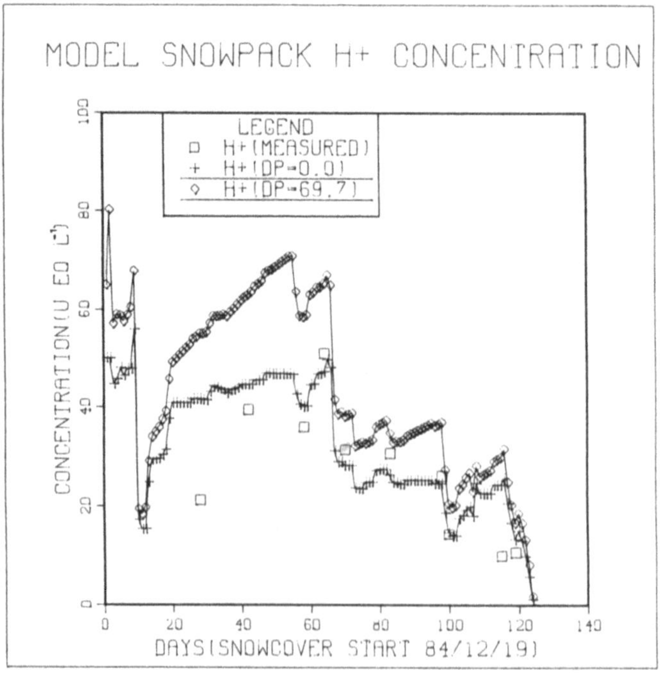
MODEL SNOWPACK H+ CONCENTRATION
LEGEND
□ H+(MEASURED)
+ H+(DP=0.0)
◇ H+(DP=69.7)
0
20
40
60
80
100
0
20
40
60
80
100
120
140
DAYS(SNOWCOVER START 84/12/19)

Fig. 4.

Figure 2 shows the fine temporal details of an early season melt event in February 1985. This melt event lasted approximately 80 hr with several melt pulses. The H^+ concentration of the melt water of the first pulse showed the expected high initial concentration and then a decrease in the receding limb of the pulse. However, the H^+ concentration of the following melt pulses did not show the same well defined decrease, but were instead rather variable. This was also observed in the other melt events over the season. The variability in melt water H^+ concentration can be affected by any short-term changes in the rainfall H^+ concentration, but only daily rainfall samples are currently analyzed. Any further investigation should consider the sampling of both melt water and rainfall samples on a shorter and more regular frequency in order to fully assess such temporal variations.

Initial modelling results are shown in Figures 3 and 4. The model ran on daily time steps for the entire snow season. Model estimates of the snowpack water equivalent compared well with snow course water equivalent measurements (Figure 3). Model estimates of the snowpack H^+ concentration were not as good partly because actual daily dry deposition values for the two field seasons were not available. In addition, the concentration constant n had to estimated at this time. A value of n = 2.94 which represented a 65% loss of pollutants in the first 30% of melt water was used. Figure 4 shows two model runs, one assuming no dry deposition input and the second assuming a mean daily dry deposition rate of 69.7 ueq m^{-2} d^{-1} which is a 5 yr mean value for the region for the November-April period. These results show the contribution of the dry deposition to the model snowpack pollutant concentration which appears to be particularly pronounced during the accumulation period. For the 1984/85 winter, the model overestimated the snowpack concentration during the snow accumulation period, but it provided reasonable H^+ concentration estimates after the first significant melt event (see Figure 4). This does put some question on the assumption that pollutants accumulate in the snowpack without neutralization over the cold season. Consideration of processes affecting the pollutant concentration in the pack during the accumulation period may also be necessary. Model estimates of the melt water H^+ concentrations were difficult to compare with the measured melt water H^+ concentrations since high variability of measured values were observed during the day while these model runs were done only in daily time steps.

Experience from the first two field seasons have provided some valuable data and experience which will be useful for future model development and field research. The snowmelt lysimeter design used during the second season generally worked well, but the measurement capacity of the system should be increased to handle melt rates of up to 6 mm hr^{-1}. The hourly melt values measured in the field provided some very detailed data on the melt events and the significant melt pulses occurring within a 24 hour period. The high hourly variability in the melt water flow and H^+ concentration suggests the need to monitor and model at hourly or at least 6 hourly time intervals, particularly for rain-on-snow events. Finally, the assumptions concerning the accumulation of pollutants in the snowpack during the cold season have to be re-examined. These problems are being considered in this continuing study.

REFERENCES

Anderson, E.A.: 1973, 'National Weather Service river forecast system. Snow accumulation and ablation model', *NOAA Tech. Memo. NWS Hydro-17*, US Dept. of Commence, Silver Springs, Maryland.

Barrie, L.A. and Sirois, A.: 1985, 'Wet and dry deposition of sulphates and nitrates in eastern Canada: 1979-82', Presented at International Symposium on Acidic Precipitation, Muskoka, Ontario, September 15-20, 1985.

Colbeck, S. C.: 1981, *Water Resour. Res.*, **17**, 1383.

Goodison, B.E.: 1977, 'Snowfall and snow cover in southern Ontario: Principles and techniques of assessment', Ph.D. thesis Univ. of Toronto, Toronto, Ontario, Canada, 403pp.

Goodison, B.E. and Metcalfe, J.R.: 1984, 'Design, instrumentation and operation of a snowmelt

runoff plot', Proceedings, Eastern Snow Conference, 41st Annual Meeting, pp 219-223.
Jeffries, D. S., Cox, C. M. and Dillon, P. J.: 1979, *J. Fish. Res. Bd. Canada*, **36**, 640.
Jeffries, D.S. and Snyder, W.R.: 1981, 'Variations in the chemical composition of the snowpack and associated melt waters in central Ontario', Proceedings, Eastern Snow Conference, 38th Annual Meeting, pp 11-22.
Jeffries, D.S. and Snyder, W.R.: 1983, 'Geology and geochemistry of the Muskoka-Haliburton study area' Ontario Ministry of Environment Data Report DR 83/2.
Johannessen, M. and Henriksen, A.: 1978, *Water Resour. Res.*, **14**, 615.
Leivestad, H., Henrey, G., Muniz, I. P. and Snekvik, E.: 1976, 'Effects of acid precipitation on freshwater organisms', pp. 87-111. In F. Braekke (ed.). *Impacts of Acid Precipitation on Forest and Freshwater Ecosystems Norway. Sur Nedbors Virkning Pa Skog og Fish Res. Rep. 6/76.*
Marmorek, D. R., Cunningham, G., Jones, M. L. and Bunnell, P.: 1984, 'Snowmelt effects related to acidic precipitation: a structured review of existing knowledge and current research activities', Final report by ESSA Ltd. (Vancouver, B.C.) to Environment Canada.
Ontario Ministry of Environment: 1983, *Handbook of Analytical Methods for Environmental Samples*, Laboratory Services Branch, OME, Toronto, Ontario.
Price, A.G.: 1984, 'Snowmelt Estimation in a Deciduous Forest', Univ. of Toronto, unpublished manuscript.
Wilson, E. E. and Barrie, L. A.: 1981 'Climatological Study of Acidic Snowmelt Shock Potential for Eastern Canada', Proceedings of the Eastern Snow Conference, 38th Annual Meeting, pp. 23-32.

PATTERNS AND IMPORTANCE OF DISSOLVED ORGANIC CARBON IN FOUR ACIDIC BROWNWATER STREAMS IN NOVA SCOTIA, CANADA

T.A. Clair
Water Quality Branch, Inland Waters Directorate
Environment Canada, P. O. Box 861
Moncton, N. B., E1C 8N6
Canada

B. Freedman
School of Resource and Environmental Studies,
and Department of Biology, Dalhousie University
Halifax, N. S., B3J 4J1
Canada

ABSTRACT. The temporal variation and importance of dissolved organic carbon (DOC) was investigated in four brownwater streams in Nova Scotia, Canada, in basins receiving approximately 20 kg/ha/yr wet deposition of sulfate. Organic anions contributed an average of 23% of total equivalents. Within each waterbody, correlations of H+ with DOC were insignificant, whereas they were significant with sulfate. DOC was consistently correlated with concentrations Ca and Mg, suggesting that they occur as organic complexes. DOC concentrations and thus organic anions were inversely correlated with stream flow.

1. INTRODUCTION

Southwest Nova Scotia receives approximately 20 kg/ha/yr of sulfate from the atmosphere (Shaw, 1979). Populations of Atlantic Salmon from this region have also exhibited reduced survival and declining populations (Watt et al., 1983; Lacroix, 1985; Lacroix et al., 1985). A high proportion of these surface waters are acidic, (Clair et al., 1982), are low in minerals and high in dissolved organic carbon (DOC) and in organic anions (Kerekes et al., 1982).

There is controversy over the role of anthropogenic acid inputs in the acidification of brown waters in Nova Scotia, because of the presence in these surface waters of organic acids leached from bogs and soils. Brownwater streams and lakes are naturally acidified by high DOC concentrations (Clymo, 1984; Gorham et al., 1984). Some researchers thus believe that most of the atmospheric acid input is neutralized by watershed soils (Krug and Frink, 1983), while others state that natural acids are not important to water acidity during high flow periods (Lazerte and Dillon, 1984).

Water, Air, and Soil Pollution **31** (1986) 139–147.

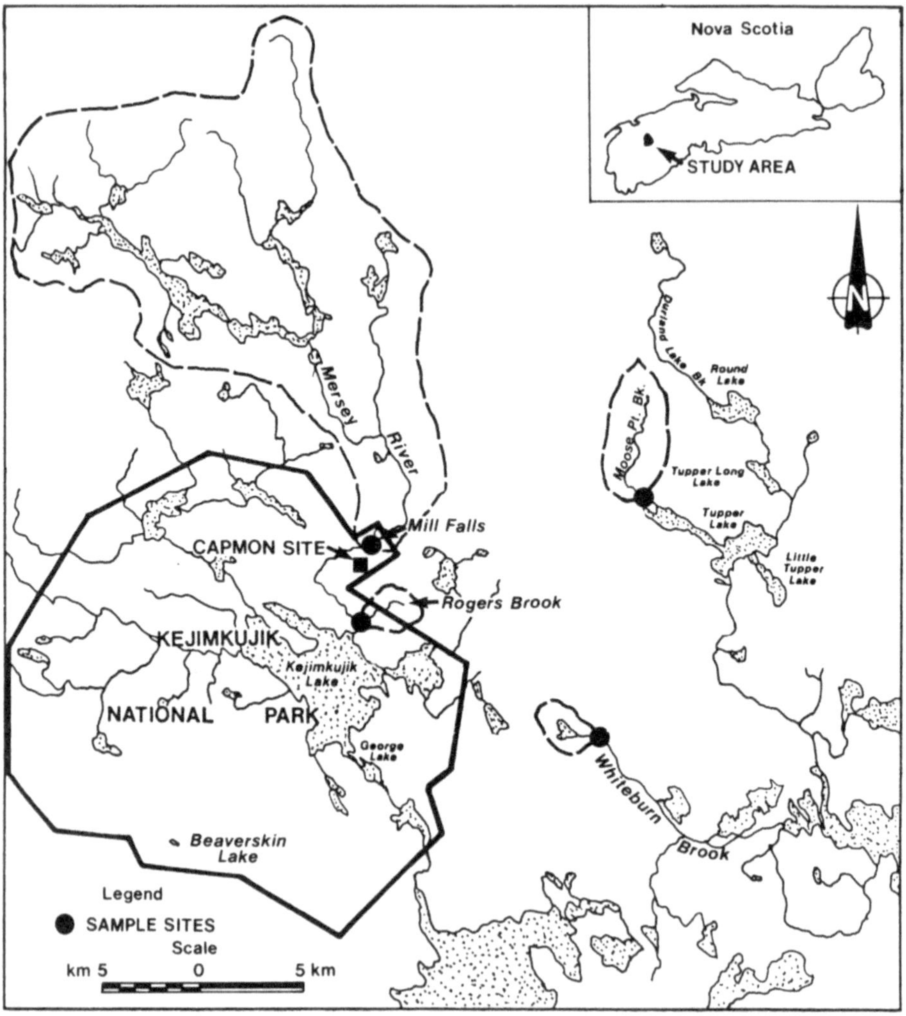

Figure 1. Location of Study Basins

The present report describes results of a study of water chemistry in four brownwater streams in Nova Scotia which were chosen to provide a range of DOC concentrations. The problems addressed include: i) the contribution of organic anions to total water anion equivalents; ii) the correlation within each stream of variations of DOC with other ions, and iii) the temporal variations of concentration and flux of DOC. We also assess the relative contributions of sulfate and organic anions, to overall acidity.

2. STUDY AREA AND METHODS

The study area is centred on Kejimkujik National Park, Nova Scotia, Canada (Figure 1). The terrestrial part of the study area is 94% covered by coniferous, deciduous, and mixed forests, with the remainder composed of wetlands, especially Sphagnum wetlands. Bedrock is variable, but mainly granite, slates with quartzite, and greywacke (Gimbarzevsky, 1975). The landscape has an almost continuous cover of shallow glacial tills, with frequent drumlin and moraine deposits. Typical surface waters in the study area are oligotrophic, and most are dark-coloured due to dissolved organic substances.

The four watersheds differ in size: the Mersey River is 295 km sq, Moose Pit Brook is 16.7 km sq, Whiteburn Brook is 9.5 km sq and Rogers Brook 8.7 km sq. All have brown, organic, acidic waters, Moose Pit Brook has a mean annual DOC = 14.1 mg/L and mean annual pH = 4.6, Mersey River, mean DOC of 9.4 mg/L and mean pH 4.9, Rogers Brook, mean DOC 7.3 mg/L, mean pH of 5.1 and Whiteburn Brook with DOC of 5.6 mg/L and pH of 5.6.

The Mersey River and Moose Pit Brook were gauged with standard Water Resources Canada automatic water level recorders. Rogers Brook and Whiteburn Brook had water level recorded manually using staff gauges, either daily or 3 to 4 times per week. Water samples were collected at least weekly in clean, 1 L Nalgene bottles. All chemical analyses were done according to standard Environment Canada NAQUADAT Procedures (Anonymous, 1984). Sulfate was analyzed by ion chromatography as erroneously high estimates of sulfate exports have resulted from the use of a methylthymol blue (MTB) technique for sulfate analysis (Kerekes et al., 1984). Organic anions (A-) were not measured, but were computed from DOC and pH following Oliver et al. (1983).

Collection of samples began in June 1980 in the Mersey River, and March 1983 for the remaining streams. In all cases, the last samples used in this analysis were collected in March 1985.

3. OBSERVATIONS AND DISCUSSION

Cation concentration in these surface waterbodies was dominated by Na (averaging 51% of total cation equivalents among the four streams), Mg (17%), and Ca (16%) (Table I). Hydrogen ion accounted for only 4% (range of less than 1 to 9%) of total cation equivalents. Anions were dominated by Cl (53% of total anion equivalents), A- (23%), and

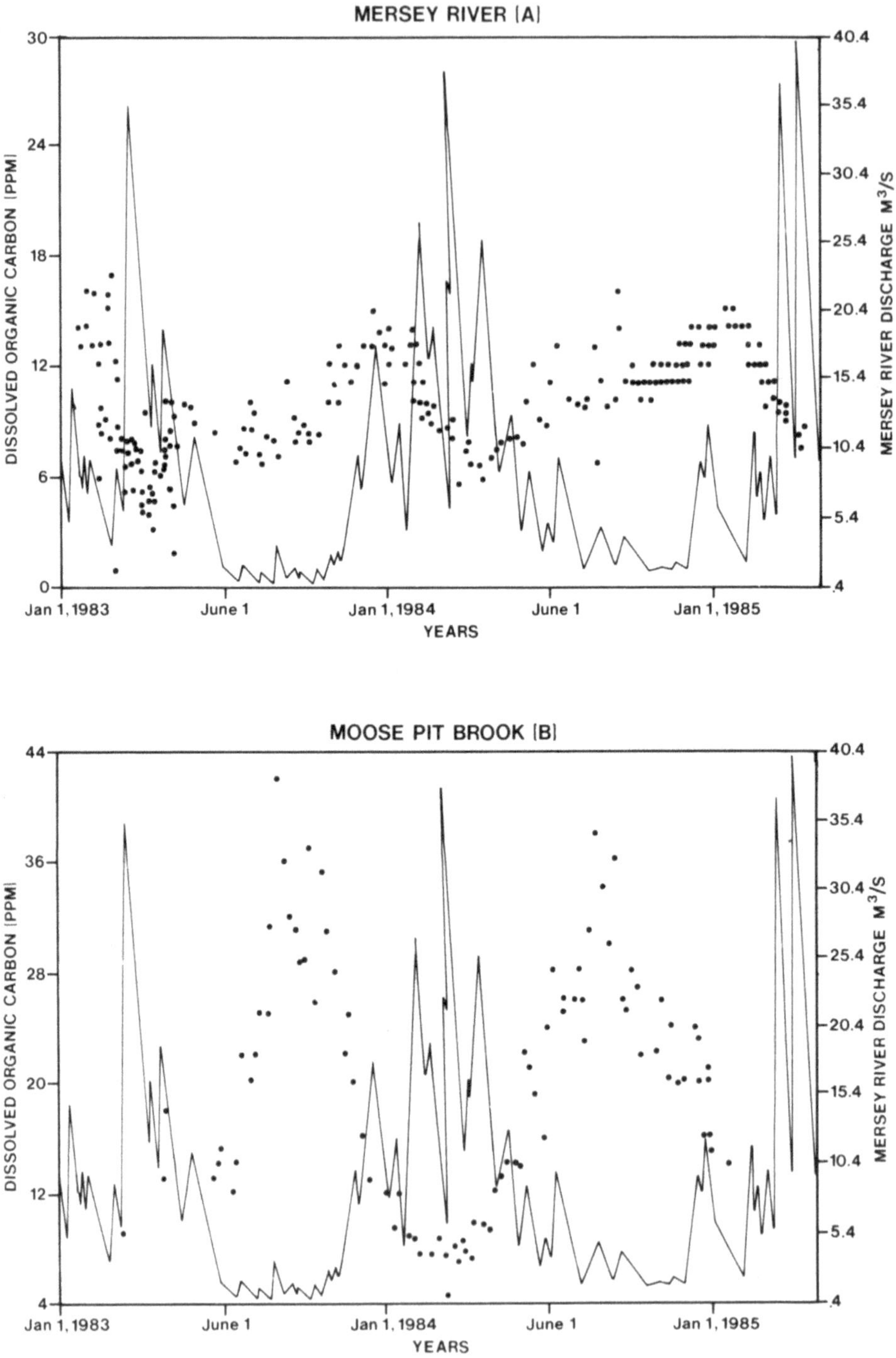

Figure 2. Dissolved Organic Carbon (DOC) Concentrations in the Four Study Streams [A], [B]

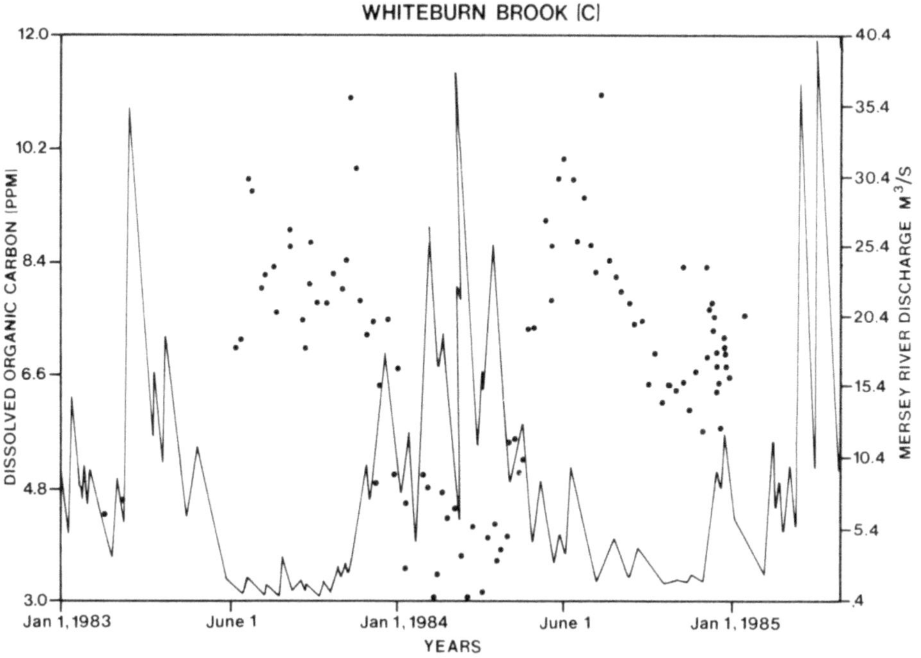

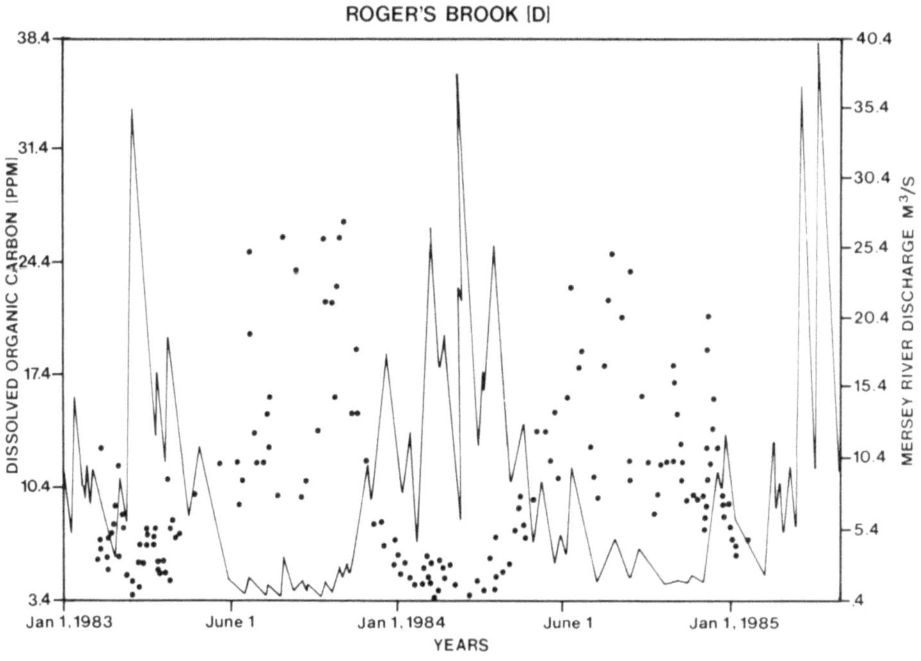

Figure 2a. Dissolved Organic Carbon (DOC) Concentrations in the Four Study Streams [C], [D]

TABLE I SELECTED WATER CHEMISTRY OF THE FOUR STUDY STREAMS - CONCENTRATION DATA ARE FLOW-WEIGHTED ANNUAL MEANS

	MERSEY RIVER	MOOSE PIT RIVER	ROGERS BROOK	WHITEBURN BROOK
Concentration (ueq/L)				
Ca	44.9	41.9	46.4	57.9
Mg	44.4	46.1	50.2	54.3
Na	141.4	118.8	137.5	197.5
H	12.0	25.0	8.1	2.3
Fe	7.5	7.2	14.3	7.2
Al	20.0	21.1	12.2	7.8
SO4	54.3	52.8	64.5	55.7
Cl	144.1	107.4	135.4	203.0
A-	69.7	98.0	47.7	37.1
Anions/cations*	0.97	0.96	0.95	0.94
DOC (mg/L)	9.4	14.1	7.3	5.6
TC (mg/L)	14.3	21.1	14.4	10.7
pH	4.9	4.6	5.1	5.6
Flux (kg/ha/yr)				
DOC	78	136	73	51
TC	118	203	150	103

* The minor constituents K and Mn are included in the cation sum, but not in the Table.

TABLE II CORRELATIONS BETWEEN SULFATE OR DOC CONCENTRATIONS IN WATER AND SELECTED CONCENTRATIONS OR FLUXES. (** = p less than 0.001; * = p less than 0.01)

	MERSEY RIVER (n=992)	MOOSE PIT BROOK (n-103)	WHITBURN BROOK (n=105)	ROGERS BROOK (n=230)
DOC X colour	0.690**	0.895**	0.816**	0.866**
X A-	0.973**	0.990*	0.948**	0.988**
X Na	0.424**	0.671**	0.179	0.525**
X Mg	0.668**	0.623**	0.311*	0.671**
X K	0.047	0.477**	0.181	0.481**
X Ca	0.533**	0.730**	0.439**	0.756**
X SO4	-0.071	-0.543**	0.475**	-0.057
X Al	0.664**	0.865**	0.400**	0.825**
X Mn	0.200**	0.631**	0.525**	0.668**
X Fe	0.395**	0.898**	0.764**	0.719**
X H	0.029	0.149	-0.122	-0.109
SO4 X H	0.799**	0.378**	0.282*	0.431**
DOC X water flow	-0.283**	-0.460**	-0.533**	-0.493**

SO_4 (21%).

Statistical relationships between DOC and other chemical constituents within each waterbody were examined by bivariate correlation analysis (Table II). As expected, DOC correlated strongly with color. The correlation with A- is also expected, since DOC is the most important variable in the calculation of A-. Other correlations with DOC were strongest for polyvalent cations, especially Ca, Mg, Al, Fe, and Mn (Table II). We interpret the correlations of these with DOC to indicate that they are occurring as organic complexes, rather than in ionic form. Clair and Komadina (1985) used the Campbell et al. (1983) chelation method and confirmed this relationship for Al in these four study basins by observing that virtually all of the total extractable Al was non-labile, and probably bound to dissolved organic matter, a form which is much less bioavailable and toxic (Baker and Schofield, 1982). This is an important difference between acidic waters high in DOC, and clear acidic waters.

Within each waterbody (as opposed to among them), variations of concentration of H+ were not correlated with variations of DOC (Table II). H+ was not consistently strongly correlated with any ionic constituent, although H+ vs. SO_4 did yield a high correlation in the Mersey River (Table II), and lower but significant correlations in the other 3 rivers. It is therefore notable that SO_4 is the most useful anion for explaining variations of H+ ion within these study streams, where organic anions play an important role in water chemistry.

Temporal variation of water discharge in all four streams is similar to that for Mersey River, though the magnitude of discharge obviously varies amongst them. The very strong seasonality in discharge occurs because of the generally higher precipitation that occurs in the fall (Ambler, 1983). Mean daily water flow remains high throughout November to mid-May, because the moderate Nova Scotian winter does not normally allow a deep snowpack accumulation, and because soils are saturated. During the growing season, low runoff is also largely due to high rates of evapotranspiration.

A strong seasonality of DOC (or A-) concentration is also apparent in all four water bodies (Figures 2a, b, c, d). In all of the streams, the tendency is for DOC concentration to be highest during the low-flow period, i.e. having the largest relative impact on water chemistry in the summer, a phenomenon also reported from the eastern United States, (Eshleman and Hemond, 1985). This trend is most marked for Moose Pit Brook (Figure 2b), Rogers Brook (Figure 2d), and Whiteburn Brook (Figure 2c). In the larger Mersey River (Figure 2a) there is a tendency for high DOC concentrations to persist into the begining of the high-flow period probably because of lag time between runoff from its first order streams to the collection site. These patterns result in significantly negative correlations between water flow and DOC concentration (Table II).

This study confirms, and expands on the work of Kerekes et al. (1984) for streams in the same area, including one waterbody which partly overlapped both studies (Mersey River). Our data also quantify the trends that Kerekes et al. (1982) have described between DOC and metal concentrations. We have also provided evidence of the importance of SO4 concentrations to the pH of waters high in DOC, supporting the work of Lazerte and Dillon (1984). Moreover, we have shown that organic acid concentrations in southwestern Nova Scotia streams are highest in the summer low flow periods in summer and/or the begining of the autumn high flow period. However, their concentrations, and thus influence are less important during high flow periods in winter.

4. ACKNOWLEDGMENTS

The authors acknowledge the help of F. Luxton, U. Prager, C. Stewart, and L. Boulter in data collection or analysis. T. Pollock critically reviewed the manuscript and the staff at Kejimkujik National Park provided access to their facilities and encouragement to the study.

REFERENCES

Ambler, D.C.: 1983, 'Surface water hydrology, Kejimkujik Calibrated Catchment Program'. Tech. Rep. IWD-AR-WRB-83-43, Environment Canada, Dartmouth, N.S.

Anonymous: 1984, 'NAQUADAT Dictionary of parameter codes'. Water Quality Branch, Environment Canada, Ottawa, 319 pp.

Baker, J.P. and Schofield, C.L.: 1982, Water, Air and Soil Pollut., 18, 289.

Campbell, P.G.C., Bisson, M., Bougie, R., Tessier, A., and Villeneuve, J.P.: 1983, Anal. Chem., 55, 2246.

Clair, T.A., Witteman, J.P., and Whitlow, S.H.: 1982, 'Acid precipitation sensitivity of Canada's Atlantic Provinces'. Tech. Bull. 124, Inland Waters Directorate, Environment Canada, Ottawa.

Clair, T.A. and Komadina, V.: 1984, 'Aluminum speciation in waters of Nova Scotia and their impact on WQB analytical and field methods'. Tech. Rep. IWD-AR-WQB-84-69. Environment Canada, Moncton, N.B.

Clymo, R.S.: 1984, Phil. Trans. R. Soc. Lond. 305, 487.

Eschleman, K.N. and Hemond, H.F.: 1985, Water Resour. Res., 21, 1503.

Gimbarzevsky, P.: 1975, 'Biophysical survey of Kejimkujik National Park'. Information report FMR-X-81. Forest Management Institute. Petawawa, Ont.

Gorham, E., Bayley, S.E., and Schindler, D.W.: 1984, Can. J. Fish. Aquat. Sci. 41, 1256.

Kerekes, J., Howell, G., Beauchamp, S., and Pollock, T.: 1982, Int.

Revue ges. Hydrobiol. 67, 679.

Kerekes, J., Howell, G., and Pollock, T.: 1984, Verb. Internat. Verein. Limnol. 22, 1811.

Krug, E.C., and Frink, C.R.: 1983, Science 221, 520.

Lacroix, G.L.: 1985, Can. J. Fish. Aquat. Sci. 42, 292.

Lacroix, G.L., Gordon, D.J., and Johnston, D.J.: 1985, Can. J. Fish. Aquat. Sci. 42, 768.

Lazerte, B.D., and Dillon, P.J.: 1984, Can. J. Fish. Aquat. Sci. 41, 1664.

Oliver, B.G., Thurman, E.M., and Malcolm, R.L.: 1983, Geochim. Cosmochim. Acta, 47, 2031.

Shaw, R.W.: 1979, Env. Sci. Technol. 13, 406.

Watt, W.D., Scott, C.D., and White, W.T.: 1983, Can. J. Fish. Aquat. Sci. 40, 462.

INTERFACING HYDROLOGICAL AND HYDROGEOCHEMICAL MODELS FOR SIMULATING STREAMWATER CHEMISTRY IN THE TURKEY LAKES WATERSHED, CANADA

D.C.L. Lam[1], S. Boregowda[2], A.G. Bobba[1], D.S. Jeffries[1] and G.G. Patry[2]

[1]National Water Research Institute
P.O. Box 5050
Burlington, Ontario, Canada L7R 4A6

[2]McMaster University
Hamilton, Ontario,
Canada L8S 4L7

ABSTRACT. The results obtained by interfacing a hydrological model with a hydrogeochemical model are presented and compared to the observed data collected from the Turkey Lakes Watershed. The most prominent feature shown in the data is the progressive increase of pH, alkalinity and Ca^{++} + Mg^{++} in the downstream direction. Analysis of the model results indicates that the proportion of groundwater input to total runoff in the headwater and downstream areas plays an important role in creating these spatial gradients. The model also reproduces many of the episodic events, both temporal and spatial, shown in the observed data.

1. INTRODUCTION

Several hydrogeochemical models (e.g. Christophersen et al., 1982; Cosby et al., 1985) have been developed to simulate the soil and water chemistry in acidified watersheds. On the other hand, hydrological models (e.g. Lam and Bobba, 1985; Seip et al., 1985) have been used to define the water pathways and contact times of the chemical ions in both the aquatic and soil regimes. Since almost all of the observations conducted in acidified watersheds pertain to daily or similar time scales, the verification of the hydrogeochemical models generally requires the interfacing with hydrological models to simulate the episodic events commonly manifested in such observations.

The purpose of this paper is twofold. The first objective is to apply an interfaced hydrogeochemical-hydrological model to a watershed small enough to receive the same sulphate loading throughout but large enough to offer several sites with different soil buffering capacity and water residence times for testing the model. The second objective is to analyze from a time-space perspective the major interactions between the hydrology and the soil and water chemistry. This type of test is necessary for understanding the processes, as most of the previous work (e.g. Seip et al., 1985) on model verification so far has been limited to one or two stations in the same watershed.

Water, Air, and Soil Pollution **31** (1986) 149–154.

2. LOCATION

The Turkey Lakes Watershed near Sault Ste. Marie, Ontario offers such a location for testing models. A distinct feature of the watershed is that it contains a series of five lakes over a total area of 10 km^2 (Figure 1). A series of sampling stations (S1-S5) are located along the main stream. The lowest elevation is 245 m and the highest, 645 m. At lower elevations, the soil depth is about 1 to 10 m and at higher elevations, about 0.2 to 1 m. Thus, the groundwater storage capacity is significantly less at the top than at the bottom of the basin. The soil is composed mainly of loam and sandy loam at higher elevations, as compared to gravelled till and fine-grained till at lower elevations. Atmospheric deposition of acid is moderate in this region, with an average pH of 4.5 and an annual sulphate load of about 20 kg SO_4 ha^{-1}.

3. MODEL INTERFACE

The hydrological model we have adopted is a three-soil-layer model which has been applied to several Canadian watersheds with satisfactory results (Bobba et al., 1986). This model has been verified with observed total runoffs, snow water equivalents and ^{18}O data. In particular, it has been verified with 3 years of observed data for each of the 5 stations (S1-S5) in Turkey Lakes (Bobba et al., 1986).

This hydrological model can be readily interfaced with several hydrogeochemical models (e.g. Cosby et al., 1985). We have chosen the model by Rustad et al. (1986). Two major modifications are needed, however. The first change is to add one more soil layer in between the top and bottom layers in the original model to make it compatible with our hydrological model. The second change to the hydrogeochemical model is to leave out the Al calculations, as they are not important for the pH range (5.5 to 7) generally observed in Turkey Lakes. Thus, the chemistry is almost reduced to the simple case of titrating Ca and $MgCO_3$ with H_2SO_4. Such a simple structure (Figure 2) is easy to follow and the interactions with the hydrology can be conveniently analyzed.

4. RESULTS

After experimenting with some model coefficients and initial values of the model variables, a calibration fit with the observed data for 1981 is obtained (Figure 3). In all five stations, it is found that some sulphate adsorption is required, as opposed to desorption in the case of the Birkenes model (Christophersen et al., 1982). By keeping the calibrated coefficients fixed, the model is then applied to the next 2 years (1982 and 1983). The results agree satisfactorily with the observed data for all five stations. However, only the detailed results for S1 and S3, as examples of headwater station and downstream station, are shown in Figures 3a and b, respectively.

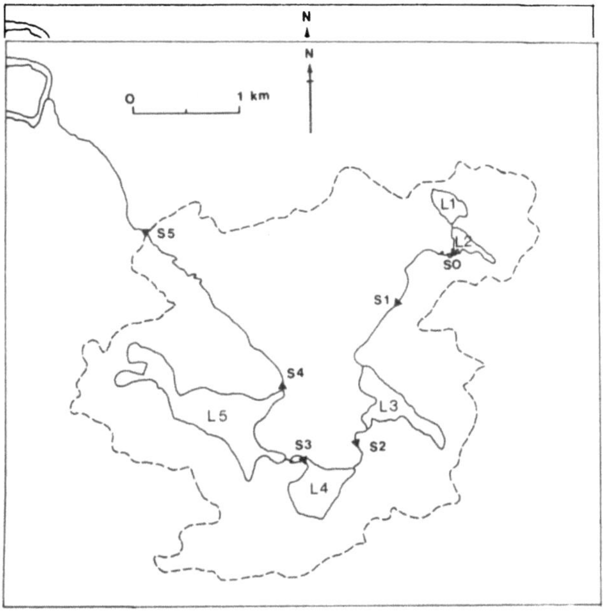

Figure 1. Location of the Turkey Lakes Watershed and the stream sampling stations, S1 - S5.

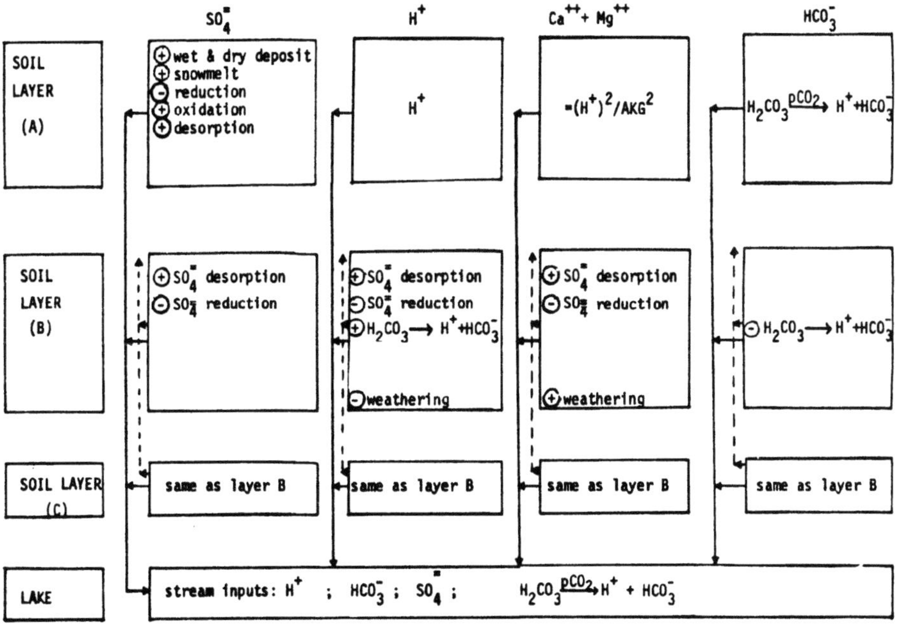

Figure 2. Interfacing hydrological and hydrogeochemical models. Arrows indicate hydrological pathways; boxes represent soil layers or water compartments in which the hydrogeochemical processes take place.

For example, snowmelt events during April and May are prominent in each year. These episodic events are simulated satisfactorily by the model and correspond to runoff peaks in these months (Bobba et al., 1986), thus underscoring the necessity for an accurate hydrological model. Indeed, the computed results tend to depart from the observed results, mainly during summertime, particularly in 1982. A possible explanation is that the low flows at summertime affect the computed concentrations strongly, because slight underestimation of the flow can produce high concentrations, and vice versa.

The spatial difference in the Ca^{++} + Mg^{++} concentration is also evident, with higher concentration at the downstream station (S3) than at the headwater station (S1). This spatial difference of concentration is consistent with the fact that higher portion in groundwater (i.e. from soil layers B and C, Figure 2) contributes to the downstream runoff than to the headwater runoff. That is, more Ca^{++} ions are dissolved from calcite at downstream areas (Figure 3b) than headwater areas (Figure 3a), as conjectured by Semkin and Jeffries (1984) and Lam et al. (1985).

5. DISCUSSION

The time-space perspective can be conveniently summarized by plotting the computed and observed results just before the snowmelt (Figure 4). Thus, the results for S1 through S5 are aligned along the "stations"-axis as well as displayed in time along the "year"-axis in Figure 4. Clearly, the $SO_4^{=}$ concentration does not change much both in time and space, but Ca^{++} + Mg^{++} and HCO_3^- do. In the case of Ca^{++} + Mg^{++}, not only is there a progressive increase in concentration downstream, but also a decline in 1983. In the case of HCO_3^-, there is a spatial gradient corresponding to that of Ca^{++} + Mg^{++}; as well, the HCO_3^- concentrations in 1982 appear to be higher than the other two years, as a result of a relatively dry year (Bobba et al., 1986). That is, drier years encourage higher groundwater portion in the stream flow, with more HCO_3^- inputs. The combined effect of Ca^{++} + Mg^{++} and HCO_3^- concentrations on the pH is also obvious, with a mild spatial gradient of pH and a slightly higher pH in 1982 (Figure 4). Detailed discussions on these spatial gradients and the pCO_2 calculations are given in Lam et al. (1985).

Of course, Figure 4 represents only one of many possible summaries of the time-space representations; other interesting perspectives can be constructed by plotting similar results at snowmelt time or summertime. As indicated on Figure 4, the computed results produce essentially similar time-space patterns as the observed. It is beyond the scope of this paper to examine the methods of evaluating the errors between the computed and observed results. As an example, however, the mean relative least-squares errors for $SO_4^{=}$, pH, HCO_3^- and Ca^{++} + Mg^{++} are 4.9%, 21.7%, 10.3% and 8.5%, respectively, for the case just before the snowmelt as given in Figure 4. These error estimates conform closely to those found by Seip et al. (1985).

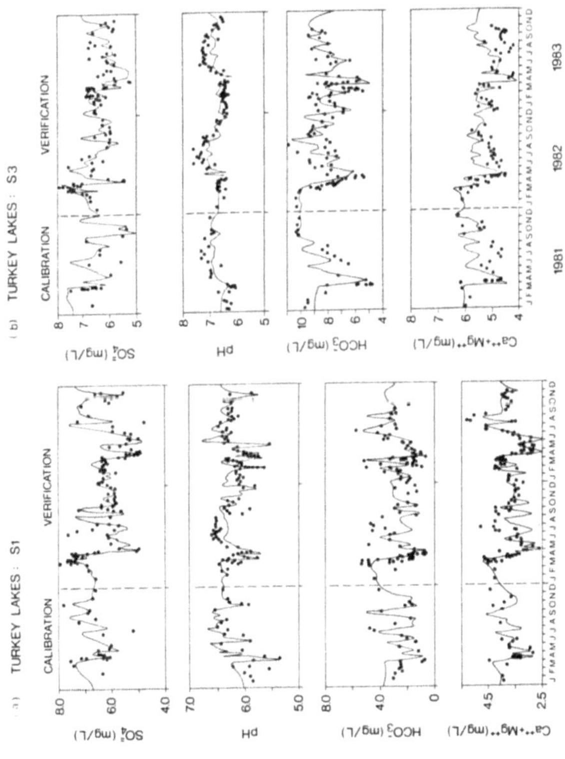

Figure 3. (a) Computed and observed results (1981-1983) for station S1; (b) for station S3.

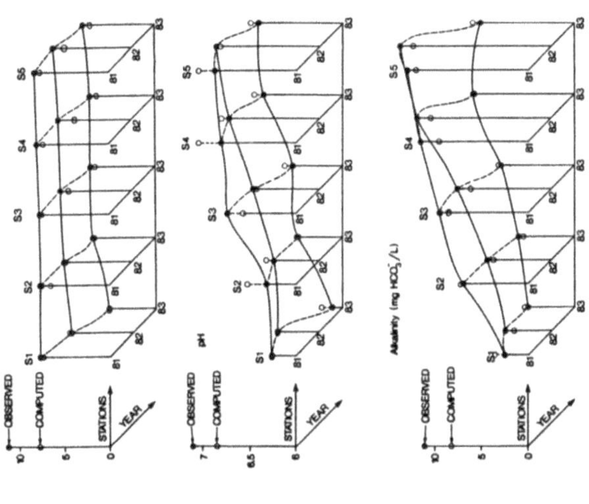

Figure 4. Computed and observed results just before snowmelt for 1981-1983 for all five stations (S1-S5), presented in a time-space perspective.

6. CONCLUSIONS

The interfaced hydrological-hydrogeochemical model satisfactorily simulates the temporal and spatial changes as observed in Turkey Lakes. In particular, the spatial gradients in pH, alkalinity and Ca^{++} + Mg^{++} are successfully reproduced by the model. At this stage of model development, it is crucial to examine the existing models by comparing the computed results to the observed data from a diversity of watersheds with different soil and water properties as well as sulphate inputs. The case of the Turkey Lakes Watershed is a step in such a direction. It is anticipated that improvements in the models will be made in such areas as Al speciation, organic acids, HNO_3 and alkalinity generation as data from ongoing studies in Canada and elsewhere become available for model calibration and verification.

ACKNOWLEDGMENTS

The authors are grateful to Drs. H.M. Seip and N. Christophersen for providing a programme listing of their model as well as many useful discussions. Discussions with F.C. Elder, M.E. Thompson and R.G. Semkin were also helpful. Two co-authors, S. Boregowda and G.G. Patry, were supported by a research contract grant from Environment Canada, No. 02SE.KW405-4-1662.

REFERENCES

Bobba, A.G., Lam, D.C.L., Jeffries, D.S., Bottomley, D., Charette, J.Y., Dillon, P.J. and Logan, L.: 1986, Modelling the hydrological regimes in acidified watersheds. In Proc. Int. Symp. on Acicic Precipitation, Muskoka Conf. '85, Ont. (submitted).

Christophersen, N., Seip, H.M. and Wright, R.F.: 1982, Water Resour. Res., **18**, 977.

Cosby, B.J., Wright, R.F., Hornberger, G.M. and Galloway, J.N.: 1985, Water Resour. Res., **21**, 51.

Lam, D.C.L. and Bobba, A.G.: 1985, Modelling watershed runoff and basin acidification. In Johansson, I. (ed.) Hydrological and Hydrogeochemical Mechanisms and Model Approaches to the Acidification of Ecological Systems, Nordic Hydrological Programme, NHP-Report No. 10, p. 205-215.

Lam, D.C.L., Bobba, A.G., Jeffries, D.S. and Kelso, J.M.R.: 1985, Relationships of spatial gradients of primary production, buffering capacity and hydrology in Turkey Lakes Watershed. ASTM Special Technical Publication, Philadelphia, Pa., STP#928, p. 42-53.

Rustad, S., Christophersen, N., Seip, H.M. and Dillon, P.J.: 1986, Can. J. Fish. Aquat. Sci., **43**, 625.

Seip, H.M., Seip, R., Dillon, P.J. and de Grosbois, E.: 1985, Can. J. Fish. Aquat. Sci., **42**, 927.

Semkin, R.G. and Jeffries, D.S.: 1984, Relationships between hydrological conditions and the ionic composition of stream waters in Turkey Lakes Watershed. In Proc. Canadian Hydrology Symposium, U. of Laval, Quebec, Canada, p. 109-122 .

MODELLING THE HYDROLOGICAL REGIMES IN ACIDIFIED WATERSHEDS

A.G. Bobba[1], D.C.L. Lam[1], D.S. Jeffries[1], D. Bottomley[2], J.Y. Charette[3], P.J. Dillon[4] and L. Logan[4]

[1]National Water Research Institute
P.O. Box 5050
Burlington, Ontario, Canada
L7R 4A6

[2]National Hydrology Research Inst.
Inland Waters Directorate
Environment Canada,
Ottawa, Ontario, Canada K1A 0E7

[3]Inland Waters Directorate, Q.R.
P.O. Box 10,100
St. Foy, P.Q., Canada G1V 4H5

[4]Ontario Ministry of Environment
Dorset Research Centre
P.O. Box #39
Dorset, Ontario, Canada P0A 1E9

ABSTRACT. Application of a hydrological model to three Canadian watersheds forms part of a verification package to test its accuracy and portability. Agreement of model results with observed data on hydrograph, groundwater flow and snowpack is reasonably good. The model results are consistent with the observed pH and with many of the episodic events that have occurred in these acidified watersheds. Contrasts have also been made on the different calibrated coefficients at several locations in these watersheds. An attempt to relate them to the geology and soil characteristics at the site has led to realistic estimation of the soil contact times.

1. INTRODUCTION

As pointed out by H.M. Seip during a recent workshop on predicting soil and water acidification (Johnson et al., 1985), a satisfactory knowledge of the hydrology is extremely important in understanding acidification. We have adopted a modeling approach (Lam and Bobba, 1985) in which both the accuracy of the hydrological model and its linkage to hydrogeochemical models are emphasized. The hydrological model we have developed (Bobba and Lam, 1984) includes these considerations.

The purpose of this paper is to report on the comparison of model results with the flow data observed in the soil layers and water compartments such as those in groundwater, snowpack and streams. Several Canadian watersheds have been chosen for confirmation purposes in order to test the general applicability of the model. In particular, we must know how the model coefficients change from watershed to watershed. Can they be related to the soil types? Do they conform to the

Water, Air, and Soil Pollution **31** (1986) 155–163.

kinematic rates measured in the laboratories or the field? It is only when questions on the accuracy of the flow predictions and versatility of the hydrological model coefficients are answered quantitatively, that we are able to link up the hydrological and hydrogeochemical models.

2. EVALUATION PROCEDURES FOR THE HYDROLOGICAL MODEL

The description of the hydrological model has been given in Bobba and Lam (1985). Briefly, the model expresses the watershed as a set of reservoirs of determinable capacities that hold water temporarily and gradually recede as their contents are diminished by infiltrations, recharge, evapotranspiration and lateral drainage.

The structure of the hydrological model, is comprised of three distinct soil regimes: upper soil zone (A), lower soil zone (B) and groundwater zone (C), corresponding to the surface runoff, interflow and groundwater flow, respectively. The sum of these three flows then constitutes the total basin runoff.

The separation of runoff into three components is a salient feature of our model, not only because of the physical considerations but also because we can conveniently associate them in the known regions of chemical processes. For example, generally speaking, humus formation may occur in the upper soil zone, cation exchange in the upper and lower soil zone and soil weathering in the groundwater soil zone.

In this paper, we aim at comparing the computed results to the observed data. As daily measurements are made on the total runoffs, the calibration and confirmation procedures mainly use such data sets. During model calibration, the data for a particular year and a particular watershed are chosen to calculate the sum of squares of the errors between the computed and observed results. Optimization procedures such as the gradient search method (Bobba and Lam, 1984) are used to find the set of parameter values for a number of watersheds (see Table I) that produces the minimal value of this sum of squares of errors. During model confirmation (Seip et al., 1985), observed data for other years are compared to results computed with the calibrated parameters. To quantify the model performance, three statistical methods are used: (a) the rank order correlation, (b) the mean relative error and (c) the coefficient of efficiency. In the rank order correlation, the correlation coefficient (r) and regression slope (s) are derived from the ordered ranks of both the computed and observed hydrographs and are purported to be good indicators of the episodic coincidences. The mean relative error (e) is given by

$$e = \frac{1}{N} \sum_{i=1}^{N} \left| q_i - \hat{q}_i \right| / q_i \qquad (1)$$

where q_i is the observed flow, $\hat{q}_i$ is the computed flow and N is the total number of data points. While the error e is intended as an averaged quantifier, it is strongly affected by low values of q_i. The coefficient of efficiency (Nash and Sutcliffe, 1970) is defined as

$$E \left[\sum_{i=1}^{N} (q_i - \bar{q})^2 - \sum_{i=1}^{N} (q_i - \hat{q}_i)^2 \right] / \sum_{i=1}^{N} (q_i - \bar{q})^2 \qquad (2)$$

where $\bar{q}$ is the mean of the observed runoffs. That is, the mean observed value, $\bar{q}$, is used as a benchmark so that E is a measure of the improvement on the variance not explainable by the mean value, but explainable by the particular model which produces the $\hat{q}_i$'s(Seip et al., 1985).

3. RESULTS

3.1 Turkey Lakes Watershed

Figure 1 shows the calibration (1981) and confirmation (1982-83) results for the total runoff at a headwater stream station, S1, at the Turkey Lakes watershed, near Sault Ste. Marie, Ontario, Canada. The computed hydrograph produces a satisfactory fit with the observed data (e.g. r=0.76 and s=0.85 for the confirmation period, see Table II). In particular, the episodic events during spring snowmelts for all 3 yr are accurately simulated as well as other episodes due to heavy rainfall (see Figure 1, e.g. June, 1981; October, 1982). The magnitudes at these peaks are also predicted reasonably well (e.g. e=11.3% with an improvement of 48% over the mean-as-model for the confirmation period). Similarly, the computed hydrograph (not shown here) for a downstream station, S3, at the same watershed produces a good fit with the observed data (e.g. r=0.89, s=0.88 for the confirmation period, see Table II). Note that the calibrated constants for infiltration (between top and middle layers) for S1 anS3 are several orders of magnitude different (Table I) and so are the calibrated constants for deep infiltration (between middle and bottom layers). These differences are due to the more compact clayey till texture, the shallower soil depths and the sometimes exposed bedrock at S1 (Table I), typical of many headwater watersheds in the area.

Also shown on Figure 1 are the observed pH data. As is well known, these pH values show depressions at the spring snowmelts around April of each year (Figure 1), because the sulphate ions accumulated in the snowpack are released during those episodes. Detailed discussions on the relationship between hydrology and the stream chemistry are given in Lam et al. (1986).

The model has been applied to several other stations in the watershed with satisfactory results. In particular, the simulated groundwater discharge has also been compared with observed groundwater data collected at yet another station (Norberg Creek). The groundwater flow data was estimated by the ^{18}O isotope method (Bottomley et al., 1984). As shown in Figure 2, the computed groundwater results conform well with the estimated.

3.2 Lac Laflamme Watershed

As in the case of Turkey Lakes simulation, model calibration is based on 1981 data and model confirmation, on 1982-83, for Lac Laflamme in

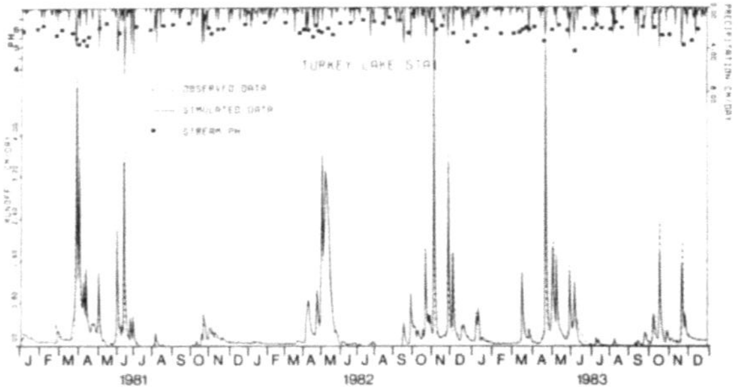

Figure 1.Simulated and observed runoff with pH at station,Turkey lakes watershed(calibration.1981, confirmation.1982-83).

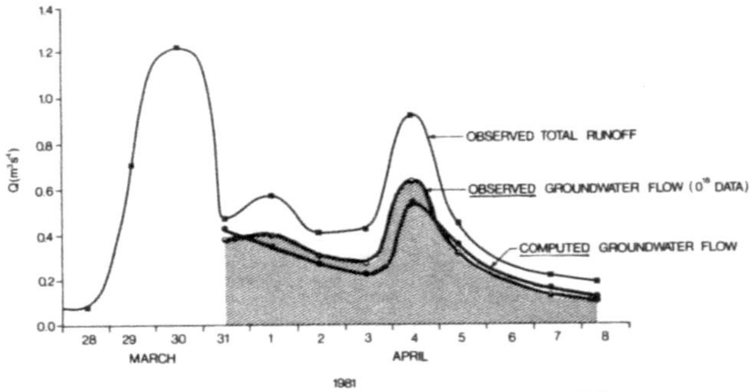

Figure 2.Observed total runoff, observed groundwater flow(O^{18}data) and computed groundwater flow at Norberg Creek,Turkey Lakes.

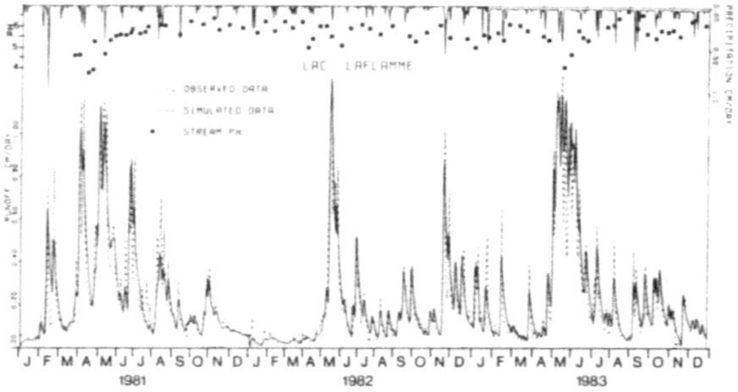

Figure 3.Simulated and observed runoff with pH at Lac Laflamme (calibration.1981,confirmation.1982-83)

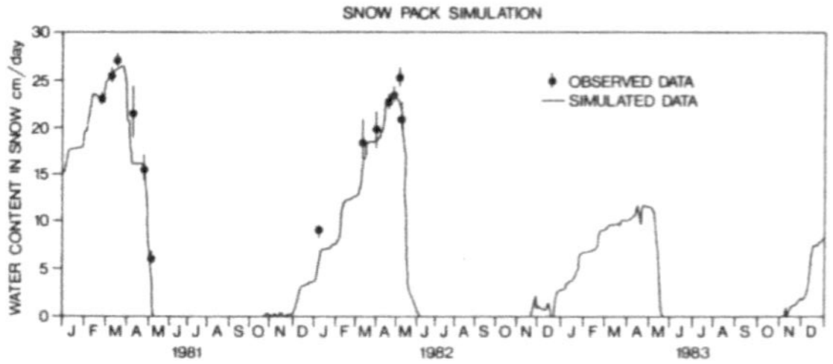

Figure 4.Simulated and observed water content in snowpack at Lac Laflamme watershed.

Quebec, Canada (Figure 3). The model simulates quite successfully the snowmelt runoff in April and May and the rain episodes in summer and fall (e.g. r=0.83, s=0.93 for the confirmation period, Table II). Again, the observed pH drops at snowmelt times. The timing and the shape of the computed hydrograph fits well with the observed (e.g. an improvement of 83% over the mean-as-model for the confirmation period, Table II). Note that there are more summer fluctuations in Lac Laflamme than in Turkey Lakes.

As part of the investigation on the accuracy of the submodel components, we have also compared the computed water equivalent of the snow pack with the observed (Figure 4). The observed snow accumulation and thawing sequences are reproduced very closely by the results of the snow accumulation and ablation submodel, with correlation coefficient of 0.99 for the calibration period and 0.96 for the confirmation period. Thus, both the snow portion and the stream portion are simulated correctly for Lac Laflamme, based on these statistical evaluations.

In general, the computation (not shown in Figure 4) shows that about 50% surface runoff contributes to the total, cf. 70% at S1 and 30% at S3 in Turkey Lakes (Lam and Bobba, 1985).

3.3 Harp Lake

Simulation of the hydrograph has been carried out for Harp Lake (station 4) by Seip et al. (1985) using a two-soil-layer model with the piston flow concept. Here, we applied our model to the same station, using 1977 data for model calibration and 1978-1982 data for model confirmation. We managed to produce, for example, for station 3A an improvement of 76% over the mean-as-model for the confirmation period (see Table II), using our three-layer model without applying the piston flow. The timing of the episodes is properly simulated, but at times (e.g. April, 1979 and April, 1980) the magnitudes show some departure (underestimated by about 40 to 60%) from the observed hydrograph and reflect a relatively larger mean relative error of e=24.6% for the confirmation period (Table II). When we rerun the model for another catchment (station 4) in the same watershed, similar shortcomings (e.g. e=25.6% for the confirmation period, Table II) occur again. Since the same kind of mismatching also occurs in Seip et al.'s model, it is possible that the input meteorological data were inaccurate or some unknown processes (e.g. beaver dams) have not been accounted for by both models. Otherwise, our computed results agree satisfactorily with the observed data (e.g. an improvement of 73% over the mean-as-model for the confirmation period at station 4, Table II).

4. DISCUSSIONS

During the calibration, the model coefficients are set by mathematically minimizing the least-squares variance between the computed and observed results for each watershed. While the procedure itself is clearly mathematical, actual observed data have been used to influence the minimization. Therefore, it is interesting to see how much realism has been instilled upon such coefficients, how they perform in the confirmation period and how they differ from watershed to watershed.

Table I lists the calibrated hydrological coefficients for the various stations in the three Canadian watersheds. For example, the infiltration coefficient which regulates the flow from the upper soil zone to the lower soil zone ranges between 1 to 2 cm day^{-1} for all locations except station S1 in Turkey Lakes. However, the deep infiltration coefficient which regulates the flow from the lower soil zone to the groundwater zone varies substantially between 0.00004 to 0.042 cm day^{-1}. Yet, these values conform to the values reported in the literature (e.g. Freeze and Cherry, 1979) for the type of geology concerned. For example, the soil zone B of station S1 in Turkey Lakes pertains to headwater glacial soil of clay-silty loam texture, whereas downstream soil such as station S3 has a larger deep infiltration coefficient because of the silty sand texture (Table I). On the other hand, the calibrated deep infiltration coefficient in Lac Laflamme conforms to the values reported for compacted till with pockets of morain and sand (Freeze and Cherry, 1979) which is more permeable than those in Turkey Lakes (Table I). In the case of Harp Lake, although both stations are covered by silty sand, the soil at station 3A is less sandy (soil zone B, Table I) and therefore less permeable than at station 4. Thus, generally speaking, the calibrated infiltration coefficients are consistent with the soil type of these watersheds.

The half-life water residence times of the three soil reservoir can be computed (Table I) using the flow coefficients. In all cases as expected, the groundwater reservoir has a longer residence time. Water tends to flow from soil zone C laterally to the stream more in Turkey Lakes than in the other two watersheds (see groundwater flow coefficients, Table I) because of the difference in soil permeability as discussed above. Thus, the headwater area (S1) in Turkey Lakes has a faster residence time and the more sandy station 4 in Harp Lake has a slower one. These estimated residence times are therefore also consistent with the geology of these watersheds.

Table II shows the detailed results of the three statistical evaluation procedures for all the five testing sites. At the outset, we must stress that it is difficult to quantify the model performance with just one number. As shown in Table II, for example, the confirmation result for Turkey Lakes S1 scores the smallest mean relative error (e=11.3%) but yet achieves the least improvement (E=0.48) over the mean-as-model among the five cases. On the other hand, the confirmation result for Harp Lake station 4 has the largest mean relative error (e=25.6%) but commands a respectable improvement (E=0.73) over the mean-as-model. This discrepancy indicates that the statistical results must be interpreted with considerable care. Thus, since the mean relative error is strongly biased towards low flow values (Equation (1)), it is not effective in quantifying the model performance at peak flows. Hence, one should rely on the rank correlation coefficient (r) or the coefficient of efficiency (E) for assessing the capability of the model for simulating episodic events. As shown in Table II, both the values of r and E indicate accurate calibration results for all five sites. More importantly, the table also shows that the model coefficients, one calibrated for each site (Table I) are able to produce results of comparable accuracy using new data for other years at the same site during the confirmation period.

Table I Geology and the calibrated model constants.

		Turkey Lakes Watershed		Lac Laflamme	Harp Lake	
		Station #S1	Station #S3		Station #3A	Station #4
Area (km^2)		2.05	3.62	0.68	0.22	1.39
Slope		steep	steep	medium	gentle	gentle
Geology:	A	clay-loam	sandy-silt loam	silty loam	silty loam	sandy loam
	B	clay-silty loam with glacial till	silty-sand with glacial till	silty loess with glacial till	silt with sand	sand with silt
	C	very compacted till with weathered bedrock	glacial till with gravel and sand	compacted till with gravel and sand	till on top of weathered bedrock	till on top of weathered bedrock
Soil Depth (m):	A	0.1	0.25	0.5	0.25	0.25
	B	0.5	1.5	0.75	0.25	0.5
	C	1.0	3.5	5.5	0.5	1.5
Calibrated Model Constants:						
Infiltration (cm d^{-1})	A to B	0.0002	1.00	1.50	1.20	2.00
Deep Infiltration (cm d^{-1})	B to C	0.00004	0.025	0.042	0.0045	0.030
Surface Flow (d^{-1})	from A	0.55	0.54	0.47	0.85	0.69
Interflow (d^{-1})	from B	0.35	0.30	0.08	0.50	0.03
Groundwater Flow (d^{-1})	from C	0.25	0.20	0.06	0.06	0.008
Computed Half-Life Residence Time (d):	A	1.3	1.3	1.5	0.8	1.0
	B	2.0	2.3	8.7	1.4	23.1
	C	2.8	3.5	11.2	11.2	86.6

(A=upper soil zone, B=lower soil zone, C=groundwater soil zone).

Table II. Statistical evaluation of computed hydrograph with observed runoff data.

Station	CALIBRATION					CONFIRMATION				
	Year	Rank-Correlation Coefficient (r)	Rank-Correlation Slope (s)	Mean Relative % Error (e)	Coefficient of Efficiency (E)	Year	Rank-Correlation Coefficient (r)	Rank-Correlation Slope (s)	Mean Relative % Error (e)	Coefficient of Efficiency (E)
Turkey Lakes (S1)	1981	0.85	0.93	18.4	0.50	1982-83	0.76	0.85	11.3	0.48
Turkey Lakes (S3)	1981	0.91	0.91	15.5	0.78	1982-83	0.89	0.88	17.2	0.71
Lac Laflamme	1981	0.94	0.92	12.8	0.85	1982-83	0.83	0.93	18.5	0.83
Harp Lake (s.3A)	1977	0.89	0.81	14.1	0.74	1978-82	0.80	0.80	24.6	0.76
Harp Lake (s.4)	1977	0.86	0.83	14.8	0.64	1978-82	0.84	0.86	25.6	0.73

Table III. Sensitivity analysis: performance of calibrated models at other sites.

	Turkey Lakes S3 Data (1981)	Lac Laflamme Data (1981)	Harp Lake s3A Data (1977)
Turkey Lakes S3 model	0.91 (0.78)	0.80 (0.62)	0.85 (0.65)
Lac Laflamme model	0.91 (0.62)	0.94 (0.85)	0.81 (0.69)
Harp Lake s.3A model	0.82 (0.60)	0.84 (0.62)	0.89 (0.74)

[The numbers are the rank correlation coefficient r and the coefficient of efficiency E (in brackets).]

Table III shows the results of a preliminary sensitivity analysis of the hydrological model. In this test, we applied the Turkey Lakes S3 model, i.e. that model which uses the five constants (see Table I) calibrated with the Turkey Lakes S3 data, to two other watersheds with similar geology (see Table I), namely Lac Laflamme and Harp Lake station 3A. Conversely, we also applied the Lac Laflamme model to Turkey Lakes S3 and Harp Lake s.3A, and the Harp Lake s.3A model to Lac Laflamme and Turkey Lakes S3. The computed results are then compared to the observed data used during the calibration (e.g. the 1981 data for Turkey Lakes S3) at the particular watershed so applied. The comparisons are made with the rank correlation coefficient r and the coefficient of efficiency E as presented in Table III. Consider these statistical coefficients in Table III as 3x3 entries of a matrix formed by the possible applications of three different calibrated models to three different watersheds. Along the diagonal of such a matrix, each model is applied to the watershed originally calibrated and hence the performance indicators are, of course, the best because of the use of the optimization procedures. The main result, however, is that the off-diagnal entries, i.e. where a model calibrated with data from one watershed is applied to another, do not indicate much worse performance either. For example, as shown in Table III, the lowest score in terms of the coefficient of efficiency occurs when the Harp Lake s.3A model is applied to Turkey Lakes S3, but, even then, the improvement over the mean-as-model is an impressive 60%. In other words, these models are quite portable in the sense that they can be used among the three sites with at least 60% improvement over the mean-as-model.

However, we do not expect these models to be so portable for sites with substantial geological differences. For example, when we apply the Lac Laflamme model to the Turkey Lakes S1 site (not shown in Table III, but see Figure 1), the rank correlation coefficient drops from the original r=0.85 (see Table II) for the model calibrated with Turkey Lakes S1 data to r=0.70 for the Lac Laflamme model. Conversely, when the Turkey Lakes S1 model is applied to Lac Laflamme, the original r=0.94 is downgraded to r=0.62 (not shown in Table III). Thus, the present hyrological model is more sensitive for sites with substantially different geology, since the calibrated model constants would be so much different (e.g. Lac Laflamme vs. Turkey Lakes S1, see Table I).

6. CONCLUSIONS

Confirmation of the hydrological model developed for acidified watersheds in Canada has produced quite encouraging results showing reasonable agreement with the observed data on stream flow, groundwater and snowpack. The model is also shown to be fairly portable for different watersheds, as long as there are sufficient and reliable observed data for model calibration. The calibrated model coefficients are fairly consistent with known geological characteristics at these watersheds.

The stream pH and chemistry are certainly affected by the hydrology, and the residence times estimated at these sites are useful for explaining the apparent differences in water chemistry and soil contact times. Most importantly, these test results have encouraged us to link the hydrological model to hydrogeochemical models for studying acidification problems in Canadian watersheds (Lam *et al.*, 1986).

REFERENCES

Bobba, A.G. and Lam, D.C.L.: 1984, Application of linearly distributed surface runoff model for watershed acidification problems, Proc. *Canadian Hydrology Symposium*, No. 15-1984, Vol. 1, pp 381-398.

Bottomley, D.J., Craig, D. and Johnston, L.M.: 1984, *Journal of Hydrology*, **75**, 1.

Freeze, R.A. and Cherry, J.A.: 1979. *Groundwater*. Prentice-Hall, Inc., Englewood Cliffs, N.J., USA, 07632.

Johnson, D.W., Nilson, I.S., Reusse, J.O., Seip, H.M. and Turner, R.S.: 1985, Predicting soil and water acidification. Proceedings of a workshop, Oak Ridge National Lab., ORNL/TM-9258, Tennessee, 47 p.

Lam, D.C.L. and Bobba, A.G.: 1985, Modelling watershed runoff and basin acidification. In Johansson, I. (ed.), *Hydrological and Hydrogeochemical Mechanisms and Model Approaches to the Acidification of Ecological Systems*, Nordic Hydrological Programme, NHP-Report No. 10, pp 205-215.

Lam, D.C.L., Boregowda, S., Bobba, A.G., Jeffries, D.S. and Patry, G.G.: 1986, Interfacing hydrological and hydrogeochemical models for simulating stream chemistry in the Turkey Lakes watershed, Canada. In *Proc. Int. Symp. on Acidic Deposition*, Muskoka Conf. 1985, Ontario (accepted).

Nash, J.E. and Sutcliffe, J.V.: 1970, River flow forecasting through conceptual models, 1. A discussion of principles. *J.Hydrol.*, **10**, 282.

Seip, H.M., Seip, R., Dillon, P.J. and de Grosbois, E.: 1985, *J. Fish. Aquatic. Science*, **42**, 927.

ORGANIC VERSUS ANTHROPOGENIC ACIDITY IN TRIBUTARIES OF THE KEJIMKUJIK WATERSHEDS IN WESTERN NOVA SCOTIA

J. Kerekes, S. Beauchamp, R. Tordon and C. Tremblay
Canadian Wildlife Service
Environment Canada
c/o Department of Biology
Dalhousie University
Halifax, Nova Scotia
B3H 4J1

T. Pollock
Water Quality Branch
Environment Canada
Post Office Box 861
Moncton, New Brunswick
E1C 8N6

ABSTRACT: Dilute waters (Ca = 0.3 to 0.8 mg L^{-1}) respond by depressed pH levels throughout the year to existing atmospheric wet deposition of sulphate (20 kg $ha^{-1}yr^{-1}$). This occurs in southwest Nova Scotia particularly during the cold, wet season when runoff is high. Colored waters of similar Ca levels receiving runoff from peaty catchments exhibit pH values one unit lower (4.7 to 4.1) than those of catchments of similar Ca levels but free from peat deposits (pH > 5.3). In colored streams sulphate and organic anions show opposing cyclic patterns while the negative gran alkalinity is the near mirror image of sulphate. Sulphate anion peaks during the high spring discharge when the organic anion concentration is lowest. Sulphate suddenly drops during the summer months during periods of high evapotranspiration, low water table and runoff, when the concentration of organic anions is highest. Both Al and Fe follow cyclic patterns similar to that of the organic anions while H^+ reflects that of the sulphate anion. Hydrogen ion concentration is always higher than that of sulphate in very colored waters, particularly during the summer months when organic anion concentrations are very high. Analysis of data in the tributaries in the Kejimkujik watersheds indicates that while considerable organic acidity is present in colored waters, anthropogenic sulphate further increases the free acidity of these waters, particularly at times of high discharge.

1. INTRODUCTION

It has been shown that dilute waters (Ca < 0.8 mg L^{-1}) in southwest Nova

Water, Air, and Soil Pollution **31** (1986) 165–173.

Scotia exhibit depressed pH levels throughout the year (pH 5.6) under current wet deposition of excess sulphate of ~20 ka $ha^{-1}yr^{-1}$ (Kerekes, 1980; Kerekes et al. 1982). Colored surface waters high in dissolved organic carbon (DOC) receiving drainage from peaty catchments show pH values approximately one unit lower than those of well drained Sphagnum-free catchments of similar Ca or Mg levels. Kerekes and Pollock (1983) and Kerekes et al. (1984) showed that organic anions, (A^{-1}), (Oliver et al., 1983) may constitute 50% of the anions in these waters. Recently, the virtual disappearance of Atlantic salmon (*Salmo salar*) was reported from several dilute, highly colored rivers in the area (Farmer et al., 1980; Watt, 1981). It could be argued that due to recent increases in soil acidification, an increase in organic acidity rather than anthropogenic acid deposition might be responsible for the decline of these fish populations in waters of high organic acidity (e.g., La Zerte and Dillon, 1984). Consequently, there is a need to elucidate the relative contribution of organic vs anthropogenic acidity in surface waters in southwest Nova Scotia, and to clarify whether a reduction of atmospheric acid deposition could reduce the acidity of these highly organic waters.

2. STUDY SITE AND STUDY PERIOD

Three streams, the outflows of two headwater lakes (Beaverskin and Pebbleloggitch) and one tributary (Atkins Brook) of Kejimkujik Lake, located in southwest Nova Scotia, are considered. The catchment of Beaverskin Outflow is essentially free from organic soils and the outflow has clear water (color = 5 Hazen units) while Pebbleloggitch Outflow (60 to 200 Hazen units) receives colored, organic water from an extensive bog which once was part of the lake, and Atkins Brook (color = 60 to 340 Hazen units) drains a peaty , organic meadow. The two lake outflows are representative of a more controlled, less fluctuating hydrological regime, while Atkins Brook,lacking any standing water body along its course, responds quickly to precipitation events.

The study area, the lakes and the climatic features are described by Kerekes et al. (1982), though the catchment of Atkins Brook, 15.0 km^2, was not previously included. While the Kejimkujik calibrated catchments have been investigated since 1978, only data covering the period of November 1981 to May 1983, which includes the June 1982 to May 1983 water year, are presented because ion chromatographic (IC) sulphate determinations are only available since November 1981. It has been pointed out by Kerekes and Pollock (1983) and Kerekes et al. (1984) and subsequently confirmed by Cheam et al. (1985) that only IC determinations give reliable values for sulphate determinations in highly colored organic waters such as those discussed here.

3. METHODS

Water samples were collected directly in polythylene bottles, and were stored until analysis for color (within a week) and ionic constituents (within 2 mo.). The analytical procedures used are given in Table I.

Table I: Analytical methods employed. Code numbers refer to the NAQUADAT dictionary of codes of detailed procedures. (Anon. 1983).

Parameter	Method	NAQUADAT Code
pH	glass electrode	10301L
Ca	atomic absorption	20110L
Mg	atomic absorption	12107L
Na	flame emission	11103L
K	flame emission	19103L
Al	atomic absorption	13305P
Fe	atomic absorption	26304P
SO_4	ion chromatograph	16309L
Cl	ion specific electrode	17205L
N- (NO_3,NO_2)	colorimetric	07110L
DOC	U.V. digestion, colorimetric	06104L
Color	Pt-Co	02011L
Gran Alkalinity		10110L
A- (Organic Anion)	Oliver *et al.* (1983)	

Dissolved organic carbon, when not measured directly was estimated from water color using the relationship (DOC mg L^{-1} = 0.115 x Color + 2.7) developed from a survey of 49 bodies of water (color range 5 to 290 Hazen units, DOC range 3.0 to 36.0 mg L^{-1}) in Kejimkujik National Park (Kerekes, 1981).

4. RESULTS

The hydrological discharge of Upper Mersey River, a major inflow of Kejimkujik Lake, can be viewed as representative of the discharge pattern of Atkins Brook (Figure 1). The annual mean monthly concen-

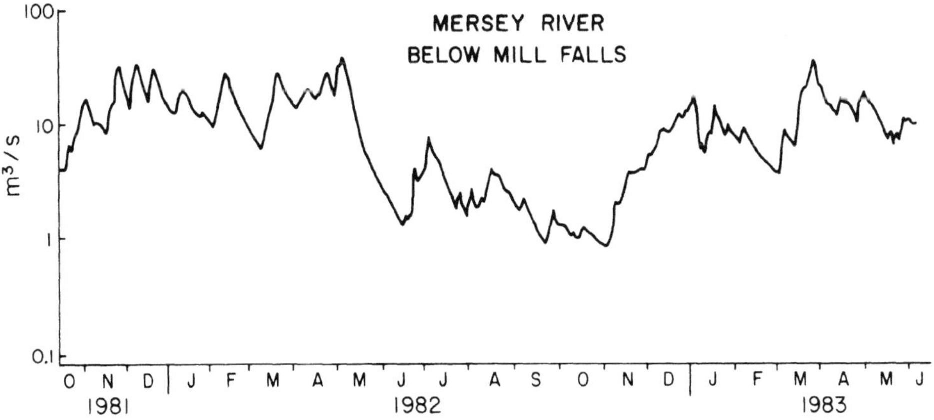

Fig.1: Measured discharge at the Upper Mersey gauging station from October 1981 to June 1983.

tration of major ions, DOC, color and gran alkalinity (not weighted for discharge) for Beaverskin and Pebbleloggitch Outflows and Atkins Brook are given in Table II. The mean Ca and Mg concentrations are very low

TABLE II: Summary of major ionic constituents, DOC, Color and gran alkalinity at 3 sites based on monthly mean values, from June 1982 to May 1983 (February and March excluded)

LOCATION	Ca	Mg	Na	K	H	Fe	Al	SO_4	SO_4*	Cl	A-	DOC	Color	Gran alk.
	ueq L^{-1}											mg L^{-1}	Hazen units	ueq L^{-1}
ATKINS BROOK														
Median	38.4	43.9	140.0	5.8	57.6	27.9	27.9	26.8	15.8	110.0	158.0	26.2	204.0	-33.2
Mean	39.2	47.4	140.0	5.2	58.5	29.0	25.4	35.2	23.4	114.0	146.0	24.7	191.0	-34.9
Std. Dev.	12.7	15.2	23.3	1.2	14.4	11.9	6.2	16.9	16.1	20.8	35.0	5.9	50.9	23.7
Coeff. of Var.	32.2	32.0	16.6	23.8	24.7	40.8	24.5	48.1	68.9	18.3	23.9	23.7	26.7	67.9
BEAVERSKIN LAKE														
Median	18.7	29.7	117.0	5.8	5.5	2.8	5.4	44.8	32.8	118.0	26.9	3.3	5.0	2.20
Mean	18.4	29.9	118.0	7.3	5.5	3.4	5.7	45.0	32.7	119.0	26.9	3.3	5.1	1.99
Std. Dev.	2.0	2.5	5.4	4.7	1.8	2.8	3.1	2.4	2.1	5.4	1.3	0.2	1.7	1.54
Coeff. of Var.	10.6	8.5	4.6	64.8	33.4	84.4	53.7	5.2	6.3	4.5	4.7	5.8	32.5	77.30
PEBBLELOGGITCH LAKE														
Median	17.3	29.6	119.0	6.6	46.4	9.8	25.8	45.8	33.9	114.0	97.3	15.3	110.0	-25.3
Mean	18.3	31.1	125.0	7.0	47.2	9.8	27.1	45.5	33.9	112.0	100.0	16.3	118.0	-26.0
Std. Dev.	2.7	5.3	15.5	1.6	8.8	1.9	6.1	6.5	5.8	9.6	13.1	2.4	20.9	14.4
Coeff. of Var.	14.8	17.0	12.4	22.7	18.5	19.5	22.4	14.3	17.2	8.6	13.1	14.7	17.7	55.5

SO_4* = Sea salt corrected sulphate
HCO_3 concentration 0 in all cases.

and similar in the two lake outflows but somewhat higher and more variable in Atkins Brook.

Time plots of monthly means of H^+, sea salt corrected (excess) sulphate (SO_4*), (A-), and Fe + Al ions in Beaverskin and Pebbleloggitch Outflows are given in Figures 2a, b and 2c, d, respectively. Calcium, Mg (not depicted) and SO_4* concentrations are similar in both lakes but (A-), Al and Fe values are consistently higher in Pebbleloggitch.

Sea salt corrected sulphate makes up ~15% of anions in Pebbleloggitch and ~20% in Beaverskin, while (A-) contributes ~40 to 50% and 15%, respectively. Hydrogen ion makes up 15 to 20% of cations in Pebbleloggitch and 1 to 3% in Beaverskin.

Sulphate and (A-) show opposing cyclic patterns while the negative gran alkalinity is a near mirror image of SO_4* in Atkins Brook (Figure 2e). Sulphate begins to drop and (A-) to suddenly increase in April when the water discharge is still very high and the peatlands on the catchment thaw out. Sulphate levels remain extremely low as a result of biological reduction during the summer low flow period when evapotranspiration is highest (see also Ogden, 1982 and La Zerte and Dillon, 1984). Sulphate increases suddenly during the late fall when the peatlands become saturated with water and discharge increases. Organic anion concentrations reach their highest levels in summer during periods of low flow while the lowest concentrations occur during the winter months when the soils are at least partially frozen. Both Al and Fe follow cyclic patterns similar to that of (A-) (Figure 2f) along with

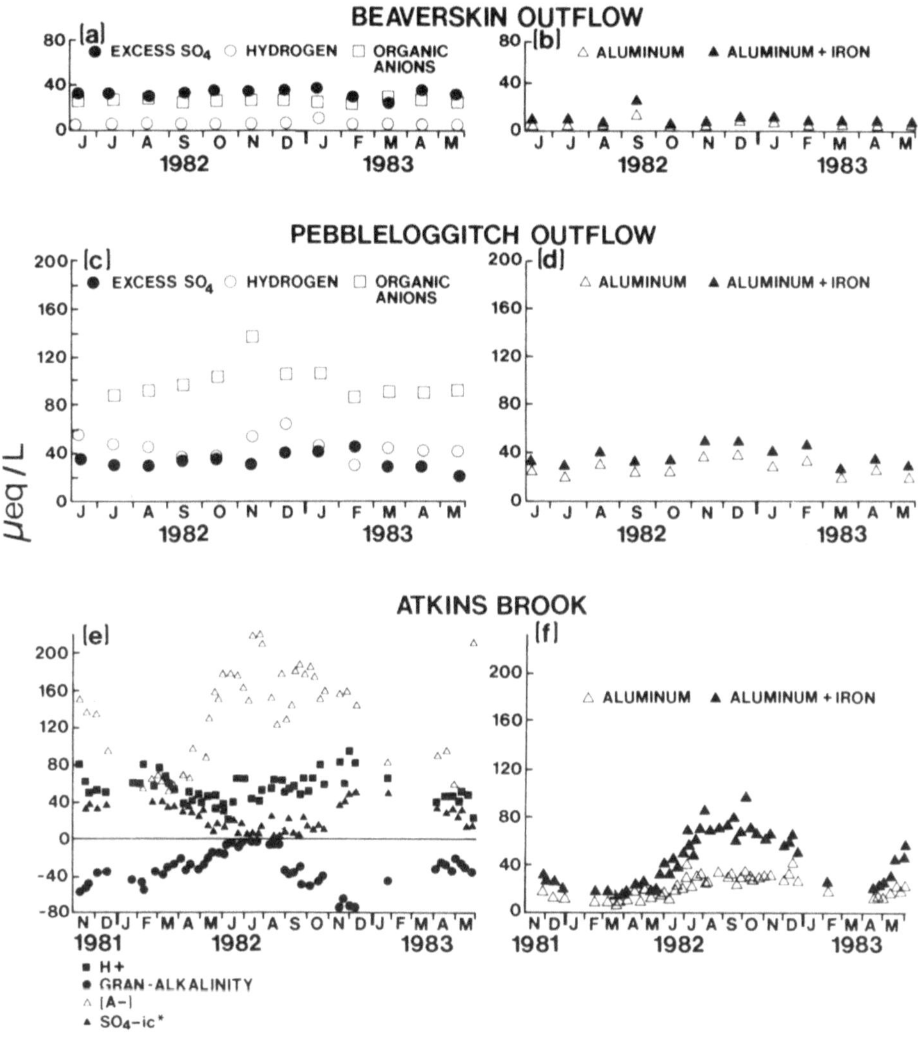

Fig. 2 (a to f): Seasonal variations in SO_4*, H^+, (A-), gran alkalinity (the latter for Atkins Brook only), Al, and Al + Fe in Beaverskin and Pebbleloggitch Outflows and Atkins Brook.

that of color and DOC (not depicted). Calcium and Mg concentrations are lowest in May at the beginning of the growing season (Figure 3). Hydrogen ion behaves similarly to SO_4* except it remains relatively high during the summer months when SO_4* values are extremely low but (A-) concentrations are high (Figure 2e). The highest H^+ ion concentrations coincide with the highest SO_4* concentrations when (A-) are low.

The contribution of SO_4* to the total anion concentration varies

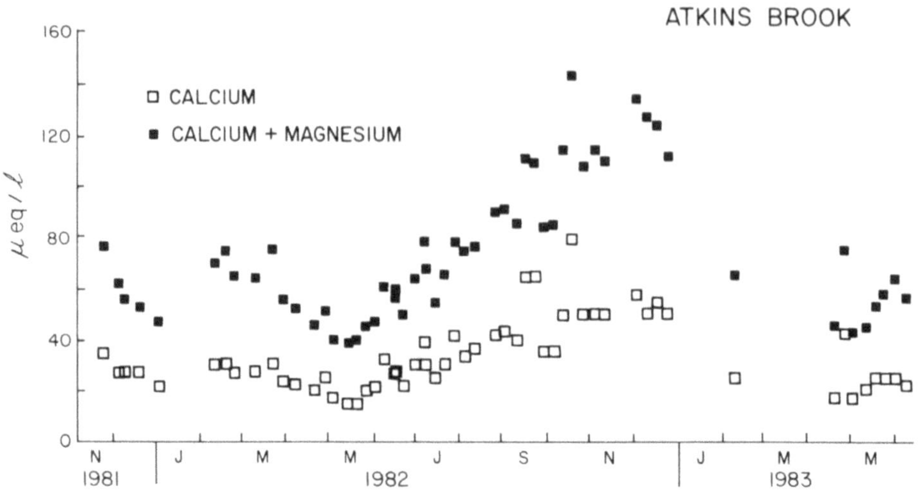

Fig. 3: Seasonal variations in Ca and Ca + Mg concentrations in Atkins Brook from November 1981 to May 1983.

between near zero or very low in summer to near 20% during the rest of the year while (A-) varies from 20% in winter to 65% in summer. Hydrogen ion makes up 12 to 22% of the total cation concentration throughout the year.

The correlation matrix for zero order correlations for selected ions and first order partial correlations between H^+ and SO_4* are given in Table III.

5. DISCUSSION

We interpret these findings to suggest that in the more dilute, clear-water headwater lake(s) in southwest Nova Scotia, the wet deposition of SO_4* of ~20 kg $ha^{-1} yr^{-1}$ is sufficient to sustain a relatively high level of free acidity or chronic pH depression throughout the year (Figure 2a, annual mean pH 5.25), while organic acids cause a further drop in pH in the colored Pebbleloggitch Outflow (Figure 2c, annual mean pH 4.34). It is equally as plausible, from the SO_4 and (A-) data presented, that the H^+ arises from organic acids as it does from H_2SO_4 in Beaverskin Outflow. However, the gran alkalinity data in Table II supports the latter. By definition the negative gran alkalinity concentration is equal to free H^+ concentrations as measured by the pH probe. If there is residual positive gran alkalinity or the negative gran alkalinity is a smaller number than the free H^+ concentration, then there are anions in solution which are capable of accepting protons at the pH of the titration. For instance, the annual mean free H^+ in Beaverskin Outflow is 5.5 ueq L^{-1} and the gran alkalinity is 2.2 ueq L^{-1}. Therefore, 7.7 ueq L^{-1} of (A-) have accepted protons at the gran titration pH of 4.1.

TABLE III: Simple and partial correlations for selected variables from Atkins Brook

Simple correlations

	H^+	SO_4*	A-	Fe + Al	Ca + Mg
H^+	1.00 --				
SO_4*	0.39 (0.02)**	1.00 --			
A-	-0.19 (0.15)	-0.60 (< 0.001)**	1.00 --		
Fe + Al	0.12 (0.27)	-0.56 (< 0.001)**	0.69 (0.001)**	1.00 --	
Ca + Mg	0.60 (< 0.001)**	0.12 (0.273)	0.20 (0.15)	0.52 (0.002)**	1.00 --

Partial Correlations

	vs SO_4*	Controlled for	vs A-	Controlled for
H^+	0.35 (0.031)**	A-	0.05 (0.39)	SO_4*
H^+	0.56 (< 0.001)**	Fe + Al	-0.38 (0.020)**	Fe + Al
H^+	0.41 (0.014)**	Ca + Mg	-0.41 (0.017)**	Ca + Mg
H^+	0.49 (0.004)**	A- and Fe + Al	-0.24 (0.11)	SO_4* and Fe + Al
H^+	0.22 (0.13)	A- and Ca + Mg	-0.20 (0.16)	SO_4* and Ca + Mg
H^+	0.30 (0.057)	Fe + Al and Ca + Mg	-0.29 (0.066)	Fe + Al and Ca + Mg
H^+	0.25 (0.10)	A- and Fe + Al and Ca + Mg (SO_4*)	-0.23 (0.121)	A- and Fe + Al and Ca + Mg (SO_4*)

SO_4* = sea salt corrected sulphate
** = significant at $p < 0.05$

If we assume that the SO_4* (33 ueq L^{-1}) is all from atmospheric deposition of H_2SO_4 the H^+ would immediately react with the excess (Ca + Mg) bicarbonate (20 ueq L^{-1}) to produce (Ca + Mg) sulphate and leave 13 ueq L^{-1} of surplus H^+. Only 5.5 ueq L^{-1} were measured thus 7.5 ueq L^{-1} of H^+ must have combined with some other base in solution.

On the other hand, if H^+ comes solely from (A-) (27 ueq L^{-1}) which will accept 7.7 ueq L^{-1} of protons on lowering the pH from 5.2 to 4.1 then the pKa of this acid has to be 3.7 and it will be 96% dissociated at pH 5.2. The H^+ thus produced will have to have been neutralized by the bases present.

The first assumption where the major H^+ contribution is from H_2SO_4 is in accord with the relationship derived by Oliver et al. (1983), while the secod assumption is not. Additional support of the first assumption may also be made from data in low Ca (10 to 20 ueq L^{-1}) clearwater lakes (DOC < 10 mg L^{-1}) in Newfoundland with lower SO_4* (20 to 25 ueq L^{-1}) but higher pH's (5.6 to 5.9) than Beaverskin (Pollock, 1985).

The relative importance of anthropogenic versus organic acidity can be seen more clearly in Atkins Brook, which exhibits opposing cyclical fluctuations of SO_4* and (A-) concentrations (Figure 2e). Organic anions make an important contribution to free acidity, particularly during the summer months when SO_4* concentrations are low and (A-) concentrations are high. However, Figure 2e suggests that SO_4* is primarily responsible for the high H^+ concentrations in Atkins Brook during fall and spring. The gran alkalinity curve demonstrates that during the summer if one were to add H_2SO_4 to the system (e.g. gran titration), the H^+ concentration would not vary directly as the SO_4 concentration but would be buffered by the (A-), whereas during the rest of the year H^+ would vary directly as the SO_4 concentration. This means that not only does the concentration of the (A-) vary, but also that its ability to accept protons varies seasonally.

Correlation analysis shows that a significant correlation exists between concentrations of SO_4* and H^+ in Atkins Brook, while the correlation between concentrations of (A-) and H^+ is not significant (Table III). Considering the significant correlation between (A-) and Fe + Al, we assume that an appreciable percentage of (A-) (~30%) are complexed with metallic ions during the summer when these ions peak in concentrations. Partial correlations also support the conclusion that SO_4* and not (A-) controls the fluctuations in H^+ concentrations in Atkins Brook. Both the correlation and the level of significance increase between SO_4* and H^+ when controlled for both (A-) and Fe + Al, variables which correlate highly. The results obtained when Ca and Mg are controlled show slightly increased correlation, suggesting that the seasonal cycling of these ions provides little buffering for H^+. The exact cycling role of Ca and Mg needs further investigation.

It should be stressed that (A-), which may be as high as 65% of the anions in summer, are important contributors to the acidity of Atkins Brook seasonally. During the summer months when SO_4* concentrations are low, (A-) can maintain elevated H^+ concentrations (pH = 4.2 to 4.5). However, it is equally important to recognize that the highest H^+ (pH ~4.1) occurs when the SO_4* concentrations are the highest and the (A-) concentrations are relatively low or declining.

We conclude that both (A-) originating from the watershed and H_2SO_4 of anthropogenic origin from atmohpheric deposition are major sources of acidity in Atkins Brook. This is somewhat different from the conclusion reached by La Zerte and Dillon (1984) in Plastic Lake in central Ontario, where anthropogenically derived H_2SO_4 from atmospheric deposition is the major source of acidity. The addition of H_2SO_4 of atmospheric origin increases tha acidity of dilute, organic waters in southwest Nova Scotia, particularly during periods of high water discharge from late fall to spring. This additional anthropogenic acidity is a plausible cause of the reported decline of Atlantic salmon in colored, organic rivers in southwest Nova Scotia (Farmer et al., 1980; Watt, 1981).

There is an urgent need to incorporate organic acidity in future models predicting H^+ concentration in surface water acidification.

REFERENCES

Anonymous: 1983, NAQUADAT Dictionary of Parameter Codes. Ion chromatographic sulphate determination No. 16309 - Environment Canada, Inland Waters Directorate, Water Quality Branch, Ottawa.

Cheam, V., Chau, A.S.Y., and Todd, S.: 1985, "Sulphate in colored waters; investigations on methodologies, data reliability and approaches for salvaging historical colorimetric data.' Environment Canada. Nat. Water Res. Inst. Rep. 85-95, 57 pp.

Farmer, G.J., Goff, T.R., Ashfield, D., and Samant, H.S.: 1980, 'Some effects of the acidification of Atlantic salmon rivers in Nova Scotia.' Tech. Rep. Fish. Aquat. Sci. No. 972. Dept. of Fisheries. St. Andrews, N. B. 13 pp.

Kerekes, J.: 1980, 'Preliminary characterization of three lake basins sensitive to acid precipitation in Nova Scotia, Canada.' in: D. Drablos and A. Tollan (eds.), Impact of Acid Precipitation, Proc. Int. Conf. Sandefjord, Norway, March 11-14, 1980: 232-233.

Kerekes, J.: 1981, Unpublished data.

Kerekes, J., Howell, G., Beauchamp, S. and Pollock, T.: 1982, Int. Revue ges. Hydrobiol., 67, 679.

Kerekes, J. and Pollock, T.: 1983, Can. J. Aquat. Sci., 40, 2260.

Kerekes, J., Howell, G., and Pollock, T.: 1984, Verh. Internat. Verein. Limnol., 22, 1811.

La Zerte, B.D. and Dillon, P.J.: 1984, Can. J. Fish. Aquat. Sci., 41, 1664.

Oliver, B.G., Thurman, E.M. and Malcolm, R.L.: 1983, Geochimica et Cosmochimica Acta, 47, 2031.

Ogden, J.G.: 1982, Water, Air and Soil Pollut. 17, 119.

Pollock, T.: 1985, Unpublished data.

Watt, W.D.: 1981, 'Present and potential effects of acid precipitation on the Atlantic salmon in Eastern Canada.' in: Sochasky, L. (ed.) Acid rain and the Atlantic salmon. The International Atlantic Salmon Foundation, St. Andrews, N. B.

OXYGEN-18 STUDY OF THE ATMOSPHERIC-AQUATIC LINKAGE IN ADIRONDACK WATERSHEDS*

Ben D. Holt and R. Kumar

Argonne National Laboratory
9700 South Cass Avenue
Argonne, Illinois 60439

ABSTRACT. Twelve monthly measurements were made of the $\delta^{18}O$ of the water and of the dissolved sulfates in inlet streams and in outlet streams of lakes in three watersheds in the Adirondack Park region of New York. The average $\delta^{18}O_{H_2O}$ of the surface waters (streams and lakes) of the three watersheds was in the typical range of seasonally varying $\delta^{18}O_{H_2O}$ of precipitation water, whereas the $\delta^{18}O_{SO_4^{2-}}$ of the surface waters was significantly lower than the typical range of seasonally varying $\delta^{18}O_{SO_4^{2-}}$ in precipitation water. Two possible causes for the apparent alteration of $\delta^{18}O$ of the sulfates during percolation of the water through various strata in the ground link between the atmosphere and the watershed lakes are: (1) bacterial redox cycling, in which the sulfate is reduced, allowing isotopic equilibration between the HSO_3^- ion and associated water, and then catalytically reoxidized to sulfate; and (2) ion exchange, in which the soil strata, containing chemically fixed sulfates, behave as a "column" that is not in sulfate-ion equilibrium with sulfates in the atmospheric recharge water.

1. INTRODUCTION

In a recent study of lake acidification processes [the Integrated Lake-Watershed Acidification Study (ILWAS), (1,2)], extensive geochemical and hydrological measurements were made on three fresh water lake watersheds in the Adirondack Park region of New York. The three watershed basins selected for that study were Panther Lake, Sagamore Lake, and Woods Lake, some pertinent data for which are

* Work performed under the auspices of the U.S. Environmental Protection Agency and the U.S. Department of Energy under Contract No. W-31-109-ENG-38.

Table I. Comparative data for the three Adirondack Lakes (1,2)

	Watershed		Lake		Water	
	Area (km^2)	Mean depth of Till (M)	Mean Depth (M)	Flushing Time (days)	*pH*	Alkalinity ($\mu eq\ L^{-1}$)
Panther	1.2	17.1	3.9	264	6.8	147
Sagamore	49.7	—	10.5	81	5.6	31
Woods	2.1	2.2	3.5	221	4.7	−10

given in Table I. Although these three basins receive similar amounts of acid precipitation, the lakes differ in ***pH***, indicating different acid neutralizing capacities among the watersheds.

We selected the same three lakes for a one-year oxygen-isotope study of sulfates in the atmospheric-aquatic linkage. (In that linkage, sulfates are transported from the atmosphere to the ground, through the soil and rock strata of the ground to inlet streams flowing into the lakes, and finally from the lakes to outlet streams.) During the period of July 1983 to June 1984, water samples were collected each month from the one major inlet and the one major outlet stream of each of the three lakes and shipped to our laboratory for isotopic analysis. Precipitation water intercepted by the three watersheds (44°N latitude) was not sampled for isotopic analysis. Instead, typical ranges of isotope ratios in precipitation water and dissolved sulfates, obtained in previous measurements at Argonne, IL, (42°N latitude) were used in this study.

The original purpose of our investigation was to apply oxygen-isotopy to test for bacterial redox cycling of sulfate (3) in each of the three lakes. It was presumed that if lake-water conditions are such that the sulfate is bacterially reduced and subsequently reoxidized by catalytic air oxidation (4), the sulfate product should have an oxygen isotope ratio, $^{18}O/^{16}O$, that is controlled by the associated lake water in which it is reoxidized. That is, if redox cycling of sulfur is proceeding at a sufficient rate in the body of the lake, the $\delta^{18}O$ [deviation in parts per thousand of the $^{18}O/^{16}O$ ratio in a sample from that of standard mean ocean water (SMOW)] of the sulfate flowing out of the lake should bear a linear relationship to the $\delta^{18}O$ of the lake water, although the $\delta^{18}O$ of the sulfate flowing into the lake might be different and unrelated to the $\delta^{18}O$ of its associated water (depending on its origin). Lloyd (5) noted that oceanic sulfate is not in oxygen isotopic equilibrium with ocean water and hypothesized that the reduction of sulfate to sulfide and the subsequent reoxidation of sulfide to sulfate exert the primary control on the oxygen isotope composition of ocean sulfate. Similarly, Cortecci (6) concluded from measurements of $\delta^{18}O$ and concentration of sulfate in surface waters of some Italian lakes that the observed seasonal variations in these qualities could be explained, not by the corresponding seasonal variations of $\delta^{18}O$ in rainwater sulfates, but rather by bacterial reduction of sulfate, followed by sulfide oxidation.

The results of our study of the three Adirondack lake watersheds suggest that bacterial redox processes control the $\delta^{18}O$ of the dissolved sulfate in the ground link of the atmospheric-ground-aquatic linkage even before the sulfate reaches the lakes. This evidence is complementary to the results of the ILWAS experiments (1, 2) which showed the importance of the ground link in the chemistries of lake-acidifying components of precipitation water.

2. EXPERIMENTAL

Four-liter samples of water were collected at sites (the main inlet and outlet streams of each of the three lakes) previously used in the ILWAS program. Twenty mg of CuCl were added to each polyethylene sample bottle to prevent bacterial decomposition of sulfate during shipment and storage before analysis.

The samples were analyzed for $\delta^{18}O$ of the water and for $\delta^{18}O$ of the dissolved sulfate, according to published procedures (7). Estimates of sulfate concentrations were obtained from measurements of the CO_2 formed from the oxygen in SO_4^{2-} in the analytical procedure.

3. RESULTS AND DISCUSSION

Isotopic results obtained at the three lakes are shown in Figure 1 for inlet streams, and in Figure 2 for outlet streams. In both figures, curve groups A and B are measured values of $\delta^{18}O_{H_2O}$ and $\delta^{18}O_{SO_4^{2-}}$, respectively, for the three lakes for each month of the collection period (July 1983–June 1984). Curves C and D represent $\delta^{18}O_{H_2O}$ and $\delta^{18}O_{SO_4^{2-}}$, respectively, for precipitation water, collected at Argonne, IL, during the corresponding months in 1977–1978 (8); they are replotted here for purposes of comparison.

Figure 3 shows the range of Adirondack lakes data in comparison to data obtained elsewhere for surface water, ground water, and precipitation water (6, 8–12). In Figure 4, monthly sulfate concentrations are plotted for the inlet and outlet streams of the three lakes.

3.1 Oxygen Isotopy of Water.

Month-to-month fluctuations in the $\delta^{18}O_{H_2O}$ of water flowing into the three lakes (curve-group A, Figure 1) are generally in agreement. These fluctuations appear to be related to stream flow rates. During the late summer and early fall when the streams were reported to be very low, the $\delta^{18}O_{H_2O}$ continued to rise, probably as a

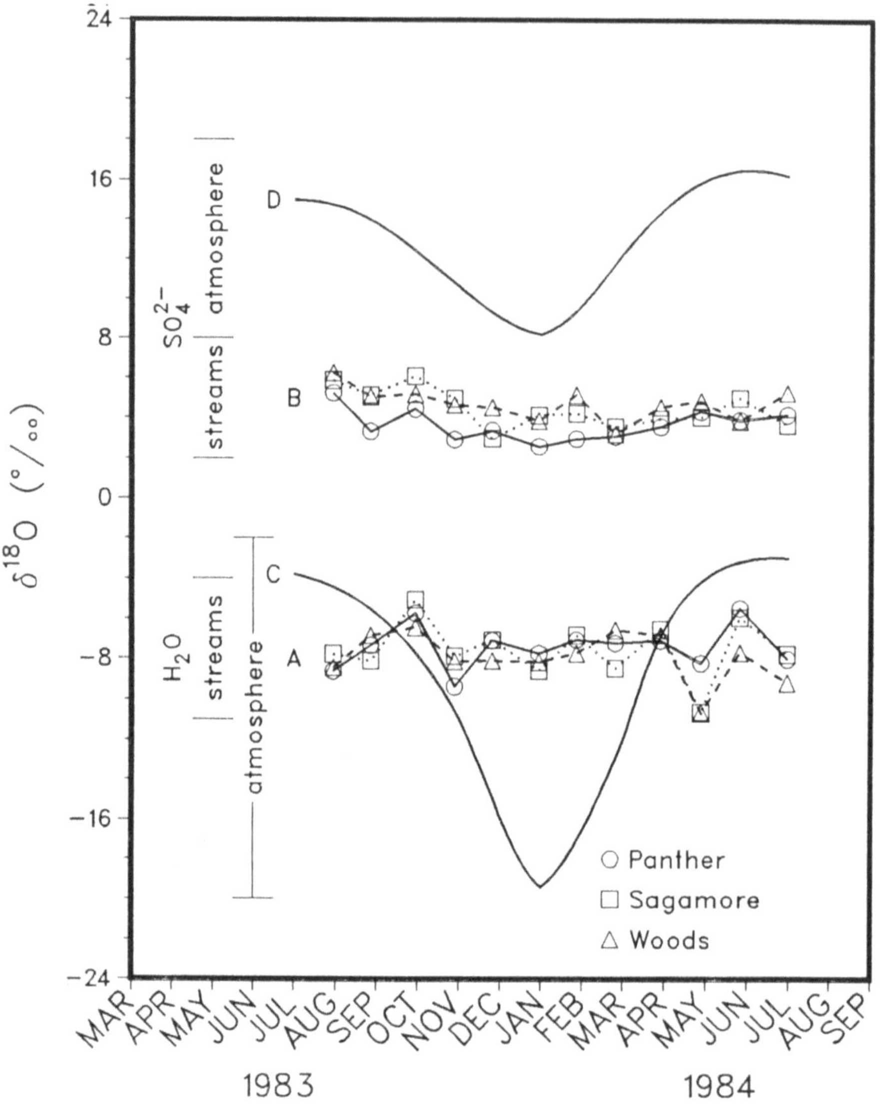

Fig. 1. Oxygen isotope data for water and dissolved sulfate in inlet streams, compared to precipitation.

consequence of ^{18}O-enrichment through evaporation; then in late fall, when heavy precipitation caused considerably higher stream flows with a higher ratio of run-off water to ground water, they sharply declined. Similarly, during snowmelting in April, the $\delta^{18}O_{H_2O}$ of the streams declined because of the characteristically low $\delta^{18}O_{H_2O}$ of winter precipitation water (8,9).

Comparison of the A curve-groups in Figures 1 and 2 shows that the fall and spring fluctuations in $\delta^{18}O_{H_2O}$ for the inlet and outlet streams were generally in phase with each other, indicating that during those periods, the water was moving fairly rapidly across the lakes with the probable consequence of relatively shallow mixing. An exception to this interlake agreement appeared to be in the outlet water of Woods Lake during the autumn months when a dam was temporarily opened for repairs. The deep mixing, caused by refilling of the lake, apparently prevented the fluctuations of $\delta^{18}O_{H_2O}$ that otherwise would have been seen in the surface water, had it not been disturbed.

None of the $\delta^{18}O_{H_2O}$ data for the inlet and outlet streams (curve-groups A and B in Figures 1 and 2, respectively) shows pronounced seasonal trends. This is in contrast to characteristic seasonal trends in $\delta^{18}O_{H_2O}$ of precipitation water (8,9). The co-plotted curve-groups A and curves C, in Figures 1 and 2, indicate that ground water flowing into the lakes is characterized by a resultant $\delta^{18}O_{H_2O}$ of precipitation waters (which recharge the ground throughout the season by percolating down to the water table and draining into the streams and lakes). The comparative display of ranges of $\delta^{18}O$ data in Figure 3 shows that our range of $\delta^{18}O_{H_2O}$ in precipitation is in agreement with data of Förstel and Hützen in West Germany (10) and of Cortecci and Longinelli in Italy (9); also our ranges of $\delta^{18}O_{H_2O}$ in the Adirondack inlet and outlet streams are not significantly different from those of ground water in West Germany (11), or in the Italian Arno and Serchio Rivers, and in Italian Corbara Lake [formed by a dam on the Tiber River (12)].

3.2 Sulfate Concentrations.

Sulfate concentration data in Figure 4 show generally higher values for the Sagamore watershed than for Panther and Woods, with no pronounced seasonal variation for any of the three watersheds. The relatively high sulfate concentrations for the Sagamore system substantiate findings of Galloway, *et al.* (2) that outputs of sulfate (draining from the lake) exceed inputs (entering the watershed as atmospheric depositions), whereas in the Panther and Woods systems, outputs approximately equal inputs.

Our results show that, unlike $\delta^{18}O_{H_2O}$, no significant changes in sulfate concentrations accompanied the drought in late summer and early fall, nor the snowmelt in April. Sulfate concentration was apparently independent of factors that affected the isotopic quality of the water. In agreement with conclusions of Galloway *et*

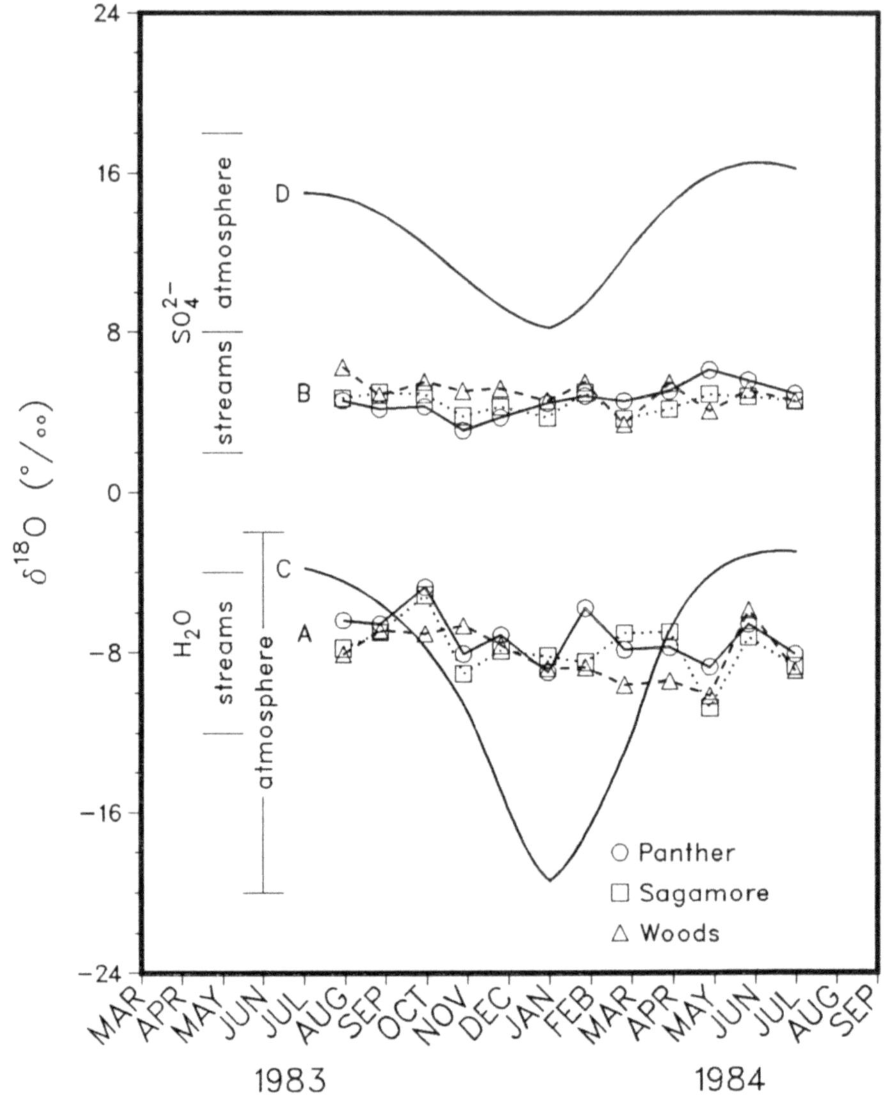

Fig. 2. Oxygen isotope data for water and dissolved sulfate in outlet streams, compared to precipitation.

OXYGEN-18 STUDY OF THE ATMOSPHERIC-AQUATIC LINKAGE IN ADIRONDACK WATERSHEDS*

Ben D. Holt and R. Kumar

Argonne National Laboratory
9700 South Cass Avenue
Argonne, Illinois 60439

ABSTRACT. Twelve monthly measurements were made of the $\delta^{18}O$ of the water and of the dissolved sulfates in inlet streams and in outlet streams of lakes in three watersheds in the Adirondack Park region of New York. The average $\delta^{18}O_{H_2O}$ of the surface waters (streams and lakes) of the three watersheds was in the typical range of seasonally varying $\delta^{18}O_{H_2O}$ of precipitation water, whereas the $\delta^{18}O_{SO_4^{2-}}$ of the surface waters was significantly lower than the typical range of seasonally varying $\delta^{18}O_{SO_4^{2-}}$ in precipitation water. Two possible causes for the apparent alteration of $\delta^{18}O$ of the sulfates during percolation of the water through various strata in the ground link between the atmosphere and the watershed lakes are: (1) bacterial redox cycling, in which the sulfate is reduced, allowing isotopic equilibration between the HSO_3^- ion and associated water, and then catalytically reoxidized to sulfate; and (2) ion exchange, in which the soil strata, containing chemically fixed sulfates, behave as a "column" that is not in sulfate-ion equilibrium with sulfates in the atmospheric recharge water.

1. INTRODUCTION

In a recent study of lake acidification processes [the Integrated Lake-Watershed Acidification Study (ILWAS), (1,2)], extensive geochemical and hydrological measurements were made on three fresh water lake watersheds in the Adirondack Park region of New York. The three watershed basins selected for that study were Panther Lake, Sagamore Lake, and Woods Lake, some pertinent data for which are

* Work performed under the auspices of the U.S. Environmental Protection Agency and the U.S. Department of Energy under Contract No. W-31-109-ENG-38.

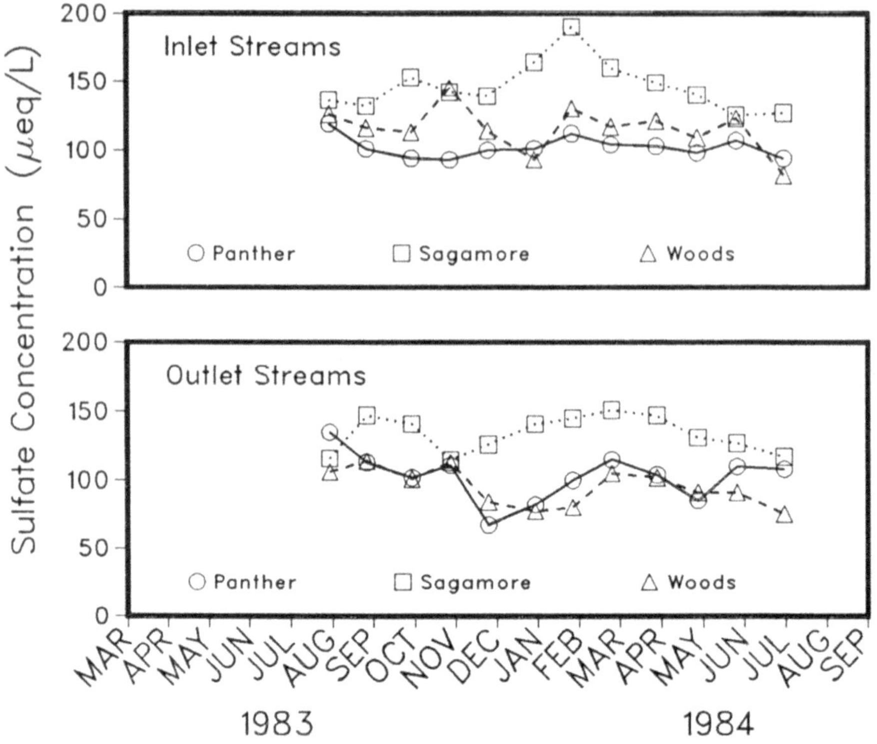

Fig. 4. Sulfate concentrations in inlet and outlet streams.

al. (2) that in the Panther and Woods systems there is no net gain or loss of sulfate as the water moves through the systems, this suggests that the output sulfate concentration was affected only by the input sulfate deposition.

3.3 Oxygen Isotopy of Sulfate.

In curve-groups B of Figures 1 and 2, the $\delta^{18}O_{SO_4^{2-}}$ values for the inlet and outlet Adirondack streams are co-plotted with curve D, which represents previously reported $\delta^{18}O_{SO_4^{2-}}$ data (8) for precipitation water at Argonne, IL. Like the $\delta^{18}O_{H_2O}$ in the inlet and outlet streams, the $\delta^{18}O_{SO_4^{2-}}$ for the dissolved sulfate did not show the pronounced seasonal variation that is characteristic of sulfate in precipitation water. The same smoothing-out effect on the $\delta^{18}O$ of precipitation water by percolation through the soil and rock strata of the ground was also seen in the $\delta^{18}O$ of the sulfate.

Unlike the $\delta^{18}O$ of the water, however, the annual range of $\delta^{18}O_{SO_4^{2-}}$ in the inlet and outlet streams was significantly lower than that of $\delta^{18}O_{SO_4^{2-}}$ in precipitation water. It may be argued that the $\delta^{18}O_{SO_4^{2-}}$ in the streams should have been compared only to the $\delta^{18}O_{SO_4^{2-}}$ of the precipitation water that was collected in the same area and during the same sampling period as that of the stream samples. However, as in the case of $\delta^{18}O_{H_2O}$ discussed above, $\delta^{18}O_{SO_4^{2-}}$ data that have been previously obtained by other workers for precipitation in the temperate zone of the Northern Hemisphere substantiate the marked difference in precipitation water and surface water. Comparative ranges of data in Figure 3 show that our range of $\delta^{18}O_{SO_4^{2-}}$ for precipitation in 1977–78 (8) was in agreement with that of Cortecci and Longinelli in 1968–69 (9); and our range of $\delta^{18}O_{SO_4^{2-}}$ in the Adirondack streams (inlets and outlets) was not too different from water in the head-water regions (sites of lowest $\delta^{18}O_{SO_4^{2-}}$) of the Italian rivers (12) and water in Corbara Lake (6).

Although it may be expected that the seasonal variation of $\delta^{18}O_{SO_4^{2-}}$ in ground water, of which the Adirondack streams are composed (except during run-off of precipitation or snow melt), has a smoothed-out pattern in comparison to that of precipitation water which recharges the ground water, an explanation is needed for why the ground water is significantly lower in $\delta^{18}O_{SO_4^{2-}}$ than that assumed for the atmospheric recharge water. Apparently, it is not caused by elution of excess soluble mineral sulfates from the soil, sufficiently low in $\delta^{18}O_{SO_4^{2-}}$ to give the observed decreased $\delta^{18}O_{SO_4^{2-}}$ in the stream water. Such would not be substantiated by the experimental evidence of Galloway, *et al.* (2) that the sulfate outputs of Panther and Woods Lakes are approximately equal to the inputs.

Ion exchange might be proposed as a mechanism by which the $\delta^{18}O$ of the sulfate that moves from soil media to the streams and lakes could be lower than that of the sulfate entering by precipitation recharge that is, if the chemically bound sulfate in the soil is lower in $\delta^{18}O$ than the incoming precipitation sulfate (experimental evidence of which we do not have). Isotope separation by column action might be expected to enrich the heavier isotope (^{18}O) on the column (the ground strata), leaving the eluted sulfate more depleted in ^{18}O than the entering sulfate. Although this is in agreement with the observed order, the magnitude of such an effect is not known. Eventual saturation of the column (ground strata), relative to the isotopic composition of the sulfate flowing through it, should allow the eluted sulfate, on the average, to be equal to the incoming sulfate in $\delta^{18}O$, and preclude the observed depletion of ^{18}O.

A more likely mechanism is bacterial reduction of sulfate, followed by aqueous air oxidation in one or more of the various ground strata (vegetation litter, organic soil, glacial till, glacial gravel, etc.). According to this mechanism, the incoming precipitation sulfate is reduced during percolation of the water through environments that are favorable to anaerobic bacterial activity, and is reoxidized by catalyzed aqueous air-oxidation as the water migrates through environments favorable to such oxidation. It is known that isotopic exchange between the HSO_3^- ion and solvent water is very rapid (4,13); consequently, as the sulfur passes through the aqueous +4-valence state (prior to reoxidation to sulfate), the bisulfite is expected to rapidly exchange with the large excess of associated water. The isotopically equilibrated HSO_3^- ions then oxidize to SO_4^{2-}, 3 oxygens of which are isotopically controlled by the associated water, regardless of the $\delta^{18}O$ of the original incoming sulfate.

Previous laboratory studies (4) have shown that when HSO_3^- is air-oxidized in an aqueous solution in the presence of a catalyst (ferric iron or elemental carbon), the isotopic relationship between the sulfate and the associated water is

$$\delta^{18}O_{SO_4^{2-}} = 0.8\ \delta^{18}O_{H_2O} + 10. \qquad (1)$$

When the relationship (1) is applied to the $\delta^{18}O_{H_2O}$ data of the curve-groups A in Figures 1 and 2, sets of calculated values of $\delta^{18}O_{SO_4^{2-}}$ are obtained for sulfates

Table II. Difference between measured and calculated $\delta^{18}O_{SO_4^{2-}}$ in inlet and outlet streams.

Sample Collection Date	$\Delta\delta^{18}O_{SO_4^{2-}} = \delta^{18}O_{SO_4^{2-}}$(measured) $- \delta^{18}O_{SO_4^{2-}}$(calculated) Panther		Sagamore		Woods	
	Inlet	Outlet	Inlet	Outlet	Inlet	Outlet
7-29-83	2.2	−0.4	2.1	1.0	3.0	2.7
8-26-83	−0.8	−0.6	1.6	0.6	0.6	0.4
9-28-83	−1.0	−1.9	0.1	−1.0	0.4	1.2
10-26-83	0.5	−0.4	1.3	1.1	1.2	0.4
11-23-83	−0.9	−0.5	−1.3	0.6	1.0	1.3
12-29-83	−1.2	1.6	1.0	0.3	0.4	1.6
1-26-84	−1.3	−0.6	−0.3	1.8	1.4	2.5
2-22-84	−1.1	0.9	0.4	−0.7	−1.5	1.1
3-28-84	−0.7	1.3	−0.8	−0.2	0.0	3.0
4-26-84	0.9	3.1	2.6	3.5	3.4	2.2
5-25-84	−1.7	0.9	−0.2	1.1	0.1	−0.2
6-29-84	0.6	1.4	−0.1	1.5	2.6	1.7
Average	−0.4	0.4	0.5	0.8	1.1	1.5

eluted into the streams after undergoing the proposed bacterial redox mechanism. The differences between these calculated values (not plotted) and the measured values (curve-groups B),

$$\Delta\delta^{18}O_{SO_4^{2-}} = \delta^{18}O_{SO_4^{2-}} \text{ (measured)} - \delta^{18}O_{SO_4^{2-}} \text{(calculated)}, \qquad (2)$$

are listed for the three watersheds in Table II. The average deviation of these differences from 0‰ are −0.4, 0.5 and 1.1‰, respectively, for inlet streams to Panther, Sagamore, and Woods Lakes; and 0.4, 0.8, and 1.5‰, respectively, for the corresponding outlet streams. These relatively small differences between measured and calculated values allow the hypothesis of either process, bacterial redox or ion exchange in the ground strata, and show that the order of relative impact of the responsible process among the three watersheds is Panther > Sagamore > Woods.

3.4 Conclusions

Oxygen isotopy verifies previous experimental observations (1, 2) that the chemistry of acidification-related solutes (*e.g.*, sulfates) in precipitation water can be significantly altered in the ground link between the atmosphere and the aquatic systems of Adirondack watersheds. Bacterial redox cycling of sulfates may be a likely mechanism in ground strata of the lake watersheds, resulting in important effects on lake water quality.

ACKNOWLEDGMENTS

We are very grateful to E. R. Altwicker, Department of Chemical and Environmental Engineering, Rensselaer Polytechnic Institute, Troy, NY, for assistance in arranging for collection of the lake water samples; to T. Suleski, Eagle Bay, NY, for collection and shipment of the samples; and to A. E. Engelkemeir and M. I. Homa, Analytical Chemistry Laboratory, for isotopic analyses.

5. REFERENCES

1. 'The Integrated Lake-Watershed Acidification Study: Proceedings of the ILWAS Annual Review Conference'; Electric Power Research Institute Rep. No. EPRI EA-2827, 1983.
2. Galloway, J. N., Schofield, C. L., Peters, N. E., Hendrey, G. R., and Altwicker, E.R.: 1983, Can. J. Fish. Aquat. Sci. **40**, 799.
3. Postgate, J. R.: 1968, 'The Sulfur Cycle,' in Inorganic Sulfur Chemistry, G. Nichless, Ed., Elsevier, New York, p. 259.
4. Holt, B. D., Kumar, R., and Cunningham, P. T.: 1981, Atmos. Environ. **15**, 557.

5. Lloyd, R. M.: 1967, Science **156**, 1228.

6. Cortecci, G.: 1973, Geochim, Cosmochim. Acta **37**, 1531.

7. Holt, B. D.: 1977, Anal. Chem. **49**, 1664.

8. Holt, B. D., Cunningham, P. T., Kumar, R.: 1981, Environ. Sci. Tech. **15**, 804.

9. Cortecci, G., Longinelli, A.: 1970, Earth Planet. Sci. Lett. **8**, 36.

10. Förstel, H. and Hützen, H.: 1978, 'The $^{18}O/^{16}O$-Ratio of Precipitation at the Jülich Nuclear Research Center from 1974 to the Middle of 1978,' Kernforschungsanlage Jülich GmbH, Institut für Chemie, Institut 2: Biophysikalische Chemie. Report Jül-1524, July 1978, translated from German by STS Incorporated for Argonne National Laboratory, Argonne, IL, ANL-TRANS-1148.

11. Förstel, H. and Hützen, H.: 1982, '$^{18}O/^{16}O$-Ratio Ground Water at the Federal Republic of Germany in Stable Isotopes, H. L. Schmidt, H. Förstel, and K. Heinzinger, Eds., Elsevier, Amsterdam.

12. Longinelli, A. and Cortecci, G.: 1970, Earth Planet. Sci. Lett. **7**, 376.

13. Eigen, M., Kustin, K., and Maass, G.: 1961, Z. Phys. Chem. **30**, 130.

WATER ACIDIFICATION IN THE FEDERAL REPUBLIC OF GERMANY PROVED BY SIMPLE CHEMICAL MODELS:

Raymund Schoen
Universität Hohenheim
Institut für Landeskultur und Pflanzenökologie
Postfach 700562
7000 Stuttgart 70

ABSTRACT. Several descriptive acidification models show the appearance of water acidification in soft water areas of the Federal Republic of Germany. The input of sulfuric and nitric acid seems to be the main reason as it is shown for example by Wright's acidification equation.

Critical situations can be found especially during snow melt. Brooks of the Black Forest, Bavarian Wood, Harz-Mountains or the Taunus are within the most endangered water courses, but we fear that all soft-water regions of Germany show already damages.

1. Introduction

The acidification of water courses and lakes in the Federal Republic of Germany was shown comperatively late (1,2). A first attempt for a regional survey was conducted during snow melt in 1983 (3,4). Symptoms of acidification were recorded in nearly all soft-water areas examined as judged from a decrease in pH, low alkalinity or even strong acidity and high concentrations of soluble Al etc. (Figure 1). Extremly high concentrations of SO_4 and NO_3 and of Al could be found. Recent research results from several authors have revealed SO_4-concentrations up to 70 mg l^{-1} (1500 µeq l^{-1}) and Al-concentrations near to 20 mg l^{-1} (2200 µeq l^{-1}) in various surface waters (not cited here).

The aim of the publication presented here is to evaluate the validity of several descriptive acidification models (8,10,11) for data (3,9,12) obtained in the Federal Republic of Germany. The models should also indicate the mechanism for the observed acidification processes such as acid deposition, natural organic acid production or ionic displacement from humic substances (5,6).

2. Results

The first model refers to the correlation between Ca + Mg and alkalinity, as described by Almer (7). In soft waters, neither influenced by acid precipitation nor by bog formation, alkalinity (generally represented

Water, Air, and Soil Pollution **31** (1986) 187–195.

by HCO_3-ions) derives mainly from weathering of Ca and Mg containing minerals caused by CO_2-production. Therefore HCO_3 is accompanied by an equivalent amount of such bivalent cations. During acidification a) the produced HCO_3 is neutralized by acids introduced into soil and surface water and b) bivalent cations are lost from the soil by exchanging against H^+ and/or NH_4. For both reasons the 1:1-ratio between Ca + Mg versus HCO_3 is completely changed. Similar effects were observed in the Federal Republic of Germany (3,4), Figure 2). A research project supported by the provinicial government of Baden-Württemberg from summer 1983 to spring 1984 revealed the same (Figure 3).

A second model uses the ionic balance and its change during acidification (8). Figure 4 shows the extremely good correlation between the "loss of alkalinity" (8) and the SO_4-content of the samples of (3); (loss of alkalinity = SO_4 = 0.91 (Ca + Mg) + H^+ + Al - HCO_3 = Δ Alk.).

This hypothesis may be a good tool to show losses of alkalinity are the result of the input of strong acids such as H_2SO_4, because all the waters tested are poor in natural SO_4-content.

Figure 5 shows the correlation between loss of alkalinity and SO_4 and the role of NO_3 in the watercourses of the Black Forest (9) during snow melt 1984. These samples, poorer in SO_4 than those from (3) (compare Figure 1), contain a relatively greater proportion of NO_3, and consequently nitrat (probably derived from input of NO_x) becomes important for the acidification equation. Nevertheless, in most cases SO_4 remains the dominant anion (compare Figure 11) on an equivalent basis, but not always on a molar basis!

Another idea - mainly derived from the view that acidification may be described as the titration of bases released from soil by acids from deposition - is the equivalent ratio between Ca + Mg vs SO_4 + NO_3. Similar suggestions have been made by Henriksen (10) and Hultberg (11), using Ca + Mg/SO_4 respectively Ca/SO_4. From our results it seems useful to take NO_3 into account (Figures 6,7,8).
Figure 6 shows that samples with a ratio above 0.9 belong to the HCO_3-buffer-system, whereas samples with a lower ratio seem to belong to the Al buffer. Figure 7 reveals a similar tendency for the data obtained in Baden-Württemberg.

The two different buffer system may also be revealed by correlation of the ratio Ca +Mg/SO_4 + NO_3 vs pH (Figure 8). Below a pH-value of 4.5, a buffer system of Al and humic substances, even Fe may be effective, above a pH of about 5.6 the water is buffered by HCO_3 (Figure 9). Therefore, we may not expect a linear relationship between the equivalent ratio of Ca + Mg/SO_4 + NO_3 and pH. Nevertheless this example shows that the model implication of an acid-base titration is reasonable as well.

Principle component analysis (PCA) - not shown here (9) - at least reveals that the equivalent ratio of Ca + Mg/SO_4 + NO_3 is of major importance for all changes and effects in the chemistry of the studied water courses, especially during snow melt, whereas this ratio depends on a lot of different factors, for example acid deposition, cation exchange capacity, age, depth and water conductivity of the soil in question. A simple example is the response upon changes in the water level (Figure 10). Rising concentrations of SO_4 and NO_3 with snow melt or rain storm events and sometimes sligth decreases in Ca- or Mg-content result in a depression of the ratio of Ca + Mg/SO_4 + NO_3 with typical acidification effects such as low pH, decrease in alkalinity and/or high values of aluminium or other metals or humic substances (Bavarian Wood/data from (12)).

PCA, and the simple quantitative analysis (Figure 11) show, that organic anions are a) a dependent variable of the Ca + Mg/SO_4 + NO_3-ratio and b) never exceed the proportion of SO_4 in the water samples from the Baden-Württemberg data set which is rather rich in samples influenced by peat (9).

3. Diskussion

Simple ion exchanged in humic substances of the soil, as suggested by Rosenqvist (6) and later by Krug and Frink (13) does not seem to be a reasonable explanation for the observed changes such as the increasing concentration of SO_4 between summer and spring flood. There is also no evidence, that soils and brooks that are influenced by road de-icing salts in the Black Forest of Baden-Württemberg (9) showed more acid water courses than do similar areas that are not influenced by salt. At least there is no evidence that the precipitation chemistry has changed during recent years to a higher salt content, except of raising parts of H^+, NH_4, SO_4 or NO_3 (14). In the contrary, the reduction of dust emissions should have led to the opposite trend.

4. Conclusions

All the results shown above may lead to the conclusion that models developed mostly for Scandinavian conditions are - except site specific differences due to the character of the soils and different acid loadings - valid for soft water courses of the Federal Republic of Germany also.

They also give good evidence that acidification of water courses in the Federal Republic of Germany follows the same rules known from other regions of the world.

The project is supported by the FEDERAL ENVIRONMENTAL AGENCY (Umweltbundesamt), Berlin, project-no.: 102 04 342. Data for (9) are from a project supported by the provincial government of BADEN-WÜRTTEMBERG.

We thank Prof. Dr. Ottow for translation, Dr. Hrebicek for data processing, Mrs. Halter for drawings and Mrs. Nies for typing the manuscript.

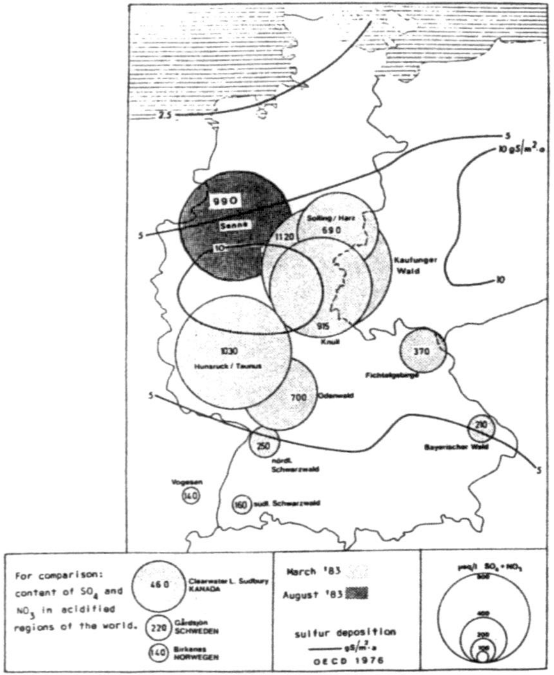

Fig. 1.

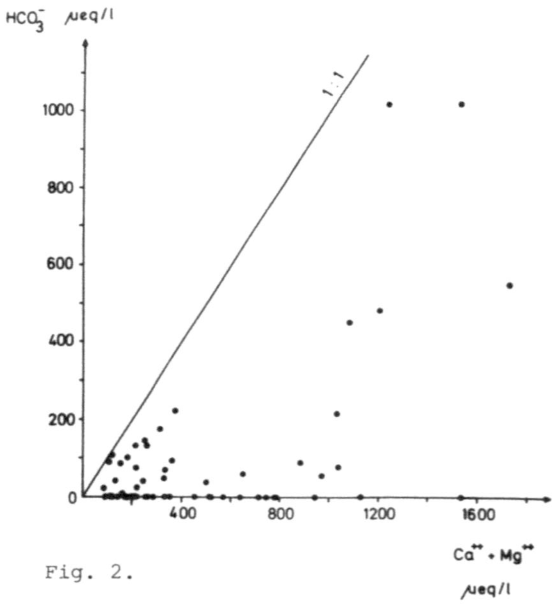

Fig. 2.

Fig. 1 Content of SO_4 and NO_3 (in µeq l^{-1}) in brooks of different soft water regions in the Federal Republic of Germany (snow melt 1983) and total S deposition (3).

Fig. 2 Relation between Ca + Mg and HCO_3 in brooks of the Federal Republic of Germany; snow melt 1983 (3). The 1:1 line indicates non-acidification (10).

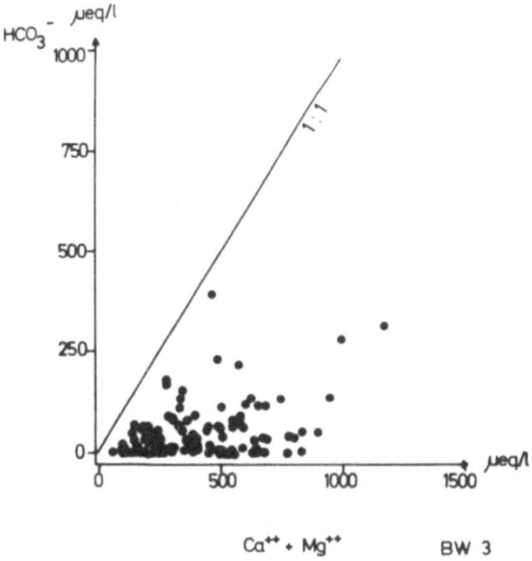

Fig. 3 Calcium and Mg vs alkalinity; Baden-Württemberg, snow melt 1984.

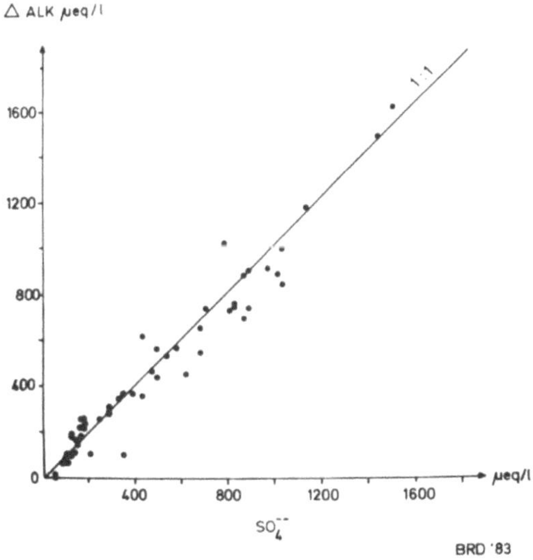

Fig. 4 Loss of alkalinity (Δalk = 0.91 (Ca + Mg) + Al - HCO_3 + H^+) vs SO_4-content (8); brooks of the Federal Republic of Germany; snow melt 1983. r - 0.967.

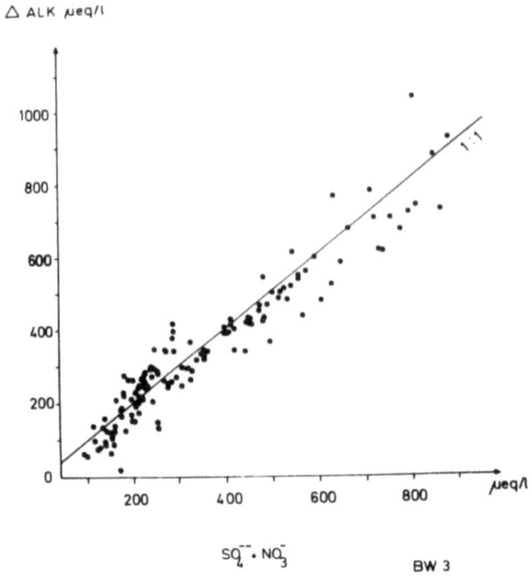

Fig. 5 Baden-Württemberg, loss of alkalinity versus SO_4 + NO_3; r = 0.969; snow melt 1984.

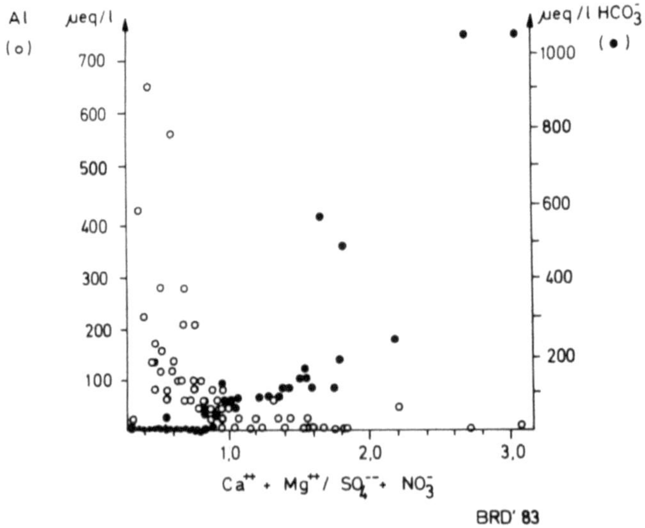

Fig. 6 Ratio of Ca + Mg/SO_4+NO_3 (all in µeq l^{-1}) vs HCO_3 and Al; FRG; snow melt 1983 (3). Notice the different scales.

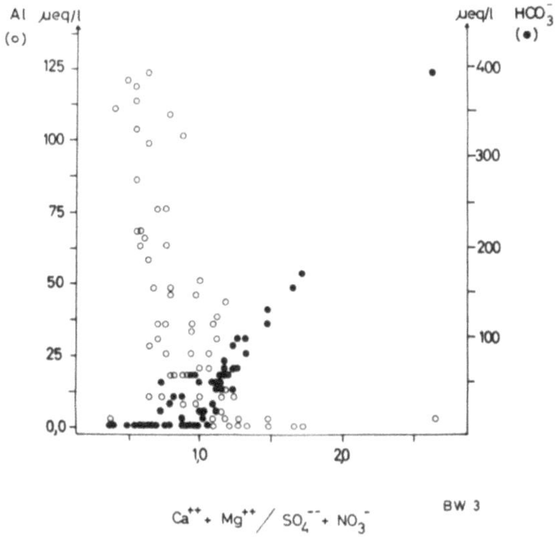

Fig. 7 Similar to Fig. 6, but data from Baden-Württemberg; snow melt 1984.

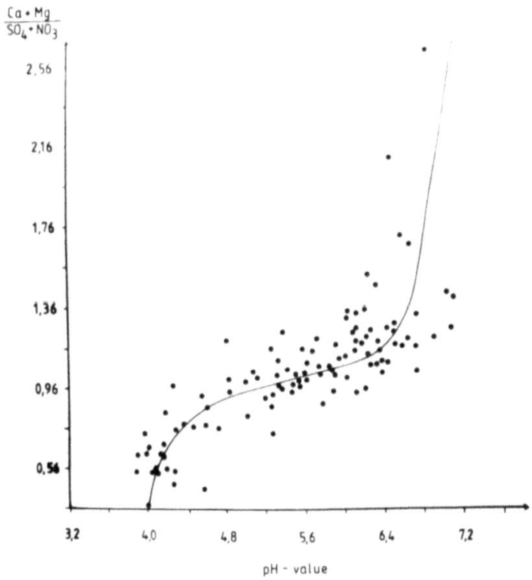

Fig. 8 Ca + Mg/SO_4 + NO_3 vs pH, Baden-Württemberg, snow melt 1984 (curve fitted by hand).

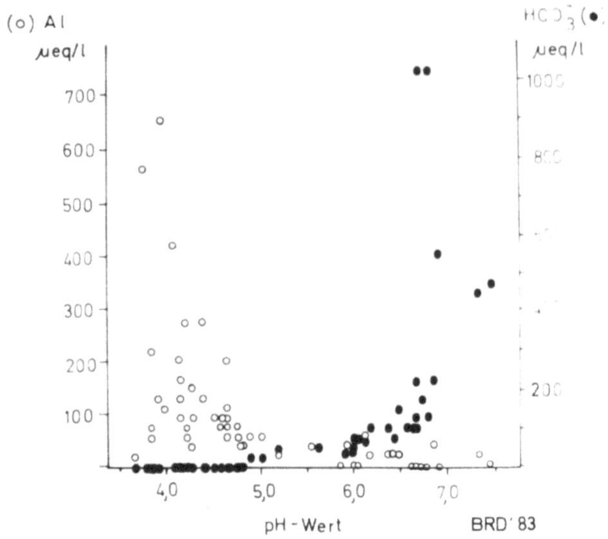

Fig. 9 Correlation between pH, Al and HCO_3; Federal Republic of Germany; snow melt 1983. Notice the scales.

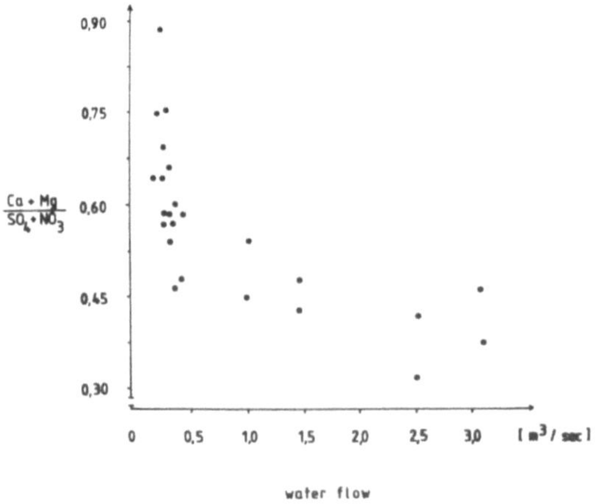

Fig. 10 Correlation between acidification and water flow of a brook in Bavaria (Große Ohe) (Hecker, 1983) in the year 1981/82.

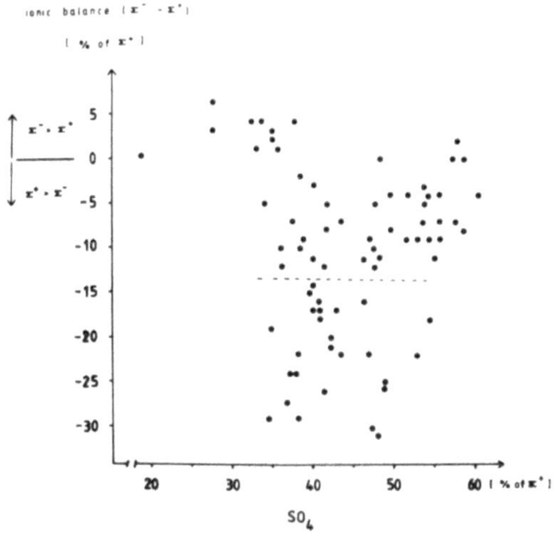

Fig. 11 Presuming that a surplus of cations is a measure for organic anions, it can be shown, that a) organic anions never exceed the proportion of SO_4 - even in peat influenced waters (samples below dashed line) - and b) that there is no direct correlation between organic anions and SO_4 (in % of Σ^+!)

References

1) Brehm, J.: 1978, Arch. Hydrobiol., 82, 49.
2) Meinel, W., Matthias, U. und Schöcke, K.: 1981, Phillipia 4,357.
3) Schoen, R., Wright, R.F. und Krieter, M.: 1984, Die Naturwissenschaften 71, 95.
4) Schoen, R., Wright, R.F. and Krieter, M.: 1983 Acid Rain Research Report 5/83 NIVA, Oslo, 15 p.
5) Overrein, L., Seip, H. and Tollan, A.: 1980, Ecological Impact of Acid Precipitation, final report of the SNSF-Projekt, OSLO, 175 p.
6) Rosenqvist, Th., Jorgensen, P. and Rueslatten, H.: 1980, in: Drablos, D./Tollan, A. (Eds); 1980: Acid Precipitation - Effects on Forest and Fish: Sandefjord-Conference; 383 p.
7) Almer, B., Dickson, W., Ekström, Ch. and Hörnström, E.: 1978, in: Nriagu, J.O. (Ed.): Sulfur in the Environment, New York, Part II, S. 273.
8) Wright, R.F.: 1983, Acid Rain Research Report, 4/83, NIVA, OSLO, 165 p.
9) Schoen, R., Neubauer, D.: 1985, unpublished results
10) Henriksen, A.: 1980, in: Drablos, D. and Tollan, A., (Eds.); Sandefjord-conference; p. 68-75.
11) Hultberg, H.: 1983, personal communication.
12) Hecker, A.: 1983, unpublished Diplomarbeit at the Ludwig-Maximilian-University München, Forstwissenschaftliche Fakultät.
13) Krug, E.C. and Frink, Ch.R.: 1983, Science 221, 520.
14) Brimblecombe, P. and Stedman, D.H.: 1982, Nature 298, 460.

RELATIONS BETWEEN LAKE ACIDIFICATION AND SULFATE DEPOSITION IN NORTHERN MINNESOTA, WISCONSIN, AND MICHIGAN

Dale S. Nichols, USDA Forest Service, N. Cent. For. Exp. Stn., 1831 Hwy 169 E, Grand Rapids MN USA 55744

Ronald E. McRoberts, USDA Forest Service, N. Cent. For. Exp. Stn., 1992 Folwell Ave., St. Paul MN USA 55108

ABSTRACT. From a level of 1 kg $ha^{-1}yr^{-1}$ in north central Minnesota, emission-related wet SO_4 deposition increases across northern Wisconsin and northern Michigan to about 18 kg $ha^{-1}yr^{-1}$ in south central Michigan. Samples taken from 82 clearwater (low color) lakes across this region in the summer of 1984 showed a pattern of acidification in proportion to deposition. We found a linear increase in the difference between alkalinity and Ca+Mg and in lake SO_4 concentration with increasing deposition. We developed a simple equation to predict the emission-related SO_4 deposition levels that will cause the alkalinity of sensitive clearwater lakes to go to zero.

1. INTRODUCTION

Many researchers have reported the acidification of surface waters in southern Scandinavia, and to a lesser extent in southeastern Canada and the northeastern United States, where strongly acidic atmospheric deposition is coincident with lakes and streams with insufficient capacity to buffer acid inputs. Many lakes with low buffer capacity are found in northern Minnesota, Wisconsin, and Michigan. However, deposition in this region is less acidic than in those areas where deposition-induced lake acidification has been documented. The purpose of our study is to determine the extent to which the lakes in northern Minnesota, Wisconsin, and Michigan are undergoing acidification in response to acid deposition.

2. METHODS

In the summer of 1984, we sampled 82 clearwater lakes--from 9 to 13 lakes in each of the seven national forests in northern Minnesota, northern Wisconsin, and northern and central Michigan (Figure 1). All had apparent color values (measured by a Hellige color comparitor) less than 30 Co-Pt color units; most values were 15 or less. For each national forest, we selected lakes to cover an alkalinity range from as low as possible to about 1000 to 1200 μeqL^{-1}. We visited each lake

Water, Air, and Soil Pollution **31** (1986) 197–206.

twice, once between mid-June and mid-July, and again between mid-August and mid-September. At each visit, samples were collected from the upper 1 m of the water column at each of six different locations on each lake, away from the littoral zone. These six samples were composited into three samples for chemical analysis. Samples were kept in iced coolers in the field and then refrigerated in the laboratory before analysis. Linear polyethylene sample bottles with caps containing conical polyethylene inserts as suggested by Schock and Schock (1982) were used to minimize gas exchange before pH and alkalinity measurements. The pH was measured in the field as well as within 24 hr of returning to the laboratory. We determined alkalinity by Gran plot titration (Gran, 1952), also within 24 hr after sample arrival at the lab. Calcium, Mg, K, and Na were measured by atomic adsorption spectrophotometry (EPA, 1983). Sulfate was determined by the automated methythymol blue colorimetric procedure described by McSwain et al. (1974). Chloride was measured by an automated ferricyanide colorimetric method (EPA, 1983).

Using data from 869 precipitation samples taken over periods of 2 to 4 years at nine National Atmospheric Deposition Program (NADP) collection sites (American Samoa; Denali, Alaska; Mead, Nebraska; Marcell, Minnesota; Fernberg, Minnesota; Wellston, Michigan; Aurora, New York; Caldwell, Ohio; and Lewiston, North Carolina), representing the entire range of precipitation chemistry found in the NADP sampling network, Verry and Harris (1984, 1986) developed a model to separate precipitation samples into groups based on chemical composition. When they stratified samples according to base cation concentration into a "low-salt" and a "high-salt" group, they found that more than 90% of the total H deposition at any particular collection site occurred in the low-salt samples, while up to 75% of the variation in the regression of H on SO_4 and NO_3 concentrations was in the high-salt group. Stratifying the low-salt sample data further, they noted that 84% of the variation in the H concentration of samples containing more than 13 μeqL^{-1} of H was related to $SO_4 + NO_3$ concentration. For samples with 13 μeqL^{-1} of H or less, $SO_4 + NO_3$ explained only 9% of this variation.

Nichols and Verry (1985) used the stratification criteria of Verry and Harris (1984, 1986) as the basis for calculating background and emission-related amounts of wet sulfate deposition in the northern Lake States. Because precipitation H is only slightly related to $SO_4 + NO_3$ at concentrations below 13 μeqL^{-1}, they considered samples in this range to represent normal background precipitation chemistry, unaffected by emissions of acid-precursor forms of S and N. Above 13 μeqL^{-1}, the H concentration is strongly correlated with levels of SO_4 and NO_3; they considered these precipitation samples to be impacted by emissions. This 13 μeqL^{-1} criterion was used to distinguish between normal and emission-affected precipitation for both low-salt and high-salt samples. They used precipitation chemistry data from eight NADP stations across the northern Lake States (Marcell and Fernberg in Minnesota; Spooner, Trout Lake, and Lake Dubay in Wisconsin; and Douglas Lake, Wellston, and Kellogg Biological Station in Michigan), collected for periods of 1 to more than 4 years through December 1982, to calculate emission-related wet SO_4 deposition in this region.

For each station Nichols and Verry (1985) used the following procedure to calculate total, background, and emission-related wet SO_4 deposition. The sample data were stratified into four groups: normal low-salt, normal high-salt, emission-related low-salt, and emission-related high-salt. For each group, SO_4 deposition was calculated by multiplying the mean volume weighted SO_4 concentration for that group by the respective precipitation volume. The sum of the four group products equals the total wet SO_4 deposition at a station. The sum of the SO_4 deposition calculated for the normal low-salt and normal high-salt groups of samples represents one portion of the normal background SO_4 deposition at that station. In addition, part of the SO_4 present in emission-affected samples is actually normal background SO_4. This portion was calculated by multiplying the mean SO_4 concentrations found in the normal low-salt and high-salt groups by the precipitation volumes for emission-related low-salt and high-salt groups. These amounts were then subtracted from the total SO_4 deposition calculated for the emission-related groups. The sum of these two remainders was considered to be the emission-related wet SO_4 deposition for that station.

From these data, Nichols and Verry (1985) constructed an isopleth map of emission-related wet SO_4 deposition across northern Minnesota, northern Wisconsin, and upper and lower Michigan (Figure 1). We estimated the level of emission-related wet SO_4 deposition at each of the 82 lakes we sampled by interpolating between isopleths.

The amounts of H in precipitation are related to both SO_4 and NO_3.

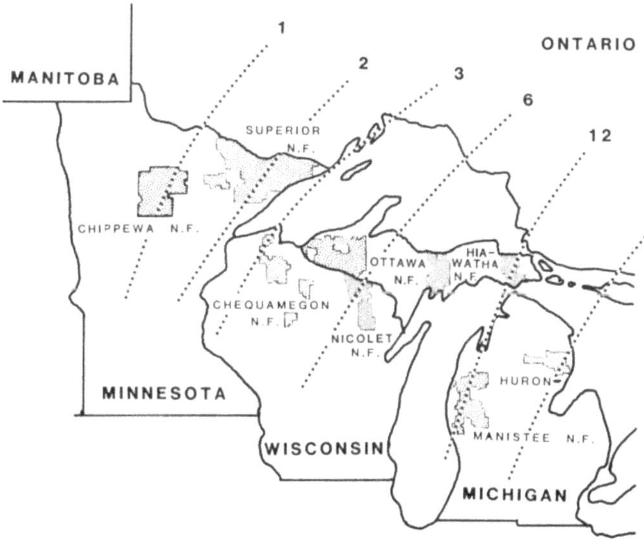

Figure 1. Emission-related wet SO_4 deposition, Kg $Ha^{-1}Yr^{-1}$, across northern Minnesota, Wisconsin, and Michigan, in relation to the seven national forests where 82 clearwater lakes were sampled.

In relating deposition to lake acidification, we consider only SO_4 in this paper. Because NO_3 is taken up rapidly by aquatic and terrestrial biota, it does not seem to promote long-term lake acidification to the extent that SO_4 does (Galloway et al., 1983; Driscoll and Likens, 1982).

3. RESULTS AND DISCUSSION

In most unpolluted surface waters, Ca and Mg together and HCO_3 constitute about 80% of the total cations and anions, respectively; the relation between the sum of Ca plus Mg and HCO_3 alkalinity is close to 1:1 (all constituents expressed in μeqL^{-1}) (Henriksen, 1980; Gorham et al., 1983). When a lake is subjected to acid deposition, HCO_3 alkalinity decreases and is replaced by SO_4, while the concentration of Ca+Mg stays the same or increases somewhat. The result is an alkalinity deficit with respect to Ca+Mg and an increase in SO_4 above normal background levels (Almer et al., 1978; Dickson, 1980; Henriksen 1980).

Alkalinity was regressed on Ca+Mg for each of six groups of lakes. The lakes sampled in each national forest were considered as one group, except that the Ottawa and Nicolet National Forests were grouped together because of their geographical proximity. The slopes of the six regression lines are all close to 1.0 and none is significantly different from any other (alpha=0.05); however, each of the intercept values is different from some or all of the others. The individual regression equations are:

Chippewa N.F.	$\widehat{alk} = -\ 25.38 + 1.024\ (Ca{+}Mg),\ R^2 = 0.996$	a
Chequamegon N.F.	$\widehat{alk} = -\ 58.92 + 1.034\ (Ca{+}Mg),\ R^2 = 0.998$	b
Superior N.F.	$\widehat{alk} = -\ 80.35 + 1.039\ (Ca{+}Mg),\ R^2 = 0.990$	c
Ottawa/Nicolet N.F.'s	$\widehat{alk} = -\ 83.75 + 1.062\ (Ca{+}Mg),\ R^2 = 0.996$	c
Hiawatha N.F.	$\widehat{alk} = -\ 89.66 + 1.009\ (Ca{+}Mg),\ R^2 = 0.998$	c
Huron-Manistee N.F.	$\widehat{alk} = -\ 127.20 + 1.015\ (Ca{+}Mg),\ R^2 = 0.996$	d

($\widehat{alk}$=estimated alkalinity).

The intercept values of equations followed by the same letter are not significantly different from each other (alpha=0.05).

Because the slopes are all close to 1.0, the Ca+Mg intercept value for each regression equation (obtained by setting alkalinity to zero and solving for Ca+Mg) represents the average amount of Ca+Mg present in low alkalinity lakes in each group that is not balanced by alkalinity. A strong gradient of emission-related wet SO_4 deposition exists from northern Minnesota to southern Michigan (Figure 1). Except for the lakes sampled in the Superior National Forest (the reason for this exception is discussed later), the Ca+Mg intercept values increase linearly with the average emission-related wet SO_4 deposition level for each group of lakes (Figure 2). This indicates a replacement of HCO_3 alkalinity by SO_4 in direct proportion to deposition. When a term for lake SO_4 concentration is added to each regression, the group intercepts are no longer different, and the alkalinity of all 82 lakes can be described by one equation:

$$\widehat{alk} = 25.79 + 1.024(Ca{+}Mg) - 1.152(SO_4). \qquad (1)$$

Alkalinity increases with increasing Ca+Mg and decreases with increasing SO_4. Both relations are essentially 1 to 1, and together account for 99.8% of the lake-to-lake variation in alkalinity.

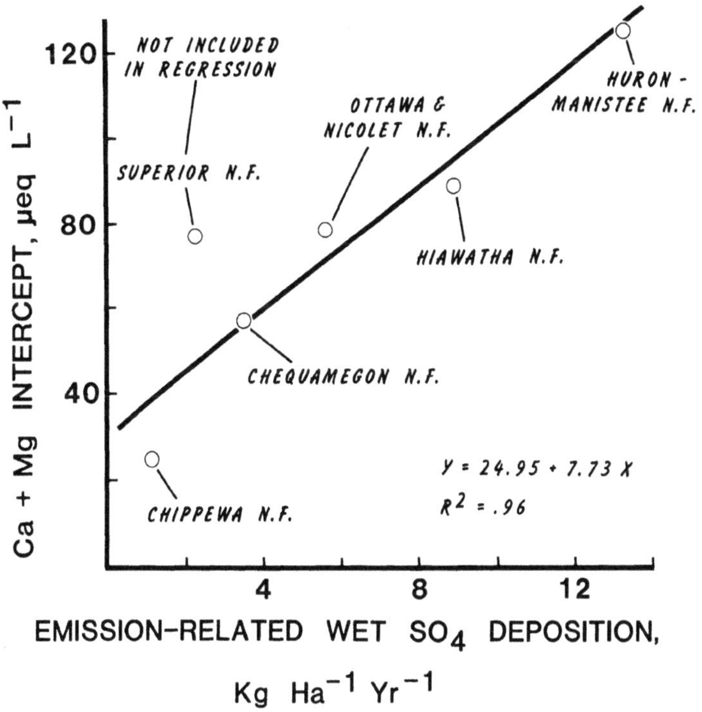

Figure 2. Alkalinity was regressed on Ca+Mg for 6 groups of national forest lakes. Ca+Mg intercepts (the concentration of Ca+Mg when alkalinity = zero) for the 6 regressions increse linearly with emission-related wet SO_4 deposition.

Lake SO_4 increases linearly with increasing emission-related wet SO_4 deposition, from north central Minnesota to central lower Michigan (Figure 3). The regression of lake SO_4 concentration on emission-related wet SO_4 deposition level is highly significant (alpha=0.01). Dry deposition of emission-related SO_4 may be as important as wet deposition; however, dry deposition data are not available. If it is assumed that dry SO_4 deposition is proportional to wet SO_4 deposition, then wet emission-related SO_4 deposition can be used as a surrogate for total emission-related SO_4 deposition in considerations of its effects on lakes.

Because of their anomolously high SO_4 concentrations, the lakes sampled in the Superior National Forest were not included in this

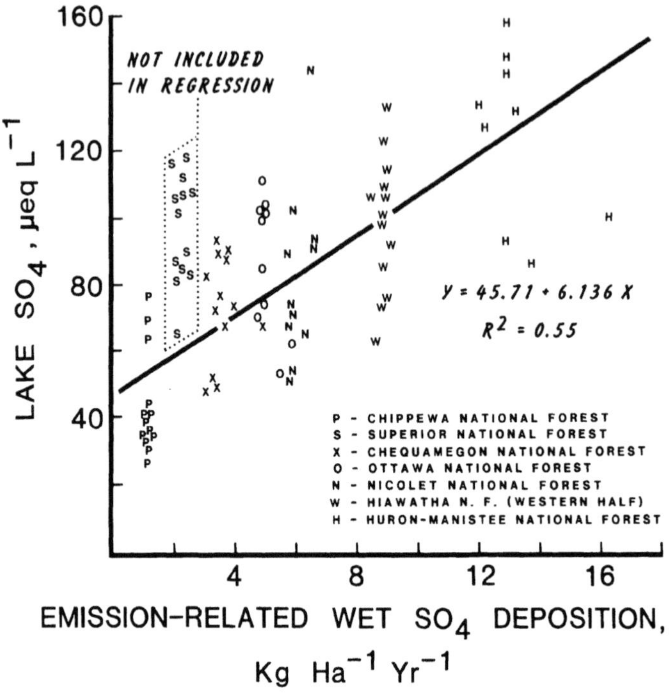

Figure 3. Sulfate concentrations in 82 clearwater lakes increase linearly with emission-related set SO_4 deposition.

regression. We used the method described by Draper and Smith (1966) to assess the lack of fit of this regression line to the data. When the Superior National Forest lakes are excluded, the lack of fit is not significant (alpha=0.05), indicating that the linear equation is adequate to describe SO_4 in the remaining lakes in terms of emission-related wet SO_4 deposition. However, when the Superior National Forest lakes are included in the regression, the lack of fit is highly significant ($p<0.001$).

We believe the high SO_4 concentrations found in the lakes sampled in the Superior National Forest are due to natural geologic sources. The other lakes sampled are all found on thick deposits of tills and outwash sands. The Superior National Forest lakes are located in or immediately adjacent to a large Upper Precambrian formation of intrusive rock called the Duluth Complex. The Duluth Complex contains extensive low-grade sulfide deposits, primarily pyrrhotite (FeS) and Chalcopyrite ($CuFeS_2$) (Sims and Morey, 1972). The oxidation of sulfide minerals such as these produces free H and SO_4 that can have the same acidifying effects on lakes as the deposition of H and SO_4 in rain and snow.

The SO_4 concentration in a lake is the result of two groups of factors: 1) anthropogenic sources of SO_4 (for the lakes we sampled, we assume that emission-related deposition is the only significant anthropogenic source), and 2) natural factors--natural sources of SO_4 such as natural deposition, SO_4 adsorption-desorption reactions, SO_4 reduction, and hydrologic processes. From the regression of lake SO_4 on emission-related wet SO_4 deposition (Figure 3), the intercept of the regression line on the lake SO_4 axis is the mean contribution of SO_4 from natural sources, while the slope of the line represents the average response of lake SO_4 to emission-related deposition. There is substantial variation of individual lake SO_4 concentrations from the values predicted by the regression line. This variation is apparently due to lake-to-lake differences in natural SO_4 sources and sinks.

Using the relations in Figure 3, we subtracted from the SO_4 concentration in each lake, the contribution due to emission-related deposition. For each lake, we attributed the remaining SO_4 to natural sources and sinks, unique to that lake. We call this the lake's natural sulfate index (NSI) concentration.

$$\widehat{NSI} = \text{lake } SO_4 - 6.14\ (ERD) \tag{2}$$

(ERD = emission-related wet SO_4 deposition).

The NSI is an index of not only the amount of SO_4 entering a lake from natural sources, but also the extent to which processes such as SO_4 adsorption-desorption and SO_4 reduction are modifying the concentration of SO_4 derived both from natural sources and emission-related deposition.

When lake SO_4 is separated into natural and emission-related deposition-derived components, lake alkalinity can be described in terms of Ca+Mg, NSI, and emission-related wet SO_4 deposition. Multiple regression yields the following equation, with an R^2 value of 0.998.

$$\widehat{alk} = 21.61 + 1.029\ (Ca{+}Mg) - 0.959\ (\widehat{NSI}) - 8.539\ (ERD). \tag{3}$$

Agreement between measured alkalinity and alkalinity predicted by Equation (3) is very good. For all 82 lakes sampled, the average difference between measured and predicted alkalinity is 10.0 μeqL^{-1}. For the most sensitive lakes, those with alkalinities below 200 μeqL^{-1}, the average difference is 6.3 μeqL^{-1}.

If the response of the 82 lakes we sampled is representative of that of other lakes, then Equation (3) appears to be useful in predicting the deposition level at which any clearwater lake will go to zero alkalinity. To make such predictions, the deposition-induced change that may occur in Ca+Mg must be estimated. In the following equations, a subscript "p" indicates present conditions, a subscript "n" indicates future (new) conditions, and a "Δ" indicates a change in concentration in response to a change in deposition. The ratio of the change in Ca+Mg to the change in lake SO_4 is represented by the letter "f."

$$f = \Delta(Ca{+}Mg)/\Delta(SO_4). \tag{4}$$

Using data from a variety of Scandinavian and North American lakes, Wright (1983) estimated "f" to be between 0 and 0.4.

From Equation (1)

$$\Delta(SO_4) = 0.89\Delta(Ca{+}Mg) - 0.87\Delta(alk). \tag{5}$$

If, in response to a future deposition level $alk_n = 0$ then:

$$\Delta alk = -alk_p. \tag{6}$$

Substituting Equation (6) into Equation (5), and solving Equations (4) and (5) simultaneously yield:

$$\Delta(SO_4) = 0.87\ alk_p/(1-0.89f). \tag{7}$$

Combining Equations (7) and (4) yields:

$$\Delta(Ca{+}Mg) = (0.87\ alk_p)(f/(1 - 0.89f)). \tag{8}$$

As deposition changes, Equation (3) becomes:

$$\widehat{alk}_n = 21.61 + 1.029((Ca{+}Mg)_p + \Delta(Ca{+}Mg)) - 0.959(\widehat{NSI}) - 8.539(ERD)_n. \tag{9}$$

We assume that the natural SO_4 index (NSI) concentration of a lake does not change significantly as deposition changes. This may not be completely true in some cases. The SO_4 derived from natural sources is not likely to change, but modifying reactions such as SO_4 adsorption-desorption may.

Substituting Equation (8) into equation 9, and solving for emission-related wet SO_4 deposition when $alk_n{=}0$ yield:

$$\widehat{ERD}_c = (21.61 + 1.029((Ca{+}Mg)_p + (0.87alk_p)(f/(1 - 0.89f))) - 0.959(\widehat{NSI}))/8.539 \tag{10}$$

($\widehat{ERD}_c$ = estimated critical emission-related wet SO_4 deposition).

The critical emission-related wet SO_4 deposition for a lake is the level of deposition that will cause that lake to go to zero alkalinity.

During the late 1970's, USDA Forest Service personnel sampled 82 clearwater lakes in the western half of the Hiawatha National Forest in Upper Michigan. Table I shows the percentage of these lakes that we estimate, using Equation (10), would have alkalinities of zero or less at various emission-related wet SO_4 deposition levels, and at f values of 0, 0.2, and 0.4. For these lakes, the value assigned to f makes little difference in the number of lakes estimated to have no alkalinity at any reasonable deposition level. If we assume that the process of acidification is reversible if deposition decreases, it appears that 1 or 2% of these lakes would have no alkalinity even with no emission-related deposition.

Table I. Estimated percentages of clearwater lakes in the western half of the Hiawatha National Forest that would have alkalinities of zero or less at various deposition levels.

Emission-related wet SO_4 deposition kg $ha^{-1}yr^{-1}$	Percent of lakes with alkalinity $\leq$ zero		
	f=0	f=0.2	f=0.4
0	0.7	1.4	2.7
3	3.8	4.0	4.3
6	4.4	4.5	4.9
9*	7.5	7.5	7.5
12	12.5	12.5	13.3
15	17.0	15.7	14.6
20	22.0	20.0	17.5
25	24.0	23.5	20.6
30	24.7	24.0	23.5

* approximate present deposition level

In other recent studies in the Upper Midwestern United States, Glass and Loucks (1985) noted a close chemical correspondence between precipitation and weakly buffered surface water composition; Schnoor et al. (1986) used a model based on mass balances for alkalinity in lakes and their watersheds to predict lake acidification at various levels of acidic deposition. In eastern Canada Lachance et al. (1985) and Thompson and Hutton (1985) found relations between lake SO_4 concentration, lake acidification, and precipitation chemistry. The findings reported in this paper fit well with the results of these studies.

4. CONCLUSIONS

A pronounced gradient of emission-related wet SO_4 deposition exists from northern Minnesota to southern Michigan. This is reflected in the alkalinity, Ca+Mg, and in SO_4 concentrations in 82 lakes sampled along this gradient. The difference between alkalinity and Ca+Mg increases as does lake SO_4 with increasing deposition. These changes can be seen even at low emission-related deposition levels. A simple equation was developed to estimate lake alkalinity from Ca+Mg concentration, an index value of natural SO_4 sources and sinks, and emission-related wet SO_4 deposition. This equation appears to have the capacity to predict the deposition level at which a clearwater lake will go to zero alkalinity, but testing is needed using an independent data set.

5. ACKNOWLEDGMENTS

These data and interpretations are a contribution to the National Atmospheric Deposition Program (NADP), and the North Central Forest Experiment Station's Aspen-Birch-Conifer (ABC) Ecosystem Program. This research has been funded as part of the National Acid Precipitation Assessment Program by the United States Forest Service.

6. REFERENCES

Almer, B., Dickson W., Ekstrom, C., and Hornstrom, E.: 1978, 'Sulfur pollution and the aquatic ecosystem', Nriagu, J. O. (ed), Sulfur in the Environment, Part II, Ecological Impacts, John Wiley and Sons, New York, p. 271.

Dickson, W.: 1980. 'Properties of acidified waters', Drablos, D., and Tollan, A. (eds), Ecological Impact of Acid Precipitation, Proceedings of an International Conference, Sandefjord, Norway, March 11-14, 1980, SNSF-Project, 1432 As, Norway, p. 75.

Draper, N. R. and Smith, Jr., H.: 1966, Applied Regression Analysis, John Wiley and Sons, New York, p. 26.

Driscoll, C. T. and Likens, G. E.: 1982, Tellus **34**, 283.

EPA: 1983, Methods for Chemical Analysis of Water and Wastes, EPA-600/4-79-020, U.S. Environmental Protection Agency, Office of Research and Development, Cincinnati, Ohio.

Galloway, J. N., Schofield, C. L., Peters, N. E., Hendrey, G. R., and Altwicker, E. R.: 1983, Can. J. Fish. Aquat. Sci. **40**, 799.

Glass, G. E. and Loucks, O. L.: 1986, Environ. Sci. Technol. **20**, 35.

Gorham, E., Dean, W. E., and Sanger, J. E.: 1983, Limnol. Oceanog. **28**, 287.

Gran, G.: 1952, Analyst **77**, 661.

Henriksen, A.: 1980, 'Acidification of freshwaters - a large scale titration', Drablos, D., and Tollan A. (eds), Ecological Impact of Acid Precipitation, Proceedings of an International Conference, Sandefjord, Norway, March 11-14, 1980, SNSF-Project, 1432 As, Norway, p. 68.

Lachance, M., Bobee, B., and Grimard, Y.: 1985, Water Air Soil Pollut. **25**, 115.

McSwain, M. R., Watrous, R. J., and Douglas, J. E.: 1974, Analytical Chem. **46**, 1329.

Nichols, D. S. and Verry, E. S.: 1985, 'Evidence for the cultural acidification of lakes in the northern Lake States', Air Pollutants Effects on Forest Ecosystems, Proceedings of a Symposium, May 8-9, 1985, St. Paul, Minnesota, The Acid Rain Foundation, St. Paul, Minnesota, p. 253.

Schnoor, J. L., Nikolaidis, N. P., and Glass, G. E.: 1986, Jour. Water Pollut. Contr. Fed. **58**, 139.

Schock, M. R. and Schock, S. C.: 1982, Water Res. **16**, 1455.

Sims, P. K., and Morey, G. B. (eds): 1972, Geology of Minnesota: a Centennial Volume, Minnesota Geological Survey, University of Minnesota, St. Paul, 632 p.

Thompson, M. E. and Hutton, M. B.: 1985, Water Air Soil Pollut. **24**, 77.

Verry, E. S. and Harris, A. R.: 1984. 'A universal relation between hydrogen and nitrate plus sulfate', Poster Summaries, International Symposium on Hydrochemical Balances of Freshwater Systems, Sept. 10-14, 1984, Uppsala, Sweden, p. 27.

Verry, E. S. and Harris, A. R.: 1986, Water Resour. Res. In press.

Wright, R. F.: 1983, Predicting Acidification of North American Lakes, Report 4/1983, Norwegian Institute for Water Research, Oslo, Norway, 163 p.

SOURCES OF SULPHATE AND ACIDITY IN WETLANDS AND LAKES IN NOVA SCOTIA

J. Kerekes, S. Beauchamp, R. Tordon
Canadian Wildlife Service
Environment Canada
c/o Department of Biology
Dalhousie University
Halifax, Nova Scotia
Canada B3H 4J1

T. Pollock
Water Quality Branch
Environment Canada
P. O. Box 861
Moncton, New Brunswick
Canada E1C 8N6

ABSTRACT. There is a declining gradient of wet SO_4 deposition from south to north in Nova Scotia with the highest values being in the south, along with a localized increase around the Halifax metropolitan area, due to local SO_4 emission. Edaphic conditions such as drainage from soils containing gypsum or drainage on disturbed rocks containing pyrite, provide additional SO_4 to surface waters. Acidity is usually absent in the former (pH > 7.0) and very high in the latter (as low as pH 3.6). By contrast peaty, organic drainages release water low in SO_4 during the growing season but they release high amounts of organic anions (A-), consequently, these waters maintain decreased pH values, usually < 4.5. A study of over 80 wetlands and lakes during the ice free period in Nova Scotia showed that sea salt corrected SO_4 concentrations range from 45 ueq L^{-1} in the south end of the province, ~30 ueq L^{-1} in the Kejimkujik area and < 17 ueq L^{-1} in the northern areas with values > 85 ueq L^{-1} in the Halifax area, reflecting the atmospheric deposition pattern of SO_4. The SO_4 concentrations may be > 2000 ueq L^{-1} in drainages containing gypsum, > 700 ueq L^{-1} in drainages over pyrite bearing rocks but < 20 ueq/L^{-1} in streams draining bogs. The SO_4 concentrations change considerably during the non-growing season when the ground is saturated with water or frozen, and the runoff is high (snow and rain often alternate in winter). Under such conditions SO_4 concentration drops in the two former cases and increases in bog drainages, accompanied with a considerable drop in (A-) concentrations. Care should be taken when interpreting SO_4 concentrations in surface waters in Nova Scotia with respect to atmospheric SO_4 deposition.

Water, Air, and Soil Pollution **31** (1986) 207–214.

1. INTRODUCTION

Atmospheric H_2SO_4 deposition is considered the primary source of acidification in surface waters, particularly in Nova Scotia where nitrate concentrations in surface waters are very low (M.O.I., 1983; Kerekes et al., 1982). Thompson and Hutton (1985) showed that SO_4 concentrations in surface waters reflect atmospheric H_2SO_4 deposition in Canada, the highest concentrations being in central Ontario at areas of highest deposition while values are declining towards east and west following the gradient of SO_4 deposition. On a much smaller scale, Underwood et al. (1986) showed that SO_4 concentrations in sensitive lakes in Nova Scotia can be attributed to SO_4 deposition. Kerekes (1984) indicated that SO_4 concentrations show a strong seasonal variation in highly colored, organic waters. Thus there is a need to examine seasonality of SO_4 concentrations in surface waters in Nova Scotia, to establish how representative water samples taken at various times of the year are with respect to annual mean SO_4 concentration. Also, there is a need to identify types of water where geological or edaphic conditions influence SO_4 concentrations.

2. METHODS

Water samples were collected directly in polyethylene bottles, and were stored until analysis for color (within a week) and ionic constituents (within 2 mo.). The analytical procedures used are given in Table I.

Table I: Analytical methods employed. Code numbers refer to the NAQUADAT dictionary of codes of detailed procedures. (Anon. 1983).

Parameter	Method	NAQUADAT Code
pH	glass electrode	10301L
Ca	atomic absorption	20110L
Mg	atomic absorption	12107L
Na	flame emission	11103L
K	flame emission	19103L
Al	atomic absorption	13305P
Fe	atomic absorption	26304P
SO_4	ion chromatograph	16309L
Cl	ion specific electrode	17205L
N- (NO_3,NO_2)	colorimetric	07110L
DOC	U.V. digestion, colorimetric	06104L
Color	Pt-Co	02011L
Gran Alkalinity		10110L
A- (Organic Anion)	Oliver et al. (1983)	

Dissolved organic carbon, when not available, was estimated from water color using the relationship (DOC mg L^{-1} = 0.115 x Color + 2.7) developed from a survey of 49 bodies of water (color range 5 to 290 hazen units, DOC range 3.0 to 36.0 mg L^{-1}) in Kejimkujik National Park (Kerekes, 1981).

3. RESULTS

The average concentration of major ions (not corrected for water discharge) for 20 water bodies in Nova Scotia (Figure 1) is given in Table II. The water bodies may be broken into three distinct groups. Those

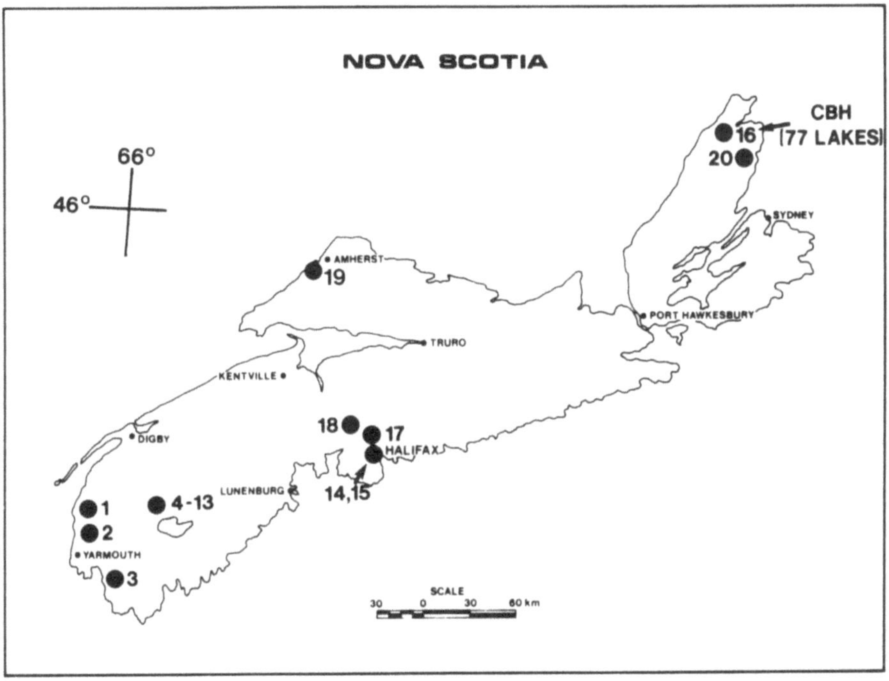

Fig. 1: Location of Nova Scotia study of lakes and wetlands.

in group 1 are lying on igneous rocks and receive SO_4 from atmospheric deposition, although many of them are highly colored and receive peatland drainage which influences the seasonality of SO_4 concentrations (Kerekes et al., 1984a). These water bodies lie along the southwest and northeast axis of the province. The "excess" (sea salt corrected) sulphate (SO_4*) concentrations are 30 to 45 ueq L^{-1} near the southwestern extremity of the Province (lakes No. 1 to 3), 21 to 31 ueq L^{-1} in the Kejimkujik area (No. 4 to 13), 90 ueq L^{-1} near Halifax (No. 14 and 15) and ~17 ueq L^{-1} on the Cape Breton Highlands on the northeast (No. 16, average of 77 lakes). The second group of lakes (No. 17 and 18) lying on disturbed slate bedrock, have high SO_4* concentrations (200

TABLE II: Major ionic constituents of various Nova Scotia wetlands bedrock substrate types.

#	LOCATION	DATE	n	COLOUR	CATIONS Ca^{++} (ueq L^{-1})	Mg^{++}	Na^{+}	K^{+}	H^{+}	Fe	Total Al	ANIONS SO_4^{--} (ueq L^{-1})	SO_4^{*--}	HCO_3^{-}	Cl-	A-	$\frac{CAT-AN}{CAT+AN}$ (% Diff)
GROUP 1 - Igneous bedrock																	
1	Boarsback L.	Summer 83-84	5	150	35.4	55.4	172	7.2	31.7	14.5	23.8	62.0	44.0	0.0	174	86.0	2.6
2	Jessie L.	Summer 83-84	4	20	60.8	75.8	210	12.2	1.5	13.2	9.7	70.7	45.9	32.5	240	42.3	-0.3
3	Sand Pond L.	Summer 83-84	6	290	30.8	47.0	174	7.4	35.2	35.9	30.9	45.8	30.3	0.0	150	163.0	0.3
4	Big Dam East L.	05/83-11/83	4	10	31.9	32.9	122	5.1	2.0	1.0	8.9	43.7	31.2	17.1	121	28.0	-1.5
5	Mountain Lake	07/83-11/83	5	< 5	19.2	27.9	112	5.1	5.0	1.9	8.2	42.4	31.0	3.1	110	18.5	-1.7
6	Beaverskin L.	06/82-05/83	66	6	18.8	29.6	115	7.2	5.5	3.0	5.7	43.3	31.2	0.0	116	26.7	0.3
7	Pebbleloggitch L.	06/82-05/83	59	110	18.5	30.1	120	6.9	52.3	8.8	24.6	47.2	35.4	0.0	113	95.0	1.2
8	Grafton L.	06/82-05/83	61	35	47.5	44.4	150	5.0	2.1	12.4	6.5	47.2	30.9	14.8	156	56.3	-1.2
9	L. Mersey R.	06/82-05/83	60	60	36.6	39.4	134	6.3	12.6	9.9	16.1	51.2	37.4	0.0	133	75.6	-1.0
10	Little R.	06/82-05/83	36	100	34.0	33.6	130	6.3	20.4	22.4	23.6	43.4	31.5	0.0	115	105.0	1.3
11	Atkins B.	06/82-05/83	52	180	33.4	39.5	130	5.5	54.9	23.9	20.9	32.9	21.5	0.0	112	133.0	2.2
12	Mt. Tom B.	06/82-05/83	35	130	34.6	38.6	129	5.6	33.3	16.3	26.5	28.2	16.5	0.0	113	123.0	0.6
13	West R.	06/82-05/83	36	150	32.4	39.6	140	6.0	42.6	20.1	33.7	37.0	25.4	0.0	111	141.0	4.3
14	Suzie L.	05/80-11/83	14	5	43.9	36.2	157	7.7	17.0	3.9	25.6	99.8	87.9	7.7	115	10.2	11.2
15	Kearney L.	05/80-11/83	17	5	115.0	52.6	335	12.2	3.0	3.6	14.5	133.0	95.7	27.9	361	21.1	-0.6
16	CBH Lakes	19/5/83	77	5-140	24.4	27.5	116	7.3	9.1	4.7	19.4	29.8	17.4	7.0	120	68.0	4.1
GROUP 2 - Disturbed pyritic slate bedrock																	
17	L. Springfield L.	10/82-08/84	39	< 5	334	214	1827	23.0	251.0	39.4	267.0	992	779	0.0	2059	2.7	-1.6
18	Lacey Mill L.	09/82-11/83	15	< 5	118	115	309	10.2	28.0	5.4	47.8	228	200	0.0	271	6.4	11.2
GROUP 3 - Gypsum bedrock																	
19	Laytons L.	Summer 83	3	10	2548	417	1699	60.3	0.0	5.5	23.4	2325	2262	633	1988	—	-2.0
20	Freshwater L.	Summer 83	1	< 5	528	107	605	17.4	0.0	0.6	2.2	360	290	239	677	18.0	1.3

* Sea salt corrected sulphate

Note: NO_3 < 3 ueq L^{-1} in all cases

to 799 ueq L^{-1}), which is the result of the biological oxydation of pyritic minerals (Kerekes et al., 1984b). Sulphate concentrations reach high levels in the third group of lakes, those lying on gypsum (No. 19 and 20), 2262 and 295 ueq L^{-1} respectively.

Water bodies in the first group may be broken down further into two categories with respect to the seasonal pattern of SO_4* concentrations:

a) clear headwater lakes with very little seasonal variation in SO_4* (Figure 2a) and colored, organic-rich lake outlets and streams which have lakes on their catchments (Figures 2b to 2e) of sufficient

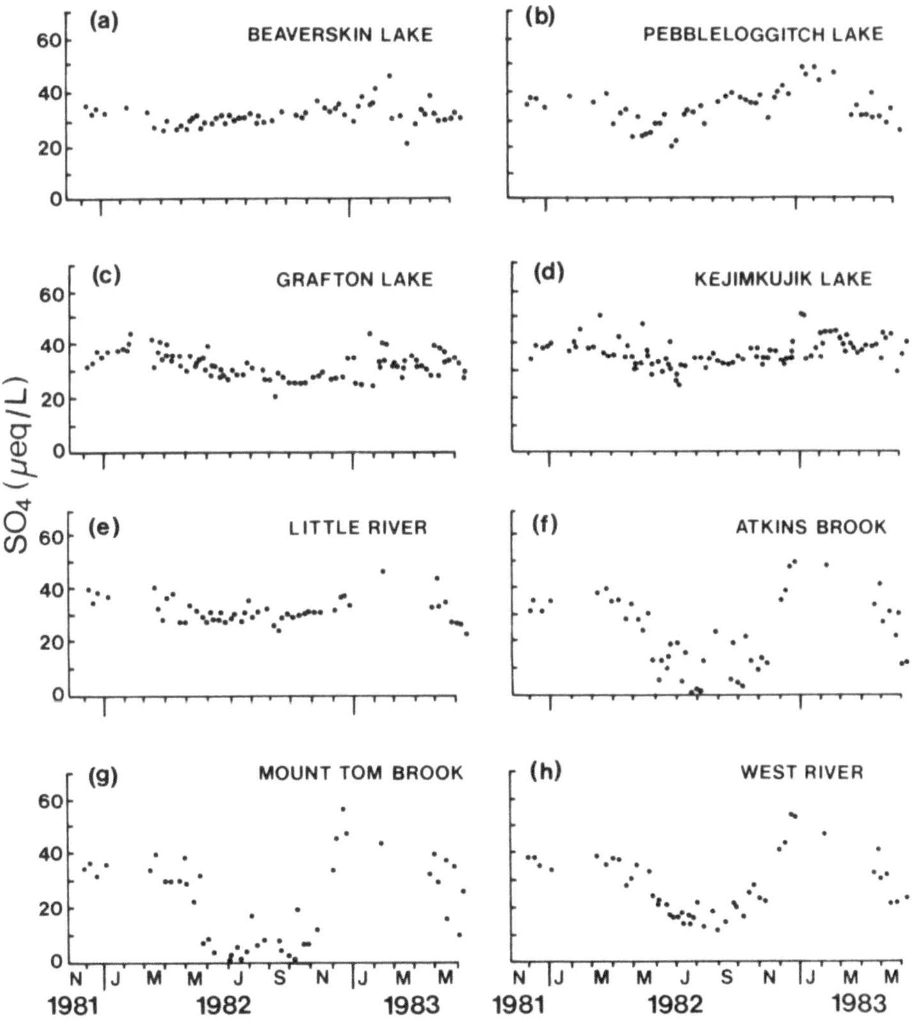

Fig. 2: Seasonal variations in excess sulphate concentration in selected waterbodies from November 1981 to May 1983.

water storage capacity to dampen the seasonality of SO_4* concentrations.

b) streams draining peatland or organic soils with strong seasonality in water discharge, where seasonal variation of SO_4* is very pronounced (0 to 60 ueq L^{-1}, Figures 2f to 2h). Sulphate and organic anions show strong opposing, cyclic patterns in these water bodies (Kerekes, 1984).

Water bodies in the first group are all acidic and exhibit a wide range of H^+ concentration (pH 4.26 to 5.82) with the highest H^+ concentrations in Atkins Brook (No. 11) where the Ca level is fairly low but organic anions are the highest. Hydrogen ion concentrations are the highest (pH 3.6 to 4.54) in the second group (pyritic bedrock) where the increase of SO_4 is not matched with similar increases in Ca and Mg concentrations although Ca and Mg levels are higher than in the first group. The third group, which includes a lake with the highest SO_4 concentration, is well buffered and the pH is greater than 7.0.

Lakes within the first group having glacial till on their catchment have higher Ca and Mg concentrations (e.g. 61 and 75 ueq L^{-1}, No. 2) than those lying on bedrock with little or no glacial till (19 and 30 ueq L^{-1}, No. 6 and 7).

4. DISCUSSION

Sulphate concentrations in water bodies lying on igneous bedrock reflect the gradient of atmospheric SO_4* deposition in Nova Scotia, which is reported as 20.0 kg $ha^{-1}yr^{-1}$ (Kerekes et al., 1982) in Kejimkujik, somewhat higher in the southwestern extremities of the Province, 41.0 kg $ha^{-1}yr^{-1}$ in Halifax and 18.3 kg $ha^{-1}yr^{-1}$ on the Cape Breton Highlands (Underwood et al., 1986). Streams draining peaty, organic soils show very strong seasonal variation of SO_4* as the result of seasonal anoxia in the bogs (Kerekes et al., 1984a). These water bodies have lower annual mean concentrations of SO_4* (not corrected for discharge) than those free from peat deposits, or those which have large water storage capacity (e.g. lakes) in their catchments. However, both of these types of water bodies are similar in terms of the amount of SO_4 exported annually. The reason for this apparent contradiction is that very low SO_4 concentrations in the organic-rich waters occur at times of low water discharge during summer; consequently they contribute very little to the total export. By contrast, the higher concentrations, which are similar to those of other water bodies, are associated with high water discharge from late fall until spring and provide the bulk of the total export (Kerekes, 1983).

The H^+ concentrations of the lakes in Table II reflect the interaction of sensitivity (e.g. Ca and Mg levels), atmospheric deposition of SO_4* (Henriksen 1980) and the amount of organic anions present. For example, lake No. 15 near Halifax with a relatively high Ca concentration, receives more than twice as much SO_4* deposition as the extremely dilute low Ca lake No. 8 in Kejimkujik, which is slightly more acidic then the former. Colored waters having high levels of organic anions exhibit higher concentrations of H^+ than clear water lakes of similar Ca, Mg and SO_4 concentrations (e.g. lakes No. 7 and 6).

Seasonal changes in SO_4 concentrations follow different patterns, depending on edaphic conditions and the nature of the runoff. Sulphate concentrations change markedly in highly colored, organic streams lying on granite. Low values persist during the summer; the highest occur in fall when the ground becomes saturated with water and runoff is high. They remain high during winter and spring until the discharge begins to fall off. However, during periods of low SO_4 levels, H^+ remains high along with organic anions in these highly organic streams (Kerekes, 1984). By contrast, SO_4 concentrations show very little seasonal variation in clear headwater lakes. Colored, organic water bodies with considerable water storage on their catchments show some seasonality with somewhat lower values during the summer months. Sulphate drops considerably from dilution at times of high water discharge in water bodies enriched with SO_4 from edaphic sources (gypsum, pyrite oxydation). For example, in an extreme case SO_4 in the major inflow of Little Springfield Lake (No. 17), drops from 1995 ueq L^{-1} (pH 2.7) to 631 ueq L^{-1} (pH 3.2) between low (summer) and high discharge (snow melt with rain) respectively (Kerekes, 1983). Consequently, care should be taken when SO_4 concentration and acidity are interpreted with respect to atmospheric SO_4 deposition in Nova Scotia water bodies not sampled at sufficient time intervals.

REFERENCES

Anonymous: 1983, NAQUADAT Dictionary of Parameter Codes. Ion chromatographic sulphate determination No. 16309 - Environment Canada, Inland Waters Directorate, Water Quality Branch. Ottawa.

Henriksen, A.: 1980, 'Acidification of freshwater - a large scale titration.' pp. 68-74 in: Ecological impact of acid precipitation. SNSF Project. Oslo, Norway.

Kerekes, J.: 1981, unpublished data.

Kerekes, J.: 1983, unpublished data.

Kerekes, J.: 1984, 'Review of the regional aquatic program,' pp. 83-85 in: Proc. Fall 1984 Workshop of Atlantic Region LRTAP Working Group. Environment Canada. Atmospheric Environment Service. Bedford, N. S.

Kerekes, J., Howell, G., Beauchamp, S., and Pollock, T.: 1982, Int. Revue ges. Hydrobiol., 67: 679.

Kerekes, J., Howell, G. and Pollock, T.: 1984a, Verh. Internat. Verein. Limnol., 22: 1811-1817.

Kerekes, J., Freedman, B., Howell, G. and Clifford, P.: 1984b, 'Comparison of the characteristics of an acidic eutrophic, and an acidic oligotrophic lake near Halifax, Nova Scotia.' Water Pollution Research Journal of Canada, 19: 1-10.

M.O.I.: 1983, U.S./Canada Memorandum of Intent Final Report, Work Group 1., Environment Canada, Downsview, Ontario.

Oliver, B.G., Thurman, E.M. and Malcolm, R.L.: 1983, Geochimica et Cosmochimica Acta, 47: 2031.

Thompson, M.E. and Hutton, M.B.: 1985, Water, Air and Soil Pollution. 24: 77.

Underwood, J.K., Odgen III, J.G., Kerekes, J.J. and Vaughan, H.H.L 1986, 'Acidification of Nova Scotia III: Atmospheric deposition of SO_4 and NO_3 and effects on rural and urban lakes.' Water, Air and Soil Pollution (in press).

STORAGE AND RELEASE OF MAJOR IONIC CONTAMINANTS FROM THE SNOWPACK IN THE TURKEY LAKES WATERSHED

R.G. Semkin and D.S. Jeffries
Environmental Contaminants Division
National Water Research Institute
P.O. Box 5050, Burlington, Ontario. L7R 4A6 Canada

ABSTRACT. The chemistry of the snowpack and snowmelt was investigated at the Turkey Lakes Watershed during the spring melt period in 1985. Ions in the snowmelt were 2 to 10 times more concentrated than those in the pre-melt snowpack and approximately 50% of the H^+, SO_4 and NO_3 were lost from the snowpack with the first 30% of the melt. Rainfall flowed directly through the snow cover even during the early stages of the melt. Four rain events, which accounted for only 18% of the flowthrough collected, were responsible for about 50% of the H^+ and SO_4 and 37% of the NO_3 ions exported from the snowpack. Sulphate to nitrate equivalent ratios in the pre-melt snowpack were generally about 0.6 but an enrichment of SO_4 relative to NO_3 in atmospheric deposition during the spring (SO_4:NO_3 > 1) resulted in snowmelt with an increasing SO_4 content relative to NO_3.

1. INTRODUCTION

Meltwater from an ablating snowpack can affect the chemistry of receiving waters in two ways. The initial production of meltwater can have an ionic composition more concentrated than that of the initial snow cover. Johannessen and Henriksen (1978) observed that 50 to 80% of the snowpack ionic load was released with the first 30% of the melt. This influx of concentrated and acidic meltwater can generate marked depressions in pH and alkalinity and peak concentrations of NO_3 and NH_4 in stream and lake waters as recorded at the Turkey Lakes (Jeffries and Semkin, 1983) and elsewhere (Galloway and Dillon, 1983). Johannesson et al. (1980) have discussed snowpack loss of ions and resultant stream chemistry observed for various locations in Norway.

Meltwater also contains low concentrations of basic cations and sulphate compared to background levels in stream and lake waters. As a consequence, runoff generated by snowmelt can effectively dilute the concentration of these ions in surface waters, particularly during the later melt stages. Jeffries and Semkin (1983) reported that Ca and SO_4 decreased to minimum values

Water, Air, and Soil Pollution **31** (1986) 215–221.

in the surface waters of Batchawana Lake and that the specific conductance dropped from a normal of 25 to 30 μS to <10 μS during high flow conditions in early May 1982.

We have conducted studies in the Turkey Lakes Watershed (TLW) during the spring of 1985 to characterize the snowpack chemistry at this location and to ascertain which chemical and physical factors influenced the quality and quantity of snowmelt. This work complements earlier snowmelt studies (Jeffries and Semkin, 1983) conducted in the TLW during 1981 and 1982.

2. STUDY AREA

The TLW is an undisturbed, forested basin located on the Precambrian Shield approximately 50 km north of Sault Ste. Marie, Ontario. Catchment area is 10.5 km^2. Annual precipitation from 1981 to 1984 averaged 1212 mm with snowfall accounting for about 30% of the total. Details on the physical and biological characteristics of the watershed are given by Jeffries and Semkin (1982).

3. METHODS

Weekly measurements of the snowpack and bulk deposition were taken at a main meteorological station located just south of the TLW. Snow samples were weighed to determine water equivalence and density and thawed at room temperature for chemical analysis. Bulk deposition was used to approximate the atmospheric loading to the basin despite the uncertainty in measuring the dry deposition component, particularly for sulphur and nitrogen species (Jeffries, 1984). A description of the bulk collector as well as the deposition chemistry is provided by Semkin and Jeffries (1985). Individual rain events during the melt were also sampled with a second bulk collector. Snowmelt was collected from a 1 m^2 plexiglass snow box installed above an insulated tunnel adjacent to the snow coring area. Data from discrete melt samples were volume-weighted to produce daily values of meltwater chemistry over the study period. All samples were analyzed for pH and specific conductivity immediately after collection. Cations and nutrients were measured in accordance with methods outlined by the Department of the Environment (1979); anions were analysed by ion chromatography.

4. RESULTS AND DISCUSSION

Prior to the first major melt, the TLW snowpack recorded a water equivalence of 228 mm (mean of 11 sites) and a density of 0.23. The chemistry of a representative snow core collected at the melt site is shown in Table I. On an equivalent basis, hydrogen ion, nitrate and sulphate accounted for approximately 70% of the ionic composition. The pH of the snowpack was consistent with measurements taken on the pre-melt snow cover since 1981 (Jeffries and Semkin, 1983).

TABLE I Snowpack chemistry at the melt site, Feb. 19/85

Parameter	Concentration (m eq L^{-1})
H	0.0246 (pH=4.61)
Ca	0.0044
Mg	0.0016
Na	0.0047
K	0.0006
NH_4	0.0081
SO_4	0.0123
Cl	0.0045
NO_3	0.0218

Atmospheric deposition in the winter months is generally enriched in nitrate relative to sulphate at the Turkey Lakes as shown by the SO_4:NO_3 equivalent ratio of ~0.6 for the snowpack. The reverse situation occurs in summer when ratio values are generally around 3.0 (Semkin and Jeffries, 1985). Jeffries (1984) reported a similar variation of SO_4 and NO_3 in bulk deposition at Sudbury.

Snowmelt was initiated by rising temperatures and rainfall on February 20, 1985. As shown in Figure 1, precipitation to the snowpack subsequent to this date could be accounted for either in the water equivalence of the snow cover or by increased melt generation. Even during the earliest phase of the melt, the snowpack appeared quite permeable with rainfall moving directly through the snow.

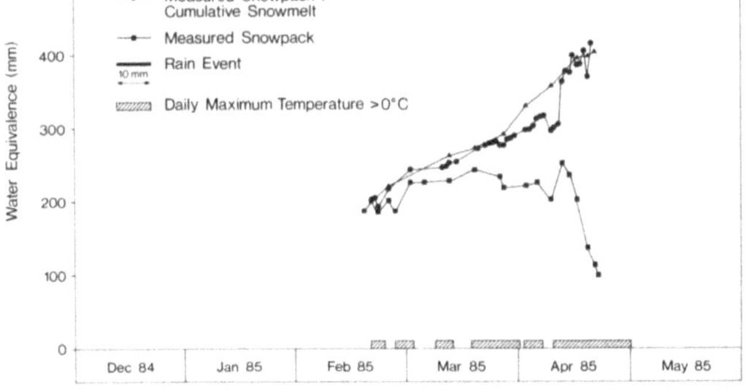

Figure 1. Water budget for bulk deposition, snowpack and snowmelt

Figure 2 illustrates that the pre-melt concentration of selected ions associated with acidity in the snowpack was relatively constant. Heavy rainfall coincident with the warmer temperatures generated snowmelt whose composition was 2 to 10 times more concentrated than that of the snowpack. Early melt waters recorded the highest ionic content and generally decreased in ionic strength as the melt progressed until concentrations approximated those in the remaining snowpack.

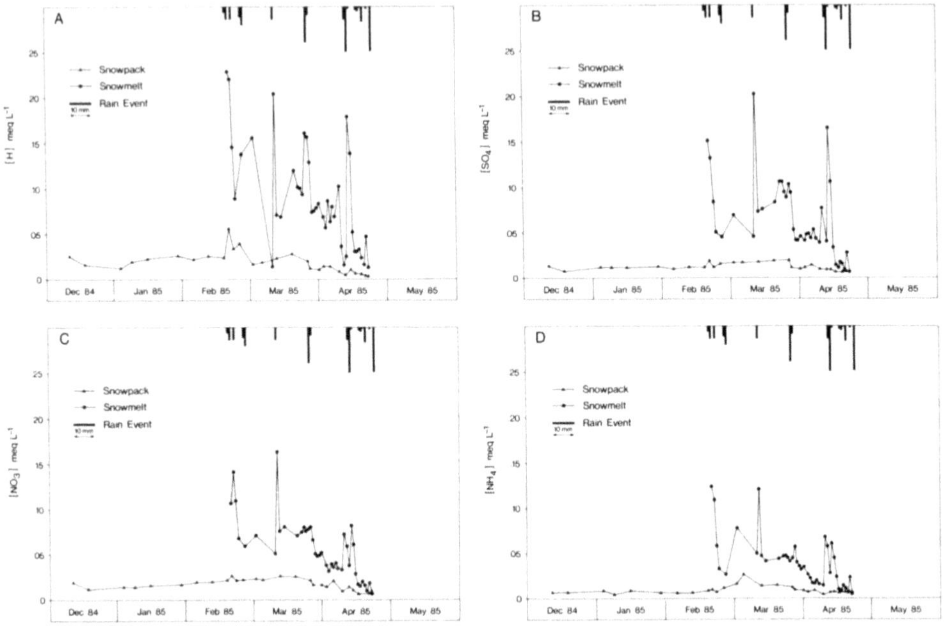

Figure 2. Concentration (meq/L) of major ions in the snowpack and in snowmelt. Note that the peak values in melt coincide with rain events. (A) H; (B) SO_4; (C) NO_3; (D) NH_4.

In terms of material exported from the snowpack, Figure 3 shows that approximately 50% of the H^+, sulphate and nitrate were lost in the first 30% of the snowmelt. All the ions behaved in the same manner with no preferential release of any particular ion.

The highest ionic concentration in the samples from the melt chamber occurred during periods of rainfall. Table II characterizes the chemistry of four major rain events during the spring.

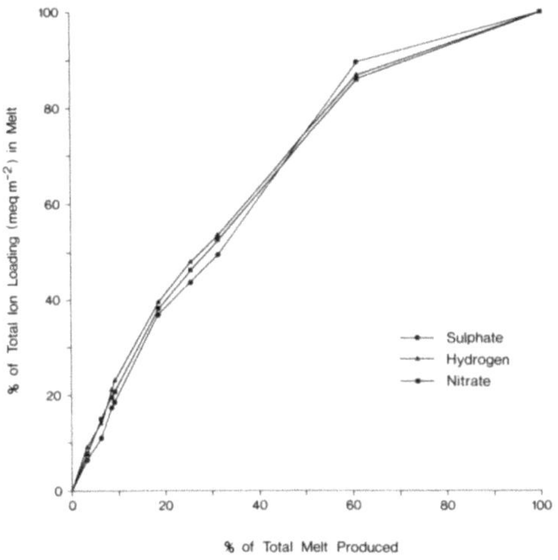

Figure 3. Material export from the snowpack as a function of snowmelt.

TABLE II. Chemistry of spring precipitation

Date	pH	SO_4 (meq L^{-1})	NO_3 (meq L^{-1})
Feb. 20-23	3.55	0.1022	0.0921
Mar. 11	4.12	0.0835	0.0835
Mar. 26-27	3.90	0.1305	0.0694
Apr. 13-14	3.62	0.1678	0.0864

The total rainfall during these episodes accounted for only 18% of the water volume collected; however, in terms of the material exported from the snowpack, 51%, 49% and 37% of the H^+, sulphate and nitrate were transported directly through the snow cover by the rain.

The chemistry of the snowmelt differed from the composition of the pre-melt snowpack as a result of changes in atmospheric deposition during the spring. This is evidenced by the pattern of SO_4:NO_3 equivalent ratios in the respective media. In the winter, nitrate generally exceeded sulphate concentrations in bulk deposition and in the snowpack such that SO_4:NO_3 equivalent ratios were less than 1.0 (Figure 4). With the meteorological conditions prevalent in March and April, sulphate became enriched relative to nitrate in atmospheric deposition and the SO_4:NO_3 ratio increased. As a consequence, SO_4:NO_3 equivalent ratios in the snowmelt were intermediate to those recorded for the snowpack and for bulk deposition during the spring.

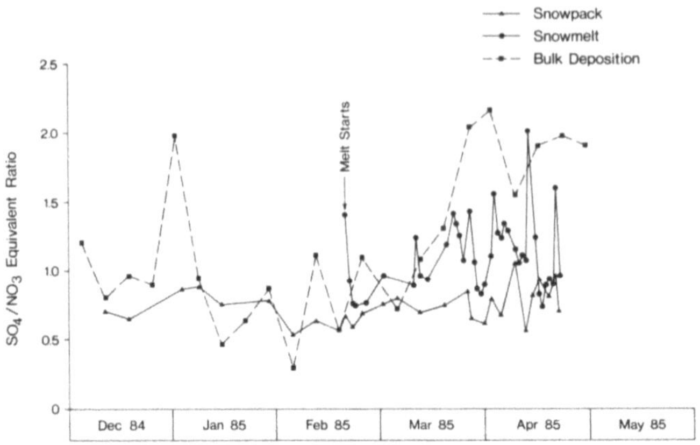

Figure 4. Sulphate/nitrate ratios in bulk deposition, snowpack and snowmelt.

If the ground beneath the snowpack is water-saturated from successive freeze/thaw events common during early melt or is too shallow to store the flux of melt waters, the flowthrough from heavy rainfall can rapidly enter the watercourses in the basin. Abrupt changes in stream and lake water chemistry may result. Observations of stream chemistry at an upland site in the watershed showed that alkalinity concentrations decreased by 40% and 57% following heavy rainfall on February 20 and March 20, 1985, respectively.

5. SUMMARY

Snowmelt at the Turkey Lakes is a complex phenomenon that varies from year to year. Runoff generated by the snowpack will be regulated not only by the energy available for the melt process but also by atmospheric deposition (rainfall) on the snow cover. With limited rainfall to the snowpack during spring, meltwater will reflect the snowpack in chemical composition, i.e. nitrate will be enriched relative to sulphate and initial fractionation of the ions during melting will lead to a short-term enrichment in the early melt-waters. However, if significant precipitation occurs at this time of the year, the runoff chemistry will be influenced by the rain composition and an increasing proportion of sulphate can be expected. The latter situation occurs more often in the TLW than the former.

The chemistry of stream and lake waters can be impacted by the input of concentrated meltwaters. Significant reductions in pH and alkalinity are recorded during periods of high stream flow at the Turkey Lakes. Even during the early stages of spring melt when the hydrologic flux through the basin is relatively low, these chemical changes can be produced by rainfall flowing directly through the snowpack.

ACKNOWLEDGEMENT

The authors thank R. Neureuther, E. Walker and M. Seymour for their technical assistance.

REFERENCES

Department of the Environment: 1979, Analytical Methods Manual. Inland Waters Directorate, Water Quality Branch, Ottawa, Canada.

Galloway, J.N. and Dillon, P.J.: 1983, Effects of acid deposition: the importance of nitrogen. In: Ecological Effects of Acid Deposition, National Swedish Envir. Protection Bd., Rept. PM 1636, Berlings Arloev, Solna, Sweden, p. 145-160.

Jeffries, D.S. and Semkin, R.G.: 1982, Basin description and information pertinent to mass balance studies of the Turkey Lakes Wateshed. Turkey Lakes Watershed Unpublished Report No. TLW-82-01, 34 pp.

Jeffries, D.S. and Semkin, R.G.: 1983, Changes in snowpack, stream and lake chemistry during snowmelt in the Turkey Lakes Watershed. Proceedings of Internat'l Symposium on Acid Precipitation. Lindau, FRG, VDI-Berichte 500, p. 377-386.

Jeffries, D.S.: 1984, Atmospheric deposition of pollutants in the Sudbury area. In: J.O. Nriagu, Ed., Environmental Impacts of Smelters. John Wiley and Sons, New York, p. 117-154.

Johannessen, M. and Henriksen, A.: 1978, Chemistry of snowmelt: changes in concentration during melting. Wat. Resourc. Res. 14, 615.

Johannessen, M., Skartvert, A. and Wright, R.F.: 1980, Streamwater chemistry before, during, and after snowmelt. Proc. Int. Conf. Ecol. Impact Acid Precip., SNSF Project, Sandefjord, Norway, p. 224-225.

Semkin, R.G. and Jeffries, D.S.: 1985, Bulk deposition of ions in the Turkey Lakes Watershed. Turkey Lakes Watershed Unpublished Report No. TLW-85-01, 24 pp.

SYSTEMATIC STUDY OF LAKE WATER ACIDITY IN QUÉBEC

J. DUPONT and Y. GRIMARD
Ministère de l'Environnement du Québec
3900, rue Marly, Sainte-Foy (Québec),
Canada, G1X 4E4.

ABSTRACT. Chemical data (pH, SO_4, Ca, Mg) from 1091 Québec lakes sampled between 1981 and 1985, and 2-yr (1982-1983) precipitation weighted mean sulphate concentrations from REPQ (Réseau d'échantillonnage des précipitations du Québec/Québec Precipitation Sampling Network) were examined to obtain relationships between precipitation sulphate, precipitation sulphate, precipitation acidity, lake sulphate, lake sensitivity and lake acidity. We found good relations between lake and precipitation sulphates, and between atmospheric sulphate and precipitation acidity. We also propose a descriptive model where lake acidity is a function of lake sensitivity and atmospheric sulphate.

1. INTRODUCTION

The concern for surface water acidification arose in the province of Québec at the end of the 1970's (Altshuller and McBean, 1979). Now we know that the largest part of Québec, which lies on the Canadian Shield, has one of the lowest buffering capacities in Canada (Kramer, 1981; Shilts et al., 1981). Moreover, the province of Québec receives precipitation with a mean annual pH ranging from 4.2 to 5.0 (Barrie and Sirois, 1982; Grimard, 1984a).

Between 1981 and 1985, federal and provincial agencies carried out several sampling surveys on water chemistry with the purpose of assessing the spatial pattern of Québec lake acidity and delineating affected areas. Some of these surveys were conducted on a provincial scale (Grimard, 1981; Bobée et al., 1982, 1983; Langlois et al., 1983) and others were made on a smaller regional basis (Jones et al., 1980; Richard, 1982; Potvin and Grimard 1983; Dupont, 1984a,b, 1985a,b). These studies have permitted identification of regions with lakes which have low pH. Affected regions are found mostly on the southern border of the Canadian Shield where precipitation acidity is the highest. Unfortunately, the lack of complete data sets for precipitation composition in Québec has hindered determination of relationships that may exist between lake and precipitation acidity. Because of this lack of information, many authors (Bobée et al., 1982, 1983; Dupont, 1984a,b)

Water, Air, and Soil Pollution **31** (1986) 223–230.

have used lake sulphate concentrations as a surrogate for atmospheric depositions, since the natural background sulphate concentration is usually negligible (Harvey et al., 1981).

In this paper, we pooled pH, sulphate, Ca and Mg data for 1091 Québec lakes taken from several surveys. We also used the recently available data on sulphate concentrations in precipitation measured by the Québec Precipitation Sampling Network (REPQ). We have studied the relationship between lake and precipitation sulphate concentrations, and have also proposed a descriptive 3 dimensional model showing relationships between lake pH, lake sensitivity (Ca+Mg) and precipitation acidity (precipitation weighted mean sulphate concentrations).

2. DATA COMPILATION

The lake water chemistry data came from several sets of field measurements made between 1981 and 1985 by le ministère de l'Environnement du Québec (Richard, 1982; Potvin and Grimard, 1983; Dupont, 1984a,b, 1985a,b) and by Environment Canada (Bobée et al., 1983). We also considered unpublished data from la Banque du Milieu Aquatique (BQMA) of le ministère de l'Environnement du Québec (MENVIQ). Lake selection criteria common to all survey data used here included the following:

1- lake watershed not significantly influenced by human activities (city, industry, agriculture, mine);

2- sample taken at the center of the ice covered lake (between January and March). Exception was made for 60 northern lakes which were sampled in August;

3- integrated sample taken between a depth of 5 m and the lake surface (or from 1 m of the bottom if lake depth was less than 6 m).

For the 933 lakes sampled by the MENVIQ, analyses were performed at the MENVIQ laboratory according to methods described by Longpré et al. (1982). Analyses for the other 158 lakes were made by the laboratory of Environment Canada with comparable methods (Bobée et al., 1983).

Automatic precipitation collectors have been in operation since June 1981 at 46 sites in Québec (Figure 1). Samples are collected on a weekly basis and sent to the MENVIQ laboratory where 15 parameters are analyzed, validated and stored in a computerized data bank (Faucher, 1982). Data used in this study are from wet deposition data published by Grimard (1984a). They also take into account information on the representativity of each sampling site (Grimard, 1984b). A 2-yr (1982-1983) precipitation weighted mean sulphate concentration ($[SO_4^{--}]_p$) was interpolated for each NTS cell (National Transverse System; scale 1:50 000). We assume that every lake associated with each one of these cells (204) is exposed to the corresponding atmospheric sulphate depositions.

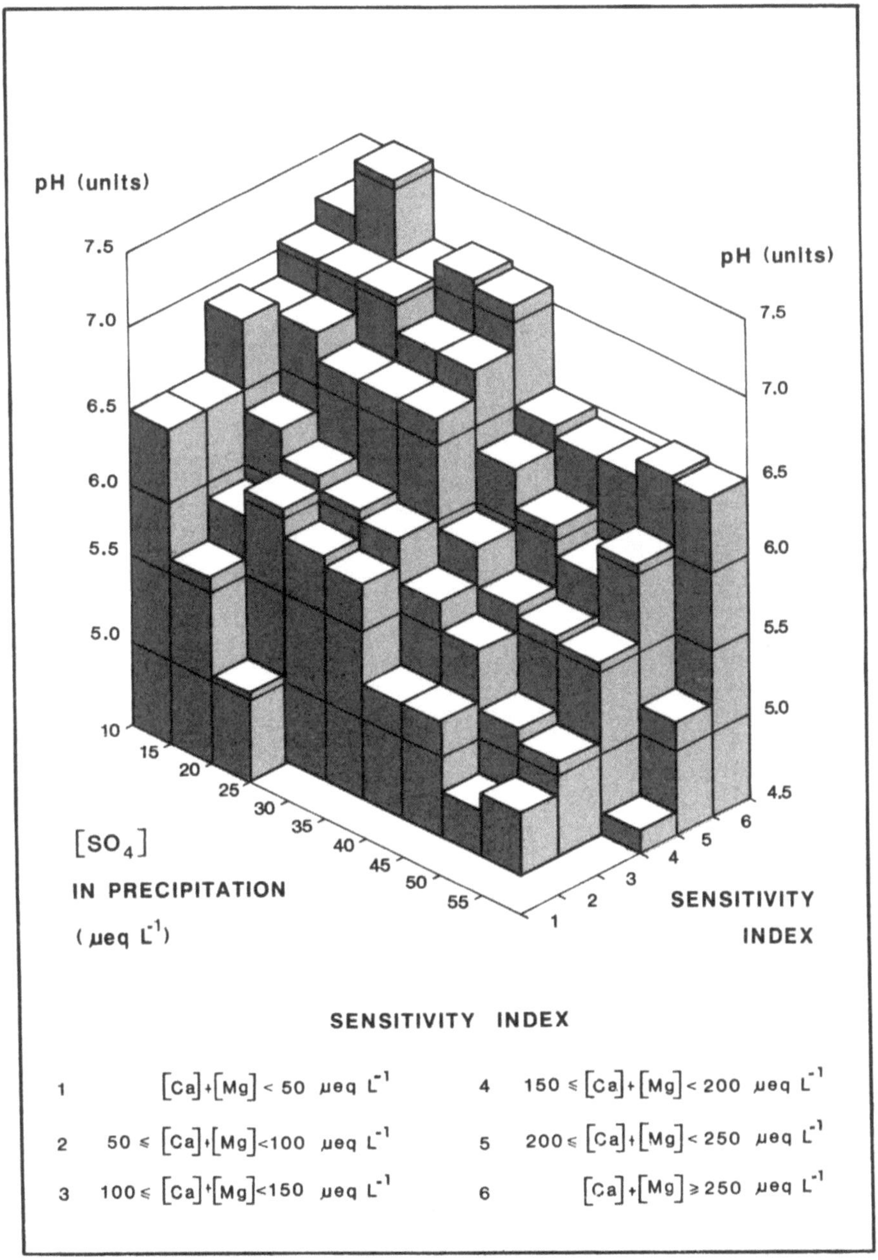

Fig. 2. Block chart relationships between lake pH, Ca + Mg (lake sensitivity) and the precipitation weighted mean sulphate concentrations (precipitation acidity) for 1982-1983.

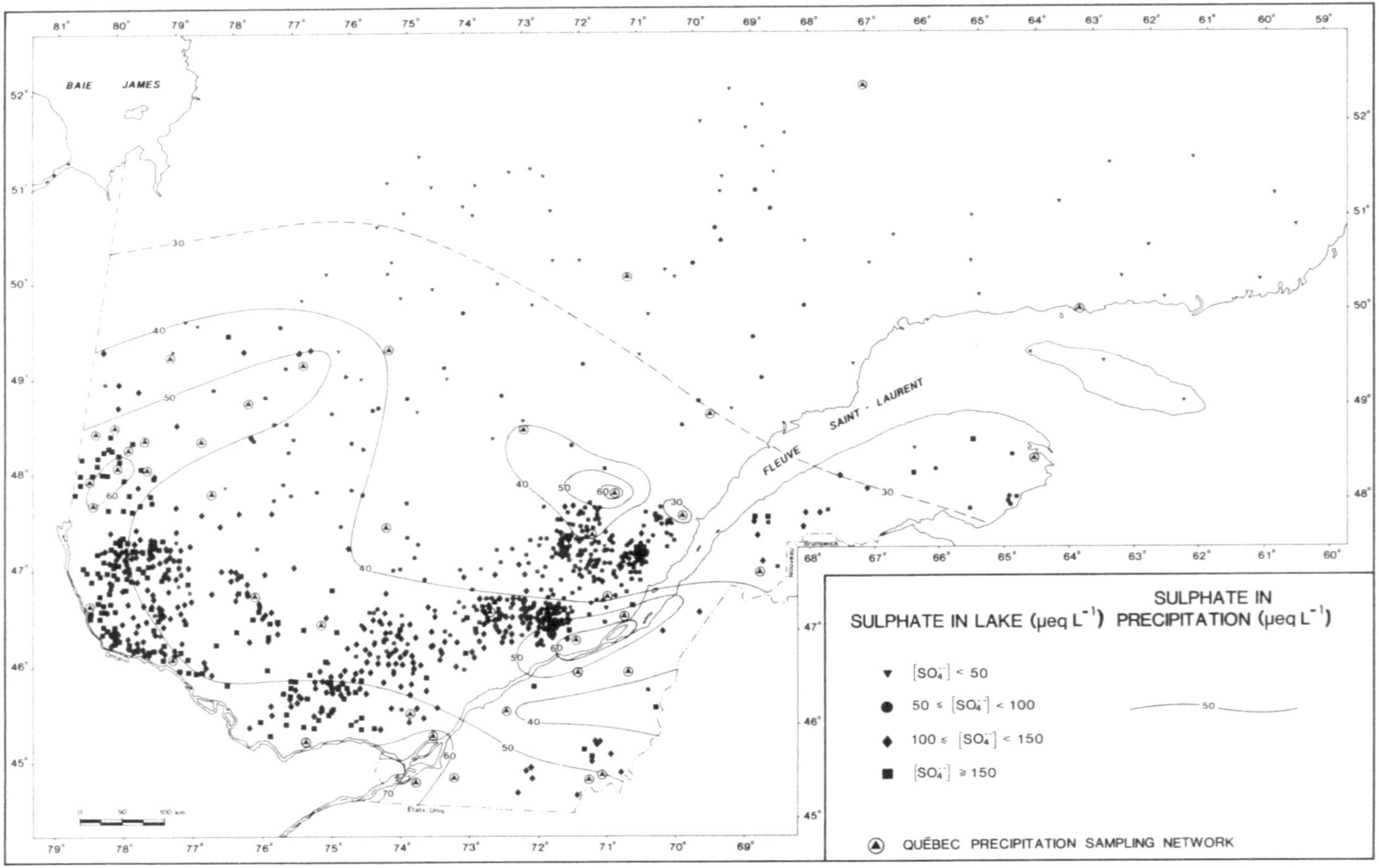

Fig. 1. Spatial distribution of sulphates (μeq L^{-1}) in Québec lakes with reference to the precipitation weighted mean sulphate concentrations (μeq L^{-1}) for 1982-1983.

3. RESULTS AND DISCUSSION

3.1. Relation between lake and precipitation sulphates

Lake sulphate concentrations were split into 4 classes (<50, 50-100, 100-150, and ⩾150 μeq L^{-1}). Occurrence of a specific lake within a given class is presented in Figure 1 along with the spatial distribution of the 2-yr precipitation weighted mean sulphate concentrations. Note that 60 lakes located near Schefferville (55°N-67°W) were not shown in this figure because of their northern position. These lakes exhibit both low lake and low precipitation sulphate concentrations.

The observed spatial distribution of lake sulphate is in good agreement with the one presented by Lachance et al. (1985), even though the two data sets are completely independant. The general pattern shows a strong decreasing gradient from the southwest to the northeast and is very similar to the one observed for sulphate in precipitation. This agreement confirms what many authors have postulated for Québec lakes, i.e. that lake sulphates are good indicators of atmospheric sulphates (Bobée et al., 1982, 1983). In fact, lake sulphate concentrations (μeq L^{-1}) are highly correlated to precipitation sulphate concentrations (both variables expressed as median by NTS cell) when considering lakes with less than 200 μeq L^{-1} of Ca + Mg ($[SO_4^{--}]_L = 3.0\ [SO_4^{--}]_p - 30$: $r = 0.85$). Only lakes with low Ca + Mg concentrations were considered in order to exclude those lakes having a high potential contribution of natural sulphate from geological sources. A relationship comparable to the one described previously has also been observed in Norway by Henriksen (1980) ($[SO_4^{--}]_L = 1.9\ [SO_4^{--}]_p - 19$: $r = 0.84$). It seems this kind of relationship is adequate for large scale studies where natural sulphate contributions are low compared to the atmospheric inputs.

Considering that atmospheric sulphates could be a good indicator of precipitation acidity, we looked at the relationship between $[SO_4^{--}]$ and $[H^+]$ in precipitation and found a high correlation of 0.81. Henriksen (1980) for Norway and Gorham et al. (1984) for the eastern United States observed similar results (0.89 and 0.92, respectively).

3.2. Descriptive model

When studying the problem of surface water acidification, 3 interrelated factors must be considered; precipitation acidity, watershed sensitivity, and the resulting lake acidity. The descriptive model that we propose here uses all 3 factors. We assume that precipitation weighted mean sulphate concentrations are a good indicator of atmospheric acid loadings; Ca + Mg is an index of the watershed sensitivity; and lake acidity is expressed as lake pH.

Figure 2 was constructed by considering 10 equidistant classes of atmospheric sulphate and 6 equidistant levels of sensitivity based on a Ca + Mg index. This arbitrary choice allows a sufficient number of lakes in each class. Afterwards, we computed the mean lake pH for each one of the 60 resulting groups and plotted the 3 dimensional block chart shown in Figure 2.

The block chart clearly shows that for any fixed level of sensitivity, an increase in atmospheric sulphate concentration is generally followed by a decrease in lake pH. An increase in lake sensitivity is also followed by a pH drop, when considering a fixed level of atmospheric sulphate. The model also shows that the steepness of the pH drop is greater when atmospheric sulphate and/or lake sensitivity is high. Analysis of variance for this model, where individual lake pH is the dependant variable and where sensitivity and atmospheric sulphate groups are independant variables, gives us a high level of signification ($P>F$: 0.01) for each explicative variable.

The correlations between lake pH and atmospheric sulphate are significant for each of the 6 sensitivity levels ($\alpha = 0.01$). Their respective correlations, beginning with the most sensitive group, are: -0.96, -0.78, -0.70, -0.50, -0.45, and -0.45. Similar results were found by Lachance et al. (1985) who examined the relation between lake pH and log $[SO_4^{--}]$ in lake for different fixed levels of Ca. For Ca-levels, their absolute values of correlation coefficients are lower than ours, but the lake density is lower ($n = 258$) and they considered lake sulphate instead of atmospheric sulphate.

Dickson (Almer et al., 1978) has found similar results in relating lake pH in Scandinavia to annual wet sulphate depositions (Kg ha^{-1} yr^{-1}) with reference to two levels of lake sensitivity. Our approach is quite similar to his, but we found that for our lakes, atmospheric sulphate concentrations showed better correlations than wet sulphate depositions which are respectively -0.93, -0.55, -0.31, -0.04, -0.18, and -0.23. The difference between these results and those presented before could be related to the fact that wet sulphate depositions do not take into account the dilution factor introduced by precipitation. This factor could be of major importance in areas where the variability of the precipitation is high, which is the case for the province of Québec.

Actually, we consider the model as a good descriptive approach for summarizing the acid precipitation impacts on Québec lakes. In the near future, we will look further into the development of this model as a predictive management tool for target atmospheric sulphate concentrations or depositions.

4. CONCLUSION

Based on a large set of data of Québec lake and precipitation chemistry, the study shows that good relationships exist between lake and atmospheric sulphate concentrations and between H^+ and sulphate concentrations in precipitation.

A model was proposed which related lake pH to atmospheric sulphate concentrations and lake sensitivity expressed as Ca + Mg. This model clearly shows the importance of atmospheric sulphate and lake sensitivity on lake pH. Thus, the more sensitive the lake, and/or the higher the atmospheric sulphate concentrations, the lower the lake pH will be.

REFERENCES

Almer, B., Dickson, W., Ekstrom, C., and Hornstrom, E.: 1978, 'Sulfur Pollution and the Aquatic Ecosystem', in Nriagu, J. (ed.), *Sulfur in the Environment*, part II, John Wiley, Toronto, pp. 271-311.

Altshuller, A.P. and McBean, G.A.: 1979, *The LRTAP Problem in North America: A Preliminary Overview*, Report of the United States - Canada research consultation group on the long-range transport of air pollutants, U.S.A. State Department and Canada Department of External Affairs, 48 p.

Barrie, L.A. and Sirois, A.: 1982, *An Analysis and Assessment of Precipitation Chemistry Measurements made by CANSAP (Canadian Network for Sampling Precipitation), 1977 to 1980*, Environment Canada, Atmospheric Environment Service, Toronto. Report AQRB82-003-T.

Bobée, B., Grimard, Y., Lachance, M., and Tessier, A.: 1982, *The Nature and Extent of Acidification in Québec Lakes*, ministère de l'Environnement du Québec, rapport PA-1, 210 p.

Bobée, B., Lachance, M., Haemmerli, J., Tessier, A., Charette, J.Y., and Kramer, J.: 1983, *Évaluation de la sensibilité à l'acidification des lacs du sud du Québec et incidences sur le réseau d'acquisition de données*, rapport scientifique no 157 pour Environnement Canada, INRS-Eau, 198 p.

Dupont, J.: 1984a, *Nature et étendue de l'acidification de 64 lacs de la région de Rouyn-Noranda*, ministère de l'Environnement du Québec, rapport PA-9, 152 p.

Dupont, J.: 1984b, *Nature et étendue de l'acidification de 97 lacs de la région de Portneuf*, ministère de l'Environnement du Québec, rapport PA-10, 113 p.

Dupont, J.: 1985a, *Étude de la physico-chimie de 22 lacs du Parc de La Vérendrye en rapport avec les processus d'acidification des eaux lacustres*, ministère de l'Environnement du Québec, rapport PA-13, 74 p.

Dupont, J.: 1985b, *Étude de la variabilité spatiale de la physico-chimie des lacs de la région de l'Outaouais en relation avec les précipitations acides*, ministère de l'Environnement du Québec, rapport PA-17, 112 p.

Faucher, M.: 1982, *Détermination des caractéristiques physico-chimiques des précipitations*, ministère de l'Environnement du Québec, rapport PA-3, 85 p.

Gorham, E., Martin, F.B., and Litzau, J.T.: 1984, *Science* **225**, 407.

Grimard, Y.: 1981, *Contribution à l'étude de la vulnérabilité des eaux lacustres québécoises face à l'acidification des précipitations*, ministère de l'Environnement du Québec, Service de la qualité des eaux, rapport no 81-6, 38 p.

Grimard, Y.: 1984a, *Réseau d'échantillonnage des précipitations du Québec: Sommaire de la qualité des eaux de précipitation 1981-1982-1983*, ministère de l'Environnement du Québec, rapport PA-11, 163 p.

Grimard, Y.: 1984b, *Réseau d'échantillonnage des précipitations du Québec: Analyse et évaluation de la qualité des sites d'échantillonnage*, ministère de l'Environnement du Québec, rapport PA-12, 77 p.

Harvey, H.H., Pierce, R.C., Dillon, P.J., Kramer, J.R. and Whelpdale, D.M.: 1981, *Acidification in the Canadian Aquatic Environment - Scientific Criteria for Assessing the Effects of Acidic Deposition on Aquatic Ecosystems*, Publication NRCC no 18475, 369 p.
Henriksen, A.: 1980, in Drablos, D. and Tollan, A. (eds.), *Ecological Impact of Acid Precipitation*, SNSF-project,pp. 68-74.
Jones, H.G., Ouellet, M., and Drakke, D.J.: 1980, in Drablos, D. and Tollan, A. (eds.), *Ecological Impact of Acid Precipitation*, SNSF-project, pp. 226-227.
Kramer, J.R.: 1981, *Calcite Saturation Index, Alkalinity, and Alkalinity Corrections to Previous Data*, McMaster University, Department of Geology, Environmental geochemistry report, Number 1981/1, 8 p.
Lachance, M., Bobée, B. and Grimard, Y.: 1985, *Water, Air and Soil Pollution* **25**, 115.
Langlois, C., Vigneault, Y., Desilets, L., Nadeau, A. and Lachance, M.: 1983, *Évaluation des effets de l'acidification sur la physico-chimie et la biologie des lacs du bouclier canadien (Québec)*, ministère des Pêches et Océans Canada, Rapport technique canadien des sciences halieutiques et aquatiques no. 1233, 129 p.
Longpré, G., Joubert, G., and Trottier, J.: 1982, *Guide d'information sur l'analyse physique, chimique, biologique et bactériologique des milieux environnementaux*, ministère de l'Environnement du Québec, Direction des laboratoires, 152 p.
Potvin, P., and Grimard, Y.: 1983, *Étude de la qualité du milieu aquatique de 67 lacs à sauvagine de la région de Schefferville, Québec*, ministère de l'Environnement du Québec, rapport PA-7, 108 p.
Richard, Y.: 1982, *Relation entre le niveau d'acidité de 158 lacs du Parc des Laurentides et l'évolution de leurs statistiques de pêche au cours des neuf dernières années*, ministère de l'Environnement du Québec, rapport PA-5, 115 p.
Shilts, W.W., Card, K.D., Poole, W.H., and Sandford, B.V.: 1981, *Sensitivity of Bedrock to Acid Precipitation: Modification by Glacial Processes*, Paper 81-14, Geological Survey of Canada, 7 p., 3 maps.

TEMPORAL VARIATION IN ALUMINUM SPECIATION AND CONCENTRATION DURING SNOWMELT

W. H. HENDERSHOT, A. DUFRESNE, H. LALANDE AND F. COURCHESNE
Department of Renewable Resources,
Macdonald College of McGill University,
Ste. Anne de Bellevue,
Québec, H9X 1C0 Canada

ABSTRACT. Variations in Al speciation, H^+ concentration and electrical conductivity in the stream and shallow groundwater wells were studied during spring snowmelt in a small watershed in the southern Laurentian Mountains of Québec. During the first major flush that coincides with high temperatures and heavy rainfall, both complexed and uncomplexed monomeric forms increased significantly in stream and groundwaters. During the following weeks when the discharge was continually high there was a distinct dilution in all the forms of Al. By late April/early May, when discharge decreased and soil temperatures increased, the values of all forms of Al began to rise towards those measured at the beginning of the snowmelt.

1. INTRODUCTION

The spring snowmelt period is widely recognized as being critical to the aquatic environment. The flush of pollutants from the terrestrial to the aquatic ecosystems can cause a sharp drop in the pH of lake or stream waters and a rise in the amount of dissolved reactive Al (Schofield and Trojnar, 1980; Charette et al., 1984; LaZerte, 1984). These chemical changes are believed to be related to sudden fish kills in lakes, even where the water chemistry during the rest of the year poses no threat to fish (Grahn, 1980; Overrein et al., 1980). The spring melt is also responsible for the recharge of surface and groundwaters in many catchment basins; hence the chemistry of meltwaters is of prime importance to the long-term degradation of fish habitats.

The objective of this paper is to show that variations in stream water chemistry are closely tied to those taking place in the soils.

Water, Air, and Soil Pollution **31** (1986) 231–237.

2. MATERIALS AND METHODS

2.1 Field Study

The research was carried out in the Lac Pin Blanc watershed. Details on the geology and vegetation can be found in Hendershot et al. (1984, 1985). Shallow groundwater wells located near the stream and pond were monitored during the snowmelt period. The wells are constructed of 5 cm diameter ABS plastic pipe, open at the bottom and pierced every 7.5 cm with two 10 mm holes on opposite sides. The water table height was measured daily and the water sampled about every five days. Stream discharge was measured continuously with a Belford water level recorder. Stream water was sampled using a Manning automatic water sampler and bulked to give four samples per day.

2.2 Laboratory Analyses

All samples were collected and stored in 1 L polyethylene bottles. Immediately on arrival in the field laboratory, the pH of the samples was measured using a Radiometer PHM 83 digital pH meter. The samples were then filtered through S & S 595 filter paper and stored at 4°C in the dark. Electrical conductivity (EC) was measured with a Radiometer CDM 83 digital meter. Aluminum speciation was determined using the 8-hydroxyquinoline (oxine) method as discussed in Lalande and Hendershot (1986). Research has shown that the first measurement (Al_1) includes mainly monomeric uncomplexed Al such as Al^{3+}, $AlOH^{2+}$, $Al(OH)_2^+$ and perhaps the dimer $Al_2(OH)_4^{2+}$. The second (Al_2) measures the same forms as Al_1 plus monomeric Al bound to fluoride or organic ligands. The third (Al_3) is considered to be the total reactive Al that, in addition to the forms measured by Al_2, includes polymeric forms of reactive Al that may be in either organic or inorganic form. In this paper Al_2-Al_1 is defined as complexed monomeric Al and Al_3-Al_2 as polymeric Al.

3. RESULTS AND DISCUSSION

Figure 1 presents the results obtained for the variations in stream discharge and water chemistry between March 21 (day 81) and May 25, 1984 (day 146). March 21 was the beginning of the period when the stream was ice-free and rapid snowmelt and high discharge began.

The stream hydrograph (Figure 1a) began with a low flow of about 7 L s^{-1}, discharge then rose considerably to form three major peaks on days 97-98, 108, and 116 with a smaller peak on day 130. These major peaks are due to the combination of warm temperatures and rainfall causing the discharge to attain average daily values of between 160 to 180 L s^{-1}.

Although data were available for six groundwater wells space limitations dictate that only one, well 7, can be discussed here. Figure 2a shows that the water table position followed a similar pattern to that of stream discharge when the decreased sampling density is taken into consideration. The water table rose to a maximum during the first major melt event (day 96 to 98) then dropped to relatively low values (day 102). A second maximum was observed on days 107-109 corresponding to the second major snowmelt event.

Figure 1b shows the variation of Al concentrations in the stream. Before day 95 the amount of total reactive Al (Al_3) is fairly high, around 14 ueq L^{-1}, with a weak relationship to stream discharge. Beginning on day 96, there is a sharp rise in Al concentration to 20 ueq L^{-1} that corresponds to the first major peak in discharge. The peak in Al concentration is the result of the flushing of soluble material from the soils aided by the rise in the water table and the generation of quickflow through surface horizons in lowlying areas. During this event the concentration of Al in well 7 peaked at about 27 ueq L^{-1}. Once this first flush has passed, the relationship of the stream water chemistry to discharge changes abruptly; by the end of the first major flush (day 98) Al concentration drops to about 10 ueq L^{-1}, in both the stream and well 7, as discharge increases. This pattern is maintained until after day 130 and is termed the dilution effect since concentration decreases as discharge increases as would be the case if a source of pure water was being added to the solution flowing through the watershed.

During the later part of the snowmelt there is a gradual rise in Al concentration from 12 to 22 ueq L^{-1} in the stream due to the increase in soil temperature and the concomitant increase in biological activity. Together they result in more rapid decomposition and solubilization of both organic and inorganic soil materials. Coupled with this is the decrease in leaching rate of the soils leading to higher concentration in the stream.

When the Al speciation in the stream is viewed as a whole it can be seen that uncomplexed monomeric Al (Al_1) and polymeric Al (Al_3-Al_2) contribute about equally to the total reactive Al concentration throughout the melt period. Complexed monomeric Al (Al_2-Al_1) is more abundant than the other two forms with its contribution increasing slightly after approximately day 125. In general the Al_1 and Al_2-Al_1 covary although there are periods when there is no apparent similarity in their patterns. On the other hand, temporal variations in Al_3-Al_2 are independent of the other two.

One of the major differences between the behavior of the stream and that of the groundwater is visible in the way the Al-speciation changes with time (Figure 2b). In the stream the peaks in total reactive Al (Al_3) were caused by increases in uncomplexed (Al_1) and complexed (Al_2-Al_1) monomeric Al. In the wells most of the variation is in Al_1 with a relatively smaller contribution from Al_2-Al_1. Considering that the most labile forms of Al are the most toxic to aquatic life, this

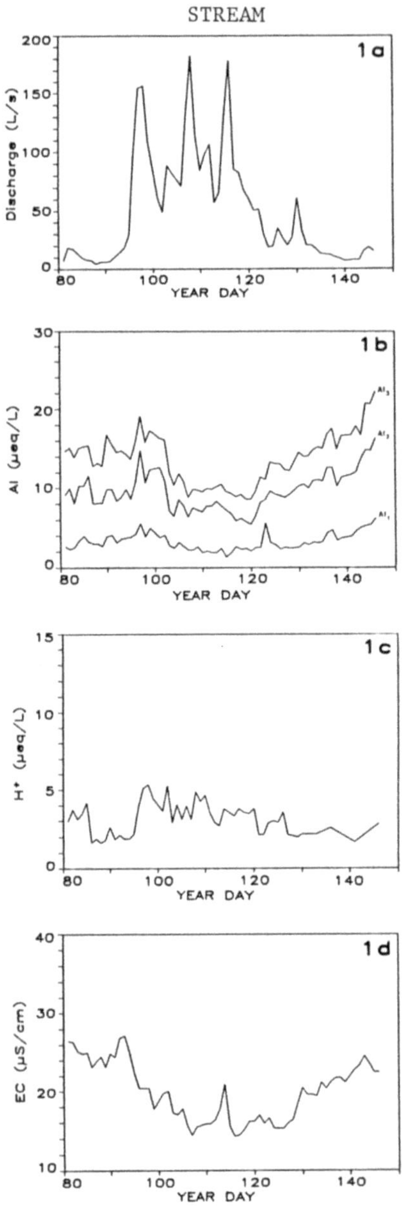

Figure 1. Variation in discharge (1a), Al-speciation (1b), H concentration (1c) and electrical conductivity (1d) in the stream during spring snowmelt period.

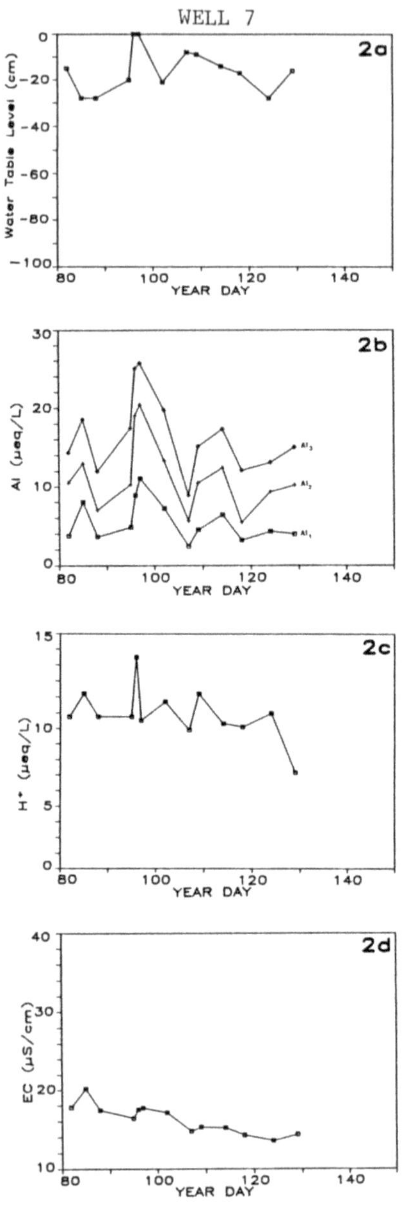

Figure 2. Variation in water table height (2a), Al-speciation (2b), H concentration (2c) and electrical conductivity (2d) in well 7 during spring snowmelt period.

transformation of uncomplexed into complexed monomeric Al, as the water moves from groundwater to the stream, is a benefitial phenomenon. In part the change in Al speciation is attributed to the higher pH in the stream that is in turn the result of water flowing through deeper till layers. These were not sampled by our shallow groundwater wells but would have higher pH.

During the initial period from day 81 to 100 there is a co-variation in H^+ and stream discharge (Figure 1c); after this date the H^+ variations do not seem to follow any such pattern. Some of the time the H^+ variations follow a similar trend to those of Al_i (monomeric uncomplexed Al) as would be expected from the pH-solubility relationship of Al^{3+}, but this is not maintained throughout.

The H^+ concentration in the well (Figure 2c) responds in much the same way as was observed in the stream. Except for the rise in H^+ concentration during the first major melt event (days 96 to 98) there is no clear correlation between H^+ and water table position. The H^+ concentration of the wells tends to be higher than that measured in the stream. Again this points to a flow of deep groundwater into the stream.

Early in the melt period the electrical conductivity (EC) of the stream is high (around 24 uS cm^{-1}) and drops rapidly with the rise in stream discharge after day 93 (Figure 1d). During the period of maximum snowmelt between days 95 and 125 the EC is relatively low, around 16 uS cm^{-1}, and is inversely related to discharge; towards the end of the snowmelt EC increases to 20 uS cm^{-1} as discharge drops. Although the intensity of the variation of EC with discharge is somewhat different from the variation of Al with discharge, the trend is the same. The results for EC and Al both show an apparent dilution of dissolved material in the stream during the period of maximum melt.

The variations in EC in the well (Figure 2d) follow the same trend as in the stream. EC is highest at the beginning of the snowmelt period and gradually drops. During the melt period the EC of well 7 dropped from 20 to 14 uS cm^{-1}. From the variations in EC in the well it is clear that the dilution effect is not just a question of stream water being diluted by pure water reaching the stream without contacting the soil; the water in the soil profile itself becomes more dilute with time.

4. CONCLUSION

The close correlation of the behavior of the groundwater well to that of the stream clearly demonstrates the importance of the terrestrial component of a watershed in controlling the stream chemistry. The temporal variations in Al-speciation, H^+ and EC all support this interpretation.

Hypothesis 1 - Flowpath

The chemistry of the stream reflects the dominant flowpath of water through the soils in the watershed. This is most clearly seen by comparing the baseflow conditions before day 95, when meltwater was mainly passing through deep soil horizons, with the first major melt event (days 95 to 98), during which quickflow through surface horizons was generated. The change in flowpath lead to the highest concentrations in dissolved Al and H^+ in both the stream and groundwater wells. The high concentration of Al, particularly the complexed monomeric form (Al_2-Al_1) provides unequivocal evidence that the meltwater is making chemically effective contact with the soil.

Hypothesis 2 - Concentration/Dilution

Early in the melt period there is a particularly large flush of many dissolved species into the stream as is seen by the peaks in Al, H^+ and EC. The high concentration of the first meltwater is due to the washing out of: a) The readily soluble products of natural decomposition of both mineral and organic soil materials; and b) the pollutants accumulated through the winter in the snowpack, intensified by the concentration and early release of ionic material as the snow recrystallizes (Johanessen and Henriksen, 1978). One important source of dissolved material is the fresh litter from the previous fall. The litter is very easily decomposed and provides a rich source of bases and Al (Hendershot et al., 1984).

Once the readily soluble material has been leached into the stream the concentration drops, even though the water continues to flow through surface soil horizons. During this period the soils are leached with relatively pure water coming from the snowpack; due to the large volume of water moving through the soil and the low soil temperatures the soil is unable to maintain as high a concentration in either the groundwater or stream.

Hypothesis 3 - Biological Activity

During the period of continuously high flow (days 100 to 130) low concentrations are maintained; when the snow has melted over most of the watershed the flow of water through the soils becomes slower and soil temperature begins to rise. The higher soil temperature promotes biological activity that increases the rate of decomposition and production of easily dissolved material. Since concentration increases as a function of a) decreased flow rate and b) increased soil temperature it is difficult to be sure which factor is dominant. However, the continued increase in concentration after day 130, during a period of relatively steady flow, suggests that soil temperature is important.

ACKNOWLEDGMENTS

This research was made possible by grants from the Natural Sciences and Engineering Research Council of Canada, Environment Canada, Fisheries and Oceans Canada, and Le Ministère de l'Education du Québec - FCAC.

REFERENCES

Charette, J.Y., Jones, H.G., Sochanska, W. and Gauthier, J.M.: 1984, 'Changes in lake water quality during spring runoff in a northern boreal forest subjected to acidic precipitation'. Proc. Can. Hydrol. Sym., Univ. de Laval, Québec, in press.

Grahn, O.: 1980, 'Fishkills due to high aluminum concentration in lake water'. Proc. Int. Conf. Ecol. Impact Acid Precip. Norway, 1980, SNSF-Project, pp. 310-311.

Hendershot, W.H., Dufresne, A., Lalande, H., and Wright, R.K.: 1985, 'Speciation of aluminum in different compartments of a drainage basin during snowmelt.' Proc. East. Snow Conf. Montreal, June, 1985 (In Press).

Hendershot, W.H., Lalande, H. and Dufresne, A.: 1984, 'Aluminum speciation and movement in three small watersheds in the southern Laurentians'. Water Poll. Res. J. Can. 19, 11.

Johannesen, M. and Henriksen, A.: 1978, 'Chemistry of snow meltwater: change in concentration during melting.' Water Resour. Res. 14, 615.

Lalande, H. and Hendershot, W. H.: 1986, 'Aluminum speciation in some synthetic systems: Comparison of the fast oxine pH 5.0 extraction and dialysis methods'. Can. J. Fish. Aquat. Sci. 43, 231.

Lazerte, B.D.: 1984, 'Forms of aqueous aluminum in acidified catchments of southern Ontario: a methodological analysis'. Can. J. Fish Aquat. Sci. 41, 766.

Overrein, L.N., Seip, H.M. and Tollan, A.: 1980, Acid precipitation effects on forest and fish. SNSF. Final Report, 175 p.

Schofield, C.L. and Trojnar, J.R.: 1980, 'Aluminum toxicity to brook trout (Salvelinus fontinalis) in acidified waters'. Polluted rain. Rochester International Conference on Environmental Toxicity, 1979. New York: Plenum Press. pp 341-366.

PREDICTED CHANGES IN STREAMWATER CHEMISTRY FOLLOWING REDUCED/ INCREASED SULPHUR DEPOSITION USING THE "BIRKENES MODEL"

H.M. Seip, N. Christophersen and S. Rustad
Center for Industrial Research
Box 350 Blindern
0314 OSLO 3
Norway

ABSTRACT. A sensitivity analysis of some important parameters in a mathematical model previously found to reproduce major short-term variations in water chemistry in some streams, has been carried out. The results are also relevant in a discussion of possible changes in stream acidity for changes in S deposition. In the simulations sulphate concentrations in streamwater were doubled or halved compared to present day values. Assuming no change in the lime potential of the soil, the peak acidity during snowmelt changed with up to 0.9 pH units; shifts from 0.4 to 0.5 units occurred in many cases. By simultaneously changing the sulphate stream concentration and the lime potential with 0.2 units shifts of about 1 pH unit were obtained in the simulations for some acidity peaks.

1. INTRODUCTION

One of the main questions related to freshwater acidification is: What changes in freshwater chemistry will result from given changes in emissions of S- and N compounds. Mathematical modeling is one useful tool in attempting to answer this question. Here we present some results obtained with the "Birkenes model" by forcing changes in stream sulphate concentrations. The lime potential and the soil CO_2 pressure have also been varied. The results may be regarded as a sensitivity analysis, but are also relevant when discussing possible changes in streamwater acidity for various deposition levels. The Birkenes model has reproduced major short-term variations in streamwater chemistry in two catchments in Norway (Christophersen et al. 1982, 1983/84, 1984) and one in Canada (Seip et al., 1985; Rustad et al., 1986). The results given here are based on the version of the model used for a tributary (Inflow # 4) to Harp lake, Ontario. The lowest pH values in this stream occur at high flow, particularly during snowmelt.

Water, Air, and Soil Pollution **31** (1986) 239–246.

2. THE MODEL

The hydrological model, which forms a basis for the chemical model, consists simply of two soil reservoirs and a snow reservoir (Figure 1). The main processes operating in the two soil reservoirs are also given in the figure. The model is based on the mobile anion concept (Seip, 1980) and stream sulphate concentrations are calculated by a submodel. Bicarbonate and organic anions are also included. The cation concentrations are determined by the charge balance and other equations describing the processes in Figure 1 (Rustad _et al._, 1986). The main equations in the upper reservoir are (all concentrations in mol L^{-1}):

$$[H^+]/[M^{2+}]^{1/2} = K_{G1} \qquad (K_{G1} = 10^{-2.27}) \qquad (1)$$

$$[Na^+]/[M^{2+}]^{1/2} = K_{G2} \qquad (K_{G2} = 10^{-2.31}) \qquad (2)$$

$$[Al^{3+}][H]^{-3} = {}^*K_{so} \qquad ({}^*K_{so} = 10^{8.1}) \qquad (3)$$

$$[HCO_3^-][H^+] = K\, P_{CO_2}^{soil} \qquad (K = 10^{-7.71}) \qquad (4)$$

where M^{2+} is the sum of Ca^{2+} and Mg^{2+} in soil water. $P_{CO_2}^{soil}$ is the CO_2 partial pressure in the soil. The numerical values given were used by Rustad _et al._ (1986). In the lower reservoir we assume Equations (3) and (4) to be valid, while cation exchange (Equations (1) and (2)) is replaced by weathering type processes. The CO_2 partial pressure in the upper reservoir was kept constant. The pressure varies seasonly in the lower reservoir but is constant during winter and snowmelt. For further details see Rustad _et al._ (1986).

3. SCENARIOS

The relationship between emissons and deposition of S compounds is not considered here. Neither do we discuss the time interval before a change in deposition is reflected in streamwater concentrations.

In this study we have assumed several scenarios which differ in concentrations of sulphate in streamwater and/or in lime potential (LP = $-\log K_{G1}$ = pH + 1/2 log $[M^{2+}]$) of the upper soil horizon. In addition to the present sulphate concentrations (approximately 200 μeq L^{-1} during snowmelt) we have used halved and doubled values obtained by halving or doubling the deposition _and_ adjusting S pools in the soil reservoirs correspondingly. Changed deposition will also, most likely, affect the lime potential of the soil. Shifts in the lime

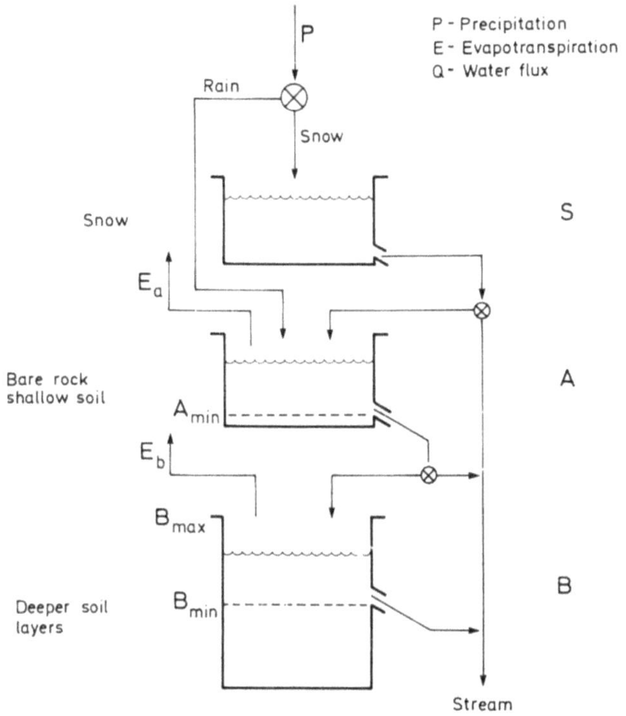

Processes Operating

	Shallow soil reservoir (A)	Deeper soil reservoir (B)
H_2O	Precipitation, evapotranspiration, infiltration to lower reservoir, discharge to stream	Infiltration, evapotranspiration, discharge to stream, piston flow
SO_4^{2-}	Wet + dry deposition, adsorption/desorption, mineralization	Adsorption/desorption, reduction during warm periods
$Ca^{2+}+Mg^{2+}$	Ion-exchange	Release by weathoring, adsorption/desorption
H^+	Ion exchange and equilibrium with gibbsite	Consumption by weathering, adsorption/desorption, equilibrium with gibbsite
Al^{3+}	Equilibrium with gibbsite	Equilibrium with gibbsite, adsorption/desorption
HCO_3^-	Equilibrium with a fixed CO_2-pressure	Equilibrium with a seasonally varying CO_2 pressure

Figure 1 Hydrological model used for Harp lake catchment and main processes operating

potential of the upper soil of 0.2 units may conceivably be caused by acid deposition (Tamm and Andersson, 1985; Christophersen *et al.*, 1985), but also by other causes such as forest growth. The lime potentials used in the simulations are LP = 1.9, 2.1, 2.3, and 2.5, cfr. Equation (1).

The results are found to be sensitive to the values for the CO_2 pressure in the soil, particularly in the lower reservoir, and this parameter has been varied. Other parameters are as given in the previous section or by Rustad *et al.* (1986).

4. RESULTS

Results are given for H^+-peaks occuring during snowmelt. In Figure 2 we show the peaks obtained for the present day sulphate level compared

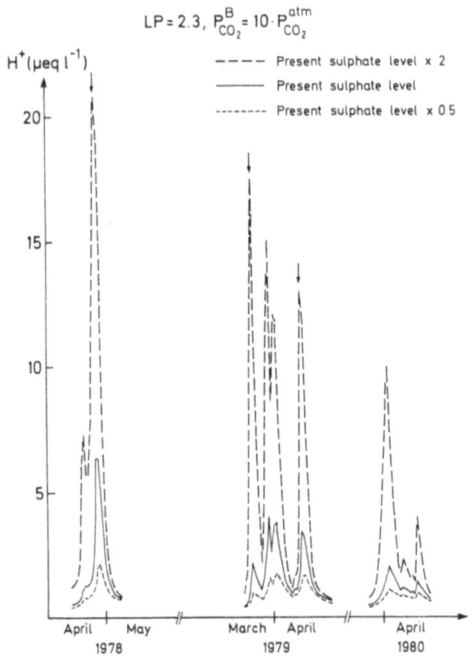

Fig. 2 H^+-concentrations during snowmelt calculated for three sulphate levels.

to those obtained by halving or doubling the level. The lime potential is equal to 2.3 in these calculations, and the CO_2 pressure during snowmelt in the lower soil equal to 10 times that in the atmosphere. The shifts vary considerably, the largest shift (0.9 pH units) occurred for the first peak during snowmelt 1979 when the present sulphate level was doubled. The effects of changing both the sulphate concentration and the lime potential are illustrated in Table I. The CO_2 pressure in lower soils was 4 times that in the atmosphere in these calculations. An increase in the CO_2 partial pressure in the lower reservoir tends to increase the pH shifts obtained by doubling the sulphate stream concentrations as seen by comparing Figure 2 with the results in Table I (lime potential 2.3). The average pH shift for the three peaks marked with arrows in Figure 2 was 0.67 units for the higher CO_2 pressure compared to 0.50 units for the lower pressure when the present sulphate concentration was doubled. The pH increases were approximately equal (0.36 and 0.39 units resp.) when the concentration was halved.

The simulated runoff is a mixture of water with bicarbonate from the lower reservoir and a somewhat more acid water from the upper reservoir. Large shifts are obtained when bicarbonate in the mixture is depleted (increased sulphate concentration) or generated (reduced sulphate concentration) (cf. Seip and Rustad, 1984). This feature is not restricted to snowmelt episodes as shown by other simulation results not discussed in this paper.

5. CONCLUSIONS

Model calculations indicated that a halving or a doubling of the sulphate concentrations in streamwater may lead to considerable changes in streamwater pH even if the lime potential of the soil is assumed fixed. Increases from about 0.4 pH units and up to 0.9 units were obtained by halving the sulphate concentration for water with pH in the range 4.6 to 5.2. Such changes are biologically very significant (Schindler et al., 1985). The pH of very acid water (pH <4.5) responded less to sulphate reduction. The model predicted pH changes in streamwater acidity peaks of 0.1 to 0.6 units for shifts in LP of 0.2 units. Simultaneous changes in both deposition (factor of 2) and lime potential (0.2 unit decrease/increase when the deposition is doubled/halved) may result in surface water pH shifts of about 1 unit.

ACKNOWLEDGEMENT

This work was supported by the Norwegian Council for Scientific and Industrial Research and by the British-Scandinavian Surface Water Acidification Programme.

TABLE I

Peak H^+– concentrations (μeq L^{-1}) for three sulphate levels and four lime potentials in the upper soil. The peaks are those marked with arrows in Figure 2. The CO_2 partial pressure in the lower soil reservoir was 4 times that in the atmosphere. P, D, H stand for present, doubled, and halved sulphate levels. ΔpH is the shift in pH going from halved to present or from present to doubled sulphate concentrations.

Peak		LP = 1.9					LP = 2.1					LP = 2.3					LP = 2.5				
		D		P		H	D		P		H	D		P		H	D		P		H
1	H^+	46.2		27.0		12.8	36.5		18.5		7.6	19.5		7.7		3.3	8.9		3.5		2.0
	ΔpH		0.23		0.32			0.30		0.39			0.40		0.37			0.41		0.24	
2	H^+	48.8		29.0		13.3	38.4		19.4		5.8	20.2		5.2		1.6	6.2		1.7		1.0
	ΔpH		0.23		0.34			0.30		0.52			0.59		0.51			0.56		0.23	
3	H^+	39.6		21.0		7.8	29.9		12.6		4.3	13.9		4.4		2.2	5.2		2.4		1.6
	ΔpH		0.28		0.43			0.38		0.47			0.50		0.30			0.34		0.18	

Christophersen, N., Seip, H.M., and Wright, R.F.: 1982, Water Resour. Res., 18, 977.

Christophersen, N., Dymbe, L.H., Johannessen, M., and Seip, H.M.: 1983/84, Ecol. Model. 21, 35.

Christophersen, N., Rustad, S., and Seip, H.M.: 1984, Phil. Trans. R. Soc. Lond. B305, 427.

Christophersen, N., Rustad, S., and Seip, H.M.: 1985, 'Modelling soil-water interactions in catchments in southern Norway'. In: Troyanowsky, C. (ed.), Air pollutants and plants, VCH Verlagsgesellschaft, Weinheim, 230.

Rustad, S., Christophersen, N., and Seip, H.M., and Dillon, P.J.: 1986, Can. J. Fish. Aquat. Sci. 43, 625.

Schindler, D.W., Mills, K.H., Malley, D.F., Findlay, D.L., Shearer, J.A., Davies, I.J., Turner, M.A., Linsey, G.A., and Cruickshank, D.R.: 1985, Science, 228, 1395.

Seip, H.M.: 1980, 'Acidification of freshwater - sources and mechanisms.' In D. Drabløs and A. Tollan (eds.). Ecological Impact of Acid Precipitation, SNSF-project, Norwegian Inst. Water Research, Oslo, 358.

Seip, H.M., and Rustad, S.: 1984, Water Air Soil Pollut. 21, 217.

Seip, H.M., Seip, R., Dillon, P.J., and de Groisbois, E.: 1985, Can. J. Fish. Aquat. Sci. 42, 927.

Tamm, C.O. and Andersson, F., 1985, 'Indirect effects of pollutants on plants and ecosystems.' In: Troyanowsky, C. (ed.), Air pollutants and plants, VCH Verlagsgesellschaft, Weinheim, 245.

WATER ACIDIFICATION BY AIR POLLUTANTS IN THE FEDERAL REPUBLIC OF Germany

Dr. Ing. Joachim Wieting
Federal Environmental Agency
Bismarckplatz 1, D-1000 Berlin 33
Federal Republic of Germany

ABSTRACT. In the Federal Republic of Germany (FRG), airborne pollutants are leading to an increased rate of acidification of aquatic systems. In particular, the upper reaches of rivers and obligotrophic lakes in low-carbonate areas are affected. These areas are a minority of the total number of waterways, but are of great importance for environmental protection. Water acidification cannot be viewed as an isolated issue. In forested areas, it is associated with damage to the entire ecosystem. In lakes and streams, one of the most significant effects of water acidification is fish mortality; several bodies of water in which fish were once plentiful are now devoid of fish. With increasing acidification of the ground water, drinking-water supply problems may also be expected. All measures which can be taken to prevent the entry of pollutants into aquatic systems are preferred to those which merely treat the symptoms of acidified aquatic systems (e.g. addition of lime). The Federal Environmental Agency therefore considers the reduction of air pollutants to be the most effective solution to the problem of acidified of aquatic systems.

1. INTRODUCTION

The problem of water acidification due to air pollutants has been known in the FRG for only a relatively short time. (Schoen et al., 1984) Until a few years ago, humic acid and leaching products from conifers in snow and snowmelt water had been suggested as possible causes of the acidification of geological subsoil. (Steidle and Pongratz in Wieting et al., 1984, Lorenz in Wieting et al., 1984) The problem was therefore thought to result from modern forestry, particularly from the planting of Spruce and monoculture on bed-rock soil (Urgesteinsboden) that was originally scantily covered by mixed forest. Recent research

Water, Air, and Soil Pollution **31** (1986) 247–256.

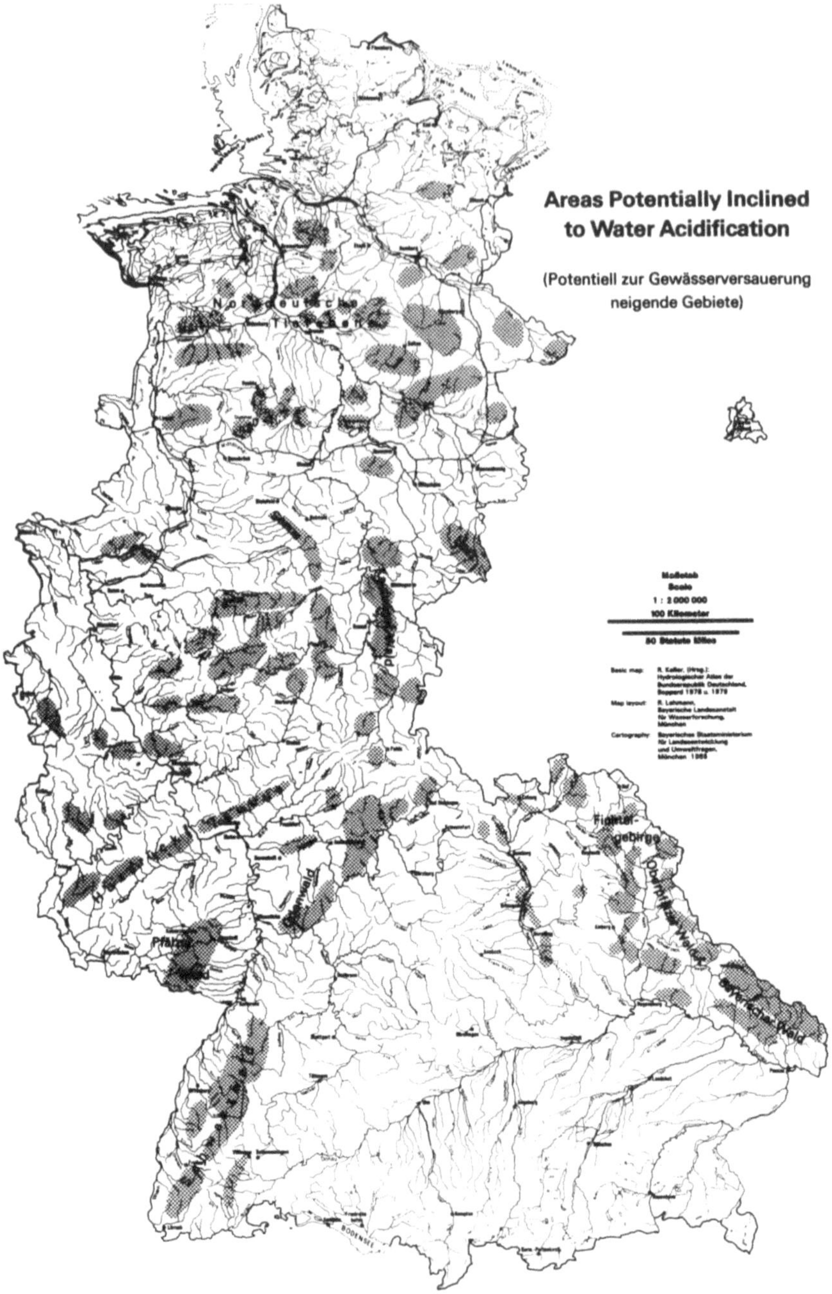

Figure 1 Areas potentially inclined to water acidification. (This map of potential geogenic risk was prepared on the basis of soil and geological maps. Maps of flowing waters, forests, and regions not subject to anthropogenic influences have been taken into account. The regional names, as indicated in the maps, represent a selection of landscape areas.)

in the FRG indicates a well-established relationship between precipitation and pH-lowering, pointing to "acid precipitation" as an additional cause. (Wieting et al., 1984) The problem of water acidification has consequences not only in Scandinavia and North America, but also in Central Europe, particularly in Germany.

The pH value of precipitation in the FRG is around 4.0 to 4.6. (Umweltbundesamt, 1983) This is clearly lower (i.e. acidity is higher) than expected under natural conditions (approximately 4.6 to 5.6 for pre-industrial times). The pH alone, however, is not sufficient to establish the acidifying power of precipitation. Together with the pH value and the amount of precipitation, the ammonium content, as a measure of acids which could be only temporarily neutralized, allows an estimate of yearly acid deposition.

2. REGIONAL OVERVIEW OF ACIDIFIED WATERS

It has been found that water systems with drainage basins composed of non-calcareous rock or sandy soil are most susceptible to acidification. Accordingly, a map showing geogenically endangered areas was developed (see Figure 1). Other areas found to be susceptible to acidification are those which experience emissions, especially where processes tending to oppose acidification (e.g. anthropogenic pollution from pastures, meadows, drainage and sewage inlets) are quantitatively unimportant. These areas are principally far from housing settlements in the upper highlands (Mittelgebirge) of West Germany.

The aspects of geological, pedological and hydrological flowing waters in the FRG were investigated using hydrochemical and hydrobiological methods. The investigation was an attempt to quantify the differences between naturally occurring and anthropogenic acidification of waterways; the results of the investigation for various landscape areas are presented in Wieting et al., (1984). All findings suggest that, in the unpolluted upper reaches of non-calcareous waterways in the FRG, acid concentrations considerably above those expected for natural conditions now exist. Critically low pH values had previously been found only in relatively short stretches of waterways (spring regions), as anthropogenic pollution due to agriculture and sewage buffers acidification over wide areas.

In the case of still waters, oligotrophic lakes with water lacking Ca are most affected by acidification. Such lakes, however, are relatively rare in the FRG. Most lakes exhibit a high buffering capacity, so acidification is not expected. Acidification has now been detected in lakes in the Bavarian Forest, Fichtelgebirge range, northern Black Forest, Lüneburg Heath and in the outwash areas of Northern Germany. In naturally acidic bog and heath ponds, additional acidification of human origin has been detected.

Since long-term measurements of lake acidification were not available, changes due to acid deposition over a long period of time were determined by paleolimnological studies of lake bottom deposits. Because the distribution of certain diatoms is strongly dependent on the pH of the water, pH values were determined retroactively using dated sediment cores. (Arzet and Steinberg in Wieting et al., 1984) A reconstruction of the course of recent pH changes in sediments revealed that acidification has increased since the 1950s, resulting in a reduction of 1 to 2 values on the pH scale (see Figure 2). This corresponds to a 10 to 100-fold increase in the acid concentration.

Only isolated results from investigations of groundwater acidification have been reported. Investigations

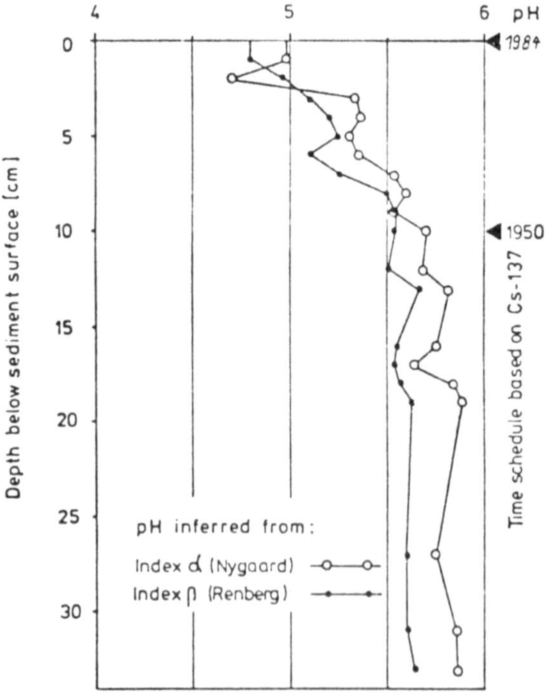

Figure 2 Down-core calculated pH values based on Index α and Index β for Lake Großer Arbersee.

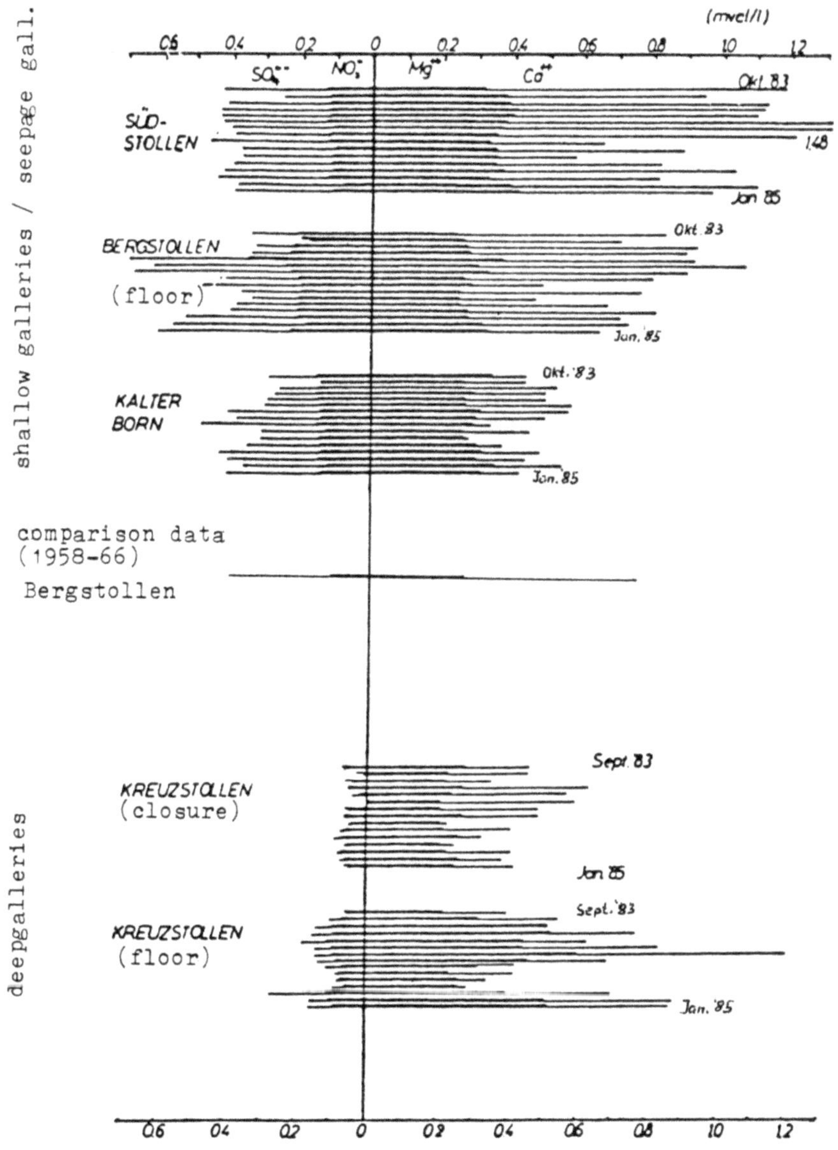

Figure 3 Comparison of Water Acidification in Shallow and Deep water Galleries in Four Areas of the Taunus Region. (Data collected over the period 1983-1985 by Krieter, in Wieting et al., 1984.)

in the Taunus, Hunsrück and Bavarian Urgebirge regions have shown that, in these non-calcareous areas, acidification already occurs in groundwater near the surface. (Krieter and Röder et al. in Wieting et al., 1984) However, due to a lack of comparable data from earlier years, it is unclear in many cases, to what extent the results can be ascribed to the effects of acid deposition on the soil, and to what extent they are ascribable to geogenic stresses on the groundwater.

In the case of the Taunus region, the risk of groundwater acidification can readily be seen, since numerous infiltration galleries permit entry of acidified water to shallow and deep groundwater. In Figure 3, acidification of shallow groundwater by significantly increased concentrations of sulfates and nitrates is compared with acidification of deep groundwater. Since the onset of acidification in the 1960s, concentrations of SO_4, NO_3 have increased up to four-fold. This can be seen by a comparison of data from the Bergstollen from the years 1958-66 and 1985.

Increased acidification of groundwaters can eventually endanger the drinking-water supply. Especially endangered are drinking-water supplies in the countryside where acidification of filter and buffer layers may lead to the washing out of acidic water with dissolved aluminium and heavy metals into the deeper groundwater. Larger waterworks which prepare drinking-water by neutralization and flocculation with Al salts can consistently guarantee hygienic water quality. Smaller waterworks which cannot afford such preparative techniques or thorough analytic facilities may not be able to supply adequate drinking-water.

3. BIOTIC AND CHEMICAL CHANGES DUE TO WATER ACIDIFICATION

The evaluation of the limnological effects of water acidification is complicated by a lack of comparable data, an incomplete documentation of aquatic faunal and floral populations in the FRG, zoogeographic differences, and other factors. Biological studies of the limnological effects of water acidification therefore yield very heterogeneous results. They range from mere suggestions that acidification of the area under examination is probable, to detailed reconstructions of the acidification process in lakes using biological ph'meters. (Brehm and Meijering, 1982; Steinberg in Wieting et al., 1984, Steinberg and Arzet in Wieting et al., 1984)

The greatest number of water acidification investigations in the FRG are carried out using a single method in official water control districts or comparable projects. The method most used, when data are available from older mappings from paleolimnological surveys, is the use of small organisms. Macrozoobenthos are taken as the indicative biocenosis. Species number and unilaterally frequent occurrence or deficiency or specific groups of organisms are used as evaluation criteria to indicate acidification. (Matthias, 1983) In mainly still waters, acidification is indicated by macrophytes. (Melzer and Rothmeyer, 1983)

Societies of diatoms (siliceous algae), Chrysophyta (golden brown algae) and Chydorides (water fleas) have been shown to be highly pH-sensitive. (Steinberg et al., 1984. Steinberg and Hartmann, 1986) The societal structure of these organisms shows characteristic changes due to acidification, and are therefore used as biological pH-meters. Since all three groups leave subfossil remnants in lake sediments, the subfossils can be used to reconstruct the entire acidification process. In the FRG, the diatom and Chydoride pH-meters are used in a wide variety of acid-sensitive geological formations. Chironomides (bloodworms), which also leave subfossil remnants in sediments, have been found to be less acid-sensitive than the groups mentioned above. (Steinberg, Meier, et al., 1984)

Fish fauna is also used on a limited basis to evaluate the extent of acidification. Since fish population and aging structures are influenced by water resources management, impoverishment of species, reproduction numbers, or changes in reproductive conditions can often be misleading. Only the total depletion of fish populations in waters can be taken as a clear indication of water acidification.

Chemical changes in still waters of the FRG is illustrated by the example of the lake, Grosser Arbersee. Figure 4 shows a plot of the distribution of alkaline earth metals, heavy metals and Al as a function of changing pH. The steadily decreasing concentration of alkaline earth metals in the upper centimeters of sediment core is a sign of water acidification. (Steinberg et al. in Wieting et al., 1984) When pH is decreased, Ca and Sr are released from carbonates and humic complexes, while Al, Zn and Pb are concentrated in the upper centimeters of the sediment core.

Chemical surveys of surface waters in the FRG are principally divided into field measurements and laboratory investigations. In the field, pH value, conductivity and temperature are measured. In parallel laboratory measurements, pH values, alkalinity, uv absorption at 254 nm, and concentrations of Ca, Mg, K, Na, NH_3, S, N, Cl, and Al (in 0.45 m filtered sample) are determined. In still waters, samples are drawn in the spring and after the

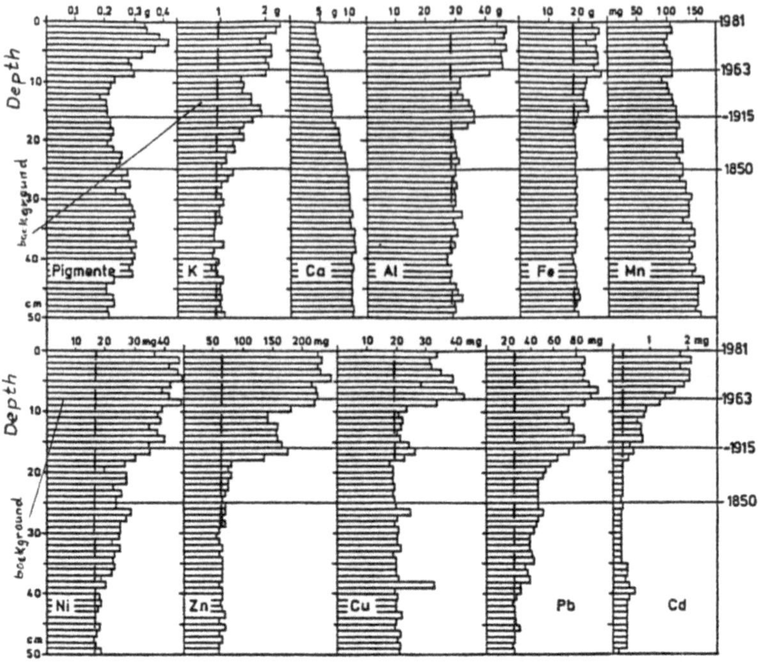

Figure 4 Vertical distribution of selected chemicals in the sediment of Lake Großer Arbersee (in kg^{-1} per dry weight). (Increased pigment content and reduced alkaline earth concentration indicated acidification near the surface. Metals such as Ca and Sn are released from carbonates. Copper, Zn and Cd build up. Aluminium, whose concentration also increased toward the surface, is released mainly by the Al buffer mechanism in the area surrounding the lake and is washed into the lake.)

end of summer stagnation. Flowing waters are measured at low-water periods and during snow melt and heavy precipitation. This minimal control program is complemented, when necessary, by measurement of additional parameters and increased measurement frequency. Extension and intensification of the investigations usually depends on the accompanying biological surveillance programs.

4. SUMMARY

- Acid deposition from the atmosphere has also been found in water systems in the Federal Republic of Germany (FRG). The relationship between acid deposition and water acidification can be considered definitely established.
- An initial inventory in the FRG indicates that acidification is, at present, limited to water systems in areas of non-calcareous rock and sandy soil. These conditions are found primarily far from habitation in the heights of the highlands (Mittelgebirge).
- Soil buffering capacity decreases with increasing acidification of the soil, resulting in increased acidification of groundwaters close to the surface. Acidification of aquifers results in the release of heavy metals and aluminium, which can cause widespread damage to drinking-water supplies.
- Severe biotic and chemical changes have been observed in areas water acidification is occurring. This can lead to fish mortality and complete desolation of waters, accompanied by the release of heavy metals on a large scale. It is expected that the cleanest waters would then suffer bilogical death.
- Acid deposition from the air into aquatic systems occurs today primarily through emissions of SO_2. Nitrogen and S compounds lead to a rapid increase in the acid concentration, above all during periods of snow melt. Although wide-range and long-term reductions in SO_2 emissions are expected in the FRG and have been recorded in some countries, the opposite trend in NO_x emissions has been observed. Increases in water acidification due to N compounds are therefore expected to occur.
- While conventional water pollution occurs through direct introduction of pollutants or by washing-out mechanisms, recent water pollution results from the indirect contact between water and air pollution which primarily results from industry and motor vehicle traffic. Environmental policies aimed at water management do not adequately consider the effects of air pollution on water systems.
- These effects can be held in check by a consistent preventative environmental policy to reduce air-borne pollutants. Maximum allowable levels for pollutants should no longer be established according to the principle of "acute toxicity for humans". Problems resulting from air pollutants must be treated with regard to environmental media such as water and soil. The reduction of air pollutants must be brought about on a national (FRG) as well an international (ECE) level.

References

Brehm, J. and Meijering, M.P.D.: 1982, <u>Arch. Hydrobiologie</u>, 95, 17.

Mathias, U.: 1983, <u>Z. Wasser-Abwasser Forschung</u>, 16, 107.

Melzer, A. and Rothmeyer, E.: 1983, <u>Bericht Bayer. Botanische Gesellschaft</u>, 54, 9.

Schoen, R., Wright, R. and Krieter, M.: 1984, <u>Naturwissenschaften</u>, 71, 95.

Steinberg, C., Arzet, K. and Krause-Dellin, D.: 1984, <u>Naturwissenschaften</u>, 71, 631.

Steinberg, C. and Hartman, H.: 1986, <u>Naturwissenschaften</u>, 73, 37.

Steinberg, C., Meier, R., Emeis-Schwarz, H., Krause-Dellin, D. and Arzet, K.: 1984, <u>Vom Wasser</u>, 63, 35.

Umweltbundesamt: 1983, <u>Jahresbericht 1983</u>, 83-88.

Umweltbundesamt: 1985, "Empfehlungen zur Überwachung von Oberflächengewässer (Gewässerversauerung durch Luftschadstoffe)", <u>Interne Empfehlungen zur Methodik zur Wahl der Entnahmestellen</u>.

Wieting, J., Lehnhart, B., Steinberg, C., Hamm, A. and Schoen, R.: 1984, "Gewässerversauerung in der Bundesrepublik Deutschland", <u>Materialien 1/84 des Umweltbundesamtes</u>, Erich Schmidt Verlag, Berlin.

Additional Literature

Lehnhart, B. and Steinberg, C.: 1984, "Limnochemische und Limnobiologische Auswirkungen der Gewässerversauerung, eine Literaturstudie", <u>Informationsberichte Bayerisches Landesamt für Wasserwirtchaft 4/84</u>.

FACTORS AFFECTING SNOWMELT STREAMWATER CHEMISTRY IN THE BLACK FOREST (WEST GERMANY)

K.H. Feger and G. Brahmer
Institute of Soil Science and Forest Nutrition
Albert-Ludwigs-University
D-7800 Freiburg i.Br., Federal Republic of Germany

ABSTRACT. Site conditions such as parent material, soils, but also vegetation cover and elevation explain the varying snowmelt streamwater chemistry in the Black Forest. The results are derived from multiple statistical analysis of a regional survey of 104 small mountain streams in the first phase of snowmelt in spring 1984. Cluster analysis classifies the snowmelt streams into three groups which are clearly linked to bedrock geology. Factor analysis finds podsolization, weathering and mineralization processes in the soils of the catchments to have most impact even under snowmelt conditions. There is no evidence that acidic atmospheric deposition directly affects the acidity of the investigated streams. However, the deposition rates are low compared to certain other regions in Central Europe. In areas with podsolic soils the organic soil layer plays a key role in the acidity and mobilization of Al and heavy metals. This is shown in the high correlations between pH, DOC, UV-extinction, color and metal concentrations. Because the concentrations of DOC are low (<10 mg.L^{-1}) and an anion deficit cannot be found, it is assumed that water acidity is not caused by dissolved humic acids, but by cations exchanged in the organic layer of acidic soils. Streamwater chemistry in areas with brown earth soil types is mainly affected by leaching of basic cations in the mineral soil horizons and mineral weathering.

1. INTRODUCTION

During the last few decades, concern has been expressed that freshwater quality in areas with siliceous bedrock geology is strongly influenced by anthropogenic acidic depositions (Overrein *et al.*, 1980; Glass *et al.*, 1982; Ulrich and Matzner, 1983). Two possible pathways have been discussed:

(1) Direct, short-term effects: It has been hypothesized that in periods of high streamflow, a high percentage of the runoff water reaches the streams and lakes chemically unchanged (Gjessing *et al.*, 1976). Thus during the wintertime, the atmospheric deposition is collected by snowpack and is released on a short-term basis with snowmelt.

(2) Indirect, soil-mediated effects: Both soils and vegetation play a major role in terrestrial to aquatic element transport (Johnson and Reuss,

Water, Air, and Soil Pollution **31** (1986) 257–265.

1984). The acidity in surface waters is often principally determined by the types of soils and the hydrological runoff patterns (Rosenqvist, 1977; Krug and Frink, 1983; Lefohn and Klock, 1985). But if soil acidity in large areas exposed to acidic depositions increases faster than is natural, it is likely also to have long-term effects on freshwater quality (Abrahamsen, 1985). Seip (1980) explained acidification of surface waters in Norway by an increase of mobile anions mainly due to increased sulphate deposition. Thus, the introduction of the mobile anion, rather than H^+ itself, is a critical factor in the terrestrial to aquatic transfer of H^+ and Al. But there is empirical and theoretical evidence for only slight effects by even high additional inputs of H_2SO_4 on surface water pH-values (Abrahamsen and Stuanes, 1980; Johnson, 1981).

Most of the examinations concerning surface water acidification in the Federal Republic of Germany (Umweltbundesamt, 1984) emphasize the direct effects of atmospheric depositions (Schoen *et al.*, 1983; Schoen and Kohler, 1984). The aim of our studies is to explain the current chemical composition of surface waters in the Black Forest. Which factors are important in determining the spatial variation of water quality? No prediction can be made about a long-term increase of water acidity (acidification), since chemical data from the past are not available. Additionally, there is a general lack of reference data from comparable forested catchments not affected by acid depositions.

2. MATERIAL AND METHODS

2.1. Sites and Field Sampling

In spring 1984, water samples were collected from 104 streams throughout the Black Forest. The catchments are almost completely forested without habitation upstream (Zöttl *et al.*, 1985a,b). Some samples were nevertheless affected by salted roads and sewage effluents and therefore not considered for statistical analysis. The final set of data consisted of n = 86 samples, one sample from each stream. According to the snowmelt progress, samples were collected successively from the lower to the higher altitudes (Brahmer, 1985).

The Black Forest, situated in SW Germany, is a largely forested mountain area of about 6000 km^2. Average annual temperatures vary between 4 and 7° C and annual precipitation between 1000 and 2000 mm due to the varying altitudes between 500 and 1500 m ab.s.l.. Bedrock geology consists of various types of gneiss and granite as well as "Buntsandstein", mesozoic quartzy sandstone extremely poor in basic cations (Geyer and Gwinner, 1968). Depending on parent material, soil forming conditions during and after the last glacial period and type of vegetation cover, a large variety of soil types has developed. On periglacial solifluction material derived from gneiss, deeply weathered brown earths ("Braunerden") prevail (Stahr, 1979). Very acidic podsols are wide-spread on Buntsandstein and some granites (Jahn, 1957; Schweikle, 1971; Stahr, 1973). At the highest altitudes, especially in the NW-area, numerous dystrophic peats with *Sphagnum* moss and mountain pine (*Pinus mugo*) occur. The original forest association, consisting mainly of common beech (*Fagus sylvatica*) and silver fir (*Abies alba*) has been changed in the course of the last few centuries. Intensified silvicultural practices causing serious soil degradation in some areas resulted in highly increased proportions of Norway spruce (*Picea abies*) (Hausrath, 1938; Hausburg, 1967). Mainly in the SW-part, the proportion of deciduous forest is higher (Brückner, 1970).

2.2. Chemical and Statistical Methods

A detailed description of sampling and analytical procedures is given by Zöttl *et al.* (1985b). Water samples were determined for conductivity, pH, HCO_3^- (alkalinity 4.3), UV-extinction (λ = 254 nm) and color (λ = 436 nm) in the laboratory on the day collected. The anions Cl^-, NO_3^- and SO_4^{2-} were analyzed by ionic chromatography, PO_4^{3-} and NH_4^+ photometrically. DOC was measured by CO_2-IR-detection after combustion at 800° C. Ca^{2+}, Mg^{2+}, Fe, Zn and Mn were measured by flame atomic absorption, Na^+ and K^+ by flame emission and Al, Pb, Cu and Cd by graphite furnace AAS-technique. Vials for trace metal analyses had been cleaned by a special procedure involving soaking in 5 % HNO_3 followed by triple rinsing with bidistilled water. After filtration through membrane filters (pore size 0.45 µm) 100-mL-aliquots for metal analyses were preserved with 0.1 mL conc. HCl suprapur. Aliquots for anion analyses were preserved with $CHCl_3$. The samples were stored in the dark at 4° C. The statistical procedures using the standard programs BMDP and SPSS are described in detail by Brahmer (1985).

3. RESULTS AND DISCUSSION

Significant relationships ($p < 0.001$) exist between H^+ and DOC (r=0.73), H^+ and color (r=0.78) as well as between DOC and color (r=0.82). Interestingly, DOC and UV-extinction relate very well (r=0.85). If UV-extinction is high, a greater part of DOC can be expected to consist of humic substances (Sontheimer and Wagner, 1977). Thus, the composition of dissolved organic matter is likely to be similar in all of our samples. There is no significant correlation between H^+ and SO_4^{2-} (r=0.07) and pH and NO_3^- (r=-0.36).

Nearly all metals show a significant relationship ($p < 0.001$) with H^+. The basic cations Ca^{2+} and Mg^{2+} correlate negatively with H^+ (r=-0.85 rsp. r=-0.69). In contrast, Al (r=0.87), Mn (r=0.82) and Fe (r=0.76) show elevated concentrations when H^+-concentrations are high. However, significant correlation coefficients ($p < 0.001$) also exist between these metals and DOC (Al r=0.83 and Fe r=0.81). Mn, Pb, Cd, and Zn behave similary. Hence, a greater part of these metals is thought to be transported by chelation with organic compounds. But this is decisively important for the ecotoxicological evaluation of metal concentrations. For this purpose, metal speciations especially for Al (Driscoll *et al.*, 1980) could be useful.

Due to the highly significant relationships between DOC and the metals Al and Fe, these cations were not considered for calculation of cation-anion balances. Since no anion deficit is evident and the absolute concentrations of DOC are low (<10 $mg.L^{-1}$, we assume that the partly low pH-values are not directly caused by humic acids. It is likely that the acidity in the investigated streams is due to protons being exchanged in the organic layer of acidic soil types. These protons form strong acids together with the mobile anions which may be derived from precipitation but also from mineralization and weathering processes. This is in accordance with a model of Rosenqvist (1977, 1983). Formation of metal-organic complexes may also be an important source of protons (Saar and Weber, 1982; van Breemen *et al.*, 1983).

To classify the 86 sampling sites, a cluster analysis was carried out. We obtained three groups clearly independent of each other. Table I shows the average values of the chemical parameters for each cluster.

TABLE I Mean values of the complete data and of the three groups of streams classified by cluster analysis, snowmelt 1984 (n = number of samples/streams).

Variable	Unit	complete data n = 86	Cluster 1 n = 22	Cluster 2 n = 41	Cluster 3 n = 23
H^+	$\mu eq.L^{-1}$	16.2	58	2.5	0.2
(pH-value)		4.8	4.2	5.6	6.7
Color (λ=436 nm)	$E.m^{-1}$	0.6	1.2	0.4	0.3
UV-extinct. (λ=254 nm)	$E.m^{-1}$	4.2	9.4	2.9	1.6
DOC	$mg.L^{-1}$	2.8	6.0	2.0	1.2
Cl^-	$\mu eq.L^{-1}$	46	49	42	50
SO_4^{2-}	$\mu eq.L^{-1}$	132	155	118	133
NO_3^-	$\mu eq.L^{-1}$	87	74	88	96
HCO_3^-	$\mu eq.L^{-1}$	72	0	51	176
Na^+	$\mu eq.L^{-1}$	62	38	61	86
K^+	$\mu eq.L^{-1}$	17	21	16	14
Ca^{2+}	$\mu eq.L^{-1}$	176	103	161	272
Mg^{2+}	$\mu eq.L^{-1}$	66	42	65	90
Al	$\mu g.L^{-1}$	234	720	93	19
Fe	$\mu g.L^{-1}$	39	98	22	11
Mn	$\mu g.L^{-1}$	29	85	15	1.4
Cd	$\mu g.L^{-1}$	0.14	0.29	0.10	0.06
Pb	$\mu g.L^{-1}$	0.58	1.15	0.39	0.36
Zn	$\mu g.L^{-1}$	9.1	18.6	4.5	2.6
Cu	$\mu g.L^{-1}$	2.1	2.9	1.8	1.7

Cluster 1 is characterized by low pH-values (3.8 to 5.0), high concentrations of DOC, Al, Mn, Fe, and other heavy metals but low concentrations for Ca^{2+}, Mg^{2+}, and HCO_3^-. Cluster 3 has higher pH-values (above 6.0), low concentrations of DOC, Al, Fe, and heavy metals, but high values of Ca^{2+}, Mg^{2+}, and HCO_3^-. Cluster 2 is intermediate. Most of the above parameters differentiate between groups. Of the anions, only NO_3^- and HCO_3^- show differences. Interestingly, Na^+ and K^+ differ considerably. The three clusters of streams are regionally distributed according to the geological map of the Black Forest (Cluster 1 Buntsandstein; Cluster 2 granite; Cluster 3 gneiss). Water samples from Buntsandstein areas have low concentrations of Na^+, Ca^{2+}, and Mg^{2+} which seems to be due to a lack of feldspars in the quartzy sandstones of Buntsandstein. The relative high concentrations of K^+ are apparently the result of weathering of the minerals mica and illite, both rich in K.

Factor analysis using 15 variables explained 81.6 % of the total variance after extraction of three factors (Figure 1). This method deduces primilary factors, which explain the observed relationships of the hydrochemical characteristics. The result can be used as a hypothetical model for the identification of causal relationships. When transfered to the natural conditions, the influence of real factors, which are responsible for the chemistry of surface water, can be estimated in part of the variance of the original data (Hötzl, 1975; Reeder *et al.*, 1972; Kalvenes *et al.*, 1979; Reid *et al.*, 1981).

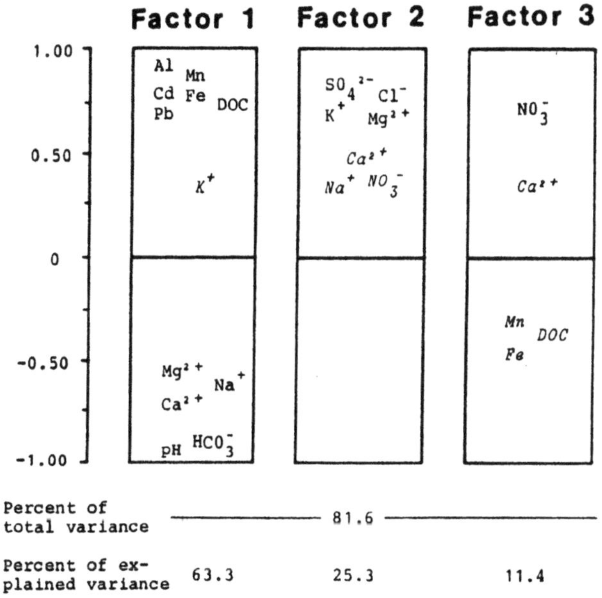

Figure 1. Diagrammatic representation of factors from varimax solution of chemical data for 86 Black Forest streams, snowmelt 1984. Variables with slight significant loadings (0.30 to 0.50) in italics. For the sake of clarity variables with loadings < 0.30 have been omitted.

Factor 1, which accounts for the largest amount of variance, reflects the influence of the soil-forming process of podsolization on the chemical composition of our water samples. Podsolization occurs typically in cool, humid climates under coniferous forest and/or heath vegetation as it is the case in the Black Forest. The degree of podsolization is decisively affected by the base status of the substrate and its permeability to leaching. Raw humus layers of podsols are effective ion exchangers highly saturated with H^+. Their acidity depends on the type of vegetation (Bergseth, 1977). The raw humus layers of soils undergoing podsolization also generate organic-rich, often brown-colored leaching solutions with a strong acidic and complexing character (Wiechmann, 1978; DeConinck, 1980). Hydrological flow pattern plays a key role, too. During periods of heavy rainfall and snowmelt, water reaches the stream by moving in the saturated upper soil layers. Soil water drainage is characteristically more rapid, because the soil hydraulic conductivities are greater and the water moves along a steeper gradient parallel to the surface slope (Lee, 1980). Due to climatic conditions, flow over frozen ground does not occur in the Black Forest. But the B_{sh} or placic horizons of podsols may act as less permeable layers as reported earlier by Jahn (1957). Stahr *et al.* (1983) found a prevailing lateral water and element transport in comparable soil types of a catchment in the Black Forest. Hence, under high flow conditions streamwater chemistry in areas with such soil types is mainly controlled by ion exchange and complexing processes in the organic and eluviated soil horizons. H^+-consuming weathering reactions are then less important for the composition of streamwater. The regional distribution of the factor scores, expressing the relative influence of a factor at each site, shows a pattern according again to the geological map. High values for Factor 1 occur in Buntsandstein-areas. This is in accordance with cluster analysis (cf. Table I, Cluster 1). The relative low factor scores in gneiss-areas may indicate that podsolization is active there to only a small degree. There is no correlation between the scores of Factor 1 and the altitudes of sampling sites (r=0.08). This indicates the overwhelming importance of substrate for podsolization. If only the streams from gneiss and granite-areas (n = 44) are correlated, a relationship to altitude (r=0.48) is more evident.

Factor 2 contributes 25.3 % of the variance. High loadings are obtained for SO_4^{2-}, Cl^-, NO_3^- as well as for Na^+, K^+, Mg^{2+}, and Ca^{2+}. If considered in the factor analysis, conductivity also plays a major role in Factor 2. There is also a significant relationship ($p < 0.001$) between factor scores and the altitudes of sampling sites (r=0.72 entire area; r=0.71 gneiss-grainite area). Thus, water samples from lower altitudes have more dissolved compounds than from higher altitudes. This may indicate a strong influence of temperature, which is a major factor controlling biogeochemical processes such as mineralization and weathering. It is suggested that Factor 2 represents the influence of a combination of various chemical processes controlling cation leaching in the brown earths of the catchments. Increased ionic strength of soil solutions cause elevated leaching of cations (Wiklander, 1974). With the present information, it is not possible to separate the influences of NO_3^- and SO_4^{2-} originating from anthropogenic atmospheric depositions and natural processes within the catchments. Various types of gneiss in the Black Forest may also have considerably high S-contents up to 2000 $mg.kg^{-1}$ (Siewers, 1974). Hence, the proportion of lithogenic S may also be important for S-turnover and varies, depending on weathering intensity. The role of anion adsorption-desorption processes and mineralization processes remains unknown, too.

Factor 3 accounts for 11.4 % of the variance. Only nitrate and also conductivity, if considered, are important in this factor. Evidently, Factor 3 describes a NO_3^--specific process (N-mineralization). The scores of Factor 3 also seem to be related to the altitudes of sampling sites especially of samples from gneiss and granite-areas (r=-0.57). If only NO_3^--concentrations are correlated with altitudes, the relationship becomes even more clear (entire area r=-0.51; gneiss-granite area r=-0.67).

Thus, even under snowmelt conditions soil and vegetation cover considerably affect water quality. Similar results were obtained from Norwegian studies where sites only consisted of bedrock partly covered with humus material (Rueslåtten and Jørgensen, 1978). Reid *et al.* (1981) investigated streamwater chemistry during storm events in an upland catchment in Scotland. They found clear evidence that nearly all water originated within the soil, whereas direct surface runoff was of minor importance. In contrast to regional surveys of surface waters in Norway (Wright *et al.*, 1977) and Scotland (Wright and Henriksen, 1980), our studies in the Black Forest provide a significant relationship between pH-values and DOC.

4. CONCLUSIONS

Spatial variation of streamwater chemistry in the Black Forest depends highly on site-specific conditions, whereas atmospheric depositions do not seem to have a differentiating influence. However, the deposition rates in the study area are low (open land 0.5 to 1 kg $H^+.ha^{-1}.a^{-1}$; 10 to 20 kg SO_4^{2-}-$S.ha^{-1}.a^{-1}$) and tend to be relatively uniform throughout the entire area (Evers, 1985; Feger, 1986). No evidence has been found for atmospheric acidic deposition affecting water quality and especially pH on a short-term basis. It still remains unclear to which proportion nitrate and sulphate in streamwater originate from atmospheric depositions or from biogeochemical processes within the catchments. Using the sum (NO_3^- + SO_4^{2-}) (Schoen *et al.*, 1983) or other simple chemical models for the identification of freshwaters acidified by anthropogenic acidic depositions seems therefore not advisable. The natural soil-forming process of podsolization is found to have a decisively important influence also on surface water chemistry. For evaluation of long-term trends in surface water acidification, it therefore seems necessary to take into consideration natural conditions and processes, cultural factors (land-use changes, silvicultural practices) as well as atmospheric deposition.

ACKNOWLEDGEMENTS

The authors would like to thank LfU Baden-Württemberg, Karlsruhe for the determination of DOC, and S. Holzmann, F. Schreiber, S. Barnert, E. Weber and F. Kurth for technical assistance. The project "Acidification and Heavy Metal Deposition in Lakes of the Black Forest" is jointly sponsored by the Commission of the European Communities and the Federal State of Baden-Württemberg (Projekt Europäisches Forschungszentrum für Maßnahmen zur Luftreinhaltung, Kernforschungszentrum Karlsruhe) grant No. 83/002/EG.

REFERENCES

Abrahamsen, G.: 1985, in: *NATO Advanced Research Workshop Effects of Acidic Depositions on Forests, Wetland and Agricultural Ecosystems*, Toronto, Springer, New York (in press).
Abrahamsen, G. and Stuanes, A.O.:, 1980, in D. Drabløs and A. Tollan (eds.), *Ecological Impact of Acid Precipitation*, pp. 152-154, SNSF, Oslo-Ås.
Bergseth, H.: 1977, *Acta Agriculturæ Scandinavica* **27**, 269.
Brahmer, G.: 1985, Diplomarbeit Institut für Bodenkunde und Waldernährungslehre, Universität Freiburg i.Br., 118 pp.
Brückner, J.: 1970, Dissertation Universität Freiburg i.Br., 174 pp.
DeConinck, F.: 1980, *Geoderma* **24**, 101.
Driscoll, C.T.; Baker, J.P.; Bisogni, J.J. and Schofield, C.L.: 1980, *Nature* **284**, 169.
Evers, F.H.: 1985, *Mitt. Ver. Forstl. Standortskunde u. Forstpflanzenzüchtung* **35**, 31.
Feger, K.H.: 1986, Dissertation Universität Freiburg i.Br., 242 pp.
Geyer, O.F. and Gwinner, M.P.: 1968, *Einführung in die Geologie von Baden-Württemberg*. E. Schweizerbart'sche Verlagsbuchhandlung, Stuttgart.
Gjessing, E.T.; Henriksen, A.; Johannessen, M. and Wright, R.F.: 1976, in: Brække, F.H. (ed.): *Impact of Acid Precipitation on Forest and Freshwater Ecosystems in Norway*, SNSF-Project, IR 6/76, Oslo-Ås, 65-85.
Glass, N.R.; Arnold, D.E.; Galloway, J.N.; Hendrey, G.R.; Lee, J.J.; McFee, W.W.; Norton, S.A.; Powers, C.F.; Rambo, D.L. and Schofield, C.L.: 1982, *Environ. Sci. Technol.* **16**, 162a.
Hausburg, H.: 1967, *Mitt. Ver. Forstl. Standortskunde u. Forstpflanzenzüchtung* **17**, 3.
Hausrath, H.: 1938, *Freiburger Universitätsreden* **26**, 23 pp.
Hötzl, H.: 1975, *Z. Dtsch. Geol. Ges.*, **126**, 121.
Jahn, R.: 1957, *Mitt. Ver. Forstl. Standortskunde u. Forstpflanzenzüchtung* **6**, 39.
Johnson, D.W.: 1981, *Water, Air, Soil Pollut.* **16**, 243.
Johnson, D.W. and Reuss, J.O.: 1984, *Phil. Trans. R. Soc. Lond.* **B305**, 383.
Kalvenes, S.; Mohn, E; Sollie, B.: 1979, SNSF-Project IR 40/79, Oslo-Ås, 71 pp
Krug, E.C. and Frinck, C.R.: 1983, *Science* **221**, 520.
Lee, R.: 1982, *Forest Hydrology*, Columbia Press, New York, 349 pp.
Lefohn, A.S. and Klock, G.O.: 1985, *JAPCA* **35**, 632.
Overrein, L.N.; Seip, H.M. and Tollan, A.: 1980, *Acid Precipitation - Effects on Forests and Fish*. Final report of the SNSF-Project 1972-1980, Oslo-Ås, 175 pp.
Reeder, S.W.; Hitchon, B. and Levinson, A.A.: 1972, *Geochim. Cosmochim. Acta* **36**, 825.
Reid, J.M.; MacLeod, D.A. and Cresser, M.S.: 1981, *J. Hydrol.* **50**, 129.
Rosenqvist, I.T.: 1977, *Sur jord - surt vann*. Ingeniørforlaget A/S, Oslo, 123 pp.
Rosenqvist, I.T.: 1983, *VDI-Berichte* **500**, 393.
Rueslåtten, H.G. and Jørgensen, P.: 1978, *Nordic Hydrology* **9**, 1.
Saar, R.A. and Weber, J.H.: 1982, *Environ. Sci. Technol.* **16**, 510a.
Schoen, R. and Kohler, A.: 1984, in: Umweltbundesamt (Ed.): *Gewässerversauerung in der Bundesrepublik Deutschland*, Materialien 1/84, E.Schmidt Verlag, Berlin, pp. 55-69.

Schoen, R., Wright, R.F. and Krieter, M.: 1983, *Acid Rain Research* **5**, NIVA, Oslo, 15 pp.
Schweikle, V.: 1971, Dissertation Universität Hohenheim, 103 pp.
Seip, H.M.: 1980, in: *Ecological Effects of Acid Precipitation*, pp. 358-366, SNSF-Project, Oslo-Ås.
Siewers, U.: 1974, Dissertation Universität Göttingen, 62 pp.
Sontheimer, H. and Wagner, I.: 1977, *Z. Wasser Abwasser Forsch.*,**10**, 77.
Stahr, K.: Hädrich, F. und Gauer, J.: 1983, *Z. Pflanzenernährung Bodenk.* **146**, 23.
Stahr, K.: 1973, *Arb. Inst. Geol. Paläont. Univ. Stuttgart*, N.F. **69**, pp 85-183.
Stahr, K.: 1979, *Mitt. Dtsch. Bodenkundl. Ges.* **29**, 649.
Ulrich, B. and Matzner, E.: 1983, *Abiotische Folgewirkungen der weiträumigen Ausbreitung der Luftverunreinigungen*, Umweltforschungsplan des Bundesministers des Innern, Luftreinhaltung Forschungsbericht 10402615, 221 pp.
Umweltbundesamt (Ed.): 1984, *Gewässerversauerung in der Bundesrepublik Deutschland*, UBA-Materialien 1/84, E. Schmidt Verlag, Berlin, 462 pp.
Van Breemen, N.; Mulder, J. and Driscoll, C.T.: 1983, *Plant and Soil* **75**, 283.
Wiechmann, A.: 1978, *Hohenheimer Arbeiten* **94**, 141 pp.
Wiklander, L.: 1974, *Grundförbättring* **26**, 155.
Wright, R.F. and Henriksen, A.: 1980, SNSF-Project IR 72/80, Oslo-Ås, 63 pp.
Wright, R.F.; Dale, T.; Henriksen, A.; Hendrey, G.R.; Gjessing, E.T.; Johannessen, M.; Lysholm, C. and Støren, E.: 1977, SNSF-Project IR 33/77, Oslo-Ås 153 pp.
Zöttl, H.W.; Feger, K.H. and Brahmer, G.: 1985a, *Berichte Projekt Europäisches Forschungszentrum für Maßnahmen der Luftreinhaltung* **2**, Kernforschungszentrum Karlsruhe, pp. 285-302.
Zöttl, H.W.; Feger, K.H. and Brahmer, G.: 1985b, *Naturwissenschaften* **72**, 268.

EFFECTS OF SULPHURIC ACID AND ACIDIFYING AMMONIUM DEPOSITION ON WATER QUALITY AND VEGETATION OF SIMULATED SOFT WATER ECOSYSTEMS

J.A.A.R. Schuurkes, I.C.C. Heck, P.L.G.M. Hesen, R.S.E.W. Leuven & J.G.M. Roelofs
Laboratory of Aquatic Ecology, Catholic University, Toernooiveld, 6525 ED Nijmegen, The Netherlands.

ABSTRACT In a greenhouse, seven identical mini-ecosystems, simulating soft water ponds, were exposed to different types of artificial rain water. The effects of rain water containing H_2SO_4 and nitrate, and rain water containing ammonium sulphate on water quality and vegetation were studied and compared. Causal relations were established between rain water quality, water chemistry and changes in floristic composition. Ammonium sulphate deposition, particularly, strongly affected water quality and vegetation development. Although ammonium sulphate deposition was only slightly acid, due to nitrification it acted as an important acid source, causing acidification to pH=3.8. Under acidified conditions, ammonium sulphate deposition lead to a luxuriant growth of *Juncus bulbosus* and *Agrostis canina*. In the mini-ecosystems, H_2SO_4 deposition with a pH of 3.5 only decreased the pH of the water to 5.1 within 1 yr. The acidification of water appeared to be coupled with changes in alkalinity, sulphate, Al, Cd, Ca, Mg, K and inorganic-N. It is concluded that in NH_3-affected regions in The Netherlands, the high atmospheric deposition of ammonium sulphate probably contributes to a large extent in the acidification, eutrophication and floristic changes of oligotrophic soft waters.

1. INTRODUCTION

In general, water acidification is ascribed to the increased atmospheric deposition of H^+, mainly as H_2SO_4 and HNO_3. Although clear correlations can be shown based on the empirical Henriksen-model (Henriksen, 1982), no causal relations have been established between the deposition of acidifying substances and their ultimate chemical and biological effects on surface waters. A few studies have documented chemical and biological responses of lake- ecosystems which have been experimentally acidified with H_2SO_4 (Schindler and Turner, 1982). Fully controlled experiments, however, have only been conducted on the effects of H_2SO_4 and HNO_3 acid rain water simulants on productivity and growth of crops and trees, or on chemical properties of soils (Drabløs and Tollan, 1980). No such experiments have been reported for aquatic systems. The present paper, however, deals with mini-ecosystems, simulating hydrologically isolated lentic soft waters, which were exposed to artificial rain solutions. Besides H_2SO_4 , ammonium sulphate containing rain was also used, as in strongly manured agricultural areas in The Netherlands this is an important component of wet deposition (Schuurkes, 1986).

Water, Air, and Soil Pollution **31** (1986) 267–272.

2. MATERIAL AND METHODS

The experiment was conducted in a greenhouse which served only as a shelter from deposition of ambient air pollutants. Light and temperature conditions were very similar to the environmental, outdoor situation. Inside the greenhouse, seven identical mini-ecosystems, comparable to small ponds, were created in small PVC-containers (2 m^2). The starting-situation of all systems simulated field-conditions of nutrient-poor soft waters, such as small lakes or moorland pools. The sediment consisted of fine mineral sand obtained from a undisturbed small soft-water system. In wintertime, plants from oligotrophic soft waters (*Littorella uniflora* (L.) Aschers; *Luronium natans* (L.) Raf. and *Pilularia globulifera* L.) and plants from acidified aquatic systems (*Sphagnum* spp.; *Juncus bulbosus* L. and *Agrostis canina* L.) were introduced to the ponds. All mini-ecosystems received a similar quantity of plant material in order to create a uniform starting situation. Each system was exposed to a different type of artifical rain which was applied by means of a watering-can and a spray-gun. The rain treatments started in November 1983. Two major types of rain solutions were used. Rain containing H_2SO_4 was applied in a staged pH gradient (treatments 2, 3 and 4), whereas rain containing ammonium sulphate (pH = 5.6) was applied in a staged concentration gradient (treatments 5, 6 and 7). Table 1 gives some chemical characteristics of the different rain applications, which are based on actual concentration and deposition figures in different parts of the Netherlands. Treatment 1 was used as a reference. In treatments 2, 3 and 4 nitrate was added, as HNO_3 is also an important constituent of acid rain. During the first year, water samples of the pools were taken monthly, while pH was measured more frequently. The above-ground biomass of some plant species was partly harvested in the late summer of 1984. Dry weight was determined by drying the plant material for 24 hr at 105°C. Water quality data were statistically compared with the Kruskal-Wallis test, $p \leq 0.05$.

TABLE I. Some chemical characteristics of the rain applications.

		theoretical concentration ($\mu mol.L^{-1}$)			measured deposition ($mol.ha^{-1}.yr^{-1}$)			
Treatment	pH	NH_4^+	SO_4^{2-}	NO_3^-	H^+	NH_4^+	SO_4^{2-}	NO_3^-
1	5.6	0	5	0	17	130	390	1
2	5.0	0	10	40	65	100	390	220
3	4.25	0	50	40	360	100	600	210
4	3.5	0	200	40	1600	80	880	210
5	5.6	250	125	0	17	1430	850	20
6	5.6	500	250	0	17	2870	1600	20
7	5.6	1500	750	0	17	8520	4650	20

3. RESULTS

3.1. Water chemistry

Striking temporal trends occurred in pH, alkalinity, ammonium, nitrate and sulphate (Figure 2). In the H_2SO_4 treatments 1, 2 and 3 the pH of the water increased, but pH remained unaffected in the water of treat-

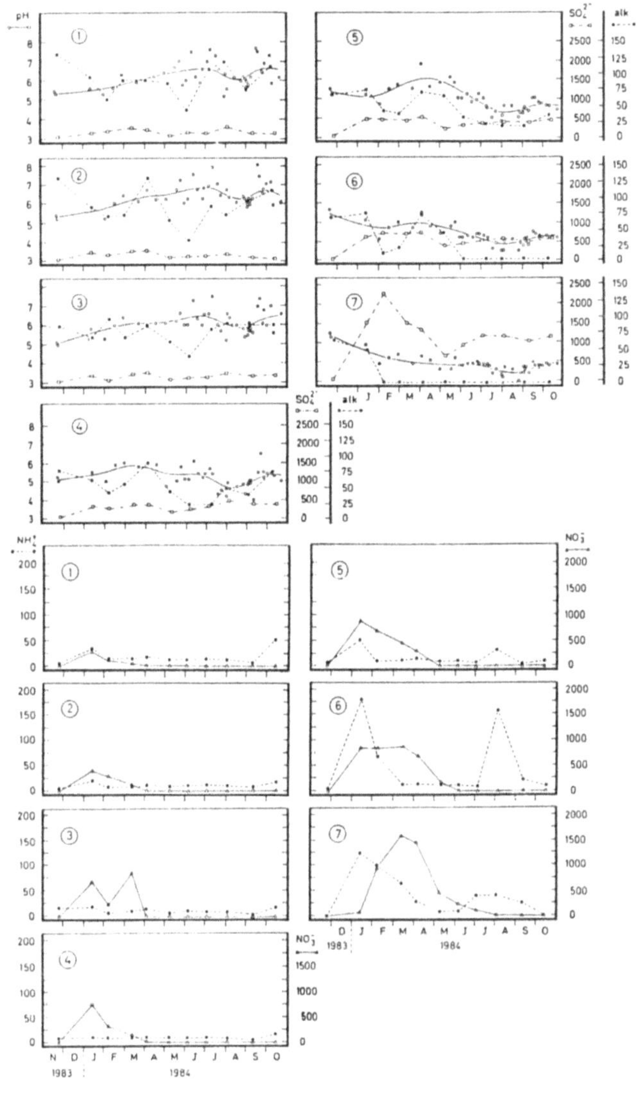

Figure 2. Time trends of pH, alkalinity, sulphate, ammonium and nitrate (µM.) in seven pools exposed to different rain solutions (Table I).

ment 4. A dramatic pH decrease occurred as a result of treatments 5, 6 and 7. Apparently, acidification occurred faster when higher ammonium sulphate concentrations were present in the rain water applied. Trends in alkalinity were very similar, although rather large fluctuations occurred. In the H_2SO_4 treatments alkalinity was only affected in pool

4. All ammonium sulphate treatments (5, 6 and 7) caused a decrease in alkalinity of the water. Alkalinity decreased most rapid in pool 7.

Ammonium and nitrate concentrations in the pools were very distinct for the different treatments. When rain containing H_2SO_4 was applied (treatments 2 to 4), ammonium concentrations were little affected, whereas the ammonium sulphate treatments caused increased levels during the first months and in late summer 1984. Also the nitrate concentrations were highest during the first months, particularly in pools 5, 6 and 7. In all pools the sulphate concentration remained rather constant after spring 1984. The absolute concentration was related to the sulphate concentration of the rain applied.

Figure 3 gives mean values of several chemical parameters in the pools of all mini-ecosystems. Values were obtained from analyses during August, September and October 1984. These figures reflect the effects of the different rain applications after one year in which biochemical processes, plant growth and other biotic factors have been interacting. The most marked differences occurred in pH, alkalinity, sulphate, Ca, Al, Cd and K. For the H_2SO_4 treatments pH was significantly affected in pool 4. Within 1 yr, the pH had decreased to 5.1. In this situation alkalinity (60µeq/L), sulphate (420 µM.), Ca (275 µM.) and Mg concentrations (110 µM.) in the water also differed significantly. Rain containing ammonium sulphate (treatments 5, 6 and 7) very clearly affected the chemical composition of the pools with pH decreases to 4.5, 4.2 and 3.8, respectively; alkalinity values were 25, 10 and 0 µeq/L respectively. In the most acidified pool (treatment 7) the mean concentrations of sulphate, Ca, Mg, K and Al were 1060, 440, 150, 44 and 209 µM., respectively. The mean Cd concentration was 48 ηM.. In pools 6 and 7, pH and concentrations of sulphate, Al, Cd and K differed significantly from all other pools. It can be concluded that water quality was most affected by rain containing ammonium sulphate.

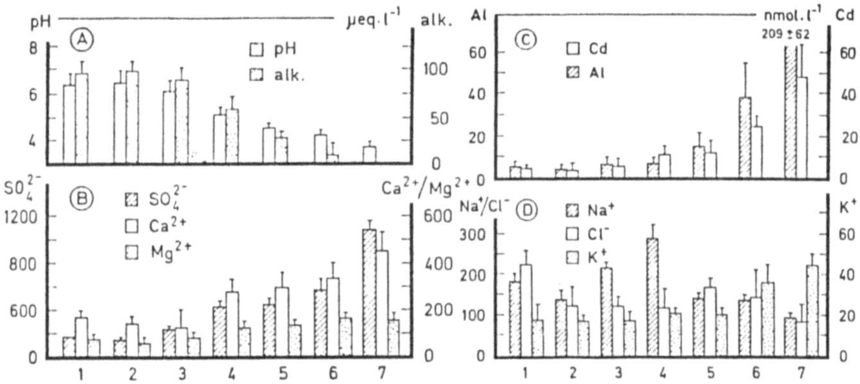

Figure 3. Some chemical characteristics of the water in seven mini-ecosystems exposed to different rain solutions as given in Table I. All concentrations are expressed as µM., except for pH, alkalinity and Cd.

3.2.Vegetation

In the pools of treatment 5, 6 and 7 the biomass of *Littorella uniflora* was higher than in the H_2SO_4 treatments (Figure 4). However, increasing ammonium sulphate deposition did not increase the biomass of *Littorella uniflora*. The biomass of both *Juncus bulbosus* and *Agrostis canina* was highest in the pools of treatment 6 and 7. Rain containing nitrate only affected *Juncus bulbosus*. The biomass of *Juncus bulbosus* in treatment 5 was similar to that of treatments 2, 3 and 4.

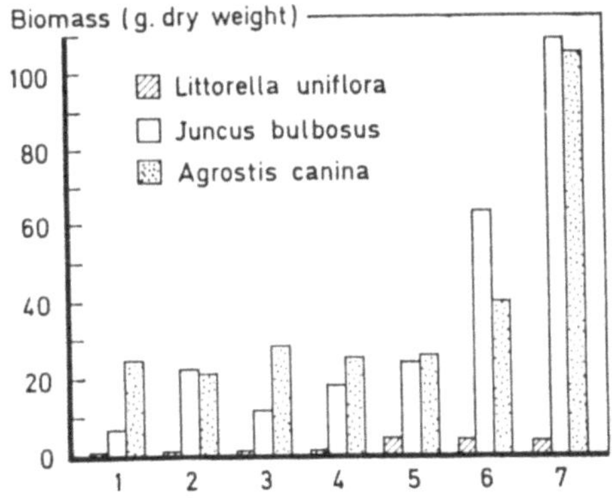

Figure 4. Biomass of *L.uniflora, J.bulbosus* and *A.canina* in seven mini-ecosystems after 10 mo exposure to the rain solutions 1 to 7.

4. DISCUSSION

The quantity of acid applied in treatments 1, 2 and 3 was not sufficient to acidify the water within one year, due to the acid neutralizing capacity of the system. Only treatment 4 (pH=3.5), equivalent to a yearly deposition of 1600 mol H^+ ha^{-1}, acidified the water significantly.

In the ammonium sulphate treatments, the temporal patterns of pH, ammonium and nitrate (Figure 2) indicate the importance of the nitrification process. This process produces 2 mol H^+ and 1 mol nitrate per mol ammonium converted. A comparison of the acidifying impact of treatments 4 and 5, in which equivalent amounts of H^+ and ammonium were applied, shows that the ultimate pH was 5.1 and 4.5 respectively. The acid production, calculated from Δ HCO_3^- and Δ H^+ in the ammonium sulphate treatment appears to be twice the acid production in the pH=3.5 treatment (105 and 50 µeq/L, respectively). Higher ammonium sulphate applications cause a stronger acidification which may reach down to pH=3.8 when the ammonium input is 8520 mol $ha^{-1}yr^{-1}$. This pH-value corresponds with the "average" pH of 98 acidified soft waters in the Netherlands (Leuven *et al.*, 1986). During the first year of the experiment, nitrification appeared to be the dominant process in determining the rate and level of acidification. However, bacterial processes and nitrogen assimilation by plants may be involved in regulating acidification during the experiment. It is obvious that the amount of ammonium sulphate applied must be taken into account with respect to the potential acidifying capacity of the rain treatments. Preliminary calculations show a relation between the amount of potential acid applied (calculated as H^+ + 2 x NH_4^+) and the degree of acidification (measured as Δ H^+ +Δ HCO_3^-).

Many other chemical constituents of the water were also changed, particularly with respect to the ammonium sulphate treatments. The concentrations of most components were not correlated with the rain water pH but appeared to be dependent on the level of acidification of the water. The increased levels of Al, Cd, Ca and Mg in pools 5, 6 and 7 will be due to both dissolution and exchange processes of the sediment adsorption complex with H^+ and ammonium ions. The latter process probably accounts for the enhanced concentrations of K. The enhanced ammonium concentrations in pools 5, 6 and 7 during summer 1984, probably resulted from inhibition of the nitrification process caused by the low pH of the water. The sulphate levels in the pools were related to the amount of sulphate applied, which indicates that it makes makes no difference whether sulphate is added as H_2SO_4 or ammonium sulphate.

The development of the vegetation is determined by the interaction between rain water composition and the subsequent water quality. Ammonium and nitrate deposition particularly, appeared to affect growth directly or indirectly. The H_2SO_4 treatments (pH 5.6 to 3.5) had no marked effect on the total biomass of the vegetation. At low pH values, the enhanced ammonium concentrations in the water (Figure 2) may be favorable for *Juncus bulbosus* and *Agrostis canina*, as the biomass of both species increased with higher ammonium sulphate deposition. *Littorella uniflora* is not able to profit from this enhanced ammonium deposition. As a consequence *Littorella uniflora* may be suppressed by a luxuriant growth of the nitrophilous species *Juncus bulbosus* and *Agrostis canina*. This conclusion is supported by eco-physiological experiments and field observations (Roelofs *et al.*, 1984; Schuurkes *et al.*, 1986).

In the Netherlands, water acidification and accompanying chemical and biological changes (Leuven *et al.*, 1986) are most pronounced in those areas where atmospheric NH_3 concentrations emitted from animal manure are very high and wet deposition is only slightly acid (Schuurkes, 1986). The results presented in this paper show that the subsequent high deposition of airborne ammonium sulphate adversely affects vulnerable soft water systems by acidification and eutrophication. Therefore, it may be expected that in NH_3 affected regions, airborne ammonium sulphate has contributed to the acidification, nitrogen-enrichment and luxuriant growth (eutrophication) of acid-tolerant plant species of oligotrophic soft waters.

Some recent papers have already focused on the possible impact of atmospheric ammonium deposition (Galloway and Dillon, 1983; van Breemen *et al.*, 1984; Roelofs *et al.*, 1984; Nihlgård, 1985; Schuurkes, 1986). Nevertheless, in general more attention should be paid to the role of ammonium sulphate deposition in the acid rain problem. Particularly in NH_3 affected areas with susceptible soils, this component may be of quantitative importance in regulating water acidification.

REFERENCES

van Breemen, N., Driscoll, C.T., and Mulder, J.: 1984, *Nature* **307**, 599.

Drabløs, D., and Tollan, A. (eds.): 1980, *Ecological impact of acid precipitation*, SNSF project, Oslo, Norway, 383 pp.

Galloway, J.N. and Dillon, P.J.: 1983, In: Ecological effects of acid deposition, NSEPB, report PM 1636, 145- 160.

Henriksen, A.: 1982, "Susceptibility of surface waters to acidification", in Johnson, R.E. (ed.), *Acid Rain/Fisheries*, Proc. Int. Symp. Acid. Precip. Fish. Imp., American Fisheries Society, Bethesda, Maryland, p. 103-121.

Leuven, R.S.E.W., Kersten, H.L.M., Schuurkes, J.A.A.R., Roelofs, J.G.M., and Arts, G.H.P.: 1986, *Water, Air & Soil Pollut*, in press.

Nihlgård, B.: 1985, *Ambio*, **14**, 2.

Roelofs, J.G.M., Schuurkes, J.A.A.R., and Smits, A.J.M.: 1984, *Aquat. Bot.*, **18**, 389.

Schindler, D.W., and Turner, M.A.: 1982, *Water, Air and Soil Pollut*, **18**, 259.

Schuurkes, J.A.A.R.: 1986, *Experientia*, **42**, 351.

Schuurkes, J.A.A.R., Kok, C.J. and den Hartog, C.: 1986, *Aquatic Botany*, **24**, 131.

Soil acidity and its relationship to root growth in declining forest stands in Germany

E. Matzner[1], D. Murach[2] and H. Fortmann[3]

[1] Research Center Forest Ecosystems/Forest Decline, University of Göttingen, Büsgenweg 2, D-3400 Göttingen, West Germany

[2] Institute for Silviculture, University of Göttingen, Büsgenweg 1, D-3400 Göttingen, West Germany

[3] Institute for Soil Science and Forest Nutrition, University of Göttingen, Büsgenweg 2, D-3400 Göttingen, West Germany

Abstract

Ingrowth core studies were used to investigate the influence of 3 different soils on root growth of Norway spruce in two stands in Northwest Germany. Root growth was significantly inhibited by the acid soil treatment. The amount of roots which had grown into the limed ingrowth cores were much higher and the percentage of dead roots was less when compared to the unlimed treatment. This is attributed to improved Ca/Al-ratios after liming. The fertilized peat-sand soil delivered optimal conditions for root growth indicated by maximum root biomass and 0% root mortality. The importance of soil acidification induced by acid deposition as the cause of the root disturbance in declining stands is addressed.

Water, Air, and Soil Pollution **31** (1986) 273–282.

I. Introduction

Because of high filtering activity of the canopy, forest soils are heavily impacted by the deposition of various air pollutants such as heavy metals (MAYER, 1981), acids (ULRICH et al., 1979) and organic compounds (MATZNER, 1984).

The present rate of acid deposition, especially in Central Europe, far exceeds the known rates of acid buffering by weathering of minerals (BACHE, 1982; MAZZARINO et al., 1983; JOHNSON et al., 1981; FÖLSTER, 1985) and thus induces losses of exchangeable basic cations and, finally, mobilization of potential toxic Al-ions. Increasing soil acidification has been documented in large forested areas in Germany during the last decades (BUTZKE, 1981; GRENZIUS, 1984; EVERS, 1983; v. ZEZSCHWITZ, 1982).

For about 70 yr it has been known that Al-ions are potentially toxic to plant roots (HUTCHINSON, 1983). Recent experiments clearly showed root damage and Al-induced Mg-deficiency and yellowing of needles related to low Ca/Al- and Mg/Al ratios in nutrient solution on Norway spruce (ROST-SIEBERT, 1985; JORNS and HECHT-BUCHHOLZ, 1985). The hypothesis that increasing levels of Al-ions in soil solution (caused by acid deposition) leading to toxic effects on tree roots may play a major role in the present forest decline in Germany has been put forward by ULRICH et al. (1979) and has been subject to much debate (e.g. REHFUESS, 1981; ZÖTTL, 1983). However, disturbance of the root system has been observed in every declined forest stand in West Germany where root systems have been studied. Symptoms of root disturbance include strong vertical gradients of the fine root biomass (MURACH, 1984), increased percentage of dead fine roots of damaged stands as compared to undamaged stands (MURACH, 1984; SCHULTE-BISPING and MURACH, 1984; LISS et al., 1984) pathological mycorrhiza (BLASCHKE, 1981; MEYER, 1985; LISS et al., 1984), damage of the endodermis (HÜTTERMANN, 1982), and changes in chemical composition (BAUCH and SCHRÖDER, 1982; MURACH, 1984). These damages place a predisposing stress on the whole tree by reducing water- and nutrient uptake.

At present, two major hypotheses are used to explain the observed root damage. One is the acid toxicity hypothesis mentioned above. SCHÜTT et al. (1983) and LICHTENTHALER and BUSCHMANN (1984) consider the direct effect of air pollutants on the leaves of the plant as the primary cause; the observed root damage is explained by a reduced supply of assimilates to the root system following the damage to the leaves.

Measurements of root biomass dynamics conducted in forest stands of North Germany contradict the hypothesis of assimilate deficiency as the primary cause of the observed root dieback: MURACH (1984) reported a strong decrease of the living fine root biomass in the top soil of the Solling spruce stand in autumn 1982 and related this finding to the drop in pH following an acidification pulse in 1982. Since fine root biomass of the deeper soil remained constant or slightly increased, MURACH excluded assimilate deficiency as cause of the observed root damage in the top soil.

Biomass inventory studies in 2 young (12 yr) Norway spruce stands (SCHULTE-BISPING and MURACH, 1984) showed that the total fine root biomass of the damaged stand is about 3 times higher as compared to a neighbouring healthy stand, while the biomass of living fine roots is about the same. Soil data indicate stronger soil acidification under the damaged stand than under the healthy stand. GEHRMANN (1984) reported damage to roots of beech seedlings on acid soils before the cotyledons appear and photoynthesis starts, which could in this case not be explained by assimilate deficiency.

In order to get more information about the importance of the chemical soil state in regulating root growth, more exact field experimentation was considerd necessary.

2. Methods

The already well documented SOLLING-site "F 1" (ELLENBERG, 1971), covered by 110 yr old Norway-spruce was chosen as one of the 2 study sites for this project. The stand is moderately damaged; yellowing of the older needles is frequent, and slight needle losses occur. For comparison, a stand of Norway-spruce (87 yr) in the Egge-mountains "E 1" was investigated which is heavily damaged, indicated by frequent yellowing of the older needles and severe (> 50%) needle losses.

Both stands grow on acidic soils derived from loess covering sandstone. Soil chemical data of the undisturbed soils are given in Table I.

Table I

The ingrowth core-method used to study root growth in this investigation was first described by PERSSON (1980a,b). In march 1984 45 corings (Ø 8cm) were taken up to a depth of 40cm, and each 15 were refilled to the top of the O-layer with 3 different soils: 1: Acid subsoil from the Solling F 1-site (20-40cm depth, see Table 1); 2: subsoil from the Solling F 1-site limed with $Ca(Mg)SiO_4$. After the experiment, the limed soil had a pH-($CaCl_2$) of about 4.3 and a base saturation of 18%; 3: fertilized 2:1 peat-sand mixture. The following concentrations of major elements in a 1:1 water extract of the peat-sand mixture were measured ($mg \cdot L^{-1}$): NH_4: 1.2; NO_3: 71; Na: 8.9; K: 16.3; Ca: 108; Mg: 6.4; Fe: 0.3; Mn: 2.7; Al: 0.9; pH 4.5. Before filling the cores, the mineral soil was dried and sieved (< 2 mm) and therefore contained no living or dead roots. Afterwards, the soil was rewetted.

Table I: Soil chemical characteristics of the experimental sites

Depth (cm)	pH ($CaCl_2$)	C %	CEC $meq \cdot kg^{-1}$	Base saturation Na+K+Ca+Mg % of CEC
Solling F1				
0-10	2.9	4.6	133	5
10-20	3.6	2.0	79	4
20-30	3.9	1.1	61	4
30-40	4.0	0.8	55	5
Egge E1				
0- 5	2.7	4.6	84	7
5-10	2.9	2.8	77	5
10-20	3.3	2.0	63	5
20-30	3.6	1.3	49	4
30-40	3.5	1.5	53	4

Table II: Vertical distribution of fine roots (< 2 mm diameter) (Values in mg/100 mL soil)

	Egge E1				Solling F1			
Depth	Vitality	$\bar{x}$	$s_{\bar{x}}$	n	Vitality	$\bar{x}$	$s_{\bar{x}}$	n
O_F	living	27	13	15	living	127	17	40
	dead	9	5	15	dead	18	4	40
O_H	living	50	13	15	living	164	19	40
	dead	11	2	15	dead	88	12	40
0- 5 cm	living	53	18	15	living	120	10	40
	dead	32	8	15	dead	109	8	40
5-10 cm	living	31	11	15	living	131	11	40
	dead	14	4	15	dead	71	7	40
10-20 cm	living	47	21	15	living	60	11	40
	dead	14	4	15	dead	28	3	40
20-30 cm	living	31	9	15	living	22	8	40
	dead	9	2	15	dead	30	10	40
30-40 cm	living	29	14	15	living	18	7	40
	dead	10	4	15	dead	25	8	40
Total living: 1800 $kg \cdot ha^{-1}$ Total dead : 640 $kg \cdot ha^{-1}$ Σ : 2440 $kg \cdot ha^{-1}$					Total living: 2700 $kg \cdot ha^{-1}$ Total dead : 2000 $kg \cdot ha^{-1}$ Σ : 4700 $kg \cdot ha^{-1}$			

The distribution and total biomass of living and dead fine roots of both stands was evaluated for background information by using the corings with the original soils from the sites. These were treated in the same way as the refilled cores.

After 7 month the refilled corings were removed again, cut into 5 layers (0 to 5, 5 to 10, 10 to 20, 20 to 30 and 30 to 40cm depth) and analyzed for the amounts of roots that had invaded. The roots were washed from the soil according to the procedure described by MURACH (1984) and subdivided into living and dead fractions by morphological criteria (for details see MURACH, 1984).

3. Results

The vertical fine root biomass distribution of both stands under undisturbed conditions is given in Table II.

Table II

The total amount of living roots found in the heavily damaged E 1-site (1800 kg $\cdot$ ha^{-1}) is much less than in the Solling F 1-site (2700 kg $\cdot$ ha^{-1}). ULRICH et al. (1984) report a decrease in fine root biomass of damaged Norway spruce as the needle losses exceeds 50% attributed it to insufficient supply of assimilates to the root system. This may also be the reason for the low root biomass found in the E 1-site and should be kept in mind when comparing the root biomass found in the refilled cores, since the initial status of the root systems of these sites is already different.

Furtheron the vertical profiles of fine root biomass differ significantly among the two sites. The F 1-site shows high fine root concentrations in the 0-horizon and in the top mineral soil, when compared to the layers > 10 cm, while the root distribution in the E 1-site was found to be more uniform, which seems to be unusual for spruce stands (MURACH, 1985).

The distribution of dead fine roots of the F 1-site indicates increasing relative amounts with increasing soil depth. This pattern not found in the E 1-site, corresponds to decreasing molar Ca/Al-ratios (0.5 to 0.1) in the F 1 soil solution (MURACH, 1984). Unfortunately, no measurements on the gradients of the Ca/Al ratio of the soil solution were made under undisturbed conditions in the E 1-site.

The total and relative amount of dead fine roots found at the F 1-site is much higher than at the E 1-site. This may indicate a higher turnover rate but since no information on root decomposition rates are available, a final conclusion on that is impossible.

Table III

Table III: Amount and distribution of fine roots (< 2 mm diameter) in ingrowth cores of different soils (Values in mg/100 ml soil, n = 15 for all treatments)

F 1 - site							
Depth (cm)	Vitality	Acid F 1 Soil		Limed F 1 Soil		Peat-Sand Soil	
		$\bar{x}$	$s\bar{x}$	$\bar{x}$	$s\bar{x}$	$\bar{x}$	$s\bar{x}$
0- 5	living	2	0.9	32	16	38	12
	dead	2	0.6	2	0.6	0	0
5-10	living	3	1.6	18	7.2	60	14
	dead	3	0.9	6	2.9	0	0
10-20	living	2	0.8	10	3.0	62	12
	dead	3	1.0	5	1.1	0	0
20-30	living	3	1.7	5	1.2	36	6.7
	dead	3	0.6	3	0.6	0	0
30-40	living	0	0	7	2.7	35	7.0
	dead	2	0.6	4	2.2	0	0
Total amount living:		75 $kg \cdot ha^{-1}$		470 $kg \cdot ha^{-1}$		1820 $kg \cdot ha^{-1}$	
Total amount dead :		105 $kg \cdot ha^{-1}$		160 $kg \cdot ha^{-1}$		0 $kg \cdot ha^{-1}$	
Σ :		180 $kg \cdot ha^{-1}$		630 $kg \cdot ha^{-1}$		1820 $kg \cdot ha^{-1}$	

E 1 - site							
Depth (cm)	Vitality	Acid F 1 Soil		Limed F 1 Soil		Peat-Sand Soil	
		$\bar{x}$	$s\bar{x}$	$\bar{x}$	$s\bar{x}$	$\bar{x}$	$s\bar{x}$
0- 5	living	0	0	11	6.2	15	8.0
	dead	2	1	2	1.7	0	0
5-10	living	0	0	2	1.3	14	5.5
	dead	2	0.7	1	0.5	0	0
10-20	living	2	0.8	7	0.4	15	4.9
	dead	2	0.5	2	0.6	0	0
20-30	living	2	0.8	6	2.2	12	3.7
	dead	3	0.6	2	0.6	0	0
30-40	living	1	0.4	1	0.6	9	3.3
	dead	3	0.9	3	0.9	0	0
Total amount living:		50 $kg \cdot ha^{-1}$		205 $kg \cdot ha^{-1}$		505 $kg \cdot ha^{-1}$	
Total amount dead :		100 $kg \cdot ha^{-1}$		80 $kg \cdot ha^{-1}$		0 $kg \cdot ha^{-1}$	
Σ :		150 $kg \cdot ha^{-1}$		290 $kg \cdot ha^{-1}$		505 $kg \cdot ha^{-1}$	

To compare the amounts of fine roots in the ingrowth cores (which are given in Table III) for specific layers with those found under undisturbed conditions (Table II), the depth of the O_{F+H}-layer must be taken into account. The O_{F+H}-layer of the F 1-site is about 5 cm high, that of the E 1-site 10 cm. Since the cores were filled to the top, the 0 to 5 cm layer of the ingrowth cores of the F 1-site and the 0 to 10 cm layer of those of the E 1-site corresponds to the O-layer under undisturbed conditions.

The different soils used influenced the growth of fine roots significantly. Only very few roots were able to grow into the acid F 1 soil and most of those present were dead after 7 month. The total amount of new fine roots in both sites is nearly equal under these conditions. The strong vertical gradient of the undisturbed fine root distribution of the F 1-site does not influence the root distribution in the ingrowth cores filled with acid soil. Liming of the soil obviously improved the root growth. The total amount of living roots increased by a factor of six in case of F 1 and of four in case of E 1 when compared to the unlimed treatment. Furthermore, the proportion of dead roots is reduced significantly to about 25% of the total root mass for both stands whereas the unlimed treatment lead to 60 to 70% of dead roots. The limed treatment resulted in a vertical gradient of the living ingrown roots for the F 1-site which is comparable to the root distribution under undisturbed conditions.

In case of liming the total amount of new fine roots found in the F 1-site exceeds that of the E 1-site more than two fold which is presumably caused by the higher amount of living roots under undisturbed conditions in F 1 as compared to E 1 (see Table II). This may also explain the difference between the sites with respect to the amounts of fine roots in the fertilized peat-sand treatment.

Optimal conditions for root growth are found in the fertilized peat-sand ingrowth cores. The amount of ingrowing roots equals 1820 kg $\cdot$ ha^{-1} for F 1 and 505 for the E 1-site with 100% living fine roots. No vertical gradients are obvious for the F 1-site which - in this case - is due to fast growing long roots that entered the ingrowth cores at the top and were able to grow into deeper layers within the time period investigated.

The different soil treatments also induced changes in root morphology. The fine roots derived from the peat-sand treatment showed normal branching pattern by numerous long secondary roots with white tips indicating high growth rates. Growth of secondary roots was found to be much less in the limed treatment when compared to the peat-sand-mixture. The branching pattern was partly abnormal (coralloid) as described by CLARKSON (1969), ROST-SIEBERT (1985) and by TISCHNER et al. (1983), who found coralloidic root branching as an effect of low Ca/Al-ratios in hydroculture experiments. The proportion of coralloidic roots increased significantly in the acid soil treatment in this study as compared to the limed one.

4. Discussion

As already stated above, the initial status of the F 1 fine root system differs from that of the E 1-site. Thus, the biomass data from the ingrowth core studies of both sites are not comparable directly. However, the relative effects of the different soil treatments on root growth are clearly evident, and correspond exceedingly well in both stands.

The root biomass, as well as the amount of new roots growing each year, are determined by environmental conditions such as soil temperature, -moisture, -mechanical properties, -chemical properties, and by the assimilate supply from the above ground parts of the tree.

Out of these parameters, soil temperature, -moisture and -mechanical properties can be excluded to explain the differing root growth rates and mortality patterns in the limed and unlimed treatment, since these parameters are the same for both. Soil mechanical properties of the peat-sand treatment might be especially favourable for root growth, and differ significantly from the mineral soil ingrowth cores. Furthermore, high nutrient content of the fertilized peat-sand mixture may enhance root growth. MEYER (1985) reported decreasing root branching and root growth with increasing mineral N-content of the soil, which was attributed to the impact of high N-levels on micorrhiza. The results from the ingrowth core studies contradict the conclusion of MEYER.

As far as we know, root ingrowth core studies on trees were only done by PERSSON (1980b). However, his experiments were designed to study root development of Pinus silvestris after irrigation and fertilization of whole stands, rather than to study effects of soil acidity. He found lower root biomass in the fertilized and irrigated stands when compared to the control-plot. No data are given on soil chemical stress parameters neither for the control nor for the fertilized and irrigated plots.

Assimilate supply may influence the amount of roots invading the ingrowth cores through sink functions resulting from favourable conditions in specific parts of the root-system. This is to be expected in case of the limed and peat-sand treatments. Therefore the data of the new build root mass detected in the cores give no information about the total fine root turnover and growth of the whole stands. High root biomass in the limed and peat-sand treatment as compared to the unlimed treatment may thus be a consequence of improved chemical (limed treatment, peat-sand), and physical (peat-sand) conditions and, furthermore, from sink functions of the cores with respect to assimilates. Improved chemical conditions imply raised pH and increased Ca/Al and Mg/Al ratios of the soil solution. Despite the fact that the improved root growth in the limed and peat-sand treatment cannot be attributed to improved soil conditions alone the data clearly indicate the inhibition of root growth by the acid soil treatment. The amounts of fine roots in the acid soil ingrowth cores are extremely low and most of the roots died. Conversely, the results from the peat-sand treatment show that unstressed fine roots of Norway spruce obviously live longer than the 7 months of investigation, as all of the fine roots were found to be alive under these conditions. In other words, the high percentage of dead roots in the acid soil treatment is caused by chemical stress, presumably by low Ca/Al-ratios of the soil solution (0.5 to 0.1, MURACH 1984) as indicated by abnormal branching morphology of the roots and by the fact that Ca/Al ratios below 1 were shown to induce root damage to Norway spruce in hydroculture (ROST-SIEBERT, 1985; JORNS and HECHT-BUCHHOLZ, 1985).

Liming the soil and raising the Ca/Al ratio and pH reduced the percentage of dead roots in both sites drastically. Since the amount of living roots increased by liming at the same time, the turnover rate of fine roots in the limed treatment has become much lower than in the unlimed one. However, the liming effect on the soil chemical status was only moderate. It led to a base saturation of about 18% and a pH ($CaCl_2$) of 4.3. A higher base saturation of this soil may be necessary to further reduce the percentage of dead roots.

5. Conclusions

The occurrence of soil-derived chemical stress caused by soil acidity to fine roots of Norway spruce is clearly emphasized by the results described above. Acid soils such as the one used for this experiment inhibit root growth and induce high root mortality. Soils of this kind are common in large forested areas of Germany. Falsification of the hypothesis of soil acidity being the major cause of root disturbances in declining stands on acid soils failed by the approach used, to the contrary, experimental quantification was possible. Since acid precipitation is the major cause of soil acidification under Middle European deposition regimes, its causal relationship to root disturbance and forest decline is obvious.

References

Bache, B.W.: 1982, in: Ecological Effects of Acid Precipitation, Swedish National Environment Protection Board, Report snv pm 1636

Bauch, J. and Schröder, W.: 1982, Forstw. Cbl. 101, 285

Blaschke, H.: 1981, Forstw. Cbl. 100, 190

Butzke, H.: 1981, Forst- u. Holzwirt 36, 542

Clarkson, D.T.: 1969, in: J.H. Rorison (ed.): Ecological aspects of the mineral nutrition of plants. Blackwell, Oxford

Ellenberg, H.(ed.): 1971, Ecol.Studies 2
Evers, F.H.: 1983, Forst- u. Holzwirt 38, 317
Fölster, H.: 1985, in: J.I. Drever (ed.): The Chemistry of Weathering, D. Reidel Publ.Comp., 197
Gehrmann, J.: 1984, Ber.d.Forschungszentr. Waldökosyst./Waldsterben d. Univ.Göttingen, Bd. 1
Grenzius, R.: 1984, Forstwiss. Cbl. 103, 131
Hüttermann, A.: 1982, Mitt.d.Landesanst.f.Ökologie, Sonderheft: Immissionsbelastungen von Waldökosystemen, Recklinghausen
Hutchinson, T.C.: 1983, in: Intern.Confer.: Heavy metals in the environment. Heidelberg, Sept. 1983, Vol. 1
Johnson, N.M., Driscoll, Ch.T., Eaton, J.S., Likens, G.E. and McDowell, W.H.: 1981, Geochim. Acta 45, 1421
Jorns, A. and Hecht-Buchholz, C.: 1985, Allgem. Forstzeitschr. 46, 1249
Lichtenthaler, H.K. und Buschmann, C.: 1984, Allgem. Forstzeitschr. 15, Heft 1, 12
Liss, B., Blaschke, H. and Schütt, P.: 1984, J.For.Path. 14, 90
Matzner, E.: 1984, Water Air and Soil Pollution 21, 425
Mayer, R.: 1981, Gött.Bodenkdl.Ber. 70, 1
Mazzarino, M.J., Heinrichs, H. und Fölster, H.: 1983, in: B. Ulrich and J. Pankrath (eds.): Effects of accumulation of air pollutants in forest ecosystems. D.Reidel Publ.Comp., Dordrecht, Netherlands
Meyer, F.H.: 1985, Allg.Forstzeitschr. 9/10, 208
Murach, D.: 1984, Gött.Bodenkdl.Ber. 72
Murach, D.: 1985, personal communication
Persson, H.: 1980a, Oikos 34, 77
Persson, H.: 1980b, Acta Phytogeographica Suecica 68, 101
Rehfuess, K.E.: 1981, Forstw. Cbl. 100, 363
Rost-Siebert, K.: 1985, Ber.d.Forschungszentr.Waldökosyst./Waldsterben d. Univ.Göttingen, Bd. 12
Schütt, P., Blaschke, H., Hoque, E., Koch, W., Lang, K.J. und Schuck, H.J.: 1983, Forstwiss. Cbl. 102, 158
Schulte-Bisping, H. und Murach, D.: 1984, Ber.d.Forschungszentr. Waldökosyst./Waldsterben d. Univ.Göttingen, Bd. 2, 207
Tischner, R., Kaiser, U. und Hüttermann, A.: 1983, Forstw. Cbl. 102, 329
Ulrich, B., Mayer, R. und Khanna, P.K.: 1979, Schriften d.Forstl. Fak.d.Univ.Göttingen, Bd. 58, Sauerländer-Verlag, Frankfurt/a.Main
Ulrich, B., Murach, D. und Pirouzpanah, D.: 1984, Forstarchiv 55, 127
v. Zezschwitz, E.: 1982, Forst- u. Holzwirt 37, 275
Zöttl, H.W.: 1983, AFZ, 206

RESPONSES OF FIVE SPECIES OF CONIFER SEEDLINGS TO ALUMINUM STRESS

Thomas C. Hutchinson, Leah Bozic and Guadalupe Munoz-Vega*
Department of Botany and Institute for Environmental Studies
University of Toronto,
Toronto, Ontario
Canada, M5S 1A1

* Now at Xochimilco Depto El Hombre y su ambiente
Universidad Autonoma Metropolitana, Mexico D.F.

ABSTRACT. Seedlings of red, white and black spruce, and white and jack pine were tested for their response to a range of Al concentrations, when grown for 12 to 14 weeks in sand culture in the greenhouse. Nutrients were supplied in solution of pH 3.8 to which Al was supplied at 0, 5, 10, 20, 40, 80 and 160 mg/L. Red and white spruce were inhibited in growth from 5 mg/L upwards, while black spruce was equally inhibited only at concentrations > 20 mg/L. In contrast, white pine was stimulated by 5 to 20 mg Al/L and was always more tolerant than the other 4 species at every Al level used. Jack pine was intermediate in its response. There was a positive linear relationship between P and Al accumulation in the shoots of red spruce. This was in contrast to the other 4 species in which P concentrations in the shoot decreased with an increase in Al concentrations. In localization studies on jack pine using a modified hematoxylin stain, Al was shown to accumulate in the root cap of root tips, and in the epidermal and outer cortical walls of older roots. EDAX-SEM analysis confirmed the tissue distribution of Al and revealed its coincidental distribution with P in roots.

1. INTRODUCTION

High elevation red spruce in parts of eastern North America are showing dieback and mortality over a wide latitudinal range, from the New England states to the Carolinas (Siccama et al.,1982; Johnson, 1983). The rapid onset of this red spruce mortality has led to speculation as to whether increased acidic deposition and acid fogs in eastern North America are primary causes. Similarly, the suggestion has been made that acidic precipitation and acid mists and fogs are involved in the European declines. Ulrich et al. (1980) have postulated that the trigger for the declines in Germany were the severe summer droughts of 1975 and 1976, which allowed oxidation of S and N compounds in the soil and which led to an acid pulse in the soil

Water, Air, and Soil Pollution **31** (1986) 283–294.

solution when the rains returned. Ulrich's group recorded both an acid pulse in the soils of the Hartz mountains at this time, and increased concentrations of Al in the soil solution in the rooting zone (Ulrich et al., 1979; Ulrich and Matzner, 1981). They suggested that the blackened and dead absorbing roots of the Norway spruce and European beech which they found at this time were caused by Al toxicity; and that this progressed via secondary bacterial infections to cause ultimate decline and dieback.

The toxicity of Al to non-adapted species on acidic soils has been reported for almost 80 yr (Hutchinson, 1983). However, many boreal forest species normally occur on acidic, podzolic soils. It might be supposed that the root systems of such plants are already well adapted to elevated levels of Al in soil solution. Indeed, both crop and native species have been shown to exhibit significant varietal and population differentiation with respect to Al tolerance (Foy et al., 1978; Rorison, 1960; Clymo, 1962). It should also be recognized that as soil pH decreases below about 4.4 to 4.2 there is a very rapid increase in Al solubility, such that Al in soil solution differs greatly between soil pH 4.5 and 3.8 (Magistad, 1925). Thus, it is possible that if soil acidity is increased so that pH declines over this range, then root systems would experience rather rapid increases in Al concentrations such that inhibition and death of sensitive roots might occur.

The general consensus concerning acidic podzolic soils has been that the increased acidity of wet and dry precipitation over the past 20 yr in the areas where tree declines are being described is unlikely to have caused significant increases in soil acidity. However, some recent evidence challanges this view. Two Swedish studies have independently reported that soil profiles sampled after 55 and 35 yr, respectively, generally showed a significant decrease in pH and that this occurred in both podzolic and gray-brown forest soils (Hallbacken and Tamm, 1986; Tyler et al., 1985). Since this decrease in pH amounted to as much as 0.7 pH units from earlier pH's between 4.00 and 4.46, the changes are exactly in the pH range and of the magnitude anticipated to cause substantial increases in Al concentrations in soil solution. They report that the soil profile pH changes also extended into the C horizon.

In view of the possibility that the dieback of red spruce and balsam fir in eastern North America might relate to Al toxicity via changes in soil acidity, and the fact that large tracts of coniferous forest in eastern Canada are now subject to acid deposition, with ambient rainfall pH's of around 4.2 and 4.3, the present study was aimed at determining the relative sensitivity of economically important eastern coniferous species to Al. Accordingly, we report on the responses of red spruce (*Picea rubens*), white spruce (*Picea glauca*), black spruce (*Picea mariana*), jack pine (*Pinus banksiana*), and white pine (*Pinus strobus*) seedlings to a range of Al concentrations in sand culture at pH 3.8. The purpose was to determine whether any of these species might be potentially sensitive to increases in soluble Al and how this might be influenced by Ca and Mg contents of the nutrient supply.

2. METHODS

Seeds of 4 species was obtained from the Ontario Ministry of Natural Resources, with the red spruce originating with the Maritimes Forest Research Centre, Fredricton, N.B.. All seeds were surface sterilized with 0.5% sodium hypochlorite. Seeds were germinated on moist filter paper in petri dishes. When the radicles were about 2 cm long the seedlings were planted in pots of a 10 cm diameter filled with acid washed sand. Nutrients were supplied to the seedlings using a modified Ingestad's solution (Ingestad, 1962), the modification being to reduce phosphate levels to 2.1 x 10^{-4} mole.L^{-1}. Aluminum was added as $AL_2(SO_4)_3 \cdot 16H_2O$, Ca as $CaCl_2 \cdot 2H_2O$ and Mg as $MgCl_2 \cdot 6H_2O$. K_2SO_4 and NaCl were added equalize the SO_4 and Cl concentrations in each treatment. There were fifteen replicate pots in a randomized block design for each of 28 treatment combinations of Al, Ca, and Mg. Aluminum concentrations used were 0, 5, 10, 20, 40, 80 and 160 mg/L, each with Ca at 4 or 16 mg/L and Mg also at 4 or 16 mg/L, for a 7x2x2 treatment arrangement. NaOH or HCl was used to adjust the solution pH to 3.8. Plants were grown for 10 to 14 weeks, depending upon species, in the greenhouse, with a supplemental fluorescent and incandescent light source and having a photoperiod of 18 hr light, 6 hr dark and a light intensity of 160 uE/m^2/s. Temperature fluctuated over the period of the experiments from 20 to 28 C, with a mean of approximately 22 C.

Leachate was collected from randomly selected pots at weekly intervals to determine whether Al and P levels were markedly changed from the nutrient solutions, and to determine leachate (drainage) pH. Approximately 70 mL of fresh nutrient solution plus Al was added per week and the pots were also watered with 70 mL distilled water twice per week. At harvest, roots and shoots were rinsed in distilled-deionized water, dried at 60 C and weighed. Al, Mg, Ca, K, Mn, Cl and S concentrations were determined by neutron activation analysis. Al, Mg and Ca were also determined using the Perkin-Elmer 460 atomic absorption sprectrophotometer after a HNO_3 digestion. Phosphorous was analyzed colorimetrically using the Molybdenum Blue method on an Auto-analyzer. Only the Al and P uptake data will be presented in this paper. The data were analyzed using ANOVA and regression analyses.

In separate experiments to determine Al and P localization in jack pine seedlings, 3 replicate plants were grown for 24 weeks in the Ingestad nutrient solution, with Al supplied at 0 and 4.0 mMol (108 mg/L). Three replicate cross sections from the root tip and 10 cm back from the tip were analyzed from each plant. These sections from fresh roots were mounted om SEM stubs and carbon coated. They were then examined in a Cambridge S180 SEM with an ORTEC 6230 Energy Dispersive X-ray Microanalysis System. X-ray spectra were recorded with a count time of 40s and an accelerating voltage of 25 kV. The Al distribution obtained from this technique was compared to a histochemical staining method. Cross sections from the same roots were sectioned fresh and stained in a modified hematoxylin stain, which in the presence of Al turns purple to indicate localization (see Havas, 1986).

3. RESULTS AND DISCUSSION

3.1 Seedling Growth

Shoot and root dry weight were measured at final harvest. Al reduced total dry matter production of the five species, but the species differed sharply in their pattern of response to the range of concentrations (Table I). The three spruces were more sensitive to Al toxicity than the two pines. Irrespective of Ca or Mg status of the nutrient solutions, red spruce and white spruce at 5 mg Al/L grew 20 and 17% less than their controls; while at 20 mg Al/L yields of the 3 spruces were reduced by more than 20%. In sharp contrast, white pine was actually stimulated at concentrations from 5 to 20 mg Al/L while jack pine showed no statistically significant decrease in growth through this range. At 40 mg Al/L, however, jack pine seemed to have exceeded an Al threshold and growth was reduced 42%, which was equivalent to the growth reductions of the 3 spruces at this Al concentration. White pine was remarkably tolerant to Al. Significant decreases in growth occurred only at 80 mg Al/L and higher. Even 160 mg Al/L did not reduce seedling growth by 50%. At this Al concentration, red spruce was extremely inhibited and sickly.

Schier (1985) found that preconditioned red spruce and balsam fir seedling shoots showed no injury when treated with up to 200 mg Al/L at pH 3.8, but that roots of red spruce were inhibited at 50 mg Al/L and above, while balsam fir was more tolerant. Rost-Siebert (1984) found that Norway spruce was much more sensitive to Al than the three North American spruces tested in this study. Norway spruce seedlings were inhibited by even 1 mg Al/L if the molar ratio of Al:Ca exceeded 1.0. However, Muller and Abrahamsen (1984) indicated that Norway spruce and silver fir can tolerate 20 to 40 mg Al/L without harmful effects.

TABLE I. Effect of Al on total dry matter production of five species. Results expressed as a percentage of the control.

Al conc. (mg/L)	White pine	Jack pine	White spruce	Black spruce	Red spruce
0	0 ab	0 a	0 a	0 a	0 a
5	+ 9 ab	4 a	20 b	3 a	17 b
10	+ 15 b	3 a	14 b	14 b	13 b
20	+ 8 ab	5 a	22 b	38 c	22 b
40	1 a	42 b	37 c	51 d	45 c
80	25 c	70 c	51 d	63 d	64 d
160	42 d	88 d	62 d	70 d	80 e

In columns, different letters indicate differences between means at the 5% level using the Newman-Keuls' multiple-range test.

Spruce showing Al inhibition had stunted brownish or blackened root systems. The needles were either yellowish or purplish in appearance. These symptoms are reminiscent of P deficiency. White pine did not show these symptoms.

Growth inhibition of red and white spruce seedlings at low Al concentrations raise the possibility that elevated Al levels in the rooting zone of natural stands may inhibit root growth. The concentrations of Al found in some Northern Ontario boreal forest soils were high enough to potentially reduce seedling growth (see section 3.4).

No significant overall effect of the Ca and Mg treatments on the response to Al was found. Higher levels of Ca and Mg would need to be used to test the capacity of these elements to ameliorate Al toxicity, especially as Rost-Siebert's (1984) work has tended to support the view of Ulrich _et al._ (1979) that the Al:Ca molar ratio in soil solution is a more determining factor for Al toxicity than is Al concentration _per se_.

3.2 Aluminum and Phosphorus Concentrations in Plants

All species except white pine were analyzed for Al concentrations in the roots and shoots. These data are summarized in Table II. White pine analyses have yet to be completed. Very high concentrations of Al accumulated in the roots of Al treated plants. For treatments of 20 mg Al/L and greater, the Al concentration in red spruce roots was significantly higher than in jack pine. White and black spruce also had higher root Al concentrations than jack pine roots at 20 to 80 mg Al/L. It is to be expected that in poorly growing plants, elemental concentrations will be increased. However, in this study, while both red spruce and jack pine growth was reduced by more than 80% at the highest Al treatment, the red spruce roots had approximately a 50% greater Al concentration than did the jack pine roots.

The highly significant correlation between red spruce seedling dry weight and the Al concentration in the shoots is shown in Figure 1, which illustrates two of the Ca + Mg treatments. Smaller seedlings have higher Al in their shoots, suggesting a possible causal relationship. This also demonstrates for red spruce that Al toxicity is not solely manifested by a root inhibition but that the Al also accumulates in the shoots at Al exposure levels which cause severe seedling inhibition.

A significant relationship between Al and P in plants has been shown by many workers (e.g. McCormick and Borden, 1974) with both an external and internal co-precipitation and co-occurence of Al and P being ascribed to the formation of an $Al\text{-}PO_4$ precipitate. As shown by the spruce in this study Al toxicity symptoms are often similar to those for P deficiency. Thus, seedlings were analyzed for P content as well as for Al. Root and shoot P concentrations are summarized in Table III. The obvious contrast in this table is the positive linear relationship between Al concentration and P in the shoots of red spruce compared with the other species, and also that the P content of the roots of red spruce was not affected by the Al concentration while

TABLE II. Mean Al concentration (mg/kg dry weight) in roots and shoots of 4 coniferous species after exposure to a range of Al concentrations.

Al conc. (mg/L)	Red Spruce	Jack Pine	White Spruce	Black Spruce
		SHOOTS		
0	29 a	39 a	41 a	40 a
5	47 a	85 ab	97 b	76 a
10	67 a	139 ab	74 b	132 ab
20	162 ab	196 ab	136 c	125 ab
40	272 b	251 bc	213 d	270 ab
80	492 c	380 c	291 e	327 b
160	772 d	988 d	237 d	704 c
		ROOTS		
0	243 a	211 a	389 a	335 a
5	446 a	411 ab	1121 ab	1470 b
10	739 a	747 abc	848 ab	1980 bc
20	1690 b	849 bc	1447 c	1898 bc
40	2212 bc	1227 c	2233 d	2333 c
80	2905 c	1744 d	2434 d	3044 d
160	5351 d	3654 e	2588 d	3529 d

In columns, root and shoot means followed by different letters are significantly different at the 5% level using the Newman-Keuls' multiple range test.

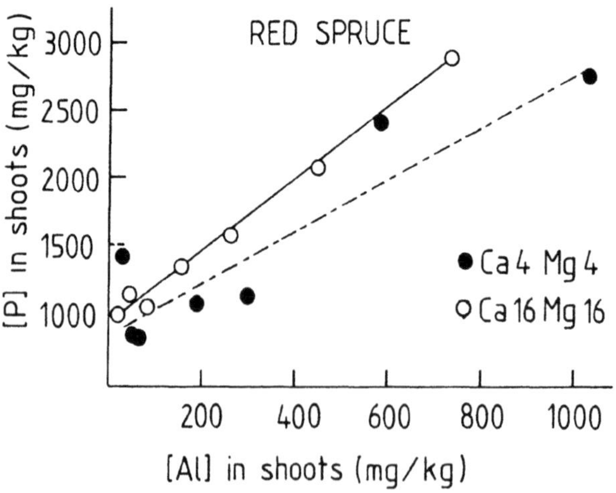

Fig. 1. The relationship between the mean seedling weight (Y) of *Picea rubens* and the Al content of the shoots (X), after exposure to the range of Al concentrations. For Ca 16 + Mg 4, $Y = 3.15X10^{-5}\ X^2 -.070X + 43.8$, $r = .99$; Ca 16 + Mg 16, $Y = 3.78x10^{-5}\ X^2 - .085X + 48.8$, $r = .98$.

TABLE III. Mean P concentration (mg/kg dry weight) of roots and shoots of 4 coniferous species after exposure to a range of Al concentrations.

Al conc. (mg/L)	Red Spruce	Jack Pine	White Spruce	Black Spruce
		SHOOTS		
0	1093 a	1287 ab	967 cd	1478 d
5	1030 a	1464 b	867 bc	1058 c
10	1010 a	1231 ab	511 a	986 c
20	1305 a	1035 a	784 bc	665 a
40	1299 a	1064 a	658 ab	709 a
80	2031 b	1006 a	905 c	836 b
160	2395 b	2132 c	1155 d	1019 c
		ROOTS		
0	2137 a	2839 ab	2859 c	4039 c
5	2062 a	2460 a	2292 abc	2676 b
10	2186 a	2834 ab	1733 a	1743 a
20	3139 a	2469 a	2088 ab	2600 b
40	3488 a	2273 a	1999 ab	2543 b
80	2923 a	2157 a	1989 ab	2353 b
160	6241 a	3346 b	2756 bc	2556 b

In columns, shoot and root means followed by different letters are significantly different at the 5% level using the Newman-Keuls' multiple-range test.

that of the other species was. Jack pine, white and black spruce all generally have the same pattern of P content in theroots and shoots; i.e., a decrease in P concentration from the 0 to 40 or 80 mg Al/L treatments, and an increase of P in the seedlings at 80 or 160 mg Al/L.

P and Al levels in shoots of jack pine and red spruce from certain Al + Mg + Ca treatments are summarized by regression analysis in figure 2. Jack pine showed a decline in P content of shoots as the Al in the shoots increased (Fig. 2a). However, at 160 mg Al/L it appears that damage to the barrier to P translocation occurred and both P and Al flooded in. The data may be interpreted as co-precipitation of Al and P occurring in shoots only at the highest Al treatment. The regressions are significant and best fit a quadratic. In striking contrast, and differing from white and black spruce also, the red spruce seedlings show a highly significant linear increase in shoot Al with increasing shoot P (Fig. 2b). Since the seedlings are severely inhibited from 20 mg Al/L upwards and appear P deficient, this can be interpreted as a continuous and persistent co-precipitation of $AlPO_4$ in the leaves as treatment Al increases.

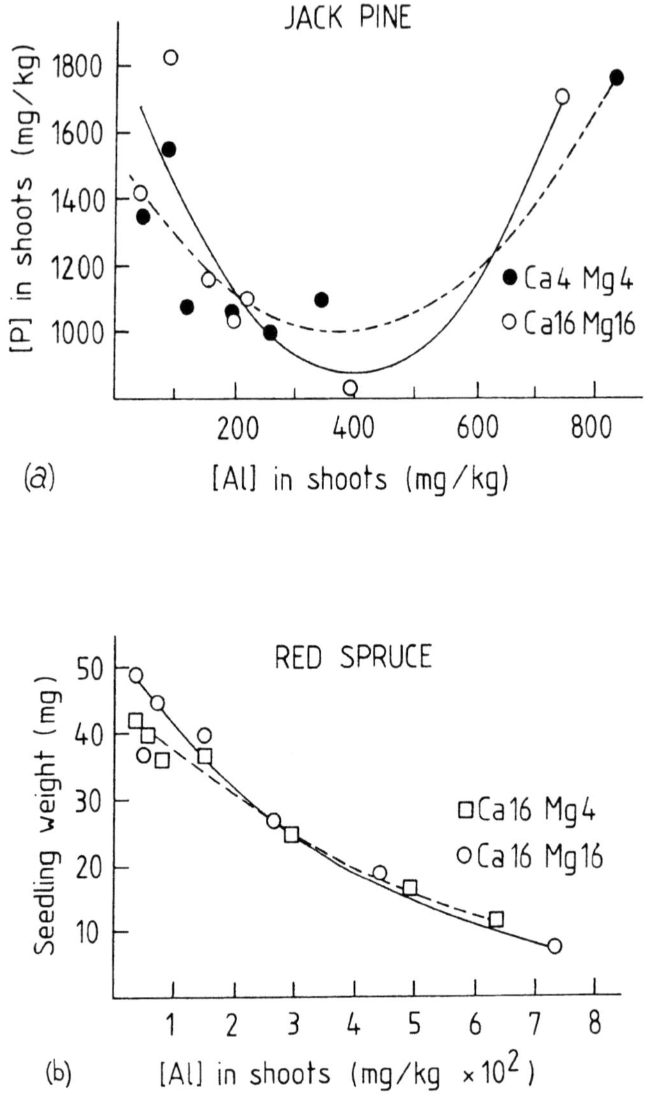

Fig. 2. The relationship between P (Y) and Al (X) in the shoots of (a.) Pinus banksiana and (b.) Picea rubens in response to Al treatments. For Pinus banksiana at Ca 4 + Mg 4, $Y = 3.76\times10^{-3} X^2 - 2.80X + 1512$, r = .89; Ca 16 + Mg 16, $Y = 6.55\times10^{-3} X^2 - 5.02X + 1838$, r = .85. For Picea rubens at Ca 4 + Mg 4, Y = 1.96X + 860, r = .92; Ca 16 + Mg 16, Y = 2.73X + 904, r = .996.

The increasing P levels in the shoots of red spruce as Al levels increase could also be interpreted as a possible consequence of P accumulation of non-translocatable P in the roots. Cumming et al. (1985) report elevated accumulations of non-translocatable P in red spruce seedling roots exposed to 10 mg Al/L. Womack and Colowick (1979) suggest that this accumulation of P in apparent free space (AFS), as a result of Al complexation or formation of an Al-ATP complex, could cause limiting levels of energetically active ATP for metabolic use.

3.3 Localization of Al and P in Jack Pine Roots

Staining of freshly cut roots in the region of the tips and at 10 cm back from the tip with hematoxylin showed substantial differences. In the region of the tip, staining for Al was intense in the root cap, the meristem, and along the outer cortical cell layers of the young root. The root cap was so densely stained it was difficult to determine localization but much of the stain may have indicated Al binding to cell walls and mucilage. The presence of stain in the meristem may indicate that inhibition of root growth is due to binding of Al in the nucleus resulting in an inhibition of cell division (Sampson et al., 1965). At the 10 cm region the stain indicated dense Al accumulation in the epidermal and cortical cell walls, with a reduced intensity towards the endodermis. There was no staining of cell walls of the vascular tissue. This could be interpreted as

TABLE IV. A comparison of peak to background ratios of Al and P in the xylem cortex and epidermis of jack pine seedling roots grown in 0 and 4 mmol/L Al. Values are means and standard deviations of three replicates.

Tissue		Al treatment conc. (mmol/L) 0	4
Xylem	Al	1.09 + 0.03 a	1.08 + 0.02 a
	P	1.18 + 0.10 a	1.11 + 0.03 a
Cortex	Al	1.06 + 0.08 a	2.28 + 0.40 b
	P	1.27 + 0.14 a	1.41 + 0.34 a
Epidermis	Al	1.04 + 0.06 a	2.87 + 0.75 c
	P	1.13 + 0.10 a	1.95 + 0.48 b

Means followed by different letters are significantly different at the 5% level using the Newman-Keuls' multiple range test.

apoplastic movement of Al across the cortex and binding to these cell walls, and that the endodermis is an effective barrier to Al.

Energy dispersive X-ray spectra were recorded in three locations of each root cross section from the 10 cm region. Data for counts from the xylem, cortex and epidermis are shown in Table IV. The data is expressed as peak to backgroud ratios of counts for Al and for P in tissues treated with 0 and 4 mmol/L Al. A significant accumulation and localization of Al occurred in the Al treated roots, in both the epidermis and in the cortex.
This localization supports the hematoxylin staining patterns described in this report. Both the Al and P concentrations in the xylem were not significantly different from the control. P localization was similar to that of Al. This coincident occurence of Al and P in the cortical and epidermal tissue indicates the capacity for cell wall material to bind Al and thus fix P (Clarkson, 1967). This Al and P distribution is supported by the scanning electron micrograph of an Al-treated root cross section, and corresponding X-ray distribution maps of Al and P (Figs. 4a, b, c). Dense localization of Al and P in the epidermis can be distinguished. The resolution of these micrographs is, however, obviously insufficient to distinguish cell wall from cytoplasmic or nuclear localization.

3.4 Extractable Al in Northern Ontario soils

To determine the potential for Al toxicity to occur in boreal forest soils and relate these levels to the tolerance of the species found in this study, we summarize data from Bozic (1986) on water-extractable Al and the pH of soils collected from 35 boreal forest sites in Northern Ontario. The soil pH's of the organic horizon ranged from 3.2 to 4.6, with a mean pH of approximately 3.7. For mineral soils the pH was consistently higher, ranging from 3.5 to 6.2 but with a mean of approximately 4.3. Thus, these podzolic soils, which support conifers, are naturally quite strongly acidic and are in the pH range in mineral horizons in which increased acidification could cause substantial increases in Al. For the organic layer the pH is already below 4.0 so Al could be expected in high concentrations. This was borne out by the extractable levels, which for the organic soils ranged from 15 to 257 mg Al/L, with a mean of just over 50 mg Al/L . These are very high levels compared with agricultural crop exposures and since the conifers at the collection sites were apparently healthy and growing well, this suggests that mature root systems, at least, are well adapted to high Al or avoid it by rooting in the B and C mineral horizon. The mineral soils had water-extractable Al levels from 2 to 40 mg Al/L and a mean of 12 mg Al/L. It seemed reasonable to choose Al concentrations used in this study considering the range of Al found in the organic horizons, since seedlings germinate and establish in the organic A horizon. Thus, the range of 0 to 160 mg Al/L was chosen for experiments. It can be expected that in the upper horizons especially, soluble Al would be largely organically bound, and that this would decrease its toxicity. Evidence for this is reported by Nilsson and Bergkvist (1983), where in Swedish podzols 83 to 97% of the dissolved Al in the A horizon, consisted of organic-bound forms, while in the mineral B horizon only 8 to 20% was organically bound.

In summary , these studies have shown that a wide range of responses to Al exist amongst the 5 conifer species studied. Red spruce and white spruce were inhibited in growth at the 5 mg Al/L, while levels greater than 20 mg Al/L were severly inhibitory. Black spruce was also severely inhibited by 20 to 160 mg Al/L. In contrast, jack pine and white pine were not affected by Al treatment concentrations of up tp 20 mg Al/L and indeed white pine was stimulated. Even 160 mg Al/L failed to reduce white pine seedling growth by 50%.

The roots of red spruce accumulated high levels of Al and both Al and P was translocated to the shoots where possible co-preciptiation occurred. Jack pine, white and black spruce differed in that increased Al treatment generally resulted in decreased P except at 160 mg Al/L.

Localization studies using SEM-EDAX and hematoxylin on jack pine seedling roots showed that Al accumulated in the root cap, root meristem and in the cells walls of the epidermis and cortex of older roots. P had a similar distribution in these older roots.

While this study is not extrapolatable directly to field conditions, it does suggest that red spruce may be an Al sensitive species, and that white pine is much more tolerant. The field situation is enormously more complex than greenhouse and sand culture experiments, not least in the addition of soil chemical complexity, spatial gradients, and mycorrhizal, pathogen and microbial interactions etc. Caution is also essential in concluding anything about mature tree responses to stress from seedlings in pots.

ACKNOWLEDGEMENTS. Financial support for this study was from a Canadian Forest Service contract, and from a Natural Sciences and Engineering Research Council operating grant to T. C. Hutchinson.

REFERENCES

Bozic, L. E.: 1986, 'The effect of Al on jack pine seedlings and its localization within the root tissue' MSc thesis, Dept. of Botany, University of Toronto.

Clymo, R. C.: 1962, J. Ecol. **50**, 707.

Cumming, J. R., Eckert, R. T., and Evans, L. S.: 1985, Can. J. Bot. **63**, 1099.

Foy, C. D., Chaney, R. C., and White, M. C.: 1978, Ann. Rev. Plant Physiol. **29**, 511.

Hallenbacken, L., and Tamm, C. O.: 1986, Water, Air and Soil Pollut., this volume.

Havas, M.: 1986, Water, Air and Soil Pollut., this volume.

Hutchinson, T. C.: 1983, 'A historical perspective on the role of aluminum in the toxicity of acid soils and waters, In: International Symp. Heavy Metals in the Environment. Heidelberg, 1983, CEP Consultants, Edinburgh, pp. 17-26.

Ingestad, T.: 1962, Medd. Stat. Skogsf. Inst. **51**, 1.

Johnson, A.: 1983, J. Air Poll. Cont. Assoc. **33**, 1049.
Krause, G., Arndt, V., Brandt, C. J., Bucher, J., Kenk, G., and Matzner, E.: 1986, Water, Air and Soil Pollut., this volume.
Magistad, O. C.: 1925, Soil Sci. **20**, 181.
McCormick, L. H. and Borden, F. Y.: 1974, Soil Sci. Soc. Am. Proc., **38**, 931.
Muller, H., and Abrahamsen, G.: 1984, personal communication.
Nilsson, S. I., and Bergkvist, B.: 1983, Water, Air and Soil Pollut., **20**, 311.
Rorison, I.: 1960, J. Ecol. **48**, 585.
Rost-Siebert, K.: 1984, 'Aluminum toxicity in seedlings of Norway spruce and beech.' In: Andersson, F. and Kelly, J. M. Aluminum Toxicity in Trees, Sveriges Lantbruksuniversitet Intern. Workshop, 49-68.
Sampson, M., Clarkson, D. T., and Davies, D.D.: 1965, Science **148**, 1476.
Schier, G.: 1985, Can. J. For. Res. **15**, 29.
Siccama, T. G., Bliss, M. and Vogelmann, H. W.: 1982, Bull. Torrey Bot. Club **109**, 162.
Tyler, G., Berggren, D., Bergvist, B., Falkengren-Grerup, U., Folkeson, L., and Ruhling, A.: 1985, 'Soil acidification and metal solubility in forests in south Sweden.' In: Hutchinson, T. C. and Meema, K. M. (Eds.) Effects of Acidic Deposition of Forests, Wetlands and Agricultural Ecosystems, NATO Advanced Research Workshop Proc. Toronto, May 12-17, 1985. Springer-Verlag, Heidelberg.
Ulrich, B., and Matzner, E.: 1983, Forschungsbericht 104 02615. UBA, Berlin.

Ulrich, B., Mayer, R., and Khanna, P. K.: 1979, Schiften aus der Forstlichen Fakultat der Universitat Gottingen **58**, 1.
Ulrich, B., Mayer, R., and Khanna, P. K.: 1980, Soil Sci. **130** 193.
Womack, F. C., and Colowick, S. P.: 1979, Proc. Nat. Acad. U.S.A. **76**, 5080.

ACTIVITIES AND RESULTS OF THE TERRESTRIAL EFFECTS PROGRAM: ACID PRECIPITATION IN ONTARIO STUDY (APIOS)

S. N. Linzon
Phytotoxicology Section
Air Resources Branch
Ontario Ministry of the Environment
Toronto, Ontario
Canada M5S 1Z8

Abstract. *Studies on the terrestrial effects of acidic precipitation in Ontario involve determining the effects on soils, crops, forests, lichens, mosses and biogeochemical systems. In the soils program, a baseline study was started in 1980 to establish a reliable and uniform data base for soils across the province, in order to identify future trends. Soil sensitivity criteria are being derived for mapping purposes. Starting in 1982, a complex, sophisticated mobile rain exclusion canopy system was constructed outdoors for controlled acid rain studies to determine the dose-response relationships of field crops. In response to numerous complaints a program was designed to determine the role that acidic precipitation is playing in the decline of sugar maple trees in Ontario. At 8 sites, tree conditions were noted, samples of foliage, bark, roots and soil were collected for chemical analysis, and increment cores and discs (from felled trees) were taken for radial growth patterns. Results from this study indicated that acidic precipitation was an additional stress to insect outbreaks and spring droughts. Intensive surveys have been conducted in selected areas in Ontario to identify the current viability and distribution of common denominator lichens and mosses. Biogeochemical studies are being conducted at four watersheds in Ontario. The different watersheds are located in progessively decreasing atmospheric deposition loadings from east to west across the province. The studies are attempting to document the role of contrasting terrestrial ecosystems in the process of lake acidification by atmospheric deposition.*

1. INTRODUCTION

Studies on the terrestrial effects of acidic precipitation in Ontario started in 1980. Initially, an annotated bibliography was prepared which included the title, author(s) and summary of literature on the effects of acidic precipitation on vegetation and soils (Linzon et al., 1981). The bibliography excluded papers dealing with precursor pollutants such as SO_2 *and* NO_x *alone unless there was a direct reference to their contribution to acidic precipitation and effects on terrestrial systems. The bibliography laid the foundation*

Water, Air, and Soil Pollution **31** (1986) 295–305.

for the development of the terrestrial effects program of the Acidic Precipitation in Ontario Study (APIOS).

Studies were established to determine the effects of acidic precipitation on vegetation (crops, ornamentals, forests, lichens and mosses), on soils and on biogeochemical systems. The studies were carried out both in the field and under controlled conditions in the laboratory. A procedures manual was prepared to ensure a standardized approach to terrestrial studies carried out by the Ontario Ministry of the Environment throughout the province (Griffin, 1983).

2. VEGETATION STUDIES

2.1. Baseline Vegetation Study

In order to document the potential effect of acidic precipitation on forest tree foliage, 76 permanent sites were established in Ontario for baseline foliar sampling. These sites for the most part were established adjacent to sites selected for the Baseline Soil Study. Foliage was collected from the trees and chemically analyzed for 30 parameters. Over the years the sites will be visited periodically and the foliage resampled in order to detect changes in elemental composition. Acidic precipitation has the potential to add or remove nutrients from trees and to change the availability of elements in the soil for uptake by the trees.

To determine if potential elemental concentration changes can be attributed to acidic precipitation, the magnitude of natural foliage variability must be established. A special site was established to assess the effects of sampling at different times of the year, in different portions of the canopy and from trees of varying age.

2.2 Experimental Studies with Simulated Acidic Rain

Studies on the effects of simulated acidic rain on ornamentals, agricultural crops and forest species are being conducted at the Phytotoxicology controlled environment facility at Brampton, Ontario. Specially designed indoor acidic rain simulation chambers were constructed in 1981 (Kuja and Enyedi, 1983). Four separate chambers were built to allow for the simultaneous application of four different rain simulants to plants being treated. The chambers are of sufficient height (7.3 m) to enable droplets released from nozzles at the top of the chambers to approach terminal velocity before impinging upon the plants set out on turn tables below.

In 1981, the Rain Acidity Interlaboratory National Study (RAINS) was organized in order to explain the discrepancies which had been reported in the literature with regard to results found in simulation experiments. It was apparent that different methodologies had a bearing on these discrepancies and thus a cooperative study was organized in which five laboratories in the USA and the Ontario Ministry of the Environment laboratory would use standardized procedures and materials. The five USA laboratories were Argonne National Laboratory, Argonne, Illinois; Boyce Thompson Institute,

Ithaca, New York; Brookhaven National Laboratory, Brookhaven, New York; Corvallis Environmental Research Laboratory, Corvallis, Oregon; and the Oak Ridge National Laboratory, Oak Ridge, Tennessee. The same cultivar of radish grown in Pro Mix BX was exposed to eight different rain simulants (pH 2.6, 3.0, 3.4, 3.8, 4.2, 4.6, 5.0 and 5.4). The SO_4^{2-} *to* NO_3^- *ratio was 2.37 and was maintained for each simulated rain. Other anions and cations were added in similar proportions as found in ambient rain. The plants were treated one hour per day for three days each week for a total of 7.6 cm of rain. After 28 days the radishes were harvested and wet and dry weights determined. Because of different latitudes of the six laboratories the growth rates varied but the responses to the simulated rain were similar. At pHs 5.0 and 4.6 there was generally a stimulation in growth, and the pooled data showed significant negative effects at pH 3.0. The data are currently being analyzed statistically for publication.*

The indoor acid rain chambers have been used to study symptomatology, sensitivities of different cultivars of crop plants, effects on growth and yield, and threshold levels of acidity causing effects on plants. Plant species tested include alfalfa, barley, cabbage, corn, cucumber, petunia, pinto bean, radish, soybean, tobacco, tomato, white birch, sugar maple, white pine and white spruce. Tests have shown that plant species vary in their sensitivity to simulated acidic rain, and that visible injury occurs on leaves, stems and flowers of plants when the pH of the simulated acidic rain treatments is 3.0 or lower. Consequently, repeated occurrences of rainfall of pH 3.0 or lower during the growing season will increase the likelihood of visible injury to plants and reductions in crop yield.

A specific pattern of acid rain injury takes place on all plant surfaces where droplets reside for a period of time. Initially, small brown pits or lesions of dead tissue develop where cells have been destroyed by hydrogen ions. Mesophyll cells within the leaf tissue can display both hypertrophy and hyperplasia (Tung et al., 1982). With repeated exposure to highly acidic rain (e.g pH 3.0), lesions will increase in size as more plant tissues in the vicinity of each lesion are affected. Eventually, leaves can become deformed or reduced in area, and their ability to photosynthesize may be inhibited.

The results of experiments with agricultural crops illustrated that plant response to simulated acidic rain treatments is not only species dependent but also strongly cultivar dependent. Rainfall pH between pH 3.0 and pH 4.2 may have a negative effect, a positive effect or no effect on plant growth. For example, certain cultivars of cucumbers and alfalfa were stimulated in growth; no effects were found on any of the cultivars of corn or barley tested; and cultivars of cabbage and radish displayed negative growth. Soybean cultivars showed all three responses, stimulation to Hodgson, no effect on Maple Arrow or Evans, and negative effects on Harcor.

This differential susceptibility is an important factor to be considered when interpreting results derived from effects studies which utilize a single test species or cultivar.

Starting in 1982, a complex, sophisticated mobile rain exclusion canopy system was constructed outdoors at Brampton, in order to conduct controlled acidic rain studies to determine the dose-response relationships of field grown crops and trees. There are three main features of the system:

(1) *Three greenhouse type canopies (19.5m x 9.2m) move automatically over field grown crops when a rain event starts, and retract after completion of the rain event. Nozzles inside the canopies apply simulated acidic rain of different pHs to 50 plots (1.5m x 1.5m) of crop plants located under each canopy.*

(2) *The amount of rain applied within the canopies matches the amount of rain falling naturally outside, as measured by a tipping bucket rain gauge.*

(3) *During dry periods the field crops are protected from elevated concentrations of gaseous pollutants, such as O_3 and SO_2 by a blower-pillow system. The blowers deliver pollutant-filtered clean air to perforated polyethylene tubing located between each row of crop plants.*

All phases of the mobile rain exclusion canopy system are computer controlled, from detection of a rain event, deployment of the canopies, mixing and application of the simulated acidic rain treatments, and retraction of the canopies. Crop plants located under each of the three canopies can be treated differently during dry periods prior to being covered by the canopies. Under the first canopy, acidic rain treatments can be applied to plants which had been exposed to ambient air during dry periods; under the second canopy, the plants can be protected from pollutants in ambient air during dry periods; and under the third canopy, increased concentrations of gaseous pollutants can be added to the blower system and delivered to plants during dry periods in order to study the interactive effects of acidic rain and elevated concentrations of O_3, SO_2 or NO_x.

The system was completed in 1983 and during the 1984 growing season radish and soybean crops were experimented with to test the system and correct any operational faults. Foliar injury was induced at pH 3.0 but there was no apparent reduction in yield of soybean var. Bicentennial as measured by the number of seeds per pod or the number of pods per plant when compared to plants treated at pH 5.6. In 1985, after certain improvements had been made to the system experiments were continued on soybean var. Hodgson.

2.3 Lichen and Moss Study

Lichens and mosses receive a significant portion of their nutrients from atmospheric deposition and may be organisms especially sensitive to acidic precipitation. They are well-known for their suceptibility to gaseous pollutants such as SO_2 and HF, and for their ability to absorb and accumulate airborne contaminants.

Intensive surveys have been conducted in selected areas of Ontario to identify the current viability and distribution of common denominator mosses and lichens. Quadrats have been established on boles of trees and standardized photographic records have been taken at three locations in Ontario. Future surveys may document changes in population structure.

Samples of selected lichen and moss species were collected in 1982 and 1983 from 45 sites across Ontario. These samples are being analyzed for a wide range of chemical constitutents. Attempts to relate lichen and moss chemistry to current precipitation chemistry is underway.

In the Thunder Bay District of northern Ontario, selected chemical constitutents of lichens and mosses which were collected in the early 1970's are being compared with analytical data from similar collections made in the early 1980's.

Metal levels in lichens are strongly influenced by their proximity to major sources of metal-contaminated particulates. The sulphur content of lichens exhibits a south-to-north decrease in concentration. The elemental composition of wet deposition can account for only part of the variability seen in the chemical content of lichens. Gaseous and airborne particulate materials, and perhaps the chemistry of the substrate, may also be required to account for the remaining variability.

3. Soil Studies

3.1 Baseline Soil Study

A baseline soil study was started in 1980 to establish a reliable and uniform data base for soils across Ontario, in order to identify future trends. Over 400 permanent soil sites on various landforms had been established by 1983. A three-volume report has been published on baseline soil data for 250 sites established during 1980 and 1981 (Griffith et al., 1985).

For each soil baseline site, the precise location was recorded and duplicate soil samples were collected from every horizon from a soil pit approximately 1 m in depth. The soil samples were analyzed for 30 different parameters including texture, pH, base saturation, cation exchange capacity, organic matter, iron, aluminum, trace metals, sulphur and nitrogen. Information such as soil colour; number and depth of roots; stones; and soil structure were recorded also in the field.

The baseline data describe the current status of soils in Ontario. Trends in soil chemistry will be monitored by resampling soil profiles over an extended period of time. Trends will vary across the province, depending on site sensitivity and atmospheric deposition loading.

Although soil surveys have been conducted in the past in Ontario, few of these are of value in the current study of the potential effects of acidic precipitation. Earlier surveys concentrated on agricultural or managed lands, and precise sampling locations were not provided. Also, because a variety of sampling and analytical methods were used, comparison with currrent results are generally unreliable. In contrast, the present APIOS baseline soil study restricted the establishment of sampling sites to unmanaged lands which offer reasonable access and long-term security. The precise locations of the soil pits are recorded as well as the exact analytical methodology used in order to make future re-sampling and analytical tehniques comparable.

In addition to providing data on the physical and chemical properties of Ontario soils for future reassessment, the soil database will provide information required for sensitivity mapping of soils throughout Ontario.

3.2 *Soil Sensitivity Mapping*

Soil sensitivity criteria are being derived from soil buffering experiments which will be applied to the soil baseline data. A soil sensitivity map for Ontario will be prepared based on the findings of all soil sensitivity experiments. Although several criteria can be used to prepare soil maps, the results of experiments indicate that soil texture, base saturation, presence or absence of carbonates and soil depth are most important, followed by soil pH and cation exchange capacity. Other criteria include organic matter content, sulphate adsorption capacity, soluble aluminum content, and vegetative cover growth. The experimental results indicate that intermediate pH soils (i.e. largely brunisolic in Ontario) are the most sensitive to change by acidic precipitation. The sensitivity criteria may vary depending on the resource one is interested in. For lake sensitivity, the parent material horizon data will be useful, whereas for forest growth sensitivity, the organic and mineral layer soil horizon data will be more useful. Thus three maps reflecting soil sensitivity may be produced reflecting the differences in soil horizon type, and the sensitivity of the resource under consideration: soils, lakes or forests.

3.3 *Soil Sensitivity Experiments*

(a) *Leaching Experiments*

1. Soil horizons of several major soil types in Ontario are being leached with simulated acidic rain over a treatment range of pH 2.0-5.6 with a chemical make-up similar to that of Ontario's rain. The soil leachate is analyzed for major cations, anions, and metals in order to predict the long-term effects of acidic precipitation on soil under controlled conditions.

2. Soil columns will be leached with heavier doses of simulated acidic solutions to accelerate and document the physical and chemical changes in soil that may occur due to acidic precipitation.

(b) *Soil Buffering Experiments*

To be able to predict possible reactions of soils to acidic precipitation, soil samples are being equilibrated with solutions of different acidity and base cation content (Dimma et al., 1983). The pH of selected soil horizons is measured after the addition of either a nitric, a sulphuric, a nitric-sulphuric or a simulated acidic precipitation solution with each being added over a pH range of 4.0, 3.5, 3.0, 2.8, and 2.5. The rate of change of soil pH is indicative of the ability of a soil to resist acidification. The study will allow the evaluation of the chemical and physical components of soil which contribute to a soil's buffering capacity.

Podzolic soils treated with rain simulants tend to adsorb added sulphate, and nitrate becomes the mobile anion in leachates. Conversely, in natural forest soil systems, nitrate tends to be conserved in the soil and sulphate becomes mobile in drainage waters.

(c) Open -Top Soil Columns

To determine the long-term chemical changes in soil exposed to ambient conditions, three different soil types (sand, silt, clay) were collected, chemically analyzed and placed into open-top columns. Several replicates of each soil type were placed in the field in the Northwestern and Northeastern Regions, and in Southern Ontario. A replicate column of each of the three soil types will be removed from the ground after 5, 10, 15 and 20 year intervals and chemically analyzed to determine changes in soil chemistry under ambient conditions.

3.4 Soil Variability Study

To better document the natural variation in some soil parameters, two existing baseline soil sites near Dorset were systematically sampled every meter on a 10 m grid to evaluate seasonal and spatial variation. These natural changes in soil properties within a sampling unit provide background information for assessing changes in soil properties due to acidic precipitation. Most soil sensitivity chemical parameters (e.g. base saturation, soluble aluminum and pH) are quite variable in the spring and fall. Summer appears to be the most stable season for sampling soil for sensitivity chemical parameters. It was found that pH can vary up to 1.4 units at different locations within a single sample plot at any given time. Thus precise site relocation is important when resampling soil for trends-over-time.

4. BIOGEOCHEMICAL STUDIES

Biogeochemical studies are being conducted by the Ontario Ministry of Environment at four watersheds in Ontario. Two are located near Dorset in central Ontario (Plastic Lake and Harp Lake) which receive wet sulphate deposition of approximately 30-35 kg/ha.yr. One of the northern sites is at High Falls, 40 km west-southwest of Sudbury and receives about 20 kg/ha.yr wet sulphate deposition. The other northern site is at Hawkeye Lake, 40 km northwest of Thunder Bay and the wet sulphate deposition rate is about 10 kg/ha.yr. The four biogeochemcal study sites are located from east to west along a northerly slope within the Great Lakes-St. Lawrence Forest Region and reside within three different zones of atmospheric deposition which will allow "gradient effects" of acidic loadings to be investigated.

The terrestrial studies at the watersheds are providing data to document the role of contrasting terrestrial ecosystems in the process of lake acidification by atmospheric deposition. In addition, the ecosystems are being examined to determine the effect of varying acidic inputs on nutrient and trace elemental cycles within these ecosystems.

The field methodology used at all four biogeochemcal sites is similar. Incident precipitation, throughfall, stemflow, and soil leachate are collected similarly and analyzed for the same chemical parameters. Litterfall traps and litter decomposition bags are used to document the flux of elements in seasonal cycles. Complete tree harvests, and dissection of trees for samples of foliage, twigs, bole discs, and roots for chemical analysis will determine the biomass on the watersheds as well as the storage of elements in the standing crop. In addition to forest inventories, soil surveys and soil pit sampling will provide information regarding total elemental reserves.

Some results to date show that throughfall is enriched in nutrient and trace elements as compared to incident precipitation. This enrichment is dependent on the species of the tree interacting with the precipitation, indicating that forest canopy type determines the flux of materials to the forest floor via precipitation/-throughfall water. Stemflow is even more enriched in nutrient ions and trace elements. As with throughfall, the species of the tree bole has an influence on the degree and nature of this enrichment. Enrichment of SO_4^{-2} *under conifers is greater than under hardwoods under similar atmospheric deposition. Enrichment of* SO_4^{-2} *in throughfall is higher in areas of higher atmospheric deposition. Shrubs contribute significantly more to the alteration of throughfall water, amplifying the role of the tree canopy in precipitation alteration. Conifers, as a rule, induce a pH depression of throughfall water while hardwood species elevate the pH. These changes are on the order of a tenth of a pH unit. Stemflow differences between conifers and hardwoods are highly pronounced. Conifer boles can depress incident precipitation pH by one-half to one pH unit. Hardwood boles have the opposite effect. Nitrogen compounds tend to be conserved by the tree canopy and bole, the reverse trend of other nutrient ions. The degree of alteration via throughfall is less during the dormant winter period.*

Soil leachate sampled at various horizons in a podzolic soil is more acidic (lower pH) than soil leachate from a brunisolic soil. Also, concentrations of SO_4^{-2} *and aluminum in the podzolic soil leachate are higher. Nitrate in soil leachate is only detectable in the early spring or late fall, corresponding to tree dormancy periods. Therefore* NO_3^- *is not an important mobile anion during the growing season. Sulphate concentrations in soil leachate are elevated following rain epiosodes which follow prolonged dry periods; possibly resulting from sulphur oxidation in the soil during the dry periods.*

Data from the biogeochemical studies will be used to develop two predictive models (1) a hydrological model aimed at predicting changes in water quality; and (2) a biogeochemical model aimed at predicting changes in forest productivity.

5. *FOREST PRODUCTIVITY AND DECLINE STUDIES*

5.1. *Sugar Maple Decline Study*

The decline of sugar maple in a rural area was first reported in Ontario in Grey County in 1952 (Griffin, 1965). Symptoms of damage included discoloured foliage and death of young twigs followed by crown dieback. The progression was slow and trees seldom died. Recovery occurred in moderately affected trees.

Over the years sugar maple decline has been observed at intervals in different parts of the Province of Ontario, with severe outbreaks occurring in 1976 and 1977 in the Parry Sound and Owen Sound districts. In this latter episode, forest tent caterpillar defoliated the forests and over 50% of the sugar maple trees were killed on 8000 ha and 500 ha in the Parry Sound and Owen Sound districts, respectively (Canadian Forestry Service, 1980). The insect population collapsed and most of the surviving sugar maple trees recovered.

In the spring of 1984, a number of maple syrup producers from the Muskoka area of Ontario complained to the Ministry of Agriculture and Food about the increase in dieback and mortality of sugar maple trees in the last six years. The producers felt that acidic precipitation may be involved and that continued sugar maple decline could jeopardize the local maple syrup industry.

The Ontario Ministry of the Environment was asked to investigate the role of acidic precipitation in this recent maple decline outbreak in Muskoka. A field study was designed by the Phytotoxicology Section, Air Resources Branch to examine the etiology of the declining sugar maple trees. Questionnaires were given to each woodlot owner to provide a history of stand management. Permanent observation plots were established in seven woodlots in the Muskoka area of south-central Ontario and in one woodlot near Thunder Bay in northwestern Ontario. Soil from the woodlots, and foliage, twigs and roots from sugar maple trees exhibiting a gradient of decline symptoms were collected for chemical analysis. Increment cores were collected from sampled trees and discs were taken from felled trees to examine chronological growth patterns. In addition, atmospheric acidic deposition rates, forest management practices, site disturbances, tree age, site quality, the presence of diseases and the history of insect defoliation and weather records were investigated at each study site location.

A report describing the field methodology and a summary of preliminary results has been prepared (McLaughlin et al., 1985).

A summary of the 1984 study results indicates that the current outbreak of sugar maple decline in Muskoka first became evident about 1978. Based on both aerial and ground observations sugar maple decline symptoms were observed on trees throughout the Muskoka region, with no consistent pattern as to topography, aspect or site. Although some degree of tree decline was observed on maples in all age classes it was most pronounced on older trees and on trees which had been tapped for maple syrup production or otherwise wounded. Site nutrient deficiencies were not implicated in the decline of sugar maple.

The soil at the Muskoka sites was acidic and contained high amounts of soluble aluminum. Declining trees in Muskoka suffered extensive root death and the fine roots had significantly higher aluminum concentrations than fine roots of healthy trees. A foliar chemical gradient was detected with reduced elemental concentrations in the tops of the tree crowns at Muskoka. Annual growth rings for both healthy and declining trees were very narrow during two years of forest tent caterpillar defoliation. Subsequent to the collapse of the insect epidemic, growth recovered in the healthy trees but not in the declining trees. Incremental growth in the declining tree population appeared to be falling relative to the healthy trees for 20 years prior to the caterpillar infestation suggesting that this group of trees may have been predisposed to decline perhaps by physiologic, genetic or environmental factors. Early season droughts during the two years of the defoliation by insects (1976 and 1977) and again in 1983 and root infections by Armillaria mellea were some of the contributing factors to the tree decline.

It is apparent that the role of acidic precipitation as a contributing factor to sugar maple decline is part of a complex system. The primary mechanism of sugar maple decline is probably site specific, and may vary between individual trees and regions. At Muskoka the soils are acidic and are classified as sensitive; aluminum is freely available to trees and therefore the potential for indirect effects to trees exists. Higher amounts of aluminum were found in the fine roots of declining trees in Muskoka compared to healthy trees. Foliar tissue leaching and reductions in incremental growth on some trees may be environmentally related. Sugar maple decline is widespread throughout the Muskoka area and the region receives high loadings of acidic deposition (about 35 kg SO_4 *in wet precipitation /ha.yr).*

From the above, it may be tentatively concluded from the first year's study of sugar maple decline at Muskoka, that acidic precipitation is an additional stress to the severe epidemic of forest tent caterpillar in the late 1970s combined with spring droughts in 1976, 1977 and 1983. Armillaria mellea, tree age and site management are also contributing factors.

Future work includes examining additional sugar maple sites in calcareous soils in southern Ontario and in areas which had not been defoliated by the forest tent caterpillar in 1976 and 1977. A province-wide survey is being conducted for the incidence and severity of hardwood decline (sugar maple, ash, beech, birch) in order to document trends over time which could be related to atmospheric deposition loadings. Also, a detailed stem analysis and dendrochronology study is being carried out in different Forest Sections of Ontario to develop regional tree growth chronologies which can be related to different soil buffering capacities and different loadings of atmospheric pollutants.

5.2 Analysis of Tree Growth

Historical records of forest stand dynamics are available for several forest woodlots in the Muskoka-Haliburton area. Three compartments in each of eight forest woodlots were measured in 1962/63 and in 1972. Two of these woodlots, a coniferous and a mixed hardwood, were selected for remeasurement of diameter growth in 1983 and 1984.

Current measurements of these compartments will provide a two-decade detailed record of tree growth and stand dynamics. In addition, tree cores and stem analysis discs are being examined in 1985 to document the development of these forest stands over an even longer period of time. Soil conditions, site, climate, stand age, and tree density are being studied to determine if these variables will be able to account for observed changes in the growth patterns or if an independent variable, such as acidic precipitation, is influencing forest growth.

6. ACKNOWLEDGMENTS

A number of staff of the Ontario Ministry of the Environment, from the Phytotoxicology Section, Air Resources Branch and from the Northeastern and Northwestern Regions, participated in the terrestrial studies described in this paper. These staff include W. I. Gizyn, D. E. Dimma, M. A. Griffith, A. L. Kuja, A. J. Enyedi, G. Tung, R. D. Jones, P. C. McGovern, D. L. McLaughlin, D. Terry, S. Rutherford, W. D. McIlveen, J. Negusanti, and H. D. Griffin. Also, the following consultants participated in the terrestrial studies: J. W. Case, F. C. Lozano, V. Timmer and D. V. Love.

7. REFERENCES

Canadian Forestry Service: 1980, "Survey Bulletin. Forest Insect and Disease Conditions in Ontario, Summer 1980", Great Lakes Forest Research Centre, Sault Ste. Marie, Ont., p.9.

Dimma, D. E., Griffith, M. A. and Linzon, S. N.: 1983, "An Experiment to Assess the Sensitivity of Soil to Acidification", Proceedings, Acid Rain and Forest Resources Conference, Quebec City, 1-19 pp.

Griffin, H. D.: 1965, "Maple Dieback in Ontario", The Forestry Chronicle **41**, 295.

Griffin, H. D. (ed.): 1983, "Procedures Manual - Terrestrial Effects", Report No. APIOS 007/83, Ontario Ministry of the Environment, Toronto, Canada, pp. 1-57.

Griffith, M. A., Spires, T. and Barclay, P.: 1985, "Ontario Soil Baseline Survey, Analytical Data 1980-1981", Three Volumes, APIOS 002/85, Ontario Ministry of the Environment, Toronto, Canada pp. 1-575.

Kuja, A. L. and Enyedi, A. J.: 1983, "Effect of Simulated Acid Rain on Agricultural Crops", Proceedings, Agrometeorological Workshop, University of Guelph, Guelph, Ontario, pp. 1-13.

Linzon, S. N. (ed.), Gizyn, W. I. and Griffith, M. A.: 1981, "An Annotated Bibliography - Terrestrial Effects of Acidic Precipitation", Report No. APIOS 003/81, Ontario Ministry of the Environment, Toronto, Canada, pp. 1-181.

McLaughlin, D. L., Linzon, S. N., Dimma, D. E. and McIlveen W. D.: 1985, "Sugar Maple Decline in Ontario", Ontario Ministry of the Environment, Toronto, Canada, pp. 1-18.

Tung, G., Kuja, A. L. and Linzon, S. N.: 1982, "Histopathology of Plant Leaf Injury Caused by Simulated Acid Rain", Proceedings, Microscopical Society of Canada, **IX**, University of Alberta, Edmonton, Alberta, pp. 64-65.

A MOBILE RAIN EXCLUSION CANOPY SYSTEM TO DETERMINE DOSE-RESPONSE RELATIONSHIPS FOR CROPS AND FOREST SPECIES

Allen Kuja, Randy Jones, and Alexander Enyedi,
Phytotoxicology Section, Air Resources Branch,
Ontario Ministry of the Environment,
880 Bay Street, 3rd floor,
Toronto, Ontario, Canada
M5S 1Z8

ABSTRACT. A fully automated rain exclusion canopy system has been designed and constructed to study the impact of simulated acid rain on the growth and yield of agricultural crops. The system consists of three, mobile greenhouse shelters which exclude ambient rainfall and apply simulated acid rain (SAR) treatments to crops established in field plots. All aspects of the field system are controlled by a microcomputer and data acquisition system. In addition, elevated ambient levels of gaseous pollutants (e.g. O_3, SO_2, NO_x) are reduced in field plots via an air exclusion system consisting of large blowers, potassium permanganate-treated alumina filters and perforated polyethylene tubes. Gaseous pollutants can be injected into the treatment plots via the blowers.

1. INTRODUCTION

Rain exclusion shelters have been utilized in drought studies to exclude ambient rain from field plots (Bruce and Shuman, 1962; Arkin et al. 1976; Upchurch et al., 1983). At present, rain exclusion canopy (REC) systems of various designs are utilized in several locations in the U.S.A. for simulated acid rain 'effects' studies (Evans et al., 1981; Irving, 1982; Heagle et al., 1983). REC systems eliminate ambient acidic ion inputs which may interfere with controlled applications of simulated acid rain on to field plots. A REC system was designed and constructed by the Ontario Ministry of the Environment to study the direct effects of simulated acid rain on crops and economically important forest species. Dose-response functions can be generated for crops or trees established in field plots and exposed to SAR.

The Ontario REC system is unique because all aspects of canopy movement, SAR preparation and SAR application are computer controlled. A gaseous pollutant reduction system was also included in the design, in lieu of open-top fumigation chambers (Heagle et. al., 1973). Filtered air is forced over test plants through perforated polyethylene tubes which run parallel to experimental crop rows (Laurence et. al., 1982; Jones et. al., 1977). Joint effects, produced by SAR and gaseous

Water, Air, and Soil Pollution **31** (1986) 307–315.

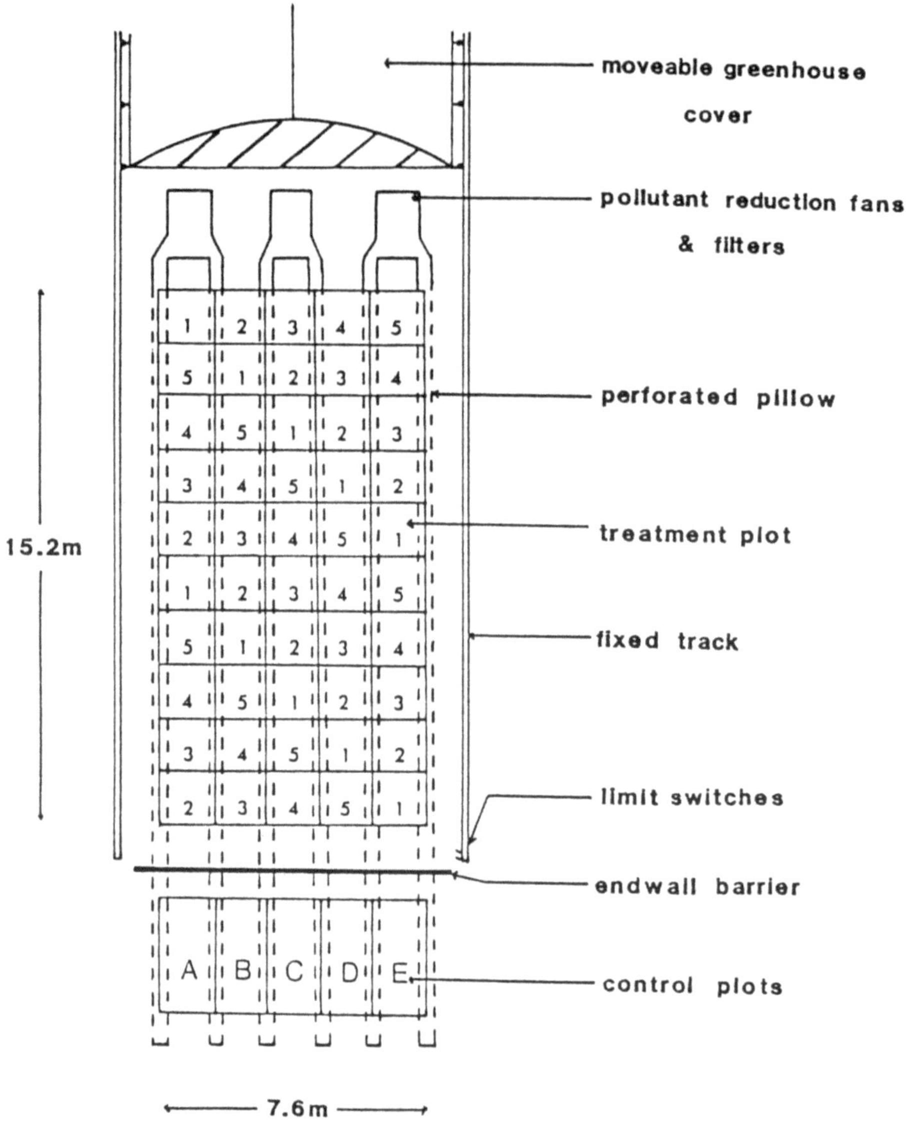

Figure 1. Precipitation Exclusion and Gaseous Pollutant Reduction System (only one system is shown for clarity).

pollutants such as O_3 on crop yield and tree growth, can be investigated.

2. DESCRIPTION OF SYSTEM

The rain exclusion canopy system consists of three, free standing greenhouse structures (dimensions: 19.5 m x 9.1 m x 4.6 m) constructed of galvanized steel and covered with polyethylene. Each greenhouse is fitted with 18 caster wheels and sits on a pair of parallel steel channels which act as tracks. Each canopy is moved on and off its respective treatment area by an electric cable drive system.

Test crops are grown on three separate treatment areas (15.2 m x 7.6 m) which are situated side by side in a west to east direction. These treatment areas are covered by the rain shelters during ambient rain events. At all other times, the shelters are stored in a position sufficiently south of the plots so as not to interfere with incident radiation to the treatment areas.

The soil in the treatment blocks consists of a tiled-drained sandy loam overlying clay. Ambient control plots which receive only natural rainfall are situated adjacent to the treatment blocks. Each treatment area is subdivided into 50 experimental plots (1.5m x 1.5m).

All control devices, treatment mixing tanks and monitors are housed in a climate controlled trailer. Adjacent to this trailer is an enclosure which houses the water deionization system and water storage tank, simulated acid rain treatment reservoir tanks, delivery pumps and electrical panels.

Ambient air concentrations of O_3, SO_2 and NO/NO_2 are measured by means of monitors situated in the control trailer (Monitor labs Inc. Model 8810 photometric O_3 analyzer, Teco Inc. Model 43 pulsed fluorescent SO_2 analyzer and Model 14B/E chemiluminescent $NO_2/NO/NO_x$ analyzer). Temperature, relative humidity, wind velocity and photosynthetic photon flux density are measured by sensors located on an 8 m tower attached to the control trailer. Ambient precipitation is detected with an electronic rain sensor mounted on the instrument tower and is measured with a tipping bucket rain gauge (Weathertronics Model 6010) situated on the control trailer roof.

Nozzles in each canopy are arranged into two 5x5 Latin Squares. Stainless steel, whirl-jet nozzles (Bete Fog Nozzles Inc., Greenfield, Mass.) are mounted to the canopy frames 2.4 m above the soil. Each nozzle is mounted in an adjustable, plexiglass levelling plate and is centred over its respective treatment plot when the canopy is positioned over test plants (Figure 1). Each nozzle emits a square pattern of droplets with a mean drop size of 1 mm diameter which is within the range of natural rain droplets (0.2 mm to 6 mm) (Laws and Parsons, 1943). Although a limited range of droplet sizes are produced by this nozzle, the whirl jet design provides total coverage of each experimental plot.

Five pumps deliver SAR from five treatment reservoirs (900 L) to the three exclusion canopies via a series of buried pvc plastic pipes and flexible vinyl hoses which are connected to the pvc lines. The flexible hoses are dragged along the ground as the canopies are moved. This arrangement allows five SAR treatment lines to continually remain attached to each mobile canopy.

A flow regulator is attached in each of five treatment lines to each canopy. Beyond the regulator, the line is separated into two lines (black polyethylene tubing) each having a PVC needle valve and pressure gauge (0 to 100 kpa) for adjusting flow rates (operating pressure is 30 kpa). One line supplies SAR to five nozzles in the first latin square array, the remaining line supplies SAR to five nozzles in the second latin square array.

The gaseous pollutant reduction system consists of 9 low pressure, horizontal draw blower fans (112,000 L min^{-1} air flow capacity), potassium permanganate-treated alumina filters and perforated polyethylene tubes. Three blower units are situated at the south end of each treatment area (Figure 1). Blower units on the west and east treatment areas are each equipped with pre-filter housings containing particulate filters in series with a filter housing containing 28 filter trays of potassium permanganate impregnated alumina to remove SO_2, O_3, NO_x, and H_2S at 90% efficiency. The blower units on the center treatment area are not equipped with filters.

Studies to determine joint effects of gaseous pollutants and simulated acid rain on crop yield can be performed using the three canopies and their respective air exclusion systems. Since the prevailing winds at the study site are from the west, filtered air is blown into experimental plots of the west canopy. Unfiltered air is forced over experimental plots of the center canopy for comparative purposes. Plots in the east treatment area can be exposed to either filtered or unfiltered air spiked with O_3 or SO_2 at controlled rates.

Air is blown into plots via 30 cm diameter perforated polyethylene tubing or 'pillows'. Each pillow has two series of holes arranged along each side of its entire length. Hole frequency and diameter were calculated according to blower static pressure specifications and to reduce pressure gradients (i.e. holes were placed 15 cm apart; hole diameters ranged from 1 cm to 1.25 cm). The pillows extend beyond the treatment areas covered by the canopies to include ambient control plots which are exposed to natural rainfall (refer to figure 1). Elevated concentrations of O_3 can be injected into the air exclusion system from an O_3 generator (Griffin Model GTC-1B) at specified rates to determine possible joint effects on simulated acid rain treatments and O_3 fumigation.

Micro-climate effects may occur when canopies are positioned over test crops during applications of simulated acid rain. However, the time required for applications accounts for less than one percent of the total growing season. In contrast, open top chamber designs placed over a test plot for the duration of a study, can create a continuous micro-climate effect.

3. COMPUTER CONTROLLED ROUTINES

All aspects of of the field system are controlled automatically by a micro-computer (Apple IIE) and data acquisition device (Cyborg Isaac 91A). The computer/controller continually monitors the precipitation sensor for ambient rain. The system status is re-checked every 5 s for: a) canopy position, b) status of all mixing and treatment application pumps, c) SAR level in mixing and storage tanks, and d) status of nine air exclusion blowers. An automated warning routine is triggered in the event of a system failure.

Meteorological and air quality data are collected 10 times per second for temperature, relative humidity, wind velocity, light intensity, and ambient concentrations of gaseous pollutants. Every 30 min, previous half hour means of all meteorological and air quality data are stored to a floppy disk and then printed for a permanent record.

Blowers are activated by the computer controller if the previous half hour mean value for a given air pollutant exceeds a specified criterion (e.g. 0.05 ppm O_3, 0.15 ppm SO_2 or 0.25 ppm NO/NO_2). Air sampling alternates in 5 min intervals among ambient air (at trailer) and each of the three canopy areas. Blowers are deactivated when precipitation occurs or the half hour mean falls below the specified criterion level.

The system has been designed to apply five SAR treatments simultaneously (Figure 2). The SAR treatments are prepared in U.V. stablized plastic mixing tanks by means of a system of limit switches, pumps and solenoid valves. Simulated acid rain is prepared in 175 L batches by dilution of concentrated acid stock solutions into deionized water on a timed basis. Tap water is deionized via a water purification system (15 mohm resistivity) and stored in a U.V. stablized polyethylene tank (7500 L) equipped with a U.V. sterilizer for maintaining water quality.

Simulated acid rain at the five acid levels is produced by the addition of 240 mL of acid/ion stock solution to 175 L of deionized water by means of an electronic metering pump (55 cm min^{-1} at 72 psi). A specific stock solution, containing mineral salts and H_2SO_4/HNO_3 acids based on ambient rainfall chemistry for southern Ontario (Chan, et al., 1983) is used (Table I). All solutions have a final SO_4NO_3 ratio of 2:1.

The SAR mixing routine is initiated on a signal from a high limit float switch located in each of five treatment reservoir tanks (900 L) made of UV stabilized polyethylene. Deionized water is then pumped into the treatment mixing tanks via large magnetic drive pumps (180 L min^{-1} at 10 psi). Thirty seconds after the filling is initiated, the appropriate acid stock solution is added to each tank by metering pump for exactly 5 min which is sufficient time to give the final pH. Mixing pumps (50 L min^{-1} at 10 psi) circulate the water in the tanks to insure mixing. The tanks are allowed to mix for an additional 30 s and are then dumped into their respective treatment reservoir tanks. The dumping procedure requires an additional 5 min to complete.

When precipitation occurs, the computer immediately moves the canopies over the plots within 50 s after rain is detected. Simulated acid rain is applied automatically to treatment plots after 0.5 mm of ambient rain has fallen. The total deposition of SAR equals the ambient rainfall deposition as determined by the tipping bucket rain gauge. For

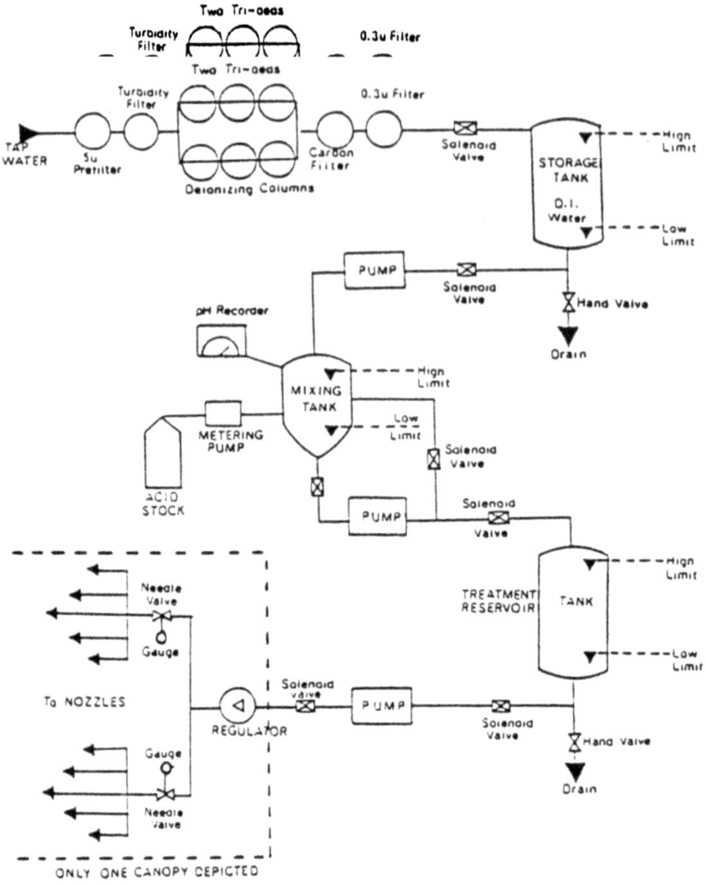

Figure 2. Simulated Acid Rain Application System

(only one treatment line is depicted for clarity)

Table I. Chemical composition of simulated acidic rain (SAR) solution

Acid levels of simulated rain solutions *

SAR	H^+ ueq L^{-1}	pH	Acid vol. mL L^{-1}	
			H_2SO_4	HNO_3
1	50	4.30	1.001E-3	8.849E-4
2	250	3.60	5.002E-3	4.424E-3
3	450	3.35	9.003E-3	7.962E-3
4	650	3.19	1.301E-2	1.151E-2
5	850	3.07	1.699E-2	1.503E-2

* All simulated rain solutions contained the following background ions (mg L^{-1}):

K - 0.063 mg L^{-1}, Ca - 0.501 mg L^{-1},

Mg - 0.103 mg L^{-1}, Na - 0.100 mg L^{-1},

NH_4 - 0.744 mg L^{-1}, Cl - 0.023 mg L^{-1},

SO_4 - 2.771 mg L^{-1}, NO_3 - 1.385 mg L^{-1},

S/N ratio - 2.0:1.0

ambient rain events of short duration (less than 0.25 mm), 0.25 mm of SAR is applied 30 min after the ambient rain detector has dried.

In addition, the computer controller can be manually programmed to apply SAR at a specified time. In the manual mode, SAR is applied in 0.8 mm. increments within 10 minute intervals (4.8 mm hr^{-1}) until the proper amount is applied. The computer maintains the canopies over the treatment plots for 30 min after manual SAR application has ceased and treatment mixing is complete. Normal program routines such as air quality monitoring and meteorological data collection continue during manual mixing and/or SAR applications.

4. SUMMARY

The rain exclusion canopy system has been designed for conducting studies on the direct impact of simulated acid rain on field grown plants. At present, experiments are being performed with the REC system to determine simulated acid rain effects on yields of soybean (*Glycine max*, cv. 'Bicentennial' and 'Hodgson') and radish (*Raphanus sativus* cv. 'Cherry Belle' and 'Comet'). The results of these studies will be reported in another paper. The REC system will also be used in future studies to investigate the joint effects of SAR and O_3 on the growth of sugar maple (*Acer saccharum* and white spruce (*Picea glauca* seedlings.

REFERENCES

Arkin, G.F., Ritchie, J.T., Thompson, M., and Chaison, R.: 1976, *Agron. J.* **68**, 429.

Bruce, R.R., and Shuman Jr., F.L.: 1962, *Trans. ASAE* **52**, 212.

Chan, W.H., Tang, A.J.S. and Lusius, M.A.: 1983, *Precipitation concentration and wet deposition fields of pollutants in Ontario September 1980 to December 1981.* Ontario Ministry of the Environment, APIOS Report ARB-61-83-ARSP, Toronto, Canada. 79 p.

Evans, L.S., Lewin, K.F., Conway, C.A., and Patti., M.J.: 1981, *New Phyt.* **89**, 459.

Heagle, A.S., Phibeck, R.B., Brewer, P.F., and Ferrell, R.E.: 1983, *J. of Envir. Qual.* **12**, 538.

Heagle, A.S., Body, D.E., and Heck, W.: 1973, *J. Envir. Qual.* **2**, 365.

Irving, P.M., Miller, J.E., Prepejchal, W., Hoff, J.,and Segers, J.: 1982, *Argonne National Laboratory Radiology and Environmental Research Division Annual Report ANL-82-65 Part III*. Argonne, Illinois, U.S.A. p. 9.

Jones, H.C., Lacasse, N.L., Liggett, W.S., and Weatherford, F.: 1977, Experimental air exclusion system for field studies of SO2 effects on crop productivity. Tenn. Valley Authority, Muscle Shoals, Alabama, EPA report 600/7-77-122.

Laurence, J.A., MaClean, D.C., Mandl, R.H., Schneider, R.E., and Hansen, K.S.: 1982, Water, Air, and Soil Pollution **17**, 399.

Laws, J.O., Parsons, D.A.: 1943, Trans.Amer.Geophys.Union **24**, 452.

Upchurch, D.R., Ritchie, J.T., and Foale, M.A.: 1983, Agron. J. **75**, 845.

ACIDIFICATION PHENOMENA IN BEECH (FAGUS SYLVATICA) FORESTS OF EUROPE

Rüdiger Wittig
Dep. of Geobotany
University of Düsseldorf
D-4000 Düsseldorf
Germany

ABSTRACT. When comparing the species composition of herb layers in loess beech forests of the Westphalian Bight (Northrhine-Westphalia, F.R. Germany) in the years 1976 and 1983, findings show that acidophilic plants tended to be favored during this period. Apart from many other factors, depositions have also been considered as causes for this. Where they should actually play a role, comparable locations having a higher acid deposition rate would have to show a stronger tendency towards acid-indicators than those having a lower one.

It is the stemflow area of the beech tree (Fagus sylvatica L.) which attracts more precipitation than the rest of the forest area, on account of its infundibular treetop. Furthermore, stemflow water from Fagus sylvatica exhibits a higher degree of acidity than rain dripping from the treetop. Thus, the stemflow area as an area with higher acidic incidence may be compared with the normal forest area displaying minor acidic incidence. Findings from all the surveys so far conducted in Middle Europe show that soil pH recorded in the stemflow area is distinctly lower, and that there are less base- and weak acid-indicators but correspondingly more (strong) acid-indicators, than in "normal" forest soil. It must be noted that the phenomena observed need not be due to acidic precipitation. They could as well be attributable to a special feature pertaining to the stemflow area of the beech as specifically natural phenomena.

In the event of natural causes being detected, these would have to be observable in the overall growth area of the beech. In Southern Europe, however, we did not find any indication whatsoever for acidification of the stemflow area. This phenomenon appears to be restricted to Central European regions affected by higher pollution.

1. INTRODUCTION

In the herb layer of beech (Fagus sylvatica L.) forests in large areas of Europe the following phenomena were observed which indicate soil acidification:

Water, Air, and Soil Pollution **31** (1986) 317–323.

- In beech forests growing in medium bufferd soils, acid indicators are increasing (Wittig *et al.*, 1985a).
- Base-indicators and even weak acid-indicators around the trunk base of beech are being replaced by strong acid-indicators (e.g. Wittig and Neite, 1983, 1985; Jochheim, 1985).

2. INCREASE IN ACID INDICATORS

2.1. Prerequisites of the study

Practically no group of organisms is so well known in terms of site requirements and thus of indicator significance as the vascular plants of Central Europe. Once the relevant species growing on a given surface are known, it is possible to derive a variety of details on the conditions prevailing there. As several investigations have shown, the findings derived from the combination of emerging species will normally correlate very well with actual conditions (e.g. Ellenberg,1979; Runge, 1983). This applies particularly of the pH of the soil (e.g.Neite and Wittig, 1985). By merely comparing former and recent specimens taken from the vegetation of one site, it is possible to gain insights into how a specific site has developed, even if there are no data existing from previous years.

2.2. Methods

Starting from the premises elaborated in 2.1., 44 surveys of the millet grass beech forest in the Westphalian Bight already documented by Burrichter and Wittig (1977) in 1976 were repeated in 1984.Subsequently, by comparing the lists of species from 1983 with those of 1976, a balance was made of the different types of acid, N, and moisture indicators (classification following Ellenberg, 1979), and the mean indicator values calculated using the Ellenberg (1979) method. Both results were checked for their significance using the Wilcoxon test (Sachs,1974).

2.3. Results

As Table I shows, both acid, N, and moisture indicators increased significantly, although far less significantly in the latter two groups. Concerning the mean indicator values only the soil reaction value shows a significant change.

2.4. Discussion

Of the two alternative methods of biological indication, the indicator value method is by far more reliable, and thus more conclusive than the consideration of separate indicator species, as in the former method it is of little significance if one or other indicator species is overlooked, an occurrence which cannot be excluded altogether even with experts (Tüxen, 1972). This has been supported by Wittig *et al.* (1985b), Ellenberg (1950) makes a similar observation: "The most reliable information on soil condiditions can be obtained if we... consider the entire plant community at that site because it represents, in a certain sense,

TABLE I: Changes in species composition and mean indicator values in the herb layer of the millet grass-beech forest from 1976 to 1983

		1976	1983	Balance (%)	Significance
Number of species	Acid indicators	41	101	+ 151	xxx
	N indicators	48	78	+ 63	xxx
	Moisture indicators	85	104	+ 22	xx
Mean average indicator values	Soil reaction	4,99	4,75	- 0,24	xxx
	N	5,25	5,23	- 0,02	n.s.
	Moisture	5,62	5,66	+ 0,04	n.s.

xxx = highly significant
xx = very significant
n.s.= non-significant

a proven local 'task force' whose constitution reflects very well the soil properties of the habitat".

As shown by investigations made by Trepl (1982), the species composition of forest communities is very conservative, i.e. it does not react to normal short-term fluctuations in habitat conditions but only to long-term changes. Thus it is possible to display a trend by comparing former and recent relevés of the same forest community.

The acidification of soils may have many separate causes or combinations of the causes, e.g. natural ageing processes in the ecological system of a forest, leaching of nutrient contents, climate, forestry, depositions("acid rain"). Whenever depositions are involved, the soil should be more acid at locations receiving more precipitation than at others receiving less. This leads us to the investigations described in the following section, related of the trunk base of beech trees.

3. ACIDIFICATION PHENOMENA IN THE TRUNK BASE AREA OF BEECH TREES

3.1. Prevailing situation

Recently, several authors have described the trunk base area of beech (Fagus sylvatica L.) as an exceptional habitat (please also refer to the bibliography in Jochheim, 1985). Most of the authors have detected for this particular area a clearly lower pH in soil solution over that prevailing at some distance from and between the trees (e.g. Glatzel *et al.*, 1983; Könies, 1982; Flückinger *et al.*, 1984; Jochheim, 1985; Neite and Wittig, 1985). This phenomen has also been recorded for the North American Fagus grandifolia (e.g. Gersper and Holowaychuk, 1971). In the Federal Republic of Germany, several instances have been recorded of an increased emergence of acid indicators in the trunk base area (Glavač *et al.*, 1970; Wittig and Neite, 1983, 1985; Jochheim, 1985), whereas base indicators tend to avoid this microhabitat as far as possible. Figure 1 and 2 show examples for these phenomena.

With respect to the exposure to immissions in this particular area,

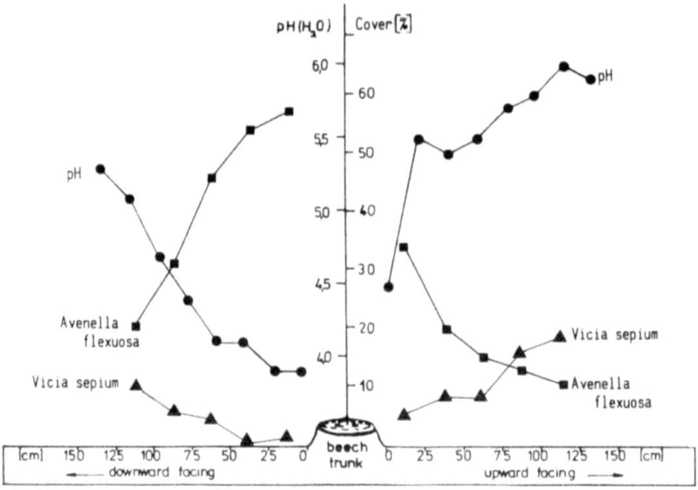

FIGURE 1: Average soil-pH (H_2O) and average distribution of Avenella flexuosa (acid-indicator) and Vicia sepium (base-indicator)in the upward and downward sloping quarter of the area around the beech trunk of a Carici-Fagetum in the Spessart (F.R. Germany)

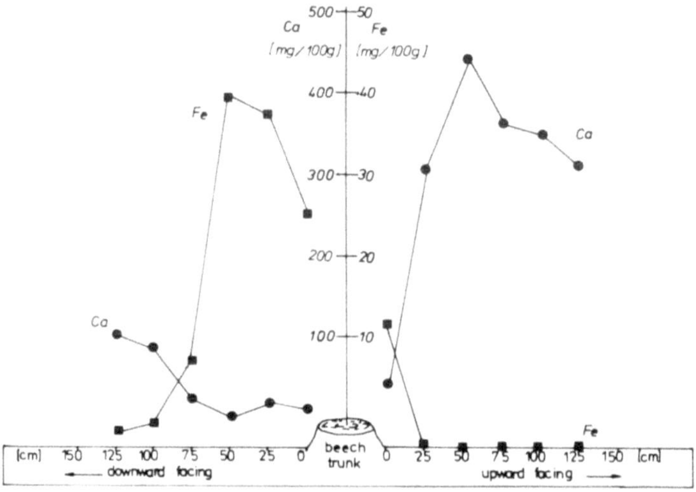

FIGURE 2: Average Ca^{2+} and Fe^{3+} content (NH_4Cl extractable) of the upper soil layer (0 - 5 cm) in the upward and downward sloping quarter of the area around the beech trunk of a Carici-Fagetum in the Spessart (F.R. Germany)

it is well-known that the bases of trunks receive some five time more rainwater than the remainder of the forest soil (Eidmann, 1959; Mayer and Ulrich, 1973), and that the water running down the trunk has a lower pH than that dripping from the crown (Jochheim, 1985: Table I.)

3.2. Alternative interpretations

The diagnosis as per 3.1. (lower pH, increase of acid indicators and decrease of base indicators in the trunk base area) could either be "from natural causes", i.e. caused by the tree itself, or be the consequence of higher acid deposition due of immissions, or stem from a combination of these causes. Certain observations by Tüken (1977) speak in favor of natural impoverishment and hence of local acidification; the author discovered that dead leaves tend to be blown away more often from the trunk base area than from the "normal" forest soil. In this area, remineralization would thus be weaker, which might lead to a gradual impoverishment of the soil. Moreover the fact that the nutrient content of soils in regions with abundant rainfall is washed out faster than in zones with less rainfall, would be one possible explanation for the phenomena observed. Finally, the tree itself may be responsible for this acidification by the excretion of acids via its bark.

If the theory of natural causes contributing to the acidification of the trunk base is correct, similar phenomena should also occur in all growth ranges of beech trees experiencing equal or higher levels of precipitation. If, however, depositions (acid rain) are the cause for these occurrences, they should only emerge in regions subjected to a certain deposition load.

In order to answer the question of causes for this particular situation in beech trunk bases, beech forests were investigated in Spain, Italy and Greece where - even without accurate measuring data to hand - one may rightly assume on the basis of the economic structure and prevailing wind directions that these Southern European countries are less exposed to acid precipitation than Central Europe.

3.3. Methods

Soil samples were recovered and herb layers mapped at 10 different locations in the Pyrenean Mountains, at 5 locations in Central and Southern Italy (one each in Foresta Umbra and Parco Nazionale della Calabria; 3 locations in Parco Nazionale d'Abruzzo), as well as at two sites in the Hungarian Bükk Mountains and at one place in the Greek National Park of Kentriki (Rhodopes). The method of soil analysis will be described in a number of our forthcoming publications.

3.4. Results

Whereas all studies from the Federal Republic of Germany, Austria and Switzerland confirm both a diminished pH and certain changes in the herb layer around the trunk base of beeches, no significant differences were observed between the trunk base area and intermediate areas either in Spain or Italy as far as the soil pH is concerned. No were any accumulations of acid indicators encountered in this area. Although the pH in the run-off area around the trunk base in the Hungarian Bükk Mountains appeared to be significantly lower, no acid indicators were detected there. Figure 3 shows the results, together with those made by other authors, in a map of Europe.

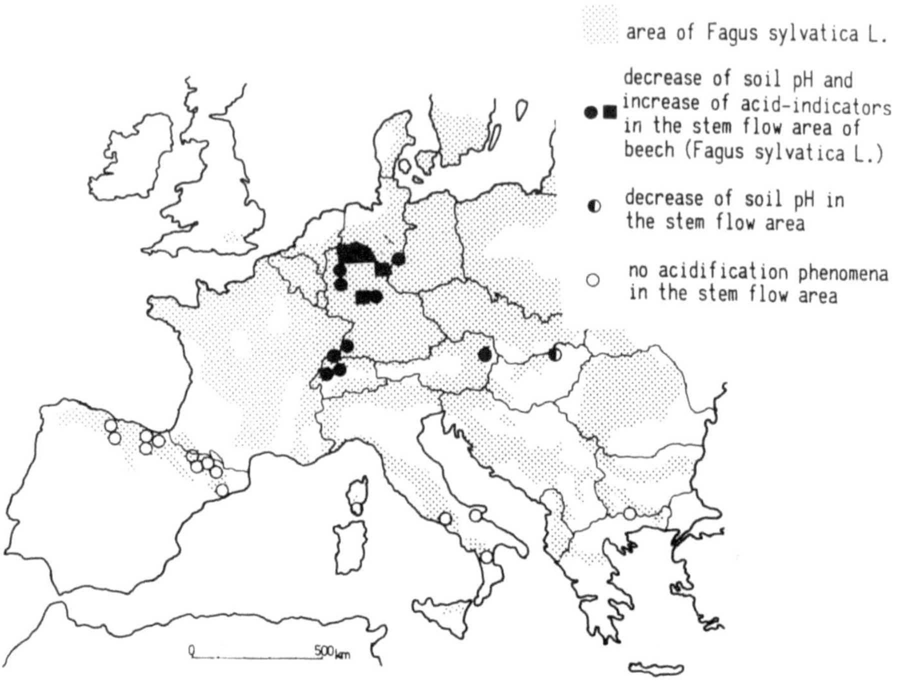

FIGURE 3: Current state of knowledge on the extent of occurrences of acidification in the stem flow area of beeches in Europe (Spain, Italy, Greece, Hungary: own investigations; Austria: Glatzel *et al.*, 1983; Switzerland: Flückinger *et al.* , 1984; Germany: own investigations, and Glavac *et al* ., 1970; Koenies, 1982; Jochheim, 1985; Corsica: Glavac *et al.*, 1985)

3.5. Conclusions

The question raised at the end of section 3.2. has thus been answered at least in part: Acidification in the stemflow area is not encountered in all of the regions in which beech trees grow, the phenomenon is quite obviously not due to the tree itself. As the regions under investigation in Southern Europe receive a comparable rate of precipitation during the vegetation period to that in Germany (see e.g. Cuatrecasas, 1932), the volume of precipitation also has to be excluded as a reason for the particular situation of the trunk base in Central European beeches. The hypothesis that immissions contribute to the phenomenon is therefore supported. In order to reconfirm or contradict such a hypothesis, it should the necessary to carry out measurements and analyses of trunk run-offs in Southern European beech forests, and to "close the geographical gaps" in the network of investigations (France, Belgium, Netherlands, Yugoslavia, Czechoslovakia, Poland, GDR).

REFERENCES

Burrichter, E. and Wittig, R.: 1977, *Mitt. Flor.-soz. Arb. gem. NF* 19/20, 377

Cuatrecasas, J.: 1932, *Veröffentl. Geobot. Inst. Rübel,* 9, 443

Eidmann, F.E.: 1959, *Union géodesique et géographique internationale* 48, 5

Ellenberg, H.: 1950, *Landwirtschaftl.Pflanzensoziologie* 1,1

Ellenberg, H.: 1979, *Scripta Geobotanica* 9 (2nd edition) 1 und 39

Flückinger, W., Braun, W. and Flückinger-Keller, H.: 1984, *Schweiz. Zeitschrift Forstwesen* 135, 389

Gersper, P. and Holowaychuk, N.: 1971, *Ecology* 52, 691

Glatzel, G., Sonderegger, E. and Kazda, M.: 1983, *Allg.Forstzeitung* 94, 111

Glavac, H., Jochheim, H., Koenies, H., Rheinstädter, R. and Schäfer, H.: 1985, *Allg. Forst. Zeitschr.* 40, 1397

Glavac, V., Krause, A. and Wolff-Straub, R.: 1970, *Schriftenreihe Vegetationskunde* 5, 187

Jochheim., H.: 1985: *Ber. Forschungszentrum Waldökosysteme/Waldsterben,* 10, 1

Koenies, H.: 1982, *Ber. Forschungszentrum Waldökosysteme/Waldsterben,*9,1

Mayer, R. and Ulrich,B.: 1973, in Ellenberg, H. (ed), *Ökosystemforschung,* Springer-Verlag, Berlin, Heidelberg, New York, pp. 165

Neite, H. and Wittig, R.: 1985, *Acta Oecologica/Oecologia Plantarum* 6, 375

Runge, M.: 1983, *Verhandl. Ges. Ökol.* 11, 339

Sachs, L.: 1974, *Angewandte Statistik,* Springer-Verlag, Berlin, Heidelberg, New York, 545 pp.

Trepl, L.: 1982, *Tuexenia, Mitt.Flor.soz.Arb.gem.NS* 2, 195

Tüxen, R.: 1972, in Tüxen, R. (ed.), *Grundfragen und Methoden in der Pflanzensoziologie,* De Junk Publishers, Den Haag, pp. 168

Tüxen, R.: 1977, in Tüxen, R. (ed.), *Vegetation und Klima,* J. Cramer, Vaduz, pp. 127

Wittig, R. and Neite, H.: 1983, *Allg.Forst.Zeitschr.*38, 1232

Wittig, R. and Neite, H.: 1985, *Vegetatio* 64, 113

Wittig, R., Ballach, H.J. and Brandt, C.J.: 1985 a, *Angewandte Botanik* 56, 219

Wittig, R., Werner W. and Neite, H.: 1985 b, *VDI Bericht* 560, 21

ASSESSMENT OF RELATIVE SENSITIVITIES DURING EARLY GROWTH STAGES OF SELECTED CROP SPECIES SUBJECTED TO SIMULATED ACIDIC RAIN

A. J. ENYEDI and A. L. KUJA
Phytotoxicolgy Section, Air Resources Branch,
Ontario Ministry of the Environment,
880 Bay Street, 3rd Floor,
Toronto, Ontario, Canada M5S 1Z8

ABSTRACT. Under greenhouse conditions, experiments were performed to assess the cultivar sensitivity (during early growth stages) to simulated acidic rain treatments of six economically important crop species. The accumulation of dry biomass was investigated for the following crops: alfalfa, barley, cabbage, corn, cucumber, and soybean. For each crop, treatments consisted of eight acidity levels in simulated rain - pH 2.6, 3.0, 3.4, 3.8, 4.2, 4.6, 5.0, and 5.6. Crops varied in response to increased rain acidity. Overall, exposure of cultivars to a range of acidities in simulated rain had very little effect on the dry biomass accumulation of shoot and root components. With the exception of cabbage, the crop species examined were highly resistant to injury and dry biomass reduction after exposure to acidic rainfall. The results also demonstrated that plant response to simulated acid rain was not only species dependent, but also cultivar dependent.

1. INTRODUCTION

Although the direct impact of acidic precipitation on agricultural crops is not yet completely understood, exposure of crops to acidic precipitation throughout the growing season has the potential to affect plant growth, development and yield (Evans, 1982). However, previous experiments designed to assess crop response to H_2SO_4 and HNO_3 in simulated rain treatments have demonstrated considerable variation in sensitivity among crop species. (Harcourt and Farrar, 1980; Cohen et al., 1981, 1982; Evans et al., 1982; Troiano et al., 1982, 1984; Plocher et al., 1985). In one study, 28 different crops were ranked into 6 groups of decreasing acidic rain sensitivity - root crops > leaf crops > cole crops > tuber crops > legumes and fruit crops (Lee et al., 1981). A comprehensive review of experimental research concluded that variable plant response to rain acidity was a result of cultivar and species influences (Irving, 1983). For example, several studies conducted with different soybean cultivars show contradictory results with increased, decreased and no net effect on soybean yield (Evans et al., 1981, 1983; Troiano et al., 1983; Irving and Miller, 1981; Heagle et al., 1983). In

Water, Air, and Soil Pollution **31** (1986) 325–335.

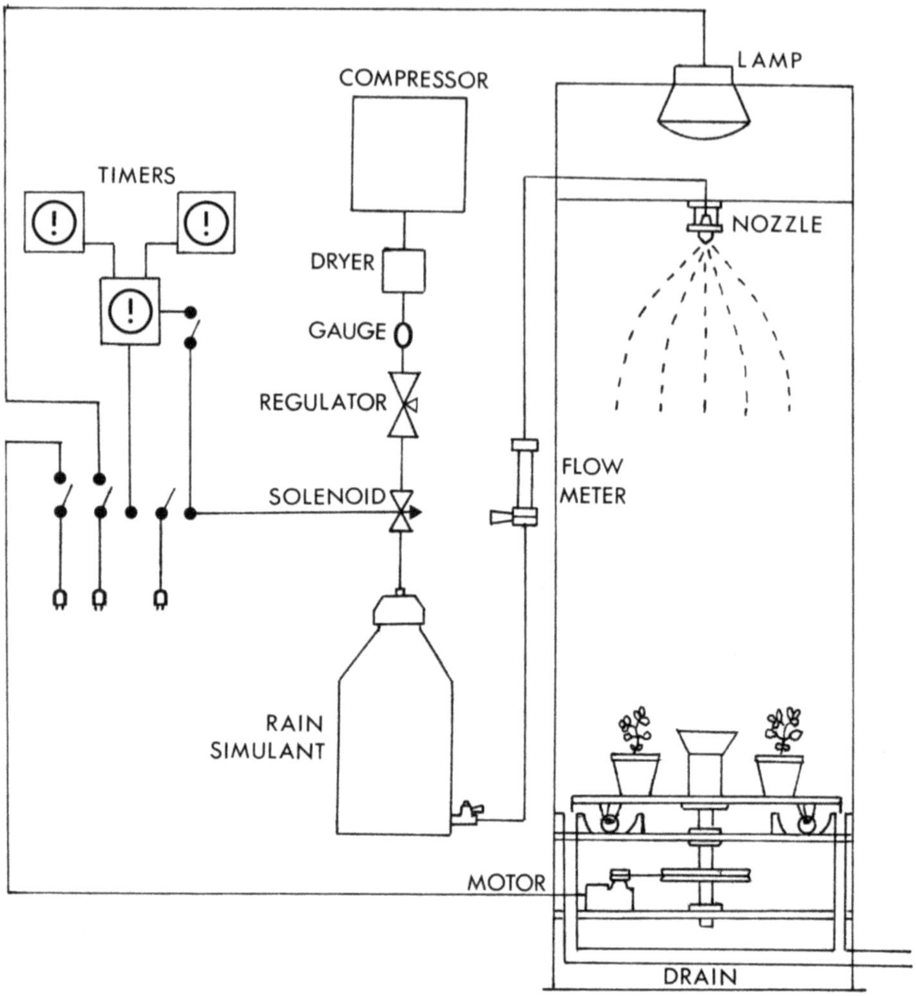

Note: The system consists of four chambers each with identical components. For clarity only one is shown here.

FIGURE I - Schematic diagram of the indoor rain simulation chambers addition, sensitivity to simulated acidic rainfall has been shown to be related to the stage of plant development when exposure first occurs (Jacobson *et al.*, 1985).

At present, the agricultural region of southern Ontario receives precipitation with an annual weighted-mean pH of 4.2 (Chan *et al.*, 1983). The purpose of this study was to investigate the relative sensitivity of six economically important agricultural crops to acidic rain. The effects of repeated simulated acidic rain application on seed germination, seedling establishment and early stages of plant growth of 24 crop cultivars were examined.

2. MATERIALS AND METHODS

2.1. Plant Culture

For this study, four cultivars of the following crop species were selected: alfalfa (*Medicago sativa*, cv. 'Titan', 'Iroquois', 'Saranac' and 'Anchor'), barley (*Hordeum vulgare*, cv. 'Perth', 'Bruce', 'Massey' and 'Mingo'), cabbage (*Brassica oleracea*, cv. 'Danish Ballhead', 'Princess', 'Pennsylvania State Ballhead' and 'Ruby Red Ballhead'), corn (*Zea mays*, cv. 'Pioneer 3925', 'Pioneer 3803', 'Pioneer 3906' and 'Pioneer 3950'), cucumber (*Cucumis sativus*, cv. 'Hymark', 'Sprint', 'Tamor' and 'Pioneer') and soybean (*Glycine max*, cv. 'Hodgson', 'Harcor', 'Maple Arrow' and 'Evans'). At the time of planting, four seeds were sown into 10 cm plastic pots filled with medium screened sandy loam soil. In each experiment, plants were maintained on elevated benches inside two adjacent greenhouses with automatic temperature and humidity control. Pot positions were blocked on each greenhouse bench and pots were moved every second day. Temperature was maintained at 22/18 °C day/night and humidity was approximately 50±20%. In addition, each greenhouse was supplied with charcoal-filtered air to eliminate background gaseous pollutants present in the ambient atmosphere. Supplementary lighting in the form of fluorescent and incandescent lamps providing a photon flux density of 350±50 uE m^{-2} s^{-1} extended the photoperiod to 16 hr. All pots received supplemental irrigation with deionized water as required.

2.2 Rain Simulation and Application

After seeding, pots were randomly assigned to one of 8 treatment groups of simulated acidic rain (SAR) for a minimum of 12 replicates per treatment level. Exposure to SAR occurred immediately after seeding and rain solutions were allowed to contact the soil. Acidity levels of SAR were pH 5.6, 5.0, 4.6, 4.2, 3.8, 3.4, 3.0, and 2.6. Solutions of SAR were made from the addition of reagent grade H_2SO_4 and HNO_3 to deionized water (Table I) and a SO_4:NO_3 mass ratio of 2:1 was maintained. Background ions typically present in ambient precipitation occurring in the agricultural region of Ontario were added as mineral salts (Table I).

To simulate rainfall, 4 rain simulation chambers were designed and constructed to expose plants to simulated acidic treatments (Figure 1).

Table I. Chemical composition of simulated acidic rain (SAR) solutions

pH of SAR	Ion Species Concentration (mg dm^{-3})†			
	H^+	$SO_4^{=}$	NO_3^-	S/N¶ Ratio
5.6	0.003	2.769	1.385	1.99:1
5.0	0.010	3.115	1.558	1.99:1
4.6	0.025	3.638	1.820	1.99:1
4.2	0.063	4.954	2.476	2.00:1
3.8	0.158	8.245	4.126	1.99:1
3.4	0.398	16.554	8.268	2.00:1
3.0	1.000	37.399	18.684	2.00:1
2.6	2.512	89.705	44.851	2.00:1

† All simulated rain solutions contained the following ions: NH_4 - 0.744 mg dm^{-3}, Ca - 0.500 mg dm^{-3}, Mg - 0.097 mg dm^{-3}, Na - 0.100 mg dm^{-3}, K - 0.063 mg dm^{-3}, Cl - 0.023 mg dm^{-3}.

¶ SO_4 to NO_3 Ratio (by Mass).

The four chamber compartments were housed within a framework of the following dimensions; height 7.3 m, width 4.9 m, and depth 1.2 m. A 1000 W mercury vapor lamp was installed at the top each chamber and illuminated plants with a photon flux density of 42+3 uE m^{-2} s^{-1} during each rain application. A turntable of 1.1 m diameter was positioned at the base of each chamber to rotate plants during rainfall exposure and ensure uniform rain drop distribution.

Simulated precipitation solutions were introduced into each chamber from a single raindrop nozzle (Bete Fog Nozzles Inc., Model SRN303) positioned 5.33 m above the center of the turntable. The nozzle produced water droplets with a median diameter of 1 mm and delivered rain solutions to the turntable surface at an intensity of 1.2 cm hr^{-1}. Rain applications were made three times each week (Monday, Wednesday, Friday) for a total of 12 events for each experiment. Each rain event was approximately 40 min in duration and 8 mm of SAR was applied per rain event (2.4 cm $week^{-1}$). The total deposition of SAR was 9.6 cm in each study.

2.3. Seed Germination and Biomass Determination

After 3 to 4 SAR events, germination of the planted seed was determined by counting emerging seedlings in each pot. All values were converted to percent germination per pot and subjected to one-way analysis of variance (Steel and Torrie, 1980) to determine significant F-ratios. All pots were then thinned to one plant per pot.

Approximately 30 days after seeding, all cultivars were harvested for dry biomass determination. Each seedling was removed from its pot and the dry biomass determined gravimetrically with either a Mettler PE3600 or a Mettler H54AR electronic balance after roots and shoots had dried for 72 hr in a forced-air oven at 80 °C. All biomass values were subjected to one-way analysis of variance to determine significant F-ratios for roots and shoots. In addition, all biomass values were subjected to regression analysis to determine dose-response functions. Finally, mean values for biomass were tested using Duncan's Multiple Range Test (Steel and Torrie, 1980) to separate treatment effects.

3. RESULTS

3.1. Seed Germination and Seedling Establishment

Simulated acidic rain had no significant effect on the germination of alfalfa, barley, cabbage, corn and cucumber cultivars. Germination of one soybean cultivar, 'Harcor', was significantly reduced at pH 2.6 - germination was 33% as compared to 62% at pH 5.6. However, the germination of 'Harcor' was significantly greater at pH 3.8 (93%) compared to pH 5.6. No significant SAR effect was seen in the remaining three soybean cultivars.

In all cultivars examined, seedling establishment and early plant growth were not impeded by repeated SAR treatment. However, foliar injury to cotyledons (when present) and primary leaves was observed in all cultivars when pH of SAR was 3.0 or less. For all 24 cultivars

Table II. Seedling shoot dry mass (g) of crop cultivars exposed to simulated acidic rain

Crop Cultivar	Treatment pH 5.6	5.0	4.6	4.2	3.8	3.4	3.0	2.6
Alfalfa								
'Titan '	0.047 a[†]	0.055 a	0.056 a	0.053 a	0.051 a	0.042 a	0.045 a	0.041 a
'Iroquois'	0.082 a	0.080 a	0.086 a	0.072 a	0.072 a	0.085 a	0.081 a	0.062 a
'Saranac'	0.081 a	0.070 a	0.059 a	0.065 a	0.068 a	0.075 a	0.062 a	0.067 a
'Anchor'	0.042 a	0.052 a	0.057 ab	0.060 ab	0.062 ab	0.063 ab	0.080 b	0.065 ab
Barley								
'Perth'	0.66 a	0.60 a	0.61 a	0.68 a	0.62 a	0.61 a	0.63 a	0.62 a
'Bruce'	0.57 a	0.52 a	0.59 a	0.49 a	0.52 a	0.63 a	0.55 a	0.62 a
'Massey'	0.88 b	0.76 ab	0.78 ab	0.67 a	0.79 ab	0.77 ab	0.83 b	0.70 a
'Mingo'	0.72 ab	0.80 b	0.77 ab	0.64 a	0.72 ab	0.68 ab	0.68 ab	0.70 ab
Cabbage								
'Princess'	0.48 bc	0.51 c	0.51 c	0.33 a	0.41 abc	0.37 a	0.40 ab	0.31 a
'Danish'	0.56 a	0.53 a	0.54 a	0.54 a	0.56 a	0.58 a	0.56 a	0.44 a
'Penn. State'	0.77 c	0.57 ab	0.66 bc	0.77 c	0.76 c	0.68 bc	0.62 abc	0.48 a
'Ruby Red'	0.88 b	0.77 ab	0.76 ab	0.72 ab	0.82 b	0.72 ab	0.69 ab	0.57 a
Corn								
'Pioneer 3925'	0.86 ab	0.79 a	0.78 a	0.75 a	0.81 a	0.89 ab	1.05 b	1.25 c
'Pioneer 3803'	0.93 a	1.12 abc	1.02 ab	1.09 abc	1.17 bc	1.18 bc	1.22 c	1.07 abc
'Pioneer 3906'	0.86 a	0.97 ab	1.04 b	0.87 a	1.04 b	0.90 ab	0.95 ab	0.86 a
'Pioneer 3950'	0.96 ab	0.91 ab	0.91 ab	0.86 ab	0.83 a	0.88 ab	1.06 b	0.85 a
Cucumber								
'Hymark'	1.21 abc	1.41 cd	1.38 bcd	1.52 d	1.53 d	1.46 d	1.07 a	1.14 ab
'Sprint'	1.82 a	2.04 a	2.05 a	2.16 a	1.99 a	2.19 a	1.94 a	1.85 a
'Tamor'	0.68 ab	0.76 abc	0.70 ab	0.64 ab	1.14 d	0.46 a	1.06 cd	0.81 bc
'Pioneer'	0.87 a	1.31 b	1.33 b	0.79 a	0.86 a	1.05 ab	1.34 b	1.35 b
Soybean								
'Hodgson'	0.46 ab	0.57 c	0.46 ab	0.45 ab	0.50 bc	0.54 bc	0.42 a	0.45 ab
'Harcor'	0.48 b	0.51 b	0.48 b	0.49 b	0.47 b	0.44 b	0.37 a	0.37 a
'Maple Arrow'	0.90 ab	0.96 b	0.87 ab	0.94 b	0.88 ab	0.93 b	0.79 a	0.87 ab
'Evans'	0.92 a	0.93 a	0.85 a	0.88 a	0.87 a	0.94 a	0.86 a	0.86 a

[†] Treatment means, within rows, followed by same letter are not significantly different at the 5% level, according to Duncan's Multiple Range Test.

Table III. Seedling root dry mass (g) of crop cultivars exposed to simulated acidic rain

Crop Cultivar	Treatment pH 5.6	5.0	4.6	4.2	3.8	3.4	3.0	2.6
Alfalfa								
'Titan '	0.024 a[†]	0.028 a	0.030 a	0.027 a	0.026 a	0.025 a	0.024 a	0.022 a
'Iroquois'	0.045 b	0.059 c	0.037 a	0.033 a	0.037 a	0.048 b	0.043 b	0.038 ab
'Saranac'	0.035 a	0.032 a	0.026 a	0.029 a	0.024 a	0.035 a	0.023 a	0.026 a
'Anchor'	0.021 a	0.025 a	0.040 a	0.026 a	0.027 a	0.028 a	0.040 a	0.028 a
Barley								
'Perth'	0.289 ab	0.267 ab	0.341 b	0.295 ab	0.293 ab	0.286 ab	0.309 ab	0.260 a
'Bruce'	0.359 ab	0.299 a	0.354 ab	0.328 ab	0.339 ab	0.353 ab	0.403 b	0.379 ab
'Massey'	0.481 a	0.415 a	0.452 a	0.392 a	0.412 a	0.429 a	0.393 a	0.401 a
'Mingo'	0.420 a	0.441 a	0.454 a	0.369 a	0.439 a	0.402 a	0.383 a	0.428 a
Cabbage								
'Princess'	0.074 ab	0.111 c	0.095 bc	0.078 ab	0.094 bc	0.066 ab	0.072 ab	0.057 a
'Danish'	0.124 a	0.121 a	0.123 a	0.108 a	0.114 a	0.114 a	0.099 a	0.101 a
'Penn. State'	0.149 b	0.129 ab	0.131 ab	0.146 b	0.146 b	0.139 ab	0.122 ab	0.107 a
'Ruby Red'	0.144 b	0.132 ab	0.146 b	0.121 ab	0.145 b	0.104 ab	0.114 ab	0.095 a
Corn								
'Pioneer 3925'	0.536 abc	0.465 a	0.487 ab	0.494 ab	0.491 ab	0.556 abc	0.618 bc	0.646 c
'Pioneer 3803'	0.581 a	0.706 a	0.615 a	0.659 a	0.714 a	0.677 a	0.724 a	0.626 a
'Pioneer 3906'	0.685 ab	0.943 c	0.811 c	0.655 ab	0.766 bc	0.664 abc	0.696 abc	0.575 a
'Pioneer 3950'	0.719 a	0.595 a	0.636 a	0.686 a	0.676 a	0.673 a	0.685 a	0.586 a
Cucumber								
'Hymark'	0.362 abc	0.476 c	0.381 abc	0.378 abc	0.331 ab	0.443 bc	0.289 a	0.293 a
'Sprint'	0.345 ab	0.377 ab	0.470 bc	0.371 ab	0.516 c	0.550 c	0.510 c	0.258 a
'Tamor'	0.161 ab	0.271 d	0.108 a	0.238 cd	0.148 ab	0.202 bcd	0.183 abc	0.175 abc
'Pioneer'	0.222 ab	0.306 bc	0.336 c	0.219 ab	0.188 a	0.268 abc	0.343 c	0.334 c
Soybean								
'Hodgson'	0.217 ab	0.298 c	0.231 b	0.219 ab	0.240 b	0.274 bc	0.176 a	0.208 ab
'Harcor'	0.259 b	0.288 b	0.275 b	0.254 b	0.245 b	0.246 b	0.199 a	0.181 a
'Maple Arrow'	0.317 b	0.339 b	0.328 b	0.326 b	0.323 b	0.336 b	0.268 a	0.309 ab
'Evans'	0.380 bc	0.392 c	0.348 ab	0.342 ab	0.336 ab	0.369 bc	0.341 ab	0.321 a

[†] Treatment means, within rows, followed by same letter are not significantly different at the 5% level, according to Duncan's Multiple Range Test.

evaluated, foliar lesions due to SAR comprised less than 5% of total leaf area. In all cases, foliar injury was typically a light-colored flecking, 1 to 2 mm in diameter, on the adaxial leaf surface.

3.2. Biomass Determinations

Results for shoot and root biomass determinations are given in Tables II and III, respectively. All comparisons have been made with reference to the least acidic treatment, pH 5.6. Results of the regression analyses showed that no significant dose-response functions could be derived for any cultivar examined. In each case, the slope of the dose-response line was not significantly different from zero at the 5% level of probability. The following are highlights for each species examined:

Alfalfa The shoot dry mass of cultivar 'Anchor' was significantly increased for plants exposed to pH 3.0 SAR as compared to pH 5.6 (Table II). The root dry mass of cultivar 'Iroquois' was significantly decreased for plants exposed to pH 4.6, pH 4.2 and pH 3.8 SAR (as compared to pH 5.6 SAR), whereas the root dry mass of plants exposed to pH 5.0 was increased (Table III). No other significant effects were observed.

Barley The shoot dry mass of cultivar 'Massey' plants exposed to SAR was decreased at pH 2.6 and pH 4.2, as compared to pH 5.6 (Table II). Root dry mass of cultivar 'Bruce' was significantly increased at pH 3.0 vs other pH 5.6 SAR (Table III). No other significant differences were found.

Cabbage Cultivar 'Danish Ballhead' did not demonstrate any biomass response (root or shoot) to SAR treatment at any level. In the remaining cultivars, ('Princess', 'Penn State' and 'Ruby Red') root and shoot dry masss were significantly altered by SAR treatment (Table II and III). For all three cultivars, the lowest biomass value (root or shoot) was recorded for plants exposed to pH 2.6 SAR, as compared to pH 5.6 SAR.

Corn The shoot dry mass for three corn cultivars was significantly increased by SAR as compared to pH 5.6 (Table II). For 'Pioneer 3925' corn plants, maximum shoot mass occurred when pH $\leq$ 3.0. For 'Pioneer 3803' corn, stimulation of dry shoot accumulation occurred over the range of pH 3.8 to pH 3.0, however shoot dry mass at pH 2.6 was not significantly diffferent from pH 5.6 (Table II). For 'Pioneer 3906', significantly greater shoot dry masss were observed at pH 4.6 and pH 3.8. No significant differences due to SAR were found in 'Pioneer 3950', compared to pH 5.6.

Cultivar 'Pioneer 3925' demonstrated greatest root dry mass when pH $\leq$ 3.4, however, this was significant from pH 5.6. Root dry mass for cultivar 'Pioneer 3906' was significantly increased at pH 5.0 and 4.6, compared to other SAR treatments. Overall, SAR had no deleterious effect on the dry biomass accumulation of all corn cultivars examined. For each cultivar, pH 2.6 SAR did not significantly reduce shoot or root dry mass, as compared to pH 5.6.

Cucumber Cultivars 'Hymark', 'Tamor' and 'Pioneer' exhibited significantly increased shoot dry mass for plants when exposed to SAR treatments in the pH 5.0 to pH 2.6 range (Table II). Root dry mass was significantly increased by SAR treatment (Table III). For 'Pioneer', root dry mass was greatest at pH 2.6, 3.0 and 4.6, as compared to pH 5.6. For 'Sprint', root dry mass was greatest at pH 3.8, 3.4 and 3.0. In cultivars 'Hymark' and 'Tamor', the stimulation of root dry mass was not as marked as in the two other cucumber cultivars. However, for both cultivars ('Hymark' and 'Tamor'), significantly greater root dry mass was observed at pH 5.0, vs pH 5.6.

Soybean The shoot dry mass of cultivar 'Harcor' exposed to SAR of pH 3.0 and 2.6 was significantly reduced, as compared to pH 5.6 (Table II). For 'Maple Arrow', 'Evans' and 'Hodgson', no signifcant difference was observed for dry shoot mass of plants exposed to pH $\leq$ 4.6 SAR vs pH 5.6 (Table II).

The root dry mass of 'Harcor' and 'Maple Arrow' were both significantly reduced at pH 3.0, and at pH 2.6 in the case of 'Harcor' (Table III). In 'Hodgson' and 'Evans' cultivars, root dry mass accumuation was stimulated by intermediate SAR levels of pH 5.0 to 3.4, however, this effect was not significant (Table III).

4. DISCUSSION

For all crops and cultivars examined, visible injury was only observed on foliage which was exposed to pH 3.0 and pH 2.6 SAR. It should be noted that the occurrence of foliar injury on seedlings had no significant relationship to other parameters determined in this study.

Overall, exposure of cultivars to a range of acidities in simulated rain had virtually no effect on the dry biomass accumulation of shoot and root components. In a few cases, dry biomass accumulation was significantly reduced for plants exposed to highly acidic rain - a "negative" effect. For other cultivars, no significant change in dry biomass accumulation was observed in response to SAR exposure. Finally, for a few cultivars, accumulation of dry biomass was significantly increased after repeated exposure to highly acidic rain - a "stimulatory" effect.

At acidity levels between pH 3.8 and pH 2.6 SAR (3 to 40 times more acidic than ambient acidic rain - pH 4.2) the shoot dry mass was unaffected in 13 cultivars, reduced in 4 cultivars and stimulated in 7 cultivars. Root dry mass was unaffected by this SAR range (pH 3.8 to 2.6) in 16 cultivars, reduced in 6 cultivars and stimulated in 2 cultivars. It should be noted that "stimulatory" or "negative" responses were often restricted to only a few of the 7 acidity levels utilized in these experiments. In addition, the dry biomass data is extremely variable and as a result, no significant dose-response functions could be determined.

With the exception of cabbage cultivars, the crop species examined are highly resistant to injury and dry biomass reduction caused by exposure to acidic rainfall during early growth stages. However, the results of this study do demonstrate that crop response to SAR is not not uniform.

These experiments with agricultural crops illustrate that plant response to simulated acidic rain treatments is not only species dependent, but can also be cultivar dependent. This differential susceptability (cultivar vs. species) is an important factor which must be considered when examining results from other SAR "effects" studies which have utilized only one crop species or cultivar in the experimental design.

REFERENCES

Chan, W.H., Tang , A.J.S., and Lusius, M.A.: 1983, Precipitation concentration and wet deposition fields of pollutants in Ontario September 1980 to December 1981. Ontario Ministry of Environment, APIOS Report ARB-61-83-ARSP, Toronto, Canada. 79 p.

Cohen, C.J., Grothaus, L.C., and Perrigan, S.C.: 1981, Effects of simulated sulphuric acid rain on crop plants. Agricultural Experimental Station, Oregon State University, Corvallis, Oregon, Special Report 619, p. 53.

Cohen, C.J., Grothaus, L.C., and Perrigan, S.C.: 1982, Effects of simulated sulphuric and sulphuric-nitric acid rain on crop plants: Results of 1980 crop survey. Agricultural Experimental Station, Oregon State University, Corvallis, Oregon, Special Report 670, p. 92.

Evans, L.S., Lewin, K.F., Conway, C.A., and Patti, M.J.: 1981, New Phytologist **89**, 459.

Evans, L.S.: 1982, Envir. Exp. Bot. **22**, 155.

Evans, L.S., Gmur, N.F., and Mancini, D.: 1982, Envir. Exp. Bot. **22**, 445.

Evans, L.S., Lewin, K.F., Patti, M.J., and Cunningham, E.A.: 1983, New Phytol. **93**, 377.

Harcourt, S.A. and Farrar, J.F.: 1980, Environ. Pollut. **22**, 69.

Heagle, A.S., Phibeck, R.B., Brewer, P.F., and Ferrell, R.E.: 1983, J. Environ. Qual. **12**, 538.

Irving, P.M. and Miller, J.E.: 1981, J. Environ. Qual. **10**, 473.

Irving, P.M.: 1983, J. Environ. Qual. **12**, 442.

Jacobson, J.S., Troiano, J., and Heller, L.: 1985, J. Exp. Bot. **36**, 159.

Lee, J.J., Neely, G.E., Perrigan, S.C., and Grothaus, L.C.: 1981, Envir. Exp. Bot. **21**, 171.

Plocher, M.D., Perrigan, S.C., Hevel, R.J., Cooper, R.M., and Moss, D.N.: 1985, Simulated acid rain on crops. Agricultural Experimental Station, Oregon State University, Corvallis, Oregon,

Special Report 739, p. 64.

Steel, R.G.D., and Torrie, J.H.: 1980, Priciples and procedures of statictics, Ed.2, McGraw-Hill, New York. p. 633.

Troiano, J., Heller, L., and Jacobson, J.S.: 1982, Environ. Pollut. (Ser.A.) **29**, 1.

Troiano, J., Colavito, L., Heller, L., McCune, D.C., and Jacobson, J.S.: 1983, Envir. Exp. Bot. **23**, 113.

Troiano, J., Jacobson, J.S., and Heller, L.: 1984, Agriculture, Ecosystems and Environment **11**, 161.

CHANGES IN SOIL pH OVER A 50-YEAR PERIOD UNDER DIFFERENT FOREST CANOPIES IN SW SWEDEN.

Carl Olof Tamm & Leif Hallbäcken
Department of Ecology and Enviromental Sciences
Swedish University of Agricultural Sciences
S-75007 Uppsala, Sweden

ABSTRACT. Reinvestigation of soil profiles, down to 70 cm depth in mineral soil, sampled for measurements in 1927 revealed a general decrease in pH for spruce and hardwood stands with 0.3 to 0.9 units. The pH-changes in the top soil can partly be explained by biological acidification, increasing with stand age. Stand age and species composition have less influence in deeper soil horizons and much of the acidification there is ascribed to acid deposition.

In 1927 157 soil profiles were described morphologically and sampled for pH at Tönnersjöheden Experimental Forest, SW Sweden. The location of all profiles was given in relation to a permanent coordinate net. Stand type and vegetation were also noted. The pH measurements were made potentiometrically with quinhydrone/calomel electrodes on field moist samples.

In 1982-84 ninety of the profiles were resampled (within about 1 m from the old profile), i.e. virtually all profiles which had not been disturbed by site preparation, road-building, etc. The pH was measured in water suspensions as in 1927, and with the same weight ratio soil:water (1:4). Additional readings with a glass electrode gave results very similar to those with the quinhydrone method.

The recent pH measurements are compared with those from 1927 in Table I (Hallbäcken and Tamm, 1986a,b). A general decrease of 0.3 to 0.7 units was found in all mineral horizons (A_2, B and C) beneath all categories of canopy (beech, oak and spruce planted during different periods. In the humus layer (A_0) decreases up to 0.9 units occurred. 25 of the 28 group differences (horizons and canopies) between pH 1927 and recent pH were statistically significant or highly significant. All mineral soil horizons are now in the aluminium buffer range (around pH 4.5), while the B and C horizons mostly were close to or above pH 5 in 1927. There are also differences in pH between hardwood sites and spruce sites, though smaller than between old and new samples. Spruce sites are typically more acid than beech and oak sites.

Both in 1927 and now there is a decrease in pH of upper soil horizons with the age of the spruce stand (Figures 1a, b and c), while pH increases in A_0 and A_2 upon clearfelling. This biological

Water, Air, and Soil Pollution **31** (1986) 337–341.

Table 1.

Measurements of pH in soil samples from Tönnersjöheden Experimental Forest in 1927 and 1982-83. All measurements made with quinhydrone electrode with calomel electrode as reference on water suspensions of field moist samples collected in late summer or early autumn. Symbols for statistical significance levels (Student's t-test): * = $p<0.05$; ** = $p<0.01$; *** = $p<0.001$.

Stand type	Median value of pH, Soil horizon A_0 1927	A_0 82-83	A_2 1927	A_2 82-83	B 1927	B 82-83	C 1927	C 82-83	Differences between arithmetic means of pH in 1927 and 1982-83, Soil horizon A_0	A_2	B	C
I Beech (Fagus silvatica), 17 profiles	4.49	3.76	4.52	4.17	4.89	4.67	5.25	4.70	0.77***	0.30**	0.27***	0.55***
II Oak (Querqus sessiliflora), 7 profiles	4.30	3.77	4.69	4.42	4.99	4.68	5.27	4.79	0.36	0.33*	0.27**	0.46***
III Spruce (Picea abies) planted 1877-90 after beech, 5 profiles	4.00	3.67	4.56	4.04	4.67	4.39	5.04	4.52	0.32	0.36	0.39*	0.51***
IV Spruce (Picea abies) planted 1890-1900 on open Calluna heath, 24 profiles	4.06	3.53	4.53	3.94	5.02	4.40	5.28	4.57	0.56***	0.65***	0.54***	0.73***
V Spruce (Picea abies) planted 1925-30 after beech, 11 profiles	4.46	3.59	4.75	4.00	4.96	4.50	5.19	4.56	0.94***	0.71***	0.48***	0.70***
VI Spruce (Picea abies) planted 1955-60 after beech, 5 profiles	4.41	3.95	4.56	3.95	5.02	4.61	5.37	4.67	0.46**	0.65*	0.56*	0.72**
VII Spruce (Picea abies) planted 1955-60 after birch, 5 profiles	4.15	3.65	4.60	4.09	5.02	4.54	5.21	4.65	0.53*	0.53*	0.51**	0.56**

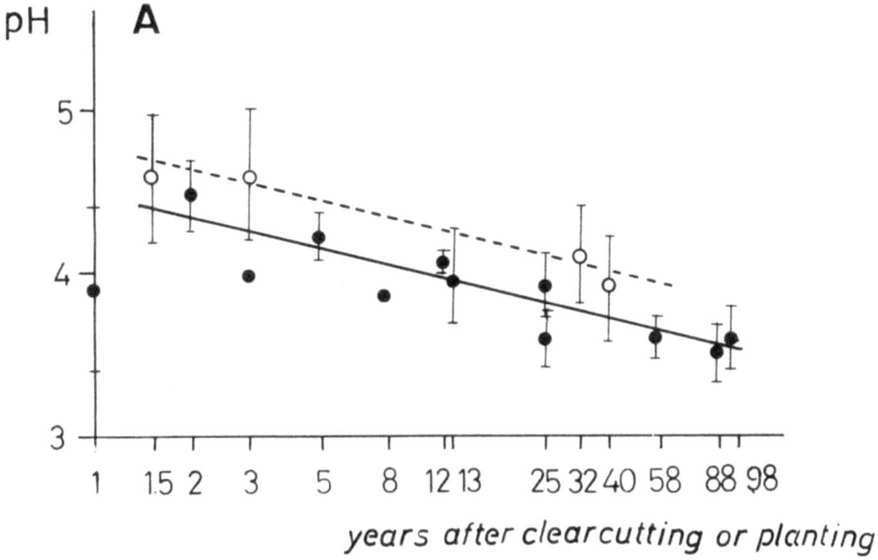
pH
A
5
4
3
1
1.5
2
3
5
8
12
13
25
32
40
58
88
98
years after clearcutting or planting

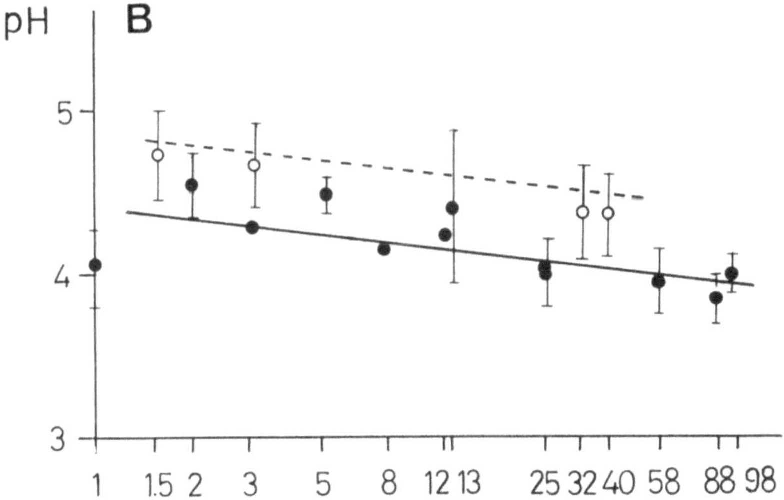
pH
B
5
4
3
1
1.5
2
3
5
8
12
13
25
32
40
58
88
98

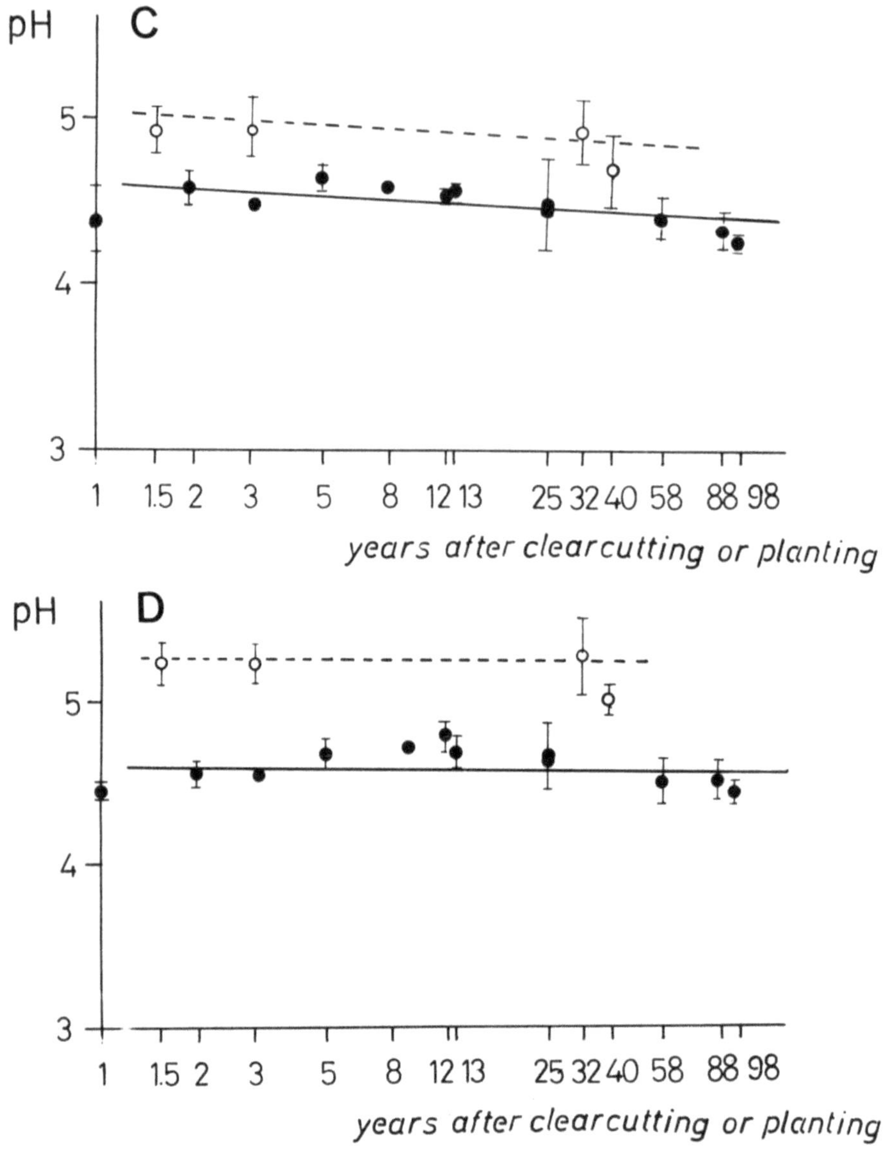

Figure 1. The pH of field moist soil samples plotted against the logarithmic age for different spruce stands at Tönnersjöheden Experimental Forest. Vertical bars represent standard deviation in measurements, where there are three or more profiles in stands of the same age.
– – – = Samples collected in 1927 ——— = Samples collected in 1984

A. Humus layer A_0 B. A_2 horizon C. B horizon D. C horizon acidification is stronger than the difference between old and new samplings in the forest floor (A_0) but not in the mineral horizons and can hardly be traced in the C horizon at 70 cm soil depth (Figure 1d). Acid deposition seems to be the only likely explanation of this deep acidification.

The acidification of the deepest horizons in the mineral soil are, when expressed in hydrogen ion concentrations, of course not so large as in the upper part of the soil profile, due to the logarithmic nature of pH. However, two important conclusions can be drawn:

1. There has been an acidification of the C horizon from pH above 5 to 4.7 or less.

2. This means important changes in soil chemical properties, e.g. in aluminium solubility which may imply an increased leaching of 'free' aluminium to soil and ground water.

References

Hallbäcken, L., and Tamm, C.O.: 1986a, Geologiska Föreningens i Stockholm Förhandlingar, In press.

Hallbäcken, L., and Tamm, C.O.:1986b, Scandinavian Journal of Forest Research, In press.

DIURNAL VARIATIONS OF AEROSOL CONCENTRATIONS INSIDE AND ABOVE A YOUNG SPRUCE STAND: MODELLING AND MEASUREMENTS

Bo L. B. Wiman
Department of Ecology, University of Lund
Ecology Building
S-223 62 Sweden.

ABSTRACT. Aerosol characteristics and aerodynamic conditions above and inside forests are strongly dynamic, especially for landscapes subject to frequent frontal activity and episodic anthropogenic aerosol inflows. Using empirical data of 1-hr time- resolution a non-stationary partial differential equation model (based on K-theory and classical filtration theory for concentration and deposition variations in a forest) is shown reasonably capable of handling such temporal complexity, which is also shown to be reflected in a temporally complex deposition pattern.

1. INTRODUCTION

Deposition models of various kinds accounting for structural and temporal details of forest landscapes are needed to couple as submodules with long-range transport models (at present forced to simplify effects of landscape heterogeneity; cf. e.g. Carmichael and Peters, 1984). Therefore, it is essential to formulate and critically evaluate approaches to such submodules. Various aspects of sink structure, efficiency and aerodynamics involved are given by e.g. Hicks *et al.* (1982), Slinn (1982), Davidson *et al.* (1982), Chamberlain *et al.* (1984), Bondietti *et al.* (1984), Wesely *et al.* (1985), Wiman *et al.* (1985), Wiman (1985). For landscapes subject to strong episodic patterns in aerosol exposure and frontal activity, determination of the resulting patterns of aerosol deposition is particularly complicated due to rapid changes in all basic systems characteristics (particle concentrations, chemical composition, size distributions, transfer and collection mechanisms). Further, complex feedbacks between canopy structure and micrometeorology are involved. Although there is, at present, no theory adequately handling all aspects of this complexity, it can be partly analysed from K-theory (particle transfer) and classical filtering theory (particle collection). This paper adapts an earlier non-stationary model (Wiman, 1985) and its handling to a situation of this strongly dynamic type and evaluates model capacity and deficiency by means of highly time-resolved field data on aerosol concentrations and aerodynamics inside and above a stand of *Picea abies* (Karst.).

2. THE MODEL

For a non-stationary situation in one space coordinate z (height above the ground surface) the partial differential equation model (pde) by Wiman and Ågren (1985) reduces to

$$\partial c/\partial t = \partial/\partial z\,(K \partial c/\partial z) + v_s\, \partial c/\partial z - S \tag{1}$$

Water, Air, and Soil Pollution **31** (1986) 343–348.

(with boundary conditions $\partial c/\partial z=0,\ z=0,\ t \geq 0;\ c\ (t,z_r)$ defined by measured concentration data; initial conditions approximated by measured concentration gradient at $t=0$) if aerosol sources within the forest are negligible; $c\ (t,z,D_p)$ is concentration at time t of particles with (equivalent) diameter D_p; $K\ (t,z)$ is K_{zz} - term in turbulent diffusivity tensor; $v_s\ (D_p)$ is particle settling velocity; $S\ (t,z,D_p)$ is a sink function due to particle collection by needles. Requirements, approximations and numerical methods for this pde are discussed by Wiman and Ågren (1985), Wiman (1985).

In contrast to an earlier study (Wiman, 1985), associated with steady large-scale weather enabling curve and surface fitting of field data for vertical leaf distribution and micrometeorology prior to input to Equation (1), interpolation routines are used here for undistorted data input. As the present study reformulates an earlier sink-submodel (Wiman and Ågren, 1985) on theoretical grounds, data on the vertical distribution of leaf surface area density ($\rho\ (z)$) from the site (Wiman, 1985) are recalculated to the equivalent cylinder area distribution (with equivalent cylinder diameter d_n = 0.8±0.2 mm) giving the leaf area index 14.6 (±5.8) ha ha^{-1}.

I will use the following description for **particle transfer** by eddy diffusion:

$$K(t,z)=(l_m(z))^2\ \partial u\,(t,z)/\partial z\ F(t,z) \tag{2}$$

where l_m is the mixing length under neutral conditions, u is x- component of the mean wind vector and F is a stability function. l_m is obtained from applying the formula by Perrier (1976) on a layer-by-layer basis in the canopy and is thus calculated directly from field data on $\rho(z)$ and from a factor w_c accounting for needle orientations (assumed random, i.e. $w_c = 0.5\ \pi^{-1}$). u (and $\partial u/\partial z$) obtains directly from field data (interpolated where needed). A stability function according to Goudriaan (1977; see also Fukui and Nakajima, 1983) is used, for simplicity here with the assumption that particles are transported as momentum (whether a mass-heat analogy would be better supported seems still debatable, cf. e.g Davidson *et al.*, 1982; Ley and Thomson, 1983; Sievering, 1986). To calculate F, Richardson's number (Ri) is evaluated from field data. However, the several F (Ri)-relationships suggested in the literature are commonly restricted to $0.04 < F < 10$, the range used here.

The **particle collection efficiency** E of various (geometrical arrangements of) objects can be formulated from pure aerosol mechanics (e.g. Kirsch and Stechkina, 1978) or obtained semiempirically (e.g. Slinn, 1982). Although both approaches are subject to limitations due to the geometrical complexity of vegetative elements (cf. Sievering and Wiman *et al.*, 1985), it is of interest to investigate the usefulness of classical filtration theory and to assume that particles will be collected by a needle (cylinder) from (i) the horizontal stream in the canopy with mean effective velocity u and (ii) vertical flow represented by $l_m\ \partial u/\partial z\ F^{1/2}$. The essential collection mechanisms are across-streamline diffusion (efficiency E_D), impaction (E_I), interception (E_R) and gravitational settling (E_G). Pich (1966), quoting classical work, gives

$$E_D = 1.71/(2-\ln(Re))^{1/3}\ Pe^{-2/3}\ ; \qquad Re < 1 \tag{3}$$

$$E_D = 4(2/\pi)^{1/2}\ Pe^{-1/2}\ ; \qquad \text{for potential flow} \tag{4}$$

$$E_I = St^3/(St^3 + 0.77St^2 + 0.22)\ ; \qquad Re = 10 \tag{5}$$

$$E_R = 1/\,(2\,(\,2-\ln\,(Re)))\{2\,(1+N_R)\ln\,(1+N_R)-N_R\,(\,2+N_R\,)\,/(1+N_R\,)\}\ ; \qquad Re < 1 \tag{6}$$

$$E_R = 1+N_R-1/(1+N_R) \qquad \text{for potential flow, } St=0,\ N_R = D_p\,/\,d_n \tag{7}$$

where Re, Pe and St are the Reynold, Peclet and Stoke numbers, respectively (related to d_n and to process (i) or (ii)). I use the formulations for potential flow to represent cases with $Re > 1$ (E_D,E_R) and (5) for all Re. A factor π^{-1} transforms from efficiency per unit length to unit area. A factor W accounts for cylinder orientations; for horizontal (vertical) flow W is w_h (w_v). I use $w_h = w_v = 0.75$ (randomly oriented needles; effects of their finite size included). A rigorous treatment must include an orientation-dependent d_n. I assume that E_D, E_I and E_R are mutually exclusive in processes (i) and (ii), respectively, but that (i) and (ii) can act simultaneously. Interference effects due to neighbouring needles are neglected for this 'loose-filter model' (cf. Pich 1966). This leads to

$$S = \{E_h \, u \, (w_h / \pi) \, \rho c\} + \{E_v \, abs(K \, (\partial c / \partial z)) \, (w_v / \pi) \, \rho + B \, m_p \, g \, w_c\} \qquad (8)$$

where E_h (E_v) refers to total efficieny of horizontal (vertical) processes and B, m_p and g are the particle mobility, particle mass and gravitational constant, respectively.

3. EXPERIMENTAL: METHODS AND BASIC RESULTS

The model will be tested versus a measurement period of 48 hr at a site (Wiman, 1985; cf. also Wiman and Lannefors, 1985) within the boreo-nemoral region of Sweden. The period (August 18-20, 1982) was subject to SE-SSE winds, strong frontal activity, rain (c. 10 mm) and relative humidity ≥80%. Data were sampled as follows: wind speed (cup anemometers, 4 levels from $z_r = 1.25\, z_c$, z_c is upper canopy limit = 6 m), temperature (thermistors, 8 levels from z_r), particle number concentrations (Climet 208C optical particle counter (OPC), classes 0.3 to 0.5 μm, 0.5 to 0.7 μm, 0.7 to 1 μm used here; 2 levels: z_r and $0.17 z_c$), mass concentrations (2-stage Nuclepore filter samplers: particles less/larger than c. 2 μm; OPC-levels; FSU-streakers (Nelson *et al.*, 1976): particles < c. 3 μm, OPC-levels;). One 12-hr impactor sample was obtained at z_r. Sampling inlets were as described by Wiman (1985).

Micrometeorological data were averaged to hourly means. OPC-data were sampled for each class 1 min every 20th min at a given height and aggregated to 3 min to represent each hour. From independent measurements with another OPC the OPC-data are concluded to represent the hour with an uncertainty <20%. Filters (12-hr sampling blocks), streaker samples (12-hr intervals) and impactor plates were analysed for fine-particle S and coarse-particle Ca with PIXE (Malmqvist *et al.*, 1982). An overview of the experimental results is given Table I, where OPC-data are aggregated to 12-hr blocks for comparison with mass concentration data.

The micrometeorological data showed the first night (N1) as fairly windy (frontal passage) with near-neutral temperature conditions in and above the canopy. During D1 and D2 temperature inversions developed within the canopy. N2 showed low wind speed and unstable to neutral temperature profiles within the canopy, tending towards stability above the canopy. Table I shows a clear trend of concentration depletion in the canopy (statistically significant in 15 out of 24 cases). Exceptions are fine-S data, streaker-S data and coarse-Ca data for D1. The factor-of-3 concentration difference showing at times between fine-S and streaker-S is to be expected from the different operational modes of the samplers (discrete and continuous, respectively). The total particle volume within all three classes, rather than any separate OPC-class, traces the S-concentrations. This indicates a fairly strong time-dependency of the fine-S particle size distribution.

4. MODEL EVALUATION AND DISCUSSION

The model was run for each OPC-class (1-hr resolution) represented by its geometrical

Table 1. Overview of experimental concentration data. Values for the 4 12-hr blocks (N1=night 1, D1=day 1, etc.) are normalized relative to absolute values (abs. v.) obtained for the N1-block. Underlined data: above canopy; not underlined: inside canopy. Errors corresponding to one standard deviation (mass concentrations) or representativity errors (OPC) given within parentheses.

	OPC .3-.5	OPC .5-.7	OPC .7-1	S_{fine}	$S_{streaker}$	Ca_{coarse}
Abs.v.:	28.0(cm^{-3})	1.0(cm^{-3})	0.29(cm^{-3})	2190(ng m^{-3})	733(ng m^{-3})	175(ng m^{-3})
N1	1.00(0.20)	1.00(0.20)	1.00(0.20)	1.00(0.11)	1.00(0.15)	1.00(0.15)
	0.67(0.13)	0.22(0.07)	0.28(0.07)	0.64(0.07)	0.79(0.09)	0.61(0.16)
D1	0.10(0.02)	0.52(0.10)	1.52(0.30)	0.22(0.02)	0.53(0.05)	<0.26
	0.08(0.02)	0.37(0.07)	0.72(0.14)	0.26(0.03)	0.59(0.06)	0.30(0.27)
N2*	0.13(0.03)	0.58(0.12)	1.41(0.28)	0.30(0.04)	0.67(0.09)	1.08(0.16)
	0.10(0.02)	0.27(0.05)	0.38(0.08)	0.22(0.02)	0.34(0.05)	0.36(0.09)
D2	0.15(0.03)	0.43(0.09)	0.79(0.16)	0.34(0.05)	0.98(0.14)	0.50(0.09)
	0.14(0.03)	0.34(0.07)	0.72(0.14)	0.29(0.04)	0.63(0.11)	0.35(0.07)

* Impactor sample above canopy giving mass median aerodynamic diameter for S (Ca) 0.3 (4) μm.

mean particle diameter (D_{pg}), with particle density set to 2 g cm^{-3}. Model predictions for concentrations at $z = 0.17 z_c$ are compared with measured data in Figure 1 where empirical errors are incorporated into in-canopy concentrations through quadratic addition of above- and in-canopy error-data. Within the size range available for evaluation the simplified approach with K - and filtration-theory produces qualitatively well-behaved predictions. Whereas 'day-predictions' are generally within a factor of 2 of test data, the quantitative deficiency of the model for nights (depletion underestimated with a factor of 3-4) may relate to their being rainy, with a milieu inside the canopy not accounted for by the K - and S -formulations (particle collection by droplets on vegetative surfaces, humid boundary layers around needles potentially implying a situation partly analogous to particle deposition to natural waters (e.g. Williams, 1982; cf. also Jenkin, 1984)). Further, particular filtration problems arise when the flow field around a collector is turbulent (Hidy and Heisler, 1978; Douglas *et al.*, 1982, 1983; see also Gutfinger and Friedlander, 1985), also suggesting an explanation to the discrepancies often found between theoretical and measured collection efficiencies (cf. Katen and Hubbe, 1985).

Model behaviour is further detailed in Figure 2 showing the full solution of Equation (1) for all 3 OPC-classes simultaneously when recalculated to a volume distribution and also the corresponding deposition patterns. High concentrations and efficient transfer and collection during N1 generate higher deposition than obtained for D1, as opposed to the more typical night-day deposition pattern examplified by N2 vs D2 (cf. Wiman, 1985; see also Hicks *et al.*, 1982). About halfway down into the canopy the non-trivial interplay between collector density, particle transfer and collection efficiency per needle optimizes to an overall-efficient deposition process.

The particle size distribution of S (and most likely Ca) was probably unstable during the period (cf. sect. 3) although the impactor sample obtained (Table I) supports the general principle with S (Ca) in accumulation (coarse) mode. Approximating the S (Ca) distribution with a stationary log-normal model with D_{pg} 0.3 to 0.5 (4 to 6) μm and a geometric standard deviation of 2 (2) gives model estimates of 0.1 to 0.2 (1 to 2) cm s^{-1} for the gross canopy deposition velocity (i.e. related to deposition rate in the stand per unit ground area, with reference concentrations at z_r) over the period. Considering the test results these values are probably underestimates.

In conclusion, operating the model with a 1-hr time resolution supports the

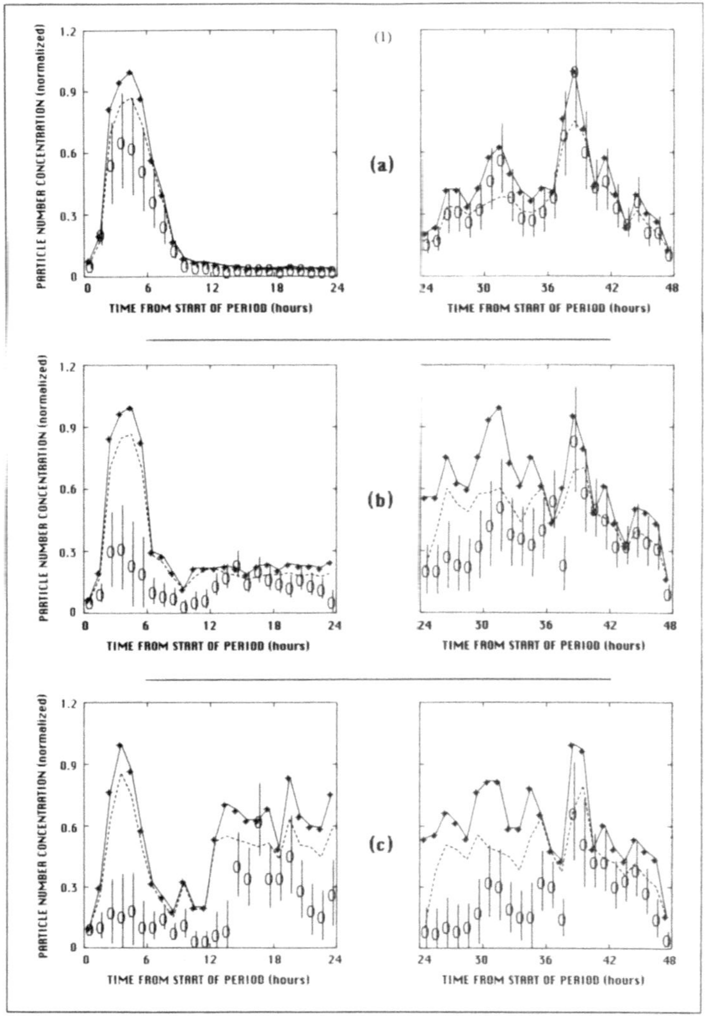

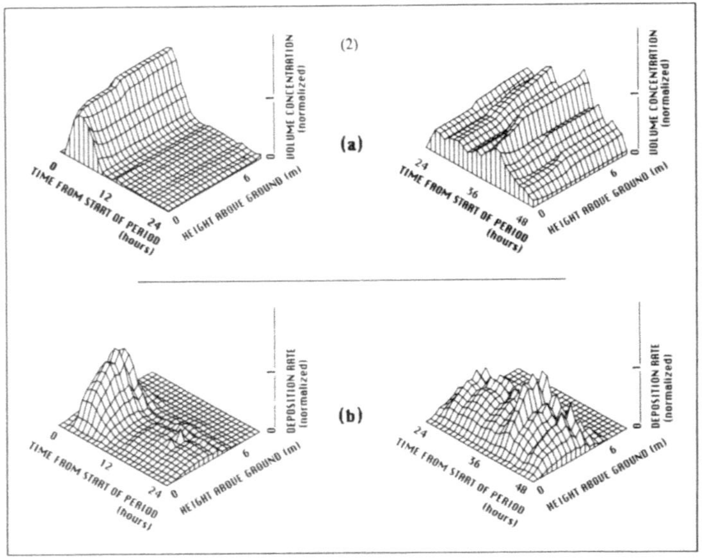

Fig. 1. Predicted (solid lines, upper line: boundary conditions) and empirical concentrations (*: above-canopy, boundary conditions; 0: in-canopy; vertical bars are representativity errors) as a function of time (t) ($t = 0$ is local time 2000 (GMT+1h)). (a): particle size class 0.3-0.5, (b): 0.5-0.7 (c): 0.7-1 μm. Maximum value in each diagram is normalized to 1.

Fig. 2. Full solution of Equation (1) for the three-classes particle volume distribution obtained from OPC-data. (a): concentration, c, as a function of t and z (height above ground surface); (b): deposition rate as a function of t and z. Maximum value in each diagram is normalized to 1.

view that (i) the model acceptably handles strongly dynamic behaviour of aerosol concentrations and aerodynamics, (ii) this behaviour, in turn, generates a strongly dynamic pattern in aerosol deposition. However, model results also suggest that detailed studies of stand architecture, wet collectors, turbulence around individual collectors and the dynamics of element-specific particle-size distributions are strongly needed to develop operative versions of aerosol deposition models. Further, whereas little seems to be gained by using higher-order closure models as an alternative to K-theory for flow descripitions above canopies (cf. Carmichael and Peters, 1984) their use inside canopies (e.g. Wilson and Shaw, 1977) may be developed and applied for aerosol deposition modelling where K-theory is only a first-order approximation of the complexity of particle transfer.

ACKNOWLEDGEMENTS. This study was sponsored by the National Swedish Environment Protection Board and the Swedish Natural Science Research Council. I thank G. I. Ågren and H.O. Lannefors for constructive support and criticism.

REFERENCES

Bondietti, E.A., Hoffman, F.O. and Larsen, I.L.: 1984, *J. Environ. Radioactivity* **1**, 4.

Carmichael, G.R. and Peters, L.K.: 1984, *Atmos. Environ.* **18**, 937-951.

Chamberlain, A.C., Garland, J.A. and Wells, A.C.: 1984, *Boundary Layer Met.* **29**, 343.

Davidson, C.I., Miller, J.M. and Pleskow, M.A.: 1982,*Wat. Air Soil Pollut.*, **18**, 25.

Douglas, P.L., Dullien, F.A.L. and Spink, D.R.: 1982, ' On the Deposition of Aerosols. A Review and Recent Theoretical and Experimental Results', *2nd Waste. Treat.Util. Proc. Int. Symp.*, 391.

Douglas, P.L., Dullien, F.A.L. and Spink, D.R.: 1983,*The Canadian Journal of Chemical Engineering* **61**, 776.

Fukui, K and Nakajima, M.: 1983,*Quart. J. R. Met. Soc.* **109**, 661.

Goudriaan,J.: 1977,*Crop Micrometeorology: A Simulation Study*, Pudoc, Wageningen.

Gutfinger, C. and Friedlander, S.K.: 1985, *Aerosol Science and Technology* **4**, 1.

Hicks, B.B., Wesely, M.L., Durham, J.L. and Brown, M.A.: 1982,*Atmos. Environ.* **16**, 2899.

Hidy, G.M. and Heisler, S.L.: 1978, 'Transport and Deposition of Flowing Aerosols', In: D.T. Shaw (ed.) *Recent Developments in Aerosol Science*, Wiley, New York, pp. 135-165.

Jenkin, M.E.: 1984, *Atmos. Environ.* **18**, 1017-1024.

Katen, P. C. and Hubbe, J.M.: 1985, *J. Geophys. Res.* **90**, D1, 2145.

Kirsch, A.A. and Stechkina, I.B.: 1978, 'The Theory of Aerosol Filtration with Fibrous Filters', In: D.T. Shaw (ed.), *Fundamentals of Aerosol Science*, Wiley, New York, pp. 165-257.

Ley, A. and Thomson, D.J.: 1983,*Quart. J. R. Met. Soc.* **109**, 847.

Malmqvist, K.G., Johansson G.I. and Akselsson, K.R.: 1982, *J. Radioanalyt. Chem.* **74**, 125.

Nelson, J.W., Jensen, B., Desadeleer, G.G., Akselsson, K.R. and Winchester, J.W.: 1976, *Adv. X-ray Anal.* **19**, 403.

Perrier, A.: 1976, 'Etude et Essai de Modélisation des Échanges de Masse et d'Énergie au Nivau des Couverts Végétaux', Ph. D. Thesis, University of Paris IV Avril, Paris (in French).

Pich, J.: 1966, 'Theory of Aerosol Filtration' by Fibrous and Membrane Filters', In: C.N. Davies (ed.), *Aerosol Science*, Academic Press, London, pp. 223-285.

Sievering, H. and Wiman, B.L.B., Ågren, G.I., Lannefors, H.O.: 1986, *Atmos. Environ.* **20**, 407.

Sievering, H.: 1986, *Atmos. Environ.* **20**, 341.

Slinn, W.G.N.: 1982, *Atmos. Environ.* **16**, 1785.

Wesely, M.L., Cook, D.R. and Hart, R.L.: 1985,*J. Geophys. Res.* **90**, 2131.

Williams, R.M.: 1982, *Atmos. Environ.* **16**, D1,1933.

Wilson, N.R. and Shaw, R.H.: 1977, *J. Appl. Meteorol.* **16**, 1197-1205.

Wiman, B.L.B. and Ågren , G.I. : 1985, *Atmos. Environ.* **19**, 335.

Wiman, B.L.B. and Lannefors, H.O.: 1985, *Atmos. Environ.* **19**, 349.

Wiman, B.L.B., Ågren, G.I. and Lannefors, H.O.: 1985, *Atmos. Environ.* **19**, 363.

Wiman, B.L.B.: 1985,*Tellus.* **37B**, 230-242.

THE EFFECT OF NICKEL AND COPPER DEPOSITION FROM A MINING AND SMELTING COMPLEX ON CONIFEROUS REGENERATION IN THE BOREAL FOREST OF NORTHERN MANITOBA.

David L. Wotton, Dale C. Jones, and S. Floyd Phillips
Environmental Management, Manitoba Department of Environment and Workplace Safety and Health
Building 2, 139 Tuxedo Avenue,
Winnipeg, Manitoba, Canada R3N 0H6

ABSTRACT. Airborne pollutants in the form of SO_2 and heavy metals have been dispersed over the boreal forest in the Thompson area of northern Manitoba since 1961. Metal deposition has been found in soils and plant material to a distance of 35 km from the Inco nickel smelter. Studies were initiated in 1979 to determine effects of Ni and Cu on coniferous regeneration of jack pine (Pinus banksiana Lamb.) and black spruce (Picea mariana (Mill) BSP.). Studies included metal accumulation in cones, effects on seed quality, seed germination and seedling response in metal contaminated soils. Nickel and Cu accumulation in cones of both species declined rapidly with distance from the smelter. Seed quality and germination potential was not related to metals in cones or air pollutant influence in general. Nickel and Cu accumulation in surface organic soils was inversely related to distance from the smelter. A significant inverse correlation was determined between seedling growth and Cu/Ni concentrations in surface organic soils. Metal accumulation in soils within 5 km of the nickel smelter inhibited root growth of conifer seedlings.

1. INTRODUCTION

Atmospheric emissions from Inco Metal Company's nonferrous Ni smelter at Thompson, Manitoba commenced in 1961. Gaseous and particulate pollutants are discharged from a 152 m stack. Principal pollutants are SO_2 and metal particulates containing Ni, Cu, Fe, Pb, and Cd.

The Thompson smelter is centered in the boreal forest ecosystem of northern Manitoba (55° 43'N, 97° 52'W). Forests of jack pine (Pinus banksiana Lamb.) and black spruce (Picea mariana (Mill.) BSP.) dominate in the area (Kotowycz et al., 1974). Air pollutants may impact surrounding forests by means of wet and dry deposition.

Emissions of SO_2 can cause vegetation injury symptoms such as foliar chlorosis, necrosis, and premature foliar loss which may ultimately alter the growth and development of vegetation (Malhotra and Blauel, 1980). Metal particulates can be directly

Water, Air, and Soil Pollution **31** (1986) 349–358.

deposited on the soil or may be physically trapped by plant surfaces and subsequently washed from the forest canopy to the soil.

In 1977 a joint federal/provincial forest study was initiated by the Canadian Forestry Service, Northern Forest Research Centre and the Province of Manitoba, Department of Environment and Workplace Safety and Health. Initial results showed a gradient of heavy metal accumulation in the soils and plant material to a distance of 35 km from the Thompson smelter (Wotton and Hogan, 1981). Nickel levels in the range of 3000 $\mu g\ g^{-1}$ were found in surface organic soils (LFH) within 5 km of the stack and decreased inversely to a background level ($\simeq$ 60 $\mu g\ g^{-1}$) at 66 km distance.

This study was carried out to investigate the effects of metal contaminated soils on coniferous regeneration in the Thompson area.

2. MATERIALS AND METHODS

Cone, seed and seedling development of jack pine and black spruce from 12 sites on two transects southeast and southwest of the Thompson smelter were studied. Each site was selected to include representative stands of jack pine and black spruce of the boreal forest in the area.

2.1. Cone and Seed Quality

In 1979 and 1980 mature cones from 5 jack pine and 5 black spruce trees were collected from each of the 12 sites. Following kiln extraction of seed the cones were analyzed for Ni and Cu by atomic absorption spectrophotometry.

Four replicates of 100 seeds per tree were germinated on moistened filter paper over sterilized silica sand in covered plastic petri dishes. Germination was carried out over a 14-day period in a Conviron (I23L) growth chamber. Incubation conditions were set at 15°C, 95% relative humidity and a 16 hr daylight period. Deionized water was used to moisten the silica as required and germination success was recorded daily. Measurements of germinant mass, length and percent germination were made after 14 days. The number of ungerminated seeds which were hollow, dormant or rancid was determined.

2.2. Seedling Growth

In 1981, seedlings were grown in LFH soils collected from the 12 study sites to evaluate the effect of various heavy metal concentrations on seedling growth. Control seed was collected from five randomly selected jack pine and black spruce trees from a background location (site 9).

The LFH material from each site was air dried, ground to a uniform texture and loaded into three 45 cell "Styrobloc'20" containers. After seeding, the soils were treated with a 2.5% solution of oxine benzoate, a fungicide, to control "damping-off" disease, and placed on floor level greenhouse benches for 24 weeks. Temperature in the greenhouse ranged from 20 to 25°C, relative humidity was 40 to 60% and lights were on for 18 hr day^{-1}. Seedlings were watered using a mist nozzle at a rate of 4 L min^{-1} for 1 min per tray every second day.

Measurements of condition, shoot length, root length, and oven dry mass were determined for each seedling at the end of the experiment. LFH soil and plant tissue samples were analyzed for Ni and Cu using an atomic absorption spectrophotometer. Correlation analysis was used to evaluate relationships between seedling growth and metal levels in LFH and seedling tissue.

3. RESULTS

3.1. Cone Analysis

The close-in sites (10, 11, 1) had elevated Ni levels in the cones of both species (Table I). For example, the Ni level in jack pine cones at site 11 was 10.8 µg g^{-1} compared with 4.00 µg g^{-1} at background (site 9). Nickel concentrations in black spruce cones were higher with 18.0 µg g^{-1} at site 10. An exponential relationship was found between Ni in cones and distance from the smelter with cones closest to the emission source having the highest metal levels.

3.2. Seed Quality

Seed germination in jack pine ranged from 75% at site 7 to 49.9% at site 4 while in black spuce it ranged from 76.3% at site 10 to 54.2% at site 5 (Table II). There were no significant relationships between metal levels and germination in either species which indicates that metal levels did not affect seed germination.

Seed mass of jack pine and black spruce were variable (Table II) but differences between sites were not significant and there was no apparent correlation with Ni and Cu concentrations in the cones. There was a trend for black spruce seed to be lighter in weight at sites which were close to the smelter. Germinant length was variable but not related to distance from the stack and thus probably not affected by smelter emissions (Table II). The precentage of hollow, dormant or rancid seeds was not correlated with Ni and Cu concentrations or to distance from the smelter.

3.3. Seedling Growth

Concentrations of Ni and Cu in LFH decreased with distance from the smelter according to an inverse curvilinear relationship (Table I). In particular, the correlation between Ni and distance (r^2=0.91) was significant at $p < 0.05$ (Figure 1) while that for Cu vs distance (r^2= 0.95) was significant at $p<0.01$. Nickel concentrations at close-in sites were up to 75 times the levels at background (2173 µg g^{-1} at site 11 vs 29 µg g^{-1} site 9) (Table I).

Nickel and Cu levels in seedling tissue were significantly correlated with distance from the stack, r^2 = 0.70 for jack pine and

TABLE I

Metal Concentration (μg g^{-1}) in Surface Organic Soil, Cones and Seedling Tissue of Jack Pine and Black Spruce

Study Site	Dist. from Smelter (km)	Cones				Surface Soils				Seedling Roots				Seedling Shoots			
		Ni		Cu		Ni		Cu		Ni		Cu		Ni		Cu	
		JP	BS	JP	BS	JP	BS	JP	BS	JP	BS	JP	BS	JP	BS	JP	BS
11	1.5	10.8	18.0	3.2	4.7	2173	1456	1210	1057	1484	259	306	212	140	23	28	18
10	2.3	9.8	16.7	2.0	4.3	1821	1813	550	925	1577	511	346	279	88	37	30	29
1	4.0	9.7	12.2	2.1	3.0	822	551	165	151	550	156	64	45	64	17	22	8
12	4.3	9.1	16.0	1.5	2.4	1233	1107	204	292	1267	317	80	67	89	54	21	20
5	9.0	6.5	11.0	1.5	2.2	370	490	54	77	837	288	49	53	79	28	24	13
4	11.0	4.6	9.0	1.3	1.6	399	451	49	61	690	234	44	40	79	24	21	11
2	12.2	5.8	8.7	1.4	2.0	253	372	27	53	280	94	29	27	51	25	29	7
6	15.7	4.9	9.8	1.1	1.7	270	288	36	48	351	162	53	34	47	19	31	13
13	18.0		3.6		1.6	47	57	12	16	22	17	26	17	7	4	30	10
7	21.8	8.3	9.5	1.5	2.1	117	197	20	48	141	65	23	28	39	32	28	31
3	31.8	2.8	6.1	1.3	2.0	79	114	12	23	68	41	16	18	23	7	20	8
9	66.0	4.0	8.4	1.8	2.2	29	24	7	9	28	14	23	16	11	4	32	10

TABLE II

Seed and Seedling Characteristics of Jack Pine and Black Spruce

Study Site	Distance from Smelter (km)	Seed and Germinants						Seedlings			
		Ger. %		Seed Mass (g/400)		Ger. Len. (mm)		Shoot Len. (cm)		Root Length (cm)	
		JP	BS	JP	BS	JP	BS	JP	BS	JP	BS
11	1.5	63.2	62.9	1.06	.334	46.7	26.3	5.2	4.3	14.1	5.3
10	2.3	73.0	76.3	1.08	.382	47.8	24.4	3.3	2.3	4.9	2.2
1	4.0	63.6	67.2	1.26	.345	41.6	30.4	4.9	8.5	20.1	17.9
12	4.3	67.8	64.6	1.03	.333	38.6	24.8	5.6	3.3	22.9	4.1
5	9.0	50.2	54.2	1.25	.308	41.4	29.6	4.8	9.0	20.3	14.9
4	11.0	49.9	69.2	1.08	.326	37.2	29.7	5.0	9.6	20.4	15.1
2	12.2	64.7	71.8	1.16	.326	37.8	31.2	8.2	19.9	27.9	20.4
6	15.7	51.3	64.3	1.30	.306	43.6	29.4	5.4	10.9	23.8	16.2
13	18.0		68.1		.355		26.2	6.5	17.6	2.9	23.3
7	21.8	75.0	73.4	1.17	.351	47.7	29.9	6.3	4.9	24.6	8.9
3	31.8	73.5	63.6	1.15	.331	49.8	29.4	6.8	13.7	30.8	19.1
9	66.0	65.7	73.1	1.18	.348	47.8	29.4	6.6	11.0	28.1	19.8

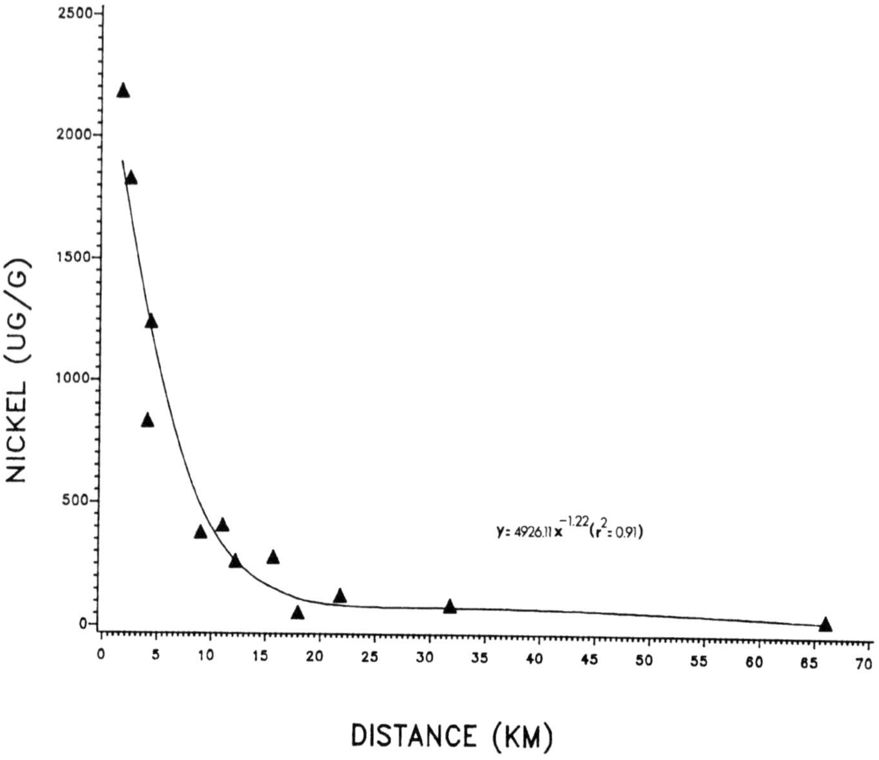

Fig. 1: Nickel in Surface Organic Soil at Jack Pine Sites vs Distance from Smelter.

$r^2 = 0.73$ for black spruce at $p < 0.05$. Significant correlations also existed for metal levels in the LFH vs seedling tissue for both species, $r^2 = 0.92$ for roots and $r^2 = 0.86$ for shoots of jackpine ($p < 0.01$) while $r^2 = 0.92$ for roots and $r^2 = 0.75$ for shoots of black spruce $p < 0.01$.

Nickel concentrations in jack pine root tissue grown in soil from sites within a 5 km radius of the stack, were up to 56 times the level at background (1577 $\mu g\ g^{-1}$ at site 10 vs 28 $\mu g\ g^{-1}$ at site 9). Background concentrations of Ni were approached at approximately 20 km (Figure 2).

Nickel and copper were also elevated in black spruce roots although concentrations were much lower than in the jack pine. We recognize that soil particles still attached to rootlets and root hairs may have contributed in part to the high levels of metals found

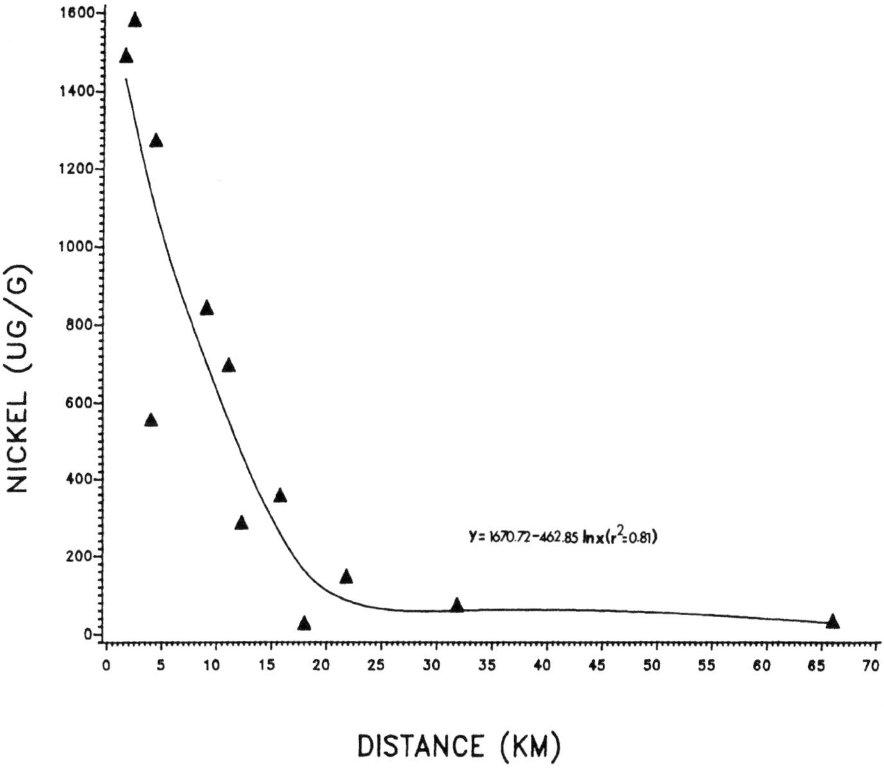

Fig. 2: Nickel in Jack Pine Roots vs Distance from Smelter.

in root tissue. However, a substantial amount of the metals were believed to have been incorporated in the root tissue since uptake to the shoots was confirmed.

Seedling growth, in particular root length at sites close to the smelter was suppressed within 5 km of the smelter and reduced at sites 11 km south and 9 km southeast (Table II). Significant correlations were found for root growth vs distance from the smelter for both jack pine and black spruce (Figures 3 and 4).

4. DISCUSSION

Nickel and Cu deposition and accumulation in jack pine and black spruce cones was related to distance from the smelter. However, no significant relationship was found between parameters of seed quality and distance from the emission source. It appears that the smelter emissions do not affect seed quality.

Metal deposition to surface organic soils of coniferous forests surrounding the smelter were elevated and decreased rapidly with distance from the smelter. Past studies of metal accumulation in soils near Thompson observed levels of Ni and Cu with the same magnitude and deposition pattern (Phillips and Slaney, 1981; Wotton and Hogan, 1981).

Sites closest to the smelter had the highest concentrations of metals in the LFH and seedling tissue and also exhibited the greatest degree of seedling growth inhibition. Growth inhibition was most pronounced in the roots; the highest concentrations of Ni and Cu were also found in the roots. In some cases root elongation was limited to the formation of rudimentary buds. Roots had very little lateral development, few or no root hairs and in some cases appeared blackened. This is typical of metal toxicity symptoms observed in

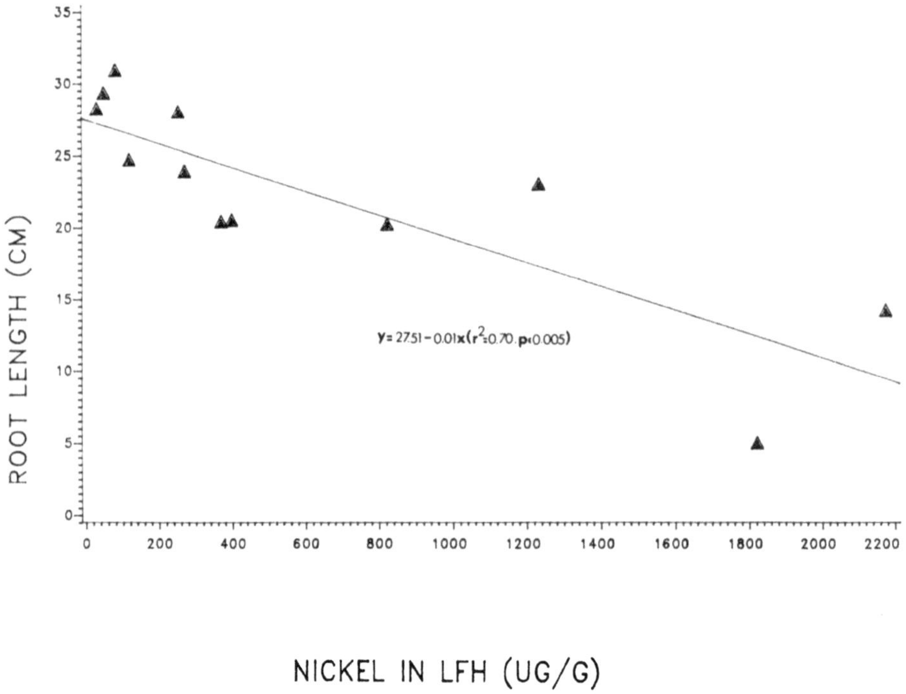

Fig. 3: Root Growth vs Ni in Surface Organic Soils at Jack Pine Sites.

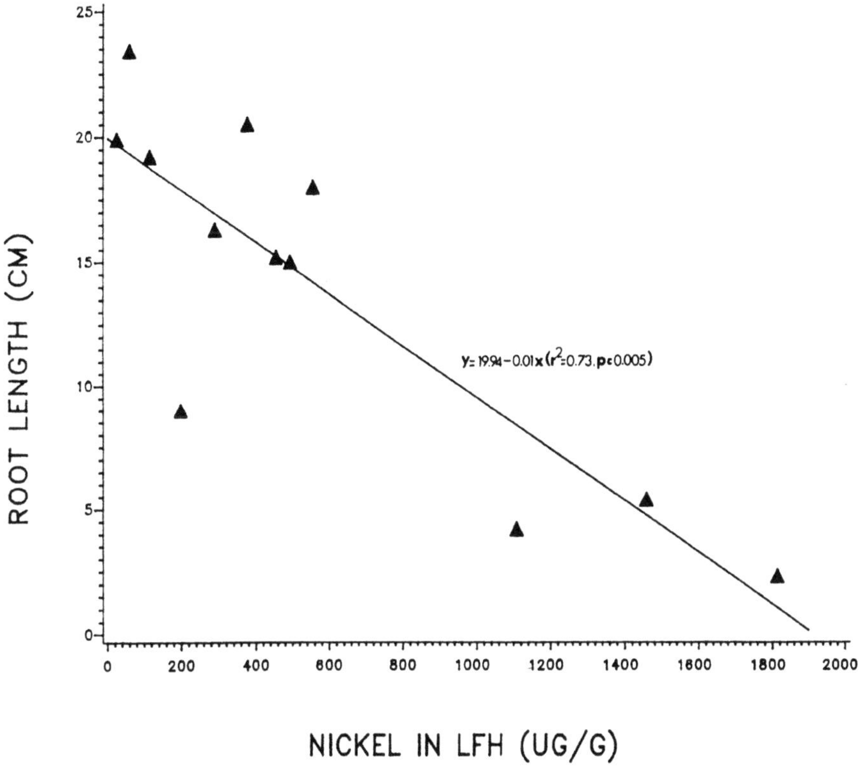

Fig. 4: Root Growth vs Ni in Surface Organic Soils at Black Spruce Sites.

other growth studies (Hutchinson and Whitby, 1973, and 1977; Whitby and Hutchinson, 1974; Malhotra and Blauel, 1980; Winterhalder, 1981).

In the present study root growth inhibition and obnormal root development was generally observed in the seedlings grown in LFH from sites within 5 km of the Inco smelter. There was a greater incidence of poor root growth at locations closest to the smelter, i.e. at 1.5 and 2.3 km (Table II). Root inhibition and reduced growth may be the result of direct or indirect toxic effects or possibly interference with plant metabolic processes such as inhibition of enzymatic activity (Malhotra and Khan, 1981; Ernst et al., 1983). It is not known whether these growth inhibition effects would persist if the seedlings were able to successfully establish their root systems in the underlying mineral soil which at present is not contaminated with heavy metals.

5. CONCLUSIONS

Metal particulates contained in Ni smelter emissions accumulated in cones of jack pine and black spruce. Nickel and Cu decreased as distance increased from the smelter. Germination and quality of jack pine and black spruce seed was not affected by smelter emissions.

Metal deposition to surface organic soils and Ni and Cu concentrations in seedling tissue were inversely correlated with distance from the smelter. Root growth of jack pine and black spruce seedlings was inhibited by metal accumulations in surface organic soils within 5 km of the Thompson smelter.

REFERENCES

Ernst, W.H.O., Verkeleij, J. A. C., and Vooijs, R.:1983. J. Environ. Monitoring and Assessment 3, 297.

Hutchinson, T. C., and Whitby, L. M.: 1973, "A study of airborne contamination of vegetation and soils by heavy metals from the Sudbury copper-nickel smelters, Canada, Inst. of Environ. Sciences and Engineering, Univ. of Toronto, Pub. No. EL-3, 16pp.

Hutchinson, T. C. and Whitby, L. M.: 1977, Water, Air, and Soil Poll. 7, 421.

Kotowycz, A. J., Kay, M, Rannard, C. D., Lamont, R. H., Jeffrey, C. A. and O'Regan, R. J.: 1974, "The Forests of Manitoba, Dept. Mines Resources and Environ. Mgm't., Manitoba, 82pp.

Malhotra, S. S., and Blauel, R. A.: 1980, "Diagnosis of air pollutant and natural stress symptoms on forest vegetation in western Canada," Environ. Can., Can. For. Serv., Nor. For. Res. Ctre., Info. Rep. NOR-X-228, 84pp.

Malhotra, S. S., and Khan, A. A.: 1981, "Effects of SO_2 and heavy metals on Pinus banksiana, Mitteilungen der Forstlichen Bundesversuchsanstatt, Wien,; 299-307.

Phillips, S. F., and F. F. Slaney, and Co. Ltd.: 1981, "The Inco Thompson environmental program. Vol. I, 300 + pp.

Whitby, L. M., and Hutchinson, T. C.: 1974, J. Environ. Conservation. 1, 191.

Winterhalder, K.: 1981, "Initiation of plant colonization of denuded, acid, metal-contaminated soils by limestone application - a case of confounded synergism; Paper presented at the XIII International Botanical Congress, Sydney, Australia, 1-9.

Wotton, D. L., and Hogan, G. D.: 1981, "The effects of atmospheric emissions from the Thompson mining and smelting industry on forest vegetation and soils, Submission to Manitoba Clean Environment Commission, 112pp.

ALUMINUM SPECIATION IN SOIL SOLUTIONS: EQUILIBRIUM CALCULATIONS

Paul A. Arp and Rock Ouimet
Department of Forest Resources
University of New Brunswick
Fredericton, New Brunswick Canada E3B 6C2

ABSTRACT. A computer simulation was done to illustrate how the equilibrium solubility and speciation of Al in well-aerated soil solutions may be affected by pH (from 2.0 to 10.0), organic acids (citric, oxalic, phthalic, and salicylic acid), metal ions (K, Mg, Ca, Al, Fe), inorganic ligands (F, OH, SO_4, PO_4, CO_3, and SiO_3), and type of Al-containing solid [kaolinite, gibbsite, or amorphous $Al(OH)_3$] thought to be present. The simulation indicated that the type of Al-oxide/hydroxide considered has a substantial influence on the inorganic and organic equilibrium composition of the soil solution, and on the occurrence (or non-occurrence) of other Al-minerals such as $KAl_3(SO_4)_2(OH)_6$ (alunite) and $Al(SO_4)(OH)\cdot 5H_2O$ (jurbanite).

1. INTRODUCTION

The purpose of this paper is to illustrate how chemical equilibrium calculations can be employed to characterize the chemical speciation of Al^{3+} in aqueous solutions resembling those found in forest soils and in soil extracts derived therefrom. Emphasis is placed on examining the functional dependencies of Al solubility and Al speciation on (i) pH, (ii) the absence or presence of a particular mix of naturally occurring organic acids, (iii) the presence of metal ions other than Al^{3+}, and (iv) the type of Al(III) solids considered to be in equilibrium with the adjacent solution.

Florence and Batley (1980) and Ouimet (1986) reviewed a number of field studies dealing specifically with the chemical speciation of Al in soil solutions. In general, one finds that much of the Al present is complexed by organic ligands (see also Hendershot et al., 1984).

2. SOIL SOLUTION: A HYPOTHETICAL EXAMPLE

Chemical components and their total concentrations are listed in Table I for a hypothetical soil solution . This solution is meant to reflect the elemental composition of soil solutions normally encounter-

Water, Air, and Soil Pollution **31** (1986) 359–366.

ed with respect to the growing of plants under field, greenhouse, or growth chamber conditions. The choice of components and concentrations followed general guidelines suggested by (e.g.) Kabata-Pendias and Pendias (1984), Stumm and Morgan (1981), Lindsay (1979), Grausten et al. (1977), Ramamoorthy and Manning (1974), Stevenson (1967), and Fried and Broeshart (1967). Excluded from this list are components such as Mn^{2+}, $B(OH)_4^-$, Zn^{2+}, Cu^{2+}, and MoO_4^{2+} because (i) the total solution concentrations of these components tend to be at trace level, (ii) the overall mass balance and speciation of Al is affected by microcompon- ponents only in a minor way, and (iii) attention is focused on Al speciation. The overall ionic strength of the solution is allowed to vary above a minimum value (I=0.01 mol) in accordance with the total amount of solutes present, and by assuming that "indifferent" ions such as Na^+, NO_3^-, NH_4^+ are also present.

The organic composition of the solution in Table I in terms of citric, oxalic, phthalic, and salicylic acid is meant to be a simplified presentation of the organic component of soil solutions which tend to contain watersoluble substances such as polyuronic acids, fulvates, and a variety of simple organic acids.

3. COMPUTER SIMULATIONS

The MINEQL computer program (Westall et al., 1976) was used to calculate the equilibrium speciation for the solution in Table I as function of pH). Together, 15 species-forming macrocomponents were explicitly considered of which 5 were metal ions, and 9 were ligands. The remaining component was H^+. The list of chemical species resulting from these components contained a total of 131 entries of which 106 were watersoluble, and 25 were solids (Ouimet, 1986). There were, however, only a few species which -- by calculation -- attained appreciable equilibrium concentrations. These are also listed in Table I. The Al^{3+}-containing oxides/hydroxides considered were (i) kaolinite, (ii) gibbsite, and (iii) amorphous $Al(OH)_3$. The Fe^{3+} solid considered to be present was amorphous $Fe(OH)_3$ as appropriate for well-aerated solutions. Not all of the solids listed in Table I can coexist at equilibrium according to the phase rule. Thus, calculations with kaolinite, gibbsite and $Al(OH)_3$(am) were done separately from one another, but each in conjunction with SiO_2 (quartz).

4. RESULTS AND DISCUSSION

The effect of pH on the most important watersoluble inorganic and organic Al species is shown in Figure 1. In general, increasing the solution pH reduces most concentrations of the inorganic Al species except for $Al(OH)_4^-$. The dominant inorganic forms of watersoluble Al at low pH are Al^{3+} and $AlSO_4^+$, followed by $AlOH^{2+}$ and $Al(OH)_2^+$ at somewhat higher pH levels. Fluoro-Al occurs as AlF^{2+}, with its maximum concentration limited by the total component concentration for F. Note that the concentrations of the inorganic Al species plotted in Figure 1

Table I. Total component concentrations and dominant Al species for the hypothetical soil solution considered in the text. Important solids are also listed.

Components	Total concentrations (mol)	Dominant Al species	Formation constants [a]	Solids [c]	Solubility products at I=0
K^+	10^{-3}	AlF^{2+}	6.63	kaolinite	-67.7 [d]
Ca^{2+}	10^{-3}	$Al(OH)^{2+}$	-4.68	gibbsite	-34.6
Mg^{2+}	10^{-3}	$Al(OH)_2^+$	-9.57	$Al(OH)_3$(am)	-31.2
Al^{3+}	10^{-1} [b]	$Al(OH)_4^-$	-23.28	H_2SiO_3(s)	-25.4
Fe^{3+}	10^{-1} [b]	$AlSO_4^+$	2.67	$Fe(OH)_3$(am)	-39.0
F^-	$5x10^{-6}$			jurbanite	-17.8
CO_3^{2-}	$10^{-3.5}$	$AlCitH^+$	24.90	alunite	-85.6
SiO_3^{2-}	10^{-1} [b]	$AlCit^o$	22.00	basaluminite	-117.7
SO_4^{2-}	10^{-3}	$AlCitOH^-$	18.69		
PO_4^{2-}	10^{-5}	$AlOx^+$	7.21		
$Phth^{2-}$	10^{-4}	$Al(Ox)_2^-$	12.61		
Ox^{2-}	10^{-4}	$AlSal^+$	15.37		
Sal^{2-}	10^{-4}				
Cit^{2-}	10^{-4}				

a All values refer to an ionic strength of I=0.01 mol.
b Total concentration in solution when all Al, Fe or Si solids are dissolved.
c Other solids included in the MINEQL calculations: $AlPO_4$, $FePO_4$, $Ca(OH)_2$, $CaSO_4$, $Ca_5(PO_4)_3OH$, $Ca_{10}(PO_4)_6F_2$, $CaHPO_4$, $Ca_4H(PO_4)_3$, $CaSiO_3$, CaF_2, CaOx, $CaCO_3$, $Ca(OH)_2$, $Mg(OH)_2$, $Mg_3(PO_4)_2$, MgOx, MgF_2.
d Solubility product refers to $Al_2(SiO_3)_2(OH)_2 \cdot H_2O$.

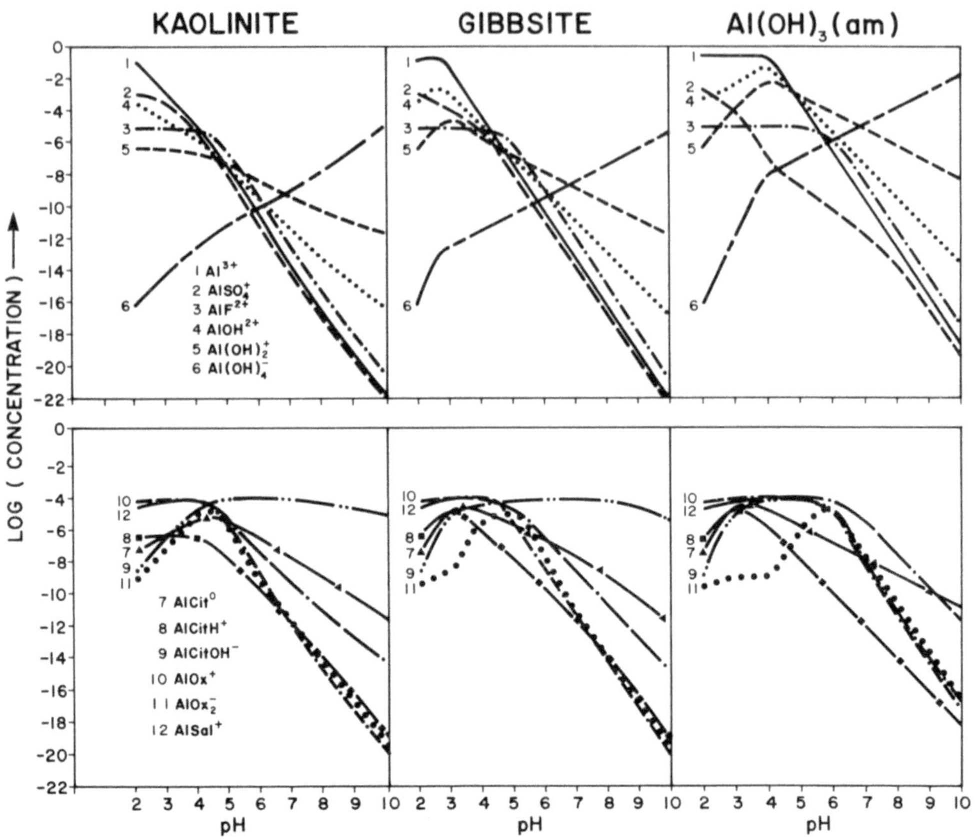

Figure 1. Equilibrium effects of pH and Al-containing solids [kaolinite, gibbsite and $Al(OH)_3$(am)] on organic and inorganic Al speciation in soil solution (Table I).

represent the inorganic Al speciation at equilibrium in the absence and presence of organic substances as long as there are finite amounts of Al solids in contact with the solution phase.

Organic Al species are to be found at maximum concentrations of ≈ 10^{-4} mol throughout the pH range from 2 to 10 (Figure 1). All the organic ligands considered contribute significantly to Al speciation except phthalate. In particular, $AlOx^+$ and $AlSal^+$ make up most of the soluble organic Al at pH<5.0. At pH>5.0, $AlCitOH^-$ dominate and represents up to 99% of the total organic Al. The decreased concentrations for all Al-Ox species at pH > 4.5 are in part due to the formation of solid calcium oxalate [CaOx(s)].

The type of Al solid considered to be in equilibrium with the aqueous solution should have a pronounced effect on Al solubility. Thus, total Al solubility (total inorganic plus total organic Al) increases in the order

$$\text{kaolinite} < \text{gibbsite} < Al(OH)_3(am)$$

as indicated in Figure 2.

The type of Al solid in equilibrium with the solution also appears to affect Fe(III) solubility (Figure 2). In particular, the formation of kaolinite should enhance the solubility of Fe due to a decreased Al^{3+} activity. The presence of gibbsite and $Al(OH)_3(am)$, in contrast, should render Al^{3+} more soluble, and therefore more competitive than Fe^{3+} for ligand binding.

Field measurements of soluble Al (mostly organic Al) in podzolic soils as reported by (e.g.) David and Driscoll (1984) and Johnson _et al_. (1968) suggest that the Al^{3+} activity in solution is possibly controlled by kaolinite. Consideration of the solubility products for various Al-SO_4 minerals, however, indicates that the solubility of soil Al should also be affected by $KAl_3(SO_4)_2(OH)_6$ (alunite) and $Al_4(OH)_{10}SO_4$ (basaluminite) (Wolt, 1981; Adams and Rawajfih, 1977) at moderate acidity, and by $Al(SO_4)(OH)\cdot 5H_2O$ (jurbanite) at strong acidity (Nordstrom, 1982). The MINEQL-calculations pertinant to the soil solution of this study revealed that alunite may indeed precipitate at (i) ≈ 3<pH<5 with gibbsite, and (ii) at ≈ 3<pH<7 with $Al(OH)_3(am)$ as dominant Al solid. In contrast, alunite may not precipitate from the solution when in equilibrium with kaolinite (Figure 3). The predicted extent of alunite formation, furthermore, would be reduced (i) by reducing the total solution concentration for SO_4 from 10^{-3} to 10^{-4} mol, and (ii) would be eliminated altogether by reducing the total solution concentration of both K and SO from 10^{-3} to 10^{-4} mol. Calculations with total SO ≥10^{-2} mol predicted the formation of alunite and also of jurbanite at ≈ 2.5≤pH≤3.5 (details are not shown here).

All of these predictions are critically dependent on the proper assessment of the corresponding solubility products. For example, resetting the solubility product for alunite from -85.6 to -83.4 (Nordstrom, 1982) in the MINEQL program predicted the formation of jurbanite instead of alunite at ≈2.5≤pH≤3.5 for the gibbsite case. In contrast, alunite was predicted to occur for the case of $Al(OH)_3(am)$ as before, but in conjunction with jurbanite precipitation at ≈ 2.5≤pH≤3.5.

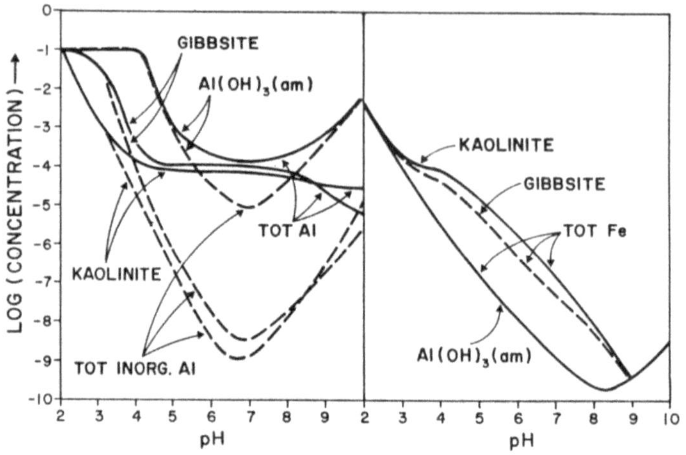

Figure 2. Equilibrium effects of pH and Al-containing solids [kaolinite, gibbsite, $Al(OH)_3$(am)] on total watersoluble Al in the presence of absence of organic ligands in the hypothetical soil solution of Table I (left). Also shown is total watersoluble Fe(III) (right) when organic ligands are present.

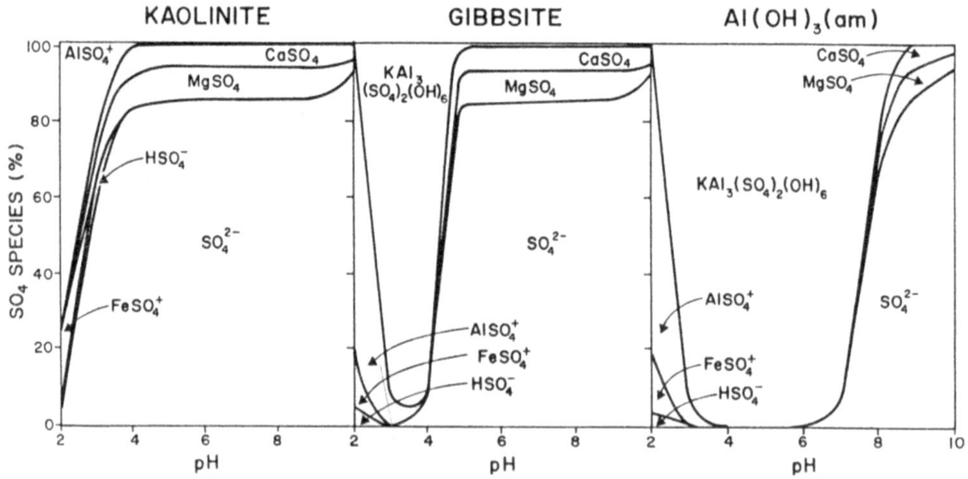

Figure 3. Equilibrium speciation of SO_4 (% of total SO_4) for the soil solution of Table I when in contact with kaolinite, gibbsite, or $Al(OH)_3$(am).

The prediction of possible alunite and jurbanite formation is further dependent on (i) the overall rates of dissolution and/or precipitation of the Al-containing solids, (ii) the rate of transformation from $Al(OH)_3(am)$ to gibbsite, (iii) and from gibbsite to kaolinite in the presence of quartz, and (iv) the occurrence of other less stable but slowly transforming Al sulfates (e.g. basaluminite to alunite). Rates of dissolution and/or precipitation of basic Al-sulfates tend to be slow often leading to over- or undersaturated solutions. For example, aluminous sulfates were not found in stream water that received Al- and SO_4- carrying water from an acid mine (pH<4.6, total SO_4>.01 mol, total Al>.0001 mol) (Nordstrom and Ball, 1986).

Overall, the eventual formation of alunite from soil solutions containing various inorganic and organic Al-complexing agents is possible, but subject to kinetic limitations. Actual formation of alunite should affect the extent of Al and SO_4 leaching from acidic soils, and may reduce the availability of K for plant uptake. These effects should be severe with $Al(OH)_3(am)$, moderately with gibbsite, and absent with kaolinite as dominant Al minerals and when the total solution concentrations for K and SO_4 range from $\approx 10^{-4}$ to 10^{-3}mol.

5. CONCLUDING REMARKS

Chemical equilibrium calculations only provide a gross picture of Al solubility in soils because soils are open heterogenous systems subject to many biological transformations and irreversible processes (Sposito, 1981). Also, the solid phases (organic or inorganic) in soils, as a rule, are impure. Furthermore, many soil reactions (in particular those leading to formation and transformation of soil minerals) proceed so slowly that a true soil equilibrium often cannot be attained. In addition, soil organics are in a constant state of flux on account of biological production and decay.

Soils, therefore, do not reflect equilibrium conditions. Yet it is possible to consider a "steady state" for some of those soil components that are known to react with one another on a short-term and nearest-neighbour basis. This steady state can be characterized by knowing (i) the soil solids that affect mineral solubility on a short-term basis, and (ii) the type and quantity of organic ligands that are present in macroamounts. For instance, the addition of, say, citric and oxalic acids to soils should (at least temporarily) transform some of the water insoluble Al and Fe into watersoluble Al and Fe (Bruckert and Jacquin, 1969). In fact, increasing (e.g.) the total citrate concentration from $1x10^{-4}$ to $2x10^{-4}$ mol for the soil solution considered here doubles most of the Al-Cit concentrations plotted in Figure 1.

Many other organic ligands (e.g. fulvates) present in soil solutions form additional complexes with Al^{3+} and Fe^{3+}. The extent of such complexation is also expected to depend on (i) the solution pH, (ii) the solubility products of the inorganic phases present, (iii) the type and concentration of organic ligands, and (iv) presence of metal ions other than Al^{3+} and Fe^{3+} which all interact with the available ligand pool. Chemical modeling offers a way to account for these factors, and allows for a general characterization of the chemical speciation and solubility of Al^{3+} and other metal ions in soil-water systems. All of this should prove to be a useful guide for assessing the actual speciation of Al and other chemical components (notably SO_4^{2-} and K^+) for field conditions as affected by soil acidification, acid precipitation, and changes in the amount and type of organic matter percolating through the soil.

ACKNOWLEDGMENTS

Financial support was received from the Canadian Forestry Service (PRUF Research Contract) and also from the Natural Science and Engineering Council of Canada (Grant # A7803).

REFERENCES

Adams, F. and Rawajfih, Z.: 1977, Soil Sci. Soc. Am. J. 41, 686.
Bruckert, S. and Jacquin, F.: 1969, Soil Biol. Biochem. 1, 275.
David, M.B. and Driscoll, C.T.: 1984, Geoderma 33, 297.
Florence, T.M. and Batley, G.E.: 1980, Crit. Rev. Anal. Chem. 9, 219.
Fried, M. and Broeshart, H.: 1967, The Soil-Plant System in Relation to Inorganic Nutrition, Academic Press, New-York and London, 358 pp.
Grausten, W.C., Cromack, Jr., K. and Sollins, P.: 1977, Sci. 198, 1252.
Hendershot, W.H., Lalonde, H. and Dufresne, A.: 1984, Water Poll. Res. J. Can. 19, 11.
Johnson, N.M., Likens, G.F., Bormann, F.H., and Pierce, R.S.: 1968, Soil Sci. Soc. Amer. Proc. 28, 381.
Kabata-Pendias, A. and Pendias, H.: 1984, Trace Elements in Soils and Plants, CRC Press, Fa., 315 pp.
Lindsay, W.L.: 1979, Chemical Equilibrium in Soils, Wiley, New York, N.Y., 448 pp.
Nordstrom, D.K.: 1982, Geochimica et Cosmochimica Acta 46, 681.
Nordstrom, D.K. and Ball, J.W.: 1986, Sci. 232, 54.
Ouimet, R.: 1986, MScF thesis, U. New Brunswick, N.B., Canada, 200 pp.
Ramamoorthy, S. and Manning, P.G.: 1974, Nucl. Chem. Letters 10, 623.
Sposito, G.: 1981, The Thermodynamics of Soil Solutions, Oxford Clarendon Press, 223 pp.
Stevenson, F.J.: 1967, 'Organic acids in soil'. Pp. 119-146 In McLaren, A.D. and G.H. Peterson, eds. Soil Biochemistry, Marcel Dekker, New York, N.Y., 509 pp.
Stumm, W. and Morgan, J.J.: 1981, Aquatic Chemistry. An Introduction Emphasizing Chemical Equilibria in Natural Waters, Wiley Interscience, 780 pp.
Westall, J.C., J.L. Zachary and Morel, F.M.M.: 1976, MINEQL: a Computer Program for the Calculation of Chemical Equilibrium Composition of Aqueous Systems, Water Quality Laboratory, Ralph M. Parsons Laboratory for Water Resources and Environmental Engineering, Dept. of Civil Eng., Mass, Inst. Technol., Tech. Note No. 18. Cambridge, Mass., 91 pp.
Wolt, J.D.: 1981, Soil Sci. Am. Soc. J. 45, 283.

UPTAKE OF AL, CA, AND P IN BLACK SPRUCE SEEDLINGS: EFFECT OF ORGANIC VERSUS INORGANIC AL IN NUTRIENT SOLUTIONS.

Paul A. Arp and Rock Ouimet
University of New Brunswick
Fredericton, NB, Canada E3B 6C2

ABSTRACT. Two-year old black spruce seedlings (Picea mariana [Mill.] B.S.P.) of greenhouse-grown paper-pot stock were subjected to chemically well-characterized nutrient solutions for 28 days to assess the elemental uptake (Al,P,Ca) of these plants in response to organic versus inorganic Al in the rooting medium (pH=3.0; 0 ≤ total Al ≤ 48 mg L^{-1}). Oxalate additions to the nutrient solutions (0 ≤ Ox ≤ 2.4 mmol) served as organic Al-complexing agent. The results indicated that the plants took up Al in proportion to the Al concentration of the rooting medium, with Al uptake from the Al-Ox treatments somewhat more extensive than the Al uptake from the inorganic Al treatments. Furthermore, root Al >> shoot Al for both cases. The pattern of P uptake was similar to that of Al uptake but for the roots only, i.e. root P was proportional to root Al. Increased root P was not associated with increased shoot P. Calcium content of the roots was slightly reduced with increased inorganic and organic Al, but increased strongly with increasing oxalate in the rooting medium.

1. INTRODUCTION

Some suggestions have been made to classify the chemical speciation of Al in soils and surface waters in terms of (i) "non-labile" Al (which is thought to be complexed with organic ligands), (ii) "labile" Al (mostly complexed with inorganic ligands), and (iii) colloidal organic and inorganic polymeric Al extractable only by strong acid digestion (Driscoll, 1984). This classification, however, is not entirely satisfactory because some organic Al may be available for plant uptake, and should, therefore, be part of the "labile" Al in the soil. For instance, VanPraag et al. (1985) reported that Al-citrate promotes Al uptake by Norway spruce beyond the Al uptake from inorganic Al in soil solutions. In contrast, uncomplexed trivalent ions for Al, Cr, Ga, and In are taken up by mustard and tomato plants, but only in trace amounts (DeKock and Mitchell, 1957). The same ions, however, are readily taken up when chelated with organic ligands. Adams and Hatchcock (1984) reported that Al toxicity of cotton occurs in some

Water, Air, and Soil Pollution 31 (1986) 367–375.

soils containing ≤ 0.4 µmol of Al in solution, but not in others with up to 11.5 µmol of "labile" Al.

The objective of the work reported in this paper is to examine the effects of watersoluble organic vs inorganic Al on the uptake of Al, P, and Ca by black spruce (Picea mariana [Mill] B.S.P.) seedlings subjected to a chemically well-characterized nutrient solution as rooting medium. The nutrient solutions were treated with various levels of Al and oxalate (the only Al binding reagent employed here) such that (i) inorganic Al, (ii) organic Al and (iii) oxalate became the dominant experimental factors.

2. MATERIALS AND METHODS

The experiment consisted of six levels of total Al (0 to 48 mg L^{-1}) and six levels of oxalate (0 to 2.4 mmol) as shown in Table I. The composition of the basic nutrient solution consisted of a modified Ingestad (1962) nutrient solution for spruce having (in mg L^{-1}): 25 $N\text{-}NH_4$, 25 $N\text{-}NO_3$, 10 P, 50 K, 40 Ca, 12 Mg, 64 S, 1 Fe-EDTA, 0.17 B, 0.002 Zn, 0.002 Cu, and 0.003 Mo.

Aluminum and oxalate were added to the nutrient solutions as $AlCl_3 \cdot 6H_2O$ and oxalic acid. The solution pH was then adjusted to 3.0 using NaOH or HCl. The experiment was conducted in a greenhouse with 2-year old seedlings of greenhouse-grown paper-pot stock (parentage not common). The seedlings were grown bare-root in non-treated nutrient solutions for 10 days following their removal from the paper-pots. Three groups of four seedlings were assigned to each treatment, with the roots of each group placed into an opaque bottle containing 250 ml of treated nutrient solution. Each treatment lasted up to 28 days, with all treatment solutions fully replaced on every second or third day.

The greenhouse conditions at the time of the experiment were variable, with hot sunny days prevailing (inside day temperatures ≈ 20 to 35 C) followed by cool nights (night temperatures ≈ 0 to 10 C).

Shoot and root clippings were taken from plants that remained intact at the end of the experiment. The plant material so collected was oven-dried (105 C for 24 hrs) and ashed (500 C for 2 hrs). The resulting ash was acid-dissolved and analyzed for Al and Ca by atomic absorption spectrophotometry, and for P by colorimetry (McKeague, 1978).

The MINEQL computer program of Westall et al. (1976) was used to calculate the organic and inorganic Al speciation in each treatment solution for a temperature of 20 C, an atmospheric pressure of 1 atm, and an ionic strength of 0.02 mol. The calculations indicated that no precipitation of Al should occur in any treatment solution. The organic Al species, in theory, were present at about 82 to 90% of total Al in the solutions treated with Al and Ox (Figure 1). The dominant Al species (see Table II) were $AlOxH_2PO_4$, AlOx, and $Al(Ox)_2$. Free Ox (Ox and HOx) ranged from 5 to 44 µmol. The Al and AlH_2PO_4 species were the dominant Al species for the Al treated solutions with no Ox.

Table I. Aluminum and oxalate levels used to assess the effects of organic and inorganic Al in nutrient solutions on black spruce seedlings.

		Total Al concentration ($mg\ L^{-1}$)					
		0	4	8	16	27	48
Total Ox concentration (mmol)	0.00	X	X	X	X	X	X
	0.15	X	X				
	0.30	X		X			
	0.60	X			X		
	1.20	X				X	
	2.40	X					X

Table II. Dominant Al species in the nutrient solutions.

Al species	Formation constants (I = 0.0)	Reference
Al^{3+}	----	----
$AlH_2PO_4^{2+}$	23.2	Larsen (1967)
$AlOx^{+}$	7.7	Martell and Smith (1976)
$AlOx_2^{-}$	13.3	Martell and Smith (1976)
$AlOxH_2PO_4^{o}$	31.6	Arp and Meyer (1985)

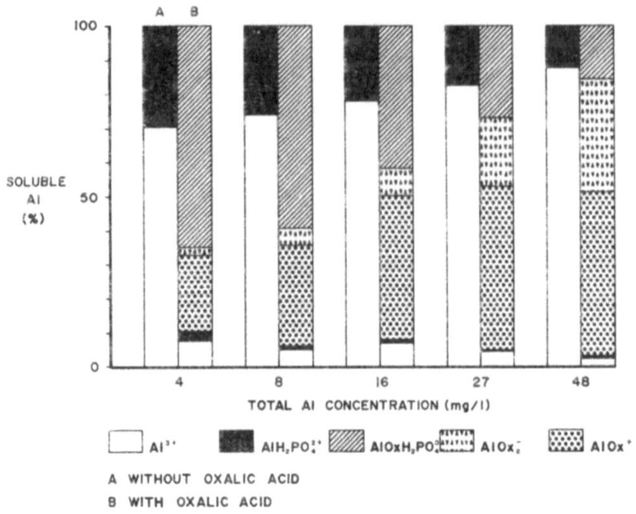

Figure 1. Chemical speciation of Al in the treatment solutions in absence (A) and in presence (B) of oxalate as determined by chemical equilibrium calculations.

Preliminary testing of the nutrient solutions revealed the following:

(1) The pH values of the solutions prior to each replacement tended to be higher than 3.0, with the oxalate treatments showing the largest increase, and the inorganic Al treatments the least.

(2) Increased pH values were related to a loss of oxalate as revealed by HPLC ion-exchange analysis. This loss was presumably due to the reaction

$$(COO^-)_2 + \tfrac{1}{2}O_2 + H_2O \rightarrow 2HCO_3^- .$$

(3) Loss of oxalate was diminished by adding Captan (a fungicide with antioxidant properties) to the treatment solutions (a one-time application)) at a rate of 12 mg L^{-1} of active ingredient.

(4) The Captan additions did not affect the growth of the seedlings. In fact, black spruce seedlings are adapted to grow on poorly-drained soils where root respiration is restricted by low rates of oxygen supply (Fowells, 1965).

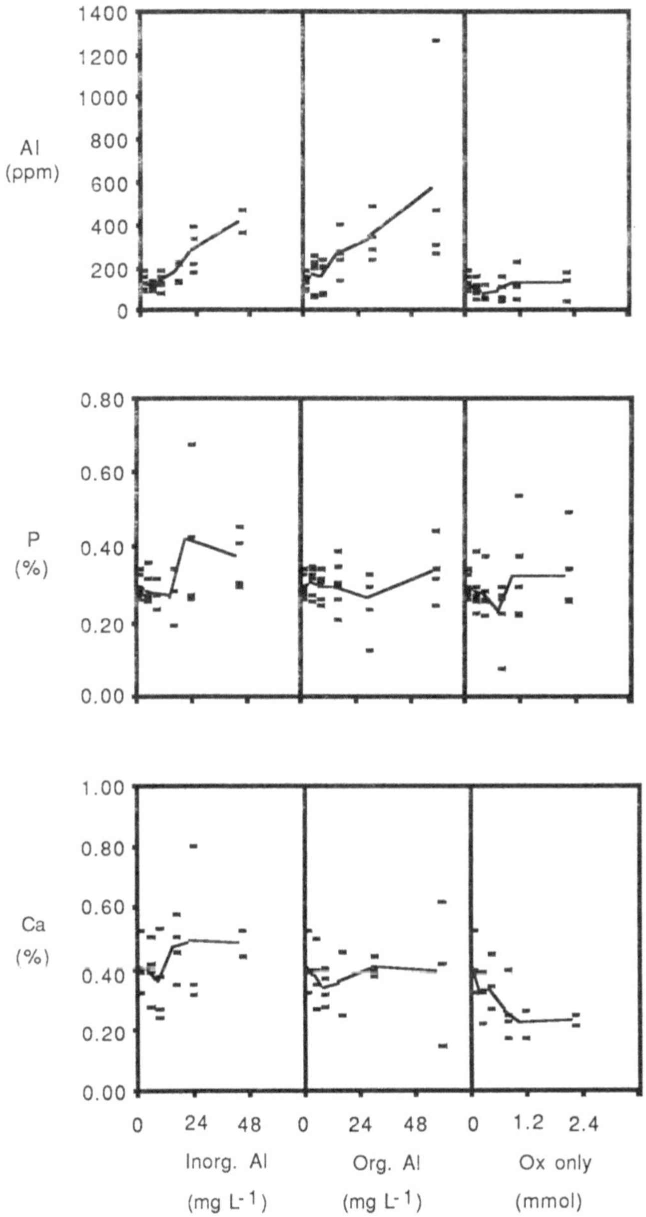

Figure 2. Effect of inorganic and organic Al and of oxalate in nutrient solutions on the uptake of Al, P and Ca in black spruce seedlings (shoots only). Actual observations are shown as dots. The plotted lines connect the means at each treatment level.

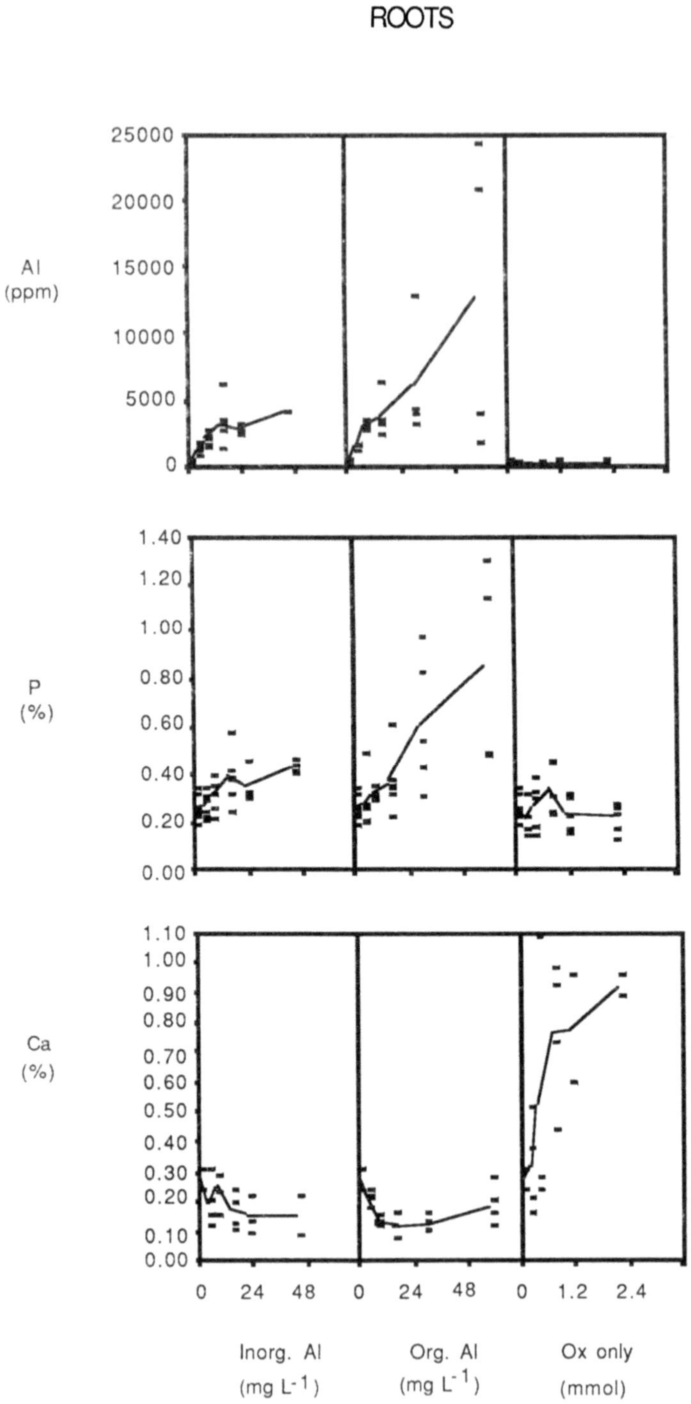

Figure 3. Effect of inorganic and organic Al and of oxalate in nutrient solutions on the uptake of Al, P and Ca in black spruce seedlings (roots only). Actual observations are shown as dots. The plotted lines connect the means at each treatment level.

3. RESULTS

Some of the plants suffered from shoot dehydration at peak greenhouse temperatures. Many of the plants so affected recovered but others -- notably those treated with high inorganic Al in the nutrient solutions -- continued to dehydrate completely.

The uptake of Al, P, and Ca in those plants that remained intact fluctuated considerably against increased intensity of the Al, Ox, and AlOx treatments as seen in Figures 2 and 3. Nevertheless, a number of trends are apparent.

(1) Aluminum uptake increased with increasing Al in the nutrient solutions, with Al uptake from the solutions containing organic Al somewhat more extensive (and also more variable) than Al uptake from the solutions containing inorganic Al.

(2) Most of the Al taken up remained in the roots, but some of it was transported to the shoots.

(3) The Al build-up in the roots lead to an increased P uptake. Most of the phosphorus so taken up, however, remained in the roots. Shoot P was unaffected by all treatments.

(4) Calcium uptake was slightly reduced by increased concentrations of inorganic and organic Al in the rooting medium. Additions of oxalate to the nutrient solutions, however, led to strong Ca binding in the roots presumably due to the precipitation of CaOx(s) on or within the root cortex. Root clippings taken from the plants treated with oxalate (or Al and oxalate) turned white after drying.

(5) Increased binding of Ca in the roots led to a slight reduction of shoot Ca.

4. DISCUSSION

The results indicate that black spruce seedlings are sensitive to inorganic as well as organic Al in the rooting medium. Hence, symptoms of Al toxicity may not only occur in soil/nutrient solutions with $pH \leq 4.2$ (when Al oxides/hydroxides become soluble) but also in soil/nutrient solutions with $pH > 4.2$ whenever organic ligands are present in appreciable concentrations. The oxalate ligand chosen here should be representative of other non-colloidal organic ligands (citrate, malonate, acetate, etc.) which may also occur in soil/nutrient solutions at comparable concentrations. The presence of colloidal organic ligands (fulvates, polyuronic acids, etc.), in contrast, is likely to reduce Al uptake by the plants on account of Al binding to a colloidal phase. Hence, one should categorize watersoluble organic Al in terms of colloidal and non-colloidal organic Al.

Similar work on spruce seedlings by Murach (1984), Persson (1983), and Rost-Siebert (1983) has shown that Al is retained by the cell walls of the endodermis, and may block the uptake of other elements. Furthermore, when Al is complexed by organic molecules, then the uptake of other elements is not blocked. This work shows that a small fraction of the Al taken up by the roots is transported to the shoots (see Figure 2 and 3). Furthermore, complexation of Al with oxalate enhances Al uptake which -- in turn -- increases root-binding of P.

The presence of Ox in the nutrient solution leads to special effects on Ca uptake due to Ca binding in or on the roots. In particular, the reducing effect of inorganic Al on Ca uptake (Clarkson and Sanderson, 1971) is probably counteracted by enhanced CaOx(s) formation in or on the roots when Ox is present. Oxalate, therefore, interferes with the competitive effect of Al on Ca uptake and may thus lead to an overall reduction of Ca transport from the roots to the shoots.

It should be remembered that the experiment represents only 28 days of plant conditioning to enhanced organic and inorganic Al in nutrient solutions. There is no doubt that some further fraction of the Al, P and Ca accumulated in the roots will be transported to the shoots due to slow but persistent plant-internal Al complexation. Furthermore, symptoms of Al toxicity (reduced root growth, reduced uptake of Ca and Mg ions, increased fixation of P in the roots) will become severe with increasing length of experimentation.

Some of the variability observed for root and shoot Al, P and Ca is most likely due to the high genetic variability of the plant material selected for experimentation. Experiments with plants with identical genes should greatly reduce the variability of the experimental results. The conclusions to be drawn from doing so, however, will not necessarily reflect the overall population response.

ACKNOWLEDGMENTS

Financial support was received from the Canadian Forestry Service (PRUF Contract). Special thanks are due to L. Cooper, J. Manasc, D. Fournier, J. Beck, W.L. Meyer, H. D'Avignon and N. Kreiberg, and to the reviewers of this paper.

REFERENCES

Adams, F and Hatchcock, P.J.: 1984, Soil Sci. Soc. Amer. J. **48**, 1305.

Arp, P.A. and Meyer, W.L.: 1985, Can. J. Chem. **63**, 3357.

Clarkson, D.C. and Sanderson, J.: 1971, Soil Sci. **22**, 267.

DeKock, P.C. and Mitchell, R.L.: 1957, Soil Sci. **84**, 55.

Driscoll, C.T.: 1984, Int. J. Environ. Anal. Chem. **16**, 267.

Fowells, H.A., 1965: 'Silvics of forst trees of the United States'. Agric. Handbook No. 271, US Dept. Agric., Washington, D.C. 20250, 762 pp.

Ingestad, T.: 1962, 'Macro-element nutrition of pine, spruce and birch seedlings in nutrient solutions', Medd. Stat. Skogsf. Inst. 51-150.

Larsen, S.: 1967, Adv. Agron. **19**, 151.

Martell, A.E. and Smith, R.M. :1976, Critical Stability Constants. Volume 3: Other Organic Ligands, Plenum Press, New York, N.Y., 495 pp.

McKeague, J.A.: 1978, 'Manual on soil sampling and methods of analysis'. 2. ed. Can Soc. Soil Sci., 211 pp.

Murach, D.: 1983,Allg. Forstztg. **38**: 639.
Persson, H.: 1983, Plant Soil **71**, 87
Rost-Siebert, K.: 1983, Allg. Forst. Zeitschr. **27**, 686.
VanPraag, H.J., Weissen, F., Sougnez-Remy, S., and Carletti, G.: 1985, Plant Soil **83**, 339.
Westall, J.C., Zachary, J.L., and Morel, F.M.M.: 1976, MINEQL: a Computer Program for the Calculation of Chemical Equilibrium Composition of Aqueous Systems Water Quality Laboratory, Ralph M. Parsons Laboratory for Resources and Environmental Engineering, Dept. of Civil Eng., Mass. Inst. Technol., Tech. Note No. 18, Cambridge, Mass., 91 pp.

POTENTIAL LEACHING OF ELEMENTS IN THREE DANISH SPRUCE FOREST SOILS

Lennart Rasmussen
Laboratory of Environmental Sciences and Ecology
Technical University of Denmark, Building 224
DK-2800 Lyngby, Denmark

ABSTRACT. On three localities in Denmark, all with 40 to 50 yr old Norway spruce plantations, but different soil chemistry, ion balance studies have been implemented in 1983. The annual variations in soil water concentrations of cations and anions are described and the potential annual leaching fluxes are determined by means of zero-tension percolation lysimeters at 2 depths (A- and B-horizon). Leaching of Al and and heavy metals was much more severe on the acidic sites, whereas especially Ca was leached from the nutrient rich site. In parallel with the high concentrations of Al and heavy metals, great needle loss was observed on the two acidic sites. The maximum leaching from the A-horizon was obtained either in the late summer, associated with the effluent of organic material, or in the winter due to large amounts of percolation water. Apart from low Ca and Mn levels of one of the sites, the findings from the two acidic localities were comparable to similar studies in Sweden.

1. INTRODUCTION

In the winter 1982/83 the first observations of a forest dieback, more severe than usually seen in Denmark under extreme climatic conditions, were reported in western Jutland. Since then, forest damage has been observed on poor soils all over the country, but still most pronounced in the western parts. However, neither data on the soil chemical status nor sufficient atmospheric chemistry data have been available from the areas in concern.

On that background ion balance studies were implemented on 3 localities, namely The Klosterhede Plantation, western Jutland; The Strødam Reserve (Gribskov Forest), northern Zealand; and The Tange Plantation, central Jutland. All 3 plantations consist of 40 to 50 yr old Norway spruce (Picea abies (L.) Karsten) growing on sandy soils, but the soil chemical status is very different for the 3 localities, ranging from very acidic to almost neutral.

In order to produce comparable results to similar studies in Sweden, West Germany, The Netherlands and USA, sample collection equipment

Water, Air, and Soil Pollution **31** (1986) 377–383.

and analytical techniques were attempted to be standardized, even in cases where it means double or triple instrumentation in the field (Bergkvist, 1983; Ulrich and Matzner, 1983; van Breemen and Jordan, 1983; Cronan, 1980).

The main objectives of this paper are to (1) describe the soil water concentrations of cations and anions on the 3 different localities, (2) describe the annual variations in soil solution concentrations, (3) estimate the potential annual leaching of cations and anions, determined by the use of lysimeters of the zero-tension percolation type, and (4) compare the findings with similar studies in Sweden.

2. STUDY AREAS

The soil characteristics of the 3 localities are presented in Table I.

TABLE I. Soil characteristics.

LOCALITY	Depth cm	pH_{H_2O}	pH_{CaCl_2}	Loss on ignition %	Clay %	Silt %	Fine sand %	Coarse sand %	Base saturation %	Ca/Al-mol in soil Water
Klosterhede	5	3.8	2.7	57					31	
	15	3.9	3.1	2.5	4.0	5.4	41.0	49.6	5	0.08
	45	4.5	4.4	0.8	2.0	4.8	35.6	57.6	14	
	55	4.6	4.6	0.4					22	0.07
	100	4.7	4.7		0.0	2.0	12.0	86.0	22	
Strødam	5	3.7	3.1	4.3					18	
	15	3.7	3.0	3.8	4.0	5.4	35.2	55.4	6	0.26
	45	4.0	3.7	3.1	2.4	8.8	36.8	52.0	6	
	55	4.1	3.8	3.0					8	0.23
	87	4.4	4.2		2.0	2.8	36.6	58.6	19	
Tange	5	4.3	3.6	5.2					36	
	15	4.4	3.7	2.6	0.0	2.0	14.6	80.4	18	13
	45	5.2	4.7	1.8	1.6	3.0	0.2	95.2	28	
	55	5.1	4.6	1.6					21	19
	110	5.0	4.5		0.0	0.4	4.0	95.6	31	

All stations are located close to weather stations and monitoring stations for atmospheric heavy metal deposition. The locality at Tange is located at the EMEP-station run by The National EPA. Mean annual SO_2 concentrations in the air here is 9 $\mu g\ m^{-3}$, and 24 hr means are occasionally over 100 $\mu g\ m^{-3}$ (Heidam, 1985). Annual mean pH in bulk precipitation at Strødam is 4.1. Ion content in precipitation and throughfall is presented by Freiesleben et al. (1986).

The Klosterhede Plantation is growing on a very nutrient poor podzolized heathplain formed during the last glacial age.

The Tange Plantation is located on Ca rich, but sandy, fluvial deposits along the river Gudenåen, which were formed during the ice melting period. No obvious B-horizon is developed.

The Strødam Reserve plantation is situated on a sandy glacial age deposit. The soil is an acidified brown soil, still containing some Ca. This site is only 5 km NW of the town of Hillerød with 34,000 inhabitants, and 30 km NW of the Copenhagen area. In contrast, the two localities in Jutland can be looked upon as background stations as regards air pollution from internal Danish emission sources.

3. MATERIAL AND METHODS

On each station 6 zero-tension percolation (soil column) lysimeters were mounted, 3 collecting soil water just under the A-horizon (25 to 30 cm) and 3 just under the B-horizon (55 to 60 cm). The percolation lysimeter consists of a plexiglass cylinder (diameter 292 mm) with a perforated bottom plate. The percolating solution was collected in a 10 L polyethylene flask, closed by a polyethylene membrane, which was perforated automatically by the drainage funnel of the lysimeter. A soil core of the same diameter as the lysimeter was punched out with a stainless steel cylinder and placed with a minimum of disturbance in the plexiglass cylinder. A more detailed description of the lysimeter installation is given by Rasmussen _et al._ (1986).

The use of this type of lysimeter prevents the natural tree root uptake of elements, a situation comparable to clear cutting of a forest stand. Without the competition from the tree roots, the forest floor grass and herb vegetation - if present - will develop better.

In this study forest floor vegetation of herbs and mosses was undisturbed. The following species were found at Klosterhede: _Chamaenerion angustifolium_ (L.) Scop., _Senecio sylvaticus_ L., _Deschampsia flexuosa_ (L.) Trin., _Dicranum scoparium_ Hedw., _Hypnum cupressiforme_ Hedw., _Pleurozium schreberi_ (Brid.) Mitt. At Tange: _Chamaenerion angustifolium_ (L.) Scop. and _Lactuca muralis_ (L.) Gaertner (Hansen, 1981; Andersen _et al._, 1976). At The Strødam Reserve no forest floor vegetation existed.

The percolates were analyzed monthly for Al, Ca, Mg, K, Na, Fe, Mn, Cd, Ni and Zn by atomic absorption techniques and for NO_3^-, Cl^- and SO_4^{2-} by ion chromatography. NH_4^+ was determined by an ammonia electrode.

4. RESULTS AND DISCUSSION

4.1. Comparison between the 3 localities

The mean soil solution concentrations of elements and pH are presented in Table II. Tange is the most nutrient rich site with relatively high values in pH, Ca and Mg, and low values in Al and heavy metals. No obvious B-horizon is developed. The area was ploughed and the upper soil turned upside down 50 yr ago before the planting of trees. This treatment is reflected in the lowest pH values in the B-horizon. However, there are no large differences in soil water concentrations between the A- and B-horizon. An exception is the low NH_4^+ concentration in the deep soil layer indicating NH_4^+ sorption and/or nitrification in the upper soil layers. The increasing NO_3^- concentration with the depth indicate nitrification. The relatively low NH_4^+ concentrations in both soil layers compared to the other localities also indicate a higher nitrification on this less acidified soil.

TABLE II. Mean soil solution concentrations and S.D. (in brackets) in A- and B-horizons. n=28-45.

		$mg \cdot L^{-1}$												$\mu g \cdot L^{-1}$		
		pH	Al	Ca	Mg	K	Na	Fe	Mn	Cl	NO_3-N	NH_4-N	SO_4-S	Cd	Ni	Zn
Klosterhede	A	3.8	6.9	0.86	2.4	7.6	23	7.8	0.03	31.5	0.42	8.3	6.6	2.9	3.3	82
		(0.2)	(2.3)	(0.52)	(1.4)	(3.3)	(8)	(5.0)	(0.01)	(14.6)	(0.66)	(3.2)	(2.9)	(2.0)	(1.6)	(41)
	B	4.5	6.0	0.63	2.8	3.4	31	0.05	0.02	54.3	1.1	7.0	11.7	7.4	2.8	9.6
		(0.2)	(2.5)	(0.75)	(1.9)	(2.0)	(13)	(0.05)	(0.01)	(31.2)	(0.8)	(1.4)	(3.4)	(4.1)	(1.7)	(11.0)
Strødam	A	3.8	11.1	4.3	1.6	7.2	9	4.3	0.10	12.4	11.6	15.3	11.6	3.0	5.4	129
		(0.2)	(3.9)	(2.1)	(0.8)	(1.9)	(2)	(4.1)	(0.05)	(3.7)	(9.5)	(6.9)	(3.4)	(1.8)	(2.3)	(52)
	B	4.0	18.2	6.1	2.8	4.4	13	0.16	0.37	18.8	22.4	6.7	18.9	8.0	12.1	295
		(0.1)	(6.7)	(1.7)	(1.0)	(1.4)	(3)	(0.13)	(0.24)	(6.4)	(11.5)	(3.2)	(5.8)	(3.3)	(3.7)	(82)
Tange	A	5.2	2.5	49	7.7	8.6	28	0.28	0.76	30.9	23.8	2.5	11.9	3.0	3.6	31
		(0.7)	(2.6)	(16)	(3.1)	(3.9)	(9)	(0.37)	(0.97)	(27.3)	(10.2)	(5.3)	(5.6)	(1.5)	(3.7)	(37)
	B	4.7	1.9	53	6.1	1.0	23	0.13	0.12	32.3	27.4	0.2	11.6	5.0	2.3	28
		(0.3)	(2.0)	(15)	(3.0)	(0.7)	(5)	(0.15)	(0.12)	(23.8)	(9.4)	(0.3)	(2.7)	(2.6)	(2.4)	(26)

In the more acidic soils at the two other sites, the levels of nutrients and pH in the soil solutions are much lower than at Tange, whereas Al and heavy metal concentrations are higher. The Klosterhede site is especially poor in Ca and Mn, and in the outflow from the B-horizon also Fe. NH_4^+ is almost the only available form of N. In comparison with Ca, the Mg concentration is remarkably high on this poor soil. This finding might be ascribed to the close distance (15 km) to the North Sea. The pH at Klosterhede is less than 4.0 in the A-horizon, but rises to about 4.5 before the lower B-horizon. At the Strødam site the mean pH in the A-horizon is 3.8, and in the B-horizon 4.0. However, in spite of the high Al concentrations, there is also some Ca in the soil solution.

In the A-horizon where soil water solutions become brown/black in summer, a major part of the elements is associated with the leaching of organic matter, especially for Al, where up to 97% organic bound Al has been reported by Nilsson and Bergkvist (1983). Most of the dissolved organic material is precipitated in the B_h-horizon in the acidic soils, associated with elements like Al, Fe and Ni (Tyler et al., 1983). From the lower B-horizon the elements are mostly leached as inorganic ions or salts, and some of them in even higher concentrations than from the A-horizon. In the case of Al and Cd at Klosterhede and Strødam, toxic levels for tree roots are reached when comparing with laboratory experiments (Rost-Siebert, 1983; Godbold and Hüttermann, 1985).

Summing up, the soil acidification at Strødam seems more advanced than at the Klosterhede site. However, there is still some Ca left in the soil at Strødam, which is almost not the case at Klosterhede. Since Mn is an easily leached element (Freiesleben and Rasmussen, 1986), the low concentrations on the acidified sites indicate a rather advanced soil acidification. This is in agreement with the fact that the most pronounced needle loss is observed at the Klosterhede site followed by the Strødam site.

4.2. Annual variation in concentrations

Since the tree root uptake of water and nutrients was prevented by the use of the percolation lysimeters, it was expected that the concentrations of nutrients in the soil water solutions would rise during the summer due to increased mineralization. This picture was obvious for most elements, especially in the upper soil. Increased mineralization would also make the pH increase, but on the other hand, the increased nitrification during the summer would make it decrease. Nitrification seemed to be important both at Tange and Strødam, whereas the very little N available at Klosterhede might be taken up by grasses and herbs in the form of NH_4^+.

Taking into account that the percolation of water is greatest in winter, the element fluxes reach a winter maximum, even in cases where soil water concentrations are relatively low.

4.3. Annual leaching of water and ions

In the sandy soils in question, with a large hydrological conductivity the annual amounts of water percolating through the soil profile (at least the B-horizon) in lysimeters are believed not to differ very much from the natural water flux. That is because most of the water is passing very fast through the soil, especially during winter, where root uptake is minimal anyhow. In the summer, both evaporation from the soil and evapotranspiration from the herbs in the lysimeters reduce water percolation to almost zero. The percentages of precipitation which percolate through the B-horizons at Klosterhede, Strødam and Tange are 55%, 55% and 43%, respectively (Table III). Tange is the site with the most developed herb vegetation, which is reflected in the smallest amount of percolation water. This observation is even more obvious in the A-horizon where the percolation at Klosterhede, Strødam and Tange is 41%, 55% and 25% respectively. Notice that the Strødam site is without forest floor vegetation and here 55% of the open field precipitation passes both the A- and B-horizon. Since the bottom plates of the zero-tension lysimeters cut off the capillary watercolumns, some water is retained for a short period just above the bottom plate. Although this situation will not produce anoxic conditions in sandy soils the result is still a greater water loss through transpiration from the A-horizon lysimeters. Therefore chemical fluxes through the B-horizon are believed to be the most representative for the <u>in situ</u> conditions.

The over all picture of leached elements on the two acidic vs. the more neutral site (Table III) is the same as reflected in the concentration values described in paragraph 4.1, but a few interesting findings can be emphasized. At Klosterhede the annual leaching of Ca and Mn is extremely poor. However, although needle analyses show low values for those elements, there should be no deficiency as an explanation to the observed needle loss. The proximity to the North Sea is reflected in high Na, Cl and Mg fluxes. The most remarkable finding at the other acidic site, Strødam, is the substantial leaching of Al and heavy metals from the B-horizon, except Fe. In contrast, Al and heavy metals are hardly mobilized in the nutrient rich Tange soil, whereas extremely high amounts of Ca are leached. Taking into account that annual throughfall of S in spruce forest in Denmark is in the range 40 to 50 kg ha^{-1} (Freiesleben et al., 1986), S seems to accumulate in the A- or upper B-horizon on the two acidic localities. On the other hand, the relatively low S concentrations in the soil solution in the A-horizon may also be due to an analytical artifact, since our tests show that in solutions with a high organic content, ion chromatography technique may give up to 50% lower values than autoanalyzer technique. Nitrification and mineralization of N-compounds seems to be most important on the base rich site at Tange with higher pH. From a watershed study Bormann and Likens (1979) reported a maximum outlet of 140 kg NO_3-N $ha^{-1}yr^{-1}$ when the forest was clearcut and any regrowth of trees and herbs was prevented with herbicide treatment. Taking the presence of a forest floor vegetation into account in this study and the smaller annual precipitation, a potential leaching of 92 kg NO_3-N $ha^{-1}yr^{-1}$ as found from the similar Tange soil seems quite comparable. In contrast, a N deficiency may most likely occur at Klosterhede, since the N demand of the trees is expected to be 20-100 kg $ha^{-1}yr^{-1}$ (Nihlgård, 1985).

TABLE III. Annual leaching of elements from the A- and B-horizon.

Locality		Precipitation mm	Percolate mm	pH mean	kg · ha^{-1}											g · ha^{-1}		
					Al	Ca	Mg	K	Na	Fe	Mn	Cl	NO_3-N	NH_4-N	SO_4-S	Cd	Ni	Zn
Klosterhede DK		788																
	A		320	3.73	20	3.3	9.7	29	83	23	0.10	80	1.6	20	15	9.3	11	305
	B		435	4.44	28	3.1	13	13	140	0.2	0.07	230	4.8	32	62	38	12	47
Strødam DK		551																
	A		303	3.89	32	13	6.2	22	28	8.5	0.32	33	38	43	32	10	17	420
	B		301	4.07	46	17	10	13	44	0.4	1.04	44	59	15	44	24	34	830
Tange DK		640																
	A		163	5.70	1.3	85	13	11	46	0.2	0.20	74	70	3.8	37	4.8	1.5	20
	B		275	4.70	4.9	135	16	1.9	73	0.2	0.25	86	92	0.9	38	12	5.7	75
Gårdsjön S	A			3.98	8.2	12	10	44	53	6.8	4.4					7.5	15	260
	B			4.51	28	20	20	34	100	3.1	9.6					7.1	36	590
Skånes Värsjö S	A			4.07	9.7	8.3	4.7	39	26	18	2.7					6.1	6.5	295
	B			4.26	68	38	14	44	62	0.5	11.7					14.5	34	1370

4.4. Comparisons between Denmark and Sweden

In similar studies at Gårdsjön, Stenungssund, SW Sweden, and Skånes Värsjö, Markaryd, S Sweden (Bergkvist, 1983), exactly the same type of lysimeters were used, making the results comparable with the present ones.

In most aspects the Swedish findings of fluxes (anions not given) were similar to the Danish Klosterhede and Strødam sites (Table III). That holds for pH and all elements except Mn and Ca. Mn was lower at the Danish sites, especially Klosterhede, and Ca was also lower at Klosterhede. By a closer examination of the figures in Table III most of the minor differences observed between the Danish and Swedish sites may be explained, when taking into consideration that 1) the sites at Klosterhede and Gårdsjön both are located relatively close to the sea, 2) Strødam and Gårdsjön are exposed to air pollution from local emission sources, and 3) the Danish and Swedish data are from 1983-84 and 1980-82, respectively.

5. ACKNOWLEDGEMENTS

This research is carried out with financial support from the Commission of the European Communities (Grant ENV-892-DK), The Danish EPA and The Ministry of Energy.

6. REFERENCES

Andersen, A.G., Boesen, D.F., Holmen, K., Jacobsen, N., Lewinsky, J., and Rasmussen, L.: 1976, Den danske mosflora, Gyldendal Copenhagen.

Bergkvist, B.: 1983, National Swedish Environment Protection Board Bulletin SNV PM 1686, 35 pp.

Bormann, F.H. and Likens, G.E.: 1979, Pattern and Process in a Forested Ecosystem, Springer-Verlag, N.Y.

Cronan, C.S.: 1980, Oikos 34, 272.

Freiesleben, N.E.v. and Rasmussen, L.: 1986, Water, Air, and Soil Pollution (this issue).

Freiesleben, N.E.v., Ridder, C., and Rasmussen, L.: 1986, Water, Air, and Soil Pollution (this issue).

Godbold, D.L. and Hüttermann, A.: 1985, Environ. Pollut. Ser. A 38, 375.

Hansen, K.: 1981, Dansk feltflora, Gyldendal Copenhagen, 757 pp.

Heidam, N.Z.: 1985, 'EMEP Bulletin 1982', Miljøstyrelsens Luftforureningslaboratorium, Risø Report MST Luft A-97, 19 pp.

Nihlgård, B.: 1985, Ambio 14, 2.

Nilsson, S.I. and Bergkvist, B.: 1983, Water, Air, and Soil Pollution 20, 311.

Rasmussen, L., Jørgensen, P. and Kruse, S.: 1986, Bulletin of Environmental Contamination and Toxicology 36, 563.

Rost-Siebert, K.: 1983, Allgem. Forstzeitschr. 27, 686.

Tyler, G., Bergkvist, B., Rühling, Å. and Wiman, B.: 1983, National Swedish Environment Protection Board Bulletin SNV PM 1692, 65 pp.

Ulrich, B. and Matzner, E.: 1983, Luftreinhaltung Forschungsbericht 104 02 615, 221 pp.

Van Breemen, N. and Jordens, E.R.: 1983, in Ulrich, B. and Pankrath, J.: Effects of accumulation of air pollutants in forest ecosystems, D. Reidel Publishing Company.

DETERMINATION OF A CHEMICAL BASIS OF AIR-POLLUTION STRESS IN WOOD OF MATURE WHITE PINE TREES IN THE SUSCEPTIVE FOREST ECOSYSTEMS.

PATHAK, S. N., LOVE, D. V., and ROY, D. N.
Faculty of Forestry, University of Toronto,
203, College Street,
Toronto, Ontario, M5S 1A1 CANADA.

ABSTRACT. This paper describes a pilot investigation dealing with changes in chemical composition of wood in pollution-stressed white pine trees, taken from two forested ecosystems of Southern Ontario, Canada. The implications of this research lie in a much needed search for a potential bioindicator of a pre-decline, incipient-stress induced (and/or contributed by) the air-pollutants.

1. INTRODUCTION

During the past 5 to 6 years, a number of phenomena which involve severe deterioration of some species of forest trees in both Europe and North America have emerged. It is the consensus of the scientists that this phenomenon and an extremely rapid overall decline in certain species of forest trees in West Germany (Ulrich, 1984), cannot possibly be explained by natural factors. The responses of trees (both mature and saplings) in forested ecosystems to stress(es) caused by natural factors, air-pollutants, and/or combinations thereof are needed.

We believe it is very difficult, if not impossible, to determine a common denominator of damage, since the decline syndromes vary from tree to tree and site to site. However, we could narrow the possibilities by focussing on identifying a pre-decline, early, pollution-stress factor(s) and then study the mechanisms involved. The present research focusses on searching for such a possible factor in apparently undamaged mature forest trees.

Several documented cases of the effects of air-pollutants onthe sites and the biochemical pathways of photosynthesis exist (Hutchinson, 1985; Huttunen, 1983, 1985). The total carbon (C) resources of the trees could effect the translocation, assimilation, and partitioning of the carbohydrates for cell wall differentiation. The consequence of all this could be seen in changes in the growth of annual rings and chemistry , structure, and functioning of wood.

Water, Air, and Soil Pollution **31** (1986) 385–392.

TABLE I

CURRENT RATES OF ACIDIC DEPOSITION (CHAN ET AL., 1985)

FROM 1981 to 1983

SITES	SULPHATES	NITRATES	PPT. WEIGHTED SO_4'' CONC.
	($gm\ m^{-2}$)		($mg\ L^{-1}$)
DORSET	9	1.51	3.5
PETAWAWA	7	1.00	2.61

$1\ gm\ m^{-2} = 10\ Kg\ hactare^{-1}$

Studies of trace elements accumulation within tree growth rings, going back as far as 100 years (Baes and McLaughlin, 1984, Legge et al. , 1984, McLaughlin et al., 1985) show that their concentrations correspond with those of known emission sources affecting the trees. Therefore, we think, that comparisons of tree growth rates and chemical composition might be one possible method of documenting cause and effect relationships. The underlying, a priori hypothesis is that the structure of wood reflects the variations of the growth in the past. The effects of air-pollutants may thus, be recorded in the variations in wood chemical composition (and/or structure-function of the wood).

The specific objective of our research was to find a measurable and quantifiable pattern of changes in the chemical components of wood and relate that to the past and present indices of pollution (mainly acidic deposition and ozone).

2. RESEARCH METHODOLOGY

The data for this pilot study are obtained from two white pine trees each from Dorset (DO) and Petawawa (PW) sites. White pine has been documented to be susceptible to the impact of gaseous pollutants (Linzon, 1950). Table I shows the sites and their acidic depositions (a major component of the air-pollutants) for the period 1981-1983. In the absence of any unpolluted "control" area (site), any relative differences were investigated in these two polluted sites (DO being more polluted than PW). Figure 1 represents a cutting- and sampling-design, adopted for the wood chemical analyses of each tree. Each of the five radial samples (H1, H2, H3, S1, S2), which comprised of ten growth rings was subjected to organic analyses (TAPPI standards) for total extractives and lignin contents, and to inorganic analyses (INAA procedure, Pathak et al., 1986) for aluminum (Al) and calcium (Ca).

3. RESULTS AND DISCUSSION

Initially, it was established that DO trees exhibited a site-specific distribution of total extractives, lignin, Al, and Ca (Pathak et al., 1985). There is no significant variability (at 95% confidence interval) within trees and within the site for these chemical components. Further, the heartwood extractives content showed a significant increase (by 5 to 6%) from the reported "normal" values (Rowell, 1984). We also reported that contrary to the documented increase in the lignin content of sapwood (by 2%) as compared to heartwood, these trees showed a decrease of about 2%. There does not seem to be any radial translocation of the chemical components studied.

Figures 2 to 5 show the patterns of distribution of the mean total extractives, lignin, Ca, and Al in both sites across the five radial sample locations. A detailed comparison of the total heartwood (H1+H2

FIGURE 1.
DESIGN FOR WOOD SAMPLES

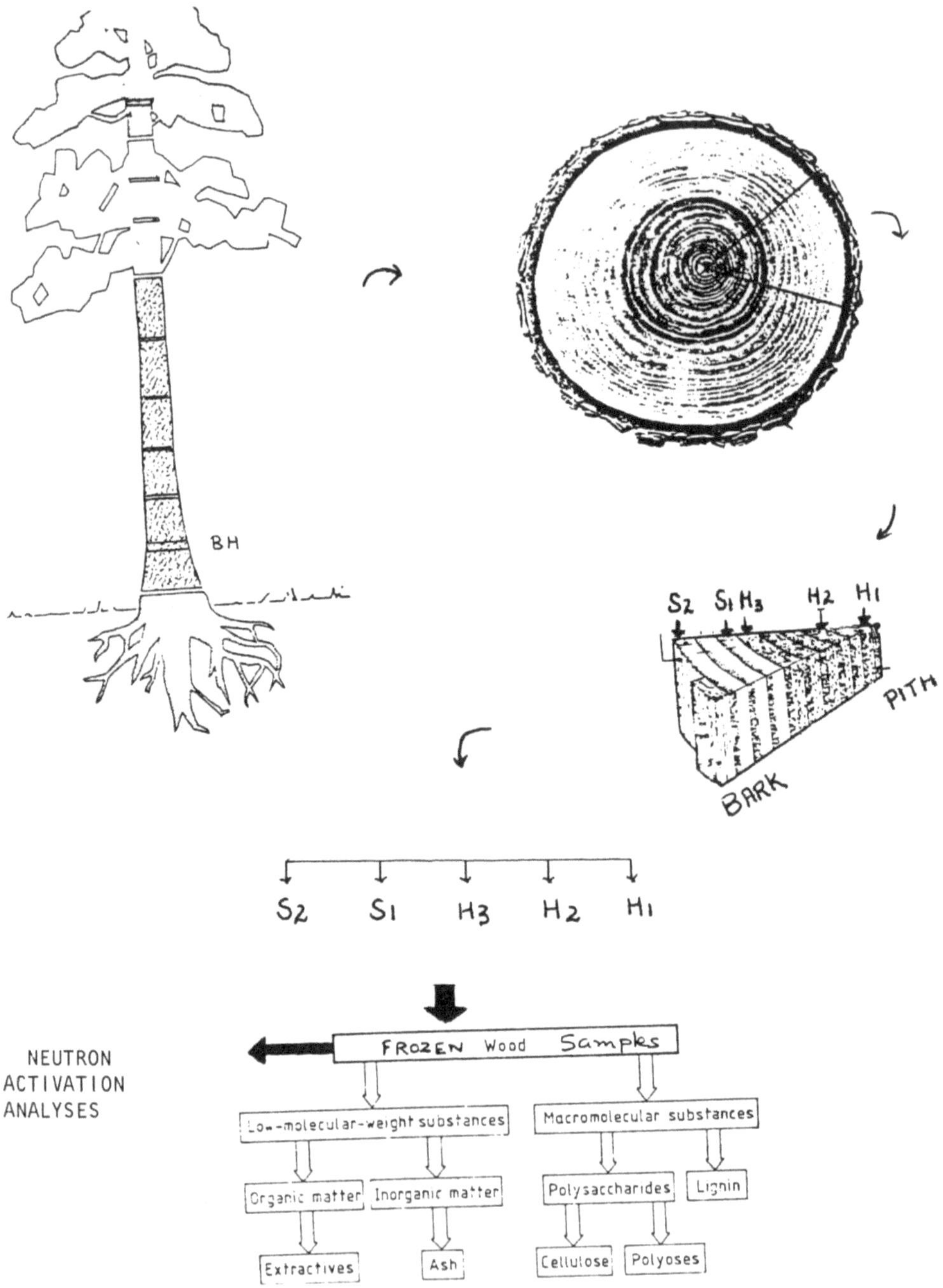

FIGURES 2-5.

DISTRIBUTION OF CHEMICAL COMPONENTS

WHITE PINE TREES

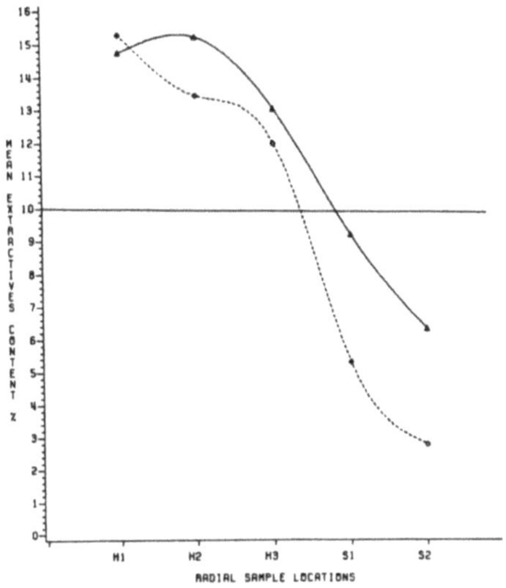

FIG. 2

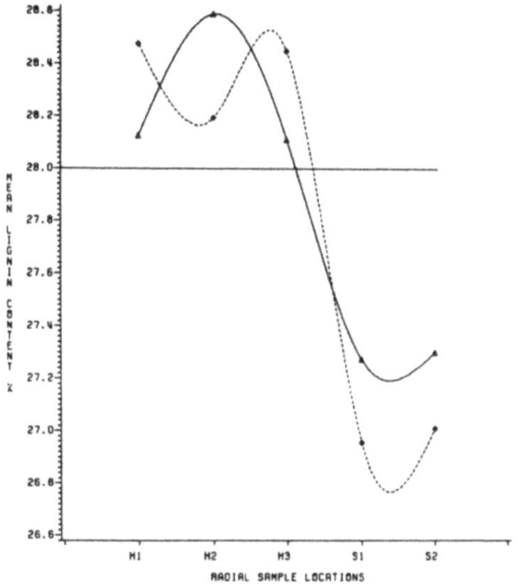

FIG. 3

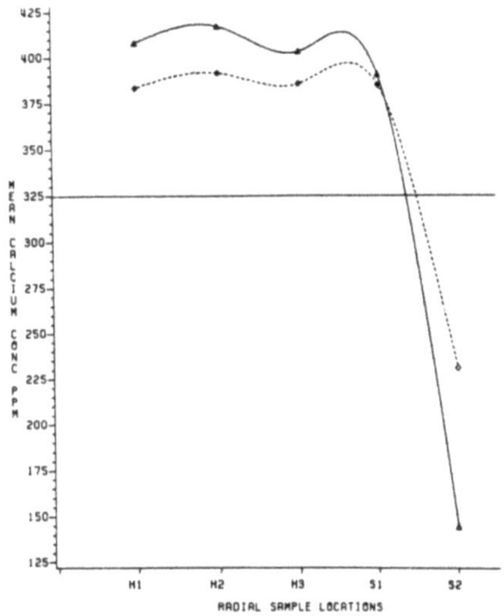

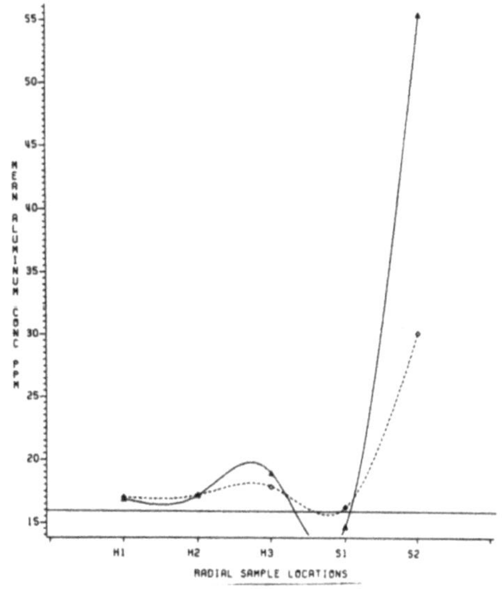

FIG. 4

LEGEND: SITE —▲—▲— DO —◆—◆— PW —— Reported Value

FIG. 5

H1=HEARTWOOD NEAR PITH
H2=MID-HEARTWOOD
S1=SAPWOOD AT HEART/SAPWOOD BOUNDARY
S2=SAPWOOD NEAR CAMBIUM

H3= Heartwood at heart/sapwood boundary

FIGURES 6-9

DISTRIBUTION OF MEAN CHEMICAL COMPONENTS

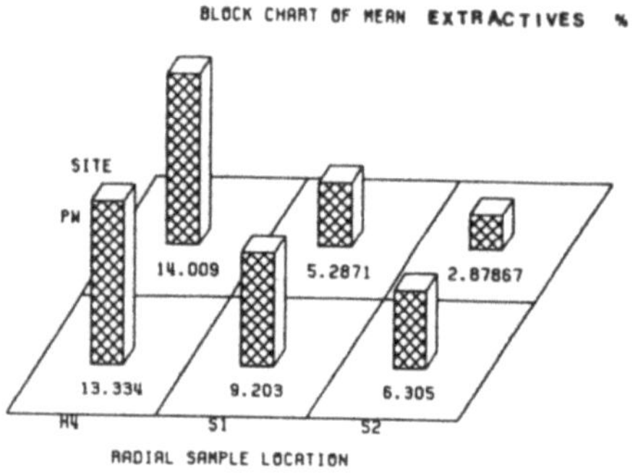

FIG. 6

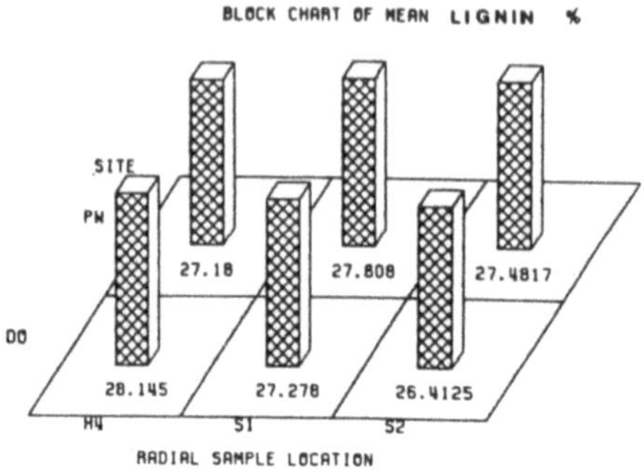

FIG. 7

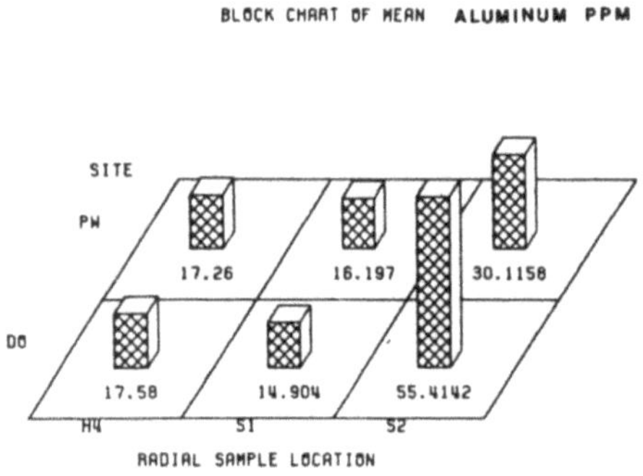

FIG. 8

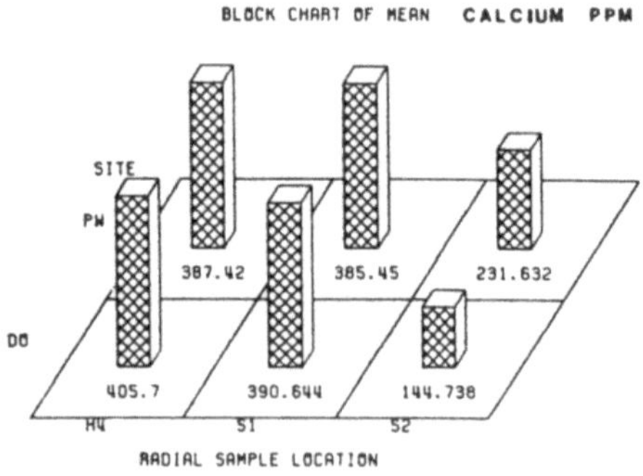

FIG.9

H4=MEAN HEARTWOOD
S1=SAPWOOD AT HEART/SAPWOOD BOUNDARY
S2=SAPWOOD NEAR CAMBIUM

+H3, that is,H4), early sapwood (S1), and latest sapwood (S2) of the total extractives content, lignin,Al, and Ca from trees of DO and PW is shown in Figures 6 to 9.

It is observed that trees at both sites show an increase in heartwood extractives content. Similar increases have been found to be induced by a herbicide, paraquat (Fengel and Wagener, 1984). It is possible that increase in the heartwood extractives is one way a tree might deal with a stressful situation. Further, the extractives content of sapwood at DO and PW differ which could be due to a differential response to different pollution regimens. The lignin content of the heartwood of both sites is much lower than the reported values, more so in PW. Ca exhibits a remarkable decline in S2 at both sites and Al concentrations increase at both sites. Since Ca and Al are antagonistic in terms of their uptake from the soil by a tree, our data support this fact. The data favor the "soil acidification" hypothesis of air-pollutant/organism interaction (Ulrich, 1983). A synchrony between drop in lignin content and decrease in Ca and increase in Al concentration in wood is evident. We believe the drop in sapwood lignin could be a direct consequence of a decreased availability of Ca, since Ca is needed in lignification of cellular polysaccharides into a pyrophenolic compound and because of its crucial role (as calcium pectate) in general cellular-communication in the middle lamella of wood parenchyma.

It is clear from this data that a specific trend of alterations in these four chemical components of wood exists in these sites. Further work exploring the relationships of such trends with indexed increments in specific volume of annual growth rings and a few predictor variables (for example, site-, climatic-, and atmospheric-variables) is in progress. It is hoped that similar trend-analyses from other polluted sites (for example, Sudbury, Saulte St. Marie) which have different pollution levels,will delineate any site-specific trends of changes in chemistry of wood of trees which have been growing under various pollution indices.

4. REFERENCES

Baes III, C. F., and McLaughlin, S. B.: 1984, 'Trace metals in tree rings: evidence of recent and historical air pollutants', Science 224, 494.

Chan, W. H., Tang, A. J., and Lusis , M. A.: 1985, 'An evaluation of the precision and accuracy of daily sampling for wet deposition of trace contaminants'. Muskoka' 85 Conference proceedings, in press.

Fengel, D., and Wagener, G.: 1984, Wood Chemistry, Ultrastructure, and Reactions, Walteridge Gruyter Publishers, Berlin. 613 pp.

Hutchinson, T. C., Adams, C., and Gaber, B.: 'The role of leaf surface in neutralization of acidic raindrops in crops and boreal forest species', Muskoka'85 Conference proceedings, in press.

Huttunen, S.: 1983, 'Air pollution induced stress and its effects on the photosynthesis of Pinus sylvestris L. in Oulu', Aquilo Ser Bot., 19, 275.

Huttunen, S.: 1985, ' Acid depsition, forest soil acidity and tree growth in Southern Finland', Muskoka'85 Conference proceedings, in press.

Legge, A. H., Kayfman, H. C., and Winchester, J. W.: 1984, 'Tree ring analyses by PIXE for a historical record of soil chemical response to acidic air pollutants', N. Inst. and methods in Physics Res. 53, 507.

Linzon, S. N.: 1950, 'Effects of Smelter fumes on white pine needle blight', A Master's thesis, University of Toronto, Ontario, Canada.

Linzon, S. N.: 1984, 'A trip report: Visit to Germany to assess symptoms and discuss cases of forest damage', personal communications.

Pathak, S. N.: 1985, 'Chemical analyses of Dorset white pine trees', MOE, semiannual meeting presentation.

Pathak, S. N.: 1986, 'Comparison of dffferent analytical techniques for wood chemicals', Physics Letters, in preparation.

Ulrich, B.: 1983, 'Effects of accumulation of air-pollutants in Forest Ecosystems', Edited by Ulrich, B., and Pankarath, J. D. Reidel Publishers, Dordecht, Holland, 133-146.

Ulrich, B.: 1984, 'Unpublished comments abridged from a lecture given at University of Toronto.

EFFECTS OF AIR POLLUTION ON THE CHEMISTRY OF SURFACE WAXES OF SCOTS PINE

J.N. Cape
Institute of Terrestrial Ecology
Bush Estate
Penicuik
Midlothian EH26 OQB
UK

ABSTRACT. Surface waxes from Scots pine (Pinus sylvestris) trees from the same provenance, but growing at a polluted and an unpolluted site, were sampled for each year class of needles at least twice during the year. There was no change in the amount of wax per unit area over the lifetime of the needles, but less wax was obtained from trees at the polluted site. Quantitative thin-layer chromatography of the wax gave several components which decreased in amount with time. They were identified as long-chain alcohols and ketones, and comprised only a small proportion of the total wax. The rate of change was greater in polluted air than in clean air, and there was a strong correlation with measured contact angles for water droplets. The degradation in structure observed by scanning electron microscopy may be associated with particular chemical components of the surface wax.

1. INTRODUCTION

The surface waxes of leaves, and in particular the epicuticular waxes, form a barrier between the leaf and its atmospheric environment. The chemistry and morphology of these waxes have been studied for many plant species; a recent review by Baker (1982) summarizes the present state of knowledge. Both the formation and subsequent development of epicuticular waxes are sensitive to the external environment of the leaf. The physical form of the wax reflects both environment and chemical composition, and may be a good indicator of environmental change. Air pollutants form part of the external environment, and several studies have explicitly considered the influence of gaseous pollutants on wax morphology as observed by scanning electron microscopy (SEM) (Bystrom et al., 1968; Grill, 1973; Percy and Riding, 1978; Huttunen and Laine, 1983; Riding and Percy, 1985; Crossley and Fowler, 1986).

There has been little or no published work on the effects of air pollution on wax chemistry. For Scots pine, the epicuticular waxes have been partially analyzed by Schütt and Schuck (1973) as a

Water, Air, and Soil Pollution **31** (1986) 393–399.

function of needle age. They observed a greater proportion of longer-chain alkanes, alkyl esters and aliphatic alcohols in older needles, but it was not possible to show whether this was caused by increases in amounts of longer-chain components or by losses of the shorter-chain components. Indirect evidence of chemical changes has come from studies of the wettability of needle surfaces, as measured by the contact angle of small water droplets (Leyton and Juniper, 1963; Cape, 1983). The study reported here is a preliminary chemical analysis of the same material collected for the contact angle measurements described by Cape (1983).

2. METHODS

Details of the 4 sites in the study are given in Cape (1983). Material from 2 of these sites was used for chemical analysis: Saltoun, a clean-air site in S.E. Scotland (annual mean SO_2 concentration ~5 μg m^{-3}) and Halifax (Gorple), a polluted site in W. Yorkshire, N. England (annual mean SO_2 concentration ~40 μg m^{-3}). The IUFRO standard for damage to conifers is in the range 25-50 μg SO_2 m^{-3}, but at the polluted site, concentrations of NO_x and suspended particles were also much greater than at the clean-air site. The Scots pine trees at both sites were from the same provenance.

Foliage samples were taken towards the end of July and again at the end of November 1981 from each year class of needles. Shoots were taken from 5 trees at each site, from the fourth or fifth whorl. At the laboratory, batches of 1,000 needles from each sample (5 for each needle age) were shaken for 10s with 200 mL of A.R. chloroform, then rapidly filtered under vacuum and the solution reduced by vacuum distillation. Wash times between 5s and 2 min had no significant effect on the amount of wax removed from the needles. The surface area of each batch of 1,000 needles was estimated from the needle dry weight. An excellent linear relationship was found for sub-samples of 25 needles from each year class between needle dry weight and surface area (calculated from measurements of needle length, breadth and thickness).

Chemical analysis of waxes was performed using an Iatroscan. Examples of the use of this method in lipid analysis are given by Ackman (1981). In these experiments 0.5 μL drops of a chloroform solution (100 μg μL^{-1}) were developed in a solvent system of chloroform: toluene: acetic acid in ratios 60: 40: 0.3. Each wax sample was analyzed in triplicate, and quantitation was achieved by using standard solutions of model compounds, with multiple calibration points. Examples of a calibration analysis and a typical wax sample are shown in Figure 1.

FIGURE 1. a) Calibration analysis (TLC/FID) of standards (0.5 μL of 100 μg μL^{-1} solution).
b) Typical analysis (TLC/FID) of wax (0.5 μL of 100 μg μL^{-1} solution).

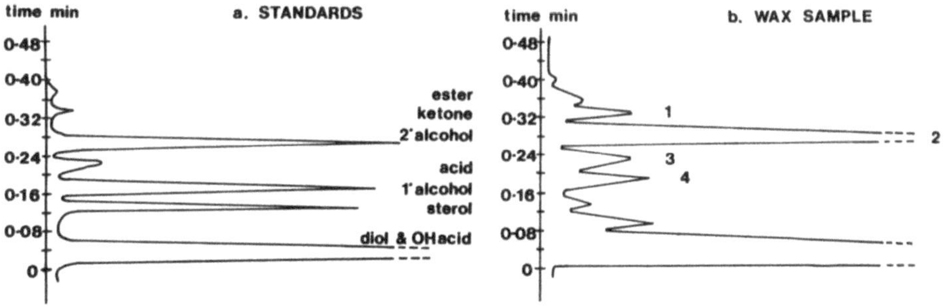

3. RESULTS

There was no significant change in the amount of surface wax removable by chloroform washing as the needles aged at either site. There was however a significant difference between the sites, with less wax at Halifax, the polluted site (Figure 2).

FIGURE 2. Variation in the amounts of wax per unit area (μg cm^{-2}) with time at Saltoun (clean air: ◆) and Halifax (polluted: ●).

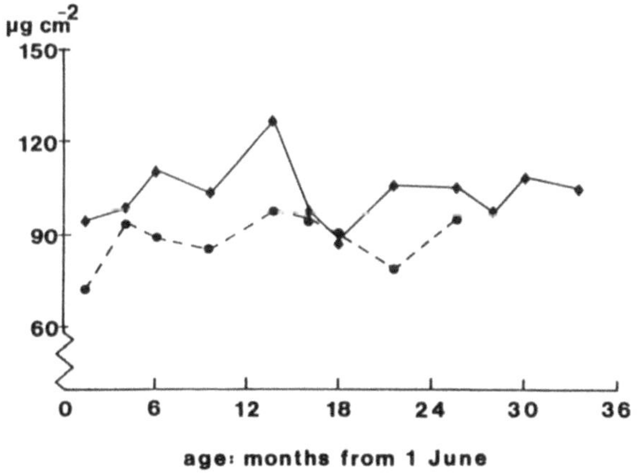

In the subsequent chemical analyses at least 4 clear peaks were recognisable (Figure 1). These have been identified on the basis of their relative retention times as representing ketones, secondary alcohols, fatty acids and primary alcohols respectively. Peaks corresponding to alkanes were also sometimes observed, but were not easily quantifiable. The amounts of the individual components identified by chromatography form only a small proportion of the total, with initial values of around 2, 3, 4 and 15 $\mu g\ cm^{-2}$ for components 1-4, respectively, compared with average total wax of 104 $\mu g\ cm^{-2}$ at Saltoun and 91 $\mu g\ cm^{-2}$ at Halifax. However, only component 4 (primary alcohol) showed a significant difference between the 2 sites over the lifetime of the needles (Figure 3(c)). Of the groups of compounds all but the third (fatty acid) showed a significant decrease with time at each site (Figure 3). There was consequently a strong correlation (Table I) between the amounts of these components (expressed as $\mu g\ cm^{-2}$) and the contact angle of water droplets (Cape, 1983) on the needle surface.

TABLE I. Correlation between amounts of wax components and contact angle (Cape, 1983) for both sites together (11 samples)

COMPONENT	CORRELATION COEFFICIENT
1 (ketone)	0.879
2 (secondary alcohol)	0.755
4 (primary alcohol)	0.768

4. DISCUSSION

Washing with chloroform is an effective way of removing surface waxes from pine needles. The observation that there was no significant decrease in the amount removed as needles aged seems at variance with observations of marked structural changes in the form of the wax (Cape and Fowler, 1981; Huttunen and Laine, 1983; Crossley and Fowler, 1986). However, the mesh of fine rods observed by SEM covers only a small fraction of the needle area, around the stomata, and may be only a small proportion of the total wax removed by chloroform washing. It is not clear whether the differences between the 2 sites in amounts of surface wax are related to air pollution or to other factors such as climate or nutrient status. However, material from the other 2 sites in the original study (Cape 1983) showed a similar pattern of more wax at the unpolluted site.

The decrease with needle age in the amounts of components 1, 2 and 4 has been treated as a first-order decay process, following the earlier analysis of contact angle data (Cape, 1983). The straight lines in Figure 3 show the relationship of form:

$$\ln (\mu g/cm^2) = C - kt$$

where C is the (natural) logarithm of the amount of wax at time zero (1 June) and k is a first-order decay constant with units mo^{-1}. The coefficients of the lines are shown in Table II. Only for component 4 is the initial amount (C) significantly different between sites. However, there is a marked difference in the decay constants (k) for all 3 components between the polluted and unpolluted sites. This suggests that the 'natural' weathering of these components is accelerated in polluted air. The relative weathering rates for components 1 and 2 (ie a factor of 1.8 greater in polluted air) is similar to the factor of 2 observed for the rates of change in contact angle at these 2 sites (Cape, 1983).

TABLE II. Equations of first-order decay process for components 1, 2 and 4 (Figure 3).

COMPONENT	SITE	INITIAL AMOUNT ($\mu g\ cm^{-2}$)	RATE (% mo^{-1})
1 (ketone)	Saltoun	1.8	-4.1
	Halifax	2.0	-7.4
2 (secondary alcohol)	Saltoun	17.1	-1.3
	Halifax	15.2	-2.3
3 (primary alcohol)	Saltoun	3.0	-4.2
	Halifax	1.9	-6.0

It has been suggested (Baker, 1982) that rods or tubes which characterize the form of the wax around the stomata in Scots pine are associated with ketones and secondary alcohols. If this is so, the present observation of a decrease in amounts of these components fits well with the structural degradation observed by SEM. The effects of these changes on the measured contact angle of water droplets would also be expected, given the dependence of this parameter on both the chemistry and structural form of the surface.

Although there is now evidence that physical changes in wax structure may be related to the loss of certain (albeit minor) components of the surface wax there is no evidence of a specific chemical effect which would identify the causal agent(s). However, the present measurements suggest that the shift towards a greater proportion of longer-chain homologues observed by Schuck (1972) and Schütt and Schuck (1973) may have resulted from the preferential loss of the shorter-chain compounds, rather than from a shift towards biosynthesis of longer-chain homologues in older needles.

FIGURE 3. Variation in the amounts of wax components with time at Saltoun (clean air: ◆) and Halifax (polluted: ⊙)
a) component 1 (ketone)
b) component 2 (secondary alcohol)
c) component 4 (primary alcohol)

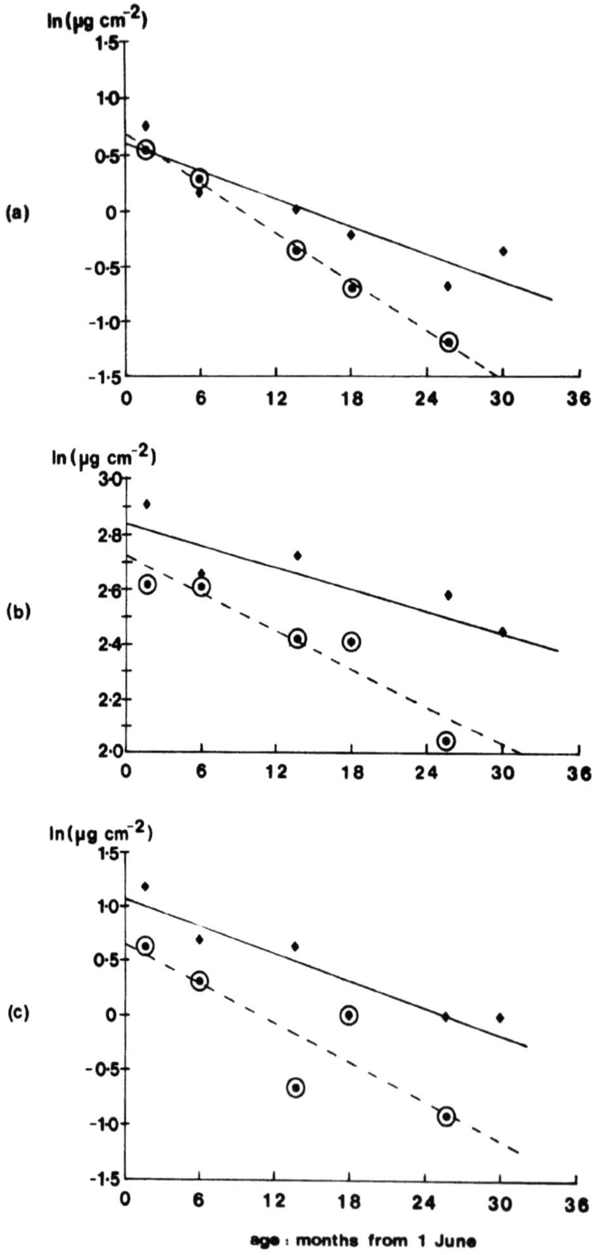

ACKNOWLEDGEMENTS

The experimental material was made available by courtesy of Mr Roger Lines (Forestry Commission). I am indebted to Dr John Sargent and colleagues at the Institute of Marine Biochemistry, Aberdeen for the use of their Iatroscan TLC/FID analyser. Finally, the efforts of Mr I D Leith (ITE Bush) and others in counting by hand some quarter million pine needles should be acknowledged.

REFERENCES

ACKMAN, R.G.: 1981, Methods in Enzymology, **72**, 205.
BAKER, E.A.: 1982, In The Plant Cuticle, Cutter, D.F., Alvin, K.L. and Price, C.E. (eds). pp 139-166. Academic Press, London.
BYSTROM, B.G., GLATER, R.B., SCOTT, F.M. and BOWLER, E.S.C.: 1968: Botanical Gazette, **128**, 133.
CAPE, J.N.: 1983, New Phytologist, **83**, 293.
CAPE, J.N. and FOWLER, D.: 1981, Silva Fennica, **15**, 457.
CROSSLEY, A. and FOWLER, D.: 1986, New Phytologist, **103**, 207.
GRILL, D.: 1973, Phytopathologische Zeitschrift, **78**, 75.
HUTTUNEN, S. and LAINE, K.: 1983, Ann. Bot. Fennici, **22**, 79.
LEYTON, L. and JUNIPER, B.E.: 1963, Nature, **198**, 770.
PERCY, K.E. and RIDING, R.T.: 1978, Canadian Journal of Forest Research, **8**, 474.
RIDING, R.T. and PERCY, K.E.: 1985. New Phytologist, **99**, 55.
SCHUCK, H.J.: 1972, Flora, **161**, 604.
SCHÜTT, P. and SCHUCK,H.J.: 1973, Flora, **162**, 206.

IN VIVO BUFFERING AND CONCENTRATION OF SIMULATED ACID RAIN DROPS ON LEAVES OF SELECTED CROPS

Bo Richter Larsen
Laboratory of Environmental Sciences and Ecology
Technical University of Denmark
Building 224, DK-2800 Lyngby
Denmark

ABSTRACT. The *in vivo* processes in simulated acid rain drops on leaf surfaces of selected crops were studied by means of a micro pH-electrode. Changes in pH were followed for 1 to 4 hr. Large differences among species in acid buffering ability and concentration due to evaporation were measured. Stressed plants showed markedly decreased buffering ability. Mathematical models suggested that the buffering mechanism was controlled by passive diffusion of leachates across the cuticle. Solid surface tensions of the leaves were determined indicating that buffering ability is related to polar forces in the cuticle. A model explaining the varying suceptibility to acid rain among plant species is proposed.

1. INTRODUCTION

Acid rain has been observed to cause foliar injury to plants and to affect growth and yield (reviewed by Evans, 1984). The susceptibility of plants to acid rain differs highly. Varying ability of plant tissue to buffer the acid has been suggested as one explanation on the basis of *in vitro* studies (Craker and Bernstein, 1984) and of field studies (Adams and Hutchinson, 1984). The present study reports on *in vivo* buffering and concentration due to evaporation of simulated rain drops on leaves of selected crops. The importance of surface tension and plant stress for the buffering ability is investigated.

2. MATERIALS AND METHODS

The following five species were selected: Bean, *Phaseolus vulgaris* L.; cress, *Lepidum sativum* L.; lettuce, *Lactuca sativa* L.; white mustard; *Sinapis alba* L. and rape, *Brassica napus* L. All plant material was seeded in plastic pots containing sphagnum and grown in growth chambers under 16 hr photoperiod at 19 ± 1 ^{o}C, 80 ± 5 %RH and 200 to 250 $\mu Em^{-2}s^{-1}$. The pots were watered as required with half concentrated Hoagland 3. At 4 to 6 weeks after sowing simulated acid rain drops of 10 μL were placed on the first trifoliolate leaf in bean, on the first leaf in lettuce and on the third leaf in cress, mustard and rape. The change in pH was measured continously with a specially designed micro pH-electrode (Radiometer Ltd. Denmark) and plots of pH vs time were recorded.

The pH-electrode consisted of a glas-electrode and a capillary 1M KCl reference-electrode both mounted inside a hypodermic needle. Before each measurement the electrode was calibrated with buffers at pH 3.0 to 7.0. The electrode has an accuracy of 0.01 pH unit and operates down to very small volumes (∿ 1 μL).

The simulated rain consisted of 50 μM aqueous NH_4NO_3 solutions adjusted to the desired pH by addition of H_2SO_4.

The measurements were performed at least in duplicate and repeated on a teflon surface as reference. Unstressed conditions were obtained in a growth chamber a 19 °C, 75 to 80 %RH and 200 to 250 $\mu Em^{-2}s^{-1}$. Drought stress was obtained by cutting off the water supply three days before measurements which were performed in the laboratory where the conditions were: 22 °C, 55 to 65 %RH and 10 to 20 $\mu Em^{-2}s^{-1}$. The soil was dry (not quantified), but the plants did not appear lax. Acid rain stress was obtained by showering the plants with simulated acid rain at pH 3.3. The measurements were carried out on a spot shielded from the rain shower while the leaves were still wet. The measuring time was 1 to 4 hr depending on when pH reached an equilibrium or when drops were totally evaporated.

The surface tension of leaves was determined as described elsewhere (Chattoraj and Birdi, 1984). Contact angles between leaf and 10 μL drops of mixtures of H_2O and n-propanol (0 to 2.5 M) were determined from photographs (x 12 magnification). The surface tensions of the mixtures were measured using the Wilhemy plate method. All measurements were done with 5 replicates and subjected to variance analyses (not shown). Significant difference between species was revealed ($p < 0.001$). The surface tension of liquids, γ_L, and solids, γ_s, can be written as a sum of apolar (dispersion), γ^D, and polar, γ^P, forces:

$$\gamma_s = \gamma_s^D + \gamma_s^P \quad (A), \qquad \gamma_L = \gamma_L^D + \gamma_L^P \quad (B).$$

The relation between the contact angle, θ, and the various tensions is given by the equation:

$$\cos\theta = -1 + 2\cdot\gamma_L^{-1}\,(\gamma_s^D\cdot\gamma_L^D)^{\frac{1}{2}} + 2\cdot\gamma_L^{-1}\,(\gamma_s^P\cdot\gamma_L^P)^{\frac{1}{2}} \quad (C).$$

Plots of $\cos\theta$ vs $(\gamma_L^D)^{\frac{1}{2}}\cdot\gamma_L^{-1}$ show linear releationships. According to the procedure delinated elsewhere (Chattoraj and Birdi, 1984) the value of γ_s^D is estimated from these plots by extrapolating to θ → zero. γ_s^D is thus given from the interception with COS θ = 1, where $\gamma_s^D \cong ((\gamma_L^D)^{\frac{1}{2}}\cdot\gamma_L^{-1})^{-2}$. The second therm of Equation (C) is estimated from plots of $(\gamma_s^D\cdot\gamma_L^P)^{\frac{1}{2}}$ vs $(\gamma_L^P)^{\frac{1}{2}}$ which show linear relationships. From the slope of these lines γ_L^P can be determined.

3. MATHEMATICAL MODELLING

The experiments suggested that three main processes were involved in the interaction of leaves with droplets: 1) Buffering, possibly due to leachates; 2) evaporation; and 3) neutralization, possibly through cellular disruption. The first process caused an increase in pH. Assuming a differential proportionality between leaching and the acticity of the hydrogen ions the following integrated equation was obtained:

$$a_{H^+}(t) = a_{H^+,start} - C_1^b \cdot (1 - \exp(-C_2^b \cdot t)) \qquad (1).$$

The evaporation caused a decrease in pH. Assuming no acid evaporated and that the volume of the drop was exponentially decreased the following integrated equation was obtained:

$$a_{H^+}(t) = a_{H^+,start} + C_1^e \cdot \exp(C_2^e \cdot t) \qquad (2).$$

Low starting pH caused a small but rapid preliminary rise in pH possibly due to disruption of cells. This rise was assumed to be instant and the following integrated equation was obtained:

$$a_{H^+}(t) = a_{H^+,start} - C_0^d + C_1^e \cdot \exp(C_2^e \cdot t) \qquad (3).$$

The constants C_0^i and C_1^i fit the curves to the starting proton activity, $a_{H^+,start}$. The constant C_2^i determines the rate of the processes. The letters 'b', 'e' and 'd' refer to buffering, evaporation and disruption respectively. The constants were estimated from 10 to 25 equidistant points of the pH-t curves by iterative least squares approximation.

4. RESULTS AND DISCUSSION

The recorded pH-t responses were higly reproducible. It appears in Figures 1 and 2 that the mathematical models can be fitted to the observations. The author recognizes that results of statistical analyses would depend on the number of data points used in calculation, hence no statistical data are presented. However, the asymptotic correlation coefficients were close to 1 with small asymptotic standard errors. The resulting estimates of the constants are shown in Tables I and II. The solid surface tension of the different plant species appear in Table III. The buffering capacity estimated from the starting pH of 3.5 is shown in Table III and appears to be correlated with γ_s^P as shown in Figure 3.

The internal pH of the plants was measured *in vivo* by inserting the pH-electrode into petioles of the leaves since the size of the electrode did not allow measurements directly in the mesophyll (data not shown). No inter-species variation was obtained ($\overline{pH} = 5.65 \pm 0.16$), but the pH decreased moving from leaf base towards leaf tip.

All species were capable of buffering acid rain drops of pH 3.5 under unstressed conditions. The relative buffering capacity was in the order: mustard > bean > cress >> lettuce > rape. The large rate constant, C_2^b, of bean indicated a changed short term buffering ability ranking: bean > mustard > cress >> lettuce > rape.

The buffering ability was decreased under dry conditions in bean, mustard and cress and was totally dominated by evaporation in lettuce, rape and stressed mustard (data for rape not shown). Increased temperature and humidity resulted in a greater buffering ability as shown for mustard (Table I).

The application of an acid rain shower at pH 3.3 immediately before the measurements caused a marked decrease in the buffering ability as shown for bean. The reversed effect occured with a shower at pH 5.6.

The plants were unable to overcome strong acidity and the evaporation became dominating at decreasing pH. The rate constant, c_2^e, in lettuce from a starting pH of 2.7 resembled that of teflon and was even greater in bean from a starting pH of 2.5 and 2.2. From a starting pH around 4 the rate constant, c_2^b, in bean and mustard was lower than from pH 3.5, but the equlibrium pH remained statistically unchanged.

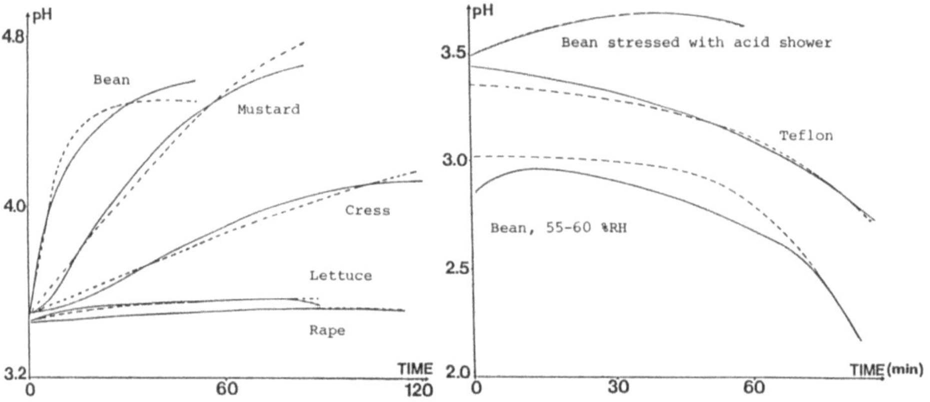

FIGURES 1 and 2. Buffering and concentration due to evaporation of acid rain drops measured under unstressed (Figure 1) or stressed (Figure 2) conditions. Measured response:(—); Fitted equations:(--).

TABLE I. The buffering and evaporation of acid rain drops on leaves. Estimated constants in Equations (1) and (2).

Species	T °C	RH %	Light $\mu E \cdot m^{-2} \cdot s^{-1}$	Soil moisture	Pre-shower	a_{H^+} start $x10^6$	c_1^b $x10^6$	c_2^b min^{-1}	c_2^e min^{-1}	c_1^e $x10^6$	N	pH start	pH stop
Bean	19	75-80	225	High	-	316	287	0.251	0	-	3	3.50	4.54
Mustard	19	75-80	225	High	-	316	303	0.048	0	-	5	3.50	4.89
Cress	19	75-80	225	High	-	316	243	0.016	0	-	4	3.50	4.14
Lettuce	19	75-80	225	High	-	355	82	0.063	0	-	2	3.45	3.56
Rape	19	75-80	225	High	-	355	66	0.018	0	-	2	3.45	3.53
Bean	22	55-60	10-20	Low	-	316	191	0.163	0	-	2	3.50	3.90
Mustard	22	55-60	10-20	Low	-	316	273	0.036	0	-	2	3.50	4.37
Cress	22	55-60	10-20	Low	-	316	101	0.017	0	-	2	3.50	3.67
Bean	22	55-60	10-20	High	-	98	74	0.211	0	-	2	4.00	4.62
Mustard	22	55-60	10-20	High	-	53	36	0.015	0	-	2	4.28	4.76
Bean	22	55-60	10-20	High	pH 5.6	316	290	0.566	0	-	2	3.50	4.59
Bean	22	55-60	10-20	High	pH 3.3	316	146	0.048	0.133	0.104	2	3.49	3.70
Mustard	24	90-95	225	High	-	316	296	0.100	0	-	2	3.50	4.70

TABLE II. The concentration due to evaporation of acid rain drops. Estimated contants in Equations (2) and (3).

Species	T °C	RH %	Light $\mu E \cdot m^{-2} \cdot s^{-1}$	Soil moisture	a_{H^+} start $x10^6$	c_1^e $x10^6$	c_2^e min^{-1}	c_o^d $x10^{-6}$	N	pH start	pH stop
Lettuce	22	55-60	10-20	High	316	44	0.041	-	4	3.50	2.73
Mustard, senescent leaf	22	40-60	5-10	Low	316	28	0.034	-	2	3.50	2.83
Lettuce	22	55-60	10-20	High	107	98	0.0033	-	2	3.97	3.77
Teflon	22	55-60	10-20	-	316	55	0.046	-	2	3.50	2.75
Teflon	22	55-60	10-20	-	78	7	0.049	-	2	4.11	3.50
Bean	22	55-60	10-20	High	933	11	0.071	100	2	3.09	2.46
Bean	22	55-60	10-20	High	1413	2	0.097	278	2	2.94	2.21

TABLE III. The solid surface tensions, γ_s, (mN/m) arising from dispersion (D) and polar forces (P) compared to buffering capacity, ΔpH, towards 10 μL rain drops of pH 3.5 measured under unstressed conditions.

Surface	γ_s^D	γ_s^P	γ_s	ΔpH
Rape	15.3	0.4	15.7	0.08
Teflon	18.9	0.0	18.9	-
Cress	21.7	10.2	31.9	0.64
Lettuce	33.9	5.7	39.6	0.11
Bean	20.1	26.3	46.4	1.04
Mustard	22.2	28.6	50.8	1.49

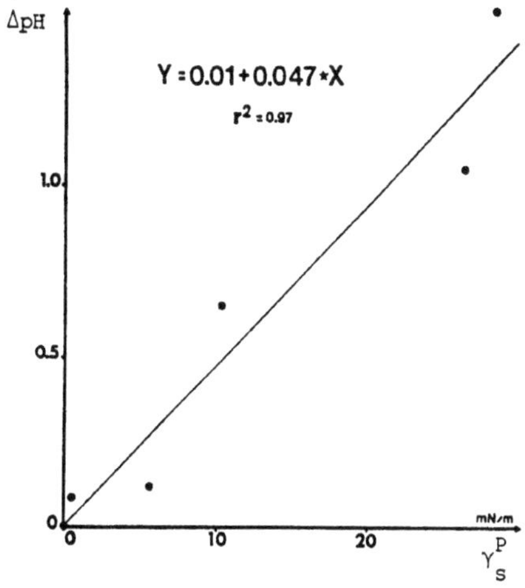

FIGURE 3. Buffering capacity, ΔpH, of unstressed plants towards 10 μL rain drops of pH 3.5 vs. surface tension from polar forces in the leaf surface, γ_s^P.

This study demonstrated that leaf tissue *in vivo* buffers acid rain. The buffering ability varied among species and was correlated with the polar forces in the leaf surfaces. This is consistent with Miller (1985), Evans *et al.* (1981) and McFarlane and Berry (1974) who report that leaf cuticles are perforated with micropores, channels of more rapid penetration of inorganic ions. It is suggested that a higher number of polar channels facilitates the leaching of buffering inorganic ions and that this higher number of micropores increase surface tensions from polar forces. Miller found that out of 32 examined taxa *Brassica* and four other do not contain transcuticular pores. The very low buffering capacity of rape is consistent herewith.

The varying amount, chemistry and ultra-structure of epicuticular wax on leaves of different species may also be importent to buffering ability through the effect on the wettability. Baker (1974) and Baker and Procopiou (1980) showed that drought stressed plants increase wax synthesis. Whitecross and Armstrong (1972) showed that increased light and/or temperature changed the ultra-structure of surface wax in rape towards increase in number and size of crystals. Holloway (1969; 1970) showed that wettability of cuticles increased with increased polarity of the wax constituents and showed that leaves become less hydrophobic when the micro-roughness is reduced *i.e.* reduced number and size of crystals.

The buffering ability was markedly decreased under dry conditions. This may be consistent with findings of Evans *et al.*(1981) who showed that nutrient leaching causing a buffering of applied rain drops was correlated with reduced adaxial leaf resistance. It is well known that leaf resistance increases under dry conditions, but the relation with nutrient leaching remains unexplained. In the same study it was shown that there was an initial temporary increase in adaxial leaf resistance following acid rain exposure. This tendency increased with acidity and may explain the results of the present study that buffering ability decreases under simultaneous apllication of acid rain of pH 3.3. It is also possible that buffering of acid rain is connected with stomatal turgor mechanisms including efflux of K^+ and malate from guard cells or influx of H^+ to guard cells (Smith and Raven, 1979). However, further evidence is needed and the change in leaf resistance and buffering ability may as well be unspecific plant stress symptoms. Investigations have recently shown, that stress-induced ethylene is developed in plants exposed to acid rain (Larsen, 1986; Chia *et al.*, 1984).

The effect of light intensity cannot be evaluated on the basis of the present study alone, since other parameters were changed simultaneously, but the marked increase in buffering ability in bean following a rain shower of pH 5.6 and in mustard at humidity around 90 %RH indicates that humidity may be much more important than light intensity.

There are differences in plant morphology between field-grown and chamber-grown plants and differences in composition and terminal velocity between artificial and natural rain. Nevertheless, the results of the present study combined with others suggest the three following main processes in acid rain drops on leaves: I) at moderate acidity and under unstressed conditions lechates or surface compounds (e.g. dust) buffer H^+. The buffering ability varies among species and is favoured by high surface tensions arising from polar forces in the leaf surface. II) at high acidity internal disruption of epidermal cells occur with release of neutralizing cell content. III) Under stressed conditions or at high acidity plants are unable to buffer acid rain sufficiently and a considerable quantity of H^+ remains in rain drops. As the water evaporates the concentration increases eventually leading to process II). The evaporation is favoured by low surface tension present in species with highly non-polar cuticles. Processes I) and II) offer protection from extensive tissue injury. Long term exposure to acid rain may cause severe nutrient loss whereby growth may be effected. Process (II) causes leaf lesions due to perturbation of upper leaf surfaces (Evans *et al.*1977) and may eventually lead to membrane damage and chlorophyll degradation (Chia *et al.*, 1984).

The differential damage picture and the varying susceptibility among species (Evans and Curry, 1979; Lee *et al.*, 1981; Irwing, 1983) can be understood in the light of the proposed processes:

Species with relatively polar surfaces (e.g. bean) strongly interact with acid rain; partly through their great number of polar micropores, partly due to the large contact area between leaf and drops caused by the polar forces. Such species are typically unaffected by acid rain directly but since they interact with the acid they become stressed and susceptible to simultaneously acting stressors, whereby they risk synergistic effects as demonstrated by Larsen (1985; 1986).

Species with higly non-polar surfaces (e.g. rape) repel rain drops. The low amount of retained acid rain and the small contact area result in weaker interactions with the rain. Due to their lower buffering ability and the slower evaporation of the drops (determined by the spherical shape of the drops on the hydrophobic surface) these species typically develop necrotic spots and visual damage. However, they are expected to be less affected by acid rain with regard to growth.

Research on the role of surface characteristics in phytotoxicity of acid rain is under progress in our laboratory.

5. ACKNOWLEDGEMENTS

The author acknowledge the cooperation of Radiometer Ltd. Denmark and is grateful for the micro pH-electrode placed at his disposal while still under development. Special thanks to Miss Susanne Bang and Mrs. Birthe Ebert for skilful technical assistance.

6. REFERENCES

Adams, C.M. and Hutcinson T.M.: 1984, *New Phytol.* 97, 463.
Baker, E.A.: 1974, *New Phytol.* 73, 955.
Baker, E.A. and Procopiou, J.: 1980, *J. Horticul. Sci.* 55, 85
Chattoraj, D.K. and Birdi, K.S.: 1984, *Adsorption and the Gibbs Excess.* Plenum Press, New York.
Chia, L.S., Mayfield, C.I. and Thompson, J.E.: 1984, *Plant Cell and Environment.* 7, 333.
Craker, L.E. and Bernstein, D.: 1984, *Environm. Poll. (A).* 36, 375.
Evans, L.S.: 1984: *Ann. Rev. Phytopatol.* 22, 397.
Evans, L.S., Gmur, N.F. and Kelsch, J.J.: 1977, *Environm. Exp. Bot.* 17, 145.
Evans, L.S. and Curry, T.M.: 1979, *Amer. J. Bot.* 66, 953.
Evans, L.S., Curry, T.M. and Lewin, K.: 1981, *New Phytol.* 88, 403.
Holloway, P.J.: 1969, *J. Sci. Fd. Agric.* 20, 124.
Holloway, P.J.: 1970, *J. Pesticide Sci.* 1, 156.
Irwing, P.: 1983, *J. Environm. Quality.* 12, 442.
Larsen, B.R.: 1985, *Ecotoxicol. Environm. Safety* 10, 228.
Larsen, B.R.: 1986, *Environm. Poll.(A).* 41, 179
Lee, J., Neely, G.E., Perrigan, S.C. and Grothaus, L.C.: 1981, *Environm. Exp. Bot.* 21, 171.
McFarlane, J.C. and Berry, W.L.: 1974, *Plant Phys.* 53, 723.
Miller, R.M.: 1985, *Annals Bot.* 55, 459.
Smith, F.A. and Raven, J.A.: 1979, *Ann. Rev. Plant. Phys.* 30, 289.
Whitecross, M.I. and Armstrong, D.J.: 1972, *Aust. J. Bot.* 20, 87.

THE EFFECT OF SIMULATED ACID RAIN ON FEATHER MOSSES AND LICHENS OF THE BOREAL FOREST.

T.C. Hutchinson, M. Dixon and M. Scott
Department of Botany, University of Toronto,
Toronto, Ontario, Canada M5S 1A1

ABSTRACT. Permanent field plots containing a dominant ground cover of feather moss (Pleurozium schreberi) and the forage lichen, (Cladina), were established in mature, boreal forest jack pine stands to monitor the effects of simulated acid precipitation. For a five-year period commencing in 1981, bimonthly sprays (pH range 2.5 to 5.6) were given throughout the growing season. The feather moss wefts were extremely sensitive to simulated rains of pH 2.5 and 3.0; but loss of cover and frond blackening were also observed at pH 3.5. The pH 2.5 treatment killed almost all of the Pleurozium, while the cover remaining in the pH 3.0 treatment after 5 years was reduced by 44%. In laboratory studies designed to compare the effects of H_2SO_4, HNO_3 and a 2:1 mixture of both, microcosms sprayed with H_2SO_4 alone (pH 3.0) were more significantly affected than fronds treated with HNO_3 alone or pH 5.6 sprays of any ratio. Although less sensitive than Pleurozium, field-sprayed lichens were also visibly damaged. At pHs less than 3.5, C. stellaris and C. rangiferina had reduced podetial height and dry weight; while C. mitis was affected by a combination of the acid rain treatment and other associated factors. While ambient rains of pH 4.2 may not in themselves be harmful to the boreal ground flora, it is apparent that the feather mosses and lichens, lacking a cuticle and true roots, are very sensitive to occasional, extremely acidic rain events.

1. INTRODUCTION

Large areas of northern Canada, the USA, Soviet Siberia, Scandinavia and upland U.K. are covered with boreal forest. In these forests, coniferous trees of the genera Picea, Pinus, Abies and Larix predominate. Much of the ground of these forests is covered by dense growths of feather mosses or, in drier, better-illuminated sites, by caribou-forage lichens (Kershaw 1977). In eastern North America, rains of mean pH 4.0 to 4.5 now fall on these forests and on the potentially acid-rain sensitive lichens and mosses. Since these lower plants largely lack a cuticle and have no true roots, their cells and chloroplasts are more exposed to direct effects of acidity and leaching. Therefore, deleterious effects on these plants may well show up before effects on higher plants. In this study, the effect of simulated acid rains on growth, survival and

Water, Air, and Soil Pollution **31** (1986) 409–416.

persistence of the feather moss Pleurozium schreberi and the common boreal lichens Cladina stellaris, C. rangiferina, and C. mitis in northern jack pine (Pinus banksiana) forests is reported.

2. METHODS

Experimental acid rain sprays were made on 5 m x 2 m plots of jack pine forest dominated in the ground flora by either Pleurozium schreberi or in drier sites by the three Cladina species listed above. Sprays equivalent to a precipitation depth of 2.5 cm were made bimonthly throughout the growing season from June - October in each of five years, 1981-85. Spray treatments of pH 2.5, 3.0, 3.5, 4.0 and 5.6 were applied to each of three replicate 5 m x 2 m plots at each of two sites. Overall, the summer sprays added approximately 30% to the annual growing season precipitation at the sites. 'Rain' was made up using deionised water, with a 2:1 molar ratio of H_2SO_4 to HNO_3 acid and sprayed on to the plots through a plastic garden hose with a plastic diffusion nozzle. In addition, a set of non-spray plots was set up. Various parameters of soil chemistry, stem-flow, throughfall and species persistence and performance have been measured and analysed on a monthly or annual basis. We report here on the data for cover of the moss and bare ground estimated in August of each year by T.C. Hutchinson. Estimates were made on the same three sub-plots, each 1 m x 1 m, at each of the three replicate plots - i.e. each cover value is the mean of nine separate estimates. The ground occupied by a vertical projection down was estimated as a percentage for each species. For the mosses only green or yellow fronds were counted in cover and blackened or dark brown shoots were considered dead. For the lichens, podetial height was measured on 30 random individuals (10 per plot), for each of the six pH treatments. Both total podetial height and height of the living, photosynthetically active portion of the podetium were measured. Measurements were also made of the length of internodes for 15 individuals, i.e. 3 replicate plots x 5 podetia per plot, selected randomly from the original 30. Podetial tips, mid-portions and bases were considered to be the combined heights of internodes 2 and 3, 4 and 5, and 6 and 7, respectively. Bases were already initiated prior to the spraying and have received the longest exposure to the simulated rain events.

A short-term field experiment was performed which compared the effect on P. schreberi of simulated sprays of pH 3.0 and 3.5 (which represent approximately 1% and 7% of the rain events in eastern Ontario) with sprays of pH 5.6. The "throughfall" solutions, consisting of deionized water plus nutrient salts to simulate throughfall of a mature jack pine canopy, were acidified with a 2:1 molar ratio of H_2SO_4 to HNO_3. Seventeen "microcosms", consisting of 30 cm diameter x 30 cm PVC pipes were inserted into the forest floor giving an intact core of moss and soil horizons H, A and B, were randomly located in an area of uniform feathermoss growth. The microcosms were sprayed twice a month from May to October 1983 and 1984. Each spray was equivalent to 0.7 cm of precipitation which, with the additional 8-10 cm of ambient rain, totalled 9.5-11.5 cm of rain per month. Cotton-polyester threads, serving as growth markers, were tied 0.5 cm from the tip of 35 fronds per microcosm at the beginning of the experiment. Chlorophyll and elemental composition data are reported elsewhere (Dixon 1986).

In addition, a 6-week laboratory experiment was performed to compare the effects of different acids on P. scherberi. Fifteen microcosms were removed from a location adjacent to the field site and taken to the University of Toronto greenhouse in October 1984. Five treatments were sprayed daily to a depth of 4.0 cm. Treatments consisted of simulated throughfall solutions acidified to pH 3.0 with H_2SO_4, HNO_3, a 2:1 molar ratio of H_2SO_4 to HNO_3 or to pH 5.6 made up with the 2:1 mixture. In addition, the last two mixtures (pH 3.0 and 5.6) were sprayed on alternate days.

Chlorophyll contents of three samples of the top one cm sections of all fronds from each microcosm were determined using a DMSO - 80% acetone extraction (Dixon 1986).

3. RESULTS AND DISCUSSION

3.1. Effects on Mosses

The acidic sprays in the 2 m x 5 m plots had rapid and severe effects on the growth and survival of Pleurozium schreberi (Table I). By the end of 1981 the cover of the moss sprayed with four, twice monthly sprays of pH 2.5 was reduced to 66.7% from approximately 85%. The fronds looked yellow or yellow-brown compared with the healthy green of the pH 3.5 or higher treatments. By August, 1982 the lowest pH treatment had had a devastating effect, with all but 1.6% of the cover destroyed. At this time, both sprays of pH 3.0 and 3.5 also had had significant effects. By 1985, pH 3.0 and pH 3.5 had caused Pleurozium to decline from 71% to 40% and from 84% to 64%, respectively. Moss treated with pH 3.0 actually showed some recovery in the very wet 1983, with survival and regeneration of certain, presumably more tolerant, individuals. Acid effects involved destruction of chlorophyll, a blackening of fronds, and total loss of structure. By 1985, in the pH 2.5 treatment it was no longer possible to recognize even decaying moss litter. Bare ground showed an inverse effect. This loss of the moss layer influenced water relations, nutrient cycling and nutrient status of the higher plants and caused changes in soil solution chemistry (described elsewhere). At pH 4.0 and 5.6 the feather moss showed a positive response, possibly due to the increased water supply.

In the field microcosm experiment, the general pattern of response was similar. As shown in Figure 1, pH 3.0 sprays had a significantly deleterious effect on frond growth, relative to pH 3.5 and 5.6, which were not significantly different from each other. The unsprayed control had significantly greater growth than any of the sprayed microcosms. Chlorophyll, Mg, Ca and to a limited extent K were reduced in the green growing tips of pH 3.0-sprayed fronds relative to pH 5.6 or unsprayed fronds.

TABLE I. Vegetative cover of Pleurozium shreberi and bare ground of Site 1 - recorded annually as % $1m^2$ plots (n = 3 x 3 per pH treatment).

	Treatment (pH spray)					
	5.6	4.0	3.5	3.0	2.5	Unsprayed
Year	% Cover					
1981	83.9 + 13.2	51.1 + 20.1	83.9 + 16.7	70.6 + 12.9	66.7 + 28.7	-
1982	86.7 + 12.7	61.1 + 23.3	68.0 + 23.8	47.2 + 20.0	1.6 + 2.5	84.4 + 11.3
1983	87.2 + 9.7	52.8 + 28.1	66.1 + 29.6	33.1 + 20.7	0.7 + 1.6	83.3 + 5.0
1984	90.0 + 7.3	74.4 + 12.8	66.9 + 24.7	50.7 + 27.6	0.2 + 0.4	76.1 + 16.4
1985	91.0 + 6.7	75.6 + 13.6	63.6 + 25.5	40.0 + 21.0	0.0	81.1 + 9.6
	% Bare Ground					
1981	3.9 + 5.5	(21.7 + 19.0)	4.1 + 7.2	9.2 + 17.3	7.9 + 9.5	-
1982	0.9 + 0.9	5.8 + 5.8	2.2 + 3.4	10.3 + 9.6	46.1 + 11.4	1.8 + 1.5
1983	2.9 + 3.2	9.8 + 12.1	5.4 + 4.5	27.9 + 16.1	64.4 + 8.5	4.8 + 2.7
1984	0.8 + 0.8	6.9 + 7.6	2.2 + 1.6	24.2 + 17.8	54.4 + 33.8	4.7 + 4.9
1985	0.9 + 1.1	4.6 + 3.4	8.0 + 10.9	25.1 + 16.2	88.0 + 14.4	3.6 + 3.8

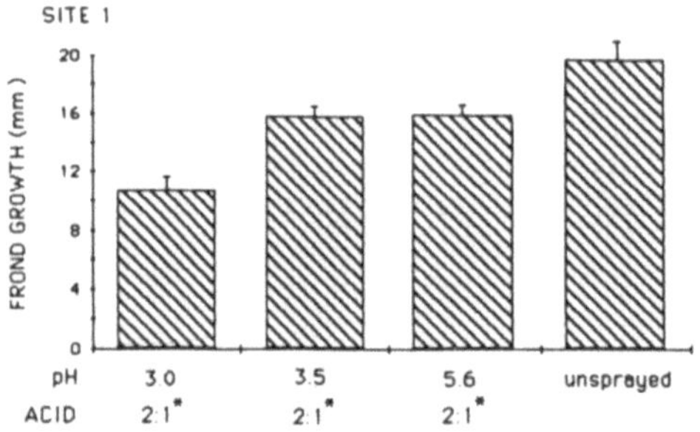

Figure 1. The growth of fronds of Pleurozium schreberi after two growing seasons of exposure to simulated acid rain of pH 3.0, 3.5 and 5.6 compared with unsprayed plants. Each histogram represents the mean and S.D. of 150 measured fronds marked at the commencement of the experiment. A one-way ANOVA and Duncan's Multiple Range test comparing mean frond growth yielded an F ratio of 27.23 ($P < 0.0001$). Means from the pH 3.5 and 5.6 groups were found to be significantly different.
* 2:1 refers to a 2:1 molar ratio of H_2SO_4 to HNO_3

In the laboratory microcosm experiment, where sprays at pH 3.0 but of different composition were used, significant differences in chlorophyll content were measured (Figure 2). The effect of acidifying the throughfall solution with H_2SO_4 or the H_2SO_4:HNO_3 mixture was significantly more destructive to chlorophyll than that of the HNO_3 acidified spray. Alternating the pH 3.0 spray with pH 5.6 ameliorated the acid influence. The HNO_3 acidified spray caused an initial greening of the fronds but after six weeks (Figure 2) the fronds had significantly less chlorophyll than the pH 5.6 sprayed fronds, as corroborated by a similar field experiment (Hutchinson 1986).

It seems that Pleurozium schreberi, a feather moss that occupies thousands of square kilometres of boreal forest, is very sensitive to acid sprays and that single rain events of as low as pH 3.0 could cause substantial damage to chlorophyll and to growth. However, at the present ambient rain pH of the sites, i.e. pH 4.2, little or no damage would be expected. A problem may arise from the occasional severely acidic events which may act selectively to change the population genetic structure of the moss, or could begin to kill or weaken the critical moss layer.

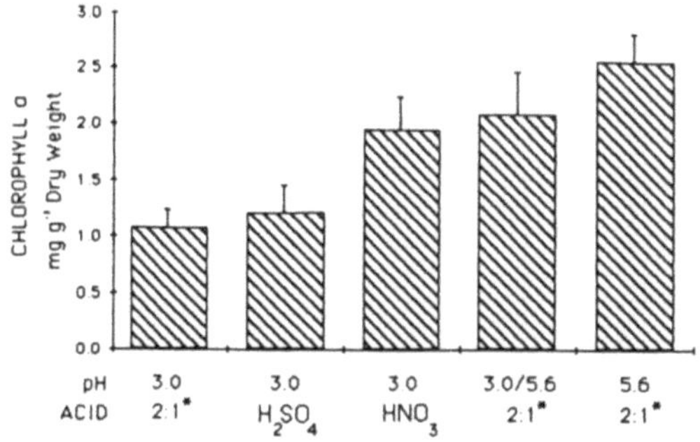

Figure 2. The chlorophyll a content of the feather moss Pleurozium schreberi when sprayed daily in greenhouse experiments with one of five spray pH treatments (see text). Values are given as means with S.D. A one-way ANOVA and Duncan's Multiple Range test comparing mean chlorophyll a content resulted in an F ratio of 35.49 ($P < 0.0001$). Two treatment pairs (2:1 and H_2SO_4 alone; HNO_3 alone and 3.0/5.6) showed significant differences.
* 2:1 refers to a 2:1 molar ratio of H_2SO_4 to HNO_3

3.2. Effects on Lichens

The responses of the three lichens to the five years of acid treatments were notably less severe than for Pleurozium. Nevertheless, by autumn 1983 browning and abnormal branching of some lichen podetia were noted. Substantial declines in both podetial height and podetial dry weight were observed at pHs less than 3.5 for all three Cladina species (Figures 3 and 4). C. stellaris showed the greatest inhibition of the three species, at pH 2.5, with a reduction in weight of 50% compared with podetia of plants sprayed with pH 3.5 (Figure 4), and a reduction in height of nearly 33%. Such declines may be the result of loss of net photosynthetic potentials at low pH (e.g., Lechowicz 1982). The vertical stratification in age of Cladina podetia is related to the degree of damage from simulated acid rain. Although an apparent growth stimulation occurs in young podetial tips sprayed with pHs less than 4.0, the corresponding basal podetial tissue shows a reduction in height. Conclusions based solely on short-term examination of podetial tips may therefore be quite misleading. In these experiments, the stunting of lower internodes (Figure 5) may be due to a) a greater number of years exposure to acid rain events and/or b) a greater susceptibility to low pH of older tissues which contain less chlorophyll and are photosynthetically inactive (Nash 1980).

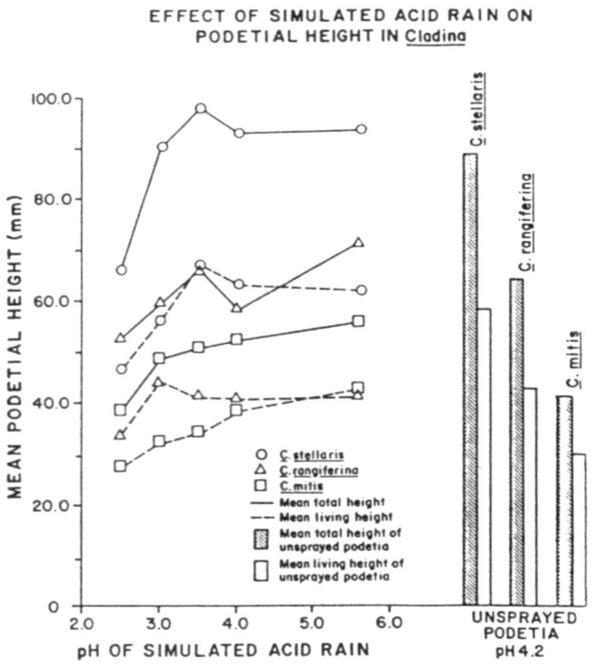

Figure 3. Effects of spraying on podetial height. A similar pattern is seen for all three species. Decreases in height are generally evident at pH 3.0 or pH 2.5. Maximum podetial height is attained at pH 3.5 for C. stellaris and C. rangiferina. The pattern of decrease in total height mirrors declines in living height; although ratios of living/total height sometimes increase at pH 2.5, possibly indicating that basal portions of the lichens are more affected than podetial tips.

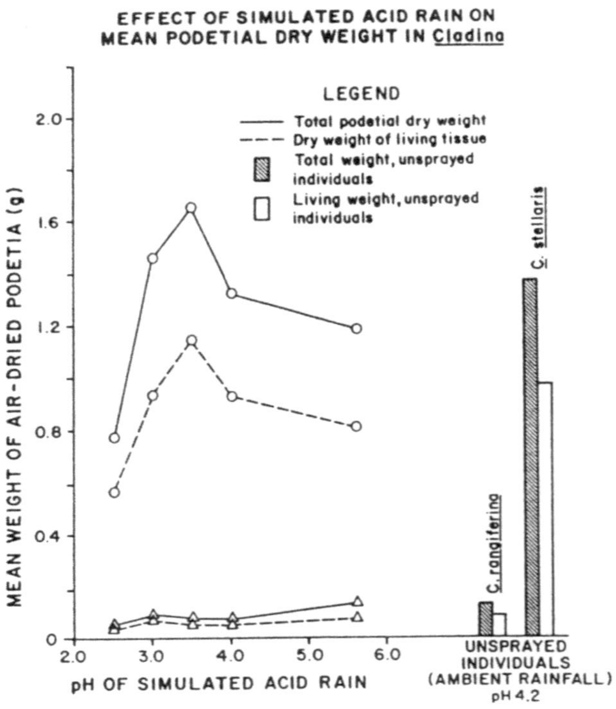

Figure 4. Effects of acidification on podetial weight. Substantial decreases in dry weight of both whole podetia and living portions of podetia are seen in pH 3.0 and pH 2.5-sprayed C. stellaris following a peak in pH 3.5-sprayed plots. A slight decline in dry matter is evident for pH 2.5-spayed C. rangiferina. Individual podetia of C. stellaris are much larger and heavier than those of C. rangiferina.

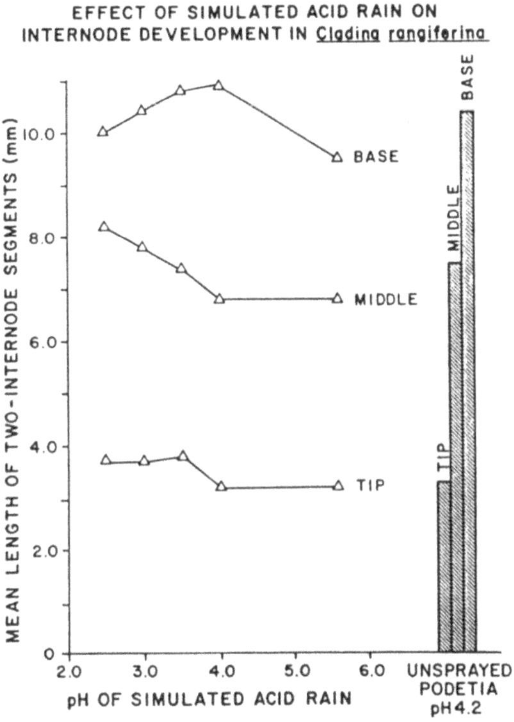

Figure 5. Living portions of podetia of C. rangiferina (n=15/treatment) were arbitrarily divided into three regions consisting of two-internode segements. The mean length (mm) of tips (internodes 2 + 3), mid-portions (internodes 4 + 5) and bases (internodes 6 + 7) were measured for pH treatments 2.5 to 5.6 and for unsprayed plots. At pH's less than 4.0, stimulation in internode lengths of tips and mid-portions occurs. Basal internodes show a reduction in height.

Separate studies at the same sites indicate that vascular perennials such as Kalmia and Vaccinium (Hutchinson, unpublished data) are tolerant of simulated rain as low as pH 2.5. In comparison, this study demonstrates that lichen and feather moss communities, which dominate huge expanses of the Canadian boreal forest, are sensitive to simulated rain events and have the potential to be affected by ambient rain episodes of less than pH 3.5. Since both Cladina and Pleurozium species intercept airborne pollutants (Moser et al. 1983; Brown and Smirnoff 1978), possible synergistic interactions between acid precipitation and gaseous pollutants (SO_2, O_3) may occur and could result in widespread devastation of boreal understory vegetation.

Although lichens are less sensitive than the feather moss, Pleurozium, it is clear from these data that rare low pH events falling on the boreal forest floor may result in a decrease in growth potential over wide geographic areas.

4. ACKNOWLEDGEMENTS

Financial assistance for the study was provided by a Natural Sciences and Engineering Council of Canada Operating Grant to T.C. Hutchinson and a NSERC Post-Doctoral Fellowship to Martha Scott. Logistic support was provided by the Ontario Ministry of Natural Resources and by funding to TCH from the Canadian National Sportsman's Fund and the National Research Council of Canada. Marilyn Feth provided invaluable field assistance with the lichen podetial measurements.

5. REFERENCES

Brown, D.H. and Smirnoff, N.: 1978, 'Observations on the effects of ozone on Cladonia rangiformis', Lichenologist, **10**, 91-94.

Dixon, M.: 1986, M.Sc. thesis, Botany Department, University of Toronto.

Hutchinson, T.C., Scott, M. and Dixon, M.: in press, Effects of experimental acidification on boreal forest ground flora lichens and mosses. In Proc. NATO Advanced Workshop on "Effects of Acidic Deposition on Forests, Wetlands and Agricultural Ecosystems". Ed. T.C. Hutchinson and K. Meema, Springer Verlag.

Kershaw, K.A.: 1977, 'Studies on lichen-dominated systems. XX. An examination of some aspects of the northern boreal lichen woodands in Canada', Can. J. Bot., 55, 393-410.

Lechowicz, M.J.: 1982, 'The effects of simulated acid precipitation on photosynthesis in the Caribou lichen Cladina stellaris (Opiz) Brodo', Water, Soil and Air Pollution, **18**, 421-430.

Moser, T.J., Nash III, T.H. and Olafsen, A.G.: 1983, 'Photosynthetic recovery in arctic caribou forage lichens following a long-term field sulfur dioxide fumigation', Can J. Bot., **61**, 367-370.

Nash III, T.H., Moser, T.J., and Link, S.O.: 1980, 'Nonrandom variation of gas exchange within arctic lichens', Can. J. Bot., **58**, 1181-1186.

EROSION EFFECTS OF AIR POLLUTION ON NEEDLE SURFACES

M.Karhu and S. Huttunen
Department of Botany
University of Oulu
SF-90570 Oulu
Finland

ABSTRACT. A scanning electron microscope was used to study the epicuticular wax of conifer needles from forests in the San Bernardino Mountains, California, the Austrian mountains near Graz and Finnish forests under local air pollution stress combined with the northern cold climate. The purpose of the study was to compare and define the role of needle erosion as a factor in tree damage, to compare the injury types in California and European countries and to characterize and further identify the possible role of O_3 in needle injuries. In all cases in this preliminary study, there were chronic erosion effects in slightly stressed trees. The different sites where trees were studied, ranging from slightly polluted to heavily polluted areas characterized by different pollutant types, yielded a variety of erosion types. A new injury type related to insect and O_3 effects in experimental and field studies was described. Additional research to confirm the results is needed.

1. INTRODUCTION

Wintertime erosion caused by air pollution of the surface of evergreen needles and its possible effects on the pathogenic sensitivity and the wintertime water economy of conifers have been studied (Huttunen and Laine, 1981, 1983; Cape and Fowler, 1981; Cape,1983). It has been noted that wintertime accumulation (mainly acid deposition of S compounds) on green surfaces results in cuticle degradation, which is widespread in trees growing in slightly polluted forests and on a larger scale in forest areas with acid deposition. Corrosion and degradation were most rapid during January and February, when the surface accumulation of aerosols and particulate substances was at its highest (Huttunen and Laine, 1981, 1983).

The erosion effects of air pollutants on needle surfaces were studied by scanning electron microscope under experimental and field conditions. The needles of the different age classes of heavily and slightly stressed trees and healthy trees were obtained from the San Bernardino Mountains, California, the Austrian mountains near Graz and Finnish forests under local air pollution stress combined with the northern cold climate. The erosion of needles was compared with O_3

Water, Air, and Soil Pollution **31** (1986) 417–423.

effects in Swiss experimental study.

2. MATERIAL AND METHODS

The material obtained is presented in Table I. The air-dried samples were covered with 25 nm gold-palladium film, using sputter equipment (Polaron E5100) and investigated with a JEOL JSM-35 Scanning electron microscope (15kV accelerating voltage, sample current 10^{-11}A, exposure 100 s). The preliminary treatment of needle surface wax structures was also studied. All injuries caused by air drying, mailing or other mechanical damage are excluded.

Table I. Micrographed needle material

AUSTRIA	
Picea abies (L.) Karst. October 1984	Site A; a clean area near Pöls-Aichfeld, 600 m Site B; a distinctly polluted area near the village of Fohnsdorf Site C; an area 1400 m a.s.l., O_3 can also be expected needle age; 1984 and 1983 needles
CALIFORNIA	
Pinus attenuata Lemon Pinus coulteri D. Don Pinus muricata Shaw March 1985	Botanic Garden, U.C.R., Riverside, 335-442 m a.s.l. hilly terrain needle age; 1984 and 1983 needles
FINLAND	
Pinus sylvestris L. Picea abies (L.) Karst. March 1985	Site A; a clean rural area about 50 km from the town of Oulu in northern Finland (<5 $\mu g\ m^{-3}a^{-1}$ SO_2) Site B; an urban industrial area in Oulu near a sulphitic cellulose plant, where the main emissions are SO_2+ dust (17 $\mu g\ m^{-3}a^{-1}$ SO_2) Site C; an urban industrial area in Oulu near a fertilizer plant, where main emissions are NO_x and SO_2 (14 $\mu g\ m^{-3}a^{-1}$ NO_2; 9 $\mu g\ m^{-3}a^{-1}$ SO_2) Site D; an industrial area near a steel mill in Raahe, where the main emissions are SO_2+ metal dust (16 $\mu g\ m^{-3}a^{-1}$ SO_2; 56 $\mu g\ m^{-3}a^{-1}$ suspended particles) and some sample areas in south-western Finland with European multiple stress disease including different stress classes needle age; 1984 and 1983 needles
SWITZERLAND	
Picea abies (L.) Karst. July 1985 after two months exposure	an experimental study with open top chambers zero (dust and active charcoal filtered), ambient air, zero +100 $\mu g\ O_3\ m^{-3}$, zero + 300 $\mu g\ O_3\ m^{-3}$ needle age; 1985 and 1984 needles

3. RESULTS

3.1 Austrian spruces

The needles of the year 1983 from slightly stressed Austrian spruce samples showed typical agglomerated surface waxes as compared with the samples from a clean rural site. Agglomerated surface waxes were seen both in a distinctly polluted area and in an area 1400 m a.s.l., where O_3 is also expected. Some samples were distinctly characterized by particles on the surfaces. The needles of the year 1983 were quite eroded (Figure 1). Agglomerated waxes in 1984 needles obtained in October 1984 were only seen in the site at 1400 m a.s.l.

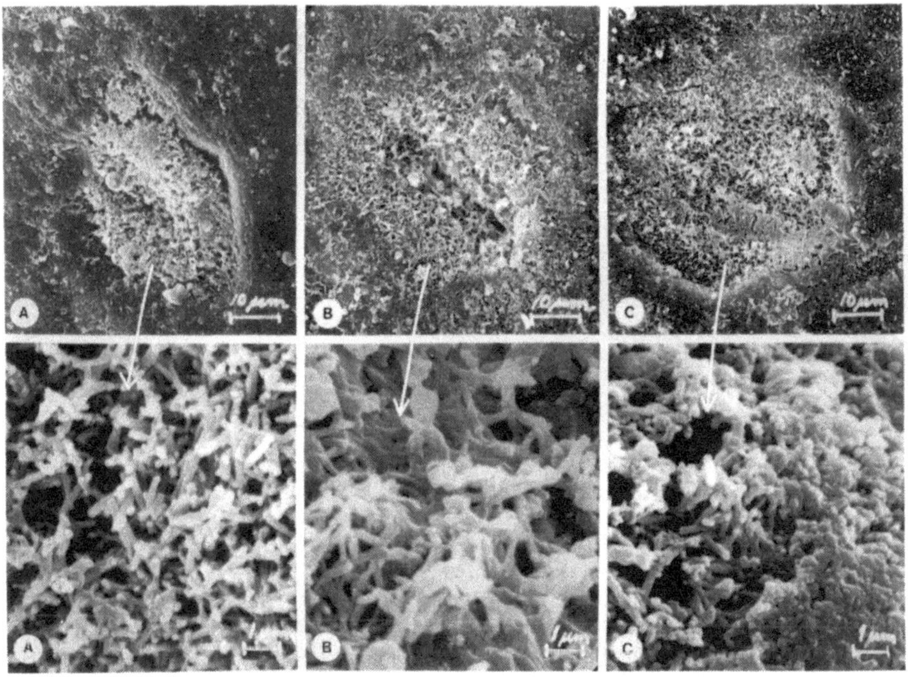

Fig. 1. Norway spruce needles from Austria near Graz. A. Needles from a clean area near Pöls-Aichfeld. Normal needle surface waxes in the 1983 needles obtained in October 1984. B. Needles from a distinctly polluted area near Fohnsdorf and C. Needles from the site 1400 m a.s.l. where one can also expect to find O_3. In both places the typical agglomerated waxes were obtained.

3.2 Californian pines

The needle surface structures of the different pine species exhibit characteristic erosion features and surface changes. A comparison of the different species growing in the same place showed these features quite clearly. For example Pinus coulteri from the Botanic Garden at Riverside is characterized by a distinct structure of stomatal areas with crystalloid waxes and a uniform ridge-like structure between stomata. Pinus attenuata at the same place had needles with occluded stomatal areas. Pinus muricata had clear holes around stomata, and a new erosion type was observed in the samples from the Botanic Garden of the University of California. This environment was observed to have distinct effects on surface structures (Figure 2).

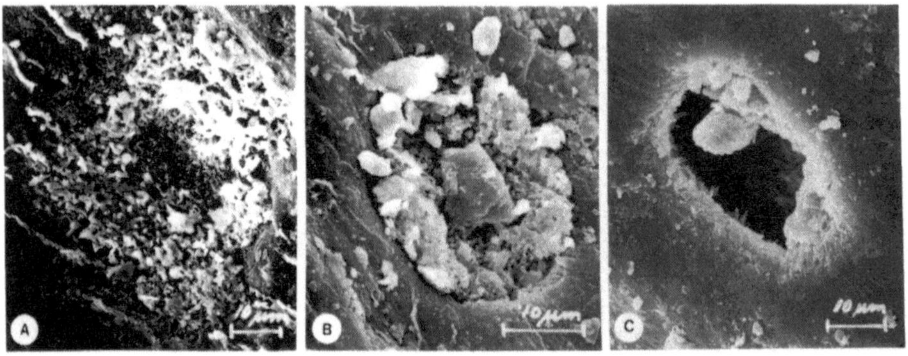

Fig. 2. Needle surface structures of the different pine species from the Botanic Garden at Riverside in California. A. Pinus coulteri. B. Pinus attenuata. C. Pinus muricata. Needles of 1984 were obtained.

3.3 Finnish spruces and pines

The needle age classes of spruce and pine needles were studied at a clean rural site in northern Finland and at two different urban industrial sites characterized by NO_x and SO_2 pollution. In spruce, needle ages up to 12 to 13 yr could be found at the rural sites, while the SO_2 polluted site was characterized by 9 needle years of growth and the NO_x polluted site by 6 needle yrs. Ruptures of the amorphous surface waxes and cuticle in the area between the stomata were the most typical changes. In addition, the change of the crystalloid starlike wax into a more curled form can be considered an ageing effect. In urban areas these changes in needles can be observed to appear much faster than in a clean rural area. In spruce needles the ageing effect cannot be noticed as easily as in pine needles.

In pine, needle ages up to 6 yr were found at rural sites; 3 needle yr were seen at the SO_2 polluted site and 4 needle yr at the NO_2 polluted site. In pine needles the well-organized structure of the crystalloid wax stomatal areas make the scanning electron microscopic monitoring of changes quite easy. The first sign of both the normal ageing of needles and the effect of air pollution is fusion of the wax fibrils. Later on all the wax fibrils are fused and eroded (Huttunen and Laine, 1981, 1983).

The different sites where trees were studied, ranging from slightly polluted to heavily polluted areas characterized by different pollutant types, yielded slightly more variations in erosion types (Figure 3).

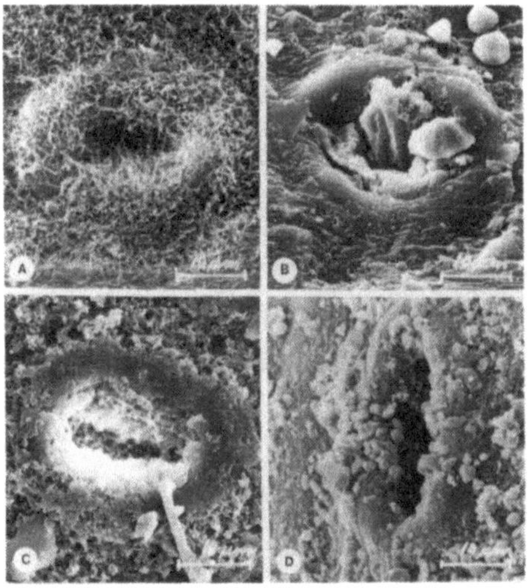

Fig. 3. The epicuticular wax structure of Scots pine from different environments. A. One-year-old needle from a rural site in Northern Finland. B. One-year-old needle from an area near the sulphite cellulose plant. C. One-year-old needle from an area near the fertilizer plant. D. One-year-old needle from an area near the steel mill.

3.4 Experimental studies with ozone

A new injury type was observed during experiments. The new injury type was characterized by holes in the epicuticular waxes and cuticle. This symptom could be easily observed because of the dark burnt-like wax material at the margins of the ruptures and the clear material in the centre (Figure 4).

All of the surface waxes from the experimental study differed somewhat from those seen in the field. The chamber conditions with air ventilation seem to give a flatter wax and surface morphology. Further studies are needed to analyze these differences. The final appearance of injuries with holes and ruptures is caused by mites or aphids, which seem to be able to find minute lesions and thinned cuticles, and are thus eat up the damaged cells. Dead bodies of mites were found on needle basis.

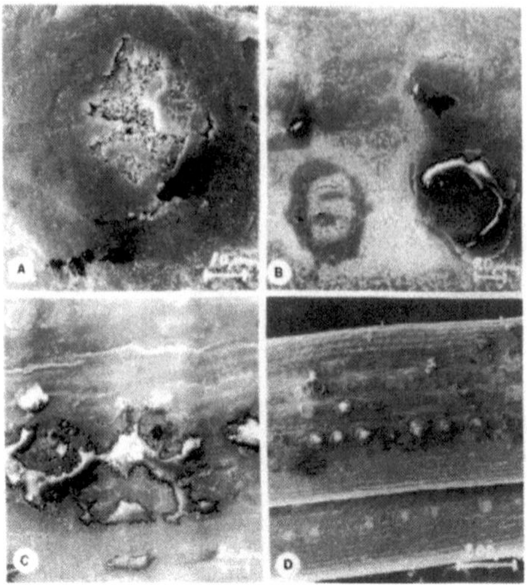

Fig. 4. Norway spruce needles from chamber experiments with ozone. A. zero (dust + charcoal filtered). B. Zürich ambient air. C. zero + 100 $\mu g\ O_3\ m^{-3}$. D. zero + 300 $\mu g\ O_3\ m^{-3}$.

4. DISCUSSION

Rehfuess (1983) and Paramesvaran *et al.* (1985) found only a minor erosion effect with an early collapse of epicuticular wax fibrils. The waxes within the epistomatal chambers are fused together, forming an occlusive encrustation. Braun and Sauter (1983) thought that this damage was caused by photochemical oxidants and SO_2 was believed to play only a synergistic role. The erosion effects of SO_2 and acidic compounds have been observed on Norway spruce needles by Grill (1973), on Scots pine needles by Fowler *et al.* (1980), Cape and Fowler (1981), and Huttunen and Laine (1981, 1983). Percy (1976) reported stomatal occlusions to be a symptom typical in conifers in polluted atmospheres and Percy and Riding (1978) have studied the erosion of *Pinus strobus* needles. Grill and Golob (1983) have demonstrated the effects of aggressive basic pollutants (e.g. the effect of magnesite dust) at very small concentrations on waxes. Bligny *et al.* (1973) have reported fluoride caused erosion effects on *Abies alba* needles. In this study erosion effects due to air pollution were observed in several different field conditions.

A new erosion type was found in O_3 gassing chamber experiments. The injury type was similar to those observed in some field samples. At the same time, the needle injury type "slight mottling" was also described in field samples. It seems possible that, in addition to acidic deposition, O_3 might have acute effects on needle surface, which might predispose the needles to insect attack. The O_3 experiment showed evidence of insect attack at the end of experiment. Additional research to confirm a cause of the new injury type is needed.

Acknowledgements. The authors wish to thank Dr. Bucher, Zürich, Dr. Bytnerowicz, Riverside, and Professor Dr. Grill, Graz, for the material for SEM studies. The scanning electron micrographies were all performed at the Institute of Electron Optics, University of Oulu, and we would especially like to thank Mrs. Sinikka Komulainen for help. The study was supported financially by the Maj and Tor Nessling Foundation.

REFERENCES

Bligny, R., Bisch, A.-M., Garrec, J.-P., and Fourcy, A.: 1973, Journal de Microscopie 17, 207.
Braun, H.J., and Sauter, J.J.: 1983, Allg. Forstzeitschr. 26/27, 656.
Cape, J.N.: 1983, New Phytol. 93, 239.
Cape, J.N., and Fowler, D.: 1981, Silva Fennica 15, 457.
Fowler, D., Cape, J.N., Nicholson, I.A., Kinnaird, J.W., and Paterson,I.S.: 1980, 'The influence of polluted atmosphere on cuticle degradation in Scots pine (Pinus sylvestris)', 146. In Drablos, D., and Tollan, A. Eds.: 1980, Ecological impact of acid precipitation. SNSF project Oslo-Ås, October 1980, p. 383.
Grill, D.: 1973, Phytopath. Z. 78, 75.
Grill, D., and Golob, P.: 1983, Aquilo Ser. Bot. 19, 255.
Huttunen, S., and Laine, K.: 1981, Archiwum Ochrony Srodowiska 2-4, 29.
Huttunen, S., and Laine, K.: 1983, Ann. Bot. Fennici 20, 79.
Paramesvaran, N., Fink, S., and Liese, W.: 1985, Eur.J.For.Path. 15, 168.
Percy, K.E.: 1976, 'Histological and histochemical effects of pulp mill emissions on eastern white pine', B.Sc.F. Thesis, University of New Brunswick, Fredericton, Mimeograph.
Percy, K.E., and Riding, R.T.: 1978, Can.J.For.Res. 8, 474.
Rehfuess, K.E.: 1983, 'Eine Arbeitshypothese über die Fichtenerkrankung in den Hochlagen des Bayerischen Waldes', 34-35. Tagungsbericht zum Symposium Sauerer Regen-Waldschäden, Jülich 27.-28.Januar 1983.

GROWTH DECLINE IN RED SPRUCE AND BALSAM FIR RELATIVE TO NATURAL PROCESSES

J.W. Hornbeck, R.B. Smith, and C.A. Federer
USDA Forest Service, Northeastern Forest Experiment Station
Durham, NH 03824 U.S.A.

ABSTRACT. Average annual growth of 3,001 red spruce (*Picea rubens* Sarg.) trees on plots randomly located across northern New England and upstate New York shows a regionally consistent decline starting soon after 1960. Depending on location, basal-area increment in 1980 was 13 to 40% less than in the peak growth year around 1960. The growth decline occurs in all size and age classes and is independent of elevation. Balsam fir (*Abies balsamea* (L.) Mill.) also shows a growth decline, but the start varies by tree age. The growth decline in both species is explainable at least in part by natural growth processes, complicating the task of evaluating the effects of atmospheric deposition.

1. INTRODUCTION

Linking atmospheric deposition to regional tree health in the Northeastern United States has proven difficult. Studies showing a narrowing of ring widths in red spruce (*Picea rubens* Sarg.) at upper elevations are one of the few consistent indications that something unusual may be occurring (Johnson, 1983; Johnson and Siccama, 1984; McLaughlin *et al.*, 1983; Roman and Raynal, 1980). In 1984 we began a tree-ring study to more fully document timing and regional and elevational extent of the red spruce decline. We also studied balsam fir (*Abies balsamea* (L.) Mill.), a common associate of red spruce.

2. METHODS

Our study is based on measurements of annual growth of 3,001 red spruce and 1,283 balsam fir trees in New England and upstate New York. The sample is composed of groups of 4 or fewer trees from several thousand randomly located plots, most of which were used in the USDA Forest Service's 10-yr inventory of forest resources in Maine, New Hampshire, and Vermont (USDA Forest Service, 1985). All sample trees were growing in forest settings and were dominant or codominant in the canopy.

Annual ring widths were measured on single increment cores

Water, Air, and Soil Pollution **31** (1986) 425–430.

extracted from each tree at about 1.4 m (breast height) above ground. Two or more cores per tree would have been preferred, but the collection has already been obtained for other objectives. Cores were mounted on wooden blocks and sanded carefully; ring widths were measured under magnification with a digital micrometer accurate to 0.01 mm. Cross-dating was done using information from Marchand (1984), and from chronologies developed during measurement.

Ring widths were converted to annual basal-area increment for comparing growth patterns (See Hornbeck and Smith {1985} for methods). This was suggested by Phipps (1983) as a means of showing whether narrowing ring widths in maturing trees represents a true growth decline. To make regional comparisons we split our sample into 5 geographic regions: (1) The Adirondack Mountains of New York, (2) Vermont, (3) New Hampshire, (4) the 5 northernmost counties of Maine which roughly correspond to the area most severly impacted by the spruce budworm, and (5) the remaining counties in southern Maine. In each region for each year we averaged the basal area increments over all trees to get a value for regional mean growth.

To determine effects of elevation, a linear regression was fitted to annual-basal area increments of each tree from 1960 through the most recent ring (1981-83). Slopes of the regression lines, indicating rates of increasing or decreasing tree growth for the period, were then compared by elevation classes.

3. GROWTH PATTERNS

Regional patterns of mean growth for red spruce are remarkably consistent across the study area (Figure 1). Except for northern Maine, where basal-area increment was suppressed from 1912-20 by budworm defoliation, growth increased steadily from 1900 to a peak in 1960 $\pm$ 2 yr. Growth then declined through the middle 1960's, recovered at varying rates until 1971, then showed further decline. By 1980, basal-area increment was 13 to 40% less than in the peak growth year around 1960, with the greatest reduction in New Hampshire and New York. This general pattern holds even when sample trees are separated by size and age classes, as shown in Figure 2 for northern Maine.

Growth patterns for balsam fir are also uniform across the region (Figure 1). However, the regional pattern differs from red spruce in that annual basal-area increment continues to increase until 1970 $\pm$ 2 yr, or 10 yr later than peak growth rates for red spruce. After 1970, annual basal-area increment in fir declined at varying rates across New England. Compared to the peak growth year around 1970, basal-area increment in 1980 ranged from 7% less in Vermont to 37% less in northern Maine. The growth pattern is independent of size class, as shown in the sample for southern Maine (Figure 3). However, separation into classes by age at breast height shows a desynchronization (Figure 3). Trees older than 60 yr had peak growth around 1960, trees 41 to 60 yr old peaked closer to 1970, and trees $<$ 40 yr old have not yet demonstrated a clear peak.

Red spruce data from New Hampshire are used to illustrate

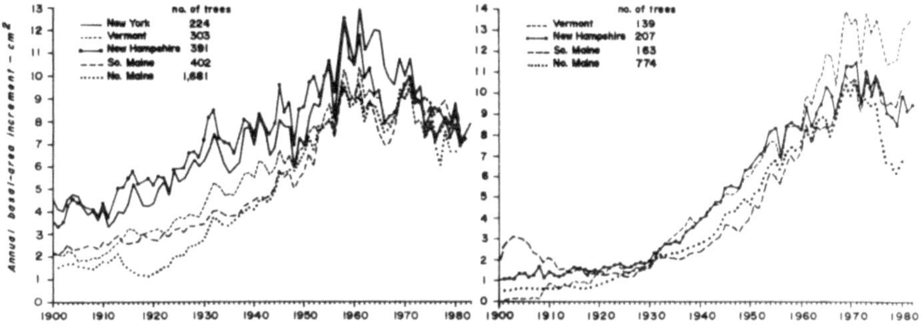

Figure 1. Mean basal-area increment for red spruce (left) and balsam fir (right).

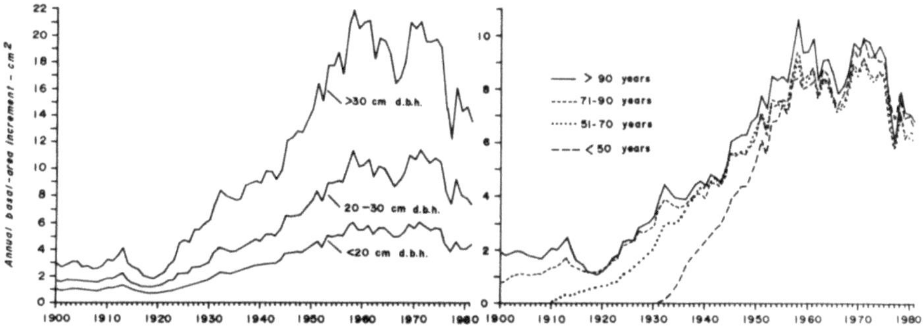

Figure 2. Mean basal-area increment for red spruce in northern Maine by size (left) and age (right) at breast height.

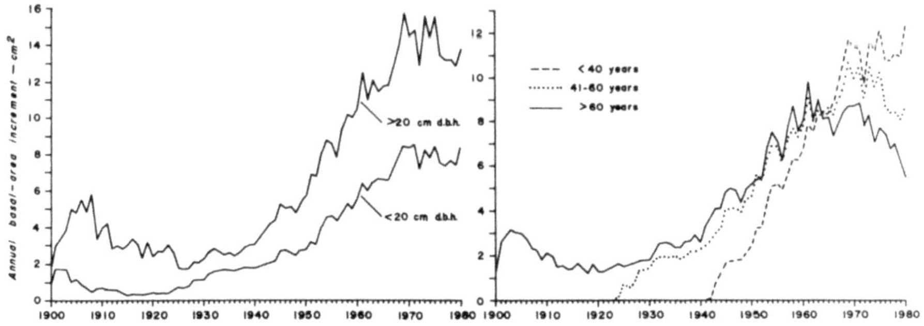

Figure 3. Mean basal area increment for balsam fir in southern Maine by size (left) and age (right) at breast height.

relationships of growth decline to elevation (Figure 4). The elevation of the 391 sample trees ranged from 100 to 1,100 m, the widest range for any of our regions. Negative regression slopes indicating declining tree growth for 1960-80 occurred for the majority of trees in all elevation classes (Figure 4), showing growth decline to be independent of elevation. These results are typical of the entire region, even low elevations in Maine where our sample included more than 500 red spruce located below an elevation of 100 m.

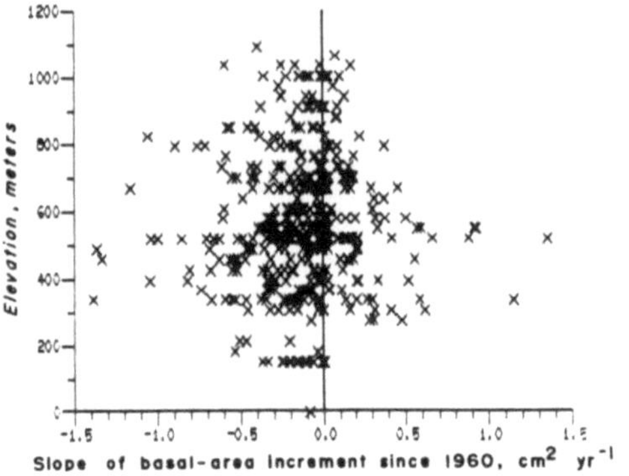

Figure 4. Relationship of growth rate for 1960-83 to elevation.

4. DISCUSSION

Documentation of the growth decline of red spruce now seems fairly complete. Our study and those either reviewed or conducted by McLaughlin (1985), Johnson and Siccama (1984), and Roman and Raynal (1980) show that growth of red spruce as determined from tree rings began a remarkably consistent decline after 1960, and that the decline extends throughout the range of red spruce within the United States. In contrast to earlier studies conducted at high elevations (Scott et al., 1984; Johnson, 1983; Siccama et al., 1982), our data and that of McLaughlin (1985) show that growth decline is unrelated to elevation, occurring from sea level to above 1,800 m.

Less information is available about the growth of balsam fir. Scott et al. (1984) found that basal area of balsam fir decreased by 35% between 1964-66 and 1982 at high elevations on Mt. Whiteface, New York, but annual growth data were not obtained. Siccama et al. (1982) reported a narrowing of ring widths in balsam fir on Camel's Hump, Vermont, similar to that for red spruce, but no data were presented. Our data substantiate growth decline for balsam fir, and show that the start of decline varies by tree age.

Several factors may be involved in growth declines; these include acid deposition, insects and diseases, climatic changes, and stand maturation (Hornbeck and Smith, 1985). Our initial investigations have focused on stand maturation, and attempt to determine if growth patterns and declines such as shown in Figure 1 differ from what would be expected as a result of natural growth processes.

Trees achieving canopy dominance normally passes through 3 phases relative to basal-area increment: (1) a steady increase through youth, (2) a leveling off during middle age, and (3) a decline with older age. Stands of mixed ages have many trees passing through each phase at a given time while even-aged stands have most trees in the same growth phase. Our data suggest that many of the red spruce in the Eastern United States are growing in stands that should be considered even-aged. As much as 80% of our trees either (1) germinated between 1900-1920, or (2) germinated between 1850 and 1900, were severly suppressed until 1900-1920, then were released and grew at rates similar to open-growing trees. The large numbers of trees either germinating or becoming free to grow between 1900-1920 result from intensive harvest around the turn of the century and from budworm activity in the early 1900's (Seymour, 1985). Thus, while tree ages may differ by as much as 70 yr, they may function as if they were even-aged, and timing of growth phases could be similar for most trees.

The phase of steadily increasing basal-area increment is roughly 1920 through the late 1950's (Figure 1). Leveling off seems to occur in the 1960's, and the decline phase is well underway in the 1970's. An approach for determining if these phases are different from that expected is to compare them with growth data that predate the present decline. Meyer (1929) contains detailed data that can be used for this purpose. His data are especially applicable, having been obtained from 201 sample plots in second growth, even-aged stands in Maine, New Hampshire, and Vermont.

Basal-area increment was derived from Meyer's data for mean stand diameter at 10-yr intervals of dominant and codominant trees on site index 60. Figure 5 shows this curve superimposed on our present-day curve for New Hampshire, the centrally located state in Meyer's study. Age zero for Meyer's data was moved forward to 1900 to coincide with initiation of many present-day red spruce stands. There is a remarkable similarity between the curves in terms of shape and timing of growth phases. Growth of present-day red spruce is following a general pattern that is predictable by Meyer's data taken in the 1920's.

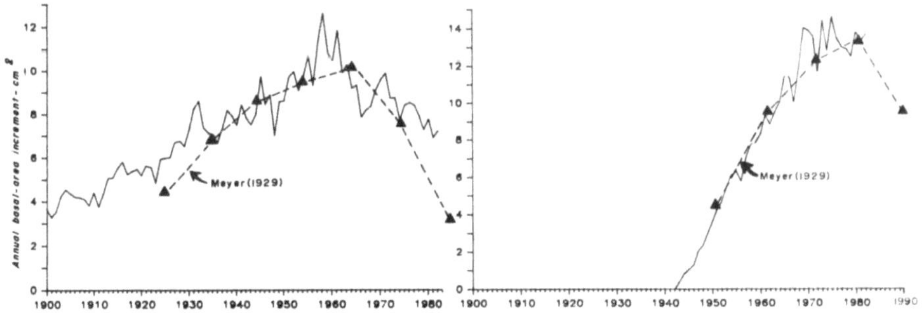

Figure 5. Basal-area increment derived from data in Meyer (1929) compared with present-day values for red spruce (left) and balsam fir (right) in New Hampshire.

The same conclusions apply to balsam fir, though complicated somewhat by age. Balsam fir matures more quickly and is shorter lived than red spruce (Westveld, 1954; Oosting and Billings, 1951), which may explain why decline occurs at different times depending on age classes (Figure 3). A curve derived from Meyer's balsam fir data matches the present-day curves of trees < 40 yr of age at breast height through at least the steadily increasing and initial leveling phases of growth (Figure 5). Growth curves for older trees (Figure 3) have the same general shape as the curve derived from Meyer, but timing and amplitude do not correspond as well as for younger trees.

5. CONCLUSION

Heavy logging and spruce budworm around the turn of the century have structured the age distribution and growth phases of large numbers of red spruce and balsam fir trees in the Northeastern United States. As a result, a decrease in basal-area increment in these species over the past 20 to 25 years is explainable at least in part by natural growth processes. The timing is coincident with increased atmospheric pollution and complicates the task of evaluating factors that cause growth decline.

6. REFERENCES

Hornbeck, J.W. and Smith, R.B.: 1985, Can. J. For. Res. **15**, 1199.

Johnson, Arthur H.: 1983, J. Air Pollut. Control Assoc. **33**, 1049.

Johnson, A.H. and Siccama, T.G.: 1984, Tappi **67**, 68.

Marchand, P.J.: 1984, Can. J. For. Res. **14**, 51.

McLaughlin, S.B.: 1985, J. Air.Pollut. Control Assoc. **35**, 512.

McLaughlin, S.B.; Blasing, T.J.; Munn, L.K.; and Duvick, D.N.: 1983, J. Air Pollut. Control Assoc. **33**, 1042.

Meyer, W.H.: 1929, Yields of second-growth spruce and fir in the northeast. USDA For. Serv. Tech. Bull. 142. 52 p.

Oosting, H.J. and Billings, W.D.: 1951, Ecology. **32**, 84.

Phipps, R.L.: 1983, In Air Pollution and the Productivity of the Forest, Izaak Walton League, Washington, DC, pp. 255-271.

Roman, J.R. and Raynal, D.J.: 1980, NYS-ERDA Rep. 80-28, Albany, NY, 4-1.

Seymour, R.S.: 1985, Forecasting growth and yield of budworm-infested forests. In Sanders, C.J., R.W. Stark, E.J. Mullins, and J. Murphy, eds., Recent advances in spruce budworm research: Proc. CANUSA spruce budworm res. symposium. Can. For. Serv. (In Press)

Scott, J.T.; Siccama, T.G.; Johnson, A.H.; and Breisch, A.R.: 1984, Bull. Torrey Bot. Club **111**, 438.

Siccama, T.G.; Bliss, M.; and Vogelmann, H.W.: 1982, Bull. Torrey Bot. Club **109**, 162.

USDA Forest Service: 1985, Forest Service Resource Inventory: An overview. USDA For. Serv., For. Resour. Econ. Res. Staff, Washington, DC. 29 p.

Westveld, M.: 1954, J. For. **52**, 11.

SNOWMELT IN A BOREAL FOREST SITE: AN INTEGRATED MODEL OF MELTWATER QUALITY (SNOQUAL1)

H.G. Jones and W. Sochanska
INRS-Eau - 2700, rue Einstein
Sainte-Foy (Québec) Canada
G1V 4C7

J. Stein, J. Roberge and
A.P. Plamondon
Université Laval
Sainte-Foy (Québec) G1K 7P4

J.Y. Charette - Environment Canada - Sainte-Foy (Québec)
G1V 4H5

ABSTRACT. An integrated model (SNOQUAL1) for the simulation of meltwater quality has been developed for a boreal forest site (Lake Laflamme, Québec). The model consists of a physical module and a chemical module; it has been validated for H^+, NO_3^- and SO_4^{2-} concentrations in discharge from the snowcover during the spring melt of 1984. Good agreement of simulated ion concentrations with those measured in the field were obtained for the strong acid anions NO_3^- and SO_4^{2-}. Greater deviations of simulated concentrations from observed concentrations were recorded in the case of H^+. These are attributed to ion exchange reactions between the ion and the organic debris which is deposited continously onto the snowcover from the forest canopy during the winter season.

1. INTRODUCTION

An integrated model for the acidity of surface waters is currently being developed by Environment Canada at the catchment area of Lake Laflamme on the Canadian Shield 80 km north of Québec City, Canada. Attention is initially being focussed on the discharge of meltwater in spring. During this period relatively high values for the concentration of H^+ in the initial discharges from the snowcover is the result of the fractionation of ionic species between the snowpack and meltwater (Johannessen and Henriksen, 1978). This process can be expressed in terms of a leaching coefficient "k" (Foster, 1978).

In the development of previous water quality models no actual field lysimeter experiments to quantify the fractionation effect have been reported. Goldstein et al. (1984), and Booty (1983) determined k by a best fit of an exponential curve to the data reported by Johannessen and Henriksen (1978). Some workers however, (Christophersen et al., 1984) have adjusted best-fit curves to the same data by using a combination of two exponential equations. The k values for these models were finally optimised during calibration runs on data from surface waters only, not on meltwater. As both the rate of melt (Seip, 1980) and snowpack composition (Jones et al., 1985) can

Water, Air, and Soil Pollution **31** (1986) 431–439.

influence the fractionation effect we have approached the development of an integrated model for snowmelt quality at Lake Laflamme, by determining k directly from data on the chemical composition of meltwater samples from the field.

The following paper describes the results of the study on the physical ablation of the snowpack and the chemical composition of the meltwaters, and their subsequent integration into a model for the quality of meltwater discharge during the spring runoff.

2. MODEL STRUCTURE

The model (SNOQUAL1, figure 1) consists of two modules; a physical module which simulates the evolution of snowcover and meltwater discharge (mm) from the pack and a chemical module which relates meltwater quality (concentration, µeq L^{-1}) to the quantity (mm) of meltwater discharge.

The basis of the physical module is the SNOW-17 model (Anderson, 1973) in which air temperature is used as an index to energy exchange across the snow-air interface. For snowmelt during non-rain periods, the following equation is utilized:

$$M = M_f \, (T_a - M_b) \tag{1}$$

For snowmelt during rain-on-snow periods, the equation becomes:

$$M = 3.67 \times 10^{-9} \, (T_a + 273)^4 - 20.4 + 0.0125 \, P_x \, T_a + 8.5 \, f(U_a) \times [(0.9 \, e_s - 6.11) + 0.00057 \, P_a \, T_a] \tag{2}$$

where M is snowmelt (mm); T_a, air temperature (°C); P_x, water equivalent of precipitation (mm); $f(U_a)$, wind function (mm mb^{-1} $6hr^{-1}$); e_s, saturated vapour pressure (mb); P_a, atmospheric pressure (mb); M_f, proportionality factor; M_b, temperature at which melt starts (°C).

The only unknowns are air temperature and precipitation, the other parameters being calculated by the program, found through calibration or as fixed input parameters.

The chemical module defines the relationship between the concentration of any ion in a meltwater sample and the quantity of meltwater discharged from the pack:

$$[C]_{i \to i+1} = \frac{[C]_o}{(H_i - H_{i+1})} \left[H_i e^{-k(H_o - H_i)} - H_{i+1} e^{-k(H_o - H_{i+1})} \right] \tag{3}$$

where:

$[C]_{i \to i+1}$ is the concentration (µeq L^{-1}) of the ion in any discrete discharge of meltwater sample ($H_i \to H_{i+1}$, mm) released during the melt period; $[C]_o$, initial concentration of the ion in the snowcover immediately before any ablation of the snowcover takes place (µeq L^{-1}); H_o, initial water equivalent of the snowcover before any ablation of

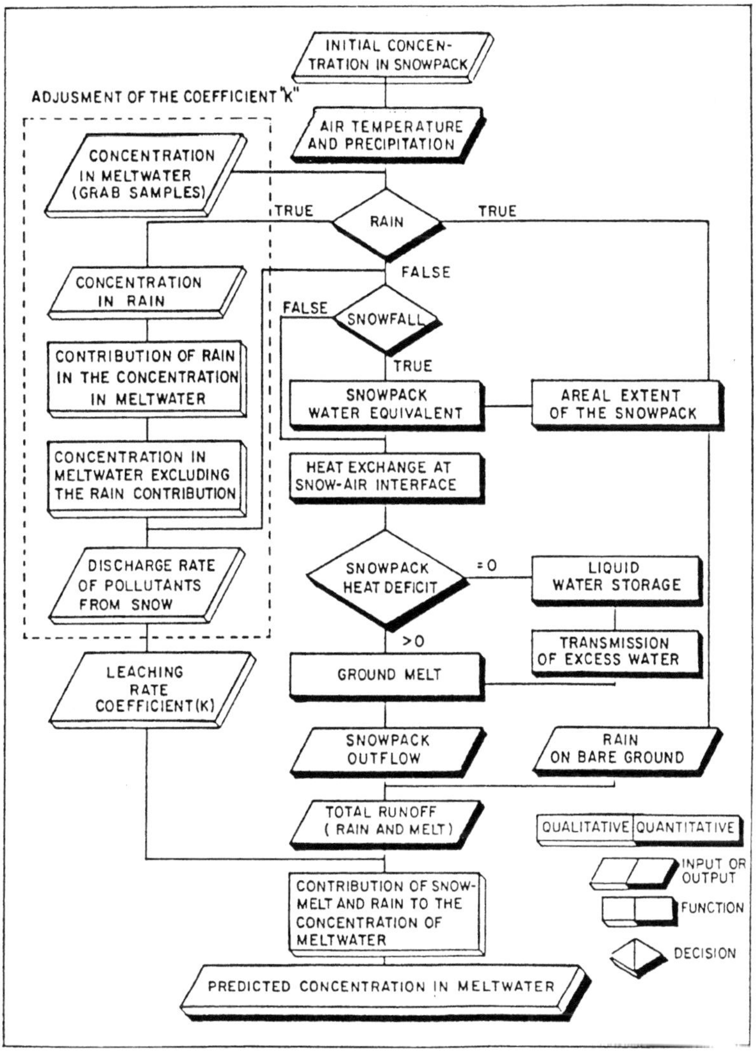

Fig.1: Integrated model for snowmelt quality (SNOQUAL1).

the snowcover takes place (mm); H_i, water equivalent of the snowcover prior to the release of the discrete meltwater sample (simulated or measured, mm); H_{i+1}, water equivalent of the snowcover after the release of the discrete meltwater sample (mm); k, leaching coefficient (mm^{-1}).

k was determined for melt periods with no precipitation by computing a best fit of equation 3 to the observed relationship between the concentrations (H^+, from pH; NO_3^-, SO_4^{2-}) and quantity of meltwater samples from the field lysimeter ($1m^2$).

For rain-on-snow episodes it is assumed that the chemical composition of all rain which passes through the pack remains unchanged. A comparison of samples from wet-only collectors and lysimeters showed that for most rain-on-snow events the relative contribution of the snowpack to the chemical content of combined meltwater-rainwater discharges, is small compared to that of rain particulary during cold (≃ 2 to 5°C) days of the melt season. The quantity of rain which can be retained in the snowcover and contribute to the next melt sequence can be estimated from the water equivalent and the density of the snowpack according to the physical module.

In cold periods with no melt, the quality of new snowfalls can be integrated into the model by recalculating a mean concentration for any ionic species in the pack.

When the physical and chemical modules are coupled the integrated model will simulate the instantaneous ionic concentration of meltwater at any time during the melt sequence.

3. SITE DESCRIPTION

The catchment area (0.68 km^2) of Lake Laflamme (0.06 km^2) is located at 47°19'N, 71°07 W in the Montmorency Forest 80 kms north of Quebec City, Canada. The area is forested; balsam fir, white spruce and white birch are the dominant species. Mean annual precipitation is 1,400 mm (34% snow), snowpack water equivalent ≃ 300 mm (max), mean temperature is 0.2°C with 132 frost-free days and the lake is ice-covered from mid-november to mid-may.

4. METHODOLOGY

All data were taken during the 1984 snowmelt season. Air temperature and surface-soil temperature were recorded by means of thermistors. Wind speed was registered at lm above the snow. Total precipitation was measured by a heated rain gauge (25 cm ID) and snowmelt from a large canvas lysimeter (20 m^2) by means of a bucket wheel. All systems were coupled to an automatic data acquisition system (CR 5, Campbell Instruments). Snowdepth and snow water equivalent were obtained with a Mount Rose sampler.

Precipitation (snow and rain) samples for chemical analysis were recovered from two Sangamo Type A wet-only collectors, one of which was installed in an open area of the forest, the other under the canopy.

Snowcover samples were obtained by coring the snowcover with a large plastic (PlexiGlas) corer having a square (25 cm x 25 cm) cutting edge. Meltwater for chemical analysis was sampled from the discharge of a field lysimeter (fiberglass tank, lm x lm x 0.5 m deep).

Precipitation samples were collected on an event basis. Lysimeter samples were collected approximately every 24 hr; the snowcover was sampled less frequently. Snow samples were kept frozen (-20°C) in polyethylene bags until melted; liquid samples were kept at 4°C up to a maximum time limit of 48 hr. pH was determined on unfiltered liquid samples by means of a pH meter (Radiometer pHM26). All liquid samples were filtered, (polycarbonate 47 mm, 0,4 μm) and the filtered samples analysed for NO_3 and SO_4 by ion chromatography (Dionex S12).

5. RESULTS AND DISCUSSION

The simulated and measured hourly runoffs for a melt sequence (Julian day 105 - 131) taken from the complete spring melt-period (Julian day 96 - 134) are presented in figure 2.

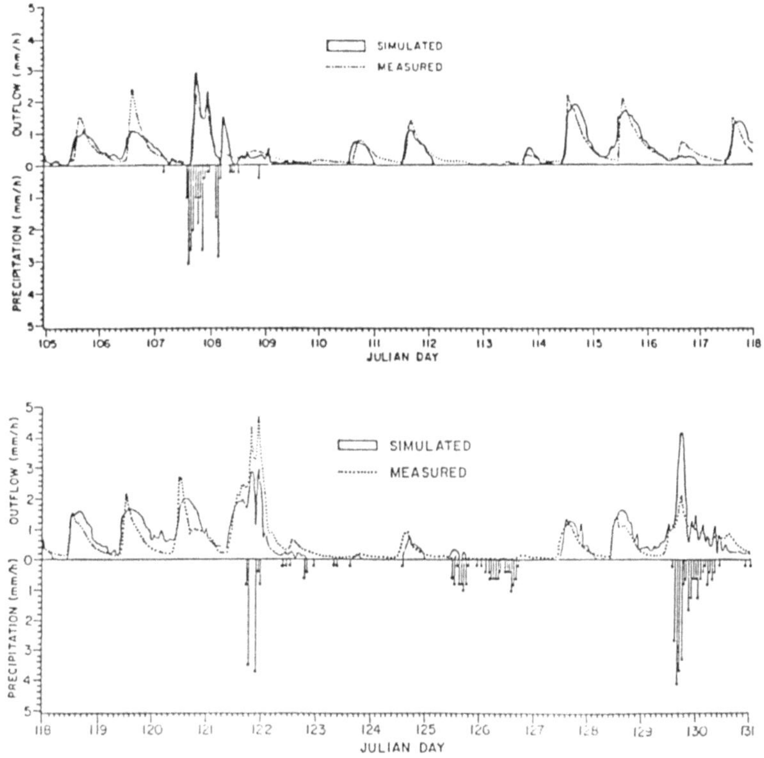

Fig.2 : Meltwater hourly outflow and total precipitation at Lake Laflamme from April 14 - May 10, 1984.

Results from the first simulation of runoff and from the sensitivity analysis of model parameters on the shape of the hydrograph led us to add two correction factors to the empirical equations describing the transmission of water in the snowpack. The time for the excess water to be released from the snowpack was shortened by 25% and its release was increased by a factor of 4.0. We attribute these correction factors to the heterogeneous nature of meltwater flow in the snowpack. The influence of rain and debris from the forest canopy leads to the formation of macropores and channels which can short-circuit some of the meltwater through the snowcover. The arithmetic mean of the errors between observed and simulated values before calibration was 0.6 mm while it was -0.1 mm after.

Using runoff data generated from the physical module, the value of k obtained from equation 3 and the initial concentration of pollutants in the snowpack, the concentration of NO_3^-, SO_4^{2-} and H^+ in meltwater discharge from the pack was simulated.

Figures 3 and 4 show the simulated and observed concentrations of NO_3 and H^+respectively, in a melt sequence that was measured from April, 24-30, 1985 (Julian day 115 - 211).

Table I reproduces the standard error of estimate (SEE) which represents the standard deviation of observed ionic concentrations from the simulated values for this period. The table also shows values for a cumulative error of estimate (CEE). CEE expresses the relative deviation (%) of the observed total ionic load discharged from the pack from that simulated by the integrated model for any specified period. SEE and CEE are complementary terms. They should be considered together as low values of CEE do not necessarily indicate satisfactory simulations.

TABLE I

Standard error of estimate (SEE, μeq L^{-1}) of observed and simulated ionic concentrations and cumulative error of estimate (CEE, %) of observed and simulated ionic loads ($L_i = \Sigma_1^n [C]_i H_i$, meq m^{-2}; L_{io} = observed L_i; L_{is} = simulated L_i; CEE = $[(L_{is} - L_{io})/L_{io}]$ 100) for the melt period April 24 - 30, 1984, Lake Laflamme, Québec

ION	SEE	L_{io}	L_{is}	CEE
NO_3^-	2.80	1.749	1.895	7.7
SO_4^{2-}	1.84	1.623	1.793	9.5
H^+	3.96	2.067	2.426	14.8

It will be noted that the model overestimates the total ionic discharges for the period by approximately 10% (7.7 - 14.8%). We believe this indicative of a satisfactory simulation considering the possible sources of error in the determination of the values of both the physical and chemical parameters required by SNOQUAL1.

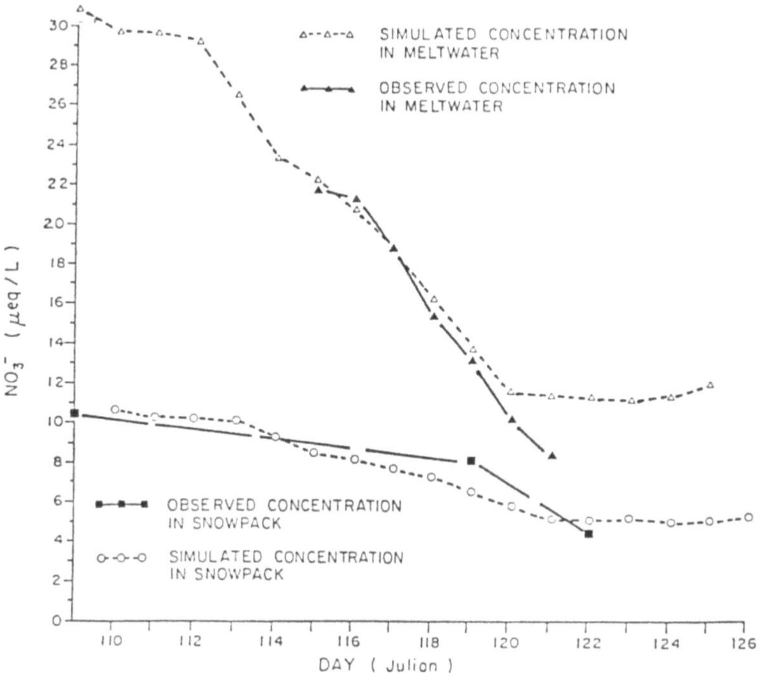

Fig.3: Concentration of NO_3^- in meltwater and snowpack, Lake Laflamme from April 18 - 30, 1984.

The computed value of k used in the simulation was 0.004 for NO_3^-, 0.005 for SO_4^{2-} and 0.0022 mm^{-1} for H^+ ions. The relatively low value of k for $H^+$is the result of ion exchange processes which take place between meltwater and organic debris deposited from the forest canopy (Jones and Sochanska, 1985). This is probably also the reason why the agreement between observed and simulated H^+concentrations is not as good as those for NO_3^- and SO_4^{2-} although errors in the determination of $[C]_o$ for H^+ from pH measurements cannot be ruled out. We have recently observed that k values change with the age and density of the snow and also vary with the species of soluble ions and the nature of the organic debris (Jones, unpublished results). The range of values for k is of two orders of magnitude (0.0002 to 0.02 mm^{-1}). At the present time the inability to determine k from physical or chemical parameters of snow samples without physically simulating the melt of whole snow cores in the laboratory is the most acute problem in the use of the model for the prediction of snowmelt quality. A series of experiments to elucidate the relationships between the value of k and the physical and chemical nature of the snowcover prior to any melt period is now being carried out.

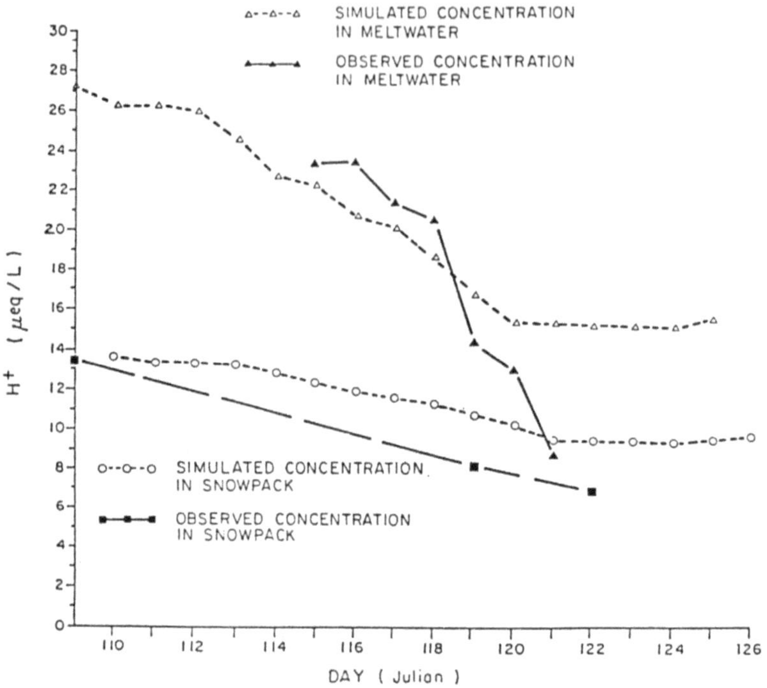

Fig.4: Concentration of H^+ in meltwater and snowpack, Lake Laflamme from April 18 - 30, 1984.

6. CONCLUSION

The integrated model simulates ion concentrations which are in good agreement with those measured in the field lysimeters for NO_3^- and SO_4^{2-}. However, the agreement between the simulated evolution for H^+ and the observed concentrations of the ion, as calculated from the pH, is not as satisfactory. To use the model as a predictive tool in the study of acid flux in snowcovers, ways to determine the leaching coefficient independantly of observed melts "in situ" or physically simulated melts in the laboratory must be found.

ACKNOWLEDGMENTS

This research was supported by the Natural Science and Engineering Research Council of Canada (operating grant); Ministère de l'enseignement supérieur et de la Science, Québec (FCAR), Environment Canada (Water Resources Research Program) and the Inland Waters Department, Québec Region, of Environment Canada (contractual research).

REFERENCES

Anderson, E.A.: 1973, 'National Weather Service River Forecast System. Snow Accumulation and Ablation Model'. NOAA Tech. Mem. NWS Hydro-17 U.S. Dept. of Commerce, Silver Spring, Md.

Booty, W.G.: 1983, 'Watershed Acidification Model and the Soil Acid Neutralization Capacity', Unpublished Ph.D. Thesis; McMaster University, Ontario.

Christophersen, N., Dymbe, L.H., Johannessen, M., and Seip, H.M.: 1984, Ecological Modelling **21**, 35.

Foster, P.: 1978, 'The Modelling of Pollutant Concentration during Snowmelt' Central Electricity Research Laboratories, U.K. Lab. note #RD/L/N 46/78.

Goldstein, R.A., Gherini, S.A., Chen, C.W., Mok, L., and Hudson, R.J.M.: 1984, Phil. Trans. R. Soc. Lond. B. **305**, 409.

Johannessen, M. and Henriksen, A.: 1978, Water Resources Research **14**: 615.

Jones, H.G. and Sochanska, W.: 1985, Annals of Glaciology, **7**: 167.

Jones, H.G., Sochanska, W., Charette, J.Y., and Stein, J.: 1985, 'Aspects Qualitatifs du Développement au Bassin du Lac Laflamme d'un Modèle Prédictif de l'Impact des Précipitations Acides: Phase de la Fonte Printanière de 1984', INRS-Eau rapport scientifique no. 182, Environment Canada, Québec Region.

Seip, H.M.: 1980, 'Acid Snow - Snowpack Chemistry and Snowmelt,' in 'Effects of Acid Precipitation on Terrestrial Ecosystems.' Hutchinson, T.C. and Havas, M. eds. Plenum Press, N.Y. pp. 77-94.

THE INFLUENCE OF MYCORRHIZAL ASSOCIATIONS ON PAPER BIRCH AND JACK PINE SEEDLINGS WHEN EXPOSED TO ELEVATED COPPER, NICKEL OR ALUMINUM

MELANIE D. JONES, MARK H.R. BROWNING and THOMAS C. HUTCHINSON
Department of Botany, University of Toronto, Toronto, Ontario, Canada M5S 1A1

ABSTRACT. Acid deposition may adversely affect northern forest ecosystems by increasing the concentration of metals in the soil solution. This study investigates the effects of ectomycorrhizal fungi on paper birch and jack pine seedlings exposed to elevated Cu, Ni, or Al in sand culture. One of four mycorrhizal fungi, _Scleroderma flavidum_, was able to reduce Ni toxicity to the birch seedlings. It did this by reducing transport of Ni to the stems. None of the fungi affected Cu toxicity in birch. In separate experiments, jack pine seedlings were exposed to combinations of Al and Ca. Infection with _Rhizopogon rubescens_ increased seedling susceptibility to Al. Seedlings inoculated with _Suillus tomentosus_ showed a greater growth stimulation by Ca than uninoculated jack pines. Thus, for both tree species, the mycorrhizal association could alter the response of seedlings to high concentrations of certain metals, although this varied with fungal species.

1. INTRODUCTION

Jack pine and paper birch are widespread and important components of the Canadian boreal forest, a region which is subject to acid rain. Aluminum is a major constituent of boreal forest soils, while Cu and Ni are serious contaminants in soils around the Sudbury smelters in Ontario. Under acid soil conditions the solubility of such metals increases, enhancing their potential availability for plant uptake.

The presence of ectomycorrhizae reduces the uptake of Zn into _Betula_ seedlings grown at potentially toxic concentrations of Zn (Brown and Wilkins, 1985). Mycorrhizae are also essential for the survival of several ericoid species in soils contaminated with Cu and Zn (Bradley _et al._, 1982).

The present study was initiated to determine: a) whether infection of either paper birch or jack pine by mycorrhizal fungi affects their growth in, and uptake of, Ni, Cu, or Al; and b) whether elevated Al concentratons affect the infectivity of two ectomycorrhizal fungi on jack pine, and whether Ca influences this effect.

Water, Air, and Soil Pollution **31** (1986) 441–448.

2. MATERIALS AND METHODS

2.1 Birch: Ni and Cu exposure

Sterile Betula papyrifera Marsh. seedlings, were grown in silica sand in 50 mL culture tubes from seeds collected from the Sudbury area. Seedlings were inoculated with agar plugs from colonies of Laccaria proxima, Lactariusm hibbardae, Lactarius rufus, or Scleroderma flavidum, or with sterile agar plugs. Following mycorrhizae formation, the seedlings were transplanted to sterile 5 cm pots which contained acid-washed silica sand, and which were topped with Al caps. Seedlings were fertilized every two weeks with 1/10 modified Ingestad's solution (Mason, 1980), at pH 3.5 with either 5 ppm Ni, as $NiCl_2$, 4 ppm Cu, as $CuCl_2$, or no metal added. After 18 weeks of treatment the seedlings were harvested, oven-dried, and digested in concentrated HNO_3. The digests were analyzed for Ni or Cu on a Perkin Elmer Model 460 atomic absorption spectrophotomer. At harvest, the degree of infection of each root system was scored as being between 0 and 1/4, 1/4 and 1/2, 1/2 and 3/4 or 3/4 and 1.

2.2 Jack pine: Al, Ca exposure

Isolates of Rhizopogon rubescens and Suillus tomentosus were cultured from basidiocarps collected in a 12 yr-old and 64 yr-old stand, respectively, of Pinus banksiana Lamb. Cultures were fragmented (Molina and Palmer, 1982), and 10 mL of inoculum applied to each seedling of P. banksiana growing in sterile chambers consisting of two nested 7.5 cm plastic pots. The upper pot contained silica sand, and was topped with an inverted translucent plastic 400 mL beaker. Four 2 cm holes in each beaker were plugged with cotton. Seedlings received one of four treatments: 1 ppm Al + 1 ppm Ca, 1 ppm Al + 20 ppm Ca, 10 ppm Al + 1 ppm Ca, or 10 ppm Al + 20 ppm Ca biweekly for 16 weeks. Al and Ca were added in a 1/4 strength modified Ingestad's solution adjusted to pH 3.8. At harvest the root area was measured using a Li-Cor LI-3100 Area Meter. The number of infected root tips on each seedling was counted under a dissecting scope. This number was divided by the seedling root area to give an indication of the degree of infection. Seedlings were then oven-dried, weighed, and digested in concentrated HNO_3. The digests were analyzed for Al and Ca using an Applied Research Laboratories 3400 Simultaneous Spectrometer.

3. RESULTS

3.1 Paper birch

Ni and Cu reduced the growth of birch by 37 and 60%, respectively, relative to controls (Figure 1a). Under Ni treatment, seedlings infected with Scleroderma flavidum weighed more than other mycorrhizal or non-mycorrhizal seedlings principally due to enhanced root growth (Figure 1b). This enhanced growth correlated with a reduced transfer of Ni to the stems of the S. flavidum-infected seedlings (Figure 2a).

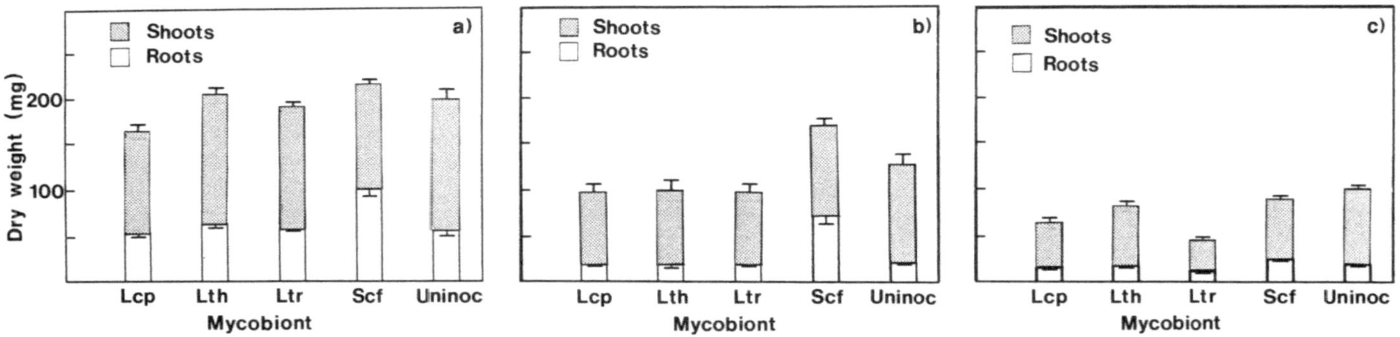

Fig. 1. Dry weights (mg) of birch seedlings grown for 18 wk a) in the absence of additional Ni or Cu, b) in the presence of 5 ppm Ni, or c) in the presence of 4 ppm Cu. Lcp = Laccaria proxima, Lth = Lactarius hibbardae, Ltr = Lactarius rufus, Uninoc = uninoculated. Means + standard errors. The significance levels for inoculation effects as detected by one-way ANOVAs are: a) shoots, P<0.05; roots, P<0.001; b) shoots, n.s.; roots, P<0.001; c) shoots, P<0.001, roots, P<0.001.

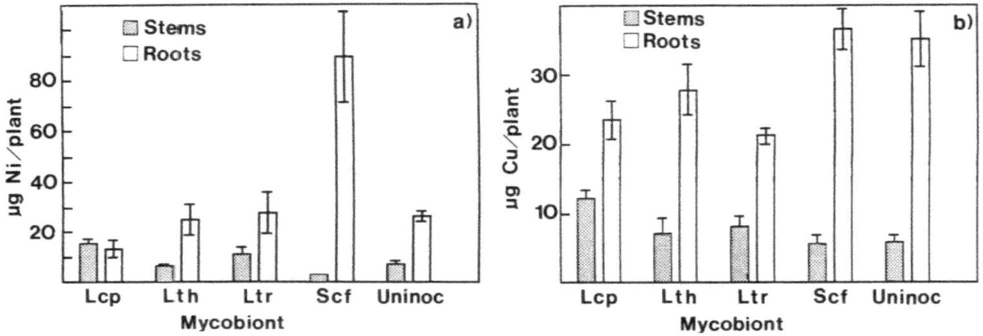

Fig. 2. The Ni or Cu content (µg per plant) of the stems and roots of birch seedlings after 18 wk of growth in the presence of a) 5 ppm Ni or b) 4 ppm Cu. The 4 mycobionts are as listed in Fig. 1. Means + standard errors. The significance levels for inoculation effects as detected by one-way ANOVAs are: a) stems, P<0.01; roots, P<0.001; b) stems, n.s., roots, P<0.05.

Other treatments resulted in at least twice as much Ni in the stems and half the amount of Ni in the roots.

None of the mycorrhizal fungi ameliorated the effect of Cu (Figure 1c). While root weights differed only slightly between inoculation treatments, shoot weights were markedly lower in Lactarius rufus-infected plants. There were no differences in the transfer of Cu to the stems of seedlings infected with different fungi (Figure 2b).

Rates of infection varied between the four fungi. Infection was greater than 1/2 in 82% of the seedlings inoculated with either Laccaria proxima or S. flavidum. Lactarius hibbardae infected only poorly with 66% of seedlings showing less than 1/4 infection. Infection by L. rufus was variable.

Neither Ni or Cu affected infection by any of the fungal species (Kruskal Wallis). Under either metal treatment, S. flavidum was the only fungus to show a significant relationship between the degree of infection and host weight ($p < 0.05$). When all seedlings were analyzed as a group, there was a positive relationship between degree of infection and root weights ($p < 0.05$) under the Ni treatment. Shoot weights were not correlated with infection for either metal.

3.2 Jack pine

Pine seedlings infected with Suillus tomentosus were smaller than uninoculated seedlings, regardless of Al or Ca treatment (Figure 4). Aluminum at 10 ppm reduced both shoot height and root weight in seedlings infected with R. rubescens. There was no corresponding effect of Al on uninoculated seedlings. Calcium at 20 ppm increased root growth in seedlings infected with S. tomentosus, especially when combined with high Al.

As well as inhibiting growth, 10 ppm Al reduced the infectivity of R. rubescens (Figure 3). There was no ameliorative effect of Ca. Aluminum did not affect the proportion of root infected by S. tomentosus.

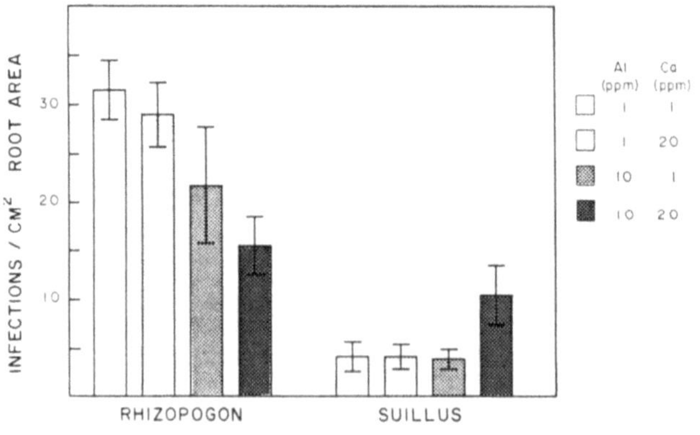

Fig. 3. The mean number of ectomycorrhizal root tips per cm2 of root area, with standard errors, for jack pine seedlings inoculated with one of two mycorrhizal fungi and treated with one of four Al and Ca combinations. The significance level for Al treatment effect on Rhizopogon rubescens infection is p + 0.005 (two-way ANOVA).

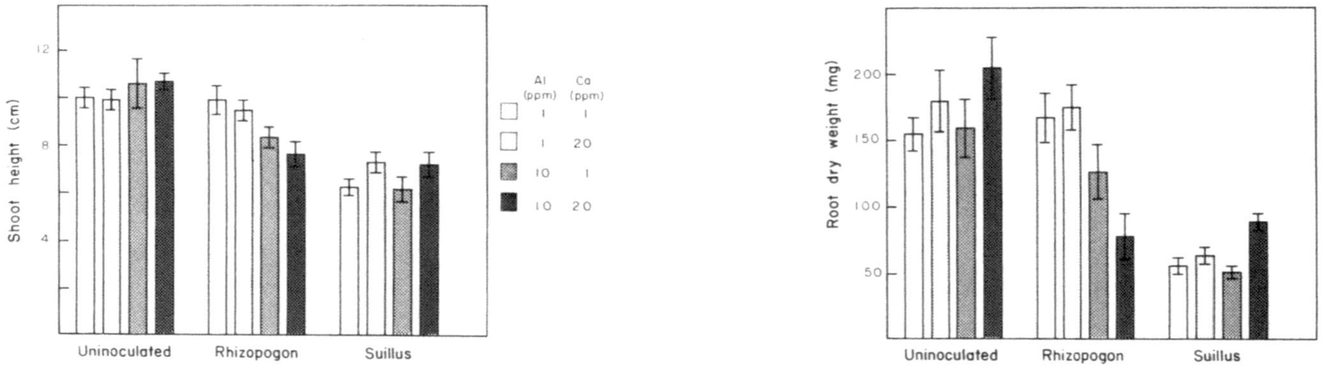

Fig. 4. The mean shoot height (cm) and root dry weight (mg) for jack pine seedlings when inoculated with one of two mycorrhizal fungi, or when not inoculated, and grown in the presence of two Al and Ca concentrations in combination. The bars indicate standard errors. The significance levels for Al treatment effects on seedlings inoculated with Rhizopogan rubescens are p=0.005 for shoot height and p=0.001 for root weight (two-way ANOVAs). Significance level for Ca treatment effect on root weight of seedlings inoculated with Suillus tomentosus is p=0.002.

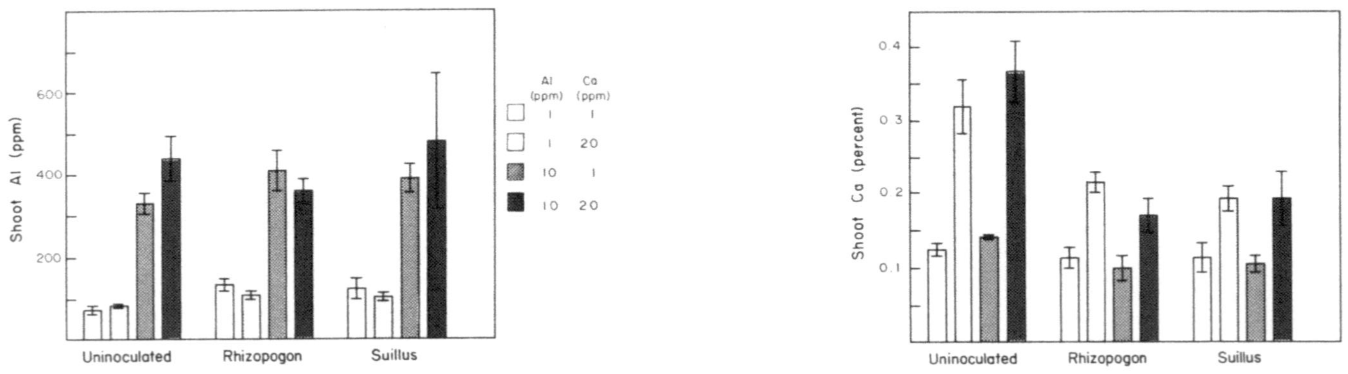

Fig. 5. The Al and Ca concentrations (ppm dry weight) in the shoots of jack pine seedlings grown in the presence of combinations of two concentrations of Al and Ca. Bars indicate standard errors. The seedlings were either uninoculated or inoculated with one of two mycorrhizal fungi. The significance level for Al treatment effect on Al concentration is $p < 0.001$ (three-way ANOVA). Significance levels for inoculum treatment and Ca treatment effects on Ca concentrations are $p < 0.001$ and $p = 0.001$, respectively.

The concentration of Al in shoots (Figure 5a) or roots of the seedlings was not influenced by fungal treatment, and thus cannot explain differences in growth between the various inoculum treatments. Calcium treatment had no effect on Al concentration in the shoots.

Shoot Ca concentration was reduced by both fungal species, especially at the high Ca level (Figure 5b).

4. DISCUSSION

Mycorrhizal and non-mycorrhizal seedlings responded differently to Al, Ca, Cu, and Ni but no generalization about the influence of ectomycorrhizal infections can be made. The differences in response of R. rubescens and S. tomentosus to Al and Ca may relate to the characteristics of the sites from which these isolates originated. The R. rubescens isolate came from a young jack pine stand. Soil water extracts (top 10 cm) from this site had low levels of both Al and Ca (Al 0.8 ppm, Ca 1.8 ppm, Browning, 1986). The S. tomentosus isolate was obtained from a mature jack pine stand where soil water extracts had higher levels of both elements (Al 40.9 ppm, Ca 70.8 ppm). It is not surprising then that R. rubescens showed a greater sensitivity to Al. In contrast, the isolates of Laccaria proxima, Lactarius hibbardae, Lactarius rufus, and Scleroderma flavidum used to infect the birch seedlings were all obtained from the same site near Sudbury, Ontario.

These results do not reflect differences in infectivity between fungi. Although infection of jack pine by S. tomentosus was lower than that of R. rubescens, it had a major effect on seedling characteristics relative to uninoculated plants. If the effect was strictly related to the degree of infection, seedlings infected by Suillus tomentosus would be expected to be more similar to uninoculated seedlings that those infected by R. rubescens. This was not observed. The same results occurred in birch seedlings which were only lightly infected by L. hibbardae. The response of birch seedlings was positively correlated with the degree of infection only for Scleroderma flavidum, the fungus which increased resistance to the metals, relative to the other fungi.

When grown in Ni, the effect of the fungi on the birch was related to their ability to control the distribution of the metal within the plant. Brown and Wilkins (1985) found that infection of Betula by either Amanita muscaria or Paxillus involutus increased Zn tolerance of the host in a similar manner, by reducing Zn translocation to the shoot. Mycorrhizae may retain metals in the polyphosphate bodies of the fungi, or in the interfacial matrix (Bradley et al., 1982; Jensen et al., 1982).

Differences in Cu tolerance were not due to differences in total uptake or translocation of Cu to the shoots. Instead, the fungi may begin to divert increasing amounts of photosynthate from the host.

Aluminum uptake, like that of Cu, was not affected by mycorrhizal formation in this study. It may be the partitioning of internal Al which differs between inoculated and uninoculated seedlings. Jack pine seedlings infected with Pisolithus tinctorius appeared to exclude Al from the nuclei of root cortical cells (Cochrane and McNabb, 1979).

Calcium has often been shown to ameliorate Al toxicity (Foy, 1974).

The reduced translocation of Ca observed in the mycorrhizal seedlings may explain why infection with R. rubescens produced Al-sensitive seedlings. The response of seedlings infected with Suillus tomentosus is complicated by the fact that this fungus is normally found on mature tree root systems. The much smaller size of all seedlings infected with S. tomentosus may indicate that this fungus has a higher carbohydrate demand than R. rubescens.

5. CONCLUSIONS

Depending on the species of mycobiont present, growth of paper birch and jack pine may be inhibited by elevated concentrations of Cu, Ni, or Al in the soil solution. This should be considered in cases where the mycobiont present can be selected for tolerance, for example, in the outplanting of containerized nursery stock to sites contaminated by metals or prone to acid deposition.

ACKNOWLEDGMENTS

Financial support for these studies was provided by Natural Sciences and Engineering Research Council of Canada grants and scholarships.

REFERENCES

Bradley, R., Burt, A.J. and Read, D.J.: 1982, New Phytol. 91, 197.

Brown, M.T. and Wilkins, D.A.: 1985, New Phytol 99, 101.

Browning, M.H.R.: 1986, M.Sc. Thesis, University of Toronto, p. 211.

Cochrane, C.E. and McNabb, H.S.: 1979, 'Aluminum and pH effects on growth and mycorrhizal relationships of Pisolithus tinctorius' In: Fourth North American Conference on Mycorrhizae, Colorado State University.

Foy, C.D.: 1974, 'Effect of aluminum on plant growth', In: The Plant Root and its Environment, Ed. W. Carson, University Press, Charlottesville, VA, pp. 601-642.

Jensen, T.E., Baxter, J., Rachlin, J.W. and Jani, V.: 1982, Environ. Poll. (A) 27, 119.

Mason, P.A.: 1980, 'Aseptic synthesis of sheathing (ecto-) mycorrhizas' In: Tissue Culture Methods for Plant Pathologists. Eds. D.S. Ingram and J.P. Hegelson, Blackwell Scientific, Oxford, pp. 173-178.

Molina, R. and Palmer, J.G.: 1982, 'Isolation, maintenance, and pure culture manipulation of ectomycorrhizal fungi'. In: Methods and Principles of Mycorrhizal Research, Ed. N.C. Schenk, American Phytopathological Society, St. Paul, pp. 115-129.

NUTRIENT SUPPLY AND FOREST DECLINE IN SOUTHWEST-GERMANY

H.W. ZOETTL and R.F. HUETTL
Institute of Soil Science and Forest Nutrition
Albert-Ludwig-University
D-7800 Freiburg i.Br.
Federal Republic of Germany

ABSTRACT. The analysis of the "high altitude Norway spruce disease" leads to the deficient nutrient supply hypothesis as the dominant predisposing stress factor for the "new type" of forest decline in the research area (Southwest Germany). Photooxidants are seen as inciting factors causing physiological and anatomical damages in the foliage. Together with acid depositions (especially fog), the photooxidants cause increased leaching of the easily mobilized elements. Compensation for increased nutrient losses is possible for the trees depending on the availability of nutrient elements in the soil. The deficiency becomes acute at the time of the highest need of nutrients, i.e. at shoot development when last year's needles turn yellow. As the soil nutrient supply is a decisive factor for the occurence of the forest damages studied an improvement of the nutrient supply by application of fertilizers generally leads to the disappearance of these disease symptoms. Fertilization is therefore a tool to mitigate forest decline when related to an insufficient or poor nutritional status. Fertilization is not an adequate way to fight the "new type" of forest decline as it is fighting the effect and not the cause. It can, however, reduce the impact that air pollution has on forest trees by revitalizing stands and ameliorating soils.

1. INTRODUCTION

The investigation of the "new type" of forest decline observed in Central Europe since the mid 70's resulted in a variety of opinions for possible causes. Phytopathologists see the phenomenon as a new "complex-disease" marked by multiple new symptoms. They believe in a sudden outbreak in wide areas which impacted various forest tree

Water, Air, and Soil Pollution **31** (1986) 449–462.

species and is independent of site differences (SCHUETT et al., 1983). Appearance and spreading of decline symptoms are considered to be related to potential stress agents or pathogens.

Another research assessment is based on forest site science. This approach takes into account those factors that differ on a regional basis and which may have positive or negative effects. In this approach effects of air pollutants fit in easily; but regional differentiation and adequate quantification of the single stress factors are substantial. Different symptoms can thus be tested on their correlations to specific factors (cf. REHFUESS, 1985). The approach of regionally varying sets of site factors can easily be transformed into the concept of the modern stress physiology that focuses on the effects of differently effective stresses. (Pre-)Disposing (mainly long-time factors), inciting (mainly short-time factors), and contributing factors are distinguished.

Our own research into the recent forest decline phenomena is focused on the role of nutrient supply of forest sites. In this context the history of the ecological development of Central Europe's commercial forests has to be considered. Forests remained often on nutrient poor substrates and frequently under unfavorable climatical conditions, whereas the better sites had been occupied for agricultural practices. Before today's management of our forests based on economical and ecological principles, manyfold devastations of the forest soils had been the rule. Forests were used as pasture land and litter utilization. The heavy biomass output resulted in decreased soil fertility. Thus, it is generally very likely that forest stands not sufficiently supplied with essential nutrients will be more susceptible to additional stress due to increased air pollution loads.

Our aim is therefore to determine the nutritional status of forest stands that reveal symptoms of the "new type" of forest decline. Another question is whether an improvement of the nutritional status of declining trees exhibiting nutritional disturbances is possible and whether this improvement would increase the resistance against further adverse impacts.

2. HIGH ALTITUDE NORWAY SPRUCE DISEASE SYMPTOMS AND SITE

Based on concepts of forest site and forest nutrition science we have started to investigate wide spread damages which can clearly be distinguished by different symptoms. The "high altitude Norway spruce disease" is characterized by a tip-yellowing of older needles and occurs independently of tree age. The yellowing starts generally at the oldest needles in the lower crown area and leads from heavy chloroses to necroses and finally to the loss of needles. The current needles stay green. Heavily diseased trees frequently show also yellowing of current needles in the fall, whereas these needles would usually turn yellow only in the following year at the time of budbreak (MIES and ZOETTL, 1985). Height growth of damaged spruce is in general not reduced. Needle loss results in the typical "sub-top-dying"-symptom. Dying trees can be found. The disease obviously started in the mid 70's at various areas at higher elevations of our central chains of mountains and at sites with very poor parent material. During recent years these damages have also spread to lower elevations. The yellowing is related to insolation (radiation). Younger spruces under the canopy of older trees do not reveal tip-yellowing even when the same aged trees at the adjacent clear-cut area are heavily damaged.

In the southern Black Forest damaged stands are mainly located at altitudes of 800 to 1200 m a.s.l. and often on skeletal rich soils derived from periglacial solifluction debris. The geological material underlaying the heavily demaged stands is frequently granite. The humus type of these stands is often mull. Raw humus layers are very seldom. The dominant soil type is the acidic, cryptopodsolic brown-earth; well developed podsols are not common. The generally high biological acitivity especially in the west of the southern Black Forest is mainly determined by the turnover activity of the very large earthworm species Lumbricus badensis (LAMPARSKI and ZOETTL, 1981).

2.1. Soil Analysis

A site close to the "Belchen" in the southern Black Forest at an altitude of 1200 m a.s.l. was examined (ZOETTL and MIES, 1983 b). Six damaged and six healthy spruce stands 12 to 20 years in age were compared. The soils are sandy-loamy acid brown-earths. Soil data of the main rooting horizons are given in Table I.

Due to the chemical composition of the parent material the fine earth has high contents of Al and low contents of Ca and Mg. The total contents of Ca and Mg show significant differences between the soils of damaged and healthy stands. In the exchangeable fraction, however, these differences are even more pronounced. Apparently the soils of damaged stands have an extremely low supply of Mg and Ca. This becomes very evident looking at the ratio of antagonistic ions. As an example the Mg:Al-ratios are listed. The base saturation is very low; pH ($CaCl_2$) values are around 3.5. The humus and N status is favorable as the C:N-ratios range around 15. No inhibition of litter decomposition could be observed.

Table I: Soil data "Belchen". Picea abies; healthy (h), damaged (d); * significant difference ($P < 0.05$)

soil depth cm		total content ($\mu mol \cdot g^{-1}$ d.m.) Al	Ca	Mg	K
0 - 10	h	2820	25*	270*	780
	d	2940	16	170	810
20 - 30	h	3100	24*	290*	870
	d	3220	16	170	830

cm		exchangeable ($\mu eq \cdot g^{-1}$ d.m.) Al	Ca	Mg	K	Mg/Al(mol)
0 - 10	h	52.2*	11.6*	5.1*	2.7	0.147
	d	75.3	4.6	2.2	3.2	0.044
20 - 30	h	43.9	6.8*	2.4*	1.6	0.082
	d	64.8	1.9	0.8	1.7	0.019

cm		pH $CaCl_2$	C%	N%	C:N
0 - 10	h	3.65*	7.53	0.46	16.0
	d	3.56	5.03	0.33	14.7
20 - 30	h	3.43	5.19*	0.33*	15.6
	d	3.81	3.41	0.23	14.9

2.2. Analysis of Needles and Roots

The nutritional status of the investigated stands is characterized by the needle analysis data in Table II. A very good supply of N, P, and K is indicated. Calcium seems sufficient, whereas the Mg contents are within the deficiency range especially for the damaged trees. Attention must be drawn to the fact that the current needles of the damaged spruces were still green. Only the older needles of the damaged spruce stands are yellowed. The microelements Mn, Fe, and Cu are within the usual range for such acidic soils. But the Zn values are rather low with significant differences between "damaged" and "healthy" trees. No differentiation between "damaged" and "healthy" can be noticed for the toxic elements. Contents are within the usual range and show the known increase with needle age. Lead contents are at a minimum with today's levels being even lower than those observed in the 70's for comparable stands and sites (ZOETTL and MIES, 1983 a). This is probably due to the reduction of Pb concentration in gasoline in 1976. Cadmium contents are also very low. It seems evident that the indicated low heavy metal depositions cannot be responsible for the recent forest decline (ZOETTL, 1985). Also the total S contents of the needles are within a range common for the southern Black Forest. They confirm the low SO_2 deposition rates discussed later on. Thus, from both soil and needle analyses the Mg deficiency of the damaged trees becomes very clear.

Table II: Needle data "Belchen". Picea abies; 1st and 4th needle year class; healthy (h), damaged (d); * significant difference ($P < 0.05$)

	N	P	K	Ca	Mg
			$mg \cdot g^{-1}$		
1 h	15.1	2.92	12.4	2.89	0.83
1 d	13.9	2.63	11.2	2.10	0.35*
4 h	10.2	2.31	8.1	3.19	0.64
4 d	10.0	2.45	9.6	3.39	0.22*

	Mn	Fe	Zn	Cu
		$\mu g \cdot g^{-1}$		
1 h	405	50	33	3.4
1 d	522	51	23*	4.8
4 h	515	99	40	2.9
4 d	784	84	24*	4.4

	Al	Pb	Cd	S
		$\mu g \cdot g^{-1}$		$mg \cdot g^{-1}$
1 h	263	0.59	0.09	1.66
1 d	216	0.91	0.07	1.43
4 h	486	1.53	0.09	1.74
4 d	471	1.86	0.07	1.56

These findings could be objected to by assuming root damages due to toxic Al ions or heavy metals like Mn. In order to test these objections damaged and healthy spruces of comparable stands and sites were analysed for element contents of all compartments. The contents of crucial elements in the current needles and in the finest roots are presented in Table III. The contents of Al and Zn confirm the well known fact that plants accumulate toxic elements such as Al and heavy metals up to high quantities in the root tissue (ZOETTL, 1983). The Mn contents are similar in the finest roots as in the current needles. Spruce is known to tolerate Mn very well, a fact that is demonstrated by the high Mn levels in needles of some best growing stands (NEBE, 1967). It is important to note that the nutritional status of damaged and healthy trees is significantly different when comparing the Zn contents of needles. The same applies for Ca and Mg: no differences in

Table III: Element contents of Picea abies at "Sirnitz". 8 yr old; healthy (h), damaged (d); n = 3 + 3; sampling fall 1983; * significant difference (no overlapping of values)

		N	P	K	Ca	Mg
		$mg \cdot g^{-1}$ d.m.				
current needles	h	14.5	2.02	6.35	2.32*	0.51*
	d	13.9	2.10	5.41	1.27	0.24
roots <1 mm ø	h	11.0	1.98	3.61	2.75	1.17
	d	11.3	1.82	2.87	2.44	1.12

		Mn	Zn	Cu	Pb	Al
		$\mu g \cdot g^{-1}$ d.m.				
current needles	h	244	28*	2.19	1.09	130
	d	218	14	2.25	0.99	121
roots <1 mm ø	h	221	146	12.8	17	6741
	d	263	148	14.0	15	7587

the roots; significant differences, however, in the current needles. The differentiated supply of these elements can only be detected via needle analysis. Root growth and mycorrhiza formation were similar for both damaged and healthy trees. Damage of the mycorrhiza is improbable due to the optimum P supply of the trees.

2.3. Atmospheric Deposition Load

The deposition load in the study area can be characterised by data from the permanent measuring station "Kaelbelescheuer" (OBLAENDER and HANSS, 1985) at the west edge of the southern Black Forest (900 m a.s.l.). The monthly mean values of SO_2 are low and vary between 2 and 6 ppb; half hour peak values reach 25 ppb. NO_2 is even considerably lower. However, O_3 values are higher (monthly mean 23 to 50 ppb) and half hour peaks up to 120 ppb are within a range where detrimental effects might be expected (PRINZ et al., 1984).

The input of S and N is also rather low: open land bulk precipitation measured by MIES (1986) from Sept. 1982 to Sept. 1984 is 15.8 $kg \cdot ha^{-1} \cdot yr^{-1}$ S (throughfall 24.8) and 15.3 $kg \cdot ha^{-1} \cdot yr^{-1}$ N (throughfall 19.7). The deposition of H^+ is 0.82 $kg \cdot ha^{-1} \cdot yr^{-1}$ in open land and 1.25 in throughfall. Open land precipitation (rain and snow) is 1563 mm (throughfall 1153 mm). An abundant growth of epiphytic lichens in this area confirms the low SO_2 air concentration.

The heavy metal deposition is well known from earlier studies at the "Baerhalde" east of the Feldberg (ZOETTL et al., 1979; ZOETTL, 1985) and also from the input measurements by MIES (1986). The input rates are relatively low: open land precipitation Kaelbelescheuer (Baerhalde) for Pb 120 (128) $g \cdot ha^{-1} \cdot yr^{-1}$, Cd 2.5 (4.3) $g \cdot ha^{-1} \cdot yr^{-1}$, Zn 225 (174) $g \cdot ha^{-1} \cdot yr^{-1}$, Cu 50 (23) $g \cdot ha^{-1} \cdot yr^{-1}$. The heavy metal contents of above ground tree compartments (RAISCH and ZOETTL, 1983) could not be linked to the recently observed forest damages.

3. HYPOTHESIS

Based on the observed decline symptoms and the analytical data the following hypothesis was derived. At higher altitudes of the southern Black Forest the chemical composition of the parent rocks and the soil development processes under the cool-humid climate led to very acid top soils with an extremely low base saturation. Especially on granites the supply of Mg, Ca, and Zn is often insufficient. Forest practices like clear cutting and removal or burning of residues have probably increased the reduction of available nutrient reserves. The impact of higher acid atmospheric deposition will increase the leaching of basic cations from the topsoil. The insufficient nutrient supply of the soil is considered as the dominant predisposing stress factor. Photooxidants (e.g. O_3) and intensive insolation, are thought as inciting stress factors. During subsequent rainy and cloudy periods acid rain and especially acid fog may enhance the leaching of the more mobile elements from the predamaged needle tissue. These are Mg, K, Mn, Zn. Since Ca in the needles is only to a small extent physiologically active and rather insoluble it is less affected. When nutrient losses from the crown increase the tree is able to compensate by increased nutrient uptake from the soil if the supply in the rooted horizons is sufficient. In the study area this is certainly possible for K and Mn, due to high reserves of K and the high mobility of Mn in the humid-acid soils. The supply of Mg (and Zn), however, is very limited. Thus, the deficiency becomes acute at the time of bud break when the trees have the highest need for nutrients. Where the soil supply is insufficient the trees are mobilizing all available reserves in older organs; especially in older needles. For Mg this mobilization process is well known. Consequently in spring time last year's green needles turn yellow (MIES and ZOETTL, 1985).

The presented hypothesis is developed from studies of Norway spruce stands showing yellow-tip-chlorosis at older needles. Those stands are frequently located at higher altitudes of the southern Black Forest on acidic soils with extremely low base saturation. The hypothesis is supported by results of BOSCH et al. (1983) from the Bavarian Forest and of ZECH and POPP (1983) from the Fichtel-Mountains. In all these cases, Mg deficiency can be seen as a predisposing factor. It depends on the parent material to what extent a Zn deficiency is involved. The triggering factors are of different origins. In the Fichtel-Mountains the SO_2 concentration in the atmosphere (annual averages 7 to 25 ppb) is significantly higher than in the Bavarian Forest or in the Black Forest. Thus, regionally different sets of stress factors do occur.

The extension of our hypothesis to stands with a different nutrient supply is possible. Extreme Mn deficiency symptoms with yellowing of the current needles is occuring on limestone sites. For those Norway spruce stands our needle analyses showed Mn contents below 10 $\mu g \cdot g^{-1}$; whilst both the supply of Ca and Mg, and K and Zn was good or at least sufficent. The insufficient supply of Mn on these dry limestone sites is due to the high fixation of this element.

On other sites such as shallow rendzina soils Scotch pine showed extreme K deficiency due to the Ca:K-antagonism. The other crucial elements were sufficiently available. Here as well initial damage was noticed in the current green needles.

If long-term predisposition in combination with insufficient nutrient supply from the soil and triggering stress factors such as moderate air pollution loads are able to cause the described forest damages, a change of the set of stresses should lead to a decrease or an increase of the decline symptoms. In extreme cases the symptoms should completely disappear or new ones should emerge. The latter is the case in the Oar Mountains (Czechoslovakia) with a very high SO_2 load. Due to permanent mean concentrations of 40 to 50 ppb SO_2 Norway spruce has died off completely within wide areas. There the symptoms are classical fumigation damages e.g. brownish tip-necrosis, especially on the current needles. No yellowing is visible. The lichen flora is heavily reduced with increasing SO_2 concentrations. Only the very tolerant species Lecanora conizaeoides is still present. Soil factors do not play a differentiating role in the development of this decline, whereas the influence of climatical factors is obvious. With SO_2 concentrations being even higher in the valleys then at higher altitudes, the damage of stands at high altitudes is progressing much faster. This can be explained by large temperature changes; e.g. frosts may occur even during the summer months.

4. EXPERIMENTAL TESTING OF THE HYPOTHESIS

To test the hypothesis we started a comprehensive program of "diagnostic fertilization trials". This experimental approach was initiated in light of the following considerations. If weak to modest air pollution loads provoke a predisposition because of an insufficient nutrient supply the improvement of the nutritional status should lead to a decrease of damage, perhaps the symptoms should even disappear. An improvement of the nutrient supply is easily possible by the application of fertilizers. For a quick response, easily soluble chemical compounds are used. The aim of the diagnostic trials is to test whether particular nutrient deficiencies determined by both needle and soil analysis can be corrected by applying the deficient nutrients. Some results from three trials are presented. They mark two deficiency situations in Southwest-Germany: Mg and K. The site conditions of the trial plots are characterised by different parent materials, soils, and climate. The level of air pollution is rather uniform. H^+, N, S deposition and SO_2, NO_x air concentrations are low; O_3 concentrations range between 30 to 50 ppb (monthly mean values).

4.1. Diagnostic Fertilization Trial Freiburg 9b

Trial Freiburg 9b is a typical example for the widely spread "Norway spruce disease at higher elevations". The 20-yr-old Norway spruce stand located at 970 m a.s.l. in the southern Black Forest was characterised by tip-yellowing of older needle year classes. The site conditions are typical for this damage type. The acidic brown-earth with mull-like humus but very low base saturation and exchangeable Mg and Ca contents (Table IV left) is derived from periglacial solifluction debris on granite.

The nutritional status of the spruce is marked by good N, very good P and K needle concentrations, normal values of Mn, Fe, and Al; but Zn and Ca contents are low, the Mg supply is deficient (< 0.7 mg·g^{-1} d.w.; see Table IV control 1984). Thus visible symptoms, data of needle and soil analysis all correspond very well. Therefore, we applied the deficient and poorly supplied nutrients using a fast release Mg trial fertilizer, Ca nitrate and Zn sulphate. Manganese was applied to test whether uptake of this element would increase even though (based on needle content) it was not currently at deficiency levels. The fertilization was carried out spring 1984. The result of this application was a greening of the needles and a reduction of needle loss compared to the control within a six-month period.

Table IV: Diagnostic Fertilization Trials

Freiburg 9 b. Picea abies 20 yr (n=5). Fertilizer applied in May 1984: Trial fertilizer (12 % K_2O, 18 % MgO, 7 % CaO) 255 kg/ha MgO; $Ca(NO_3)_2$ 650 kg/ha; $ZnSO_4$ 7 kg/ha Zn; $MnSO_4$ 10 kg/ha Mn.

Freiburg 11 b. Abies alba 10 yr (n=10). Fertilizer applied in May 1984: 650 kg/ha $Mg(NO_3)_2$.

Ebnat 3. Picea abies 64 yr (n=5). Fertilizer applied in May 1984: Patentkali (28 % K_2O, 8 % MgO) 350 kg/ha K_2O.

Chemical soil properties: Control, 20 - 30 cm soil depth

	Freiburg 9 b	Freiburg 11 b	Ebnat 3
exchangeable cations	----------	meq·100 g^{-1} d.m.	----------
K^+	0.18	0.17	0.06
Ca^{2+}	0.40	0.32	0.14
Mg^{2+}	0.13	0.08	0.05
Mn^{2+}	0.07	0.03	0.08
Fe^{2+}	0.01	0.01	0.01
Al^{3+}	5.20	3.87	4.66
H^+	0.87	0.58	0.65
pH ($CaCl_2$)	3.9	4.1	3.9
K/Ca (mol)	0.9	1.063	0.857
Mg/Al (mol)	0.038	0.031	0.016
Ca/Al (mol)	0.116	0.12	0.38
base saturation (%)	10.5	11.4	4.5

Status of nutrition: Current needles, sampling fall 1984

	Freiburg 9 b control	Freiburg 9 b fertil.	Freiburg 11 b control	Freiburg 11 b fertil.	Ebnat 3 control	Ebnat 3 fertil.
	----------	----------	mg·g^{-1}	d.m.	----------	----------
N	13.3	15.0	14.3	16.2	16.4	15.3
P	1.8	1.7	2.3	2.4	1.6	1.3
K	7.7	7.8	8.7	10.9	2.6	5.1
Ca	1.5	3.3	4.1	4.6	4.2	6.8
Mg	0.60	0.87	0.54	0.87	1.29	1.07
	----------	----------	µg·g^{-1}	d.m.	----------	----------
Mn	890	1270	430	360	2415	2390
Zn	20	42	28	34	31	26
Fe	44	43	53	51	35	36
Al	105	125	385	405	<50	<50

The needle analysis data of the fertilized spruce reveal a highly improved nutritional status: the N content had increased, P and K stayed at optimum, the Ca content was doubled, Mg is well above the borderline value. Application of Zn and Mn had stimulated the uptake of these elements considerably. Zinc is now optimum (40 to 50 $\mu g \cdot g^{-1}$ d.w.). Thus, fertilization helped to significantly revitalize the declining trees on this site.

4.2. Diagnostic Fertilization Trial Freiburg 11b

Occasionally yellow-tip-symptoms can also be observed on silver fir (HUETTL and ZOETTL, 1985), the second most important coniferous tree species in the Black Forest. Damage on silver fir occurs under the same site conditions as described for Norway spruce. Thus, trial Freiburg 11b can be considered another example of the above mentioned damage typ. The aim of this diagnostic trial was to test the hypothesis for another coniferous tree species.

The site of Freiburg 11b has an acidic brown-earth formed on gneiss at an elevation of 600 m a.s.l. In fall 1983 yellowing of older needles and 25 to 30% needle loss were noticed for the 10-yr-old firs. The chemical soil data (Table IV middle) are similar to Freiburg 9b. Again the Al concentration in the soil solution is high impeding Mg uptake (Mg:Al-ratio). Needle and soil analytical data correspond well indicating a strong Mg deficiency which coincides with the visible discoloration symptoms. It has to be noted that threshold values for Norway spruce and silver fir cannot be considered the same. Silver fir has in general higher P, K, Ca, Mg, Mn, and Al needle contents (based on dry weight) whereas N levels are at the same level for both species (HUETTL, 1985). Also in this trial N, P, and K are rather high or at optimum levels. Calcium and Zn seem sufficiently supplied. Manganese, Fe, and Al are revealing values within the expected range. In the spring of 1984 dissolved Mg nitrate was applied to improve the Mg supply. Referring to agricultural findings (PFAFF and BUCHNER, 1958) the anion NO_3^- serves as a cation carrier, and should stimulate the Mg^{2+} uptake. In the fall of 1984 most symptoms had disappeared. The trees looked healthy and were vigorously growing (height growth). Also this treatment resulted in a considerable increase of the Mg concentration (Table IV). In addition the low Zn content was somewhat improved.

Furthermore the increased N uptake and the better Mg supply resulted in a well balanced status of nutrition according to findings of INGESTAD (1967). He suggests that for optimum nutrition all essential nutrients have to be taken up in relative proportions of N (see Table V). TAMM (1977) and INGESTAD (1967) consider needle contents between 17 and 25 $mg \cdot g^{-1}$ d.w. an optimum N supply, whereas threshold values for a sufficient N supply (depending on age and coniferous tree species) range from 12 to 15 $mg \cdot g^{-1}$ d.w. (ZOETTL, 1964; BAULE and FRICKER, 1967; KERN and MOLL, 1971). Due to the nutrient supplement the silver fir of trial Freiburg 11b fulfill these requirements (Table V).

Table V: Trial Freiburg 11b. Picea abies; element content (current needles) in relative proportions of N content; required values after INGESTAD (1967)

Element	required	control	fertilized
N	100	100	100
P	≥10	16.1	14.8
K	≥50	60.8	67.3
Ca	≥5	28.7	28.4
Mg	≥5	3.8	5.4

4.3. Diagnostic Fertilization Trial Ebnat 3

Recently needle analysis data of Norway spruce stands on loamy soils derived from either jurassic limestone or moranic sediments in Southwest Germany revealed K deficiency and often low Zn supply (HUETTL, 1985). It also has been shown by comparing earlier data of comparable stands and sites that a historical change had occured causing a 60 to 70% reduction of K contents within the last two decades (ZOETTL and HUETTL, 1985). The typical symptoms for this region are moderate needle loss (25 to 30%), "silver tinsel"-syndrom, and a heterogenous needle discoloration pattern dominated by a lemon green color.

An example for this damage type is the 64-yr-old Norway spruce stand at Ebnat 3. There jurassic limestone is covered by a layer of acid loam (Feuersteinlehm). Base saturation is low (Table IV right). The pH is also very low. The site is located at 600 m a.s.l. and receives 800 mm mean annual precipitation, about 30 to 50% less than the Black Forest sites.

Needle analysis (Table IV right) shows a clear K deficiency for the control plot (<4.0 $mg \cdot g^{-1}$ d.w.) whereas the other elements are at least sufficiently supplied. Due to the acid-humid soil the Mn needle content is high; but no Mn toxicity symptoms could be found. The observed lemon green discoloration symptoms conform to the K deficiency. The cause for the deficient K supply is unclear. Soil analysis data show only moderate differences of the exchangeable K^+ content for this site and the Freiburg area, where an optimum K nutrition was found via needle analysis. Also the K:Ca antagonism cannot fully account for the insufficient K uptake. However, the mineral composition of the parent rocks is different. The crystalline rocks (granite, gneiss) are very rich in K-feldspars. Accelerated soil acidification will lead to higher K leaching from the rooted horizons which at the Ebnat site can obviously not be replaced sufficiently. These results do also indicate that the K supply cannot be estimated sufficiently by soil analysis. Needle analysis is a much safer method.

To improve the vitality of these spruces 350 $kg \cdot ha^{-1}$ K_2O were applied. Once again in fall 1984 the trees on the fertilized plot looked revitalized. They had dark green needles and vigorous shoots. The K supplement produced a significant rise in K content (Table IV right). Due to the well known antagonistic interaction of K and Mg, the Mg uptake was reduced in favor of the deficient nutrient element.

5. SUMMARY

The results of the diagnostic fertilization trials show that nutrient deficiencies play a key role in many cases of the recently observed forest decline in Southwest-Germany. The deficiency type varies due to different site conditions. Therefore the causes for the damages can only be detected by precise differentiation of damage symptoms and distinguished analysis of potential sets of stress factors.

The hypothesis derived from the "high altitude Norway spruce disease" considers a deficient nutrient supply (Mg, K, Zn) as a decisive predisposing stress factor. Inciting factors such as photooxidants are presumed to lead to physiological and anatomical damage of the needle tissue. In combination with acid precipitation (fog) increased leaching of the easily mobilized elements (K, Zn, Mg, Mn) from the canopy may occur. The tree's reaction to compensate for the nutrient losses depends on the site properties. For trees on granite it will be possible to increase K and Mn uptake, but Mg and Zn supply is short. The frequently observed Mg (and Zn) deficiencies at these sites are thus not surprising. The deficiency situation becomes acute at the time of highest need for nutrients i.e. at budbreak. The improvement of the nutrient supply by nutrient supplements leads to a reduction or the disappearance of damage symptoms. The nutrient supply is hence a crucial factor for the development of the investigated forest damages.

6. ACKNOWLEDGEMENTS

We would like to thank the Bundesministerium für Forschung und Technologie (BMFT), the Deutsche Forschungsgemeinschaft (DFG) and the Ministerium für Ernährung, Landwirtschaft, Umwelt und Forsten Baden-Württemberg (MELUF) for financial support.

7. REFERENCES

BAULE, H. and FRICKER, C.: 1967, Die Düngung von Waldbäumen. BLV Verlag München.
BOSCH, C., PFANNKUCH, E., BAUM, U., REHFUESS, K.E.: 1983, Forstwiss. Cbl. 102, 167.
HÜTTL, R.F. and ZÖTTL, H.W.: 1985, Allg. Forstz. 40, 1011.
HÜTTL, R.F.: 1985, Freiburger Bodenkundl. Abh. 16, 1.
INGESTAD, T.: 1967, Proc. 14th IUFRO Congress, Section 22, 265.
KERN, K. G. and MOLL, W.: 1971, Allg. Forst- u. Jagdz. 142, 127.
LAMPARSKI, F. and ZÖTTL, H.W.: 1981, Mitt. Dtsch. Bodenkundl. Ges. 32, 499.
MIES, E.: 1986, Ph. D. Thesis University Freiburg (in prep.).
MIES, E. and ZÖTTL, H.W.: 1985, Forstwiss. Cbl. 104, 1.
NEBE, W.: 1967, Archiv Forstwesen 16, 109.
OBLÄNDER, W. and HANSS, A.: 1985, LfU-Bericht 97/85.
PFAFF, C. and BUCHNER, A.: 1958, Z. Pflanzenern., Düng., Bodenk. 81, 102.
PRINZ, B., KRAUSE, G.H.M., JUNG, K.-D.: 1984, Düsseldorfer Geobot. Kolloq. 1, 11.
RAISCH, W. and ZÖTTL, H.W.: 1983, Mitt. Dtsch. Bodenkundl. Ges. 38, 399.
REHFUESS, K.E.: 1985, Düsseldorfer Geobot. Kolloq. 2, 25.
SCHÜTT, P., KOCH, W., BLASCHKE, H., LANG, K.J., SCHLUCK, H.J., SUMMERER, H.: 1983, So stirbt der Wald. BLV Verlag München.
TAMM, C.O.: 1977, Skogs- o. Lantbr. Akad. Tidskr. Suppl. 11, 7.
ZECH, W. and POPP, E.: 1983, Forstwiss. Cbl. 102, 50.
ZÖTTL, H.W.: 1964, Cbl. f. ges. Forstw. (Wien) 81, 1.
ZÖTTL, H.W.: 1983, Allg. Forstz. 38, 206.
ZÖTTL, H.W.: 1985, Experientia 41, 1104.
ZÖTTL, H.W. and HÜTTL, R.F.: 1985, Allg. Forstz. 40, 197.
ZÖTTL, H.W. and MIES, E.: 1983a, Allg. Forst- u. Jagdz. 154, 110.
ZÖTTL, H.W. and MIES, E.: 1983b, Mitt. Dtsch. Bodenkundl. Ges. 38, 429.
ZÖTTL, H.W. STAHR, K., HÄDRICH, F.: 1979, Mitt. Dtsch. Bodenkundl. Ges. 29, 569.

FSC
www.fsc.org
MIX
Papier aus verantwortungsvollen Quellen
Paper from responsible sources
FSC® C105338

If you have any concerns about our products,
you can contact us on
ProductSafety@springernature.com

In case Publisher is established outside the EU,
the EU authorized representative is:
Springer Nature Customer Service Center GmbH
Europaplatz 3, 69115 Heidelberg, Germany

Printed by Libri Plureos GmbH
in Hamburg, Germany

ACIDIC PRECIPITATION

ACIDIC PRECIPITATION

Proceedings of the
International Symposium on Acidic Precipitation,
Muskoka, Ontario, September 15–20, 1985

Edited by

HANS C. MARTIN
Atmospheric Environment Service, Downsview, Ontario

PART 2

Reprinted from Water, Air, and Soil Pollution, Volume 31 (1986)

Springer-Science+Business Media, B.V.

Library of Congress Cataloging-in-Publication Data

CIP

International Symposium on Acidic Precipitation
(1985 : Muskoka, Ont.)
Acidic precipitation.

'Reprinted from Water, air, and soil pollution, volumes 30 and 31 (1986).'
Includes bibliographies and indexes.
1. Acidic precipitation (Meteorology)—Environmental aspects—Congresses. 2. Acid deposition—Environmental aspects—Congresses. 3. Acid pollution of river, lakes, etc.—Congresses. 4. Air—Pollution—Environmental aspects—Congresses. I. Martin, H. C. II. Title.
QC924.5.I57 1985 363.7'386 86-33883
ISBN 978-94-010-8019-4 ISBN 978-94-009-3385-9 (eBook)
DOI 10.1007/ 978-94-009-3385-9

Originally published by D. Reidel Publishing Company, Dordrecht, Holland in 1987
Softcover reprint of the hardcover 1st edition 1987

ACIDIC PRECIPITATION

Proceedings of the
International Symposium on Acidic Precipitation
Muskoka, Ontario, September 15–20, 1985

Guest Editor

H. C. MARTIN

Atmospheric Environment Service, Downsview, Ontario

PART 2

ACIDIC PRECIPITATION

Proceedings of the
International Symposium on Acidic Precipitation
Muskoka, Ontario, September 15–20, 1985

Guest Editor

H. C. MARTIN

Atmospheric Environment Service, Downsview, Ontario

PART 1

REGIONAL PATTERN OF ACID DEPOSITION AND FOREST DECLINE ALONG A CROSS SECTION THROUGH EUROPE

M. Hauhs and R.F. Wright
Norwegian Institute for Water Research, Box 333,
0314 Oslo, Norway

ABSTRACT. Acid deposition has caused acidification of waters and is implicated in forest dieback in Europe. Norway spruce is most widely affected. Air pollution is clearly related to this new dieback, but no single pollution factor (acid deposition, SO_2 or O_3) can explain the regional pattern. The symptoms of dieback fall into two general types - loss of brown and green needles, and needle yellowing. We propose a multiple-cause hypothesis involving the predisposing factor soil acidification which when combined with one or more inciting factor gives rise to either loss of brown and green needles (drought or frost) or needle yellowing (O_3 episodes). Forest dieback causes changes in water chemistry. Nitrate in streamwater may provide an early warning of incipient forest damage.

1. INTRODUCTION

Deposition of strong acids from the atmosphere has caused acidification of surface waters in Europe and has been implicated in the recent widespread forest dieback. Soil chemical processes mediate water acidification (REUSS and JOHNSON,1985), and a similar indirect link through the soil is incorporated in ULRICH's (1984) hypothesis for the recent forest decline in central Europe. Others such as PRINZ et al. (1984), however, suggest that the recent forest dieback is due to direct effects of gaseous air pollutants such as O_3. In either case air pollution emerges as the only factor that has changed dramatically over the last decades and thereby fits the temporal pattern of the damage.

Water, Air, and Soil Pollution **31** (1986) 463–474.

Two distinct types of damage symptoms accompany the dieback of these coniferous forests. Both types have developed during the past 10 yr at Lange Bramke, a calibrated catchment in West Germany at which data for deposition, soil, soil solution and runoff chemistry are available (HAUHS,1985a). We use these data here to develop a multiple-cause hypothesis for forest dieback and then examine the ability of this hypothesis to explain regional patterns of forest dieback, air pollution and acid deposition at 11 sites across a south-to-north gradient through Europe. We consider only Norway spruce (<u>Picea abies Karst.</u>) as this species is present at all sites and is most widely affected by recent forest decline in Europe.

2. THE MULTIPLE-CAUSE HYPOTHESIS

The calibrated catchment at Lange Bramke (West Germany) has been studied intensively since 1977 (HAUHS,1985a). During the course of these investigations the trees began to exhibit several types of visible symptoms that are typical of the widespread dieback of Norway spruce (<u>Picea abies Karst.</u>) in West Germany. We thus have the rare situation in which continuous measurements are available before and after forest effects occur.

During the 8-yr period 1977-84 the soils of the catchment sustained a net loss of Ca and Mg (HAUHS,1985a). This loss is caused by the 42 kg S ha^{-1} yr^{-1} deposition and subsequent sulfate flux through the catchment. The ecosystem is currently unstable as the remaining available base cations that are within the reach of the root system (0-60 cm) are equivalent to only about 10 years of uptake.

The decrease in capacity factors such as base saturation at Lange Bramke caused concurrent changes in soil solution chemistry (HAUHS,1985a). The Ca/Al molar ratio in soil solution at 80 cm depth has dropped by about 50% during the 8 years.

These observed changes in intensity factors in soil solution coincided with the development of forest decline symptoms at Lange Bramke. In 1982 older, sunlight-exposed needles began to yellow and this symptom spread within the spruce stand. Needles of affected trees showed Mg-deficiency; needles from two year-old shoots had only 0.17 -0.43 mg Mg g^{-1} and 1.0-2.6 mg Ca g^{-1}. Yellowing at Lange Bramke first struck the dominating trees. Affected trees show growth reductions since 1982 (basal area) and 1983 (height) (HAUHS,1985a).

In 1984 these trees began to lose needles from the center of the crown. This is often described as subtop dying. As the needle chlorosis is the most dominant symptom and can also be indentified from infrared aerial photography we will term this symptom class as "yellowing (type I)".

Table I: Hypothesis of two types of forest dieback.

	Type I Mg-deficiency	Type II Water stress
Deposition Kg ha^{-1} yr^{-1}	S < 50 N ≈ 25	S > 50 N ≈ 25
	Needles: - Chlorosis caused by Mg-deficiency	Needles: - Loss of needles due to water stress - Reduced excess bases in needle dry matter
	↑ H_2O: full water uptake ↑ NO_3: ecosystem is a sink for N-Deposition	↑ NO_3: ecosystem indicates N-saturation ↑ H_2O: physiologic water stress
	Roots: - functioning with respect to N- and water uptake	Roots: - reduced fine root biomass in mineral horizons due to toxic metal conc.
	Soil capacity factor - Decrease in base saturation Soil intensity factor: - decrease in conc. of base cations	Soil capacity factor: - Decrease in base saturation Soil intensity factor: - increase in conc. of pot. toxic ions.
	- dominant trees affected first - no increase in nitrate leaching prior to visible symptoms	- suppressed trees affected first - changes in seepage and runoff water may proceed visible symptoms (volume of runoff, NO_3, alk.).

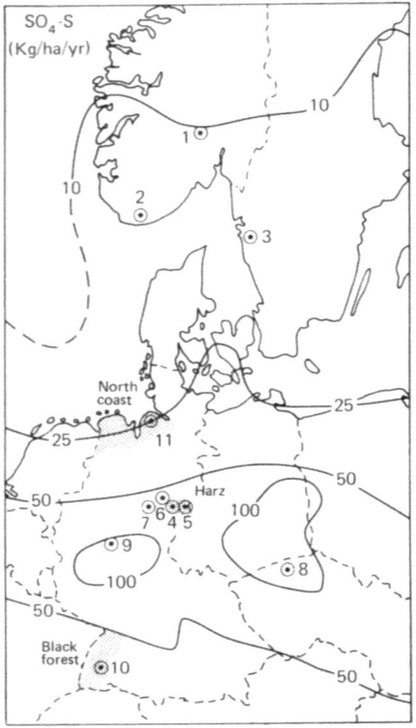

Figure 1: Location of the sites used in this comparison, the estimated pattern of total S deposition-Kg S ha^{-1} yr^{-1}) (from OECD, 1976) and the three regions within West Germany shown in Table III.

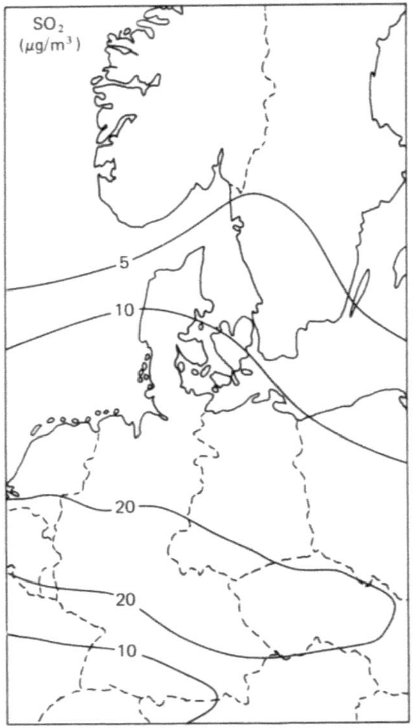

Figure 2: Mean SO_2 concentration over Europe based on rural stations for the period Jan.78 - Dec.82 (from EMEP, 1984).

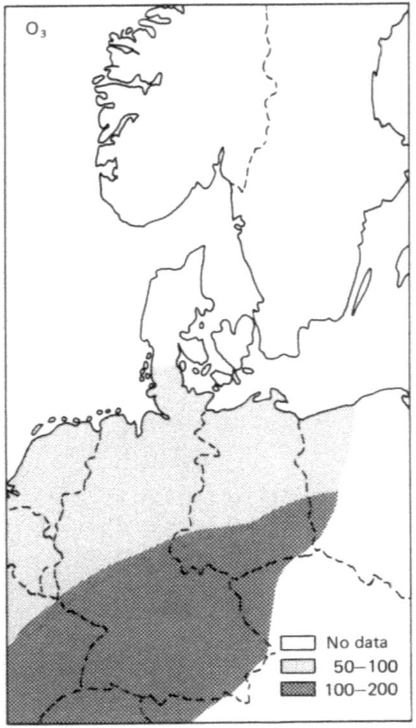

Figure 3: Frequency of episodes with elevated ozone levels in central Europe. Shown are hours per year in which ozone exceeds 75 ppbv (from Fowler et al., 1986).

At Lange Bramke the spruce also exhibits a second type of dieback symptom. At sites on the ridges and north slope loss of green and brown needles occurs. This is not preceeded by yellowing. The affected trees are mainly suppressed and growth reductions started in 1981 (basal area) and 1982 (height) (HAUHS,1985a). These trees also had reduced fine root biomass and reduced transpiration (HAUHS,1985a).

Loss of brown and green needles starts at the outer parts of the crown (ridges) or is evenly distributed throughout the crown (north slope). Needles are dropped when brown or often with a light green colour. As the needle loss proceeds in a different pattern compared to the loss of chlorotic needles, it is possible to separate these two types by infrared aerial photography (HARTMANN <u>et al.</u>, 1985). We term this class of symptoms "brown/green (type II)".

The observations in the Lange Bramke catchment on which the multiple-cause hypothesis is based are summarized in Table I. Al toxicity cannot be excluded as the critical intensity factor predisposing for both types of damages. Variations in the initial soil chemistry and in the rates of internally-produced acidification of the growing spruce stand may determine whether forest decline becomes visible as Mg-deficiency (type I) or as water stress (type II) (HAUHS,1985a).

This hypothesis implies that:

(1) Type II damage (needle loss) is preceeded by an increased nitrate leaching and

(2) Type I damage (yellowing) occurs only at moderate rates of soil acidification.

This can be tested from other sites in Europe.

3. REGIONAL PATTERNS IN EUROPE

3.1. Air pollution

The three stress factors: acid deposition, SO_2 and O_3 show distinctly different patterns in Europe. A belt of high acid deposition occurs over central Europe due to the location of the SO_2-emission sources and the prevailing wind directions (Figure 1). Both S and N compounds contribute to acid deposition.

The acid deposition belt has its largest extension in the west-east direction (Figure 1). Our 11 sites are positioned along a north-south transsect through this belt. Total S deposition (wet and dry) varies among these sites between 10 and 108 kg S ha^{-1} yr^{-1} (Table II). Sites with low or medium deposition occur both north and south of the high deposition belt.

The regional pattern in SO_2 concentration is similar to that of total S-deposition (Figure 2). The SO_2 levels affect the dry fraction of the total S-deposition. Dry deposition is a major part of the total S-load in the belt over central Germany. In southern Germany and Scandinavia, however, wet deposition comprises the major part of the S input (FOWLER et al.,1986). Four sites (4-7 in Table II) located in the same region (Figure 1) differ in their total S-deposition probably due to local differences in dry deposition and interception of cloud water. Site 6 (Hils, West Germany) is exposed to the prevailing westerly winds whereas site 4 (Lange Bramke) represents a sheltered position.

High deposition of N compounds to forests is reported from areas with intensive agriculture such as the Netherlands (ROELOFS et al., 1986) and site 11 Northern Germany (BÜTTNER et al., 1986). Most of the N-deposition to forests in those areas comes as dry deposition of ammonium sulfate particles (ROELOFS et al., 1986).

The occurrence of episodes with high O_3 levels shows a NW-SE gradient (FOWLER et al., 1986) (Figure 3). For central Europe this gradient cuts across the belt of high S deposition and high SO_2 levels. Prelimnary data from Scandinavia indicate that the pattern of O_3 concentrations follows the S-deposition with maxima along the south coast of Norway and the west coast of Sweden (GRENNFELT and SCHJOLDAGER, 1985).

FOWLER et al. (1986) distinguish among three types of pollution climate along these gradients.

Region 1: O_3 at high altitudes as the dominant pollutant with significant inputs of S and N by mainly wet deposition (example: southern Germany)

Region 2: SO_2 and NO_2 as primary pollutants, high S deposition mainly by dry deposition and a number of significant O_3 episodes (example: central Germany)

Region 3: Moderate S deposition mainly by wet deposition and a number of significant O_3 episodes (example: south coast of Norway and west coast of Sweden)

3.2. Forest dieback

Recent decline of Norway spruce is widespread in several countries in central Europe. National surveys of forest decline are based on total needle losses and do not generally distinguish between the several types of damage symptoms. Along our cross section we therefore restrict the classification to the criterium whether needle losses of more than

20% have been reported and which type of symptom dominates in these cases.

Table II: Forest dieback and surface water status at 11 sites in Europe.

Site	S-Dep. (kg S ha^{-1} yr^{-1})	Sensitivity to water acidification	Type of symptom[a]	Runoff chemistry ($\mu eq\ L^{-1}$) pH	SO_4	NO_3	Reference
1. Langtjern (F?)	10	high	none	4.9	80	3	Wright (1983)
2. Birkenes	23	high	none	4.5	132	7	Christophersen et al. (1982)
3. Gårdsjön (F1)	25	high	?	4.2	275	2	Hultberg (1985a)
4. Lange Bramke	42	low	I+II	6.1	230	49	Hauhs (1985a)
5. Eckertal	48	high	I+II	4.8	327	63	Groth (1984); Hartmann et al.(1985)
6. Hils	73	medium	II	4.0	1030	510	Raben (1985) Wiedey (1986)
7. Solling F1	85	low	II	4.0[b]	1460[b]	206[b]	Ulrich (1984)
8. X-14	108	medium-low	(dead)	4.9	1392	186	Paces (1985)
9. Elberndorfer Creek	69	low	?	6.1	396	73	Görtz et al.(1984) Block and Bartels (1984)
10. Black Forest[c]	23	high-medium	I	4.2-6.7	118-155	74-96	Zöttl et al.(1985) Huettl (1986)
11. Wingst	63	medium	II	4.0[b]	2103[b]	659[bd]	Büttner et al. (1986)

a Type I - Chlorotic needles (yellowing)
Type II - Loss of brown or green needles

b Lysimeter water from depth 1.0 m.

c this site includes a whole region where the distribution of dieback symptoms (Schöpfer and Hradetzky,1984) and surface water chemistry is available

d NH_4-N at 1.0 depth is 247 $\mu eq\ L^{-1}$ in seepage water.

No forest decline has yet been reported at either of the Norwegian sites (no.1 Langtjern and no.2 Birkenes) (Table II). On the Swedish west coast at Gårdsjön (no.3) prelimnary information suggests needle loss (Hultberg, 1986b), but it is yet unclear whether this is the same type of decline as in central Europe. At the site in northwest Czechoslovakia (no.8), on the other hand, the forest of Norway spruce was completely destroyed by 1975 (PACES, 1985). This site is situated in an area with high SO_2 levels (yearly mean of 112 µg m^{-3}) and extremely high S-deposition (Table II). The high SO_2 concentrations stem from local sources and are not reflected in the regional pattern (Figure 2).

Although the national surveys do not yet provide quantitative comparison of the different symptoms, most reports of the yellowing typical of Mg-deficiency (type I) come from high altitude sites in southern Germany (BOSCH et al., 1983; ZECH and POPP, 1983; ZÖTTL and MIES, 1983) (site 10, Table II). Infrared aerial photography indicates that the needle yellowing (type I) is also widely distributed in the Harz Mountains (central Germany) where it is concentrated on north- and east-facing slopes (HARTMANN et al., 1985). The Harz area has high S deposition (Figure 1, Table II), and cloud interception is probably a major deposition pathway. Total deposition differs substantially between exposed (southwest-facing) and sheltered (northeast-facing) slopes (Table III). At the sheltered site 4 (Lange Bramke, Harz) deposition is relatively low and comparable to site 10 (Black Forest, West Germany). Loss of brown or green needles (type II) occurs at exposed slopes in the Harz Mountains and at the north coast (Table III).

Table III: Comparison of the pollution climate and total damage at three areas with forest decline in West Germany.

Region	Pollution climate	Dominant type of damage	symptom-free Norway spruce (%)	severely damaged Norway spruce (%)
North coast	3 - moderate S-deposition, high N-dep, moderate O_3	II*	19[a]	33[a]
Harz exposed	2 - high S-deposition (cloud water interception), moderate O_3	II	20[a]	32[a]
" sheltered	3 - moderate S-deposition, moderate O_3	I+		
Black Forest	1 - moderate S-deposition, high O_3	I	13[b]	36[b]

* II = loss of green or brown needles
\+ I = yellowing

a) from HARTMANN and VORETZSCH (1985)
b) from HRADETZKY and SCHÖPFER (1984)

At three widely separated areas the surveys based on infrared air photography in 1983 showed that about 1/3 of the older (>60 yrs.) Norway spruce trees exhibited severe dieback symptoms (HARTMANN and VORETZSCH, 1985; HRADETZKY and SCHÖPFER, 1984). These areas are the north coast, the Harz Mountains, and the Black Forest and are examples of the most affected regions in Germany (Table III). Although the total damages are about the same, the type of damage symptom differs. Furthermore these regions are characterized by differences in pollution climate (Table III).

4. DISCUSSION

4.1. Links to air pollution

A multiple-cause hypothesis is necessary to explain the regional pattern of Norway spruce damages in Europe. Although very high levels of a single pollutant can explain damages at individual sites, no single pollution factor alone can account for the observed regional pattern. At the site 8 (north-west Czechoslovakia), for example, the extremely high concentrations of SO_2 alone are sufficient to explain the death of Norway spruce. The distribution of the type of damages at the other sites, however, is in general agreement with the multiple-cause hypothesis (Table I). The exposure effect in the Harz area found by HARTMANN <u>et al.</u> (1985) for type I damages (yellowing) can be explained by a lower input of acid deposition on these sheltered north-east slopes. At the west slopes exposed to interception deposition type II (loss of brown or green needles) is the dominant symptom. From the Black Forest, where the S-deposition is generally lower (Figure 1 and Table II), an exposition effect such as in the Harz area was not revealed by infrared aerial photography (SCHÖPFER and HRADETZKY,1984). PRINZ <u>et al.</u> (1984), however, mention that needle yellowing (type I) in the Black Forest occurs predominantly on west and southwest slopes; these are also the more wind-exposed. In areas that have similar S and N deposition as in the sheltered Lange Bramke valley in the Harz, Type I becames the dominant decline symptom.

Coastal areas with high Mg-inputs from cycled seasalts should not show type I damages (yellowing and Mg-deficiency). At a given S-input a lower Ca-deposition than is typical for coastal sites will result in a lower Ca/Al ratio in the soil and thereby increase the probability of type II damages. The coastal site in northern Germany (Wingst, site 11) receives a high input of NH_4 of which most is leached as nitrate. Such an ammonium sulfate deposition can have a much higher acidifying affect than the S component alone.

The two Scandinavian sites at the coast (sites 2 and 3) receive about the same acid deposition and have similar bedrock as the Black Forest (site 10). The effects on surface waters in these areas are also comparable (ANDERSSON and OLSSON,1985; ZÖTTL et al., 1985). The difference in the status of the forest cannot be explained by a hypothesis that is based on acid deposition alone. Site 2 (Birkenes,Norway) does not show decline symptoms whereas site 10 (Black Forest, West Germany) is a center of dieback. These sites differ in O_3 levels (Figure 3), and it may be this additional inciting stress factor that causes forest dieback in areas already predisposed by soil acidification. An alternative explanation for the differences between Birkenes (Norway) and the Black Forest area (West Germany) can be the contribution of internal soil acidification caused, for example, by net losses of organic-bound N from the soil at the Black Forest.

A future increase in O_3 levels in southern Scandinavia could trigger type I damages at inland sites at which the predisposing factor of soil acidification is already well advanced. As Mg-deficiency seems to be a necessary condition for damages that result in type I symptoms (yellowing), these are unlikely to occur at the coast. At coastal sites such as Birkenes further acid deposition may be required before type II symptoms will appear following climatic stress (e.g. a dry summer).

3.2. Links to water acidification

Forest dieback and water acidification are linked in three independent ways as exemplified by the data from Lange Bramke, West Germany (site 4): (1) trends in volume of runoff, (2) trends of nitrate in soil solution and runoff, and (3) changes in flux of bicarbonate in streamwater (HAUHS, 1985b). These parameters are related to those elements that show large turnover rates in forest ecosystems (water, N and C). Changes in the state of the forest such as fine root damages that affect these turnover rates are likely to have consequences for the seepage and runoff water.

The proposed link between forest dieback and water status also holds on the regional scale. Most of the 11 sites in our comparison do not have comparable long-term data on runoff. We can, however, compare nitrate levels in streamwaters at these sites (Table II).

All sites with symptoms of the new forest dieback have high nitrate levels in runoff water (sites 4 to 11 in Table II). If this represents a causual relation, surface water chemistry may provide an early warning of forest dieback (HAUHS, 1985b). A test of this hypothesis is possible at Gårdsjön, Sweden (site 3) where symptoms similar to type II have recently been observed. If these damages belong to the same decline as observed in middle Europe, they should be accompanied by an increase in nitrate exports in runoff water within several years. A similar increase in nitrate levels in streamwater is expected if forest decline starts at Birkenes, Norway (site 2).

5. CONCLUSIONS

We have examined information about long range transport of air pollutants as a cause of water acidification and forest dieback along a cross section from central to northern Europe. No single air pollutant stress factor (acid deposition, SO_2, O_3) can by itself explain forest dieback in Europe. A multiple-cause hypothesis based on the combination of predisposing factors in form of soil acidification and inciting factors such as O_3 episodes or climatic stress is sufficient to account for the distribution of different forest damage symptoms on several scales: (1) on the local scale at Lange Bramke (HAUHS,1985a), (2) on the regional scale (Harz) and (3) on the large scale along a cross section from southern Germany to central Scandinavia. The hypothesis is supported by empirical evidence for increased nitrate leaching at sites with symptoms of forest dieback.

Our hypothesis provides a basis for prediction of future forest damage symptoms to be expected in areas not yet affected. The links between acid deposition, soil acidification and acidification of surface waters are much better understood than those involved in forest dieback. Surface water chemistry may offer an early warning of incipient forest decline. One of the most promising variables may be nitrate.

ACKNOWLEDGMENTS

We thank G. Büttner, M. Dobertin, J. Gehrmann, H. Hultberg, G. Raben, J. Schjoldager and G. Wiedey for providing unpublished data.

REFERENCES

BLOCK, J., and BARTELS, U.: 1984, 'Pilotprojekt "Saure Niederschläge"'. LÖLF Mittlg. 9, 14 (Landesanstalt für Ökologie, Landespfl. und Forstpl., Düsseldorf, West Germany),

BOSCH, C., PFANNKUCH, E., BAUM, U., and REHFUESS, K.E.: 1983, Forstw. Cbl. 102, 167.

BÜTTNER, G., LAMERSDORF, N., SCHULTZ, R., and ULRICH, B.: 1986, personal communication.

CHRISTOPHERSEN, N., SEIP, H.M., and WRIGHT, R.F.: 1982, Water Resour. Res. 18, 977.

EMEP: 1984, Summary report from the chemical Co-ordinating Centre for the second phase of EMEP. Report 2/84 (Norwegian Institute for Air Research, Lillestrøm, Norway), 120 pp.

FOWLER, D., CAPE, J.N., JOST, D., and BEILKE, S.: in press, Water Air Soil Pollut.

GÖRTZ, W., MAASFELD, W., and ANNA, H.: 1984, Wasser und Boden, 36, 538.

GRENNFELT, P., and SCHJOLDAGER, J.: 1984, Ambio, 13, 61.

GROTH, P.: 1984, 'Aufbereitung eines huminsauren Oberflächenwassers im Trinkwasserwerk Eckertalsperre', p.327-331, In: Materialien 1/84, (Umweltbundesamt, E.Schmidt Verlag, Berlin, FRG).

HARTMANN, G., and VORETZSCH, A.: 1985, Forst- u. Holzwt., 40, 271.

HARTMANN, G., UEBEL, R. and STOCK, R.: 1985, Forst- u. Holzwt., 40,286.

HAUHS, M.: 1985a, 'Wasser- und Stoffhaushalt im Einzugsgebiet der Langen Bramke (Harz)', Berichte des Forschungszentrums Waldökosysteme/Waldsterben, Bd.17 (3400-Göttingen Büsgenweg 2, West Germany), 206pp.

HAUHS, M.: 1985b, Zeitschr. Dt. Geol. Gesell., 136, 585.

HRADETZKY, J., and SCHÖPFER, W.: 1984, Allg. Forstzeitschr., 39, 1080.

HUETTL, R.: (1986) Research methods and prelimnary results in the Black Forest of Germany, Proceedings of the Mid-South Symposium on Acid Deposition, Little Rock AR (in press)

HULTBERG, H.: 1985a, Ecol. Bull. (Stockholm), 37, 133.

HULTBERG, H.: 1985b, personal communication.

OECD: 1976, Longe-range transport of air pollutants, (Organisation for Economic Cooperation and Development, 2 Rue Andre, Paris).

PACES, T.: 1985, Nature, 315, 31.

PRINZ, B., KRAUSE, G.H.M., and JUNG, K.D.: 1984, Düsseldorfer Geobotanische Kolloquien, 1, 25.

RABEN, G.: 1985, personal communication.

REUSS, J.O., and JOHNSON, D.W.: 1985, J. Environ. Qual., 14, 26.

ROELOFS, J.G.M., KEMPERS, A.J., HOUDIJK, A.L.F.M. and JANSEN, J.: 1986, Plant and Soil, 84, 45.

SCHÖPFER, W. and HRADETZKY, J.: 1984, Forstw. Cbl., 103, 231.

ULRICH, B.: 1984, Atmos. Environ., 18, 621.

WIEDEY, G.: 1986, personal communication.

WRIGHT, R.F.: 1983, Hydrobiologia, 101, 1.

ZECH, W., and POPP, E.: 1983, Forstw. Cbl., 102, 55.

ZÖTTL, H.W., and MIES, E.: 1983, Allg. Forstzeischr., 38, 110.

ZÖTTL, H.W., FEGER, K.H., and BRAHMER, G.: 1985, Naturwissensch., 72, 268.

NEUTRALIZATION OF ACIDIC RAINDROPS ON LEAVES OF AGRICULTURAL CROP AND BOREAL FOREST SPECIES

T. C. Hutchinson, C. M. Adams and B. A. Gaber
Institute for Environmental Studies
University of Toronto
Toronto, Ontario, Canada M5S 1A4

ABSTRACT. The abilities of foliage of selected agricultural crop and native boreal forest species to neutralize acidic raindrops were compared. The species differed widely in their responses. Neutralization was influenced to a large extent by leaf wettability and was poorly related with species' susceptibility to foliar injury from acid rain sprayings. Little neutralization of pH 3.0 droplets occurred on very waxy leaves, e.g. cabbage (*Brassica oleracea* L.), due to the small contact area between the leaf surface and raindrops. In contrast, on sunflower (*Helianthus annuus* L.) and radish (*Raphanus sativus* L.) leaves, which are pubescent and easily wettable, neutralization was considerable. For all agricultural crop species examined, the pH of droplets drying on cotyledons was consistently higher than on the leaves. The pH values of raindrops were also higher when the foliage was injured by the acid rain, probably due to leakage of cellular contents. Among boreal forest species examined, bunchberry (*Cornus canadensis* L.) was particularly good at neutralizing natural acid rain, increasing the pH from 3.9 to 6.6 after 9 hr of foliar contact, while the response of other boreal species ranged from a final pH of 4.8 to 5.7 under the same conditions. Simulated raindrops on wild sarsaparilla (*Aralia nudicaulis* L.) were never neutralized but increased in acidity as they evaporated. Chemical analyses of droplets collected from foliage showed calcium (Ca) and potassium (K) to be the major cations entering the neutralized droplets. Neutralization of acidic raindrops appears to occur through two processes: solubilization of alkaline dusts and exudates on the leaf surface, and ion exchange removal of H^+ by the foliage.

1. INTRODUCTION

The most direct influence of rain on plants occurs at the leaf surface. Raindrops bounce off leaves, wash leaves, drip from them or accumulate as droplets. Rain contains a great diversity of material including essential minerals for plant growth, trace metals, organic compounds and acidity. Particulates from the air are also present.

Recently concern has developed over the potential for acid precipi-

Water, Air, and Soil Pollution **31** (1986) 475–484.

tation to decrease plant growth. There is relatively little knowledge available, however, on the interactions (both physical and chemical) between acidic raindrops and leaf surfaces as the droplets dry on the foliage. Exposure of plants to simulated rain with a pH below 3.5 has been shown to injure the foliar epidermis and underlying palisade and mesophyll cells (Evans and Curry, 1979; Adams *et al.*, 1984) and accelerate leaching of foliar cations (Wood and Bormann, 1975; Evans *et al.*, 1981; Scherbatskoy and Klein, 1983). Throughfall and stemflow beneath forest canopies have been shown to be substantially different from the rain above the canopy, as a consequence of leaf and bark interactions (e.g., Cronan and Reiners, 1983; Richter *et al.*, 1983). Cations such as calcium (Ca), potassium (K) and magnesium (Mg) on exchange sites on the cuticle and especially in cell walls of leaves are readily exchanged with hydrogen ions (H^+) in acid precipitation (Tukey and Morgan, 1963). Accumulations of dry deposited particulates, together with salts exuded from the plants through glandular trichomes and hydathodes, solubilize in raindrops on the leaf surface. Both processes may result in significant neutralization of acidic raindrops on the leaf surface (Adams and Hutchinson, 1984). In our earlier study we found considerable variation in the ability of the leaves of three plant species to neutralize acidic droplets, which was consistent with the amount of foliar damage they developed from sprayings with simulated acid rain. Wormwood (*Artemisia tilesii* Ledeb.), which was extremely tolerant to both SO_2 exposures and acidic rain, had the greatest ability to neutralize pH 3.0 rain. Spinach (*Spinacea oleracea* L.) and bean (*Phaseolus vulgaris* L. var. 7 Pencil Pod Wax) had poorer abilities to neutralize acidic rain and showed greater foliar injury. This suggested that the ability of leaves to neutralize acidic raindrops may be a protective mechanism. In the present study the objectives were to (1) compare the abilities of a range of agricultural crop and boreal forest species to neutralize acidic rain, (2) to evaluate the hypothesis that species which are able to effectively neutralize acid rain have a higher tolerance to acid rain damage, and (3) to determine changes in raindrop chemistry after contact with leaves of a number of plant species. Droplet pH measurements were made with a microelectrode after sprayings with simulated rain solutions in the greenhouse and in the boreal forest, and after a natural acidic rain in the boreal forest.

2. MATERIALS AND METHODS

Measurements of rain droplet pH were made on leaves of a range of agricultural crop and boreal forest species following applications of simulated rain. The chemical composition of rain used in the crop experiments approximated that of southern Ontario rain and is published elsewhere (Hutchinson and Adams, submitted for publication), while the rain sprayed on the boreal forest species contained the same ions in lower concentrations, representative of rain in northern Ontario (based on CANSAP data 1979-81). Simulated rains from pH 3.0 to 5.6 were used. Rain solutions were acidified using a 2:1 equivalent ratio of H_2SO_4 and HNO_3.

Experiments on crop species were conducted in a greenhouse and plants were sprayed in an enclosed rain chamber. The following 10 varieties were used: sunflower (*Helianthus annuus* L.), radish (*Raphanus sativus* L.), sugar beet (*Beta vulgaris* L.), cabbage (*Brassica oleracea* L.), broccoli (*Brassica oleracea* L.), brussel sprouts (*Brassica oleracea* L.), kohlrabi (*Brassica oleracea* L.), wax bean (*Phaseolus vulgaris* L.), garden beet (*Beta vulgaris* L.) and tomato (*Lycopersicum escultanum* L.). Rain nozzles produced droplets with an average diameter of 480 μm (Bete Fog Nozzle, model 1/8WL, 3/4, 120SQ). Rain showers were either 15 or 30 min in duration, delivering 0.45 and 0.90 cm of rain, respectively. pH measurements were made on droplets as the plants dried in the rain chambers. Large sample sizes of droplets were necessary due to the variability in droplet pH even on a single leaf. Values given represent the mean pH of drops on several leaves of 5 replicate plants for each of the species.

Experiments on boreal forest species were carried out *in situ* in plots established at a northern Ontario site near Kirkland Lake. A plastic backpack sprayer was used to wet foliage thoroughly. Rain applications were made to the two herbaceous species, bunchberry (*Cornus canadensis* L.) and wild sarsaparilla (*Aralia nudicaulis* L.), simultaneously, while paper birch (*Betula papyrifera* Marsh.) was sprayed separately. These experiments were carried out in July 1984. Raindrops were randomly measured on many leaves of 15 replicate plants of bunchberry and 5 replicates of the two other species. In addition, raindrop pH was measured on a range of tree and herbaceous species at this site following a natural rainfall in August 1983. These measurements were made on replicate samples of raindrops collected from the leaves using micro-pipettes. pH measurements were made with a Lazar micro-electrode (DJM-146 or 1210) in conjunction with a Markson Digital pH/millivolt meter. Parafilm surfaces were used as controls in all of the experiments.

Chemical analyses for Ca, K and Mg in raindrops collected using microelectrodes from leaves of agricultural crop species and from parafilm were made with an atomic absorption spectrophotometer.

3. RESULTS AND DISCUSSION

The acidity of rainwater changed during its exposure to a leaf surface (Figures 1 and 2, Table I). Rain with an initial pH of 5.6 was affected much less by the leaf surface than the more acidic rain (pH 3.8 and 3.0). The largest increase in pH, i.e. from 5.6 to 7.0 (Table 1), represents a decrease of only 2 μeq/L H^+, compared with a decrease of 937 μeq/L H^+ when the pH was increased from 3.0 to 4.2 (Figure 1b). There was a wide range of response of the various leaf surfaces to acidic rain. Species generally fell into 3 categories: (1) those which neutralized the low pH rain (sunflower, radish, cabbage c.v. "Salarite", broccoli, brussel sprouts, garden beet, wax bean, tomato, bunchberry and paper birch) (2) those which had no effect on rain acidity at the low pH (cabbage c.v. 'Viking Golden Acre', kohlrabi and sugar beet) and (3) one species whose raindrops acidified (wild

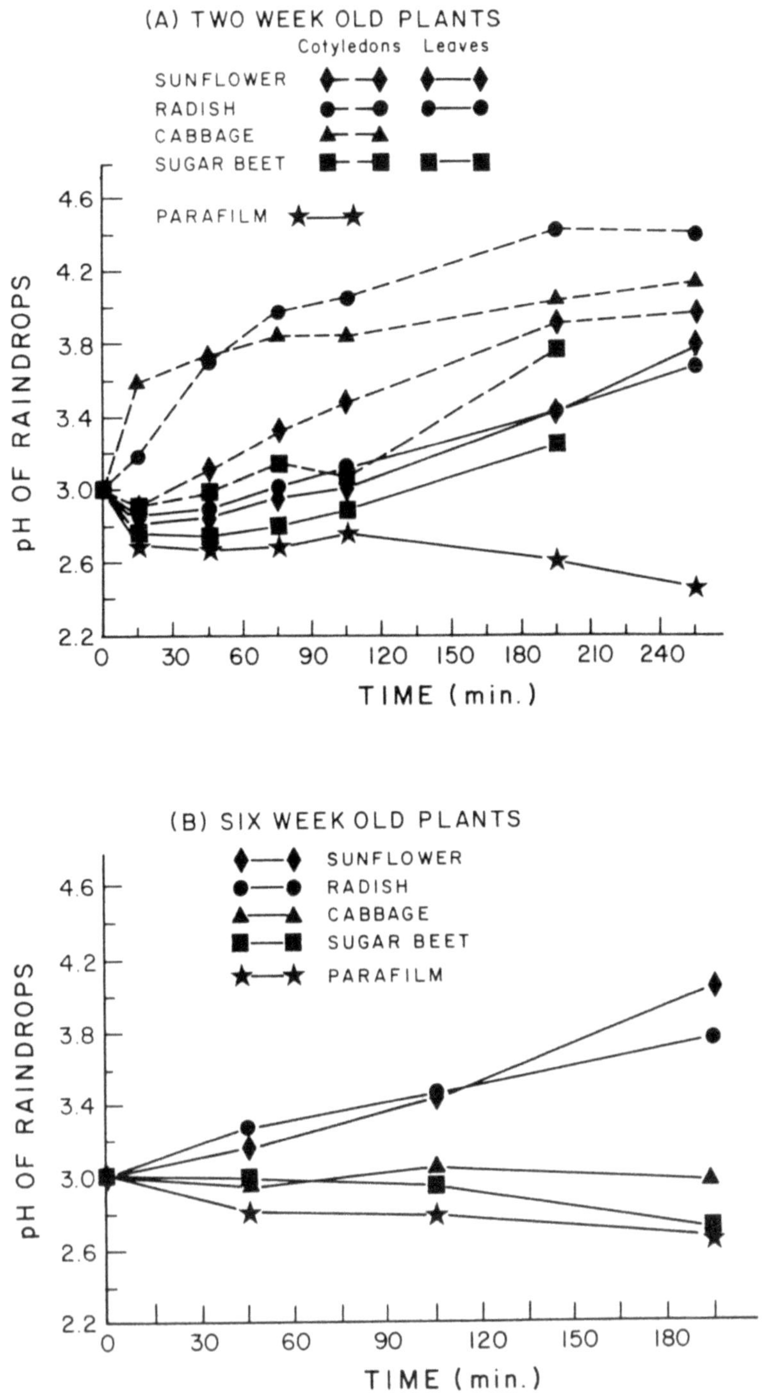

Fig. 1. Changes in the pH of raindrops on foliage of four crop species following a 15 min pH 3.0 simulated rain treatment. (A) 2 wk old plants, cotyledons and leaves, (B) 6 wk old plants, leaves.

Table I. pH of raindrops on leaves of several crop species 45 min after a 30 min duration 'rain shower' of either pH 5.6 or 3.0.

	pH 5.6 rain	pH 3.0 rain		
Species	mean pH (n)	pH over uninjured leaf tissue mean pH (n)	pH over lesions mean pH (n)	% of leaves injured
Broccoli	6.2 (20)	3.17 (29)	3.9* (3)	13
Brussel sprouts	6.5 (16)	3.14 (19)	3.3* (5)	10
Cabbage				
"Viking Golden Acre"	6.2 (31)	3.06 (24)		6
Cabbage "Salarite"	6.2 (36)	3.20 (30)	4.7* (7)	13
Garden beet	6.4 (43)	3.16 (39)		-
Kohlrabi	6.2 (14)	3.07 (14)		12
Radish	6.5 (52)	3.34 (60)	3.9* (2)	52
Sugar beet	6.1 (73)	3.06 (53)		30
Tomato	6.6 (66)	3.36 (69)	4.3* (7)	87
Wax Bean 1° Leaf	6.9 (32)	3.15 (49)		47
Trifoliate	7.0 (3)	3.69 (25)		33
Parafilm surface	5.4 (10)	3.06 (10)		-

Three week old seedlings were used in this experiment. Plants had received three previous 30 min pH 3.0 'rains' prior to this spraying.
* $p < 0.01$, as determined by a t-test, comparing droplet $[H^+]$ over injured and uninjured leaf tissue.

sarsaparilla). In a greenhouse experiment using four agricultural crop species, sunflower and radish leaves were found to have a strong neutralizing capacity (Figure 1b), with sunflower increasing the pH of droplets from an initial 3.0 to 4.0 after 195 min of contact. In contrast, pH 3.0 raindrops were not neutralized on leaves of cabbage and sugar beet. Cotyledons had a significantly greater neutralizing ability than true leaves of these species (Figure 1a). Based on the results of a second greenhouse experiment in which 10 agricultural crop varieties were sprayed with a low pH (3.0) rain, we have not found a good relationship between the neutralizing ability of leaves and their susceptibility to injury from acid rain (Table I). Leaves of cabbage, broccoli, kohlrabi and brussel sprouts (all brassicas) generally had a poor neutralizing ability, but showed less injury than the other species. These leaf surfaces are densely covered with epicuticular wax (Baker, 1982; Caporn and Hutchinson, 1986) which makes them extremely hydrophobic. Because less rain remains on the surface of these leaves, they show less injury. However, raindrops that are retained have large contact angles (i.e. a very small area of contact) between the leaf and the drops (Caporn and Hutchinson, 1986). This limits the area over which neutralization of the acidic drops can take place and therefore lesions may develop at these sites. Thus, the degree to which leaves are wettable is a major determinant of a plant's susceptibility to acid rain in that it determines both how much rainwater is retained as well as the amount of interaction (including neutralization) that can take place between the leaf surface and individual acidic raindrops.

pH values for raindrops sitting directly over lesions were 1 to 2 pH units higher than the pH of raindrops elsewhere on the same leaves (Table I). These higher pH's strongly suggest that cations and other neutralizing substances are released from leaking, damaged cells. Tukey and Morgan (1963), using a distilled water mist, found that various factors which caused injury to foliage (e.g., nutrient deficiencies, low temperature and mechanical injury), increased leaching losses of metabolites. The higher pH values measured on the cotyledons of the crop species (Figure 1a) may also be due to accelerated leaching, which may result from a poorly developed cuticle and less epicuticular wax. The surface of cotyledons showed quite extensive injury in both greenhouse experiments (15 to 100% of the cotyledon surface); injury was apparent even before the spraying (pH 3.0) was finished. Leaves of the same species had < 5.0% of the surface area injured (data not shown).

The raindrops were affected in two ways when they came in contact with the surface of a leaf. First, as the drops evaporated the concentration of H^+ increased. This was substantial when the rain was very acidic as shown by the parafilm controls on which pH 3.0 raindrops decreased to around pH 2.5 when they were nearly dry (Figures 1a and b). It was notable that droplets on wild sarsaparilla foliage dried quickly in comparison to the other boreal forest species studied (and the parafilm). This was evidenced in the decrease in pH of the raindrops from 3.8 to almost 3.3 (Figure 2b). Secondly, changes occurred in the chemical composition of the raindrops. Results of chemical analyses of the raindrops showed concentrations of Ca, Mg and K to be 5 to 50 times greater after a 75 min contact with leaves of sunflower, radish, sugar beet and cabbage (Table II). Ca was the prevalent cation entering raindrops on cabbage and sugar beet leaves, whereas K was the predominant cation in raindrops on radish and sunflower.

The increase in the concentrations of these cations in the raindrops on leaves suggests they are important in neutralizing acidic raindrops. Neutralization of acidic raindrops may be brought about when H^+ ions in the raindrop are exchanged for cations on exchange sites in the cell walls of the leaf or by the solubilization of alkaline dusts and exudates on the leaf surface. Studies have shown that cations penetrate the cuticle through pores (McFarlane and Berry, 1974) and considerable quantities of salts may also be deposited onto the leaf surface from within by processes such as exudation and guttation. In this study, we cannot judge the contribution from solubilization of dry deposited material on the leaf surface. Data which we have obtained in recent greenhouse experiments comparing the neutralizing ability of plants grown in filtered air with those grown in unfiltered air suggest a large component of the neutralization is from solubilized alkaline dust (Hutchinson and Adams, submitted). In these experiments, plants from the filtered air treatment showed only a very small increase in droplet pH in the first 60 min of drying, followed by a continuous gradual increase in pH over the remainder of the drying period (approx.

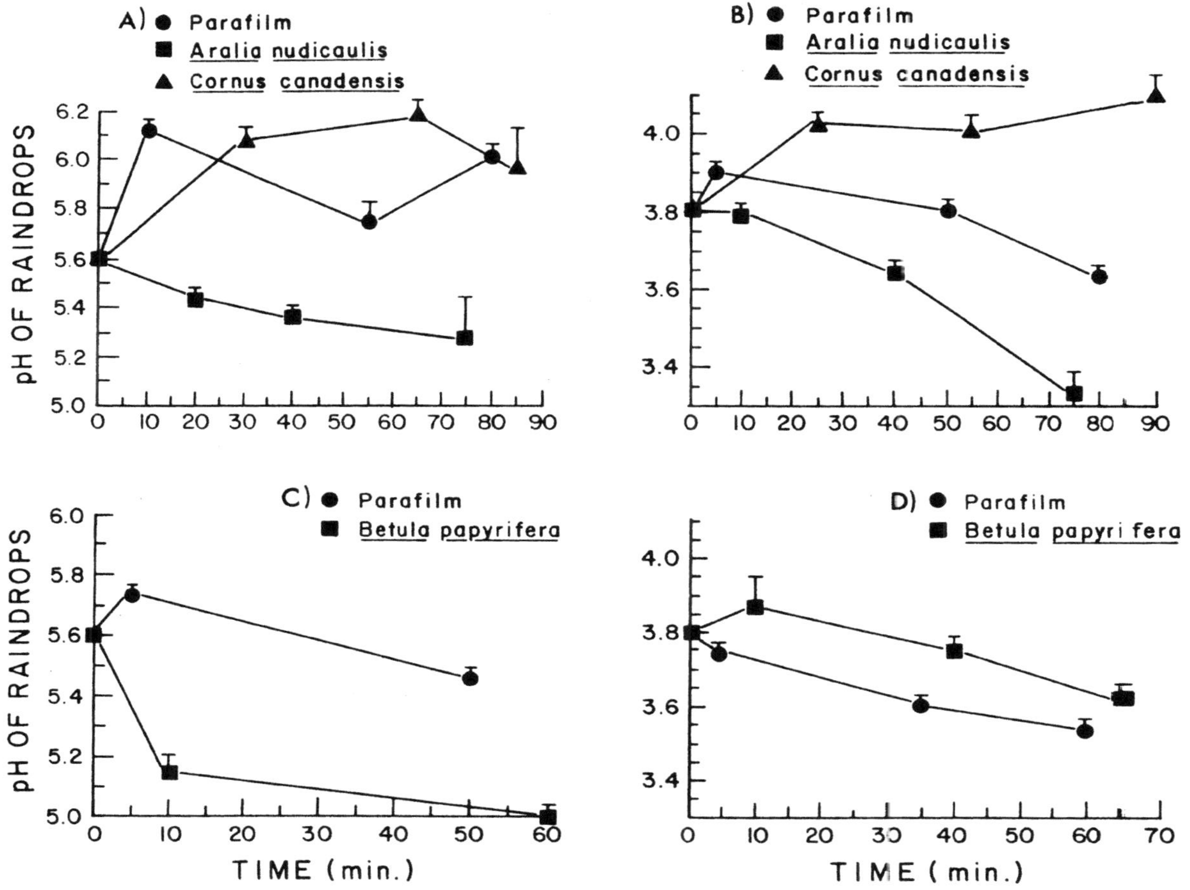

Fig. 2. Changes in the pH of raindrops on foliage of three boreal forest species following a simulated rain of pH 5.6 (A and C) or pH 3.8 (B and D).

Table II. Change in rain pH and base metal composition after droplet contact with leaves of four crop species. Concentrations are expressed in mg/L. (Droplets were collected 75 min after a five min rain shower.)

	pH	Ca	K	Mg
Simulated rain	3.00	1.9	0.44	0.51
Parafilm droplets	2.90	2.4	0.54	0.66
Sunflower droplets	3.10	6.3	23.30	1.70
Sugar beet droplets	3.00	7.1	3.60	2.70
Radish droplets	3.10	11.8	21.50	3.00
Cabbage droplets	3.10	19.3	1.90	2.80

Values are the mean of 8 samples for rain and parafilm, 2 samples (each from 6 plants) for cabbage and 4 samples (each from 3 plants) for the other species. Drops were collected after a simulated rain when plants were 4 wk old except for cabbage. On cabbage, drops were collected later when plants were 7 wk old.

3 hr). In contrast, plants grown in unfiltered air showed a very rapid initial increase in pH, followed by a continuous gradual pH increase for the rest of the drying period. These results suggest that the initial chemical and pH changes in raindrops on foliage are mainly due to solubilization of material already present on the leaf surface, while ion exchange processes become important in droplet pH changes later in their drying. Thus, in the present study, the chemical analyses of raindrops collected after 75 min of foliar contact may largely reflect changes due to solubilization of exudates and dusts on the leaf surfaces. In a study by Lovett (1984) in a mixed oak forest, dry deposition dominated contributions of Ca^{2+} to throughfall while canopy exchange dominated that for K^+. Leaf surface characteristics, such as the presence of trichomes and the amount and structure of epicuticular waxes, as well as leaf orientation are important features determining the amount of particulates retained on the leaves of different plants species.

Changes in the pH of rain on foliage of boreal forest herbs and shrubs also occurred following a natural acidic rain. There was an increase, the magnitude of which was species dependent, in the pH of rainwater following a 15 hr natural rain of pH 3.9 at our boreal forest site (Figure 3). The pH of raindrops on leaves 1 hr after the rain ranged from 4.1 on false honeysuckle (Diervilla lonicera Mill.) to 5.0 on bunchberry. Nine hr after the rain, considerably higher pH values for raindrops were found on all the species, e.g. raindrops on bunchberry had a pH of 6.6. Clearly, the amount of neutralization varies not only between species, but also with the drying conditions following a rain event. The rain ended in the late evening and raindrops did not dry until after sunrise the following day, by which time large increases in pH occurred. Under sunny conditions where the leaves dry rapidly, less neutralization would be expected and this might increase the risk of foliar injury from rains having a low pH. Interestingly, the plants reported in Figure 3 maintained this high neutralizing ability despite a very heavy drenching of over 15 hr and indeed after heavy rains for the previous few days. This raises interesting questions about the regeneration of neutralizing ability.

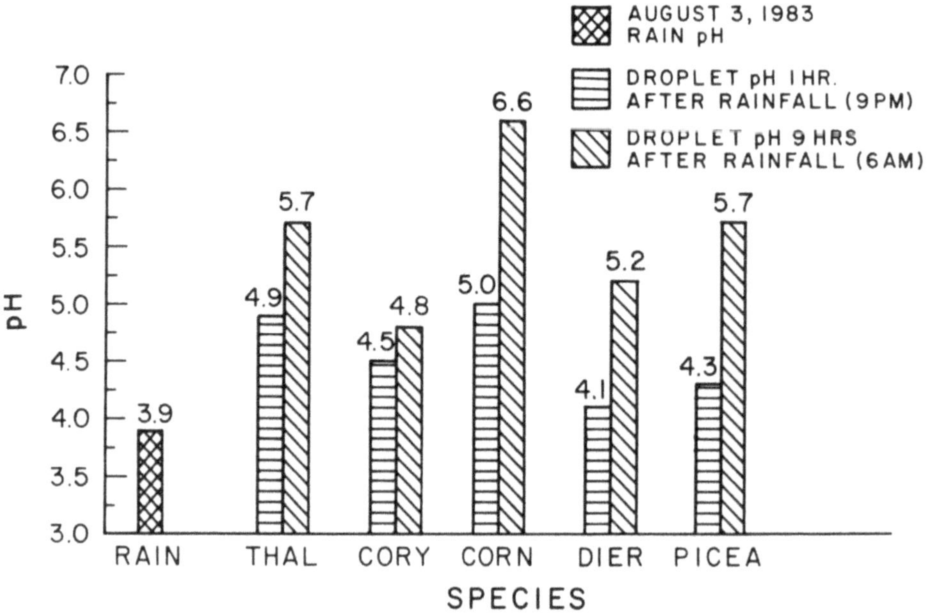

Fig. 3. Comparison of the pH of natural rainfall and the pH of rain droplets on leaves of 5 boreal forest species 1 hr and 9 hr after the rain ceased. Values are the mean of 2 or 3 samples (each approximately 4 mL) of raindrops. Species: THAL = Thalictrum polygamum, CORY = Corylus cornuta, CORN = Cornus canadensis, DIER = Diervilla lonicera, PICEA = Picea glauca.

In conclusion, our studies show that foliar neutralization of acid rain differs widely between species. The data do not support the hypothesis that foliar neutralization of acid rain alone acts as a protective mechanism from damage, although it may still be important for some species. Other factors such as leaf wettability, environmental conditions, and leaf surface morphology play determining roles in the outcome of the leaf-raindrop interactions and need to be investigated further.

4. ACKNOWLEDGEMENTS

The funding for this study was provided through a Strategic grant from the Natural Sciences and Engineering Research Council of Canada to T.C. Hutchinson and R.L. Jefferies, from support funding of the Canadian Sportsman's Fund, and from an NSERC Scholarship and financial assistance from FCAC to B. Gaber.

5. REFERENCES

Adams, C.M., Dengler, N.G., and Hutchinson, T.C.: 1984, 'Acid rain effects on foliar histology of Artemisia tilesii', Can. J. Bot. **62**, 463-474.

Adams, C.M., and Hutchinson, T.C.: 1984, 'A comparison of the ability of leaf surfaces of three species to neutralize acidic rain drops', New Phytol **97**, 463-478.

Baker, E.A.: 1982, 'Chemistry and morphology of plant epicuticular waxes', In: Cutler, D.F., Alvin, K.L., and Price, C.E. (eds.), The Plant Cuticle, Academic Press, London, pp. 139-165.

CANSAP DATA Summary: 1979, 1980, 1981, Canadian Network for Sampling Precipitation, Atmospheric Environment Service, Environment Canada.

Caporn, S.J.M., and Hutchinson, T.C.: 1986, 'The contrasting response to simulated acid rain of leaves and cotyledons of cabbage (Brassica oleracea L.)', New Phytol., in press.

Cronan, C.S., and Reiners, W.A.: 1983, ' Canopy processing of acidic precipitation by coniferous and hardwood forests in New England', Oecologia **59**, 216-223.

Evans, L.S, and Curry, T.M.: 1979, 'Differential responses of plant foliage to simulated acid rain', Am. J. Bot. **66**, 953-962.

Evans, L.S., Curry, T.M., and Lewin, K.F.: 1981, 'Responses of leaves of Phaseolus vulgaris to simulated acidic rain', New Phytol. **88**, 403-420.

Lovett, G.M.: 1984, 'Dry deposition and canopy exchange in a mixed oak forest as determined by analyses of throughfall', J. of Ecology **21** (3), 1013-1027.

McFarlane, J.C., and Berry, W.L.: 1974, 'Cation penetration through isolated cuticles', Plant Physiol **53**, 723-727.

Richter, D.D., Johnson, D.W., and Todd, D.E.: 1983, 'Atmospheric sulfur deposition, neutralization and ion leaching in two deciduous forest ecosystems', J. Environ. Qual. **12**, 263-270.

Scherbatskoy, T., and Klein, R.M.: 1983, 'Response of spruce and birch foliage to leaching by acidic mists', J. Environ. Qual. **12**, 189-195.

Tukey, H.B. Jr., and Morgan, J.V.: 1963, 'Injury to foliage and its effect upon the leaching of nutrients from above-ground plant parts', Physiol. Plant. **16**, 557-564.

Wood, T., and Bormann, F.H.: 1975, 'Increases in foliar leaching caused by acidification of an artificial mist', Ambio **4**, 169-171.

SIMULTANEOUS EXPOSURE OF FOREST TREES TO POLLUTANTS AND CLIMATIC STRESS

Hans-D. Payer, Lutz W. Blank, Christof Bosch, Gerhard Gnatz,
Wolfgang Schmolke, Peter Schramel
Gesellschaft fuer Strahlen- und Umweltforschung Muenchen,
Ingolstädter Landstr. 1, D-8042 Neuherberg
W.-Germany

ABSTRACT. This paper describes a facility designed to simultaneously generate complex climatic and pollutant conditions for research into the effects of climate and air pollution on plants. The system, commissioned in 1985, consists of a set of 4 environmental chambers which are each equipped with four identical subchambers. These 4 subchambers are subjected to identical climatic conditions, but are equipped with independent gas supplies. The chambers are designed to operate within a temperature range of -20 to +40° C at a relative humidity of 20 to 95 % with an illumination system providing high light intensities at a spectrum very close to sunlight. A computer controlled fumigation system allows the injection of various gases singly and in combination at concentrations between 0.005 to 1,000 ppm (SO_2, NO_x, O_3, CO_2, PAN).

1. INTRODUCTION

Large forest areas in Germany and other countries are threatened by an unprecedented large-scale decline. More than 50% of the trees in Germany are affected (BML 1983, 1984, 1985). The currently favored explanatory hypothesis for damage at higher altitudes proposes that O_3(and other photooxidants), acid mist, mineral deficiencies and extreme climatic conditions are of primary importance (Blank, 1985). Additionally, biological factors (e.g. pathogens) may play a role.

Research into the causes of the decline is using various experimental approaches: e.g. field fumigation systems (McLeod et al., 1985), open top systems (Krause und Höckel, 1986), modified greenhouses (Roberts, 1981), and a variety of laboratory studies. To verify hypotheses, however, only few phytotrons are available which are capable of simulating a wide range of realistic and reproducible climatic and pollutant conditions simultaneously (Aiga et al., 1984; Guderian, 1984). The new environmental chambers at the GSF are probably the technically most advanced facility of this type (Payer et al., 1985). This paper describes the technical and performance characteristics of the chambers during the first experiment, the results of which will be published separately (Bosch and Rehfuess, 1986).

Water, Air, and Soil Pollution **31** (1986) 485–491.

2. DESIGN OF THE NEW PHYTOTRON

2.1 General design characteristics

The technical design of this new phytotron, shown in Figure 1, follows the concept of a chamber-in-chamber facility (van Haut et al., 1979). Four large environmental chambers were commissioned in 1985. The chambers are constructed as twin units with a common monitoring room. To avoid contaminations and corrosion by acid gases the chambers were constructed of the following materials only: stainless steel; glass; PTFE; polypropylene; polyethylene; and silicone. Each chamber, with a floor area of approx. 10 m^2 at a height of 2.5 m, can be operated independently. Cooling and heating is carried out indirectly using a brine system in the experimental room as well as in an air jacket. With this technique, temperature gradients and condensing of water during climatic transients and the resulting sorption of gases to the wall can be kept to a minimum.

The environmental performance characteristics of an empty chamber are as follows (mean and standard error indicated):

Temperature: -20 to +40° C ± 0.4° C without illumination, ± 1° C at 100 klux; gradient per 3 m distance: 0.4° without illumination, 2° at 100 klux (details see below).

Relative humidity (above -6°C): 20 to 95 ± 10 %; (gradient depending on temperatures); 100 % rel. humidity during periods of misting

Light: 500 to 120,000 ± 3,000 lux; horizontal gradient ± 10 %; quality of daylight spectrum (details see below).

Wind: 0.05 to 0.7 m s^{-1} horizontal air flow corresponding to a circulation rate of 3,000 to 23,000 ± 500 m^3 hr^{-1} (equivalent to 1 to 12 air-changes min^{-1}).

Fresh air supply: 200 to 750 ± 20 m^3 hr^{-1} prefiltered air (equivalent to 0.1 to 0.4 air-changes min^{-1}).

Alternatively, four smaller subchambers placed within each large chamber may be used. These subchambers with a floor area of approx. 0.6 m^2 and a height of 1.1 m are constructed of stainless steel and perspex. Light and climatic conditions are identical for the four subchambers placed in the same large chamber, but rain, mist and gas supplies may be varied individually. The performance characteristics of the subchamber are approximately the same as those of the large chambers. The ventilation, however, differs markedly. In the subchamber fresh air is drawn in from the large chamber resulting in the following aeration characteristics: Wind: 0.03 to 0.1 m s^{-1} vertical air stream downward; there are 0.5 to 4 air-changes min^{-1} (100 % fresh air; once-through system).

Both, the large and the subchambers can be equipped with root compartments of 0.35 m depth providing independently controlled root temperatures between -15° and +25°C (±2°).

There is a microcomputer-based central monitoring and control system. Back-ups for all essential systems (e.g. ventilation, pumps, cooling, humidifying, data registration) are provided (Walter und Payer, 1985).

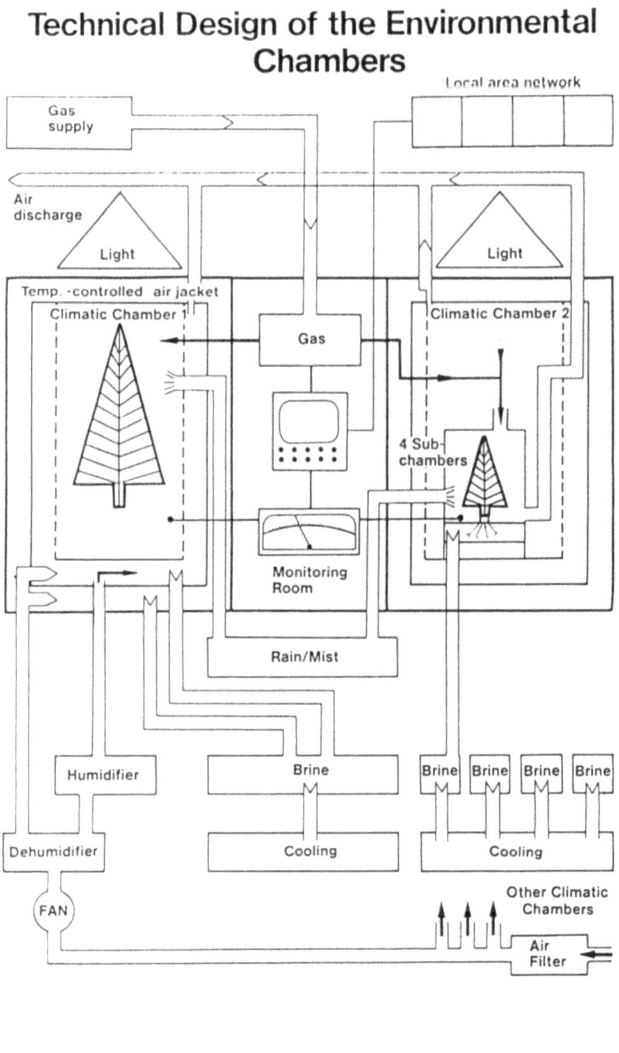

Fig. 1: Schematic outline of the Environmental Chambers

2.2 The light system

As light is a major factor in the development of plants, great care was taken to provide a light source producing light intensities and a spectrum close to realistic environmental conditions as found at field sites. The light system consists of three types of lamps: i) Metalhalide (ultra) lamps and ii) Xenon lamps are simultaneously in operation when standard daylight is required, and iii) Krypton lamps are triggering the biological rhythm of plants and are used during dawn and emergency situations. The lamps are arranged in four units per large chamber and provide an acceptable distribution even at high light intensities in both the large chamber and the respective four subchambers. Distinct peaks in the transmission spectra of both the Xenon and Ultra lamps are levelled out by combining both lamp types, resulting in an acceptable spectrum close to natural sunlight as necessary for plant development (Schauz, 1981). The CIE standard sunlight spectrum D-65 served as a reference (Figure 2).

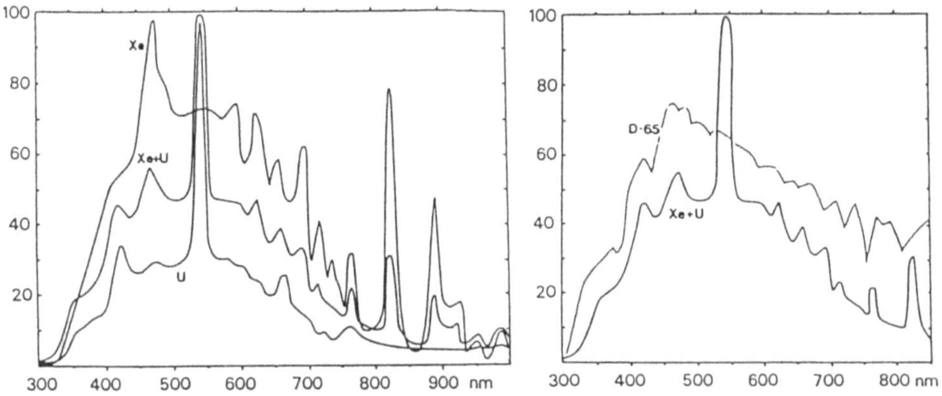

Fig. 2 a: Transmission spectra of Xenon and Ultra-lamps and filters

Fig. 2b: Transmission spectrum of the mixed lamp system, compared to CIE D-65 standard sunlight spectrum. Abscissa: wave length; Ordinate: relative units of the light intensity.

Action spectra of some photobiological reactions were used to evaluate the corresponding segments of the artificial spectra (Figure 3). The results, related to the value of the full range of the transmission spectrum, show the mixture of two lamp types to be better than Xenon or Ultra lamps on their own. Other types of light sources do not provide the required light intensities of up to 120 000 lux and lack UV and far-red emissions.

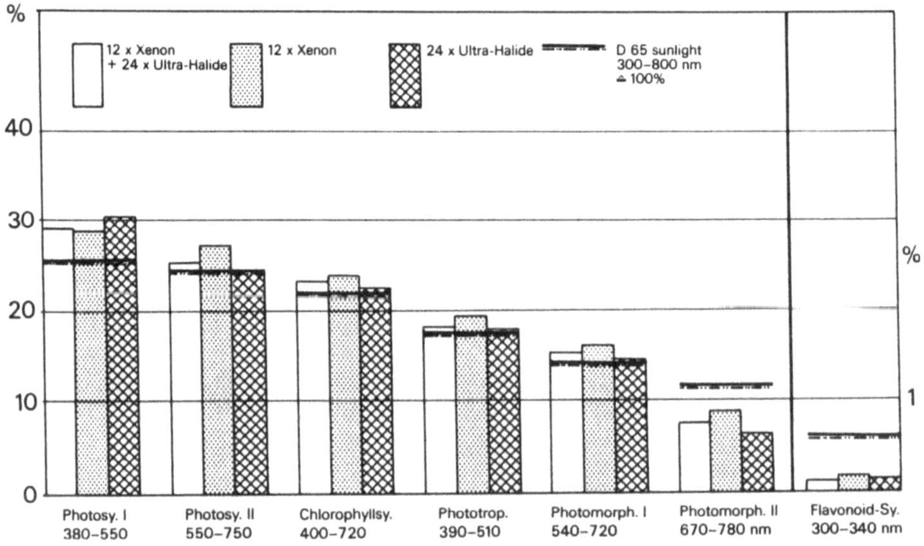

Fig. 3: The spectral quality of the light sources evaluated with reference to photobiological action spectra. Ordinate: relative biological efficiency (basis: 300 - 800 nm = 100 %); Abscissa: wave length ranges

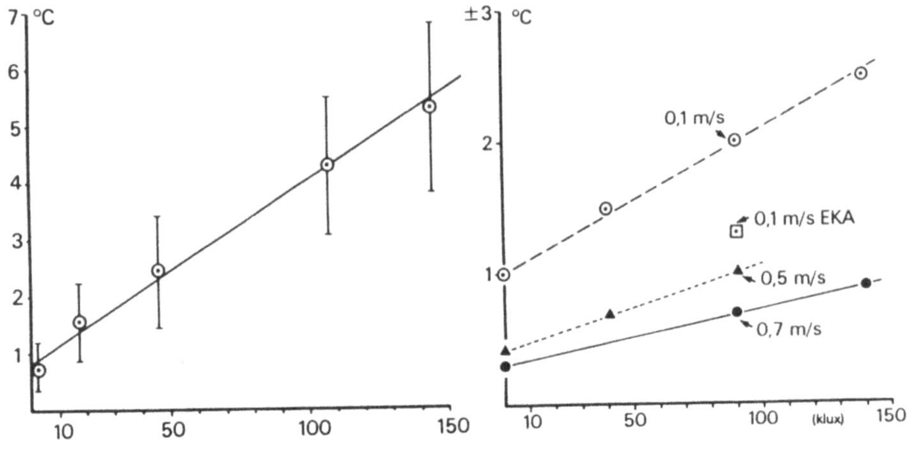

Fig. 4: Impact of the light system on the accuracy and precision of the temperature of the environmental chambers. Abscissa: Illuminance (klux)

a) left ordinate: temperature of the subchamber exceeding the set value of the surrounding large chamber (overheating)

b) right ordinate: standard error of the temperature of the large chamber at 20° C (reproducibility and gradients).

Whilst most of the heat produced by the illumination system is absorbed or removed before entering the actual experimental chamber, penetrating visible irradiation may reach an energy input of up to 1 kW m^{-2} with the resulting caloric energy affecting the accuracy as well as the precision of the temperature regime (Figure 4). The systematic shift of the temperatures, most pronounced in the subchambers, is corrected by a control program. At high light intensities an increased air exchange and circulation rate is required.

2.3 Aeration, gas dosing and analysis

The fresh-air supply for the phytotron has a maximum capacity of 4,000 m^3 hr^{-1}. Ambient air, pre-conditioned to a relative humidity between 40 and 70 %, is subjected to O_3 (0.8 mg m^{-3}) in order to oxidize NO to NO_2. This air is then passed to a central filter system (fine-particle filters; permanganate; activated carbon) before the required pollutant gases are injected into the individual chambers. The filter system retains most pollutant gases, resulting in a purified airstream with the following residuals:
HC < 50 µg m^{-3}; NO < 10 µg m^{-3}; NO_2, SO_2, O_3, < 5 µg m^{-3}. Episodes of elevated NO concentrations in the ambient air (up to 1 ppm), however, result in a peak NO concentration of 30 µg m^{-3}.

The pollutant gases to be injected into the chambers through PTFE tubes are diluted with filtered air to a preliminary concentration not acutely toxic to the plants. Pre-mixing and dosing of the diluted gases are carried out by computer-controlled mass-flow controllers and may be varied to follow diurnal patterns. For safety reasons there is no control loop from analytical system to dosing system. Gas dosing is automatically stopped in case of emergency, including a failure of the climatic control. The gas dosing system can accomodate a variety of gases and has so far been used for the following pollutants: Peroxiacetylnitrate (PAN): 7 ± 0.5 µg m^{-3} (Russi and v.Tiedemann, 1986); SO_2, NO, NO_2, O_3: 10 ± 5 µg m^{-3} to 2,000 ± 40 µg m^{-3} (Payer et al., 1986); CO_2: 200 to 2,000 mg m^{-3} ± 2 %.

Air samples, taken through PTFE tubes every 5 to 80 min, are analyzed using the following methods: O_3 by UV-absorption; SO_2 by UV-fluorescence; NO_x by chemoluminescence; hydrocarbons by flame ionization; CO_2 (which serves as a substrate for photosynthesis) by infra red absorption. Current analytical limits of O_3, SO_2, NO_x are 5 µg m^{-3}; limits of routine measurements of CO_2 and HC are approx. 0.5 mg m^{-3} and 0.05 mg m^{-3}, respectively.

3. RESEARCH IN THE CHAMBERS

The environmental chambers are used for long- and short-term studies of the effect of pollutants and climatic stresses on plants, predominantly forest trees. Multidisciplinary research projects are jointly carried out by member institutes of the Munich Working Group on Air Pollutants (MAGL). The first long-term project started in January 1985 and was successfully carried out for 150 days. In this experiment, environmental conditions as found in the Bavarian Forest, where a distinct type of forest decline is observed (Bosch et al:, 1983) were simulated. The factorial design focussed on the effects of: i) frost; ii) O_3; iii) acid mist; iiii) nutrient supply singly and in combination, on a young clone of Norway spruce. The results of this first project will be published in the "Forstwissenschaftliches Centralblatt", 105, (Bosch and Rehfuess, 1986). This first project will in 1986/87 be followed by a one year study with a similar experimental design but with various clones of Norway spruce.

References

Aiga, Ichiro, Kenji Omasa and Shigeru Matsumoto: 1984, Research Report NIES Japan, 66, 133.

Blank, L.W.: 1985, Nature 314, 311

BML: (Der Bundesminister für Ernährung, Landwirtschaft und Forsten): 1983, 'Ergebnisse der Waldschadensinventur'. Bonn 1983

BML: (Der Bundesminister für Ernährung, Landwirtschaft und Forsten): 1984, 'Ergebnisse der Waldschadensinventur'. Bonn 1984

BML: (Der Bundesminister für Ernährung, Landwirtschaft und Forsten): 1985, 'Ergebnisse der Waldschadensinventur'. Bonn 1985

Bosch, C., Pfannkuch, E., Baum, U., und Rehfuess, K.E.: 1983, Forstwiss. Cbl 102, 167.

Bosch, C., und Rehfuess, K.E.,: 1986, Forstwiss. Cbl., 105, in press

Guderian, R.,: 1984, Personal communication

v. Haut, H., Prinz, B., und Höckel, F.E.: 1979, Schriftenreihe der LIS Essen 49, 29.

Krause, G.H.M. und Höckel, F.E.: 1986, Staub Reinhaltung Luft, in press

McLeod, A.R., Fackrell, J.E., and Alexander, K.: 1985, Atmospheric Environm. 19, 1639-1649.

Payer, H.D. et al.,: 1986, 'Beschreibung der Expositionskammern und der Versuchsbedingungen bei der Belastung von Pflanzen mit Luftschadstoffen und Klimastreß'. Forstwiss. Cbl. 105, in press

Payer, H.D., Schramel, P., Schmolke, W., Gnatz, G., Bosch, C., und Blank, L.W.: 1985, 'Neue Umweltsimulationskammern der GSF zur Erforschung von Waldschäden'. 14. Jahrestagung der Gesellschaft f. Umweltsimulation (GUS), Pfinztal, 14.3.1985

Roberts, T.M.: 1981, 'Effects of stack emissions on agriculture and forestry'. CEGB Research 12, 11-14.

Russi, H., and v.Tiedemann, A.: 1986, in preparation

Schauz, K.: 1981, Angew. Botanik 55, 57.

Walter, A. and Payer, H.D.: 1985, 'Expositionskammern, Technik und Ausführung'.Den Wald ins Labor holen, GSF Neuherberg, 20

TRACE ELEMENT DISTRIBUTION WITHIN THE TREE PHYTOMASS AND FOREST FLOOR OF A TOLERANT HARDWOOD STAND, ALGOMA, ONTARIO

Ian K. Morrison and Gary D. Hogan
Government of Canada
Canadian Forestry Service
P.O. Box 490
Sault Ste. Marie, Ontario, Canada
P6A 5M7

ABSTRACT. An examination of tree phytomass and trace metal concentrations (w/w) and pools in the tree stratum and forest floor of a sugar maple-yellow birch forest was carried out at Turkey Lakes Watershed, Algoma District, Ontario. Estimated aboveground tree phytomass (167,500 kg ha^{-1}) was dominated by stemwood, branches > 2 cm, and stem bark. Highest trace metal concentrations were found in foliage (Cu, Fe, Mn, Ni) and stem bark (Cd, Pb, Zn). Concentrations of essential trace metals found in all sugar maple components followed the expected sequence of Mn>Fe>Zn>Cu. Lead and Ni concentations were always higher than those of Cd. Concentrations of essential elements in foliage and other components were comparable to those reported in the literature for other localities in North America. There was no indication that availability of essential trace elements (e.g., Cu) to vegetation had been increased as a result of increased atmospheric deposition. Lead, Ni and Cd levels in vegetation and forest floor were lower than those reported for similar forested areas of the northeastern United States.

1. INTRODUCTION

In studies concerning effects of acidic deposition on forests and forest watersheds, it is often assumed that acidic inputs lead to changes in the distribution of elements within the system. These changes could be related to the leaching of ions from aboveground components of the forest, or from the soils, or to the mobilization of metals present within or deposited onto the soil surface (Morrison, 1984).

This hypothesized interaction between trace elements and acidic deposition formed the basis for the present study. Specifically addressing the possibility that atmospheric deposition may have increased concentrations of trace metals in vegetation and forest floor materials in general, we conducted an examination of trace metals in, and losses from, the aboveground tree component of a mature sugar maple *(Acer saccharum* Marsh.) - yellow birch - *(Betula alleghaniensis* Britt.) stand at Turkey Lakes Watershed (TLW), Algoma District,

Water, Air, and Soil Pollution **31** (1986) 493–500.

Ontario. In view of the fact that, over the years, Turkey Lakes Watershed may have been exposed to some of the lowest metal deposition rates in North America (Lazrus *et al.*, 1970), we believed that a study of metal distribution at this site would serve as a useful comparison with more heavily contaminated sites in northeastern North America.

2. MATERIALS AND METHODS

2.1 Study Area

The Turkey Lakes Watershed (Lat. 47°03'N; Long. 84°25'W) is located within the Algoma Section of the Great Lakes - St. Lawrence Forest Region (Rowe, 1972). Annual additions to the watershed via precipitation were approximately 0.5, 20.4 and 25.5 kg ha^{-1} for H^+, NO_3^- and SO_4^{2-}, respectively; average annual volume-weighted pH of precipitation was 4.5 (Foster and Nicolson, 1984).

2.2 Tree Phytomass

Four 0.1-ha circular permanent sample plots were established in the study area. Diameter at breast height (DBH) was recorded to the nearest 0.1 cm for all trees > 5 cm DBH. Sugar maple, yellow birch and other species of trees from across a range of diameters were felled and component relationships were determined according to a method adapted from Hegyi (1972). These relationships were used in conjunction with the aforementioned stand table to estimate aboveground tree phytomass by components for each of the four plots.

For purposes of the present study, trace metals by components were determined in a subset of 15 sugar maple trees. Samples of foliage, live and dead branches, stem wood and stem bark were collected from all sample trees during August 1981. Samples were removed to the laboratory, and oven dried prior to chemical analysis. Bioelement contents per unit area were obtained for each component by multiplying phytomass per unit area of each component by concentration of metal found.

2.3 Forest Floor

Mean forest floor bulk and organic weights were determined by quantitatively excavating 80 (20 per 0.1-ha plot) 0.1-m^2 samples and separating them in the field into L, F, and H horizons. Dry weights of bulk samples were determined after drying to constant weight at 70°C. Organic matter was determined by applying organic weight/bulk weight conversions after determination of total organic matter content by loss-on-ignition (LOI) at 500°C in a muffle furnace. Samples for chemical analysis were ground in a laboratory knife mill. Amounts of element per area were obtained by multiplying element concentration by area organic weight.

2.4 Metal Analysis

Analysis for Cu, Fe, Mn, and Zn was carried out by flame AAS after overnight dry ashing at 500°C and $HCl:HNO_3$ extraction. Cadmium, Pb and Ni were determined by flame and graphite furnace AAS after wet digestion in nitric-perchloric acid according to the method of Blanchar *et al.* (1965).

3. RESULTS AND DISCUSSION

3.1 Tree Phytomass and Metal Content

Our estimated total aboveground biomass figure of 167,750 kg ha^{-1} (Table I) indicates that the stand is typical for temperate deciduous forest (Rodin and Bazilevich 1965). As expected, the aboveground phytomass is dominated, in order of importance, by stem wood, branches >2 cm, stem bark, branches <2 cm, foliage and dead branches (Table I).

Table I. Tree phytomass and trace metal pools in a sugar maple-yellow birch stand.

Component	Mass[a]	Metal						
		Fe	Mn	Zn	Cu	Pb	Ni	Cd
Foliage	3300	207	2775	99	32	2.9	21.8	0.89
Branches >2 cm[b]	7150	328	2938	178	43	65.2	15.0	2.14
Branches <2 cm[b]	31700	1553	8368	412	114	152.2	37.7	6.97
Stem bark	15600	826	10654	561	117	223.1	101.4	10.45
Stem wood	108700	2065	7500	1087	369	32.6	54.3	20.65
Dead branches	1300	91	348	27	5	16.9	1.5	0.36
Total	167750	5070	32583	2364	680	492.9	231.7	41.46

[a] kg ha^{-1}, all others g ha^{-1}
[b] diameter

Concentrations of Fe and Mn ranged from highest to lowest, respectively, in the following materials: foliage, bark, branches (both living and dead), and stem wood (Table II). Other metals followed a similar pattern, with the exception of Pb and Cd, which were found in highest concentrations in bark. Dead branches displayed metal levels similar to those found in bark, and this suggests that these metals have an affinity for non-living woody tissues.

Table II. Trace metal levels in sugar maple components.

		Metal[a]						
Component		Fe	Mn	Zn	Cu	Pb	Ni	Cd
Foliage	$\overline{X}$	63	841	30	9.8	0.9	6.6	0.27
	SD	6	292	6	2.3	0.3	2.1	0.12
Branches >2 cm[b]	$\overline{X}$	46	411	25	6.1	9.8	2.1	0.30
	SD	15	126	7	1.5	2.8	0.7	0.01
Branches <2 cm[b]	$\overline{X}$	49	264	13	3.6	4.8	1.2	0.22
	SD	15	99	3	0.8	3.7	0.2	0.05
Stem bark	$\overline{X}$	53	683	36	7.5	14.3	6.5	0.67
	SD	11	157	10	1.0	1.3	0.4	0.04
Stem wood	$\overline{X}$	19	69	10	3.4	0.3	0.5	0.19
	SD	5	28	4	0.8	0.1	0.2	0.04
Dead branches	$\overline{X}$	69	268	21	4.0	13.0	1.2	0.28
	SD	37	112	16	2.3	5.0	0.1	0.06

[a] All values expressed as mg kg^{-1}, for Fe, Mn, Zn and Cu, n = 15; for Pb, Ni and Cd, n = 5.

[b] diameter

For comparative purposes, analytical data are available in the literature mainly for foliage and essential elements. Fewer comparable data are available for the other components, or for non-essential elements (Pb, Ni, and Cd). Exceptions include studies by Van Hook *et al.* (1977); Siccama and Smith (1978); Heinrichs and Mayer (1977, 1980); Friedland and Johnson (1985). Our previously published (Morrison 1985) foliar analysis for essential metals in sugar maple followed the expected sequence Mn > Fe > Zn > Cu, and compares well with the published foliar data of Lea *et al.* (1979) for Huntington Forest, New York. This suggests a similar uptake of metals at both sites. There is a strong likelihood, then, on the basis of "expected" levels of the foliar trace metals, that neither site has been subjected to an increased *availability* of any of these metals through soil reactions or through deposition to leaf surfaces, even though Lazrus *et al.* (1970) demonstrated substantially higher trace metal depositions to sites in the northeastern United States than to the Algoma site.

Comparison of our data for essential trace metals in tree components other than foliage, with other available data (e.g., Young and Guinn, 1966) shows no obvious differences. Young and Guinn's (1966) data for red maple (*Acer rubrum* L.) are comparable with ours for every element except Zn, for which they are higher. If these data are typical of maples in general, there is no indication that the availability of these metals to sugar maples is increased under the influence of high loading.

The situation for Pb differs from that observed for the essential metals. Comparison of our results with those obtained by Smith and Siccama (1981) shows that Pb levels in various tree components are lower at TLW than at Hubbard Brook Experimental Forest (HB), New Hampshire. This was expected, since estimated levels of Pb deposited annually at TLW are less than one-tenth those deposited at the New Hampshire site (Foster and Nicolson, 1986). With a deposition figure for HB of 266 g Pb ha^{-1} yr^{-1}, it is surpising that Pb levels in all components are not higher. Comparison of the two data sets (TLW and HB) shows that Pb levels at HB are three to four times higher in foliage and bark, and one and a half times higher in branches; on the other hand, there is no apparent enrichment in stem wood. It suggests that deposition onto exposed plant surfaces is an important route for Pb entry into association with the organic matter cycle. This observation has been made not only for HB (Smith and Siccama, 1981) but also for the Pine Barrens Region of New Jersey (Turner *et al.*, 1985; Friedland and Johnson, 1985). The lack of stem wood Pb enrichment at heavily contaminated sites in the northeastern United States suggests that Pb transport from the soil is low and that Pb deposited on the soil surface is not readily available to the tree. The results obtained for Cd in all components are comparable to those obtained for red maple in eastern Tennessee (Van Hook *et al.*, 1977). Our results and those of Van Hook *et al.* (1977) reflect the fact that Cd deposition in North America is lower over all than that of Pb. Atmospheric inputs of Cd at the Tennessee site were estimated to be about 20 g ha^{-1} yr^{-1} (Van Hook *et al.*, 1977). At the present time we have been unable to obtain a reliable estimate of atmospheric Cd input to the TLW but it is conceivable that it could be as much as an order of magnitude less.

In the tree phytomass, the largest pools of trace metals (Table I) are in stem bark (Mn, Pb, and Ni) or stem wood (Fe, Zn, Cu and Cd), followed by those in small and large branches, and foliage and dead branches, respectively. Over 70% of the Cu, Zn, Ni, and Cd found within the tree stratum was contained within the stem wood and stem bark. Generally variation in pool size was more influenced by component mass than by element concentration (Table II).

3.2 Forest Floor Mass, Metal Concentration and Amount

Iron concentrations in the forest floor (Table III) were strongly influenced by mineral matter present. Total Fe in HF digests of the underlying eluviated horizon was 7000 to 11,000 mg kg^{-1}. Therefore, it appears that some Fe was solubilized from mineral soil particles during wet ashing. This pattern was not present for other metals and was consistent with the fact that a number of metals are retained within the upper organic layers, where they are usually found at higher concentrations than in the mineral soil. Lead, Ni and Cd concentrations were lower in the forest floor at TLW than they were in deciduous stands in Vermont (Reiners *et al.*, 1975; Friedland *et al.*, 1984), though not radically different from those reported by Van Hook *et al.* (1977) for eastern Tennessee. (Strict comparisons are difficult, however, because of differences between Canadian and American systems of classification of organic horizons.) Levels of Cu were similar to those reported by Friedland *et al.* (1984) but levels of Zn were higher at TLW. These higher Zn levels might be related to higher levels of Zn available for cycling and derived from a greater component of litter from yellow birch, a known accumulator of Zn.

Table III. Trace metal levels in the forest floor[a].

Horizon		Metal Fe	Mn	Zn	Cu	Pb	Ni	Cd
L	$\bar{X}$	1010	872	116	12.8	44.0	3.84	0.5
	SD	263	181	39	2.4	8.3	1.41	0.2
F	$\bar{X}$	5147	531	139	15.8	101.8	7.05	0.5
	SD	3219	313	54	1.9	14.2	1.38	0.4
H	$\bar{X}$	11198	372	71	10.9	68.1	8.40	0.4
	SD	4179	479	20	3.2	15.0	1.30	0.1

[a] mg kg^{-1} oven dried. (L, F, and H layers contained 91, 80 and 34% volatile matter, respectively.) For Fe, Mn, Zn and Cu, n = 19. For Pb, Ni and Cd, n = 10.

Our results on pool size within the forest floor are not strictly comparable with those of Friedland *et al.* (1984) because of a difference in the weight of the forest floor (96,000 kg ha^{-1} for Friedland *et al.* (ibid.) versus 32,482 kg ha^{-1} in the present study) and in manner of expression (mg kg^{-1} forest floor and mg kg^{-1} organic matter, respectively). Pool size estimates for Pb, Cu, Zn, Ni and Cd of 8450, 1310, 8580, 1400 and 160 g ha^{-1}, respectively, found by Friedland *et al.* (1984) were much larger than those observed in the present study (Table IV). In spite of difficulties in making comparisons between the two sites, it is apparent that a major portion of this difference could be attributed to differences in rate and duration of metal deposition.

Values for the forest floor in Table IV indicate that this is a considerable sink for metals in general. For example, 85% of the Pb in the system is contained within the forest floor layers. In all cases (metal x horizon) the F horizon represented the largest pool of metal within the forest floor.

Table IV. Soil organic horizon biomass and trace metal pools.

Horizon	Organic matter[a]	Metal						
		Fe	Mn	Zn	Cu	Pb	Ni	Cd
L	5642	5698	4919	654	72	248	21.6	2.8
F	15280	78646	8113	2123	241	1555	107.7	8.3
H	11560	129449	4300	820	126	787	97.1	4.6
TOTAL	32482	213793	17332	3597	439	2590	226.4	15.7

[a] kg ha^{-1}, all others as g ha^{-1} based on organic matter, n = 80.

3.3 Metal Inputs By Atmospheric Deposition

Estimated metal inputs by atmospheric processes (g ha^{-1} yr^{-1}) are as follows: Fe 647, Mn 45, Cu 41, Pb 20, Ni 14[1] and Cd no data (Foster and Nicolson, 1986). Comparable values for the eastern United States (likewise in g $ha^{-1} yr^{-1}$) are: Fe 360, Mn 60, Zn 918, Cu 222, Pb 432, Ni 36 calculated from Lazrus *et al.* (1970) for New York and Maine sites (5-mo data extrapolated to annual deposition) and Cd 90 (Schlesinger *et al.*, 1974). Comparison of these data sets shows that TLW is relatively uncontaminated, by North American standards, with respect to atmospheric deposition of anthropogenic metal emissions.

4. SUMMARY

Atmospheric depositions of most metals to the TLW are small in comparison with those in the northeastern United States. Levels of most of the metals examined, both essential and non-essential, within the various aboveground components were within the range of values reported in the literature. There was no indication, after direct comparison of results, that tree components at TLW were subject to increased availability of trace metals brought about by present levels of acidic deposition. Interpretation of any current data set, however, is hampered by the lack of pre-industrial revolution data. Most metals tended to accumulate within the forest floor, e.g., 85% of the Pb within the forest system examined was retained by these horizons. Comparisons with data in the eastern United States, with TLW used as a lower limit, show that ill-defined relationships existed between the level of metal deposited from the atmosphere and that in foliar material. It

[1] estimated from Barrie and Vet (1984).

was assumed that this was because of immobilization of the metal within the forest floor.

5. ACKNOWLEDGMENTS

Field work and data analysis related to this study were supervised by D. Ropke. Sample preparation and chemical analysis were carried out by S. Beith, L. Chartrand and M. Claxton.

6. REFERENCES

Barrie, L.A. and Vet, R.J.:1984, *Atmosph. Environ.* **18**, 1459.
Blanchar, R.W., Rehm, G., and Caldwell, A.G.:1965, *Soil Sci. Soc. Proc.* **29**, 71.
Foster, N.W. and Nicolson, J.A.:1984, Proc. For. Congr., Quebec City, P.Q., 337.
Foster, N.W. and Nicolson, J.A.:1986, *Water, Air, Soil Pollut.* (this volume).
Friedland, A.J., and Johnson, A.H.:1985, *J. Environ. Qual* **14**, 332.
Friedland, A.J., Johnson, A.H., and Siccama, T.G.:1984, *Water, Air, Soil Pollut.* **21**, 161.
Hegyi, F.:1972, *Forest. Chron.* **48**, 193.
Heinrichs, M., and Mayer, R.:1977, *J. Environ. Qual.* **9**, 402.
Heinrichs, M., and Mayer, R.:1980, *J. Environ. Qual.* **9**, 111.
Lazrus, A.L., Lorange, E., and Lodge, J.D.:1970, *Environ. Sci. and Technol.* **4**, 55.
Lea, R., Tierson, W.C., Bickelhaupt, D.H., and Leaf, A.L.:1979, *Plant and Soil* **51**, 535.
Morrison, I.K.:1984, *Commonw. For. Bur., For. Abstr.* **45**, 483.
Morrison, I.K.:1985, *Can. J. Forest Res.* **15**, 179.
Reiners, W.A., Marks, R.H., and Vitousek, P.M.:1975, *Oikos* **26**, 264.
Rodin, L.E. and Bazilevich N.I.:1965, *Production and mineral cycling in terrestrial vegetation.* Oliver and Boyd, London.
Rowe, J.S.:1972, Can. For. Serv., Ottawa, Ont., Publ. 1300.
Schlesinger, W.H., Reiners, W.A., and Knopman, D.S.:1974, *Environ. Pollut.* **6**, 39.
Siccama, T.G. and Smith, W.H.:1978, *Environ. Sci. and Technol.* **12**, 593.
Smith, W.H. and Siccama, T.G.:1981, *J. Environ. Qual.* **10**, 323.
Turner, R.S., Johnson, A.H., and Wang, D.:1985, *J. Environ. Qual.* **14**, 305.
Van Hook, R.I., Harriss, W.F., and Henderson, G.S.:1977, *Ambio* **6**, 281.
Young, H.E. and Guinn, V.P.:1966, *Tappi* **49**, 190.

TRACE ELEMENTS IN THE HYDROLOGIC CYCLE OF A TOLERANT HARDWOOD FOREST ECOSYSTEM

N.W. Foster and J.A. Nicolson
Government of Canada
Canadian Forestry Service
Great Lakes Forestry Centre
Sault Ste. Marie, Ont.
Canada
P6A 5M7

ABSTRACT. The concentrations and annual fluxes of Fe, Al, Mn, Cu and Pb were measured during 1983 in bulk precipitation, throughfall, stemflow, forest floor percolate, mineral soil solution below the root zone and streamflow in a maple-birch stand on an acid podzolic soil at the Turkey Lakes Watershed (TLW), Ontario. Inputs of metals to TLW in precipitation were small in comparison with those in the eastern United States and Europe. Considerable loss of Mn and Cu from the vegetation during both the growing and the dormant (leafless) periods was observed and presumed to be due to leaching. The enrichment in soil solution of all metals examined, in relation to throughfall, was greatest for Al (7X) and least for Cu (1.2X). Aluminum was mobilized in both the forest floor and the mineral soil, the latter possibly in association with SO_4^{2-}. Copper was solubilized in the lower forest floor or the mineral soil. Surface soil contents of Al and Cu were reduced by Al and, to a lesser extent, Cu leaching beyond the effective rooting zone. Iron, Mn and Pb were mobilized largely in the F horizon of the forest floor, most likely by organic acids. Leaching of Fe, Mn and Pb was reduced by metal accumulation in vegetation, the lower forest floor, or mineral soil within the effective rooting zone of the vegetation. Most (80 to 99%) of the metals leached from the rooting zone were retained in the watershed and did not appear in streamwater.

1. INTRODUCTION

Trace elements in the hydrologic cycle of forest ecosystems have been studied largely either in systems under the influence of point source pollution (e.g., Tyler, 1978) or in systems receiving very high levels of acid deposition (e.g., Heinrichs and Mayer, 1977). Trace element movement in temperate forests remote from pollution has received less attention (e.g. Smith and Siccama, 1981).

Water, Air, and Soil Pollution **31** (1986) 501–508.

We report on the deposition to and movement of trace metals within a maple (*Acer* spp.) -birch (*Betula* spp.) forest at Turkey Lakes Watershed (TLW), a remote catchment receiving moderate levels of acid deposition (mean annual H^+ content 0.41 kg ha^{-1} (Foster *et al.*, 1986). A watershed analysis approach was used to examine the concentrations and fluxes of Fe, Al, Mn, Cu and Pb above and below the forest canopy, through the soil, and in streamwater.

2. METHODS

Data were collected in 1983 from a 1-ha study site in an unevenaged 'old-growth' sugar maple (*Acer saccharum* Marsh.)-dominated stand with a lesser component of yellow birch (*Betula alleghaniensis* Britton) from within basin 31 of the Turkey Lakes Watershed, Ontario (Foster, 1985). Precipitation and throughfall were collected in continuously open (bulk) collectors (plastic funnel-bottle sets), each 20 cm in diameter. Precipitation was collected with four gauges at two locations, adjacent to and 3 km distant from the study site, and throughfall was obtained from 12 collectors within the study site largely by discrete storm events or occasionally by composite events. Eighteen sugar maple trees (DBH 8 to 53 cm) were fitted with 2.5-cm plastic collars at 1.5 m above ground level for stemflow sampling by event. Soil water was sampled by event below the F (Oe) horizon (5-2 cm depth) by 24 tension-free lysimeters and weekly at the bottom of the effective tree rooting (60 cm) by three ceramic cups under 20 kPa tension. Greater replication of lysimeters was required to estimate forest floor solution chemistry with the same precision as the other processes. Four throughfall and forest floor (bulk) samples per collection were taken for chemical analysis. Streamflow was monitored continuously and sampled every two weeks at the outlet of the basin. Snowfall was sampled every second week from three "snowboards" outside and four within the forest; soil water and streamflow were sampled every three days during the snowmelt period in the spring.

Total trace metal concentrations were determined on filtered solutions (Whatman #41) by graphite furnace atomic absorption spectrophotometry (AAS) in all samples except forest floor percolate which were determined by flame AAS. Precipitation, throughfall and stream samples were analyzed as collected. Stemflow, forest floor percolate and soil solution samples were digested with HNO_3 prior to analysis. Ion fluxes were calculated from the product of concentration and estimated volume. Estimated volumes for mineral soil water were calculated according to the water balance equation (Thornthwaite and Mather, 1957). Ion fluxes in stemflow were determined as the sum, over all diameter classes, of the products of the estimated flux/tree and the number of trees/hectare. Unweighted mean seasonal ion concentrations for each variable were compared by means of Tukey's multiple comparison test.

3. RESULTS

In precipitation Fe and Al were the dominant trace metals, with mean annual volume weighted concentrations of 52 and 28 μg L^{-1}, respectively (Figures 1 and 2). Precipitation was greatly enriched in Mn, and to a lesser extent Cu, as it passed through the canopy, and was further enriched in all metals by contact with the forest floor during the dormant season (Table I) and growing season (Table II). The mean annual concentrations of Mn and Fe were greatest in stemflow and forest floor percolate and decreased in mineral soil solution. Mean annual Al and Cu concentrations were greatest in mineral soil solution. Lead was detectable (>1 μg L^{-1}) only in stemflow and forest floor percolate. Trace metal concentrations were considerably higher in soil solution than in streamwater.

Mean trace metal concentrations (unweighted) in precipitation below the canopy and in solution within the soil, during the growing season were generally equal to or greater than those during the dormant period (Table II). Stream Fe concentrations were highest during the growing season. The mean Al and Mn concentrations in streamwater were similar for the two periods.

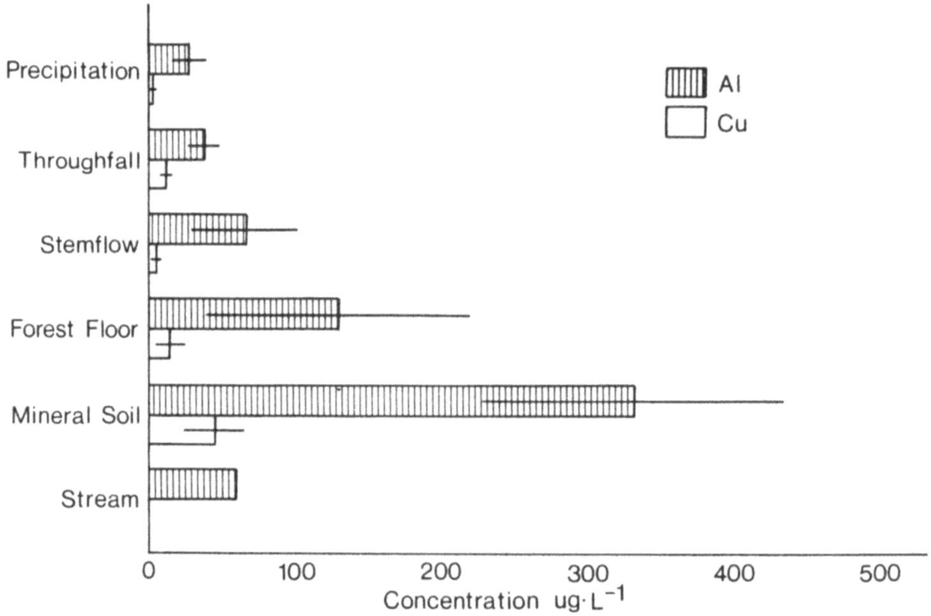

Fig. 1. Mean annual volume weighted Al and Cu concentrations at Basin 31 in 1983 (mean and standard deviation).

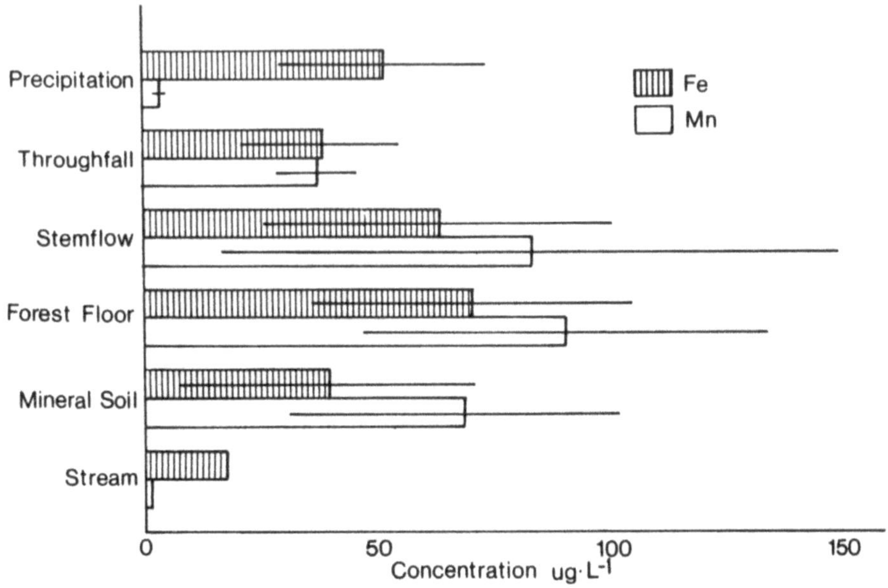

Fig. 2. Mean annual volume weighted Fe and Mn concentrations at Basin 31 in 1983 (mean and standard deviation).

Table I. Dormant season unweighted concentrations (μg L^{-1}) of metals in solutions sampled within Basin 31 (mean ± standard deviation.)

Variable	Fe	Al	Mn	Cu	Pb	Number** of samples
Precipitation	60 ± 15^{ac}	31 ± 9^{a}	5 ± 1^{a}	3 ± 2^{a}	N.D.	72
Throughfall	43 ± 11^{ab}	46 ± 11^{a}	32 ± 10^{a}	9 ± 3^{b}	N.D.	153
Stemflow	83 ± 38^{c}	90 ± 27^{b}	152 ± 89^{c}	5 ± 3^{ab}	1 ± 1	98
Forest floor	70 ± 37^{cd}	134 ± 93^{c}	85 ± 45^{b}	17 ± 12^{c}	6 ± 5	195
Mineral soil	45 ± 33^{ad}	380 ± 105^{d}	47 ± 22^{a}	52 ± 19^{d}	N.D.	21
Stream	15^{b}	52^{a}	1^{a}	N.D.*	N.D.	21

* Not detectable in 1981 and 1982 observations.
** Number of samples = replication x sample dates.
Non-corresponding letters (columns) significant at .05 level.

Table II. Growing season unweighted concentrations ($\mu g\ L^{-1}$) of metals in solutions sampled within Basin 31 (mean ± standard deviation.)

Variable	Fe	Al	Mn	Cu	Pb	Number** of samples
Precipitation	80 ± 32^{a}	33 ± 11^{a}	6 ± 2^{a}	4 ± 2^{a}	N.D.*	93
Throughfall	57 ± 21^{b}	51 ± 12^{a}	60 ± 12^{b}	23 ± 5^{b}	N.D.	276
Stemflow	71 ± 42^{ab}	72 ± 44^{b}	69 ± 66^{b}	5 ± 3^{a}	1 ± 1	245
Forest floor	119 ± 37^{c}	183 ± 87^{c}	181 ± 56^{c}	11 ± 7^{c}	19 ± 13	228
Mineral soil	36 ± 33^{b}	357 ± 103^{d}	84 ± 48^{b}	41 ± 20^{d}	N.D.	33
Stream	24^{b}	54^{ab}	1^{a}	N.D.*	N.D.	9

* Not detectable in 1981 and 1982 observations.
** Number of samples = replication x sample dates.
Non-corresponding letters (columns) significant at .05 level.

In precipitation the annual fluxes of trace metals increase in the order Pb < Cu = Mn < Al < Fe (Table III). In throughfall the annual fluxes of Fe, Al and Mn were the same order of magnitude and three times greater than the flux of Cu. The flux of all metals in stemflow was extremely small. The flux of Al, Mn and Cu from the rooting zone exceeded their input from precipitation. Metal fluxes in streamwater were considerably lower than those in soil water collected below the rooting zone. A comparison of precipitation and streamflow annual fluxes indicated net ecosystem gains for all trace metals.

Table III. Annual fluxes ($\mu g\ ha^{-1}$) of metals in solutions sampled within Basin 31 (mean ± standard deviation.)

Variable	Fe	Al	Mn	Cu	Pb
Precipitation	647 ± 265	351 ± 135	45 ± 12	41 ± 20	25
Throughfall	433 ± 181	416 ± 119	423 ± 100	142 ± 35	23
Stemflow	6 ± 1	6 ± 1	7 ± 1	1	1
Forest floor	785 ± 353	1458 ± 960	785 ± 465	166 ± 110	79 ± 61
Mineral soil	355 ± 279	2933 ± 843	551 ± 320	393 ± 124	6
Stream	66	223	5	4	2

4. DISCUSSION

Inputs of trace metals to TLW in precipitation were much lower than those reported for other sites remote from industrial and urban centres in Europe (e.g., Semb, 1978) and eastern North America. For example, inputs of Pb in precipitation were estimated to be at least one order of magnitude lower than the 200 g ha^{-1} estimated by Reiners *et al.* (1975) the 266 g ha^{-1} reported by Smith and Siccama (1981) for the White Mountains of New Hampshire, the 140 to 340 g ha^{-1} documented by Turner *et al.* (1985) for the New Jersey Pine Barrens, and the 300 g ha^{-1} in precipitation and 400 g ha^{-1} in cloud water interception estimated by Friedland and Johnson (1985) for the Green Mountains of Vermont. Lead levels in precipitation, throughfall, mineral soil solution and streamwater at TLW represent upper limits since only those concentrations that were above the limit of detection were used to calculate the Pb flux. Manganese Cu and Pb concentrations in precipitation at TLW were one order of magnitude lower, and Fe concentrations were similar to values reported for West Germany. Mayer *et al.* (1980) reported that unweighted mean half-year trace metal concentrations in the Solling Mountains of West Germany were 31 $\mu g\ L^{-1}$ (Mn), 38 $\mu g\ L^{-1}$ (Pb), 53 $\mu g\ L^{-1}$ (Cu) and 110 $\mu g\ L^{-1}$ (Fe) in the winter and 39 $\mu g\ L^{-1}$ (Mn), 21 $\mu g\ L^{-1}$ (Pb), 14 $\mu g\ L^{-1}$ (Cu), and 43 $\mu g\ L^{-1}$ (Fe) in the summer. The mean weighted Mn and Cu concentrations at TLW during the dormant period were similar to mean unweighted December-April bulk precipitation Mn (2.8 $\mu g\ L^{-1}$) and Cu (5.0 $\mu g\ L^{-1}$) in a eucalyptus (*Eucalyptus* spp.) forest in California (McColl, 1981), but in California Fe (12.9 $\mu g\ L^{-1}$) was considerably lower.

The ratios of the concentrations of Fe, Al, Mn, Cu and Pb in throughfall and precipitation at TLW were 0.7, 1.5, 6.0, 3.0 and 1.0, respectively, for the dormant period and 0.7, 1.5, 10.0, 5.7 and 1.0 for the growing season. The Pb ratio is based solely on data from the few occasions on which Pb values in precipitation and throughfall were detectable. It is assumed that ions are concentrated by adsorption and evaporation of water from the canopy. (The ratio of throughfall to precipitation was 0.9.) The difference between the growing season and dormant period ratio for Mn suggests that leaching of vegetation is responsible for a major portion of the Mn in throughfall. The concentrations of Mn in maple and birch foliage in this stand (~900 to 1500 $\mu g\ g^{1}$) were much higher than those of Fe (~70 to 90 $\mu g\ g^{-1}$) and Cu (9 to 11 $\mu g\ g^{-1}$) (Morrison, 1985). A similarity in Mn concentrations in sugar maple bark and foliage at the TLW has been observed (Hogan, 1985). The importance of Mn leaching from bark and branches is revealed by the high Mn concentrations in stemflow during the dormant period when the trees are leafless. McColl (1981) reported that increases in Mn in throughfall within a eucalyptus forest in California were related largely to leaching of the vegetation. The washoff of Mn in atmospherically derived particulates deposited in the forest canopy has been cited by Mayer *et al.* (1980) and McColl (1981) as contributing to Mn fluxes in German and California forests, respectively. The contribution of wet and dry deposition of Mn to throughfall fluxes in TLW forests in unknown, but under investigation.

In contrast to Mn and Cu removal from the canopy Fe was retained in both seasons, but more so during the growing season. This suggests storage of Fe in leaves and other plant parts. Heinrich and Mayer (1977) reported Fe and Cu retention by a Solling beech canopy and suggested that adsorption on leaf surfaces was involved in metal retention. There was no net Al interaction with the vegetation since no significant differences between throughfall and precipitation concentrations were observed in either season. The fact that stemflow transported a very shall quantity of trace metals to the forest floor reflects its minor contribution (~2%) to below-canopy water flux.

Interaction with the forest humus and mineral soil resulted in the greatest alterations in the quantity of trace metals that migrated through the ecosystem in solution. Leaching and mineralization of decomposing organic matter in the forest floor enhanced the content of metal ions and organic acids in solutions infiltrating the soil. Carbon concentrations in the forest floor percolate (20 to 50 mg L^{-1}) greatly exceeded those in precipitation (0 to 5 mg L^{-1}). Fulvic acids attack and degrade minerals, releasing trace metals into the soil solution (Schnitzer, 1978). The formation and migration of water-soluble metal chelates in solution (Stevenson, 1972) may explain the observed increase in flux of trace metals from the forest floor to the mineral soil.

Positive input-output balances for the rooting zone indicate that accumulation of Fe, Mn and Pb is occurring in the vegetation, the lower forest floor or the mineral soil. The reason for any accumulation in soil has not yet been determined, but may be related to free ions forming secondary minerals, and/or precipitation of metal-fulvic acid complexes whose solubility decreases as their metal/fulvic acid ratio increases (Schnitzer, 1978). Low C concentrations in mineral soil percolate (2 to 10 mg L^{-1}) were evidence that fulvic acids were precipitated within the rooting zone. Complexation with fulvic and humic acids is a major mechanism for metal retention in soil (Schnitzer, 1978). Uptake by vegetation and metal accumulation in roots may contribute also to the positive balances.

Negative input-output balances for the rooting zone indicate that Al, and to a lesser extent Cu, were leaching beyond the rooting zone thereby reducing soil contents. Nilsson and Bergkvist (1983) have suggested that Al concentrations in the lower soil profile of acid podzolic forest soils may be controlled by equilibria with aluminum sulphates. Sulphate was the major anion in soil solution collected below the effective rooting zone at TLW (Foster *et al.*, 1986) which supports the Al sulphate equilibria mechanism hypothesis.

The low trace metal fluxes in streamflow at TLW, relative to estimated metal fluxes from the rooting zone, probably reflects the reduced solubility of free trace metal ions in streamwater, which has a considerably higher mean annual pH (pH 6.7) than does soil solution (pH 4.4). Most of the metals (90 to 99% of the Al, Mn, Cu and Pb and/or 80% of the Fe) leached from the effective rooting zone of the soil are therefore retained in the watershed and do not reach the surface waters.

5. ACKNOWLEDGMENT

The assistance of Mr. P. Hazlett and Mr. W. Johns with sample collection and analysis of the data and Mrs. L. Chartrand with chemical analysis of samples is gratefully acknowledged.

6. REFERENCES

Foster, N.W.: 1985, *For. Ecol. Manage.* **12,** 215.
Foster, N.W., Morrison, I.K., and Nicolson, J.A.: 1986, *Water, Air, Soil Pollut.* In press.
Friedland, A.J. and Johnson, A.H.: 1985, *J. Environ. Qual.* **14,** 332.
Heinrichs, H. and Mayer, R.: 1977, *J. Environ. Qual.* **6,** 402.
Hogan, G.: 1985, personal communication.
Mayer, R., Heinrichs, H., Seekamp, G. and Fabbender, H.W.: 1980, *Z. Pflanzenernähr. Bodenkd.* **143,** 221.
McColl, J.G.: 1981, *Plant Soil* **62,** 337.
Morrison, I.K.: 1985, *Can. J. For. Res.* **15,** 179.
Nilsson, I. and Bergkvist, B.: 1983, *Water, Air, Soil Pollut.* **20,** 311.
Reiners, W.A., Marks, R.H. and Vitousek, P.M.: 1975, *Oikos* **26,** 264.
Schnitzer, M.: 1978, 'Humic Substances: Chemistry and Reactions.' In M. Schnitzer and S.V. Khan (Ed.) *Soil Organic Matter,* Developments in Soil Science 8. Elsevier Scientific Publishing Com pany, New York, 319 p.
Semb, A.: 1978, 'Deposition of Trace Metals from the Atmosphere in Norway.' SNSF Project, Res. Rep. 13/78, Norway, 28 p.
Smith, W.H. and Siccama, T.G.: 1981, *J. Environ. Qual.* **10,** 323.
Stevenson, F.J.: 1972, *Bio Science* **22,** 643.
Thornthwaite, C.W. and Mather, J.R.: 1957, *Climatology* **3,** 185.
Turner, R.S., Johnson, A.H. and Wang, D.: 1985, *J. Environ. Qual.* **14,** 305.
Tyler, G.: 1978, *Water, Air, Soil Pollut.* **9,** 137.

THE UPTAKE AND TOXICITY OF MERCURY AND LEAD TO SPRUCE (PICEA ABIES KARST.) SEEDLINGS

D.L Godbold and A. Hüttermann
Forstbotaniches Institut, University of Göttingen,
Büsgenweg 2, 3400 Göttingen, F.R.G.

ABSTRACT. Inhibition of root elongation rate was used to estimate the toxicity of methyl-Hg, Hg and Pb to spruce seedlings in nutrient solution. Methyl-Hg was found to be considerably more toxic than $HgCl_2$, this was found to be due to its intrinsic toxicity and not a greater uptake into the roots. The toxicity of Pb was dependent upon the composition of the nutrient solution, a greater inhibition of root elongation was found in a nutrient poor artificial soil solution than in a full nutrient solution. The implications of these studies to spruce decline are discussed.

1. INTRODUCTION

Beyond all reasonable doubt atmospheric pollution has been shown to be the cause of the present forest decline in West Germany (BMFT, 1985). SO_2 and NO_x emitted from the combustion of coal and oil are thought to be the primary agents causing decline through soil acidification and direct action on tree foliage. Associated with trees showing the symtoms of decline are high heavy metal contents of soils, roots and needles (Mayer, 1981, 1983), but the importance of this to forest decline has not been fully assessed. Apart from the emission of SO_2 and NO_x, the combustion of coal and oil results in the emission of Pb and other heavy metals as aerosols (Pacyna, 1983), and Hg in gaseous and particulate forms (Rogers, 1976) . About 60% of the total anthropogenic emission of Pb results from the combustion of leaded petrol. Ninety-seven percent of all Hg contained in coal is released in the flue gas (Anderson and Smith, 1977), leading to localized and long range Hg contamination. The dominent portion of Hg is homogenously distributed in the atmosphere as Hg^0, which is oxidized to soluble forms, possibly by photochemical oxidants, and precipitated with wet or dry deposition (Jernelöv et al., 1984). Ionic forms of Hg travel great distances in acidic aerosols and clouds (Tomlinson et al., 1980).

In the work presented here the effects of Hg, methyl-Hg and Pb on root elongation of spruce have been investigated, and the possible implications of these studies to spruce decline discussed.

Water, Air, and Soil Pollution **31** (1986) 509–515.

2. MATERIALS AND METHODS

2.1. Culture of plants.

All plant material was grown from seed under sterile conditions. Seeds were surface sterilized in 1% w/v $Ca(OCl)_2$ and germinated on 1% w/v water agar. After 3 weeks seedlings were transferred to sterile nutrient solutions based on either Ingestad solution (Ingestad, 1959) of the following composition: (µmol) $Ca(NO_3)_2$ 450, $MgSO_4$ 315, KH_2PO_4 250, NH_4NO_3 225, $FeCl_2$ 5, $MnSO_4$ 1, H_3BO_3 5, Na_2MoO_4 0.1, $CoSO_4$ 0.02, $ZnSO_4$ 0.1, $CuSO_4$ 0.1, pH 4.5, or an artifical soil solution of the following composition: (µmol) $Ca(NO_3)_2$ 68, $CaCl_2$ 62, $MgSO_4$ 82, KH_2PO_4 16, KNO_3 334, $NaNO_3$ 174, $FeCl_2$ 5, $MnSO_4$ 3, H_3BO_3 5, Na_2MoO_4 0.1, $ZnSO_4$ 0.1, $CuSO_4$ 0.1, pH 4.0 and grown for a further 7 days. Nutrient solutions were constantly aerated with sterile air. All work was carried out at 23/21°C day/night temperatures, 100 µE m^{-1} s^{-1} light intensity (Osram L 18 W/25 lamps), and 16 h daylength.

2.2. Estimation of root elongation.

Seedlings were transferred to non-sterile nutrient solutions and allowed to equilibrate for 2 days, then transferred to nutrient solutions containing additions of $PbCl_2$, $HgCl_2$ or methyl-HgCl for 1 to 7 days. Root elongation was measured as previously described (Godbold and Hüttermann, 1985).

The Hg content in roots was estimated by atomic absorption spectrophotometry using the method of Hatch and Ott (1968).

3. RESULTS

The rate of root elongation was significantly inhibited by 0.05 µmol $HgCl_2$ and methyl-Hg within 1 to 2 days of Hg exposure (Figure 1). At 0.05 µmol $HgCl_2$ the degree of inhibition remained relatively constant, whereas at 0.05 µmol methyl-Hg root elongation completely ceased after 3 days exposure. At 0.01 and 0.001 µmol methyl-Hg (Figure 2) the rate of root elongation was inhibited after 2 and 4 days treatment, respectively. After 7 days exposure to 0.01 µmol methyl-Hg root elongation almost completely ceased. The data show that at low levels both $HgCl_2$ and methyl-Hg are extremely toxic to spruce, with methyl-Hg being the more toxic of the two forms.

In short term experiments, at 0.1 µmol methyl-Hg the rate of root elongation was inhibited within 4 hr of exposure, whereas 0.1 µmol $HgCl_2$ root elongation was not inhibited until 8 hr (Table I). Comparision of the amounts of Hg taken up into the roots, show that after 4 and 8 hr exposure similar amounts of Hg as $HgCl_2$ and methyl-Hg had been accumulated in the root tissue. Only after 8 hr was a greater accumulation of methyl-Hg than $HgCl_2$ found.

Table I Hg contents and percentage root growth (control 100 %) of roots of Picea abies seedlings maintained for 16 hr in nutrient solutions containing 0 (control), 0.1 µmol $HgCl_2$ or 0.1 µmol methyl-HgCl. Standard error in parenthesis.

Duration of treatment (hr)	Hg contents µmol g^{-1} Dwt Control	$HgCl_2$	Me-Hg	% root growth $HgCl_2$	Me-Hg
4	0	0.45 (0.01)	0.46 (0.07)	116.0	69.8
8	0	0.76 (0.05)	0.73 (0.06)	86.4	18.9
16	0	0.83 (0.07)	1.03 (0.02)	63.2	13.8

Lead was also found to inhibit root elongation of Picea abies seedlings. The degree of inhibition was dependent upon the Pb supply and the ionic composition and pH of the surrounding medium (Table II). A greater inhibition of root elongation was found in the artificial soil solution than in the full nutrient solution at 0.05 µmol Pb. In the artificial soil solution the severity of inhibition increased as the Pb concentration in the medium increased.

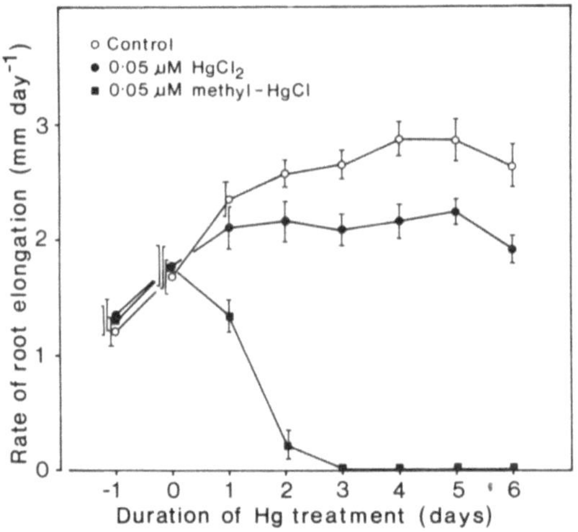

Fig. 1 Root elongation rate of Picea abies seedlings grown for 6 days in nutrient solutions containing 0 (control), 0.05 µmol $HgCl_2$ or 0.05 µmol methyl-HgCl. Bars indicate standard error.

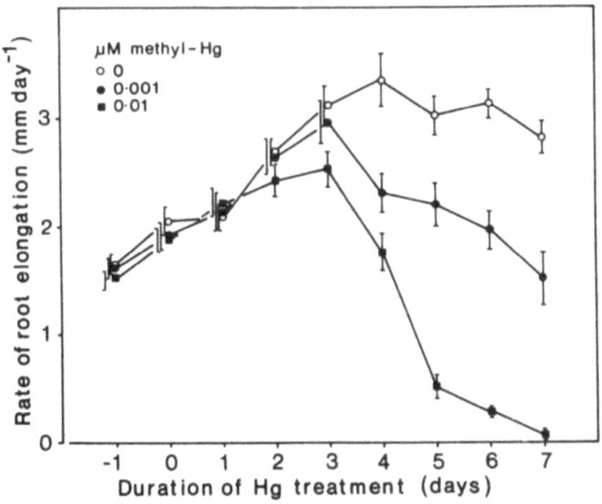

Fig. 2 Root elongation rate of Picea abies seedlings grown for 7 days in nutrient solutions containing a range of methyl-HgCl concentrations (0 - 0.01 µmol). Bars indicate standard error.

Table II Percentage inhibition of root elongation rate (mm day^{-1}) of Picea abies seedling after 6 days growth in two different nutrient solutions containing a range Pb concentrations.

Pb (µmol)	Full Nutrient solution	Artificial soil solution
0.1	----	36.8
0.5	34.1	57.9
2.0	55.5	71.1

4. DISCUSSION

Inhibition of root elongation has been shown to be a sensitive parameter for assessing the toxicity of heavy metals (Hassett et al., 1976; Godbold et al., 1984). The greater toxicity of methyl-Hg compared to $HgCl_2$, has often been credited to a more rapid uptake of methyl-Hg than $HgCl_2$ (Boudan et al., 1983). In the data presented here a greater inhibition of root elongation was induced by methyl-Hg than by $HgCl_2$ at similar tissue concentrations. Indicating that in spruce roots, the greater toxicity of methyl-Hg is due to a higher toxicity to root metabolism of methyl-Hg than $HgCl_2$, rather than a greater uptake. Inhibition of root elongation by Pb was considerably influenced by the ionic composition and pH of the surrounding medium. Inhibition of root growth of spruce by Al has been found to be mediated by Ca availability (Rost-Siebert, 1985), similarly Ni toxicity symptoms in oats was strongly influence by soil Mg level (Procter and McGowan, 1976). But, as the two nutrient solutions used in these studies vary in pH, Ca, Mg and PO_4 concentration and N source, further studies are required to fully estimate which factors influence Pb toxicity. However the data indicate that Pb toxicity is strongly influence by nutrient availability and small changes in pH.

To estimate the toxicity of heavy metals to Picea abies seedlings, their effect on root elongation were estimated in solution culture. Direct comparison of the concentrations of metals in nutrient and soil solutions is difficult as the activity and speciation of the metals in the soil solution are not known. In forest soils in Sweden 36% of total Pb and 95% of total Zn were found to be present as cationic species (Tyler et al., 1985). In heavily polluted soils Pb and Cd were present mainly as chlorides (Tills and Alloway, 1983a,b). When the levels of metals shown to inhibit root growth in nutrient solution are compared to those in the soil and soil solution under stands of Picea abies showing decline symptoms (Table III), it can be seen that the levels of Pb and Hg are sufficiently high to influence root growth at the Solling and Wohldorf. The level of Pb shown to inhibit root elongation is similar to that in the soil solution at Solling, and below that at Wohldorf. A similar level of Pb as that shown to inhibit root elongation has been found in soil solution directly in contact with spruce roots, at forest sites near Hamburg (Bauch, 1985). At Bärhalde, Pb concentrations are ten fold less than those found to inhibit root elongation, however these samples were collected at a soil depth of 30 cm, whereas the principal rooting zone for spruce is in the humus layer (Ulrich, 1983) where considerably higher Pb concentrations would be expected.

Table III. Pb concentrations in soil solution (µmol) and Hg content (µmol Kg^{-1}) of some forest soils in West Germany.

	Bärhalde[a] Black Forest Pb	Wohldorf[b] Hamburg Pb	Solling[c] Pb	Solling[c] Hg
mean	0.009	1.05	0.14	4.6
min	-----	0.36	0.07	4.1
max	-----	3.52	0.30	4.8

Source: a) Trüby and Zöttl (1984)
b) Rastin (1985)
c) Matzner (1985)

For Hg, the concentration in the soil solution is not known, however , if only 1 % of the total Hg in the humus layer were plant available as Hg^{2+} this would be sufficient to influence root growth. Mercury methylation in soils has been shown to be principally an abiological process (Rogers, 1977) mediated by humic and fulvic acids (Nagase et al., 1982). The rate of methylation is increased by low pH and an increase in Hg concentration. Conversely loss of methyl-Hg via demethylation and conversion to more volatile dimethyl-Hg is a biological process (Rogers, 1976). Hence factors associated with soil acidification, decrease in soil pH, an increase in the depth of the humus layer and a decrease in microbial activity (Chet et al., 1984), would favour the production of methyl-Hg. In esturine sediments methyl-Hg constitutes about 0.01 % of the total Hg (Andren and Harriss, 1973). If 0.01 % of the total Hg in the humus layer was in the form of methyl-Hg, a similar level would be found to that of the lowest methyl-Hg level (0.001 µmol) found to inhibit root elongation in nutrient solution.

The accumulation of heavy metals in the humus layer is due principally to the high affinity of humic and fulvic acids for heavy metals. Thus the majority of the total metals in the humus layer will be bound to organic material. However (Kerndorf, 1980) has shown that even when humic and fulvic acids are in great excess all the metals available are not complexed. The percentage of metals bound was found to be pH dependent, a decrease in pH decreased the percentage of bound metals. Heavy metals are readily released from humic acids by addition of H^+ and SO_4 ion (Thanabalasingham and Pickering, 1985), similarly a decrease in soil solution pH of 0.2 units, between pH 4.0 to 4.5, results in an increased Cd concentration by a factor of 2 or 3 (Tyler et al., 1985). Hence any sudden decrease in the pH of the soil for example due to an acidification push (Hüttermann and Ulrich, 1984) is likely to release bound heavy metals. The rapidity in which metals are taken up by Picea abies and inhibition of growth is induced, suggest that such occurences are likely to be significant for root growth and turnover.

While it is unlikely that heavy metal toxicity per se is the dominent causitive factor in forest decline, the level of metals are sufficiently high at the Solling and Hamburg sites that they cannot be excluded as a possible contributing factor. Over long term exposures of years, growth reduction of roots, due to heavy metals and other factors (Hüttermann and Ulrich, 1984), may be sufficient to critically reduce nutrient and water uptake.

ACKNOWLEDGEMENTS

We thank the Royal Society (London) and the European Community (Grant ENV-726-D (B)) for financial support.

REFERENCES

Anderson, B.J. and Smith, K.E.: 1977, Environ. Sci. Technol., **11** , 75.
Andren, A.W. and Harriss, R.C.: 1973, Nature London, **235** , 256.
Bauch, J.: 1975, Personal communication.
BMFT report 3: 1985, Umweltforsch. z. Waldschäden.

Boudou, A., Desmazes, J.P., Gavach, C. and Georgescauld, D.: 1983, Int. Con. Heavy Metals Environ., **1** , 382.
Chet, I., Merg, G. and Hüttermann, A.: 1984, In: Ulrich, B. (ed) Berichte des Forschungszentrums Waldökosysteme/Waldsterben, vol 3 , Göttingen, p 135 (1959)
Godbold, D.L., Horst, W.J., Collins, J.C., Thurman, D.A. and Marschner, H.: 1984 J. Plant Physiol., **116** , 59.
Godbold, D.L. and Hüttermann, A.: 1985, Environ. Pollut., **38** , 375.
Hassett, J.J., Miller, J.E. and Koeppe, D.E.: 1976, Environ. Pollut., **11** , 297.
Hatch, R.W. and Ott, W.L.: 1968, Anal. Chem., **40** , 2085.
Hüttermann, A. and Ulrich, B.: 1984, Phil. Trans. R. Soc. Lond. **B305** , 353.
Ingestad, T.: 1959, Physiol. Plant., 12 , 568.
Jernelöv, A., Johansson, K., Lindkvist, O. and Rodhe,H.: 1984, In, Acidification research in Sweden, p 1.
Kerndorff, H.: 1980 Ph.D thesis, Universität Mainz, F.R.G.
Mayer, R.: 1981, Göttinger Bodenkdl. Ber., **71** , 1.
Mayer, R.: 1983, In, Ulrich, B. and Pankrath, J. (eds) Accumulating air pollutants in forest ecosystems, De Reidel, Dortrecht, p 47.
Matzner, E.: 1975, Personal communication.
Nagase, H., Ose, Y., Sato, T. and Ishikawa, T.: 1982, Sci. Total Environ., **24** , 133.
Pacyna, J.M.: 1983, Int. Con. Heavy Metals Environ., **1** , 178.
Proctor, J. and McGowan,I.D.: 1976, Nature, **260** , 134.
Rastin, N.: 1975, Personal communication.
Rogers, R.D.: 1976, J. Environ. Qual., **5** , 454.
Rogers, R.D.: 1977, J. Environ. Qual., **6** , 463.
Rost-Siebert, K.: 1985, Ph.D. thesis, University Göttingen, F.R.G.
Thanabalasingham, P. and Pickering, W.F.: 1985, Environ. Pollut., **9** , 267.
Tills, A.R. and Alloway, B.J.: 1983a, J. Soil Sci., **34** , 769.
Tills, A.R. and Alloway, B.J.: 1983b, Environ. Tech. Letts., **4** , 529.
Tomlinson, G.H., Brouzes, R.J.P., McClean, R.A.N. and Kadlecek, J.: 1980, Proc. Int. conf. ecol. impact acid precip., Norway, SNSF project. p 134.
Trüby, P. and Zöttl, H.W.: 1984, Angew. Botanik., **58** , 39.
Tyler, G., Berggren, D., Bergkvist, B., Falkengren-Grerup, U., Folkeson, L. and Rühling, A.:1985, In. Hutchinson, T.C. (ed), Effects of acidic deposition on forest, wetland and agricultural ecosytems, Springer, Toronto, In press.
Ulrich, B.: 1983, LÖLF-Mitt. Sonderh. **9** , Recklinghausen, F.R.G.

EFFECTS OF NH_4^+ ON THE GROWTH AND K^+ (^{86}Rb) UPTAKE OF VARIOUS ECTOMYCORRHIZAL FUNGI IN PURE CULTURE.

Boxman, A.W., Sinke, R.J. and Roelofs, J.G.M.
Department of Aquatic Ecology, Catholic University,
Toernooiveld, 6525 ED Nijmegen, The Netherlands.

ABSTRACT. As a result of considerable deposition of ammonium sulphate on vegetation and soil the NH_4/K ratios in the soil water extracts of pine forests in the Netherlands are increasing. Increasing the NH_4/K ratio from 1 to 40 in the solid nutrient medium of isolated ectomycorrhizal fungi will in general result in a 60 to 80% inhibition of the lateral growth. On increasing the NH_4/K ratio the biomass of most of the fungi examined shows a maximum at ratios of approximately 10 to 20, although the colony diameter at those ratios remains rather small. NH_4^+ also exerts a pronounced effect on the uptake rate of Rb^+ in *Rhizopogon luteolus*. At low Rb^+ concentrations the inhibitory effect of NH_4^+ is maximal, whereas on increasing the Rb^+ concentration the inhibitory effect of NH_4^+ is abolished. This means that NH_4^+ competitively inhibits the Rb^+ uptake in *R. luteolus*.

1. INTRODUCTION

In the Netherlands ammonium, besides NO_x and SO_2, is one of the major components of atmospheric deposition. Due to the very intensive agricultural activities , large emissions of NH_3 from animal manure take place. According to van Breemen (1982) NH_3 from the manure reacts with SO_2 from the atmosphere, causing wet and dry deposition of $(NH_4)_2SO_4$ on vegetation and soil. The average total deposition of acid or acidifying substances in the Netherlands is approximately equivalent to 6 kmol $H^+.ha^{-1}.yr^{-1}$ of which 80% is ammonium sulphate (Aalst van, 1984). It has been postulated by Roelofs *et al.* (1985) and Nihlgård (1985) that the deposition of NH_4^+ is probably one of the major factors for the forest dieback in the Netherlands and parts of Europe. It has been shown by Roelofs *et al.* (1985) that needles of the Corsican pine, *Pinus nigra* var. *maritima* (Ait.) Melville, absorp NH_4^+ in exchange for K^+, Mg^{2+} and Ca^{2+}. As a result, NH_4^+ absorption may lead to K^+ and Mg^{2+} deficiency, causing chlorotic yellow-brown needles. In the forest soil, usually nutrient poor and acidic, K^+, Mg^{2+} and Ca^{2+} are readily exchanged by NH_4^+ resulting in increased ratios of NH_4^+ to K^+, Mg^{2+} and Ca^{2+}. Roelofs *et al.* (1985) clearly demonstrated a correlation between these increased ratios and the condition of the pine forests. Increased NH_4^+ to K^+ and Mg^{2+} ratios may strongly depress the uptake of K^+ (Smith and Epstein, 1964) and of Mg^{2+} (Jacob, 1958). Furthermore a reduction in growth rate of *Chrysanthemum* (Nelson and Hsieh, 1971), induction of necrotic spots in the same genus and stem lesions in tomato plants (Maynard *et al.*, 1969) have been observed and ascribed to an increased NH_4/K ratio in the nutient solution.

Water, Air, and Soil Pollution **31** (1986) 517–522.

In healthy stands of *P. nigra* and *Pseudotsuga menziesii* (Mirb.) Franco in the Netherlands the mean ratio of NH_4/K in the soil water extract is approximately 4 (0.5 to 14), whereas it is 9 (1 to 36) in moderately damaged stands and 14 (2 to 64) in severely damaged stands (Roelofs *et al.* , 1985; Roelofs, 1986).

Particularly in areas with intensive stockbreeding the total number of symbiotic fungi has been decreasing (Jansen, 1985;Arnolds, 1985). Ectomycorrhizas play an important role in the nutrient supply of the host and generally enhance nutrient uptake by plant roots (Harley and Smith, 1984). Therefore, the role of NH_4/K ratios (from 1 to 40) on the growth of 14 ectomycorrhizal fungi (Trappe, 1962) is examined. The role of NH_4^+ on the uptake of K^+has been examined by studying the uptake of $^{86}Rb^+$ (as a tracer for K^+) in the fungus *Rhizopogon luteolus* Fr. in order to determine whether there are direct effects of NH_4^+ on the Rb^+ uptake in this fungus.

2. MATERIALS AND METHODS

Mycelial cultures of ectomycorrhizal fungi were grown on a solid MMN medium according to Marx (1969), except that the salts were added at an one-fourth strength and except that $(NH_4)_2HPO_4$ was replaced by an equivalent amount of NH_4Cl. The NH_4/K ratios were varied by increasing the NH_4^+ concentrations and keeping the K^+ concentration constant. For growth experiments mycelial discs of 6 mm in diameter, cut from the edge of a freshly grown mycelial mat, were transferred to Petri dishes (ϕ 9 cm) containing 15 mL of agar medium of pH 5.5 (final pH after sterilization). Three replicates for each treatment were used. It was assumed that fungal growth was concentric. Two perpendicular lines were drawn on the Petri dishes and at regular intervals the diameter of the surface covered with fungal mycelium was measured, along the predrawn lines. The cultures were incubated at 20 ^{0}C in the dark. At the end of the experiments fungal dry weight was measured. The mycelial mats were transferred into a beaker filled with 200 mL of bidistilled water. Then the beakers were incubated at 121 ^{0}C at 15 psi for 15 min to dissolve the agar. Next, the mycelium was filtered on pre-weighed filters (Schleicher and Schull; 602H) and washed three times with bidistilled water of approximately 60 ^{0}C. The filters were dried at 75 ^{0}C for 48 hr and weighed again.

For the experiments on the uptake of Rb^+, three mycelial discs of *R. luteolus* were grown in 250 mL of the same medium used for the growth experiments except that the agar was omitted. After three or four weeks of growth in an incubator shaker at 125 rpm in the dark the cultures were used in the uptake experiments. At that time the fungus had been growing into spheres of approximately 5 to 10 mg dry weigth. The fungal spheres were washed three times with bidistilled water, suspended in 50 mM KCl and aerated overnight in order to exhaust endogeneous substrate. After starvation the fungal spheres were washed again three times with bidistilled water, placed into 25 mL test tubes and pre-incubated with 14 mL (final volume) of 45 mM Tris-succinate buffer pH 4.0 and 3% (w/v) glucose for 60 minutes at 20 ^{0}C. Oxygen was bubbled through the suspension continuously. The uptake of Rb^+ was started by addition of $^{86}Rb^+$ as a tracer, as described earlier (Theuvenet and Borst-Pauwels, 1976). After 60 min of incubation the fungal spheres were filtered on pre-weighed filters by suction and washed with respectively 50 mM ice-cold MgCl (to remove extracellularily bound Rb^+) and bidistilled water. The filters were dried (75 ^{0}C ,48 hr) and weighed, after which the radioactivity on the filters was determined by means of liquid scintillation. In a preliminary study it was found that the uptake of Rb^+ was linear with time throughout the incubation (data not shown).

The fungi used in this study were provided by Dr. D.Marx (USDH, Athens, USA): *Laccaria laccata* (Scop. ex Fr.) Cooke #254, *Pisolithus tinctorius* (Pers.) Coker and Couch #298, *Thelephora terrestris* Ehrh. ex Fr. #223 ATCC-38058 and Dr. T.Limonard (Agricultural University of Wageningen, The Netherlands): *Laccaria proxima* (Boud.) P.D. Orton, *Pisolithus tinctorius* #293 and *Suillus bovinus* (L. ex Fr.) O. Kuntze or obtained from CBS (Baarn, the Netherlands): *Amanita muscaria* (L. ex Fr.) Hook. CBS236.76, *Amanita rubescens* (Pers. ex Fr.) S.F. Gray CBS107.34, *Cenococcum geophilum* Fr. CBS883.68, *Hebeloma crustuliniforme* (Bull. ex St.Amans) Quel., *Rhizopogon luteolus* Fr. CBS304.37, *Suillus granulatus* (L. ex Fr.) O.Kuntze CBS114.42, *Suillus luteus* (L. ex Fr.) S.F.Gray CBS136.36, *Suillus variegatus* (Sow. ex Fr.) O.Kuntze CBS118.42 .

3. RESULTS

With most of the fungi examined growth occurred in three defined stages: an initial lag phase, a logarithmic phase and a stationary phase. Only the slow growing fungus *C. geophilum* did not reach the stationary phase. The lag period appeared to be approximately one-fourth of the total growth period (see Table I).

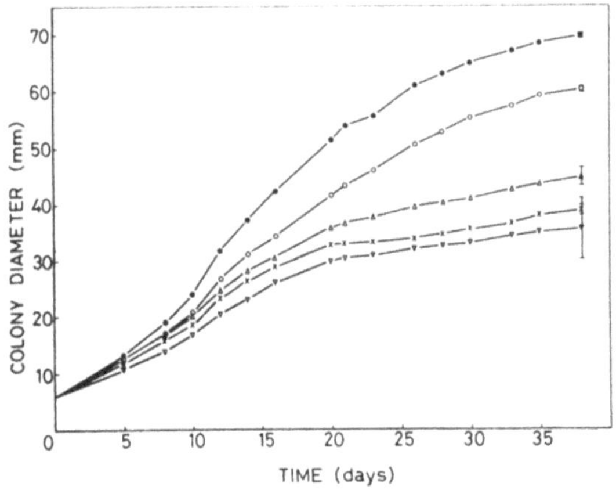

Fig 1. The effect of NH_4/K ratios on the lateral growth of *R. luteolus* at pH=5.5. •, control $NH_4/K=1$ o, $NH_4/K=5$; Δ, $NH_4/K=10$ x, $NH_4/K=20$; ∇, $NH_4/K=40$. Mean of triplicate experiments.

A typical example of the effect of increasing NH_4^+ concentrations on the growth of *R. luteolus* is shown in figure 1. At an NH_4/K ratio of 1 lateral growth was maximal, whereas on increasing this ratio, growth became increasingly inhibited. Table I shows, that with all of the fungi examined it was found that the lateral growth was inhibited, although the degree of inhibition varied with the species. For *C. geophilum* and *H. crustuliniforme* the value of the NH_4/K ratio was very critical for growth. On increasing this ratio from 1 to 5, a severe inhibition of the lateral growth occurred, whereas a further increase of this ratio had little effect. *S. bovinus* and *S. variegatus* showed only inhibition of lateral growth at ratios of more than 5 and 10 respectively. With the other fungi growth decreased more or less gradually with increasing NH_4/K ratios. Two strains of *P. tinctorius* reacted very similarly to increased NH_4/K ratios.

Table I. The effect of increasing NH_4/K ratios on the lateral growth and biomass production of various ectomycorrhizal fungi at pH=5.5. Upper row: lateral growth expressed as % of the lateral growth at an NH_4/K ratio of 1. Lower row: fungal biomass (mg dry weight) at the end of the growth period. (-): not determined. Mean of triplicate experiments.

	growth period (days)	NH_4/K ratio 1	5	10	20	40
Amanita muscaria	42	100	45	19	24	26
		7.5	13.8	10.1	16.8	12.3
Amanita rubescens	49	100	32	19	22	19
		7.7	10.7	16.5	15.8	14.8
Cenococcum geophilum	63	100	25	23	20	25
		26.8	27.1	27.0	34.2	28.6
Hebeloma crustuliniforme	42	100	7	6	8	9
		9.2	7.2	7.6	7.8	7.5
Laccaria laccata	25	100	38	33	28	22
		-	-	-	-	-
Laccaria proxima	42	100	57	42	30	29
		5.5	20.9	24.0	22.7	16.6
Pisolithus tinctorius S-293	28	100	73	60	53	49
		10.1	27.7	29.3	25.4	21.3
Pisolithus tinctorius S-298	28	100	73	64	49	45
		10.8	34.0	29.2	24.4	18.8
Rhizopogon luteolus	35	100	85	53	45	41
		7.5	20.8	26.8	24.9	21.1
Suillus bovinus	26	100	97	78	76	65
		-	-	-	-	-
Suillus granulatus	35	100	52	36	36	34
		8.4	15.9	25.5	16.7	15.4
Suillus luteus	28	100	44	29	29	34
		20.8	33.7	22.9	25.3	24.2
Suillus variegatus	42	100	111	99	51	27
		13.2	30.0	33.4	17.9	18.5
Thelephora terrestris	23	100	73	69	-	24
		11.2	23.5	11.3	-	14.1

Table I further shows the biomass of the fungi at the end of the growth period. With *H. crustuliniforme* a slight decrease in biomass was observed on increasing the NH_4/K ratio. With the other fungi the lowest biomass was found at an NH_4/K ratio of 1. On increasing this ratio an maximum in biomass was found at ratios of approximately 10 to 20, whereas a further increase in NH_4/K ratio resulted again in a decrease in biomass.

Table II. The effect of increasing NH_4^+ concentrations on the uptake rate of Rb^+ at a Rb^+ concentration of 100 μM by *R. luteolus* at pH=4.0. Mean of duplicate experiments.

NH_4^+ (μM)	NH_4/K	uptake rate %
0	0	100
100	1	59
200	2	27
1000	10	11
5000	50	3

The effect of increasing amounts of NH_4^+ on the uptake of Rb^+ is shown in Table II. With increasing NH_4^+ concentrations the uptake of Rb^+ was inhibited increasingly. It was also examined whether NH_4^+ inhibited the Rb^+ uptake rate at other Rb^+ concentrations. The effect of NH_4^+ on the kinetics of the Rb^+ uptake is shown in Table III. At low Rb^+ concentrations the greatest inhibition in Rb^+ uptake was found on adding NH_4^+. The inhibition in Rb^+ uptake rate was abolished when the Rb^+ concentration was raised.

Table III. The effect of NH_4^+ on the kinetics of Rb^+ uptake by *R. luteolus* at pH=4.0. v: uptake rate in $\mu mol.g^{-1}$ dry weight.min^{-1} ; %: % of control. Mean of triplicate experiments.

Rb^+ (mM)	control v	+ 100 μM NH_4^+ v	%	+ 1000 μM NH_4^+ v	%
0.001	0.0081	0.004	49	0.0005	6
0.010	0.075	0.040	54	0.0047	6
0.025	0.19	0.09	50	0.011	6
0.10	0.50	0.31	61	0.043	9
0.25	0.97	0.68	71	0.088	9
1.0	1.87	1.57	84	0.44	24
10.0	4.66	4.04	86	3.60	77

4. DISCUSSION

In large parts of the Netherlands, particularly in areas with intensive stockbreeding, the total number of symbiotic fungi is decreasing (Jansen, 1985, Arnolds, 1985). In order to study whether high depositions of NH_4^+ in such forests soils may be involved in this decline, the effect of NH_4^+ on the growth of various ectomycorrhizal fungi has been examined in pure culture. This study shows that the lateral growth of all of the fungi examined is inhibited with increasing NH_4/K ratios. For two of the examined fungi the value of the NH_4/K ratio is rather critical, for the other fungi the inhibition of the lateral growth is more gradual. In general the inhibition of the lateral growth is 60 to 80 % at NH_4/K ratios of 10 or more as compared to the control with an NH_4/K ratio of 1. This means that at low NH_4/K ratios the fungi can cover a great area with their mycelium. The lateral growth of fungi is of great importance for the host, since one of the main symbiotic functions of these fungi is to expand the absorbing area of the root and thus facilitating nutrient and water uptake of the host (Bowen, 1973). Increasing the NH_4/K ratio might thus lead to a decreased nutrient uptake.

Although the lateral growth is inhibited at increased NH_4/K ratios, fungal biomass initially increases with increasing NH_4/K ratios. Plassard *et al.* (1982) showed, with the fungus *S. luteus* that addition of NH_4^+ led to an increased biomass production, resulting in an increased chitin content. Addition of NO_3^- as N-source had no effect on the chitin content. This might indicate that the fungi protect themselves at high NH_4^+ concentrations by forming more chitin and remaining rather small. In this way the absorption capacity of the fungi will be decreased, which enables the cells to survive at higher NH_4^+ levels. At very high NH_4/K ratios the biomass production decreases again which could mean that the fungi are not capable to detoxify all of the offered NH_4^+ anymore.

The results, presented here are in good agreement with the inventarization studies on the changes in macro fungi in forests stands (Janssen, 1985; Arnolds, 1985). An important question remains whether experiments under laboratory conditions may be extrapolated to the field situation. However, in greenhouse experiments, in which pines on a Podzolic soil are being treated with artificial rain, containing an NH_4^+ gradient, a positive correlation has been found between the NH_4^+ concentration in the soil water extract and the disappearence of the ectomycorrhizas (van Dijk, personal communication).

NH_4^+ also exerts a pronounced effect on the Rb^+ uptake in *R. luteolus*. At low Rb^+ concentrations the inhibitory effect of NH_4^+ is maximal, whereas on raising the Rb^+ concentration this inhibitory effect is abolished. This indicates that NH_4^+ competitively inhibits the Rb^+ uptake in *R. luteolus*.

Normally rather low concentrations of K^+ in the soil water extracts of pine forests are found (up to approximately 100 µmol.L^{-1}), indicating that at a relatively low NH_4^+ input the uptake of K^+ might already be decreased (see Table II). Not only the NH_4/K ratio is of importance but also the absolute value of the soil K^+ content. NH_4^+ therefore exerts a mutual effect on the soil. Due to leaching the K^+ content of the soil will decrease (Roelofs *et al.*, 1985) and due to this decreased K^+ content NH_4^+ will inhibit the K^+ uptake to a relative greater extent. Ultimately high depositions of ammonium may inhibit fungal growth and K^+ uptake.

5. REFERENCES

Arnolds, E.: 1985, *Veranderingen in de paddestoelen flora (Mycoflora)*. Wetenschappelijke mededelingen K.N.N.V. no. 167.

Aalst, R.M. van: 1984, Depositie van verzurende stoffen in Nederland. In: *Zure regen, oorzaken, effecten en beleid*, p. 66. Proc. Symp. Zure regen, 's-Hertogenbosch. Eds. E.H. Adema and J. van Ham, Pudoc, Wageningen.

Bowen, G.D.: 1973, Mineral nutrition of ectomycorrhizae. In: *Ectomycorrhizae*, p. 15. Eds. G.C. Marks and T.T. Kozlowski. Acad. Press, New York and London.

Breemen, N. van, Burrough, P.A., Velthorst, E.J., Dobben, H.F. van, Wit, T. de, Ridder, T.B. and Reynders, H.F.R.: 1982, *Nature*, **229**, 548.

Harley, J.L. and Smith, S.E,: 1984, *Mycorrhizal Symbiosis*. Acad. Press, New York and London.

Jacob, A.: 1958, *Magnesia, der funfte Pfanzenhauptnahrstoff.*, p. 159, Ferdinand Enke Verlag, Stuttgart.

Jansen, P.J.: 1985, *Coolia*, **28** , 13.

Marx, D.H.: 1969, *Phytopath.*, **59** , 153.

Maynard, D.N., Barker, A.V. and Lachman, W.H.: 1969, *Proc. Am. Soc. Hort. Sci.*, **92** , 537.

Nelson, P.V. and Hsieh, K.H.: 1971, *Comm. in Soil Science and Plant Analysis*, **2** , 439.

Nihlgård, B.: 1985, *Ambio*, **14** , 2.

Plassard, C.S., Moussain, D.G. and Salsac, L.E.: 1982, *Phytochemistry*, **21**, 345.

Roelofs, J.G.M., Kempers, A.J., Houdijk, A.L.F.M. and Jansen, J.: 1985, *Plant and Soil*, **84** , 45.

Roelofs, J.G.M.: 1986, *Experientia*, **42** , 372.

Smith, R.C. and Epstein, E.: 1964, *Plant Physiol.*, **39** , 992.

Theuvenet, A.P.R. and Borst-Pauwels, G.W.F.H.: 1976, *Biochim. Biophys. Acta*, **426** , 745.

Trappe, J.M.: 1962, *Bot. Rev.*, **28** , 538.

THE GEOGRAPHICAL DISTRIBUTION AND TEMPORAL VARIATIONS OF ACIDIC DEPOSITION IN EASTERN NORTH AMERICA

Peter W. Summers
Atmospheric Environment Service
Environment Canada
4905 Dufferin Street
Downsview, Ontario M3H 5T4, Canada

Van C. Bowersox & Gary J. Stensland
Illinois State Water Survey
2204 Griffith Drive
Champaign, Illinois, 61820
U.S.A.

ABSTRACT. Data from two national precipitation chemistry monitoring networks, and several regional air and precipitation chemistry networks are used to describe some broad-scale features of acidic deposition in eastern North America. In northeastern North America, the coefficient of variation is shown to increase from 10-16% for annual averages to nearly 100% for daily values. There is a strong annual cycle in H^+, $SO_4^=$ and NH_4^+ deposition and some of the other ions although these cycles are not all in phase. The wet NO_3^- deposition contributes relatively more than $SO_4^=$ to the acidity of snow as compared to rain. Wet deposition is highly "episodic" with about 50% to 70% of the total annual deposition of $SO_4^=$ and NO_3^- accumulating in the highest 20% of the days. Estimates made in various ways indicate that, over eastern North America as a whole, dry deposition is approximately equal to wet for both $SO_4^=$ and NO_3^-. Dry may exceed wet in the high emissions zone but drops to about 20% of the total deposition in more remote areas. Deposition via fog or low cloud impaction is an important input to high elevation forests, but more data are required to quantify the magnitude and regional extent of this.

1. INTRODUCTION

The deposition of chemical species from the atmosphere to the earth's surface is the end result of a complex chain of physical and chemical processes in the atmosphere and is an input to another complex chain of processes that takes place in terrestrial and aquatic ecosystems. Atmospheric deposition has traditionally been separated into two distinct mechanisms - wet deposition and dry deposition. However, a third mechanism is now receiving increased attention, namely the deposition via fog and cloud droplets. It is the purpose of this paper

Water, Air, and Soil Pollution **31** (1986) 523–535.

to use recently available data to update previous results on the temporal and spatial variations in, and relationships between, the major ions contributing to acidic deposition over eastern North America. This will be done from the perspective of providing information relevant to the study of effects of deposition on ecosystems, rather than the understanding of atmospheric processes.

The spatial variations of the major ions in precipitation in the eastern United States for the year 1979 were discussed by Wilson and Mohnen (1982), and recently published information on deposition patterns in eastern North America have concentrated on the year 1980 (National Academy of Sciences, 1983; Barrie and Hales, 1984). These analyses emphasized annual averages. Now that another three or four years of data are available, other time scales of possible importance for studying ecological effects can be discussed, together with some indication of the variability and representativeness of the measurements. Since $SO_4^=$ is the ion of most importance when considering aquatic effects (Newcombe, 1985), it will be emphasized in this paper and used to provide examples.

For this paper eastern North America (ENA) is defined as the region roughly east of the Mississippi Valley and the Manitoba - Ontario border and south of latitude 60°N.

2. THE DATABASE

Many precipitation chemistry monitoring networks operate in ENA but only two national networks provide coverage of the whole area. These are the United States National Atmospheric Deposition Program/ National Trends Network (NADP/NTN) in the United States and the Canadian Network for Sampling Precipitation (CANSAP). Note that after 1983 CANSAP was replaced by the Canadian Air and Precipitation Chemistry Monitoring Network (CAPMoN) in the east. Details of the operation and an assessment of these networks have been given by Stensland et al. (1980) and Barrie and Sirois (1982), with a brief summary in Barrie and Hales (1984). The spatial resolution of these networks is of the order 100 to 500km - enough to indicate the large-scale patterns of wet deposition in ENA. Several research networks are also operated, e.g. the Canadian Air and Precipitation Monitoring Network (APN), the Utility Acid Precipitation Study Program (UAPSP) and the Multi-State Atmospheric Power Production Pollution Study (MAP3S) networks. In addition to precipitation chemistry, some of these networks also report ambient air concentrations of selected S and N compounds. Since there are no direct measurements of dry deposition on the scale of ENA, fluxes are calculated from the product of these ambient air measurements and the appropriate deposition velocity estimates. Data from all these sources will be used as appropriate to illustrate various points.

3. WET DEPOSITION

3.1. Annual Patterns and Variability

The inter-annual variability of wet deposition over ENA is illustrated in Figure 1 by maps displaying the spatial distribution of sulfate for 1981, 1982, and 1983. These maps are based on data from the Acid Deposition System (ADS) - a management and reporting system for North American deposition monitoring data (Watson and Olsen, 1984). Screening criteria were applied so that each site used in the preparation of the maps would have valid precipitation data and valid precipitation chemistry data covering a minimum fraction of the year,

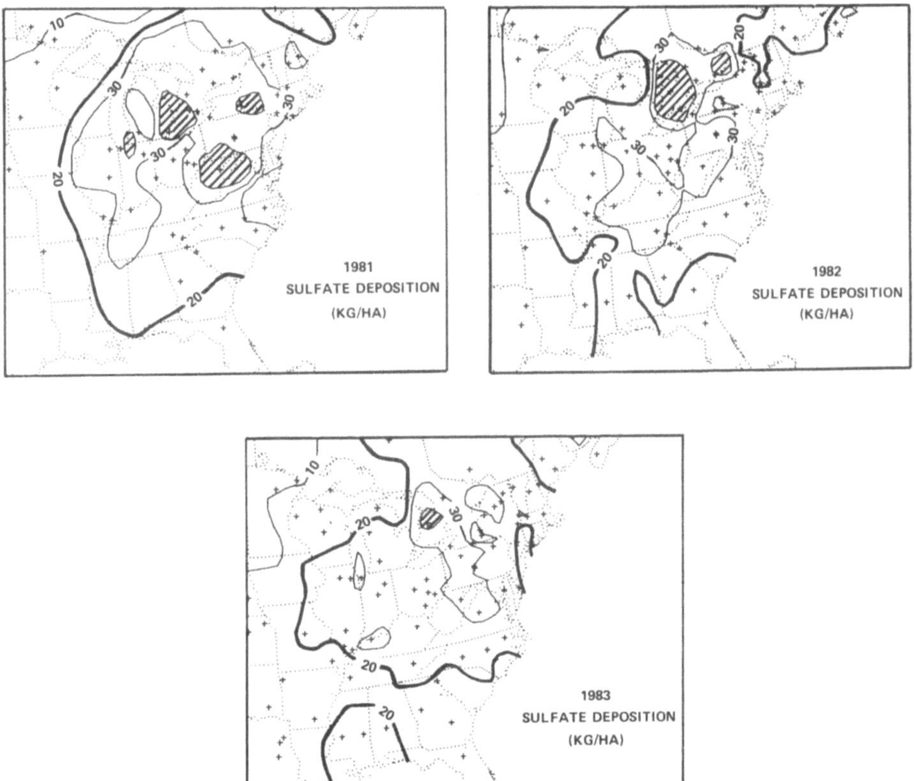

Figure 1. The annual sulphate deposition (kg $SO_4^=$ ha^{-1} yr^{-1}) in eastern North America for three recent years. The 20 kg $SO_4^=$ ha^{-1} yr^{-1} contour is emphasized and the region with deposition exceeding 40 kg $SO_4^=$ ha^{-1} yr^{-1} is cross-hatched.

the year, following the general approach of Voldner et al. (1984). These criteria were explicitly defined by Lindberg et al. (1985), but were relaxed slightly to accommodate the undercatch of snow in areas where there was enough snow to seriously reduce the annual average catch efficiency. Annual wet deposition values were calculated, from the product of the annual weighted concentration and the measured annual precipitation total. The comparability of data between the networks or from individual sites within the networks was not critically evaluated, but no apparent discontinuities were detected at the boundaries.

Spatial patterns were based on the sites meeting the selection criteria for each year as indicated in Figure 1. No seasalt adjustment was applied to the data. The weighted annual deposition data were objectively analyzed using a computerized scheme to contour meteorological variables (Barnes, 1964; Barnes, 1973; Achtemeier et al., 1977).

From 1981 through 1983 the region contained within the 20 kg ha^{-1} contour line remained fairly constant but the area with sulfate deposition > 30 kg ha^{-1} decreased. Very high deposition areas (> 40 kg ha^{-1} tended to move around from year to year and were frequently the result of values for only a small number of sites. The 1980 sulfate deposition pattern (Barrie and Hales, 1984) is similar to those for 1981 and 1982. The high deposition area was considerably smaller in 1983.

From the effects standpoint, it is clear that much of ENA received wet sulfate fluxes in excess of 20 kg ha^{-1} yr^{-1} for the 3 yr of data analyzed. Areas where fluxes exceeded 30 kg ha^{-1} yr^{-1} were within and near source regions and, compared to the 20 kg ha^{-1} region, these areas exhibited much greater spatial and temporal variability.

3.2. Seasonal Variability

Summarizing atmospheric deposition on an annual basis obscures the strong seasonal cycle in the deposition of sulfate and many of the other ions. These variations are apparently related to the seasonal cycle of the meteorological and chemical processes, such as solar radiation, temperature, precipitation type and amount, SO_2 transformation rates, etc.

The seasonal variations of wet deposition at a single station, are often erratic because of the small number of contributing observations. By spatially averaging data over geographical areas with similar characteristics (such as climate, topography, relationship to pollutant source regions), a larger number of data points is used in the calculations and a smoother and statistically more meaningful annual cycle results. Such an analysis was done for monthly CANSAP data covering the years 1977-80 by Barrie and Sirois (1982). A similar analysis was done for NADP data grouped into regions and averaged over seasonal quarters by Bowersox and Stensland (1981).

In this paper, analyses were done for four regional groupings of stations located as defined in Figure 2 in ecosystems identified as being sensitive to acid deposition - two in the U.S.A. and two in Canada. Data from these regions were averaged over a 5-mo warm period (May through September) and a 5-mo cold period (November through March) representing, respectively, the growing and dormant seasons. Because of the strong latitudinal variation of the growing season, April and October were treated as transition months and left out of this analysis. In eastern Canada the cold period is also the time of snowpack accumulation. Ratios of warm to cold period deposition were calculated to show the intra-annual or seasonal variations in wet deposition. These are presented in Table I. For the U.S., maps showing the distributions of these ratios for major ion concentrations indicated a sulfate maximum in the Northeast U.S.A. (see Bowersox and Stensland, 1985). For the regions identified in Figure 2, maxima in the northeast U.S.A. occur for the deposition ratios of $SO_4^=$, as well as for H^+ and NH_4^+. Ratios for these three ions are greater than 1.3 in all 4 regions, thus depositions are clearly higher in the biologically active warm period than in the cold period. In the central and maritime Canadian regions, approximately equal amounts of nitrate are deposited in the cold and warm periods. The observation that nitrate depositions in the Canadian regions exhibit little to no variations, while sulfate depositions are 1 1/2 to 2 times higher in the warm than the cold periods, points to the stronger relative importance of NO_3^- to snowpack acidity. Indeed Bowersox and dePena (1980) concluded that NO_3^- was a principal contributor to the free acidity of snow while sulphate was a principal contributor to rain at an eastern U.S. site. The deposition of Na^+ and Cl^- are clearly out of phase with the major acidic species. Deposition ratios in all regions are <1. The ratios in maritime Canada are the lowest, probably owing to higher wintertime airborne seasalt concentrations. Finally the soil-related cations (Ca^{++}, Mg^{++}, and K^+) exhibit a mixed pattern of ratios, some above and some below 1. Values <1 in maritime Canada are consistant with the seasalt influence discussed for Na^+ and Cl^-. Elsewhere, the patterns reflect the somewhat more complicated influence of highly variable sources on precipitation chemistry.

3.3. Individual Events

On the time scale of a day, or events (episodes) lasting a few days, the wet deposition is highly variable, because it is the product of two highly variable quantities, namely the precipitation amount and the species concentration within the precipitation. Data from the APN site at Chalk River, Canada shows day-to-day variations in H^+ deposition of one to two orders of magnitude (Whelpdale and Barrie, 1982). Though infrequent, extreme wet deposition episodes can account for a relatively large fraction of the annual deposition that impacts ecosystems. When events were ranked in order of H^+ or $SO_4^=$ deposition, Barrie and Sirois (1986) found that about 20% of the daily events accounted for about 50% of the total wet deposition in ENA. Kurtz et

Table I: Intra-annual variability of the wet deposition of the major ions expressed as the mean and coefficient of variation for the warm (May to September) to cold (November to March) period ratio for the four groupings of stations indicated in Figure 2.

	Ratio of warm/cold period deposition				
	Southeast USA	Northeast USA	Central Canada	Maritime Canada	Coeff. of Variation
H^+	1.80	2.32	1.78	1.67	
$SO_4^=$	1.63	2.60	2.05	1.59	18-25%
NO_3^-	1.58	1.54	1.04	0.99	18-25%
NH_4^+	1.86	2.94	2.09	1.40	18-25%
Ca^+	1.30	1.83	1.25	0.78	
Mg^{++}	0.98	1.48	1.46	0.27	35-45%
K^+	1.34	2.22	1.07	0.42	35-45%
Na^+	0.74	0.90	0.60	0.18	50-70%
Cl^-	0.62	0.94	0.78	0.17	50-70%

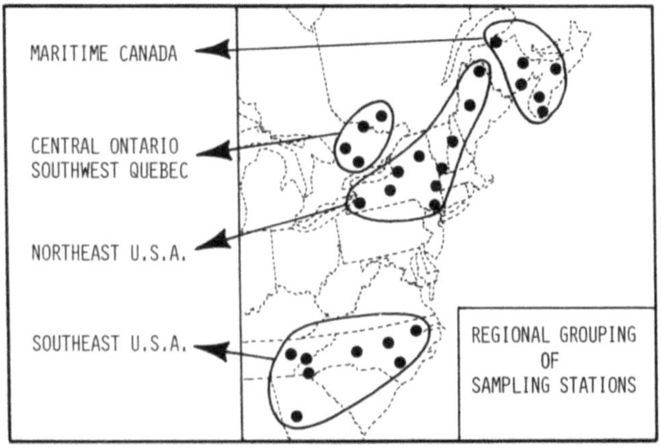

Figure 2: The regional groupings of monitoring stations in ecologically sensitive parts of eastern North America used to calculate the monthly deposition values.

al. (1984) reported on one extreme event in 1981 at Dorset, Ontario where during an 8-day period 100mm of rain (12% of the 1981 total) resulted in a $SO_4^=$ deposition of 7 kg ha^{-1}, 28% of the 1981 total. The observed pH depression in one nearby lake was well over 1.0 pH unit and equivalent in magnitude to that typically observed as a result of spring snowmelt.

Table II: Dependence of the Coefficient of Variation as percent (standard deviation divided by the mean) on averaging time for precipitation $SO_4^=$ concentrations from various networks.

Averaging Time	MAP3S[a] (8 sites)	Eastern CANSAP (10 sites)	NE NADP network (15 sites)	Central Ontario (single site)
Year		13.2	15.4	12.5
Month	50			44
Week	74			
Day	88			90[b]

a) MAP3S/RAINE Research Community (1982)
b) Munn et al., (1986)

3.4. Variability and Averaging Time

There has not been a systematic analysis of precipitation chemistry data to investigate the relationship between variability and averaging time. For $SO_4^=$ concentrations, published results together with additional analyses, are summarized in Table II. These clearly show a systematic decrease in the coefficient of variation (standard deviation divided by the mean) with increasing averaging time. Munn et al. (1986) present data indicating an inverse relationship between C.O.V. of the daily values and the annual mean concentration over eastern Canada.

For longer averaging times the C.O.V. for point annual averages of $SO_4^=$ deposition does not show a regional pattern but varies randomly between 7.7 and 31.6%, with average network values between 13 and 16%.

4. DRY DEPOSITION

Very little is known about the geographical distribution of dry deposition fluxes in eastern North America because, at present, there is no method to routinely measure them; thus, no monitoring networks exist. As an alternative, estimates can be made using the physically based concept of deposition velocity, analogous to an exchange coefficient, which, when multipled by the ambient air concentration, gives the rate of deposition to a surface.

The dry deposition velocity is dependent on both the meteorological conditions and the underlying surface characteristics. In general, it tends to be low over snow and ice surfaces and highest over wet vegetated surfaces. This dependence has been established at only a few research sites, but can be extrapolated to generate fields of dry deposition velocity over larger areas. The approach suggested by Wesley and Hicks (1977), and Sheih et al. (1979) has been expanded by Voldner and Sirois (1986) to produce maps of the monthly average deposition velocity field east of the Rockies for SO_2, NO_x, HNO_3 and particulate $SO_4^=$ and NO_3^-. Rural air monitoring data in the U.S.A. are available for sites operated during the Sulphate Regional Experiment (SURE) in 1978 and for other specialized research sites. In Canada ,the APN network was set up specifically to obtain concurrent daily air and precipitation chemistry data at rural sites. In both countries data at more sites are needed to establish geographic patterns. When the APN stations are ranked according to distance from a major pollution source, the ambient air concentrations of SO_2 decrease with increasing distance, as shown in Table III. Ambient $SO_4^=$ and NO_3^- particulate levels fall off rapidly at first then stay nearly constant in part because the rate of production by oxidation from the gaseous phase is balanced by the deposition rate. If the underlying surface and meteorological conditions were the same at all sites, then the dry deposition flux would decrease at the same rate as the air concentrations. Estimates made by Chan et al. (1986) using the air quality data from a regional scale network of 36 stations operated in Ontario since 1980 show an order of magnitude drop in the dry deposition flux from about 0.4g S $m^{-2}yr^{-1}$ (12 kg $SO_4^=$ $ha^{-1}yr^{-1}$) along the north shore of Lake Erie to less than 0.05g S $m^{-2}yr^{-1}$ in northern Ontario.

Ecosystems respond to both wet and dry deposition but sometimes in quite different ways, thus it is useful to have some indication of the relative magnitude of wet and dry deposition. The mass budget approach, based on a careful combination of emissions, ambient air and precipitation measurements and atmospheric flow fields (Galloway and Whelpdale, 1980; Galloway et al., 1984) indicated that over the whole of eastern North America wet deposition slightly exceeds dry deposition of S. Over eastern Canada, generally farther from the sources, wet deposition exceeded dry by a factor 2.5 to 1, but in the eastern United States dry exceeded wet by about 30%. Logan (1984) used a similar approach to develop a nitrogen budget for eastern North America from which the fraction dry deposited can be calculated. More recent estimates from the APN network (Barrie and Sirois, 1986) averaged over 6 stations in eastern Canada, and from the APIOS network in Ontario (derived from data in Chan et al., 1986), are summarized together with the budget studies in Table IV. Both the APN and APIOS data showed a decrease of the percentage due to dry deposition from about 40% on the north shore of Lake Erie to less than 20% in more remote areas.

Table III: Average air concentrations at the APN sites as a function of distance (D) from the centre of Lake Erie.

Station	D (km)	SO_2	$SO_4^=$	NO_3^-
			$\mu g\ m^{-3}$	
Long Point	100	13.3	4.9	2.6
Chalk River	400	6.5	2.6	0.5
Algoma	600	3.3	2.6	0.6
Montmorency Forest	950	2.1	1.8	0.3
Kejimkujik	1200	1.7	2.4	0.4
ELA - Kenora	1300	1.3	1.6	0.5
Baie d'Espoir	2000	0.6	1.3	0.2

Table 4. The dry deposition expressed as a percentage of the total estimated deposition obtained from several studies.

REGION	$SO_4^=$	NO_3^-
BUDGET STUDIES		
Eastern North America	45	50
Eastern U.S.A.	57	59
Eastern Canada	29	23
NETWORK STATIONS		
APN (Eastern Canada)	22	30
APIOS (Ontario)	16	38*

* not including NO_2 contribution, but also assumes all NO_3^- as HNO_3 which leads to an overestimate.

5. DEPOSITION BY FOG AND LOW CLOUD

Although it has long been known that "fog drip" can provide a substantial fraction of the water input from the atmosphere to high elevation forest ecosystems, chemical deposition via this route had not been considered especially important until more recent sampling of fogs and low cloud indicated high concentrations of acidic species, often leading to extremely low pH values.

Some of the higher elevation forests in ENA such as the Green Mountains of Vermont, are bathed in cloud water about 10 to 20% of the time (Siccama, 1974). The resultant "fog-drip" can account for up to 50% of the total annual moisture input to the forest. A model to simulate this collection by a balsam fir canopy has been developed by Lovett (1984) and indicates water deposition rates on the order of tenths of a millimeter per hour. An exposure of 1000 hours per year, which is typical of the high elevations, in the NE would thus produce water deposition equivalent to a few tens of centimeters.

Observations of fog and low cloud chemistry are very limited, originating from isolated research projects rather than from any systematic regional monitoring efforts. The most extensive series of mountain fog/low cloud measurements have been made atop Whiteface Mountain, New York, at an elevation of 1480 m. Early measurements of pH values in non-precipitating clouds ranged from 3.0 to 4.5, with occasional values below 3.0 (Falconer and Falconer, 1980; Castillo et al., 1983). More recent measurements, reported by Castillo et al. (1985), showed that in the summers of 1981 and 1982 the mean pH of cloud water was 3.5 and 3.9, respectively. In August with polluted southwesterly air flow the mean pH was 2.6 and 3.0 in 1981 and 1982 respectively, with the mean sulphate ion concentration of 32 mg ℓ^{-1} for the two Augusts. When the water flux is multiplied by the concentrations of H^+ and $SO_4^=$, cloud impaction becomes a significant deposition mechanism which in some cases could exceed that from either precipitation or dry deposition. Deposition by this mechanism is highly dependent on elevation, and the extent of the affected areas in eastern North America has yet to be determined.

6. SUMMARY

The area receiving in excess of 20 kg $SO_4^=$ ha^{-1} yr^{-1} wet deposition was similar over the three years 1981 to 1983, but smaller regions of higher deposition (>40 kg $SO_4^=$ ha^{-1} yr^{-1}) exhibited large year-to-year variations in location and areal extent. The spatial variability of wet deposition depends strongly on the time over which the data are averaged. The annual wet deposition of $SO_4^=$ had a network average coefficient of variation between 12 and 16%. At individual stations the coefficient of variation of the annual average was as high as 36%. For daily measurements the coefficient of variation was of the order of 100%.

A strong seasonal variation in wet deposition is observed in eastern North America. The warm (growing) period deposition of all major ions except Na^+ and Cl exceeded the cold (dormant) period deposition in the region of highest deposition. The H^+, $SO_4^=$ and NH_4^+ ions had the largest and most consistent warm to cold period variations. Deposition also tends to be highly episodic in the sense that only a few events contribute to most of the deposition over a year.

The large-scale patterns of dry deposition cannot yet be well defined due to the lack of suitable method for routine direct monitoring. Based on measured ambient air concentrations and estimated dry deposition velocities (extrapolated from measurements at a few research sites), the calculations suggest that within or close to the major emission source regions the dry deposition of sulfur and nitrogen is greater than the wet. In more remote areas wet deposition predominates. Over eastern North America as a whole, wet deposition of sulfur slightly exceeds dry. A rural air monitoring network is now being established in the U.S.A. and more sites will gradually be added to the Canadian networks. These will eventually provide information on the regional patterns of dry deposition.

There is increasing evidence that substantial deposition can occur via fog and low clouds, especially to the higher elevation forests in the Appalachians, New England and southern Quebec. In order to establish the magnitude and areal extent of this input, a co-operative project between the U.S.A. and Canada to monitor the composition of mountain fogs at several sites, began in 1985.

7. ACKNOWLEDGMENTS

The analyses of deposition data from the U.S. and Canadian networks were based on information obtained from the Acid Deposition System (ADS) data-base operated by Pacific Northwest Laboratory funded as part of the National Acid Precipitation Assessment Program by the U.S. Environmental Protection Agency. The contribution of two of the authors (G.J.S. and V.C.B.) was supported as part of the National Acid Precipitation Assessment Program by the Department of Energy, Office of Health and Safety Research Division under Contract DE-AC02-76EV01199.

The authors wish to acknowledge the computer programming assistance of Randall K. Stahlhut at the Illinois State Water Survey and of Alain Sirois and Balbir Saini at the Atmospheric Environment Service.

8. REFERENCES

Achtemeier, G.L., Hildebrand, P.H., Schickedanz, P.T., Ackerman, B., Changnon, Jr., S.A., and Semonin, R.G.,: 1977, 'Illinois Precipitation Enhancement Program Design and Evaluation Techniques for High Plains Cooperative Program', Ill. State Water Survey Final Report on Contract 14-06-D-7197, Appendix B.

Barnes, S.L.: 1964, J. Appl. Meteorol., 3, 396.

Barnes, S.L.: 1973, 'Mesoscale objective map analysis using weighted time-series observations'. NOAA Tech. Memo. 62, NSSL, 60 pp.

Barrie, L.A. and Sirois, A.: 1982, 'An Analysis and Assessment of Precipitation Chemistry Measurements made by CANSAP (The Canadian Network for Sampling Precipitation): 1977-1980'. Report AQRB-82-003-T, Atmospheric Environment Service, Toronto, Canada, 163pp.

Barrie, L.A. and Sirois, A.: 1986, J. Water, Air and Soil Poll., 22, in press.

Barrie, L.A. and Hales, J.M.: 1984, Tellus., 36B, 333.

Bowersox, V.C. and dePena, R.G.:1980, J. Geophys. Res., 85, 5614.

Bowersox, V.C. and Stensland, G.J.: 1981, 'Seasonal Patterns of Sulfate and Nitrate in Precipitation in the United States'. Proc. 74th Annual Meeting of APCA, Philadelphia, PA, June 1981.

Bowersox, V.C. and Stensland, G.J.: 1985, 'Seasonal Variations in the Chemistry of Precipitation in the United States', Proc. 78th Annual Meeting, Air Pollution Control Assoc. Detroit, Mich., June 1985.

Castillo, R.A., Juisto, J.E., and McLaren, E.: 1983, Atmos. Env., 17, 1497.

Castillo, R.A., Kadlecek, J. and McLaren, S.: 1985, J. Water, Air and Soil Poll., 24, 323.

Chan, W.H., Tang, A.J.S., Chung, D.H.S., Kirk, R.W. and Lusis, M.A.: 1986, Atmos. Env., 20, in press.

Falconer, R.E. and Falconer, P.D.: 1980, J. Geophys. Res., 85, 7465.

Galloway, J.N. and Whelpdale, D.M.: 1980, Atmos. Env., 14, 409.

Galloway, J.N., Whelpdale, D.W., and Wolff, G.T.: 1984, Atmos. Env., 18, 2595.

Kurtz, J., Tang, A.J.S., Kirk, R.W. and Chan, W.H.: 1984, 'Analysis of an Acid Deposition Episode at Dorset, Ontario', Atmos. Env., 18, 387-394.

Lindberg, S., Bigelow, D., Bowersox, V., Knapp, W. and Olsen, T.: 1985, 'NADP Annual Data Summary, Precipitation Chemistry in the United States, 1982'. Chapter II - Data Selection Criteria, Natural Resource Ecology Laboratory, Ft. Collins, Co., pp. 3-8.

Logan, J.A.: 1984, J. Geophys. Res., 88, 10, 785.

Lovett, G.M.: 1984, Atmos. Env., 18, 361.

MAP3S/RAINE Research Community.: 1982, Atmos. Env., 16, 1603.

Munn, R.E., Whelpdale, D.M., Oehlert, G.W., and Summers, P.W.: 1986, Atmos. Env., 20, in press.

National Academy of Sciences: 1983, 'Acid Deposition: Atmospheric Processes in Eastern North America', National Academy Press, Washington, D.C., 375 pp.

Newcombe, C.P.: 1985, Env. Management, 9, 277.

Sheih, C.M., Wesley, M.L. and Hicks, B.B.: 1979, Atmos. Env., 13, 1361.

Siccama, T.G.: 1974, Ecol. Mon., 44, 325.

Stensland, G.J., Semonin, R.G. Peden, M.E., Bowersox, V.C. McGurk, F.F., Skowron, L.M., Slater, M.J. and Stahlhut, R.K.: 1980, 'NADP Quality Assurance Report - Central Analytical Laboratory, 1979 National Atmospheric Deposition Program, Illinois State Water Survey, Champaign, Ill., U.S.A.

Voldner, E.C., Sirois, A. and Clark, T.L.: 1984, 'Data Screening and Calculation Procedures for the North American Precipitation Chemistry Data to be Used in the International Sulfur Deposition Model Evaluation'. Proceedings of the APCA/ASQC Specialty Conference on Quality Assurance in Air Pollution Measurements, Air Pollution Control Assoc., Pittsburgh, PA, 13 pp.

Voldner, E. and Sirois, A.: 1986, J. Water Air and Soil Poll., 22, in press.

Watson, C.R. and Olsen, A.R.: 1984, 'Acid Deposition System (ADS) for Statistical Reporting: System Design and User's Code Manual'. EPA-600-8-84-023, U.S. Environmental Protection Agency, Research Triangle Park, N.C.

Wesley, M.L. and Hicks, B.B.: 1977, J. Air Poll. Control Assoc., 27, 1110.

Whelpdale, D.M. and Barrie, L.A.: 1982, J. Water, Air and Soil Poll., 18, 7.

Wilson, J.W., and Mohnen, V.A.: 1982, J. Water, Air and Soil Poll., 18, 199.

EFFECTS OF GASEOUS POLLUTANTS ON FORESTS IN EASTERN NORTH AMERICA

Samuel N. Linzon
Phytotoxicology Section
Air Resources Branch
Ontario Ministry of the Environment
Toronto, Ontario
Canada M5S 1Z8

Abstract. The purpose of this paper is to summarize the current state of knowledge concerning direct forest damage caused by atmospheric gaseous pollutants. The major gaseous pollutants affecting forest growth are O_3, SO_2, and HF. Ozone is the most prevalent phytotoxic air pollutant on a regional scale in eastern North America. It has been demonstrated to cause foliar injury and growth effects on trees in both laboratory and field studies. Both SO_2 and HF have been found to cause damage to forests primarily in the vicinity of strong point sources. Case histories of forest damage are reported in the paper which describe foliar and radial growth effects, with associated volume losses. Threshold levels of the gaseous pollutants causing effects on forest growth are provided.

1. INTRODUCTION

Ozone is not emitted directly by any stationary or mobile source but is formed in the atmosphere from a photochemical reaction involving precursor pollutants, NO_x and reactive hydrocarbons. The two precursor pollutants are emitted mainly by automobiles. Since NO_x are emitted by high temperature combustion processes, fossil fuel electric generating stations are also a large source, and hydrocarbons are emitted also by various petroleum operations from manufacturing, storage to sale by gasoline stations. Since it takes time for O_3 to be formed in the atmosphere, high concentrations are usually found downwind of the sources, with O_3 affecting agricultural crops or forests often hundreds of kilometers from the sources of the precursor pollutants.

In contrast, SO_2 and HF are emitted directly by point sources and high concentrations usually cause damage to vegetation in the vicinity of the industries, with the distance of the damage from the source depending on the magnitude of the emissions. Sulphur dioxide effects on forests have been found as far as 112 km northeast of Ni smelters in the Sudbury area of Ontario (Linzon, 1958), while HF effects on forests have been documented as far as 12 km northeast of a P manufacturing plant in the Long Harbour area of Newfoundland (Sidhu, 1979).

Water, Air, and Soil Pollution **31** (1986) 537–550.

Nitrogen oxides are prevalent in the atmosphere but there is little evidence of ambient levels of NO_x alone causing direct damage to forests. Nitrogen deposition may indirectly affect forest health by delaying dormancy and making trees more susceptible to frost (Friedland et al., 1985). Nitrogen oxides are important in that by combining with other pollutants they can lead to the formation of acid rain and O_3, which are both long range transported pollutants.

It is well documented that O_3 is responsible for widespread and costly damage to agricultural crops both in the United States (Heck et al., 1982) and in Canada (Linzon et al., 1984) and that O_3 has caused extensive damage to forests in California (Miller et al., 1982). However, this paper will discuss only the direct effects of the major gaseous air pollutants on forests and trees primarily in eastern North America. The forest effects will include not only tree damage, but also impacts on soils and lichens. This paper will not review the effects of acidic precipitation on forests by either direct or indirect routes.

2. OZONE

Many reviews have been written on the effects of O_3 on vegetation, which discuss dose-response relationships, the symptoms of injury, sensitivity of plant species, and the reduction in growth and yield of crops and trees (Hill et al., 1970; Linzon et al., 1975; and Jacobson, 1977). Accordingly, trees listed as sensitive to O_3 include white ash, trembling aspen, catalpa, honey locust, silver maple, eastern white pine, ponderosa pine and Virgina pine.

Ozone injury can result in four general types of foliar lesions. These include pigmentation, bleaching, chlorosis, and bifacial necrosis (Hill et al., 1970). In pigmentation, the palisade mesophyll cells take on brown, black, red or purple colorations. Bleached lesions usually occur on upper surfaces of leaves due to the collapse of the palisade cells and sometimes the epidermal cells. Chlorotic areas may also occur on the upper surface of the leaves, often coalescing to present a mottled appearance. Pine needles which do not possess differentiated mesophyll tissue may develop the chlorotic mottling on any leaf surface. Bifacial necrotic lesions may range from ivory to orange-red in coloration.

Studies on needle blights of eastern white pine have been conducted in several locations in eastern North America starting in Canada in 1959 and in the USA in 1961 (Linzon and Costonis, 1971). In fumigation experiments in North Carolina, Berry and Ripperton (1963) induced a tip necrosis on newly developing immature needles of eastern white pine with 128 $\mu g\ m^{-3}$ (0.065 ppm) O_3 for 4 hr. (At 25°C and 760 mm Hg, 1 ppm O_3 = 1963 $\mu g\ m^{-3}$). Costonis and Sinclair (1969) in New York State showed that O_3 acting on the new needles of certain sensitive strains of eastern white pine could cause a wider range of symptoms from silvery and chlorotic flecks through chlorotic mottling to tip necrosis. In artificial fumigation experiments this syndrome was induced by controlled doses of O_3 of 137 $\mu g\ m^{-3}$ (0.07 ppm) for 4 hr or 59 $\mu g\ m^{-3}$ (0.03 ppm) for 48 hr. At Chalk River, Ontario, in experimental fumigations Linzon (1973) induced chlorotic flecks on new needles of eastern white pine with concentrations of 196 $\mu g\ m^{-3}$ (0.10 ppm) O_3 for 6 hr. In field studies at Chalk River, Linzon (1973) found *chlorotic flecks developing on new needles of eastern white pine ramets (clonal grafts) which had been exposed to ambient air containing* O_3 *concentrations of 59*$\mu g\ m^{-3}$ *(0.03 ppm) or over for periods of 10 to 15 hr, while no fleck symptoms developed on ramets which had been covered with polyethylene bags. Blanchard et al., (1979) monitored oxidant levels in New Hampshire and found new needles of eastern white pines were injured when oxidant levels above 59*$\mu g\ m^{-3}$ *(0.03 ppm) were recorded for 56 hr in 1974, 36 hr in 1975 and 34 hr in 1976.*

Ozone damage to forests in eastern North America has been extensively investigated in the Blue Ridge Mountains of Virginia starting in the early 1970's (Skelly et al., 1983). Elevated concentrations of O_3 *were measured frequently from April to October, with monthly 8 hr average* O_3 *concentrations ranging from 69 to 128*$\mu g\ m^{-3}$ *(0.035 to 0.065 ppm) and peak hourly concentrations from 157 to 255*$\mu g\ m^{-3}$ *(0.08 to 0.13 ppm)* O_3*. The most intense* O_3 *episode occurred in July 1977, with a peak hourly concentration of 326*$\mu g\ m^{-3}$ *(0.166 ppm)* O_3 *with 36% of the total monthly hours exceeding 157*$\mu g\ m^{-3}$ *(0.08 ppm)* O_3 *(Galloway and Skelly, 1978). Eastern white pine was the first tree species to exhibit symptoms of* O_3 *injury, which included reduced needle length and needle retention, foliar chlorosis and reduced radial growth. Other native tree species exhibiting oxidant air pollutant induced symptoms included tulip poplar, green ash, hickory, black locust, table mountain pine, Virginia pine, pitch pine and hemlock. Open top chambers in which* O_3 *was filtered from ambient air entering the chambers which enclosed several coniferous tree species resulted in an increased height growth of up to 26% compared to trees growing in the open. Ozone concentrations in the air were found to increase with elevation throughout the study region and foliar injury on sensitive tree species was greater at the highest elevations (Skelly et al., 1983).*

A number of studies have been conducted on the effects of O_3 *on various tree species under controlled conditions. Davis et al., (1981) artificially fumigated several trees species in Pennsylvania with 393*$\mu g\ m^{-3}$ *(0.20 ppm)* O_3 *for 5 hr periods and found black cherry to be the most sensitive, exhibiting 27% foliar injury. In fact, visible injury was produced on black cherry foliage at 196*$\mu g\ m^{-3}$ *(0.10 ppm)* O_3 *for 4 hr and 373*$\mu g\ m^{-3}$ *(0.19 ppm)* O_3 *for 2 hr.*

Karnosky (1976) in New York State produced injury on foliage of trembling aspen in artificial fumigations with 98$\mu g\ m^{-3}$ *(0.05 ppm)* O_3 *for 3 hr.*

Eighteen species of coniferous tree seedlings were exposed to 196$\mu g\ m^{-3}$ *(0.10 ppm)* O_3 *for 8 hr or to 491*$\mu g\ m^{-3}$ *(0.25 ppm)* O_3 *for 4 to 8 hr in Pennsylvania by Davis and Wood (1972). The six most sensitive tree species in descending order of sensitivity were Virginia pine, jack pine, European larch, Austrian pine, Scots pine and eastern white pine. Chlorotic mottle and tip necrosis or complete needle necrosis were the most commonly observed symptoms.*

Sugar maples were exposed to O_3 *under laboratory and field conditions in New York State by Hibben (1969). The threshold level for visible injury to occur on maple foliage was 393 to 589*$\mu g\ m^{-3}$ *(0.20 to 0.30 ppm)* O_3 *for 2 hr. Extended intermittent doses of 196*$\mu g\ m^{-3}$ *(0.10 ppm)* O_3 *for 2 hr day*$^{-1}$ *for 2 wk caused slight injury. In the field, understory saplings in the shade were not injured by high doses of* O_3*.*

Nine forest tree species were exposed to 589 $\mu g\ m^{-3}$ (0.30 ppm) O_3 for 5 mo in Ohio by Jensen (1973). Significant reductions in growth were found in sugar maple, silver maple, and sycamore. Jensen and Dochinger (1974) in Ohio reported from 50 to 62% significant reductions in dry weight of new shoots of hybrid poplar cuttings exposed to 294 $\mu g\ m^{-3}$ (0.15 ppm) O_3 for 8 hr day^{-1}, 5 days wk^{-1}, for 6 wk.

Effects on growth and yield of tree species are usually associated with visible injury to the photosynthetically functional foliage. In fumigation experiments, Kress and Skelly (1982) in Virginia found significant growth reductions (height and total dry weight) for sugar maple, sweet gum, American sycamore, green ash, loblolly pine and pitch pine exposed to 196 $\mu g\ m^{-3}$ (0.10 ppm) and/or 294 $\mu g\ m^{-3}$ (0.15 ppm) O_3 for 6 hr day^{-1} for 28 days. Foliar biomass of native ground vegetation was reduced by 30% in ambient air compared to filtered-air chambers without the occurrence of foliar symptoms of O_3 injury on most plant species (Skelly et al., 1982).

Radial growth studies on eastern white pine in the Blue Ridge Mountains indicated that ambient O_3 concentrations reduced the growth of sensitive individuals by 30 to 50% annually during the period 1955 to 1978 (Benoit et al., 1982). Mann et al., (1980) found similar results for eastern white pines in eastern Tennessee.

Studies conducted in Ohio on the chlorotic dwarf disease of eastern white pine by Dochinger (1968) demonstrated that air pollution, of which O_3 forms a part, is the primary causal agent of the disease. Artificial fumigation studies (Dochinger et al., 1970) showed that sensitive eastern white pine trees exposed to a mixture of 196 $\mu g\ m^{-3}$ (0.10 ppm) O_3 and 260 $\mu g\ m^{-3}$ (0.10 ppm) SO_2 for 4 to 8 hr daily for 2 to 4 wk could develop chlorotic mottling symptoms similar to those seen in the field. It was suggested that the two pollutants acted synergistically since the degree of injury produced by either O_3 or SO_2 alone was much less than that produced by the mixture.

There is little information on the effects of O_3 on lichens in eastern North America. Sigal and Nash (1983) reported that lichens in southern California were adversely affected if the cumulative oxidant dose level exceeded 589 mg m^{-3} (300 ppm) hr^{-1} yr^{-1}.

Atmospheric O_3 does not penetrate soil to any appreciable depth. Experiments were conducted by Tingey (1978) which showed that O_3 diffused less than 20 mm into soil and thus could not have any direct effect on root growth. The indirect nature of O_3 inhibition on root growth was confirmed in an experiment in which plants were either enclosed in plastic bags or not, and then exposed to O_3. Root growth was reduced only in those plants in which the O_3 could penetrate the foliage, which reduced the photosynthate material translocated to the roots.

With regard to limiting values for O_3 effects on vegetation, Jacobson (1977) reviewed the available scientific data and estimated that 196 $\mu g\ m^{-3}$ (0.10 ppm) O_3 for 1 hr and 79 $\mu g\ m^{-3}$ (0.04 ppm) O_3 for 4 hr were the lowest concentrations and exposure periods for causing visible injury on the most sensitive crop species. For the most sensitive trees, 393 $\mu g\ m^{-3}$ (0.20 ppm) O_3 for 1 hr and 118 $\mu g\ m^{-3}$ (0.06 ppm) O_3 for 4 hr were required.

3. SULPHUR DIOXIDE

Several reviews on the effects of SO_2 on vegetation have been published which include those by Barrett and Benedict, 1970; Heck and Brandt, 1977; Linzon, 1978a; and the U.S. Environmental Protection Agency, 1982. Accordingly, trees listed as sensitive to SO_2 include white ash, trembling aspen, white birch, catalpa, Chinese elm, Douglas fir, western larch, mulberry, eastern white pine, jack pine and black willow.

One of the earliest documented cases of vegetation damage caused by SO_2 in eastern North America occurred in the early 1900s at Ducktown, Tennessee in the vicinity of copper smelters. Acute injury to forests extended 24 km to the north of the source and coniferous trees were found to be much more susceptible to damage than hardwood trees (Hedgcock, 1914). Although forest damage had been observed since the turn of the century in the vicinity of Ni and Cu ore roasting in the Sudbury area of Ontario, detailed investigations on the effects of SO_2 were not started until the 1940's (Katz, 1949; Linzon, 1958).

Sulphur dioxide is emitted from various industrial sources and the adverse effects on vegetation in the vicintiy of these sources have been documented. Some of the industrial sources are Zn, Pb, Ni, Cu and Ag smelters, Fe concentrators, petroleum refineries, pulp and paper mills, and fossil fuel thermal generating stations (Linzon, 1978a).

The first visible evidence of SO_2 injury to plants is discernible in the foliage. The stems, bud, and reproductive parts of plants are visibly more resistant to SO_2 than foliar parts. Acute injury on broad leaves takes the form of bifacial lesions, which usually occur between veins. In some cases, injury can occur on the margins of the leaves. Young leaves rarely display necrotic markings, whereas fully expanded leaves are most sensitive to this type of SO_2 injury. In conifers, acute injury usually appears as a bright orange-red terminal necrosis on current-year needles, often with a sharp line of demarcation between the injured tips and the normally green bases. Occasionally the injury may occur as bands in apical, medial, or basal locations on the needles. In conifers, chronic injury on older needles appears as a yellowish-green color, then changes to reddish-brown, starting at the tips and developing basipetally (Linzon, 1978a).

It is important to stress the differences between acute, chronic, and subtle effects of SO_2 on plant life. Acute injury is caused by a rapid accumulation of bisulphite and sulphite. When the oxidation product, sulphate, accumulates beyond a threshold value that the plant cells can tolerate, chronic injury occurs. It is estimated that sulphate is about 30 times less toxic than sulphite (Thomas, 1951).

3.1 Acute Effects of SO_2 on Trees

Some extremely low doses of SO_2 have been reported in the literature to cause acute effects on very sensitive clones of eastern white pine in fumigation experiments. These were 78 $\mu g\ m^{-3}$ (0.03 ppm) SO_2 for 1 hr (slight injury) and 78 $\mu g\ m^{-3}$ (0.03 ppm) SO_2 for 3 hr (severe injury) in North Carolina (Costonis, 1971), and 65 $\mu g\ m^{-3}$ (0.025 ppm) SO_2 for 6 hr in Wisconsin (Houston, 1974). (At 25°C and 760 mm Hg, 1 ppm SO_2 = 2600 $\mu g\ m^{-3}$). These *very sensitive strains of eastern white pine are not considered representative of the forest population. Berry (1967) reported injuring foliage of eastern white pine in fumigation experiments conducted in North Carolina with SO_2 at 650 $\mu g\ m^{-3}$ (0.25 ppm) for 1 hr. In later experiments conducted in Georgia, Berry (1974) injured seedlings of Virginia, shortleaf, slash and loblolly pines with SO_2 at 650 $\mu g\ m^{-3}$ (0.25 ppm) for 2 hr. Karnosky (1976) in New York State produced acute injury on foliage of trembling aspen in artificial fumigations with 910 $\mu g\ m^{-3}$ (0.35 ppm) SO_2 for 3 hr.*

Measurements of atmospheric SO_2 and observations of foliage on trees in natural forests in the vicinity of Ni smelters at Sudbury, Ontario revealed that the following doses were required to cause acute injury: 2470 $\mu g\ m^{-3}$ (0.95 ppm) SO_2 for 1 hr, 1430 $\mu g\ m^{-3}$ (0.55 ppm) SO_2 for 2 hr, 910 $\mu g\ m^{-3}$ (0.35 ppm) SO_2 for 4 hr and 650 $\mu g\ m^{-3}$ (0.25 ppm) SO_2 for 8 hr (Dreisinger, 1965). For acute injury to occur other environmental and plant factors are important. These include sunlight, moderate temperature, high relative humidity, adequate soil moisture, and plant genotype and stage of growth. If these factors are not conducive to injury, the plants will escape harm even in the presence of doses two or three times higher than those noted above. Conversely, if the predisposing factors are especially conducive to plant injury, the SO_2 doses could be reduced by about 25% to 1820 $\mu g\ m^{-3}$ (0.70 ppm) for 1 hr, 1040 $\mu g\ m^{-3}$ (0.40 ppm) for 2 hr, 676 $\mu g\ m^{-3}$ (0.26 ppm) for 4 hr and 468 $\mu g\ m^{-3}$ (0.18 ppm) for 8 hr (Dreisinger and McGovern, 1970).

Based on available evidence of acute effects of SO_2 on vegetation (tree and crop species) from both controlled fumigation experiments and field data, McLaughlin (1981) suggested the following air quality standards for the eastern half of the USA where there is a high density of power plants: 1300 $\mu g\ m^{-3}$ (0.50 ppm) SO_2 for 1 hr and 780 $\mu g\ m^{-3}$ (0.30 ppm) for 3 hr. He concluded that the current U.S. standard of 1300 $\mu g\ m^{-3}$ (0.50 ppm) SO_2 for 3 hr was not adequate to protect vegetation from visible injury or yield effects.

3.2 Chronic Effects of SO_2 on Trees

Banfield (1972) reported injuring foliage of eastern white pine in artificial fumigations in Massachusetts with 182 $\mu g\ m^{-3}$ (0.07 ppm) SO_2 for three continuous days. Roberts (1976) placed two-year-old grafted clones of sensitive eastern white pine in the field in Ohio where annual ambient SO_2 levels were 91 $\mu g\ m^{-3}$ (0.035 ppm). Both foliar injury and reduced leaf growth were observed after a 5 mo period.

Sulphur dioxide was found to cause effects on forests in an area of about 1619 ha located 30 km SW of Bathurst, New Brunswick (Van Sickle, 1973). Large quantities of waste sulphide materials from a heavy metal mine which had been buried in an underground mined-out section of another mine oxidized in 1971 and the resulting SO_2 was vented to the surface. Concentrations of SO_2 over 2600 $\mu g\ m^{-3}$ (1.0 ppm) were monitored at a distance of 2.4 km from the source. Sulphur dioxide damage occurred on eastern white pine, white birch, eastern white cedar, balsam fir, black spruce, American beech and red maple. Injury on eastern white pine was found up to a distance of 13 km NE of the source. The emissions of SO_2

decreased gradually over the years and by 1976 only eastern white pine near the source exhibited symptoms of SO_2 toxicity. Possible changes in the chemical properties of the soil were investigated by Mahendrappa (1977). Only a very small decrease in the pH values of mineral soil horizons was observed. The pH values of the upper organic horizons decreased because of the SO_2 contamination. Areas severely affected by SO_2 had organic soil pH as low as 2.2 with an average of 3.1 compared to an average pH of 4.6 in unaffected areas.

In the Sudbury area of Ontario, three large nickel smelters had discharged prior to 1972 approximately 7000 tonne of SO_2 per day into the surrounding atmosphere. Forest effects in the area were excessive with severe injury on trees occurring up to 40 km northeast of Sudbury. Based on studies of over 6000 eastern white pine trees on 42 sample plots during a 10-yr period, (1953-1963) the Sudbury area was segregated into three fume zones: Inner, Intermediate and Outer (Linzon, 1966). In the Inner Fume zone, an area of about 1860 km^2, eastern white pine trees displayed both acute and chronic foliar injuries which resulted in reduced radial and volume growth and in excessive tree mortality. In the Intermediate Fume Zone, and area of about 4140 km^2, chronic SO_2 injury was present, while in the Outer Fume Zone, atmospheric contamination was too dilute to cause visible injuries.

The greatest increase in chronic foliar injuries during the growing season was found to occur on the one-year-old needles of eastern white pine trees. The continuous development of injuries on the older needles was reflected in early abscission of the oldest foliage, and reduced radial and volume growth and premature death of the trees. Over five times as many eastern white pine trees died annually in the Inner Fume Zone than in a control area located at a distance from Sudbury. The eastern white pine forests in the Inner Fume Zone were found to exhibit a net average annual negative growth in volume, since the trees that died during the investigation period exceeded the volume added by the surviving trees (Linzon, 1971). Radial increment growth for the period 1940 to 1960 as measured on cores extracted in 1963 from 20 dominant living eastern white pine trees on each of 42 sample plots showed a gradual decline in the Inner Fume Zone, whereas a constant pattern was maintained in the other areas (Linzon, 1973).

It has been observed in the Sudbury area that other biological effects can occur also as a result of atmospheric contamination. The disease blister rust (Cronartium ribicola) of white pine was found to be practically non-existent in areas close to the source of the S fumes and the incidence of weevil insect injuries was less than in control areas (Linzon, 1958).

Epiphytic lichens, which are perennial and evergreen are extremely sensitive to SO_2 because they are continually exposed to the gases in a polluted environment. In studies conducted on the occurrence of lichens in the Sudbury area Leblanc et al., (1972) found that the number of epiphytes growing on poplar trees was drastically reduced in zones where the growing season mean levels of SO_2 were over 52 $\mu g\ m^{-3}$ (0.02 ppm) and slightly reduced in zones where the mean levels of SO_2 were approximately 26 $\mu g\ m^{-3}$ (0.01 ppm).

Abeles, et al., (1971) observed that soil was able to reduce an SO_2 concentration of 260,000 $\mu g\ m^{-3}$ (100 ppm) to 20,000 $\mu g\ m^{-3}$ (8 ppm) in 15 *min. Autoclaving the soil prior to the experiment reduced the SO_2 concentration to 52,000 $\mu g\ m^{-3}$ (20 ppm) after 15 min. This evidence indicates that the primary absorption effect is chemical. However, soil microorganisms appear to play a role in the absorption of SO_2 by soils.*

Sulphur depositions in the Sudbury area resulted in decreased pH of the soils. Hutchinson and Whitby (1974) reported that the pH values of surface soils up to 7.4 km from the Coniston smelter were generally less than 3.0, while the pH of the soils in unpolluted areas in the Sudbury area is in the range of 3.8 to 4.8.

In another study conducted on the degree of S and heavy metal contamination of soil and vegetation in the Sudbury area, McGovern and Balsillie (1973) found that within 8 km of the three smelters there was an intense zone of contamination by S, Cu, Fe and Ni. Between 8 and 16 km some reduction in the level of contamination was found, and from 16 to 32 km there was a major decrease in the levels of these elements. Snow sampled during the winter of 1972 showed that S levels in snow were elevated over an area of about 2400 km^2, whereas levels of Cu, Fe and Ni were high in an area of about 640 km^2.

It was observed that chronic effects on eastern white pine trees in the forest were prominent where SO_2 air concentrations for the growing season averaged 44 $\mu g\ m^{-3}$ (0.017 ppm), the arithmetic mean for the 10-yr (1954-1963) measurement period, and chronic effects were slight where SO_2 concentrations averaged 21 $\mu g\ m^{-3}$ (0.008 ppm) (Linzon, 1971). It is important to note that the forest damage which occurred in the Sudbury area prior to 1972 was primarily the result of exposure to atmospheric SO_2. Other pollutants, such as O_3 are extremely low in the area, and the SO_2 monitoring stations were located sufficiently close to the forest plots under observation so that both were subjected to the same SO_2 fumigations emanating from the Sudbury smelters. In addition, substantial reductions in the emissions of SO_2 from the Sudbury smelters after 1972 resulted in concomitant reductions in injury to forests in the area.

Starting in 1972, several abatement activities were undertaken by the Ni-producing industries in the Sudbury area. The existing three smokestacks at the Copper Cliff smelter were replaced with one tall stack (380 m) in 1972 which reduced the concentrations of SO_2 in the vicinity of Sudbury, but contributed further to long range transported pollutants. The smelter at Coniston and the pyrrhotite plant at Falconbridge were shut down permanently in 1972. Other abatement activities included the operation of sulphuric acid manufacturing plants at both Copper Cliff and Falconbridge and the construction of a new smelter at Falconbridge in 1978. Emissions of SO_2 from the smelters in the Sudbury area were reduced from over 7000 tonne day^{-1} in the 1960s to about 2400 tonne day^{-1} currently (a 66% reduction). Further reductions in emissions are planned. The reductions in emissions of SO_2 in the Sudbury area have resulted in little or no acute damage occurring on vegetation in recent years. The fact that the drastic reductions in SO_2 emissions resulted in dramatic improvements in forest conditions in the Sudbury area demonstrate that SO_2 was the prime pollutant responsible for the severe forest damage observed earlier. Starting in 1978 the Regional Municipality of Sudbury became engaged in reclaiming the damaged land (Lautenbach, 1985). By the end of 1984, over 2,636 ha of land had been limed and fertilized, and grass and 387,580 trees had been planted.

3.3 *Effects of SO_2 in Combination with O_3*

There have been a number of studies on the effects of combined SO_2 and O_3 mixtures on eastern white pine. Costonis (1973) in North Carolina found an antagonistic response during a 2 hr exposure of 130 μg m^{-3} (0.05 ppm) SO_2 + 98 μg m^{-3} (0.05 ppm) O_3. Houston (1974) in Wisconsin found a synergistic response during a 6 hr exposure of 65 μg m^{-3} (0.025 ppm) SO_2 and 98 μg m^{-3} (0.05 ppm) O_3. In both of the above studies 98 μg m^{-3} (0.05 ppm) O_3 alone did not injure the eastern white pines being tested.

Neilsen et al., (1977) in Ohio fumigated Scots pine seedlings with 650 μg m^{-3} (0.25 ppm) SO_2 and either 274 μg m^{-3} (0.14 ppm) O_3 or 568 μg m^{-3} (0.29 ppm) O_3 for 6 hr day^{-1} for varying time periods and found an antogonistic response.

Karnosky (1976) studied the effects of SO_2 and O_3 on trembling aspen and found the response to a mixture of the two gases to be synergistic. Whereas it required either 910 μg m^{-3} (0.35 ppm) SO_2 alone for 3 hr or 98 μg m^{-3} (0.05 ppm) O_3 alone for 3 hr to injure the foliage of trembling aspen, a combination of 520 μg m^{-3} (0.20 ppm) SO_2 and 98 μg m^{-3} (0.05 ppm) O_3 for 3 hr caused more injury than the sum of the individual pollutants.

It is apparent that a number of different responses can occur when plants are exposed to a mixture of gaseous air pollutants. Some investigators report synergistic (greater than additive) responses, others report antagonistic (less than additive) responses and others have reported no change from the additive response of the indivudual pollutants. In some experiments, changing the concentrations of the different pollutants can result in different responses. In order for gaseous pollutant mixtures to be considered in future derivation of air quality standards the results of experiments will have to be more uniform for a diverse range of plant species under varying conditions.

4. *HYDROGEN FLUORIDE*

Major reviews on the effects of HF on vegetation have been written by Thomas and Alther (1966), Treshow and Pack (1970), and Weinstein (1977). Accordingly, trees listed as sensitive to HF include apricot, box elder (Manitoba maple) Douglas fir, western larch, peach, eastern white pine, loblolly pine, lodgepole pine, Mugho pine, Scots pine, ponderosa pine, plum and blue spruce.

One of the earliest reports of fluoride injury to trees in North America was in the vicinity of an Al plant in California (De Ong, 1946). English walnut and apricot trees within one-half mile of the plant were severely injured with apricot leaves found to contain between 247 and 403 μg g^{-1} fluoride (based on dry weight). Normal concentrations of fluoride in plant foliage growing in areas remote from industrial sources is in the range between 2 to 20 μg g^{-1}, on a dry-weight basis (Weinstein, 1977). Some trees are accumulators of fluoride, for example, bitternut hickory accumulated up to 1000 μg g^{-1} fluoride in foliage without showing visible injury (Linzon, 1970).

Fluorides are released into the atmosphere from a number of sources, including the manufacture of Al, steel, ceramics, P chemicals and fertilizers (Treshow and Pack, 1970). Fluorides are among the most phytotoxic gases *of the major air pollutants and may cause injury to susceptible plant species at concentrations 10 to 1000 times lower than the others, (HF being toxic at concentrations less than 1 ppb) (Weinstein, 1977). (At 25 ^{o}C and 760 mm Hg, 1 ppb HF = 0.82 $\mu g\ m^{-3}$). In addition, fluoride accumulates in plant tissue and can be harmful to livestock consuming contaminated forage.*

Hydrogen fluoride enters plants primarily through the stomata of foliage, and moves with the transpiration stream towards the leaf tips and margins where it accumulates and results in chlorotic or necrotic injury (Treshow and Pack, 1970). Intercostal lesions and chlorotic areas can result when HF is absorbed at higher concentrations than can be translocated. Marginal necrotic tissue on broadleaves often develop an abscission layer next to the healthy portions of the leaf and the injured portion breaks away leaving an irregular leaf margin. This is quite noticeable on apricot and plum trees. Fluoride injury symptoms on coniferous tree needles consist of necrosis which begins at the tip of the current year needles and progresses towards the base (Treshow and Pack, 1970).

Adams et al., (1956) fumigated both ponderosa pine and gladiolus plants and induced foliar injury with 0.98 $\mu g\ m^{-3}$ (1.2 ppb) HF over a 24 hr period. Adams et al., (1957) exposed various tree species to 1.23 $\mu g\ m^{-3}$ (1.5 ppb) HF for 8 hr daily, 5 days per week until foliage was injured. Elm, western larch, mulberry, ponderosa pine, lodgepole pine, peach and willow accumulated up to 265, 53, 273, 83, 39, 92 and 144 $\mu g\ g^{-1}$ fluoride, respectively, prior to injury.

A major study of fluoride effects on forests has been conducted in the vicinity of a P manufacturing plant at Long Harbour, Newfoundland (Sidhu, 1979). Foliage of balsam fir, black spruce, larch, and white birch displayed severe symptoms of fluoride injury up to a distance of 8 km from the source, moderate symptoms between 8 to 9 km, light symptoms from 9 to 12 km, and no symptoms beyond 13 km. At the end of the growing season of 1976, maximum fluoride conentrations in foliage of balsam fir, black spruce, larch and white birch were 265, 96, 411 and 357 $\mu g\ g^{-1}$. Beyond 12 km the fluoride concentrations in foliage of the same tree species were less than 20 $\mu g\ g^{-1}$. Air monitoring was conducted using Na - formate plates and the calculated maximum fluoride level in the air as was about 5 $\mu g\ m^{-3}$ (6.10 ppb) at a distance of 0.7 km from the source. Beyond 12 km the calculated level of fluoride in the air was 0.20 $\mu g\ m^{-3}$ (0.24 ppb) or less.

Studies were started in 1969 on Cornwall Island, Ontario to document the effects of fluorides emitted from an Al reduction plant located immediately to the south in New York State (Linzon, 1970). At a sampling station on Cornwall Island about 1.6 km northeast of the Al plant, eastern white pine needles accumulated 135 $\mu g\ g^{-1}$ fluoride and displayed severe injury. Many dead eastern white pine trees were present in the forest stand at this sampling station. Trembling aspen trees had 495 $\mu g\ g^{-1}$ fluoride in injured foliage. The eastern white pines displayed an orange-red terminal necrosis on needles; the trembling aspen foliage displayed black necrotic marginal injury on leaves; and choke cherry leaves displayed reddish-brown marginal lesions. At a control location about 6.4 km northeast of the source, eastern white pine and trembling aspen foliage displayed no injury and contained less than 20 $\mu g\ g^{-1}$ fluoride content. Lime candle measurments showed over 4000 μg F/100 cm^2 in the air near the eastern white pine stands located 1.6 km from the source. A control order to reduce fluoride emissions was served to the Al company by New York State. The Al company began operations in 1959 and emitted 139 kg F hr^{-1} until 1972 when abatement measures reduced the emissions to about 34 kg F hr^{-1}. The eastern white pine stands on Cornwall Island located 1.6 km from the source demonstrated a remarkable recovery in the radial growth of the trees following the abatement measures taken by the Al company in New York State. In addition, in the years following 1972 little or no fluoride injury occurred on the current year's needles of the eastern white pine trees (Linzon, 1978b).

An investigation was conducted in the vicinity of an oil refinery in Ontario regarding the sudden appearance of injury on Scots pine trees to a distance of 5 km from the refinery (Linzon and Tung, 1976). In addition to Scots pine, eastern white pine, spruce, larch, plum, and Manitoba maple trees displayed foliar injury. On the broadleaf trees, the foliage displayed both marginal necrosis and bifacial lesions in the central portions of the leaves. Chemical analysis showed higher levels of fluoride in foliage close to the source compared to control non-injured trees. A histopathological study was conducted on the injured Scots pine needles. Cellular damage was found to include collapsed mesophyll cells, hypertrophied transfusion parenchyma cells in the stele, a compressed phloem and collapse of the endodermal ring of cells surrounding the stele. The combination of symptomatology, resistance and sensitivity of different tree species, the chemical analysis results, and the histopathological findings confirmed that the injuries had been caused by an acute HF fumigation.

Soil normally contains between 20 and 500 $\mu g\ g^{-1}$ fluoride (Weinstein, 1977), and the amount of fluoride normally accumulated from the soil by plants is small. However, Davis and Barnes (1973) found adverse growth effects (height, root and stem dry weight) on both loblolly pine and red maple by watering soil that the seedlings were growing in with a solution containing 2×10^{-2} M fluoride which was estimated to increase fluoride content in the soil by about 1000 $\mu g\ g^{-1}$ at the end of the experiment. Studies conducted in the vicinity of an Al reduction plant in Ohio 18 yr after operations began, indicated an increase of up to 180 $\mu g\ g^{-1}$ in surface soils 3 km northeast of the source with a decreasing trend with distance to background levels at about 10 km downwind (McClenahen, 1976). In another experiment in which soil had been amended previously by additions of up to 974 $\mu g\ g^{-1}$ fluoride, McClenahen and Weidensaul (1977) found decreases in dry weight of leaves, stems and roots of shagbark hickory, black locust and red maple.

Leblanc et al., (1971) studied the effects of HF on epiphytic lichens and mosses in the vicinity of an Al reduction plant at Arvida, Quebec. Lichen and moss-bearing bark discs cut from trees in an unpolluted area were fixed to trees at various sites near the Al plant and left for periods of 4 or 12 mo. At a distance of 1 km from the source lichens accumulated 990 $\mu g\ g^{-1}$ fluoride in 4 mo compared to 70 $\mu g\ g^{-1}$ in a control location 40 km distant. Mosses accumulated 570 $\mu g\ g^{-1}$ fluoride in 4 mo compared to 20 $\mu g\ g^{-1}$ in the control area.

With regard to limiting values for HF effects on vegetation, McCune et al., (1974) suggested 8.20 to 4.10 $\mu g\ m^{-3}$ (10 to 5 ppb) for 2 to 4 hr periods, 3.28 to 0.82 $\mu g\ m^{-3}$ (4 to 1 ppb) for daily mean concentrations depending upon the magnitude of diurnal fluctuations, and 0.49 to 0.25 $\mu g\ m^{-3}$ (0.6 to 0.3 ppb) as mean concentrations for periods of 30 to 60 days.

5. CONCLUSIONS

From the foregoing it may be concluded that of the major gaseous air pollutants directly affecting forests in eastern North America, O_3 currently is the most prevalent phytotoxic pollutant on a regional scale. Nitrogen oxides have caused damage to trees in the immediate vicinity of point sources, but are more important regionally as precursor pollutants in the formation of both O_3 and acid rain. Sulphur dioxide has caused damage to forests to distances of over 100 km from major point sources, but on a regional scale it is more important as one of the precursor pollutants leading to the formation of acid rain. Finally, HF has been extremely phytotoxic to forests in the immediate vicinity of point sources out to distances about 12 km downwind.

6. REFERENCES

Abeles, F. B., Craker, L. E., Forrence, L. E. and Leather, G. R.: 1971, Science **173**, 914.
Adams, D. F., Hendrix, J. W. and Applegate, H. G.: 1957, Agric. Food Chem. **5**, 108.
Adams, D. F., Shaw, C. G. and Yerkes, W. D. Jr.: 1956, Phytopath. **46**, 587.
Banfield, W. M.: 1972, Phytopath. **62**, 493.
Barrett, T. W. and Benedict, H. M.: 1970, In: Recognition of Air Pollution Injury to Vegetation: A Pictorial Atlas" (J. S. Jacobson and A. C. Hill, eds.), Air Pollut. Control Assoc., Pittsburgh, Pa., pp. C1-C17.
Benoit, L. F., Skelly, J. M., Moore, L. D. and Dochinger, L. S.: 1982, Can. J. For. Res. **12**, 673.
Berry, C. R.: 1967, Phytopath. **57**, 804.
Berry, C. R.: 1974, Phytopath. **64**, 207.
Berry, C. R. and Ripperton, L. A.: 1963, Phytopath. **53**, 552.
Blanchard, R. O., Baas, J. and Van Cotter, H.: 1979, Plant Dis. Reporter **63**, 177.
Costonis, A. C.: 1971, Phytopath. **61**, 717.
Costonis, A.C.: 1973, Eur. J. For. Path. **3**, 50.
Costonis, A. C. and Sinclair, W. A.: 1969, Phytopath. **59**, 1566.
Davis, D. D. and Wood, F. A.: 1972, Phytopath. **62**, 14.
Davis, D. D., Umbach, D. M. and Coppelino, J. B.: 1981, Plant Dis. **65**, 904.
Davis, J. B. and Barnes, R. L.: 1973, Environ. Pollut. **5**, 35.
De Ong, E. R.: 1946, Phytopath. **36**, 469.
Dochinger, L. S.: 1968, J. Air Pollut. Control Assoc. **18**, 814.
Dochinger, L. S., Bender, F. W., Fox, F. L. and Heck, W. W.: 1970, Nature **225**, 476.
Dreisinger, B. R.: 1965, 58th Annual Meeting, Air Pollut. Control Assoc., Toronto, Ontario, Paper No. 65-121. pp. 1-21.
Dreisinger, B. R. and McGovern, P. C.: 1970, in: Proceedings Impact of Air Pollution on Vegetation (S. N. Linzon ed.), APCA Specialty Conference, Toronto, Ontario, pp. 11-28,
Friedland, A. J., Hawley, G. J. and Gregory, R. A.: 1985, "Investigations of nitrogen as a possible contributor to red spruce (Picea rubens Sarg.) decline", Presented at Effects of Air Pollutants on Forest Ecosystems Symposium, Minneapolis, Mn.

Galloway, J. M. and Skelly, J. M.: 1978, Virginia's Climate Advisory **3**, 4.
Heck, W.W. and Brandt, C. S.: 1977, In Air Pollution (A. C. Stern, ed.), Academic Press, New York, pp. 157-229.
Heck, W. W., Taylor, O. C., Adams, R., Bingham, G., Miller, J., Preston, E. and Weinstein, L.: 1982, J. Air Pollut. Control Assoc. **32**, 353.
Hedgcock, G. G.: 1914, J. of the Washington Academy **4**, 70.
Hibben, C. R.: 1969, Phytopath. **59**, 1423.
Hill, A. C., Heggestad, H. E. and Linzon, S. N.: 1970, In: Recognition of Air Pollution Injury to Vegetation: A Pictorial Atlas (J. S. Jacobson and A. C. Hill, eds.). Air Pollut. Control Assoc., Pittsburgh, Pa., pp. B1-B22.
Houston, D. B.: 1974, Can. J. For. Res. **4**, 65.
Hutchinson, T. C. and Whitby, L.M.: 1974, Environ. Conservation **1**, 123.
Jacobson, J. S.: 1977, In: Ozone und Begleitsubstanzen im Photochemischen Smog. VD1 Berlin 270, 163.
Jensen, K. F.: 1973, Plant Dis. Reporter **57**, 914.
Jensen, K. F., and Dochinger, L. S.: 1974, Envir. Pollut. **6**, 289.
Karnosky, D. F.: 1976, Can. J. For. Res. **6**, 166.
Katz, M.: 1949, Ind. Eng. Chem. **41**, 2450.
Kress, L. W. and Skelly, J. M.: 1982. Plant Dis. **66**, 1149.
Lautenbach, W. E.: 1985, "Land reclamation program 1978-1984 - Regional Municipality of Sudbury", Vegetation Enhancement Technical Advisory Committee, pp. 1-65.
Leblanc, F. Comeau, G. and Rao, D. N.: 1971, Can. J. Botany **49**, 1691.
Leblanc, F., Rao, D. N. and Comeau, G.: 1972, Can. J. Bot. **50**, 519.
Linzon, S. N.: 1958, "The influence of smelter fumes on the growth of white pine in the Sudbury region", Joint Publication, Ontario Dept. Lands and Forests and Ontario Dept. Mines., Toronto, Ontario, pp. 1-45.
Linzon, S. N.: 1966. J. Air Pollut. Control Assoc. **16**, 140.
Linzon, S. N.: 1970, In: Proceedings, Second. Int. Clean Air Congress, Washington, D.C. pp. 277-289.
Linzon, S. N.: 1971, J. Air Pollut. Control Assoc. **21**, 81.
Linzon, S. N.: 1973. "The effects of air pollution on forests" Presented to Fourth Joint Chemical Engineering Conference, Vancouver, British Columbia, pp. 1-17.
Linzon, S. N.: 1978a, In: Sulfur in the Environment: Part II, Ecological Impacts (J. O. Nriagu, ed.), John Wiley & Sons Inc., New York, pp. 109-162.
Linzon, S.N.: 1978b, In: Proceedings, 25th Ontario Industrial Waste Conference, Toronto, Ontario. pp. 105-121.
Linzon, S. N. and Costonis, A. C.: 1971, In: Proceedings, 1971 Congress, International Union of Forest Research Organizations, Gainsville, Florida, and in Mitteilungen Der Forestlichen Bundes - Versuchsanstalt Wien **92**, 71.
Linzon, S. N., Heck, W. W. and Macdowell, F. D. H.: 1975, In: Photochemical Air Pollution: Formation, Transport and Effects, National Research Council of Canada, Public. No. NRCC 14096, pp 89-142.
Linzon, S. N., Pearson, R. G., Donnan, J. A. and Durham, F. N.: 1984, "Ozone effects on crops in Ontario and related monetary values", Report No. ARB-13-84-Phyto, Ontario Ministry of the Environment, pp. 1-70.
Linzon, S.N. and Tung, G.: 1976, in: Proceedings, American Phytopath. Society **3**, 308.

Mahendrappa, M. K.: 1977, Bi-monthly Research Notes, Can. Forestry Service **33**, 32.
Mann, L. K., McLaughlin, S. B. and Shriner, D. S.: 1980, Environ. Exp. Bot. **20**, 99.
McClenahen, J. R.: 1976. J. Environ. Qual. **5**, 472.
McClenahen, J. R. and Weidensaul, T. C.: 1977, "Geographic distribution of airborne fluorides near a point source in southeast Ohio", Research Bulletin 1093, Ohio Agric. Research and Devel. Centre, Wooster, Ohio, pp. 1-29.
McCune, D. C., Temple, P. J. and Witherspoon, A. M.: 1974, 67th Annual Meeting of the Air Pollut. Control Assoc. Reprint No. 74-226, pp. 1-12.
McGovern, P. C. and Balsillie, D.: 1973, "Sulphur dioxide (1972) - heavy metal (1971) levels and vegetation effects in the Sudbury area", Ontario Ministry of the Environment, Northeastern Region, pp. 1-50.
McLaughlin, S. B.: 1981, In: The Proposed SO_x and Particulate Standard, APCA Specialty Conference, Atlanta, Georgia, pp. 62-83.
Miller, P. R., Taylor, O. C. and Wilhour, R. G.: 1982, "Oxidant air pollution effects on a western coniferous forest ecosystem". E.P.A - 600/D-82-276. US E.P.A., Corvallis, Oregon.
Nielson, D.G., Terrell, L.E. and Weidensaul, T.C.: 1977, Plant Dis. Reporter **61**, 699.
Roberts, B.R.: 1976, Envir. Pollut **11**, 175.
Sidhu, S. S.: 1979, J, Air Pollut. Control Assoc. **29**, 1069.
Sigal, L. L. and Nash, T. H. III.: 1983, Ecology **64**, 1343.
Skelly, J. M., Chevone, B. L. and Yang, Y-S.: 1982, In: Proceedings - International Symposium in Hydrometeorology, Amer. Water Research Association, pp. 69-73.
Skelly, J. M., Yang, Y-S., Chevone, B. I. and Long, S. J.; 1983, In: Proceedings - Air Pollution and the Productivity of the Forest Symposium, Washington, D.C., pp. 143-159
Thomas, M. D.: 1951, Ann. Rev. Plant Physiology, pp. 293-322.
Thomas, M. D. and Alther, E. W.: 1966, in: Handbook of Experimental Pharmacology, Vol. 20, Part 1, Pharmacology of Fluorides, (F.A. Smith, ed.) Springer - Verlag, New York. pp. 231-306.
Tingey, D. T.: 1978, California Air Environment, Univ. of California **7**, 5
Treshow, M. and Pack, M. R.: 1970, In: Recognition of Air Pollution Injury to Vegetation: A Pictorial Atlas (J. S. Jacobson and A. C. Hill (eds.), Air Pollut. Control Assoc., Pittsburgh, Pa, pp. D1-D17.
U.S. Environmental Protection Agency.: 1982, In: Air Qualtiy Criteria for Particulate Matter and Sulfur Oxides, Volume III, Chapter 8, pp. 1-101.
Van Sickle, G. A.: 1973, Bi-monthly Research Notes, Can. Forestry Service, **29**, 32.
Weinstein, L. H.: 1977, J. of Occupational Medicine **19**, 49.

REGIONAL CHEMICAL CHARACTERISTICS OF LAKES IN NORTH AMERICA: PART I - EASTERN CANADA

D.S. Jeffries
National Water Research Institute
P.O. Box 5050, Burlington, Ontario, Canada L7R 4A6

D.L. Wales
Fisheries Research Branch
Ontario Ministry of Natural Resources
Queen's Park, Toronto, Ontario, Canada M7A 1W3

J.R.M. Kelso
Great Lakes Fisheries Research Branch
1219 Queen Street East
Sault Ste Marie, Ontario, Canada P6A 5M7

R.A. Linthurst
U.S. EPA, EMSL MD 75
Research Triangle Park, NC 27511, U.S.A.

ABSTRACT. Data defining the major ion chemistry of lakes located in eastern Canada have been compiled for the purpose of evaluating the current status of surface water quality in relation to acidic deposition. A companion paper for lakes in the eastern United States (i.e. Part II, Linthurst et al., 1986) has been prepared also. Data sources in Canada included the National Inventory Survey, the Ontario Lake Sensitivity data set, and the National Aquatic Data base which provided an overall data base of approximately 5700 lakes. Only recently collected data (largely 1980 or later) were used in the analysis. Frequency distribution statistics were obtained for pH, acid neutralizing capacity (ANC), SO_4 and organic anion (A^-) concentrations. Acidic and low ANC waters in eastern Canada occur in a pattern explained by a combination of biogeochemical factors and atmospheric deposition. Nova Scotia contained the highest proportion of acidic and ultralow ANC lakes of any region surveyed in eastern North America; since this region receives approximately 20 $kg.ha^{-1}.yr^{-1}$ wet SO_4 deposition, the proposed target loading may be too high to protect the highly sensitive waters of Maritime Canada. Compared to the rest of eastern Canada, lakes in Ontario have relatively high ANCs due to the influence of $CaCO_3$ contained in the glacial till of the area. Variation in the SO_4 concentration of lakes approximately follows expected gradients in wet SO_4 deposition. Naturally occurring organic acids do not play a dominating role in the acidification of eastern Canadian lakes.

Water, Air, and Soil Pollution **31** (1986) 551–567.

1. INTRODUCTION

Within the past 5 years, the relationship between acidification of aquatic ecosystems and deposition has been extensively analyzed, reviewed, and debated (NAS, 1981; Harvey et al., 1981; U.S./Canada, 1983). Since the North American publication of Beamish and Harvey (1972) linking observed chemical aquatic responses to acidic deposition, a number of trend analyses have been conducted in various areas of Canada and the United States (Schofield, 1976; Dillon et al., 1979; Hendry et al., 1980; Thompson et al., 1980; and Burns et al., 1981). The conclusion of all analyses was that acidic deposition was instrumental in effecting surface water acidification (i.e. decreases in pH and/or ANC).

Nevertheless, what remains unresolved is the geographical extent to which North American surface waters as a whole are threatened by acidic deposition. Statistical uncertainty in extrapolating the results of specific studies to a national or continental scale exists due to: the unknown degree to which the study sites represent the larger regional population, absence of data for many of the key chemical variables required for a critical assessment of effects, difficulty in inter-study comparisons of data collected by different methodologies, and variable documentation of quality assurance protocols.

Even if the limitations of extant data are ignored, evaluating effects of acidic deposition on surface waters is extremely complex due to a number of confounding and inter-related physical, chemical, and biological factors which collectively control the response of the aquatic ecosystem. Although alkalinity or ANC has often been used as an index of surface water sensitivity, sensitivity is dependent on the buffering capacity in the entire watershed and the ecosystem's ability to regenerate that capacity.

A currently important question is whether the influence of naturally occurring organic acids can account for observed widespread acidification. Some lakes and streams with low pH are, or have become, acidic because of natural, internally driven acidification processes (Patrick et al., 1981). The in-situ production of dissolved organic carbon compounds, a considerable portion of which is organic acids, results from microbially-mediated decomposition of organic matter. Organic acids are functionally important in some surface waters for 2 reasons: they can contribute significantly to the acidity (Dickson, 1980; Oliver et al., 1983), and they chelate various metals, thereby increasing total metal solubility or decreasing the concentrations of biologically available metals (Beck et al., 1974; Reuter and Perdue, 1977). The relative contribution of organic acids to the acidification of lakes and the sensitivity of organic acid lakes to mineral acid inputs have not been examined on a regional or national basis.

The objective of this paper is to re-examine hypotheses about effects of SO_4 deposition on lakes in eastern Canada using extensive, recently acquired data. Within different subregions in Canada:

a) regional characterization of the distributions of pH, ANC, SO_4 and organic anion (A^-) are presented;
b) the relationship between SO_4 concentrations and ANC is investigated; and
c) the relationship between A^- concentrations and ANC is discussed.

A parallel paper (i.e. Part II, Linthurst et al., 1986) has been prepared for lakes located in the eastern United States. Reference to and comparison with information presented in Part II will be made, where appropriate, throughout this paper.

2. METHODS

2.1 Regions/Subregions

Three major geographical regions within eastern Canada were selected: Ontario, Quebec-Labrador, and the Maritimes (Figure 1). Within Ontario, 3 subregions were delineated, Northwestern (NW ONT), Northeastern (NE ONT), and South-Central (SC ONT). Quebec (QUE) and Labrador (LAB) were analyzed as subregions. Newfoundland (NF), Nova Scotia (NS), and New Brunswick (NB) were designated as subregions within the Maritime region. Provincial boundaries were used to delineate subregions in all regions except Ontario. Within Ontario, NE ONT and NW ONT were divided arbitrarily at the 85° 20' longitude and SC ONT was situated south of the 45° 50' latitude.

Of the subregions in eastern Canada, NE ONT is particularly variable in character since it contains both geologically sensitive and insensitive terrain, and also the very large emission source at Sudbury. The QUE subregion is extremely large and therefore experiences a large gradient in sulphate deposition; however, the data presented in this paper are for lakes located almost exclusively in the southern, higher deposition portion of the subregion.

2.2 Data Base Descriptions

The Canadian data base is a compilation of results from extensive surveys conducted largely between 1980 and the present. Data contained in the National Inventory Survey (NIS) were collected during 1980-1983 by the Department of Fisheries and Oceans from Ontario, NF, NB, NS, QUE, and LAB. The Ontario Sensitivity Data Base (SENS1) is a compendium of data sets collected in Ontario from 1978 to the present by the Ontario Ministry of Natural Resources, the Ontario Ministry of the Environment, the University of Toronto and Ontario Hydro. Over 84% of the data in SENS1 were collected since 1980. The National Water Quality Data Base (NAQUADAT) contains data collected from LAB during the summer and fall of 1980 by the Canadian Wildlife Service (CWS), and from QUE collected by the Inland Waters Directorate (IWD) during January-March 1982. NAQUADAT also contains data collected by another group within IWD from the Maritimes and LAB as

Figure 1 Subregions of eastern Canada used in this paper.

part of a long-term monitoring program. NAQUADAT data prior to 1980 were not considered in this paper. Depending on parameter, the entire Canadian data set is comprised of approximatley 5700 lakes of which 4500 are located in Ontario.

2.3 Lake Selection

The data base presented in this paper was compiled from a number of separate surveys and, therefore, was not derived from lakes selected by using a single statistical design (see Part II). Because of this, the statistics calculated below are given as sample estimates. Kelso et al. (1986) concluded that water chemistry found in the predominantly headwater NIS data base was not different from that found in more diverse regional data sets. Specifically, they found that the key indicators of acidification (pH and ANC) were not significantly correlated with lake size, and thus, the sampled chemical conditions were assumed to be representative of the total resource. This evidence plus the fact that the data base used here is both very much larger and does not contain such a strong bias toward low order lakes (as the NIS), suggests that sample estimates given below do reflect population estimates.

As noted above, most of the lakes samples by the NIS in Ontario, NF, NS,and NB were headwater lakes (100, 97, 85, and 88% respectively). In the QUE and LAB surveys, only 75 and 50%

respectively were headwater lakes. Lakes within the Ontario SENS1 data base were moderately biased toward important, accessible sport fisheries lakes; lakes $<$ 10 ha in surface area were undersampled. Lakes contained within SENS1 were excluded from this analysis when pH $<$ 4 (indicating acid mine drainage) and when Na $>$ 2.0 $mg.L^{-1}$ (indicating road salt effects). Many lakes present in SENS1 have been sampled on more than 1 occasion; in these cases, the multiple sets of data were pre-screened and the "best" sample for each lake used in the following analysis (based on sampling time, number of parameters analyzed, laboratory, etc.).

Lakes in NAQUADAT sampled by CWS (LAB) were biased towards those waters which contained suitable waterfowl habitat and permitted floatplane access. Generally, these waters would not be headwater lakes (W. Whitman, pers. comm.). The remaining data in NAQUADAT collected by IWD are biased towards headwater systems but have a reasonably even geographic distribution over the Maritime provinces, LAB, and southern QUE. Based on sampling site descriptions, lakes contained in NAQUADAT were eliminated from this analysis if local anthropogenic influences were probable. Many lakes in NAQUADAT have data from multiple sampling occasions; in this event, the median value for each parameter was used in the following analysis. Finally, latitudes and longitudes for each lake were cross-checked between each data base to ensure that no duplication occurred.

2.4 Sampling Methodologies

Five meter integrated tube samples were collected during the NIS; lakes $<$ 5 m deep were only sampled to 1 m from the bottom. Eighty-three percent of the data contained in SENS1 were also derived from 5 or 10 m composite tube samples. Data obtained by IWD were similarly tube composites while the CWS survey in LAB collected surface grab samples.

2.5 Analytical Methodologies

H^+ concentration (from pH) was measured electrometrically throughout. ANC was measured using the Gran technique. SO_4 analyses were conducted either by automated colorimetry (Methyl thymol blue = MTB procedure) and/or by ion chromatography (IC). IC determined SO_4 data were used over MTB values whenever possible. When only MTB SO_4 data were available, these data were included in the following analysis only when sample colour was $\leq$30 Hazen units and/or dissolved organic carbon was less than 6 $mg.L^{-1}$. The elimination of MTB sulphate data collected from high organic waters is necessary due to the analytical interferences present for this type of sample (Kerekes et al., 1984). Howell and Pollock (in press) suggest that differences in SO_4 concentrations between the 2 analytical methodologies are minimal when sample colour is $<$20 Hazen units and it is possible to extend this threshold level to 40 units. Finally, all SO_4 data for the Maritime region, LAB, and QUE were sea salt corrected.

Dissolved organic carbon (DOC) was determined for the NIS by infrared (IR) spectrophotometry and in the SENS1 and NAQUADAT data bases by IR or automated colorimetry. A^- concentrations were calculated using DOC and pH according to the method described by Oliver et al. (1983). The A^- concentration is a surrogate for the acidity associated with naturally occurring organic substances.

2.6 Quality Assurance

Almost all lake samples used to develop the data base reported here were analyzed at one of the large federal or provincial water quality laboratories. Hence, data reliability was maintained through operation of their comprehensive in-house quality assurance programs. Moreover, all of these laboratories also take part in the extensive LRTAP round robin program conducted by the Quality Assurance and Methods Section, National Water Research Institute, Burlington, Ontario (K. Aspila, pers. comm.) in order to identify and correct between-laboratory bias and variability. A few of the samples were analyzed by either university laboratories (U. of Toronto, 8% of the SENS1 data; U. of Minnesota, part of the CWS LAB survey) or contract laboratories. In both cases, their quality assurance procedures were modelled after or monitored directly by the federal or provincial laboratories. It should be noted that the predominating use of the composite tube technique to give an approximate whole-lake integrated sample can be expected to reduce substantially data variability caused by natural fluctuations in near surface water quality. The simplicity of the tube methodology also tends to minimize sampling errors.

3. RESULTS

3.1 Regional Characterization - pH and ANC

Median pH values ranged from a low of 5.20 in NS to a high of 7.10 in NW ONT (Table I). Minimum ANCs occurred in the Maritimes while the highest median values were observed in northern Ontario. In eastern Canada, with the exception of NS, the proportions of sampled lakes characterized by pH $\leq$ 5.0 were relatively low; however, a majority of the lakes in all subregions except NE ONT and NW ONT had ANCs $\leq$ 200 $\mu eq.L^{-1}$.

Within the Ontario region, SC ONT had the lowest median pH (6.29) and ANC (65 $\mu eq.L^{-1}$). NW ONT had the highest median pH (7.10) and ANC (254 $\mu eq.L^{-1}$). These values for NW ONT are the highest for any subregion in eastern Canada. The largest proportion of acidic lakes (ANC < 0) and lakes with pH $\leq$ 5.0 were found in NE ONT; however, the highest percentages of lakes with pH $\leq$ 6.0 and ANC $\leq$ 200 $\mu eq.L^{-1}$ occurred in SC ONT.

Median pH in QUE and LAB were the same (6.33 and 6.36 respectively) and similar to that in SC ONT. However, ANCs were different (QUE = 36 and LAB = 46 $\mu eq.L^{-1}$), both being substantially

Table I Minimum, median, and Q_4* values for pH and ANC ($\mu eq.L^{-1}$) in Canadian lakes. Percentages of lakes sampled which fall into 3 pH and 3 ANC classes ($\mu eq.L^{-1}$) are given. All statistics are sample estimates.

Subregion	pH				Percentage			ANC Concentration				Percentage		
	Sample n	Min.	Med.	Q_4	≤5.0	≤5.5	≤6.0	Sample n	Min.	Med.	Q_4	≤0	≤40	≤200
NW ONT	1080	5.03	7.10	7.70	0	1	4	1078	- 3	254	762	<1	4	40
NE ONT	1820	4.00	6.85	7.70	6	10	20	1805	-99	165	839	9	24	51
SC ONT	1619	4.22	6.29	6.89	2	8	28	1578	-66	63	159	3	32	83
QUE	434	4.40	6.33	6.89	3	12	30	429	0	36	100	1	55	90
LAB	198	4.84	6.36	6.67	<1	2	20	182	2	46	88	0	43	95
NB	84	4.51	6.34	6.94	7	15	36	81	-22	38	98	11	51	96
NS	232	4.20	5.20	6.00	39	63	82	198	-82	0	21	51	93	99
NF	270	4.84	6.17	6.62	3	14	44	176	-16	28	62	7	68	94

*The maximum value expected to characterize 80 percent of the lakes sampled.

less than Ontario values. Few lakes were characterized by pH $\leq$ 5.0 or ANC $\leq$ 0 $\mu eq.L^{-1}$, but a considerable proportion had pH $\leq$ 6.0 and the majority of all lakes in both subregions had ANC $\leq$ 200 $\mu eq.L^{-1}$. Fifty-five percent of lakes in QUE and 43% in LAB had ANC $\leq$ 40 $\mu eq.L^{-1}$.

Median pH in NB and NF were similar to those observed in SC ONT and QUE-LAB; however, NS had the lowest median pH (5.20) of all subregions in eastern Canada. Similarly, median ANCs were comparable to QUE-LAB except for NS where the median lake was, in fact, acidic (i.e. ANC = 0 $\mu eq.L^{-1}$). Within eastern Canada, the Maritime region as a whole had the highest proportion of lakes with pH $\leq$ 6.0 and ANC $\leq$ 200 $\mu eq.L^{-1}$.

3.2 Regional Characterization - SO_4

Median SO_4 concentrations were quite variable in eastern Canada, ranging from 24 (LAB) to 158 $\mu eq.L^{-1}$ (SC ONT); see Table II. NE ONT and SC ONT had the highest proportion of high SO_4 lakes; 46 and 63% respectively had values $\geq$ 150 $\mu eq.L^{-1}$. Minimum concentrations were $\leq$ 10 $\mu eq.L^{-1}$ in all subregions except SC ONT and NB (41 and 40 $\mu eq.L^{-1}$ respectively).

In Ontario, the highest median SO_4 concentrations occurred in SC ONT (158 $\mu eq.L^{-1}$) where only 37% of the lakes had values $\leq$ 150 $\mu eq.L^{-1}$. NW ONT had the lowest median (75 $\mu eq.L^{-1}$) and 98% of the lakes were $\leq$ 150. NE ONT was intermediate in both instances.

The median concentration of SO_4 in LAB was 24 $\mu eq.L^{-1}$ and in QUE, 82 $\mu eq.L^{-1}$. In LAB only 3% of the lakes had concentrations $\geq$ 50 $\mu eq.L^{-1}$, whereas QUE had 82% in this category.

The highest median SO_4 in the Maritimes occurred in NB (77 $\mu eq.L^{-1}$) followed by NS and NF (46 and 31 $\mu eq.L^{-1}$ respectively). While very few lakes in any subregion of the Maritimes had concentrations $\geq$ 100 $\mu eq.L^{-1}$, the proportion in the lowest SO_4 class ($\leq$ 50 $\mu eq.L^{-1}$) varied greatly, i.e. NB=7%, NS=54%, and NF=91%.

3.3 Regional Characterization - A^-

The highest minimum, median, and fourth quintile A^- concentration occurred in NW ONT (see Table II). Median values for the remaining subregions of eastern Canada ranged from 38 (SC ONT) to 56 (NF) $\mu eq.L^{-1}$. Most lakes had A^- concentrations $\leq$ 100 $\mu eq.L^{-1}$.

The lakes sampled in NW ONT were characterized by higher A^- concentrations (minimum = 25, median = 94 $\mu eq.L^{-1}$) than those in NE ONT or SC ONT. Forty-two percent of the lakes in NW ONT had A^- > 100 $\mu eq.L^{-1}$; in contrast, NE ONT had 19% and SC ONT had 0%.

QUE and LAB exhibited similar A^- distributions; 73 and 78% respectively had A^- concentrations between 30 and 100 $\mu eq.L^{-1}$. While the distribution of lakes among A^- categories was somewhat more variable in the Maritimes compared to QUE-LAB, the range of median concentrations was similar (44 - 56 $\mu eq.L^{-1}$). Only 3% of the lakes in NB had values > 100 $\mu eq.L^{-1}$ while 19% in NF were in this class.

Table II Minimum, median, and Q_4* values for lake SO_4 and A^- concentrations ($\mu eq.L^{-1}$) in Canada. Percentages of lakes sampled which fall into 3 SO_4 and 3 A^- classes ($\mu eq.L^{-1}$) are given. All statistics are sample estimates.

Subregion	SO_4 Concentration				Percentage			A^- Concentration				Percentage		
	Sample n	Min.	Med.	Q_4	≤ 50	≤ 100	≤ 150	Sample n	Min.	Med.	Q_4	≤ 3	≤ 30	≤ 100
NW ONT	339	10	75	98	15	82	99	47	25	94	126	0	4	58
NE ONT	813	0	148	201	13	34	54	652	<1	45	86	<1	27	81
SC ONT	750	41	158	178	<1	4	37	747	7	38	53	0	26	100
QUE	246	8	82	124	18	59	90	287	10	48	71	0	27	97
LAB	180	4	24	32	97	100	100	198	8	47	66	0	22	98
NB	52	40	77	91	7	96	100	75	16	44	64	0	23	97
NS	130	5	46	65	54	98	100	167	4	48	78	0	14	90
NF	176	2	31	42	91	99	100	253	3	56	97	4	43	81

*Maximum value expected to characterize 80 percent of the lakes sampled.

3.4 Regional Characterization - Relationship between ANC and SO_4

Figure 2 shows the distribution of only those lakes having ANC $\leq$ 200 $\mu eq.L^{-1}$ arranged by SO_4 class for regions in eastern Canada. In addition, subregions are arranged within region by order of decreasing median SO_4, i.e. from left to right. Five SO_4 classes were derived from the cumulative frequency distribution of SO_4 using the entire data sets from both Canada and the U.S. Hence, the SO_4 concentration classes in Figure 2 ($\leq$ 47, 48-77, 78-117, 118-163,and >163 $\mu eq.L^{-1}$) were based on the first through fifth quintiles, respectively; that is, where 20, 40, 60, 80, and 100% of the sample (Canada) and population (U.S.) would be found. Within each SO_4 class, lakes characterized by ANCs $\leq$ 0, 0-40, 41-100, and 101-200 $\mu eq.L^{-1}$ are shown by differing shading of the vertical bars.

The distribution of SO_4 concentrations for lakes in SC ONT and NE ONT (i.e. those with the highest medians) is skewed toward the highest SO_4 classes; in contrast, most lakes in NF and LAB are found in the lowest classes. The greatest variability in SO_4 distribution occurred in QUE while the least was in LAB and NF. The relative proportion of lakes with ultralow ANC ($\leq$ 40 $\mu eq.L^{-1}$) was generally fairly constant across SO_4 classes except for NE ONT where the majority of lakes with ANCs $\leq$ 40 $\mu eq.L^{-1}$ were found in the highest SO_4 class.

3.5 Regional Characterization - Relationship between ANC and A^-

Figure 3 shows the distribution of only those lakes having ANCs $\leq$ 200 $\mu eq.L^{-1}$ by A^- class. Figure 3 was prepared in a comparable manner to the ANC within SO_4 Figure 2. Thus the A^- class values which characterize 20, 40, 60, 80, 100% of the lakes sampled in Canada and the target populations in the U.S. were <26, 26-37, 38-50, 51-73, and >73 $\mu eq.L^{-1}$.

While the subregions within a region have been ranked in Figure 3 by decreasing median A^- (i.e. in an analogous manner to the ranking by median SO_4 in Figure 2), there is no analogous skewing of the A^- distribution. Within Ontario for example, the median A^- in NE ONT was higher than in SC ONT (45 vs 38 $\mu eq.L^{-1}$ respectively), but the distribution of A^- was more skewed to the lowest A^- class in NE ONT than in SC ONT. In QUE and LAB, A^- medians and distributions were similar, while in the Maritimes, NF had then highest median concentration but most of the lakes with ANC $\leq$ 200 $\mu eq.L^{-1}$ were characterized by A^- < 26 $\mu eq.L^{-1}$, i.e. were in the lowest A^- class.

The proportion of lakes with ultralow ANCs within each subregion generally showed no trend of occurring in the highest A^- classes; in fact, the reverse situation occurs in both NE ONT and NF. The distribution of lakes in NS with negative ANCs was fairly uniform across the upper 4 A^- classes.

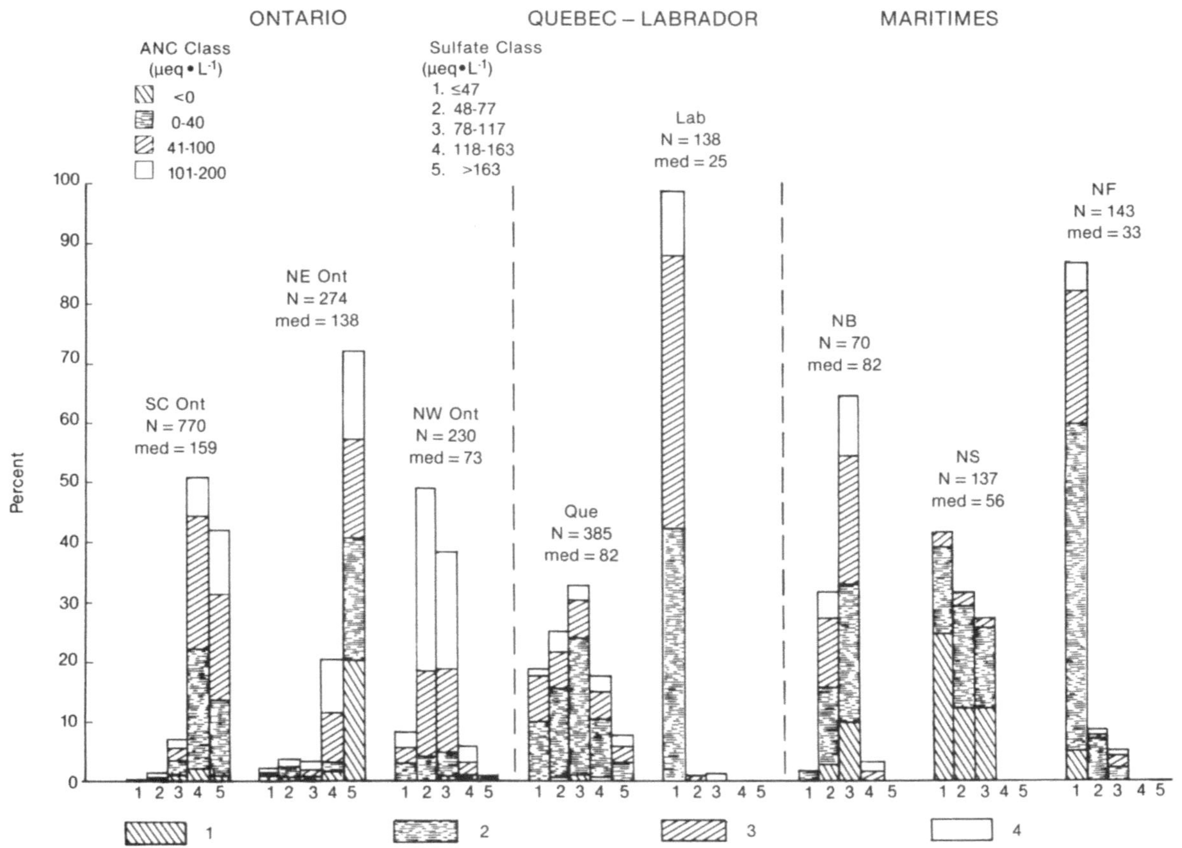

Figure 2 The distribution of lakes with ANC $\leq$ 200 μeq.L^{-1} across 5 SO_4 classes for 3 regions (8 subregions) across eastern Canada. Within each SO_4 class, the proportion of lakes characterized by each of 4 ANC classes are given. Subregions are arranged within region in order of decreasing median SO_4 concentration.

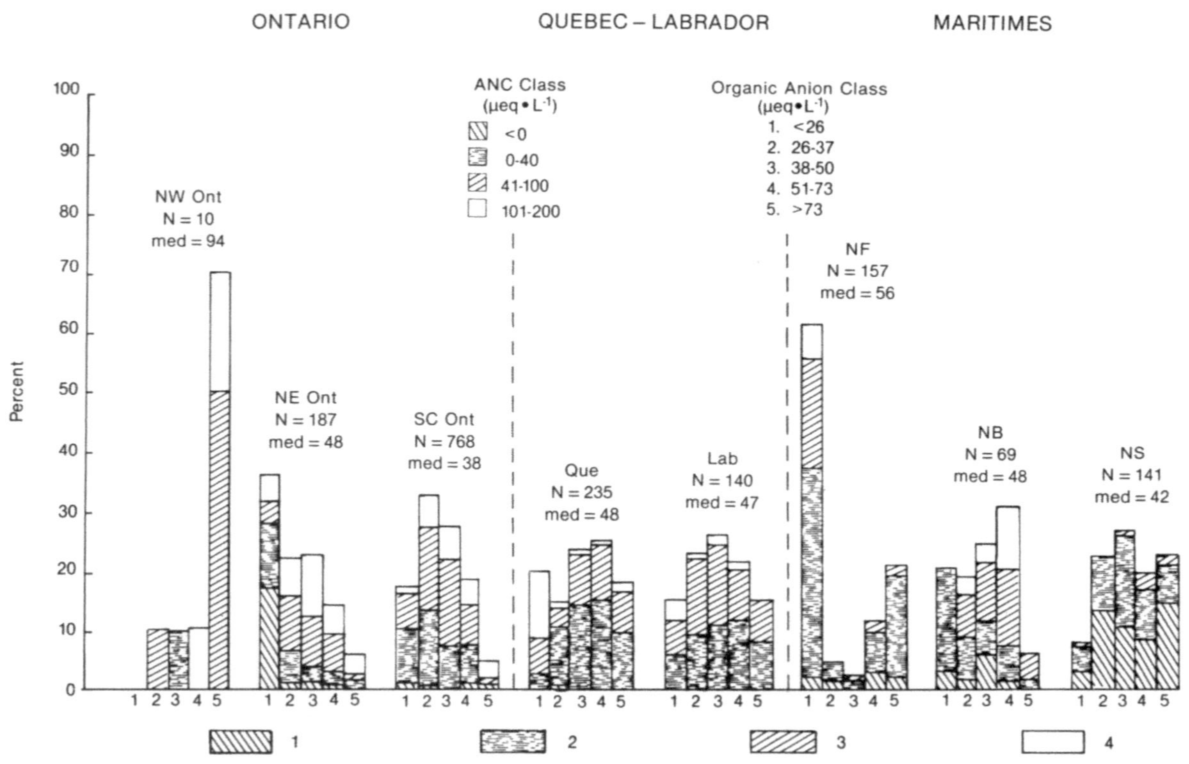

Figure 3 The distribution of lakes with ANC $\leq$ 200 μeq.L^{-1} across 5 A^- classes for 3 regions (8 subregions) across eastern Canada. Within each A^- class, the proportion of lakes characterized by each of 4 ANC classes are given. Subregions are arranged within region in order of decreasing median A^- concentration.

4. DISCUSSION

Variations in the regional and subregional distribution of pH, ANC, and SO_4 in lakes can be explained from the combined influence of surrounding terrain characteristics and atmospheric deposition. Excluding NW ONT and NE ONT, most of the lakes sampled in subregions of eastern Canada may be considered sensitive to acidic deposition (i.e. 83-99% of the lakes have ANC < 200 $\mu eq.L^{-1}$; median ANCs range from 0 to 165 $\mu eq.L^{-1}$). In contrast, lake populations in subregions of the eastern U.S. are generally less sensitive (e.g. maximum percent sensitive = 71 and median ANCs range from 84 to 802 $\mu eq.L^{-1}$; see Linthurst et al., 1986).

The extreme sensitivity of Maritime region lakes arises from their widespread occurrence on silicate bedrock which is overlain by extremely thin soils and/or wetlands. In particular, our data showed that 51% of the lakes surveyed in NS are currently acidic and an additional 42% have ultralow ANC concentrations < 40 $\mu eq.L^{-1}$. It may be argued that the extreme sensitivity apparent for NS waters is an artifact of sampling bias towards headwater lake systems. We do not believe that this is the case. First, as noted above, Kelso et al. (1986) found that the NIS data set (most headwaters and a major component of our NS data) was chemically representative of the regional situation. Second, Underwood et al. (1986) conducted a complementary survey of 234 NS lakes which specifically included lakes from these regions having carbonate bedrock. They found a median pH of 5.37 and a mean ANC of 48 $\mu eq.L^{-1}$ (mean biased high since negative values for 37 acidic lakes not included). These data for lakes plus earlier documentation of the acidification of some rivers with concurrent loss of salmonid fisheries (Watt et al., 1983) confirm NS waters as among the most affected systems in eastern North America.

The occurrence of acidic and acidified waters in terrain receiving approximately the target loading of 20 $kg.ha^{-1}.yr^{-1}$ wet SO_4 deposition (U.S./Canada, 1983) suggests that this target may be too high to protect sensitive NS systems. Even the NF subregion which receives 10-20 $kg/ha^{-1}.yr^{-1}$ wet SO_4 deposition showed 7% of the 176 lakes surveyed to be in the acidic (ANC < 0) class. While lakes in LAB also exhibit high sensitivity (80% have ANC < 88 $\mu eq.L^{-1}$), this subregion to the north of NF which receives approximately 10 $kg.ha^{-1}.yr^{-1}$ deposition (or less) contained no lakes in the acidic class and a median ANC = 46 $\mu eq.L^{-1}$.

The Ontario subregions exhibited the highest median ANCs (63-254 $\mu eq.L^{-1}$) in eastern Canada; ANC increases to the north and west. As discussed by Shilts (1981), this pattern follows $CaCO_3$ distribution in the glacial tills which overlay the silicate bedrock of the region. The source area for the carbonate found in the tills is the limestone basin located on the western side of Hudson Bay. Since no similar source area is present on the eastern side, glacial till in QUE contains little carbonate material and lakes in that subregion have lower ANCs.

NE ONT does contain some extremely sensitive terrain such as the Killarney area (Beamish and Harvey, 1972) and areas which historically received very high deposition from major point sources at Sudbury and Wawa (see various papers in Nriagu, 1984). Most of the 9% of lakes surveyed in NE ONT with ANC < 0 are associated with such areas. Coupled with the influence of the tills noted above, the lakes in NE ONT therefore exhibit an extremely wide range of ANC (20% have ANC > 839 $\mu eq.L^{-1}$).

The geographic differences in median SO_4 concentration can be related to atmospheric deposition. As a rule, soils in eastern Canada do not adsorb SO_4; most "calibrated basin" studies show a small excess SO_4 output over input measured as bulk or wet deposition (Harvey et al., 1981). Hence, when subregions of eastern Canada are listed in order of median SO_4 concentrations (i.e. LAB < NF < NS < NW ONT ≃ NB < QUE < NE ONT < SC ONT), it is not surprising that they follow a general gradient of increasing SO_4 deposition as presented by U.S./Canada (1983). Similar observations have been made by Thompson and Hutton (1985). The shift in distribution of SO_4 concentration toward higher quintiles in areas of higher deposition is clearly apparent in Figure 2. From Figure 2 it is also clear that there is no simple relationship between ANC distribution (when ANC < 200 $\mu eq.L^{-1}$) and SO_4 distribution and therefore by inference, SO_4 deposition. The only direct SO_4 - ANC relationship observed in our data occurs in NE ONT, and, as above, is explained by the combined influence of the large point sources in this subregion, localized areas of high sensitivity, and perhaps the relatively high sampling frequency present in these areas. Overall, there is no indication that low ANC lakes are associated necessarily with high SO_4 waters. This result makes geochemical sense, since there are several terrestrial and within-lake sources that contribute ANC to surface water systems, and the magnitude of the sources can vary greatly between lakes.

A current issue in the debate over lake acidification is the role of natural organic acids produced by the microbial degradation of plant and animal matter. Contribution of H^+ by these organic acids to the acidity of water systems can be assessed from the distribution of A^- in the waters. Highest median A^- occurred in NW ONT (Table II, 94 $\mu eq.L^{-1}$) while all the remaining subregions fell within the remarkably narrow range of 38-56 $\mu eq.L^{-1}$ despite the fact that the geographic extent of wetlands (a major source of organic acids) varies greatly among the subregions. The high A^- concentrations in NW ONT are consistent with those found in the Boundary Waters Canoe Area of northern Minnesota (see Part II). NS, the subregion which contains the highest proportion of acidic lakes and the common occurrence of coloured lakes and rivers, has a fourth quintile concentration (78 $\mu eq.L^{-1}$) only slightly greater than the other areas. Furthermore, Figure 3 shows that low ANC classes are spread, often fairly uniformly, across all classes of A^-. The exception to the uniform distribution of ANC across A^- classes is in NE ONT (Figure 3) where waters which are acidic tend to have very low A^- concentrations - the "classical" clear acidic lakes. If the occurrence of acidic lakes is primarily related to the presence of organic acids, then one would expect the low ANC lakes to be primarily associated with the high A^- classes. The fact that this is not the case refutes the argument for a predominating influence of natural sources of organic acidity in acidic lakes in eastern Canada.

5. CONCLUSIONS

From compilation and assessment of recent lake composition data, we conclude the following:

a) Acidic and ultralow ANC waters in eastern Canada occur in a pattern explained by a combination of biogeochemical factors and atmospheric deposition.
b) Nova Scotia contains the highest proportion of acidic and ultralow ANC lakes of any subregion surveyed in eastern North America despite the fact that it receives atmospheric deposition near the target loading of 20 $kg.ha^{-1}.yr^{-1}$ wet SO_4.
c) While Newfoundland has the next most sensitive waters, occurrence of acidic lakes in this subregion is much smaller probably due to a lower level of deposition.
d) The target loading of 20 $kg.ha^{-1}.yr^{-1}$ wet SO_4 deposition may be too high to protect the highly sensitive waters of Maritime Canada.
e) Relatively high median ANCs in Ontario occur in response to greater terrain modification of the higher deposition levels received in this area. Most acidic lakes in Ontario occur in small areas of high sensitivity and/or near major point sources.
f) Variation in the subregional median SO_4 concentration of lakes in eastern Canada follows expected gradients in SO_4 deposition; however, there is no simple relationship between lake SO_4 and ANC.
g) Natural organic acid contribution to lake waters was highest for Northwestern Ontario, and remarkably uniform across other subregions. Moreover, low ANC waters were found across all classes of A^-.
h) The hypothesis that most acidic lakes in eastern Canada are naturally acidic due to the presence of organic acids is refuted.

6. ACKNOWLEDGEMENTS

Many individuals were involved in the collection of the data used here: NIS - C.K. Minns, Y. Vigneault, D.A. Scruton, R.H. Peterson, and W. Watt; SENS1 - W. Keller, L. Maki, M. Johnson, J. Beaver, and P.J. Dillon; NAQUADAT - R. Lemieux, B. Bobee, M. Papineau, W. Whitman, T. Clair, and J. Kerekes. P. Kanciruk took the major responsibility for the data processing required for the paper, and P. Kellar and B. Emmel were involved in the organization of the information. We acknowledge the substantial contribution made by all these people.

7. REFERENCES

Beamish, R.J. and Harvey, H.H.: 1972, J. Fish. Res. Board Can., **29**, 1131.

Beck, K.C., Reuter, J.H., and Perdue, E.M.: 1974, Geochim. Cosmochim. Acta, **38**, 341.

Burns, D.A., Galloway, J.N., and Hendry, G.R.: 1981, Wat. Air Soil Pollut., **16**, 277.

Dickson, W.: 1980, 'Properties of acidified waters', IN: Ecological Impact of Acid Precipitation, Proc. Internat. Conf., Sandefjord, Norway, 75.

Dillon, P.J., Jeffries, D.S., Snyder, W., Reid, R., Yan, N.D., Evans, D., Moss, J., and Scheider, W.A.: 1979, J. Fish. Res. Board Can., **35**, 809.

Harvey, H.H., Pierce, R.C., Dillon, P.J., Kramer, J.R., and Whelpdale, D.M.: 1981, 'Acidification in the Canadian environment: scientific criteria for assessing the effects of acidic deposition on aquatic ecosystems', National Research Council Canada, Rep. No. 18475, 369 pp.

Hendry, G.R., Galloway, J.N., Norton, S.A., Schofield, C.L., Schaffer, P.W., and Burns, D.A.: 'Geological and hydrochemical sensitivity of the eastern United States to acid precipitation', EPA-600/3-80/024, 100 pp.

Howell, G.D., and Pollock, T.L.: in press, 'Sulphate, water colour, and dissolved organic carbon relationships in organic waters of Atlantic Canada', IN: Proc. Workshop on Statistical Aspects of Water Quality Monitoring, Burlington, Ontario, Canada.

Kelso, J.R.M., Minns, C.K., Gray, J.E., and Jones, M.L.: 1986, 'Acidification of surface waters in eastern Canada and its relationship to aquatic biota'. Can. Spec. Publ. Fish. Aquat. Sci., **87**, 42 pp.

Kerekes, J., Howell, J., and Pollock, T.: 1984, Verh. Internat. Verein. Limnol., **22**, 1811.

Linthurst, R.A., Landers, D.H., Eilers, J.M., Kellar, P.E., Brakke, D.F., Overton, W.S., Crowe, R., Meier, E.P., Kancirak, P., and Jeffries, D.S.: 1986, Wat. Air Soil Pollut., this volume.

National Academy of Sciences: 1981, 'Atmosphere-biosphere interactions: towards a better understanding of the ecological consequences of fossil fuel combustion', National Academy Press, Washington, D.C., 263 pp.

Nriagu, J.O. (ed): 1984, Environmental Impact of Smelters, Advances in Environmental Science and Technology, **15**, J. Wiley and Sons, New York, 608 pp.

Oliver, B.G., Malcolm, R.L., and Thurman, E.M.: 1983, Geochim. Cosmochim. Acta, **47**, 2031.

Patrick, R., Binwtti, V.P., and Hatterman, S.F.: 1981, Science, **211**, 446.

Reuter, J.H. and Perdue, E.M.: 1977, Geochim. Cosmochim. Acta, **41**, 325.

Schofield, C.L.: 1976, 'Lake acidification in the Adirondack Mountains of New York: causes and consequences', IN: Proc. First Internat. Conf. Acid Precip. and the For. Ecosystem, L.S. Dochinger and T.A. Seliga (eds), U.S.D.A. For Serv. Tech. Rep. NE-2, 477.

Shilts, W.W.: 1981, 'Sensitivity of bedrock to acid precipitation: modification by glacial processes', Geol. Survey Can. Paper 81-14, 7 pp.

Thompson, M.E., Elder, F.C., Davis, A.R., and Whitlow, S.: 1980, 'Evidence of acidification of rivers in eastern Canada', IN: Ecological Impact of Acid Precipitation, Proc. Internat. Conf., Sandefjord, Norway, 244.

Thompson, M.E. and Hutton, M.B.: 1985, Wat. Air Soil Pollut., **24**, 77.

Underwood, J.K., Ogden, J.G. III, and Smith, D.L.: 1986, Wat. Air Soil Pollut., this volume.

U.S./Canada: 1983, Memorandum of Intent on Transboundary Air Pollution, Impact Assessment, Working Group I.

Watt, W.D., Scott, C.D., and White, W.T.: 1983, Can. J. Fish. Aquat. Sci., **40**, 462.

METALS AND ACIDIFICATION: AN OVERVIEW

Bruce LaZerte
Dorset Research Centre
Ontario Ministry of the Environment
P.O. Box 39
Dorset, Ontario Canada POA 1EO

0. Abstract

Both upland soils and lake sediments appear to retain atmospherically deposited trace metals (e.g. Pb) even under acid conditions. The abundant, local, mineralogically derived metals (e.g. Al and Mn) are exported from upland soils under acidic conditions, but are usually retained by lake sediments. Acidification, however, reduces the extent of retention in lake sediments and soils, potentially inducing elevated metal levels in lake water. Only under conditions of extreme acidification and with the more mobile metals do lakes become net exporters of metals. Surface depletion of metals in sediment cores may not be the result of recent acidification.

1. Introduction

A recent detailed review pertaining to the effects of acid deposition on the geochemical cycling of many trace elements is available (Campbell et al., 1985) and will not be repeated here. The authors conclude that Al, Mn, Zn and possible Cd and Ni can be mobilized from soils and that their concentrations can increase in acidified aquatic systems. Conversely, Pb and perhaps Cu are not expected to be mobilized. Other trace metals probably fall somewhere in between these extremes but there was less evidence available. In particular, little is known about the geochemical mobility of Hg, although increased Hg burdens in fish are apparently associated with acidification (Suns et al., 1980; Lindqvist et al., 1984). I will attempt here to provide only a brief overview of the subject, concentrating more on potential metal sources and sinks within the landscape that might be influenced by acidic deposition. Lakes and well-drained upland soils will be emphasized; streams and wetlands will be ignored.

Water, Air, and Soil Pollution **31** (1986) 569–576.

2. Potential Sources and Sinks

Acid deposition already contains large amounts of anthropogenically derived metals. In addition, its acidity can extract local, mineralogically derived metals from soils and sediments. As it passes "downstream", the water will eventually lose part of its acidity and metal load. The importance of particular hydrologic pathways in this process cannot be overestimated.

2.1 Upland, well-drained soils

In forested terrestrial systems, atmospheric deposition is usually the only possible source of anthropogenic metals. Freidland et al., (1984) show that recent accumulations of Pb and Cu in organic forest soil layers agree well with that expected from independent estimates of deposition rates. Also, comparison of current Pb burdens in the forest floor of the North Eastern United States with old, buried organic horizons indicates that at least 80 to 90% of the Pb burden is of recent anthropogenic origin (Johnson et al., 1982).

In addition, metals are accumulated within terrestrial biomass such that, for some metals such as Cd, 27% of the total metal burden is actually found in the woody tissues (Heinrichs and Mayer, 1980; van Hook et al., 1977). For local, mineralogically derived metals such as Al and Mn, cycling through the trees and leaf litter provides one path for accumulation in the upper soil organic layers (Cronan, 1980), but burrowing and mixing between the mineral and organic horizons may also be important.

Thus, there is in the surface organic layers a large burden of metals potentially susceptible to leaching by increasingly acidic deposition. It appears that Pb is so tightly bound by organics that it is not released under the usual acidic conditions (pH>3). Zinc, Cd and Ni are leached to a greater extent (Tyler, 1978; Biddapa et al., 1982) although organic horizons may not have as great a burden of these metals as of Pb (e.g. Friedland et al., 1984). Local mineralogically derived metals (Al and Mn) are also easily leached from these layers (Cronan, 1980; Tyler, 1978; Biddapa et al., 1982) and the canopy (Cronan, 1980).

Friedland et al., (1984) also show that recent accumulations of Zn (as well as Pb and Cu) agree well with independent estimates from atmospheric deposition. Although the Pb and perhaps the Cu results are to be expected, the Zn is not, based on the above leaching studies, as deposition in their study areas is quite acidic. Perhaps there are some additional mineralogical sources of Zn not accounted for; Conroy et al., (1975) argue that in the vicinity of the Sudbury, Ontario smelter complex, local sources of Zn and Cd are more important than atmospheric deposition. Other studies have not found any consistent relationship between Zn deposition and increased burden in organic soil horizons (Reiners et al., 1975; Siccama et al., 1980).

It is expected that under the normal course of podzolization, organically complexed Pb, Al and Fe (Tyler, 1981) and perhaps colloidal Al-silicates (Farmer et al., 1980; Childs et al., 1983) would be precipitated in the mineral horizons. Large amounts of ammonium - oxalate extractable amorphous aluminum (perhaps aluminum silicates) can be found in these layers and is a fraction easily leached by dilute acid solutions (Cronan, 1985).

Which of these potential soil sources and sinks is actually operative at any given time depends, in part, on local hydrology as well as the current acidity. Cronan (1980) and Hendershot et al., (1985) argue that surface organic horizons are important sources of Al during spring melt in parts of North America. Similar results may be expected for other metals.

2.2 Lake sediments

Like the surface organic layer of upland soils lake sediments have accumulated atmospherically transported metals (directly deposited onto the lake surface and/or onto the surrounding catchment and subsequently eroded), as well as mineralogically derived metals eroded from their surrounding catchment. Sediment profiles permit the separation of recent anthropogenic from natural contributions (e.g. Kemp and Thomas, 1976; Robbins, 1978; Dillon and Evans, 1982; Evans et al., 1983) to sediment metal burdens. However, only in pristine watersheds with no catchment development can the anthropogenic pathway to lake sediments be clearly assigned to atmospheric deposition. For example, David and Galloway (1981) assigned 70% of the Zn and 80 to 90% of the Pb burden in two relatively undisturbed lakes to anthropogenic sources obtained via atmospheric deposition, whereas in Lake Erie large amounts of anthropogenic metals are derived from upstream sources and sewage treatment plants as well (Nriagu et al., 1979).

If a lake were to acidify in isolation from its catchment, the more mobile metals would be expected to migrate from the sediments (e.g. Al, Mn and perhaps Zn, Schindler and Turner, 1982; but not Pb, Davis et al., 1982), suggesting that the lake may act as a source of metals. However, as acid deposition impacts the whole catchment, an increased external load of the mobile metals to the lake (Campbell et al., 1985) is also expected; in addition, lake sediments can be substantial sources of alkalinity, increasing the pH of the incoming water and pore water (Oliver and Kelso, 1983; Cook et al., 1986) and inducing the precipitation or adsorption of these metals (White and Driscoll, 1985). Consequently, lake sediments may still act as a NET sink for metals, even while acidifying.

Aluminum budgets for acidifying lakes in South-Central Ontario (LaZerte, 1986) and New York (Driscoll and Schafran, 1983) indicate that the lakes are an Al sink. Mn and possibly Zn (Dillon, 1986; fewer data are available for Zn) are also retained in Plastic Lake, Ontario (mean pH of 5.6). In five lakes near the Sudbury, Ontario smelter complex, Al, Mn,

Cu, Ni and Zn were mostly retained (but occasionally in balance, +- 9% retention) in all lakes, acid or not. Only in the most acidic lake (Clearwater, pH of 4.2) was Mn lost, with a retention of -54% (Dillon, 1986).

This scenario contradicts the original interpretation of sediment metal profiles in acidifying lakes, whereby a surface decrease in metal is taken as evidence for acid leaching from the sediment (Reuther et al., 1983; Norton et al., 1981). A decreased rate of deposition to the sediment surface of acidifying lakes (Evans et al., 1983; Kahl et al., 1984) is another possible explanation for this surface decrease that is in better agreement with Dillon's (1986) budget results.

However there are other potential explanations, and increasing lake acidity may not always be responsible. Many such surface depleted profiles (particularly of Pb, Cd and Zn) have been found in South-Central Ontario lakes (including Plastic Lake, Ontario, Evans et al., 1983; Smith, 1983), but with no apparent relationship to the alkalinity of the lakes and with a wide variability from core to core within each lake. There has also been a dramatic decline in lead deposition rates since the 1970's (Eisenreich et al., 1986) which could be important.

Carignan and Tessier (1985) have proposed an explanation consistent with such profiles that invokes a diffusive accumulation of Zn in the sediments rather than net loss. This mechanism is seemingly independent of lake water acidification, but may be facilitated by catchment acidification and the consequent elevated levels of metals and sulphate in the lake waters. Carignan and Nriagu (1985) have also suggested that Cu and Ni may behave similarly in the heavily metal polluted lakes in the Sudbury area. Interestingly, they found that Mn was the only metal studied that was exported from the sediments of Clearwater lake, in agreement with the budgets of Dillon (1986).

Nonetheless, acid lakes do appear to accumulate less anthropogenic Zn and Cd in their sediments than do neutral lakes (Evans et al., 1983), while one might expect them to have higher loads from their catchments. There is also less Pb-210 accumulating in the sediments of acid lakes (Dillon et al., 1986).

Conversely, White and Driscoll (1985) hypothesize that lakes in acidifying catchments may be INCREASING the burden of Pb and presumably other trace metals in their sediments. They suggest that because these lakes have a higher load of dissolved Al, much of which precipitates within the lake (Driscoll and Schafran, 1983), there could be greater trace metal scavenging than in a less acid lake. There is, however, little evidence to support this hypothesis other than the fact that Al precipitation appears to be correlated with Pb deposition in acidic Dart Lake, New York. In addition, even if there were less precipitating material available at higher pH's, one would expect it to have a higher affinity for trace metals. Nonetheless, the hypothesis cannot be completely ignored, and points out the need for further studies on metal cycling in acidifying systems.

I tentatively conclude that acidified lakes usually remain NET metal sinks but retain less of the metals (Al, Mn, Zn, Pb and possibly Cd and Ni) than more neutral lakes, as suggested by Dillon (1982; Dillon and Smith, 1984) for the Sudbury, Ontario lakes. Except for Pb, this proposal is identical to that suggested for the upland organic soil horizons (see above).

2.3 Lake water

Turning to the lake water itself, one would expect to see increases in metal concentrations in low pH lakes (when compared to more alkaline lakes having the same metal burdens), not necessarily because of positive net leaching from the sediments but because of reduced net losses to the sediments and increased leaching from the surrounding catchment. Most large scale surveys addressing this question have considered lakes with a gradient in pH due primarily to a gradient in acid deposition. Unfortunately this confounds increased atmospheric metal deposition with increased atmospheric acid deposition (e.g. Henriksen and Wright, 1978). However, Borg's (1983) study appears to confirm this expectation for Al, Mn, Zn, Cd and even Pb. The Pb levels are probably not elevated because of increased leaching from the surrounding catchment (Campbell et al., 1985) or from the sediments (Davis et al., 1982), but because of decreased sedimentation of Pb deposited directly onto the lake's surface (Dillon et al., 1986).

As with upland soils, hydrology is important. Under-ice flow paths in the spring can permit large amounts of snowmelt to bypass the body of the lake and be shunted to its outflow (LaZerte, 1984; Charette et al., 1986), reducing a lake's potential for neutralizing acidity and stripping metals out of the water. Also, lake surveys must minimize differences in lake residence time in order to detect the impact of other factors on lake retention, as it has been shown that residence time has a significant impact on the retention of Pb-210 (Cornett et al., 1986) and stable Pb (White and Driscoll, 1985), and probably most other metals.

References

Biddappa, C.C., Chino, M., and Kumazawa, K.: 1982, Plant and Soil **66**, 299.

Borg, H.: 1983, Hydrobiologia **101**: 27-34.

Campbell, P. G., Stokes, P. M., and Galloway, J. N.: 1985, Acid deposition: effects on geochemical cycling and biological availability of trace elements. Subgroup on metals of the tri-academy committee on acid deposition. National Academy Press, Washington D.C. 83p.

Carignan, R., and Tessier, A.: 1985, Geochem. Cosmochim. Acta 49, 1753.

Carignan, R., and Tessier, A.: 1985, Science 228, 1524.

Charette, J.Y., Jones, H.G., Sochanska, W. and Gauthier, J.M.: 1986, Atmosphere and Oceans, In Press.

Childs, C.W., Parfitt, R.L., and Lee, R.: 1983, Geoderma 29, 139.

Cook, R.B., Kelly, C.A., Schindler, D.W., and Turner, M.A.: 1986, Limnol. Oceanogr. 31, 134.

Conroy, N., Hawley, K., Keller, W., and Lafrance, C.: 1975, Influences of the atmosphere on lakes in the Sudbury area. Proc. First Specialty Symposium on Atmospheric Contribution to the Chemistry of Lake Waters. Internat. Assoc. Great Lakes Res. Sept. 28-Oct. 1, 1975.

Cornet, R.J., Chant, L. and Link, D.: 1986, Water Pollut. Res. J. Canada, In Press.

Cronan, C.S.: 1980, Oikos 34, 272.

Cronan, C.S.: 1985, ALBIOS preliminary data report. July 1985. Unpublished manuscript. Univ. of Maine.

Davis, A.O. and Galloway, J.N.: 1981, Atmospheric lead and zinc deposition into lakes of the eastern United States. Chapt. 19. In S.J. Eisenreich (ed) Atmospheric Pollution in Natural Waters, Ann Arbor Publishers, Ann Arbor, Michigan.

Davis, A.O., Galloway, J.N., and Nordstrom, D.K.: 1982, Limnol. Oceanogr. 27, 163.

Dillon, P.J.: 1982, Mass balance models: an explanation of the observed chemistry of lakes near Sudbury, Ontario. Final Limnological Report. Ont. Min. Environ. Rep. SES 009/82. Dorset, Ontario.

Dillon, P.J.: 1986, unpublished results.

Dillon, P.J. and Evans, R.D.: 1982, Hydrobiologia 91, 121.

Dillon, P.J. and Smith, P.J.: 1984, Trace metal and nutrient accumulation in the sediments of lakes near Sudbury, Ontario, p. 375-426. In J.O. Nriagu (ed.) Environmental impact of smelters, Adv. Environ. Sci. Ser., John Wiley.

Dillon, P.J., Smith, P.J., and Evans, H.E.: 1986, J. Env. Geol. & Wat. Sci., In Press.

Driscoll, C.T. and Schafran, G.C.: 1981, The chemistry, transport and fate of aluminum in dilute acidified lake systems. USEPA/NCSU Final report APP-0094-1981.

Eisenreich, S.J., Metzer, N.A., Urban, N.R., and Robbins, J.A.: 1986, Environ. Sci. Technol. **20**, 171.

Evans, H.E., Smith, P.J. and Dillon, P.J.: 1983, Can. J. Fish. Aquat. Sci. **40**, 570.

Farmer, V.C., Russell, J.D. and Berrill, M.L.: 1980, J. Soil Sci. **31**, 673.

Friedland, A.J., Johnson, A.H. and Siccama, T.G.: 1984, Wat. Air Soil Pollut. **21**, 161.

Heinrichs, H. and Mayer, R.: 1980, J. Environ. Qual. **9**, 111.

Hendershot, W.H., Dufresne, A., Lalande, H. and Wright, R.K.: 1985, Speciation of aluminum in different compartments of a drainage basin during snowmelt. Proc. Eastern Snow Conference, Montreal, Quebec.

Henriksen, A. and Wright, R.F.: 1978, Water Research **12**, 101.

Johnson, A.H., Siccama, T.G. and Frieland, A.J.: 1982, J. Environ. Qual **11**, 577.

Kahl, J.S., Norton, S.A., and Williams, J.S.: 1984, Chronology, magnitude and paleolimnological record of changing metal fluxes related to atmospheric deposition of acids and metals in New England. Chapt. 2. In O.P. Bricker [ed] Geological aspects of acid deposition, Butterworth Publishers, Boston.

Kemp, A.L.W. and Thomas, R.L.: 1976, Wat. Air and Soil Pollut. **5**, 469.

LaZerte, B.D.: 1984, Can. J. Fish. Aquat. Sci. **41**, 766.

LaZerte, B.D.: 1986, unpublished results.

Lindqvist, O., Jernelov, A., Johansson, K. and Rodhe, H.: 1984, Mercury in the Swedish environment: global and local sources. Report SNV PM 1816. National Swedish Env. Protect. Bd. 105 p.

Norton, S.A., Hess, C.T. and Davis, R.B.: 1981, Rates of accumulation of heavy metals in pre- and post-European sediments in New England lakes. Chapter 20. In S.J. Eisenreich (ed.) Atmospheric Pollution in Natural Waters. Ann Arbor Publishers, Ann Arbor, Michigan.

Nriagu, J.O., Kemp, A.L.W., Wong, H.K.T. and Harper, N.: 1979, Geochim. Cosmochim. Acta **43**, 247.

Oliver, B.G. and Kelso, J.R.: 1983, Wat. Air Soil Pollut. **20**, 379.

Reiners, W.A., Marks, R.H. and Vitousel, P.M.: 1975, Oikos **26**, 264.

Reuther, R., Andersson, J.B. and Forstner, U.: 1983, Effect of pH changes on the distribution and chemical forms of heavy metals in sediment cores from Swedish lakes, p. 868-871. In Heavy Metals in the Environment Vol. 2., International Conference, Heidelberg. 1983.

Robbins, J.A.: 1978, Geochemical and geophysical applications of radioactive lead, p. 285-408. In J.O. Nriagu (ed.) The biogeochemistry of lead in the environment. Elsevier, Amsterdam.

Schindler, D.W. and Turner, M.A.: 1982, Wat. Air Soil Pollut. **18**, 259.

Siccama, T.G., Smith, W.H. and Mader, D.H.: 1980, Environ. Sci. Technol. **14**, 54.

Smith, P.J.: 1983, Sediment chemistry of lakes in the Muskoka-Haliburton study area. Data report DR 83/7. Ont. Min. Environ. Dorset, Ontario.

Suns, K., Curry, C. and Russell, D.: 1980, The effects of water quality and morphometric parameters on mercury uptake by yearling yellow perch. Tech. Report LTS 80-1. Ont. Min. Env. Rexdale, Ontario. 16p.

Tyler, G.: 1978, Wat. Air Soil Pollut. **9**, 137.

Tyler, G.: 1981, Wat. Air Soil Pollut. **15**, 353.

Van Hook, R.I., Harris, W.F. and Henderson, G.S.: 1977, Ambio **6**, 281.

White, J. and C.T. Driscoll: 1985, Environ. Sci. Technol. **19**, 1182.

REGIONAL CHEMICAL CHARACTERISTICS OF LAKES IN NORTH AMERICA

PART II: EASTERN UNITED STATES

by

R. A. Linthurst[1], D. H. Landers[2], J. M. Eilers[3], P. E. Kellar[4],

D. F. Brakke[5], W. S. Overton[6], R. Crowe[7], E. P. Meier[7]

P. Kanciruk[8] and D. S. Jeffries[9]

[1]U. S. EPA, Environmental Monitoring Systems Laboratory, MD-39, Research Triangle Park, NC 27111, USA

[2]State University of New York, King Hall, 300 Washington Blvd., Oswego, NY 13126 USA Present Address: U.S. EPA, Environmental Research Laboratory, 200 S. W. 35th Street, Corvallis, OR 97333, USA

[3]Northrop Services, Inc., 200 S. W. 35th Street, Corvallis, OR 97333, USA

[4]Radian Corporation, P.O. Box 13000, Research Triangle Park, NC 27709, USA

[5]Western Washington University, Institute for Watershed Studies, Bellingham, WA 98225, USA

[6]Oregon State University, Department of Statistics, Kidder Hall, Corvallis, OR 97333, USA

[7]U.S. EPA, Environmental Monitoring Systems Laboratory, Las Vegas, NV 89114, USA

[8]Environmental Sciences Division, Oak Ridge National Laboratory, Oak Ridge, TN 37830, USA

[9]National Water Research Institute, Box 5050, Burlington, Ontario L7R 4A6, Canada

Water, Air, and Soil Pollution **31** (1986) 577–591.

ABSTRACT. A survey of 1592 lakes (>4 ha and $\leq 2{,}000$ ha) was conducted in Fall, 1984. The study area was three regions of the eastern United States (Northeast, Upper Midwest, and Southeast) expected to contain the most lakes with acid neutralizing capacity (ANC) ≤ 400 µeq/L. The design of the survey provides statistically reliable estimates of the number, location and chemical characteristics of lakes in the study area. Frequency distribution statistics are presented for pH, ANC, sulfate and organic anion. The highest percentages and numbers of acidic lakes occurred in Florida, the Adirondacks, and the Upper Peninsula of Michigan. The highest percentages and numbers of lakes with high sulfate concentrations occurred in the Poconos/Catskills, Southern New England and Florida. The highest percentages and numbers of lakes with high organic anion concentrations occurred in Northeastern Minnesota, the Upper Great Lakes Area and Florida. The percentage of acidic lakes by sulfate concentration class revealed no discernable pattern across subregions. In contrast, the acidic lakes occur with the highest frequency in the lowest organic anion concentration class. The complex distribution of low ANC lakes cannot be explained simply by sulfate concentrations or organic acids.

1. INTRODUCTION

Despite several attempts to quantify change related to acidic deposition, the geographical extent of lakes and streams potentially at risk, or acidic, and the degree or severity to which North American surface waters as a whole are threatened by acidic deposition, have remained unresolved. Considerable statistical uncertainty in extrapolating the results of specific studies to a regional or national scale has continued to exist largely due to: the unknown degree to which the individual study sites represented the larger regional population; bias in lake selection; the absence of measurements of key chemical variables which are considered critical in assessing effects; the difficulty in inter-study comparisons of data collected by different investigators; non-standard methodological techniques; and the variable documentation of quality assurance protocols.

The objective of this paper is to explore the chemical relationships among lakes receiving sulfate loading from atmospheric deposition in the eastern United States, using a uniformly generated data base that was subjected to rigorous quality assurance protocols (Linthurst et al., 1986; Overton et al.,1986; Kanciruk et al., 1986). Specifically, the questions posed here are: 1) do pH and concentrations of ANC, sulfate and organic anion in lakes show patterns that can be distinguished regionally? 2) are sulfate concentrations in lakes inversely related to ANC concentrations? 3) are increased organic anion concentrations directly related to ANC concentrations? These are addressed with the assumption that atmospheric deposition of strong acids, principally sulfuric acid, can produce acidic lakes. Of interest in our analysis is to determine what patterns are regional in nature, what relationships are not valid on a regional scale, and to define the specific sub populations of lakes where acidic deposition effects are observed.

2. METHODS

Detailed methods are given in Linthurst et al. (1986). Methods bearing on interpretations are summarized below.

2.1. Population Sampled

Lakes were selected using a stratified design with equal allocation of sample lakes among strata (Linthurst et al., 1986). Three stratification factors were used: region, subregion and alkalinity map class. Lakes were selected in each stratum by systematic sampling of an ordered list following a random start.

Three geographical regions in the eastern United States (Figure 1) were chosen for study during the first phase of the Eastern Lake Survey: the Northeast, Upper Midwest, and Southeast. The regions were chosen on the basis of alkalinity maps generated from historical data by Omernik and Powers (1983). The lakes within these regions were expected to be characterized predominantly by ANC ≤ 400 µeq/L. Subregions (Figure 1) were delineated on the basis of similarities in surrounding physical geography, land use characteristics, and available water quality information.

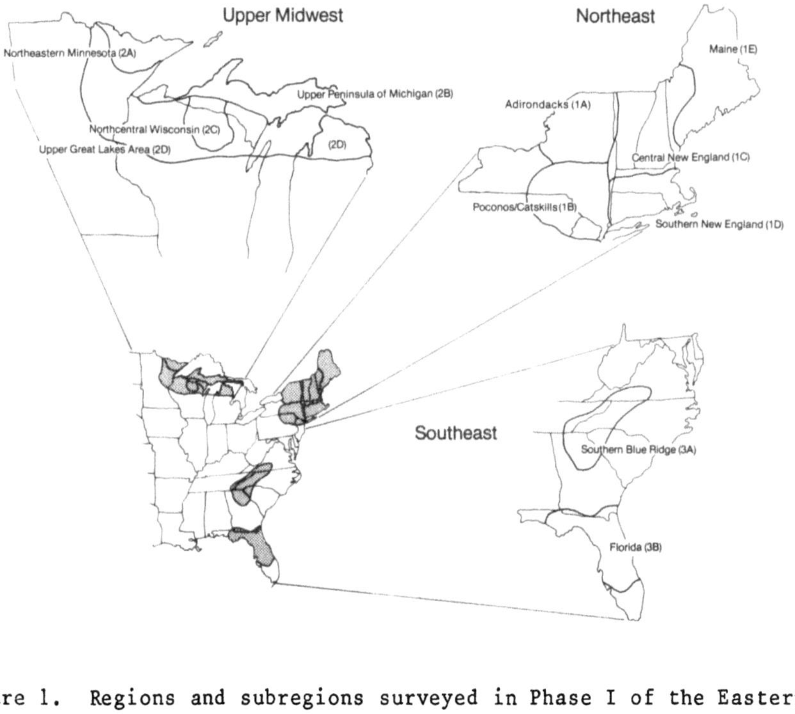

Figure 1. Regions and subregions surveyed in Phase I of the Eastern Lake Survey.

Each subregion was further divided into three smaller areas using the alkalinity maps of Omernik et al. (1985), Omernik and Kinney (1985) and Omernik and Griffith (1985). Areas within each subregion were delineated such that three classes of lakes with expected alkalinities of <100, 100 to 200 and > 200 μeq/L were distinguished. The lakes to be sampled within each alkalinity class were drawn from a preliminary map population list so that each had a known probability of being chosen for sampling. An approximately equal number of lakes (about 50) was chosen from each class.

The target population within the three regions was based on 1:250,000 scale U.S. Geological Survey (USGS) maps. Because such maps in the eastern U.S. usually do not include lakes with surface area less than approximately 4 ha, these smaller lakes are not represented in the target population. The target population was refined before sampling using large-scale (e.g., 1:24,000) USGS maps, during sampling using observations by the sampling teams and following sampling based on evaluation of the data. The final target population excluded lakes which were <4 ha, non-lakes (dry or flowing water), wetlands, high conductance (>1500 μS/cm), impacted by anthropogenic sources, inaccessible or frozen. All population estimates reported in this paper refer to the target population.

2.2. Sampling Methods

Sampling was conducted using helicopters equipped with floats (Morris et al., 1986), during fall turnover when most of the lakes were unstratified. Sampling periods were October 8 to November 8, 1984 in the Northeast, October 17 to November 13, 1984 in the Upper Midwest, and November 17 to December 14, 1984 in the Southeast. In most lakes discrete samples from 1.5 m were taken from sites generally located in the deepest part of the lake. In a few lakes that were too shallow to permit obtaining a clean sample from 1.5 m, samples were collected at 0.5 m. Lakes greater than 2000 ha were sampled but have been excluded from discussion in this paper.

2.3. Analytical Methodologies

Preliminary sample processing, including filtration and preservation, was conducted at base sites in specially equipped mobile field laboratories usually within 12 h of collection. Samples were then shipped by overnight air to one of four analytical laboratories where analyses were completed within pre-established holding times specific to each chemical variable. A total of 25 analytical measurements as made in situ, in the field laboratory and in the analytical laboratories (Hillman et al. 1986). Results for pH, ANC, sulfate and organic anion are presented here.

An extensive quality assurance program was implemented during the ELS -I to ensure that all sampling and analytical protocols were standardized (Best et al., 1986). The data were subjected to two primary quality assurance checks--verification and validation--conducted by the EPA Environmental Monitoring Systems Laboratory at Las Vegas, NV

(Drouse et al., 1986) and the EPA Environmental Research Laboratory at Corvallis, Oregon (Eilers et al., 1986).

pH was determined electrometrically in the field laboratory on samples collected in gas-tight syringes using a special chamber to avoid exchange of dissolved carbon dioxide with air (Hillman et al., 1986). ANC was determined in an analytical laboratory using a modification of the Gran technique (Hillman, ibid). Sulfate was determined using ion chromatography. Sulfate data were not corrected for sea salt because of the influence of road salt in some areas.

Dissolved organic carbon (DOC) was determined by flame ionization detector spectrophotometry. Organic anion concentrations were calculated using DOC and H^+ concentrations according to the model equations described in Oliver et al. (1983). This model was developed using data from colored waters and assumes that most of the organic carbon is humic material. The organic anion values for this paper were calculated using DOC data from both colored and clear lakes and, consequently, may be overestimates.

3. RESULTS

3.1. Regional Characterization - pH and ANC

Within the eastern U.S., median pH values ranged from a low in Subregion 3B (Florida) of 6.6 to a high of 7.4 in the Upper Great Lakes Area (2D) in the Upper Midwest (Table I). The lowest Q_1 values were observed in the Adirondacks subregion (1A) and Florida. The highest percentages of lakes with pH ≤ 5.0 occurred in Adirondacks subregion, the Upper Peninsula of Michigan (2B) and Florida (3B). Relatively few lakes in the target population in the Eastern U.S. were characterized by having low pH (≤ 5.0).

In the Northeast, median pH values were fairly similar among subregions. The median values ranged from 6.7 in Subregion 1A to 7.0 in Subregion 1B (Poconos/Catskills), and the Q_4 values ranged from 7.1 in Subregions 1A and 1D to 7.4 in 1B. In Maine (1E) 66.8% of the lakes had ANC ≤ 200 µeq/L but only 0.5% had negative ANC and pH ≤ 5.0.

In the Upper Midwest, the lowest median pH (6.7) and ANC (93.9 µeq/L) occurred in Northcentral Wisconsin (2C). Over 98% of the lakes in Northeastern Minnesota (2A) which contains the Boundary Waters Canoe Area, had pH values > 6.0. The highest percentage (nine) of low pH lakes (≤ 5.0) occurred in the Upper Peninsula of Michigan (2B) as did the highest percentage of lakes with ANC ≤ 0 µeq/L (9.8%). As in Subregion 2A, nearly all the lakes in Subregion 2D had pH values ≤ 6.0 and less than one-third had ANC ≤ 200 µeq/L.

Florida (3B) had the lowest median pH and ANC of any subregion in the U.S. All of the lakes in the Southern Blue Ridge (3A) had pH values ≥ 5.5; no lakes with ANC ≤ 0 µeq/L were sampled. In contrast, 12.4% of the Florida lakes had pH values ≤ 5.0 and 22.0% had negative ANC.

Table I

Median, Q_1[a] and Q_4[a] values for pH (Table I) and ANC (Table II) in the eastern United States. Percentages of lakes sampled which fall into three pH (Table I) and ANC (Table II) classes are given. All values are target[b] population estimates. The 95% upper confidence limit for the percentage is given in parentheses.

		pH Values			Population Estimates(%)[c]		
	$\hat{N}$	Q_1	Med.	Q_4	≤5.0	≤5.5	≤6.0
1A	1290	5.5	6.7	7.1	10.0(14.0)	19.9(25.1)	26.6(32.4)
1B	1479	6.6	7.0	7.4	0.8(1.3)	5.7(9.9)	7.8(12.1)
1C	1483	6.3	6.8	7.3	1.7(3.2)	7.8(11.2)	12.9(17.2)
1D	1318	6.2	6.8	7.1	5.0(8.7)	10.0(14.7)	14.6(19.8)
1E	1526	6.6	6.9	7.3	0.5(1.3)	1.6(3.0)	4.8(7.2)
1	7096	6.3	6.9	7.3	3.4(4.5)	8.6(10.3)	12.9(14.9)
2A	1457	6.6	6.9	7.3	0.0(-)	0.0(-)	1.4(3.0)
2B	1050	6.1	7.1	7.8	9.4(14.6)	13.6(19.4)	17.7(24.1)
2C	1480	5.8	6.7	7.5	2.1(3.7)	11.1(14.7)	27.7(33.0)
2D	4515	6.6	7.4	8.1	0.0(-)	0.1(0.2)	4.5(8.8)
2	8501	6.4	7.1	7.9	1.5(2.2)	3.6(4.5)	9.6(10.0)
3A	258	6.7	7.0	7.3	0.0(-)	0.0(-)	0.4(0.5)
3B	2098	5.4	6.6	7.4	12.4(18.4)	20.6(27.3)	32.7(41.8)

Table II

		ANC (μeq L^{-1})			ANC (μeq L^{-1}) Population Estimates (%)[c]		
	$\hat{N}$	Q_1	Med.	Q_4	≤0	≤40	≤200
1A	1290	14.0	111.8	250.8	10.7(14.8)	31.3(37.4)	70.5(77.2)
1B	1479	102.2	297.4	530.5	5.3(9.5)	12.2(18.1)	38.7(48.6)
1C	1483	52.0	119.9	387.7	2.4(4.3)	16.4(21.1)	67.6(74.2)
1D	1318	39.4	161.8	402.6	5.0(8.1)	19.6(25.5)	57.3(66.9)
1E	1526	72.1	148.4	298.4	0.5(1.3)	7.6(10.7)	66.8(73.4)
1	7096				4.6(6.0)	17.0(19.3)	60.0(63.6)
2A	1457	98.2	184.9	403.4	0.0(-)	1.9(4.1)	57.0(64.5)
2B	1050	55.5	283.6	1284.4	9.8(15.0)	18.3(24.6)	41.7(51.0)
2C	1480	16.8	93.9	600.7	3.1(5.0)	37.0(43.1)	56.7(63.3)
2D	4515	104.0	801.7	2021.9	0.0(-)	3.9(7.1)	31.3(40.8)
2	8501				1.7(2.4)	11.1(13.3)	41.4(46.9)
3A	258	149.1	250.2	421.4	0.0(-)	0.8(1.1)	34.3(41.9)
3B	2098	-7.5	83.5	713.0	22.0(29.3)	32.9(37.5)	55.1(67.3)

[a] It is estimated that 20% of the lakes have values less than or equal to Q_1 and 80% are less than Q_4.

[b] Target population size (≤ 2000 ha) was estimated by weighting sample n for the probability of the lake being sampled.

[c] (-) = undefined

3.2 Regional Characterization - Sulfate Concentrations in Lakes

In general, median sulfate concentrations in lakes were higher in the Northeastern U.S. than in the Upper Midwest (Table III). In the Southeast, the median concentration in Florida was within the range observed in the Northeast while the Q_4 concentration (272 μeq/L) was the highest observed in all three subregions. The lowest median sulfate concentration (32 μeq/L) occurred in the Southern Blue Ridge (3A).

Within the Northeast, the subregion containing the Poconos/Catskills Subregion (1B) had the highest median sulfate concentration while Maine (1E) had the lowest. Only 1% of the lakes in Maine had sulfate concentrations >150 μeq/L while 12.8% of the lakes in Subregion 1A had sulfate in excess of 150 μeq/L. The highest percentage of lakes with sulfate concentrations >150 μeq/L occurred in Subregion 1B, followed by Southern New England (1D).

The highest sulfate concentration in the Upper Midwest occurred in the Upper Peninsula of Michigan. The lowest Q_1 and median occurred in Subregion 2D (the Upper Great Lakes Area). In Subregions 2A and 2C nearly all the lakes were characterized by sulfate concentrations >150 μeq/L. Subregion 2D had the highest proportion of lakes with sulfate concentrations >150 μeq/L followed by Subregion 2B.

The Southern Blue Ridge had the lowest median, Q_1 and Q_4 sulfate concentrations and the lowest percentage of lakes with concentrations >50 μeq/L of any U.S. subregion. In contrast, the median concentration in Florida was nearly three times that in Subregion 3A and within the range observed in the Northeast. The Q_4 concentration in Florida was the highest observed in the eastern U.S. Sixty-eight percent of the lakes in Subregion 3B had concentrations >50 μeq/L, while 40.3% of the lakes had concentrations >150 μeq/L.

3.3 Regional Characterization - Organic Anion Concentrations

Generally, the highest organic anion concentrations were observed in the Upper Midwest and Florida (Table IV). The greatest proportions of lakes with organic anion >80 μeq/L occurred in Subregions 2A, 2D and 3B where 49% or more of the lakes fell into this category. The lowest percentage of lakes with high organic anion concentrations was found in the Southern Blue Ridge (3A), where only 2.4% of the lakes had organic anion >80 μeq/L.

Maine had the highest median and Q_4 organic anion concentrations in the Northeast. Five percent or fewer of the lakes in the subregions containing the Adirondacks and the Poconos/Catskills had organic anion >80 μeq/L.

Northeastern Minnesota (2A) had the highest median organic anion concentration in the Upper Midwest. The lowest median concentration occurred in Subregion 2C, which also had the lowest percentage of lakes with organic anion concentrations >80 μeq/L.

In the Southern Blue Ridge (Subregion 3A) the median organic anion concentration was 19.7 μeq/L compared to that in Florida which was 78.2 μeq/L. Only 34.4% of the lakes in 3A had organic anion concentrations >30 μeq/L while in Subregion 3B, 81.6% of the lakes occurred in

Table III

Median, Q_1[a] and Q_4[a] values for lake sulfate (Table III) and organic anion (Table IV) in the eastern United States. Percentages of lakes sampled which fall into three sulfate classes are given. All values are target[b] population estimates. The 95% upper confidence limit for the percentage is given in parentheses.

		Sulfate Concentrations (μeq L^{-1})			Sulfate (μeq L^{-1}) Population Estimates (%)[c]		
	$\hat{N}$	Q_1	Med.	Q_4	> 50	>100	> 150
1A	1290	104.0	118.7	141.9	99.4 (100)	84.1(91.0)	12.8(16.8)
1B	1479	118.1	159.3	221.0	99.8 (100)	90.6 (100)	56.1(67.9)
1C	1483	77.9	101.2	136.1	99.5 (100)	51.2(58.1)	15.7(20.2)
1D	1318	108.2	141.1	190.3	100 (100)	87.0(98.6)	45.8(55.9)
1E	1526	56.4	74.6	87.8	88.5 (95.8)	9.3(12.8)	1.1(2.3)
1	7096	79.4	115.4	164.9	97.3 (100)	63.0(66.8)	26.0(29.3)
2A	1457	46.6	62.5	83.9	73.7 (82.7)	9.7(14.4)	1.9(4.0)
2B	1050	49.1	77.7	103.5	79.1 (90.8)	26.3(35.5)	6.6(11.8)
2C	1480	41.9	56.9	76.4	69.5 (76.9)	6.4(9.6)	1.0(2.6)
2D	4515	29.0	50.1	85.4	50.1 (62.0)	16.5(24.9)	11.0(18.0)
2	8501	34.5	57.1	87.8	61.1 (67.9)	14.8(19.5)	7.1(10.9)
3A	258	19.5	31.8	59.6	23.1 (31.7)	14.6(21.6)	8.5(13.9)
3B	2098	28.4	93.7	272.4	68.1 (82.0)	49.4(61.5)	40.3(51.8)

Table IV

		Organic Anion Concentrations (μeq L^{-1})			Organic Anion (μeq L^{-1}) Population Estimates (%)[c]		
	$\hat{N}$	Q_1	Med.	Q_4	>10	>30	>80
1A	1290	26.1	37.3	53.3	96.6 (100)	73.1(80.2)	4.3(6.9)
1B	1479	28.4	37.3	57.5	97.3 (100)	72.6(84.4)	84.6(8.8)
1C	1483	29.9	42.9	65.4	98.3 (100)	80.0(87.1)	10.9(14.9)
1D	1318	22.3	37.9	65.7	96.0 (100)	62.9(73.9)	11.0(16.5)
1E	1526	28.2	51.5	80.8	98.9 (100)	72.0(84.4)	23.2(28.5)
1	7096	27.3	41.2	64.8	97.5 (100)	73.4(77.5)	11.1(13.1)
2A	1457	54.8	89.0	121.4	100.0 (100)	98.8 (100)	58.5(67.3)
2B	1050	32.0	66.6	111.0	99.6 (100)	83.9(95.5)	33.7(43.4)
2C	1480	29.7	45.6	77.5	98.5 (100)	78.7(85.8)	19.7(25.6)
2D	4515	49.8	81.8	128.0	100.0 (100)	96.3 (100)	52.4(64.3)
2	8501	42.6	72.7	120.2	99.7 (100)	92.1(98.5)	52.1(45.4)
3A	258	11.1	19.7	38.0	85.3 (93.2)	34.4(44.8)	2.4(5.1)
3B	2098	35.7	78.2	130.5	91.1 (100)	81.6(97.6)	49.0(62.4)

[a] It is estimated that 20% of the lakes have values less than or equal to Q_1 and 80% are less than Q_4.

[b] Target population size (< 2000 ha) was estimated by weighting sample n for the probability of the lake being sampled.

[c] (-) = undefined

this category.

3.4 Regional Characterization - Relationship Between ANC and Sulfate

The distribution of lakes with ANC $\leq$ 200 µeq/L by sulfate class is shown for the Northeast (Figure 2A) and Upper Midwest and Southeast (Figure 2B). Data are arranged within region by increasing median sulfate for all lakes sampled within the subregion. Five sulfate classes, based on the first through fifth quintiles, were derived from the cumulative frequency distribution of sulfate using the entire data sets from Canada (see Jeffries, et al., 1986) and the U.S. The quintile values represent the maximum concentration below which 20, 40, 60, 80, and 100% of the sample (Canada) and estimated target population (U.S.) would be found. Within each sulfate class, lakes characterized by ANC $<$ 0, 0 to 40, 41 to 100, and 101 to 200 µeq/L are shown. In subregions with the highest overall median sulfate, the distribution of lakes with ANC $\leq$ 200 µeq/L is skewed toward the higher sulfate classes.

In the Northeast, all the lakes with ANC $\leq$ 200 µeq/L within the Poconos and Catskills (1B) and Southern New England (1D) were found in the upper three sulfate classes. In Subregion 1E, which had the lowest population median sulfate concentration in the Northeast, most of the lakes with ANC $\leq$ 200 µeq/L were found in the lower three sulfate classes. The largest percentage of lakes with ANC $\leq$ 40 µeq/L were not consistently found in the highest sulfate classes within each subregion. The distributions of lakes among sulfate classes were more variable in Subregions 1A, 1C and 1E than in 1B and 1D.

In the Upper Midwest, lakes with sulfate concentrations in the fifth class ($>$ 163 µeq/L) occurred only in Northcentral Wisconsin (2C). Very few lakes in Subregions 2A, 2B and 2C occurred within the fourth class. Most of the lakes in the Upper Midwest occurred in the lower three sulfate classes. A greater percentage of lakes with ANC $\leq$40 µeq/L was found in sulfate class 2 than in sulfate class 3 in both Subregions 2B and 2C, both of which contained the highest percentages of low ANC lakes in the Upper Midwest.

In the Southeast, lakes in Florida (3B) were distributed across all sulfate classes. The greatest percentage of lakes with negative ANC was found in the lowest sulfate class, but lakes with ANC $\leq$ 40 µeq/L were found among all sulfate classes. The highest percentage of low ANC lakes was found in the lowest and not in the highest sulfate class. In the Southern Blue Ridge, most of the lakes occurred in the lowest sulfate class with the remaining few occurring in sulfate classes 2 and 5. Those lakes in sulfate class 5 had ANC in the 101 to 200 µeq/L range.

3.5 Regional Characterization - Relationship Between Organic Anion and ANC

The distributions of lakes with ANC $\leq$ 200 µeq/L by organic anion class are shown for the Northeast and the Upper Midwest and Southeast (Figure 3B). Data are arranged within region by increasing median organic anion concentration for all lakes sampled within the subregion. Five organic anion categories were defined in the same manner as the sulfate classes.

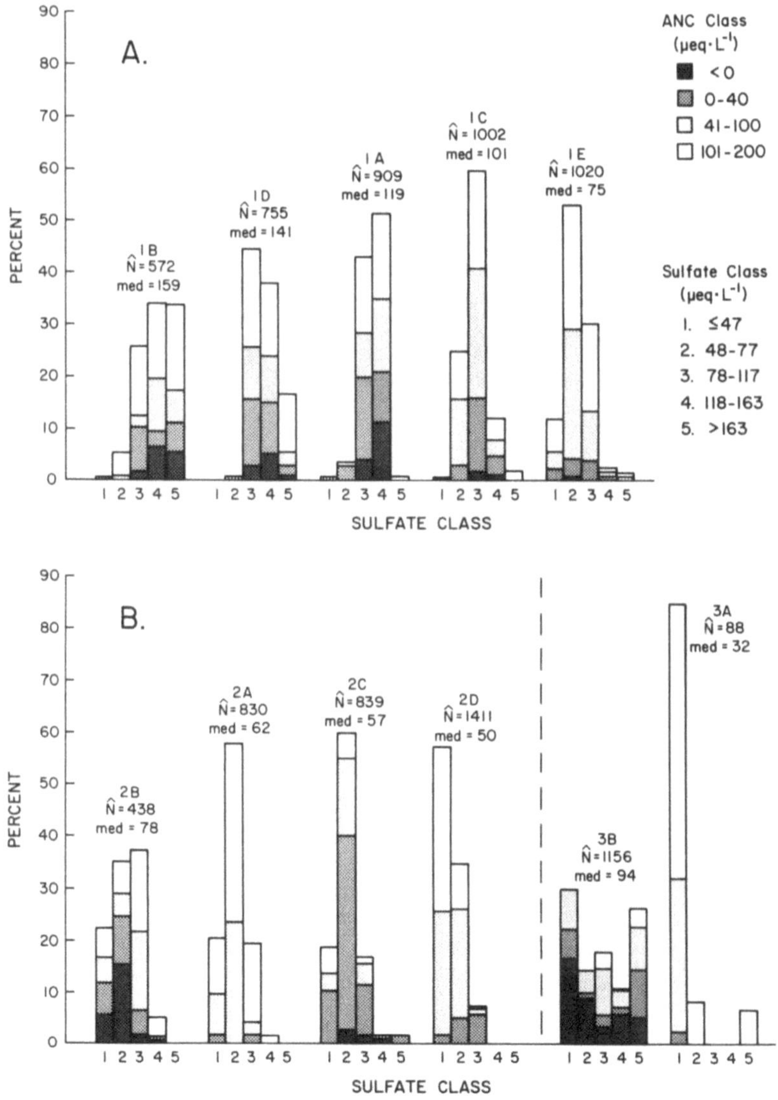

Figure 2. The distribution of lakes with ANC≤200 μeq/L across 5 sulfate classes for the Northeast (A) and the Upper Midwest and Southeast (B). Within each sulfate class, the proportions of lakes characterized by one of four ANC concentration classes are given. Subregions are arranged in order of decreasing overall population median estimates of sulfate concentration.

In the Northeast (Figure 3A) as the overall median organic anion concentration (including lakes with ANC $\leq$ 200 µeq/L) decreased, the distributions generally tended to shift to the lower organic anion classes. The distributions within Maine (1E) and Central New England (1C) were similar with a slightly higher percentage of lakes occurring in the lowest organic anion class in Subregion 1C, which had a lower overall median organic anion concentration. The distributions within Subregions 1B, 1A, and 1D were somewhat different despite the fact that the overall median values were ~37 µeq/L in each of these subregions. In each of these three subregions, proportionally more lakes were found in the lower organic anion classes than in Subregions 1E and 1C.

The majority of lakes with negative ANC were found in the lowest organic anion class. In general, no relationship between increasing organic anion concentration and decreasing ANC concentration was observed; that is, as organic anion increased, the percentages of lakes characterized by ANC $\leq$ 40 µeq/L did not increase. In fact, with increasing organic anion, ANC tended to increase.

In the Upper Midwest and Southeast, as the overall median organic anion decreased, the percentages of lakes with ANC $\leq$ 200 µeq/L generally shifted from the upper to the lower organic anion classes. The distributions in Northcentral Wisconsin (2C) and Florida (3B) were bimodal with relatively few lakes in the middle organic anion class (38-50 µeq/L). In Northeastern Minnesota (2A), which had the highest overall median organic anion concentration, most of the lakes with ANC $\leq$ 200 µeq/L were characterized by >73 µeq/L organic anion. Most of the lakes with ANC $\leq$ 200 µeq/L in the Southern Blue Ridge (3A) were characterized by organic anion concentrations >26 µeq/L. As in the Northeast, lakes with negative ANC were generally characterized by organic anion concentrations in class 1 and class 2. In the Upper Peninsula of Michigan (2B) and in Florida (3B), a small percentage of lakes with negative ANC were found in organic anion classes 4, and 4 and 5, respectively.

4. DISCUSSION

Concentrations of ANC, sulfate, and pH in lakes of the eastern U.S. exhibit high variability within and among subregions. This regional heterogeneity in the eastern U.S. underscores the difficulty in drawing direct inferences between the occurrence of acidic lakes and atmospheric deposition. For example, the Upper Peninsula of Michigan (2B) and the Upper Great Lakes Area (2D) lie along a similar depositional gradient, and yet 2D contains virtually no lakes with pH $\leq$ 5.0 whereas 9.4% of the lakes in 2B have pH $\leq$ 5.0. In this case, terrigenous properties of the watersheds strongly influence lake chemistry. Certain patterns are apparent, however, that suggest a distribution in lake chemistries among subregions that may be related to acidic deposition. For example, the Northeast, Southern New England, the Poconos/Catskills, and the Adirondacks subregion receive high loadings of sulfate deposition, whereas northern New England and most of Maine receive substantially less (U.S./Canada, 1983). The subregional sulfate concentrations correspond with this pattern in sulfate deposition for the Northeast and Upper Midwest.

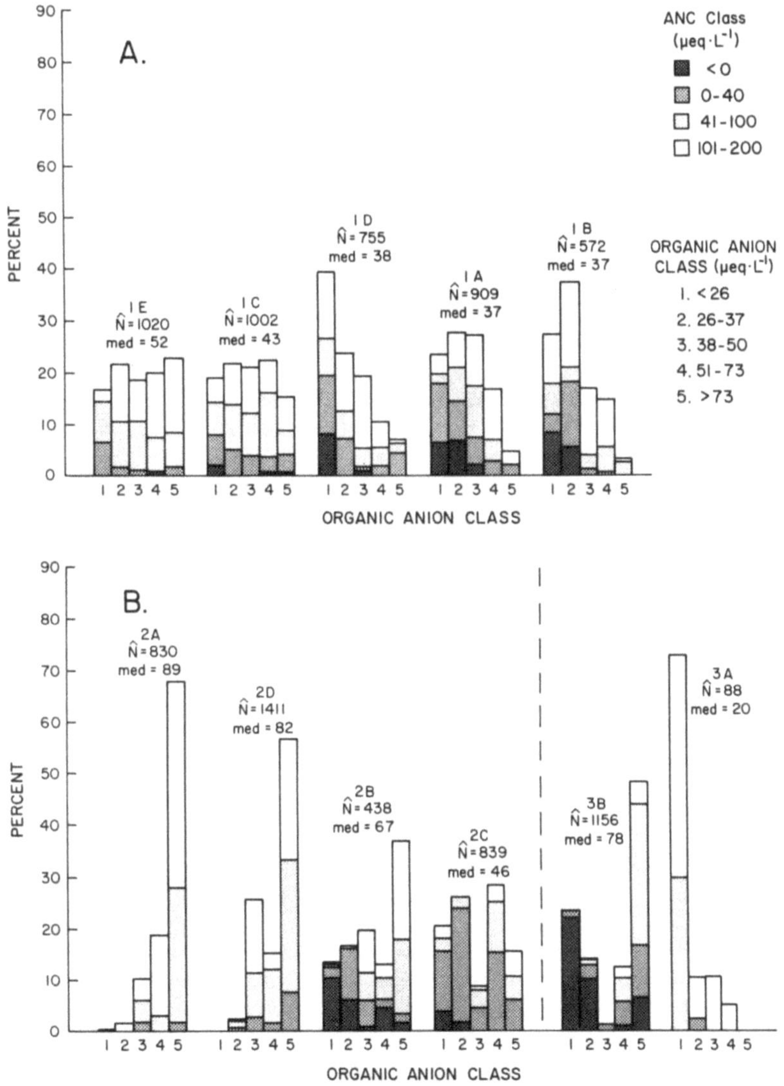

Figure 3. The distributions of lakes with ANC≤200 µeq/L across each of 5 organic anion classes for the Northeast (A) and the Upper Midwest and Southeast (B). Within each organic anion class, the proportions of lakes characterized by one of four ANC classes are given. Subregions are arranged in order of decreasing overall population median estimates of organic anion.

The data in this survey and the companion Canadian compilation (Jeffries et al., 1986) indicate that: (1) in-lake ANC concentrations (or pH) are not related directly to sulfate concentrations; and (2) most acidic lakes are not caused by naturally occurring organic acids. The absence of an obvious relationship between sulfate and ANC within the subregions reflects a variety of non-conservative watershed processes and multiple sulfate sources influencing lake concentrations, as well as differences in sources of ANC. These processes include sulfate adsorption by soils, biological uptake by vegetation, groundwater contributions from sulfide minerals or marine deposits, sea salt contributions, in-lake sulfate reduction, and alkalinity generation. Further stratification of the lakes by relevant ecological criteria may reveal a stronger relationship between ANC and sulfate (e.g., in clearwater lakes within uniform geological terrain). The apparently complex relationship between ANC and sulfate in lakes requires further analysis within individual watersheds and from a regional scale perspective.

Another significant finding is the small number of low pH, negative ANC lakes with appreciable concentrations of organic anions. Lakes having negative ANC and also associated with the highest organic anion class were found only in the Upper Peninsula of Michigan and the subregion containing Florida and Georgia. The only significant number of high organic anion, low pH lakes sampled occurred in Subregion 3B in the Okefenokee Swamp, Georgia, where lakes were among the most acidic sampled; the minimum pH was 3.81. Overall, in Subregion 3B, 20% of the lakes had pH $\leq$ 5.4 and ANC $\leq$ -7.5 µeq/L. This demonstrates the potential for organic acids to serve as naturally acidifying sources in some cases. However, a much higher percentage of lakes in the two lowest organic anion classes had negative ANC than in the two highest organic anion classes. In Northeastern Minnesota, which had the highest median organic anion concentration of all subregions 98.6% of the lakes had pH $>$ 6.0 and no lakes had ANC $\leq$ 0 µeq/L. Although other processes, such as watershed sources of pyrite, can cause a lake to be naturally acidic, the widespread distribution of acidic lakes suggests that a more general mechanism such as atmospheric deposition is likely to be responsible.

The analysis of the relationships between sulfate and ANC, and organic acids and acidic lakes yields similar results from both the U.S. and Canadian surveys. Both countries have regions in which lakes are characterized by high and low acid neutralizing capacity, often coincident with observed patterns of sulfate deposition. However, the lakes in the eastern U.S. generally have considerably higher ANC and pH. This difference can be attributed to the dominant influence of the Precambrian Shield in Canada in contrast with the deeper soils, deeper glacial tills, and smaller aerial extent of igneous bedrock in the eastern U.S. Lakes in the U.S. also exhibit greater regional heterogeneity than those in Canada. For example, bimodal distributions in organic acid classes were observed in Northcentral Wisconsin, Florida, the Upper Peninsula of Michigan and the Upper Great Lakes Area, but not in the Canadian regions.

5. CONCLUSIONS

The ELS-I provides a consistent data base that permits population estimates to be computed for the U.S. regions containing the majority of low ANC lakes. The initial observations are:

a) Acidic and low ANC lakes occur in subregions receiving a considerable range of SO_4 deposition levels. However, the highest percentages of acidic lakes in the Northeast and Upper Midwest are associated with areas receiving the highest sulfate deposition. This association is consistent with the hypothesis that elevated sulfate deposition has caused chemical alterations and acidification in some areas.

b) The large differences in the percentages of low pH and low ANC lakes among subregions in the eastern U.S. receiving similar levels of sulfate deposition indicate that the patterns of atmospheric deposition alone are insufficient to explain fully the distribution of acidic waters without characterizing the underlying sensitivity of the areas.

c) Florida contains by far the highest percentage of acidic lakes in the eastern U.S., followed by the Adirondacks, Southern New England, and the Upper Peninsula of Michigan. The acidic lakes are generally associated with low organic anion concentrations. An adequate explanation is currently unavailable for the high percentage of acidic lakes in Florida.

d) The high percentages of acidic lakes with low concentrations of organic anions is consistent with the observation in Canada that naturally acidic lakes caused by organic acids comprise a small percentage of the total number of acidic lakes. It will be important to identify and define the conditions causing natural acidification.

e) Based on the data presented in this paper, the percentage of low pH and low ANC lakes in the eastern U.S. is far lower than in Canada.

6. ACKNOWLEDGEMENTS

Lockheed Engineering Management Services Co. and Northrop Services, Inc. contributed to the successful completion of the Survey. We would also like to acknowledge the sampling support of EPA Regions 1 (Boston, MA), 2 (Edison, NJ), 3 (Philadelphia, PA), 4 (Atlanta, GA), 5(Chicago, IL), 9 (San Francisco, CA) 10 (Seattle, WA) of the U.S. EPA.

REFERENCES

Best, M.D., L.W. Creelman, S.K. Drouse and D.J. Chaloud. EPA-600/4-86-011, U.S. Environmental Protection Agency, Las Vegas, Nevada, 1986.

Drouse, S.K., D.C.J. Hillman, L.W. Creelman, J.F. Potter, and S.J. Simon. EPA-600/4-86-008, U.S. Environmental Protection Agency, Las Vegas, Nevada, 1986.

Eilers, J.M., D.J. Blick, Jr., M.D. DeHaan. U.S. Environmental Protection Agency, Corvallis, Oregon. 1986.

Hillman, D.C.J., J. F. Potter and S.J. Simon. EPA-600/4-86-009, U.S. Environmental Protection Agency, Las Vegas, Nevada, 1986.

Jeffries, D.S., Walco, D., Kelso, J.R.M. and Linthurst, R.A.: 1986, Wat. Air Soil Poll., this volume.

Kanciruk, P., J.M. Eilers, R.A. McCord, D.H. Landers, D.F. Brakke and R.A. Linthurst. EPA-600/4-86-007C, U.S. Environmental Protection Agency, Washington, D.C., 1986, 439 pp.

Linthurst, R.A., D.H. Landers, J.M. Eilers, D.F. Brakke, W.S. Overton, E.P. Meier and R.E. Crowe. EPA-600/4-86-007A, U.S. Environmental Protection Agency, Washington, D.C., 1986, 275 pp.

Morris, F.A., D.V. Peck, M.B. Bonoff and K.J. Cabble. EPA-600/4-86-010, U.S. Environmental Protection Agency, Las Vegas, Nevada, 1986.

Oliver, B.G., Malcolm, R.L., and Thurman, E.M.: 1983, Geochim. Cosmochim. Acta 47, 2031.

Omernik, J.M., Kinney, A.J., Griffith, G.E. and Yeaple, A.L.: 1985, U.S. Environmental Protection Agency, Corvallis, OR (in press).

Omernik, J.M., Griffith, G.E.: 1985, U.S. Environmental Protection Agency, Corvallis, OR. EPA-600lD-85-043, June 1985.

Omernik, J.M., Kinney, A.J.: 1985, U.S. Environmental Protection Agency, Corvallis, OR. EPA-600lD-84-216, June 1985.

Overton, W.S., P. Kanciruk, L. A. Hook, J.M. Eilers, D.H. Landers, D.F. Brakke, D.J. Blick, Jr., R.A. Linthurst, M.D. DeHaan, and J.M. Omernik. EPA-600/4-86-007B, U.S. Environmental Protection Agency, Washington, D.C., 1986, 374 pp.

United States/Canada: 1983, Memorandum of Intent of Transboundary Air Pollution. Impact Assessment, Working Group I.

THE IMPACT OF ACIDIFICATION ON AQUATIC BIOTA IN NORTH AMERICA: A COMPARISON OF FIELD AND LABORATORY RESULTS

G. Mierle
Ontario Ministry of the Environment
Dorset Research Centre
P.O. Box 39
Dorset, Ontario, Canada POA 1E0

K. Clark
Biota Environmental Contractors
P.O. Box 21
Dorset, Ontario, Canada POA 1E0

R. France
Institute of Environment Studies
University of Toronto
Toronto, Ontario, Canada M5L 1A4

ABSTRACT. Acidification of lakes and streams can lead to adverse effects on aquatic biota, and the recognition of widespread atmospheric deposition of strong acids in eastern North America has led to considerable concern over its existing and potential impact on th flora and fauna of sensitive (i.e. low alkalinity) systems. While most of the concern has been directed toward the impact on the fisheries resource, it is well documented that all levels of biota are affected (Almer et al., 1974; Likens et al., 1983; Dillon et al., 1984), and full appreciation of the impact of acidification requires examination of all living components of the aquatic ecosystem. Although extensive work has already been conducted in the Scandinavian countries, there are numerous differences in species between those countries and North America, and a sound appraisal of the threat to North American biota requires comparable work for this continent. Large scale surveys and monitoring programs are under development or are in the early phases of execution by various institutions, but much has already been learned from laboratory and field acidification experiments, and from comparisons of acidic and non-acidic lakes and streams. The purpose of this paper is to compare the results of these studies for several dominant species from several major groups of organisms. We hope to focus attention on the nature and quality of information about the sensitivity of organisms with known or apparent significant ecological roles, identify gaps or weaknesses, and provide insight into the merits and limitations of the various methods of study.

Water, Air, and Soil Pollution **31** (1986) 593–604.

1. ALGAE

This group of primary producers has usually been treated in terms of group properties (e.g. primary production, chlorophyll *a* concentration, total biomass) or in terms of community properties such as species richness, various diversity indices or relative proportions at the level of order. The results of these efforts have been reviewed by others (NRCC, 1981; Dillon *et al*., 1984). Briefly, primary production and biomass are not significantly affected even down to pH 4.2. Species richness declines at low pH, and there is a marked change in community composition as lakes acidify. The size distribution of algae determines their edibility by zooplankton, and shifts towards large algae can have severe impacts on energy transfer efficiency (e.g. Yan and Strus, 1980). This subject, however, has been poorly investigated in the context of acidification.

There is a remarkable dearth of matching survey and experimental data on the effects of acidification on individual species of algae. A number of algal species have been studied in the laboratory (Moss, 1973; Bowen and Ward, 1978; Gensemer and Kilham, 1984), but with the exception of *Asterionella formosa* these species include none of the chrysophytes and diatoms that commonly dominate the phytoplankton and periphyton of circumneutral lakes (Dillon *et al*., 1984; Kling and Holmgren, 1972; Nakamoto *et al*., 1983; Stockner, 1971; Fox *et al*., 1969).

A notable exception to the above comments is the work on *Chrysochromulina breviturrita* Nicholls. Blooms of this alga occur in moderately acidic lakes (pH 5.5 to 6.2) and produce lakewide, obnoxious odors (Nicholls *et al*., 1982). In laboratory studies of its pH tolerance optimum yield was obtained between pH 5.5 and 6.8 (Wehr *et al*., 1985). This species also requires Se, and it has been suggested that pulses of atmospheric deposition of Se (which co-occurs with sulfate deposition) may explain the unpredictable nature of *Chrysochromulina* blooms (Wehr and Brown, 1985).

Although quantitative survey data are sparse, a number of investigators have observed the development of dense growths of green or blue-green filamentous algae in the littoral zone of acidic lakes (Stokes, 1981; Singer, 1984; Jackson, 1985). In large enclosure experiments by Muller (1980) lowering the pH from 6.5 to 6.0 produced a dramatic increase in *Mougeotia* spp. Similarly, acidification of Lake 223 led to blooms of *Mougeotia* (Schindler *et al*., 1985). In one laboratory study optimum growth of *Mougeotia* was observed at pH 4.5 (Jackson, 1985), but others have found significant differences between strains (Turner and Stokes, 1985).

Equally problematic is the lack of species specific survey data. Most of the data that are available (Conroy *et al*., 1975; Kwiatkowski and Roff, 1976; Yan, 1979; Crisman *et al*., 1980; Conway and Hendrey, 1982; Blouin *et al*., 1984) have been summarized at the level of class which obviously precludes comparison to species specific laboratory studies. Because different algal species can play very different ecological roles, it is unfortunate that more experimental and field work has not been performed to establish the tolerance of the dominant algal species of soft water oligotrophic lakes.

2. PLANKTONIC CRUSTACEANS

This is the only group of aquatic organisms other than fish for which extensive survey data is available, both in the context of general zoogeographic research (Carter et al., 1980) and lake acidification (Sprules, 1975; Keller, 1981; Joubert and Tousignant, 1981; Confer et al., 1983). All of these surveys have revealed a decline in species richness and diversity as pH declines, and in very acid lakes (pH <5.0) one or two species (*Diaptomus minutus* and *Bosmina longirostris*) often account for nearly the whole community.

Experimental work with planktonic crustaceans has been rather modest, but most of the results agree with the patterns in the survey data. During acidification of Lake 223 *Daphnia galeata mendotae* declined from its dominant status, disappeared at about pH 5.5 and was replaced by *D. catawba* (Schindler et al., 1985), a pattern consistent with the survey data. The results of laboratory experiments with *D. pulex* (with which Dodson (1981) has lumped *D. catawba*) are also consistent with the occurrence of *D. catawba* at moderately low pH. In 4- and 21-day exposures survivorship was not reduced at pH 4.3 and 5.0, respectively (Walton et al., 1982). However, reproduction was delayed and the intrinsic rate of natural increase was reduced at pH 5.0 and below. These effects on reproduction suggest that tolerance based on survivorship alone might be a poor predictor of population success in a natural competitive environment. In another study acute toxicity tests indicated that *D. catawba* would be significantly impaired at pH 5.0 and below (Havas and Likens, 1985), but high mortality in the control treatments of these experiments suggests an additional unidentified stress, probably starvation.

Acidification of Lake 223 to 5.0 led to an increase in *Holopedium gibberum* (Schindler et al., 1985) which agrees well with the survey results. In three-day exposures mortality of this species was not significantly different than controls (pH 6.5) down to pH 4.0 (Havas and Likens, 1985), which suggests that in the short term at least it is fairly tolerant of low pH.

Given the ease with which many zooplankton species are cultured and handled, especially the cladocerans, it is unfortunate that greater advantage of the experimental possibilities has not been taken with these animals. Chronic exposures, transfer experiments, and enclosure studies are all feasible. The last technique may be especially useful for unravelling biotic and abiotic interactions. For example, in acidified enclosures the abundance and mean body size of *Bosmina longirostris* increased relative to the controls, a result attributed to the decline of its predator, *Mesocyclops edax* (Havens and DeCosta, 1985). These types of experiments coupled with manipulations of the initial densities of competitors and predators could put the analysis and interpretation of field data on much firmer ground.

3. BENTHIC CRUSTACEANS

Until relatively recently benthic crustaceans, in particular mysids, amphipods and decapods, had received little attention, but it is becoming apparent that many members of this group are very acid sensitive. The extinction of *Mysis relicta* at about pH 5.8 during acidification of Lake 223 (Nero and Schindler, 1983) was the first experimental evidence of the sensitivity of this organism, although an early survey by Dadswell in North America and the success of stocking attempts in Swedish lakes of various pHs also suggests a minumum pH of 5.7 to 5.9 for survival (cited by Nero and Schindler, 1983).

More recently *Hyalella azteca*, probably the most common and abundant amphipod in North America, has been shown to be quite sensitive. In a survey of 79 lakes in Ontario *H. azteca* was found in 69 of 70 lakes with positive alkalinities and in none of 9 lakes which are presently or were historically acidified, i.e. with negative alkalinities (Stephenson and Mackie, 1986). Apparent reproductive failure was observed in one lake with very low alkalinity, delayed reproduction in another, but normal populations and recruitment in four other lakes with moderate alkalinities. Four- and ten-day LC50's were estimated to be pH 4.4 and 4.5, respectively, although 8-day LC50's estimated from the extensive bioassay work by France and Stokes (1986) were slightly higher (4.8 to 5.0). The latter authors also found that small animals and postmolt stages were the most sensitive forms, and that pH resistance was very dependent on the history of exposure. Since spring pH depression in the littoral zones of the study lakes were well below these short term tolerance values, acute mortality along with reproductive impairment may be affecting this species. In a stream acidification experiment *H. azteca* was nearly eliminated at pH 5.0 over a 1-mo period (Zischke *et al.*, 1983), but was not affected at pH 6.0.

Three common species of crayfish, *Orconectes virilis*, *O. propinquus* and *O. rusticus* are extremely sensitive to low pH. A variety of stresses on *O. virilis* developed below pH 5.7 during acidification of Lake 223: Ca uptake was inhibited, hardening of the exoskeleton was delayed (Malley, 1980), parasitism by *Thelohenia* greatly increased (France and Graham, 1985), egg mortality increased either from direct H^+ stress, improper adhesion or fungal infestation (France, 1985, 1986) and the young of the year failed to recruit in three successive years (Davies, 1986). In stream transplant experiments, softwater of pH 5.6 to 6.1 was toxic to the young of *O. propinquus* and *O. rusticus*. Similarly, in laboratory constituted softwater at pH 5.4 to 5.7 all but one brood died within 8 days, whereas no mortality was observed at pH 6.0 to 6.3. Survey and experimental results indicated that two other species, *Cambarus robustus* and *C. bartoni*, are tolerant to low pH (Berrill et al., 1985). Thus, 3 of the 5 common species of crayfish of softwaters appear to be very sensitive to acidification.

4. MOLLUSCS

Although huge surveys in Norway of snails and clams have been conducted (Okland, 1980a,b), work in North America has been more modest. Mackie and Flippance (1983) surveyed 53 lakes in Ontario over a wide range of hardness and alkalinity, and many species appeared to be restricted to high alkalinity, circumneutral lakes. In a more detailed survey of 6 low alkalinity lakes several species were found at lower alkalinities than previously recorded, and certain *Pisidium* species had highest densities in the lowest alkalinity lakes (Rooke and Mackie, 1984a). However, the common and usually abundant snail, *Amnicola limnosa*, displayed a significant decrease in density with decreasing alkalinity. Cohort growth patterns revealed decreased growth with decreasing alkalinity, and apparent reproductive failure was observed in Heney Lake, one of the lakes with lowest alkalinity (Rooke and Mackie, 1984b). Reduced growth was also observed with young *Amnicola* that were transplanted to the lowest alkalinity lakes, although no mortality was observed. Acute toxicity tests indicated that young and adult animals were quite tolerant to low pH; four-day LC50s were well below pH 4.0, and mortality from spring pH depressions appeared unlikely (Servos, 1983). Embryonic development, however, was delayed at pH 5.5 and total developmental failure occurred at pH ≤5.0 (Servos *et al*., 1985). This last result may explain the reproductive failure that was noted by Rooke and Mackie (1984b) in Heney Lake. Recent sampling suggests that *A. limnosa* is now extinct in Heney Lake. A qualitative survey of 32 lakes by Servos (1983) suggests that other populations of *Amnicola* may also have been lost. Its frequency of occurrence was 100% in lakes with early summer pHs at or above pH 6.3, but it dropped to 75% in the 16 lakes between 5.3 and 6.2.

Three other snails occur fairly frequently in low alkalinity, circumneutral lakes and may be susceptible to acidification. *Campeloma decisum*, *Gyraulus parvus* and *Helisoma anceps* were found in about 65% of the lakes above pH 6.3 in Servos' survey, but their respective frequency of occurrences were 50, 25 and 19% in the lakes between pH 5.5 and 6.2. Laboratory and field transplant studies on these species would be quite useful.

Pisidiid clams appear to be less susceptible than snails to acidification (Rooke and Mackie, 1984b; Servos *et al*., 1985), probably because they spend most of their time burrowed in the sediments where the pH level of the interstitial water may be substantially higher than the lake, and because embryo development occurs within the shell of the mother. More survey, laboratory and transplant results would be helpful to better define the tolerance of these and other (e.g. *Elliptio complanata*) clams to low pH.

5. INSECTS

A major portion of the animal biomass of the benthos is comprised of insect larvae and nymphs, and they form the majority of the diet of several species of fish in soft water, oligotrophic lakes (Post and Cucin, 1984). Despite their obvious importance in lake ecosystems they have been relatively poorly studied. There are few quantitative survey data, and complimentary survey and experimental data at the species level are almost nonexistent.

Studies in Scandinavia indicated that *Chaoborus* spp. were tolerant of low pH (Nilssen, 1974) and in North America the tolerance of *C. punctipennis* (Say) and *C. flavicans* (Meigen) has been confirmed by the only survey which has considered the relationship of the occurrence of particular species to the pH of lakes. Yan *et al*. (1985) sampled 33 lakes ranging from pH 4.55 to 7.42, and marked differences in *Chaoborus* densities were observed. There were significant correlations of density with nutrient levels and color, but not with pH. Laboratory studies also indicate that *C. punctipennis* is very tolerant of low pH; no increase in mortality was observed at pH 3.5 relative to controls (Havas and Likens, 1985).

The tolerance of more than 30 species of insect larvae and nymphs have been determined in laboratory studies (summarized by Singer, 1984), and these data demonstrate three important phenomena. Firstly, 4-day exposures can underestimate long term (30- to 90-day) tolerance by more than two pH units. Secondly, emergence appears to be the critical life stage, although a comparison of the sensitivity of early instars and emergence has not been performed. Thirdly, there is great variation in sensitivity between species. For example, at one extreme the median tolerance limits of *Brachycentrus americanus* (a caddisfly) for 4-day survival, 30-day survival, and emergence were pH 1.5, 2.5 and 4.0, respectively, whereas for *Ephemerella subvaria* (a mayfly) the corresponding values were 4.6, 5.4 and 5.9. Unfortunately, survey data for these species and others that have been tested in the laboratory are not available.

A small scale stream survey of insects has been performed, and these data suggest that a number of genera of mayflies and stoneflies are sensitive to low pH (Mackay and Kersey, 1985). Data from stream acidification experiments (Hall *et al*., 1980) suggest that a number of species respond to low pH with increased drift, and treated areas were found to have significantly lower emergence. However, given the lack of species specific data and the corresponding large scale survey data the significance of this information is unclear. Much work remains to be done in this area.

6. AMPHIBIANS

The adult stage of most amphibians is primarily terrestrial, and consequently direct impact will probably occur only on the strictly aquatic phases, eggs and larvae. The sensitivity of several species to low pH has been studied, and generally they appear to be relatively tolerant. For example, in field transplants by Dale *et al*. (1985) with *Rana sylvatica* eggs mortality was not significantly different between pH 5.7 and 4.1, although survivorship was only 7% at pH 4.1. The lack of statistical significance in this instance may reflect the high variance within replicates at the control site rather than a true lack of impact. Clark and Hall (1985) also performed field transplants and conducted *in situ* exposures, and they observed no effect down to pH 4.3. In laboratory studies *R. sylvatica* was not affected at pH 4.8 but hatching success was reduced slightly at pH 4.1 (Clark and LaZerte, 1985). Pierce *et al*. (1984) observed a large number of developmental deformities at pH 4.07, but no effect on hatching success. These results agree well with field observations of successful hatching between pH 4.3 to 7.8 (Dale *et al*., 1985; Clark, 1986).

Bufo americanus is somewhat more sensitive to H^+ ion than *R.*

sylvatica. In field transplants Dale *et al*. (1985) observed 100% mortality of eggs at pH 4.1, and Clark and Hall (1985) detected increased mortality at pH ≤4.6 although some hatching (11 to 75%) was observed even at pH 4.2. In laboratory experiments reduced hatching was observed at pH ≤4.3 (Clark and LaZerte, 1985).

Ambystoma maculatum is the most sensitive amphibian that has been tested, and laboratory and field experiments suggest that its eggs are stressed at pH <5.0 (Pough, 1976; Pough and Wilson, 1977). Pough (1976) observed >65% mortality of embryos in ponds with pH <5.0. In a census of 20 ponds hatching success was positively correlated with pond pH (Clark, 1986). In field transplants slightly reduced hatching success occurred in ponds with pH <5.0 (Clark and Hall, 1985; Clark, 1986a), and Dale *et al*. (1985) observed complete mortality of embryos at pH 4.1. Similar results have been obtained in a laboratory experiment where hatching success at pH 4.0 was 0 to 45% compared to 62 to 94% at pH 7.0 (Pough and Wilson, 1977).

Although amphibians appear to be relatively tolerant to direct H^+ stress, two additional factors must be considered for a realistic assessment of the impact of acid deposition. Firstly, the extent of acidification of the preferred habitat of many amphibians (small ponds, temporary pools, etc.) may be much greater than lakes because of their extremely high flushing rates. Unfortunately, we have little reliable chemistry survey data for small ponds. In two large surveys (Dale *et al*., 1985; Glooshenko *et al*., 1985) of small water bodies, ponds were located by listening for calling males. This procedure obviously biases the data set towards suitable habitat. Also, the region studied by Glooshenko *et al*. (1985) was within the influence of smelter emissions, and metal contamination can obscure the effects of H^+. Secondly, there is little abundance data, and that which is available suggests that some populations are reduced. Clark (1986a,) found significant positive correlations of the densities of adult *Hyla crucifer*, *Rana catesbiana*, small *R. clamitans* and *R. sylvatica* egg masses with pond alkalinity. Because the pH of the ponds was within the tolerance ranges of these species, the results were attributed to indirect effects such as changes in food resources, habitat structure and predator-prey relationships. These data clearly indicate that as yet we cannot accurately appraise the impact of acid deposition on amphibians. Hopefully, future work will quantify indirect effects and the chemical status of potential breeding sites so that a realistic assessment can be made.

7. GENERAL DISCUSSION

To optimize the choice of an acceptable level of deposition of strongly acidic substances we must have an accurate understanding of the responses of aquatic biota to acidification of their habitat. No one research method will yield the required accuracy, for each has its own limitations. For example, the basic premise of the survey approach is that occurrence indicates tolerance and absence indicates intolerance to low pH. There are three major weaknesses in this reasoning, however. Firstly, there are often multiple correlations (e.g. Clark and Hall, 1985) which obscure cause-effect relationships. Secondly, presence/absence data does not take into account changes in abundance or productivity which may be affected long before a species is eliminated. Thirdly, complete adjustment of the biotic community to acid stress probably requires a long period of time, particularly where sublethal toxic effects are involved. For example, delayed reproduction has been reported for *Daphnia pulex* and *Hyalella azteca* (Walton *et al*., 1982; Stephenson and Mackie, 1985), and this in itself may lower the replacement rate below one. The change may be small, e.g. 10 to 20%, but nonetheless may lead to the extinction of the population over several tens of generations. For small animals (e.g. zooplankton) a few years may be required, but for larger animals (e.g. crayfish, amphibians) the process may require several decades. Since most acidification in North America has occurred within the last 3 or 4 decades, it is likely that many low alkalinity lakes contain species which will fail in the years to come even if the status quo is maintained. Clearly, surveys based on presence/absence data will tend to underestimate the effect of acidification.

Laboratory experiments can be used to establish cause-effect relationships, and to define with reasonable precision the dose-response relationship. However, the conditions in such experiments are often highly unnatural. Physical factors are usually maintained very constant, and other stresses that may arise in the field (e.g. parasitism) are eliminated or minimized in the lab. Consequently, laboratory results will tend to underestimate the sensitivity of organisms to low pH.

Field experiments may give information about tolerance under more realistic conditions, and in addition they may yield insights into various indirect effects, in particular biotic interactions. However, the multiplicity of changes that accompany reduced pH in the field often makes data interpretation difficult and firm conclusions may be elusive.

Although each strategy has its limitations, when all the results converge (or anomalies explained) we may feel confident that the tolerance of an organism has been defined, and that those results are suitable for setting standards. This goal is near for several species, e.g. *Amnicola limnosa*, *Hyallela azteca* and several species of crayfish.

There are several generalizations which emerge from the lab and field studies which merit comment. Firstly, different life stages have different sensitivities, and usually the smallest forms, either the eggs or very young juveniles, are most sensitive (e.g. crayfish and amphipods). Secondly, acute toxicity tests (4-day) usually underestimate the long term tolerance of an organism by 1 to 1.5 pH units, and extrapolation of such data to the field will normally not be justified. However, spring pH depressions of 1 to 1.5 pH units are not unusual, and application of the results of acute toxicity tests may be warranted if they are based on the life stage that occurs during that time. Some littoral species (e.g. *Hyalella azteca*) may fall into this category. Thirdly there are marked differences in tolerance between species even within closely related taxonomic groups. These generalizations clearly indicate that tolerance must be determined on a species specific basis, and that size and life history phenomena must be carefully considered in its determination.

Several groups of organisms and areas of investigation have not been considered in any detail but deserve mention. Two recent papers on rotifers indicate that there are marked alterations in community structure (Siegfried et al., 1984) and community biomass (Yan and Geiling, 1985). These effects may in part be the result of biotic interactions; laboratory toxicity tests with these organisms would be most interesting. A detailed understanding of the physiological basis hydrogen intoxicity (e.g. disruption of ion regulation) would be valuable, for it may provide tools for predicting how other ions (e.g. Na^+, Ca^{2+}) will affect pH sensitivity, and possibly a means of measuring sublethal effects.

The direct effects of H^+ toxicity on individual species, have been stressed in this paper because these impacts are most easily quantified and the results can be generalized to all lakes. Severe indirect impacts may also occur, however, and they should not be overlooked. Although a full discussion of this topic is not possible here, two examples will illustrate the point. Yan and Strus (1980) estimated the energy transfer efficiency from phytoplankton to zooplankton in an acidified, metal-contaminated lake and concluded that it was much lower than in circumneutral lakes, perhaps by a factor of 10. Clearly, higher trophic levels will experience diminished food supply and reduced production even in the absence of direct H^+ stress.

This phenomenon may already be manifesting itself among aquatic dependent wildlife. Many species of ducks, in particular the ducklings, are obligate insectivors, and reproductive failure of ducks in acid stressed areas of Ontario has been linked to the availability of prey (McNicoll et al., in prep.). In Quebec normal growth of ducklings did not occur in acid lakes which contained competing fish species (Des Granges, 1985). Although these impacts are difficult to quantify and may vary from site to site, they must be considered carefully in evaluating the impacts of acidification.

In summary, surveys, laboratory experiments and field studies indicate that several species of crustaceans and probably several molluscs are extremely sensitive to low pH and may be eliminated before the alkalinity of a lake is exhausted. A number of insect species may also fall in this group but much more work at the species level is required before the extent of susceptibility in this group is revealed. Given the widespread nature of acidification we should endeavour to fill the gaps in our understanding of biotic responses as quickly as possible.

ACKNOWLEDGMENTS

The authors would like to thank N. Yan, R. Hall, K. Fisher and V. Glooshenko for providing assistance, advice and constructive criticisms during the preparation of this manuscript.

REFERENCES

Almer, B., Dickson, W., Ekstrom, C. and Hornstrom, E.: 1974, Ambio 3, 30.
Berrill, M., Hollett, L., Margosian, A., and Hudson, J.: 1985, Can. J. Zool. **63**, 2586.
Blouin, A.C., Lane, P.A., Collins, T.M. and Kerekes, J.J.: 1984, Int. Revue ges Hydrobiol. **69**, 819.
Bowen, M.S., and Ward, H.B.: Micro Lett. **6**: 77.
Carter, J.C.H., Dadswell, M.J., Roff, J.C. and Sprules, W.G.: 1980, Can. J. Zool. **58**, 1355.
Clark, K.L.: 1986a, Can. Field Natur., in press.
Clark, K.L.: 1986b, 'Distributions of anuran populations in central Ontario to habitat acidity. Int. Symp. on Acidic Precipitation, Sept 15-20, 1985.
Clark, K.L. and Hall, R.J.: 1985, Can. J. Zool. **63**, 116.
Clark, K.L. and LaZerte, B.D.: 1985, Can. J. Fish. Aquat. Sci. **42**, in press.
Confer, J.L., Kaaret, T. and Likens, G.E.: 1983, Can. J. Fish. Aquat. Sci. **40**, 36.
Conroy, N., Hawley, K., Keller, W. and LaFrance, C.: 1975, 'Influences of the atmosphere on lakes in the Sudbury area, in: Proc. First Specialty Symp. on Atmospheric Contribution to the Chemistry of Lake Waters. Internat. Assoc. Great Lakes Res., 28 Sept. - 1 Oct. 1975, 146.
Conway, H.L. and Hendrey, G.R.: 1982, 'Ecological effects of acid precipitation on primary producers', in: Acid Precipitation, Effects on Ecological Systems (F.M. D'Itr., ed.) Ann Arbor Science, Ann Arbor, 506 pp.
Crisman, T.L., Schulze, R.L., Brezonik, P.L., and Bloom, S.A.: 1980, 'Acid precipitation: the biotic responses in Florida lakes', in: Ecological Impact Acid Precipitation (Drablos, D. and Tollan, A., eds.) SNFS Project, Oslo.
Dale, J.M., Freedman, B. and Kerekes, J.: 1985, Can. J. Zool. **63**, 97.
Davies, I.J.: 1986, Can. J. Fish. Aquat. Sci. **43**, in press.
DesGranges, J.L.: 1985, 'Why do some ducks avoid acid lakes while others don't?', Int. Symp. on Acidic Precipitation, Sept 15-20, 1985. p. 146 (Abstract)
Dillon, P.J., Yan, N.D., and Harvey, H.H.: 1984, CRC Crit. Rev. Environ. Control **13**, 167.
Dodson, S.I.: 1981, 'Morphological variation of Daphnia pulex Leydig (Crustacea: Cladocera) and related species from North America'.
Fox, J.L., Odlaug, T.O. and Olson, T.A.: 1969, The Ecology of Periphyton in Western Lake Superior, Wat. Resources Res. Center Bull. 14, Univ. of Minn., 127 pp.
France, R.L.: 1985, Bull. Environ. Contam. Toxicol., in press.
France, R.L.: 1986, Can. J. Fish. Aquat. Sci., in press.
France, R.L. and Graham, L.: 1985, Water, Air, Soil Pollut. **26**, 129.
France, R.L. and Stokes, P.M.: 1986, Can. J. Fish. Aquat. Sci., in press.

Gensemer, R.W. and Kilham, S.S.: 1984, Can. J. Fish. Aquat. Sci. 41, 1240.
Glooshenko, V., Weller, W. and Eastwood.: 1985, 'Amphibian abundance and distribution with respect to environmental contaminants and pH stress adjacent to Sudbury, Ontario', Technical Report, Wildlife Branch, Ont. Min. of Nat. Res.
Hall, R.J., Likens, G.E., Fiance, S.B., and Hendrey, G.R., 1980; Ecol. 61, 976.
Havas, M. and Likens, G.E.: 1985, Can. J. Zool. 63, 1114.
Havens, K. and DeCosta, J.: 1985, Hydrobiol. 122, 153.
Jackson, M.: 1985, 'Filamentous algae in Ontario softwater lakes', Int. Symp. on Acidic Precipitation, Sept 15-20, 1985. p. 117-118 (Abstract)
Joubert, G. and Tousignant, L.: 1981, The Effect of pH on the Distribution of Crustacean Zooplankton in 158 Quebec Lakes, Manuscript rep., Ministere de l'Environement du Quebec. 24 pp.
Keller, W.: 1981, Planktonic Crustacea in Lakes of the Greater Sudbury Area, Ontario Ministry of the Environment. Tech. Rep. 33 pp. plus appendices.
Kling, H.J., and Holmgren, S.K.: 1972, 'Species composition and seasonal distribution of phytoplankton in the Experimental lakes Area, Northwestern Ontario', Fish. Res. Bd. Canada, Tech. Rep. 337, 56 pp.
Kwiatkowski, R.E. and Roff, J.C.: 1976, Can. J. Bot. 54, 2546.
MacKay, R.J. and Kersey, K.E.: 1985, Hydrobiol. 122, 3.
Mackie, G.L. and Flippance, L.A.: 1983, Freshwat. Invertebr. Biol. 2, 48.
Malley, D.F.: 1980, Can. J. Fish. Aquat. Sci. 37, 237.
Moss, B.: 1973, J. Ecol. 61, 157.
Muller, P.: 1980, Can. J. Fish. Aquat. Sci. 37, 355.
Nakamoto, L., Heintsch, L. and Nicholls, K.: 1983, 'Phytoplankton of lakes in the Muskoka-Haliburton area', Ontario Ministry of the Environ. Data Report DR 83/8, 44 pp.
Nero, R.W. and Schindler, D.W.: 1983, Can. J. Fish. Aquat. Sci. 40, 1905.
Nicholls, K.H., Beaver, J.L. and Estabrook, R.H.: 1982, Hydrobiol. 96, 91.
Nilssen, J.P.: 1974, Norsk ent. Tidsskv. 21, 37.
NRCC: 1981, Acidification in the Canadian Aquatic Environment: Scientific Criteria For Assessing the Effects of Acidic Deposition on Aquatic Ecosystems', Publications No. 18475 of the National Research Council of Canada, Ottawa, 369 pp.
Okland, J.: 1980a, 'Environment and snails (Gastropoda): Studies of 1,000 lakes in Norway', in: Ecological Impacts of Acid Precipitation (Drablos, D. and Tollan, A., eds.), SNSF Project, Oslo.
Okland, K.A.: 1980b, 'Mussels and crustaceans: studies of 1,000 lakes in Norway', in: Ecological Impacts of Acid Precipitation (Drablos, D. and Tollan, A., eds.), SNSF Project, Oslo.
Pierce, B.A., Hoskins, J.B. and Epstein, E.: 1984, J. Herpet. 18, 159.

Post, J.R. and Cucin, D.: 1984, Can. J. Fish. Aquat. Sci. 41, 1496.
Pough, F.H.: 1976, Sci. 192, 68.
Pough, F.H. and Wilson, R.E.: 1977, Air and Soil Poll. 7, 307.
Roff, J.C., Sprules, W.G., Carter, J.C.H. and Dadswell, M.J.: 1981, 'The structure of crustacean zooplankton communities in glaciated eastern North America'.
Rooke, J.B. and Mackie, G.L.: 1984a, Can. J. Fish. Aquat. Sci. 41, 777.
Rooke, J.B. and Mackie, G.L.: 1984b, Can. J. Zool. 62, 1474.
Schindler, D.W., Mills, K.H., Malley, D.F., Findlay, D.L., Shearer, J.A., Davies, I.J., Turner, M.A., Lindsey, G.A., Cruikshank, D.R.: 1985, Sci. 228, 1395.
Servos, M.R.: 1983, The Effect of Short and Long Termed Acidification on Selected Molluscs of South-Central Ontario. M.Sc. thesis, University of Guelph, Guelph, Ont.
Servos, M.R., Rooke, J.B. and Mackie, G.L.: 1985, Can. J. Zool. 63, 511.
Siegfried, C.A., Sutherland, J.W., Quinn, S.O. and Bloomfield, J.A.: 1984, Verh. Interat. Verein. Limnol. 22, 549.
Singer, R.: 1984, 'Benthic organisms', In The Acidic Deposition Phenomenon and Its Effects, Vol. II Effects Sciences (Altshuller, A.P. and Linthurst, R.A., eds.) U.S. Envir. Prot. Agency Report EPA-600/8-83-016BF.
Sprules, G.W.: 1975, J. Fish. Res. Board Can. 32, 389.
Stephenson, M. and Mackie, G.L.: 1986, Can. J. Fish. Aquatic Sci., in press.
Stockner, J.G.: 1971, J. Fish. Res. Bd. Can. 28, 265.
Stokes, P.M.: 1981, 'Benthic algal communities in acid lakes'. in Effects of Acid Precipitation on Benthos (Singer, R., ed.). North America Benthological Society, Hamilton, N.Y.
Turner, P.A.E. and Stokes, P.M.: 1985, 'The growth of filamentous algae (Mougeotia spp.) isolated from acidic lakes grown over a pH gradient from 4.5 to 6.9', Int. Symp. on Acidic Precipitation, Sept. 15-20, 1985. p. 120 (Abstract)
Walton, W.E., Compton, S.M., Allan, J.D., Daniels, R.E.: 1982, Can. J. Zool. 60, 573.
Wehr, J.D., and Brown, L.M.: 1985, Can. J. Fish. Aquat. Sci. 42, in press.
Wehr, J.D., Brown, L.M. and O'Grady, K.: 1985, Can. J. Bot. in press.
Yan, N.D.: 1979, Wat., Air, Soil Pollut. 11, 43.
Yan, N.D., and Geiling, W.: 1985, Hydrobiol. 120, 199.
Yan, N.D., Nero, R.W., Keller, W. and Lasenby, D.C.: 1985, Holarctic Ecol. 8, 93.
Yan, N.D. and Strus, R.: 1980, Can. J. Fish. Aquat. Sci. 37, 2282.
Zischke, J.A., Arthur, J.W., Nordie, K.J., Hermanutz, R.O., Standen, D.A., and Henry, T.: 1983, Water Res. 17, 47.

EVIDENCE OF FISH POPULATION RESPONSES TO ACIDIFICATION IN THE EASTERN UNITED STATES

Terry A. Haines
U.S. Fish and Wildlife Service
Columbia National Fisheries Research Laboratory
Field Research Station
Department of Zoology, University of Maine
Orono, ME 04469 USA

Joan P. Baker
Kilkelly Environmental Associates
P.O. Box 31265
Raleigh, NC 27622 USA

ABSTRACT. The hypothesis that acidification has reduced or eliminated fish populations in certain areas of the eastern United States was investigated by examining present and historical fishery survey records. The causes of acidification (e.g., atmospheric deposition) were not specifically considered, although instances of obvious alternative explanations (e.g., acid mine drainage, organic acids) were avoided. The number of usable data sets located was small. Trend analyses are severely limited by the lack of high quality historical data. The strongest evidence for fisheries declines associated with acidification is provided by data for the Adirondack Mountain region of New York. In some lakes, fish populations have declined or disappeared; lakes experiencing fishery declines are now acidic. Alternative explanations for changes in fish communities over time were examined. In 49 lakes, some or all fish populations have apparently been lost with no available explanation other than acidification. Extrapolation of these data to the entire Adirondack Mountain region suggests that perhaps 200 to 400 lakes may have lost fish populations as a result of acidification. Streams in Pennsylvania and Massachusetts also had documented declines in fish populations that were associated with acidity; however, the data are fewer and less complete than those for New York. Acidification effects on fishery resources in other regions of the eastern United States are apparently minimal. The extent of the damage to date appears small relative to the total resource.

Water, Air, and Soil Pollution **31** (1986) 605–629.

1. INTRODUCTION

The most widely reported consequence of acidic deposition is the reduction or elimination of fish populations in acidified waters (Overrein et al., 1980; Haines, 1981; Altshuller and Linthurst, 1984; Dillon et al., 1984). That acidification (below pH 5 to 6, depending on fish species) adversely affects fish and fish populations has been clearly demonstrated in laboratory and field experiments. In the laboratory, chemical conditions characteristic of acidified waters (low pH, elevated Al concentration, and low Ca concentration) decrease fish survival, growth, and reproductive potential (Baker and Schofield, 1982; Brown, 1983). Fish transferred into acidic lakes and streams, or held in cages during episodic acidification typically experience osmoregulatory stress and high mortality, occasionally 100% (Leivestad and Muniz, 1976; Johnson et al., 1986). The experimental acidification of Lake 223 in western Ontario (Schindler, et al., 1985) confirmed that acidification results in significant population effects - most notably recruitment failure and population extinction.

The issue, therefore, is not cause-and-effect. If acidification occurs, fish populations will be lost. However, difficult questions remain:

What levels of acidity (low pH, elevated Al, or both) result in fish population decline and loss?

How many fish populations have been lost or adversely affected as a result of acidic deposition?

What fraction of the resource has been lost or is likely to be lost in the near future?

To demonstrate conclusively that fish population trends in a water body are related to acidic deposition, certain criteria concerning temporal changes in surface water chemistry and fish populations are required. It must be shown that (1) the water body formerly supported a viable fish population, either self-sustaining or hatchery-maintained, (2) one or more fish species formerly present have been reduced or eliminated, (3) the water body is more acidic now than when fish were present or more abundant, (4) the increased acidity did not result from local factors, and (5) other factors have not adversely affected the fish population. Few available data sets meet these criteria, and most have not been published in the peer-reviewed literature.

Presumably, each region in the U.S. with surface waters potentially acidified by atmospheric deposition may have experienced loss of fisheries resources. Geographic areas of concern are (1) the Adirondack region of New York, (2) northern New England (Vermont, New Hampshire, Maine), (3) parts of Massachusetts and Rhode Island, (4) the Catskills/Pocono/Kittatinny Ridge regions of New York, New Jersey, and Pennsylvania, (5) the central Appalachian mountain region (Southern Blue Ridge Province) of western North Carolina, eastern Tennessee, and northern Georgia, and (6) northern Florida.

The following discussions briefly summarize our current understanding of the status of fish populations in these areas. The potential effects of acidic deposition and acidification on fisheries resources in each of these regions have not been thoroughly evaluated. The causes of acidification (e.g., acidic deposition) are not specifically considered, although instances of obvious alternative explanations (e.g., acidic mine drainage, organic acids) are avoided.

Some of the best documentation for loss of fish populations as a result of acidification involves historic records on water quality and fisheries status in the La Cloche Mountain region of Ontario (Harvey and Lee, 1982), in Nova Scotian salmon rivers (Watt et al., 1983), and in southern Norway (Muniz, 1984). These data are well known, have been widely published, and consequently are not repeated here.

2. CASE STUDIES

2.1. New York

The first reports of acidic deposition-induced acidification of lakes and concomitant declines in North American fish populations concerned the Adirondack Mountain region (Schofield, 1965, 1973, 1976a,b). Several reports (Pfeiffer and Festa, 1980; Colquhoun et al., 1984) summarized recent survey data concerning water chemistry and fish populations in lakes within this region. Schofield (1976b) surveyed 40 lakes that had been surveyed in the 1930s. In the 1930s, three lakes had pH <5.0 and no fish; one lake with pH between 6.0 and 6.5 was also fishless. In 1975, 19 of the 40 lakes had pH <5.0 and were fishless. Two additional lakes with pH between 5.0 and 5.5 also lacked fish. Thus 17 lakes apparently lost fish populations between the 1930s and 1975. Pfeiffer and Festa (1980) estimated that at least 180 lakes in the region had lost fish populations as a result of acidification; however the data supporting their conclusion were not published.

We recently reviewed all available fish population and water chemistry data for the Adirondack Mountain region (Baker and Harvey, 1984). The objectives of the study were to evaluate fish population status, scrutinize explanations for decline or loss of fish populations, and analyze statistical associations between fish population decline or loss and chemical variables indicative of acidity. Because fish sampling procedures have been highly variable over time, quantitative indicators of fish population status, such as catch per unit effort or population presence/absence, could not be used. To evaluate temporal changes in fish population status, discrete indicators of population status were developed. These indicators, known as ordered classifications, are used in cases where different levels of some factor can be recognized but not quantified with a standard unit of measurement on a continuous scale (Kleinbaum and Kupper, 1978). Population status was evaluated for 14 fish species common to the area and considered vulnerable to capture with the gear used. Evaluations were based on results from field surveys, including number, size, age, and sex of fish caught; sampling technique, e.g., gear used, gear size, effort, and depth fished; stocking records; and physical data on lake area, depth, and elevation. Ratings were based on a series of objective criteria to ensure consistency. All evaluations were conducted by the same person

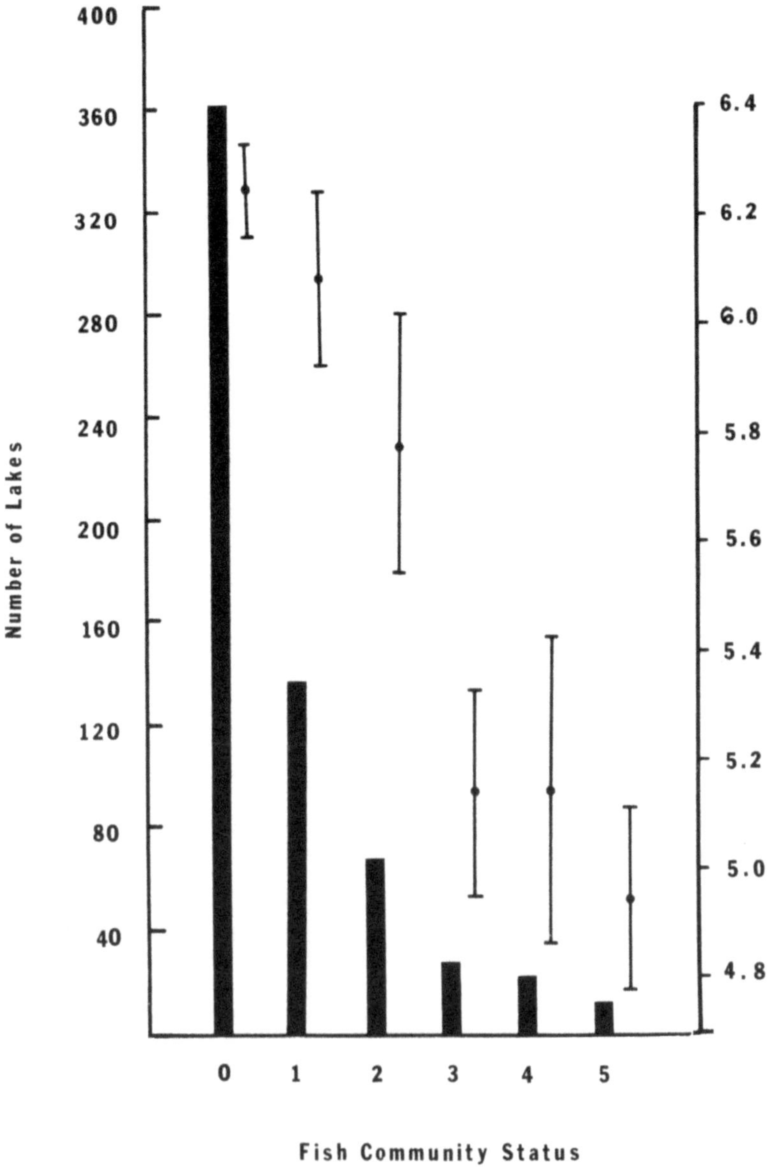

Figure 1. Number (solid bar) and fish community status rating for Adirondack lakes with mean 5-yr pH for the period centered on the year of the last fishery survey (dot). Vertical bar represents 95% confidence interval.

and based solely on fish population information, without knowledge of lake identity or acidity status.

Three ratings were assigned to each species population on a scale of probability from 0 (low) to 9 (high). Ratings were assigned for each of three probabilities: (1) that the fish population was constant or increased over time, (2) that the population declined over time, and (3) that the population was lost. For each population assigned a probability rating above 0 for decline or loss over time, the likelihood that the decline or loss resulted from factors other than acidification was evaluated. These evaluations were based on survey results, stocking records, and comments by field biologists regarding angler pressure, yield, presence of beavers, or other factors that could affect fish populations. Again, the evaluations were made without reference to lake identity or chemical status. The objective was to evaluate alternative explanations for fish population declines, such as reclamation, change in stocking practices, and introduction of new species. The rating scale ranged from 0 to 9, such that ratings 0 to 3 indicated that factors other than acidification were responsible for the change, ratings 4 to 5 were marginal, and ratings 6 to 9 represented increasing probabilities that the change resulted from acidification (i.e., no apparent alternative explanation). In addition, a single overall fish community status rating was assigned to each lake (Table I). After ratings were assigned, the statistical association between the ratings and lake chemical status was examined.

Table I. Fish community status ratings applied to New York lakes.

Rating	Community Status
0	Normally diverse, apparently healthy fish community, only random changes.
1	1 or 2 species declined or disappeared, but not acid-sensitive species. Population decline or loss apparently due to factors other than acidification.
2	1 or 2 species declined or disappeared, <u>may</u> be acid-sensitive species. Cause of population decline or loss uncertain.
3	1 or 2 acid-sensitive species declined or disappeared; no apparent explanation other than perhaps acidification.
4	Most or all species disappeared; no apparent explanation other than perhaps acidification. Any remaining species are acid-tolerant.
5	All species disappeared; no apparent explanation other than perhaps acidification. Confirmed fishless.

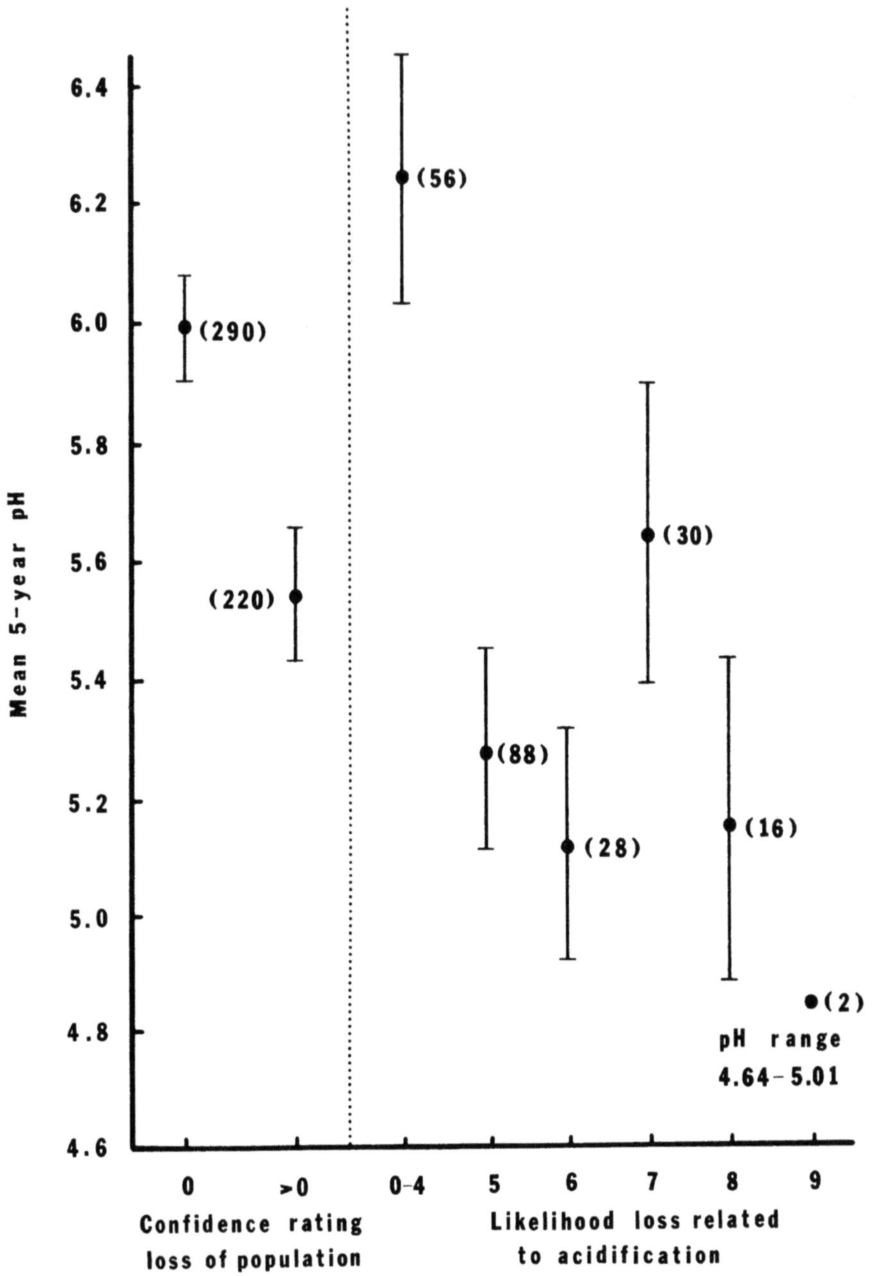

Figure 2. Adirondack brook trout population confidence ratings for the likelihood that the population was lost, and the likelihood that the loss was related to acidification, as compared to mean 5-yr pH. The number of lakes in each rating class is given in parentheses. (0 = low probability; 9 = high probability). Vertical bar represents 95% confidence interval.

Of the 2759 lakes (total area 99,666 ha) in the Adirondack Mountain region, chemical data were located for 1543 (95,304 ha) and fish survey data for 1461 (90,897 ha). Recent (1974-1983, excluding 1979) chemical data including electrometric pH measurements were available for 693 lakes (47,982 ha), and recent fish survey data for 427 (57,495 ha). Large lakes were overrepresented in the data set for lakes having both chemical and fish population data. Of the 693 lakes with recent chemical data, 171 (24.7%) had pH <5. Fish population data were sufficient for 604 lakes to assign fish community summary ratings; 49 (8.1%) were rated 3 to 5 and thus had a high probability that fish populations had actually declined or disappeared with no available explanation other than acidification (Figure 1). Both fish population and recent pH data were available for 396 lakes; 38 (9.6%) of these received summary fish community status ratings of 3 to 5. A series of *t*-tests indicated that lakes rated 3 to 5 had significantly ($p < 0.05$) lower pH than lakes rated 0 to 2 (Figure 1). Many lakes in the Adirondack Mountain region are classified as brook trout lakes (Pfeiffer, 1979), including 220 of the 396 lakes discussed above. Of these, 30 (13.6%) received fish community summary ratings of 3 to 5.

Temporal change in pH was evaluated for lakes with both fish and water chemistry data. Lakes with summary fish community status ratings of 0 to 2, which were not likely to have populations that declined or disappeared, had a mean pH increase of 0.035 unit/decade. Lakes with summary ratings of 3 to 5 had a mean pH decrease of 0.117 unit/decade, which differed significantly from the 0 to 2 group (*t*-test, $p < 0.01$); this suggests that pH had declined in lakes that probably had lost fish populations. However, the rate of change in pH over time was highly correlated with present pH.

Similar analyses were conducted for each of the 14 fish species considered. All species had populations that were rated as likely to have been lost (Table II). The proportion of populations lost was substantial for all species, ranging from 18% for white sucker to 67% for lake chub. Two species, lake chub and chain pickerel, had no populations for which population loss was rated as likely to have resulted from acidification (rating >5). The remaining 12 species received at least some probability ratings above 5, indicating that population losses had occurred with no apparent explanation other than acidification. The proportion of the populations for these 12 species that had disappeared probably because of acidification ranged from <1% for yellow perch to 12 to 14% for brook trout and lake trout. For all species, the proportion of populations lost for reasons other than acidity exceeded that lost apparently because of acidity. For these 12 species, analysis of variance indicated that lake pH was not significantly different among those rated 0 to 4, but was significantly ($p < 0.05$) lower for those rated 6 to 9. Examples for three common species are presented in Figures 2 to 4.

Eight lakes having chemical and fishery data have also been sampled for evaluation of sediment diatoms and trace metals, allowing examination of multiple sources of information on chemical and biological trends (Table III). Sagamore and Panther lakes exhibit little or no evidence of acidification. Fish populations reflect only

Table II. Number and percent of fish species populations (with adequate data for evaluation of trends) rated as likely to have been lost (rating >0), and number and percent with no apparent explanation for population loss other than acidification (rating >5).

Species	Number of Populations	Likely to Have Been Lost		Likely Caused by Acidification	
		Number	Percent	Number	Percent
Brook trout (*Salvelinus fontinalis*)	707	298	42.1	98	13.9
Lake trout (*Salvelinus namaycush*)	111	43	38.7	8	11.8
Rainbow trout (*Salmo gairdneri*)	90	36	40.0	2	2.2
White sucker (*Catostomus commersoni*)	412	76	18.4	10	2.4
Brown bullhead (*Ictalurus nebulosus*)	520	100	19.2	13	2.5
Pumpkinseed (*Lepomis gibbosus*)	351	111	31.6	7	2.0
Golden shiner (*Notemigonus crysoleucas*)	326	113	34.7	12	3.7
Creek chub (*Semotilus atromaculatus*)	254	107	42.1	7	2.8
Lake whitefish (*Coregonus clupeaformis*)	53	25	47.2	1	1.9
Smallmouth bass (*Micropterus dolomieui*)	136	44	32.4	2	1.5
Largemouth bass (*Micropterus salmoides*)	62	13	21.0	1	1.6
Chain pickerel (*Esox niger*)	41	12	29.3	0	-
Yellow perch (*Perca flavescens*)	267	82	30.7	2	0.7
Lake chub (*Couesius plumbeus*)	18	12	66.7	0	-

local management activities and are otherwise unchanged. The chemistry of Woodhull Lake is ambiguous. The pH has fluctuated between 5.0 to 6.5 since 1954. Diatom stratigraphy indicates no acidification, and sediment chemistry demonstrates possible slight acidification. The fish population in Woodhull Lake may be experiencing acid stress as demonstrated by the disappearance of smallmouth bass and brook trout, and a recent decline in lake trout abundance. However, eight fish species are still present, including lake trout, cisco, lake whitefish, and white sucker, and declines in

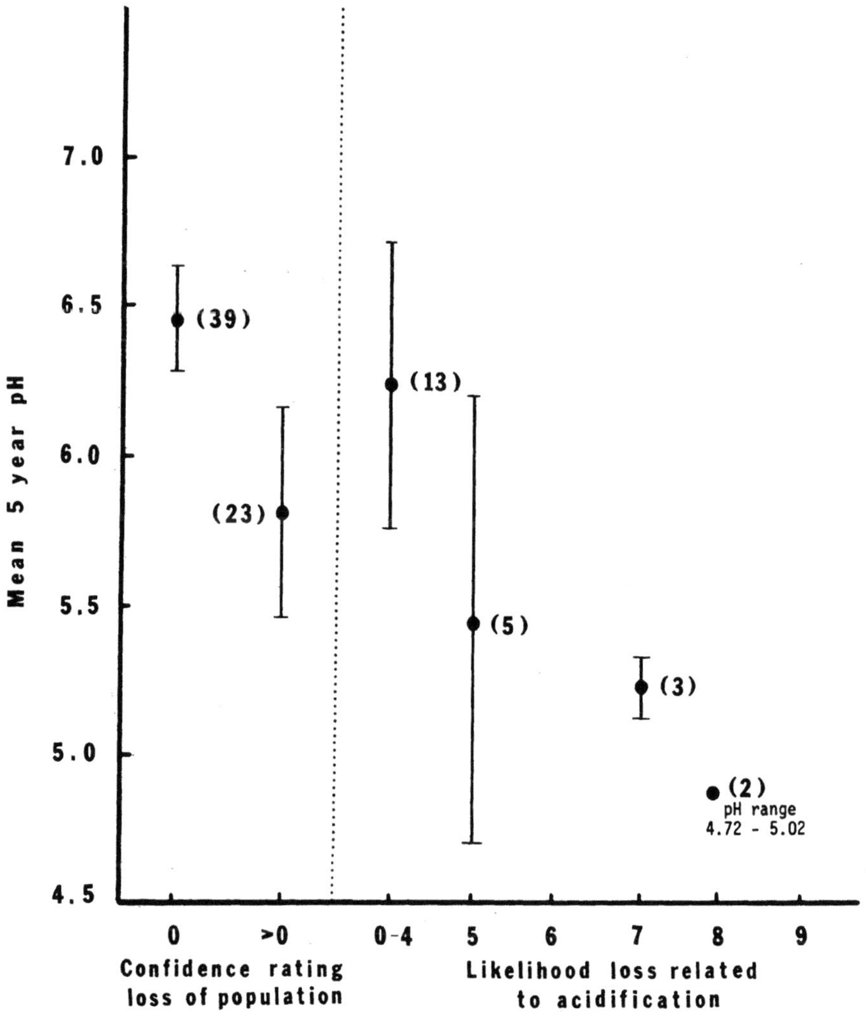

Figure 3. Adirondack lake trout population confidence ratings for the likelihood that the population was lost, and the likelihood that the loss was related to acidification, as compared to mean 5-yr pH. The number of lakes in each rating class is given in parentheses. (0 = low probability; 9 = high probability). Vertical bar represents 95% confidence interval.

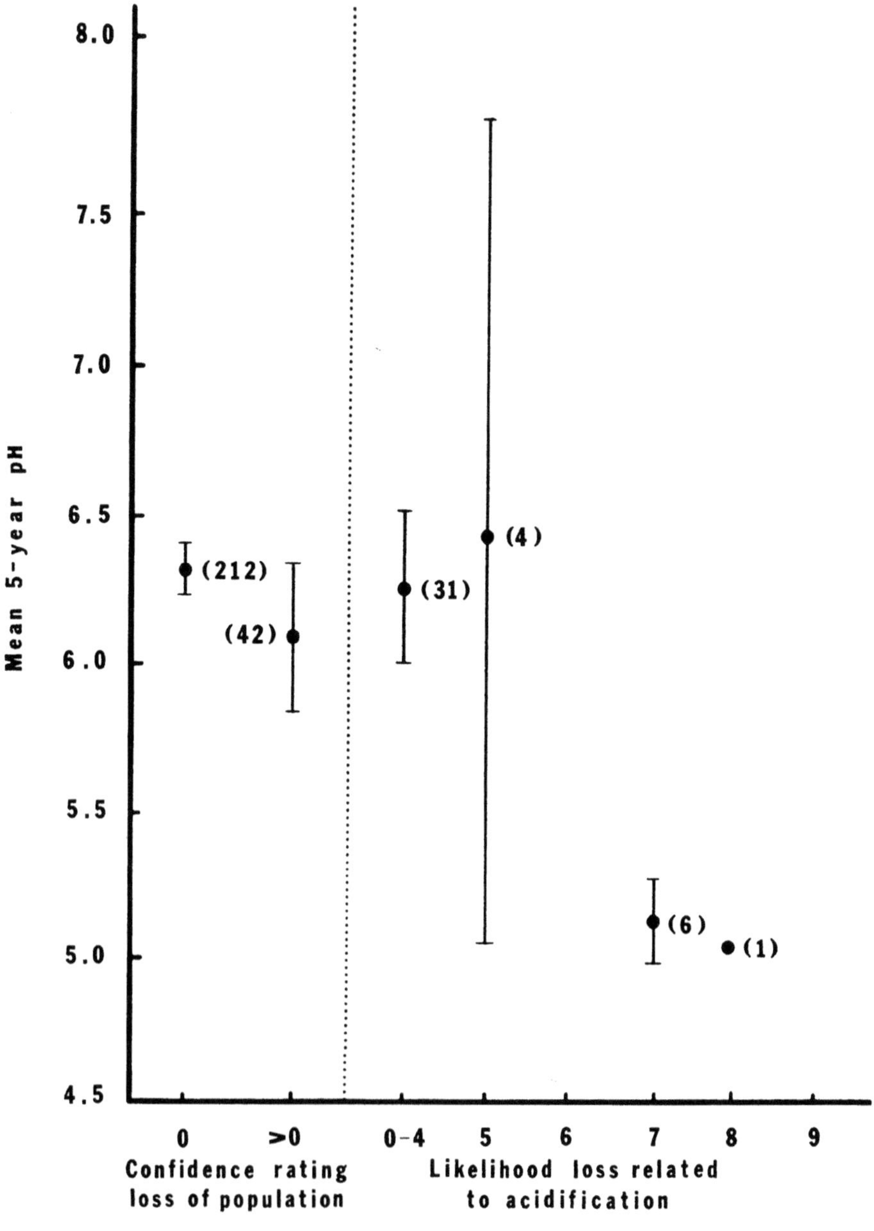

Figure 4. Adirondack white sucker population confidence ratings for the likelihood that the population was lost, and the likelihood that the loss was related to acidification, as compared to mean 5-yr pH. The number of lakes in each rating class is given in parentheses. (0 = low probability; 9 = high probability). Vertical bar represents 95% confidence interval.

Table III. Water chemistry, sediment diatom, sediment chemistry, and fish population changes over time in eight New York Lakes. Diatom data were provided by Donald Charles, Indiana University, and sediment chemistry data were provided by Stephen Norton, University of Maine.

Measured pH Lake	Year	pH	Diatom Inferred pH Present/pre-1980	Sediment Chemistry	Fish Survey Data
Panther	1975	6.88	6.1/6.4 Steady decrease toward surface from 5-10 cm in core.	No acidification indicated.	Fish population eradicated in 1951 and then restocked; no changes in any species other than that caused by direct action.
	1978-80	6.2			
Sagamore	1933	7.0	6.3/6.1 No overall significant trend.	No acidification indicated.	Nine species in 1933, 8 species remaining in 1976; no changes in any major species.
	1961	5.6			
	1976	6.0			
	1978-80	5.6			
Woodhull	1954	5.0	6.0/6.1 No obvious trend.	Dating poor. Possible slight acidification indicated by decline in Zn in surface sediment.	6 species in 1931; 8 species collected plus 3 reported in 1954; 8 species in 1974-81; brook trout last collected 1954 and smallmouth bass last collected 1974, but lake trout and lake whitefish still present.
	1969	6.5			
	1974	6.0			
	1975	5.8			
	1976	5.1-5.3			
	1978	5.7			
	1981	5.4			
	1982	5.6			
Big Moose	1948	6.0	4.9/5.8 Marked diatom-inferred pH decrease and floristic change beginning about 1950.	Acidification indicated by declining relative concentration of Zn, Ca, and Mn as well as declining deposition rates relative to TiO2.	Ten species in 1948, 5-6 species since 1962; smallmouth bass and lake whitefish last collected in 1948; lake trout recruitment failure in late 1940s, last collected in 1966.
	1960	5.2			
	1965	5.2			
	1971	5.5			
	1974	4.9			
	1978-79	4.6-5.0			
	1981	5.3			
	1982	4.9			

Table III continued - page 2.

Honnedaga	1959-60 1975 1976 1981 1982	3.9-6.0 4.91 4.7-4.8 4.92 4.84	5.2/6.1 Insufficient data to determine pattern of change; major floristic change from bottom to top of core.	Not sampled.	Lake trout became rare and emaciated in early 1950s, last collected in 1954; brook trout declined in mid-1960s and were very rare by the mid-1970s.
Upper Wallface	1963 1968 1975 1978 1979	5.6 4.9 4.9 5.0 5.0	4.7/5.1 Most rapid pH and floristic change starting in 1945-55.	Slight acidification indicated by relative decline of deposition rates for Ca and Zn.	Brook trout moderately abundant in 1963-68, absent in 1975.
Woods	1968 1973 1975 1976 1978-80 1982	5.0 6.0 5.2 5.8 4.7	4.8/5.2 Dating information not available. More low pH indicators at top of core.	Dating poor. Only trace metal data available. Acidification not indicated by Zn. Ca and Mn data not available.	Lake trout reported present previously, but were never collected; formerly known for large brook trout, brook trout abundant in 1961, but scarce and all of hatchery origin since 1966; golden shiner last collected in 1966.
Deep	1954 1968 1978 1979	4.70 4.70 4.62-4.76 4.62-4.72	4.8/5.0 Decrease in diatom-inferred pH and major floristic changes beginning 1940-50.	Slight acidification indicated by relative decline in Ca, Mn, and Zn concentrations and deposition rates.	Brook trout present, moderately abundant in 1954, absent in 1964; no survival of stocked fish since 1964.

Table IV. Recent and historical pH and fish species composition of 20 New Hampshire lakes. Historical data on fish species composition includes reports and surveys. (Source: Singer and Boylen, 1984).

Lake	pH Recent	pH Historical (year)	Fish Species Recent	Fish Species Historical	Reclaimed
Cone	4.5	5.2 (51)	-[a]	CP, BB, YP	yes
Rocky	4.9	- (52)	-[a]	BT[b], BB	no
Caldwell	5.1	6.2 (48)	BT	BT[b]	yes
Loon	5.3	6.1 (51)	-	BT[b]	no
Upper Hall	5.8	5.6 (60)	BT	BT[b], LT, LD	no
Hemingway	5.8	- (50)	BB	BB	no
Crane	5.8	6.4 (62)	BB, CP, YP, LB, GS,	BT[b], CP, YP, BB, S	no
Kiah	5.9	5.7 (51)	BT, BB	BT[b], BB	no
Beaver	5.9	6.2 (53)	BB	BT[b]	no
Bean	6.2	6.0 (52)	BT, WS, GS, LC	BT[b]	no
East	6.3	6.5 (51)	BT	BT[b]	no
Russell	6.3	6.6 (38)	BT	BT[b], GS, BD, BB	yes
Moody	6.5	6.4 (37)	WS, GS, LC	BR[b]	no
Blue	6.6	6.4 (54)	BT, BB, RS	BT[b]	yes
Munn	6.7	6.7 (48)	BT	BT[b]	yes
Long	6.8	5.7 (52)	BT, WS	BT[b], RT[b], BB	no
Sessions	6.8	6.6 (50)	BT, FF, CC	BT[b]	yes
Phillips	6.9	6.5 (52)	BT, LS	BT[b]	yes
Echo	6.9	6.9 (39)	BT	BT[b], RT[b], RS[b]	yes
Round	7.3	7.2 (39)	BT, WS, RD	BT[b], S	yes

[a] Stocked with brook trout; stocking discontinued.
[b] Stocked.

Fish Species Codes:
BB = brown bullhead, BD = blacknose dace, BR = brown trout,
BT = brook trout, CC = creek chub, CP = chain pickerel, FF = fallfish,
GS = golden shiner, LB = largemouth bass, LC = lake chub,
LD = longnose dace, LS = longnose sucker, LT = lake trout,
RD = northern redbelly dace, RS = rainbow smelt, RT = rainbow trout,
S = "shiners", WS = white sucker, YP = yellow perch

Table V. Present pH and historical evidence of fish survival in four south-central Pennsylvania streams. Source: Sharpe et al. (1984).

Stream	Present Ph	Date	Event
Wildcat	5.6-7.0	1900s	Brook trout routinely caught
		1980-81	Brook trout and rainbow trout present and reproducing
McGinnis	4.4-5.7	1910	Fisherman caught 60 brook trout
		1977	Brook trout stocking fails
		1981	No fish present in stream
Card Machine	4.6-6.0	1920s	Brook trout routinely caught
		1960s	Brook trout nursery established
		1966	Brook trout mortality at nursery
		1972	Brook trout mortality at nursery
		1978	Brook trout mortality at nursery
		1981	No fish present in stream
Linn	4.6-5.6	1932	Brook trout stocked, no mortality
		1951	Brook trout nursery established
		1960	Brook trout mortality at nursery
		1970	Brook trout mortality in stream 2 days after stocking
		1976	Brook trout mortality after stocking
		1977	Brook trout mortality after stocking
		1981	No fish present in stream

lake trout abundance may reflect the cessation of hatchery introductions. Big Moose and Honnedaga lakes exhibit pronounced acidification, and all sources of information are consistent. Trends for Upper Wallface, Woods, and Deep lakes are less apparent. All show declines in pH and fish populations, although disappearance of fish generally lags behind pH decline by a few years. However, the change in pH in these lakes is apparently quite small. The consistency of the multiple and independent sources of information leads us to conclude that acidification has occurred. A possible explanation for the loss of fish populations over such a small pH change may be provided by the recent finding that total organic carbon declined in two Norwegian acidic lakes coincident with a decline in pH from ca 4.9 to 5.1 to ca 4.4 to 4.7 (Davis et al., 1985). Thus these lakes may have changed from organic weak acid systems to mineral strong acid systems with a concomitant increase in concentrations of toxic Al species.

2.2. Vermont

The Vermont Agency of Environmental Conservation has completed a fairly systematic assessment of the current status of fish populations in Vermont lakes relative to acidity (Langdon, 1983, 1984). Initially, potential susceptible areas were identified based on maps of bedrock geology, and 205 lakes in these areas were surveyed for chemistry in 1979-1984. Six of these (representing 0.2% of the total lake surface area in the state, excluding large border waters) had alkalinity <0 μeqL^{-1}; 60 (10.8% of the total lake surface area) had alkalinity <100 μeqL^{-1}.

The 36 Vermont lakes with the lowest alkalinity were selected for long-term chemical monitoring. Twenty-nine of these were surveyed to determine fisheries status. Two lakes (0.1% of the total lake surface area) were fishless, presumably due to low pH or high Al. The remaining 27 (about 9% of the total lake surface area) contained 1 to 10 fish species (Langdon, 1983, 1984).

Historical information, occasionally dating back to the 1880s, has been located for 19 of these lakes in Vermont Agency of Environmental Conservation files. Eighteen lakes now have annual mean pH 5.5 or above, with 2 to 10 species per lake. Thirteen of these had historical fisheries data. Although some shifts in species composition have occurred, major fisheries declines or loss of acid-sensitive species, which would be suggestive of acidification, have not.

Eleven lakes now have annual mean pH below 5.5, with 0 to 3 species per lake. Six of these had some historical information. Results from trend analysis are somewhat equivocal. For example, Little and Haystack ponds contained no fish in 1982-83. Little Pond once supported a naturally reproducing population of brook trout, reported as early as 1889 and as late as 1958 to be in good condition (Langdon, 1983). The surface pH was 5.2 in both 1958 and 1982. Haystack Pond (pH 4.6) may never have sustained a naturally-reproducing fish population. The earliest reports (1881), stated that no fish inhabited the lake and that none could live there. Two other lakes (pH 4.7 and 5.0) have been stocked with brook trout with no angler success. Both lakes now support only brown bullhead. Finally, historical data for two additional waters (pH 4.7 and 5.2) suggest that "shiner" populations may have been lost. Brook trout populations in these lakes are currently maintained by stocking.

Given that these 29 lakes are those most likely to have been affected, the overall impact of acidification on Vermont's lakes to date has apparently been minimal (Langdon, 1984). The 29 lakes together represent <10% of the total surface area of lakes in Vermont.

2.3. New Hampshire

Singer and Boylen (1984) surveyed 20 low alkalinity lakes in New Hampshire, ranging in pH from 4.5 to 7.3. Historical fisheries data were located for all lakes, and historical pH data for all but two (Table IV). Three lakes were fishless, and all three formerly supported fish. However, the fish population of Loon Pond (pH 5.3) was maintained by stocking, which was discontinued in the 1950s. No historical pH data were located for Rocky Pond (pH 4.9). Cone Pond (pH 4.5) has apparently been acidic for at least 30 yr; however, pH may have declined slightly. The lake was chemically reclaimed and stocked with brook trout for a number of years, but the stocking was discontinued following reports of poor fishing. Rocky Pond (pH 4.9) contained brown bullhead in 1952 and was stocked with brook trout for a number of years. It was surveyed in 1972, and only one brook trout was caught. The only other lake where fish populations may have declined is Upper Hall Pond (pH 5.8), where lake trout and longnose dace (*Rhinichthys cataractae*) may have disappeared. Russell Pond (pH 6.3), where golden shiner, blacknose dace, and brown bullhead may have disappeared, has been reclaimed. The interpretation of fish population changes in relation to water chemistry in New Hampshire is complicated by extensive fishery management, including chemical reclamation and hatchery introductions.

2.4. Maine

We have surveyed 88 Maine lakes for chemical and fishery status. Of these, 9 (10.2%) had pH <5.0 and were fishless or contained only hatchery-maintained populations of brook trout. There were no historical fishery data available.

Maine has 5770 lakes of 0.5 ha surface area or larger. Water chemistry and fisheries survey data were obtained for 1459 lakes from the Maine Department of Inland Fisheries and Wildlife. Fisheries survey records were screened to identify lakes that had been surveyed two or more times. About half the records were reviewed and 312 lakes were identified that had suitable survey data. The data for species and number of individuals collected were reviewed without reference to lake chemistry. Lakes for which species disappeared were investigated for evidence of reclamation, stocking, level of collecting effort, type of sampling gear used, or other factors that might explain the absence of species in the catch. We identified 12 lakes (3.9%) that had lost one or more species of fish with no obvious cause. Chemical data were unavailable for two of the lakes. The remaining 10 lakes had colorimetric pH and occasional methyl orange alkalinity data available. Two lakes (Pleasant Lake, MIDAS #1100, and Scraggly Lake, MIDAS #4264) had lost lake trout populations and were acidic (pH 4.4 to 4.8). The remaining lakes had lost brook trout, white sucker, or yellow perch populations, but generally had pH $\geq$5.5. However, much of the pH data were old and predated the most recent fish sample. Therefore, although temporal shifts in species composition occurred in some Maine lakes, there was no clear association between decline or disappearance of fish populations and lake chemistry.

2.5. Massachusetts

A state-wide survey of the chemical status of lakes in Massachusetts showed that 46 of 900 (5.1%) had pH <5.0 and 159 of 829 (19.2%) had alkalinity $\leq$40 $\mu eq L^{-1}$ (Godfrey *et al.*, 1985). The acidic lakes were located in southeastern (Cape Cod and Buzzards Bay watersheds) and central Massachusetts (Millers, Chicopee, Quinebaug, and Farmington watersheds). Comprehensive surveys of fisheries status in Massachusetts have not been completed. There have been reports of declining fish populations in the Cape Cod region, and in Quabbin Reservoir in central Massachusetts (Halliwell, 1985a), but definitive evidence linking these declines to acidification is lacking.

A temporal association study of water chemistry and fish species distribution was conducted in tributaries to the Millers River, Massachusetts (Halliwell, 1985a,b). Sixteen streams that had been surveyed in 1953 and 1967 were surveyed again in 1983. Fish surveys were conducted with fish toxicant (rotenone) in 1953 and by electrofishing in 1967 and 1983. Species were identified but individuals were not enumerated in 1967. Stream pH was measured colorimetrically in 1953 and 1967, and electrometrically in 1983. Streams that had pH ≥ 6.0 in 1983 generally had relatively stable fish communities and pH over time. Most of the streams with pH <6.0 in 1983 had apparently lost three or more species. Two streams had pH <5.0 in 1983 and had lost all fish species. These streams were of different character than the others surveyed, being of lower gradient, higher color (organic acids), and containing primarily warmwater fish species.

2.6. Rhode Island

Fish kills involving stocked rainbow trout and attributed to low pH were reported for Wallum Lake, a large clear-water oligotrophic lake on the Rhode Island-Massachusetts border. The lake was subsequently neutralized by addition of ground limestone (Guthrie and Stolgitis, 1977). A recent survey noted an apparent loss of the native brook trout population in Cedar Swamp Brook. The stream had pH 6.8 in 1962 and contained trout. In 1983 pH was 4.2 to 5.0 and trout were absent (DeMaine, 1983). No other studies of fishery status in relation to acidity have been conducted in the state.

2.7. Catskill/Pocono/Kittatinny Ridge regions of New York, Pennsylvania and New Jersey

Although some surveys of water quality and aquatic biota have been completed for these regions (e.g., Colquhoun et al., 1981, 1984; Faust and McIntosh, 1983; Berg and Bradt, 1984; Bradt et al., 1984), assessments of current fisheries status or trends relative to potential effects of acidification are unavailable. Colquhoun et al. (1984) noted that most of 50 Catskill streams surveyed had acceptable pH for fish. Well known trout streams (e.g., Willowemoc and Beaverkill) have apparently not been affected. Several small lakes on Kittatinny Ridge in New Jersey are clear-water and moderately to highly acidic (Faust and McIntosh, 1983). Two, with measured pH ranging 3.6 to 4.7 and 3.8 to 4.9, reportedly support only yellow perch, a relatively acid-tolerant species (Brennan et al., 1985). Historical trends have not been evaluated.

2.8. Pennsylvania

A spatial survey of fish populations and water chemistry was conducted in 51 streams in south-central Pennsylvania (Sharpe, 1985). Reproducing fish populations were present in 37 streams, three streams contained only hatchery fish, and 11 streams were fishless. Fish population status was correlated with water chemistry; many of the fishless streams had pH below 5.0 to 5.5 with high levels of Al. Anecdotal information concerning historical trends in fish populations was located for four streams, three of which were included in the above survey (Table V). All three streams that are presently acidic (minimum pH 4.4) and fishless formerly supported fish. Periodic fish mortalities were first reported in the 1960s, and stocking failures were reported in the 1970s. No such mortalities have been reported for the stream that is not acidic and presently supports fish. An *in situ* fish toxicity bioassay was conducted on one acidic (McGinnis Run) and the non-acidic stream (Wildcat Run)(Sharpe *et al*., 1983). Trout fry survived only 4 to 9 days in the acidic stream, but survived for the duration of the experiment (36 days) in the non-acidic stream. Inasmuch as these streams formerly did support viable populations (either self-sustained or long-term survival of hatchery fish), it is likely that acidification caused the demise of fish in these streams.

Kimmel (1985) reviewed the stream fisheries data files of the Pennsylvania Fish Commission for evidence of fisheries declines or fish kills possibly associated with acidification. Only streams meeting the following criteria were considered: (1) current alkalinity $\leq 200\ \mu eqL^{-1}$, (2) presence of native or hatchery-reared salmonids or historical record of same, (3) no past or present mining or drilling in watershed, (4) stream order 1 to 3, and (5) undeveloped forested watershed. Of the 344 streams that met the above criteria, 124 (36% totalling 550 km) had records indicating one or more of the following: (1) an absence of fish, (2) a loss of wild salmonid populations, (3) fish kills, particularly following stocking, or (4) a change in fish stocking policy (e.g., elimination of pre-season stocking, replacement of brown or rainbow trout with brook trout, termination of all stocking), presumably due to high acidity and poor survival. This information should be interpreted cautiously. It is based solely on written comments filed by fisheries biologists and presumed effects of acidification. Although these records are inconclusive, they do provide a preliminary indication of a problem. The 124 streams represent 6.6% of the managed trout resource and 2% of the total managed and unmanaged trout resource in Pennsylvania.

2.9. West Virginia

In 1983, the state of West Virginia initiated a survey of 82 streams to obtain baseline data and evaluate potential effects from acidic deposition (Menendez, 1985). About 10% of the streams surveyed (focusing on those areas most likely to be sensitive) in fall 1984 had pH <5.0, and 24% had alkalinity $<40\ \mu eqL^{-1}$. Unfortunately, associated fisheries surveys have not been completed.

In West Virginia, as in Pennsylvania and other states of the central and southern Appalachians, acidic streams are common, due to acidic mine drainage or naturally acidic bedrock. Thus, effects of acidic deposition are often difficult to distinguish. In addition, prior stream surveys focused on the more fertile streams. Little historical data are available for streams with low alkalinity. However, changes in fisheries and water chemistry that cannot be attributed to logging, mining, or other on-site activity have apparently occurred in some watersheds. For example, in 1935 the Cranberry River was an excellent trout stream (McGavock and Davis, 1935). In the 1960s, surveys in the North Fork of the Cranberry River found only sparse trout populations confined to a few tributaries. In the 1980s, trout populations have consisted principally of stocked fish and springtime pH frequently falls below 5.0 (Menendez, 1985).

Evidence for effects of acidic deposition and acidification on West Virginia fishery resources is thus not definitive. Moreover, the extent of effects or sensitive aquatic systems is not known.

2.10. Southern Blue Ridge Province

Headwater streams in the Southern Blue Ridge Province (SBRP), which encompasses parts of northern Georgia, eastern Tennessee, and western North Carolina, typically have low alkalinity (<100 μeqL^{-1}) with baseflow pH above 6.0 (Winger et al., 1985). There is no documentation of losses or declines of fish populations in streams, rivers, or reservoirs in the region due to acidic deposition. Fowler et al. (1984) surveyed 12 high elevation (>500 m) first and third order streams and found no relationship between fish assemblage characteristics and baseflow pH, alkalinity, or other chemical measurements. Jones et al. (1983) investigated fish kills in fish-rearing facilities near Cherokee, North Carolina, coincident with storm events and pH depressions (pH <5.0) in the tributary stream, Raven Fork. They concluded that the pH depressions resulted from organic acids draining from the forest-soils complex rather than acidic deposition. Monitoring of other streams in the SBRP during storms and high stream flow suggests that Raven Fork is atypical for the region. In five streams sampled during 20 storm events, pH levels never dropped below 5.3, and most (75%) remained above pH 6.0 (Olem, 1985).

Virtually all of these streams have low alkalinity and therefore may be susceptible to future effects of acidic deposition. These streams provide valuable salmonid habitat in this region, but fish communities have historically been affected by logging, overfishing, stocking, and other factors (Larson and Moore, 1985), complicating detection of acidification effects. In addition, fishery survey data are sparse. The available evidence, however, suggests no substantial effects to date. Further, soils in the region generally retain sulfate strongly, which may delay or mitigate aquatic effects of atmospheric inputs (Altshuller and Linthurst, 1984).

2.11. Florida

Results from the U.S. National Lake Survey indicate that Florida had more acidic (pH <5.0) lakes than any other region in the eastern United States, including the Adirondack Mountain region of New York (Linthurst et al., 1986). Yet no effects on fishery resources have been reported. Surveys of fish populations in a small number of Florida lakes having pH ranging 3.7 to 8.9 reveal no loss of major fish populations (Keller, 1984; Canfield et al., 1985). Canfield et al. (1985) surveyed five acidic (pH 3.7 to 4.9) clear-water and three acidic (pH 4.1 to 4.6) colored lakes. All contained populations of largemouth bass. Intensive mark-recapture studies in Cue Lake (clear-water, oligotrophic lake; pH ca. 4.0) demonstrated that all fish species present were reproducing and growing normally (Canfield, 1985). The number and biomass of largemouth bass in Cue Lake are within the range reported for oligotrophic lakes in Florida.

Predictions of effects of acidification on fisheries resources in Florida lakes are uncertain. The reasons for the large number of acidic lakes are not clear. Hypotheses proposed to explain the apparent tolerance of Florida fishes to acidic conditions include low concentrations of Al (Hendry and Brezonik, 1984; Canfield et al., 1985), and long-term adaptation to natural acidity (Canfield, 1985).

3. DISCUSSION

Considering the interest in the effects of acidification on fishery resources, the number of available data sets showing a relation between surface water acidification and fish population status is remarkably small. This may reflect, in part, a lack of historical data for the generally small, remote lakes that are vulnerable to acidification, and of variability in fish census methods and record keeping. The strongest evidence of adverse effects of acidification on fish populations is provided by the Adirondack Mountain data. Rigorous evaluation of the fishery and chemical data for the Adirondacks clearly indicates both that fish populations have declined in some lakes and that lakes where fish populations have declined are now acidic. Alternative reasons for the decline in fish populations in these lakes are not evident. There was a high probability that fish communities had declined as a result of acidification in 49 lakes, and there was marginal evidence for such a decline in 64 additional lakes. Extrapolating this estimate to the entire Adirondack Mountain region indicates that fish communities in 200 to 400 lakes may have been adversely affected by acidification. This is nonetheless a crude estimate, and the error associated with it cannot be quantified.

Pennsylvania and Massachusetts streams also have documented fish population declines associated with acidity; however the data are fewer and less complete than those for New York and the analyses less rigorous. That is not to say that effects have been less severe; rather, the data are not sufficient to document this. Lake surveys in New Hampshire and Vermont identified very few lakes where fish populations may have declined because of acidification, and analysis of fish survey records in Maine failed to identify any such lakes. More so than the other states surveyed, New Hampshire relies on an extensive fish stocking program to manage coldwater fisheries. Such hatchery-maintained fisheries may be less susceptible to decline from acidification. Although acidic, fishless lakes exist in Maine, it is unknown whether these lakes ever supported fish. The lakes having historical fishery data have apparently not been greatly affected by acid deposition. Minimal historical fisheries data were obtained for Rhode Island, the Catskill/Pocono/Kittatinny Ridge region, West Virginia, the Southern Blue Ridge Province, and Florida. Available evidence suggests that effects of acidic deposition on fishery resources in these regions are minimal.

Alternative factors that could reduce or eliminate fish populations have been raised (Retzsch et al., 1982). These factors include use of chemical pesticides, changes in fish hatchery production, changing angler pressure, and increased beaver activity. Direct effects of human activities (e.g., obstruction of fish migrations by dams, degradation of water quality by agricultural and urban run-off, municipal sewage, and industrial wastes) greatly reduced fish populations in accessible waters in the colonial and post-civil war periods (Thompson, 1970). However, such factors primarily affected lakes near population centers; remote lakes were not directly affected. Lumbering and subsequent burning of the watersheds undoubtedly affected fish populations in less accessible lakes, but these factors generally were most important in the late 1800s and early 1900s, and thus predated the acidity-related population declines documented here. For example, Coolidge (1963) tabulated the area of Maine forests burned from about 1900 to 1960. The area was greatest before 1912 and declined substantially to 1960. Smith (1972) stated that wood harvest in Maine was greatest in 1890-1910. Thereafter, harvest gradually declined, reaching a minimum just before World War II. The harvest then increased and in 1969 was about double the previous maximum. In the Adirondack Mountain region, Charles (1984) failed to find a correlation between lake pH inferred from sediment diatom species assemblage and logging or fire in the Big Moose Lake watershed. Similarly, Bradbury (1986) failed to detect any influence of fire in the watershed on sediment diatom species composition in a Minnesota lake.

Early fishery scientists were unaware of the inherently low productivity of high elevation, oligotrophic lakes and believed that supplementing the sparse fish populations with hatchery fish would enhance species diversities and standing crops to levels comparable to those in more productive, low elevation lakes. However, these attempts generally failed and consequently fell into disfavor. For example, Smallwood (1918) documented the failure of stocking to supplement the fish population in Lake Clear, New York. Lake Clear presently has a pH >7.0 and contains at least five species of fish (Colquhoun et al., 1984).

Beaver activity can either enhance or degrade fish populations, depending on the particular circumstances. Beaver were reintroduced into the Adirondack Mountains of New York in 1905. By the 1920s population densities had increased enough to raise concern about beaver damage to various resources (Johnson, 1927). Therefore, beaver activity also apparently predates the acidity-related fish population declines. Aerial photographs from 1968 and 1978 were used to identify Adirondack Mountain region headwater lakes with beaver activity. Fish community status ratings were unrelated to beaver activity (Hunsaker, 1985).

Chemical pesticides have been used in remote areas of northeastern North America to control spruce budworm and blackfly, with detrimental effects on fish (Burdick et al., 1964; Anderson and Everhart, 1966; Elson, 1967; Kerswill and Edwards, 1967; Locke and Havey, 1972). Organochlorine compounds are generally no longer used for these purposes, and most formerly affected fish populations have recovered (Dean et al., 1979). Analyses of brook trout from remote lakes of varying pH in northern New England revealed low organochlorine residues (Haines, 1983); concentrations were comparable to those in Antarctic fish and apparently represent world-wide background levels.

It is impossible to rule out factors other than acidification as important in the decline or loss of fish populations except by intensive, case-by-case studies. Such studies have seldom been made, and the quantity of data for estimating trends in fish populations is exceedingly sparse. However, a few such data sets do exist; in the cases cited herein factors other than acidification apparently were not involved with the decline of fish populations. The extent of the resource that has been damaged is apparently small, however, and the extent is geographically limited. We believe the major focus for future research efforts should be the determination of possible trends in fish communities subjected to present or future acidic deposition scenarios.

ACKNOWLEDGMENTS

The following persons graciously provided us with unpublished data and assessments, without which we could not have prepared this report: Daniel E. Canfield, Jr., University of Florida; David Halliwell, Massachusetts Division of Fisheries and Wildlife; William Kimmel, California University of Pennsylvania; Richard Langdon, Vermont Agency of Environmental Conservation; Raymond Menendez, West Virginia Department of Natural Resources; Harvey Olem, Tennessee Valley Authority; William Sharpe, Pennsylvania State University; Michael Van Den Avyle, U. S. Fish and Wildlife Service, University of Georgia. Financial support for this work was provided by the U. S. Environmental Protection Agency, U. S. Fish and Wildlife Service, and the National Research Council.

REFERENCES

Altshuller, A. and Linthurst, R. (editors): 1984, The acidic deposition phenomenon and its effects: critical assessment review papers. Vol. II. Effects sciences. U. S. Environmental Protection Agency EPA-600/8-83-016BF.

Anderson, R. and Everhart, W.: 1966, Trans. Am. Fish. Soc. 95, 160.

Baker, J. and Havey, H.: 1984, Critique of acid lakes and fish population studies in the Adirondack region of New York state. Report to Environmental Research Laboratory, U.S. Environmental Protection Agency, Corvallis, OR.

Baker, J. and Schofield C.: 1982, Water Air Soil Pollut. 18, 289.

Berg, M. and Bradt, P.: 1984, Macrozoobenthos of three lakes with varying sensitivities to acidification. Page 7 in Abstracts for the 47th Annual Meeting, American Society of Limnology and Oceanography, Vancouver, BC.

Bradbury, J.: 1986, Arch. Hydrobiol. 106, 203.

Bradt, P. Barasso, D. and Dudley, J.: 1984, Phytoplankton communities in three lakes with different sensitivities to acidification. Page 11 in Abstracts for the 47th Annual Meeting, American Society of Limnology and Oceanography, Vancouver, BC.

Brennan, E., Ehrenfeld, J., Faust, S., Fritz, R., Graedel, T., Kish, G., Lord, D., McIntosh, A., Morgan, M., Nelson, D., Reiss, N., Riley, D., Spatola, J. and Tedrow J.: 1985, Acidic deposition in New Jersey: a report to the Governor and Legislature of New Jersey. Trenton, NJ.

Brown, D.: 1983, Bull. Environ. Contam. Toxicol. 30, 582.

Burdick, G., Harris, E., Dean, H., Walker, T., Skea, J. and Colby D.: 1964, Trans. Am. Fish. Soc. 93, 127.

Canfield, D.: 1985, personal communication.

Canfield, D., Maceina, M., Nordlie, F. and Shireman J.: 1985, Trans. Amer. Fish. Soc. 114, 423.

Charles, D.: 1984, Verh. Internat. Verein. Limnol. 22, 559.

Colquhoun, J., Symula, J. Pfeiffer, M. and Feuer J.: 1981, Preliminary report of stream sampling for acidification studies 1980. New York State Dept. Environmental Conservation, Albany, NY, Tech. Rep. 81-2.

Colquhoun, J., Kretzer, W. and Pfeiffer M.: 1984, Acidity status of lakes and streams in New York State. New York State Dept. Environmental Conservation, Report WM P-83(6/84).

Coolidge, P.: 1963, History of the Maine woods. Furbush-Roberts Printing Co., Bangor, ME.

Davis, R., Anderson, D. and Berge F.: 1985, Nature 316, 436.

Dean, H., Skea, J., Colquhoun, J. and Simonin, H: 1979, N.Y. Fish Game J. 26, 188.

DeMaine, A.: 1983, Pond and stream surveys: acidification susceptibility. Performance Report for Project F-20-4-24. Rhode Island Division of Fish and Wildlife, West Kingston, RI.

Dillon, P., Yan, N. and Harvey, H.: 1984, CRC Critical Reviews in Environmental Control 13, 167.

Elson, P.: 1967, J. Fish. Res. Bd. Can. 24, 731.

Faust, S. and McIntosh, A.: 1983, J. Environ. Sci. Health A18, 155.
Fowler, D., Van Den Avyle, M. and Hudy, M.: 1984, Fish assemblage characteristics of acid sensitive streams in the Southern Blue Ridge Province. Page 31 in Abstracts of the 114th Annual Meeting of the Am. Fish. Soc., Ithaca, NY.
Godfrey, P., Ruby, A. and Zojicek, O.: 1985, The Massachusetts Acid Rain Monitoring Project A.R.M.: Phase I. Publication No. 147, Water Resources Research Center, University of Massachusetts, Amherst, MA.
Guthrie, R. and Stolgitis, J.: 1977, Fisheries investigations and fish management in Rhode Island lakes and ponds. Fisheries Report No. 3, Rhode Island Division of Fish and Wildlife, Providence, RI.
Haines, T.: 1981, Trans. Am. Fish. Soc. 110, 669.
Haines, T.: 1983, Water Air Soil Pollut. **20**, 47.
Halliwell, D.: 1985, Massachusetts Wildlife 36, 10.
Halliwell, D.: 1985b, personal communication.
Harvey, H. and Lee, C.: 1982, Historical fisheries changes related to surface water pH changes in Canada. Pages 45-55 in R. Johnson, editor, Acid Rain/Fisheries. American Fisheries Society, Bethesda, MD.
Hendry, C. and Brezonik, P.: 1984, Water Resour. Bull. 20, 75.
Hunsaler, C.: 1985, personal communication.
Johnson, C.: 1927, Roosevelt Wild Life Bull. 4, 501.
Johnson, D., Simonin, H., Colquhoun, J. and Flack, S.: 1986, Biogeochemistry, in press.
Jones, H., Naggle, J., Young, R., Kelly, J., Olem, H., Ruane, R., Pasch, R., Hyfantis, G. and Parkhurst W.: 1983, Investigations of the cause of fishkills in fish-rearing facilities in Raven Fork watershed. TVA/ONR/WR-83/9.
Keller, A.: 1984, Fish communities in Florida lakes: relationship to physico-chemical parameters. M.S. Thesis, University of Florida, Gainesville, FL.
Kerswill, C. and Edwards, H.: 1967, J. Fish. Res. Bd. Can. **24**, 709.
Kimmel, W.: 1985, personal communication.
Kleinbaum, D. and Kupper, L.: 1978, Applied regression analysis and other multivariable methods. Duxbury Press, Boston, MA.
Langdon, R.: 1983, Fisheries status in relation to acidity in selected Vermont lakes. Vermont Agency of Environmental Conservation Report, Montpelier, VT.
Langdon, R.: 1984, Fishery status in relation to acidity in selected Vermont lakes, 1983. Report of the Dept. of Water Resources and Environmental Engineering, Agency of Environmental Conservation, Montpelier, VT.
Larson, G. and Moore, S.: 1985, Trans. Am. Fish. Soc. 114, 195.
Leivestad, H. and Muniz, I.: 1976, Nature **259**, 391.
Linthurst, R., Landers, D., Eilers, J., Keller, T., Brakke, D. and Jeffries, D.: 1986, Water Air Soil Pollut. this volume.
Locke, D. and Havey, K.: 1972, Trans. Am. Fish. Soc. 101, 638.
McGavock, A. and Davis, H.: 1935, A stream survey of the waters of the Monongahela National Forest. Dept. of Commerce, Bureau of

Fisheries, Washington, DC.
Menendez, R.: 1985, personal communication.
Muniz, I.: 1984, Phil. Trans. R. Soc. Lond. B305, 517.
Olem, H.: 1985, personal communication.
Overrein, L., Seip, H. and Tollan, A.: 1980, Acid precipitation - effects on forest and fish. Research Report 19, Acid Precipitation - Effects on Forest and Fish Project, Aas, Norway.
Pfeiffer, M.: 1979, A comprehensive plan for fish resource management within the Adirondack Zone. FW-P142 (12/9), New York State Department of Environmental Conservation, Albany, NY.
Pfeiffer, M. and Festa, P.: 1980, Acidity status of lakes in the Adirondack region of New York in relation to fish resources. New York State Department of Environmental Conservation Report FW-P168 (10/80), Albany, NY.
Retzsch, W., Everett, A., Duhaime, P. and Northwanger, R.: 1982, Alternative explanations for aquatic ecosystem effects attributed to acidic deposition. Report prepared for Utility Air Regulatory Group by Everett and Associates, Rockwell, MD.
Schindler, D., Mills, K., Malley, D., Findlay, D., Shearer, J., Davies, I., Turner, M., Linsey, G. and Cruikshank, D.: 1985, Science 228, 1395.
Schofield, C.: 1965, Trans. Am. Fish. Soc. 94, 227.
Schofield, C.: 1973, The ecological significance of air-pollution-induced changes in water quality of dilute-lake districts in the northeast. Trans. N.E. Fish Wildl. Conf. 98-111
Schofield, C.: 1976a, Lake acidification in the Adirondack Mountains of New York: causes and consequences (abstract). Page 477 in L. Dochinger and T. Seliga, editors. Proceedings of the First International Conference on Acid Precipitation and the Forest Ecosystem. USDA Forest Service, Gen. Tech. Rep. NE-23.
Schofield, C.: 1976b, Ambio 5, 228.
Sharpe, W.: 1985, personal communication.
Sharpe, W., Kimmel, W., Young, E. and DeWalle, D.: 1983, Northeast Environ. Sci. 2, 171.
Sharpe, W., DeWalle, D., Leibfried, R., Dinicola, R., Kimmel, W. and Sherwin L.: 1984, J. Environ. Qual. 13, 619.
Singer, R. and Boylen, J.: 1984, Biological field surveys of northeastern acidified lakes. Final Report to EPA/NCSU Acid Deposition Program, North Carolina State University, Raleigh, NC.
Smallwood, W.: 1918, Amer. Naturalist 52, 322.
Smith, D.: 1972, A history of lumbering in Maine 1861-1960. University of Maine Press, Orono, ME.
Thompson, P.: 1970, The first fifty years - the exciting ones. Pages 1-12 in N. Benson, editor. A Century of Fisheries in North America. Special Publication 7, American Fisheries Society, Bethesda, MD.
Watt, W., Scott, C. and White. W.: 1983, Can. J. Fish. Aquat. Sci. 40, 462.
Winger, P., Lasier, P., Hudy, M., Fowler, D. and Van Den Avyle M.: 1985, J. Environmental Qual., Submitted.

SENSITIVE SURFACE WATERS - A UK PERSPECTIVE

David E Wells*
Department of Agriculture and Fisheries For Scotland
Freshwater Fisheries Laboratory, Pitlochry, Perthshire
Scotland, PH16 5LB

Alun S Gee
Welsh Water Authority, Llanelli, Dyfed, South Wales

and

Richard W Battarbee
Palaeoecology Research Unit, Department of Geography
University College, London

ABSTRACT. This overview of sensitive surface waters in the United Kingdom has focused on palaeolimnological studies, the estimation of acidification values according the Henriksen, and correlation with non-marine sulphate values of available chemical data and trends in salmonid populations. The palaeolimnological studies of lakes with unafforested catchments in Galloway led to the conclusion that changes in water acidity, since the 1800's, have been caused by acid deposition. Significant differences in stream chemistry between mature conifer forests and adjacent moorland catchments, in both Wales and Scotland, suggest that afforestation as well as acid deposition have contributed to the decline and, in some areas, the elimination of the salmonid population.

1. INTRODUCTION

Since the 1980 Sandefjord Conference (Drablos and Tollan, 1980) and the 1982 Stockholm Conference (Anon, 1982) there has been a substantial increase in reports on acid deposition and effects on surface waters in the UK (Last, 1982; Buckley-Golder, 1984; Farmer, 1984; Anon, 1984; HMSO, 1984; Seip, 1985). In view of the questions raised by the public, politicians and media alike concerning the effects of acid waters it is appropriate that these data be drawn together to produce a synoptic picture of the available information for the UK as a whole.

Maps of sensitive ground waters, based on solid geology of the UK, prepared by the British Geological Survey (Kinniburgh and Edmunds, 1984) have been used in conjunction with exisiting water chemistry data to identify the regions of sensitive surface waters. A condensed form of these maps and the current study areas are given in Figure 1. Such

* Author for all correspondence

Water, Air, and Soil Pollution **31** (1986) 631–668.

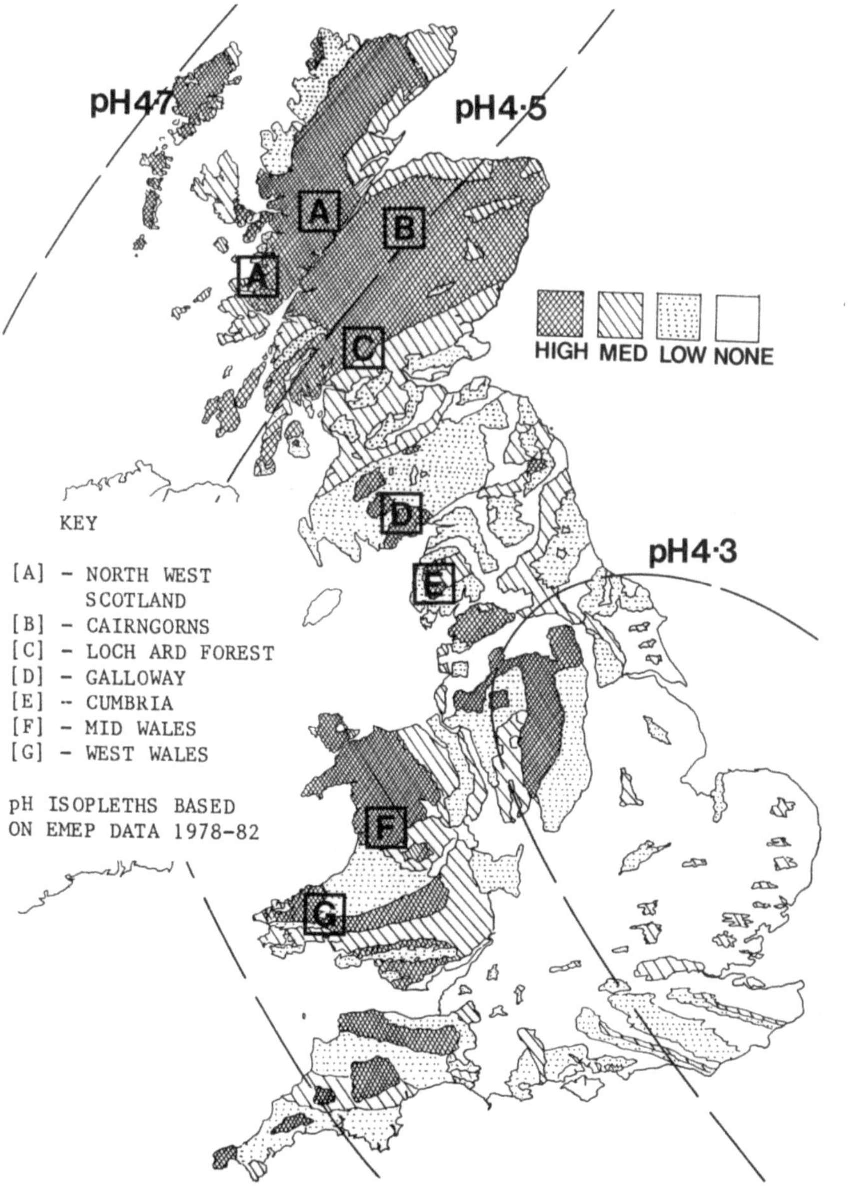

Figure 1. Areas of sensitive groundwaters [reconstructed from maps prepared by Kinniburgh and Edmunds for the British Geological Survey (1984)] and surface waters study sites in the UK.

overview maps are only indicative of the general location of vulnerable catchments and in specific cases local enrichment, such as a small calcareous outcrop, can often ameliorate the poorer water chemistries.

With the possible exception of some sandstone areas in S.E. England, sensitive waters are predominantly confined to upland Britain, comprising ~15% of the total land mass in England, and over 50% in Scotland and Wales. Most of these upland areas contain important salmonid spawning and nursery streams.

Stream chemistry patterns are affcted by (1) N. Atlantic storms producing a 'sea salt effect' (Harriman and Wells, 1985) (2) a more temperate climate resulting from the influence of the Gulf Stream on a smaller land mass, reducing incidents of snowmelt-related acid flushes experienced in Scandanavia and N. America and (3) enrichment by the lowland geology which reduces the threat, in terms of vulnerability of returning adult salmon. Two exceptions to this have been reported from the River Duddon and Esk in Cumbria (Prigg, 1983), where kills of adult and juvenile salmonids have been attributed to acid flushes.

2. TRENDS FROM LAKE SEDIMENT STUDIES

Long-term postglacial acidification of upland catchments has been reported by Pennington (1984) in Cumbria. These studies clearly indicate changes in catchment vegetation and in the character of surface soil horizons over the last 14,000 yr. However, the speculation that "these acidification processes had, by c 1800 AD, produced a situation where no further threshold of change was crossed as a result of burning fossil fuels" (Pennington, 1984) is unsubstantiated and assumes that soil acidification necessarily leads to surface water acidification.

Of the upland lakes in Cumbria, data on recent (post-1800 AD) changes are available only for Burnmoor Tarn (Haworth, 1985). Here, however changes in diatoms which could be associated with acidification were not observed; which is in keeping with the present pH which ranges between 6.0 and 7.0. Although the lake has a granite catchment overlain by peat it is probably enriched by small calcite veins. Further work by Haworth on more acidic upland lakes, such as Scoat Tarn (pH $<$5) is underway.

The clearest evidence on post-18th C trends in acidification has come from sites in Galloway, SW Scotland where diatom analysis and 210-Pb dating, soot particle and pollen analyses, and geochemical analysis for trace heavy metals have been carried out. (Flower and Battarbee, 1983; Battarbee, 1984, Battarbee et al , 1985). The specific objectives of this work were to characterise the recent (last 200 yr) ecology of six lakes in relation to the commencement, rate, extent and cause of any detectable acidification. The lochs studied are either entirely or partly situation on granite bedrock; in addition three have ploughed and afforested catchments and three have un-afforested and undisturbed catchments.

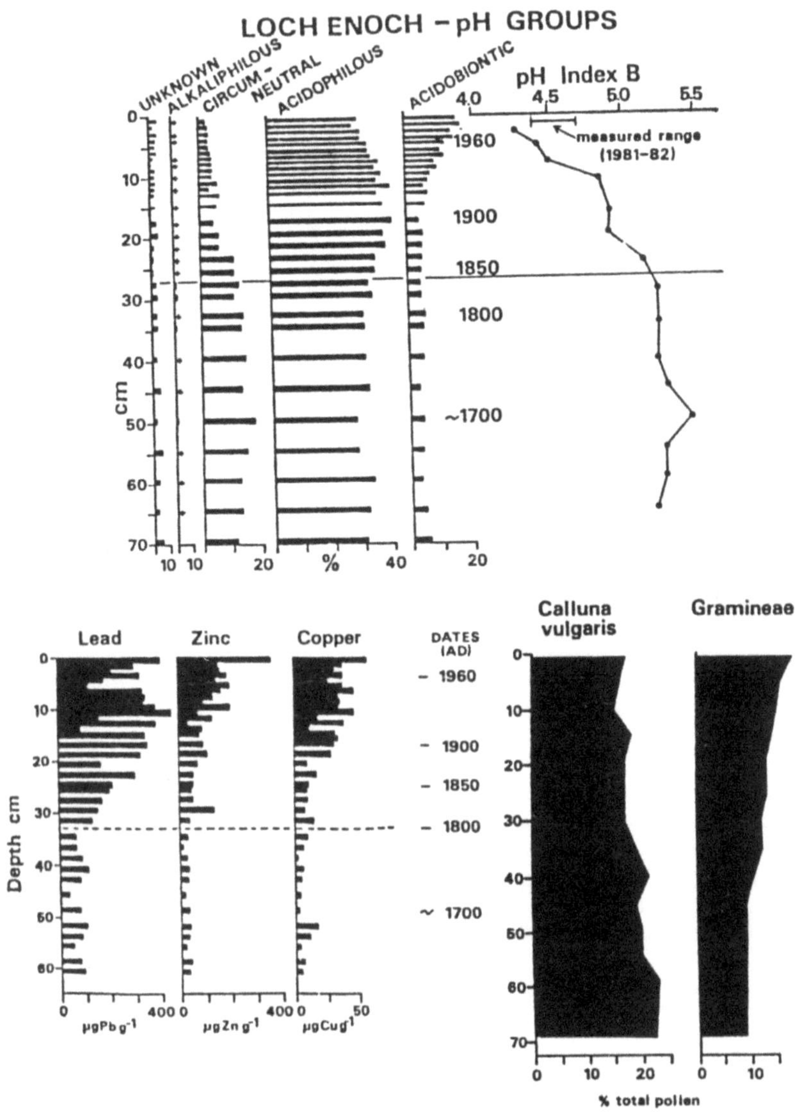

Figure 2. Summary of diatom,heavy metal and pollen data for L Enoch,SW Scotland showing changes correlated with pH through time.

TABLE I

CHANGES IN LAKE ACIDITY DETERMINED BY DIATOM ASSEMBLAGE INDEX AND ACIDIFICATION VALUES IN SOME GALLOWAY LOCHS, S.W.SCOTLAND

SITE	ALT (m)	CATCHMENT AREA (ha)	LAKE (ha)	FOREST %	PREDICTED pH 1850-PRESENT	CURRENT pH*	ACIDIFICATION INDEX**
L.ENOCH	493	50.1	3.7	-	5.2 to 4.3	4.4	42
L.VALLEY	320	34.1	1.9	-	5.5 to 4.5	4.6	44
ROUND L.OF GLENHEAD	299	12.6	8.6	-	5.7 to 4.8	4.65	45
L.DEE	225	100.0	14.6	20.8	6.1 to 5.6	5.0	62
L.GRANNOCH	210	114.3	12.3	69.5	5.5 to 4.4	4.55	69
L.FLEET	344	173	6.2	21.8	-	4.33	92
L.SKERROW	126	50.7	30.1	14.8	no change	5.05	42

All cores cover at least the period from 1840 to the present. For the sites with undisturbed catchments on the L. Doon intrusion L. Enoch, L. Valley, and Round Loch of Glenhead the diatom-based inferences indicate pre-acidification pH values of between 5.3 and 6.0 for the centuries before c. 1850. However from 1850 progressive acidification occurred, accelerating in the mid 20th century as the acidobiontic taxa Tabellaria binalis and T. quadriseptata increased, and the circumneutral species decreased. A decline in pH of about one unit is indicated for these three lakes and the reconstructed pH values from the surface sediment diatom assemblages are in good agreement with the measured pH values of the loch (Table I). Trend in lakes with afforested catchments are less clear, less marked or confused by the effect of soil erosion. In the case of L Grannoch, acidification occurred somewhat later than the other sites, but the floristic change was the same resulting in a diatom assemblage dominated by T. binalis and T. quadriseptata by 1960. However the catchment ploughing since 1960 has led to acclerated soil erosion and the probable inwash of catchment diatoms to the extent that errors are introduced in the pH reconstruction (Battarbee and Flower, 1984). For L. Dee and L. Skerrow acidification is less marked since both lakes have substantial parts of their catchment on non-granitic rock. Both sites, however, lost their Cyclotella plankton at the end of the 19th century and the pH of L. Dee, prior to the beginning of liming in 1980, had decreased from approximately pH 6.1 to 5.6.

In an attempt to evaluate alternative explanations for the acidification of non-forested sites, pollen and geochemical analysis of the core from L. Enoch, the most isolated and acidic site, was carried out. The explanation that relates acidification to land use changes (Rosenquist, 1978), involving a decline in grazing and increase in Calluna heathland, cannot be sustained since the evidence indicates a decline in Calluna in recent decades and a reciprocal expansion of Gramineae (Figure 2); a trend that agrees with local contemporary practises in upland grazing, and heather burning in the West of Scotland.

Sections of the sediment core were also analysed for Cu, Zn and Pb as indicators of atmospheric input from industrial sources. The increased concentrations above background levels found are accounted for by increased deposition of the metals from the atmosphere (Figure 2). These increases are paralleled by increases in the concentration of carbonaceous particles ('soot') in the sediment (Darly, 1985) and together provide substantial evidence of the increasing pollutant load of airstreams over Galloway in the nineteenth and twentieth centuries.

3. TRENDS IN WATER QUALITY

3.1 Scotland

Although the association between acid deposition and acid waters in upland areas has been made for some time (Gorham, 1955), it was Harriman and Morrison (1982) (HMSO, 1981) who first reported that water quality was additionally affected by afforestation. These studies in the Loch

TABLE II

SUMMARY OF CHEMICAL DATA FOR SENSITIVE WATERS IN THE UNITED KINGDOM

SITE	N	pH MIN	pH MAX	Ca MIN	Ca MAX	EXSO4 MIN	EXSO4 MAX	Al MIN	Al MAX	REFERENCE
SCOTLAND										
L.Ard										
Rainfall		4.4		6		72		-		Harriman and
Streams(NF)	4	5.3	6.0	67	91	118	131	9	13	Morrison, 1982
Streams(F)	4	4.3	5.1	56	73	148	188	18	28	
Galloway										
Rainfall		4.4		12		70		-		Harriman
Streams(NF)	10	4.7	8.0	30	132	76	266	10	240	et al ., 1986
Streams(F)	9	4.3	6.7	40	165	112	201	36	384	
Lochs (NF)	18	4.4	5.8	15	80	96	136	58	265	
Lochs (F)	4	4.9	5.7	35	90	106	153	87	145	
Cairngorm										
Rainfall										DAFS, 1
Streams(NF)	28	4.6	8.1	19	525	39	174	0	14	
Lochs (NF)	28	4.5	7.6	13	247	21	121	0	110	
North West Scotland										
Rainfall		5.44		43		39		-		
Streams(NF)	4	6.2	6.5	54	94	73	88	30	50	
Streams(F)	3	6.1	6.6	39	193	90	130	52	97	Harriman and Wells, 1985
WEST WALES										
Rainfall		4.9		35		54		64		Stoner et
Streams(NF)	7	5.3	6.8	84	273	75	135	81	342	al ., 1984
Streams(F)	6	4.7	5.4	62	100	92	126	25	450	
R.Camddw(NF)	1	5.3		68		54		126		Stoner and
R.Tywi (F)	1	4.8		69		102		234		Gee, 1985
MID WALES										
Lakes (NF)	2	5.0	5.4	54	70	122	135	81	108	ibid
Lakes (F)	3	4.7	5.1	47	100	138	235	180	360	
CUMBRIA										
Upland lakes	22	4.4	6.3	28	220	60	167	-	-	Carrick and
Lowland lakes	21	4.7	6.9	43	542	70	216	-	-	Sutcliffe,1982
Streams(NF)*	15	4.8	6.7	51	205	44	202	71	283	Selected sites see below

*Site Nr.M3,M5,M6,M7,M14,T9,T24,T41,T46,T49,T50 Crawshaw,1984
3,11,e & f Carrick & Sutcliffe , and 1983
N=Number of sites. Ca, ExSO4 as ueqL-1 , Al as ugL-1 .

Ard Forest, Central Scotland (Figure. 1), an area receiving relatively high acid deposition, (Table II) showed highly significant differences in acidity and excess sulphate between streams draining open moorland and mature conifer forests [>50% catchment cover, >15 yr old]. However the correlation between acidity and sulphate was highly significant for both the forested and non-forested catchments (Figure 3c). There were no significant difference between the streams in young plantations and the open moorland suggesting that the increased acidity is not accounted for, in the long term [> 2yr], by ploughing and planting. Streams in the forest were permanently acid, with a total loss of alkalinity, while the moorland streams exhibited a strong seasonal trend being temporally acid [pH <5.6] from autumn through to spring.

In 1978-9 and 1983-4 water quality was measured in the bulk precipitation, and in 22 lochs and 57 streams in Galloway, S W Scotland (Harriman et al ., 1986). The area is also subjected to episodes of highly acidic deposition (Table II), and drains catchments of granite or metamorphosed shales and schists overlain with well leached soils giving, in most cases, permanently acid clear waters. Sixteen of the 22 lochs and 12 of the streams had a mean pH < 5.0; all lochs and 20 streams had a mean [Ca] < 100 ueqL-1 and only 5 streams in the Lower Fleet catchment had a mean [Alk] >50 ueqL-1. Aluminium concentrations were high; 6 to 924 ugL-1 labile Al, 87 to 1034 ugL-1 total monomeric Al. As in the Loch Ard Forest streams, drainage from mature forest catchments were more acidic and had substantially elevated [Al] (Table II and Figure 3c).

A series of samples has also been analysed from 22 lochs and 14 streams in the Cairngorms, most of which are non-forested or have a low density of indigenous Pinus sylvestris . In N. W. Scotland samples have been taken from 6 streams in moorland and intensely forested catchments. This area has a lower concentration of acidity in precipitation relative to either Galloway or L.Ard (Table II).

3.2 Wales

In West Wales Stoner et al . (1984) obtained data during 1981/2 for three rainwater sites and 13 streams in the Upper Tywi catchment. The episodic nature of acid deposition, and the mean acidity in precipitation, is generally similar to that in other remote areas of the UK (Gee and Stoner 1985). The pH values in each of the streams had a wider seasonal variability than the Scottish streams, typically pH 5.4 to 8.0 for a non-forested catchment and pH 3.8 to 7.2 for a forested catchment. Superimposed on this were short term episodic events, related to storm events with low pH 3.8 to 4.3, and high [Al] 900 to 1000 ugL-1.

The differences in stream chemistry between forested and non-forested catchments reported by Harriman and Morrison (1982) for the L.Ard area were confirmed by Stoner et al . (1984) and Stoner and Gee (1985) in the Tywi streams and Mid-Wales Lakes [n=5] (Table II). The greater

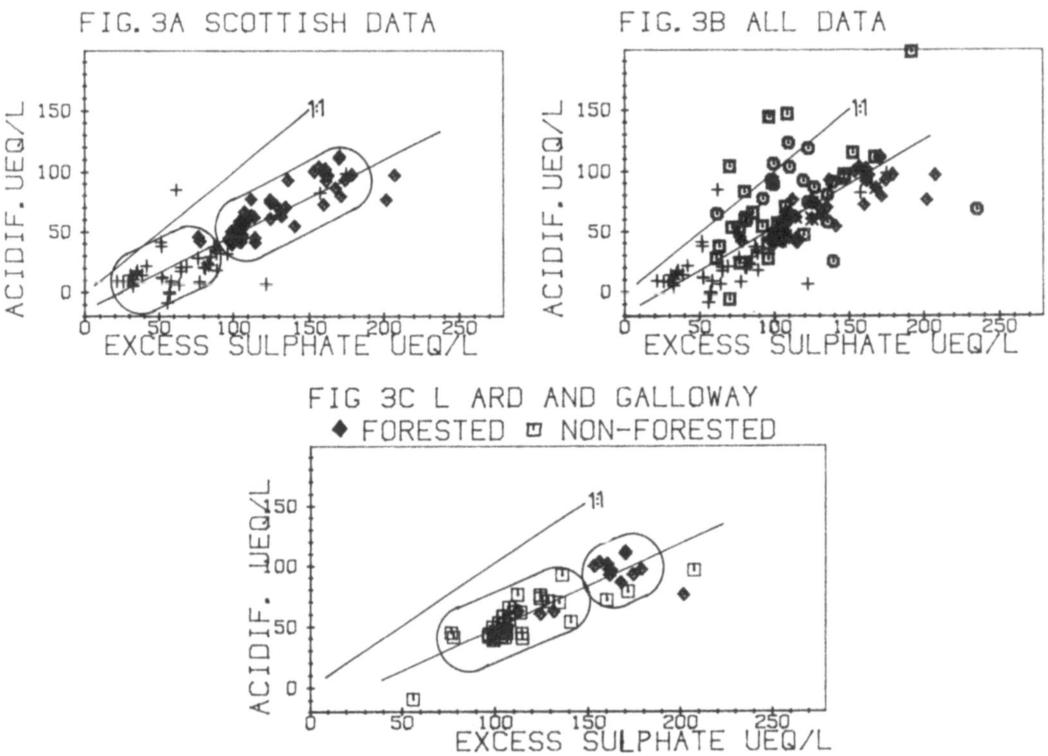

Figure 3. Correlation of Acidification Values (Henriksen, 1982) and excess sulphate for some sensitive surface waters in the UK. [A]= Scottish data only. (+) NW, Cairngorm,(◆) L.Ard and Galloway sites, [B] = All data including (◻) Wales and (◎) Cumbria. See Table III for correlation data, [C] (◆) L Ard and (◻) Galloway sites comparing forested and non-forested catchment.

TABLE III

CORRELATION OF ACIDIFICATION VALUES AND EXCESS SULPHATE FOR UK SUFFACE WATERS

DATA SETS	n	m	c	r[p 0.05]
A	92	0.61	-14.4	0.87
A+B	108	0.57	-6.1	0.76
A+C	107	0.52	0.2	0.72
A+B+C	123	0.48	-0.1	0.64
A+B+C+D	150	0.52	0.2	0.61

See Table II for data and references.
A=SCOTLAND[N=92] B=WELSH[N=16] C=DUDDON(NWWA+FBA)[N=15] D=CUMBRIA(FBA)[N=27]

hydrological buffer of these lakes moderates the short-term seasonal changes found in the streams, but the difference between forested and non-forested sites remained highly significant for pH, [Al] and excess [SO4]. Using these data sets streams were classified on the basis of 'hardness' (expressed as CaCO3) and land-use. Stoner et al .(1984) have attributed the current stream chemistry to the acid geology in an area which receives episodes of highly acidic precipitation and to the changes in land use by afforestation.

3.3 Cumbria

Data from Cumbria have been collected by the Freshwater Biological Association (Carrick and Sutcliffe 1982) primarily for the interpretation of biological problems unconnected with acidification, although recently these data have been reviewed in relation to acid deposition. Sutcliffe et al . (1982) and Sutcliffe, (1983) conclude that "in seeking evidence for acidification they failed to find it. " However it may be more appropriate to conclude that despite the large data set for the eutrophic lakes there was insufficient data from the unproductive upland lakes and tarns in this region to allow any assessment of acidification trends.

For these less productive lakes, Sutcliffe comments that "there are insuffícent data to draw detailed conclusions, except that pH and Alk have not altered markedly in the last 50 yr". However there are no continuous data sets available (Pearsall, 1930; Mortimer, 1941 - 42; Mackereth et al . 1957 - no pH data) and from the report of bulked data for the upland lakes (Carrick and Sutcliffe, 1982) it is not possible to trace the true history of the water quality during this time.

Any conclusion from these sites must await the results of palaeolimnological analysis.

Data from the Duddon; 20 sites between 1971 and 1974, and the Windermere catchments; 31 sites between 1975-78, (Carrick and Sutcliffe, 1983), is quite extensive. However there are only 6 sites where pH, Alk, Ca, Mg, Cl & SO4 were collected simultaneously; for most data there was no information on alk or sulphate which precludes the calculation of acidification values and excess sulphate.

Crawshaw (1984) reports on detailed investigations of the water quality on the Rivers Esk and Duddon, and other Cumbrian catchments following fish mortality incidents on the Esk in 1980 and 1982. Chemical data were collected from 100 potentially sensitive sites, but Na was not included in the analysis, and chloride data were limited. As a result the marine salt contribution could only be deduced from the data for 11 of the sites where the mean pH was < 6.5.

3.4 Acidification Values

All water quality data where the determinands pH, Alk, Ca, Mg, Cl and

SO_4 were available were used to calculate the acidification values for UK surface waters, according to Henriksen's definition (1982) and to determine the correlation against excess sulphate (Figure 3). Despite some criticism which has been levelled at this empirical approach it remains a valuable tool in such an overview analysis.

The Scottish data set (Figure 3a) [n=92] covers the widest range of excess sulphate, acidification values and rainfall acidity (Table II).The regression ($Y = -14.4 + 0.61X$, $r = 0.87$) shows two distinct clusters between areas occurring above and below the pH 4.5 acid deposition isopleth, similar to the Norwegian lakes data (Henriksen, 1982) ($Y = -12 + 0.98X$). The Cairngorms and NW Scotland, are areas of lowest UK acid deposition (Barrett, 1983), have low acidification values, and excess sulphate and are still bicarbonate controlled, despite the catchment soil acidity. The L.Ard and Galloway areas have higher acidification values, and excess sulphate and cover an area receiving significantly higher levels of acid deposition. Within each cluster forested and non-forested catchments form a second division. The difference in equivalence is accounted for by metal ions such as Al which are leached from the catchment. Al has not been added because the speciation and the equivalence of the Al for much of the data are unknown.

The addition of the Welsh data [n=16], Duddon [n=15] and particularly the Cumbrian upland lakes data [n=27] reduced the correlation possibly because (a) these data were, in most cases, single values and (b) there is a wider range of natural sulphate and magnesium associated with bicarbonate from the more diverse geology (Figure 3b).

4. TRENDS IN FISH POPULATIONS

Both long and short term changes in water quality can seriously affect fish populations, in particular salmonid species, over and above variations caused by changes in land use, climate, habitat and stocking.

4.1 Scotland

Harriman and Morrison (1982) (HMSO, 1981) found that most of the L.Ard Forest streams sampled contained only eel, while the adjacent moorland streams maintained a sparse brown trout [Salmo trutta L] [~5/100 m^2] population. Attempts to hatch eyed or newly fertilized salmon eggs in the forest streams were only successful when limestone chips were added to the egg boxes. Streams were also stocked in the spring with salmon fry, but none were later recovered in the forested streams. Eggs (54 to 92%) and fry survived in the moorland streams.

In Galloway (Figure 1) Harriman *et al*. (1986, HMSO, 1981) have reported changes in fish populations in 22 lochs and 26 streams. The absence of migratory fish was accounted for by natural and artifical barriers. However the spatial and temporal distribution of the brown trout could not be explained in terms of altitude, access, suitable spawning grounds, overfishing or food availability. Of the 22 lochs gill-netted, 7 were fishless and of the 26 streams electrofished 7 were also fishless. There was clear evidence from historical records of declines in fish catches in L. Riecawr, Grannoch and Fleet and the fishing effort for Riecawr indicated a significant [$p<0.05$] drop in the catch per unit effort between the '50s and '70s. The main cause of the declines and in some catchments, failure of the fisheries in these clear waters has been attributed to the water quality associated with forestry practices and the high levels of acid deposition in this area.

Streams currently being studied in N.W. Scotland and in the Cairngorms all have a population of brown trout [10 to 30/100 m^2] below the normal altitude limits of 500 to 600m and where the stream substrate gives reasonable cover, and a salmon population where there are no physical barriers.

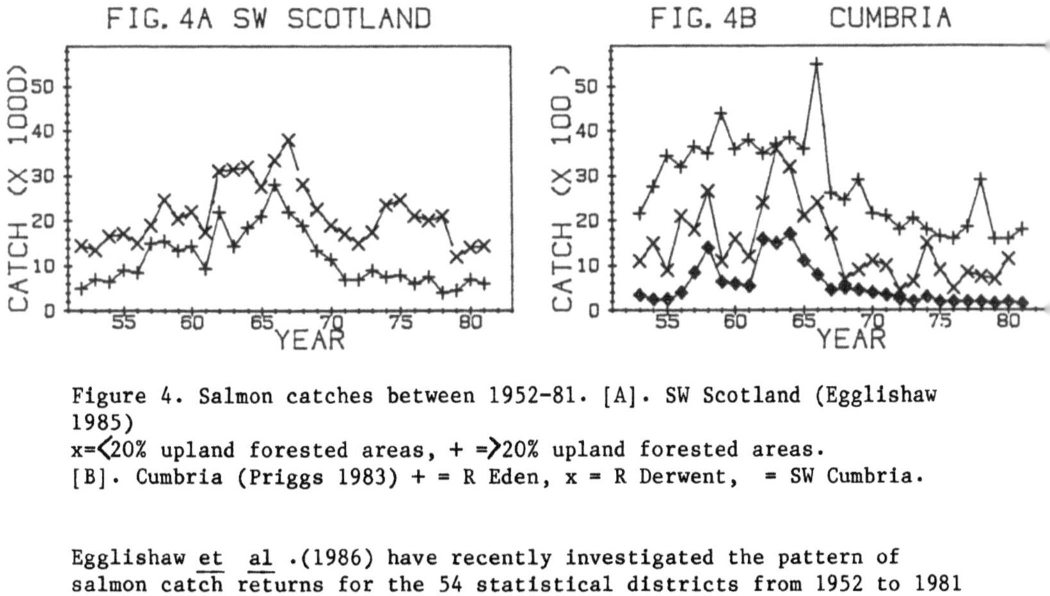

Figure 4. Salmon catches between 1952-81. [A]. SW Scotland (Egglishaw 1985)
x=<20% upland forested areas, + =>20% upland forested areas.
[B]. Cumbria (Priggs 1983) + = R Eden, x = R Derwent, = SW Cumbria.

Egglishaw *et al*.(1986) have recently investigated the pattern of salmon catch returns for the 54 statistical districts from 1952 to 1981 and found a highly significant correlation [Chi Sq=25. 45, $p \ll 0.001$] between the decline in catch and districts with 20% upland conifer forests; the extent of the decline increasing as the forest cover increases. Such a highly significant correlation outweighs many of the other possible causes such as the variation in fishing effort and variability of such records. In S.W. Scotland both forested and non-forested areas showed marked and continuous declines in catches implicating other factors, in addition to forestry (Figure 4a). Similar catch statistics are not currently available for the whole of England and Wales.

4.2 Wales

4.2 Wales

In Wales Gee and Stoner (1985) investigated the effects of seasonal variation in water quality on salmonid survival in forested and non-forested streams. Although only a small number (5,0+) of trout were used in each stream they confirmed a significant correlation between mortality (LT50) and Al concentration for 2 autumn periods, Oct'81 and Nov'82, such that
Mortality (LT50) = 0.0038 [Al] - 0.018 (r = 0.95, n=13, p 0.001).

In contrast all fish survived in the same streams during July'82, except for the heavily forested catchments where the pH and Ca levels remained low.

The effects of short term variations was also investigated in Nov'82 in the Tywi catchments where 0+ trout and salmon were caged in each of the streams of poorer water quality (5 forested, 1 non-forested) and richer stream (pH 5.8 to 7.8 , [Ca] 60 to 560 ueqL-1) as control. Following rainfall (135mm) the [Ca] rose 100 ueqL-1 with pH, and [Al] (50 to 100 ueqL-1) remained relatively constant; giving a typical sea salt effect. Fish in the non-forested streams survived while those in the forest died.

4.3 Cumbria

Studies on water quality (Crawshaw, 1984) and fish stocks (Prigg 1983) on the Esk and Duddon and other selected lakeland catchments in Cumbria were undertaken following reports of declining fish stocks (Figure 4b) and adult and juvenile salmonid mortalities in the lower reaches of the Esk during a prolonged summer storm following the drought in 1980. The Esk is a clear water river, even during such spates, and in the course this episode the pH dropped to 5.0. This, and the elevated levels of Al [200 to 500 ugL-1] and low Ca [100 ueq L-1] values were suggested as probable causes of the incident. Further mortalities occurred in 1983 on the Duddon and the Esk where the pH dropped to 4.3 and remained between pH 5.0 to 5.5 for several days.
Such events may be due partly to the reduction [80 to 90%] of lime application to the lower areas of the catchment as a result of the cessation of the government subsidy in 1976 (Robinson, 1984). Although this may have aided the adaptation of the fresh run salmonids to freshwater conditions during low pH/high flow conditions, the streams in the upper catchments would not be affected. In these streams Prigg (1983) found that the fish populations declined rapidly below pH 5.6 to 5.8 and particularly [Ca] $<$ 100 ueqL-1. Streams in the upper catchments with low pH and [Ca] were either fishless or maintained a sparse population with missing year classes. Fish population and chemical data from Galloway (Harriman et al., 1985) showed a similar trend. Such data are indicative of sensitive catchments where the intermittent variability of the stream chemistry can have a fatal effect on fish stocks particularly at either the spawning or egg stages of development.

5. CONCLUSIONS.

[1]. Sensitive surface waters in the UK are confined primarily to upland catchments in West and Mid Wales, some catchments in Cumbria, particularly the R. Duddon and Esk, and the Southern Uplands & Highlands of Scotland (Figure 1).
[2] The strongest evidence for acidification of sensitive waters since the 1800s comes from palaeolimnological studies in SW Scotland; where the decline in pH of the water in non-afforested sites has been attributed to acid deposition related to human activity. This is supported by declines in brown trout populations in the Galloway lochs and streams.
[3] There is a significant correlation between acidification values and excess sulphate and acid deposition, particularly for the Scottish data.

[4] Salmonid populations in streams in West Wales, Cumbria, Galloway and L.Ard have been correlated with mean values for pH, Ca & Al. Mortality is most likely at the egg and fry stages during stream episodes where acidity and Al levels are highest.
[5] Significant differences in pH, excess sulphate and elevated Al were found between forested and non-forested catchments in L.Ard, Galloway, West and Mid Wales. The correlation between declines in salmon catch returns and areas of mature conifer plantations that occupy >20% of the catchment in Scotland is also highly significant.

6. REFERENCES

Anon: 1982, Ecological Effects of Acid Deposition. Report and Background Papers. Stockholm Conference on the Acidification of the Environment. National Swedish Environment Protection Board Report No. 1636.
Anon: 1984, Acid Rain Report No 14, The Watt Committee on Energy , London.
Barrett, C. F. et al : 1983, Acid Deposition in the United Kingdom, Stevenage, Warren Springs Laboratory.
Battarbee, R. W. and Flower, R. J.: 1984, Limnol. Oceanog ., 29, 1325.
Battarbee, R. W., Flower, R. J., Stevenson, A. C. and Rippey, B.: 1985, Nature 314, 350.
Battarbee. R. W.: 1984, Phil. Trans. R. Soc. Ser B ., 305, 451.
Battarbee, R. W. and Flower, R. J.: 1985 Working Paper No 8, Paleoecology Research Unit, Univ. Coll. , London.
Buckley-Golder, D. H.: 1984, Acidity in the Environment. Energy Techn. Support Unit, Harwell, Oxfordshire, 181 pp HMSO.
Carrick, T. R. and Sutcliffe, D. W.: 1982, Occ. Publs. Freshwat. Biol. Assoc. No. 16.
Carrick, T. R. and Sutcliffe, D. W.: 1983, Occ. Publs. Freshwat. Biol. Assoc. No. 21
Crawshaw, D. H.: 1984, North West Water Authority Report,(UK) TSN 84/3
Darley, J.: 1985, Particulate soot in Galloway Lake Sediments:its application as an indicator of environmental changes and as a technique

for dating recent sediments. BSc Dissertation, University College, London.
Drablos, D. and Tollan, A. (Eds): 1980, "Ecological Impact of Acid Precipitation" Proc. Int. Conf. Sandefjord , Norway.
Egglishaw, H. J., Gardiner, J., Foster, J.: 1985, Scottish Geograph. Mag, 102, 57
Farmer, P.: 1984, Acid Rain & the Environment 1980 - 84; A selected bibliography. Technical Communication ISBN 0-94665-04-9.
Flower, R. J. and Battarbee, R. W.: 1983, Nature 305, 130.
Gee, A. S. and Stoner, J. H.: 1985, Environ. Poll . in press.
Gorham, E.: 1955, Geochem. Cosmochim. Acta 7, 321.
Harriman, R. and Morrison, B. R. S. 1982, Hydrobiologia 88, 251.
Harriman, R. and Wells, D. E.: 1985, J. Inst. Water. Poll. Cont. 84, 215.
Harriman, R., Morrison, B. R. S., Caines, L. A., Collen, P. and Watt, A. W.: 1986, Water, Air & Soil Poll ., in press.
Haworth, E. Y.: 1985, 53rd Annual Report of the Freshwat. Biolog. Assoc. 60-79.
Henriksen, A.: 1982, In Acid Rain/Fisheries (Ed Johnson R. E.) Cornell Univ., Ithaca, NY p103.
HMSO 1981, DAFS Freshwater Fisheries Laboratory Triennial Review of Research 1979-1981, 35-38.
HMSO 1984, Acid Rain:Fourth Report from the Environmental Committee:House of Commons.
Kinniburgh, D. G. and Edmunds, W. M.: 1984, The Susceptibility of UK Groundwaters to Acid Deposition. Report to the Dept. of the Environ. from the British Geological Survey (Wallingford, UK) .
Last, F. T.: 1982, Agric. & Environ. I. , 299.
Mackereth, F. J. H, Lund J. W. G. and Macan, T. T.: 1957, Proc. Linn. Soc. Lond. 167, 157.
Mortimer, C. H.: 1941-2, J. Ecol ., 29, 280, 30, 147.
Pennington, W., Haworth, E. Y., Bonny, A. P., and Lishman, J. P.: 1972, Phil. Trans. R. Soc., (B) 264, 191.
Pennington, W.: 1984, 52nd Annual Report of the Freshwat. Biolog. Assoc. 28-64. Pearsall, W. H.: 1930, J. Ecol . 18, 306.
Prigg, R. F.: 1983, North West Water Authority Report BN 77-2-83.
Rosenquist, I. T.: 1978, Sci. Tot. Environ . 10, 39.
Robinson, J. F.: 1984, North West Water Authority Report (UK) October
Seip, H. M.: 1985, Acid Precipitation-Literature Review. S1-80701-4-1984(English) , Min. of Environ. (Also S1-810701-1, Summer 1980-October 1981(Norwegian) , S1-80701-2 Autumn 1981-Autumn 1982(Norwegian), S1-80701-3, Autumn 1982-Autumn 1983(English))
Stoner, J. H. Gee, A. S. and Wade, K. R.: 1984, Environ. Poll. (A) 35, 125.
Stoner, J. H. and Gee, A. S.: 1985, Inst. Water. Eng. & Sci . 39, 27.
Sutcliffe, D. W., Carrick, T. R., Heron, J., Rigg, E., Talling, J. F., Woof, and C., Lund, J. W. G.: 1982, Freshwater Biol ., 12, 451.
Sutcliffe, D. W.: 1983, 51st Annual Report of the Freshwat. Biolog. Assoc. 30-62.

FOREST DECLINE IN EUROPE: DEVELOPMENT AND POSSIBLE CAUSES

G.H.M. KRAUSE (1), U. ARNDT (2), C.J. BRANDT (3), J. BUCHER (4), G. KENK (5), and E. MATZNER (6)

1) Landesanstalt für Immissionsschutz des Landes NRW, Wallneyer Straße 6, 4300 Essen 1, FRG; 2) Universität Hohenheim, Fachbereich Pflanzenökologie, Postfach 700 562, 7000 Stuttgart 70, FRG; 3) Battelle, Pacific Northwest Laboratories, Richland, Wash., 99352, USA; 4) Eidgenössische Anstalt für das Forstliche Versuchswesen, 8903 Birmensdorf/ZH, Switzerland; 5) Forstliche Versuchs- und Forschungsanstalt Baden-Württemberg, Sternwald- straße 16, 7800 Freiburg, FRG; 6) Forschungszentrum Waldökosysteme - Waldsterben, Hüskenweg 2, 3400 Göttingen, FRG.

ABSTRACT. Symptoms of recent forest decline were first observed on silver fir (Abies alba) in the early 1970s in "clean-air" areas of southern Germany; they have subsequently been described in the early 1980s for Norway spruce (Picea abies) and recently broad-leaved trees such as European beech (Fagus sylvatica). The history of forest decline is discussed with emphasis on symptomatology, development and distribution. After a discription of results of recent surveys of tree injury, indicating an increase in the occurrence of decline, attempts will be made, by exploiting knowledge of pollutant emissions and deposition, to deduce the possible causes of European forest decline.

1. INTRODUCTION

In the early 1970s, silver fir (Abies alba Mill.) showed first symptoms of decline in rural and subalpine areas of southern Germany. Since silver fir is known to be quite

---

1) to whom requests for reprints should be addressed

Water, Air, and Soil Pollution **31** (1986) 647–668.

sensitive to stress, as observed during previous periods of decline over the last 150 yr (Wachter, 1978), observed symptoms, like needle loss, needle discoloration and general reduction in vitality, i.e. growth increment, were initially attributed to climatic stress situations, mainly water stress. Later in the 1970s and early 1980s, Norway spruce (*Picea abies* Karst), the economically most important tree species in German forestry, showed similar symptoms of decline. Symptomatology in general was comparable over all affected areas, although with regional differentiation, and was very often associated with nutrient deficiencies. Especially after the dry summers of 1982 and 1983, deciduous trees such as beech (*Fagus sylvatica*) and oak (*Quercus* spp.) seemed to be affected as well. Since the late 1970s, air pollution has been considered increasingly as the major causative factor (SR-U, 1982). Acid rain (Ulrich, 1981a), SO_2 (Wentzel, 1982a) and photooxidants, mainly O_3 (Arndt *et al.*, 1982; Bucher, 1982; Prinz *et al.*, 1982), were made responsible for decline of spruce and fir as well as natural environmental factors (Kandler, 1985). Not only in FRG, but also in parts of France, Italy, Switzerland, Austria, Belgium, The Netherlands as well as, GDR, USSR, Czechoslovakia and northern Jugoslavia, tree decline became apparent in a manner similar to the "Neuartigen Waldschäden" of FRG (Bucher, 1985a; Wentzel, 1985). The term, "novel forest decline" is an attempt to differentiate symptomatology from that of known decline-causing factors such as climatic and biotic stresses and also from "classical smoke injury" observed since the beginning of the industrial revolution. However, in the latter cases, a true cause-effect relationship can be defined. Today we are dealing in central Europe in most regions with the problem of an obvious effect but with no obvious cause. This paper gives an overview of distribution and development of decline as well as of possible causes of injury in central Europe.

2. SYMPTOMATOLOGY

Symptoms of the "novel forest decline" syndrome are observed on an ecosystem level as well as on individual trees. Those occuring at the ecosystem level include soil acidification, damage to the decomposer chain and damage to ground vegetation. Symptoms on leaves and needles of individual trees are most obvious and have been described with regional variation in many publications (e.g. Rehbock, 1982; Schütt, 1982; Prinz *et al.*, 1982, Rehfuess, 1983; VDI, 1983, 1985). However, fungal diseases, insect infestation as well as climatic and edaphic factors, and silviculture may vary the picture considerably.

In the following, only symptoms considered specific for novel forest decline are briefly described for tree species affected most:

Silver fir (*Abies alba* Mill.): Injury is characterised by:

- chlorosis and yellowing of needles (mainly upper side of older needles)
- casting of older needles (mostly starting from the base of the crown and from the inner parts of branches)
- premature reduction in height increment, i.e. stork's nest formation already found in relatively young trees.

It should be mentioned that several episodes of fir dieback have been reported over the last two centuries (Wachter, 1978; Seitscheck, 1981) with some similarities in symptomatology, resulting in the difficulty to differentiate between "old" and "new" dieback. Increment reduction, frequently observed (Schweingruber *et al.*, 1983; Kenk *et al.*, 1984), started in the early 1950s in some silver fir stands in Baden-Württemberg and in Switzerland now showing severe injury; first rather slow, later in greater pulses, related to climatic influences (Kenk *et al.*, 1984).

Norway spruce (*Picea abies* (L.) Karst.): Injury shows the general symptoms:

- yellow needles and chlorosis of older needles, partly also reddish-brown
- casting of older needles, delayed in comparison to fir (mostly starting from the base of the crown and from the inner parts of branches)

In contrast to silver fir, a corresponding loss in increment growth is not observed to start in spruce until needle loss has reached 30 to 50 % (Kenk, 1985).

It is nearly always observed that the upper sides of needles show more intense yellowing than the lower sides. Furthermore a clear shading, or shielding, effect is observed i.e. overlapping branches have yellow foliage, while those immediately below remain green (Prinz *et al.*, 1982). There is a strong correlation between the key symptom of yellow needles and nutrient deficiencies in needles as well as with low nutrient status of the soil, mainly Mg, Ca, probably K and Zn (Krause *et al.*, 1983 a; Rehfuess and Rodenkirchen, 1984; Zöttl and Hüttl, 1985). Evaluation of causative factors have therefore focussed on this phenomenon. However, yellow needles and chlorosis are not the predominant symptoms in all countries with novel forest decline.

The main symptom on pine (*Pinus austriaca nigra*, *Pinus sylvestris*) is again the yellowing of the older needles as well as needle loss. However, for more than 20 yr, pine dieback has been observed throughout Europe and is partly related to rapid and extreme temperature changes between summer and winter and to water stress as, for example, in eastern parts of Austria (Tomiczek, 1982; SR-U, 1983), and is in most other cases related to poor site conditions.

Symptoms on other tree species are not so clear cut. However, common beech (*Fagus sylvatica* L.) seems to have been affected for more than 10 yr, showing reduced shoot growth and dieback of lateral branches (Roloff, 1985). This is enhanced, and even probably triggered, by dry summers like 1976, 1982, 1983. Premature leaf drop, also seen as a symptom of "novel forest decline" by Schütt *et al.* (1983b, a.o.), seems to be more related to biotic factors (Roloff, 1985). The overall symptomatology, however, is too unspecific to be related to specific causes, including the air pollution factor (Halbwachs, 1971; Bucher, 1985a; Donaubauer, 1985).

3. DEVELOPMENT AND DISTRIBUTION OF NOVEL FOREST DECLINE

Since vitality of silver fir and Norway spruce deteriorated rapidly for no apparent cause in Baden-Württemberg (Schröter, 1983) and similar damage was reported from Bavaria (Schütt, 1982), the German Forest Services conducted their first detailed survey on forest decline in 1982, showing that the decline phenomenon was not restricted to southern parts of Germany (SR-U, 1982). Since then, more detailed and sophisticated inventories were carried out annually. Neighboring countries such as Austria, Switzerland, The Netherlands, as well as France and Italy started similar inventories in 1983 or 1984. However, results are comparable only since 1984, when inventories in Federal Republic of Germany (FRG) and other countries were carried out on the basis of random sample plots with single tree evaluation. As a whole, a region of central Europe between 53°N and 46°N now seems to be affected. The following gives a detailed description of the distribution of forest decline in central Europe.

3.1. Federal Republic of Germany

According to the inventory carried out by the Federal Ministry for Agriculture and Forestry (BML) in 1984 (Breloh, 1985), 3.7 Mha showed injury, equivalent to 50 % of the total forested area. Based on injury classes, 0 to 10 % needle loss for "not affected", 11 to 25 % "slight injury", 26 to 60 % "moderate injury", and over 60 % "severe injury", 33 % of the forested area showed "slight injury", 16 % were "moderately injured" but only 1,5 % are "severely" affected, including dead trees. Large-scale death of trees has not been observed. However, a shift from lower to higher injury classes has become apparent. In relation to the forested area, 51 % of spruce, 59 % of pine, 87 % of fir, as well as 50 % and 43 % of common beech and oak showed injury in 1984. Age classes above 60 yr are affected most, but recently, younger stands are deteriorating as well (Breloh, 1985). In many stands, dominant spruce trees are affected preferrentially and trees at the edges of stands are more frequently injured than those within a stand (Prinz et al., 1982). Similar observations were made during the Swiss inventory in 1984 (Schmid-Haas, 1985).

Regional differentiation shows that southern parts of Germany are still affected most, for example 66 % in Baden-Württemberg, 57 % in Bavaria, compared to 27 % in Schleswig-Holstein and 36 % in Lower Saxony. This is due to the prevailing higher elevations and frequent distribution of silver fir in Bavaria and Baden-Württemberg.

Most injury was classified as slight (11 to 25 % needle loss), leaving a total of about 17 % moderate to severe injury. Injury of fir and spruce, which first appeared at higher elevations of about 800 to 1000 m in southern Germany, is currently observed between 500 to 600 m in the Black Forest area on poor granite and sandstone soils and 350 to 500 m in the mountainous areas north of the river Main. West and southwest facing slopes are usually affected first (Mies and Zöttl, 1982; Reichelt, 1983; Prinz, 1983). However, this can not be regarded as a strict rule, since injury is also observed to some extent in flatlands (near the German-Dutch border) and valley floors (Prinz et al., 1982; Genßler, 1985).

3.2. Switzerland

The first Swiss inventory of forest decline was carried out in 1984 using the same criteria of injury assessment, and similar methods of evaluation as in FRG (Schmid-Haas, 1985, Schöpfer and Hradetzky, 1984). But, while in FRG most changes from normal were attributed to novel forest decline phenomena and included in the data presented above, the Swiss inventory excluded injury due to site condition, and known biotic and abiotic causes. Of all trees surveyed, 34 % were considered to be injured. Again, the highest percentage was found in injury class 1 (11 to 25 %), while 1 % was regarded as severely injured or dead. Pine is affected most (50 %), but one third of all coniferous species and some deciduous trees such as oak (30 %) and beech (25 %) also showed symptoms of injury. Damage has been observed in all regions, but seems to focus on the inner alpine valleys and high altitudes. Similar to observations in FRG, forest decline is clearly related to site characteristics such as tree age, canopy density, and species composition of stands. No correlation has been found so far between acidity of underlying bedrock and occurrence of injury (Schmidt-Haas, 1985).

3.3. Austria

Not all federal states of Austria were included in the 1984 inventory which is based on a 4 x 4 km grid with a somewhat different classification than that used in Switzerland and FRG. Furthermore, the survey was limited to the main tree species, spruce and pine. According to this first program, roughly 70 % of the trees assessed were not affected, while 30 % showed injury symptoms; pine was more severely injured than spruce (Pollanschütz, 1985).

3.4. Italy

Using a similar method of assessment as in Austria and the classification of FRG, the 1984 inventory was executed only in the autonomus Italian province of South Tyrol; 16 % of spruce, 35 % of fir, and 6 % of pine showed injury of unknown origin, mostly as needle loss, in injury class 1 (11 to 25 %) (Minerbi, 1985).

3.5. The Netherlands

On the basis of random sampling, 85 % of the total forested area in The Netherlands was surveyed in 1984. Vitality was characterized primarily by needle loss and needle discoloration. Of all tree species, about 43 % of conifers, and 67 % of deciduous trees were considered healthy; 44 % and 28 %, respectively, showed initial reduced vitality and 13 % of the conifers as well as about 5 % of the broad-leaved trees were considered to be severely injured or dying (den Boer and van den Tweel, 1984).

3.6. France

Forest damage was assessed in 1984 in the region of Alsace-Lorraine, adjacent to the Black Forest in FRG and Franche-Comte, in the French Jura. Based on the system of injury evaluation in the FRG, but using a slightly different system of classification for needle loss assessment, about 65 % of all trees surveyed are considered healthy. Around 24 % were found to be "slightly", and 10 to 13 % "severely" injured in both regions. Injury to fir was greatest, followed by pine and spruce (Bucher, 1985b).

3.7. General Remarks

Forest decline has been observed in various other countries of non-Nordic-Europe (Seitscheck, 1984). However, it is questionable, whether this injury can be included under "novel forest decline". In many areas of Czechoslovakia, GDR and Poland, for example decline can be related to classical effects of SO_2 (Wenzel, 1982b).

Reflecting the results of the various European inventories carried out so far, novel forest decline seems to be ubiquitous in spruce, pine and fir forests as well as to some extent in areas with beech, oak and other deciduous tree species. However, it should be mentioned that, according to the 1984 inventories, just 13 % of conifers in Alsace-Lorraine and Franche-Comte (France), 17 % in FRG, 13 % in South Tyrol (Italy), 13 % in The Netherlands, and 8 % in Switzerland showed needle loss exceeding 25 %. Since needle loss below 25 % is quite difficult to estimate, errors cannot be excluded. Natural variation of site characteristics, tree provenances, and other growth factors may be just as responsible for such needle loss as any anthropogenic factors (Ballach and Brandt, 1983, 1985). The new type of European forest decline is definitely present but should not be overemphasized in light of reports of previous decline episodes (Wachter, 1978; VDI, 1985).

4. POSSIBLE CAUSES OF NOVEL FOREST DECLINE

The wide distribution of forest decline in Europe, the wide range of varying factors in respect to soil, climate, and, last but not least, the lack of perceiving the "air pollution environment" (Last et al., 1984) makes it difficult to attribute this phenomenon to a defined cause in a pure and simple way, and thus many hypothesis have been presented to explain the new type of forest decline. Prinz (1983) has postulated the following three criteria which, theoretically, must be met for a conclusive explanation of the new decline syndrome.

- It must be possible to relate specific symptoms of injury to the causal factor in question.
- Temporal development of injury must coincide with temporal development of the causal factor in question, including accumulation effects and delayed action of the factor.
- Spatial distribution of injury must largely coincide with spatial distribution of the factor in question.

4.1. Possible Natural and Silvicultural Factors

Considering the experience of forestry in previous centuries, extreme climatic events, particularly drought, have the most significant impact on tree vitality and hence on stand or even ecosystem composition. Since of all economically important tree species, silver fir, Norway spruce, and common beech are relatively drought sensitive, these species were likely to be affected during past dry periods and also show the most extensive signs of novel forest decline. The last important dry period in Europe was experienced in the 1970s and the early 1980s probably indicating that the climate in some areas of Europe is shifting toward dry, continental summer-type conditions with all possible ecological implications (Asthalter, 1984). Schwerdtfeger (1981) points out that long lasting dry periods are the most important predisposing factors for diseases of biotic or abiotic origin. Recent dry periods increased root and leaf injury and were followed by periods of epidemic spreading of bark beetles or pathogens such as Oligonychus ununguis Jac., Lophodermium spp., Chrysomyxa spp., Rhizosphaera kalkhoffii affecting spruce and, to some extent, fir or Rhynchaenus fagi L., Cryptococcus fagi Bärspr., and Apiognomonia errabunda (Rob.) affecting beech (Rehfuess, 1983; Donaubauer, 1985; Minerbi, 1985; Roloff, 1985). However, other aspects of indirect effects of drought are given by Ulrich (1981b) and Matzner and Cassens-Sasse (1984) who observed increased soil acidification due to so called "acid pulses" in some regions, and by Arndt and Lindner (1981) as well as Prinz et al., (1982) who pointed out in '82 the increased impact of photochemical oxidants on vegetation during warm periods with intense solar radiation. Furthermore, high solar radiation together with water stress and nutrient imbalances may lead to photoinhibition and reduction in assimilate production (Powles, 1984; Lichtenthaler and Buschmann, 1984). Besides drought, periods of severe frost, especially with rapid and extreme temperature changes as observed in FRG for example during the winter of 1978/79 and the spring of 1981, may increase susceptibility of trees to action of other biotic and abiotic stressors (Ernst, 1934; Wentzel, 1956; Materna, 1979; Cramer, 1984; Cramer and Cramer-Middendorf, 1984), resulting in considerable needle loss (Rehfuess, 1983; Friedland et al., 1984) or retarded development of leaves (Cramer-Middendorf, 1984).

Viruses or virus-like organisms were considered as a possible cause for novel forest decline, due to the apparent spreading of the disease from more or less specific areas of origin. Viruses of the Potex group have been observed in beech and oak and virus-like particles have been found in terminal buds of spruce and fir (Frenzel, 1985; Nienhaus, 1985). However, Nienhaus (1985a) considers viruses as an important predisposing factor but not as a primary cause, based on most recent research.

Besides these predisposing factors, local site conditions are also important for the evaluation of possible causes, since they strongly influence water and nutrient status. The key symptom of yellow needles is due to a "sudden" deficiency of Mg and Ca (Prinz et al., 1982; Rehfuess, 1983a; Zech and Popp, 1983). Although the new type of forest decline appears on many different soil types, including the calcarious soils of the Swabian alps, which does not favor soil as a primary causative factor, decline seems to be related to poor soil conditions with generally low nutrient availability (Pollanschütz, 1985; Rehfuess, 1985; Schmid-Haas, 1985; Zöttl, 1985;) partly explaining regional differences in extent and development of decline (Schmidt and Hartmann, 1984; Evers, 1985; Rehfuess, 1985). It is reasonable to assume that soil conditions are very significant predisposing factors.

One factor difficult to discuss extensively here, includes problems associated with silviculture. As forest science developed methods for producing more timber on the basis of sustained yield, natural habitats of forests were altered dramatically in many areas of Europe. Stands of deciduous trees have been diminished in favor of more economically important coniferous trees, mostly Norway spruce, sometimes without regard for natural requirements of the tree for stand site and climate. Another problem is associated with the original provenance of the tree and hence its genetic adaptation to a given environment and its propagation outside this environment. Forest management practices are also considered as a predisposing factor (Manion, 1981) but does not seem to have a major causal effect on the occurrence of forest decline. Injury is not limited to unsuitable sites or to particular forest types, but is also found on optimal sites and in well managed mixed stands.

Considering only natural factors, it is possible to accept the hypothesis of Kandler (1985) that nature is again showing her capriciousness and forest decline is a disease affecting natural systems, as has happened many times before. However, studies at the permanent observation plots of the Forestry Research Institute (FVA) of Baden-Württemberg (Schröter et al., 1985) where health of specific trees has been evaluated over many years for the development of specific, ubiquitous symptoms of decline, such as yellowing of needles, make it apparent that causes of decline presently observed in central Europe are not due to natural and silvicultural factors alone.

4.2. Air Pollutants as a Possible Cause of Forest Decline

Recent research indicates that the impact of air pollutants is a major factor contributing to the development of the disease (SR-U, 1982; Breloh, 1985), and, for various reasons, secondary air pollutants today probably pose the greatest threat to forests (Smith, 1984). However, the most striking phenomenon is that symptoms of forest decline were first observed, and became most severe, in areas quite remote from industrial and urban agglomerations. If pollutants are considered to be a dominant cause of injury, special attention must be paid to those pollutants with the greatest impact in such areas. Two major pathways regarding impact of air pollutants on forest trees are discussed:

- Direct effect of gaseous air pollutants, including interaction with wet deposition on above-ground plant parts with indirect effects on the root system.
- Accumulation of dry and wet deposited substances in soils, changes in soil properties, and their effects on plants via the root system.

4.2.1. Impact of Gaseous Air Pollutants

When considering the classical air pollutants and their possible effects on vegetation, SO_2 has to be named first. However, SO_2-emissions have been decreasing since 1970 (Sartorius, 1984). SO_2-concentrations in remote areas with widespread decline symptoms range between 15 and 40 µg m^{-3}. In southern Black Forest, southern Bavarian forests, and parts of Switzerland with higher elevation annual means are well below 10 µg m^{-3} (Bucher, 1985b; Obländer and Hauß, 1985). Increased concentrations are usually assigned to definite source areas. Peak values are about 500 µg SO_2 m^{-3} over several hours and are associated with high wind speeds and certain wind directions. Based on the IUFRO-resolution (IUFRO, 1983) these values are well below reccommended ambient standards for protection of trees. However, IUFRO-standards do not consider interaction with other air pollutants present (Wentzel, 1982b) like O_3, known to increase the phytotoxic potential. In general, there is neither the necessary spatial nor temporal correlation between affected areas and SO_2-concentrations, apart from some regions in eastern countries or parts of Bavaria bordering Czechoslovakia, where annual means exceed 70 µg m^{-3} and peaks are around 1800 µg m^{-3} (Materna, 1983; Dässler and Lux, 1984), SO_2 can be eliminated as an overall cause for novel forest decline.

Emissions of NO_x however, have increased considerably during the 1970s in most areas of Europe (Sartorius, 1984) e.g. in FRG by 50 % between 1966 and 1982. Gaseous N-compounds have a low phytotoxic potenial (Taylor et al., 1975) and since concentrations in rural areas are about 5 to 20 µg m^{-3} (calculated as NO_2) with maximum daily average values of about 60 to 90 µg m^{-3}, NO_x alone can also be eliminated as a possible cause. Even more so, since NO_x concentrations are much higher in industrial areas (Pfeffer, 1985), where novel forest decline is not prevalent.

Ozone, as a secondary air pollutant is formed under the influence of UV-light in photochemical reactions involving NO_x and reactive hydrocarbons. Chemical processes in the atmosphere are more or less well understood (Finlayson and Pitts, 1976; EPA, 1978; Bruckmann, 1983) and lead to a mixture of oxidizing air pollutants, of which O_3 is the major component. Unlike the other pollutants in the discussion, O_3 is the only component which, because of its specific mechanisms of formation and decomposition (Becker et al., 1983), has been found in higher concentrations in areas of greater altitude (500 to 1500 m above sea level) remote from urban and industrial sources (Fricke, 1983; Ashmore et al., 1985; Bucher, 1985a; Pfeffer, 1985). In some regions of southern Germany and the German Democratic Republic (GDR), O_3 concentrations increased between 1967 and 1982 by 50 to 80 % and as much from 1981 to 1982 as in the two preceeding 7-yr-periods (Warmbt, 1979; Attmannspacher et al., 1984). In the past 25 yr O_3 concentrations have been found to increase even during winter (Feistner and Warmbt, 1984). In 1980 the monthly mean O_3 concentration at the station Schauinsland in southern Black Forest was e.g. 113 $\mu g\ m^{-3}$ and in the range of concentrations measured in the United States (Ashmore et al., 1985). In severely affected mountainous areas of Northrhine-Westphalia, O_3 concentrations are generally higher by a factor of 2.0 to 2.4 compared to the Ruhr area, and can reach monthly means during the summer of 100 $\mu g\ m^{-3}$, as in August 1984 (Pfeffer, 1985). High O_3 concentrations, with peak values over 600 $\mu g\ m^{-3}$ have been registered under favorable weather conditions in 1976 (Becker et al., 1983). Ozone-maxima in 1976, 1980, 1982, and 1983 correspond more or less to the increase in forest injury observed in most areas of FRG (Arndt et al., 1982; Prinz et al., 1982).

In elevated areas of FRG O_3 concentrations are high enough to injure O_3 sensitive plants species (Guderian, 1984; Arndt, 1985) as shown many times also in other parts of Europe by active monitoring techniques (Posthumus and Tonneijck, 1982; Skärby and Sellden, 1984; Schwela and Radermacher, 1985, Arndt et al., 1986). Besides acute injury symptoms to sensitive plants, even more resistent tree species such as spruce and fir may be affected on the cellular level and in combination with other pollutants as well as site factors (Arndt et al., 1982; Bucher, 1982; Prinz et al., 1985; Ashmore et al., 1985). The primary site of action for O_3 are the cell membranes of the plant, leading to changes in membrane permeability, as shown by changes in fluxes of inorganic ions and organic plant metabolites (Heath and Frederick, 1979; Heath, 1980; Tingey and Taylor, 1982; Keitel and Arndt, 1983). Although comparable symptoms to those found in the field have not yet been reproduced in laboratory experiments, enhancement of ion leaching, including Mg, by O_3 occured in combined fumigation and misting experiments with young spruce trees (Krause et al., 1983b; Krause et al., 1985; Rehfuess, 1986). Reduced tree vitality due to nutrient deficient soils enhances this mode of action. These findings could be in agreement with changes in the membrane system of chloroplasts found in needles of yellow trees from various sites in FRG (Parameswaran et al., 1985) especially since similar morphological effects were observed in O_3 fumigation experiments with spruce (Guderian and Vogels, 1986). However, the link between these phenomena and the impact of O_3 is still controversely discussed (Fink, 1985). Furthermore O_3 is associated with a reduction in assimilate production (Miller et al., 1969; Tingey and Taylor, 1982; Arndt and Kaufmann, 1985) and this may lead in consequence to reduced growth of fine roots before leaf area is affected (Tingey et al., 1976; Tingey, 1977; Blum and Heck, 1980). Similar effects were observed with other pollutants and pollutant mixtures (Freer-Smith, 1983). Based on these experiments, Krause et al. (1983 a) outlined a probable way of how trees are affected by O_3 in combination with acid mist under consideration of accompaning factors, like nutrient deficiency, chlorophyll reduction, photosensitivity, and reduced fine root growth. However, so far field experiments which would adopt laboratory experiments to real world conditions are lacking, but are under current research (Seufert and Arndt, 1985; Krause and Höckel, 1986; Rehfuess, 1986; Kenk, 1986).

Of all pollutants present in areas of forest decline, O_3 probably has the highest degree of phytotoxicity based on concentration and duration of exposure. Distribution and formation of O_3 in the atmosphere also correlates well with spatial and temporal development of forest decline, so that it is reasonable to assume that O_3 is the major contributing factor of all air pollutants involved.

4.2.2. Impact of Wet and Dry Deposition on Forests via Soils

Dry and wet deposition are the two major sinks for emissions from natural and anthropogenic sources, whereby the rate and distribution of wet deposition is determined largely by the precipitation pattern (Georgii et al., 1982). Analysis of rain shows the presence of cations such as H^+, NH_4^+, Ca^{2+}, Mg^{2+}, and Na^+ and anions such as SO_4^{2-}, NO_3^-, and Cl^-. The relative contribution of anionic equivalents forming strong acids with H^+ are about 60 % for SO_4, 30 % for NO_3 and 10 % for Cl (Beilke, 1983). Contrary to sulfate deposition, nitrate deposition seems to increase with increasing distance from sources (Specovius, 1985) but the processes are complex (VDI, 1985). Although, often anticipated, pH of rainwater has not changed greatly in the past 50 yr, having reached a point of "acid saturation" in 1940 at a pH of about 4.2 (Winkler, 1982). Since that time, the affected area has expanded considerably, although few spatial differences in pH-values throughout central Europe are observed (Winkler, 1983), apart from Scandinavia.

Due to the large relative surface area and long residence time of fog droplets, this form of wet deposition usually has a much greater ionic load in comparison to rain (Schrimpff et al., 1984; Specovius, 1985) and pH values below 3 are observed frequently (Fuzzi and Orsi, 1983; Kaminski and Winkler, 1983). Since fog episodes are much more frequent at higher elevations, impact on vegetation and soil have often been anticipated (Prinz et al., 1982; Ulrich and Matzner, 1983). Because of the high filtering capacity of the canopy, forest soils are heavily impacted by deposition of various air pollutants such as heavy metals (Mayer, 1981), acids (Matzner and Ulrich, 1984; Evers, 1985) and organic compounds (Matzner, 1984). Since the buffering capacity of most forest soils is limited, (Bache, 1982; Johnson et al., 1981; Mazzarino et al., 1983) high rates of acid deposition can ultimately lead to soil acidification. This may be followed by release of compounds into the soil solution which are potentially toxic to roots, such as H^+, Al^{3+} and heavy metal ions. Forest soil acidification during the previous decades has been reported in numerous papers (e.g. Butzke, 1981; Grenzius, 1984; v. Zezschwitz 1982; Blume 1981). H^+-budgets reveal that atmospheric deposition represents the major load of strong acidity to forest soils (Ulrich et al., 1979; Matzner and Ulrich, 1984; Driscoll and Likens, 1982). The effect of acid deposition on soils is enhanced by natural fluctations in chemical characteristics of soil i.e., HNO_3 accumulation during warm and dry seasons. The effect of these acidification pulses on soil organisms and plant roots depends upon the ability of the soil to buffer the increasing amounts of protons during this period. Here, the interrelationship between natural acidification pulses and acid deposition and other processes leading to soil acidification (e.g., logging etc.) becomes clear, since acid deposition continuously stresses and slowly exhausts the buffer capacity of the soil. The occurrence of acidification pulses had been postulated on a theroretical basis (Ulrich, 1981), and was recently quantified by measurements (Matzner and Thoma, 1983; Matzner and Cassens-Sasse, 1984). Silvicultural pactices appear to have an important influence on H^+ input to forest soils due to the high filtering effect of coniferous stands (i.e. Evers, 1985). This input is of particular concern for soils with low natural base-saturation capacity.

Root damage has been observed frequently in declining forest stands in FRG (Bauch and Schröder, 1982; Blaschke, 1981; Schütt et al., 1983b; Murach, 1984), and is attributed to toxic concentrations of H^+, Al^{3+} and other ions in the soil solution caused by soil acidification due to acid deposition and natural acidification pulses. Laboratory experiments indicate that low levels of Al in the soil solution reduces Ca and Mg uptake (Evers, 1983; Junga, 1984; Hüttermann, 1983), and hence symptoms related to Mg and Ca deficiency in needles can also be explained by antagonistic effects of Al and H^+ ions during uptake. The Ca/Al ratio that induced root damage (Rost-Siebert, 1985; Junga, 1984) and Ca and Mg uptake in waterculture experiments are found in the soil solution of declining stands such as the Solling Forests (Murach, 1984; Ulrich et al., 1984a).

However, the hypothesis that soil acidification causes root damage due to high concentrations of H^+ or Al^{3+} is subject to controversal discussion (Rehfuess, 1981; Bauch, 1983; Johnson et al., 1983; Zöttl, 1983; a.o.). Parameters for evaluation of the risk of soil acidification for root damage have been outlined by Ulrich et al. (1984b). Observations on soils where, due to soil reaction, excessive mobilization of toxic metal ions is not to be expected, lead to the assumption that Al-toxicity can not explain the forest decline syndrome at every site, but is surely of importance on acid soils.

Increased input of N compounds into forest ecosystems in recent years has become the subject of considerable concern (Nihlgard, 1985), although the high N input of up to 40 kg $ha^{-1}yr^{-1}$ has often been responsible for the increase in increment growth frequently observed especially in southern Germany (Kenk, 1985). Input of nitrate can lead not only to a depletion of base saturation in soils under certain conditions, but combined with acid deposition may also lead to reduced microbial activity (Härtl and Cerny, 1981) and nitrification with increased ammonium fixation (Hüttermann and Ulrich, 1984) will take place. This may eventually lead to reduced mycorrhizal propagation, with implications for tree vitality (Meyer, 1985).

5. CONCLUSIONS

Taking the European inventories, the criteria on which they are based, such as symptomatology, and the different methods of evaluations, it becomes very difficult to determine possible causes of novel forest decline. Since questions like what is "old", what is "new", what "is" and what "is not" forest decline still remain unanswered and are not unequivocally accepted (Bucher, 1985b), it is absolutely necessary to overcome this substantial deficit in establishing generally aceptable criteria attributable to novel forest decline. This lack is probably the reason for the changing emphasis of possible causes in discussions, from acid rain, to SO_2, photooxidants, and N input, apart from natural factors. To avoid an undefined broth of possible causes, only sufficient differentiation of the observed phenomena will lead to a higher degree of perception. It has been repeatedly pointed out that air pollutants should be more closely examined as possible causes of forest damage, and that large-scale epidemiological studies should be conducted (Last et al., 1984; Bucher, 1985a,b). So far such a study has been carried out only in Switzerland in 1983, where comparison of the results with maps of forest damage and maps of pollutant distribution based on needle analyses, showed that the sites with increased needle content of S, Cl^-, F^- and Fe were located in those areas with the greatest forest damage (Bucher et al., 1984; Kaufmann et al., 1984; Landolt et al., 1984).

However, no clear statement of a true cause-effect relationship in respect to novel forest decline in Europe can be made at the present time. It is indeed very questionable if this will ever be possible in the strict sense and hence the only concept of proof is that of highest probability (Prinz et al., 1985). Forest decline is obviously related to nutrient status (Rehfuess, 1981; Rehfuess, 1983; 1985; a.o.), but since trees are able to grow on extremely poor sites when adapted to these conditions from the beginning, novel forest decline seems to be more an imbalance of the source-sink relation between the crown and the root system of the tree (Prinz et al., 1985). This imbalance is expressed in the general independent appearance of yellow needles associated with nutrient deficiencies. Leaching due to acidic deposition and enhanced by the interaction with photooxidants leads to nutrient depletion and impact on the root system by reduced assimilate production or direct effects of acid deposition on soil stability. Through these processes, nutrient uptake is disturbed, inciting a spiral of steadily decreasing vitality. Climate, may be a triggering factor, since the dry summers of 1976, 1980, 1982, and 1983 coincided with a considerable increase in the development of novel forest decline with all its variation. Furthermore, pathogens interact with the primary attack of air pollutants, after having found most favourable conditions in the weakened host. This scenario is surely influenced by regional and temporal variation but seems to be the only conceivable explanation according to the current state of research.

REFERENCES

Arndt, U.: 1985, In: Nießlein, E., and Voss, G. (eds): Was wir über das Waldsterben wissen. Deutscher Instituts Verlag GmbH, Köln, ISBN 3-602-14158-6, p 160.

Arndt, U., and Lindner, G.: 1981, Staub-Reinhalt. Luft 41, 349.

Arndt, U., Seufert, G., and Nobel, W.: 1982, Staub-Reinhalt. Luft 42, 243.

Arndt, U., Obländer, W., König, E., and Maier, W.: 1983, VDI-Berichte 500, 249.

Arndt, U., and Kaufmann, M.: 1985, Allg.Forst Z. 40, 19.

Arndt, U., Barttolmeß, H., and Schlüter, Chr.: 1986, Allg. Forst Z. 41, 3.

Ashmore, M., Bell, N., and Rutter, J.: 1985, Ambio 14, 81.

Asthalter, K.: 1984, Allg.Forst Z. 39, 549.

Attmannspacher, W., Hartmannsgruber, R., and Lang, P.: 1984, Meterol.Rdsch. 37, 1983.

Bache, B.W.: 1982, In: Ecological Effects of acid precipitation, Swedish National Environment Protection Board, Report SUV pm 1636.

Ballach, H.-J., and Brandt, C.J.: 1983, Staub-Reinhalt. Luft 43, 448.

Ballach, H.-J., and Brandt, C.J.: 1985, Staub-Reinhalt. Luft 45, 1.

Bauch, J., and Schröder, W.: 1982, Forstw. Cbl. 101, 285.

Bauch, J.: 1983, In: Ulrich, B., and Pankrath, J.: Effects of accumulation of air pollutants in forest ecosystems. D. Reidel Pbl. Co., Dordrecht, pp. 377-386.

Becker, K.H., Löbel, J., and Schurath, U.: 1983, Umweltbundesamt Berichte 5/83, p 3.

Beilke, S.: 1983, VDI-Berichte 500, 35.

Blaschke, H.: 1981, Fortsw. Cbl. 100, 190.

Blum, U., and Heck, W.W.: 1980, Environm. Experim. Prof. 20, 73.

Blume, H.P.: 1981, Berliner Naturschutz, 715.

den Boer, W.M.J., van den Tweel, Ir. P.A.: 1981, The condition of the Dutch Forests in 1984, Internal Report of the National Forest Service, Division Forestry-Development, PO Box 20020, 3502 LA Utrecht, The Netherlands, 20 p.

Breloh, P.: 1985, In: Die Waldschadenssituation in der Bundesrepublik Deutschland. VDI Kommission Reinhaltung der Luft, Düsseldorf, Bd. 1 Schriftenr., pp. 5-20.

Bruckmann, P.: 1983, VDI-Berichte 500, 21.

Bucher, J.B.: 1982, In: Waldschäden durch Immissionen?, Gottlieb Duttweiler Institut, 8803 Rüschlikon, Zürich, pp. 91-109.

Bucher, J.B.: 1984, Forstw. Cbl. 103, 16.

Bucher, J.B.: 1985a, In: IUFRO, Intern. Symp. Human Impacts on forests, Strasbourg, France, Sept. 16.-22., 1984. INRA Les Colloques de l' INRA 30, 107.
Bucher, J.B.: 1985b, In: Nato Research Workshop: Effects of Acidic Deposition and Air Pollution on Forests, Wetlands and Agricultural Ecosystems, May 13-17, 1985, Toronto, Canada, in press.
Bucher, J.B., Kaufmann, E., and Landolt, W.: 1984, Schweiz. Z. Forstw. 135, 271.
Butzke, H.: 1981, Forst- und Holzwirt 36, 542.
Cramer, H.H.: 1984, Pflanzenschutznachrichten Bayer 37, 97.
Cramer, H.H., and Cramer-Middendorf, M.: 1984, Pflanzenschutznachrichten Bayer 37, 208.
Dässler, H.-G., and Lux, H.: 1984, Hereynia N.F., Leipzig, 21, 73.
Donaubauer, E.: 1985, Allg. Forst Z. 40, 164.
Driscoll, C.T., and Likens, G.E.: 1982, Tellus 34, 283.
EPA: 1978, Air quality criteria for ozone and other photochemical oxidants. EPA 600/8-78-004, 341 p.
Ernst, F.: 1934, Forstw. Cbl., pp. 86-102.
Evers, F.H.: 1981, Forstw. Cbl., 100, 253.
Evers, F.H.: 1983, Forst- und Holzwirt, 38, 305.
Evers, F.H.: 1985, Mitteil. Vereins Forstl. Standortskd. Pflanzenzüchtung H 31, 31.
Fink, S.: 1985, Bundesminister für Forschung und Technologie, Forschungsbericht 03 7330.
Finlayson, B.J., and Pitts Jr., J.N.: 1976, Science 192, 111.
Feistner, U., and Warmbt, W.: 1984, In: Proc. Quadr. Internat. Ozone, Smyp., 3.-7. September 1984, Halkidiki, Greece.
Freer-Smith, P.H.: 1983, Chronic pollution injury to some tree species in response to SO_2 and NO_2 mixtures. Ph.D. Thesis, Univ. of Lancaster, U.K.
Frenzel, B.: 1985, In: Waldschäden '85, Kolloq. Energie-Versorgung Schwaben, Stuttgart-Hohenheim 24.-25.01.1985 in press.
Fricke, W.: 1983, VDI-Berichte 500, 55.
Friedland, A.J., Gregory, R.A., Kärenlampi, L., Johnson, A.H.: 1984, Can.J.For.Res. 14, 963.
Fuzzi, S., and Orsi, G.: 1983, J. Aerosol Sci. 14, 135.
Genßler, H.: 1985, Die Waldschadenssituation in der Bundesrepublik Deutschland, VDI-Kommissions Reinh. Luft, Schriftenr. Bd. 1, 75.
Georgii, H.-W., Perseke, C., Rohbock, E.: 1982, Umweltbundesamt, Berlin, Forschungsprojekt 10402600, 205 p.
Grenzius, R.: 1984, Forstw. Cbl. 103, 131.
Guderian, R.: 1984, XIII. Arbeitstagung forstlicher Rauchschadenssachverständiger (IUFRO), MOST, CSSR, 27.8.-1.9.1984, in preparation.

Guderian, R., and Vogels, K.: 1986, Proceedings of the 7th World Clean Air Congress, Sidney, 25.-29.8.1986, in preparation.
GVSt: 1982, Saurer Regen und Forstschäden: Eine Dokumentation. Gesamtverband des Deutschen Steinkohlenbergbaus, 33 p. + Literaturdokumentation.
GVSt: 1983, Saurer Regen und Forstschäden: Eine Dokumentation, 2. neubearbeitete Auflage; Gesamtverband des Deutschen Steinkohlenbergbaus, 146 p.
Halbwachs, G.: 1971, Mitt. Forstl. BVA, 92, 33.
Härtl,O., and Cerny, M.: 1981, Mitt. Forstl. BVA Wien, 137/II, 233.
Heath, R.L., and Frederick, P.: 1979, Plant Physiol. 64, 455.
Heath, R.L.: 1980, Ann. Rev. Physiol. 31, 395.
Hüttermann, A.: 1983, Allg. Forst Z., 38, 663.
Hüttermann, A., and Ulrich, B.: 1984, Phil. Trans. R. Soc. London, B 305, 353.
IUFRO: 1983, Aquilo Ser. Bot. 19, 175.
Johnson, A.H., Siccama, T.G., Turner, R.S., and Lord, D.G.: 1983, Science,
Johnson, N.M., Driscoll, C.T., Eaton, J.S., Likens, G.E., and McDowall, W.H.: 1981, Geochim. Acta 45, 1421.
Junga, U.: 1984, Ber. d. Forschungszentrum Waldökosysteme Univ. Göttingen, Bd. 5, pp. 1-174.
Kandler, O.: 1985, In: Waldschäden '85, Kolloquium der Energieversorgung Schwaben, Universität Stuttgart Hohenheim vom 24.-25.01.1985, in press.
Kaminski, U., and Winkler, P.: 1983, Ann. Meteorol. 20, 149.
Kaufmann, E., Bucher, J.B., Landolt, W., Jud, B., and Hoffmann, Ch.: 1984, Schweiz. Z. Forstw. 135, 817.
Keitel, A., and Arndt, U.: 1983, Angew. Bot. 57, 193.
Kenk, G.: 1985, Pers. Communication.
Kenk, G.: 1986, Pers. Communication.
Kenk, G., Kremer, W., Brandl, H., and Burgbacher, H.: 1984, Allg. Forst Z. 39, 692.
Krause, G.H.M., Jung, K.-D., and Prinz, B.: 1983 a, VDI-Berichte 500, 257.
Krause, G.H.M., Jung, K.-D., and Prinz, B.: 1983 b, In: Davis, D.D., and Dochinger, L. (eds) Proceedings of the Symposium Air Pollution and the Productivity of the Forest. Izaak Walton League, Washington D.C., Oct. 04.-05. 1983, pp. 297-332.
Krause, G.H.M., Jung, K.-D., and Prinz, B.: 1985, In: Waldschäden - Einflußfaktoren und ihre Bewertung. VDI-Kolloquium 18.-20.06.1985, Goslar, Harz.
Krause, G.H.M., and Höckel, F.E.: 1986, Open-Top-Kammeranlage Versuchsstation Eggegebirge - Ein Beitrag zur Kausalanalyse der neuartigen Waldschäden, Staub-Reinh. Luft, in press.

Landolt, W., Bucher, J.B., and Kaufmann, E.: 1984, Schweiz. Z. Forstw. 135, 637.
Last, F.T., Fowler, D., and Freer-Smith, P.H.: 1984, Forstw. Cbl. 103, 28.
Lichtenthaler, H.K., and Buschmann, C.: 1984, Allg. Forst Z., 39, 12.
Linthurst, R.: 1983, VDI-Berichte 500, 175.
Manion, T.D.: 1981, Tree disease concepts, Prentice Hall, 324.
Materna, J.: 1979, Zbornik 1, 341.
Materna, J.: 1983, Aquilo Sev. Bot. 19, 147.
Matzner, E.: 1984, Water, Air, and Soil Pollut. 21, 425.
Matzner, E., and Thoma, E.: 1983, Allg. Forst Z., 38, 677.
Matzner, E., and Cassens-Sasse, E.: 1984, Ber. d. Forschungszentrum Waldökosysteme Univ. Göttingen, Bd. 2, 50.
Matzner, E., and Ulrich, B.: 1984, Z. Pflanzenernähr. Bodenkunde 147, 290.
Mazzarino, M.J., Heinrichs, H., and Fölster, H.: 1983, In: Ulrich, B., and Pankrath, J. (eds.) Effects of accumulation of air pollutants in forest ecosystems, D. Reidel Pbl. Co., Dordrecht, NL.
Meyer, F.H.: 1981, Forschungserg. z. Förderung d. forstl. Erzeugung III, Hiltrup b. München, pp. 24-62.
Meyer, F.H.: 1985, Allg. Forst. Z. 40, 208.
Mies, E., and Zöttl, H.W.: 1982, Allg. Forst. Z. 37, 1296.
Miller, P.R., Parmeter, Jr., J.R., Flick, B.H., and Martinez, C.W.: 1969, JAPCA 19, 435.
Minerbi, St.: 1985, Ergebnisse der Waldschadenserhebung 1984 in Südtirol. Informationsschrift des Landesforstinspektorates Bozen. LFI 1/85: 26 p.
Mülder, D.: 1983, Möglichkeiten der Forstbetriebe, sich Immissionsbelastungen waldbaulich anzupassen bzw. deren Schadwirkung zu mildern. Beitrag zum Gutachten "Waldschäden und Luftverunreinigungen", see SR-U, 1982.
Murach, D.: 1984, Gött. Bodenkdl. Berichte, 72.
Nihlgard, B.: 1985, Ambio 14, 2.
Nienhaus, F.: 1985, Allg. Forst Z., 40, 119.
Nienhaus, F.: 1985a, Pers. Communication.
Obländer, W., Wörth, R., König, E., Braunger, H., and Schröter, H.: 1983, Allg. Forst-Jagd Z. 154, 175.
Obländer, W., and Hauß, A.: 1985, Landesanstalt für Umweltschutz Baden-Württemberg, Bericht 97/85.
Parameswaran, N., Liese, W., and Schmitt, U.: 1985, Bundesminister für Ernährung, Landwirtschaft und Forsten, Forschungsvorhaben 83 HS 014.
Pfeffer, U.: 1985, Staub-Reinhalt. Luft 45, 287.
Pollanschütz, J.: 1985, Schriftenr. d. Forstl. BVA, Wien, 8, 29 p.
Posthumus, A.C., and Tonneijk, A.E.G.: 1982, In: Steubing, L., and Jäger, H.J. (eds): Monitoring of air pollutants by plants. W. Junk Publ., The Hague, NL.

Powles, St.B.: 1984, Ann. Rev. Plant Physiol. 35, 15.
Prinz, B.: 1983, Forst- und Holzwirt 38, 460.
Prinz, B., Krause, G.H.M., and Stratmann, H.: 1982, LIS-Berichte Nr. 28, 154 p., Landesanstalt für Immissionsschutz des Landes NRW, Essen, FRG.
Prinz, B., Krause, G.H.M., and Jung, K.-D.: 1985, In: Nato Advanced Research Workshop: Effects of Acidic Deposition and Air Pollution on Forests, Wetlands and Agricultural Ecosystems; May 13-17, 1985, Toronto, Canada, in press.
Rehbock, N.: 1982, Allg. Forst. Z. 37, pp. 1179, 1182-1183, 1186-1187.
Rehfuess, K.E.: 1981, Forstw. Cbl. 100, 363.
Rehfuess, K.E.: 1983, Allg. Forst. Z. 38 (24), 601.
Rehfuess, K.E.: 1983a, VDI-Berichte 500, 273.
Rehfuess, K.E.: 1985, Düsseldorfer Geobot. Kolloq. 2, 25.
Rehfuess, K.E.: 1986, Pers. Communication.
Rehfuess, K.E., and Rodenkirchen, H.: 1984, Forstw. Cbl. 103, 248.
Reichelt, G.: 1983, Allg. Forst und Jagd Z. 154, 66.
Roloff, A.: 1985, Allg. Forst Z. 40, 157.
Rost-Siebert, K.: 1985, Ber. d. Forschungszentr. Waldökosyst./Waldsterben, Bd. 12, Göttingen, FRG.
Sartorius, R.: 1984, In: Kolloquium Waldsterben-Diagnose und Therapie, FGN Berlin e.V. 02.-03.04.1984.
Schmid-Haas, P.: 1985, Schweiz. Z. Forstwes. 136, 251.
Schmidt, A., and Hartmann, R.: 1984, Allg. Forst. Z. 39, 552.
Schöpfer W., and Hradetzky, J.: 1984, Forstw. Cbl. 103, 231.
Schrimpff, E., Klemm, O., Eiden, R., Frevert, T., and Hermann, R.: 1984, Staub-Reinhalt. Luft, 44, 72.
Schröter, H.: 1983, Allg. Forst- und Jagdz. 154, 123.
Schröter, H., Achstetter, L., and Holzapfel, W.: 1985, Allg. Forst- und Jagdz. 156, 123.
Schröter, H., and Aldinger, E.: 1985, Allg. Forst Z. 40, 438.
Schütt, P.: 1982, Bild der Wissenschaft 12, 86.
Schütt, P., Blaschke, H., Hoque, E., Koch, W., Lang, K.-J., and Schuck, H.J.: 1983a, Forstw. Cbl. 102, 158.
Schütt, P., Koch, W., Blaschke, H., Lang, K.J., Schuck, H.J., and Summerer, H.: 1983b, So stirbt der Wald - Schadbilder und Krankheitsverlauf. BLV Verlagsgesellschaft München, Wien, Zürich, 1983, ISBN 3-405-12844-7.
Schweingruber, F.H., Lontic, R., and Winkler-Seifert, A.: 1983, Ber. der Eidg. Anstalt für das forstl. Versuchsw. 253, 29 p.
Schwela, D., and Radermacher, L.: 1985, Staub Reinh. Luft, 45, 284.
Schwerdtfeger, F.: 1981, Die Waldkrankheiten, Verlag Paul Parey, Hamburg, Berlin.

Seitscheck, O.: 1981, Forstwiss. Cbl. 100, 138.
Seitscheck, O.: 1984, Forst- und Holzwirt 12, 299.
Seufert, G., and Arndt, U.: 1985, Allg. Forst. Z. 1/2, 13.
Skärby, L., and Sellden, G.: 1984, Ambio 13, 68.
Smith, W.H.: 1984, Forstw. Cbl. 103, 48.
Specovius, J.: 1985, Pers. Communication.
SR-U: 1983, Waldschäden und Luftverunreinigungen. Der Rat von Sachverständigen für Umweltfragen. Sondergutachten, März 1983, 371 p.
Taylor, O.C., Thompson, C.R., Tingey, D.T., Reinert, R.A.: 1975, In: Mudd, J.B., and Kozlowski, T.T., (eds.), Responses of Plants to Air Pollution. Academic Press, New York, pp. 121-139.
Tingey, D.T., Wilhour, R.G., and Standley, C.: 1976, Forest Sci, 22, 234.
Tingey, D.T.: 1977, In: Dimitrades, B. (ed.): Proceedings of an Internat. Conf. on Photochemical Oxidant Pollution and it Control. pp. 601-609, E817/600/3-77-0016, US Environmental Protection Agency, Research Triangle Park, NC.
Tingey, D.T., and Taylor, Jr., G.E.: 1982, In: Unsworth, M.H., and Ormrod, D.P. (eds.): Effects of gaseous air pollution in agriculture and horticulture. London: Butterworth, Scientific, pp 113-138.
Tomiczek, Ch.: 1982, Ursachen des Kiefernsterbens in Niederösterreich. Dissertation Universität für Bodenkultur, Wien, 206 p.
Ulrich, B.: 1981 a, Forstw. Cbl. 100, 228.
Ulrich, B.: 1981 b, Forst- und Holzwirt 36, 525.
Ulrich, B., Meyer, R., and Khanna, P.K.: 1979, Schriftenr. a.d. Forstl. Fak. der Univ. Göttingen u.d. Niedersächsischen Forstl. Versuchsanstalt, Band 58.
Ulrich, B., and Matzner, E.: 1983, VDI-Berichte 500, 289.
Ulrich, B., Murach, D., and D. Pirouzpanah: 1984a, Forstarch. 55, 127.
Ulrich, B., Meiwes, K.J., and Khanna, P.K.: 1984b, Forst- und Holzwirt, 39, 278.
VDI: 1983, Säurehaltige Niederschläge-Entstehung und Wirkungen auf terrestrische Ökosysteme, VDI-Kommission Reinhalt. Luft, Düsseldorf, 277 p.
VDI: 1985, Acid precipitation - Formation and Impact on Terrestrial Ecosystems, VDI-Kommission Reinhalt. Luft, Düsseldorf, in press.
Wachter, A.: 1978, Z. Pflanzenkrankh. Pflanzensch. 85, 361.
Warmbt, W.: 1979, Z. Meteorologie 29, 24.
Wentzel, K.-F.: 1956, Allg. Forst Z., 11, 541.
Wentzel, K.-F.: 1982a, Forst- und Holzw., 37, 410.
Wentzel, K.-F.: 1982b, Aquilo Ser. Bot., 19, 167.
Wentzel, K.-F.: 1983, Waldschäden durch Immissionen, Gottlieb-Duttweiler Inst., Rüschlikon/Zürich, Switzerland, p. 7-17.

Wentzel, K.-F.: 1985, Forstarchiv, 56, 51.
Winkler, P.: 1982, Z. Pflanzenernährung Bodenkd. 145, 576.
Winkler, P.: 1983, VDI-Berichte 500, 141.
Zech, W., and Popp, E.: 1983, Forstw. Cbl. 102, 50.
v. Zezschwitz, E.: 1982, Forst- und Holzwirt 37, 275.
Zöttl, H.W.: 1983, Allg. Forst. Z. 38, 206.
Zöttl, H.W.: 1985, Düsseldorfer Geobot. Kolloq. 2, 31.
Zöttl, H.W., and Hüttl, R.: 1985, Allg. Forst Z., 40, 197.

EFFECTS OF LIMING ON CRAYFISH AND FISH IN SWEDEN

P. Nyberg*, M. Appelberg** and E. Degerman*

* Nat.Board of Fish.
Inst. of Freshw.Res
S-170 11 Drottningholm

** Inst. of Limnol.
Univ. of Uppsala
Box 557
S-751 22 Uppsala

ABSTRACT. The effects of lime treatment on crayfish (Astacus astacus and Pacifastacus leniusculus) populations in 17 lakes and fish populations in 47 lakes and 7 rivers within the trial period 1976-82 have been evaluated. An increase in the catch of crayfish per unit effort was observed in 7 lakes, although significantly in one lake only. The varying results in the other lakes indicate that factors other than pH may be of greater importance for the development of crayfish populations after liming. Recruitment of fish improved in waters where liming resulted in pH >5.5. In lakes with pH <5.5 before and pH >5.5 after treatment, there was a significant increase in the number of fish caught, from 12 to 34 per unit effort. Due to improved recruitment the individual average weight was smaller and hence the catch in weight per unit effort was about the same as before liming. After lime treatment in streams which resulted in a stable pH of >5.5, the abundance of juvenile salmonids increased to numbers found in non-acidified streams. In other streams acid spates reduced the positive influence of liming on the abundance of juvenile salmonids.

1. INTRODUCTION

The extent of acidification and its effects on fish populations in fresh waters in Sweden is well documented (e.g. Almer, 1972; Almer et al., 1978; Johansson and Nyberg, 1981). In order to try and protect waters of special value for fisheries, nature conservancy or recreational uses, a trial period of liming measures was performed in Sweden during 1976-82, with the support of Government grants. During this period about 450 projects, often including several lakes, were started. In every project different kinds of follow-up studies were performed (data at the National Swedish Board of Fisheries). In most projects studies were restricted to analyses of pH, alkalinity, water color and the sum of Ca and Mg twice a year. In 158 of the projects some kind of special study, such as gill-net surveys, electrofishing or crayfish fishing, was performed.

Water, Air, and Soil Pollution **31** (1986) 669–687.

After 1982 and onwards no biological studies have been carried out within the projects. Instead test fishing has been performed by the Institute of Freshwater Research using standardized methods.

The ecological effects of liming in waters within the Swedish liming program have earlier been reported by Appelberg (1984a), Eriksson et al. (1982, 1983), The National Swedish Board of Fisheries and the National Swedish Environment Protection Board (1981) and Nyberg (1984a).

2. MATERIALS AND METHODS

All projects with data sets conforming with evaluation requirements and where at least 2 yr had elapsed since liming have been evaluated. In lakes, the effects on fish and crayfish populations have only been related to changes in water quality, and specific liming methods have not been taken into consideration. Moreover, as complete chemical analyses were not included in most projects, the pH increase is taken as an indication of an overall improvement in water quality, e.g. decreased concentrations and toxicity of potentially toxic metals. In all waters pH and alkalinity have been measured twice a year, usually in spring and autumn.

Test fishing for the crayfish species *Astacus astacus* and *Pacifastacus leniusculus* was performed at least once in each of the 32 lakes studied. Before liming, all lakes had an alkalinity less than 0.12 $meq \cdot l^{-1}$ and could be characterized as oligotrophic. In 17 lakes fishing efforts were carried out either the year before or during the year of liming, and 2 to 5 yr after liming. In most cases, fish-baited cylindrical traps with two entrances, and occasionally cages with one entrance at the top were used. One effort is synonymous with one night's fishing using one trap or cage. In 15 of the lakes studied data from both crayfish and fish catches were available.

It was assumed that the relative density of crayfish, estimated from the catch per unit effort (c.p.u.e.), always had a normal, or an approximately normal distribution. This may not be true in a strict sense, but skewness was regarded to have little effect on the results. Linear regression models were used to study the development of crayfish populations both before and after lime treatment. Multiple linear regression and backward elimination was used to select the factors affecting crayfish catches in lakes of low alkalinity (SAS Inst.Inc., 1982).

When evaluating the effects of liming on fish populations in lakes, the lakes have been pooled into three classes according to the pH before liming and the pH during 2 to 4 yr after treatment. The three classes are:

I) Lakes with pH <5.5 both before and after liming, i.e. unsuccessful treatment.
II) Lakes with pH <5.5 before liming and pH >5.5 after treatment.
III) Lakes with pH >5.5 both before and after treatment, i.e. mildly acidified waters.

Hence these classes are consequently referred to as "lake category I, II and III", respectively. The limit of pH 5.5 has been chosen because at this pH the buffering capacity is very low and acid episodes may cause severe drops in pH. Moreover, a pH of 5.5 has been found to be critical for the recruitment of roach (Rutilus rutilus) (Almer et al., 1978; Milbrink and Johansson, 1975), which is a fairly common fish species in the studied lakes.

Test fishing in lakes was usually performed 0 to 2 yr before liming and 2 to 4 yr after. All test fishing was performed during April-December and most occasions were during June-September. The benthic gill-nets used were 36 m long and 1.5 m deep, and consisted of 12 different mesh sizes (10 to 75 mm) (Filipsson 1972). The smallest fishes caught in the gill-nets were 7 to 8 cm perch (Perca fluviatilis) and 8 to 9 cm roach.

For age determination of perch, the opercular bone was used. Total lengths of perch were back-calculated in 14 lakes and the data were pooled in such a way that fish born in a certain year before, during or after the year of liming were grouped together. The average length during 5 to 8 yr was calculated for each age class and the length in certain years before and after liming was expressed as a percentage of the average.

Electrofishing in running waters was carried out the year before, or during the year of liming and up to 3 to 8 yr after liming. The investigated stretches consisted of salmonid reproduction areas and were fished 2 to 4 consecutive times using standard electric fishing gear with pulsating direct current (300 to 700 V). The total fish population was estimated by the removal method (Bohlin, 1981; Zippin, 1956). The age classes 0+, I+ and >II+ were distinguished using length frequency distributions. Brown trout (Salmo trutta) in the investigated streams consisted mainly of anadromous populations, but one lake-migrating and two stream-resident populations were also present.

3. RESULTS

3.1. Effects on crayfish population

In two of the lakes information on the yearly catch of crayfish was available for 8 and 10 yr prior to liming (Figure 1). The mean c.p.u.e. in both lakes decreased significantly during this period (linear regression, r^2=0.52, p <0.001). Though after liming the decrease was inhibited.

Changes of crayfish abundance was insignificant in 16 of the 17 lakes after liming. Only one lake, which was limed repeatedly for 16 yr, showed a significant increase in c.p.u.e. (p <0.05). In the remaining waters the c.p.u.e. had increased in 6, decreased in 4, and was about the same in 6 lakes 1 to 5 years after liming, although, no significant changes was observed.

Factors which may influence crayfish populations in acidified waters were elucidated using multiple linear regression, and c.p.u.e. of crayfish the year before or within the year of liming as the de-

Table I. Correlation matrix for independent variables used in the multiple linear regression.

	pH	Alkalinity		Conduc-	Water	Lake	Roach		Perch		Abund.	No. of
	autumn	spring	autumn	tivity	color	area	c.p.u.e.	weight	c.p.u.e.	weight	of eel	fish species
pH, spring	0.60	0.82	0.72	0.30	-0.22	-0.13	-0.10	-0.27	0.19	-0.14	-0.22	0.14
pH, autumn		0.63	0.69	0.32	-0.25	-0.02	-0.10	-0.40	0.03	0.08	-0.26	0.23
Alkalinity, spring			0.82	0.53	-0.17	-0.21	-0.15	-0.28	0.09	-0.18	-0.12	0.34
Alkalinity, autumn				0.56	-0.09	-0.28	-0.29	-0.35	-0.12	-0.26	-0.05	0.53
Conductivity					-0.19	0.03	0.23	-0.25	-0.19	-0.01	-0.27	0.86
Water color						-0.19	-0.42	0.54	-0.42	-0.01	0.30	-0.28
Lake area							0.53	0.13	0.33	0.93	-0.25	0.07
Catch/effort, roach								0.08	0.31	0.51	-0.44	0.21
Average weight, roach									-0.02	0.13	-0.33	-0.46
Catch/effort, perch										0.31	-0.38	-0.24
Average weight, perch											-0.13	0.00
Abundance of eel												0.15

Table II. Parameters and estimate describing catch of crayfish per unit effort in 15 lakes using a multiple linear regression model.

Parameter	Estimate	Probability
Intercept	-23.08	< 0.009
Median pH in autumn	+ 4.72	< 0.003
Lake area (ha)	+ 0.006	< 0.003
Average weight of perch (g)	- 0.103	< 0.02

Table III. Average lengths and weights of roach and perch before and after liming. (n.s. = not significant difference, p >0.05.)

	Lake category		
	I (n=1)	II (n=26)	III (n=18)
Roach			
Average length (mm)			
Before liming	274	258	188
After liming	253	182	174
Average weight (g)			
Before liming	69	156	56
After liming	95	71	43
p (sign test)	n.s.	<0.01	n.s.
	(n=3)	(n=26)	(n=18)
Perch			
Average length (mm)			
Before liming	183	187	153
After liming	190	154	151
Average weight (g)			
Before liming	125	119	44
After liming	59	50	42
p (sign test)	n.s.	<0.001	n.s.

pendent variable (Table I). The independent variables were pH, alkalinity, conductivity, water color, relative catches and mean weights of perch and roach, number of fish species, abundance of eel (*Anguilla anguilla*), and lake area. The lowest measured values of pH and alkalinity in spring, as well as the median pH and mean alkalinity in autumn during the 3 yr period preceding crayfish catches were used. A correlation matrix for independent variables showed, as expected, that pH was correlated to both alkalinity and conductivity. A high correlation could also be noted between the mean weight of perch and lake surface area (Table I).

Catch per unit effort of crayfish was found to be positively dependent on median autumn pH ($p < 0.05$), when c.p.u.e. and the chemical parameters in 32 lakes were tested. This model explains 17% of the variation (r^2=0.169) which indicates that the pH in autumn was of importance to crayfish populations in these kind of lakes, but that other factors probably played a more important role. In the 15 lakes for which both chemical and biological variables were available, the best model, which included three variables, accounted for 73% of the variation (r^2=0.725, $p < 0.002$, Table II).

3.2. Effects on fish populations

3.2.1. Effects on recruitment. After liming, the size distribution of roach changed in all of the lakes studied except one. The anomaly belongs to category I, i.e. the pH is constantly <5.5 and it was one of the two lakes in this category inhabited by roach. In this lake the smallest roach caught were ca. 25 cm in length both before and after treatment (Table III). In lakes from category II, the average length decreased from 258 to 182 mm and there was a significant decrease in the average weight from 156 g before liming to 71 g after treatment. In lakes where the pH was always >5.5, i.e. category III, liming resulted in an increase in the number of roach in the size class 10 to 15 cm. The average lengths decreased from 188 to 174 mm and the average weight decreased from 56 to 43 g, though the latter change was not significant. For all of the investigated lakes the average weight of roach was significantly higher in lakes with pH <5.5 than in lakes with pH >5.5 (Mann-Whitney U-test $p < 0.01$).

Perch populations in lake category I had high average lengths, high average weights and abnormally low numbers of small fish both prior and after liming. In lakes of category II the situation was similar before liming, but the size distribution changed after treatment, viz. average lengths decreased from 187 to 154 mm. The latter length was almost identical to the average length found in lakes of category III, both before and after treatment. In lakes of category II the average weight decreased significantly, from 119 to 50 g after liming, while in category III, the change was very small (Table III).

The recruitment of one summer old (0-group) brown trout, increased considerably in limed streams. In the year of successful liming and during the following 1 to 3 yr 0+ fishes averaged 80 and 58%, respectively, of the populations. This was similar to the situation in waters with a naturally high alkalinity in the same region (Figure 2).

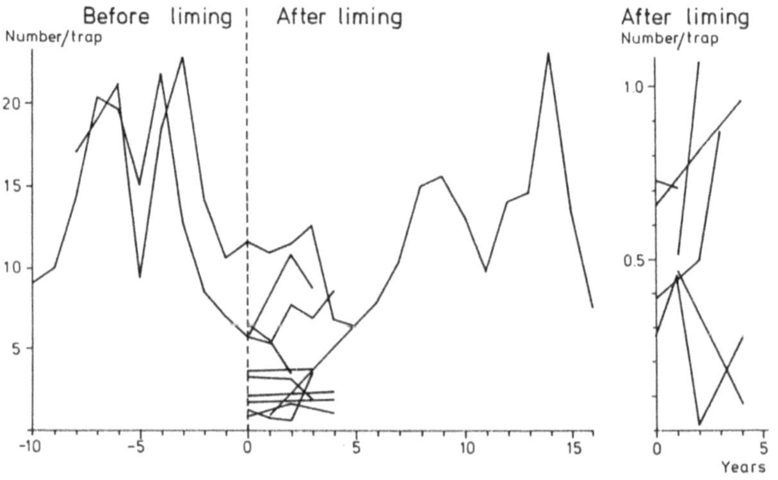

Figure 1. The catch of crayfish per unit effort before and after lime treatment in 17 waters where test fishing was performed on at least two occasions.

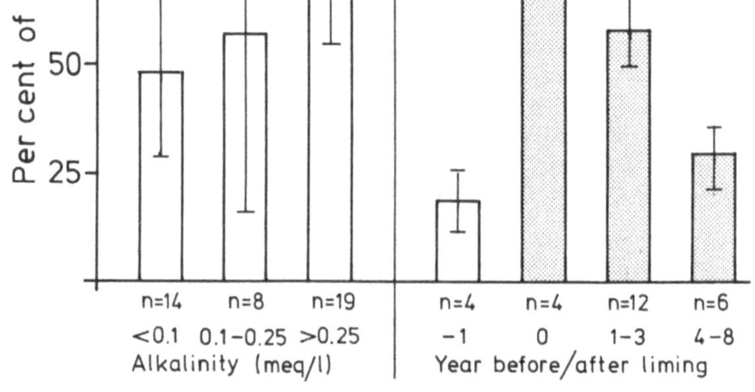

Figure 2. The proportion of 0+ salmonid parr in waters of different alkalinities. Non-limed running waters (left) and siz limed streams before and after liming (right). The mean and standard error are given.

The proportion of 0+ fish up to 3 yr after successful liming was significantly higher when compared to the situation before liming or more than three years after liming (Mann-Whitney U-test, $p < 0.001$).

3.2.2. Effects on growth. Perch growth was studied in 14 lakes where data was available for a number of years before and after liming. None of these lakes have been seriously affected by acidification and therefore recruitment may be regarded as intact. The effect of liming on perch after their first growth season (I) was slight and the average length was at most 102% of the normal length, 1 and 2 yr after treatment (Figure 3). The effect on the lengths of II and III year olds was more marked and 4 yr after treatment the average length was 105% of the normal value. The effect of liming on the growth of IV and V year old fish was about the same as for younger fish, while VI year old perch were 11% longer 4 yr after liming than the average. The increase in the average lengths after liming was significant when all ages were compared (Mann-Whitney U-test, $p < 0.05$), but the material varied and the actual differences in some age groups, especially age I, were small.

3.2.3. Effects on catch per unit effort in lakes and abundance in streams. In the lakes of category I, the number of fish caught per unit effort averaged 22 to 24 and did not alter after liming. In lakes of category II there was a marked and significant increase in c.p.u.e. from 12 to 34 fish (sign test, $p < 0.01$). In category III the c.p.u.e. increased, though not significantly, from 18 to 26 fish on average (Figure 4).

There was no significant change in the c.p.u.e. as expressed in weight in any lake category after liming. When all the lakes were pooled, there was, however, a significant increase in the c.p.u.e. (sign test, $p < 0.05$).

After successful liming a rapid increase in the abundance of brown trout was noted. However, depending on the liming technique and the intervals between treatments, somewhat different results were obtained. When lakes were limed by applying a single dose, an immediate positive effect on the abundance of brown trout was obtained downstream (Figure 5a), though the positive effect usually diminished as reacidification occurred. When liming was repeated or performed on a more or less yearly basis, however, the positive effect endured (Figure 5b, c). The same result has been obtained for Atlantic salmon (*Salmo salar*) by combining lake liming with the use of continuous lime-dosing equipment in the streams (Figure 5d, e). In both of the last cases presented, the abundances of salmonids have increased following treatment but have not yet returned to the levels observed before or at the beginning of acidification. In the Vammsjöarna lakes the brown trout population in two investigated streams increased in abundance after the first liming occasion, but after a second liming the increase was of short duration (1 yr) in one of the streams while the population decreased in the other (Figure 5f). Since this system was limed with a lime slag (CaO, MgO, SiO_2), which is highly soluble, high pH values resulted. In July 1983, which was half a year after the second liming occasion, pH values above 9 were measured in the surface water.

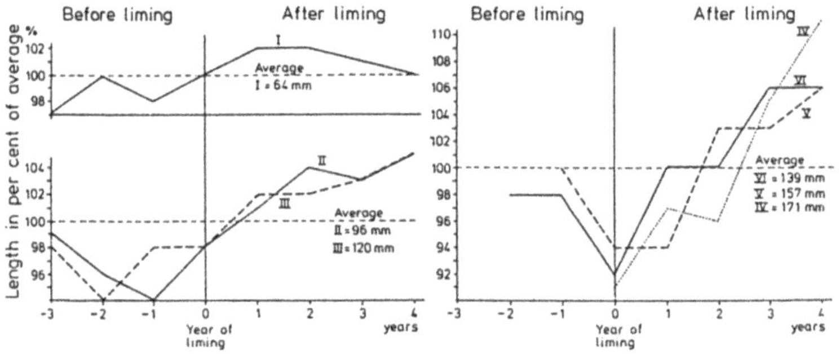

Figure 3. Length of perch (Perca fluviatilis) from their first to sixth (I-VI) growth season, given as length in per cent of average length after each growth season, before and after liming.

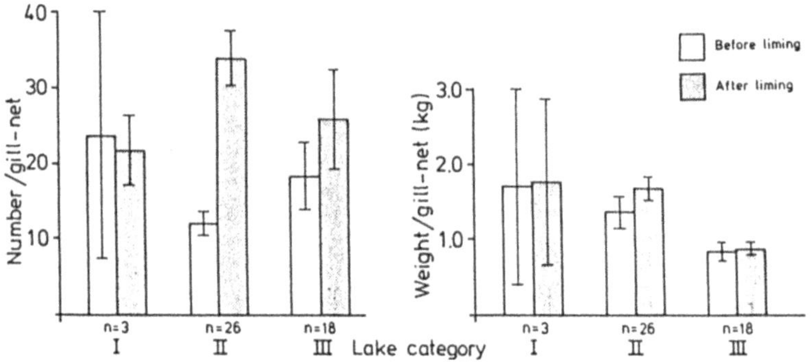

Figure 4. Average catch per unit effort before and after liming in the three lake categories. (Number (left) and weight (right) per gill-net.) The mean and standard error are given.
(I: pH <5.5 before as well as after treatment. II: pH <5.5 before and pH >5.5 after treatment. III: pH >5.5 before as well as after liming.)

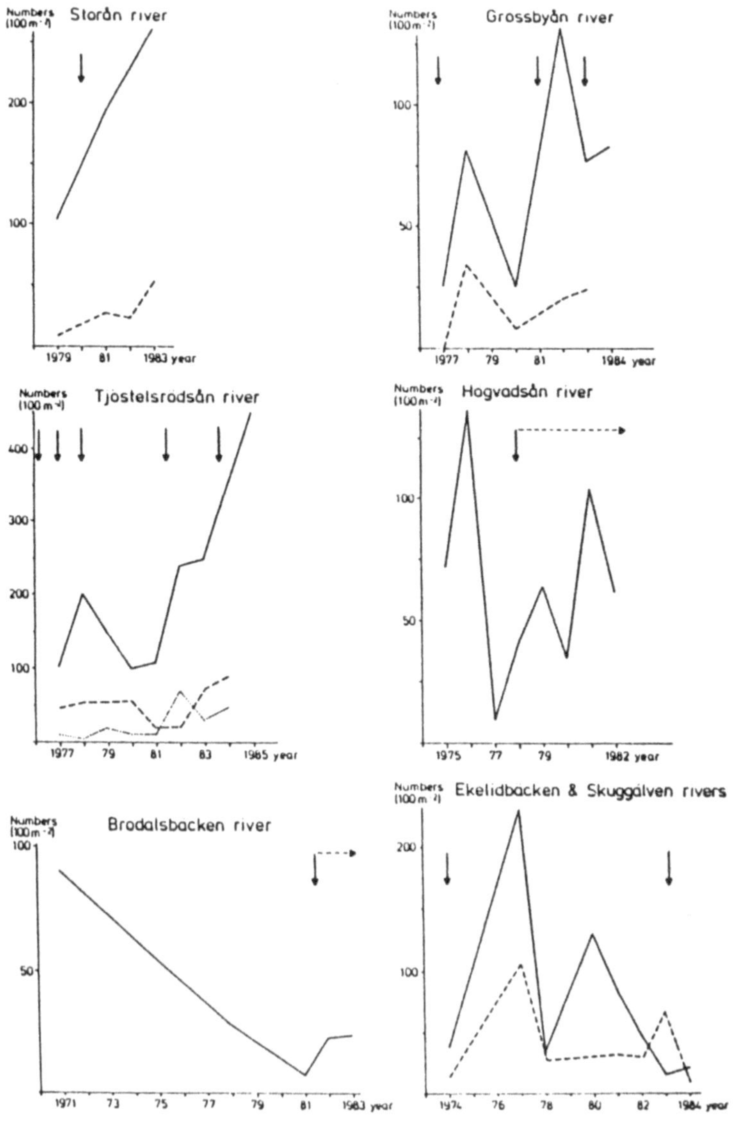

Figure 5. Abundance of salmonid parr per 100 m^2 at different stations in seven limed streams on the west coast. A vertical arrow indicates a single liming occasion in an upstream lake and a dotted horizontal arrow indicates continuous stream liming.

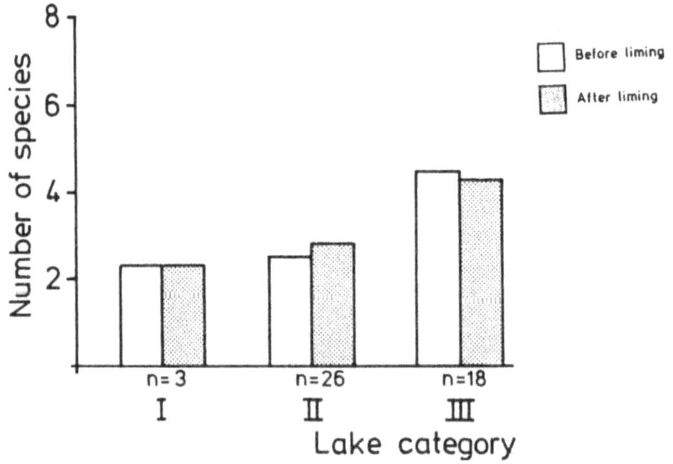

Figure 6. The average number of fish species caught in gill-nets before and after liming in the three lakes categories. The mean and standard error are given.

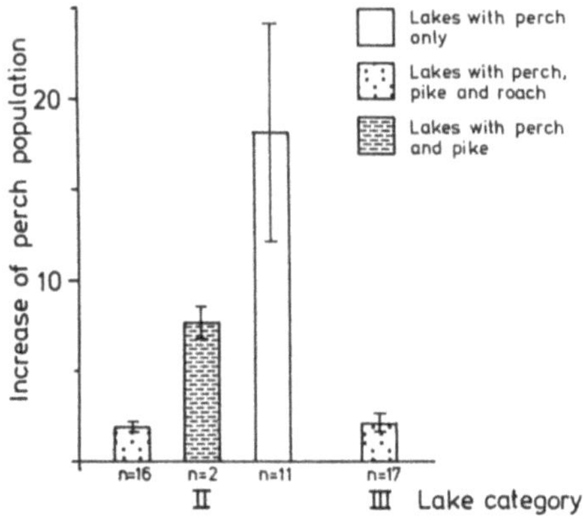

Figure 7. Increase in numbers, expressed as a proportion of the numbers before treatment, of the perch populations in lakes of categories II and III. The mean and standard error are given.

Sparse data exists on species other than salmonids. In the River Storån minnows (*Phoxinus phoxinus*) and brown trout dominated the fish fauna. At 7 localities studied by electrofishing before and 2 to 3 yr after successful liming, the abundance of minnows increased by a factor of 2.2 on average, but the increase was not significant as the abundances fluctuated between years.

3.2.4. Effects on species composition. Species richness increased with lake category and the difference was especially pronounced between categories I-II and category III (Figure 6), although no statistically significant changes in the number of species caught were noted after liming.

Significant changes in species composition occurred in only two of the 47 lakes after liming (chi-square, $p < 0.05$). In both of these lakes the changes were due to a greater catch of roach.

The most striking differences between lakes, as regards the development of fish populations after liming, were observed in perch populations. In lakes of category III, perch coexisted with both pike (*Esox lucius*) and roach and in these lakes the perch populations increased on average by a factor of 2 after treatment (Figure 7). The increase was almost identical in lakes of category II with the same species composition. In two lakes in this category perch coexisted with pike only and the population increased by a factor of almost 8. In lakes inhabited by perch only, however, the population increases were remarkably high, and were a factor of 18 on average. The increase in the perch populations were significantly different between lakes with perch only and lakes where perch coexisted with pike and roach (Mann-Whitney U-test, $p < 0.01$).

Follow-up studies of liming measures in running water have been conducted predominantly in salmonid waters. During the acidification phase the salmonids decreased in abundance and were in some cases wiped out. On no occasion has any other fish species been observed to take advantage of the absence of salmonids and increase in abundance. In the Storån river the abundances of brown trout parr and minnows were significantly positively correlated (Spearman Rank Correlation, $p < 0.05$) with one another. The abundance of minnows fluctuated greatly, however.

In a few running waters where brown trout was the only salmonid observed prior to liming, electrofishing after treatment has revealed parr of Atlantic salmon to be present.

4. DISCUSSION

4.1. Effects on crayfish population

The results do not solely support the hypothesis that lime treatment improves the catch of crayfish. There may, however, be several possible explanations for the differences in the development of crayfish populations after liming. It is obvious that some of the lakes were only slightly acidified before treatment, and therefore liming only had a minor effect on the water quality. No dramatic changes in the development of the crayfish populations would then be expected. Disturbance of the reproductive cycle due to low pH, which has been reported for several crayfish species (Appelberg, 1984b; France, 1983), is probably one of the most important factors affecting crayfish populations in acidified waters. However, improvements in recruitment may not be seen during the first 4 to 5 yr after treatment, since the traps and cages used do not normally catch crayfish less than 4 to 5 yr old.

Changes in the structure of the ecosystem due to either low pH or lime treatment have been shown in several studies (e.g. Almer *et al.*, 1978; Eriksson *et al.*, 1983). It is probable that such changes affect crayfish populations both before and after liming, regarding some of the factors that have been shown to be of importance to crayfish populations in acidified lakes (Table II).

The seasonal variation in the pH is generally less in large lakes than in small ones. Moreover, in small lakes the crayfish are restricted to litoral areas which may be more prone to the effects of acid spates. In larger lakes on the other hand, the crayfish are often distributed over larger areas, which might be an explanation for the positive dependence on lake area. The negative relationship to mean weight of perch is possibly an effect of predation from this fish species. Large perch prey on crayfish, especially when the availability of other food objects is reduced (Appelberg and Odelström, 1984; Morales and Appelberg, 1984). Since the composition of the fish fauna changes during the acidic phase (for example declines in the roach populations, as shown above), this probably indirectly affects crayfish populations.

Despite the lack of fundamental variables and the limited size of the material, the results indicate that the factors affecting crayfish density are of a complex nature. An increase in pH and Ca will ameliorate the primary effects of H^+ concentration on crayfish physiology and recruitment (Appelberg, 1984b, 1985; Malley, 1980; McMahon and Morgan, 1983), as well as mitigate the effects of acid-soluble metal ions (Rosseland *et al.*, 1984). However, as indicated above, other factors such as changes in predation pressure may affect crayfish density. A combination of the long life history of both crayfish species as well as a possible counter-action from other parts of the ecosystem may conceal or/and mitigate the effect of liming on *A. astacus* and *P. leniusculus* within the first 4 to 5 yr after treatment.

4.2. Effects on fish populations

The main reason for the reduction in numbers or extinction of fish populations in acidified waters is disturbed or totally failed recruitment, due to the mortality of eggs and larvae in particular (Daye and Garside, 1979; Muniz, 1981).

The noted changes of length distributions, decreased average lengths and weights indicates improved recruitment after liming. Recruitment of roach has recommenced in all of the studied lakesfrom categories I and II except one. The decreased average length and the significant decrease in the average weight, clearly indicate improved recruitment following liming. In lakes from category III the recruitment of roach had only been disturbed to a minor extent before liming. This is also in agreement with the results of Almer *et al.* (1978) and Milbrink and Johansson (1975). In spite of the almost intact recruitment, liming resulted in an increase in the number of roach in the size class 10 to 15 cm and a small decrease in the average length and weight. The results from the different categories of lakes are expected, as a pH of around 5.5 is critical for roach recruitment.

Perch is one of the most common and acid-tolerant fish species in Sweden (Almer *et al.*, 1978). Only in a few lakes had the recruitment of perch ceased before lime application started. The high average lengths and weights in lakes with pH <5.5, i.e. lake category I both pre and post liming and category II before liming, indicate disturbed recruitment. This is also shown by the fact that the average length and weight in lakes of category II changed after treatment and was almost identical to what was found in lakes of category III (pH >5.5), both before and after liming.

The recovery or improvement of perch, roach and brown trout recruitment in most waters where liming gave pH >5.5, must reflect a reduction in the toxicity of the water. An increase in pH implies less stress on fish (Leivestad *et al.*, 1976) and also initiates the precipitation of toxic metals such as Al. In addition increased concentrations of Ca improve the survival of fish at low pH values (Brown, 1981) and high concentrations of Al (Rosseland *et al.*, 1984).

Typical predatory species such as pike often display stunted growth in very acidic water, due to a lack of prey fish. Perch and roach, on the other hand, may have a very fast growth rate in populations where recruitment is disturbed or has ceased (Alenäs *et al.*, 1981; Nyberg, 1984a). After liming in this kind of lake the fast growth rates of planktivorous and benthivorous species decreased (*op.cit.*), due to increased competition for food in the growing populations. The growth rates of predatory species were favored by the recovered recruitment of the prey species (*op.cit.*). The observed increases of the total lengths of perch in the present study after each growth season were relatively small and were probably caused by an increased food supply.

Liming has generally been shown to result in increased abundances of zooplankton (Eriksson *et al.*, 1983; Nyberg, 1984) and benthic invertebrates (Eriksson *et al.*, 1983; Raddum *et al.*, 1984), i.e. suitable food organisms. This should favor the growth rate of perch smaller than ca. 15 cm, i.e. age groups I-IV. The improvement in food conditions and water quality may also reduce the mortality of young perch. The increase in the size of older and piscivorous perch (VI) after liming may thus have been due to a greater supply of prey.

Larger and deeper lakes in general have a more diverse fish fauna than smaller lakes. There were, however, no statistical differences between lakes from categories II and III with regard to surface area, altitude or maximum depth. This leads to the conclusion that the smaller number of species in lakes with pH <5.5 before treatment, i.e. category I and II was probably due to the disappearance of certain acid-sensitive species.

In lakes from category II, however, where the pH was critically low (<5.5) for acid-sensitive species before liming and >5.5 after treatment, an increase in the number of species would have been expected.

The very slight increase in the number of species after liming in this lake category may be explained by the fact that the most sensitive species, such as roach and other cyprinids, had already been eradicated before liming and probably are not able to recolonize treated lakes by natural means.

The effects of liming on the c.p.u.e. (number) were to be expected, i.e. there was no effect in acid lakes where liming did not result in any increase in pH (I), there was a marked increase in acid lakes after successful treatment (II) and a smaller increase in lakes that were only slightly acid, i.e. pH >5.5 before treatment (III).

C.p.u.e. (weight) were highest in lakes with pH <5.5 before treatment (I, II), while lowest in category III. This is due to the fact that the lakes are situated in different geographical regions of Sweden. Most of the lakes in categories I and II, viz. severely acidified lakes, are found in the southwestern part of the country, while the lakes from category III are situated in central and inland Sweden. In the former lakes perch, which is the dominating species, seldom becomes stunted and catches are normally about twice as high as in oligotrophic waters in inland areas.

The explanation for the very slight increase in the weight of the catch in category II lakes after liming, despite the significant increase in numbers, is that the mean weight of perch and roach decreased from 119 to 50 g and from 156 to 71 g, respectively.

The community structure due to interactions between fish species in neutral lakes may be altered in an acidic environment. Liming, which improves the water quality for all fish species, may thus affect the species composition of acidified lakes. Moreover, roach is a very dominant species (Svärdson, 1976). The populations of roach were expected to increase and presumably affect other species, such as perch, after liming. An increase of c.p.u.e. of roach resulting in significant altered species composition was, however, only noticed in two lakes. The influence of roach on perch populations is, however, most pronounced in shallow lakes which are rich in nutrients (Lessmark, 1983). Hence the waters investigated are probably too oligotrophic to permit dominance of roach.

Changes in species composition following liming have been reported from a few other lakes. In a lake with perch and Arctic char (Salvelinus salvelinus), for example perch seemed to be favored by mild acidification, which only affected the char population and the condition was reversed after liming (Nyberg, 1984a). In another lake the cisco (Coregonus albula) population declined following liming and an increase in the roach population was observed (Alenäs, 1985). This lake was, however, rather shallow, and thus not a typical habitat for cisco.

Marked differences in the increase of perch populations between lakes with perch only and perch coexisting with pike only or pike and roach, strongly indicate the importance of predation and competition for the development in fish populations after liming.

It has been proposed that in running waters in southern Sweden with high summer alkalinities (>0.25 meq·l^{-1}) the total abundance of anadromous populations of salmonid parr (brown trout and Atlantic salmon) should exceed 100 fish/100 m^2 (Degerman et al., 1985). Such abundances have been obtained in limed streams in some of the investigated localities.

At stations at higher altitudes with poorer bedrock or soil and in other streams, these abundances have not always occurred after liming. In the River Högvadsån it has been suggested that the absence of 1+ Atlantic salmon in the years immediately before and after liming (1977-78), despite the fact that 0+ fish were found during previous years, depended on low pH values and elevated levels of Al (Nyberg, 1984a). The Al complexes precipitate slowly in cold water (Dickson, 1983) and may affect a long stretch of stream water. The optimum pH-range for Al precipitation is between pH 6 and 7, whereas below pH 5 and above pH 9 the dissolution of Al from the sediment is considerable. In the Vammsjöarna area (Figure 5f) thus higher concentrations of Al (>500 $ug \cdot l^{-1}$) occurred than before liming (Dickson, pers.comm.), probably leading to negative effects on the fish fauna.

The positive correlation between brown trout parr and minnows in the Storå river, indicate that the improved water quality after liming favoured both species. Alenäs (1985) reports that the populations of bullhead and Alpine bullhead (Cottus gobio and C. poecilopus) increased in two small salmonid streams on the Swedish west coast after liming, whilst there was little or no increase in the salmonid populations. However, it is probable that these liming operations did not improve the water quality sufficiently. Although the water chemistry data is meager it can be concluded that in these small streams the summer alkalinities are too low (0.2 $meq \cdot l^{-1}$) and hence acidic spates may induce salmonid mortality (Degerman et al., 1985). The bullheads spawn in March-May and the eggs hatch after 3 to 4 weeks. Thus the bullheads may in some years show good recruitment while the autumn-spawning salmonids may suffer diminshed recruitment.

The appearence of Atlantic salmon in a few brown trout streams, which also has been reported by Alenäs (1985), indicate that the improved water quality following liming has enhanced the survival of the former species, which is more acid sensitive (Leivestad et al., 1976).

5. CONCLUSIONS

After lime treatment resulting in pH >5.5, the characteristics of the fish fauna were similar to those found in non-acidified waters. Species composition and interactions among fish species, e.g. perch, pike and roach, were of importance for the increase in perch populations after liming. As regards the development of crayfish populations after treatment, factors other than the water quality, such as changes in predation pressure, seem to be of great importance. No negative effects have been observed after liming with $CaCO_3$ in surface waters. However, in spite of liming acid spates in connection with high concentrations of Al probably have negative effects on fish populations in running waters.

6. REFERENCES

Alenäs, I.: 1985, Negativa effekter av kalkning med avseende på omstrukturering av befintliga fiskpopulationer. Swed.Wat.Air Pollut. Res.Inst., Gothenburg. 22 p. (In Swedish.)

Alenäs, I., Hultberg, H. and Andersson, I.: 1981, Kalkningsprojekt Härskogen. Swed.Wat.Air Pollut.Res.Inst., Gothenburg. 51 p. (In Swedish.)

Almer, B.: 1972, The effect of acidification on fish stocks in lakes on the west coast of Sweden. Inform.Inst.Freshw.Res., Drottningholm (12). 47 p. (In Swedish with English summary.)

Almer, B., Dickson, W., Ekström, C. and Hörnström, E.: 1978, Sulfur pollution and the aquatic ecosystem. p. 272-311. In Sulfur in the environment. II. Ecological impact. Ed.: J.O. Nriago. John Wiley, New York.

Appelberg, M.: 1984a, Effekter av försurning och kalkning på populationer av flod- och signalkräftor. LIU 1984:B7. Inst.Limnol., Univ. Uppsala, Sweden. 130 p. (In Swedish.)

Appelberg, M.: 1984b, Early development of the crayfish Astacus astacus L. in acid water. Rep.Inst.Freshw.Res., Drottningholm 61, 48.

Appelberg, M.: 1985, Hydrobiologia 121, 19.

Appelberg, M. and Odelström, T.: 1984, Betydelsen av predation och näring för kräftpopulationer i kalkade vatten. LIU 1984:B8. Inst. Limnol., Univ. Uppsala, Sweden. 35 p. (In Swedish.)

Bohlin, T.: 1981, Rep.Inst.Freshw.Res., Drottningholm 59, 5.

Brown, D.J.A.: 1981, J.Fish.Biol. 18, 31.

Daye, P.G. and Garside, E.T.: 1979, Can.J.Zool. 57, 1713.

Degerman, E., Fogelgren, J.-E., Tengelin, B. and Thörnelöf, E.: 1985, Occurrence of brown trout, Atlantic salmon and eel in small acidified water courses on the west coast of Sweden. Inform.Inst.Freshw. Res., Drottningholm (1). 84 p. (In Swedish with English summary.)

Dickson, W.: 1983, Vatten, Lund, Sweden 39, 400.

Eriksson, F., Hörnström, E., Mossberg, P. and Nyberg, P.: 1982, Ecological effects of lime treatment of acidified lakes and rivers. Inform.Inst.Freshw.Res., Drottningholm (6). 96 p. (In Swedish with English summary.)

Eriksson, F., Hörnström, E., Mossberg, P. and Nyberg, P.: 1983, Hydrobiologia 101, 145.

Filipsson, O.: 1972, Sötvattenslaboratoritets provfiske- och provtagningsmetoder. Inform.Inst.Freshw.Res., Drottningholm (16). 24 p. (In Swedish.)

France, R.L.: 1983, Response of the crayfish Orconectes virilis to experimental acidification of a lake with special reference to the importance of calcium. p. 98-111. In Freshwater crayfish V. Pap.Fifth Int.Symp.Freshw.Crayfish, Davis, Calif., USA 1981. Ed.: C.R. Goldman. AVI Publ.Company, Westport, Conn.

Johansson, K. and Nyberg, P.: 1981, Acidification of surface waters in Sweden - effects and extent 1980. Inform.Inst.Freshw.Res., Drottningholm (6). 118 p. (In Swedish with English summary.)

Leivestad, H., Hendry, G., Muniz, I.P. and Snekvik, E.: 1976, Effects of acid precipitation on freshwater organisms. p. 87-111. In Impact of acid precipitation on forest and freshwater ecosystems in Norway. Ed.: F. Braekke. S.N.S.F.-Project. FR 6/76. Oslo-Ås.

Lessmark, O.: 1983, Competition between perch (Perca fluviatilis) and roach (Rutilus rutilus) in south Swedish lakes. Ph.D. Thesis, Inst.Limnol., Univ.Lund, Sweden. 172 p.

Malley, D.F.: 1980, Can.J.Fish.Aquat.Sci. 37, 364.

McMahon, B.R. and Morgan, D.O.: 1983, Acid toxicity and physiological responses to sub-lethal acid exposures in crayfish. p. 71-85. In Freshwater crayfish V. Pap.Fifth Int.Symp.Freshw.Crayfish, Davis, Calif., USA 1981. Ed.: C.R. Goldman. AVI Publ.Company, Westport, Conn.

Milbrink, G. and Johansson, N.: 1975, Rep.Inst.Freshw.Res., Drottningholm 54, 52.

Morales, J. and Appelberg, M.: 1984, Studier av en signalkräftpopulation före och efter en sjökalkning. LIU 1984:B13. Inst.Limnol., Univ. Uppsala, Sweden. 20 p. (In Swedish.)

Muniz, I.P.: 1981, Int.Atlantic Salmon Found.Spec.Publ.Ser. 10, 65.

National Swedish Board of Fisheries and the National Swedish Environment Protection Board: 1981, Liming of lakes and rivers 1977-1981 in Sweden. Inform.Inst.Freshw.Res., Drottningholm (4). 201 p. (In Swedish with English summary.)

Nyberg, P.: 1984a, Phil.Trans.R.Soc., Lond. B 305, 549.

Nyberg, P.: 1984b, Rep.Inst.Freshw.Res., Drottningholm 61, 154.

Raddum, G.G., Hagenlund, G. and Halvorsen, G.A.: 1984, Rep.Inst.Freshw. Res., Drottningholm **61**, 167.

Rosseland, B.O., Skogheim, O.K., Abrahamsen, H. and Matzow, D.: 1984, Survival and reduction of physiological stress of Atlantic salmon (*Salmo salar* L.) smolts in an acid river through slurry liming. Kalkningsprosjektet Rapp. 17:84. 24 p. Fylkeshuset, 4800 Arendal, Norway.

SAS Institute Inc.: 1982, SAS user's guide: statistics. 1982 Edition. Cary, N.C. SAS Institute Inc. 384 p.

Svärdson, G.: 1976, Rep.Inst.Freshw.Res., Drottningholm **55**, 144.

Zippin, C.: 1956, Biometrics **12**, 163.

A SCANDINAVIAN MODEL USED TO PREDICT THE REACIDIFICATION OF LIMED LAKES IN NOVA SCOTIA AND ONTARIO, CANADA

H. U. Sverdrup
Dept. of Chemical Engineering, Chemical Center, LNTH
P.O.Box 124
S-221 00 Lund
Sweden

ABSTRACT: A Scandinavian reacidification model was used to predict the reacidification of three limed lakes in the Sudbury area, Ontario, and Sandy Lake in Nova Scotia. A prediction was made also for the future reacidification of Bowland Lake, 70 km north of Sudbury, limed in 1983. The reacidification of a neutralized lake is dependent on the continuing dilution of the dissolved calcium carbonate, and for a limited period of time, the dissolution of calcite from the bottom. The test of the model on the Canadian laked shows that the model may be a useful tool for lake liming design and planning.

1. INTRODUCTION

A larger number of lakes and rivers in Scandinavia and North America located on granitic bedrock or surrounded by soils with a slow weathering mineral component are suffering from the effects of acid rain.

The mitigative measure, most frequently taken, is liming of the river or lake affected. This is in fact the only large scale remedy available until emissions of SO_2 and NO_x are reduced drastically.

Earlier, we have developed models for predicting the reacidification of limed lakes (Sverdrup , 1985) and these have been thoroughly tested against data from actual lake limings in Sweden (Sverdrup , 1985; Sverdrup and Warfvinge, 1986)

The reacidification model is a powerful tool for lake liming planning and management as well as for cost optimization. Due to the usefulness of the model in Scandinavia, it was decided to try the reacidification model on some limed lakes in Canade.

2. MODEL DESCRIPTION

The hydrology of most lakes varies substantially with the seasons. The two main reasons for this are the seasonal variations in the tributary flow rate and the thermal stratification of lakes.

Water, Air, and Soil Pollution **31** (1986) 689–707.

The stratification may in shallow lakes reach such a depth that the volume under the thermocline, the hypolimnion, becomes small as compared to the total lake volume. In such a case the stratification of the lake will have little influence on residence time variations in relation to the influence caused by the tributary flow rate variations. Such a system may well be modelled as a stirred tank reactor with variable flow rate.

The residence time distribution will be influenced by both the stratification and the flow variations in deeper lakes with short residence times. A two tank reactor model will be able to describe the system in which the two tanks are merged at every turnover, and the flow rate varies with the seasons.

In deeper lakes where the residence time is considerably longer than each stratification period, the lake may be considered to be fairly well mixed during one residence time period. Accordingly, the lake may be modelled with a one tank model with variable flow rate.

Stagnant zones or separated basins generally contain a very small fraction of the lake volume or the water exchange is not sufficiently restricted over the actual time periods involved.

On basis of the considerations discussed above, a simple model was designed in order to model the lake reacidification process after liming. The tank is modeled as one stirred tank with variable flow rate.

The dissolution of calcite in dilute acidic aqueous solution can be described as a heterogeneous solid-liquid reaction controlled by diffusion. The chemical reactions involved are:

$$CaCO_3(s) + H^+ \qquad Ca^{2+} + HCO_3 \qquad (I)$$

$$CaCO_3(s) + H_2O + CO_2 \qquad Ca^{2+} + 2HCO_3 \qquad (II)$$

$$CaCO_3(s) + H_2O \qquad Ca^{2+} + HCO_3 + OH^- \qquad (III)$$

In natural waters with pH-values below pH 5.5 and a carbon dioxide partial pressure less than 0.1 atm, reaction (I) will dominate and the dissolution rate will be proportional to the hydrogen ion concentration while at higher pH-values reaction (III) will dominate. If the carbon dioxide partial pressure exceeds 0.1 atm, the reaction (II) will contribute to the dissolution rate and it may be increased as much as one order of magnitude (Plummer et al 1978).

In natural lakes, acidic or limed, the carbon dioxide partial pressure will usually be much less than 0.1 atm and the expression for the dissolution rate at pH-values below pH 5.5 may be simplified to:

$$R = K1\ (H^+) + KW - KB(Ca^{2+})(HCO_3^-) \qquad (kmol/m^2{\cdot}s)$$

where the last term takes the backward reaction at high alkalinities into account and KW accounts for the effect of the reaction with water and carbon dioxide. In open systems this may be approximated as a constant factor.

Several coupled mechanisms will influence the dissolution rate of calcite from the bottom: The dissolution of calcite and a subsequent increase in the pH-value due to the neutralization of the acid. Due to the increase in pH-value, precipitation of humates and metal complexes on the calcite surfaces will occur and the dissolution of calcite will be inhibited. Due to the increase in the calcium concentration in the water column, calcium will enter into ion exchange with hydrogen ions in the sediments.

Calcite is consumed on the bottom according to the equation:

$$(2) \quad -(dm/dt) = R\ A\ B(t) \qquad (kmol/s)$$

where R is the reaction rate and A the available mineral surface area for dissolution and B(t) the deactivation function. The area of exposed calcite on the bottom is expressed as a fraction of the lake bottom area:

$$(3) \quad B(t) = e^{-KH\ t} + KO$$

KH is the deactivation rate and was in the calculations set at the value KH=0.6 yr corresponding to a diffusional boundary layer with a thickness of 6 cm above the bottom. For each tank in the model structure, a complete mass balance is made for the components to be calculated.

The inlet flux from the tributaries plus the amount dissolved or neutralized, (dm/dt), is balanced by the accumulation in the lake volume V, plus the amount sorbed to the sediment (dS/dt), plus the outlet flux.

$$(4) \quad (Q_i\ C_i) + (dm/dt) = V(dC/dt) + (dS/dt) + Q_u\ C \quad (kmol/s)$$

where the sum of the discharge of n tributaries is assumed to be equal to the outlet flow: The lake volume is assumed to be constant which is nearly true for deeper lakes and more approximate for shallow lakes. Each tank is assumed to be well mixed and the concentration in the lake equal to the concentration at the outlet.

The seasonal runoff variations may be divided into different classes. We assumed two types of runoff regions: 1) Variations typical of coastal regions (Nova Scotia); 2) Variations typical of inland regions (Ontario). Both types are shown in Figure 1.

3. THE DISSOLUTION OF CALCITE IN LAKES

The reacidification rate of a limed lake was modeled considering several chemical processes in the lake system:

- The dissolution of calcite and a subsequent increase in the pH-value due to the neutralization of the acid.
- Due to the increase in pH-value, precipitation of huates and metal complexes on the calcite surfaces will occur and the dissolution of calcite will be inhibited.
- Due to the increase in the Ca concentratior in the water column, Ca will enter into ion exchange with H^+ in the sediments.

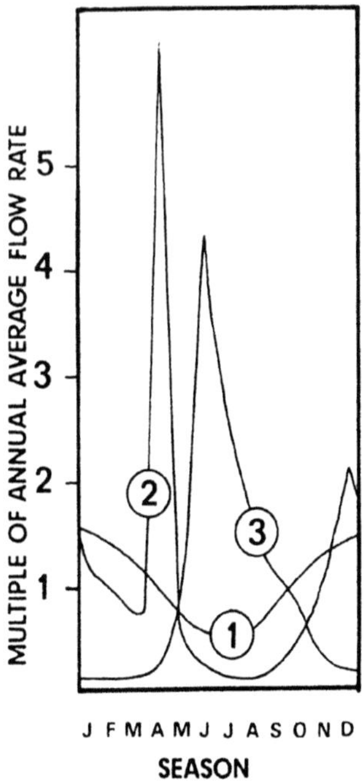

Figure 1. The three seasonal variation types used in the model is a simplification of classifications introduced by Gottschalk et al. (1979). They apply for 1: Coastal regions, 2: Inland regions and 3: Mountain regions.

TABLE I

Lake	Year	$Ca(OH)_2$ metric tons	$CaCO_3$ metric tons
Middle	1973	20.0	13.6
Lohi	1973	18.2	-
	1974	4.1	-
	1975	5.0	15.0
Total		27.3	15.0
Hannah	1975	13.0	7.5

Amounts of neutralizing agent added to the Sudbury lakes

For each tank in the model structure, a complete mass balance was made for H and Ca.

The inlet flux from the tributaries plus the amount dissolved or neutralized is balanced by the accumulation in the lake volume, plus the amount sorbed to the sediment, plus the outlet flux.

It was chosen to work with the mass balances for H and Ca and calculate the alkalinity from a titration curve. It was assumed that the carbonate system was approximately at equilibrium in the water column and the alkalinity could be considered as dependent on the concentrations of the H and Ca. The lake was also assumed to be an open system to the atmosphere with respect to CO_2.

4. NUMERICAL METHODS AND DATA REQUIRED

A computer code was written to solve the differential equations numerically with the Runge-Kutta algorithm and calculate the reacidification of a limed lake. The code was written for a small personal computer to facilitate model availability.

The simple model structure will need the following input data which are among those usually provided in Swedish lake liming projects, or may be easily estimated from these:

1.	Initial pH in the lake after liming	
2.	Annual average pH in the tributary	
3.	Initial Ca concentration after liming	(mg l^{-1})
4.	Average Ca concentraiton in the tributary	(mg l^{-1})
5.	Amount of calcite on the bottom of the lake	(metric tons)
6.	Fraction of the bottom covered by calcite	(%)
7.	Mean lake depth	(m)
8.	Lake volume	(mill m^3)
9.	Average residence time	(yr)

The amount on the bottom may be estimated from the total amount distributed and the amount instantly dissolved. The fraction of the bottom fully covered by calcite may be estimated from the spatial distribution of calcite powder over the lake surface and the amount on the lake bottom (Sverdrup 1984; Sverdrup, 1985).

In addition, the tributary flow variation type is indicated with a number that can be read from Figure 1.

10. Tributary flow variation type.

The deactivation rate for calcite on the bottom was estimated internally in the model and may assume a value between 0.6 and 0.8 yr. The deactivation rate has been determined from large amounts of lake liming data (Sverdrup et al 1984, Sverdrup,1985) as well as the deactivation process which has been studied in detail (Warfvinge, 1986).

TABLE II

Name of the lake	Sandy Lake	Hannah Lake	Middle Lake	Lohi Lake	Bowland Lake
pH in the lake after liming	6.9	7.3	8.0	7.0	6.6
Average pH in the tributary	4.9	4.6/5.2/5.4	4.6/5.2	4.5	5.0
Ca conc after liming (mg l^{-1})	6.5	21	18	9	4.5
Ca conc in the tributaries (mg l^{-1})	3.6	13	10	6	2.3
Lake volume (10^6 m^3)	5.1	1.1	1.75	2.5	8.3
Average depth (m)	6.9	4	6.2	6.2	7.6
Average hydr.retention time (yr)	0.43	2.5	1.4	1.1	3.0
Calcite bottom cover (P)	0.03	0.03	0.05	0.02	0.05
Amount calcite on the bottom (tonne)	100	2	6	6	3
Runoff regim	Costal	Inland	Inland	Inland	Inland
Contrib. from upstream lake	-	-	37 %	-	-

Indata used for the reacidification calculation. The quthor was informed by Norman Yan that the deposition of acid in the Sudbury area probably declined somewhat during the years 1975 and onwards. Accordingly, runs were made where the yearly average pH of the tributaries to Hannah Lake varied with time; pH 4.6 was used from 1973-1977 and after 1979 pH 5.2 was used. In the Sudbury Lakes (Lohi, Hannah, Middle), virtually all addad base material dissolved instantly, leaving a very small residual amount on the bottom. The amount given in the table includes this amount and the equivalent amount calcite absorbed in the sediments as estimated by the authors. The average tributary pH and calcium was estimated from the pre-liming data for the lakes.

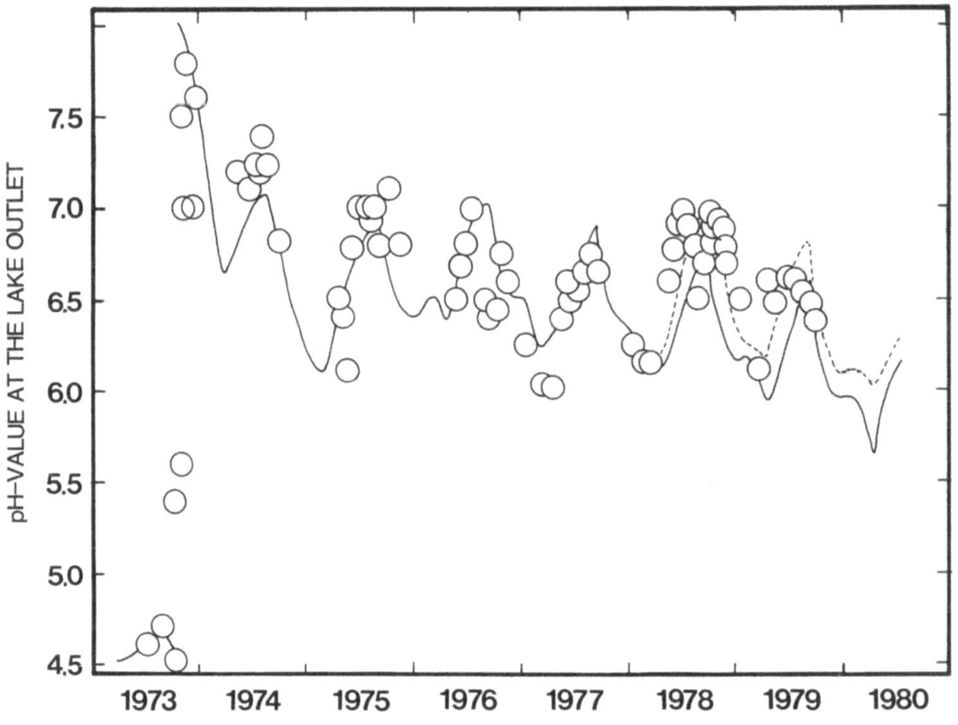

Figure 2. The predictions of the reacidification model compared to the observed pH-values in Middle Lake. The drawn line corresponds to an average pH in the tributaries of pH 4.6 (——), the dotted line pH 5.2 (---).

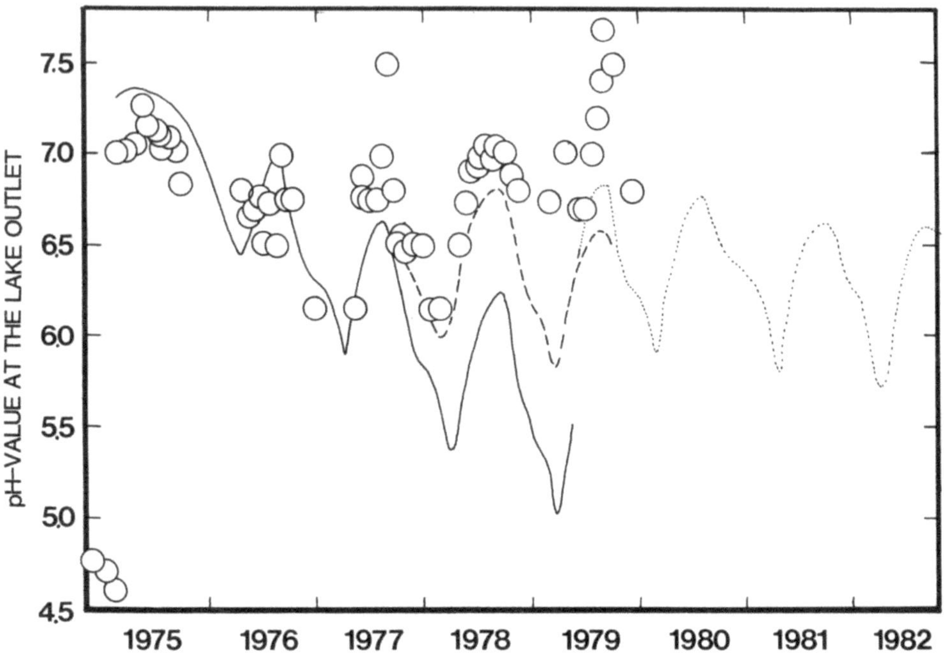

Figure 3. The predictions of the reacidification model compared to the observed pH values in Hannah Lake. It is assumed that the deposition of acid in the lake watersheds has decreased over the years. The drawn line corresponds to an average pH-value in the tributaries of pH 4.6 (——), in 1977 this was assumed to change to pH 5.2 (---) and in 1979 to 5.4 (···).

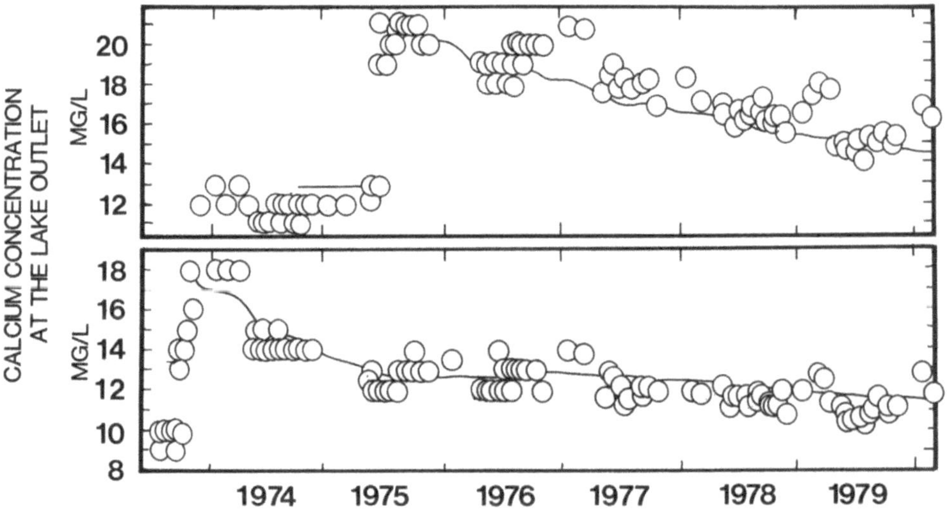

Figure 4. The predicted Ca concentrations in Hannah Lake (A) and Middle Lke (B). 37 % of the tributary water to Middle Lake originates in Hannah Lake.

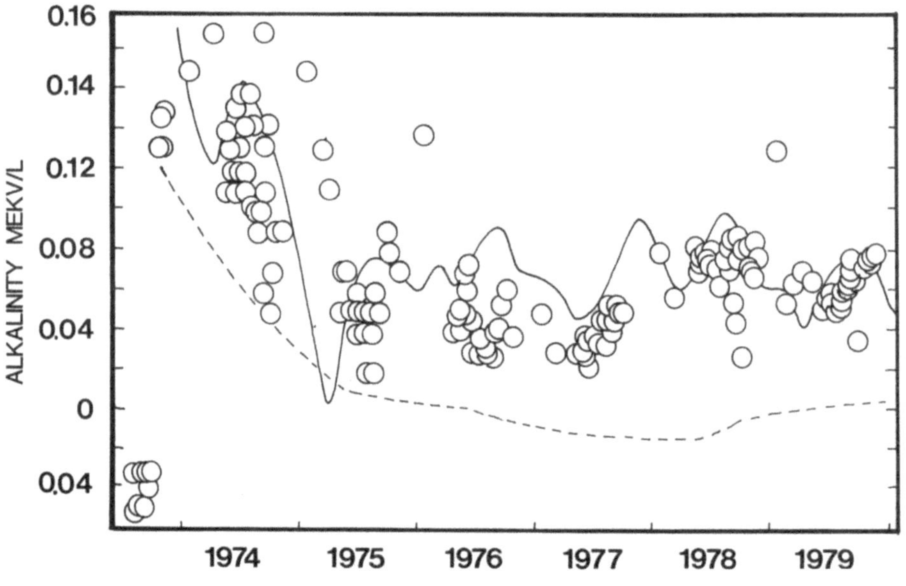

Figure 5. Prediction of alkalinity in Middle Lake. The drawn line represents the predictions of our model (——), the dotted line the model of Dillon and Scheider(1982), Dillon et al (1984) (---).

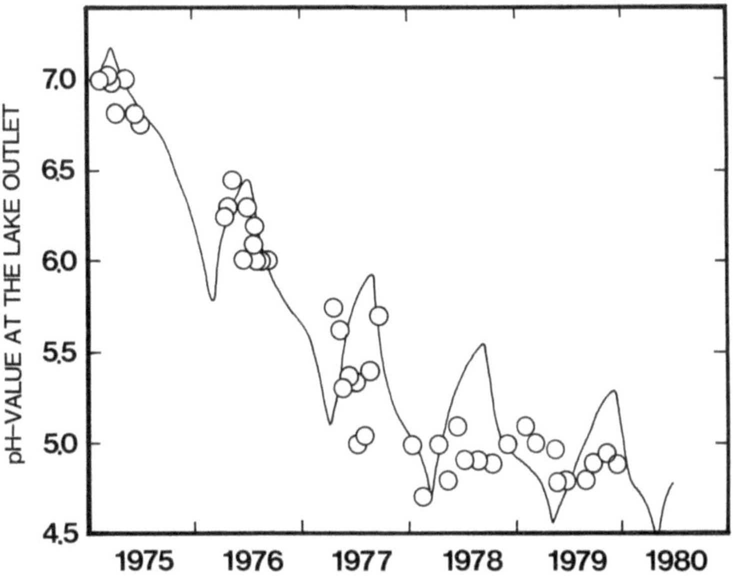
pH-VALUE AT THE LAKE OUTLET
7.0
6.5
6.0
5.5
5.0
4.5
1975
1976
1977
1978
1979
1980

Fig. 6a.

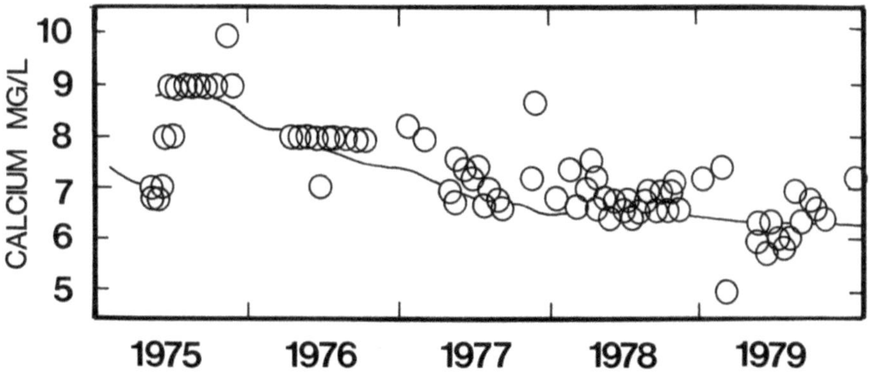
CALCIUM MG/L
10
9
8
7
6
5
1975
1976
1977
1978
1979

Fig. 6b.

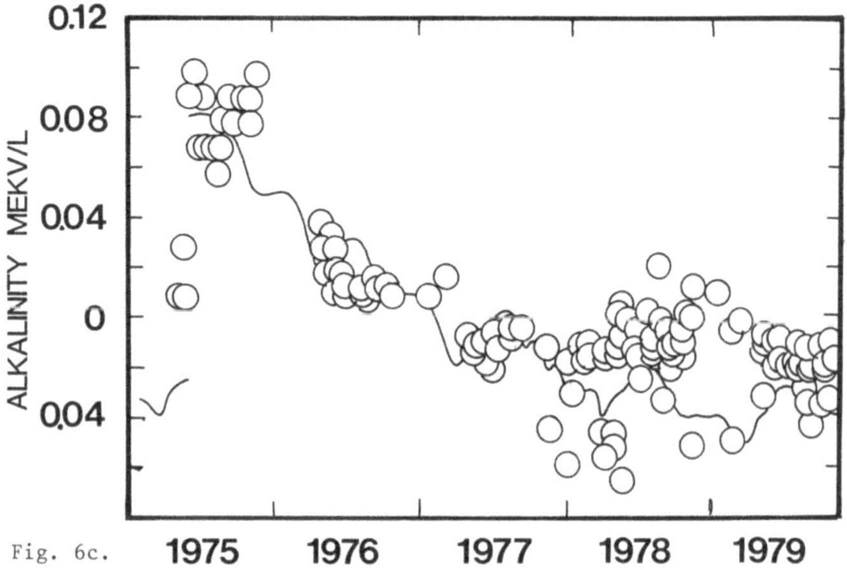

Fig. 6c.

Figure 6. Model prediction for the reacidification of Lohi Lake. Lohi Lake was not fertilized in any way during this period.

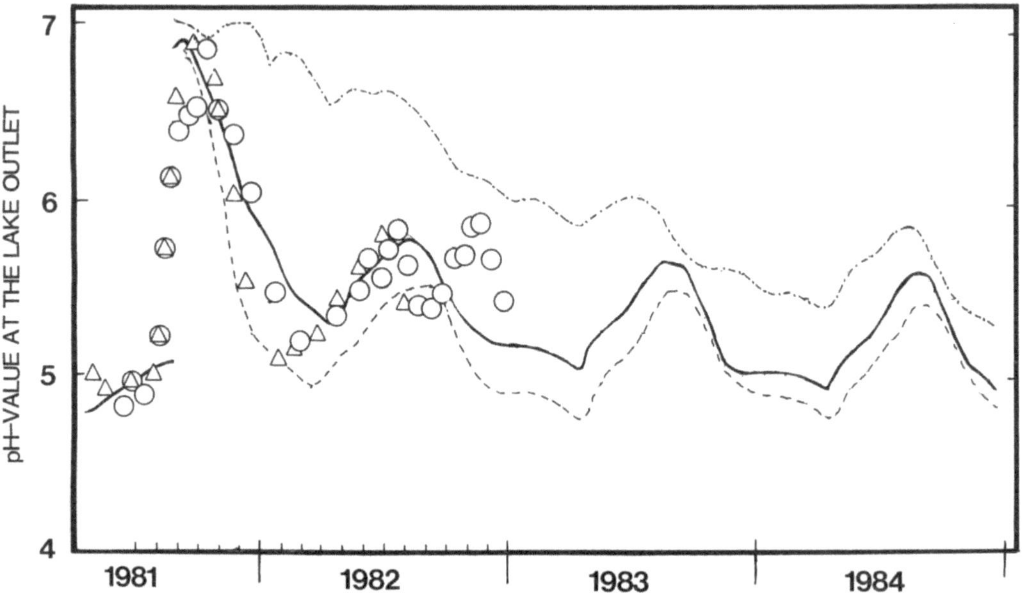

Figure 7. Model prediction for the reacidification of Sandy Lake. Circles represent surface values, trianles bottom values. The lower dotted line represents the case with no calcite on the bottom, the upper dotted line a hypthetical case with 400 tonne on the bottom.

5. THE NEUTRALIZATION OF THE SUDBURY LAKES AND SANDY LAKE

Sheider et al (1975) described the method used for addition of the neutralizing chemicals to the lakes. Briefly, surface water was pumped into a slurrying chamber on a moving boat where the water was combined with dry chemicals supplied via a hopper. The slurry was discharged into the backwash of the boat while traversing the lake. Approximately 8 metric tons of dry chemical were added per day.

Enough $Ca(OH)_2$ was added to Lohi and Middel lakes to raise the pH to pH 8.0 (determined by titration), a level that was considered adequate for subsequent neutralization of the entire water column and the top few cm of lake sediments (Scheider et al. 1975). Calcium carbonate was simutaneously added to Middle Lake, with the intent of providing a residual buffering system. The amount of both chemical compunds added to Hannah Lake per unit volume was identical to that added to Middle Lake. A detailed schedule of additions is given by Ontario Ministry of the Environment (Dillon and Scheider 1982)(Table 1).

Middle Lake was fertilized with annual additions of P in 1975 to 1978 and Hannah was fertilized annually with P in 1976-1978. Hannah Lake is located upstream from Middle Lake.

White et al (1984) have described the method for liming Sandy Lake. It was limed in 1981 with 135 metric tons of calcite. The calcite powder was distributed from a barge and slurried with water in a tank before application.

6. RESULTS AND DISCUSSION

The model was used to calculate the reacidification for the Sudbury lakes and Sandy Lake. The data required by the model was obtained from the available information, the data are listed for all lakes in Table II. Bowland Lake was included as well. It was limed in 1983 and it is not yet reacidified.

Hannah and Middle Lake were modeled as two lakes in series, and the calculation seems to predict the observed pH, Ca and alkalinity well for Middle Lake (Figures 2, 4 and 5). For Hannah Lake the production follows the observed data when it is assumed that the acid tributary pH was set at pH 4.6 for the period 1975-1977, atpH 5.2 between 1977 and 1979 and at pH 5.5 after 1979. The calculated pH values are compared to the observed pH values in Figure 5. The calculated concentrations of Ca are shown in Figure 6 together with the observed values. It can be seen that the calculations follow the observations. Dillon and Sgheider (1982), reports that they included additions of Ca containing road salt in their model of the Ca budget. We did not include this in our calculation and still obtain an answer that correlates better with the observed values. This may also be seen in the calculation of the alkalinity for Middle Lake. For comparison the curve predicted by Dillon and Scheider (1982), Dillon (1984) is included in the diagram shown in Figure 5. The inclusion of the amounts of Ca containing road salt in the model, will cause the model to predict a higher Ca concentration than observed.

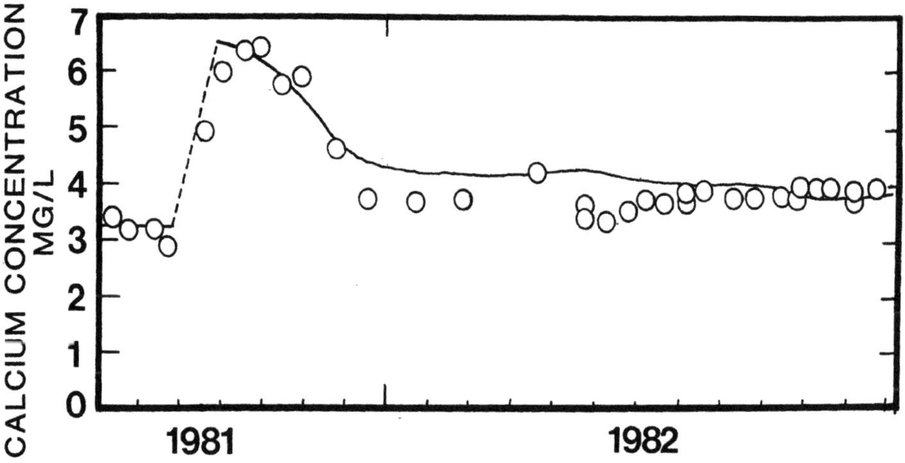

Figure 8. The predicted calcium concentration for Sandy Lake compared to the actual observation by White et al.(1984)

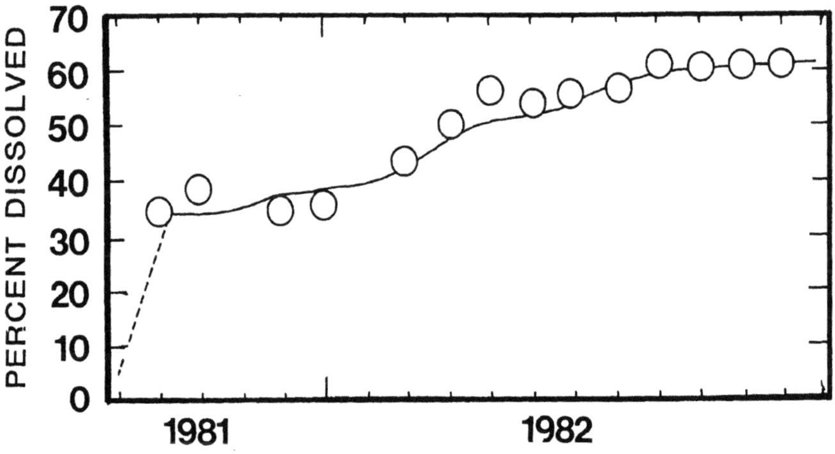

Figure 9. The predicted long term dissolution for Sandy Lake compared to the determination by White et al.(1984)

TABLE III

Water column dose	Amount on the bottom	Areal dose	Reacidification time to pH 6.0	Average amount required	Material cost	
tonne	tonne	tonne ha^{-1}	mo	tonne yr^{-1}	\$ $tonne^{-1}$	\$ yr^{-1}
35	0	0	3	140	65	9 100
35	100	1.4	4	410	30	12 300
35	200	2.7	14	202	25	5 050
35	400	5.4	17	310	20	6 200
35	700	9.5	25	350	20	7 000
35	1500	20.3	26	710	15	10 650
Stream liming in a main tributary				180	35	6 500

Cost analysis of hypothetical lake limings in Sandy Lake. It was assumed that the lake could be limed in different ways such as using different grinding grades or place smaller or larger fractions of the addid amount on very shallow water, to obtain very different amounts on the bottom. If it is assumed that the lake cannot be limed more than once every year, a minimum of 200 tonne of calcite is neede on the bottom, or approximately 200 tonne yr^{-1} for the lake to be kept constantly above pH 6.0. The required amount for a powder doser to keep the lake above pH 6.0 would be 180 tonne yr^{-1} if 50 % dissolution efficiency is assumed.

TABLE IV

Water column dose	Bottom dose	Bottom calcite cover	Reacidification time	Yearly required amount
tonne	tonne	%	mo	tonne yr^{-1}
35	100	3	4	410
35	100	10	6	273
35	100	20	13	126
35	100	30	13	126
35	100	40	12	137
35	100	100	7	234

Cost analysis of hypothetical lake limings in Sandy Lake. It was assumed that the lake was limed in such a way that 35 tonne dissolved in the water column and 100 tonne would settle on the bottom. The reacidification time was calculated for different degrees of calcite cover on the bottom. The analysis indicates a cover of 25 % as optimal in terms of amounts required per year.

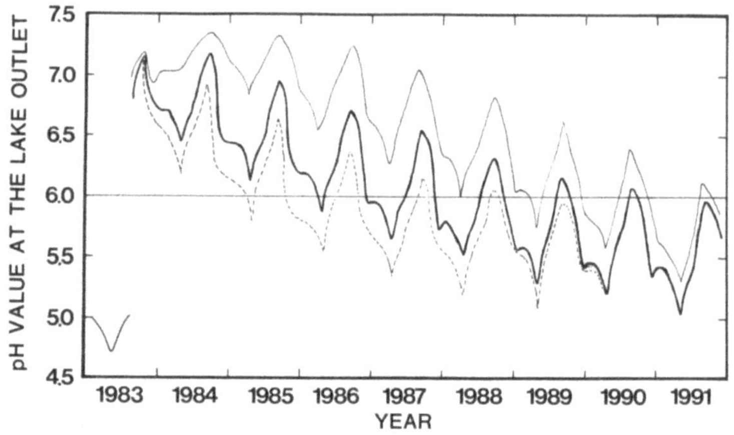

Figure 10. The predictions of the model for the reacidification of Bowland Lake after liming (——). For comparison two hypothetical cases are shown as well; Bowland Lake if there was no calcite on the bottom (---) and if there had been 8 times the actual amount of calcite on the bottom (——). The model predicts that Bowland Lake will reacidify during 1987 and 1988, indicating a duration of approximately 4-5 yr for the neutralization treatment.

The lakes were fertilized with P for several years. The model does not take this into account, but still seems to be able to describe the observed pH values and alkalinities. This leaves the impression that if the tributary acid load for these lakes have declined over the past years, this may as well explain the trends observed in pH values and alkalinities.

The numerical value for the average tributary pH and tributary Ca was for the Sudbury lakes taken to be equal to the values of pH and Ca in the lakes prior to liming.

In this way generation of alkalinity in the lake is incorporated in the model as a constant source.

The good correlation between the model predictions and the observed data strongly indicates that the internal generation of alkalinity of the lake was not changed by the changes induced by liming.

Lohi Lake is also located in the Sudbury area and was limed in 1975, but has not received any fertilization. The model seems to predict the reacidification of this lake fairly well(Figures 6 a-e).

The liming of Sandy Lake and the observed results are described in an accurate report by White et al (1984) and the data was used to test the model. The calculated pH value is compared to the observed values in Figure 7. Even if the lake has a short retention time and accordingly is difficult to model, the calculations and the observations do correlate very well. The model also predicts the Ca concentration (Figure 8) and the long term dissolution of calcite very accurately when compared to the reported data (Figure 9).

With the model we may investigate some hypothetical questions such as "could the reacidification have been delayed by adding more calcite on the bottom"? If the answer is yes,"is it cost efficient to do so"?

In a series of runs we calculated the reacidification for Sandy Lake with increasing amounts of calcite on the bottom, ranging from 0 ton ha^{-1}, the actual dose of 1.4 metric tons ha^{-1}, continuing with 2.4 metric tons ha^{-1}, 9.5 metric tons ha^{-1} and 20 metric tons ha^{-1}. The result of the calculation is shown in Table III.

As the amount on the bottom increases, the reacidification time increases towards a limit. We may now analyse this information in different ways. The cost efficeincy may either be defined in terms of amounts of calcite the yearly amount required to keep the lake above pH 6.0.

Amount per year required = Total amount/Reacidification time (ton yr^{-1})

or it may be defined in terms of costs. This is usefull if different calcite powders are used. We then define the cost per year to keep the lake above the reacidification limit, pH 6.0:

Cost per year = Total project cost/Reacidification time (Skr yr^{-1})

For our four alternatives we calculate the amount per year required and list them in table III. The same costs for transportation and application labor etc. was used for all examples calculated. However further study with the model reveals a more efficient way to delay reacidification. In the actual liming, the bottom had a small fraction of the bottom covered with calcite. Instead of increasing the added amount on the bottom, an equivalent delay of the reacidification may be achieved by increasing the bottom calcite cover, with a substantial increase in cost efficiency. The result of the calculation is shown in Table IV.

7. PREDICTION FOR THE REACIDIFICATION OF BOWLAND LAKE

We have used data available from Bowland Lake to predict reacidification as well as to predict reacidification in some hypothetical alternatives.

The predicted pH for Bowland Lake until 1991 is shown in Figure 10. Four different cases were considered.

- A hypothetical case with no calcite on the bottom, but water column chemistry as in the actual case (P = 0.0, 45 metric tons dissolved)
- The actual case with 35 metric tons of calcite on the bottom, the p-factor was set at 7.5 % bottom cover (P = 0.075)
- A hypothetical case with 200 metric tons of calcite on the bottom and a bottom cover of 20 % (P = 0.2). Water column chemistry as in the actual case
- A hypothetical case with 1000 metric tons of calcite covering 80 % of the bottom. Water column chemistry as in the actual case.

The model predicts that Bowland Lake will require reliming in 1987/88.

8. DISCUSSION

The study shows that the reacidification of Canadian limed lakes may be predicted with reasonable accuracy in a simple way.

The model runs for the lakes in the Sudbury region and indicate that if the assumptions on a gradual reduction in acid load to the lakes are correct, then fertilization of those lakes did not change the long term net generation of alkalinity in the lakes.

The model was originally designed to be used in the Swedish Operational Liming Programme. Accordingly, it was kept simple in order to work on the limited amount of data available in each project. More sophisticated models have been developed and tested by the authors, taking such things as thermal stratification into account. However, the increased requirement for data to run the model, largely limits the use of such models in operational liming. Accordingly, it was decided to test the simple model, the management tool, rather than the academicly more correct model.

The model in its present version works without calibration to any specific data set. All data required are physically or chemically observable.

The deactivation rate for calcite may seem dubious, but a fixed rate is used based on studies of the actual process (Sverdrup et al 1983, Warfvinge, 1986). The fraction of the bottom with calcite cover is equally a physical parameter, however the most difficult to estimate.

A fixed value may be calculated based on the fraction of the lake covered in the liming operation, the degree of wet slurrying and the particle size distribution.

Finally, the study shows how the model may be used for neutralization cost optimization in a liming project. Cost optimization is generally without much interest in research projects, but becomes very important in operational liming programs such as the Swedish or Norwegian Program.

The model may be obtained by writing to the authors, and is available for IBM computers.

LIST OF SYMBOLS

A	Available mineral surface	m^2
B(t)	Reactivation function	
C	Concentration	$kmol\ m^{-3}$
C_i	Concentration in the tributary i	$kmol\ m^{-3}$
KI	Rate constant for the reaction with the hydrogen ion	$m\ yr^{-1}$
KW	Rate of hydrolysis	$kmol\ yr^{-1}\ m^{-1}$
KB	Backward reaction rate	$m^3\ kmol^{-1}\ s^{-1}$
KH	Reactivation rate	yr^{-1}
KO	infinite deactivation limit	
M	Mass	kmol
n	Number of tributaries	
Q_i, Q	Tributary flow rate, flow rate	$m^3\ s^{-1}$
P	Fraction of the bottom with full calcite cover	
R	Raction rate	$kmol/m^2\ s$
S	Sorbed amount	kmol
t	Time	s
V	Lake volume	m^3

REFERENCES

Bengtson, B. et al:1981. Kalkning av sjöar och vattendrag 1977-1981. Information från Sötvattenslaboratoriet, Drottningholm 1981:4.

Dillon, P.J.:1984, The use of mass balances and mass balance models for the quantification of the effects of anthropogenic activities on lakes near Sudbury, Ontario. In Environmental impacts by Smelters. J. Nriagu (Ed.), J. Wiley & Sons Inc.

Dillon P.J., Scheider, W.A. 1982. Modelling the reacidification rates of neutralized lakes near Sudbury, Ontario. 185th National meeting, American Chemical Society, Las Vegas, Nevada, March 28 - April 2 1982.

Hultberg, H. and Andersson, I.;1982. Water, Air and Soil Poll.18;311.

Keller, W. and Pitblads, J.R.: 1984, Water Quality changes in Sudbury area lakes 1974-1975 to 1981-1983. APIDS No: APIO 20/84. Ontario Ministry of the Environment, Toronto, Canada.

Molot, L.,Hamilton, J. and Booth,G.: 1985, The experimental lake neutralization program: Technical Report No. 1; Bowland Lake before and during liming. Report prepared for the Ontario Ministries of the Environment and Natural Resources. Toronto, Canada.

Plummer, L.N.,Wigley, T.M.L. and Parkhurst, B.L.:1978, Amer. I. Sci. 273;179.

Sheider, W.A., Alamski, J. and Paylor, M.:1975, Reclamation of acidified lakes near Sudbury, Ontario. Ontario Minist. Environ., Toronto, Canada.

Sverdrup, H.:1985, Calcite dissolution kinetics and lake neutralization. Technical Doctorate Thesis, Dept. Chem. Eng., Lund. Institute of Technology, Lund, Sweden. 170 pp.

Sverdrup, H., Rasmussen, R., and Bjerle, I.:1984, Chemica Scripta 23;119.

Sverdrup, H. and Warfvinge, P.:1984, Vatten 40;369.

White, W. J., Watt, W. D., and Scott, C.D.: 1984, Fischeries 9;25.

Warfvinge, P. G.,1986, Neutralization of soil systems. Licentiate thesis. Dept. Chem. Eng. Lund Institute of Technology, Lund, Sweden. 270 pp.

LIMING IN ONTARIO: SHORT-TERM BIOLOGICAL AND CHEMICAL CHANGES

G.M. Booth, J.G. Hamilton, L.A. Molot
Booth Aquatic Research Group Inc.
532 Queen Street East
Toronto, Ontario, M5A 1V2
Canada

ABSTRACT. The short-term results of a 5-yr experimental lake neutralization program involving whole-lake liming of Trout and Bowland Lakes are reviewed. Whole-lake liming did not prevent spring melt pulses of acid and Al in a zone situated under the ice nearshore. Liming did improve whole-lake water quality in a previously acidified lake (Bowland Lake) so that lake trout (Salvelinus namaycush) could be restocked successfully. In situ bioassays showed a statistically significant improvement in survival of overwintering lake trout eggs after liming. Lake trout fry exposed during spring melt also showed increased survival rates after liming except at one site in Trout Lake where liming did not prevent the occurrence of 100% mortality. The resident yellow perch (Perca flavescens) population in Bowland Lake exhibited short-term increases in survival and in recruitment the first year following neutralization. Growth rate responses differed among age cohorts. A preliminary shoal liming experiment in Laundrie Lake suggested that nearshore spawning shoals which are not necessarily protected by whole-lake liming may be protected by treatment with fine calcite gravel.

INTRODUCTION

In 1981 the Government of Ontario initiated a 5-yr program to assess the role of liming in the management of acid stressed lakes. The principal objectives included both the documentation of the effects of whole-lake liming on the ecosystem and the testing of alternatives to whole-lake liming. This paper provides an overview of the program. Future publications are planned to provide detailed accounts of each experiment.

Two lakes, Bowland and Trout, were selected for whole-lake neutralization and detailed monitoring. A third lake, Laundrie Lake, was used for shoal liming experiments. In this report we review some of the preliminary findings regarding nearshore water chemistry, in situ bioassays and fish ecology.

Bowland Lake ($47^{o}05'N$, $80^{o}50'W$) is an acidic lake located on the

Water, Air, and Soil Pollution **31** (1986) 709–720.

granitic Canadian shield 70 km north of Sudbury, Ontario. It is a 109 ha lake with a mean depth of 7.6 m. At one time it supported a lake trout (*Salvelinus namaycush*) population. By 1980 the lake had acidified to pH 5 and had total Al concentrations of 130 to 150 $\mu g\ L^{-1}$. The former lake trout population had disappeared and restocking attempts had repeatedly failed, leaving only a population of yellow perch (*Perca flavescens*). The lake was limed in August, 1983, with dry calcite powder (Molot *et al.*, 1986). Our goal in liming Bowland Lake was to reestablish a self-sustaining lake trout population.

Trout Lake (45°35'N 80°10'W), the second study lake, is a 290 ha lake north of Parry Sound, Ontario. In contrast to Bowland Lake, this lake supports a complex fish community of over 16 species, including lake trout and smallmouth bass (*Micropterus dolomieui*). While summer epilimnetic pH levels were about 6, during spring melt nearshore zones were invaded by acidic, Al rich, meltwater. The lake was limed with a calcite slurry in May, 1984 (Molot *et al.*, 1986). The goal in liming Trout Lake was to protect the fish community, including nearshore lake trout spawning habitat, without producing negative effects on the ecosystem.

Laundrie Lake (47°07'N 80°52'W) is a 364 ha lake, 70 km north of Sudbury, Ontario, that receives the outflow from Bowland Lake. At the time of the experiment, water chemistry was intermediate to Bowland and Trout Lakes with pH values 5.5 to 6.0 and total Al levels about 100 $\mu g\ L^{-1}$. Only a relic lake trout population of older fish remained from the originally healthy fishery. Stocking of hatchery-reared lake trout had been successful, but there was no evidence of natural recruitment. The goal for Laundrie Lake was to test whether shoal liming could be used as a means of improving water quality for lake trout egg and fry development.

In both Bowland and Trout Lakes, liming successfully raised the pH and alkalinity to over 6.5 and 80 $\mu eq\ L^{-1}$, respectively (Molot *et al.*, 1986). In the six months following liming, the alkalinity ranged from 80 to 90 $\mu eq\ L^{-1}$ in both lakes. Total Al levels in Bowland Lake dropped from 130 to 80 $\mu g\ L^{-1}$ 1 mo after liming. Al apparently precipitated out of solution and settled to the bottom.

NEARSHORE WATER QUALITY DURING SPRING MELT

The quality of water in the nearshore zone during spring runoff is important, because it is in this zone that lake trout eggs are spawned (McMurtry, 1985). Lake trout sac fry occupy the interstitial spaces in the spawning shoals during the spring melt. The sac fry stage has been shown to be particularly sensitive to acidic water (Holtze and Lee, 1982; Peterson *et al.*, 1982; Hutchinson and Holtze, 1985).

A monitoring program was designed to document the chemistry of nearshore water before and after liming in Trout and Bowland Lakes. It was of particular interest whether whole-lake liming could protect the ecologically important nearshore zones during acid spates. Nearshore water was sampled in both study lakes at the surface, 1 and 2 m depths below the water-ice interface along several transects from shore to

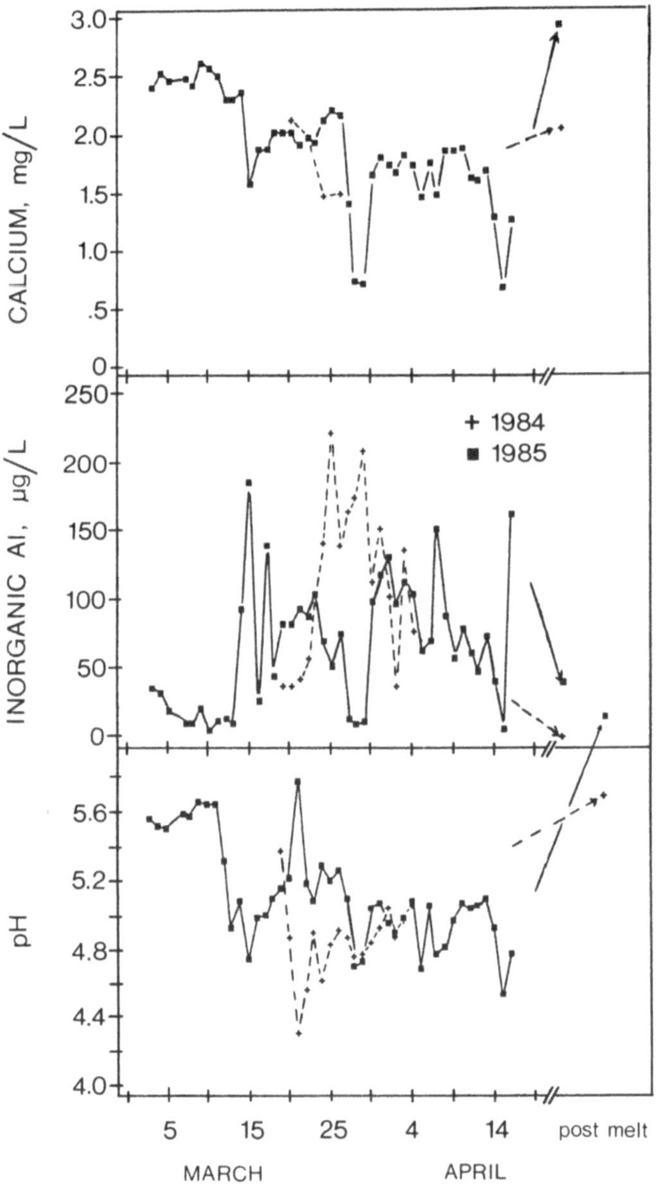

Figure 1. Fluctuations of pH, inorganic Al and Ca immediately below the ice 0 m offshore at Site 5 in Trout Lake in 1984, before liming, and 1985, after liming, showing peaks in inorganic Al coinciding with depressions in pH and Ca levels.

30 m offshore. Additional samples were taken interstitially from the shoal substrates, using the method of Gunn and Keller (1984).

Figure 1 illustrates the fluctuations in pH, Ca and inorganic Al which occurred before (1984) and after (1985) liming at the water-ice interface at Site 5, 2 m offshore in Trout Lake. Both before and after liming there were substantial inflows of potentially toxic meltwater. During the spring pulses, the Al laden, acidic meltwater contained little Ca to mitigate its toxicity and mixed little with whole-lake water, thereby being potentially damaging to biota restricted to the nearshore region. Toxic meltwater was not, however, found at all sites around the lake.

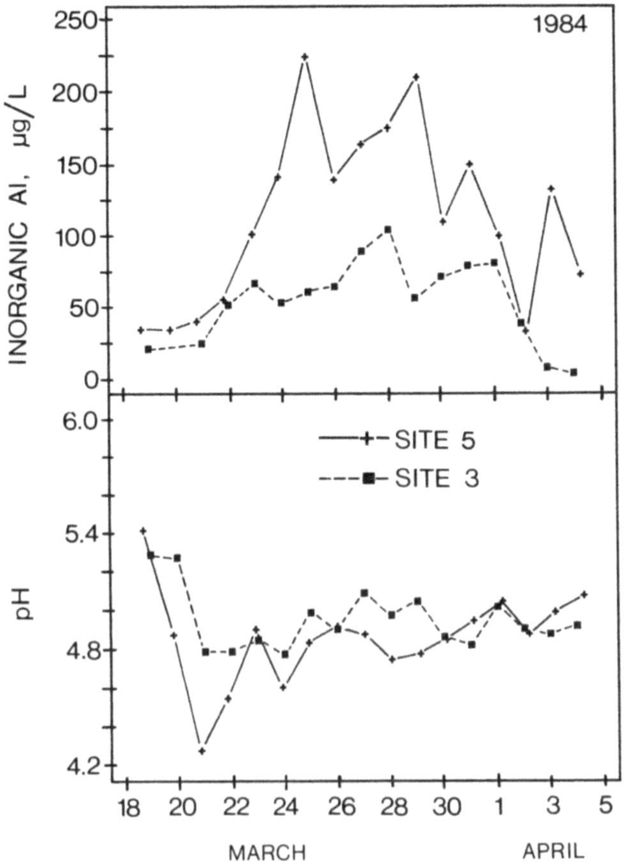

Figure 2. Fluctuations in pH and inorganic Al immediately below the ice 0 m offshore at 2 sites in Trout Lake before liming. The variability between sites was primarily a function of inorganic Al levels.

Large inter-site variation in water quality was observed. This may be due to a combination of factors including slope, watershed size and snow accumulation (Keller and Gunn, 1985). Figure 2 illustrates the differences which occurred in the spring before liming, at two sites, 2 m offshore in Trout Lake. While Site 3 had low pH levels, these did not coincide with high inorganic Al levels as they did at Site 5 where similarly low pH levels occurred together with a peak in inorganic Al. This indicates that the toxicity of nearshore water differs at various sites around the lake.

BIOASSAY OF LAKE TROUT

Hatchery-Reared Fry: _In situ_ bioassays of lake trout fry were used to determine the toxicity of near-surface, inshore water during spring pulses in Trout and Bowland Lakes and in a circumneutral control lake (Fowke Lake, $45^{o}40$'N, $79^{o}40$'W). Six 20 L baskets containing 20 hatchery-reared fry each were held at 3 sites on each lake during spring melt (except 1983 in Bowland Lake - see below) for 1 week in 1983 and 2 weeks in 1984 and 1985. The baskets were suspended immediately below the ice in less than 1.5 m of water within 10 m of shore. One to 2 g fry from North Bay Hatchery (pH 7.2, conductivity 100 μmhos cm^{-1}) were allowed to acclimate to within 1.5^{o}C of lake temperature before commencing the exposure.

There was less than 10% mortality in the control lake in all years. Figure 3 shows the mortality at each site in Trout and Bowland Lakes. In 1983, the Bowland Lake exposure was carried out one week after the ice melted in early May. The lake was not receiving spates of acidic snow melt at that time. Even so, more than 50% of the fish died within 7 days. In the 2 yr after liming, less than 12% of the fish died each year after two weeks exposure.

In Trout Lake there was considerable inter-site and inter-year variation. In 1983 few fish died during the one week exposure. In 1984 only one site, Site 5, had higher mortality than the control lake (28% after 2 weeks). The 1984 water chemistry at Site 5 is illustrated in Figure 2. After liming Trout Lake in early summer 1984, the bioassay in March, 1985 resulted in 100% mortality at Site 5 compared to 4% and 18% at other sites.

It should be noted that lake trout fry would not be found naturally in these nearshore surface waters during spring melt, thus the bioassay does not purport to exactly simulate any ecologically significant exposure. It is intended only as a relative measure of toxicity of water in the nearshore surface zone. Nevertheless, the high mortality at Site 5 does demonstrate that major pulses of toxic water can occur in the nearshore surface water, unmitigated by whole-lake liming. The ecological significance of this is currently unknown.

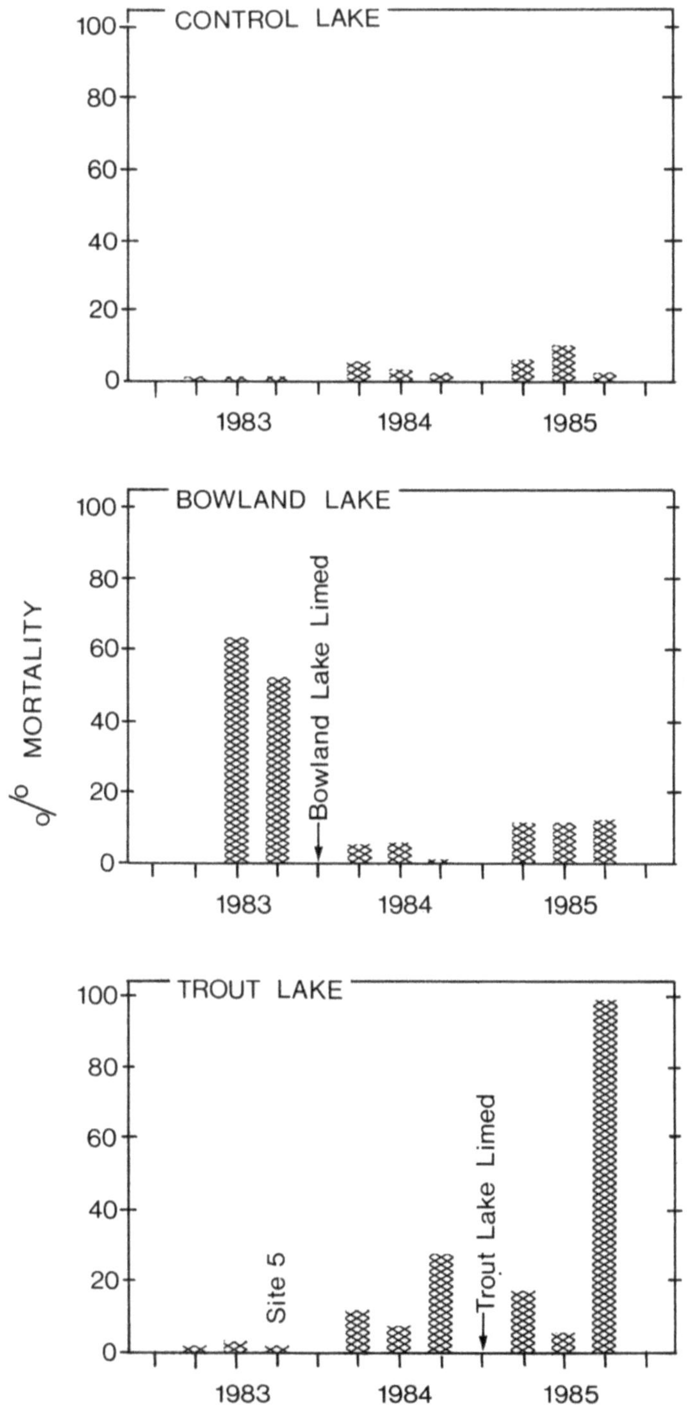

Figure 3. Mortality of hatchery-reared lake trout fry exposed for 2 weeks at 3 sites in Trout, Bowland and a control lake, showing decrease in toxicity of Bowland Lake after liming and large inter-site variability in Trout Lake with 100% mortality at Site 5 after liming.

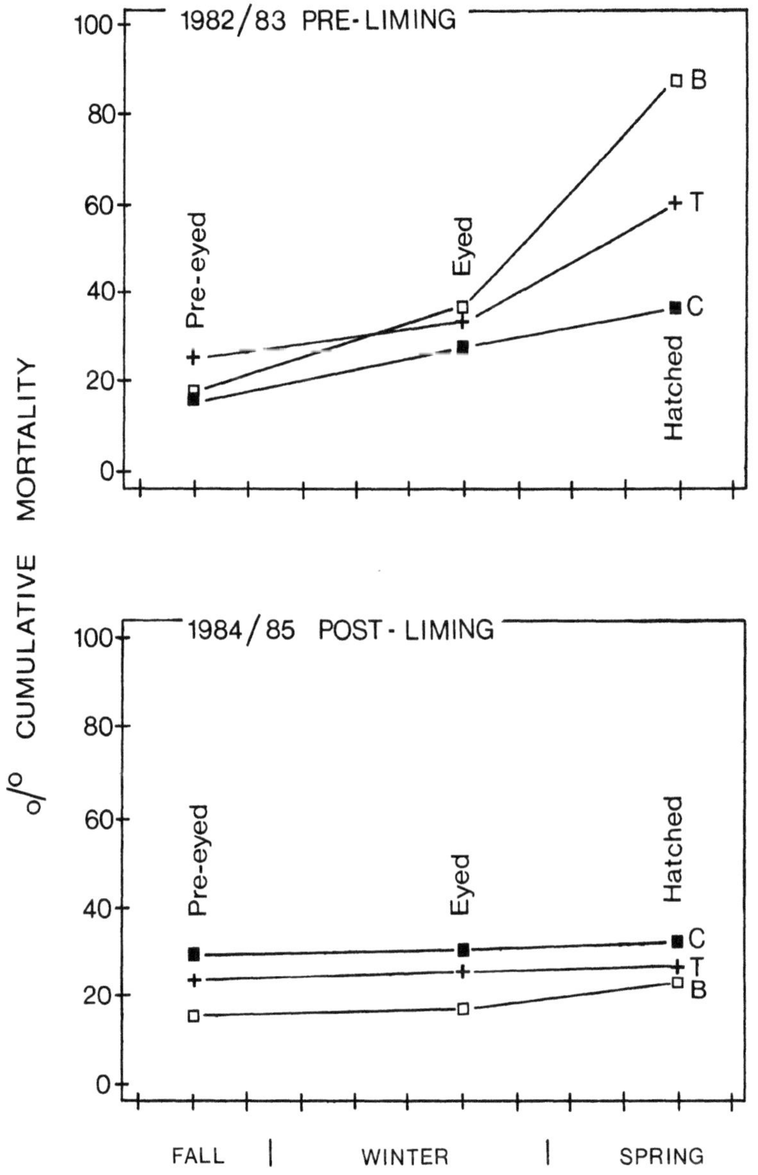

Figure 4. Mortality of lake trout eggs in Bowland (B) and Trout (T) Lakes before and after liming, and a control lake (C), showing decrease in mean mortality levels (5 sites per lake) during spring melt in the limed lakes.

Egg Incubations: In natural populations, the fry hatch in late winter and then remain on the shoals until after ice-out. It is the hatched sac fry, therefore, that are potentially exposed to the spring runoff.

Twenty incubators containing 50 newly fertilized eggs each were situated in 1 m of water on nearshore shoals in Bowland, Trout and a circumneutral control lake (Fowke Lake). The eggs were left from late October until ice-out in late April/early May. The time of death was inferred both from periodic sampling over the winter and from stage of egg or sac fry development at death. In contrast to the extreme variability between sites in the surface water exposures of fry, lake trout eggs incubated on the surface of nearshore shoals showed less inter-site variability. Figure 4 shows that prior to liming there was a high rate of mortality at the hatched sac fry stage in Bowland Lake and a lower rate in Trout Lake. The early pre-eyed and eyed mortality was not very different among the three lakes. After liming, there was a decrease in mortality of hatched fry in both Trout and Bowland Lakes compared to the control. Both these decreases were found to be significant ($P<0.05$) when tested using ANOVA.

There are several differences which should be noted between the egg bioassays and the hatchery-reared fry bioassays. The eggs were in deeper water (1 m) directly on the bottom and were surrounded by granite rubble. While the hatchery-reared fry were exposed to runoff directly beneath the ice, the eggs were somewhat more protected from direct runoff effects. In addition, there is evidence to show that eggs, sac fry and fry have different responses to pH and Al (Holtze and Lee, 1982). It has been suggested that Al may mitigate low pH effects in eggs while increasing the toxicity to fry (Baker and Schofield, 1982; Brown, 1983).

FISH COMMUNITY RESPONSES IN BOWLAND AND TROUT LAKES

Fish populations were sampled with various entrapment gear (1.22 m trap nets, plexiglass minnow traps and baited minnow traps). Data are reported from the intensive spring samplings of 1983, 1984 and 1985 and the fall fishing periods in 1983 and 1984. Fish collected in the spring were marked with distinctive fin clips and released. Subsamples were measured for length and weight, and scale samples were taken to determine the age structure of the population at time of annuli formation. In the fall, recovery of marked fish allowed Petersen population estimates to be determined in 1983 and 1984.

Our studies in Trout Lake have not detected any short-term changes in the fish community or in any individual populations which might be attributed to liming. The changes in the Bowland Lake fish community were more dramatic.

Figure 5 shows the percentage frequency distributions for the yellow perch population in Bowland Lake the year before liming and for two years after liming. There is a progression from a fairly erratic distribution of year class strength to a more natural distribution as the younger year classes become dominant. Population estimates of perch $\geq$ age I show a fourfold increase the spring after liming, due to high overwintering survival. Preliminary estimates for 1985 show an

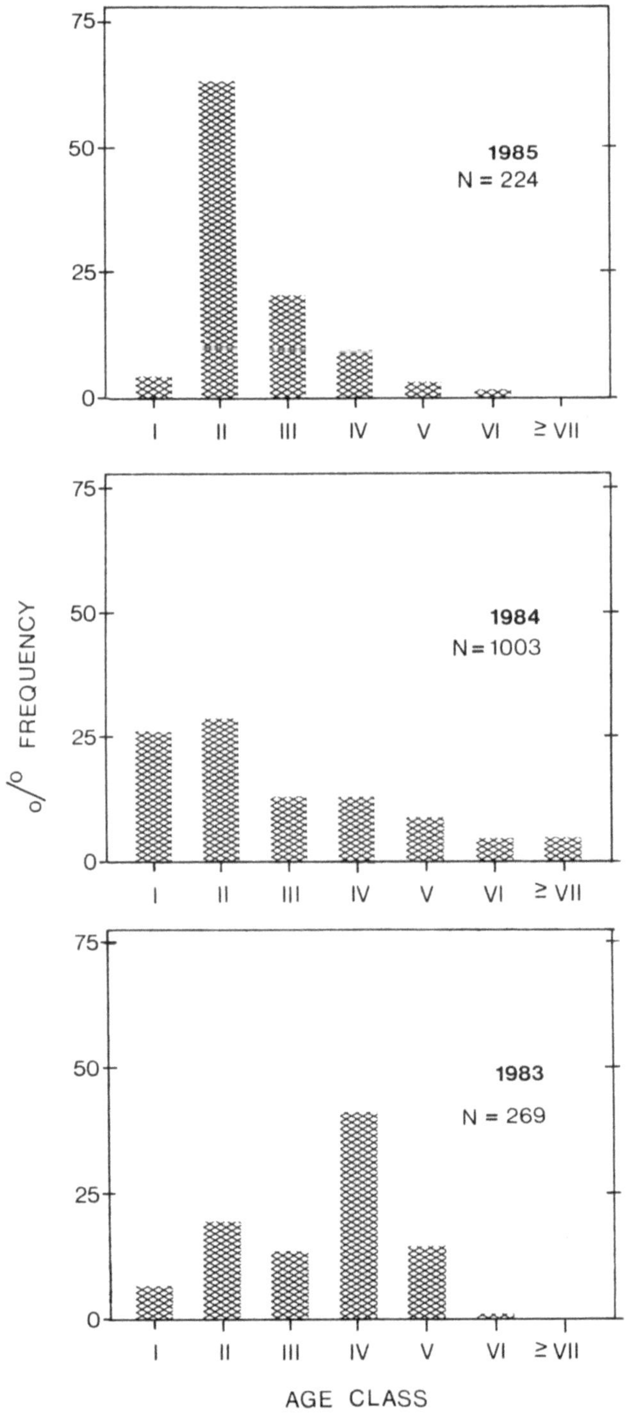

Figure 5. Percent frequency distribution for age classes of yellow perch in Bowland Lake at time of annulus formation, 1983 (pre-neutralization), 1984 and 1985 (post-neutralization).

even greater increase in population strength consistent with observations by Nyberg et al. (1985).

Size-at-age estimates are presented for the cohorts before and after neutralization (Figure 6). In the first season after liming (the last increment for each age group on the graph) the age I cohort grew similarly to cohorts prior to neutralization. The age II and III cohorts grew more slowly, while older age IV, V and VI fish grew faster following neutralization. This change in growth is a possible result of reduced stress (evidenced by increased growth of older cohorts) and changes in food supply and competition (evidenced by decreased size of younger cohorts) due to a changed community structure, all of which are probably related to the treatment of the lake with calcite. The lake was stocked shortly after liming with lake trout (age 10 mo and 22 mo) that on subsequent recapture appear to be thriving. Adult and age III lake trout were also stocked and these fish will be monitored to determine whether there is successful reproduction and recruitment of new generations of lake trout in Bowland Lake. In the fall of 1985, fish were observed spawning on 5 shoals in Bowland Lake and fertile eggs were collected. It is not yet known whether these eggs will survive to yearling fish.

SHOAL LIMING

Whole-lake liming is expensive. It is also questionable whether it provides adequate protection in the nearshore environment where lake trout eggs, for example, are spawned. For these reasons it was important to test alternative methods of protecting spawning habitat. Experiments by Gunn and Keller (1980) had shown that incubation of eggs in limestone could enhance survival in acidic conditions. Shoal liming, if proved to be effective, could be used in a limed lake to provide additional nearshore protection. It could also be useful in a lake where water quality was adequate most of the year but spring melt caused toxic water to invade the spawning shoals.

A small-scale experiment was conducted on Laundrie Lake in 1984/85 involving bioassays of lake trout eggs in limestone plots. Although preliminary, the results are promising and it appears possible that lake trout eggs and fry may be protected in a moderately stressed lake by treating the spawning shoals with calcite. Experiments are now under way to confirm the preliminary results and to determine how long the treatments remain effective. The behavioral response of lake trout to treated shoals is also being investigated.

SUMMARY

No short-term detrimental effects of liming were seen in Trout Lake. However, whole-lake liming does not necessarily protect the nearshore shallow zones of the lake where lake trout sac fry are found during spring melt. In situ toxicity studies suggest that both low pH and high inorganic Al are major contributors to ecological damage. We have some promising preliminary results from the shoal liming experiment which may be important to fisheries managers. The ecological community

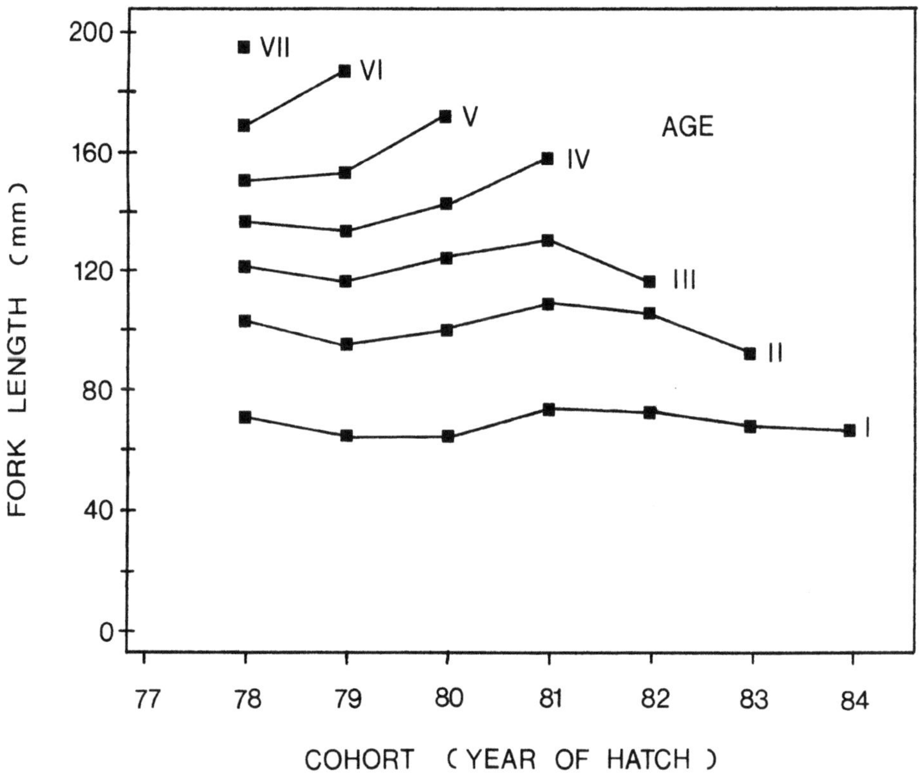

Figure 6. Fork length (mm) for age classes of yellow perch in Bowland Lake based on back-calculated growth rates (pre-1983) and measured lengths at time of annulus formation in 1984 and 1985. (Note: last interval for each age class represents growth during first growing season following neutralization.)

responses in Bowland Lake are difficult to attribute directly to liming because of the complexities of annual variation and interactions involving competition and changing food supply. However, a more normal size/age distribution is appearing and population size of the perch is increasing dramatically. The stocked lake trout are also surviving and spawning.

It remains to be seen whether there are any long-term effects of liming and whether alternative techniques, such as shoal liming, will be required to achieve or sustain recruitment in the lake trout population.

ACKNOWLEDGMENTS

The Experimental Lake Neutralization Program was sponsored by the Ontario Ministries of the Environment and Natural Resources. We are especially grateful to the scientists at these ministries who have provided direction and encouragement, to Bill Keller for helpful criticism, to Alan Burt, Gary Bowen, Greg Pope, Dave A. Ross and Ed Trippel, among others, for data collection and preparation, and finally to Rosemary Simmons for preparation of the manuscript.

REFERENCES

Baker, J. and Schofield, C.: 1982, Water, Air and Soil Poll. 18, 289.

Brown, D.J.A.: 1983, Bull. Environ. Cont. Tox. 30, 582.

Gunn, J.M. and Keller, W.: 1985, 'Effects of Acidic Meltwater on Chemical Conditions at Nearshore Spawning Sites'. Abstracts from the International Symposium on Acidic Precipitation, Muskoka, Ontario, Sept. 15-20, 1985.

Gunn, J.M. and Keller, W.: 1984, Can. J. Fish. Aquat. Sci. 41, 319.

Gunn, J.M. and Keller, W.: 1980, Can. J. Fish. Aquat. Sci. 37, 1522.

Holtze, K.E. and Lee, J.: 1982, Technology Transfer Conference #3. Ontario Ministry of the Environment, Toronto, Canada.

Hutchinson, N.J. and Holtze, K.E.: 1985, Abstracts from the International Symposium on Acidic Precipitation, Muskoka, Ontario, Sept. 15-20, 1985.

McMurtry, M.J.: 1985, 'Susceptibility of Lake Trout (Salvelinus namaycush) Spawning Sites in Ontario to Acidic Meltwater'. Ontario Ministry of Natural Resources manuscript. 14 pp.

Molot, L.A., Hamilton, J.G. and Booth, G.M.: 1986, Wat. Res. 20, 757.

Nyberg, P., Degerman, E. and Appelberg, M.: 1985. Abstracts from the International Symposium on Acidic Precipitation, Muskoka, Ontario, Sept. 15-20, 1985.

Peterson, R.H., Daye, P.G., LaCroix, G.L. and Garside, E.T.: 1982, In Acid Rain/Fisheries. Proceedings of an International Symposium on Acidic Precipitation and Fishery Impacts in Northeastern North America. August 2-5, 1981. 177-196.

LIMING THE ACID LAKE HOVVATN, NORWAY: A WHOLE-ECOSYSTEM STUDY

G.G. Raddum[1], P. Brettum[2], D. Matzow[3], J.P. Nilssen[4], A. Skov[4], T. Sveälv[4], and R.F. Wright[2]

[1] Zoological Museum, University of Bergen, 5014 Bergen, Norway
[2] Norwegian Institute for Water Research, Box 333, 0314 Oslo, Norway
[3] Directorate for Game and Freshwater Fish, Box 55, 4801 Arendal, Norway
[4] Zoological Institute, University of Oslo, 0316 Oslo, Norway

ABSTRACT. Hovvatn, a 1-km^2 chronically-acidified lake in southernmost Norway, was treated with 200 tonne of powdered limestone in March 1981. An additional 40 tonne were added to a 0.046 km^2 pond (Pollen) draining into Hovvatn. The lakes were stocked with brown trout in June 1981 and in each subsequent year. At ice-out pH rose from 4.4 to 6.3 (Hovvatn) and 7.5 (Pollen), Ca and alkalinity increased, and total Al decreased by 120 $\mu g\ L^{-1}$. None of the other major ions exhibited significant changes in concentration. Total organic C and P increased after liming. The phytoplankton community was dominated by chrysophytes and did not change significantly following liming. The zooplankton community was typical of acid lakes prior to liming. There was a clear succession in species dominance following treatment, although no new species immigrated to the lakes. Zoobenthos changed from a community characterized by low abundance and reduced number of species to increased abundances of oligochaetes, mayflies and chironomids. Hovvatn and Pollen were barren of fish prior to stocking. The stocked fish showed remarkably high growth rate during the first years. Liming apparently improved conditions for zoobenthos, enhancing the processing of fine detritus which in turn resulted in elevated levels of TOC and P in the lakewaters during the first year after liming. The "oligotrophication" process typical of acid lakes was temporarily reversed by liming. The interactions between groups of organisms in Hovvatn and Pollen indicates that many years are required before a new steady-state can be attained following liming.

1. INTRODUCTION

Acid precipitation, acidification of freshwaters and loss of fish populations have affected thousands of lakes and streams in southern Norway (Jensen and Snekvik, 1972; Overrein et al., 1980). Fish populations are extinct in a 13000 km^2 area and damaged populations are found in an additional 20 000 km^2 (Sevaldrud et al., 1980).

Acid waters are characterized by the absence of bicarbonate alkalinity, elevated levels of inorganic Al species toxic to fish and generally low levels of P (Dickson, 1980; Henriksen, 1980). These acidified ecosystems exhibit reduced number of species at all tropic levels, reduced rates of decomposition and reduced rates of nutrient cycling (Almer et al., 1978; Davis and Berge, 1980; Hobæk and Raddum, 1980; Harvey, 1980; Grahn et al, 1974; Raddum and Fjellheim, 1984).

Liming acidic lakes is a whole ecosystem manipulation that changes ecosystem structure and function. Fish again occupy the top predator position. Phytoplankton, zooplankton and zoobenthos exhibit changes (Eriksson et al., 1983; Hultberg and Andersson, 1982).

In 1980 the Norwegian government started a 5-yr research project with the goal to document to what extent neutralization (liming) provides a responsible biological, technical, and economic means to reduce the damage caused by acidification (Matzow, 1982). Lake Store Hovvatn, (here called Hovvatn), a chronically-acidic lake in southernmost Norway, was chosen as one of the sites to be studied. The chemistry and biology of Hovvatn had been investigated during 1974-1980 as part of the Norwegian SNSF-project (Acid precipitation-effects on forest and fish) (Overrein et al., 1980).

After a year collection of background data Hovvatn and an adjoining pond (Pollen) were limed in March 1981. We report here the changes in water chemistry, phytoplankton, zooplankton, zoobenthos and fish during the first 3 yr after liming.

2. MATERIAL AND METHODS

2.1. Site description

Hovvatn, a 1.1 km^2 lake with a 5.8 km^2 terrestrial catchment, is located at 600 m above sea level, 50 km from the south coast (Figure 1). The catchment is of granitic bedrock with thin to non-existent poorly-developed podzolic and mor soils. Open forests of spruce, pine

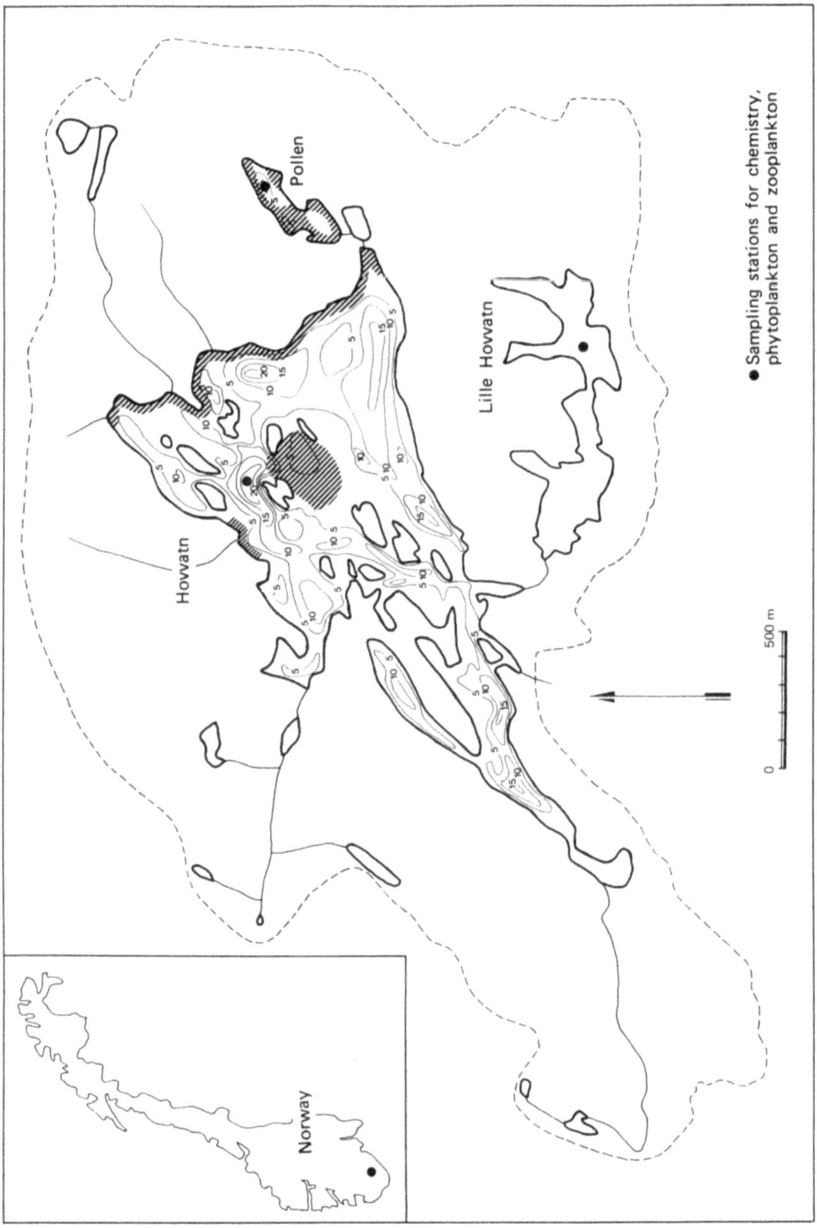

Figure 1. Hovvatn and its catchment. Pollen is the small pond to the east, and Lille Hovvatn (unlimed control) is the lake to the south. Limestone powder was spread on the areas marked by shading. Bathymetry is given in meters.

and birch are interspersed with peaty areas. Hovvatn has a mean water residence time of about 0.8 yr. Shore morphology is complicated with numerous bays and islands (Figure 1). A small 4.6 ha pond (called Pollen) is attached to the eastern arm of the lake by a 1-m wide shallow channel.

The liming experiment at Hovvatn comprised 2 parts: liming of Hovvatn itself and liming of Pollen. Lille Hovvatn, a small lake draining into Hovvatn from the south, was used as the untreated "control".

Prior to liming Hovvatn and Lille Hovvatn were acidic (pH 4.4), oligotrophic and barren of fish. Hovvatn has no tributaries suited for spawning and the original brown trout population which went extinct in the late 1940's presumably spawned in the central part of the lake.

2.2 Liming

In March 1981 200 tonne of powdered limestone were spread on the snow in a 10-m wide 3-km long strip along the eastern shore of Hovvatn and around the islands (Figure 1). An additional 40 tonne were spread on the ice over the entire Pollen.

The liming doses for Hovvatn and Pollen were 0.27 and 1.3 tonne ha^{-1} terrestrial catchment, respectively, which correspond to 25 g m^{-3} and 210 g m^{-3} lakewater. The dose for Hovvatn was chosen to achieve pH 6 to 7 and alkalinity of 50 to 100 µeq L^{-1} after liming and to neutralize an additional 1-yr of acidic runoff. It was assumed that only 50% of the applied limestone would dissolve. The higher dose for Pollen was chosen with the aim of maintaining pH levels above 5.5 for 5 yr.

2.3 Fish stocking

In June and October 1981 a total of 4800 brown trout (Salmo trutta L.) of 3 year-classes (0.1 and 2) were stocked at two locations in Hovvatn and in Pollen (Figure 2). The fish were marked with different fin clipping codes for each location. To compensate for catch, natural mortality, and assumed absence of reproduction, an additional 2800 fin-clipped trout, age group 0+ and 1+, were stocked at the same three locations in October 1982. In 1981 the stocking density in Pollen was 244 fish ha^{-1}, considerably higher than in Hovvatn, 32 fish ha^{-1}.

Fish in Hovvatn and Pollen were caught with bottom gill nets at four stations and with pelagic gill nets at one station (Figure 2) in

June, August and October 1982 and in May 1983. In addition trout were caught with gill nets at station SG in November 1982. A comparative sample of the parent population from Lake Byglandsfjorden was collected in November 1982 by bottom gill nets at three spawning grounds. All tributaries and the outlet of Hovvatn were fished with electrical fishing in October - November 1982. Uncontrolled fishing in Hovvatn and Pollen was negligible.

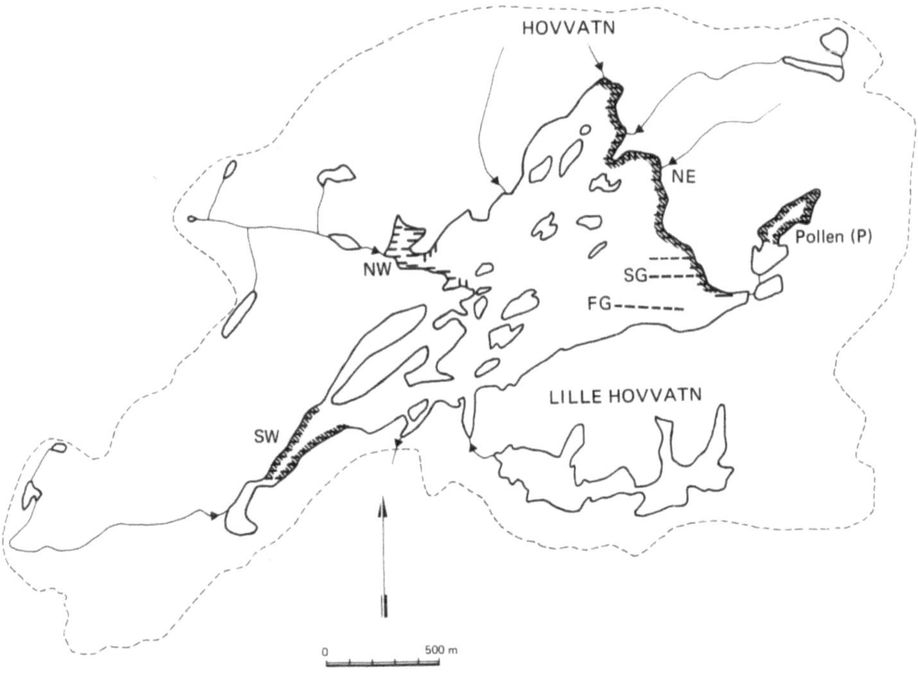

Figure 2. Areas stocked with brown trout (shaded) and gill-net stations in Pollen (P) and Hovvatn (NE, SW, NW, FG and SG) (from Sveälv, 1985).

Food selection was examined in stomach contents of 479 fish representing 10 fishes (if possible) of each year class, lake location, and sampling date. The quantity of main food groups were estimated using a point scale based on dry weights. Details are given by Sveälv (1985).

We investigated fish reproduction by incubating fertilized trout eggs from the parent population in Lake Byglandsfjorden into Hovvatn during the winter 1982-83. The eggs were placed in small plastic boxes one egg to each of 24 separated chambers 16 mm in diameter with holes in the bottom and top of each. The boxes were placed on the bottom, 2 m depth in the central part of Hovvatn in November 1982. Possible spawning grounds at site SG were inspected by a diver in March 1983.

2.4 Water chemistry

Water samples for chemical analysis were collected about 8 times a year at 2 to 5 depths from 4 (later 3) stations in Hovvatn (Wright, 1984). We use here the results only from one station in Hovvatn, one in Pollen (beginning in 1983) and one in Lille Hovvatn (Figure 1). The major inflowing stream to Hovvatn (from Lille Hovvatn) and the outlet of Hovvatn were sampled more frequently.

The samples were collected in polyethylene bottles and stored cool and in the dark during transport to the laboratory at the Norwegian Institute for Water Research. Analysis is by routine methods for softwaters.

In addition to this routine sampling and analysis program, occasional samples were collected in acid-washed glass vials and analyzed for Fe, Mn, Pb, Cd, Cu, and Zn by atomic absorption spectrophotometry and graphite furnace. The complete analytical methods are listed in Wright (1984).

2.5 Phytoplankton

Quantitative samples for analysis of phytoplankton were collected in Hovvatn, Pollen, and Lille Hovvatn. The samples analyzed are composite of the uppermost 10 m (Hovvatn), 8 m (Pollen), and 8 m (Lille Hovvatn), respectively. The volume of each species and the total volume of phytoplankton in each samples were estimated using Utermöhl's (1958) method.

Measurements of <u>in situ</u> primary productivity were carried out in May-September 1982 at all three sites. The method uses light and dark bottles at six different depths after addition of 0.2 mL $NaH^{14}CO_3$ with an activity of 4 µCi. The exposure time was 4 hr. Carbon fixation was determined by scintillation counting. CO_2 analysis was carried out using Hongve's (1982) method. Samples for chlorophyll analysis were collected simultaneously and analysed fluorimetrically.

2.6 Zooplankton

Zooplankton was sampled with a 90 μm net with opening 27 cm and length 1 m. The net was drawn vertically from 8 , 10 and 8 m to the surface in Pollen, Hovvatn, and Lille Hovvatn, respectively. Zooplankton species abundance (no. m^{-2}), fecundity (eggs per total females), and fertility (eggs per egg-bearing female) were calculated.

2.7 Benthic invertebrates

Zoobenthos was collected in spring, summar and autumn each year along 2 fixed transects in Hovvatn and 1 in Pollen (Figure 2). A core sampler of 77-mm diameter was used to collect 6 or 10 quantitative samples at each of depths 0.5, 2, 5 and 9 to 12 m. The material was sieved through a 0.225 mm net, the samples preserved in alcohol, and sorted beneath the microscope in the laboratory.

3. RESULTS

3.1 Water chemistry

As expected liming resulted in dramatic increases in pH, alkalinity and Ca and large declines in total-Al concentrations (Table I) (Figures 3 to 5) (Wright, 1985). Following ice-out in May 1981 pH, Ca and alkalinity in Hovvatn and Pollen increased immediately as the limestone came in contact with the water. The levels of total-Al in unfiltered water, however, declined slowly over the next several weeks (Figure 5). Furthermore, Al concentrations exhibited a marked vertical gradient in the water column with higher concentrations in the bottom waters. Slow removal of Al after liming acid water has also been reported from Sweden (Swedish National Board of Fisheries, 1981).

In both Hovvatn and Pollen, alkalinity and Ca concentrations increased further during the first summer after liming to maximum levels in early September, prior to the onset of autumn rains. This indicates that limestone continued to dissolve during the first summer.

Reacidification of Hovvatn and Pollen began with the onset of autumn rains in September 1981 (Wright, 1985). Precipitation and runoff from the terrestrial catchments were, of course, acidic and unaffected by the lime treatment. The effect of the liming decreased in both water bodies as the high pH, limed water was back-titrated and flushed out by incoming acidic runoff and precipitation. By 25 May

Table I. Changes in water chemistry as a result of liming and acidification in Hovvatn and Pollen (1 m depth). Units: $\mu eq\ L^{-1}$. Δalk and ΔCa are changes relative to pre-liming conditions. Δalk is defined as $\Delta HCO_3 - \Delta H^+ - \Delta Al$ (from Wright, 1985).

	HOVVATN				POLLEN			
	before	- - - - after - - - -			before	- - - - after - - - - -		
		0 yr	1 yr	2 yr		0 yr	1 yr	2 yr
	810318	810519	820525	830513	810318	810519	820525	830513
H^+	32	0	3	12	30	0	0	4
(pH)	4.49	6.42	5.57	4.91	4.53	7.43	6.38	5.39
Na	44	41	42	-	37	40	40	-
K	4	5	4	-	3	7	4	-
Ca	24	88	79	60	22	609	139	81
Mg	17	14	16	-	12	17	14	-
Al^A	30	16	10	12	36	22	18	19
NH_4	3	4	-	-	2	8	-	-
NO_3	16	16	18	-	8	11	15	-
SO_4	75	75	75	-	73	73	67	-
Cl	56	48	45	-	51	54	45	-
HCO_3	0	18	8	0	0	543	64	9
Δalk		64	56	38		587	112	52
ΔCa		64	55	36		588	116	59
TOC mg L^{-1}	3.1	2.8	3.6	2.7	3.9	4.2	4.8	3.1
tot-P $\mu g\ L^{-1}$	5	7	5	-	5	10	8	-
tot-N $\mu g\ L^{-1}$	440	430	470	-	300	610	440	-
Fe $\mu g\ L^{-1}$	120	100	70	-	140	740	460	-
Mn "	14	12	6	-	11	6	8	-
Cu "	4	6	3	-	6	9	3	-
Zn "	20	10	20	-	19	20	10	-
Cd "	0.6	0.3	0.2	-	0.6	0.9	0.3	-
Pb "	4	1	4	-	2	6	2	-

[A] measured as total-Al but expressed as $\mu eq\ L^{-1}$ with valence 3.

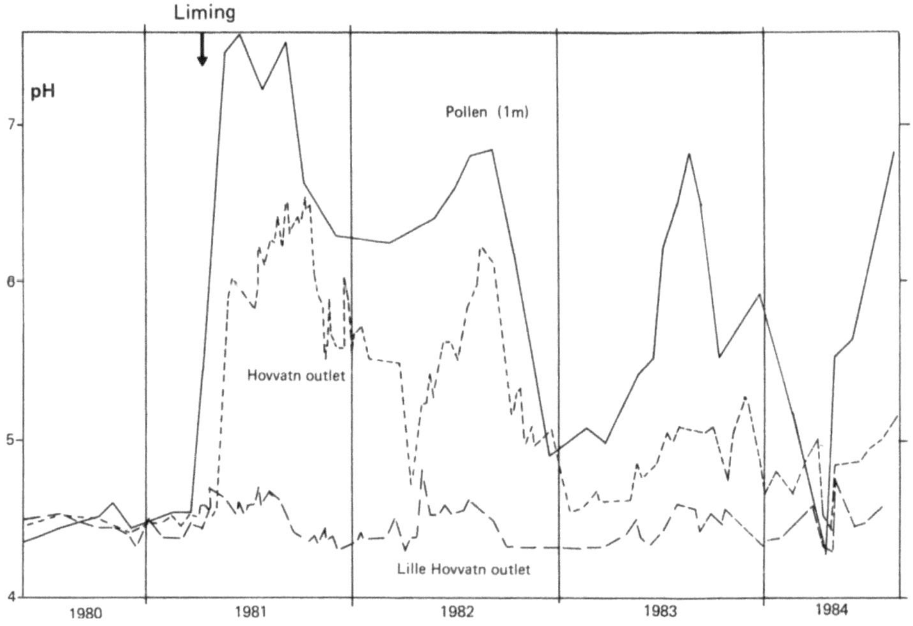

Figure 3. pH levels in the untreated Lille Hovvatn (outlet), the limed Hovvatn (outlet) and limed Pollen (1 m) over the period May 1980 - August 1984 (from Wright, 1985).

1982, 1 yr after liming, the pH in Hovvatn had decreased to 5.6 (from a high of 6.7 in September 1981); in Pollen the pH was 6.4 (from a maximum 7.5). Ca and bicarbonate levels had also decreased. After two full years, 13 May 1983, pH in Hovvatn was 4.9 and 5.4 in Pollen (Table I) (Figure 3).

Liming did not affect the concentrations of the other major ions (Na, K, Mg, NH_4, NO_3, SO_4 and Cl), or the trace metals (Cu, Zn, Cd, Pb, Fe) (Table I) (Wright, 1985). Mn concentrations appeared to decrease following liming in both Hovvatn and Pollen from concentrations of 10 to 20 µg L^{-1} before liming to 5 to 10 µg L^{-1} by the end of the first summer after liming.

Measurements of the dissolved O_2 concentration in Hovvatn in April

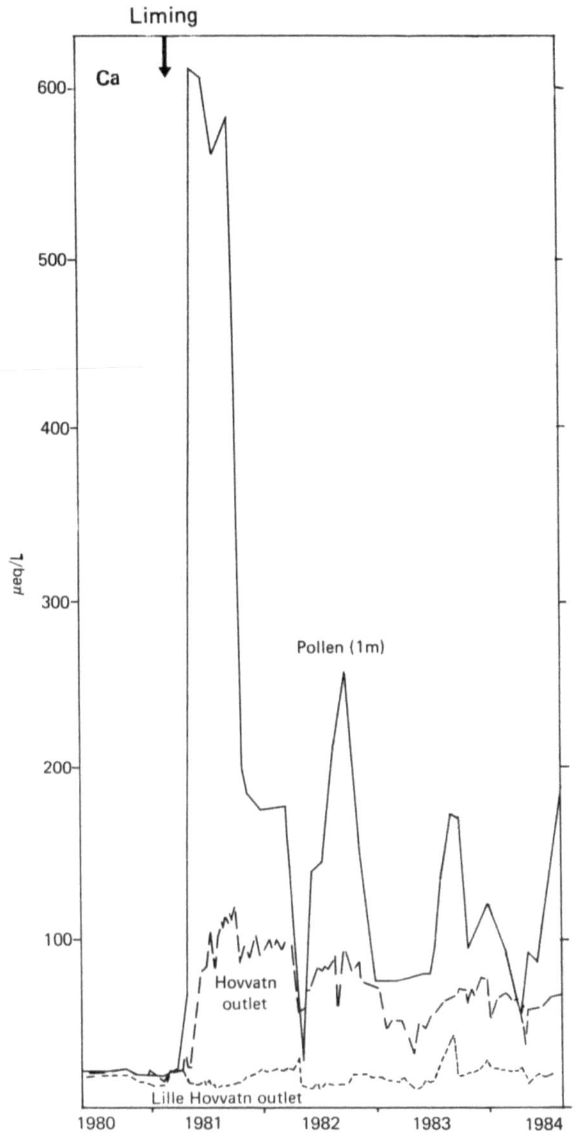

Figure 4. Concentrations of Ca (µeq L^{-1}) in the untreated Lille Hovvatn (outlet), the limed Hovvatn (outlet) and limed Pollen (1 m) over the period May 1980 - August 1984 (from Wright, 1985).

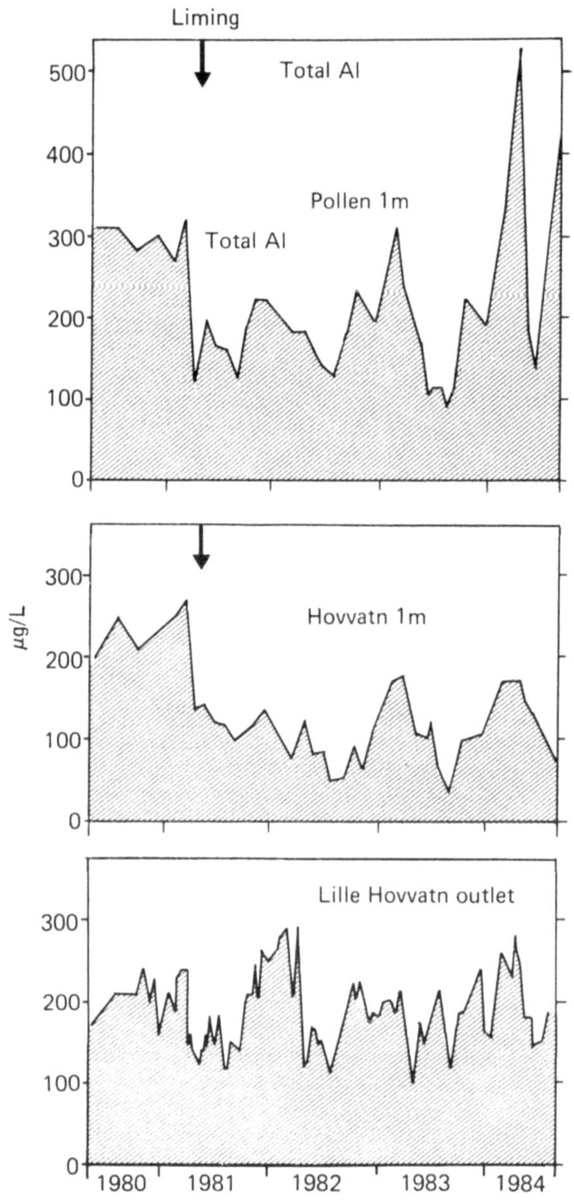

Figure 5. Concentrations of total Al (μg L^{-1}) in Pollen (1 m), Hovvatn (1 m) and Lille Hovvatn (outlet) over the period May 1980 - August 1984 (from Wright, 1985).

1982, after the winter-long stagnation and just prior to ice-out, showed > 50% saturation at all water depths.

Total organic C (TOC), total P (tot-P) and possibly also total N (tot-N) were affected by the limestone treatment (Wright, 1985). TOC levels increased from about 2 to 3 mg C L^{-1} before liming to about 4 mg L^{-1} and 5 to 8 mg L^{-1} in Hovvatn and Pollen, respectively, during the first and second year after treatment. Data for the first half of 1983 indicate that TOC levels had decreased to pre-liming concentrations (Figure 6). Concentrations of tot-P in Hovvatn and Pollen increased with liming (Figure 6). In Hovvatn the increase was from about 4-5 µg L^{-1} to 7 µg L^{-1} and was only of a few months duration. In Pollen the increase was larger - to 10 to 12 µg L^{-1} and was evident during both 1981 and 1982.

3.2 Phytoplankton

At all 3 sites (Hovvatn, Pollen and Lille Hovvatn) the total number of algae was low both before and after liming (Table II), and the composition was typical of oligotrophic lakes in Norway. The phytoplankton community at all 3 sites was dominated by Chrysophyceae species (mainly small forms) and "µ-algae", a term for all small forms not identified to species (Figure 7). Several of the samples collected in August and September contained relatively large numbers of Chlorophyceae and Dinophyceae. Occurrence of Cryptophyceae species was sporadic.

Table II. Mean volume of phytoplankton in the ice-free season at 3 sites before and after liming. Units: mm^3 m^{-3}. Number of samples is given in parentheses.

	1980	1981	1982	1983
Hovvatn (limed)	29 (2)	57 (5)	95 (7)	72 (7)
Pollen (limed)	79 (2)	59 (5)	129 (7)	91 (7)
Lille Hovvatn (control)	-	-	103 (7)	148 (7)

At both the limed sites, Hovvatn and Pollen, the mean total volume of phytoplankton during the ice-free season doubled from 1981 to

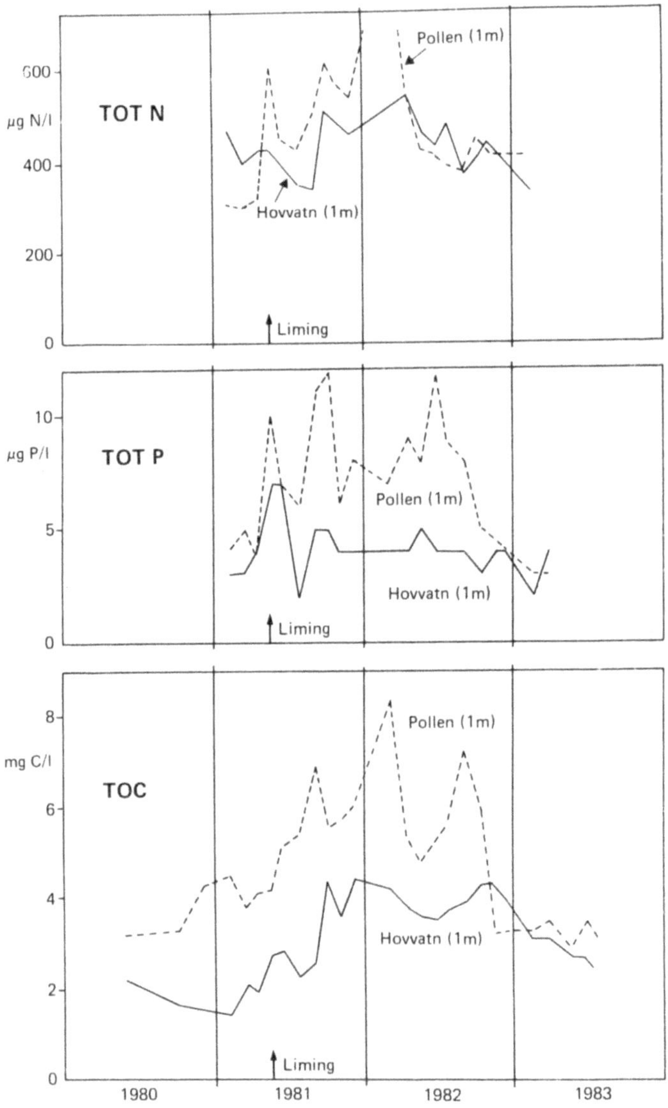

Figure 6. Concentrations of tot-N, tot-P and total organic C in surface waters of Pollen and Hovvatn over the period May 1980 - August 1983 (from Wright, 1984).

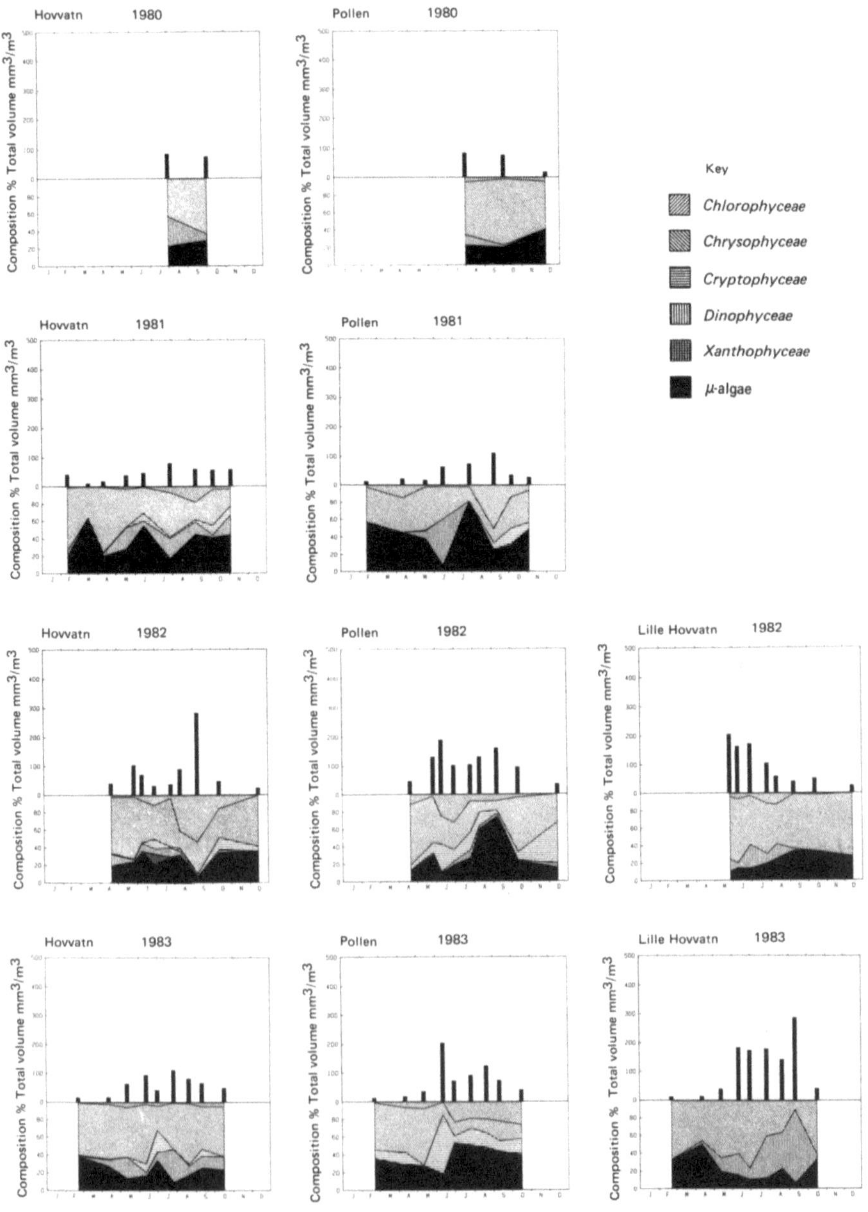

Figure 7. Variation in total volume and composition of phytoplankton in Hovvatn (limed), Pollen (limed) and Lille Hovvatn (control) 1980-83.

1982 but decreased again in 1983 (Table II). At the reference locality Little Hovvatn, there was a marked increase of phytoplankton from 1982 to 1983. These data were in the range observed in Hovvatn in 1982. The year-to-year fluctuations observed in Hovvatn after liming may therefore be natural. The pre-liming samples are insufficient to establish the range of natural fluctuations.

There were no clear-cut changes in species composition of phytoplankton following liming (Table III). An exception might be the Chlorphyceae which showed a relative increase in volume. In September 1982 a species similar to Coccomyxa sp. comprised about 25% of the total phytoplankton volume in Hovvatn. Similarily a species of the genus Gloeocystis made up about 40% of the total volume in Pollen in September 1981, and a coccoid green algae about 20% in June 1982.

Common species of the Chrysophyceae were Dinobryon crenulatum and chrysomonads of different sizes. Especially in September 1982 there was a relatively large number of individuals of D. crenulatum in Hovvatn when the highest total volume, 280 mm^3 m^{-3} was observed. Bitrichia chodatii and Chrysoikos skujai were both common species in the lakes. The former is common in acid lakes (Raddum et al., 1980).

A general feature of the phytoplankton in Hovvatn is that the species of Chrysophyceae such as Dinobryon crenulatum, D. sociale v. americanum, Bitrichia spp., and Kephyrion spp. had a lesser specific volume than usually observed in oligotrophic lakes in Norway.

Peridinium inconspicuum is a typical abundant alga in oligotrophic and acid lakes (Schindler and Holmgren, 1971; Hörnström et al., 1973; Eriksson et al., 1983). This dinoflagellate was common and often dominant during the midsummer in both Hovvatn and Lille Hovvatn.

Primary productivity measurements in 1982 show relatively high maximum daily production rates of 270, 310 and 270 mg C m^{-2} day^{-1} at Hovvatn, Pollen and Little Hovvatn, respectively. The corresponding seasonal mean daily production was 110, 120 and 140 (Figure 8). The yearly primary production amounts to 20 g C m^{-2} in Hovvatn, 23 g C m^{-2} in Pollen and 25 g C m^{-2} in Lille Hovvatn. These values are high given the very low algal biomass in the lakes.

The efficiency of primary production increased in Little Hovvatn in August and September, a period in which the total volume of phytoplankton was decreasing. Lowest efficiency occurred when algal biomass was at a minimum. This same trend was also found in Hovvatn and Pollen, but here the efficiency also decreased in the late summer and autumn along with the decrease in algal biomass (Figure 8).

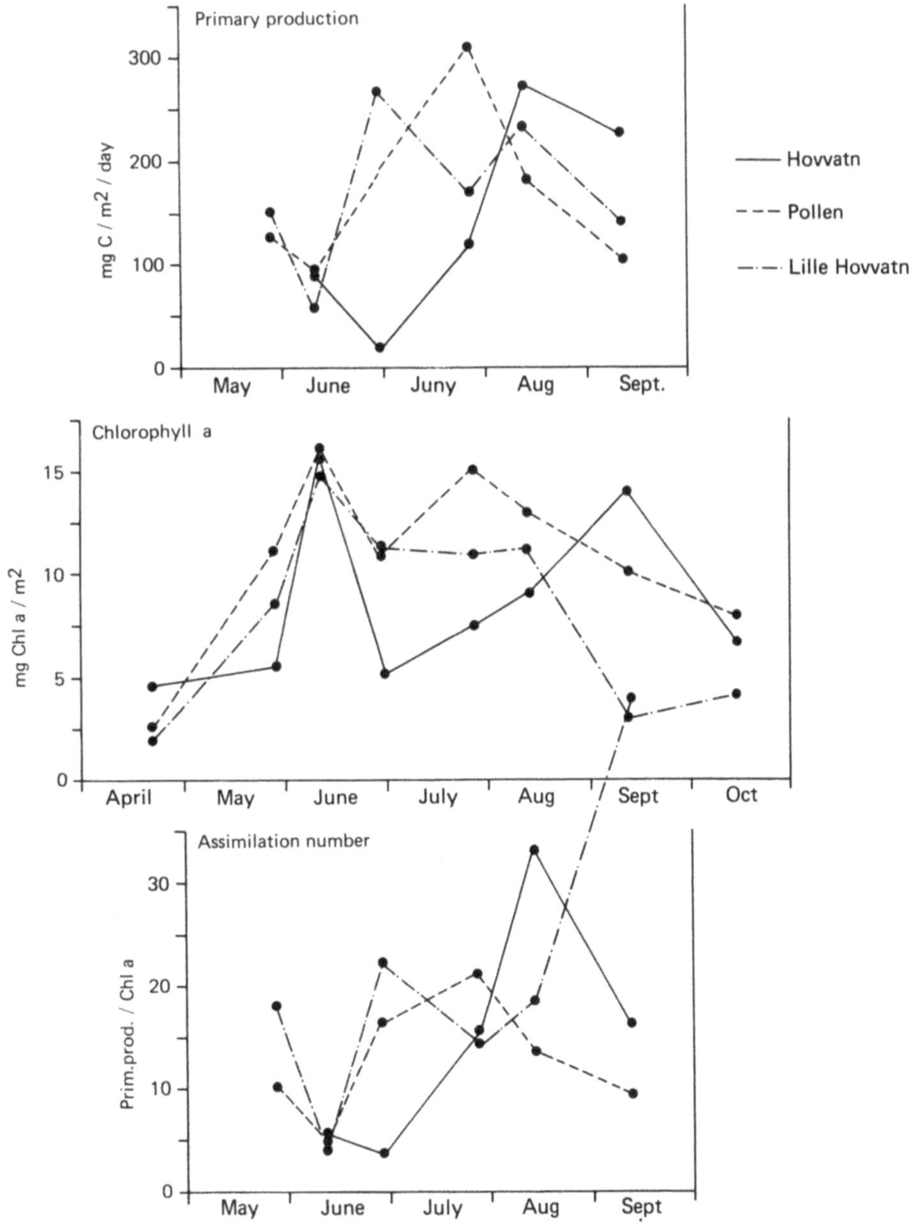

Figure 8. Primary production, chlorophyll a and assimilation number in Hovvatn (limed), Pollen (limed, and Lille Hovvatn (control) in 1982.

Table III. Phytoplankton species found in quantitative samples from Hovvatn, Pollen, and Lille Hovvatn in 1981, 1982 and 1983.

	HOVVATN			POLLEN			LILLE HOVVATN	
	81	82	83	81	82	83	82	83
Chlorophyceae								
Botryococcus braunii Kütz	x							
Chlamydocapsa planctonica (West & West) Fott		x						
Chlamydomonas spp.	x	x	x	x	x	x	x	x
Coccoid monads	x	x	x	x	x	x	x	
Coccomyxa sp. (cf.)		x						
Cosmarium sp.					x	x		
Elakatothrix gelatinosa Wille	x	x		x	x	x		
Euastrum sp.			x		x	x	x	x
Gloeocystis sp.	x			x		x		
Green flagellate	x		x	x			x	
Monomastix sp.			x			x	x	
Monoraphidium dybowskii (Wolosz.) Hind. & Kom.-Legn.		x	x		x	x		x
Oosystis lacustris Chod.	x	x	x		x	x	x	x
Oosystis submarina v. variabilis Skuja	x	x	x	x	x	x	x	x
Paramastix conifera Skuja	x		x	x	x	x	x	x
Scenedesmus spinosus Chod.					x			
Scourfieldia cf. cordiformis Takeda						x	x	x
Sphaerocystis schroederi Chod.					x			
Tetraedron minimum (A.B.) Hansg.	x				x		x	
Tetraedron minimum v. tetralobatum Reinsch.					x	x		
Chrysophyceae								
Bicosoeca planctonica Kiss.		x			x	x	x	
Bitrichia chodatii (Rev.) Chod.	x	x	x			x		
Bitrichia ollula (Fott) Bourr.							x	
Bitrichia phaseolus (Fott) Bourr.			x				x	x
Chrysoikos skujai (Nauw.) Willen	x	x	x	x		x	x	x
Chrysolykos planctonicus Mack		x						
Chrysomonads (d<7)	x	x	x	x	x	x	x	x
Chrysomonads (d>7)	x	x	x	x	x	x	x	x
Chrysophyceae, cysts	x	x	x	x	x	x		x
Chrysophyceae, indet	x	x	x	x	x	x	x	x
Craspedomonads	x	x	x	x	x	x	x	x
Dinobryon spp.			x					x
Dinobryon, cysts					x			
Dinobryon crenulatum West & West	x	x	x		x	x	x	x
Dinobryon korschikovii (Korsch.) Matv.								x
Dinobryon sertularia Ehr.		x				x		
Dinobryon sociale v. americanum (Brunnth.) Bachm.	x		x	x			x	x
Kephyrion spp.	x	x	x		x	x	x	
Phaester aphanaster (Skuja) Bourr.				x	x		x	
Psuedokephyrion angulosm Hill.	x							
Psuedokephyrion sp.	x	x						
Psuedokephyrion taeniatum Nicholls			x			x	x	x

Table III (cont.)

	HOVVATN 81	HOVVATN 82	HOVVATN 83	POLLEN 81	POLLEN 82	POLLEN 83	LILLE HOVVATN 82	LILLE HOVVATN 83
Bacillariophyceae								
Achnanthes sp. (l=15-25)	x		x					
Synedra sp. (l=30-40)				x				
Synedra sp. (l=40-70)				x				
Tabellaria fenestrata (Lyngb.) Kütz.			x					
Cryptophyceae								
Cryptaulax vulgaria Skuja				x				
Cryptomonas marssonii Skuja	x	x	x		x	x		
Cryptomonas sp. (l=15-18)					x	x	x	x
Cryptomonas sp. (l=20-22)					x	x		
Cryptomonas sp. (l=24-28)					x	x		
Cryptophyceae, indet	x	x	x	x	x	x		x
Katablepharis ovalis Skuja	x	x	x	x	x	x		
Rhodomonas lacustris Pasch. & Ruttn.	x		x		x			
Dinophyceae								
Dinophyceae, indet	x	x	x	x	x	x	x	x
Gymnodinium cf. lacustre Schill.	x	x	x	x	x	x	x	x
Gymnodinium sp. (l=14-15)	x		x		x			x
Gymnodinium sp. (l=17-18)								x
Gymnodinium sp. (l=28-33)								x
Peridinium inconspicuum Lemm.	x	x	x	x	x	x	x	x
Peridinium sp.								x
Xanthophyceae								
Isthomochloron trispinatum (West & West) Skuja	x	x	x				x	
μ-algae	x	x	x	x	x	x	x	x

3.3 Zooplankton

Prior to liming Hovvatn had a low number of zooplankton species (Table IV) (Hobæk and Raddum, 1980). The dominant rotifer species was Kellicottia longispina (Figure 9). Liming did not affect this species the first year, 1981. In 1982 K. longispina decreased in number both in Hovvatn and Pollen, while in the reference lake, Lille Hovvatn, it remained abundant. The rotifer Conochilus unicornis increased the first year after liming, and became the dominant species in Hovvatn in 1982 and in Pollen in both 1982 and 1983. In 1983 the rotifers increased further in the limed sites, with densities ten times higher than the two previous years.

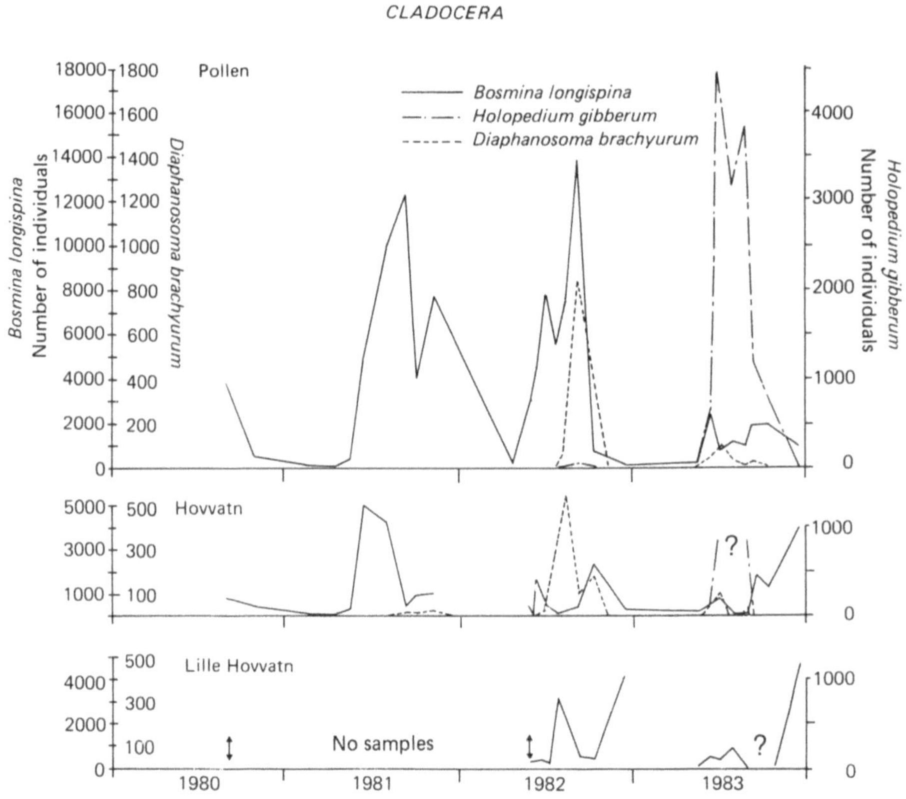

Figure 10. Numbers of the 3 dominant species of cladocera found in the zooplankton of Pollen (limed), Hovvatn (limed), and Lille Hovvatn (control) in 1980-83.

Table IV. Zooplankton found at Hovvatn, Pollen and Lille Hovvatn.
[A] species with mainly littoral habitat.
[B] species found in both Hovvatn and Lille Hovvatn.

ROTATORIA	COPEPODA	CLADOCERA
Kellicottia longispina [B]	Eudiaptomus gracillis [B]	Bosmina longispina [A B]
Keratella hiemalis [B]	Megacyclops gigas [A B]	Holopedium gibberum
K. serrulata [B]	Diacyclops bicuspidatus [A B]	Diaphanosoma brachyurum
Lecane sp. [A B]	D. nanus [A B]	Chydoridae sp. [A B]
Polyarthra minor [B]	Heterocopa saliens [B]	
Synchaeta kitina/oblonga		
Collatheca pelagica		
Conochilus unicornis [B]		
Ascomorpha sp. [B]		

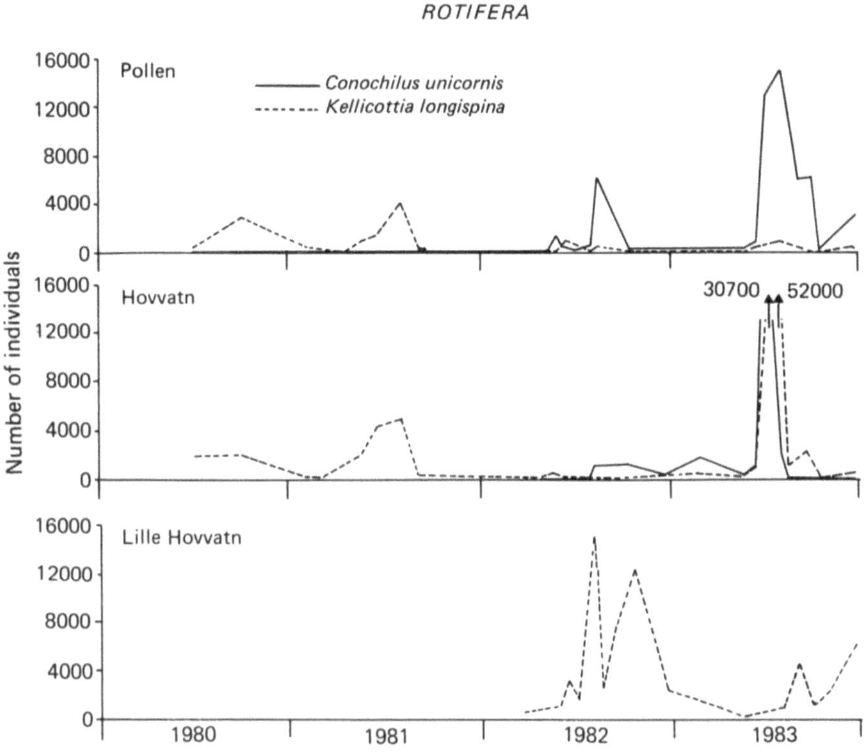

Figure 9. Number of rotifers found in the zooplankton of Pollen (limed), Hovvatn (limed), and Lille Hovvatn (control) in 1980-83.

Bosmina longispina was the only cladoceran found in Hovvatn in measurable numbers prior to liming (Figure 10) (Hobæk and Raddum, 1980). The first season after liming, this species increased to relatively high numbers. Another cladoceran, Diaphanosoma brachyurum, which was not found before liming appeared in Hovvatn with a few individuals in late summer 1981. During 1982 D. brachyurum increased further while B. longispina decreased.

In Pollen B. longispina increased the first and second year after liming, but decreased in 1983. D. brachyurum appeared in high number in 1982 but was reduced in 1983. Further, a third species, Holopedium gibberum, appeared in Pollen the second year of liming. This species increased to high numbers in 1983. At Lille Hovvatn the only cladoceran found was B. longispina, and no marked change from 1982 to 1983 was observed.

At all 3 sites the two copepods, Eudiaptomus gracilis and Heterocope saliens, were common during the entire period. E.gracilis was to some extent reduced in number in all localities in Hovvatn and Pollen the second and third year after liming.

3.4 Benthic invertebrates

Before liming the highest density of invertebrates was found in the littoral zone (0.5 to 2 m depth), while the density at 5 and 10 m depth was very low (Table V).

Liming of Hovvatn and Pollen resulted in a successive increase in abundance of benthic animals over the 3 yr. The oligochaets were relatively abundant prior to liming at 0.5 m depth on the sandy shore of Hovvatn. Oligochaets were not caught at this site the second year after the liming. The third year, however, they again occurred at this depth and at higher abundance than observed earlier. Excess lime was present along the shore in this part of Hovvatn at least the first year after liming.

Oligochaets were not found at 5 and 10 m depth and only with a few individuals at 2 m depth in Hovvatn before liming. The first year after the treatment (1981) these animals were not found at any of these depths, but in 1982 the organisms were present at all three depths. During 1983, the abundance of oligochaets increased further to a relatively high number.

In Pollen oligochaets had about the same depth distribution as in Hovvatn before liming (Table V). In contrast to Hovvatn the

Table V. Abundance of 3 groups of benthic invertebrates in samples collected in autumn at 4 depths before and after liming. Unit: individuals m^{-1}. oligo = Oligochaeta; ephem = Ephmeroptera; chiro = Chironomidae

Depth m	Group	6 acid lakes	HOVVATN 1977	HOVVATN 1981	HOVVATN 1982	HOVVATN 1983	POLLEN 1980	POLLEN 1981	POLLEN 1982	POLLEN 1983
	oligo	828	1029	-	0	1645	336	567	441	455
0.5	ephem	0	0	-	0	385	0	2030	1435	700
	chiro	2925	8001	-	7070	23253	3948	11060	15960	31647
	oligo	56	0	0	140	844	0	525	110	2722
2	ephem	0	0	0	804	525	0	175	1190	735
	chiro	2376	1722	2457	6405	13300	3612	4872	12746	79160
	oligo	0	0	-	252	350	0	147	483	4510
5	ephem	0	0	-	252	455	0	35	315	175
	chiro	455	814	-	5005	15436	1974	840	8855	29824
	oligo	15	0	0	35	50	35	70	0	0
10	ephem	0	0	0	0	0	35	70	0	0
	chiro	412	240	1008	660	2527	714	777	770	7623

oligochaets become present or increased in abundance at all depths in Pollen in 1981. The numbers increased further during 1982 and 1983 at 2 and 5 m with 2700 and 4500 individuals m^{-2}, respectively. At 10 m depth, however, the organisms disappeared in autumn 1982 and were not found in the samples taken in 1983.

After liming mayflies larvae became frequent in the samples. This was most pronounced in Pollen where more than 2000 individuals m^{-2} were recorded at 0.5 m depth in 1981. In Hovvatn larvae showed the highest densities at 2 and 5 m depth. The increase in abundance here was delayed by at least one season as compared with Pollen (Table V). Pollen had the highest number of mayflies in 1982, while in Hovvatn mayflies continued to increase in density in 1983.

After liming the density of chironomidae larvae increased at most depths in both Hovvatn and Pollen (Table V, Figure 11). The greatest increase was found in Pollen, where relative to pre-liming conditons the abundance increased by 2.8 times at 0.5 m depth the first autumn after liming (1981). At other depths no significant response was noted in 1981. The second autumn (1982) a 3.5 to 4.5 fold increase of chironomids was found in the depth-zone 0.5 to 5 m, while at 10 m depth no effect was recorded. The third autumn (1983) the increase was enormous, especially at 2 m depth with densities 22 times that prior to liming (nearly 80 000 larvae m^{-2}). At 5 and 10 m depth the increase was 15 and 11 times, respectively, while 0.5 m depth had 8 times. A similar pattern of response was found in Hovvatn, but here the 5-m depth showed the highest relative increase.

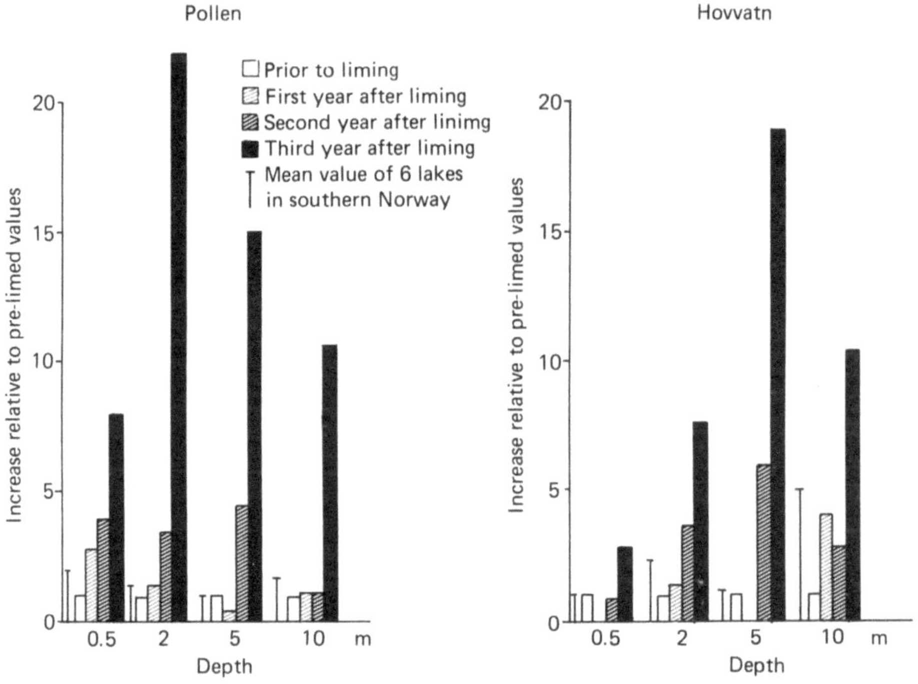

Figure 11. Density of chironomids in the benthos of Pollen and Hovvatn before liming (1980), and the three years after liming (1981, 1982, 1983), expressed as increase relative to 1980. Also shown are the densities found in 6 other acid lakes in southern Norway (Raddum, 1980).

Before liming Pollen had 20 chironomid species and Hovvatn 21 species (Table VI). After liming 22 species were found in both lakes. Species that increased in number after liming are shown in Figure 12. The second year after liming the increase generally accelerated except for the species Sergentia sp., Polypedilum breviantennatum and Chironomus spp. which decreased in Pollen. This reduction continued during the third year.

In Hovvatn Sergentia sp. and Chironomus spp. were sparse, while P. breviantennatum was common. P. breviantennatum showed a reduction in abundance the third year after liming.

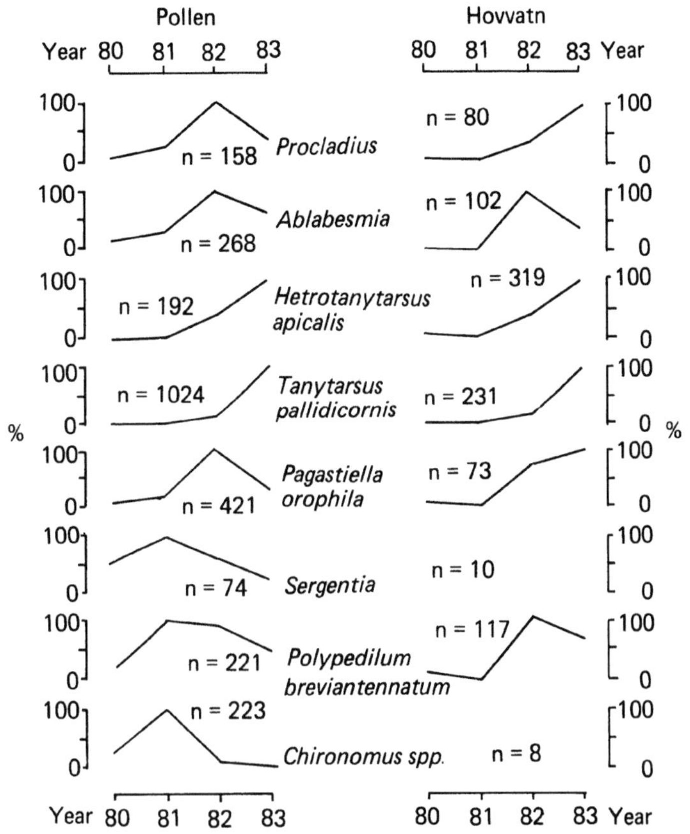

Figure 12. Relative changes in abundances of several chironomid species in the benthos of Pollen and Hovvatn in 1980-83.

Table VI. Occurrence of species and groups of Chironomids in Hovvatn and Pollen before and after liming. 1=rare; 2=uncommon; 3=common; 4=frequent; 5=dominant

	HOVVATN		POLLEN	
	before	after	before	after
Macropelopia goetghebueri	2	2	2	2
Procladius	2	3	2	3
Ablabesmyia longistyla	2	3	2	4
Protanypus mori		1		1
Heterotanytarsus apicalis	2	4	2	4
Heterotrissocladius marcidus	2	2	2	2
H. grimshawi	1	2		1
Psectrocladius sordidellus grp.	2	2	1	1
P. septentrionalis	1	2	1	1
P. spp.			1	1
Mesocricotopus thienemanni		1		
Cricotopus	1			
Thienemannia	3			
Parakiefferiella	4	2	1	1
Tanytarsus pallidicornis type	1	4	1	5
Tanytarsini indet		2	1	1
Cladotanytarsus vanderwulpi			1	3
Pagastiella orophila	1	3	1	4
Sergentia	1	1	2	2
Stictochironomus	3			
Polypedium breviantennatum type	2	3	2	3
Dicrotendipes tritomus	2	3	1	1
D. nervosus		1		
Chironomus spp.	1	1	3	1
Cladopelma	1	1	2	2
Cryptochironomus	1	1	1	1
Demicryptochironomus vulneratus	1	1	1	1
Number of species and groups	21	22	20	22

In Pollen several other species decreased in density the third year (Pagastiella orophila, Procladius sp. and Ablabesmyia spp.). Ablabesmyia spp. also showed the same trend in Hovvatn, while the two others continued the increase the third year in Hovvatn. The third year the dominating species in number was Tanytarsus pallidicornis in Pollen and Heterotanytarsus apicalis in Hovvatn.

The species Protanypus morio and D. nervosus were only recorded after liming in a very low number. On the other hand the Thienemannia sp. and Stictochironomus sp. disappeared after liming. A strong reduction of Parakiefferiella was also noted after the treatment. The other species listed in Table VI were not affected by the liming.

3.5 Fish

No fish were caught in the fish fence at the outlet of Hovvatn or with electrical fishing in the inflowing streams and outlet of Hovvatn.

The trout mainly reside along the shore in Hovvatn and the catch was approximately the same throughout 1981, 1982 and 1983.

The migration of trout from stocking locations increased from 1982 to 1983 In Pollen the trout were mainly stationary. Twenty %, however, were recaptured at other stations in May 1983,and both age groups 1 and 2+3 showed significant increases in distribution. The migration from station NE and SW to other stations was most pronounced, and increased significantly through May 1983 for age groups 1 and 2+3 at station NE and for age group 2+3 at station SW.

Growth rate of the stocked trout differed among stations. Growth differences were highest in age group 1 and lowest in age group 3. Generally, growth was highest at station NW where no fish were stocked, and lowest at SW, the deepest station. The condition factor varied between 0.98 to 1.16 and was lowest in age group 1 and highest in age group 3. The variation in condition factor was small among the four stations. Sveälv and Matzow (1985) give more information about migration and growth.

Trout stocked in Hovvatn grew significantly faster ($p < 0.001$), than the parent trout of the same strain in Lake Byglandsfjorden (Figure 13). The trout in Lake Byglandsfjorden grow asymptotically to 34 cm (Sveälv, 1985), the same length the trout in Lake Hovvatn reached in age group 3. Trout in age group 2 in Lake Hovvatn and Byglandsfjorden showed the greatest annual growth at 15 and 6 cm, respectively.

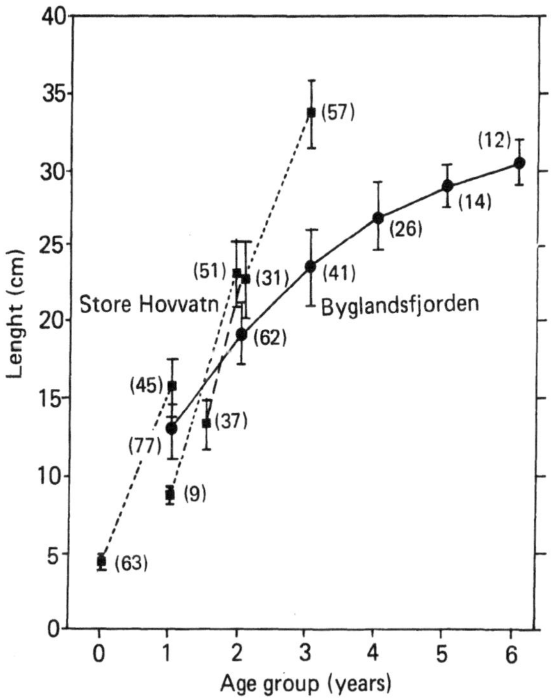

Figure 13. Growth of trout (mean and standard deviation) stocked in Hovvatn after liming and in lake Byglandsfjorden, their native habitat. Numbers of fish are indicated (from Sveälv, 1985).

Yield increased with age, and in age group 3 it was higher than the production (Figure 14) (Sveälv, 1985). The yield in age group 1 (8 kg) was approximately 25% of the yield in age group 2 (33 kg) which in turn constituted about 25% of yield in age group 3 (144 kg).

Natural reproduction of the stocked trout was indicated by the presence of two spawning redds at 3.7 m depth in March 1983 observed by diver. The eggs lay just under sediment surface. Of 54 eggs collected, only one was fertilized (Sveälv, 1985). Eighty-seven % of the trout eggs in "hatching boxes" placed in the central part of Hovvatn survived to hatching in May 1983.

Insects were the bulk food category for the fish at all locations (Figure 15). Larvae and imagines of Coleoptera, mainly Dytiscidae, were main food items throughout the year at all stations. Ephemeroptera were

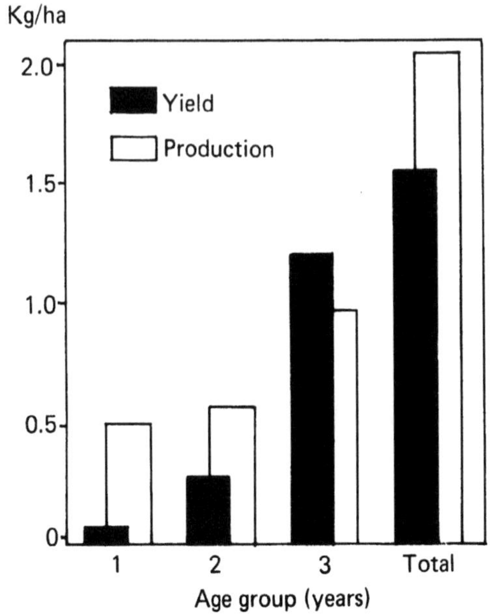

Figure 14. Production and yield of fish in Hovvatn (total and for each of 3 age classes) during 1982 (from Sveälv, 1985).

the dominant trout food in May and October. They also occurred in June, especially in Pollen, but in August they were of minor importance. Nymphs and imagines of Corixidae were most important in August, especially at station NE. Larvaes of the Megalopteran _Sialis lutaria_ were an important food object under all fishing periods, except in October. Trichoptera larvae and Odonata nymphs occurred in trout diet at all periods and stations but never as main food item. Diptera occurred infrequently as trout food in spite of their importance in the benthic fauna. Zooplankton played a minor role as fish food, The only planktonic crustacean of importance was the large cladoceran _Bychotrephes longimanus_ which occurred in August at all catch stations. The benthic cladoceran _Eurycercus lamellatus_ is mainly littoral, and was eaten to some extent in August.

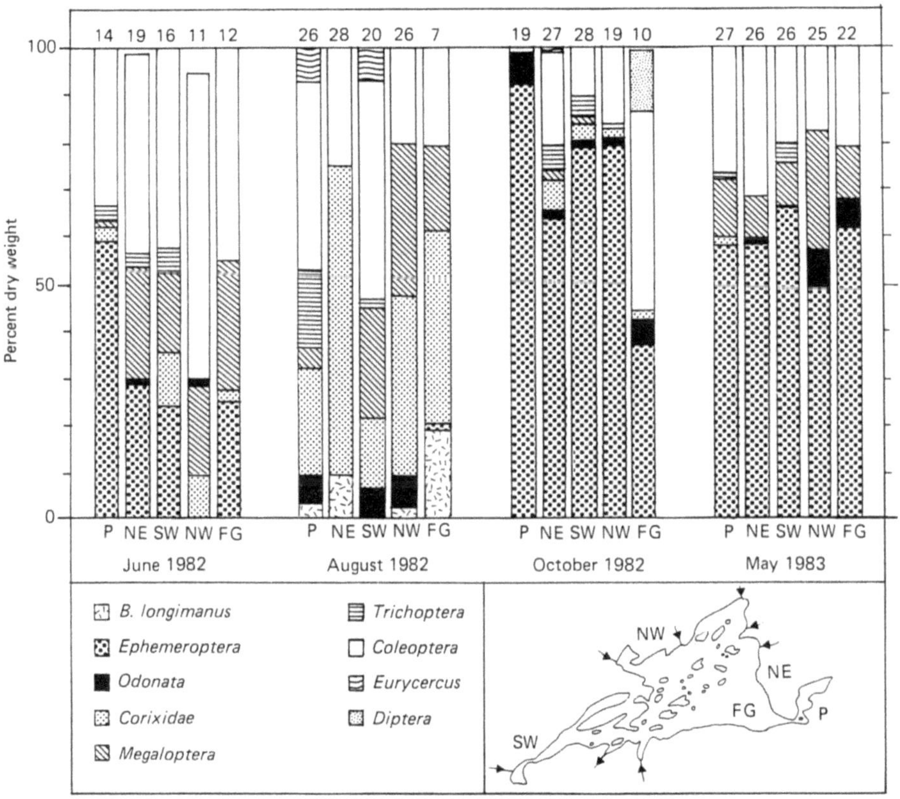

Figure 15. Food choice of trout at several net stations in Pollen (P) and Hovvatn (NO, SV, NV, FG) during June 1982 - May 1984. Numbers of fish are given above the bars (from Sveälv, 1985).

4. DISCUSSION

4.1 Effect of zoobenthos on water chemistry

Many studies have shown that liming acidic lakes causes changes in TOC, color and transparency. Liming with high doses of $CaCO_3$ can result in decreases in TOC and color (transparency increases) apparently due to precipitation of humic substances. Several lakes in Sweden responded in this manner (Swedish National Board of Fisheries, 1981; Hultberg and Anderson, 1982) as did a lake in the Adirondack Mountains, NY (Driscoll et al., 1982). On the other hand, liming can cause increases in TOC and color and decrease transparency (Dickson, 1979). Mechanisms here include increased phytoplankton growth, increased respiration in the sediment with release of organic substances to the overlying water, and simple pH-dependent light absorption by humic substances (Hultberg and Andersson, 1982; Yan, 1983). In Hovvatn, as in several other limed lakes in Ontario (Dillon et al., 1979; Yan , 1983), changes in TOC, color and transparency cannot be explained by changes in the phytoplankton.

The changes in TOC, color and transparency with liming and reacidification are variable and probably due to several factors. As both Hultberg and Andersson (1982) and Hasselrot et al. (1984) point out, such changes in transparency influence light penetration, heat balance and mixing in lakes and may be of biotic significance.

Increases in P could come from dissolution of trace amounts of P contained in the limestone or from release of P from the sediments as a result of increased respiration. P in a Swedish limestone amounts to about 0.5 to 1 µg P/l per mg Ca/l dissolved (Swedish National Board of Fisheries, 1981). In Hovvatn Ca concentrations increased about 2 mg L^{-1}. Using the Swedish data an increase of 1 to 2 µg P L^{-1} would be expected, approximately the increase observed. In Pollen the high dose gave a Ca increase of about 12 mg L^{-1} which would correspond to 6 to 12 µg P L^{-1}, again approximately as observed.

The increased P concentrations may, however, be due to release from the lake sediments. Several lines of evidence support this mechanism. (1) The high increase of benthic animals combined with the long duration of increased P levels in Pollen, suggests a continued supply rather than a one-time addition with the limestone. (2) The elevated P levels roughly parallel the elevated TOC levels which returned to pre-liming levels in 1983. The activity of micro- and macroorganisms in the fine sediments of Hovvatn can be responsible for the elevated levels of both P and TOC.

This activity is apparently due to the increase of oligochaets, mayflies and chironomids after liming. The invertebrates continuously break down organic material from larger to smaller particles (Cummins, 1973; Peterson and Cummins, 1974). During each step in this process dissolved organic material is released to the water. The increase in number of animals dwelling in the fine detritus and probable elevated level of bacterial activity after liming may explain the elevated TOC and P levels.

Whereas liming accelerates the prosessing of fine detritus, the effect on decompostion of coarse detritus was minor. An experiment in which birch leaves were placed in nylon-mesh bags on the sediment in Hovvatn, Pollen and Lille Hovvatn revealed no significant differences in decomposition rate between localities (Fjellheim and Raddum, 1984). The litterbags were, as expected, occuppied by shredders which did not increase significantly.

We have not measured the respiration or the level of microorganisms at either Hovvatn or Pollen. The succession of the zooplankton to species consisting mainly of filter-feeders suggests that microorganisms increased in abundance following liming. Phytoplankton production apparently had only minor influence on the zooplankton succession.

Together with the benthic animal development this demonstrates that an increased rate of decomposition is a main effect of the liming and affects water chemistry as well as the lake biota. Similar effect of lime treatment on the microbial activity of a Swedish lake sediment were observed by Gahnström et al. (1980).

The return of TOC concentrations to pre-liming levels the third year is more difficult to explain. Possibly the easily-decomposed organic matter accumulated during the acid condition was exploited during the first 2 yr after liming and by the third year the decomposition rate reached equilibrium with the deposition rate. The high density of microfiltrators the third year can also have influenced the TOC levels.

Other explanations can be related to the reacidification. A portion of the TOC can have flocculated (Gjessing, 1976) as the pH levels decreased the third year. Increased growth of bacteria or fungi utilizing dissolved material directly can also have cut down the TOC level.

4.2 Effect of liming on phytoplankton

Liming at Hovvatn and Pollen did not result in major changes in phytoplankton species composition and abundance. The total algal volume was higher in 1982 relative to 1981, but then again decreased in 1983. Compared with the acid lake Lille Hovvatn no decrease was expected. The decrease may be due to the fluctutation in pH of Hovvatn after liming and reacidification: a circum-neutral community of phytoplankton probably was developing in 1982 and then the water turned acid again in 1983. Lille Hovvatn, with a stable low pH, has developed an acid "maximum" flora without disturbances in water chemistry.

In Swedish liming experiments Eriksson et al. (1983) report increased phytoplankton biomass in two lakes the second year after liming. This is consistent with surveys of lakes in Ontario in which the very acid lakes had lowest phytoplankton biomass (Conroy et al., 1976). In contrast, the synoptic surveys in Sweden conducted by Hörnström et al. (1973) revealed no apparent relationship between the phytoplankton biomass and lake pH. Yan and Stokes (1978) conclude that the number of phytoplankton species and community structure were more sensitive to lake acidification than phytoplankton biomass. This was also the conclusion of Eriksson et al. (1983). Phytoplankton biomass in oligotrophic lakes is related primarily to P level and secondarily to pH. Low pH lakes commonly have low P levels and thus low phytoplankton biomass (Hörnström et al. 1973; Conroy et al. 1975). Liming at Hovvatn increased total P content the first years, but apparently this increase was insufficient to cause significant increase in phytoplankton biomass.

Our data from Hovvatn indicate no changes in phytoplankton species number or community structure caused by the lime treatment. An exception may be the appearance of small green algae, which could be an effect of the liming. Chlorophyceae do not generally dominate the phytoplankton community in acid lakes in southern Norway (Raddum et al., 1980) indicating that the presence of green algae in Hovvatn and Pollen is due to liming.

The primary production measured in 1982 showed no significant differences among Hovvatn, Pollen and Lille Hovvatn. The increase in number of the cladoceran species Holopedium gibberium in Pollen could be an effect of increased production of phytoplankton, but it is much more likely a result of increased amount of bacteria and other organic material in the water column caused by the liming.

4.3 Effects of liming on zooplankton

The low number of zooplankton species found in Hovvatn, Pollen and Lille Hovvatn before liming is characteristic of clear acidic lakes (Hobæk and Raddum, 1980). Throughout the season, the acid community usually is dominated by one species, either *Eudiaptomus gracilis* or *Bosmina longispina*.

The dominant species of rotifer during the acid condition in Hovvatn was *Kellicottia longispina*. No increase in this species was found in 1981, which supports earlier findings that the rotifers do not react the year of liming (Eriksson *et al*., 1983; Hillbricht-Ilkowska *et al*., 1977). In 1982 the number of *K. longispina* was reduced. This may be caused by the increase of *Diaphanosoma brachyurum* which is an effective competitor (microfiltrator) of rotifers (Geller and Müller, 1981). In 1983 the *D.brachyurum* decreased, and the rotifers increased.

In the reference lake Lille Hovvatn the rotifers showed a high and stable abundance in the absence of *D. brachyurum* during all three years. This supports the competition theory between *K. longispina* and *D. brachyurum*.

Conochilus unicornis increased in density during 1982 and 1983 both in Pollen and Hovvatn. This species feeds on detritus and bacteria (Pourriot, 1977) and competes better with *D. brachyurum* during periods with improved nutrition supply. Increasing amounts of bacteria in the water after liming acid lakes have been found by Scheider and Dillon (1976) and Hillbricht-Ilkowska *et al*. (1977).

The cladoceans exhibited a marked response to liming. In the acid condition the community was dominated by one species, either *Eudiaptomus gracilis* or *Bosmina longispina*. In Hovvatn the first year after liming *B. longispina* increased to high numbers, probably because of improved food production of detritus and bacteria. Larsson (1978) found that *B. longispina* mainly feed on such diet in a oligotrophic mountain lake. At Hovvatn *B. longispina* could expand on unexploited resource in 1981 because of little competition from other cladocerans. During 1982, however, the cladoceran system became more differentiated with increased numbers of the microfiltrator *D. brachyurum*.

In Pollen *Holopedium gibberum* appeared in small numbers in 1982, indicating that qualitative changes had occured in the food resource (phytoplankton, bacteria or humic substances). *H. gibberum* increased further in 1983 both in Pollen and Hovvatn and partly displaced the two other cladoceran species.

The pattern observed in the cladoceran system has also been

recorded in other lake liming experiments (Hultberg and Andersson, 1982; Eriksson et al., 1983) In enclosure experiments the r-selected species D. brachyurum rapidly built up high abundances, although it was not present in detectable numbers at the start of the experiments (Brettum et al., 1984).

The abundance of the copepod Eudiaptomus gracilis was not influenced by the liming and fish introduction in Hovvatn. The observed decrease in 1983 in Hovvatn might be caused by predation from the increasing population of Heterocope saliens and competition from the increasing numbers of cladoceran species.

Fecundity (clutch-size) of E. gracilis, however, increased after liming, but the patterns were irregular. Fertility increased immediately after liming perhaps as a direct physiological effect; subsequent decrease to pre-lime conditions could be due to indirect biotic conditions. The same pattern was exhibited by other species, especially B. longispina and H. gibberum where fertility decreased from 1982 to 1983. Shortage of food may explain the decrease in 1983, as TOC levels had begun to decrease.

4.4 Effect of liming on zoobenthos

The strength and duration of the effect of liming on zoobenthos apparently depends on the amount of unexploited organic material and the lime dose. The key factor in the sediments appears to be the bacterial activity, as many of the sediment-dwelling animals utilize and are dependent on microbial conditioning of the sediment particles as food (Brittain, 1972; Cummins, 1973; Petersen and Cummins, 1974). Oligochaets are typical animals in the detritus system, and their increase the second and third year after liming indicates that accumulated food has become available. Further, physiological effects due to elevated Ca concentrations in the water might also have stimulated the oligochaets (Havas, 1981).

The difference in oligochaet abundances between Hovvatn and Pollen after liming may be due to differences in bottom substrate. The sediment in Pollen consists of soft organic material even in the uppermost part of the littoral zone. The shallow sediment at Hovvatn was limestone mixed with sand, which may be deleterious for the oligochaets the first season after liming. Limestone mixed with organic material in Pollen did not have this effect.

Similar response of oligochaet populations was found by Raddum et al. (1984) in the limed lake Søndre Boksjø. Oligochaets increased in abundance the first season after liming and then decreased. Reduction to the pre-limed level occurred rapidly at the site with the lowest content of organic material, while the reduction in number was slower at the site with higher content of detritus, indicating that amount of available food is the most important factor.

The mayflies (Leptophlebia sp.) present in acid lakes prefer shallow bays with high content of organic material and presence of macrovegetation. The samples in Hovvatn and Pollen were from sites not rich in vegetation, which explains the absence or low occurrence of the mayfly larvae prior to liming. After liming, however, the larvae occurred frequently in the bottom samples from the depth zone 0.5 to 5 m (Table V). As these larvae are able to survive in acid water the observed increase is not of physiological cause.

Fine detritus is the most common food for mayflies, but the material requires processing by bacteria and fungi before utilization (Brittain, 1972). Kjellberg (1972) found that Leptophlebia sp. feed on both algae and detritus. Increased bacterial activity in lake sediment after lime treatment as observed by Gahnström et al. (1980) improves nutrient supply to the larvae. This can explain the immigration of the mayflies to the soft bottom both in Hovvatn and Pollen after the treatment. A similar development of distribution was observed in lake S. Boksjø (Raddum et al., 1984).

The difference in the population development of mayflies observed between Hovvatn and Pollen after liming may be due to the initial populations in the two localities. Hovvatn had much lower content of organic material in the littoral zone and relatively less macrovegetation than Pollen. The initial population density of Leptophlebia sp. was propably higher in Pollen than in Hovvatn. The delay in population increase observed in Hovvatn seems to be caused by this relationship.

The development of chironomid community after liming apparently is the result of physiological factors, food availability, and intraspecific competition. The increase of Tanytarsus pallidicornis can be a result of both physiological effects and food availability. Another species of this genus has low tolerance to acid water (pH < 5.5) during emergence (Bell, 1970). Decline in Tanytarsus during acidification in Swedish lakes was observed by Henrikson et al. (1982), which also underlines the sensitivity of Tanytarsus species to acidification. The rare occurrence of the species prior to liming and the increase observed the second and third year can therefore be due to the elevated pH of the water.

The enormous increase of the larvae, however, can not occur without available food resources. Gut contents of larvae in Pollen from 10 m depth (fine detritus) and 5 m depth (zone of dying sphagnum) indicated that both detritus and dead sphagnum can be used as food. Sphagnum death after liming is well known from Sweden (Eriksson et al., 1983). Sphagnum is rich in nutrient cations and can give valuable inputs of salts to the fauna living in this area.

In Hovvatn no sphagnum zone occurred at the sampling site. Here fine detritus was the only substrate, and *Heterotanytarsus apicalis* dominated over *T. pallidicornis*. This shift in dominance between Hovvatn and Pollen could be due to differences in the food quality at the two sites as *T. pallidicornis* seems to take advantage in the sphagnum zone in Pollen. The sphagnum also increases the surface of the bottom, which allow increased density of larvae. Other species of chironomids living in fine detritus are *Pagastiella orophila*, *Sergentia* sp. and *Chironomus* spp. In Pollen *Sergentia* sp. and *Chironomus* spp. increased the first season after liming, but decreased in number the following two years. *P. orophila* increased in abundance the second year and was reduced the third year.

The development of *Sergentia* sp. and *P. orophila* is probably not due to physiological effects as these species can sustain high abundance both in acid and less acid waters (Wiederholm and Eriksson, 1977; Raddum and Sæther, 1981). The changes in abundance of *Chironomus spp.* can be due to physiological effects in addition to interspecific effects. The species are common in oligotrophic acid waters (Raddum and Saether, 1981; Henrikson et al., 1982), and rapid pH change can probably damage this population.

Chironomus prefer increasing amounts of organic matter (Bruce, 1962), a pattern which occurs in acid lakes due to reduced microbiological activity (Hendrey et al. 1976; Gahnström et al. 1980). The sharp reduction of *Chironomus* in Pollen the second and third year after liming could be due to decline in the pool of organic matter accumulated during the acid period.

The effect of liming on the zoobenthos in Hovvatn largely parallels the effect observed in other limed lakes such as lake Søndre Boksjø and the lakes in Gjerstad (Raddum et al., 1984, Kroglund and Raddum, 1985). Together these studies show that accumulated organic material becomes available as food to different kinds of bottom animals after liming, with accelerated biological activity and increasing density of animals for at least 2-3 years.

The response to liming starts at the shallower parts of the lakes and moves deeper each season. This succession is probably a result of the decreasing temperature and light with increasing depth which slows down the biological aactivity and parallels in general the steps during acidification. The delayed effect in the deepest parts may also be due to precipitation and slow removal of Al-hydroxides in the deeper parts after liming. Al is most toxic during the Al hydrolysis (Driscoll et al., 1980; Baker, 1981). Deleterious concentrations of Al to some bottom animals can have occurred in the deeper parts both in Hovvatn and Pollen thus counteracting a positive effect of the liming.

Acidification leads to reduced microbiological activity (Hendrey et al. 1976; Gahnström et al., 1980). This results in a reduction in turnover-rate and recycling of nutrients which in turn leads to accumulation of organic matter. This was first described by Grahn et al. (1974) as the oligotrophication process.

The high biological activity in the sediments observed in Hovvatn and Pollen after liming points to a reversal of the oligotrophication process. The accumulated organic material appears to be utilized over a short period and the localities have been in a state of "eutrophication" with rapid increase and high activity of sediment-dwelling animals with the increase of both P and TOC the first and second year after treatment. The reacidification of Hovvatn will probably soon turn the "eutrophication" to "oligotrophication" again, and we expect to return to an acid lake with initially low pool of organic matter.

4.5 Fish

The increased emigration by trout from Pollen to Hovvatn during winter 1982-83 is difficult to explain, but can be due to "acid shocks" and toxic water quality for fish in the upper layer. Stressed fish select new habitats (Muniz et al., 1978). Other migration of fish between stations seems to be due to factors not directly caused by the liming or reacidification.

Differences in growth of trout among bottom net stations in October 1982 are due to fish density, bottom-, depth- and food-condition rather than water quality. The best growth was generally exhibited by immigrants to station NW, where no trout were stocked. The condition factor of the trout was very good and increases with age as a result of a higher fraction of mature trout.

Trout from Lake Byglandsfjorden exhibit significantly slower growth during hatchery rearing than under natural conditions after stocking in Hovvatn, due to low stocking density, high abundance of easily-available and energy-rich food, and low annual mortality The growth increment was among the best found in natural brown trout populations (Sømme, 1941). The growth rate decreased, however, already the second year after stocking (Sveälv and Matzow, 1985).

High growth during the first years after stocking of brown trout into barren lakes is a well-known phenomenon (Sømme, 1941; Nilsson, 1972) and has also been reported after stocking of brook trout (Salvelinus fontinalis M.) into fishless acid lakes (Grande et al., 1980).

The total yield is within the range of rich natural brown trout lakes (Sømme, 1941) and consists mainly of age group 3. With higher stocking rate of more catchable trout from the beginning in Hovvatn the yield probably could be raised considerably, as the carrying capacity after liming was apparently much higher than expected.

Spawning occurred in Hovvatn and in spite of low fertilization, experiments with eggs indicate high survival when the eggs were laid in suitable substratum where the water could circulate.

The stomach contents of trout in Hovvatn was dominated by large sized, mobile, littoral species, typical of acid lakes with no fish (Raddum, 1980; Hultberg and Andersson, 1982). The preferences of the fish exert a high selection pressure on the invertebrate community in Hovvatn which may lead to alterations in species composition (Henrikson et al., 1980). Due to the great abundance of easily-available littoral animals the trout seldom fed on chironomids and zooplankton. The similarity of food composition in trout stomachs from different catch stations in Hovvatn indicate that the abundance of preferred animals was not appreciably reduced the first year.

Liming of Hovvatn and Pollen caused changes in most major aspects of ecosystem function and structure. Biological changes following liming is manifest as short-term transient response such as the explosion of Chironomid populations, as well as long-term response such as the species succession observed in the zooplankton. The long-term responses were not complete at Hovvatn before reacidification set in. The interactions between groups of organisms in Hovvatn and Pollen indicates that more than 3 yr are required before a new steady-state can be attained following liming.

5. CONCLUSIONS

1. Liming of Hovvatn and Pollen increased the pH from 4.4 to 6.3 and 7.5, respectively. Reacidification started the first season after liming, and the pH was brought back to about 5.0 (Hovvatn) and 5.5 (Pollen). High fluctuations in pH were observed in Pollen the third year, depending on the inflow of acid water.

2. Ca and alkalinity increased and Al decreased after liming. Only small changes in concentrations of the other ions occurred.

3. Phytoplankton biomass and production did not change significantly after liming.

4. Zooplankton showed a succession in the species composition. *Bosmina longispina* dominated before liming and the first year after liming. *Diaphanosoma bracyurum* increased to high numbers the second year of liming, while *Holopedium gibberum* dominated the third year. All species are acid tolerant, and the succesion results from food competition and food utilization and rather than from reinvasion or physiological effects.

5. Detritus dwelling invertebrates increased to very high abundance after liming. All the species or groups are acid tolerant. The increase is probably due to the microbial conditioning of detritus as food for the invertebrates.

6. Fish grow very fast after liming, and their diet consists of the fauna typical of acid lakes (coleoptera, corixidae and mayflies). This community was strong enough to sustain the predation pressure, and the fish were not forced to feed on less-valuable food organisms. Fish density was suitable for the food resources.

7. The interactions between groups of organisms indicates that more than 3 yr are required before a new steady-state is attained following liming.

ACKNOWLEDGMENTS

We thank A. Jacobsen, M. Johannessen, E. Kleiven, G. Myklebostad, and A. Prytz for help in these investigations.

REFERENCES

Almer, B., Dickson, W., Ekström, C., and Hörnström, E.: 1978, 'Sulfur pollution and the aquatic ecosystem', In Nriagu, J.O. (ed.): Sulfur In the Environment, Part II, Ecological Impacts, John Wiley & Sons, p. 271-311.

Baker, J.: 1981, 'Aluminium toxicity to fish as related to acid precipitation and Adirondack surface water quality', Thesis, Cornell University, Ithaca, N.Y.

Bell, H.L.: 1970, Can. Ent. 102, 636.

Brettum, P., Kroglund, F., Nilssen, J.P., Sandøy, S., Skov, A., and Wærvågen, S.B.: 1984, 'Eksperimentelle innhegningsforsøk i Gjerstad, Aust-Agder. Et forsøk på alternativ behandling av forsurede vann', Report 16/84, (Kalkingsprosjektet, DNV, Fylkeshuset, Arendal, Norway), 38pp.

Brittain, J.E.: 1972, Freshw. Biol. 2, 271.

Bruce, D.: 1962, Trans Soc. Br. Ent. 15, 41.

Conroy, N., Hawley, K., Keller, W., and Lafrance, C.: 1976, J. Great Lakes Res. 2(Suppl.1), 145.

Cummins, K.W.: 1973, Ann.Rev.Entomol. 18, 183.

Davis, R.B., and Berge, F.: 1980, 'Atmospheric deposition in Norway during the last 300 years as recorded in SNSF lake sediments. II. Diatom stratigraphy and inferred pH', p. 270-271, In: Drabløs, D. and Tollan, A. (eds.): Ecological Impact of Acid Precipitation, (SNSF-project, NISK, 1432-Ås, Norway).

Dickson, W.: 1979, Skogs- o. Lantbra.-akad. Tidsk. Suppl. 13, 38.

Dickson, W.: 1980, 'Properties of acidified waters', p.77-83, In: Drabløs, D. and Tollan, A. (eds.): Ecological Impact of Acid Precipitation, (SNSF-project, NISK, 1432-Ås, Norway).

Dillon, P.J., Yan, N.D., Schneider, W.A., and Conroy, N.: 1979, Arch. Hydrobiol. Beih. Ergeb. Limnol., 13, 317.

Driscoll, C.T., White, J.R., Schafran, G.C., and J.D. Rendall: 1982, J. Environ. Eng. Div. ASCE 108, 1128.

Driscoll, C.T., Baker, J., Bisogni, J., and Schofield, C.: 1980, Nature 284, 161.

Eriksson, F., Hörnström, E., Mossberg, P., and Nyberg, P.: 1983,

Hydrobiologia 101, 145.

Fjellheim, A., and Raddum, G.G.: 1984, 'Nedbryting av organisk materiale i forbindelse med kalking av sure sjøer', p.23-34, In: S. Bosheim and M. Nicholls (ed.), Interaksjoner Mellom Trofiske Nivåer i Ferskvann (Norsk limnologforen., Box 86, 0313 Oslo, Norway).

Gahnström, G., Andersson, G., and Fleischer, S.: 1980, 'Decomposition and exchange processes in acidified lake sediment', p. 306-307, In: Drabløs, D. and Tollan, A. (eds.): Ecological Impact of Acid Precipitation, (SNSF-project, NISK, 1432-Ås, Norway).

Geller, W., and Müller, H.: 1981, Oecologia, 49, 316.

Gjessing, E.T.: 1976, Physical and Chemical Characteristics of Aquatic Humus, (Ann Arbor Science Publishers, Ann Arbor, Michigan, USA), 120 pp.

Grahn, O., Hultberg, H., and Landner, L.: 1974, Ambio 3, 93.

Grande, M., Andersen, S., and Sevaldrud, I.: 1980, 'Forsøk med utsetting av bekkerøye (Salvelinus fontinalis Mitchill) i sure innsjøer 1975-1978', Internal Report 68/80 (SNSF-project, NISK, 1432 Ås, Norway), 88 pp.

Harvey, H.: 1980, 'Widespread and diverse changes in the biota of North American lakes and rivers coincident with acidification' p. 93-98, In: Drabløs, D. and Tollan, A. (eds.): Ecological Impact of Acid Precipitation, (SNSF-project, NISK, 1432-Ås, Norway).

Hasselrot, B., Andersson, I., and Hultberg, H.: 1984, Rep. Inst. of Freshw. Research Drottningholm, 61, 78.

Havas, M.: 1981, 'Physiological response of aquatic animals to low pH', p.49-67, In: R. Singer (ed.): Effects of Acid Precipitation on Benthos, (North American Benthological Soc., Illinois EPA, Springfield, IL).

Hendrey, G.R., Baalsrud, K., Traaen, T.S., Laake, M., and Raddum, G.: 1976, Ambio 5, 224.

Henriksen, A.: 1980, 'Acidification of freshwater - a large-scale titration', p. 68-74, In: Drabløs, D. and Tollan, A. (eds.): Ecological Impact of Acid Precipitation, (SNSF-project, NISK, 1432-Ås, Norway).

Henrikson, L., Oscarson, H.G., and Stenson, J.A.C.: 1980, 'Does the change of predator system congtribute to the biotic development in acidified lakes' p. 316, In: Drabløs, D. and Tollan, A. (eds.): Ecological Impact of Acid Precipitation, (SNSF-project, NISK, 1432-Ås, Norway).

Henrikson, L., Olofsson, J.B., and Oscarson, H.G.: 1982, Hydrobiologia,

86, 223.
Hillbricht-Ilkowska, A., Rybak, J.I., Kajak, Z., Dusoge, K., Ejsmont-Karabin, J., Spodniewska, I., Weglenske, T., and Godlewske-Lipowa, W.A.: 1977, Ecol.Polska 25, 379.
Hobæk, A., and Raddum, G.G.: 1980, 'Zooplankton communities in acidified regions of South Norway', Internal Report 75/80, (SNSF-project, NISK, 1432 Ås, Norway), 132 pp.
Hongve, D.: 1982, Limnos(Oslo) 3, 3.
Hultberg, H., and Andersson, I.B.: 1982, Water Air Soil Pollut., 18, 311.
Hörnström, E., Ekström, C., Miller, O., and Dickson, W.: 1973, 'Försurningens inverkan på västkustsjöar', Publication 7/1973, (Statens naturvårdsverk, Solna, Sweden), 97 pp.
Jensen, K.W., and Snekvik, E.: 1972, Ambio 1, 223.
Kjellberg, G.: 1972, Ent. Tidsskr. 9, 1.
Kroglund, F., and Raddum, G.G.: 1985, 'Effekter av kalking på bunnfaunaen i Gjerstad', Report 23/84 (Kalkingsprosjektet, DNV, Fylkeshuset, Arendal, Norway), 92pp.
Larsson, P.: 1978, Holarct. Ecol. 1, 162.
Matzow, D.: 1982, Vatten 38, 419.
Muniz, I.P., Leivestad, H., and Rosseland, B.O.: 1978, 'Stressmålinger på fisk i sure vassdrag, p. 233-247, In: Publication No.2 (Nordforsk, Miljøvårdssekretariatet, Stockholm, Sweden).
Nilsson, N.A.: 1972, J.Fish.Res.Bd. Canada, 29, 693.
Overrein, L., Seip, H.M., and Tollan, A.: 1980, 'Acid precipitation - effects on forest and fish'. Research Report 19/80, (SNSF-project, NISK, 1432 Ås, Norway), 175 pp.
Peterson, R.C., and Cummins, K.W.: 1974, Freshw. Biol. 4, 343.
Pourriot, R.: 1977, Arch.Hydrobiol. Beih.Ergebn.Limnol. 8, 243.
Raddum, G.G.: 1980, 'Comparison of benthic invertebrates in lakes with different acidity' p. 330-331, In: Drabløs, D. and Tollan, A. (eds.): Ecological Impact of Acid Precipitation, (SNSF-project, NISK, 1432-Ås, Norway).
Raddum, G.G., Lømsland, E.R., and Johnsen, T.M.: 1980, 'Klorofyll a, feopigment og artssammensetning av fytoplankton i vann med ulik pH', Technical Note 56/80 (SNSF-project, NISK, 1432 Ås, Norway), 60 pp.
Raddum, G.G., and Sæther, O.A.: 1981, Verh.int. Ver. Limnol. 21, 367.
Raddum, G.G., and Fjellheim, A.: 1984, Verh.int. Ver. Limnol. 22, 1973.
Raddum, G.G., Hagenlund, G., and Halvorsen, G.A.: 1984, Rep. Inst. Freshw. Res. Drottningholm, 61, 167.
Scheider, W., and Dillon, P.J.: 1976, 'Neutralization and fertilization

of acidified lakes near Sudbury, Ontario'. Proc. 11th Canadian Symp.Water Poll. Research, 11, 93.

Schindler, D.W. and Holmgren, S.K.: 1971, J.Fish.Res.Bd. Canada, 28, 189.

Sevaldrud, I.H., Muniz, I.P., and Kalveness, S.: 1980, 'Loss of fish populations in Southern Norway. Dynamics and magnitude of the problem' p.350-351, In: Drabløs, D. and Tollan, A. (eds.): Ecological Impact of Acid Precipitation, (SNSF-project, NISK, 1432-Ås, Norway).

Sveälv, T., 1985, 'Utplantering av öring i det partiellt kalkade Store Hovvatn, Södra Norge. Spridning och populationsdynamik samt jämförande beståndskaraktäristik med öring i Byglandsfjorden', Thesis, Univ. Oslo.

Sveälv, T., and Matzow, D.: 1985, 'Studium av öring utplanterad i det partiellt kalkade Store Hovvatn, Aust-Agder', Report 25/85 (Kalkingsprosjektet, DVF, Fylkeshuset, Arendal, Norway), 81 pp.

Swedish National Board of Fisheries: 1981, 'Liming of lakes and rivers 1977-1981 in Sweden', (Nat'l.Bd.Fisheries, Box 2565, 40317 Gothenburg, Sweden), 210 pp.

Sømme, I.D.: 1941, Ørretboka, (Jacob Dybwads forlag, Oslo), 591 pp.

Utermöhl, H.: 1958, Mitt.int.Ver.theor.angew Limnol. 9, 1.

Wiederholm, T., and Eriksson, L.: 1977, Oikos 29, 261.

Wright, R.F.: 1984, 'Changes in the chemistry of Lake Hovvatn, Norway, following liming and reacidification', Acid Rain Res. Rept. 6/1984, (Norwegian Inst. Water Res., Box 333, 0314 Oslo 3, Norway), 68 pp.

Wright, R.F.: 1985, Can.J. Fish. Aquat. Sci., 42, 1103.

Yan, N.D.: 1983, Can. J. Fish. Aquat. Sci., 40, 621.

Yan, N.D., and Stokes, P.: 1978, Environ. Conserv., 5, 93.

WATER QUALITY IMPROVEMENTS IN THE SUDBURY, ONTARIO, CANADA AREA RELATED TO REDUCED SMELTER EMISSIONS

W. Keller[1], J. Roger Pitblado[2], and N. I. Conroy[1]

[1]Ministry of the Environment, 199 Larch Street
Sudbury, Ontario, Canada, P3E 5P9
[2]Geography Department, Laurentian University
Sudbury, Ontario, Canada, P3E 2C6

ABSTRACT. Data from water quality studies conducted in the Sudbury, Ontario, Canada area indicate that substantial decreases in the acidity of surface waters have accompanied reductions in SO_2 emissions from the Sudbury smelting industry since 1977. On average, acidic lakes in the Sudbury area showed a decrease in H^+ of $\sim$ 50% between 1974-76 and 1981-83, and the severity of springtime pH depressions in streams decreased. Although many Sudbury area surface waters remain highly acidic, general decreases in acidity appear to be continuing. The results demonstrate that reductions in emissions of acids and acid precursors result in concomitant improvements in water quality.

1. INTRODUCTION

In recent decades, the effect of acidic deposition on freshwater systems has emerged as a major environmental concern, with reported adverse effects occuring at various aquatic trophic levels (see reviews by Haines, 1981; Harvey et al., 1981; Dillon et al., 1984). Mitigation of the effects of acidic deposition by the addition of neutralizing agents, usually powdered limestone, has been conducted in several areas of the world including Sweden (Bengtsson et al., 1980; Hasselrot and Hultberg, 1984), Norway (Rosseland and Skogheim, 1984; Wright, 1985) the USA (Krester and Colquhoun, 1984) and Canada (Molot et al., 1984; White et al., 1984; Yan and Dillon, 1984). However, despite the successes reported for most neutralization efforts, the application of neutralizing agents remains a temporary measure, which must be periodically repeated for continuing protection. It is generally recognized that abatement of emissions of acids and acid precursors to the atmosphere is the only ultimate solution to the problem of acidic deposition.

The Sudbury, Ontario, Canada area provides a unique opportunity to assess water quality responses associated with recent reductions in SO_2 emissions. Lake acidification related to smelter emissions has long been recognized in the area (Gorham and Gordon, 1960; Beamish and Harvey, 1972). During the period 1960-1972, estimated annual SO_2 emissions from Sudbury area metal smelters averaged between 4240 and 7034 tonne day^{-1} while during 1973-77 average emissions declined to

Water, Air, and Soil Pollution **31** (1986) 765–774.

3663-4038 tonne day^{-1}. Since 1977, SO_2 emissions were further reduced (average annual emissions of 1065 to 2562 tonne day^{-1} for the period 1978-84) as a result of smelting process changes, pollution abatement measures and extended strike and shutdown periods in the Sudbury metal recovery industry (Figure 1; summarized in OCTF, 1982).

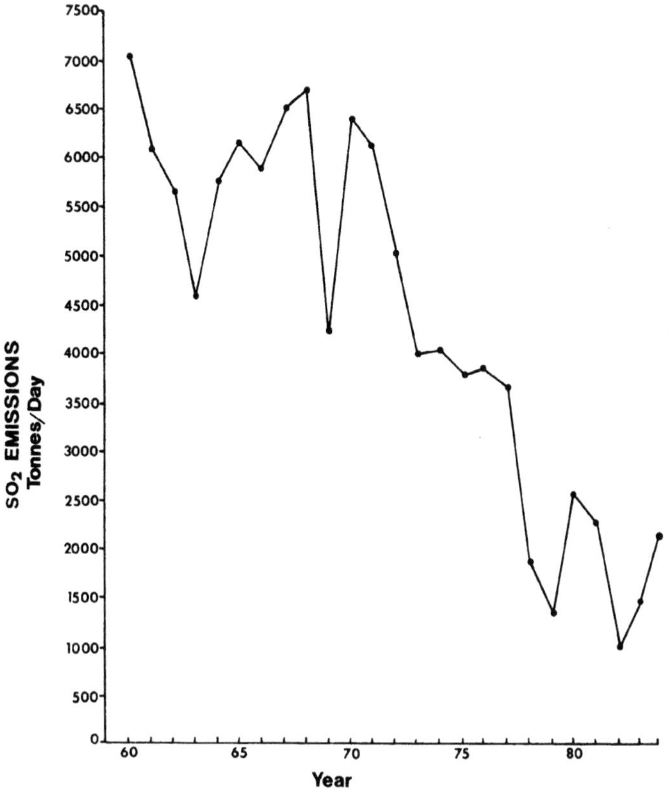

Fig. 1. Annual emissions of SO_2 (tonne day^{-1}) from the Sudbury area smelters (data from OCTF, 1982 and records on file with the Ministry of the Environment).

In this paper, we assemble and examine data for Sudbury area surface waters and assess trends in water quality in light of reductions in SO_2 emissions.

2. METHODS

All data presented herein were collected by the Ontario Ministry of the Environment (MOE). From 1973-1978, sampling was conducted as part of the Sudbury Environmental Study (SES) designed to document and evaluate regional water quality problems related to emissions from the Sudbury smelting industry. From 1979 on, data were obtained under the Acidic Precipitation in Ontario Study (APIOS), an expanded program implemented to examine the effects of acidic deposition on a provincial scale.

Lake samples were collected at a central location in each study lake. Prior to 1979, lake samples were taken at surface (0.5 m depth). Commencing in 1979, to conform with sampling procedures used under the overall APIOS study, samples were collected as non-volume-weighted water column composites by the Tygon tube method (MOE, 1979). For the parameters considered herein, comparison of data from simultaneously collected surface and composite samples for 43 lakes did not reveal any differences in values attributable to differing sampling methodologies (Keller and Pitblado, 1986). Stream samples were collected by hand from a well mixed reach of each stream, holding containers completely below the water surface.

In all years, pH was measured by meter, after appropriate buffer calibration over the range pH 4.0 to 7.0. Analyses for SO_4 were done by the methylthymolblue (MTB) colorimetric method before 1981, and by ion chromatography (IC) from 1981 on. Only data for waters with low color (< 20 Hazen units) are considered herein, since the MTB method overestimates SO_4 concentrations in colored waters, giving higher values than IC measurements (Kerekes and Pollock, 1983; Watt *et al.*, 1983).

3. RESULTS AND DISCUSSION

3.1 Synoptic Surveys

During the period 1974-76, sampling of 209 Sudbury area lakes (within ∿ 250 km) was conducted to investigate the spatial extent of water quality impacts related to emissions from the Sudbury area smelters (Conroy *et al.*, 1978). That study identified a large zone (∿ 650 km^2 of lake surface area) of low pH (<5.5) lakes centred around Sudbury (Conroy *et al.* 1978; Keller *et al.*, 1980; Pitblado *et al.*, 1980). Re-sampling of the same lakes in 1981-83 (Keller and Pitblado, 1986) after a decrease of ∿ 50% in Sudbury area SO_2 emissions (Figure 1) demonstrated general increases in pH (decreases in H^+) among the acidic lakes (Figure 2). On average, the decrease in H^+ within the lakes with pH <5.5 in 1974-76 was ∿ 50%. Correspondingly, decreases in lake water SO_4 concentrations (Figure 3) accompanied reduced SO_2 emissions. Decreases in SO_4 concentrations were significantly correlated (r=0.64; P=<0.01) with distance from Sudbury.

Annual data for 37 low pH (<5.5) Sudbury area lakes were collected during five years (1981-1985 inclusive). Examination of these data (Table I) showed a general increase in pH and decrease in SO_4 concentrations over the study period, on average; but overall changes were not statistically significant (ANOVA, P > 0.1). On an individual lake basis, however, 14 and 20 lakes showed continuing significant (P < 0.05) pH increases and SO_4 decreases, respectively, based on linear regression of annual values against sampling year. It is possible that because of lag time in lake responses due to hydrology and watershed effects, some of the observed water quality improvements may partially reflect emission reductions in the early

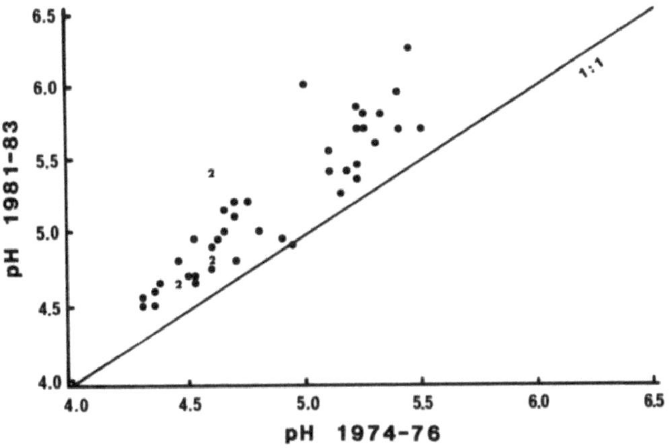

Fig. 2. Average pH in 1974-76 compared to average pH in 1981-83 for 45 study lakes with pH<5.5 in 1974-76. Numbers indicate coincident points.

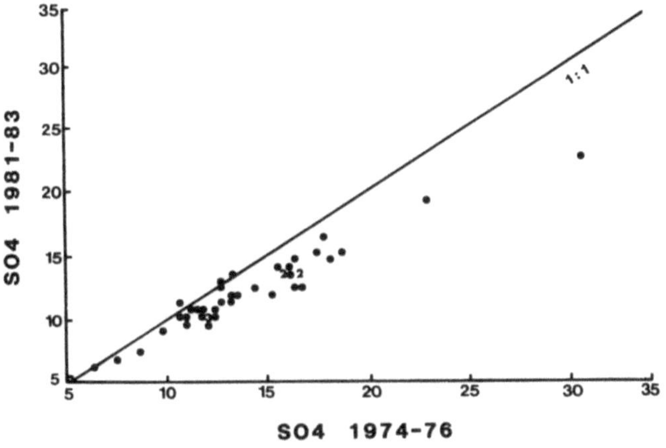

Fig. 3. Average concentrations of SO_4 (mg L^{-1}) in1974-76 compared to average concentrations in 1981-83 for 45 acidic study lakes.

1970's (Figure 1). However, monitoring data for Sudbury area lakes for the period 1973-76 (Dillon, 1982; Keller and Pitblado, 1986) did not show decreases in concentrations of SO_4 or H^+ during that period, indicating that the effects of these emission reductions were at least largely realized before commencement of the studies reported herein.

TABLE I

Average annual pH and SO_4 concentrations (1981-85) for 37 Sudbury area lakes with pH <5.5 in 1974-76. Data are based on a single sampling of each lake during each study year.

	1981	1982	1983	1984	1985	Significance[a]
pH	4.74	4.80	4.79	4.82	4.93	NS
SO_4(mg L^{-1})	13.9	13.1	13.3	12.5	11.9	NS

[a] based on a one way ANOVA; NS = no significant difference ($P>0.1$) over the study period.

3.2 Intensive Surveys

3.2.1 Lakes. Between 1979 and 1982, 29 northeastern Ontario lakes were sampled five to six times per year. The lakes covered a broad geographical area and ranged widely in average pH (Figure 4). Linear regression analysis of pH against the date of sampling showed small, but significant ($P<0.05$) increases in pH over the sampling period for ten lakes, all of which are located relatively close (within ∿ 120 km) to Sudbury. Examples of the temporal pH patterns in four of these lakes are shown in Figure 5. Based on our other observations, substantial changes in the pH of some of the study lakes may have occurred before initiation of sampling in 1979.

Other temporally intensive data on Sudbury area lakes are scarce. Comparative data for two acidic lakes (Laundrie, 80 km north and Florence, 105 km north) showed a significant (*t* test, $P<0.01$) increase in pH and decrease in SO_4 concentrations between 1974-76 and 1979-83 (Table II). LaZerte and Dillon (1984) and Dillon *et al.* (1986) documented substantial increases in pH and decreases in SO_4 concentrations in two other lakes near Sudbury over the same general time period (Clearwater, 16 km southwest; Swan, 15 km southwest). Decreases in acidity related to reduced atmospheric deposition have also been reported for other Sudbury area lakes (Alice and Baby Lakes, ∿ 10 km southeast; Hutchinson and Havas, 1986) and for lakes in Sweden (Forsberg *et al.*, 1985).

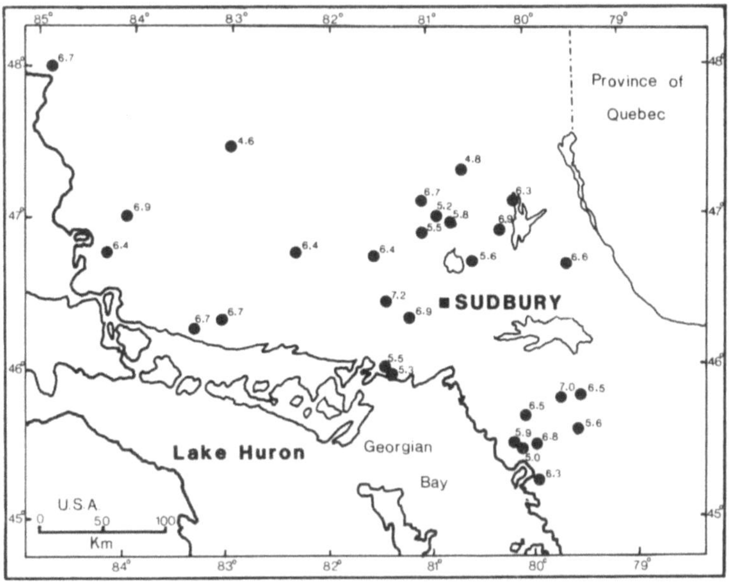

Fig. 4. Location and average pH of 29 northeastern Ontario lakes monitored during 1979-82.

TABLE II

Average pH and SO_4 concentrations (mg L^{-1}) for two Sudbury area lakes during 1974-76 and 1979-83.

Parameter	Laundrie Lake		Florence Lake	
	$\bar{x}^a$	N	$\bar{x}^a$	N
pH				
(1974-76)	4.7	21	4.6	10
(1979-83)	5.2	16	4.8	6
SO_4				
(1974-76)	12.5	21	14.2	10
(1979-83)	9.7	6	12.2	6

[a] all period means significantly different ($P<0.01$) based on a t test.

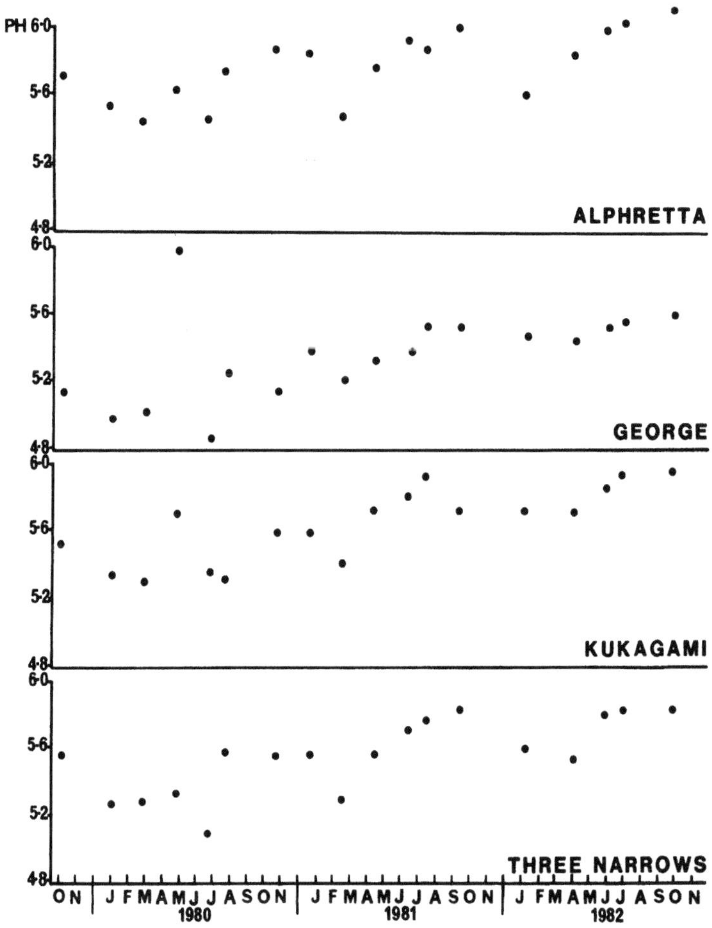

Fig. 5. Temporal patterns in the pH of four Sudbury area lakes, 1979–1982.

3.2.2 Streams. Data, usually collected weekly, are available for four stream systems located in the Killarney area. This locale, ∿ 50 km southwest of Sudbury, exhibits widespread lake acidification (Beamish and Harvey, 1972).

Three of the waters, the Chikanishing, Mahzenazing and West Mahzenazing rivers, drain acidic lakes (George, Tyson and Johnnie lakes, respectively). Sampling points are located ∿ 1.2, 4.0 and 2.5 km, respectively, downstream from the lake outlets. The fourth stream, an unnamed creek, originates at the base of the LaCloche Mountains and drains through a large marshy area followed by wooded terrain, to the sampling point, ∿ 3.0 km downstream (Keller and Gunn, 1982).

Monitoring data (Figure 6) showed that these water-courses were subjected to annual depressions in pH coinciding with snowmelt and high spring runoff. However, the severity of springtime pH decreases appeared generally reduced in 1981 and 1982 in comparison with 1979 and 1980, likely reflecting changes in atmospheric deposition. Unfortunately, hydrological data are not available for these streams, thus, the influence of natural variations in runoff quantity on the observed pH depressions cannot presently be assessed.

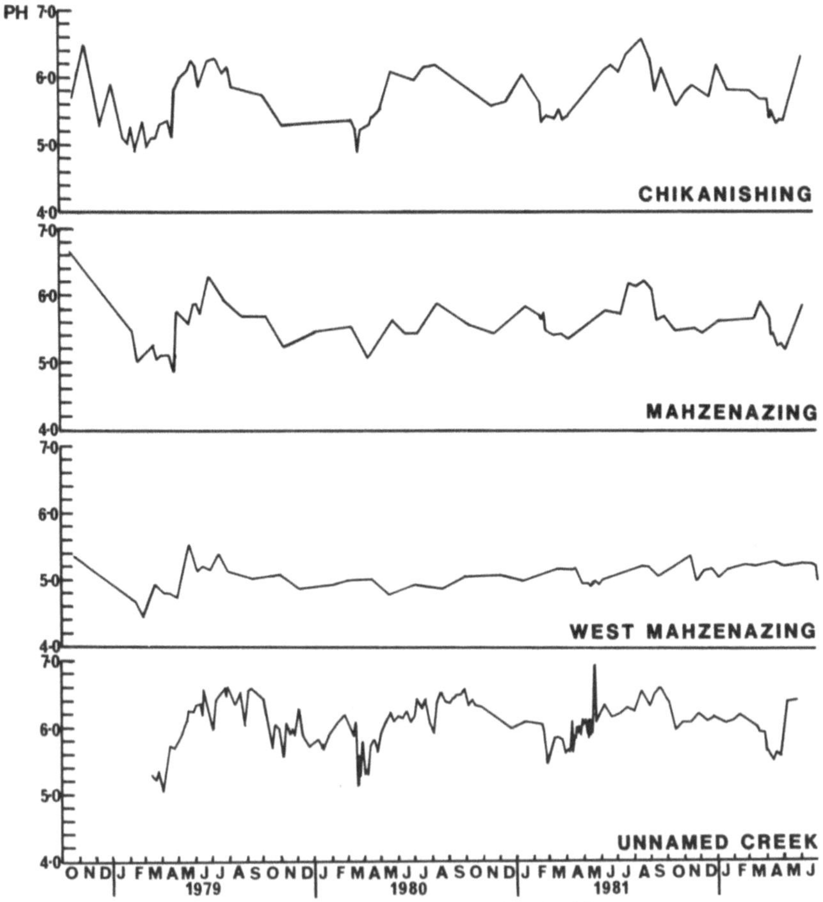

Fig. 6. Temporal patterns in the pH of four Sudbury area streams, 1978-1982.

4. SUMMARY AND CONCLUSIONS

Water quality data for the Sudbury, Ontario, Canada region, including the results of synoptic lake surveys and temporally intensive lake and stream monitoring were examined for the period 1974 to 1985. During this period, SO_2 emissions from the Sudbury area metal smelters were greatly reduced, declining from an average of ∿ 3,900 tonne day^{-1} during 1973-77 to ∿ 1800 tonne day^{-1} during 1978-84.

Available water quality data indicate that substantial increases in pH and decreases in SO_4 concentrations occurred during 1977-81, which corresponded with the period of substantial emission reductions in the smelting industry. Although some improvement in water quality appears to be continuing, many waters remain highly acidic.

Observations in the Sudbury area clearly demonstrate that reductions in SO_2 emissions lead to rapid decreases in the acidity of impacted surface waters.

ACKNOWLEDGEMENTS

We thank Peter Dillon, Donna Wales and Frank Tomassini for constructive reviews of the manuscript, Peggy Gale for drafting figures and Cheryl Cluff for typing. Ray Labbé (Laurentian University) provided photographic reductions of the figures. This work was conducted under the Acidic Precipitation in Ontario Study (APIOS).

REFERENCES

Beamish, R.J. and Harvey, H.H.: 1972, J. Fish. Res. Board Can. **29**, 1131.
Bengtsson, B., Dickson, W., and Nyberg, P.: 1980, Ambio **9**, 34.
Conroy, N.I., Hawley, K., and Keller, W.: 1978, Extensive Monitoring of Lakes in the Greater Sudbury Area, 1974-76, Ontario Ministry of the Environment Tech. Rep., 40 p. plus appendices.
Dillon, P.J.: 1982, 'Mass Balance Models: An Explanation of the Observed Chemistry of Lakes Near Sudbury, Ontario'. In Studies of Lakes and Watersheds Near Sudbury, Ontario: Final Limnological Report, Ontario Ministry of the Environment Tech. Rep. SES 009/82, 133 p. plus figures.
Dillon, P.J., Yan, N.D., and Harvey, H.H.: 1984, CRC Critical Reviews in Environmental Control **13**, 167.
Dillon, P.J., Reid, R.A., and Girard, R.: 1986, Water, Air and Soil Pollut. (this issue).
Forsberg, C., Morling, G., and Wetzel, R. G.: 1985, Ambio **14**, 164.
Gorham, E. and Gordon, A. G.: 1960, Can. J. Bot. **38**, 477.
Haines, T.A.: 1981, Trans. Am. Fish. Soc. **110**, 669.
Harvey, H.H., Pierce, R.C., Dillon, P.J., Kramer, J.R., and Whelpdale, D.M.: 1981, Acidification in the Canadian Aquatic Environment, Publication NRCC No. 18475 of the Environmental Secretariat, National Research Council of Canada.
Hasselrot, B. and Hultberg, H.: 1984, Fisheries **9**, 4.
Hutchinson, T.C. and Havas, M.: 1986, Water, Air and Soil Pollut. (in press).

Keller, W., Gunn, J.M., and Conroy, N.I.: 1980, 'Acidification Impacts on Lakes in the Sudbury, Ontario, Canada Area' In D. Drablos and A. Tollan (eds.) Ecological Impacts of Acid Precipitation, Proc. Int. Conf. Ecol. Impacts Acid Precip., Sandefjord, Norway, pp. 228-229.

Keller, W. and Gunn, J.M.: 1982, Experimental Neutralization of a Small Seasonally Acidic Stream Using Crushed Limestone, Ontario Ministry of the Environment Tech. Rep. APIOS 004/82, 24 p.

Keller, W. and Pitblado, J.R.: 1986, Water, Air and Soil Pollut. (in press).

Kerekes, J. and Pollock, T.: 1983, Can. J. Fish. Aquat. Sci. **40**, 2260.

Kretser, W.A. and Colquhoun, J.R.: 1984, Fisheries **9**, 36.

LaZerte, B.D. and Dillon, P.J.: 1984, Can. J. Fish. Aquat. Sci. **41**,1664.

MOE (Ministry of the Environment): 1979, Determination of the Susceptibility to Acidification of Poorly Buffered Surface Waters, Tech. Rep., 21 p.

Molot, L.A., Hamilton, J.G., and Booth, G.M.: 1984, 'Ontario's Experimental Lake Neutralization Project: Calcite Additions and Short Term Changes in Lake Chemistry', In Lake and Reservoir Management, EPA 440/5/84-001, pp. 356-359. U.S. Environmental Protection Agency, Washington, D.C.

OCTF (Ontario/Canada Task Force): 1982, Report of the Ontario / Canada Task Force for the Development and Evaluation of Air Pollution Abatement Options for Inco Limited and Falconbridge Nickel Mines Limited in the Regional Municipality of Sudbury, Ontario, Intergovernmental Task Force Report, Toronto, Ontario, 255 p.

Pitblado, J.R., Keller, W., and Conroy, N.I.: 1980, J. Great Lakes Res. **6**, 247.

Rosseland, B.O. and Skogheim, O.K.: 1984, Fisheries **9**, 10.

Watt, W.D., Scott, C.D., and White, W.J.: 1983, Can. J. Fish. Aquat. Sci. **40**, 2261.

White, W.J., Watt, W.D., and Scott, C.D.: 1984, Fisheries **9**, 25.

Wright, R.F.: 1985, Can. J. Fish. Aquat. Sci. **42**, 1103.

Yan, N.D. and Dillon, P.J.: 1984, 'Experimental Neutralization of Lakes Near Sudbury, Ontario', In J. Nriagu (ed.) Environmental Impacts of Smelters, John Wiley and Sons Inc., pp. 417-456.

THE CASE FOR LIMING SOME NOVA SCOTIA SALMON RIVERS

Walton D. Watt
Fisheries Research Branch
P.O. Box 550
Halifax, B3J 2S7
CANADA

ABSTRACT. Long range transport of H_2SO_4 has caused many Nova Scotian rivers to decline in pH to the point where their Atlantic salmon (Salmo salar) stocks have been destroyed or much diminished. Chemical records show a declining pH trend in N.S. rivers since the early 1950's. The annual variation in hydrogen ion activity is positively correlated with sulphate and Al and negatively correlated with organic anions. It is technically feasible to restore the acidified salmon habitat by the addition of limestone, and the total cost of mounting a liming program to restore the lost habitat has been calculated. The pre-acidification Atlantic salmon production capacity was estimated from physical habitat surveys and tag return data. The estimated costs of the liming program are much greater than the anticipated economic benefits of the salmon restoration. Present plans are for a small liming program to establish a series of refuges for the preservation of nuclei of native salmon stocks.

1. THE CHEMISTRY OF ACIDIFICATION

Reports of declining pH's of precipitation reported for eastern Canada led the Department of Fisheries and Oceans to examine surface water data in the Maritime Provinces for indications of acidification that might impact on the freshwater fisheries resource. A resurvey by Watt et al. (1979) of 22 Nova Scotian lakes which were first examined in 1955 by Gorham (1957) showed that 16 of these lakes were still in a relatively undisturbed state except for atmospheric input and that all 16 had significantly lower pH's. During 1978-80, a survey of water chemistry in the Atlantic salmon rivers of the Maritimes area (Farmer et al., 1980; Watt 1981) revealed that the only severely acidified area in the Maritimes was the Southern Upland zone of Nova Scotia (Figure 1). This zone comprises approximtely one-half of the total area of the province, and is a region of shallow soils and poor drainage, underlain by granites and metamorphic rocks lacking in basic minerals. The area

Water, Air, and Soil Pollution **31** (1986) 775–789.

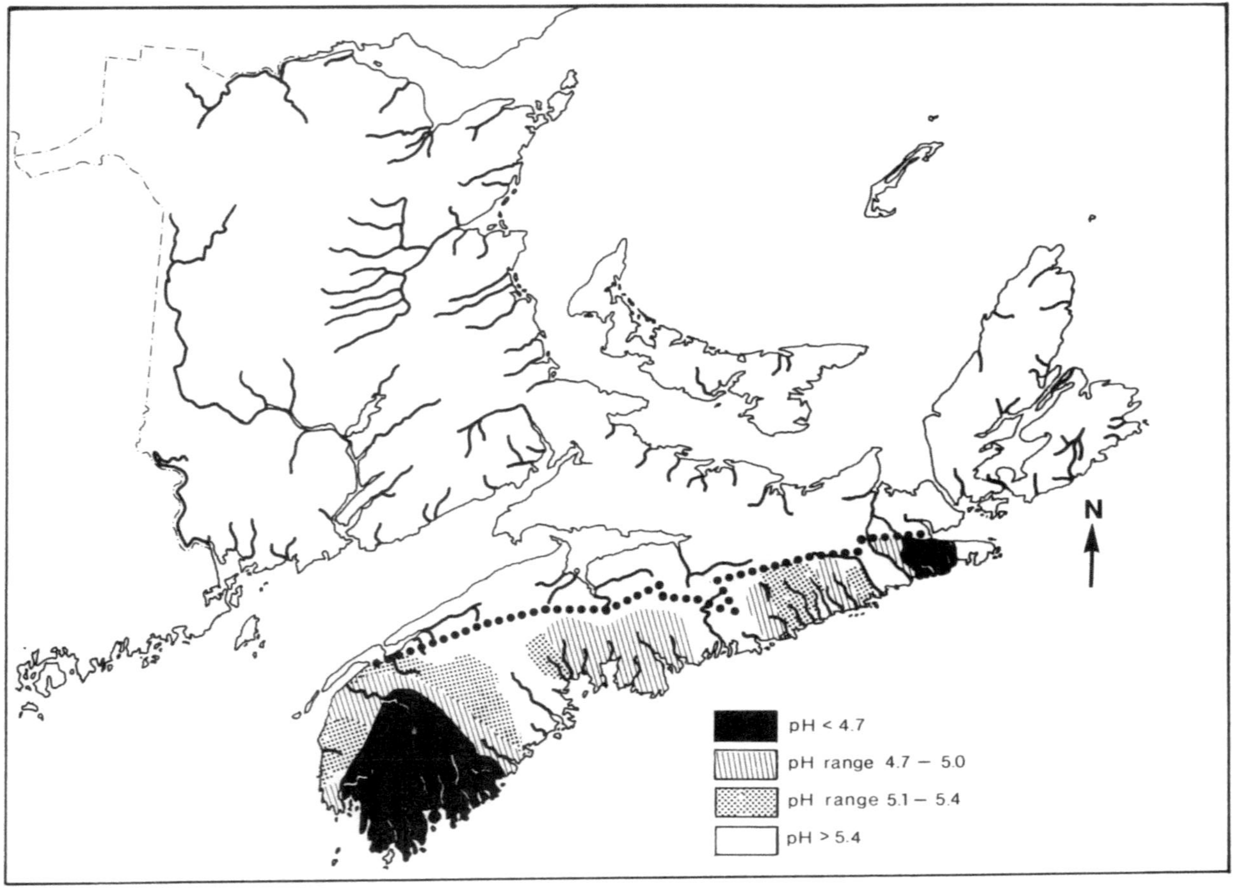

Fig. 1. Surface water surveys (1979-85 data) show that the mean annual pH of Atlantic salmon rivers in the Maritime Provinces are above 5.4 (considered the danger threshold for salmon), except in Nova Scotia's Southern Upland, which lies south of the dotted line.

Table I

The Southern Upland rivers listed below contain habitat that is physically suited to the rearing of juvenile Atlantic salmon, and have been taken into consideration in this report. In category 1 rivers the native salmon runs are extinct. In category 2 rivers only remnant populations survive in one or two higher pH tributaries. In category 3 rivers salmon stocks have been depleted by acidification of some of the smaller tributaries, but in the main stem and most tributaries salmon production rates are normal. In the category 4 rivers Atlantic salmon stocks have not yet been affected by acidification.

Category	Mean pH (main stem)	N	Rivers	
1	<4.7	13	Argyle	Patterson
			Barrington	Pennant
			Broad	Roseway
			* Clyde	Sable
			Jordan	* Salmon (Echo)
			Larrys	Tidney
			* Nine Mile	
2	4.7-5.0	18	Cole Harbour	* ** Mersey
			* East (Lunenburg)	* Middle
			* East (Sheet Harbour)	Rocky Run
			East (St. Margarets)	Sackville
			** Indian (Halifax)	Salmon (Major)
			* Ingram	** Sissiboo
			* Issacs Harbour	* Tangier
			* Liscomb	Three Fathom
			Little East	* Tusket
3	5.1-5.4	13	Annis	* Moser
			Chezzetcook	New Harbour
			* Gaspereau	* Salmon (Digby)
			* Gold	Salmon (Jeddore)
			* Kirby	* Salmon (Dufferin)
			* LaHave	* West
			* Medway	
4	>5.4	16	Belliveau	Mushamush
			Boudreau	* Musquodoboit
			Chegoggan	Necum Teuch
			Country Harbour	* Petite
			* Ecum Secum	* Quoddy
			Indian (Gysborough)	* Ship Harbour
			Martins	* St. Marys
			** Meteghan	Taylor

* For these rivers angling records are available from 1936 to the present.

** In these rivers most of the Atlantic salmon habitat is inaccessible due to impassable dams. Inacessible portions of these drainages were excluded from the analysis in estimating habitat availability and limestone requirements.

contains 60 rivers known to contain habitat physically suitable for Atlantic salmon (Salmo salar) (Table I) of which 13 have mean annual pH's less than 4.7, 18 have pH's in the range 4.7 to 5.0, 13 have pH 5.1 to 5.4 and 16 are of pH greater than 5.4 (1979-85 data).

Data from the daily sampling of East River, N.S. have confirmed the seasonal pattern of river pH variation as reported by Watt (1981) and Watt et al. (1983a) from monthly sampling. Minimal pH's typically occur in winter and then rise gradually throughout spring and summer until the onset of the autumn freshet when pH falls rapidly. The range the range of seasonal variation is always less than 1 pH unit. The total range of pH variation in East River in 1983 was 0.65 pH units from the low of pH 4.77 on March 5 to 5.42 on November 1. The steepest rate of decline was in the period November 1-12 when pH fell 0.32 units (average decline of 0.03 units per day). The greatest 24 hr variations observed were a fall of 0.17 pH units on November 10-11, 1983 and a rise of 0.20 pH units which occurred on April 21-22, 1984. In 1984 the minimum pH (4.63) occured in January, and the pH maximum (5.18) in October. No evidence has been encountered in monthly or daily data of significant episodic pH depressions under snowmelt conditions.

Previous data (Watt et al., 1983a) were inadequate to decipher the relationships between pH and flow beyond that of a simple straight line regression model. Daily pH data (1983-84) from East River, N.S. vs instantaneous flow rates at the time of sampling now reveal that this curve is actually an "S" shaped polynomial. At higher flows the H^+ concentration approaches an upper limit, and so very high episodic flows do not cause extreme pH depressions. For the East River data the highest flows occurred in March and April (1984) after heavy rainfall. Under these conditions the H^+ ion concentration remained at about 20 μeq L^{-1} (pH 4.7). This limit was probably set by the pH of the early spring rainfall which was also near 4.7. Similarly, under low flow conditions, the decline in H^+ activity is not linear, but appears to be approaching a lower limit at extreme low flows. The lower limit, which is near pH 5.3 to 5.4, appears to be representative of the pH of groundwater discharge.

For the East River, N.S. data the H^+ is correlated (at the less than 0.001 significance level) positively with excess sulphate and total aluminum and negatively with organic acid anions. A multiple regression model on excess sulphate (XSS04), total Al and organic acid anion (OA) concentrations will account for 80.4% of the H^+ variation (concentrations in eq L^{-1}):

$$[H^+] = 6.64 + 0.0975\ [XSSO_4] + 0.223\ [Al] - 0.379\ [OA].$$

The natural causes behind this regression are readily apparent: the strong positive regression of H^+ on excess sulphate is because sulphate is the major source of the strong acid and the positive regression on total Al is due to mobilization of that metal by high levels of strong acid; the negative regression on organic acid anions (calculated as per Oliver, 1982; Oliver et al., 1983) is due to reduced dissociation of the relatively weak organic acids.

For five rivers (Roseway, Medway, Mersey, LaHave, and Musquodoboit, see Table I), the monthly data collected in 1980-81 can be compared to similar monthly data collected by Thomas (1960) in 1954-55. All five show a fall in pH over this 26 year period (Watt *et al.*, 1983a), though only for the four lower pH rivers is the decline statistically significant (p less than 0.01). Unfortunately, two points in time are not adequate to prove a trend and for most sites that is all that we have at present. By a fortunate coincidence, however, Thomas' (1960) site on the Medway River was also sampled monthly during the period 1965-78 by Environment Canada as part of a surface water quality monitoring program, and the combined data clearly show a declining pH trend (Watt *et al.*, 1983a).

The chemical observations from Nova Scotia's Southern Upland rivers appear to support the hypothesis of acidification by long range transport of H_2SO_4. Krug and Frink (1983) have suggested that some of the observed acidification of surface waters could result from increased production of organic acids in the soils as a result of changes in land use, most notably reforestation. There are several objections to the organic acids hypothesis as an explanation for the acidification of Nova Scotia's rivers:

1. Watt *et al.*, (1983a,b) report color values in N.S. rivers significantly lower than Thomas' (1960) values for the same sampling sites. Color is very well correlated with dissolved organic C concentrations (Watt *et al.*, 1983b) and color is still measured with the same procedure employed in 1954-55 by Thomas (1960). The only anion increase noted by Watt *et al.*, (1983a) was in sulphate.
2. Daily chemical data from East River (N.S.) show a negative correlation between H^+ and organic anions, i.e., organic anion concentrations tend to be highest during times of highest pH.
3. In all chemically monitored N.S rivers the lowest pH's occur after winter rainstorms when the ground is frozen and runoff is direct. Under these conditions the runoff pH is at or below that of the precipitation. When the ground is not frozen, the runoff pH is always higher than that of the precipitation.
4. Reforestation is not a significant phenomenon in the Southern Upland, though heath formation (deforestation), due to forest fires in the late nineteenth and early twentieth centuries, has been noted in the Clyde and Roseway drainages.

2. ATLANTIC SALMON TOXICITY

Watt *et al.* (1983a) reported that in their electrofishing survey of 19 rivers in the Southern Upland, juvenile Atlantic salmon were never found in waters of pH less than 5.0. In waters of pH 5.0 to 5.4, juvenile populations were low, but above pH 5.4 juveniles were present at normal population densities.

In a laboratory study of the effect of low pH's on salmon embryos, Daye (1980) reported a lethal level below pH 4.5 The hatching success of Atlantic salmon embryos incubated in the natural substrate of acidic rivers (pH 4.5 to 5.0) is positively correlated with the pH of the interstitial water percolating through the substratum, and Lacroix (1985a) reported an LL50 at pH 4.7. Emergence of salmon fry from natural redds is also reduced at pH levels below 4.8, indicating a low survival of alevins after hatching.

Studies reported by Farmer et al. (1980) and Lacroix et al. (1985) have shown that the most acid-sensitive stage in the life cycle of the Atlantic salmon is the fry emergence stage, when feeding is initiated. At this stage, in acid rivers of pH 5.0, mortality is about 50% (range 30 to 70%). The fry appear to die of starvation due to inability to initiate exogenous feeding at low pH's. Responses to pH stress at this phase of the life cycle also include decreased swimming and feeding activity, severe emaciation, growth retardation, and decreased body ionic content. An increase in the pH to 6.0 by treatment of the river water with limestone removed all adverse effects and reduced mortality to 5% (Lacroix et al., 1985).

Physiological responses determined in juvenile salmon inhabiting acidic rivers of Nova Scotia at pH levels of about 5.0 are increased hematocrit level and decreased plasma Na and chloride concentrations. Laboratory "models" of acid exposure in soft water (Lacroix, 1985b) indicated that salmon smolts are more sensitive to low pH than parr, and young parr (0+ age class) are more sensitive than older parr. The overwintering mortality rate of young parr is generally much higher in acidic rivers where pH levels are below 5.0 from December to April. Smoltification is impaired at pH levels below 5.0, and mortalities are high when salmon smolts are chronically exposed to pH levels below 4.7 (Saunders et al., 1983; Johnston et al., 1984).

Total Al levels are typically high (more than 100 $\mu g\ L^{-1}$) in acidified Nova Scotian rivers, but Al toxicity is not yet a common problem because most rivers in Nova Scotia still have concentrations of dissolved organic matter in excess of 5 mg L^{-1}, which chelates the ionic aluminum and so detoxifies it (Lacroix, 1985a,b; and Lacroix et al., 1985). Hutcheson and Odense (1984) report that, under conditions of high total Al and low organic matter such as occurred in Sandy Lake, N.S. prior to liming (White et al., 1984), Atlantic salmon fry died of Al toxicity at pH 5.5.

With adult fish, depressed levels of the androgenic hormones testosterone and 11-ketotestosterone were found in male Atlantic salmon caught in the acidified Westfield River at pH 4.7 (Freeman et al., 1983), whereas levels of these hormones in salmon caught in the Medway River at pH 5.6 were much higher. Freeman and Sangalang (1985) report that metabolism of these and other steroid hormones was abnormal in caged salmon held in the acidified Westfield River but was normal in a similar group of salmon caged in the Medway River. The abnormalities in hormone metabolism in salmon from acidified waters resembled those seen in other vertebrates subjected to stress. Eggs taken from salmon caged in the more acidic river were less viable than eggs obtained from fish held in the Medway, although both groups were incubated at a pH of 5.6 (Freeman and Sangalang, 1985).

All of the foregoing toxic effects of acidification on Atlantic salmon appear to be fully reversible simply by raising the pH of the rivers.

3. N.S. SALMON ANGLING TRENDS

There are 60 rivers flowing through the Southern Upland area of Nova Scotia that have the physical potential to support Atlantic salmon stocks. Watt et al. (1983a) have described the angling trends for 22 of the rivers for which angling data is available since 1936 and whose drainage basins still remain very similar to their condition as it was 50 years ago. The major industry is still forestry and, though the frequency of cutting has increased due to a change over from lumber to pulp as the final product (which has resulted in younger forests), the extent of forest cover is still much the same. Agriculture has never been a significant industry on the impoverished Southern Upland soils, and industrialization and urbanization are coastal phenomena occurring near the rivers' mouths.

The rivers can be divided into four groups based on their present (1985) mean annual pH; three of pH less than 4.7, seven of pH 4.7-5.0, six of pH 5.1-5.4 and six of pH greater than 5.4. For rivers presently of pH less than 4.7, the angling record ends in the 1970's. Electrofishing in ten of the rivers, where pH is less than 4.7, has failed to discover any surviving juvenile salmon. We have concluded that for rivers at this pH level the salmon runs are now extinct. This conclusion accounts for 13 of the 60 rivers in the Southern Upland (Table 1). For rivers in the pH range 4.7 to 5.0 the angling catch has declined to about 10% of levels prevalent during the 1936 to 50 period. Electrofishing data from twelve of the 18 rivers in this pH range indicates the presence of small populations of salmon juveniles, commonly in higher pH tributaries which function as natural refuges. The decline in salmon returns was simultaneous in the rivers of pH less than 4.7 and in the pH 4.7 to 5.0 rivers. Between 1950 and 1960 both of these lower pH river groupings suffered a 60% decline in angling removals. The simultaneous decline in adult salmon returns suggests a large increase in acid deposition during the 1945-55 decade (the generation time of Atlantic salmon in this area of Nova Scotia is 4 to 5 yr) sufficient to overwhelm acid neutralization and buffering capacity in these rivers. Although several of the lowest pH rivers are exceptionally rich in organic acids which may provide some slight buffer capacity, pH's in these rivers were probably already (in 1945) marginal for Atlantic salmon survival, hence only a slight drop in pH was required to create lethal conditions. The rivers presently within the pH range 4.7 to 5.0, have slightly greater acidification resistance, attributable to cation exchange capacity in the somewhat less boggy soils of their drainages.

The rivers in pH categories 5.1 to 5.4 and greater than 5.4 show no sign, yet, of an impact of acidification on angling returns, though for 13 rivers in pH range 5.1 to 5.4 there is electroseining and toxicity evidence of acidification impacts limiting juvenile salmon production in the lower pH tributaries (pH near 5.0) of these rivers (Lacroix, 1985a).

4. SALMON PRODUCTION CHANGES DUE TO ACIDIFICATION

The available data for an estimate of pre-acidification production potential are:

(1) Forty-eight years of angling returns from 22 undisturbed (physically) rivers flowing through the Southern Upland.

(2) Twelve years of tag returns for hatchery smolts stocked in the LaHave River, from the various distant and local fisheries, and from fishway traps (Cutting and Gray, 1984).

(3) Total physical habitat estimates for 14 of the rivers (Anon., 1978; 1985) including four rivers, with pH's greater than 5.0, that have not been physically altered over the past 50 years, and for which angling data are available from 1936 to the present. The physical habitat potential of the other rivers can be estimated from comparisons with similar nearby rivers where habitat surveys have been conducted.

The LaHave tag return data are from the the years 1973-84, and from these it is possible to divide the total stock into proportions accounted for by the various fisheries. Total stock is defined here as the sum of the Greenland, Newfoundland, and Nova Scotian commercial fisheries catches, the angling fishery, poaching, and the spawning escapement. The total breakdown (12 yr average) is Greenland 8.4%, Newfoundland 9.7%, Nova Scotia commercial 29.8%, river angling 13.3%, and spawning escapement and poaching 38.8%.

The Medway, Petite, Gold and St. Marys rivers have mean annual pH's greater than 5.0; and, although there is acidification impact in some tributaries of the Medway and Gold, this has not yet reached the level where it is significantly impacting the angling catch. The total angling catch for these four rivers shows no trend over the period 1936-83. From the estimates of total salmon habitat area in these drainages, and using the average angling catch for 1936-83 to which we can apply the LaHave River value that the angling fishery accounts on average for 13.3% of the total stock, we can estimate a total stock yield per unit of habitat for unacidified rivers. For these rivers the estimated average total stock yield was 2,200 fish (salmon + grilse) yr^{-1} km^{-2} of salmon rearing habitat.

The total accessible habitat area physically suitable for Atlantic salmon in the Southern Upland rivers is estimated at 17.17 km^2. Of this amount, 9.91 km^2 has pH's greater than 5.0 and can be expected to yield an average of 2,200 fish yr^{-1} km^{-2}. There are 3.78 km^2 of habitat in rivers with mean annual pH's in the 4.7 to 5.0 range and salmon production is estimated at 220 yr^{-1} km^2 (angling data indicate production at about 10% of the pre-acidification rate). The remaining 3.48 km^2 of habitat are in rivers with pH's below 4.7, and salmon production is zero (evidence from angling and electrofishing).

Given that the above analysis is approximately correct, then the total annual salmon and grilse production from the Southern Upland rivers should presently be about 22,700 fish yr^{-1}. The total physical potential of the 17.17 km^2 of salmon habitat (if none of it were acidified) is 37,800 fish yr^{-1}.

As a check on the accuracy of the above calculations, the calculated angling fishery for the total Southern Upland may be compared to the actual angling catch. The angling fishery yield from wild stocks in the Southern Upland rivers should be approximately 13.3% of 22,700 which is 3,019 angled fish yr^{-1}. The recorded angling catch from this area (Salmon Management Zones 7 and 9) averaged 3,558 fish yr^{-1} for the period 1970-83. (O'Neil and Swetnam, 1984; Swetnam and O'Neil, 1984). The actual angling catch includes returns from hatchery stocking which during the period 1976-79 contributed an average of 478 fish per year to the angling catch (McPhee et al, 1981). The corrected actual angling catch is thus 3,080 fish yr^{-1} of wild fish from the Southern Upland rivers, which compares favorably with the 3,019 estimate.

Taking the ratio of salmon habitat area to drainage area, and expressing it as the number of m^2 of salmon habitat available per km^2 of drainage (third row of the following tabulation); the data indicate an anomalously low value for habitat per unit of drainage in the lowest pH category.

pH Category:	< 4.7	4.7-5.0	5.1-5.4	> 5.4
km^2 Salmon Habitat:	3.48(6.86)	3.78	5.38	4.53
km^2 Drainage:	6,240	3,510	4,820	4,250
m^2 Habitat/km^2 Drainage:	560	1,080	1,120	1,070

It is difficult to escape the suspicion that, over the years in the absence of detailed river survey information, fisheries biologists have compensated for the lack of salmon production in what we now know to be acidified rivers by underestimating the habitat availability. On this basis, it is quite plausible to argue that the true ratio of habitat area to drainage area is approximately 1,100 m^2 of habitat per km^2 of drainage throughout the Southern Upland. From this assumption, a corrected value for salmon habitat in what are now the lowest-pH rivers is 6.86 km^2. This correction has no effect on the estimate of current production level (22,700 fish yr^{-1}); but it does raise the estimate of total production potential, in the absence of acidification to 45,200 fish yr^{-1} (from 37,800) from the 20.6 km^2 of available salmon habitat.

5. LIMING FEASIBILITY STUDIES

The Department of Fisheries and Oceans has undertaken to test the feasibility of deacidifying Atlantic salmon habitat by addition of limestone or other substances to lakes or streams (Watt et al., 1984; White et al., 1984).

Table II

Liming doses (as multiples of whole lake acidity and as mass per unit area) and deacidification responses from five Nova Scotian headwater lake treatments. All five lakes have mean residence times in the 4 to 6 mo range, and median particle diameter of the limestone was 35 μm in all cases.

Lake Treatment #	1*	2*	3*	4**	5**
Dose (x acidity)	0.7	2.0	2.7	3.3	8.2
Dose (tonnes/ha)	0.2	0.6	0.4	1.7	2.2
% dissolved after 2 yr	62	50	47	59	45
Pre-liming pH	4.6	4.4	4.6	4.8	4.5
	Outlet Stream pH				
One week after liming	5.3	5.6	6.5	6.9	6.7
After 6 months	4.4	4.5	5.5	5.2	6.4
After 1 year	4.5	4.5	4.7	5.4	5.7
After 2 years	4.5	4.4	4.5	5.1	5.2

* Treatments 1 to 3 were with dolomitic limestone (11% Mg) slurried with lake water and sprayed evenly over the lake surface.

** Treatments 4 and 5 were with calcitic limestone (1.5% Mg), sprayed as a slurry in treatment 4, and dry powder spread over the ice in treatment 5.

The experiments conducted to date indicate that the pH of salmon streams can be adjusted to satisfactory levels (pH above 5.0) by liming, but that fresh limestone must be added annually and in some cases, twice annually. Different liming methods have been tested and estimates have been made of the relative costs and effectiveness. The two liming approaches most thoroughly investigated to date are the use of instream limestone gravel deposits, and the liming of headwater lakes.

Watt *et al.* (1984) reported on 3 yr of experiments with instream limestone gravel. The effectiveness of instream limestone gravel is inversely related to flow and is significantly reduced at low temperatures by lower dissolution rates. Under winter conditions, the tonnages of limestone that would theoretically be required to ensure satisfactory pH levels are so high as to be impractical.

Satisfactory levels of pH can be achieved in salmon streams if the headwater lakes are treated with powdered limestone doses of more than 3 times the lake acidity (Table II). Because of the low mean residence times in most Nova Scotian lakes, retreatment would be required on an annual basis. A major problem with this approach, as was noted by White *et al.* (1984) and Watt *et al.* (1984) has been the advent of midwinter rainstorms, the runoff from which accumulates as a low pH surface layer on the limed lakes and delivers a low pH shock to the salmon juveniles in the outlet streams. During the winter of 1983-84, this problem was overcome by spreading a layer of powdered limestone over the ice (Table II). When heavy rainstorms occurred, the first runoff entering the lake was from off of the ice, and this in turn caused the surface pH of the lake and the pH of the outlet stream to rise and remain up in spite of very low pH's in the inlet streams.

This has also led to the discovery that winter liming with tractors is faster and cheaper than summer liming from boats, through further monitoring is required to ascertain how much the dissolution efficiencies (near 60% with summer liming) are affected.

The headwater lake liming technique has proven feasible in the field, but this technology has only been tried on a pilot scale, and its practicality has yet to be proven in a full-scale demonstration project. Only about 70% of the acidified salmon habitat in Nova Scotia could be treated with this liming technique, because of the requirement for suitable headwater lakes with road access. To do even 70% of the acidified habitat with this technique will require considerable seasonal road construction to achieve lake access.

Additional access road requirements were estimated for five rivers (Clyde, Medway, East(Lunenburg), Ingram and Tangier). An average value for the five comes to 4.1 km^2 of seasonal road per 100 km^2 of drainage. Seasonal road construction (including stream crossings) presently costs about $4,000 km^{-1}. An alternative to road construction would be to deliver the lime to the lakes by air, which has been found to cost about twice as much as liming from a boat when road access is available.

For the approximately 30% of acidified habitat that is not amenable to headwater lake liming there is another alternative. Most of this habitat could be treated by continuous delivery of limestone to the river from a storage facility (silo). The new road construction for this operation would be about the same number of kilometers per 100 km^2 as for headwater lake liming, but these would have to be all weather roads, at an approximate cost of \$9,000 km^{-1}. In addition, there would be a requirement for upgrading of existing seasonal roads to all-weather status at a cost of approximately \$5,000 km^{-1}. The road upgrading requirement would be for approximately 12 km of seasonal road per 100 km^2 of treated drainage. Also, for the operation of a silo system, electric power would be required at the site; hence about 20 km of power line construction will be required for each 100 km^2 of treated drainage. Silo construction cost for a unit capable of delivering 200 tonnes yr^{-1} (at varying seasonal rates) would be about \$70,000. This would likely be the most practical size for the average installation.

In addition to the actual construction costs estimated above; there will be annual maintenace costs for roads, powerlines and equipment (snowplows, etc.), and in the case of the silos, a requirement for replacement after about 10 yr.

The total quantity of limestone per square kilometer of drainage that would be required for de-acidification (pH's above 5.0) has been worked out for the same five representative rivers examined for access road requirements, and average values from these have been applied to all rivers within the pH category (Table I).

The total drainage area with accessible habitat wherein surface water pH's average less than 4.7 and salmon runs are extinct, is 6,340 km^2. These rivers would require, on average, about 6 tonnes of limestone per km^2 of drainage per year to de-acidify them (maintain pH greater than 5.1), so that Atlantic salmon could again utilize this habitat. The total tonnage requirement is 37,400 tonnes per year. The salmon restoration would require an extensive restocking operation. The shortage of suitable donor stocks would probably dictate a large-scale hatchery program.

The drainage area of accessible salmon habitat in rivers with average pH's in the range 4.7 to 5.0 is 3,510 km^2. Rivers in this pH range would require about 4 tonnes of limestone km^{-2}, for a total requirement of 14,000 tonnes yr^{-1}. Most of these rivers would recover by natural reproduction of the surviving remnants of native runs that are still present. If left to themselves, however, a full stock recovery would require at least four generations of natural reproduction, or about 20 yr. It might be more economical to hasten the recovery by initiating a restocking program (perhaps hatchery based), and/or prohibiting salmon fishing during the recovery period.

Over the next 20 yr it is likely that most rivers presently in the pH range 5.1 to 5.4 will require some liming to maintain present production levels in the face of continuing high levels of acid input. There are 4,820 km^2 of drainage in this pH category and the necessary maintenance dosage would probably average about 1 tonne km^{-2} of drainage, thus requiring about 4,800 tonne of limestone yr^{-1}. This treatment should probably be phased in gradually over the 20 yr period. This assumes that there will be no dramatic increases (or decreases) in the present levels of acid input.

6. TOTAL COSTS AND BENEFITS OF LIMING

The estimated cost of road construction to provide seasonal access for headwater lake liming and all-weather access for resupplying the silos is $\$17 \times 10^6$ (1984 \$ Can.). The estimated cost for limestone trucking and spreading in the headwater lake treatment is $\$1.1 \times 10^6\ yr^{-1}$. The capital cost of lime dispensing silos on rivers where headwater lake treatment is impractical is $\$16 \times 10^6$, and the operating cost would be about $\$0.7 \times 10^6\ yr^{-1}$. Replacement time for silos is estimated at 10 years. Project management and scientific monitoring (biological and chemical) are estimated at $\$0.5 \times 10^6\ yr^{-1}$.

The total estimated cost for a 20 year project of deacidifying the Atlantic salmon habitat of the Southern Upland of Nova Scotia is $\$95 \times 10^6$ (1984 \$ Can.). This total figure includes the capital cost for road construction, silo construction and replacement, annual limespreading, silo operation costs, and the costs of a modest monitoring program.

The aim of the production effort would be to bring the Atlantic salmon production level of the Southern Upland back up to the pre-acidification level; which is estimated at 45,200 adult salmon (total stock) per year. The actual annual Canadian catch would be about 24,000 salmon (Newfoundland and Nova Scotia commercial, plus angling). This represents an enhancement of 12,000 salmon above the present average catch, at an average annual cost over 20 yr of $\$4.75 \times 10^6$. These costs amount to approximately \$400 per restored salmon. The value per landed salmon to the eastern Canadian economy is, on average, less than \$100 per fish (McPhee et al., 1981), hence the liming operation cannot be justified on economic grounds.

7. LIMING FOR GENETIC PRESERVATION

A large area of Nova Scotia has already been rendered barren of Atlantic salmon, and more is expected to be lost (most of what is presently in the pH range 4.7 to 5.0) over the next 20 yr. After the acid rain problem is brought under control, it will be necessary to mount a salmon restoration program for the barren habitat.

Each river has its genetically unique stocks of Atlantic salmon. The loss of a native salmon stock is irreversible. Salmon can be restored to empty habitats, but the process is lengthy (about 10 yr) and the probability of success is not high unless a suitable nearby donor stock is available. The farther afield the donor stock comes from, the lower are the chances for a successful transplantation. The job of restoration would be easier and entail a higher probability of success if a number of nuclei of local native wild stocks were available to provide a genetically diverse selection of potential donor stocks.

To achieve this gene bank, a small number (three to six) of the stocks currently threatened with extinction should be preserved by creating de-acidified refuges. This goal could be accomplished by applying the headwater lake liming technique to a selected tributary in each river. The average refuge would comprise about 50 square kilometers of drainage.

The costs of establishing (and maintaining) such a refuge would be similar (on an area basis) to those entailed in the wholesale liming operation discussed previously, though some economies of scale would be absent. Since the object of this program would be to maintain genetic diversity in the wild salmon stocks of the Southern Upland of Nova Scotia, it cannot be evaluated on the basis of a simple cost benefit estimate, and it cannot be compared with enhancement projects.

REFERENCES

Anon: 1978, Harvesting the Resource, Biological Conservation Subcommittee Report, prepared for the Atlantic Salmon Review, Department of Fisheries and Oceans, Halifax, N.S., 194 pp.

Anon: 1985, Anadromous, Catadromous and Freshwater Fisheries Subcommittee report to the Canadian Atlantic Fisheries Scientific Advisory Committee, CAFSAC Subcommittee Report 85/2, Department of Fisheries and Oceans, Halifax, N.S., 12 pp.

Cutting, R.E. and Gray, R.W.: 1984, Assessment of the status of the Atlantic salmon stocks of the LaHave River, Nova Scotia. CAFSAC Research Document 84/40. Department of Fisheries and Oceans, Halifax, N.S., 44 pp.

Daye, P.G: 1980, Can. J. Fish. Aquat. Sci. 37, 1,035.

Farmer, G.J., Goff, T.R., Ashfield, D.O., and Samant, H.S.: 1980, Some effects of the acidification of Atlantic salmon rivers in Nova Scotia. Can. Tech. Rep. Fish. Aquat. Sci. No. 972: 13 pp.

Freeman, H.C., Sangalang, G.B., Burnes, G., and McMenemy, M.,: 1983, Sci. Tot. Env. 32: 87.

Freeman, H.C., and Sangalang, G.B: 1985, (in press), The effects of an acid river, caused by acid rain, on weight gain, steroidogenesis, and reproduction in the Atlantic salmon (Salmo salar). Aquatic Toxicology and Hazard Assessment: Eighth Symposium, ASTM STP891, R.C. Balmer and D.J. Hansen, Eds., American Society for testing Materials, Philadelphia, p. 339-49.

Gorham, E.: 1957. Limnol. Oceanog. 2: 12.

Hutcheson, M.S., and Odense, R.B: 1984, A study of the toxicity mechanisms to Atlantic salmon of acidic lake waters with high aluminum concentrations. Report submitted to Department of Fisheries and Oceans under Contract# 1SC83-00163. Atlantic Oceanics Co. Ltd., Dartmouth, N.S., 25 pp.

Johnston, C.E., Saunders, R.L., Henderson, E.B., Harmon, P.R., and Davidson, K: 1984, Chronic effects of low pH on some physiological aspects of smoltification in Atlantic salmon (Salmo salar). Can. Tech. Rep. Fish. Aquat. Sci. No. 1294, 7 pp.

Krug, E.C. and Frink, C.R.: 1983, Science 221: 520.

Lacroix, G.L., 1985a. Can. J. Fish. Aquat. Sci. 42, 292.

Lacroix, G.L. 1985b. Can. J. Zool. 63, 2,254-2,261.

Lacroix, G.L., Gordon, D.J., and Johnston, D.L.: 1985. Can. J. Fish. Aquat. Sci. 42: 768.
McPhee, R.D., Ruggles, C.P., and Farmer, G.: 1981, Program Evaluation: Fisheries Resource Enhancement Program. Evaluation and Audit Branch, Department of Fisheries and Oceans, Ottawa, Ont., 198 pp.
Oliver, B.G.: 1982, The organic contribution to the acidity of natural waters, and the acid neutralization capacity of sediments. Proc. Workshop. Nov. 18. 1981. Kejimkujik Calibrated Catchments Program, J. Kerekes, Ed. Atlantic Reg. LRTAP Monitoring and Effects Working Group. Environment Canada, Halifax, N.S., p. 23-24.
Oliver, B.G., Malcolm, R.L., and Thurman, E.M.: 1983, Geochem. Cosmochim. Acta 47: 2,031.
O'Neil, S.F., and Swetnam, D.A.B.: 1984, Collation of Atlantic salmon sport catch statistics, Maritime Provinces, 1970-79. Can. Data Rept. Fish. Aquat. Sci. No. 481., 297 pp.
Saunders, R.L., Henderson, E.B., Harmon, P.R., Johnston, C.E., and Eales, J.G.: 1983, Can. J. Fish. Aquat. Sci. 40, 1,203.
Swetnam, D.A., and O'Neil, S.F.: 1984, Collation of Atlantic salmon sport catch statistics, Maritime Provinces, 1980-83. Can. Data Rept. Fish. Aquat. Sci. No. 450, 194 pp.
Thomas, J.F.J: 1960, Industrial water resources of Canada, Water Survey Report No. 11, the Atlantic Provinces and the Saint John River Drainage Basins in Canada, 1954-56. Department of Mines and Technical Surveys, Ottawa. 143 pp.
Watt, W.D: 1981, Present and potential effects of acid precipitation on the Atlantic salmon in Eastern Canada. In Acid Rain and the Atlantic Salmon. Int. Atl. Salmon Found. Spec. Pub. Ser. 10: 39-45. Proceedings of a conference held November 22-23, 1980 in Portland, Maine.
Watt, W.D., Farmer, G.J., and White, W.J.: 1984. Studies on the use of limestone to restore Atlantic salmon habitat in acidified rivers. In Lake and Reservoir Management. Proceedings of the third annual conference of the North American Lake Management Society (Knoxville, Tennessee, October 18-20, 1983).
Watt, W.D., Scott, D., and Ray, S.: 1979, Limnol. Oceanogr. 24, 1134.
Watt, W.D., Scott, C.D., and White, W.J.: 1983a. Can. J. Fish. Aquat. Sci. 40, 462.
Watt, W.D., Scott, C.D., and White, W.J.: 1983b. Can. J. Fish. Aquat. Sci. 40: 2,260.
White, W.J., Watt, W.D., and Scott, C.D.: 1984, Fisheries, 9, 25.

NATURAL CHANGES IN THE PLANKTONIC ROTIFERA OF A SMALL ACID LAKE NEAR SUDBURY, ONTARIO FOLLOWING WATER QUALITY IMPROVEMENTS

Hugh J. MacIsaac[1], W. Keller[2], T.C. Hutchinson[1,3] and N.D. Yan[4]

1 Department of Botany, University of Toronto, Toronto, Ontario, Canada M5S 1A1

2 Ministry of the Environment, 199 Larch St., Sudbury, Ontario, Canada P3E 5P9

3 Institute for Environmental Studies, Haultain Building, University of Toronto, Toronto, Ontario, Canada M5S 1A4

4 Ministry of the Environment, P.O. Box 39, Dorset, Ontario, Canada P0A 1E0

ABSTRACT. Swan Lake is a small, acidic, metal-contaminated lake located near Sudbury, Ontario. During the past 8 yr, the lake has experienced a substantial increase in pH, together with significant reductions in the concentrations of heavy metals and base cations. These changes were observed only after acid and metal emissions from Sudbury area smelters were reduced. The composition of the planktonic Rotifera in the lake has changed concurrent with improvements in water chemistry. Dominance of the rotifer community by the acidophile *Keratella taurocephala* has been sharply reduced in recent years, while significant increases in the densities of *Polyarthra* spp., *Chromogaster ovalis*, *Conochiloides natans* and *Trichocerca similis* have been observed.

1. INTRODUCTION

A large number of studies have documented the deleterious impact of acidification on aquatic ecosystems. In Ontario, synoptic (Sprules, 1975; Keller and Pitblado, 1984) and experimental (Schindler *et al*., 1985) investigations have revealed extensive modifications of zooplankton communities concurrent with lake acidification. However, most studies to date have been confined to examining acidification effects on macrozooplankton communities. Since rotifers often contribute significantly to zooplankton production (Makarewicz and Likens, 1979) and abundance (Roff and Kwiatkowski, 1977) and since species abundances are often affected by pH (see review in Pennak, 1978), rotifers are of particular interest to studies attempting to assess acidification-induced changes in zooplankton communities.

In the Sudbury, Ontario region many lakes have been acidic for decades owing to emissions from large-scale smelting operations. However, Sudbury SO_2 and metal emission levels have decreased dramatically since 1972 (Keller and Pitblado, 1986), due primarily to the implementation of emission abatement equipment and a general reduction in smelting activities. Some lakes in the region have already responded

Water, Air, and Soil Pollution **31** (1986) 791–797.

chemically to the decreased inputs of acid and metals (Hutchinson and Havas, 1986; Keller and Pitblado, 1986).

While the biological responses of acidic Sudbury area lakes to experimental liming are known (Yan and Dillon, 1984; Hamilton et al., 1985), natural responses of lakes following improvements in water quality have not yet been described. In this study we attempt to characterize the chemical and biological responses which have occurred since 1977 in Swan Lake, located near Sudbury. Emphasis is placed on changes which have occurred in the planktonic rotifer community. Rotifers have previously been demonstrated to comprise an important component of the Swan Lake zooplankton (Yan and Geiling, 1985).

2. MATERIALS AND METHODS

Swan Lake is located 15 km southwest of Sudbury. The lake has an area of 5.78 ha, a volume of 1.62 $m^3 * 10^5$ and a maximum depth of 8.8 m (MOE, 1982). Whole lake water samples for chemical analyses were collected from a central station on a twice monthly basis during the ice-free season, and were volume-weighted according to lake morphometry. Chemical analyses were conducted following the methods described by Ontario Ministry of the Environment (MOE, 1981).

Planktonic rotifer samples were collected bi-weekly in 1977, 1982, 1983 and 1984 with a 34 L Schindler-Patalas trap. The trap was outfitted with 76 um mesh Nitex netting in 1977, and 35 um netting during other years. Samples were collected from June through September at depths of 1, 4 and 7 m. All samples were enumerated using a 1 ml Sedgewick-Rafter cell, except 1983 samples which were counted using sedimentation tubes. Median rotifer densities were derived from monthly mean values, which were calculated from mean depth densities. Rotifer biomass was calculated assuming dry weight to represent 7% of wet weight. Dry weight data are summarized in MacIsaac (1986).

Parametric and non-parametric (Kruskal-Wallis) one-way analysis of variance (ANOVA) tests were performed on chemical parameters and rotifer species densities respectively, to determine the significance of differences between years.

3. RESULTS

The water chemistry of Swan Lake has changed considerably since 1977, concurrent with the observed reductions in emissions from Sudbury area smelters. For example, mean lake pH has risen from 4.0 in 1977 to 5.1 in 1984 (Table I; also see: Dillon et al., 1986). The greatest changes in pH were observed between 1977 and 1982, although mean values have continued to rise slowly.

The concentration of sulphate in the lake decreased significantly ($P < 0.001$) from 583 ueq L^{-1} in 1977 to 182 ueq L^{-1} in 1984 (Table I). A dramatic reduction was also observed in the conductivity of the lake ($P < 0.001$); Table I). Similarly, significant decreases in concentration occurred between 1977 and 1984 for Cu, Ni, Al, Mn, Zn ($P < 0.001$; Table I), Ca, Mg, K and Na. The decrease in concentrations of heavy metals were of greater magnitude than those observed for the base cations.

Decreases in concentrations of Cu, Ni and Al were especially great.

In addition to dramatic chemical changes, both lake transparency and color have changed significantly ($P < 0.05$; Table I). Secchi disc transparency values decreased from 6.7 m in 1977 to 4.1 m in 1984. Lake color increased from 8.1 to 16.5 Hazen units between 1982 and 1984.

Table I.
Selected chemical characteristics of Swan Lake between 1977 and 1984.

Parameter	1977 $\bar{X} \pm SD$	n	1982 $\bar{X} \pm SD$	n	1983 $\bar{X} \pm SD$	n	1984 $\bar{X} \pm SD$	n
pH	4.0 (3.8-4.1) *	19	4.8 (4.7-5.0)	12	5.0 (4.7-5.6)	13	5.1 (4.6-5.7)	12
Conductivity (μS/cm)	98 + 4	18	51 + 2	12	46 + 2	13	46 + 2	12
Sulphate (μeq/L)	583 + 17	10	223 + 17	11	202 + 16	13	182 + 16	11
Total Phosphorus (μg/L)	11 + 3	16	10 + 5	11	12 + 8	13	11 + 4	11
Colour (Hazen)	–	–	8.1 + 4.5	11	6.0 + 4.9	13	16.5 + 8.5	11
Secchi Disc (m)	6.7 + 1.3	18	4.5 + 0.4	12	5.3 + 0.9	13	4.1 + 0.8	12
Cu (μg/L)	64 + 12	17	17 + 6	11	12 + 5	13	9 + 3	11
Ni (μg/L)	301 + 30	18	82 + 11	11	80 + 12	13	58 + 22	11
Zn (μg/L)	36 + 5	17	17 + 12	11	11 + 3	13	12 + 11	11
Al (μg/L)	289 + 51	11	76 + 17	11	67 + 38	13	47 + 17	11
Mn (μg/L)	213 + 10	11	116 + 6	11	100 + 3	13	99 + 3	11

* range given for pH

The composition of the planktonic rotifer community in Swan Lake has changed significantly since 1977, when it was dominated by two species. During 1977, *K. taurocephala* had a median density of 250.0 individuals L^{-1} and comprised 94.3% of the total rotifer abundance (Table II), and 59.5% of the biomass (Table III). A large species of *Synchaeta* had a 1977 median density of 12.6 individuals L^{-1} (Table II) and contributed an additional 4.8% and 39.0% to community abundance and biomass (Table III), respectively. Thus, *K. taurocephala* and *Synchaeta* together accounted for greater than 98% of both rotifer abundance and biomass during 1977. In contrast, *K. taurocephala* had median densities of 140.0, 62.0 and 68.7 individuals L^{-1} in 1982, 1983 and 1984, respectively (Table II). These densities correspond to 52.0%, 11.0% and 31.2% of the total rotifer abundances for those years (Table II), and only 33.4%, 5.8% and 16.8% of the biomass (Table III). However, since the within-year variability in species density was great, the overall density differences between years were not significantly ($P > 0.05$) different for *K. taurocephala*. Similarly, the density of *Synchaeta* did not differ significantly ($P > 0.05$) between study years, although a general decline appears to have occurred since 1977 (Table II).

Densities of *Polyarthra* (*P. remata* and *P. vulgaris*), *C. natans*, *T. similis* and *C. ovalis* differed significantly ($P < 0.05$ or less) between 1977 and 1984. Each species was very rare in 1977, although by 1984 they accounted for 55.8% of the total rotifer density (Table II) and 55.6% of community biomass (Table III). The increase in importance of *Polyarthra*

Table II.
Median density and % of total abundance for planktonic rotifers in Swan Lake between 1977 and 1984.

	YEAR							
	1977		1982		1983		1984	
	Median Density (ind./L)	% Total Abundance	Median Density (ind./L)	% Total Abundance	Median Density (ind./L)	% Total Abundance	Median Density (ind./L)	% Total Abundance
Keratella taurocephala	250.0	94.3	140.0	52.0	62.0	11.0	68.7	31.2
Polyarthra spp.	<0.1	<0.1	87.0	32.3	345.5	61.5	81.9	37.1
Synchaeta spp.	12.6	4.8	7.8	2.9	22.2	4.0	5.5	2.5
Trichocerca similis	<0.1	<0.1	5.1	1.9	107.3	19.1	10.2	4.6
Conochiloides natans	<0.1	<0.1	<0.1	<0.1	13.6	2.4	30.3	13.7
Chromogaster ovalis	<0.1	<0.1	1.3	0.5	7.7	1.4	1.0	0.4
other rotifers	2.4	0.8	27.9	10.4	3.6	0.6	23.0	10.5

between 1977 and 1982 was especially great, and corresponded to the greatest changes in lake chemistry. T. similis also increased substantially between 1977 and 1982, although another large increase was observed in 1983 (Table II). In contrast, C. natans and C. ovalis did not show demonstrable increases in density until 1983.

Other species which have increased in abundance since the lake's chemistry has 'improved' include Gastropus stylifer, Keratella crassa and Pleosoma truncatum.

Table III.
Median biomass and % of total biomass for planktonic rotifers in Swan Lake between 1977 and 1984.

	YEAR							
	1977		1982		1983		1984	
	Median Biomass (μg/L)	% Total Biomass	Median Biomass (μg/L)	% Total Biomass	Median Biomass (μg/L)	% Total Biomass	Median Biomass (μg/L)	% Total Biomass
Keratella taurocephala	5.00	59.5	2.8	33.4	1.2	5.8	1.4	16.8
Polyarthra spp.	<0.1	<0.1	3.0	35.3	11.8	55.0	2.8	34.5
Synchaeta spp.	3.3	39.0	2.0	24.3	5.8	27.1	1.4	17.7
Trichocerca similis	<0.1	<0.1	<0.1	1.0	1.6	7.5	0.2	1.8
Conochiloides natans	<0.1	<0.1	<0.1	<0.1	0.7	3.3	1.6	19.4
Chromogaster ovalis	<0.1	<0.1	<0.1	0.2	0.1	0.7	<0.1	0.2
other rotifers	.0.1	1.4	0.6	5.7	0.2	0.6	0.7	9.1

4. DISCUSSION

Acidic deposition and heavy metal contamination have impacted both terrestrial (Freedman and Hutchinson, 1980) and aquatic (Keller and Pitblado, 1984, and the references cited therein) ecosystems in the Sudbury region. Swan Lake has been strongly affected, likely because it is underlain with Precambrian Shield bedrock, lies along a prevailing wind direction from Sudbury, and is in close proximity to the Sudbury smelters.

However, recent reductions in the input of acid and metals into the lake have been manifested as significant improvements in water quality. Such changes have also been demonstrated for other lakes in the Sudbury area (Hutchinson and Havas, 1986; Keller and Pitblado, 1986). Thus, the observed natural response of Swan Lake is part of a more general chemical recovery of Sudbury area lakes.

The community composition of the planktonic Rotifera in Swan Lake has also changed significantly since 1977. In particular, the relative contributions of K. taurocephala and Polyarthra have changed dramatically. Three hypotheses may be constructed to account for the decline of K. taurocephala. First, the observed differences do not reflect actual density differences between years, but rather reflect differences in the sampling protocols between the study years. Second, seasonal and annual variation in rotifer species abundance is often striking (Nauwerck, 1978; Makarewicz and Likens, 1979), and may be sufficient to account for the observed pattern. Third, K. taurocephala is a species highly adapted for living in acidic waters, but it is virtually excluded from high pH waters either due to physiological intolerance or species interactions.

The first hypothesis may be tested by considering the density of K. taurocephala for the years 1977 and 1982 (Table II). The median density of the species in 1977 was 250.0 individuals L^{-1} when collected with a 76 um mesh Schindler-Patalas trap, and 140.0 individuals L^{-1} in 1982 when collected with a 35 um mesh Schindler-Patalas trap. Yan and Geiling (1985) estimated that a 76 um mesh trap would under-estimate K. taurocephala densities by a factor of two. Thus, even if its densities were under-estimated in 1977, K. taurocephala densities were still much greater than in years when finer grades of net were used. Thus, the data in Table II do not support the first hypothesis. However, the marked increases in abundance of C. ovalis, Polyarthra, C. natans and T.similis indicates that enumeration procedures may have affected some 1983 densities. This problem does not detract from the overall results of the study, however, as it is not sufficient to explain the shift in community importance from K. taurocephala to Polyarthra.

The second hypothesis appears untenable on the grounds that the density of K. taurocephala in a nearby acidic lake (Clearwater), has remained very stable during the periods of 1976-1978 (Yan and Miller, 1984) and 1981-1985 (Yan, 1986). In each year, the species has accounted for more than 90% of the total abundance.

Evidence has accrued in the literature which provides limited support for the third hypothesis. A number of studies (Siegfried et al., 1984; Carter et al., 1986; MacIsaac, 1986) have demonstrated a preference by K.

taurocephala for acidic waters. Indeed, McCormick *et al*. (1985) reported that the species displayed extensive acid tolerance in field bioassays, while Chengalath and Koste (1983) described the species as an acidobiont. Thus, although the species appears highly adapted to acidic waters, it should be emphasized that the reason(s) for the decline in abundance of *K. taurocephala* with increasing pH in this study is not clear.

Hamilton *et al*. (1985) found that after an acidified Ontario lake was neutralized with calcite, *K. taurocephala* was replaced by *Polyarthra* and a number of other species, an observation which coincides with the natural changes which have occurred in Swan Lake. Moreover, a survey of 47 Ontario lakes of varying acidity revealed that members of the genus *Polyarthra* were present only in lakes of pH greater than 4.65 (MacIsaac, 1986), hence the virtual absence of *Polyarthra* in Swan Lake during 1977 was not unexpected. Similarly, the occurrence of *Synchaeta* in Swan Lake during 1977 was not surprising, as members of the genus have been recorded in other highly acidic Sudbury area lakes (MOE, 1982).

Conochiloides natans was very rare in Swan Lake while the lake was highly acidic, but has increased since 1983 (Table II). This observation accords with its geographic distribution in other Sudbury area lakes (MacIsaac, 1986). Furthermore, the species demonstrated a distinct preference for cool hypolimnetic waters (Keller, 1986), agreeing with its classification as a cold stenotherm (Stemberger 1979).

In summary, the changes which have occurred in the planktonic Rotifera in Swan Lake are likely the result of many interacting factors. Certainly pH is a contributing factor; however, caution should be exercised in over-emphasizing pH alone, since it is difficult, if not impossible, to separate its effects from changes in food resources, crustacean competitors, invertebrate predators and direct metal effects. It appears likely, however, that the decrease in SO_2 and metal emissions from Sudbury smelters initiated both the chemical and biological changes.

ACKNOWLEDGEMENTS

We thank W. Geiling for enumerating 1977, 1982 and 1984 plankton samples and R. Chengalath for identifying difficult specimens. K. Nicholls and an anonymous rewiewer provided helpful comments. This research was supported by the Natural Sciences and Engineering Research Council of Canada (to TCH and HJM) and by a Canadian Wildlife Federation scholarship to HJM.

REFERENCES

Carter, J.C.H., Taylor, W.D., Chengalath, R. and Scruton, D.A.: 1986, *Can. J. Fish. Aquatic Sci*. 43, 444.

Chengalath, R. and Koste, W.: 1983, *Hydrobiol*. 104, 49.

Dillon, P.J., Reid, R.A. and Girard, R.: 1986, *Water, Air and Soil Poll*., (this issue).

Freedman, B. and Hutchinson, T.C.: 1980, *Can. J. Bot*. 58, 2123.

Hamilton, J.G., Molot, L.A. and Booth, G.M.: 1985, 'Changes in biological communities in an acidified lake neutralized with calcite'. Abstracts, International Symposium on Acidic Precipitation. Muskoka, Ontario.

Hutchinson, T.C. and Havas, M.: 1986, Water, Air and Soil Poll. (in press).
Keller, W.: 1986, unpublished data.
Keller, W. and Pitblado, J.R.: 1984, Water, Air and Soil Poll. 23, 271.
Keller, W. and Pitblado, J.R.: 1986, Water, Air and Soil Poll. (in press).
MacIsaac, H.J.: 1986, An analysis of zooplankton communities from northern Ontario lakes of varying acidity, with emphasis on planktonic Rotifera, M.Sc. thesis, University of Toronto, 210 p.
Makarewicz, J.C. and Likens, G.E.: 1979, Ecol. Monogr. 49, 109.
McCormick, J.H., Shepard, B. and Eaton, J.G.: 1985, 'Toxicological studies of fish and zooplankton from Little Rock Lake'. American Society of Limnology and Oceanography, Minneapolis, Minnesota.
MOE (Ontario Ministry of the Environment): 1981, Outlines of analytical methods, Rexdale, Ontario, 246 p.
MOE.: 1982, Studies of the lakes and watersheds near Sudbury, Ontario. Final limnological report. SES 009/82, Supplementary volume, Rexdale, Ontario.
Nauwerck, A.: 1978, Archiv fur Hydrobiol. 84, 269.
Pennak, R.W.: 1978, Freshwater Invertebrates of the United States, 2nd Edition, J. Wiley and Sons, Toronto, Ontario, 803 p.
Roff, J.C. and Kwiatkowski, R.E.: 1977, Can. J. Zool. 55, 899.
Schindler, D.W., Mills, K.H., Malley, D.F., Findlay, D.L., Shearer, J.A., Davies, I.J., Turner, M.A., Linsey G.A. and Cruikshank, D.R.: 1985, Science 228, 1395.
Seigfried, C.A., Sutherland, J.W., Quinn, S.O. and Bloomfield, J.A.: 1984, Verh. Internat. Verein. Limnol. 22, 549.
Sprules, W.G.: 1975, J. Fish. Res. Board Can. 32, 389.
Stemberger, R.S. 1979. A guide to rotifers of the Laurentian Great Lakes, U.S. E.P.A. 000/4-79-021. Cincinnati, Ohio. 177 p.
Yan, N.D. and Dillon, P.J.: 1984, 'Experimental neutralization of lakes near Sudbury, Ontario'. In: J. Nriagu (ed.), Environmental Impacts of Smelters, J. Wiley and Sons, Toronto, pp. 417-456.
Yan, N.D. and Miller G.E.: 1984, 'Effects of deposition of acids and metals on chemistry and biology of lakes near Sudbury, Ontario'. In: J. Nriagu (ed.), Environmental Impacts of Smelters, J. Wiley and Sons, Toronto, pp. 243-282.
Yan, N.D. and Geiling, W.: 1985, Hydrobiol. 120, 199.
Yan, N.D.: 1986, unpublished data.

LABORATORY AND FIELD TESTING OF A DOSE CALCULATION MODEL FOR NEUTRALIZATION OF ADIRONDACK ACIDIC LAKES

Richard D. Scheffe, Joseph V. DePinto and Kurt R. Bilz
Department of Civil and Environmental Engineering, Clarkson University, Potsdam, NY 13676 USA

Abstract. A deterministic lake dose calculation model has been applied to estimate the quantity of Ca carbonate-based material needed to achieve a desired initial water chemistry in a whole lake treatment of two Adirondack acid lakes. The study lakes were treated with calcite during May 30-31, 1985.

The model accurately simulated laboratory column water chemistry profiles (spatially and temporally) and dissolution efficiencies; however, the model predicted conservative efficiencies for the study lakes. Time-series water chemistry profiles of the lakes suggest that epilimnetic mixing extended suspension time of fine calcite particles, and that atmospheric carbon dioxide intrusion rate was greater than expected. The combination of these factors enhanced dissolution efficiency.

1. INTRODUCTION

As part of the Electric Power Research Institute (EPRI) funded Lake Acidification Mitigation Project (LAMP), a deterministic dose calculation model has been developed, tested in a laboratory column, and applied to estimate the quantity of Ca carbonate-based material needed to achieve a desired initial water chemistry in a whole lake treatment of two Adirondack acidic lakes. The two lakes, located in the southwestern portion of the Adirondack Park, were treated with a very high quality calcite powder (OMYA, Inc., Proctor, VT) on May 30 and 31, 1985. Prior to treatment Woods Lake was a highly acidic system with pH ranging seasonally from 4.4 to 5.2. It has a surface area of 25 ha, a mean depth of 3.2 m, and a mean hydraulic retention time of approximately 300 d. Cranberry Pond had a similar chemistry to Woods Lake; however, it is a smaller lake (7.2 ha, 2.9 m mean depth) with a shorter mean hydraulic retention time of approximately 100 d. An intensive, post-liming transition monitoring program was conducted to, among other things, provide initial conditions for the reacidification predictions presented by Scheffe et al. (1985) and to evaluate the dose calculation model.

Two materials of very different particle size and carbonate purity were evaluated for use in the LAMP lakes: 1) Nautex SL (Omya, INC.), a

98% pure Ca carbonate material with a mean particle diameter of 2 μm (top cut of 10μm), and 2) an agricultural limestone (Eastern Rock Products, INC.) of 82% Ca carbonate with a mean particle diameter of 45 μm. Both materials were used in column settling experiments to obtain data for comparison to theoretical model predictions; the actual lake neutralization used the Nautex SL product. The contents of this paper include a description of the model development, laboratory testing and comparison to actual lake neutralization results.

2. MODEL DEVELOPMENT - LABORATORY TESTING

2.1. Model Description - Assumptions

The deterministic lake dose model has been designed to predict calcite dissolution efficiencies as well as temporal and spatial water column chemistry. The model is based on Sverdrup's (1983) sinking-dissolution model with major refinements incorporating chemical equilibrium considerations.

Calcite added to an acidic water column experiences two dominant driving forces - dissolution and settling. Conceptually, the model considers calcite powder with a range of particle sizes sinking through a water column. As the larger particles settle and dissolve they increase the pH; and, accordingly, successively smaller particle sizes will encounter higher pH values until the finest particles have passed through the column.

The assumptions applied in developing the model are as follows:

1) The lake (or a specified basin within the lake) is viewed as a cylindrical column subdivided into incremental strata through which successively smaller particles settle. Also, vertical and horizontal dispersion in the water column is assumed to be a negligible transport mechanism during the treatment period - vertical transport is governed entirely by Stoke's settling. Allowance is made, however, for a decrease of individual particle radii (and hence an increase in settling time) due to dissolution throughout the settling period.
2) The lake is assumed to be a closed system with respect to atmospheric CO_2 exchange during the time that calcite is settling in the water column.
3) The neutralization material is spread uniformly over the treated portion of the lake.
4) Calcite dissolution is assumed to be transport controlled at low pH (below 6) and reaction controlled at higher pH.
5) Acid-base reactions resulting from calcite dissolution are assumed to be very rapid relative to calcite dissolution itself and particle transport; therefore, a thermodynamic equilibrium exists between dissolved calcite and other water column chemical species.

The dissolution of solid calcite in an acidic aqueous solution is described by three heterogeneous reactions (Plummer et al., 1978):

$$CaCO_3 + H^+ = Ca^{2+} + HCO_3^- \qquad [1]$$

$$CaCO_3 + H_2O + CO_2 = Ca^{2+} + 2HCO_3^- \quad [2]$$

$$CaCO_3 + H_2O = Ca^{2+} + HCO_3^- + OH^-. \quad [3]$$

Reaction (1) dominates at pH-values less than about 6. Kinetic studies have shown that the reaction rate is transport limited in this pH range and can be appropriately modeled by application of a film mass transport model in which H^+ diffuses through a boundary layer surrounding the particle (Sverdrup and Bjerle, 1982).

At pH values above 6, reactions (2) and (3) tend to dominate. In addition to the forward reactions, an empirical backward rate expression has been suggested by Plummer et al. (1978) and is included in the overall rate expression for this model. Incorporation of these reaction terms with the transport term yields the following overall dissolution rate expression:

$$dM/dt = D([H^+]-HSAT)/h + k_2[H_2CO_3] + k_3 - k_4[Ca^{2+}][HCO_3^-] \quad [4]$$

where M is the mass of undissolved calcite, D is the diffusion coefficient for H^+, h is the boundary layer thickness surrounding a sinking calcite particle, HSAT is the surface H^+, k_2 and k_3 represent rate constants for reactions (2) and (3), respectively, and k_4 is a backward reaction rate constant.

Since the water column pH and other water chemistry define the driving force for calcite dissolution, a suitable method for updating the water chemistry in each lake stratum is required in the model. In order to eliminate the empiricism inherent in the Sverdrup titration curve approach, a chemical equilibrium calculation submodel similar to MINEQL (Westall et al., 1976) is used to calculate the water chemistry change resulting from a predetermined quantity of dissolved calcite. In addition to the carbonate system, the charge balance includes inorganic Al species (sulfate, flouride and hydroxyl complexes) as well as dissolved organic carbon (DOC). DOC is treated as a simple monoprotic acid, with an empirical pK_a and H^+ site-measured DOC concentration relationship as determined by Driscoll et al. (1982).

2.2. Comparison with Experimental Data

A 3.66 m high by 15 cm diameter column with sample ports located at several depths provided a means of testing the model (details in Bilz, 1986). Two materials at various doses (Table I) were used in the column experiments. Predetermined doses of the material were applied as a slurry evenly at the top of the column (filled with synthetic lake water with pH of about 4.4 and other chemical species concentrations similar to the study lakes), and aliquots were extracted at various time intervals throughout the column. A summary of the overall efficiencies (based on the total change in dissolved Ca compared to total applied Ca) for both the column runs and the model predictions are listed in Table I. For these trials the model results deviated very little from the column observations.

A comparison of dissolved Ca profiles between a column and model run is shown in Figures 1 and 2, respectively. The discrepancies (time and space) between observed and simulated values are small; however, these data show a significant limitation inherent in the model. Since the model does not allow for dispersion, discontinuous chemistry profiles are possible and flattening of the profiles does not occcur (eg., compare the 8 hr profiles).

TABLE I

Comparison of Dissolution Efficiencies between Experimental Column Data and Lake Dose Model

Material Dose (mgL^{-1})	Dissolution Efficiency (%) Column Data (after 48 hr)	Model Prediction
Eastern Rock Products		
20	54	48
50	34	35
Nautex SL		
4	95	100
8	100	100
16	78	84
31.4	42	43

3. CALCITE APPLICATION DESCRIPTION - FIELD OBSERVATIONS

3.1. Field Application Procedure

On May 30 and 31, 1985 a Ca carbonate slurry was applied to Woods Lake and Cranberry Pond. The slurry (Nautex SL, a product of Omya, Inc., Proctor, VT) was a mixture of a very pure (98% $CaCO_3$ by weight) calcite material in water at a concentration of 71% solids (dry weight). In order to maintain adequate fluid properties in this concentrated slurry, it also contained a sodium polyacrylate dispersant (Omyasperse, Omya, Inc.) at a concentration of 0.15 % (by weight). The applied slurry doses for Woods Lake and Cranberry Pond were 31.8 (22.7 dry) and 10 (7.26 dry) tonne, respectively. The material was delivered in two pressurized tank trucks to a staging area in Big Moose, NY located approximately 13 km from the lakes.

The slurry was applied via helicopter supplied by Helicopter Systems, Inc., Scottdale, PA. The helicopter, a Bell 204B, was equipped with a 1271 kg capacity slurry holding tank and a dispensing unit called a Hydro-Spyder (Union Carbide). This system was capable of delivering

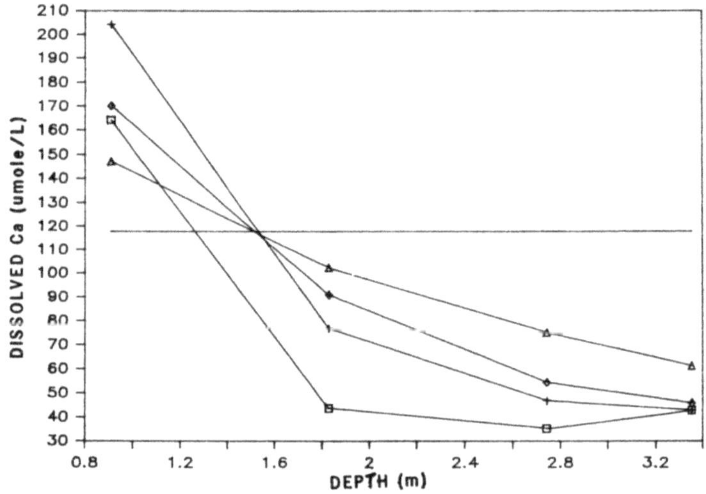

Figure 1. Time-series profiles of dissolved Ca for a typical column run (Eastern Rock Aglime, 20 mgL^{-1}. □ 1/2 hr, + 2 hr, ◇ 4 hr, △ 8 hr.

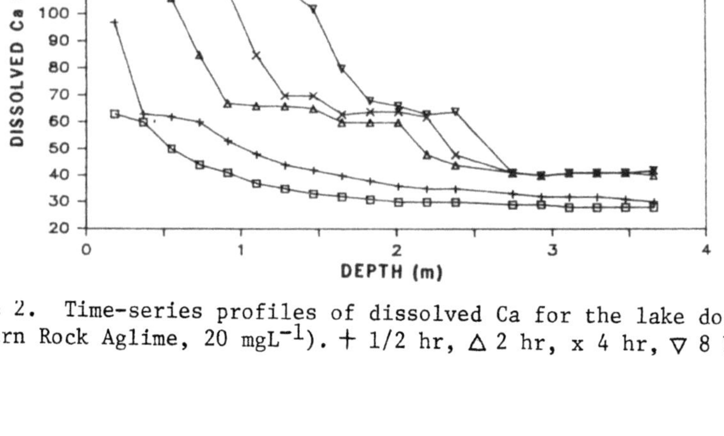

Figure 2. Time-series profiles of dissolved Ca for the lake dose model (Eastern Rock Aglime, 20 mgL^{-1}). + 1/2 hr, △ 2 hr, x 4 hr, ▽ 8 hr.

0.61 m^3 (1.1 metric tons) of slurry per trip. An average round trip, including flight to and from the lake, slurry loading at the staging area, and slurry application at an average rate of 0.7 m min , took 10 minutes; this converts to a slurry delivery rate of 6.54 tonne (4.63 dry) hr^{-1}.

The lakes were treated on a volumetric basis by dividing the surface into grids (13 grids for Woods Lake and 7 grids for Cranberry Pond) of known volume and dosing each grid with an amount of slurry necessary to attain a uniform whole lake concentration. This resulted in applied doses of 29.6 mgL^{-1} and 26.5 mgL^{-1} for Woods Lake and Cranberry Pond, respectively. The calcite doses applied to these lakes exceeded $CaCO_3$ saturation for a closed system. Under these conditions, calcite dissolution efficiency was expected to suffer at the expense of undissolved $CaCO_3$ being deposited in the lake sediments. This was one of the goals of the liming demonstration and, hence, the reason for using a supersaturation dose.

3.2. Results of Woods Lake and Cranberry Pond Treatment

An intensive, transition monitoring program (conducted by LAMP personnel) during and for approximately three weeks after the treatment of Woods Lake and Cranberry Pond provided an excellent data set for field testing the dose estimation model presented above. Following is a summary of the short-term chemical response of the lakes to the treatment, and a comparison of the actual response to model predictions.

Dose Model Predictions. Based on the pre-liming chemical data and the known treatment doses for Woods Lake and Cranberry Pond, the dose calculation model described above has been applied to predict *a priori* dissolution efficiency and resulting water chemistry for both lakes. A summary of the model predictions is presented in Table II.

Two deterministic model predictions are shown for each lake, one assuming equilibrium between Al(III) and gibbsite and the other without the equilibrium assumption. Under the temporal and dose circumstances, it is unlikely that $Al(OH)_4^-$ will actually form as pH is increased; hence, the prediction without Al(III) equilibrium is probably more valid.

Comparison with field observations. The system responses during the transition period (Figures 3 to 5) indicate that, relative to the dose model predictions, higher dissolution efficiencies were obtained in both lakes by the end of the transition phase (day 9).

There appear to be two major factors, when considered together, that have contributed to the above discrepancies. First, it is apparent that these lakes were responding to considerable CO_2 invasion while the calcite was still in the water column in order to produce such high ANC values (in excess of closed-system equilibrium) within a week after treatment. Analysis of the DIC and Calcium data in Figures 4 and 5 revealed average CO mass transfer rates of 2.2 and 2.9 md^{-1} for Woods and Cranberry, respectively. Second, it was observed that undissolved

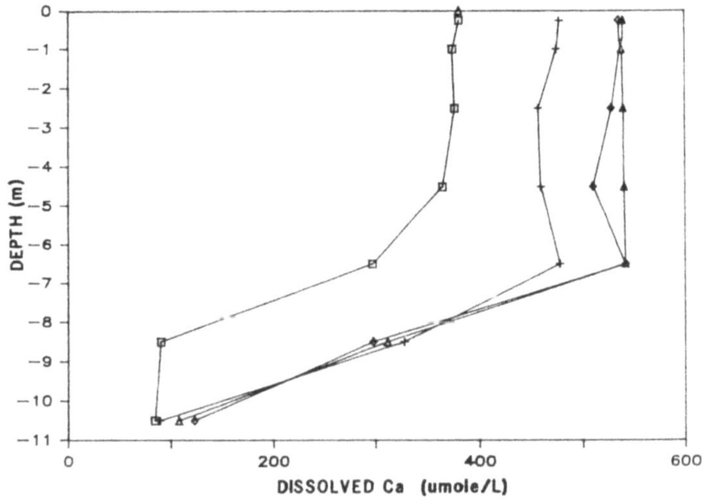

Figure 3. Time-series profiles of dissolved Ca for Woods Lake during transition monitoring period. □ 6/1/85, + 6/5/85, ◇ 6/9/85, △ 6/19/85.

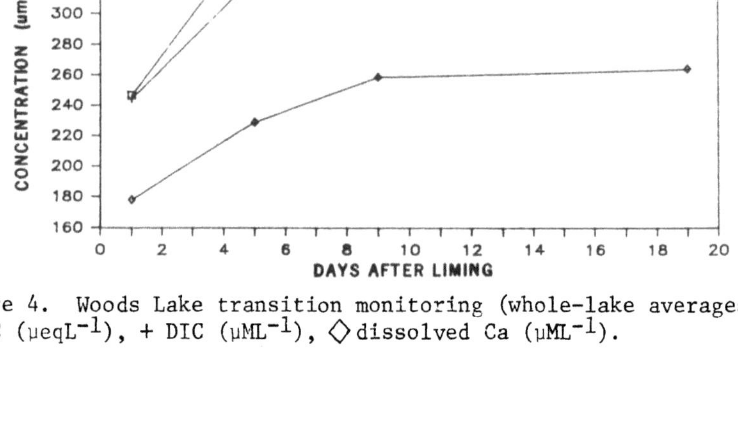

Figure 4. Woods Lake transition monitoring (whole-lake averages). □ ANC (μeqL^{-1}), + DIC (μML^{-1}), ◇ dissolved Ca (μML^{-1}).

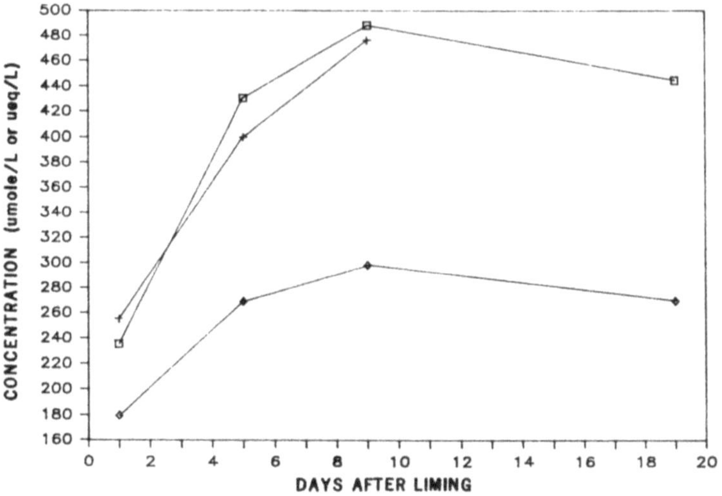

Figure 5. Cranberry Pond transition monitoring (whole-lake averages). □ ANC (μeqL^{-1}), + DIC (μML^{-1}), ◇ dissolved Ca (μML^{-1}).

calcite remained in the water column longer than expected. Apparently, vertical mixing in the upper waters of both lakes reduced net settling velocities (relative to Stokes velocity) of suspended particles. Futhermore, in Woods Lake suspension times were prolonged by an inability of the undissolved calcite to penetrate the thermocline (Figure 3); this is not a major problem since less than 5% of the Woods Lake volume is below 7.5 m. However, the combination of these processes is probably the cause for the high dissolution efficiencies and, consequently, lower than expected sediment doses.

TABLE II

Dose Calculation Predictions for Woods Lake and Cranberry Pond Based on Actual Application of Nautex SL

Description	ANC	DIC	pH	Ca	Dissolution Efficiency (%)
WOODS LAKE (dose = 29.6 mgL^{-1} dry basis)					
Pre-Liming Conditions	-6.7	48.8	4.88	46.1	
PREDICTIONS					
Lake Dose Model:					
with Al(III) equilibrium	402	199	9.59	196	53
without Al(III) equilibrium	248	174	9.73	171	44
Open System Equilibrium Model	573	584	8.88	336	100

Description	ANC	DIC	pH	Ca	Dissolution Efficiency (%)
CRANBERRY POND (dose = 26.5 mgL^{-1} dry basis)					
Pre-Liming Conditions	-15.9	72.6	4.57	40.1	
PREDICTIONS					
Lake Dose Model:					
with Al(III) equilibrium	407	253	9.43	220	70
without Al(III) equilibrium	295	228	9.56	390	59
Open System Equilibrium Model	535	546	8.02	300	100

(all units $\mu moleL^{-1}$, except ANC - μeqL^{-1})

Acknowledgments

The authors wish to thank K.A. West of Clarkson University and C.T. Driscoll of Syracuse University for providing water chemistry data. This paper was presented as a poster session at the International Symposium on Acidic Precipitation, Muskoka, Ontario, Canada, September, 1985. The research described in this article has been funded by the Electric Power Research Institute through a subcontract agreement with General Research Corporation and is part of the Lake Acidification Mitigation Project (LAMP), project number RP 2337. (Clarkson, Cornell, Indiana and Syracuse Universities are participating in the field portion of LAMP).

References

Bilz, K.R.: 1986, Empirical Evaluations and Application Strategies for Neutralizing Substances in Acidified Waters. M.S Thesis, Clarkson University, Potsdam, NY.

Driscoll, C.T.: 1982, Calcium Carbonate Neutralization of Acidified Surface Waters. In: Proceedings of the American Society of Civil Engineers, 108(EE6):1128-1145.

Plummer, L.N., Wigley, T.M.L., and Parkhurst, D.L.: 1978, American Journal of Science, **(279)**, 537.

Scheffe, R.D., Booty, W.G., and DePinto, J.V.: 1985, Development of Methodology for Predicting Reacidification of Calcium Carbonate Treated Lakes. International Symposium on Acidic Precipitation. Muskoka, Ontario, Canada.

Sverdrup, H.: 1983, Chemica Scripta, **(22)**, 8.

Sverdrup, H., and Bjerle, I.: 1982, Vatten **(38)**, 59.

Westall, J.C., Zachary,J.L., and Morel F.M.M.: 1976, Technical Note 18, R.M. Parsons Laboratory for Water Resources and Hydrodynamics, M.I.T., Cambridge, MA.

LIMING IN SWEDEN

Olof Lessmark
National Board of Fisheries
P.O. Box 2565
S-403 17 Göteborg, Sweden

Eva Thörnelöf
National Environmental Protection Board
P.O. Box 1302
S-171 25 Solna, Sweden

ABSTRACT. Air-borne pollutants have acidified 90,000 km of running water and at least 18,000 of Sweden's 90,000 lakes. In 1977 a national liming program to mitigate the effects of acidification in surface waters started with governmental support. Governmental subsidies usually amount to 85 % of the total liming cost but can in rare cases be extended to 100 % when waters of great national interest are threatened. The main liming agent is limestone of the size 0 to 0,2 mm. In lakes lime is spread by pontoon boats, helicopters and other vehicles. In running waters dosers and diversion wells are used. By 1985, the national liming program has been able to lime approximately 3,000 lakes and 100 streams.

1. INTRODUCTION

Due to air-borne pollutants approximately 18,000 lakes and 90,000 km of running water in Sweden have pH values at levels where damage to living organisms in the water can be expected (Bengtsson et al., 1980, 1981; Persson, 1982). In 1977 a national liming program with governmental support was started by the Swedish National Board of Fisheries to mitigate the effects of acidification in lakes and running waters. The results presented here refer to experiences from liming operations performed with governmental subsidies since 1977.

2. OBJECTIVE

The purpose of the liming program is to restore acidified lakes and streams and to maintain natural ecosystems in waters threatened by acidification. Fishing interests often are the driving force behind the local liming projects. However, environmental conservation is also an important purpose. To avoid serious biological damage, the target of the

Water, Air, and Soil Pollution **31** (1986) 809–815.

liming program is to start mitigation when a waters's alkalinity has decreased below 0.05 meq L^{-1} and/or when the pH has dropped below 6.0, and increase the alkalinity to > 0.1 meq L^{-1}.

3. ORGANIZATION

Since 1982 the National Environmental Protection Board has been responsible for the program in cooperation with the National Board of Fisheries. These authorities are also responsible for research, monitoring, information and guidelines for liming.

The County Administrations plan the liming operations and allocate national grants. They also run regional monitoring programs and control the effects of the individual liming projects. Regional liming plans, rolling over five years, are renewed every year. Based on these The National Environmental Protection Board makes a division among the County Administrations of the governmental subsidies.

Local authorities and fishing organizations are generally responsible for the local liming projects and their execution. A local liming project may include just one single lake or a total water system. The local organizations, often in cooperation with the County Administrations or consultants, prepare the detailed local liming programs which are then executed by companies specialised in liming. At present a total of about 1200 local liming projects are running.

4. FINANCING

The governmental subsidies to surface-water liming have increased from US$ 1.0 million in 1977 to US$ 8.9 millions in 1985. Governmental subsubsidies are given to a liming project only if the water has a pH below 6 or an alkalinity below 0.05 meq L^{-1}, and provided a detailed liming plan is presented. Furthermore, waters with easy public access are favored.

The national subsidies normally cover 85 % of the cost of a liming project. Projects that are especially difficult to finance or have national importance can have a 100 % governmental funding.

5. LIMING TECHNIQUES

A multitude of liming techniques and liming agents have been developed and tested within the framework of the national liming program. The aim when selecting the liming agent is to obtain the best biological and chemical result at the lowest cost. The main liming agent for lakes is limestone of the recommended size-fraction 0 to 0.2 mm, which gives both a good instantaneous and a long-term dissolution effect in relation to cost. Recommendations on doses are based on practical full-scale liming operations (Table I). Several other liming agents have been tested, e.g. burnt lime (CaO), slaked lime ($CaOH_2$), lye (NaOH) and soda (Na_2CO_3). However, these have not become commonly used in full-scale liming operations because of high price and small or undesired effects.

pH	Dose ($g m^{-3}$) as $CaCO_3$
4.0-4.5	30
4.6-5.0	20
5.1-5.5	15
5.6-6.0	10

Table I. Recommended doses of limestone for surface waters of different pH to increase alkalinity to > 0.1 meq L^{-1}.

The highest instantaneous rate of limestone in water is obtained at low pH, small dose per volume of water, even distribution of lime, finely ground fractions, large depth, vigorous stirring and separation of lime-particles and when the lime dose necessary for a lake is spread over the total water surface in proportion to depth. However, surplus amounts of lime are frequently spread on shallow littoral areas, where water turbulence agitates lime-particles. Thereby a continuous dissolution of lime is attained and the long-term effect of the liming agent is increased. The best continuous dissolution is obtained in clear water lakes, and the most prolonged duration of a liming operation, expressed as number of lake turnover times, is obtained in lakes with short retention times, due to the fact that humus and metals are gradually deposited on the lime particles, thereby inactivating them (Bengtsson et al., 1981). In most lakes, instantaneous plus continuous dissolution amount to 50 to 70 % of the total lime dose before reacidification (Bengtsson op cit.).

In lakes the bulk of lime is spread by various kinds of pontoon boats (Table II). Helicopters are used when there are no roads to the lakes. Spreading is also performed on the ice and along the shores by trucks and other vehicles. Lime is spread dry or wet mixed with water. Advantages by wet dosing are that small lime particles are separated from bigger ones, resulting in a better dissolution of the lime before it has sunk to the bottom (Warfvinge et al., 1984) and that no lime dust, unpleasant to people working with lime-spreading, is formed. Lime can also be mixed with water and pumped in a tube from the shore to a small boat mobile over the lake.

The final choice of liming method and liming agent for a lake is determined by the kind of biological result that is wanted, water quality, hydrology, accessibility to the lake from roads and cost. Lakes with shorter retention time than 0.2 ys and lakes where most of the inflowing water is concentrated to one inlet can successfully be limed with the same methods as running waters.

The duration of the positive biological and chemical effects obtained by a liming operation are mainly determined by the lake's flushing rate (Figure 1). Lakes with retention times of about 1 ys have to be relimed about every second or third year, and lakes with retention times of

Liming method	Proportion of lime spread,%	Cost per ton limestone (0-0.2 mm), U.S.$. Purchase	Transport	Application
Boat	72	12.9-16.5	4.7-11.8	5.3-20.6
Vehicle	22	"	"	2.9-35
Helicopter	2	"	"	50.0-79.4
				Investment
Doser				
dry/wet	4	"		2300-35000
slurry		65 (0-15 um)		18000-35000
Diversion well	0.4	10.5-16.1 (>3 mm)		1800-47000

Table II. Proportions of total amount of lime spread by different methods and actual costs for purchase, transport and application including man hours. (1 U.S. Dollar = 8.5 SKr).

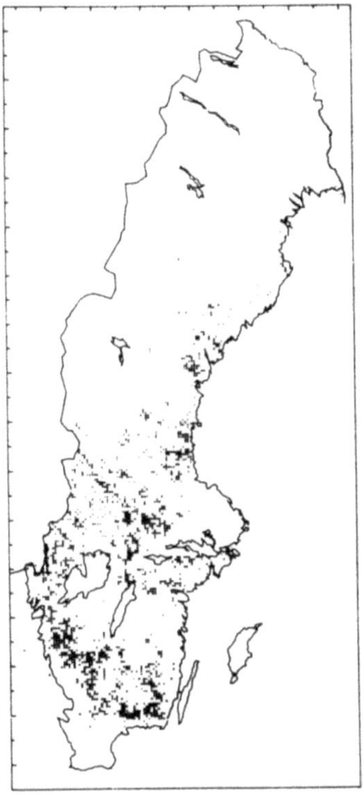

Figure 1. Relationship between the time required for a lake to reacidify to pH < 6 and/or alkalinity < 0.05 meq/l and the lakes theoretical retention time.

2 to 3 years have to be relimed every 3rd to 4th year. Most acidified lakes are shallow, and 63 % of acidified and limed lakes have retention times shorter than one year (Table III). Most lakes have to be relimed rather frequently. Reliming is done at the appropriate time, i.e. as soon as the alkalinity drops below 0.05 meq L^{-1}.

	median	range
Volume ($x\ 10^6\ m^3$)	0.7	0.5-703
Area (km^2)	0.22	0.01-49
Average depth (m):	3.3	0.2-22
Maximum depth (m):	9.0	0.5-62
Theoretical retention time (years):	0.66	0-15
Color (mg Pt/l):	60	5-1000

Table III. Data characterizing 1112 acidified, limed Swedish lakes.

The liming cost of a conventional lake liming operation by boat amounts to approximately US$ 32 to 47 $tonne^{-1}$ lime: costs for lime US$ 13 to 16, transportation US$ 5 to 12 and for application US$ 5 to 21 (Table II).

In running waters the best biological result is obtained when it is possible to avoid acid episodes by liming of upstream lakes (Engblom and Lingdell, 1985; Degerman, 1986, Nilsson and Johansson 1985). If it is not sufficient or possible to lime upstream lakes, running waters are limed by various kinds of dosers and diversion wells. Soil liming on seepage areas around streams are practised in order to increase the pH and decrease the outflow of toxic metals from the soil, especially Al. This method is generally used for small streams where other methods are not suitable. Lime doses on these specific areas have to be 2 to 20 tonne ha^{-1} to obtain the object desired (Dickson and Nyberg, 1985; Bogelius, 1985). When seepage areas are limed the yearly outflow of lime to the water amounts to approximately 10 % per year during the first 2 to 3 ys. The corresponding value if lime is spread on dry land is a maximum of 1 to 2 % (Bengtsson et al., 1981).

Dosers working with water-power have been developed for small streams with a water-flow of up to 1 to 3 $m^{3}s^{-1}$. For larger streams electric power dosers are used. Installation costs for dosers vary over a great range depending on size of the water course, distance to nearest electric power supply and kind of constructions that have to be build at the stream (Table II).

The liming agent for dosers is generally of the size-fraction 0 to 0.2 mm, resulting in approximately 50 % instantaneous dissolution, but finer fractions are occasionally used. Slurry dosers work with very small lime particles (0 to 15 μm) suspended in water, resulting in a high dissolution of dosed lime (90 to 100 %)(Tideström and Moberg, 1984). However, slurry lime is more expensive than 0 to 0.2 mm lime powder (Table II), and slurry dosers are not common.

Liming wells operate with coarse crushed limestone which is ground by the water turbulence to smaller particles, which are then dissolved in the water. To give acceptable water quality the stream shall have a pH above 5, and at least 30 % of the stream has to pass through the well (Tideström op cit.). The wells are only operative in streams with relatively uniform water flow, since if the velocity becomes too low the lime particles are not agitated and ground to smaller particles and if the velocity becomes too high the lime is washed out from the well.

6. RESULTS AND REMAINING PROBLEMS

By the national liming program we have been able to lime approximately 3000 lakes and more than 100 streams (Figure 2). Fish populations quickly respond positively, and even when just a few old fishes remain, reproduction has recommended and resulted in normal or more than normal survival of offspring in the following years (Nyberg et al., 1986). Insects, zoo- and phytoplankton also quickly respond positively to liming, whereas snails, mussels and other invertebrates with low mobility need more time to recolonize (Ericzon et al., 1982). No negative effects of liming have been observed on biota.

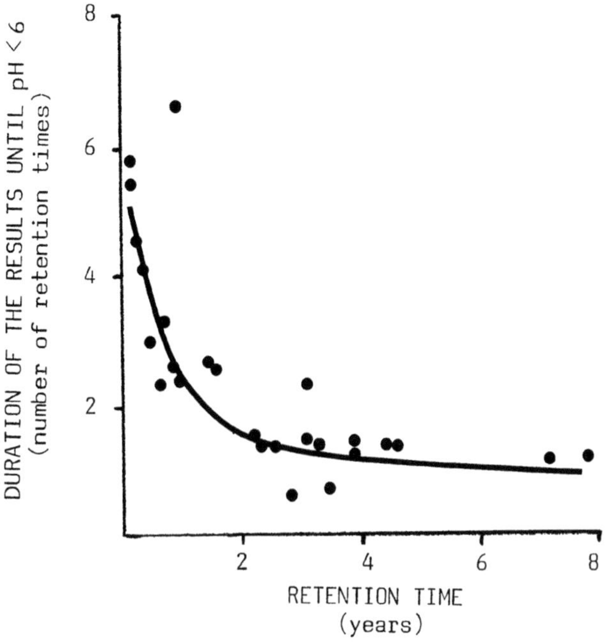

Figure 2. Limed lakes in Sweden. Each dot represents a 5 x 5 km square with one or more limed lakes.

A remaining problem is the inflow to ice-covered lakes of cold, acid water during snow melt which may be negative to crayfish and other acid-sensitive organisms dwelling in shallow littoral zones.

Only a minor part of acidified streams have been treated, and hence reproduction areas for salmonid fish have been lost. The combination of low pH and high concentration of toxic metals leaching out from the soil is a problem which can not successfully be solved by lime dosers. Although liming dosers improve the environment for salmonid parr, the abundance seldom reaches the same levels as in similar not acidified waters (Nyberg et al., 1986).

Lime dosers are also afflicted with many technical problems and frequently collapse because of mechanical, electronic failures, freezing, shortage of lime due to roads being unaccessible during winter and spring. A short acid episode when liming dosers stop can cause devasting effects on biological conditions (Engblom and Lingdell, 1985). This means that only a small proportion of acidified streams can be handled successfully with the technique available today. In the future more effort will be spent on work directed to improve the liming technique for streams including development of better lime dosers and improvement of the soil liming technique.

However, in conclusion, liming is a successful method to maintain fish and many other organisms in acidified lakes and streams until the amount of air-borne acidifing substances have decreased to a tolerable level.

REFERENCES

Bengtsson, B., Dickson, W., Nyberg, P.:1980, Ambio 9, 34.
Bengtsson, B. (ed.):1981, Liming of lakes and streams. Information från Sötvattenslaboratoriet. No 4. (In Swedish).
Bogelius, A. (ed.):1985, Liming of lakes and streams. Statens Naturvårdsverk rapport nr 1919. Fiskeristyrelsen meddelanden nr.1:1985. (In Swedish).
Degerman, E.:1986, Water, Air, and Soil Pollution.
Dickson, W. and Nyberg, P.:1985, personal communication.
Ericxon, F., Hörnström, E., Mossberg, P., Nyberg, P.:1982, Ecological effects of liming in lakes and streams. Information från Sötvattenslaboratoriet. No 6. (In Swedish).
Engblom, E. and Lingdell, P.E.:1985, How effect lime dosers the bottom fauna? Statens Naturvårdsverk rapport nr.1994. (In Swedish).
Nilsson and Johansson:1985, A comparative study of the benthos of limed and non-limed streams: effects of different liming methods. Information från Sötvattenslaboratoriet. No. 11. (In Swedish).
Nyberg, P., Appelberg, M., Degerman, E.:1986, Water, Air and Soil Pollution.
Persson, G. (ed.):1982, Acidification today and tomorrow. Swedish Ministry of Agriculture. Environment '82 Comittee.
Tideström, H. and Moberg, P.O.:1984, Liming running waters. Machine characteristics assessed. SNV PM 1873 or Board of Fisheries, No. 2 1984.
Warfvinge, P., Sverdrup, H. and Bjerle. I.:1984, Chemica Scripta 24, 67.

LOCH FLEET: TECHNIQUES FOR ACIDITY MITIGATION

G.D. Howells and D.J.A. Brown
Present address: Department of Applied Biology
University of Cambridge
Central Electricity Research Laboratories
Leatherhead, Surrey KT22 7SE
UK

ABSTRACT. At a small upland acid lake in southwest Scotland, deposition, water quality and the lake catchment have been characterized over a two year period. In Spring 1986 powdered limestone was applied to three sectors of the lake's catchment. This paper outlines the scope, timescale and objectives of the project and gives some data on pre-intervention conditions.

1. INTRODUCTION

Lake acidification implies the progressive loss of neutralizing bases from catchment soils, and addition of Ca or other minerals is an essential step to restoring the fisheries of affected waters. Since both Ca (beneficial) and Al (detrimental) are derived principally from within the catchment, altering catchment yields of these elements, rather than treating stream or lake water directly, seems preferable, especially if retention time is short. While an extensive literature (see Fraser and Britt, 1982) reports surface water liming techniques, there is less experience on catchment interventions, field trials are few, and efficacy over time largely unknown. Less conventional techniques involving e.g. hydrological manipulations, and reducing, or removing potential proton sources, are scarcely tested. Techniques of soil/catchment treatment, if effective and economic, could be focussed on affected areas/lakes on a much shorter time scale than could be envisaged for a programme of emission control. This strategy can also reduce acidification due to natural processes, and has a more certain and immediate, though not permanent, outcome.

A small (17 ha), remote, lake in south west Scotland (Figure 1) was chosen as the site for field investigation. The lake is acid (pH $\sim$4.5), it has lost its trout population in the recent past (1965), and it provides suitable ecological and other conditions. The site is mostly rough moorland, but a part (12%) was planted with conifers in 1960/61. Its characteristics are rather extreme (344 m altitude, >2 m yr^{-1} rainfall, granitic bedrock, poor peaty soils (Bown and Heslop, 1979)) in relation to upland areas elsewhere in Britain.

Water, Air, and Soil Pollution **31** (1986) 817–825.

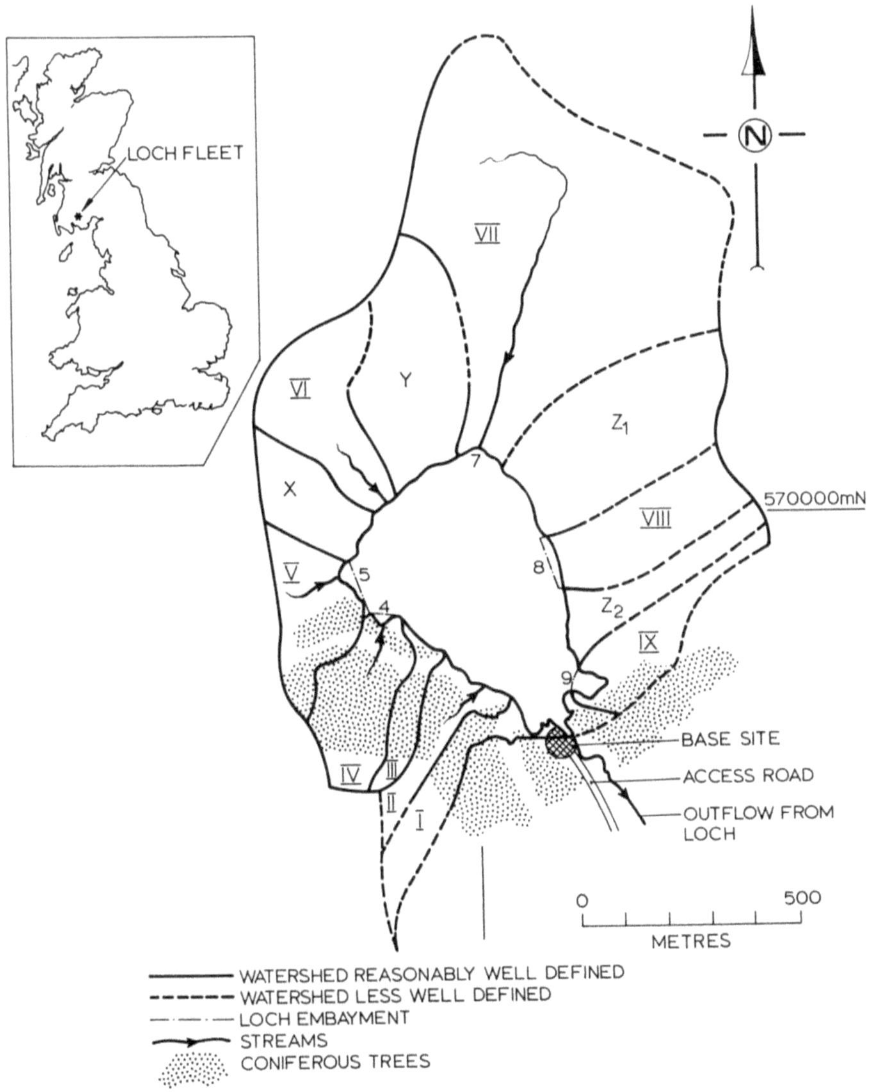

Fig. 1. Loch Fleet Catchment.
Showing the location of subcatchments associated with Loch embayments as numbered.

Nevertheless, techniques of acidity mitigation developed there should be applicable to the more typical, less acid, waters and catchments of upland Britain.

2. OBJECTIVES AND STEPS IN RESEARCH

The underlying assumption is that fish were lost because of a recent change in water quality; it follows that amended water quality would allow fish re-establishment. The principal objectives of the project are:-

(a) to demonstrate that the water chemistry of a lake can be brought into a range suitable for trout by treatment of the catchment or waters, including addition of basic minerals to soils and/or waters, manipulation or hydraulic contact with natural or added minerals, manipulation of the ion exchange system, e.g. by burning, and

(b) once suitable chemical conditions have been achieved, to demonstrate the suitability of the water for a self-sustaining trout population.

It is expected that practical land, lake or stream management procedures can be developed and will be applicable elsewhere. Alternative treatments will be applied in independent sectors, and their effects monitored in soils, soil percolates, and run-off. Semi-permeable Fabrene "curtains" provide "macrosamplers" to catch the run-off and seepage water from catchment sectors (Figure 1, Table I).

TABLE I: LOCH FLEET AND CATCHMENT DATA

	Area (ha)	% Forest	Rain av (mm)	Flow av (L s^{-1})
Whole catchment (incl. loch)	128	10	2153*	71.0+
Subcatchments**				
Sector I	3.8	29	2200	2.1
II	3.2	56	2190	1.7
III	2.0	75	2180	1.0
IV	5.6	88	2170	2.8
V	5.6	45	2150	2.9
VI	7.3	0	2180	4.1
VII	37.4	0	2225	22.2
VIII	7.0	0	2160	3.9
IX	4.3	23	2150	2.4

* 30 yr mean 1941-70
\+ Calculated from rainfall and area
** As indicated in Fig. 1

The changed water quality will be tested against brown trout survival (both long- and short-term). In parallel, the trout spawning potential of the lake and tributary waters will be assessed, (and physical improvements made if necessary) and an estimate made of the lake productivity and potential sustainable yield.

Eventually the whole catchment and tributary streams will be tested by the method(s) of choice and when water quality is favorable, brown trout will be restocked.

The project thus encompasses: an initial phase of collecting and collating existing and baseline information; a second phase when drainage "macrosamplers" are installed, and sectoral treatments initiated; a third phase for additional treatments and tests of changed water quality; and a final phase for treatment the whole catchment, and restocking.

The project is collaborative (Central Electricity Generating Board, North of Scotland Hydro Board, South of Scotland Electricity Board, National Coal Board) and substantial components are undertaken by independent research groups.

3. INITIAL FINDINGS

3.1. Rain and Water Quality

The baseline chemistry of rainfall, tributary and loch waters is summarized in Table II. Flow or volume weighted data are calculated on the basis of 1985 samples.

About half the solutes in both rain and surface waters are attributable to NaCl, demonstrating a strong maritime influence. Acidity is 2 to 3 times greater in the lake and stream waters than in rainfall, suggesting significant catchment sources. Neutralizing bases (Ca and Mg) are enhanced in the drainage waters, presumably derived from rocks or soils. Drainage from the forested Sector (IV) compared with that from moorland (Sector VII) is somewhat more acid, and

TABLE II: MAJOR ELEMENT EONCENTRATIONS
μeq L^{-1}, January 1985 to December 1985, vol. weighted means

	H^+	Na^+	Ca^{2+}	Mg^{2+}	$SO_4{}^{2-}$-S	$NO_3{}^-$-N	Cl^-	$NH_4{}^+$
Rain	14	59	8	14	46	11	67	21
Loch inflows:								
Sector VII	34	198	56	48	120	18	224	
Sector IV	42	250	48	51	113	14	264	
Loch outflow	36	210	96	66	143	11	288	

enhanced in sea salts (Na, Cl, Mg) but sometimes also has higher Ca concentrations. Aluminium (total) is also higher from Sector IV (0.36 mg L^{-1} mean) than from Sector VII (0.27 mg L^{-1}). Typically forest drainage is coloured and has high organic carbon content (>8 mg L^{-1}). The major drainage is from moorland, however, and outflow concentrations do not differ greatly from those of Sector VII.

Time series data for inflow (Site 7) and outflow (Site 0) indicate the range of fluctuations through the critical winter and ice/snowmelt period - acidity varied from pH 4 (prior to snow or ice melt), to pH 4.9 (at least one peak occurring over a thaw). Calcium was minimal (0.1 mg L^{-1}) in early winter with other "troughs" at times of heavy rain or thaw; maximum values were $\sim$1.8 mg L^{-1}, marginal for brown trout survival. Aluminium (total) was also variable, with lowest values (0.075 mg L^{-1}) found unexpectedly during thaw, and peaks (e.g. 0.47 mg L^{-1}) unassociated with acid episodes. (In May, 17.5% of total Al was present as the labile monomeric fraction.) While this pattern is inconsistent with the current conception of acid "episodes", water quality in the major inflow stream is clearly inhospitable for brown trout spawning or as a nursery for fry.

Although the project is not designed as a catchment study, budgets for major elements will be calculated when flows and dry deposition estimates are available. Ambient SO_2 is at the level of detection (2 $\mu g\ m^{-3}$) (Bennett et al., 1985) and estimated fluxes suggest no net annual retention or release of sulphur; dry deposition of atmospheric S is estimated to be negligible. Retention (reduction to $S^{=}$) or release (reoxidation) is possibly important seasonally however, and will be studied in more detail, since Al mobilization has been linked with anion fluxes.

3.2. Soils and Sediments

3.2.1. Soils have been sampled on a 50 × 50 m grid. Soil parent materials are of three principal types: organic deposits, colluvial deposits, and weathering rock or till. The till deposits are thin and widely scattered; they are mainly derived from local granites, but some graywacke is present. Despite the rocky appearance of the terrain, outcrops or surface boulders constitute only 2% of the area. The most notable features of the soils are their peaty surface horizons and their shallowness; no soils with mineral horizons have been found. The principal soil type is the peaty ranker (39%), with peaty gleys (24%) and deep (>50 cm) peat (33%). Soil profiles and topsoil samples are being analyzed to establish the level of exchangeable bases and the 'lime requirement'.

3.2.2. Lake sediments are often neglected, yet may be important sources or sinks for acidic or neutralising materials. At Loch Dee, nearby, limestone applications brought to change in pH, alkalinity or calcium, possibly due to sediment adsorpton (Burns et al., 1984). The extent and characteristics of Loch Fleet sediments have been established from surface sediments and deep cores. The deeper, northerly, part of the Loch has substantial deposits of an organic (>50% LOI) nature, while the shallow, southerly, half has rock outcrops with overlying sands, silts and gravels. Elemental analysis further distinguishes the organic sediments as much richer in Zn and Pb, and the mineral sediments richer in Al, Mg and particularly Mm (Table III).

TABLE III: SELECTED ELEMENTS IN SEDIMENTS
(g/100 g air dried weight, means)

	% Ash	Ca	Mg	Al	Zn	Pb	Mn
"Organic"* sediments (n = 14)	48	0.33	0.19	3.0	0.24	0.74	0.28
"Mineral" sediments (n = 18)	86	0.37	0.40	7.8	0.07	0.27	1.61

* Defined as >50% LOI

3.3. Aquatic Biology Programme

Fisheries investigations at Loch Fleet aim to discover why fish are not now present, and to identify the remedial action needed to re-establish them. Historic trout catches (from 1936) have been obtained (Figure 2); although not strictly quantitative, a dramatic decline in fish caught between 1950 and 1960 is evident. Stocking of fish was rarely reported and so presumably the fishery prior to 1960 was sustained by natural spawning. A qualitative survey employing gill nets and traps in the lake, and electrofishing in the streams, confirm that the lake is fishless, and that trout are not found above a waterfall (5 m height) 7 km downstream. In general the water quality below the fall is "better", possibly associated with a more diverse geology and a sparse deciduous tree cover. A pH of ~7 has been recorded there during low flow periods but during critical spring high flows pH was 4.1 (cf. Loch Fleet pH 4.1).

During the 1984/85 winter, potential spawning areas in the Fleet system were identified, and >10% survival of artificially implanted freshly fertilised eggs was observed at sites considered accessible to a resident population. Survival of eyed eggs was <90%, however. Early mortality was almost certainly due to poor water quality (low pH, low Ca, high Al) (Howells, Brown and Sadler, 1983), as well as to freezing at some sites. Spawning potential is thus minimal at present, and recruitment failure can account for the loss of trout populations.

Baseline information on invertebrate and planktonic organisms is also being collected. Data on primary productivity and invertebrate standing crop, and their changes in response to altered water quality, will be used to estimate the size of fishery that could be supported.

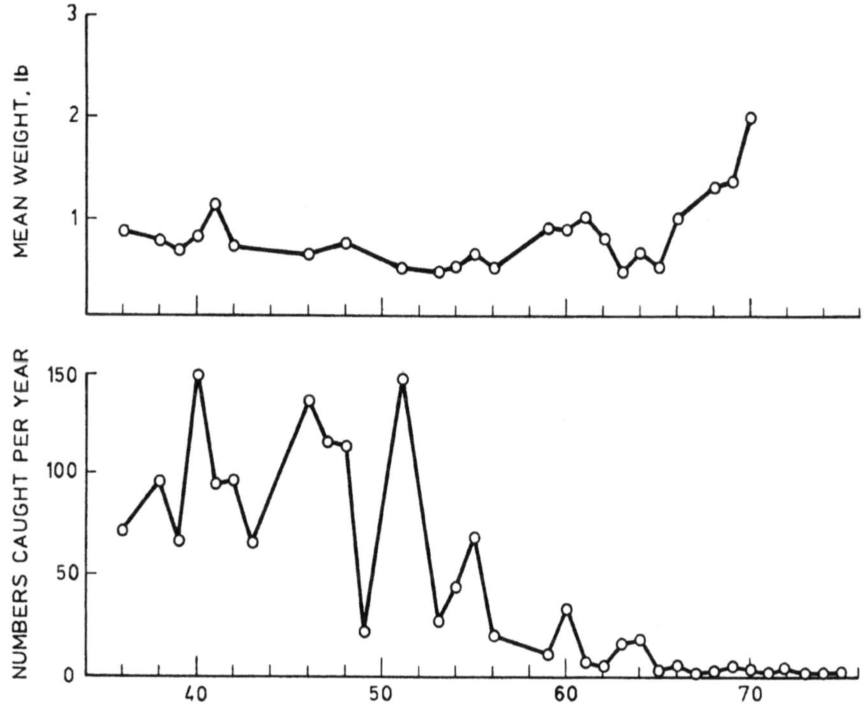

Fig. 2. Numbers of Brown trout caught by rod and line, and their mean weight, in Loch Fleet, 1936-1975.

4. MANIPULATIVE OPTIONS

The principal objective is to increase the base saturation and to recharge the cation exchange system within the soils. The options available include:-

- application of limestone or chalk materials on or in the soil, at different loading rates;
- improved access of drainage water to weathering mineral materials and bed rock;
- packing limestone materials the path of drainage waters;
- burning of living/dead organic materials to reduce biological acid production and the accretion of unsaturated exchange sites; and
- installation of flow or pH activated lime dosing systems in streams.

Soil treatments have first been tested in the laboratory on replicated soil cores; using limestone or chalk, of a range of size fractions. Finely ground limestone at a rate of 20 tonnes ha^{-1} has been applied dry to Sector VI, and as a slurry to Sector IV. Lime dust has also been loaded (10 tonnes ha^{-1} of wetland areas) to Splagnum beds in the upper part of Sector VII. Water quality changes are being monitored and initial and subsequent values for soil pH, BS and CEC will be obtained.

5. PRELIMINARY ASSESSMENT AND CONCLUSIONS

5.1.

Loch Fleet is a highly acid aquatic system and its present drainage water quality is unsuitable for brown trout, being acid (pH $\leq$ 4.5), low in Ca (approx. 1 mg L^{-1}), and possibly too high in Al (approx. 200 μg L^{-1}, total, 20 to 30% as "labile".

5.2.

The Loch is significantly more acid than its atmospheric deposition, so that substantial acidity must also derive from vegetation and soils within the catchment. Although the underlying bedrock (granite) is poor in base minerals, some Ca and Mg are mobilized, although in quantities insufficient to bring the pH/alkalinity into a range suitable for trout.

5.3.

Changes in land use: planting a part (12%) of the catchment with coniferous forest, reduced grazing by sheep and goats, reduced upland management by burning, may all have played some part in influencing water quality over the past 50 yr. However, no specific contemporaneous change in land use is recorded in the period 1950 to 1960 when fish were lost.

5.4.

This oligotrophic loch can never have been very productive. Further, the present spawning capacity is limited by the availability of suitable spawning beds, suggesting that major physical changes to the stream beds may have occurred.

5.5.

A range of manipulative treatments, designed to increase the base saturation and to saturate the cation exchange system of soils, sediments and bogs is planned, so as to bring the water quality into the range defined from bioassay studies as suitable for brown trout survival and maintenance, from the early sensitive, newly fertilized eggs through to adult spawning.

5.6.

It is envisaged that effective and economic treatments will be applied to the lake and its catchment, and that improved water quality will result leading to the successful restocking of a trout.

ACKNOWLEDGEMENTS

Many people have contributed information which has been used to illustrate this paper. In particular, we wish to thank J.J. Smith (SSEB), analytical chemistry; G. Glover (NCB), lake sediments; J. Bibby (Macaulay Institute), soil characteristics; A.W.H. Turnpenny (CEGB) fishery studies; R.W. Bell (EAU, Liverpool), manipulative treatments. The project is supported by the CEGB, SSEB, NSHEB and the NCB and this paper is published with permission of the Central Electricity Generating Board.

REFERENCES

Bennett, M., Chown, M., and Gwyn, B.: 1985, Measurements of SO_2, NO_x and O_3 at Loch Fleet, 24-28 June 1985, CEGB Report (pers. comm).

Bown, C.J., and Heslop, R.E.F.: 1979, The Soils of the Country Round Stranraer and Wigton, Macaulay Institute for Soil Research.

Burns, J.C., Coy, J.S., Tervet, D.J., Harriman, R. Morrison, B.R.S., and Quine, C.P.: 1984, Fish. Management, 15, 145.

Fraser, J.E. and Britt, D.L.: 1982, Liming of Acidified Waters: A Review of Methods and Effects on Aquatic Ecosystems, FWS/OBS-80/40.13, US Fish and Wildlife Service.

Howells, G.D., Brown, D.J.A., and Sadler, K.: 1983, J. Sci. Food Agric., 34, 559.

THE DISSOLUTION EFFICIENCY FOR DIFFERENT STREAM LIMING METHODS

H. U. Sverdrup
Dept. of Chemical Engineering, Chemical Center, LNTH
P.O. Box 124
S-221 00 Lund
Sweden

ABSTRACT: The dissolution efficiency for different stream liming methods is shown to depend upon the following parameters: Particle size distribution; Initial pH-value; Dose intensity; and Degree of wet slurrying. The particle size distribution and the degree of wet slurrying are the only free variables and when they are determined, the others are determined by the conditions at the dosing site. The way to influence the dissolution efficiency, and accordingly the cost of neutralization, will be to vary the particle size distribution of the powder used for dosing. This way the cost optimum may be located.

1. INTRODUCTION

In Sweden, the first serious effects of acidification started to appear in our lakes during the 1960s. By then several decades of acidic deposition had changed the upper soil layers in regions with low weathering, and the runoff could no longer be completely neutralized during its passage through the soil profile.

Today, some 20 000 lakes of a total of 100 000 in Sweden will have a pH-value below pH 5.5 during a substantial part of the year. Every year new lakes and rivers will become acidic as the soil in their watersheds continue to acidify.

Liming is in no way a solution to the acidification problem. Still, it is the only large scale remedy available for the postponing of some of the effects of acidification until emissions of SO_2 and NO_x are reduces.

For the fiscal year 1985-86, the Swedish Parliament has budgeted 90 million SEK for acidificaiton mitigation measures ($10 million). In addition, the local authorities in countries and provinces will spend additional money as financial support of projects, as well as make personnel available adding cost of another 15 to 20 million SEK.

The large program results in an increased demand for competent selection of acidified objects to be considered, as well as efficient administration and planning, competent technical planning and cost efficient liming operations.

Water, Air, and Soil Pollution **31** (1986) 827–837.

2. STREAM LIMING TECHNOLOGIES

In Sweden, calcite powder dosers have been extensively used for neutralizing acid streams generally in order to protect fish spawning grounds. By the end of 1984, approximately 75 large calcite dosers and an equal number of small units had been installed, consuming an estimated amount of 20 000 ton of calcite powder each year.

The economic performance of different technical solutions may be very different and this is closely related to the utilization of the calcite powder.

The dissolution efficiency may be defined as the total utilization of the calcite seen from the target area, regardless of where it was added to the stream.

The calcite is purchased and paid for per bulk tonne, but the neutralization downstream will only be effected by the actual amount dissolved int the water.

2.1. Important parameters

It is known from theoretical studies and from lake liming (2.4.5.6.7) that among the factors influencing the dissolution efficiency, a few are dominating. The strongest factor by far is the grinding grade of the powder used, usually defined by the particle size distribution curve. In addition, the initial pH-value of the water, the dose intensity and the degree of wet slurrying before dosing, will include the main governing parameters.

Accordingly, we will try to relate the following parameters, ranked after their relative importance: Grinding grade; Initial pH-value; Dose intensity; and Degree of wet slurrying, as independently as possible to dissolution efficiency.

The settling depth, which is rather important in lake liming, has a more marginal influence in rivers. Calcites used for liming in Scandinavia are very low in Mg carbonate content and therefor the effect of the dolomite content is not considered in this study.

2.2. Confounding factors

Several confounding factors are encountered during the analysis of the material.

In Sweden, most calcite powders used in liming are marked under commercial names. Names like 0 to 0.2 mm lake liming powder, 0 to 0.5 mm, 0 to 0.1 mm aglime are often used, giving an impression of describing the average particle size of the material. However, only a wet sieved particle size distribution, in the form of a diagram or table correctly describes the true distribution of particle sizes contained in a powder. As an example, the powder sold as 0 to 0.2 mm from the Ignaberga Quarry, has approximately the same particle size distribution curve as 0 to 0.125 mm from the Gåsgruvan Quarry. The origin of the calcite powder used at a specific site is not regularly reported and this information had to be collected to enable the evaluation of some of the cited material (3).

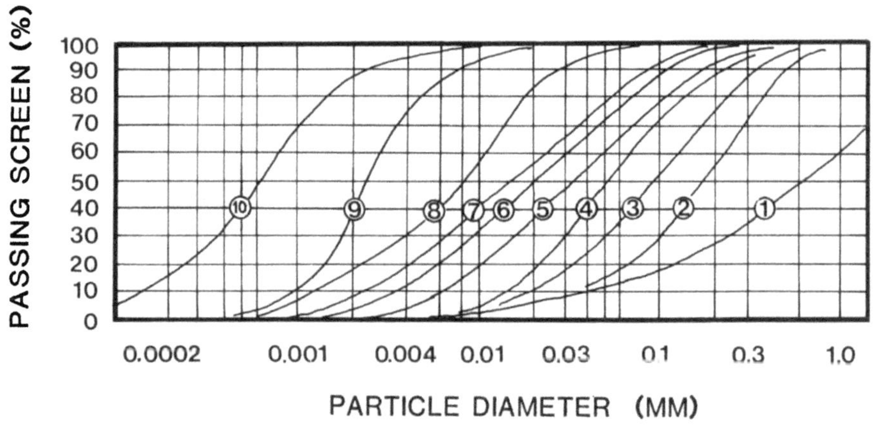

Figure 1. Particle size distribution of calcite powders used for liming. Frquently used commercial names are 1-4: 0-3 mm aglime, 2-4: 0-2 mm aglime, 4-5: 0-1 mm aglime, 6: 0-0.5 mm liming powder, 7: 0-0.2 mm, 0-0.125 mm, 0-0.1 mm liming powder, 8: 0-0.05 mm, 0-0.044 mm, 9-10: filler slurry.

Figure 1 shows the particle size distribution curves for some Swedish calcite powders used for liming, together with their commercial names.

The degree of wet slurrying before dosing is important as it separates agglomerated particles in the powder. In a normal dry powder, a substantial fraction of all smaller particles will adhere to the larger ones and without thorough wet slurrying it will behave as a coarser powder (2,4).

The degree of wet slurrying may be mechanically achieved by special arrangements on the doser or by turbulent mixing in the stream itself. Laboratory experiments show that very vigorous mixing is needed (4). That implies that very turbulent flow downstream a doser is required in order to have a significant effect on the dissolution efficiency.

In many streams there is a small but significant rise in the Ca concentration down the stream. If the dissolution efficiency is measured between points widely separated this may show up as a systematic error if not considered and measured.

2.3. Confounding factors in Tideström's survey

Tideström, in his extensive survey (3), comissioned by the Swedish Environmental Protection Board, was careful to consider the possible effect of a slowly rising Ca concentration downstream without dosing. He also records whether the dosers were equipped with special wet slurrying units or not, as well as the dose intensity and the pH-value prior to liming.

However, in the survey of some 30 dosers, he only gives the commercial names for the calcite powders used, and the quarry that delivered the calcite is not listed.

By contacting all the local authorities or responsible persons involved, the origin of the calcite powders used could be established. The results are given in Table I.

TABLE I

Commercial name	Quarry	Curve No.	Doser type	Commercial name
0- 3 mm	Ignaberga	1	DW	Tryssjo
5- 9 mm	Gotland	-	DW	Mover
0- 2 mm	Ignaberga	4	DW	Sundtorpsaan
0- 0.5 mm	Igna berga	6	DW	Sundtorpsaan
6- 14 mm	Gaasgruvan	-	DW	Tobybrunn
12-30 mm	Gaasgruvan	-	DW	Tobybrunn
0-0.2 mm	Ignaberga	7	DD	Hällefors
0-0.2 mm	Ignaberga	7	DD	Borlänge
0-0.125 mm	Gaasgruvan	7	DD	Borlänge
0-0.2 mm	Ignaberga	7	DD	Tomal
0-0.125 mm	Gaasgruvan	7	DD	Gnosjö
0-0.2 mm	Ignaberga	7	WSD	Boliden
0-0.2 mm	Ignaberga	7	WSD	Movab
0-0.2 mm	Kopinge	7	WSD	Boliden
0-0.2 mm	Kopinge	7	WSD	Movab
0-0.2 mm	Nordcen	7	WSD	Movab
0-0.015 mm	Malmökrita	9	SD	Sydkraft
0-0.015 mm	Malmökrita	9	SD	Movab
0.005> mm	Hustadmarmor	10	SD	Norw.Lim.Proj.

The table shows the commercial names of the calcite powders reported in Tideström's survey, related to their true particle size distribution curves as shown in Figure 1. DW signifies diversion well, DD Dry applicaiton doser, WSD stands for wet slurried doser, SD for slurry doser.

3. DISSOLUTION

3.1. Dissolution efficiency vs initial pH

From the cited references (2,3) information concerning the dissolution efficiency may be extracted as a function of different initial pH-values. To sort out the possible effects of other parameters, we will consider only one grinding grade, No. 7, and low dosing intensities, below 20 g m^{-3}. To avoid one of the confounding factors, only data from wet-slurried limings were used. The resulting diagram, when initial pH-value is plotted versus the dissolution efficiency, is seen in Figure 2.

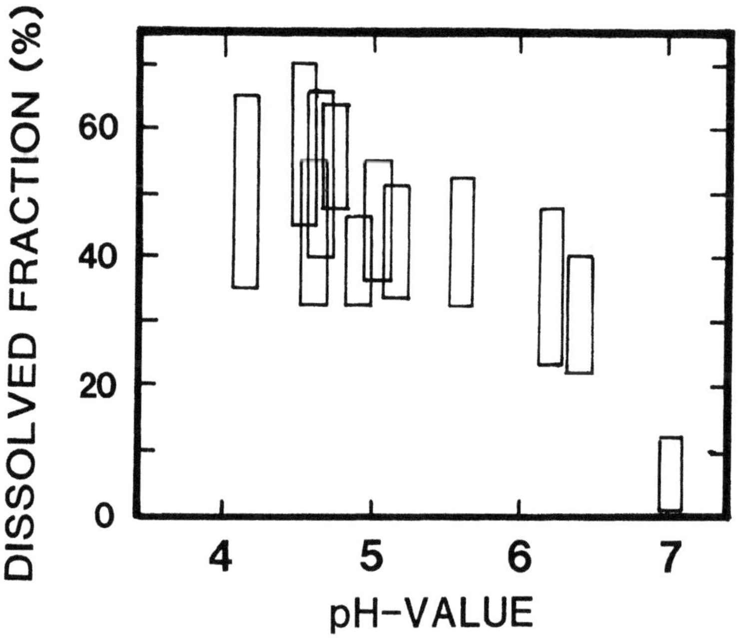

Figure 2. The diagram shows the relation between initial pH-value and the dissolution efficiency calcite powder No. 7, called 0-0.2 mm at low dosing intensities, 0-20 g m^{-3} when added wet slurried. The bars represent the accuracy of the measurements as reported by Tideström in his report (3).

3.2. Dissolution efficiency vs dose intensity

To be able to sort out the effect of dose intensity on the dissolution efficiency, it is necessary to have data from different dose intensities, but all at the same initial pH, grinding grade and degree of wet slurrying. If the data available are screened with these criteria, we end up with 8 datapoints for the powder No. 7 at an initial pH-value around pH 4.6, and three points at pH 5.6 all wet slurried.

In Figure 3, the first set of data is shown related to the dissolution efficiency over a wide range of dose intensities, 10 to 50 g m^{-3}.

The diagram seems to indicate what may be deduced from theoretical considerations (4). The dosing intensity will have little effect at

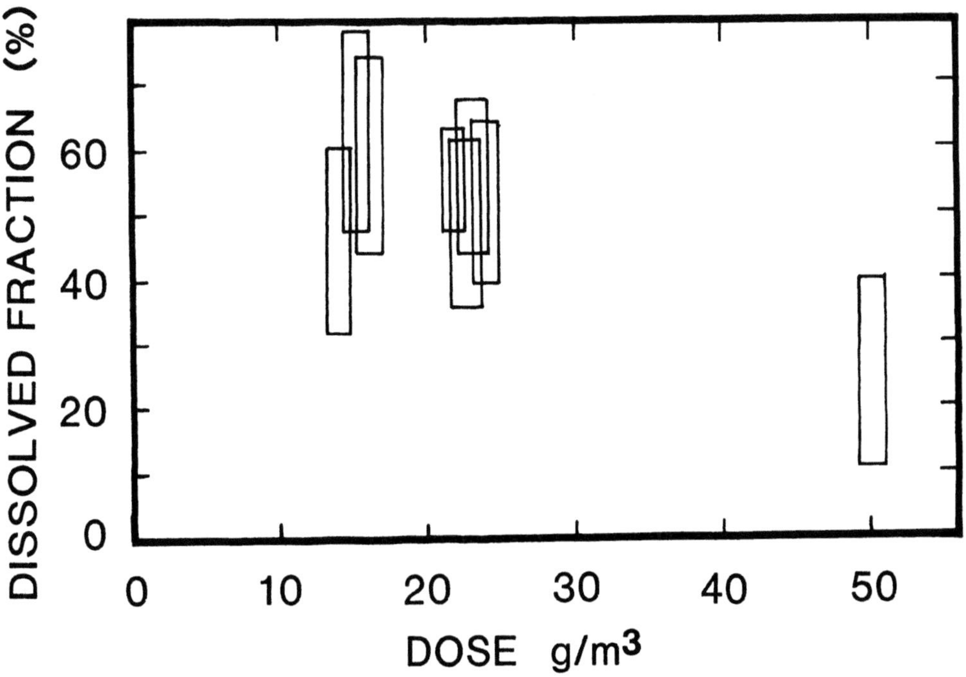

Figure 3. The dose intensity related to the dissolution efficiency for powder No. 7, 0-0.2 mm, at an initial pH-value around pH 4.5-4.7. The length of the bars represents the accuracy of the measurements.

doses less than 20 g m^{-3}, but it increases as high alkalinities are produced by large doses. Then the backward rate of dissolution will come to balance the forward rate, and the particles may sink wihtout dissolving.

3.3 Degree of wet slurrying vs dissolution efficiency

Several dosers on the Swedish market have no special arrangements for wet slurrying of the calcite powder. One is the "Borlänge", which uses electric vibrators to achieve dosing; the other is "Hällefors", which is a purely mechanical waterdriven device. Some old prototypes of the larger wet slurried dosers of Boliden, still distribute the powder dry (Högvadsån).

Data from the "Hällefors" are available and if we make a plot of the dissolution efficiency versus the initial pH, we arrive at Figure 4.

It can be seen that the dissolution falls below what may be expected from wet slurried dosing. The powder used was No. 7 or 0 to 0.2 mm. A similar diagram may be produced with data from the "Borlänge" device, with powder No. 7, under the name of 0 to 0.125 mm.

The observed reduction in dissolution efficiency for dry application is in good agreement with laboratory experiments (4).

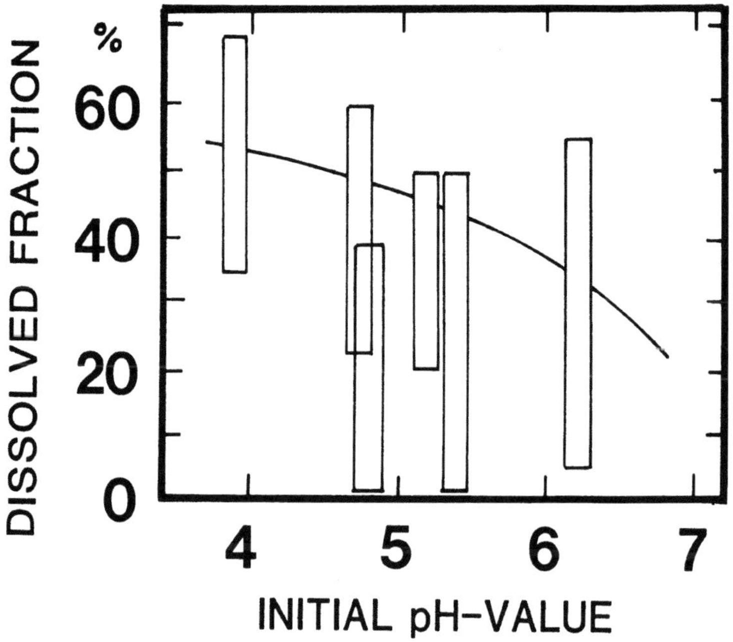

Figure 4. The dissolution efficiency of dry applied powder No. 7 or 0-0.2 mm by the "Hällefors" at low dose intensities. The line (--) represents the expected dissolution with wet slurrying.

3.4. Conclusions on dissolution efficiency

The effect of the main parameters grinding grade, initial pH, dose intensity and degree of wet slurrying on the dissolution efficiency may be interpreted from the available data, in a simple diagram. For three different grinding grades dissolution diagrams may be drawn for low dose intensities, less than 20 g m^{-3}. Such a diagram is shown in Figure 5.

The dagram applies to wet slurried dosing only of the powders No. 6, 7 and 9, commercially called 0 to 0.5 mm, 0 to 0.2 mm or 0.to 0.125 mm filler. The confidence limits for the line representing powder No. 6 are quite wide.

Dry application dosing usually implies a reduction in the observed dissolution efficiency in relation to wet slurried dosing. The reduction is approximately 25 to 50 % of wet slurried dissolution.

The effect of the dose intensity could be investigated for two grinding grades, powder No. 7 and No. 9. Powder No. 9 is so finely ground that it will always produce very near saturation concentration when applied in large doses.

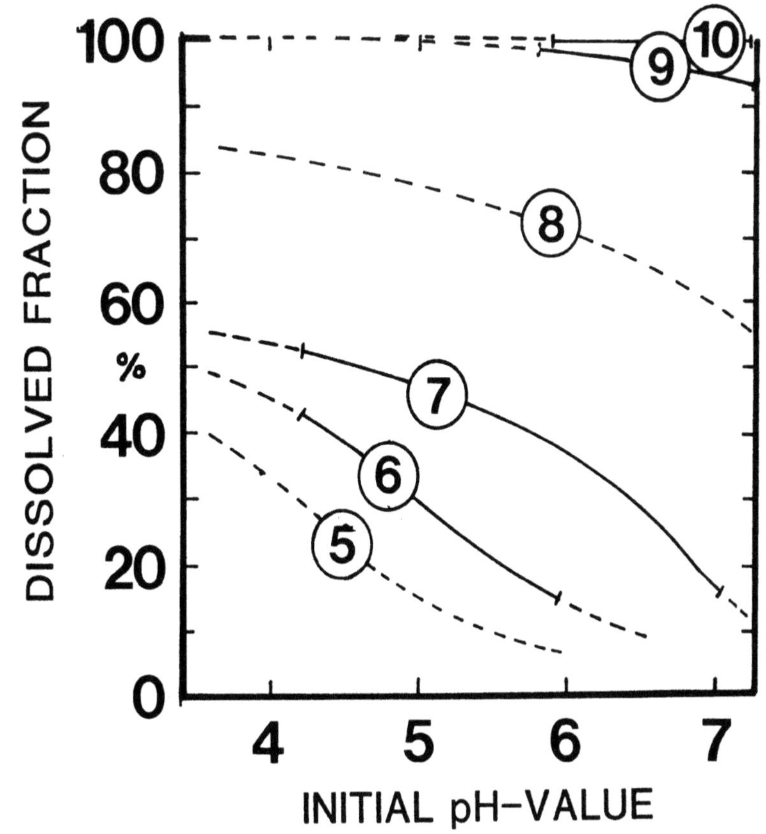

Figure 5. A diagram that my be used to estimate the expected dissolution when a calcite powder is added to a stream. The drawn lines (—) represent what is indicated by the data, the dotted lines (---) are extrapolations. The numbers correspond to the grinding grades shown in Figure 1. The average particle surface related diameters of the powders are: No. 6; 14 m, No. 7; 12 m, No. 8; 7.5 m, No. 9; 3.5 m. Powders coarser than No. 7 will become relatively cost ineffective in stream liming.

The data investigated are summarized in Figure 6 where the effect of dose intensity is expressed as a reduction of the low dose dissolution efficiency. The diagram applies to wet slurried conditions only.

It can be seen that the dose intensity has little or no effect at doses lower than 25 g m^{-3}.

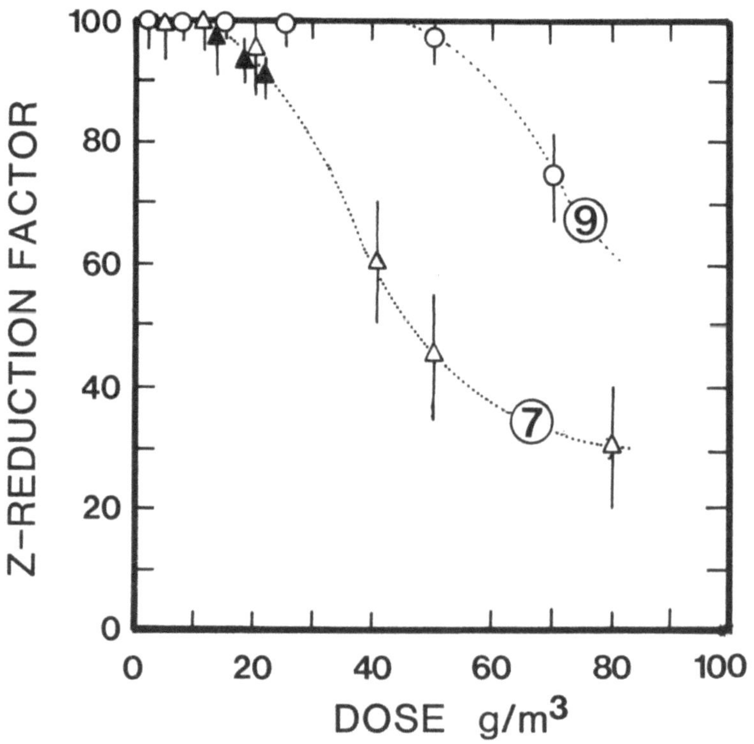

Figure 6. The reduction of the dissolution efficiency as percent of the limiting dissolution at very low dose intensity. The upper line is for powder No. 9 represented by circles (o), the lower for powder No. 7, marked with crosses (+). The length of the bars represents the accuracy of the measurements.

3.5. The effect of dissolution efficiency on the neutralization cost

As mentioned earlier, the dissolution efficiency will have a large impact on the neutralization cost, as the neutralization is only accomplished with the dissolved part of the calcite. One way to compare different stream liming methods is then to look at the cost effectiveness, which we may measure in cost per tonne? dissolved calcite, the neutralization cost, NC. In this measure we would also have to include the cost of capital investment in the dosing equipment used. The expression becomes:

$$NC = \frac{\text{(Total amount calcite)(brutto price)+(Capital cost)/(Total amount)}}{\text{Dissolution efficiency}}$$

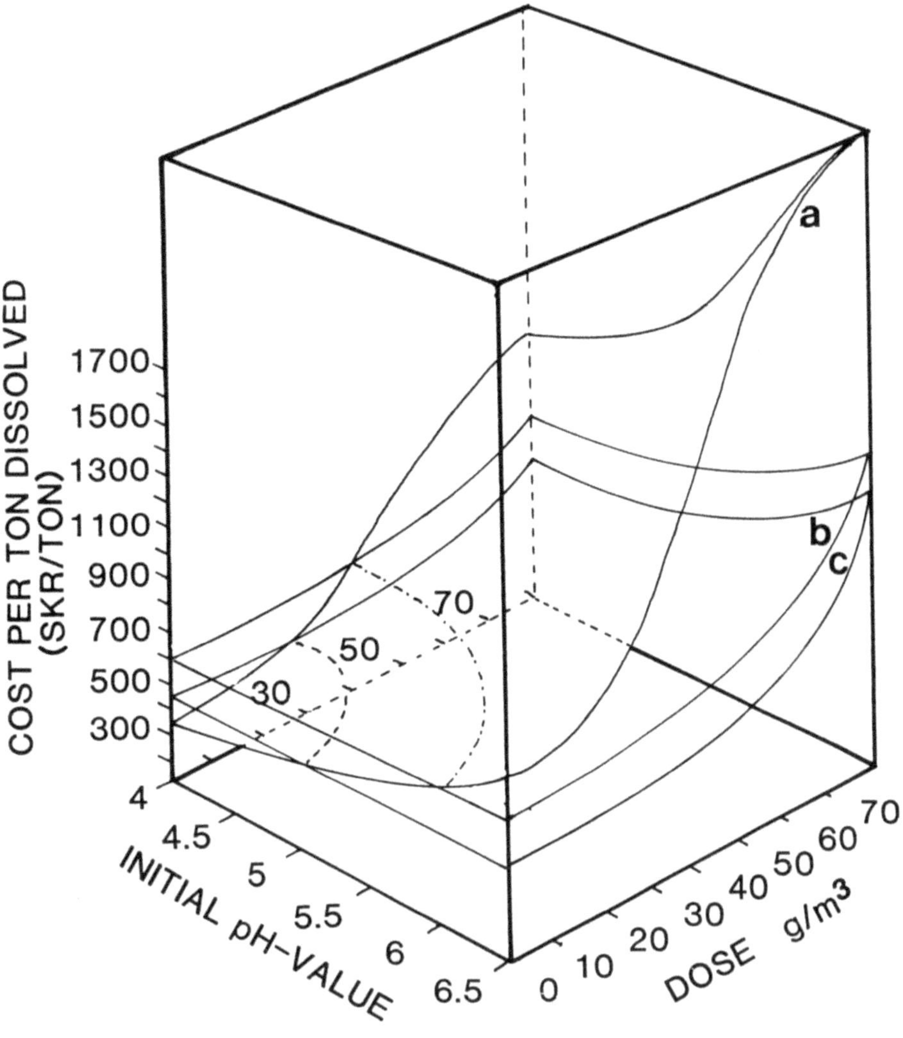

Figure 7. The response surfaces for the calcite powder No. 0-0.2 mm or 0-0.125 mm and the powder No. 9: filler. It is assumed that the cases imply the same capital costs, that both are dosed wet slurried and that the calcite costs are 190 Skr ton^{-1} and 450 Skr ton delivered at the doser site. Dosers usually operate in the range pH 4.5-6.0 and at doses from 5-50 g m^{-3}. a represents the responce surface for powder No. 7, o is the surface for powder No. 9. b is the responce surface for powder No. 9 as well, but the price was assumed to be 600 Skr ton^{-1} instead of 450 Skr ton^{-1}.

where NC has the dimension Swedish kronor per ton dissolved (SEK/ton)

As we have four strongly acting parameters which influence the neytalization cost, as well as the price for delivered calcite powder at the site, which is partly dependent on the grinding grade and the cost of capital investment, the picture becomes very complex. There is a problem with at least 6 independent variables.

4. CONCLUDING REMARKS

For well designed dosers, the neutralization cost efficiency achieved is not as much a function of the apparatus as it is of the characteristics of the neutralization agent and the chemical and physical conditions at the doser site.

It is possible to give bounds for the neutralization cost for different techniques, indirectly implying certain grinding grades or degrees of wet slurrying etc. but at a specific site cost optimization must be carried out specificly if desired, as no universal answers are realistic.

REFERENCES

1. Abrahamsen H.: 1984, Miljöverntekníkk 5;23, 26.

2. Sverdrup, H., Bjerle, I., Warfvinge, P., Östensson, P.:1983, Kalkning av rinnande vatten. Proceedings to symposium at Alvestad, Sweden, September 1984. CODEN: LUTKDH/TKKT-7015/1-112/1983.

3. Tideström, H.: 1984, Kalkning av rinnande vatten. Utverdering av kalkdoseringsanleggningar. Statens Naturvaardsverk SNV-PM 1873.

4. Warfvinge, P., and Sverdrup, H.: 1984, Chemica Scripta 24, 67.

5. Sverdrup, H.: 1984, Chemica Scripta 22, 12.

6. Sverdrup, H, and Bjerle, I.: 1982, Vatten 38, 59.

7. Plummer, L.N., Wigley, T.M.L., Parkhurst, D.L.: 1978, Amer. Jour. Sci. 278, 179.

CHARACTERIZATION AND NEUTRALIZATION REQUIREMENTS OF TWO ACIDIC LAKE SEDIMENTS

Thomas C. Young, James R. Rhea, and Genevieve McLaughlin
Department of Civil and Environmental Engineering, Clarkson University; Potsdam, New York 13676 USA

ABSTRACT. Preparations for neutralization of two acidic lakes in the southwest Adirondacks included determination of base neutralization capacity for sediments as well as the water column. For sediments this involved assessing vertical and horizontal gradients in sediment properties related to sediment-bound acidity. The analyses showed significant variance vertically (depth within the sediment profile) and horizontally (due mainly to depth of overlying water) in both lakes. Base requirements, equivalent to the acidity present in 2 cm of sediments ranged between 0.011 and 0.022 kg $CaCO_3$ m^{-3} for the two lakes. Preliminary observations indicate pH values and Ca content increased in the surficial sediments of both lakes after treatment.

1. INTRODUCTION

Chemical mitigation of the effects of acidic precipitation on aquatic ecosystems may be accomplished through the addition of limestone or other basic material. The Lake Acidification Mitigation Project (LAMP), sponsored by the Electric Power Research Institute, was initiated in the fall of 1983 to study technical questions concerning such base addition to acidic lakes. Project objectives have focused on assessing the scientific feasibility of acid lake neutralization from a comprehensive evaluation of the chemical and biological effects of treating three lakes in the Adirondack region of New York State. Part of the LAMP activities has involved determination of base neutralization requirements of the study lakes. The sediment-bound acidity has been included in this determination, since it represents base neutralizing capacity that may prevent achievement of treatment objectives (water column pH and alkalinity) and, further, may accelerate reacidification after treatment, if it is not part of base addition requirements.

We present a description of work performed for estimation of the sediment neutralization requirements for two of the LAMP study lakes and include an assessment of early data on changes in sediment pH and Ca content resulting from treatment. Estimation of the amount of base for neutralization of sediment acidity in each of the LAMP lakes required

Water, Air, and Soil Pollution **31** (1986) 839–846.

estimation of sediment acidity, sediment bulk density, and dependence of these quantities on lake bathymetry. Development of the whole-lake dose and logistics of base application for treatment of the LAMP lakes has been described elsewhere (Scheffe et al. 1985a,b; EPRI 1985).

2. METHODS OF STUDY

The two lakes selected for neutralization, Cranberry Pond (43° 51' 52.4" N, 75° 58' 34.7" W; $A_o = 7.2 \times 10^4$ m^2, z_{max} = 7.2 m) and Woods Lake (43° 52' 11.3" N, 75° 57' 8.8" W; $A_o = 2.5 \times 10^5$ m^2, z_{max} = 12 m), are situated in the southwestern Adirondack region of New York State (Big Moose quadrangle). Details of the hydrology and morphometry of the two lakes have been reported elsewhere (Scheffe et al. 1985a,b; EPRI 1984).

Finely-powdered, high-purity $CaCO_3$ (Nautex; Omya, Inc.) was applied to the two lakes as a slurry by aerial methods on 30-31 May 1985. The amount of calcite added to each lake was 22.6×10^3 and 7.2×10^3 kg, respectively. These amounts were nearly double the rates of application necessary to neutralize the water column alone. The excess base amounted to approximately 10.5×10^3 kg and 3.5×10^3 kg for Woods Lake and Cranberry Pond, respectively. The water column limestone excess also was greater than that required to achieve the sediment acidity treatment goal, as will be developed in this paper. The excess lime was retained for the treatment, however, for several reasons, partly economic, partly as a contingency, but mainly to test the value of having a reserve of unreacted limestone to counteract post-treatment reacidification. Thus, the actual treatments appeared more than adequate for neutralization of significant amounts of sediment acidity in the two lakes.

The design of the sediment sampling plan made allowances for potential spacial heterogeneity of sediment properties, expected to be largely a function of overlying water column depth, but also dependent on location relative to specific reference points (ie. inlets, outlets, etc.). To accomplish this a stratified sampling program was employed which included water column depth and spatial orientation in selection of sampling locations.

Sediments were collected from the two lakes before and after treatment by gravity coring methods using a corer with a 5 cm diameter plexiglas liner. Sediment cores were segmented into 2 to 3 cm increments to a depth of 12 cm each. All cores were sliced in the field immediately upon collection.

Physical and chemical variables quantitated for the core segments included moisture content, organic matter by loss on ignition, bulk density, pH, exchangeable acidity, total Ca, and total Mg. Moisture, loss on ignition, and bulk density determinations were performed according to standard soil test procedures (Page 1982). Sediment acidity was estimated using a modification of the rapid method for exchangeable hydrogen ion determination described by Brown (1943). Total Ca and Mg were determined by flame atomic absorption spectroscopy according to manufacturer's recommendations (Perkin-Elmer) on

HNO_3-digested sediment aliquants in an aqueous matrix containing 1% HNO_3/1% $LaCl_2$ (v/v). Interstitial pH of sediment cores was determined by direct insertion of a temperature compensated, calibrated, combination pH electrode into intact sediment cores at 2 cm intervals.

3. RESULTS AND DISCUSSION

Pre-treatment sediment data, summarized in Tables I and II, were analyzed by analysis of variance and covariance to determine differences among lakes and spacially within lakes in order to permit appropriate weighting of dose calculations for systematic heterogeneity in sediment acidity.

A small but significant difference existed in average moisture content between lakes ($p<0.05$), due mainly to the lower moisture content of sediments from shallow regions of Woods Lake compared with other sediment samples from Woods Lake or Cranberry Pond. Minor differences ($0.05<p<0.1$) were detected between lake-wide averages of sediment bulk densities when corrected for varying moisture content (adjusted by regression to the mean moisture content). The average moisture-corrected bulk density for shallow water sediments in Woods Lake was significantly greater ($p<0.05$) than that observed for shallow water samples from Cranberry Pond and seemed to be due to relative enrichment by sand in the former samples compared to relative enrichment by organic matter in the latter.

Exchangeable sediment-bound acidity, though quantitated per unit of dry sediment weight, depended strongly on sediment moisture content ($p<0.05$). Moisture content alone would not contribute directly to sediment-bound exchangeable acidity. Consequently, this relationship likely reflects the effects of other sediment variables not included in the analysis, such as amount of organic matter or hydrous metal oxides. Significant differences did not occur between lake-wide averages of moisture-corrected exchangeable acidity in the bottom sediments ($p>0.1$), but within each lake the levels of exchangeable acidity fluctuated with the magnitude of overlying water depth ($p<0.05$). In both lakes the shallow water sediments contained the highest amounts of exchangeable acidity.

From the spacial distribution of sediment acidity and bulk density revealed by these analyses, calculations were performed for estimation of the amount of $CaCO_3$ required to neutralize all of the exchangeable acidity in 2 cm of sediment in both lakes. The estimates were based on the acidity and bulk density of either (1) the top 2 cm core segments or (2) the 12 cm core profile averages. Selection of 2 cm as the depth of sediment for treatment was somewhat somewhat arbitrary, but it was felt that particulate limestone would not penetrate to a depth greater than 2 cm.

Calculation of sediment dose was based on the assumption that the reaction of limestone with exchangeable sediment acidity followed a 1:1 equivalent stoichiometry and could be represented such simple neutralization and exchange reactions as:

Table I. Pre-liming values for characteristics of Woods Lake sediment as a function of water column depth and depth in sediment core.*

Water Column Depth Range (m)	Characteristic	Units	Depth in sediment (cm) 1	3	5	7	10
Less Than 1.5	Moisture Content	%	94.72 (0.54)	89.85 (2.94)	85.63 (6.89)	84.96 (6.64)	83.91 (7.03)
	Bulk Density	g(dry)/mL(wet)	0.05 (.00)	0.10 (0.03)	0.15 (0.08)	0.16 (0.08)	0.17 (0.08)
	Interstitial pH		4.78 (–)	5.60 (–)	6.20 (–)	6.62 (–)	7.00 (–)
	Exchangeable Acidity	meq/g	0.27 (0.06)	0.24 (0.08)	0.23 (0.10)	0.23 (0.11)	0.23 (0.10)
	Organic Matter	%	36.31 (2.81)	28.05 (8.19)	23.95 (11.86)	22.59 (12.11)	21.22 (12.55)
	Total Ca	mg Ca/g	0.54 (0.11)	0.34 (–)	0.37 (–)	0.31 (–)	0.54 (–)
	Total Mg	mg Mg/g	0.44 (0.08)	0.45 (–)	0.39 (–)	0.38 (–)	0.43 (–)
1.5–3.9	Moisture Content	%	96.68 (0.81)	94.51 (2.05)	93.82 (2.04)	91.92 (2.70)	89.63 (2.25)
	Bulk Density	g(dry)/mL(wet)	0.03 (0.01)	0.06 (0.02)	0.06 (0.02)	0.08 (0.03)	0.11 (0.02)
	Interstitial pH		6.22 (–)	6.38 (–)	6.43 (–)	6.43 (–)	6.47 (–)
	Exchangeable Acidity	meq/g	0.25 (0.04)	0.26 (0.04)	0.27 (0.04)	0.26 (0.05)	0.21 (0.04)
	Organic Matter	%	45.09 (4.13)	42.60 (5.76)	40.24 (6.06)	35.74 (5.57)	31.62 (3.78)
	Total Ca	mg Ca/g	0.72 (0.14)	0.76 (–)	0.53 (–)	0.51 (–)	0.97 (–)
	Total Mg	mg Mg/g	0.45 (0.06)	0.50 (–)	0.55 (–)	0.54 (–)	0.48 (–)
Greater Than 3.9	Moisture Content	%	96.29 (0.44)	93.81 (0.40)	93.17 (0.62)	91.63 (0.88)	90.82 (1.16)
	Bulk Density	g(dry)/mL(wet)	0.04 (.00)	0.06 (.00)	0.07 (0.01)	0.09 (0.01)	0.09 (0.01)
	Interstitial pH		6.03 (–)	6.14 (–)	6.21 (–)	6.21 (–)	6.21 (–)
	Exchangeable Acidity	meq/g	0.26 (0.06)	0.27 (0.04)	0.30 (0.03)	0.27 (0.04)	0.25 (0.04)
	Organic Matter	%	40.91 (3.59)	40.54 (2.06)	38.95 (0.92)	35.75 (1.62)	35.00 (1.87)
	Total Ca	mg Ca/g	0.83 (0.15)	1.14 (–)	0.82 (–)	0.98 (–)	1.19 (–)
	Total Mg	mg Mg/g	0.62 (0.04)	0.65 (–)	0.69 (–)	0.65 (–)	0.54 (–)

* Values are reported as means (and standard deviations) for duplicate analyses of triplicate samples (N=6), or representative values for unreplicated samples.

Table II. Pre-liming values for characteristics of Cranberry Pond sediment as a function of water column depth and depth in sediment core.*

Water Column Depth Range (m)	Characteristic	Units	Depth in sediment (cm) 1.5	4.5	7.5	10.5
Less Than 1.5	Moisture Content	%	95.96 (2.40)	95.85 (1.05)	92.61 (3.25)	89.89 (3.57)
	Bulk Density	g(dry)/mL(wet)	0.04 (0.02)	0.04 (0.01)	0.08 (0.02)	0.10 (0.03)
	Interstitial pH		6.00 (-)	6.10 (-)	5.30 (-)	6.30 (-)
	Exchangeable Acidity	meq/g	0.31 (0.01)	0.32 (0.03)	0.30 (0.07)	0.28 (0.06)
	Organic Matter	%	57.04 (1.16)	53.60 (4.33)	48.85 (9.55)	41.47 (9.75)
	Total Ca	mg Ca/g	0.75 (0.07)	0.51 (-)	0.60 (-)	0.70 (-)
	Total Mg	mg Mg/g	0.33 (0.03)	0.34 (-)	0.35 (-)	0.34 (-)
Greater Than 1.5	Moisture Content	%	95.51 (1.63)	94.86 (1.79)	92.62 (0.22)	91.77 (0.10)
	Bulk Density	g(dry)/mL(wet)	0.04 (0.02)	0.05 (0.02)	0.07 (.00)	0.07 (0.01)
	Exchangeable Acidity	meq/g	0.27 (0.01)	0.27 (0.02)	0.26 (0.01)	0.26 (0.01)
	Organic Matter	%	43.76 (2.52)	38.11 (0.54)	36.52 (0.28)	35.64 (0.81)
	Total Ca	mg Ca/g	1.16 (0.20)	1.09 (-)	0.82 (-)	1.11 (-)
	Total Mg	mg Mg/g	0.51 (0.02)	0.53 (-)	0.56 (-)	0.44 (-)

* Values are reported as means (and standard deviations) for duplicate analyses of triplicate samples (N=6), or representative values for unreplicated samples.

$$2R\text{-}H + CaCO_3 = 2\text{-}R^- + Ca^{2+} + CO_2 + H_2O$$

$$R\text{-}H_2 + CaCO_3 = R\text{-}Ca + CO_2 + H_2O.$$

Sediment dose calculations indicated calcite requirements of 2.71 X 10^3 and 0.87 X 10^3 kg for Woods Lake and Cranberry Pond, respectively, for treatment of the top 2 cm of each lake. For 2 cm of profile-averaged sediment, a more conservative treatment, the required treatments amounted to 5.55 X 10^3 and 1.29 X 10^3 kg of calcite for the two lakes. Calculation details for the sediment calcite requirements of the two lakes have been presented in detail elsewhere (EPRI 1985). Compared to the amounts of limestone used in the whole-lake treatments, the latter quantities are about one-half the excess limestone that was noted previously as being part of the whole-lake treatments. Thus, the dose for each lake should have been adequate for neutralization of several cm through the sediment profile.

Preliminary data collected during the first two months after treatment show increased sediment pH and Ca levels in upper sediment layers (Figure 1). The sediment pH profiles, determined approximately

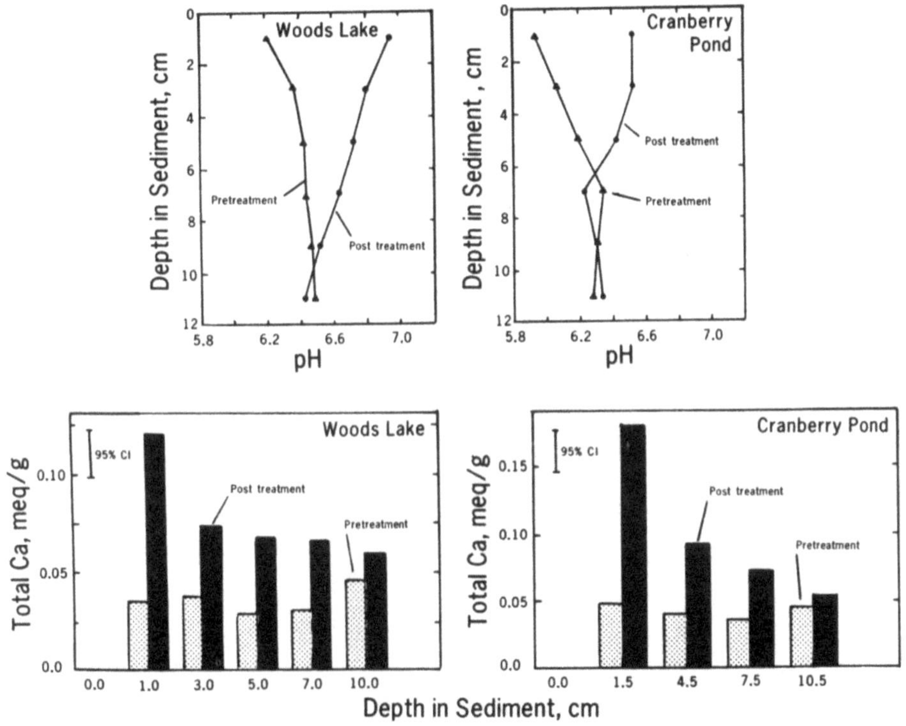

Figure 1. Effects of Lime Treatment on Sediment pH and Ca_T.

three weeks before and from 3 to 6 weeks after treatment, indicate the effects of treatment in a reasonably straightforward manner which might be anticipated from knowledge of the high water content of the lake sediments. The pH near the sediment-water interface followed the change in pH of the overlying lake water rather closely in both lakes; consequently, a reversal in the sign of the vertical pH gradient occurred after application of limestone to the water column of the lakes. The extent to which the limestone neutralized sediment-bound acidity has yet to be determined. The Mg content of the limestone used for treatment was quite low (1% w/w) and did not change through the sediment profile as a result of treatment ($p>0.1$). On the other hand, the Ca content of the surficial sediments in both lakes did increase significantly ($p<0.05$) above the levels present before treatment (Figure 1). It remains to be determined whether the extent to which the increase in sediment Ca may be associated with such factors as exchange of Ca for sediment-bound acidity, settlement losses and burial of solid phase $CaCO_3$, or adsorption of Ca cations from lake and sediment pore water; however, research on LAMP is in progress to answer these questions.

4. ACKNOWLEDGEMENTS

The authors wish to acknowledge the Electric Power Research Institute (Palo Alto, CA) for supporting this research by a grant to Clarkson University through the General Research Corporation (McLean, VA). The research has been part of the Lake Acidification Mitigation Project (LAMP, EPRI Project No. RP 2337). We also wish to acknowledge the assistance of our co-investigators at Clarkson: J.V. De Pinto and R. Scheffe and that of the other institutions who are participating in the field portion of LAMP: Cornell, Indiana, and Syracuse Universities, and the US Geological Survey.

5. LITERATURE CITED

Brown, I.C.: 1943. 'A rapid method of determining exchangeable hydrogen and total exchangeable bases of soils.' Soil Sci. **56**, 353-357.

EPRI: 1984. Lake Acidification Mitigation Project: First Annual Report. Electric Power Research Institute, Palo Alto, CA.

EPRI: 1985. Lake Acidification Mitigation Project: Second Annual Report. Electric Power Research Institute, Palo Alto, CA.

Page, A.L.: 1982. Methods of Soil Analysis. Amer.Soc. Agron., Madison, WI.

Scheffe, R.D., Bilz, K., and De Pinto, J.V.: 1985a. Laboratory and Field Testing of Dose Calculation Methods for Neutralization of Adirondack Acid Lakes. International Symposium on Acidic Precipitation. Muskoka, Ontario, Canada.

Scheffe, R.D., Booty, W., and De Pinto, J.V.: 1985b. Development of Methodology for Predicting Reacidification of Calcium Carbonate Treated Lakes. International Symposium on Acidic Precipitation. Muskoka, Ontario, Canada.

SULFATE CONTENT AND ADSORPTION IN SOILS OF TWO FOREST WATERSHEDS IN SOUTHERN NORWAY

Bal Ram Singh *
Department of Soil Fertility and Management
Agricultural University of Norway, Box 28, N-1432
Ås, Norway

Dale W. Johnson
Environmental Science Division
Oak Ridge National Laboratory
P.O. Box X
Oak Ridge, TN 37831 USA

*Current Address: Misamfu Regional Research Station,
Soil Productivity Research Programme,
P.O. Box 410055,
KASAMA, Zambia.

ABSTRACT

Laboratory studies were conducted to determine the sulfate content and sulfate adsorption properties of soils from two watersheds in Southern Norway, which show no net sulfate retention in the field (input - output). Soils from both watersheds had generally high soluble sulfate contents, especially in surface horizons and in marsh soils. With one notable exception - an Fe podzol at Birkenes - adsorbed insoluble sulfate contents were low or negligible, even in cases in which free Fe oxides were fairly high. Aluminium oxide in the Fe podzol at Birkenes seemed to lend it considerable ability to adsorb sulfate. With a few exceptions - Fe-humus podzol (B-2) and Fe podzol (B-3) at Birkenes - there were minor changes in water soluble or adsorbed sulfate contents of the soils collected in early automn or in spring (after snow melt). Only one of the soils from Storgama (an Fe-humus podzol subsoil) and two of the soils from Birkenes (subsoils from an Fe-humus podzol and an Fe podzol) adsorbed additional sulfate with CaSO4 treatments. Most soils from these watersheds have limited ability to retain sulfate, a result which is consistent with the sulfate budgets at these sites. However, the real extent of the sulfate-adsorbing soils at both sites will strongly affect the rapidity with which watershed sulfate output responds to changes in atmospheric sulfate loading at these sites.

1. INTRODUCTION

It is now generally recognized that sulfate mobility is an important factor determining the effect of acid deposition on ion leaching from soils. This applies not only to the potential loss of base cations from soils (Johnson and Cole, 1977; Singh et al.,1980) but also to the transport of H^+ and Al^{3+} to aquatic ecosystems (Seip, 1980).

Water, Air, and Soil Pollution **31** (1986) 847–856.

A knowledge of a soil's ability to adsorb sulfate is essential to employ an appropriate anion model for assessing acid deposition effects on ion leaching from soils. Soil sulfate adsoption effectively reduces the ability of atmospherically deposited S tò cause ion leaching from soils simply by reducing the concentration of sulfate in soil solution (Johnson and Cole, 1977; Johnson et al., 1979; Singh et al., 1980; Lee and Webber, 1982). Sulfate adsorption generally occurs on hydrous Fe and Al oxides in subsoils, whereas organic matter has a blocking effect in surface soils and in B2hir horizons of spodosols (Harward and Riesenauer, 1966; Singh, 1980; Johnson and Told, 1983; Singh, 1984). Thus, very acid soils high in organic matter such as histosols and spodosols, should have low sulfate adsorption capacities and be conducive to the transfer of H^+ and Al^{3+} to surface waters. In this paper, we describe the results of some Fe, Al, and sulfate analyses as well as sulfate adsorption tests on soils from two forested watersheds: Birkenes and storgama.

2. MATERIALS AND METHODS

2.1. Site Description

Birkenes is a 4100-ha watershed located at 8°.30'E 58°.48'N at an elevation of 200 to 300 m, 20 km north of Kristiansand, on the south coast of Norway. Vegetation consists primarily of Scots pine (Pinus sylvestris) and Norway spruce (Picea abies), with some birch (Betula pubescens). Understroy consists of blueberry-bracken community (Eu-Piceetum myrtilletosum/dryopteridestosum) and pine community (Barbilophozia-Pinetum). Soils are primarily Spodozols and Inceptisols derived from glacial till overlying a biotite-granite and about half of the watershed has a soil depth of less than 20 cm (Frank, 1980). The profiles B-1, B-2, B-3, B-4, and B-5 were classified into Histic Humaquept, Lithic Haplorthod, Lithic Haplorthod, Haplaquod, and Typic Haplorthod, respectively (Frank, 1980). The areal extent of these soils has not been worked out.

The Storgama watersheds are located 8°E 59°N at an elevation of 500 to 700 m in Telemark, Norway. These watersheds are characterized by barren granitic bedrocks, peatland, and shallow outwash moraines (Braekke, 1981). Vegetation consists primarily of heather (Calluna vulgaris), with bog wortheberry (Vaccinium uliginosum) and mountain heather (Phyllodoce coerulea) (Braekke, 1981).

The soils have not been classified according to the USDA series, but the Norwegian classification for profiles S-1, S-2, S-3, S-4, and S-5 was Fe Podzol, Humus podzol, Marshy Soil, Fe-Humus podzol, and Marshy soil, respectively.

2.2. Soil Chemical Analysis

Soil samples were collected in August 1980 from five profiles, each from Birkenes and Storgama watersheds. The profiels at each site was exposed, and the samples were collected by horizons. These profiles were marked with wooden poles to locate them the following season.

These sites were sampled for a second time (new pits) in May 1981 in the same way. The samples collected in May 1981 were analyzed only for sulfate content to ascertain the variation in sulfate content after automn leaching and spring snow melting. All other analytical data are from samples collected in August 1980. All soil samples were airdried and passed through a 2-mm sieve prior to analysis. Sulfate extractions were conducted according to the methods described by Johnson and Todd (1983). Initial sulfate content was determined by sequential extractions of triplicate 10g of soil with 0.2 L of water (soluble sulfate) followed by 0.05 L of 0.16 M NaH_2PO_4 (adsorbed sulfate). Total sulfate was defined as the sum of sulfate extracted by H_2O and NaH_2PO_4. Sulfate in H_2O and Na_2SO_4 leaching was analyzed by ion chromatography. Sulfate content was also determined at the Norwegian Forest Research Institute (NISK), and the samples collected in May 1981 were analyzed only at NISK. Sulfate extracted by distilled water and 0.10 M ($Ca(H_2PO_4)2$) solution in 1:5 soil to solution (volume basis) ratio was determined by autoanalyzer as per procedure described by Ogner and Haugen (1977).

Free Fe and Al fractions were determined by separate (not sequentials) extractions with (1) 0.1 M $Na_4P_2O_7$, (2) 0.2 M ammonium oxalate adjusted to pH_2, and (3) the dithionite-citrate bicarbonate reagent according to standard soil survey methods (USDA, 1972). It is generally believed that extraction 1 removes organically bound Fe (Fe p); extraction 2 removes organically bound plus inorganic amorphous Fe (Fe o); and extraction 3 removes organically bound, inorganic amorphous, and crystalline Fe (Fe d) (McKeaque et all.,1971). Crystalline Fe (Fe c) is designated as Fe d - Fe o, and amorphous inorganic Fe (Fe a) is designated as Fe o - Fe p. The same general scheme can be applied to Al, but the extractions are less specific and include more overlap (McKeaque et al., 1971). Percent C was determined by combustion in an EC-12 LECO carbon analyzer.

2.3. Sulfate Adsorption

Sulfate adsorption was determined on selected profiles, based on results of initial sulfate analyses. Adsorption tests were conducted by placing 10g of air-dried soil in soil columns in a mechanical vacuum extractor and subjecting them to 5-m L aliquots of $CaSO_4$ at 4, 8, 16, and 32 Mg L^{-1} of sulfate-S. Interstitial water volume was calculated from volume recovery of the first aliquot, and a correction factor was applied to each subsequent aliquot to separate dilution effects of interstitial water from soil sulfate adsorption as a cause of loss of sulfate from solution.

3. RESULTS AND DISCUSSION

3.1. Soil Sulfate, Iron, Aluminium, and Organic Carbon Contents

Water soluble sulfate was highest in surface horizons of all soils from both Birkenes and Storgama (Figures 1 and 2). Water soluble sulfate contents from analyses at the Norwegian Forest Research Institu-

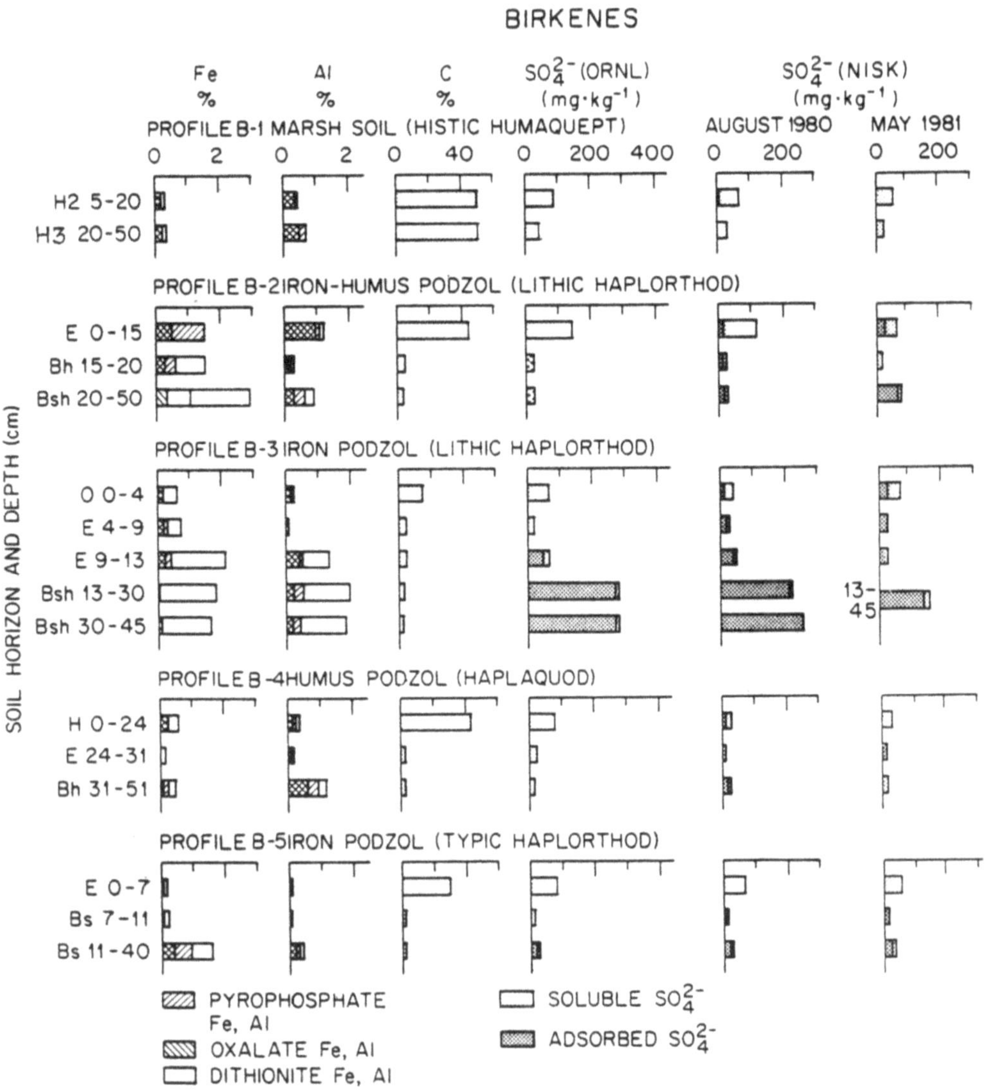

Fig. 1. Iron, Al, organic C, and sulfate (measured at ORNL and NISK) in soils from Birkenes Watershed

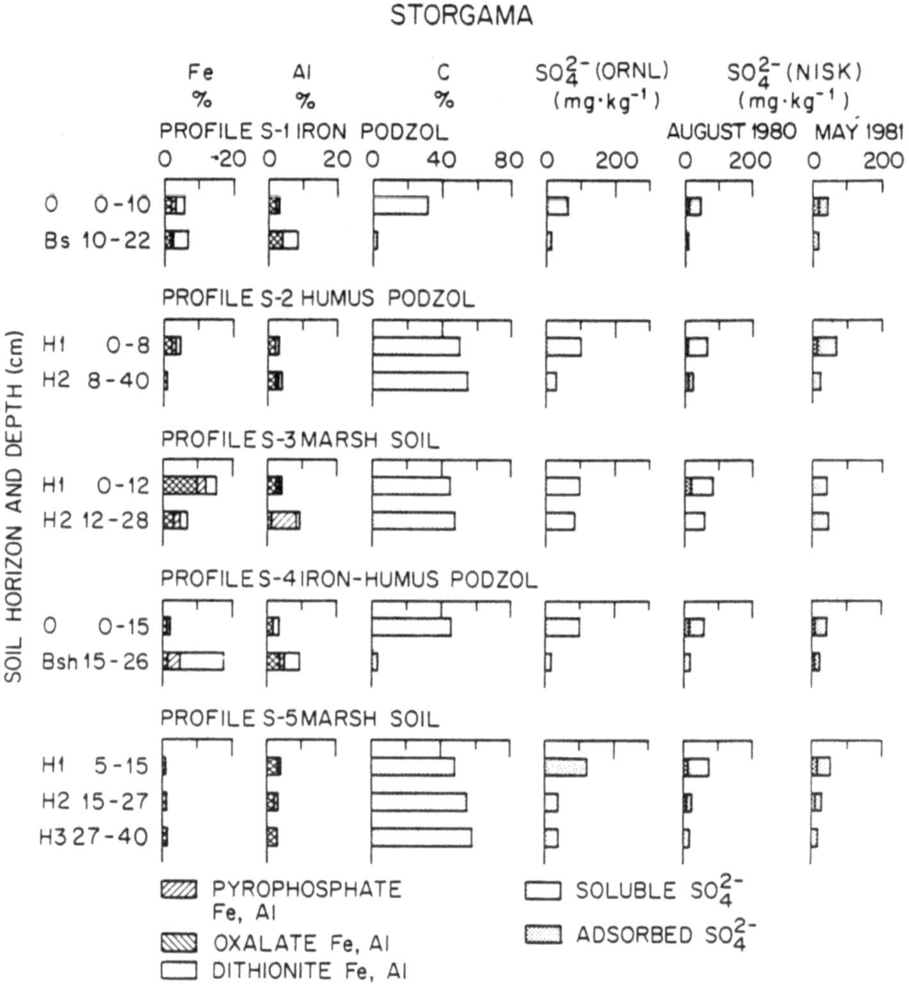

Fig. 2. Iron, Al, organic C, and sulfate (measured at ORNL and NISK) in soils from Storgama Watersheds

te (NISK) were generally lower than those determined at Oak Ridge National Laboratory (ORNL), but patterns were basically the same. These differences in analytical values may results, to some extent, from the different soil-to-solution ratios employed. At ORNL, the soil-to-solution ratio was 1:20 on a weight basis, but at NISK, it was 1:5 on a volume basis, thus varying from about 1:4.9 for mineral horizons to about 1:25 for organic horizons on weight basis.

The high soluble sulfate values are clearly associated with high percent carbon (Table I) as noted previously in other soils (Johnson and Todd, 1983), possibly reflecting a large pool of mineralizable S in these soils.

Adsorbed, insoluble sulfate was either absent or very low (< 20 mg kg^{-1} in all horizons of all soils except for the Fe podzol (profile B-3) at Birkenes (Figure 1). The Bsh horizons of this profile contained adsorbed sulfate values in excess of any measured by Johnson and Todd (1983), including a tropical soil from Costa Rica, and they were also in excess of those measured for other Norwegian soils (Singh, 1980).

The presence of large amounts of adsorbed sulfate in these horizons was clearly associated with high Fe c and Al c values and low percentage C values. Fe c was quite high in the Bsh horizon of the Fe-humus podzol (profile B-2) at Birkenes and the 15 to 26 cm layer of the Fe-humus prodzol (profile S-4) at Storgama, also; yet adsorbed sulfate was low or absent in the latter soils (Figures 1 and 2). It appears, then that Al was most responsible for the high adsorbed sulfate in profile B-3 at Birkenes. Correlation coefficients among all soils also reflect the importance of Al, as both Al c and Al d showed highly significant correlation with adsorbed sulfate (Table I) and Al correlated better than Fe in these soils. These results are in contrast to those reported by Johnson and Todd (1983), but corroborate those reported by Singh (1980).

Sulfate contents of samples collected in May 1981 after snow melt at Birkenes was generally lower than those collected in early automn. In the surface horizon of profile B-3, an apparent increase in water soluble was observed after snow melt (Figure 1). But, in the surface horizon of profile B-2, there was a decrease in water soluble sulfate from 1980 to May 1981. At the same time there was a large increase in adsorbed sulfate in Bsh horizon of this profile, suggesting that water soluble sulfate in the surface horizons was leached down to subsurface horizons by snow melting. Adsorbed sulfate in Bsh horizon of B-3 profile decreased from 237 mg kg^{-1} in 1980 to 147 mg kg^{-1} in May 1981. 1981, thus releasing a large amount of sulfate to lower layers or ground water. There were minor changes in adsorbed sulfate in other profiles from Birkenes and only minor changes in either water soluble or adsorbed sulfate of all profiles collected at Storgama from August 1980 to May 1981.

TABLE I

Correlation coefficients (r) between soil sulfate and Fe and Al fractions and organic C

	Soluble sulfate		Adsorbed sulfate	
	ORNL[a]	NISK[b]	ORNL[a]	NISK[b]
%C	0.71[c]	0.66[c]	-0.34[d]	-0.34[d]
Fe_d	-0.24	-0.18	0.40[d]	0.40[e]
Fe_o	0.39[c]	0.48[c]	-0.23	-0.21
Fe_p	0.24	0.34	-0.18	-0.14
Fe_c	-0.39	-0.34	0.51[e]	0.51[e]
Fe_a	0.06	-0.17	0.10	0.06
Al_d	-0.15	-0.14	0.78[c]	0.78[c]
Al_o	0.31	0.35[d]	-0.10	-0.07
Al_p	-0.25	0.30	0.12	0.14
Al_c	-0.31[d]	-0.33	0.89[c]	0.88[c]
Al_a	0.03	-0.06	-0.35[d]	-0.35[d]

a Oak Ridge National Laboratory

b Norwegian Forest Research Institute

c 99% level.

d Significant at the 95% level.

e 90% level.

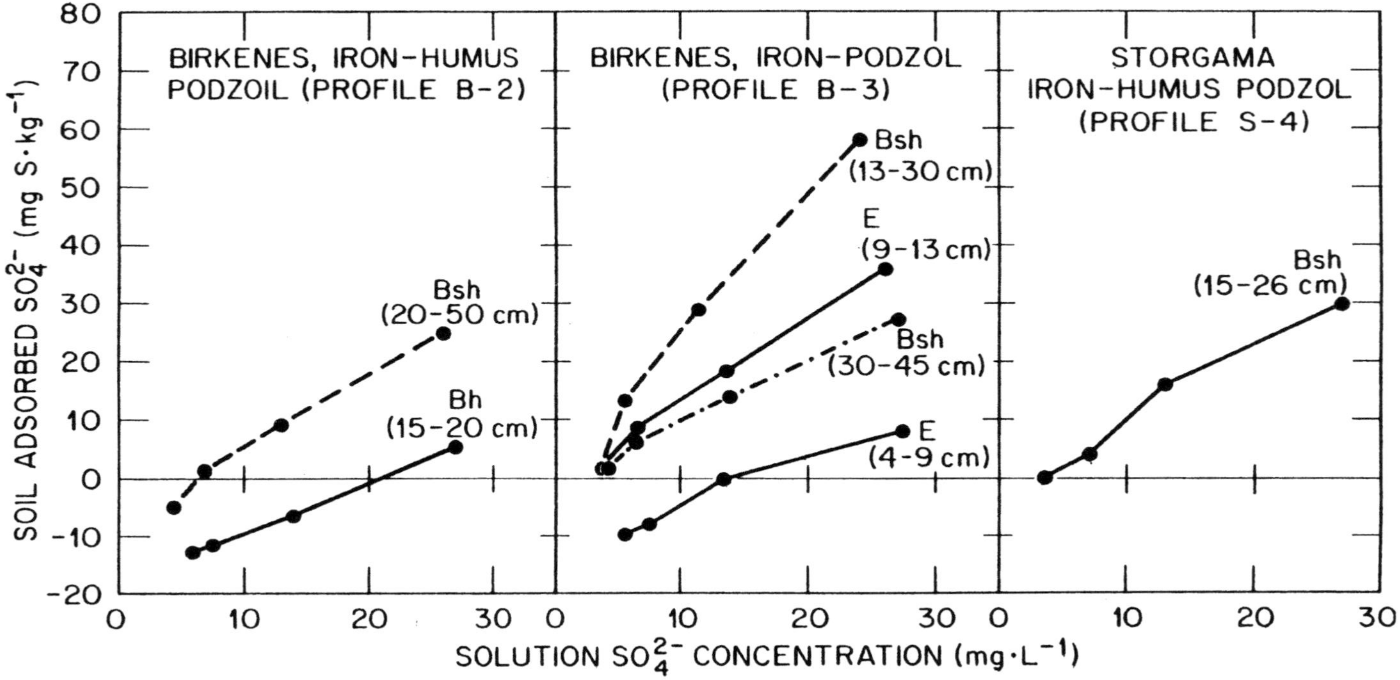

Fig. 3. Sulfate adsorption isotherms for selected soil horizons from the Birkenes and Storgama Watersheds.

3.2. Sulfate Adsorption by Soils.

Profiles B-2 and B-3 from Birkenes and profiles S-2, S-3, and S-4 from Storgama were subjected to sulfate adsorption tests as outlined in the methods. Of the five profiles tested, only subsoils from the Fe-humus podzol and Fe podzol (profiles B-2 and B-3) from Birkenes and subsoil from the Fe-humus podzol (profile S-4) at Storgama adsorbed any additional sulfate (Figure 3). Surface horizons from Birkenes and all but one horizon at Storgama lost sulfate even when treated with 32 mg L^{-1} of S, in $CaSO_4$ solutions.

In general, those soils that showed positive sulfate adsorption were high in Fe c, Al c, or both, and low in organic matter. Apparently the lack of sulfate adsorption in other soils was the result of higher organic matter content; this is especially obvious in the Storgama soils in which the main distinguishing feature of the one soil that did adsorb sulfate (profile 4, Bsh horizon) was its low organic matter content (Figure 2). It is interesting to note that the high initial sulfate contents of the Bsh horizons of profile 3 at Birkenes did not preclude further sulfate adsorption in column studies (Figures 2 and 3).

This is consistent with the concept that sulfate adsorption is concentration-dependent and, thus, to describe a soil as "saturated" with sulfate is misleading. The sulfate adsorption capacity of a soil is dependent on its basic properties such as Fe, Al, and organic matter content, but the net gain or loss of sulfate from any soil is dependent on the concentration of sulfate entering the soil relative to the soils, the equilibrium sulfate concentration as prescribed by the soils sulfate content, and its adsorption isotherm.

ACKNOWLEDGMENTS

The authors thank Drs. N. Christopherson and R. F. Wright for their help in sample collection and Dr. A. Stuanes for constructive suggestions in the preparation of this manuscript. Research sponsored by the Office of Health and Environmenal Research, U.S. Department of Energy, under Contract No. DE-AC05-84OR21400 with Martin Marietta Energy Systems, Inc. Publication No.____, Environmental Sciences Division, ORNL.

REFERENCES

Braekke, F. H. 1981, Medd. Nor. Inst. Skogforsk. 36, 6.

Frank, J. : 1980, Soil Survey at Birkenes, a small catchment in Augst-Agder County, Southern Norway (in Norwegian) SNSF-Project. TN 60/80.

Harward, M. E., and Reisenauer, H. M.: 1966, Soil Science 101, 326.

Johnson, D. W.: 1980, Site susceptibility to leaching by H_2SO_4 in acid rainfall, in Hutchinson, T. C., and Havas, M., (eds.) Effects of acid precipitation on terrestrial ecosystems. Plenum Press, New York/Toronto, P. 525.

Johnson, D. W., and Cole, D. W.: 1977, Water Air Soil Pollution 7, 489.

Johnson, D. W., Cole, D. W., and Gessel, S. P.,: 1979, Biotropica. 11,38
Johnson, D. W., and Todd, D. E.: 1983, Soil Sci. Soc. Amer. J, 47, 792.
Lee, J. J., and Weber, D. E.: J. Environ. Qual. 11, 57.
McKeaque, J. A., Brydon, J. E., and Miles, N. M.: 1971. Soil Sci. Am. Proc. 35, 33.
Ogner, G., and Haugen, A.: 1977, Analyst 102, 453.
Seip, H. M.: 1980, Acidification of fresh waters: Sources and mechanisms in Drablos, D. and Tollon, A. (eds.) Ecological impact of acid precipitation. Johs. Grefslie Trykkeri A/S, Mysen, Norway. P. 358.
Singh, B. R.: 1980, Acta Agric. Scand. 28, 313.
Singh, B. R., Abrahamsen, G., and Stuanes, A. O.: 1980, Soil Sci. Soc. Am. J. 44, 75.
Singh, B. R.: 1984, Soil Sci. 138, 294.
USDA, (U.S. Department of Agriculture): 1972, Soil Survey Laboratory methods and procedures for collecting soil samples. Soil Survey Investigations Report No. 1. USDA. U.S. Govt. Print. Off., Washington, D.C.

DEVELOPMENT OF METHODOLOGY FOR PREDICTING REACIDIFICATION OF CALCIUM CARBONATE TREATED LAKES

Richard D. Scheffe, William G. Booty and Joseph V. DePinto
Department of Civil and Environmental Engineering
Clarkson University, Potsdam, NY 13676 USA

ABSTRACT As part of the EPRI funded Lake Acidification Mitigation Project(LAMP), an ongoing study to evaluate the feasibility of acid lake liming, two models (WAM and ALaRM) have been tested for two calibrated Adirondack Mountain study lakes(Woods and Cranberry). A priori predictions of reacidification rates based on Ca carbonate application and historical hydrological data are presented. WAM (Watershed Acidification Model) is a deterministic model that is capable of simulating the movement of water and chemical constituents through a watershed. WAM generates indata in the form of hydrologic and alkalinity to ALaRM (Acid Lake Reacidification Model), a general mass balance model developed for the temporal prediction of the principal chemical species in both the water column and sediment pore water of small lakes and ponds.

A matrix of runs to determine model sensitivity to input loadings, kinetic coefficients, and sediment dosage levels indicate that reacidification of the water column is most sensitive to variation in hydrologic loading followed by variation in sediment dosage levels. Baseline estimates (initial water column alkalinity between 400 to 500 μeqL^{-1}) indicate that reacidification to near zero alkalinity occurs after a time period equivalent to approximately two to three average hydraulic retention times.

1. INTRODUCTION

As part of the Electric Power Research Institute (EPRI) funded Lake Acidification Mitigation Project (LAMP) -- an ongoing study to evaluate the feasibility of acid lake liming -- two mathematical models (WAM and ALaRM) have been applied to the prediction of reacidification rates following the neutralization of two Adirondack Park (New York) study lakes (Woods Lake and Cranberry Pond). The two lakes, located in the southwestern portion of the Adirondack Park (Scheffe et al., 1985), were treated with a very high quality calcite powder (OMYA, Inc., Proctor, VT) on May 30 and 31, 1985 (Scheffe et al., 1985). Woods Lake was a highly acidic system with a surface area of 25 ha, a mean depth of

Water, Air, and Soil Pollution **31** (1986) 857–864.

3.2 m, and a mean hydraulic retention time of approximately 10 mo. Cranberry Pond had a similar chemistry to Woods Lake; however, it is a smaller lake (7.2 ha, 2.9 m mean depth) with a shorter mean hydraulic retention time of approximately 3.5 mo. Both lakes were chemically, biologically, and hydrologically monitored by the LAMP team for pre-liming conditions from January, 1984 through May, 1985. These data have been used to calibrate the models. An intensive, post-liming transition monitoring program was conducted to, among other things, provide initial conditions for the reacidification predictions presented in this paper.

The reacidification rate determinations require the sequential application of two models: WAM (Watershed Acidification Model) and ALaRM (Acid Lake Reacidification Model). A generalized schematic of the flow of information to and from the two models is shown in Figure 1. WAM is a deterministic model that is capable of simulating the movement of water and chemical constituents through a watershed based solely on meteorological input information, precipitation chemistry, and basic watershed characteristics. WAM is used in this project to provide the time-variable hydrology and principal state variable (DIC, ANC, and Ca) loadings that are necessary inputs to ALaRM. ALaRM then combines these inputs with measured lake characteristics and initial conditions (either measured or calculated post-liming water and sediment chemistry) to calculate the temporal profile of reacidification for the lake. The ultimate goals of this reacidification determination are:

1) to optimize the management of acidic lakes via base addition by providing information on the duration of treatment as a function of such things as dose and type of material used; and,
2) to provide a methodology for predicting the time when reapplication is necessary, hence aiding in the planning of long-term acidic lake neutralization programs either on an individual lake or a regional basis.

Scandanavian based reacidification models have been successfully applied to Swedish and Norwegian liming projects (Sverdrup et al., 1984; Sverdrup, 1985). Those models involve surface water chemistry and loading routines based on empirical relations gleaned from watersheds of vastly different character from Adirondack basins; and, consequently, they are not readily transferable to our study systems.

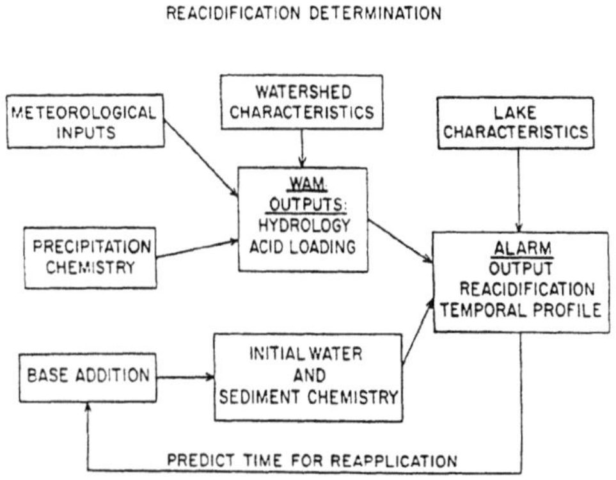

Figure 1. Schematic of WAM-ALaRM sequential application for reacidification prediction.

2. INPUT ACID LOADINGS (WATERSHED ACIDIFICATION MODEL)

The overall structure of WAM is shown in Figure 2. WAM is a deterministic model that is capable of simulating the movement of water and chemical constituents through a watershed. The hydrologic portion of the model calculates 1) interception by and evapotranspiration from the forest canopy, 2) evaporation from the soil horizons, 3) snowpack accumulation and snowmelt volumes, 4) percolation and interflow in the soil horizons, 5) groundwater flow, 6) streamflow and, 7) lake water balance. The chemical portion of the model calculates 1) attenuation and enrichment of ions in the throughfall, 2) ion accumulation and leaching in the snowpack, 3) attenuation and enrichment of ions passing through each soil horizon, and 4) mixing of waters and equilibrium with atmospheric CO_2 in the streams.

In the model, the watershed is divided into homogeneous sub-basins. The sub-basins are defined by surface area, ground slope, percentage open, forested and exposed bedrock areas as well as vegetation-type areas. For each sub-basin a 'typical' soil profile is defined by horizon thickness, porosity, field capacity, saturated hydraulic conductivity, soil and soil water chemistry. The model operates on a daily time step and includes seasonal controls that determine which processes are active.

2.1 Development of ALaRM Input Data

Seasonal inflows and outflows for Woods Lake and Cranberry Pond, along with loadings of ANC, DIC, and Ca have been generated by WAM. Figure 5a shows 1984 ANC loadings that were generated for the ALaRM calibration. Three different annual watershed meteorological loading scenarios have been developed for Woods Lake and Cranberry Pond Watersheds: 1) maximum 2) minimum and 3) average. Daily precipitation volumes for the period April 14, 1978 - December 27, 1981 were obtained from the ILWAS project (Valentini, 1985) for the study area. Using these data, monthly total precipitation volumes were calculated. A maximum 1 year data set was generated by selecting a maximum monthly volume for each month from the 4 years of measured data for that particular month. The same procedure was followed in generating a minimum and an average 1 year data set. Further documentation concerning WAM and generation of input loadings to the study lakes is available (Booty, 1983; EPRI, 1985).

3. ACID LAKE REACIDIFICATION MODEL (ALaRM)

ALaRM is a general mass balance model designed for the temporal prediction of the principal chemical species in both the water column and sediment pore water of neutralized water bodies. The model features five-segment partitioning; the water column is divided into epilimnion and hypolimnion, and the underlying sediments are divided into an active layer adjacent to the water column and an underlying inactive layer. The active sediment layer is further partitioned (horizontally) based on the sediment surface area fraction exposed to the epilimnion. Schematic

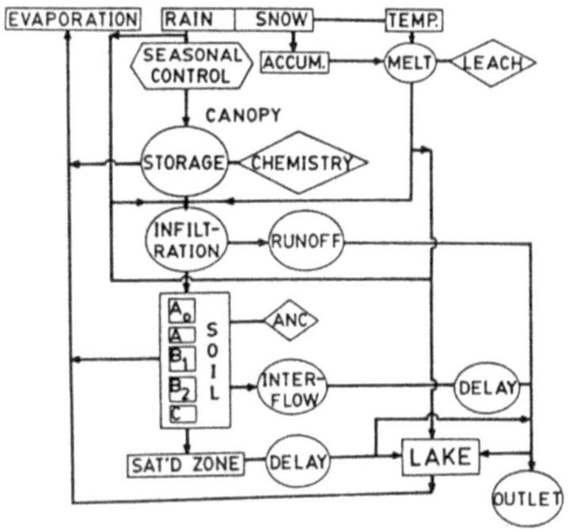

Figure 2. WAM process flow diagram.

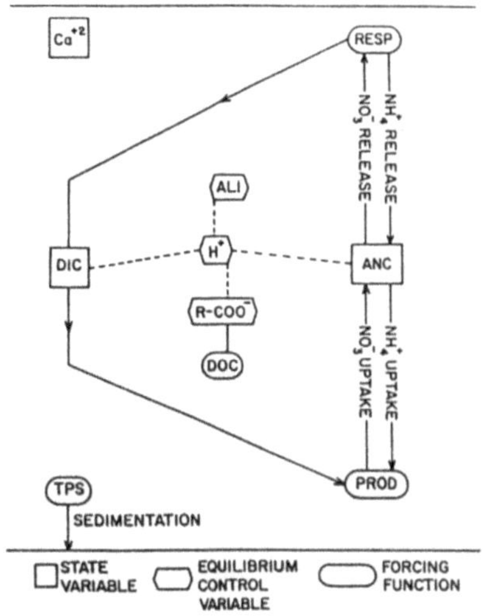

Figure 4a: Water column transformations in ALaRM.

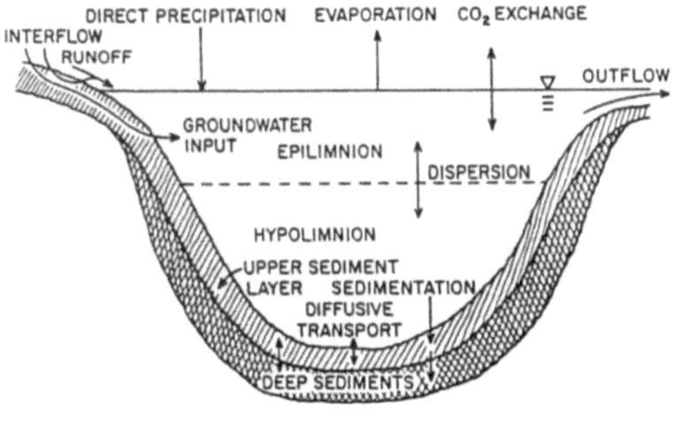

Figure 3. ALaRM mass transport diagram.

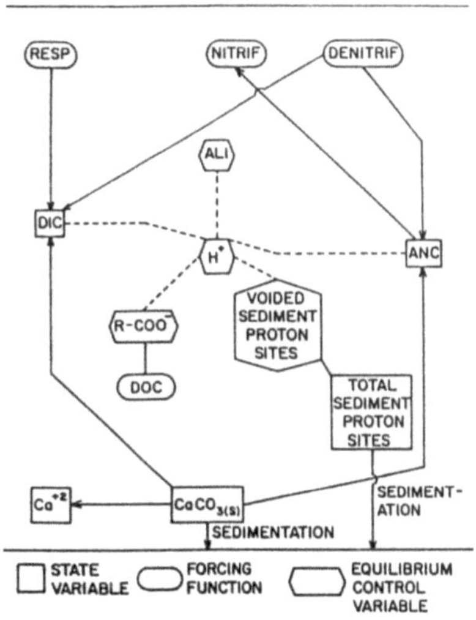

Figure 4b. Sediment transformations in ALaRM.

diagrams illustrating segmentation, transformations and transfer paths are shown in Figures 3 and 4.

Input loadings (generated by WAM) of DIC, ANC and dissolved Ca are introduced to the epilimnion. These three variables are free from thermodynamic constraints and are treated as time dependent state variables in all segments. The remaining state variables in the sediment segments include undissolved calcite and total sediment proton sites. The model incorporates internal sources for DIC and ANC inputs in the form of biological forcing functions: phytoplankton respiration and productivity, sediment respiration and nitrification-denitrification. Hydrogen ion concentration is governed by equilibrium with carbonates, inorganic Al, dissolved organic C and, for the sediments, unoccupied sediment proton sites.

All principal anions (carbonate, bicarbonate, hydroxyl) and cations (H^+, Ca) can diffuse between each segment; this transport subsequently affects DIC and ANC levels in the segments involved (Figures 3 and 4). The model runs on a half-day time step; a system of twenty-one first order coupled ordinary differential equations are solved (fourth order Runge-Kutta) and the new values for each state variable (ANC, DIC) define pH in each segment (H^+ is solved by iteration via a Newton-Raphson routine). In addition to the carbonate system, the proton balance includes inorganic Al hydroxyl species (gibbsite equilibrium is assumed, $pK_{so} = -8.11$) as well as dissolved organic carbon (DOC). DOC is treated as a simple monoprotic acid, with an empirical pK and hydrogen ion site versus measured DOC concentration relationship as determined by Driscoll (1982). Once pH is established in each segment, speciation of DIC, Al hydroxyl complexes, and DOC is determined. It is assumed that the relevant proton reactions are very fast relative to species transport.

3.1 Sediment Chemistry

Sediments can act as either a source or sink for ANC (relative to the water column). Undissolved calcite can dissolve in the sediments and ANC can subsequently diffuse into the water column; or ANC from the water column can diffuse to acidic (relative to the water column) sediments. Loss of potential water column neutralization from undissolved calcite occurs by direct neutralization of the sediments and burial of the upper sediment layer at the rate of 1 $mmyr^{-1}$.

The interpretation of ANC in the sediment layers is based on the assumption that equilibrium exists between interstitial protons and sediment bound protons, and this equilibrium is modeled as a Freundlich isotherm (sediment bound proton is modeled by a simple power law function and increases with interstitial proton concentration) using sediment titration data (EPRI, 1985). The total number of potential sediment proton sites (TPS) is the sum of sediment bound protons and dissociated sites,

$$TPS = BP + BPN,$$

where BP and BPN refer to the number of sites occupied and void of protons, respectively. From this relation and the equilibrium relation between BP and interstitial proton [H^+], it is clear that BPN is a function of TPS and interstitial pH. Furthermore, we assume that one mole of BPN (potential sites for proton removal from the pore water) represents the contribution of one equivalent of ANC. Finally, the proton balance for sediment layers includes BPN in addition to the dissolved species included in the water column proton balance.

3.2 Calibration

To date, post treatment water chemistry data are not available; hence, the calibration presented here (Figure 5b) reflects an adjustment of various biological forcing functions. Values for the intersegment exchange coefficients, the calcite dissolution rate constant and the atmospheric exchange coefficient for CO_2 are within ranges reported in the literature; details can be found in the LAMP interim report (EPRI, 1985).

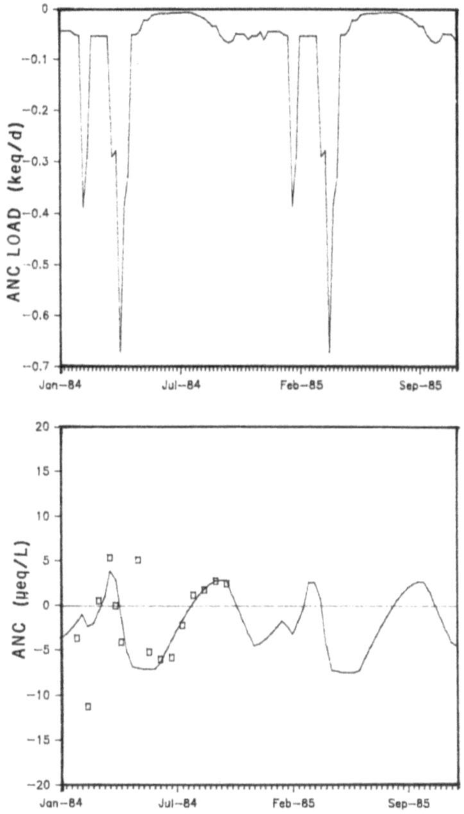

Figure 5. a(top)-WAM generated ANC loadings for 1984; b-ALaRM calibration of Woods Lake using 1984 observations and the above loadings.

4. RESULTS AND DISCUSSION

Initial conditions for DIC, ANC and pH are those observations following neutralization of both lakes (Scheffe et al., 1985). Following neutralization and an initial transition dissolution period, the whole-lake average ANC for both lakes was between 400 and 500 μeqL^{-1}. The quantity of undissolved calcite reaching the sediments (0.026 tonne ha^{-1}, Woods Lake; 0.00 tonne ha^{-1}, Cranberry Pond) was calculated by subtracting the observed whole-lake average dissolved Ca increase from the applied Ca dosage.

The baseline predictions (Figure 6) determined with average hydrologic loadings indicate that reacidification to near zero ANC will occur during the spring runoffs of 1987 and 1986 for Woods Lake and Cranberry Pond, respectively. The results for Woods Lake, however, suggest that reacidification to below zero ANC may hold off until the spring of 1988.

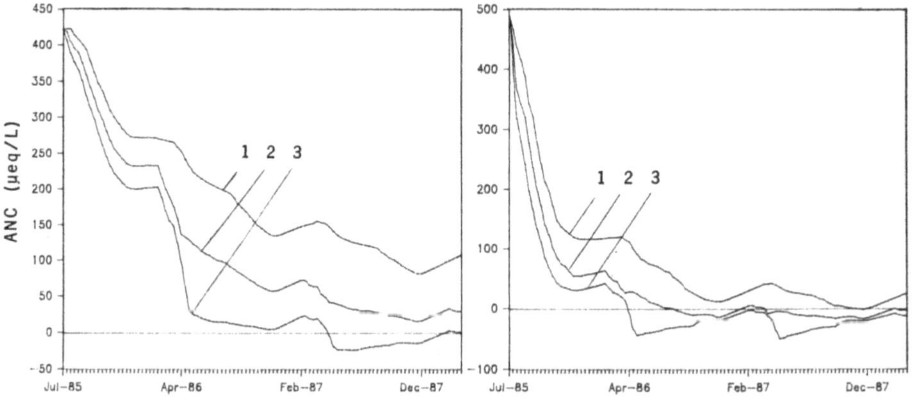

Figure 6. Reacidification as a function of watershed loadings: 1-minimum flow, 2-base flow, 3-maximum flow.

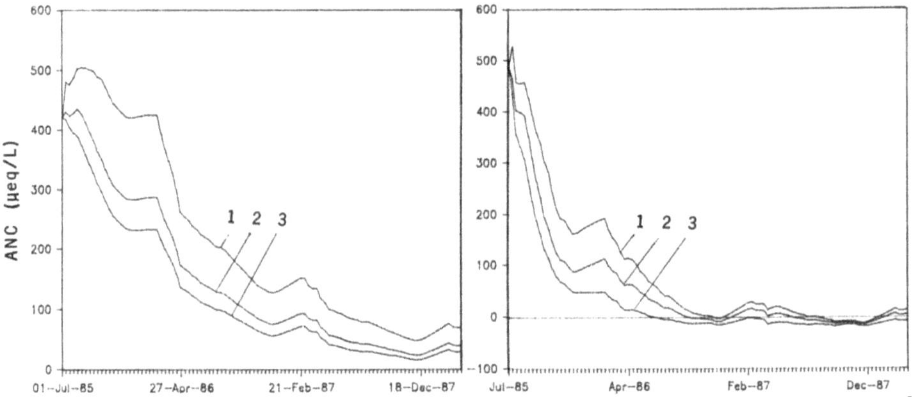

Figure 7. Reacidification as a function of sediment dosing: 1-0.10 kgm^{-2}; 2-0.05 kgm^{-2}; 3-0.026 kgm^{-2} (Woods Lake), 0 kgm^{-2} (Cranberry Pond).

Reacidification is most sensitive to hydrologic variation and flushing behavior (Figures 6 and 7). Sufficient snowpack could affect reacidification in both lakes as early as the first spring following summer treatment; and, Cranberry Pond with a mean hydraulic retention time of 3.5 months reacidifies before Woods Lake (mean hydraulic retention time of 10 months) in all scenarios.

Sediment dosing (Figures 7) appears to provide residual neutralization capability; again, Woods Lake receives a more effective sediment feedback due to the lower flushing rate. It should be pointed out, however, that the effect of sediments on water column chemistry is not a simple function of mean hydraulic retention time; rather, one must consider the lumped effect of both retention time and average water column depth. For a given retention time, the sediments in a relatively shallow lake will impart greater influence on water column chemistry because the surface area (sediment-water column interface) to water column volume ratio is relatively large.

A matrix of runs to determine sensitivity to various model coefficients has been performed (EPRI, 1985). Relative to variations of hydrologic loading and sediment dosing, reacidification rate is less sensitive to intersegment exchange coefficients.

ACKNOWLEDGMENTS

The authors wish to thank T.C. Young of Clarkson University for sediment titration data and C.T. Driscoll of Syracuse University for providing water chemistry data. This paper was presented as a poster session at the International Symposium on Acidic Precipitation, Muskoka, Ontario, Canada, September, 1985. The research described in this article has been funded by the Electric Power Research Institute through a subcontract agreement with General Research Corporation and is part of the Lake Acidification Mitigation Project (LAMP), project number RP 2337. (Clarkson, Cornell, Indiana and Syracuse Universities are participating in the field portion of LAMP).

REFERENCES

Booty, W.G.: 1983, Watershed Acidification Model. Ph.D. Dissertation, McMaster University, Ontario, Canada.

Driscoll, C.T.: 1982, Calcium Carbonate Neutralization of Acidified Surface Waters. In: Proceedings of the American Society of Civil Engineers, 108(EE6):1128-1145.

EPRI: 1985, Annual Report for Lake Acidification Mitigation Project. Electric Power and Research Institute, Palo Alto, California.

Scheffe, R.D., DePinto, J.V., and Bilz, K: 1985, Laboratory and Field Testing of Dose Calculation Methods for Neutralization of Adirondack Acid Lakes. International Symposium on Acidic Precipitation. Muskoka, Ontario, Canada.

Sverdrup, H., Rasmussen, R., and Bjerle, I.: 1984, Chemica Scripta, **(24)** 137.

Sverdrup, H.: 1985, Calcite Dissolution Kinetics and Lake Neutralization. Ph.D. Dissertation, Lund Institute of Technology, Lund, Sweden.

Valentini, J.L.: 1985, Personal communication. Tetra Tech, INC., Lafayette, California.

LYSIMETER STUDY OF EFFECTS OF ACID DEPOSITION ON PROPERTIES AND LEACHING OF GLEYED DYSTRIC BRUNISOLIC SOIL IN NORWAY

Gunnar Abrahamsen and Arne O. Stuanes
Division of Forest Ecology
Norwegian Forest Research Institute
1432 Aas-NLH, Norway.

ABSTRACT. The effects of artificial rain of various acidities were studied in a lysimeter experiment. Lysimeters, 29 cm in diameter, and 40 cm deep contained a Gleyed Dystric Brunisol. Natural structure, stratification and original vegetation were maintained. Artificial rain was produced from groundwater with "high" concentrations of neutral salts and from rainwater with lower concentrations. pH levels of 6, 4, 3 and 2 were established by adding H_2SO_4. Effects of dilution with given amounts of acid were examined by comparing the effects of 50 mm "rain" month^{-1} of pH 3 and pH 2 with 500 mm "rain" month^{-1} of pH 4 and pH 3, respectively. The study continued for 5 yr. Totals of 1250 or 12500 mm "rain" were applied in addition to a natural input of 2773 mm. Increased input of SO_4^{2-} increased the output of SO_4^{2-} but, as concentration increased, sorption of SO_4^{2-} in the soil also increased. Concentrations of base cations in the leachate increased parallel to the output of SO_4^{2-}. However, significant effects on leaching of base cations and on the content of exchangeable cations in the soil, was only found in the pH 2 treatment with 1250 mm of "rain" and in the pH 4 and pH 3 treatment with 12500 mm of "rain".

1. INTRODUCTION

Atmospheric acid deposition is likely to increase the acidity of many forest soils, but the degree of acidification will vary with soil properties. Acid deposition includes both H_2SO_4 and HNO_3. In areas where N is in surplus in the soil/plant system, increased soil leaching and acidification has been measured (e.g. van Breemen et al., 1983; van Miegroet and Cole, 1984, 1985). In general however, the deposition of S is more important for leaching and soil acidification than the deposition of N (e.g. Abrahamsen, 1980a). This is due to the higher requirement for N than for S in many forest ecosystems. Due to the electrochemical balance of anions and cations in the percolating water, increased leaching of anions results in increased leaching of cations. If more base cations are leached from the soil than the incoming quantity, soils may become more acidic. However, the weathering of

Water, Air, and Soil Pollution **31** (1986) 865–878.

soil minerals which release base cations may compensate for increased leaching. Due to the variability of SO_4^{2-} sorption, the content of base cations, hydrologic conditions, rate of weathering among soils etc., it is often difficult to predict the long term effect of acid deposition.

Experiments have been carried out in Norway and elsewhere to estimate the long term effect of acid deposition on soil acidity (e.g. Abrahamsen and Tveite, 1983). These experiments are based on applications of artificial rain with higher concentrations of acid and with different contents of other electrolytes than are found in natural rainwater. The present lysimeter study was carried out to examine the effect on leaching of electrolytes and on soil properties of: 1) different concentrations of acid in artificial rain, 2) given amounts of acid applied in dilute and concentrated form and 3) two different levels of neutral salts in artificial rain. Preliminary results from this study have been described previously (Abrahamsen, 1986a).

2. MATERIALS AND METHODS

The lysimeters were of the bucket type, being 40 cm high and 29 cm wide polyethylene cylinders. "Undisturbed" soil monoliths were obtained by pressing the cylinders into the soil and, at the same time, digging around the cylinders. Natural soil stratification and structure were maintained and the vegetation was not significantly altered. Before the bottom was welded on, a layer of small, inert, polyethylene pellets were placed underneath the soil to prevent water saturation and anaerobic conditions.

Lysimeters were established in two soil types: a Gleyed Dystric Brunisol and a richer Orthic Melanic Brunisol. In this paper only results of the former type will be described. Properties of this soil are given in Table I.

Twenty lysimeters were placed in the field and exposed to natural precipitation in addition to artificial rains of various acidities.

Ten different treatments were established, with each treatment replicated twice (Table II). The treatments were organized to compare 1) the effects of different acidities of artificial rain on leaching, 2) the effects of given H^+ and SO_4^{2-} depositions applied in different concentrations, and finally 3) the effects of different concentrations of neutral salts in the artificial rain. Different levels of acidity in the artificial rains were obtained by adding H_2SO_4.

The effect of acid concentration was examined by comparing applications of 50 mm of "rain" of pH 3 and 2 with applications of 500 mm of "rain" of pH 4 and 3, respectively. In this way effects of a given H^+ and SO_4^{2-} deposition of one concentration were compared to a 10 times larger water volume with 1/10 of the former concentration.

Effects of different neutral salt contents were examined by using groundwater with "high" concentrations of salts compared with the low concentrations in collected rainwater. The least acidic treatment (pH 6) involved a comparison between groundwater and rainwater with no adjustment of pH. The effect of groundwater of pH of 6.1 was compared

Table I. Some properties of the Gleyed Dystric Brunisol (Umbric Dystrochrept, U.S.D.A.) used in the lysimeter experiment (From Stuanes and Sveistrup, 1979).
* Extracted by 1 M NH_4Cl.

Soil horizon	Soil depth	Texture %			Soil	Exchangeable cations* $mmol(+)kg^{-1}$				CEC*	Base*	Al*
	cm	Clay	Silt	Sand	$pH_{(H_2O)}$	Na	K	Mg	Ca	$mmol(+) kg^{-1}$	saturation %	$mmol kg^{-1}$
Ah	0-7	9	36	55	4.2	0.5	2.2	5.8	19.1	116	24	22
Bw	7-31	8	33	59	4.3	0.3	0.6	0.4	0.9	51	4	14
Cw	31-	5	34	61	4.6	0.3	0.5	0.2	0.8	18	10	4

Table II. Experimental design of the lysimeter experiment. (Names used for the different treatments are in brackets. The experiment had two replicates and a total of 20 lysimeters. Lysimeters were established in autumn of 1975, but treatment with artificial acid rain began in June 1977. pH was adjusted with H_2SO_4.)

Effect of different concentrations of neutral salts in the artificial rain		Effect of "concentrated" acid compared to dilute acid - same H^+ deposition	
50 mm monthly			
pH of "rain" Groundwater	pH of "rain" Rainwater	pH of 50 mm rainwater mo^{-1}	pH of 500 mm rainwater mo^{-1}
pH 6 (GWC-50)	pH 4.4 (RWC-50)		
pH 4 (GW4-50)	pH 4 (RW4-50)		
pH 3 (GW3-50)	pH 3 (RW3-50)	pH 3 (RW3-50)	pH 4 (RW4-500)
pH 2 (GW2-50)	pH 2 (RW2-50)	pH 2 (RW2-50)	pH 3 (RW3-500)

to the effect of rainwater of pH of 4.3.

Groundwater was taken from a forested area approximately 40 km north of Oslo where the lysimeters were located (Abrahamsen et al., 1976). Rainwater was collected by means of a 4 by 4 m polyethylene plate connected to a buried 3000 L plastic tank. The groundwater is richer in Cl^-, Na^+, K^+, Mg^{2+} and Ca^{2+} and poorer in H^+, NH_4^+ and NO_3^- than the rainwater. The differences in H^+, Mg^{2+} and Ca^{2+} cause different lime potentials in the groundwater and the rainwater (Table III).

In addition, the lysimeters were exposed to natural precipitation. Precipitation was monitored at the plot and the average chemical composition is shown in Table III. Rainwater collected for lysimeter application had slightly different chemical composition than average rain, possibly related to contamination of the polyethylene plate.

The artificial rains were applied monthly during the 5-mo frostfree period. The artificial rains were applied in small quantities every other hour over one day for the 50 mm treatments and over two days for the 500 mm treatments.

The lysimeters were placed in the field in the autumn of 1975. During 1976, 50 mm of groundwater was applied to all lysimeters 5 times. The treatments (Table II) began in June 1977 and continued until September 1981. In June 1982 the lysimeters were disassembled.

Leachate was collected after each application of artificial rain and in conjunction with natural precipitation. pH, conductivity and content of Cl^- were measured immediately after sampling. Remaining

Table III. Chemical characteristics of natural rainwater, and of rainwater and groundwater used in the experiment (μM)

	Rainwater	Rainwater in collector	Groundwater 1976-1981
pH	4.3	4.4	5.9
Cl^-	20	25	73
SO_4^{2-}	34	31	48
NO_3^-	34	42	3
NH_4^+	32	26	1
Na^+	15	21	82
K^+	6	6	12
Mg^{2+}	2	4	23
Ca^{2+}	6	15	63
Lime potential*	1.75	2.0	3.9

* Lime potential = pH + 1/2 log $[Ca^{2+} + Mg^{2+}]$

water was preserved with HCl and monthly samples were prepared by taking samples proportional to the amount of leachate produced at different times during the month. These samples were analysed by standard methods as described by Ogner et al., (1984).

When the lysimeters were disassembled the vegetation was analyzed, clipped, dried at 60°C and milled before chemical analysis. Lysimeter walls were cut in half to remove the soil. The litter layer was removed, dried at 60°C and milled before analyzing. The upper 10 cm of the soil was cut in 2.5 cm slices and the lower part in 5 cm slices. Each soil slice was weighed, dried at 25°C and passed through a 2 mm sieve before analyzing. Exchangeable cations were determined after extraction with 1 M NH_4OAc at pH 7. Water plus phosphate extractable SO_4^{2-} was determined after shaking the soil for 1 hr with 0.01 M $Ca(H_2PO_4)_2$. Soil pH was measured in a 1:2.5 soil-water suspension. For further descriptions of analytical methods, see Ogner et al., (1984).

3. RESULTS

Results are restricted to input/output budgets of water, SO_4^{2-} and the base cations: Na^+, K^+, Mg^{2+} and Ca^{2+}. Variation in leachate pH over time is also described. The description of effects on soil include content of extractable SO_4^{2-}, exchangeable base cations and pH.

3.1. Input and output of water

The leachate included in this study was collected during the period May 26, 1977 to June 1, 1982. The May 26, 1977 sample included snow melt from the winter of 1976/77. During the period Jan. 1977 to June 1982 precipitation totalled 4067 mm of which 38% was snow. However, because of frozen soil, most of the snow melt did not percolate through the lysimeters, but ran off outside the lysimeters. Since the edges of the lysimeters were about 5 cm above the soil surface, snow melt has been estimated to be 50 mm each spring. Therefore total input of natural precipitation into the lysimeters amounted to 2773 mm. Artificial rain was additional. For the 50 mm treatment it amounted to 1250 mm. The total input of water to these lysimeters was therefore 4023 mm (Table IV) which was close to the natural amount of precipitation during the period. For the 500 mm treatment, the total amount of artificial rain was 12500 mm and the total input of water 15273 mm (Table IV) or, on average, 3050 mm per year. This was almost four times the natural input during the period.

The average output of water from the lysimeters for the two replicates varied with treatment (Table IV). The lowest was 1428 mm from the GWC-50 treatment, and the greatest was 2379 mm from the RW4-50 treatment. Evapotranspiration amounted to 41 to 65% for the lysimeters supplied with 50 mm water per watering. Lysimeters supplied with the 10-fold amount of water had average evapotranspiration of 21 and 23%.

The large variation in output of water among treatments is not easily explained. When disassembling the lysimeters signs of leakage were found in only one of two lysimeters of the GW3-50, RW4-50 and RW4-500 treatments. Since the outputs of water from these treatments were not particularily low, leakage does not explain the large variation in water flow. Water output in relation to differences in vegetative cover and height does not account for variation either.

3.2. Leaching of chemical compounds

SO_4^{2-} was the major anion in the leachate from all lysimeters except one where NO_3^- dominated. Excluding this lysimeter, SO_4^{2-} represented between 68 and 82% of the total anion flow (negative charge basis) from the control (GWC-50 and RWC-50) lysimeters. Dominance of SO_4^{2-} increased with increasing input. Cl^- was the next important anion representing 10 to 17% of the total anion flow from the GWC-50 and RWC-50 lysimeters. NO_3^- varied between 0.3 and 20%. The anomalous lysimeter, had NO_3^- representing 67% of the total anion flow. Phosphate was negligible in comparison with other anions.

In the leachate from control lysimeters, Ca and Al were equally

Table IV. The budgets of water, sulphate and base cations for the different treatments. Mean of two replicates.

	GROUNDWATER				RAINWATER					
	GWC-50	GW4-50	GW3-50	GW2-50	RWC-50	RW4-50	RW3-50	RW2-50	RW4-500	RW3-500
Water, l m^{-2}										
Input	4023	4023	4023	4023	4023	4023	4023	4023	15273	15273
Output	1428	1976	1788	2202	2112	2379	1682	2053	11775	12033
Out/in	0.35	0.49	0.44	0.55	0.52	0.59	0.42	0.51	0.77	0.79
Sulphur, mmol(-) m^{-2}										
Input	309	495	1689	13190	266	319	1569	13190	1489	13990
In veg. & litter	123	95	101	147	75	90	91	162	75	121
In soil	636	755	1071	1617	634	814	875	2305	490	1231
Output	563	820	1294	10702	849	900	1097	9901	3109	9231
Out/In	1.82	1.66	0.77	0.81	3.19	2.82	0.70	0.75	2.09	0.66
Base cations, mmol(+) m^{-2}										
1) Input	435	435	435	435	184	184	184	184	916	916
2) In veg. & litter	858	752	545	521	596	452	521	479	541	250
3) In soil	2511	3660	2619	1841	3152	2098	2537	1158	2850	998
4) Output	470	810	886	4061	638	1003	675	2538	2060	4576
Out/In	1.1	1.9	2.0	9.3	3.5	5.5	3.7	14.0	2.3	5.0
2+3+4-1	3404	4787	3615	5988	4202	3369	3549	3991	4535	4908

important if the mean valence of Al was assumed to be two. Both elements represented about 21 to 32% of the total positive charge. Na amounted to 15 to 21%, Mg 9 to 14%, K 5 to 8% and H^+ 3 to 7% of the total cation leaching. With the increasing acidity of input water, Al dominance increased.

Assuming that all cations and anions are accounted for, the balance between cations and anions can be used to estimate the valence of Al. Balance is obtained for the different lysimeters if the valence varies between 0.7 and 2.9. As expected lower valences were found for the control and pH 4 lysimeters whereas high valences were found for the most acidified lysimeters.

Limited space does not permit a detailed description of the leaching pattern over time. However, variation in leachate pH is of interest when discussing the total leaching of base cations. Figure 1 shows the annual average pH for the different treatments from 1976 to 1982. Only results from the lysimeters supplied with rainwater are given. The mean annual volume-weighted pH is calculated by transforming pH to H^+ concentration and weighting this with the volume of water before retransformation to pH. The variation in average annual pH in 1976, before treatment, was from 4.01 to 4.66. At the end of the experiment variation was between 4.00 to 5.49. The figure shows that rainwater of pH 3 and below reduced pH in the leachate but few effects can be observed among the other treatments. The figure also shows a considerable annual variation.

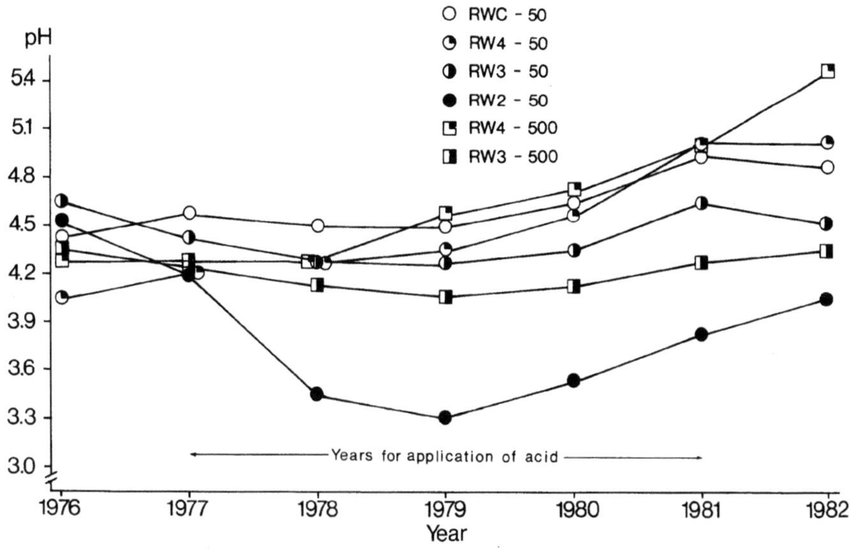

Fig. 1. Variation in mean annual volume-weighted pH of leachate from the lysimeters supplied with rainwater.

3.3. Input and output of sulphate

Input of SO_4^{2-} varied from 266 mmol (-) m^{-2} in the least acidic treatment to more than 13000 mmol (-) m^{-2} in most acidic treatments (Table IV). The total output of SO_4^{2-} varied from 563 to almost 11000 mmol (-) m^{-2}. In the two least acidic treatments, the output was considerably larger than the input. This means that a large proportion of SO_4^{2-} came from the soil itself with a lesser proportion due to the input. When the input SO_4^{2-} concentration increased to pH 3 and pH 2 there was a retention of SO_4^{2-} amounting to 19 to 34% of the input. This effect was observed for both groundwater and rainwater and for both "rain" quantities. Relative increase in sorption of SO_4^{2-} with increased concentration in the input demonstrates the concentration effect on SO_4^{2-} sorption.

3.4. Input and output of base cations

Input of base cations (Na^+, K^+, Mg^{2+} and Ca^{2+}) to the lysimeters supplied with groundwater amounted to 435 mmol (+) m^{-2} (Table IV). To those supplied with 50 mm mo^{-1} of rainwater, the input represented 184 mmol (+) m^{-2} and to those receiving the 10-fold amount of rain water, the input was 916 mmol (+) m^{-2}. The output from the soil was almost in balance for the GWC-50 treatment. For all other treatments, output was significantly larger than input - for the most acidic treatments about 10 times larger. Variability among the less acidic treatments makes it impossible to find significant difference in the output from these treatments. However, looking at concentrations of base cations instead of total amounts, increased concentrations were observed with increasing acid load.

3.5. Chemical content in vegetation and litter

Total amount of S and base cations in vegetation and litter are shown in Table IV. Compared to input and to content of sorbed SO_4^{2-} in soil, the amount of S in vegetation and litter was relatively low. If there is an treatment effect, it is seen only in the most acidic treatment. Of the total S content in vegetation and litter, 13 to 40% was in the vegetation and the rest was in the litter.

Content of base cations in vegetation and litter was slightly larger than the input to the lysimeters supplied with groundwater and represented 20 to 40% of the store of exchangeable cations in the soil. Obvious effects of treatment appears only to be found in the lysimeters supplied 500 mm mo^{-1} of rainwater of pH 3.

3.6. Chemical content of the soil

The total amount of extractable SO_4^{2-} in soil increased with increasing input (Table IV). Except for the GWC-50 treatment, total amount leached was greater than the extractable amount. Comparison of the treatment RW2-50 with RW3-500 and RW3-50 with RW4-500 clearly shows the concentration effect for the sorption of SO_4^{2-}. Decreasing the

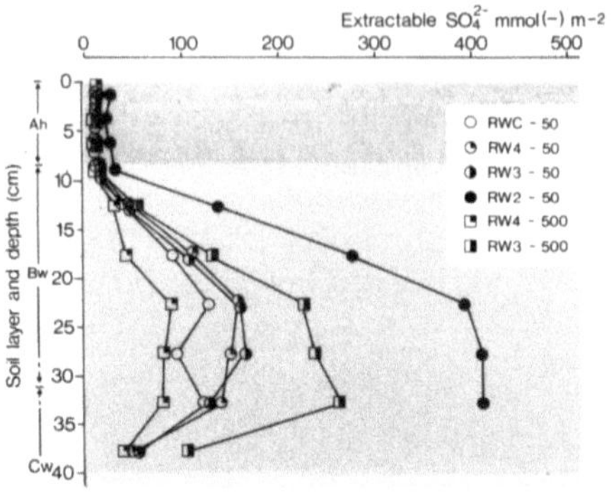

Fig. 2. Vertical variation in total extractable SO_4^{2-} in the soil of lysimeters supplied with rainwater.

concentration by a factor of ten and maintaining the total input resulted in a decrease in the sorbed amount by a factor of two.

The distribution of extractable SO_4^{2-} with depth for the rainwater treatments is shown in Figure 2. The highest amounts were sorbed in the lower Bw-horizon. There were clear differences in amounts of sorbed SO_4^{2-} between the treatments. In the Ah-horizon, no difference between the treatments could be found except for the pH 2 treatment. The distribution of SO_4^{2-} shown in Figure 2 agrees with available knowledge of soil profil development. Content of organic matter gradually decreases with soil depth giving the highest amount of positive sorption sites in the lower Bw-horizon.

The sum of exchangeable base cations Na^+, K^+, Mg^{2+} and Ca^{2+} for the different treatments is shown in Table IV. Except for the pH 2 and the RW3-500 mm treatments, there was no clear effect of treatment on the exchangeable amount. The exchangeable amount for the two pH 2

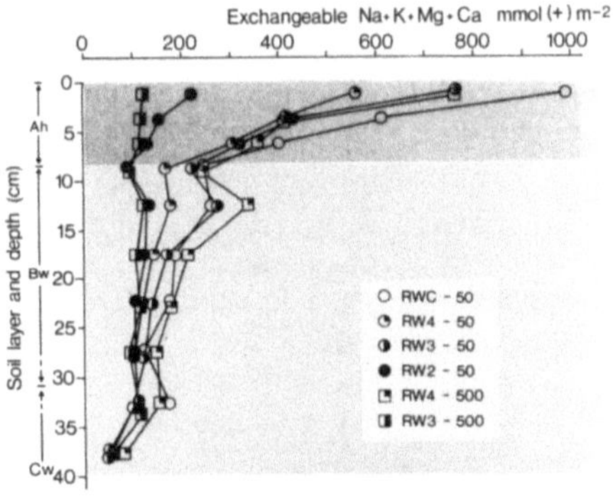

Fig. 3. Vertical variation in exchangeable base cations in the soil of lysimeters supplied with rainwater.

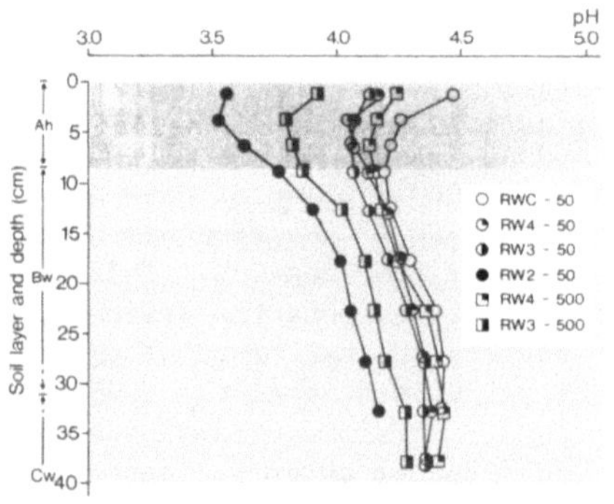

Fig. 4. Vertical variation in soil $pH_{(H_2O)}$ in lysimeters supplied with rainwater.

treatments are 63 and 45% of the mean for the other treatments. Rainwater of pH 2 produced lower amounts of exchangeable cations than groundwater of pH 2.

Variation in exchangeable base cations with soil depth is shown in Figure 3. Differences between treatments RWC-50, RW4-50 and RW3-50 on the one hand and the RW2-50 and RW3-500 on the other hand, were clear in the upper 20 cm of the soil. Below 20 cm, however, small differences were observed even between these treatments.

Figure 4 shows the variation in soil pH (H_2O) with soil depth. There were clear differences between lysimeters receiving the highest acid input and those receiving lower input. Differences were most obvious in the upper parts of the soil and they decreased with increasing soil depth. Unlike the content of exchangeable base cations, there appeared to be differences, although small, in soil pH even between the least acidic treatments.

4. DISCUSSION AND CONCLUSION

The lysimeters described in this study were dug out from an apparently homogenous area of less than 100 m^2. In the same plot a soil profile was described and sampled. Some of the characteristics of this profile are shown in Table I. However, when analysing the leachate and the soil of the individual lysimeters significant random variation was found. Variation in leaching can partly be ascribed to variation in water flow through the lysimeters. On the average, for the lysimeters supplied with 50 mm artificial rain mo^{-1}, evapotranspiration amounted to 414 mm yr^{-1}. This is close to the potential evapotranspiration for this area, which has been estimated at 400 mm yr^{-1} (Werner Johannessen, 1970).

Leaching of SO_4^{2-} increased with increasing H_2SO_4 input. However, increased sorption of SO_4^{2-} in the lysimeters receiving the most acidic rain reduced the differences in SO_4^{2-} leaching between treatments.

When the pH of the rain was reduced to pH 2 or pH 3 in the lysimeters supplied with 500 mm water mo^{-1}, SO_4^{2-} leaching increased drastically.

The sharp increase in SO_4^{2-} leaching when pH of the "rain" dropped from 3 to 2 or the rain quantity increased from 50 to 500 mm mo^{-1}, was accompanied by a sharp increase in the leaching of base cations. For less acidic treatments, variation in SO_4^{2-} leaching was not reflected in a similar variation in leaching of base cations. This result is probably due to a large random variation among the lysimeters. When looking at concentrations in leachate instead of total amounts, a slight increase in base cations was evident when the acidity of the "rain" increased from pH 6 to pH 3. It seems clear, however, that leaching increases more when the input of water is increased by a factor of ten than when the input of acid is increased by a factor of ten.

In general, the results are consistant, but less clear,when compared with results from a similar lysimeter study on an Eluviated Dystric Brunisol (Abrahamsen, 1980b, 1983; Bjor and Teigen, 1980). In addition, Morrison and Foster (1986) observed the same pattern in their lysimeter studies with an Orthic Humo Ferric Podzol and a Dystric Brunisol.

When considering the total amount of exchangeable cations in the soil, the conclusions are the same as those for leaching. The only significant effect that can be observed is from the pH 2 lysimeters and the pH 3 500 mm mo^{-1} lysimeters.

Calculating total amounts of base cations in the vegetation, litter, soil and leachate and allowing for input, did not reveal any significant effect among the three least acidic treatments (Table IV). However, at pH 2 and at the higher "rain" inputs the total amount may be enlarged. This indicates that the release of base cations from weathering of minerals or mineralization of organic material has increased.

The vertical pattern in sorbed SO_4^{2-} and exchangeable base cations is also consistant with previous results, but again the effects of the least acidic treatments are less clear. In a similar study with an Eluviated Dystric Brunisol small, but clear reductions in the amount of exchangeable Mg and Ca were found in the O and E layers after application of 1100 mm of pH 3 "rain" (Stuanes, 1980; Abrahamsen, 1983). The vertical distribution of SO_4^{2-} described in the present study is also supported by previous studies (Singh, 1980a,b). Singh (1984) found the SO_4^{2-} sorption maximum of the Bw-horizon of the reported soil by using Langmuir equation to be 8.5 mmol kg^{-1}. The soil which received the highest acid input in the present study had a maximum of 3.3 mmol sorbed SO_4^{2-} kg^{-1}.

In accordance with a number of other studies conducted in Norway the results presented here do not indicate that acid deposition in quantities found in Norway is likely to have rapid and large effects on soil properties. Therefore, these studies do not support the hypothesis that acid deposition has drastically changed soil pH even in the C horizon of Norwegian soils (See Abrahamsen, 1986b for references).

REFERENCES

Abrahamsen, G., Bjor, K. and Teigen, O.: 1976, 'Field experiments with simulated acid rain in forest ecosystems'. SNSF-Project, Norway FR 4/76 p.15.

Abrahamsen, G.: 1980a, 'Acid precipitation, plant nutrients and forest growth'. In Drabløs, D. and Tollan, A. (eds.), Proc. Int. Conf. Ecol. Impact Acid Precip., SNSF-Project, Norway 1980, 58.

Abrahamsen, G.: 1980b, 'Impact of atmospheric sulphur deposition on forest ecosystems'. In Schriner, D.S., Richmond, C.R. and Lindberg, S.E. (eds.), Atmospheric sulfur deposition. Environmental Impact and Health Effects. Ann Arbor Science Publishers, Inc. Ann Arbor, Michigan, 397.

Abrahamsen, G.: 1983, 'Sulphur pollution: Ca, Mg and Al in soil and soil water and possible effects on forest trees'. In Ulrich, B. and Pankrath, J. (eds.), Effects of accumulation of air pollutants in forest ecosystems, 207.

Abrahamsen, G.: 1986a, 'Assessment of the long-term effect of acid deposition on leaching from forest soil. A methodical study'. Paper presented at the International Symposium on "Air Pollution and Stability of Coniferous Forest Ecosystems", Ostravice/Czechoslovakia, Oct. 1-5, 1985, in press.

Abrahamsen, G.: 1986b, 'Air pollution and soil acidification'. Paper presented at Nato Advanced Research Workshop "Effects of acidic deposition on forests, wetlands and agricultural ecosystems", Toronto, May 12-17, 1985, in press.

Abrahamsen, G., Tveite, B.: 1983, 'Effects of air pollutants on forest and forest growth'. In Ecological effects of acid deposition. National Swedish environment protection board - Rep. PM, 1636, 199.

Bjor, K. and Teigen, O.: 1980, 'Effects of acid precipitation on soil and forest. 6. Lysimeter experiment in greenhouse'. In Drabløs, D. and Tollan, A. (eds.), Proc., Int. Conf. Ecol. Impact Acid Precip. SNSF-Project, Norway, 200.

Morrison, I.K. and Foster, N.W.: 1986, 'Limits of cation leaching of weakly-podzolized forest soils: An emperical evaluation'. Paper presented at NATO Advanced Research Workshop "Effects of acidic deposition on forest, wetlands and agricultural ecosystems", Toronto, May 12-17, 1985, in press.

Ogner, G., Haugen, A., Opem, M., Sjøtveit, G. and Sørlie, B.: 1984, 'The chemical analysis program at The Norwegian Forest Research Institute, 1984', Norwegian Forest Research Institute, Aas, Norway.

Singh, B.R.: 1980a, Acta Agric. Scand., 30, 357.

Singh, B.R.: 1980b, 'Effect of acid precipitation on soil and forest. 3. Sulfate sorption by acid forest soils'. In Drabløs, D. and Tollan, A. (eds.), Proc., Int. Conf. Ecol. Impact Acid Precip. SNSF-Project, 194.

Singh, B.R.: 1984, Soil Science, 138, 189.

Stuanes, A.O.: 1980, 'Effects of acid precipitation on soil and forest. 5. Release and loss of nutrients from a Norwegian forest soil due to artificial rain of varying acidity'. In Drabløs, D. and Tollan, A. (eds.), Proc., Int. Conf. Ecol. Impact Acid Precip. SNSF-Project, Norway, 198.

Stuanes, A. and Sveistrup, T.E.: 1979, 'Field experiments with simulated acid rain in forest ecosystems. 2. Description and classification of the soils used in field, lysimeter and laboratory experiments'. SNSF-Project, Norway FR 15/79, p. 35.

van Breemen, N., van Grinsven, J.J.M. and Jordens, E.R.: 1983, 'H^+ budgets and nitrogen transformation in woodland soils in the Netherlands influenced by high inputs of atmospheric ammonium sulphate'. VDI-Berichte Nr. 500, 345.

Van Miegroet, H. and Cole, D.W.: 1984, J. Environ. Qual., 13, 586.

Van Miegroet, H. and Cole, D.W.: 1985, Soil Sci. Soc. Am. J., 49, 1274.

Werner Johannessen, T.: 1970, 'The climate of Scandinavia'. In Landsberg, H.E. (ed.), World survey of climatology 5, 23.

ACID DEPOSITION AND ION LEACHING FROM A PODZOLIC SOIL UNDER HARDWOOD FOREST

N.W. Foster, I.K. Morrison and J.A. Nicolson
Government of Canada
Canadian Forestry Service
Great Lakes Forestry Centre
Sault Ste. Marie, Ontario
Canada
P6A 5M7

ABSTRACT. The contribution of atmospheric acids to cation leaching from a podzolic soil under mature maple-birch forest in central Ontario was examined during 1983. The movement of base cations was associated largely with NO_3^-, SO_4^{2-} and organic acid anions in surface soil horizons, with SO_4^{2-} and NO_3^- below the effective rooting zone, and SO_4^{2-} and HCO_3^- in streamflow. Mineral soil horizons could adsorb little additional SO_4^{2-} or associated cations at current soil solution SO_4^{2-} concentrations. Therefore, it is concluded that the soil *in situ* lacks a strong affinity for SO_4^{2-}. Current annual inputs to the forest of SO_4^{2-} and NO_3^- in bulk precipitation (26.4 and 18.2 kg ha^{-1}, equivalent to 8.8 kg S and 4.1 kg N ha^{-1}, respectively) contributed significantly to cation leaching from the soil. In order to maintain exchangeable cations in soil at current levels, a rate of weathering yielding 29.6, 5.0, 4.4 and 2.2 kg ha^{-1} yr^{-1} of Ca^{2+}, Na^+, Mg^{2+} and K^+, respectively, would be required.

1. INTRODUCTION

Base cations percolate through forest soils in association with mobile anions that are produced by microbially mediated or geochemical reactions in soil or from atmospheric deposition. Anions controlling cation leaching may be organic acid (fulvic acids) (Ugolini *et al.*, 1977), bicarbonate (HCO_3^-) (McColl and Cole, 1968), chloride (Cl^-) (Johnson, 1981), sulphate (SO_4^{2-}) (Cronan, 1980) or nitrate (NO_3^-) (Van Miegroet and Cole, 1984). Anthropogenic SO_4^{2-} and NO_3^-, by ultimately raising anion levels in soil solution, may increase the rate of cation leaching through the soil. The objective of this paper is to report on the extent that acid precipitation is contributing to base cation leaching from an old-growth forest on an acid podzolic soil at the Turkey Lakes Watershed (TLW), Algoma District, Ontario.

Water, Air, and Soil Pollution **31** (1986) 879–889.

2. METHODS

Data were collected in 1983 from a 1.0-ha study site within an uneven-aged 'old growth' sugar maple *(Acer saccharum* Marsh.) dominated stand with a lesser component of yellow birch *(Betula alleghaniensis* Britton), located in Basin 31 of the TLW (lat. 47°03'N, long. 84°24'W), Ontario. A site description and methods for solution and soil collection and chemical analysis are reported in Foster (1985) and Foster and Nicolson (1984).

Nitrogen and S mineralization were determined, using a method adapted from Foster *et al.* (1985), on soil collected in 1983 from a 0.05-ha sampling area within the study site. Forest floor weights were determined by quantitatively excavating 80 samples, each 0.1 m^2, and drying them to a constant weight at 70°C. Mineral soil was collected by horizons from four soil pits within the study site. Sulphate adsorption was determined by sequential leaching with Na_2SO_4, according to a method adapted from Johnson and Henderson (1979). Adsorbed SO_4^{2-} was extracted from soil, with and without Na_2SO_4 treatment, by means of NaH_2PO_4 containing 500 ppm P, and was followed by analysis of the extract for SO_4^{2-} by ion chromatography.

Ion contents of precipitation, forest floor percolate and streamflow were calculated as the product of concentration and measured volume. The amount of soil leachate at the 60-cm depth was estimated by means of the water balance equation (Thornthwaite and Mather, 1957).

3. RESULTS

The forest floor horizons were enriched in exchangeable base cations in contrast with soil lower in the profile (Table I). The forest floor, therefore, had a considerable potential for neutralizing the H^+ inputs in precipitation (Table II). Base cations replaced H^+ in solution as precipitation interacted with the forest canopy and forest floor. Sulphate and NO_3^- anions provided almost half the counter ions required to maintain electrical neutrality in forest floor solutions and to balance the increase in cation concentration over precipitation. A potential for even greater accumulation of NH_4^+, NO_3^- and SO_4^{2-} was observed when forest floor materials were incubated in the laboratory in the absence of plant uptake (Table III).

The Bhf_2 horizon contained the most adsorbed SO_4^{2-} (Table IV) and had the greatest capacity to adsorb additional SO_4^{2-} (Figure 1). The content of adsorbed SO_4^{2-} in the other horizons was Bf > Bhf_1 >IIBC > Ae (Table IV). The ability of these horizons to adsorb additional SO_4^{2-} was Bf > Bhf_1 (not shown) > Ae >IIBC (Figures 1 and 2).

In the TLW soil, tree rooting was restricted largely to the 50 to 60 cm of loose ablation till by the presence of a dense basal till below. Basin 31 received 1,307 mm of precipitation in 1983. We estimated, by using the method of Thornthwaite and Mather (1957), that 40% of the precipitation (530 mm water) percolated beyond the effective rooting zone. The chemical composition of water collected below the effective rooting

TABLE I

Soil properties related to cation retention in Basin 31 (mean ± standard deviation for four profiles).

Horizon	Depth (cm)	pH	Ca	Mg	K	Na	CEC	Base saturation (%)
			Exchangeable cations (NH_4OAc, pH 7.0) c mol $(p^+)\cdot kg^{-1}$					
L	7-5	5.1	31.23 ± 7.98	5.00 ± 1.04	2.31 ± 0.23	0.05 ± 0.04	65.39 ± 10.26	59
F	5-2	4.7	18.80 ± 7.08	3.65 ± 1.04	2.13 ± 0.38	0.13 ± 0.10	66.27 ± 10.18	37
H	2-0	4.4	5.17 ± 2.67	1.68 ± 0.81	0.56 ± 0.15	0.07 ± 0.06	32.37 ± 10.44	23
Ae	0-4	4.1	1.56 ± 0.40	0.31 ± 0.07	0.15 ± 0.07	0.21 ± 0.15	9.68 ± 2.58	23
Bhf_1	4-7	4.4	2.60 ± 1.60	0.53 ± 0.35	0.14 ± 0.03	0.16 ± 0.04	19.48 ± 3.42	18
Bhf_2	7-15	4.5	0.66 ± 0.45	0.17 ± 0.14	0.10 ± 0.03	0.20 ± 0.13	18.52 ± 1.60	6
Bf	15-57	4.7	0.19 ± 0.10	0.02 ± .01	0.04 ± 0.02	0.27 ± 0.14	9.78 ± 2.71	5
IIBC	57-82	5.0	0.11 ± 0.05	0.02 ± .01	0.02 ± 0.01	0.32 ± 0.28	4.83 ± 1.93	10
IIC	82+	5.1	0.08 ± 0.05	0.01 ± .01	0.01 ± .01	0.13 ± 0.06	1.49 ± 0.11	15

TABLE II

Annual fluxes (kg ha^{-1}) of ions in solutions sampled within watershed 31 in 1983 (mean ± standard deviation).

Variable	H^+	Ca^{2+}	Mg^{2+}	K^+	Na^+	NH_4^+-N	SO_4^{2-}-S	Cl^-	NO_3^--N	HCO_3^-
Precipitation	0.52 ± 0.04	2.9 ± 0.3	0.7 ± 0.1	1.3 ± 0.3	1.3 ± 0.3	2.7 ± 0.6	8.8 ± 0.6	3.1 ± 0.9	4.1 ± 0.3	0.8 ± 1.4
Forest floor	0.27 ± 0.09	32.6 ± 11.1	5.2 ± 1.6	24.0 ± 7.2	2.2 ± 1.0	8.1 ± 3.9	12.2 ± 3.3	4.7 ± 2.6	13.9 ± 5.6	n.d.
Mineral soil	0.25 ± 0.14	24.5 ± 4.6	4.6 ± 0.5	1.4 ± 0.6	6.2 ± 0.6	0.2 ±0.1	14.7 ± 2.6	4.2 ± 0.9	13.1 ± 4.4	0.8 ± 2.5
Stream	0.01	19.5	2.3	1.9	2.9	0.3	7.5	1.4	2.6	42.6

TABLE III

Nitrogen and S mineralized in soil held for 20 weeks at constant 20°C and 33 kPa moisture.

Horizon	NH_4^+-N (mM kg^{-1})	NO_3^--N (mM kg^{-1})	SO_4^{2-}-S (mM kg^{-1})	NH_4^+-N (kg ha^{-1})	NO_3^--N (kg ha^{-1})	SO_4^{2-}-S (kg ha^{-1})
L-H*	†95.4 ± 10.9	86.1 ± 5.3	4.06 ± 2.20	69.4	37.6	6.1
Ae	15.3 ± 0.9	3.1 ± 0.8	0.56 ± 0.13	39.1	8.0	3.2
Bhf_1	3.6 ± 2.9	6.6 ± 1.6	0.14 ± 0.05	7.9	14.4	0.7
Bhf_2	2.4 ± 0.4	0.1 ± 0.1	0.19 ± 0.08	13.9	0.9	2.5
Bf	1.2 ± 0.3	0.1	0.09 ± 0.05	43.1	1.7	7.0
IIBC	0.2 ± 0.1	0.1	0.03 ± 0.01	5.7	0.5	2.2

* Estimated from mineralization in F horizon
† mean ± standard deviation

TABLE IV

Soil properties related to N and S retention in Basin 31 (mean ± standard deviation for four profiles).

Horizon	N	C	S	Sulphate - S Soluble	Sulphate - S Adsorbed	Fe	Al
	g kg^{-1}			mg kg^{-1}		g kg^{-1}	
Ae	2.41 ± 0.94	28 ± 11	0.17 ± 0.07	21.4 ± 2.2	9.0 ± 2.3	2.6 ± 1.9	1.5 ± 0.4
Bhf_1	4.04 ± 0.47	54 ± 5	0.43 ± 0.07	23.2 ± 3.6	26.1 ± 2.5	15.2 ± 8.3	4.8 ± 2.6
Bhf_2	3.18 ± 0.22	51 ± 5	0.38 ± 0.01	9.1 ± 1.9	56.0 ± 13.1	17.2 ± 5.8	12.3 ± 2.8
Bf	1.64 ± 0.37	26 ± 6	0.18 ± 0.05	6.0 ± 1.6	37.2 ± 19.7	3.9 ± 1.8	7.8 ± 2.0
IIBC	0.50 ± 0.24	11 ± 4	0.06 ± 0.03	4.9 ± 0.5	14.4 ± 5.6	1.4 ± 0.4	4.4 ± 0.9
IIC	0.16 ± 0.03	3 ± 1	----	2.7 ± 1.3	----	0.5 ± 0.1	1.8 ± 0.3

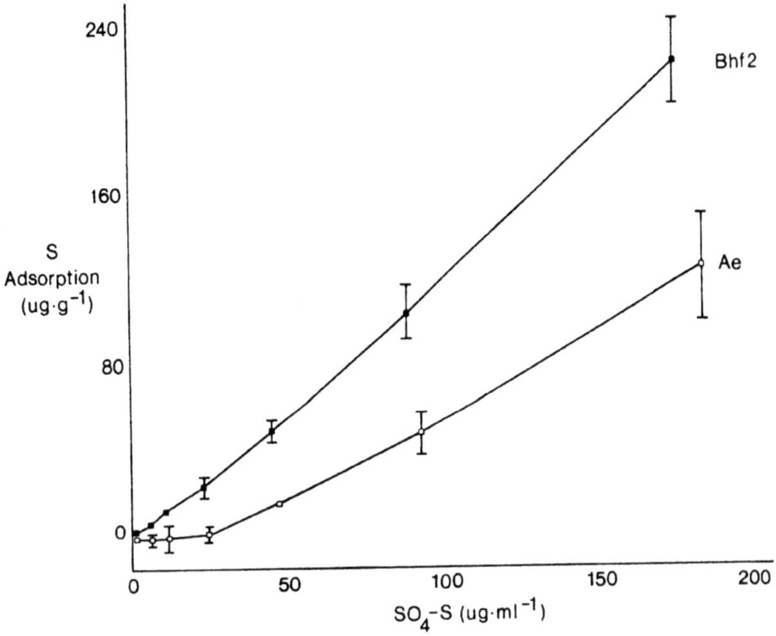

Fig. 1. Adsorption of SO_4^{2-}-S from equilibrium solutions by Bhf_2 and Ae soil horizons under a maple-birch forest (mean ± standard deviation).

zone consisted of less H^+ and NH_4^+ and more base cations, SO_4^{2-} and NO_3^- than did precipitation (Table II).

Although the total base content of the soil solution water exceeded that of precipitation, individual ions exhibited variability in their susceptibility to leaching. Base cations were leached from the

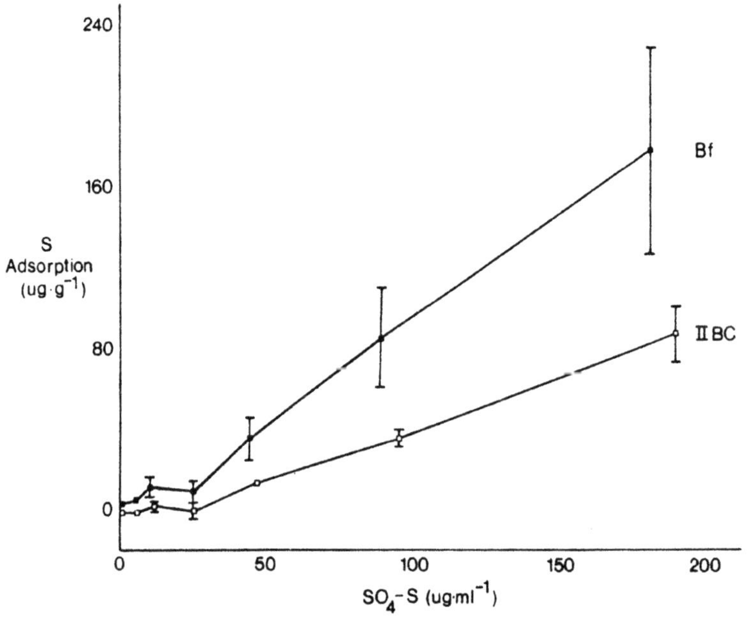

Fig. 2. Adsorption of SO_4^{2-}-S from equilibrium solutions by Bf and IIBC soil horizons under a maple-birch forest (mean ± standard deviation).

soil in association with SO_4^{2-} and NO_3^-. The combined NO_3^- and SO_4^{2-} concentrations in mineral soil water were equal to 80% of the total base cation concentrations on an equivalent weight basis. We estimate that leaching (rooting zone ion output minus precipitation ion input) was 21.6 Ca^{2+}, 4.9 Na^+, 3.9 Mg^{2+} and 0.1 K^+ kg ha^{-1} yr^{-1} in 1983. Outputs of all base cations in streamflow exceeded inputs in precipitation, whereas most of the H^+ and NH_4^+ was retained in the terrestrial ecosystem and did not appear in streamwater (Table II).

4. DISCUSSION

4.1. Cation Fluxes

Morrison (1986) has reported that incorporation into the forest vegetation at TLW was 8.0 Ca^{2+}, 0.1 Na^+, 0.5 Mg^{2+} and 2.2 K^+ kg ha^{-1} yr^{-1}. Weathering rates in the TLW soil, therefore, would have to be sufficient to yield approximately 29.6 Ca^{2+}, 5.0 Na^+, 4.4 Mg^{2+} and 2.2 K^+ kg ha^{-1} yr^{-1} to balance net losses to incorporation and leaching. Soils vary considerably, both in their content of mineral elements and in the rate of weathering of minerals. The soils of the Precambrian Shield are relatively young, being derived from the Wisconsin glaciation ∿ 10,000 yr ago. The TLW soil contained, within the main rooting zone to a depth of 60 cm, 286 kg Ca^{2+}, 74 kg K^+ and 33 kg Mg^{2+} ha^{-1} that were exchangeable with 1M NH_4OAc (pH 7.0). The rate of change of base saturation in the TLW soil has not yet been determined. If cations removed from the exchange sites by vegetation and leaching are not replaced by the weathering of primary minerals the supply of exchangeable base cations would be rapidly depleted (e.g., ∿10 yr for Ca^{2+}). The TLW soil contains considerable mineral nutrients as reserves in the unweathered mineral component. It is more likely then that base cation losses are being sustained by cation release from primary minerals. The net change in base cation storage on exchange sites in the TLW soil, then, is likely to be small. Net changes, however, even small ones, are cumulative, acting to lower the base saturation of the soil over time.

Schmoldt *et al.* (1985) have reported that better growth of sugar maple dominated forest in northern Wisconsin and upper Michigan was positively correlated with soil Ca^{2+} and Mg^{2+} that is exchangeable with NH_4OAc (pH 7.0). They found that the best growth was associated with plots containing 2000 kg Ca^{2+} ha^{-1} and 400 kg Mg^{2+} ha^{-1} in the surface 15 to 20 cm of soil, while the poorest growth was associated with 700 kg Ca^{2+} and 200 kg Mg^{2+} ha^{-1}. The upper 20 cm of soil at TLW contained 200 kg Ca^{2+} ha^{-1} and 30 kg Mg^{2+} ha^{-1}. The TLW soil appears, therefore, to contain less Ca^{2+} and Mg^{2+} than required for optimum growth of sugar maple in the northern part of its range.

4.2. Nitrogen Fluxes

According to McColl and Cole (1968), cation leaching in soils is limited by the supply of anions mobile in the soil. Mineralization of N and, to a lesser extent, SO_4^{2-} in the forest floor (Table III) increased anion concentrations in water that infiltrated the organic horizons and significantly contributed to the leaching of cations from the forest floor to the mineral soil. The anion deficit in the forest floor percolate was presumably made up largely of organic acid anions released from decomposing plant residues. Evidence for this was carbon concentrations in the forest floor percolate (20 to 50 mg L^{-1}) that exceeded those in precipitation (0 to 5 mg L^{-1}).

Low C concentrations in mineral soil solution (2 to 10 mg L^{-1}) in comparison with those in forest floor percolate suggest that organic acid anions were retained within the effective rooting zone. Chloride ions were mobile in the soil and constituted, on an equivalent weight basis, ∿5% of the total anion flux from the rooting zone. Base cation leaching, therefore, was associated with a mix of anions (organic acid, SO_4^{2-}, NO_3^- and Cl^-) in the forest floor. Fluxes of SO_4^{2-}-S and NO_3^--N in soil solution were similar in magnitude (Table II). They were considerably greater than those observed in both precipitation and streamwater. The total N concentrations in surface silty-loam mineral soil horizons were much higher than those in deeper sandy loam materials (Table IV). If the inorganic N produced from decomposing organic matter is in excess of microbial and higher plant needs, NH_4^+ and/or

NO_3^- will accumulate in the soil. We have estimated that up to 85 kg ha^{-1} of inorganic N may accumulate in Ae, Bhf_1 and Bhf_2 over 20 weeks, under optimum temperature and moisture conditions (Table III). This potential N mineralization was determined in the laboratory in the absence of plant N uptake. Potential N mineralization in the forest floor horizons may be as high as 100 kg N ha^{-1} (Table III).

Mineralization in the field, where temperature and moisture are seldom maintained at optimum levels, is unlikely to approach potential levels. The measured annual inorganic N flux through the forest floor was 22 kg ha^{-1} (Table II). Our estimate for N turnover between the vegetation and soil for this stand is 45 kg ha^{-1} yr^{-1} (Foster *et al.*, 1986). Much of the mineral N, therefore, cycles between vegetation and soil and is retained within the forest ecosystem.

The relatively small NH_4^+ fluxes in streamflow can be attributed to nitrification in soil and effective N uptake by the vegetation. Appreciable amounts of NH_4^+ were nitrified in the soil both in the laboratory (Table III) and in the field (Table II). Nitrification in acid organic and mineral horizons from tolerant hardwood forest has been widely reported (e.g., Krause, 1982; Federer, 1983; Klein *et al.*, 1983). Most of the NO_3^- in the Turkey Lakes soil was produced in organic, Ae, and Bhf_1 horizons rather than at greater soil depths. The estimated annual NO_3^--N flux from the effective rooting zone was 13 kg ha^{-1}. Nitrate made a significant contribution to base cation leaching from the soil, especially during the dormant period (Foster, 1985).

4.3. Sulphur Fluxes

Morrison (1986) has estimated that the annual turnover of S between the maple-birch forest and soil at the TLW is 3 to 4 kg ha^{-1}. The fluxes of SO_4^{2-}-S through the soil, 12 to 15 kg ha^{-1} yr^{-1}, were larger than estimated S uptake by the stand. The cycle of N and S between the vegetation and soil is generally in proportion to the mole ratio of organic forms of the two elements (33:1) in plant material. An N:S ratio of 21 in maple foliage at TLW suggests the presence of SO_4^{2-}S in foliage. Sulphate appears, therefore, to be present in the plant and soil in excess of the demands of the vegetation.

It is generally recognized that control of SO_4^{2-} leaching occurs through adsorption by Fe and Al sesquioxides in the soil (e.g., Johnson and Todd, 1983), rather than by plant uptake as was the case with N. The TLW soil contains considerable Fe and Al sesquioxides throughout and below the rooting zone, with highest concentrations in the Bhf horizons (Table IV).

Adsorbed SO_4^{2-} in the soil and adsorption of additional SO_4^{2-} by the soil were similar to those in podzols beneath temperate forest in the northeastern United States (Fuller *et al.*, 1985). Giles *et al.* (1974) classified adsorption isotherms according to initial slope. All the TLW soil horizons exhibited a C or constant partition isotherm type. The C isotherm types are indicative of lower SO_4^{2-} adsorption, at low equilibrium solution concentrations, than is the case with the L type (Langmuir) isotherms that characterized the podzols examined by Singh (1984). Although all mineral horizons of the TLW soil can adsorb additional SO_4^{2-}, the equilibrium concentration that had to be reached before SO_4^{2-}-S was initially retained was 12 mg L^{-1} for Bhf_1 and 25 mg L^{-1} for Ae and IIBC horizons (Figures 1 and 2). Sulphate, however, was adsorbed from all concentrations of the equilibrium solutions for Bhf_2 and Bf horizons. The mean SO_4^{2-}-S concentrations of water entering the mineral soil (1.1 mg L^{-1}) and soil solution (60 cm) (1.8 mg L^{-1}) were below the equilibrium concentrations that must be attained for adsorption by Bhf_1, Ae, and IIBC horizons.

Fuller *et al.* (1985) concluded that the podzols they examined were in equilibrium with soil solution concentrations and were not currently adsorbing SO_4^{2-}. Mollitor and Raynal (1982) have reported, for a podzolic soil under a hardwood canopy, that the SO_4^{2-} concentration in water leaving the B horizon did not differ from that observed in the A horizon. The flux of SO_4^{2-} beyond the rooting zone in the TLW soil was roughly in balance with inputs to the mineral soil, and this also suggests a lack of adsorption of SO_4^{2-} by the soil. Potential mineralization of SO_4^{2-}-S in the mineral soil horizons (Table III) was roughly equal to annual inputs from the forest floor (12-13 kg ha^{-1}). If actual mineralization in the field approaches potential mineralization, then considerable SO_4^{2-} must be adsorbed. If actual mineralization rates are low then little adsorption of SO_4^{2-} must take place.

The concentration of soluble SO_4^{2-}-S in the TLW soil was highest in the Ae and Bhf_1 horizons and decreased with depth (Table IV). Mineralization of SO_4^{2-} by microbial activity in the mineral soil over 20 weeks was greatest in the Ae horizon (Table III). The surface mineral horizons, therefore, may be a source of SO_4^{2-}. The Bhf_2 and Bf horizons may act as a sink, retaining some of the SO_4^{2-} mineralized in the upper profile and that from the atmosphere. Additional information, which is currently being collected on SO_4^{2-} fluxes through individual horizons, will help sort out the extent to which potential mineralization rates are being realized. This information in time will provide insight into adsorption of SO_4^{2-} by soil *in situ*.

Bulk deposition of SO_4^{2-}-S at Turkey Lakes was 8.8 kg ha^{-1} yr^{-1} in 1983 (Table IV). The TLW currently receives 10 to 12 kg ha^{-1} yr^{-1} of total SO_4^{2-}-S (Barrie, 1985) of which 8 to 9 kg ha^{-1} yr^{-1} are estimated to be wet deposition. The TLW falls within the deposition zone of 4 kg ha^{-1} yr^{-1} of *anthropogenic* wet deposition of SO_4^{2-}-S (Nicols and Verry, 1985). Nicols and Verry (ibid.) calculated the wet SO_4^{2-} deposition that was related to anthropogenic emissions on the basis of the relationship of H^+ to SO_4^{2-} in precipitation. Anthropogenic dry deposition of SO_4^{2-}-S would be 1.5 kg ha^{-1} yr^{-1} if we assume the ratio of anthropogenic to natural S in dry deposition is similar to that in wet deposition. In the worst possible situation, if we assume no interaction of SO_4^{2-} with vegetation or soil, all atmospheric SO_4^{2-} might contribute to cation leaching from soil. Anthropogenic SO_4^{2-} would represent 37% of the total SO_4^{2-} leached from the soil. The 5.5 kg ha^{-1} of anthropogenic SO_4^{2-}-S (wet plus dry) could leach 330 eq ha^{-1} of base cations from the soil in the worst situation, and this would represent 21% of the current estimated base leaching. An additional 10% of the current base leaching would be related to anthropogenic NO_3^-, if we assume that the ratio of anthropogenic NO_3^- to natural NO_3^- in precipitation is similar to the SO_4^{2-} ratio, and no anthropogenic NO_3^- is removed by soil and vegetation. In the worst situation, therefore, base cation leaching from the TLW soil would be increased significantly by current levels of acid precipitation. Natural rainfall (∿2 mg L^{-1} SO_4^{2-}-S) and acidified natural rainfall (∿6 mg L^{-1} SO_4^{2-}-S) have been applied to the TLW soil during the growing season. The added SO_4^{2-} was not retained by the soil (Hern *et al.*, 1985) and Ca^{2+} and Mg^{2+} leaching was increased three times (Rutherford *et al.*, 1985). It is reasonable to assume, therefore, that with current or increased additions of anthropogenic SO_4^{2-}, little of the SO_4^{2-} will be adsorbed by the soil, as discussed previously, and the worst possible situation will be approached.

Annual inputs in precipitation and outputs in streamflow for SO_4^{2-} are roughly in balance (Table II). Since the flux of SO_4^{2-} in streamwater is only half of that below the rooting zone, SO_4^{2-} must be retained by the subsoil, perhaps by adsorption. Another possibility is that precipitation of SO_4^{2-} with Al (Meiwes *et al.*, 1980; Nilsson and Bergkvist, 1983) may also contribute to SO_4^{2-} retention in TLW subsoils. Foster and Nicolson (1985) have observed that fluxes of Al in the mineral soil solution below the tree rooting zone at TLW are higher than those in throughfall, forest floor percolate and streamflow, which supports the aluminum sulphate precipitation hypothesis. The dominance of Ca^{2+} and HCO_3^- in streamwater is evidence that cation leaching from this watershed is regulated by the dissolution of $CaCO_3$ in the subsoil.

5. SUMMARY

Inputs to the TLW in bulk precipitation in 1983 were 8.8, 4.1 and 0.5 kg ha^{-1} yr^{-1} for SO_4^{2-}-S, NO_3^--N and H^+, respectively. The soil lacks a strong affinity for SO_4^{2-} at current soil solution concentrations. Present levels of SO_4^{2-} deposition from the atmosphere, therefore, have raised SO_4^{2-} levels in soil solution and increased cation leaching. Mineralization of NO_3^- in soil exceeded inputs to the ecosystem in precipitation, and that of SO_4^{2-} was equal to such inputs. Sulphur and N mineralization rates were greatest in the forest floor and upper mineral soil horizons (0 to 10 cm). Base cation movement was associated with NO_3^-, SO_4^{2-} and organic acid anions in surface soil horizons, largely with NO_3^- and SO_4^{2-} below the effective rooting zone and with SO_4^{2-} and HCO_3^- in streamflow. Calcium and Mg^{2+} are being leached from the soil of this maple-birch forest, therefore, in association with both anthropogenic and naturally derived SO_4^{2-} and NO_3^-.

Even though an accelerated rate of cation leaching is observed as a result of the action of anthropogenically derived anions, we do not know the rate at which reserves of exchangeable base cations in the soil are being depleted. The estimated net annual base cation losses to leaching would rapidly deplete the supply of exchangeable base cations in the soil in the absence of weathering of primary minerals. No estimates of weathering rates for this particular soil are available; but it is likely that a considerable proportion of the cation losses are being sustained by cation release from primary minerals. Weathering would have to approximate 29.6 Ca^{2+}, 5.0 Na^{+}, 4.4 Mg^{2+} and 2.2 K^{+} kg ha^{-1} yr^{-1} to completely replace losses to leaching and storage in the vegetation. Observations over longer periods and additional information on fluxes between individual soil horizons are needed to clarify the effect of leaching on soil base cation reserves.

6. ACKNOWLEDGMENT

The assistance of P. Hazlett, W. Johns, and D. Ropke with sample collection and analysis of the data, and of D. Kurylo, J. Longlade, G. Hauer, C. Vaillancourt, A. Spadafora, L. Chartrand and A. Juliani with chemical analysis of samples is gratefully acknowledged.

7. REFERENCES

Barrie, L.A.:1985, Unpublished data.
Cronan, C.S.:1980, *Oikos* **34**, 272.
Federer, C.A.:1983, *Soil Sci. Soc. Am. J.* **47**, 1008.
Foster, N.W.:1985, *For. Ecol. Manage.* **12**, 215.
Foster, N.W. and Nicolson, J.A.:1984, 'Acid Precipitation and Water Quality within a Tolerant Hardwood Stand and Soil', in: Speeches and Papers. *Forest Resources Management, The Influence of Policy and Law.* Internat'l For. Congr., Quebec City, p. 337.
Foster, N.W. and Nicolson, J.A.:1985, Unpublished data.
Foster, N.W., Beauchamp, E.G. and Corke, C.T.:1985, *Soil Biol. Biochem.* **17**, 699.
Foster, N.W., Nicolson, J.A. and Morrison, I.K.:1986, 'Acid Deposition and Element Cycling in Eastern North American Forests', in: Proc. *Acid Rain and Forest Resources Conference,* Quebec, Canada. 14-17 June 1983 (in press).
Fuller, R.D., David, M.B., and Driscoll, C.T.:1985, *Soil Sci. Soc. Am. J.* **49**, 1034.
Giles, C.H., Smith, D., and Huitson, A.:1974, *J. Colloid Interface Sci.,* **47**, 755.
Hern, J.A., Rutherford, G.K. and Van Loon, G.W.:1985, *Can. J. For. Res.* **15**, 839.
Johnson, D.W.:1981, *Water, Air, Soil Pollut.* **16**, 243.
Johnson, D.W. and Henderson, G.S.:1979, *Soil Sci.* **128**, 34.
Johnson, D.W. and Todd, D.E.:1983, *Soil Sci. Soc. Am. J.* **47**, 792.
Klein, T.M., Kreitinger, J.P. and Alexander, M.:1983, *Soil Sci. Soc. Am. J.* **47**, 506.
Krause, H.H.:1982, *Can. J. For. Res.* **12**, 922.
McColl, J.G. and Cole, D.W.:1968, *NW. Sci.* **42**, 134.
Meiwes, K.J., Khanna, P.K. and Ulrich, B.:1980, *Z. Pflanzenern. Boderkd.* **143**, 402.
Mollitor, A.V., and Raynal, D.J.:1982, *Soil Sci. Soc. Am. J.* **46**, 137.

Morrison, I.K.:1986, 'Biomass and Macroelements in a Tolerant Hardwood Stand, Turkey Lakes Watershed, Ontario', in: Proc. *Acid Rain and Forest Resources Conference*, Quebec, Canada, 14-17 June 1983 (in press).

Nichols, D.S. and Verry E.S.:1985, 'Evidence for the Cultural Acidifica tion of Lakes in the Northern Lake States', in: Proc. Symp. *Effects of Air Pollutants on Forest Ecosystems*, St. Paul, MN, 8-9 May 1985(in press).

Nilsson, S.I. and Bergkvist, R.:1983, *Water, Air, Soil Pollut.* **20**, 311.

Rutherford, G.K. Van Loon, G.W., Mortensen, S.F. and Hern, J.A.:1985, *Can. J. For. Res.* **15**, 848.

Schmoldt, D.L., Martin, G.L. and Bockheim, J.G.:1985, *For. Sci.* **31**, 209.

Singh, B.R.: 1984, *Soil Sci.* **138**, 189.

Thornthwaite, C.W. and Mather, J.R.:1957, *Climatology* **3**, 185.

Ugolini, F.C., Minden, R., Dawson, H. and Zachara, J.:1977, *Soil Sci.* **124**, 291.

Van Miegroet, H. and Cole, D.W.:1984, *J. Environ. Qual.* **13**, 586.

THE EFFECT OF FIVE YEARS ACID TREATMENT ON LEACHING, SOIL CHEMISTRY AND WEATHERING OF A HUMO-FERRIC PODZOL

R.A. Skeffington and K.A. Brown,
Central Electricity Research Laboratories, CEGB
Kelvin Avenue
Leatherhead
Surrey, KT22 7SE, UK

ABSTRACT. This paper describes the effect of treating a nutrient-poor forest soil in monolith lysimeters with H_2SO_4, pH 3.0, for 4.75 yr. The lysimeters were instrumented with porous cup probes to distinguish processes occurring in each soil horizon. In the A horizon base cation exchange and sulphate absorption were the principal proton- consuming processes whereas lower down the profile Al^{3+} dissolution from hydrous oxides dominated. Acid treatment thus reduced the amount of amorphous Al in the lower horizons, but exchangeable Al was unaffected. Sulphate absorbtion was positively correlated with the distribution of Al hydrous oxides. High rates of nitrification reduced the differences between acid and control monoliths, but acid treatment significantly reduced soil pH down to 75 cm and reduced the levels of exchangeable base cations in the litter and A horizons. Acid treatment increased the leaching rates of base cations and Al. Consideration of the total base cation content shows that acid treatment increased the rate of weathering by 0.7-1.4 k eq ha^{-1} yr^{-1}. The results should be useful in modelling more realistic rates of acid input to similar soils.

1. INTRODUCTION

The effects of acid precipitation on soil have been the subject of a great deal of research (Bache, 1980; Drabløs and Tollan, 1980; Ulrich and Pankrath, 1983), but there is still disagreement about the likely effects of current rates of acid deposition. It has been suggested that natural soil acidification processes are likely to outweigh atmospheric acid inputs as far as soil acidification (Krug and Frink, 1983) and consequent water acidification (Rosenqvist, 1978) are concerned, though these views are hotly disputed by others (e.g. Johnson et al., 1984). With such disagreement about fundamentals, experimental studies of soil acidification are clearly required. Because soils are very variable, it is important to identify and quantify the effects of acidification on soil processes, so that the effects of acidification on soil types other than the one studied may

Water, Air, and Soil Pollution **31** (1986) 891–900.

be calculated. This paper reports on the response of a podzolic forest soil to experimental acidification.

The approach adopted was the long-term treatment of a large undisturbed core of forest soil with sulphuric acid, pH 3. The soil chosen was a very base-poor humo-ferric podzol and the rate of acid input was chosen to give a high acid load (about 30 times the average UK H^+ deposition) without subjecting the soil to concentrations of acid which would not occur in the natural environment. As soils are layered structures and the processes which occur in each layer may be very different, soil probes were installed in intermediate horizons to study the leachate chemistry down the profile. Such experiments as these can be used to answer a wide range of questions relating the reaction of the soil to externally-applied acidity. Here we give a first assessment of some of the results. The effects of acid treatment on soil composition have been reported elsewhere (Brown, 1985, 1986).

2. MATERIALS AND METHODS

2.1. The Monolith Lysimeters

Four undisturbed cores of soil and overlying litter (monoliths) were obtained in fibreglass cylinders 0.8 m in diameter and 1.35 m deep (Belford, 1979). The collection site was a clearing in woodland near Headley Park, Hants. (Latitude 51°N, 1°W). The ground vegetation was largely bracken (Pteridium aquilinium (L.) Kuhn). The soil was a sandy humo-ferric podzol of the Shirrell Heath series: a typic haplorthod in the USDA soil taxonomy and an orthic podzol in the FAO system. More details of the soil can be found in Rippon (1984).

The lysimeters were installed in a pit outside the Biology Laboratory on 6 April 1976. They were each fitted with 4 acid-washed ceramic cup probes of the type pioneered by Wagner (1962), which were inserted from the side of the lysimeters to correspond with horizons Ah, Bs, C1 and C2. Every 2 weeks a water sample was extracted through the probes, and water draining from the base of each lysimeter was also collected. This description of lysimeter operation is necessarily abbreviated - for a full description see Rippon et al. (1985).

2.2. Treatments

The lysimeters were watered every Monday, Wednesday and Friday in 4 batches of 1.25 L d^{-1} per lysimeter, giving a nominal annual total of 1500 mm rain per lysimeter. Distilled water was applied to all lysimeters at this rate from 12 April 1977, being replaced by 0.5 mM H_2SO_4 pH 3.0, from 5 October 1977 on two of the lysimeters (A1 and A2). The lysimeters were covered by a greenhouse partially open at both ends during the experimental period. The monoliths were kept free of vegetation, but fresh litter from the sampling site was added each winter to a density comparable with that in the field. On 6 July 1982 after 1735 days (4.75 yr) of treatment, the lysimeters were dismantled in 3 cm layers and the soil and litter analysed.

2.3. Chemical Analysis

Soil and litter samples were analysed according to methods in Allen (1974) unless otherwise stated. Exchangeable Na, K, Ca, Mg and Mn were extracted with 1M ammonium acetate, pH 7, using air dried, 2 mm-sieved samples. Al and Fe hydrous oxides were extracted with citrate-dithionite (Hesse, 1971). Exchangeable Al was extracted with 1.0 M KCl and analysed for monomeric and polymeric hydroxy complexes and organic complexes (Bache and Sharp, 1976). Adsorbed plus readily-soluble SO_4^{2-} was extracted with 0.032 M KH_2PO_4 solution, and adsorbed SO_4^{2-} was calculated by subtracting the amount of SO_4^{2-} extracted with 2.5% acetic acid (readily- soluble SO_4^{2-}). Total S in dried soil was measured after digestion with $Mg(NO_3)_2$-HNO_3 (full references in Brown, 1986).

Metal cations in extracts and soil probe water were analysed by inductively-coupled plasma emission spectrometer, and SO_4^{2-}, NH_4^+, NO_3^- and Cl^- by autoanalyser methods (Skeffington, 1984a).

3. RESULTS

3.1. Fluxes

The fluxes of most of the major cations and anions in one control and one acid lysimeter are shown diagramatically in Figure 1. Other ions were also measured but their contribution to the ion balance was generally small. Inspection of the fluxes in the acid lysimeters (right-hand side of Figure 1) shows a progressive retention of SO_4^{2-} down the soil profile, which occurs mostly in the 'A' horizon. This should reduce the acid load on the soil (Johnson, 1980), but the appearance of NO_3^- in the soil solution, due presumably to nitrification, will increase it again as nitrification is an acidifying process (Reuss, 1977). The A horizon is a strong absorber of protons in the acid lysimeters. The processes that absorb them are in order of importance: the release of base cations, SO_4^{2-} retention, and Al release. In the control lysimeters there is a net production of H^+ in the 'A' horizon. Nitrification is the major H^+-producing process, and the H^+-consuming processes are base cation and Al release as in the acid lysimeters, except that the former is relatively more important. Below the 'A' horizon, nitrification continues to produce H^+, but the $[H^+]$ in both acid and control lysimeters progressively declines. In the B and upper C horizons the H^+-consuming process is almost exclusively Al release. In the lower C horizon there is once again some base cation release in the acid lysimeters, but the flux of the Al continues to increase even deep in the soil profile.

3.2. Effect of Acid Treatment on the Monoliths

3.2.1. Sulphate and pH. The applied H_2SO_4 (6.6 equivalents m^{-2}) acidified the soil, and some of the added S was retained in the profile (see Figure 1). The effect of acid treatment on pH ($CaCl_2$) was

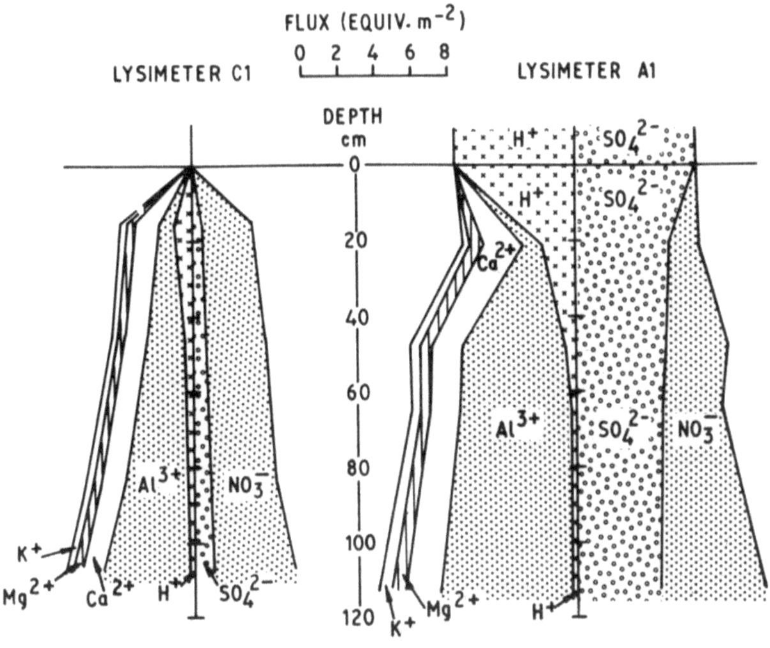

Fig. 1 Fluxes of major cations over the whole experimental period, in two lysimeters, control (C1) and acid (A1) treated. Lysimeters C2 and A2 showed similar results. The width of any sector at a given depth represents the flux of an ion which passed that depth during the 4.75 years of the experiment.

greatest in the litter (Figure 2A), but the soil was significantly acidified down to 75 cm. The total S reserve increased throughout the mineral soil profile due to acid treatment, but an increase could not be distinguished in the litter and A horizon where more than 90% of the total S was present as organic compounds. The reserves of readily soluble and absorbed SO_4^{2-} (Figure 2B) also increased down the profile, and could account for all the increase in total S in the B and C horizons of the acid lysimeters. See Brown (1986) for further detail.

3.2.2. Organic matter and cations. Acid treatment reduced the content of all exchangeable cations in the litter (Brown, 1985) and of exchangeable Ca^{2+}, Mg^{2+} and Mn^{2+} in the mineral soil down to about 25 cm depth (Brown, 1986). See these papers for a detailed discussion.

3.2.3. Aluminium and iron. Al hydrous oxides (extracted with the dithionite-citrate reagent, Figure 2C) were significantly lower in the acid-treated lysimeters. The distribution of Al extractable with 2.5% acetic acid was very similar to that in Figure 2C, although the concentrations were about one-tenth those of the dithionite-citrate

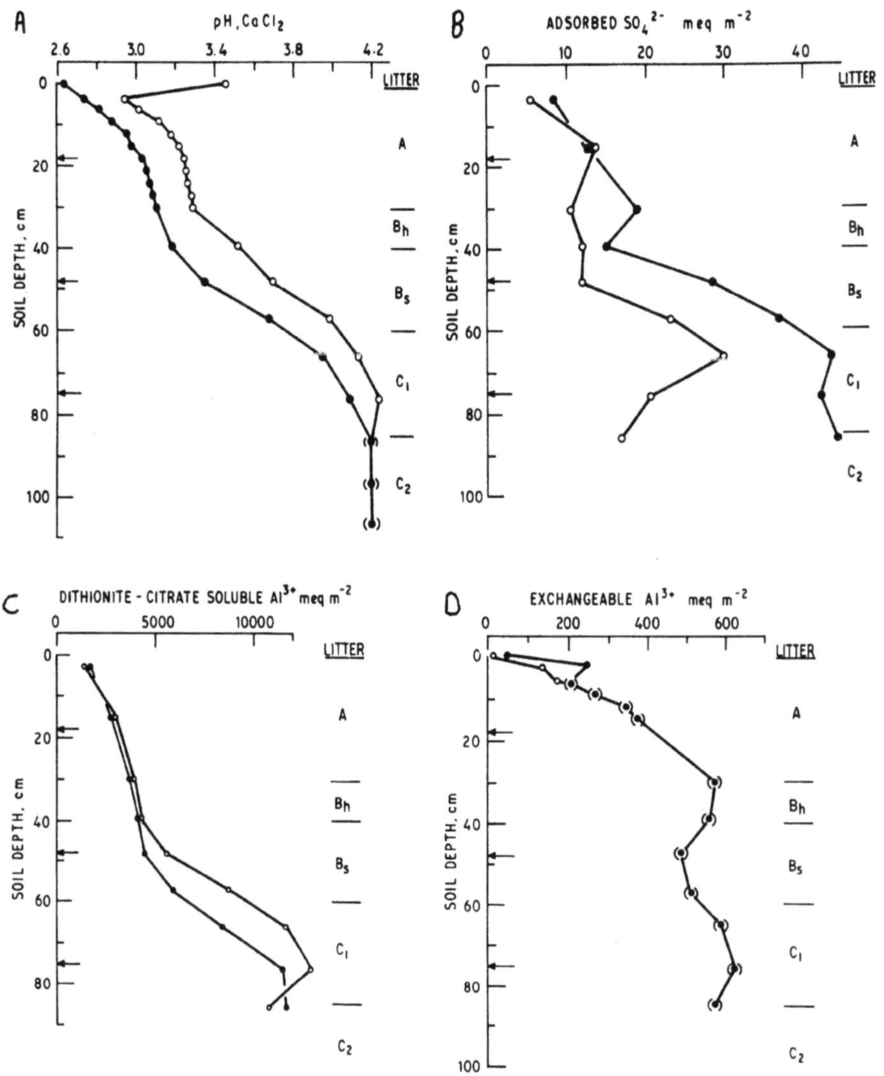

Fig. 2A Depth profile of geometric mean pH ($CaCl_2$) in control (o) and acid-treated (●) lysimeters. Where duplicate values overlapped between treatments, the mean value for all four lysimeters is given in parentheses (●).

2B Depth profile of mean absorbed SO_4^{2-} (KH_2PO_4 extractable less 2.5% acetic acid extractable).

2C Depth profile of mean dithionite-citrate soluble Al.

2D Depth profile of mean exchangeable (1M KCl) Al.

extractable Al. The distribution of exchangeable Al (soluble in 1M KCl) was different from the hydrous oxides (Figure 2D). Only in the litter and the top 3 cm of the 'A' horizon was there a significant increase due to acid treatment (Figure 2D). Essentially all the

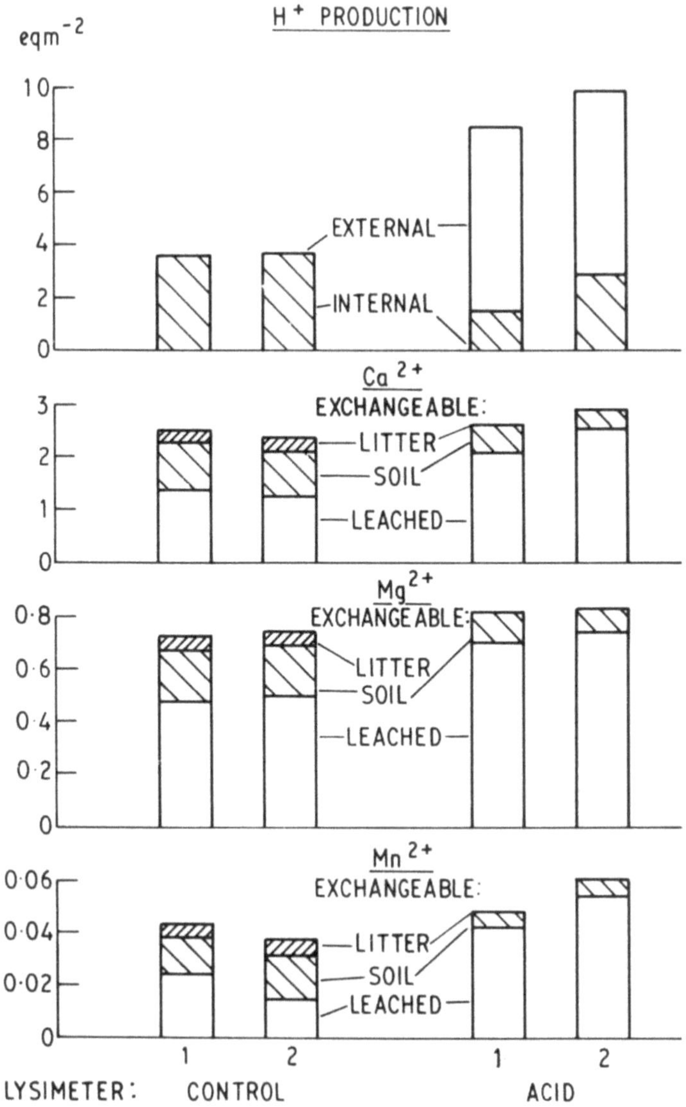

Fig. 3 Proton inputs, internal generation, and weathering, 'A' horizon. Weathering is calculated as the difference in height between the lowest 'control' bar and the two 'acid' bars, making slight corrections for probe depth.

exchangeable Al in the mineral profile, and more than 90% of that in the litter, was in a monomeric form as measured by the method of Bache and Sharp (1976). At the pH of the soil solution (~4.3) this implies that Al^{3+} was the dominant Al species.

No effect of acid treatment was observed on acetic-acid extractable Fe or Fe hydrous oxides.

3.3. Weathering

Weathering in this context is defined as release of base cations other than by ion exchange processes. In principle the amount of weathering which took place in the monoliths can be calculated by the formula:

Weathering = Leaching - (Exchangeable cations before treatment
Exchangeable cations after treatment).

In practice, it proved impossible to apply this formula because the flux of water applied to the lysimeters (twice natural rainfall) moved organic matter (the principal source of cation exchange capacity) down the profile, because the exchangeable cation contents below the top few layers were close to the detection limits of the methods originally used, and because the lysimeters were subjected to chemically unquantified rain and dry deposition in the period before acid treatment. It is possible however to compare the acid and control lysimeters. Figure 3 shows that proton input and internal production in the 'A' horizon of the acid monoliths was roughly double that in the control monoliths. More Ca^{2+} and Mg^{2+} were leached from the acid monoliths than the control monoliths, but less exchangeable ($\underline{M}$ NH_4 OAC) Ca^{2+} and Mg^{2+} remained behind. However, the sum of the exchangeable and leached Ca^{2+} and Mg^{2+} is greater in the acid (A1 and A2) than in the control monoliths (C1 and C2), demonstrating an extra source of base cations in the acid monoliths. This is an increase in weathering (as defined above) due to acid treatment: rates are given in Table I. These are minimum rates as they assume no weathering in the control lysimeters. In the B and C horizons this method of calculation shows no increase in weathering in the acid monoliths.

Table I: Calculated Increase in Weathering Rates Due to Acid Treatment: 'A' Horizon

Ion	Lysimeter	
	A1	A2
Ca^{2+}	0.42	1.1
Mg^{2+}	0.25	0.27
Mn^{2+}	0.02	0.04
Total	0.69	1.41

Figures in keq ha^{-1} yr^{-1}

4. DISCUSSION

4.1. Fluxes

The flux data are generally explicable by well-known concepts of soil science. Cation exchange laws appear prima facie to be obeyed: the mobile anion concept (Johnson and Cole, 1980) holds good. It had been suggested that if SO_4^{2-} deposition were increased then transport of cations through the profile by organic anions would decrease in almost exact compensation (Krug and Frink, 1983) - this is clearly not the case here. An unexpected finding was the large amount of nitrification which occurred (127 to 203 kg N ha^{-1} yr^{-1}), as when we started this experiment it was generally held that nitrification does not occur in acid soils. These high rates are probably an artefact of the absence of vegetation from the lysimeters, as there is then no sink for the NH_4^+ produced by organic matter decomposition. However, the data show that nitrification can occur in acid soils. Nitrification reduced the difference between the acid fluxes in the acid and control lysimeters.

Aluminium in the probe leachates was not speciated but should be largely Al^{3+} at the pHs involved if Al is all in inorganic monomeric forms. In the 'A' horizon consideration of the ion balance indicates that some (~10%) of the Al in leachates was probably organically complexed. This is a smaller proportion than in some studies (Nilsson and Bergkvist, 1983), though the pattern of Al dynamics is similar. There is no net Al deposition in the lower soil horizons, as occurs in podzols, but Al deposition due to immobilisation of organic complexes may still occur - the effect would be swamped because H^+ produced by nitrification in the lower horizon is absorbed by Al dissolution.

4.2. Effects on Soil Monoliths

Acid treatment reduced soil pH to a depth of 75 cm. This is not surprising because of the lack of weatherable minerals to neutralise the considerable flux of protons coming from the horizons above and those generated internally by nitrification. Only in the lower C horizon, where nearly all the H^+ in the downcoming flux had been exchanged for Al, was the acidification not significant.

The distribution of exchangeable Al (Figure 2) shows little effect of acid treatment in the lower horizons, since most of the Al released apparently originated from the dissolution of Al^{3+} from the solid phase of poorly-ordered hydrous oxides, the exchange sites being already saturated with Al. At the top of the profile, however, some dissolution of Al from organic complexes appears to have occurred, and the Al has displaced base cations from the exchange complex, leading to the observed increase in exchangeable Al (Figure 2D). As Al is bound more tightly to exchange sites than are base cations, there was little tendency for displaced base cations to exchange with Al on exchange sites lower down the profile, which remained almost completely Al saturated. Absorption of SO_4^{2-} occurs on Al and Fe hydrous oxides (Chao et al., 1964; Johnson, 1980). In this soil the absorbed SO_4^{2-} correlated quite well with the Al hydrous oxides and not the Fe hydrous oxides (Figure 2). The soil retained about 34% of the applied SO_4^{2-} (Figure 1): 65% of that retained was retained in the top 15 cm of the soil profile. Only about 30% of SO_4^{2-} retained by the litter and 'A' horizons was recovered in the soil analysis. The cause of this discrepancy is not known but may be related to the low precision in estimating total S reserves in the upper horizons due to variations in organic matter content between the monoliths. In the B and C horizons the SO_4^{2-} retained as shown by the flux measurements was consistent with the differences between total S content in the soils. There was no significant increase in cation exchange capacity which might indicate the presence of specific absorption of SO_4^{2-}, (Johnson, 1980), and no evidence for $AlOHSO_4$ precipitation, suggested as a mechanism of SO_4^{2-} absorbtion by Prenzel (1983), and Nilsson and Bergkvist (1983).

4.3. Weathering

Stuanes (1980) calculated weathering rates in a forest soil subjected to three different acid treatments, and found that weathering increased with acid input. This is true in this experiment also - the rate of weathering in the acid lysimeters exceeded that in the control by 0.7-1.4 keq ha^{-1} yr^{-1}. In Stuanes' (1980) experiment some of the weathering term may have originated from organic matter decomposition and plant uptake rather than mineral weathering whereas in this experiment there was no plant uptake and litter decomposition was slightly reduced by acid treatment. The estimates given are thus minimum estimates, as mineral weathering would also have occurred in the control treatments. The values obtained are consistent with other studies, e.g. Ulrich (1983). No weathering could be demonstrated in the B and upper C horizons, but in the lower C horizon some base cation release occurred (Figure 1). Weathering, as the principal process neutralising acid production in soil, deserves much further study.

5. CONCLUSIONS

The acid application rates used in this study were high (about 14 keq ha^{-1} $year^{-1}$ of H^+) compared to rates commonly observed in the natural environment (up to 1 keq ha^{-1} yr^{-1} in wet deposition: Table 7 in Skeffington, 1984b). Even in throughfall under mature trees in polluted areas the highest rate observed was 6.2 keq ha^{-1} yr^{-1} (Skeffington, 1983). Moreover, rain and throughfall always contain cations, which increases the lime potential of these solutions, (Bache, 1980) whereas the maximum lime potential of the treatments used in this study was -0.4 based on the analytical detection limits for Ca and Mg. Any exchanger in contact with this solution would inevitably acidify however low its base saturation. At the same time the acid treatment used was not so extreme as to produce shock effects due to the high acid concentration (Abrahamsen, 1986), and significant results, such as that the source of Al release is amorphous rather than exchangeable Al, have been obtained. The main use of such data should be to understand the processes whereby soil reacts to applied acidity and what controls their rates. The results can be incorporated into models which will allow extension into different soil types and more realistic acid loadings. We hope to be able to develop this approach in future.

6. ACKNOWLEDGEMENTS

The lysimeter experiment was designed and operated by Dr J.E. Rippon. Many people contributed to the experiment but enormous contributions were made by Derek Foster, Dave Haggar and Mike Russell.

7. REFERENCES

Abrahamsen, G.: 1986, Water, Air and Soil Pollut., this issue
Allen, S.E.: 1974, Chemical Analysis of Ecological Materials, Blackwell Scientific Publications, Oxford
Bache, B.W.: 1980, 'Acidification of Soils'. In Effects of Acid Precipitation on Terrestrial Ecosystems (T.C. Hutchinson and M. Havas, Eds.), pp. 183-202, Plenum Press, New York
Bache, B.W. and Sharp, G.S.: 1976, Geoderma, 15, 91.
Belford, R.K.: 1979, Journal of Soil Science, 30, 363.
Brown, K.A.: 1985, Soil Biology and Biochemistry, 17, 31.
Brown, K.A.: 1986, Water, Air and Soil Pollution submitted
Chao, T.T., Harward, M.E. and Fang, S.C.: 1964, Soil Science Society of America Proceedings, 28, 632.
Drabløs, D. and Tollan, A.: 1980, Ecological Impact of Acid Precipitation, SNSF-Project, Oslo.
Hesse, P.R.: 1971, Textbook of Soil Chemical Analysis, p. 355, John Murray, London
Johnson, D.W.: 1980, 'Site susceptibility to leaching by H_2SO_4'. In Effects of Acid Precipitation on Terrestrial Ecosystems (T.C. Hutchinson and M. Havas, Eds.), pp. 525-535, Plenum Press, N.Y
Johnson, D.W. and Cole, D.W.: 1980, Env. Internat., 3, 79.
Johnson, N.M., Likens, G.E., Feller, M.C. and Driscoll, C.T.: 1984, Science (NY), 225, 1424.
Krug, E.C. and Frink, C.R.: 1983, Science (NY), 221, 520-525.
Nilsson, S.I. and Bergkvist, B.: 1983, Water, Air, Soil Pollut. 20, 311.
Prenzel, J.: 1983, in Ulrich and Pankrath (1983) q.v., pp. 157-170.
Reuss, J.O.: 1977, Water, Air, Soil Pollut., 7, 461.
Rippon, J.E.: 1984, CERL Lab. Note TPRD/L/2668/N84.
Rippon, J.E., Foster, D.J. and Skeffington, R.A.: 1985, CERL Lab. Note TPRD/L/2847/N85, Leatherhead, England.
Rosenqvist, I.T.: 1978, Sci. Tot. Environ., 10, 39.
Skeffington, R.A.: 1983, in Ulrich and Pankrath (1983) q.v., p. 219.
Skeffington, R.A.: 1984a, Atmospheric Environment, 18, 1683.
Skeffington, R.A.: 1984b, Atmospheric Environment, 18, 1695.
Stuanes, A.O.: 1980, In Drabløs and Tollan (1980) q.v., pp. 198-199.
Ulrich, B. and Pankrath, J., eds.: 1983, Effects of Accumulation of Air Pollutants in Forest Ecosystems, D. Reidel, Dordrecht.
Ulrich, B.: 1983, in Ulrich and Pankrath (1983) q.v., pp. 1-29
Wagner, G.H.: 1962, Soil Science, 94, 379.

LEACHING OF METALS FROM A SPRUCE FOREST SOIL AS INFLUENCED BY EXPERIMENTAL ACIDIFICATION

Bo Bergkvist
Department of Ecology (Plant Ecology)
University of Lund
Östra Vallgatan 14
S-223 61 LUND
Sweden

ABSTRACT. Acidified (H_2SO_4+HNO_3, 3:1) throughfall waters (pH 3.16 and 3.40 as volume weighted means or control (untreated throughfall water, pH 3.72) were applied for 3.5 yr by an automatic irrigation device to lysimeters containing podzolized spruce forest soils of 0-5, 0-15 and 0-35 cm soil depth. The total volume of the leachates was measured together with their pH and total content of DOC, Na, K, Ca, Mg, Fe, Mn, Al, Cu, Zn, Cd and Pb and the initial amounts of metals and H in the soil. The main part of H^+ added with the throughfall waters was retained within the soil. Concentrations and fluxes of Mg, Ca, Mn, Zn and Cd in the soil were significantly increased by addition of acidified throughfall waters; K was less affected. As a consequence of lowered flux of DOC in the A horizon as acid input increased, Fe, Al, Cu, and Pb fluxes also decreased. The mobility of these metals in the A horizon was shown to be regulated mainly by the formation of water-soluble organic compounds rather than directly by pH variations. Compared to the control, the additional annual loss of Mg from the soil profile in the most acid treatment was c. 10% of the currently exchangeable amount.

1. INTRODUCTION

The impact of increased atmospheric deposition on forest soils is being paid much attention, especially since damage to forest trees has become a widespread phenomenon. The risk of the pool of necessary nutrients diminishing was pointed out by Mayer and Ulrich (1977) who suspected that the supply of Mg was becoming deficient for the vegetation. Zöttl (1985) considered the deficient nutrient supply as the dominant stress factor causing the forest damage. It has been shown that an improvement of the nutrient supply (especially of Mg) by application of fertilizers can lead to the disappearance of these disease symptoms. In a recent litterature review (Nilsson, 1985) it was concluded that a decrease of the pool of exchangeable cations in podzolized forest soils is probably taking place in S. and SW. Sweden. Root damage due to high concentrations of toxic metals is also likely

Water, Air, and Soil Pollution **31** (1986) 901–916.

to develop. It has been stated that a successive decrease of exchangeable Ca and a concomitant increase of available Al can cause damage to tree roots (Ulrich et al., 1979; Ulrich 1983):

This study is part of a larger project, on cycling of metals in forest ecosystems financed by the Swedish Environmental Protection Board. The objectives of the study are to evaluate the extent to which an accelerated but moderate acidification of a podzolized forest soil can cause high concentrations of metals in soil solutions; to quantify metal losses from the soil; and to relate these losses to the soil pool.

2. MATERIALS AND METHODS

A mature mixed-coniferous forest situated in southwest Sweden was chosen for the experiment. The main tree species is Norway spruce (*Picea abies*); some Scots pine (*Pinus silvestris*) also occurs, but does not influence the studied soil plots. The study area (Värsjö) is situated about 130 km NNE of Lund. The stand is mainly even-aged, the spruce being c. 80 yr. A few species dominate the vegetation of the forest floor: the grass *Deschampsia flexuosa*, the dwarf-shrub *Vaccinium myrtillus*, the mosses *Ptilium crista-castrensis* and *Pleurozium schreberi*. Areas devoid of vegetation are found beneath the tree crowns. No shrub layer is usually present. The topography is characterized by low hills of sandy-silty glacial till, originating from siliceous rocks, poor in carbonates, surrounded by mires. The soil is generally very deep and freely drained. It may be characterized as an Fe podzol, with a 7-10 cm purely organic litter + mor horizon ($A_{00}+A_0$), a 5-12 cm greyish mineral horizon (A_2), and a 25-35 cm illuvial horizon (B_1). The vertical distribution of organic matter, clay content, pH and metal content in the soil profile is shown in Tables I and II.

The mean monthly temperature of the warmest month is c. 16°C, of the coldest month c. -2°C. 1980 and 1981 were colder than normal; 1982, 1983 and 1984 were warmer than normal. Mean annual precipitation is c. 785 mm according to data from four meteorological stations within 15 km (Ellesson, 1985). The main part of the precipitation falls in late autumn and winter. Snow cover is seldom stable throughout the winter. However, the two winters 1980/1981 and 1981/1982 were characterized by a considerable snow cover (20-50 cm) of above-normal duration.

The lysimeters were constructed in plexiglass. Soil cores were taken close together beneath the spruce crowns. They were lacking any vascular vegetation. The cores were taken with steel cylinders and transferred to the lysimeters with a minimum of disturbance. Then the lysimeters were installed in the soil at the original positions. They were equipped with irrigation devices and Erlenmeyer flasks (Duran glass) for the collection of soil water. For a more detailed description of lysimeter design and installation see Tyler (1981) or Nilsson and Bergkvist (1983).

Three different soil depths were chosen for each set of

TABLE I

Content of organic matter as organic C and clay (percent of the dry weight) and pH-H_2O and pH-KCl in the studied soil.

Soil horizon	Soil depth cm	Organic matter	Clay	pH H_2O	pH KCl
$A_{oo} + A_o$	0- 5	48	1.1	3.64	2.81
A_2	5-15	9	1.6	3.76	2.99
B_1	15-35	2.9	1.0	4.19	4.00

TABLE II

Content of metals and H in the studied soil (g m^{-2})

	Acid digestable ("total")*				Extractable**			
	0-5	5-15	15-35	Σ0-35	0-5	5-15	15-35	Σ0-35
Na	5.1	34	60	9	0.96	2.3	1.9	5.2
K	4.8	36	106	147	7.5	3.8	4.7	16
Mg	11	70	232	313	3.7	2.9	1.4	8
Ca	53	242	674	969	9.7	18	4.8	33
Al	44	492	3040	3580	0.24	9	96	105
Fe	74	1260	5010	6340	0.15	2.9	12	15
Mn	5.9	28	87	121	0.28	1.7	0.7	2.7
Cu	0.2	0.2	0.6	1.1	0.04	0.03	0.04	0.11
Zn	1.1	2.5	6.7	10.3	0.83	1.4	1.3	3.5
Cd					0.007	0.009	0.009	0.03
Pb	2.3	2.1	2.9	7.2	1.7	1.7	1.3	4.7
H					0.30	1.04	1.3	2.6

* Conc. HNO_3:$HClO_4$ (4:1)

** Na - EDTA: Cu, Zn, Cd, Pb

Acid NHO_4Ac (pH 4.8): Na, K, Mg, Ca, Al, Fe, Mn

KCl: H

lysimeters. They represented approximately the A_{00} + A_0 horizons (depth 0-5 cm), the A horizon including most of A_2 (depth 0-15 cm), and the A horizon + the B_1 horizon (depth 0-35 cm).

The diameters of the lysimeters were 290 mm (0-5 and 0-15 cm) and 190 mm (0-35 cm). Three different sets of lysimeters were installed which makes one lysimeter for each treatment at a given soil depth. Throughfall water was collected in 15 systematically distributed plastic funnels (diam. 30 cm) connected to plastic bottles. Plastic nets in the bottom of these funnels trapped the litter fall which was also collected. During winter an equal number of plastic buckets was used (diam. 30 cm). Polyethylene was used in all plastic equipment.

The throughfall waters were filtered (filter papers Munktell OOK), mixed in a large plastic barrel and divided into three identical aliquots. One aliquot was acidified by H_2SO_4 and HNO_3 (3:1) to pH 3.30, the second one to pH 3.00 and the third one was left untreated as a control. The waters were then placed in the irrigation device and automatic irrigation of the lysimeters continued during 5 weeks after each refill. In each irrigation device the water was put into a 5-litre plastic bottle and flowed through a solenoid valve to an irrigation vessel of plexiglass placed 10 cm above the lysimeter as a shelter against natural rain. The perforated (Ø 1.5 mm perforation, 2.5 cm distance) bottom was covered by a filter paper (Munktell OOH). The battery-operated solenoid valve opened two times a week and each time c. 6 mm water was added to the soil.

The irrigation device was only operating from March/April to November/December. One set of lysimeters (0-5, 0-15 and 0-35 cm) received the pH 3.00 throughfall, a second set the pH 3.30 and a third set received the untreated througfall. During the winter they received snow and rain in ambient amounts and properties. Appropriate amounts of litterfall were given to the lysimeter soils at each sampling time during the sheltered period.

The lysimeters were installed in August 1980. The experiment started on October 1, 1980 and ceased on April 7, 1984. The collection interval was about 5 weeks.

Total concentrations of 11 metals in the drainage waters from the lysimeters were measured by AAS tehcnique: Na, K, Ca, Mg, Fe, Mn, Al, Cu, Zn, Pb and Cd. Furthermore, pH and dissolved organic C (DOC) were determined, DOC by infrared technique using a Beckman 915B carbon analyser. Soil parameters (organic matter, clay content, pH and content of metals) are from a budget study in the immediate vicinity (Bergkvist, 1986). Detailed descriptions of field sampling, sample pretreatment and analysis are found in Tyler (1981) and Nilsson and Bergkvist (1983).

3. RESULTS AND DISCUSSION

3.1 General Considerations and Hydrogen Flux into and Out of The Soil

Data are presented in the following ways: (1) Mean concentrations of metals and DOC for the whole period, 3.5 yr; (2) Extractable and

TABLE III

Total cumulated fluxes of metals, H and DOC during three years of the experiment (g m^{-2} 3 yr^{-1}).

Treatment soil depth cm	H	DOC	Na	K	Mg	Ca	Fe	Mn	Al	Cu	Zn	Cd	Pb
Input to soil:													
Control	0.36		10.4	7.2	3.3	6.4	0.24	1.6	0.39	0.007	0.12	0.0008	0.024
pH 3.40	0.73												
pH 3.16	1.3												
Output from soil:													
Control													
0- 5	0.21	122	10.2	8.5	1.3	3.1	0.35	0.43	0.41	0.006	0.04	0.0005	0.019
0-15	0.41	195	9.5	9.1	1.3	2.7	1.5	0.39	2.4	0.006	0.06	0.0012	0.023
0-35	0.10	20	14.1	9.5	1.7	2.5	0.24	0.67	8.6	0.001	0.10	0.0019	0.002
pH 3.40													
0- 5	0.28	86	9.8	9.1	2.6	5.6	0.24	0.93	0.59	0.004	0.07	0.0008	0.015
0-15	0.46	167	9.5	10.0	1.9	2.9	0.73	0.21	2.0	0.004	0.08	0.0015	0.017
0-35	0.32	110	16.3	10.4	2.4	4.0	0.47	0.43	8.1	0.002	0.13	0.0018	0.013
pH 3.16													
0- 5	0.57	81	9.5	8.8	3.1	9.5	0.29	1.3	0.37	0.004	0.13	0.0014	0.016
0-15	0.67	84	11.5	12.0	3.0	4.5	0.37	0.40	1.43	0.002	0.17	0.0022	0.012
0-35	0.19	60	15.6	12.5	4.4	5.8	0.73	0.92	10.6	0.003	0.16	0.0028	0.004

TABLE IV

Test of the differences between the means in Figure 1 (t-test, $p < 0.01$). a = between the control and the pH 3.40 treatment, b = between the control and the pH 3.16 treatment, c = between the two acidification treatments. (+) means a significant increase and (-) a decrease. n = 33.

	0-5	0-15	0-35
Na			
K			
Ca	+b		+b
Mg	+a +b	+b +c	+b +c
Fe		-c	+a +b -c
Mn	+c		
Al			
Cu	-a	-b -c	
Zn	+a +b +c	+b +c	
Cd	+b +c	+b	+c
Pb		-b	+a -c
DOC		-b -c	+a +b -c
pH	-b -c	-b -c	-a -b +c

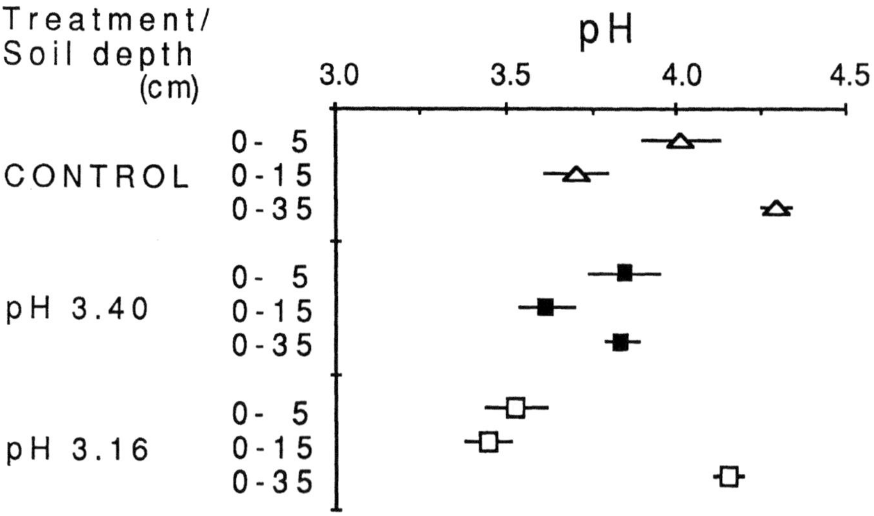

Fig. 1a.

"total" amounts of metals in the soils; (3) Calculations of metal budgets of the soils; (4) Stepwise regression (multiple) analysis of variable interrelation.

Concentrations are calculated as mg l^{-1} for DOC, Na, K, Ca, Mg, Fe, Mn and Al; as ug l^{-1} for Cu, Zn, Pb and Cd. Stores of metals in soils are calculated as g m^{-2}; the flow of metals as g $m^{-2}3yr^{-1}$.

The average annual precipitation during the experiment was 950 mm. Throughfall is c. 60% of the precipitation, more during years with above normal precipitation sums. As the lysimeters were irrigated by throughfall water collected in the study area, the actual amounts of precipitation water were governing the amount added to the lysimeters. The yearly mean sum of throughfall added to the lysimeters (including ambient throughfall during winter months) was 620 mm or 65% of the precipitation. As the lysimeters received artificially acidified water only during the irrigation period, c. 8 mo yr^{-1}, the resulting weighted mean pH for the whole volume added during three years was 3.16 for the 3.00 treatment and 3.40 for the 3.30 treatment. The pH of the total volume added to the control lysimeters was 3.72. The different treatments are in the following text called "pH 3.16", "pH 3.40" and "control" (= pH 3.72). In each step the added amounts of H ions were approximately doubled (Table III).

In a spruce stand in S Sweden, due to the acid deposition in the region, the throughfall pH varies between 3.20 and 4.20 during the year, with extremes of 3.00. The mean pH of the control (3.72) is a normal yearly mean value. Stemflow is mostly below 3.20 (Bergkvist, 1986; Bergkvist et al., 1986). The experimental treatments must therefore be considered realistic, avoiding shock effects on the biological system in the soil.

Acid supplied with the leaching solutions was partly neutralized by the soil, though acidity in soil leachates was markedly increased when acidity of throughfall inputs increased (Table III). The outflow from the uppermost topsoil (0-5 cm) is less than half of the inflow; after passage of the B horizon (0-35 cm) most of the acidity added was retained within the soil. A t-test of the means of the 3.5 yr in Figure 1a shows that the lowering of pH in soil leachates from the A horizon is highly significant when the most acid treatment is tested against the other acid treatment and against the control (Table IV).

Unexpected fluctuations of the depth of the transition zone between the A_2 and B_1 horizons produced differences in that part of the B_1 horizon which was inlcuded in each 0-35 cm lysimeter. In a podzol, dissolved organic matter precipitates in the upper part of the B horizon (see e.g. Petersen, 1976) and only minor quantities pass through the deeper B horizon. An expected concentration of DOC in soil leachates was found in the control at this soil depth, but not in the two acid treated lysimeters. In both of these, most probably due to fluctuations of the depth of the transition zone, DOC concentrations were higher than in the control. In particular, the pH 3.40 treated lysimeter deviates. All comparisons of the 0-35 lysimeters must therefore be made with caution.

Due to the use of lysimeter technique (root uptake by the trees was excluded) concentrations and flow of certain elements and water in soil leachates from the lysimeter soils may be over-estimations of the actual situation in situ.

3.2 Leachability of Metals

The leachability as affected by an increase in acidity of the leaching solutions differs greatly among metals. It is evident from Figure 1b and Tables III and IV that Ca, Mg, Zn and Cd throughout the soil profile and Mn in the mor layer (0-5 cm) have higher concentrations in soil leachates and greater outflow from soil when more acid is added to the soil. This is in accordance with the findings of Tyler (1978), in a laboratory study of the leaching of heavy metals from metal-polluted and unpolluted mor layers. No significant increase in soil leachate concentrations was found for Na, indicating that weathering rate was not increased as the acid input increased. Nor was K concentrations significantly increased, though the trend for K is toward higher output from soil as leaching solutions become more acid.

The influence of acid precipitation on the leaching of K, Ca and Mg has attracted special interest in the discussion of acid rain impact on terrestrial ecosystems. It has been shown experimentally that an increased acidity of artificial rain most significantly increases the Ca and Mg loss from soil; K appears to be less sensitive (Abrahamsen et al., 1977; Tamm et al., 1977; Abrahamsen, 1980; Abrahamsen and Stuanes, 1980; Bjor and Teigen, 1980; Cronan, 1980; Farrell et al., 1980; Stuanes, 1980; Morrison, 1983). In an experimentally acidified mountain stream in the Hubbard Brook Experimental Forest, USA, (mean pH 4.0) streamwater concentrations of Al, Ca, Mg, K, Mn, Fe and Cd increased with acidity (Hall and Likens, 1980). In a study where lysimeter soils were irrigated with waters of different acidity, only in the most acid treatment was an effect on soil chemistry shown. A decrease in extractable amounts of Ca and Mg, decreased base saturation and soil pH were limited to the uppermost 15 cm. At 15-30 cm depth extractable amounts of Ca and Mg were increased compared to control, indicating transport of these cations from the soil surface (Lee, 1985).

The influence of acidification on soil budgets of trace elements has not attracted as much concern as the macronutrients but some studies are reported. The mobility of Cd and Zn was shown to increase with soil acidity (Esser and El Bassam, 1981; Brümmer and Herms, 1983; Scokart et al., 1983) and the same was also shown for Mn (Bjor and Teigen, 1980; Norton et al., 1980).

In the present study, the release of Al, Fe, Cu and Pb did not seem to be enhanced at all by the acidity added to the soil. The release in the A horizon was in fact lowered by the addition of more acid to the soil. The release of these metals was closely related to the release of organic matter. The DOC content of soil leachates was significantly lowered in the A horizon when acid was added to the soil and so were the concentrations of Fe, Cu and Pb in the most acid treatment. Strong acidification seems to reduce the litter decomposition rate. It was demonstrated in an early study that the decomposition of humus and cellulose declined in direct proportion to

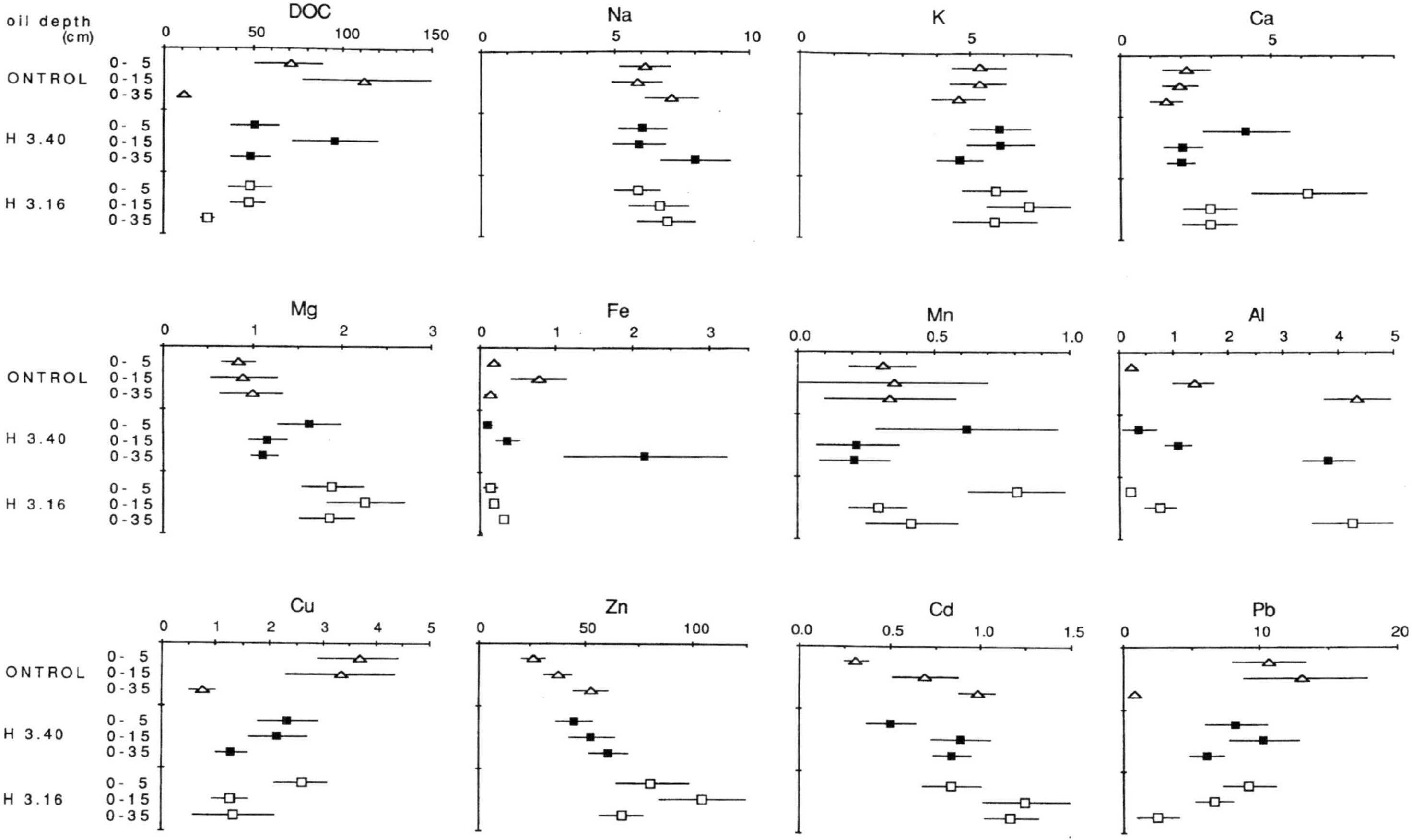

Fig. 1b.

Fig. 1. Mean concentrations, and the 95% confidence interval, in the leachates from the three lysimeter sets (see text) irrigated with throughfall waters of different acidity. N=33. (a) pH; (b) DOC, Na, K, Ca, Mg, Fe, Mn and Al as mg l^{-1}; Cu, Zn, Cd and Pb as μg l^{-1}.

the pH decrease caused by the addition of elemental sulphur to a mineral soil (White et al., 1934). Later studies have verified these findings. Tamm et al. (1977) obtained data suggesting that the addition of H_2SO_4 to soil reduced the C mineralization rate in humus. Irrigation with pH 2 water reduced birch litter decomposition, while pH 3 did not (Hågvar and Kjøndal, 1981). In an acid "rain" experiment with a pine forest soil (Bååth et al., 1979), pH 2.0 water caused a reduction in soil respiration but with pH 3.0 no statistically significant effect was noted.

The high correlation of Fe, Cu, Pb and Al with DOC in the A horizon disappears in the B_1 horizon. At least for Al this is in accordance with a speciation study of Al reported by Nilsson and Bergkvist (1983), where organic and inorganic Al species in soil leachates from different soil horizons in a podzol were determined. Almost all Al in the A horizon soil leachates was organically bound, compared to only c. 20% in the lower B horizon. According to these findings no pH-sensitivity of Al is to be expected in the A horizon. No sensitivity of Al in the mineral part of the soil was detected. This lack of pH-sensitivity of Al was also found by Cronan (1980). But in other acidification experiments an increase of Al concentrations with acidity was found (Abrahamsen and Stuanes, 1980; Bjor and Teigen, 1980; Norton et al., 1980). The Al release from the mineral soil seems to be a very fast process below a critical pH value. If the Al concentrations of the soil solutions from the deeper B horizons of a spruce forest soil are plotted against the solution pH, the curve bends very sharply upward at a pH below 4.4 (Tyler et al., 1985). A pH decrease of 0.2 units from pH 4.4 results in an increase of the Al concentration from 2 to 10 mg l^{-1}. In the upper part of the soil the upward bend of the curve starts at a lower pH, indicating a gradual diminution of the exchangeable and easily weatherable pools in the soil. In this study only the upper B horizon is included in the lysimeters. The added amounts of acid during the experiment were not sufficient to increase the Al mobility. A dramatic increase in Al concentrations was demonstrated by Teigen et al. (1976) and Abrahamsen (1983) when "rain" acidity increased from pH 3.0 to pH 2.0; Al concentration in the effluent increased from an average of 3 to 48 mg l^{-1}.

Tyler (1978) showed in the above mentioned laboratory experiment with mor layers that Cu and Pb were also released when enough acid had been added to the soil. In fact, a weak tendency toward increased Cu output from the B_1 horizon is indicated in the strongest acid treatment in the present study (Table III and Fig. 1b).

Great differences between the leachability of Mg and Cd on the one hand and Pb on the other and the role of humus are evident from Figure 2. The cumulative sums, during 3.5 yr, of metals and DOC in gross outflow from the A horizons (0-15 cm) at the different acid treatments and the control are plotted against the cumulative sums of H added to the soils during the same period. Mg and Cd are released from the A horizon almost proportionately to the input of acid. The amount of Mg outflow during 3.5 yr in the pH 3.16 treatment corresponds to a 10-yr outflow from the control soil. In Cd the corresponding time is 7 yr. Interestingly, the outflow of Pb was lowered

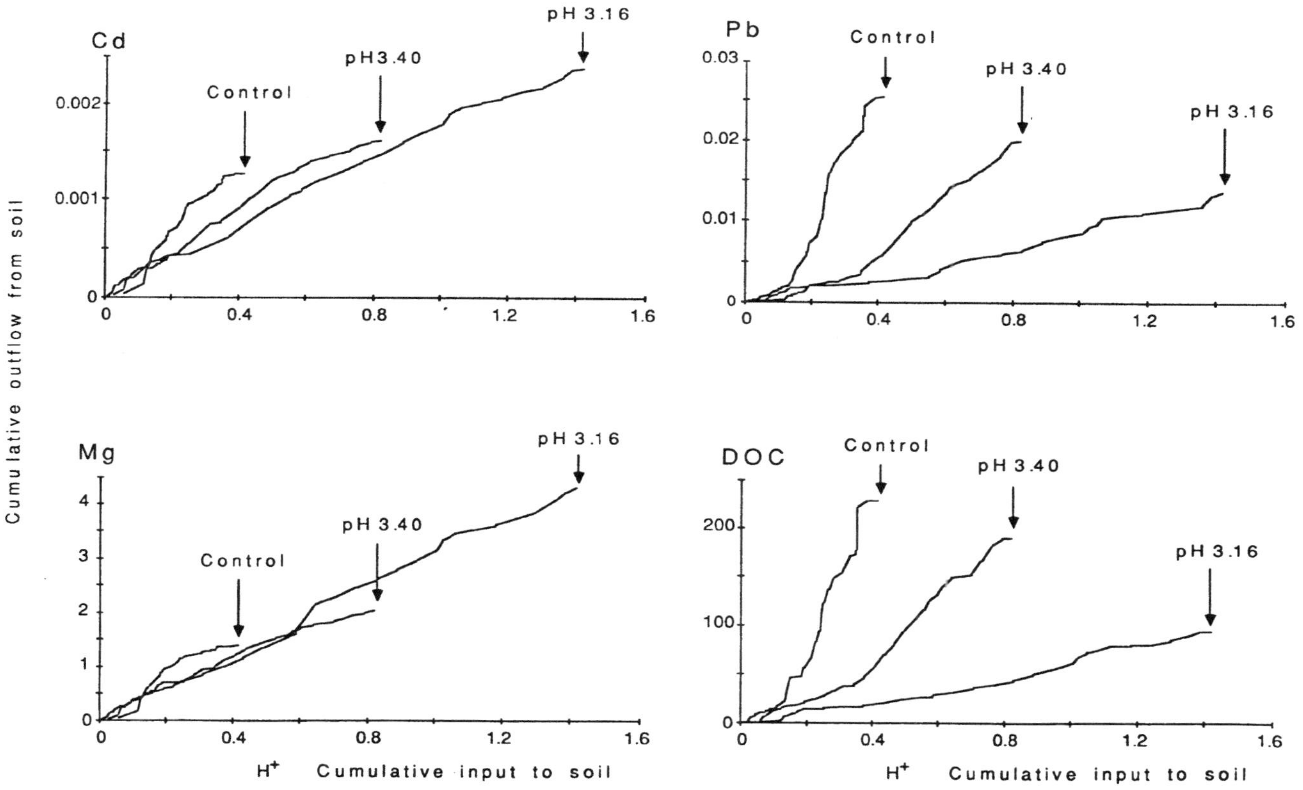

Fig. 2. Cumulative sum in outflow (gross, Y-axis) of Mg, Cd, Pb and DOC in the leachates from the three 0-15 cm lysimeters from each of the different acid treated lysimeter sets, during the whole experiment (3.5 yr) and on the X-axis, cumulative sum of H^+ added to the lysimeter soils by irrigation with throughfall waters and in ambient amounts during the winters (g m^{-2}). The arrows indicate the amounts of H^+ added during 3.5 yr by each treatment.

TABLE V

Multiple relations between metal concentrations of lysimeter solutions and certain variables, according to stepwise regression analysis of the pH 3.16 treatment. Independent variables: pH in throughfall, pH, DOC content, Σ base cations (Na^+, K^+, Ca^{2+}, Mg^{2+}), and Mn concentrations of soil (lysimeter) solutions. a after value of simple and partial r indicates a highly significant ($p < 0.01$) correlation, 100 r^2 the percentage variability accounted for by the variables entered. n = 33.

Step no., variables entered	multiple 100 r^2	simple r	partial r	Step no., variables entered	multiple 100 r^2	simple r	partial r
Zn (0-5 cm)	(87)			**Zn** (0-15 cm)	(72)		
1. pH soil sol.	72	-0.85a	-0.44	1. pH soil sol.	51	-0.72a	-0.65a
2. Mn	83	0.81a	0.61a	2. Mn	68	0.21	0.44
3. DOC	86	0.07	-0.45a	3. Σ base cations	72	0.65a	0.30
4. Σ base cations	87	0.79a	0.23	4. DOC	72	0.30	-0.12
5. pH throughfall	87	-0.77	-0.22	5. pH throughfall	72	-0.61a	-0.04
Cd (0-5 cm)	(75)			**Cd** (0-15 cm)	(80)		
1. pH soil sol.	66	-0.81a	-0.28	1. pH soil sol.	72	-0.85a	-0.76a
2. Σ base cations	72	0.77a	0.29	2. Σ base cations	79	-0.61a	0.51a
3. pH throughfall	74	-0.79a	-0.28	3. DOC	79	0.29	-0.18
4. Mn	75	0.68a	0.18	4. pH throughfall	80	-0.65a	0.18
Cd (0-35 cm)	(55)			**Cu** (0-35 cm)	(63)		
1. pH throughfall	31	-0.56a	-0.25	1. Mn	47	0.69a	0.76a
2. DOC	42	-0.50a	-0.52a	2. pH soil sol.	59	0.052	0.49a
3. Σ base cations	48	0.43	0.38	3. pH throughfall	63	0.33	0.28
4. pH soil sol.	54	0.17	0.28				
Mg (0-5 cm)	(77)			**K** (0-15 cm)	(60)		
1. pH soil sol.	69	-0.83a	-0.38	1. pH throughfall	46	-0.67a	-0.58a
2. Mn	74	0.71a	0.29	2. Σ base cations	51	0.56a	0.30
3. pH throughfall	76	-0.78a	-0.22	3. pH soil sol.	58	-0.32	0.37
4. Σ base cations	77	0.71a	0.19	4. DOC	60	0.48a	0.23

in the acid treatments. The Pb graph (Figure 2) is almost a reflection of the DOC graph (Figure 2) and the release of organic compounds governs the Pb release entirely. These close relationships between Pb and humus release and between the acidity of the soil system and release of Mg and Cd were also shown by Tyler (1981) at ambient soil pH variations.

3.3 Stepwise Regression Analysis

As an attempt to find the variables regulating the mobility of the metals studied, simple correlation, partial correlation, and stepwise regression analysis were performed on the material of the pH 3.16 treatment (Table V). pH in added throughfall water, and pH, DOC content, Σ base cations (Na, K, Ca and Mg) and Mn concentrations of soil solutions have been treated as independent variables with respect to other metals (inorganic anions (eg. sulphate, nitrate and chloride) were not measured, though they probably to a certain extent influence the transport of metals in soils). The analysis shows for each metal the share of dependent variability which might be explained by the variability of one, two or several of the independent variables. Only if the variability accounted for was greater than c. 60% is the metal included in Table V.

Zinc shows a close inverse relation to pH, and a positive correlation to Mn both in the mor layer and in the entire A horizon. Σ base cations also shows a close positive relation to Zn. With the included variables, 87 and 72%, respectively, of the variability is accounted for in the two horizons.

In the mor layer and in the entire A horizon, pH shows a close inverse relation to Cd concentration and accounts for 66 and 72% of the Cd variability, respectively. At the level of the upper B horizon, pH of the added throughfall water accounts for only 31% though if the other variables are also included 54% of the Cd variability is accounted for.

Manganese is the variable first entered when Cu is the dependent variable, and accounts for 47% of the Cu variability in the soil profile down to the upper B horizon.

The pH of the system shows a close inverse relation to Mg concentrations and accounts for 69% of the variability of Mg in the mor layer.

pH of the added throughfall shows a close inverse relation to the variability of K. If Σ base cations and pH in soil solution are also included 58% of the variability is accounted for.

In the pH 3.16 treated soils the acidity of the system seems to be very important for the mobility of these metals. Interestingly, Mn appears to be very important for the mobility of of Zn, Cu and Mg. It is known (Taylor and McKenzie, 1966; Sidhu et al., 1976) that manganiferous nodules in soils contain varying amounts of other metals, eg. Zn, Cu and Mg. Factors affecting the solubility of these concretions also affect the mobility of the associated metals.

4. CONCLUSIONS

It seems reasonable to conclude that increased soil acidification by acid rain may increase the concentrations of several metals in the soil solution, eg. Ca, Mg, Mn, Zn, Cd and (as has been shown elsewhere) eventually also Al. Because of increased losses of nutrients, and because there are indications that soil weathering rate does not keep pace with these losses, there is a risk that exchangeable stores will decrease, with consequences for plant uptake. Compared to the control, the additional annual loss of Mg in the most acid treatment was c. 10% of the currently exchangeable amount in the soil. Biological recycling of nutrients by decomposition and uptake may be impeded. The concentrations of eg. Cd and (as has been shown elsewhere) Al are increased to such an extent that risks of root injury cannot be disregarded.

REFERENCES

Abrahamsen, G: 1980, in Drabløs, D. and Tollan, A. (eds.), Proc. Int. Conf. Ecol. Impact Acid Precip., Norway 1980, SNSF Project, p 196.

Abrahamsen, G.: 1983, in Ulrich, B. and Pankrath, J. (eds.), Effects of Accumulation of Air Pollutants in Forest Ecosystems, D. Reidel Publishing Company, Dordrecht, p 127.

Abrahamsen, G., Horntvedt, R. and Tveite, B.: 1977, Water, Air, and Soil Pollut. **20**, 311.

Abrahamsen, G. and Stuanes, A.O.: 1980, in Drabløs, D. and Tollan, A. (eds.), Proc. Int. Conf. Ecol. Impact Acid Precip., Norway 1980, SNSF Project, p 152.

Bergkvist, B.: 1986, Water, Air, and Soil Pollut. submitted.

Bergkvist, B., Folkeson, L. and Olsson, K.: 1986, this volume.

Bjor, K. and Teigen, O.: 1980, in Drabløs, D. and Tollan, A. (eds.), Proc. Int. Conf. Ecol. Impact Acid Precip., Norway 1980, SNSF Project, p 200.

Brümmer, G. and Herms, U.: 1983, in Ulrich, B. and Pankrath, J. (eds.), Effects of Accumulation of Air Pollutants in Forest Ecosystems, D. Reidel Publishing Company, Dordrecht, p 233.

Bååth, E:, Lundgren, B. and Söderström, B.: 1979, Bull. Environ. Contam. Toxicol. **23**, 737.

Cronan, C.S:: 1980,Plant and Soil **56**, 301.

Ellesson, J.: 1985, personal communication.

Esser, J. and El Bassam, N.: 1981, Environ. Pollut. (Ser. A) **26**, 15.

Farrel, E.P., Nilsson, I., Tamm, C.O. and Wiklander, G.: 1980, in Drabløs, D. and Tollan, A. (eds.), Proc. Int. Conf. Ecol. Impact Acid Precip., Norway 1980, SNSF Project, p 186.

Hall, R.J. and Likens, G.E.: 1980, in Drabløs, D. and Tollan, A. (eds.), Proc. Int. Conf. Ecol. Impact Acid Precip., Norway 1980, SNSF Project, p 375.

Hågvar, S. and Kjøndal, R.: 1981, Pedobiologia **22**, 232.

Lee, J.J.: 1985, Water, Air, and Soil Pollut. **25**, 185.

Mayer, R. and Ulrich, B.: 1977, Water, Air, and Soil Pollut. **7**, 409.

Morrison, I.K.: 1983, in Ulrich, B. and Pankrath, J. (eds.), Effects of Accumulation of Air Pollutants in Forest Ecosystems, D. Reidel Publishing Company, Dordrecht, p 195.

Nilsson, S.I.: 1985, Swedish Environ. Protection Board PM **1979**, Solna, p 61.

Nilsson, S.I. and Bergkvist, B.: 1983, Water, Air, and Soil Pollut. **20**, 311.

Norton, S.A., Hanson, D.W. and Campana, R.J.: 1980, United States Department of Agriculture, Forest Service, General Technical Report PSW-43, Berkeley California, p 152.

Petersen, L.: 1976. Podzols and Podzolisation, DSR Forlag, Copenhagen.
Scokart, P.O., Meeus-Verdinne, K. and de Borger, R.: 1983, Water, Air, and Soil Pollut. **20**, 451.

Sidhu, P.S., Sehgal, J.L. and Sinha, M.K.: 1976, J. Indian Soc. Soil Sci. **24**, 330.

Stuanes, A.O.: 1980, in Drabløs, D. and Tollan, A. (eds.), Proc. Int. Conf. Ecol. Impact acid Precip., Norway 1980, SNSF Project, p 198.

Tamm, C.O., Wiklander, G. and Popovic', B.: Water, Air, And Soil Pollut. **8**, 75.

Taylor, R.M. and McKenzie, R.M.: 1966, Aust. J. Soil Res. **4**, 29.

Teigen, O., Abrahamsen, G. and Haugbotn, O.: 1976, The Norwegian interdisciplinary research programme, SNSF project, IR 26/76, Oslo-Ås, p 45.

Tyler, G.: 1978, Water, Air, and Soil Pollut. **9**, 137.

Tyler, G.: 1981, Water, Air, and Soil Pollut. **15**, 353.

Tyler, G., Berggren, D., Bergkvist, B., Falkengren-Grerup, U., Folkeson, L. and Rühling, Å.: 1985, NATO workshop on effects of Air Pollutants, especially Acid Deposition, on Forest, Agriculture and Wetlands, Toronto, May 12-17.

Ulrich, B.: 1983, in Ulrich, B. and Pankrath, J. (eds.), Effects of Accumulation of Air Pollutants in Forest Ecosystems, D. Reidel Publishing Company, Dordrecht, p 127.

Ulrich, B., Mayer, R. and Khanna, P.K.: 1979, Deposition von Luftverunreinigungen und ihre Auswirkungen in Waldökosystemen im Solling, J.D: Saurländers Verlag, Frankfurt, p 291.

White, J.W., holben, F.J. and Jeffries, C.D.: 1934, Soil Sci. **37**, 1.

Zöttl, H.W.: 1985, Düsseldorfer Geobot. Kolloq. **2**, 31.

SOLUBLE ORGANIC ACIDS AND THEIR CHELATING INFLUENCE ON Al AND OTHER METAL DISSOLUTION FROM FOREST SOILS.

J. G. McColl and A. A. Pohlman
Department of Plant and Soil Biology
108 Hilgard Hall
University of California
Berkeley, CA. 94510 USA

ABSTRACT. The effects of mineral acid deposition on chemical properties of leachates of three granitic forest soils and pine O_1-litter were investigated under controlled laboratory conditions. Of the array of organic acids released from litter and SOM as detected by HPLC analysis, nine compounds were identified and six compounds have been quantified. Organic acids possessing certain functional group arrangements, such as di- and tri-carboxylic acids containing β-hydroxyl groups, and phenolic acids with ortho-hydroxyls were about three to ten times faster in dissolving Al than similar organics with other functional group combinations or mineral acids at comparable pH and concentration levels. There was little correlation between the initial pH of chelating organic acid input solution and rate of metal dissolution. On a mole metal-leached per gram basis, dissolution of metals from Shaver soil followed the order Al>Mn>Fe in both organic and mineral acids.

1. INTRODUCTION

A major concern about the terrestrial effects of acidic atmospheric deposition is the leaching of toxic metals and cations from forest soil, and the effects on forest growth and on the aquatic systems. Soils most sensitive to acidic atmospheric deposition generally occur in mountainous areas on young, shallow soils of granitic parent material and which are high in organic matter content. Although a number of studies have been conducted in reference to cation leaching from forest floors (Johnson et al., 1983), and mobility and transport of Al (Cronan and Schofield, 1979; Johnson et al., 1981) as affected by acid deposition, only few papers have addressed the question of the role of in situ soil organic matter (SOM) and litter on the mobility of Al and other metals and cations (Driscoll et al., 1985; DeWalde et al., 1985; Cronan and Aiken, 1985). Evidence indicates that soluble organic acids released from litter and soil organic matter by mineral acid leaching (DeWalde et al., 1985) affect the chemistry of leachates from

Water, Air, and Soil Pollution **31** (1986) 917–927.

forest floors. Soil organic matter can also weather primary minerals (Schalascha *et al.*, 1967; Iskander and Syers, 1972; Schnitzer and Kodama, 1976), chelation being the main process of metal removal from both solid and aqueous phases. Organic acids such as citric, malic, tartaric, salicylic, gallic and tannic have been shown to complex Al from aqueous solution (Toy *et al.*, 1973) from minerals such as micas (Razzaghe-Karimi and Robert, 1975), and from forest soils (Pohlman and McColl, 1985).

This paper presents the effect of acidity on the release of dissolved organic carbon (DOC) and organic acids from forest litter and soils. The chelating influences of organic acids are discussed in relationship to the relative rates of Al, Mn, and Fe dissolution from soil constituents.

2. MATERIALS AND METHODS

2.1. Soils and Litter

The A-horizons of three soil series were sampled from the Sierra Nevada in Fresno County, California, along an elevational transect on granitic parent material. Chiquito (Ruptic Lithic Cryumbrept) is a very shallow soil at 2,835 m elevation, near the tree-line at Kaiser Pass; Shaver (Pachic Ultic Xerumbrept) is a deeper, more developed soil supporting a mixed-conifer forest at 1,710 m near Shaver Lake; and Vista (Typic Xerochrept) supports oak-savanna at 250 m northeast of Fresno. *Pinus ponderosa* O_1 litter was collected from Blodgett Forest Research Station in El Dorado county, California.

Soils were mechanically sieved and the < 200 μm fraction of Chiquito and Vista soils, and < 500 μm fraction of Shaver soil were used for leaching studies. The pine O_1 litter was air-dried and ground; no particle-size fractionation was made.

2.2. Leaching

Samples were leached using a mechanical extraction method described by Holmgren *et al.* (1977). Triplicate 5.0 g samples of soil were leached with about 50 mL of 10^{-2}M and 10^{-4}M acid solution at a constant rate of 0.40 mL min^{-1}. Kinetic studies were conducted on 3.0 g samples at 0.70 mL min^{-1} with volume inputs ranging from 35 to 160 mL. Litter samples (1.0 g) were leached in quadruplicate, then two duplicate samples were combined for solvent extraction and HPLC analysis. Prior to leaching, samples were washed with about 60 mL of demineralized water using gravity flow and light suction. Leachates were filtered through a 0.4 μm nucleopore filter, or centrifuged (kinetic studies) for 15 min at 15,000 rpm. Sub-samples were acidified to pH 2.5 and stored at 10°C for analysis.

2.3. Analytical

Aluminum concentrations in leachates were measured by the Aluminon

Method following digestion of samples in 30% H_2O_2 and concentrated HNO_3 (Jayman and Sivasubramanian, 1974). Iron and Mn were determined by atomic absorption spectrophotometry. Dissolved organic carbon (DOC) determinations were performed on acidified (pH 2) leachate samples using Dohrman (Xertex) carbon analyzer with an IR detector. Organic acids were extracted from the leachates by three successive additions of redistilled ethyl acetate. The solvent was evaporated to dryness at 55°C under vacuum, and the samples were dissolved in 0.05 mL of 0.008 N H_2SO_4 containing 10% acetonitrile. Organic acid mixtures were separated on Aminex HPX-87H 300 x 7.8 mm column using 5 to 30% acetonitrile/0.008 N H_2SO_4 gradient solution at 0.8 ml min^{-1} and 45°C column temperature. Organic compounds were detected simultaneously at 210 and 254 nm using Hewlett Packard HP-1040A HPLC system with a diode-array detector (DAD). Peak areas and retention times (t_R) were obtained with a computing data processing unit (DPU) integrator. Organic acids were identified by comparing retention times, and by matching spectra from 190 nm to 350 nm of unknowns to pure organic acids. The retention times of identified compounds along with molar absorption data at 210 and 254 nm are listed in Table I.

TABLE I. Retention times (t_R)† and molar absorptivities at 210 nm and 254 nm of aliphatic and phenolic acids.

Compound	t_R (min)	Molar Absorptivity (ε)* 210 nm	254 nm
1. Oxalic	4.82	740	32
2. Citric	5.71	190	ND
3. Aconitic	6.58	7,300	1,700
4. Succinic	8.10	65	ND
5. Fumaric	9.24	7,500	930
6. Protocatechuic	20.0	14,000	7,800
7. p-Hydroxybenzoic	26.1	10,000	12,000
8. Vanillic	27.4	15,000	8,500
9. Salicylic	29.8	13,000	370

† t_R = average of three runs, measured at 210 nm.
* Molar absorptivities in water.
ND = not detected to 10^{-2} M.

3. RESULTS AND DISCUSSION

Typical leaching curves for dissolved organic carbon (DOC), Al, Mn, and Fe, over a wide range of H_2SO_4 and HNO_3 acid input for two soils are listed in Figure 1 (A,B,C, and D). The ratio of DOC to Al dissolved remained fairly constant and equal to about 50:1 up to about 10^{-3} M H^+ input, irrespective of the actual amounts of these constituents dissolved; for example, about 48 μmol g^{-1} DOC and 1.0 μmol g^{-1} Al dissolved from Chiquito soil, and about 11 μmol g^{-1} DOC and 0.2 μmol g^{-1} Al dissolved from Vista soil, over this range of acid input. Similarly, the amounts of Mn and Fe dissolved from both

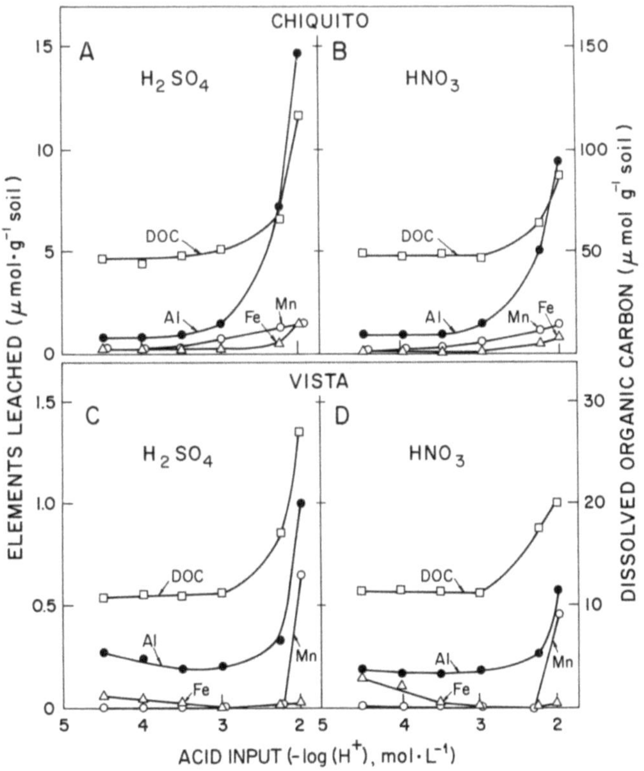

Fig. 1. Dissolved organic C, Al, Mn and Fe in leachate of: A) Chiquito soil leached with H_2SO_4 solutions, B) Chiquito soil leached with HNO_3 solutions, C) Vista soil leached with H_2SO_4 solutions, D) Vista soil leached with HNO_3 solutions.

soils also remained fairly constant over this acid range. This consistency in pattern between DOC and dissolved metals for two quite distinct soils suggests that leaching over the range of acid input up to about 10^{-3} M H^+ represents "background" leaching. With increasing H^+ ion input up to 10^{-2} M H^+, about 2.5-fold more DOC was dissolved from the two soils and 15- and 5-fold more Al was dissolved from Chiquito and Vista soils respectively. At higher acid input DOC was probably controlled by H^+ hydrolysis of bound organics. Such hydrolytic process along with mass-flow action of percolating solution releases greater quantities of soluble organics and metals than the "background" leaching. The actual quantities of metals released appeared to be related to DOC, and the organic carbon content of the soils.

To gain greater insight into the chemistry and functional characteristics of DOC, we subjected the leachates to high pressure liquid chromatography (HPLC) analysis. Typical liquid chromatography of a wide array of organic acids extracted in 10^{-4} M H^+ and 10^{-2} M H^+ from pine O_1 litter-horizon and Chiquito A-horizon are shown in Figure 2 (A,B,C, and D). These chromatograms demonstrate that reasonably good peak separations for both aliphatic and phenolic acids

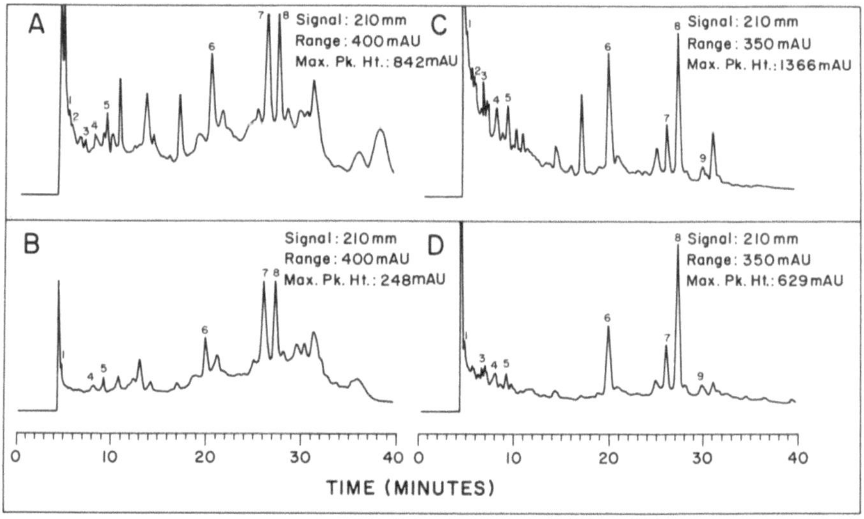

Fig. 2. Typical HPLC chromatograms showing the array of organic acids leached as affected by acid input from: A) ponderosa pine O_1 litter leached with 10^{-2} M H^+, B) ponderosa pine O_1 litter leached with 10^{-4} M H^+, C) Chiquito soil leached with 10^{-2} M H^+. Numbers refer to identified compounds listed in Table II.

can be obtained in a short analysis time. The numbered peaks have been identified and the retention data for the aliphatic and phenolic acids are listed in Table I. The amounts of individual organic acids expressed as nmol g^{-1} soil are listed in Table II. Of the total organic acids quantified, 376 nmol g^{-1} were leached from O_1-litter in 10^{-2} M H^+ compared to 238 nmol g^{-1} in 10^{-4} M H^+.

Similarly, about two-fold more organic acids were leached from Chiquito soil at the higher acid input. The quantities of organic acids leached from Vista soil were negligible in comparison, even at 10^{-2} M H^+ input. These organic acids can be classified as "free" form and do not constitute total amount of aliphatic and phenolic acid content of litter or soil. Aliphatic acids such as oxalic, citric, and succinic, were not quantified because of inferences from tailing of unknown materials, and/or partial resolution of peaks. Such aliphatic acids may constitute a large fraction of the "free" organic acids in leachates, since their molar absorptivities (ε) at 210 nm are several

TABLE II. The effect of pH of input solution on the amounts of organic acids leached.

Organic Acids Identified	Pine O_1 Litter		Chiquito Soil		Vista Soil	
	pH of H_2SO_4 Input Solution					
	2.0	4.0	2.0	4.0	2.0	4.0
	n mol g^{-1}					
Oxalic	*	*	*	*	0.0	0.0
Citric	*	*	*	*	0.0	0.0
Aconitic	17.0	9.0	8.4	2.6	*	*
Succinic	*	*	*	*	*	*
Fumaric	35.0	18.0	8.9	3.2	0.6	0.0
Protocatechuic	91.0	49.0	17.0	7.3	0.0	0.0
P-Hydroxybenzoic	150.0	100.0	7.1	5.0	0.0	0.8
Vanillic	83.0	62.0	11.0	9.3	0.0	0.4
Salicylic	0.0	0.0	*	*	0.3	0.6
Total Organic Acids	376.0	238.0	52.4	27.4	0.9	1.8

* Detected but not quantified.

orders of magnitude lower than ε_{210} for the organic compounds quantified (Table I).

A number of organic acids identified in leachate exhibit metal chelating properties. Pohlman and McColl (1985) have shown that di- and tri-carboxylic acids containing β-hydroxyl functional groups, and phenolic acids containing ortho-OH groups dissolve Al, Mn, and Fe from soils faster than similar organic compounds with other functional group combinations. The faster dissolution of metals by such compounds is assumed to be due to the interaction of functional groups with metals that results in the formation of stable 5- or 6-membered ring structures (Martell and Calvin, 1952). For example, malic acid can form two types of ring structures with a metal through the interaction of one acidic and one coordinating group as depicted in Figure 3 (structures I and II). In contrast, fumaric acid, because of the position of its functional groups, cannot form such structures and consequently is not considered to be a chelating agent. Similarly, citric acid, but not aconitic acid, can form 2-ring structures (III and IV), while protocatechuic acid, but not p-hydroxybenzoic acid, can form one 5-membered ring structure (V) through the participation of two phenolic -OH groups. Thus, compounds such as oxalic and salicylic acids can also be classified as possessing chelating properties based on the above criteria.

The effectiveness of Al, Mn, and Fe dissolution from Shaver soil by chelating acids compared to non-complexing acids and NH_4OAc is depicted in Figures 4 and 5. The average rate of Al dissolution in 10^{-3} M malic acid was 33 x 10^{-3} μmol g^{-1} min^{-1} compared to average rates of 9 x 10^{-3} μmol g^{-1} min^{-1} in 10^{-3} M fumaric acid. The dissolution of Al in the chelating acid was about four-fold faster, even though fumaric acid was a stronger acid. Similarly, the average rate of Al dissolution in the chelating citric acid was 56 x

10^{-3} μmol g^{-1} min^{-1} compared to 14 x 10^{-3} mol g^{-1} min^{-1} in aconitic acid, or four times faster (Figure 4, A, B, C and D). The rate of Al dissolution was slower in phenolic acids with an average rate of 18 x 10^{-3} μmol g^{-1} min^{-1} for protocatechuic acid compared to 4.7 x 10^{-3} μmol g^{-1} min^{-1} for the non-complexing p-hydroxybenzoic acid (Figure 5, A and B). The slower rate of Al dissolution in non-complexing organic acids was assumed to be by H^+

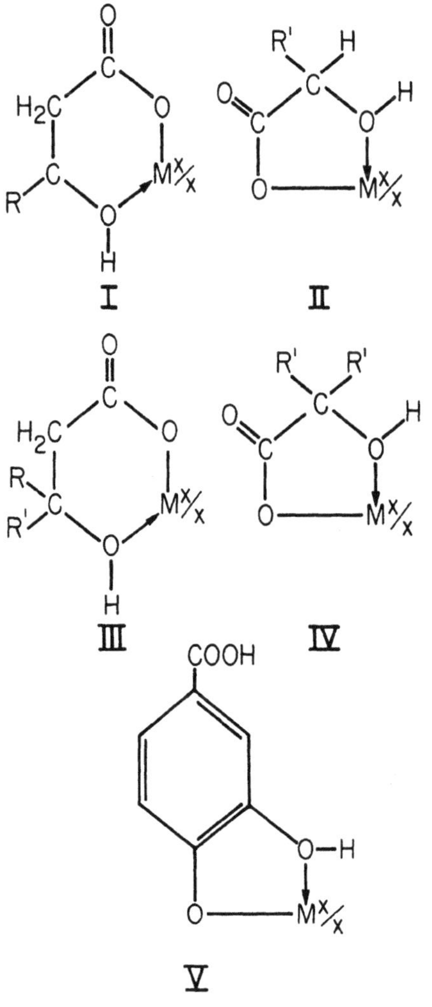

Fig. 3. Suggested ring structures involved in metal chelation. Di-carboxylic acids with β-hydroxyl showing; I) 6-membered ring structure; II) 5-membered ring structure. Tri-carboxylic acid with a β-hydroxyl showing; III) 6-membered ring structure; IV) 5-membered ring structure; V) Phenolic acid with ortho-hydroxyls showing a 5-membered ring structure.

ion exchange. Such rates would be afected by a degree of H^+ ion dissociation, or pK of the organics.

In addition, initial pH of input solution was not readily correlated with the degree of metal dissolution in a chelating acid (Table III). Average rates of Al, Mn, and Fe decreased only slightly as the initial pH of the citric acid solution was increased from pH 3.25 to pH 5.65. In comparison to citric acid, the rates of Al dissolution in the three mineral acids were an order of magnitude slower and ranged from 6.8 x 10^{-3} to 2.5 x 10^{-3} μmol g^{-1} min^{-1}. The average rates of metal dissolution in both organic and mineral acid followed the order Al>Mn>Fe on a mole per gram soil basis.

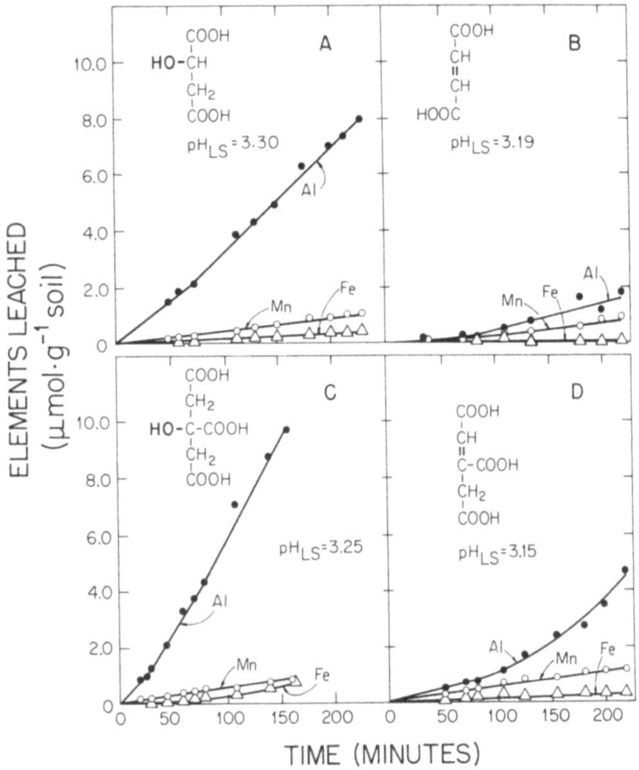

Fig. 4. The effect of chemical structure of di- and tri-carboxylic acid on rates of Al, Mn and Fe dissolution from Shaver soil. Dissolution in: A) 10^{-3} M malic, B) 10^{-3} M fumaric, C) 10^{-3} M citric, and D) 10^{-3} M aconitic acid.

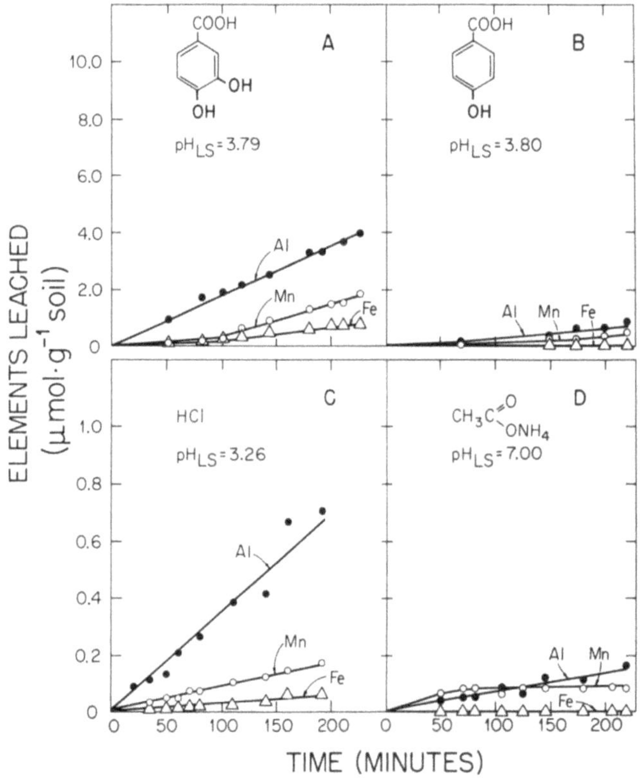

Fig. 5. Rates of Al, Mn and Fe dissolution from Shaver soil in phenolic acids are compared to dissolution in hydrochloric acid and ammonium acetate. Dissolution in A) 10^{-3} M protocatechuic, B) 10^{-3} M p-hydroxybenzoic, C) 5.5 x 10^{-4} M HCl, and D) 1 M NH_4OAc.

TABLE III. The effect of initial concentrations and pH of citric and mineral acids on the rate of Al, Mn and Fe dissolution from Shaver soil.

Acid	Conc. 10^{-3}M	pH_{LS}	Average Rates of Dissolution Al	Mn	Fe
			μmol g^{-1} min^{-1} 10^{-3}		
Citric	5.0	2.78	240.0	16.0	13.0
	1.0	3.25	56.0	6.7	4.2
	1.0	4.92	46.0	5.7	3.0
	1.0	5.65	41.0	5.6	2.6
	0.2	3.75	29.0	2.4	1.5
Hydrochloric	1.6	2.79	2.5	2.5	0.3
	0.55	3.26	3.6	0.8	0.3
Nitric	1.0	3.00	4.8	2.1	0.2
Sulfuric	1.0	3.00	6.8	2.2	0.2

4. SUMMARY AND CONCLUSIONS

Soluble organic acids are released from pine O_1 litter and Chiquito soil under mild acid conditions and amounts rapidly increase with acidity, but release from Vista soil was negligible. Such differences in leachates suggest that DOC is not homogenous but varies between soils and litter types.

Release of certain types of organic acids such as di- and tri-carboxylic acids containing β-hydroxyl groups and phenolic acids containing ortho-hydroxyls can leach Al and other metals faster than mineral acids at comparable pH and concentrations. Such organic acids possess metal chelating capabilities by forming stable 5- and 6-membered ring structures with metals. Organic acids such as oxalic, malic, citric, protocatechuic and salicylic can be classified as chelating organic acids. The rate of release of Al in non-chelating organic acids is the same order of magnitude as in mineral acids suggesting that H^+ ions released by carboxylic acid group dissociation play a major role in metal dissolution. On a mole metal leached per gram soil basis, dissolution of metals from Shaver soil followed the order Al>Mn>Fe in both organic and mineral acids.

Although total identity of free organic acids in DOC and their weight contributions in soil and litter leachates was not determined, this study shows the importance of soluble organics on the dissolution of three metals. Greater attention should be directed to the importance of soil organic matter and its role in regulating the effects of acidic deposition. Organic acid leaching could be important in shallow granitic soils that are high in organic matter and covered with a litter layer.

5. ACKNOWLEDGMENTS

This study was supported by the National Science Foundation, grant BSR

84-16661, and the Agricultural Experimental Station, University of California, Berkeley, Hatch Project CA-B-3664-H. We thank R. Forsstrom for his technical assistance.

6. REFERENCES

Cronan, C. S. and Schofield, C. L.: 1979, 'Aluminum Leaching Response to Acid Precipitation: Effects on High Elevation Watersheds in the North-East', Science 204, 304.

Cronan, C. S. and Aiken, G. R.: 1985, 'Chemistry and Transport of Soluble Humic Substances in Forested Watershed of the Adirondack Park, New York', Geochim. et Cosmochim. Acta 49, 1697.

DeWalde, D. R., Ribblett, G. C., Helvey, J. D., and Kochenderfer, J.: 1985, 'Laboratory Investigation of Leachate Chemistry from Six Appalachian Forest Floor Types, Subjected to Simulated Acid Rain', J. Environ. Qual. 14, 234.

Driscoll, C. T., van Breeman, N., and Mulder, S.: 1985, 'Aluminum Chemistry in a Forested Spodosol', Soil Sci. Soc. Am. J. 49, 437.

Holmgren, G. G. S., Juve, L., and Geschwender, R. C.: 1977, 'A Mechanically Controlled Variable Rate Leaching Device', Soil Sci. Soc. Am. J. 41, 1207.

Iskander, I. K. and Syers, J. K.: 1972, 'Metal-Complex Formation by Lichen Components', J. Soil Sci. 23, 254.

Jayman, T. C. Z. and Sivasubramanian, S.: 1974, 'The Use of Ascorbic Acid to Eliminate Interference from Iron in the Aluminon Method for Determining Aluminum in Plant and Soil Extracts', Analyst 99, 256.

Johnson, N. M., Driscoll, C. T., Eaton, J. S., Likens, G. E., and McDowell, W. H.: 1981, 'Acid Rain, Dissolved Aluminum and Chemical Weathering at the Hubbard Brook Experimental Forest, New Hampshire', Geochim. Cosmochim. Acta 45, 1421.

Johnson, D. W., Richter, D. D., van Miegroet, H., and Cole, D. W.: 1983, 'Contributions of Acid Deposition and Natural Processes to Cation Leaching from Forest Soils: A Review', J. Air Pollution Control Assoc. 33, 1036.

Martell, A. E. and Calvin, M.: 1952, Chemistry of the Metal Chelate Compounds., Prentice-Hall, Inc., N. Y.

Pohlman, A. A. and McColl, J. G.: 1985, 'Kinetics of Metal Dissolution from Forest Soils by Soluble Organic Acids', J. Env. Qual. 15, 86.

Razzaghe-Karimi, J. and Robert, M.: 1975, 'Alteration des Micas et Geochimie de L'aluminum: Role de la Configuration de la Molecule Organique sur la Complexation', C. R. Acad. Sci., Paris 2800, 2645.

Schnitzer, M. and Kodama, H.: 1976, 'The Dissolution of Micas by Fulvic Acid', Geoderma 15, 381.

Schalascha, E. B., Appelt, H. and Schatz, A.: 1967, 'Chelation as a Weathering Mechanism: I. Effect of Complexing Agents in the Solubilization of Iron from Minerals and Gianodiorite', Geochim. Cosmochim. Acta 31, 587.

Toy, A. D., Smith, T. D., and Pilbrow, S. R: 1973, 'Al-27 Nuclear Magnetic Resonance in Aqueous Solutions of its Chelates with Hydroxy-Carboxylic Acids', Aust. J. Chem. 27, 1889.

EFFECTS OF FOREST VEGETATION ON SPATIAL VARIABILITY OF SURFACE MINERAL SOIL PH, SOLUBLE ALUMINUM AND CARBON

Susan J. Riha, Gail Senesac, and Eric Pallant
Department of Agronomy
Cornell University
Ithaca, NY 14853
USA

ABSTRACT. The purpose of this study was to examine spatial variability in forest soils at several levels including variability due to soil structure, to the presence of individual trees and to populations of different species of trees. Both classic statistical and geostatistical methods were used. Soil chemical properties measured include pH and Al and C in solution which was in equilibrium with the surface mineral soil. Results indicated that soil cores 1.8 cm in diameter were as effective as larger cores in incorporating variability in surface mineral soil pH. There was no spatial correlation in soil pH in samples separated by a distance of 20 to 360 cm. The presence of individual Norway spruce and red pine trees affected soil in their vicinity. Soil pH was depressed and soluble Al elevated in soil near the base of the tree compared to soil 120 cm from the tree, independent of direction. In addition, in soil sampled at least 100 cm from the base of trees, pH was lower in Norway spruce compared to sugar maple plantations and soluble C was greater in red pine compared to Norway spruce plantations. It is concluded that in less than 50 yr the presence of individual trees and populations of different tree species can affect chemical properties of surface mineral soils. These effects should be considered in the design and interpretation of experiments.

1. INTRODUCTION

There is general agreement that patterns of spatial variation in soils in general, and forest soils in particular, are complex. Some of the patterns are so complex that only a statistical description of them can be attempted. Other patterns may be at least partially predictable from a mechanistic understanding of forest ecosystem processes. In either case, the description of spatial variation in forest soil properties can be of value in 1) elucidating underlying mechanisms of importance in forest ecosystems, 2) specifying areas of critical or special importance when perturbations of forest ecosystems occur, 3) designing field experiments, and 4) constructing

Water, Air, and Soil Pollution **31** (1986) 929–940.

and validating models of field processes.

The purpose of this study was to describe and compare different scales of spatial variability in soils using both classic statistical and geostatistical methods. Examples will be further discussed in terms of how they relate to our understanding of the effects of acid precipitation on forest ecosystems.

2. METHODS

2.1 Site Description

The experiments reported here were conducted in tree plantations located in Dryden, New York, USA (latitude 42°27'0", longitude 76°25'0", elevation 430 m). The site includes one-acre experimental plantations of Norway spruce (*Picea abies* (L.) Karst), sugar maple (*Acer saccharum* Marsh.), and red pine (*Pinus resinosa* Ait.) which were established by R.F. Chandler between 1939 and 1941 as part of a long term study of the effects of forest species on soil properties. Previous to plantation establishment, the land had been under cultivation for approximately 100 yr. Trees in the plantations studied were planted at about 1.5 by 1.5 m spacing and received minimal management after planting.

The soils associated with the plantations studied are developed from silt-enriched glacial (Wisconsin) till derived primarily from local sandstones and siltstones. The soils are well drained due to topographic position and are very acid. They are classified as coarse-loamy, mixed mesic Typic Fragiochrepts or Dystrochrepts. Depth of soil development is limited to about 1 m or less in both types of soil either by the presence of a fragipan or bedrock.

2.2 Sampling

Several different sampling designs were used, depending on the level of variability that was being investigated. To examine variability associated with soil structure, a 4 m transect was established in the sugar maple plot. All points along the transect were at least 100 cm from the base of any tree. Soils were sampled along the transect to a depth of 5 cm in the surface mineral horizon using three different size corers: 1.8 cm, 4.5 cm, and 10 cm in diameter. Twenty samples of each size were collected from points along the transect in the pattern presented in Figure 1. Soil was sampled in June of 1985.

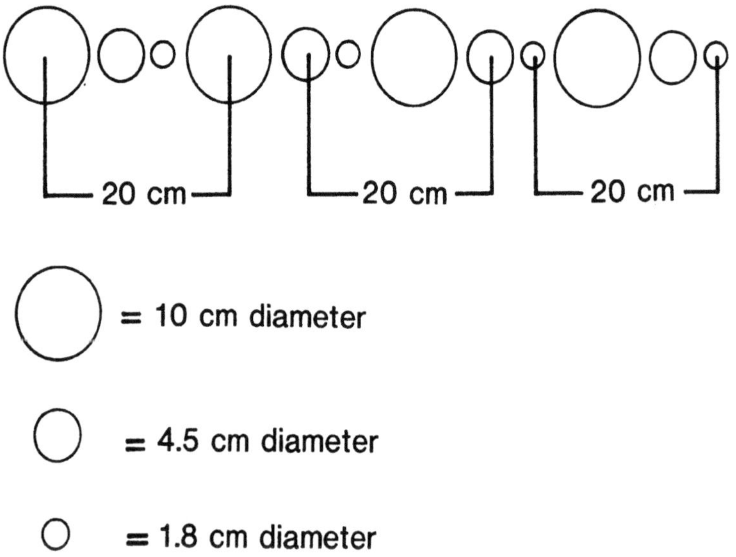

Figure 1. Sampling design used to study forest soil variability associated with soil structure.

To examine soil variability associated with the presence of individual trees, cluster samples were taken in each plot around two trees, chosen to avoid edge effects. Each cluster consisted of a series of samples taken at the base of a tree and radiating out from the tree at 30 cm intervals in the 4 cardinal directions for 120 cm. This resulted in a total of 20 samples per cluster. In the sugar maple plot, the cluster samples were collected only in the N, S, and E directions. Cluster samples were made between April and May of 1984 using a corer 4.5 cm in diameter. Cores were taken to a depth of approximately 10 cm.

To examine soil variability associated with the presence of populations of different species of trees, plots were sampled every 5 m along 60 m transects oriented in a north-south direction. Transect samples were taken between April and May of 1984 using a soil corer 4.5 cm in diameter. Cores were taken to a depth of approximately 10 cm.

All soil cores were placed in polyethylene bags and stored at 4°C within a few hours after sampling. The mineral soil of each core was not subdivided into horizons but was carefully separated from the surface organic horizons, passed through a plastic 4 mm sieve, and thoroughly mixed.

2.3 Measurements

Soil sample pH was determined using a 2:1 solution (10 mM $CaCl_2$) to soil (dry weight basis) dilution. Suspensions were allowed to settle for 20 min. after which pH was measured using a flat surface electrode. In order to analyze for soluble Al and C, field moist samples equivalent to 5.0 g dry soil were shaken with 25 mL 0.05 M NH_4NO_3 for 3 hr on a reciprocating shaker, centrifuged and filtered through 0.45 µm PVC membranes. Filtrates were then analyzed for labile and total Al concentration using the 8-hydroxyquinoline method of James et al. (1983) and for oxidizable C with a modified Mebius procedure using 42 mM (0.25N) $K_2Cr_2O_7$ and 50 mM $Fe(NH_4)_2(SO_4)_2$ (Nelson and Sommers, 1982).

2.4 Statistical Analyses

Classical statistical analyses were performed on an IBM 4341 using the General Linear Models program of the Statistical Analysis System (SAS). Least significant differences were calculated between means when a probability of significance at the 0.05 level existed. All geostatistical analyses were performed on an IBM personal computer (XT) using GEOSTAT (Knighton and Wagenet, 1985).

3. RESULTS AND DISCUSSION

3.1 Small Scale Soil Spatial Variability

Spatial variability of forest soil chemical properties can be considered to occur on several different levels or scales. The first level of variability considered here is assumed to arise primarily form mesoscopic and sometimes macroscopic differences in soil structure. Much of the variability at this level is due to the fact that soils are composed of populations of grains of various sizes, shapes and chemical properties which are in turn arranged to create pores of different sizes and shapes. Earthworm, insect, and root channels further contribute to pore development. Depending on the goal of the particular study, it may be important to sample a large enough volume of soil to represent the pore size distribution of the plot in general. For example, if a large percentage of acid precipitation were moving through a few large pores and the chemical composition of the solution in these pores did not change much in comparison to the change in the solution moving through smaller pores, a method to sample soil solution that did not include these large pores would not be representative of the soil solution of larger areas.

The problem of how to determine the volume of soil to sample in order to include this type of variability has been approached both theoretically and empirically. Two empirical approaches are used here to examine how soil pH, soluble C and soluble total and labile Al vary over short distances. The first was to take increasingly larger samples of soil, thoroughly mix each sample and then take a subsample for soil chemical measurement. Table I presents such data

Table I. Soil chemical properties as a function of sample volume.

		Small	Medium	Large	PR>F+
Soil pH	$\bar{x}$	3.73^{a}	3.75^{a}	3.71^{a++}	0.2075
	SD	0.08	0.05	0.07	
Soluble C	$\bar{x}$	6.1^{a}	11.2^{b}	3.5^{a}	0.0063
($\mu g\ g^{-1}$)	SD	5.9	10.8	3.1	
Total Al	$\bar{x}$	11.7^{a}	13.1^{a}	8.7^{b}	0.0002
($\mu g\ g^{-1}$)	SD	3.1	2.9	3.4	
Labile Al	$\bar{x}$	12.9^{a}	13.8^{a}	8.1^{b}	0.0001
($\mu g\ g^{-1}$)	SD	3.4	3.3	3.1	

+Probability level at which treatment is not significant.
++Values sharing same letters within horizontal rows not significantly different at p = 0.05.

for surface mineral horizon pH, soluble C and soluble total and labile Al. Increasing the sample size from 13 to 393 cm^3 changed neither the mean pH nor standard deviation. The coefficient of variation was also relatively small. This suggests that mesoscopic variability with respect to soil pH is sufficiently represented in the smallest volume sampled. It also implies that, for the plots studied, a relatively few number of small soil cores could be used to adequately describe the surface mineral pH, assuming other larger levels of spatial variability are not imposed on the system.

However, mean soluble C, labile and total Al varied significantly with soil core size. Both labile and total Al were lower in filtrates from subsamples of large core soil then filtrates from subsamples of small and medium core soil. Differences in mean C and Al among core sizes may have been due to differences in core sampling techniques, which could have resulted in less surface organic material being included in the large core sample. In addition, increasing core size did not appear to systematically decrease variability. This suggests that mixing of large soil samples does not homogenize the sample with respect to a property such as soluble carbon and that significant variability exists on a scale of the subsample size.

Another approach to examining spatial variability is to devise a sampling grid and use a statistical method that is appropriate for an analysis of variation in sample values with respect to their spatial location. Geostatistical methods allow us to determine if samples taken close to one another are more likely to be similar, with respect to a given property, than samples taken further apart. The specific technique used is to calculate a semivariogram, which is a measure of the variance of estimation of the sample value x at a location (z+h) using the sample value measured at location (z), where h is the lag distance between samples. The semivariogram is therefore defined as (Rendu, 1978)

$$\gamma(h) = 1/2\ E\{[x(z) - x(z+h)]^2\}.$$

The sampling grid presented in Figure 1 is designed to examine spatial structure with respect to surface mineral horizon with lag distances of 20 to 360 cm. The resulting semivariograms for soil pH for the three different soil volumes sampled are presented in Figure 2. The general variances for the small, medium, and large cores are

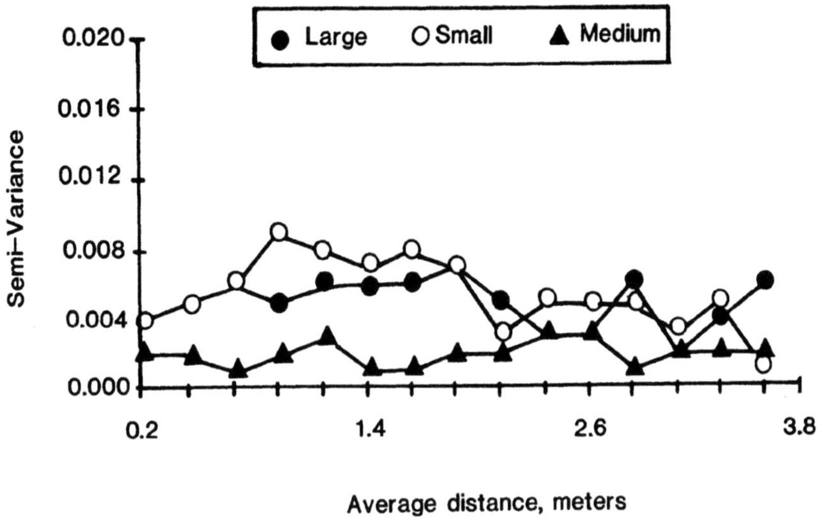

Figure 2. Semivariogram of surface mineral soil pH with lag distances of 20 to 360 cm, used to study small-scale spatial variability.

0.006, 0.002, and 0.005, respectively. For all size cores, over the distance examined, there appears to be no spatial correlation with respect to pH between samples at shorter distances. If there were, this would be indicated by the semivariance decreasing below the general variance and approaching zero or some constant as lag distances decreased (moving left on the abcissa). The lack of spatial correlation for surface mineral horizon pH on a scale of 20 to 360 cm is not surprising given the low coefficient of variability for this property in these plots. However, it is conceivable that in forest soils that were never cultivated, there would be considerably more variability in surface mineral horizon pH on this scale due to factors such as pit and mound topography. There is also no spatial correlation with respect to soluble C over these same lag distances (20 to 360 cm). However, there was some indication for labile and total Al that samples less than 10 m apart were increasingly more highly correlated than samples more than 10 m apart.

3.2 Soil Spatial Variability due to Individual Trees

A larger scale of variability than that previously discussed is variability imposed by the presence of an individual tree. Researchers have found, for example, that pH decreases near the bole of a tree (Gersper and Holowaychuk, 1971; Skeffington, 1983). Differences in soil chemical properties as a function of distance and direction from a tree are believed to be related to the effect a tree has on the distribution and composition of precipitation and organic matter.

In the study presented here, individual trees were selected in 3 different monospecific plantations (Norway spruce, sugar maple and red pine) that were in close proximity to each other. Surface mineral horizons underneath these trees were then sampled as a function of distance and direction from the tree. An analysis of variance was used to determine if surface mineral horizon pH, soluble total and labile Al and soluble C varied significantly as function of distance and direction. Significant interactions between distance and direction were also tested. Results indicated that in general the effects of direction and direction by distance interactions were not highly significant for any of the species studied. The effects of distance were dependent on the species of tree present (Table II).

Table II. Differences in soil chemical properties as a function of distance from tree.

	Norway Spruce	Sugar Maple	Red Pine
pH(0-120 cm)	3.5-3.7	3.8-3.9	3.6-3.9
PR>F+	0.0716	0.6092	0.0001
Labile Al, $\mu g\ g^{-1}$ (0-120 cm)	18-16	15-15	15-11
PR F	0.0758	0.9354	0.0003
Total Al, $\mu g\ g^{-1}$ (0-120 cm)	23-20	18-19	19-14
PR>F	0.0887	0.5221	0.0001
Soluble C, $\mu g\ g^{-1}$ (0-120 cm)	68-34	------	184-207
PR>F	0.5793		0.9317

+Probability level at which treatment is not significant.

The distance of a soil sample from a sugar maple had no effect on pH, total or labile Al. In comparison, the surface mineral horizon pH near the base of red pine was significantly lower and soluble labile and total Al higher than further from the tree. Soluble carbon however, did not appear to vary as a function of distance from tree (Table II). Soil chemical properties under individual Norway spruce trees appeared to vary in a manner similar to red pine, but the changes were not as significant (Table II). Using the same sampling scheme, surface mineral soil was collected under Norway spruce trees of similar age growing in a different, less acid soil type 2.1 km away. The results were similar to those for Norway spruce and red pine presented in Table II. The soil pH was 4.1 at the base of the tree as compared to 4.4 at 120 cm (PR>F=0.0033). Labile and total soluble Al were 1.4 and 4.3 ppm, respectively, near the base of the tree and decreased to 0.4 and 2.0 ppm at 120 cm (PR>F=0.0456 and 0.0275, respectively). Unlike the trees in Table II, there was a significant (PR>F=0.0001) effect of distance on soluble C (11.4 ppm at the base of the tree compared to 1.4 ppm at 120 cm).

There are several implications of these experiments for acid precipitation studies. In some cases, a soil sampling design may wish to account for or even focus on the variability in soil chemical properties imposed by individual trees. For example, roots and microorganisms growing in soil near the base of spruce and pine trees may be most susceptible to further effects of acid precipitation. Also, if a relatively large percentage of precipitation is moving through a relatively small area of soil at the base of the tree, failure to account for the difference in soil chemical properties as a function of distance from tree would result in an unrealistic model of the effects of soil chemical properties on water leaving the surface horizons. However, if the presence of a small volume of more acid soil was thought to have little effect on total soil solution composition or on root and microbe populations, sampling near the tree might be avoided.

3.3 Soil Spatial Variability due to Populations of Different Tree Species

It is clear that the presence of an individual tree can impose a pattern of variability in soil chemical properties such as surface mineral soil pH and soluble Al. At a larger scale than individual trees, it has long been considered that populations of different species of trees can differentially affect the chemistry of the soil. For example, soil under conifers is thought to become more acid than the same soil under many hardwoods. There could be differences in soil chemical properties under populations of different species of trees due to such factors as differences in amount, composition and incorporation of litter and differences in root distribution and ion uptake.

Tree species effects on soil chemical properties are not easy to study because it is often not possible to control covariants such as soil type, climate, aspect, slope, and tree age while varying only tree species. In the plots available for this study, it was initially assumed that none of these factors were significant covariates. Accepting this assumption, soil cores at least 100 cm from individual trees within each of the three vegetation types were sampled and analyzed. The soil chemical data was then evaluated to assure that values were randomly distributed. A simple t-test using a pooled variance was employed to compare means (Table III). With

Table III. Differences in soil chemical properties as a function of tree species.

	Norway Spruce	Sugar Maple	Red Pine
pH (PR>F = 0.0001)	3.76^a	4.00^b	3.83^{c+}
Labile Al ($\mu g\ g^{-1}$) (PR>F = o.0031)	15.9^a	13.2^b	11.7^b
Total Al ($\mu g\ g^{-1}$) (PR>F = 0.307)	20.1^a	16.9^{ab}	14.9^b
Soluble C ($\mu g\ g^{-1}$)	52^a	---	206^b

+Values sharing same letters within horizontal rows not significantly different at p = 0.05.
++Probability level at which treatment is not significant.

this approach, one concludes that surface mineral soil pH beneath spruce and pine is more acid and soluble labile Al concentration higher than soil beneath maple. Also, it appears that there is more soluble total Al and less soluble carbon in surface soil under Norway spruce compared to red pine (Table III).

One major objection to the conclusions reached above is that there is no replication of the plots and therefore no way of comparing between treatment (tree species) variation to between plot variation. If soil properties are varying between these three plots regardless of the presence of vegetation, differences due to this type of variation will be erroneously assigned to treatment differences. However, in this study, like many other forest soil studies, there are not appropriate treatment replications. A geostatistical approach to this problem is presented here which does not rely on plot replication to delineate treatment differences.

A sampling transect was used where the spatial relationship of all samples were known. In this case, a sampling scheme was devised that transected three plots of monospecific populations of different trees. Variability in soil pH, soluble Al and C were then studied by generating a semivariogram which included all three plots for each of these soil chemical properties studied (Figures 3, 4 and 5). The increase in the semivariance above the general variance is attributed to trend, i.e. the presence of some factor imposing systematic variability on the area being studied. In these studies, where trend was present, the data was then detrended by subtracting the mean plot soil chemical property (for example, red pine pH) from the values for that plot (Riha et al., 1986; McBratney and Webster, 1981; Webster and Cuanalo, 1975). If the resulting semivariogram showed no trend and the semivariance fluctuated (randomly) around the general variance, then it was assumed that the systematic change (trend) of the soil chemical property in space coincided with the change in tree species. Conclusions reached using this approach were that differences in soil pH under spruce compared to maple existed due to the presence of vegetation, but differences in soil pH between maple and pine or spruce and pine could not be clearly attributed to vegetation (Figure 3). There were no obvious differences due to vegetation in labile or total Al (Figure 4) between any of the plots. However, there was a clear difference in soluble C (Figure 5) in soil under spruce and pine due to the presence of the different tree species.

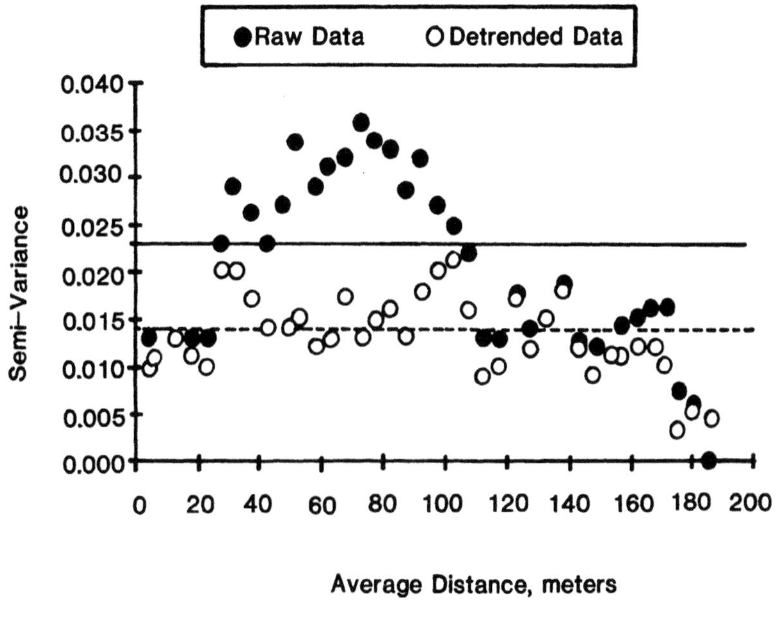

Figure 3. Semivariance of soil pH as a function of distance between samples.

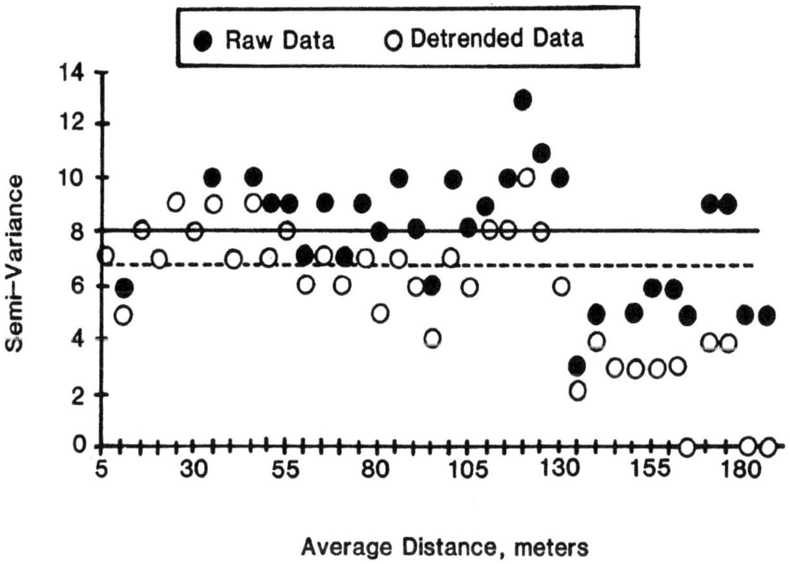

Figure 4. Semivariance of total soluble Al as a function of distance between samples.

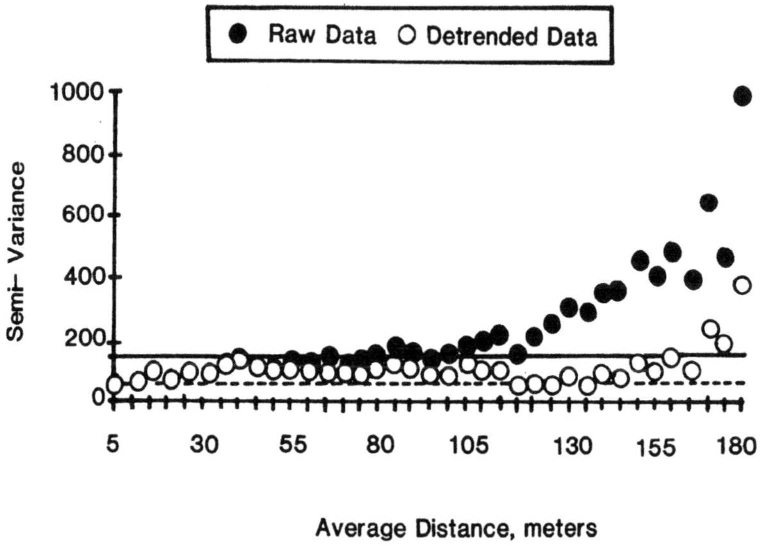

Figure 5. Semivariance of total soluble C as a function of distance between samples.

These results indicated that various species of trees can differentially affect surface mineral horizon chemical properties even in a relatively short time of less that 50 yr. It may be necessary to consider the species of tree growing in the soil as well as the general soil type in order to evaluate the past effects of acid precipitation and predict future effects.

4. CONCLUSIONS

The major conclusion of this study is that starting with a previously cultivated soil, in less than 50 yr the presence of individual trees and different species of trees affected the chemistry of the surface mineral soil. In considering soil pH and soluble Al, trees imposed patterns of variability on the chemistry of the soil that appeared greater than small scale variability due to soil structure or other mesoscopic features. However, in considering soluble C, much variability appeared to occur on a mesoscopic scale. Such variability needs to be considered when designing experiments and interpreting their results.

REFERENCES

Gersper, P.L. and Holowaychuk, N.: 1971, Ecology 52, 691.

James, B.R., Clark, C.J., and Riha, S.J.: 1983, Soil Sci. Soc. of Am. J. 47, 893.

Knighton, R. and Wagenet, R.J.: 1985, Theory and user's manual for GEOSTAT: A geostatistical analysis computer program for auto- and cross-correlation, semi- and cross-variograms and kriging. Agron. Mimeo 85-11.

McBratney, A.B. and Webster, R.: 1981, Geoderma 26, 63.

Nelson, D.W. and Sommers, L.E.: 1982, Total carbon, organic carbon, and organic matter. In A.L. Page, R.H. Miller, and D.R. Keeney (eds.). Methods of soil analysis. Part 2. Chemical and microbiological properties. 2nd edn. pp. 539-580. Am. Soc. Agron., Madison, WI.

Rendu, J.M.: 1978, An introduction to geostatistical methods of mineral evaluation. South African Institute of Mining and Metallurgy. Johannesburg.

Riha, S.J., James, B.R., Senesac, G.P. and Pallant, E.: 1986, Soil Sci. Soc. of Am. J. 50 (In press).

Skeffington, R.A.: 1983, Soil properties under three species of tree in southern England in relation to acid deposition in throughfall. In B. Ulrich and J. Pankrath (eds.). Effect of accumulations of air pollutants in forest ecosystems. pp. 219-231. Reidel Pub. Co.

Webster, R. and de la Cuanalo, H.E.: 1975, J. Soil Sci. 26, 176.

ALKALINITY PRODUCTION IN EPILIMNETIC SEDIMENTS: ACIDIC AND NON-ACIDIC LAKES.

S.L. Schiff and R.F. Anderson
Lamont-Doherty Geological Observatory
Department of Geological Sciences, Columbia University
Palisades, NY 10964 USA

ABSTRACT. Interstitial water profiles in epilimnetic sediments of lakes with varying water column alkalinities were collected to assess the origin and importance of sedimentary alkalinity in freshwater lakes. Release of Ca^{2+} and NH_4^+, and consumption of SO_4^{2-} are the most important contributors to alkalinity production in sediments of non-acidic lakes. In acidic lakes, Fe^{2+} and Mn^{2+} replace Ca^{2+} as the dominant cation contributors to alkalinity production. The sedimentary alkalinity flux is an important component of the acid neutralizing capacity of freshwater lakes. However, the presence of large alkalinity gradients in sediment porewaters does not necessarily indicate a large source of alkalinity for the lake, as a significant portion of the alkalinity is associated with the formation of Fe^{2+}, Mn^{2+} and NH_4^+. Oxidation of Fe^{2+} and Mn^{2+} at the anoxic-oxic interface and biological removal of NH_4^+ in the overlying water column results in consumption of the co-diffusing alkalinity.

1. Introduction

Acid neutralization processes are important components of models which predict temporal trends in acidification of freshwater lakes. Researchers have noted that within lake sediments, the interstitial pH typically increases above the water column value, even in acidic lakes with pH < 4.5 (Dickson, 1980; Tolonen and Jaakkola, 1983; Carignan, 1985; Rudd et al., 1986). The production of alkalinity in anoxic hypolimnia due to sedimentary reactions has been well documented (e.g. Mortimer, 1941) and related to the neutralization of acidic precipitation (Kelly et al., 1982). Recently, epilimnetic sediments have also been identified as an important site of alkalinity production (Kelly and Rudd, 1984; Schiff and Anderson, 1986). Porewater alkalinity in epilimnetic and hypolimnetic sediments has been found to increase with depth (Emerson 1976, Kuivila and Murray 1984, Carignan 1985) indicating a flux of alkalinity to the overlying water column. In situ alkalinity production can be the most important source of alkalinity in lakes with inherently low water column alkalinity (Schindler et al., 1986). In this study, porewater profiles in epilimnetic sediments of lakes with varying water column alkalinities were collected to assess the origin and importance of this alkalinity flux in freshwater lakes.

2. Sampling locations and methods:

In order to span a range of water column alkalinites and temporal exposure to acidic precipitation, eight lakes were chosen. Three lakes; L239, L223, and L114, are part of the Experimental Lakes Area (ELA) in NW Ontario. L239 is the ELA control lake. L223 has received artificial additions of H_2SO_4 for 8 years, sufficient to lower the water column pH by 0.25 pH units per year to a low of 5.0 in 1980 (Schindler et al., 1986). L114 has received artificial additions of H_2SO_4 equivalent to an average annual rainfall of pH 4.0 rain

Water, Air, and Soil Pollution **31** (1986) 941–948.

since 1979. Batchewana and Little Turkey lakes are part of the Turkey Lakes Watershed near Sault Ste. Marie, Ontario, which receives precipitation with an average annual pH of 4.3 (Semkin and Jeffries, 1985). Bowland lake, located 70 km north of Sudbury, Ontario was sampled immediately prior to a whole lake liming experiment in August 1983. Mohonk and Awosting lakes are located in the Shawangunk Mountains, a ridge of quartzite conglomerate approximately 145 km north of New York City. Awosting lake has had a pH of less than 4.5 since before 1931 (Smiley and Huth, 1983). Mohonk has been protected from the effects of acidic precipitation by a shale contact within the lake drainage basin.

The epilimnetic sediments at 1 to 2 metres water depth in these lakes differ in physical characteristics. L114, Batchewana and Mohonk have flocculent organic-rich sediments with surficial porosities of 0.98 to 0.99. Bowland, L239, L223, Little Turkey and Awosting have sediments of lower organic content and consequently lower porosity. The sediments of Bowland lake were covered with moss to a depth of 4 cm.

A porewater dialysis sampler (Hesslein, 1976) with a vertical resolution of 1 cm and fitted with a 0.2 μm Nucleopore membrane, was used to collect the interstitial porewater profiles. To maintain the redox integrity of the samples, the sampling syringes were pre-filled with N_2 or He. All of the subsequent analyses were performed by subdividing the sample from one dialysis compartment (≈4 mL). Alkalinity was determined by Gran titration with 0.01 N HCl, in the pH range of 4.2 to 3.5. The cation subsample was acidified with 10 or 50 μL of ultrapure acid and analyzed by IC plasma emission spectroscopy for Ca, Mg, Na, K, Fe, Mn, Al, Sr and Si. SO_4^{2-}, Cl^- and F^- were analyzed by ion chromatography. NH_4^+ was analyzed by the phenolhypochlorite method. Procedural blanks showed no significant contamination of any ions.

3. Calculations:

Alkalinity can be expressed as the excess of strong basic cations over strong acidic anions:

$$ALK = Ca^{2+} + Mg^{2+} + NH_4^+ + Na^+ + K^+ + Fe^{2+} + Mn^{2+} - SO_4^{2-} - Cl^- - F^- - NO_3^- \quad (1)$$

where all quantities are expressed in equivalents. Fe and Mn passing the dialysis membrane are assumed to be in the +2 oxidation state. The contribution of each of the ions on the right hand side of Equation (1) to a change in alkalinity can be assessed by computing the change in concentration of these ions from a reference level (chosen here as the overlying water column) to the depth of interest in the sediments. Diffusion coefficients (Li and Gregory, 1974) used in the flux calculations are corrected for temperature, porosity, tortuosity, and cross-coupling (McDuff and Ellis, 1979).

Assumptions inherent in these calculations are that these porewater profiles are at steady-state, and that adsorption and advection are negligible. As well, the D of alkalinity is assumed to be equal to the D of HCO_3^- even though a significant portion of alkalinity measured by Gran titration may be composed of organic acid anions (Herczeg and Hesslein, 1984) which would have a much lower D.

4. Results and discussion:

An increase in porewater pH with depth in the sediments (Fig. 1) was observed in all lakes with a water column pH of less than 6.5. Except for lake Awosting, the sediment porewater pH seemed to approach a constant pH with sediment depth in the pH range of 6.0 to 6.8, regardless of water column pH or SO_4^{2-}, or sediment organic matter content.

Porewater titration alkalinities increase to greater than one order of magnitude above the water column values at some depth in the sediments of all lakes studied (Fig. 2). The increase in alkalinity begins within 2 cm of the sediment-water interface except in Bowland and Awosting lakes. Although Awosting and Bowland have water column pH's of 4.2 and

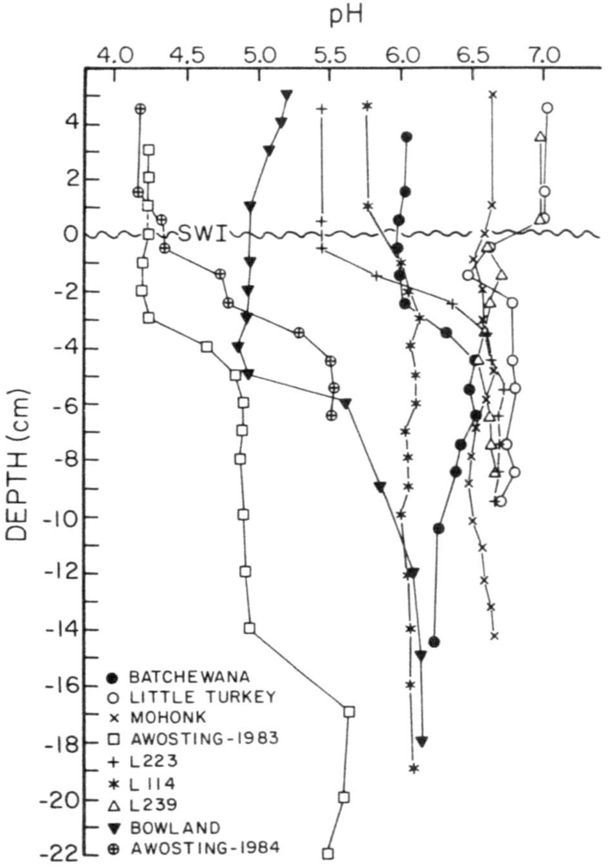

Figure 1: Sediment porewater pH versus depth above the sediment-water interface.

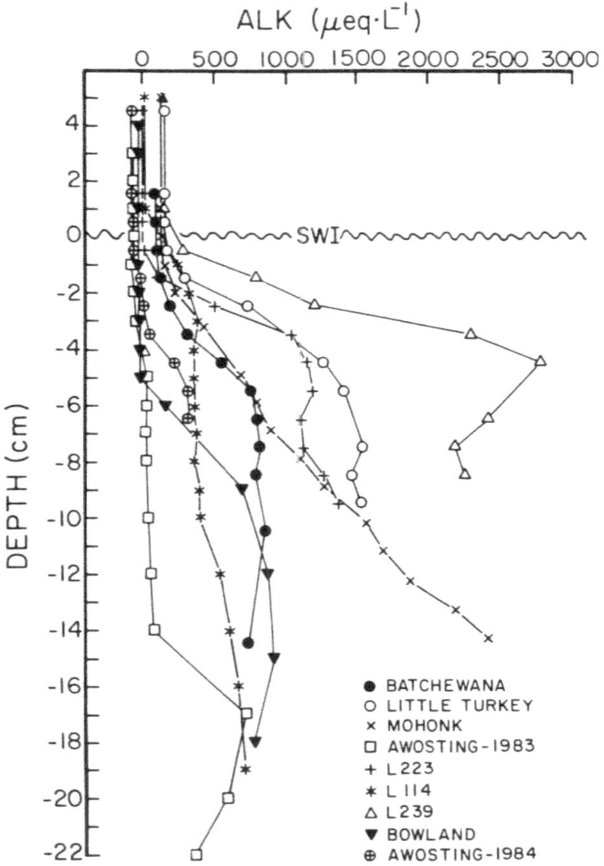

Figure 2: Gran titration alkalinity in sediment porewaters versus depth above the sediment-water interface.

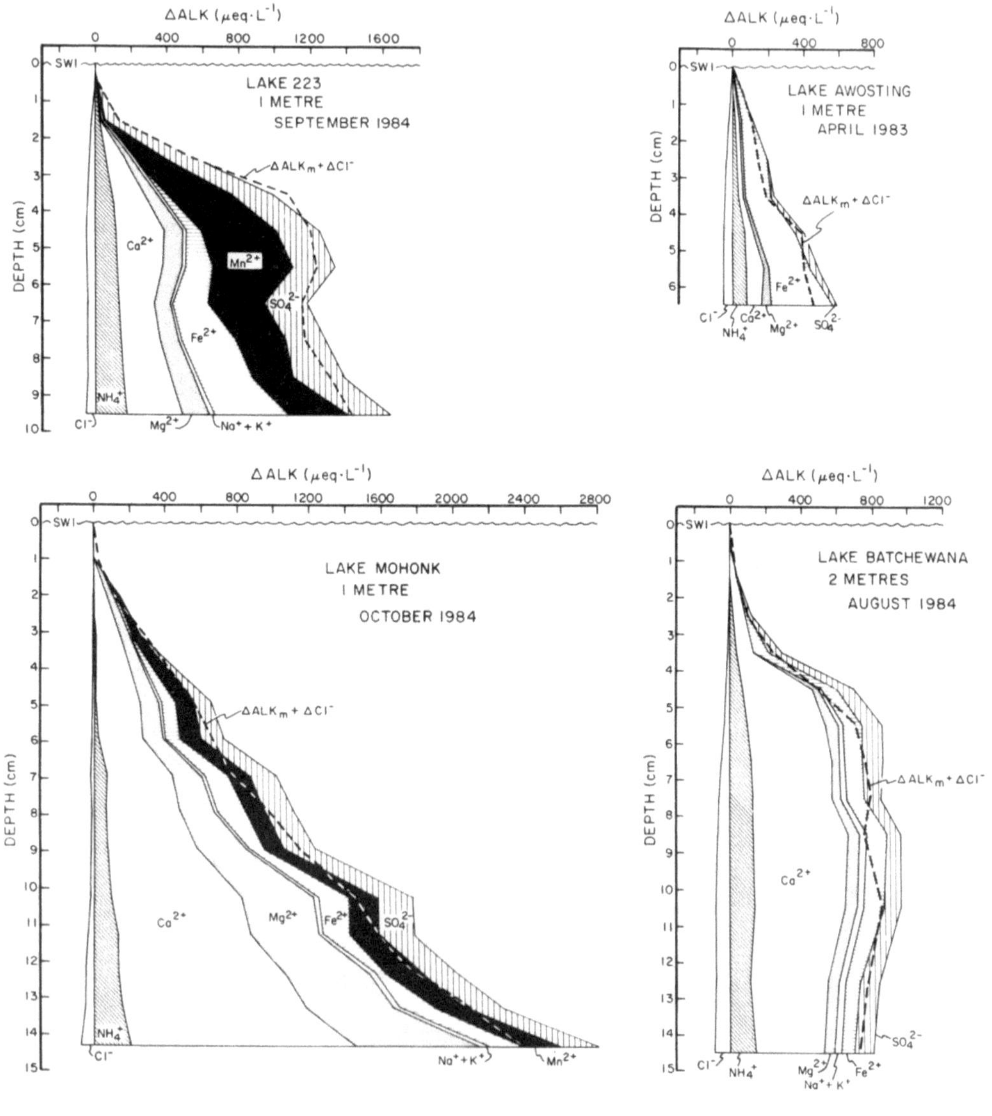

Figure 3: Gran titration alkalinity (ΔALK_m) and independantly calculated ionic contributions to alkalinity with depth below the sediment-water interface in: a) L223, b) Mohonk, c) Batchewana and d) Awosting 1983. Since the Cl^- concentration increases with depth, and the contribution of ΔCl^- to ΔALK is negative, ΔCl^- has been added to ΔALK_m to allow the direct comparison between the sum of the individual positive contributions to alkalinity and the Gran titration alkalinity. ΔCl^- has also been plotted so that the contribution of ΔCl^- to ($\Delta ALK_m + \Delta Cl^-$) is apparent.

5.2, respectively, both lakes have considerable positive alkalinity at depth in the sediments. Alkalinity gradients are in general steeper in non-acidic lakes with lower porosity sediments. In all lake sediments, except lake Mohonk, the porewater alkalinity reaches a maximum at some depth below which the alkalinity either increases slowly or decreases.

To assess the origin of the alkalinity production in the sediments, the individual ionic contributions to alkalinity were calculated at each depth as described above (Fig. 3):

$$\Delta ALK_{calc} = \Delta NH_4 + \Delta Ca + \Delta Mg + \Delta Na + \Delta K + \Delta Fe + \Delta Mn - \Delta SO_4 - \Delta Cl \qquad (2)$$

ΔALK_m is the difference in measured Gran alkalinity between the water column and the given depth. Since the porewater Cl^- concentration increases with depth, the contribution of ΔCl^- to ΔALK_{calc} is negative. If all the ionic contributions to ΔALK are accounted for, $\Delta ALK_m + \Delta Cl^-$ should match the sum of the calculate positive individual ionic contributions to alkalinity:

$$\Delta ALK_m + \Delta CL = \Delta NH_4 + \Delta Ca + \Delta Mg + \Delta Na + \Delta K + \Delta Fe + \Delta Mn - \Delta SO_4 \qquad (3)$$
$$= \Delta ALK_{calc} + \Delta Cl^-$$

In general the agreement between $\Delta ALK_m + \Delta Cl^-$ and the sum of the positive ionic contributions is good, within 10 to 15%, ruling out other significant contributors to porewater alkalinity production.

4.1 Contributions to alkalinity production in the sediments:

Calcium is the most important single contributor to alkalinity production in the sediment porewater profiles of Batchewana, L239, Mohonk and Little Turkey. None of these lakes have a water column pH of < 6.0. L223, Bowland and Awosting all have a water column pH of less than 5.5 and all have Fe^{2+} (Bowland and Awosting) or Fe^{2+} + Mn^{2+} (L223) contributing more to alkalinity production than Ca^{2+} in the shallower sediment intervals, suggesting that Ca^{2+} has been leached from the sediments of acidic lakes. In L114, most of ΔALK is associated with NH_4^+ production, but $\Delta Ca^{2+}+\Delta Mg^{2+}+\Delta Na^+ + \Delta K^+$ exceeds ΔFe^{2+} at all sediment depths.

Magnesium, Na^+, and K^+ also contribute to ΔALK but are consistently less than the contribution of Ca^{2+}. The order $Ca^{2+} > Mg^{2+} > Na^+ \approx K^+$ is in agreement with the observed order of release of exchangeable cations in lake sediments (Oliver and Kelso, 1983, Schiff and Anderson, 1986).

Sulfate is reduced to low concentrations of 10 to 30 $\mu eq \cdot L^{-1}$ by 5 cm in sediments of all lakes except Bowland and Awosting where SO_4^{2-} consumption takes place deeper in the sediments. The ΔSO_4^{2-} contribution to alkalinity is more pronounced in lakes with higher water column SO_4^{2-}. In Little Turkey, Batchewana, Awosting, Bowland and L239, $\Delta Cl^- \approx -\Delta SO_4^{2-}$, and the contributions of these two ions to ΔALK_{calc} cancel each other.

Iron increases with depth in all porewater profiles. In Bowland, Batchewana, L239, Mohonk, and Little Turkey, Fe^{2+} reaches a maximum within the sediments and then decreases with increasing depth. This decrease of Fe^{2+} with an increase in ALK suggests that the control on Fe solubility may be siderite ($FeCO_3$) formation. However, solubility calculations indicate that supersaturation with respect to siderite is reached only in the sediment porewaters of L239 and L223. Mn^{2+} can be an important contributor to alkalinity and exceeds the contribution of Fe^{2+} or Ca^{2+} in L223 at depths < 7 cm. All the porewaters studied are undersaturated with repect to $MnCO_3$ and $CaCO_3$. NH_4^+ is an important contributor to alkalinity production in all lakes except Bowland Lake.

4.2 Alkalinity flux out of the sediments:

Table I contains three estimates of the alkalinity flux from the sediments to the overlying water column, calculated by applying Fick's first law to: 1) the measured Gran alkalinity profiles, 2) all the calculated ionic contributions to alkalinity, and 3) only the

TABLE 1: ALKALINITY FLUX OUT OF THE SEDIMENTS[1]

lake	water column pH	water column SO_4^{2-} (μeq.L^{-1})	alkalinity flux(2) ΔALK_m (meq.m^{-2}.yr^{-1})	alkalinity flux(3) ΔALK_{calc} (meq.m^{-2}.yr^{-1})	alkalinity flux(4) ΔALK_{calc}-ΔNH_4^+-ΔFe^{2+}-ΔMn^{2+} (meq.m^{-2}.yr^{-1})
Lake 223	5.4	255	60	60	35
Lake 114	5.8	84	510	520	70
Lake 239	7.0	106	290	280	220
Little Turkey [5]	7.0	175	60	---	---
Batchewana	6.0	130	70	80	60
Bowland [6]	5.2	231	0.4	---	---
Mohonk	6.6	235	240	240	200
Awosting 1983	4.2	166	20	20	5
Awosting 1984	4.2	174	-15	-30	-3

1. All alkalinity fluxes have been calculated by applying Fick's first law to the porewater gradients in 0-2 cm of the sediment column, assuming ALK=HCO_3^-. Diffusion coefficients were corrected for an annual average temperature of 9 °C, cross-coupling, porosity and tortuosity.
2. Alkalinity flux calculated from the measured Gran alkalinity porewater gradient.
3. Alkalinity flux calculated from the sum of all the ionic contributions to alkalinity.
4. Alkalinity flux calculated from the sum of the contributions of Ca^{2+}, Mg^{2+}, Na^+, K^+, SO_4^{2-}, Cl^- to alkalinity.
5. In Little Turkey, the alkalinity fluxes of (3) and (4) were not calculated as Ca^{2+} in the first sediment interval is inordinately high.
6. In Bowland Lake the calculated alkalinity fluxes ((3) and (4)) are not given as there is no significant gradient in the 0-3 cm of the porewaters within the errors of the cation analyses.

calculated contributions of Ca^{2+}, Mg^{2+}, Na^{+}, K^{+}, SO_4^{2-} and Cl^- to alkalinity. Oxidation of Fe^{2+} and Mn^{2+} at the anoxic-oxic interface in the upper few cm of the sediments results in consumption of alkalinity. Thus the alkalinity co-diffusing with Fe^{2+} and Mn is not released to the overlying oxic water column. In addition to Fe^{2+} and Mn^{2+}, the flux of alkalinity associated with the NH_4^+ flux in the porewaters is not a source of alkalinity for the lake. NH_4^+ does not accumulate in the water columns of any of these lakes, and the associated alkalinity is lost when NH_4^+ is assimilated by phytoplankton or oxidized by nitrifiers.

4.3 Importance of the sedimentary alkalinity flux in neutralizing acidic precipitation:

The third method of calculating the alkalinity flux (Table I) gives the true supply of alkalinity to the overlying water column. This is the alkalinity production term important in the prediction of temporal trends of lake acidification. Driscoll and Likens (1982) estimate the annual H^+ loading in wet and dry deposition in NE U.S.A. to be 132 $meq.m^{-2}.yr^{-1}$. The sedimentary alkalinity flux by method (3) in Mohonk and L239 clearly exceeds the estimated H^+ load. In Batchewana, L114 and Little Turkey, the alkalinity flux is approximately 1/2 of the estimated NE U.S.A. H^+ deposition rate, and it appears that these presently non-acidic lakes would be sensitive to an increase in their present H^+ load. The alkalinity fluxes in Table I, however, cannot be used to quantitatively predict lake sensitivity as the SO_4^{2-} load will increase with the increase in H^+ load. Since SO_4^{2-} is reduced to constant low levels in the sediments of all lakes studied, an increase in SO_4^{2-} consumption would result in an increase in alkalinity production.

Although L223 has been artificially acidified for 8 years, a substantial alkalinity flux out of the epilimnetic sediments was observed due to Ca^{2+} release and SO_4^{2-} consumption.

In the two most acidic lakes, Bowland and Awosting, the flux of alkalinity is very low. In both the Awosting profiles, oxidation of reduced S resulted in an increase of SO_4^{2-} in the upper 2 cm of the porewaters and a concomittant decrease in alkalinity. This oxidation of reduced S cannot be a steady-state feature but must be re-oxidation of reduced S which was formed during warmer summer months when the redox potential of the surficial sediments was more reducing. In addition, even though SO_4^{2-} is consumed in the sediments, the increased depth of the SO_4^{2-} consumption zone in both lakes, compared to the non-acidic lakes studied, decreases the resultant alkalinity flux out of the sediments. Since the alkalinity gradient in Awosting and Bowland sediments is associated primarily with Fe^{2+} and NH_4^+, oxidation of Fe^{2+} at the oxic-anoxic interface and uptake of NH_4^+ in the overlying water column will consume the co-diffusing alkalinity. The large increase in porewater alkalinity with sediment depth in these acidic lakes does not necessarily indicate a large capacity to neutralize acidic precipitation.

Acknowledgments:

M. English, P. Huth, L. Molot, D. Schindler, R. Semkin, and D. Smiley, provided logistical support. M. English and P. Froelich critically read the manuscript. This research was funded by NSF DEB 80-17639, the Charles A. Lindberg Fund and an NSERC Postgraduate Scholarship to S.S. This is LDGO contribution # 4033.

References:

Carignan, R.: 1985, Nature **317** 158.
Dickson, W.: 1980, **In** Drablos, D. and A. Tollan (eds). Ecological Impact of Acid Precipitation. Symp. Proc. SNSF Proj., Oslo, Norway. pp 75-83.
Driscoll, C.T. and Likens, G.E.: 1982, Tellus **34** 282.
Emerson, S.: 1976, Geochim. Cosmochim. Acta **40** 925.
Herczeg, A. and Hesslein, R.: 1984, Geochim. Cosmochim. Acta **48** 837.
Hesslein, R.H.: 1976, Limnol. Oceanogr. **21** 912.
Kelly, C., Rudd, J., Schindler, D., and Cook, R.,: 1982, Limnol. Oceanogr. **27** 610.
Kelly, C. and Rudd, J.: 1984, Biogeochem. **1** 63.
Kuivila, K.M. and Murray, J.W.: 1984, Limnol. Oceanogr. **29** 1218.
Li, Y.-H. and Gregory, S.: 1974, Geochim. Cosmochim. Acta **38** 703.
McDuff, R.E. and Ellis, R.A.: 1979, Am. J. Sci. **279** 666.
Mortimer, C.H.: 1941, J. Ecol. **29** 280.
Oliver, B.G. and Kelso, J.R.M.: 1983, Water Air Soil Poll. **20** 379.
Rudd, J., Kelly, C., Flett, R., and Schindler, D.: 1985, Limnol. Oceanogr. (in press)
Schiff, S.L. and Anderson, R.F.: 1986, Can. J. Fish. Aquatic Sci. (in press).
Schindler, D., Turner, M., Stainton, M., and Linsey, G.: 1986, Science **232** 844.
Semkin, R. and Jeffries, D.: 1985, Turkey Lakes Watershed. Data report: 85-01 24 pp.
Smiley, D. and Huth, P.: 1983, Mohonk Preserve Inc. Research Report. March. 43 pp.
Tolonen, K. and Jaakkola, T.: 1983, Ann. Bot. Fennici. **20** 57.

EFFECT OF LAKE ACIDIFICATION ON THE ADSORPTION OF PHOSPHORUS BY SEDIMENTS

T. Mayer
Aquatic Ecology Division
National Water Research Institute
Canada Centre for Inland Waters
P.O. Box 5050
Burlington, Ontario, Canada L7R 4A6

J.R. Kramer
McMaster University
Geology Department
Hamilton, Ontario, Canada L8S 4M1

ABSTRACT. The retention properties of acidic and non-acidic lake sediments were determined in order to assess the effects of lake acidification on the immobilization of P from solution by sediments. The adsorption of P by solids was described by the Langmuir model which was used to determine the sorption parameters, e.g. sorption maxima and equilibrium constant of adsorption. The pH of solution and the chemical and mineralogical characteristics of sediments affect mainly the magnitude of adsorption maxima. The binding strength of the adsorbed complex is similar for all the investigated sediments (ΔG=-25.3 to -28.5 kJ mol^{-1}) and it is affected little by variation in pH or by chemical and mineralogical composition of sediments. The results indicate that the magnitude of P removal is determined more by sediment chemistry and mineralogy (amorphous Al/Fe oxy-hydroxides, carbonate content) than by pH of the water.

1. INTRODUCTION

The role of sediments in regulating dissolved inorganic P concentration has long been recognized. Hence, sediments are important components of the whole-lake mass balances of P. Whereas numerous studies have been concerned with uptake of P by sediments of neutral lakes, little information is available on P immobilization by sediments of acidic lakes.

The purpose of this investigation is to evaluate the effect of pH on the sorption of P by lacustrine sediments. The retention properties of sediments are assessed by investigating the reversibility of adsorption.

Water, Air, and Soil Pollution **31** (1986) 949–958.

Sampling was carried out to allow comparison of sediments from lakes of different geology and pH. The samples include sediments from acidic and neutral Shield lakes situated in Sudbury area (north-eastern Ontario). The Shield lakes are located in terrain underlain by non-calcareous bedrock of low solubility with thin glacial overburden. The lakes are oligotrophic with low alkalinity and conductivity. Their limnology and geological setting have been described in detail (Ontario Ministry of Environment, 1978, 1982). Two sediments (E-30; WB) from depositional basins of Lake Ontario, which is a mesotrophic hardwater lake were included for comparison.

2. MATERIALS AND METHODS

Surficial sediments from Shield lakes were obtained in the summer of 1981 using an Ekman dredge. Surface sediments from Lake Ontario locations were collected in May 1975 with a double-Shipek. The samples were frozen and subsequently freeze dried. Sediment particles less than 250 μm in diameter were used for all experiments.

Methods of determining the chemical properties of sediments included: total elemental composition by X-ray fluorescence; total and organic C (Kemp, 1971); total and organic P (Mayer and Williams, 1981); and oxalate extractable Fe and Al (McKeague, 1978).

The sorption of inorganic P was evaluated by equilibrating sediment suspensions with various levels of P for 72 hr at 21°C. 0.1 N or 10^{-4} N NaCl was used to maintain constant ionic strength and facilitate particle settling during centrifugation. Bacterial uptake of P was eliminated by addition of 147 μmol $HgCl_2$. On equilibration, the aqueous and solid phase were separated and the filtrate was analyzed colorimetrically for phosphate (Harwood et al., 1969). The solid residue was retained for subsequent desorption analysis. Measurement of P desorption from the sediment involved addition to the residue of 40 mL 0.1 N NaCl and 72-hr equilibration. The subsequent separation and P determination were carried out as above. The pH of the experiments varied. The details of the experimental procedures can be found in Mayer (1984).

Native exchangeable P, determined from isotope dilution (Li et al., 1973) was used as an estimate of the native sorbed P. This value was added to all values of added P sorbed used in the analysis of P sorption data.

3. RESULTS

A summary of chemical characteristics of the investigated sediments is given in Table I. Oxalate extractable Fe, which measures Fe associated with amorphous oxy-hydroxides constituted a similar proportion of total Fe in all investigated sediments, except in the Windy Lake sediment. This Fe rich sediment contained unusually high

TABLE I Water chemistry data and chemical characteristics of sediments

Sample	Water		Sediment						
	pH*	Alkalinity*	Total Fe	Oxalate extract. Fe	Total Al	Oxalate extract. Al	Organic C	Organic P	Total P
		meq L^{-1}	mg g^{-1}	mg g^{-1}	mg g^{-1}	mg g^{-1}	%	µg g^{-1}	µg g^{-1}
Wavy	4.5	-0.03	26.7	4.69	34.3	16.0	12.84	742	1845
Clearwater	4.4	-0.04	41.7	6.96	36.2	8.8	9.68	535	1142
Windy	6.4	0.09	71.7	44.37	48.3	7.7	6.53	460	1863
L.O. E-30	8.5	1.7	49.2	12.19	76.4	1.1	2.48	155	1108
L.O. WB	8.5	1.7	48.5	13.76	78.9	2.3	2.72	200	1191

*Data from Ontario Ministry of the Environment (1978; 1982) and Dobson (1980)

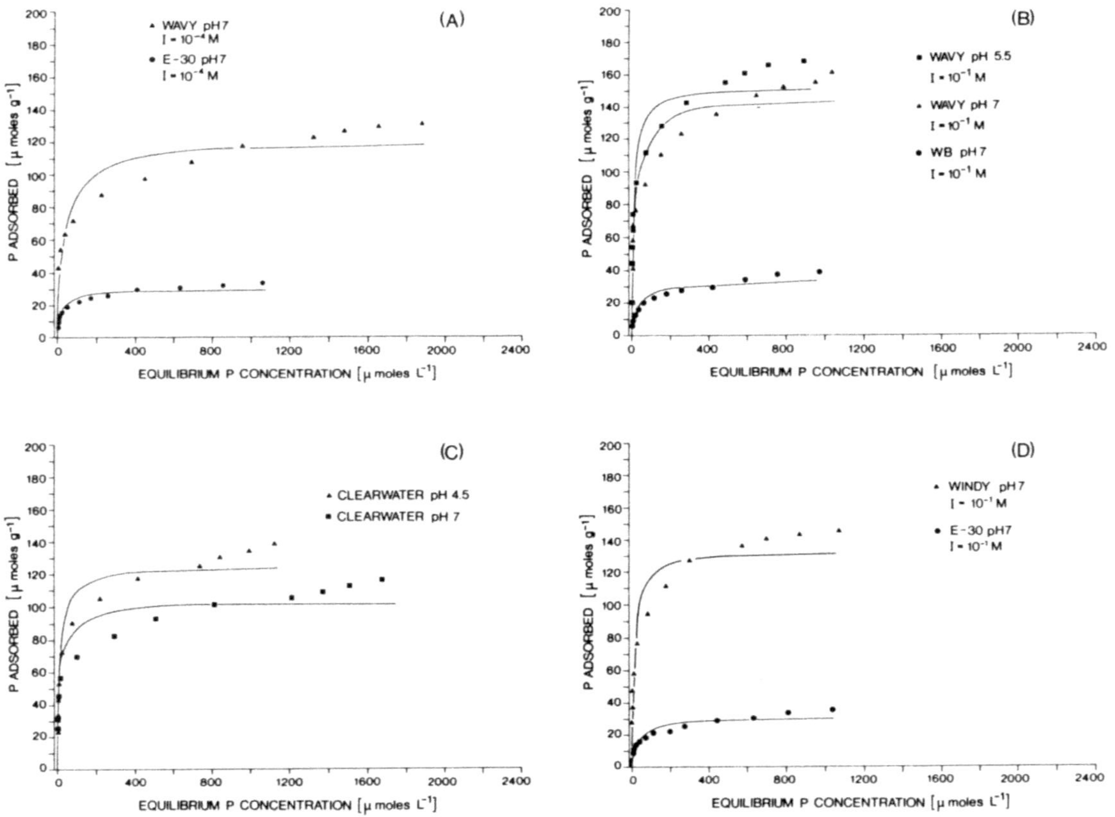

Fig. 1. Adsorption isotherms for (a) Wavy Lake and Lake Ontario E-30 sediments, matrix solution 10^{-4} M CaCl; (b) Wavy Lake and Lake Ontario WB sediments, matrix solution 10^{-2} M NaCl, (c) Clearwater Lake sediments, matrix solution 10^{-1} M CaCl; (d) Windy Lake and Lake Ontario E-30 sediments, matrix solution 10^{-1} M CaCl. Solid lines reflect the Langmuir fit.

TABLE II Langmuir parameters and the values of free energy of adsorption

Sediment Sample	pH*	I** (M)	K L μmol^{-1}	Γ_M μmol g^{-1}	R^2	ΔG kJ mol^{-1}	
Wavy	5.5	10^{-1}	0.1137	151.8	0.8735	-28.5	
Wavy	7	10^{-1}	0.0612	145.2	0.8460	-26.9	
Wavy	7	10^{-4}	0.0329	118.2	0.8365	-25.4	
Clearwater	4.5	10^{-1}	0.0895	124.5	0.9189	-27.9	
Clearwater	7	10^{-1}	0.0690	102.9	0.8622	-26.9	
Windy	7	10^{-1}	0.0842	132.7	0.9004	-27.7	
L.O. E-30	7	10^{-1}	0.0308	31.3	0.8941	-25.3	
L.O. E-30	7	10^{-4}	0.0578	29.7	0.8838	-26.8	
L.O. WB	7	10^{-1}	0.0311	33.9	0.8952	-25.3	
Lake sediments		dist. water +KH_2PO_4		27.7-48.4		-11.5	Kuo and Lotse (1974)
Lake sediments	4.8-7.0	dist. water	0.033-1.036	36.3-71.6			Ku *et al.* (1978)
New Zealand soils	7	10^{-1}		10-104		-28.5- -30.1	Ryden *et al.* (1977)

* pH of matrix solution.
** I = Ionic strength of matrix solution (NaCl).

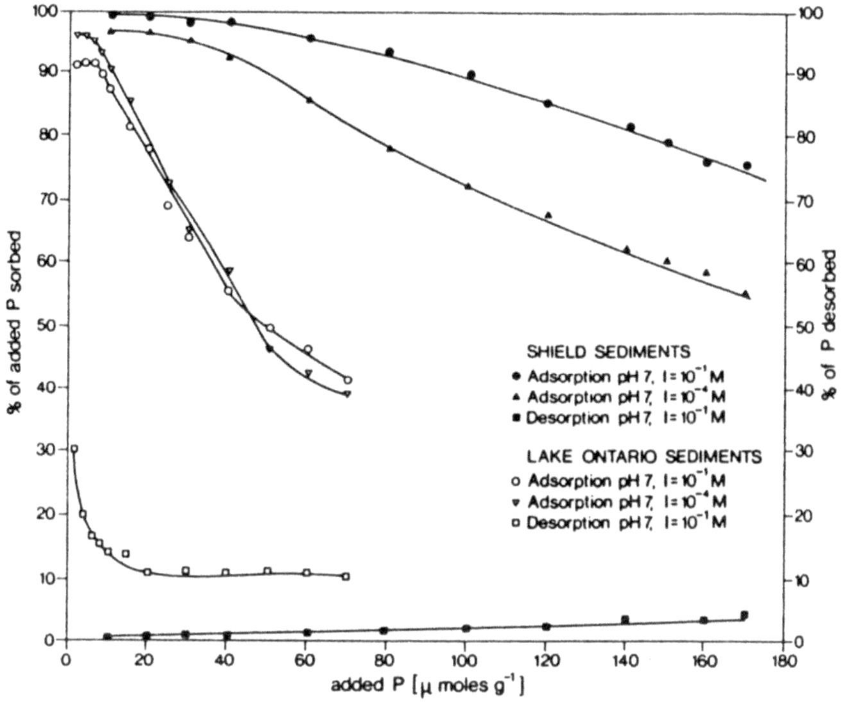

Fig. 2. Proportions of P adsorbed/desorbed at different levels of added P.

proportions of oxalate extractable Fe (Table I). Oxalate extractable Al was, however, significantly lower in sediments from Lake Ontario than that in Shield sediments (Table I). The low values of oxalate extractable Al in Lake Ontario sediments suggest that most of the Al in these sediments is associated with clay minerals. The carbonate content of Lake Ontario sediments was also higher (Mayer, 1984) than that of Shield sediments, reflecting different lithology of lake basins.

The Langmuir model was used to describe the adsorption data. The adsorption isotherms (Figure 1) for given experimental data sets were obtained by least squares fit to the standard non-linear form of the Langmuir equation (Stumm and Morgan, 1981).

Values of the Langmuir parameters (equilibrium constant adsorption, K, and adsorption capacity, Γ_M) and R^2 values are summarized in Table II, along with results from other studies for comparison.

The experiments with different lake sediments show that the sediments which adsorb more added P release less P in subsequent desorption step. The retention properties of sediments are indicated by the proportion of added inorganic P adsorbed and the proportion of P desorbed in subsequent extraction (Figure 2).

4. DISCUSSION

The adsorption data indicate discrepancy between the experimental data and the curve obtained by the least squares fit to Langmuir equation. The discrepancy may be attributed to the fact that the Langumuir model takes no account of surface charges (Parfitt, 1977; Barrow, 1978; Posner and Bowden, 1980). Thus, it describes the adsorption process better at lower concentrations. Despite this limitation, the model can be used to give an empirical description of the curves and so summarize them with two parameters which then allow ready comparison of adsorption characteristics between the sediments. The concept of an upper limit of adsorption implied from saturation of adsorption sites is a definite advantage of this model.

The results of the adsorption experiments indicate that the type and properties of sediments are reflected only in the magnitude of the sorption maxima Γ_M (Table II). The noncalcareous sediments exhibited significantly higher adsorption capacity than did the calcareous Lake Ontario sediments. This suggests that more surface sites are available for adsorption in the former than in the latter. At the phosphate concentration typical of that found in the lake, the Shield sediments adsorbed nearly 100% of added P. This is consistent with the field observations of Yan and Lafrance (1982), who reported between 72 to 86% loss of added P to the Shield sediments. At higher phosphate solution concentrations, the Shield sediments exhibited a continued linear adsorption (Figure 2), which is thought to indicate the presence of more than one type of adsorption site (Parfitt, 1977). Results with model systems containing Al and Fe oxides suggest that the phosphate is adsorbed very strongly at low solution concentration and therefore it is difficult to distinguish between two types of sites (Parfitt, 1977). At higher phosphate solution concentration, the isotherms on Al containing surfaces often show continued linear adsorption in contrast to the well defined plateau on Fe oxide surfaces (Parfitt, 1977). Thus, it is reasonable to assume that continued linear adsorption observed at higher P concentration on the Shield sediments is due to the presence of Al containing surfaces. The results of oxalate extraction (Table I) support this conclusion.

Values of the free energy of adsorption (ΔG) are remarkably similar for all the investigated sediments (Table II). Such similarity strongly suggests that the chemical and mineralogical characteristics of sediments do not affect the nature of the adsorption process. Neither do the experimental conditions (e.g., variation of pH and ionic strength I) affect the nature of the adsorption reaction. The relatively large negative values of ΔG suggest that the nature of the P uptake is strong chemisorption.

Comparison of Γ_M values (Table II and Figure 1) reveals that more P is adsorbed by the same sediment at lower pH. For example, in the case of Wavy sediment an increase in adsorption of 4% coincided with a decrease in pH of 1.5 units. For Clearwater sediment, an adsorption increase ~17% coincided with an experimental decrease in pH of 2.5 units. The observed increase in magnitude of phosphate uptake with decreasing pH is consistent with the adsorption model.

The results also show that an 18% increase in adsorption occurred in Wavy sediment when the ionic strength of solution was increased by three orders of magnitude. A smaller and perhaps insignificant increase occurred in Lake Ontario E-30 sediments. The increase in adsorption resulting from the increase in ionic strength is consistent with the electric double layer (EDL) theory.

The reversibility of adsorption evaluated from the amount of added P desorbed, indicates that calcareous Lake Ontario sediments have a lower retention ability than noncalcareous Shield sediments. Between 10 to 30% of added sorbed P was desorbed from Lake Ontario sediments, whereas only 0.3 to 4% of added P was released from the Shield lake sediments. Hence, the sediments which adsorbed the most added P tended to release the least during the subsequent desorption (Figure 2). Similar findings were reported on sediments from Wisconsin lakes by Williams et al. (1970).

In general, the relatively low desorption of added inorganic P suggests that the adsorption process is not readily reversible, provided the sediments remain oxic. This is to be expected if removal of phosphate from solution takes place by specific adsorption.

The results of the absorption study can be summarized as follows: Sediments from three Shield lakes tend to have considerably higher adsorption capacities and better retention abilities than do Lake Ontario sediments. The difference in the magnitude of adsorption caused by the shifts of pH or ionic strength is relatively small when compared with the difference in adsorption due to the chemical and mineralogical characteristics of sediments. Therefore, the sediment chemistry and mineralogy appears to be more important in determining the adsorption capacity of sediments than the changes in pH of the lakes.

ACKNOWLEDGMENTS

The assistance of Drs. R.A. Bourbonniere, D.S. Lean and B.K. Burnison is gratefully acknowledged. The authors thank Dr. P.G. Manning and Mr. M. Charlton for their review of the initial manuscript. The work formed a basis for M.Sc. thesis submitted to McMaster University, with DOE support.

REFERENCES

Barrow, N.J.: 1978, J. Soil Sci. **29** , 447.
Dobson, H.F.H.: 1980, Great Lakes Focus on Water Quality, **6**, (1), International Joint Commission, Windsor, Ontario.
Harwood, J.E., van Steenderen, R.A., and Kuhn, A.L.: 1969, Water Res. **3** , 417.
Ku, W.C., DiGiano, F.A., and Feng, T.H.: 1978, Water Res. **12** , 1069.
Kuo, S. and Lotse, E.G.: 1974, Soil Sci. Soc. Amer. Proc. **38** , 50.
Kemp, A.L.W.: 1971, J. Sediment. Petrol. **41**, 537-548.
Li, W.C., Armstrong, D.E., and Harris, R.F.: 1973, Environ. Sci. Technol. **7** , 454.
Mayer, T., and Williams, J.D.H.: 1981, Modified Procedures for Determining the Forms of Phosphorus in Freshwater Sediments. Technical Bulletin No. 119, National Water Research Institute, Inland Waters Directorate, Canada Centre for Inland Waters, Burlington, Ontario.
Mayer, T.: 1984, Phosphorus Mobility in Lacustrine Sediments upon Lake Acidification. M.Sc. Thesis, McMaster University, Hamilton.
McKeague, J.A. (ed): 1978, Manual on Soil Sampling and Methods of Analysis. Canadian Society of Soil Science.
Ontario Ministry of the Environment: 1978, Extensive Monitoring of Lakes in the Greater Sudbury Area, 1974-1976. Sudbury Environmental Study.
Ontario Ministry of the Environment: 1982, Studies of Lakes and Watershed near Sudbury, Ontario. Sudbury Environmental Study. Supplementary Volume to SES 009/82.
Parfitt, R.L.: 1977, Sco. Am. J. **41** , 1064.
Posner, A.M. and Bowden, J.W.: 1980, Journal of Soil Science. **31** , 1-10.
Ryden, J.C., McLaughlin, J.R. and Syers, J.K.: 1977, J. Soil Sci. **28** , 72.

Stumm, W. and Morgan, J.J.: 1981, Aquatic Chemistry, 2nd ed'. Wiley-Interscience, New York. 780 p.

Williams, J.D.H., Syers, J.K., Harris, R.F. and Armstrong. D.E.: 1970, Environ. Sci. Technol. **4** , 517.

Yan, N.D. and Lafrance, C.: 1982, in Studies of Lakes and Watersheds Near Sudbury, Ontario: Final Limnological Report SES 009/82, Ontario Ministry of the Environment.

SIMULATED ACID RAIN EFFECTS ON ACIDIFICATION OF A LEXINGTON SILT LOAM

B. L. Lilieholm and S. E. Feagley
Agronomy Department
Louisiana Agricultural Experiment Station
Louisiana State University Agricultural Center
Baton Rouge, LA 70803

ABSTRACT

Intact soil cores were collected to a depth of 15 cm from a Lexington silt loam. Simulated precipitation with adjusted pH values of 3.7, 4.7 and 5.7 (control), was applied to the cores in increments of 500 mL day^{-1} until totals which approximated 10, 20, 40, and 80 yr of effective rainfall in Louisiana were reached. The exchangeable acidity and Al and H_2O and KCl pH were measured from 2.5 cm sections of the cores after treatment. Only the 3.7 treatment at the 80 yr volume significantly affected the soil pH and exchangeable acidity and Al.

1. INTRODUCTION

The determination of the effects of acid precipitation on soils is a complex problem due to the numerous factors such as the type of vegetative cover, amount of past acid precipitation input, natural rates of acid formation and parent material minerology (Johnson et al., 1981; Cronan, 1980; Mollitor and Raynal, 1982). Some effects that have been shown are soil acidification, changed nutrient availability, accelerated leaching of exchangeable bases, and increased solubility of Al and other phytotoxic substances (Johnson et al., 1981; Lodhi, 1982; Cronan, 1980; Abrahamsen, 1980).

Acidity has been found to be a major factor in solution reactions in several ecosystems (Hoffman et al., 1980). Acidification of forest and aquatic ecosystems have shown reduced productivity and decomposition of litter (Overrein, 1980). Studies suggest reduced productivity of forests may occur due to depletion of nutrients caused by long term acid precipitation in the U.S. and Canada (Norton, 1976; Cronan, 1980; Mollitor and Raynal, 1982; Roberts, 1984).

Several studies have shown acidification of the soil surface over less than 2 yr under natural conditions close to SO_2 emittors (Baker et al., 1976) and under simulated conditions (Strayer and

Water, Air, and Soil Pollution **31** (1986) 959–964.

Alexander, 1981; Bitton and Baylan, 1985). Other studies have found increased leaching of Ca, Mg and Al with increased acidity of treatment solutions (Cronan and Schofield, 1971; Singh et al., 1980; Schier, 1984). The order of mobility for metal ions has been determined by to be K>Mg>Mn>Al upon application of simulated acid rain (Hovland et al., 1980).

The objective of this research was to determine the effects of acid precipitation on acidification of a Lexington silt loam (fine-silty, mixed, thermic Typic Paleudalf). This particular soil is used for agriculture and forestry in Louisiana. Therefore, its potential acidification due to acid rain is extremely important to especially the timber industry of Louisiana.

2. PROCEDURE

The Lexington silt loam is a deep, well-drained soil formed in loess 0.7 to 1.0 m thick above the underlying sandy and loamy Coastal Plains sediment. Soil cores were taken under a stand of loblolly pine at the Idlewild Research Station in Clinton, Louisiana, by placing a 5.0 cm ID PVC sleeve inside the barrel of a truck mounted Giddings hydraulic probe to a depth of 20 cm.

The cores were attached to a rack and simulated acid rain treatments of pH 3.7, 4.7 and 5.7 were applied. The treatment solutions had the same chemical composition (except for Cl and Na) as the naturally occurring rainfall of Louisiana as determined by National Atmospheric Deposition Program data from 1983 (Feagley, 1984). These ionic concentrations and the H^+ added due to the pH treatment and volume of solution are listed in Table I. The pH of the solutions was adjusted using NaOH and/or HCl.

The total volume of simulated acid rain applied to the soil cores was calculated by subtracting evapotranspiration for Clinton, 59 cm yr^{-1} (Muller and Larimore, 1975), from the average annual rainfall in Louisiana, 132 cm yr^{-1} (Feagley, 1984). Effective rainfall (surplus rainfall minus evapotranspiration) is estimated as the amount of rainfall that percolates through the soil. Treatments, replicated 3 times, were applied 5 days a week at increments of 500 mL day^{-1} until total volumes of application approximating the volume of effective rainfall in 10, 20, 40, and 80 yr had been reached (Table I). Cores were discarded if they flooded.

Each set of soil cores was sectioned into 2.5 cm increments after the total volume was applied. Samples were taken from the sections and measured for pH (1:2 in H_2O and 1:2 in 1N KCl), exchangeable acidity and Al (Soil Survey Staff, 1972).

The data was analyzed using a General Linear Model. The design was completely randomized with a split plot at depth. Duncan's Multiple Range was used for mean separation.

RESULTS AND DISCUSSION

The statistical analyses of H_2O and KCl pH and exchangeable

Table I. Ionic concentrations and volume to represent years of application of simulated precipitation leached through soil cores.

Ion*	Concentration	Yr	Volume	pH	Added H^+
	$mg\ l^{-1}$		L		$cmol(+)kg^{-1}$**
Ca^{2+}	1.36×10^{-2}	10	14.061	3.7	5.6×10^{-3}
				4.7	5.6×10^{-4}
Mg^{2+}	0.19×10^{-2}			5.7	5.6×10^{-5}
		20	28.121	3.7	11.2×10^{-3}
K^+	0.10×10^{-2}			4.7	11.2×10^{-4}
				5.7	11.2×10^{-5}
NH_4^+	1.30×10^{-2}	40	56.242	3.7	22.5×10^{-3}
				4.7	22.5×10^{-4}
NO_3^-	4.04×10^{-2}			5.7	22.5×10^{-5}
		80	112.485	3.7	45.0×10^{-3}
SO_4^{2-}	13.83×10^{-2}			4.7	45.0×10^{-4}
				5.7	45.0×10^{-5}

* Na^+ and Cl^- also in solutions. Concentrations of these species were not constant since NaOH and HCl used to adjust pH of solutions.
** $cmol\ kg^{-1}$ = centimole of charge kg^{-1}

acidity (EA) and Al (EAl) of the soil cores showed that the volume of simulated acid rain applied to the soil, pH of treatment and depth of measurement were significant for all four factors (α=0.01). Interactions were all significant except for effective acidity for the yr by pH, yr by depth and yr by pH by depth interaction (α = 0.05). Caution must be used when interpreting the H_2O and KCl pH data since all interactions except yr by pH by depth were significant. Significance of interactions may cause significant differences among the main effects when real significance does not exist.

Listed in Table II are means of H_2O and KCl pH and EA and EAl averaged over all parameters. For example, the H_2O pH means listed by yr are averaged over all pH treatments and depth, etc. The 80 yr volume of application caused the lowest pH value for H_2O and KCl pH. There was not a linear decrease in pH with treatments from 0 to 80 yr volume of application as expected since pH is a log function. The data indicated that this soil was affected even at the 10 yr volume of application, which was significantly the same as the 40 yr volume. This soil has the potential to be affected by conditions in this experiment at a 10 yr volume of simulated acid rain application, and greatly affected at a 80 yr volume. The 20 yr volume of application caused no difference in soil pH, which may be due to suttle changes in buffering capacity, mineral weathering, cation exchange capacity (CEC), base saturation, etc. in the soil.

The pH treatments caused significant differences in the data (α

Table II. Means of H_2O pH, KCl pH and exchangeable acidity and Al at volume, pH and depth.

H_2O PH						KCL PH					
VOL	MEAN	PH	MEAN	DEPTH	MEAN	VOL	MEAN	PH	MEAN	DEPTH	MEAN
yr				cm		yr				cm	
C	5.04a**	C	5.04b	2.5	4.70e	C	4.18a	C	4.18b	2.5	3.80a
10	4.82b	3.7	4.60d	5.1	4.70e	10	4.01b	3.7	3.94d	5.1	3.87b
20	5.08a	4.7	5.25a	7.6	4.77d	20	4.19a	4.7	4.26a	7.6	3.98c
40	4.77b	5.7	4.94c	10.2	4.91c	40	4.01b	5.7	4.05c	10.2	4.11d
80	4.65c			12.7	4.98b	80	3.93c			12.7	4.26e
				15.2	5.05a					15.2	4.34f

EXCHANGEABLE ACIDITY $cmol(+)kg^{-1}$						EXCHANGEABLE AL $cmol(+)kg^{-1}$					
VOL	MEAN	PH	MEAN	DEPTH	MEAN	VOL	MEAN	PH	MEAN	DEPTH	MEAN
yr				cm		yr				cm	
C	1.00c	C	1.00c	2.5	1.07b	C	0.87c	C	0.87c	2.5	0.78e
10	1.13b	3.7	1.25a	5.1	1.29a	10	0.94b	3.7	1.06a	5.1	1.05b
20	0.93d	4.7	0.76d	7.6	1.31a	20	0.81d	4.7	0.60d	7.6	1.11a
40	0.97c	5.7	1.08b	10.2	1.09b	40	0.83d	5.7	0.91b	10.2	0.95c
80	1.34a			12.7	0.92c	80	1.13a			12.7	0.85d
				15.2	0.81d					15.2	0.72f

* C stands for the control which received no treatment.
** Means with the same letter are not significantly different using Duncan's Multiple Range ($\alpha = 0.05$).

= <0.001). The pH 3.7 treatment had the lowest value and 4.7 the highest. Again, the decrease in soil pH was not linear from the control to the 3.7 pH treatment, as would be expected. The pH 3.7 treatment, however, did cause the greatest acidification of the soil.

At varying depths of the soil, the KCl pH was significantly different (α = <0.001). The pH increased steadily from 2.5 to 15.2 cm, which showed that the surface of the soil was acidified. This was mainly due to the 3.7 pH 80 yr volume treatment. The effect of treatment decreased with each increment measured. The results for H_2O pH indicated the same pattern as KCl pH except the soil pH increased from 5.1 cm.

The EA and EAl had the greatest values at the 80 yr volume of application. The EA and EAl had also been affected at the 10 yr volume treatment, which corresponded to the H_2O and KCl pH treatment effects. The 3.7 pH treatment caused the greatest EA and EAl, which was inversely correlated to the H_2O and KCl pH values as expected. It is well documented in basic soils textbooks that as pH decreases, the EA increases. The data shown at varying depths, however, was different from the H_2O and KCl data. The levels increased from the surface to 7.6 cm, then decreased from there down, slightly deeper than the H_2O and KCl pH. The H_2O and KCl pH decreased in the 0 to 7 cm section for the pH 3.7 treatment, although averages and mean separation may not have shown this. This indicated downward movement and accumulation of EA and EAl in the soil profile. This accumulation is probably due to a textural change, slight increase in clay amount, at about 7 cm, therefore a higher sink for EA. It is probable that the increase in EAl was responsible for generation of H^+ due to ionization of Al species to a depth of 7 cm, thus lowering the pH.

4. CONCLUSION

It must be stressed that the data presented in this paper cannot be directly extrapolated to situations in the environment. The experiment was designed to be an accelerated study similar to mineralogical weathering studies, to evaluate the potential effect of acid precipitation on a soil having an initial acid pH (5 to 6) and a medium CEC (6 to 8 cmol(+)kg^{-1}) and base saturation (60 to 70%). Several important components of this experiment were held as constant as possible such as organic matter content, nutrient cycling, mineral weathering and plant growth, so that a small portion of this complex environmental issue could be studied more efficiently. It is, however, recognized that these factors play a major role in soil chemical changes and mineralogical weathering, which in turn contribute to soil acidification.

The data suggested that the Lexington silt loam has the potential to be acidified at the surface of the soil upon receiving a volume of rainfall approximating 10 to 80 yr of effective rainfall at pH 3.7. This acidification occurred to a depth of 7 cm due to downward movement and accumulation of EA and EAl.

REFERENCES

Abrahamsen, G.: 1980, 'Impact of Atmospheric S Deposition on Forest Ecosystems', in D. S. Shriner, C. R. Richmond and S. E. Lindberg (eds.), Atmospheric S: Environmental Impact and Health Effects, Ann Arbor Science, Ann Arbor, MI.
Baker, J., Hocking, D., and Nyborg, M.: 1976, 'Acidity of Open and Intercepted Precipitation in Forests and Effects on Forest Soils in Alberta, Canada', Proceedings of the First International Symposium on Acid Precipitation and the Forest Ecosystem, Gen. Tech. Rep. NE-23, USDA Forest Service, Washington, D.C.
Bitton, G., and Boylan, R. A.: 1985, J.E.Q. 14, 66.
Cronan, C. S., and Schofield, C. L.: 1972, Sci. 204, 304.
Cronan, C. S.: 1980, 'Consequences of Sulfuric Acid Inputs to a Forest Soil', in D. S. Shriner, C. R. Richmond, and S. E. Lindberg (eds.), Atmospheric Sulfur Deposition, Ann Arbor Science, Ann Arbor, MI.
Feagley, S. E.: 1984, La. Agric. 27, 24.
Hoffman, W. A. Jr., Lindberg, S. E., and Turner, R. R.: 1980, J.E.Q. 9, 95.
Hovland, J., Abrahamsen, G., and Ogner, G.: 1980, Plant and Soil 56, 365.
Johnson, A. H., Siccama, T. G., Wang, D., Turner, R. S., and Barringer, T. H.: 1981, J.E.Q. 10, 421.
Lodhi, M. A. K.: 1982, Am. J. of Botany 69, 474.
Mollitor, A. V. and Raynal, D. J.: 1982, Soil Sci. Soc. Am. J 46, 137.
Muller, R. and Larimore, P. B. Jr.: 1975, Publications in Climatology 28, 31.
Norton, S. A.: 1976, 'Changes in Chemical Processes in Soils Caused by Acid Precipitation', Proceedings of the First International Symposium on Acid Precipitation and the Forest Ecosystem, Gen. Tech. Rep. NE-23, USDA Forest Service, Washington, D.C.
Overrein, L. N.: 1980, 'Acid Precipitation - Effects on Forest and Fish', Final Report of SNSF Project, Norweigan Council for Science and Industrial Resources Research, Oslo, Norway.
Roberts, L.: 1984, Bioscience 34, 291.
Schier, G. A.: 1984, 'Growth of Pitch Pine Seedlings in Soil Cores from the New Jersey Pine Barrens Treated with Simulated Acid Rain', Proceedings of North American Forest Biology Workshop, Logan, UT.
Singh, B. R., Abrahamsen, G., and Stuanes, A.: 1980, Soil Sci. Soc. Am. J. 44, 75.
Soil Survey Staff. 1972, 'Soil Survey Laboratory Methods and Procedures for Collecting Soil Samples', Soil Survey Investigation Report No. 1, USDA Soil Conservation Service, Washington, D.C.
Strayer, R. F. and Alexander, M.: 1981, J.E.Q. 10, 960.

EFFECTS OF ACID RAIN ON ION LEACHING IN A DANISH FOREST SOIL

N. E. v. Freiesleben and L. Rasmussen
Laboratory of Environmental Sciences and Ecology
Technical University of Denmark
Building 224, DK 2800 Lyngby
Denmark

ABSTRACT. The mobility of major cations (H^+, ammonium, Al, Ca, Na, Mg, K, Fe), heavy metals (Mn, Zn, Ni, Cd) and anions (chloride, sulphate and nitrate) was studied in the laboratory in an acidified brown soil from a Norway spruce forest. Lysimeters containing undisturbed soil columns of the A-horizon and the A- plus B-horizon were watered with 540 mm of throughfall precipitation collected *in situ*, either directly (pH 3.6) or adjusted to pH 3.3 or 2.8. The pH 3.3 treatment increased leaching of Mn and Cd from the B-horizon. The pH 2.8 treatment increased leaching of ammonium, Na, Ca, Mg, K, Mn, Zn and Cd from the A-horizon and ammonium, Al, Na, Ca, Mg, K, Mn, Zn and Cd from the B-horizon. Fe leaching from the A-horizon was decreased by both acidic treatments, and the pH of the leachates was not significantly affected. Sulphate retention was 138-161 meq m^{-2} yr^{-1} by all treatments. Due to experimental conditions nitrate leaching was observed in all lysimeters.

1. INTRODUCTION

Acid precipitation accelerates the export of cations from terrestrial ecosystems and increases soil mineral weathering, especially so in natural forest soils not exposed to liming and fertilization (Tyler, 1978; Singh *et al*., 1980; Lee and Weber, 1982). The risk of depletion of the nutrient pool was pointed out by Mayer and Ulrich (1977), who especially stressed Mg deficiency as a possible cause of forest die-back. Zöttl (1985) considered the deficient nutrient supply as the dominant stress factor causing forest damage. Root damage due to high concentrations of Al has also been reported (Ulrich, 1983; Matzner and Murach, 1986).

The objective of the present study was to characterize short term effects on the ion mobility in an acidic Danish forest soil exposed to increased acid precipitation applicated at natural precipitation rates.

2. MATERIALS AND METHODS

The soil was an acidic, slightly podzolized brown soil developed on a

Water, Air, and Soil Pollution **31** (1986) 965–968.

low hill of sandy glacial till at the Strødam-Reserve, Gribskov Forest, Zealand, Denmark. The soil was deep and freely drained. The forest plantation consisted of even-aged 40 yr old Norway spruce (*Picea abies*), which for research purposes has been without forest management for the last 20 yr. There was no forest floor herb vegetation. Soil characteristics are shown in Table I. The soil horizons were as follows: A_o (0 to 3 cm) mostly litter, A_1 (3 to 11 cm) black, A_2 (11 to 23 cm) grayish, B_1 (23 to 28 cm) dark brown, B_2 (28 to 55 cm) yellowish-red, C (55- cm) yellowish.

TABLE I
Soil characteristics.

Depth	(cm)	5	15	45	55
pH, H_2O		3.7	3.7	4.0	4.1
pH, $CaCl_2$		3.1	3.0	3.7	3.8
Loss on ignition	(%)	4.3	3.8	3.1	3.0
Clay	(%)	3.9	4.0	2.4	3.6
Silt	(%)	6.7	5.4	8.8	1.1
Fine sand	(%)	41.2	35.2	36.8	34.4
Coarse sand	(%)	48.2	55.4	52.0	60.9
CEC	(meq/100g dry soil)	42.0	27.4	33.1	26.1
Base saturation	(meq/100g dry soil)	18.0	5.6	5.7	7.9

Soil cores were taken vertically from the surface with a steel cylinder and transferred to zero-tension column lysimeters without disturbing the stratification. The lysimeters were constructed in plexiglass in order to avoid contamination with heavy metals. The inside diameter was 290 mm and the length was either 23 cm, comprising the A_o, A_1 and A_2 horizons (9 lysimeters), or 55 cm comprising the A plus the B_1 and B_2 horizons (9 lysimeters). The soil core rested on a polyethylene net and a perforated plexiglass bottom plate. The leachate was led to a polyethylene collecting flask through a polyethylene funnel. The lysimeters were wrapped in aluminium foil to prevent light induced microbial activity. The experiment was run at 20^oC. The applicated natural throughfall was collected throughout the year on large plastic sheets mounted between the trees in Gribskov Forest. It was applied either directly or adjusted to pH 3.3 or 2.8 with a mixture of H_2SO_4 and HNO_3 (1:1, molar basis). For each treatment and both soil depths three replications were run. The apparatus for water application was composed of a cylindrical plexiglass vessel of the same diameter as the lysimeter (Freiesleben *et al.*, 1987). Throughfall water was supplied by a peristaltic pump and polyethylene tubing, giving a slight over-pressure forcing the water through 90 small polyethylene tublets mounted in the bottom plate of the vessel. 540 mm of throughfall was applicated during 9 mo as daily showers of 15 mm during 4 successive days once every month. In the intervening period the lysimeters were subjected to natural evaporation. The mean annual precipitation (open field) at the study site is 550 mm, of which 350 mm reach the forest floor as throughfall (Freiesleben *et al.*, 1986). The mean leachate volume was 67% (A-horizon) and 64% (A+B-horizon) of the applicated throughfall volume, the rest being lost by evaporation between the treatments. The water fluxes were constant throughout the experiment. Leachates from the A-horizon were slightly colored (brown) whereas those from the A+B-horizon were clear and colorless.

Analysis of Al, Na, Ca, Mg, K, Fe, Mn, Ni, Cd and Zn was performed by AAS. Chloride, nitrate and sulphate was determined by ion chromatography (Dogan and Hærdi, 1981) and ammonium was analyzed for using an ammonia electrode.

3. RESULTS AND DISCUSSION

The data in Table II represent the mean annual fluxes calculated from the whole study period. The ion budgets of the soil when natural unadjusted throughfall was applied (pH 3.6, volume weighted mean), showed a significant accumulation of H^+, K, Mn and SO_4 in the A-horizon. In the B-horizon H^+, NH_4, K and SO_4 was accumulated. The ion budgets of the soil were negative, showing a loss of NH_4, Al, Fe, Cd and NO_3 from the A-horizon, and of Al, Ni, Cd and NO_3 from the B-horizon. The remaining ion fluxes did not show significant changes by passage of the soil horizons (Na, Ca, Mg, Zn and Cl).

The experimental acidification (pH 3.3 and 2.8) in this short-term study caused significant changes in ion leaching, even though the applied volume of water only corresponded to 18 mo of the natural throughfall.

The mean proton fluxes of both the A-horizon and the B-horizon percolates did not change significantly by the three different application treatments. For all treatments the pH decreased during the experiments, but this decrease was equivalent to the formation of H^+ from nitrification processes. However, long-term experimental application of acid rain to soil column lysimeters has by others been shown to cause significant decrease in soil water pH (Stuanes, 1980; Bjor and Teigen, 1980). In this study the increased input of H^+ was completely consumed in the A-horizon. The following proton consuming processes must be taken into account: Mineralization of cations in organic matter, weathering of inorganic minerals, protonation and precipitation of anions, reduction reactions and cation exchange.

Increased leaching of base cations as an effect of acid precipitation has been shown in numerous studies (Abrahamsen, 1980; Abrahamsen and Stuanes, 1980; Cronan, 1980; Farrell et al., 1980; Morrison, 1983; Bergkvist, 1986). The mobility of Zn, Mn, Cd and other trace metals has been reported to increase with soil acidity as well (Esser and El Bassam, 1981; Brümmer and Herms, 1983; Scokart et al., 1983; Bjor and Teigen, 1980). Even minor pH decreases have been shown to increase the leaching of Zn, Cd, Ni and Mn significantly from both metal-polluted and unpolluted mor layers (Tyler, 1978).

In the present study the pH 3.3 treatment, which is only slightly more acidic than natural throughfall (pH 3.6), caused a significant decrease in Fe leaching from the A-horizon, but also a significant increase in Mn and Cd mobility in the B-horizon. Increased leaching from both soil horizons were observed for Na, Ca, Mg, K, Zn, Mn and Cd by the pH 2.8 treatment. The cation leaching increase from the A-horizon was evi-

TABLE II
Mean yearly ion fluxes in applicated throughfall and lysimeter leachates.

	No pH adjustment			pH 3.3			pH 2.8		
	Throughfall	A-horizon percolate	B-horizon percolate	Throughfall	A-horizon percolate	B-horizon percolate	Throughfall	A-horizon percolate	B-horizon percolate
H^+	96	44	28	175	49	25	555	57	29
$NH4^+$	96	350	32	96	260	38	96	516*	64*
Al^{3+}	21	174	446	21	177	375	21	174	553*
Na^+	125	124	128	125	112	133	125	162*	195*
Ca^{2+}	114	96	116	114	77	106	114	142*	201*
Mg^{2+}	73	63	66	73	52	60	73	96*	122*
K^+	82	58	13	82	49	19	82	74*	34*
Fe^{3+}	4.6	41	1.2	4.6	15*	1.2	4.6	19*	1.5
Mn^{2+}	8.0	3.4	3.5	8.0	4.3	6.7*	8.0	8.5*	9.1*
Zn^{2+}	1.4	1.3	1.6	1.4	1.2	1.2	1.4	3.0*	4.0*
Ni^{2+}	0.076	0.057	0.15	0.076	0.059	0.14	0.076	0.073	0.17
Cd^{2+}	0.009	0.013	0.029	0.009	0.017	0.051*	0.009	0.026*	0.062*
Total cations	621	955	835	700	797	765	1080	1252	1213
Cl^-	169	165	142	169	150	141	169	178	166
$NO3^-$	80	335	488	120	254	373	309	374	657
$SO4^{2-}$	321	291	183	361	267	200	551	501	397
Total anions	570	791	813	650	671	714	1029	1053	1220
Anion deficit	51	164	22	50	126	51	51	199	-7

Fluxes: meq $m^{-2}yr^{-1}$ (1 yr correspond to 350 mm of throughfall precipitation).
Standard deviations were 5-40% (ammonium s.d. was 10-80%).
*: Significantly different from unadjusted value ($p<0.01$, Students T-test).

dent from the very first sampling after treatment start, indicating that fast processes were involved, but both cation exchange capacity (CEC) and base saturation was unaffected by the acidification treatment. Probably cation exchange processes were involved, and base saturation was restored by weathering processes. Soil pH was below 4.0 which means that soil buffering processes included dissolution of Al-rich clay minerals or Al-salts (Al_2O_3, $AlOHSO_4$, $Al(OH)_3$) (Ulrich, 1983). Al was the dominant cation in the percolates from the B-horizon. The leaching of Al increased by the most acidic treatment (pH 2.8), and increased also during the course of the experiment due to internal acidification processes.

Fe and Al are known to form stable complexes with dissolved organic compounds in the A-horizon. These are precipitated in the B_1-horizon (Petersen, 1976). The high organic content of the A-horizon percolates was reflected by their brown color and the large anion deficit. The decrease in Fe mobility in the A-horizon caused by the acidification treatments may be due to changes in organic complexation.

Retention of SO_4 in the soil was observed by all treatments, and was most pronounced in the B-horizon. Quantitatively the retention was 138 to 161 meq m^{-2} yr^{-1} by all treatments, but with increasing SO_4-load a smaller proportion was retained, 45 and 28% for the pH 3.3 and 2.8 treatment, respectively. A similar constancy of accumulation has been observed in both less and more acidified catchments in Central Europe, where limits of variation were 79 to 99 meq m^{-2} yr^{-1} (Paces, 1985).

The soil column lysimeter technique means cutting off all living tree roots. Root uptake is thus excluded and the pool of elements in the dead organic material will be mineralized and exposed to potential leaching. This situation is similar to clear cutting of a forest stand under natural conditions. The result is high organic contents in the soil solutions and nitrification causing an additional internal acid input. In this case 408, 253 and 348 meq m^{-2} yr^{-1} for the three treatments respectively. In an evaluation of soil column experiments, it is necessary to take these experimental artifacts into account. Especially the nitrogen compounds are highly affected by biological activities. Microbial decomposition of organic material produces NH_4 in the A-horizon, but soil sorption, microbial assimilation and nitrification result in very low B-horizon output. Nitrification and lack of soil retention cause the NO_3 flux to increase with soil depth. By the pH 2.8 treatment only a minor part of the NH_4 in the A-horizon was nitrified. This was reflected in a high NH_4 flux, and, in comparison with the high NO_3 input with throughfall, a relatively low NO_3 output from the A-horizon. The nitrification inhibition might be caused by aluminium toxicity.

Summarizing, Mn and Cd seemed in the present study to be the elements most succeptible to leaching by increased proton input. Mn mobility might, however, have been generally increased in this study because of the lacking tree root uptake and by lowering of the redox potential in water saturated micro sites or by high microbial activity (Tyler, 1981). However, the rapid mobility changes of Mn and Cd might be used as an early warning of acidification changes in the soil chemical status.

4. ACKNOWLEDGEMENTS

This research was carried out with financial support from the Commission of the European Communities (Grant ENV-892-DK), The Danish EPA and The Ministry of Energy.

REFERENCES

Abrahamsen, G.: 1980, in Drabløs, D. and Tollan, A. (eds.), Proc. Int. Conf. Ecol. Impact Acid Precip., Norway 1980, SNSF project, p 196.

Abrahamsen, G. and Stuanes, A. O.: 1980, in Drabløs, D. and Tollan, A. (eds.), Proc. Int. Conf. Ecol. Impact Acid Precip., Norway 1980, SNSF Project, p 152.

Bergkvist, B.: 1986, Water, Air, and Soil Pollut. (in Press).

Bjor, K. and Teigen, O.: 1980, in Drabløs, D. and Tollan, A. (eds.),Proc. Int. Conf. Ecol. Impact Acid Precip., Norway 1980, SNSF Project, p 200.

Brümmer, G. and Herms, U.: 1983, in Ulrich, B. and Pankrath, J. (eds.), Effects of Accumulation of Air Pollutants in Forest Ecosystems, D. Reidel Publishing Company, Dordrecht, p 233.

Cronan, C. S.: 1980, Oikos 34, 272.

Dogan, S. and Hærdi, W.: 1981, Chimia 35, 339.

Esser, J. and El Bassam, N.: 1981, Environ. Pollut. (Ser. A) 26, 15.

Farrell, E. P., Nilsson, I., Tamm, C. O. and Wiklander, G.: 1980, in Drabløs, D. and Tollan, A. (eds.), Proc. Int. Conf. Ecol. Impact Acid Precip., Norway 1980, SNSF Project, p 186.

Freiesleben, N. E. v., Ridder, C. and Rasmussen, L.: 1986, Water, Air, and Soil Pollut. (in press).

Freiesleben, N. E. v., Rasmussen, L. and Jørgensen, P.: 1987, Geoderma (in prep.).

Lee, J. J. and Weber, D. E.: 1982, J. Environ. Qual. 11, 57.

Matzner, E. and Murach, D.: 1986, Water, Air, and Soil Pollut. (in press)

Mayer, R. and Ulrich, B.: 1977, Water, Air and Soil Pollut. 7, 409.

Morrison, I. K.: 1983, in Ulrich, B. and Pankrath, J. (eds.), Effects of Accumulation of Air Pollutants in Forest Ecosystems, D. Reidel Publishing Company, Dordrecht, p 195.

Paces, T.: 1985, Nature 315, 31.

Petersen, L.: 1976, Podzols and Podzolisation, DSR Forlag, Copenhagen.

Scokart, P. O., Meeus-Verdinne, K. and de Borger, R.: Water, Air, and Soil Pollut. 20, 451.

Singh, B. R., Abrahamsen, G. and Stuanes, A.: 1980, Soil Sci. Soc. Am. J. 15, 137.

Stuanes, A. O.: 1980, in Drabløs, D. and Tollan, A. (eds.), Proc. Int. Conf. Ecol. Impact Acid Precip., Norway 1980, SNSF Project, p 198.

Tyler, G.: 1978, Water, Air, and Soil Pollut. 9, 137.

Tyler, G.: 1981, Water, Air, and Soil Pollut. 15, 353.

Ulrich, B.: 1983, in Ulrich, B. and Pankrath, J. (eds.), Effects of Accumulation of Air Pollutants in Forest Ecosystems, D. Reidel Publishing Company, Dordrecth, p 127.

Zöttl, H. W.: 1985, Düsseldorfer Geobot. Kolloq. 2, 31.

GEOCHEMISTRY OF AQUATIC AND TERRESTRIAL SEDIMENTS, PRECAMBRIAN SHIELD OF SOUTHEASTERN ONTARIO

E.H.W. Hornbrook, I.M. Kettles, W.W. Shilts
Geological Survey of Canada
601 Booth Street, Ottawa, Ontario
Canada K1A 0E8

ABSTRACT

Lake water and sediment samples from approximately 2200 lakes and glacial sediment (sub-solum) samples from about 1800 sites were collected throughout a 38000 km^2 rectangular area extending from Georgian Bay east to the Ottawa and St. Lawrence Rivers, Ontario, Canada. Lake water alkalinity and pH patterns are similar to the distribution of carbonate components in glacial drift. Carbonate-rich drift derived from the Paleozoic limestone terrain on the northeast flank of the Precambrian Frontenac Arch has been dispersed in a southwestward direction across a variety of non-calcareous metasedimentary and igneous rocks of the Canadian Shield, providing a buffering capacity to lakes situated in granitic terrain. The distribution patterns of mobile trace and minor elements are influenced by geochemical processes associated with subaerial weathering, ground and surface water transport, and the geochemical environment within the lakes themselves. Although composition of the drift is generally reflected by lake geochemistry, these post depositional processes can cause significant variations between patterns derived from the two sample types. Anions and cations such as $SO_4^=$, Cl^-, Na^+, and F^- exhibit concentration patterns thought to reflect both anthropogenic inputs and natural variations due to differences in the geology. All regional geochemical patterns may show evidence of local enhancement caused by high concentrations of chemically distinctive minerals in drift or nearby bedrock.

1. INTRODUCTION

Water and sediment samples from approximately 2,200 lakes and glacial sediment (sub-solum) samples from about 1800 sites (Figure 1) were collected from the Precambrian Frontenac Arch, and from an adjacent 100 km wide strip of the Canadian Shield, extending from Thousand Islands on the St. Lawrence River to Georgian Bay (Figure 2).

Water, Air, and Soil Pollution **31** (1986) 969–979.

These samples were analyzed to determine concentrations of over 40 trace, minor and major chemical components including the content of clay-sized particles and carbonate minerals with potential buffering capacity. The objectives of the project were to map the regional distribution of environmentally critical components of the aquatic and terrestrial systems in order to: 1) determine to what extent patterns are related to the complex bedrock and glacial geology of the region; 2) establish regional trends that may be related to anthropogenic input as opposed to natural variations; 3) determine to what extent the composition of the aquatic elements is influenced by the composition of the terrestrial elements; 4) determine to what extent the aquatic chemistry is controlled by local atmospheric and hydrogeological factors; and 5) determine to what extent the chemical components of the sample media reflect known and probable mineralization in the bedrock and overlying glacial material.

2. METHODS

Glacial sediments, primarily till, were sampled from below the postglacial solum at easily accessible man-made exposures along public and private roads and from natural exposures along streams and rivers. In the Ottawa River valley and the St. Lawrence River valley, marine silty clay and proglacial lacustrine silty clay, which form an almost continuous surficial cover in lower areas, were sampled. The clay (<2μm) size fraction was removed from each of these samples by centrifugation. The silt-clay (<64μm) fraction was removed by sieving from a separate, dried sample split. Trace and minor element concentrations were determined on the clay fraction by Atomic Absorption using a hot HCl/HNO_3 leach (Cu, Pb, Zn, Co, Ni, Ag, Cr, Fe, Mn, Cd), colorimetric methods (As), and fluorimetric methods (U). Carbonate determinations were made on the <64μm fraction using a Leco Carbon Analyzer. The difference between C present in leached and unleached splits of the same sample was converted to $CaCO_3$ equivalent weight % using a formula modified from Foscolos and Barefoot (1970).

Open File 947 lists some of the geochemical data for till and other surficial sediments in this area. Compositonal data similar to those described in Open File 947 are now routinely collected in regional surveys of drift composition, most of which are carried out primarily for mineral exploration purposes.

Lake sediment and water samples were collected utilizing helicopter support at an average density of one sample site per 13 km^2. Water samples were collected at least 25 cm below the water surface. Lake sediments were collected from profundal lake basins using a hollow pipe-like sampling apparatus and the upper 10 to 15 cm of the sediment core were discarded. The water samples were analyzed without any preparation but the field dried lake sediment samples were ball milled and sieved to minus 177μm. Appropriate quality control standards, field and analytical duplicates were incorporated into the analytical units.

The lake sediment and water data for this area was published in the form of 1:250,000 scale geochemical value maps in Geological Survey of Canada (GSC) Open Files 899 and 900. Each of these files consists of 42 maps of geochemical variables: 28 for lake sediments (Zn, Cu, Pb, Ni, Co, Ag, Mn, As, Mo, Fe, Hg, LOI (500°C), U, CO_2, C (organic), S, Sr, Cr, SiO_2, Al_2O_3, Fe_2O, CaO, MgO, K_2O, P_2O_5, TiO_2, BaO and LOI (900°C)) and 14 for lake waters (U, F, pH (laboratory), Na, K, Ca, Mg, C (organic), Cl, NO_3, SO_4, alkalinity (equivalent $CaCO_3$), conductivity and dissolved O_2). These open files also include one sample site location map and a text of field measurements, analytical and statistical data. Similar surveys with 15 to 20 variables have been carried out across Canada totalling approximately 1,400,000 km^2, primarily in support of the mineral exploration industry and mineral commodity assessment studies, but the data include a number of environmentally significant parameters such as Hg, U, F, As, pH etc.

3. RESULTS AND DISCUSSION

Regional distribution of buffering components, such as carbonate minerals (Figure 3), in samples of glacial drift is closely mimicked by patterns of alkalinity, pH (Figures 4 and 5) and other carbonate-related parameters in both lake waters and lake sediments. The overall pattern of concentrations of components related to carbonate minerals reflects both the overprint of southwestward glacial dispersal of debris from the Paleozoic limestone terrain that flanks the Frontenac Arch along its northeastern side and the more limited dispersal of marble debris from outcrops on the Frontenac Arch. High alkalinities measured in lakes underlain by bedrock that generally provides few components capable of buffering the effects of acid rain, demonstrate how glacial processes can distort the mineralogical/chemical signature of local bedrock by transporting carbonate-rich debris over areas much larger than those of the source outcrops. It is important to recognize that aquatic and terrestrial systems in areas thought to be sensitive to acidic precipitation on the basis of bedrock geology alone, cannot necessarily be evaluated without considering the composition and distribution of the drift cover.

Comparison of concentration patterns for trace and minor elements between the clay-sized (-2μm) fraction of drift and the -177μm fraction of lake sediments demonstrates that element mobility is related to a complex mix of influences of drift mineralogy/chemistry, geochemical processes associated with overland or groundwater transport, and local geochemical environment of lakes from which samples were collected. For example, U distribution in the clay size fraction of drift samples (Figure 6) bears only vague resemblance to its distribution in lake waters (Figure 7). Comparison of patterns of regional alkalinity, drift carbonate content, and concentration of U in lake waters suggests high solubility and hydromorphic transport of U in high bicarbonate waters. Elevated concentrations of U in lake waters mimmick almost exactly regional patterns of 1) high alkalinity, 2) elevated pH in surface lake waters, and 3) high carbonate content in drift.

Uranium in lake bottom sediments (not shown here), on the other hand, more closely relates to regional distribution of U in drift. This indicates that U is not so mobile in non-carbonate regions and that it is being fixed in lake bottom sediments which are composed largely of organic material with minor contribution of silt and clays from surrounding glacial deposits. However, where bicarbonate waters are present, little U is fixed in lake sediment and it may be transported great distances from its source.

Arsenic, which is somewhat less affected by these processes, has concentration patterns that are broadly similar for both drift and lake bottom samples (Figures 8 and 9). The As map for drift clearly delineates strong anomalies over the meta-sedimentary and metavolcanic rocks that host the arsenical Au of the old Ontario gold mining belt. Lake sediment As reflects the eastern limb of the belt. The source of major lake sediment anomalies in the Tweed area is presently unknown, but may in part reflect contamination by arsencial byproducts of Au ore processing formerly carried out in the region.

Comparison of patterns of concentrations of trace and minor elements between terrestrial sediments and aquatic sediments and waters shows that the less mobile elements such as As (see Figures 8 and 9) tend to be concentrated in lakes in patterns corresponding to those for the terrestrial system. Conversely, the more mobile elements, Fe and Mn particularly (not shown), are concentrated in lakes in patterns that reflect local hydrochemical factors. A third group of anions and cations (such as $SO_4^=$ (shown in Figure 10)) and Cl^-, Na^+, and F^- (not shown), exhibit anomalous concentration patterns that may reflect anthropogenic inputs from acid precipitation or road salt, as well as natural variation due to differences in geology. For example, many of the areas characterized by higher sulphate concentrations are underlain by sulphide-rich bedrock.

4. CONCLUSION

In this necessarily brief and preliminary discussion, we have attempted to emphasize the dominant role played by bedrock and glacial geology in producing variations in geochemical and mineralogical patterns in environmentally significant surficial deposits of the Canadian Shield. We have demonstrated how both natural post or syndepositional processes and anthropogenic inputs may alter or otherwise influence the primary geological compositional signatures of the landscape. It is our opinion that the effects of the former processes must be evaluated more carefully than has heretofore been done with reference to the primary geological characteristics of any landscape for which an assessment of anthropogenic effects is being carried out.

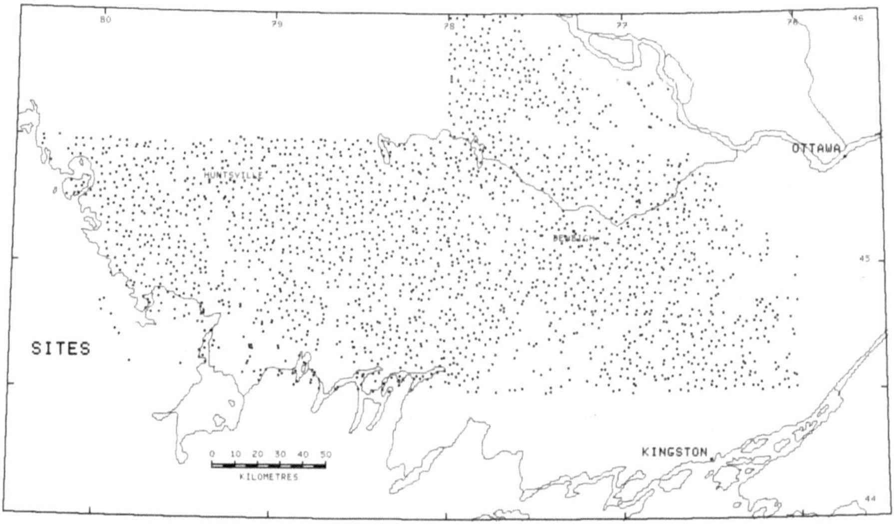

Figure 1 Map of lake bottom sediment and lake water sample sites. Over 1800 till sammples were collected within the area outlined with a dotted line.

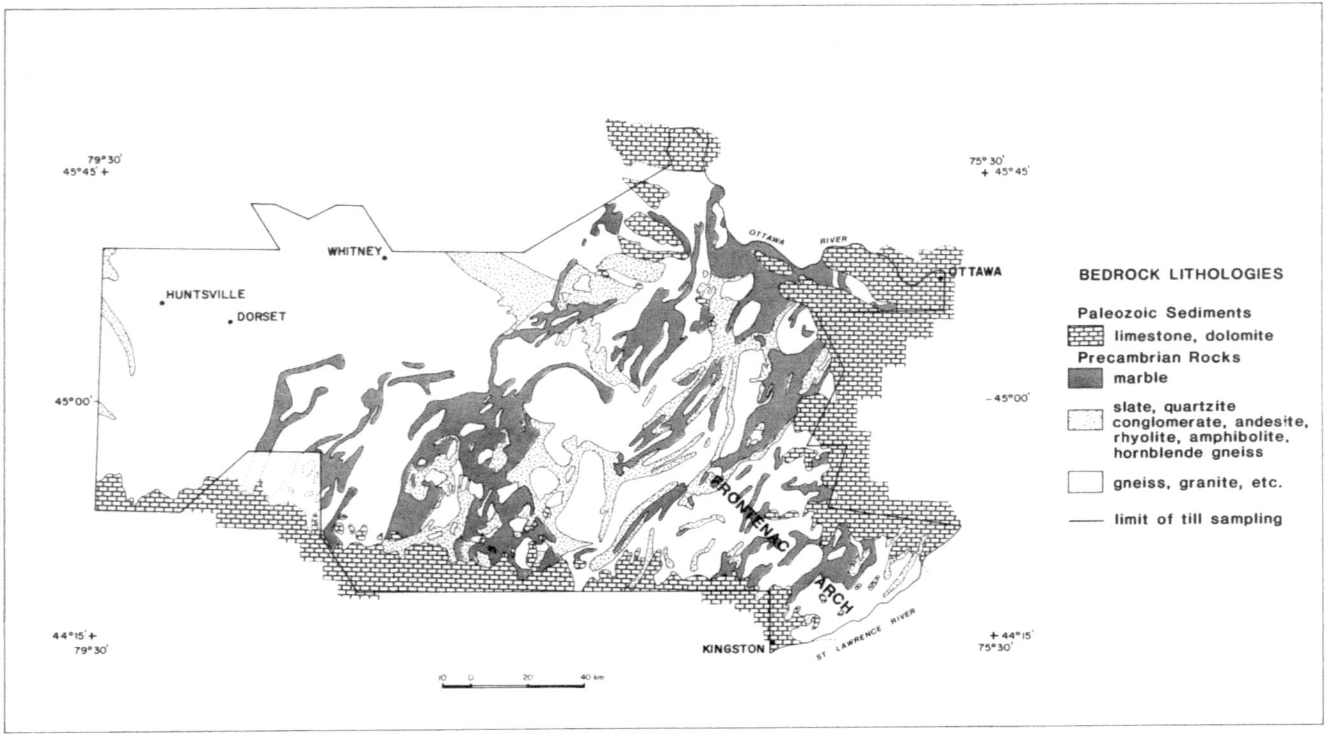

Figure 2 Map of bedrock geology of the Frontenac Arch area, Ontario, Canada.

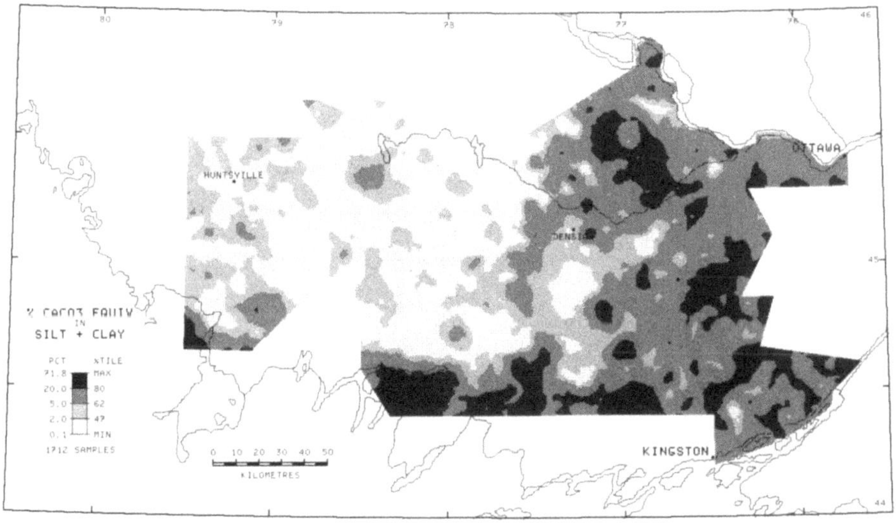

Figure 3 Map of carbonate concentrations (%$CaCO_3$ equivalent in $\leq$ 63 µm fraction) in till and other glacial sediments.

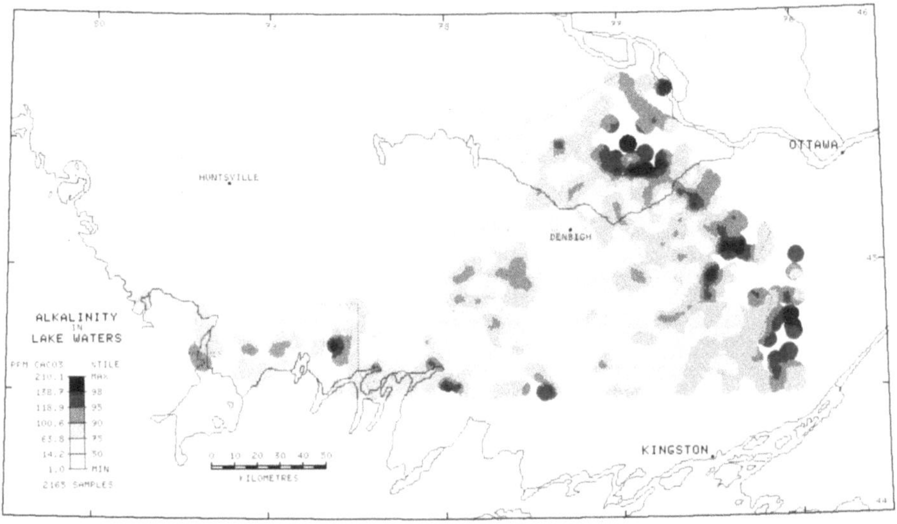

Figure 4 Map of alkalinity (as ppm $CaCO_3$) in lake waters.

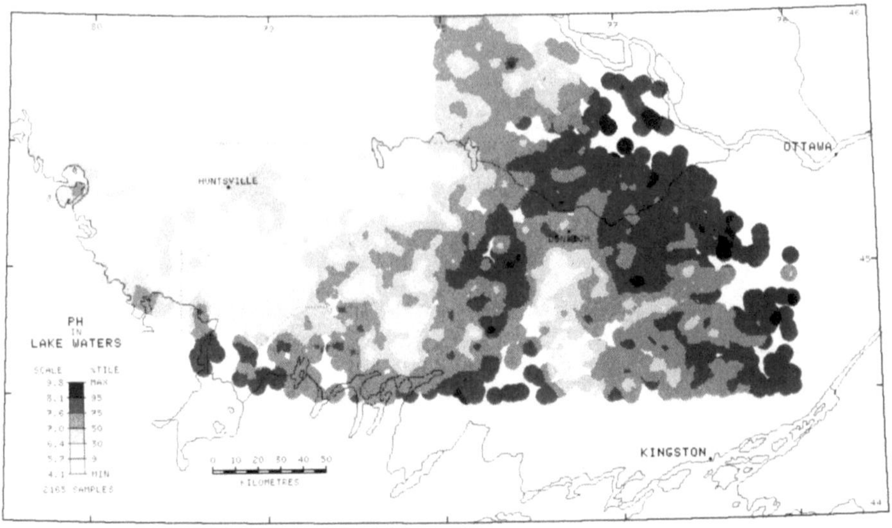

Figure 5 Map of pH levels in lake waters.

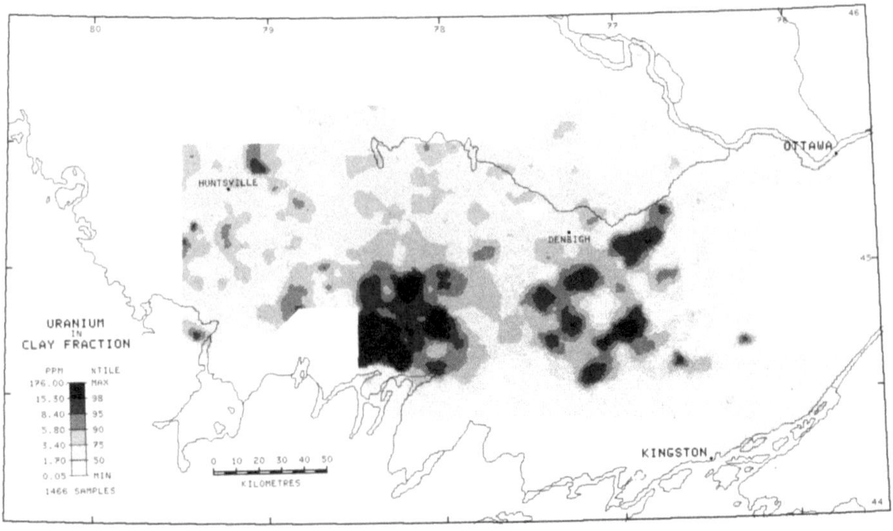

Figure 6 Map of U concentrations in till and other glacial sediments ($\leq$ 2 μm fraction).

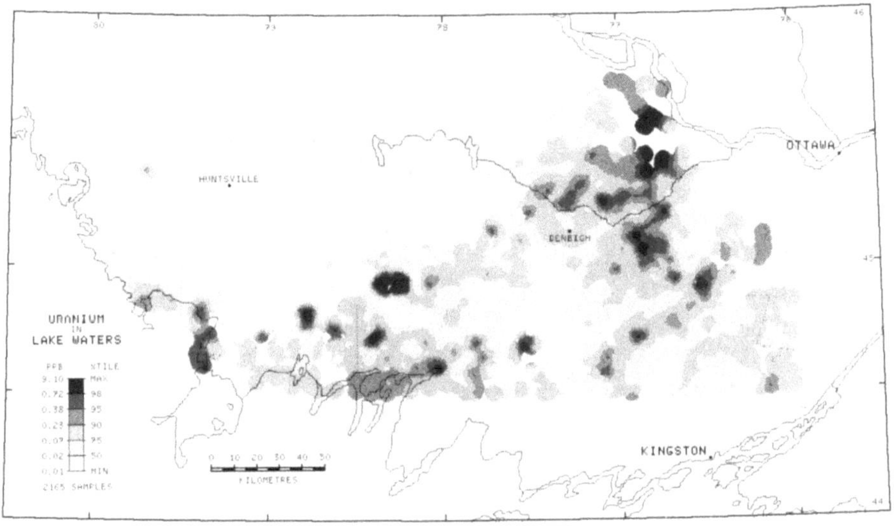

Figure 7 Map of U concentrations in lake waters.

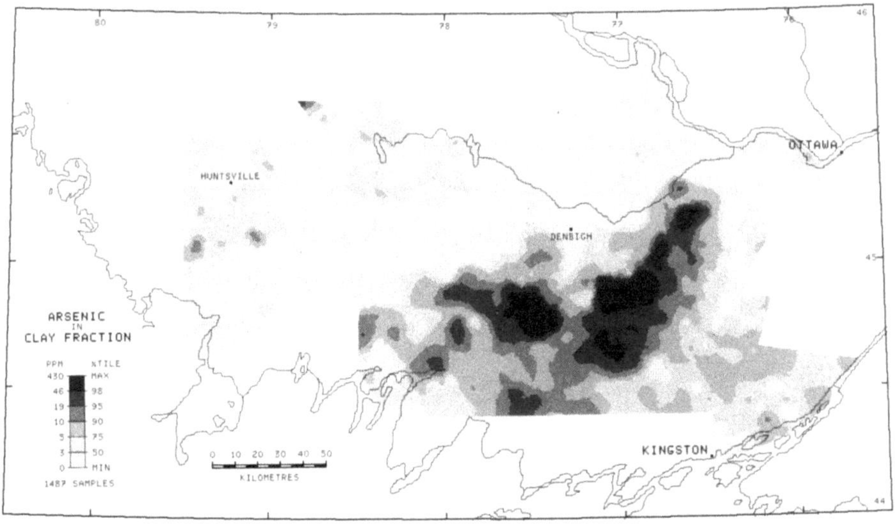

Figure 8 Map of As concentrations in till and other glacial sediments ($\leq$ 2 μm fraction).

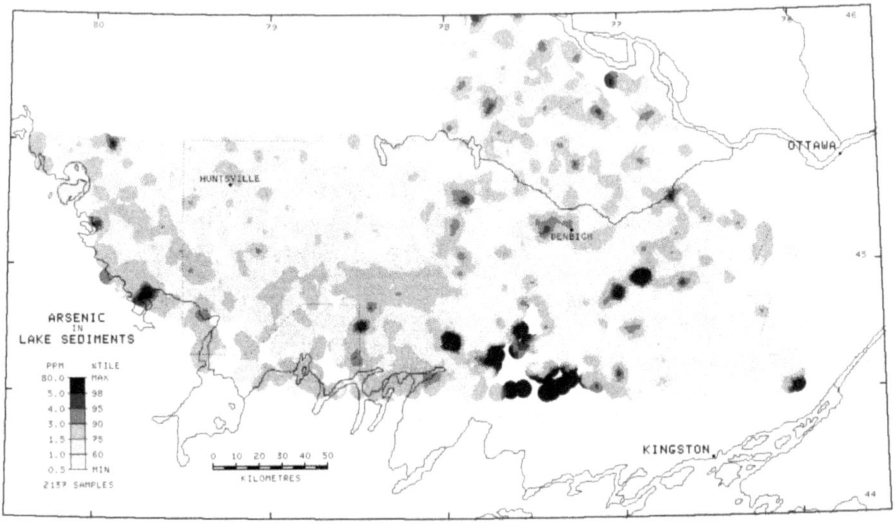

Figure 9 Map of As concentrations in centre lake bottom sediments ($\leq$ 177 µm fraction).

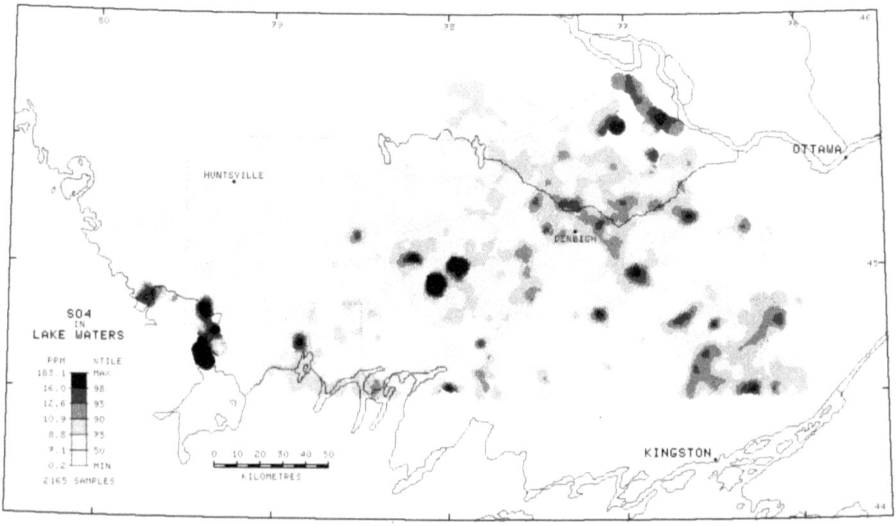

Figure 10 Map of $SO_4^=$ concentrations in lake waters.

REFERENCES

Foscolos, A.E. and Barefoot, R.R.: 1970, A rapid determination of total organic and inorganic carbon in shales and carbonates; Geological Survey of Canada, Paper 70-11.

Hornbrook, E.H.W., Lund, N.G. and Lynch, J.J.: 1984, Regional lake sediment and water geochemical reconnaissance data, Ontario 1982, parts of NTS 31C(S/2) and 31F (Ontario only); Geol. Surv. Can., Open File 899.

Hornbrook, E.H.W., Lund, N.G. and Lynch, J.J.: 1984, Regional lake sediment and water geochemical reconnaissance data, Ontario, 1982, parts of NTS 31D, 31E, and 41H; Geol. Surv. Can., Open File 900.

Kettles, I.M. and Shilts, W.W.: 1983, Reconnaissance geochemical data for till and other surficial sediments, Frontenanc Arch and surrounding area, Ontario; Geol. Surv. Can., Open File 947.

KINETICS AND MECHANISM OF MINERAL DISSOLUTION IN A SOIL AT pH VALUES BELOW 4

J.J.M. van Grinsven, G.D.R. Kloeg and W.H. van Riemsdijk
Department of Soil Science and Geology and the Department of Soil Science and Plant Nutrition, Agricultural University
P.O. Box 8005
6700 EC Wageningen
The Netherlands

ABSTRACT. Kinetics of H^+ consumption by mineral dissolution in a surface soil and a subsoil of an acid woodland soil were studied by means of batch type experiments and stationary pH titrations. Mineral dissolution in the subsoil could effectively be described by congruent dissolution of illite ($K_{0.6}Mg_{0.25}Al_{2.3}Si_3O_{10}(OH)_2$) followed by an incongruent dissolution stage. In the incongruent stage the concentration of Al decreased while concentrations of K and Mg continued to increase. Congruent dissolution rates of illite could be predicted with power type relationships from transition state theory. In the incongruent dissolution stage the activities of H and Al appeared to be controlled by precipitation of $Al(OH)_3$. Results of this kind of experiments could be helpful to understand the behavior of Al and base cations in acid soils and soils with high atmospheric loadings of acid. Comparison of the results of batch experiments for a surface soil and a subsoil showed that initially proton consumption in the surface soil is faster and more effective due to the presence of a relatively high cation exchange capacity and base saturation value. However with progressive proton consumption and increasing reaction times proton consumption in the subsoil becomes faster than in the surface soil, presumably due to higher rates of mineral dissolution in the subsoil. Lower dissolution rates in the surface soil may result from lower contents of easily weatherable minerals, due to excessive leaching in the past which is reflected in a 17% lower acid neutralizing capacity than in the subsoil.

1. INTRODUCTION

Ultimately consumption of protons in acid soils always originates from the dissolution of alumino-silicates. Exchange of base cations and dissolution of oxides are faster buffer processes than dissolution of alumino-silicates, however the origin, and therefore also the capacity of fast buffer mechanisms lies in continued dissolution of alumino-silicates. The difference between the rates of mineral dissolution and

Water, Air, and Soil Pollution **31** (1986) 981–990.

atmospherically derived H^+ production controls the long-term changes in soil chemistry. If the H^+ input exceeds the dissolution rate of base cations, acidity levels (H + Al) both in solution and on the exhange complex will increase. If dissolution rates of base cations can keep up H^+ production rates, acidity levels will not increase and no adverse effects on biota will result. Generally in acid soils Al is the main product of mineral dissolution.

In order to evaluate longterm sensitivity of acid soils to harmful effects of acid atmospheric deposition by means of simulation models such as ILWAS (EPRI, 1983), dissolution of alumino-silicates must be described in terms of rate (r_1), acid neutralizing capacity (ANC) and stoichiometry. Present knowledge of the kinetics of dissolution of alumino-silicates applies mainly to pure mineral systems (Helgeson et al., 1984; Busenberg and Clemency, 1976; Lasaga, 1983), and not (quantitatively) to soils, which are complex mixtures of various minerals and organic substances.

A budget study for an acid woodland soil (Van Breemen et al., 1984) showed that atmospheric H^+ loads of 4 kmol ha^{-1} yr^{-1} are largely consumed between 0 and 10 cm depth, producing almost equivalent amounts of Al.

Batch type experiments were carried out on the surface soil and the subsoil of this woodland to study the rates and mechanisms of mineral dissolution.

2. MATERIALS AND METHODS

Soil samples (sandy mixed acid mesic Umbric Dystrochrept (USDA, 1975)) were taken every 10 cm, air dried and mixed well in bulk samples for 0 to 20 and 60 to 100 cm depth. Prior to all batch experiments soils were prewetted 24 hr with demineralized water. In experiment I 30 mL of a 1:1 mixture of 0.001 M HNO_3 and 0.0005 M H_2SO_4 (pH 3) were added per 10 g of dry soil. Experiments were carried out in duplicate and with one blank, in which soil was shaken with demineralized water. Batches were shaken, end over end, during 0.5, 2, 8, 23, 47, 96, 191, 336 and 672 hr. In experiment II 20 g of subsoil and 120 mL of demineralized water were titrated with a 1:1 mixture of 0.1 M HNO_3 and 0.05 M H_22SO_4 at an automatically adjustable rate in order to maintain a constant pH (Radiometer, ABU 80 autoburette and TTT 60 titrator). Sediment and solution were stirred at respective rates of 4 and 500 rpm in order to minimize abbrasion of the soil grains that may result in an artificial increase of reactive surface. Titrations were carried out at pH values of 3.0, 3.2, 3.5 and 3.8. Experiment III was comparable to experiment I except that the batch solution was replaced five times by a fresh acid solution. The time intervals between the replacements were 1, 24 and 168 hr.

Elemental analysis of soil extracts and the solid phase was carried out according to procedures as described by Begheijn (1980). Ca and Mg were measured by atomic absorption spectrometry, Na and K by atomic emission spectrometry, Al by colorimetry (pyrocathecholoviolet), SO_4 by ion chromatography and H_4SiO_4 by colorimetry (blue silica molybdic acid complex). The cation exchange complex was analyzed by an adapted unbuffered $BaCl_2$ extraction. Exchangeable H and Al were separated by titration of the extract up to a pH of 8.2, followed by an independent analysis of Al. Elemental analyses were done by X-ray fluorescence spectrometry.

3. RESULTS AND DISCUSSION

3.1. Solid phase characteristics

The acid neutralizing capacity (ANC) was estimated by the component composition (Van Breemen et al., 1984) according to $(6(Al_2O_3)+2(CaO)+2(MgO)+2(K_2O)+2(Na_2O))$. For the surface soil ANC was 1365, (371 in the clay fraction) and for the subsoil 1623 mmol kg^{-1} (dry soil). In the surface soil 30% of Al, 50% of Mg and 6% of K is present in the clay fraction. The CEC of the surface soil ranged from 60 to 100 mmol kg^{-1} and is mainly occupied by H (50%) and Al (25%) (on ionic equivalent basis). CEC and clay content in the subsoil are negligible. X-ray diffraction showed the presence of both illite and expandable three-layer clay minerals. The surface soil and the subsoil contain 20% (mass fraction) of K- and Na-feldspars.

3.2. A mechanistic interpretation of experiment I.

Mineral dissolution kinetics were studied first in the subsoil because due to the small CEC cation exchange reactions caused by acid additions will be unimportant. Figure 1 shows concentrations of Al, Mg, K, H and SO_4 for experiment I. Initial concentrations of H and SO_4 were 1 and 0.25 mmol L^{-1}, respectively. All concentration changes in the blank experiment were small compared to those in the experiment with acid addition except for Si. Therefore Si was not furher considered.
Apparent is the the occurrence of a maximum in the Al concentration at 47 hr. Concentrations of K and Mg increase throughout the experiment. In contrast concentrations of Ca and Na (not shown) hardly change after 2 hr.
The experimental results may be explained by an initial, congruent dissolution of a silicate mineral, releasing Al, K and Mg until the solution becomes oversaturated with respect to a secondary Al-mineral. Up to 47 hr the ratio of the concentration changes of H and Al is -8/2.2, which agrees well with the theoretical ratio of -8/2.3 for congruent illite $(K_{0.6}Mg_{0.25}Al_{2.3}Si_3O_{10}(OH)_2)$ dissolution:

$$\text{Illite} + 8H^+ + 2H_2O \Longleftrightarrow 0.6K^+ + 0.25Mg^{2+} + 2.3\,Al^{3+} + 3.5H_4SiO_4 \qquad \log K_{so} = 10.34. \quad (1)$$

Furthermore the batch solutions gradually become saturated with respect to illite (Figure 2), and after 191 hr remain slightly undersaturated. The decrease of [Al] after 47 hr suggests the formation of a secondary Al-mineral. At 47 hr log(Al)+3log(H) is 8.42 and fluctuates between 8.68 and 8.21 afterwards (Figure 2). These values suggest the formation

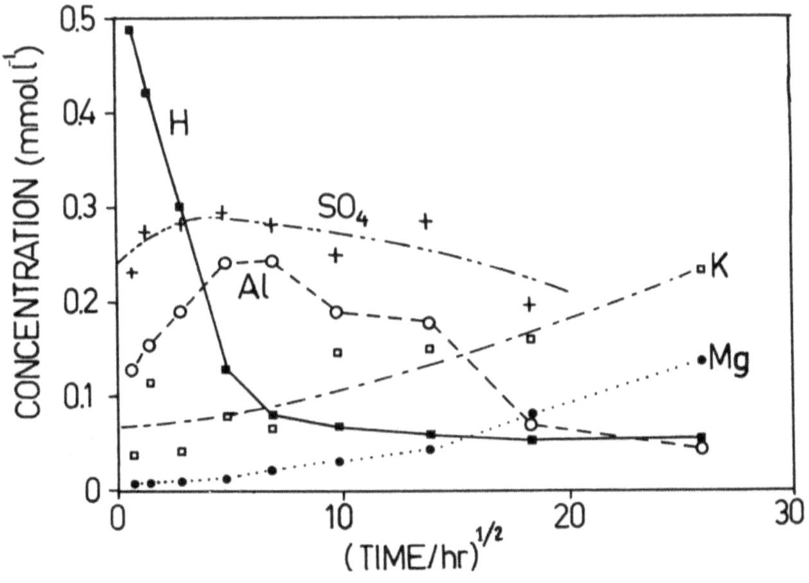

1. Concentrations of Al, K, Mg, H and SO_4 for experiment I, after addition of 30 mL of a 1:1 mixture 0.001 M HNO_3 and 0.0005 M H_2SO_4 to 10 g of subsoil.

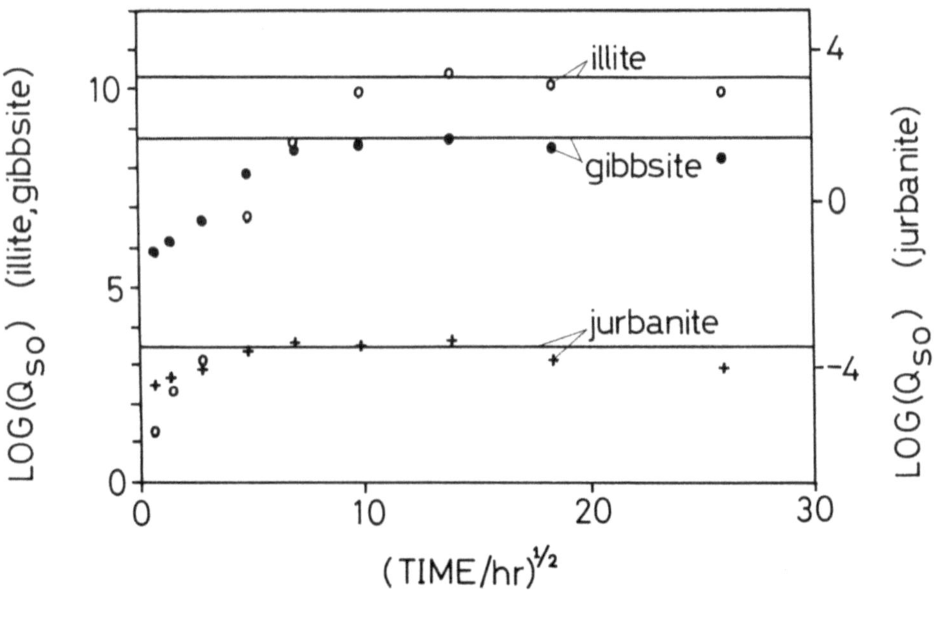

2. Saturation of batch solutions in experiment I with respect to illite ($\log(K_{so})=10.34$), natural gibbsite ($\log(K_{so})=8.77$) and jurbanite ($\log(K_{so})=-3.50$).

of a precipitate of $Al(OH)_3$ intermediate between synthetic ($\log(K_{so}) = 8.11$) and natural gibbsite ($\log(K_{so}) = 8.8$). After 47 hr the net release of Al from illite stops. Because the pH will still slowly increase due to continued, incongruent dissolution of illite, Al released earlier will precipitate as $Al(OH)_3$ and supply the major part of the H^+ for further release of K and Mg from illite. The incongruent dissolution reaction for illite is:

$$\text{Illite} + 1.1H^+ + 8.9\ H_2O \Longleftrightarrow 0.6K^+ + 0.25Mg^{2+} + 2.3Al(OH)_3 + 3.5H_4SiO_4 \qquad \log K_{so} = -9.9 \text{ to } -14.5. \quad (2)$$

The observed decrease of SO_4 after 191 hr (Figure 1) and the presence of some correspondence between activities of H, Al and SO_4 and stability lines for jurbanite ($Al(SO_4)(OH).5H_2O$) (Figure 2) suggest the precipitation of jurbanite. Yet jurbanite was not further considered as its formation remains obscure (Mulder et al., 1986) and the balance between the concentration changes of H and Al does not agree with the stoichiometry of jurbanite precipitation.

3.3. A kinetic illite dissolution model for experiment I.

To illustrate the validity of the proposed dissolution mechanism for experiment I, a simple simulation model was developed. The required kinetic parameters and the equilibrium constant for gibbsite were derived from the experiment itself.

A general expression for the rate of mineral dissolution in terms of the transition theory (Aagaard and Helgeson, 1982) is:

$$r^* = K^* (H^+) (1-\exp(A/\sigma RT) \qquad (3)$$

where r^* stands for the dissolution rate of the mineral (mol cm^{-2} s^{-1}), (H^+) represents the proton activity, A stands for the chemical affinity of the overall reaction (J mol^{-1}), R is the gas constant (J mol^{-1} K^{-1}), T represents the temperature (^oK), σ is the ratio between the rate of decomposition of the activated complex and the overall reaction, and K^* is the rate constant (mol cm^{-2} s^{-1}). Equation (3) can be rewritten as:

$$r = K (H^+) (1 - (Q/K_{so})^p) \qquad (4)$$

where r and K are the dissolution rate and rate constant in mol kg^{-1} hr^{-1} (the reactive surface area was not measured), Q is the activity product for the dissolution reaction, which is $\prod_i a_i^n$, where n stands for the reaction coefficient for the ith reactant species, K_{so} is the equilibrium constant for overall dissolution of the mineral and p is $1/\sigma$. According to Aagaard and Helgeson (1982) the last term of Equation (4) is negligble far from equilibrium. At equilibrium Q is equal to K_{so} and r is equal to 0.

The non-linear optimization package OPTPAC (Kilsdonk, 1977) was used to determine parameters for Equation (4) for the congruent dissolution stage. The function to be minimized was:

$$\sum_{i=1}^{n} ((r'_i - r_i)/r'_i)^2 \qquad (5)$$

in which r'_i is the experimentally found mineral dissolution rate, which is calculated from the rate at which ions are released to solution assuming congruent illite stoichiometry, and n is the number of observations. The standard error of estimate and prediction accuracy (SEE) was calculated by dividing the result of Equation (5) by (n-2). Values for p and C were calculated for rates of release of H alone (n=5) and of H, Al and Mg (n=15). Equation (4) could not be fitted to the incongruent dissolution stage because Q essentially remained constant (Figure 2) and the changes in the rates of release of K and Mg did not show a trend. Results are shown in Table I.

n	K (mol g^{-1} hr^{-1})	p	K_{so} (illite)	SEE
5	$-3.06\ 10^{-2}$	$-4.45\ 10^{-2}$	$2.19\ 10^{10}$	0.480
5	$-2.63\ 10^{-1}$	$-6.51\ 10^{-3}$	$1.00\ 10^{11}$	0.407
15	-2.02	$-1.51\ 10^{-1}$	$2.19\ 10^{10}$	0.565
15	-1.12	$-1.68\ 10^{-1}$	$1.00\ 10^{11}$	0.569

TABLE I. Optimized kinetic parameters for congruent dissolution of illite according to Equation (4).

In contradiction with transition state theory values of p are negative, which makes the last term of Equation (4) not negligeable. Allthough apparently transition state theory does not apply Equation (4) still fits the rate data.

Simulation of the congruent stage of experiment I using values for K and p, based on the consumption rates for H, gave the best prediction of the starting point of the incongruent stage. Exactly at 47 hr saturation was predicted with gibbsite (log(K_{so}=8.40). From this point several combinations of [H] and [Al] in equilibrium with $Al(OH)_3$ were calculated iteratively with the Newton-Raphson procedure for increasing values of H^+ consumption. The resulting concentrations of K and Mg were calculated from the total proton consumption and Equation (2). The reaction times to attain the newly calculated sets of concentrations in the incongruent stage were calculated both from Equation (4) (Figure 3) and by assuming a constant dissolution rate after equilibrium is attained with gibbsite (Figure 4). The simulation results show that the congruent dissolution stage is decribed well. In the incongruent stage the observed decrease of [Al] and increase of [Mg] are predicted best when assuming a constant illite dissolution rate. However, the latter assumption causes oversaturation with respect to illite after 230 hr (Figure 5). After 191 hr the predicted concentrations of K are too high, probably due to neglecting the observed decrease of [SO_4]. The fact that Equation (4) does not apply

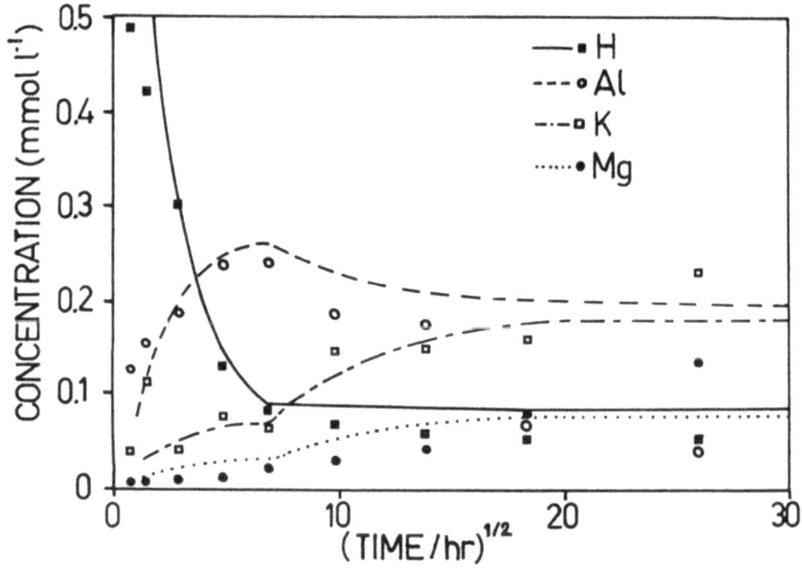

3. Simulated and measured concentrations for experiment I, assuming congruent dissolution of illite until saturation with gibbsite ($\log(K_{so})=8.40$) occurs at 47 hr. Illite dissolution rates are a function of the degree of saturation with respect to illite.

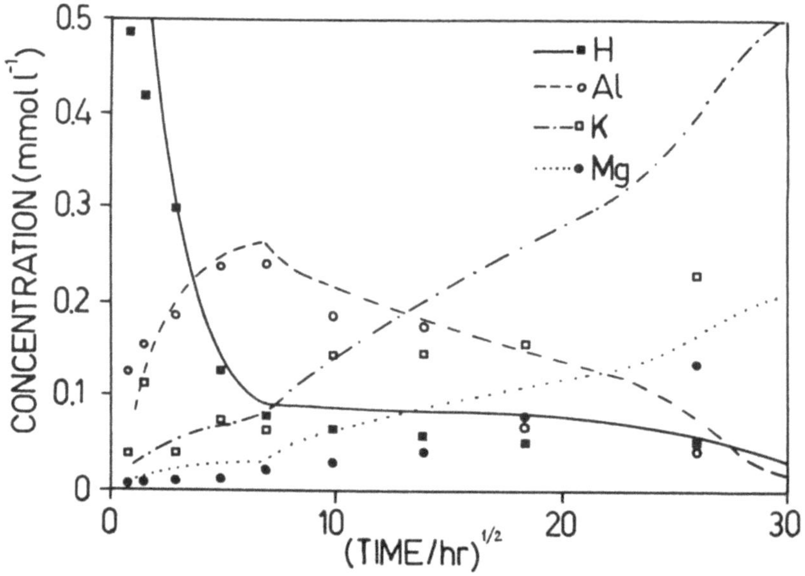

4. Simulated and measured concentrations for experiment I, assuming congruent dissolution of illite until saturation with gibbsite ($\log(K_{so})=8.40$) occurs at 47 hr. Before 47 hr illite dissolution rates are a function of the degree of saturation with respect to illite. After 47 hr illite dissolution rates are assumed constant.

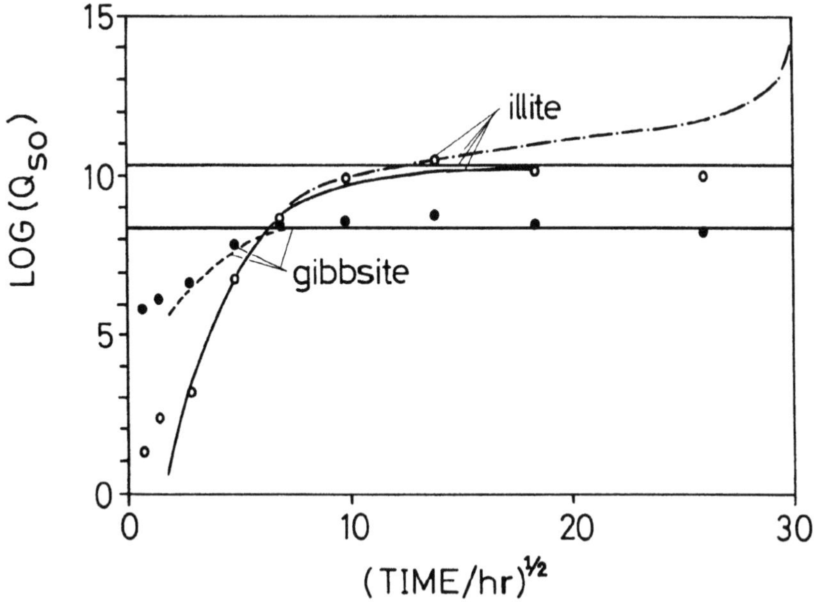

5. Simulated and observed activity products (Q_{so}) for illite and gibbsite. After saturation with gibbsite Q_{so} for illite is calculated assuming the illite dissolution rate to be a function of Q_{so} (——) or to remain constant (—·—).

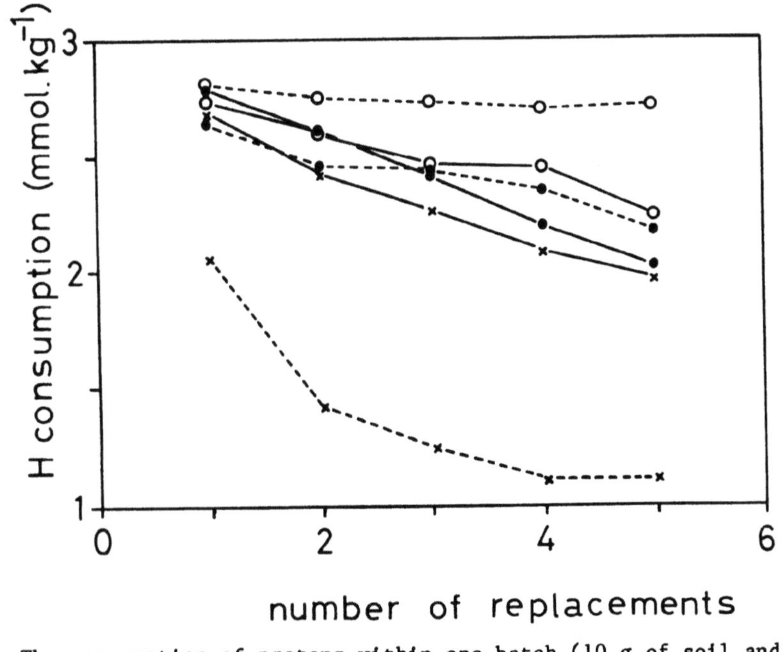

6. The consumption of protons within one batch (10 g of soil and 30 mL of solution) as a function of the number of times the solution is replaced by a fresh acid solution (0.5 mmol L^{-1} HNO_3 and 0.25 mmol L^{-1}), for surface soil (——) and subsoil (----), equilibrating for 1 hr (X), 1 day (●) or 1 week (O).

to the incongruent stage may also be caused by a too low equilibrium constant for illite. If we assume higher values for $\log(K_{so})$, e.g. 11 (Table I), undersaturation is maintained longer, but predicted illite dissolution rates in the incongruent stage remain too low.

3.4. Dissolution kinetics in experiment II.

Only the final composition of the batches in experiment II was available. The overall reaction stoichiometry of the stationary pH titrations, i.e. $\Delta H:\Delta Al:\Delta Mg$ was -10.8:2.5:0.25 and is close to illite stoichiometry. Stationary pH titrations were intended to establish a simple relationship between (H^+) and r far from equilibrium:

$$r = K' \, (H^+)^s. \qquad (6)$$

As effects of ANC are not considered in Equation (6), r values for H were compared at Δ ANC fixed at 12.5, 17.5, 22.5 and 27.5 mmol kg^{-1}.

Linear regression on the logaritmic form of Equation (6) gave 0.8 for s (correlation coefficient 0.95). For experiment I, using r values from H, Al and Mg, a value of 2.4 (correlation coefficient 0.76) was found. Disagreement between both s values most likely is caused by not considering effects of Δ ANC, amounting to 3 mmol kg^{-1} in experiment I and to 40 mmol kg^{-1} in experiment II, or to artefacts from comparing a shaken to a stirred batch experiment.

3.5. Comparison of H^+ consumption rates for the surface soil and the subsoil in experiment III.

Figure 6 shows H^+ consumption per batch experiment as a function of the number of times the acid solution was replaced. At the same time the effect of reaction time per batch experiment is illustrated. In all cases H^+ consumption within one batch decreases with increasing number of acid solution replacements, but the effect is most clear in the subsoil for the shortest reaction time. This cannot be explained by Equation (4) as undersaturation and (H^+) tend to increase with repeated acid additions. More likely decreased H^+ consumption rates are a result of a decreasing ANC. The decrease of H^+ consumption rates with decreasing reaction time and increasing number of replacements is smallest in the surface soil, because here the main neutralizing process will be fast exchange of base cations. Observations for the subsoil for 1 hr reaction time, might agree with diffusion type dissolution mechanisms (Busenberg and Clemency, 1976): A step change in acid input and the resulting drop of pH will cause an extension of the diffusion layer, which will be expressed in a peak increase of r_i. Proton consumption in the surface soil was largely balanced by mobilization of base cations and in the subsoil by dissolution of Al. Total H^+ consumption after 5 solution replacements in the surface soil is 12.5 mmol kg^{-1}, which corresponds to 112% of the exchangeable bases. As a result it can be expected that during the last acid additions H^+ consumption due to mineral dissolution will also be important in the surface soil. This seems to be reflected by the observation that for a reaction time of 1 week H^+ consumption in the subsoil becomes increasingly higher than in the surface soil possibly due to higher mineral dissolution rates. It should be noted that ANC for the surface soil is 16% lower than for the subsoil. As parent material for the surface soil and the subsoil is similar this might indicate removal of reactive silicate minerals from the surface soil.

4. CONCLUSIONS

For a full description of mineral dissolution in acid soils we need to consider effects of saturation with respects to the dissolving minerals in addition to effects of a decreasing ANC. Experiment I and II clearly illustrate that the dissolution rate is very sensitive to small changes of ANC, which does not seem to be caused by exhaustion of a specific reactive mineral but more likely by a fast decrease of reactive spots such as dislocations and etch pits (Lasaga, 1983) on the mineral surfaces.

ACKNOWLEDGEMNTS

This research was supported in part by the Netherlands Directorate General for Science.

REFERENCES

Aagaard, P. and Helgeson, H.C.: 1982, Amer. J. Sci. **282**, 237.
Begheijn, L.Th.: 1980, Methods of chemical analysis for soils and waters. Agricultural University, Wageningen, The Netherlands.
Busenberg, E. and Clemency, C.V., 1976. Geoch. et Cosmoch. Acta. **40**, 41.
EPRI: 1983, The Integrated Lake-Watershed Acidification Study. Volume 1. Model principles and application procedures. Report EA-3221. TETRA Tech. Inc., Lafayette Ca, USA.
Helgeson, H.C., Murphy, W.M. and Aagaard, P.: 1984, Geoch. et Cosmoch. Acta. **48**, 2405.
Kilsdonk, A.C.M.: 1977, Non-Linear Optimization package OPTPAC 3. Program description and reference manual. ISA-DSA Scientific Computer Application, no UDV-DSA/SCA/TU/77/054/ap, N.V. Philips, Eindhoven, The Netherlands.
Lasaga, A.C.: 1983, Kinetics of silicate dissolution. Proc. 4th Int. Conf. Water Rock Interaction. p 269-274.
Mulder, J., Van Grinsven, J.J.M. and Van Breemen, N.: 1986, Hydrochemical budgets of woodland soils affected by atmospheric deposition. III. Aluminum Chemistry. (submitted to S.S.S.A.J.)
U.S. Department of Agriculture, Soil Conservation Service: 1975, Soil Taxonomy, US Government Printing Office, Washington D.C..
Van Breemen, N., Driscoll, C.T. and Mulder, J.: 1984, Nature **307**, 599.

SIMULATED MOBILIZATION OF METALS FROM SEDIMENTS IN RESPONSE TO LAKE ACIDIFICATION

Nabil Arafat & Jerome O. Nriagu
National Water Research Institute
Box 5050
Burlington, Ontario, Canada L7R 4A6

ABSTRACT. The regeneratable forms of pollutant metals in lake sediments were estimated using different chemical extractants and the effects of lake acidification on the mobilization of such metal fractions studied in laboratory microcosm. The release of each metal was found to increase exponentially below a threshold pH value, which was about 4.0 for Cu, Ni, Zn, Cd and Fe. The data suggest that a large percentage of pollutant metals are mobile as determined by chemical extraction techniques and long-term release experiments.

1. INTRODUCTION

The most recent layers of lake sediments in many parts of Canada show highly elevated levels of metals which can be related to the increased contributions from anthropogenic sources. Since some lakes which are sensitive to acid rain have shown a marked drop in their pH values in recent years (Likens and Bormann, 1974), there has therefore been some concern that the increasing acidification of the lakes will liberate metals from the sediments and consequently will adversely affect the quality of the overlying lake water. The objectives of this study are two-fold: (a) using different chemical extractants, to determine the leachable fractions of metals in highly contaminated lake sediments; and (b) to evaluate the mobilization of the available metals in the sediments as a function of pH during an extended period of time. Previous studies of metal release from soils and sediments have generally been of short-term duration (Khalid, 1980; Patchineelam and Forstner, 1983).

Water, Air, and Soil Pollution **31** (1986) 991–998.

2. METHODOLOGY

The highly contaminated lake sediments used in the study were obtained from Ramsey Lake (45°25'N, 81°00'W) located about 8 km from the tall smelter stack at Sudbury, Ontario. The lake has an area of about 7.904×10^6 m^2 and a mean depth of 12.2 m. Its metal input comes mostly from the atmosphere so that the chemical composition of metal-bearing aerosols falling into the lake is, presumably, similar to that falling into many lakes in Ontario (Nriagu _et al._, 1982); The pH of the Ramsey Lake water was 6.8, an important consideration in a study aimed at understanding the potential effects of acid stress on lake sediments.

Large compositive sediment samples were collected using Birge-Eckman grab sampler at deepest part of each of the two basins at depths of 9 and 21 m. The sediment-water interface was well-defined and only cores with the interface interact were used in the study. In the laboratory, about 30 L of the homogenized wet sediments were put into each of two identical polyethylene containers measuring 60x60x75 cm. To the sediment which covered a depth of about 10 cm, 110 L of distilled water was added and the mixture was allowed to settle. A steady stream of air was passed through the water column to provide mixing and aerobic conditions without disturbing the sediment-water interface. The tanks were maintained at the temperature of 22+2°C, and the pH of the mixture progressively lowered over a period of 12 mo by adding H_2SO_4.

Three sediment cores were also collected at three different locations using a Benthos gravity corer. An aliquot (2.0 g) of the freeze-dried sample was leached with dithionate-citrate-bicarbonate (DCB) mixture while another aliquot (1.0 g) was treated with 20 mL of 0.5 N H_2SO_4. The sample extraction was done in plastic tubes which were shaken gently for 24 hr and the leachate was then removed by high speed centrifugation. Total metal concentration in each sediment sample was determined according to the method of Agemian and Chau (1976); organic C in the sediments was determined by dry combustion in a Leco furnace (Kemp, 1971). Trace metals were determined by atomic absorption spectrometry according to the Analytical Methods Manual (1979).

3. RESULTS AND DISCUSSION

The total metal concentrations in Ramsey Lake sediment cores (Table I) show marked enrichment of Cu, Pb, Ni and Fe in the surficial sediments, especially in the 0-5 horizon. The eastern basin (Stn E1) sediments generally show slightly lower levels of metals in the most recent layers. From the sediment accumulation rate of 0.9 mm yr^{-1} (185 g $m^{-2}yr^{-1}$) (Nriagu and Coker, 1980) in the western basin (Stns W1 and W2), the onset of accelerated metal deposition (7 to 9 cm below sediment-water interface) can be dated to the beginning of mining and smelting operations in Sudbury around 1890 (Nriagu _et al._, 1982).

TABLE I. Average Total metal concentrations $\mu g\ g^{-1}$ in Ramsey Lake sediments.

Depth (cm)	Fe	Ni	Cu	Zn	Pb	Cr
Station W1						
0 - 2	61200.	1880.	1320.	198.	122.0	132.
2 - 3	30600.	370.	130.	139.	18.5	139.
3 - 5	37900.	656.	236.	174.	35.9	133.
5 - 7	29500.	114.	40.	114.	16.2	114.
Station W2						
0 - 2	60600.	4340.0	2930.	384.	253.0	131.
2 - 3	54400.	989.0	593.	198.	98.9	109.
3 - 5	59900.	118.0	56.	138.	31.4	128.
5 - 7	44600.	78.5	32.	136.	24.2	145.
Station E1						
0 - 2	35700.	1740.0	1060.0	212.	81.0	82.
2 - 3	26500.	397.0	204.0	122.	50.9	77.
3 - 5	25700.	119.0	40.3	110.	12.8	101.
5 - 7	26900.	76.7	27.9	120.	12.0	88.

In general, the profiles of DCB extractable metals tend to parallel those of total metals with marked enrichment of this phase in the surficial sediments (Figure 1). This implies that the smelters at Sudbury are a major contributor of DCB extractable metals in the lake sediments or that this fraction is precipitated diagenetically in lake sediments (Carignan and Nriagu, 1985). It is generally believed that the DCB treatment liberates metals adsorbed or coprecipitated with oxides and hydroxides of Fe and Mn and some of the Al present as coatings on clays and other sediment particles (Mehra and Jackson, 1960; Farmer et al., 1980. The observed build-up of Fe in the surficial sediments can be attributed to (a) the fact that Fe input from anthropogenic sources may be more mobile or (b) the precipitation of labile Fe derived diagenetically from the older sediments.

It is generally accepted that extraction with reducing DCB reagent (pH, about 7) gives a rough estimate of that portion of trace metals bound to metal oxides. A comparison of data in Table I and Figure 1 shows that 25 to 35% of the total Ni, 1 to 4% of the Cu, 25 to 30% of the Zn and 10 to 20% of the total Cr in the surficial sediments are associated with metal oxides. The values for Pb are in the range of 15 to 30%. These levels of trace metal association with the reducible oxide phases are comparable to what have been reported for polluted sediments in other areas (Forstner and Wittman, 1979; Salomons and Forstner, 1984).

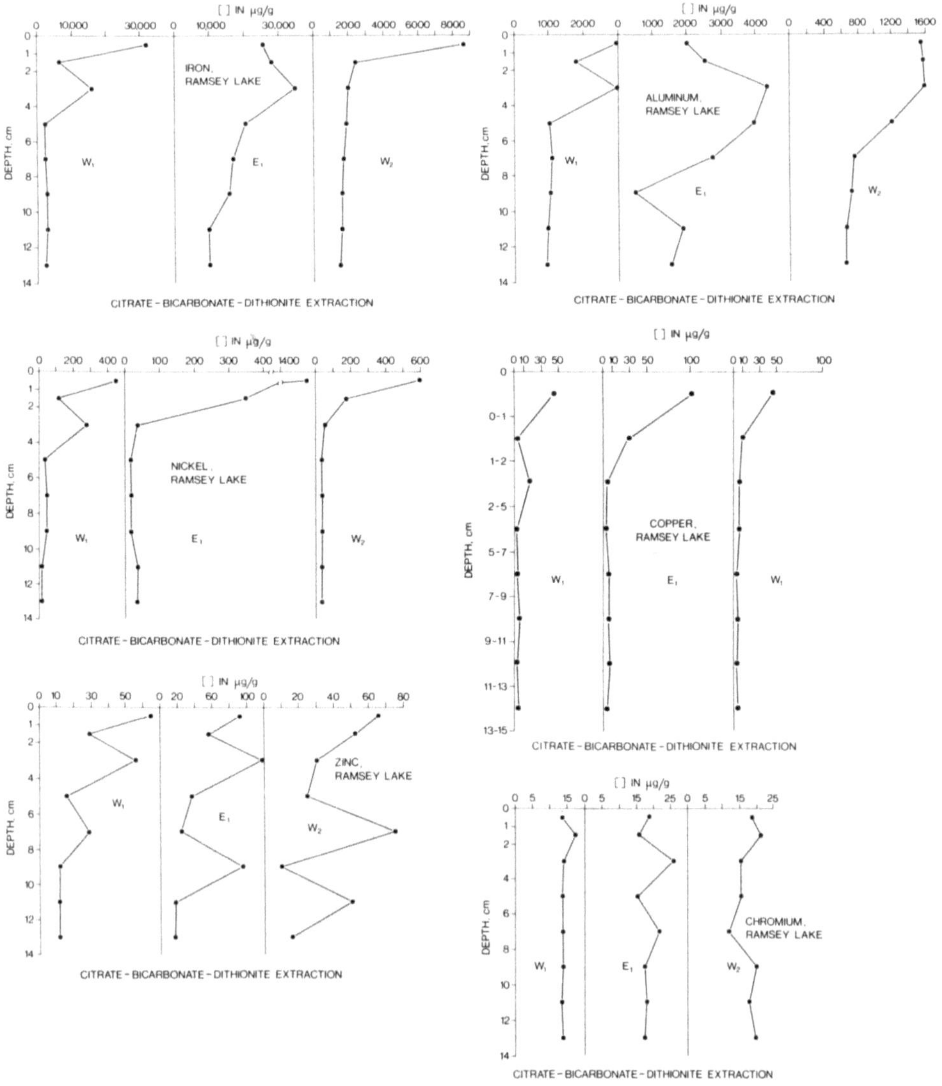

Fig. 1 Profiles of citrate-bicarbonate-dithionite (CDB) extractable forms of metals in the lake sediment cores.

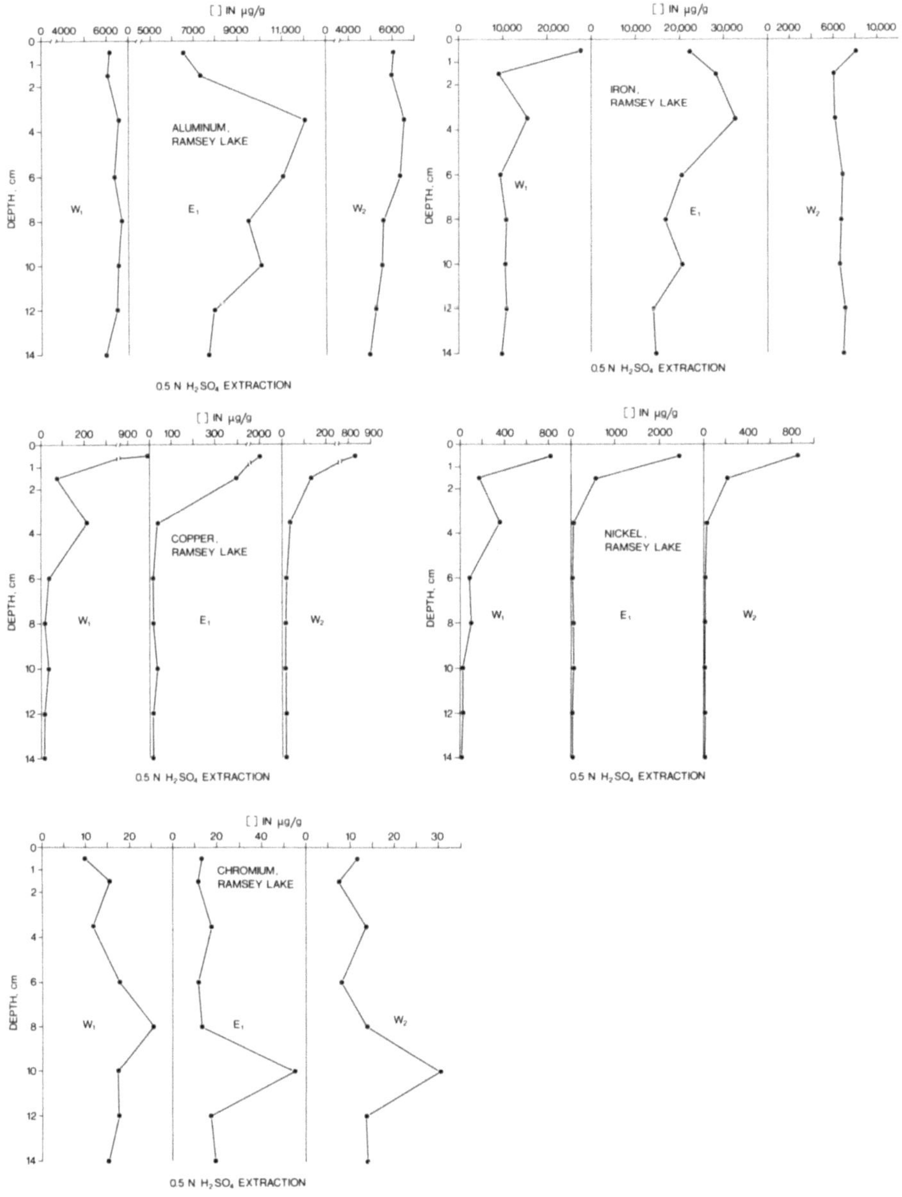

Fig. 2 Profiles of 0.5N H_2SO_4 extractable forms of metals in the lake sediment cores.

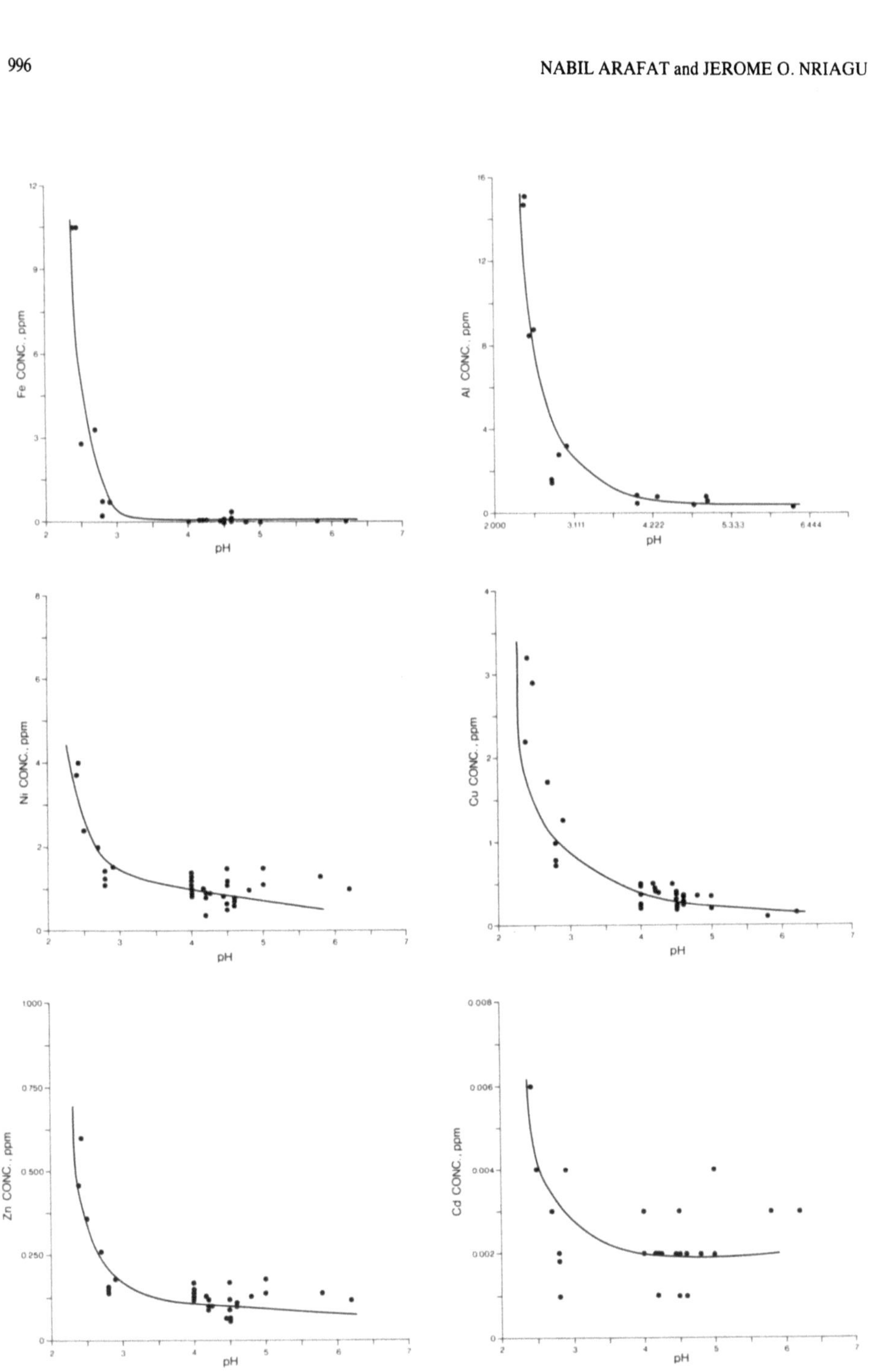

Fig. 3 Metal release as function of pH in sediments under laboratory conditions.

The profiles of metals extracted using dilute H_2SO_4 are shown in Figure 2. For Fe, this reagent extracts about 80 to 90% of the Fe-DCB in the surficial sediments. By contrast, the dilute acid extracts considerably (2- to 3-fold) more Al than DCB (compare Figures 1 and 2). The acid treatment, however, liberates 45 to 55% of the total Ni and 70 to 80% of the total Cu in the surficial sediments (compare Figure 2 and Table I). The implication with reference to Figure 1 is that less than 10% and 55 to 70% of the potentially mobile Cu and Ni in the surficial sediments are tied up with the oxyhydroxides. Copper has a marked affinity for organic matter in lacustrine sediments (Thornton, 1979; Nriagu and Coker, 1980). It would thus appear that the acid treatment has resulted in the release of organically bound Cu which apparently accounts for about 50 to 60% of the total burden of this metal in the sediments. The dissolution of metal sulfides can be discounted in the aerobic surficial sediments used in the present experiments.

The long-term, pH-induced release of metals from the sediments to the overlying water is depicted in Figure 3. Without exception, the release of each metal increases exponentially with a decrease in pH below a threshold value. For Cu, Ni, Zn and Cd, the threshold value is in the range of 4 to 3 pH units, which is similar to the breakthrough values for the release of metals from sewage sludge (Adams and Sanders, 1984). The actual amount of metals liberated from the sediments is significant. At the pH of 2.8, it is estimated that 120 $\mu g\ cm^{-2}$ of Ni, 110 $\mu g\ m^{-2}$ of Cu and 18 $\mu g\ m^{-2}$ of Zn are released by the sediments. Assuming that the 10 m water depth at the sampling station is completely mixed, such an acid stress would be expected to add 120 $\mu g\ L^{-1}$ Ni, 110 $\mu g\ L^{-1}$ Cu and 18 $\mu g\ L^{-1}$ Zn to the overlying water. Such potential contributions from sediments may be compared with the current Ni, Cu and Zn concentrations in the water of 220, 16 and 8.0 $\mu g\ L^{-1}$, respectively (Nriagu <u>et al</u>., 1982).

It should be emphasized that the present study was designed to test specifically the role of pH alone on the release of metals under static aerobic conditions. Extrapolation of the results to field situations must be done in a circumspect manner. For example, the resuspension and bioturbation of sediments can strongly affect the amount of metals released, and marked enrichment of the upper levels is common throughout "Canadian" lakes (Brannon et al., 1980; Semkin and Kramer, 1976; Wong et al., 1984). Furthermore, a reduction in Eh of the sediments can have a dramatic effect on the rate of metal release as documented by the DCB extraction and by field observations already referred to.

Ramsey Lake is probably typical of many oligotrophic lakes where an oxidized surficial layer overlies the generally reduced sediments (Mortimer, 1941; Wetzel, 1983). As the redox boundary is diagenetically moved upward with sediment accretion, some of the trace metals may be expected to be mobilized. This raises the possibility that the metal enrichment in the oxidized surficial sediments of some oligotrophic lakes could be diagenetically mediated. The function of the oxidized layer as a barrier in the transfer of mobilized metals to the overlying water has certainly not received the attention it deserves.

REFERENCES

Adams, T. McM. and Sanders, J.R.: 1984, Environ. Pollut. (Series B) **8**, 85.

Agemian, H. and Chau, A.S.Y.: 1976, Analyst. **101**, 761.

Analytical Methods Manual: 1979, Water Quality Branch, Inland Waters Directorate, Ottawa, Ontario, Canada.

Brannon, J.M., Plumb, R.H. and Smith, I.: 1980, In "Contaminants and Sediments". Baker, R.A., Ed., Ann Arbor, Science Publishers: Ann Arbor, Michigan, Vol. 2, 221.

Carignan, R. and Nriagu, J.O.: 1985, Geochim. Cosmochim. Acta. **49**, 1753.

Farmer, J.G., Swan, D.S. and Baxter, M.S.: 1980, Sci. Total Environ. **16**, 131.

Farmer, V.C., Russell, J.D. and Smith, B.F.L.: 1983, J. Soil Sci. **34**, 571.

Forstner, U. and Wittman, G.T.W.: 1979, "Metal Pollution in the Aquatic Environment". Springer: Berlin.

Kemp, A.L.W.: 1971, J. Sed. Petrol. **41**, 537.

Khalid, R.A.: 1980, In "Cadmium in the Environment". Nriagu, J.O., Ed. and Wiley: New York, Vol. 1, 257.

Likens, G.E. and Bormann, F.H.: 1974, Science. **184** , 1176.

Mehra, O.P. and Jackson, M.L.: 1960, Clays and Clay Minerals. **7**, 317.

Mortimer, C.H.: 1941, 1942, J. Ecol. **29** , 280; **30**, 147.

Nriagu, J.O. and Coker, R.D.: 1980, Environ. Sci. Technol. **14**, 443.

Nriagu, J.O., Wong, H.K.T. and Coker, R.D.: 1982, Environ. Sci. Technol. **16**, 551.

Patchineelam, S.R. and Forstner, U.: 1983, In "Proc. International Conference Heavy Metals in the Environment". Heidelberg, Vol. 2, pp. 860-863.

Salomons, W. and Forstner, U.: 1984, "Metals in the Hydrocycle". Springer-Verlag: Heidelberg.

Semkin, R.G. and Kramer, J.R.: 1976, Can. Mineral. **14**, 73.

Thornton, I.: 1979, In "Copper in the Environment". Nriagu, J.O., Ed. and Wiley: New York, Vol. 1, 171.

Wetzel, R.G.: 1983, "Limnology". 2nd Ed.; W.B. Saunders: Philadelphia.

Wong, H.K.T., Nriagu, J.O. and Coker, R.D.: 1984, Chemical Geology. **44**, 187.

WHAT FRACTION OF THE TOTAL METAL FLUX INTO LAKES IS RETAINED IN THE SEDIMENTS?

Jerome O. Nriagu & Henry K.T. Wong
National Water Research Institute
Box 5050, Burlington, Ontario L7R 4A6
Canada.

Abstract. The concentrations of, and deposition rates for, the pollutant metals (Pb, Cu, Zn, Ni, and Cd) and the rare earth elements are reported for the sediments and settling particulates in a large, remote lake (Lake Opeongo) in the Algonquin Provincial Park, Ontario. Using a mass conservation model based on the rare earth elements, it is estimated that only 40 to 60% of pollutant metals getting into the lake is retained by the sediments. The low retention anada rate may explain the widely reported intra-basin and intra-lake differences in metal contents of sediments and casts some doubt on the use of pollutant metal profiles in retrospective assessment of the human contribution to the metal flux into many lake basins.

1. Introduction.

Sediments provide an historical record of past changes in the loading of metal pollution into lakes (Forstner and Wittmann, 1981; Salomons and Forstner, 1984; MARC, 1985). The reading of such a record occasionally presupposes that most of the metal inputs into a lake is retained in the sediments. Such an assumption has not been validated by actual field measurements. The rare earth (RE) elements as a group represent a conservative non-pollutant in lakes and a study of their mass conservation can potentially provide some insight into the retention of metals by the sediments. In this report, we compare the behaviors of the RE and pollutant metals in a large softwater lake (Lake Opeongo) in Central Ontario and conclude that the pollutant metals are more mobile than has hitherto been realized. Increased lake acidification thus is expected to accentuate the mobilization of pollutant metals stored in sediments of such lakes.

Water, Air, and Soil Pollution 31 (1986) 999–1006.

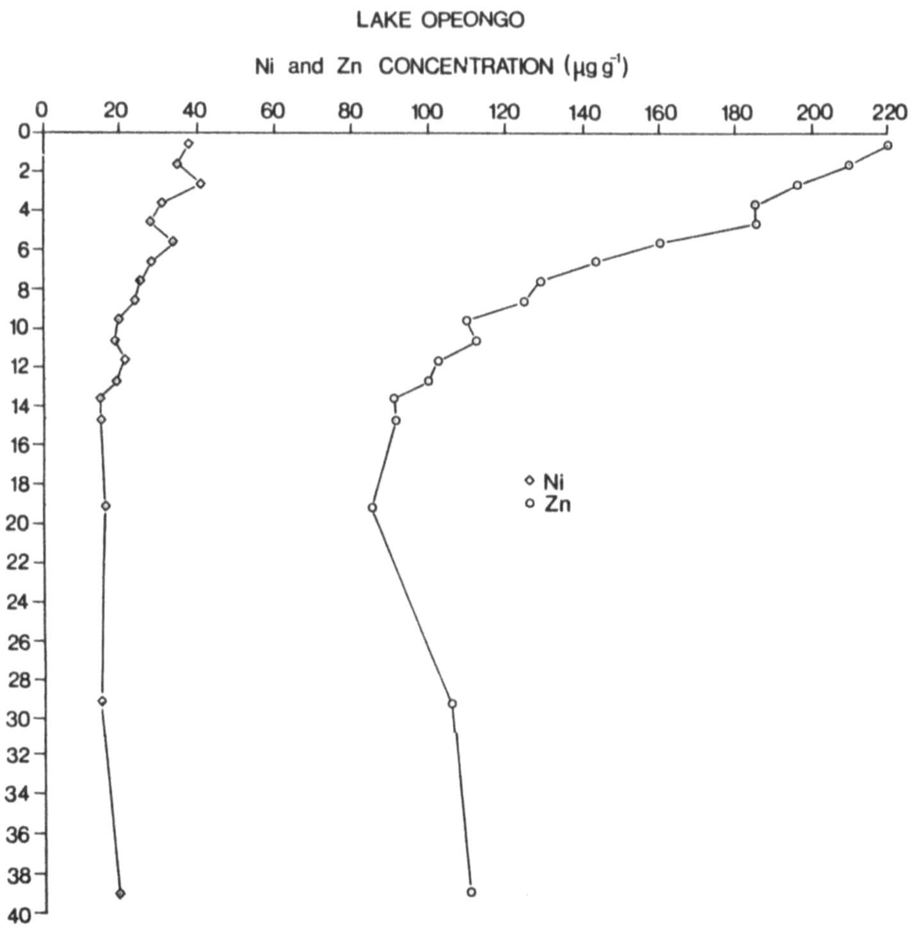

Figure 1. Profiles of Ni and Zn in Lake Opeongo sediments showing recent changes in rates of pollutant metal input.

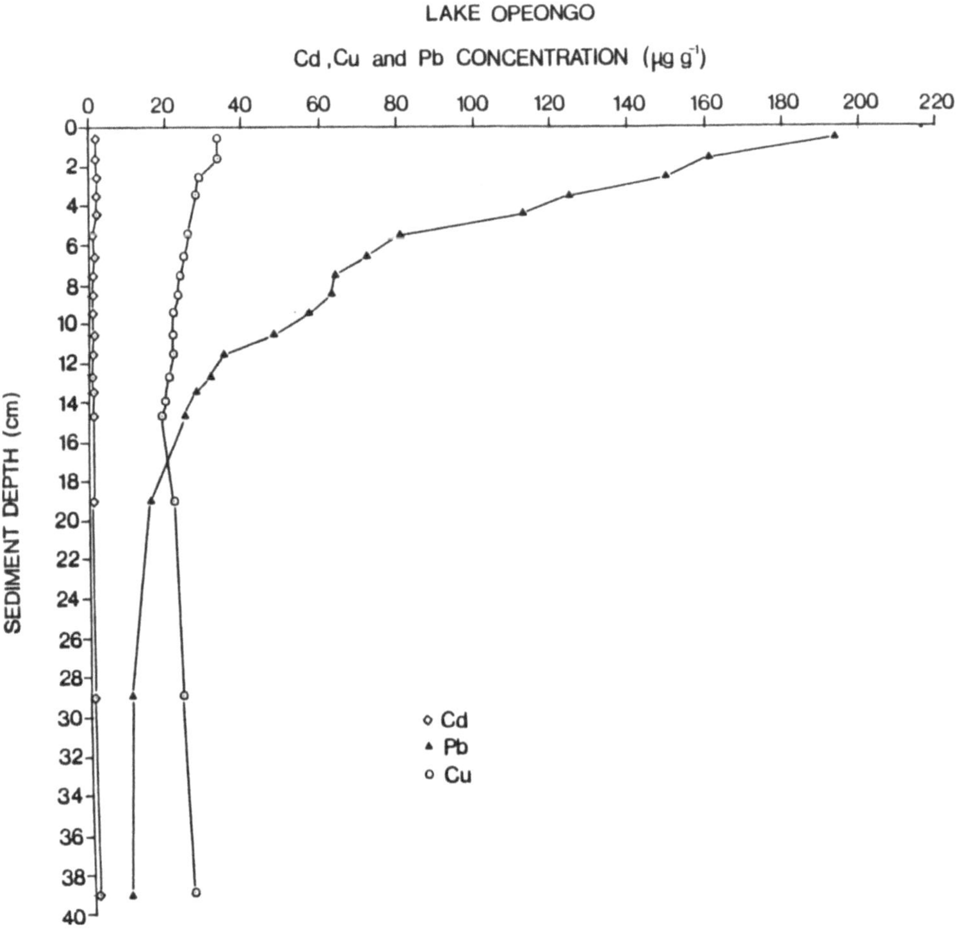

Figure 2. Profiles of Cd, Pb and Cu in Lake Opeongo sediments showing recent changes in rates of pollutant metal input.

2. Materials and Method

Lake Opeongo is a large softwater,oligotrophic lake with a maximum depth of 50m and mean depth of 15m. The major physical properties and chemical characteristics of this lake lake are summarized in Table I; additional data are presented elsewhere (Wong et al., 1984; Wong and Nriagu, 1985).

The settling particulates in the water column were collected using sediment traps or particle interceptors (Wong and Nriagu, 1985). The traps were deployed at depths of 3.0 m. below the air-water interface and about 3.0 m. above the sediment-water interface at each of the sampling stations in the lake. The traps were tapped every 28 days in June, July, August and September, 1983 and the seston flux was calculated from the solid matter in the freeze-dried samples and the quantity of material intercepted per unit time.

The sediments were sampled by the lightweight coring device (Williams and Pashley, 1979). The cores were obtained at the same sampling sites as the sediment traps. These were located at the deepest parts of the four basins in the lake, each of which has a relatively flat bottom as determined by echo sounding. Four traps were deployed at different depths across one of the basins (South Arm), and the data presented below represent the averages for the stations. The sediment accumulation rate was determined by Pb-210 geochronology (Robbins, 1978; Nriagu et al., 1979). Total metal concentrations were determined by atomic absorption spectrometry following the wet digestion of the samples with aqua regia and $HClO_3$(Nriagu, et al., 1982; Wong et al., 1984). The rare earth elements were measured by neutron activation while the standard dry combustion method (Perkin Elmer Elemental Analyzer) was used in determining the carbon and nitrogen contents of the freeze-dried samples.

3. Results and Discussion

The lakewide average concentrations of Cu, Ni, Pb, Zn, organic C and N in suspended particulates collected between May and October, 1983 are shown in Table II. A notable feature is the high concentration of Pb in the seston; lakes in Central Ontario are known to receive highly elevated inputs of anthropogenic Pb (Evans and Rigler, 1985). The high particulate Zn content may reflect the high background levels in this watershed, the average Zn concentration in the precolonial sediment layers being about 100 $\mu g\ g^{-1}$ (Figure 2). By contrast, the average Cu, Ni and Pb concentrations in the precolonial layers are only 21, 16 and 12 $\mu g\ g^{-1}$, respectively.

The average concentrations of the rare earths (RE) in the seston are also shown in Table II. One can assume that most of the RE in seston are derived from the surrounding bedrock and, indeed, the data compare favorably with the average abundance (in $\mu g\ g^{-1}$) of the RE in the Precambrian shield of Canada: La=32, Ce=65, Nd=26, Sm=4.5, Eu=0.94, Tb=0.48, Yb=1.5, and Lu=0.23 (Shaw *et al*., 1976). Notice also that the relative ratios of the RE in the seston are very similar to those of the shield rocks suggesting that there is no unusual behavior or supply of any particular RE element to the watershed (e.g., see Crocket and Kabir, 1981; Olmez and Gordon, 1985).

The downcore profiles (Figure 1) show the characteristic enrichment of the pollutant metals in the most recent sediments. Accumulation of excess metal burden in the surficial sediments has been reported in many parts of Europe and North America and is generally attributed to the recent influx of metal pollution into lakes in these areas (see MARC, 1985 for a recent review). By contrast, there is no discernable enrichment of the RE in the most recent sediment (Table III); industrial activities apparently do not now release large quantities of these elements into the environment (Olmez and Gordon, 1985).

The ratios of the concentrations of individual RE elements in the sediments to their concentrations in the suspended particulates ($R_{s/p}$) vary from 2.4 to 3.1 (Table II) and average about 2.7+0.26. In other words, the ratios are relatively constant. Such an enrichment of the RE in sediments in relation to the seston can be regarded as a natural lake phenomenon which may be attributed primarily to the biodegradative removal of the sestonic biomass. This point is emphasized by the $R_{s/p}$ for particulate organic C which is 0.35, or roughly the reciprocal of that for the RE. The organic matter content itself varies from well over 50% in the seston (Table II) to under 20% in the surficial sediments (Wong and Nriagu, 1985). Other factors which can also contribute to the relative enhancement of the RE in sediments include (a) addition of weathered and fractionated material from the bluff and bedrock, and (b) diagenetic processes which pump the mobile components out of the sediments. The relative importance of the latter two processes is unkown at the present time. It is, however, worth noting that $R_{s/p}$ values above 2.0 seem to be characteristic of other conservative components in these lakes. For example, Al has an $R_{s/p}$ value of 2.6, which is similar to the RE values, while the expectably more mobile Fe has a slightly lower $R_{s/p}$ of 2.2.

TABLE I

WATER CHEMISTRY DATA FOR LAKE OPEONGO

Ca (mg L^{-1})	3.3	pH	6.8
Mg	1.2	Alkalinity(mg L^{-1}) as $CaCO_3$	4.6
Na	1.1	Conductivity ($\mu S\ cm^{-1}$)	42
K	0.62	Lake Area (ha)	5900
SO_4	8.6	Maxium depth (m)	43
Al	0.018	Lake volume ($m^3 \times 10^6$)	856
		Watershed area (ha)	34609

* Ontario Ministry of Natural Resources Fisheríes Acidification Program data (1982)

TABLE II

PARTICULATE METAL CONCENTRATIONS AND ENRICHMENT FACTORS IN LAKE OPEONGO

Element	Concentration in seston*	$R_{s/p}$**	Element	Concentration in seston*	$R_{s/p}$**
Cu	38±14 (n=41)	1.3	Th	2.5 (n=4)	2.4
Ni	31±10 (n=43)	1.2	La	18 (n=4)	2.7
Pb	108±24 (n=46)	1.6	Ce	31 (n=4)	2.6
Zn	146±43 (n=44)	1.4	Nd	16 (n=4)	3.0
Fe	1.6±0.6%(n=44)	2.2	Sm	3.1 (n=4)	3.1
Al	1.7±0.4%(n=53)	2.6	Eu	0.75(n=4)	2.4
Org. C	28 %(n=45)	0.35	Tb	0.32(n=4)	2.9
			Yb	0.82(n=4)	3.0
			Lu	0.14(n=4)	2.6

*Concentrations are in μg g-1 (dry wt.) unless otherwise stated; n= the number of data points averaged; the ± error function shown is the standard deviation.

** $R_{s/p}$ is the ratio of concentration in sediments to the concentration in suspended particulates.

TABLE III

CONCENTRATION OF THE RARE EARTH ELEMENT IN LAKE OPEONGO SEDIMENTS

Depth (cm)	U	Th	La	Ce	Nd	Sm	Eu	Tb	Yb	Lu
	($\mu g\ g^{-1}$)									
0- 1	1.1	5.4	42	72	40	7.9	1.6	0.8	2.2	0.30
1- 2	1.4	5.6	42	75	38	7.9	1.7	0.8	2.2	0.31
3- 4	0.9	6.5	46	84	50	9.0	1.6	0.8	2.4	0.35
6- 7	1.3	6.6	45	81	49	8.8	1.8	0.9	2.4	0.35
9-10	1.2	6.3	49	81	49	9.6	1.7	0.9	2.5	0.37
14-15	1.1	6.7	50	84	48	9.8	2.2	1.2	2.5	0.38
20-22	1.4	5.8	53	81	50	11.0	2.0	0.9	2.7	0.40
28-30	1.1	5.5	54	77	53	11.0	1.7	1.0	2.7	0.40
43-45	1.4	6.0	56	87	54	11.0	1.7	1.1	2.8	0.42

The $R_{s/p}$ values for the pollutant metals are also fairly constant but with a range of only 1.2 to 1.6 (Table II) and a mean of 1.4. In other words, there is also a fractionation of the pollutant metals between the sediments and the seston. The interesting feature however is the marked difference in the $R_{s/p}$ values of the pollutant metals and those of the conservative elements. The RE and Al in the sediments come from the runoff and river inputs, the weathering of bluff and bedrock materials, and from atmospheric fallout into the lakes. On the other hand, the pollutant metals are believed to come primarily from the atmosphere, and one can envisage a complex source-driven model to explain the observed disparity in the $R_{s/p}$ values. We do, however, prefer the simple explanation which follows.

Pending strong evidence to the contrary, we believe that the reduced $R_{s/p}$ for pollutant metals is due to their leakage out of the sediments. If the pollutant metals were totally conserved in the sediments, just like the RE and Al, the $R_{s/p}$ values close to 2.6 would also be expected. A deviation from this value should give a rough estimate of the regenerative loss of the pollutant metals from the lake sediments. On the basis of this assumption, it is estimated that only 50 to 65% of the total metal input into Lake Opeongo is retained by its sediments. The low retention figure is contrary to the general belief that sediments are the sink for most of the pollutant metal flux into a lake (see Salomons and Forstner, 1984; MARC, 1985). This partial retention/loss may explain some of the common (peculiar) features of trace metal distribution in lake sediments, including the inter-basin and inter-lake differences in the rates of metal accumulation (Forstner, 1977; Nriagu et al., 1979, 1982; Norton et al., 1981; Evans and Dillon, 1983; Dillon and Smith, 1984). The low retentivity also emphasizes the need for caution in the use of trace metal profiles in retrospective assessment of recent changes in rates of pollutant input into a given watershed.

Possible mechanisms involved in the release of pollutant metals from sediments are not difficult to find. Several studies show that the downward flux of seston (and particulate metals) increases exponentially with depth towards the sediment-water interface of many lakes (Chambers and Eadie, 1981; Rosa et al., 1983, Charlton, 1983). This implies that the turnover time for particles at the sediment-water interface is only a few days. The massive resuspension of sediments, orchestrated by bioturbation and bottom currents, represents the basic physical process that drives the metal release. In addition, the dissolution or desorption of sediment-bound metals results in elevated levels of the pollutant metals in the porewaters (Carignan and Tessier, 1985; Carignan and Nriagu, 1985). The continuous out-flux of the metal-enriched porewaters (due to diffusion and bioturbation) also contributes to the self-cleansing of metal pollution from the sediments. Studies of porewater profiles for trace metals indeed show that the regeneration of pollutant metals is a common feature in sediments of many lakes in Ontario (Carignan and Tessier, 1985; Carignan and Nriagu, 1985).

References

Carignan, R. and Nriagu, J.O.: 1985, *Geochim. Cosmochim. Acta* **49**, 1753.

Carignan, R. and Tassier, A.: 1985, *Science* **228**, 1524.

Chambers, R.L. and Eadie, B.J.: 1981, *Sedimentology* **28**, 439.

Charlton, M.N.: 1983, *J. Great Lakes Res.* **9**, 201.

Crocket, J.H. and Kabir, A.: 1981, *J. Great Lakes Res.* **7**,455.

Dillon, P.J. and Smith, P.J. :1984,' *Environmental Impact of Smelters* ',J.O. Nriagu(ed.), Wiley, New York, pp. 375-416.

Evans, H.E. and Dillon, P.J.: 1983, *Can. J. Fish. Aquatic Sci.* **40**, 570.

Evans, R.D. and Rigler, F.H.: 1985, *Water, Air & Soil Pollut.* **24**, 141.

Forstner, U.: 1977, *Arch. Hydrobiol.* **80**, 172.

Forstner, U. and Wittman, G.T.W.: 1981,' *Metal Pollution in the Aquatic Environment* ', Springer-Verlag, Berlin, 486 pp.

MARC: 1985, *Historical Monitoring* , MARC Report No. 31, Monitoring and Research Assessment Centre, University of London, 320 pp.

Norton, S.A., Hess, C.T. and Davies, R.B.:1981,' *Atmospheric Pollutants in Natural Waters* ' S.J. Eisenrich (ed.), Ann Arbor Sci. Inc., Michigan, p. 409.

Nriagu, J.O., Kemp, A.L.W., Wong, H.K.T. and Harper, N.: 1979, *Geochim. Cosmochim. Acta* **43**, 247.

Nriagu, J.O., Wong, H.K.T. and Coker, R.D.: 1982, *Environ. Sci. Technol.* **16**, 551.

Olmez, I. and Gordon, G.E.: 1985, *Science* **229**, 966.

Robbins, J.A.: 1978,' *The Biogeochemistry of Lead in the Environment* 'J.O. Nriagu (ed.), Elsevier, Amsterdam, p 285.

Rosa, F., Nriagu, J.O. and Wong, H.K.T.: 1983, *Chemosphere* **12**, 1345.

Salomons, W, and Forstner, U.: 1984,' *Metals in the Hydrophere* ' Springer-Verlag, Berlin, 340 pp.

Shaw, D.M., Dostal, J. and Keays, R.R.: 1976, *Geochim. Cosmochim. Acta* **40**, 73.

Williams, J.D.H. and Pashley, A.E.:1979, *J. Fish Res. Board Can.* **36**, 241.

Wong, H.K.T., Nriagu, J.O. and Coker, R.D.: 1984, *Chem. Geol.* **44**, 187.

Wong, H.K.T. and Nriagu, J.O.: 1985, Proc.' *Trace Substances and Environmental Health* ' **19**, 388.

ATMOSPHERIC DEPOSITION AND FOREST DECLINE

Thomas D. Crocker
Economics Department
University of Wyoming
Laramie, WY
82070 U.S.A.

Bruce A. Forster
Economics Department
University of Guelph
Guelph, Ontario
Canada, N1C 2W1

ABSTRACT. This paper discusses the economic implications of alternative scientific explanations of observed forest decline. The various explanations can be grouped into two categories. One category posits direct damage to the forest canopy that is reversible. The other category posits indirect damage via soil quality changes that may be irreversible or slow to recover. The paper argues that the decision maker should take into account all of the explanations in designing a control strategy for acid deposition. It would be costly to wait until a single explanation has emerged victorious. If the irreversible decline explanation turns out to be correct, the forest damages will already have been sustained with no recovery possible. Assuming that the impact of acid deposition is to reduce forest productivity by 5% per annum then the annual losses to the commercial timber industry in Canada is estimated to be \$197 million (in 1981 dollars). To this an annual loss of \$1.29 billion (in 1981 dollars) should be added for disruption to recreation and wildlife habitat values.

1. INTRODUCTION

This paper provides an economic assessment of the possible implications of atmospheric deposition (acidic deposition and other air pollutants) for the forest sector with particular reference to Canada. Interest in forestry impacts of acid deposition in particular has been spurred in part by the rapid spread in forest die-back that has occurred in Germany. It has been estimated that the area of forest damage in West Germany has increased from 8% in 1982 to 35% in 1984 (Cowling, 1984). Consensus has been achieved that large acreages of forests in the eastern United States and Canada have been declining over the past two or three decades.

There are six hypotheses: gaseous pollutant (Tomlinson, 1983), soil acidification - heavy metal toxicity (Ulrich et al, 1980), drought

Water, Air, and Soil Pollution **31** (1986) 1007–1017.

(Johnson, 1983), general stress (Johnson, et al, 1984), mg deficiency (Abrahamson, 1980), and, excess N (McLaughlin, 1985). The pollution damage hypotheses can be grouped into two categories. One category posits direct damage to the forest canopy that is reversible. The other category posits indirect damage via changes in soil quality that are either irreversible or slow to recover.

This study formalizes the economic implications of these alternative scientific explanations by incorporating them into dynamic bioeconomic models of forest growth. The losses involved when decisions about acid deposition control policies fail to acknowledge the existence of competing explanation are demonstrated. Finally, the paper presents rough estimates of the economic value of damages thought to be occurring to Canadian forests as a result of acid deposition.

2. ATMOSPHERIC DEPOSITION IMPACTS AND FOREST DECLINE

As mentioned the various competing explanations can be grouped into roughly two categories - those for which there are direct impacts to the forest canopy and those for which there are indirect impacts to the forest soils.

Consider first the case in which there is direct damage to the forest canopy as a result of ambient concentrations of pollutants say W. Denoting forest biomass by R, then the forest dynamics may be described by a pollution version of the Lotka growth process

$$\dot{R} \equiv \frac{dR}{dt} = F(R, W) \quad , \quad R(0) = R_0 \tag{1a}$$

where $\frac{\partial F}{\partial R} \gtrless 0$, $\frac{\partial F}{\partial W} < 0$.

A more specific form of this growth process common to the literature would be

$$\dot{R} = R(a-bR) \quad , \quad R(0) = R_0 \tag{1b}$$

where b is a constant but $a = a(W)$, $a' < 0$. The steady-state forest biomass is given by

$$R^W = a(W)/b < R_0 = a(0)/b. \tag{2}$$

Increases in W will reduce a(W) and hence the biological maximum sustainable forest biomass. This may be illustrated in the following simple phase diagram, Figure 1, for the general form (1a).

If the pollution damage is direct to the canopy and if W is reduced to 0, the growth function immediately returns to its original position and the forest grows to R_0.

Consider the case in which the forest impacts are not direct due

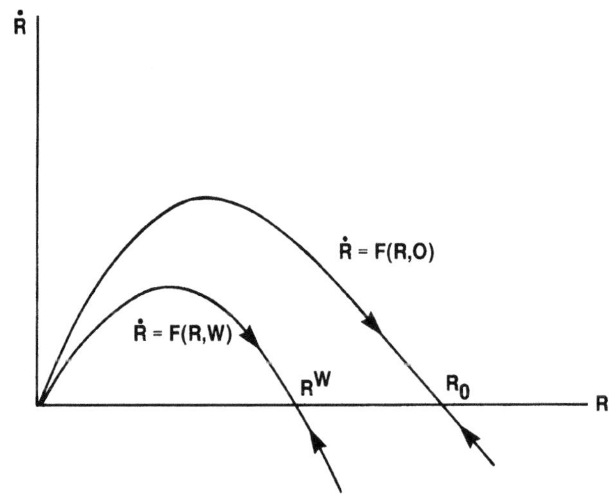

Figure 1: Forest Growth Process with Direct Impacts

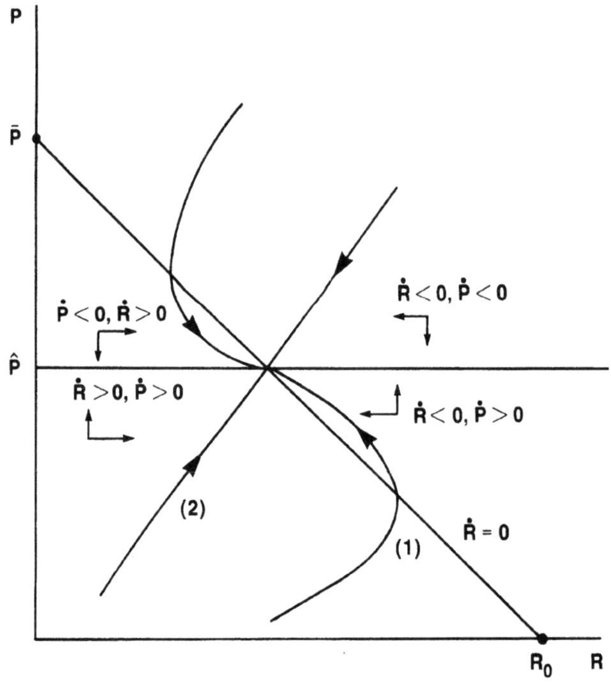

Figure 2: Phase Diagram for Pollution and Forest Biomass

to ambient air concentrations but rather are indirect due to changes in soil quality that result from continued deposition from the air.

Let P denote the accumulated stock of soil pollution (such as Al) and suppose the stock accumulates according to

$$\dot{P} = g(W) - \alpha P \qquad (3)$$

The function g(W) summarizes the geochemical reactions that generate P as a result of inputs of W. It is assumed that the stock of pollution degrades naturally at a rate α. If $\alpha = 0$ then the stock of pollution never decays and the consequent reduction in soil quality is completely irreversible.

The forest dynamics may be re-written as

$$\dot{R} = F(R,P) \qquad (4)$$

where Equation (4) obeys the same conditions as Equation (1) with W replaced by P. However, P may change through time even if W is held constant. In Equation (4) it is assumed that F is independent of W. An alternative assumption is simply that the ambient concentration W is below no-damage threshold levels which is consistent with Tomlinson's (1983a) discussion of the recent German experience.

Assuming a constant ambient concentration W then the forest dynamics are determined by the solution of the dynamical system (5), (6)

$$\dot{R} = F(R,P) \quad , R(0) = R_0 \qquad (5)$$

$$\dot{P} = g(W) - \alpha P \quad , P(0) = P_0 \qquad (6)$$

This two-dimensional system of differential equations may be analyzed qualitatively using phase diagram techniques.

In Figure 2 the locus of (R,P) combinations that yield steady-states for forest biomass ($\dot{R}=F(R,P)=0$) is negatively sloped since from the above assumptions increases in P will reduce the value of R for which F(R,P)=0. Above the curve $\dot{R} < 0$ and the forest declines while the reverse is true below the curve. R_0 is the pollution-free biomass level for the forest, while $\bar{P}$ is the choke pollution level that implies an ultimate complete dessication of the forest.

The locus of steady-states for P is horizontal at $\hat{P} = g(W)/\alpha$. $\dot{P} < 0$ above $\hat{P}$ and $\dot{P} < 0$ below $\hat{P}$. The region below $\hat{P}$ appears to be more relevant for current discussions.

Figure 2 has been constructed so that an equilibrium $(\hat{R},\hat{P})$ exists. This requires that the $\dot{R}=0$ locus intersects the $\dot{P}=0$ locus. It is not difficult to demonstrate that $(\hat{R},\hat{P})$ is locally stable. In Figure 2, two representative time paths for R and P are shown. Path (1) is typical of paths for which the forest continues to grow while the pollution level is rising. However, in finite time the path hits the $\dot{R}=0$ locus and the forest ceases to grow as the P variable has choked off further growth. The pollution stock continues to accumulate and the forest enters a decline phase while moving to the equilibrium. The warning in this is that monitoring forest growth in the short-run may be very misleading except for very mature stands with biomass close to R_0. One could conclude the absence of damage even though a decline is about to set in. This may explain the sudden decline observed in German forests - the damage has been occurring but the actual decline is more recent. In fact for those forests now showing visible signs of decline, tree ring analyses indicate that the decline was initiated 2 to 3 decades earlier (McLaughlin, 1985). Along paths such as (2) decline is not observed, but rather the forest always exhibits growth but to lower biomass than could have been achieved without the pollution impacts.

In this case if W were reduced to zero, the stock of P is still present although it would decay exponentially, $\dot{P} = -\alpha P$. Thus the forest dynamics would be explained by

$$\dot{R} = F(R, P_T e^{-\alpha t})$$

where P_T is the pollution level in existence when W is reduced to zero. The F function shifts up - but only over time as P decays towards zero. This may take a very long time - the smaller is α the slower is the recovery of the soil. Even when W has been decreased the forest may continue to decline for a period of time.

There are two possibilities that produce even more pessimistic results. If $\hat{P} > \overline{P}$ then no meaningful equilibrium exists and all paths show a short term growth phase followed by the complete collapse of the forest. This collapse may be offset if W is reduced sufficiently so that $\hat{P} < \overline{P}$ results but the forest may still have a growth phase followed by a decline phase.

If $\alpha = 0$ then there is no equilibrium with $W > 0$. In this case the pollution of the soil is completely irreversible. The forest can never recover to its pollution free biomass even if W is reduced to zero. Once P reaches $\overline{P}$ the forest is no longer a replenishable resource, but rather a "wasting" resource like coal or oil.

3. DECISION THEORY AND THE CHOICE OF COMPETING HYPOTHESES

Alternative hypotheses must be studied because acid deposition policymakers are trying to decide upon some course of action; they are concerned with measures of environmental and economic consequences rather than with the scientific validity of particular stories. Fundamentally, their problem is to establish a probability distribution for these consequences as a function of their policy decision variables. In economic efficiency terms, their decisions will be improved within the Bayesian framework if they use all variable information rather than only a subset [Raiffa and Schlaifer (1961)].

Presume that over any particular time interval the policymaker is interested in maximizing the prospective net economic benefits of forest growth. This prospective forest growth, Y, could be influenced by the future levels, A, of atmospheric deposition. At this time, the policymaker has observations on the past behavior of forest growth, Y, and past and present atmospheric deposition, A. He also is confronted with six explanations, $S_1,\ldots,S_6$, each of which purports to explain the observed empirical association between Y and A. Each of the six explanations is represented by its parameter vector, $\beta_1,\ldots,\beta_6$; the vectors do not necessarily have the same dimensionality among explanations. For example, the parameters for the drought explanation differ greatly from the mg and ca foliage leaching explanation. In effect, each of the explanations specifies a density function $f(Y|S_i,\beta_i,A)$, $i = 1,\ldots,6$, between past observations on forest growth and acid deposition.

Given the historical observations, the policymaker attaches probabilities $p(S_i|Y,A)$ to each of the explanations. For each explanation, he will also have a density function, $g(\beta_i|S_i,Y,A)$, for its parameters. With this information, the policymaker can construct a mixed marginal or predictive distribution:

$$f(\hat{Y}|Y,A,\hat{A}) = \sum_{i=1}^{6} p(S_i|Y,A)\int_{\beta_i} f(\hat{Y}|S_i,\beta_i,\hat{A})g(\beta_i|S_i,Y,A)d\beta_i \ . \qquad (7)$$

As Zellner (1971) shows, this expression reduces to:

$$f(\hat{Y}|Y,A,\hat{A}) = \sum_{i=1}^{6} p(S_i|Y,A)f(\hat{Y}|S_i,Y,A,\hat{A}) \ . \qquad (8)$$

In effect, Equation (7) is simply a mixture of the individual model marginal distributions where the mixing weights are the individual model probabilities, $p(S_i|Y,A)$. Note that the distribution in Equation (2) incorporates all the policymaker's information.

Now suppose the policymaker must select an atmospheric deposition control action, $\hat{A}$. If $R(\hat{A}, \hat{Y})$ is the policymaker's payoff function and U is his expected utility, then the optimal $\hat{A}$ will solve the problem

$$\max_{\hat{A}} \int_{\hat{Y}} U(R(\hat{A},\hat{Y}))f(\hat{Y}|Y,A,\hat{A})d\hat{Y} \ . \qquad (9)$$

Suppose now that instead of using the mixed predictive distribution in (8), the policymaker first selects a particular model, S_i, and then employs the individual predictive distribution of S_i to select an $\hat{A}$. Let $\hat{A}^*$ solve (9), and let $\hat{A}_i^*$ solve that version of (9) which occurs when $f(\hat{Y}|S_i,Y,A,\hat{A})$ is used in place of (8). In the absence of differences in computational cost, the solution A_i^* can be no better than A^* since more information is always valuable; that is

$$\int_{Y} \left\{ U(R(\hat{A}^*,\hat{Y}))-U(R(\hat{A}_i^*,\hat{Y})) \right\} f(\hat{Y},A,\hat{A})d\hat{Y} \geq 0 \qquad (10)$$

Thus a decision strategy which insists that forest decline be attributed to a single explanation prior to taking action fails to account for model and parameter uncertainties. This strategy cannot improve outcomes and may make them even worse. In the case of atmospheric deposition impacts upon forests, the strategy is myopic since some of the explanations imply path-dependent intertemporal effects that we have earlier termed irreversibilities. Tesfatsion (1980) has shown that the opportunity losses associated with myopic optimization become greater as the degree of uncertainty about future outcomes increases and as the degree of positive correlation among period-by-period returns declines. Intuition suggests that current decisions into which possible irreversible atmospheric deposition impacts upon forests do not enter will increase future uncertainties and/or reduce the degree of correlation among period-by-period returns.

4. ECONOMIC VALUATION OF FOREST IMPACTS

Despite the concern over negative impacts of atmospheric deposition to forests, the forest products industry in North America has remained skeptical about any "acid rain" link to observed, or possible future, damages (Popovich, 1984). It is easy to understand why foresters and the forestry sector have not yet reacted strongly in Canada to the issue of atmospheric deposition. Given the highly complex and speculative nature of the scientific debate, this issue would seem to pall into insignificance when compared to known losses to the forest sector from so-called "natural" causes. In 1980-81 the average annual loss of commercial timber due to fire, insects and disease was 100 Mm^3 (Forestry Chronicles, June 1983). The annual net production of commercial timber is roughly 150 Mm^3 with a possible expansion to 200 Mm^3 (McDougall 1984). Thus known losses are 50 to 66% of net production levels.

The annual forest growth in Eastern Canada exposed to various levels of sulphate deposition (as a surrogate for atmospheric or acidic deposition) are shown in Table I.

In their study of socioeconomic impacts of LRTAP in Canada, DPA (1984) decided that "Without appropriate dose-response functions for the forestry sector, biophysical and socioeconomic impacts cannot be usefully estimated at this time." As a counter approach we follow that expressed recently by Crocker and Regens (1985) that there is no good reason not to attribute observed reductions in forest growth to acid deposition. The United States National Academy of Sciences ascribed a 5% annual forest growth reduction to acid deposition. According to Loucks (1985) forest productivity has been reduced 5 to 15% below historic levels; however, only part of this reduction is attributable to acidic inputs (Loucks, 1985). Fraser (1985) reports that based upon Delphi survey techniques, the estimated productivity decline for forests in Atlantic Canada was 3.5 and 5% for Ontario and Quebec during 1974-1984 and 5 and 7%, respectively, for 1954-1984.

Assuming that all eastern Canadian forest growth occurring in regions receiving at least 10 kg $ha^{-1}.yr^{-1}$ of sulphate deposition are

TABLE I

ANNUAL FOREST GROWTH ($m^3.10^3$) BY PROVINCE RECEIVING GREATER THAN 10 kg $ha^{-1}.yr^{-1}$ SULPHATE DEPOSITION

	Annual Growth $m^3.10^3$		
	Sulphate Deposition kg $ha^{-1}.yr^{-1}$		
	10-20	20-40	> 40
Ontario	31,065	12,544	998
Quebec	47,363	39,659	4,438
New Brunswick	0	11,474	0
Nova Scotia	0	7,077	0
Prince Edward Island	0	443	0
Newfoundland	1,734	3,456	0

Source: Memorandum of Intent, 1983.

suffering an average reduction of 5%, we estimate the economic value of the losses for each province by simply taking the percentage loss of the current value of growth. The results are summarized in Table II.

This estimate should be used cautiously since it relies on the assumed 5% loss coefficient from the National Academy of Sciences although this seems to be in the right ball park. Even if the 5% is scientifically correct, the estimate at best provides an "order of magnitude" assessment. It is less clear what "economic welfare" implications lie behind it since it does not allow for price adjustments that might occur.

The estimate of an annual loss of \$197 million (in real terms) presumes that the reduction is a constant 5% per annum. This is possible only if the result of deposition is to produce a loss in sustainable yield. If the impact of the deposition however is to cause a widespread die-back that is either getting progressively worse or irreversible then the estimates of loss will be substantially higher than the above. It is worth pointing out that while agricultural areas may be able to cope with air pollution and other stresses by adapting new crop species, frequently this option is not available to forest areas (Smith, 1985, 88). Forests respond to stress by altering their composition to less diverse communities (McLaughlin, 1985).

To the estimated commercial losses we must add the non-timber losses due to the disruption of recreational areas of wildlife

TABLE II

FOREST GROWTH, VALUE AND ESTIMATED ECONOMIC LOSSES DUE TO ACID DEPOSITION

Province	Total Annual Yield (000s) m^3	Value \$000 (\$1981 Cdn)	Estimated Economic Losses Due to Acid Deposition (\$000) (\$1981 Cdn)
Quebec	81,460	2,138,326	106,916
Ontario	44,607	1,170,934	58,547
New Brunswick	11,474	301,193	15,060
Nova Scotia	7,077	185,771	9,289
Newfoundland	5,190	136,238	6,812
Totals	149,808	3,932,462	196,624

Source: Memorandum of Intent, 1983.

habitats. Drawing on results in Crocker et al. (1980), a 5% reduction in non-timber values produces a 1981 loss of C$13.68 ignoring differences in population density and accessibility between the U.S. and Canada. There are 94 x 10^6 ha of forests in eastern Canada receiving at least 10 kg of sulphate ha^{-1}. Assuming that this entire area is subject to the 5% loss in values, this loss amounts to $1.29 billion in 1981. Adding the timber and non-timber losses produces an annual estimated loss of $1.49 billion in 1981 dollars. These estimates which are very crude should be treated as "ball-park" estimates that may be useful in the rank-ordering of receptor categories. The estimate for the United States in 1978 dollars was U.S. $1.75 billion of which roughly U.S. $600 million was for commercial timber losses (Crocker and Regens, 1985).

5. CONCLUSIONS

The alternative scientific explanations of observed forest decline can be grouped into two categories that have potentially quite different long-run implications. Those that posit direct damage to the forest canopy imply that the forest can fully recover if atmospheric deposition is stopped. Those that posit indirect damage to forest soils imply that such a recovery may not be feasible or will occur much more slowly than in the direct damage scenarios. In planning acid deposition control strategies the decision maker should resist the temptation to wait until an explanation has emerged with well-defined parameters. The decision maker should use all the available information. Planning should consider a wide range of cases.

ACKNOWLEDGMENTS

The authors thank Dr. Billy M. McCormac and an anonymous referee for helpful suggestions in preparing this paper.

REFERENCES

Abrahamson, G.: 1980, "Acid Precipitation, Plant Nutrients, and Forest Growth,"in D. Drablos, and A. Tollan, eds., Ecological Impact of Acid Precipitation, Proceedings of an International Conference, SNSF Project, Oslo, Norway, 58-63.

Cowling, E.B.: 1984, The Environmental Forum, 7.

Crocker, T.D. and Regens, J.L.: 1985, Environmental Science Technology 19, 112.

Crocker, T.D., Tschirhart, J.T., Adams, R.M., and Forster, B.A.: 1980, Methods Development for Assessing Acid Precipitation Control Benefits, a report to the United States Environmental Protection Agency.

DPA Group, Inc.: 1984, LRTAP in Canada: Socioeconomic Resources at Risk and Preliminary Damage Estimates, a report submitted to the Environmental Strategies Directorate, Environment Canada.

Fraser, G.A.: 1985, "Issues in the Evaluation of Long Range Air Pollution Damage to Canadian Forests," paper presented to the meetings of the Mid-West Forest Economists.

Johnson, A.H.: 1983, Journal of the Air Pollution Control Association 33, 1049.

Johnson, A.H. Siccama, T.G., Turner, R.S., and Lord, D.G.: 1984, "Assessing the Probability of a Link between Acid Precipitation and Decreased Growth Rates of Trees in the Northeastern United States," in R.A. Sinthurst, ed., Direct and Indirect Effects of Acidic Deposition on Vegetation, Boston, Mass.: Butterworth Publishers, 81.

Loucks, O.L.: 1985, "Biological Productivity," draft manuscript.

McDougall, F.W.: 1984, The Forestry Chronicle, 175.

McLaughlin, S.B.: 1985, Journal of the Air Pollution Control Association 35, 1985.

Popovich, L.: 1984, Journal of Forestry, 278.

Raiffa, H., and Schlaifer, R.: 1961, Applied Statistical Decision Theory, Cambridge, Mass.: The M.I.T. Press.

Smith, W.H.: 1985, Journal of Forestry 83, 82.

Tesfatsion, L.: 1980, Journal of Economic Dynamics and Control 2, 135.

Tomlinson, G.H.: 1983a, Environmental Science and Technology 17, 246A.

Tomlinson, G.H.: 1983b, "Die-back of Red Spruce, Acid Deposition, and Changes in Soil Nutrient Status - A Review," in B. Ulrich and J. Pankrath, eds., Effects of Accumulation of Air Pollutants in Forest Ecosystems, Boston, Mass.: D. Reidel Publishing Co., 331.

Ulrich, B., Mayer, R., and Khanna, P.K.: 1980, Soil Science 130, 193.

Zellner, A.: 1971, An Introduction to Bayesian Inference in Econometrics, New York: John Wiley & Sons.

ECONOMIC VALUATION OF ACIDIC DEPOSITION DAMAGES: PRELIMINARY RESULTS FROM THE 1985 NAPAP ASSESSMENT

J. M. Callaway
R. F. Darwin
R. J. Nesse
Pacific Northwest Laboratory
Richland, WA 99352 USA

ABSTRACT. This paper presents estimates of the monetary value of the biophysical damages caused by acid deposition for the National Acid Precipitation Assessment Program's 1985 Assessment. Economic damages are estimated for four effect areas: commercial agriculture, forests, recreational fishing and selected types of materials. The preliminary nature of the value of the estimated damages in each area is emphasized. The estimated values should be interpreted with appropriate regard for the uncertainties about the physical effects of acid deposition and the assumptions in the economic analysis.

1. INTRODUCTION

The objectives of this paper are to discuss methods used to estimate the monetary value of the biophysical damages caused by acid deposition for the 1985 Assessment being conducted by the National Acid Precipitation Assessment Program (NAPAP) and to present the preliminary estimates of economic damages from the Assessment.

2. OVERVIEW OF ECONOMIC ANALYSIS

The primary goal of the economic analysis contained in the 1985 Assessment was to estimate the monetary value of the biophysical damages currently being caused by acid deposition. An important aspect of the damages-oriented focus of the 1985 Assessment is that the preliminary damage estimates reported in this study may not be the same as the benefits of control policies for regulating the precursors of acid deposition. The estimated damage values contained in the

---

This research was conducted for NAPAP by Energy and Resource Consultants, Mathtech, by Drs. R. W. Adams and B. A. McCarl at Oregon State University and by Pacific Northwest Laboratory under a Related Services Agreement with the U.S. Department of Energy under Contract DE-AC06-76RLO 1830.

Water, Air, and Soil Pollution **31** (1986) 1019–1034.

1985 Assessment assume fairly pristine environmental conditions. Control policies would probably not achieve these conditions. In general, information about the incremental benefits of various control options needs to be integrated with information about incremental control costs to determine economically efficient levels of damage reduction.

Four types of receptor systems, sensitive to acid deposition, have been included in NAPAP's economic analysis. These include: commercial agriculture, commercial forests, recreational fishing, and common building, cultural and infrastructure materials. Specific details about market coverage and geographic focus in each of these areas are dealt with in Sections 3.1-3.4 of this study.

Several comments about the extent of coverage in each of these areas are necessary. First, the treatment of economic damages in each of these areas is confined largely to the economic agents who actually use the services provided by the resources at risk in a particular receptor system, such as fishermen, timber owners, anglers and homeowners. The analyses do not take into account damages to non-users who would, in fact, be willing to pay some amount of money to preserve resources at risk for their own future use (Cicchetti and Freeman, 1971) or for future generations (Krutilla and Fisher, 1975). Second, only damages in the United States were evaluated. Finally, the ancillary effects which the precursors of acid deposition may have on health and visibility have also been omitted from coverage in the 1985 Assessment. This latter decision was based on the fact that these effects are covered by other programs within the U.S. Environmental Protection Agency, while NAPAP's focus is on acid deposition.

Two different approaches can be used to value the biophysical damages of acid deposition (Smith, 1982). The first is the damage function approach. It uses dose-response functions to translate changes in acid deposition into physical damages. Physical damages are then multiplied by a unit economic value, such as a market price, to measure economic damages. This approach is used in the estimation of economic damages to recreational fishing and materials. There are two basic problems with this technique (Crocker, 1982). First, it does not take into account possible changes in the prices of resources at risk as a result of acid deposition-induced reductions in the supplies of these resources. Second, it ignores substitutions which individuals and firms can make in their consumption and production decisions due to changes in prices and technology. As a result, extremely unrealistic assumptions have to be made about the preferences of individuals, the technology of firms, and the structure of the markets being analyzed for the damage function approach to yield damage measures which are consistent with economic theory.

A second approach overcomes these shortcomings by using mathematical representations (models) of supply and demand curves in appropriate markets to characterize the behavior of consumers and producers. Dose-response information is used to shift the relevant supply or demand curves to simulate the effects of acid deposition on key physical parameters. Appropriate damage/benefit measures can then be evaluated from these simulated shifts in supply and demand curves. This approach is used to value the damages of acid deposition to commercial agriculture, commercial forests and recreational fishing (along with the damage function approach).

In these three areas, the damages born by resource users and consumers are approximated (Willig, 1976) by changes in consumer surplus, while the damages experienced by producers and resource owners are measured by changes in producer surplus. Consumer surplus is conventionally defined as the difference between the maximum amount of money an individual is willing to pay for a good rather than do without it and the amount actually paid. Producer surplus represents the difference between the amount of money which a producer receives from the sale of a good and the minimum amount he would accept rather than forego selling it (Just, Hueth, Schmitz 1982).

3. METHODS AND DATA

3.1 Commercial Agriculture

The agricultural portion of the 1985 Assessment examines the direct and indirect effects of increases in acid deposition on consumer and producer surplus in the U.S. agricultural sector. For this assessment, the direct effects of acid deposition consist of changes in soybean yields in the 28 leading soybean producing states* due to increases in H^+ concentrations in wet deposition. Indirect effects consist of changes in farm production costs for all crops grown in the same area as a result of the passive fertilization of cropland by acid deposition (Hoeft et al., 1972), and increases in the intensity of liming applications on more acidic soils (Reuss, 1977).

The model used in this analysis was previously employed by Adams et al. (1984) to estimate the economic effects of O_3-induced crop yield reductions on U.S. agriculture. This model is a multi-regional, non-linear programming model of the U.S. agricultural sector (Baumes, 1978; Chattin et al., 1983). It maximizes the sum of producer and consumer surplus in regional markets for 12 field crop and 11 livestock commodities, 12 processed agricultural commodities, and three national inputs-land, labor and fertilizer. The constraints in the model consists of a series of regional crop-mix and input use combinations consistent with profit maximizing behavior of individual farmers (McCarl, 1982). Regional crop yields can be modified to reflect information about the magnitude of acid deposition-induced yield changes. The model solves for market clearing commodity and input

*These states are: Alabama, Arkansas, Delaware, Florida, Georgia, Illinois, Indiana, Iowa, Kansas, Kentucky, Louisiana, Maryland, Michigan, Minnesota, Mississippi, Missouri, Nebraska, New Jersey, North Carolina, North Dakota, Ohio, Oklahoma, South Carolina, South Dakota, Tennessee, Texas, Virginia, and Wisconsin.

prices in all markets simultaneously. It also provides information about the magnitude and distribution of producer and consumer surplus for each market and region.

Estimation of the economic value of the damages due to the direct effects of acid deposition was performed as follows. The agricultural model was first solved using 1982 soybean production (yield) data and the resulting values of consumer and producer surplus for the U.S. agricultural sector were recorded. Next, soybean yields in the 28 state region were modified to reflect the physical effects of acid deposition on soybean yields. These modifications were based on simulated changes in yields due to differences between current (1982) H^+ concentrations and more pristine base conditions, including 10, 30 and 50% reductions in H^+ from the 1982 figure. The sector model was then solved using this yield information and the new values for consumer and producer surplus in the agricultural sector were recorded. The value of the economic damages associated with a given set of yield modifications was calculated as the difference in total economic surplus, with and without acid deposition.

The same general approach was adopted to incorporate the indirect effects of acid deposition into the analysis. For this part of the analysis the assumed soybean yield changes for the 50% case were combined with reductions in fertilizer costs of $1.19 to $3.21 per hectare and increases in liming costs of $0.17 to $0.52 per hectare. These cost changes were distributed spatially based on the analysis contained in Moskowitz et al. (1985) regarding the effects of acid deposition on soil amendment practices.

Soybeans were selected for the direct effects analysis because this is the only major, commercially-important crop which has shown a consistent negative yield response to increases in H^+ concentrations in wet deposition (Irving, 1983). Dose-response functions were developed by Moskowitz et al. (1985) using data from field experiments on four varieties of soybeans by Evans et al. (1984, 1985) and Banwart et al. (1985). County-level yield changes for the 28 states were estimated by evaluating these dose-response functions using state-level estimates of 1982 soybean production and simulated changes in concentrations of H^+.

3.2 Commercial Forests

The forests section of the 1985 Assessment addresses the impacts of hypothetical changes in annual tree growth in the Northeast and Southeast timber supply regions* on regional stumpage and primary forest product markets in the United States as a whole. No empirical relationships between variations in wet and dry deposition and

*The Northeast includes Connecticut, Delaware, Maine, Maryland, Massachusetts, New Hampshire, New Jersey, Pennsylvania, Rhode Island, Vermont, and West Virginia; the Southeast includes Florida, Georgia, North Carolina, South Carolina, and Virginia.

recently-observed episodes of forest dieback or decreases in forest growth (Siccama _et al._, 1982; Johnson and Siccama, 1983) have been found. Thus, this portion of the _Assessment_ does not estimate the economic damages of current levels of acid deposition or of any form of air pollution. The analysis is designed to show, instead, the range of economic effects that could occur due to a wide range of growth slowdowns, regardless of the source of those slowdowns.

The economic effects of hypothetical reductions in tree growth were estimated using existing softwood and hardwood forest sector market models, TAMMSOFT (Adams and Haynes, 1980, 1985) and TAMMHARD (Lange, 1983), which are physically linked to the inventory projection model, TRAS (Larson and Goforth, 1970, 1974). TAMMSOFT and TAMMHARD are multi-market spatial equilibrium models of the softwood and hardwood stumpage and primary forest products markets in the United States. Both models are spatially disaggregated into supply and demand regions. Regional supply and demand equations interact to determine market prices, production and consumption, as appropriate, for lumber and/or plywood. Aggregate stumpage demand is derived directly from product supply by a series of conversion factors. Stumpage supply equations in the models reflect the harvest response of private owners to changes in stumpage prices and inventories. Hardwood and softwood inventory volumes are supplied to both models directly from TRAS. The market clearing harvest levels are used in TRAS to update the softwood inventories in each region. The outputs of the market models include information about the size and distribution of consumers' and producers' surpluses in regional stumpage and product markets.

Estimation of the value of the damages due to hypothetical decreases in tree growth was performed as follows. First, base case values for consumer and producer surplus in all markets was established by running the above system of models for the period 1976-2030. Total economic surplus in the U.S. forest sector was calculated annually from 1985 to 2030 as the sum of the economic surpluses of timber owners, plywood and lumber producers and consumers of lumber and plywood in all regions. Next, the radial growth rates in TRAS used to calculate the average annual grown increments for both hardwoods and softwoods were reduced simultaneously in the Northeast and Southeast by 10, 15 and 20% to reflect hypothetical decreases in tree growth. The system of models was rerun using this information and the new values for consumer and producer surplus in the U.S. forest sector were recorded and aggregated. Finally, the annual value of the economic damages for each hypothetical growth reduction was calculated as the difference between the total surplus values, with and without the growth change. The present value of these damages for the period 1985-2030 was calculated and annualized (i.e., levelized) using a 10% discount rate.

3.3 Recreational Fishing

This section of the 1985 Assessment focuses on the economic consequences of acid deposition damages to recreational fishing in the Adirondack region of New York. This was done because of the public attention this particular area has received and the region's importance as a center of freshwater game fishing both to New York residents and out-of-state visitors. The focus of the analysis was further narrowed to include only damages to Adirondack Lakes due to the lack of information on the extent of physical damages to streams in the Adirondack region.

To provide checks on the estimated value of the damages, two methods were used to estimate the economic damages associated with Adirondack Lake fishing: a participation model and a travel cost model. The participation model was a regression model which related variation in the time which individuals spent fishing at each of 24 multiple-lake sites to variation in selected characteristics of these sites, including fishable area and catch rates. Using information about the effects of acid deposition on these characteristics, the model was used to forecast changes in the number of fishing days. Estimates of economic damages at each site were obtained by multiplying the simulated change in fishing days at the site due to acid deposition by the value of a fishing day ($30), obtained from the travel cost model and checked against Vaughn and Russell (1982a). Total damages were estimated by aggregating the damages from each of the 24 sites.

The travel cost approach is based on the premise that the demand curves for recreation at a site can be derived by observing how people respond to differences in the costs of traveling to particular recreation sites. As the distance from the point of origin to a recreation site increases, the time and distance costs of travel increase and visitation rates decrease. The travel cost method used in this analysis employed a procedure based on the work of Vaughn and Russell (1982b). It involved estimating the parameters of 24 site-specific regression models, relating the number of fishing days at a site by an individual to travel distance, income and fishing experience. The intercept and distance parameters of the previous models were then regressed against fishable area and catch rates. Next, estimates of the effects of acid deposition on fishable area and catch rates were used to shift the first-stage travel cost demand equation. Changes in consumer surplus were evaluated from these equations for each of the sites and aggregated over all sites to obtain an estimate of total annual damages.

The participation model and travel costs models did not use individual pond or lake data. Individual lakes and ponds were aggregated into 24 sites, based on the location of the pond or lake's outlet. The basis for this aggregation scheme and for the site characteristics data was the 1984 version of the Adirondack Lake and Pond Survey, made available through the New York State Department of Environmental Conservation. Information on individual anglers, their travel expenses and their catches was obtained from the New York Angler Survey for 1976-1977 (Kretser and Klatt, 1981).

The physical effects information used in this analysis was based on a recent NAPAP study of the status of Adirondack fish populations by Baker and Harvey (1984), who estimated that approximately 8.1% of the region's lakes had lost some or all fish populations. Another 10.6% of these lakes evidenced marginal damages due to acidification. These losses translate into estimates of lost fishable area of 3.2 and 10.0%, respectively. To examine how sensitive the damage estimates might be to changes in catch rates, economic damages were calculated for a 3.2% loss of fishable area with no change in catch rates and then with a 3.2% decrease in catch rates. The analysis was repeated for a 10% reduction in fishable area first with no change and then a 10% decrease in catch rates, for a total of four damage scenarios.

3.4 Materials

The materials section of the 1985 Assessment covers damages to selected common building, cultural and infrastructure materials due to increases in SO_2 and H^+ concentrations in wet deposition within cities in the northeastern quadrant of the United States.* Common building materials included in the analysis are steel, Cu, Al, paint, marble and mortar used in commercial and residential buildings. Cultural materials included bronze, marble and limestone statutes and marble and limestone trim on historic buildings. Coverage of infrastructure materials was limited to transmission towers.

The methodologies used to estimate the economic value of acid deposition damages in all of the materials categories were quite similar. For brevity, only the method used to estimate economic damages to common building materials is presented here. It consisted of the following steps. First, an inventory of exposed materials for four sample cities (New Haven, Connecticut; Portland, Maine; Pittsburgh, Pennsylvania; and Cincinnati, Ohio) was developed to determine the exposed surface area for each of the selected materials. Predictor equations were developed to project exposed surface areas from the sampled areas to a total group of 117 cities in the 17 state region covered by the assessment. Next, damage functions for paint, Zn, stone and mortar were used in conjunction with information about local changes in SO_2 and H^+ concentrations to calculate the decrease in the maintenance and replacement interval for each type of material in the projected inventory of exposed materials. Unit repair/replacement costs were developed for each component of the projected inventory of exposed materials. Annual economic damages were computed for each material component as the product of: cost per unit surface area,

*The area covered includes Connecticut, Delaware, Illinois, Indiana, Kentucky, Maine, Maryland, Massachusetts, New Hampshire, New Jersey, New York, Ohio, Pennsylvania, Rhode Island, Vermont, Virginia, District of Columbia, and West Virginia.

exposed surface area in the inventory and the estimated change in the maintenance/repair interval. These damages were aggregated over each type of inventory component to obtain city-wide and then area-wide damage estimates for the 117-city group. These future value damage estimates were converted to present value terms and then annualized present value damage estimates using a 10% real discount rate.

The inventory of common construction materials was based on data from a ground-based survey, conducted by the U.S. Army Corps of Engineers (Merry and LaPotin, 1985). The inventory of historic buildings was based on the National Register of Historic Places. The inventory of statues and monuments was developed from an inventory of outdoor statues and monuments in the eastern United States by Panhorst (1985). Transmission towers were inventoried through contacts with individual electric utilities.

Damage (mass loss) functions for Zn, Cu, and Al were derived by Lipfert et al. (1985a) from extant atmospheric corrosion data. Damage (mass loss) functions for carbonate and silicate paint were developed by Haynie (1985). Damage (dissolution) functions for marble and limestone were developed using data from three NAPAP test sites (Reddy et al., 1985). A damage function for mortar was developed from the limestone damage function based on the assumption that mortar can be represented by an inert matrix composed of one-third calcium carbonate (lime).

The unit costs of repairing/replacing common building materials was based on industry norms. Replacement costs of statues and monuments were based on original costs as reported in Monument News (1870-1930). Rehabilitation costs for bronze statues were developed from case study data (Society for the Preservation of New England Antiquities, 1985). Replacement costs for historic buildings with carved or plain stone trim were derived by the U.S. Park Service.

Economic damage estimates were calculated based on changes in the physical integrity of relevant materials due to changes in SO_2 and H^+ concentrations from base to current conditions. Base conditions were assumed to be an SO_2 concentration of 0.5 μgm^{-3} and pH of 5.2. Current concentrations of SO_2 and H^+ were simulated in each of the 117 cities using an existing long-range transport model, ASTRAP (Shannon, 1981), for SO_2 sources outside each metropolitan area. A parameterized air quality model, PAQMAN (Lipfert et al., 1985b), was developed to simulate locally-generated concentrations of SO_2 and H^+ in each city.

4. RESULTS

4.1 Commercial Agriculture

A summary of the major preliminary results from the Assessment is presented in Table I. The first three rows highlight the results from the direct effects analysis. The annual decrease in the total value of economic surplus (e.g., damages) for these cases ranges from about

TABLE I. Agriculture: Estimated Annual Changes in Economic Surplus Due to Simulated Changes in Crop Yields, Passive Fertilization and Liming (Millions of 1984 Dollars)

Case	Change in Economic Surplus		
	Consumers	Producers	Total
Direct Effects			
-10%*	-17.2	-3.2	-20.4
-30%*	-66.5	-8.6	-75.1
-50%*	-184.6	-32.2	-152.4
Direct† plus Indirect effects	-109.5	161.0	51.5

*Indicates reference levels for H^+ concentrations in relation to 1982 environmental conditions
†Yield reduction consistent with -50% case

$20 million for the 10% case to about $152 million in the 50% case. These damages are a small fraction (never more than 0.002) of the estimated $144 billion total surplus value for the agricultural sector in 1982. These damages are also significantly lower than $2.4 billion welfare loss estimated by Adams et al. (1984) for a 25% increase in tropospheric O_3 concentrations. In addition, the analysis shows that these damages are not borne evenly by consumers and producers: higher commodity prices caused by reduced production hurt consumers more than producers. The final row of Table I shows the combined effects of acid deposition-induced yield and cost changes. Consumers are made worse off in this last case due to higher prices, while producers tend to benefit more from free fertilizer than they are damaged by small increases in liming costs. The net result is a small increase in the value of total surplus, suggesting that economic agents in the U.S. agricultural sector may be benefited, rather than damaged, by current levels of acid deposition.

4.2. Commercial Forests

Table II presents preliminary estimates for the annualized changes in the net present value of economic surplus cumulated over the period 1985-2030 for the three reduced growth cases. Estimated total damages, shown in the last column of Table II ranged from about $340 to $510 million annually depending on the severity of the hypothetical growth reduction. The corresponding changes in the net present value of damages over the entire period (not shown) are roughly $3.4, $4.4 and $5.0 billion.

TABLE II. Commercial Forests: Estimated Annualized Changes in the Present Value* of Economic Surplus Due to Hypothetical Reductions in Radial Growth for Selected Years (Millions of 1984 Dollars)

Case	Change in Economic Surplus			
	Timber Owners	Producers	Consumers	Total
-10%†	63.4	-63.4	-342.4	-342.4
-15%†	109.9	-40.4	-516.3	-446.8
-20%†	220.7	148.0	-878.8	-510.1

*For the period 1985-2030.
†Indicates percent reduction in radial growth rates in Northeast and Southeast.

The distributional impacts of the assumed growth reductions are shown in the first three columns of Table II. In general, timber owners as a group benefit from reductions in radial growth due to their ability to pass along higher prices to the producers of primary wood products. The annualized benefit received by timber owners was estimated to range from about $60 to $220 million. The picture for producers of primary wood products is mixed. As a group, they experience annual damages in the 10 and 15% reduced growth cases of $60 and $40, million but receive roughly $150 million in annual benefits when radial growth is reduced by 20%. Of the three groups, consumers of wood products bear the brunt of the damages due to simulated reductions in radial growth. The decreases in economic surplus of this group ranged from about $340 to $880 million annually. Moreover, a substantial portion of these losses represent transfers of economic surplus from consumers to producers and timber owners as physical damage levels were increased.

Table II does not present a geographic breakdown of economic damages. However, timber owners in the Southeast suffer economic damages due to all growth slowdowns. In addition, producers in that region are hurt considerably more by these changes in growth than producers in the Northeast, who actually experience slight benefits. This suggests a relative (but not absolute) improvement in the competitive position of the Northeast over the Southeast due to decreased tree growth in both regions.

At least three caveats must be applied to the results of Table II. First, the economic surplus changes shown in these tables are short-run estimates which have not been adjusted to include changes in investment in plant capacity in lumber and plywood production. While the market models used in the analysis do simulate changes in production capacity, the costs associated with these changes are not calculated. Second, the possibility of incorporating more intensive forest management practices to counteract slower tree growth through improved forest management was not explored. The capability exists to do this in TAMM (Adams et al., 1980); however, there is limited information about specific investments to curtail recently-observed declines in forest growth in the Northeast and Southeast. Finally, it needs to be emphasized, again, that the growth changes used in the analysis are strictly hypothetical.

4.3. Recreational Fishing

The preliminary results of the above simulations are presented for both the participation and travel cost approaches in Table III. The two approaches produce comparable results. Estimates of the annual value of economic damages range from \$0.7 to \$12.0 million per year. The damage estimates associated with a 3.2% reduction in fishable area and no change in catch rates are of particular interest. Recently, Mullen and Menz (1985) reported a \$1.1 million loss in net economic value in the same region given a 5% reduction in fishable area, very close to the \$0.7 million loss obtained from the travel cost model at a slightly lower level of physical damage.

There are several basic caveats attached to the damage estimates in Table III. First of all, the estimates apply only to Adirondack lakes and do not include streams for which key data were unavailable. Second, the aggregation procedure used to combine sites may not have captured important localized impacts, such as in high altitude lakes. Third, the damage estimates do not include changes in the non-user values as discussed in Section 2.1. Fourth, the treatment of fishing

TABLE III. Adirondack Lakes: Estimated Annual Differences in Economic Surplus Due to Simulated Reductions in Fishable Area and Catch Rates (Millions of 1984 Dollars)

Model Type	Case	Physical Effect 3.2% Damage	Physical Effect 10% Damage
Participation	A	-1.7	- 5.2
	B	-3.2	-10.2
Travel Cost	A	-0.7	- 4.6
	B	-4.8	-12.0

Case A Change in fishable area only.
Case B Change in fishable area and average catch.

behavior in the travel cost model also does not account for the ability of anglers to switch to substitute fishing locations, which would presumably result in lower total economic damages. Finally, as mentioned earlier, the estimated changes in fishable area and catch rates used in the analysis contain unquantified uncertainties that have not been incorporated into the economic damage estimates.

4.4. Materials

The preliminary damage estimates for cultural materials and transmission towers are shown in Table IV. The ranges in values for cultural materials reflect the use of real discount rates of 5 and 10%. The annualized estimated value of the total damages for the broad material components shown in Table IV ranges from $24 to $114 million per year. This is close to the estimated annual damages for the agricultural sector; however, these estimates do not include damages to other types of infra-structure materials, such as automobiles and bridges. The damage estimates for cultural materials also do not account for the willingness of individuals to pay to prevent the visual/aesthetic changes to these materials which may take place as a result of increases in SO_2 and H^+ concentrations.

Table IV does not include a preliminary damage estimate for common building materials in the 17-state study region. While this analysis has been conducted, the results are currently undergoing review to deal with several important methodological problems. First, the paint damage functions can be criticized on both theoretical and empirical grounds. These functions predict mass loss, but not loss of adhesion or delamination which may be more important causes of repainting. Also, the data for estimating the parameters of the paint damage functions were based on experiments with aluminum and stainless

TABLE IV. Materials: Estimated Differences in Annualized Maintenance Costs Due to Simulated Changes in SO_2 and H^+ for Cultural Materials and Transmission Towers in the Northeast (Millions of 1984 Dollars)

Material Component	Annualized Change in Maintenance Cost
Cultural Materials	
Statues and Monuments	5 - 7
Historical Buildings	17 - 100
Transmission Towers	2

steel substrates and are difficult to generalize to other surfaces - wood, brick, stucco, etc. These problems are important since paint damage accounted for about 75% of the estimated total damages to common building materials in the preliminary analysis.

A second set of problems relates to questionable economic and behavioral assumptions. For example, the painting budgets include the costs of activities such as sanding, scraping and stripping which are only needed when paint surfaces are subject to peeling or delaminating. Also, the methodology does not account for changes in behavior to mitigate or avoid economic losses, such as changes in paint technology and paint selection. Different painting budgets which are more consistent with actual practice could reduce the nature of total damages to common building materials by approximately 40%. Finally, systematic biases may be present within the inventory protocol which classified all wood and aluminum surfaces as painted.

5. CONCLUDING COMMENTS

The primary objectives of this paper have been to identify the methods used to estimate the monetary value of the biophysical damages caused by acid deposition for the NAPAP 1985 Assessment and to present the preliminary damage estimates from the Assessment. Methods and damages estimates were presented in four different economic effect areas: commercial agriculture and forests, recreational fishing, and building, cultural and infrastructure materials. In all but the last area, methods were used which incorporated the behavioral responses of individuals and firms to simulated physical damages to resources at risk. The analysis of materials' damages was performed using the more traditional economic damage function approach.

The results presented in this paper should be interpreted for policy purposes with caution, for several reasons. First, the estimated damage values were calculated assuming fairly pristine base case conditions. It is unlikely that these conditions could be achieved with control policies to reduce the precursors of acid deposition. Thus, these damages estimates should not be construed as estimates of the benefits of reducing acid deposition. Furthermore, the damage estimates presented in this study do not cover the full range of economic losses that may occur within the four selected areas. Finally, there is a great deal of uncertainty about the physical effects of acid deposition in the two areas in which economic value of damages could conceivably be large enough to warrant concern, namely: commercial forests and common building materials.

6. REFERENCES

Adams, R. M., Hamilton, S. A., and McCarl, B. A.: 1984, The Economic Effects of Ozone on Agriculture. EPA-600/3-84-090. U.S. Environmental Protection Agency, Corvallis, Oregon.

Adams, D. M. and Haynes, R. W.: 1980, Supplement to Forest Science **26**, 1.

Baker, J. and Harvey, T.: 1984, "Critique of Acid Lakes and Fish Population Studies in the Adirondack Region of New York State." Report prepared for EPA project E3-25. North Carolina State University, Raleigh, North Carolina.

Banwart, W. L.: 1985, "Sensitivity of Soybean Cultivars to Simulated Acid Rain." In Agricultural Effects Task Group F Peer Review, held at Brookhaven national Laboratory, February 4-7, 1985. North Carolina State University, Raleigh, North Carolina.

Baumes, H.: 1978, "A Partial Equilibrium Sector Model of U.S. Agriculture Open to Trade: A Domestic Agricultural and Agricultural Trade Policy Analysis." Unpublished Ph.D. Thesis, Purdue University, Lafayette, Indiana.

Chattin, B., McCarl, B. A., and Baumes, H.: 1983, "Users Guide and Documentation for a Partial Equilibrium Sector Model of U.S. Agriculture." Agricultural Experiment Station Bulletin No. 313, Purdue University, Lafayette, Indiana.

Cicchetti, C. J. and Freeman, A. M. III.: 1971, Quarterly Journal of Economics **85**, 528.

Crocker, T. D.: 1982, "Pollution Damage to Managed Ecosystems." In Economic Assessments in Effects of Air Pollution on Farm Commodities, J. S. Jacobson and A. A. Miller (eds.), Izaak Walton League of American.

Evans, L. S., Thompson, R. F., and Patti, M. J.: 1984, New Phytologist **96**, 207.

Evans, L. S., Hendrey, G. R., and Lewin, K. F.: 1985, "Comparison of Yields of Several Cultivars of Field-Grown Soybeans Exposed to Simulated Acidic Rainfalls." In Agricultural Effects Task Group F Peer Review at Brookhaven National Laboratory, 4-7 February 1985. North Carolina State University Acid Deposition Program, Releigh, North Carolina.

Haynes, R. W. and Adams, D. M.: 1985, Simulations of the Effects of Alternative Assumptions on Demand-Supply Determinants on the Timber Situation in the United States. USDA Forest Service, Washington, D.C.

Haynie, F. H.: 1985, "Atmospheric Damage to Paints." EPA Environmental Research Brief, EPA/600/M-85/019, June 1985.

Hoeft, R. G., Keeney, D. R., and Walsh, L. M.: 1972, Journal of Environmental Quality **1**, 203.

Irving, P. M.: 1983, Journal of Environmental Quality **12**, 442.

Johnson, A. H. and Siccama, T. G.: 1983, "Spruce Decline in the Northern appalachians: Evaluating Acid Deposition as a Possible Cause." In Proceedings of the Annual Meeting of Technology Assessment of the Pulp and Paper Industry. March 7-11, Atlanta, Georgia.

Just, R. E., Hueth, D. L., and Schmitz, A.: 1982, Applied Welfare Economics and Public Policy. Prentice-Hall, Inc., Englewood Cliffs, New Jersey.

Kretser, W. A. and Klatt, L. E.: 1981, "1976-1977 New York Angler Survey-Final Report." New York Department of Environmental Conservation, May 1981.

Krutilla, J. V. and Fisher, A. C.: 1975, The Economics of Natural Environments: Studies in the Valuation of Commodity and Amenity Resources. The Johns Hopkins University Press, Baltimore, Maryland.

Lange, W. J.: 1983, "An Econometric model of Hardwood Lumber and Stumpage Markets in the United States." Unpublished Ph.D. Dissertation. Oregon State University, Corvallis, Oregon.

Larson, R. W. and Goforth, M. H.: 1970, "TRAS--A Computer Program for the Projection for Timber Volume." Agricultural Handbook No. 377, USDA Forest Service, Washington, D.C.

Larson, R. W. and Goforth, M. H.: 1974, "TRAS--A Timber Volume Projection Model." Technical Bulletin No. 1508, USDA Forest Service, Washington, D.C.

Lipfert, F. W., Bernarie, M., and Daum, M. L.: 1985a, Derivation of Metallic Corrosion Damage Functions for Use in Environmental Assessments. Brookhaven National Laboratory, Upton, New York.

Lipfert, F. W., Dupuis, L. R., and Schaedler, J. S.: 1985b, Methods for Mesoscale Modeling for Materials Damage Assessment. Brookhaven National Laboratory, Upton, New York.

McCarl, B. A.: 1982, American Journal of Agricultural Economics **64**, 768.

Merry, C. and Le Potin, P.: 1985, A Description of the New Haven, Connecticut Building Material Data Base. U.S. Army Cold Regions Research and Engineering Laboratory, Hanover, New Hampshire. Forthcoming.

Moskowitz, P. D., Maderos, W. H., Oden, N. C., and Thode, H. C. Jr.: 1985, Effects of Acid Deposition on Agricultural Production. BNL 51889. Brookhaven National Laboratory, Upton, New York.

Mullen, J. K. and Menz, F. C.: 1985, American Journal of Agricultural Economics **67**, 112.

National Acid Precipitation Program (NAPAP): 1983, The Acidic Deposition Phenomenon and Its Effects: Critical Assessment Review Papers. 2 Vols., EPA-600/8-83-016B. United States Environmental Protection Agency, Washington, D.C.

Panhorst, M.: 1985, Distribution of Cultural Properties: Outdoor Sculpture and Public Monuments Data Base. National Park Service, Washington, D.C.

Reddy, M. S., Sherwood, S., and Doe, S.: 1985, "Limestone and Marble Dissolution by Acid Rain." Proceedings of 5th International Congress on Deterioration and Conservation of Stone, June 1985.

Reuss, J. O.: 1977, Water, Air, and Soil Pollution **7**, 461.

Shannon, J. D.: 1981, Atmospheric Environment **15**, 689.

Siccama, J. G., Bliss, M., and Vogelman, H. W.: 1982, Bulletin of the Torrey Botany Club **109**, 162.

Smith, V. R.: 1982, Climatic Change 4, 5.

Vaughn, W. S. and Russell, C. S.: 1982a, Freshwater Recreational Fishing. Johns Hopkins University Press. Baltimore, MD.

Vaughn, W. S. and Russell, C. S.: 1982b, Land Economics **58**, 450.

Willig, R. D.: 1976, American Economic Review **66**, 589.

COST EFFECTIVE MANAGEMENT OF WET SULFATE DEPOSITION

A. Castel **C. Griffith**
Corporate Policy and Planning
Ontario MOE
135 St. Clair Avenue West
Toronto, Ontario
Canada M4V 1P5

ABSTRACT. The purpose of the paper is to present the results of a study that assesses the cost and environmental implications of management options for controlling acid rain. Specifically, the paper illustrates the application of a mathematical programming model which identifies sulphur dioxide emission reductions in eastern North America such that total abatement costs are minimized subject to achieving a prescribed acidic deposition objective. The approaches to reduce acidic deposition considered are emission rollback and least cost targetted. The analysis shows that control strategies developed from a least cost targetted perspective are far less costly and no less effective than emission rollback management approaches. A sensitivity analysis of key input parameters and the effect on management options will also be discussed.

1. INTRODUCTION

Recent years have seen increasing resources dedicated to the study of "Acid Rain", and impressive progress has been made in understanding the causes, effects and technologies available for reducing the emissions found to be responsible for acid rain production (Memorandum of Intent On Transboundary Air Pollution 1982/83). Possession of this information has enabled Canada to set environmental protection objectives, and make for example, a commitment to a lowering of wet sulphate deposition in sensitive areas to no more than 20 kg $ha^{-1}yr^{-1}$.

Mathematical programming to determine cost-effective abatement strategies have been done. Applications include: to achieve compliance with ambient standards (Atkinson, 1983) (Seskin et al., 1983); to minimize control costs within an industry (Rodman and Keenan 1984-85); and to set emission charges to bring about desired reductions (Anderson, 1982). While employing similar analytical principles, the present analysis differs from previous applications by its focus on SO_2 as a long range pollutant, use of deposition rates as the key parameter and incorporation of sources from different industrial sectors.

The purpose of this paper is to describe an analytical procedure

Water, Air, and Soil Pollution **31** (1986) 1035–1045.

which was developed in response to the policy maker's dilemma. Specifically, a linear programming model was developed and used to limit the number of possible policy options, by identifying those strategies which are both efficient in terms of minimizing control costs and effective in terms of achieving desired deposition objectives. Moreover, the model also provides an opportunity to assess the relative cost savings associated with different policy options or strategies.

2. OVERVIEW OF MODEL STRUCTURE

The Screening Model was designed expressly to analyze acid deposition management decisions. It relies on linear programming, a mathematical procedure that makes a systematic search through a large set of possible management options in order to identify one which is optimal from the point of view of specified criteria. Criteria can be formulated either as constraints that must be met or as an objective to be achieved. In the context of acid deposition, minimization of emission control costs was formulated as the modeling objective. This objective was achieved subject to constraints requiring that deposition in sensitive areas be at, or below, a prescribed maximum allowable level.

The SO_2 abatement levels at each source are adjusted during operation of the model in order to satisfy both the objective and the constraints. They are the planning parameters about which decisions are being made within the model and thus are called decision variables.

Total abatement costs are measured as the sum of removal costs incurred at every controllable source. This value is minimized by a search procedure whereby an initial feasible solution that satisfies the constraints is obtained and then, by successive adjustments, lower cost configurations of abatement are found.

In searching for the minimum value of the objective (lowest total abatement costs) the linear programming routine progresses through two stages. First, costs are ignored and the model merely adjusts the emission control removal levels until constraints such as those describing allowable deposition rates are all satisfied.

For example, the linear programming routine will increase the SO_2 emission control removal efficiencies until total deposition rates at each receptor are no greater than the allowable rates. When this is achieved, the collection of source reduction requirements is called a "feasible solution." If the desired deposition level at any receptor cannot be achieved given the SO_2 control technologies used in the analysis then no feasible solution exists.

If a feasible solution can be found, the next stage in the linear programming procedure commences. This entails an iterative procedure to identify where, and by how much, emission control removal levels can be relaxed without violating deposition constraints. In this step, all of the many source/receptor interactions must be considered simultaneously.

3. DATA INPUT REQUIREMENTS

The starting point for assembling the requisite data inputs is a network of sources and receptors. For the current model applications this network encompasses areas of Canada and the U.S. east of the Rocky Mountains. Twenty locations are currently employed within the model to assess deposition rates in areas that are most susceptible to damage from acid rain (Figure 1). Additional receptors can be included as needed.

The deposition criteria at these receptors are expressed in terms of maximum allowable deposition rates at a given receptor area. These rates, which represent wet deposition only, are imposed as constraints within the model.

Emission sources are characterized either as large point sources or as aggregate area sources. A total of 235 large point sources have currently been selected on the basis that their 1980 emissions exceeded 19,000 tonne yr^{-1} of SO_2. Included in this group are Canadian and American power plants and non-power plants as follows:

	Power Plants	Non-Power Plants	Total
Canadian	9	7	16
American	178	41	219
Total	187	48	235

Overall, these large sources represent approximately 67% of total man-made emissions of SO_2 that originated within the modeling area in 1980. Detailed estimates of SO_2 abatement technologies, removal efficiencies and associated cost information were developed only for these large sources.

The remaining emission sources, consisting of a few large point sources, point sources emitting less than 19,000 tonne yr^{-1}, and diffuse sources, have been aggregated by state and province into area sources. There are 92 such area sources, including:

	Power Plants	Non-Power Plants	Total
Canadian	9	9	18
American	32	42	74
Total Number	41	51	92

These area sources may be introduced to the model by assuming, for example, the desulfurization of heating oil at refineries and the resulting combustion of lower S fuels at points throughout the area source. However, since abatement cost functions have not yet been estimated for area sources their emissions are currently fixed.

In addition to the emissions that originate from point and area sources, there are emissions both from natural sources and from man-made sources that lie outside of the boundaries of the study area. These sources are accounted for in the Screening Model by means of a

MAP OF RECEPTOR AREAS

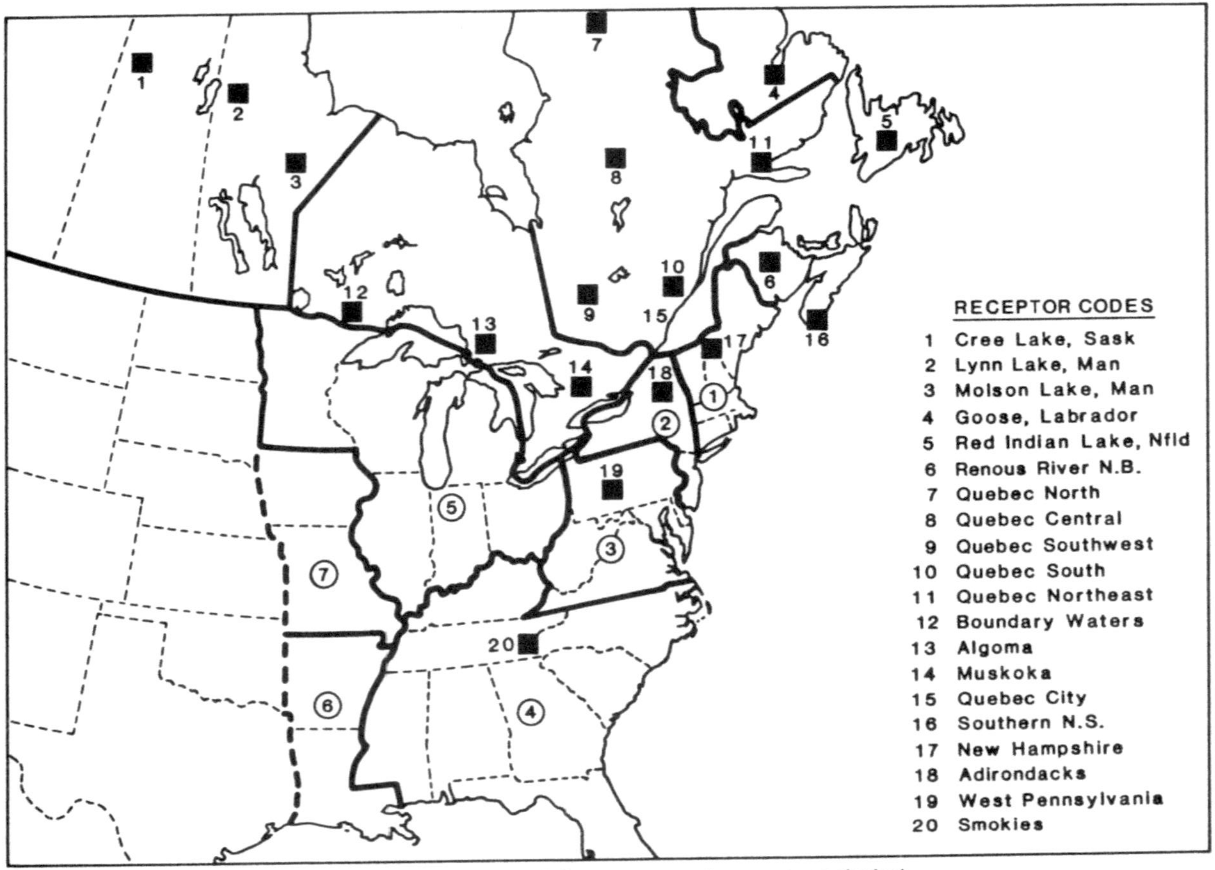

NOTE: Numbers for U.S. source regions are encircled

"background deposition level" at each receptor. Currently, the background deposition levels have been set for all receptors at 2 kg $ha^{-1}yr^{-1}$ of wet-deposition S, (6 kg $ha^{-1}yr^{-1}$). These values can be individually assigned for the various receptors and, therefore, can be varied if necessary.

The linkage between a source of SO_2 emissions and receptor areas deposition rates is provided by the Ministry of the Environment's statistical long-range transport model (Ontario Ministry of the Environment, 1981). Specifically, the source-receptor relationship is captured in a transfer coefficient. Transfer coefficients represent kg $ha^{-1}yr^{-1}$ S at the receptor that can be attributed to an annual emission of one tonne of SO_2 at a source. Since a transfer coefficient is needed for each source/receptor pair, 6,540 coefficients are required for the model.

4. ABATEMENT COST FUNCTIONS

For each of the 235 large emission sources in the screening model, SO_2 emission levels are specified along with estimated removal efficiencies and costs for new emission control techniques.

Emission data are based on 1980 estimates of emissions prepared by the Ontario Ministry of the Environment. These estimates reflect production activity levels for that year and the impact of control technologies that were in place at the time.

For a limited number of sources all within Ontario, detailed cost estimates were available for levels of SO_2 emission control.

For the majority of sources, namely all those in the U.S.A. and Canadian power plants outside Ontario use have been made of abatement cost functions in the form of generic cost curves which do not account for site specific conditions.

Control technologies considered in the analysis of power plants include:

i) physical coal cleaning (25% removal of coal S);
ii) coal substitution down to 1.4% S; and
iii) flue gas desulfurization using limestone scrubbing (up to 90% removal of SO_2).

Fuel S content control options such as coal cleaning are handled at the plant level. For flue-gas and off-gas desulfurization techniques, removal efficiencies and control costs are determined for individual boiler and production lines that are found at each plant. Thus, a power plant with four coal fired boilers could have limestone scrubbers assigned to each boiler separately as well as coal cleaning for all boilers. The alternative emission control options at that plant would then consist of coal cleaning with or without the four scrubbers taken separately or in combination. Maximum reductions would occur with all four scrubbers installed plus coal cleaning.

For the industrial or non-power plant sources, a variety of technologies were considered, depending on the industry and/or production processes uses at each source:

Technology Process	Removal	Application
Limestone FGD	90%	- weak gas streams from boilers and various production processes
Acid Plant	95%	- strong gas streams from non-ferrous smelters
Wellman Lord FGD	90%	- H_2SO_4 plant off-gases
Vacuum Carbonate Process with Claus S Recovery	93%	- cooking oven off-gas in the iron and steel industry

For each of the control techniques identified above, separate expressions were obtained that related variables such as fuel S concentrations and flue gas flow rates to capital cost and operation and maintenance costs. These expressions are based on published cost estimates for the various technologies.

Total costs for a technology applied at an emission point in a plant were calculated as the sum of capital costs and the present value of annual operating and maintenance costs incurred over a 20 yr period. Capital costs include all expenditures incurred to put a control technology in place.

Costs in the Screening Model do not include the possibility of production costs savings or by-product revenues. For the non-power plant sources these could be significant and an upward bias in certain costs could therefore occur. On the other hand, the generic cost functions used in the model could be low for a number of sources. These two sources of bias would be expected to cancel each other out when costs are aggregated.

5. ABATEMENT STRATEGIES: A COMPARATIVE ANALYSIS

To illustrate how efficiency and effectiveness can vary with policy approach three SO_2 control strategies are compared. The strategies are:

1. Uniform 50% reduction
2. Least Cost 50% Reduction To Achieve a Deposition Objective of 25 kg $ha^{-1}yr^{-1}$.
3. Least Cost Reduction To Achieve a Deposition Objective of 25 kg $ha^{-1}yr^{-1}$.

Option 1: Uniform 50% Reduction

This strategy was analysed to show the magnitude and distribution of control costs, the distribution of abatement requirements, and environmental effects of adopting what is generally conceded to be simplest management approach. As displayed in Table I, the annualized cost of control to Canadian and U.S. sources is $415 million and

TABLE I

50% ACROSS THE BOARD: EACH SOURCE REDUCES EMISSIONS BY 50 PERCENT

Province Target	Base Emissions (1980 ktonne)	Reduced Emissions (ktonne)	Remaining Emissions (ktonne)	Annualized Cost (U.S. $ Millions)[1]	Wet Sulphate Deposition Expected For Sensitive Area[2]
Manitoba	492.6	231.3	261.3	15	7 - Lynn Lake
Ontario	1,848.2	732.5	1,115.7	297	22 - Muskoka
Quebec	1,155.5	305.2	850.3	38	22 - South
New Brunswick	215.4	53.6	161.8	31	15 - Renous River
Nova Scotia	218.0	23.1	194.9	24	15 - Southern
Newfoundland	67.4	10.0	57.4	12	14 - Red Indian Lake
Eastern Canada	3,997	1,355	2,641	415	21 - New Hampshire
Eastern U.S.	19,772	6,447	13,324	6,900	20 - Adirondacks

1 20-yr amortization period
2 Deposition in wet sulphate - kg $ha^{-1}yr^{-1}$ in sensitive areas in 1980

TABLE II

LEAST COST 50% REDUCTION OF CANADIAN AND U.S. SOURCES TO ACHIEVE A DEPOSITION TARGET OF 25 KG (SO_4) PER HECTARE YEAR

Province Target	Base Emissions (1980 ktonne)	Reduced Emissions (ktonne)	Remaining Emissions (ktonne)	Annualized Cost (U.S. $ Millions)[1]	Wet Sulphate Deposition Expected For Sensitive Area[2]
Manitoba	492.6	257.4	235.2	23	7 - Lynn Lake
Ontario	1,848.2	653.1	1,195.1	196	21 - Muskoka
Quebec	1,155.5	407.3	748.2	48	22 - South
New Brunswick	215.4	29.7	185.7	15	16 - Renous River
Nova Scotia	218.0	-	218.0	-	15 - Southern
Newfoundland	67.4	-	64.4	-	14 - Red Indian Lake
Eastern Canada	3,997	1,347	2,646	282	21 - New Hampshire
Eastern U.S.	19,772	7,138	12,633	2,066	20 - Adirondacks

1 20-yr amortization period
2 Deposition in wet sulphate - kg $ha^{-1}yr^{-1}$ in sensitive areas in 1980

TABLE III

LEAST COST REDUCTION OF CANADIAN AND U.S. SOURCES TO ACHIEVE A DEPOSITION TARGET OF 25 KG (SO_4)

Province Target	Base Emissions (1980 ktonne)	Reduced Emissions (ktonne)	Remaining Emissions (ktonne)	Annualized Cost (U.S. $ Millions)[1]	Wet Sulphate Deposition Expected For Sensitive Area[2]
Manitoba	492.6	-	492.6	-	8 - Lynn Lake
Ontario	1,848.2	539.4	1,308.8	167	25 - Muskoka
Quebec	1,155.5	339.3	816.2	40	20 - South
New Brunswick	215.4	29.7	185.7	15	17 - Renous River
Nova Scotia	218.0	-	218.0	-	17 - Southern
Newfoundland	67.4	-	67.4	-	15 - Red Indian Lake
Eastern Canada	3,997	908	3,088	222	24 - New Hampshire
Eastern U.S.	19,772	2,273	17,488	471	25 - Adirondacks

1 20-yr amortization period
2 Deposition in wet sulphate - kg $ha^{-1}yr^{-1}$ in sensitive areas in 1980

TABLE IV

SUMMARY OF SCREENING MODEL RUNS WITH ALTERNATIVE TRANSFER COEFFICIENTS

		U.S. Sources		Canadian Sources		Ontario Sources	
Deposition Target	Transfer Coefficients	Removal of SO_2 (5)	Cost ($$10^9$)	Removal of SO_2 (%)	Cost ($$10^9$)	Removal of SO_2 (%)	Cost ($$10^9$)
25	STAT	4	9	49	20	73	17
	LAG80	9	24	47	18	66	13
	LAG79	28	97	48	19	66	13
	LAG78	29	100	51	26	71	20

NOTE: STAT = Statistical model
LAG80 = Lagrangian model, 1980
LAG79 = Lagrangian model, 1979
LAG78 = Lagrangian model, 1978

$6,900 million respectively over 20 yr. Wet sulphate deposition for selected sensitive areas is also shown in Table I. The major deficiency of this emissions-driven approach is that abatement is required at all sources independent of their contribution to acidic deposition and their abatement costs.

Option 2: Least Cost 50% Reduction

Table II shows the implications of retaining the 50% emission reduction in total, but allowing sources in each jurisdiction to use the most efficient means of achieving it. By abandoning the constraint that every source remove 50% of their emissions, the annualized cost to Canada decreases by $133 million to a cost of $282 million per year over 20 yr. Moreover, under this option some provinces are not required to reduce SO_2 emissions at all, while deposition in all receptor areas will be reduced to 25 kg $ha^{-1}yr^{-1}$ or less. Cost savings in the United States are even more dramatic; annualized costs are reduced by almost $5 billion to about $2 billion yr^{-1}.

Substantial savings are achieved and deposition levels are reduced compared to those resulting from the 50% uniform approach. Since, however, under this approach some sources do not have to abate at all, it may be questioned on the grounds that, in appearance it is "unequitable".

Option 3: Least Cost Targetted

Option 3 is intended to show the deposition and the cost consequences of a joint U.S./Canadian program to achieve a specified deposition target. As can be seen in Table III, by elimination of the 50% total emission reduction constraint and focusing only on finding the least cost configuration to achieve a deposition objective (i.e. 25 kg $ha^{-1}yr^{-1}$) further cost savings can be achieved. Adopting a least-cost targetted management approach annual abatement costs for the United States would be three times lower than the control costs under Option 2. Although not as significant as those in the United States, potential cost savings to Canada reach $60 million yr^{-1}.

Programs can thus be designed in which lower or equivalent deposition rates can be achieved at a lower cost than options defined solely in terms of emission reductions. Moreover, joint Canada/U.S. abatement programs were found to be beneficial to both countries in terms of reducing control costs and protecting the environment.

Robustness Under Uncertainty

Results from the screening model have highlighted that abatement configurations can be cost-effective so long as economics and deposition objectives are an <u>ex ante</u> design feature of policy formulation and not left as a consequence of the abatement program.

The detailed results concerning specific reduction requirements and their costs are, however, not absolute since input data are subject to variability. What is more, data refinements may never totally suppress the uncertainties. Consequently, a critical policy issue in assessing abatement allocations is how reduction requirements stipulated by the model respond to changes in the input parameters. A preliminary sensitivity analysis has indicated that abatement configurations were not especially sensitive to variations in control costs and emission levels. For these reasons, emphasis was placed on determining the sensitivity of screening model results to changes in transfer coefficients.

Screening model applications, to date, have been conducted using transfer coefficient data from the "statistical model". Transfer coefficients are now available from other long-range atmospheric transport models including a Lagrangian model developed by the Ontario Ministry of the Environment (Ellenton et al., 1985). The Lagrangian model differs from the statistical model in it is a trajectory model (versus a grid model) and makes far greater use of date describing meteorologic conditions. The Lagrangian differs from the statistical model in it is a trajectory model (versus a grid model) and makes far greater use of date describing meteorologic conditions. The Lagrangian model transfer coefficients were generated for 1980, 1979 and 1978 and together with the 1980 data from the statistical model, allowed for a point of entry into testing the sensitivity of abatement configuration to transfer coefficient variability.

The screening model was run to achieve a deposition target of 25 kg $ha^{-1} yr^{-1}$ at least cost for each set of transfer coefficients. SO_2 removal requirements are summarized for American, Canadian and Ontario sources in Table IV. For American sources, removal requirements demonstrate considerable variability, especially between the statistical and Lagrangian data sets. Removal requirements for Ontario and Canadian sources however remained fairly constant across the transfer coefficient data sets.

The increase in American removal requirements when Lagrangian coefficients were used suggests that the statistical model understates the need for American removal efforts at the expense of Canadian emission sources.

6. CONCLUDING COMMENTS

The findings have shown that the approach to policy development can have substantial effects on abatement cost and environmental protection achieved. And, specifically, that the least cost targeted model solutions are less expensive and no less effective than the often proposed so-called uniform policy approach.

The screening model has the weaknesses of any simplification but it is the best available at this time, and the current state of policy analysis is much superior to that of only a few years ago. It must be stressed, however, that the least cost targeted rationale is intended to be no more than a guide for decision making, and not necessarily the decision rule. There are many other socio-economic, engineering and political issues that come into play when trying to put together a program for controlling acid rain. During a time when both public and private sector funds for pollution control are scarce, however, the policy options which can accomplish objectives at the lowest cost deserve consideration and discussion. There may be compelling reasons for diverging from a least cost approach but we must at least identify those least cost programs in order to make informed policy comparisons and choices.

REFERENCES

Anderson, R.J. (Jr.): 1983, "Using Mathematical Programming Models for Cost-Effective Management of Air Quality". Mathematical Models for Planning and Controlling Air Quality, pp. 59-74. Pergamon Press, Oxford, England.

Atkinson, S: 1983, "Nonoptimal Solutions Using Transferable Discharge Permits: The Impication of Acid Rain Deposition". Land Economics Monographs Number 6, pp. 55-70. The University of Wisconsin Press, Madison, Wisconsin.

Ellenton, G., Ley, B., and Misra, P.K.: 1985, Atmospheric Environmental in press.

Ontario Ministry of the Environment: 1981, "Documentation of the Ontario Ministry of the Environment Statistical LRT Model OME-LRT.

Ontario Ministry of the Environment: 1983, "Linear Programming Screening Model for Development and Evaluation of Acid Rain Abatement Strategies". Prepared by A.E. McBean, J.H. Ellis and M. Fortin.

Rodman, K. and Keenan, J.D.: 1984-85, "Economic Efficiency of SO_x Standards". Journal of Environmental Systems, Volume 14(2), pp. 333-346.

Seskin, E.P., Anderson, R.J. and Reid, R.O.: 1983, "An Empirical Analysis of Economic Strategies for Controlling Air Pollution". Journal of Environmental Economics and Management, 10 pp. 112-124.

TRENDS IN PUBLIC OPINION ON ACID RAIN: A COMPREHENSIVE REVIEW OF EXISTING DATA

Keith Neuman
Corporate Policy and Planning
Ontario Ministry of the Environment
135 St. Clair Avenue West
Toronto, Ontario, CANADA M4V 1P5

ABSTRACT. Questions concerning the value society places on the benefits and costs of control measures for acidic deposition cannot be fully achieved without considering the views held by the public-at-large. While numerous opinion polls have been conducted on acid rain, to date no systematic review of these data has been undertaken to identify underlying trends. This paper summarizes a comprehensive analysis of existing public opinion data in Canada and the U.S. from 1979 to the present, with particular attention given to identifying longitudinal trends, comparisons across jurisdictions, demographic group differences and comparisons with other issues. It is concluded that over the past five years, the North American public has become aware of acid rain as an environmental issue and, despite a limited understanding of the problem, believe it to be a serious one. Concern about acid rain has by now reached the level where most believe concrete actions must be taken, and a majority are willing to help pay for abatement programs. Canadians and residents of regions susceptible to acid rain damage have typically been more informed and concerned about acid rain, but such regional and national differences have diminished over time as the issue has become more widely recognized. These findings are discussed in terms of implications for policy and education strategies.

INTRODUCTION

Accurate and comprehensive information is critical to effectively addressing major environmental problems such as acid rain. A wide range of data are required from a variety of scientific disciplines to develop an understanding of causes and effects as well as viable options for control. As with most problems of this magnitude, however, acid rain cannot be solved through scientific discovery alone, but encompasses a substantial political challenge as well. In addition to the data provided by biologists, chemists and economists, consideration must also be given to the diverse and often conflicting interests and values present in our society.

One important political force to be considered is the general

Water, Air, and Soil Pollution **31** (1986) 1047–1059.

public. Though rarely well-articulated in comparison with organized interest lobbies, public opinion has increasingly become a decisive factor in influencing and often shaping public policy. With respect to environmental hazards, public perceptions are often a critical determinant of government response, sometimes irrespective of scientific opinion (e.g. the Three Mile Island nuclear accident, PCB spills in northern Ontario). Policy-makers must therefore have a clear understanding and appreciation of public perceptions and beliefs as important input in making decisions.

As acid rain emerged as a high profile issue in North America, it has commanded attention in public opinion polls and survey studies posing questions about public awareness and concern. Such polls, first published in 1980, have indicated an apparent growth in concern about acid rain in North America, undoubtedly a factor in keeping this issue on the main policy agenda at both the national and regional levels.

As useful as such polls are, as individual studies their value is limited because they capture the views of a specific population on specific questions at a specific point in time. Findings can rapidly become dated as events transpire and public sentiment shifts. The broader significance of a specific poll may be discernible only in the context of the whole body of research data. A systematic analysis of these studies has yet to be undertaken, however, leaving a fragmented picture of public attitudes and beliefs about acid rain.

This paper addresses this gap by presenting a brief overview of a comprehensive analysis of public opinion data on acid rain from Canada and the U.S. Particular attention is paid to comparisons across jurisdictions (regional as well as national), trends over time, differences among demographic groups and comparisons with other issues. The results of this analysis provide a basis for drawing conclusions about public opinion on acid rain as well as implications for policy and education strategies.

CHARACTERISTICS OF RESEARCH ACTIVITY

Between 1979 and 1985, 36 public opinion studies have been conducted in Canada and the U.S. which included at least one question on acid rain and have been published, referenced or made publicly available (a list of these studies is presented in Table I). Most of this research has been conducted by professional polling organizations (e.g., Gallup, Roper), whose clients typically include industry, government, media and environmental groups. These organizations utilize established research methods and collect data that in most cases can be considered valid and representative of the populations studied.

Two broad areas of acid rain issues have been addressed in these surveys. The majority of questions (60% out of a total of 200 questions) addressed aspects of acid rain as a problem, pertaining to respondents': a) awareness of acid rain (15% of all questions); b) knowledge about acid rain (18%); and c) level of concern about the problem (27%). The remaining 40% of questions considered issues concerning potential responses to the problem, including:

TABLE I U.S. and Canadian Public Opinion Studies on Acid Rain, 1979 - 1985

Polling Organization (Sponsor)	Survey Date	Population Studied
Canadian Studies		
1. Canadian Gallup Poll Ltd. (Canadian Coalition on Acid Rain)	1985	Canada
2. Canadian Gallup Poll Ltd.	1985	Canada
3. Decima Research Ltd.	1985	Canada
4. Canadian Gallup Poll Ltd. (Canadian Coalition on Acid Rain)	1984	Canada
5. Centre de Recherche Opinion Publique (Environment Canada)	1984	Canada
6. Canadian Facts (Tricil Ltd.)	1983	Southern Ontario
7. Centre de Recherche Opinion Publique	1983 (Nov.)	Canada
8. Canadian Gallup Poll Ltd. (Toronto Star)	1983	Canada
9. Centre de Recherche Opinion Publique	1983 (Feb.)	Canada
10. Decima Research Ltd.	1982	Canada
11. Centre de Recherche Opinion Publique	1981	Canada
12. ARA Consultants (Ontario Ministry of the Environment)	1981	Selected Ontario sub-populations
13. Canadian Gallup Poll Ltd.	1980	Canada
14. Centre de Recherche Opinion Publique	1980	Canada
15. Decima Reseach Ltd. (Ontario Ministry of the Environment)	1979	Ontario
16. Atlantic Opinion Research Centre (Environment Canada)	1979	Atlantic Provinces
U.S. Studies		
1. Louis Harris and Associates	1985	U.S.
2. Kentucky Survey Research Centre (Kentucky Energy Cabinet)	1984	Kentucky
3. St. Paul Pioneer Press - Dispatch/WCCO-TV/WCCO-Radio	1984	Minnesota
4. Louis Harris and Associates (Business Week)	1984	U.S.
5. New England Conqressional Caucus	1984	New England constituency groups
6. Parker, Julie (MA Thesis), St. Lawrence University	1984	Ogdensburg, N.Y. and Prescott, Ont.
7. Louis Harris and Associates	1983	U.S.
8. Temporary State Commission on Tug Hill and N.Y. State Cooperative Extension	1983	Tug Hill area (upstate New York)
9. Gordon Black Associates (N.Y. State Dept. of Environmental Conservation and the Centre for Financial Management, SUNY, Albany	1983	New York State
10. Kentucky Survey Research Centre	1983	Kentucky
11. Adirondack Museum Blue Mountain Lake, N.Y.	1983	Museum visitors
12. ABC News-Washington Post	1983	U.S.
13. CBS News-New York Times	1983	U.S.
14. Roper Organization	1983	U.S.
15. Cambridge Reports Inc. (Missouri Union Electric Co.)	1983	Missouri
16. Roper Organization	1982 (July)	U.S.
17. Roper Organization	1982 (Jan)	U.S.
18. Institute for Social Inquiry, University of Connecticut	1982	Connecticut
19. University of North Carolina (Chapel Hill) School of Journalism (National Wildlife Federation)	1982	North Carolina
20. Resources for the Future (U.S. Council on Environmental Quality)	1980	U.S.

d) satisfaction with government actions (11%); and e) support for policy options for controlling acid rain (30%).

Comparisons between survey research in Canada and the U.S. reveal differences which reflect how acid rain issues have evolved and been viewed in each country. The number of studies undertaken is roughly equivalent but a greater proportion of U.S. studies have focused exclusively on acid rain, overall accounting for two-thirds of the total number of survey questions. This difference is most pronounced with awareness and knowledge questions, reflecting the U.S. emphasis on defining the problem, in contrast to the Canadian context in which the problem has been more quickly acknowledged and the focus has been placed on possible solutions. Canadian studies were more likely to include questions concerning the government's role in addressing acid rain.

National differences are also evident in the fact that Canadian studies more commonly surveyed at the national level (71% versus 45% of U.S. studies), whereas acid rain has been considered primarily a regional issue by Americans. The rapid emergence of acid rain as a major issue in the U.S. in the early 1980s is reflected in the significant growth in U.S. research activity over time, in both the number of studies conducted as well as the number and breadth of questions addressed. For example, only one knowledge question was included in 1980 studies, but this increased to 9 in 1982 and 12 in 1984. The level of Canadian activity, on the other hand, has remained stable, with 1 or 2 studies conducted each year.

Overall, there has been a substantial amount of opinion research on acid rain over the past six years, more than has been done on any other specific environmental issue. While data are not available for all populations in all time periods, comparisons can be made in many cases to assess both the consistency in responses within specific populations, time periods and subgroups as well as significant differences between them.

ANALYSIS OF SURVEY DATA

Awareness of Acid Rain

Over the past five years most of the North American public has became aware of acid rain as an environmental issue. This awareness developed sooner in Canada, where by 1981-82 over 90% of the public expressed an awareness of acid rain at a simple recognition level (i.e. responding affirmatively to questions such "Are you aware of acid rain?" or "Have you heard of acid rain?"). By comparison, public awareness in the U.S. was initially much lower, but grew steadily over time so that by 1985 it was consistent with awareness levels in Canada (30% in 1980, 63% in 1983, 94% in 1985). These data reflect the fact that widespread government and media attention on acid rain in Canada predated that in the U.S. by several years.

In both countries, awareness is greatest in those areas most susceptible to acid rain damage (e.g., Ontario, upstate New York). But awareness appears to be almost as great in certain other regions less vulnerable to such damage, notably the prairie provinces in Canada and the midwestern and western portions of the U.S. In 1983, for example, awareness levels were 75% in the northeastern U.S. (where damages have received the greatest attention) but 84% in the midwest (where major sources of acid gas emissions are located) and 80% in the west.

The fact that awareness of acid rain is high in most areas of the continent indicates that the national and regional media is likely playing a major role in raising such awareness. This impact is clearly illustrated in Atlantic Canada, where high profile media coverage of regional acid rain impacts during the summer of 1979 boosted awareness over a six week period from 25% to 50%.

Media impact may be particularly prominent in the case of acid rain because it is a form of environmental damage which differs on an experiential level from others which are more localized and visible. Without noxious fumes or muddied water to provide clear evidence of pollution, the public must rely almost exclusively on media reports for information on the existence and impact of acid rain.

Across North America, people are somewhat more likely to be aware of acid rain if they are male, of higher socio-economic status (more education, higher income, white collar or professional employment) and older (up to 60 years of age). While this pattern is consistent across virtually all studies, the magnitude of differences is not great (e.g., awareness among Kentucky males in 1984 was 76% compared to 61% for females). These findings on demographic group differences are generally consistent with previous studies of public concern about environmental issues, although recent evidence indicates that environmental concern is becoming universal across all sectors of the public (see below).

Knowledge of Acid Rain

While awareness of acid rain has grown considerably in recent years, the public's understanding of what it is remains superficial. In 1980, for example, only 15% of Canadians aware of acid rain were able to correctly define it (based on making reference to either SO_2 or NO_2), with 65% giving partially correct answers (less specific, e.g., pollution killing lakes) and 20% either wholly mistaken or unable to give an answer. In a U.S. national study that same year (using even more lenient criteria for correct responses), only 32% of Americans sampled gave a correct or partially correct answer and 58% were either unable to give any response or unaware of acid rain. The limited extent of people's understanding of acid rain is underscored by the fact that most survey questions tapped knowledge at a very basic level, typically providing multiple choice alternatives or, if open-ended, using lenient criteria for correct responses.

The primary focus of knowledge-oriented questions tapped people's ability to identify effects of acid rain. Relying on these data to assess the public's understanding is problematic, however, because the considerable scientific uncertainty about effects makes it difficult to definitively classify responses as correct or incorrect. These findings do provide a clear picture of what the public believes to be the effects of acid rain, uncovering some common misconceptions.

Overall, the public is most likely to mention acid rain damage to aquatic bodies (lakes, rivers, fish), the class of effects with the most scientific documentation. Other correctly identified effects commonly mentioned include damage to forests and property (e.g. buildings, automobiles). In numerous studies, however, a sizeable proportion of the public cited effects which have not (or had not at the time) been documented scientifically. The most common of these is damage to human health, likely a function of people confusing acid rain with other forms of air pollution, as well as the widespread belief that all forms of pollution are hazardous to health.

Canadians appear to be somewhat more certain than Americans about damages caused by acid rain. By 1985, the relative priority of acid rain effects given by Canadians is generally consistent with current scientific evidence (aquatic, forests, fishing). Americans have typically exhibited greater uncertainty and were more likely to attribute incorrect effects to acid rain, such as the death of Lake Erie.[1] Over time, the U.S. public has expressed less uncertainty and more frequent mention of all effect categories (including misconceptions), suggesting that Americans may be growing more concerned (as is discussed in the next section) but not necessarily more knowledgeable.

Significant regional differences on knowledge of effects did not emerge in either country, although variations in the emphasis given to certain types of effects reflect predictable regional concerns (e.g. fishing effects in Atlantic Canada, crop damage in the U.S. midwest).

A number of U.S. studies indicate that, in comparison to knowledge of effects, Americans have a somewhat better understanding of acid rain sources (of which there is considerably more scientific certainty). The most frequent responses to source questions were generally the correct ones (e.g., coal-burning power plants, 58-80%; automobile emissions, 29-80%). As with effects, popular misconceptions were also evident, such as aerosols destroying atmospheric ozone (22-25%), Mt. Saint Helens (15-26%), nuclear power (19%) and industrial water pollution (18-25%). Regional differences indicate that residents of those areas with major acid rain sources (midwest) and those particularly vulnerable to damage (upstate New York) tend to be most knowledgeable of what causes acid rain.

Knowledge about acid rain is positively associated with socio-economic status (SES), although not as consistently as has been found with awareness. U.S. data indicate that middle-aged people (higher SES) are most knowledgeable about acid rain sources, while younger people are more cognizant of effects.

People's limited understanding of acid rain as portrayed in these studies is typical of the depth of public knowledge of most political and social issues of the day, particularly those for which media is the primary source of information. But these findings also raise the question about how well the relevant facts are being communicated. In one survey of relatively well-informed residents (in the Tug Hill area in upstate New York), 70% expressed dissatisfaction with the available information on acid rain, finding it either not sufficiently specific or too difficult to understand. Such dissatisfaction undoubtedly reflects the considerable scientific uncertainty about acid rain, but it may also indicate the need for more effective strategies for educating the public.

Concern About Acid Rain

Along with a growing awareness, there has been increasing public concern that acid rain is a serious problem. Across Canada, a clear majority of the population has increasingly viewed acid rain as a serious problem, rising from 69% in 1980 to 86% in 1984. Consistent with findings on awareness and knowledge, people in Ontario exhibited the greatest degree of concern: as early as 1979, 50% of the adult population in that province considered acid rain a serious problem, rising to 75% by 1981. Elsewhere in Canada, acid rain was more likely to be seen as serious and of concern in Quebec and British Columbia than in Atlantic Canada or the West, although these regional differences diminished over time.

In comparison to Canadians, Americans have historically been less concerned and more uncertain about acid rain. But this difference has almost disappeared over the past couple years as public concern exhibited sharp growth, particularly in key receptor areas (upper New York State, Minnesota). Sixty percent of the U.S. public in 1981 believed acid rain to be a serious problem, increasing to 90% in 1983 and 76% in 1985. Regional differences do not reveal a consistent pattern with respect to perceptions of seriousness, but do indicate that level of personal concern about acid rain is somewhat higher in receptor areas. These findings suggest that in comparison to more general perceptions of seriousness, levels of personal concern about acid rain is partly a function of the extent to which one anticipates or experiences direct personal impacts (e.g. health effects, loss of fishing). In general, Americans are more likely to view acid rain as a serious problem than express personal concern about it, suggesting that personally-felt impacts are not yet an issue in most parts of the country.

The absence of multiple studies in the U.S. covering the same populations prevents an analysis of trends in level of concern or seriousness over time. It is interesting to note, however, that in both Kentucky (1984) and Tug Hill (1983), close to one-half of those sampled felt acid rain was getting worse in their area, most likely without the benefit of any direct evidence of changes in damages or deposition levels.

In both the U.S. and Canada, concern about acid rain varied somewhat by demographic groups, but not as consistently as with awareness or knowledge. There was a trend toward higher SES groups expressing greater concern (and less uncertainty), but this was consistent across all studies only for education. These findings reinforce evidence from U.S. studies indicating that concern for the environment is becoming widespread across all segments of the general population (Anthony, 1982; Van Liere and Dunlap, 1980).

While the above findings indicate that the public views acid rain as a major problem, it is not considered to be as serious or of as much concern as other pressing social and environmental issues of the day, such as unemployment or toxic waste leaks. Canadians in 1982, for example, were less apt to view acid rain as important an issue as air, water or waste pollution; only oil spills, loss of parkland and noise were seen to be less important than acid rain.

Moreover, while concern about acid rain has increased over the last several years, such concern does not appear to have increased relative to other environmental problems. Acid rain must compete for the public's limited attention with other issues, such as toxic dump leaks and contaminated drinking water, problems with which the perceived risks are much greater and more vividly imaged. These findings may be explained in terms of how people perceive the risks associated with different kinds of environmental hazards (cf. Slovic et al, 1980). The extent to which acid rain is perceived as personally or directly threatening is diminished because its consequences are chronic (rather than catastrophic) and delayed (versus immediate) in comparison with other more dreaded hazards. Nevertheless, acid rain has become a major public concern in part because it is global (versus localized) in scope, unfamiliar and poorly understood, involves involuntary exposure placing innocent parties at some risk, and has received considerable media attention.

Satisfaction with Government Actions

A small set of questions (primarily in Canada) have tapped the public's attitudes about how well governments have addressed the acid rain problem. Canadians in the past have generally expressed satisfaction and optimism with government efforts to curb acid rain, particularly in comparison with the perceived lack of action by the U.S. In 1983, 58% of Canadians felt their country was doing at least its fair share to reduce acid rain (versus 27% who felt Canada had done less than its fair share), while only 16% could say the same about the U.S.

Recent polls indicate, however, that Canadians have become increasingly dissatisfied with what is viewed as a lack of progress in addressing acid rain. Seventy-four percent of those polled in early 1985 felt little or no progress was being made, and at mid-year two of three Canadians polled did not believe their country was doing enough to deal with the problem. Ontario residents have been the most critical of federal and provincial government efforts, perhaps as a result of being the most informed and concerned about acid rain. Elsewhere, Atlantic Canadians have been the least critical of both U.S. and Canadian efforts. Three studies of upstate New York suggest that residents there have been as dissatisfied as Canadians with U.S. efforts on acid rain.

While public dissatisfaction can be traced to a lack of visible policy action on acid rain, there has been growing dissatisfaction in North America with the handling of environmental issues in general. In 1984, for example, only 38% of Canadians felt their federal government was spending enough on environmental protection, while in 1985, the majority of Americans were critical of the environmental record of governments at all levels.

While Canadians are more likely to blame the U.S. for acid rain, most place the responsibility for addressing the problem on both countries. U.S. studies, on the other hand, have focused exclusively on domestic responsibility. Most New York State residents in 1983 (72%) believed that acid rain was a problem for state and national governments to solve together. In Minnesota, however, residents were most likely to believe that action should be taken at the national level. This difference might be explained by the fact that acid rain is not considered to be as serious a problem in Minnesota itself and therefore less of a regional priority.

Limited data on demographic differences reveals a weak trend toward higher SES individuals being more critical and dissatisfied with government efforts, as well as more likely to see a broader jurisdictional responsibility for addressing acid rain (i.e., bi-national in Canada, national in the U.S.). This finding again suggests that those more informed about acid rain have higher expectations for government action.

Support for Acid Rain Controls

A major focus of these studies has been on assessing public support for possible policy approaches to dealing with acid rain, including the key issue of who will bear the costs of abatement. The majority of Americans accept emission controls as either necessary or the best means of reducing acid rain. For example, 61% of North Carolina residents in 1982 agreed that tighter controls were needed on eastern U.S. power plants. Yet there is a widespread belief that more research is also necessary before such controls are imposed, even in key receptor areas such as Tug Hill, N.Y. where 60% in 1983 agreed that more research was probably needed before controls were applied. Support for more research undoubtedly reflects a great deal of public uncertainty about acid rain effects and the best means of control, particularly given the substantial economic implications of abatement options.

The public in all regions of the U.S. appears to be sensitive to the cost of acid rain abatement, with support for controls diminishing noticeably when presented in terms of hypothetical economic implications. In Missouri, where several major sources of acid rain emissions are located, 62% of registered voters in 1983 supported a then current national abatement strategy (8 million ton reduction in SO_2 emissions in 31 eastern states, not including Missouri), but only 47% maintained this support when told the strategy would cost $7 billion a year. New York State residents that same year overwhelmingly (95%) favoured controls if cost were not a factor, but only 47% accepted such controls if they resulted in an average increase of $25 per month in consumers' utility bills.

By contrast, Canadians unequivocally favour additional controls on polluting industries. In 1983, 86% of Canadians nation-wide supported restrictions on Canadian industry, with little variation across regions. The unanimity of this view most likely explains why this issue has not received more attention in Canadian surveys. A more pressing issue in recent years has been how Canada should deal with the lack of U.S. commitment on acid rain abatement. In 1985, 80% of the Canadian public felt Canada should proceed with reductions unilaterally, without waiting for U.S. cooperation.

The high cost of acid rain controls raises the question of who should bear the burden of this cost. Americans and Canadians alike place most of this burden on polluting industries, but also appear to recognize their own responsibility for contributing toward reducing acid rain. Sixty percent of Americans polled in 1983 believed it was fair for electric consumers nationwide to help pay for needed abatement controls. In Canada, 75% in 1984 and again in 1985 felt that polluting industries should receive financial assistance from government if need could be shown. Canadians of higher SES status were more likely to support government assistance, although not at a higher level of assistance; level of government assistance was in fact inversely related to education level and income.

Since extensive abatement controls may ultimately require public subsidies and/or higher prices for certain commodities, the political viability of major abatement initiatives may depend on the extent to which the public is willing to help pay for them. In both countries, a small majority of those polled expressed a willingness to pay extra dollars in either higher prices or taxes to help reduce acid rain damage. Data collected from 1980 through 1985 indicate that roughly 50% in Canada and 70% in Ontario would pay something, with half of these groups willing to pay over $100 per year. An even greater proportion indicated in 1984 a willingness to donate a day's work toward acid rain abatement (69% in Canada, 73% in Ontario).

U.S. data on this topic is less definitive but generally consistent with Canadian results. One-half of New York state residents expressed a willingness to pay an extra $25 per month ($300 per year) in higher utility bills, with another 25% willing to pay a smaller amount. In Missouri, on the other hand, only 9% would pay $300 per year, although 58% would pay $50. These findings indicate that, as in Canada, residents of receptor regions are more willing to pay for acid rain abatement. Willingness-to-pay values are in fact more similar among receptor areas on both sides of the border than to either national average, suggesting that regional distinctions may be more significant on this issue than national ones.

Across studies, support for abatement controls and willingness to pay for them was positively related to income and education. Younger respondents were more willing to pay something while those of middle-age were willing to pay more. It is to be expected that higher income people express a greater willingness to pay since they have more disposable income with which to pay for environmental protection. Education may also play an important role here, although its collinearity with income makes it difficult to assess its effect separately.

A lack of similar data on willingness-to-pay for other types of pollution abatement limit comparisons with acid rain. The 1981 Ontario study did indicate, however, that people's general willingness to pay for environmental damage was partly a function of the extent to which: a) the problem was natural versus man-made in origin; and b) the polluter could not be clearly identified. Respondents in this study were more willing to pay for acid rain than for chemical waste pollution, but less so than for natural disasters. Acid rain falls somewhere in the middle of this dynamic: it is clearly man-made, but the source is not easily identified.

CONCLUSION AND IMPLICATIONS

Over the past several years, acid rain has entered the consciousness of the North American public. While most people's understanding of the problem is limited, the general level of concern has reached the point where most now believe that concrete steps must be taken to address the problem. Canadians have typically been more informed and concerned about acid rain than Americans, but this gap has narrowed considerably in the last couple of years. Moreover, while residents of areas most clearly at risk from acid rain have understandably been more concerned about it, regional as well as national differences have diminished over time as the issue has become more widely recognized.

Socio-economic status appears to have some bearing on the importance given to the acid rain issue, but the impact of socio-economic variables on attitudes toward acid rain was not significant. Research suggests that other factors, such as personal belief and value structures, may be more important determinants of how individuals view environmental matters (Milbrath, 1984; Van Liere and Dunlap, 1980).

The growing public expectation for action warrants serious attention by policy-makers. In Canada, dissatisfaction with a lack of action is likely to grow until the public becomes aware of new initiatives which are seen to be significant enough to match the magnitude of the problem. In the U.S., survey data do not reveal the extent of regional divisiveness echoed in the political arena and indicate that Americans nationwide support acid rain abatement. Regional differences which do exist are likely to diminish as acid rain impacts are increasingly felt nationwide, resulting in greater political pressure for action at a national level.

The public's willingness to help finance abatement programs is particularly significant considering that this support is widespread across all regions, providing governments with key political support for undertaking major initiatives. But the public's willingness to pay may depend in part on the extent to which the costs are seen to be fairly distributed between the public purse and emitting industries. Canadians would appear to accept cost-sharing on an equal basis, but further study is needed to more accurately identify what the public views as fair. One possible approach would be to finance abatement through a clearly identified acid rain tax or surcharge (on tax returns or utility bills), thereby distinguishing it from other rate increases and highlighting people's personal contribution to solving the problem.

The combined efforts of environmental interests, media and government to raise public awareness of acid rain have clearly been successful. The challenge now is to expand the public's limited understanding. Given the limits to the level of scientific information which the general public is capable of grasping, programs aimed at a broad audience must focus on communicating basic facts and clearing up misconceptions. Simple concepts typically have the greatest impact, as was demonstrated by a successful Dutch campaign which vividly portrayed the progressive damage to a statue of the madonna using three photographs labelled "yesterday", "today", and "tomorrow?"

Communication programs must also address those questions most relevant to target audiences, such as how acid rain is likely to affect them personally. Risks to health and property (e.g. automobiles, historic buildings) are likely to be of greater concern to most urban residents than damages to distant aquatic ecosystems. Further efforts are needed to identify more clearly what the public wants to know about acid rain.

As useful as these studies have been in providing an indication of the public's views on acid rain, these data are neither wholly conclusive nor without limitations. Comparisons across jurisdictions and time were necessarily limited by the absence of comparable studies and questions. Variations in question wording and format, for instance, will often elicit very different responses. In some studies, questions were framed in ways that either prompted certain responses or limited responses to an identified range. At a more fundamental level, because survey research techniques tend to simplify issues to ensure they are understood by a broad audience, these data may be presenting an oversimplified picture of the public's beliefs and perceptions.

A more comprehensive and accurate assessment of public opinion is warranted and could be developed in part by standardizing question format, thereby providing a firm basis for identifying trends and group or regional differences. Ensuring consistent coverage of relevant populations may require that national or regional surveys be carried out on a regular basis (annual or semi-annual).

Future studies must also focus on developing a more in-depth understanding of how the public perceives acid rain and possible policy options. This might entail several strategies, including:

a) conducting more sophisticated analyses of survey data to identify important relationships between different issues (e.g. how support for controls is affected by level of knowledge);
b) utilizing relevant theories from communications and the social sciences (e.g., current research on hazard perception) to identify key factors affecting beliefs and attitudes about acid rain; and
c) utilizing complementary research methods, such as focus group interviews, designed to tap people's opinions in greater depth than is possible with sample surveys. These techniques may be particularly useful in identifying public acceptability of policy options for financing abatement programs.

Knowing what the public thinks will not in itself provide the answers to solving the acid rain problem. But such information is very likely essential if society is to reach some consensus on how this problem is to be addressed.

NOTES

[1]An important methodological caveat to these findings is that the frequency of incorrect responses is somewhat inflated due to the prevalent use of closed-format questions, in which the response choices (correct and incorrect) are provided, making it easier and therefore more likely for incorrect responses to be given.

REFERENCES

Anthony, R.: 1982, Polls, Pollution and Politics: Trends in Public Opinion on the Environment. Environment, **24**, 4, 14-20, 33-34.

Milbrath, L.: 1984, Environmentalists: Vanguard for a New Society, SUNY Press, Buffalo.

Slovic, P., Fischoff, B., and Lichtenstein, S.: 1980, Facts and Fears: Understanding Perceived Risk. In Schwing, R.C. and Albers, W.A. (Eds.) Societal Risk Assessment: How Safe is Safe Enough? Plenum, New York, 181-214.

Van Liere, K.D. and Dunlap, R.E.: 1980, The Social Basis of Environmental Concern: A Review of Hypotheses, Explanations and Empirical Evidence. Public Opinion Quarterly, **44**, 2, 181-197.

A COMPARATIVE REGIONAL ANALYSIS OF THE STATUS OF AQUATIC RESOURCES WITH RESPECT TO ACID DEPOSITION.

John L. Malanchuk, David A. Bennett,
Patricia A. Mundy, and Gregory J. Mallon
Acid Deposition Assessment Staff (RD-680)
U.S. Environmental Protection Agency
Washington, DC 20460

ABSTRACT. A national scale assessment of the effects of acid deposition on aquatic resources is currently unavailable for the United States. A more limited assessment has been performed in three potentially sensitive geographical regions: the Adirondack Mountains of New York, the Southern Blue Ridge Province of North Carolina, Tennessee and Georgia, and the Upper Midwest of Minnesota, Wisconsin and Michigan. A series of questions organized the assessment, and historical and current data from each region were employed to address the questions. Where possible, independent data sets were used singly and in combination within a given case study area, to respond to the questions and to provide an estimate of confidence. Conclusions among regions were also compared. In general, the impact of acid deposition on aquatic resources is difficult to detect, but positive correlations between atmospheric deposition and effects do exist. Thus, there is evidence to suggest that acid deposition is at least partially responsible for the acidification of aquatic resources. The extent of acidification, however, varies from region to region. The Adirondacks appear to represent the region of greatest impact, but independent studies indicate that the area of acidified lakes is a small percentage of the total resource. Despite the observation of minimal damage, application of damage estimates from the Adirondacks to other regions would be inappropriate. Future Assessments will seek to predict the rate of acidification of watersheds with other regional characteristics under alternative loading scenarios.

1. INTRODUCTION

In the Aquatic Effects chapter of 1985 Assessment*, prepared by the National Acid Precipitation Assessment Program, three component levels are considered 1) national, 2) multi-state (New York and New England) and 3) regional.

*The draft 1985 Assessment, described in this paper, is a contribution to the First Interim Assessment that will be available to the public at the end of 1986.

Water, Air, and Soil Pollution **31** (1986) 1061–1068.

TABLE I

Questions to be addressed in the 1985 Assessment of Aquatic Effects

---

1. What is the extent of low pH, low alkalinity surface waters?

2. What percent of the low pH, low alkalinity waters have been acidified (at least partially) due to natural processes?

3. What characteristics of watersheds are most closely related to surface water alkalinity and pH?

4. Have there been changes in the chemical status of surface waters?

5. What is the rate of acidification of surface waters?

6. What is the relationship between deposition and water chemistry?

7. a. What are the effects of acidification?

   b. Can these effects be mitigated; what are the economics of mitigation?

8. What are the physical and economic impacts of acid deposition on fisheries?

9. Is there a statistical relationship between water chemistry and biological effects that can be used to predict future impacts?

10. What are the future chemical and biological impacts at current deposition rates, and at other deposition rates?

Regional case studies have been prepared for the Adirondack Mountains of New York, the Southern Blue Ridge Mountains, the upper Midwest, and headwater streams in Pennsylvania. The Adirondack Regional Case Study (ARCS) is the most complete and provides the focus for this paper. The other case studies are included in the 1985 Assessment.

To explore the extent of current understanding of aquatic issues related to acid deposition, a series of questions of interest to policy-makers, scientists and the public was raised, and, where possible, answers were provided. See Table I.

2. ADIRONDACK REGIONAL CASE STUDY

Historical data from the Adirondacks spanning the period between approximately 1932 to 1980 were used in the analysis of changes in lake chemistry (Baker and Harvey, 1984). Upon analysis, several trends were apparent; there was no significant change in lake pH over time when all lakes were considered. However, significant decreases in lake pH were found among high elevation (>610 m) lakes (mean rate of 0.12 ± 0.05 pH units per decade). Due to the nature of the historical data used in this analysis the uncertainty in these estimates may be high. The Baker and Harvey (1984) data set, Fish Information Network (FIN), used in the historical analysis includes the New York State Department of Environmental Conservation (NYSDEC) data set. These data sets also were employed to evaluate the current chemical status of Adirondack lakes and compared to the results from the National Surface Water Survey (NSWS) (Table II). The media lake area for 2,427 lakes in the Adirondacks for which areas are available is 5.3 ha. The mean for these lakes is 41 ha. The NSWS provides unbiased estimates of lake pH in its target population, which did not include lakes <4 ha. Biases in the other two data sets used do exist and corrections for some known biases were applied. Row 2b of Table II attempts to correct for lake area and elevation biases in the FIN data base.

Multiple regression analysis was performed on data taken from FIN, and on data contained within the Multi-State Analysis, another 1985 Assessment component. These analyses were performed to address assessment question 3 concerning watershed characteristics important in predicting the chemical status of a lake; at this stage of our analysis, important characteristics include: cover type, slope, watershed area, area in rock outcropping, elevation, soil cation exchange capacity and bedrock buffering capacity. In addition to lake chemistry, the fish population status in Adirondack lakes was investigated using FIN.

2.1. Physical and Economic Impacts of Acid Deposition on Fisheries

Central to the estimation of the extent of impact on fisheries was the development of a method to describe the fish population status of lakes. The following considerations are relevant to these results: 1) the population of sampled lakes used in the analysis is not a random or representative sample of the entire population of Adirondack lakes; 2) the ratings of fish community status assigned to the lakes have not been verified for experimenter bias; 3) the results do not prove how fish communities declined in the Adirondacks, but rather suggest that for 8% to 18% of these lakes (the percent depending upon the definition of "adequate evidence"), acidification may have caused the loss of fish populations; 4) the effect of uncertainty in the data, bias in sample selection, and bias in the ratings of fish community status have not been assessed.

As indicated in Figure 1, there are significant differences in the mean pH between those lakes rated 1 (decline of one or two not particularly acid-sensitive species) and those rated 2 (declines in one or two acid-sensitive species) as well as those between those rated 2 and those rated 3 (declines of several acid-sensitive species). Figure 2 shows the number of individual species populations rated as having adequate evidence for loss due to acidification to be small relative to the total number of populations in the sampled lakes. Brook trout show the strongest evidence of population declines due to acidification.

The economic dimension of the impact of acidification upon fish communities in the Adirondacks was investigated. The economic damages presented are based on current levels of resource damage, and are not intended as estimates of future economic damage or of the economic benefits of reducing acid deposition. Economic damages to Adirondack anglers were measured using two common approaches for valuing environmental damages: a participation model and a travel cost model. Data for these models were derived from the 1976 New York State Angler Survey (NYSDEC 1981) coupled with recent information from Baker and Harvey (1984). These models were then used to estimate current levels of economic damage. Damage estimates ranged from 0.7 to 12 million dollars per year.

3. CONCLUSIONS

The effects of acidification were analyzed using historical data, fish community status information, and economic models. The results are uncertain due to the nature of the data utilized. Chemical data for Adirondack lakes suggest that significant change in pH has occurred in high elevation (>610 m) lakes and in currently acidic lakes. However, when all Adirondack lakes are considered in the analysis, no significant trend in lake pH is observed. Multiple regression analysis was performed on each set of data, and specific watershed variables -- such as elevation, bedrock buffering capacity, and percent watershed in conifers -- thought to be important in predicting lake chemistry were identified for each set of data. The specific variables selected depend on the data set used. Analysis of fish community status showed that 8 to 18% of lakes in the Adirondacks may have lost populations as a result of acidification. Economic damage attributable to this level of fish effect was predicted using a participation model and a travel cost model. Analyses based on these models indicate damages from 0.7 to 12 million dollars per year.

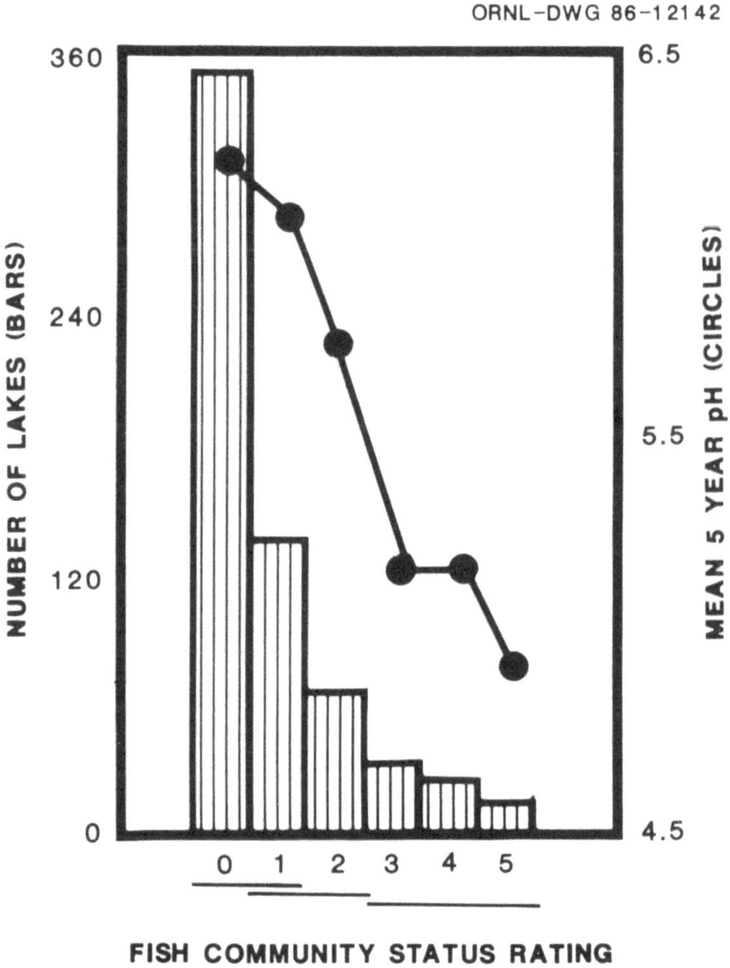

Figure 1. Fish community status vs lake pH (Baker and Harvey, 1984). A distribution of sample lakes with different ratings of fish community status, and mean 5-year pH of those lakes is shown. The pH levels of lakes rated 2 are significantly lower than the pH levels for lakes rated 1; the pH levels of lakes rated 3 are significantly lower than for lakes rated 2. The pH levels for lakes rated 0 and 1, as well as 3, 4, and 5, are not significantly different from each other. (0 = healthy, 1 = declines unlikely due to acidification, 2 = marginal evidence of decline, 3 = evidence of decline resulting from acidification, 4 = majority of acid sensitive species have disappeared, 5 = all species have disappeared).

TABLE II

Comparison of Current Lake Chemical Status between New York State Department of Environmental Conservation, Fish Information Network, and National Surface Water Survey Data Bases

Data set name (# lakes sampled)	Median lake area[a]	Min. lake area[a]	% of lakes			% of lake area		
			pH <5	pH 5-6	pH >6	pH <5	pH 5-6	pH >6
1. NYSDEC[b] (1074)	13.0 ha[d]	none	19	25	56	3.9	10	86
2. FIN[c] (a) (693)	10.5 ha	none	24.7±2.8	23[d]	52[d]	4.2±1.3	22[d]	73[d]
(b)	10.5 ha	none	20-25			1.4±6.8		
3. NSWS[e] (127)	20.9 ha	4 ha	10	17	73	2	8	90

[a]Biases may exist because, historically, lakes with fish were more likely to be sampled, and, recently, lakes suspected of damage have been preferentially sampled.

[b]Colquhoun et al. (1984). New York State Department of Environmental Conservation (NYSDEC).

[c]Baker and Harvey (1984). Fish Information Network (FIN).

[d]Estimated from graphs in Baker and Harvey (1984).

[e]National Surface Water Survey (NSWS) in Kanciruk et al. (1986).

ORNL-DWG 86-12143

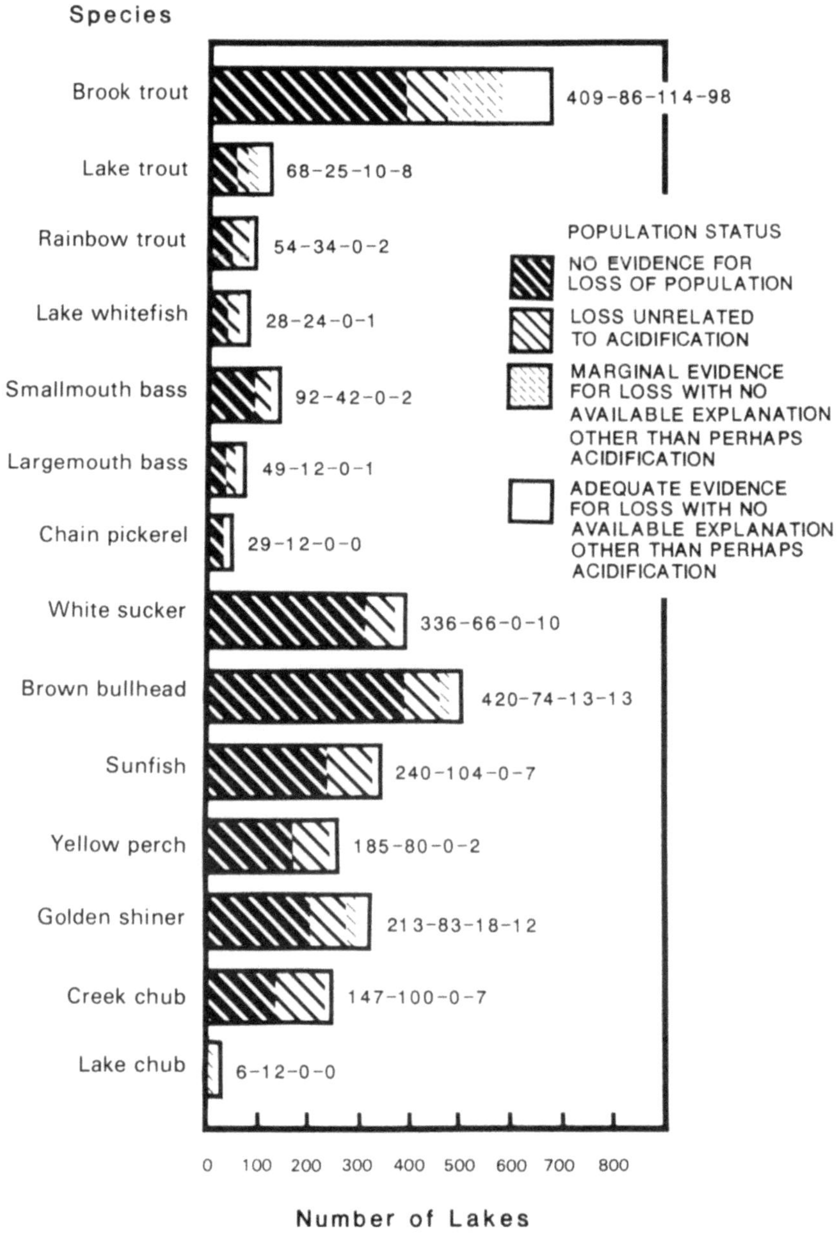

Figure 2. Classification of population status among sample lakes (Baker and Harvey, 1984).

REFERENCES

Baker, J.P. and Harvey, T.B.: 1984, Critique of acid lakes and fish population status in the Adirondack Region of New York state, draft. Acid Deposition Program, North Carolina State University, Raleigh, North Carolina.

Colquhoun, J., Kretser, W., and Pfeiffer, M.: 1984, Acidity status update of lakes and streams in New York State. New York State Department of Environmental Conservation. 140 pp.

New York State Department of Environmental Conservation: 1981, Final Report -- New York Angler Survey, 1976-1977. NYSDEC FY-P177(7/81) Ray Brook, NY. 12977. 214 pp.

U.S. Environmental Protection Agency: 1984, National Surface Water Survey National Lake Survey -- Phase I. NAPAP project reference number E-23. Office of Research and Development, Washington, DC.

AQUATIC CHARACTERIZATION FOR RESOURCES AT RISK IN EASTERN CANADA

Andrew S. Fraser
Aquatic Physics and Systems Division
National Water Research Institute
Burlington, Ontario, Canada

ABSTRACT. To ascertain quantitative estimates of resources at risk for eastern Canada, linkage must be made between geological terrain sensitivity to acid deposition and water chemistry. An evaluation has been made of watershed areas principally in the Province of Quebec to identify and characterize factors related to LRTAP effects. The watershed areas are geographically located within the bounds of the high sulphate deposition zone of the continental plume. To identify specific test watersheds for detailed analysis, alkalinity frequency distributions were computed. A selection of watersheds has been made that spans the Pre-Cambrian shield region of the Laurentian highlands, the St. Lawrence lowlands and the Gaspé Peninsula. There is evidence that areas in the far eastern areas of Quebec, removed from strong anthropogenic sources may be considered as approaching critical levels of alkalinity. Ionic composition for selected watersheds display similar chemical characteristics for regions of diverse geology and soil characteristics. These differences have been assessed in light of the common effects of sulphate deposition.

1. INTRODUCTION

The region of eastern Canada south of latitude 52°N lies within the proscribed region of the continental pollution plume. The southern part of the Province of Quebec receives an estimated sulphate deposition, when considering both wet and dry contributions of between 10 and 30 kg ha^{-1} yr^{-1} (Barrie and Hales, 1984). Characterization of the aquatic resources within this region will lead to projected estimates, with confidence intervals of the total aquatic resource at risk.

Investigations into the aquatic effects of acidic deposition in southern Quebec have centred on the documentation and clarification of the chemical relationships of importance in rivers and lakes under stress from acid deposition (Jones et al.,

Water, Air, and Soil Pollution **31** (1986) 1069–1078.

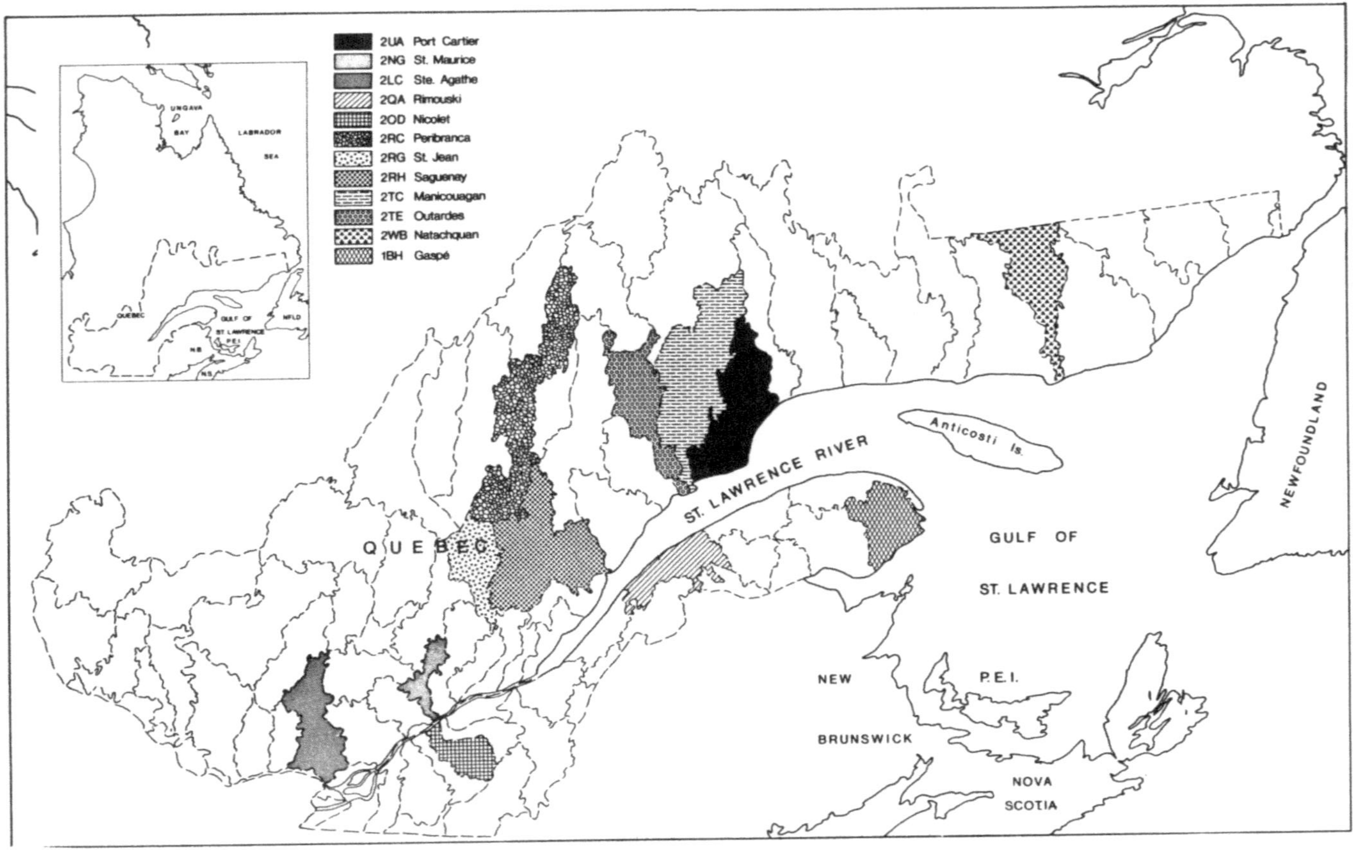

Figure 1 Watershed regions of southern Quebec with key regions highlighted.

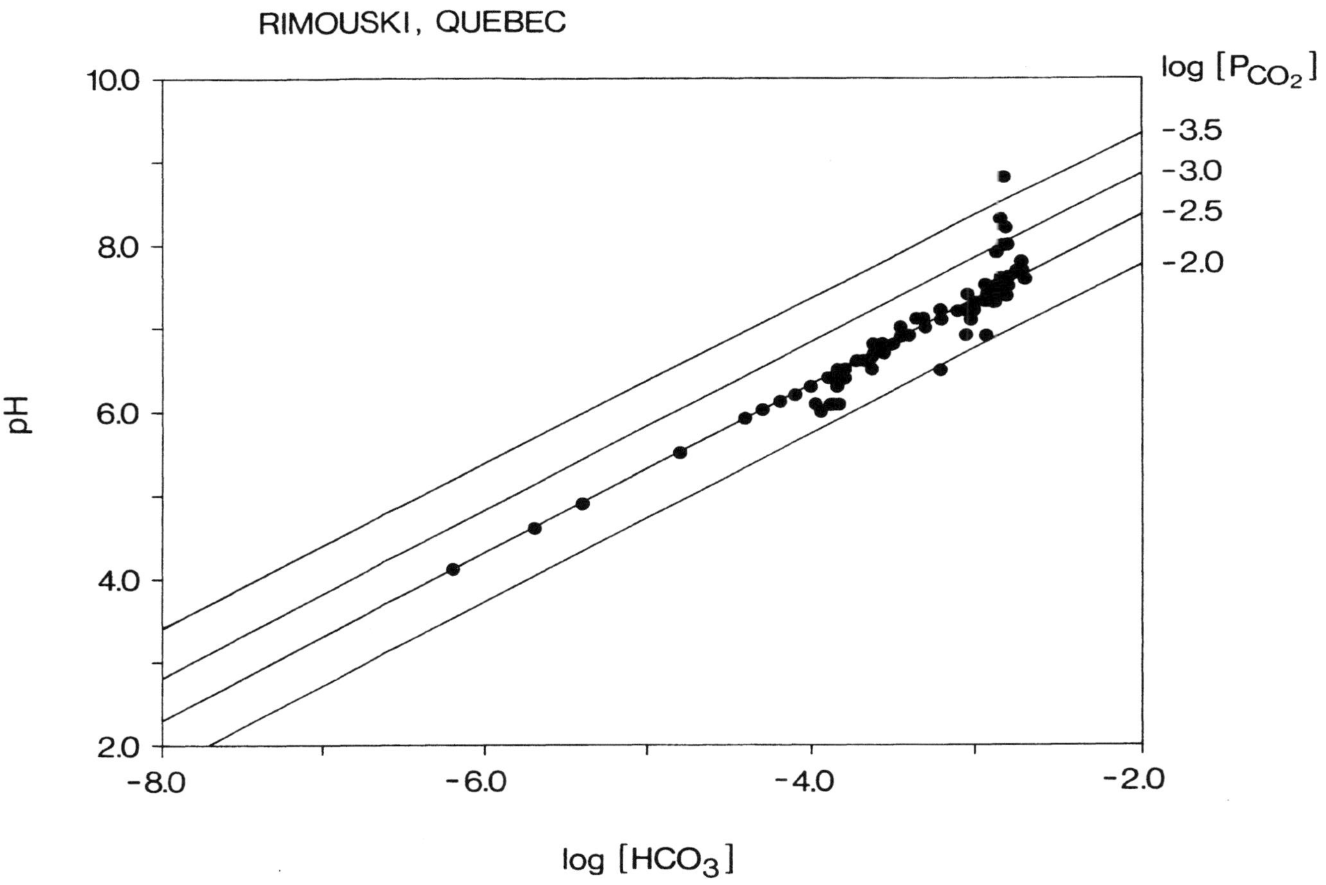

Figure 2 Bicarbonate/pH relationship to the partial pressure of CO_2 in Quebec surface waters.

1983). Significant insights into the chemical dynamics and of chemical relationships in 177 lakes in the region between between the Ottawa River and the Saguenay River have been produced (Bobée et al., 1983).

In light of the common problems experienced in regions impacted by large scale deposition of SO_x and NO_x, as described by Henriksen (1980), Lachance et al. (1985), and the general sensitivity of the bedrock and derived soils of the Laurentian Shield discussed by Shilts (1981), a detailed analysis of the entire available data set for the Province of Quebec to clearly identify in quantitative terms the aquatic risk has been undertaken.

2. MATERIALS AND METHODS

The basic geographic unit of the study was the watershed designation based upon the Water Survey of Canada map of the Province of Quebec hydrometric stations (Figure 1). Data were subgrouped into watershed areas and all further analyses were conducted under the watershed restriction.

The data base used in this study is contained on the NAQUADAT system maintained by Environment Canada (Demayo, 1970). A subset of all the available data for lakes between 1967 and 1982 which were of interest were evaluated for data precision and representativeness. Numerous problems were encountered, particularly with reference to missing data. The primary variable which was missing was the alkalinity. To solve the data availability situation an editing process of alkalinity estimation was devised.

To estimate alkalinity the average value of two methods was selected. The first method was to compute the alkalinity in μeq L^{-1} based upon an ion balance in which alkalinity was the only missing variable (Equation (1)).

$$[Ca^{+2}] + [Mg^{+2}] + [Na^{+}] + [K^{+}] - [SO_4^{-2}] - [Cl^{-}] - [HCO_3^{-}]=0 \quad (1)$$

(units in μeq L^{-1}).

The second method was to compute the alkalinity as in Equation (2).

$$\log [HCO_3^{-}] = -7.8 + \log [P_{CO_2}] + pH . \quad (2)$$

This latter equation is dependent upon $[P_{CO_2}]$, the partial pressure of CO_2 in surface waters. An examination of the pH/CO_2 relationship in surface waters showed considerable variance but in general a value of $10^{-2.5}$ atm. was acceptable (Figure 2). This is in good agreement with the findings of the cation denudation rate model proposed by Thompson (1982).

To begin the characterization of the available lake data, a method to represent large amounts of data in an easy to read form was selected. Ion rosettes which graphically depict the relative contribution to the total ionic content of a sample for each of the major cations and anions were chosen. This technique, first developed by Maucha (1932), was applied to all 57 watersheds under study for the period of record.

It should be noted that some of the regions are located near marine systems and as such sea salt ratio corrections have been applied where appropriate.

3. RESULTS AND DISCUSSION

Regionally, the ion rosettes represent five distinct geographical and geological regions (Figure 3). Mt. Tremblant displays an ionic composition which is well buffered with calcium bicarbonate dominating the system. However, a high level of sulphate (SO_4^{-2}) shows the presence of acidic stress upon the system. This can be compared to the Natachquan, a system under marine influence located in the far eastern region of Quebec on the north shore of the St. Lawrence River. Note the similarity in the ionic composition. Although the Natachquan basin is a relatively soft water system, the dominance of the sulphate component of the ion balance clearly shows that the system is under stress. The St. Maurice watershed shows an episode of strong acid load wherein the bicarbonate component has been overloaded and eliminated through normal stoichiometric processes. This system, however, has sufficient buffering capacity to re-establish a more normal ionic composition through weathering processes over time. Nevertheless, such acid shocks can have disasterous effects upon the biotic communities within such systems (Jefferies et al., 1979).

Geographical factors dominate the aquatic effects observed in the Province of Quebec. On the south shore of the St. Lawrence River the geology is dominated by limestones. Ion rosettes for the Nicolet basin and the area around Rimouski, Quebec clearly show calcium bicarbonate systems which easily override the presence of the other major ions despite the level of sulphate deposition which is comparable to that deposited on the more sensitive north shore of the St. Lawrence River (Thompson, 1983).

To further classify the 57 regions studied frequency distributions of alkalinity were prepared. Presented here are the results for five of the watershed areas showing again the diversity of the region (Figure 4). Alkalinity reported as less than zero represents an acid system and is resultant from the Gran analysis technique for alkalinity or is an estimate computed from pH and ion balance procedures discussed previously. The diverse geology of the Mt. Tremblant watershed, which is made up primarily of pre-Cambrian shield in the north and glacial till with some limestone in the south, contributes to the diversity of the

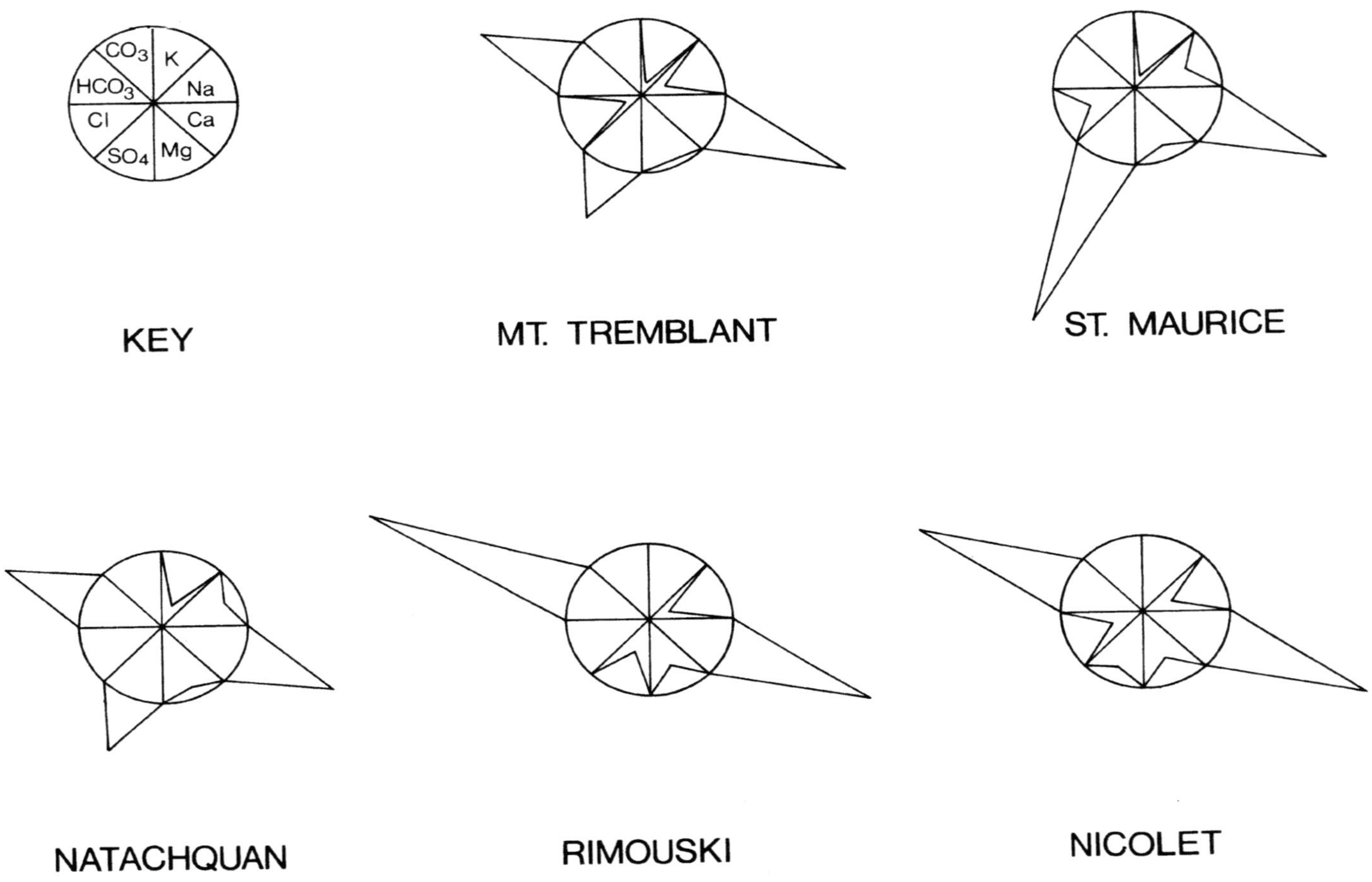

Figure 3 Ion rosettes representing five key regions in southern Quebec.

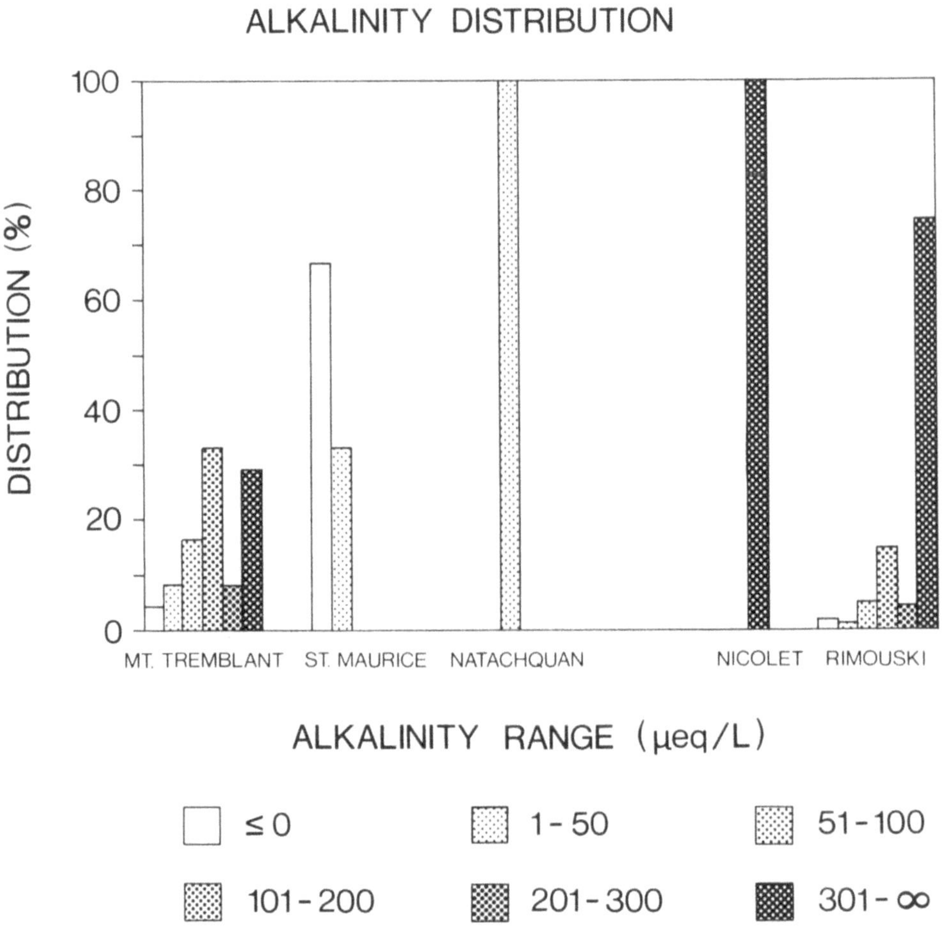

Figure 4 Alkalinity frequency distributions for key regions in southern Quebec.

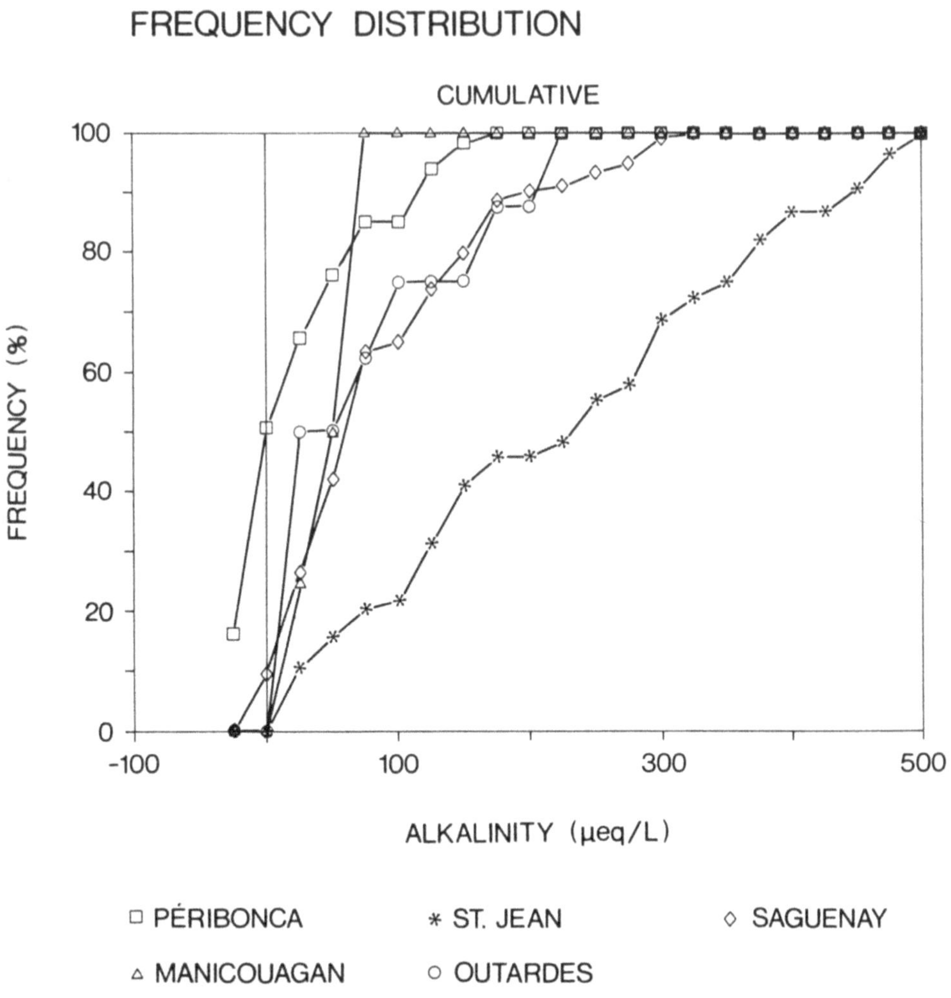

Figure 5 Cumulative frequency distributions for alkalinity in key regions of southern Quebec.

distribution, whereas regions such as the Natachquan and Nicolet are distinctly univariate. These data can be reassessed in terms of the cumulative frequency distribution. Such a plot indicates the percentage of data gathered within each watershed as a function of the basin susceptibility (Figure 5). Regions located on the north shore of the St. Lawrence River within the region of high sulphate deposition display 30 to 50% alkalinity below zero and 80 to 100% below 100 $\mu eq\ L^{-1}$.

4. CONCLUSIONS

Diversity of the geologic factors, particularly between the north and south shores of the St. Lawrence River, is the paramount element controlling the system response to acidic loadings from the atmosphere. A common history of sulphate deposition diminishing from west to east links the two shores to the long range transport issue. There is now sufficient quantitative information available on the 57 watersheds to identify key regions as sites for model formulation and development bringing together components of the aquatic and geological chemistry. Investigations to bring this next phase of study to completion are currently underway.

ACKNOWLEDGMENTS

The author wishes to express gratitude to F.C. Elder whose encouragement and guidance fostored this effort D.C.L. Lam and M.E. Thompson who offered constructive criticism in the progress of the work and also to the many unmentioned individuals who gather, analyze and store the data required for such large scale studies.

REFERENCES

Barrie, L.A. and Hales, J.M.: 1984, Tellus, 36B, 333.

Bobée, B., Lachance, M., Haemmerli, J., Tessier, A., Charette, J. and Kramer, J.: 1983, Evaluation de la sensibilité à l'acidification des lacs du soud du Québec et incidences sur le réseau d'acquisition de données. Scientifique rapport no. 157, Institute national de la recherche scientifique, Environnement Canada.

Demayo, A.: 1970, A storage and retrieval system for water quality data. Dept. Energy, Mines and Resources. IWD Report No. 9, Ottawa, Ontario.

Henriksen, A.: 1980, Acidification of freshwaters - a large scale titration. Proceedings international conference ecological impact of acid precipitation, Oslo, pp. 68-74.

Jefferies, D.S., Cox, C.M. and Dillon, P.J.: 1979, J. Fish. Res. Bd. Canada, 36, 640.

Jones, M.L., Greig, L.A. and Marmorek, D.R.: 1983, Methods and results of an analysis of chemical and biological data collected by the national inventory survey program. Essa Environmental and Social Systems Analysts Ltd.

Lachance, M., Bobée, B. and Grimard, Y.: 1985, Water, Air and Soil Pollution. 25, 115.

Mauncha, R.: 1932, Hydrochemische methoden in der limnologie. Die Binnengewasser, 12. Stuttgart, 173 pp. 557-558.

Shilts, W.W.: 1981, Sensitivity of bedrock to acid precipitation: modification by glacial processes. Geological Survey of Canada, Paper 81-14.

Thompson, M.E.: 1982, Water, Air and Soil Pollution. 18, 215.

Thompson, M.E.: 1983, Sulfur yields of Quebec rivers. Nat. Wat. Res. Institute, Environment Canada, No. 83-22.

Estimates of existing and potential impact of acidification on the freshwater fishery resources and their use in eastern Canada.

Charles K. Minns
Fisheries & Oceans
Great Lakes Fisheries Research Branch
867 Lakeshore Rd, P.O. Box 5050
Burlington, Ontario L7R 4A6
Canada

and

John R.M. Kelso
Great Lakes Fisheries Research Branch
1219 Queen Street East
Sault Ste. Marie, Ontario P6A 5M7
Canada

ABSTRACT. There are over 700,000 lakes in Canada, east of the Ontario-Manitoba border and south of latitude 52. This region supports much of Canada's freshwater fishery resources, mostly in acid-sensitive habitat. Of these lakes, half have alkalinities less than 50 $\mu eq.L^{-1}$ and 150,000 have a pH less than 6. There are 14,000 acidic lakes i.e., with a pH less than 4.7.

Models of lake-watershed systems have been developed to predict the future response of fishery resources in eastern Canada. Despite considerable uncertainties in key parameters, the predicted range of losses is not great. The models predict that with current acid deposition, up to 9 % of all lakes will ultimately have a pH less than 5. Reduction of deposition rates, either by 40 % everywhere or to a maximum of 12 $kg.ha^{-1}SO_4$, greatly reduces the predicted losses. The models give conservative predictions of impact as inclusion of loss of watershed acid neutralizing capacity and fish mortality due to episodic pH depression would indicate greater losses. The freshwater fishery resources of eastern Canada are important economically accounting for 42 x 10^6 angler-days and \$1.4 x 10^9 yr^{-1} per year. Up to 10 % of this sector of the Canadian economy could be at risk.

1. INTRODUCTION

Here we present an overview of attempts to assess the current and potential impact of acidic deposition on lakes and fish resources in eastern Canada. Resource and economic data are integrated to assess

Water, Air, and Soil Pollution **31** (1986) 1079–1090.

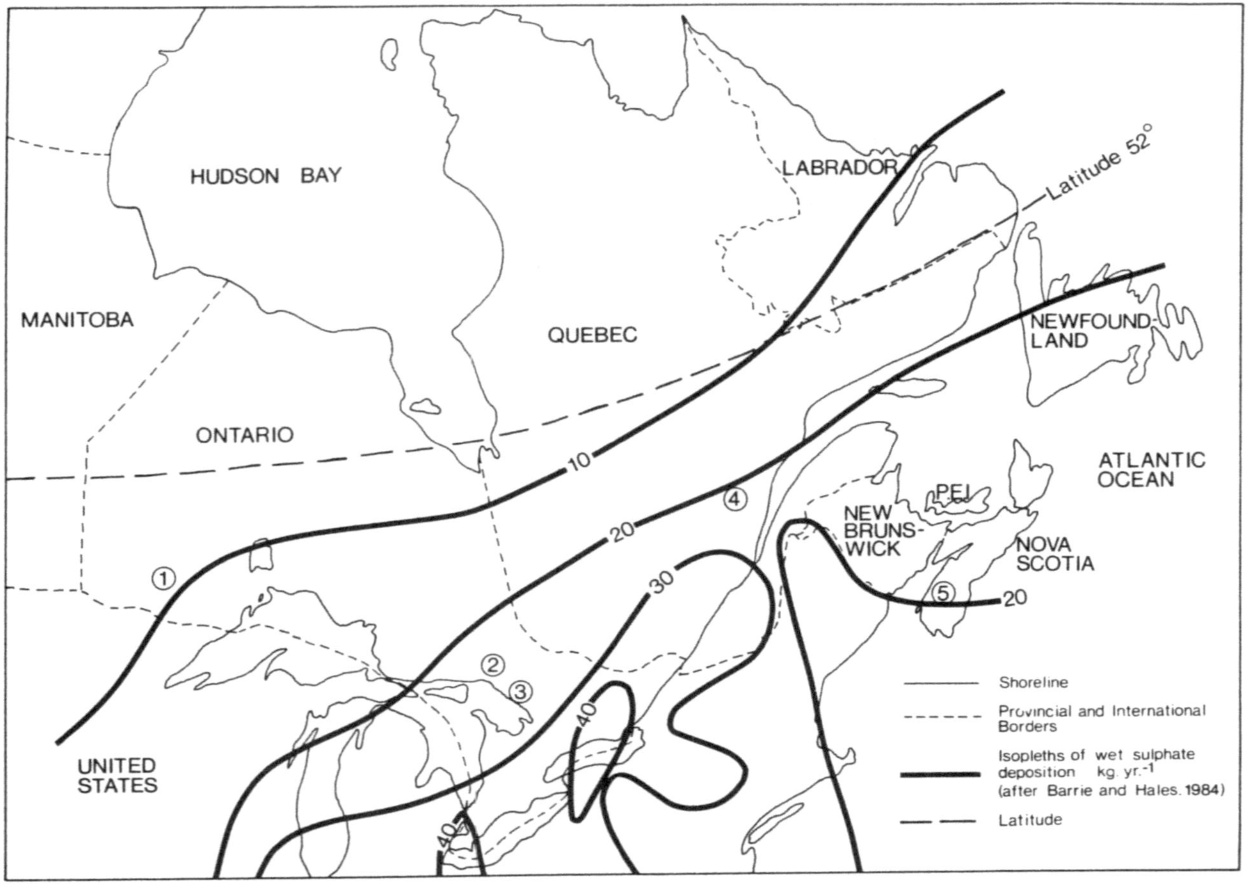

Figure 1 A map of eastern Canada showing the provinces, the region south of latitude 52°, and approximate isopleths of wet sulphate deposition in 1980 (after Barrie and Hales, 1984).

the socioeconomic value of recreational fishing. Much of the analysis of freshwaters focuses on chemistry, but links are made which show the fisheries implications.

2. CURRENT STATUS

Kelso et al. (1986) presented an extensive analysis of the results of lake surveys conducted throughout eastern Canada. Though the sample was small (≃1000 lakes), the chemical data were considered to be representative of the freshwaters in that region; average conditions were not correlated with lake size or position in the drainage network. They found evidence of the impact of sulphate deposition on the chemistry of lakes and rivers. Biota (phytoplankton, zooplankton, benthos and fish) consistently showed reduced diversity in waters with a pH below 6.0. There were many lakes which showed elevated Al and Hg levels, considered harmful to fish.

Kelso et al. (1986) used available lake counts and measures data from the different regions to extrapolate the survey results by province to the area affected in eastern Canada west of the Ontario-Manitoba border and south of latitude 52° assuming the survey data was a random sample of the total resource (Table I). This is the region of Canada most affected by acidic deposition (Figure 1) according to the 1980 results reported by Barrie and Hales (1984). Kelso et al. estimated there are ≃720,000 lakes in the specified region. Nearly half of these have an alkalinity less than 50 $\mu eq.L^{-1}$, considered to be a threshold for impact and almost 150,000 have a pH less than 6. The results also indicated there are already more than 14,000 acid lakes (pH<4.7). Many of those in the Atlantic region are dystrophic and may therefore be naturally acidic. In Ontario and Quebec, the acid lakes are most likely to be casualties of acidic deposition, as these are the areas receiving the greatest loadings (Figure 1). Most of the area south of latitude 52° is receiving acid deposition above background levels (considered to be less than 10 kg wet SO_4 $ha^{-1}.yr^{-1}$). Therefore, most of the acid-sensitive lakes must be considered at risk.

Important inland fishery resources are at risk in these regions. Johnson et al. (1977) and Minns (1981) have shown that much of Ontario's lake trout and brook trout resources are found in acid-sensitive waters. In Quebec, Legendre and Legendre (1984) showed that many of the preferred game fish species are distributed in areas receiving elevated acid deposition. In the Atlantic region, brook trout are an important inland fishery resource. Also, there are the Atlantic salmon resources of Quebec and the Maritimes to be considered. Some Nova Scotian salmon rivers have lost their stocks because of acidification (Watt et al., 1983).

3. PREDICTIONS OF FUTURE LAKES STATUS

Since 1979 a series of studies (Minns, 1981, Minns et al., 1986) have

Table I: **Estimates of lake resources and recreational angling by province in eastern Canada.**

Category	Ontario[c]	Quebec	Newfoundland	New Brunswick	Nova Scotia	Total
Numbers of lakes[a]:						
Total	126,278	441,752	138,711	3,596	6,585	716,922
-Alkalinity <50 μeq.L^{-1}	40,662	229,711	75,181	1,718	6,144	353,417
-pH<6	28,918	72,006	41,475	1,269	4,853	148,521
Acid (pH<4.7)	2,020	6,185	2,497	554	3,115	14,371
Angling[b]:						
Fishermen 10^{-3}	2,372.0	989.2	102.3	105.6	69.0	3,638.1
Angler days 10^{-3}	23,414.0	12,471.0	1,916.0	1,637.0	1,503.0	40,941.0
Expenditures $M	776.3	533.9[d]	37.5	47.1	26.5	1,417.9

[a] After Kelso *et al.* (1986), lakes south of latitude 52°.
[b] After DFO (1980), direct expenditures and major purchases (1980 Canadian dollars).
[c] Great Lakes excluded.
[d] Estimated from the average expenditure per angler in the rest of Canada.

been conducted with the purpose of using models of lake acidification to predict the future state of freshwater resources throughout eastern Canada. Earlier efforts described in Minns et al. (1986) concentrated on the development of a methodology for prediction. Recently, a predictive model (Marmorek et al., 1986; Jones et al., 1986) has been developed and used.

The chemical predictions are largely based on the empirical models of Henriksen (1979), Thompson (1982) and Wright (1983). Henriksen (1979) presented the concept of whole lake titration with acid whereby the acid neutralizing capacity (ANC) of a lake is consumed. Henriksen (1982) introduced the concept of the F-factor, that part of the acid falling on the watershed of a lake that is exchanged for base cations so that not all the acid deposition causes depletion of lake ANC. Wright (1983) presented an extensive application of the chemical balance approach taking into account the F-factor and using lake sulphate levels as an indicator of deposition.

These approaches were concerned with the sums of major anion and cation concentrations. Thompson (1982) offered a budget approach where the output of major ions by watersheds could be computed with allowance for changes in runoff and acid deposition, but not the F-factor. Marmorek et al. (1986) have generalized these approaches to allow the budgeting of major cations (Ca and Mg), sulphate and alkalinity with explicit consideration of the F-factor, the role of the watershed to lake area ratio (r = Aw/Ao), and the levels of background sulphate i.e., non-marine sulphate not due to anthropogenic sources. A major change from the calculations of Wright (1983), is the use of sulphate deposition (wet plus dry) as the determinant of lake chemistry rather than vice versa. The Marmorek et al. (1986) site model combines two approaches: back calculation of watershed acid neutralizing capacity, and back calculation of original alkalinity given background sulphate and cation exchange; to predict the original (pre-acidic deposition) lake alkalinity.

A simple statement of the alkalinity mass balance, assuming no retention, is combined with the premise that there is a proportionality between alkalinity and major ions, which is altered by sulphate loadings and cation exchange. The resulting equations can be rearranged to allow the prediction of steady state Alkalinity, Alk^{∞}, for a level of sulphate deposition, SD_t:

$$Alk^{\infty} = \alpha C_t^* - \left[\alpha\left(\frac{r}{r+1}\right)F_W\right](SO_4^* - \beta) - \left[1 - \left(\frac{r}{r+1}\right)F_W\right]\frac{SD_t}{P.WR} \qquad (1)$$

where

r = Aw/Ao

Aw = watershed area excluding the lake area, m^2

Ao = lake area, m^2

P = annual precipitation $L.m^{-2}.yr^{-1}$

WR = fraction of water remaining after evapotranspiration

α = the proportionality between Alkalinity and C_t after Wright (1983)

C_t^* = current concentration of non-marine Ca+Mg, $eq.L^{-1}$

F_W = empirical factor accounting for cation exchange with sulphate after Henriksen (1982) and Wright (1983)

SO_4^*, β = current and original (background) concentrations of non-marine sulphate, $eq.L^{-1}$, respectively

SD_t = sulphate deposition (negative alkalinity) $eq.m^{-2}.yr^{-1}$.

The first term in the equation represents original alkalinity with no cation exchange i.e., F_W is zero; the second term corrects for cation exchange; and the third term reflects the titration of the original alkalinity due to changes in sulphate deposition. Since the level of background sulphate (β) strongly determines the proportionality (α) of alkalinity and cations (Wright, 1983), the most important factors in determining a lake's response to increased sulphate deposition are β, the background sulphate and F_W, the cation exchange factor. Wright (1983) and Henriksen (1982) suggested that F_W may have a range from 0 to 0.5 in acid sensitive areas. Estimates of range from 20 to 100 $\mu eq.L^{-1}$ in different remote areas of Canada. Consequently, simulations to predict Alk∞ have been run using the four combinations of the two most uncertain parameters.

The above site model is embedded in a regionalization model (Jones et al., 1986). For each secondary watershed unit in the area east of the Ontario-Manitoba border south of latitude 52° on the Canadian shield, frequency distributions of all measured parameters (C_t^*, r, SO_4^*, SD_t, P,WR) have been developed using available lake survey data e.g., Kelso et al. (1986). In addition, the number and area of lakes in each secondary watershed unit have been estimated. By assuming that each parameter is orthogonal to all others, or where necessary allowing for correlations, the frequency distribution of lake Alk∞'s can be computed. Via a standard pH-alkalinity curve, the frequency distribution of final lake pHs can be predicted for a given pattern of sulphate deposition. Overall, the model is used to predict the number of acid ($pH < 5$) lakes. Few fish species inhabit lakes with a pH <5 (Harvey et al., 1981).

The model has been run for current (1980) sulphate deposition using the four combinations of the key unknown parameter extremes. The model has also been run for two different scenarios of acid deposition control: (i) proportional reductions (or increases) in all areas; and, (ii) threshold controls which target the areas receiving the highest deposition. Despite the wide range of parameter assumptions, the range of predicted numbers of lakes with a pH <5 is not very large (Figure 2) when considered against total sulphate deposition (Figure 2). The central curve is the geometric mean of the four predictions. The upper bound results from assuming a high background sulphate and an F_W of zero. The lower bound results from assuming that F_W is 0.5. Acidification of 25,000 to 60,000 lakes given current deposition less the 14,000 (Table I) already considered acid, would represent a large loss for Canada.

A 40 % reduction of deposition everywhere would reduce the worst case predictions of the number of acid lakes to less than the current estimate (Figure 2). Establishing a ceiling of 20 kg sulphate $ha^{-1}.yr^{-1}$ deposition would still predict a further worst case loss up to 15,000 lakes in the future (Figure 2).

These predictions must be considered conservative for at least three reasons. First, the model assumes that the acid neutralizing capacity of watersheds is not reduced by acidic deposition. Some authorities (Arp, 1983; Booty and Kramer, 1984) suggest that there is a reservoir of ANC which is depletable. If this were the case, once depleted, regardless of whether or not cation exchange occurs in the short term, many shield lakes would become acid permanently even if deposition rates were reduced to low levels. Thus, the predictions presented here may represent a lower bound of possible acidification. Secondly, episodic pH depression, whether during spring snowmelt or heavy rain events, may have serious consequences to fisheries. This phenomenon is expected to occur (Harvey et. al., 1981) on a widespread basis throughout eastern Canada and has been shown (Lievestad and Muniz, 1976; Gunn and Keller, 1984) to cause mortality to fish. Third, there is evidence of biotic effects where the pH is between five and six (Kelso et al., 1986).

4. SOCIOECONOMICS

In eastern Canada, there are 3.7×10^6 anglers spending 41×10^6 days angling and $\$1.4 \times 10^9$ in direct and indirect expenditures (Table I) according to a 1980 survey (DFO, 1980). This effort is directed almost entirely at freshwater lakes and rivers. In Ontario a wide range of fish species is caught including salmonids, yellow perch, walleye, bass, esocids and smelt. In the maritime provinces, the catches are dominated by brook trout and then smelt, with other salmonids accounting for most of the rest.

The current and projected resource assessments given earlier can be linked to the socioeconomic data if a number of assumptions are

Table II: Estimated value of recreational angling in acid sensitive and acidified freshwater resources in eastern Canada south of latitude 52°.

Province	Angler - days, 10^{-3}			Expenditures, $\$M.yr^{-1}$		
	Alkalinity $<50\ \mu eq.L^{-2}$	pH <6	pH <4.7	Alkalinity $<50\ \mu eq.L^{-1}$	pH <6	pH <4.7
Ontario	7,281	5,068	381	241.4	168.0	12.6
Quebec	6,400	1,885	177	274.0	80.7	7.6
Newfoundland	1,022	548	35	20.0	10.7	0.7
New Brunswick	626	385	298	18.0	11.1	8.6
Nova Scotia	1,312	753	1,349	23.1	13.3	23.8
Total	16,642	8,639	2,240	576.6	283.8	53.2

Table III: Estimated value of projected recreational angling losses in eastern Canadian freshwater resources under current (1980) acidic deposition rates based on four combinations of the two least certain model parameters.

Region	Angler-days, 10^{-3}			Expenditures, $\$M.yr^{-1}$		
	Geom.Mean	Lower Bound	Upper Bound	Geom.Mean	Lower Bound	Upper Bound
Ontario	1,485	1,020	2,905	49.2	33.8	96.3
Quebec	96	17	602	4.1	0.7	25.8
Maritimes	289	123	1,012	6.4	2.7	22.2
Total	1,870	1,169	4,519	59.7	37.6	144.3

Figure 1: Plot of Observed vs Predicted Alkalinity Concentration (ueq L^{-1}) for Lakes in the Adirondack Park, New York

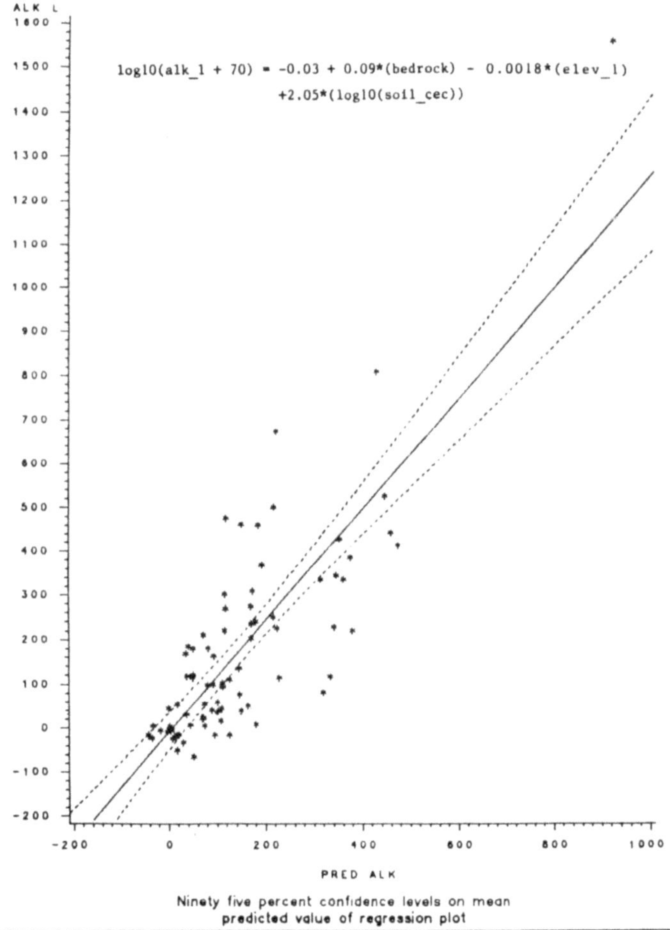

Figure 2 Predicted relationship between the number of lakes with pH <5 and total sulphate deposition in eastern Canada south of latitude 52°.

made. If resource assessments are matched with economic statistics, the value of existing and expected losses due to acidic deposition can be estimated. We have assumed that there is no change in the price or value per unit of resource with successive reductions in the supply of resource and that the anglers who exploited previously non-acidic resources, cease to angle. We have assumed that all recreational angling takes place south of latitude 52°. We have assumed that the 1980 angling activity was applied proportionately to all non-acidified (Table I) resources. It is more likely that a disproportionately large share of the angling is directed to Shield lakes which are the more acid-sensitive. Thus, this assumption will tend to underestimate the economic importance of impacts. We have assumed that all angling is directed to lakes. Obviously some angling takes place in rivers. However, the lake numbers can be considered to represent rivers, since the resource estimates were based on total freshwater resources by area. They are merely represented as lakes since they dominate the landscape. River resources can be classified by the same acid-sensitivity categories and expressed in terms of lengths of river as was done by Loucks et al. (1984) for the eastern United States.

The total resource less the acidified lakes (Table I) is assumed to account for all angling activity (Table II) by province. The value of resource at risk is calculated on a proportional basis (Table II). Lakes which have an alkalinity less than 50 $\mu eq.L^{-1}$, account for 16.6 x 10^6 angler-days, worth \$576.6 x 10^6 yr^{-1}. Lakes with a pH less than 6, but not less than 4.7, are worth \$283.8 x 10^6 yr^{-1} in angling activity. Assuming that acidified lakes have an equivalent value, we estimate that the more than 14,000 acidic lakes represent a value of \$53.2 x 10^6 yr^{-1} or 2.24 x 10^6 angler-days.

The projected losses (Figure 2), given current deposition rates, have a moderate range of values (Table III). The geometric mean of projections is close to the estimate of currently acidic resources. At the upper bound of predictions, we could expect to lose over 2 x 10^6 angler-days and \$90 x 10^6 yr^{-1} in addition to existing acidified losses (Table II), mostly in Ontario and Quebec.

5. DISCUSSION

These results represent our best overall assessment of the current status and potential future status of freshwater fishery resources in eastern Canada. Given current acid deposition rates, we might expect up to 65,000 lakes to become acidic (pH <5) compared to an estimated current 14,000. This would represent 9 % of the freshwater lakes in eastern Canada south of latitude 52°, and 10.2 % of the recreational angling value.

The various deposition control scenarios indicate that neither a 30 % reduction in deposition everywhere or limitation of 20 kg SO_4 $.ha^{-1}.yr^{-1}$, (current control objectives) would reduce the worst case results below the existing level of acidification. A 40 % reduction everywhere or ≈ a 12 kg SO_4 $.ha^{-1}.yr^{-1}$ limitation would achieve that result. Loucks and Glass (1984) indicated that a limit of 12 $kg.ha^{-1}.yr^{-1}$ was necessary to protect the most sensitive freshwaters. Nonetheless, the current control objective (20 kg $SO_4.ha^{-1}.yr^{-1}$) would bring about some improvement.

Given that the cost of emission control is expected to be high, amelioration of the acidification effects by chemical treating of lakes is a possible alternative. Some authorities (cf. Forster, 1984) have suggested that liming could be carried out for as little as $50 $.ha^{-1}.yr^{-1}$. Even assuming the logistics of such an effort were solved and all worries about the side effects of liming were allayed, treatment of the projected 65,000 lakes could run to $\$75 \times 10^6$ $.yr^{-1}$. If episodic events were to be addressed in the treatment scenario, the figure could be much higher. For instance, if all lakes with a pH <6 were treated, the cost would be $\$172 \times 10^6$ $.yr^{-1}$.

Projected freshwater fishery impacts are only part of the overall acid deposition problem. In Canada, natural resource industries are extremely important sectors of the economy. Forestry, fisheries, and tourism sectors must be protected. Loss of natural resources in one area because of inadequate pollution control measures in another, exemplifies the need for an adjustment in the relationship between man and the biosphere.

REFERENCES

Arp, P.A.: 1983, Ecol. Modelling, 19,105.

Barrie, L.A. and Hales, J.M.: 1984, Tellus 36B,333.

Booty, W.G. and Kramer, J.R.: 1984, Phil. Trans. R. Soc. Lond. B, 305,441.

DFO: 1980, Survey of sportfishing in Canada. Department of Fisheries and Oceans, Ottawa, Canada.

Forster, B.A.: 1984, Economic impact of acid precipitation: A Canadian perspective, In Crocker, T.D. (ed.). Acid precipitation series vol. 8. Economic perspectives on acid deposition control. Ann Arbor Science. 180p.

Gunn, J.M. and Keller, W.: 1984, Can. J. Fish. Aquat. Sci., 41,319.

Harvey, H.H., Pierce, R.C., Dillon, P.J., Kramer, J.R., and Whelpdale, D.M.: 1981, Acidification in the Canadian aquatic environment. Publ. NRCC No. 18475. Environmental Secretariat, National Research Council of Canada, Ottawa, Ont. 369p.

Henriksen, A.: 1979, Nature, 278,542.

Henriksen, A.: 1982, Changes in base cation concentrations due to freshwater acidification. Acid Rain Res. Rep. 1/1982, NIVA, Oslo, Norway.

Johnson, M.G., Leach, J.H., Minns, C.K., and Olver, C.H.: 1977, J. Fish. Res. Board Can., 37,1592.

Jones, M.L., Marmorek, D.R., and Minns, C.K.: 1986, Can. J. Fish. Aquat.Sci., in prep.

Kelso, J.R.M., Minns, C.K., Gray, J.E., and Jones, M.L.: 1986, Can. Spec. Publ. Fish. Aquat. Sci., 87,42p.

Legendre, P. and Legendre, V.: 1984, Can. J. Fish. Aquat. Sci., 41,1781.

Lievestad, H. and Muniz, I.P.: 1976, Nature 1259,391.

Loucks, O.L. and Glass, G.E.: 1984, U.S./Canada aquatic impacts assessment: Integration of experimental studies, monitoring and modelling of acidic deposition effects. In Schnoor, J.L. (ed.). Acid precipitation series vol. 9. Modelling of total acid precipitation impacts. Ann Arbor Science. 222p.

Loucks, O.L., Miller, R.W., and Armentano, T.V.: 1984, Northeast. Environ. Sci., 3,8.

Marmorek, D.R., Jones, M.L., and Minns, C.K.: 1986, Can. J. Fish. Aquat.Sci., in prep.

Minns, C.K., Kelso, J.R.M., and Johnson, M.G.: 1986, Can. J. Fish. Aquat. Sci. 43,900.

Minns, C.K.: 1981, Acid rain: a preliminary estimate of the risk to Ontario's inland fisheries. Can. MS Rep. Fish. Aquat. Sci., 1622:vi + 36p.

Thompson, M.E.: 1982, Water Air Soil Pollut., 18,215.

Watt, W.D., Scott, C.D., and White, W.J.: 1983, Can. J. Fish. Aquat. Sci., 40,462.

Wright, R.F.: 1983, Predicting acidification of North American lakes. Acid Rain Res. Rep. 4/1983, NIVA, Oslo, Norway. 165 p.

LAKE RESOURCES AT RISK TO ACIDIC DEPOSITION IN THE EASTERN UNITED STATES

Jerald L. Schnoor, Sijin Lee, Nikolaos P. Nikolaidis, D.R. Nair
Department of Civil and Environmental Engineering
University of Iowa
Iowa City, IA 52242 USA

ABSTRACT. Watershed descriptors have been obtained or compiled for 1439 watersheds in the northeastern and upper midwestern United States. A methodology, which combines multiple linear regression procedures with a simple deterministic model for alkalinity shows promise as a tool for acid precipitation assessements. Mean absolute errors in predicted lake alkalinity concentrations of approximately ± 100 $\mu eq\ L^{-1}$ were obtained with no significant difference (at the 0.05 significance level) between predicted and observed alkalinity histograms. Estimates of the lake resources-at-risk across the depositional gradient from Minnesota to the Adirondack Park of New York were established.

1. INTRODUCTION

The northeastern and upper midwestern portions of the U.S. contain thousands of lakes which are sensitive to acidic deposition, defining sensitivity as acid neutralizing capacity less than 40 $\mu eq\ L^{-1}$ measured by Gran titration. The United States Environmental Protection Agency (1985) has recently completed a survey of lakes (greater than 4 ha, less than 200 ha in surface area) and reported about 5% of the Northeastern lakes, and 2% of the Upper Midwest lakes to be acidic. Schnoor et al. (1986) used a simple steady state model to make projections of lake resources at risk to increased deposition in northcentral Wisconsin and the eastern portion of the Upper Peninsula of Michigan. The purpose of this paper is to extend the analysis to a ten state area. Multiple regression statistical analysis based on watershed descriptors for 1439 lakes were combined with a steady state deterministic model to estimate the resources-at-risk at various loadings of acidic deposition.

2. STUDY AREAS

Portions of ten states were included in this study. Moving from west to east, they are: northeastern Minnesota, northcentral Wisconsin, the eastern part of the Upper Peninsula of Michigan, the Adirondack Park of New York, and the New England States of Connecticut, Massachusetts, Rhode Island, New Hampshire, Vermont, and Maine.

Data for the Upper Midwest (Minnesota, Wisconsin, and Michigan) were collected in cooperation with the Wisconsin Department of Natural

Water, Air, and Soil Pollution **31** (1986) 1091–1101.

Table I Number and Areal Extent of Lakes in Study Area

Name of Sampled States	No. of Acid Lakes with pH < 5.0	No. of Lakes Sampled	Total Number of Lakes in the Region	Surface Area of Acid Lakes (pH < 5.0), ha	Total Surface Area of Lakes in the Region, ha
Adirondacks	208	1,124	2,877	3,786	114,000
Maine	18	752	4,000	335	590,000
Vermont	6	232	525	80	88,000
New Hampshire	17	560	1,088	754	74,000
Rhode Island	8	101	360	94	41,000
Massachusetts	16	624	2,443	260	120,000
Connecticut	1	225	1,218	3	38,000
Michigan (U.P)*	10	53	900	243	25,000
Minnesota (N.E.)*	0	342	2,200	0	100,000
Wisconsin (N.C.)*	18	427	1,600	673	81,000

* The Number of Acid Lakes for Michigan (U.P), Minnesota (N.E.) and Wisconsin (N.C.) was evaluated as Alkalinity < 0.0 ueq/L

Resources (Rhinelander District Office) and the U.S. Environmental Protection Agency (Environmental Research Laboratory-Duluth). The data are compiled in the U.S. EPA "ERLD-UMD Acid Rain Deposition Gradient Susceptibility Database" by Glass (1985). Data for the northeastern states were obtained from several sources including Haines and Akielaszek (1983), ACID database by Hendrey et al. (1983), Pfeiffer and Festa (1980, 1984), Norton et al. (1982) bedrock maps, and the GEOECOLOGY database at Oak Ridge National Laboratory (Olson et al., 1980). These references were compiled into one database at The University of Iowa (UI database).

Table I provides a summary of the number of acid lakes (less than pH 5.0) in each of the study areas. As one moves from west to east, following a gradient of decreasing precipitation pH, the percentage of acid lakes that were sampled increases. In northeastern Minnesota, there were no acid lakes found, while 4.2% were observed in northcentral Wisconsin, followed by 18.9% in the eastern part of the Upper Peninsula of Michigan, and 18.5% in the Adirondack Park of New York. In New England, the percentage of acid lakes sampled ranged from 0.4% in Connecticut to 7.9% in Rhode Island. The number of acid lakes sampled in the combined Upper Midwest and Northeast was less than 7%, based on Table I, and the surface area of these acid lakes represented less than 1% of the total surface area of lakes in the regions.

Since the sampling of lakes in each region was not random nor entirely representative, the percentage of acid lakes is not definitive, but it does illustrate a trend among these study areas. The study areas are all relatively sensitive to acid deposition (crystalline bedrock and/or acidic soils). Among these sensitive regions, the greater is the acidity of precipitation, generally, the greater the percentage of acid lakes. Where the volume-weighted average precipitation pH is 4.6 and the precipitation rate is 100 cm yr^{-1}, the acid deposition is 250 eq ha^{-1} yr^{-1} (wet only). It seems that most lakes in these sensitive areas can generate greater than 250 eq ha^{-1} yr^{-1} of alkalinity to neutralize the acid deposition.

3. METHODOLOGY

Utilizing the existing databases, the objectives of this study were to: 1) determine the watershed factors which contribute to lake sensitivity, and 2) predict the number of lakes which would be acidic under various acid deposition loadings.

The UI database consisted of four water quality parameters (alkalinity, pH, conductivity, and color) and 23 watershed descriptors (lake elevation, lake surface area, watershed area, ratio of watershed area to lake area, bedrock type, hydrologic type, number of inlets, number of outlets, drainage order, total watershed relief, number of cabins in the watershed, annual runoff, annual precipitation, annual potential evapotranspiration, annual H-ion loading (eq m^{-2} yr^{-1}), average soil pH, soil percent base saturation, soil cation exchange capacity, and the percentage of the watershed as wooded swamp, marsh, woodlands, open land, or developed land).

Multiple linear regression analyses were used to determine the watershed variables which correlated well with lake alkalinity in each region. Other investigators have related lake alkalinity to watershed descriptors using multivariate statistics (Rapp et al., 1986; Eilers et al., 1983; Haines et al., 1983; Hunsaker et al., 1985; Baker and Harvey, 1984). The following statistical procedures were utilized in this research.

1) Obtain descriptive statistics (mode, median, mean, kurtosis, and skewness) for each variable in the database.
2) Perform a log-transformation on variables to obtain a near-normal distribution as necessary.
3) Correlate all variables with each other by constructing a matrix of correlation coefficients from linear, least-squares regressions.
4) Use exploratory linear stepwise multiple regression analyses with selected watershed descriptors to determine the best fit (maximum r^2) models with 2-6 independent variables and with alkalinity as the dependent variable.
5) Select the best multiple regression model for each state or region, making sure that the variable associations were logical and reasonably independent. Check the statistical significance for each of the models.
6) Verify the multiple regression model for each state or region by predicting the alkalinity concentrations for lakes using only watershed descriptors and withholding alkalinity data. Estimate the prediction error.
7) Use the steady state deterministic mass balance model to predict alkalinity concentrations under various acid loading conditions.

4. MASS BALANCE MODEL FOR ALKALINITY

The primary drawback to multiple regression models is that they cannot be used to predict alkalinity concentrations under different acid loading scenarios due to the fact that they are dealing with only watershed descriptors. A combination of multiple regression techniques and a deterministic steady state model (Schnoor et al., 1984, 1986; Stumm et al., 1983a) were used to predict the number and areal extent of lake resources-at-risk at double and half the present-day acid loadings. The Adirondack region in New York State was chosen for analysis because the multiple regression model worked best (r^2 = 0.53) for this region and because the Adirondack region has the greatest number of acid lakes. The Minnesota and Wisconsin areas were also chosen for analysis because the regression models produced statistically significant predictions of alkalinity based only on watershed variables.

The steady-state alkalinity concentration in the model depends on only four factors: the evapocencentration factor (I/Q), the amount of precipitation (R), the weathering or chemical neutralization rate (W), and the acidity of precipitation (L_{acy}). W includes all neutralization processes such as chemical weathering, ion exchange, and reduction reactions. The mass balance equation is given below.

$$Alk = \frac{I}{Q}\left\{\frac{W}{R(0.1)} - L_{acy}\right\} \tag{1}$$

where Alk = alkalinity concentration, μeq L^{-1}
I = inflow volume per time, $m^3\ yr^{-1}$
L_{acy} = total acidity of precipitation (dry and wet), μeq L^{-1}
Q = outflow or seepage rate, $m^3\ yr^{-1}$
R = precipitation rate, cm yr^{-1}
W = weathering and neutralization rate, eq $ha^{-1}\ yr^{-1}$
0.1 = units conversion factor.

A mixed kinetic model (with a fractional-order rate and a zero-order rate) has been determined in studies of mineral weathering by Stumm et al. (1983b). The fractional order term represents H^+ attack on mineral surfaces, and the zero-order term represents chemical weathering due to attack of organic ligands. Schnoor et al. (1986) have used an analogous mixed kinetic model to represent neutralization in the watershed, and the following form of the equation is applicable.

$$Alk = \frac{I}{Q}\left\{\frac{k_h\left(\frac{L_{acy}}{L_{ref}}\right)^m + k_o}{R(0.1)} - L_{acy}\right\} \tag{2}$$

where $W = k_h\left(\frac{L_{acy}}{L_{ref}}\right)^m + k_o$ = weathering or neutralization rate
k_h = hydrolysis rate constant for H^+ attack on minerals, eq ha^{-1} yr^{-1}
k_o = ligand weathering rate constant for organic ligands and CO_2, eq $ha^{-1}\ yr^{-1}$
L_{ref} = reference precipitation acidity, μeq L^{-1}
m = fractional order weathering rate constant, dimensionless.

One difficulty in the use of the steady state model is how to determine the rate constants, m, k_h, and k_o. Theory dictates that m, the fractional-order hydrolysis rate constant, must be between 0 to 1.0 if the reaction is truly controlled by the rate of dissolution at the surface of minerals. Laboratory results in the literature on acid hydrolysis with a wide variety of minerals indicates that the range of m for most crystalline rock primary minerals is 0.3 to 0.7. A sensitivity analysis was performed as a part of this study, and it indicated that m was best chosen as 0.6 for Adirondacks and Wisconsin, and 0.5 for Minnesota. This optimum value of m for lakes in the three study areas ensured that calibrated rate constants were positive and that alkalinity predictions increased as m increased. Results were not very sensitive to the values of m which were chosen.

Verification of the model (Equation (2)) was performed while withholding measured lake alkalinity field data. Thus, a true prediction error could be obtained. In order to verify the model, a multiple regression equation was used to generate estimated alkalinity and rate constant k_o values from watershed descriptors alone. Then, the model calibration consisted of an inverse procedure to obtain k_h from Equation (2). Finally, Equation (2) was used to predict the alkalinity concentration of each lake in the data set, and the predicted alkalinity concentration was compared to the measured value.

5. RESULTS AND DISCUSSION

5.1. Regression Models

Linear regression correlations were developed between the alkalinity concentration of the lakes (dependent variable) and each watershed descriptor. The strongest relationships were indicated by correlation coefficients with lake area, lake elevation, annual rainfall, bedrock type, watershed area, soil pH and soil cation exchange capacity (CEC). Lakes in the Northeast (New England states and the Adirondack Park), the Adirondack Park alone, northeastern Minnesota, and northcentral Wisconsin were investigated. A positive correlation coefficient indicates that lake alkalinity increases as the value of the watershed descriptor increases, and a negative correlation coefficient signifies that alkalinity decreases as the watershed descriptor increases. Norton's (1982) bedrock scale was strongly positively correlated with alkalinity of lakes in the Northeast region ($r = 0.395$) and the Adirondack Park, N.Y. (0.369), while annual rainfall and lake elevation were highly negatively correlated with alkalinity in the Northeast region (-0.403, -0.254), and the Adirondack Park, N.Y. (-0.350, -0.650). In the Midwest region, the bedrock scale (from 1 to 10) of Rapp et al. (1986) was utilized, which goes in the opposite direction of Norton's (1982) bedrock scale. Low numbers on the Rapp et al. (1986) bedrock scale indicate non-sensitive bedrock geology, such as limestone or dolomite. As expected, it correlated negatively with lake alkalinity in Minnesota (-0.262) and Wisconsin (-0.085). The percentage of the immediate watershed area in developed land (agriculture, residential, etc.) was correlated with higher alkalinity concentrations.

In Wisconsin, very strong positive correlations were determined between alkalinity with the lake area (0.450) and hydrologic type (0.723). Hydrologic type was scaled by Eilers et al. (1983) from 1 to 4 according to the potential degree of groundwater influence on the lake. Lakes of hydrologic type 1 are precipitation-dominated seepage lakes, and they represent a class of low alkalinity systems that were clearly discriminated by linear regression analysis. Approximately 50% of the variation in alkalinity among lakes in Wisconsin was explained by the watershed descriptor hydrologic type. H^+ deposition was used as a watershed variable (county-averaged from GEOECOLOGY), and it did not vary enough within a region to be statistically significant in most cases. Figure 1 gives the multiple linear regression equation which was obtained for lakes in the Adirondack Park of New York. The coefficient of determination (r^2) was 0.53, so 53% of the variance in the observed alkalinity concentration values was explained by only three watershed variables (bedrock type, elevation of the lake, and soil CEC). The mulitple linear regression model resulted in an absolute error residual of ± 110 μeq L^{-1} and little bias in prediction (Figure 1).

Figure 1: Plot of Observed vs Predicted Alkalinity Concentration (ueq L^{-1}) for Lakes in the Adirondack Park, New York

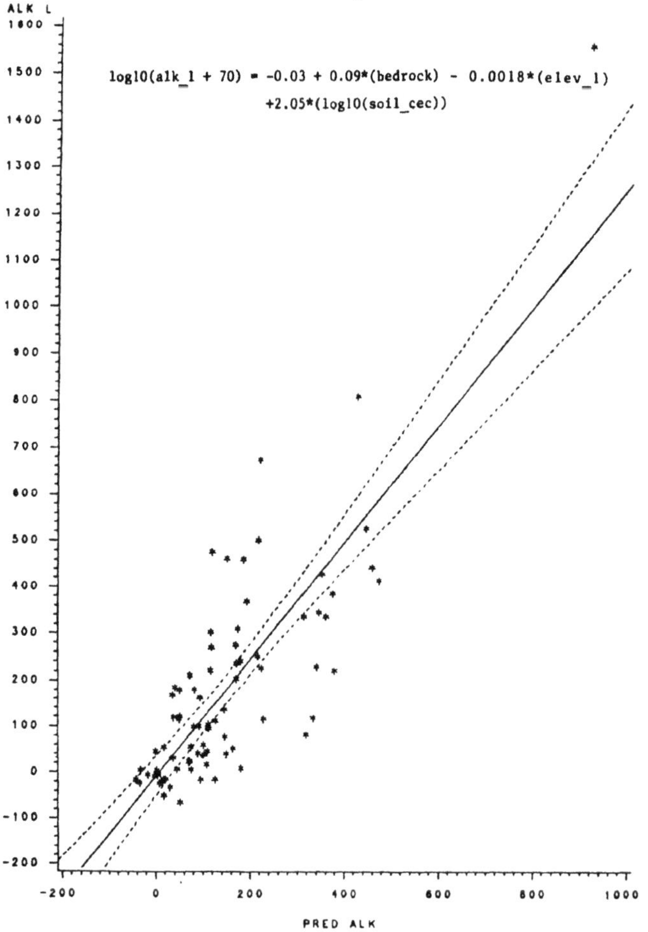

Ninety five percent confidence levels on mean predicted value of regression plot

Figure 2: Verification of Trickle-Down Model for Lakes in the Adirondack Park, New York

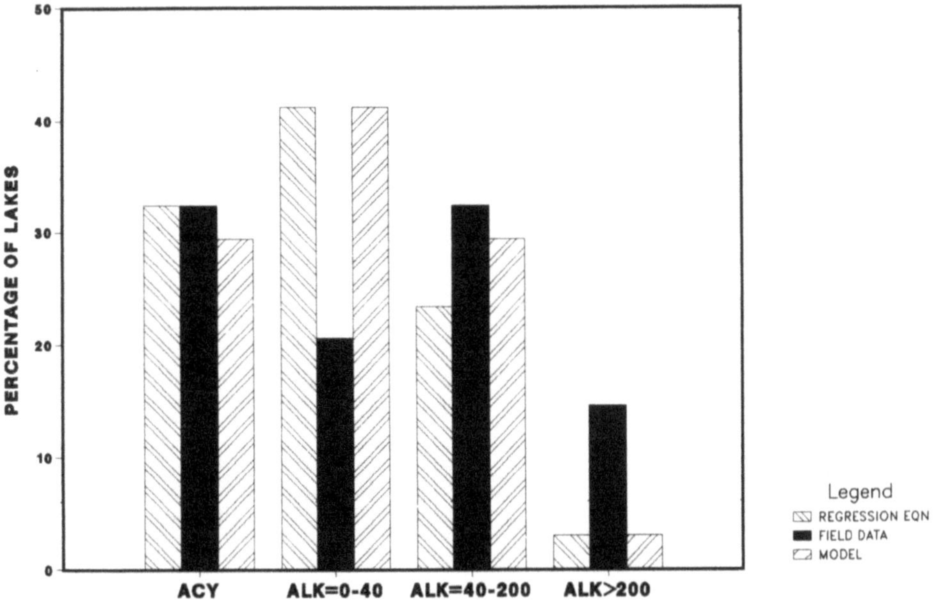

Figure 3: Lake Resources at risk at Varying Deposition

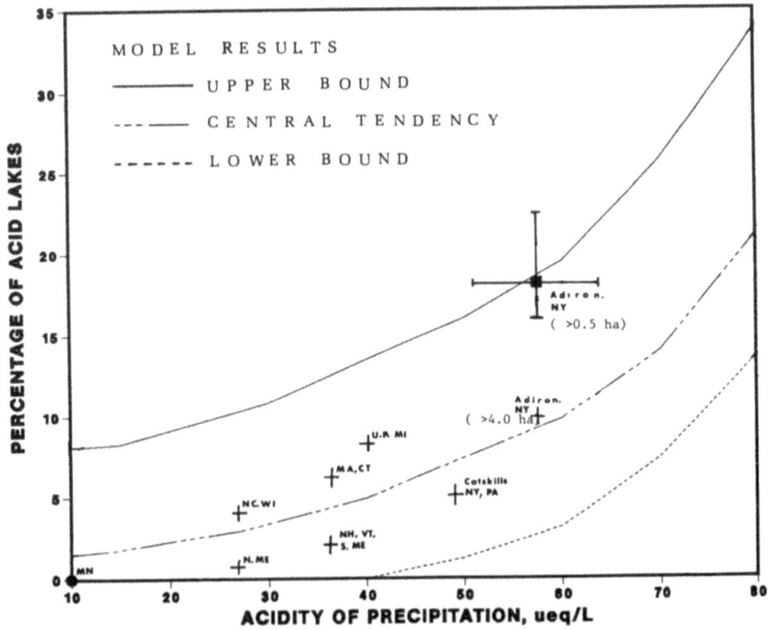

5.2. Steady-State Model

The steady-state deterministic model has sometimes been called the "Trickle-Down Model" by Schnoor et al. (1986) and Stumm et al. (1983a). Results of the model verification are presented in Figure 2 for four alkalinity categories (Alk < 0,0 ≤ Alk < 40,40 ≤ Alk < 200, and Alk ≥ 200 μeq L^{-1}). Approximately 32% of the verification set of lakes (m = 34) were measured to be acidic (Alk < 0 μeq L^{-1}), 32% were predicted to be acidic by the multiple linear regression model, and 29% were predicted to be acidic by the steady-state model. The regression equation and the steady-state model gave similar results (which is logical since the steady-state model used the regression equation in its calibration procedure). The models did a poor job of predicting the percentage of lakes in the high alkalinity category (Alk ≥ 200 μeq L^{-1}), but this is not very critical to the risk assessment.

The Steady-state model was used to predict the alkalinity concentration of all lakes in the Adirondack Park at double and half-loading of H^+ deposition. If the loading doubled, the percentage of acid lakes in the Adirondacks would be predicted to increase from 16 to 46%. At half the present acid loading, approximately 6% of the acid lakes in the Adirondacks would recover. These lakes are sensitive to changes, either increases or decreases in acid deposition.

The three lines (upper bound, central tendency, and lower bound) of Figure 3 summarize the results of the study. The crosses indicate the percentage of acid lakes (Gran alkalinity ANC < 0 μeq L^{-1}) in each region according to field data from EPA (1985) on lakes greater than 4.0 ha surface area. The large cross indicated for the Adirondacks includes the range and estimate for all lakes greater than 0.5 ha from this study. The central tendency of all model results is given by the semi-dashed line. In Figure 3, the boundary lines were calculated from Equation (2) using total acidity of precipitation (wet + dry), but the crosses represent available field data based on wet-only precipitation acidity. The central tendency corresponds closely to the model results for northcentral Wisconsin with model coefficients ranging from k_o = 5 - 4778 eq ha^{-1} yr^{-1}, k_h = 144-230 eq ha^{-1} yr^{-1}, m = 0.6, and L_{acy} = 5 μeq L^{-1} background acidity. The solid line corresponds to an upper bound for the most sensitive region, the Adirondack Park of New York, where the model coefficients were k_o = 0 - 7487 eq ha^{-1} yr^{-1}, k_h = 273-613 eq ha^{-1} yr^{-1}, m = 0.6, and L_{acy} = 10 μeq L^{-1}. The lower bound represents the least sensitive region, and it is shown by the dashed line. It corresponds closely to the model results for northeastern Minnesota with model coefficients ranging from k_o = 228-5841 eq ha^{-1} yr^{-1}, k_h = 55-59 eq ha^{-1} yr^{-1}, m = 0.5 and L_{acy} = 1 μeq L^{-1}. The uncertainty of the model results is large, but results are consistent with the field measurements from EPA (1985) and they can be used to estimate (in a preliminary assessment) the lake resources-at-risk to acid deposition.

6. SUMMARY

Multiple regression models and the steady state version of the Trickle-

Down model were used to predict the lake resources at risk under different acid loadings in the Adirondack Park. Error estimates were determined based on a procedure of model calibration, verification (cross-validation), and prediction. A database (UI database) was compiled from existing data sources, including four water quality parameters and 23 watershed descriptors. Watershed descriptors which were significantly correlated with lake alkalinity concentrations were bedrock type, elevation of the lake, annual rainfall, watershed area, lake surface area, and soil pH and CEC.

The Trickle-Down model was coupled with multiple linear regression models. Mean absolute prediction errors were on the order ± 100 μeq L^{-1} alkalinity. The model tended to underpredict the number of high alkalinity lakes (> 200 μeq L^{-1}) and to overpredict the number of lakes in the alkalinity range from 0 to 40 μeq L^{-1} for the Adirondack Park. Application of the Trickle-Down model worked at least as well as the multiple regression equation to predict alkalinity concentrations of the lakes. An estimate of the number of acid lakes in the Adirondack Park under current loading conditions is 16%, but the number would increase to 46% if acid deposition were to double, and it would decrease to 10% at half-loading.

A risk assessment methodology has been developed to estimate the number of acid lakes in any region under current conditions. The prediction error (± 100 μeq L^{-1}) is rather large for an individual lake, but it may be acceptable for estimating distributions of lakes by alkalinity classes. Final verification of the model under different loading conditions is not possible except for cases of whole-lake or whole-watershed manipulation.

7. ACKNOWLEDGMENTS

This project was supported by a cooperative agreement with the U.S. Environmental Protection Agency (CR-811257-01). Advice and assistance came from M. Robbins Church (EPA), Joseph Eilers (Northrop), George Hendrey (Brookhaven National Laboratory), Nicholas Gmur (Brookhaven National Laboratory), Gary Glass (EPA), Kathy Webster (Wisconsin Department of Natural Resources), John Malanchuk (EPA), Richard Olson (Oak Ridge National Laboratory), Webb Van Winkle (Oak Ridge National Laboratory), and George Rapp (University of Minnesota-Duluth).

8. REFERENCES

Baker, J.P. and Harvey, T.B.: 1984, Critique of Acid Lakes and Fish Population Status in the Adirondack Region of New York State, Final Report, USEPA, Corvallis Environ. Res. Lab., Corvallis, Oregon.

Eilers, J.M., Glass, G.E. and Webster, K.E.: 1983, Canadian Journal of Fisheries and Aquatic Sciences, 40, 1896.

Glass, G.E.: 1985, ERLD-UMD Acid Rain Deposition Gradient Susceptibility Database, USEPA, Duluth Environ. Res. Lab., Duluth, Minnesota, Unpublished data.

Haines, T.A. and Akielaszek, J.: 1983, A Regional Survey of Chemistry of Headwater Lakes and Streams in New England, US Fish and Wildlife Service, Orono, Maine.

Hendrey, G.R., Hoogendyk, C.G. and Gmur, N.S.: 1983, Establishing the Acidification Chemistry Information Database (ACID), Brookhaven Nat. Lab., Upton, New York.

Hunsaker, C.T., Olson, R.J., Christensen, S.W., Turner, R.S., Rosen, A.E., Beauchamp, J.J., Durfee, R.C., coleman, P.R., Wilson, D.L. and Latham, F.E.: 1985, Adirondack Regional Case Study: Relationships Between Watershed Attributes and Water Chemistry, International Symposium on Acidic Precipitation, Sept. 15-20, Muskoka, Canada.

Norton, S.A., Akielaszek, J.J., Haines, T.A., Stromberg, K.L. and Longcore, T.R.: 1982, Bedrock Geologic Control of Sensitivity of Aquatic Ecosystems in the United States to Acid Deposition, Nat. Atm. Dep. Program, USEPA, Corvallis Environ. Res. Lab., Corvallis, Oregon.

Olson, R.J., Emerson, C.J. and Nungesser, M.K.: 1980, Geoecology Database, US Dept. of Energy, Oak Ridge Nat. Lab., Oak Ridge, Tennessee.

Pfeiffer, M.H. and Festa, P.J.: 1980, Acidity Status of Lakes in the Adirondack Region of New York in Relation to Fish Resources, New York State Dept. of Environ. Conservation, New York, revised in 1984.

Rapp, G., Jr., Allert, J.D., Liukkonen, B.W., Ilse, J.A., Loucks, O.E. and Glass, G.E.: 1986, Environ. International, 11, 425.

Schnoor, J.L., Palmer, W.D., Jr. and Glass, G.E.: 1984, in Modeling of Total Acid Precipitation Impacts, Schnoor, J.L. (ed.), Butterworth, Boston, Mass., 155.

Schnoor, J.L., Nikolaidis, N.P. and Glass, G.E.: 1986, Lake Resources at Risk to Acidic Deposition in the Upper Midwest, J. Water Pollut. Control Fed., 58, 139.

Stumm, W., Morgan, J.J. and Schnoor, J.L.: 1983a, Naturwissenschaften, 70, 216, (in German).

Stumm, W., Furrer, G. and Kunz, B.: 1983b, Croat. Chem. Acta, S6, 585.

US Environ. Prot. Agency: 1985, National Surface Water Survey, Eastern Lake Survey, Phase I - Synoptic Chemistry, International Symposium on acidic Precipitation, Sept. 15-20, Muskoka, Canada.

AUTHOR INDEX

PARTS 1 AND 2

PART 1

AUTHOR INDEX

PARTS 1 AND 2

PART 2

SUBJECT INDEX

PARTS 1 AND 2

PART 1

SUBJECT INDEX

PARTS 1 AND 2

PART 2

MIX
Papier aus verantwortungsvollen Quellen
Paper from responsible sources
FSC® C105338

If you have any concerns about our products,
you can contact us on
ProductSafety@springernature.com

In case Publisher is established outside the EU,
the EU authorized representative is:
Springer Nature Customer Service Center GmbH
Europaplatz 3, 69115 Heidelberg, Germany

Printed by Libri Plureos GmbH
in Hamburg, Germany